Habits of a Successful Math Student

We care about your success, and throughout this text we will continue to support your efforts to achieve your goals. In addition to our opening Chapter S, within each chapter, we highlight a characteristic of a successful math student and provide tips and strategies to think about throughout your course. Some chapter openers focus on skills and others attitudes. When you need a little extra motivation, scan through these chapters as well as our dedicated *Chapter S, Strategies to Succeed in Math,* for tips and suggestions to guide you in your course.

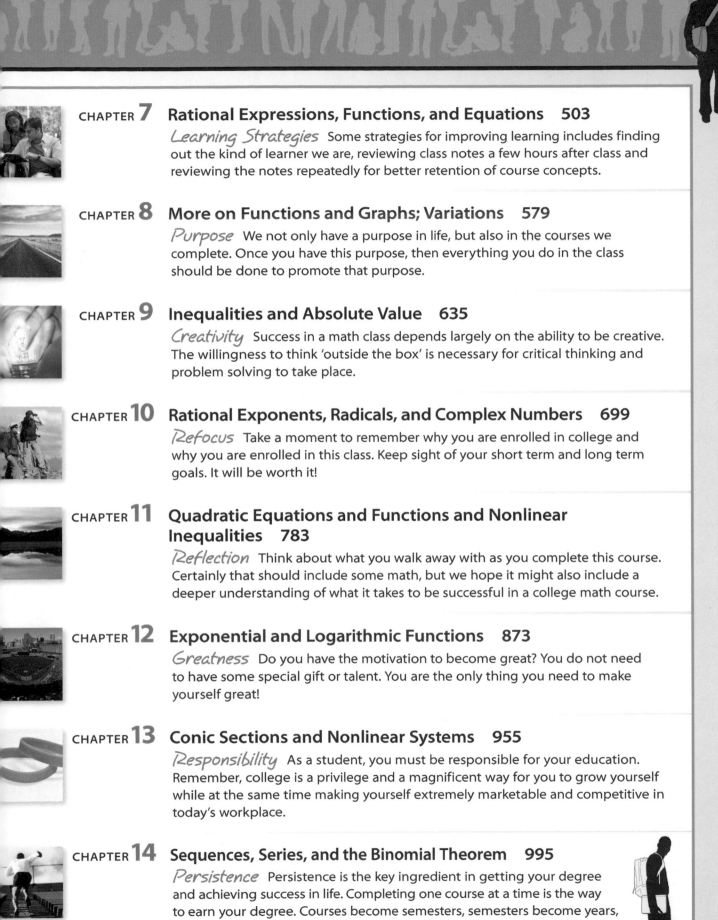

Beginning & Intermediate Algebra

Andrea Hendricks
Georgia Perimeter College

Oiyin Pauline Chow
Harrisburg Area Community College

Connect
Learn
Succeed™

BEGINNING & INTERMEDIATE ALGEBRA

Published by McGraw-Hill, a business unit of The McGraw-Hill Companies, Inc., 1221 Avenue of the Americas, New York, NY 10020. Copyright © 2013 by The McGraw-Hill Companies, Inc. All rights reserved. Printed in the United States of America. No part of this publication may be reproduced or distributed in any form or by any means, or stored in a database or retrieval system, without the prior written consent of The McGraw-Hill Companies, Inc., including, but not limited to, in any network or other electronic storage or transmission, or broadcast for distance learning.

Some ancillaries, including electronic and print components, may not be available to customers outside the United States.

This book is printed on acid-free paper.

6 7 8 9 10 11 QVS/QVS 21 20 19 18 17

ISBN 978–0–07–338453–5
MHID 0–07–338453–4

ISBN 978–0–07–729704–6 (Annotated Instructor's Edition)
MHID 0–07–729704–0

Vice President, Editor-in-Chief: *Marty Lange*
Vice President, EDP: *Kimberly Meriwether David*
Senior Director of Development: *Kristine Tibbetts*
Editorial Director: *Stewart K. Mattson*
Executive Editor: *Dawn R. Bercier*
Sponsoring Editor: *Mary Ellen Rahn*
Director of Digital Content Development: *Emilie J. Berglund/Nicole Lloyd*
Developmental Editor: *Emily Williams*
Marketing Manager: *Peter A. Vanaria*
Senior Project Manager: *Vicki Krug*
Senior Buyer: *Sherry L. Kane*
Senior Media Project Manager: *Sandra M. Schnee*
Senior Designer: *Laurie B. Janssen*
Cover Illustration: *Imagineering Media Services Inc.*
Senior Photo Research Coordinator: *Lori Hancock*
Photo Research: *Danny Meldung/Photo Affairs, Inc*
Compositor: *Cenveo Publisher Services*
Typeface: *10.5 Times LT Std*
Printer: *Quad/Graphics*

All credits appearing on page or at the end of the book are considered to be an extension of the copyright page.

Library of Congress Cataloging-in-Publication Data

Hendricks, Andrea.
    Beginning & intermediate algebra / Andrea Hendricks, Oiyin Pauline Chow.
       p. cm.
    Includes index.
    ISBN 978–0–07–338453–5—ISBN 0–07–338453–4 (hard copy : alk. paper) 1. Algebra. I. Chow, Oiyin Pauline. II. Title. III. Title: Beginning and intermediate algebra.
    QA152.3.H3345 2013
    512—dc23
                                                                                2011032563

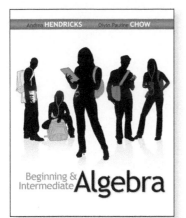

In developmental math, the focus needs to be about the "whole student" and providing students with "more than just the math." As Andrea and Pauline say, we want students to know that we care about their success. Therefore, we chose to include students on the covers to show how interested we are in their pursuit to succeed and persist through their math courses. We also wanted to visually represent today's students in their current environments. Our authors have made this their focus too by providing an entire chapter of robust resources for teaching and learning success strategies beyond just the math. And, no matter the course format, Andrea and Pauline have provided purposeful examples and exercises, current and relevant applications, and critical thinking exercises to reach today's whole student. Our hope is that students will want to envision themselves as the successful and confident math students on these covers.

Dedications

David and Judy McMullen—my wonderful parents, who taught me that with commitment and perseverance, I can achieve my goals and dreams. Thanks for your love and support.
Todd, Andy, Charlie, and Cory—my family. Thanks for being with me on this journey.

   —Andrea Hendricks

Chow Cheung—my father, who recognized the importance of education and sent all seven of his children abroad to pursue a college education in the 70s.
Wong Shing—my mother, who constantly provides us with love and nurture.
Michael, Amy, and Andrew—my family, pride and joy.

   —Oiyin Pauline Chow

Hendricks & Chow

Developmental Math Hardcover Series

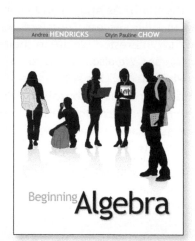

Beginning Algebra

Andrea Hendricks and Oiyin Pauline Chow, ©2013

ISBN: 978-0-07-338427-6
MHID: 0-07-338427-5

Intermediate Algebra

Andrea Hendricks and Oiyin Pauline Chow, ©2013

ISBN: 978-0-07-338426-9
MHID: 0-07-338426-7

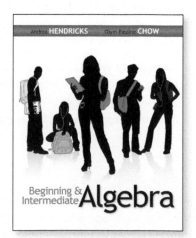

Beginning & Intermediate Algebra

Andrea Hendricks and Oiyin Pauline Chow, ©2013

ISBN: 978-0-07-338453-5
MHID: 0-07-338453-4

About the Authors

Andrea Hendricks I am an Associate Professor of Mathematics at Georgia Perimeter College (GPC), Online Campus. I have been teaching at the college level since 1992 and have taught the full range of math courses. In 2008, I joined the online campus of GPC and teach exclusively online. Prior to joining the online campus, I taught traditional face-to-face classes, hybrid classes, and online classes. During my tenure at GPC, I have served as Assistant Department Chair for one of the ground campuses and am currently Assistant Department Chair for the online math department, which gives me responsibility for the part-time faculty members and for managing and overseeing the developmental math courses. In addition, I have chaired and served on various curriculum committees, the Faculty Senate, Peer Review Committees, Promotion and Tenure Panels, and the Math Conference Committee. I have received the NISOD Teaching Excellence Award, the Faculty Teaching and Service Award, and the GPC Collegiality Award. I particularly enjoy teaching developmental math classes so that I can help students overcome their fear and anxiety toward learning math. I look forward to sharing my strategies and teaching moments on a larger scale within this Developmental Math series. My husband, Todd, is also a professor of mathematics at GPC, and we have three growing boys under the age of 13: Andy, Charlie, and Cory. In additon to teaching, authoring, and being a mom, I enjoy golf, playing the piano, and am involved with my church through teaching a 4-year-old Sunday school class and also AWANA classes.

Oiyin Pauline Chow As a Senior Professor and Chair of the Mathematics and Computer Science Department at Central Pennsylvania's Community College (HACC), I have approximately 30 years of teaching experience, and I have taught a full curriculum—from Developmental Math to Calculus, Linear Algebra, and Differential Equations in both traditional and online formats. Currently, I serve as Mathematics Department Chair at HACC's five regional campuses and their virtual campus. I also participate in many active roles on campus to support the Faculty Council and various developmental education and department committees. My other interests include serving on various state and national math organizations: the executive boards of AMATYC (secretary) and three Pennsylvania math organizations, PCTM (president), PADE (member at large), and PSMATYC (president). I have received the NISOD Teaching Excellent Award, The PADE Exemplary Teaching in Developmental Education award, and the PCTM Outstanding Contribution to Mathematics Education Award. The greatest joy in this profession is being able to work with students, empower them to take ownership in learning, and help them see their own accomplishments. In particular, at the Developmental Math level, I take pride if I can change students' outlook on mathematics, help them achieve an appreciation of math that they never had, and gain a better understanding of math. My husband Mike and I are experiencing an official empty nest, but we enjoy visiting and keeping tabs on our son, Andrew, at his software developer job in Palo Alto California and our daughter, Amy, in New York City at a digital marketing firm.

> "Students want to know that someone is interested in their success."
> —Andrea Hendricks and Pauline Chow

Our Development Story

From our years of teaching developmental math students, we have observed that more and more students need a review of study skills and how these can be applied to equal success in college. So, when we teach, we teach more than just the math—we teach life skills. Accordingly, more and more schools are integrating study skills into their developmental math classes to help retain students and improve their success rates. Students want to know that someone is interested in their success and can make the math attainable without sacrificing the integrity of the course. Therefore, we felt that it is our role to provide a new, purposeful developmental math text series. We have thoroughly enjoyed partnering in this quest of authoring and have formed a great team with Andrea creating the organization and framework for the text and writing the narrative and explanations/examples while Pauline manages the exercise sets and oversees the digital content. Together, we have focused on the following three key course needs to provide an accessible, relevant, and motivating series that will help drive students to succeed.

Success Strategies For Today's Students

Since student success is critical, we incorporated a separate chapter on student success strategies including topics like time management, note taking, and preparing for tests. We integrated these topics throughout the text by highlighting each chapter opener with a characteristic of a successful student and a motivational quote. Therefore, students will find new resources needed to be successful at their fingertips while studying math.

Student-Centered Examples and Exercise Sets

Today, we must address and serve a diverse population of developmental math students with various learning styles. When students realize the relevance of math in their lives, they are motivated to learn the material and are prepared to move into various academic disciplines. We have worked diligently to ensure that our clear, concise explanations, exercises, and applications mean something to today's students. Our numerous worked examples are organized by objectives with detailed, step-by-step procedures wherever possible. We also have worked to provide multiple ways for students to learn—troubleshooting their own mistakes, writing about the math, advancing through various levels of problems, identifying other's errors, and critically taking ownership of the material that they are learning.

Digital Solutions For Every Teaching Environment

Because of the growth of online learning, hybrid classes, and course redesigns, it has become even more important to create materials that help students grow into successful, independent learners. This book is designed not only to motivate and inspire students but students can also learn from it in any type of classroom environment. From our experience teaching online, it is clear to us that students depend on the textbook more so than in a traditional classroom. We also worked to take an active role in the creation of the digital content, and while developing the organization and progression of our exercise sets, we considered how a student works through online homework for additional practice. Each and every exercise has a purpose both in the text and online to support the objectives. We were involved in the process of ensuring that the guided solutions in the online homework represent the same author voice and narrative as the printed book to eliminate any inconsistencies.

Andrea M. Hendricks   and

Contents

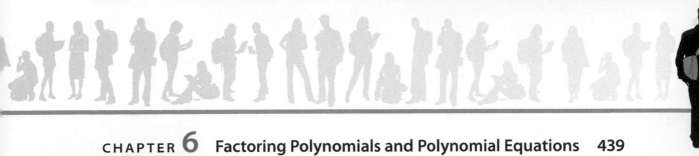

Success Strategies for Today's Students

Today's students need more guidance than ever before on how to succeed in both their current math course and future courses.

▶ Do your students come to class prepared to study, take good notes, make time for homework, and study for tests?

▶ How do you cover all of the math curriculum and teach these success strategies?

The Hendricks/Chow series provides materials dedicated to student success strategies that are integrated within both the text and the instructor and student supplements.

▶ How do you incorporate study skills or success strategies into your course?

An entire chapter, entitled **Strategies to Succeed in Math**, is dedicated to study skills, such as time management, test taking, note taking, and tips to succeed in online/hybrid courses.

A **Learning Style Inventory Quiz** will help students identify if they are a visual, auditory, or kinesthetic learner and then provides specific math strategies to help support the various learning styles.

Instructors and students will benefit from several helpful strategies to use while in these courses.

CHAPTER S

Strategies to Succeed in Math

Success

It is with great delight that we welcome you to this course and the materials we have provided you. You are embarking on perhaps the most rewarding experience of your life. College is a series of challenges that will prepare you for a lifetime of learning and accomplishments. At the end of your college education, you will be awarded a degree—a degree that establishes your ability to learn and to persevere. It is a statement that you can work to accomplish a goal, no matter what the obstacles.

At this point, you must evaluate the reason you are here. Are you here to fulfill your dreams or the dreams of someone else? To successfully complete this journey, you must be here to fulfill your own desires, not those of a friend or family member. It will be most difficult to withstand the trials of college if you do not have a personal desire to see it through.

A semester or quarter can be very overwhelming. Focus on one day at a time and not on everything that you must learn throughout the entire course. Before you know it, the course will be over. We know that it is very easy to get distracted from your goals. Stay committed and motivated by remembering your ultimate reason for attending school.

We wish you success in this course and in your future educational endeavors. It is our hope that you are successful, not only this semester or quarter, but every term until your ultimate goal is achieved. This course will pave the way for that success. It is the door to achieving your dreams.

Mrs. Andrea Hendricks and *Mrs. Pauline Chow*

Andrea M. Hendricks *Pauline Chow*

Question for Thought: What do you dream about doing? How does attending college make that dream possible? How does this course help you meet your goals? What is your biggest obstacle in being successful in this course? What can you do to overcome that obstacle? What is your plan for succeeding in this course?

Chapter S will introduce some important strategies that can enhance your performance in this course.

Chapter Outline

Section S.1 Time Management and Goal Setting

Section S.2 Learning Styles

Section S.3 Study Skills

Section

Section

ACTIVITY 1   Complete the following survey to determine your math learning style.

Math Learning Styles Survey

Answer each question with the number "3" if you agree most of the time, "2" if you agree sometimes, and "1" if you agree rarely or do not agree.

_____ 1. When there is talking or noise in class, I get easily distracted.

_____ 2. If a problem is written on the board, I have difficulty following the steps unless the teacher verbally explains the steps.

_____ 3. I find it easier to have someone explain something to me than to read it in my math book.

_____ 4. If I know how the math is used in real life, it is easier for me to learn.

_____ 5. To remember formulas and definitions, I need to write them down.

_____ 6. I prefer listening to the lecture rather than taking notes.

_____ 7. Using manipulatives, hands-on activities, or games helps me learn math concepts.

_____ 8. When solving a problem, I try to picture working it out in my mind first.

_____ 9. Quiet places are the best places to study math.

SECTION S.3   Study Skills

▶ OBJECTIVES

As a result of completing this section, you will be able to

1. Identify habits that good math students employ.
2. Read the textbook more effectively and efficiently.
3. Take quality notes.
4. Organize materials in a notebook or portfolio.
5. Complete homework in ways that maximize retention.
6. Establish an action plan.
7. Troubleshoot common errors.

Objective 1 ▶

Identify habits that good math students employ.

Study skills are the skills students need to improve their learning capacity and to acquire new knowledge. No two students are going to employ the same set of study techniques, though there are some that every student should utilize when studying math. This section will address some of the skills that will enhance the success of all students.

Habits of Successful Math Students

Good math students study with purpose. When you underline something, it should be because it is important. If you write out a note card, it should be because it is something you want to remember long term. When you work a homework problem, it should lead to a better understanding of the process used to solve that problem. If you can't answer "Why am I doing this?" then what you are doing may not be a productive use of your time. You should always have goals to achieve when you sit down to study and everything you do during your study time should work toward those goals. The following suggestions should help you approach your math class with purpose.

1. *Successful students are responsible.*
   ▶ Attend every class meeting. Try to be a few minutes early so that your materials are out and you are ready for class to begin.
   ▶ Find an accountability partner in class. Encourage one another and use each other as a resource if one of you is absent from class.
   ▶ Adhere to all deadlines for assignments.

2. *Successful students make the most of their time in class.*
   ▶ When you are in class, "be in class." Try not to worry about other things going on in your life. Focus your attention on the topics being discussed. (No texting, surfing, or doing other homework.)
   ▶ Either take notes in class or record the lecture so that you can refer to it later. Some instructors even post their class notes on their website.
   ▶ Actively participate in class. Follow along with the instructor. Ask questions

Each chapter also contains additional materials on success strategies, such as motivational quotes and chapter openers dedicated to a characteristic of a successful student.

CHAPTER

1

Real Numbers and Algebraic Expressions

Time Management

As you begin this chapter, we want you to focus on how you are managing your time. Your success in this course depends on your ability to devote the appropriate time to learning and practicing the material.

* The general rule of thumb is that you should spend 2 hours studying for each hour you are in class.
* Prepare a calendar of your week and schedule an appointment with yourself for study time.
* Review how you are using your time through the week and make an honest assessment of whether you have the time to make this class a priority.
* See the Preface for additional resources concerning time management.

❝Imagination is more important than knowledge. For while knowledge defines all we currently know and understand, imagination points to all we might yet discover and create.❞

— Albert Einstein (Mathematician and scientist)

Coming Up...

In Section 1.3, we will learn that the number of viewers (in millions) for the season premiere of Fox Network's *American Idol* for Seasons 1 to 8 can be approximated by the expression $0.1x^3 + 18.4x - 2.54x^2 - 4.6$, where x is the number of seasons aired. We will learn how to use this expression to estimate the number of viewers for future seasons.

1

❝The key is not to prioritize what's on your schedule, but to schedule your priorities.❞

—Stephen Covey

Resources Available in Success Strategies Manual

* Overcoming Math Anxiety Strategies
* Note Taking and Homework Strategies
* Test Preparation Worksheets
* Time Management Activities
* Goal Management

* How to Navigate a Math Textbook
* Organizing a Portfolio
* Printable Math Study Sheets
* Error Analysis Activities
* Math Term Glossaries

In addition to **Sucess Strategies** and the chapter opener materials, more resources can be found in our robust **Success Strategies Manual,** with both Student and Instructor versions available. The manual will provide additional materials geared toward success strategies, including worksheets, tips, handouts and templates.

❝This is a GREAT chapter. All math classes should start with this information!❞

—Elise Price, *Tarrant County College*

Relevant Examples and Exercises for Today's Students

Every example and exercise has a purpose! The Hendricks/Chow series offers unique types of examples and exercises to capture students' interest and help them build different skill sets as they move throughout the chapter.

▶ How Relevant are the applications in your text?

SECTION 1.2 **Fractions Review**

▶ **OBJECTIVES**

As a result of completing this section, you will be able to

1. Write the prime factorization of a number.
2. Define and write fractions.
3. Simplify fractions.
4. Multiply fractions.
5. Divide fractions.
6. Add or subtract fractions with common denominators.
7. Add or subtract fractions with unlike denominators.
8. Troubleshoot common errors.

Objective 1 ▶

Write the prime factorization of a number.

As of 2009, there were approximately 309 million people living in the United States and approximately 116 million Facebook users in the United States. Write a fraction that represents the part of the United States population that were Facebook users in 2009. (Source: http://flavorwire.com/82308/awesome-infographic-facebook-vs-the-united-states)

In this section, we will learn how to use a fraction to represent such information. We will also learn how to simplify fractions and perform operations with fractions.

Factoring Numbers

To express a number in *factored form* is to write the number as a product of two or more numbers. For example, in the statement $4 \cdot 5 = 20$, the numbers 4 and 5 are called **factors** and 20 is the *product*. Factors divide a number evenly without any remainder.

The numbers 2, 3, 5, 7, 11, 13, 17, and 19 have something in common. The factors of these numbers are only 1 and the number itself. These types of numbers are called *prime numbers*. The following prime numbers are written in factored form.

$$2 = 1 \cdot 2, \ 3 = 1 \cdot 3, \ 5 = 1 \cdot 5$$

> Each section opens with a relevant application that is then revisited in the worked examples.

93. The table shows the U.S. unemployment rates between 2001 and 2010. (Source: Bureau of Labor Statistics)

Years After 2001	0	1	2	3	4	5	6	7	8	9
Unemployment Rate (percent)	4.7	5.8	6.0	5.5	5.1	4.6	4.6	5.8	9.3	9.6

a. Write the ordered pairs (x, y) that correspond to the data in the table, where x is the years after 2001 and y is the unemployment rate.

b. Interpret the meaning of the first and last ordered pairs in the context of the problem.

c. In what year was the unemployment rate the highest? The lowest?

... scatter plot of the data.

29. The top 10 U.S. Internet search providers processed a total of approximately 9,200,000,000 search requests, during August 2010. Google processed about 6,000,000,000 and Yahoo! processed about 1,210,000,000 search requests. Write fractions that represent the portion of Google search requests

... lists the approximate salary of Peyton ... quarterback for the Indianapolis Colts, for ... 9. (Source: http://content.usatoday.com/sportsdata/ ...salaries/player/Peyton-Manning)

03	0	1	2	3	4	5	6
ns)	$11.3	$35	$0.7	$10	$11	$11.5	$14

84. The total number of individual songs purchased digitally in 2008 and 2009 was 2.231 billion. The total number of individual songs purchased digitally in 2008 was 0.089 billion less than the total number in 2009. Let x = total number of individual songs purchased digitally in 2008 (in billions) and let y = total number in 2009 (in billions). (Source: http://www.ritholtz.com)

a. Write an equation using x and y that represents the combined total number of individual songs purchased digitally in 2008 and 2009.

b. Write an equation that relates the total number of songs purchased digitally in 2008 to the total number of songs purchased digitally in 2009.

... pairs that ... ond to the ... the table, ... is years ... 03 and y is Manning's salary (in millions of dollars).

b. Interpret the meaning of first and last ordered pairs in the context of the problem.

c. In what year was his salary the highest? The lowest?

d. Make a scatter plot of the data.

> How often do you find the applications in your current text irrelevant and stale? In order to promote active learning, the authors have worked to provide current and relevant applications that are interesting to students and contain subjects that are not always in math books, such as social media, smart phones, consumer topics, and current events.

> " I like the application problems included here—very appropriate for this level student: connected to their real-world future occupations. Kudos to you!"
>
> —Vicki Schell, *Pensacola State College*

How do your current resources help your students progress to become critical thinkers?

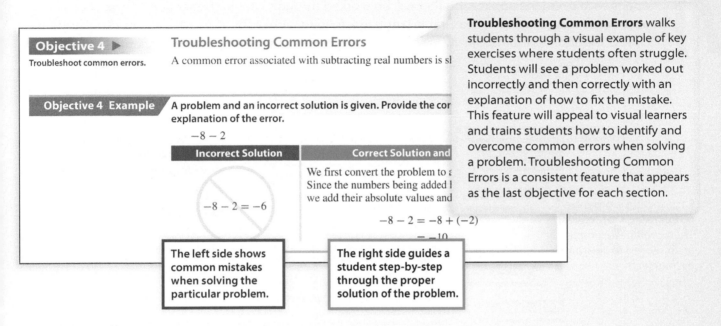

Objective 4 ▶
Troubleshoot common errors.

Troubleshooting Common Errors

A common error associated with subtracting real numbers is s[...]

Objective 4 Example
A problem and an incorrect solution is given. Provide the cor[...] explanation of the error.

$$-8 - 2$$

Incorrect Solution	**Correct Solution and**
$-8 - 2 = -6$	We first convert the problem to [...] Since the numbers being added [...] we add their absolute values and [...] $-8 - 2 = -8 + (-2)$ $= -10$

The left side shows common mistakes when solving the particular problem.

The right side guides a student step-by-step through the proper solution of the problem.

Troubleshooting Common Errors walks students through a visual example of key exercises where students often struggle. Students will see a problem worked out incorrectly and then correctly with an explanation of how to fix the mistake. This feature will appeal to visual learners and trains students how to identify and overcome common errors when solving a problem. Troubleshooting Common Errors is a consistent feature that appears as the last objective for each section.

You Be the Teacher! exercises reinforce and build critical thinking skills by teaching students to constantly reevaluate their work. In these exercises, students are asked to check a student's work and make necessary corrections to ensure the problem is answered correctly. Like Troubleshooting Common Errors, this feature emphasizes Error Analysis and trains students to take ownership and identify their own mistakes.

 **You Be the Teacher!**

Correct the student's errors, if any.

103. Determine if the following relations are functions.
  a. $\{(3, 1), (7, 5), (1, 2), (5, 2), (0, -2)\}$
  b. $\{(1, 3), (5, 7), (2, 1), (2, 5), (-2, 0)\}$
  Vivian's work: **a.** a function and **b.** not a function
104. Find $f(-3)$ if $f(x) = 12x - 5$.
  William's work:
  $f(-3) = -3(12x - 5) = -36x + 15$

❝The Troubleshooting Common Errors examples make this text stand out among other books. I use this concept in my classes now and would continue to reinforce the material with this feature. Critical thinking is so important for students and this idea helps students to think more about process rather than obtaining the 'right' answer.❞

—Carol Ann Poore, *Hinds Community College, Ramond Campus*

Chapter Walkthrough & Organization

When developing the organization and framework for this series, Andrea considered her teaching methodologies. **She organized each section with a five-step process, just as she presents her lectures: (1) lead-in, (2) objectives, (3) lesson, (4) check for understanding, and (5) summary.** This five-step process has served as the framework of each section of the textbook. Most every section opens with a real-life application that will be worked through later in the section. The section's objectives, stated in clear, measurable language, follows with many worked examples and student checks for understanding. Finally each section closes with a summary of key concepts.

SECTION 3.5 · Writing Equations of Lines

▶ **OBJECTIVES**

As a result of completing this section, you will be able to

- Write the equation of a line given its slope and y-intercept.
- Write the equation of a line given a point and a slope.
- Write the equation of a line given two points.
- Write the equation of a line given a point and a relationship to another line.
- Solve application problems.
- Troubleshoot common errors.

The average price of a movie ticket in the United States in 1996 was \$4.42. The average price of a movie ticket in the United States in 2010 was \$7.89. Write a linear equation that models the average price of a movie ticket. To answer this question, we need to know how to use two points to write the equation of a line. (Source: http://www.natoonline.org/statisticstickets.htm)

In Sections 3.1 to 3.4, we were given the equation of a line and we had to find certain information about it. In this section, we will be given specific information about a line and will write the equation of the line that satisfies the given conditions. The answers to these types of problems will be linear equations of the form $y = mx + b$ or $Ax + By = C$.

> Each section is organized by **Learning Objectives.** Then, multiple examples are provided to illustrate each objective.

Use the Slope and y-Intercept to Write an Equation of a Line

Objective 1 ▶

Write the equation of a line given its slope and y-intercept.

The first case we consider when writing equations of lines is when we are given a line's slope and y-intercept. Recall that the slope-intercept form of a line is $y = mx + b$ where m is the slope and b is the y-intercept.

So, if we know the slope and y-intercept, we will substitute these values into slope-intercept form to obtain the equation.

> **Procedure boxes** are included whenever possible to help verbalize the math steps and processes.

> **Procedure: Writing the Equation of a Line Given its Slope and y-Intercept**
>
> The line's slope is the value of m and the y-coordinate of the y-intercept is the value of b.
>
> So, the equation is $y = \_\_\_ x + \_\_\_$.

Objective 1 Examples · Write the equation of the line that satisfies the given information.

> Each objective is followed by thorough and **multiple examples,** numbered exactly as the objective, to reinforce the concepts and show the stepped out solutions.

1a. $m = 4$ and the y-intercept is $(0, 5)$ **1b.** $m = -\dfrac{1}{2}$ and the y-intercept is $(0, -4)$

1c. $m = 0$ and the y-intercept is $(0, 1)$

1d.

Solutions

1a. The slope is $m = 4$ and the value of $b = 5$. So, the equation is
$$y = 4x + 5$$

1b. The slope is $m = -\dfrac{1}{2}$ and the value of $b = -4$. So, the equation is
$$y = -\dfrac{1}{2}x - 4$$

> The examples provide a high level of **step-by-step** detail so that students do not lose track of the various steps.

After each set of examples for an objective, there is a **Student Check** feature that provides additional exercises for students to try on their own, similar to the examples shown.

☑ **Student Check 2** Write the equation of the line that has the given slope and passes through the given point. Write the answer in both slope-intercept form and standard form.

a. $m = -5; (3, 1)$ **b.** $m = -\frac{2}{7}; (4, 5)$

c. $m = 0; (-6, 9)$ **d.** undefined slope; $(8, 1)$

Note: Using method 1 or 2 produces the same linear equation. When using the slope-intercept form, we must substitute the values of m and b into $y = mx + b$ to obtain the equation. When using the point-slope form, the equation is obtained through the process.

Notes from the author appear throughout the text to give students valuable tips and information.

The end of each section contains the **Answers to the Student Checks**, along with a **Summary of Key Concepts** and most include **Graphing Calculator Skills** for those instructors who incorporate calculator use.

ANSWERS TO STUDENT CHECKS

Student Check 1 **a.** $y = -2x + 9$ **b.** $y = \frac{7}{3}x - 1$
c. $y = \frac{2}{3}$ **d.** $y = \frac{3}{2}x + 1$
Student Check 2 **a.** $y = -5x + 16$ **b.** $y = -\frac{2}{7}x + \frac{43}{7}$
c. $y = 9$ **d.** $x = 8$
Student Check 3 **a.** $y = -5x + 8$ **b.** $y = -\frac{4}{3}x + \frac{2}{3}$
c. $y = -1$ **d.** $x = -8$

Student Check 4 **a.** $y = \frac{1}{2}x - 4$ **b.** $y = -2x + 1$
c. $x = 2$ **d.** $y = -3$
Student Check 5 **a.** **i.** $y = -3000x + 30,000$
ii. \$18,000 **iii.** 7 yr
b. **i.** $y = 0.54x + 15.3$ **ii.** 21.78 million **iii.** 2024

SUMMARY OF KEY CONCEPTS

1. If the slope and y-intercept of an equation are known, writing the equation that satisfies this information is immediate. The value of m and b are substituted into the slope-intercept form $y = mx + b$.
2. There are three other situations that provide enough information for an equation of a line to be written. They are:
   - a point and a slope
   - two points
   - a point and a line parallel or perpendicular
   In each of these situations, the slope must be determined. If it is not given, use the slope formula or the relationship to a given line to find it. After the slope is found, use it

with one of the points in either the point-slope form or the slope-intercept form to write the equation of the line.
3. If a line is described as vertical or with undefined slope, the equation of the line will be of the form $x = h$, where h is the x-coordinate of the given point.
4. If a line is described as horizontal or with zero slope, the equation of the line will be of the form $y = k$, where k is the y-coordinate of the given point.
5. In application problems, we will either know the slope (how the values change) and y-intercept (initial value) from the problem or we will be given two points that enable us to write the equation.

GRAPHING CALCULATOR SKILLS

The graphing calculator has the ability to calculate the equation of the line if provided enough information. At this point, it is more beneficial to use the calculator to check our work instead of allowing it to do the work for us.

Example: Use the calculator to verify that $y = \frac{3}{2}x + 6$ is the equation of the line that goes through the points $(-4, 0)$ and $(4, 12)$.

> " I think the Hendricks/Chow series will improve the way I teach. The authors offer techniques I've thought about but wasn't sure how to implement. The resources and tips are very helpful and will enhance my teaching! "
> —Edward Ennels, *Baltimore City Community College*

Exercise Sets for Today's Whole Student

The Exercise Sets contain a wide quantity and variety of exercise types that purposefully help students persist from basic skills through differentiation of topics, to practicing critical thinking and evaluating.

Write About It! exercises help students practice their vocabulary and verbal skills.

Mix 'Em Up! is another layer of practice that mirrors what students might see on a test and also mixes together exercises from various objectives when possible.

Think About It! exercises asks students to think critically about conceptual problems.

Group Activity features provide multistep projects for groups of students to complete. This feature appears at the end of each chapter.

SECTION 3.5 / EXERCISE SET

Write About It!

Use complete sentences in your answer to each question.

1. How can you determine the equation of a line if you know its slope and y-intercept?
2. How can you determine the equation of a line if you know two points on the line?
3. Which method do you prefer to find the equation of a line—using the slope-intercept form or using the point-slope form? Why?

Practice Makes Perfect!

Write the equation of the line that satisfies the given information. (*See Objective 1.*)

9. $m = 4$, $b = 5$
10. $m = -2$, $b = 1$
11. $m = 3$, passes through $(0, -4)$
12. $m = \frac{1}{2}$, passes through $(0, 7)$

$(0, -10)$

Mix 'Em Up!

Determine if the ordered pair is a solution of the given inequality.

43. $x + 3y < 6$; $(0, 0)$
44. $x - 7y < 1$; $(10, 3)$
45. $y > \frac{1}{2}x + 8$; $(8, -5)$
46. $y < -\frac{1}{3}x - 11$; $(-9, 0)$
47. $x > 12y$; $(1, 0)$
48. $y > 6x$; $(0, -1)$
49. $y - 11 < 4$; $(-3, 25)$
50. $x + 5 \geq 1$; $(-2, 4)$
51. $6x - y \leq 1$; $(0.2, -0.5)$
52. $x + 8y > 3$; $(-1.4, 0.5)$

Graph the solution set of eac
variables.

53. $x + 3y < 2$
55. $5x - 2y > 0$
57. $x + 3 > 0$

at least $350 per week.

64. If Brasil wants to make at least $450 per week, (a) write a linear inequality in two variables that represents this situation, (b) graph the linear inequality, and (c) give three possible combinations of hours that Brasil could work at each job to at least $450 per week.

You Be the Teacher!

Correct each student's error, if any.

65. Graph $4x + y < 8$. Jamie's work: The boundary line is $4x + y = 8$ and my test point is $(0, 0)$.
$4(0) + 0 < 8$
$0 < 8$

You Be the Teacher! exercises have students correct another students' work and help students analyze common errors.

Calculate It!

109. Each table shows the points on the graph of a given line. Determine which graph contains the points $(0, 2)$ and $(-1, -3)$.

a.

b.

Think About It!

73. Give an example of a linear inequality for which the point $(0, 0)$ cannot be used as the test point to determine which half-plane to shade.
74. Is it possible for a linear inequality in two variables to have one point as a solution? Explain.
75. Is it possible for a linear inequality in two variables to have only its boundary line as a solution? Explain.
76. Graph the inequalities in parts a–d and use your results to answer parts e–g.

g. How does the graph of $y \geq mx + b$ differ from the graph of $y > mx + b$?

77. Graph the inequalities in parts a–d and use your results to answer parts e–g.

a. $y < -2x + 4$ b. $y < \frac{1}{3}x - 1$
c. $y < 4x$ d. $y < 1$
e. Which half-plane was shaded in each of these inequalities?
f. What can you conclude about the graph of the solution of $y < mx + b$?
g. How does the graph of $y \leq mx + b$ differ from the graph of $y < mx + b$?

PIECE IT TOGETHER / SECTIONS 3.1–3.3 Review

Determine algebraically if the ordered pair is a solution of the equation. (*Section 3.1, Objective 1.*)

1. $(6, 2)$; $y = |x - 4|$ 2. $(7, -4)$; $8x - 2y = 62$

Identify the quadrant or axis where the point is located. (*Section 3.1, Objective 2.*)

3. $(-8.9, -1)$ 4. $\left(\frac{7}{8}, -\frac{5}{6}\right)$
5. $(-5, -1)$ 6. $(9, 4)$

Use the given graph of an equation to determine if each ordered pair is a solution of the equation. (*Section 3.1, Objective 4.*)

7. $(0, 3)$

Graph each equation. (*Section 3.2, Objectives 2–4.*)

9. $y = 4x + 2$ 10. $y = -5x + 5$ 11. $4x - y = 0$
12. $y = -7x$ 13. $3x + 8 = 0$ 14. $2y - 11 = 0$

Use the slope formula to determine the slope of the line between each pair of points. (*Section 3.3, Objective 1.*)

15. $(-2, 3)$ and $(4, 7)$ 16. $(-2, 8)$ and $(5, 8)$

Write the equation in slope-intercept form, if necessary. State the slope and y-intercept of the line from its equation. Write the y-intercept as an ordered pair. Explain how the x- and y-values change with respect to one another. (*Section 3.3, Objective 2.*)

17. $3x - 2y = 6$

At an appropriate mid-point in the chapter, a review, **Piece It Together,** allows students to review and show mastery of the concepts presented thus far in the chapter.

GROUP ACTIVITY / Use a Linear Equation to Model Population Growth

1. Go to http://www.google.com/publicdata?ds=uspopulation and select a state or expand the state to select a specific county.
2. After you make your selection, a graph will appear in the main window. Use your cursor to trace the graph to see the population for each year. Write two ordered pairs of the form (year, population) for the state or county that you have selected.
3. Use the two ordered pairs from step 2 to write a linear equation that models the population of the selected state/county.
4. What is the slope of the model? What does it mean in the context of the problem?
5. What is the y-intercept of the model? What does it mean in the context of the problem?
6. Use the model to predict the population of your selected state/county in the year 2020.

Review Material

The Review Material has been designed purposefully for student interaction. Students are provided a list of key terms, formulas, and properties from the chapter, as well as a Chapter Summary that students can quickly complete to review the main concepts that were presented. This section ends with Chapter Review Exercises, a Chapter Test, and Cumulative Review Exercises.

What's the big idea? helps students understand the "why" behind what they learned in that particular chapter.

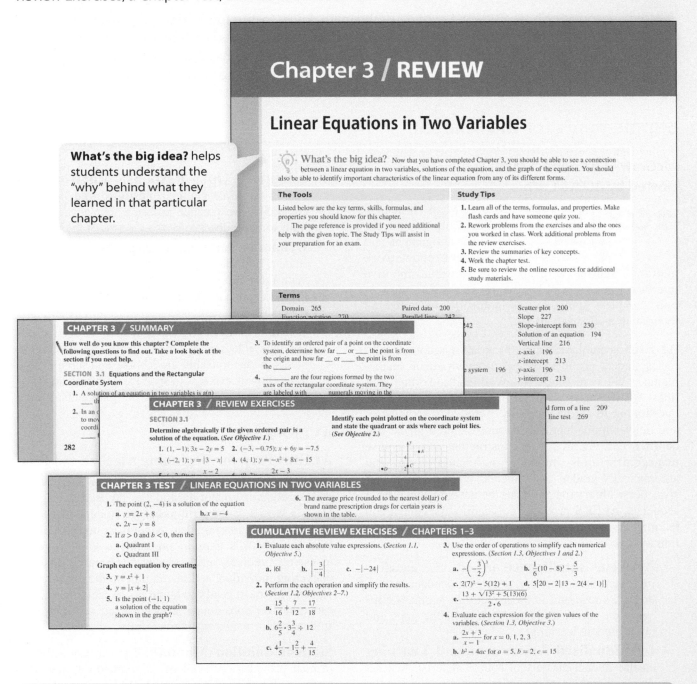

Chapter 3 / REVIEW

Linear Equations in Two Variables

What's the big idea? Now that you have completed Chapter 3, you should be able to see a connection between a linear equation in two variables, solutions of the equation, and the graph of the equation. You should also be able to identify important characteristics of the linear equation from any of its different forms.

The Tools

Listed below are the key terms, skills, formulas, and properties you should know for this chapter.

The page reference is provided if you need additional help with the given topic. The Study Tips will assist in your preparation for an exam.

Study Tips

1. Learn all of the terms, formulas, and properties. Make flash cards and have someone quiz you.
2. Rework problems from the exercises and also the ones you worked in class. Work additional problems from the review exercises.
3. Review the summaries of key concepts.
4. Work the chapter test.
5. Be sure to review the online resources for additional study materials.

Terms

Domain 265	Paired data 200	Scatter plot 200
Function notation 270	Parallel lines 242	Slope 227
		Slope-intercept form 230
		Solution of an equation 194
		Vertical line 216
		x-axis 196
	system 196	x-intercept 213
		y-axis 196
		y-intercept 213

CHAPTER 3 / SUMMARY

How well do you know this chapter? Complete the following questions to find out. Take a look back at the section if you need help.

SECTION 3.1 Equations and the Rectangular Coordinate System

1. A solution of an equation in two variables is a(n)

2. In an

282

3. To identify an ordered pair of a point on the coordinate system, determine how far ___ or ____ the point is from the origin and how far ___ or ____ the point is from the ____.

4. _____ are the four regions formed by the two axes of the rectangular coordinate system. They are labeled with ____ numerals moving in the

... form of a line 209
... line test 269

CHAPTER 3 / REVIEW EXERCISES

SECTION 3.1

Determine algebraically if the given ordered pair is a solution of the equation. (*See Objective 1.*)

1. $(1, -1)$; $3x - 2y = 5$ 2. $(-3, -0.75)$; $x + 6y = -7.5$
3. $(-2, 1)$; $y = |3 - x|$ 4. $(4, 1)$; $y = -x^2 + 8x - 15$

Identify each point plotted on the coordinate system and state the quadrant or axis where each point lies. (*See Objective 2.*)

CHAPTER 3 TEST / LINEAR EQUATIONS IN TWO VARIABLES

1. The point $(2, -4)$ is a solution of the equation
   a. $y = 2x + 8$ **b.** $x = -4$
   c. $2x - y = 8$

2. If $a > 0$ and $b < 0$, then the
   a. Quadrant I
   c. Quadrant III

Graph each equation by creating

3. $y = x^2 + 1$
4. $y = |x + 2|$
5. Is the point $(-1, 1)$ a solution of the equation shown in the graph?

6. The average price (rounded to the nearest dollar) of brand name prescription drugs for certain years is shown in the table.

CUMULATIVE REVIEW EXERCISES / CHAPTERS 1–3

1. Evaluate each absolute value expressions. (*Section 1.1, Objective 5.*)
   a. $|6|$ **b.** $\left|-\dfrac{3}{4}\right|$ **c.** $-|-24|$

2. Perform the each operation and simplify the results. (*Section 1.2, Objectives 2–7.*)
   a. $\dfrac{15}{16} + \dfrac{7}{12} - \dfrac{17}{18}$
   b. $6\dfrac{2}{5} \cdot 3\dfrac{3}{4} \div 12$
   c. $4\dfrac{1}{5} - 1\dfrac{2}{3} + \dfrac{4}{15}$

3. Use the order of operations to simplify each numerical expressions. (*Section 1.3, Objectives 1 and 2.*)
   a. $-\left(\dfrac{3}{2}\right)^3$ **b.** $\dfrac{1}{6}(10 - 8)^3 - \dfrac{5}{3}$
   c. $2(7)^2 - 5(12) + 1$ **d.** $5[20 - 2|13 - 2(4 - 1)|]$
   e. $\dfrac{13 + \sqrt{13^2 + 5(13)(6)}}{2 \cdot 6}$

4. Evaluate each expression for the given values of the variables. (*Section 1.3, Objective 3.*)
   a. $\dfrac{2x + 3}{x - 1}$ for $x = 0, 1, 2, 3$
   b. $b^2 - 4ac$ for $a = 5, b = 2, c = 15$

> **"** This is a very good book and I would like to use it to teach my courses as soon as possible! **"**
> —Zakia Ibaroudene, *Northeast Lakeview College*

Supplements

Comprehensive Resources for Every Teaching and Learning Environment

Teaching to today's whole student, beyond just the math, forces instructors to find more resources and tools to implement both inside and outside of class. As virtual classrooms continue to evolve, these types of resources become even more necessary. The Hendricks/Chow series has a number of quality, relevant supplements designed to save instructors preparation time and motivate students to learn and succeed.

Supplements for the Student

McGraw-Hill Connect Math
Hosted by ALEKS Corp.

 Connect Math Hosted by ALEKS Corp. is an exciting, new assignment and assessment ehomework platform. Starting with an easily viewable, intuitive interface, students will be able to access key information, complete homework assignments, and utilize an integrated, media-rich eBook.

ALEKS is a unique, online program that dramatically raises student proficiency and success rates in mathematics, while reducing faculty workload and office-hour lines. ALEKS uses artificial intelligence and adaptive questioning to assess precisely a student's knowledge, and deliver individualized learning tailored to the student's needs. ALEKS offers instructors robust course management tools, including automated reports and automatically-graded assignments. With a comprehensive course library that includes the developmental math sequence, ALEKS provides a dynamic assessment and learning system that can be used for a variety of instructional purposes.

- **Artificial Intelligence** and adaptive questioning determine precisely what each student knows, doesn't know, and is most ready to learn. ALEKS can then successfully target knowledge gaps and guide student learning.

- **Individualized Assessment and Learning** ensure student mastery of course material. ALEKS delivers highly individualized instruction on the exact topics each student is most **ready to learn**. The student is periodically reassessed to fill knowledge gaps and ensure long-term retention.

- **Adaptive, Open-Response Environment** avoids multiple-choice questions and includes comprehensive practice problems, explanations, and immediate feedback.

- **Dynamic, Automated Reports** track detailed student and class progress toward course mastery. With these reports, instructors can effectively direct instruction by identifying what students know and are ready to learn.

- **Robust Course Management Tools** include textbook integration, automatically-graded assignments, a customizable gradebook, and more. These tools allow instructors to spend less time on administrative tasks and more time directing student learning.

ALEKS Prep

ALEKS Prep for Beginning Algebra and **Prep for Intermediate Algebra** focus on prerequisite and introductory material, and can be used during the first six weeks of the term to ensure student success in Beginning and Intermediate Algebra courses. ALEKS Prep quickly fills gaps in prerequisite knowledge by assessing precisely each student's preparedness and delivering individualized instruction on the exact topics students are most **ready to learn**. As a result, instructors can focus on core course concepts and see improved student performance with fewer drops.

Student Solution Manual The student solution manual provides comprehensive, worked-out solutions to the odd-numbered exercises in the section exercises, review exercises, piece-it-together exercises, chapter tests, and the cumulative review. The steps shown in the solutions match the style of solved examples in the textbook.

Lecture and Exercise Videos   Online lecture and exercise videos will feature Andrea Hendricks and other instructors guiding students through the learning objectives, examples, and also exercises using the same methodology from the text. Additionally, Andrea Hendricks developed a subset of **Troubleshooting Common Errors videos** that show students how to learn from and overcome common mistakes. These videos support the last objective of each section. All videos are available online as part of Connect Math Hosted by ALEKS Corp. or within ALEKS 360. Other supplemental videos include eProfessor videos, which are animations based on examples in the book. The videos are closed-captioned for the hearing impaired, and meet the Americans with Disabilities Act Standards for Accessible Design.

Student Success Strategies Manual

To support the Chapter S materials in the text, Kelly Jackson from Camden County College, has developed a practical manual of activities, worksheets, tips, and strategies to support Chapter S (*Success Strategies*). This manual is available online as a resource for Connect Math Hosted by ALEKS Corp.

Guided Student Workbook

Developmental Math students often struggle with taking quality notes while in class or listening to a lecture. To support the Hendricks & Chow text, this guided workbook provides a template of a lecture for the students to fill-in the important topics, terms, and procedures so that they spend less time creating notes and more time engaged in class. Also, in addition to the notes sections, students can then practice with additional student check exercises, additional problems similar to those in the text, and extra You Be the Teacher! problems. The Guided Student Workbook is an excellent companion that instructors and students can use to support the main text. This workbook is fully editable and available online as part of Connect Math Hosted by ALEKS Corp. or also available for custom packages. Students can download and print as needed.

Supplements for the Instructor

McGraw-Hill Connect® Math
Hosted by ALEKS Corp.

Connect Math Hosted by ALEKS Corp. is an exciting, new assignment and assessment ehomework platform. Instructors can assign an AI-driven ALEKS Assessment to identify the strengths and weaknesses of each student at the beginning of the term rather than after the first exam. Assignment creation and navigation is efficient and intuitive. The grade, based on instructor feedback, has a straightforward design and allows flexibility to import and export additional grades.

Instructor Success Strategies Manual   This manual is an excellent resource of additional materials for full- and part-time instructors to incorporate into their courses. This robust manual includes teaching strategies, extra classroom activities, concept reviews, and activities that support the materials in Chapter S to help teach study skills including note taking, test preparation, and studying, etc. This manual is downloadable at Connect Math Hosted by ALEKS Corp.

Annotated Instructor's Edition   In the Annotated Instructor's Edition (AIE), answers to exercises, review, and tests appear adjacent to each exercise set, in a color used only for annotations. Instructors will also find helpful hints and notes within the margins to consider while teaching.

Instructor's Solution Manual   The instructor's solution manual provides comprehensive, worked-out solutions to all exercises in the section exercises, review exercises, piece-it-together exercises, chapter tests, and the cumulative review. The steps shown in the solutions match the style and methodology of solved examples in the textbook.

Instructor's Testing and Resource Online   Among the supplements is a computerized test bank utilizing Brownstone Diploma algorithm-based testing software to create customized exams quickly. This user-friendly program enables instructors to search for questions by topic, format, or difficulty level; to edit existing questions, or to add new ones; and to scramble questions and answer keys for multiple versions of a single test. Hundreds of text-specific, open-ended, and multiple-choice questions are included in the question bank. Sample chapter tests are also provided. CDs are available upon request.

ALEKS 360: A Total Course Solution

With *eBook* Integration

Connect
Learn
Succeed™

A cost-effective total course solution: fully integrated, interactive eBook combined with ALEKS individualized assessment and learning.

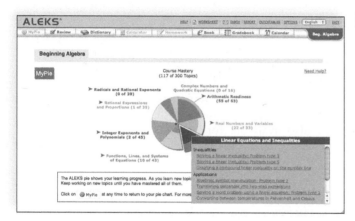

Individualized Learning

- The ALEKS Pie summarizes a student's current knowledge and provides individualized learning on the exact topics the student is **ready to learn**

- Artificial intelligence successfully targets gaps by assessing precisely a student's knowledge and periodically reassessing for long-term retention

- Adaptive, open-response environment avoids multiple-choice and includes problems, explanations, and realistic answer input tools

Interactive eBook

- eBook access provides worked examples, videos, and additional support

- Robust virtual features include highlighting, bookmarking, and note-taking capabilities

- Students can easily access the eBook, multimedia resources, and their notes from within their ALEKS Student Accounts

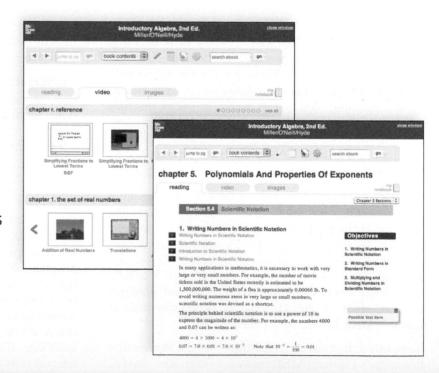

Learn More: www.aleks.com/highered/math/aleks360

ALEKS is a registered trademark of ALEKS Corporation.

ALEKS Course Management Tools

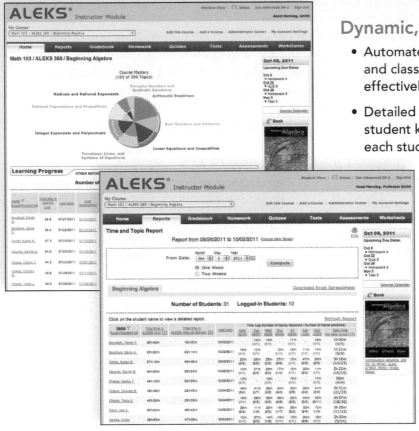

Dynamic, Automated Reporting

- Automated reports dynamically track student and class learning progress so instructors can effectively direct classroom instruction

- Detailed reports identify precisely what each student knows, and more importantly, what each student is ready to learn next

- Time and Topic Report offers up-to-the-minute daily progress, including time logged, topics attempted, and topics mastered

Course Control and Customization

- Align ALEKS topics with a textbook or course syllabus

- Create and customize course objectives and modules

- Set due dates for course objectives to pace student progress

- Assign automatically-graded homework, quizzes, and tests

- Seamlessly track and adjust student scores with the customizable gradebook

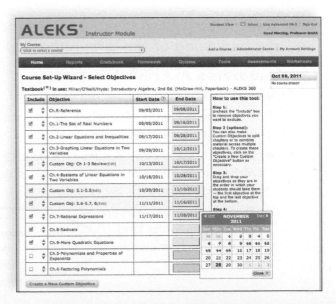

McGraw Hill connect®
|MATH

Hosted by **ALEKS Corp.**

Connect Math Hosted by ALEKS Corporation is an exciting, new ehomework platform combining the strengths of McGraw-Hill Higher Education and ALEKS Corporation. Connect Math Hosted by ALEKS Corporation is the first platform on the market to combine an artificially-intelligent, diagnostic assessment with an intuitive ehomework platform designed to meet your needs.

Connect Math Hosted by ALEKS Corporation is the culmination of a one-of-a-kind market development process involving full-time and adjunct Math faculty at every step of the process. This process enables us to provide you with a solution that best meets your needs.

Connect Math Hosted by ALEKS Corporation is built by Math educators for Math educators!

1 *Your students want a well-organized homepage where key information is easily viewable.*

Modern Student Homepage

▶ This homepage provides a dashboard for students to immediately view their assignments, grades, and announcements for their course. (Assignments include HW, quizzes, and tests.)

▶ Students can access their assignments through the course Calendar to stay up-to-date and organized for their class.

Modern, intuitive, and simple interface.

2 *You want a way to identify the strengths and weaknesses of your class at the beginning of the term rather than after the first exam.*

Integrated ALEKS® Assessment

▶ This artificially-intelligent (AI), diagnostic assessment identifies precisely what a student knows and is ready to learn next.

▶ Detailed assessment reports provide instructors with specific information about where students are struggling most.

▶ This AI-driven assessment is the only one of its kind in an online homework platform.

Recommended to be used as the first assignment in any course.

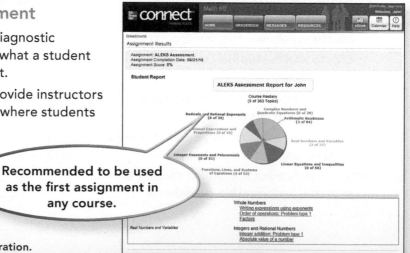

ALEKS is a registered trademark of ALEKS Corporation.

Resources for Online Homework

3 *Your students want an assignment page that is easy to use and includes lots of extra help resources.*

Efficient Assignment Navigation

▶ Students have access to immediate feedback and help while working through assignments.

▶ Students have direct access to a media-rich eBook for easy referencing.

▶ Students can view detailed, step-by-step solutions written by instructors who teach the course, providing a unique solution to each and every exercise.

Students can easily monitor and track their progress on a given assignment.

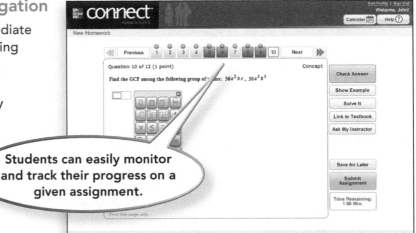

4 *You want a more intuitive and efficient assignment creation process because of your busy schedule.*

Assignment Creation Process

▶ Instructors can select textbook-specific questions organized by chapter, section, and objective.

▶ Drag-and-drop functionality makes creating an assignment quick and easy.

▶ Instructors can preview their assignments for efficient editing.

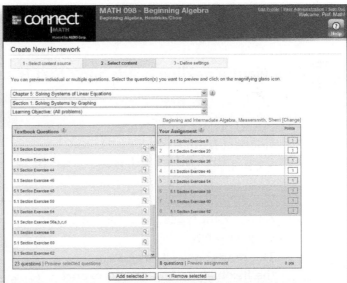

Connect
Learn
Succeed™

www.connectmath.com

5 Your students want an interactive eBook with rich functionality integrated into the product.

Hosted by **ALEKS Corp.**

Integrated Media-Rich eBook

▶ A Web-optimized eBook is seamlessly integrated within ConnectPlus Math Hosted by ALEKS Corp. for ease of use.

▶ Students can access videos, images, and other media in context within each chapter or subject area to enhance their learning experience.

▶ Students can highlight, take notes, or even access shared instructor highlights/notes to learn the course material.

▶ The integrated eBook provides students with a cost-saving alternative to traditional textbooks.

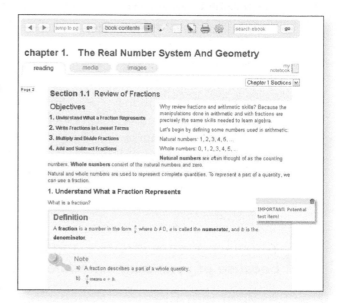

6 You want a flexible gradebook that is easy to use.

Flexible Instructor Gradebook

▶ Based on instructor feedback, Connect Math Hosted by ALEKS Corp.'s straightforward design creates an intuitive, visually pleasing grade management environment.

▶ Assignment types are color-coded for easy viewing.

▶ The gradebook allows instructors the flexibility to import and export additional grades.

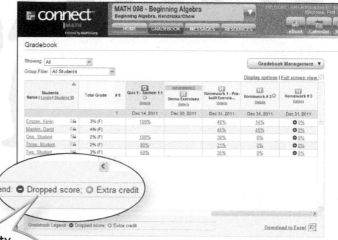

Instructors have the ability to drop grades as well as assign extra credit.

Built by Math Educators for Math Educators

You want algorithmic content that was developed by math faculty to ensure the content is pedagogically sound and accurate.

Digital Content Development Story

As the usage of online homework progresses and evolves, McGraw-Hill understands the need to have author involvement and author approval of the digital content to ensure that what students see in the online homework system is consistent with what they see in their textbooks. For this new developmental math series, co-author Pauline Chow has not only been closely involved with writing exercises for the text but also has overseen and led the creation of the digital content to ensure a seamless transition from print to digital offerings.

The development of McGraw-Hill's Connect Math Hosted by ALEKS Corporation content involved collaboration between McGraw-Hill, our authors, experienced instructors, and ALEKS Corporation, a company known for its high-quality digital content. The result of this process, outlined below, is accurate content created with your students in mind. It is available in a simple-to-use interface with all the functionality tools needed to manage your course.

1. McGraw-Hill partnered with author Pauline Chow to lead and oversee the digital content development.
2. Pauline Chow selected the textbook exercises to be included in the algorithmic content to ensure appropriate coverage of the textbook content.
3. McGraw-Hill auditioned and selected experienced instructors to work as digital contributors and represent the author's voice.
4. These digital contributors created detailed solutions for use in the Guided Solution and Solve It features, matching the voice of authors Andrea Hendricks and Pauline Chow.
5. Pauline and the digital contributors provided detailed instructions for authoring the algorithm specific to each exercise to maintain the original intent and integrity of each unique exercise.
6. Each algorithm was reviewed by Pauline and the contributors, then went through a detailed quality control process by ALEKS Corporation before being copyedited and posted live.

Solutions in Connect Math Hosted by ALEKS Corp. match the procedure and language of the text.

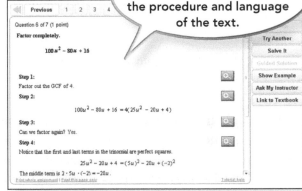

RESULT = Truly Vetted, Consistent Digital Content That Is Approved by the Authors and Supported by ALEKS Corporation.

Author and Lead Digital Contributor, O. Pauline Chow, *Harrisburg Area Community College*

Lead Digital Contributor, Amy Naughten

Digital Contributors
Dihema Ferguson, *Georgia Perimeter College*
Marianne Rosato, *Massasoit Community College*
Chris Yarrish, *Harrisburg Area Community College*
Allison Williams, *Georgia Perimeter College*
Eric Bennett, *Lansing Community College*
Katy Cryer
Christi Verity

McGraw Hill — *Connect Learn Succeed™*

Market Development

Our Commitment to Market Development and Accuracy

McGraw-Hill's Development Process is an ongoing, never-ending, market-oriented approach to building accurate and innovative print and digital products. We begin developing a series by partnering with authors that desire to make an impact within their discipline to help students succeed. Next, we share these ideas and manuscript with instructors for review for feedback and to ensure that the authors' ideas represent the needs within that discipline. Throughout multiple drafts, we help our authors adapt to incorporate ideas and suggestions from reviewers to ensure that the series carries the same pulse as today's classrooms. With any new series, we commit to accuracy across the series and its supplements. In addition to involving instructors as we develop our content, we also utilize accuracy checks through our various stages of development and production. The following is a summary of our commitment to market development and accuracy:

1. 3 drafts of author manuscript
2. 5 rounds of manuscript review
3. 2 focus groups
4. 1 consultative, expert review
5. 3 accuracy checks
6. 3 rounds of proofreading and copyediting
7. Towards the final stages of production, we are able to incorporate additional rounds of quality assurance from instructors as they help contribute towards our digital content and print supplements

This process then will start again immediately upon publication in anticipation of the next edition. With our commitment to this process, we are confident that our series has the most developed content the industry has to offer, thus pushing our desire for quality and accurate content that meets the needs of today's students and instructors.

Acknowledgements

Paramount to the development of *Beginning and Intermediate Algebra* was the invaluable feedback provided by the instructors from around the country that reviewed the manuscript or attended a market development event over the course of the several years the text was in development.

A Special Thanks To All of The Event Attendees Who Helped Shape Beginning and Intermediate Algebra.

Focus groups and symposia were conducted with instructors from around the country to provide feedback to editors and the authors and ensure the direction of the text was meeting the needs of students and instructors.

Mihaela Blanariu, *Columbia College Chicago*
Eddie Ennels, *Baltimore City Community College*
Dihema Ferguson, *Georgia Perimeter College–Decatur Campus*
Stephanie Fernandez, *Lewis and Clark College*
Cathy Hoffmaster, *Thomas Nelson Community College*
Joe Howe, *St. Charles Community College*
Kelly Jackson, *Camden County College*
Jason King, *Moraine Valley Community College*
Rob King, *Harrisburg Area Community College*
Michael Kirby, *Tidewater Community College*
Viktoriya Lanier, *Middle Georgia College*
Cindy Light, *Indiana University–Southeast*
Catherine Moushon, *Elgin Community College*
Sandi Nieto, *Santa Rosa Junior College*
Toni Parise, *Southern Maine Community College*
Mari Peddycoart, *Lone Star College–Kingwood*

David Price, *Tarrant County College*
Elise Price, *Tarrant County College*
Amber Rust, *University of Maryland*
Mark Schwartz, *Southern Maine Community College*
Andrew Stephan, *St. Charles Community College*
Brad Stetson, *Schoolcraft College*
Richard Watkins, *Tidewater Community College*
Karen Watson, *Cypress College*
Carol White, *Tarrant County College–Southeast Campus*
Joanna Wilson, *Georgia Perimeter College*

Manuscript Review Panels

Over 200 instructors reviewed the various drafts of manuscript to give feedback on content, design, pedagogy, and organization. Their reviews were used to guide the direction of the text.

Ricki Alexander, *Harrisburg Area Community College*
Marie Aratari, *Oakland Community College*
Dr. Eric Aurand, *Mohave Community College*
Chris Barker, *San Joaquin Delta College*
Scott Barnett, *Henry Ford Community College*
Disa Beaty, *Rose State College*
David Behrman, *Somerset Community College*
Sandra Belcher, *Midwestern State University*
Monika Bender, *Central Texas College*
Francois Bereaud, *Miramar College*
Teresa Betkowski, *Gordon College*
John Beyers, *University of Maryland*
Katrina Bishop, *Craven Community College*
Tammy Bishop, *Wayne Community College*
Bret Black, *Oxnard College*
Gregory Bloxom, *Pensacola State College*
Stacey Boggs, *Allegany College of Maryland*
Karen Bond, *Pearl River Community College*
Anthony Bottone, *Arizona Western College*
Cynthia Box, *Georgia Perimeter College*
Susan P. Bradley, *Angelina College*
Chandra Breaux, *Georgia Perimeter College*
Carmen Buhler, *Minneapolis Community & Technical College*
Kirby Bunas, *Santa Rosa Junior College*
Julia Burch, *Central Michigan University*
Rebecca Burkala, *Rose State College*
David Busekist, *Southeastern Louisiana University*
Susan Byars, *Gordon College*
Nick Bykov, *Delta College*
Lynn Cade, *Pensacola State College*
Kimberley Caldwell, *Volunteer State Community College*
Edythe Carter, *Amarillo College*
Yungchen Cheng, *Missouri State University*
Kim Clark, *Wayne Community College*
Adam Cloutier, *Henry Ford Community College*
Delaine Cochran, *Indiana University Southeast*
David Cooper, *Wake Technical Community College*
Wendy Davidson, *Georgia Perimeter College*
Carlos de la Lama, *San Diego City College*
Marlene Dean, *Oxnard College*
Robert Diaz, *Fullerton College*
Paul Diehl, *Indiana University Southeast*
David Dillard, *Patrick Henry Community College*
Michael Dubrowsky, *Wayne Community College*
Scott M. Dunn, *Central Michigan University*
John Edwards, *University of Oklahoma*
Cheryl Eichenseer, *St. Charles Community College*
Mark Ellis, *Central Piedmont Community College*
Edward Ennels, *Baltimore City Community College*
Marcos Enriquez, *Moorpark College*
Paul Farnham, *Fullerton College*
Dale Felkins, *Arkansas Tech University*
Dihema Ferguson, *Georgia Perimeter College*
Jacqui Fields, *Wake Technical Community College*

Elise Fischer, *Johnson County Community College*
Rhoderick Fleming, *Wake Technical Community College*
Cynthia Fletcher, *Pulaski Technical College*
Donna Flint, *South Dakota State University*
Cindy Fowler, *Central Piedmont Community College*
Dorothy French, *Community College of Philadelphia*
Dick Frost, *Rose State College*
John Fulk, *Georgia Perimeter College*
Jenine Galka, *Moraine Valley Community College*
Angela Gallant, *Inver Hills Community College*
Sunshine Gibbons, *Southeast Missouri State University*
Sharon L. Giles, *Grossmont Community College*
Suzette Goss, *Lone Star College*
Kathleen Grigsby, *Moraine Valley Community College*
Kathryn Gunderson, *Three Rivers Community College*
Jin Ha, *Northeast Lakeview College*
Shawna Haider, *Salt Lake Community College*
Mark Harbison, *Sacramento City College*
Cynthia Harrison, *Baton Rouge Community College*
Jennifer Hastings, *Northeast Mississippi Community College*
Dr. Annette Hawkins, *Wayne Community College*
Alan Hayashi, *Oxnard College*
Kristy Hill, *Hinds Community College*
Kayana Hoagland, *South Puget Sound Community College*
Irene Hollman, *Southwestern College*
Teresa Houston, *East Mississippi Community College*
Heidi Howard, *Florida State College at Jacksonville*
Steven Howard, *Rose State College*
Joe Howe, *St. Charles Community College*
Susan Howell, *University of Southern Mississippi*
Denise Hum, *Canada College*
Zakia Ibaroudene, *Northeast Lakeview College*
Sally Jackman, *Madisonville Community College*
Marilyn Jacobi, *Gateway Community Technical College*
Yvette Janecek, *Blinn College*
Tina Johnson, *Midwestern State University*
Nancy Johnson, *State College of Florida*
Linda Jones, *Vincennes University*
Diane Joyner, *Wayne Community College*
Dynechia Jones, *Baton Rouge Community College*
Paul Jones, *University of Cincinnati*
Diane Joyner, *Wayne Community College*
Laura Kalbaugh, *Wake Technical Community College*
Krystyna Karminska, *Thomas Nelson Community College*
Susan Kautz, *Lone Star College*
Edward Kavanaugh, *Schoolcraft College*
Pallavi Ketkar, *Arkansas Tech University*
Kerin Keys, *City College of San Francisco*
Rob King, *Harrisburg Area Community College*
Jason King, *Moraine Valley Community College*
Jeff Koleno, *Lorain County Community College*
Lee Koratich, *Hillsborough Community College*
Jacek Kostyrko, *San Joaquin Delta College*
Eugene Kramer, *University of Cincinnati; Raymond Walters College*

Stan Kubicek, *Blinn College*
Jan La Turno, *Rio Hondo College*
Jason Lachowicz, *Patrick Henry Community College*
Marsha Lake, *Brevard Community College*
Debra Landre, *San Joaquin Delta College*
Carol Lanfear, *Central Michigan University*
Betty J. Larson, *South Dakota State University*
Lonnie Larson, *Sacramento City College*
Sungwook Lee, *University of Southern Mississippi*
Cynthia Light, *Indiana University*
Lisa Lindloff, *McLennan Community College*
Barbara Little, *Central Texas College*
Wanda J. Long, *St. Charles Community College*
Francine Long, *Edgecombe Community College*
Mike Long, *Shippensburg University of Pennsylvania*
Yixia Lu, *South Suburban College*
Joanne Mann, *Lone Star College*
Amy Marolt, *Northeast Mississippi Community College*
Dorothy S. Marshall, *Edison College*
Abbas Masum, *Houston Community College & Alvin
    Community College*
Barabara Maurice, *Three Rivers Community College*
Julie Mays, *Angelina College*
Carlea Mcavoy, *South Puget Sound Community College*
Toni McCall, *Angelina College*
Roger McCoach, *County College of Morris*
Michael McComas, *Marshall Community & Technical College*
Mikal McDowell, *Cedar Valley College*
Leslie Mcginnis, *Blinn College*
Bridget Middleton, *Santa Fe Community College*
Edward Migliore, *University of California*
Shahnaz Milani, *Blinn College*
Bronte Miller, *Patrick Henry Community College*
Phillip Miller, *Indiana University Southeast*
Jon David Miller, *Lone Star College*
Dennis Monbrod, *South Suburban College*
Roya Namavar, *Rogers State University*
Martha Nega, *Georgia Perimeter College*
Cao Nguyen, *Central Piedmont Community College*
Kevin Olwell, *San Joaquin Delta College*
Priti Patel, *Tarrant County College*
Curtis Paul, *Moorpark College*
Mari Peddycoart, *Lone Star College*
Karen Pender, *Chaffey College*
Vic Perera, *Kent State University*
Kathy Perino, *Foothill College*
Michele Poast, *Dixie State College of Utah*
Carol Ann Poore, *Hinds Community College*
Carol Poore, *Hinds Community College–Raymond*
Richard Porter, *Mercer County Community College*
David Price, *Tarrant County College*
Elise Price, *Tarrant County College*
Jill Rafael, *Sierra College*
Cynthia Reed, *Moorpark College*
Pamelyn Reed, *Lone Star College*
Lynn Rickabaugh, *Aiken Technical College*
Christopher Riola, *Moraine Valley Community College*
Dianne Robinson, *Ivy Tech Community College*
Cosmin Roman, *The Ohio State University*

Jody Rooney, *Jackson Community College*
Elaine Russel, *Angelina College*
Amber Rust, *University of Maryland*
Kristina Sampson, *Lone Star College*
Robyn Scalzo, *Baton Rouge Community College*
Vicki Schell, *Pensacola State College*
Laura Schoppmann, *Seton Hall University*
Mark Schwartz, *Southern Maine Community College*
Daniel Seaton, *University of Maryland Eastern Shore*
Jerry Shawyer, *Florida State College at Jacksonville*
Jenny Shotwell, *Central Texas College*
Jean Shutters, *Harrisburg Area Community College*
Craig Slocum, *Moraine Valley Community College*
Jennifer Smeal, *Wake Technical Community College*
Donna Spikes, *Minneapolis Community & Technical College*
Brad Stetson, *Schoolcraft College*
Mark Stigge, *Baton Rouge Community College*
Daniela Stoevska-Kojouharov, *Tarrant County College*
Panyada Sullivan, *Yakima Valley Community College*
Sharon L. Sweet, *Brevard Community College*
Marcia Swope, *Santa Fe Community College*
Nader Taha, *Kent State University*
M. Kaye Tanner, *Linn Benton Community College*
Linda Tansil, *Southeast Missouri State University*
Carolyn Thomas, *San Diego City College*
Mary Thurow, *Minneapolis Community & Technical College*
Tejan Tingling, *Community College Baltimore County–Essex*
Robyn Toman, *Anne Arundel Community College*
Lee Topham, *Lone Star College–Kingwood*
Scott Travis, *Lone Star College*
Cameron Troxell, *Mt San Antonio College*
Barbara Jo Tucker, *Tarrant County College*
Laura Tucker, *Central Piedmont Community College*
Chris Turner, *Pensacola State College*
Jewell Valrie, *Wake Technical Community College*
Terry R. Varvil Jr., *Hillsborough Community College*
Mansoor Vejdani, *University of Cincinnati*
Mildred Vernia, *Indiana University Southeast*
Carol Walker, *Hinds Community College*
Jimmy Walker, *Hill College*
Sherry Wallin, *Sierra College*
Jane Wampler, *Housatonic Community College*
Michelle Watts, *Lone Star College, Tomball*
Pamela Webster, *Texas A & M University Commerce*
Gail Whitaker, *Wharton County Junior College*
Robert White, *Allan Hancock College*
Ernest Williams, *Calhoun Community College*
Suzanne Williams, *Central Piedmont Community College*
Nathan Wilson, *Saint Louis Community College*
Olga Cynthia Wilson Harrison, *Baton Rouge Community College*
Jackie Wing, *Angelina College*
Rick Woodmansee, *Sacramento City College*
Grethe Wygant, *Moorpark College*
Tzu-Yi Alan Yang, *Columbus State Community College*
Mina Yavari, *Allan Hancock College*
Patty Zachary, *Lone Star College*
Carol Zavarella, *Hillsborough Community College*
Loris Zucca, *Lone Star College*

Acknowledgments

This first edition would not have been possible without the encouragement and support of our families. There really are not words that convey our thanks and appreciation for their understanding as we worked seemingly night and day on the manuscript these last few years. We love you Todd, Andy, Charlie, and Cory Hendricks and Michael, Amy, and Andrew Ko.

Just as it takes a village to raise children, it takes a village to write a book. There are some important members of this village we would like to personally thank. We first extend our thanks to our dear friend and first sponsoring editor at McGraw-Hill, David Millage, who had the foresight to bring us together for this exciting journey. We also thank our Sponsoring Editor, Mary Ellen Rahn, for her support in this project. The resources, guidance, and insight you have provided through this project have been invaluable. We extend our gratitude to our Developmental Editors, Adam Fischer and most recently Emily Williams, for keeping us on task and coordinating all of the helpful feedback we have received through this process. We also truly appreciate the many hours that the production team has spent with us to make this series come together. We thank Vicki Krug for teaching us the world of production and we thank Laurie Janssen for the beautiful, current and creative design. Many thanks also goes to Emilie Berglund and Nicole Lloyd, Directors of Digital Content for overseeing the process of making the book come alive through Connect Math Hosted by ALEKS. Thanks for your countless hours and dedication to this book. We would like to thank every other member of the McGraw-Hill team for their excitement and enthusiasm in this project.

Some of the key contributors that we would like to thank are Calandra Davis, Kelly Jackson, Andrew Stephan, Joe Howe, Lisa Collette, Pat Steele, and Bea Sussman. Thanks, Calandra, for being there and offering your support to get this project going. Thanks, Kelly, for your wonderful insight and enhancements for Chapter S. You really helped our vision for this chapter become a reality. Thanks, Andy and Joe, for your contribution of Chapter 12 for Intermediate Algebra. Your work provided a great addition to the text. Thanks, Lisa, for your priceless input and suggestions for the final manuscript of this text. Your detailed comments were very helpful. Thanks, Pat and Bea, for your keen eyes and thorough review of the final manuscript. Your suggestions have provided the polishing touch to the book for which we are most appreciative.

We also thank all of the supplements and digital content contributors that helped complete this series. Thanks, Emily Whaley, for your dedicated and detailed efforts on the Solutions Manuals. Also, we must thank all of the several instructors that have reviewed this series throughout various stages of manuscript. We appreciate your guidance, feedback and support to help this series come to life. We look forward to you seeing the completed project to see how your suggestions have shaped this series.

Lastly, we extend our thanks to all of the students we have taught in our collective 45 years of teaching for making us the teachers we are today.

Application Index

Strategies to Succeed in Math

S

Success

It is with great delight that we welcome you to this course and the materials we have provided you. You are embarking on perhaps the most rewarding experience of your life. College is a series of challenges that will prepare you for a lifetime of learning and accomplishments. At the end of your college education, you will be awarded a degree—a degree that establishes your ability to learn and to persevere. It is a statement that you can work to accomplish a goal, no matter what the obstacles.

At this point, you must evaluate the reason you are here. Are you here to fulfill your dreams or the dreams of someone else? To successfully complete this journey, you must be here to fulfill your own desires, not those of a friend or family member. It will be very difficult to withstand the trials of college if you do not have a personal desire to see it through.

A semester or quarter can be very overwhelming. Focus on one day at a time and not on everything that you must learn throughout the entire course. Before you know it, the course will be over. We know that it is very easy to get distracted from your goals. Stay committed and motivated by remembering your ultimate reason for attending school.

We wish you success in this course and in your future educational endeavors. It is our hope that you are successful, not only this semester or quarter, but every term until your ultimate goal is achieved. This course will pave the way for that success. It is the door to achieving your dreams.

Mrs. Andrea Hendricks and Mrs. Pauline Chow

Andrea M. Hendricks   *Pauline Chow*

Question for Thought: What do you dream about doing? How does attending college make that dream possible? How does this course help you meet your goals? What is your biggest obstacle in being successful in this course? What can you do to overcome that obstacle? What is your plan for succeeding in this course?

Chapter S will introduce some important strategies that can enhance your performance in this course.

Chapter Outline

"Continuous effort—not strength or intelligence—is the key to unlocking our potential."

Winston Churchill

SECTION S.1 — Time Management and Goal Setting

▶ OBJECTIVES

As a result of completing this section, you will be able to

1. Set realistic grade and attendance goals.
2. Establish a schedule for studying math.
3. Establish an action plan.
4. Troubleshoot common errors.

A student might have great study skills but without good utilization of time, those skills are a wasted treasure. Time is like money; we should know how every moment is spent so that we do not waste it, and we should spend it wisely. The great news is that everyone has exactly the same amount of time. The bad news is that once the time has passed, it can never be recaptured. In this section, we will learn some helpful ways to manage our time and to set appropriate goals that make the best use of our time.

Setting Realistic Goals

Objective 1 ▶

Set realistic grade and attendance goals.

As we begin a new course, we have a "clean slate." No grades are in the grade book, no absences have been recorded, and no assignments are late or outstanding. We can control our behavior in such a way that we maximize our chance of success in this course. The beginning of a semester is an optimal time to set realistic goals that we can achieve by the end of the semester. Setting goals provides us with a sure focus and clarifies the direction in which we are headed. Activity 1 will provide us a chance to think about the goals we have for this course and for college in general.

ACTIVITY 1   Answer each question.

1. What was the last math course you took? When did you take it? Was it a successful experience? Why or why not? _____
   _____

2. What grade would you like to earn in this course? How many hours a week do you think you will need to devote outside of class to meet this goal?
   _____

3. How many absences do you think could put your goal at risk? What other activities could put your goal at risk?_____
   _____

4. What is your short-term goal? What do you hope to accomplish this semester?
   _____

5. What is your long-term goal? Why are you in college?_____
   _____

6. What are some things you can do that will enable you to reach your goals?
   _____

7. Are there any other activities that you need to avoid or limit that would prevent you from reaching your goals?_____
   _____

Note: *Keep in mind that being late or leaving class early can be just as harmful as missing class completely. Attending class regularly, punctually, and for the entire time is a goal that will lead us to success in math.*

Plan Study Time

Before we can establish study time, we need to know how we currently use and manage our time. One way to do this is by recording our activities in a calendar or planner. The key is to record all of our activities, including time for driving, sleeping, and eating, as shown in Figure S.1. When we do not take the time to record our activities, we are more apt to waste time and to spend time on things that are not aligned with our goals. Once we understand and realize how we use our time, we can be more deliberate in planning a schedule that is beneficial to our goals.

Figure S.1

Time	Monday, 9/22
7:00	Shower/get dressed
7:30	Breakfast
8:00	Leave for school
8:30	
9:00	Math 1001 (9–10:15)
9:30	
10:00	
10:30	Engl 1101 (10:30–11:30)
11:00	
11:30	
12:00	Lunch
12:30	
1:00	Drive to work
1:30	Work (1:30–6)
2:00	
2:30	
3:00	
3:30	
4:00	
4:30	
5:00	
5:30	
6:00	Drive home
6:30	Dinner
7:00	Watch TV
7:30	
8:00	Study
8:30	
9:00	Watch TV
9:30	
10:00	Exercise
10:30	
11:00	Go to bed

Following are some suggested guidelines for planning study time.

1. Make a daily appointment to study for this class.
   a. A general guideline is to spend 2 hr studying for each hour in class. For example, a 3-credit-hour class will require an average of 6 hr of study time each week.

   b. Devote some time each day to studying math. Do not cram the suggested hours into one day.

   c. The best time to study is the hour immediately following class time. If this is not doable, study time should be as soon after class as possible.

2. Multitask when feasible.

   a. When waiting in line or waiting for an appointment, review your class notes.

   b. When riding a bus, listen to a lecture or review note cards.

3. Prioritize tasks. Differentiate between the things that must be done and the things that you just want to do.

4. Maintain a balanced schedule by not overcommitting your time. Say no when you don't have the time to devote to a task.

5. Make a habit of writing things down in a calendar or a to-do list.

Note: *Doing something right the first time takes less time in the long run than doing it poorly and having to redo it later.*

ACTIVITY 2   Answer each question.

1. Without using a calendar or planner, estimate how much time you spend sleeping, eating, watching TV, playing on the computer, working, socializing with friends, sitting in classes, studying, taking care of children, running errands, traveling from one place to another, exercising, and so on.

   a. What did you discover about yourself and how you spend your time? _____

   _____

   b. Is there anything that you want to spend more time doing? _____

   _____

   c. Is there anything you want to spend less time doing?_____

   _____

2. Create a planned time chart for next week using one similar to Figure S.1.

   a. First record large blocks of time commitments (class time, work time, driving time, church activities, and so on)._____

   _____

   b. Schedule necessary activities like sleeping and eating._____

   _____

   c. Schedule regular activities such as grocery shopping, going to the gym, cleaning house, putting kids to bed, and the like._____

   _____

   d. Based on suggested guidelines, you should study 2 hr for each hour you are in class. If you are enrolled in 12-credit hours, you should have 24 hr of study time. Distribute this time over the week in reasonable blocks of time. Do not set yourself up for cram sessions. _____

   _____

   e. Schedule yourself some fun time and time to decompress._____

   _____

   f. Allow flexibility in your schedule. Things will come up, so you shouldn't have every moment scheduled._____

   _____

3. Are there things in your life that need to be removed to improve your chances of success?_____

   _____

Action Plan

Objective 3 ▶

Establish an action plan.

1. Utilize the print and digital resources provided to you that accompany your textbook. Work with your instructor to access the available Success Strategies Manual. You will find things such as
   - a time tracker.
   - a tip sheet for getting to know your teacher.
   - a shopping list for math success that includes a list of supplies you might need.
   - positive affirmations that you can recite to help motivate you.
   - other resources to get organized and to help you have a productive semester.

2. Set some specific goals for this course.
   - I will get a grade of _____ in this course.
   - Three things I can do to ensure that I meet this goal are
     1. _____
     2. _____
     3. _____
   - I plan to spend _____ hours per week outside of class for this course.
   - Three things I can do to be sure that I have enough time to devote to math are
     1. _____
     2. _____
     3. _____
   - At most I will miss _____ classes this semester.
   - Three things I can do to be sure I attend each class meeting are
     1. _____
     2. _____
     3. _____
   - What could interfere with my ability to meet these goals? _____
   - What can I do to overcome or prevent this challenge? _____

Troubleshooting Common Errors

Objective 4 ▶

Troubleshoot common errors.

Some of the common errors associated with time management are shown in the following table along with a more appropriate behavior.

Poor Time Management Behaviors	Better Time Management Behaviors
▶ Student waits until Sunday afternoon to do his homework for the week.	▶ Student sets aside 1 hr per night for math and gets a little bit done each day.
▶ Student arrives late to class each day.	▶ Student arrives on time and is ready for class to start.
▶ Student has no idea when the next test is scheduled.	▶ Student uses a planner to record due dates and exam dates.
▶ Student plans a vacation starting a week before the semester ends.	▶ Student checks the Academic Calendar for the college before making travel plans.

"You can't control how your instructor teaches, but you can control what you do to learn the material. Use strategies that emphasize your strengths and de-emphasize your weaknesses."

Kelly Jackson, *Camden County College*

SECTION S.2 / Learning Styles

OBJECTIVES

As a result of completing this section, you will be able to

1. Identify your dominant learning style(s).
2. Implement strategies to maximize success in math based on your learning style(s).
3. Establish an action plan.
4. Troubleshoot common errors.

People learn in many different ways. The way a person best gathers, processes, organizes, and remembers information refers to their **learning style.** There are three basic learning styles—auditory, visual, and kinesthetic. If you do not know how you best learn, it will be helpful for you to complete a learning styles inventory. One is included in this section, but there are many inventories available online.

Objective 1 ▶

Identify your dominant learning style(s).

Learning Styles

The following survey has been created to identify a student's learning style as it pertains to mathematics. Following the survey is a tally sheet for your responses. The category with the highest percentage represents your dominant math learning style.

ACTIVITY 1   Complete the survey.

Math Learning Styles Survey

Answer each question with the number "3" if you agree most of the time, "2" if you agree sometimes, and "1" if you agree rarely or do not agree.

_____ 1. When there is talking or noise in class, I get easily distracted.

_____ 2. If a problem is written on the board, I have difficulty following the steps unless the teacher verbally explains the steps.

_____ 3. I find it easier to have someone explain something to me than to read it in my math book.

_____ 4. If I know how the math is used in real life, it is easier for me to learn.

_____ 5. To remember formulas and definitions, I need to write them down.

_____ 6. I prefer listening to the lecture rather than taking notes.

_____ 7. Using manipulatives, hands-on activities, or games helps me learn math concepts.

_____ 8. When solving a problem, I try to picture working it out in my mind first.

_____ 9. Quiet places are the best places to study math.

_____ 10. Talking myself through a math problem helps me solve it.

_____ 11. When I write things down I remember them better.

_____ 12. I have to do a math problem myself to learn it. I can't really know it by watching someone else do it.

_____ **13.** If someone lectures about math without writing things down, I find it difficult to follow.

_____ **14.** When I take a math test, I recall more of what was said to me than what I read in my notes or in my math book.

_____ **15.** When I take a math test I can picture my notes or problems on the board in my head.

_____ **16.** I can do math problems sometimes, but can't verbally explain what I did.

_____ **17.** Reading math makes my eyes feel tired or strained.

_____ **18.** When I study math I need to take a lot of breaks.

_____ **19.** I am good at using my intuition to know how to solve a math problem.

_____ **20.** I can learn math fast when someone explains it to me.

_____ **21.** Puzzles and games are a good way to practice math.

_____ **22.** When I work a math problem I say the numbers I am working with to myself.

_____ **23.** I try to write down everything and take a lot of notes in math class.

_____ **24.** I do best with math if I just roll up my sleeves and work on problems.

Developed by Kelly Jackson, Camden County College

Math Learning Styles Tally Sheet

Carefully copy your responses (1, 2, or 3) to each question on the survey into the spaces provided.

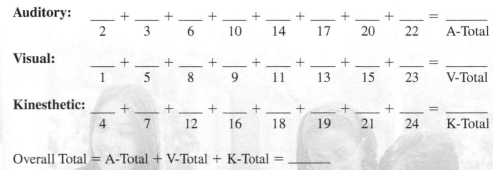

Auditory: $\underset{2}{\rule{1.5em}{0.4pt}} + \underset{3}{\rule{1.5em}{0.4pt}} + \underset{6}{\rule{1.5em}{0.4pt}} + \underset{10}{\rule{1.5em}{0.4pt}} + \underset{14}{\rule{1.5em}{0.4pt}} + \underset{17}{\rule{1.5em}{0.4pt}} + \underset{20}{\rule{1.5em}{0.4pt}} + \underset{22}{\rule{1.5em}{0.4pt}} = \underset{\text{A-Total}}{\rule{1.5em}{0.4pt}}$

Visual: $\underset{1}{\rule{1.5em}{0.4pt}} + \underset{5}{\rule{1.5em}{0.4pt}} + \underset{8}{\rule{1.5em}{0.4pt}} + \underset{9}{\rule{1.5em}{0.4pt}} + \underset{11}{\rule{1.5em}{0.4pt}} + \underset{13}{\rule{1.5em}{0.4pt}} + \underset{15}{\rule{1.5em}{0.4pt}} + \underset{23}{\rule{1.5em}{0.4pt}} = \underset{\text{V-Total}}{\rule{1.5em}{0.4pt}}$

Kinesthetic: $\underset{4}{\rule{1.5em}{0.4pt}} + \underset{7}{\rule{1.5em}{0.4pt}} + \underset{12}{\rule{1.5em}{0.4pt}} + \underset{16}{\rule{1.5em}{0.4pt}} + \underset{18}{\rule{1.5em}{0.4pt}} + \underset{19}{\rule{1.5em}{0.4pt}} + \underset{21}{\rule{1.5em}{0.4pt}} + \underset{24}{\rule{1.5em}{0.4pt}} = \underset{\text{K-Total}}{\rule{1.5em}{0.4pt}}$

Overall Total = A-Total + V-Total + K-Total = _____

Auditory Percentage = (A-Total ÷ Overall Total) × 100 _____

Visual Percentage = (V-Total ÷ Overall Total) × 100 _____

Kinesthetic Percentage = (K-Total ÷ Overall Total) × 100 _____

Note: _Very few people learn math only one way. We hear things, see things, and do things that help us understand math better. One style may dominate the others though._

Strategies to Enhance Your Learning

Objective 2 ▶

Implement strategies to maximize success in math that are specific to your learning style(s).

Now that you have identified your math learning style, some suggestions for things you can do to increase your success in math class based on which style(s) suits you best are provided. A summary of each type of learner and strategies to apply are shown next. General success strategies that apply to all students will be presented in a later section of this chapter.

> **Definition:** An **auditory learner** learns best through listening to lectures, having discussions, talking things through, and listening to what others have to say. These learners often benefit from reading the text aloud and using a recording device.

Strategies to Assist Auditory Learners

✔ Sit near the front of the class so you can hear your instructor.

✔ Sit away from auditory distractions (air conditioner, door, window, and so on).

✔ Participate in class discussions. Ask and answer questions.

✔ Listen carefully to what the teacher says, their tone of voice, inflection, and volume will give cues about what is important.

✔ Record lectures to aid in taking notes.

✔ Read the text out loud.

✔ Use computer tutorials, online sites, or videos with an audio track.

✔ Use songs, rhymes, and other auditory memory devices.

✔ Discuss your ideas verbally.

✔ Dictate to someone while they write down your thoughts.

✔ Study with a classmate, which allows you to talk about and hear the information.

✔ Recite out loud the information you want to remember several times.

✔ Make recordings of important points you want to remember and listen to them repeatedly.

✔ Ask your teacher to repeat or restate something, as needed.

✔ Verbalize your goals for completing your assignments and say your goals out loud each time you begin work on a particular assignment.

> **Definition:** A **visual learner** learns best through seeing. These learners prefer to sit at the front of the classroom to avoid visual obstructions and do best when the teacher provides detailed notes. They generally think in terms of pictures and learn best from visual displays, such as diagrams, illustrated textbooks, videos, PowerPoint slides, and handouts. During a lecture or classroom discussion, visual learners often prefer to take notes to help them grasp the material.

Strategies to Assist Visual Learners

✔ Use visual materials such as pictures, charts, graphs, and the like.

✔ Have a clear view of your teacher when she is speaking so you can see body language and facial expressions.

✔ Use colored markers, Post-it notes, or highlighters to identify important points in the text.

✔ Illustrate your ideas as a picture or brainstorming bubble before writing them down.

✔ Use multimedia (e.g., computers and videos) learning tools.

✔ Use different colors and pictures in your notes, exercise books, and the like.

✔ Study in a quiet place away from auditory disturbances.

✔ Visualize information as a picture to aid memorization.

✔ Write down things that you want to remember.

✔ Take many notes and write down lots of details.

✔ Learn new material by writing out notes, cover your notes then rewrite them.

✔ Write your goals down and read them as you complete your assignments.

✔ Take advantage of study guides, handouts, and workbooks that include worked out solutions.

✔ Choose a seat away from visual distractions and close to the front of the class.

✔ Prepare flashcards and review them often.

> **Definition:** A **kinesthetic learner** learns best through moving, doing, and touching. These students learn best through a hands-on approach. It is difficult for these students to sit still for long periods.

Strategies to Assist Tactile/Kinesthetic Learners

✔ Practice, practice, practice. Because you prefer a hands-on approach, you need to work as many problems as possible.

✔ Take a 3- to 5-min study break every 15–25 min.

✔ Use manipulatives that can help you investigate the topic you are studying (algebra tiles, base 10 blocks, and the like).

✔ Work in a standing position.

✔ Chew gum while studying.

✔ Use bright colors to highlight what you are reading.

✔ If you wish, listen to music while you study (be sure it is not distracting though).

✔ Make or use a model.

✔ Pace or walk around while reciting to yourself or using flashcards or notes.

✔ If the opportunity arises, volunteer to go to the board to work problems.

✔ Study while sitting in a comfortable lounge chair or on cushions or a beanbag.

✔ Cover your desk with your favorite colored construction paper or even decorate your area to help you focus.

✔ Memorize information by closing your eyes and writing the information in the air, try to picture and hear the words in your head as you are doing this.

✔ Make flashcards, card games, floor games, and the like to help you process information.

✔ When working with someone, after they show you a problem, ask if you can try one.

✔ Make a graphic organizer showing the connections between topics.

✔ Get a study group together in a classroom where you can write on the board, walk around, talk about the math, but do problems.

✔ Use applets or computer tutorials with guided solutions in which you have to enter information throughout.

> **Note:** *There is no right or wrong learning style. Determine what you have a tendency toward and use strategies to enhance your performance in class. You cannot control the way your instructor delivers information, but you can control what you do to best receive the information.*

Action Plan

Objective 3 ▶

Establish an action plan.

1. My dominant learning style(s) is(are) dominant _____

_____.

2. Five strategies that I will implement immediately to improve my chances of success in math based on my learning style(s) are

     a. _____.

     b. _____.

     c. _____.

     d. _____.

     e. _____.

3. Does it seem like my teacher's style of delivery matches my style or is it a mismatch? (Example: you prefer to learn visually but your teacher does not write on the board a lot)_____

_____.

Troubleshooting Common Errors

Objective 4 ▶

Troubleshoot common errors.

One common error is how students deal with the situation in which their teacher's teaching style does not match their learning style. Students can control their own behavior but not those of others. Following are some suggestions on how to deal with this situation.

Learning Style/Teaching Style Mismatch	What Can You Do?
Visual learner with a teacher who talks a lot but doesn't write down a lot.	**1.** Record the lesson. Later go back and listen to the lesson and fill in your notes with anything you missed the first time. **2.** Read the section in the book prior to class so you know the topic that will be discussed and write down some of the vocabulary and steps. **3.** Visit your instructor or a tutor in a setting where you can ask questions and have time to write things down.
Kinesthetic learner with a teacher who does not use activities, does not have you work problems in class, and does not use manipulatives.	**1.** Rework the problems that the teacher did in class, check your work with your instructors, and then try some similar problems on your own. **2.** Use the Explain or Show me features in your online homework system that will ask you to enter each step. **3.** Meet with your instructor or a tutor. Each time they show you a problem ask, "Can I try one now?"
Auditory learner in a blended or online class, with limited access to "lectures."	**1.** Take advantage of lecture and exercise videos that accompany your textbook or from popular Internet sites that have great educational videos. **2.** Make audio notes for yourself, reading important rules, definitions, and processes into a recorder. Listen to your recordings as often as possible. **3.** Have a study group or tutoring session where you can discuss the math concepts with someone.

SECTION S.3 — Study Skills

 OBJECTIVES

As a result of completing this section, you will be able to

1. Identify habits that good math students employ.
2. Read the textbook more effectively and efficiently.
3. Take quality notes.
4. Organize materials in a notebook or portfolio.
5. Complete homework in ways that maximize retention.
6. Establish an action plan.
7. Troubleshoot common errors.

Study skills are the skills students need to improve their learning capacity and to acquire new knowledge. No two students are going to employ the same set of study techniques, though there are some that every student should utilize when studying math. This section will address some of the skills that will enhance the success of all students.

Habits of Successful Math Students

 Objective 1

Identify habits that good math students employ.

Good math students study with purpose. When you underline something, it should be because it is important. If you write out a note card, it should be because it is something you want to remember long term. When you work a homework problem, it should lead to a better understanding of the process used to solve that problem. If you can't answer "Why am I doing this?" then what you are doing may not be a productive use of your time. You should always have goals to achieve when you sit down to study and everything you do during your study time should work toward those goals. The following suggestions will help you approach your math class with purpose.

1. *Successful students are responsible.*
   ▶ Attend every class meeting. Try to be a few minutes early so that your materials are out and you are ready for class to begin.
   ▶ Find an accountability partner in class. Encourage one another and use each other as a resource if one of you is absent from class.
   ▶ Adhere to all deadlines for assignments.

2. *Successful students make the most of their time in class.*
   ▶ When you are in class, "be in class." Try not to worry about other things going on in your life. Focus your attention on the topics being discussed. (No texting, surfing, or doing other homework.)
   ▶ Either take notes in class or record the lecture so that you can refer to it later. Some instructors even post their class notes on their website.
   ▶ Actively participate in class. Follow along with the instructor. Ask questions when you don't understand and be prepared to answer questions that you know.

3. *Successful students dedicate an appropriate amount of time to math outside of class.*
   ▶ Immediately after class, take a few minutes to write down a summary of the key concepts that you remember. (It has been shown that students who write down what they know retain 1½ times as much as those who don't, 6 weeks later.)
   ▶ Review your class notes as soon after class as possible.
   ▶ Begin your homework assignment as soon as you can.
   ▶ Review notes from previous classes on a regular basis so that you do not forget material covered earlier in the course.
   ▶ Devote some time every day to your math class.

4. *Successful students aren't afraid to ask for help.*

▶ Math does not need to be a solo activity. Ask your instructor, a tutor, or a classmate about problems you do not understand. Form a study group.

▶ Take advantage of your college's tutoring center and your instructor's office hours.

▶ Use self-help books, other texts, online websites, computer programs, DVDs, or any other outside sources you can find to supplement your course materials.

5. *Successful students are persistent.*

▶ If you get "stuck" on a problem, refer to your notes, the book, or other help resources. Rework the problem until you can do it without referring to these things.

▶ Be patient with yourself in learning the material. Don't get frustrated that other students may seem to learn more quickly than you. Every student comes to class with a different mathematical background. Some classmates just graduated from high school, while others may be returning after several years.

▶ Be willing to try new approaches to solving a problem. If your first attempt fails, continue trying other possible strategies. Often there is more than one way to get to the solution.

Study skills consist of the things you do in class as well as the things you do outside of class. We will take a closer look at four of these topics: reading a math textbook, taking notes, organizing your notebook/portfolio, and completing homework.

Reading a Math Textbook

Objective 2 ▶

Read the textbook more effectively and efficiently.

The textbook is a great resource to aid in the mastery of material presented in class. Many students pay a lot of money for a textbook but do not take the time to read it and often have never been shown how to read a math textbook. Before you attempt homework problems, it is important that you carefully read the relevant sections of your math textbook. The examples should be studied and definitions, properties, and formulas should be learned.

Reading a math textbook is very different than reading a psychology or history text and especially different from reading a novel or a newspaper. Math textbooks generally do not have a whole lot of prose. Textbooks are typically organized around new definitions, new procedures, and worked examples that illustrate these concepts. Graphs, tables, diagrams, equations, formulas, and notations are used to visualize some of the concepts, but these are no more and no less important that the worded passages themselves. Together, they explain the complete math concept you are reading about. When reading a page in your math textbook, you will be reading from left to right, right to left, up and down (think tables), diagonally (think graphs), and from the inside out (think equation solutions).

Some guidelines for reading a math textbook are

1. Skim a new section briefly to identify new vocabulary or processes. Identify the objectives to be presented in the section, which are located in the headings and subheadings.

2. Read the section a second time, using more time and concentration. Try to connect your prior knowledge with the new material you are reading. You will need to read slowly, reread sections, and constantly ask yourself if you understand.

3. When you get to the worked examples, use your own paper to work the problems in detail, making sure that you understand how each step follows from the previous one.

4. Make a list of concepts, formulas, and vocabulary you do *not* understand and seek help.

5. Make a list of formulas and vocabulary that you *do* know and understand.

6. Make sure you understand and use the mathematically correct definition for each vocabulary word. Often words are "borrowed" from everyday language and the meaning can be the same or different in mathematics.
7. You must be able to decode and understand the math notation along with the vocabulary.
8. Learn the formal definitions and properties. It is a good idea to be able to paraphrase into your own words but you don't want to create "new" math rules that may not always work.
9. Get help if you do not understand the reading material.
   - Go back and review the previous section to see if it might be related.
   - Review your instructor's notes on the material.
   - Use the resources that came with your book (videos, CDs, animations, etc.).
   - Ask a classmate, a tutor, or your instructor. Be specific with what you do not understand.

ACTIVITY 1   Get to know the features of this book.

1. Where are the answers located and which answers are available?_____
   _____

2. Where is the chapter review? What is included in the review?_____
   _____

3. What do the different icons used in the book mean?_____
   _____

4. What colors are used to set off definitions, formulas, tips, and other important information?_____
   _____

Taking Notes

Objective 3 ▶

Take quality notes.

Effective note taking begins when you enter the classroom. After you get to your seat, prepare to take notes by taking out your paper, pencil, and book. Taking notes enables you to record how your instructor explains processes, to record examples that are worked, and to identify important class information, such as homework assignments and test dates. Note taking also helps you listen more attentively and to be actively engaged in the learning process. Studies show that people may forget 50% of a lecture within 24 hr, 80% in 2 weeks, and 95% within 1 month if they do not take notes.

Some guidelines for taking notes are as follows.

1. To be an effective note taker, it is important that you attend class.
   Don't be late.
   Don't leave early.
2. To be an effective note taker, you must be a good listener.
   a. Be actively involved in the lecture so that you stay focused on what is being presented.
   b. Sit near the front of the class so that distractions are minimized.
   c. Try to relate the new topics to the material that you have already learned.
   d. Ask your instructor for clarification, if you do not understand. For example, "I don't understand how you got from step 2 to step 3…", "I don't know what the symbol ____ means.", and "What is the difference between ____ and ____?"
   e. Listen for words that signal important information.
   f. Notice how your teacher uses formal math vocabulary when speaking.

3. To be an effective note taker, you must actually take notes.

4. Bring pencils and paper to class and take math notes in pencil, not pen.

    a. Each day start a new set of notes on a new page. Date and label them with the chapter (and section) that is being discussed.

    b. Copy down *everything* that the instructor writes on the board. If the instructor takes the time to write something, it is important.

    c. Take notes, even though your understanding may not be complete.

    d. Leave a space where you may have missed steps or have questions so that you can get these filled in after class. Then actually get them filled in by reading your book, going to a tutor, visiting your instructor's office, or asking a study buddy.

    e. Develop a good note-taking system. Ideas for note-taking systems can be found on the Web or also available with the optional Success Strategies Manual.

5. To be an effective note taker, you should review your notes.

    a. Review and reorganize your notes as soon as possible after class. Fill in any steps you missed.

    b. Write clearly and legibly so you can understand what you have written later.

    c. Rewrite ideas in your own words.

    d. Highlight important ideas, examples and issues with colored pens, pencils, or highlighters.

    e. Review your class notes before the next class period.

    f. Ask questions during office hours or the next class period if there are items that are unclear.

    g. Review all of your notes at least once each week to get a perspective on the course.

The following is an example of a page of effective notes.

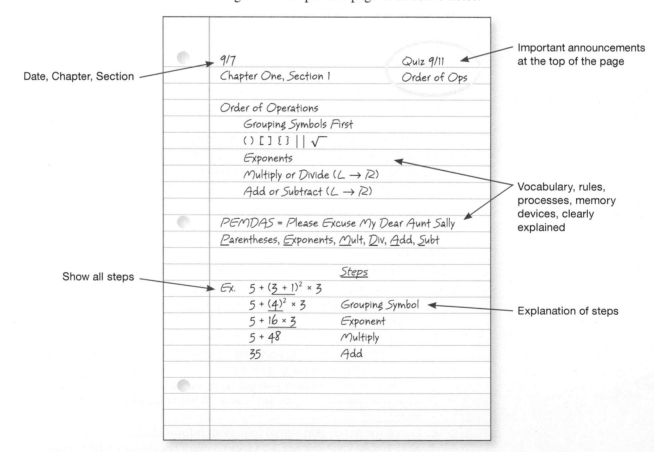

Organizing a Notebook

Objective 4 ▶

Organize materials in a
notebook or portfolio.

You attend class, take notes, get handouts, complete homework assignments, and so on. All of this information can be overwhelming if it is not kept in an orderly fashion. The purpose of organizing your course materials is so that you can use them as tools to prepare for tests and other assignments. There are many different ways that you can organize your materials. The main thing is that you keep them all together and in a logical manner.

Some organization suggestions are presented next. Use what seems useful to your situation and adapt them as necessary. The key is to have a system that works for you.

1. Keep materials in a three-ring binder with pockets.
2. At the front of the binder, have a calendar and a copy of your syllabus, a list of homework assignments for the term, and other handouts your instructor gives you on the first day of class.
3. In the front pocket, keep a to-do list of what needs to be done in class.
4. Use dividers with tabs to divide the materials into different sections. Have a section for class notes and handouts, homework, study guides and test reviews, and tests and quizzes.

   - Class notes should be ordered by date. If your instructor provides a handout with additional notes or instructions for a section, keep this with your class notes for the day. (Use a three-hole punch on handouts so they fit in the binder.)
   - Homework should be written out neatly and brought to class each day.
   - If your instructor provides a study guide or test reviews, keep these together.
   - Finally, be sure to keep all of your tests and any other graded assignments to assist you in preparing for the final exam.

Completing Homework

Objective 5 ▶

Complete homework in ways that maximize retention.

Have you heard the saying, "Math is not a spectator sport?" What this means is that you cannot expect to watch someone else do it and master the material. A colleague used the following example with her class. One summer, she attended 37 Atlanta Braves baseball games. By the end of the summer, her ability to watch and enjoy baseball improved greatly. However, her ability to perform baseball did not improve at all, because she did not play baseball; she only observed it.

While math is not a baseball game, the same general rule applies. Attending class every day is definitely helpful (and fun!), but it is not the only thing needed to successfully master the subject. You must practice. This is where homework comes in. Homework is not just a necessary "evil." This is your opportunity to apply the new definitions, formulas, and procedures to gain deeper understanding.

Some guidelines for completing your homework are as follows.

1. Start your homework as soon after class as possible. The longer the time between class and homework, the more difficult it will be to remember everything you learned in class.

2. Complete your homework neatly and in order. It might look like this.

9/20

Chapter 1, Section 5
Order of Operations
Pg. 59   #2, 6, 10, 14, 18, 22

  2. $10 + \underline{5 \times 2}$
        $10 + 10$ multiply first
  ✓ $\boxed{20}$ add

  6. $20 \div \underline{5 \times 2}$
        $20 \div 10$ multiply first   *Ask why this is wrong*
    $\boxed{2}$ divide

  10. $(\underline{3 + 2})^2$
       $(5)^2$ grouping symbols
      5×5 exponent
  ✓ $\boxed{25}$ multiply

  14. $4 + 2 \times \underline{3^2}$
      $4 + \underline{2 \times 9}$ exponent first
      $4 + 18$ multiply
  ✓ $\boxed{22}$ add

3. If you cannot complete a problem after 10 minutes, skip the problem and continue with your assignment. If, after returning to the problem, you still cannot work it, get help from someone.

4. If you cannot even get started on the homework assignment, review your class notes and rework the problems the instructor worked in class.

5. At the top of the first page of your homework exercises, list the problems that you have questions on. At the next class meeting or during your instructor's office hours, ask for help with these exercises.

6. After completing your homework assignment, compare your answers with those in the back of the book. Most books contain answers to the odd-numbered problems. Rework any problems that you missed.

7. After completing your homework, answer the following questions.
   a. What do I understand clearly?
   b. What do I not understand?
   c. What new definitions, rules, or formulas did I apply?
   d. What types of mistakes did I make on problems I worked incorrectly?
      i. Secretarial: miscopied, misread handwriting, misaligned
      ii. Computational: arithmetic mistake (like $8 \times 7 = 54$)
      iii. Procedural: missed a step, steps out of order, stopped too soon
      iv. Conceptual: no idea how to start, wrong method or formula used

8. Review your homework each day until class meets again so you do not forget your newly acquired skills.

Note: *If your instructor requires you to complete online homework, print a copy of the homework and work it offline. Most systems will allow you to log back in and enter your answers. There really is no substitute for working problems by hand. As your brain directs the motor skills involved in writing the exercises out, a connection is made that impacts your ability to remember the academic concept.*

Action Plan

Objective 6 ▶

Establish an action plan.

Check out additional homework strategies in the supplemental Success Strategies Manual for examples of a "Homework Cover Sheet," a "Chapter Preview" note-taking tool for use when you read your textbook, a tip sheet for "Making Note Cards," "Creating Memory Devices and Mnemonics," "Getting the Most Out of Tutoring," and several other study aids.

List one new strategy under each heading that you will implement immediately to improve your chances of success in your math course:

1. I will make it a habit to _____
   _____

2. From now on when I read my book I will _____
   _____

3. My notes would be better if I _____
   _____

4. I will make my notebook more organized by _____
   _____

5. When I do my homework I will _____
   _____

Troubleshooting Common Errors

Think about the behaviors on the left compared with those on the right. Which habits seem more likely to lead to success in your course?

Poor Habits	Better Habits
A student shows up 10 minutes late for class with a latte in hand but no pen or paper.	A student shows up 10 minutes early for class and has his supplies out and has some questions ready for the instructor.
A student texts friends during class and surfs the net.	A student records the lesson, noting in her notebook the counter number that goes with each problem she is working on.
A student emails her instructor with a message that looks like a text… "i missed ur class pls send HW"	Good Morning Dr. Jones, I missed the 10:00 Algebra class this morning. I got the homework and notes from a classmate. Would it be possible for you to email me a copy of the handout that you distributed in class?
A student comes to class and asks, "I wasn't here for the last class; what did I miss?"	A student misses class and contacts a classmate prior to the next class meeting to get the notes, HW assignment, and reads the book to try to learn what she missed.
A student starts his homework and feels stuck. He closes the book and decides to ask his teacher about it next class.	A student starts his homework and feels stuck. He looks to his notebook for a similar problem; he checks the textbook examples to see if he can find a similar problem. He checks the tutoring schedule to see when he can get in for some help.

SECTION S.4 | Test Taking

▶ OBJECTIVES

As a result of completing this section, you will be able to

1. Prepare for tests and quizzes.
2. Maximize success during the test-taking process.
3. Analyze mistakes on a test.
4. Establish an action plan.
5. Troubleshoot common errors.

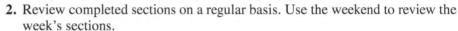

Tests are used to demonstrate mastery or knowledge. Tests can show that we can apply what we have learned. With adequate preparation, tests should simply be an extension of homework. Test taking is only successful if you have employed successful study strategies prior to the test. As the saying goes, "practice makes perfect."

Objective 1 ▶

Prepare for tests and quizzes.

Before the Test

1. Complete your homework assignments regularly.
2. Review completed sections on a regular basis. Use the weekend to review the week's sections.
3. Create a practice test from the problems your teacher worked in class. Work the problems without referring to your notes or books so that you simulate a real test environment. Include problems at all levels of difficulty on the practice test. Check the answers by reviewing your notes or have a classmate check your work.
4. Know which formulas will be provided and which ones must be memorized.
5. Understand all formulas and definitions you need to know for the test. Flashcards are very helpful with this task. Use the front of an index card for the name of the formula or definition and write the formula or definition on the back of the card. Be sure you know what each symbol in a formula represents.
6. Begin preparing a week before an exam. Use your weekly planner to assist with study time. Record the date of the exam on your calendar and then schedule test preparation time beginning a week earlier.
7. Find out as much information about the test as you can:
   a. How many questions will be on the test?
   b. What types of questions (multiple choice, free response) will be included?
   c. How long will you be given to complete the test?
   d. What materials are you allowed to use?
   e. Will there be bonus or extra credit problems?
   f. What chapter/sections are covered?

During the Test

Objective 2 ▶

Maximize success during the test-taking process.

1. Arrive at class early with all the necessary supplies and any aids you are permitted to use, and be ready to begin.
2. When you get your exam, write down all of the formulas on the top or sides of the first page. If your test is computerized, use scratch paper to write down the formulas.
3. Read the instructions carefully.
4. Review the test and complete the problems you know how to do first. This will build your confidence and bring to mind other things you have learned.
5. Keep an eye on the time. Do not spend too much time on one problem (especially if it is a low-points question).

6. Check your answers to make sure they are reasonable. For example, if you are solving for a length, a negative answer would not make sense.

7. Show all of your work neatly and clearly.

8. If you have time, review your work. Don't change any answers unless you have good reason. Often times, first instincts are correct. Double check your signs and arithmetic.

9. Use all of the allotted time to take your test. There is no prize for finishing first nor is there a penalty for being the last one to turn in the exam.

10. Mark up the test paper, if you are allowed, with information that will help you save time.

    a. Circle what you are looking for.

    b. With multiple-choice questions, if you know an answer is impossible, cross it out.

    c. If you have checked an answer, put a check mark next to the problem.

    d. If you skipped a problem but want to come back to it, put a plus sign next to it.

    e. If you see a problem that you don't recognize at all, put a minus sign next to it and come back to it last.

    f. If you have eliminated some options in a multiple-choice question, put down how many options are left. When running out of time, go back to the ones with the fewest options first.

Here is what your test might look like after the first time through.

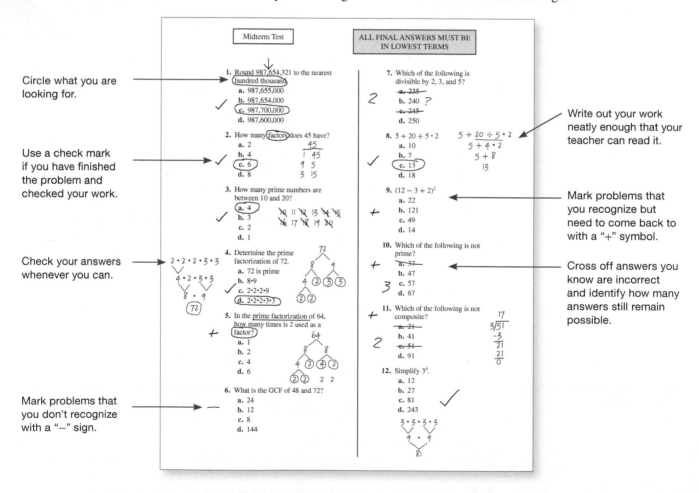

After the Test

Objective 3 ▶

Analyze mistakes on a test.

1. Look through the test for the problems you missed. Rework these problems until you get them correct.

2. Keep your test corrections with the exam (if it is returned to you) for future studying.

3. Make an appointment to meet with your instructor to review any questions you cannot figure out.

4. Most likely, the material on your test will be covered on your final, so it is important that you keep the exam if your instructor allows you to do so. File it in your notebook so you can review it later in the course.

5. Here is a strategy for test corrections:

   a. Divide a piece of paper in thirds.

   b. On the left side, write out the original problem with your original work.

   c. In the middle, write out the correct answer to the problem.

   d. On the right side, work a similar problem to ensure that you can do this type of problem correctly.

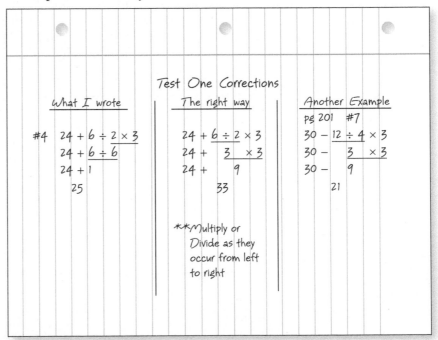

Action Plan

Objective 4 ▶

Establish an action plan.

Check out additional homework strategies in the supplemental Success Strategies Manual for a "Test Preparation Worksheet," a "Post-Test Debriefing" form, suggestions for "How to Reduce Test Anxiety," and tips for "Problem-Solving Strategies."

1. Two things that I will commit to doing before a test to improve my chances of success:

   _____

   _____

2. Two strategies that I will employ during a test to keep my anxiety down:

   _____

   _____

3. After a test I will do the following to learn from my mistakes:

   _____

   _____

Troubleshooting Common Errors

After you check your homework or take a quiz or test, you should review it to determine the types of mistakes you made and think about how you can fix them. It is important to understand whether the errors you make are conceptual, procedural, computational, or secretarial.

A **conceptual error** shows little or no understanding of the underlying concept or procedure.

In a **procedural error**, the correct process is used but there is a mistake made with the steps.

In a **computational error,** an arithmetic mistake is made.

In a **secretarial error**, a miscopy of some kind is made.

Evaluate	
$24 + 6 \div 2 \times 3$	
$\underline{24 + 6} \div 2 \times 3$	Add
$\underline{30 \div 2} \times 3$	Divide
15×3	Multiply
45	

The student shows no understanding of the concept of "order of operations." He just works straight through from left to right. This is an example of a conceptual error.

Evaluate	
$24 + 6 \div 2 \times 3$ PEMDAS	
$24 + 6 \div \underline{2 \times 3}$	Multiply
$24 + \underline{6 \div 6}$	Divide
$24 + 1$	Add
25	

The student clearly identifies that the order of operations is being used as she uses the acronym PEMDAS. She knows the right process. However, she thinks that she must multiply then divide, rather than the correct rule: Multiplication and division are done as they occur, from left to right. This is an example of a procedural error.

Evaluate	
$24 + 6 \div 2 \times 3$ PEMDAS	
$24 + \underline{6 \div 2} \times 3$	Divide
$24 + \underline{3 \times 3}$	Multiply
$24 + 9$	Add
32	

The student clearly identifies that "order of operations" is being used as he uses the acronym PEMDAS. He does all of the right steps, but then adds incorrectly. This is an example of a computational error.

A fourth type of mistake is *simply secretarial.* This type of mistake includes miscopying any part of the problem, misreading your handwriting, misaligning the problem, or any other non-math-related issue with how you write out your work.

How do I correct these errors?

A conceptual error needs to be corrected before the test begins. This mistake is typically about lack of preparation for the test.

> **Note:** *Remember P^5*
>
> *Proper Preparation Prevents Poor Performance*

▶ If you have not been doing any homework, do some. If you have been doing some, do more. If you have been doing all of the homework and are still struggling with the concepts, you may need some tutoring.

▶ Sometimes the issue is problem recognition. Perhaps you can do all of the problems when they are organized into separate sections in the book, but not when they are jumbled together on a test.

- A great way to mimic this situation is to mix up your notecards.
- Don't study everything in the same order each time.
- Mix problems from different sections and chapters together.
- Work any cumulative reviews in the text.
- Review old homework each time you complete a new homework assignment to compare how the new assignment is the same or different from the old topics.

A procedural mistake usually requires only a minor fix. You understand the concept; you know what the steps are, but have confused them in some way. This mistake is also about preparation; the steps should be natural by the time you get to a test. If you have repeated a process enough times, you will not even think about the steps any more, you will just know what to do. Whether it is playing music, throwing a baseball, making a pie, or doing a math problem, practice makes perfect!

Computational and secretarial mistakes both have similar fixes. Check your work and slow down. These mistakes are typically made on problems that you are very confident with and are working through quickly. You do not want quickly to mean *carelessly*!

▶ For many problem types, you can check your answers. When this is possible you should always do so. Check the solution of a problem with the original problem. In this way even a miscopy can be found.

▶ Estimate your answer before you begin your work. If your answer is not near the estimate, investigate why.

▶ Think about the reasonableness of your answer. If you made a secretarial error with a decimal point and have an answer where a car is driving 500 mph, it should be clear to you that this type of answer is not reasonable.

▶ If you are permitted to use a calculator, check hand calculations with the calculator. Also, check the calculator output with your reasonableness criteria. If you press the wrong buttons, a calculator can still give you a wrong answer.

▶ Use grid paper instead of lined paper to help with misalignment issues. You can also turn lined paper horizontally to create columns.

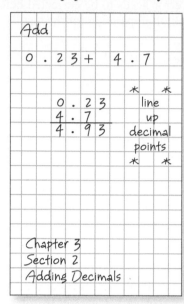 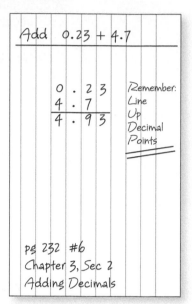

 Note: *If you can't read it, you can't review or study it. If your teacher can't read it, you may get it marked wrong.*

Be Neat!

Procedure: Determining the Type of Mistake on a Test or Quiz

Step 1: Conceptual: Left it blank, wrong process, mis-memorized rule
Step 2: Procedural: Performed steps out of order, missed a step, or couldn't finish all steps
Step 3: Computational: Added, subtracted, multiplied, or divided wrong
Step 4: Secretarial: Miscopy, omission, misalignment, misread handwriting

After a test, review each problem that you got wrong and evaluate what went wrong. What types of mistakes did you make? Is there a trend in which type of mistakes you make? What will you do to eliminate these mistakes in the future? Also, reflect about whether the test is helping you meet your grade goal or harming you in making your grade goal. After each test, you should evaluate your progress toward meeting your grade goal.

"Organization and Commitment go hand in hand. For each section, organize the study plan and make a commitment to follow the schedule."

Pauline Chow, *Harrisburg Area Community College*

SECTION S.5   Blended and Online Classes

OBJECTIVES

As a result of completing this section, you will be able to:

1. Differentiate between blended, hybrid, and online classes.
2. Use techniques to maximize your success in blended and online settings.
3. Establish an action plan.
4. Troubleshoot common errors.

The desire for flexibility and convenience in scheduling classes has led to a continued increase in blended and online classes.

Identifying Different Delivery Methods

Objective 1 ▶

Differentiate between blended, hybrid, and online classes.

Blended classes, or **hybrid** classes, incorporate traditional classroom experiences with online learning activities. Students in these types of classes are required to attend a specific number of on-campus class meetings. The remaining portion of the course is delivered online, thereby reducing the amount of time a student spends in class. In summary, blended classes strive to combine the best elements of traditional face-to-face instruction with the best aspects of learning at a distance.

Online classes, unlike blended classes, deliver entire courses at a distance. Some may have required meetings for exams, but typically all aspects of the course are provided online.

With a portion of a course or an entire course delivered at a distance, the roles of the teacher and student change. The teacher becomes more of a facilitator and the student becomes more like a teacher in these environments. While in a traditional classroom, the student's learning is his or her responsibility, this responsibility becomes even more important in blended and online courses. The information provided in the previous sections certainly applies to students in any type of class, but there are some additional tips for students in blended and online classes.

Tips for Blended and Online Classes

Objective 2 ▶

Use techniques to maximize your success in blended and online settings.

1. The first day of class is extremely important. If you are in a blended class, be sure to attend the first on-campus meeting. If you are in an online class, be sure to log into the course the first day the course is available.
2. Review the syllabus and course policies to determine what is expected of you.
   a. How many required on-campus meetings are there?
   b. Are there proctored exams?
   c. Upon what activities is your grade based (e.g., online homework, discussion board postings, online quizzes and/or tests, group projects, and so on)?
3. Familiarize yourself with the course resources.
   a. Does your instructor post notes or videos, solutions for quizzes/tests, or other material?
   b. Does your instructor hold online office hours through chat, email, or a live classroom? If so, when?
   c. What learning materials (videos, practice problems, ebook, and the like) are available with your textbook?

4. Make a calendar for the semester (one may be available by your instructor) and write down due dates for homework, quizzes, tests, discussion posts, and other assignments.

5. Determine a time each week to have class with yourself.

   a. If the face-to-face version of the class meets, for example, 2 days a week for 1 hr and 45 min, then you should set aside this same amount of time to "have class" on your own.

   b. Class time is time for you to watch videos, to read the textbook, to do what is necessary to learn the material, and to complete quizzes, tests, and discussions. As you watch videos or read the textbook, take notes as you would if you were attending a traditional class.

6. Determine a time each week to have study time. In addition to "class," you need study time.

   a. For each credit or contact hour for the course, you should have three hours of class plus study time (3 credit class = 9 hr, 3 hr for "class" and 6 hr of study time).

   b. Study time is time for you to complete homework. So that you do not get dependent on the help resources online, we suggest that you print the homework assignment, work it offline, and then log back in to enter your answers. For the problems that you answered incorrectly, use the help features to assist you.

7. Remember that you are not alone in the online classroom. Interact with your instructor and other classmates through discussion boards, email, and other features that are available in your class.

8. Ask your instructor for help when you do not understand the material or the technology.

9. Maintain your notes, homework, printed copies of assignments, and any other printables in your notebook.

10. Check in frequently, daily if possible, for announcements, changes to the calendar or assignments, and for any updates you need to know.

Action Plan

Objective 3 ▶

Establish an action plan.

Set some goals for how you will be successful in a class with a nontraditional delivery method.

1. I will "have class with myself" _____ times a week for _____ minutes each time.

2. If I find I can't do it alone, I will get help by _____
   _____.

3. Five things that I am committed to do to ensure that I maximize my success are
   _____.
   _____
   _____
   _____
   _____

Troubleshooting Common Errors

Objective 4 ▶
Troubleshoot common errors.

Blended and online classes present some different challenges than face-to-face classes. Here are some suggestions to overcome those challenges from students who have taken blended and/or online classes.

Study Skills	Suggestions
Homework	"Plan ahead and don't procrastinate in completing your homework. And try to do the take home assignments without looking at notes . . . it will help a lot at test time!" "Follow the study guide, do all the online homework and take-home work. Try the practice tests to help with confidence if you get nervous taking tests."
Textbook	"Read the book first, then do book questions, then do all online homework until a score of 100 percent is reached. Review all materials again before an exam. You have to be disciplined in getting into a routine just like a classroom based course."
Communication	"I would recommend keeping in constant contact with the instructor. Monitor your progress and make sure you are constantly evaluating your study habits. Reviewing feedback is another great way to improve your understanding."
Time Management	"You will need a lot of time to succeed in this course. Just because it's an online class doesn't mean you can sign on for 5 minutes do some work and fly by. Time is essential to succeeding in this course." "This course requires a good bit of time, and dedication. Had I not spend much time on the course at the beginning, I would have not been able to perform well for the remainder of the course." "It is very flexible but students have to be prepared to discipline themselves enough to do the work on time and not wait until the last minute. Don't try to do all the problems at one time, it is very overwhelming. It's not always possible but try to treat it like a normal class and set aside an hour a day or every other day, to work on it. Sometimes you have a busy week and that's what's nice about having an online class, the flexibility to work on the material when you have time." "In order to succeed you have to put the time into this class as if it were on campus. I feel that if individuals apply themselves to this course they will succeed!!!!"

Strategies to Succeed in Math

 What's the big idea? Chapter S provides many strategies for maximizing your success in math class. No one will implement all of these ideas, but using even some will improve your chances of success. The more you implement, the better your chance of success.

The Tools

Connect Hosted by ALEKS has several additional Student Resources to go along with the tips in this chapter.

You can learn math!

Summary of Success Strategies

- ▶ Set goals.
- ▶ Manage your time.
- ▶ Adopt habits of successful math students.
- ▶ Read your book with purpose.
- ▶ Take meaningful notes.
- ▶ Keep an organized notebook/portfolio.
- ▶ Do homework to ensure deep understanding.
- ▶ Adopt good test-taking skills before, during, and after an exam or quiz.
- ▶ If you are in a blended or online class, adapt your strategies.

What's Next?

Throughout this text we will continue to support your efforts to succeed. With each chapter we will revisit one of the characteristics of success discussed in this opening chapter. Some chapter openers will focus on skills and others attitudes. When you need a little extra motivation, scan through this chapteror through the chapter openers to find suggestions.

The topics for each chapter opener are as follows.

Real Numbers and Algebraic Expressions

Time Management

As you begin this chapter, we want you to focus on how you are managing your time. Your success in this course depends on your ability to devote the appropriate time to learning and practicing the material.

- The general rule of thumb is that you should spend 2 hours studying for each hour you are in class.

- Prepare a calendar of your week and schedule an appointment with yourself for study time.

- Review how you are using your time through the week and make an honest assessment of whether you have the time to make this class a priority.

- See the Preface for additional resources concerning time management.

Question For Thought: Of the activities you participate in, which are essential? Can you give up any of these activities to provide more time for your studies?

Chapter Outline

Coming Up...

In Section 1.3, we will learn that the number of viewers (in millions) for the season premiere of Fox Network's *American Idol* for Seasons 1 to 8 can be approximated by the expression $0.1x^3 + 18.4x - 2.54x^2 - 4.6$, where x is the number of seasons aired. We will learn how to use this expression to estimate the number of viewers for future seasons.

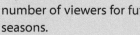

*"*The key is not to prioritize what's on your schedule, but to schedule your priorities.*"*

—Stephen Covey

SECTION 1.1   The Set of Real Numbers

Chapter 1 lays the foundation of algebra. In this chapter, we will explore the set of real numbers, operations with real numbers, and numerical and algebraic expressions. The skills learned in this chapter will be applied in every chapter of this text. These are the skills that will enable us to solve equations, simplify expressions, and solve application problems.

▶ **OBJECTIVES**

As a result of completing this section, you will be able to

1. Classify a number as a natural number, whole number, integer, rational number, or irrational number.
2. Graph real numbers on a real number line.
3. Compare the value of two real numbers.
4. Find the opposite of a real number.
5. Find the absolute value of a real number.
6. Troubleshoot common errors.

How would you classify the following numbers?

    12 hr: the number of semester hours to be a full-time student

    $-70°F$: the coldest recorded temperature in the continental United States (1971)

    $7\frac{1}{2}$ ft: the height of one of the tallest NBA players, Yao Ming

    π: the distance around a circle divided by its diameter

In this section, we will learn the different types of real numbers and will classify numbers such as the ones shown.

The Set of Real Numbers

Objective 1 ▶

Classify a number as a natural number, whole number, integer, rational number, or irrational number.

Numbers are used to represent so many things in our lives—the amount of money in a bank account, hours in a day, temperature, weight, height, number of credit hours, GPA, salaries, measurements, and so on. In addition, computers store all of their data as numbers. The world as we know it would basically cease to exist without numbers. So, it seems logical to begin with a review of the different types of numbers.

We first need to define a few terms related to sets. A **set** is a collection of objects. Each object in a set is called a **member** or an **element**. A set is written in braces, { }, and is usually denoted with a capital letter. Sets can be either **finite** or **infinite**. A finite set has a specific number of elements. An example of a finite set is $A = \{1, 2, 3, 4, 5\}$. An infinite set has infinitely many elements. An example of an infinite set is $B = \{1, 3, 5, 7, 9, \dots\}$. The three dots at the end are called an **ellipsis** and indicate that the set continues indefinitely.

In sets A and B, the elements of the sets are listed explicitly. This method of listing each element of the set is called the **roster method**. Another method, the **set-builder notation**, states the conditions the elements must satisfy to be included in the set. Set-builder notation is written in the form:

$$\{x \,|\, \text{condition } x \text{ must satisfy}\}$$

This is read as "the set of x such that _____." In the blank, we insert the condition that x must satisfy. The letter x is a **variable** and represents some unknown number.

An example of set-builder notation is $C = \{x \,|\, x \text{ is a positive odd number}\}$. The given condition tells us that x can be 1, 3, 5, 7, 9, ... Note that set C is the same as set B.

To denote that a number is or is not a member of a set, we use the notation shown in the following table.

Symbol	Meaning	Example	Verbal Statement
$\in$	Element of	$4 \in A$	"4 is an element of A." or "4 is a member of A."
$\notin$	Is not an element of	$6 \notin A$	"6 is not an element of A." or "6 is not a member of A."

Numbers belong to certain sets. The next table defines these sets.

Natural numbers (or counting numbers)	$\mathbb{N} = \{1, 2, 3, 4, 5, \dots\}$
Whole numbers	$\mathbb{W} = \{0, 1, 2, 3, 4, 5, \dots\}$
Integers	$\mathbb{Z} = \{\dots, -5, -4, -3, -2, -1, 0, 1, 2, 3, 4, 5, \dots\}$
Rational numbers	$\mathbb{Q} = \left\{\dfrac{p}{q} \,\middle|\, p, q \in \mathbb{Z} \text{ with } q \neq 0\right\}$
Irrational numbers	$\mathbb{I} = \{\text{numbers that are not rational}\}$
Real numbers	$\mathbb{R} = \{\text{rational or irrational numbers}\}$

The set of **rational numbers** is given in set-builder notation. It is read as "the set of numbers p over q, such that p and q are **integers** with q not equal to 0." That is, the values of p and q can be replaced with any integer except that q cannot be replaced with zero. So, a rational number is the quotient, or ratio, of two integers whose denominator is not zero. It is impossible to list all of the rational numbers, since there are infinitely many of them. Some examples of rational numbers are

$$\frac{10}{1} = 10, \quad \frac{37}{7} = 5\frac{2}{7}, \quad -\frac{3}{5} = -0.6, \quad \frac{1}{3} = 0.3333\dots = 0.\overline{3}$$

Rational numbers include **integers, mixed numbers, decimals that terminate,** and **repeating decimals.** Note all integers are rational numbers since each integer can be written as its value divided by 1, as illustrated by $10 = \dfrac{10}{1}$.

The set of **irrational numbers** arose out of the study of geometry, specifically triangles and circles. A simple definition of an irrational number is a number that is not rational. This means that the number cannot be written as a quotient of two integers.

Note: *Irrational numbers are numbers whose values are nonterminating and nonrepeating decimals.*

The number π (pi) may be the most commonly known irrational number. We often use the value 3.14 for π but this is only a decimal approximation. The exact value of π is 3.1415926535 . . . . This value continues indefinitely with no repeating pattern. Another irrational number used in advanced math classes is the number e, Euler's number. It is used in many important formulas, such as those involving population growth. Its value is approximately 2.72.

Other examples of irrational numbers involve square roots. The **square root** of a number is the number that must be multiplied by itself to get the original number. For instance, $\sqrt{9} = 3$ since $3 \cdot 3 = 9$. The square root of 9 is a rational number since its value is equivalent to 3, which can be expressed as a quotient of two integers, $\dfrac{3}{1}$. However, not all square roots are rational numbers.

We can use the calculator to approximate square roots to determine if the number is rational or irrational. If the value on the calculator is a decimal that repeats or terminates, then the number is rational. If the value on the calculator is a decimal that is nonterminating and nonrepeating, then the number is irrational. Some irrational numbers are

$$\sqrt{3} = 1.7320508075\ldots$$
$$\sqrt{8} = 2.8284271247\ldots$$

A real number is either rational or irrational. It cannot be both. The next diagram illustrates how the sets of numbers relate to one another.

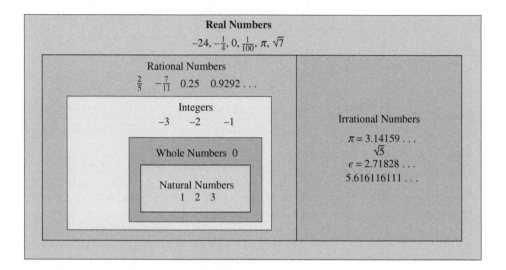

This diagram shows us the following facts.

- All natural numbers are whole numbers, integers, and rational numbers.
- All whole numbers are integers and rational numbers.
- All integers are rational numbers.
- Natural numbers, whole numbers, integers, and rational numbers are *not* irrational numbers.
- Natural numbers, whole numbers, integers, rational numbers, and irrational numbers are all real numbers.

In Chapter 10, another set of numbers, *complex numbers*, is introduced. This set contains all of the real numbers.

Procedure: Classifying a Real Number

Step 1: If the number is 1, 2, 3, . . . , then the number is a natural number.

Step 2: If the number is 0, 1, 2, 3, . . . , then the number is a whole number.

Step 3: If the number is . . . , −3, −2, −1, 0, 1, 2, 3, . . . , then the number is an integer.

Step 4: Determine if the number is rational or irrational.
    a. If the number is equivalent to a decimal that terminates or repeats, then the number is rational.
    b. If the number is equivalent to a decimal that does not terminate or repeat, then the number is irrational.

Step 5: If the number is rational or irrational, it is a real number.

Objective 1 Examples

Classify each number as a natural number, whole number, integer, rational number, irrational number, and/or real number. If the number is a rational number, write it in fractional form. If the number is irrational, approximate its value to two decimal places.

1a. 0.75 **1b.** $-3\frac{1}{2}$ **1c.** $\sqrt{11}$ **1d.** 0

1e. $\sqrt{25}$ **1f.** $\dfrac{11}{16}$ **1g.** -7

Solutions

	Number	Natural	Whole	Integer	Rational	Irrational	Real
1a.	$0.75 = \dfrac{75}{100} = \dfrac{3}{4}$				X		X
1b.	$-3\dfrac{1}{2} = -\dfrac{7}{2}$				X		X
1c.	$\sqrt{11} = 3.316624\ldots$ ≈ 3.32					X	X
1d.	$0 = \dfrac{0}{1}$			X	X		X
1e.	$\sqrt{25} = 5 = \dfrac{5}{1}$	X	X	X	X		X
1f.	$\dfrac{11}{16}$				X		X
1g.	$-7 = -\dfrac{7}{1}$			X	X		X

✓ Student Check 1

Classify each number as a natural number, whole number, integer, rational number, irrational number, and/or real number. If the number is a rational number, write it in fractional form. If the number is irrational, approximate its value to two decimal places.

a. 0.4 **b.** $-6\dfrac{4}{7}$ **c.** $\sqrt{36}$ **d.** 10

e. $\sqrt{21}$ **f.** $\dfrac{5}{6}$ **g.** -5 **h.** $1.232232223\ldots$

Objective 2 ▶

Graph real numbers on a real number line.

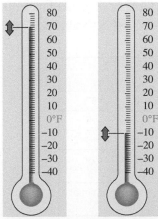

Figure 1.1   Figure 1.2

The Real Number Line

Graphing or plotting real numbers on a number line is a skill that will be used in later sections when we solve inequalities and graph equations. The act of reading a thermometer can assist us in plotting real numbers. Figure 1.1 illustrates a thermometer reading of 70°F and Figure 1.2 illustrates a thermometer reading of −10°F.

If we rotate the thermometer, we get a number line. If we put a dot at 70°F, we have graphed the real number 70. If we put a dot at −10°F, we have graphed the real number −10, as shown.

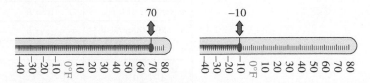

The **real number line** is similar to this rotated thermometer. It is a horizontal line drawn with arrows on both ends to indicate that the real numbers are infinite. Tick marks are used to divide the number line into equal segments. Positive numbers are located to the right of 0 and negative numbers are located to the left of 0. 0 is neither positive nor negative.

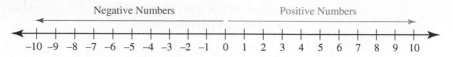

> **Procedure: Graphing or Plotting Real Numbers on a Number Line**
>
> **Step 1:** If the number is not an integer, approximate its value to two decimal places.
> **Step 2:** Place a dot on its position on the number line. Make the dot easy to distinguish from the number line itself.

Objective 2 Examples   **Graph each number on a real number line.**

2a. 5   **2b.** -3 **2c.** $\dfrac{2}{3}$ **2d.** $-4\dfrac{1}{2}$ **2e.** 6.1   **2f.** $\sqrt{2}$

Solutions
2a. 5   Graph at the appropriate tick mark.
2b. -3 Graph at the appropriate tick mark.
2c. $\dfrac{2}{3} \approx 0.67$ Graph this value between 0 and 1 but closer to 1.
2d. $-4\dfrac{1}{2} = -4.50$ Graph this value exactly halfway between -4 and -5.
2e. 6.1   Graph this value between 6 and 7 but closer to 6.
2f. $\sqrt{2} \approx 1.41$ Graph this value close to halfway between 1 and 2.

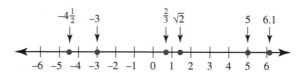

✓ Student Check 2   Graph each number on a real number line.

a. -3 **b.** -1 **c.** $-\dfrac{5}{2}$ **d.** $1\dfrac{1}{4}$ **e.** -5.2 **f.** $\sqrt{24}$

Ordering Real Numbers

Objective 3 ▶

Compare the value of two real numbers.

We can express the relationship between two real numbers a and b using symbols of equality and inequality. The equality symbol is the equals sign, $=$. The inequality symbols are $<$, $>$, $\leq$, $\geq$, or $\neq$. The equality and inequality symbols can be used to write mathematical statements. The statement is either true or false. Some examples of true statements are shown.

Verbal Statement	Mathematical Statement	Example	Read as
a is equal to b	$a = b$	$3 = 3$	"3 is equal to 3"
a is less than b	$a < b$	$3 < 4$	"3 is less than 4"
a is greater than b	$a > b$	$3 > 2$	"3 is greater than 2"
a is less than or equal to b	$a \leq b$	$3 \leq 4$	"3 is less than or equal to 4"
a is greater than or equal to b	$a \geq b$	$3 \geq 2$	"3 is greater than or equal to 2"
a is not equal to b	$a \neq b$	$1 \neq 3$	"1 is not equal to 3"

To determine how the numbers -10 and -20, for example, relate to one another, we will use a thermometer. As we move up a thermometer, the temperatures get larger. As we move down a thermometer, the temperatures get smaller. Since -20 lies below -10 on the thermometer, -20 is less than -10. That is, $-20 < -10$.

When the thermometer, or number line, is rotated horizontally, we see that -20 is to the left of -10.

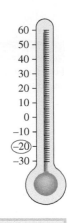

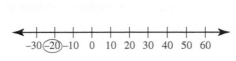

Property: Order of Real Numbers

If a number a lies to the left of a number b on a real number line, then $a < b$ or $a \leq b$.

If a number a lies to the right of a number b on a real number line, then $a > b$ or $a \geq b$.

If a number a lies in the same location as a number b on a real number line, then $a = b$.

Objective 3 Examples | Use the number line to compare the numbers. Use a $<$, $>$, or $=$ symbol to make a true statement.

Problems	Solutions
3a. -3 ___ -4	![number line from -5 to 1] $-5\ -4\ -3\ -2\ -1\ 0\ 1$ -3 lies to the right of $-4 \Rightarrow -3 > -4$.
3b. -2 ___ 0	![number line from -5 to 1] $-5\ -4\ -3\ -2\ -1\ 0\ 1$ -2 lies to the left of $0 \Rightarrow -2 < 0$.
3c. π ___ $\sqrt{5}$	Note that $\pi \approx 3.14$ and $\sqrt{5} \approx 2.24$. $\sqrt{5}\quad \pi$![number line from 0 to 6] $0\ 1\ 2\ 3\ 4\ 5\ 6$ π lies to the right of $\sqrt{5} \Rightarrow \pi > \sqrt{5}$.
3d. 3.5 ___ $\dfrac{7}{2}$	Note that $\dfrac{7}{2} = 3.5$. The two numbers are equal, so $3.5 = \dfrac{7}{2}$.

✓ Student Check 3 | Use the number line to compare the numbers. Use a $<$, $>$, or $=$ symbol to make a true statement.
   a. -5 ___ -4 **b.** $\sqrt{2}$ ___ $\dfrac{1}{2}$ **c.** 1.25 ___ $\dfrac{5}{4}$

Note: *The inequality statements in Example 3 can also be written by reversing the inequality symbol. In either case, the arrow points to the smaller number.*

The statement $-3 > -4$ can also be written as $-4 < -3$.

The statement $-2 < 0$ can also be written as $0 > -2$.

The statement $\pi > \sqrt{5}$ can also be written as $\sqrt{5} < \pi$.

Opposites of Real Numbers

A number line not only provides us a way to order real numbers, it also helps us visualize the distance between numbers. Two numbers that lie equal distances from zero and lie on opposite sides of zero are *opposites* or *additive inverses* of one another.

> **Definition: Opposite of a Real Number**
>
> The **opposite** or **additive inverse** of a real number a is the number that has the same distance from 0 but lies on the opposite side of 0. The opposite of a real number a is denoted as $-a$.

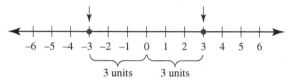

This number line illustrates that -3 and 3 are opposites of one another since they are the same distance from zero but on opposite sides of zero. We state these facts as shown in the table.

Verbal Statement	Mathematical Statement
The *opposite of* negative three is three.	$-(-3) = 3$
The *opposite of* three is negative three.	$-(3) = -3$

> **Facts:**
> - *The opposite of a positive real number is a negative number.*
> - *The opposite of a negative real number is a positive number.*

Find the opposite of each number. Write a mathematical statement to represent the answer.

	Solution	
Problem	**Opposite**	**Mathematical Statement**
4a. -2	The *opposite* of -2 is 2.	$-(-2) = 2$
4b. 16	The *opposite* of 16 is -16.	$-(16) = -16$
4c. $\dfrac{3}{4}$	The *opposite* of $\dfrac{3}{4}$ is $-\dfrac{3}{4}$.	$-\left(\dfrac{3}{4}\right) = -\dfrac{3}{4}$
4d. $-\dfrac{9}{5}$	The *opposite* of $-\dfrac{9}{5}$ is $\dfrac{9}{5}$.	$-\left(-\dfrac{9}{5}\right) = \dfrac{9}{5}$
4e. π	The *opposite* of π is $-\pi$.	$-(\pi) = -\pi$
4f. -0.23	The *opposite* of -0.23 is 0.23.	$-(-0.23) = 0.23$

☑ **Student Check 4**   Find the opposite of each number. Write a mathematical statement to represent the answer.

a. -7 b. 4 c. $\dfrac{1}{2}$ d. $-\dfrac{25}{6}$ e. $\sqrt{6}$ f. 8.2

Absolute Value

We just defined numbers that are opposites as numbers that have the same *distance* from zero on a number line. A number's distance from zero is called its *absolute value*.

Definition: The **absolute value** of a real number a is the distance between 0 and a on the real number line. The absolute value of a is denoted $|a|$.

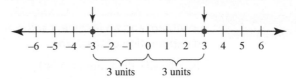

3 units   3 units

From the graph, we see that the distance between 3 and 0 is 3 units and the distance between -3 and 0 is 3 units. We can state these facts as shown in the table.

Verbal Statement	Mathematical Statement
The *absolute value* of three is three.	$|3| = 3$
The *absolute value* of negative three is three.	$|-3| = 3$

Because absolute value refers to distance, the absolute value of a number is always greater than or equal to zero. The following property summarizes this fact.

Property: Absolute Value

If $a \geq 0$, then $|a| = a$.

If $a < 0$, then $|a| = -a$.

This property states that if a number is greater than or equal to zero, its absolute value is equal to that number. For instance, $5 \geq 0$ and 5 is 5 units from zero, so $|5| = 5$. If a number is less than zero, its absolute value is equal to the opposite of the number. For instance, $-2 < 0$ and -2 is 2 units from zero, so $|-2| = 2$. Note that $2 = -(-2)$.

Procedure: Finding the Absolute Value of a Number

Step 1: Find the number's distance from zero.
Step 2: The absolute value of a number is always greater than or equal to 0.

Objective 5 Examples Simplify each absolute value expression.

Problem	Solution
5a. $|6|$	$|6| = 6$, since 6 is 6 units from 0.
5b. $|-8|$	$|-8| = 8$, since -8 is 8 units from 0.
5c. $|0|$	$|0| = 0$, since 0 is 0 units from 0.
5d. $\left|-\dfrac{1}{2}\right|$	$\left|-\dfrac{1}{2}\right| = \dfrac{1}{2}$, since $-\dfrac{1}{2}$ is $\dfrac{1}{2}$ unit from 0.
5e. $-|2|$	First, find the absolute value of 2, which is 2. Then find the opposite of 2, which is -2. $$-|2| = -(2) = -2$$
5f. $-|-3|$	First, find the absolute value of -3, which is 3. Then find the opposite of 3, which is -3. $$-|-3| = -(3) = -3$$

✓ Student Check 5 Simplify each absolute value expression.

a. $|10|$ **b.** $|-6|$ **c.** $|\pi|$ **d.** $\left|-\dfrac{2}{5}\right|$ **e.** $-|4|$ **f.** $-|-7|$

Objective 6 ▶
Troubleshoot common errors.

Troubleshooting Common Errors

Some common errors associated with the objectives in this section are shown next.

Objective 6 Examples A problem and an incorrect solution are given. Provide the correct solution and an explanation of the error.

6a. Classify the number $\sqrt{49}$.

Incorrect Solution	Correct Solution and Explanation
$\sqrt{49}$ is an irrational number since it is a square root.	$\sqrt{49} = 7 = \dfrac{7}{1}$, so this number is a rational number, an integer, a whole number, and a natural number. The square root of a number is irrational only if the number has a decimal value that is nonrepeating and nonterminating.

6b. Plot the number $\dfrac{3}{4}$ on a real number line.

Incorrect Solution	Correct Solution and Explanation
	$\dfrac{3}{4} = 0.75$ So, the plot of the point is between 0 and 1. 

6c. Simplify $-|-10|$.

Incorrect Solution	Correct Solution and Explanation
$-|-10| = 10$	This problem means to find the opposite of the absolute value of -10. So, we first find the absolute value of -10 and then take its opposite. $-|-10| = -(10) = -10$

ANSWERS TO STUDENT CHECKS

Student Check 1 **a.** $0.4 = \dfrac{4}{10}$ rational, real

b. $-6\dfrac{4}{7} = -\dfrac{46}{7}$ rational, real

c. $\sqrt{36} = 6 = \dfrac{6}{1}$ rational, integer, whole, natural, real

d. $10 = \dfrac{10}{1}$ rational, integer, whole, natural, real

e. $\sqrt{21} \approx 4.58$ irrational, real

f. $\dfrac{5}{6}$ rational, real

g. $-5 = \dfrac{-5}{1}$ integer, rational, real

h. $1.232232223\ldots$ irrational, real

Student Check 2

Student Check 3 **a.** $-5 < -4$ **b.** $\sqrt{2} > \dfrac{1}{2}$

c. $1.25 = \dfrac{5}{4}$

Student Check 4 **a.** $-(-7) = 7$ **b.** $-(4) = -4$

c. $-\left(\dfrac{1}{2}\right) = -\dfrac{1}{2}$ **d.** $-\left(-\dfrac{25}{6}\right) = \dfrac{25}{6}$

e. $-\left(\sqrt{6}\right) = -\sqrt{6}$ **f.** $-(8.2) = -8.2$

Student Check 5 **a.** 10 **b.** 6 **c.** π

d. $\dfrac{2}{5}$ **e.** -4 **f.** -7

SUMMARY OF KEY CONCEPTS

1. Real numbers consist of rational numbers and irrational numbers. Rational numbers are made up of numbers that can be written as the ratio of two integers. All of the integers are elements of the set of rational numbers. The integers are positive and negative whole numbers. Whole numbers are the natural numbers combined with zero. The natural numbers are counting numbers.

2. Every real number can be graphed on a real number line.

3. For any two numbers a and b, $a = b$, $a < b$, or $a > b$. To determine the order of numbers, place them on the number line. The number to the right is the larger number.

4. The numbers a and $-a$ are opposites of one another. These numbers have the same distance from zero and are on opposite sides of zero.

5. The absolute value of a number measures the number's distance from zero on the real number line. The absolute value of a number is always zero or positive. Vertical bars, $| \ |$, denote absolute value.

GRAPHING CALCULATOR SKILLS

There are three basic skills that the calculator can be applied to in this section: approximating an irrational number, finding the opposite of a number, and finding the absolute value of a number. The keystrokes correspond to a TI-83 Plus or TI-84 Plus. Similar keystrokes can be used with other calculators as well.

1. Approximate the values of $\sqrt{3}$ and π.

Solution: Enter the expressions on the calculator.

So, $\sqrt{3} \approx 1.73$ and $\pi \approx 3.14$.

2. Find the opposite of -6.

Solution: The symbol $(-)$ denotes the opposite of a number and is also used to enter a negative number. So, we must use it twice to find the opposite of -6.

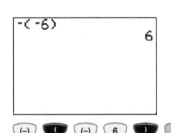

So, $-(-6) = 6$.

3. Determine the values of $|-5|$ and $-|-7|$.

Solution: The absolute value operation is found under the MATH menu. Press the MATH menu and then arrow right to go to the NUM menu. The first option is abs, which denotes absolute value.

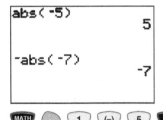

So, $|-5| = 5$ and $-|-7| = -7$.

SECTION 1.1 / EXERCISE SET

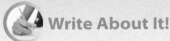 **Write About It!**

Use complete sentences to explain the meaning of each term or phrase.

1. Rational number

2. Irrational number

3. The opposite of a number

4. The absolute value of a number

Determine if the statement is true or false. If a statement is false, provide an example that contradicts the statement.

5. Every rational number is also an integer.

6. All real numbers can be classified as either a rational number or an irrational number.

7. All whole numbers are natural numbers.

8. All irrational numbers are real numbers.

9. A decimal that doesn't terminate is an irrational number.

10. The sum of a number and its additive inverse is 1.

Practice Makes Perfect!

Classify each number as a natural number, whole number, integer, rational number, irrational number, and/or real number. If the number is a rational number, write it in fractional form. If the number is irrational, approximate its value to two decimal places. (*See Objective 1.*)

11. -7

12. -15

13. 3

14. 10

15. 2.5

16. 7.3

17. $1\frac{7}{8}$

18. $-4\frac{1}{7}$

19. 0.1

20. 2.1

21. $\sqrt{81}$

22. $\sqrt{49}$

23. $\sqrt{15}$

24. $\sqrt{21}$

25. $-\pi$

26. $-\sqrt{7}$

27. 365 days

28. 52 weeks

29. $-80°F$: the coldest recorded temperature in Prospect Creek Camp, Alaska, on January 23, 1971

30. $-28°F$: the coldest recorded temperature in Caesars Head, South Carolina, on January 21, 1985

31. $109°F$: the hottest recorded temperature in Monticello, Florida, on June 29, 1931

32. $134°F$: the hottest recorded temperature in Greenland Ranch, California, on July 10, 1913

Give an example of a real number that satisfies each condition. (*See Objective 1.*)

33. A rational number that is not an integer

34. A rational number that is an integer

35. An irrational number between 2 and 3

36. An irrational number between -2 and -1

37. A rational number between -3 and -4

38. A rational number between 4 and 5

39. An integer between $\sqrt{3}$ and $\sqrt{6}$

40. An integer between $\sqrt{5}$ and $\sqrt{10}$

Graph each number on a real number line. (*See Objective 2.*)

41. -4

42. -1

43. $\frac{1}{2}$

44. $\frac{8}{3}$

45. 3.5

46. 1.5

47. -4.5

48. -2.5

49. $1.\overline{3}$

50. $0.\overline{29}$

51. $\frac{\pi}{2}$

52. 2π

53. $\sqrt{8}$

54. $\sqrt{14}$

55. $\sqrt{16}$

56. $-\sqrt{25}$

Compare the values of each pair of numbers. Use a $<$, $>$, or $=$ symbol to make the statement true. (*See Objective 3.*)

57. -7 ___ -8

58. -3 ___ -2

59. $\frac{2}{5}$ ___ 0.4

60. $-\frac{7}{4}$ ___ -1.75

61. $\sqrt{3}$ ___ 1.75

62. $\sqrt{11}$ ___ 3

63. $-(-2)$ ___ 2

64. $-(4)$ ___ 4

65. 0 ___ -2

66. $-(-6)$ ___ 0

67. π ___ 3.1

68. 1.4 ___ $\sqrt{2}$

Find the opposite of each real number. (*See Objective 4.*)

69. 6

70. 14

71. -1

72. -5

73. $\frac{1}{4}$

74. $-\frac{10}{3}$

75. -5.1

76. 10.5

77. $-\pi$

78. $\sqrt{13}$

79. $-1.\overline{3}$

80. $2.\overline{23}$

Simplify each absolute value expression. (*See Objective 5.*)

81. $|5|$

82. $|-3.8|$

83. $|-7.5|$

84. $|6.7|$

85. $\left|\frac{1}{3}\right|$

86. $-\left|\frac{5}{4}\right|$

87. $-\left|-\frac{9}{11}\right|$

88. $\left|-\frac{4}{7}\right|$

89. $-|7|$

90. $-|-14|$

Mix 'Em Up!

Classify each real number as a natural number, whole number, integer, rational number, irrational number, and/or real number. If the number is a rational number, write it in fractional form. If the number is irrational, approximate its value to two decimal places. (*See Objective 1.*)

91. 12.5

92. -17.5

93. $-\frac{11}{3}$

94. $\frac{11}{5}$

95. $96 billion: the cost of the damage from hurricane Katrina in Louisiana in 2005

96. 4160 mi—the length of the Great Wall of China, the world's longest man-made structure

97. 6.3

98. -3.47

99. $\sqrt{12}$

100. $-\sqrt{36}$

101. 1.37%: interest rate for a 12-month certificate of deposit at Ally bank

102. 5.75%: a 30-yr mortgage rate

103. $-\pi$

104. $\sqrt{31}$

105. -1293 ft: elevation of the Dead Sea

106. -86 m: elevation of Death Valley

Graph each number on a real number line. (*See Objective 2.*)

107. $\left\{-5.2, -2\frac{1}{2}, 1, \sqrt{6}\right\}$

108. $\left\{-\sqrt{18}, -3\frac{1}{3}, 1.5, 2\pi\right\}$

109. $\left\{|-5|, -1.5, \frac{4}{3}, \sqrt{9}\right\}$ **110.** $\left\{-|-8|, -\frac{\pi}{2}, \sqrt{4}, 5\frac{1}{2}\right\}$

Compare the value of each pair of numbers. Use a <, >, or = symbol to make the statement true. (*See Objective 3.*)

111. $\frac{99}{70}$ ___ $\sqrt{2}$ **112.** $\frac{21}{7}$ ___ π

113. $-|-9|$ ___ $\sqrt{81}$ **114.** $\sqrt{16}$ ___ $-|-4|$

115. $-\left|-\frac{10}{3}\right|$ ___ $\frac{\pi}{2}$ **116.** $\frac{12}{6}$ ___ $\sqrt{4}$

Find the opposite of each real number. (*See Objective 4.*)

117. $\frac{14}{3}$ **118.** $-\frac{3}{11}$

119. $4.\overline{29}$ **120.** $-1.\overline{5}$

Simplify each absolute value expression. (*See Objective 5.*)

121. $-|6|$ **122.** $|16|$

123. $|-9.5|$ **124.** $-\left|-\frac{9}{8}\right|$

 You Be the Teacher!

Correct each student's errors, if any.

125. Is the opposite of a number always negative?

     Betty's work: Yes, if x is a number, the opposite is $-x$.

126. Graph the number on the real number line: -1.99.

     Gloria's work:

          $-2 \quad -1 \quad 0 \quad 1$

127. Fill in the blank with the appropriate symbol to make the statement true: $\frac{5}{7}$ ___ $\frac{7}{10}$.

     Enefiok's work: $\frac{5}{7} < \frac{7}{10}$

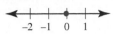 **Calculate It!**

Approximate the value of each real number to the nearest hundredth. Specify if the number is rational or irrational.

128. $\frac{\pi}{4}$ **129.** $\frac{\pi}{6}$

130. $1 + \sqrt{5}$ **131.** $2 + \sqrt{3}$

132. $2 + \frac{5}{6}$ **133.** $4 + \frac{2}{9}$

134. $6 + \sqrt{81}$ **135.** $9 - \sqrt{4}$

136. $|-142|$ **137.** $-|23|$

| SECTION 1.2 | **Fractions Review** |

▶ OBJECTIVES

As a result of completing this section, you will be able to

1. Write the prime factorization of a number.
2. Define and write fractions.
3. Simplify fractions.
4. Multiply fractions.
5. Divide fractions.
6. Add or subtract fractions with common denominators.
7. Add or subtract fractions with unlike denominators.
8. Troubleshoot common errors.

As of 2009, there were approximately 309 million people living in the United States and approximately 116 million Facebook users in the United States. Write a fraction that represents the part of the United States population that were Facebook users in 2009. (Source: http://flavorwire.com/82308/awesome-infographic-facebook-vs-the-united-states)

In this section, we will learn how to use a fraction to represent such information. We will also learn how to simplify fractions and perform operations with fractions.

Objective 1 ▶

Write the prime factorization of a number.

Factoring Numbers

To express a number in *factored form* is to write the number as a product of two or more numbers. For example, in the statement $4 \cdot 5 = 20$, the numbers 4 and 5 are called **factors** and 20 is the *product*. Factors divide a number evenly without any remainder.

The numbers 2, 3, 5, 7, 11, 13, 17, and 19 have something in common. The factors of these numbers are only 1 and the number itself. These types of numbers are called *prime numbers*. The following prime numbers are written in factored form.

$$2 = 1 \cdot 2, 3 = 1 \cdot 3, 5 = 1 \cdot 5$$

> **Definition:** A number p is a **prime number** if its whole number factors are only 1 and the number p itself.

A number that is not prime is called **composite**. This means that the number has factors other than 1 and itself. Examples of composite numbers are 4, 6, and 16, and their factored forms are shown.

Number	Factored Forms
4	$1 \cdot 4$ or $2 \cdot 2$
6	$1 \cdot 6$ or $2 \cdot 3$
16	$1 \cdot 16$ or $2 \cdot 8$ or $4 \cdot 4$

The number 1 is neither prime nor composite. Every number, other than 1, can be expressed as a product of prime factors. This factorization is called the **prime factorization** of the number. We can find the prime factorization of 60, for example, using a **factor tree**. We begin with any two factors of 60, such as 6 and 10 or 2 and 30. We find factors of any number that is not prime and continue the process until all factors are prime numbers. When we find a prime factor, we circle it for easy reference, as shown.

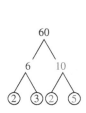
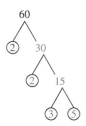

So, the prime factorization of 60 is $2 \cdot 3 \cdot 2 \cdot 5$ or $2^2 \cdot 3 \cdot 5$ because the factors 2, 3, and 5 are prime numbers. Note that it doesn't matter which two factors we choose to begin the factor tree, because we will always obtain the same factorization except for the order of the factors.

> **Procedure: Writing the Prime Factorization of a Number**
>
> **Step 1:** Write the number as a product of any two of its factors.
> **Step 2:** If both factors are prime, then the factorization is complete.
> **Step 3:** If either of the factors is not prime, rewrite that factor as a product. If these factors are prime, then the factorization is complete.
> **Step 4:** Continue this process until all factors are prime.

Objective 1 Examples   **Write the prime factorization of each number.**

1a. 42   **1b.** 120

Solutions   1a.

Write 42 as a product of any two of its factors. Circle the prime factor, 7.

Rewrite 6 as $2 \cdot 3$.

The prime factorization of 42 is $2 \cdot 3 \cdot 7$.

1b.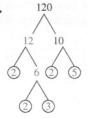

Write 120 as a product of any two of its factors.

Rewrite 12 as 2 · 6 and 10 as 2 · 5. Circle the prime factors 2, 2, and 5.

Rewrite 6 as 2 · 3.

The prime factorization of 120 is 2 · 2 · 3 · 2 · 5 or $2^3 \cdot 3 \cdot 5$.

☑ **Student Check 1**   Write the prime factorization of each number.
  a. 24   **b.** 128

Fractions

Objective 2 ▶

Define and write fractions.

In Section 1.1, we defined a rational number as a number that can be written as the quotient of two integers. A *fraction* is a quotient of two real numbers.

> **Definition:** A **fraction** is a number of the form $\dfrac{a}{b}$, where a and b are real numbers with $b \neq 0$. The number a is called the **numerator** and b is called the **denominator**.

A fraction represents a division of a whole into parts. The denominator of a fraction represents the "total number of parts in the whole" while the numerator represents the "number of parts chosen." For example, the fraction $\dfrac{3}{4}$ represents 3 parts out of 4. We can visualize this as shown in Figure 1.1.

Figure 1.1

It is important to note that a fraction may or may not be a rational number. For example,

$\dfrac{3}{4}$	This fraction is a rational number since it is a quotient of two integers.
$\dfrac{\pi}{2}$	This fraction is not a rational number since π is not an integer.

If the numerator of a fraction is less than the denominator, the fraction is called a **proper fraction**. Some examples are $\dfrac{1}{2}$ and $\dfrac{4}{15}$. If the numerator is greater than the denominator, the fraction is called an **improper fraction**. Some examples are $\dfrac{5}{3}$ and $\dfrac{17}{2}$. An improper fraction can also be written as a **mixed number**. For instance,

$$\frac{5}{3} = 1\frac{2}{3} \quad \text{and} \quad \frac{17}{2} = 8\frac{1}{2}$$

Objective 2 Examples   **Write a fraction that represents each quantity.**

2a. Prior to the 2010 gubernatorial elections, 23 Republicans, 1 Independent, and 26 Democrats held the office of governor in the United States. Write fractions that represent the portion of the U.S. governors who were Republican, Independent, and Democratic.

Solution   **2a.** The whole is the total number of governors in the United States, 50. The part is the number of governors from each party. So, the fraction of United States governors for each party is

Republican	$\dfrac{23}{50}$
Independent	$\dfrac{1}{50}$
Democrat	$\dfrac{26}{50}$

2b. As of 2009, there were approximately 309 million people living in the United States and approximately 116 million Facebook users in the United States. Write a fraction that represents the portion of the U.S. population who were Facebook users in 2009. (Source: http://flavorwire.com/82308/awesome-infographic-facebook-vs-the-united-states)

Solution   **2b.** The whole is the population of the United States, 309 million. The part is the number of Facebook users, 116 million. So, the fraction of the U.S. population who were Facebook users is

$$\frac{116{,}000{,}000}{309{,}000{,}000}$$

✔ **Student Check 2**   Write a fraction that represents each quantity.

a. Thirty-nine states elected a governor during the 2010 gubernatorial elections. Write a fraction that represents the portion of states that went through a gubernatorial election. (Source: http://statehouserock.com/)

b. The National Center for Education Statistics published the number of bachelor's degrees conferred by major for the 2007–2008 academic year. The pie graph to the right shows this information. Write a fraction that represents the portion of Education degrees that was awarded.

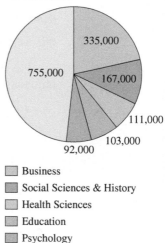

- ☐ Business
- ☐ Social Sciences & History
- ☐ Health Sciences
- ☐ Education
- ☐ Psychology
- ☐ Other

Writing Fractions in Simplified Form

Objective 3 ▶

Simplify fractions.

A fraction is **simplified** or in **lowest terms** if the numerator and denominator share no common factors other than 1. The following property provides us a method to write a fraction in lowest terms.

Property: The Fundamental Property of Fractions

If $\dfrac{a}{b}$ is a fraction, where $b \neq 0$ and c is a nonzero real number, then

$$\frac{a \cdot c}{b \cdot c} = \frac{a}{b}$$

The reason the fundamental property works is that $\frac{c}{c} = 1$.

$$\frac{a \cdot c}{b \cdot c} = \frac{a}{b} \cdot \frac{c}{c}$$

$$= \frac{a}{b} \cdot 1$$

$$= \frac{a}{b}$$

Procedure: Simplifying Fractions

Step 1: Write the numerator and denominator with their prime factorizations.
Step 2: Divide the numerator and denominator by their common factors.
Step 3: Multiply the remaining factors.

Objective 3 Examples Simplify each fraction.

3a. $\frac{28}{30}$ **3b.** $\frac{144}{99}$ **3c.** $\frac{116,000,000}{309,000,000}$

Solutions **3a.**

$$\frac{28}{30} = \frac{2 \cdot 2 \cdot 7}{2 \cdot 3 \cdot 5}$$ Write the prime factorization of 28 and 30.

$$= \frac{\cancel{2} \cdot 2 \cdot 7}{\cancel{2} \cdot 3 \cdot 5}$$ Divide out the common factor of 2.

$$= \frac{14}{15}$$ Multiply the remaining factors.

Note: *It is not necessary to write the prime factorization of the numerator and denominator of a fraction to simplify it. For instance, observe that both 28 and 30 are divisible by 2. So, we can divide them each by 2 to obtain the simplified form of the fraction.*

$$\frac{\overset{14}{\cancel{28}}}{\underset{15}{\cancel{30}}} = \frac{14}{15}$$

3b.

$$\frac{144}{99} = \frac{2 \cdot 2 \cdot 2 \cdot 2 \cdot 3 \cdot 3}{3 \cdot 3 \cdot 11}$$ Write the prime factorization of 144 and 99.

$$= \frac{2 \cdot 2 \cdot 2 \cdot 2 \cdot \cancel{3} \cdot \cancel{3}}{\cancel{3} \cdot \cancel{3} \cdot 11}$$ Divide out the common factors.

$$= \frac{16}{11}$$ Multiply the remaining factors.

3c.

$$\frac{116,000,000}{309,000,000} = \frac{116,000,000 \div 1,000,000}{309,000,000 \div 1,000,000}$$ Divide the numerator and denominator by their common factor of 1,000,000.

$$= \frac{116}{309}$$ Simplify.

✓ Student Check 3 Simplify each fraction.

a. $\frac{20}{32}$ **b.** $\frac{180}{200}$ **c.** $\frac{104,000}{624,000}$

Multiplying Fractions

Objective 4 ▶

Multiply fractions.

We can multiply fractions by multiplying the numerators of the fractions and by multiplying the denominators of the fractions. We always want to express the answers in lowest terms.

> **Property: Multiplying Fractions**
>
> For $b, d \neq 0$,
>
> $$\frac{a}{b} \cdot \frac{c}{d} = \frac{a \cdot c}{b \cdot d}$$

Objective 4 Examples   **Multiply the fractions and write each answer in lowest terms.**

4a. $\dfrac{3}{4} \cdot \dfrac{5}{2}$ **4b.** $\dfrac{10}{3} \cdot \dfrac{18}{25}$ **4c.** $2\dfrac{2}{3} \cdot 5\dfrac{3}{4}$

Solutions   **4a.** $\dfrac{3}{4} \cdot \dfrac{5}{2} = \dfrac{3 \cdot 5}{4 \cdot 2}$ Multiply the numerators and denominators.

$= \dfrac{15}{8}$ Simplify the products.

4b. $\dfrac{10}{3} \cdot \dfrac{18}{25} = \dfrac{10 \cdot 18}{3 \cdot 25}$ Multiply the numerators and denominators.

$= \dfrac{2 \cdot 5 \cdot 2 \cdot 3 \cdot 3}{3 \cdot 5 \cdot 5}$ Write the prime factorization of each number.
$\quad 10 = 2 \cdot 5, \ 18 = 2 \cdot 3 \cdot 3, \ 25 = 5 \cdot 5$

$= \dfrac{2 \cdot 2 \cdot 3}{5}$ Divide out the common factors of 5 and 3.

$= \dfrac{12}{5}$ Multiply the remaining factors.

4c. $2\dfrac{2}{3} \cdot 5\dfrac{3}{4} = \dfrac{8}{3} \cdot \dfrac{23}{4}$ Convert the mixed fractions to improper fractions.
$\quad 2\dfrac{2}{3} = \dfrac{8}{3} \quad$ and $\quad 5\dfrac{3}{4} = \dfrac{23}{4}$

$= \dfrac{\overset{2}{\cancel{8}} \cdot 23}{3 \cdot \underset{1}{\cancel{4}}}$ Multiply the numerators and denominators. Divide out the common factor of 4 from the numerator and denominator.

$= \dfrac{46}{3}$ Multiply the remaining factors.

✔ **Student Check 4**   Multiply the fractions and write each answer in lowest terms.

a. $\dfrac{5}{8} \cdot \dfrac{3}{7}$ **b.** $\dfrac{4}{9} \cdot \dfrac{15}{14}$ **c.** $1\dfrac{3}{7} \cdot 2\dfrac{1}{10}$

Dividing Fractions

Objective 5 ▶

Divide fractions.

Before we can divide fractions, we must discuss the concept of reciprocals. Two numbers are *reciprocals* if their product is 1. Some examples are shown.

Number	Reciprocal	Their Product
6	$\dfrac{1}{6}$	$6 \cdot \dfrac{1}{6} = \dfrac{6}{1} \cdot \dfrac{1}{6} = 1$
$\dfrac{4}{5}$	$\dfrac{5}{4}$	$\dfrac{4}{5} \cdot \dfrac{5}{4} = 1$

> **Definition:** Two numbers are **reciprocals** if their product is 1. The reciprocal of $\frac{c}{d}$ is $\frac{d}{c}$ since
>
> $$\frac{c}{d} \cdot \frac{d}{c} = \frac{c \cdot d}{d \cdot c} = 1$$

Dividing fractions is related to multiplying fractions. For example, we know

$$30 \div 6 = 5, \text{ but we also know } 30 \cdot \frac{1}{6} = \frac{30}{1} \cdot \frac{1}{6} = \frac{30}{6} = 5$$

Therefore,

$$30 \div 6 = 30 \cdot \frac{1}{6}$$

This illustrates that dividing by a number is the same as multiplying by the reciprocal of the number.

> **Property: Dividing Fractions**
> For $b, c, d \neq 0$,
>
> $$\frac{a}{b} \div \frac{c}{d} = \frac{a}{b} \cdot \frac{d}{c}$$

Objective 5 Examples Divide the fractions and write each answer in lowest terms.

5a. $\frac{8}{7} \div \frac{12}{14}$ **5b.** $\frac{3}{5} \div 9$ **5c.** $3\frac{3}{7} \div \frac{7}{15}$

Solutions **5a.** $\frac{8}{7} \div \frac{12}{14} = \frac{8}{7} \cdot \frac{14}{12}$ Multiply by the reciprocal of $\frac{12}{14}$, which is $\frac{14}{12}$.

$= \frac{8 \cdot 14}{7 \cdot 12}$ Multiply the numerators and denominators.

$= \frac{2 \cdot 2 \cdot 2 \cdot 2 \cdot 7}{7 \cdot 2 \cdot 2 \cdot 3}$ Rewrite each number as a product of prime factors. $8 = 2 \cdot 2 \cdot 2, 14 = 2 \cdot 7, 12 = 2 \cdot 2 \cdot 3$

$= \frac{2 \cdot 2}{3}$ Divide out the common factors of 2, 2, and 7.

$= \frac{4}{3}$ Multiply the remaining factors.

5b. $\frac{3}{5} \div 9 = \frac{3}{5} \cdot \frac{1}{9}$ Multiply by the reciprocal of 9, which is $\frac{1}{9}$.

$= \frac{3 \cdot 1}{5 \cdot 9}$ Multiply the numerators and denominators.

$= \frac{3 \cdot 1}{5 \cdot 3 \cdot 3}$ $9 = 3 \cdot 3$. Rewrite each number as a product of prime factors.

$= \frac{1}{5 \cdot 3}$ Divide out the common factor of 3.

$= \frac{1}{15}$ Multiply the remaining factors.

5c. $3\frac{3}{7} \div \frac{7}{15} = \frac{24}{7} \cdot \frac{15}{7}$ Write $3\frac{3}{7}$ as $\frac{24}{7}$ and multiply by the reciprocal of $\frac{7}{15}$, which is $\frac{15}{7}$.

$= \frac{24 \cdot 15}{7 \cdot 7}$ Multiply the numerators and denominators.

$= \frac{360}{49}$ or $7\frac{17}{49}$ Simplify the products.

The answer can be written as either an improper fraction or as a mixed number.

☑ **Student Check 5**   Divide the fractions and write each answer in lowest terms.

a. $\frac{12}{15} \div \frac{30}{24}$ **b.** $\frac{6}{7} \div 12$ **c.** $5\frac{2}{3} \div 4\frac{1}{9}$

Adding and Subtracting Fractions with Like Denominators

Objective 6 ▶

Add or subtract fractions with like denominators.

One common method for adding or subtracting fractions requires that the denominators of the fractions be the same. If the denominators are the same, we simply add or subtract the numerators and place this number over their common denominator.

For example,

$$\frac{2}{7} + \frac{1}{7} = \frac{2+1}{7} = \frac{3}{7}$$

> **Property: Adding or Subtracting Fractions with Common Denominators**
>
> $$\frac{a}{b} + \frac{c}{b} = \frac{a+c}{b}, \quad \text{for } b \neq 0$$
>
> $$\frac{a}{b} - \frac{c}{b} = \frac{a-c}{b}, \quad \text{for } b \neq 0$$

Objective 6 Examples   **Add or subtract the fractions and simplify the result.**

6a. $\frac{5}{8} + \frac{2}{8}$ **6b.** $1\frac{1}{3} + 6\frac{2}{3}$ **6c.** $\frac{8}{9} - \frac{2}{9}$ **6d.** $5\frac{1}{7} - 4\frac{3}{7}$

Solutions   **6a.** $\frac{5}{8} + \frac{2}{8} = \frac{5+2}{8}$ Add the numerators and place over the common denominator.

$= \frac{7}{8}$ Simplify the numerator.

6b. $1\frac{1}{3} + 6\frac{2}{3} = \frac{4}{3} + \frac{20}{3}$ Write each mixed number as an improper fraction.

$= \frac{4+20}{3}$ Add the numerators and place over the common denominator.

$= \frac{24}{3}$ Simplify the numerator.

$= 8$ Simplify the fraction.

6c. $\frac{8}{9} - \frac{2}{9} = \frac{8-2}{9}$ Subtract the numerators and place over the common denominator.

$= \frac{6}{9}$ Simplify the numerator.

$= \frac{2 \cdot 3}{3 \cdot 3}$ Factor the numerator and denominator.

$= \frac{2}{3}$ Divide out the common factor of 3.

6d. $5\dfrac{1}{7} - 4\dfrac{3}{7} = \dfrac{36}{7} - \dfrac{31}{7}$ Write each mixed number as an improper fraction.

$\qquad\qquad = \dfrac{36 - 31}{7}$ Subtract the numerators and place over the common denominator.

$\qquad\qquad = \dfrac{5}{7}$ Simplify the numerator.

✓ **Student Check 6**   Add or subtract the fractions and simplify the result.

a. $\dfrac{4}{5} + \dfrac{2}{5}$ **b.** $\dfrac{3}{8} + \dfrac{1}{8}$ **c.** $\dfrac{4}{5} - \dfrac{1}{5}$ **d.** $9\dfrac{2}{9} - 7\dfrac{4}{9}$

Adding and Subtracting Fractions with Different Denominators

Objective 7 ▶

Add or subtract fractions with unlike denominators.

If the fractions we need to add do not have the same denominator, we must first rewrite each fraction as an *equivalent fraction* with the same denominator, or a *common denominator*.

Equivalent fractions are fractions that represent the same quantity. For example, $\dfrac{3}{4}$ and $\dfrac{6}{8}$ are equivalent fractions as shown.

We can obtain equivalent fractions by multiplying a given fraction by 1. Recall multiplying by 1 does not change the value of the fraction. When we multiply a fraction by 1, we are applying the Fundamental Property of Fractions. This property not only enables us to simplify fractions, but also enables us to multiply the numerator and denominator by the same nonzero number—that is, by a form of 1. For example, to rewrite $\dfrac{3}{4}$ as a fraction with a denominator of 8, we can multiply the numerator and denominator by 2 as shown.

$$\frac{3}{4} = \frac{3}{4} \cdot 1 = \frac{3}{4} \cdot \frac{2}{2} = \frac{3 \cdot 2}{4 \cdot 2} = \frac{6}{8}$$

If we want to add $\dfrac{4}{9}$ and $\dfrac{3}{2}$, for example, we must first convert these fractions to equivalent fractions with the same, or common denominator. If we find the *least common denominator*, that will make the process somewhat easier.

The **least common denominator (LCD)** is the smallest number that all denominators divide into evenly. A *common denominator* is a number that all the denominators divide evenly but may not be the smallest such number. So, if the denominators are 9 and 2, the LCD is the smallest number that both 9 and 2 divide into—that is, the number 18.

If the LCD is not easily identified, we can use the prime factorizations of the denominators to determine it. For example, the least common denominator of 6 and 20 is 60. We know this by their prime factorizations as shown.

$$\begin{array}{l} 6 = 2 \cdot ③ \\ 20 = ②\cdot②\cdot⑤ \end{array} \qquad \longrightarrow \qquad \text{LCD} = 2 \cdot 2 \cdot 3 \cdot 5 = 60$$

Notice that 6 has a prime factor of 2 and 20 has two prime factors of 2. Since 2 appears twice in the prime factorization of 20, we use two factors of 2 in the LCD. Since 3 and 5 appear only once in the prime factorizations of 6 and 20, respectively, we include one factor of each of them in the product of the LCD.

Procedure: Finding the LCD

Step 1: Write the prime factorization of each denominator.
Step 2: For each of the different factors, circle the largest number of occurrences of this factor found in the factorizations.
Step 3: The product of the circled factors is the LCD.

Fact: *The LCD must include all of the different factors found in the denominators. The factor must repeat the largest number of times it appears in any of the denominators.*

Procedure: Adding or Subtracting Fractions with Unlike Denominators

Step 1: Determine the LCD.
Step 2: Convert each fraction to an equivalent fraction with the LCD as its denominator. Multiply the numerator and denominator by the number that is needed to make the denominator the LCD.
Step 3: Add or subtract the fractions.
Step 4: Simplify the result, if necessary.

Objective 7 Examples   Add or subtract the fractions and simplify the result.

7a. $\dfrac{2}{3} + \dfrac{1}{4}$ **7b.** $\dfrac{5}{12} - \dfrac{3}{8}$ **7c.** $2 + \dfrac{5}{6}$ **7d.** $4\dfrac{1}{12} - 1\dfrac{9}{14}$

Solutions   **7a.**
$$\dfrac{2}{3} + \dfrac{1}{4} = \dfrac{2}{3} \cdot \dfrac{4}{4} + \dfrac{1}{4} \cdot \dfrac{3}{3}$$
Multiply each fraction by a form of 1 to obtain the LCD, 12.

$$= \dfrac{8}{12} + \dfrac{3}{12}$$
Simplify each product.

$$= \dfrac{8 + 3}{12}$$
Add the numerators and place over the common denominator.

$$= \dfrac{11}{12}$$
Simplify the numerator.

7b. The LCD of 12 and 8 can be found as follows.

$$12 = 2 \cdot 2 \cdot ③$$
$$8 = ②\cdot 2 \cdot 2$$
$$\longrightarrow \qquad \text{LCD} = 2 \cdot 2 \cdot 2 \cdot 3 = 24$$

$$\dfrac{5}{12} - \dfrac{3}{8} = \dfrac{5}{12} \cdot \dfrac{2}{2} - \dfrac{3}{8} \cdot \dfrac{3}{3}$$
Multiply each fraction by a form of 1 to obtain the LCD, 24.

$$= \dfrac{10}{24} - \dfrac{9}{24}$$
Simplify each product.

$$= \dfrac{10 - 9}{24}$$
Subtract the numerators and place over the common denominator.

$$= \dfrac{1}{24}$$
Simplify the numerator.

7c.
$$2 + \dfrac{5}{6} = \dfrac{2}{1} + \dfrac{5}{6}$$
Write 2 as $\dfrac{2}{1}$.

$$= \dfrac{2}{1} \cdot \dfrac{6}{6} + \dfrac{5}{6}$$
Multiply the first fraction by a form of 1 to obtain the LCD, 6.

$$= \frac{12}{6} + \frac{5}{6}$$ Simplify the product.

$$= \frac{12 + 5}{6}$$ Add the numerators and place over the common denominator.

$$= \frac{17}{6}$$ Simplify the numerator.

7d. We first convert each mixed number to an improper fraction, $4\frac{1}{12} = \frac{49}{12}$ and $1\frac{9}{14} = \frac{23}{14}$. We find the LCD, as follows.

$$14 = 2 \cdot \boxed{7}$$
$$12 = \boxed{2 \cdot 2} \cdot \boxed{3}$$ $\longrightarrow$ $$LCD = 2 \cdot 2 \cdot 3 \cdot 7 = 84$$

$$4\frac{1}{12} - 1\frac{9}{14} = \frac{49}{12} - \frac{23}{14}$$ Write each fraction as an improper fraction.

$$= \frac{49}{12} \cdot \frac{7}{7} - \frac{23}{14} \cdot \frac{6}{6}$$ Multiply each fraction by a form of 1 to obtain the LCD, 84.

$$= \frac{343}{84} - \frac{138}{84}$$ Simplify each product.

$$= \frac{343 - 138}{84}$$ Subtract the numerators and place over the common denominator.

$$= \frac{205}{84}$$ Simplify the numerator.

$$= 2\frac{37}{84}$$ Convert to a mixed number.

 **Student Check 7**   Add or subtract the fractions and simplify the result.

    a. $\frac{2}{7} + \frac{1}{3}$ **b.** $\frac{5}{24} - \frac{1}{30}$ **c.** $7 + \frac{1}{3}$ **d.** $6\frac{3}{16} - 4\frac{9}{20}$

Objective 8 ▶

Troubleshoot common errors.

Troubleshooting Common Errors

Some common errors associated with fractions are shown next.

Objective 8 Examples   **A problem and an incorrect solution are given. Provide the correct solution and an explanation of the error.**

8a. Simplify $\frac{2}{3} \cdot \frac{1}{5}$.

Incorrect Solution	Correct Solution and Explanation
$$\frac{2}{3} \cdot \frac{1}{5} = \frac{2}{3} \cdot \frac{5}{5} \times \frac{1}{5} \cdot \frac{3}{3}$$ $$= \frac{10}{15} \times \frac{3}{15}$$ $$= \frac{30}{225}$$ $$= \frac{2}{15}$$	We do not need to find the LCD when multiplying fractions, we can simply multiply the numerators and denominators. $$\frac{2}{3} \cdot \frac{1}{5} = \frac{2 \cdot 1}{3 \cdot 5} = \frac{2}{15}$$

8b. Simplify $\dfrac{4}{9} + \dfrac{1}{9}$.

Incorrect Solution	Correct Solution and Explanation
$\dfrac{4}{9} + \dfrac{1}{9} = \dfrac{4+1}{9+9} = \dfrac{5}{18}$	To add fractions with the same denominator, we add their numerators and place this sum over the common denominator. $$\dfrac{4}{9} + \dfrac{1}{9} = \dfrac{4+1}{9} = \dfrac{5}{9}$$

ANSWERS TO STUDENT CHECKS

Student Check 1   **a.** $24 = 2 \cdot 2 \cdot 2 \cdot 3$
  b. $128 = 2 \cdot 2 \cdot 2 \cdot 2 \cdot 2 \cdot 2 \cdot 2$

Student Check 2   **a.** $\dfrac{39}{50}$ **b.** $\dfrac{103,000}{1,563,000}$

Student Check 3   **a.** $\dfrac{5}{8}$ **b.** $\dfrac{9}{10}$ **c.** $\dfrac{1}{6}$

Student Check 4   **a.** $\dfrac{15}{56}$ **b.** $\dfrac{10}{21}$ **c.** 3

Student Check 5   **a.** $\dfrac{16}{25}$ **b.** $\dfrac{1}{14}$ **c.** $\dfrac{51}{37}$

Student Check 6   **a.** $\dfrac{6}{5}$ **b.** $\dfrac{1}{2}$ **c.** $\dfrac{3}{5}$ **d.** $\dfrac{16}{9}$ or $1\dfrac{7}{9}$

Student Check 7   **a.** $\dfrac{13}{21}$ **b.** $\dfrac{7}{40}$ **c.** $\dfrac{22}{3}$ **d.** $\dfrac{139}{80}$ or $1\dfrac{59}{80}$

SUMMARY OF KEY CONCEPTS

1. The prime factorization of a number is a way to write a number as a product of prime numbers. Prime numbers are 2, 3, 5, 7, 11, 13, and so on.
2. A fraction represents a part of a whole and is written as $\dfrac{a}{b}$.
3. Fractions can be simplified by dividing the numerator and denominator by their common factor.
4. Multiply fractions by multiplying their numerators and their denominators together.
5. Divide fractions by multiplying by the reciprocal. The reciprocal of the fraction $\dfrac{a}{b}$ is $\dfrac{b}{a}$ for a and b not zero.
6. We can only add and subtract fractions with a common denominator.
7. If the given fractions do not have the same denominator, convert each fraction to an equivalent fraction with the LCD as its denominator.

GRAPHING CALCULATOR SKILLS

Practice operations with fractions by hand before relying on the calculator to perform the work for you. At this point, use the calculator only as a check of your work. The calculator will display all answers as decimals. To convert an answer to a fraction, use the **MATH** menu.

1. Simplify $\dfrac{12}{18}$.

```
12/18
          .6666666667
Ans▶Frac
                  2/3
```

2. $\dfrac{1}{4} \cdot \dfrac{12}{15}$

```
1/4*12/15
                    .2
Ans▶Frac
                   1/5
```

3. $\dfrac{8}{7} \div \dfrac{12}{14}$

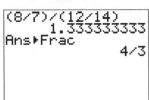

4. $\dfrac{5}{12} - \dfrac{3}{8}$

SECTION 1.2 — EXERCISE SET

Write About It!

Use complete sentences to explain the meaning of the given term or process.

1. Fraction

2. Simplifying a fraction to lowest terms

3. Prime number

4. Composite number

5. Prime factorization of a number

6. (a) Equivalent fractions

    (b) Converting a fraction to an equivalent fraction

7. (a) Least common denominator

    (b) Finding the LCD between two fractions

8. Multiplying fractions

9. Dividing fractions

10. (a) Adding fractions

    (b) Subtracting fractions

Determine if the statement is true or false. If a statement is false, explain why.

11. The fraction $\dfrac{2+3}{5+3}$ simplifies to $\dfrac{2}{5}$.

12. All odd numbers are prime numbers.

13. The only even prime number is the number 2.

14. The LCD of $\dfrac{3}{4}, \dfrac{5}{6}$, and $\dfrac{1}{12}$ is 24.

15. The number 0 has a reciprocal.

16. $4 \div \dfrac{1}{8} = \dfrac{1}{4} \times \dfrac{1}{8}$

17. $\dfrac{1}{2} + \dfrac{1}{5} = \dfrac{2}{7}$

18. $\dfrac{3}{7} - \dfrac{1}{3} = \dfrac{2}{4} = \dfrac{1}{2}$

Practice Makes Perfect!

Write the prime factorization of each number. (*See Objective 1.*)

19. 12   **20.** 27   **21.** 80

22. 84   **23.** 144   **24.** 210

Write a fraction that represents the given quantity. (*See Objective 2.*)

25. As of June 30, 2010, there were approximately 3835 million people living in Asia and approximately 825 million Internet users in Asia. Write a fraction that represents the portion of Internet users in Asia. (Source: http://internetworldstats.com/america.htm)

26. As of June 30, 2010, there were approximately 344 million people living in North America and approximately 226 million Internet users. Write a fraction that represents the portion of Internet users in North America. (Source: http://internetworldstats.com/america.htm)

27. California had a total of 53 members of the U.S. House of Representatives serving in the 111th Congress and 34 representatives were Democrats. Write a fraction that represents the portion of Democrats who served as representatives for California. (http://www.house.gov)

28. Georgia had a total of 13 members of the U.S. House of Representatives serving in the 111th Congress and 7 representatives were Republicans. Write a fraction that represents the portion of Republicans who served as representatives for Georgia. (http://www.house.gov)

29. The top 10 U.S. Internet search providers processed a total of approximately 9,200,000,000 search requests, during August 2010. Google processed about 6,000,000,000 and Yahoo! processed about 1,210,000,000 search requests. Write fractions that represent the portion of Google search requests and Yahoo! search requests among the top 10 U.S. providers. (http://en-us.nielsen.com/content/nielsen/en_us/insights/rankings/internet.html)

30. Senators in Class III took office in 2005 and their terms expired in 2011. With a total of 34 senators in Class III, 16 are Democrats and 18 are Republicans. Write fractions that represent the portion of Democrats and the portion of Republicans in Class III. (http://www.senate.gov)

Simplify each fraction. (*See Objective 3.*)

31. $\dfrac{8}{24}$ 32. $\dfrac{12}{60}$ 33. $\dfrac{64}{8}$

34. $\dfrac{70}{35}$ 35. $\dfrac{192,000}{256,000}$ 36. $\dfrac{126,000}{432,000}$

Multiply the fractions and simplify each result. (*See Objective 4.*)

37. $\dfrac{1}{2} \cdot \dfrac{3}{5}$ 38. $\dfrac{4}{7} \cdot \dfrac{2}{11}$ 39. $\dfrac{2}{3} \cdot \dfrac{9}{28}$

40. $\dfrac{4}{15} \cdot \dfrac{5}{24}$ 41. $\dfrac{3}{5} \cdot \dfrac{5}{3}$ 42. $\dfrac{11}{10} \cdot \dfrac{10}{11}$

43. $\dfrac{5}{4} \cdot 20$ 44. $12 \cdot \dfrac{3}{2}$ 45. $2\dfrac{2}{3} \cdot 3\dfrac{3}{4}$

46. $3\dfrac{1}{2} \cdot 1\dfrac{1}{7}$

Divide the fractions and simplify each result. (*See Objective 5.*)

47. $\dfrac{1}{2} \div \dfrac{3}{5}$ 48. $\dfrac{4}{7} \div \dfrac{2}{11}$

49. $\dfrac{2}{3} \div \dfrac{9}{28}$ 50. $\dfrac{7}{15} \div \dfrac{14}{36}$

51. $\dfrac{1}{3} \div 3$ 52. $7 \div \dfrac{1}{7}$

53. $20 \div \dfrac{5}{4}$ 54. $12 \div \dfrac{3}{5}$

55. $6\dfrac{2}{3} \div 1\dfrac{1}{9}$ 56. $7\dfrac{1}{2} \div 2\dfrac{1}{2}$

Add or subtract the fractions and simplify each result. (*See Objective 6.*)

57. $\dfrac{1}{2} + \dfrac{7}{2}$ 58. $\dfrac{2}{3} + \dfrac{10}{3}$ 59. $\dfrac{3}{7} + \dfrac{1}{7}$

60. $\dfrac{5}{10} + \dfrac{1}{10}$ 61. $\dfrac{9}{10} - \dfrac{1}{10}$ 62. $\dfrac{5}{12} - \dfrac{3}{12}$

63. $7\dfrac{3}{4} + 1\dfrac{3}{4}$ 64. $6\dfrac{1}{5} - 3\dfrac{4}{5}$

Find the least common denominator (LCD) of the fractions. (*See Objective 7.*)

65. $\dfrac{2}{3}, \dfrac{3}{4}$ 66. $\dfrac{1}{5}, \dfrac{1}{2}$ 67. $\dfrac{5}{8}, \dfrac{1}{2}$

68. $\dfrac{3}{5}, \dfrac{7}{20}$ 69. $\dfrac{3}{18}, \dfrac{4}{27}$ 70. $\dfrac{5}{6}, \dfrac{2}{21}$

71. $\dfrac{1}{2}, 8$ 72. $\dfrac{4}{7}, 5$

Add or subtract the fractions and simplify each result. (*See Objective 7.*)

73. $\dfrac{1}{2} + \dfrac{1}{7}$ 74. $\dfrac{1}{4} + \dfrac{1}{5}$ 75. $\dfrac{3}{8} + 4$

76. $\dfrac{6}{11} + 2$ 77. $\dfrac{7}{2} - \dfrac{7}{3}$ 78. $\dfrac{6}{4} - \dfrac{6}{5}$

79. $5 - \dfrac{1}{2}$ 80. $4 - \dfrac{2}{3}$ 81. $\dfrac{3}{14} - \dfrac{1}{12}$

82. $\dfrac{4}{10} - \dfrac{2}{15}$ 83. $2\dfrac{1}{6} + 3\dfrac{2}{3}$ 84. $6\dfrac{1}{4} - 3\dfrac{1}{2}$

Mix 'Em Up!

Simplify each fraction.

85. $\dfrac{243}{405}$ 86. $\dfrac{11400}{60800}$

87. $\dfrac{336}{384}$ 88. $\dfrac{144}{464}$

Perform the operation and simplify each result.

89. $\dfrac{4}{15} + \dfrac{13}{30}$ 90. $\dfrac{1}{3} + \dfrac{5}{12}$ 91. $4\dfrac{1}{5} \cdot 1\dfrac{5}{7}$

92. $3\dfrac{3}{10} \cdot 1\dfrac{2}{3}$ 93. $\dfrac{5}{3} - \dfrac{1}{5}$ 94. $\dfrac{7}{2} - \dfrac{1}{3}$

95. $6\dfrac{7}{8} \div 2\dfrac{3}{4}$ 96. $1\dfrac{2}{7} \div 6\dfrac{3}{7}$ 97. $9\dfrac{1}{4} - 6\dfrac{2}{3}$

98. $7\dfrac{2}{5} + 8\dfrac{4}{5}$ 99. $\dfrac{5}{18} \cdot \dfrac{2}{15}$ 100. $\dfrac{2}{3} \cdot \dfrac{21}{16}$

101. $10\dfrac{2}{5} - 8\dfrac{4}{5}$ 102. $2\dfrac{5}{6} - 1\dfrac{7}{8}$

You Be the Teacher!

Correct each student's errors, if any.

103. Find the numerator that makes the fractions equivalent: $\dfrac{3}{8} = \dfrac{}{16}$.

Mary's work:

$\dfrac{3}{8} = \dfrac{3+8}{8+8} = \dfrac{11}{16}$. So, $\dfrac{3}{8} = \dfrac{11}{16}$.

104. Multiply and simplify the result: $\dfrac{2}{5} \times \dfrac{3}{10}$.

Jennifer's work:

$\dfrac{2}{5} \times \dfrac{3}{10} = \dfrac{2 \times 2}{5 \times 2} \cdot \dfrac{3}{10} = \dfrac{4}{10} \cdot \dfrac{3}{10} = \dfrac{12}{10} = \dfrac{2 \times 6}{2 \times 5} = \dfrac{6}{5}$

105. Multiply and simplify the result: $2\frac{1}{3} \times 3\frac{3}{4}$.

Scott's work:

$$2\frac{1}{3} \times 3\frac{3}{4} = 6\frac{3}{12} = 6\frac{1}{4} = \frac{25}{4}$$

106. Subtract and simplify the result: $\frac{3}{8} - \frac{1}{12}$.

Pauline's work: $8 \times 12 = 96$

$$\frac{3}{8} - \frac{1}{12} = \frac{3 \times 12}{8 \times 12} - \frac{1 \times 8}{12 \times 8} = \frac{36}{96} - \frac{8}{96}$$

$$= \frac{28}{96} = \frac{4 \times 7}{4 \times 24} = \frac{7}{24}$$

 **Calculate It!**

Use a calculator to perform the operation. Verify the answer by performing the operation by hand.

107. $\frac{3}{7} + \frac{12}{11}$ **108.** $\frac{8}{13} \cdot 6$ **109.** $3 - \frac{4}{3}$

110. $\frac{12}{35} \cdot \frac{14}{3}$ **111.** $\frac{3}{14} \cdot 3$ **112.** $5 + \frac{6}{5}$

The Order of Operations, Algebraic Expressions, and Equations

▶ **OBJECTIVES**

As a result of completing this section, you will be able to

1. Evaluate exponential expressions.
2. Use the order of operations to simplify numerical expressions.
3. Evaluate algebraic expressions.
4. Determine if a value is a solution of an equation.
5. Express relationships mathematically.
6. Solve application problems.
7. Troubleshoot common errors.

While we may not simplify numerical expressions on a daily basis, many of the things we encounter in life are based on this skill. One example involves saving money. There is an equation that can be used to calculate how much money we will accumulate if we invest a specific amount of money in a savings account. For example, if we invest $5000 in a savings account that earns 6% annual interest for 3 yr, the total amount saved is given by the equation

$$\text{Amount saved} = 5000(1.06)^3$$

To find the value of this expression, we must know the order in which to perform the operations. In this section, we will learn how to simplify numerical expressions using the accepted order of operations. We will also learn how to apply this skill to evaluate algebraic expressions and equations.

Exponential Expressions

The equation shown in the section opener contains an expression of the form $(1.06)^3$. This is an example of an *exponential expression*.

Objective 1 ▶

Evaluate exponential expressions.

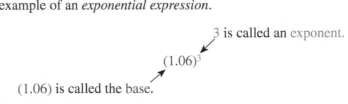

3 is called an exponent.

$(1.06)^3$

(1.06) is called the base.

The exponent implies that we use the base as a repeated factor 3 times. That is,

$$(1.06)^3 = (1.06)(1.06)(1.06) \approx 1.19$$

Definition: Exponential Notation

For b a real number and n a natural number,

$$b^n = \underbrace{b \cdot b \cdot b \cdots b}_{n \text{ times}}$$

The number b is called the **base** of the **exponential expression** and the number n is called the **exponent**.

When the number n is a natural number, it indicates the number of times b is multiplied by itself or used as a factor. Some examples are illustrated next.

Phrase	Mathematical Expression
3 squared	$3^2 = 3 \cdot 3 = 9$
5 cubed	$5^3 = 5 \cdot 5 \cdot 5 = 125$
6 to the 4th	$6^4 = 6 \cdot 6 \cdot 6 \cdot 6 = 1296$

Procedure: Evaluating an Exponential Expression

Step 1: Identify the base and the exponent.
Step 2: Rewrite the expression as repeated multiplication as indicated by the exponent.
Step 3: Simplify the result.

Objective 1 Examples Complete the chart by identifying the base, the exponent, and then evaluate the expression.

Problems	Base	Exponent	Evaluate
1a. 2^5	2	5	$2^5 = 2 \cdot 2 \cdot 2 \cdot 2 \cdot 2 = 32$
1b. 4.2^3	4.2	3	$4.2^3 = (4.2)(4.2)(4.2) = 74.088$
1c. $\left(\dfrac{2}{3}\right)^4$	$\dfrac{2}{3}$	4	$\left(\dfrac{2}{3}\right)^4 = \dfrac{2}{3} \cdot \dfrac{2}{3} \cdot \dfrac{2}{3} \cdot \dfrac{2}{3} = \dfrac{16}{81}$
1d. -6^2	6	2	We must find the *opposite* of 6 squared. $-6^2 = -(6)(6) = -36$

✓ **Student Check 1** Identify the base, the exponent, and then evaluate the expression.

    a. 10^3 **b.** 1.5^2 **c.** $\left(\dfrac{6}{7}\right)^3$ **d.** -8^4

The Order of Operations

Objective 2 ▶

Use the order of operations to simplify numerical expressions.

The expression $5000(1.06)^3$, found in the section opener, is an example of an expression involving a combination of operations. This expression involves both multiplication and exponents. To simplify this expression, we need to know the order in which to perform the operations. Without an order to perform the operations, we would get different values for the same expression.

If we multiply 5000 and 1.06 first in the expression $5000(1.06)^3$, we get

$$[5000(1.06)]^3 = (5300)^3 = 148{,}877{,}000{,}000$$

If we evaluate the exponential expression $(1.06)^3$ first, we get

$$5000(1.06)^3 = 5000(1.191016) = 5955.08$$

Sometimes expressions include *grouping symbols* such as parentheses ( ), brackets [ ], braces { }, absolute value symbols | |, square roots $\sqrt{\ }$, or the fraction bar. Depending on where the grouping symbols are placed in an expression, different results may be obtained.

To avoid confusion and errors, mathematicians developed an order for which we perform operations when simplifying expressions. This is called the *order of operations* and it is stated next.

> **Procedure: Order of Operations**
>
> When simplifying a numerical expression, perform the operations in the following order:
>
> **Step 1:** Simplify expressions inside grouping symbols first, starting with the innermost set. If there is a fraction bar, simplify the numerator and denominator separately.
> **Step 2:** Simplify any exponential expressions.
> **Step 3:** Perform multiplication or division in order from left to right.
> **Step 4:** Perform addition or subtraction in order from left to right.

Objective 2 Examples   Use the order of operations to simplify each expression.

2a. $9 + 5(6)$ **2b.** $4 \cdot 5 \div 2 + 3 \cdot 8$ **2c.** $\dfrac{2}{3}(6-2)^2 - \dfrac{1}{3}$

2d. $\dfrac{6 - \sqrt{6^2 - 4(5)(1)}}{2(1)}$ **2e.** $4[7 - |9 - 4(5-3)|]$

Solutions   **2a.** $9 + 5(6)$

$9 + 30$ Multiply 5 and 6.

39 Add.

2b. $4 \cdot 5 \div 2 + 3 \cdot 8$

$20 \div 2 + 3 \cdot 8$ Multiply 4 and 5.

$10 + 3 \cdot 8$ Divide 20 by 2.

$10 + 24$ Multiply 3 and 8.

34 Add the numbers.

2c. $\dfrac{2}{3}(6-2)^2 - \dfrac{1}{3}$

$\dfrac{2}{3}(4)^2 - \dfrac{1}{3}$ Simplify inside parentheses.

$\dfrac{2}{3}(16) - \dfrac{1}{3}$ Simplify 4^2.

$\dfrac{32}{3} - \dfrac{1}{3}$ Multiply: $\dfrac{2}{3}\left(\dfrac{16}{1}\right) = \dfrac{32}{3}$.

$\dfrac{31}{3}$ Subtract the resulting fractions.

2d. $\dfrac{6 - \sqrt{6^2 - 4(5)(1)}}{2(1)}$

$\dfrac{6 - \sqrt{36 - 4(5)(1)}}{2}$ Simplify 6^2. Multiply the numbers in the denominator.

$\dfrac{6 - \sqrt{36 - 20}}{2}$ Multiply: $4(5)(1) = 20(1) = 20$.

$\dfrac{6 - \sqrt{16}}{2}$ Subtract the numbers in the square root.

$\dfrac{6 - 4}{2}$ Simplify $\sqrt{16}$.

$\dfrac{2}{2}$ Subtract the numbers in the numerator.

1 Divide the numbers.

2e. $4[7 - |9 - 4(5 - 3)|]$

$4[7 - |9 - 4(2)|]$ Simplify inside parentheses.

$4[7 - |9 - 8|]$ Multiply 4 and 2 in the absolute value.

$4[7 - |1|]$ Subtract the numbers in the absolute value.

$4[7 - 1]$ Simplify the absolute value of 1.

$4(6)$ Subtract the numbers in the brackets.

24 Multiply.

✔ Student Check 2   Use the order of operations to simplify each expression.

a. $3 + 2(9)$ **b.** $8 \cdot 3 \div 6 - 2 \cdot 2$ **c.** $\dfrac{1}{5}(8 - 6)^3 - \dfrac{3}{5}$

d. $\dfrac{4 + \sqrt{8^2 - 4(7)(1)}}{2(7)}$ **e.** $2[11 - 4|8 - 3(5 - 3)|]$

Evaluating Algebraic Expressions

Objective 3 ▶

Evaluate algebraic expressions.

We will now apply the order of operations to evaluating algebraic expressions. An **algebraic expression** is an expression that involves variables and/or numbers joined by arithmetic operations. Recall a *variable* is a letter or symbol that represents some unknown number. Examples of algebraic expressions are

$$2x \qquad 4y - 5 \qquad 3x^2 - 5x + 7$$

When we evaluate an algebraic expression, we assign a specific value for each of the variables and determine the value of the resulting expression.

> **Procedure: Evaluating an Algebraic Expression**
>
> **Step 1:** Replace the variable(s) with the given number(s).
> **Step 2:** Use the order of operations to simplify the resulting expression.

Objective 3 Examples

Evaluate each expression for the given values.

3a. Find the value of $2x + 3$ when $x = 0$, $\dfrac{1}{2}$, 1, and 2.

3b. Find the value of $\dfrac{x - 1}{x}$ for $x = 1$, 1.5, 2, and 3.

3c. Find the value of $b^2 - 4ac$ when $a = 3$, $b = 5$, and $c = 2$.

Solutions

3a. Since we are evaluating the same expression for more than one value, we can organize the information in a chart.

x	Value of $2x + 3$	
0	$2(0) + 3 = 0 + 3 = 3$	Replace x with 0.
$\dfrac{1}{2}$	$2\left(\dfrac{1}{2}\right) + 3 = 1 + 3 = 4$	Replace x with $\dfrac{1}{2}$.
1	$2(1) + 3 = 2 + 3 = 5$	Replace x with 1.
2	$2(2) + 3 = 4 + 3 = 7$	Replace x with 2.

3b.

x	Value of $\dfrac{x-1}{x}$	
1	$\dfrac{1-1}{1} = \dfrac{0}{1} = 0$	Replace x with 1.
1.5	$\dfrac{1.5-1}{1.5} = \dfrac{0.5}{1.5} = \dfrac{1}{3}$	Replace x with 1.5.
2	$\dfrac{2-1}{2} = \dfrac{1}{2}$	Replace x with 2.
3	$\dfrac{3-1}{3} = \dfrac{2}{3}$	Replace x with 3.

3c. $b^2 - 4ac = (5)^2 - 4(3)(2)$ Replace b with 5, a with 3, and c with 2.

$\qquad\qquad = 25 - 4(3)(2)$ Simplify the exponential expression.

$\qquad\qquad = 25 - 12(2)$ Multiply 4 and 3.

$\qquad\qquad = 25 - 24$ Multiply 12 and 2.

$\qquad\qquad = 1$ Subtract 25 and 24.

✔ **Student Check 3**   Evaluate each expression for the given values.

    a. Find the value of $3x + 5$ for $x = 0$, 1, and 2.

    b. Find the value of $\dfrac{x-4}{x+1}$ for $x = 4$, 5, 5.2, and 6.

    c. Find the value of $b^2 - 4ac$ when $a = 6$, $b = 8$, and $c = 2$.

Solutions of Equations

Objective 4 ▶

Determine if a value is a solution of an equation.

A statement that shows two expressions are equal is an **equation**. The equals sign "$=$" is used to denote this equality. We will study many different kinds of equations throughout this text. Some examples of equations are

$$4x + 5 = 9, \qquad x^2 - x = 6, \qquad \sqrt{y + 4} = 2y, \qquad P = 2l + 2w$$

One of the goals in algebra is to **solve an equation**. To solve an equation means to find the value(s) of the variable(s) that satisfies the equation, that is, makes the equation true. Each such value is called a **solution of an equation**. For instance,

$x = 1$ is a solution of $4x + 5 = 9$: $x = 2$ is not a solution of $4x + 5 = 9$:

$\qquad\qquad 4(1) + 5 = 9 \qquad\qquad\qquad\qquad\qquad 4(2) + 5 = 9$

$\qquad\qquad\quad 4 + 5 = 9 \qquad\qquad\qquad\qquad\qquad\quad 8 + 5 = 9$

$\qquad\qquad\qquad\quad 9 = 9 \qquad\qquad\qquad\qquad\qquad\qquad 13 = 9$

$\qquad\qquad\qquad\quad$ True $\qquad\qquad\qquad\qquad\qquad\qquad\quad$ False

---

Procedure: Determining if a Value is a Solution of an Equation

Step 1: Replace the variable(s) with the given values.

Step 2: Simplify each side of the equation.

Step 3: If the resulting equation is true, then the value is a solution. If the resulting equation is false, then the value is *not* a solution of the equation.

| **Objective 4 Examples** | Determine if the given value is a solution of the equation. |

4a. $4y + 5 = y^2$; $y = 5$ **4b.** $3x - 2(x + 1) = x + 5$; $x = 4$

Solutions   4a.

$$4y + 5 = y^2$$
$$4(5) + 5 = (5)^2 \qquad \text{Replace } y \text{ with 5.}$$
$$20 + 5 = 25 \qquad \text{Multiply 4 and 5 and simplify } 5^2.$$
$$25 = 25 \qquad \text{Add.}$$

Since $y = 5$ makes the equation true, it is a solution of $4y + 5 = y^2$.

4b.
$$3x - 2(x + 1) = x + 5$$
$$3(4) - 2(4 + 1) = 4 + 5 \qquad \text{Replace } x \text{ with 4.}$$
$$12 - 2(5) = 9 \qquad \text{Multiply 3 and 4. Add in parentheses and on the right side..}$$
$$12 - 10 = 9 \qquad \text{Multiply 2 and 5.}$$
$$2 = 9 \qquad \text{Simplify.}$$

Since $x = 4$ makes the equation false, it is *not* a solution of $3x - 2(x + 1) = x + 5$.

| ✔ **Student Check 4** | Determine if the given value is a solution of the equation. |

a. $a^2 - a + 6 = a - 2$; $a = 4$ **b.** $7b - 5(b + 2) = b + 3$; $b = 13$

Translate Expressions into Symbols

| **Objective 5** ▶ |
| Express relationships mathematically. |

Now that we know how to evaluate an algebraic expression, we will turn our focus on how to express or write mathematical relationships given certain phrases or sentences. The following table shows some common phrases for the basic mathematical expressions.

Addition	$a + b$	sum of a and b a increased by b b more than a a added to b a plus b
Subtraction	$a - b$	difference of a and b b subtracted from a b less than a a minus b a decreased by b from a, subtract b
Multiplication	ab	product of a and b a times b a multiplied by b a of b
Division	$\dfrac{a}{b}$	a divided by b quotient of a and b ratio of a to b
Equation	$a = b$	a is b a is equal to b a yields b The result of a is the same as b

Many of the problems that we encounter in math require us to have the ability to express mathematical relationships correctly. Example 5 practices this skill. When one of the numbers is not known, a variable is used to represent this unknown number.

Objective 5 Examples   Translate each phrase into an algebraic expression, equation, or inequality. Use the variable *x* to represent the unknown number.

Problems	Solutions
5a. A number increased by 7	Since "increase" means add, the expression is $$x + 7$$
5b. Five less than a number	Since "less than" means subtract, the expression is $$x - 5$$
5c. The difference of a number and 2	Since "difference" means subtract, the expression is $$x - 2$$
5d. The product of 5 and a number	Since "product" means multiply, the expression is $$5 \cdot x \text{ or } 5x$$
5e. The quotient of a number and 3	Since "quotient" means divide, the expression is $$\frac{x}{3}$$
5f. The sum of twice a number and 4	Since "twice a number" means multiply and "sum" means add, the expression is $$2x + 4$$
5g. The difference of three times a number and 1	Since "three times" means multiply and "difference" means subtract, the expression is $$3x - 1$$
5h. Twice the sum of a number and 6	Since "sum" means add and "twice" means multiply, we have $$2(x + 6)$$
5i. Three less than one-half of a number	Since "less than" means subtract and "$\frac{1}{2}$ of" means multiply we have $$\frac{1}{2}x - 3$$
5j. A number less than 8 is the same as three times the number.	Since "is the same as" means equal to, the translation is an equation of the form $$8 - x = 3x$$
5k. A number is less than 3.	Since "is less than" denotes inequality, we have $$x < 3$$

✓ Student Check 5   Translate each phrase or sentence into an algebraic expression, equation, or inequality. Use the variable *x* to represent the unknown number.

a. A number decreased by 2

b. Three more than a number

c. The sum of three times a number and 5

d. The quotient of a number and 2

e. The difference of a number and 4

f. The product of a number and 6

g. The difference of twice a number and 9

h. Three times the sum of a number and 1

i. Five less than one-third of a number

j. One more than four times a number is 15.

k. Twice a number is less than 4.

Applications

Objective 6 ▶

Solve application problems.

The skills of simplifying numerical expressions, evaluating algebraic expressions or equations, and expressing relationships mathematically will be used in the context of real-life situations.

One of the specific applications deals with calculating the *perimeter* of polygons. Polygons are two-dimensional closed shapes made up of straight lines, such as triangles, squares, rectangles, trapezoids, and the like.

> **Definition: Perimeter of a Polygon**
>
> The **perimeter** of a polygon is the total distance around the outside of the polygon.

For example, the perimeter of a rectangular yard with width 100 ft and length 250 ft is

$$P = 100 + 250 + 100 + 250$$
$$P = 700$$

So, the perimeter is 700 ft.

Objective 6 Examples | **Solve each problem.**

6a. If $5000 is invested in a savings account that earns 6% annual interest for 3 yr, the total amount saved is given by the equation.

$$\text{Amount saved} = 5000(1.06)^3$$

Find the amount that would be saved.

Solution **6a.** Simplifying the expression on the right side of the equation gives us that

$$\text{Amount saved} = 5955.08$$

So, in 3 yr, a total of $5955.08 will be in the account.

6b. After renovations in 2010, the Michigan Stadium at the University of Michigan in Ann Arbor, Michigan, became the largest football stadium in the United States with a seating capacity of 108,000. The football field is in the shape of a rectangle. The perimeter P of a rectangle is the sum of 2 times its length and 2 times its width.

   i. Write an equation that denotes the perimeter of a rectangle, where l is the length and w is the width.

   ii. If the football field has a length of 360 ft and a width of 160 ft, what is the field's perimeter?

Solution **6b.**   **i.** The equation that represents the perimeter is $P = 2l + 2w$.

    ii. $P = 2l + 2w$ Use the relationship defined in part (i).

        $P = 2(360) + 2(160)$ Replace *l* with 360 and *w* with 160.

        $P = 720 + 320$ Simplify each product.

        $P = 1040$ Add.

The perimeter of the football field is 1040 ft.

6c. The number of viewers (in millions) for the season premiere of Fox Network's *American Idol* for Seasons 1 to 8 is shown in the table.

Season	Viewers (in millions)
1 (2002)	9.9
2 (2003)	26.5
3 (2004)	28.56
4 (2005)	33.58
5 (2006)	35.53
6 (2007)	37.7
7 (2008)	33.4
8 (2009)	30.4

 i. The number of viewers can be approximated by the expression $0.1x^3 + 18.4x - 2.54x^2 - 4.6$, where x is the number of seasons aired. Write an equation that represents the approximate number of viewers, where v is the number of viewers in millions.

 ii. Use the equation to estimate the number of viewers for Season 9 (Jan. 2010).

Solution **6c.** **i.** The equation that represents the number of viewers, in millions, is
$$v = 0.1x^3 + 18.4x - 2.54x^2 - 4.6.$$

 ii. $v = 0.1x^3 + 18.4x - 2.54x^2 - 4.6$

  $v = 0.1(9)^3 + 18.4(9) - 2.54(9)^2 - 4.6$ Replace the variable x with 9.

  $v = 0.1(729) + 18.4(9) - 2.54(81) - 4.6$ Evaluate the exponential expressions.

  $v = 72.9 + 165.6 - 205.74 - 4.6$ Add or subtract from left to right.

  $v = 28.16$

The expression estimates that there were approximately 28.16 million, or 28,160,000, viewers for Season 9.

6d. The basal metabolic rate (BMR) is the number of calories a person burns to sustain life. It is the number of calories burned even if a person stayed in bed all day. It is based on a person's height, weight, and age. The BMR for women is 655 plus 4.35 times a person's weight in pounds plus 4.7 times a person's height in inches minus 4.7 times a person's age.

 i. Write an equation that represents a woman's BMR, where w is the woman's weight (in pounds), h is the woman's height (in inches), and a is the woman's age.

 ii. If Susan is a 35-yr-old who weighs 145 lb and is 5 ft 2 in. tall, what is her BMR? Round to the nearest integer.

Solution **6d.** **i.** The equation that represents a woman's basal metabolic rate is given by

BMR	is	655	plus	4.35 times weight	plus	4.7 times height	minus	4.7 times age

$$\text{BMR} = 655 + 4.35w + 4.7h - 4.7a$$

 ii. To find Susan's BMR, we must make appropriate substitutions for the variables. We are given that $w = 145$, $h = 62$, and $a = 35$. So, her BMR is

$$\text{BMR} = 655 + 4.35w + 4.7h - 4.7a$$

$$\text{BMR} = 655 + 4.35(145) + 4.7(62) - 4.7(35)$$

$$\text{BMR} = 655 + 630.75 + 291.4 - 164.5$$

$$\text{BMR} = 1412.65$$

Susan's BMR is approximately 1413 calories.

✓ **Student Check 6**   Solve each problem.

a. Suppose $2000 is invested in a certificate of deposit, CD, account that earns 5% annual interest for 4 yr. The amount of money in the account after 4 yr is given by Amount = $2000(1.05)^4$. Simplify this expression to determine the amount of money in the account in 4 yr.

b. The Singapore Flyer is the world's largest Ferris wheel. The Ferris wheel is in the shape of a circle. The distance around a circle is called its *circumference*. The circumference of a circle is twice the product of π and the radius of the circle, where r is the radius of the circle.

    i. Write an equation that represents the circumference of the circle.

    ii. If the radius of the Singapore Flyer is 75 m, find its circumference. Use 3.14 for π.

c. The number of bachelor's degrees conferred in the Computer and Information Sciences (CIS) majors from 1970 to 2005 are shown in the table. The number of degrees conferred in this field can be approximated by the expression $0.3x^4 - 18x^3 + 308.4x^2 + 273.3x + 1096.6$, where x is the number of years after 1970.
(Source: National Center for Education Statistics)

Years after 1970	Degrees Conferred
0	2388
5	5652
10	15121
15	42337
20	25159
25	24506
30	44142
35	47480

    i. Write an equation that can be used to approximate the number of CIS degrees conferred, where d is the number of degrees.

    ii. Use the equation to estimate the number of degrees conferred in 2010. Round the answer to the nearest integer.

d. The basal metabolic rate (BMR) for men is 66 plus 6.23 times a person's weight in pounds +12.7 times a person's height in inches minus 6.8 times a person's age in years.

    i. Write an equation that represents a man's BMR, where w is the man's weight (in pounds), h is the man's height (in inches), and a is the man's age.

    ii. If Todd is a 43-yr-old who weighs 250 lb and is 6 ft 4 in. tall, what is his BMR? Round to the nearest integer.

Objective 7 ▶

Troubleshoot common errors.

Troubleshooting Common Errors

Some of the common errors associated with numerical expressions and algebraic expressions are shown next.

Objective 7 Examples   **A problem and an incorrect solution are given. Provide the correct solution and an explanation of the error.**

a. Simplify $3(4)^2$.

Incorrect Solution	Correct Solution and Explanation
$3(4)^2$ 12^2 144	Based on the order of operations, we must evaluate the exponent first and then multiply. $3(4)^2$ $3(16)$ 48

b. Translate the phrase: 6 less than a number.

Incorrect Solution	Correct Solution and Explanation
The translation is $6 - x$.	The phrase "6 less than a number" is translated as $x - 6$. The order of this translation is very important. Think of how we find 6 less than 9—that is, $9 - 6$ or 3. Note that the expression $6 - x$ is the translation of "6 less a number" or "the difference of 6 and a number."

ANSWERS TO STUDENT CHECKS

Student Check 1   **a.** 1000   **b.** 2.25   **c.** $\dfrac{216}{343}$ **d.** -4096

Student Check 2   **a.** 21   **b.** 0   **c.** 1   **d.** $\dfrac{5}{7}$ **e.** 6

Student Check 3   **a.** 5, 8, 11   **b.** $0, \dfrac{1}{6}, \dfrac{6}{31}, \dfrac{2}{7}$ **c.** 16

Student Check 4   **a.** no   **b.** yes

Student Check 5   **a.** $x - 2$ **b.** $x + 3$ **c.** $3x + 5$ **d.** $\dfrac{x}{2}$

e. $x - 4$ **f.** $6x$ **g.** $2x - 9$ **h.** $3(x + 1)$
i. $\dfrac{1}{3}x - 5$ **j.** $4x + 1 = 15$ **k.** $2x < 4$

Student Check 6   **a.** \$2431.01   **b.** $C = 2\pi r$; 471 m
c. $d = 0.3x^4 - 18x^3 + 308.4x^2 + 273.3x + 1096.6$; 121,469 degrees conferred
d. $BMR = 66 + 6.23w + 12.7h - 6.8a$; 2296 cal

SUMMARY OF KEY CONCEPTS

1. An exponent is a way of writing repeated multiplication. The exponent indicates the number of times to repeat the base as a factor.

2. The order of operations is used to simplify numerical expressions. The order is: parentheses (or any grouping), exponents, multiplication/division in order from left to right, and addition/subtraction in order from left to right.

3. To evaluate an algebraic expression, replace the variable with the given value and use the order of operations to simplify the resulting expression. It is helpful to put parentheses around the number being substituted into the expression.

4. An equation is a statement that two expressions are equal. To determine if a number is a solution of an equation, substitute it into the equation. If the resulting statement is true, the number is a solution. If it is not true, the number is not a solution.

5. Be familiar with the common phrases for addition, subtraction, multiplication, division, and equality. Use the phrases to express relationships mathematically.

6. When solving application problems, we must be able to write an appropriate equation that represents the problem and then use the equation to determine values when given specific information.

GRAPHING CALCULATOR SKILLS

We can use the calculator for simplifying exponential expressions, simplifying numerical expressions, and evaluating algebraic expressions. The most important part of entering a numerical expression is "telling" the calculator where the grouping symbols occur in the problem. We use parentheses to indicate grouping symbols on the calculator. If parentheses are not used correctly, an incorrect order of operations is applied.

Example: Use the calculator to find the value of the numerical expressions.

1. $\left(\dfrac{2}{3}\right)^4$

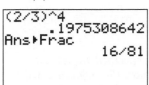

2. $4 \cdot 5 \div 2 + 3 \cdot 8$

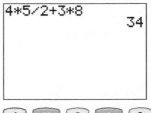

3. $\frac{2}{3}(6-2)^2 - \frac{1}{3}$

4. $\frac{6 - \sqrt{6^2 - 4(1)(5)}}{2(1)}$

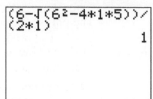

5. $|9 - 4(5-3)|$

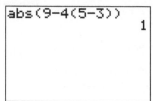

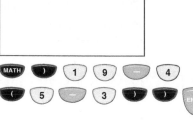

SECTION 1.3   EXERCISE SET

 Write About It!

Use complete sentences in your answers to the following exercises.

1. Explain how to apply the order of operations to simplify numerical expressions.

2. How does a numerical expression differ from an algebraic expression?

3. Explain how to apply the order of operations to simplify the expressions $(4 + 3)^2$ and $4^2 + 3^2$.

4. Explain how to apply the order of operations to simplify the expressions $\frac{6+4}{3+2}$ and $\frac{6}{3} + \frac{4}{2}$.

5. Explain what it means to evaluate an algebraic expression.

6. What is the first step in translating phrases into algebraic expressions?

7. Explain how to translate an English phrase involving "more than" into an algebraic expression.

8. Explain how to translate an English phrase involving "subtracted from" into an algebraic expression.

Determine if the statement is true or false. If false, explain the error.

9. The phrase "four less than three times a number" can be translated into $4 - 3x$.

10. When the expression $x^2 - 2x$ is evaluated for $x = -4$, the result is -8.

Practice Makes Perfect!

Identify the base, the exponent, and then evaluate the expression. (See Objective 1.)

11. 4^4 12. 5^4 13. -3^3

14. -7^3 15. 2.5^3 16. 1.2^3

17. -0.6^3 18. -0.5^3 19. $-\left(\frac{3}{5}\right)^2$

20. $-\left(\frac{4}{7}\right)^2$ 21. $\left(\frac{2}{3}\right)^5$ 22. $\left(\frac{1}{4}\right)^3$

Use the order of operations to simplify each expression. (See Objective 2.)

23. $5 + 4(3)$

24. $6 + 2(10)$

25. $4(7) - (3)(2)$

26. $11(4) - (7)(5)$

27. $12 \cdot 4 \div 8 - 2 \cdot 2$

28. $21 \cdot 4 \div 6 - 5 \cdot 2$

29. $\frac{1}{7}(6-4)^3 - \frac{2}{5}$

30. $\frac{1}{4}(8-5)^3 - \frac{5}{2}$

31. $\frac{16 - 11}{5 - 1}$

32. $\frac{5 + 2}{4 - 1}$

33. $2(4)^2 - 5(4) + 7$

34. $5(2)^2 + 4(2) - 1$

35. $4 \cdot 3^2 - 5 \cdot 3 + 1$

36. $6 \cdot 5^2 - 7 \cdot 5 + 2$

37. $(4 \cdot 2)^2 - 8 \div 2 \cdot 7$

38. $(2 \cdot 3)^2 - 4 \cdot 2 \div 8 + 3$

39. $(5)^2 - 4(2)(3)$

40. $3 \cdot 7^2 - 4(5)(6)$

41. $2[44 - 2|35 - 5(6 - 2)|]$

42. $5[18 - 6|14 - 2(9 - 3)|]$

43. $\frac{13 - \sqrt{13^2 - 4(4)(9)}}{2 \cdot 9}$ 44. $\frac{10 - \sqrt{10^2 - 4(2)(8)}}{2 \cdot 8}$

Evaluate each expression for the specified values of x. Organize the information in a chart. (See Objective 3.)

45. $3x + 2$ for $x = 0, \frac{1}{3}, 2,$ and 4

46. $4x + 1$ for $x = \frac{1}{4}, 1, 2,$ and 4

47. $\frac{3x - 2}{x}$ for $x = 1, 2, 3,$ and 4

48. $\dfrac{5x - 2}{x}$ for $x = 1, 2, 3,$ and 4

49. $\dfrac{x + 3}{x - 1}$ for $x = 2, 3, 4,$ and 5

50. $\dfrac{x + 1}{x - 2}$ for $x = 3, 4, 5,$ and 6

Evaluate each expression for the given values of the variables. (See Objective 3.)

51. $2x + 3y$ for $x = 0$ and $y = 2$

52. $4x + y$ for $x = 2$ and $y = 0$

53. $\dfrac{y_2 - y_1}{x_2 - x_1}$ for $x_1 = 2, x_2 = 5,$ $y_1 = 4,$ and $y_2 = 6$

54. $\dfrac{y_2 - y_1}{x_2 - x_1}$ for $x_1 = 0, x_2 = 3,$ $y_1 = 2,$ and $y_2 = 7$

55. $\dfrac{1}{2}bh$ for $b = 4$ and $h = 5$

56. $\dfrac{1}{2}bh$ for $b = 3$ and $h = 6$

57. $2l + 2w$ for $l = 8$ and $w = 10$

58. $2l + 2w$ for $l = 4$ and $w = 12$

59. $\sqrt{(x_1 - x_2)^2 + (y_1 - y_2)^2}$ for $x_1 = 1,$ $x_2 = 4, y_1 = 12, y_2 = 16$

60. $\sqrt{(x_1 - x_2)^2 + (y_1 - y_2)^2}$ for $x_1 = 0,$ $x_2 = 6, y_1 = 3, y_2 = 11$

61. $b^2 - 4ac$ for $a = 2, b = 5, c = 3$

62. $b^2 - 4ac$ for $a = 1, b = 6, c = 2$

Determine if the given value is a solution of the equation. (See Objective 4.)

63. $2x^2 - 3x - 1 = x + 5; x = 2$

64. $y^2 + 3y - 6 = y + 9; y = 3$

65. $a^2 - 2a + 3 = 2a + 8; a = 5$

66. $3b^2 - 6b - 10 = 5b - 6; b = 4$

67. $9x - 6(x - 1) = 4x + 2; x = 3$

68. $4z - 2(z + 1) = z + 2; z = 5$

69. $6y - 2(y - 3) = y + 9; y = 1$

70. $8z - 4(z + 1) = 2z + 8; z = 6$

Translate each phrase into an algebraic expression. Let x represent the unknown number. (See Objective 5.)

71. The sum of a number and 4

72. The sum of a number and 12

73. 2 more than three times a number

74. 6 more than four times a number

75. The difference of a number and 9

76. The difference of a number and 15

77. 3 less than four times a number

78. 3 decreased by twice a number

79. The product of a number and 10

80. The product of a number and 7

81. The product of 3 and the sum of a number and 15

82. The product of 4 and a number increased by 6

83. The sum of four times a number and 3

84. The sum of twice a number and 4

85. Twice the difference of a number and 7

86. Twice the difference of a number and 3

87. The quotient of three times a number and 4

88. The quotient of four times a number and 5

Write an algebraic expression that represents each unknown quantity. (See Objective 6.)

89. Shanika invested money in two different accounts. She invested a total of $5000. If x represents the amount she invested in the first account, write an expression that represents the amount she invested in the second account.

90. David invested money in two different accounts. He invested a total of $2000. If x represents the amount he invested in the first account, write an expression that represents the amount he invested in the second account.

91. In a rectangle, the length is three less than twice the width. If w represents the width, write an expression that represents the length of the rectangle.

92. In a rectangle, the width is four more than the length. If l represents the length, write an expression that represents the width of the rectangle.

93. In 2011, Oprah Winfrey and Tiger Woods were both among the 100 highest paid celebrities. Oprah Winfrey was paid $215 million more than Tiger Woods. If x represents the amount Tiger Woods was paid in millions, write an algebraic expression that represents how much Oprah Winfrey was paid. (Source: http://www.forbes.com)

94. In 2011, pop stars Lady Gaga and Justin Bieber were among the 100 highest paid celebrities. Justin Bieber was paid $37 million less than Lady Gaga. If x represents the amount Lady Gaga was paid in millions, write an algebraic expression that represents how much Justin Bieber was paid. (Source: http://www.forbes.com)

Solve each problem. (See Objective 6.)

95. Find the perimeter of the Lincoln Memorial Reflecting Pool in Washington, D.C. The reflecting pool is approximately 2029 ft long and 167 ft wide.

96. The perimeter of a pentagon is five times the length of a side of the pentagon. If each outside wall of the Pentagon is 921 ft long, what is the perimeter of the Pentagon?

97. The expression $P(1 + r)^t$ represents the amount of money in an account when P dollars is invested at an annual interest rate of r (in decimal form) for t yr. Find the amount of money that will be in an account at the end of 5 yr if $1000 is invested at 5% annual interest. (Hint: Convert 5% to a decimal value before substituting this value in the expression.)

98. The expression $P(1 + r)^t$ represents the amount of money in an account when P dollars is invested at an annual interest rate of r (in decimal form) for t yr. Find the amount of money that will be in an account at the end of 3 yr if $5000 is invested at 3% annual interest. (Hint: Convert 3% to a decimal value before substituting this value in the expression.)

99. In 2011, the *U.S. News & World Report Guide to America's Best Colleges* ranked Princeton University as one of the top national universities. Yearly tuition and fees at Princeton  University is represented by the expression $1100c + 9200$, where c is the number of credit hours taken in a year. What is the yearly tuition and fees if 30 credit hours are taken in a year?

100. In 2011, the *U.S. News & World Report Guide to America's Best Colleges* ranked the University of Pennsylvania as the top Business School in the United States. Yearly tuition and fees at University of Pennsylvania is represented by the expression $3105c + 9804$, where c is the number of credit units taken each year. What is the yearly tuition and fees if 11 credit units are taken in 1 yr?

101. According to the Bureau of Labor Statistics, the average hourly rate of registered nurses in the state of Illinois is $25. The expression $25x + 37.5y$ represents the average weekly earnings of a registered nurse in Illinois, where x is the number of regular hours worked in a week and y 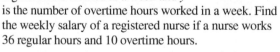 is the number of overtime hours worked in a week. Find the weekly salary of a registered nurse if a nurse works 36 regular hours and 10 overtime hours.

102. According to the Bureau of Labor Statistics, airplane pilots and navigators earned an average of $95.80 per hour. The expression $95.80x$ represents the average weekly earnings of an airplane pilot and navigator, where x is the number of hours worked in a week. What are the weekly earnings for an airplane pilot who works 30 hr per week?

103. The height of a baseball hit upward with an initial velocity of 100 ft/sec from an initial height of 2.5 ft is represented by the expression $-16t^2 + 100t + 2.5$, where t is the number of seconds after the ball has been hit. What is the height of the ball after 4 sec?

104. The USS *Constitution* is the oldest commissioned warship afloat in the world. The height of a cannonball fired from the USS *Constitution* with an initial velocity of 388 ft/sec from a height of 24 ft is represented by the expression $-16t^2 + 388t + 24$, where t is the number of seconds after the cannonball has been fired. What is the height of the cannonball after 12 sec?

 Mix 'Em Up!

Use the order of operations to simplify each expression.

105. 6^3

106. -3^2

107. $|-12(7) + 8(4)|$

108. $|-6(-13) + 8(-7)|$

109. 1.4^3

110. $-\left(\dfrac{3}{4}\right)^3$

111. $5[20 - |3 \cdot 8 - 2(11 - 7)|]$

112. $6[18 - 7|4 \cdot 3 - 2(6 - 1)|]$

113. $\dfrac{25 + \sqrt{25^2 - 4(16)(9)}}{2 \cdot 9}$

114. $\dfrac{25 - \sqrt{25^2 - 4(8)(18)}}{2 \cdot 18}$

115. $\dfrac{1}{15}(45) - 1\dfrac{7}{15}$

116. $\dfrac{1}{19}(76) - 2\dfrac{17}{19}$

117. $\dfrac{21 - 13}{3 \cdot 4 - 7}$

118. $\dfrac{18 \div 3 + 9}{27 + 13}$

119. $10 \cdot 3^2 - 28 \div 4 \cdot 7 + 1$

120. $3 \cdot 5^2 - 32 \div 4 \cdot 8 + 6$

121. $\dfrac{1}{6}(14 - 11)^3 - 12 \div 3$

122. $(1 \cdot 3)^2 + 6 \cdot 4 \div 2 + 10$

Evaluate each expression for the given values of the variables. When multiple values are given, organize the information in a chart.

123. $5x - 3$ for $x = 1, 2, 3,$ and 4

124. $\dfrac{3x + 2}{x + 1}$ for $x = 0, 1, 2,$ and 3

125. $7x - 6y$ for $x = 3$ and $y = 2$

126. $\dfrac{y_2 - y_1}{x_2 - x_1}$ for $x_1 = 2, x_2 = 14, y_1 = 13,$ and $y_2 = 22$

127. $\dfrac{1}{2}(b_1 + b_2)h$ for $b_1 = 10, b_2 = 6, h = 3$

128. $2l + 2w$ for $l = 9$ and $w = 17$

129. $\sqrt{(x_1 - x_2)^2 + (y_1 - y_2)^2}$ for $x_1 = 2, x_2 = 14, y_1 = 17, y_2 = 22$

130. $b^2 - 4ac$ for $a = 2, b = 10, c = 5$

Determine if the given value is a solution of the equation.

131. $x^2 + 3x - 6 = x + 9; x = -3$

132. $6y + 2(5y - 3) = y + 9; y = 1$

Translate each phrase into an algebraic expression. Let x represent the unknown number.

133. Five times the sum of a number and 11

134. Three times the sum of a number and 6

135. The difference of six times a number and 18

136. 6 less than four times the sum of a number and 1

137. Twice the product of a number and 15

138. The product of 18 and the sum of a number and 4

139. The product of 14 and a number increased by 9

140. The quotient of four times a number and 6

Write an algebraic expression that represents each unknown quantity.

141. Norma mixes a 5% salt solution with a 40% solution to get a 25% solution. Let x oz represent the volume of the 5% salt solution. If Norma wants 50 oz of 25% salt solution, write an expression that represents the volume of the 40% salt solution.

142. David mixes a 10% alcohol solution with a 50% solution to get a 20% solution. Let x mL be the volume of the 50% alcohol solution. If David wants to get 30 mL of 20% solution, write an expression that represents the volume of the 10% solution.

143. In a rectangle, the length is two more than three times the width. If w represents the width, write an expression that represents the length of the rectangle.

144. In a rectangle, the width is four less than twice the length. If l represents the length, write an expression that represents the width of the rectangle.

Solve each problem.

145. The expression $P(1 + r)^t$ represents the amount of money in an account when P dollars is invested at an annual interest rate of r (in decimal form) for t yr. Find the amount of money that will be in an account at the end of 5 yr if $1500 is invested at 2% annual interest. (Hint: Convert 2% to a decimal value before substituting this value in the expression.)

146. The expression $P(1 + r)^t$ represents the amount of money in an account when P dollars is invested at an annual interest rate of r (in decimal form) for t yr. Find the amount of money that will be in an account at the end of 2 yr if $3200 is invested at 4% annual interest. (Hint: Convert 4% to a decimal value before substituting this value in the expression.)

147. The height of a baseball hit upward with an initial velocity of 120 ft/sec from an initial height of 6 ft is represented by the expression $-16t^2 + 120t + 6$, where t is the number of seconds after the ball has been hit. What is the height of the ball after 5 sec?

148. The height of a baseball hit upward with an initial velocity of 95 ft/sec from an initial height of 4.5 ft is represented by the expression $-16t^2 + 95t + 4.5$, where t is the number of seconds after the ball has been hit. What is the height of the ball after 4 sec?

149. At Central Pennsylvania Community College, in-state-resident tuition and fees for the academic year 2010–2011 were calculated by the expression $211c$, where c is the number of credit units taken per semester. If Angela took 13 credits in Spring 2011, what was her tuition and fees?

150. At Los Angeles Valley College, in-state-resident tuition and fees for academic year 2010–2011 were calculated by the expression $26c + 42$, where c is the number of units taken per semester. If Goran took 15 units in Spring 2011, what was his tuition and fees?

 You Be the Teacher!

Correct each student's errors, if any.

151. Evaluate the expression $3x^2 - 2x + 1$ for $x = 2$.

Keith's work:

$3 \cdot 2^2 - 2 \cdot 2 + 1 = 6^2 - 4 + 1 = 36 - 4 + 1 = 33$

152. Evaluate the expression $-\dfrac{2}{3}x + 5$ for $x = 3$.

Frankie's work:

$$-\frac{2}{3} \cdot 3 + 5 = -\frac{2}{3} \cdot 8 = -\frac{2}{3} \cdot \frac{24}{3} = -\frac{48}{9}$$

153. Evaluate the expression $30 - 2x^2$ for $x = 3$.

Lestine's work:

$$30 - 2(3)^2 = 28(3^2) = 28(9) = 252$$

154. Evaluate the expression $|15 - 4x|$ for $x = 3$.

Ian's work:

$$|15 - 4(3)| = |11(3)| = |33| = 33$$

155. Simplify the expression $-3 \cdot 2^2 + 28 \div 7 \cdot 4 + 36$.

Basil's work:

$$-3 \cdot 2^2 + 28 \div 7 \cdot 4 + 36 = -6^2 + 28 \div 28 + 36$$
$$= -36 + 1 + 36$$
$$= 1$$

156. Simplify the expression $38 - 2^2 - 45 \div 5 \cdot 3$.

Abul's work:

$$38 - 2^2 - 45 \div 5 \cdot 3 = 38 - 4 - 45 \div 15$$
$$= 38 - 4 - 3$$
$$= 31$$

 Calculate It!

Use a calculator to evaluate each algebraic expression.

157. $x^3 - 2x + 1$ for $x = 0$, 2.5, and 3.2

158. $\dfrac{5x + 1}{x - 1}$ for $x = 1.5$, 1.8, and 4

159. $\sqrt{x^2 - 9}$ for $x = 3$, 5, and 9

160. $|500 - 3x^2|$ for $x = 8.5$, 10, and 12

| | SECTION 1.4 | Addition of Real Numbers |

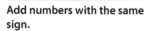

▶ **OBJECTIVES**

As a result of completing this section, you will be able to

1. Add numbers with the same sign.
2. Add numbers with different signs.
3. Solve real-life problems.
4. Troubleshoot common errors.

The operating income for Delta Airlines for the years 2002 to 2009 is shown in the table. Find the total operating income for the years 2002–2009. (Source: http://www.airlinefinancials.com/uploads/2002–2009_Delta _mainline.pdf)

Year	Operating Income (in millions)
2002	–$1309
2003	–$ 786
2004	–$5985
2005	–$1605
2006	–$1139
2007	$ 78
2008	–$ 58
2009	$ 605

To answer this question, we must know how to perform operations with real numbers. In this section, we will examine addition of real numbers.

Adding Numbers with the Same Signs

| **Objective 1** ▶ |

Add numbers with the same sign.

The real number line can help us visualize the process of adding real numbers. Adding two positive numbers is a skill that we encounter on a daily basis. For example, if a student enrolls in 6 hr of classes and then adds 4 more hours during the drop/add period, the student enrolls in a total of $6 + 4 = 10$ hr of classes.

Adding two negative numbers is not something that we encounter as often as adding two positive numbers. Suppose we owe a friend $5 and then we borrow $2 more. At this point, we owe our friend a total of $7. Mathematically, the total we owe can be represented as

$$-5 + (-2) = -7$$

On the number line, we can visualize this situation as beginning at 0, moving to the left 5 units, and moving left again 2 units until we arrive at -7.

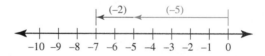

While number lines provide us the ability to visualize the operations on the numbers, using them every time we add two numbers can be very time consuming. From the previous examples, we can conclude that adding two positive numbers results in a positive number and adding two negative numbers results in a negative number.

> **Procedure: Adding Numbers with the Same Sign**
>
> **Step 1:** Add the absolute values of the numbers.
> **Step 2:** The sign of the result is the same as the sign of the numbers in the original expression.

| **Objective 1 Examples** | **Add. For parts (a) and (b), use a number line to add the numbers.** |

1a. $7 + 2$ **1b.** $(-1) + (-4)$ **1c.** $(-2) + (-7)$ **1d.** $(-1.5) + (-3)$

1e. $\dfrac{1}{5} + \dfrac{3}{10}$ **1f.** $(-7) + \left(-\dfrac{2}{3}\right)$ **1g.** $(-8) + (-2) + (-4)$

Solutions   **1a.** We begin at 0, move right 7 units, and then right 2 more units.

So, $7 + 2 = 9$.

1b. We begin at 0, move left 1 unit, and then left 4 more units.

So, $(-1) + (-4) = -5$.

1c. $(-2) + (-7) = -9$

Add the absolute values.
$|-2| + |-7| = 2 + 7 = 9$
Use the sign of the original numbers.

1d. $(-1.5) + (-3) = -4.5$

Add the absolute values.
$|-1.5| + |-3| = 1.5 + 3 = 4.5$
Use the sign of the original numbers.

1e. $\dfrac{1}{5} + \dfrac{3}{10} = \dfrac{2}{10} + \dfrac{3}{10}$

$= \dfrac{5}{10}$

$= \dfrac{1}{2}$

Convert $\dfrac{1}{5}$ to a fraction with a
denominator of 10, $\dfrac{1}{5} = \dfrac{2}{10}$.

Add the absolute values.

$\left|\dfrac{2}{10}\right| + \left|\dfrac{3}{10}\right| = \dfrac{2}{10} + \dfrac{3}{10} = \dfrac{5}{10}$

Keep the sign of the original numbers.
Simplify the result.

1f. $(-7) + \left(-\dfrac{2}{3}\right) = \left(-\dfrac{7}{1}\right) + \left(-\dfrac{2}{3}\right)$

$= \left(-\dfrac{21}{3}\right) + \left(-\dfrac{2}{3}\right)$

$= -\dfrac{23}{3}$

Rewrite -7 as a fraction, $-\dfrac{7}{1}$, and
convert it to a fraction with a denominator
of 3, $-\dfrac{7}{1} = -\dfrac{21}{3}$.

Add the absolute values of the two
numbers.

$\left|\dfrac{-21}{3}\right| + \left|\dfrac{-2}{3}\right| = \dfrac{21}{3} + \dfrac{2}{3} = \dfrac{23}{3}$

Use the sign of the original numbers.

1g. $(-8) + (-2) + (-4) = -14$

Add the absolute values.
$|-8| + |-2| + |-4| = 8 + 2 + 4 = 14$
Use the sign of the original numbers.

☑ **Student Check 1**   Add. For parts (a) and (b), use a number line to add the numbers.

    a. $3 + 5$ **b.** $(-4) + (-3)$ **c.** $(-2.6) + (-1.3)$ **d.** $\dfrac{2}{3} + \dfrac{1}{6}$

    e. $\left(-\dfrac{4}{5}\right) + \left(-\dfrac{7}{15}\right)$ **f.** $(-5) + \left(-\dfrac{1}{2}\right)$ **g.** $(-5) + (-1) + (-6)$

Adding Numbers with Different Signs

Objective 2 ▶

Add numbers with different signs.

Adding real numbers with different signs occurs in many different situations. An every-day occurrence deals with temperature changes. Suppose the temperature at 7:00 A.M. in Duluth, Minnesota, one winter morning was 2°F below zero and the temperature rose 5°F

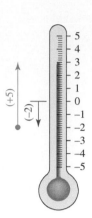

after 3 hr. The temperature at 10 A.M. can be found by simplifying the expression $-2°F + 5°F$. To perform this operation, we will use a thermometer. We begin at 0°F and go down to $-2°F$ and then go up by 5°F. This takes us to 3°F. So, we have

$$-2 + 5 = 3$$

We can also obtain the same result by using a number line as well.

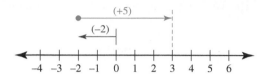

Suppose the temperature is 5°F at 12:00 A.M. and the temperature fell 8°F in 2 hr. To determine the temperature at 2:00 A.M., we need to find the result of $5°F + (-8)°F$. The thermometer can assist us again. In this case, we begin at 0°F, and go up by 5°F and then go down 8°F. This takes us to $-3°F$. So, we have

$$5 + (-8) = -3$$

Again, we can obtain the same result by using a number line.

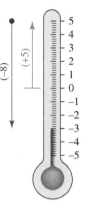

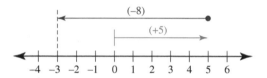

From these examples, we conclude that adding numbers with opposite signs can result in either a positive or negative number. The final sign is actually determined by the sign of the number with the larger absolute value. In the first example, 5 is the number with the larger absolute value and the sign of the $-2 + 5$ is positive. In the second example, -8 is the number with the larger absolute value and the sign of $5 + (-8)$ is negative.

Procedure: Adding Numbers with Different Signs

Step 1: Subtract the absolute values of the numbers.
Step 2: The sign of the result is the same as the sign of the number with the larger absolute value.

Objective 2 Examples Add. For parts (a) and (b), use a number line to add the numbers.

2a. $3 + (-2)$ **2b.** $(-7) + 4$ **2c.** $20 + (-35)$ **2d.** $-14 + 15.5$

2e. $-7 + 7$ **2f.** $\left(-\dfrac{1}{4}\right) + \dfrac{2}{3}$ **2g.** $(-9) + 5 + (-2)$

Solutions **2a.** We begin at 0 and move right 3 units and then left 2 units.

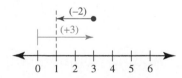

So, $3 + (-2) = 1$.

2b. We begin at 0 and move left 7 units and then right 4 units.

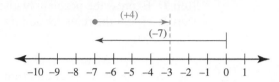

So, $(-7) + 4 = -3$.

2c. $20 + (-35) = -15$

Subtract the absolute values of the numbers.
$|20| = 20, |-35| = 35, 35 - 20 = 15$
Since $|-35| > |20|$, the sign of the answer is negative.

2d. $-14 + 15.5 = 1.5$

Subtract the absolute values of the numbers.
$|-14| = 14, |15.5| = 15.5, 15.5 - 14 = 1.5$
Since $|15.5| > |-14|$, the sign of the answer is positive.

2e. $\quad -7 + 7 = 0$

Subtract the absolute values of the numbers.
$|-7| = 7, |7| = 7, 7 - 7 = 0$
Since the difference of the absolute values is zero, the sign of the answer is neither positive nor negative.

2f. $\left(-\dfrac{1}{4}\right) + \dfrac{2}{3} = -\dfrac{3}{12} + \dfrac{8}{12}$

Rewrite each fraction with a denominator of 12.
$-\dfrac{1}{4} = -\dfrac{3}{12}$ and $\dfrac{2}{3} = \dfrac{8}{12}$

$\qquad\qquad\qquad = \dfrac{5}{12}$

Subtract the absolute values of the numbers.
$\left|-\dfrac{3}{12}\right| = \dfrac{3}{12}, \left|\dfrac{8}{12}\right| = \dfrac{8}{12}, \dfrac{8}{12} - \dfrac{3}{12} = \dfrac{5}{12}$
Since $\left|\dfrac{8}{12}\right| > \left|-\dfrac{3}{12}\right|$, the sign of the answer is positive.

2g. $(-9) + 5 + (-2)$

Add the first two numbers. Their signs are different so we subtract their absolute values and keep the sign of the number with the larger absolute value.

$\qquad = -4 + (-2)$

$\qquad = -6$

$|-9| = 9, |5| = 5, 9 - 5 = 4$
$|-9| > |5| \rightarrow (-9) + 5 = -4$
Add the result and the last number. Their signs are the same, so we add their absolute values and keep the sign of the original numbers.
$|-4| = 4, |-2| = 2, 4 + 2 = 6$
Since the numbers are negative, the result is negative.

✓ **Student Check 2**   Add the numbers.

    a. $2 + (-10)$ **b.** $-4 + 8$ **c.** $(-16) + 4$ **d.** $18.4 + (-6.2)$

    e. $3 + (-3)$ **f.** $\dfrac{3}{8} + \left(-\dfrac{5}{4}\right)$ **g.** $12 + (-2) + (-6)$

Applications

Objective 3 ▶

Solve real-life problems.

Addition of real numbers occurs in many different aspects of real life. We find adding real numbers when balancing checkbooks, determining net worth, determining values of stocks, calculating golf scores, calculating total gain/loss in yardage for football plays, changes in temperature, changes in elevation, and so on.

> **Procedure: Solving Applications of Adding Signed Numbers**
>
> **Step 1:** Express the problem mathematically.
> **Step 2:** Apply the rules for adding signed numbers.

Objective 3 Examples

Write a mathematical equation that models the given situation and then find the result.

The operating income for Delta Airlines for the years 2002 to 2009 is shown in the table. A negative number represents a loss in operating income. (Source: http://www.airlinefinancials.com/uploads/2002–2009_Delta_mainline.pdf)

3a. Find the total operating income (TOI) for the years 2002–2005.

3b. Find the average operating income (AOI) for the years 2007–2009.

Year	Operating Income (in millions)
2002	−$1309
2003	−$ 786
2004	−$5985
2005	−$1605
2006	−$1139
2007	$ 78
2008	−$ 58
2009	$ 605

Solutions

3a. We add the values for the 4 yr. Since the numbers all have the same sign, we add their absolute values and keep the sign.

$$\text{TOI} = (-1309) + (-786) + (-5985) + (-1605)$$
$$\text{TOI} = -9685$$

So, Delta Airlines had a total loss of $9685 million during the years 2002 to 2005.

3b. We add the values for the given years and divide by the number of years, 3.

$$\text{AOI} = \frac{78 + (-58) + 605}{3}$$
$$\text{AOI} = \frac{625}{3}$$
$$\text{AOI} = 208.33$$

So, Delta Airlines had an average operating income of $208.33 million during the years 2007 to 2009.

✓ Student Check 3

Write a mathematical equation that models the given situation and then find the result.

The table shows the average monthly low temperature for one of the northernmost points in the United States: Barrow, Alaska. Find the average yearly low temperature.

Month	°F	Month	°F
Jan.	−20	July	34
Feb.	−22	Aug.	34
Mar.	−22	Sept.	28
Apr.	−7	Oct.	10
May	15	Nov.	−6
June	30	Dec.	−16

Objective 4 ▶

Troubleshoot common errors.

Troubleshooting Common Errors

Some common errors associated with adding real numbers are shown next.

Objective 4 Examples | A problem and an incorrect solution are given. Provide the correct solution and an explanation of the error.

4a. $7 + (-11)$

Incorrect Solution	Correct Solution and Explanation
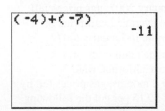 $7 + (-11) = 4$	Since the signs are different, we should subtract the absolute values of the numbers and take the sign of the number with the larger absolute value. The difference of the absolute values is 4 but -11 has the larger absolute value, so the sign of the answer should be negative. $$7 + (-11) = -4$$

4b. $-5 + (-4)$

Incorrect Solution	Correct Solution and Explanation
$-5 + (-4) = -1$	Since the signs are the same, we should add the absolute values of the numbers and keep the sign the same. So, the correct solution is $$-5 + (-4) = -9$$

ANSWERS TO STUDENT CHECKS

Student Check 1   **a.** 8   **b.** -7 **c.** -3.9 **d.** $\dfrac{5}{6}$
  e. $-\dfrac{11}{15}$ **f.** $-\dfrac{11}{2}$ **g.** -12

Student Check 2   **a.** -8 **b.** 4   **c.** -12 **d.** 12.2
  e. 0   **f.** $-\dfrac{7}{8}$ **g.** 4
Student Check 3   4.83°F

SUMMARY OF KEY CONCEPTS

1. To add numbers with the same sign, add their absolute values and keep the sign.

2. To add numbers with different signs, subtract the absolute values of the numbers. The sign of the answer is the sign of the number with the larger absolute value.

3. When solving real-life applications, look for the key phrases to determine how to solve the problem.

GRAPHING CALCULATOR SKILLS

In this section, the calculator will help you confirm your work. You should be able to enter operations with signed numbers.

Example: Simplify $(-4) + (-7)$.

```
(-4)+(-7)
            -11
```

Very important: Please note the difference between a negative sign and the subtraction sign. Use the negative sign $(-)$ to denote the sign of a number and the subtraction sign to perform an operation with numbers.

SECTION 1.4   EXERCISE SET

 Write About It!

Use complete sentences to explain how to perform the given operation.

1. Add two negative numbers

2. Add a positive number and a negative number

Practice Makes Perfect!

Perform the indicated operation and simplify. (*See Objective 1.*)

3. $1.95 + 10.26$

4. $1.37 + 2.56$

5. $-5 + (-4)$

6. $-7 + (-9)$

7. $-1.68 + (-6.77)$

8. $-4.33 + (-11.86)$

9. $\frac{9}{10} + \frac{3}{5}$

10. $\frac{13}{14} + \frac{3}{4}$

11. $-2 + \left(-\frac{3}{5}\right)$

12. $-1 + \left(-\frac{5}{7}\right)$

13. $\left(-\frac{1}{10}\right) + \left(-\frac{3}{4}\right)$

14. $\left(-\frac{1}{3}\right) + \left(-\frac{3}{8}\right)$

15. $\left(-3\frac{1}{4}\right) + \left(-1\frac{1}{4}\right)$

16. $\left(-2\frac{1}{5}\right) + \left(-1\frac{3}{5}\right)$

17. $-4 + (-12) + (-9)$

18. $-13 + (-8) + (-14)$

19. $-5.5 + (-2.4) + (-3.1)$

20. $-6.2 + (-4.2) + (-5.8)$

Perform the indicated operation and simplify. (*See Objective 2.*)

21. $1 + (-3)$

22. $4 + (-7)$

23. $-6 + 10$

24. $-30 + 15$

25. $6 + (-10)$

26. $30 + (-18)$

27. $(-1.5) + 3.6$

28. $(-4.1) + 0.8$

29. $1.6 + (-4.2)$

30. $3.7 + (-1.9)$

31. $\frac{2}{3} + \left(-\frac{4}{5}\right)$

32. $\frac{4}{7} + \left(-\frac{1}{14}\right)$

33. $1\frac{1}{3} + \left(-2\frac{2}{3}\right)$

34. $3\frac{3}{8} + \left(-4\frac{5}{8}\right)$

35. $15 + (-9) + (-2)$

36. $18 + (-8) + (-5)$

37. $-21 + 9 + (-3)$

38. $-19 + 11 + (-6)$

Solve each problem by first writing a mathematical expression that represents the situation. (*See Objective 3.*)

39. Daimler reported a net income of -332 million euros in 2008, -4765 million euros in 2009 and 5399 million euros in 2010. What was Daimler's total net income for 2008 to 2010? (Source: www.daimler.com)

40. United Airlines reported a total operating income of $1037 million in 2007, $-$4438 million in 2008, $-$161 million in 2009, and $976 million in 2010. What was United's total operating income for 2007 to 2010? (Source: www.united.com)

 Mix 'Em Up!

Perform the indicated operation and simplify. (*See Objectives 1 and 2, 4 and 5.*)

41. $-1.5 + (-3.6)$

42. $(-4.1) + (-0.8)$

43. $1.27 + (-3.73)$

44. $1.98 + (-3.19)$

45. $-\frac{10}{3} + \frac{1}{6}$

46. $\left(-\frac{3}{2}\right) + \left(-\frac{10}{3}\right)$

47. $2 + \left(-\frac{3}{7}\right)$

48. $\left(\frac{1}{2}\right) + \left(-\frac{5}{6}\right)$

49. $\left(-3\frac{1}{2}\right) + 1\frac{1}{4}$

50. $\left(-1\frac{2}{5}\right) + 2\frac{4}{5}$

51. $(-6) + 5$

52. $(-9) + 4$

53. $5.6 + (-6.2)$

54. $1\frac{5}{8} + \left(-2\frac{3}{8}\right)$

Translate each phrase into a mathematical expression and then simplify the result.

55. 2 more than -5

56. 6 more than -1

57. The sum of -3 and -4

58. The sum of 7 and -10

59. 12 increased by -4

60. -7 increased by 3

61. 9 added to -2

62. -4 added to -1

Write the expression needed to solve each problem and then answer the question. (*See Objective 3.*)

63. According to an article in the *Washington Post* in April 2005, Apple Computer Inc. had a profit of $46 million in 2004. In 2005, the profits of Apple were $244 million more than the profit in 2004. This boost in Apple's profits was due to the increased popularity of the iPod. What was the profit of Apple Computer in 2005?

64. On August 1, 2006, the average retail price of regular gas in Atlanta, Georgia, was $2.93 per gal. Over the next 2 months, gas prices fell by $0.93 per gal. Gas prices then rose by $0.21 per gal during the following 2 months. By January 2007, gas prices fell $0.25 per gal. What was the average retail price of regular gas in Atlanta, Georgia, in January 2007?

65. On August 1, 2006, the average retail price of regular gas in San Francisco, California, was $3.15 per gal. Over the next 2 months, gas prices fell by $0.58 per gal. Prices fell by $0.15 per gal the following months. By January 2007, gas prices rose by $0.25 per gal. What was the average retail price of regular gas in San Francisco, California, in January 2007?

66. Rhonda has $152.35 in her checking account. She deposits her paycheck of $423.45. Rhonda writes a check for daycare for $135, for groceries for $93.58, and then for rent for $350. What is Rhonda's checking account balance?

67. Mohammad has $52.75 in his checking account. He writes a check for groceries for $48.21. He writes a check for books for $475 and then deposits his student loan check for $750. What is Mohammad's checking account balance?

68. American Airlines reported an operating income of $965 million in 2007, −$1889 million in 2008, −$1004 million in 2009, and $308 million in 2010. What was American's total operating income for 2007 to 2010? (Source: www.aa.com)

69. Ford Motor Company reported a net income of −$14,766 million in 2008, $2717 million in 2009, and $6561 million in 2010. What was Ford's total net income for 2008 to 2010? (Source: http://corporate.ford.com/investors)

 You Be the Teacher!

Answer each student's question.

70. Benjamin: *I am trying to solve the following problem:*

   Simplify the following expression: $-\frac{3}{4} + \left(-\frac{1}{2}\right)$.

   Please check my work and tell me if I did it correctly.

Since $-\frac{3}{4}$ and $-\frac{1}{2}$ are the same sign, I just need to add their absolute values together. I know that $\left|-\frac{3}{4}\right| = \frac{3}{4}$ and $\left|-\frac{1}{2}\right| = \frac{1}{2}$.

So, $-\frac{3}{4} + \left(-\frac{1}{2}\right) = \frac{3}{4} + \frac{1}{2} = \frac{3}{4} + \frac{1 \times 2}{2 \times 2} = \frac{3}{4} + \frac{2}{4} = \frac{5}{4}$.

71. Renae: *I am trying to solve the following problem:*

   Simplify the following expression: $\frac{1}{4} + \left(-\frac{1}{2}\right)$.

   I'm having trouble visualizing what this means in terms of the number line. Can you explain this problem to me using a number line?

 Calculate It!

Use a calculator to simplify each expression. Express each answer in fractional form, when applicable.

72. $4.56 + (-8.27)$ **73.** $-3.12 + (-10.89)$

74. $-\frac{1}{3} + 12$ **75.** $-4\frac{1}{2} + \frac{3}{5}$

76. $5\frac{1}{7} + \left(-2\frac{3}{7}\right)$ **77.** $\left(-4\frac{1}{4}\right) + \left(-6\frac{1}{5}\right)$

PIECE IT TOGETHER / SECTIONS 1.1–1.4

Graph each number on a real number line and classify the number as natural, whole, integer, rational, irrational, and/or real. (*Section 1.1, Objectives 1 and 2*)

1. $\left\{-\frac{15}{2}, -\sqrt{3}, 2.5, 6\frac{1}{2}\right\}$

Compare the values of each pair of numbers. Use a <, >, or, = symbol to make the statement true. (*Section 1.1, Objective 3*)

2. -5 _____ $-(-5)$ **3.** 7 _____ $-(-7)$

Perform the operation and simplify each result. (*Section 1.2, Objectives 4–6*)

4. $\frac{2}{3} \cdot \frac{6}{10}$ **5.** $\frac{7}{12} \cdot \frac{2}{7}$ **6.** $\frac{1}{3} \div 3$

7. $7 \div \frac{1}{7}$ **8.** $\frac{9}{10} - \frac{1}{10}$ **9.** $\frac{5}{18} - \frac{3}{12}$

10. $\frac{7}{12} - \frac{1}{6}$ **11.** $5 - \frac{1}{2}$

Identify the base, the exponent, and then simplify the expression. (*Section 1.3, Objective 1*)

12. 2.1^2 **13.** -3^4

Use the order of operations to simplify each expression. (*Section 1.3, Objective 2*)

14. $7[40 - 2|5 + 3(15 - 11)|]$ **15.** $\dfrac{3 + \sqrt{3^2 - 4(2)(1)}}{2 \cdot 2}$

16. $2(16) - 7(2)$

Evaluate each expression for the specified values of x. When multiple values are given, organize the information in a chart. (*Section 1.3, Objective 3*)

17. $\frac{2}{5}x + 3$ for $x = 0, 1, 5,$ and 10

18. $2x - y$ for $x = 3$ and $y = 1$

Translate each phrase into an algebraic expression. Let x represent the unknown number. (*Section 1.3, Objective 5*)

19. The quotient of a number and 6

20. Twice the sum of a number and 4

21. 3 more than one-fifth of a number

22. 8 less than twice a number

Perform the indicated operation and simplify. (*Section 1.4, Objectives 1 and 2*)

23. $(-8) + (-2)$ **24.** $-19 + 11 + (-6)$

25. $\left(-\frac{3}{5}\right) + \frac{19}{15}$

/ **Subtraction of Real Numbers**

As a result of completing this section, you will be able to

1. Subtract real numbers.
2. Apply the order of operations.
3. Solve real-life problems.
4. Troubleshoot common errors.

The world's coldest recorded temperature is $-129°F$ in Vostok, Antarctica. The world's hottest recorded temperature is $134°F$ in Death Valley, California. What is the difference between the hottest and coldest temperatures?

To answer this question, we must know how to subtract real numbers.

Subtracting Real Numbers

Objective 1 ▶

Subtract two real numbers.

To understand how to subtract two real numbers, consider the following expressions.

$$7 - 2 = 5 \qquad 7 + (-2) = 5$$

These two expressions are equivalent, which indicates that subtraction is really a form of addition. Subtraction is equivalent to adding the opposite of a number.

> **Property: Subtraction of Real Numbers**
>
> For all real numbers a and b,
>
> $$a - b = a + (-b)$$

> **Procedure: Subtracting Real Numbers**
>
> **Step 1:** Rewrite the expression by adding the opposite of the second number.
> **Step 2:** Use the rules for addition to simplify the expression.

Objective 1 Examples   **Perform the indicated operation.**

1a. $-6 - 2$ **1b.** $3.5 - (-2.5)$ **1c.** $5 - 12$

1d. $-100 - (-25)$ **1e.** $-\dfrac{1}{2} - \dfrac{3}{5}$

Solutions   **1a.** $-6 - 2 = -6 + (-2)$ Rewrite as adding the opposite of 2, which is −2.

$= -8$ Add the numbers.

1b. $3.5 - (-2.5) = 3.5 + (2.5)$ Rewrite as adding the opposite of −2.5, which is 2.5.

$= 6$ Add the numbers.

1c. $5 - 12 = 5 + (-12)$ Rewrite as adding the opposite of 12, which is −12.

$= -7$ Add the numbers.

1d. $-100 - (-25) = -100 + (25)$ Rewrite as adding the opposite of −25, which is 25.

$= -75$ Add the numbers.

1e. $-\dfrac{1}{2} - \dfrac{3}{5} = -\dfrac{1}{2} + \left(-\dfrac{3}{5}\right)$ Rewrite as adding the opposite of $\dfrac{3}{5}$, which is $-\dfrac{3}{5}$.

$\qquad\qquad = -\dfrac{5}{10} + \left(-\dfrac{6}{10}\right)$ Convert the fractions to equivalent fractions with their common denominator of 10.

$\qquad\qquad = -\dfrac{11}{10}$ Add the numbers.

✓ **Student Check 1**   Perform the indicated operation.

  a. $1 - 4$ **b.** $11.3 - (-4.3)$ **c.** $-75 - 15$

  d. $-32 - (-8)$ **e.** $-\dfrac{2}{3} - \left(-\dfrac{1}{4}\right)$

Note: *Instead of rewriting a subtraction problem as adding the opposite, we observe that, after any double negative signs are eliminated, the operation connecting the numbers is equivalent to the sign of the number being added. This enables us to use the rules for addition to simplify the expression.*

For instance,

$-6 - 2 \rightarrow$ *Both numbers are negative, so we get* -8.

$6 - 10 \rightarrow$ *These numbers have opposite signs, so we get* -4.

$-1 - (-5) = -1 + 5 \rightarrow$ *These numbers have opposite signs, so we get* 4.

Also, note that $-(-a) = a$.

More about the Order of Operations

Objective 2 ▶
Apply the order of operations.

In Section 1.3, we learned the order of operations. We will use that again to simplify expressions, but this time, we will apply the rules for adding and subtracting signed numbers.

> **Procedure: Order of Operations**
>
> When simplifying a numerical expression, perform the operations in the following order.
>
> **Step 1:** Simplify expressions inside grouping symbols first, starting with the innermost set. If there is a fraction bar, simplify the numerator and denominator separately.
> **Step 2:** Evaluate any exponential expressions.
> **Step 3:** Perform multiplication or division in order from left to right.
> **Step 4:** Perform addition or subtraction in order from left to right.

Objective 2 Examples   Simplify each expression.

2a. $-4 - (-3) + 5 - 7$ **2b.** $-4^2 + 6 - |-10 + 3|$
2c. $5 - \{4 - [6 - (1 - \sqrt{2} - (-2))]\}$

Solutions   **2a.** $-4 - (-3) + 5 - 7 = -4 + 3 + 5 + (-7)$ Rewrite.

$\qquad\qquad\qquad\qquad = -1 + 5 + (-7)$ Add the first two numbers, $-4 + 3 = -1$.

$\qquad\qquad\qquad\qquad = 4 + (-7)$ Add the first two numbers, $-1 + 5 = 4$.

$\qquad\qquad\qquad\qquad = -3$ Add the final numbers.

2b. $-4^2 + 6 - |-10 + 3|$

$= -4^2 + 6 - |-7|$ Add inside the absolute value.

$= -16 + 6 - (7)$ Simplify 4^2 and $|-7|$.

$= -16 + 6 - 7$

$= -10 - 7$ Add from left to right.

$= -17$

2c. $5 - \{4 - [6 - (1 - \sqrt{2 - (-2)})]\}$ Rewrite the subtraction inside the square root symbol as addition.

$= 5 - \{4 - [6 - (1 - \sqrt{2 + 2})]\}$ Add inside the square root.

$= 5 - \{4 - [6 - (1 - \sqrt{4})]\}$ Simplify the square root.

$= 5 - \{4 - [6 - (1 - 2)]\}$ Subtract the numbers in parentheses.

$= 5 - \{4 - [6 - (-1)]\}$ Rewrite the expression inside the brackets as addition.

$= 5 - \{4 - [6 + 1]\}$ Add the numbers inside the brackets.

$= 5 - \{4 - [7]\}$ Subtract the numbers inside the braces.

$= 5 - \{-3\}$ Rewrite the subtraction as addition.

$= 5 + 3$ Add.

$= 8$

✓ Student Check 2   Simplify each expression.

a. $-2 - 1 - (-3) + 5$ **b.** $-3^3 + 15 - |4 - (-2)|$

c. $-8 - \{2 - [7 - (3 - \sqrt{25 - 16})]\}$

Applications

Objective 3 ▶

Solve real-life problems.

Subtracting real numbers occurs in real life when we need to find changes in temperatures, elevation, the stock market, bank accounts, and so on. It is important to be able to express the given change mathematically. We will use the skills from Section 1.3 to express mathematical relationships together with the skills for adding and subtracting real numbers.

    Other types of applications come from the study of geometry. There are some special angle relationships that are helpful to know.

Definition: The sum of the measure of the angles in a *triangle* is 180°.

$$a + b + c = 180°$$

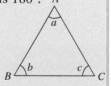

Definition: Complementary Angles

Complementary angles are two angles whose sum is 90°. Complementary angles form a **right angle** when they are adjacent to one another. A right angle is an angle whose measure is 90°. If a and b are complementary angles, then the sum of their measures is 90°.

$$a + b = 90°$$

Some examples of complementary angles are shown in the following table.

a	b	$a+b$
20°	70°	$20° + 70° = 90°$
30°	60°	$30° + 60° = 90°$
45°	45°	$45° + 45° = 90°$
50°	40°	$50° + 40° = 90°$
$x°$	$(90 - x)°$	$x° + (90 - x)° = 90°$

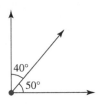

Definition: Supplementary Angles

Supplementary angles are two angles whose sum is 180°. Supplementary angles form a **straight angle** when they are adjacent to each other. A straight angle is an angle whose measure is 180°. If a and b are supplementary angles, then the sum of their measures is 180°.

$$a + b = 180°$$

Some examples of supplementary angles are shown in the following table.

a	b	$a+b$
20°	160°	$20° + 160° = 180°$
30°	150°	$30° + 150° = 180°$
45°	135°	$45° + 135° = 180°$
50°	130°	$50° + 130° = 180°$
$x°$	$(180 - x)°$	$x° + (180 - x)° = 180°$

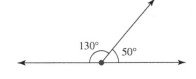

Objective 3 Examples

Vostok, Antarctica

Write a mathematical expression that represents the given situation and then find the result.

3a. The world's coldest recorded temperature is $-129°$F in Vostok, Antarctica. The world's hottest recorded temperature is 134°F in Death Valley, California. What is the difference between the hottest and coldest temperatures?

3b. Find the measure of the unknown angle of the given triangle.

3c. Find the complement and supplement of an angle whose measure is 25°.

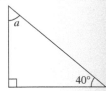

Solutions

3a. We must find the difference between the hottest and coldest temperatures. This can be translated as "hottest − coldest."

Since the hottest temperature is 134°F and the coldest is $-129°$F, we need to find

$$134 - (-129) = 134 + 129 = 263°F$$

Therefore, the difference between the hottest and coldest temperatures is 263°F.

3b. The two known angles have measures of 90° and 40°. So, the sum of the known angles is $90° + 40° = 130°$.

Since the total of the three angles must equal 180°, the remaining angle is the difference of 180° and the sum of the known angles. So,

$$a = 180° - 130° = 50°$$

The measure of the third angle is 50°.

3c. The complement of a 25° angle is
$$90° - 25° = 65°$$

The supplement of a 25° angle is
$$180° - 25° = 155°$$

✓ **Student Check 3**   Write a mathematical expression that represents the given situation and then find the result.

a. The highest point on Earth is the peak of Mount Everest in the Himalaya Mountains. The peak is 29,028 ft above sea level. The lowest point on land is the Dead Sea at 1312 ft below sea level. What is the difference in these altitudes?

b. Find the measure of the unknown angle of the given triangle.

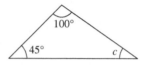

c. Find the complement and the supplement of an angle whose measure is 85°.

Objective 4 ▶
Troubleshoot common errors.

Troubleshooting Common Errors

A common error associated with subtracting real numbers is shown.

Objective 4 Example   **A problem and an incorrect solution is given. Provide the correct solution and an explanation of the error.**

$$-8 - 2$$

Incorrect Solution	Correct Solution and Explanation
$-8 - 2 = -6$	We first convert the problem to an addition problem. Since the numbers being added have the same signs, we add their absolute values and keep the sign. $$-8 - 2 = -8 + (-2)$$ $$= -10$$

ANSWERS TO STUDENT CHECKS

Student Check 1   **a.** -3 **b.** 15.6 **c.** -90
   d. -24 **e.** $-\dfrac{5}{12}$

Student Check 2   **a.** 5 **b.** -18 **c.** -3
Student Check 3   **a.** $30,340$ **b.** $35°$ **c.** $5°, 95°$

SUMMARY OF KEY CONCEPTS

1. Subtracting real numbers is equivalent to adding the opposite of a number.

2. We can apply the order of operations to simplify expressions that involve adding and subtracting real numbers.

3. We should be able to solve real-life applications by first expressing the situation mathematically and then simplifying the expression. Some applications involve triangles and angles. The sum of the measure of the three angles in a triangle is 180°. Complementary angles have a sum of 90°; supplementary angles have a sum of 180°.

GRAPHING CALCULATOR SKILLS

When using the graphing calculator for subtraction, it is important to distinguish between the negative sign, $(-)$, and the subtraction sign, $-$. Use the negative sign to denote the sign of a number and the subtraction sign to perform an operation with numbers.

Example: $-4 - (-7)$

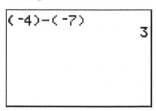

SECTION 1.5 / EXERCISE SET

 **Write About It!**

Use complete sentences in your answer to each exercise.

1. Explain how to subtract two real numbers.

2. Explain the difference between complementary angles and supplementary angles.

Determine if the statement is true or false. If a statement is false, provide an example with real numbers that contradicts the statement.

3. $a - (-a) = 0$ for all real numbers a.

4. $a - a = 0$ for all real numbers a.

 Practice Makes Perfect!

Perform the indicated operation and simplify. (See Objective 1.)

5. $2 - 9$

6. $5 - 11$

7. $2.1 - (-4.7)$

8. $3.7 - (-9.4)$

9. $-2 - 6$

10. $-3 - 9$

11. $-43 - 35$

12. $-15 - 19$

13. $2 - (-6)$

14. $3 - (-9)$

15. $-2 - (-6)$

16. $-3 - (-9)$

17. $\dfrac{2}{3} - \dfrac{7}{3}$

18. $\dfrac{3}{5} - \dfrac{3}{10}$

19. $-\dfrac{2}{3} - \dfrac{7}{3}$

20. $-\dfrac{3}{5} - \dfrac{3}{10}$

21. $\dfrac{2}{3} - \left(-\dfrac{7}{3}\right)$

22. $\dfrac{3}{5} - \left(-\dfrac{3}{10}\right)$

23. $-\dfrac{2}{3} - \left(-\dfrac{7}{3}\right)$

24. $-\dfrac{3}{5} - \left(-\dfrac{3}{10}\right)$

25. $1\dfrac{1}{5} - 6\dfrac{4}{5}$

26. $3\dfrac{1}{7} - 4\dfrac{6}{7}$

27. $-1\dfrac{1}{5} - 6\dfrac{4}{5}$

28. $-3\dfrac{1}{7} - 4\dfrac{6}{7}$

29. $1\dfrac{1}{5} - \left(-6\dfrac{4}{5}\right)$

30. $3\dfrac{1}{7} - \left(-4\dfrac{6}{7}\right)$

31. $-1\dfrac{1}{5} - \left(-6\dfrac{4}{5}\right)$

32. $-3\dfrac{1}{7} - \left(-4\dfrac{6}{7}\right)$

33. $-2.1 - (-4.7)$

34. $-3.7 - (-9.4)$

Simplify each expression. (See Objective 2.)

35. $1 - (-5) + (-3) - 7$

36. $-2 + 4 - 5 - (-6)$

37. $-8 - 9 + 4 - 5$

38. $3 - (-7) - 6 - 4$

39. $5 - 15 - (-12) + 25$

40. $-8 - (-12) - (-6)$

41. $-1.9 - (-0.6) - 2.8$

42. $5.8 - (-2.2) + (-1.3)$

43. $-\dfrac{1}{2} + \dfrac{3}{4} - \left(-\dfrac{3}{8}\right)$

44. $\dfrac{3}{5} + \left(-\dfrac{1}{5}\right) - \left(-\dfrac{7}{10}\right)$

45. $-4^2 + 13 - |7 - (-2)|$

46. $-7^2 + 11 - |8 - (-5)|$

47. $(-3)^2 - (-6) - |2 - 11|$

48. $(-6)^2 - (-10) - |3 - 15|$

49. $-12 - \{4 - [1 - (6 - \sqrt{19 - 10})]\}$

50. $-10 - \{7 - [2 - (3 - \sqrt{29 - 13})]\}$

Solve the problem by first writing a mathematical expression that represents the situation. *(See Objective 3.)*

51. The lowest recorded temperature in Vermont was $-46°$C at the Bloomfield weather station on December 30, 1933. The highest recorded temperature was $41°$C at the Vernon weather station on July 4, 1911. What is the difference between the highest and lowest temperatures?

52. The lowest recorded temperature in Alabama was $-33°$C at the New Market weather station on January 30, 1966. The highest recorded temperature was $44°$C at the Centerville weather station on Sept. 5, 1925. What is the difference between the highest and lowest temperatures?

53. The highest point on the North American continent is the peak of Mount McKinley. The peak is 20,320 ft above sea level. The lowest point on land in Syria is near Lake Tiberias with an elevation of 656.2 ft below sea level. What is the difference between these altitudes?

54. The highest point on the Antarctica continent is the peak of Vinson Massif. The peak is 4897 meters above sea level. The lowest point  on land in Russia is the Caspian Sea shoreline 28 m below sea level. What is the difference between these altitudes?

55. The table shows the annual net income, in thousands, for AirTran Airways for 2005–2009. Find the change in net income for each year. (Source: http://investor.airtran.com, 2009 Annual Report)

Year	Net Income	Change in Net Income
2005	$ 9,320	n/a
2006	$ 14,494	
2007	$ 50,545	
2008	-$266,334	
2009	$134,662	

56. The operating profit of General Motors (GM) Corporation was $10,705 million in December 2005. By March 2006, the operating profit  decreased by $5904 million. In June 2006, the operating profit decreased another $4990 million. By September 2006, the operating profit increased by $5583 million. What was the operating profit of GM in September 2006? (Source: CNNMoney.com)

Find the measure of the unknown angle of the given triangle.

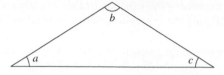

57. $a = 74°$, $b = 38°$ **58.** $b = 114°$, $c = 21°$

59. $a = 23°$, $b = 97°$ **60.** $b = 61°$, $c = 53°$

61. Find the complement and the supplement of an angle whose measure is 52°.

62. Find the complement and the supplement of an angle whose measure is 26°.

63. Find the complement and the supplement of an angle whose measure is 73°.

64. Find the complement and the supplement of an angle whose measure is 34°.

 Mix 'Em Up!

Perform the indicated operation and simplify.

65. $(-3.6) - 5.1$ **66.** $(-5.78) - 7.15$

67. $-\dfrac{5}{13} - \dfrac{1}{2}$ **68.** $-\dfrac{9}{14} + \dfrac{1}{2}$

69. $14 - 8 + (-5) - (-7)$ **70.** $22 + (-5) - (-10) - 1$

71. $-4 - \left(-\dfrac{5}{4}\right)$ **72.** $3\dfrac{2}{3} - \left(-6\dfrac{1}{2}\right)$

73. $-2^2 + 11 - |-8 - (-4)|$

74. $-3^2 - 16 - |-8 - (-2)|$

75. $4.5 - (-1.8) + 3.2 - 0.1$

76. $2.9 + (-1.1) - 3.6 - 0.5$

77. $18 - \{10 - [4 - (7 - \sqrt{30 - 5})]\}$

78. $20 - \{6 - [5 - (1 - \sqrt{20 - 4})]\}$

79. $-2.6 - (-1.2) - 2.7$ **80.** $9.5 - 2.9 + (-0.1)$

81. $-\dfrac{5}{6} + \dfrac{1}{2} - \left(-\dfrac{2}{3}\right)$ **82.** $\dfrac{5}{7} + \left(-\dfrac{3}{4}\right) - \left(-\dfrac{9}{14}\right)$

83. $(-1.5)^2 - (-4.5) - |3.2 - 7.2|$

84. $(-2.5)^2 + (-6.4) - |5.7 - 11.7|$

85. $1 - 18 - (-10) - 34$ **86.** $-6 - (-17) - 1$

Write an expression needed to solve each problem and then answer the question.

87. The northernmost city in the United States is Barrow, Alaska. The southernmost city in the Continental United States is Key West, Florida. The coldest recorded temperature in Barrow, Alaska, is $-47°$F.

The coldest recorded temperature in Key West, Florida, is 41°F. What is the difference between the coldest recorded temperature in Key West and the coldest recorded temperature in Barrow?

88. The record low temperature in New York City is 1 degree Fahrenheit below zero. The record low temperature in Palm Springs, California, is 23°F. What is the difference between the record low temperatures of New York City and Palm Springs?

Find the measure of the unknown angle of the given triangle.

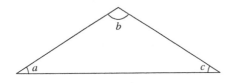

89. $a = 41°$ $b = 99°$

90. $b = 82°$, $c = 27°$

91. Find the complement and the supplement of an angle whose measure is 69°.

92. Find the complement and the supplement of an angle whose measure is 18°.

 You Be the Teacher!

Correct each student's errors, if any.

93. Simplify the following expression: $\frac{1}{5} - \left(-1\frac{3}{4}\right)$.

Elvernice's work:

$$\frac{1}{5} - \left(-1\frac{3}{4}\right) = \frac{1}{5} - \left(-\frac{7}{4}\right) = \frac{1}{5} - \frac{7}{4} = \frac{1 \times 4}{5 \times 4} - \frac{7 \times 5}{4 \times 5}$$

$$= \frac{4}{20} - \frac{35}{20} = -\frac{31}{20}.$$

94. Simplify the following expression: $-\frac{1}{4} - (-3)$.

Sherry's work:

$$-\frac{1}{4} \times 3 = -\frac{1}{4} \times \frac{3}{1} = -\frac{3}{4}.$$

 Calculate It!

Use a calculator to simplify each expression. Express answers in fractional form, when applicable.

95. $25.19 - 17.26$

96. $-2.45 - 5.68$

97. $-\frac{5}{6} - 5$

98. $3\frac{2}{5} - \frac{3}{5}$

SECTION 1.6	**Multiplication and Division of Real Numbers**

▶ **OBJECTIVES**

As a result of completing this section, you will be able to

1. **Multiply real numbers.**
2. **Simplify exponential expressions.**
3. **Divide real numbers.**
4. **Evaluate algebraic expressions.**
5. **Solve real-life problems.**
6. **Troubleshoot common errors.**

Objective 1 ▶

Multiply real numbers.

Ryan decides to play some poker. To play a hand, he must bet $5. He plays seven hands and does not win any of them. How much money did Ryan lose? To solve this problem, we need to know how to multiply positive and negative numbers together. If Ryan lost $5 for each of seven hands, he lost a total of $35. This can be determined mathematically by finding the product of 7 and −5.

In this section, we will discuss multiplying and dividing real numbers along with real-life illustrations.

Multiplying Real Numbers

Multiplication of real numbers is just repeated addition. For example, 2(3) means that 3 is added two times or $3 + 3 = 6$. Similarly, 3(2) means that 2 is added three times or $2 + 2 + 2 = 6$. Using this concept, we can find the product $2(-3)$. This expression means that −3 is added two times or

$$2(-3) = (-3) + (-3) = -6$$

Now we need to find out what happens when two negative numbers are multiplied. Consider the product $(-2)(-3)$.

$$(-2)(-3) = -(2)(-3)$$
$$= -[2(-3)]$$
$$= -(-6)$$
$$= 6$$

| Rewrite negative 2 as the opposite of 2. |
| Rewrite as the opposite of 2 times −3. |
| The product of 2 and −3 is −6. |
| The opposite of −6 is 6. |

Procedure: Multiplying Real Numbers

Step 1: Determine the sign of the product.
     a. The product of two real numbers with the *same* sign is positive.
     b. The product of two real numbers with *different* signs is negative.
Step 2: Multiply the numbers.

Objective 1 Examples   **Multiply the real numbers.**

1a. $7(-5)$ **1b.** $(-4)(-6)$ **1c.** $\left(-\dfrac{3}{5}\right)\left(\dfrac{5}{3}\right)$ **1d.** $(-4)\left(-\dfrac{1}{8}\right)$

1e. $(-2.35)(10)$ **1f.** $(-4)(-8)(-3)$ **1g.** $(-5)(-4)(-2)(-1)$

Solutions   **1a.** $\qquad\qquad 7(-5) = -35$ The product of two numbers with different signs is negative.

1b. $\qquad\qquad (-4)(-6) = 24$ The product of two numbers with the same sign is positive.

1c. $\qquad \left(-\dfrac{3}{5}\right)\left(\dfrac{5}{3}\right) = -\dfrac{3(5)}{5(3)}$ The product of numbers with different signs is negative.

$\qquad\qquad\qquad\qquad = -\dfrac{15}{15}$ Multiply the numerators and denominators.

$\qquad\qquad\qquad\qquad = -1$ Simplify.

1d. $\qquad (-4)\left(-\dfrac{1}{8}\right) = \left(-\dfrac{4}{1}\right)\left(-\dfrac{1}{8}\right)$ Rewrite -4 as $-\dfrac{4}{1}$.

$\qquad\qquad\qquad\qquad = \dfrac{4(1)}{1(8)}$ The product of numbers with the same sign is positive.

$\qquad\qquad\qquad\qquad = \dfrac{4}{8}$ Simplify.

$\qquad\qquad\qquad\qquad = \dfrac{1}{2}$

1e. $\qquad (-2.35)(10) = -23.5$ The product of numbers with different signs is negative.

1f. $\qquad (-4)(-8)(-3) = 32(-3)$ Multiply from left to right. The product of the first two numbers is positive.

$\qquad\qquad\qquad\qquad = -96$ The final product is negative.

1g. $(-5)(-4)(-2)(-1) = (20)(-2)(-1)$ Multiply from left to right. The product of the first two numbers is positive.

$\qquad\qquad\qquad\qquad = (-40)(-1)$ The product of the numbers is negative.

$\qquad\qquad\qquad\qquad = 40$ The final product is positive.

☑ Student Check 1   Multiply the real numbers.

a. $(-5)(3)$ **b.** $(-12)(-2)$ **c.** $\dfrac{2}{3}\left(-\dfrac{3}{2}\right)$ **d.** $(-6)\left(-\dfrac{1}{18}\right)$

e. $-5.71(100)$ **f.** $(-7)(-3)(-4)$ **g.** $(-8)(-9)(-2)\left(-\dfrac{1}{2}\right)$

Note: *Example 1(f) contains a product of three negative numbers and their product is negative.*
Example 1(g) contains a product of four negative numbers and their product is positive.

The product of an even number of negative numbers is positive.
The product of an odd number of negative numbers is negative.

Simplifying Exponential Expressions

Objective 2 ▶

Simplify exponential
expressions.

In Section 1.3, we defined an exponent. Recall the following definition of an exponent.

Definition: Exponents

For b a real number and n a natural number,

$$b^n = \underbrace{b \cdot b \cdot b \cdots b}_{n \text{ times}}$$

where b is the base and n is the exponent.

We will now examine what happens when a negative base is raised to an even or odd exponent.

Negative Base Raised to an Even Exponent (2, 4, 6, . . .)	Negative Base Raised to an Odd Exponent (1, 3, 5, . . .)
$(-5)^2 = (-5)(-5)$	$(-4)^3 = (-4)(-4)(-4)$
$\quad\;\; = 25$	$\quad\;\; = 16(-4)$
	$\quad\;\; = -64$
$(-2)^4 = (-2)(-2)(-2)(-2)$	$(-3)^5 = (-3)(-3)(-3)(-3)(-3)$
$\quad\;\; = 4(-2)(-2)$	$\quad\;\; = 9(-3)(-3)(-3)$
$\quad\;\; = -8(-2)$	$\quad\;\; = -27(-3)(-3)$
$\quad\;\; = 16$	$\quad\;\; = (81)(-3)$
	$\quad\;\; = -243$
A negative base raised to an even exponent is *positive*.	A negative base raised to an odd exponent is *negative*.

Identifying the base of an exponent incorrectly can cause the result to be incorrect. The error most often occurs with expressions of the form -3^2 and $(-3)^2$. While these expressions look very similar, they represent two very different things.

Expression	Verbal Representation	Mathematical Representation	Base
-3^2	Opposite of 3 squared	$\begin{aligned}-3^2 &= -(3 \cdot 3)\\ &= -(9) \text{ or } -9\end{aligned}$	3
$(-3)^2$	Negative 3 squared	$\begin{aligned}(-3)^2 &= (-3)(-3)\\ &= 9\end{aligned}$	-3

The examples in the table illustrate the following facts.

1. If a negative sign is in front of an exponential expression, the negative sign indicates the opposite of the value of the exponential expression. In other words, the negative sign is not part of the base of the exponent. That is,

$$-b^n = -\underbrace{(b \cdot b \cdot b \cdots b)}_{n \text{ times}}$$

The base b is repeated as a factor n times.

2. If there are parentheses around a negative number raised to an exponent, then the exponent is applied to the negative number. The negative number is repeated as a factor. That is,

$$(-b)^n = \underbrace{(-b)(-b)(-b) \cdots (-b)}_{n \text{ times}}$$

The base $-b$ is repeated as a factor n times.

Procedure: Simplifying an Exponential Expression

Step 1: Identify the base and exponent.
Step 2: Rewrite the expression as repeated multiplication.
Step 3: Simplify the result.

Objective 2 Examples   Identify the base, determine if the negative sign repeats in the product, and simplify each exponential expression.

Problems	Base	Does negative sign repeat?	Evaluate
2a. $(-4)^3$	-4	yes	$\begin{aligned}(-4)^3 &= (-4)(-4)(-4)\\ &= -64\end{aligned}$
2b. -5^4	5	no	$\begin{aligned}-5^4 &= -(5 \cdot 5 \cdot 5 \cdot 5)\\ &= -625\end{aligned}$
2c. $\left(-\dfrac{1}{2}\right)^2$	$-\dfrac{1}{2}$	yes	$\begin{aligned}\left(-\dfrac{1}{2}\right)^2 &= \left(-\dfrac{1}{2}\right)\left(-\dfrac{1}{2}\right)\\ &= \dfrac{1}{4}\end{aligned}$
2d. $-(-3)^5$	-3	yes	We must find the *opposite* of -3 raised to the fifth power. $\begin{aligned}-(-3)^5 &= -[(-3)(-3)(-3)(-3)(-3)]\\ &= -(-243)\\ &= 243\end{aligned}$

✔ **Student Check 2** Identify the base, determine if the negative sign repeats in the product, and simplify each exponential expression.

a. $(-8)^2$ **b.** -8^2 **c.** $\left(-\dfrac{2}{3}\right)^5$ **d.** $-(-0.5)^3$

Dividing Real Numbers

Objective 3 ▶

Divide real numbers.

As discussed earlier in this chapter, division by a number is equivalent to multiplying by the *reciprocal* of the number. For example,

$$4 \div 2 = 2 \text{ is equivalent to } 4 \cdot \frac{1}{2} = 2$$

> **Definition: Division of Real Numbers**
>
> For real numbers a and b, with $b \neq 0$,
>
> $$a \div b = \frac{a}{b} = a \cdot \frac{1}{b}$$
>
> The value a is called the **dividend** and the value b is called the **divisor**.

Since the quotient $a \div b$ can be written as the product $a \cdot \dfrac{1}{b}$, the sign rules for multiplication also apply to division.

> **Procedure: Dividing Real Numbers**
>
> **Step 1:** Determine the sign of the quotient.
>   **a.** The quotient of two numbers with the *same* sign is positive.
>   **b.** The quotient of two numbers with *different* signs is negative.
> **Step 2:** Simplify the quotient.

Recall that division can always be checked by multiplication. For example,

$$\frac{-6}{3} = -2 \text{ because } 3(-2) = -6$$

Recall also that $\dfrac{0}{6} = 0$ because $0 \cdot 6 = 0$. This result is stated below.

> **Property: Zero as Dividend**
>
> The quotient of 0 and any nonzero real number, b, is 0.
>
> $$\frac{0}{b} = 0, b \neq 0$$

If we try to find $\dfrac{6}{0}$, we would have to find a number that, when multiplied by 0, is 6. There is no such number since 0 times any real number is 0. Therefore, we say that this quotient is undefined.

> **Property: Zero as Divisor**
>
> The quotient of any nonzero real number, b, and 0 is *undefined*.
>
> $$\frac{b}{0} \text{ is undefined}$$

One last fact about negative signs and fractions is important to note. Consider the following four quantities.

$$-\frac{4}{2} = -2 \qquad \frac{-4}{2} = -2 \qquad \frac{4}{-2} = -2 \qquad \frac{-4}{-2} = 2$$

So,

$$-\frac{4}{2} = \frac{-4}{2} = \frac{4}{-2} \qquad \text{but} \qquad -\frac{4}{2} \neq \frac{-4}{-2}$$

The placement of the negative sign in a fraction can be done in three ways. The negative sign can be placed in the numerator of the fraction, *or* the denominator of the fraction, but *not* both, *or* in front of the fraction.

Property: Negative Signs in Fractions

For real numbers a and b with $b \neq 0$,

$$-\frac{a}{b} = \frac{-a}{b} = \frac{a}{-b}$$

Objective 3 Examples	Perform the indicated operation.

Problems	**Solutions**	
3a. $\dfrac{24}{-6}$	$\dfrac{24}{-6} = -4$	
3b. $\dfrac{-40}{0}$	$\dfrac{-40}{0}$ is undefined	
3c. $\dfrac{-35}{-7}$	$\dfrac{-35}{-7} = 5$	
3d. $\dfrac{-6}{-7}$	$\dfrac{-6}{-7} = \dfrac{6}{7}$	
3e. $\dfrac{0}{-3}$	$\dfrac{0}{-3} = 0$	
3f. $20 \div \left(-\dfrac{2}{5}\right)$	$20 \div \left(-\dfrac{2}{5}\right) = 20 \cdot \left(-\dfrac{5}{2}\right)$	Rewrite as multiplying by the reciprocal.
	$= \dfrac{20}{1} \cdot -\dfrac{5}{2}$	Rewrite 20 as a fraction.
	$= -\dfrac{100}{2}$	Multiply the numerators and denominators.
	$= -50$	Simplify.
3g. $-\dfrac{3}{4} \div \dfrac{5}{8}$	$-\dfrac{3}{4} \div \dfrac{5}{8} = -\dfrac{3}{4} \cdot \dfrac{8}{5}$	Rewrite as multiplying by the reciprocal.
	$= -\dfrac{24}{20}$	Multiply the numerators and denominators.
	$= -\dfrac{4 \cdot 6}{4 \cdot 5}$	Factor the numerator and denominator.
	$= -\dfrac{6}{5}$	Divide out the common factor of 4.

☑ **Student Check 3**   Perform the indicated operation.

a. $\dfrac{30}{-5}$ **b.** $\dfrac{0}{-4}$ **c.** $\dfrac{-60}{-12}$ **d.** $\dfrac{3}{5} \div (-9)$ **e.** $\dfrac{-11}{-15}$ **f.** $\left(-\dfrac{1}{2}\right) \div \left(-\dfrac{9}{10}\right)$

Algebraic Expressions

Objective 4 ▶

Evaluate algebraic expressions.

Now that we know how to perform operations with signed numbers, we will revisit evaluating algebraic expressions. Recall that we evaluate an algebraic expression by replacing the variable(s) with the given values. Then we use the order of operations to simplify the resulting expression.

Objective 4 Examples **Evaluate each expression at the given values.**

4a. $\dfrac{-b + \sqrt{b^2 - 4ac}}{2a}$, for $a = 3, b = -11, c = -20$

4b. $3x^2 - x + 5$, for $x = -2, -1, 0, 1, 2$

Solutions **4a.** $\dfrac{-b + \sqrt{b^2 - 4ac}}{2a}$

$= \dfrac{-(-11) + \sqrt{(-11)^2 - 4(3)(-20)}}{2(3)}$ Let $a = 3, b = -11, c = -20$.

$= \dfrac{11 + \sqrt{121 - 4(3)(-20)}}{6}$ Simplify the opposite of -11 and the exponential expression. Simplify the denominator.

$= \dfrac{11 + \sqrt{121 - 12(-20)}}{6}$ Multiply from left to right inside the square root.

$= \dfrac{11 + \sqrt{121 + 240}}{6}$ Multiply -12 and -20.

$= \dfrac{11 + \sqrt{361}}{6}$ Add the numbers inside the square root.

$= \dfrac{11 + 19}{6}$ Simplify the square root.

$= \dfrac{30}{6}$ Add the numerators.

$= 5$ Divide the numbers.

4b. Make a chart to organize the information. Make the appropriate substitution for the variable. Simplify the expression.

x	$3x^2 - x + 5$	
-2	$3(-2)^2 - (-2) + 5 = 3(4) - (-2) + 5$ $= 12 + 2 + 5$ $= 19$	Replace x with -2. Simplify the exponent. Multiply. Add or subtract from left to right.
-1	$3(-1)^2 - (-1) + 5 = 3(1) - (-1) + 5$ $= 3 + 1 + 5$ $= 9$	Replace x with -1. Simplify the exponent. Multiply. Add or subtract from left to right.
0	$3(0)^2 - (0) + 5 = 3(0) - (0) + 5$ $= 0 - 0 + 5$ $= 5$	Replace x with 0. Simplify the exponent. Multiply. Add or subtract from left to right.
1	$3(1)^2 - (1) + 5 = 3(1) - (1) + 5$ $= 3 - 1 + 5$ $= 7$	Replace x with 1. Simplify the exponent. Multiply. Add or subtract from left to right.
2	$3(2)^2 - (2) + 5 = 3(4) - (2) + 5$ $= 12 - 2 + 5$ $= 15$	Replace x with 2. Simplify the exponent. Multiply. Add or subtract from left to right.

✓ **Student Check 4**   Evaluate each expression at the given values.

a. $\sqrt{(x_1 - x_2)^2 + (y_1 - y_2)^2}$, for $x_1 = -3$, $x_2 = 2$, $y_1 = 8$, $y_2 = -4$

b. $\dfrac{|x - 2|}{x + 2}$, for $x = -2, -1, 0, 1, 2$

Applications

Objective 5 ▶

Solve real-life problems.

Multiplication and division of real numbers occur in real life through various types of problems. Multiplication is needed to calculate the *area* and *volume* of geometric shapes. Division is needed to calculate the *average* of a group of numbers.

> **Definition:** The **average** of a group of two or more values is the sum of the values divided by the number of values.

For example, the average of 50 and 100 is $\dfrac{50 + 100}{2} = \dfrac{150}{2} = 75$.

Division is also needed to find the percent change in a value.

> **Property: Percent Change**
>
> If an original value has taken on a new value, then the percent change is calculated as
> $$\frac{\text{new value} - \text{original value}}{\text{original value}} \quad \text{or} \quad \frac{\text{change in value}}{\text{original value}}$$

For example, if the price of a $30 item is on sale for $24, then the percent change in the price is

$$\frac{\text{new} - \text{original}}{\text{original}} = \frac{24 - 30}{30} = \frac{-6}{30} = -0.20 \quad \text{or} \quad \text{a decrease of } 20\%$$

Objective 5 Examples   Solve each problem.

5a. The area of a rectangle is length times width. Write an equation that represents the area of a rectangle, where l is the length and w is the width of the rectangle. Use the equation to find the area of an iPad which has a length of 9.56 in. and a width of 7.47 in. Round to two decimal places.

Solution   **5a.** The area of a rectangle can be written as $A = (l)(w)$ or $A = lw$.

To find the area of an iPad, replace l with 9.56 and w with 7.47.

$$A = lw$$
$$A = (9.56)(7.47)$$
$$A = 71.41$$

So, the area of an iPad is 71.41 in.^{2}

5b. The following chart shows the closing price of Apple's stock for August 23 to August 27, 2010. (Source: http://www.google.com/finance/historical?q=NASDAQ:AAPL)

Date	Closing Price	Daily Change in Price	Percent Change in Price
8/27/2010	$241.62		
8/26/2010	$240.28		
8/25/2010	$242.89		
8/24/2010	$239.93		
8/23/2010	$245.80	n/a	n/a

i. Complete the chart by showing how the price of the stock changed from the previous day to the next and by calculating the percent change in the price for each day.

ii. After completing the chart, find the average daily change in price.

Solution   5b.   i. The daily change in price is the difference in one day's closing price and the previous day's closing price. The percent change is the daily change divided by the previous day's closing price.

Date	Closing Price	Daily Change in Price	Percent Change in Price
8/27	$241.62	$241.62 - 240.28 = \$1.34$	$\dfrac{1.34}{240.28} = 0.0056$ or 0.56%
8/26	$240.28	$240.28 - 242.89 = -\$2.61$	$\dfrac{-2.61}{242.89} = -0.0107$ or -1.07%
8/25	$242.89	$242.89 - 239.93 = \$2.96$	$\dfrac{2.96}{239.93} = 0.0123$ or 1.23%
8/24	$239.93	$239.93 - 245.80 = -\$5.87$	$\dfrac{-5.87}{245.80} = -0.0239$ or -2.39%
8/23	$245.80	n/a	n/a

ii. The average daily change in the closing price is the sum of the daily changes divided by the number of days.

$$\frac{-5.87 + 2.96 + (-2.61) + 1.34}{4} = \frac{-4.18}{4} = -1.045$$

On average between August 23 and August 27, the closing price decreased by approximately $1.05 each day.

✓ **Student Check 5**   Solve each problem.

a. The volume of a right circular cylinder is the square of the radius, times the product of π and the height of the cylinder. Write an equation that represents the volume of a right circular cylinder, where r is the radius and h is the height. Then use the equation to find the volume of a fuel oil storage tank that is a right circular cylinder with radius 24 in. and height 72 in. If there are 231 in.^{3} in a gallon, how many gallons will this tank hold?

b. The following chart shows the closing price of Callaway Golf Company's stock for August 9 to August 13, 2010. (Source: http://www.google.com/finance/historical?q=NYSE:ELY)

Date	Closing Price	Daily Change in Price	Percent Change in Price
8/13/2010	$6.41		
8/12/2010	$6.64		
8/11/2010	$6.68		
8/10/2010	$7.07		
8/9/2010	$7.25	n/a	n/a

i. Complete the chart by showing how the price of the stock changed from the previous day to the next and by calculating the percent change in the price for each day.

ii. After completing the chart, find the average daily change in price.

Troubleshooting Common Errors

Objective 6 ▶
Troubleshoot common errors.

Some common errors associated with multiplying and dividing real numbers are shown next. One of the main errors arises when negative signs are involved in exponential expressions. Other errors occur when division with zero is involved.

Objective 6 Examples   A problem and an incorrect solution are given. Provide the correct solution and an explanation of the error.

6a. Simplify -7^2.

Incorrect Solution	Correct Solution and Explanation
$-7^2 = 49$	The base of the exponential expression is 7. The negative sign indicates that we find the opposite of squared. $$-7^2 = -(7 \cdot 7) = -49$$

6b. Simplify $\dfrac{-2}{0}$.

Incorrect Solution	Correct Solution and Explanation
$\dfrac{-2}{0} = 0$	When we divide by zero, the expression is undefined. So, $\dfrac{-2}{0}$ is undefined.

6c. Simplify $\dfrac{0}{-2}$.

Incorrect Solution	Correct Solution and Explanation
$\dfrac{0}{-2}$ is undefined	When zero is divided by any nonzero number, the result is zero. So, $$\dfrac{-2}{0} = 0$$

ANSWERS TO STUDENT CHECKS

Student Check 1   **a.** -15 **b.** 24   **c.** -1 **d.** $\dfrac{1}{3}$ **e.** -571 **f.** -84 **g.** 72

Student Check 2   **a.** 64   **b.** -64 **c.** $-\dfrac{32}{243}$ **d.** 0.125

Student Check 3   **a.** -6 **b.** 0   **c.** 5   **d.** $-\dfrac{1}{15}$ **e.** $\dfrac{11}{15}$ **f.** $\dfrac{5}{9}$

Student Check 4   **a.** 13   **b.** undefined, 3, 1, $\dfrac{1}{3}$, 0

Student Check 5 **a.** $V = \pi r^2 h$; $V = 130{,}288$ in.^{3} or approximately 564 gal

b. i. in table below **ii.** The average daily change in price is $-\$0.21$.

Date	Closing Price	Daily Change in Price	Percent Change in Price
8/13/2010	$6.41	−$0.23	− 0.0346 or − 3.46%
8/12/2010	$6.64	−$0.04	− 0.0060 or − 0.60%
8/11/2010	$6.68	−$0.39	− 0.0552 or − 5.52%
8/10/2010	$7.07	−$0.18	− 0.0248 or − 2.48%
8/9/2010	$7.25	n/a	n/a

SUMMARY OF KEY CONCEPTS

1. The product or quotient of two real numbers with the same sign is positive. The product or quotient of two real numbers with different signs is negative.

2. The product of an even number of negative numbers is positive. The product of an odd number of negative numbers is negative.

3. An exponent indicates how many times a number is multiplied by itself or used as a factor.

   a. A negative number raised to an odd exponent is negative.

   b. A negative number raised to an even exponent is positive.

   c. When a negative sign is in the base of an exponential expression, it should be repeated in the multiplication.

   d. When a negative sign is not in parentheses of an exponential expression, we find the opposite of the value of the exponential expression. The negative sign is not repeated.

4. Division by zero is undefined but 0 divided by a nonzero number is 0.

5. Evaluate an algebraic expression by replacing the variable with the given number. Simplify using the order of operations.

6. To solve real-life problems, express the problem mathematically and then simplify.

GRAPHING CALCULATOR SKILLS

The calculator can assist us with multiplying and dividing real numbers as well as evaluating exponential expressions. The caret symbol  is used with exponents. To enter the power of 2, you may either press the  x^2 key or you may use  and 2. If parentheses are given in the problem, you must enter them on the calculator.

Examples:

1. $(-4)(-5)$

2. $-\dfrac{60}{15}$

3. $(-6)^2$

4. -6^2

5. $(-4)^3$

```
(-4)(-5)
              20
-60/15
              -4
```

```
(-6)²
              36
-6²
             -36
(-4)^3
             -64
```

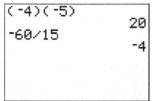

SECTION 1.6   EXERCISE SET

 Write About It!

Use complete sentences to explain the process or problem.

1. Multiply signed numbers

2. Simplify an exponential expression

3. Divide signed numbers (where the denominator is nonzero)

4. Explain why division by zero is undefined.

Determine if the statement is true or false. If a statement is false, provide an example with real numbers that contradicts the statement.

5. The expression $(-b)^n$, where n is a natural number, is always positive.

6. The expression $-b^n$ is equivalent to multiplying $-b$ by itself n times.

7. The product of an odd number and a negative number is negative.

8. A negative base raised to an odd exponent is always negative.

9. The product of a nonzero real number and its additive inverse is 1.

10. Zero divided by a nonzero real number is always 0.

 Practice Makes Perfect!

Perform the indicated operation and simplify.
(See Objective 1.)

11. $8(-6)$

12. $(-3)(5)$

13. $\frac{1}{3}(-9)$

14. $\left(-\frac{1}{4}\right)(20)$

15. $\left(-\frac{1}{3}\right)(-9)$

16. $\left(-\frac{1}{4}\right)(-20)$

17. $\left(\frac{2}{5}\right)\left(-\frac{4}{7}\right)$

18. $\left(-\frac{3}{8}\right)\left(\frac{4}{9}\right)$

19. $\left(-\frac{2}{5}\right)\left(-\frac{4}{7}\right)$

20. $\left(-\frac{3}{8}\right)\left(-\frac{4}{9}\right)$

21. $5(-2.3)$

22. $(-6)(1.4)$

23. $-3.24(1000)$

24. $4.98(-100)$

25. $-5\left(-\frac{1}{5}\right)$

26. $\left(-\frac{4}{5}\right)\left(-\frac{5}{4}\right)$

27. $(-1)(-2)(3)$

28. $(-4)(5)(-6)$

29. $-7(-4)(-9)\left(\frac{2}{3}\right)$

30. $-3(-4)(-7)\left(\frac{1}{14}\right)$

31. $(-9)(-8)(0)(-1)$

32. $(-4)(0)(-5)(3)$

33. $(-8.2)(10)$

34. $(-7.25)(100)$

Simplify each exponential expression. (See Objective 2.)

35. 8^2

36. 12^2

37. 5^3

38. 7^3

39. $(-9)^2$

40. $(-11)^2$

41. $(-4)^3$

42. $(-5)^3$

43. $(-2)^4$

44. $(-1)^4$

45. -13^2

46. -2^3

47. $-(-5)^4$

48. $-(-10)^4$

49. $\left(\frac{1}{4}\right)^2$

50. $\left(-\frac{3}{8}\right)^2$

51. $-\left(-\frac{3}{2}\right)^3$

52. $-\left(-\frac{5}{4}\right)^3$

53. $(0.8)^2$

54. $(0.5)^2$

55. $-(-1.2)^3$

56. $-(-1.5)^3$

Perform the indicated operation and simplify.
(See Objective 3.)

57. $\frac{-12}{3}$

58. $\frac{-40}{10}$

59. $\frac{14}{-2}$

60. $\frac{18}{-9}$

61. $\frac{-16}{-2}$

62. $\frac{-1}{-7}$

63. $-5 \div \frac{1}{5}$

64. $-7 \div \frac{1}{7}$

65. $\frac{4}{7} \div \left(-\frac{1}{21}\right)$

66. $\frac{5}{6} \div \left(-\frac{1}{30}\right)$

67. $\left(-\frac{2}{3}\right) \div \left(-\frac{2}{3}\right)$

68. $\left(-\frac{3}{5}\right) \div \left(-\frac{3}{5}\right)$

69. $\frac{0}{-7}$

70. $\frac{0}{-3}$

71. $\frac{-5}{0}$

72. $\frac{-9}{0}$

Evaluate each expression for the given values.
(See Objective 4.)

73. $\frac{-b + \sqrt{b^2 - 4ac}}{2a}$ for $a = 2, b = -9, c = 10$

74. $\frac{-b + \sqrt{b^2 - 4ac}}{2a}$ for $a = 3, b = 14, c = 8$

75. $\frac{-b - \sqrt{b^2 - 4ac}}{2a}$ for $a = 2, b = -1, c = -28$

76. $\frac{-b - \sqrt{b^2 - 4ac}}{2a}$ for $a = -6, b = -5, c = 4$

77. $2x^2 - 5x + 1$ for $x = -2, 0, \frac{1}{2}, 2$

78. $3x^2 + x - 2$ for $x = -2, 0, \frac{1}{2}, 2$

79. $-2x^2 + 4x + 3$ for $x = -2, -1, \dfrac{1}{2}, 2$

80. $-x^2 - 3x + 4$ for $x = -2, -1, \dfrac{1}{2}, 2$

81. $\sqrt{(x_1 - x_2)^2 + (y_1 - y_2)^2}$ for $x_1 = -1$, $x_2 = -4$, $y_1 = -3$, $y_2 = 1$

82. $\sqrt{(x_1 - x_2)^2 + (y_1 - y_2)^2}$ for $x_1 = -7$, $x_2 = 13$, $y_1 = 9$, $y_2 = -6$

83. $\dfrac{|x - 2|}{x + 1}$ for $x = -2, -1, 0, 1, 2$

84. $\dfrac{|x - 2|}{x + 2}$ for $x = -2, -1, 0, 1, 2$

85. $\dfrac{y_2 - y_1}{x_2 - x_1}$ for $x_1 = -2$, $x_2 = 8$, $y_1 = -4$, and $y_2 = -6$

86. $\dfrac{y_2 - y_1}{x_2 - x_1}$ for $x_1 = 5$, $x_2 = -3$, $y_1 = -7$, and $y_2 = 2$

87. $b^2 - 4ac$ for $a = 3, b = -5, c = -6$

88. $b^2 - 4ac$ for $a = -1, b = 6, c = -3$

89. The volume of an inverted cone is the square of the radius, times the product of π and $\dfrac{1}{3}$ and the height of the cone. Write an equation that represents the volume of an inverted cone, where r is the radius and h is the height. Then use the equation to find the volume of a liquid storage tank that is an inverted cone with radius 30 in. and height 48 in. Round to the nearest integer.

90. The volume of a rectangular box is the product of the length, width, and height. Write an equation that represents the volume of a rectangular box, where l is the length, w is the width, and h is the height. Then use the equation to find the volume of a liquid storage tank that is a rectangular box with the length 24.5 ft, width 36.5 ft, and height 56 ft. Round to the nearest integer.

91. The chart shows the closing daily price of Hershey Company's stock for October 4 to October 8, 2010. (Source: http://www.google.com/finance/historical?q=NYSE:ELY)

Date	Closing Price	Daily Change in Price	Percent Change in Price
10/8/2010	$48.75		
10/7/2010	$47.88		
10/6/2010	$48.17		
10/5/2010	$48.10		
10/4/2010	$47.63	n/a	n/a

a. Complete the chart by showing how the price of the stock changed from the previous day to the next and by calculating the percent change in the price for each day.

b. After completing the chart, find the average daily change in price.

92. The following chart shows the closing weekly price of Walmart Stores' stock for September 10 to October 8, 2010. (Source: http://www.google.com/finance/historical?q=NYSE:ELY)

Date	Closing Price	Weekly Change in Price	Percent Change in Price
10/8/2010	$54.41		
10/1/2010	$53.36		
9/24/2010	$54.08		
9/17/2010	$53.01		
9/10/2010	$51.97		

a. Complete the chart by showing how the price of the stock changed from the previous week to the next and by calculating the percent change in the price for each week.

b. After completing the chart, find the average weekly change in price.

 Mix 'Em Up!

Perform the indicated operation and simplify.

93. $(-7)(5)$

94. $(-13)(-11)$

95. $(-3.03)(10)$

96. $(-7.36)(-1000)$

97. $(-55)\left(-\dfrac{7}{11}\right)$

98. $-4\left(\dfrac{1}{20}\right)$

99. 3^4

100. 6^4

101. $\left(\dfrac{4}{7}\right)\left(-\dfrac{3}{16}\right)$

102. $\left(-\dfrac{10}{7}\right)\left(-\dfrac{14}{25}\right)$

103. $\dfrac{-63}{-7}$

104. $\dfrac{-44}{11}$

105. -5^4

106. 10^3

107. $\left(\dfrac{2}{5}\right)^3$

108. $\left(\dfrac{3}{7}\right)^3$

109. $(-4)^4$

110. $-(-3)^5$

111. -0.2^4

112. $-(-0.5)^3$

113. $(4.2)(-3.5)$

114. $(-1.2)(-5.2)$

115. $\left(-\dfrac{2}{3}\right)\left(-\dfrac{3}{2}\right)$

116. $\left(-\dfrac{7}{5}\right)\left(-\dfrac{5}{7}\right)$

117. -14^2

118. -6^3

119. $\dfrac{0}{-2}$

120. $\dfrac{-2}{0}$

121. $2 \div 0$

122. $0 \div (-9)$

123. $\dfrac{10}{-9} \div (-25)$

124. $\dfrac{-12}{5} \div (-10)$

125. $\left(-\dfrac{15}{19}\right) \div \left(-\dfrac{6}{19}\right)$

126. $\left(-\dfrac{8}{21}\right) \div \left(\dfrac{12}{21}\right)$

127. $\left(-\dfrac{5}{6}\right)^3$ **128.** $-\left(-\dfrac{8}{11}\right)^2$

129. $6(10)(-1)$ **130.** $10(-6)(-9)$

Evaluate each expression for the given values.

131. $\sqrt{(x_1 - x_2)^2 + (y_1 - y_2)^2}$ for $x_1 = 17$, $x_2 = -3$, $y_1 = -11$, $y_2 = 4$

132. $\sqrt{(x_1 - x_2)^2 + (y_1 - y_2)^2}$ for $x_1 = -10$, $x_2 = 2$, $y_1 = 3$, $y_2 = -2$

133. $2l + 2w$ for $l = 11$ and $w = 15$

134. $2l + 2w$ for $l = 3$ and $w = 8$

135. $b^2 - 4ac$ for $a = -4, b = -7, c = -9$

136. $b^2 - 4ac$ for $a = -7, b = -8, c = 2$

137. $\dfrac{1}{2}bh$ for $b = 4$ and $h = 5$

138. $\dfrac{1}{2}bh$ for $b = 3$ and $h = 6$

139. $\dfrac{-b - \sqrt{b^2 - 4ac}}{2a}$ for $a = 5, b = -7, c = -6$

140. $\dfrac{-b + \sqrt{b^2 - 4ac}}{2a}$ for $a = 5, b = -7, c = -6$

141. $\dfrac{y_2 - y_1}{x_2 - x_1}$ for $x_1 = -4, x_2 = 13, y_1 = -1$, and $y_2 = -9$

142. $\dfrac{y_2 - y_1}{x_2 - x_1}$ for $x_1 = 2, x_2 = -3, y_1 = -6$, and $y_2 = -11$

143. $\dfrac{|x - 2|}{x + 3}$ for $x = -3, -2, 0, 1, 2$

144. $\dfrac{|2x - 3|}{x}$ for $x = -2, -1, 0, 1, 2$

Solve each problem.

145. The volume of a rectangular box is the product of the length, width, and height. Write an equation that represents the volume of a rectangular box, where l is the length, w is the width, and h is the height. Then use the equation to find the volume of a liquid storage tank that is a rectangular box with the length 32.4 in., width 26.6 in., and height 64.5 in. If there are 231 in.^{3} in a gallon, how many gallons will this tank hold? Round to the nearest integer.

146. The following chart shows the closing weekly price of Dow Jones Industrial Average prices for September 10 to October 8, 2010. (Source: http://www.google.com/finance/historical?q=NYSE:ELY)

Date	Closing Price	Weekly Change in Price	Percent Change in Price
10/8/2010	$11,006.48		
10/1/2010	$10,829.68		
9/24/2010	$10,860.26		
9/17/2010	$10,607.85		
9/10/2010	$10,462.77	n/a	n/a

a. Complete the chart by showing how the price of the stock changed from the previous week to the next and by calculating the percent change in the price for each week.

b. After completing the chart, find the average weekly change in price.

 You Be the Teacher!

Correct each student's errors, if any.

147. Determine if $\dfrac{0}{a}$ or $\dfrac{a}{0}$ is undefined.

Marcia's work:

$\dfrac{0}{a} =$ undefined and $\dfrac{a}{0} = 0$

148. Simplify $-(-12)^2$.

Basil's work:

$-(-12)^2 = 12^2 = 144$

149. Simplify $\dfrac{12}{15} \div \left(-\dfrac{10}{9}\right)$.

Holly's work:

$\dfrac{12}{15} \div \left(-\dfrac{10}{9}\right) = -\dfrac{12}{15} \cdot \dfrac{10}{9} = -\dfrac{\overset{4}{\cancel{12}}}{\underset{5}{\cancel{15}}} \cdot \dfrac{\overset{2}{\cancel{10}}}{9} = -\dfrac{8}{9}$

 Calculate It!

Use a calculator to simplify each expression. Express each answer in fractional form, when applicable.

150. $(-0.4)^5$ **151.** $(-1.2)^2$

152. $-\left(-\dfrac{2}{3}\right)^5$ **153.** $\left(-\dfrac{3}{4}\right)^2$

154. $\dfrac{7}{30} \div \left(-\dfrac{1}{21}\right)$ **155.** $-\dfrac{15}{8} \div \left(-\dfrac{20}{16}\right) \cdot 12$

SECTION 1.7   **Properties of Real Numbers**

▶ **OBJECTIVES**

As a result of completing this section, you will be able to

1. Apply the identity and inverse properties.
2. Apply the commutative and associative properties.
3. Apply the distributive property.

The set of real numbers contains some interesting properties. We use some of the properties without even thinking about them. These properties are very important to the study of algebra, so we will state them explicitly in this section.

The Identity and Inverse Properties

The first property that we will discuss is the *identity property*. An **identity element** is a number which leaves another number unchanged when an operation is performed on it.

- When we add numbers, the only number that can be added to another number without changing its value is *zero*.
- When we multiply numbers, the only number that can be multiplied to another number without changing its value is *one*.

Objective 1 ▶

Apply the identity and inverse properties.

	Identity Property	Identity Element	Example ($a = 6$)
Addition	For all real numbers a, $a + 0 = 0 + a = a$	Zero is the **additive** identity.	$6 + 0 = 0 + 6 = 6$
Multiplication	For all real numbers a, $a \cdot 1 = 1 \cdot a = a$	One is the **multiplicative** identity.	$6 \cdot 1 = 1 \cdot 6 = 6$

A related property is the *inverse property*. An **inverse** is a number which produces the identity element when an operation is performed on it.

- The *additive inverse* of a number is its opposite since adding a number and its opposite results in 0.
- The *multiplicative inverse* of a number is its reciprocal since multiplying a number and its reciprocal results in 1.

	Inverse Property	Inverse Element	Example ($a = 5$)
Addition	For all real numbers a, $a + (-a) = (-a) + a$ $= 0$	$-a$ is the **additive inverse** (or opposite) of a.	The opposite of 5 is -5. $5 + (-5) = (-5) + 5$ $= 0$
Multiplication	For all real numbers $a \neq 0$, $a \cdot \dfrac{1}{a} = \dfrac{1}{a} \cdot a$ $= 1$	$\dfrac{1}{a}$ is the **multiplicative inverse** (or reciprocal) of a, $a \neq 0$.	The reciprocal of 5 is $\dfrac{1}{5}$. $5\left(\dfrac{1}{5}\right) = \left(\dfrac{1}{5}\right)(5)$ $= 1$

Objective 1 Examples   Find both the additive and multiplicative inverses of each number. Assume any variables are nonzero.

Problems	Solutions	
	Additive Inverse, $-a$	Multiplicative Inverse, $\dfrac{1}{a}$
1a. -6	$-(-6) = 6$	$\dfrac{1}{-6} = -\dfrac{1}{6}$
1b. $\dfrac{3}{4}$	$-\left(\dfrac{3}{4}\right) = -\dfrac{3}{4}$	$\dfrac{1}{\frac{3}{4}} = 1 \cdot \dfrac{4}{3} = \dfrac{4}{3}$
1c. $2x$	$-(2x) = -2x$	$\dfrac{1}{2x} = \dfrac{1}{2x}$
1d. $-3y$	$-(-3y) = 3y$	$\dfrac{1}{-3y} = -\dfrac{1}{3y}$
1e. $\dfrac{x}{7}$	$-\left(\dfrac{x}{7}\right) = -\dfrac{x}{7}$	$\dfrac{1}{\frac{x}{7}} = 1 \cdot \dfrac{7}{x} = \dfrac{7}{x}$

✔ **Student Check 1**   Find both the additive and multiplicative inverses of each number. Assume any variables are nonzero.

a. -10 **b.** $\dfrac{7}{8}$ **c.** $4y$ **d.** $-9b$ **e.** $\dfrac{a}{3}$

The Commutative and Associative Properties

Objective 2 ▶

Apply the commutative and associative properties.

Additional properties of the real numbers relate to how we add and multiply the numbers. These properties form the foundation of how we work with algebraic expressions.

The *commutative property* of the real numbers states that the order in which we add real numbers or multiply real numbers doesn't change the result.

	Commutative Property	**Examples ($a = 2, b = 3$)**
Addition	For all real numbers a and b, $a + b = b + a$	$2 + 3 = 3 + 2 = 5$
Multiplication	For all real numbers a and b, $a \cdot b = b \cdot a$	$2 \cdot 3 = 3 \cdot 2 = 6$

The *associative property* of the real numbers states that the way numbers are grouped when they are added or multiplied doesn't change the outcome.

	Associative Property	**Examples ($a = 2, b = 3, c = 4$)**
Addition	For all real numbers a, b, and c, $a + (b + c) = (a + b) + c$	$2 + (3 + 4) = (2 + 3) + 4$ $2 + (7) = (5) + 4$ $9 = 9$
Multiplication	For all real numbers a, b, and c, $(a \cdot b) \cdot c = a \cdot (b \cdot c)$	$(2 \cdot 3) \cdot 4 = 2 \cdot (3 \cdot 4)$ $(6) \cdot 4 = 2 \cdot (12)$ $24 = 24$

Objective 2 Examples   Apply the commutative and associative properties to rewrite each expression and simplify the result.

2a. $2 + y + 7$ **2b.** $2(y)(7)$ **2c.** $(x + 6) + 4$ **2d.** $4(6x)$

Solutions   **2a.** $2 + y + 7 = y + 2 + 7$ Apply the commutative property of addition.
$= y + 9$ Add the numbers.

2b. $2(y)(7) = 2(7)y$ Apply the commutative property of multiplication.
$= 14y$ Multiply the numbers.

2c. $(x + 6) + 4 = x + (6 + 4)$ Apply the associative property of addition.
$= x + 10$ Add the numbers.

2d. $4(6x) = (4 \cdot 6)x$ Apply the associative property of multiplication.
$= 24x$ Multiply the numbers.

✔ **Student Check 2**   Apply the commutative or associative properties to rewrite each expression and simplify the result.

a. $3 + x + 5$ **b.** $3(x)(5)$ **c.** $(b + 2) + 9$ **d.** $2(9b)$

Note: *Subtraction and division are* not *commutative. We cannot change the order of the numbers being subtracted or divided and obtain the same result.*

For instance, $5 - 3 = 2$ *but* $3 - 5 = 3 + (-5) = -2$.

Also, $\frac{6}{3} = 2$ *but* $\frac{3}{6} = \frac{1}{2}$.

The Distributive Property

Objective 3 ▶

Apply the distributive property.

The distributive property is a property that is used extensively in algebra. It provides a way for us to multiply a group of numbers by a number—that is, it enables us to rewrite a product as a sum or a sum as a product.

To illustrate, we will simplify the two expressions below using the order of operations.

$$3(4 + 5) = 3(9) \qquad\qquad 3(4) + 3(5) = 12 + 15$$
$$= 27 \qquad\qquad\qquad\qquad = 27$$

These two expressions are equivalent—that is,

$$3(4 + 5) = 3(4) + 3(5)$$

The preceding example illustrates that when a number is multiplied by a sum, it is equivalent to the sum of the products.

Also, consider a difference multiplied by a number.

$$3(4 - 5) = 3(-1) \qquad\qquad 3(4) - 3(5) = 12 - 15$$
$$= -3 \qquad\qquad\qquad\qquad = -3$$

These two expressions are equivalent, so

$$3(4 - 5) = 3(4) - 3(5)$$

The preceding example illustrates that when a number is multiplied by a difference, it is equivalent to the difference of the products.

These properties make up the distributive property over addition and over subtraction.

Property: Distributive Property over Addition

For all real numbers a, b, and c,

$$a(b + c) = ab + ac$$

Property: Distributive Property over Subtraction

$$a(b - c) = ab - ac$$

When asked to apply the distributive property, our goal is to multiply the factor outside of the parentheses by each of the terms inside the parentheses.

Because multiplication is commutative, we can also distribute from the left and write the distributive property as follows.

Property: Alternate Form of the Distributive Property

$$(b + c)a = ba + ca$$

An illustration of this alternate form is $(2 + 6)4 = 2(4) + 6(4) = 8 + 24 = 32$.

The distributive property can be applied when there are more than two terms inside parentheses. We just multiply the factor by each term in parentheses.

Objective 3 Examples	**Apply the distributive property to rewrite each expression. Simplify the result.**

3a. $2(x + 4)$ **3b.** $3(x - 5)$ **3c.** $4(2x + 7)$ **3d.** $-5(x + 2)$

3e. $-3(x - 7)$ **3f.** $8(2x + 3y - 5)$ **3g.** $-(6x - 9)$ **3h.** $10\left(\dfrac{1}{5}x + \dfrac{1}{2}\right)$

Solutions

3a.
$$2(x + 4) = 2(x) + 2(4)$$ Apply the distributive property.
$$= 2x + 8$$ Simplify.

3b.
$$3(x - 5) = 3(x) - 3(5)$$ Apply the distributive property.
$$= 3x - 15$$ Simplify.

3c.
$$4(2x + 7) = 4(2x) + 4(7)$$ Apply the distributive property.
$$= 8x + 28$$ Simplify.

3d.
$$-5(x + 2) = -5(x) + (-5)(2)$$ Apply the distributive property.
$$= -5x - 10$$ Simplify.

3e.
$$-3(x - 7) = -3(x) - (-3)(7)$$ Apply the distributive property.
$$= -3x + 21$$ Simplify.

3f.
$$8(2x + 3y - 5) = 8(2x) + 8(3y) - 8(5)$$ Apply the distributive property.
$$= 16x + 24y - 40$$ Simplify.

3g.
$$-1(6x - 9) = -1(6x) - (-1)(9)$$ Distribute -1 to each term.
$$= -6x + 9$$ Simplify.

 Note: *Multiplying by* -1 *changes the signs of the original terms in parentheses.*

3h.
$$10\left(\frac{1}{5}x + \frac{1}{2}\right) = 10\left(\frac{1}{5}x\right) + 10\left(\frac{1}{2}\right)$$ Apply the distributive property.
$$= 2x + 5$$ Simplify.

✓ **Student Check 3**	Apply the distributive property to rewrite each expression. Simplify the result.

a. $5(x + 6)$ **b.** $7(x - 3)$ **c.** $2(3x + 1)$ **d.** $-3(x + 2)$

e. $-6(x - 4)$ **f.** $9(4x + 5y - 3)$ **g.** $-(8x + 2)$ **h.** $15\left(-\dfrac{1}{3}x + \dfrac{2}{5}\right)$

Objective 4 ▶

Troubleshoot common errors.

Troubleshooting Common Errors

Some common errors associated with the properties of real numbers are shown next.

Objective 4 Examples / A problem and an incorrect solution are given. Provide the correct solution and an explanation of the error.

a. Simplify $(2x + 6) + 3$.

Incorrect Solution	Correct Solution and Explanation
$(2x + 6) + 3$ $6x + 18$	To simplify this expression, we must apply the associative property of addition. The distributive property does not apply since 3 is connected by addition, not multiplication. $(2x + 6) + 3$ $2x + (6 + 3)$ $2x + 9$

b. Simplify $4(6x + 7)$.

Incorrect Solution	Correct Solution and Explanation
$4(6x + 7)$ $24x + 7$	We must distribute 4 to each term in parentheses, not just the first term. $4(6x + 7)$ $24x + 28$

ANSWERS TO STUDENT CHECKS

Student Check 1   **a.** $10, -\dfrac{1}{10}$ **b.** $-\dfrac{7}{8}, \dfrac{8}{7}$ **c.** $-4y, \dfrac{1}{4y}$

d. $9b, -\dfrac{1}{9b}$ **e.** $-\dfrac{a}{3}, \dfrac{3}{a}$

Student Check 2   **a.** $x + 8$ **b.** $15x$ **c.** $b + 11$ **d.** $18b$

Student Check 3   **a.** $5x + 30$ **b.** $7x - 21$ **c.** $6x + 2$
   d. $-3x - 6$ **e.** $-6x + 24$ **f.** $36x + 45y - 27$
   g. $-8x - 2$ **h.** $-5x + 6$

SUMMARY OF KEY CONCEPTS

1. The identity element is the number that leaves another number unchanged when an operation is performed on it.
   a. For instance, 0 is the identity element for addition since any number added to zero is the original number.
   b. Also, 1 is the identity element for multiplication since any number multiplied by 1 is the original number.
2. An inverse is a number which produces the identity element when an operation is performed on it.
   a. The additive inverse is the opposite of a number.

   b. The multiplicative inverse is the reciprocal of a number.
3. Addition and multiplication are both commutative and associative. We can change the order and the grouping of the numbers being added or multiplied and obtain the same result.
4. The distributive property enables us to multiply a group of numbers by another number. The outer number distributes to all terms inside parentheses.

SECTION 1.7 / EXERCISE SET

 Write About It!

Use complete sentences to explain the term or the process.

1. Identity element for addition
2. Identity element for multiplication
3. Additive inverse

4. Multiplicative inverse
5. Commutative property for addition
6. Commutative property for multiplication
7. Associative property for addition
8. Associative property for multiplication

9. Distributive property
10. How to apply distributive property to simplify an expression

 Practice Makes Perfect!

Find the additive inverse and the multiplicative inverse of each number. Assume all variables are nonzero. (*See Objective 1.*)

11. -7 12. 8 13. 25

14. -12 15. $\dfrac{3}{5}$ 16. $\dfrac{7}{4}$

17. $-\dfrac{4}{9}$ 18. $-\dfrac{3}{10}$ 19. $3x$

20. $12y$ 21. $-6a$ 22. $-5b$

23. $\dfrac{x}{6}$ 24. $\dfrac{2y}{5}$

25. $-\dfrac{7a}{2}$ 26. $-\dfrac{4b}{3}$

Apply the commutative and associative properties to rewrite each expression and simplify the result. (*See Objective 2.*)

27. $4 + x + 7$ 28. $9 + y + 5$

29. $-12 + a - 1$ 30. $-15 + b - 7$

31. $-6 + x + (-10)$ 32. $-18 + y + (-12)$

33. $3(x)(9)$ 34. $5(y)(15)$

35. $(-1)(a)(-16)$ 36. $(-1)(b)(-24)$

37. $5(x)(-14)$ 38. $-9(y)(4)$

39. $(c - 6) + (-3)$ 40. $(d - 12) + (-9)$

41. $[x - (-2)] - 12$ 42. $[y - (-28)] - (-15)$

43. $-2(3x)$ 44. $-8(2y)$

45. $-7(-3a)$ 46. $-4(-5b)$

47. $\left(-\dfrac{1}{2}\right) + x + \dfrac{3}{4}$ 48. $\dfrac{2}{3} + y - \dfrac{5}{12}$

49. $\left(a + \dfrac{5}{7}\right) + 2$ 50. $\left(b - \dfrac{3}{5}\right) + 2$

51. $12(x)\left(\dfrac{1}{6}\right)$ 52. $18(y)\left(\dfrac{5}{6}\right)$

53. $-6\left(\dfrac{1}{4}x\right)$ 54. $-5\left(\dfrac{7}{10}y\right)$

Apply the distributive property to rewrite each expression. Simplify the result. (*See Objective 3.*)

55. $4(x - 7)$ 56. $5(x - 9)$

57. $3(x + 5)$ 58. $6(x + 2)$

59. $11(3x - 4)$ 60. $12(5x - 3)$

61. $6(2x + 3y - 5)$ 62. $3(5x - 2y + 10)$

63. $-(-5x + 2y - 9)$ 64. $-11(3x + 4y - 6)$

65. $-(16x - 3)$ 66. $-(23x - 12)$

67. $12\left(-\dfrac{1}{2}x + \dfrac{1}{3}\right)$ 68. $10\left(-\dfrac{1}{5}x + \dfrac{1}{2}\right)$

69. $-12\left(\dfrac{3}{2}x - \dfrac{1}{4}\right)$ 70. $-24\left(\dfrac{1}{6}x - \dfrac{3}{8}\right)$

 Mix 'Em Up!

Find the additive inverse and the multiplicative inverse of each number. Assume all variables are nonzero.

71. $15x$ 72. $-2a$ 73. $\dfrac{6x}{7}$

74. $\dfrac{3c}{5}$ 75. -11 76. 21

77. $\dfrac{9}{5}$ 78. $-\dfrac{6}{5}$

Apply the commutative, associative, and/or distributive properties to rewrite each expression and simplify the result.

79. $8 + x + 12$ 80. $-12 + b - 4$

81. $2(x)(14)$ 82. $3(y)(19)$

83. $-8 + x + (-28)$ 84. $-4 + y + (-22)$

85. $(-1)(a)(-18)$ 86. $-7(y)(14)$

87. $[x - (-5)] - 19$ 88. $[y - (-17)] - (-1)$

89. $\left(-\dfrac{3}{5}\right) + a + \dfrac{1}{2}$ 90. $-\dfrac{1}{3} + b - \dfrac{5}{6}$

91. $4(2.1a)$ 92. $3(1.2b)$

93. $\left(a - \dfrac{7}{8}\right) + 1$ 94. $\left(b - \dfrac{7}{10}\right) + 1$

95. $-(-x - 12y + 5)$ 96. $8(2x - 3y + 1)$

97. $-2(7x)$ 98. $-9(-5b)$

99. $5(-0.3x + 0.2)$ 100. $2(-0.9x + 0.1)$

101. $-5(2x + 3)$ 102. $-(12x + 5)$

103. $12(x)\left(-\dfrac{1}{4}\right)$ 104. $-5\left(-\dfrac{3}{10}y\right)$

105. $6\left(-\dfrac{1}{6}x + \dfrac{1}{3}\right)$ 106. $0.4(1.5x - 0.3)$

107. $0.8(0.7x - 1.2)$ 108. $5\left(-\dfrac{1}{10}x + \dfrac{2}{5}\right)$

 You Be the Teacher!

Correct each student's errors, if any.

109. Simplify $3(2a)$.

Josh's work:
$$3(2a) = 3(2)(3a) = 6(3a) = 18a$$

110. Simplify $3(6x + 2y - 4)$.

Grace's work:
$$3(6x + 2y - 4) = 18x + 2y - 4$$

111. Simplify $-(2 - x + 3y)$.

Charlotte's work:
$$-(2 - x + 3y) = -2 - x + 3y$$

112. Simplify $-6(-4b)$.

Mary's work:
$$-6(-4b) = -6(-4)(-6b) = 24(-6b) = -144b$$

SECTION 1.8 / Algebraic Expressions

▶ OBJECTIVES

As a result of completing this section, you will be able to

1. Identify terms and coefficients of algebraic expressions.
2. Identify like terms.
3. Combine like terms.
4. Simplify algebraic expressions.
5. Troubleshoot common errors.

Objective 1 ▶

Identify terms and coefficients of algebraic expressions.

In the previous sections, we have evaluated algebraic expressions and also expressed relationships mathematically as algebraic expressions. We will now discuss these types of expressions in more detail and learn some terminology that we use to describe them.

Terminology Dealing with Algebraic Expressions

A **term** is a number or a product of a number and variables raised to exponents. Terms are separated by addition or subtraction signs. Algebraic expressions consist of **variable terms** and **constant terms**. Variable terms are terms that contain a variable. Constant terms do not contain variables. Each variable term has a numerical **coefficient**. The coefficient is the number that is multiplied by the variable.

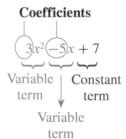

Procedure: Identifying Terms and Coefficients of an Algebraic Expression

Step 1: Determine how many expressions are connected by an addition or subtraction sign. This gives us how many terms the expression contains.
  a. The terms with variables are the variable terms.
  b. The terms with numbers only are the constant terms.
Step 2: The coefficients of the variable terms are the numbers multiplied by the variable. Be careful to identify the sign of the coefficient. If the variable term is connected by a subtraction sign, then the coefficient of that term is a negative value. Remember that subtraction is adding the opposite. Also, remember the Identity Property of Multiplication: $x = 1 \cdot x$.

Objective 1 Examples / Complete the table to identify the number of terms, the variable terms, the constant terms, and the coefficients of the variable terms of each expression.

Expression	Total Number of Terms	Variable Term(s)	Constant Term(s)	Coefficients of Each Variable Term
1a. $2x$	1	$2x$	None	2
1b. $-x$	1	$-x = -1x$	None	-1
1c. $x + 2$	2	$x = 1x$	2	1
1d. $\dfrac{y}{2} - 5$	2	$\dfrac{y}{2} = \dfrac{1y}{2} = \dfrac{1}{2}y$	-5	$\dfrac{1}{2}$
1e. $3x^2 - 5x + 7$	3	$3x^2, -5x$	7	$3, -5$
1f. $-2y^3 + 3y^2 - y - 6$	4	$-2y^3, 3y^2, -y$	-6	$-2, 3, -1$

☑ **Student Check 1** Complete the table to identify the number of terms, the variable terms, the constant terms, and the coefficients of the variable terms of each expression.

Expression	Total Number of Terms	Variable Term(s)	Constant Term(s)	Coefficients of Each Variable Term
a. $7a$				
b. $-y$				
c. $x - 7$				
d. $\dfrac{x}{3} - 4$				
e. $6y^2 - y + 3$				
f. $4y^3 - y^2 + 2y - 1$				

Like Terms

Objective 2 ▶

Identify like terms.

Terms that have the same variables with the same exponents are **like terms**.
Terms that do not have the same variables with the same exponents are **unlike terms**.

Like Terms	Unlike Terms
$2x, -5x$	$2x, -5$
$y^2, 4y^2$	$y^2, 4y$
$-3xy, 7yx$	$-3xy, 7x^2y$

> **Procedure: Identifying Like Terms**
>
> **Step: a.** If the variables and exponents are the same, the terms are *like*.
> **b.** If the variables or exponents are different, the terms are *unlike*.

Objective 2 Examples Determine if the terms are like or unlike and explain why.

Problems	Like Terms	Unlike Terms	Why?
2a. $-9y$ and -9		X	One term contains the variable y and the other term does not contain a variable.
2b. x and $-8x$	X		Each term contains the variable x.
2c. $2ba, 3ab, ab$	X		Each term contains the variables ab. Recall $ba = ab$ by the commutative property of multiplication.
2d. $6xy^2$ and $2x^2y$		X	These terms are unlike terms because the variables have different exponents.

 **Student Check 2** Determine if the terms are like or unlike and explain why.

a. $4h$ and 4 **b.** $2y$ and y **c.** $6rh$ and $-6hr$ **d.** x^2y and xy^2

Combining Like Terms

Objective 3 ▶

Combine like terms.

Algebraic expressions that contain like terms can be simplified by combining their like terms. There are two ways to think about combining like terms. Consider the algebraic expression $4x + 2x$. The terms $4x$ and $2x$ are like terms.

To simplify this expression, we can think of the problem in this way.

$$4x + 2x = x + x + x + x + x + x = 6x$$

We can also use the distributive property to combine like terms.

$$4x + 2x = (4 + 2)x = 6x$$

In either method, the result is the same, $4x + 2x = 6x$.

> **Procedure: Combining Like Terms**
>
> **Step 1:** Apply the distributive property to determine the coefficient of the like terms.
> **Step 2:** Simplify. Keep the variable component the same.

Objective 3 Examples Simplify each expression by combining like terms, if possible.

3a. $2x + x$ **3b.** $8x - x$ **3c.** $4y^2 + 3y^2$

3d. $3 + 5x$ **3e.** $\dfrac{x}{2} + 5x$ **3f.** $-2a^2 - 4a^2 + 5a - 3a$

Solutions   **3a.** $2x + x = 2x + 1x$ Recall $x = 1x$.

                          $= (2 + 1)x$ Apply the distributive property.

                          $= 3x$ Simplify.

It is not necessary to show the distributive property step. We can simply add the coefficients of the like terms mentally to get

$$2x + x = 2x + 1x$$
$$= 3x$$

3b. $8x - x = 8x - 1x$ | $8x - x = 8x - 1x$

                $= (8 - 1)x$ | $= 7x$

                $= 7x$ |

3c. $4y^2 + 3y^2 = (4 + 3)y^2$ | $4y^2 + 3y^2 = 7y^2$

                           $= 7y^2$ |

3d. The terms are unlike and therefore cannot be combined.

3e. $\dfrac{x}{2} + 5x = \dfrac{1}{2}x + 5x$ Recall that $\dfrac{x}{2} = \dfrac{1x}{2} = \dfrac{1}{2}x$.

                    $= \left(\dfrac{1}{2} + 5\right)x$ Apply the distributive property.

                    $= \left(\dfrac{1}{2} + \dfrac{10}{2}\right)x$ Write 5 as $\dfrac{10}{2}$.

                    $= \dfrac{11}{2}x$ Add.

3f. There are two sets of like terms. Apply the distributive property for each set of like terms to get

$$-2a^2 - 4a^2 + 5a - 3a = (-2 - 4)a^2 + (5 - 3)a$$
$$= -6a^2 + 2a$$

Adding the coefficients of the like terms and keeping the variables of the like terms the same gives us

$$-2a^2 - 4a^2 + 5a - 3a = -6a^2 + 2a$$

✔ **Student Check 3**   Simplify each expression by combining like terms, if possible.

   a. $11y + y$ **b.** $10b - b$ **c.** $2a^2 + 9a^2$

   d. $6 - 2x$ **e.** $\dfrac{a}{3} + 4a$ **f.** $-3x^2 - 4x^2 + 7x - 6x$

Simplifying Algebraic Expressions

Objective 4 ▶

Simplify algebraic expressions.

Simplifying algebraic expressions is a skill that we use in most aspects of algebra. Simplifying algebraic expressions involves clearing parentheses and combining any like terms.

> **Procedure: Simplifying Algebraic Expressions**
>
> **Step 1:** Clear any parentheses by applying the distributive property.
> **Step 2:** Combine like terms.

Objective 4 Examples   **Simplify each algebraic expression.**

4a. $3y - 5 + 8y - 1$ **4b.** $3(4x + 2) - 8$

4c. $2(4x - 5) + 3(2x - 1)$ **4d.** $7 - 2(8x + 6)$

Solutions   **4a.**

$$3y - 5 + 8y - 1 = 3y + 8y - 5 - 1$$ Apply the commutative property.

$$= 11y - 6$$ Combine like terms.

4b.

$$3(4x + 2) - 8 = 12x + 6 - 8$$ Apply the distributive property.

$$= 12x - 2$$ Combine like terms.

4c.

$$2(4x - 5) + 3(2x - 1) = 8x - 10 + 6x - 3$$ Apply the distributive property.

$$= 8x + 6x - 10 - 3$$ Apply the commutative property.

$$= 14x - 13$$ Combine like terms.

4d.

$$7 - 2(8x + 6) = 7 - 16x - 12$$ Apply the distributive property.

$$= -16x + 7 - 12$$ Apply the commutative property.

$$= -16x - 5$$ Combine like terms.

✓ **Student Check 4**   Simplify each algebraic expression.

a. $4y - 8 - 2y$ **b.** $5(6x + 3) - 4$

c. $6(3x + 1) + 4(5x - 3)$ **d.** $5 - 3(2x + 4)$

Objective 5 ▶

Troubleshoot common errors.

Troubleshooting Common Errors

Some common errors associated with simplifying algebraic expressions are shown next.

Objective 5 Examples   **A problem and an incorrect solution are given. Provide the correct solution and an explanation of the error.**

5a. Combine like terms: $5 + 2x$

Incorrect Solution	Correct Solution and Explanation
$5 + 2x$ $7x$	The terms 5 and $2x$ are not like terms. Therefore, they cannot be combined. The problem would have to be $5x + 2x$ to get an answer of $7x$.

5b. Combine like terms: $x^2 + x^2$

Incorrect Solution	Correct Solution and Explanation
$x^2 + x^2$ x^4	The terms are like. To add them, we must add their coefficients, not their exponents. Recall $x^2 = 1x^2$. $$x^2 + x^2$$ $$1x^2 + 1x^2$$ $$2x^2$$

5c. Simplify $4 - 3(x - 2)$.

Incorrect Solution	Correct Solution and Explanation
$4 - 3(x - 2)$ $1(x - 2)$ $x - 2$	The distributive property must be applied first because, according to the order of operations, multiplication comes before addition or subtraction. After distributing, we can combine like terms. $4 - 3(x - 2)$ $4 - 3x + 6$ $-3x + 4 + 6$ $-3x + 10$

ANSWERS TO STUDENT CHECKS

	Expression	Total Number of Terms	Variable Term(s)	Constant Term(s)	Coefficients of Each Variable Term
Student Check 1	**a.** $7a$	1	$7a$	None	7
	b. $-y$	1	$-y$	None	-1
	c. $x - 7$	2	x	-7	1
	d. $\frac{x}{3} - 4$	2	$\frac{1}{3}x$	-4	$\frac{1}{3}$
	e. $6y^2 - y + 3$	3	$6y^2, -y$	3	$6, -1$
	f. $4y^3 - y^2 + 2y - 1$	4	$4y^3, -y^2, 2y$	-1	$4, -1, 2$

Student Check 2 **a.** unlike **b.** like **c.** like **d.** unlike

Student Check 3 **a.** $12y$ **b.** $9b$ **c.** $11a^2$

d. can't be combined **e.** $\frac{13}{3}a$ **f.** $-7x^2 + x$

Student Check 4 **a.** $2y - 8$ **b.** $30x + 11$

c. $38x - 6$ **d.** $-6x - 7$

SUMMARY OF KEY CONCEPTS

1. A term is a number or a product of a number and a variable raised to exponents. The terms in an algebraic expression are separated by addition or subtraction. The coefficient of a variable term is the number multiplied by the variable. Remember that a variable by itself is understood to have a coefficient of 1 since $1 \cdot x = x$.

2. Like terms are terms with the same variables and same exponents.

3. To combine like terms, combine the coefficients of the like terms and keep the variable the same.

4. To simplify an algebraic expression, clear any parentheses by applying the distributive property and then combine like terms.

SECTION 1.8 / EXERCISE SET

 Write About It!

Use complete sentences to explain each term or process.

1. Difference of a variable term and a constant term

2. Numerical coefficient

3. Terms and coefficients of an algebraic expression

4. Like terms

5. How to apply the distributive property to combine like terms

6. How to simplify an algebraic expression

7. Determine if the following statement is true or false: An algebraic expression must have a variable term.

8. Determine if the following statement is true or false; if false, explain the error: $-2(3x - 4) = -6x - 8$

Practice Makes Perfect!

For each algebraic expression, complete the chart by identifying the total number of terms, the variable terms, the constant terms, and the coefficients of the variable terms. (*See Objective 1.*)

Expression	Total Number of Terms	Variable Term(s)	Constant Term(s)	Coefficient of Each Variable Term
9. $-5x$				
10. $6x$				
11. $2y - 5$				
12. $-y + 4$				
13. $x^2 - 5x + 3$				
14. $-x^2 + 6x - 7$				
15. $-3x^4 - x^2 + 4x - 1$				
16. $6x^3 - 2x^2 + 4x - 5$				
17. $\dfrac{x}{2}$				
18. $\dfrac{x}{4} - 2$				

Determine if the terms are like or unlike.
(*See Objective 2.*)

19. $5k$ and k

20. h and $-3h$

21. $3r^2h$ and r^2h

22. $2\pi rh$ and πrh

23. $5xy$ and $-6x^2y$

24. $3ab^2$ and a^2b

25. $\dfrac{1}{2}c^3$ and $-c^3d$

26. $\dfrac{4}{3}\pi r^3$ and $4\pi r^2$

Simplify each expression by combining like terms, if possible. (*See Objective 3.*)

27. $12x + x$

28. $20y + y$

29. $12a - a$

30. $23b - b$

31. $14x - 26x$

32. $-9y - 12y$

33. $4x^2 + 15x^2$

34. $5y^2 + 8y^2$

35. $5x^2 - 23x^2$

36. $-11y^2 + 18y^2$

37. $2x - 3y$

38. $4a + 6b$

39. $9xy - 3y$

40. $4ab + 5b$

41. $\dfrac{x}{4} + 2x$

42. $\dfrac{2y}{5} + 3y$

43. $2a - \dfrac{3a}{5}$

44. $3b - \dfrac{5b}{7}$

45. $-4x^2 + 5x + 2x^2 - 8x$

46. $6y^2 - 9y - 8y^2 + 7y$

47. $6x^2 - 21x - 18x^2 - 4x$

48. $-9y^2 + 12y - 7y^2 - 2y$

Simplify each expression. (*See Objective 4.*)

49. $8x - 7 - 6x$

50. $12x - 3 - x$

51. $5(3x + 4) - 10$

52. $6(10x + 5) - 12$

53. $-5(2x + 4) + 3(4x - 1)$

54. $-6(7x + 1) + 2(3x - 1)$

55. $4(x - 3) + 2(-3x - 5)$

56. $-3(4x + 1) - 7(5x + 6)$

57. $8\left(\dfrac{1}{4}x - 2\right) - 3x$

58. $-9\left(\dfrac{1}{3}x + 2\right) + 5x$

59. $18 - 3(-5x + 4)$

60. $7 - 2(3x + 6)$

61. $-12\left(\dfrac{5}{6}y - \dfrac{1}{12}\right) - 10\left(\dfrac{1}{5}y + \dfrac{1}{2}\right)$

62. $-10\left(\dfrac{1}{10}x - \dfrac{1}{10}\right) + 6\left(\dfrac{5}{6}x - \dfrac{2}{3}\right)$

 Mix 'Em Up!

For each algebraic expression, complete the chart by identifying the total number of terms, the variable terms, the constant terms, and the coefficients of the variable terms.

Expression	Total Number of Terms	Variable Term(s)	Constant Term(s)	Coefficient of Each Variable Term
63. $-x + 12$				
64. $6x - 13$				
65. $-3x^2 + 7x - 2$				

Expression	Total Number of Terms	Variable Term(s)	Constant Term(s)	Coefficient of Each Variable Term
66. $4x^2 - x + 8$				
67. $-x^4 + 3x^2 - 5x + 6$				
68. $-6x^3 - 7x^2 + x - 1$				
69. $\dfrac{3x}{2} - \dfrac{5}{4}$				
70. $\dfrac{7x}{2} + \dfrac{1}{3}$				
71. $0.5x^2 - 1.8x - 2.1$				
72. $-0.1x^2 + 2.5x + 3.6$				

Simplify each algebraic expression.

73. $-5(x + 6) + 12$

74. $-2(b + 7) + 9$

75. $-7(x - 1) + 3(2x - 5)$

76. $-2(6x + 5) - 3(4x - 1)$

77. $1.8 - 3(-0.2x + 1.6)$

78. $2.7 - 0.5(1.3x + 6)$

79. $14\left(\dfrac{1}{2}x - \dfrac{3}{7}\right) + 8$

80. $20\left(-\dfrac{2}{5}x + \dfrac{3}{4}\right) - 24$

81. $-18\left(\dfrac{2}{9}y - \dfrac{1}{2}\right) + 6\left(-\dfrac{2}{3}y + \dfrac{1}{2}\right)$

82. $-21\left(\dfrac{1}{3}x + \dfrac{1}{7}\right) + 4\left(\dfrac{1}{2}x - \dfrac{3}{4}\right)$

83. $0.5(2x - 6) + 0.4(x + 4) - 1.5$

84. $-0.6(1.5x - 2) + 0.7(3x + 6) + 2.9$

85. $10(0.5x - 2) + 10(0.4)$

86. $100(0.6x + 0.25) + 100(0.35x)$

 You Be the Teacher!

Correct each student's errors, if any.

87. Simplify $5 - 3(2 + 4)$.

Brian's work:
$5 - 3(2 + 4) = 2(2 + 4) = 2(6) = 12$

88. Simplify $3a - a$.

Shirl's work:
$3a - a = 3a$

89. Simplify $x^2 + x^2$.

Molly's work:
$x^2 + x^2 = x^4$

90. Simplify $3(6x + 2) - 4$.

Isaac's work:
$3(6x + 2) - 4 = 18x + 2 - 4 = 18x - 2$

91. Simplify $4 + 2(5 + x)$.

Warren's work:
$4 + 2(5 + x) = 6(5 + x) = 30 + x$

92. Simplify $6 - (2 + x)$.

Ashley's work:
$6 - (2 + x) = 4 - 2 + x = 2 + x$

93. Simplify $-6b - 4b$.

Sue's work:
$-6b - 4b = 2b^2$

94. Simplify $9a - 2(a + 5) + 7$.

Rob's work:
$9a - 2(a + 5) + 7 = -19a - 10 + 7 = -19a - 3$

 GROUP ACTIVITY   **Mathematics of Being Fit**

1. One method of determining if you are at a healthy weight for your height is by finding your body mass index (BMI). The BMI is 703 times the quotient of a person's weight (in pounds) and the square of a person's height (in inches). Write an equation that represents BMI. Let w represent weight and h represent height.

2. Use the equation from Step 1 to determine your BMI. Use the chart to determine your weight status. (You do not have to share this information with your group members.)

BMI	Weight Status
Below 18.5	Underweight
18.5–24.9	Normal
25.0–29.9	Overweight
30.0 and above	Obese

3. In Section 1.3, the basal metabolic rate was presented. (See pages 35 and 36.) Recall this calculates the number of calories that your body needs each day to perform essential functions. Calculate your BMR now, if you have not already done so. Based on the level of activity you perform each day, your body needs additional calories to perform its basic functions and the activities in which you participate.

Activity Level	Calories Needed
Sedentary (little or no exercise, desk job)	1.2 times BMR
Lightly active (light exercise/sports 1–3 days/week)	1.375 times BMR
Moderately active (moderate exercise/sports 3–5 days/week)	1.55 times BMR
Very active (hard exercise/sports 6–7 days/week)	1.725 times BMR
Extra active (hard daily exercise/ sports and physical job or training for a marathon)	1.9 times BMR

What activity level describes your lifestyle? Calculate the number of calories you need to consume for your body to sustain this activity level and to maintain your current weight.

4. If you want to lose weight, you must take in fewer calories than what you found in Step 3. One pound of body fat is equal to 3500 cal. So, if you want to lose 1 lb in a week, you will need to decrease your intake by 3500 cal or by an average of 500 cal per day. If you want to lose 2 lb in a week, you will need to decrease your intake by 1000 cal per day. Assume you want to lose 1 lb a week, how many calories should you consume each day?

5. In addition to decreasing caloric intake, you can also increase your physical activity to burn off additional calories. List a number of physical activities you do during a week, such as walking, jogging, or climbing stairs.

6. Research the Internet to find estimates for the number of calories you burn per minute for each of these activities.

7. Develop an expression for the estimated number of calories you burn doing each of these activities in a week, using a variable for the number of minutes you do each activity.

8. Evaluate the expression you found in Step 3 to determine the amount of calories burned by engaging in these activities for 30 min, 1 hr, 2 hr, 3 hr, 4 hr, 5 hr, and 6 hr per week.

9. How long would you have to engage in these activities to burn 3500 cal in a week?

Real Numbers and Algebraic Expressions

What's the big idea? Chapter 1 provides us with the skills to simplify numerical expressions by applying the rules for signed numbers, the properties of real numbers, and the order of operations. These rules and properties also provide the framework for us to simplify and evaluate algebraic expressions. We also learned how to express relationships in mathematical form. These skills form the foundation for algebra and will be used in every section of this text. Success in this course depends on mastering these skills.

The Tools

Listed below are the key terms, skills, formulas, and properties you should know for this chapter.

The page reference is provided if you need additional help with the given topic. The Study Tips will assist in your preparation for an exam.

Study Tips

1. Learn all of the terms, formulas, and properties. Make flash cards and have someone quiz you.
2. Rework problems from the exercises and also the ones you worked in class. Work additional problems from the review exercises.
3. Review the summaries of key concepts.
4. Work the chapter test.
5. Be sure to review the online resources for additional study materials.

Terms

Absolute value 9	Factor 13	Multiplicative identity 71	Right angle 52
Additive identity 71	Factor tree 14	Multiplicative inverse 71	Roster method 2
Additive inverse 71	Finite 2	Natural number 3	Set 2
Algebraic expression 30	Fraction 15	Numerator and denominator 15	Set-builder notation 2
Average 64	Identity element 71	Opposite 8	Simplified 16
Base 27	Improper fraction 15	Perimeter 34	Solve an equation 31
Coefficient 77	Infinite 2	Prime factorization 14	Solution of an equation 31
Composite 14	Integer 3	Prime number 14	Square root 3
Constant terms 77	Inverse 71	Proper fraction 15	Straight angle 53
Dividend 61	Irrational number 3	Rational number 3	Term 77
Divisor 61	Least common denominator (LCD) 21	Real number 3	Unlike terms 78
Equation 31	Like terms 78	Real number line 5	Variable 2
Ellipsis 2	Lowest terms 16	Reciprocal 19	Variable term 77
Equivalent fractions 21	Member or element 2		Whole number 3
Exponential expression 27	Mixed number 15		

Formulas and Properties

- Associative property of addition 72
- Associative property of multiplication 72
- Average 64
- Commutative property of addition 72
- Commutative property of multiplication 72
- Complementary angles 52
- Distributive property 73
- Fundamental property of fractions 16

- Identity property of addition 71
- Identity property of multiplication 71
- Inverse property of addition 71
- Inverse property of multiplication 71
- Order of operations 29, 51
- Percent change 64
- Supplementary angles 53

CHAPTER 1 / SUMMARY

How well do you know this chapter? Complete the following questions to find out. Take a look back at the section if you need help.

SECTION 1.1 The Set of Real Numbers

1. A(n) ___ is a collection of objects.

2. The set of _____ numbers is $\{1, 2, 3, \ldots\}$.

3. The set of _____ numbers is $\{0, 1, 2, 3, \ldots\}$.

4. The set of _____ is the set of positive and negative whole numbers.

5. A(n) _____ number is a number that can be written as the quotient of integers.

6. A(n) _____ number is a number whose decimal form continues indefinitely without a repeating pattern.

7. The statement $a < b$ is read as a _____ b. The statement $a > b$ is read as a _____ b.

8. The _____ of a number is a number with the same distance from zero but lies on the other side of zero on the number line.

9. The _____ of a number is the distance the number is from zero on the real number line.

SECTION 1.2 Fractions Review

10. A number with factors of 1 and the number itself is a(n) _____ number.

11. A fraction is a(n) _____ of ___ numbers. The top is called the _____ and the bottom is called the _____.

12. A fraction is in _____ ____ if the top and bottom of the fraction do not have any common factors.

13. The _____ _____ of _____ enables us to divide the top and bottom of the fraction by their common factor. $\frac{a \cdot c}{b \cdot c} = _____$.

14. The numbers a and $\frac{1}{a}$ are _____.

15. To divide fractions, _____ by the _____ of the second fraction.

16. We can add or subtract fractions with _____ denominators.

17. The ____ _____ _____ is the number that all denominators divide into evenly.

SECTION 1.3 The Order of Operations, Algebraic Expressions, and Equations

18. A(n) _____ indicates repeated multiplication.

19. In the expression b^n, b is called the ____ and n is called the _____ or _____.

20. The ____ of _____ provides us a way to simplify numerical expressions. First, simplify what is in _____ symbols, then evaluate _____, _____ or _____ from left to right, and finally ___ or _____ from left to right.

21. A(n) _____ represents an unknown number and is represented by a ____.

22. A(n) _____ _____ is an expression that involves variables and/or numbers.

23. To evaluate an algebraic expression, replace the _____ with the ____ ____ and simplify.

24. A statement that two expressions are equal is a(n) _____.

25. A(n) _____ of an equation is a value that makes the equation true.

SECTION 1.4 Addition of Real Numbers

26. To add numbers with the same sign, add their _____ _____ and keep the ____ the same.

27. To add numbers with different signs, subtract their _____ _____. The sign of the answer has the same sign as the number with the ____ _____ ____.

SECTION 1.5 Subtraction of Real Numbers

28. Subtracting real numbers is the same as _____ the _____ of a number. $a - b = _____$.

SECTION 1.6 Multiplication and Division of Real Numbers

29. The product of real numbers with the same signs is _____. The product of real numbers with opposite signs is _____.

30. A negative base raised to an even power is _____.

31. A negative base raised to an odd power is _____.

32. To divide real numbers is the same as _____ by the _____. $a \div b = \_\_\_\_$.

33. Zero divided by a nonzero number is ____. A number divided by zero is _____.

SECTION 1.7 Properties of Real Numbers

34. The _____ property of addition states that changing the order of the things being added doesn't change the result. $a + b = \_\_\_\_$.

35. The _____ property of addition states that the grouping of the things being added doesn't change the result. $a + (b + c) = _____$.

36. ____ is the identity element of addition.

37. The _____ property of addition states that adding ____ to a number doesn't change its value. $a + \_ = a$.

38. The _____ property of addition states that adding a number and its _____ is zero. $a + \_\_\_ = 0$.

39. The _____ property of multiplication states that changing the order of the things being multiplied doesn't change the result. $ab = \_\_$.

40. The _____ property of multiplication states that the grouping of the things being multiplied doesn't change the result. $a(bc) =$ _____.

41. ____ is the identity element of multiplication.

42. The _____ property of multiplication states that multiplying a number by ___ doesn't change its value. $a \cdot \_ = a$.

43. The _____ property of multiplication states that multiplying a number and its _____ is one.

$a \cdot \_ = 1$.

44. The _____ property enables us to multiply a number by a sum or difference. $a(b + c) =$ _____.

45. When a(n) _____ number is distributed to a sum, the signs of each of the terms _____.

SECTION 1.8 Algebraic Expressions

46. A(n) ____ is a number or a product of a number and a variable raised to powers. The _____ in an algebraic expression are separated by addition or subtraction.

47. A(n) _____ ____ is a term that contains a variable.

48. A(n) _____ ____ is a term that does not contain a variable.

49. The _____ of a variable term is the number multiplied by the variable.

50. Terms that have the same variables with the same powers are ____ terms. Terms that do not have the same variables or same powers are _____ terms.

51. To combine like terms, add their _____ and keep the _____ the same.

52. To simplify algebraic expressions, clear any _____ and then combine ___ _____.

CHAPTER 1 / REVIEW EXERCISES

SECTION 1.1

Classify each number as a natural number, whole number, integer, rational number, irrational number, and/or real number. If the number is an irrational number, approximate its value to the nearest hundredth. (*See Objective 1.*)

1. $5\frac{1}{3}$

2. $7.1\overline{9}$

3. 10π

4. $\sqrt{51}$

Give an example of a real number that satisfies each condition. (*See Objective 1.*)

5. An integer that is not a whole number.

6. An irrational number that is between 2 and 3.

7. A rational number that is between $\frac{3}{2}$ and 2.

8. An integer that is not a natural number.

Graph each number on a real number line. (*See Objective 2.*)

9. $\left\{ -2.\overline{3}, 2\frac{1}{4}, 5 \right\}$

10. $\left\{ -\pi, \sqrt{7}, \frac{20}{3} \right\}$

Compare the values of each pair of numbers. Use a $<$, $>$, or $=$ symbol to make the statement true. (*See Objectives 3 and 5.*)

11. $\frac{22}{7}$ _____ π

12. $\sqrt{49}$ _____ $-|-7|$

Find the opposite of each real number. (*See Objective 4.*)

13. $1.\overline{12}$

14. 5.2

15. $-3\frac{1}{5}$

16. $\frac{7}{9}$

Simplify each absolute value expression. (*See Objective 5.*)

17. $|-5.8|$

18. $-|1.9|$

19. $-\left| -\frac{6}{13} \right|$

20. $-\left| \frac{5}{4} \right|$

SECTION 1.2

Write the prime factorization of each number. (*See Objective 1.*)

21. 90

22. 560

23. 945

24. 200

Simplify each fraction to lowest terms. (*See Objective 3.*)

25. $\frac{30}{45}$

26. $\frac{168}{288}$

27. $\frac{336}{378}$

28. $\frac{126}{360}$

Perform the indicated operation. Express answers in lowest terms, when applicable. (*See Objectives 4 and 5.*)

29. $\frac{7}{12} + \frac{11}{28}$

30. $\frac{1}{2} - \frac{3}{10}$

31. $3\frac{1}{5} \cdot 1\frac{1}{4}$

32. $2\frac{1}{7} \div 6\frac{3}{7}$

33. $\frac{7}{12} \cdot \frac{9}{14}$

34. $6\frac{1}{5} - 7\frac{3}{5}$

SECTION 1.3

Use the order of operations to simplify each expression. (*See Objectives 1 and 2.*)

35. $4[(-11) - 9](-2)$

36. $2.15(-1)^2 + (-3.16)(0)$

37. $[-5 - (-5)][-6 + (-2)]$

38. $-\frac{3}{14}(-12) + \frac{10}{7}$

39. $\frac{12 + (-4)}{-21 + (-9)}$

40. $-1.5(-4)^2 + 3.2(2) + 6.5$

41. $(-1 \cdot 6)^2 - 5 \cdot 12 \div 6 - (-3.6)$

42. $(-6)^2 - (-15) \div 5 \cdot 3 - 7$

43. $|-8(-12) + 2(-13)|$

44. $\sqrt{(14-10)^2 + (-18 + 15)^2}$

Translate each phrase into a mathematical expression and then simplify the resulting expression, if applicable. Use the variable x for any unknown number. (See Objective 5.)

45. 7 less than -18

46. The difference of 22 and -10

47. 6 subtracted from -11 **48.** -10 decreased by 4

49. 9 less than the sum of -13 and 7

50. 21 more than the difference of 18 and -4

51. 5 more than -2 **52.** The sum of 12 and -25

53. 15 increased by -3 **54.** -19 added to 16

55. 7 less than -18

56. The product of -1, -5, and 2

57. Twice -35 **58.** Three times 25

59. 30% of -12 **60.** One-half of 15

61. The ratio of 28 and 42   **62.** 54 divided by -2

63. The quotient of -32 and -48

64. Four times the sum of a number and 34

65. The difference of seven times a number and 12

66. 2 less than five times the sum of a number and 6

67. Twice the product of a number and 45

68. The product of 12 and the sum of a number and -7

69. Six times the quotient of a number and -9

Solve each problem. (See Objective 6.)

70. The expression $P(1 + r)^t$ represents the amount of money in an account when P dollars is invested at an annual interest rate of r (in decimal form) for t years. Find the amount of money in the account at the end of 5 yr if $1250 is invested at 1.8% annual interest.

71. The height of a baseball hit upward with an initial velocity of 128 ft/s from an initial height of 16 ft is represented by the expression $-16t^2 + 128t + 16$, where t is the number of seconds after the ball has been hit. What is the height of the ball after 3 sec?

Write an algebraic expression that represents the unknown quantity. (See Objective 6.)

72. Norma mixes a 6% salt solution with a 35% solution to get a 22% solution. Let x oz represent the volume of the 6% salt solution. If Norma wants 60 oz of 22% salt solution, write an expression that represents the volume of the 35% salt solution.

73. In a rectangle, the length is two more than four times the width. If w represents the width, write an expression that represents the length of the rectangle.

SECTION 1.4

Perform the indicated operation and simplify. (See Objectives 1 and 2.)

74. $-5.6 + (-2.1)$ **75.** $3.12 + (-2.85)$

76. $-\dfrac{11}{12} + \dfrac{17}{18}$ **77.** $\left(-\dfrac{7}{3}\right) + \left(-\dfrac{13}{6}\right)$

78. $\left(\dfrac{3}{4}\right) + \left(-\dfrac{2}{5}\right)$ **79.** $\left(-5\dfrac{5}{6}\right) + 3\dfrac{1}{4}$

SECTION 1.5

Perform the indicated operation. (See Objectives 1 and 2.)

80. $(-13.8) - 9.4$ **81.** $2.62 - (-1.79)$

82. $3 - \dfrac{8}{5}$ **83.** $1\dfrac{3}{5} - \left(-3\dfrac{4}{5}\right)$

84. $19 - 3 + (-7) - (-11)$

85. $13 + (-10) - (-9) - 5$

86. $-4 + (-14) - (-6)$

87. $-9 + (-16) - (-1)$

88. $3.2 - (-2.7) + 1.9 - 0.6$

89. $5.1 + (-3.4) - 4.2 - 0.1$

Write the mathematical expression needed to solve each problem and then answer the question asked. (See Objective 3.)

90. The highest point on the South American continent is the peak of Mount Aconcagua. The peak is 22,834 ft above sea level. The lowest point on land in the United States is in Death Valley at 282.2 ft below sea level. What is the difference between these elevations?

91. The Hershey Company reported an annual net sales of $5671 million, total cost and expenses of $4766 million, interest expense of $96 million, and income taxes of $299 million in 2010. What was Hershey Company's total net income in 2010? (Source: http://www.thehersheycompany.com/investors/financialreports.aspx)

92. The lowest recorded temperature in Illinois was $-36°F$ at the Congerville weather station on January 5, 1999. The highest recorded temperature was $117°F$ at the East St. Louis weather station on July 14, 1954. What is the difference between the highest and lowest temperatures?

93. Wayne has $163.23 in his checking account. He deposits his paycheck of $545.85. Wayne writes a check for phone bill for $85.55, for groceries for $113.43, and then for rent for $455. What is Wayne's checking account balance?

SECTION 1.6

Perform the indicated operation and simplify. (See Objectives 1, 2, and 3.)

94. $(-12)(-20)$ **95.** $(-2.5)(0.8)$

96. $(-6)\left(\dfrac{7}{20}\right)$ **97.** $\left(-\dfrac{5}{12}\right)\left(-\dfrac{16}{25}\right)$

98. $\dfrac{35}{-14}$ **99.** -6^4

100. $-(-2)^4$ **101.** $-\left(-\dfrac{2}{3}\right)^3$

102. $\dfrac{0}{5}$ **103.** $6 \div 0$

104. $\dfrac{28}{-6} \div (-4)$ **105.** $\left(-\dfrac{42}{18}\right) \div \left(\dfrac{8}{14}\right)$

106. $9(-12)(-2)$ **107.** $-6(-3)(-5)$

Evaluate each expression for the specified values of *x*. Organize the information in a chart. (*See Objective 4.*)

108. $-2x + 7$ for $x = -4, -3, -2, -1,$ and 0

109. $\dfrac{2x + 1}{x - 3}$ for $x = -2, -1, 0, 1,$ and 2

110. $-x^2 + 5$ for $x = -2, -1, 0, 1,$ and 2

111. $x^2 - 2$ for $x = -2, -1, 0, 1,$ and 2

112. $x^2 - x + 1$, for $x = -2, -1, 0, 1$ and 2

113. $|5x - 1|$ for $x = -4, -2, 0, 2,$ and 4

Evaluate each expression for the given values of the variables. (*See Objective 4.*)

114. $2x - 5y$ for $x = -1$ and $y = 2$

115. $\dfrac{y_2 - y_1}{x_2 - x_1}$ for $x_1 = -1, x_2 = -9, y_1 = 1, y_2 = -15$

116. $2l + 2w$ for $l = 23$ and $w = 12$

117. $\dfrac{1}{2}(b_1 + b_2)h$ for $b_1 = 6, b_2 = 4, h = 7$

118. $\sqrt{(x_1 - x_2)^2 + (y_1 - y_2)^2}$ for $x_1 = -6,$
$x_2 = 6, y_1 = -5, y_2 = 0$

119. $b^2 - 4ac$ for $a = -4, b = 8, c = -3$

SECTION 1.7

Identify the property of addition or multiplication applied in each problem. (*See Objectives 1 and 2.*)

120. $0 + (-8) = -8$

121. $\dfrac{1}{6} + \left(-\dfrac{3}{10}\right) = \left(-\dfrac{3}{10}\right) + \dfrac{1}{6}$

122. $\left[2 + \left(-\dfrac{1}{3}\right)\right] + \dfrac{4}{5} = 2 + \left(-\dfrac{1}{3} + \dfrac{4}{5}\right)$

123. $12 + (-7 + (-5)) = 0$

124. $\left(-\dfrac{4}{3}\right)\left(-\dfrac{3}{4}\right) = 1$

125. $-24(6 \cdot 7) = (-24 \cdot 6)(7)$

126. $-1 + 1 = 0$

127. $-6\left(5 \cdot \dfrac{1}{5}\right) = -6$

128. $(-1.7)(0.8) = (0.8)(-1.7)$

129. $(-4 \cdot 5)(9) = -4(5 \cdot 9)$

Use the distributive property to rewrite each expression and simplify the result. (*See Objective 3.*)

130. $-7(x - 12)$ **131.** $-5(3x - 4y + 2)$

132. $0.5(1.6x - 0.4)$ **133.** $9\left(-\dfrac{1}{3}x + \dfrac{1}{6}\right)$

SECTION 1.8

For each algebraic expression, complete the chart by identifying the total number of terms, the variable terms, the constant terms, and the coefficients of the variable terms. (*See Objective 1.*)

Expression	Total Number of Terms	Variable Term(s)	Constant Term(s)	Coefficient of Each Variable Term
134. $-5x + 11$				
135. $-20x^4 + 13x^2 + 5$				
136. $\dfrac{5x}{2} - \dfrac{3}{2}$				
137. $-0.2x^2 + 3.2x + 3.6$				

Determine if the terms are like or unlike. (*See Objective 2.*)

138. $1.9h, 2.4h,$ and $-h$ **139.** $5x^3y$ and $-10x^2y$

140. $\dfrac{1}{4}cd$ and $-\dfrac{1}{8}c^2d$ **141.** $\pi r^2 h$ and $3\pi r^2$

148. $4\left(\dfrac{1}{8}x - \dfrac{3}{4}\right) + 5$

149. $-15\left(\dfrac{2}{3}y - \dfrac{1}{5}\right) + 8\left(\dfrac{1}{4}y - \dfrac{1}{2}\right)$

Simplify each algebraic expression. (*See Objectives 3 and 4.*)

142. $-(1 - 5x) + (6x + 4)$ **143.** $-4(b)(7)$

144. $10(3x - 7) - 26x$ **145.** $4x - 3(1 - 4x)$

146. $-3(x - 5) - 2(4x + 5)$ **147.** $2.6 - 0.2(-0.3x + 1.5)$

150. $5(2x^2 - 1) - 3(x^2 + 4)$ **151.** $7x^2 + 2(x + 3x^2) - 2x$

152. $0.3(3x - 1) + 0.2(x + 5) - 2.6$

153. $2.8 + 0.3(4x - 1)$ **154.** $3.2 - 0.4(2x + 1.5)$

155. $3(0.7x - 6) + 20(0.3)$

CHAPTER 1 TEST / REAL NUMBERS AND ALGEBRAIC EXPRESSIONS

1. The number that is not a natural number is

    a. $(-2)^2$ **b.** $\sqrt{25}$ **c.** $\dfrac{9 - 3}{4 - 2}$ **d.** $3 - 4(2)$

2. Provide an example of a rational number that is also an integer.

3. The number that is an irrational number is

    a. $\sqrt{6^2 + 8^2}$ **b.** 4π

    c. $\dfrac{4 + 5(-2)}{3^2}$ **d.** $\dfrac{0}{-4}$

4. Place an appropriate symbol that makes the statement true.

   a. -3 __ -5 **b.** $\sqrt{2}$ __ π

5. Simplify each expression.

   a. $-\left(-\dfrac{7}{2}\right)$ **b.** $-(2-5)$

   c. $|4-5(3^2)|$ **d.** $|(-4)^3|$

6. Find the prime factorization of 180.

7. Use the fundamental property of fractions to write each fraction in lowest terms.

   a. $\dfrac{124}{200}$ **b.** $\dfrac{-56}{-14}$

8. Perform each operation. Write answers in lowest terms, if appropriate.

   a. $\dfrac{12}{25} \cdot -\dfrac{5}{3}$ **b.** $32 \div \dfrac{1}{4}$

c. $-\dfrac{5}{9}+\left(-\dfrac{4}{9}\right)$ **d.** $\dfrac{1}{4}-\dfrac{2}{7}$

e. $\dfrac{6}{11}+\left(-\dfrac{1}{2}\right)\left(\dfrac{2}{5}\right) \div 3$ **f.** $(-4)+(-5)(2)$

g. $(9-12)-(7-9)$ **h.** $(-2)^3 - 4 \div 2 \cdot 6$

i. $\sqrt{(1-(-4))^2 + (-4-8)^2}$

j. $\dfrac{6-3(-2)}{4-2^2}$

9. Identify the property that is illustrated.

   a. $5(6+2)=(6+2)5$ **b.** $4 \cdot \dfrac{1}{4}=1$

   c. $3+(4+7)=(3+4)+7$

   d. $\dfrac{1}{5}+\left(-\dfrac{1}{5}\right)=0$ **e.** $3 \cdot 1 = 3$

   f. $-9+0=-9$ **g.** $4+(-3)=(-3)+4$

10. Complete the table for the given algebraic expressions.

Expression	Total Number of Terms	Variable Term(s)	Constant Term(s)	Coefficients of Each Variable Term
a. $-3p$				
b. $-a$				
c. $y-1$				
d. $\dfrac{x}{2}+3$				
e. $-2y^2+y-7$				
f. $9y^3-y^2+8y-4$				

11. Simplify each expression.

   a. $4(3x+7)$ **b.** $8(6x-5)$

   c. $-6\left(\dfrac{1}{3}x+\dfrac{1}{2}\right)$ **d.** $x+2+5x-4$

   e. $2(5x-3)-(7x-9)$

   f. $2x^2-8x+1+x^2-x-3$

   g. $x+\dfrac{3}{2}x$ **h.** $5+4(3x+2)$

12. Evaluate the expression $4x^2-2x+1$ for $x=-3$.

13. Translate each phrase into a mathematical expression.

   a. Two more than a number

   b. The difference of a number and 5

   c. The sum of three times a number and 4

   d. Twice the sum of a number and 7

   e. The ratio of a 4 times a number and 3

   f. Six less than one-fourth of a number

Solve each problem.

14. The country with the highest rate of decrease in the natural birth rate is the Ukraine. Its population is currently 46.8 million. It is expected to be 33.4 million in 2050. What is the difference between the expected population and the current population? (Source: http://geography.about.com/od/populationgeography/a/zero.htm)

15. The following chart shows the closing price of Google's stock for October 26 to November 2, 2009. (Source: http://finance.yahoo.com/q/hp?s=GOOG&a=09&b=19&c=2009&d=10&e=3&f=2009&g=d)

Date	Closing Price of Google Stock	Daily Change in Price
11/2/2009	$533.99	
10/30/2009	$536.12	
10/29/2009	$551.05	
10/28/2009	$540.30	
10/27/2009	$548.29	-5.92
10/26/2009	$554.21	n/a

   a. Complete the chart by showing how the price of the stock changed from the previous day to the next.

   b. After completing the chart, find the average daily change in price.

16. If $5000 is invested in a savings account that earns 4.25% annual interest for 3 yr, the amount in the account at the end of the 3 yr is given by $5000(1.0425)^3$. Simplify this expression to determine how much is in the account after 3 yr.

17. The volume of a sphere is given by $\dfrac{4}{3}\pi r^3$, where r is the radius of the sphere. Find the volume of a basketball that has a radius of 4.7 in. Round the answer to the nearest hundredth.

Linear Equations and Inequalities in One Variable

Organization

Organization is critical for your success in college. Good organizational skills will assist you in

- meeting assignment deadlines.
- keeping you focused on your tasks.
- memory skills.
- note taking.
- test preparation and test taking.
- writing.

Organization is needed for your personal study space and your course information. Your personal study space should be organized so that it is

- free from clutter.
- free from distractions.
- easy to access class materials.

Organize the course information with a

- binder with dividers for notes, exams, homework, handouts, and course information.
- different color binder for each course.
- course calendar placed at the front of each binder.

There are many websites that contain helpful information on how to get organized. Take a few moments to review these if you need assistance.

Question For Thought: Do you consider yourself organized? If not, what is one area that could use improvement?

Chapter Outline

Coming Up...

In Section 2.7, we will learn how to use linear inequalities in one variable to determine the number of attendees a couple can have at their wedding reception.

> "In reading the lives of great men, I found that the first victory they won was over themselves . . . self-discipline with all of them came first."
>
> —Harry S. Truman (U.S. President)

SECTION 2.1 Equations and Their Solutions

In this chapter, we will learn how to solve equations and inequalities using two important properties—the addition property of equality and the multiplication property of equality. We will also learn real-life applications of these equations and inequalities. We will rely heavily on the skills we obtained from Chapter 1.

▶ **OBJECTIVES**

As a result of completing this section, you will be able to

1. Identify an equation and expression.
2. Determine if a number is a solution of an equation.
3. Express statements as mathematical equations.
4. Set up equations for application problems.
5. Troubleshoot common errors.

From June 2009 to June 2010, Oprah Winfrey and James Cameron, director of the movies *Avatar* and *Titanic*, were the two highest paid celebrities. They earned a combined total of $525 million. Winfrey earned $105 million more than Cameron. How much did each of them earn? (Source: http://www.forbes.com/lists/)

In this section, we will learn how to express this type of information as a mathematical equation. We will learn how to solve these types of equations in Section 2.2.

Equations and Expressions

In Chapter 1, we worked with numerical expressions and algebraic expressions. In this section, we turn our attention to equations. An *equation* is a mathematical statement that two expressions are equal. Equations can be numerical or algebraic. Some examples of each are shown in the following table.

Objective 1 ▶

Identify an equation and expression.

Numerical Equations	Algebraic Equations
$2 + 3 = 5$	$x + 4 = 2$
$-7 + 6 = 1$	$x^2 = x$

Numerical equations are either true or false. In the above example, $2 + 3$ equals 5, so this equation is true. The next example is false since $-7 + 6$ does not equal 1. *Algebraic equations* are neither true nor false until we assign a value to the variable. The value(s) of the variable that makes the equation true is called a **solution of the equation**.

It is helpful to understand the difference between an equation and an expression. Equations are solved but expressions are simplified.

Definition: Equations and Expressions

1. An **expression** consists of terms that are combinations of letters and numbers. Recall from Section 1.8 that terms are connected by addition or subtraction signs.
2. If two expressions are connected by an equals sign, then it is an **equation**.

Objective 1 Examples Determine if each problem is an expression or equation.

Problems	Solutions
1a. $2x + 4 - 6x + 3$	$2x + 4 - 6x + 3$ is an **expression** because it does not contain an equals sign.
1b. $2x + 4 = 6x + 3$	$2x + 4 = 6x + 3$ is an **equation** because it contains an equals sign.
1c. $4y^2 + 2y = 1$	$4y^2 + 2y = 1$ is an **equation** because it contains an equals sign.
1d. $a^2 + (a + 1)^2 - 6$	$a^2 + (a + 1)^2 - 6$ is an **expression** because it does not contain an equals sign.

☑ **Student Check 1**   Determine if each problem is an expression or equation.

    a. $-4y + 2 = 3y + 1$ **b.** $-4y + 2 - 3y + 1$

    c. $-2x^2 + 4x = 3x$ **d.** $5r^2 - r - 4$

Solutions of Equations

Objective 2 ▶

Determine if a number is a solution of an equation.

As we have learned, a *solution*, or *root*, of an equation is a value of the variable that makes the equation true. The process of finding the value(s) of the variable that makes the equation true is called *solving* an equation for the given variable.

    The process of solving equations is one of the most important skills in algebra. Two fundamental properties of equation solving will be presented in Section 2.2. Before we actually solve equations, we will review how to check if a number is a solution of an equation. We did this in Section 1.3 but it is worth repeating.

> **Procedure: Determining if a Number Is a Solution of an Equation**
>
> **Step 1:** Evaluate each side of the equation for the given value.
> **Step 2:** Determine if the two values are equal.
>   **a.** If the two values are equal, then the given number makes the equation true and is a solution of the equation.
>   **b.** If the two values are not equal, then the given number makes the equation false and is *not* a solution of the equation.

Objective 2 Examples   Determine if the given numbers are solutions of the equation.

    2a. $-4x + 2 = 2x - 1; x = -2, 4,$ or $\dfrac{1}{2}$ **2b.** $x^2 = x; x = -0.5, 0, 1$

Solutions   **2a.**

x	$-4x + 2 = 2x - 1$
-2	$-4(-2) + 2 \overset{?}{=} 2(-2) - 1$ $8 + 2 \overset{?}{=} -4 - 1$ $10 \neq -5$
4	$-4(4) + 2 \overset{?}{=} 2(4) - 1$ $-16 + 2 \overset{?}{=} 8 - 1$ $-14 \neq 7$
$\dfrac{1}{2}$	$-4\left(\dfrac{1}{2}\right) + 2 \overset{?}{=} 2\left(\dfrac{1}{2}\right) - 1$ $-2 + 2 \overset{?}{=} 1 - 1$ $0 = 0$

Since $10 \neq -5$, $x = -2$ is *not* a solution.

Since $-14 \neq 7$, $x = 4$ is *not* a solution.

Since $0 = 0$, $x = \dfrac{1}{2}$ is a solution.

Of the given values, $x = \dfrac{1}{2}$ is a solution of the equation $-4x + 2 = 2x - 1$.

2b.

x	$x^2 = x$
-0.5	$(-0.5)^2 \overset{?}{=} -0.5$ $0.25 \neq -0.5$
0	$(0)^2 \overset{?}{=} 0$ $0 = 0$
1	$(1)^2 \overset{?}{=} 1$ $1 = 1$

Since $0.25 \neq -0.5$, $x = -0.5$ is *not* a solution.

Since $0 = 0$, $x = 0$ is a solution.

Since $1 = 1$, $x = 1$ is a solution.

Of the given values, $x = 0$ and $x = 1$ are solutions of the equation $x^2 = x$.

☑ **Student Check 2**   Determine if the given numbers are solutions of the equation.

a. $1 - 2x = -5x - 8$; $x = -3, \frac{1}{4}$, or 2   **b.** $x^2 = 2x$; $x = \frac{1}{2}, 0$, or 2

Translating Phrases into Mathematical Equations

Objective 3 ▶

Express statements as mathematical equations.

In Chapter 1, we learned how to translate phrases into mathematical expressions and equations. The phrases for equality are shown in the following table.

Phrases for " = "
is equal to
the result is
is the same as
equals
is

In many sections of this book, we will be required to solve application problems. The most difficult part of solving an application problem is setting up the equation correctly. Example 3 illustrates the process of setting up an equation to solve an application problem. We will only set up the equation. Solving the equations will come later.

> **Procedure: Expressing Statements as Mathematical Equations**
>
> **Step 1:** Read the problem carefully.
> **Step 2:** Determine the unknown and assign a variable to it. (If there is more than one unknown, the other unknowns will be represented in terms of the same variable initially chosen.)
> **Step 3:** From the given statement, determine the phrase that represents the equals sign. The expression before this phrase is the left side of the equation. The expression after the phrase is the right side of the equation. Translate both expressions and set them equal to obtain the equation.

Objective 3 Examples   **For each problem, define a variable and write an equation that can be used to solve the problem.**

3a. Twice the sum of a number and 3 is the same as 4 less than the number.

Solution   **3a.** What is unknown? The number is unknown. Let n represent the number. What is known? Twice the sum of a number and 3 is the same as 4 less than the number.

The phrase "is the same as" represents the equals sign.
"Twice the sum of a number and 3" represents the left side of the equation.
"Four less than the number" represents the right side of the equation.

Twice the sum of a number and 3	is the same as	4 less than the number
$2(n + 3)$	$=$	$n - 4$

So, the equation used to solve this problem is $2(n + 3) = n - 4$.

3b. The difference of three times a number and 6 is the same as the quotient of the number and 9.

Solution   **3b.** What is unknown? The number is unknown. Let n represent the number.
What is known? The difference of three times a number and 6 is the same as the quotient of the number and 9.

The phrase "is the same as" represents the equals sign.

"The difference of three times a number and 6" represents the left side of the equation.

"The quotient of the number and 9" represents the right side of the equation.

The difference of three times a number and 6	is the same as	the quotient of the number and 9
$3n - 6$	$=$	$\dfrac{n}{9}$

So, the equation used to solve this problem is $3n - 6 = \dfrac{n}{9}$.

✓ **Student Check 3**   For each problem, define a variable and write an equation that can be used to solve the problem.

   a. Three times the sum of a number and five is the same as four times the number.

   b. The sum of twice a number and one is equal to two more than the number.

Representing Applications Algebraically

Objective 4 ▶

Set up equations for application problems.

Setting up equations for application problems is very similar to the previous objective. The main difference is that we may have to rely on key words in the problem that define the relationships implicitly. For instance, we might be told the perimeter of an object or that two angles are complementary. In these cases, we have to know how to calculate the perimeter of the object or what it means for two angles to be complementary to set up the equation.

Objective 4 Examples   **For each problem, define a variable and write an equation that can be used to solve the problem.**

   4a. One angle is 30° more than another angle. The angles are complementary. Find the measure of each angle.

Solution   **4a.** What is unknown? The measure of the two angles is unknown.

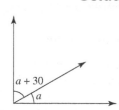

   Let a represent the measure of one of the angles.

   The other angle is 30° more, so $a + 30$ represents the measure of the other angle.

What is known? The angles are complementary, so their sum is 90°.

The phrase "is" represents the equals sign.

"Their sum" represents the left side of the equation.

"90°" represents the right side of the equation.

Their sum	is	90°
$a + a + 30$	$=$	90°

So, the equation used to solve this problem is $a + a + 30 = 90°$.

   4b. From June 2009 to June 2010, Oprah Winfrey and James Cameron, director of the movies *Avatar* and *Titanic*, were the two highest paid celebrities. They earned a combined total of $525 million. Winfrey earned $105 million more than Cameron. How much did each of them earn (in millions)? (Source: http://www.forbes.com/lists/)

Solution **4b.** What is unknown? The earnings of each celebrity are unknown.

Let j represent James Cameron's earnings. Winfrey's earnings are $105 million more than Cameron's, so $j + 105$ represents Winfrey's earnings.

What is known? They earned a combined total of $525 million.

The phrase "of" represents the equals sign.

"They earned a combined total" represents the left side of the equation.

"525" represents the right side of the equation.

$$\text{They earned a combined total} \quad \text{of} \quad \mathbf{525}$$
$$j + j + 105 \qquad\qquad = \quad \mathbf{525}$$

So, the equation used to solve this problem is $j + j + 105 = 525$.

☑ Student Check 4

For each problem, define a variable and write an equation that can be used to solve the problem.

a. One angle is 20° less than another angle. The angles are supplementary. Find the measure of each angle.

b. The total number of U.S. deaths from World War I and World War II was approximately 522 thousand. The number of deaths from World War I was 288 thousand less than the number of deaths from World War II. How many deaths occurred in each world war? (Source: http://www.fas.org/sgp/crs/natsec/ RL32492.pdf)

Objective 5 ▶

Troubleshoot common errors.

Troubleshooting Common Errors

Some common errors associated with determining if a value is a solution of an equation and translating expressions are shown next.

Objective 5 Examples | **A problem and an incorrect solution are given. Provide the correct solution and an explanation of the error.**

5a. Is $4x - 2 - 6x + 4$ an equation or expression?

Incorrect Solution	Correct Solution and Explanation
$4x - 2 - 6x + 4$ is an equation.	$4x - 2 - 6x + 4$ does not contain an equals sign, so this is an expression, *not* an equation.

5b. Determine if $x = -5$ is a solution of $3 - 2(x + 4) = 5$.

Incorrect Solution	Correct Solution and Explanation
$3 - 2(x + 4) = 5$ $3 - 2(-5 + 4) = 5$ $3 - 2(-1) = 5$ $1(-1) = 5$ $-1 = 5$ Since this is a false statement, $x = -5$ is not a solution.	The error was made in evaluating the left side of the equation. After simplifying what is in parentheses, we should multiply before adding. $3 - 2(x + 4) = 5$ $3 - 2(-5 + 4) = 5$ $3 - 2(-1) = 5$ $3 + 2 = 5$ $5 = 5$ Since $x = -5$ makes the equation true, it is a solution.

c. Three less than a number is twice the number. Find the number. Write an equation to solve this problem.

Incorrect Solution	Correct Solution and Explanation
Let x be the number. $3 - x \cancel{=} x^2$	Three less than a number is expressed as $x - 3$ and twice a number is $2x$. So, the equation is $x - 3 = 2x$.

ANSWERS TO STUDENT CHECKS

Student Check 1   a. equation   **b.** expression   **c.** equation
d. expression

Student Check 2   a. $x = -3$ is a solution.   **b.** $x = 0$ and $x = 2$ are solutions.

Student Check 3   a. $3(n + 5) = 4n$ **b.** $2n + 1 = n + 2$

Student Check 4   a. $a + a - 20 = 180$
b. $x + x - 288 = 522$

SUMMARY OF KEY CONCEPTS

1. An equation is a statement that two expressions are equal. Equations can be solved, whereas expressions can be simplified or evaluated.

2. The number(s) that make an equation a true numerical statement are called solutions of the equation.

3. Expressing statements as mathematical equations takes a lot of practice. It is very important to recognize the key phrases for the different operations and for the equals sign.

4. Writing equations that can be used to solve application problems requires us to know some essential information about the situation. This essential information can be gathered from key words in the problem—perimeter, complementary, supplementary, and the like.

GRAPHING CALCULATOR SKILLS

The graphing calculator can be used to determine whether a number is a solution of an equation. There are several ways to use the calculator to perform this skill.

Example: Determine if $x = -2$ is a solution of $-4x + 2 = 2x - 1$.

Method 1: Use the calculator to evaluate the left and right sides of the equation.

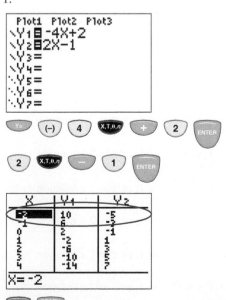

Since the left side, 10, does not equal the right side, -5, the value $x = -2$ is not a solution of the equation.

Method 2: Enter the left side and right side of the equation in the equation editor and use the table to determine the value of each side.

When $x = -2$, the values in the columns for Y_1 and Y_2 are not equal. This shows that $x = -2$ is not a solution.

SECTION 2.1 / EXERCISE SET

 Write About It!

Use complete sentences in your answer to the following exercises. Provide a specific example of each term or process.

1. Explain the meaning of an equation.

2. Explain what it means for a number to be the solution of an equation.

3. Explain the process of verifying if a number is a solution of an equation.

4. Explain the difference between an expression and an equation.

5. Explain the difference between a numerical equation and an algebraic equation.

6. Explain the difference between the two statements: "4 less than a number" and "4 minus a number".

 Practice Makes Perfect!

Determine whether each of the following is an expression or an equation. (*See Objective 1.*)

7. $3x + 2$

8. $4y + 6 = 3$

9. $2x - 1 = -5$

10. $2x - 1$

11. $-7x$

12. $7x = 0$

13. $x^2 - x - 6 = 0$

14. $x^2 - x - 6$

15. $3(x - 8) = 4$

16. $3(x - 8)$

Determine if the given numbers are solutions of the equation. (*See Objective 2.*)

17. Is $x = -2$, $x = -1$, $x = 0$, $x = 1$, or $x = 2$ a solution of $x - 1 = -3$?

18. Is $x = -2$, $x = -1$, $x = 0$, $x = 1$, or $x = 2$ a solution of $x + 1 = 0$?

19. Is $x = -2$, $x = -1$, $x = 0$, $x = 1$, or $x = 2$ a solution of $3x = 0$?

20. Is $x = -2$, $x = -1$, $x = 0$, $x = 1$, or $x = 2$ a solution of $-2x = -4$?

21. Is $x = -2$, $x = -1$, $x = 0$, $x = 1$, or $x = 2$ a solution of $x^2 - x - 2 = 0$?

22. Is $x = -2$, $x = -1$, $x = 0$, $x = 1$, or $x = 2$ a solution of $x^2 + x - 2 = 0$?

23. Is $x = -2$, $x = -1$, $x = 0$, $x = 1$, or $x = 2$ a solution of $x^2 = 1$?

24. Is $x = -2$, $x = -1$, $x = 0$, $x = 1$, or $x = 2$ a solution of $x^2 = 2x$?

25. Is $x = -2$, $x = -1$, $x = 0$, $x = 1$, or $x = 2$ a solution of $x - 2 = x$?

26. Is $x = -2$, $x = -1$, $x = 0$, $x = 1$, or $x = 2$ a solution of $3x - 8 - x = 2(x - 4)$?

For each exercise, define a variable and write an equation that can be used to solve the problem. (*See Objective 3.*)

27. The sum of a number and 4 is -3.

28. The sum of a number and -6 is 8.

29. The difference of a number and 7 is 4.

30. The difference of a number and 10 is -2.

31. Twice a number is 18.

32. Three times a number is -12.

33. The quotient of a number and 4 is 8.

34. The quotient of a number and 2 is -5.

35. The sum of twice a number and 7 is the same as one less than the number.

36. Two more than three times a number equals the number decreased by 9.

37. Three divided by a number gives the same result as one-third of the number.

38. The ratio of a number to 5 equals twice the number.

39. Four less than three times a number is six more than the number.

40. Six less than five times a number is two more than the number.

41. If three times a number is added to four times the same number, the result is the same as seven less than five times the number.

42. If twice a number is added to three times the same number, the result is the same as seven more than four times the number.

43. Four times the sum of a number and -2 is the same as twice the number.

44. Six times the difference of a number and -10 is the same as four times the number.

45. The difference of four times a number and -12 is equal to 50 more than the number.

46. The difference of twice a number and -7 is equal to 18 more than the number.

47. The sum of twice a number and 18 is equal to 27 less than the number.

48. The sum of eight times a number and 24 is equal to 32 less than the number.

For each problem, define a variable and write an equation that can be used to solve the problem. (*See Objective 4.*)

49. One angle is $13°$ less than another angle. Their sum is $90°$. Find the measure of each angle.

50. One angle is $16°$ more than another angle. Their sum is $180°$. Find the measure of each angle.

51. From June 2008 to June 2009, Carl Icahn and Larry Page were the two of the 31 richest Americans. Their combined net worth was $25.8 billion. Carl Ichan's net worth was $4.8 billion less than Larry Page. What was the net worth of each person (in billions of dollars)? (Source: http://www.forbes.com/lists/)

52. From June 2008 to June 2009, William Gates III and Jacqueline Mars were the two of the 31 richest Americans. Their combined net worth was $61 billion. William Gate's net worth was $39 billion more than Jacqueline Mars. What was the net worth of each person (in billions of dollars)? (Source: http://www.forbes.com/lists/)

53. The perimeter of a rectangular garden is 32 ft. The width of the garden is 4 ft less than the length. Let l represent the length of the garden and let w represent the width of the garden.

 a. Write an equation in terms of l and w that represents the perimeter of the garden.

 b. Write an equation that relates the length of the garden to the width of the garden.

54. The perimeter of a regulation size basketball court is 288 ft. The length of the court is 44 ft more than the width. Let l represent the length of the court and let w represent the width of the court.

 a. Write an equation in terms of l and w that represents the perimeter of the court.

 b. Write an equation that relates the length of the court to the width of the court.

55. The Dixie Chicks reached number 1 on the Billboard 200 charts for their albums *Taking the Long Way* (2006) and *Home* (2002). The two albums sold a total of 1.31 million albums. The *Taking the Long Way* album sold 0.254 million albums less than the *Home* album. Let l represent the number of *Taking the Long Way* albums sold (in millions) and let h represent the number of *Home* albums sold (in millions). (Source: www.billboard.com)

 a. Write an equation in terms of l and h that represents the combined sales of the two albums.

 b. Write an equation that relates the number of *Taking the Long Way* albums sold to the number of *Home* albums sold.

56. In 2008 and 2009, a total of 802.2 million albums were sold in the United States. The number of albums sold in 2009 was 54.4 million less than the number of albums sold in 2008. Let x represent the number of albums sold in 2008 (in millions) and let y represent the number of albums sold in 2009 (in millions). (Source: http://washingtonpost.com)

 a. Write an equation using x and y that represents the total number of albums sold in 2008 and 2009.

 b. Write an equation that relates the number of albums sold in 2009 to the number of albums sold in 2008.

57. There are about 3.1 million elementary and middle school teachers in the United States. Elementary and middle

school teachers make up about half of all teachers in the United States. Let e represent the number of elementary teachers (in millions) and let m represent the number of middle school teachers (in millions).

 a. Write an equation using e and m that represents the number of elementary and middle school teachers in the United States.

 b. If t represents the total number of teachers in the United States, write an equation that relates the total number of teachers in the United States to the number of elementary and middle school teachers in the United States. (Source: U.S. Census)

58. Financial planners often recommend that a monthly house payment be no more than about $\frac{1}{3}$ of your monthly income. If x represents a person's monthly income and y represents the maximum recommended monthly house payment, write an equation that relates x and y.

 Mix 'Em Up!

Determine whether each of the following is an expression or an equation.

59. $6z - 8 = 8z - 6$ **60.** $-2r^2 + 12 - 13r - 46$

61. $3x + 10 - 8y + 5$ **62.** $5y^2 - 20 = 18y - 35$

63. $-6r^2 + 5 = 11r + 4$

64. $-7y + 16 + 11y + 14$

Determine if the given numbers are solutions of the equation.

65. Is $x = -4$, $x = -1$, $x = 0$, $x = 4$, or $x = \frac{5}{14}$ a solution of $8x - 8 = -6x - 3$?

66. Is $x = -7$, $x = -2$, $x = 0$, $x = 1$, or $x = -\frac{11}{4}$ a solution of $-5x + 10 = -9x - 1$?

67. Is $x = -4$, $x = -2$, $x = 0$, $x = 5$, or $x = \frac{1}{2}$ a solution of $x^2 = x + 20$?

68. Is $x = -2$, $x = -1$, $x = 0$, $x = 2$, or $x = \frac{1}{3}$ a solution of $x^2 = -3x - 2$?

For each exercise, define a variable and write an equation that can be used to solve the problem.

69. Twice the sum of a number and 11 is the same as five times the number.

70. Five times the difference of a number and 20 is the same as three times the number.

71. The sum of four times a number and −10 is equal to 20 more than the number.

72. The difference of twice a number and −4 is equal to 14 less than the number.

73. Four more than twice a number equals three times the number decreased by 5.

74. The ratio of a number to 4 gives the same result as one-half of the number plus 3.

75. If five times a number is subtracted from four, the result is the same as six less than twice the number.

76. If three times a number is added to ten, the result is the same as six more than five times the number.

77. One angle is 25° more than another angle. Their sum is 180°. Find the measure of each angle.

78. One angle is 40° less than another angle. Their sum is 90°. Find the measure of each angle.

79. One angle is 19° less than another angle. Their sum is 90°. Find the measure of each angle.

80. One angle is 37° more than another angle. Their sum is 180°. Find the measure of each angle.

81. From June 2008 to June 2009, Alice Walton and Lawrence Ellison were two of the 31 richest Americans. Their combined net worth was $46.3 billion. Alice Walton's net worth was $7.7 billion less than Lawrence Ellison. What was the net worth of each person (in billions of dollars)? (Source: http://www.forbes.com/lists/)

82. From June 2008 to June 2009, Christy Walton and Forrest Edward Mars were two of the 31 richest Americans. Their combined net worth was $32.5 billion. Christy Walton's net worth was $10.5 billion more than Forrest Edward Mars. What was the net worth of each person (in billions of dollars)? (Source: http://www.forbes.com/lists/)

83. The combined total sales of a corporation in 2009 and 2010 was $69.1 billion. The total sales in 2009 was $1.5 billion less than the total sales in 2010. Let x = total sales in 2009 (in billions of dollars) and let y = total sales in 2010 (in billions of dollars).

    a. Write an equation using x and y that represents the combined total sales in 2009 and 2010.

    b. Write an equation that relates the total sales in 2010 to the total sales in 2009.

84. The total number of individual songs purchased digitally in 2008 and 2009 was 2.231 billion. The total number of individual songs purchased digitally in 2008 was 0.089 billion less than the total number in 2009. Let x = total number of individual songs purchased digitally in 2008 (in billions) and let y = total number in 2009 (in billions). (Source: http://www.ritholtz.com)

    a. Write an equation using x and y that represents the combined total number of individual songs purchased digitally in 2008 and 2009.

    b. Write an equation that relates the total number of songs purchased digitally in 2008 to the total number of songs purchased digitally in 2009.

 You Be the Teacher!

Anita and Maria are studying together. They are solving problems requiring them to translate phrases into mathematical statements. However, the wording of the problem is ambiguous so that both of their answers are correct. Change the wording of each statement so that only Anita's answer is correct.

85. The square of a number subtracted by 3 is 10.

    Anita: $x^2 - 3 = 10$

    Maria: $(x - 3)^2 = 10$

86. The absolute value of a number added to 7 is 14.

    Anita: $|x| + 7 = 14$

    Maria: $|x + 7| = 14$

87. Half of a number plus 2 is 6.

    Anita: $\frac{1}{2}x + 2 = 6$

    Maria: $\frac{1}{2}(x + 2) = 6$

88. Twice a number subtracted from 12 is 23.

    Anita: $2(12 - x) = 23$

    Maria: $12 - 2x = 23$

 Calculate It!

Use a graphing calculator to determine if the given numbers are solutions of the equation. Specify the equation(s) that were entered into the calculator.

89. $3x^2 - x = 2x;\ x = -1, 0, 1$

90. $|x - 3| = 5;\ x = -2, 0, 2$

91. $4x = x + \frac{3}{2};\ x = 0, \frac{1}{2}, 1$

92. $x + \frac{1}{2} = x^2 - \frac{1}{4};\ x = -1, -\frac{1}{2}, \frac{1}{2}$

 Think About It!

93. Write a statement that corresponds to the equation $2(x - 5) = x + 6$.

94. Write a statement that corresponds to the equation $2x - 5 = 6x$.

95. Write an equation that has $x = 3$ as its solution.

96. Write an equation that has $x = -1$ as its solution.

The Addition Property of Equality

▶ **OBJECTIVES**

As a result of completing this section, you will be able to

1. Define and recognize a linear equation.

2. Use the addition property of equality to solve a linear equation.

3. Solve equations that require more than one step.

4. Solve applications of linear equations.

5. Troubleshoot common errors.

The selling price of a new car is $24,891. The selling price is the base price plus fees and options. The base price is $22,985. What is the total of the fees and options?

In this section, we use the skills from Section 2.1 to write an equation that represents the situation and to solve the equation using the addition property of equality.

Linear Equations

In Section 2.1, we defined a mathematical equation. In the remaining sections of this chapter, we focus on one type of equation, a *linear equation*.

Objective 1 ▶

Define and recognize a linear equation.

> **Definition:** A **linear equation in one variable** is an equation that can be written in the form $ax + b = c$, where a, b, and c are real numbers and $a \neq 0$.

The main characteristic of a linear equation in one variable is that the largest exponent of the variable is one. Recall that $x = x^1$. The exponent of one is called the *degree* of the equation. Linear equations are also called *first-degree equations*. As we will learn in later chapters, the degree denotes the maximum number of solutions of an equation. Therefore, a linear equation will have one solution except for two special cases, which we will solve in Section 2.4. Some examples of linear equations in one variable are

$$x + 3 = -5 \qquad 2x - 5 = 4x + 1 \qquad 3(x + 4) = 8$$

The last two of these linear equations are not in the form $ax + b = c$. However, we will learn methods that enable us to write these equations in this form.

Examples of equations that are not linear are

$$x^2 + 5x - 6 = 0$$
$$x^3 = 8$$

The largest exponent of the variable is 2, not 1.
The largest exponent of the variable is 3, not 1.

> **Procedure: Recognizing a Linear Equation**
>
> **Step 1:** Identify the largest exponent on the variable.
> **Step 2: a.** If this exponent is one, then the equation is a linear equation.
>   **b.** If this exponent is not one, then the equation is not linear.

Objective 1 Examples   Determine if the equation is a linear equation in one variable.

Problems	Solutions
1a. $x = 5$	The equation $x^1 = 5$ is a linear equation since the largest exponent of the variable is 1.
1b. $\dfrac{2}{5} - x = \dfrac{3}{4}(x + 2)$	The equation $\dfrac{2}{5} - x^1 = \dfrac{3}{4}(x^1 + 2)$ is a linear equation since both occurrences of the variable x have an exponent of 1.
1c. $3y^2 - 4y = 1$	The equation $3y^2 - 4y = 1$ is not a linear equation because the largest exponent of the variable y is 2.

✔ **Student Check 1**   Determine if the equation is a linear equation in one variable.

a. $2y - 3 = 7$ **b.** $a - \dfrac{1}{2} = \dfrac{3}{2}$ **c.** $y^4 + 8y^2 = 9$

The Addition Property of Equality

We have learned that a solution of an equation is the value of the variable that makes a true statement. It would be too tedious to use evaluating an equation as a method for solving it. The goal of this section is to determine the solution of a linear equation by performing a series of steps that yield simpler equations that are *equivalent* to the original equation.

Equivalent equations are equations that have the same solution set. The process of producing equivalent equations will enable us to isolate the variable to one side of the equation. These steps will ultimately produce an equation of the form

$$x = \text{some number} \qquad \text{or} \qquad \text{some number} = x$$

Since two sides of an equation are equal, or equivalent, to each other, we must perform the same operations to each side of the equation to keep the equation balanced.

- It is helpful to think of a balance scale to see this relationship.
- If each side of the equation represents some weight, these two weights must be the same since they are set equal to one another.
- If we add something to one side of the equation or "scale," we must add it to the other side to keep the scales balanced.
- If we subtract something from one side of the equation or "scale," we must subtract it from the other side to keep the scales balanced.

Consider the equation, $x + 2 = 5$. The following diagram shows each side of the equation on the balance scale. Note the scales are balanced since the two sides are equal.

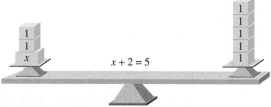

To isolate the variable x on the left side of the scale, we must take 2 units away from the left side of the scale. To maintain balance, we must also take 2 units away from the right side of the scale.

This leaves us with x on the left side of the scale and 3 on the right side of the scale. This action leads us to the solution of the equation, $x = 3$.

The equations $x + 2 = 5$ and $x = 3$ are called equivalent equations since they have the same solution set. We obtained the second equation by subtracting two from each side of the original equation, that is, $x + 2 - 2 = 5 - 2$ is equivalent to $x = 3$. This leads to the addition property of equality.

> **Property: Addition Property of Equality**
>
> If $a = c$, then
> $$a + b = c + b \qquad \text{and} \qquad a - b = c - b.$$

This property tells us that if two expressions are equal, we can

- add the same number to both expressions and the expressions will remain equal.
- subtract the same number from both expressions and the expressions will remain equal.

Example 2 illustrates the most basic use of the addition property of equality. We will use this property to solve for the variable in one step.

Procedure: Using the Addition Property of Equality to Solve an Equation

Step 1: Determine the operation that will isolate the variable on one side of the equation. Perform this operation on each side of the equation. Remember the inverse property for addition: $a + (-a) = -a + a = a - a = 0$.

Step 2: Simplify each side of the equation, as necessary. The result should be of the form

$$x = \text{some number} \quad \text{or} \quad \text{some number} = x$$

Step 3: Check the solution by substituting the value into the original equation.

Step 4: Write the solution in set notation.

Objective 2 Examples Solve each equation using the addition property of equality. Check each answer.

2a. $x + 12 = 10$

2b. $y - 7 = -13$

2c. $b + \dfrac{1}{2} = -\dfrac{3}{2}$

2d. $1.8 = 2.1 + r$

Solutions

2a.

$$x + 12 = 10$$

$$x + 12 - 12 = 10 - 12 \qquad \text{Subtract 12 from each side.}$$

$$x = -2 \qquad \text{Simplify.}$$

Check:

$$x + 12 = 10 \qquad \text{Original equation}$$

$$-2 + 12 = 10 \qquad \text{Replace } x \text{ with } -2.$$

$$10 = 10 \qquad \text{Simplify.}$$

Since $x = -2$ makes the equation true, the solution set is $\{-2\}$.

2b.

$$y - 7 = -13$$

$$y - 7 + 7 = -13 + 7 \qquad \text{Add 7 to each side.}$$

$$y = -6 \qquad \text{Simplify.}$$

Check:

$$y - 7 = -13 \qquad \text{Original equation}$$

$$-6 - 7 = -13 \qquad \text{Replace } y \text{ with } -6.$$

$$-13 = -13 \qquad \text{Simplify.}$$

Since $y = -6$ makes the equation true, the solution set is $\{-6\}$.

2c.

$$b + \frac{1}{2} = -\frac{3}{2}$$

$$b + \frac{1}{2} - \frac{1}{2} = -\frac{3}{2} - \frac{1}{2} \qquad \text{Subtract } \frac{1}{2} \text{ from each side of the equation.}$$

$$b = -\frac{4}{2} \qquad \text{Simplify each side.}$$

$$b = -2 \qquad \text{Simplify the fraction.}$$

Check: $b + \dfrac{1}{2} = -\dfrac{3}{2}$ Original equation

$$-2 + \dfrac{1}{2} = -\dfrac{3}{2}$$ Replace b with −2.

$$-\dfrac{4}{2} + \dfrac{1}{2} = -\dfrac{3}{2}$$ Write −2 with the LCD of 2 and simplify.

$$-\dfrac{3}{2} = -\dfrac{3}{2}$$ Simplify.

Since $b = -2$ makes the equation true, the solution set is $\{-2\}$.

2d. Since the variable r is on the right side, we will isolate the variable on the right side by subtracting 2.1 from each side.

$$1.8 = 2.1 + r$$
$$1.8 - 2.1 = 2.1 + r - 2.1 \quad \text{Subtract 2.1 from each side.}$$
$$-0.3 = r \quad \text{Simplify each side of the equation.}$$
$$r = -0.3$$

While the equation is solved when r is isolated on the right side, note that we can write the final answer with the variable on the left since $a = b$ is equivalent to $b = a$.

Check: $1.8 = 2.1 + r$ Original equation

$$1.8 = 2.1 + (-0.3) \quad \text{Replace } r \text{ with −0.3.}$$
$$1.8 = 1.8 \quad \text{Simplify.}$$

Since $r = -0.3$ makes the equation true, the solution set is $\{-0.3\}$.

✔ **Student Check 2**   Solve each equation using the addition property of equality. Check each answer.

a. $y - 5 = -2$ **b.** $y + 10 = -9$

c. $x - \dfrac{1}{3} = \dfrac{5}{3}$ **d.** $4 = 2.4 + x$

Multistep Equations

Objective 3 ▶

Solve equations that require more than one step.

The equations we solved in Example 2 were set up to immediately apply the addition property of equality since each side of the equation was as simplified as possible and because each equation contained a single variable term. This is not the case for most equations. Most equations require us to do some work to get them in the form that enables us to apply the addition property of equality.

Procedure: Using the Addition Property of Equality

Step 1: Simplify each side as much as possible by combining any like terms, removing parentheses, and the like. Remember that each side must be treated as a separate expression.

Step 2: If there are variable terms on both sides of the equation, use the addition property of equality to remove the variable term from one side of the equation.

Step 3: If there is a term still attached to the variable, use the addition property of equality to isolate the variable.

Step 4: Simplify each side of the equation. The equation should be in the form "$x =$ some number" or "some number $= x$."

Step 5: Check the solution by substituting the value into the original equation.

Step 6: Write the solution set in set notation.

| Objective 3 Examples | Solve each equation. |

3a. $4 + x - 3 = -2$ **3b.** $1.5y - 3 - 4 - 0.5y = 0$

3c. $6a - 3 = 5a + 1$ **3d.** $3(x + 2) - (2x - 5) = 0$

3e. $3(x + 6) = 4(x - 1)$

Solutions   **3a.**

$4 + x - 3 = -2$	
$x + 1 = -2$	Combine like terms on the left.
$x + 1 - 1 = -2 - 1$	Subtract 1 from each side of the equation.
$x = -3$	Simplify.

Check:

$4 + x - 3 = -2$	Original equation
$4 + (-3) - 3 = -2$	Replace x with -3.
$-2 = -2$	Simplify.

Since $x = -3$ makes the equation true, the solution set is $\{-3\}$.

3b.

$1.5y - 3 - 4 - 0.5y = 0$	Combine like terms on the left side. Note that
$y - 7 = 0$	$1.5y - 0.5y = 1y = y.$
$y - 7 + 7 = 0 + 7$	Add 7 to each side.
$y = 7$	Simplify.

Check:

$1.5y - 3 - 4 - 0.5y = 0$	Original equation
$1.5(7) - 3 - 4 - 0.5(7) = 0$	Replace y with 7.
$10.5 - 3 - 4 - 3.5 = 0$	Simplify each product.
$0 = 0$	Simplify.

Since $y = 7$ makes the equation true, the solution set is $\{7\}$.

3c.

$6a - 3 = 5a + 1$	
$6a - 3 - 5a = 5a + 1 - 5a$	Subtract $5a$ from each side.
$a - 3 = 1$	Simplify each side.
$a - 3 + 3 = 1 + 3$	Add 3 to each side.
$a = 4$	Simplify.

Check:

$6a - 3 = 5a + 1$	Original equation
$6(4) - 3 = 5(4) + 1$	Replace a with 4.
$24 - 3 = 20 + 1$	Find each product.
$21 = 21$	Simplify each side.

Since $a = 4$ makes the equation true, the solution set is $\{4\}$.

3d.

$3(x + 2) - 1(2x - 5) = 0$	Recall $-(2x - 5) = -1(2x - 5)$.
$3x + 6 - 2x + 5 = 0$	Apply the distributive property.
$x + 11 = 0$	Combine like terms.
$x + 11 - 11 = 0 - 11$	Subtract 11 from each side.
$x = -11$	Simplify.

Check: $3(x + 2) - (2x - 5) = 0$ Original equation

$3(-11 + 2) - [2(-11) - 5] = 0$ Replace x with -11.

$3(-9) - (-22 - 5) = 0$ Simplify the left side using the

$-27 - (-27) = 0$ order of operations.

$-27 + 27 = 0$ Add.

$0 = 0$ Simplify.

Since $x = -11$ makes the equation true, the solution set is $\{-11\}$.

3e. $3(x + 6) = 4(x - 1)$

$3x + 18 = 4x - 4$ Apply the distributive property.

$3x + 18 - 3x = 4x - 4 - 3x$ Subtract $3x$ from each side.

$18 = x - 4$ Simplify.

$18 + 4 = x - 4 + 4$ Add 4 to each side.

$22 = x$ Simplify.

Check: $3(x + 6) = 4(x - 1)$ Original equation

$3(22 + 6) = 4(22 - 1)$ Replace x with 22.

$3(28) = 4(21)$ Simplify each side.

$84 = 84$ Multiply.

Since $x = 22$ makes the equation true, the solution set is $\{22\}$.

✓ Student Check 3   Solve each equation.

a. $-3t + 5 + 4t = 0$ **b.** $0.2y + 6 - 3 + 0.8y = 0$

c. $5t - 1 = 4t + 4$ **d.** $2(x - 5) - (x + 3) = 5$

e. $5(x + 2) = 4(x - 1)$

Applications

Objective 4 ▶

Solve applications of linear equations.

The key to solving applications is setting up the equation correctly. In Section 2.1, we practiced setting up the equation but not solving them. Feel free to go back and review this material as needed. The only difference in what we are doing in this section is that we will solve the equation we set up.

Objective 4 Examples   **For each problem: (1) assign a variable to the unknown, (2) write an equation that represents the situation, (3) solve the equation, and (4) answer the question using complete sentences.**

4a. Four less than twice a number has the same result as eight more than a number. Find the number.

Solution   **4a.** What is the unknown? The number is unknown. Let n be a number.

What is known? Four less than twice a number has the same result as eight more than a number.

From the given information, we can write the equation.

Four less than twice a number	has the same result as	eight more than a number.

$$2n - 4 = n + 8$$
$$2n - 4 - n = n + 8 - n \qquad \text{Subtract } n \text{ from each side.}$$
$$n - 4 = 8 \qquad \text{Simplify.}$$
$$n - 4 + 4 = 8 + 4 \qquad \text{Add 4 to each side.}$$
$$n = 12 \qquad \text{Simplify.}$$

Check:
$$2n - 4 = n + 8 \qquad \text{Original equation}$$
$$2(12) - 4 = 12 + 8 \qquad \text{Replace } n \text{ with 12.}$$
$$24 - 4 = 20 \qquad \text{Simplify each side.}$$
$$20 = 20 \qquad \text{Simplify.}$$

The solution, $n = 12$, checks in the original equation. So, the number that satisfies the given relationship is 12.

4b. The selling price of a new car is $24,891. The selling price is the base price plus fees and options. The base price is $22,985. What is the total of the fees and options?

Solution **4b.** What is unknown? The total fees and options are unknown. Let x represent the fees and options.

What is known?
- The selling price is the base price plus fees and options.
- The selling price is $24,891.
- The base price is $22,985.

From the given information, we can write the equation.

Selling price	is	base price plus fees and options.

$$24,891 = 22,985 + x$$
$$24,891 - 22,985 = 22,985 + x - 22,985 \qquad \text{Subtract 22,985 from each side.}$$
$$1906 = x \qquad \text{Simplify.}$$

The total of the fees and options is $1906. We can check by replacing x with 1906. With application problems, we also want to check to make sure the answer is reasonable. It wouldn't make sense to have a negative value for this answer nor would it make sense for the fees and options to be more than the selling price.

✔ **Student Check 4**

For each problem: (1) assign a variable to the unknown, (2) write an equation that represents the situation, (3) solve the equation, and (4) answer the question using complete sentences.

a. Twice the sum of a number and 3 is the same as 4 less than the number. Find the number.

b. The tuition for 15 credit hours for a new in-state student at Georgia Perimeter College, a community college in Atlanta, Georgia, is $1235 per semester for students entering in Fall 2012. Tuition plus mandatory fees is $1710 per semester. How much are the fees the student pays each semester? (Source: http://www.usg.edu/fiscal_affairs/documents/tuition_and_fees/FY2012_Tuition_Rates.pdf)

Objective 5 ▶
Troubleshoot common errors.

Troubleshooting Common Errors

Some common errors associated with the addition property of equality are shown next.

Objective 5 Examples
A problem and an incorrect solution are given. Provide the correct solution and an explanation of the error.

5a. Solve $4 - x = 6$.

Incorrect Solution	Correct Solution and Explanation
$4 - x = 6$ $4 - x - 4 = 6 - 4$ $x = 2$ The solution set is $\{2\}$.	The first step is correct in that we should subtract 4 from both sides. The resulting equation is incorrect. It should be $$4 - x = 6$$ $$4 - x - 4 = 6 - 4$$ $$-x = 2$$ This tells us that the opposite of x is 2. Therefore, x must be the opposite of 2, which is -2. The solution set is $\{-2\}$. The equation $-x = 2$ could also be solved by adding x to both sides and subtracting 2 from each side. $$-x = 2$$ $$-x + x = 2 + x$$ $$0 = x + 2$$ $$0 - 2 = x + 2 - 2$$ $$-2 = x$$

5b. Solve $3(x - 2) + 4 = 2x - 5$.

Incorrect Solution	Correct Solution and Explanation
$3(x - 2) + 4 = 2x - 5$ $3x - 2 + 4 = 2x - 5$ $3x + 2 = 2x - 5$ $3x + 2 - 2x = 2x - 5 - 2x$ $x + 2 = -5$ $x + 2 - 2 = -5 - 2$ $x = -7$ The solution set is $\{-7\}$.	Three was not distributed to the second term in the parentheses. $$3(x - 2) + 4 = 2x - 5$$ $$3x - 6 + 4 = 2x - 5$$ $$3x - 2 = 2x - 5$$ $$3x - 2 - 2x = 2x - 5 - 2x$$ $$x - 2 = -5$$ $$x - 2 + 2 = -5 + 2$$ $$x = -3$$ The solution set is $\{-3\}$.

ANSWERS TO STUDENT CHECKS

Student Check 1 **a.** linear equation   **b.** linear equation
  c. not linear

Student Check 2 **a.** $\{3\}$ **b.** $\{-19\}$ **c.** $\{2\}$
  d. $\{1.6\}$

Student Check 3 **a.** $\{-5\}$ **b.** $\{-3\}$ **c.** $\{5\}$ **d.** $\{18\}$
  e. $\{-14\}$

Student Check 4 **a.** The number is -10. **b.** The student fees are $475.

SUMMARY OF KEY CONCEPTS

1. A linear equation is an equation that can be written in the form $ax + b = c$, where a, b, and c are real numbers and $a \neq 0$. The largest exponent of the variable term is one.

2. The addition property of equality enables us to add or subtract the same number from both sides of an equation. The key is to isolate the variable to one side of the equation.

3. To solve an equation, first simplify each side of the equation as much as possible. Then get variables on one side of the equation and constants on the other using the addition property of equality.

4. The key to solving applications is in setting up the equation that represents the given situation. Review Section 2.1 to assist you with this process.

GRAPHING CALCULATOR SKILLS

Solving equations is a critical skill that we need in algebra. At this level, the calculator should be only used to check our work. Two methods were shown in Section 2.1 as to how the calculator can be used to verify solutions. Another method is shown next.

Example: Verify that $x = -14$ is a solution of $x + 4 = -10$.

Solution: Store the value of x into the calculator. Enter the equation and press enter. If "1" is displayed, the statement is true for the stored value of x and this value is a solution of the equation. If "0" is displayed, the statement is false for the stored value of x and this value is *not* a solution of the equation.

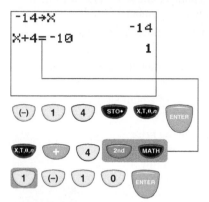

Because a 1 is displayed, $x = -14$ is a solution of $x + 4 = -10$.

SECTION 2.2   EXERCISE SET

 Write About It!

Use complete sentences to explain the meaning of the given term or process.

1. Linear equation in one variable

2. The addition property of equality

3. Solving a linear equation using the addition property of equality

4. The solution of an equation

Determine if each statement is true or false. If a statement is false, explain why.

5. To isolate the variable term in the equation $7 + x = -4$, add 4 to both sides of the equation.

6. To isolate the variable term in the equation $-5x = 2$, add 5 to both sides of the equation.

7. In the equation $-5x = -6x + 4$, the variable term on the right side of the equation can be eliminated by adding $6x$ to both sides.

8. In the equation $7x = 9x - 14$, the variable term on the right side of the equation can be eliminated by adding $9x$ to both sides.

Practice Makes Perfect!

Decide if each equation is linear. If the equation is not linear, explain why. (*See Objective 1.*)

9. $2x - 5 = 0$

10. $3(4x + 5) - 7 = 2(x + 1)$

11. $2x^2 - x + 1 = 3$

12. $-4x^2 + 3x + 5 = 0$

13. $\dfrac{1}{5}x - 3 + \dfrac{4}{5}x = 7$

14. $\dfrac{1}{2}x - 5 = -\dfrac{1}{2}x + 2$

Use the addition property of equality to solve each equation. Remember to check each solution. (*See Objective 2.*)

15. $x + 3 = -5$

16. $y + 8 = -1$

17. $b - 7 = -3$

18. $a - 1 = -10$

19. $x - 3 = 3$

20. $x + 12 = 12$

21. $-3 = x - 4$

22. $10 = 5 + x$

23. $x + \dfrac{6}{7} = -\dfrac{1}{7}$

24. $x - \dfrac{3}{4} = \dfrac{5}{4}$

25. $y + 2.5 = .5$

26. $b - 5.4 = 5$

27. $6.5 + c = 2.4$

28. $-4.2 + a = -7.3$

Solve each equation. (*See Objective 3.*)

29. $5 - x + 2x - 4 = 7$

30. $-3y + 7 + 4y - 10 = -3$

31. $-9 - 6y - 7 + 7y = 10 - 13$

32. $8b + 11 - 7b - 12 = -7 - 5$

33. $3x + 6 = 2x - 5$

34. $5x - 7 = 4x - 6$

35. $-2x - 3 = -3x + 4$

36. $-5y + 1 = -6y + 2$

37. $\dfrac{4}{3}x - 5 = \dfrac{1}{3}x + 2$

38. $\dfrac{7}{6}x - 7 = \dfrac{1}{6}x + 2$

39. $1.5a + 4 = 0.5a + 2$

40. $2.4y + 3 = 1.4y - 7$

41. $4(2x + 6) = 7x - 4$

42. $3(4x - 5) = 11x + 20$

43. $3(2x - 1) - 5(x + 3) = 0$

44. $7(4x + 2) - 9(3x - 1) = 0$

45. $\dfrac{6}{5}(x - 10) = \dfrac{1}{5}(x + 5)$

46. $\dfrac{5}{4}(y - 8) = \dfrac{1}{4}(y + 8)$

For each problem: **(1) assign a variable to the unknown, (2) write an equation that represents the situation, (3) solve the equation, and (4) answer the question using complete sentences.** (*See Objective 4.*)

47. The sum of twice a number and 4 is equal to 19 more than the number. Find the number.

48. Five times the sum of a number and 6 is the same as six times the number.

49. The difference of twice a number and -7 is equal to 38 less than the number.

50. Five times the difference of a number and -9 is the same as six times the number.

51. Many jewelers mark up the price of engagement rings by as much as 300%! One jeweler buys a ring for $1250 but sells it for $5000. (The selling price is the cost to the jeweler plus the markup.) What is the markup that this jeweler adds to their cost of the ring?

52. Many college bookstores mark up the price of a textbook by 30%. If a math textbook costs the bookstore $80, but the bookstore sells it for $104, what is the markup amount?

53. The price of a roundtrip airline ticket from Atlanta, Georgia, to San Francisco, California, is $576.10. This price includes the cost of the ticket plus taxes and fees. What are the taxes and fees if the cost of the ticket is $497.21?

54. Many mortgage companies set up escrow accounts to pay a homeowner's property taxes and insurance. With such a setup, a person's mortgage payment includes the home loan payment plus the escrow payment. If Bailey's monthly mortgage payment is $1210.80 and $199.61 of it is his escrow payment, what is Bailey's monthly home loan payment?

55. A department store sells a stand mixer for $299.99. This price includes a $50 discount. What is the original price of the mixer?

56. A department store sells a gourmet single cup home brewing system for $105.96. This price includes a $6 sales tax. What is the price of the brewing system before taxes?

57. The 2010 Forbes 400 Richest Americans list included Facebook founder, Mark Zuckerberg, and Facebook co-founder, Dustin Moskovitz. The difference in Zuckerberg's net worth and Moskovitz's net worth was $5.5 billion. If Zuckerberg's net worth was $4.1 billion more than twice Moskovitz's net worth, what was the net worth of each person? (Source: http://www.forbes.com/lists/)

58. The 2010 Forbes 400 Richest Americans list included Apple founder, Steve Jobs, and Dell founder, Michael Dell. The difference in Dell's net worth and Job's net worth was $7.9 billion. If Dell's net worth was $1.8 billion more than twice Job's net worth, what was the net worth of each person? (Source: http://www.forbes.com/lists/)

 Mix 'Em Up!

Solve each equations.

59. $3x + 1 = 2x + 1$

60. $7x + 7 = 6x - 12$

61. $3 - x = 11 - 2x$

62. $10 - 9x = 14 - 10x$

63. $\frac{3}{4}x + 1 = 6 - \frac{1}{4}x$

64. $\frac{5}{8}x - 8 = 3 - \frac{3}{8}x$

65. $\frac{2}{7}x - 1 = 5 + \frac{9}{7}x$

66. $-\frac{7}{12}x + 2 = \frac{5}{12}x - 6$

67. $3(x - 1) + 5 = 4(x - 2) - 3$

68. $10(2x + 3) - 16 = 19x - 15$

69. $5.68x + 1.02 = 4.68x + 3.88$

70. $4.19x + 3.64 = 3.19x - 1.45$

71. $5.1x - 1.8 + 2.3x = 6.4x - 2.7$

72. $4.1x - 2.1 + 0.4x = 3.5x - 1$

73. $14 - 11x = 14 - 12x$

74. $1.2 - 2.5x = 1.2 - 1.5x$

For each problem: **(1) assign a variable to the unknown, (2) write an equation that represents the problem situation, (3) solve the equation, and (4) answer the question asked using complete sentences.**

75. The difference of a number and 3 is -10. Find the number.

76. The difference of a number and 10 is -55. Find the number.

77. Four less than a number is -7. Find the number.

78. Eight less than a number is -19. Find the number.

79. Twice the sum of a number and 4 is equal to one more than the number. Find the number.

80. Three times the difference of a number and 8 equals twice the number. Find the number.

81. On average, Americans spent $6129 a year on food at home and away from home in 2010. If the average expenditures on food at home was $3624, what was the average expenditures on food away from home? (Source: http://www.bls.gov/news.release/cesan.nr0.htm)

82. In 2010, the difference in the average income before taxes and the average annual expenditures in the United States was $14,372. If the average income before taxes in 2010 was $62,481, what was the average annual expenditures in 2010? (Source: http://www.bls.gov/news.release/cesan.nr0.htm)

83. In 2010, Forbes Celebrity 100 list included the two highest paid female athletes, Serena Williams and Maria Sharapova. The difference in Williams's income and Sharapova's income was $-\$5$ million. If Williams's income was $30 million less than twice Sharapova's income, how much did each of them earn? (Source: http://www.forbes.com/lists/)

84. In 2010, Forbes Celebrity 100 list included the two of the highest paid teen stars, Taylor Swift and Miley Cyrus. The difference in Swift's income and Cyrus's income was $-\$3$ million. If Swift's income was $51 million less than twice Cyrus's income, how much did each of them earn? (Source: http://www.forbes.com/lists/)

You Be the Teacher!

Correct each student's errors, if any.

85. Solve: $2(x - 1) + 3 = x - 4$

Blanche's work:

$$2(x - 1) + 3 = x - 4$$
$$2x - 1 + 3 = x - 4$$
$$2x + 2 = x - 4$$
$$\underline{\quad -2 \qquad -2\quad}$$
$$2x \quad\; = x - 6$$
$$\underline{-x \qquad -x\quad}$$
$$x = -6$$

86. Solve: $5(x - 13) + 3 = 4x - 14$

Darcy's work:

$$5(x - 13) + 3 = 4x - 14$$
$$5x - 65 + 3 = 4x - 14$$
$$5x - 62 = 4x - 14$$
$$\underline{-62 \qquad -62\quad}$$
$$5x \quad\; = 4x - 76$$
$$\underline{-4x \qquad\;\; -4x\quad}$$
$$x = -76$$

87. Solve:

$$\frac{4}{5}x + 5 + \frac{1}{5}x = 12$$

Charlie's work:

$$\frac{4}{5}x + 5 + \frac{1}{5}x = 12$$
$$\frac{4}{5}x + \frac{1}{5}x + 5 = 12$$
$$\frac{5}{5}x + 5 = 12$$
$$x + 5 = 12$$
$$\underline{-5 \quad -5\quad}$$
$$x = 7$$

88. Solve:

$$\frac{1}{6}x - 4 = 10 - \frac{5}{6}x$$

Wanda's work:

$$\frac{1}{6}x - 4 = 10 - \frac{5}{6}x$$
$$\underline{\quad +4 \quad +4\qquad}$$
$$\frac{1}{6}x \quad\; = 14 - \frac{5}{6}x$$
$$\underline{+\frac{5}{6}x \qquad\quad +\frac{5}{6}x\quad}$$
$$\frac{6}{6}x = 14$$
$$x = 14$$

Calculate It!

Solve each equation. Then use a graphing calculator to check the solution.

89. $6(x - 1) + 3 = 5x - 9$

90. $3 + 2(x - 1) = x - 3$

91. $\frac{3}{4}x + 2 = 5 - \frac{1}{4}x$

92. $\frac{1}{2}x + 4 + x = \frac{1}{2}x - 7$

93. $1.2(x - 2.5) - 2.2(x + 10.8) = 4.6$

94. $2.3(x + 3.1) - 1.3(x - 2.4) = 6.8$

Think About It!

95. According to a national retail organization, the average person spent \$116.21 on traditional Valentine's Day merchandise in 2011 compared to an average of \$103.00 in 2010. Use this information to write a word problem.

96. The following pie chart shows the total number of retailers by type in the state of Georgia in 2010. Use the information to write a word problem that relates two of the quantities. (Source: http://www.nrf.com)

Types of Retailers in the State of Georgia

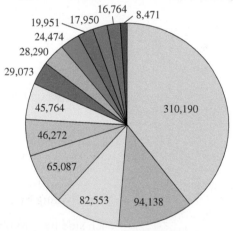

- ☐ Restaurants & bars
- ☐ Grocery & liquor stores
- ☐ Clothing & clothing accessories stores
- ☐ Health & personal care stores
- ☐ Miscellaneous store retailers
- ☐ Sporting goods, hobby, book & music stores
- ☐ Catalog, Internet & mail-order retailers
- ☐ Department stores, warehouse clubs & super stores
- ☐ Motor vehicle & parts dealers
- ☐ Building material & garden supply stores
- ☐ Gasoline stations
- ☐ Furniture & home furnishings stores
- ☐ Electronics & appliance stores

The Multiplication Property of Equality

▶ **OBJECTIVES**

As a result of completing this section, you will be able to

1. Use the multiplication property of equality to solve linear equations.
2. Apply both the addition and multiplication properties of equality to solve linear equations.
3. Solve application problems.
4. Solve consecutive integer problems.
5. Troubleshoot common errors.

To rent a moving truck for one day, a company charges $44.95 plus $0.45 per mile. If a family budgets $300 for a moving truck, how many miles will they be able to drive the truck? To answer this question, we have to solve the equation

$$44.95 + 0.45x = 300,$$

where x is the number of miles driven.

In this section, we will learn how to find the solution of this type of equation where the variable term has a coefficient other than one.

The Multiplication Property of Equality

The addition property of equality, presented in Section 2.2, only enables us to solve for the variable when its coefficient is 1. We need another property to solve equations which have coefficients other than 1 on the variable. Some examples of equations of this form are

$$3x - 4 = 5 \qquad \frac{2}{3}x = 6$$

Consider how we can solve the equation $2x = 6$ using our reasoning skills.

$2x = 6 \rightarrow$ This equation means 2 times some number is 6. We know that 2 times 3 is 6, so the solution of this equation is $x = 3$.

The equations $2x = 6$ and $x = 3$ are equivalent equations. To derive the first equation from the second equation, we must divide each side of the equation by 2.

$$2x = 6$$
$$\frac{2x}{2} = \frac{6}{2}$$
$$x = 3$$

Objective 1 ▶

Use the multiplication property of equality to solve linear equations.

Note dividing by 2 is the same as multiplying by $\frac{1}{2}$, so we could have also multiplied each side by $\frac{1}{2}$ to obtain the same result. That is,

$$2x = 6$$
$$\frac{1}{2}(2x) = \frac{1}{2}(6)$$
$$x = 3$$

This leads us to another property of solving equations.

Property: Multiplication Property of Equality

If $a = c$, and $b \neq 0$, then

$$ab = cb$$

and

$$\frac{a}{b} = \frac{c}{b}$$

This property tells us that if two expressions are equal or equivalent, we can

- multiply both expressions by the same nonzero number and the expressions will remain equal.
- divide both expressions by the same nonzero number and the expressions will remain equal.

Example 1 illustrates how we solve an equation that requires us to use only the multiplication property of equality. In the next objective and in Section 2.4, we will solve equations that require both the addition and multiplication properties of equality.

> **Procedure: Using the Multiplication Property of Equality to Solve Equations of the Form $ax = b$**
>
> **Step 1:** Determine what operation will isolate the variable on one side of the equation. Remember the inverse property for multiplication:
>
> $a \cdot \dfrac{1}{a} = \dfrac{a}{a} = 1$, that is, the product of reciprocals is 1.
>
> **Step 2:** Perform this operation on each side of the equation.
> **Step 3:** Simplify each side of the equation, as necessary. The result should be of the form
>
> $$x = \text{some number} \qquad \text{or} \qquad \text{some number} = x$$
>
> **Step 4:** Check the solution by substituting the value into the original equation.
> **Step 5:** Write the solution in set notation.

Objective 1 Examples   Use the multiplication property of equality to solve each equation.

1a. $3x = -12$ **1b.** $0.05y = 40$ **1c.** $-x = 6$

1d. $\dfrac{3}{2}a = 12$ **1e.** $\dfrac{x}{4} = -8$

Solutions   **1a.**

$$3x = -12$$

$$\frac{3x}{3} = \frac{-12}{3} \qquad \text{Divide each side by 3.}$$

$$x = -4 \qquad \text{Simplify. Recall } \frac{3x}{3} = \frac{3}{3}x = 1x = x.$$

Check:

$$3x = -12 \qquad \text{Original equation}$$

$$3(-4) = -12 \qquad \text{Replace } x \text{ with } -4.$$

$$-12 = -12 \qquad \text{Simplify.}$$

Since $x = -4$ makes the equation true, the solution set is $\{-4\}$.

1b.

$$0.05y = 4$$

$$\frac{0.05y}{0.05} = \frac{4}{0.05} \qquad \text{Divide each side by 0.05.}$$

$$y = 80 \qquad \text{Simplify. Recall } \frac{0.05y}{0.05} = \frac{0.05}{0.05}y = 1y = y.$$

Check:

$$0.05y = 4 \qquad \text{Original equation}$$

$$0.05(80) = 4 \qquad \text{Replace } y \text{ with } 80.$$

$$4 = 4 \qquad \text{Simplify.}$$

Since $y = 80$ makes the equation true, the solution set is $\{80\}$.

1c.

$$-x = 6$$

$$-1x = 6 \qquad \text{Rewrite } -x \text{ as } -1x.$$

$$\frac{-1x}{-1} = \frac{6}{-1} \qquad \text{Divide each side by } -1.$$

$$x = -6 \qquad \text{Simplify.}$$

Check:

$$-x = 6 \qquad \text{Original equation}$$

$$-(-6) = 6 \qquad \text{Replace } x \text{ with } -6.$$

$$6 = 6 \qquad \text{Simplify.}$$

Since $x = -6$ makes the equation true, the solution set is $\{-6\}$.

1d. $\dfrac{3}{2}a = 12$

$\dfrac{2}{3}\left(\dfrac{3}{2}a\right) = \dfrac{2}{3}(12)$ Multiply each side by the reciprocal of $\dfrac{3}{2}$, $\dfrac{2}{3}$.

$1a = \dfrac{24}{3}$ Simplify each side.

$a = 8$ Simplify.

Check: $\dfrac{3}{2}a = 12$ Original equation

$\dfrac{3}{2}(8) = 12$ Replace a with 8.

$\dfrac{24}{2} = 12$ Simplify each side.

$12 = 12$ Simplify.

Since $a = 8$ makes the equation true, the solution set is $\{8\}$.

1e. $\dfrac{x}{4} = -8$

$\dfrac{1}{4}x = -8$ Rewrite as $\dfrac{x}{4} - \dfrac{1}{4}x$.

$4\left(\dfrac{1}{4}x\right) = 4(-8)$ Multiply each side by the reciprocal of $\dfrac{1}{4}$, 4.

$x = -32$ Simplify each side.

Check: $\dfrac{x}{4} = -8$ Original equation

$\dfrac{-32}{4} = -8$ Replace x with -32.

$-8 = -8$ Simplify.

Since $x = -32$ makes the equation true, the solution set is $\{-32\}$.

✔ **Student Check 1**   Use the multiplication property of equality to solve each equation.

a. $2x = -14$ **b.** $0.4x = 24$ **c.** $-x = -8$ **d.** $\dfrac{5}{4}a = -20$ **e.** $\dfrac{y}{3} = -15$

Solving Linear Equations with the Addition and Multiplication Properties of Equality

Objective 2 ▶

Apply both the addition and multiplication properties of equality to solve linear equations.

Most linear equations cannot be solved using only the addition property of equality or only the multiplication property of equality; they require the use of both properties. The addition property of equality is used to isolate the variable on one side of an equation and the multiplication property of equality is used to get a coefficient of one on the variable.

Example 2 illustrates the use of both the addition property and multiplication property of equality to solve linear equations.

Objective 2 Examples   Solve each equation.

2a. $2x + 1 = -5$ **2b.** $4y + 3 = -2y - 9$

2c. $5x - 3 - 2x = 8x - 1$ **2d.** $8(2x - 1) = 10x + 7$

Solutions   **2a.** We need to isolate the variable term, $2x$, on one side of the equation. One way to do this is to subtract 1 from both sides.

$$2x + 1 = -5$$
$$2x + 1 - 1 = -5 - 1 \qquad \text{Subtract 1 from each side.}$$
$$2x = -6 \qquad \text{Simplify.}$$
$$\frac{2x}{2} = \frac{-6}{2} \qquad \text{Divide each side by 2.}$$
$$x = -3 \qquad \text{Simplify.}$$

Check: $\quad 2x + 1 = -5 \qquad$ Original equation
$$2(-3) + 1 = -5 \qquad \text{Replace } x \text{ with } -3.$$
$$-6 + 1 = -5 \qquad \text{Simplify.}$$
$$-5 = -5 \qquad \text{Simplify.}$$

Since $x = -3$ makes the equation is true, the solution set is $\{-3\}$.

2b. We need to remove a variable term from either the left side or the right side of the equation. One way to do this is to add $2y$ to both sides.

$$4y + 3 = -2y - 9$$
$$4y + 3 + 2y = -2y - 9 + 2y \qquad \text{Add } 2y \text{ to each side.}$$
$$6y + 3 = -9 \qquad \text{Simplify.}$$
$$6y + 3 - 3 = -9 - 3 \qquad \text{Subtract 3 from each side.}$$
$$6y = -12 \qquad \text{Simplify.}$$
$$\frac{6y}{6} = \frac{-12}{6} \qquad \text{Divide each side by 6.}$$
$$y = -2 \qquad \text{Simplify.}$$

Check: $\quad 4y + 3 = -2y - 9 \qquad$ Original equation
$$4(-2) + 3 = -2(-2) - 9 \qquad \text{Replace } y \text{ with } -2.$$
$$-8 + 3 = 4 - 9 \qquad \text{Simplify.}$$
$$-5 = -5 \qquad \text{Simplify.}$$

Since $y = -2$ makes the equation true, the solution set is $\{-2\}$.

Note: *We would have obtained the same solution of this equation if we had subtracted 4y from both sides.*
$$4y + 3 = -2y - 9$$
$$4y + 3 - 4y = -2y - 9 - 4y \qquad \text{Subtract } 4y \text{ from each side.}$$
$$3 = -6y - 9 \qquad \text{Simplify.}$$
$$3 + 9 = -6y - 9 + 9 \qquad \text{Add 9 to each side.}$$
$$12 = -6y \qquad \text{Simplify.}$$
$$\frac{12}{-6} = \frac{-6y}{-6} \qquad \text{Divide each side by } -6.$$
$$-2 = y \qquad \text{Simplify.}$$

2c. $5x - 3 - 2x = 8x - 1$
$$3x - 3 = 8x - 1 \qquad \text{Combine like terms on the left.}$$
$$3x - 3 - 3x = 8x - 1 - 3x \qquad \text{Subtract } 3x \text{ from each side.}$$
$$-3 = 5x - 1 \qquad \text{Simplify.}$$
$$-3 + 1 = 5x - 1 + 1 \qquad \text{Add 1 to each side.}$$
$$-2 = 5x \qquad \text{Simplify.}$$
$$\frac{-2}{5} = \frac{5x}{5} \qquad \text{Divide each side by 5.}$$
$$-\frac{2}{5} = x \qquad \text{Simplify.}$$

Check: $5x - 3 - 2x = 8x - 1$ Original equation

$$5\left(-\frac{2}{5}\right) - 3 - 2\left(-\frac{2}{5}\right) = 8\left(-\frac{2}{5}\right) - 1$$ Replace x with $-\frac{2}{5}$.

$$-\frac{10}{5} - 3 + \frac{4}{5} = -\frac{16}{5} - 1$$ Multiply.

$$-\frac{10}{5} - \frac{15}{5} + \frac{4}{5} = -\frac{16}{5} - \frac{5}{5}$$ Convert each number to a fraction with the common denominator of 5.

$$-\frac{21}{5} = -\frac{21}{5}$$ Add.

Since $x = -\frac{2}{5}$ makes the equation true, the solution set is $\left\{-\frac{2}{5}\right\}$.

2d. $8(2x - 1) = 10x + 7$

$16x - 8 = 10x + 7$ Apply the distributive property.

$16x - 8 - 10x = 10x + 7 - 10x$ Subtract $10x$ from each side.

$6x - 8 = 7$ Simplify.

$6x - 8 + 8 = 7 + 8$ Add 8 to each side.

$6x = 15$ Simplify.

$$\frac{6x}{6} = \frac{15}{6}$$ Divide each side by 6.

$$x = \frac{5}{2}$$ Simplify. Recall $\frac{15}{6} = \frac{5}{2}$.

Check: $8(2x - 1) = 10x + 7$ Original equation

$$8\left[2\left(\frac{5}{2}\right) - 1\right] = 10\left(\frac{5}{2}\right) + 7$$ Replace x with $\frac{5}{2}$.

$8(5 - 1) = 25 + 7$ Multiply.

$8(4) = 32$ Subtract.

$32 = 32$ Simplify.

Since $x = \frac{5}{2}$ makes the equation true, the solution set is $\left\{\frac{5}{2}\right\}$.

✔ **Student Check 2**   Solve each equation.

   a. $-3x + 2 = -7$ **b.** $2y - 6 = -9y + 1$

   c. $7x + 4 + 5x = 14x + 3$ **d.** $6(x - 2) = 9x + 8$

Applications

Objective 3 ▶

Solve application problems.

As we have already noted, expressing relationships mathematically is a key component to successfully solving word problems. While there may be some problems that we can solve by reasoning, it is important that we be able to justify how we came up with our answer and why we know it is valid. The ability to speak, think, and write mathematically equips us with the tools needed to provide such justification. The more we practice the skills at this level, the easier it is for us to extend them to more difficult concepts.

Procedure: Solving Word Problems

Step 1: Read the problem and determine what is unknown. Assign a variable for the unknown quantity.

> **Step 2:** Read the problem and determine what is given. Use this information to write an equation that models the situation.
>
> **Step 3:** Solve the equation using the addition and multiplication properties of equality.
>
> **Step 4:** Verify that the solution is reasonable.
>
> **Step 5:** Write the answer to the problem and attach appropriate units, if needed.

Objective 3 Examples **Write an equation that represents each situation. Solve the equation and explain the answer using a complete sentence.**

3a. One less than three times a number is -13. Find the number.

Solution **3a.** What is known? The number is unknown. Let n represent the number.

What is known? One less than three times a number is -13.

So, the equation is

$$3n - 1 = -13$$

We solve the equation to find the number.

$$3n - 1 = -13$$
$$3n - 1 + 1 = -13 + 1 \qquad \text{Add 1 to each side.}$$
$$3n = -12 \qquad \text{Simplify.}$$
$$\frac{3n}{3} = \frac{-12}{3} \qquad \text{Divide each side by 3.}$$
$$n = -4 \qquad \text{Simplify.}$$

The number is -4.

3b. How many quarters does it take to make $20.00?

Solution **3b.** What is unknown? The number of quarters is unknown. Let q represent the number of quarters.

What is known? The total value of the quarters is $20.00. To find the value of an unknown number of quarters, we use our knowledge of how we calculate the value of a specific number of quarters.

Number of Quarters	Total Value
1	$(0.25)(1) = \$0.25$
2	$(0.25)(2) = \$0.50$
3	$(0.25)(3) = \$0.75$
q	$(0.25)(q) = 0.25q$

> The total value of a collection of quarters is the value of a single quarter times the number of quarters.

So, to determine how many quarters we need to make $20.00, we can use the equation

$$0.25q = 20$$

We solve the equation to find the number of quarters.

$$0.25q = 20$$
$$\frac{0.25q}{0.25} = \frac{20}{0.25} \qquad \text{Divide each side by 0.25.}$$
$$q = 80 \qquad \text{Simplify.}$$

So, it takes 80 quarters to make $20.00.

3c. To rent a moving truck for one day, a company charges $44.95 plus $0.45 per mile. If a family budgets $300 for a moving truck, how many miles will they be able to drive the truck?

Solution

3c. What is unknown? The number of miles the truck is driven is unknown. Let m represent the miles driven.

What is known? The total cost of renting the truck is \$44.95 plus \$0.45 per mile and the total budgeted is \$300.

To find the cost of renting the truck, we calculate the cost of renting the truck for a specific number of miles and use this to determine an expression for the cost of driving the truck m miles.

Number of Miles	Cost of Renting
100	$44.95 + 0.45(100) = \$\ 89.95$
250	$44.95 + 0.45(250) = \ \ 157.45$
400	$44.95 + 0.45(400) = \ \ 224.95$
m	$44.95 + 0.45m$

> The cost of renting the truck is the flat fee of \$44.95 plus a mileage fee. We multiply the miles driven by \$0.45 and add this to the flat fee.

The equation comes from setting the cost of renting equal to the amount budgeted, \$300.

$$44.95 + 0.45x = 300$$

Solving the equation enables us to determine how many miles the truck can be driven.

$$44.95 + 0.45x = 300$$
$$44.95 + 0.45x - 44.95 = 300 - 44.95 \qquad \text{Subtract 44.95 from each side.}$$
$$0.45x = 255.05 \qquad \text{Simplify.}$$
$$\frac{0.45x}{0.45} = \frac{255.05}{0.45} \qquad \text{Divide each side by 0.45.}$$
$$x = 566.78 \qquad \text{Simplify.}$$

So, the family can drive approximately 567 miles for a cost of \$300.

✓ Student Check 3

Write an equation that represents each situation. Solve the equation and explain the answer using a complete sentence.

a. Four more than three times a number is -5. Find the number.

b. How many nickels does it take to make \$30?

c. The cost to rent a car for one day is \$13.99 plus \$0.20 per mile. How many miles can be driven for \$50? Round to the nearest mile.

Consecutive Integer Problems

Objective 4 ▶

Solve consecutive integer problems.

Consecutive means "successive" or "following one after another without interruption." So, **consecutive integers** are integers that follow one another. Some examples are listed in the table.

	Specific Example	Variable Representation
Consecutive integers	4 5 6 (+1) (+1)	$x, x + 1, x + 2, \ldots$
Consecutive even integers	10 12 14 (+2) (+2)	$x, x + 2, x + 4, \ldots$
Consecutive odd integers	11 13 15 (+2) (+2)	$x, x + 2, x + 4, \ldots$

Note that **consecutive odd** and **consecutive even integers** are represented in the same way since the difference between these types of integers is two units. The value of x determines the numbers that will be generated from the expressions $x + 2$ and $x + 4$.

If $x = 15$, then $x + 2 = 15 + 2 = 17$ and $x + 4 = 15 + 4 = 19$. So, the numbers generated are 15, 17, and 19.

If $x = 22$, then $x + 2 = 22 + 2 = 24$ and $x + 4 = 22 + 4 = 26$. So, the numbers generated are 22, 24, and 26.

Objective 4 Examples | **Write an equation that represents each problem. Solve the equation and explain the answer a complete sentence.**

4a. The sum of two consecutive integers is -21. Find the integers.

4b. Three times the smallest of three consecutive odd integers is 17 more than the sum of the other two integers. Find the integers.

Solutions | **4a.** What is unknown? Two consecutive integers are unknown. Let x and $x + 1$ represent the two integers.

What is known? The sum of the two integers is -21.

We use this statement to write the equation.

The sum of the
two integers is -21.

$x + (x + 1) = -21$	Express the relationship.
$2x + 1 = -21$	Combine like terms.
$2x + 1 - 1 = -21 - 1$	Subtract 1 from each side.
$2x = -22$	Simplify.
$\dfrac{2x}{2} = \dfrac{-22}{2}$	Divide each side by 2.
$x = -11$	Simplify.

Since $x = -11$, the integers are -11 and $-11 + 1 = -10$.

4b. What is unknown? Three consecutive odd integers are unknown. Let x, $x + 2$, and $x + 4$ represent the three odd integers.

What is known? Three times the smallest of three consecutive odd integers is 17 more than the sum of the other two integers.

We use this statement to obtain the equation.

Three times the
smallest of three   17 more than the
consecutive odd   sum of the other
integers is two integers

$3(x) = (x + 2) + (x + 4) + 17$	Express the relationship.
$3x = 2x + 23$	Combine like terms.
$3x - 2x = 2x + 23 - 2x$	Subtract $2x$ from each side.
$x = 23$	Simplify.

Since $x = 23$, the integers are 23, $23 + 2 = 25$, and $23 + 4 = 27$.

☑ **Student Check 4** | Write an equation that represents each problem. Solve the equation and explain the answer using a complete sentence.

a. The sum of two consecutive integers is -137. Find the integers.

b. Five times the smallest of three consecutive even integers is two more than twice the sum of the other two integers. Find the integers.

Troubleshooting Common Errors

Some common errors associated with solving linear equations are shown next. It is very important to be able to distinguish when the addition property of equality applies and when the multiplication property of equality applies. We use the addition property of equality to remove terms from one side of the equation. The multiplication property of equality is used to obtain a coefficient of one on the variable.

Objective 5 Examples **A problem and an incorrect solution are given. Provide the correct solution and an explanation of the error.**

5a. Solve $-2x = 30$.

Incorrect Solution	Correct Solution and Explanation
$-2x = 30$ $-2x + 2 = 30 + 2$ $x = 32$ The solution set is $\{32\}$.	The expression $-2x + 2$ cannot be combined since they are not like terms. So, adding 2 to both sides will not isolate x. Note that -2 is the coefficient of x. So, we divide both sides by -2 to obtain a coefficient of 1 on the variable. $$\frac{-2x}{-2} = \frac{30}{-2}$$ $$x = -15$$ The solution set is $\{-15\}$.

5b. Solve $2x + 8 = 4x + 10$.

Incorrect Solution	Correct Solution and Explanation
$2x + 8 = 4x + 10$ $4x - 2x + 8 = 4x - 4x + 10$ $2x + 8 = 10$ $2x + 8 - 8 = 10 - 8$ $2x = 2$ $x = 1$ The solution set is $\{1\}$.	The first step of subtracting $4x$ from both sides was applied incorrectly on the left side of the equation. We should subtract $4x$ at the end of the terms on the left, not insert it at the beginning. $2x + 8 = 4x + 10$ $2x + 8 - 4x = 4x - 4x + 10$ $-2x + 8 = 10$ $-2x + 8 - 8 = 10 - 8$ $-2x = 2$ $$\frac{-2x}{-2} = \frac{2}{-2}$$ $$x = -1$$ The solution set is $\{-1\}$.

ANSWERS TO STUDENT CHECKS

Student Check 1 **a.** $\{-7\}$ **b.** $\{60\}$ **c.** $\{8\}$ **d.** $\{-16\}$
 e. $\{-45\}$

Student Check 2 **a.** $\{3\}$ **b.** $\left\{\dfrac{7}{11}\right\}$ **c.** $\left\{\dfrac{1}{2}\right\}$

 d. $\left\{-\dfrac{20}{3}\right\}$

Student Check 3 **a.** The number is -3.
 b. It takes 600 nickels to make $30.
 c. The car can be driven for 180 miles for a cost of $50.

Student Check 4 **a.** The integers are -69 and -68.
 b. The integers are 14, 16, and 18.

SUMMARY OF KEY CONCEPTS

1. The multiplication property of equality enables us to multiply or divide both sides of an equation by the same nonzero number.

2. Most equations are going to require use of both the addition and multiplication properties of equality to solve them. The addition property of equality is used to eliminate terms from one side of the equation. The multiplication property of equality is used to make the coefficient of the variable one.

3. Word problems are solved by expressing the given relationships as a mathematical equation.

4. Consecutive integers are $x, x + 1, x + 2, \dots$. Consecutive even and odd integers are $x, x + 2, x + 4, \dots$. The equation used to solve these problems will be a result of direct translation. Refer to the key phrases learned in Chapter 1 and this chapter for assistance.

GRAPHING CALCULATOR SKILLS

The calculator should only be used to check that an equation was solved correctly at this point.

Example: Verify that $x = -\dfrac{2}{5}$ is a solution of $5x - 3 - 2x = 8x - 1$.

Method 1: Evaluate each side of the equation at the proposed solution. If each side produces the same number, then the solution is correct.

Method 2: Store the value of x into the calculator. Enter the equation and press enter. If a 1 is displayed, the statement is true for the stored value of x. If a 0 is displayed, the statement is false for the stored value of x.

$$5\left(-\frac{2}{5}\right) - 3 - 2\left(-\frac{2}{5}\right) \stackrel{?}{=} 8\left(-\frac{2}{5}\right) - 1$$

```
5(-2/5)-3-2(-2/5
)
            -4.2
8(-2/5)-1
            -4.2
```

```
-2/5→X
            -.4
5X-3-2X=8X-1
             1
```

Since each side of the equation produces the same result when evaluated at $x = -\dfrac{2}{5}$, the solution is correct.

Since 1 is displayed, the solution is correct.

SECTION 2.3   EXERCISE SET

 Write About It!

Use complete sentences in your answers to the following exercises.

1. Explain the multiplication property of equality.

2. Give an example of an equation that requires use of the addition property of equality to solve it, but not the multiplication property of equality. Explain your answer.

3. Give an example of an equation that requires use of the multiplication property of equality to solve it, but not the addition property of equality. Explain your answer.

4. Give an example of an equation that requires use of both the additional and multiplication properties of equality to solve it. Explain your answer.

Determine if each statement is true or false. If a statement is false, explain why.

5. The equation, $-p = 5$, is solved for p.

6. To solve the equation $-3x = 6$, add three to both sides.

7. The solution of $\dfrac{1}{4}x = 3$ is $x = 12$.

8. To make the coefficient of $-\frac{3}{5}x$ equal 1, we must multiply the expression by $\frac{3}{5}$.

9. To represent consecutive odd integers, use x, $x + 1$, $x + 3$, $x + 5$, etc.

10. To represent consecutive even integers, use x, $2x$, $4x$, $6x$, etc.

Practice Makes Perfect!

Solve each equation using the multiplication property of equality. (*See Objective 1.*)

11. $2x = 10$

12. $4x = 48$

13. $5x = -15$

14. $8x = -16$

15. $-0.3x = 12$

16. $-0.7x = 49$

17. $-x = 11$

18. $-x = 10$

19. $-4x = -24$

20. $-6x = -30$

21. $\frac{x}{5} = 2$

22. $\frac{a}{9} = -5$

23. $\frac{2}{3}x = 6$

24. $\frac{7}{2}y = 14$

25. $-\frac{5}{6}a = -10$

26. $-\frac{3}{8}b = -27$

27. $0.25y = 8$

28. $3.5a = -7$

Solve each equation. (*See Objective 2.*)

29. $6x + 2 = 14$

30. $4x + 5 = 21$

31. $5x - 10 = 20$

32. $7y - 6 = 36$

33. $-2a + 4 = -20$

34. $-9y + 6 = 6$

35. $-1.4x + 3.7 = 5.38$

36. $0.2b - 2.8 = -1.18$

37. $3x + 7 = x - 9$

38. $11y - 8 = 8y - 5$

39. $-5a - 6 = -2a + 3$

40. $-4x - 3 = 2x + 9$

41. $2x + 3 - 4x = 7x - 8$

42. $8a + 1 - 4a = 13 + 7a$

43. $3(2z + 6) = 4(z + 5)$

44. $5(4x - 1) = 2(3x + 5)$

45. $-4.5x + 10.5 = -3.1x + 11.06$

46. $-0.1b + 7 = 3.3b - 1.84$

47. $0.3(1.2z + 5.2) = -0.4(2.5z - 9.0)$

48. $-0.6(1.5x - 4.2) = 0.5(-x + 6.96)$

Write an equation that represents each situation. Solve the equation and explain the answer using a complete sentence. (*See Objective 3.*)

49. The cost to rent a luxury car for one day is $75 plus $0.25 per mile. This daily cost is represented by the expression $75 + 0.25x$, where x is the number of miles driven. How many miles has the car been driven if the cost of the rental car is $112.50?

50. The cost to rent a banquet facility for a wedding reception is $1500 plus $35 per person for the food. This is represented by the expression $1500 + 35x$, where x is the number of attendees. If a couple budgets $5000 for their reception, how many people can attend the reception?

51. How many nickels does it take to make $8.75?

52. How many nickels does it take to make $6.15?

53. How many quarters are needed to make $19.50?

54. How many quarters are needed to make $12.25?

Write an equation that represents each situation. Solve the equation and explain the answer using a complete sentence. (*See Objective 4.*)

55. The sum of two consecutive even integers is -78. Find the integers.

56. The sum of three consecutive odd integers is -309. Find the integers.

57. Four times the largest of three consecutive integers is 47 more than the sum of the other two integers. Find the integers.

58. The sum of the two smallest of three consecutive even integers is the same as three times the largest integer. Find the integers.

59. The sum of the two largest of three consecutive odd integers is the same as 5 more than the smallest integer. Find the integers.

60. The lengths of the sides of a triangle are consecutive integers. If the perimeter of the triangle is 12 ft, what are the lengths of the three sides of the triangle?

Mix 'Em Up!

Solve each equation.

61. $5x = 10$

62. $2x + 5 = 11$

63. $4y - 1 = 15$

64. $12y + 3 = 51$

65. $-4x + 16 = 20$

66. $-3x + 15 = 21$

67. $\frac{a}{4} = -2$

68. $-\frac{3a}{2} = \frac{1}{2}$

69. $10z + 3 - 5z = 12 - 4z$

70. $-3z + 1 - 2z = 10 + 3z + 7$

71. $-7x + 2 - 5x = 12 + 4x - 6$

72. $4y - 9 - y = 14 - 7y - 8$

73. $-\frac{1}{3}(x + 3) + 2 = x + 1$

74. $\frac{1}{4}(x - 8) = -2x + 3$

75. $-1.2b + 4 = 22$

76. $-2.1b + 1.1 = 7.4$

77. $-0.7y + 13.5 = 7y + 31.98$

78. $0.2a - 8.4 = 5.9a + 22.38$

79. $3.9(1.6x - 0.6) = -0.4(4.2x - 2.07)$

80. $-2(-0.7x + 3) = -0.5(0.2x + 16.8)$

81. $8(x - 3) = 2x - 42$

82. $-5(x + 11) = 3x + 1$

83. $-6(2x - 5) = 4x - 2$

84. $7(4x - 9) = 6x + 3$

Write an equation that represents each situation. Solve the equation and explain the answer in a complete sentence.

85. Twice a number is equal to -18. Find the number.

86. One-half of a number is equal to -10. Find the number.

87. Three-fifths of a number is 15. Find the number.

88. Two-sevenths of a number is -14. Find the number.

89. The quotient of a number and 7 is -21. Find the number.

90. The quotient of a number and 3 is 12. Find the number.

91. 30% of a number is 9. Find the number.

92. 15% of a number is 1.5. Find the number.

93. Two less than five times a number is -7. Find the number.

94. Four more than twice a number is -8. Find the number.

95. How many dimes will it take to make \$15?

96. How many half-dollars are needed to make \$75?

97. The cost to rent a midsize car for one day is \$63 plus \$0.28 per mile. This daily cost is represented by the expression $63 + 0.28x$, where x is the number of miles driven. How many miles has the car been driven if the cost of the rental car is \$121.80?

98. The cost to rent a midsize car for one day is \$56 plus \$0.32 per mile. This daily cost is represented by the expression $56 + 0.32x$, where x is the number of miles driven. How many miles has the car been driven if the cost of the rental car is \$134.40?

99. The sum of four consecutive integers is -154. Find the integers.

100. The sum of three consecutive odd integers is 51. Find the integers.

101. Three times the largest of three consecutive even integers is 2 more than the sum of the other two integers. Find the integers.

102. Three times the largest of three consecutive odd integers is 13 more than the sum of the other two integers. Find the integers.

 You Be the Teacher!

Correct each student's errors, if any.

103. Solve: $\frac{x}{2} + 3 = 7$

Michael's work:

$$\frac{x}{2} + 3 - 3 = 7 - 3$$
$$\frac{x}{2} = 4$$
$$\frac{x}{2} - 2 = 4 - 2$$
$$x = 2$$

104. Solve: $-\frac{2}{3}x + 3 = x - 2$

Gabriel's work:

$$-\frac{2}{3}x + 3 - 3 = x - 2 - 3$$
$$-\frac{2}{3}x = x - 5$$
$$-\frac{2}{3}x - x = x - 5 - x$$
$$-\frac{2}{3}x - \frac{3}{3}x = -5$$
$$-\frac{5}{3}x = -5$$
$$-\frac{5}{3}x + \frac{5}{3} = -5 + \frac{5}{3}$$
$$x = -\frac{10}{3}$$

105. Translate: Five less than twice a number equals four times the number plus ten.

Kristen's answer:

Let x be the number. $5 - 2x = 4x + 10$

106. Translate: Half the sum of 4 and a number equals the number subtracted from 6.

Lauren's answer:

Let x be the number. $\frac{1}{2}(4 + x) = x - 6$

107. Three times the largest of three consecutive odd integers is 87 more than the sum of the other two integers. Find the integers.

Cynthia's work: Let $x, x + 1, x + 2$ be the three consecutive odd integers.

$$3(x + 2) = 87 + x + (x + 1)$$
$$3x + 6 = 88 + 2x$$
$$x = 82$$

108. The sum of the two largest of three consecutive even integers is 30 less than four times the smallest integer. Find the integers.

Ron's work: Let $x, x + 2,$ and $x + 4$ be the three consecutive even integers.

$$(x + 2) + (x + 4) = 30 - 4x$$
$$2x + 6 = 30 - 6x$$
$$8x = 24$$
$$x = 3$$

 Calculate It!

Solve each equation. Then use a graphing calculator to check the solution.

109. $8x + 20 = 2x - 15 + x$

110. $2(x + 10) - 5 = 4x + 18$

111. $\frac{5}{6}x - x = 8 - \frac{3}{2}x$

112. $-3.5(-1.6x + 2.4) = 0.8(-2.1x - 41.44)$

SECTION 2.4   More on Solving Linear Equations

▶ **OBJECTIVES**

As a result of completing this section, you will be able to

1. Solve linear equations with fractions.
2. Solve linear equations with decimals.
3. Solve linear equations with no solution.
4. Solve linear equations with infinitely many solutions.
5. Troubleshoot common errors.

Sections 2.2 and 2.3 laid the groundwork for us to solve linear equations. In this section, we will solve linear equations that require a few more steps than what we have encountered so far and we will also explore equations that have special solution sets.

Linear Equations Containing Fractions

We have solved equations with fractions in earlier sections in which we simply worked with the fractions to obtain the solution. Working with fractions can often be cumbersome, so we can use the multiplication property of equality to first clear fractions from the equation and then solve it. Consider the expression $\frac{3}{4}x - \frac{1}{6}$. The least common denominator (LCD) of the fractions in this expression is 12. If we multiply the expression by 12, we get

$$12\left(\frac{3}{4}x - \frac{1}{6}\right) = \frac{12}{1}\left(\frac{3}{4}x\right) - \frac{12}{1}\left(\frac{1}{6}\right)$$

$$= \frac{36}{4}x - \frac{12}{6}$$

$$= 9x - 2$$

When the expression is multiplied by its LCD, the fractions are cleared and the expression completely changes. But if we perform this operation to both sides of an equation with fractions, the new equation is equivalent to the original equation.

So, if we solve an equation with fractions, we can clear fractions before we apply the steps to solve the equation. These steps are presented next in a general strategy for solving linear equations in one variable. These steps can be applied to solving any linear equation in one variable.

Objective 1 ▶

Solve linear equations with fractions.

> **Procedure: Solving Linear Equations in One Variable—A General Strategy**
>
> **Step 1:** Write an equivalent equation that doesn't contain fractions or decimals, if necessary.
>   **a.** Multiply both sides of the equation by the LCD to clear fractions in the equation if they occur.
>   **b.** If decimal numbers appear in the equation, multiply both sides of the equation by the power of 10 that eliminates the number with the most decimal places.
> **Step 2:** Apply the distributive property to remove any parentheses if they occur.
> **Step 3:** Use the addition property of equality to get all the variable terms on one side of the equation and all constants on the other side of the equation.
> **Step 4:** Isolate the variable by applying the multiplication property of equality.
> **Step 5:** Check the proposed solution by substituting it into the *original* equation.

Objective 1 Examples   Solve each equation by first clearing fractions. Check each answer.

1a. $\dfrac{3x}{4} - 5 = \dfrac{x}{4}$ **1b.** $\dfrac{a}{6} - \dfrac{1}{8} = \dfrac{a}{12}$

1c. $-\dfrac{2}{3}(x - 6) = \dfrac{4}{5}(x + 10)$ **1d.** $\dfrac{y - 5}{4} - \dfrac{2y}{9} = \dfrac{1}{6}$

Solutions **1a.**
$$\frac{3x}{4} - 5 = \frac{x}{4}$$

$$4\left(\frac{3x}{4} - 5\right) = 4\left(\frac{x}{4}\right)$$ Multiply each side by the LCD, 4.

$$4\left(\frac{3x}{4}\right) - 4(5) = x$$ Apply the distributive property. Simplify.

$$3x - 20 = x$$ Simplify.

$$3x - 20 - x = x - x$$ Subtract x from each side.

$$2x - 20 = 0$$ Simplify.

$$2x - 20 + 20 = 0 + 20$$ Add 20 to each side.

$$2x = 20$$ Simplify.

$$\frac{2x}{2} = \frac{20}{2}$$ Divide each side by 2.

$$x = 10$$ Simplify.

Check:
$$\frac{3x}{4} - 5 = \frac{x}{4}$$ Original equation

$$\frac{3(10)}{4} - 5 = \frac{10}{4}$$ Replace x with 10.

$$\frac{30}{4} - 5 = \frac{10}{4}$$ Simplify.

$$\frac{30}{4} - \frac{20}{4} = \frac{10}{4}$$ Subtract the numbers on the left by converting 5 to $\frac{20}{4}$.

$$\frac{10}{4} = \frac{10}{4}$$ Simplify.

Since $x = 10$ makes the equation a true statement, the solution set is $\{10\}$.

1b.
$$\frac{a}{6} - \frac{1}{8} = \frac{a}{12}$$

$$24\left(\frac{a}{6} - \frac{1}{8}\right) = 24\left(\frac{a}{12}\right)$$ Multiply each side by the LCD, 24.

$$24\left(\frac{a}{6}\right) - 24\left(\frac{1}{8}\right) = 2a$$ Apply the distributive property. Simplify.

$$4a - 3 = 2a$$ Simplify.

$$4a - 3 - 2a = 2a - 2a$$ Subtract $2a$ from each side.

$$2a - 3 = 0$$ Simplify.

$$2a - 3 + 3 = 0 + 3$$ Add 3 to each side.

$$2a = 3$$ Simplify.

$$\frac{2a}{2} = \frac{3}{2}$$ Divide each side by 2.

$$a = \frac{3}{2}$$ Simplify.

Check:
$$\frac{a}{6} - \frac{1}{8} = \frac{a}{12}$$ Original equation

$$\frac{\frac{3}{2}}{6} - \frac{1}{8} = \frac{\frac{3}{2}}{12}$$ Replace a with $\frac{3}{2}$.

$$\frac{3}{12} - \frac{1}{8} = \frac{3}{24}$$ Simplify. Recall: $\frac{3}{2} \div 6 = \frac{3}{2} \cdot \frac{1}{6} = \frac{3}{12}$.

$$\frac{3}{2} \div 12 = \frac{3}{2} \cdot \frac{1}{12} = \frac{3}{24}$$

$$\frac{6}{24} - \frac{3}{24} = \frac{3}{24}$$ Write equivalent fractions with LCD, 24.

$$\frac{3}{24} = \frac{3}{24}$$ Simplify.

Since $a = \frac{3}{2}$ makes the equation true, the solution set is $\left\{\frac{3}{2}\right\}$.

1c.
$$-\frac{2}{3}(x - 6) = \frac{4}{5}(x + 10)$$

$$15\left[-\frac{2}{3}(x - 6)\right] = 15\left[\frac{4}{5}(x + 10)\right]$$ Multiply each side by the LCD, 15.

$$\left[15\left(-\frac{2}{3}\right)\right](x - 6) = \left[15\left(\frac{4}{5}\right)\right](x + 10)$$ Apply the associative property.

$$-10(x - 6) = 12(x + 10)$$ Multiply the coefficients.

$$-10x + 60 = 12x + 120$$ Apply the distributive property.

$$-10x + 60 - 12x = 12x + 120 - 12x$$ Subtract $12x$ from each side.

$$-22x + 60 = 120$$ Simplify.

$$-22x + 60 - 60 = 120 - 60$$ Subtract 60 from each side.

$$-22x = 60$$ Simplify.

$$\frac{-22x}{-22} = \frac{60}{-22}$$ Divide each side by -22.

$$x = -\frac{30}{11}$$ Simplify.

Check:
$$-\frac{2}{3}(x - 6) = \frac{4}{5}(x + 10)$$ Original equation

$$-\frac{2}{3}\left(-\frac{30}{11} - 6\right) = \frac{4}{5}\left(-\frac{30}{11} + 10\right)$$ Replace x with $-\frac{30}{11}$.

$$-\frac{2}{3}\left(-\frac{30}{11}\right) - \left(-\frac{2}{3}\right)(6) = \frac{4}{5}\left(-\frac{30}{11}\right) + \frac{4}{5}(10)$$ Distribute.

$$\frac{20}{11} + 4 = -\frac{24}{11} + 8$$ Simplify.

$$\frac{20}{11} + \frac{44}{11} = -\frac{24}{11} + \frac{88}{11}$$ Add the fractions.

$$\frac{64}{11} = \frac{64}{11}$$ Simplify.

Since $x = -\frac{30}{11}$ makes the equation true, the solution set is $\left\{-\frac{30}{11}\right\}$.

1d.
$$\frac{y-5}{4} - \frac{2y}{9} = \frac{1}{6}$$

$$36\left(\frac{y-5}{4} - \frac{2y}{9}\right) = 36\left(\frac{1}{6}\right) \qquad \text{Multiply each side by the LCD, 36.}$$

$$36\left(\frac{y-5}{4}\right) - 36\left(\frac{2y}{9}\right) = 6 \qquad \text{Apply the distributive property. Simplify.}$$

$$9(y-5) - 4(2y) = 6 \qquad \text{Simplify each product.}$$

$$9y - 45 - 8y = 6 \qquad \text{Multiply the remaining factors.}$$

$$y - 45 = 6 \qquad \text{Simplify each side.}$$

$$y - 45 + 45 = 6 + 45 \qquad \text{Add 45 to each side.}$$

$$y = 51 \qquad \text{Simplify.}$$

Check:
$$\frac{y-5}{4} - \frac{2y}{9} = \frac{1}{6} \qquad \text{Original equation}$$

$$\frac{51-5}{4} - \frac{2(51)}{9} = \frac{1}{6} \qquad \text{Replace } y \text{ with 51.}$$

$$\frac{46}{4} - \frac{102}{9} = \frac{1}{6} \qquad \text{Simplify the numerators.}$$

$$\frac{414}{36} - \frac{408}{36} = \frac{6}{36} \qquad \text{Write equivalent fractions with LCD, 36.}$$

$$\frac{6}{36} = \frac{6}{36} \qquad \text{Simplify.}$$

Since $y = 51$ makes the equation true, the solution set is $\{51\}$.

✓ **Student Check 1** Solve each equation by first clearing fractions. Check each answer.

a. $\dfrac{7x}{6} - 4 = \dfrac{x}{6}$

b. $\dfrac{y}{4} - \dfrac{1}{12} = \dfrac{y}{8}$

c. $\dfrac{3}{4}(x - 8) = \dfrac{2}{7}(x + 14)$

d. $\dfrac{a+7}{3} - 4 = \dfrac{a}{6}$

Note: *When we multiply an equation by the LCD of all the fractions in the equation, the resulting equation should not have any fractions. If the resulting equation still contains fractions, then we either simplified incorrectly or calculated the LCD incorrectly.*

Linear Equations Containing Decimals

Objective 2 ▶

Solve linear equations with decimals.

As with equations that contain fractions, equations containing decimals can also be tedious to solve by hand. So, we will clear the decimals from the equation.

Consider the following products. Recall that multiplying a number by a power of 10 moves the decimal point to the right.

$$10(\underline{3.2}\,y) = (10)(3.2)y = 32y$$

One decimal place

Multiplying an expression with one decimal place by 10^1 or 10 clears the decimal from the expression.

$$100(\underbrace{0.05x}) = (100)(0.05)x = 5x$$

Two decimal places

> Multiplying an expression with two decimal places by 10^2 or 100 clears the decimal from the expression.

$$1000(\underbrace{2.34a + 12.567})$$
$$= 1000(2.34a) + 1000(12.567)$$
$$= 2340a + 12{,}567$$

Two and three decimal places

> Multiplying an expression with three decimal places by 10^3 or 1000 clears the decimal from the expression.

Note that when we multiply the preceding expressions by an appropriate power of 10, the decimals are cleared from the expression. We will apply this same concept to remove decimals from an equation; that is, we will multiply each side of the equation by a power of 10 (10, 100, 1000, and so on) that will eliminate the decimal from the number with the most decimal places.

To solve equations with decimals, we apply the general strategy that was stated at the beginning of the section.

Objective 2 Examples | Solve each equation by clearing the decimals. Check each answer.

2a. $x + 0.15x = 36.80$ **2b.** $0.04x + 0.05(500 - x) = 23$

Solutions **2a.**

$$x + 0.15x = 36.80$$
$$100(x + 0.15x) = 100(36.80)$$ Multiply each side by 100.
$$100(x) + 100(0.15x) = 3680$$ Distribute and simplify.
$$100x + 15x = 3680$$ Simplify.
$$115x = 3680$$ Combine like terms.
$$\frac{115x}{115} = \frac{3680}{115}$$ Divide each side by 115.
$$x = 32$$ Simplify.

Check:

$$x + 0.15x = 36.80$$ Original equation
$$32 + 0.15(32) = 36.80$$ Replace x with 32.
$$32 + 4.8 = 36.80$$ Simplify.
$$36.8 = 36.80$$ Add.

Since $x = 32$ makes the equation true, the solution set is $\{32\}$.

Note: *We could also solve the equation without removing decimals by combining like terms on the left.*

$$1x + 0.15x = 36.80$$ Recall that $x = 1x$.
$$1.15x = 36.80$$ Combine like terms.
$$\frac{1.15x}{1.15} = \frac{36.80}{1.15}$$ Divide each side by 1.15.
$$x = 32$$ Simplify.

2b.
$$0.04x + 0.05(500 - x) = 23$$

$100[0.04x + 0.05(500 - x)] = 100(23)$	Multiply each side by 100.
$100(0.04x) + 100(0.05)(500 - x) = 2300$	Distribute and simplify.
$4x + 5(500 - x) = 2300$	Simplify each product.
$4x + 2500 - 5x = 2300$	Apply the distributive property.
$-x + 2500 = 2300$	Combine like terms.
$-x + 2500 - 2500 = 2300 - 2500$	Subtract 2500 from each side.
$-x = -200$	Simplify.
$\dfrac{-x}{-1} = \dfrac{-200}{-1}$	Divide each side by −1.
$x = 200$	Simplify.

Check:		
	$0.04x + 0.05(500 - x) = 23$	Original equation
	$0.04(200) + 0.05(500 - 200) = 23$	Replace x with 200.
	$8 + 0.05(300) = 23$	Simplify.
	$8 + 15 = 23$	Multiply.
	$23 = 23$	Add.

Since $x = 200$ makes the equation true, the solution set is $\{200\}$.

 **Student Check 2** Solve each equation by first clearing the decimals. Check each answer.
a. $x + 0.25x = 90$ **b.** $0.05x + 0.07(800 - x) = 54$

Linear Equations with No Solution

Objective 3 ▶

Solve linear equations with no solution.

All of the linear equations we have encountered so far have had one solution. When solving these equations, we were able to isolate the variable on one side of the equation and the constant on the other side. These types of equations are called *conditional equations*.

> **Definition:** A **conditional equation** is an equation that is true for some values of the variable and not true for other values of the variable.

We will now turn our attention to a special type of linear equation, one that has no solution. Consider the equation $x = x + 1$. The left and right sides of this equation will never be equal since the right side of the equation is always one more than the left side of the equation as shown in the table.

x	$x = x + 1$	
-2	$-2 = -2 + 1$ $-2 = -1$	False
-1	$-1 = -1 + 1$ $-1 = 0$	False
0	$0 = 0 + 1$ $0 = 1$	False
1	$1 = 1 + 1$ $1 = 2$	False
2	$2 = 2 + 1$ $2 = 3$	False

So, $x = x + 1$ is an example of an equation with no solution. It is called a *contradiction*.

> **Definition:** An equation that is not true for any value of the variable is called a **contradiction**. A contradiction has no solution. The solution set is the empty set, or null set, $\varnothing$.

Contradictions do not initially look any different from conditional equations. The difference becomes apparent in the equation-solving process. As we solve the equation, the variable terms are eliminated from both sides of the equation and a false statement (e.g., $5 = 7$) remains.

   If we solve an equation and obtain a false statement, then the equation is a contradiction and has no solution. The solution set is the empty set, $\varnothing$.

Objective 3 Examples   Solve each equation.

3a. $x + 3 = x + 2$ **3b.** $4x + 5 - x = 3(x + 2)$

Solutions   **3a.**
$$x + 3 = x + 2$$
$$x + 3 - x = x + 2 - x \qquad \text{Subtract } x \text{ from each side.}$$
$$3 = 2 \qquad \text{Simplify.}$$

The resulting equation, $3 = 2$, is a contradiction, or false statement. So, the solution set is the empty set, $\varnothing$.

3b. $4x + 5 - x = 3(x + 2)$ Apply the distributive property on the right side and
$$3x + 5 = 3x + 6 \qquad \text{combine like terms on the left side.}$$
$$3x + 5 - 3x = 3x + 6 - 3x \qquad \text{Subtract } 3x \text{ from each side.}$$
$$5 = 6 \qquad \text{Simplify.}$$

The resulting equation, $5 = 6$, is a contradiction, so the solution set is the empty set, $\varnothing$.

✓ Student Check 3   Solve each equation.

   a. $2x - 4 = 2x + 1$ **b.** $5(2x - 4) = 3(3x + 1) + x$

> **Note:** *Once an equation is simplified on each side, we know we have a contradiction when the variable term on each side of the equation is the same but the constant terms are different.*

Linear Equations with Infinitely Many Solutions

Objective 4 ▶

Solve linear equations with infinitely many solutions.

We will now turn our attention to solving linear equations that have infinitely many solutions. Consider the equation $x + 1 = x + 1$. The left and right sides of this equation will always be equal since the right side of the equation is the exact same as the left side of the equation as shown in the following table.

x	$x + 1 = x + 1$	
-2	$-2 + 1 = -2 + 1$ $-1 = -1$	True
-1	$-1 + 1 = -1 + 1$ $0 = 0$	True
0	$0 + 1 = 0 + 1$ $1 = 1$	True
1	$1 + 1 = 1 + 1$ $2 = 2$	True
2	$2 + 1 = 2 + 1$ $3 = 3$	True

So, $x + 1 = x + 1$ is an example of an equation with infinitely many solutions. It is true for every value of the variable. This type of equation is called an *identity*.

> **Definition:** An **identity** is an equation that is true for all values of the variable. An identity has infinitely many solutions. The solution set is all real numbers, denoted $\mathbb{R}$.

Again, this type of equation looks no different from a conditional equation. Through the equation-solving process, we obtain a true statement (e.g., $5 = 5$ or $x = x$).

If we solve an equation and a true statement remains, then the equation is an identity and has infinitely many solutions of the equation. The solution set is the set of all real numbers, denoted by $\mathbb{R}$.

| **Objective 4 Examples** | Solve each equation. |

4a. $2x + 3 = 2(x - 2) + 7$ **4b.** $4(3x + 5) - (x - 3) = 5(3x + 5) - (4x + 2)$

Solutions   **4a.**

$$2x + 3 = 2(x - 2) + 7$$
$$2x + 3 = 2x - 4 + 7 \qquad \text{Apply the distributive property.}$$
$$2x + 3 = 2x + 3 \qquad \text{Combine like terms.}$$
$$2x + 3 - 2x = 2x + 3 - 2x \qquad \text{Subtract } 2x \text{ from each side.}$$
$$3 = 3 \qquad \text{Simplify.}$$

The resulting equation, $3 = 3$, is a true statement, so this equation is an identity. The solution set is all real numbers, or $\mathbb{R}$.

4b. $4(3x + 5) - (x - 3) = 5(3x + 5) - (4x + 2)$

$$12x + 20 - x + 3 = 15x + 25 - 4x - 2 \qquad \text{Apply the distributive property.}$$
$$11x + 23 = 11x + 23 \qquad \text{Combine like terms.}$$
$$11x + 23 - 23 = 11x + 23 - 23 \qquad \text{Subtract 23 from each side.}$$
$$11x = 11x \qquad \text{Simplify.}$$

This resulting equation, $11x = 11x$, is always true, so this equation is an identity. The solution set is all real numbers, or $\mathbb{R}$.

| ✓ **Student Check 4** | Solve each equation. |

a. $2(6x + 4) = 4(3x + 2)$ **b.** $7x + 3(2x - 1) = 4(3x - 1) + x + 1$

> **Note:** *Once an equation is simplified on each side, we know we have an identity when the variable terms on each side of the equation are the same and the constant terms on each side of the equation are the same.*

| **Objective 5** ▶ | **Troubleshooting Common Errors** |
| Troubleshoot common errors. | Some common errors associated with solving linear equations are shown next. |

Objective 5 Examples   A problem and an incorrect solution are given. Provide the correct solution and an explanation of the error.

5a. Solve $\dfrac{3x}{2} + 5 = \dfrac{x}{2}$.

Incorrect Solution	Correct Solution and Explanation
$$\dfrac{3x}{2} + 5 = \dfrac{x}{2}$$ $$2\left(\dfrac{3x}{2} + 5\right) = 2\left(\dfrac{x}{2}\right)$$ $$3x + 5 = x$$ $$2x = -5$$ $$x = -\dfrac{5}{2}$$	The error was made in not applying the distributive property to each term on the left side of the equation. $$\dfrac{3x}{2} + 5 = \dfrac{x}{2}$$ $$2\left(\dfrac{3x}{2} + 5\right) = 2\left(\dfrac{x}{2}\right)$$ $$3x + 10 = x$$ $$2x = -10$$ $$x = -\dfrac{10}{2}$$ $$x = -5$$ The solution set is $\{-5\}$.

5b. Solve $2x - 4(x - 1) = -2(x + 2)$.

Incorrect Solution	Correct Solution and Explanation
$$2x - 4(x - 1) = -2(x + 2)$$ $$2x - 4x + 4 = -2x - 4$$ $$-2x + 4 = -2x - 4$$ $$\underline{+2x - 4 \quad\quad +2x - 4}$$ $$0x = -8$$ The solution set is $\{-8\}$.	The error was made in not recognizing that $0x = 0$. The resulting statement should be $0 = -8$, which is a contradiction. So, the solution set is the empty set, $\varnothing$.

ANSWERS TO STUDENT CHECKS

Student Check 1   a. $\{4\}$ **b.** $\left\{\dfrac{2}{3}\right\}$ **c.** $\left\{\dfrac{280}{13}\right\}$ **d.** $\{10\}$ **Student Check 3   a.** $\varnothing$ **b.** $\varnothing$

Student Check 2   a. $\{72\}$ **b.** $\{100\}$ **Student Check 4   a.** $\mathbb{R}$ **b.** $\mathbb{R}$

SUMMARY OF KEY CONCEPTS

1. To solve an equation with fractions, multiply both sides by the LCD of all fractions in the equation. This will create an equation that does not have any fractions. If the resulting equation has fractions, then the LCD was not correct or the equation was simplified incorrectly.

2. To solve an equation with decimals, multiply both sides by a power of 10 that will eliminate decimals in the number with the largest number of decimal places.

3. Equations that produce a false statement have no solution. These are called contradictions and are denoted by the null set, $\varnothing$.

4. Equations that produce a true statement have infinitely many solutions and are called identities. The solution set is the set of all real numbers, $\mathbb{R}$.

GRAPHING CALCULATOR SKILLS

In previous sections, we learned how to check solutions of equations on the calculator by using the main screen, the store feature, and the table feature. We will now use the table feature to verify solutions when the proposed solution is large or a fraction. We can change the table settings in one of two ways as shown next.

Example: Determine if $y = 51$ is a solution of $\dfrac{y - 5}{4} - \dfrac{2y}{9} = \dfrac{1}{6}$.

Solution:

Method 1: Enter each side of the equation in the equation editor and view the table. Since the calculator accepts only the variable x, we will use x in place of y in the equation.

Instead of scrolling to $x = 51$ in the table, we can change the starting value of the table.

When we view the table we see the x-value of 51 and find that the corresponding y-values are equal, which confirms $x = 51$ is a solution of the equation.

Method 2: We can also change the table settings to enable us to enter the value of x that we want to check. Go to the Table Setup and change the Indpnt setting from Auto to Ask. Move the cursor to Ask in the Indpnt row and press enter.

Now go to the table and enter the value for x and press enter.

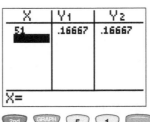

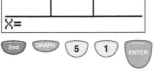

Since the y_1 and y_2 values are equal for $x = 51$, the solution is correct.

Note: *If the equation is an identity, the* Y_1 *and* Y_2 *values are equal for every value of x. If the equation is a contradiction, the* Y_1 *and* Y_2 *values are not equal for any value of x.*

SECTION 2.4 / EXERCISE SET

Write About It!

Use complete sentences to explain the meaning of the given term.

1. Conditional equation   2. Contradiction   3. Identity

Use complete sentences to answer each question.

4. How can you eliminate fractions from an equation?

5. How can you eliminate decimals from an equation?

6. When solving an equation, how will you know if the solution is the empty set?

7. When solving an equation, how will you know if the solution is all real numbers?

Determine if each statement is true or false. If a statement is false, explain why.

8. The solution of the equation $2x = 5x$ is the empty set.

9. If the solution of the equation $x + a = x + b$ is the empty set, then $a \neq b$.

10. If the solution of the equation $x + a = x + b$ is all real numbers, then $a = b$.

11. An equivalent form of the equation $\frac{y}{4} + 3 = \frac{y}{8}$ is $2y + 3 = y$.

12. An equivalent form of the equation $-5(2x - 3) = 2x - 9$ is $-10x - 3 = 2x - 9$.

Practice Makes Perfect!

Solve each equation by first clearing fractions.
(See Objective 1.)

13. $\frac{x}{2} + \frac{3}{2} = \frac{1}{2}$

14. $\frac{y}{3} - \frac{4}{3} = -\frac{10}{3}$

15. $\frac{a}{6} + 5 = \frac{a}{3}$

16. $\frac{x}{5} - 2 = \frac{x}{10}$

17. $\dfrac{3x}{5} + \dfrac{1}{3} = \dfrac{2x}{15}$ **18.** $\dfrac{4y}{7} + \dfrac{1}{2} = \dfrac{5y}{14}$

19. $2 - \dfrac{x}{6} = \dfrac{x}{3}$ **20.** $4 - \dfrac{y}{2} = \dfrac{y}{8}$

21. $\dfrac{2}{9}a - \dfrac{3}{18} = \dfrac{5}{18}a + \dfrac{4}{9}$ **22.** $\dfrac{7}{3}y + \dfrac{5}{6} = \dfrac{1}{6}y - \dfrac{2}{3}$

23. $-3 = \dfrac{9}{2} + b$ **24.** $5 = -\dfrac{4}{7} + b$

25. $\dfrac{3}{2}(x + 4) = \dfrac{1}{3}(x + 9)$ **26.** $-\dfrac{6}{7}(x - 14) = \dfrac{2}{3}(x + 6)$

27. $\dfrac{x + 4}{2} - 5 = \dfrac{x - 1}{6}$ **28.** $\dfrac{x - 2}{3} + 1 = \dfrac{x + 4}{8}$

Solve each equation by first clearing decimals. (See Objective 2.)

29. $x + 0.3x = 65$ **30.** $y + 0.1y = 11.88$

31. $a - 0.25a = 48.75$ **32.** $x - 0.95x = 6.25$

33. $0.06x + 0.05(4000 - x) = 235$

34. $0.03x + 0.045(10{,}000 - x) = 390$

35. $0.6x + 0.25(10) = 0.5(x + 10)$

36. $0.15x + 0.75(40) = 0.3(x + 40)$

Solve each equation. If the equation is a contradiction, write the solution as ∅. If the equation is an identity, write the solution as ℝ. (See Objectives 3 and 4.)

37. $x - 7 = x + 4$ **38.** $y + 2 = y - 1$

39. $2x + 3 = 2x - 3$ **40.** $8a + 2 = 8a - 2$

41. $3(2a + 5) = 2(3a - 4)$ **42.** $5(x + 2) = 3x + 10 + 2x$

43. $4x + 1 = 4(x - 2) + 9$

44. $2 - 5x = 7(1 - x) + 2x - 5$

45. $-2(y - 7) - (y + 3) = -3(y - 4) - 1$

46. $-7(2 - a) - (a + 3) = -3(5 - 2a) - 4$

47. $3(2a - 1) - (a + 4) = 5(a - 6) - 8$

48. $12(y - 4) + 3 - y = 5(2y - 9) + y$

 Mix 'Em Up!

Solve each equation. If the equation is a contradiction, write the solution as ∅. If the equation is an identity, write the solution as ℝ.

49. $2x + 3 = 3$ **50.** $6x - 7 = -7$

51. $3(y + 1) - (y - 4) = 5(y + 2)$

52. $4(x + 6) - (3x - 2) = 2(x + 3)$

53. $\dfrac{4x}{5} - 3 = \dfrac{x}{10}$ **54.** $\dfrac{7}{3}y + 1 = \dfrac{5}{9}y$

55. $x + 0.8x = 900$ **56.** $x - 0.30x = 332.5$

57. $\dfrac{3}{4}b + \dfrac{7}{10} = \dfrac{2}{5}b - \dfrac{1}{4}$ **58.** $\dfrac{5}{6}c - \dfrac{2}{3} = \dfrac{1}{2}c - \dfrac{5}{12}$

59. $0.021x + 0.045(2400 - x) = 91.2$

60. $0.016x + 0.02(8800 - x) = 158.4$

61. $\dfrac{y - 2}{7} - 3 = \dfrac{y - 11}{4}$

62. $\dfrac{y + 5}{6} - 2 = \dfrac{y - 3}{10}$

63. $0.046x + 0.05(6900 - x) = 339.4$

64. $0.066x + 0.08(9300 - x) = 727.2$

65. $\dfrac{2}{3}(x - 1) = \dfrac{1}{6}(4x - 4)$

66. $\dfrac{1}{4}(2x + 3) = \dfrac{1}{2}(x - 3)$

67. $0.81x + 2.5(-1.06) = 0.2(x + 2)$

68. $0.31x + 0.5(-6.54) = 0.3(x - 9)$

69. $3(a + 4) + 2a = 5(a + 2)$

70. $7(2x - 3) - 5x = 3(3x - 7)$

71. $\dfrac{3}{2}(x - 4) = \dfrac{1}{4}(5x + 1)$

72. $\dfrac{7}{12}(3x + 5) = \dfrac{5}{6}(x - 2)$

73. $0.26x + 0.8(73.5) = 0.75x$

74. $0.66x + 2.5(-6.448) = 0.2(x - 7)$

75. $\dfrac{3x - 8}{6} - 1 = \dfrac{x + 3}{4}$

76. $\dfrac{2x - 7}{10} + 1 = \dfrac{x + 3}{2}$

 You Be the Teacher!

Correct each student's errors, if any.

77. Solve: $\dfrac{x}{5} = \dfrac{1}{2} - \dfrac{x}{3}$

Susan's work:

$$\dfrac{x}{5} = \dfrac{1}{2} - \dfrac{x}{3}. \text{ LCD} = 30$$

$$30\left(\dfrac{x}{5}\right) = 30\left(\dfrac{1}{2} - \dfrac{x}{3}\right)$$

$$\dfrac{30x}{5} = \dfrac{30}{2} - \dfrac{x}{3}$$

$$6x = 15 - \dfrac{x}{3}$$

$$3(6x) = 3\left(15 - \dfrac{x}{3}\right)$$

$$18x = 45 - \dfrac{3x}{3}$$

$$18x + x = 45 - x + x$$

$$19x = 45$$

$$\dfrac{19x}{19} = \dfrac{45}{19}$$

$$x = \dfrac{45}{19}$$

78. Solve: $\frac{5}{6}(x + 12) = \frac{1}{8}(2x - 9)$

Weston's work:

$$\frac{5}{6}(x + 12) = \frac{1}{8}(2x - 9). \text{ LCD} = 24$$

$$24\left(\frac{5}{6}(x + 12)\right) = 24\left(\frac{1}{8}(2x - 9)\right)$$

$$\left(24 \cdot \frac{5}{6}\right)(x + 12) = \left(24 \cdot \frac{1}{8}\right)(2x - 9)$$

$$20(x + 12) = 3(2x - 9)$$

$$20x + 12 - 6x = 6x - 27 - 6x$$

$$14x + 12 - 12 = -27 - 12$$

$$14x = -39$$

$$x = -\frac{39}{14}$$

79. Solve: $3x - 5 = 3(x - 5)$

Katie's work:

$$3x - 5 = 3(x - 5)$$

$$3x - 5 = 3x - 5$$

$$3x - 5 - 3x = 3x - 5 - 3x$$

$$-5 = -5$$

Since this is true, the solution is $\mathbb{R}$.

80. Solve: $-2(x - 1) + x - 3 = x - 1$

Mitchell's work:

$$-2(x - 1) + x - 3 = x - 1$$

$$-2x + 2 + x - 3 = x - 1$$

$$-x - 1 = x - 1$$

$$-x - 1 - x = x - 1 - x$$

$$-2x - 1 = -1$$

$$-2x - 1 + 1 = -1 + 1$$

$$-2x = 0$$

$$\frac{-2x}{-2} = \frac{0}{-2}$$

$$x = 0$$

Therefore, the solution is $\varnothing$.

 Calculate It!

Solve each equation. Then use a graphing calculator to check the solution.

81. $5x - 1 - (x + 3) = 4x - 4$ **82.** $3x - 4 = 2(x + 1) + x$

83. $\frac{2}{3}(x - 6) = x + 3$ **84.** $\frac{x}{2} + \frac{3}{2} = x - \frac{x}{2} + \frac{3}{2}$

85. $0.86x + 1.25(3.344) = 0.7(x + 11)$

86. $0.56x + 0.4(-31.3) = 0.25(x + 2)$

 Think About It!

87. Write an equation that has no solution.

88. Write an equation that has infinitely many solutions.

89. Write an equation that has 0 as a solution.

 PIECE IT TOGETHER / **SECTIONS 2.1–2.4**

Determine whether the following is an expression or an equation. (*Section 2.1, Objective 1*)

1. $2x - 16 = 0$ **2.** $3x + 19$

Decide if the equation is linear. If the equation is not linear, explain why. (*Section 2.2, Objective 1*)

3. $-x^2 + 4x - 12 = 0$ **4.** $\frac{3}{2}x - 4 + \frac{5}{6}x = 11$

Solve each equation. If the equation is a contradiction, write the solution as $\varnothing$. If the equation is an identity, write the solution as $\mathbb{R}$. (*Sections 2.2–2.4*)

5. $y - 5 = 5$

6. $8.6 + c = 1.4$

7. $2(x - 3) = 3x$

8. $9(a - 7) = 4(2a + 3)$

9. $\frac{x}{3} = 9$

10. $-3x + 7 = -14$

11. $-5.2x + 13.5 = -3.6x + 14.46$

12. $3(2z - 6) = 4(z + 5)$

13. $2(x - 3) = 3x$

14. $5(3a + 1) = 7(2a - 3)$

15. $\frac{2}{3}(x + 6) = \frac{1}{2}(x - 4)$

16. $\frac{x + 7}{5} - 2 = \frac{x - 2}{2}$

17. $0.6x + 4.2 = 0.3(x + 24)$

18. $x - 2 + 3x = 2(2x - 1)$

Write an equation that represents each situation. Solve the equation and explain the answer using a complete sentence. (*Section 2.3, Objectives 3 and 4*)

19. The sum of three times a number and 13 is the same as two less than the number. Find the number.

20. The sum of the two smaller of three consecutive odd integers is the same as 82 less than six times the largest integer. Find the integers.

Formulas and Applications from Geometry

▶ **OBJECTIVES**

As a result of completing this section, you will be able to

1. Evaluate formulas.
2. Use perimeter, area, and circumference formulas.
3. Solve formulas for a specified variable.
4. Solve problems involving complementary and supplementary angles.
5. Solve problems involving straight and vertical angles.
6. Solve problems involving triangles.
7. Troubleshoot common errors.

The Columbus Circle in New York City, built in 1905, was the first traffic circle in the United States. It was renovated a century later in 2005. The distance around the inner circle measures approximately 673 ft. The distance around the outer circle is approximately 1364 ft. Knowing these values, we can find the radius of each circle and also the area of each circle. The formulas that enable us to solve this problem are provided in this section. (Source: http://en.wikipedia.org/wiki/Columbus_Circle)

Formulas

A mathematical **formula** is an equation that expresses the relationship between two or more quantities. These quantities are represented by variables. Mathematical formulas enable us to find valuable information if we know some specific values of some of the variables in the formula. In Section 1.3, we evaluated algebraic expressions; some of these were mathematical formulas.

Some of the commonly used formulas are for calculating the area and perimeter of geometric shapes, converting degrees Fahrenheit to degrees Celsius, and calculating the amount of money in a savings account. Some of these formulas are listed within this section. Evaluating formulas is exactly like evaluating algebraic expressions.

Objective 1 ▶

Evaluate formulas.

> **Procedure: Evaluating Formulas**
>
> **Step 1:** Substitute known values in place of the appropriate variables.
> **Step 2:** Simplify the resulting expression or solve the resulting equation.

Objective 1 Examples   **Use the formula and the given values to solve for the missing variable.**

1a. The formula $A = P(1 + r)^t$ is used to calculate the amount of money in an account at the end of t years if P dollars are initially invested at an annual interest rate, r. Find the total value, A, if \$10,000 is invested for 5 yr at 4.5% annual interest.

Solution   **1a.**

$A = P(1 + r)^t$	State the formula.
$A = 10,000(1 + 0.045)^5$	Let $P = 10,000$, $r = 4.5\% = 0.045$, and $t = 5$.
$A = 10,000(1.045)^5$	Add the expression in parentheses.
$A = 10,000(1.246182)$	Evaluate the exponential expression.
$A = \$12,461.82$	Multiply.

In 5 yr, the total value in the account will be \$12,461.82.

1b. The formula $F = \dfrac{9}{5}C + 32$ is used to convert degrees Celsius (C) to degrees Fahrenheit (F). Find the value of C if $F = 63°$.

Solution   **1b.**

$F = \dfrac{9}{5}C + 32$	State the formula.
$63 = \dfrac{9}{5}C + 32$	Replace F with 63.
$5(63) = 5\left(\dfrac{9}{5}C + 32\right)$	Multiply each side by the LCD, 5.
$315 = 5\left(\dfrac{9}{5}C\right) + 5(32)$	Simplify and apply the distributive property.
$315 = 9C + 160$	Simplify.

$$315 - 160 = 9C + 160 - 160 \qquad \text{Subtract 160 from each side.}$$
$$155 = 9C \qquad \text{Simplify.}$$
$$\frac{155}{9} = \frac{9C}{9} \qquad \text{Divide each side by 9.}$$
$$17.2 = C \qquad \text{Simplify.}$$

So, 63°F is equivalent to 17.2°C.

1c. The formula $d = rt$ calculates the distance, d, traveled by an object going at an average rate r for time t. A cruise ship travels from Fort Lauderdale, Florida, to San Juan, Puerto Rico, in 45.5 hr at an average speed of 24 knots (about 27.6 mph). What is the distance the ship traveled?

Solution   **1c.**
$$d = rt$$
$$d = (27.6)(45.5) \qquad \text{Replace } r \text{ with 27.6 and } t \text{ with 45.5.}$$
$$d = 1255.8 \qquad \text{Multiply.}$$

The cruise ship traveled 1255.8 mi.

☑ **Student Check 1**   Use the formula and the given values to solve for the missing variable.

a. The formula $I = Prt$ is used to calculate simple interest for an investment of P dollars at an annual interest rate, r, for t years. Find the value of I if $P = \$1000$, $r = 5\%$, and $t = 3$.

b. The formula $C = \dfrac{5}{9}(F - 32)$ is used to convert degrees Fahrenheit (F) to degrees Celsius (C). Find the value of F if $C = 50°$.

c. The world's fastest passenger train, the Harmony Express in China, opened in December 2009. It travels at an average speed of 217 mph from Wuhan, China, to Guangzhou, China. It takes 3 hr to complete the trip. Use the formula $d = rt$ to calculate the distance between the cities of Wuhan and Guangzhou. (Source: http://www.telegraph.co.uk/news/worldnews/asia/china/7230137/China-steams-ahead-with-worlds-fastest-train.html)

Perimeter, Area, and Circumference

Objective 2 ▶

Use perimeter, area, and circumference formulas.

As presented in Chapter 1, the **perimeter** of a polygon is the distance around the outside of the figure. In other words, it is the sum of the lengths of the sides of the polygon. The distance around a circle is called the **circumference** of the circle.

The **area** of a figure is the number of square units it takes to cover the inside of the figure. (A *square unit* is a square whose length and width are both 1 unit long.)

The formulas for some common shapes are provided.

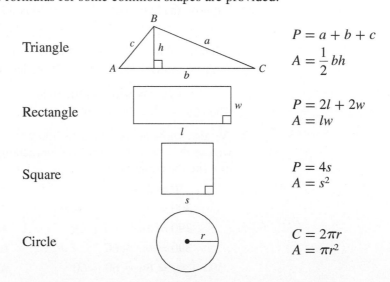

Triangle
$$P = a + b + c$$
$$A = \frac{1}{2}bh$$

Rectangle
$$P = 2l + 2w$$
$$A = lw$$

Square
$$P = 4s$$
$$A = s^2$$

Circle
$$C = 2\pi r$$
$$A = \pi r^2$$

Objective 2 Examples   Use the perimeter, area, or circumference formulas to solve each problem.

2a. Assume that the perimeter of the *Mona Lisa* by Leonardo da Vinci is 260 cm. If the length of the painting is 77 cm, find its width. Then find the area of the painting.

Solution   2a.

$P = 2l + 2w$	State the perimeter formula.
$260 = 2(77) + 2w$	Replace P with 260 and l with 77.
$260 = 154 + 2w$	Simplify.
$260 - 154 = 154 + 2w - 154$	Subtract 154 from each side.
$106 = 2w$	Simplify.
$\dfrac{106}{2} = \dfrac{2w}{2}$	Divide each side by 2.
$53 = w$	Simplify.

The width of the painting is 53 cm. Since we know both the length and width of the painting, we can calculate the area using the formula $A = lw$.

$A = lw$	State the area formula.
$A = (77)(53)$	Replace l with 77 and w with 53.
$A = 4081$	Multiply.

So, the area of the painting is 4081 cm².

2b. What is the height of a triangle whose area is 16 ft² and whose base is 8 ft?

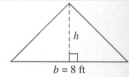

Solution   2b.

$A = \dfrac{1}{2}bh$	State the area formula.
$16 = \dfrac{1}{2}(8)h$	Replace A with 16 and b with 8.
$16 = 4h$	Simplify.
$\dfrac{16}{4} = \dfrac{4h}{4}$	Divide each side by 4.
$4 = h$	Simplify.

So, the height of the triangle is 4 ft.

2c. Suppose the length of a soccer field is 30 yd less than two times its width. If its perimeter is 390 yd, find the dimensions of the soccer field.

Solution   2c. What is unknown? The length and width of the soccer field are unknown.

What is known? The length is 30 yd less than two times its width.

Let $w =$ width of the field.
Then $2w - 30 =$ length of the field.

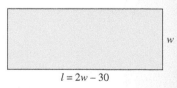

We also know the perimeter is 390 yd. So, we use the perimeter formula of a rectangle to write the equation.

$P = 2l + 2w$	State the perimeter formula.
$390 = 2(2w - 30) + 2w$	Replace l with $2w - 30$.
$390 = 4w - 60 + 2w$	Apply the distributive property.
$390 = 6w - 60$	Combine like terms.
$390 + 60 = 6w - 60 + 60$	Add 60 to each side.

$$450 = 6w \qquad \text{Simplify.}$$

$$\frac{450}{6} = \frac{6w}{6} \qquad \text{Divide each side by 6.}$$

$$75 = w \qquad \text{Simplify.}$$

So, the width of the soccer field is 75 yd. The length is
$2(75) - 30 = 150 - 30 = 120$ yd.

2d. The circumference of the inner circle of the Columbus Circle in New York City measures approximately 673 ft, and the circumference of the outer circle is approximately 1364 ft. Find the radius of the inner and outer circles. Find the area of the inner circle. Round all values to the nearest hundredth.

Solution **2d.** Since we know the circumference of each of the inner and outer circles, we use the circumference formula to find the radii.

Circumference of inner circle   Circumference of outer circle

$$C = 2\pi r \qquad \text{State the circumference formula.} \qquad C = 2\pi r$$

$$673 = 2\pi r \qquad \text{Replace the value of } C \qquad 1364 = 2\pi r$$

$$\frac{673}{2\pi} = \frac{2\pi r}{2\pi} \qquad \begin{array}{l}\text{with the given circumference.} \\ \text{Divide each side by } 2\pi.\end{array} \qquad \frac{1364}{2\pi} = \frac{2\pi r}{2\pi}$$

$$\frac{673}{2\pi} = r \qquad\qquad\qquad \frac{1364}{2\pi} = r$$

$$107.11 \approx r \qquad \text{Simplify.} \qquad 217.09 \approx r$$

The radius of the inner circle   The radius of the outer circle
is approximately 107.11 ft.   is approximately 217.09 ft.

Now that we know the radius of the inner circle, we can find its area.

$$A = \pi r^2 \qquad \text{State the area formula.}$$

$$A = \pi (107.11)^2 \qquad \text{Replace } r \text{ with 107.11.}$$

$$A = \pi (11{,}472.5521) \qquad \text{Simplify } (107.11)^2.$$

$$A = 36{,}042.09 \qquad \text{Multiply by } \pi.$$

So, the area of the inner circle is approximately 36,042.09 ft².

✓ **Student Check 2** Use the perimeter, area, or circumference formulas to solve each problem.

a. The largest painting at the Louvre Museum is the *Wedding at Cana* by Veronese. Its perimeter is 1304 in. and its width is 262 in. Find its length and then the area of the painting.

b. Find the length of the base of a triangle whose area is 10 ft² and whose height is 4 ft.

c. The length of the White House is 3 ft less than twice its width. If the perimeter of the White House is 507 ft, find its length and width.

d. Find the area and circumference of the clock face on top of Big Ben in London if the diameter of the clock face is 23 ft.

Rewriting Formulas

Objective 3 ▶

Solve formulas for a specified variable.

The Fahrenheit-Celsius formula illustrated in Example 1b can be used to find Fahrenheit or Celsius values. When the Celsius temperature was given, we evaluated the formula to determine the corresponding Fahrenheit value because the formula was already solved for F. But when the Fahrenheit temperature was given, we had to solve the equation for C to find the equivalent Celsius value. An alternate way of dealing with this is to first rewrite the Fahrenheit formula so that the variable C is isolated on one side of the equation. Then we can substitute the known Fahrenheit value.

When working with formulas and algebraic equations containing more than one variable, it is sometimes necessary for us to rewrite the formula or equation so that a different variable is isolated.

Procedure: Solving a Formula for a Specified Variable

Step 1: Multiply both sides of the equation by the LCD to clear fractions from the equation, if necessary.

Step 2: Apply the distributive property to clear parentheses as needed.

Step 3: Apply the addition property of equality to move all terms containing the specified variable to one side of the equation and all other terms to the other side of the equation.

Step 4: Apply the multiplication property of equality to obtain a coefficient of 1 on the specified variable.

Objective 3 Examples   Solve each formula for the specified variable.

3a. The formula $A = lw$ is used to find the area of a rectangle. Solve for w.

3b. The formula $P = 2l + 2w$ gives the perimeter of a rectangle. Solve for l.

3c. The formula $F = \dfrac{9}{5}C + 32$ converts degrees Celsius to degrees Fahrenheit. Solve for C.

3d. Solve $3x - y = 6$ for y.

3e. Solve $2x + 3y = 12$ for y.

Solutions   **3a.**

$$A = lw$$ 　　　　　Highlight the variable to isolate.

$$\frac{A}{l} = \frac{lw}{l}$$ 　　　　　Divide each side by l.

$$\frac{A}{l} = w$$ 　　　　　Simplify.

So, $w = \dfrac{A}{l}$.

3b.

$$P = 2l + 2w$$ 　　　　　Highlight the variable to isolate.

$$P - 2w = 2l + 2w - 2w$$ 　　　　　Subtract $2w$ from each side.

$$P - 2w = 2l$$ 　　　　　Simplify.

$$\frac{P - 2w}{2} = \frac{2l}{2}$$ 　　　　　Divide each side by 2.

$$\frac{P - 2w}{2} = l$$ 　　　　　Simplify.

So, $l = \dfrac{P - 2w}{2}$.

 Note: $\dfrac{P - 2w}{2} = \dfrac{P}{2} - \dfrac{2w}{2} = \dfrac{P}{2} - w.$ It is not equal to $P - w$.

3c.
$$F = \frac{9}{5}C + 32 \qquad \text{Highlight the variable to isolate.}$$

$$5(F) = 5\left(\frac{9}{5}C + 32\right) \qquad \text{Multiply each side by the LCD, 5.}$$

$$5F = 5\left(\frac{9}{5}C\right) + 5(32) \qquad \text{Apply the distributive property.}$$

$$5F = 9C + 160 \qquad \text{Simplify.}$$

$$5F - 160 = 9C + 160 - 160 \qquad \text{Subtract 160 from each side.}$$

$$5F - 160 = 9C \qquad \text{Simplify.}$$

$$\frac{5F - 160}{9} = \frac{9C}{9} \qquad \text{Divide each side by 9.}$$

$$\frac{5F - 160}{9} = C \qquad \text{Simplify.}$$

So, $C = \dfrac{5F - 160}{9}$.

3d.
$$3x - y = 6 \qquad \text{Highlight the variable to isolate.}$$

$$3x - y - 3x = 6 - 3x \qquad \text{Subtract 3x from each side.}$$

$$-y = -3x + 6 \qquad \text{Note the term containing } y \text{ has a coefficient}$$

$$-1(-y) = -1(-3x + 6) \qquad \text{of } -1, \text{ so multiply each side by } -1.$$

$$y = 3x - 6 \qquad \text{Apply the distributive property.}$$

So, $y = 3x - 6$.

3e.
$$2x + 3y = 12 \qquad \text{Highlight the variable to isolate.}$$

$$2x + 3y - 2x = 12 - 2x \qquad \text{Subtract 2x from each side.}$$

$$3y = -2x + 12 \qquad \text{Simplify.}$$

$$\frac{3y}{3} = \frac{-2x + 12}{3} \qquad \text{Divide each side by 3.}$$

$$y = \frac{-2x}{3} + \frac{12}{3} \qquad \text{Divide each term by 3.}$$

$$y = -\frac{2}{3}x + 4 \qquad \text{Simplify.}$$

So, $y = -\dfrac{2}{3}x + 4$.

 Note: *Parts (d) and (e) will be presented more fully in Chapter 3 when we discuss linear equations in two variables.*

☑ **Student Check 3**   Solve each formula for the specified variable.

    a. The formula $d = rt$ calculates distance for a given rate and time. Solve for t.

    b. The formula $V = \dfrac{1}{3}bh$ calculates the volume of a regular pyramid, given the length of its base and height. Solve for h.

    c. The formula $A = \dfrac{h}{2}(b_1 + b_2)$ calculates the area of a trapezoid. Solve for b_1.

    d. Solve $4x - y = 8$ for y.

    e. Solve $4x + 5y = -20$ for y.

Complementary and Supplementary Angles

Recall from Chapter 1 that **complementary angles** are angles whose sum is 90° and **supplementary angles** are angles whose sum is 180°.

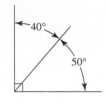

A 40° angle and 50° degree angle are complementary because their sum is 90°. The two angles together form a *right angle*. A **right angle** is an angle whose measure is 90°.

A 40° angle and 140° degree angle are supplementary because their sum is 180°. The two angles together form a *straight angle*. A **straight angle** is an angle whose measure is 180°.

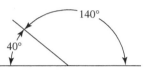

Some other examples of complementary and supplementary angles are listed in the table.

Complementary Angles	Supplementary Angles
$20°, 90° - 20° = 70°$	$20°, 180° - 20° = 160°$
$34°, 90° - 34° = 56°$	$34°, 180° - 34° = 146°$
$45°, 90° - 45° = 45°$	$45°, 180° - 45° = 135°$
$73°, 90° - 73° = 17°$	$73°, 180° - 73° = 107°$
$x°, (90 - x)°$	$y°, (180 - y)°$

In Section 2.1, we practiced setting up equations given information about complementary and supplementary angles. In this section, we will go through the entire process: setting up the equation and solving it.

Objective 4 Examples   **Find the measure of each unknown angle. Write an equation that represents the situation and solve it.**

4a. Find the measure of an angle whose complement is 15° less than twice the measure of the angle.

4b. Find the measure of an angle whose supplement is 40° more than the measure of the angle.

4c. The supplement of an angle is 10° more than twice its complement. Find the measure of the angle.

Solutions   **4a.** What is unknown? The measure of the angle and its complement are unknown.

$$\text{Let } a = \text{the measure of the angle.}$$
$$\text{Then } 90 - a = \text{measure of the complement.}$$

What is known? The complement is 15° less than twice the measure of the angle. We use this statement to write the equation that we will use to solve the problem.

Complement   is   15° less than twice the measure of the angle.

$90 - a = 2a - 15$	Express the relationship.
$90 - a + a = 2a - 15 + a$	Add a to each side.
$90 = 3a - 15$	Simplify.
$90 + 15 = 3a - 15 + 15$	Add 15 to each side.
$105 = 3a$	Simplify.
$\dfrac{105}{3} = \dfrac{3a}{3}$	Divide each side by 3.
$35 = a$	Simplify.

So, the measure of the angle is 35°.

4b. What is unknown? The measure of the angle and its supplement are unknown.

$$\text{Let } a = \text{the measure of the angle.}$$

$$\text{Then } 180 - a = \text{the measure of the supplement.}$$

What is known? The supplement is 40° more than the measure of the angle. We use this statement to write the equation that we will use to solve the problem.

$$\underset{\text{Supplement}}{} \text{ is } \underset{\substack{40° \text{ more than the} \\ \text{measure of the angle.}}}{}$$

$180 - a = a + 40$	Express the relationship.
$180 - a + a = a + 40 + a$	Add a to each side.
$180 = 2a + 40$	Simplify.
$180 - 40 = 2a + 40 - 40$	Subtract 40 from each side.
$140 = 2a$	Simplify.
$\dfrac{140}{2} = \dfrac{2a}{2}$	Divide each side by 2.
$70 = a$	Simplify.

So, the measure of the angle is 70°.

4c. What is unknown? The measure of the angle, its complement and supplement are unknown.

$$\text{Let } a = \text{the measure of the angle.}$$

$$\text{Then } 90 - a = \text{the measure of the complement}$$

$$\text{and } 180 - a = \text{the measure of the supplement.}$$

What is known? The supplement of an angle is 10° more than twice its complement. We use this relationship to write the equation to solve the problem.

$$\underset{\text{Supplement}}{} \text{ is } \underset{\substack{10° \text{ more than twice} \\ \text{its complement.}}}{}$$

$180 - a = 2(90 - a) + 10$	Express the relationship.
$180 - a = 180 - 2a + 10$	Apply the distributive property.
$180 - a = 190 - 2a$	Combine like terms on the right.
$180 - a + 2a = 190 - 2a + 2a$	Add $2a$ to each side.
$180 + a = 190$	Simplify.
$180 + a - 180 = 190 - 180$	Subtract 180 from each side.
$a = 10$	Simplify.

So, the measure of the angle is 10°.

✓**Student Check 4**   Find the measure of each unknown angle. Write an equation that represents the situation and solve it.

    a. Find the measure of an angle whose complement is 10° more than three times the angle.

    b. Find the measure of an angle whose supplement is 60° more than the angle.

    c. The supplement of an angle is 30° more than twice its complement. Find the measure of the angle.

Straight and Vertical Angles

Objective 5 ▶

Solve problems involving straight and vertical angles.

As mentioned in Objective 4, straight angles are angles whose measure is 180°. Straight angles form a straight line.

When two lines intersect one another, four angles are created. Angles opposite from one another are **vertical** or **opposite angles**. These angles have equal measure.

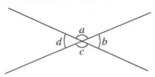

In this figure, angles a and c are vertical angles and angles b and d are vertical angles.

$$a = c \qquad \text{and} \qquad b = d$$

Note:

Angles a and b form a straight angle. So, $a + b = 180°$.
Angles b and c form a straight angle. So, $b + c = 180°$.
Angles c and d form a straight angle. So, $c + d = 180°$.
Angles d and a form a straight angle. So, $d + a = 180°$.

Objective 5 Examples   **Find the measure of each unknown angle.**

5a.

5b.

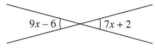

Solutions   **5a.** The two angles form a straight angle and, therefore, have a sum of 180°.

$5a + a = 180$	Express the relationship.
$6a = 180$	Combine like terms.
$\dfrac{6a}{6} = \dfrac{180}{6}$	Divide each side by 6.
$a = 30$	Simplify.

Since a is 30°, the other angle is 5(30) or 150°.

5b. The two angles are vertical angles and, therefore, have equal measure.

$9x - 6 = 7x + 2$	Express the relationship.
$9x - 6 - 7x = 7x + 2 - 7x$	Subtract 7x from each side.
$2x - 6 = 2$	Simplify.
$2x - 6 + 6 = 2 + 6$	Add 6 to each side.
$2x = 8$	Simplify.
$\dfrac{2x}{2} = \dfrac{8}{2}$	Divide each side by 2.
$x = 4$	Simplify.

Since $x = 4$, the measures of the angles are $9(4) - 6 = 36 - 6 = 30°$ and $7(4) + 2 = 28 + 2 = 30°$.

✓ Student Check 5   Find the measure of each unknown angle.

a.

b.

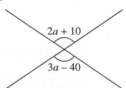

Triangles

A **triangle** is one of the basic geometric shapes. It has three sides and the sum of its angles is 180°.

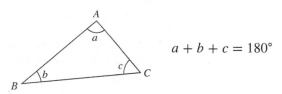

$$a + b + c = 180°$$

We will use this relationship to find the measures of the angles in a triangle.

Objective 6 Examples **Find the measure of each angle in the triangle.**

6a.

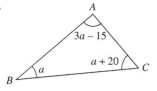

Solution **6a.** $a + a + 20 + 3a - 15 = 180$ Express the relationship.

$$5a + 5 = 180$$ Combine like terms.

$$5a + 5 - 5 = 180 - 5$$ Subtract 5 from each side.

$$5a = 175$$ Simplify.

$$\frac{5a}{5} = \frac{175}{5}$$ Divide each side by 5.

$$a = 35$$ Simplify.

Since $a = 35$, the measures of the angles are 35°, 35° + 20° = 55°, and 3(35°) − 15° = 90°. Note that the sum of these measures is 35° + 55° + 90° = 180°.

6b. In a triangle ABC, angles B and C have the same measure. The measure of angle A is four times the measure of either of the other angles. Find the measure of each angle in the triangle.

Solution **6b.** Draw a diagram to represent the three angles of the triangle. Let a be the measure of angle B. Since angle B and angle C have the same measure, a is also the measure of angle C. Then the measure of angle A is $4a$.

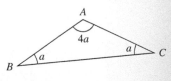

The sum of the measures of the angles in a triangle is 180°, so we have the following equation.

$$a + a + 4a = 180$$ Express the relationship.

$$6a = 180$$ Combine like terms.

$$\frac{6a}{6} = \frac{180}{6}$$ Divide each side by 6.

$$a = 30$$ Simplify.

The measure of angles B and C is 30°. The measure of angle A is 4(30) = 120°.

✓ Student Check 6 Find the measure of each angle in the triangle.

a.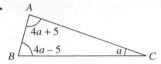

b. In a triangle ABC, the measure of angle A is twice the measure of angle C. The measure of angle B is 20° less than the measure of angle C. Find the measure of each angle.

Objective 7 ▶

Troubleshoot common errors.

Troubleshooting Common Errors

Some common errors associated with formulas and geometry applications are illustrated next.

Objective 7 Examples A problem and an incorrect solution are given. Provide the correct solution and an explanation of the error.

7a. Solve $d = rt$ for t.

Incorrect Solution	Correct Solution and Explanation
$d = rt$ $d - r = t$	The variable r is connected to t by multiplication. Therefore, we must divide both sides by r to isolate t. $d = rt$ $\dfrac{d}{r} = \dfrac{rt}{r}$ $\dfrac{d}{r} = t$

7b. The complement of an angle is six degrees less than seven times the measure of the angle. Find the measure of the angle.

Incorrect Solution	Correct Solution and Explanation
Let a be the measure of the angle. Then $a - 90$ is the measure of the complement. $a - 90 = 7a - 6$ $-84 = 6a$ $\dfrac{-84}{6} = \dfrac{6a}{6}$ $-14 = a$ The angle is 14°.	The error was made in representing the complement of the angle. If a is the measure of the angle, then $90 - a$ is the measure of the complement. Also, it is unreasonable to have a negative value for the measure of an angle. $90 - a = 7a - 6$ $96 = 8a$ $\dfrac{96}{8} = \dfrac{8a}{8}$ $12 = a$ The angle is 12°.

ANSWERS TO STUDENT CHECKS

Student Check 1 **a.** $150 **b.** $F = 122°$ **c.** 651 mi

Student Check 2 **a.** The length of the painting is 390 in. Its area is 102,180 in.². **b.** The base of the triangle is 5 ft.

c. The dimensions of the White House are 168 ft by 85.5 ft. **d.** The circumference of the clock is 72.3 ft. The area is 415.5 ft².

Student Check 3   **a.** $t = \dfrac{d}{r}$ **b.** $h = \dfrac{3V}{b}$ **c.** $b_1 = \dfrac{2A - b_2 h}{h}$

     d. $y = 4x - 8$ **e.** $y = -\dfrac{4}{5}x - 4$

Student Check 4   **a.** The angle measures 20°.   **b.** The angle measures 60°.   **c.** The angle measures 30°.

Student Check 5   **a.** The measures of the angles are 56° and 124°.   **b.** The angle measure is 110°.

Student Check 6   **a.** The angles in the triangle have measures of 85°, 75°, and 20°.   **b.** The angles in the triangle have measures of 100°, 30°, and 50°.

SUMMARY OF KEY CONCEPTS

1. Evaluating formulas is similar to evaluating algebraic expressions. Substitute the given values for the appropriate variables and simplify. Use the order of operations to simplify the resulting expression. The goal in this objective is to find the value of the missing variable. To find this value, we will either simplify an expression or solve an equation.

2. Perimeter, area, and circumference formulas should be memorized for the basic shapes (triangle, square, rectangle, and circle). If the dimensions of the figure are given, we can find the perimeter, area, or circumference by substituting the given dimensions. We can also find the dimensions of the figure if the perimeter, area, or circumference are provided as well.

3. To solve a formula for a specified variable is to rewrite the equation so that a different variable is isolated to one side. No substitutions are made in this process. We simply apply the addition and multiplication properties of equality to solve for a different variable.

4. Complementary angles and supplementary angles are two angles whose sum is 90° and 180°, respectively. The most common error is misrepresenting the unknown form of these angles. If a is the measure of the angle, then $90 - a$ is the measure of the complement and $180 - a$ is the measure of the supplement.

5. Straight angles measure 180°. Vertical angles are formed when two lines intersect. Vertical angles are equal in measure.

6. The sum of the measures of the angles in a triangle is 180°.

GRAPHING CALCULATOR SKILLS

We can use the calculator to approximate expressions involving π.

1. $\dfrac{673}{2\pi}$

So, $\dfrac{673}{2\pi}$ is approximately equal to 107.11.

2. $\pi(107.1)^2$

So, $\pi(107.1)^2$ is approximately equal to 36,035.36.

SECTION 2.5   EXERCISE SET

Write About It!

Use complete sentences to explain the meaning of each term.

1. Formula
2. Perimeter
3. Area
4. Circumference
5. Complementary angles
6. Supplementary angles
7. Straight angle
8. Vertical angles

Determine if each statement is true or false. If a statement is false, explain why.

9. When $8x - y = 16$ is solved for y, we get $y = -8x$.

10. If the measure of an angle is a, then the measure of the angle's supplement is $a - 180$.

11. If the radius of a circle is 10 ft, the area of the circle is 100π ft^2.

12. If angles with measures $(4a - 5)°$ and $(2a + 20)°$ are vertical angles, then $4a - 5 + 2a + 20 = 180$.

 **Practice Makes Perfect!**

The formula $A = P(1 + r)^t$ is used to calculate the amount of money in an account at the end of t years if P dollars are invested at an annual interest rate, r. (*See Objective 1.*)

13. Find the value of A if $8000 is invested for 4 yr at 2.5% annual interest.

14. Find the value of A if $6000 is invested for 3 yr at 1.3% annual interest.

15. Find the value of A if $12,000 is invested for 2 yr at 2% annual interest.

16. Find the value of A if $15,000 is invested for 2 yr at 1.8% annual interest.

The formula $d = rt$ calculates the distance, d, traveled by an object traveling at an average rate r for time t. (*See Objective 1.*)

17. Find d if $r = 60$ mph and $t = 5$ hr.

18. Find d if $r = 75$ mph and $t = 2$ hr.

19. Find r if $d = 440$ mi and $t = 8$ hr.

20. Find r if $d = 243$ mi and $t = 4.5$ hr.

21. Find t if $d = 80$ mi and $r = 40$ mph.

22. Find t if $d = 156$ mi and $r = 65$ mph.

The formula $V = lwh$ calculates the volume of a box with length l, width w, and height h. (*See Objective 1.*)

23. Find V if $l = 10$ in., $w = 5$ in., and $h = 2$ in.

24. Find V if $l = 3$ ft, $w = 4$ ft, and $h = 6$ ft.

25. Find w if $V = 28$ yd³, $l = 7$ yd, and $h = 2$ yd.

26. Find h if $V = 960$ m³, $l = 12$ m, and $w = 10$ m.

Use the perimeter, area, or circumference formula to find the indicated quantity. (*See Objective 2.*)

27. Find the area of a triangle whose base is 10 in. and whose height is 5 in.

28. Find the area of a triangle whose base is 9 ft and whose height is 14 ft.

29. Find the height of a triangle whose area is 20 in.² and whose base is 5 in.

30. Find the base of a triangle whose area is 35 cm² and whose height is 7 cm.

31. Find the perimeter of a rectangle whose length is 6 ft and whose width is 8 ft.

32. Find the perimeter of a rectangle whose length is 12 in. and whose width is 20 in.

33. Find the length of a rectangle if the perimeter is 24 m and the width is 7 m.

34. Find the width of a rectangle if the perimeter is 180 ft and the length is 40 ft.

35. If the area of a rectangle is 32 ft² and the width is 8 ft, find the perimeter of the rectangle.

36. If the area of a rectangle is 50 in.² and the length is 10 in., find the perimeter of the rectangle.

37. Find the circumference and area of a circle with radius 8 cm. Use 3.14 for the value of π to approximate answers to two decimal places.

38. Find the circumference and area of a circle with radius 12 in. Use 3.14 for the value of π to approximate answers to two decimal places.

39. If the circumference of a circle is 157 ft, find the radius of the circle and the area of the circle. Use 3.14 for the value of π to approximate answers to two decimal places.

40. If the circumference of a circle is 200 in., find the radius of the circle and the area of the circle. Use 3.14 for the value of π to approximate answers to two decimal places.

Solve each formula for the specified variable. (*See Objective 3.*)

41. $A = lw$ for l

42. $d = rt$ for t

43. $P = a + b + c$ for c

44. $P = 2l + 2w$ for w

45. $x + 2y = 4$ for x

46. $x + 2y = 4$ for y

47. $4x - y = 8$ for x

48. $4x - y = 8$ for y

49. $y = \frac{1}{4}x + 5$ for x

50. $y = \frac{1}{2}x - 3$ for x

51. $7x - 3y = -21$ for x

52. $3x - 5y = -15$ for y

53. $A = P + Prt$ for r

54. $A = P(1 + rt)$ for P

Find the measure of each unknown angle. (*See Objective 4.*)

55. Find the measure of an angle whose complement is twice the measure of the angle.

56. Find the measure of an angle whose complement is eight times the measure of the angle.

57. Find the measure of an angle whose complement is 10° more than three times the measure of the angle.

58. Find the measure of an angle whose complement is 15° more than four times the measure of the angle.

59. Find the measure of an angle whose supplement is eight times the measure of the angle.

60. Find the measure of an angle whose supplement is five times the measure of the angle.

61. Find the measure of an angle whose supplement is 16° less than six times the measure of the angle.

62. Find the measure of an angle whose supplement is 12° less than twice the measure of the angle.

63. The supplement of an angle is 20° less than three times its complement. Find the measure of the angle.

64. The supplement of an angle is 18° less than four times its complement. Find the measure of the angle.

Find the measure of each angle labeled in the figure. (*See Objective 5.*)

65.

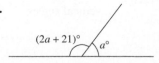

66.

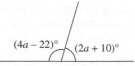

$(4a - 22)°$ $(2a + 10)°$

67.

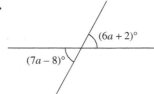

$(6a + 2)°$

$(7a - 8)°$

68.

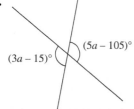

$(5a - 105)°$

$(3a - 15)°$

Find the measure of each angle in the triangle that is illustrated or described. (*See Objective 6.*)

69.

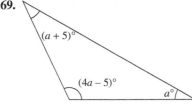

$(a + 5)°$

$(4a - 5)°$

$a°$

70.

$(6a - 1)°$

$(5a + 1)°$

$a°$

71. In triangle *ABC*, the measure of angle *B* is 8° more than the measure of angle *A*. The measure of angle *C* is 37° more than the measure of angle *A*. Find the measure of each angle in the triangle.

72. In triangle *ABC*, the measure of angles *A* and *B* are the same. The measure of angle *C* is 105° larger than the each of angle *A* and *B*. Find the measure of each angle in the triangle.

**Mix 'Em Up!**

Find the requested information.

73. Use the formula $A = P(1 + r)^t$ to find the value of *A* if $5000 is invested for 2 yr at 4% annual interest.

74. Use the formula $A = P(1 + r)^t$ to find the value of *A* if $13,500 is invested for 5 yr at 5.5% annual interest.

75. If the area of a rectangle is 1290 m² and the width is 43 m, find the perimeter of the rectangle.

76. If the area of a rectangle is 240 mm² and the length is 40 mm, find the perimeter of the rectangle.

77. The revenue *R*, unit price *p*, and sale level *x* of a product are related by the formula $R = xp$. Use this formula to find *R* if $p = \$26$ and $x = 120$.

78. The revenue *R*, unit price *p*, and sale level *x* of a product are related by the formula $R = xp$. Use this formula to find *R* if $p = \$15$ and $x = 65$.

79. The formula to find the volume of a cylinder with radius *r* and height *h* is $V = \pi r^2 h$. Use $\pi \approx 3.14$ to find *V* if $r = 3$ in. and $h = 2$ in.

80. The formula to find the volume of a cylinder with radius *r* and height *h* is $V = \pi r^2 h$. Use $\pi \approx 3.14$ to find *V* if $r = 2$ ft and $h = 7$ ft.

81. Use the formula $d = rt$ to find *d* if $r = 45$ mph and $t = 3.2$ hr.

82. Use the formula $d = rt$ to find *r* if $d = 244.8$ mi and $t = 5.1$ hr.

83. Solve $y = -\dfrac{3}{2}x + 3$ for *x*.

84. Solve $y = \dfrac{7}{2}x - 21$ for *x*.

85. The formula to find the volume of a cylinder with radius *r* and height *h* is $V = \pi r^2 h$. Use $\pi \approx 3.14$ to find *h* if $V = 1384.74$ m³ and $r = 4.2$ m.

86. The formula to find the volume of a cylinder with radius *r* and height *h* is $V = \pi r^2 h$. Use $\pi \approx 3.14$ to find *h* if $V = 615.44$ yd³ and $r = 3.5$ yd.

87. Use the formula $A = P(1 + r)^t$ to find the value of *P* if $23,500 is in an account at the end of 8 yr at 5% annual interest.

88. Use the formula $A = P(1 + r)^t$ to find the value of *P* if $16,500 is in an account at the end of 4 yr at 5.9% annual interest.

89. Solve $R = xp$ for *x*.

90. Solve $V = IR$ for *I*.

91. Solve $0.4x - y = 100$ for *x*.

92. Solve $x + 0.2y = 12$ for *y*.

93. Find the measure of an angle whose complement is 12° less than the measure of the angle.

94. Find the measure of an angle whose supplement is 40° more than three times the measure of the angle.

95. Solve $C = 6l + 4w$ for *w*.

96. Solve $P = 2l + 2w$ for *l*.

97. Find the measure of each angle labeled in the figure.

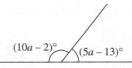

$(10a - 2)°$ $(5a - 13)°$

98. Find the measure of each angle labeled in the figure.

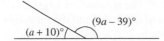

$(9a - 39)°$

$(a + 10)°$

99. In triangle ABC, the measure of angles A and B are the same. The measure of angle C is $27°$ larger than the measure of angle A. Find the measure of each angle in the triangle.

100. In triangle ABC, the measure of angles A and B are the same. The measure of angle C is $6°$ smaller than the measure of angle A. Find the measure of each angle in the triangle.

101. Find the height of a triangle whose area is 537.5 m^2 and whose base is 25 m.

102. Find the height of a triangle whose area is 693 cm^2 and whose base is 33 cm.

103. Solve $0.02x + 0.05y = 80$ for y.

104. Solve $0.05x + 0.01y = -50$ for x.

105. Find the radius and area of a circle with circumference 119.32 cm. Use 3.14 for the value of π to approximate answers to two decimal places.

106. Find the radius and area of a circle with circumference 175.84 in. Use 3.14 for the value of π to approximate answers to two decimal places.

107. Find the measure of each angle labeled in the figure.

108. Find the measure of each angle labeled in the figure.

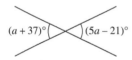

109. Solve $5x - y = -10$ for x.

110. Solve $3x + 2y = 12$ for x.

111. The supplement of an angle is $24°$ more than four times its complement. Find the measure of the angle.

112. The supplement of an angle is $19°$ less than five times its complement. Find the measure of the angle.

113. Find the measure of each angle labeled in the figure.

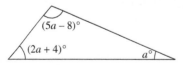

114. Find the measure of each angle labeled in the figure.

115. In triangle ABC, the measure of angle B is $9°$ less than the measure of angle A. The measure of angle C is $27°$ less than the measure of angle A. Find the measure of each angle in the triangle.

116. In triangle ABC, the measure of angle B is $16°$ more than the measure of angle A. The measure of angle C is $2°$ more than the measure of angle A. Find the measure of each angle in the triangle.

117. The circumference of the inner circle of a washer measures 50.3 in., and the circumference of the outer circle measures 370.7 in. Find the radius of the inner and outer circles. Find the area of the inner circle. Use 3.14 for the value of π. Round the values to two decimal places.

118. The circumference of the inner circle of a washer measures 88.0 mm, and the circumference of the outer circle measures 157.1 mm. Find the radius of the inner and outer circles. Find the area of the inner circle. Use 3.14 for the value of π. Round the values to two decimal places.

 **You Be the Teacher!**

Correct each student's errors, if any.

119. Find the measure of each angle labeled in the figure.

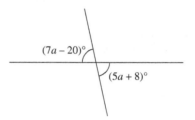

Emma's work:

$$(7a - 20) + (5a + 8) = 180$$
$$12a - 12 = 180$$
$$12a = 192$$
$$a = 16$$
$$7a - 20 = 7(16) - 20 = 92$$
$$5a + 8 = 5(16) + 8 = 88$$

120. Solve: $F = \dfrac{9}{5}C + 32$ for C.

Cynthia's work:

$$F = \frac{9}{5}C + 32$$
$$5 \cdot F = 5 \cdot \frac{9}{5}C + 32$$
$$5F = 9C + 32$$
$$5F - 32 = 9C$$
$$C = \frac{5F - 32}{9}$$

121. Solve: $A = 2\pi r^2 + 2\pi rh$ for h.

Kimberly's work:

$$A = 2\pi r^2 + 2\pi rh$$
$$A - 2\pi r^2 = 2\pi rh$$
$$\frac{A - 2\pi r^2}{2\pi rh} = \frac{2\pi rh}{2\pi rh}$$
$$\frac{A - 2\pi r^2}{2\pi rh} = h$$

122. The supplement of an angle is 14° less than six times its complement. Calculate the measure of the angle.

Francis' work:

$$180 - x = 14 - 6(x - 90)$$
$$180 - x = 14 - 6x + 540$$
$$5x = 374$$
$$x = 74.8$$

Calculate It!

Solve each equation. Then use a graphing calculator to check the answer.

123. $0.03(x + 5.4) = 0.02(x - 1.8) + x$

124. $180 - x = 5(90 - x) - 24.8$

125. $5x - 32.6 + 3(x + 4.2) = 180$

126. The formula to find the volume of a cylinder with radius r and height h is $V = \pi r^2 h$. Use $\pi = 3.14$ to find V if $r = 15.6$ ft and $h = 6.5$ ft.

| SECTION 2.6 | **Percent, Rate, and Mixture Problems** |

▶ OBJECTIVES

As a result of completing this section, you will be able to

1. Solve percent applications.
2. Solve simple interest applications.
3. Solve mixture applications.
4. Solve distance, rate, and time applications.
5. Troubleshoot common errors.

According to Oprah's Debt Diet, how a family should spend their monthly income is based on this pie chart. Because the pie chart illustrates how a family should allocate all of their monthly income, the percents add to 100%. If a family following this "diet" spends $1400 on housing each month, what is their monthly income? (Source: www.oprah.com)

We will learn how to solve this problem using a linear equation.

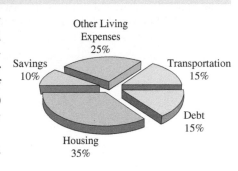

Percent Applications

Objective 1 ▶

Solve percent applications.

Many real-life situations involve percents—paying taxes, sale discounts, investments, commissions, and so on. In this section, we will explore applications of linear equations involving percents. Recall that percent means per hundred. For example, $5\% = \dfrac{5}{100} = 0.05$. Percents can be converted to decimals by dropping the percent sign and moving the decimal two places to the left.

Following are four specific types of applications of percents.

Type 1: General Percent Problem

An example of a general percent problem is to answer the question "30 is 15% of what number?" To answer this question, we let n represent the unknown number. Recall that 15% of a number is equivalent to multiplying the number by 0.15. So, we have that

$$30 = 0.15n$$
$$\frac{30}{0.15} = \frac{0.15n}{0.15}$$
$$200 = n$$

The number is 200.

Type 2: Discount Problem

Suppose a sweater that is regularly priced at $36 is on sale for 25% off. To find the sale price, we must first find the amount to be taken off the original price, which is the discount amount.

$$\text{Discount} = 25\% \text{ of } \$36$$
$$= 0.25(36)$$
$$= 9$$

$$\text{Sale price} = \text{original price} - \text{discount}$$
$$= \$36 - \$9$$
$$= \$27$$

Type 3: Markup Problem

Suppose a bookstore pays $80 for a textbook. Then the bookstore has a markup of 40% to sell to the customer. What is the selling price of the book?
To find the price of the book, we must find the amount of the markup.

$$\text{Markup} = 40\% \text{ of } \$80$$
$$= 0.40(80)$$
$$= 32$$

$$\text{Selling price} = \text{cost} + \text{markup}$$
$$= \$80 + \$32$$
$$= \$112$$

Type 4: Percent Increase or Decrease

If a salary increased from $30,000 to $35,000, what percent increase does this represent? To find the percent increase, we must find the change in the values and divide this by the original value.

$$\frac{\text{New amount} - \text{original amount}}{\text{original amount}} = \frac{35,000 - 30,000}{30,000}$$
$$= \frac{5000}{30,000}$$
$$= 0.1667$$

So, the salary increased by 16.67%.

| Objective 1 Examples | Write an equation that represents each situation. Solve the equation and answer the question in a complete sentence. |

1a. According to Oprah's Debt Diet, how a family should spend their monthly income is based on this pie chart. If a family following this "diet" spends $1400 on housing each month, what is their monthly income?
(Source: www.oprah.com)

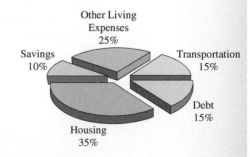

Solution **1a.** What is unknown? The monthly income is unknown.

$$\text{Let } x = \text{monthly income.}$$

What is known? 35% of the monthly income should be spent on housing. The amount spent on housing is $1400.

35% of monthly   amount spent on
income **is**   housing

$$0.35x = 1400 \qquad \text{Express the relationship.}$$

$$\frac{0.35x}{0.35} = \frac{1400}{0.35} \qquad \text{Divide each side by 0.35.}$$

$$x = 4000 \qquad \text{Simplify.}$$

So, the monthly income of this family is $4000.

1b. An electronics store has a sale price of $1559.99 on a 50-in. plasma HDTV. This price is 40% off the original selling price. What is the original selling price of the television?

Solution   **1b.** What is unknown? The original selling price of the TV is unknown.

Let x = the original selling price of the TV.

What is known? The sale price is $1599.99.

The sale price is 40% off the original selling price.

To write the equation, we use the following relationship.

Sale price = original price − discount amount

$$1559.99 = x - 0.40x \qquad \text{Express the relationship. The discount amount is } 0.40x.$$

$$1559.99 = 1x - 0.40x \qquad \text{Write } x \text{ as } 1x.$$

$$1559.99 = 0.60x \qquad \text{Combine like terms: } 1x - 0.40x = (1 - 0.40)x = 0.60x$$

$$\frac{1559.99}{0.60} = \frac{0.60x}{0.60} \qquad \text{Divide each side by 0.60.}$$

$$2599.98 = x \qquad \text{Simplify.}$$

So, the original selling price of the plasma HDTV was $2599.98.

1c. Nanette receives a 5% raise at work. She now makes $26,250 per year. What was her annual salary before the raise?

Solution   **1c.** What is unknown? The salary before the raise is unknown.

Let x = Nanette's original salary.

What is known? She received a 5% raise. So, her raise is $0.05x$.

This problem is similar to the markup illustration. So, we use the following relationship to write the equation.

Original salary + raise = new salary

$$x + 0.05x = 26{,}250 \qquad \text{Express the relationship.}$$

$$1x + 0.05x = 26{,}250 \qquad \text{Write } x \text{ as } 1x.$$

$$1.05x = 26{,}250 \qquad \text{Combine like terms.}$$

$$\frac{1.05x}{1.05} = \frac{26{,}250}{1.05} \qquad \text{Divide each side by 1.05.}$$

$$x = 25{,}000 \qquad \text{Simplify.}$$

So, Nanette's salary before her raise was $25,000.

1d. In fall 2000, enrollment in public preK–12th grade nationwide was 47.2 million. In fall 2010, the enrollment in preK–12th grade was 50 million. What was the percent increase in enrollment from 2000 to 2010?

Solution   **1d.** What is unknown? The percent increase in enrollment between 2000 and 2010 is unknown.

Let x = percent increase in enrollment in schools.

What is known? The original enrollment (in 2000) was 47.2 million. The new enrollment (in 2010) was 50.0 million.

To solve this problem, we use the following relationship.

$$\text{Percent increase or decrease} = \frac{\text{new amount} - \text{original amount}}{\text{original amount}}$$

$$x = \frac{50 - 47.2}{47.2} \qquad \text{Replace the new amount with 50 and the original amount with 47.2.}$$

$$x = \frac{2.8}{47.2} \qquad \text{Simplify the numerator.}$$

$$x \approx 0.059 \qquad \text{Divide.}$$

So, the enrollment increased by 5.9% from fall 2000 to fall 2010.

✓ **Student Check 1**   Write an equation that represents each situation. Solve the equation and answer the question in a complete sentence.

    a. Using the percentages from Oprah's Debt Diet, what is a family's monthly income if they spend $450 on transportation each month?

    b. A shirt is on sale for $21. This is 30% off the original price. What is the original selling price of the shirt?

    c. Aisha's salary after her raise is $55,640. Prior to the raise, her salary was $52,000. What percent raise did she receive?

    d. The national debt was $10.6 trillion on the day President Obama took office. The Obama Administration's 4-year estimate shows that by the end of September 2012, the debt will reach $16.2 trillion. What is the projected percent increase in the national debt from 2009 to 2012? (Source: http://www.cbsnews.com/blogs/2009/03/17/politics/politicalhotsheet/entry4872310.shtml)

Simple Interest Problems

Objective 2 ▶

Solve simple interest applications.

Interest is the amount of money collected for investing or borrowing money. Banks pay its customers interest when they invest money into their bank. Customers pay banks interest when they borrow money from the bank.

One type of interest is simple interest. Simple interest is interest earned on the amount that was initially invested or borrowed. Simple interest is computed as follows.

Property: Simple Interest Formula

$$\text{Interest} = \text{principal} \times \text{rate} \times \text{time} \qquad \text{or} \qquad I = prt$$

Principal, p, is the amount initially invested or borrowed.

Rate, r, is the annual interest rate (must be converted to a decimal).

Time, t, is the years that the money was invested or borrowed.

Objective 2 Examples	Use the simple interest formula to solve each problem.

2a. Marcus invested $10,000 in two accounts. He invested $2000 in a savings account that pays 3.5% annual interest and the remaining amount in a CD that pays 6% annual interest. How much interest did he earn from the two accounts in one year?

Solution **2a.** What is unknown? The total interest earned from the two accounts is unknown. What is known? A total of $10,000 was invested, $2000 invested at 3.5% and $8000 invested at 6%.

Account	Principal	×	Rate	× Time =	Interest
Savings	2000		$3.5\% = 0.035$	1 yr	$(2000)(0.035)(1) = 70$
CD	$10{,}000 - 2000 = 8000$		$6\% = 0.06$	1 yr	$(8000)(0.06)(1) = 480$

The total interest earned from the two accounts is given by

$$\text{Total interest} = (2000)(0.035)(1) + (8000)(0.06)(1)$$
$$= 70 + 480$$
$$= 550$$

So, Marcus earned a total of $550 in interest from the two accounts.

2b. Rosa has $5000 to invest. She invests part of her money in a savings account that earns 3% annual interest and the rest in a CD that pays 5% annual interest. If the total interest earned in 1 yr is $230, how much did she invest in each account?

Solution **2b.** What is unknown? The amount invested in each of the accounts is unknown.
Let $x =$ amount invested in the savings account.
Then $5000 - x =$ amount invested in the CD.

What is known? The total invested is $5000 and the total interest earned from the two accounts is $230.

Account	Principal	×	Rate	× Time =	Interest
Savings	x		$3\% = 0.03$	1 yr	$(x)(0.03)(1)$ or $0.03x$
CD	$5000 - x$		$5\% = 0.05$	1 yr	$(5000 - x)(0.05)(1)$ or $0.05(5000 - x)$
Total	$5000				$230

The sum of the interest from the savings account and the interest from the CD account is the total interest earned.

$0.03x + 0.05(5000 - x) = 230$	Express the relationship.
$0.03x + 250 - 0.05x = 230$	Apply the distributive property.
$-0.02x + 250 = 230$	Combine like terms.
$-0.02x + 250 - 250 = 230 - 250$	Subtract 250 from each side.
$-0.02x = -20$	Simplify.
$\dfrac{-0.02x}{-0.02} = \dfrac{-20}{-0.02}$	Divide each side by -0.02.
$x = 1000$	Simplify.

So, Rosa invested $1000 in the 3% account and $5000 - $1000 = $4000 in the 5% account.

✓ **Student Check 2**   Use the simple interest formula to solve each problem.

a. Allison invested a total of $2000 in two accounts. She invested $1500 in a retirement account that pays 5.5% annual interest and the rest in a savings account that pays 4% annual interest. How much interest did she earn in 1 yr from these two accounts?

b. Huynh has $8000 to invest. She invests part of the money in a money market account that pays 6.5% annual interest and the rest of the money in a CD that pays 5% annual interest. If she earns $490 interest in 1 yr, how much did she invest in each account?

Mixture Problems

Objective 3 ▶

Solve mixture applications.

Mixture applications involve combining two substances together to obtain a third substance. For example, combining different types of coins or dollar bills to get a monetary total or mixing two different types of liquid solutions together to make a third type of liquid solution are all mixture problems.

The main point to remember when solving mixture problems is that we cannot add unlike quantities together. We must find a way to convert the different quantities to a like quantity.

When working with monetary mixture problems, we convert the given number of coins or bills to its total value. We discussed this in Section 2.3 but it is worth repeating here.

Coins/Bills	Value of Coins/Bills
5 quarters	$5(0.25) = \$1.25$
8 dimes	$8(0.10) = \$0.80$
4 $10 bills	$4(10) = \$40$
x nickels	$x(0.05) = \$0.05x$

To find the value of a collection of coins, multiply the number of coins by the monetary worth of the coin.

When working with liquid mixture problems, we must determine the amount of "pure" substance in each solution. To illustrate this concept, consider the percent alcohol in beer, wine, and liquor. Most law enforcement agencies consider each of the following equivalent to one drink for the purposes of calculating blood alcohol concentration. (Source: http://www.bloodalcoholcontent.org/alcoholinformation.html)

Solution	Amount of Alcohol
1.5 oz of 80 proof liquor (40% alcohol)	$1.5(0.40) = 0.6$ oz of alcohol
12 oz of regular beer (5% alcohol)	$12(0.05) = 0.6$ oz of alcohol
5 oz of table wine (12% alcohol)	$5(0.12) = 0.6$ oz of alcohol

Though these three alcoholic beverages vary in quantity and percent alcohol, they each contain the same amount of alcohol. If we combine the beverages, the total amount of alcohol is the sum of the alcohol found in each beverage. This is true not only when dealing with alcoholic beverages, but for any combination of solutions with different concentrations.

Note: *To find the amount of substance in a solution, multiply the given amount by the strength of the solution.*

Objective 3 Examples   **Write an equation that can be used to solve each problem. Solve the equation and answer the question using a complete sentence.**

3a. Teresa has a collection of dimes and quarters. She has 40 less dimes than quarters. The value of her collection is $20.50. Find the number of dimes and quarters she has in her collection.

Solution   **3a.** What is unknown? The number of dimes and quarters is unknown.
Let q = number of quarters.
Then $q - 40$ = number of dimes.

What is known? The total value of the collection is $20.50.

Type of Coin	Number of coins	×	Value of coin	=	Total value of the coins
Quarters	q		$0.25		$0.25q$
Dimes	$q - 40$		$0.10		$0.10(q - 40)$

The total value of the collection is the value of the quarters plus the value of the dimes.

$0.25q + 0.10(q - 40) = 20.50$	Express relationship.
$0.25q + 0.10q - 4 = 20.50$	Apply the distributive property.
$0.35q - 4 = 20.50$	Combine like terms.
$0.35q - 4 + 4 = 20.50 + 4$	Add 4 to each side.
$0.35q = 24.50$	Simplify.
$\dfrac{0.35q}{0.35} = \dfrac{24.50}{0.35}$	Divide each side by 0.35.
$q = 70$	Simplify.

So, Teresa has 70 quarters and $70 - 40 = 30$ dimes in her collection.

3b. Louisa is a chemist and she needs to conduct an experiment with a solution that is 30% copper sulfate. She does not have a solution with this concentration but she does have 40 mL of a 25% copper sulfate solution and she also has a 60% copper sulfate solution. How many milliliters of the 60% copper sulfate solution does she need to mix with the 40 mL of the 25% solution to obtain a 30% copper sulfate solution?

Solution   **3b.** What is unknown? The amount of the 60% copper sulfate solution is unknown.
Let x = the amount of the 60% solution.

What is known? She needs a solution that is 30% copper sulfate. She has 40 mL of a 25% copper sulfate solution.

25% solution 40 mL
60% solution x mL

$\left.\begin{array}{l} x\ \text{mL} \\ 40\ \text{mL} \end{array}\right\} 40 + x$

30% solution

Type of Solution	Number of mL ×	Solution strength =	Total mL of copper sulfate
25%	40 mL	25% or 0.25	0.25(40)
60%	x	60% or 0.60	$0.60x$
30%	$40 + x$	30% or 0.30	$0.30(40 + x)$

The amount of copper sulfate in the 25% solution and the 60% solution equals the amount of copper sulfate in the 30% solution. This gives us the following equation.

$$0.25(40) + 0.60(x) = 0.30(40 + x)$$ Express the relationship.
$$10 + 0.60x = 12 + 0.30x$$ Simplify.
$$10 + 0.60x - 0.30x = 12 + 0.30x - 0.30x$$ Subtract 0.30x from each side.
$$10 + 0.30x = 12$$ Simplify.
$$10 + 0.30x - 10 = 12 - 10$$ Subtract 10 from each side.
$$0.30x = 2$$ Simplify.
$$\frac{0.30x}{0.30} = \frac{2}{0.30}$$ Divide each side by 0.30.
$$x = \frac{20}{3} \text{ or } 6\frac{2}{3}$$ Simplify.

Louisa needs to add $6\frac{2}{3}$ mL of the 60% copper sulfate solution to the 40 mL of 25% copper sulfate solution to obtain a 30% copper sulfate solution.

3c. A specialty coffee shop wants to make an After Dinner Blend from two of their coffees, French Roast and Java. The French Roast sells for $14 per pound and Java sells for $8 per pound. How many pounds of French Roast should be mixed with 10 lb of Java to produce an After Dinner Blend that sells for $12 per pound?

Solution   3c. What is unknown? The pounds of French Roast are unknown.

Let x = pounds of French Roast.

What is known? French Roast sells for $14 per pound, Java sells for $8 per pound, and the blend sells for $12 per pound. There are 10 lb of Java.

Type of Coffee	Pounds of coffee	× Price per pound	= Total price of coffee
French Roast	x	$14	$14x$
Java	10	$ 8	8(10)
After Dinner Blend	$x + 10$	$12	12(x + 10)

The combined price of the French Roast and Java equals the total price of the After Dinner Blend. This gives us the following equation.

$$14x + 8(10) = 12(x + 10)$$ Express the relationship.
$$14x + 80 = 12x + 120$$ Simplify and distribute.
$$14x + 80 - 12x = 12x + 120 - 12x$$ Subtract 12x from each side.
$$2x + 80 = 120$$ Simplify.
$$2x + 80 - 80 = 120 - 80$$ Subtract 80 from each side.
$$2x = 40$$ Simplify.
$$\frac{2x}{2} = \frac{40}{2}$$ Divide each side by 2.
$$x = 20$$ Simplify.

To make the blend, 20 lb of French Roast should be mixed with 10 lb of Java.

✓ **Student Check 3**   Write an equation that can be used to solve each problem. Solve the equation and answer the question using a complete sentence.

a. Damien has a collection of nickels and dimes. He has twice as many dimes as nickels. The value of his collection is $5.25. Find the number of nickels and dimes in his collection.

b. Todd needs to make 4 gal of a 40% antifreeze solution for his truck. How much pure antifreeze and how much water should he mix together to get 4 gal of a 40% antifreeze solution?

c. The Nut Company mixes pecans that sell for $8.25 per pound and walnuts that sell for $6.45 per pound to make 50 lb of a mixture that sells for $7.15 per pound. How much of each kind of nut should be put in the mixture?

Distance Problems

Objective 4 ▶

Solve distance, rate, and time applications.

As illustrated in Section 2.5, Example 1c, distance traveled depends on one's speed and the time traveled.

> **Property: Distance Formula**
>
> $$\text{Distance} = \text{rate} \times \text{time} \qquad \text{or} \qquad d = rt$$
>
> Distance is measured in miles, feet, kilometers, and the like.
>
> Rate is measured in miles per hour, feet per second, kilometers per hour, and the like.
>
> Time is measured in hours, minutes, seconds, and the like.

We can use this formula to determine how far a car travels if it is driven at a speed of 65 mph for 2 hr.

$$d = (65 \text{ mph})(2 \text{ hr}) = 130 \text{ mi}$$

For these word problems, there will be two people/things that are traveling. So, there will be two distances, rates, and times. One of these quantities will be given for each people/things. The other quantity will be unknown. The remaining quantity will be obtained using the distance formula. It is helpful to organize the information in a chart.

Objective 4 Examples   **Use the distance formula to write an equation that will solve each problem. Solve the equation and answer the question using a complete sentence.**

4a. Tatiana and Blair live 390 mi apart. They leave their homes at the same time and drive towards one another until they meet. Tatiana is traveling at 60 mph and Blair is traveling at 70 mph. How long will they drive until they meet one another?

Solution   **4a.** What is unknown? The time that Tatiana and Blair travel is unknown. Since they leave at the same time and travel until they meet, their times are the same.

Let $t =$ the time Tatiana and Blair travel.

What is known? They live 390 mi apart. Tatiana travels at a rate of 60 mph. Blair travels at a rate of 70 mph.

	Rate	×	Time	=	Distance
Tatiana	60		t		$60t$
Blair	70		t		$70t$

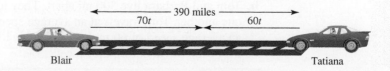

Since Blair and Tatiana are traveling toward each other, their combined distance traveled is 390 mi. We use this to write the equation.

Tatiana's distance + Blair's distance = 390

$$60t + 70t = 390 \qquad \text{Express the relationship.}$$

$$130t = 390 \qquad \text{Combine like terms.}$$

$$\frac{130t}{130} = \frac{390}{130} \qquad \text{Divide each side by 130.}$$

$$t = 3 \qquad \text{Simplify.}$$

So, they both traveled for 3 hr until they met. Note that Tatiana drove $60(3) = 180$ miles in 3 hr and Blair drove $70(3) = 210$ miles in 3 hr.

4b. Highway inspectors are examining a highway. Two inspectors start from the same point but one travels north and the other travels south. The inspector heading north leaves at 8:00 A.M. and travels at an average speed of 40 mph. The inspector heading south leaves at 10:00 A.M. the same day and travels at an average speed of 50 mph. When will the inspectors finish examining a 260-mi portion of the highway?

Solution   **4b.** What is unknown? The time that each inspector travels is unknown. The inspector heading south leaves 2 hr after the inspector heading north.

Let t = time for the inspector heading north and
$t - 2$ = time for the inspector heading south.

What is known? The total distance covered is 260 mi. The north inspector has a rate of 40 mph and the south inspector has a rate of 50 mph.

	Rate	×	Time	=	Distance
North	40		t		$40t$
South	50		$t - 2$		$50(t - 2)$

Since their total distance is 260 mi, the equation is as follows.

$$40t + 50(t - 2) = 260 \qquad \text{Express the relationship.}$$

$$40t + 50t - 100 = 260 \qquad \text{Apply the distributive property.}$$

$$90t - 100 = 260 \qquad \text{Combine like terms.}$$

$$90t - 100 + 100 = 260 + 100 \qquad \text{Add 100 to each side.}$$

$$90t = 360 \qquad \text{Simplify.}$$

$$\frac{90t}{90} = \frac{360}{90} \qquad \text{Divide each side by 90.}$$

$$t = 4 \qquad \text{Simplify.}$$

The inspector heading north will be done in 4 hr and the inspector heading south will be done in $4 - 2 = 2$ hr. So, the job will be complete at 12:00 P.M.

✓ **Student Check 4**   Use the distance formula to write an equation that will solve each problem. Solve the equation and answer the question using a complete sentence.

a. Two cars leave the same town heading in opposite directions. One car travels at 50 mph and the other at 70 mph. How long will it take for the cars to be 480 mi apart?

b. Tom and Barbara live 500 mi apart. They leave their homes and travel toward one another. Tom travels at an average speed of 65 mph and Barbara travels at an average speed of 60 mph. Tom leaves 1 hr before Barbara. How long will they each travel before they meet?

Objective 5 ▶

Troubleshoot common errors.

Troubleshooting Common Errors

The most common errors in solving word problems occur in setting up the equation incorrectly. Example 5 illustrates some of these errors.

Objective 5 Examples / **A problem and an incorrect solution are given. Provide the correct solution and an explanation of the error.**

5a. A golf club is originally priced at $224.99 and is on sale for $149.99. What is the discount rate?

Incorrect Solution	Correct Solution and Explanation
The original − discount = sale price. $$224.99 - x = 149.99$$ $$224.99 - x - 224.99 = 149.99 - 224.99$$ $$-x = -75$$ $$x = 75$$ So, the discount rate is 75%.	The discount amount is the discount rate times the original amount. So, the equation is $$224.99 - 224.99x = 149.99$$ $$-224.99x = -75$$ $$\frac{-224.99x}{-224.99} = \frac{-75}{-224.99}$$ $$x \approx 0.3333$$ So, the discount rate is 33.3%.

5b. Jackson has $20,000 in a retirement account that he wants to invest in two different ways. He invests some of the money in a money market account that earns 5% annual interest and the other in a high risk stock fund that yields 10% annual interest. If he earned $1600 interest in a year, how much did he invest in each account? Write the equation that can be used to solve the problem.

Incorrect Solution	Correct Solution and Explanation
Let x = amount invested in 5% account. Then $x - 20,000$ = amount invested in 10% account. $$0.05x + 0.10(x - 20,000) = 1600$$	The error is the amount remaining. If x = amount invested in the 5% account, then $20,000 - x$ is the amount invested in the 10% account. $$0.05x + 0.10(20,000 - x) = 1600$$

ANSWERS TO STUDENT CHECKS

Student Check 1   **a.** $3000   **b.** $30   **c.** 7%   **d.** 52.8%
Student Check 2   **a.** $102.50   **b.** $6000 in the 6.5% account and $2000 in the 5% account

Student Check 3   **a.** 21 nickels and 42 dimes   **b.** 1.6 gal of pure antifreeze and 2.4 gal of water   **c.** 19.44 lb of pecans and 30.56 lb of walnuts
Student Check 4   **a.** 4 hr   **b.** Barbara travels for 3.5 hr and Tom travels for 4.5 hr.

SUMMARY OF KEY CONCEPTS

1. Percent applications arise in many situations. Convert percents to decimals before using them in an equation.
   - Discount problems: Original − discount = sale price
     The discount amount is a percentage of the original amount.
   - Markup: Original + markup = new price
   - Percent increase or decrease: $\dfrac{\text{new amount} - \text{original amount}}{\text{original amount}}$
   - Recall "of" means multiplication.
2. Simple interest is given by $I = Prt$. When a total amount of money is given that is split between two accounts,

let x represent one amount and the remaining amount is represented as the "total amount − x."

3. Mixture problems are set up by adding the values/amounts from collections/solutions to equal a total value/amount. Remember that the quantities being added must be equivalent.

4. Distance = rate × time. Organize the information in a chart by using variables for one of the columns, given amounts for another column, and then the formula to derive the third column. The expressions obtained in the last column will be used in the equation.

GRAPHING CALCULATOR SKILLS

There are no new calculator skills needed for this section.

SECTION 2.6 / EXERCISE SET

Write About It!

Use complete sentences to explain the meaning of each term.

1. Percent   2. Simple interest   3. Mixture applications

Determine if each statement is true or false. If a statement is false, explain why.

4. The price of an item increased from $50 to $75. To find the percent increase, we must compute $\dfrac{75 - 50}{75}$.

5. If $6000 is split between two different investments, we can represent the two investment amounts as x and $x - 6000$.

6. James and Terry live 300 mi apart. They leave their homes at the same time traveling toward one another. James travels at 60 mph, and Terry travels at 72 mph. To find the time when James and Terry meet, we solve the equation $\dfrac{300}{60} = \dfrac{x}{72}$ for x.

Practice Makes Perfect!

Solve each problem involving percents. Round answers to two decimal places, if needed. (*See Objective 1.*)

7. A computer store has a deal on a notebook computer. The computer is on sale for $699.99. This is 20% off the original price. What is the original price of the computer?

8. A suit is on clearance at a department store for $60. If this price is 75% off the original price, what is the original price of the suit?

9. A digital camera with a printer dock is regularly priced at $329.99. The camera is on sale for 15% off the regular price. What is the sale price of the camera?

10. A leather sofa is regularly priced at $999.99. The furniture store is running a sale in which all furniture is 20% off. What is the sale price of the sofa?

11. A campus bookstore sells a graphing calculator for $104. This is a 30% markup on the cost the bookstore pays for the calculator. What does the bookstore pay for the calculator?

12. The price of an MP3 player is $200. This is a markup of approximately 122% from the actual cost to make the player. What is the cost to make the MP3 player?

13. The price of a luxury car plus 7% sales tax is $45,582. What is the price of the car without tax?

14. Real estate commission is a percentage of the selling price of a house that is paid to realtors for services rendered. Michael is a real estate agent assisting a family in buying a new home. If Michael's commission rate is 7% of the selling price of the home and his total commission is $13,020, what is the selling price of the home?

15. Selena receives a 5% raise at work, making her salary $31,500. What was her salary before the raise?

16. Raj receives a 4% raise at work. He now makes $15.60 per hour. What was his hourly wage before the raise?

The pie chart shows a suggested breakdown for one's monthly income. Use the pie chart to answer each question.

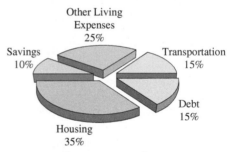

Source: www.oprah.com

17. The Donnellys' monthly income is $5000. According to the pie chart, what should they spend on their housing each month?

18. The Andersons' monthly income is $4200. According to the pie chart, what should they spend on their transportation each month?

19. Sonja is a single mom and is trying to save money. If she saves $250 each month, what should her monthly income be according to the pie chart?

20. If Brandon spends $550 each month on other living expenses, what should his monthly income be according to the pie chart?

According to a national education organization, the average time it takes for students to complete their bachelor's degrees varies based on the type and number of colleges they attend, as shown in the following table. Use this information for Exercises 21 and 22.

Students attending only one institution	51 months
Students attending two institutions	59 months
Students attending three or more institutions	67 months
Students beginning at public 2-yr colleges	71 months
Students attending only one private institution	51 months
Students attending only one public 4-yr college	55 months

21. What is the percentage increase in the time it takes to complete a bachelor's degree for students attending one public 4-yr college and students attending three or more institutions? Round to the nearest hundredth of a percent.

22. What is the percentage increase in the time it takes to complete a bachelor's degree for students attending only one private institution and students beginning at public 2-yr colleges? Round to the nearest hundredth of a percent.

23. New York City is the most populated city in the United States. The 2000 census reported that NYC had a population of 8,008,278. In 2009, the population  was estimated to be 8,391,881. By what percent did the population increase from 2000 to 2009? Round to the nearest hundredth of a percent. (Source: www.census.gov)

24. China is the world's most populated country. In 2004, the population of China was 1,288,307,000. In 2010, the population was estimated to be 1,330,141,295. By what percent did the population increase from 2004 to 2010? Round to the nearest hundredth of a percent. (Source: http://internetworldstats.com)

For each simple interest problem: (1) use a chart to organize the information, (2) write an appropriate equation, (3) solve the equation, and (4) write the answer in complete sentences. (*See Objective 2.*)

25. Rob earns a $20,000 bonus for the year. He invests this money in two accounts. One account is highly risky but earns 11% annual interest. The other account is less risky and earns 5% annual interest. If Rob earns a total of $1300 in interest for the year, how much did he invest in each account?

26. Marissa has saved $4000 over the last several years. She invests part of her money in a CD that earns 6% annual interest and the rest in a savings account that pays 4% annual interest. If Marissa earns $210 in interest for the year, how much did she invest in each account?

27. Isaiah invests some money into a money market account that earns 10% annual interest. He also invests $3000 less than the amount invested in the money market account in a CD that earns 5.5% annual interest. If he earns $1385 in interest for the year, how much did he invest in each account?

28. Georgia receives an inheritance of $100,000. She invests this money in two different accounts. She invests part of the money in the stock market and earns an average of 12% annual interest. She invests the rest of the money in a money market account that earns 7% annual interest. If Georgia earns $11,000 in interest for the year, how much did she invest in each account?

For each mixture problem: (1) use a chart to organize the information, (2) write an appropriate equation, (3) solve the equation, and (4) write the answer in complete sentences. (*See Objective 3.*)

29. Thomas has a collection of nickels and quarters. He has three times as many quarters as nickels. If the collection is worth $24, how many nickels and quarters does he have?

30. Samantha has a total of 300 coins, which consist of dimes and quarters. If her collection is worth $46.50, how many of each coin does she have?

31. A bank teller has a collection of $10 bills and $20 bills. He has fifteen more $20 bills than $10 bills. The value of his collection is $900. How many of each bill does he have?

32. A department store clerk has only $5 bills and $10 bills in her cash register. She has a total of 30 bills and the value of the collection is $215. How many of each bill does she have?

33. A lab technician needs a 40% alcohol solution. He has 10 gal of a 35% alcohol solution. How many gallons of a 50% alcohol solution need to be mixed with the 10 gal of the 35% solution to obtain a 40% alcohol solution?

34. A chemist has a 50% acid solution. She needs 30 L of a 15% acid solution. How many liters of water and how many liters of the 50% acid solution should be mixed together to obtain 30 L of a 15% acid solution?

35. A dairy farmer has two types of milk. How many gallons of a 3% milk fat solution and how many gallons of a 4.5% milk fat solution should be mixed to obtain 750 gal of a 4% milk fat solution?

36. How many liters of a 47% hydrochloric acid solution should be mixed with 4 L of a 30% hydrochloric acid solution to obtain a 35% hydrochloric acid solution?

For each problem involving distance, rate, and time: (1) use a chart to organize the information, (2) write an appropriate equation, (3) solve the equation, and (4) write the answer in complete sentences. (*See Objective 4.*)

37. Andy and Brittany live in towns 420 mi apart. They leave their homes at the same time and begin traveling toward one another. Andy travels at 65 mph and Brittany travels at 55 mph. How long will it take before they meet?

38. Two planes leave Chicago's Midway airport. One plane travels east at 450 mph and the other plane travels west at 600 mph. How long will it take for the planes to be 2310 mi apart?

39. Charlie leaves his home for a meeting traveling at an average rate of 50 mph. After his business meeting is over, he travels back home at an average rate of 70 mph. If his total driving time is 9 hr, how long does he drive going to and from his meeting? What is the total distance Charlie travels?

40. Mohammad goes on a leisurely drive on a beautiful day. He leaves his home traveling at an average rate of 55 mph. He travels until he is tired and stops for lunch. After lunch, he decides to drive back home. He travels back home at an average speed of 65 mph. If his total driving time is 4 hr, how long did he drive each way? What is the total distance he traveled?

41. Sheree leaves her job and heads east at 55 mph. One hr later, Janis leaves the same workplace and heads east at 75 mph. How long will it take Janis to catch Sheree?

42. A plane leaves Atlanta heading to San Francisco at 9 A.M. traveling at 400 mph. Another plane leaves Atlanta heading to San Francisco at 10 A.M. the same day traveling at 550 mph. How long will it take for the second plane to catch up with the first?

 Mix 'Em Up!

For each problem: (1) write an appropriate equation, (2) solve the equation, and (3) write the answer in complete sentences.

43. A 46-in. LED-LCD HDTV is on sale for $1220.99. This is 45% off the original price. What is the original selling price of the TV?

44. A 40-in. LCD-HDTV is on sale for $1614.99. This is 15% off the original price. What is the original selling price of the TV?

45. Lynette's salary after her raise is $62,829. Prior to the raise, her salary was $58,300. What percent raise did she receive?

46. Tyrone's salary after his raise is $51,450. Prior to the raise, his salary was $50,000. What percent raise did he receive?

47. A bank teller has a collection of $1 bills and $5 bills. He has seven more $1 bills than $5 bills. The value of his collection is $229. How many of each bill does he have?

48. A department store clerk has only $5 bills and $10 bills in her cash register. She has a total of 200 bills and the value of the collection is $1,565. How many of each bill does she have?

Use the percentages from Oprah's Debt Diet, to answer Exercises 49 and 50.

Oprah's Debt Diet	Percentages
Housing	35%
Debt	15%
Saving	10%
Transportation	15%
Other Living Expenses	25%

Source: www.oprah.com

49. The Basils' monthly income is $3000. According to the table, what should they spend on their debt each month?

50. The Morans' monthly income is $1714.29. According to the table, what should they spend on their housing each month?

51. Jim invests $10,200 in two accounts. He invests $4850 in a mutual fund that pays 2.2% annual interest and the remaining amount in a money market fund that pays 3.3% annual interest. How much interest does he earn from the two accounts in 1 yr?

52. Adrianne invests $4250 in two accounts. She invests $1650 in a mutual fund that pays 1.1% annual interest and the remaining amount in a money market fund that pays 5.3% annual interest. How much interest does she earn from the two accounts in 1 yr?

53. Two planes leave Wittman Regional Airport in Oshkosh. One plane travels west at 355 mph and the other plane travels east at 340 mph. How long will it take for the planes to be 3683.5 mi apart?

54. Two cars leave the same town heading in opposite directions. One car travels at 69 mph and the other at 62 mph. How long will it take for the cars to be 275.1 mi apart?

55. A dairy farmer has two types of milk. How many gallons of a 2% milk fat solution and how many gallons of a 4.5% milk fat solution should be mixed to obtain 500 gal of a 2.9% milk fat solution?

56. Michael needs to make 243 L of a 58% alcohol solution. How much pure alcohol and how much 37% alcohol solution should be mixed together to obtain 243 L of a 58% alcohol solution?

57. Brandon and Kimberly live 320 mi apart. They leave their homes at the same time and drive toward one another until they meet. Brandon travels at 55 mph and Kimberly travels at 73 mph. How long will it take before they meet?

58. Karen and Evan live 416 mi apart. They leave their homes at the same time and drive towards one another until they meet. Karen travels at 68 mph and Evan travels at 62 mph. How long will it take before they meet?

59. John earns a $9100 bonus for the year. He invests part of this money in two accounts. One account is highly risky but earns 5.7% annual interest. The other account is less risky and earns 2.1% annual interest. If John earns a total of $367.50 in interest for the year, how much did he invest in each account?

60. Maria has $11,750 to invest. She invests part of her money in a stock that earns 2.9% annual dividends and the rest in a mutual fund account that pays 5.5% annual dividends. If Maria earns $553.95 in dividends for the year, how much did she invest in each account?

61. Kevin has a collection of dimes and quarters. He has 10 more dimes than quarters. If the collection is worth $98.30, how many dimes and quarters does he have?

62. Joan has a collection of dimes and nickels. She has 38 fewer dimes than nickels. If the collection is worth $7, how many dimes and nickels does she have?

63. A sweater is on sale for $20.35. If this price is 45% off the original price, what is the original price of the sweater?

64. A jacket is on sale for $39.60. If this price is 55% off the original price, what is the original price of the jacket?

65. Alison needs to make 68 L of a 56% hydrochloric acid solution. How much 8% hydrochloric acid solution and how much 72% hydrochloric acid solution should she mix to get 68 L of a 56% hydrochloric acid solution?

66. Yarrish has a 78% acid solution. He needs 143 L of a 24% acid solution. How many liters of water and how many liters of the 78% acid solution should he mix together to obtain 143 L of a 24% acid solution?

67. The passenger volume at a major international airport was 41.89 million in 2005 and 47.81 million in 2010. What is the percent increase in the passenger volume from 2005 to 2010? Round to one hundredth of a percent.

68. The passenger volume at an airport in Michigan was 36.39 million in 2005 and 35.14 million in 2011. What is the percent decrease in the passenger volume from 2005 to 2011? Round to one hundredth of a percent.

69. A campus bookstore sells an elementary algebra text for $93.60. This is a 30% markup on the cost the bookstore pays for the text. What does the bookstore pay for the text?

70. A campus bookstore sells a chemistry text for $125.35. This is a 15% markup on the cost the bookstore pays for the chemistry text. What does the bookstore pay for the text?

71. Jeanette leaves her job and heads west at 48 mph. Thirty minutes later, Warren leaves the same workplace and heads west at 68 mph. How long will it take Warren to catch up with Jeanette?

72. A plane leaves Newark heading to Las Vegas at 7 A.M. traveling at 320 mph. A second plane leaves Newark heading to Las Vegas at 8:18 A.M. the same day traveling at 580 mph. How long will it take for the second plane to catch up with the first plane?

 You Be the Teacher!

Correct each student's errors, if any.

73. The price of a 2010 SUV plus 8% sales tax is $52,629.48. Find the price of the car without tax.

Emma's work:
sales tax $= 52{,}629.48(8\%)$
$= 52{,}629.48(0.08)$
$= 4210.3584$
≈ 4210.36

Price before tax $= 52{,}629.48 - 4210.36$
$= 48{,}419.12$

74. A 32-in. LCD-HDTV is on sale for $209.99. This is 40% off the original price. Find the original selling price of the TV.

Cynthia's work:

Original selling price $= \dfrac{209.99}{40\%}$

$= \dfrac{209.99}{0.4}$

$= 524.975$

≈ 524.98

75. The passenger volume an international airport in Florida was 30.17 million in 2004 and 34.06 million in 2011. Find the percent increase in the passenger volume from 2004 to 2011.

Ralpha's work:

Averge percent $= \dfrac{34.06 - 30.17}{34.06}$

$= \dfrac{3.89}{34.06}$

$= 0.114210$

$\approx 11.4\%$

76. Haazim has $13,850 to invest. He invests part of the money in a stock that pays 4.5% annual dividends and the rest of the money in a mutual fund that pays 3.8% annual dividends. If he earns $581.95 in dividends in 1 yr, find the amount invested in each account.

Haazim's work:

Let x be the amount invested in the stock and $x - 13{,}850$ in the mutual fund account.

$4.5x + 3.8(x - 13850) = 581.95$
$4.5x + 3.8x - 52630 = 581.95$
$8.3x = 53211.95$
$x \approx 6411.08$

77. Shan needs to make 210 L of a 60% acid solution. Calculate the number of liters of pure acid and the number of liters of 40% solution Shan needs to mix together to get 210 L of a 60% acid solution.

Shan's work:

Let x be the number of liters of pure acid and $x - 210$ be the number of liters of 40%.

$0x + 0.40(x - 210) = 0.60(210)$
$0.40x - 84 = 126$
$0.40x = 210$
$x = 210/0.40$
$x = 525$

Linear Inequalities in One Variable

▶ **OBJECTIVES**

As a result of completing this section, you will be able to

1. Graph the solution set of an inequality.

2. Write the solution set of an inequality in interval notation.

3. Write the solution set of an inequality in set-builder notation.

4. Solve linear inequalities.

5. Solve applications of linear inequalities.

6. Troubleshoot common errors.

As math classes come to an end, a prevailing thought of many students is, "What grade do I need to make on the final to pass this class, or to keep my A?" In this section, we will look at one example of how to determine the grade needed on a final exam to achieve a certain grade in a class. This is an application of linear inequalities.

Graphs of Inequalities

Until now, we have solved only linear equations in one variable. We now turn our focus to solving linear inequalities in one variable. Linear equations and linear inequalities are similar except that a linear inequality has an inequality symbol instead of an equals sign. In Chapter 1, inequality symbols were introduced.

> **Definition:** A **linear inequality in one variable** is an inequality of the form $ax + b < c$, where a, b, and c are real numbers with $a \neq 0$. Note that a linear inequality can have any of the inequality symbols: $<$, $\leq$, $>$, or $\geq$.

Objective 1 ▶

Graph the solution set of an inequality.

Some examples of linear inequalities in one variable are

$$2x + 3 < -7 \qquad 7x - 2 \geq 6x + 4 \qquad 4(3y - 1) < -2(y - 9)$$

Like linear equations, linear inequalities are neither true nor false until a value is assigned to the variable. Our job is to determine the values of the variable that make the inequality true. Each such value is a **solution of a linear inequality**.

Linear inequalities generally have infinitely many values that make the statement true. Because we will not be able to list all of these values, we graphically represent the solution set of all the solutions on a number line. So, before we actually solve a linear inequality, we will examine the graphs of possible solution sets.

> **Procedure: Graphing the Solution Set of a Linear Inequality**
>
> **Step 1:** Draw a number line.
> **Step 2:** Shade the portion of the number line that satisfies the inequality.
> **Step 3:** Put the appropriate symbol on the endpoint.
>   **a.** If the endpoint satisfies the inequality, then it is included in the solution set as denoted by a bracket. A closed circle can also be used in place of the bracket.
>   **b.** If the endpoint does not satisfy the inequality, then it is not included in the solution set as denoted by a parenthesis. An open circle can also be used in place of the parentheses.

Objective 1 Examples   **Graph the solution set of each inequality on a number line.**

1a. $x > 3$ **1b.** $x \leq 4$ **1c.** $\dfrac{1}{2} < x$ **1d.** $-2 < x \leq 5$

Solutions   **1a.** To graph the solutions of $x > 3$, we find the values of x that are larger than 3. We know the integers 4, 5, 6, and so on are larger than 3, but we must not forget that there are numbers between 3 and 4 that are also larger than 3 (e.g., 3.9, 3.54, 3.0002, etc.). The number 3 itself is not a solution of this inequality since 3 is not greater than itself. So, we shade the portion of the number line that is to the right of 3. To indicate that 3 is not included in the solution set, put a parenthesis on it.

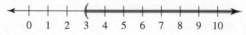

1b. To graph the solution of $x \leq 4$, we shade the portion of the number line that corresponds to values less than 4 or equal to 4. Numbers less than 4 are numbers to the left of 4. A bracket is used to show that 4 is included in the solution set since 4 equals 4.

1c. The inequality $\frac{1}{2} < x$ means that $\frac{1}{2}$ is smaller than all of the solutions, or that all of the solutions are larger than $\frac{1}{2}$. So, $\frac{1}{2} < x$ is equivalent to $x > \frac{1}{2}$. Therefore, we shade the portion of the number line to the right of $\frac{1}{2}$ and place a parenthesis on $\frac{1}{2}$ since it is not included in the solution set.

 Note: *The inequality $c < x$ is equivalent to $x > c$.*

1d. To graph the solutions of $-2 < x \leq 5$, we find the values that are between -2 and 5, including 5. So, we shade the portion of the number line that lies between these endpoints. The number -2 is not included since -2 is not less than -2.

✓ **Student Check 1** Graph the solution set of each inequality on a number line.

a. $x > -2$ **b.** $x \leq -4.2$ **c.** $\frac{5}{3} > x$ **d.** $-4 \leq x < 2$

Examples 1a–1c illustrate a **simple inequality**. The solutions of simple inequalities have to satisfy one only inequality.

Example 1d illustrates a **compound inequality**. The solutions of compound inequalities must satisfy two inequalities, not just one. For instance,

$-2 < x \leq 5$ means that "$-2 < x$ and $x \leq 5$"   or   "$x > -2$ and $x \leq 5$"

Interval Notation

Objective 2 ▶

Write the solution set of an inequality in interval notation.

Solutions of inequalities can also be represented in a concise way that represents the smallest and largest values in the solution set. This notation is called **interval notation**.

When the graph of an inequality extends to the right indefinitely, we say that the numbers in the set approach ∞ (infinity). This means that the numbers continue getting larger without bound. When the graph of an inequality extends to the left indefinitely, we say that the numbers in the set approach $-\infty$ (negative infinity). This means that the numbers continue getting smaller without bound.

Interval notation makes use of parentheses and brackets like graphing solution sets of inequalities. If the endpoint is included in the solution set of the inequality, then a bracket is used with the endpoint. If the endpoint is not included, then a parenthesis is used.

Positive and negative infinity are always written with parentheses because they don't represent a specific number. Two examples are shown next.

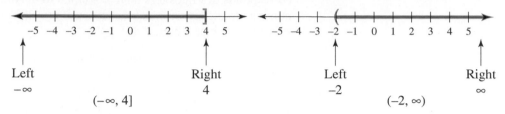

	Left	Right		Left	Right
	$-\infty$	4		-2	∞
		$(-\infty, 4]$			$(-2, \infty)$

Objective 2 Examples   Complete the table by graphing each inequality and writing the interval notation that corresponds to the solution set.

Inequality	Graph	Interval Notation
2a. $x > 3$		$(3, \infty)$
2b. $x \le 1.5$		$(-\infty, 1.5]$
2c. $\dfrac{1}{2} < x$		$\left(\dfrac{1}{2}, \infty\right)$
2d. $-2 < x \le 5$		$(-2, 5]$

✓ Student Check 2   Graph each inequality and express the solution set in interval notation.

    a. $x > -2$ **b.** $x \le -4.2$ **c.** $\dfrac{5}{3} > x$ **d.** $-4 \le x < 2$

Set-Builder Notation

Objective 3 ▶

Write the solution set of an inequality in set-builder notation.

Set-builder notation is another way to represent the solution set of an inequality. Set-builder notation was briefly discussed in Chapter 1. **Set-builder notation** is used to state conditions that the solutions must satisfy. An example is shown.

Set-builder Notation	Verbal Statement
$\{x \mid x \le 4\}$	The set of all x such that x is less than or equal to 4.

Objective 3 Examples   Graph each inequality and express the solution set in set-builder notation.

Inequality	Graph	Set-Builder Notation
3a. $x > 3$		$\{x \mid x > 3\}$
3b. $x \le 1.5$		$\{x \mid x \le 1.5\}$
3c. $\dfrac{1}{2} < x$		$\left\{x \mid x > \dfrac{1}{2}\right\}$
3d. $-2 < x \le 5$		$\{x \mid -2 < x \le 5\}$

✓ **Student Check 3** Graph each inequality and express the solution set in set-builder notation.

a. $x > -2$ **b.** $x \le -4.2$ **c.** $\dfrac{5}{3} > x$ **d.** $-4 \le x < 2$

Linear Inequalities

Objective 4 ▶

Solve linear inequalities.

Solving linear inequalities is very similar to solving linear equations. The goal is the same: to isolate the variable to one side of the inequality. The properties that enable us to do this are the addition and multiplication properties of inequality. While only one inequality symbol is used in the statement of the following properties, the properties work with all inequality symbols.

> **Property: Addition Property of Inequality**
> If $a < b$ and c is a real number, then
> $$a + c < b + c \quad \text{and} \quad a - c < b - c.$$

The property illustrates that adding or subtracting a number from both sides of an inequality produces an equivalent inequality. For an illustration of this property, let $a = -5, b = 2,$ and $c = 4$.

$$
\begin{array}{ccc}
a < b & a + c < b + c & a - c < b - c \\
-5 < 2 & -5 + 4 < 2 + 4 & -5 - 4 < 2 - 4 \\
\text{True} & -1 < 6 & -9 < -2 \\
 & \text{True} & \text{True}
\end{array}
$$

Adding or subtracting a number from both sides of an inequality maintains the inequality relationship.

> **Property: Multiplication Property of Inequality**
>
> **1.** If $a < b$ and $c > 0$, then $ac < bc$ and $\dfrac{a}{c} < \dfrac{b}{c}$.
>
> **2.** If $a < b$ and $c < 0$, then $ac > bc$ and $\dfrac{a}{c} > \dfrac{b}{c}$.

This property states the following:

1. Multiplying or dividing by a positive number produces an equivalent inequality.
2. Multiplying or dividing by a negative number produces an equivalent inequality only if the inequality symbol is reversed.

For an illustration of the property, let $a = -6$ and $b = 12$.

$$
\begin{array}{cccc}
\text{Let } c = 3. & a < b & ac < bc & \dfrac{a}{c} < \dfrac{b}{c} \\[2mm]
 & -6 < 12 & -6(3) < 12(3) & \dfrac{-6}{3} < \dfrac{12}{3} \\[2mm]
 & \text{True} & -18 < 36 & -2 < 4 \\
 & & \text{True} & \text{True}
\end{array}
$$

Multiplying both sides of an inequality by a positive number maintains the inequality relationship.

Let $c = -1$. $a < b$ $ac > bc$ $\dfrac{a}{c} > \dfrac{b}{c}$

$-6 < 12$ $-6(-1) > 12(-1)$ $\dfrac{-6}{-1} > \dfrac{12}{-1}$

True $6 > -12$ $6 > -12$

True   True

Multiplying both sides of an inequality by a negative number requires us to *reverse* the inequality symbol to maintain the inequality relationship.

Procedure: Solving a Linear Inequality

Step 1: Clear any parentheses from the equation by applying the distributive property.

Step 2: Remove any fractions by multiplying by the LCD.

Step 3: Use the addition property of inequality to collect all variable terms on one side and all constant terms on the other side.

Step 4: Use the multiplication property of inequality to get a coefficient of 1 on the variable. Remember that multiplying or dividing by a negative number, reverses the inequality symbol.

Step 5: If the inequality is a compound inequality of the form $a < x < b$, then the variable must be eliminated in the middle. Any operation that is required to isolate the variable must be done to all three parts of the inequality.

Step 6: Graph the solution set.

Step 7: Write the solution set in interval notation or set-builder notation.

Objective 4 Examples Solve each inequality. Graph the solution set and write the solution set in interval and set-builder notation.

4a. $x + 4 < -1$ **4b.** $-4x \leq 4$

4c. $3y - 7 > 2$ **4d.** $3(2a + 1) - a \geq 2a - 7$

4e. $\dfrac{1}{2}b - \left(b + \dfrac{3}{4}\right) > \dfrac{1}{4}b + \dfrac{1}{2}$ **4f.** $2(4x - 3) > 2x + 3(2x + 1)$

4g. $x + 5(x - 2) < 3(2x + 1) + 7$ **4h.** $-3 < 2x + 1 < 7$

Solutions **4a.** $x + 4 < -1$

$x + 4 - 4 < -1 - 4$ Subtract 4 from each side.

$x < -5$ Simplify.

The graph of the solution set consists of all numbers less than -5; that is, the numbers to the left of -5 but not including -5.

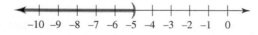

Interval notation: $(-\infty, -5)$

Set-builder notation: $\{x \mid x < -5\}$

4b.
$$-4x \leq 4$$

$$\frac{-4x}{-4} \geq \frac{4}{-4} \qquad \text{Divide each side by } -4 \text{ and reverse the inequality symbol.}$$

$$x \geq -1 \qquad \text{Simplify.}$$

The graph consists of all values greater than or equal to -1; that is, numbers to the right of -1 and including -1.

Interval notation: $[-1, -\infty)$

Set-builder notation: $\{x | x \geq -1\}$

4c.
$$3y - 7 > 2$$

$$3y - 7 + 7 > 2 + 7 \qquad \text{Add 7 to each side.}$$

$$3y > 9 \qquad \text{Simplify.}$$

$$\frac{3y}{3} > \frac{9}{3} \qquad \text{Divide each side by 3.}$$

$$y > 3 \qquad \text{Simplify.}$$

The graph consists of all values greater than 3, or to the right of 3, but not including 3.

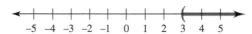

Interval notation: $(3, \infty)$

Set-builder notation: $\{y | y > 3\}$

4d. $3(2a + 1) - a \geq 2a - 7$

$$6a + 3 - a \geq 2a - 7 \qquad \text{Apply the distributive property.}$$

$$5a + 3 \geq 2a - 7 \qquad \text{Combine like terms.}$$

$$5a + 3 - 2a \geq 2a - 7 - 2a \qquad \text{Subtract } 2a \text{ from each side.}$$

$$3a + 3 \geq -7 \qquad \text{Simplify.}$$

$$3a + 3 - 3 \geq -7 - 3 \qquad \text{Subtract 3 from each side.}$$

$$3a \geq -10 \qquad \text{Simplify.}$$

$$\frac{3a}{3} \geq \frac{-10}{3} \qquad \text{Divide each side by 3.}$$

$$a \geq -\frac{10}{3} \qquad \text{Simplify.}$$

The graph consists of all numbers to the right of $-\dfrac{10}{3}$, including $-\dfrac{10}{3}$.

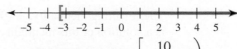

Interval notation: $\left[-\dfrac{10}{3}, \infty\right)$

Set-builder notation: $\left\{a \middle| a \geq -\dfrac{10}{3}\right\}$

4e. $\dfrac{1}{2}b - 1\left(b + \dfrac{3}{4}\right) > \dfrac{1}{4}b + \dfrac{1}{2}$

$\dfrac{1}{2}b - b - \dfrac{3}{4} > \dfrac{1}{4}b + \dfrac{1}{2}$ Distribute -1 to $\left(b + \dfrac{3}{4}\right)$.

$4\left(\dfrac{1}{2}b - b - \dfrac{3}{4}\right) > 4\left(\dfrac{1}{4}b + \dfrac{1}{2}\right)$ Multiply each side by the LCD, 4.

$2b - 4b - 3 > b + 2$ Simplify.

$-2b - 3 > b + 2$ Combine like terms.

$-2b - 3 - b > b + 2 - b$ Subtract b from each side.

$-3b - 3 > 2$ Simplify.

$-3b - 3 + 3 > 2 + 3$ Add 3 to each side.

$-3b > 5$ Simplify.

$\dfrac{-3b}{-3} < \dfrac{5}{-3}$ Divide each side by -3. Remember to reverse the inequality symbol.

$b < -\dfrac{5}{3}$ Simplify.

The graph consists of all numbers to the left of $-\dfrac{5}{3}$ but not including $-\dfrac{5}{3}$.

Interval notation: $\left(-\infty, -\dfrac{5}{3}\right)$

Set-builder notation: $\left\{ b \middle| b < -\dfrac{5}{3} \right\}$

4f. $2(4x - 3) > 2x + 3(2x + 1)$

$8x - 6 > 2x + 6x + 3$ Apply the distributive property.

$8x - 6 > 8x + 3$ Combine like terms.

$8x - 6 - 8x > 8x + 3 - 8x$ Subtract $8x$ from each side.

$-6 > 3$ False.

The resulting inequality, $-6 > 3$, is always false. Therefore, there is no solution of this inequality. The solution set is the empty set, or $\varnothing$. The graph is a blank number line as shown.

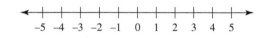

4g. $x + 5(x - 2) < 3(2x + 1) + 7$

$x + 5x - 10 < 6x + 3 + 7$ Apply the distributive property.

$6x - 10 < 6x + 10$ Combine like terms.

$6x - 10 - 6x < 6x + 10 - 6x$ Subtract $6x$ from each side.

$-10 < 10$ True.

The resulting inequality, $-10 < 10$, is always true. Therefore, any real number is a solution of this inequality. So, the solution set is all real numbers, or $\mathbb{R}$.

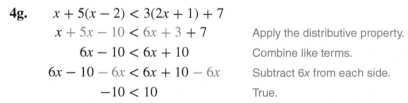

Interval notation: $(-\infty, \infty)$

Set-builder notation: $\{x \mid x \text{ is a real number}\}$

4h. This is a compound inequality. We must isolate the variable in the middle.

$$
\begin{array}{ccc}
3 < & 2x + 1 & < 7 \\
-3 - 1 < 2x + 1 - 1 & < 7 - 1 & \quad \text{Subtract 1 from each part.} \\
-4 < & 2x & < 6 \quad \text{Simplify.} \\
\dfrac{-4}{2} < & \dfrac{2x}{2} & < \dfrac{6}{2} \quad \text{Divide each part by 2.} \\
-2 < & x & < 3 \quad \text{Simplify.}
\end{array}
$$

Since $-2 < x < 3$ is equivalent to $x > -2$ and $x < 3$, the graph consists of all the points greater than -2, but not including -2, and less than 3, but not including 3. This is equivalent to all points between, but not including, -2 and 3.

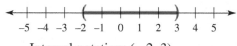

Interval notation: $(-2, 3)$

Set-builder notation: $\{x \mid -2 < x < 3\}$

✓ **Student Check 4** Solve each inequality. Graph the solution set and write the solution set in interval and set-builder notation.

a. $y + 3 < -2$ **b.** $-x \le 2$

c. $7y - 1 > 6$ **d.** $3(a + 2) - 7 \ge -4a + 10$

e. $\dfrac{1}{3}y - 2\left(y + \dfrac{1}{2}\right) > \dfrac{2}{3}y + \dfrac{1}{6}$ **f.** $4(x - 3) + 1 \ge 5(x + 2) - x$

g. $7x - 2(4x + 3) < 3(x + 5) - 4x$ **h.** $7 \le 6x - 3 \le 15$

Applications

Objective 5 ▶

Solve applications of linear inequalities.

There are a few key phrases that we need before we can solve applications relating to inequalities.

Phrase	Mathematical Statement
a is less than b	$a < b$
a is less than or equal to b	
a is no more than b	$a \le b$
a is at most b	
a is greater than b	$a > b$
a is greater than or equal to b	
a is no less than b	$a \ge b$
a is at least b	

These phrases together with the ones presented in Chapter 1 enable us to write an inequality to solve application problems.

Procedure: Solving Applications of Linear Inequalities

Step 1: Read the problem and determine the unknown. Assign a variable to the unknown value.

Step 2: Read the problem and determine the given information.

Step 3: Find the statement in the problem that states the inequality relationship, looking for key phrases that were listed in the chart above.

Step 4: Use the statement in Step 3 to write the inequality.

Step 5: Apply the addition and multiplication properties of inequalities to solve the inequality.

Step 6: Answer the question with a complete sentence.

Objective 5 Examples   **Write an inequality that models each situation. Solve the inequality and answer the question in a complete sentence.**

5a. Juanita has math test scores of 73, 85, and 80. What score does she need on her fourth math test to have a test average of at least 80?

Solution   **5a.** What is unknown? The score on the fourth test is unknown. Let x represent the score on the fourth math test.

What is known? The first three test scores are 73, 85, and 80.

To obtain the inequality, we use the following statement.

The average of the is at
four test scores least 80.

$$\frac{73 + 85 + 80 + x}{4} \geq 80$$ Recall the average is $\frac{\text{the sum of all items}}{\text{the number of items}}$.

$$\frac{238 + x}{4} \geq 80$$ Simplify the numerator of the fraction.

$$4\left(\frac{238 + x}{4}\right) \geq 4(80)$$ Multiply each side by 4.

$$238 + x \geq 320$$ Simplify.

$$238 + x - 238 \geq 320 - 238$$ Subtract 238 from each side.

$$x \geq 82$$ Simplify.

Juanita needs at least an 82 on her fourth test to have at least an 80 test average.

5b. Jamaal's final grade in his math class is based on a weighted average. The weights are as follows:

Homework average 10%   Quiz average 15%

Test average 50%   Final exam 25%

If Jamaal's homework average is 85, his quiz average is 80, and his test average is 75, what grade does he need on his final exam to have a final grade of at least 70 in the class? final grade of at least 80 in the class?

Solution   **5b.** What is unknown? The final exam grade is unknown. Let x = final exam grade.

What is known? The final grade is calculated by

10% (home work average) + 15% (quiz average) + 50% (test average) + 25% (final exam) = 0.10 (home work average) + 0.15 (quiz average) + 0.50 (test average) + 0.25 (final exam) = 0.10(85) + 0.15(80) + 0.50(75) + 0.25x

To obtain the inequalities, we use the following statements.

Final grade is at least 70.	Final grade is at least 80.
$0.10(85) + 0.15(80)$ $+ 0.50(75) + 0.25x \geq 70$	$0.10(85) + 0.15(80)$ $+ 0.50(75) + 0.25x \geq 80$
$8.5 + 12 + 37.5 + 0.25x \geq 70$	$8.5 + 12 + 37.5 + 0.25x \geq 80$
$58 + 0.25x \geq 70$	$58 + 0.25x \geq 80$
$58 + 0.25x - 58 \geq 70 - 58$	$58 + 0.25x - 58 \geq 80 - 58$
$0.25x \geq 12$	$0.25x \geq 22$
$\dfrac{0.25x}{0.25} \geq \dfrac{12}{0.25}$	$\dfrac{0.25x}{0.25} \geq \dfrac{22}{0.25}$
$x \geq 48$	$x \geq 88$

| Jamaal needs at least a 48 on the final exam to have a grade of at least 70 in the class. | Jamaal needs at least an 88 on the final exam to have a grade of at least 80 in the class. |

5c. The cost to rent a banquet facility for a wedding reception is $1500 plus $35 per person for the food. What is the number of attendees the couple can have at the reception if they have at most $5000 budgeted for the reception?

Solution   **5c.** What is unknown? The number of attendees is unknown. Let x represent the number of people attending the reception.

What is known? The cost of the reception is $1500 plus $35 per person or $1500 + 35x$.

$$\underset{\text{The cost}}{} \quad \underset{\substack{\text{is at} \\ \text{most}}}{} \quad \underset{\$5000.}{}$$

$1500 + 35x \le 5000$	Express statement mathematically. "at most" is $\le$.
$1500 + 35x - 1500 \le 5000 - 1500$	Subtract 1500 from each side.
$35x \le 3500$	Simplify.
$\dfrac{35x}{35} \le \dfrac{3500}{35}$	Divide each side by 35.
$x \le 100$	Simplify.

The couple can have at most 100 people attending their reception to have a cost of no more than $5000.

✔ **Student Check 5**   Write an inequality that models each situation. Solve the inequality and answer the question in a complete sentence.

a. Darryl has scores of 90, 85, 83, and 92 on his first four tests in his math class. What score does he need on the fifth test to have a test average of at least a 90?

b. Rosita's final grade in her math class is based on a weighted average. The weights are as follows:

Homework average 10%   Test average 45%
Quiz average 15%   Final exam 30%

If Rosita's homework average is 100, her quiz average is 75, and her test average is 80, what does she need on her final exam to have a final grade of at least 70 in the class? a final grade of at least 80 in the class?

c. The cost of the Verizon Wireless Nationwide Select 450 is $59.99 per month. The plan includes 450 anytime minutes and unlimited text, picture, video and instant messaging to anyone on any network in the United States. Additional anytime minutes cost $0.45 each. How many additional minutes can Lisa talk so that her total monthly cell phone bill is at most $75? Round to the nearest minute. (Source: www.myrateplan.com)

Objective 6 ▶

Troubleshoot common errors.

Troubleshooting Common Errors

Some common errors associated with linear inequalities are shown next.

Objective 6 Examples   **A problem and an incorrect solution are given. Provide the correct solution and an explanation of the error.**

6a. Write the interval for the graph.

Incorrect Solution	Correct Solution and Explanation
The interval is $[-2, -\infty)$.	The interval must be written in order from smallest to largest or from left to right. Since the graph continues indefinitely to the left, the interval begins at $-\infty$ and ends at the right with -2. So, the interval is $(-\infty, -2]$.

6b. Solve $-6x + 2 > 8$.

Incorrect Solution	Correct Solution and Explanation
$$-6x + 2 > 8$$ $$-6x + 2 - 2 > 8 - 2$$ $$-6x > 6$$ $$\frac{-6x}{-6} > \frac{6}{-6}$$ $$x > -1$$ The solution is $(-1, \infty)$.	The only error made was that the inequality symbol was not reversed when both sides were divided by a negative number. $$-6x + 2 > 8$$ $$-6x + 2 - 2 > 8 - 2$$ $$-6x > 6$$ $$\frac{-6x}{-6} < \frac{6}{-6}$$ $$x < -1$$ The solution is $(-\infty, -1)$.

ANSWERS TO STUDENT CHECKS

		Graph	Interval Notation	Set-Builder Notation
Student Check 1	a.			
Student Check 2	a.	$-5\ -4\ -3\ -2\ -1\ \ 0\ \ 1\ \ 2\ \ 3\ \ 4\ \ 5$	$(-2, \infty)$	$\{x \mid x > -2\}$
Student Check 3	a.			
Student Check 1	b.			
Student Check 2	b.	$-6\ -5\ -4\ -3\ -2\ -1\ \ 0\ \ 1\ \ 2\ \ 3\ \ 4$	$(-\infty, -4.2)$	$\{x \mid x \le -4.2\}$
Student Check 3	b.			
Student Check 1	c.			
Student Check 2	c.	$-6\ -5\ -4\ -3\ -2\ -1\ \ 0\ \ 1\ \ 2\ \ 3\ \ 4$	$\left(-\infty, \frac{5}{3}\right)$	$\left\{x \mid x < \frac{5}{3}\right\}$
Student Check 3	c.			
Student Check 1	d.			
Student Check 2	d.	$-5\ -4\ -3\ -2\ -1\ \ 0\ \ 1\ \ 2\ \ 3\ \ 4\ \ 5$	$[-4, 2)$	$\{x \mid -4 \le x < 2\}$
Student Check 3	d.			
Student Check 4	a.	$-7\ -6\ -5\ -4\ -3\ -2\ -1\ \ 0\ \ 1\ \ 2\ \ 3$	$(-\infty, -5)$	$\{y \mid y < -5\}$
	b.	$-7\ -6\ -5\ -4\ -3\ -2\ -1\ \ 0\ \ 1\ \ 2\ \ 3$	$[-2, \infty)$	$\{x \mid x \ge -2\}$
	c.	$-5\ -4\ -3\ -2\ -1\ \ 0\ \ 1\ \ 2\ \ 3\ \ 4\ \ 5$	$[1, \infty)$	$\{y \mid y > 1\}$
	d.	$-5\ -4\ -3\ -2\ -1\ \ 0\ \ 1\ \ 2\ \ 3\ \ 4\ \ 5$	$\left[\frac{11}{7}, \infty\right)$	$\left\{a \mid a \ge \frac{11}{7}\right\}$
	e.	$-5\ -4\ -3\ -2\ -1\ \ 0\ \ 1\ \ 2\ \ 3\ \ 4\ \ 5$	$\left(-\infty, -\frac{1}{2}\right)$	$\left\{y \mid y < -\frac{1}{2}\right\}$
	f.	$-5\ -4\ -3\ -2\ -1\ \ 0\ \ 1\ \ 2\ \ 3\ \ 4\ \ 5$	$\varnothing$	$\varnothing$
	g.	$-5\ -4\ -3\ -2\ -1\ \ 0\ \ 1\ \ 2\ \ 3\ \ 4\ \ 5$	$(-\infty, \infty)$	$\{x \mid x \text{ is a real number}\}$
	h.	$0\ \ 1\ \ 2\ \ 3\ \ 4\ \ 5\ \ 6\ \ 7\ \ 8\ \ 9\ \ 10$	$\left[\frac{5}{3}, 3\right]$	$\left\{x \mid \frac{5}{3} \le x \le 3\right\}$

Student Check 5   a. Darryl must make a 100 or better on his fifth test to have a test average of at least a 90.

b. Rosita needs at least a 42.5 on her final exam to have a final grade of at least 70. She needs at least an 75.8 on her final exam to have a final grade of at least 80.

c. Lisa can talk no more than 33 additional minutes to have a monthly bill less than or equal to $75.

SUMMARY OF KEY CONCEPTS

1. The graph of the solution set of an inequality is a picture of all real numbers that make the inequality a true statement. A parenthesis (used with < or >) on a number indicates the number is not included in the solution set but every number very close to it is included. A bracket (used with ≤ or ≥) on a number indicates the number is included in the solution set.

2. Interval notation is a concise way to represent the solution set of an inequality. It represents the interval of real numbers in which solutions lie. The interval notation always begins with the left bound of the solution set and ends with the right bound of the solution set. When a set continues indefinitely to the right, the right bound is represented by ∞. When a set continues indefinitely to the left, the left bound is represented by $-\infty$. A parenthesis is always used with ∞ or $-\infty$.

3. Set-builder notation is another way to represent the solution set of an inequality. It is written using { }. We write {variable|final inequality}. Example: $\{y \mid y < 5\}$.

4. Linear inequalities are solved using the addition and multiplication properties of inequality. The most important thing to remember is that when you multiply or divide by a negative number, you must also reverse the inequality symbol.

5. Applications are solved by translating the given statements into appropriate inequalities. Key phrases are "is at least" and "is at most." A good way to remember the translations is to think of money. For example, if you have at least $10, you would have $10 or more; if you have at most $10, you would have $10 or less.

GRAPHING CALCULATOR SKILLS

The calculator can be used as a reference to check the work we have done by hand.

Example: $x - 5 > 2$

By adding 5 to both sides, we find the solution of $x - 5 > 2$ to be $x > 7$. The graph is

To check our work on the calculator, we should determine if the inequality is true for a number larger than 7, if it is false for a number less than 7, and what happens at 7.

We will use the store feature to test values. We first store in a value for x using the STO> command. Then we enter the original inequality in the calculator and press Enter. The result will either be 1 or 0. If the result is 1, then the stored value satisfies the inequality and is a solution; if the result is 0, then the stored value does not satisfy the inequality and is not a solution.

Check: $x = 8$ (a value larger than 7).

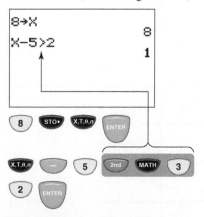

Check: $x = 6$ (a value less than 7).

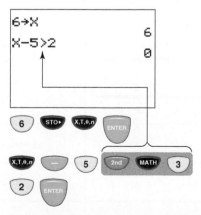

Check: $x = 7$.

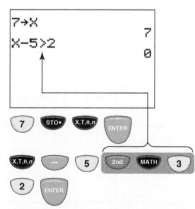

The result of 1 confirms that $x = 8$ is a solution of the inequality. The shaded portion of the graph should contain 8 (to the right of 7).

The result of 0 means that $x = 6$ is not a solution of the inequality. The shaded portion of the graph should not include the side with 6.

The result of 0 means that $x = 7$ is not a solution of the inequality. The graph should not include 7, so we put a parenthesis on 7.

SECTION 2.7 / EXERCISE SET

 Write About It!

Use complete sentences in your answer to the following exercises.

1. Define an inequality in one variable.

2. When do you use brackets and when do you use parentheses when graphing solutions of inequalities?

3. What operation do you perform to both sides of an inequality that requires you to reverse the inequality symbol?

4. Explain how to write the interval notation for the solution set of an inequality. Use an example.

5. Explain how to write the interval notation for the solution set of a compound inequality, $a \leq x \leq b$. Use an example.

6. Explain how to write the interval notation for the solution set of a compound inequality, $a < x \leq b$. Use an example.

Determine if each statement is true or false. If the statement is false, explain why.

7. The value $x = 2$ is a solution of the inequality $x - 3 > -1$.

8. The solution set of $-2x \geq 4$ is $[-2, \infty)$.

9. The inequality $5 > x$ is equivalent to $x < 5$.

10. The inequality $-3 \leq x \leq 8$ is equivalent to $8 \geq x \geq -3$.

11. The inequality $-6 \leq -3x \leq 12$ is equivalent to $2 \leq x \leq -4$.

12. Solving a linear inequality is no different from solving a linear equation.

 **Practice Makes Perfect!**

Fill in the missing information. (*See Objectives 1–3.*)

Inequality	Graph	Interval Notation	Set-Builder Notation
13. $x > -2$			
14. $y \geq -5$			
15. $a < 1$			
16. $x \leq 8$			
17.			
18.			
19.			
20.			
21.		$(-\infty, -3]$	
22.		$(2, \infty)$	
23.		$\left(-\infty, \dfrac{1}{4}\right]$	

Inequality	Graph	Interval Notation	Set-Builder Notation
24.		$\left[\dfrac{3}{5}, \infty\right)$	
25.			$\{x \mid x < 10\}$
26.			$\{y \mid y > 1\}$
27.			$\left\{x \mid x \le -\dfrac{5}{4}\right\}$
28.			$\{y \mid y \ge -1\}$
29. $-2 \le x \le 5$			
30. $-6 < y \le 8$			
31. $3 \le x < 10$			
32. $0 < a < 12$			
33.		$[-7, 3)$	
34.		$(1, 9]$	
35.			$\left\{x \mid -\dfrac{7}{3} \le x \le 8\right\}$
36.			$\left\{y \mid -\dfrac{5}{2} < y < \dfrac{1}{2}\right\}$

Solve each inequality. Graph the solution set and write the solution set in interval notation and in set-builder notation. (*See Objective 4.*)

37. $x - 3 > -7$

38. $y - 6 \ge -15$

39. $x + 1 \ge 8$

40. $a + 9 \ge -2$

41. $y - 4 < -4$

42. $a + 5 > 5$

43. $x + 10 \le 6$

44. $y - 2 \le -3$

45. $2 < y + 3$

46. $-4 \ge a - 7$

47. $-3x < 9$

48. $7a \ge -14$

49. $\dfrac{y}{4} > 1$

50. $\dfrac{x}{5} \ge -2$

51. $-a > 8$

52. $-y \le -4$

53. $-6a \ge -12$

54. $-9a < 27$

55. $-\dfrac{2}{3}x > 18$

56. $-\dfrac{4}{5}a \le -16$

57. $2y - 4 < -8$

58. $6a + 1 > -5$

59. $-5a + 3 \ge -7$

60. $4 - 7x \le -10$

61. $6x - 5 \le 2x + 8$

62. $-9x + 3 > 8x - 1$

63. $4 - 3(x - 5) > -2x + 1$

64. $8 - (7x + 9) \ge 10x + 4$

65. $\dfrac{1}{2}y - 3 \le \dfrac{1}{6}$

66. $\dfrac{3}{4}a + 2 < \dfrac{5}{8}$

67. $75 + 0.25x < 100$

68. $30 + 0.15a > 51$

69. $-9 < a + 4 < 12$

70. $-3 \le b - 7 \le 6$

71. $-13 \le 3x - 7 \le 8$

72. $-18 < 5y - 8 < 12$

73. $18 \le 6 - 4x < 34$

74. $-11 < 3 - 2y \le 25$

75. $5 \le -6a - 1 \le 17$

76. $-13 < -9b + 5 < 32$

77. $\dfrac{1}{8} < \dfrac{3}{4}a + 1 < \dfrac{5}{8}$

78. $-\dfrac{5}{12} \le \dfrac{2}{3}b - 1 \le \dfrac{7}{6}$

79. $-5.2 < 1.2x - 2.8 < 5.6$ **80.** $1.5 \le 1.8 - 0.3x \le 7.5$

81. $4(x - 2) - 9 \ge -7x + 16$

82. $2(x + 5) - 14 \le -9x + 18$

83. $3(x - 1) - 5 < 8(x + 2) - x$

84. $4(x - 1) + 10 > 9(x + 3) - 2x$

85. $\dfrac{2}{5}y - 2\left(y - \dfrac{3}{10}\right) < \dfrac{1}{2}y + \dfrac{7}{10}$

86. $\dfrac{3}{4}y - 2\left(y + \dfrac{1}{3}\right) > \dfrac{1}{2}y - \dfrac{5}{6}$

87. $12x - 5(2x + 3) > 6(x + 7) - x$

88. $16y - 7(y + 4) > 4(y - 2) + y$

Write an inequality that models each situation. Solve the inequality and answer the question in complete sentences. (*See Objective 5*.)

89. Mykel scored 85, 78, 72, and 81 on four math tests. What score does she need on the fifth test to have at least an 80 average?

90. Lucas scored 77, 78, and 75 on three math tests. What score does he need on the fourth test to have at least an 80 average?

91. Shanika's final grade in her math class is based on a weighted average. Homework counts as 5%, quizzes count as 10%, projects count as 10%, tests count as 50%, and the final exam counts as 25%. If she has a homework average of 85, a quiz average of 78, a project average of 90, and a test average of 73, what score does she need on her final exam to have at least an 80 for her final grade?

92. Hassan's final grade in his math class is based on a weighted average. Homework counts as 10%, quizzes count as 20%, participation counts as 5%, tests count as 40%, and the final exam counts as 25%. If he has a homework average of 88, a quiz average of 82, a participation grade of 100, and a test average of 80, what score does he need on his final exam to have at least an 80 for his final grade?

93. Skype is a software program that allows users to call any number in the world with their computer. Skype offers 12 months of unlimited phone calls in the United States for $29.95 per year. Sergeant Walker is stationed in Iraq for one year. Skype charges 37.2¢ per minute plus 3.9¢ per international call. If Sergeant Walker's family calls him once a week for the year, the cost of the phone service is given by $C = 29.95 + 0.372m + 0.039(52) = 31.978 + 0.372m$ where m is the number of minutes. How many minutes can the family talk in a year so that their cost in phone calls for the year is at most $1000? (Source: http://www.skype.com)

94. The cost to rent a car for a week from a luxury rental agency is $350 plus 44 cents per mile. How many miles can be driven if you have at most $500 to rent the car for a week? Round to the nearest whole number.

 Mix 'Em Up!

Solve each inequality. Graph the solution set and write the solution set in interval notation and in set-builder notation.

95. $-x + 12 < -12$ **96.** $-y + 16 \ge 14$

97. $9x - 20 > -2$ **98.** $-10y + 19 \le 39$

99. $0.2a - 7 \le -1.5$ **100.** $-0.32b + 4 < 10.4$

101. $\dfrac{y}{8} - 2 \le 6$ **102.** $-\dfrac{1}{3}x + 5 > \dfrac{4}{3}$

103. $-\dfrac{5}{8}x - 3 \le \dfrac{3}{8}$ **104.** $\dfrac{1}{4}y + 3 \ge \dfrac{11}{4}$

105. $-11 < 2a - 1 < 17$ **106.** $15 < 6b - 3 \le 21$

107. $-5 \le 4x + 3 \le 19$ **108.** $2 \le 5 - 3y < 26$

109. $-9 < -2a + 5 < 5$ **110.** $-6 \le 4b + 10 < 2$

111. $12 - 2(x - 3) > -2x + 1$ **112.** $6x - (3x - 2) > 3x + 1$

113. $7x + 2(x + 5) < 9x + 8$ **114.** $x - (4x - 2) > 5 - 3x$

115. $\dfrac{5}{2}x + 3\left(x + \dfrac{3}{2}\right) \ge \dfrac{9}{2}x - \dfrac{7}{2}$

116. $\dfrac{8}{3}y - 2\left(y - \dfrac{2}{11}\right) \ge \dfrac{5}{3}y - \dfrac{2}{11}$

117. $-7 \le -3(x - 3) + 2 \le -1$

118. $-3 < -2(y + 4) + 7 < 5$

119. $-\dfrac{7}{2} < \dfrac{1}{3}x - 1 < \dfrac{13}{6}$ **120.** $-\dfrac{4}{5} \le \dfrac{1}{2}y + 2 \le \dfrac{11}{5}$

121. $-3.8 < 0.25x - 2.8 < 13.2$

122. $2.5 \le 0.12y + 3.1 < 12.7$

Write an inequality that models each situation. Solve the inequality and answer the question in complete sentences.

123. Azziz scored 57, 97, and 85 on three math tests. What score does he need on the fourth test to have at least an 80 average?

124. Elaine scored 84, 60, and 63 on three chemistry tests. What score does she need on the fourth test to have at least an 70 average?

125. Derek's final grade in his math class is based on a weighted average. Homework counts as 10%, quizzes count as 20%, tests count as 50%, and the final exam counts as 20%. If he has a homework average of 67, a quiz average of 65, and a test average of 89, what score does he need on his final exam to have at least an 80 for his final grade?

126. Sue's final grade in her math class is based on a weighted average. Homework counts as 15%, quizzes count as 10%, tests count as 55%, and the final exam counts as 20%. If she has a homework average of 91, a quiz average of 42, and a test average of 82, what score does she need on her final exam to have at least an 80 for her final grade?

127. The cost to rent a banquet facility for a wedding reception is \$1050 plus \$36 per person for the food. What is the number of attendees the couple can have at the reception if they have at most \$9400 budgeted for the reception?

128. The cost to rent a banquet facility for a wedding reception is \$1030 plus \$25 per person for the food. What is the number of attendees the couple can have at the reception if they have at most \$4900 budgeted for the reception?

129. The cost to rent a car for a week from Convenient Rental is \$304 plus 30 cents per mile. How many miles can be driven if you have at most \$960 to rent the car for a week? Round to the nearest whole number, if necessary.

130. The cost to rent a car for a week from Economy Rental is \$257 plus 30 cents per mile. How many miles can be driven if you have at most \$410 to rent the car for a week? Round to the nearest whole number, if necessary.

 You Be the Teacher!

Correct each student's errors, if any.

131. Solve the inequality: $-2x + 5 < 11$.

Mike's work:

$$-2x + 5 < 11$$
$$\underline{\quad -5 \quad -5 \quad}$$
$$-2x \quad\quad < 6$$
$$\frac{-2x}{-2} < \frac{6}{-2}$$
$$x < -3$$

132. Solve the inequality $-9 < 3 - 2x < 13$.

Edna's work:

$$-9 < 3 - 2x < 13$$
$$-12 < -2x < 10$$
$$6 < x < -5$$

133. Solve the inequality: $3(2x - 5) + 9 > 2(3x + 1)$.

Gabriel's work:

$$3(2x - 5) + 9 > 2(3x + 1)$$
$$6x - 5 + 9 > 6x + 1$$
$$6x + 4 > 6x + 1$$
$$4 > 1$$

The solution is $\mathbb{R}$.

134. Solve the in equality: $5(x - 1) + 2x < 7(x + 3) - 10$.

Kristen's work:

$$5(x - 1) + 2x < 7(x + 3) - 10$$
$$5x - 1 + 2x < 7x + 3 - 10$$
$$7x - 1 < 7x - 7$$
$$-1 < -7$$

The solution is $\varnothing$.

 Calculate It!

Solve each inequality. Then use a graphing calculator to check the solution.

135. $8x + 20 < 2x - 15 + x$

136. $2(x + 10) - 5 \geq 4x + 18$

137. $\dfrac{5}{6}x - x < 8 - \dfrac{3}{2}x$

138. $-3.5(-1.6x + 2.4) \geq 0.8(-2.1x - 41.44)$

 GROUP ACTIVITY   **The Growth of Cell Phones and Cell Sites**

Group Project for Chapter 2

1. The number of U.S. wireless subscriber connections, w, in millions is approximately 19.35 more than 17.83 times the number of years after 1995. If x is the number of years after 1995, write an equation that approximates the number of wireless subscriber connections.

2. Use the equation in Step 1 to determine the number of wireless subscriber connections for the given years. Round to the nearest tenth.

Year	Years After 1995, x	Number of Connections, w
1995		
2000		
2005		
2010		

3. In 2010, the number of wireless subscriber connections was 93% of the population of the United States. Use the previous information to approximate the U.S. population. Show the equation that was used to solve the problem.

4. Research the Internet to determine the current U.S. population. Round the population to the nearest million. Use the equation in step 1 to determine when the number of wireless subscriber connections will reach the current U.S. population.

5. The number of cell sites can be given by $c = 15{,}552.28x + 19{,}662.9$, where x is the years after 1995. Complete the table to determine the number of cell sites for the given years.

Year	Years After 1995, x	Number of Cell Sites, c
1995		
2000		
2005		
2010		

6. Determine the percent increase in wireless subscriber connections and cell sites for each 5-yr period. Does the growth of cell sites meet the growth of subscribers?

5-yr Period	Percent Increase in Wireless Subscriber Connections	Percent Increase in Cell Sites
1995–2000		
2000–2005		
2005–2010		

Source: www.ctia.org

Linear Equations and Inequalities in One Variable

-ᓂᑯ- **What's the big idea?** Chapter 2 provides us with the skills to solve equations and inequalities in one variable. Knowing how to solve these equations and inequalities gives us the foundation we need to solve some word problems. The properties that enable us to solve these types of problems also provide the framework for us to solve more difficult equations, which will be encountered in later chapters.

The Tools

Listed below are the key terms, skills, formulas, and properties you should know for this chapter.

The page reference is provided if you need additional help with the given topic. The Study Tips will assist in your preparation for an exam.

Study Tips

1. Learn all of the terms, formulas, and properties. Make flash cards and have someone quiz you.
2. Rework problems from the exercises and also the ones you worked in class. Work additional problems from the review exercises.
3. Review the summaries of key concepts.
4. Work the chapter test.
5. Be sure to review the online resources for additional study materials.

Terms

Area 137	Equivalent equations 102	Perimeter 137
Circumference 137	Expression 92	Right angle 142
Complementary angles 142	Formula 136	Set-builder notation 168
Compound inequality 167	Identity 131	Simple inequality 167
Conditional equation 129	Interest 154	Solution of a linear inequality 166
Consecutive even integers 119	Interval notation 167	Solution of the equation 92
Consecutive integers 118	Linear equation in one	Straight angle 142
Consecutive odd integers 119	variable 101	Supplementary angles 142
Contradictions 130	Linear inequality in one variable 166	Triangle 145
Equation 92	Opposite angles 144	Vertical angles 144

Formulas and Properties

- Addition property of equality   102
- Addition property of inequality   169
- Discount problem   152

- Distance formula   159
- General percent problem   151
- Markup problem   152
- Multiplication property of equality   112

- Multiplication property of inequality   169
- Percent increase or decrease   152
- Simple interest formula   154

CHAPTER 2 / SUMMARY

How well do you know this chapter? Complete the following questions to find out. Take a look back at the section if you need help.

SECTION 2.1 Equations and Their Solutions

1. A(n) _____ is a statement in which two expressions are equal.

2. A(n) _____ is a value of the variable that makes the equation true.

3. When translating phrases, the expressions _____, _____, _____, _____, and __ represent the equals sign.

SECTION 2.2 The Addition Property of Equality

4. A linear equation in one variable is an equation that can be written in the form _____. An example of a linear equation in one variable is _____ _____.

5. To solve a linear equation in one variable, the goal is to produce an equation of the form _____ or _____.

6. _____ equations are equations with the same solution set.

7. The addition property of equality states that we can _____ or _____ the same number from both sides of an equation and not change the solution set.

8. To solve an equation, we must _____ each side as much as possible. Then we must isolate the _____ to one side and the _____ to the other.

SECTION 2.3 The Multiplication Property of Equality

9. The multiplication property of equality enables us to _____ or _____ both sides of an equation by the same number.

10. If the coefficient of the variable is a fraction, we can multiply both sides of the equation by its _____ to obtain a coefficient of 1 on the variable.

11. _____ integers are integers that follow one another. An example of consecutive integers is _____, an example of consecutive odd integers is _____, and an example of consecutive even integers is _____. If x is an integer, then _____ and _____ represents the next two consecutive integers.

SECTION 2.4 More on Solving Linear Equations

12. To solve equations with fractions, we can multiply both sides of the equation by the _____ to obtain an equivalent equation without fractions.

13. To solve equations with decimals, we can multiply both sides of the equation by the _____ that eliminates the decimals.

14. A(n) _____ equation is an equation that is true for some values of the variable but not true for others.

15. A(n) _____ is an equation that is not true for any values of the variable. The solution set for these equations is _____.

16. A(n) _____ is an equation that is true for all values of the variable. The solution set for these equations is _____.

SECTION 2.5 Formulas and Applications from Geometry

17. A(n) _____ is an equation that expresses the relationship between two or more variables.

18. To evaluate a formula, we _____ the given values in place of the variable and simplify.

19. The perimeter of a polygon is the _____ around the figure.

20. The distance around a circle is its _____.

21. The _____ of a figure is the number of square units it takes to cover the inside of the figure.

22. The sum of the measures of complementary angles is ____. The sum of the measures of supplementary angles is _____.

23. A right angle has measure _____ and a _____ angle has measure of 180°.

24. If x represents the measure of an angle, _____ represents the measure of its complement and _____ represents the measure of its supplement.

25. Vertical angles, or _____ angles, are formed by intersecting lines. These angles are _____ in measure.

26. The sum of the measures of the angles in a triangle is _____.

SECTION 2.6 Percent, Rate, and Mixture Problems

27. Percent means _____. So, 30% = ____ = ____.

28. _____ is the amount of money collected for using or borrowing money. The _____ _____ _____ is $I = Prt$.

29. When working with mixture problems, we must find the amount of _____ substance in each solution. We do this by multiplying the strength of the solution by the _____ of the solution.

30. To find the value of a collection of coins, multiply the number of coins by the _____ of each coin.

31. Distance traveled is the ___ multiplied by the ____ traveled. In symbols, $d = $ __.

SECTION 2.7 Linear Inequalities in One Variable

32. A linear inequality in one variable is an inequality of the form _____.

33. A(n) _____ of a linear inequality is a value that makes the inequality true.

34. The picture of the solution set of an inequality is the _____ of the solution set.

35. If an endpoint of a solution set of an inequality is included in the solution, a _____ is used. If an endpoint of a solution set is not included in the solution set, a _____ is used.

36. A concise way to express the solution set of an inequality is _____ _____.

37. The symbol __ indicates that the solution of an inequality continues to the right indefinitely. The symbol ___ indicates that the solution of an inequality continues indefinitely to the left. A _____ is always used with these notations.

38. The _____ notation is also used to represent the solution set of an inequality. This notation states the conditions the solution must satisfy.

39. When solving linear inequalities, we can ___ or _____ the same number from both sides of an inequality and not change the relationship between the two expressions.

40. When solving linear inequalities, we can multiply or divide both sides by a _____ number and not change the relationship between the two expressions.

41. When solving linear inequalities, we can multiply or divide both sides by a _____ number, but we must

also _____ the inequality symbol to maintain the relationship between the two expressions.

42. The phrase "is at most" can be translated by the symbol __. The phrase "is at least" can be translated by the symbol __.

CHAPTER 2 / REVIEW EXERCISES

SECTION 2.1

Determine whether each of the following is an expression or an equation. (*See Objective 1.*)

1. $7x - 9 = 2x - 12$

2. $-5r^2 + 11 - 12r - 24$

3. $5x - 11 - 2x + 3$

4. $6y^2 - 10 = 12y - 5$

Determine if the given numbers are solutions of the equation. Organize the information in tables. (*See Objective 2.*)

5. Is $x = -4$, $x = -1$, $x = 0$, $x = 4$, or
$x = -\dfrac{4}{5}$ a solution of $3x - 1 = -2x - 5$?

6. Is $x = -6$, $x = -2$, $x = 0$, $x = 3$, or
$x = \dfrac{1}{2}$ a solution of $x^2 + 3x = 18$?

Assign a variable to the unknown quantity and then write an equation that represents each statement. (*See Objective 3.*)

7. Four times the sum of a number and -17 is the same as twice the number. Find the number.

8. The sum of a number and 18 is the same as seven times the number. Find the number.

9. The difference of four times a number and -3 equals 45 less than the number. Find the number.

10. One angle is 6° more than another angle. Their sum is 90°. Find the measure of each angle.

11. One angle is 17° less than another angle. Their sum is 180°. Find the measure of each angle.

12. From June 2008 to June 2009, John Mars and Larry Page were the two of the 31 richest Americans. Their combined net worth was $26.3 billion. John Mars's net worth is $4.3 billion less than Larry Page's. Find the net worth of each person. (Source: http://www.forbes.com/lists/)

SECTION 2.2

Solve each equation. (*See Objectives 2 and 3.*)

13. $5x + 6 = 2x + 18$

14. $3 - x = 11 - 5x$

15. $\dfrac{3}{5}x + 2 = -6 - \dfrac{2}{5}x$

16. $\dfrac{4}{7} - x = 5x - \dfrac{3}{7}$

17. $5(x - 2) + 13 = 3(x - 2) - 9$

18. $1.24x + 3.25 = 2.94x - 1$

19. $3.5x - 2.6 + 1.9x = 7.3x - 13.24$

20. $32 - 10x = 32 - 15x$

For each problem: (1) assign a variable to the unknown, (2) write an equation that represents the situation, (3) solve the equation, and (4) answer the question using complete sentences. (*See Objective 4.*)

21. The sum of a number and -18 is the same as five times the number. Find the number.

22. Three times the sum of a number and 4 is the same as twice the number. Find the number.

23. Twice the difference of a number and -5 is the same as four times the number. Find the number.

24. Six more than a number is -1. Find the number.

25. The difference of twice a number and -19 is equal to 47 more than the number. Find the number.

26. One angle is 24° more than another angle. Their sum is 90°. Find the measure of each angle.

27. From June 2008 to June 2009, Christy Walton and Philip Knight were the two of the 31 richest Americans. Their combined net worth was $31 billion. Christy Walton's net worth is $12 billion more than Philip Knight's. Find the net worth of each person. (Source: http://www.forbes.com/lists/)

28. From June 2008 to June 2009, Dr. Phil McGraw and Oprah Winfrey were the two of the top 100 celebrities. Their combined earnings were $355 million. Dr. Phil McGraw earned $195 million less than Oprah Winfrey. How much did each of them earn? (Source: http://www.forbes.com/lists/)

SECTION 2.3

Solve each equation. (*See Objectives 1 and 2.*)

29. $5y - 7 = 23$

30. $-2x + 30 = 4x$

31. $\dfrac{a}{2} = -8$

32. $8z + 5 - 2z = 15 - 3z$

33. $-10x + 9 - 3x = 18 + 5x - 16$

34. $-\dfrac{1}{2}(x + 6) + 8 = x + 5$

35. $-1.8b + 6 = 42$

36. $-0.6y + 15.2 = 2.7y + 24.44$

37. $1.2(5.2x - 3.2) = -0.6(2.5x - 0.05)$

38. $12(x - 2) = 2(6x - 1)$

For each problem: (1) assign a variable to the unknown, (2) write an equation that represents the situation, (3) solve the equation, and (4) answer the question using complete sentences. (*See Objectives 3 and 4.*)

39. Three times a number is equal to -54. Find the number.

40. Two-fifths of a number is 12. Find the number.

41. The quotient of a number and 3 is -12. Find the number.

42. 60% of a number is 9. Find the number.

43. The sum of a number and 15 is the same as three more than twice a number. Find the number.

44. How many nickels will it take to make $8.50?

45. The sum of two consecutive integers is -97. Find the integers.

46. The cost to rent a midsize car for one day is $55 plus $0.32 per mile. The daily cost is represented by the expression $55 + 0.32x$, where x is the number of miles driven. How many miles has the car been driven if the cost of the rental car is $167?

SECTION 2.4

Solve each equation. If the equation is a contradiction, write the solution as ∅. If the equation is an identity, write the solution as ℝ. (See Objectives 1–4.)

47. $2x - 5 = 5$

48. $4(y + 1) - (y - 5) = 3(y + 1)$

49. $\dfrac{2}{5}x - 4 = \dfrac{3}{10}x$

50. $1.2x + 0.8x = 900$

51. $\dfrac{x - 8}{14} + 1 = \dfrac{x + 5}{7}$

52. $0.032x + 0.013(540 - x) = 13.67$

53. $\dfrac{5}{6}(x - 2) = \dfrac{1}{6}(5x - 8) - \dfrac{1}{3}$

54. $\dfrac{3}{4}a - \dfrac{5}{8} = \dfrac{1}{2}a + \dfrac{1}{4}$

55. $0.3(a - 1) + 0.5a = 0.4(2a + 9)$

56. $0.64x + 1.6(-4.5) = 0.4(x - 17.1)$

SECTION 2.5

The formula $A = P(1 + r)^t$ is used to calculate the amount of money in an account at the end of t years if P dollars are invested at an annual interest rate, r. (See Objective 1.)

57. Find the value of A if $1200 is invested for 2 yr at 1.5% annual interest.

58. Find the value of P if $2521.50 is expected to be accumulated at the end of 2 yr at 2.5% annual interest.

The distance d, rate r, and time t of an object in motion can be related by the formula $d = rt$. (See Objective 1.)

59. Find d if $r = 62$ mph and $t = 2.5$ hr.

60. Find t if $d = 306$ mi and $r = 68$ mph.

The revenue R, unit price p, and sale level x of a product are related by the formula $R = xp$. (See Objective 1.)

61. Find R if $p = \$125$ and $x = 56$.

62. Find p if $R = \$1392$ and $x = 96$.

The formula to find the volume of a cylinder with radius r and height h is $V = \pi r^2 h$. Use $\pi = 3.14$. (See Objective 1.)

63. Find V if $r = 6$ in. and $h = 15$ in.

64. Find h if $V = 157$ m³ and $r = 2.5$ m.

Use the perimeter, area, or circumference formula to find the indicated quantity. (See Objective 2.)

65. Find the height of a triangle whose area is 84 m² and whose base is 12 m.

66. Find the perimeter of the rectangle whose area is 432 in.² and whose length is 27 in.

67. Find the area of a rectangle if the perimeter is 74 m and whose width is 20 m.

68. Find the radius and area of a circle with circumference 150.72 cm. Use 3.14 for the value of π to approximate answers to two decimal places.

Solve each formula for the specified variable. (See Objective 3.)

69. $R = xp$ for p

70. $C = 3l + 4w$ for l

71. $7x + 3y = 21$ for y

72. $y = \dfrac{1}{4}x - 1$ for x

73. $0.05x + 0.02y = 35$ for x

74. $0.36x - 0.2y = 216$ for y

Find the measure of each unknown angle. (See Objective 4.)

75. Find the measure of an angle whose complement is 25° less than the measure of the angle.

76. Find the measure of an angle whose supplement is 54° more than the measure of the angle.

77. Find the measure of an angle whose supplement is 20° less than three times the measure of the angle.

78. The supplement of an angle is 30° more than three times its complement. Find the measure of the angle.

Find the measure of each angle labeled in the figure. (See Objectives 5 and 6.)

79.

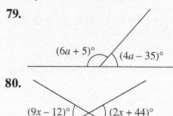

80.

81.

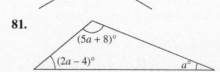

82. In triangle ABC, the measure of angle B is 18° less than the measure of angle A. The measure of angle C is 27° more than the measure of angle A. Find the measure of each angle in the triangle.

SECTION 2.6

Write an equation that represents each situation, solve the equation, and write the answer in complete sentences. (*See Objectives 1–5.*)

83. A 46-in. LED–LCD HDTV is on sale for $1170. This is 35% off the original price. What is the original selling price of the TV?

84. A sweater is on sale for $18.90. If this price is 65% off the original price, what is the original price of the sweater?

85. A campus bookstore sells an intermediate algebra text for $101.25. This is a 25% markup on the cost the bookstore pays for the text. What does the bookstore pay for the text?

86. Bev's salary after her raise is $44,285. Prior to the raise, her salary was $42,500. What percent raise did she receive?

87. The passenger volume at an international airport in Florida was 34.13 million in 2005 and 35.66 million in 2011. What is the percent increase in the passenger volume from 2005 to 2011? Round to one-tenth of a percent.

88. Kevin invests $8600 in two accounts. He invests $5350 in a mutual fund that pays 3.4% annual interest and the remaining amount in a money market fund that pays 2.6% annual interest. How much interest did he earn from the two accounts in 1 yr?

89. Stephanie has a collection of half-dollars and quarters. She has 40 more half-dollars than quarters. If the collection is worth $50, how many half-dollars and quarters does she have?

90. Joan has a collection of $20 bills and $1 bills. She has 29 more $20 bills than $1 bills. If the value of her collection is $1756, how many of each bill does she have?

91. Rachael needs to make 72 L of a 33.5% hydrochloric acid solution. How much 79% hydrochloric acid solution and how much 27% hydrochloric acid solution should she mix to get 72 L of a 33.5% hydrochloric acid solution?

92. A dairy farmer has two types of milk. How many gallons of a 2.5% milk fat solution and how many gallons of a 4.4% milk fat solution should be mixed to obtain 380 gal of a 2.9% milk fat solution?

93. Kyle and Bevin live 1026 mi apart. They leave their homes at the same time and drive toward one another until they meet. Kyle travels at 63 mph and Bevin travels at 72 mph. How long will it take before they meet?

94. Two planes leave an international airport in Tennessee at the same time. One plane travels west at 460 mph and the other plane travels east at 375 mph. How long will it take for the planes to be 4258.5 mi apart?

95. The sum of the two smaller of three consecutive even integers is the same as 300 less than nine times the largest integer. Find the integers.

96. A plane leaves Atlanta heading to Seattle at 6 A.M. traveling at 340 mph. A second plane leaves Atlanta heading to Seattle at 8 A.M. the same day traveling at 510 mph. How long will it take for the second plane to catch up to the first plane?

SECTION 2.7

Solve each inequality. Graph the solution set and write the solution set in interval notation and in set-builder notation. (*See Objectives 1–4.*)

97. $a + 5 \geq -9$

98. $4x - 18 > -6$

99. $0.3a - 7.2 \leq -2.4$

100. $\frac{2}{7}x + 1 \geq \frac{5}{7}$

101. $-1 < 3a + 2 < 23$

102. $-2 \leq 4x + 10 \leq 26$

103. $17 - 2(x - 5) < -6x + 3$

104. $6x + 2(x + 12) < 9x + 15$

105. $\frac{5}{3}y + 3\left(y + \frac{1}{6}\right) \geq \frac{2}{3}y - \frac{1}{2}$

106. $-19 \leq -2(x + 1) + 5 \leq -1$

107. $-3.5 < 0.5y + 1.4 < 1.2$

108. $-1.6 < 0.3x - 2.8 < 12.2$

Write an inequality that will solve each problem. Solve the inequality and answer the question in complete sentences. (*See Objective 6.*)

109. Baziz scored 67, 85, and 80 on three math tests. What score does he need on the fourth test to have at least an 80 average?

110. Shirl's final grade in her math class is based on a weighted average. Homework counts as 10%, quizzes count as 20%, tests count as 50%, and the final exam counts as 20%. If she has a homework average of 72, a quiz average of 70, and a test average of 84, what score does she need to make on her final exam to have at least a 80 for her final grade?

111. The cost to rent a banquet facility for a wedding reception is $1250 plus $45 per person for the food. What is the number of attendees the couple can have at the reception if they have at most $9600 budgeted for the reception?

112. The cost to rent a car for a week is $280 plus $0.32 per mile. How many miles can be driven if you have at most $842 to rent the car for a week? Round to the nearest whole number.

CHAPTER 2 TEST / LINEAR EQUATIONS AND INEQUALITIES IN ONE VARIABLE

1. An example of a linear equation in one variable is
   a. $3x + 5 - 7x$ b. $3x + 5 = 7x$

2. A solution of $x^2 = 4x$ is
   a. $x = -4$ b. $x = -2$
   c. $x = 2$ d. $x = 4$

3. The statement "The sum of three times a number and 6 is the same as 4 less than the number." can be written as

a. $3(n + 6) = n - 4$ **b.** $3(n + 6) = 4 - n$

c. $3n + 6 = n - 4$ **d.** $3n + 6 = 4 - n$

4. The equation that can be solved by subtracting 3 from each side is

a. $3x = 21$ **b.** $4 = 3 + x$

c. $x - 3 = 5$ **d.** $\dfrac{1}{3}x = 4$

5. In the Korean and Vietnam Wars, the United States lost a total of approximately 112,499 people. There were about 4007 fewer deaths in the Korean War than in the Vietnam War. How many U.S. deaths were there in each of these wars? If x represents the number of deaths in the Vietnam War,

a. write an equation that can be used to solve this problem.

b. solve the equation and explain the answer.

Solve each equation. Write each solution in a solution set.

6. $7x - 3(2x + 1) = 0$ **7.** $4(x + 3) = 5(2 - x)$

8. $3 - x = -7$ **9.** $\dfrac{3}{2}b = 9$

10. $4 + 3(2x - 5) = 7x + 6 - 2x$

11. $\dfrac{y}{6} + \dfrac{1}{2} = \dfrac{7}{6}$ **12.** $-\dfrac{1}{5}(x - 5) = \dfrac{2}{5}(x + 10)$

13. $0.05x + 0.03(10,000 - x) = 440$

14. $9y - 3(4y + 2) = 2y - 4 - 5y - 2$

15. $-2(b - 5) = 4(b + 1) - 6b$

16. Solve the equation $4x + 3y = -12$ for y.

17. Evaluate the formula $C = \dfrac{5}{9}(F - 32)$ if $F = 59°$.

18. Explain how solving linear equations and solving linear inequalities are the same. Explain how they are different.

19. The graph of $-4 \geq x$ can be represented as

a.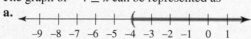

b.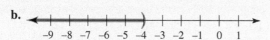

c.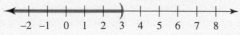

d.
```
  ◄──┼──┼──┼──┼──┼──┼──]──┼──┼──┼──┼──►
    -9 -8 -7 -6 -5 -4 -3 -2 -1  0  1
```

20. The interval notation for the following graph is
```
  ◄──┼──┼──┼──┼──┼──┼──)──┼──┼──┼──┼──►
    -2 -1  0  1  2  3  4  5  6  7  8
```

a. $(-\infty, 3)$ **b.** $(3, -\infty)$

c. $(\infty, 3)$ **d.** $(3, \infty)$

Solve each inequality. Graph the solution set. Write the solution in interval notation and set-builder notation.

21. $4x - 5(x - 2) \geq -3$

22. $\dfrac{2}{3}(x - 1) < x + \dfrac{4}{3}$

Define a variable to represent the unknown quantity. Write an equation (or inequality) that represents each situation. Solve the equation and answer the question in a complete sentence.

23. A taxicab authority in Nevada has set standard fares and fees. To pick up a passenger, the cab company charges a fee of $3.30. For each mile driven, there is a charge of $2.40.

a. If you have at most $75, how far could you travel in the cab? Round your answer to the nearest tenth.

b. The distance from your hotel to the Hoover Dam is 40.8 mi. Do you have enough money to travel there by taxicab?

24. The measure of the complement of an angle is 6° less than three times the measure of the angle. Find the measure of the angle.

25. The final grade in Joann's math class is based upon a weighted average. Tests count as 50%, quizzes count as 15%, homework counts as 10%, and the final exam counts as 25%. If she has a test average of 84, a quiz average of 86, and a homework average of 100, what does she need on the final exam to have at least an 80 for her final grade?

26. Find the value of a and the measures of the vertical angles shown.

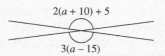

27. Two of the angles in a triangle have equal measure. The third angle is 20 less than six times the measure of the two equal angles. Find the measure of each angle in the triangle.

28. An architect designs a circular stadium. If the footprint of the stadium has a perimeter of 180π ft, what is the area of the footprint of the stadium? Use 3.14 for the value of π. Round your answer to the nearest integer.

29. David has $3000 to invest. He invests part of the money in a savings account that earns 4% annual interest and the remaining amount in a CD that earns 6% annual interest. If he earns $164 in interest in 1 yr, how much money did he invest in each account?

30. A department store advertises that they donate 5% of their weekly profit to various charities. Their donation is approximately $3 million each week. What is the department store's weekly profit?

31. Dr. Cleves is a pharmacist who needs a 15% glycerol solution. He has 10 mL of a 50% glycerol solution. How much water does he need to mix with the 10 mL of the 50% solution to obtain a solution that is 15% glycerol? Round to the nearest tenth.

32. The sum of the two smallest of three consecutive odd integers is the same as one less than three times the third consecutive odd integer. What are the three odd integers?

33. Gloria and Richard go out for an afternoon jog. They begin running at an average speed of 9 mph. After some time, they turn around and walk back home at an average speed of 4 mph. If they were gone for a total of 2 hr, how long did they jog before turning around?

CUMULATIVE REVIEW EXERCISES / CHAPTERS 1 AND 2

1. Classify each number as a natural number, whole number, integer, rational number, irrational number, and/or a real number. If the number is a rational number, write it in fractional form. If the number is an irrational number, approximate its value to two decimal places. (*Section 1.1, Objective 1*)

    a. $\dfrac{9}{11}$

    b. $\sqrt{13}$

    c. $-47°F$: the coldest recorded temperature in Elkader, Iowa on February 31, 1996

2. Give an example of a real number that satisfies the given conditions. (*Section 1.1, Objective 1*)

    a. An integer that is not a whole number.

    b. An irrational number that is between 3 and 4.

    c. A rational number that is between 5/2 and 3.

    d. An integer that is not a natural number.

3. Graph each number on a real number line. (*Section 1.1, Objective 2*)

    a. $\left\{-4.1\overline{1}, -1, \sqrt{7}, 5\dfrac{1}{2}\right\}$ b. $\left\{-2\pi, -3, \sqrt{11}, \dfrac{40}{3}\right\}$

4. Compare the values of each pair of numbers. Use a $<$, $>$, or $=$ symbol to make the statement true. (*Section 1.1, Objective 3*)

    a. π ____ 3.3   b. $\sqrt{16}$ ____ $-|-4|$

5. Find the opposite of each real number. (*Section 1.1, Objective 4*)

    a. $1.3\overline{3}$ b. 7.5 c. $-1\dfrac{2}{3}$

6. Simplify each absolute value expression. (*Section 1.1, Objective 5*)

    a. $|5|$ b. $\left|-\dfrac{4}{7}\right|$ c. $-|-14|$

7. Write the prime factorization of each number. (*Section 1.2, Objective 1*)

    a. 180   b. 98   c. 500

8. For the academic year 2009–2010, it was reported that there were 6896 higher education institutions in the United States and other jurisdictions; 2853 were 4-yr, 2259 were 2-yr, and 1784 were less-than-2-yr institutions. Write a fraction that represents the portion of less-than-2-yr institutions. (*Section 1.2, Objective 2*)

9. Simplify each fraction. (*Section 1.2, Objective 2*)

    a. $\dfrac{36}{90}$ b. $\dfrac{250}{516}$

10. Perform each operation and simplify the result. (*Section 1.2, Objectives 2–7*)

    a. $\dfrac{13}{18} + \dfrac{25}{42}$ b. $\dfrac{7}{12} - \dfrac{3}{10}$ c. $4\dfrac{1}{5} \cdot 3\dfrac{1}{3}$

    d. $2\dfrac{5}{6} \div 2\dfrac{1}{6}$ e. $5\dfrac{1}{5} - 4\dfrac{3}{5}$

11. Use the order of operations to simplify each numerical expression. (*Section 1.3, Objectives 1 and 2*)

    a. 3.5^2 b. $-\left(-\dfrac{2}{7}\right)^3$

    c. $\dfrac{1}{6}(9-7)^4 - \dfrac{5}{12}$ d. $2(6)^2 - 5(3) + 7$

    e. $7\big[28 - 3|15 - 2(7-3)|\big]$

    f. $\dfrac{5 + \sqrt{5^2 + 4(3)(8)}}{2 \cdot 3}$

12. Evaluate each expression for the given values of the variables. (*Section 1.3, Objective 3*)

    a. $\dfrac{3x - 5}{x + 1}$ for $x = 0, 1, 2, 3$

    b. $4x + 6y$ for $x = 2$ and $y = 3$

    c. $b^2 - 4ac$ for $a = 2, b = 6, c = 3$

13. Determine if the given value is a solution of the equation. (*Section 1.3, Objective 4*)

    a. $x^2 - 5x - 12 = 2x + 6$; $x = 9$

    b. $5y - (y - 3) = y + 9$; $y = 4$

14. Translate each phrase into an algebraic expression. Let x represent the unknown number. (*Section 1.3, Objective 5*)

    a. 16 more than three times a number

    b. The difference of a number and 12

    c. 13 less than five times a number

    d. The sum of four times a number and 8

15. The height of a baseball hit upward with an initial velocity of 96 ft/sec from an initial height of 6 ft is represented by the expression $-16t^2 + 96t + 6$, where t is the number of seconds after the ball has been hit. What is the height of the ball after 3 sec? (*Section 1.3, Objective 6*)

16 The expression $P(1 + r)^t$ represents the amount of money in an account when P dollars is invested at an annual interest rate of r (in decimal form) for t yr. Find the amount of money that will be in an account at the end of 4 yr if \$2500 is invested at 1.2% annual interest. (*Section 1.3, Objective 6*)

17. At a community college, tuition and fees are calculated by the expression $135c + 86.75$, where c is the number of credit hours taken per semester. If Bianca takes 10 credit hours next semester, what are her tuition and fees? (*Section 1.3, Objective 6*)

18. Perform the indicated operation and simplify. (*Sections 1.4 and 1.5, Objectives 1 and 2*)

a. $5.16 + (-3.97)$ **b.** $\left(-\dfrac{5}{4}\right) + \left(-\dfrac{3}{2}\right)$

c. $5 - 9 + (-10) - (-15)$

d. $5.4 - (-3.2) + (-2.1) + 2.8$

e. $\dfrac{3}{4} - \dfrac{5}{8} - \left(-\dfrac{1}{2}\right)$

f. $28 - \left\{9 - \left[5 - (1 - \sqrt{30 - 14})\right]\right\}$

g. $-6^2 - (-15) - |2 - 10|$

For Exercises 19–22, write a mathematical expression needed to solve each problem and then answer the question. (*Sections 1.4 and 1.5, Objective 3*)

19. The highest point on the Australian continent is the peak of Mount Kosciuszko. The peak is 7310 ft above sea level. The lowest point on land in China is the Turpan Pendi, which is 505.2 ft below sea level. What is the difference between these altitudes?

20. If the lowest recorded temperature in New Mexico is $-50°F$ and the highest recorded temperature is $122°F$, what is the difference between the highest and lowest temperatures?

21. Dave has $245.75 in his checking account. He deposits his paycheck of $635.25. Dave writes a check for his phone bill for $65.43, for groceries for $125.78, and then for rent for $525. What is Dave's checking account balance?

22. Find the measure of the unknown angle of the given triangle.

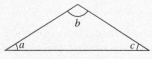

a. $a = 29°, b = 87°$
b. $b = 103°, c = 43°$

23. Find the complement and the supplement of an angle whose measure is $78°$.

24. Perform the indicated operation and simplify. (*Section 1.6, Objectives 1–3*)

a. $(-3.2)(1.5)$ **b.** $(-18)\left(\dfrac{-7}{24}\right)$

c. $-(-3)^4$ **d.** $\dfrac{0}{6}$

e. $10 \div 0$ **f.** $8(-25)(-3)$

g. $\left(-\dfrac{5}{12}\right) \div \left(-\dfrac{10}{9}\right)$ **h.** $-8 \div \dfrac{20}{7}$

25. Evaluate each expression for the given values of the variables. (*Section 1.6, Objective 4*)

a. $\sqrt{(x_1 - x_2)^2 + (y_1 - y_2)^2}$ for
$x_1 = -2, x_2 = 1, y_1 = 7, y_2 = 3$

b. $\dfrac{|2x + 3|}{x - 2}$ for $x = -2, -1, 0, 1, 2$

26. Find the additive and multiplicative inverses of each number. Assume all variables are nonzero. (*Section 1.7, Objective 1*)

a. $-16x$ **b.** $\dfrac{5}{8}$ **c.** $6a$

27. Apply the commutative, associative, and/or distributive properties to rewrite each expression and simplify the result. (*Section 1.7, Objectives 2 and 3*)

a. $-18 + c - (-15)$ **b.** $-2(x - 10y - 7)$

c. $\left(b - \dfrac{3}{10}\right) + 2$ **d.** $10\left(-\dfrac{2}{5}x + \dfrac{3}{2}\right)$

28. Determine if the terms are like or unlike. (*Section 1.8, Objective 2*)

a. $8ab^2$ and $-34ab^2$ **b.** $\dfrac{1}{3}cd^2$ and $-\dfrac{5}{7}c^2d$

c. $\dfrac{4}{3}\pi r^3$ and $\pi r^2 h$

29. Simplify each algebraic expression. (*Section 1.8, Objectives 3 and 4*)

a. $-5(x)(14)$ **b.** $15(2x - 1) - 36x$

c. $-(x - 12) - 2(5x + 7)$ **d.** $3.7 - 0.8(-0.1x + 2.5)$

e. $-12\left(\dfrac{1}{3}a - \dfrac{3}{4}\right) + 14\left(\dfrac{3}{7}a - \dfrac{1}{2}\right)$

f. $2x^2 - 4(x - 3x^2) - x$

30. Determine whether the following is an expression or an equation. (*Section 2.1, Objective 1*)

a. $6x - 19 = 4x - 11$ **b.** $7x - 15 - 3x + 4$

31. Determine if each number is a solution of the equation. (*Section 2.1, Objective 2*)

a. $x = -2, x = 0$, or $x = 2$; $5x - 2 = -2x - 16$

b. $x = -2, x = 0$, or $x = 2$; $8x - 12 = -3x + 10$

32. For each problem, define the variable, write an equation, and solve the problem. (*Section 2.1, Objectives 3 and 4*)

a. Five times the sum of a number and -26 is the same as three times the number. Find the number.

b. The sum of a number and 28 is the same as five times the number. Find the number.

c. The difference of twice a number and -6 equals 32 less than the number. Find the number.

d. One angle is $16°$ more than another angle. Their sum is $180°$. Find the measure of each angle.

e. One angle is $12°$ less than another angle. Their sum is $90°$. Find the measure of each angle.

f. One angle is $6°$ more than another angle. Their sum is $90°$. Find the measure of each angle.

33. Solve each equation. (*Section 2.2, Objectives 2 and 3; Sections 2.3 and 2.4, Objectives 1 and 2*)

a. $7(x - 3) + 18 = 2(x - 6) - 26$

b. $2.46x + 12.25 = 5.28x + 8.02$

c. $4.5x - 3.6 + 2.9x = 8.2x - 5.04$

d. $10(3 - x) - 5x = 20 + 5(2 - 3x)$

e. $4z + 32 - z = 5 - 6z$

f. $-\dfrac{2}{3}(6x + 12) + 10 = x + 5$

g. $-0.8y + 18.4 = 3.7y + 5.8$

h. $4(3x - 1) = 4(x - 1) + 8x$

For each problem: (1) assign a variable to the unknown, (2) write an equation that represents the situation, (3) solve the equation, and (4) answer the question using complete sentences. (*Section 2.2, Objective 4; Section 2.3, Objectives 3 and 4*)

34. Twice the sum of a number and 6 is the same as three times the number. Find the number.

35. The cost to rent a midsize car for one day is $58 plus $0.35 per mile. The daily cost is represented by the expression $58 + 0.35x$, where x is the number of miles driven. How many miles has the car been driven if the cost of the rental car is $250.50?

36. The sum of three consecutive even integers is 102. Find the integers.

37. The sum of the two smaller of three consecutive even integers is the same as 40 less than three times the largest integer. Find the integers.

38. Solve each equation. If the equation is a contradiction, write the solution as $\varnothing$. If the equation is an identity, write the solution as $\mathbb{R}$. (*Section 2.4, Objectives 1–4*)

a. $3(2y + 3) - 2(y + 2) = 4(y + 1) + 1$

b. $\dfrac{3}{5}x - 4 = \dfrac{7}{10}x$

c. $0.6x + 1.8x = 960$

d. $0.021x + 0.012(840 - x) = 15.12$

e. $\dfrac{5}{6}(x - 2) = \dfrac{1}{6}(5x - 8) + \dfrac{1}{3}$

39. The formula $A = P(1 + r)^t$ calculates the amount of money in an account at the end of t years if P dollars invested at an annual interest rate, r. (*Section 2.5, Objective 1*)

a. Find the value of A if $4800 is invested for 2 yr at 1.5% annual interest.

b. Find the value of P if $5181.62 is expected to be accumulated at the end of 2 yr at 1.8% annual interest.

40. The distance d, rate r, and time t of an object in motion can be related by the formula $d = rt$. (*Section 2.5, Objective 1*)

a. Find d if $r = 65$ mph and $t = 3.2$ hr.

b. Find t if $d = 203$ mi and $r = 58$ mph.

41. The revenue R, unit price p, and sale level x of a product are related by the formula $R = xp$. (*Section 2.5, Objective 1*)

a. Find R if $p = 112.50 and $x = 60$.

b. Find p if $R = 2249.10 and $x = 90$.

42. Use the perimeter, area, or circumference formula to find the indicated quantity. (*Section 2.5, Objective 2*)

a. Find the perimeter of a rectangle if its area is 400 in.^{2} and its length is 16 in.

b. Find the area of a rectangle if its perimeter is 92 m and its width is 18 m.

43. Solve each formula for the specified variable. (*Section 2.5, Objective 3*)

a. $A = lw$ for l **b.** $C = 5l + 6w$ for w

c. $3x - 8y = 24$ for y **d.** $0.02x + 0.05y = 4.5$ for x

44. Find the measure of each unknown angle. (*Section 2.5, Objective 5*)

a. Find the measure of an angle whose complement is 20° less than the measure of the angle.

b. Find the measure of an angle whose supplement is 64° more than the measure of the angle.

c. The supplement of an angle is 42° more than three times its complement. Find the measure of the angle.

45. Find the measure of each angle labeled in the figure. (*Section 2.5, Objectives 4 and 5*)

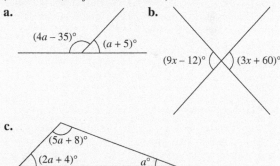

a. $(4a - 35)°$ $(a + 5)°$

b. $(9x - 12)°$ $(3x + 60)°$

c. $(5a + 8)°$ $(2a + 4)°$ $a°$

46. In triangle ABC, the measure of angle B is 18° less than the measure of angle A. The measure of angle C is 45° more than the measure of angle A. Find the measure of each angle in the triangle. (*Section 2.5, Objective 5*)

Solve each problem. Write an appropriate equation, solve the equation, and write the answer in complete sentences. (*Section 2.6, Objectives 1–5*)

47. A 46-in. LED-LCD HDTV is on sale for $975. This is 35% off the original price. What is the original selling price of the TV?

48. A sweater is on sale for $31.50. If this price is 65% off the original price, what is the original price of the sweater?

49. Kye invested $2500 in two accounts. He invested $1500 in a mutual fund that pays 2.4% annual interest and the remaining amount in a money market fund that pays 1.6% annual interest. How much interest did he earn from the two accounts in 1 yr?

50. Richard needs to make 72 L of a 25% hydrochloric acid solution. How much 40% hydrochloric acid solution and how much 10% hydrochloric acid solution should be mixed to get 72 L of a 25% hydrochloric acid solution?

51. Solve each inequality. Graph the solution set and write the solution set in interval notation and in set-builder notation. (*Section 2.7, Objectives 1–4*)

   a. $0.4a - 7.7 \leq -2.1$

   b. $-6 < 4a + 6 < 26$

   c. $8x + 3(x + 12) < 9x + 16$

   d. $-3.5 \leq 0.5y - 1.4 \leq 1.2$

Write an inequality that can be used to solve each problem. Solve the inequality and answer the question in complete sentences. (*Section 2.7, Objective 5*)

52. Myra's final grade in her math class is based on a weighted average. Homework counts as 10%, quizzes count as 20%, tests count as 50%, and the final exam counts as 20%. If she has a homework average of 72, a quiz average of 68, and a test average of 82, what score does she need on her final exam to have at least an 80 for her final grade?

53. The cost to rent a car for a week is $320 plus $0.35 per mile. How many miles can be driven if you have at most $740 to rent the car for a week? Round to the nearest whole number, if necessary.

3

Linear Equations in Two Variables

Commitment and Perseverance

Commitment and perseverance are necessary to be successful in any college classroom. Some courses may require more commitment and perseverance than others, especially if you find the material somewhat difficult to conquer. Your willingness to stick to your goals is far more important than your mathematical abilities. We truly believe that, given the right resources and the right amount of time, you can pass this class. You cannot compare yourself to other students in the classroom since college brings together students of all backgrounds—some may have seen the material last year, some 5 years ago, and some may never have seen the material. So do not get frustrated when others seem to learn the concepts faster than you. This does not mean that you will not learn them; it just means you need to have a little more perseverance until you do.

When we believe we can do something and believe that what we are doing is important, we will be committed to doing it. Know that what you are doing in the classroom is important. It is one step in achieving your college education. Try to remember the big picture as you take these small steps to reach your goal. Find a person that is supportive of your goals and allow them to hold you to the commitment of being successful in this class.

Question For Thought: How would you evaluate your ability to see things through to the end? Do you typically give up on your commitments or do you see them through to the end?

Chapter Outline

Coming Up...

In Section 3.5, we are going to learn how to write a linear equation that models the average price of a movie ticket given that the average price of a movie ticket in the United States in 1996 was $4.42 and the average price of a movie ticket in the United States in 2010 was $7.89. (Source: http://www.natoonline.org/statisticstickets.htm)

Equations and the Rectangular Coordinate System

In Chapter 2, we studied linear equations in one variable. In this chapter, we will focus on linear equations in two variables. We will also study real-life applications of these equations and introduce the concept of a function.

▶ **OBJECTIVES**

As a result of completing this section, you will be able to

1. Determine algebraically if an ordered pair is a solution of an equation.
2. Plot ordered pairs and identify quadrants.
3. Graph the solutions of an equation.
4. Determine graphically if an ordered pair is a solution of an equation.
5. Solve application problems.
6. Troubleshoot common errors.

The following line graph shows the average cost of a 30-sec Super Bowl advertisement for 2002 to 2010. This is an example of a graph that we will learn how to read to obtain specific information.

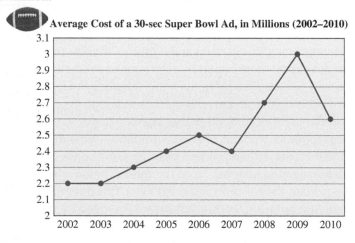
Average Cost of a 30-sec Super Bowl Ad, in Millions (2002–2010)

Verifying Solutions of Equations Algebraically

Objective 1 ▶

Determine algebraically if an ordered pair is a solution of an equation.

In Chapter 2, we studied linear equations with one variable and learned how to find solutions of these equations. We found that a linear equation in one variable has one solution with the exception of identities and contradictions. Identities are satisfied by all real numbers and contradictions do not have a solution.

In this chapter, our attention turns to equations with two variables. Some examples of equations in two variables are

$$x + y = 3 \qquad y = 4x - 1 \qquad 2x - 3y = 6 \qquad y = x^2 + 2x - 3$$

Equations containing two variables have pairs of numbers that satisfy the equation. This pair of numbers is called an **ordered pair** and is denoted by (x, y), where x and y are the variables in the equation. A **solution of an equation** containing two variables is an ordered pair (x, y) that satisfies the equation.

Consider the equation $x + y = 3$. Solutions of this equation must satisfy the fact that the sum of the x and y values is 3. Some solutions are $(1, 2)$, $(3, 0)$, $(-1, 4)$, and $(5, -2)$. We know these ordered pairs are solutions because they make the equation true when we replace the variables with their given values as shown in the following table.

(x, y)	$x + y = 3$	Solution?
$(1, 2)$	$1 + 2 = 3$ $3 = 3$	Yes
$(3, 0)$	$3 + 0 = 3$ $3 = 3$	Yes
$(-1, 4)$	$-1 + 4 = 3$ $3 = 3$	Yes
$(5, -2)$	$5 + (-2) = 3$ $3 = 3$	Yes

There are, in fact, infinitely many solutions of an equation in two variables. Just as there are infinitely many solutions of this type of equation, there are infinitely many ordered pairs that do not satisfy the equation $x + y = 3$. Any ordered pair whose values do not add to 3 is not a solution of $x + y = 3$. Some ordered pairs that are not solutions are $(1, 5)$, $(3, -3)$, and $(-1, -2)$.

(x, y)	$x + y = 3$	Solution?
$(1, 5)$	$1 + 5 = 3$ $6 \neq 3$	No
$(3, -3)$	$3 + (-3) = 3$ $0 \neq 3$	No
$(-1, -2)$	$-1 + (-2) = 3$ $-3 \neq 3$	No

Procedure: Determining if an Ordered Pair Is a Solution of an Equation in Two Variables

Step 1: Replace the values of x and y with the numbers given in the ordered pair.
Step 2: Simplify each side of the equation.
Step 3: If the resulting equation is true, then the ordered pair is a solution of the equation. If it is not true, then the ordered pair is not a solution.

Objective 1 Examples   Determine if the ordered pair is a solution of the equation.

1a. $(3, -1)$; $2x - y = 7$ **1b.** $\left(\dfrac{3}{2}, 0\right)$; $y = 2x + 3$ **1c.** $(0, 5)$; $y = x^2 - 3x + 5$

Solutions   **1a.**
$$2x - y = 7$$
$$2(3) - (-1) = 7 \qquad \text{Let } x = 3 \text{ and } y = -1.$$
$$6 + 1 = 7 \qquad \text{Simplify.}$$
$$7 = 7 \qquad \text{True}$$

Since $(3, -1)$ makes the equation a true statement, it is a solution of $2x - y = 7$.

1b.
$$y = 2x + 3$$
$$0 = 2\left(\frac{3}{2}\right) + 3 \qquad \text{Let } x = \frac{3}{2} \text{ and } y = 0.$$
$$0 = 3 + 3 \qquad \text{Simplify.}$$
$$0 = 6 \qquad \text{False}$$

Since $\left(\dfrac{3}{2}, 0\right)$ makes the equation a false statement, it is *not* a solution of $y = 2x + 3$.

1c.
$$y = x^2 - 3x + 5$$
$$5 = (0)^2 - 3(0) + 5 \qquad \text{Let } x = 0 \text{ and } y = 5.$$
$$5 = 0 + 5 \qquad \text{Simplify.}$$
$$5 = 5 \qquad \text{True}$$

Since $(0, 5)$ makes the equation a true statement, it is a solution of $y = x^2 - 3x + 5$.

✓ Student Check 1   Determine if the ordered pair is a solution of the equation.

a. $(-2, 6)$; $3x - y = 0$ **b.** $\left(\dfrac{1}{4}, 0\right)$; $y = 4x - 1$ **c.** $(-3, 5)$; $y = |x - 2|$

The Rectangular Coordinate System

Objective 2 ▶

Plot ordered pairs and identify quadrants.

As mentioned, there are infinitely many solutions of equations in two variables. Our ultimate goal is to visualize the solutions of these types of equations. Recall that we use a real number line to visualize solutions of equations and inequalities in one variable. To graph solutions of equations in two variables, we need a special system, called the *rectangular coordinate system* or the *Cartesian coordinate system*.

The **rectangular coordinate system** consists of two real number lines intersecting at right angles. The horizontal number line is referred to as the ***x*-axis** and the vertical number line is referred to as the ***y*-axis**. The point where the two number lines intersect is called the **origin**.

Every point on the coordinate system has an "address." This address is denoted by an ordered pair (x, y). Each point on the plane is determined by knowing how far left or right the point is from the origin and how far up or down the point is located from the *x*-axis.

In an ordered pair (x, y), the value x is called the *first coordinate* or *x-coordinate*. The *x*-coordinate tells us how far left (if x is negative) or right (if x is positive) to move from the origin. The value y is called the *second coordinate* or *y-coordinate*. The *y*-coordinate tells us how far up (if y is positive) or down (if y is negative) to move from the *x*-axis.

For example, in the ordered pair $(4, 5)$,

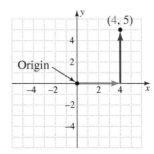

4 is the *x*-coordinate → move from the origin 4 units right

5 is the *y*-coordinate → move from the position on the *x*-axis 5 units up

Note that the ordered pair $(5, 4)$ is different than $(4, 5)$. For the point $(5, 4)$, we move 5 units to the right and 4 units up.

Procedure: Plotting an Ordered Pair

Step 1: From the origin, move left or right to the given *x*-value in the ordered pair.
Step 2: From this *x*-value, move up or down to the given *y*-value in the ordered pair.
Step 3: The point is located at the result of the two movements.

The two numbers lines that form the Cartesian coordinate system divide the plane into four regions called **quadrants**. We label these four quadrants with Roman numerals I, II, III, and IV beginning in the upper right quadrant and rotating counterclockwise.

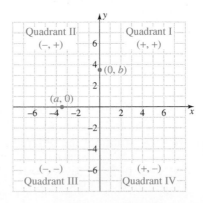

The signs of a point's coordinates determine where on the plane the point is located.

- If both the *x*- and *y*-coordinates are positive, the point lies in Quadrant I.
- If the *x*-coordinate is negative and the *y*-coordinate is positive, the point lies in Quadrant II.
- If both the *x*- and *y*-coordinates are negative, the point lies in Quadrant III.
- If the *x*-coordinate is positive and the *y*-coordinate is negative, the point lies in Quadrant IV.
- If both the *x*- and *y*-coordinates are 0, the point lies on both the *x*- and *y*-axes (or at the origin).
- If the *x*-coordinate is 0 and the *y*-coordinate is nonzero, the point lies on the *y*-axis.
- If the *x*-coordinate is nonzero and the *y*-coordinate is 0, the point lies on the *x*-axis.

Objective 2 Examples   Plot each ordered pair on the Cartesian coordinate system and state the quadrant or axis where each point is located.

2a. $(4, 3)$ **2b.** $(-2, 5)$ **2c.** $(5, -2)$ **2d.** $(0, 4)$ **2e.** $(-3, 0)$ **2f.** $\left(-\dfrac{3}{2}, -\dfrac{8}{2}\right)$

Solutions   **2a.** To plot $(4, 3)$, move 4 units right from the origin and 3 units up. $(4, 3)$ is located in Quadrant I.

2b. To plot $(-2, 5)$, move 2 units left from the origin and 5 units up. $(-2, 5)$ is located in Quadrant II.

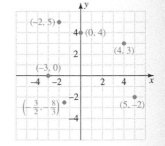

2c. To plot $(5, -2)$, move 5 units right from the origin and 2 units down. $(5, -2)$ is located in Quadrant IV.

2d. To plot $(0, 4)$, do not move left or right, move 4 units up. $(0, 4)$ is located on the *y*-axis.

2e. To plot $(-3, 0)$, move 3 units left from the origin and do not move up or down. $(-3, 0)$ is located on the *x*-axis.

2f. The point $\left(-\dfrac{3}{2}, -\dfrac{8}{3}\right) = (-1.5, -2.7)$. Move 1.5 units left from the origin and 2.7 units down. $\left(-\dfrac{3}{2}, -\dfrac{8}{3}\right)$ is located in Quadrant III.

 Note: *The points $(-2, 5)$ and $(5, -2)$ illustrate the fact that the order is important. Changing the order of the numbers in the ordered pairs changes the location of the point.*

✔**Student Check 2**   Plot each ordered pair on the Cartesian coordinate system and state the quadrant or axis where each point is located.

a. $(-5, -1)$ **b.** $(3, -4)$ **c.** $(-4, 3)$ **d.** $(0, -2)$ **e.** $(1, 0)$ **f.** $\left(\dfrac{1}{2}, \dfrac{5}{4}\right)$

Graphing Equations

Objective 3 ▶

Graph the solutions of an equation.

The graph of an equation is a picture of all the ordered pairs that satisfy the equation. When an equation contains two variables, there are infinitely many ordered pairs that satisfy the equation. It is impossible to find every solution of an equation in two variables. So, we plot several ordered pairs (solutions) to get an idea of the graph's shape. Recall in Chapter 1 that we evaluated algebraic expressions and recorded our results in a table. We use the table again to record the values for *x* and *y*.

Procedure: Graphing Solutions of an Equation

Step 1: Determine at least three solutions of the equation. Use a table to show the solutions.

    a. Substitute any value for x.

    b. Solve the resulting equation or simplify the resulting expression to find the value of y.

    c. The ordered pair (x, y) is a solution of the equation.

Step 2: Plot the solutions on a coordinate system.

Step 3: Connect the points and use their pattern to sketch the graph.

 Note: *We will study graphing special types of equations in later sections and chapters. This method is a basic introduction to graphing.*

Objective 3 Examples   Graph each equation by completing a table of solutions.

3a. $y = x + 3$ **3b.** $y = x^2 - 1$ **3c.** $y = |x + 2|$

Solutions   **3a.**

x	$y = x + 3$	(x, y)
-2	$y = -2 + 3 = 1$	$(-2, 1)$
-1	$y = -1 + 3 = 2$	$(-1, 2)$
0	$y = 0 + 3 = 3$	$(0, 3)$
1	$y = 1 + 3 = 4$	$(1, 4)$
2	$y = 2 + 3 = 5$	$(2, 5)$

Now plot the solutions and connect them to form the graph.

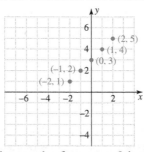

Notice the graph of $y = x + 3$ is a line.

3b.

x	$y = x^2 - 1$	(x, y)
-2	$y = (-2)^2 - 1 = 4 - 1 = 3$	$(-2, 3)$
-1	$y = (-1)^2 - 1 = 1 - 1 = 0$	$(-1, 0)$
0	$y = (0)^2 - 1 = 0 - 1 = -1$	$(0, -1)$
1	$y = (1)^2 - 1 = 1 - 1 = 0$	$(1, 0)$
2	$y = (2)^2 - 1 = 4 - 1 = 3$	$(2, 3)$

Now plot the solutions and connect them to form the graph.

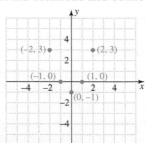 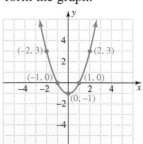

Notice the graph of $y = x^2 - 1$ has a "U-shape" pattern.

3c.

x	$y = |x + 2|$	(x, y)
-4	$y = |-4 + 2| = |-2| = 2$	$(-4, 2)$
-3	$y = |-3 + 2| = |-1| = 1$	$(-3, 1)$
-2	$y = |-2 + 2| = |0| = 0$	$(-2, 0)$
-1	$y = |-1 + 2| = |1| = 1$	$(-1, 1)$
0	$y = |0 + 2| = |2| = 2$	$(0, 2)$

Now plot the solutions and connect them to form the graph.

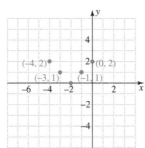 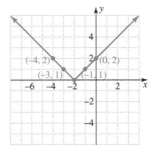

Notice the graph of $y = |x + 2|$ has a "V-shape" pattern.

✓ **Student Check 3**   Graph each equation by completing a table of solutions.

   a. $y = x - 1$ **b.** $y = x^2$ **c.** $y = |x - 1|$

Verifying Solutions Graphically

Objective 4 ▶

Determine graphically if an ordered pair is a solution of an equation.

In Objective 1, we determined if an ordered pair is a solution of an equation by substituting the values of x and y in the equation. We can also examine the graph of an equation to determine if an ordered pair is a solution of an equation. If a point lies on the graph, then it is a solution of the equation. If a point does not lie on the graph, then it is not a solution of the equation.

Objective 4 Examples   The graph of the equation $y = x^2 - 4$ is provided. Use the graph to determine if each ordered pair is a solution of the equation.

4a. $(-4, 0)$ **4b.** $(0, -4)$

4c. $(2, 0)$ **4d.** $(0, 2)$

4e. $(-3, 5)$

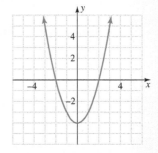

Solutions   After plotting the points, we find that $(0, -4)$, $(2, 0)$, and $(-3, 5)$ lie on the graph; thus, they are solutions of the equation. The points $(-4, 0)$ and $(0, 2)$ are not solutions of the equation because they do not lie on the graph.

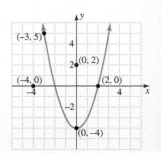

✔ **Student Check 4**   The graph of the equation $y = |x| - 3$ is provided. Use the graph to determine if each ordered pair is a solution of the equation.

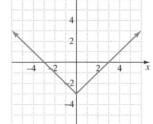

a. $(3, 0)$ b. $(0, 3)$

c. $(0, -3)$ d. $(4, -1)$

e. $(-4, 1)$

Applications

Objective 5 ▶

Solve application problems.

The applications in this section require us to read graphs to obtain information. Data that can be represented as an ordered pair are called **paired data**. For example, a person's weight for a given year can be written as an ordered pair in the form (year, weight).

When we are given real-life data in tables, bar graphs, or line graphs, we can convert the data to ordered pairs and use them to plot the data. The plot of this data is called a **scatter plot** or *scatter diagram*. Scatter plots can be used to look for patterns or trends that occur in paired data.

Objective 5 Examples   Use the table or graph to answer each question.

5a. The following table provides the number of cell phone subscribers in the United States for the given year. Write an ordered pair for each year, where x is the number of years after 2000 and y is the number of cell phone subscribers (in millions).

Years After 2000	0	1	2	3	4	5	6	8	9
Subscribers (in millions)	109	128	141	159	182	208	233	263	286

(Sources: www.infoplease.com and http://www.ctia.org/media/industry_info/index .cfm/AID/10323)

   i. Write the ordered pairs that correspond to the data in the table.

  ii. Interpret the meaning of the first and last ordered pairs.

 iii. Plot the ordered pairs to form a scatter plot of the given data.

Solution   **5a.**   **i.** The ordered pairs that correspond to the data in the table are (0, 109), (1, 128), (2, 141), (3, 159), (4, 182), (5, 208), (6, 233), (8, 263), and (9, 286).

  ii. The first ordered pair (0, 109) means that 0 yr after 2000, or in 2000, there were approximately 109 million cell phone subscribers in the United States. The last ordered pair (9, 286) means that 9 yr after 2000, or in 2009, there were approximately 286 million cell phone subscribers in the United States.

 iii. The scatter plot of the ordered pairs is

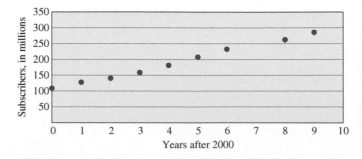

5b. The following bar graph shows the amount of money spent, in millions, on Google ads for the top 10 companies in June 2010. (Source: www.adage.com)

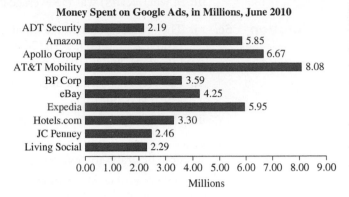

Money Spent on Google Ads, in Millions, June 2010

i. Write ordered pairs for this bar graph, where *x* is the company and *y* is the money spent on advertising, in millions.

ii. Which company spent the most money on Google ads in June 2010? How much was spent?

iii. Which company spent the least money in Google ads in June 2010? How much was spent?

iv. What was the total amount of money spent on Google ads for these companies in June 2010?

Solution **5b.** **i.** The ordered pairs corresponding to the given data are (ADT Security, 2.19), (Amazon, 5.85), (Apollo Group, 6.67), (AT&T Mobility, 8.08), (BP Corp, 3.59), (eBay, 4.25), (Expedia, 5.95), (Hotels.com, 3.30), (JC Penney, 2.46), and (Living Social, 2.29).

ii. The company that spent the most on Google ads was AT&T Mobility, since the horizontal bar is the longest for this company. AT&T Mobility spent $8.08 million.

iii. The company that spent the least on Google ads was ADT Security since this horizontal bar is the shortest. ADT Security spent $2.19 million.

iv. The total spent on Google ads for these 10 companies in June 2010, was $2.19 + 5.85 + 6.67 + 8.08 + 3.59 + 4.25 + 5.95 + 3.30 + 2.46 + 2.29 = $44.63 million.

5c. The line graph shows the average cost, in millions, of a 30-sec Super Bowl ad for the years 2002–2010. (Sources: http://www.huffingtonpost.com/2010/01/11/super-bowl-commercial-pri_n_418245.html and www.marketingcharts.com)

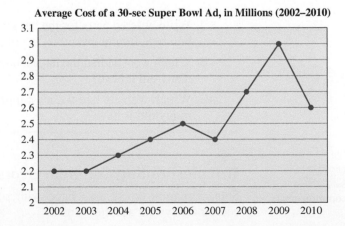

Average Cost of a 30-sec Super Bowl Ad, in Millions (2002–2010)

i. Write an ordered pair for the points on the graph.

ii. What year had the greatest average ad cost? What was it?

iii. In what year(s) was the average cost of an ad $2.4 million?

Solution   **5c.**   i. The ordered pairs on the graph correspond to the points (2002, 2.2), (2003, 2.2), (2004, 2.3), (2005, 2.4), (2006, 2.5), (2007, 2.4), (2008, 2.7), (2009, 3), and (2010, 2.6).

ii. The average cost of a Super Bowl ad was greatest in the year 2009. The cost was $3 million.

iii. The average cost of an ad was $2.4 million in the years 2005 and 2007.

✓ **Student Check 5**   Use the table or graph to answer each question.

a. The table provides information on the number of murder victims by juvenile gang killings for the given year. (Source: http://www.fbi.gov/ucr)

Years After 2000	0	1	2	3	4
Number of Victims	653	862	911	819	804

i. Write an ordered pair for each year, where x is the number of years after 2000 and y is the number of victims.

ii. Interpret the first and last ordered pairs.

iii. Plot the ordered pairs to form a scatter plot of the data.

b. The bar graph shows the number of unique visitors, in millions, for the given websites for August 2010. (Source: http://www.compete.com)

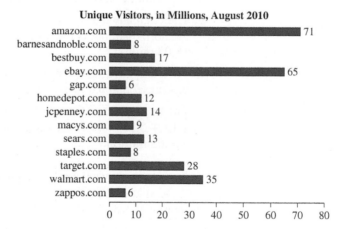

Unique Visitors, in Millions, August 2010

i. Write ordered pairs for the bar graph, where x is the website and y is the number of unique visitors, in millions.

ii. Which website had the largest number of visitors in August 2010? How many did it have?

iii. Which website had the least number of visitors in August 2010? How many did it have?

Objective 6 ▶

Troubleshoot common errors.

Troubleshooting Common Errors

Some common errors associated with solutions of equations and the rectangular coordinate system are shown next.

| Objective 6 Examples | A problem and an incorrect solution are given. Provide the correct solution and an explanation of the error. |

6a. Is the ordered pair $(4, -2)$ a solution of $y = 2x - 10$?

Incorrect Solution	Correct Solution and Explanation
$y = 2x - 10$ $4 = 2(-2) - 10$ $4 = -4 - 10$ $4 = -14$ Since the point makes the equation false, it is not a solution.	In the ordered pair $(4, -2)$, the x-value is 4 and the y-value is -2. $$y = 2x - 10$$ $$-2 = 2(4) - 10$$ $$-2 = 8 - 10$$ $$-2 = -2$$ Since the point makes the equation true, it is a solution of the equation.

6b. Plot the points $(0, 3)$ and $(-2, 0)$.

Incorrect Solution	Correct Solution and Explanation
The points are shown in the graph. 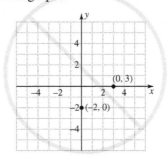	When the first coordinate is 0, there is no movement left or right from the origin. When the second coordinate is 0, there is no movement up or down from the x-axis. The correct plot is shown here. 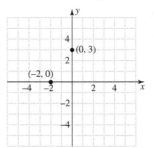

6c. Graph $y = |x + 3|$.

Incorrect Solution	Correct Solution and Explanation
The table that corresponds to this graph is	While the points in the table are solutions of the equation, we cannot assume that the graph continues in this manner. It is always a good idea to input values for x that make the expression inside the absolute value negative. So, we find a couple more points to get

x	$y = |x + 3|$	(x, y)
-2	$y = |-2 + 3| = 1$	$(-2, 1)$
-1	$y = |-1 + 3| = 2$	$(-1, 2)$
0	$y = |0 + 3| = 3$	$(0, 3)$
1	$y = |1 + 3| = 4$	$(1, 4)$
2	$y = |2 + 3| = 5$	$(2, 5)$

x	$y = |x + 3|$	(x, y)
-4	$y = |-4 + 3| = |-1| = 1$	$(-4, 1)$
-3	$y = |-3 + 3| = 0$	$(-3, 0)$

So, the graph is

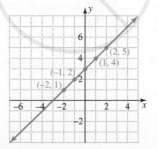

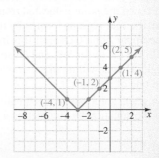

ANSWERS TO STUDENT CHECKS

Student Check 1   **a.** no   **b.** yes   **c.** yes

Student Check 2   **a.** Quadrant III

    b. Quadrant IV

    c. Quadrant II

    d. y-axis

    e. x-axis

    f. Quadrant I

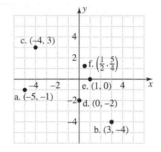

Student Check 3   **a.**   **b.**

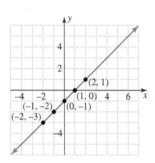

    c.

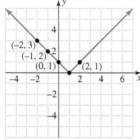

Student Check 4   **a.** yes   **b.** no   **c.** yes   **d.** no   **e.** yes

Student Check 5   **a. i.** $(0, 653), (1, 862), (2, 911), (3, 819),$
$(4, 804)$

    ii. The first point means that in 0 yr after 2000, in the year 2000, there were 653 murder victims by juvenile gang killings. The last point means that in 4 yr after 2000, in the year 2004, there were 804 murder victims by juvenile gang killings.

    iii. The scatter plot is

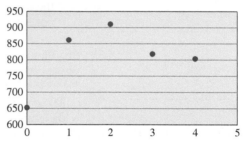

b. i. (amazon.com, 71), (barnesandnoble.com, 8), (bestbuy.com, 17), (ebay.com, 65), (gap.com, 6), (homedepot.com, 12), (jcpenney.com, 14), (macys.com, 9), (sears.com, 13), (staples.com, 8), (target.com, 28), (walmart.com, 35), (zappos.com, 6)

    ii. The website with the largest number of visitors in August 2010 was amazon.com. The site had 71 million visitors.

    iii. The websites with the least number of visitors in August 2010 were gap.com and zappos.com, each with 6 million visitors.

SUMMARY OF KEY CONCEPTS

1. A solution of an equation in two variables is an ordered pair that makes the equation a true statement.

2. Ordered pairs can be plotted on a Cartesian coordinate system. The x-value determines the movement from the origin on the x-axis. The y-value determines how far up or down the point is located from the x-axis. Quadrants are the four regions formed by the axes. Quadrant I is in the upper right, Quadrant II is in the upper left, Quadrant III is in the lower left, and Quadrant IV is in the lower right. Points will lie in either one of the four quadrants or on one of the axes.

3. To graph solutions of an equation, find several solutions of the equation by completing a table. Plot the solutions

and then connect them to form the graph. There are infinitely many solutions of an equation in two variables. The graph is a picture of these solutions.

4. We can determine if an ordered pair is a solution of an equation if it lies on the graph of the equation.

5. For application problems, it is important to understand what the x- and y-coordinates represent. If the data are given in a table, bar graph, or line graph, the set of ordered pairs is called paired data. Graphing paired data gives us a scatter plot of the data.

GRAPHING CALCULATOR SKILLS

The calculator can be used to make a scatter plot and to graph equations. These skills will come in handy in later math courses. At this point, know the features are available, but do not rely on the calculator to plot points or draw graphs.

Example 1: Plot the points $(3, -2)$, $(0, 4)$, and $(-2, 0)$.

To plot points on the calculator, the x- and y-values must be entered into lists. Press STAT, 1 to access the list feature. Then enter the values in the appropriate columns, using L1 as the x-values and L2 as the y-values.

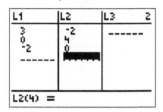

Once the points are entered, turn the STAT PLOT feature ON and graph.

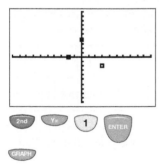

Example 2: Graph $y = x + 3$.

To graph an equation, enter the equation into the equation editor by pressing Y = and then press GRAPH to view the graph.

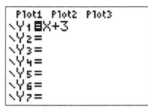

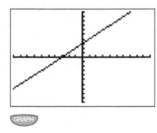

To view a table of specific solutions, access the TABLE feature by pressing 2nd GRAPH.

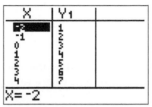

More detail will be provided in Section 3.2 on graphing with the calculator.

SECTION 3.1 / EXERCISE SET

 Write About It!

Use complete sentences to explain the meaning of the given term or process.

1. Solution of an equation in two variables

2. Plotting a point on a rectangular coordinate system

3. Identifying a quadrant or axis where a point is located

4. Graphing the solutions of an equation in two variables

Determine if each statement is true or false. If the statement is false, provide an explanation.

5. The ordered pair $(0, -4)$ is a solution of the equation $x - y = -4$.

6. The point $(3, 5)$ does not lie on the graph of the equation $y = 2x + 1$.

7. Any point located on the x-axis has an x-value of 0.

8. Any point located on the y-axis has a y-value of 0.

9. If $a > 0$ and $b < 0$, the point (a, b) is located in Quadrant II.

10. If $a > 0$ and $b < 0$, the point $(-a, -b)$ is located in Quadrant III.

Practice Makes Perfect!

Determine algebraically if the ordered pair is a solution of the equation. (See Objective 1.)

11. $(2, 0)$; $3x + y = 6$ 12. $(-8, 0)$; $2x - 3y = 16$

13. $(0, -3)$; $4x - 2y = -6$ 14. $(0, 5)$; $x - 2y = -10$

15. $(5, -1)$; $7x + y = 36$ 16. $(-4, -2)$; $5x - 6y = -8$

17. $\left(\frac{1}{2}, -\frac{3}{2}\right)$; $2x - 4y = 7$ 18. $\left(-\frac{3}{5}, \frac{1}{4}\right)$; $10x + 4y = 5$

19. $\left(-\frac{2}{5}, 0\right)$; $y = 5x + 2$ 20. $\left(\frac{4}{7}, 0\right)$; $y = 7x + 4$

21. $(0, 2)$; $y = x^2 - 5x + 2$ 22. $(3, 4)$; $y = x^2 - 5x + 2$

23. $(-3, 8); y = 2x^2 + 3x - 1$ **24.** $(0, 1); y = 2x^2 + 3x - 1$

25. $(0, -4); y = |x - 4|$ **26.** $\left(-\dfrac{1}{2}, 3\right); y = |4x - 1|$

27. $(4, 0); y = \dfrac{x - 4}{x + 2}$ **28.** $(-2, 0); y = \dfrac{x - 4}{x + 2}$

Plot each ordered pair on the Cartesian coordinate system and state the quadrant or axis where each point is located. (*See Objective 2.*)

29. $(1, 5)$ **30.** $(5, 1)$ **31.** $(-6, 2)$

32. $(2, -6)$ **33.** $(3, -4)$ **34.** $(-4, 3)$

35. $\left(\dfrac{7}{2}, -\dfrac{1}{2}\right)$ **36.** $\left(\dfrac{1}{2}, \dfrac{7}{2}\right)$ **37.** $(-6.2, -2.4)$

38. $(-3.5, 7.2)$ **39.** $(0, -5)$ **40.** $(-3, 0)$

41. $(2, 0)$ **42.** $(0, -4)$ **43.** $(\pi, -6)$

44. $(4, \pi)$

Identify each point plotted on the coordinate system and state the quadrant or axis where each point is located. (*See Objective 2.*)

45. A

46. B

47. C

48. D

49. E

50. F

51. G

52. H

53. I

54. J

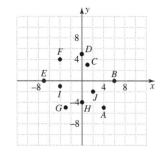

Identify the quadrant or axis where each point is located. (*See Objective 2.*)

55. $(5, -3)$ **56.** $(-2, -6)$ **57.** $(4, 10)$

58. $(-9, 15)$ **59.** $(0, -10)$ **60.** $(15, 0)$

61. $(1.2, 3.7)$ **62.** $\left(-\dfrac{1}{5}, 80\right)$ **63.** $\left(-\dfrac{1}{4}, 0\right)$

64. $(0, 20)$

Graph each equation by completing a table of solutions. (*See Objective 3.*)

65. $y = x + 2$ **66.** $y = x - 3$ **67.** $y = -2x + 4$

68. $y = -3x - 6$ **69.** $y = x^2 + 2$ **70.** $y = x^2 - 3$

71. $y = -x^2 + 4$ **72.** $y = -x^2 - 1$ **73.** $y = |x - 3|$

74. $y = |x + 2|$

Use the given graph of an equation to determine if each ordered pair is a solution of the equation. (*See Objective 4.*)

75. $(-2, -3)$

76. $(0, -1)$

77. $(3, 5)$

78. $(-2, 2)$

79. $\left(\dfrac{1}{2}, 0\right)$

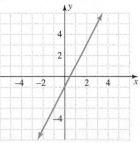

80. $(0, 2)$

81. $(-2, 0)$

82. $(3, 1)$

83. $(-4, -2)$

84. $(-6, 4)$

85. $(-5, 3)$

86. $(3, 5)$

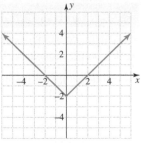

Use the graph to complete the ordered pair that is a solution. (*See Objective 4.*)

87. $(0, \underline{\quad})$

88. $(-1, \underline{\quad})$

89. $(\underline{\quad}, 0)$

90. $(\underline{\quad}, 2)$

91. $(3, \underline{\quad})$

92. $(\underline{\quad}, -6)$

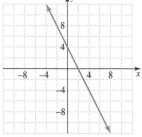

Use the table or graph to answer each question. (*See Objective 5.*)

93. The table shows the U.S. unemployment rates between 2001 and 2010. (Source: Bureau of Labor Statistics)

Years After 2001	0	1	2	3	4	5	6	7	8	9
Unemployment Rate (percent)	4.7	5.8	6.0	5.5	5.1	4.6	4.6	5.8	9.3	9.6

   a. Write the ordered pairs (x, y) that correspond to the data in the table, where x is the years after 2001 and y is the unemployment rate.

   b. Interpret the meaning of the first and last ordered pairs in the context of the problem.

   c. In what year was the unemployment rate the highest? The lowest?

   d. Make a scatter plot of the data.

94. The table lists the approximate salary of Peyton Manning, quarterback for the Indianapolis Colts, for 2003–2009. (Source: http://content.usatoday.com/sportsdata/football/nfl/salaries/player/Peyton-Manning)

Years After 2003	0	1	2	3	4	5	6
Salary (in millions)	$11.3	$35	$0.7	$10	$11	$11.5	$14

   a. Write the ordered pairs that correspond to the data in the table, where x is years after 2003 and y is Manning's salary (in millions of dollars).

   b. Interpret the meaning of first and last ordered pairs in the context of the problem.

   c. In what year was his salary the highest? The lowest?

   d. Make a scatter plot of the data.

95. The graph shows the national average hourly salary for registered nurses between 1998 and 2005. (Source: Bureau of Labor Statistics)

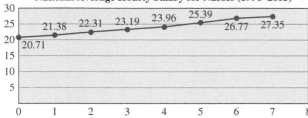

National Average Hourly Salary for Nurses (1998–2005)

a. Write an ordered pair for each point labeled on the graph.

b. What was the national average hourly salary for nurses in 1998?

c. What was the national average hourly salary for nurses in 2002?

d. How much did the average hourly salary increase from 1998 to 2005?

e. Based on the graph, how would you describe the average hourly salaries of nurses?

96. The bar graph shows the number of unique visitors, in millions, for the given websites for August 2010. (Source: http://www.compete.com)

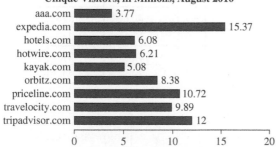

Unique Visitors, in Millions, August 2010

a. Write ordered pairs for this bar graph, where x is the website and y is the number of unique visitors, in millions.

b. Which website had the largest number of visitors in August 2010? How many did it have?

c. Which website had the least number of visitors in August 2010? How many did it have?

 Mix 'Em Up!

Determine algebraically if the given ordered pair is a solution of the equation.

97. $(1, -3)$; $9x - y = 12$

98. $(5, 4)$; $3x + 2y = 25$

99. $(-0.75, -3)$; $6x + y = -7.5$

100. $(0.7, 0)$; $-5x + 10y = 6.5$

101. $(-2, 5)$; $y = 2x^2 + 9x + 16$

102. $(10, 4)$; $y = -x^2 + x + 94$

103. $\left(\frac{1}{2}, -1\right)$; $y = -2x^2 + 9x - 5$

104. $\left(\frac{2}{3}, -2\right)$; $y = -3x^2 - 7x + 4$

105. $(-0.8, 7.4)$; $y = |3x - 5|$ **106.** $(0.6, 3.4)$; $y = |4x + 2|$

107. $(3, 0)$; $y = \dfrac{10x - 9}{x - 3}$ **108.** $(1, 3)$; $y = \dfrac{5x + 4}{-6x + 9}$

Plot each ordered pair on the Cartesian coordinate system and state the quadrant or axis where each point is located.

109. $\left(-\dfrac{3}{2}, -5\right)$ **110.** $\left(0, -\dfrac{1}{2}\right)$ **111.** $(6.5, 0)$

112. $\left(3, -\dfrac{5}{2}\right)$ **113.** $(-4, 5)$ **114.** $(0, 3)$

115. $(2.5, 4.5)$ **116.** $(-2, 1.5)$

Identify each point plotted on the coordinate system and state the quadrant or axis where each point is located.

117. A

118. B

119. C

120. D

121. E

122. F

123. G

124. H

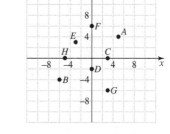

Graph each equation by completing a table of solutions.

125. $y = 3 - x$ **126.** $y = 1 - 2x$ **127.** $y = |2 - x|$

128. $y = 4 - x^2$ **129.** $y = 1 - |x|$ **130.** $y = 1 - x^2$

Use the given graph of an equation to determine if each ordered pair is a solution of the equation.

131. $(3, 0)$

132. $(0, -3)$

133. $(4, 6)$

134. $(-2, 4)$

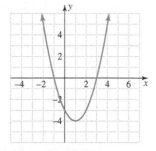

Use the given graph to complete each ordered pair that is a solution.

135. $(\underline{\hspace{1cm}}, -3)$

136. $\left(-\dfrac{1}{2}, \underline{\hspace{1cm}}\right)$

137. $(\underline{\hspace{1cm}}, 1)$

138. $(\underline{\hspace{1cm}}, -8)$

139. $(1, \underline{\hspace{1cm}})$

140. $(3, \underline{\hspace{1cm}})$

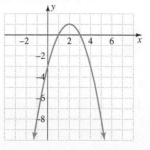

Use the graphs to answer each question.

141. The graph shows the total expenditures for airline transportation in the United States between 1998 and 2007. (Source: http://www.bts.gov/publications/national_transportation_statistics/)

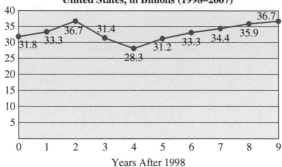

Yearly Expenditures for Airline Transportation in the
United States, in Billions (1998–2007)

a. Write an ordered pair for each point labeled on the graph.

b. In what year were the expenditures the greatest? The least?

c. Between what consecutive years did the expenditures decrease? By how much?

d. Based on the graph, how would you describe expenditures for U.S. airline transportation?

142. The graph shows the gross public debt amount in the United States between 2000 and 2010. (Source: http://www.treasurydirect.gov/NP/BPDLogin?application=np)

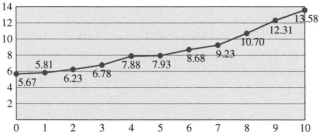

Gross Public Debt Amount, in Trillions (2000–2010)

a. Write an ordered pair for each point labeled on the graph.

b. In what year was the public debt amount the greatest? The least?

c. Between what consecutive years did the debt increase most? By how much?

d. Based on the graph, how would you describe public debt amount for the United States?

 You Be the Teacher!

Answer each student's question or correct the errors, if any.

143. Andrea: I always mix up the location of the four quadrants. Can you give me a way to remember which quadrant is which?

144. Graph the equation $y = |x - 4|$.

Bernadette's work:

x	$y = |x - 4|$	(x, y)
-1	-5	$(-1, -5)$
0	-4	$(0, -4)$
1	-3	$(1, -3)$
2	-2	$(2, -2)$
3	-1	$(3, -1)$

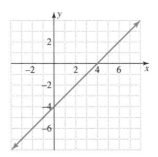

145. Graph the equation $y = -x^2 + 1$.

Valerie's work:

x	$y = -x^2 + 1$	(x, y)
-2	5	$(-2, 5)$
-1	2	$(-1, 2)$
0	1	$(0, 1)$
1	2	$(1, 2)$
2	5	$(2, 5)$

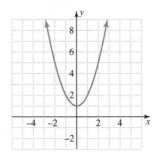

 **Calculate It!**

Use a graphing calculator to graph the points or equations. Sketch the graph displayed on the calculator.

146. $(-3, 9), (-2, 4), (-1, 1), (0, 0), (1, 1), (2, 4), (3, 9)$

147. $(-3, -5), (-2, -3), (-1, -1), (0, 1), (1, 3), (2, 5), (3, 7)$

148. $y = 4 - 2x$

149. $y = x^3 - 2x^2 + 4$

SECTION 3.2 — Graphing Linear Equations

▶ OBJECTIVES

As a result of completing this section, you will be able to

1. Identify a linear equation in two variables.
2. Plot points to graph a linear equation in two variables.
3. Use intercepts to graph linear equations in two variables.
4. Recognize and graph horizontal and vertical lines.
5. Solve application problems.
6. Troubleshoot common errors.

The median annual income for men with a bachelor's degree is given by the linear equation $y = 1443x + 38{,}843$, where x is the number of years after 1990. In this section, we will graph this equation and use the equation to find important information.

Linear Equations in Two Variables

In Chapter 2, we defined a linear equation in one variable as an equation of the form $Ax + B = C$, where A, B, and C are real numbers. The defining characteristic is that the exponent of the variable is 1. A linear equation in two variables is defined in a similar manner.

> **Definition:** An equation of the form $Ax + By = C$, where A, B, and C are real numbers with A and B not both zero is a **linear equation in two variables**.

Objective 1 ▶

Identify a linear equation in two variables.

> **Note:** *The exponent of the variables x and y is 1.*

When the values of A, B, and C are integer values with $A > 0$, the linear equation is considered to be written in the **standard form of a line**, or $Ax + By = C$.

Some examples of linear equations in two variables are

$$y = 2x - 1 \qquad \frac{x}{4} + 7y = 1 \qquad 3x - 5y = 15 \qquad x + 4y = -8$$

$$\text{(standard form)} \qquad \text{(standard form)}$$

The last two equations are written in standard form, while the first two are not.

Objective 1 Examples

Determine if the equation is a linear equation in two variables. If the equation is a linear equation in two variables, write it in standard form and identify the values of A, B, and C.

1a. $x - y = 6$ **1b.** $y = 3$ **1c.** $3x^2 + y = 9$ **1d.** $y = \dfrac{3}{2}x - 2$

Solutions

1a. This equation is a linear equation in two variables because the exponents of the variables are 1. It is also written in standard form.

$$x - y = 6$$
$$1x + (-1)y = 6$$
$$A = 1, B = -1, C = 6$$

1b. This equation is a linear equation in two variables. The term with x is missing, so it is understood to have a coefficient of 0.

$$y = 3$$
$$0x + 1y = 3$$
$$A = 0, B = 1, C = 3$$

> **Note:** *It is not necessary to write the coefficients of 0 or 1. They are written here for emphasis.*

1c. The equation $3x^2 + y = 9$ is *not* a linear equation since the exponent of x is 2.

1d. The equation is a linear equation because the exponents of the variables are both 1. It is, however, not written in standard form. To write it in standard form, we need to clear fractions and get the variable terms on one side of the equation and the constant on the other.

$$y = \frac{3}{2}x - 2$$

$$2(y) = 2\left(\frac{3}{2}x - 2\right) \qquad \text{Multiply each side by the LCD, 2.}$$

$$2y = 3x - 4 \qquad \text{Apply the distributive property and simplify.}$$

$$2y - 3x = 3x - 4 - 3x \qquad \text{Subtract } 3x \text{ from each side.}$$

$$-3x + 2y = -4 \qquad \text{Rewrite the equation with the } x\text{-term first.}$$

$$-1(-3x + 2y) = -1(-4) \qquad \text{Multiply each side by } -1.$$

$$3x - 2y = 4 \qquad \text{Simplify.}$$

So, $A = 3$, $B = -2$, and $C = 4$.

✓ **Student Check 1**   Determine if the equation is a linear equation in two variables. If the equation is a linear equation in two variables, write it in standard form and identify the values of A, B, and C.

a. $x - 2y = -4$ **b.** $x = 4$ **c.** $x^2 + y^2 = 25$ **d.** $y = \frac{2}{3}x + 4$

Graphing Linear Equations in Two Variables

Objective 2 ▶

Plot points to graph a linear equation in two variables.

In Section 3.1, we learned how to graph equations in two variables. In this section, we will specifically discuss how to graph linear equations in two variables.

Consider the equation $x + y = 5$. The solutions of this equation are ordered pairs (x, y) such that the sum of the x-coordinate and y-coordinate is 5. There are many ordered pairs that satisfy this condition. Some of the solutions are listed next.

x	y	(x, y)
5	0	$(5, 0)$
4	1	$(4, 1)$
3	2	$(3, 2)$
2	3	$(2, 3)$
1	4	$(1, 4)$
0	5	$(0, 5)$
-1	6	$(-1, 6)$
-2.5	7.5	$(-2.5, 7.5)$

$$5 + 0 = 5$$
$$4 + 1 = 5$$
$$3 + 2 = 5$$
$$2 + 3 = 5$$
$$1 + 4 = 5$$
$$0 + 5 = 5$$
$$-1 + 6 = 5$$
$$-2.5 + 7.5 = 5$$

This list of solutions continues indefinitely. Since we cannot possibly list all of the solutions of this equation, we plot the points and try to recognize a pattern that enables us to connect the points and extend the graph.

Notice that the graph of this equation is a *line*. This is the reason the equation is called a *linear* equation. Any point on the graph of this line is a solution of the equation $x + y = 5$. Note that we put arrows on the ends of the graph to indicate there are infinitely many solutions.

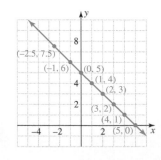

Note: *The graph of any linear equation in two variables is a line.*

Procedure: Graphing a Linear Equation in Two Variables by Plotting Points

Step 1: Make a table of at least three ordered pairs that satisfy the equation. Only two ordered pairs are required to graph a line, but a third point is good to use as a check.

    a. To get the solutions, choose a value for x, substitute this in the equation and solve for y. (A good idea is to use a negative value, zero, and a positive value.)

    b. We can also choose a value for y, substitute this in the equation, and solve for x.

Step 2: Plot the three ordered pairs. These three points should lie in a straight line. If they do not, an error has been made. Recheck your work.

Step 3: Draw the line that contains the three points. The line should extend beyond the points and have arrows at both ends to indicate that there are infinitely many solutions of the equation.

Objective 2 Examples **Graph each linear equation.**

2a. $y = -x + 3$ **2b.** $y = \dfrac{4}{3}x - 4$ **2c.** $x - 2y = 4$

Solutions **2a.** Let $x = -1, 0,$ and 1. We will use the table to organize our work to obtain the solutions.

x	$y = -x + 3$	(x, y)
-1	$y = -(-1) + 3$ $y = 1 + 3$ $y = 4$	$(-1, 4)$
0	$y = -(0) + 3$ $y = 0 + 3$ $y = 3$	$(0, 3)$
1	$y = -(1) + 3$ $y = -1 + 3$ $y = 2$	$(1, 2)$

Plot the points $(-1, 4)$, $(0, 3)$, and $(1, 2)$. Since these points lie on a line, we draw a line through the points to obtain the graph of $y = -x + 3$.

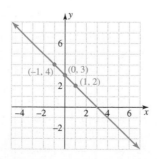

2b. Because the coefficient of x is $\dfrac{4}{3}$, our calculations for y will be easier if the values of x are divisible by 3. So, we will let $x = -3$, 0, and 3.

x	$y = \dfrac{4}{3}x - 4$	(x, y)
-3	$y = \dfrac{4}{3}(-3) - 4$ $y = -4 - 4$ $y = -8$	$(-3, -8)$
0	$y = \dfrac{4}{3}(0) - 4$ $y = 0 - 4$ $y = -4$	$(0, -4)$
3	$y = \dfrac{4}{3}(3) - 4$ $y = 4 - 4$ $y = 0$	$(3, 0)$

Plot the points $(-3, -8)$, $(0, -4)$, and $(3, 0)$. Since the points lie in a line, we draw the line through the points to obtain the graph of $y = \dfrac{4}{3}x - 4$.

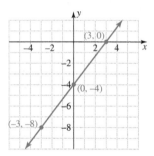

2c. Let $x = -2, 0, 1$.

x	$x - 2y = 4$	(x, y)
-2	$-2 - 2y = 4$ $-2 - 2y + 2 = 4 + 2$ $-2y = 6$ $\dfrac{-2y}{-2} = \dfrac{6}{-2}$ $y = -3$	$(-2, -3)$
0	$0 - 2y = 4$ $-2y = 4$ $\dfrac{-2y}{-2} = \dfrac{4}{-2}$ $y = -2$	$(0, -2)$
1	$1 - 2y = 4$ $1 - 2y - 1 = 4 - 1$ $-2y = 3$ $\dfrac{-2y}{-2} = \dfrac{3}{-2}$ $y = -\dfrac{3}{2}$	$\left(1, -\dfrac{3}{2}\right)$

Plotting these points and connecting them gives us the graph of $x - 2y = 4$.

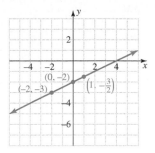

 Student Check 2   Graph each equation.

a. $y = x - 2$ **b.** $y = -\dfrac{1}{3}x + 6$ **c.** $x + 3y = 9$

Finding Intercepts

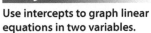 **Objective 3** ▶

Use intercepts to graph linear equations in two variables.

Another method of graphing a linear equation in two variables involves finding two important points that lie on the graph of the line. These points are called the *intercepts* of the graph.

> **Definition:** The **x-intercept** is the point on the graph where the graph intersects the x-axis. The **y-intercept** is the point on the graph where the graph intersects the y-axis.

In the graph from Example 2b, the x-intercept is $(3, 0)$ and the y-intercept is $(0, -4)$. Notice the x-intercept has a y-coordinate of zero since every point on the x-axis has a y-coordinate of zero. Similarly, the y-intercept has an x-coordinate of zero since every point on the y-axis has an x-coordinate of zero.

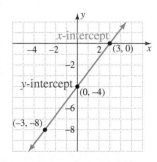

> **Procedure: Finding the Intercepts**
>
> **Step 1:** To find the x-intercept, replace y with 0 and solve for x. The point will always be of the form $(a, 0)$.
> **Step 2:** To find the y-intercept, replace x with 0 and solve for y. The point will always be of the form $(0, b)$.

> **Note:** While we can graph a line knowing only two points, it can be helpful to find a third point as a checkpoint.

Objective 3 Examples Find the *x*- and *y*-intercepts for each equation. Use these points to graph each line.

3a. $2x + 3y = 6$ **3b.** $y = 2x - 4$ **3c.** $2x - y = 0$

Solutions **3a.**

	x	y	(x, y)
x-intercept	$2x + 3(0) = 6$ $2x = 6$ $x = 3$	0	$(3, 0)$
y-intercept	0	$2(0) + 3y = 6$ $3y = 6$ $y = 2$	$(0, 2)$
Checkpoint	1	$2(1) + 3y = 6$ $2 + 3y = 6$ $3y = 4$ $y = \dfrac{4}{3}$	$\left(1, \dfrac{4}{3}\right)$

Plotting the points $(3, 0)$, $(0, 2)$, and $\left(1, \dfrac{4}{3}\right)$ gives us the graph of $2x + 3y = 6$.

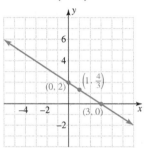

Notice that the checkpoint also lies on the graph of the line, so we know our work is correct.

3b.

	x	y	(x, y)
x-intercept	$0 = 2x - 4$ $0 + 4 = 2x - 4 + 4$ $4 = 2x$ $\dfrac{4}{2} = \dfrac{2x}{2}$ $2 = x$	0	$(2, 0)$
y-intercept	0	$y = 2(0) - 4$ $y = 0 - 4$ $y = -4$	$(0, -4)$
Checkpoint	3	$y = 2(3) - 4$ $y = 6 - 4$ $y = 2$	$(3, 2)$

Plotting the points $(2, 0)$, $(0, -4)$, and $(3, 2)$, gives us the graph of $y = 2x - 4$.

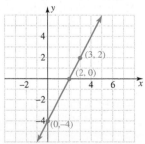

Notice that the checkpoint lies on the graph of the line, so we know our work is correct.

3c.

x	y	(x, y)
$2x - 0 = 0$ $2x = 0$ $\dfrac{2x}{2} = \dfrac{0}{2}$ $x = 0$	0	$(0, 0)$
0	$2(0) - y = 0$ $-y = 0$ $\dfrac{-y}{-1} = \dfrac{0}{-1}$ $y = 0$	$(0, 0)$
2	$2(2) - y = 0$ $4 - y = 0$ $4 - y - 4 = 0 - 4$ $-y = -4$ $\dfrac{-y}{-1} = \dfrac{-4}{-1}$ $y = 4$	$(2, 4)$

The x- and y-intercepts for this graph are the same point, the origin. So, we need to find another point on the graph to have enough information to draw the line. Choose any value for x and solve for y.

Plotting $(0, 0)$ and $(2, 4)$ gives us the graph of $2x - y = 0$.

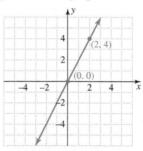

 Note: *Equations of the form $Ax + By = 0$ have graphs that go through the origin since $(0, 0)$ is always a solution of this type of equation.*

 Student Check 3 Find the x- and y-intercepts for each equation. Use these points to graph each line.

a. $4x - y = -8$ **b.** $y = 3x - 3$ **c.** $4x + y = 0$

Horizontal and Vertical Lines

Objective 4 ▶

Recognize and graph horizontal and vertical lines.

Horizontal and vertical lines are special cases of linear equations in two variables. Consider the graph of the *horizontal line* shown next.

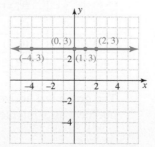

Some of the points on the line are $(0, 3)$, $(1, 3)$, $(2, 3)$, and $(-4, 3)$. The only points that this graph contains are points whose y-value is 3. A key characteristic for an ordered pair to be a solution of the equation of this line is that the y-value equals 3. Therefore, the equation of this line is $y = 3$. The equation $y = 3$ is equivalent to $0x + y = 3$. Any value of x substituted in this equation is multiplied by 0, so the solutions only depend on the value of y being 3.

> **Definition:** The graph of any equation of the form $y = k$, where k is a real number, is a **horizontal line** through k on the y-axis. The point $(0, k)$ is the y-intercept of the graph.

Now, we will examine the graph of a *vertical line,* as follows.

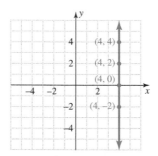

Some of the points on the line are $(4, 4)$, $(4, 2)$, $(4, 0)$, and $(4, -2)$. The only points that this graph contains are points whose x-value is 4. A key characteristic for an ordered pair to be a solution of the equation of this line is that the x-value equals 4. Therefore, the equation of this line is $x = 4$. The equation $x = 4$ is equivalent to $x + 0y = 4$. Any value of y substituted in this equation is multiplied by 0, so the solutions only depend on the value of x being 4.

> **Definition:** The graph of any equation of the form $x = h$, where h is a real number, is a **vertical line** through h on the x-axis. The point $(h, 0)$ is the x-intercept of the graph.

Objective 4 Examples

Graph each linear equation.

4a. $y = 2$ **4b.** $x = -3$ **4c.** $2x + 5 = 0$

Solutions **4a.** The equation $y = 2$ can be written as $0x + y = 2$. Notice that for any x-value chosen, y is 2. So, any ordered pair whose y-coordinate is 2 is a solution. If $x = 3, 0$, and -1, the ordered pair solutions are

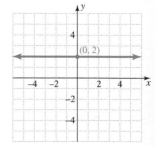

x	y	(x, y)
3	2	$(3, 2)$
0	2	$(0, 2)$
-1	2	$(-1, 2)$

The graph is a horizontal line with a y-intercept of $(0, 2)$. Note the graph does not have an x-intercept because the y-coordinate is never zero.

4b. The equation $x = -3$ can be written as $x + 0y = -3$. Notice that for any y-value chosen, x is -3. So, any ordered pair whose x-coordinate is -3 is a solution. If $y = 0$, 1, and -1, the ordered pair solutions are

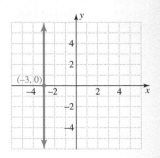

x	y	(x, y)
-3	0	$(3, 0)$
-3	1	$(3, 1)$
-3	-1	$(-3, -1)$

The graph is a vertical line with an x-intercept of $(-3, 0)$. Note the graph does not have a y-intercept because the x-coordinate is never zero.

4c. The equation $2x + 5 = 0$ can be written as $2x + 0y = -5$ or as $x + 0y = -\dfrac{5}{2}$. Notice that for any y-value chosen, x is $-\dfrac{5}{2}$. So, any ordered pair whose x-coordinate is $-\dfrac{5}{2}$ is a solution. If $y = 1$, -4, and 3, the ordered pair solutions are

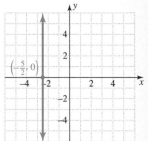

x	y	(x, y)
$-\dfrac{5}{2}$	1	$\left(-\dfrac{5}{2}, 1\right)$
$-\dfrac{5}{2}$	-4	$\left(-\dfrac{5}{2}, -4\right)$
$-\dfrac{5}{2}$	3	$\left(-\dfrac{5}{2}, 3\right)$

The graph is a vertical line with x-intercept $\left(-\dfrac{5}{2}, 0\right)$. Notice the line has no y-intercept since the x-coordinate is never zero.

✔ **Student Check 4**   Graph each linear equation.

    a. $y = -1$ **b.** $x = 5$ **c.** $3y + 2 = 0$

Applications

Objective 5 ▶

Solve application problems.

In the application problems encountered in this section, we will be given a linear equation that models a real-world situation. We will construct the graph of the equation and extract information from the equation.

Objective 5 Examples   **Use the equation to answer each question.**

5a. The median annual income for men with a bachelor's degree can be modeled by the equation $y = 1443x + 38{,}843$, where x is the number of years after 1990.
(Source: www.infoplease.com)

    i. Is this equation a linear equation in two variables? Why or why not?

    ii. Find the y-intercept of the equation and interpret its meaning in the context of this problem.

    iii. Find the x-intercept of the equation and interpret its meaning in the context of this problem.

    iv. Find the median annual income for the years 2000 and 2010.

Solution   **5a.**   **i.** The equation $y = 1443x + 38{,}843$ is a linear equation in two variables since each variable has a power of 1.

ii. The y-intercept of $y = 1443x + 38{,}843$ is found by replacing x with 0 and solving for y.

$$y = 1443x + 38{,}843$$

$$y = 1443(0) + 38{,}843 \qquad \text{Substitute 0 for } x.$$

$$y = 0 + 38{,}843 \qquad \text{Multiply.}$$

$$y = 38{,}843 \qquad \text{Add.}$$

The y-intercept is the ordered pair $(0, 38{,}843)$. This means that in 0 yr after 1990, or 1990, the median annual income for men with a bachelor's degree was $38,843.

iii. The x-intercept of $y = 1443x + 38{,}843$ is found by replacing y with 0 and solving for x.

$$y = 1443x + 38{,}843$$

$$0 = 1443x + 38{,}843 \qquad \text{Substitute 0 for } y.$$

$$0 - 38{,}843 = 1443x + 38{,}843 - 38{,}843 \qquad \text{Subtract 38,843 from each side.}$$

$$-38{,}843 = 1443x \qquad \text{Simplify.}$$

$$\frac{-38{,}843}{1443} = \frac{1443x}{1443} \qquad \text{Divide each side by 1443.}$$

$$-26.9 \approx x \qquad \text{Simplify.}$$

The x-intercept is the ordered pair $(-26.9, 0)$. This point does not make sense in the context of the problem since we cannot have a negative value for x.

iv. The median annual income for 2000 is found by replacing x with 10 ($2000 - 1990 = 10$) and the income for 2010 is found by replacing x with 20 ($2010 - 1990 = 20$).

Let $x = 10$:

$$y = 1443(10) + 38{,}843$$

$$y = 14{,}430 + 38{,}843$$

$$y = 53{,}273$$

In 2000, the median annual income was $53,273.

Let $x = 20$:

$$y = 1443(20) + 38{,}843$$

$$y = 28{,}860 + 38{,}843$$

$$y = 67{,}703$$

In 2010, the median annual income was $67,703.

5b. Straight-line depreciation is the most common method of depreciating business assets. Mr. Rowell's surveying company purchased a truck for their business. On the company's tax records, Mr. Rowell must report the depreciated value of the truck. The truck's value is given by $y = -4000x + 32{,}000$, where x is the age of the truck in years.

   i. Find the x- and y-intercepts and interpret their meaning in the context of this problem.

   ii. Find the value of the truck after 2 yr and after 4 yr.

   iii. Use the ordered pairs found in parts (i) and (ii) to draw the graph of the equation.

Solution   **5b.**   **i.** The x- and y-intercepts of $y = -4000x + 32{,}000$ are shown next.

x	y	Meaning
$0 = -4000x + 32{,}000$ $4000x = 32{,}000$ $\dfrac{4000x}{4000} = \dfrac{32{,}000}{4000}$ $x = 8$	0	The x-intercept is $(8, 0)$. It means that when the truck is 8 yr old, its value is \$0.
0	$y = -4000(0) + 32{,}000$ $y = 32{,}000$	The y-intercept is $(0, 32{,}000)$. It means that when the truck is 0 yr old, or when it is new, its value is \$32,000.

ii. The value of the truck after 2 yr and after 4 yr is shown next.

x	y	Meaning
2	$y = -4000(2) + 32{,}000$ $y = -8000 + 32{,}000$ $y = 24{,}000$	The point is $(2, 24{,}000)$. It means that in 2 yr, the value of the truck is \$24,000.
4	$y = -4000(4) + 32{,}000$ $y = -16{,}000 + 32{,}000$ $y = 16{,}000$	The point is $(4, 16{,}000)$. It means that in 4 yr, the value of the truck is \$16,000.

iii. The graph of the equation is found by plotting the points $(8, 0)$, $(0, 32{,}000)$, $(2, 24{,}000)$, and $(4, 16{,}000)$. Because the y-values range from 0 to 32,000, we let each tick mark on the y-axis represent 4000 units. Also, the graph is drawn only in Quadrant I since it doesn't make sense for the years to be negative or for the value of the truck to be negative.

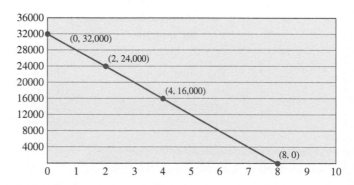

✔ **Student Check 5**   The median annual income for women with a bachelor's degree is given by the equation $y = 1100x + 27{,}645$, where x is the number of years after 1990. (Source: www .infoplease.com)

   a. Find the y-intercept of the equation and interpret its meaning in the context of this problem.

   b. Find the median annual income for the years 2000 and 2010.

   c. Use the information in parts (a) and (b) to graph the equation.

Objective 6 ▶

Troubleshoot common errors.

Troubleshooting Common Errors

Some common errors associated with graphing linear equations in two variables are illustrated next.

Objective 6 Examples A problem and an incorrect solution are given. Provide the correct solution and an explanation of the error.

6a. Graph the equation $5x + y = 0$.

Incorrect Solution	Correct Solution and Explanation
The x-intercept is $(0, 0)$ and the y-intercept is $(0, 0)$. So, the graph is 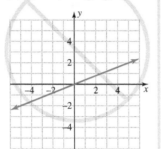	We must have two points to draw the graph of this line. To find another point, make a substitution for x. If $x = 1$, then $$5x + y = 0$$ $$5(1) + y = 0$$ $$5 + y = 0$$ $$y = -5$$ We can graph the equation using the points $(0, 0)$ and $(1, -5)$. 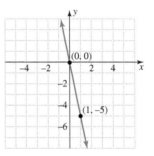

6b. Find the x- and y-intercepts of $8x - y = -16$.

Incorrect Solution	Correct Solution and Explanation
$$8(0) - y = -16$$ $$y = -16$$ $$8x - 0 = -16$$ $$8x = -16$$ $$x = -2$$ The x-intercept is $(0, -16)$ and the y-intercept is $(-2, 0)$.	The first error is in solving the first equation. It should be $$8(0) - y = -16$$ $$-y = -16$$ $$y = 16$$ The other error was in assigning the points. The point $(0, 16)$ is the y-intercept and the point $(-2, 0)$ is the x-intercept.

ANSWERS TO STUDENT CHECKS

Student Check 1 **a.** yes; $x - 2y = -4$; $A = 1, B = -2, C = -4$ **b.** yes; $x + 0y = 4$; $A = 1, B = 0, C = 4$
c. no **d.** yes; $2x - 3y = -12$; $A = 2, B = -3, C = -12$

Student Check 2

a. $y = x - 2$

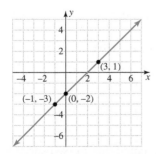

b. $y = -\dfrac{1}{3}x + 6$

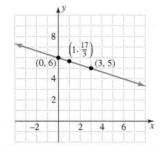

c. $x + 3y = 9$

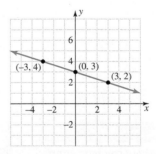

Student Check 3

   a. intercepts: $(-2, 0)$ and $(0, 8)$ **b.** intercepts: $(1, 0)$ and $(0, -3)$ **c.** intercepts: $(0, 0)$

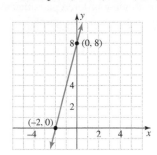 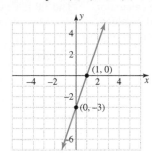 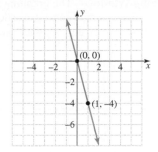

Student Check 4

   a. $y = -1$ **b.** $x = 5$ **c.** $3y + 2 = 0$

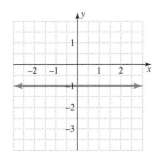 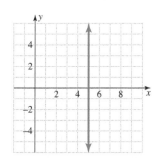 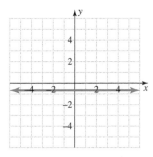

Student Check 5   **a.** The y-intercept is $(0, 27{,}645)$. This means that in 1990, the median annual income for women with a bachelor's degree is \$27,645.   **b.** The median income in 2000 is \$38,645 and was \$49,645 in 2010.

  c.

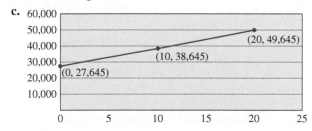

SUMMARY OF KEY CONCEPTS

1. A linear equation in two variables is an equation of the form $Ax + By = C$, where A, B, and C are real numbers with A and B not both zero. The exponent of the variables is one. If the values of A, B, and C are not fractions and $A > 0$, this form is called the standard form. It is important to be able to write equivalent forms of a linear equation in two variables.

2. To graph an equation by plotting points, select at least three different values for x or y, substitute into the equation, and solve for the corresponding value. These three points should lie in a straight line. If they do not, a mistake was made in your calculations or in graphing.

3. A linear equation can also be graphed by finding the x- and y-intercepts.

   a. To find the x-intercept, replace y with zero and solve for x. This is a point of the form $(a, 0)$.

   b. To find the y-intercept, replace x with zero and solve for y. This is a point of the form $(0, b)$.

   c. If both the x- and y-intercepts are $(0, 0)$, then the graph goes through the origin. Another point is needed to graph the line.

4. Horizontal lines are represented by an equation of the form $y = k$. These lines have a y-intercept of $(0, k)$. Equations of the form $x = h$ have graphs that are vertical lines. These lines have an x-intercept of $(h, 0)$.

5. When a linear equation is given that represents a real-life situation, the x- and y-intercepts have meaning. It is important to be able to state the meaning of these points. Finding other points and the graph of these models is also an important skill.

GRAPHING CALCULATOR SKILLS

Graphing Window—The graphing window on the calculator comes with a standard setting, as shown. The *x*- and *y*-values range between −10 and 10, with each tick mark representing 1 unit.

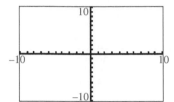
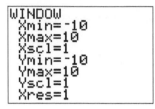

Equation Editor—The equation editor is what we use to enter equations into the calculator. This is accessed by pressing [Y=].

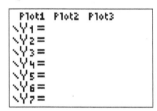

Example: Graph $x - 2y = 4$ on the calculator.

We must first solve the equation for *y*, because this is the only way equations can be entered into the calculator.

$$x - 2y = 4$$
$$x - 2y - x = 4 - x \qquad \text{Subtract } x \text{ from both sides.}$$
$$-2y = -x + 4 \qquad \text{Simplify.}$$
$$\frac{-2y}{-2} = \frac{-x + 4}{-2} \qquad \text{Divide both sides by } -2.$$
$$y = \frac{1}{2}x - 2 \qquad \text{Simplify.}$$

We can now graph the equation using a graphing calculator.

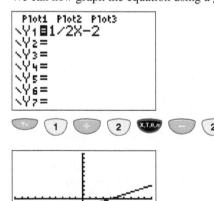

The Table Feature enables us to view ordered pairs that satisfy the equation and are, therefore, points on the graph.

With some equations, we may have to adjust the viewing window to see the graph on the graphing calculator as this next example illustrates.

Example: Graph $y = -4000x + 32{,}000$.

We enter it into the equation editor. When graphed in the standard window, the graph almost looks like a vertical line (which is not the case).

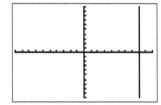

We must change the view of the window on the calculator to see the correct graph.

To change the view on the calculator, we should find the intercepts by hand to know how to change the window. From example 5b, we know the *x*-intercept is (8, 0) and the *y*-intercept is (0, 32,000). We also know that negative *x*- and *y*-values do not make sense in this problem.

To view a complete graph, we must set the window to see the intercepts. Since the graph crosses the *x*-axis at (0, 8), we need the maximum *x*-value to be larger than 8. Let's use 10.

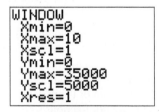

Since the graph crosses the *y*-axis at (0, 32,000), we need the maximum *y*-value to be larger than 32,000. Let's use 35,000. The value for Yscl represents how many units each tick mark on the *y*-axis represents. If we were drawing this by hand, we might let each tick mark be 5000 units.

This is the graph of the equation in the new window.

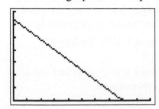

SECTION 3.2 / EXERCISE SET

Write About It!

Use complete sentences to explain the meaning of each term or process.

1. Graph a line by point-plotting.

2. Graph a line by using the x-intercept and y-intercept.

3. Which method is easier to use with the equation $y = -4x + 2$? Which method is easier to use with the equation $4x - 3y = -12$?

Determine if each statement is true or false. If a statement is false, provide an explanation.

4. The equation $y = 3x - 4$ is equivalent to $3x - y = -4$.

5. The graph of the equation $7x - y = 14$ passes through the point $(0, 2)$.

6. The x-intercept of $y = 8x - 24$ is $(0, -24)$.

7. The graph of the equation $x = -2$ has a y-intercept at $(0, -2)$.

8. The graph of the equation $y = 3$ has no x-intercept.

9. The graph of the equation $y = 5x$ passes through the origin.

10. The graph of the equation $2x + 3y = 0$ passes through the origin.

Practice Makes Perfect!

Determine if the equation is a linear equation in two variables. (*See Objective 1.*)

11. $4x + 5y = 6$
12. $3x - 2y = -12$
13. $x^2 + y = 4$
14. $x + y^2 = 9$
15. $x = 2y$
16. $y = -\dfrac{4}{3}x$
17. $\dfrac{x}{2} - \dfrac{y}{4} = \dfrac{1}{8}$
18. $x^2 + y^2 = 4$
19. $x = 3$
20. $y = -1$

Make a table of at least three ordered pairs that are solutions of the equation and use these points to graph each equation. (*See Objective 2.*)

21. $x + 4y = 8$
22. $x - 6y = 12$
23. $3x + y = -6$
24. $5x + y = 10$
25. $y = -x + 1$
26. $y = -x - 3$
27. $y = 2x - 4$
28. $y = -2x + 3$
29. $y = \dfrac{2}{3}x - 6$
30. $y = -\dfrac{2}{5}x + 4$
31. $y = -\dfrac{1}{4}x + 2$
32. $y = \dfrac{5}{6}x - 1$

Use the intercepts to graph each equation. Label the intercepts as ordered pairs on the graph. (*See Objective 3.*)

33. $x - 2y = 6$
34. $x + 6y = 6$
35. $7x - y = -7$
36. $2x + y = -4$
37. $y + 2x = 0$
38. $y - 2x = 0$

39. $y = -x + 4$
40. $y = -x - 7$
41. $y = 4x - 5$
42. $y = -6x + 3$
43. $y = \dfrac{1}{2}x + 5$
44. $y = -\dfrac{3}{4}x + 6$

Graph each equation. (*See Objective 4.*)

45. $y = 6$
46. $y = -5$
47. $x = -7$
48. $x = 4$
49. $8y - 2 = 0$
50. $6y + 3 = 0$
51. $2x + 4 = 0$
52. $3x - 9 = 0$
53. $4y + 3 = 0$
54. $2y - 7 = 0$
55. $\dfrac{1}{2}x + 5 = 0$
56. $-\dfrac{3}{4}x + 8 = 0$
57. $\dfrac{1}{3}y - 2 = 0$
58. $-\dfrac{3}{4}y + 1 = 0$

Use the equation to answer each question. (*See Objective 5.*)

59. The median annual income for men with a master's degree is given by the equation $y = 1873x + 49{,}217$, where x is the number of years after 1991. (Source: www.infoplease.com)

   a. Find the y-intercept and interpret its meaning.

   b. Find the median annual income for men in 2001.

   c. Find the expected median annual income for men in 2012.

60. The median annual income for women with a master's degree is given by the equation $y = 1335x + 35{,}660$, where x is the number of years after 1991. (Source: www.infoplease.com)

   a. Find the y-intercept and interpret its meaning.

   b. Find the median annual income for women in 2001.

   c. Find the expected median annual income for women in 2012.

61. The average retail price, in cents per kilowatt-hour, of electricity for residential use can be modeled by the equation $y = 0.395x + 8.154$, where x is the number of years after 2001. (Source: www.eia.gov)

   a. Find the y-intercept and interpret its meaning.

   b. Find the average retail price of electricity for residential use in 2010.

   c. Find the expected average retail price of electricity for residential use in 2015.

62. The average domestic first purchase price of crude oil, in dollars per barrel, can be modeled by the equation $y = 6.679x + 16.173$, where x is the number of years after 2000. (Source: www.eia.gov)

   a. Find the y-intercept and interpret its meaning.

   b. Find the average domestic first purchase price of crude oil in 2010.

   c. Find the expected average domestic first purchase price of crude oil in 2013.

Mix 'Em Up!

Graph each equation. Identify at least two points on the graph.

63. $9x - 3y = 6$ **64.** $4x - 2y = 8$ **65.** $y - 1 = 0$

66. $x + 3 = 0$ **67.** $y = 3x$ **68.** $y = -2x$

69. $y = -\dfrac{1}{2}x + 5$ **70.** $y = \dfrac{1}{3}x - 2$ **71.** $y = -0.4x + 1$

72. $y = 0.25x - 1$ **73.** $y = 2.4x - 1.8$ **74.** $y = -1.5x - 4.5$

75. $x = 0$ **76.** $y = 0$ **77.** $3x - 2y = 0$

78. $4x + 5y = 0$ **79.** $y = -\dfrac{1}{5}x$ **80.** $y = \dfrac{5}{6}x$

81. $2y + 5 = 0$ **82.** $5y - 2 = 0$ **83.** $2x - 10 = 0$

84. $3x + 3 = 0$ **85.** $4x - 5 = 0$ **86.** $6x - 9 = 0$

87. $0.01x - 0.02y = 1.2$ **88.** $0.03x + 0.02y = 0.6$

89. $\dfrac{2}{3}x - \dfrac{1}{2}y = 1$ **90.** $\dfrac{3}{4}x + \dfrac{1}{3}y = -2$

Use the equation to answer each question.

91. The percent of high school students who are users of cigarettes is on the decline. The percent of students who are users can be modeled by the equation $y = -2.5x + 38$, where x is the number of years after 1997. (Source: National Cancer Institute)

  a. Find the y-intercept of the equation and interpret its meaning.

  b. What percent of high school students used cigarettes in 2000?

  c. What percent of high school students used in cigarettes in 2010?

  d. Use the intercepts to graph the equation.

92. The expenditures (in billions) for public elementary and secondary education in the United States can be modeled by the equation $y = 20x + 285$, where x is the number of years after 1997. (Source: http://nces.ed.gov/programs/projections/projections2018/tables/table_34.asp)

  a. Find the y-intercept of the equation and interpret its meaning.

  b. What were the expenditures for public elementary and secondary education in the United States in 2005?

  c. What were the expenditures for public elementary and secondary education in the United States in 2012?

  d. Use the information obtained in parts (a) to (c) to graph the equation.

93. The percent of men aged 18 yr and older who are current cigarette smokers can be modeled by the linear equation $y = -0.33x + 25.64$, where x is the number of years after 1998. (Source: National Cancer Institute)

  a. Find the x-intercept of the equation and interpret its meaning. Is this realistic?

  b. Find the y-intercept of the equation and interpret its meaning.

  c. What percent of men aged 18 yr and older were cigarette smokers in 2005? How close is the answer to the actual estimate of 23.39%?

  d. What percent of men aged 18 yr and older were cigarette smokers in 2012?

  e. Use the intercepts to graph the equation.

  f. The Healthy People Campaign targets to reduce the percent of smokers. Find the year when the percent of men aged 18 yr and older who are current cigarette smokers will be reduced to 12%.

94. The percent of women aged 18 yr and older who are current cigarette smokers can be modeled by the linear equation $y = -0.53x + 22.12$, where x is the number of years after 1998. (Source: National Cancer Institute)

  a. Find the x-intercept of the equation and interpret its meaning. Is this realistic?

  b. Find the y-intercept of the equation and interpret its meaning.

  c. What percent of women aged 18 yr and older were cigarette smokers in 2005? How close is the answer to the actual estimate of 18.38%?

  d. What percent of women aged 18 yr and older were cigarette smokers in 2012?

  e. Use the intercepts to graph the equation.

  f. The Healthy People Campaign targets to reduce the percent of smokers. Find the year when the percent of women aged 18 yr and older who are current cigarette smokers will be reduced to 12%.

95. The percent of U.S. adults aged 20 yr and over who are obese can be modeled by the linear equation $y = 0.71x + 19.83$, where x is the number of years after 1997. (Source: National Center for Health Statistics)

  a. Find the y-intercept of the equation and interpret its meaning.

  b. What percent of U.S. adults were obese in 2008? How close is the answer to the actual estimate of 27.6% in 2008?

  c. What percent of U.S. adults were obese in 2012?

  d. Use the information obtained in parts (a) to (c) to graph the equation.

  e. What does the graph indicate about the percent of U.S. adults aged 20 yr and over who are obese?

96. The owner of a limousine rental company purchases a stretch limousine for $110,000. For tax purposes, the owner uses straight-line depreciation for reporting the value of the limousine. The depreciated  value of the limousine is given by $y = -22,000x + 110,000$, where x is the age of the limousine in years.

  a. Find the x-intercept and interpret the meaning.

  b. Find the y-intercept and interpret the meaning.

  c. When will the depreciated value of the limousine be $66,000?

d. What will be the depreciated value of the limousine when it is 3 yr old?

e. Use the information obtained in parts (a) to (c) to graph the equation.

You Be the Teacher!

Answer each student's question.

97. What is the minimum number of points that you must find to graph a straight line? Why?

98. For the equation $y = \frac{3}{7}x + 4$, how can I choose x-values to plug into the equation so that I get integer values for y?

99. How do you find the x-intercept of the equation $x = 7$?

100. How do you find the y-intercept of the equation $y = -10$?

101. I am trying to graph the equation $4y + x = 0$. When I find the intercepts, I can only find one point $(0, 0)$. How can I find another point to graph the equation?

102. I am trying to graph the equation $3x - 5y = 15$. When I find the intercepts, I only get one point $(5, -3)$. How can I find another point to graph the equation?

Calculate It!

Use a graphing calculator to complete each exercise.

103. Draw, by hand, the coordinate system that is defined by these window settings. Label the x-axis and y-axis with the appropriate tick marks.

```
WINDOW
 Xmin=0
 Xmax=100
 Xscl=1
 Ymin=0
 Ymax=500
 Yscl=50
 Xres=1
```

104. Draw, by hand, the coordinate system that is defined by these window settings. Label the x-axis and y-axis with appropriate tick marks.

```
WINDOW
 Xmin=-20
 Xmax=20
 Xscl=5
 Ymin=-50
 Ymax=50
 Yscl=10
 Xres=1
```

105. Determine the ordered pairs for the x-intercept and y-intercept of the graph if each tick mark represents one unit.

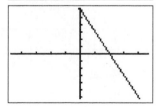

106. Determine the ordered pairs for the x-intercept and y-intercept of the graph if each tick mark on the x-axis represents one unit and each tick mark on the y-axis represents five units.

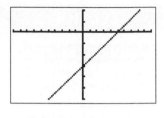

107. Graph $y = -2x + 20$ in the standard viewing window.

a. Sketch the graph shown on your calculator.

b. Which intercept(s) can be seen on your calculator?

c. Find the x-intercept and y-intercept by hand.

d. What viewing window will show the complete graph that contains both intercepts? State the values that you would use for Xmin, Xmax, Xscl, Ymin, Ymax, and Yscl.

e. Sketch the graph shown on your calculator using the window settings from part d.

108. Graph $y = 0.5x + 40$ in the standard viewing window.

a. Sketch the graph that is shown on your calculator.

b. Which intercept(s) can be seen on your calculator?

c. Find the x and y-intercepts by hand.

d. What viewing window will show the complete graph that contains both intercepts? State the values you would use for Xmin, Xmax, Xscl, Ymin, Ymax, and Yscl.

e. Sketch the graph shown on your calculator using the window settings from part d.

Think About It!

109. Make a table of solutions for the equation $y = 2x + 4$. Use x-values of $-2, -1, 0, 1, 2$, and 3. Do you notice anything about how the y-values are changing in the table? How does this relate to the equation?

110. Make a table of solutions for the equation $y = -3x + 1$. Use x-values of $-2, -1, 0, 1, 2$, and 3. Do you notice anything about how the y-values are changing in the table? How does this relate to the equation?

The Slope of a Line

As a result of completing this section, you will be able to

1. Use the slope formula to determine the slope of a line.

2. Use the slope-intercept form of a line to find its slope and y-intercept.

3. Graph a line given its slope and y-intercept.

4. Troubleshoot common errors.

In Section 3.2, we learned how to graph lines such as $y = 2x$ and $y = \frac{1}{2}x$, which are shown in the following graphs. In this section, we will discuss a property of lines. This property will answer the question, "How do the following graphs of the two lines compare to one another?"

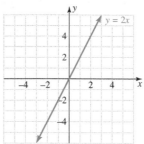
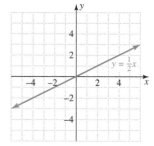

Observing the lines, we can see that the line on the left is steeper than the line on the right. A line's steepness, or *slope*, can be used to describe its graph. In this section, we will explore the slope of a line and how we can determine it.

The Slope Formula

Objective 1 ▶

Use the slope formula to determine the slope of a line.

In Section 3.2, we graphed the line $y = 2x - 4$. Let's explore some of its solutions and how they relate to the graph of the line.

x	y	Change in x	Change in y	$\dfrac{\text{Change in } y}{\text{Change in } x}$
0	$2(0) - 4 = -4$			
1	$2(1) - 4 = -2$	$1 - 0 = 1$	$-2 - (-4) = 2$	$\dfrac{2}{1} = 2$
2	$2(2) - 4 = 0$	$2 - 1 = 1$	$0 - (-2) = 2$	$\dfrac{2}{1} = 2$
3	$2(3) - 4 = 2$	$3 - 2 = 1$	$2 - 0 = 2$	$\dfrac{2}{1} = 2$

The y-values of the solutions in the table increase by 2 units every time the x-values of the solutions increase by 1 unit. This can be seen from the graph in that as we move 1 unit to the right, the graph rises 2 units.

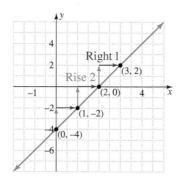

The ratio of these changes, $\dfrac{\text{Change in } y \text{ (vertical change)}}{\text{Change in } x \text{ (horizontal change)}}$, is called the slope of the line. So, the slope of the line $y = 2x - 4$ is 2.

$$\text{Slope} = \frac{\text{change in } y}{\text{change in } x} = \frac{2}{1} = 2$$

This means that for each 2 units of change in the *y*-coordinates, there is a corresponding change of 1 unit in the *x*-coordinates.

Note: *The slope of the line is the same for every pair of points on the line. If the slope is not the same, then the graph is not a line.*

Definition: The **slope** of a line is the ratio of the change in *y* to the change in *x* between two points on a line. The slope is denoted by the letter *m*. It is often called "vertical change over horizontal change" or "rise over run."

$$m = \frac{\text{Change in } y}{\text{Change in } x} = \frac{\text{vertical change}}{\text{horizontal change}} = \frac{\text{rise}}{\text{run}}$$

Using a table of values to find the slope of a line can become quite cumbersome. We can also find the slope of a line by using the slope formula.

Property: The Slope Formula

If (x_1, y_1) and (x_2, y_2) are two points that lie on a line with $x_1 \neq x_2$, then the slope is

$$m = \frac{\text{change in } y}{\text{change in } x} = \frac{y_2 - y_1}{x_2 - x_1} = \frac{y_1 - y_2}{x_1 - x_2}$$

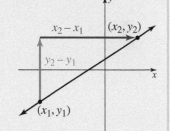

If $x_1 = x_2$, then the line through the points is vertical and the slope is undefined.

Procedure: Using the Slope Formula to Find the Slope between Two Points

Step 1: Label one point as (x_1, y_1) and the otherpoint as (x_2, y_2).
Step 2: Substitute the values into the formula

$$m = \frac{y_2 - y_1}{x_2 - x_1}$$

Step 3: Simplify the numerator and denominator and reduce the fraction, if necessary.

Objective 1 Examples   Use the slope formula to determine the slope between each pair of points.

1a. $(-1, -2)$ and $(4, 5)$ **1b.** $(5, -3)$ and $(-4, 2)$
1c. $(2, -3)$ and $(2, 5)$ **1d.** $(0, -2)$ and $(4, -2)$
1e.

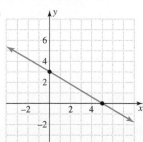

Solutions

1a. $m = \dfrac{y_2 - y_1}{x_2 - x_1}$ State the slope formula.

$m = \dfrac{5 - (-2)}{4 - (-1)}$ Let $(x_1, y_1) = (-1, -2)$ and $(x_2, y_2) = (4, 5)$.

$m = \dfrac{7}{5}$ Simplify.

Notice from the graph that if we move 7 units up from $(-1, -2)$ and 5 units to the right, we will be at the point $(4, 5)$.

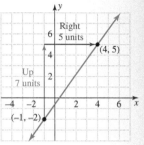

1b. $m = \dfrac{y_2 - y_1}{x_2 - x_1}$ State the slope formula.

$m = \dfrac{2 - (-3)}{-4 - (5)}$ Let $(x_1, y_1) = (5, -3)$ and $(x_2, y_2) = (-4, 2)$.

$m = \dfrac{5}{-9}$ Simplify.

$m = -\dfrac{5}{9}$

Notice from the graph that if we move 5 units down from $(-4, 2)$ and 9 units to the right, we will be at the point $(5, -3)$.

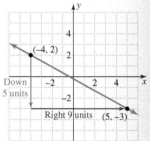

1c. $m = \dfrac{y_2 - y_1}{x_2 - x_1}$ State the slope formula.

$m = \dfrac{5 - (-3)}{2 - (2)}$ Let $(x_1, y_1) = (2, -3)$ and $(x_2, y_2) = (2, 5)$.

$m = \dfrac{8}{0}$ undefined   Simplify.

Notice from the graph that if we move up 8 units from the point $(2, -3)$, we will be at the point $(2, 5)$ since there is no movement left or right.

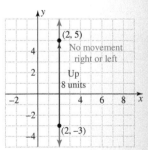

1d. $m = \dfrac{y_2 - y_1}{x_2 - x_1}$ State the slope formula.

$m = \dfrac{-2 - (-2)}{4 - (0)}$ Let $(x_1, y_1) = (0, -2)$ and $(x_2, y_2) = (4, -2)$.

$m = \dfrac{0}{4}$ Simplify.

$m = 0$ Simplify.

Notice from the graph that if we move right 4 units from the point $(0, -2)$, we will be at the point $(4, -2)$ since there is no movement up or down.

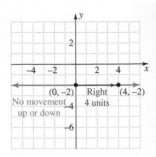

1e. Since we are given the graph of a line, we can find the slope by identifying two points on the graph and substituting their values in the slope formula. We can also find the ratio of the vertical change to the horizontal change.

Method 1: Two points on the graph are $(5, 0)$ and $(0, 3)$. So,

$$m = \frac{y_2 - y_1}{x_2 - x_1}$$

$$m = \frac{3 - 0}{0 - 5}$$

$$m = \frac{3}{-5}$$

$$m = -\frac{3}{5}$$

Method 2: To move from $(0, 3)$ to $(5, 0)$, we go down 3 units and right 5 units.

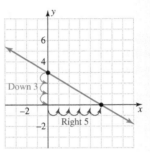

So, the slope is

$$m = \frac{\text{vertical change}}{\text{horizontal change}} = \frac{-3}{5} = -\frac{3}{5}$$

Notice that as we examine the lines in Example 1, we find that

- *The line in 1a "goes up" from left to right or has positive slope.*
- *The line in 1b "goes down" from left to right or has negative slope.*
- *The line in 1c is "vertical" and has an undefined slope.*
- *The line in 1d is "horizontal" and has a zero slope.*

So, we have four different possibilities for the slope of a line.

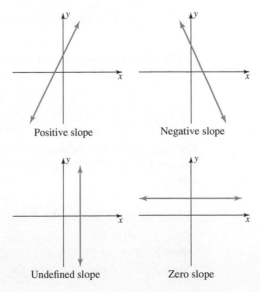

Student Check 1   Use the slope formula to determine the slope of the line between each pair of points.

a. $(1, -6)$ and $(7, -6)$ **b.** $(-2, 5)$ and $(3, 2)$

c. $(4, 7)$ and $(2, -1)$ **d.** $(-4, 2)$ and $(-4, -1)$

e.

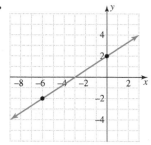

The Slope-Intercept Form of a Line

Objective 2 ▶

Use the slope-intercept form of a line to find its slope and y-intercept.

In the previous objective, we found that the slope of the line $y = 2x - 4$ is $m = \dfrac{2}{1} = 2$. We discovered this by examining the ratio of the vertical change to the horizontal change using the table on page 226. We can also confirm this by substituting two points that lie on the line into the slope formula. For instance, $(0, -4)$ and $(3, 2)$ lie on the line.

$$m = \frac{y_2 - y_1}{x_2 - x_1} \qquad \text{State the slope formula.}$$

$$m = \frac{2 - (-4)}{3 - 0} \qquad \text{Let } (x_1, y_1) = (0, -4) \text{ and } (x_2, y_2) = (3, 2).$$

$$m = \frac{6}{3} \qquad \text{Simplify.}$$

$$m = 2 \qquad \text{Simplify.}$$

The ordered pair $(0, -4)$ is the y-intercept of the graph of $y = 2x - 4$ since its x-coordinate is zero.

Note that the value of the slope and the y-intercept are provided in the equation, $y = 2x - 4$. The coefficient of x, 2, is the same as the slope of the line. The constant term, -4, is the same as the y-coordinate of the y-intercept of the line. The equation $y = 2x - 4$ is in what we call the *slope-intercept form* of a line.

Property: The Slope-Intercept Form of a Line

The equation $y = mx + b$ is the **slope-intercept form** of a linear equation. The value m is the slope of the line and the point $(0, b)$ is the y-intercept.

$$y = \underline{mx} \; + \; \underline{b}$$
$$\qquad\;\; \uparrow \qquad\quad \uparrow$$
$$\text{Slope} \quad y\text{-intercept}$$

The key difference between the standard form of a linear equation and the slope-intercept form of a linear equation is that the slope-intercept form is the equation solved for y.

Procedure: Finding the Slope and y-Intercept of a Linear Equation

Step 1: Solve the linear equation for y, if necessary.
Step 2: The coefficient of x is the slope of the line.
Step 3: The constant term is the y-coordinate of the y-intercept of the line.

| **Objective 2 Examples** | Write the equation in slope-intercept form, if possible. State the slope and y-intercept of the line from its equation. Explain how the x- and y-values change with respect to one another. |

2a. $y = 4x + 1$ **2b.** $y = -\dfrac{2}{3}x$ **2c.** $6x - y = 2$

2d. $4x + 3y = 24$ **2e.** $y = 3$ **2f.** $x = 2$

Solutions

2a. The equation $y = 4x + 1$ is in slope-intercept form since it is solved for y.

The coefficient of x is 4, so the slope is $m = 4$.

The constant term is 1, so the y-intercept is $(0, 1)$.

The slope, $m = 4 = \dfrac{4}{1}$, means that as the y-values increase by 4 units, the x-values increase by 1 unit.

2b. The equation, $y = -\dfrac{2}{3}x + 0$, is in slope-intercept form since it is solved for y.

The coefficient of x is $-\dfrac{2}{3}$, so, the slope is $m = -\dfrac{2}{3}$.

The constant term is 0, so the y-intercept is $(0, 0)$.

The slope $m = -\dfrac{2}{3} = \dfrac{-2}{3}$ or $\dfrac{2}{-3}$. This means that as the y-values decrease by 2 units, the x-values increase by 3 units. An equivalent way to state this is that as the y-values increase by 2 units, the x-values decrease by 3 units.

2c. We must solve the equation for y to write it in the slope-intercept form.

$$6x - y = 2$$

$6x - y - 6x = 2 - 6x$	Subtract 6x from each side.
$-y = -6x + 2$	Simplify.
$-1(-y) = -1(-6x + 2)$	Multiply each side by −1.
$y = 6x - 2$	Simplify.

The coefficient of x is 6, so the slope is $m = 6$.

The constant term is -2, so the y-intercept is $(0, -2)$.

The slope $m = 6 = \dfrac{6}{1}$. This means that as the y-values increase by 6 units, the x-values increase by 1 unit.

2d. We must solve the equation for y.

$$4x + 3y = 24$$

$4x + 3y - 4x = 24 - 4x$	Subtract 4x from each side.
$3y = -4x + 24$	Simplify.
$\dfrac{3y}{3} = \dfrac{-4x + 24}{3}$	Divide each side by 3.
$y = -\dfrac{4}{3}x + 8$	Simplify.

The coefficient of x is $-\dfrac{4}{3}$, so the slope is $m = -\dfrac{4}{3}$.

The constant term is 8, so the y-intercept is $(0, 8)$.

The slope $m = -\dfrac{4}{3} = \dfrac{-4}{3}$ or $\dfrac{4}{-3}$. This means that as the y-values decrease by 4 units, the x-values increase by 3 units. An equivalent way to state this is that as the y-values increase by 4 units, the x-values decrease by 3 units.

2e. The equation $y = 3$ is the equation of a horizontal line. It is in slope-intercept form and is equivalent to $y = 0x + 3$.

The coefficient of x is 0, so the slope is $m = 0$.
The constant term is 3, so the y-intercept is $(0, 3)$.

The slope $m = 0 = \dfrac{0}{1}$. This means that there is no change in the y-values as the x-values increase by 1 unit.

2f. The equation $x = 2$ represents a vertical line through the point $(2, 0)$. It cannot be written in slope-intercept form since there is no y-variable. From Example 1, we know that the slope of a vertical line is undefined. There is no y-intercept since the graph doesn't cross the y-axis.

Note that we can use the slope formula to find the slope using two points on the line, such as $(2, 0)$ and $(2, 3)$.

$$m = \frac{y_2 - y_1}{x_2 - x_1} \qquad \text{State the slope formula.}$$

$$m = \frac{3 - 0}{2 - 2} \qquad \text{Let } (x_1, y_1) = (2, 0) \text{ and } (x_2, y_2) = (2, 3).$$

$$m = \frac{3}{0} \qquad \text{Simplify.}$$

m is undefined.

✔ **Student Check 2**   Write the equation in slope-intercept form, if possible. State the slope and y-intercept of the line from its equation. Explain how the x- and y-values change with respect to one another.

a. $y = 7x + 8$ **b.** $y = \dfrac{1}{5}x$ **c.** $9x - y = 5$

d. $x + 2y = 10$ **e.** $x = -4$ **f.** $y = -2$

> **Property: Slopes of Vertical and Horizontal Lines**
>
> The slope of the equation $x = h$, where h is a real number, is undefined.
> The slope of the equation $y = k$, where k is a real number, is zero.

> **Note:** *To remember the slopes of horizontal and vertical lines, we can think of the following situation: If we can walk on the line, then the line has a slope. If we can't walk on the line, then the slope is undefined. Since we cannot "walk" on a vertical line, it has a slope that is undefined. We walk on horizontal "lines" every day, so the slope of a horizontal line is a real number, the real number 0.*

Graph a Line Using Its Slope and y-Intercept

In Section 3.2, we graphed a line by finding at least two points on the line. We can also graph a line if we know its slope and y-intercept.

Objective 3 ▶

Graph a line given its slope and y-intercept.

- The y-intercept $(0, b)$ is one point on the line.
- Another point can be obtained from the slope.

Since the slope is the ratio of the change in y to the change in x, it tells us how to move from the y-intercept to another point on the line. Recall the numerator of the slope corresponds to the change in y, or the vertical movement between the two points, and the denominator of the slope corresponds to the change in x, or the horizontal movement between the two points.

Procedure: Graphing a Line Given Its Slope and y-Intercept

Step 1: Plot the y-intercept $(0, b)$.

Step 2: Use the slope to determine another point on the line.

    a. The numerator of the slope tells us how many units to go up (if positive) or down (if negative) from the y-intercept.

    b. The denominator tells us how many units to move to the right (if positive) or left (if negative).

Step 3: Draw the graph of the line through these two points.

Note: $y = mx + b \rightarrow b$ *tells us where to begin and* m *tells us how to move.*

Objective 3 Examples / **Graph the line using its slope and y-intercept.**

    3a. $y = 3x + 2$ **3b.** $y = -\dfrac{2}{3}x + 2$

Solutions   **3a.** In the equation, $y = 3x + 2$, the slope $m = 3$ and the y-intercept is $(0, 2)$.

Step 1: Plot the y-intercept $(0, 2)$.

Step 2: Since $m = 3 = \dfrac{3}{1}$, we move up 3 units as we move right 1 unit to get to another point.

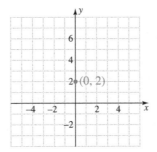 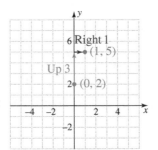

Step 3: Connect the points to obtain the graph of $y = 3x + 2$.

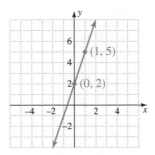

3b. In the equation $y = -\frac{2}{3}x + 2$, the slope $m = -\frac{2}{3}$ and the y-intercept is $(0, 2)$.

Step 1: Plot the
y-intercept $(0, 2)$.

Step 2: Since $m = -\frac{2}{3} = \frac{-2}{3}$, we move down
2 units as we move right 3 units to get to
another point.

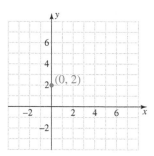

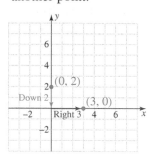

Step 3: Connect the points to obtain the graph of $y = -\frac{2}{3}x + 2$.

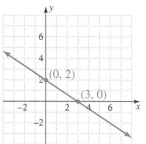

✓ **Student Check 3**   Graph the line using its slope and y-intercept.

    a. $y = \frac{1}{4}x + 3$ **b.** $y = -3x + 6$

Objective 4 ▶
Troubleshoot common errors.

Troubleshooting Common Errors

Some common errors associated with the concept of slope are shown next.

Objective 4 Examples   **A problem and an incorrect solution are given. Provide the correct solution and
an explanation of the error.**

4a. Find the slope between $(3, -2)$ and $(1, 7)$.

Incorrect Solution	Correct Solution and Explanation
The slope is $$m = \frac{3-1}{-2-7}$$ $$m = \frac{2}{-9}$$ $$m = -\frac{2}{9}$$	The values were substituted incorrectly. The numerator is the change in y and the denominator is the change in x. So, the slope between these points is $$m = \frac{y_2 - y_1}{x_2 - x_1}$$ $$m = \frac{7 - (-2)}{1 - (3)}$$ $$m = \frac{9}{-2} = -\frac{9}{2}$$

4b. Find the slope of $5x + y = 10$.

Incorrect Solution	Correct Solution and Explanation
The coefficient of x is 5, so the slope $m = 5$.	The equation is not in slope-intercept form, so we cannot conclude the slope is 5. We must first solve the equation for y. $$5x + y = 10$$ $$5x + y - 5x = 10 - 5x$$ $$y = -5x + 10$$ The coefficient of x is -5. So, the slope $m = -5$.

4c. Graph the line $y = \frac{1}{2}x - 1$.

Incorrect Solution	Correct Solution and Explanation
Begin at $(0, -1)$ and then move right 1 and up 2 since the slope is $\frac{1}{2}$. 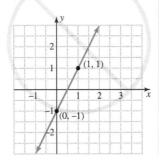	The slope was used incorrectly to obtain the graph. Since the slope of $y = \frac{1}{2}x - 1$ is $\frac{1}{2}$, we move up 1 unit as we move right 2 units. 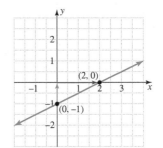

ANSWERS TO STUDENT CHECKS

Student Check 1   **a.** $m = 0$ **b.** $m = -\dfrac{3}{5}$

   c. $m = 4$ **d.** undefined   **e.** $m = \dfrac{2}{3}$

Student Check 2   **a.** $m = 7$ (y increases by 7 units

as x increases by 1 unit); $(0, 8)$ **b.** $m = \dfrac{1}{5}$ (y increases

by 1 unit as x increases by 5 units); $(0, 0)$ **c.** $y = 9x - 5$;

$m = 9$ (y increases by 9 units as x increases by 1 unit);

$(0, -5)$ **d.** $y = -\dfrac{1}{2}x + 5$; $m = -\dfrac{1}{2}$ (y decreases by

1 unit as x increases by 2 units); $(0, 5)$ **e.** can't be

written in slope-intercept form, slope is undefined

f. $y = 0x - 2$; $m = 0$ (y doesn't change as x increases

by 1 unit); $(0, -2)$

Student Check 3

a.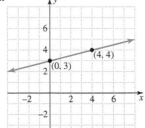

b.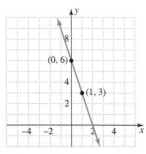

SUMMARY OF KEY CONCEPTS

1. The slope of a line is a measure of the steepness of the line. It is defined as the ratio of the change in y to the change in x.

2. The slope of a line can be found using its equation, using a table of solutions, using its graph, or using the slope formula.

 • To find the slope given its equation, the equation must be in slope-intercept form $y = mx + b$. The coefficient of x is the slope.

 • From a table of solutions, we can find the change in the y-values and the change in the x-values between the points. The ratio of these changes is the slope.

 • From a graph, we can find the slope by determining the vertical change and the horizontal change between two points on the graph. The ratio of the vertical change to the horizontal change is the slope.

- If given two ordered pairs, we can find the slope by substituting the x and y coordinates of the two points into the formula, $m = \dfrac{y_2 - y_1}{x_2 - x_1}$, and simplifying.

3. The slope tells us the direction of the line.
   - A positive slope indicates the line goes up from left to right.

- A negative slope indicates the line goes down from left to right.
- A zero slope indicates the line is horizontal.
- An undefined slope indicates the line is vertical.

4. The graph of a line can be obtained using the slope and y-intercept. Plot the y-intercept and use the slope to move to another point on the line.

GRAPHING CALCULATOR SKILLS

The graphing calculator can be used to verify our graphs as was illustrated in the previous section. We can also verify the slope of a line by examining the table of solutions and the graph of an equation.

Example: Graph $y = -\dfrac{2}{3}x + 2$ using its slope and y-intercept.

From the graph, we can verify that the slope is $-\dfrac{2}{3}$ since the movement between points is down 2 and right 3.

Examining the table also confirms the slope of the line. As the y-values decrease by 0.66667, the x-values increase by 1 unit or as the y-values decrease by 2 units, the x-values increase by 3 units.

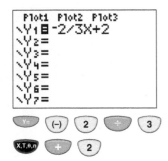

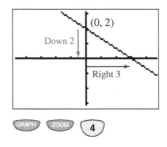

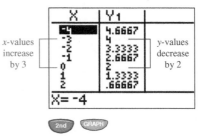

SECTION 3.3   EXERCISE SET

 ### Write About It!

Use complete sentences to explain the meaning of each process.

1. Finding the slope of line between two points
2. Finding the slope of line from a linear equation
3. Finding the slope of a horizontal line
4. Finding the slope of a vertical line

Determine if each statement is true or false. If a statement is false, provide an explanation.

5. The slope of the line $2x + 5y = 10$ is 2.

6. If the slope of a line is $-\dfrac{4}{3}$, we would move down 4 units and left 3 units to get to another point on the line.

7. The slope of the line $x = 5$ is 0.

8. The slope of the line $y = 4$ is undefined.

9. The slope of the line through $(3, 5)$ and $(4, -1)$ is $-\dfrac{1}{6}$.

10. The slope of the line through $(-2, -3)$ and $(2, -1)$ is undefined.

 ### Practice Makes Perfect!

Use the slope formula to determine the slope of the line between each pair of points. (*See Objective 1.*)

11. $(-2, 3)$ and $(4, 5)$ 12. $(1, -7)$ and $(-3, 7)$

13. $(-6, -2)$ and $(0, 4)$ 14. $(5, 0)$ and $(-4, 2)$

15. $(-4, 3)$ and $(-4, 5)$ 16. $(3, 7)$ and $(3, -2)$

17. $(1, -8)$ and $(4, -8)$ 18. $(-2, 6)$ and $(5, 6)$

19. $\left(-\dfrac{1}{6}, \dfrac{1}{4}\right)$ and $\left(\dfrac{1}{3}, -\dfrac{5}{4}\right)$ 20. $\left(-\dfrac{1}{3}, -\dfrac{5}{2}\right)$ and $\left(\dfrac{2}{3}, \dfrac{3}{2}\right)$

21. $\left(\dfrac{7}{3}, \dfrac{1}{2}\right)$ and $\left(-\dfrac{2}{3}, -\dfrac{3}{2}\right)$

22. $\left(\dfrac{1}{2}, -\dfrac{2}{5}\right)$ and $\left(-\dfrac{1}{4}, \dfrac{4}{5}\right)$

23.

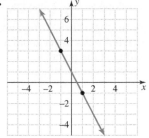

24.

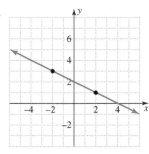

25.

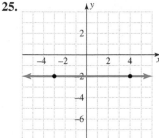

26.

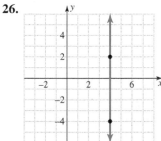

27.

X	Y₁
-3	-2
-2	-1.667
-1	-1.333
0	-1
1	-.6667
2	-.3333
3	0

X=3

28.

X	Y₁
-2	11
-1	7
0	3
1	-1
2	-5
3	-9
4	-13

X=4

Write each equation in slope-intercept form, if necessary. State the slope and y-intercept of the line from its equation. Write the y-intercept as an ordered pair. Explain how the x- and y-values change with respect to one another. (*See Objective 2.*)

29. $y = 4x - 3$ **30.** $y = 3x + 5$ **31.** $y = -2x - 1$

32. $y = -x - 4$ **33.** $y = \frac{1}{2}x + \frac{3}{2}$ **34.** $y = -\frac{7}{3}x + \frac{1}{3}$

35. $y = 5x$ **36.** $y = -3x$ **37.** $4x + y = 8$

38. $8x + y = -24$ **39.** $7x + 4y = 0$ **40.** $9x + 2y = 0$

41. $x = -2$ **42.** $y = 7$ **43.** $y = -3$

44. $x = 1$ **45.** $6y - 8 = 0$ **46.** $3x + 5 = 0$

47. $2x - 1 = 0$ **48.** $7y + 3 = 0$

Graph each line using its slope and y-intercept. Label two points on the graph. (*See Objective 3.*)

49. $y = x - 2$ **50.** $y = x + 3$ **51.** $y = -x + 4$

52. $y = \frac{2}{3}x - 2$ **53.** $y = -\frac{3}{5}x + 4$ **54.** $y = -\frac{1}{3}x + 3$

55. $2x - 4y = -8$ **56.** $4x + 5y = 20$ **57.** $x - 2y = 0$

58. $2x + y = 0$

Use the given slope of a line and the point on the line to find one additional point on the line. Graph the line using the two points. (*See Objective 3.*)

59. $m = -4$ and $(3, 4)$ **60.** $m = -2$ and $(-4, -1)$

61. $m = \frac{3}{2}$ and $(0, 2)$ **62.** $m = \frac{4}{5}$ and $(0, -3)$

63. $m = -\frac{6}{7}$ and $(0, 7)$ **64.** $m = -\frac{3}{8}$ and $(0, -8)$

65. $m = 0$ and $(2, -3)$ **66.** $m = 0$ and $(-2, 5)$

67. m is undefined and $(-1, 4)$

68. m is undefined and $(-5, -2)$

 Mix 'Em Up!

Find the slope of each line given its equation, two points on the line, its graph, or a table of solutions.

69. $y = \frac{1}{2}x + 6$

70. $y = -\frac{1}{3}x - \frac{1}{4}$ **71.** $(0, -3)$ and $(3, 5)$

72. $(-1, 5)$ and $(1, -5)$ **73.** $y = 14$

74. $x = -10$ **75.** $(11, -4)$ and $(2, 2)$

76. $(-5, 1)$ and $(1, -14)$ **77.** $\left(\frac{1}{3}, 3\right)$ and $\left(2, \frac{1}{2}\right)$

78. $\left(\frac{3}{2}, -\frac{1}{4}\right)$ and $\left(\frac{1}{2}, -\frac{3}{4}\right)$ **79.** $3x - 5y = 4$

80. $8x + 3y = 7$ **81.** $3x + 1 = 0$

82. $4y + 3 = 0$ **83.** $0.2x + 0.5y = 3.2$

84. $0.45x - 0.3y = 4.8$ **85.** $(-3.5, 12.6)$ and $(1.5, 5.6)$

86. $(-7.5, -0.6)$ and $(2.5, -11.6)$

87.

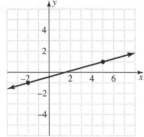

88.

89.

90.

Graph each described line. Label two points on the line.

91. $y = 4x - 3$ **92.** $y = -2x + 5$

93. $m = 1$, passes through $(0, 3)$

94. $m = -5$, passes through $(2, -1)$

95. $m = 3$, $b = 6$ **96.** $m = \frac{1}{2}$, $b = 0$

97. $m =$ undefined, passes through $(3, 4)$

98. $m = 0$, passes through $(3, 4)$

99. $m = -\frac{1}{4}$, passes through $(0, 0)$

100. $m = \frac{3}{2}$, passes through $(0, 0)$

 You Be the Teacher!

Each student is trying to solve the same problem: If a line that passes through the point $(-1, 1)$ has a slope of $-\frac{3}{4}$, specify one other point on the line. Correct each student's errors, if any.

101. Lisbelle: I can rewrite the slope as $\frac{3}{-4}$. I also know slope = rise/run. This means that if I rise, or go up, 3 units, then I have to go left 4 units to get another point on the line. So, starting at the point $(-1, 1)$, I end up at the point $(-1, 4)$ if I rise 3 units. When I go left 4 units, I end up at the point $(-5, 4)$. So, another point on the line is $(-5, 4)$.

102. Florida: I can write the slope as $\frac{3}{-4}$. I know slope = rise/run. This means that if I rise, or go right 3 units, then I have to go down 4 units to get another point on the line. So, starting at the point $(-1, 1)$, I end up at the point $(2, 1)$ when I go right 3 units. When I go down 4 units, I end up at $(2, -3)$. So, another point on the line is $(2, -3)$.

103. Lewis: I can rewrite the slope as $-\frac{3}{4}$. So if I start at the point $(-1, 1)$, I must go down 3 units and right 4 units to get another point on the line. This means that $(3, -2)$ is another point on the line.

104. James: I can rewrite the slope as $-\frac{3}{4}$. So, if I start at the point $(-1, 1)$, I must go left 3 units and up 4 units to get another point on the line. This means that $(-4, 5)$ is another point on the line.

 Calculate It!

Determine the slope of the line depicted in each graphing calculator screen shot.

105.

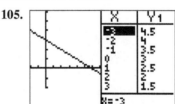

106.

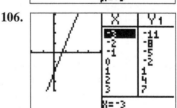

 Think About It!

Draw a graph of a line with the following characteristics, where m is the slope of the line and b is the y-value of the y-intercept of the line.

107. $m > 0$ and $b < 0$ **108.** $m > 0$ and $b = 0$

109. $m > 0$ and $b > 0$ **110.** $m < 0$ and $b < 0$

111. $m < 0$ and $b = 0$ **112.** $m < 0$ and $b > 0$

113. Slope is undefined.   **114.** $m = 0$ and $b > 0$

PIECE IT TOGETHER   / SECTIONS 3.1–3.3

Determine algebraically if the ordered pair is a solution of the equation. (*Section 3.1, Objective 1*)

1. $(6, 2)$; $y = |x - 4|$ **2.** $(7, -4)$; $8x - 2y = 62$

Identify the quadrant or axis where each point is located. (*Section 3.1, Objective 2*)

3. $(-8.9, -1)$

4. $\left(\dfrac{7}{8}, -\dfrac{5}{6}\right)$

5. $(-5, -1)$

6. $(9, 4)$

Use the given graph of an equation to determine if each ordered pair is a solution of the equation. (*Section 3.1, Objective 4*)

7. $(0, 3)$

8. $(2, -3)$

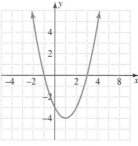

Graph each equation. (*Section 3.2, Objectives 2–4*)

9. $y = 4x + 2$ **10.** $y = -5x + 5$ **11.** $4x - y = 0$

12. $y = -7x$ **13.** $3x + 8 = 0$ **14.** $2y - 11 = 0$

Use the slope formula to determine the slope of the line between each pair of points. (*Section 3.3, Objective 1*)

15. $(-2, 3)$ and $(4, 7)$ **16.** $(-2, 8)$ and $(5, 8)$

Write each equation in slope-intercept form, if necessary. State the slope and y-intercept of the line from its equation. Write the y-intercept as an ordered pair. Explain how the x- and y-values change with respect to one another. (*Section 3.3, Objective 2*)

17. $3x - 2y = 6$

18. $5x + 3y = -15$

Graph each line using its slope and y-intercept. Label two points on the graph. (*Section 3.3, Objective 3*)

19. $y = \dfrac{1}{4}x - 1$ **20.** $x - 7y = 14$

SECTION 3.4 / More About Slope

▶ OBJECTIVES

As a result of completing this section, you will be able to

1. Interpret the meaning of the slope and y-intercept in real-world applications.

2. Determine the slope in real-world applications.

3. Determine if two lines are parallel or perpendicular.

4. Graph parallel or perpendicular lines if given the equation of one line and a point on the line parallel or perpendicular to it.

5. Troubleshoot common errors.

The depreciated value of a truck is given by $y = -4000x + 32{,}000$, where x is the age of the truck. How does the truck's value change each year? What is the initial value of the truck?

In this section, we will learn how to interpret the meaning of the slope and y-intercept in the context of applications. We will also examine the relationships between parallel and perpendicular lines.

Interpret the Meaning of the Slope and y-Intercept

When a linear equation in two variables represents a real-world situation, the slope and y-intercept have practical meanings that relate to the situation. In the equation $y = mx + b$, we know that

- The point $(0, b)$ is the y-intercept. Therefore, the value b is the *beginning* or **initial value**.
- The slope m represents a **rate of change**. The slope tells us how the y-values change with respect to a change in x. When dealing with lines, the slope is constant. So, lines are used to describe quantities that change at a constant rate.

Objective 1 ▶

Interpret the meaning of the slope and y-intercept in real-world applications.

Objective 1   Examples / **Interpret the meaning of the slope and y-intercept in the context of each situation. Use the slope and y-intercept to create a table of three ordered pairs that satisfy the given equation.**

1a. Jose's salary as an engineer can be given by $y = 3500x + 45{,}000$, where x is the number of years he has worked for the company.

1b. A truck's value is given by $y = -4000x + 32{,}000$, where x is the age of the truck in years.

1c. The life expectancy for men in the United States can be approximated by $y = 0.2x + 66$, where x is the years after 1960. (Source: http://www.cdc.gov/nchs/data/nvsr/nvsr56/nvsr56_09.pdf)

Solutions

1a. The x-value represents years worked and y represents salary in dollars. In $y = 3500x + 45{,}000$, the slope is $m = 3500 = \dfrac{3500}{1} = \dfrac{\text{change in salary}}{\text{change in years}}$ and the y-intercept is $(0, 45{,}000)$. So, Jose's initial salary is \$45,000 and his salary increases by \$3500 per year.

From this information, we can obtain the following table of solutions.

x	y
0	45,000
$0 + 1 = 1$	$45{,}000 + 3500 = 48{,}500$
$1 + 1 = 2$	$48{,}500 + 3500 = 52{,}000$

1b. The x-value represents the age of the truck in years and y represents the value of the truck. In $y = -4000x + 32{,}000$, the slope is $m = -4000 = \dfrac{-4000}{1} = \dfrac{\text{change in value}}{\text{change in years}}$ and the y-intercept is $(0, 32{,}000)$. So, the initial value of the truck is \$32,000 and its value decreases by \$4000 per year.

From this information, we can obtain the following table of solutions.

x	y
0	32,000
$0 + 1 = 1$	$32{,}000 - 4000 = 28{,}000$
$1 + 1 = 2$	$28{,}000 - 4000 = 24{,}000$

1c. The x-value represents the years after 1960 and y represents life expectancy in years. In $y = 0.2x + 66$, the slope is $m = 0.2 = \dfrac{2}{10} = \dfrac{1}{5} = \dfrac{\text{change in life expectancy}}{\text{change in years after 1960}}$ and the y-intercept is $(0, 66)$.

So, the life expectancy in 1960 was 66. Life expectancy increases 1 yr for every 5 yr after 1960. We can also say that life expectancy increases by 0.2 yr (2.4 months) for every year after 1960.

From this information, we can obtain the following table of solutions.

x	y
0	66
$0 + 5 = 5$	$66 + 1 = 67$
$5 + 5 = 10$	$67 + 1 = 68$

✓ Student Check 1

Interpret the slope and y-intercept in the context of each situation. Use the slope and y-intercept to create a table of three ordered pairs that satisfy the given equation.

a. The median annual income for women with a bachelor's degree is given by the equation $y = 1100x + 27{,}645$, where x is the number of years after 1990.

b. A car's value is given by $y = -2500x + 15{,}000$, where x is the age of the car.

c. The daily cost to rent a car is $y = 0.50x + 40$, where x is the number of miles driven.

Objective 2 ▶

Determine the slope in
real-world applications.

Slope in Real-Life Settings

Slope has many meaningful applications in real life. Slope is used to construct stairs, wheelchair ramps, highways, and roofs. It is also used to express the difficulty level of ski slopes. In these physical situations, the slope is often described in terms of rise and run.

For other applications, slope is considered a rate. For instance, it can be used to represent the growth in population, a tax rate, and other rates of change. In these situations, we must state the slope as the change in the *y*-values over the change in the *x*-values.

Objective 2 Examples ▸ **Determine the slope described in each situation.**

2a. The guidelines for constructing a wheelchair ramp state that the maximum slope should be 1 in. vertical distance for every 12 in. of horizontal distance. Express the maximum slope of a wheelchair ramp.

2b. The *grade* of a highway is the measure of the incline or steepness of the road. The grade is generally expressed as a percentage. The steepest road in the United States is Canton Avenue in Pennsylvania. The road rises 37 ft of vertical distance for each 100 ft of horizontal distance. What is the grade of the road? (Source: http://www.geographylists.com/list17y.html)

2c. The population of a town increases by 3000 people every year.

Solutions **2a.** The slope of the wheelchair ramp is

$$m = \frac{\text{vertical change}}{\text{horizontal change}} = \frac{\text{rise}}{\text{run}} = \frac{1}{12}$$

2b. The road rises 37 ft vertically for each 100 ft of horizontal change. So, the slope is

$$m = \frac{\text{vertical change}}{\text{horizontal change}} = \frac{\text{rise}}{\text{run}} = \frac{37}{100} = 37\%$$

2c. In real-life situations, time is generally used for the variable *x*. So, in this situation, population represents the variable *y*. Therefore, the slope is

$$m = \frac{\text{change in } y}{\text{change in } x} = \frac{3000}{1} = 3000$$

☑ **Student Check 2** Determine the slope described in each situation.

a. The *pitch* of a roof is defined as the ratio of the vertical increase of the roof to its horizontal increase. If a roof rises 8 in. for each 12 in. it extends horizontally, what is the pitch?

b. Ski Dubai is the first indoor ski resort in the Middle East. It has five ski runs with differing difficulty levels. The longest run is 400 m with a fall of approximately 60 m. What is the slope of this run?

c. The value of a car decreases by $5000 each year.

Parallel and Perpendicular Lines

Objective 3 ▶

Determine if two lines are parallel or perpendicular.

We will now turn our discussion to the comparison of two lines. Two lines are *parallel* if they never intersect. For lines to be parallel, they must have the same slope but different *y*-intercepts. Consider the graphs of the equations $y = \frac{3}{4}x - 1$ and $y = \frac{3}{4}x + 3$. These lines have the same slope, $\frac{3}{4}$. The lines have different *y*-intercepts, so the lines are parallel.

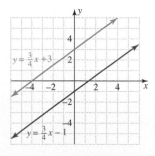

Note that if lines have the same slope and the same y-intercept, the lines are the same.

Definition: Parallel Lines

Two nonvertical lines $y_1 = m_1 x + b_1$ and $y_2 = m_2 x + b_2$ are parallel if they have the same slope (that is, $m_1 = m_2$) and different y-intercepts.

Vertical lines of the form $x = a_1$ and $x = a_2$ are parallel if $a_1 \neq a_2$.

Two nonvertical lines are *perpendicular* to each other if the lines form a right angle (90°) at their point of intersection. For this to happen, the graph of one line must be increasing (have positive slope) and the other line must be decreasing (have negative slope). Also, the rise of one line must be the run of the other line and vice versa. Mathematically, this means that the slopes of the lines are negative reciprocals of each other.

Consider the graphs of the equations $y = \dfrac{3}{4}x + 3$ and $y = -\dfrac{4}{3}x + 1$.

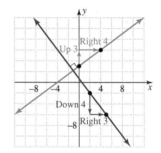

One line rises 3 units vertically and runs 4 units horizontally, so its slope is $m = \dfrac{3}{4}$.

The other line falls 4 units vertically and runs 3 units horizontally, so its slope is $m = -\dfrac{4}{3}$.

These lines have slopes that are negative reciprocals and are perpendicular to one another.

Definition: Perpendicular Lines

Two nonvertical lines $y_1 = m_1 x + b_1$ and $y_2 = m_2 x + b_2$ are perpendicular if their slopes are negative reciprocals—that is, if

$$m_1 = -\frac{1}{m_2}$$

Any vertical line of the form $x = a$ is perpendicular to any horizontal line $y = b$.

Perpendicular lines may or may not have the same y-intercept.

Note: *The negative reciprocal means that the signs must be different and the fractions must be reciprocals of each other.*

Procedure: Determining if Two Lines Are Parallel or Perpendicular

Step 1: Write each equation in slope-intercept form.
Step 2: Find the slope of each line.
Step 3: Compare the slopes of the lines.
     a. If the slopes are the same and the y-intercepts are different, then the lines are parallel.
     b. If the slopes are negative reciprocals of one another, then the lines are perpendicular.

Objective 3 Examples / Determine if the lines are parallel, perpendicular, or neither.

3a. $y = x + 4$ and $y = -x - 3$ **3b.** $y = 2x + 3$ and $y = \dfrac{1}{2}x - 4$

3c. $2x + y = 4$ and $4x + 2y = 6$ **3d.** $4x - y = 8$ and $x + 4y = -4$

Solutions **3a.** Both equations are in slope-intercept form.

The slope of $y = x + 4$ is $m = 1$.

The slope of $y = -x + 3$ is $m = -1$.

The slopes are not the same, so the lines are not parallel. The slopes, however,

are negative reciprocals of one another since $1 = -\left(\dfrac{1}{-1}\right)$. So, the lines are perpendicular.

3b. Both equations are in slope-intercept form.

The slope of $y = 2x + 3$ is $m = 2$.

The slope of $y = \dfrac{1}{2}x - 4$ is $m = \dfrac{1}{2}$.

The slopes are not the same, so the lines are not parallel. The slopes are reciprocals but not opposites. Therefore, the lines are not perpendicular to one another. So, the lines are neither parallel nor perpendicular.

3c. We must first write each equation in slope-intercept form.

$$\begin{aligned} 2x + y &= 4 \\ 2x + y - 2x &= 4 - 2x \\ y &= -2x + 4 \\ m &= -2 \end{aligned} \qquad \begin{aligned} 4x + 2y &= 6 \\ 4x + 2y - 4x &= 6 - 4x \\ 2y &= -4x + 6 \\ \frac{2y}{2} &= \frac{-4x + 6}{2} \\ y &= -2x + 3 \\ m &= -2 \end{aligned}$$

The slope of each line is $m = -2$. The lines have different y-intercepts. Therefore, the lines are parallel.

3d. We must first write each equation in slope-intercept form.

$$\begin{aligned} 4x - y &= 8 \\ 4x - y - 4x &= 8 - 4x \\ -y &= -4x + 8 \\ -1(-y) &= -1(-4x + 8) \\ y &= 4x - 8 \\ m &= 4 \end{aligned} \qquad \begin{aligned} x + 4y &= -4 \\ x + 4y - x &= -4 - x \\ 4y &= -x - 4 \\ \frac{4y}{4} &= \frac{-x - 4}{4} \\ y &= -\frac{1}{4}x - 1 \\ m &= -\frac{1}{4} \end{aligned}$$

The slopes of the lines are negative reciprocals of one another, so the lines are perpendicular.

✔ Student Check 3 Determine if the lines are parallel, perpendicular, or neither.

a. $y = -5x - 1$ and $y = \dfrac{1}{5}x + 2$ **b.** $y = 3x - 4$ and $y = -3x + 5$

c. $3x + 2y = 6$ and $9x + 6y = 10$ **d.** $2x + 3y = 6$ and $3x - 2y = 12$

Graph Parallel and Perpendicular Lines

Objective 4 ▶

Graph parallel or perpendicular lines if given the equation of one line and a point on the line parallel or perpendicular to it.

In Section 3.5, we will determine equations of lines that go through particular points that are parallel or perpendicular to a given line. It is helpful to visualize this situation before writing their equations.

> **Procedure: Graphing a Line through a Point That Is Parallel or Perpendicular to a Given Line**
>
> **Step 1:** Determine the slope of the given line.
> **Step 2:** Determine the slope of the parallel or perpendicular line.
> **a.** The slope of the parallel line is equal to the slope of the given line.
> **b.** The slope of the perpendicular line is equal to the negative reciprocal of the slope of the given line.
> **Step 3:** Graph the given line using its slope and y-intercept.
> **Step 4:** Plot the given point and use the slope in step 2 to locate another point to the line. Draw the line between these two points.

Objective 4 Examples   **Graph the line that satisfies each condition.**

4a. Graph the line $3x + y = -3$ and the line that is parallel to it that passes through the point $(1, 5)$.

4b. Graph the line $y = \dfrac{2}{3}x + 2$ and the line perpendicular to it that passes through $(-2, 1)$.

Solutions   **4a.** The slope of $3x + y = -3$ is found by writing the equation in slope-intercept form.

$$3x + y = -3$$
$$3x + y - 3x = -3 - 3x$$
$$y = -3x - 3$$

The slope of the given line is $m = -3$. So, the slope of a line parallel to it is also $m = -3$ since parallel lines have the same slope.

Graph the given equation, $y = -3x - 3$, by plotting the y-intercept $(0, -3)$ and then moving down 3 units vertically and right 1 unit horizontally to $(1, -6)$.

Graph the line parallel to $y = -3x - 3$ and through $(1, 5)$, by plotting $(1, 5)$. From this point, move down 3 units vertically and right 1 unit horizontally to $(2, 2)$.

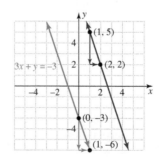

4b. The line $y = \dfrac{2}{3}x + 2$ has slope $m = \dfrac{2}{3}$. The line perpendicular to it has slope $m = -\dfrac{3}{2}$ since perpendicular lines have slopes that are negative reciprocals.

Graph the line, $y = \dfrac{2}{3}x + 2$, by plotting the y-intercept $(0, 2)$ and then rising 2 units vertically and moving right 3 units horizontally to get to $(3, 4)$.

The line perpendicular to the given line has slope of $-\dfrac{3}{2}$. So, we graph this line by plotting the point $(-2, 1)$ and then move down 3 units vertically and right 2 units horizontally to get to $(0, -2)$.

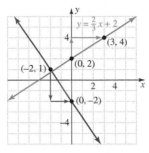

✓ Student Check 4   Graph the line that satisfies each condition.

a. Graph the line $2x - y = -4$ and the line that is parallel to it that passes through $(2, -1)$.

b. Graph the line $y = 5x - 5$ and the line that is perpendicular to it that passes through $(5, 1)$.

Objective 5 ▶
Troubleshoot common errors.

Troubleshooting Common Errors

A common error associated with slope and parallel and perpendicular lines is shown next.

Objective 5   Examples   **A problem and an incorrect solution are given. Provide the correct solution and an explanation of the error.**

Are the lines $y = 4x + 6$ and $4x + y = 0$ parallel, perpendicular, or neither?

Incorrect Solution	Correct Solution and Explanation
The coefficient of each equation is 4, so the slopes are the same and the lines are parallel.	The second equation is not in slope-intercept form. We must rewrite $4x + y = 0$ in this form to identify its slope. $$4x + y = 0$$ $$y = -4x$$ The slope of $y = 4x + 6$ is $m = 4$. The slope of $y = -4x$ is $m = -4$. The slopes are not the same, so the lines are *not* parallel. The slopes are not negative reciprocals of one another either, so the lines are *not* perpendicular.

ANSWERS TO STUDENT CHECKS

Student Check 1   a. The median annual income for women in 1990 was $27,645 and it increased by $1100 per year after 1990. Three points are (0, 27,645), (1, 28,745), and (2, 29,845).   **b.** The initial value of the car was $15,000 and its value decreased by $2500 each year. Three points are (0, 15,000), (1, 12,500) and (2, 10,000)   **c.** The cost of renting the car is $40 plus $0.50 per mile driven. Three points are (0, 40), (1, 40.50), and (2, 41).

Student Check 2   a. $\dfrac{8}{12} = \dfrac{2}{3}$ **b.** $-\dfrac{60}{400} = -\dfrac{3}{20}$ **c.** $-\dfrac{5000}{1} = -5000$

Student Check 3   a. perpendicular   **b.** neither   **c.** parallel   **d.** perpendicular

Student Check 4   a. 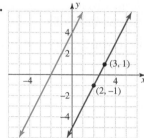   b.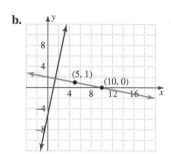

SUMMARY OF KEY CONCEPTS

1. The slope and y-intercept have meaning when the linear equation represents a real-life situation. The y-intercept represents the initial value and the slope is a rate of change indicating how the y-values change as the x-values change.

2. Slope has many physical applications, such as the grade of a road, wheelchair ramps, and roofs. Slopes are also used to denote rates of change, such as a population rate.

3. Lines that have the same slope and different y-intercepts are parallel to one another. Lines that have slopes that are negative reciprocals are perpendicular to one another.

GRAPHING CALCULATOR SKILLS

The calculator can help us determine if lines are parallel or perpendicular. It is often helpful to graph the lines in the ZDecimal or ZSquare format so that the graphs are not distorted. Because of the possible distortion, it is best to compare the line's slopes to determine how they relate to each other.

Example: Are $y = 2x - 3$ and $y = \dfrac{1}{2}x + 1$ parallel, perpendicular, or neither?

 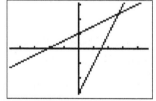

By graphing the lines on the calculator, we can verify that the lines are neither parallel nor perpendicular.

Example: Are $y = 2x - 3$ and $y = -\dfrac{1}{2}x + 1$ parallel, perpendicular, or neither?

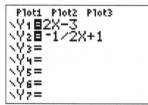 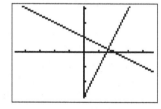

The graphs show that the lines are perpendicular to one another.

SECTION 3.4 / EXERCISE SET

Write About It!

Interpret the meaning of the slope and y-intercept in the context of each situation. Use the slope and y-intercept to create a table of three ordered pairs that satisfy the given equation. (*See Objective 1.*)

1. The expenditure per student in fall enrollment for public elementary and secondary education in the United States can be approximated by the equation $y = 374.15x + 7672.73$, where x is the number of years after 2001. (Source: http://nces.ed.gov/programs/projections/projections2018/tables/table_34.asp)

2. The total student enrollment (in thousands) in all degree-granting institutions (2-yr and 4-yr colleges and universities) can be approximated by the equation $y = 351x + 16{,}121$, where x is the number of years after 2001. (Source: http://nces.ed.gov/programs/projections/projections2018/tables/table_10.asp)

3. The percentage of high school students who regularly smoke cigarettes is on the decline. The percentage of students who are smokers is given by the equation $y = -2.5x + 38$, where x is the number of years after 1997. (Source: National Cancer Institute)

4. The percent of U.S. adults aged 20 yr and over who were obese can be modeled by the linear equation $y = 0.71x + 19.83$, where x is the number of years after 1997. (Source: National Center for Health Statistics)

5. The owner of a limousine rental company purchases a stretch limousine for $110,000. For tax purposes, the owner uses straight-line depreciation for reporting the value of the limousine. The depreciated value of the limousine is given by $y = -22,000x + 110,000$, where x is the age of the limousine in years.

6. The number of U.S. households with cable television is approximated by $y = 2.3x + 17.3$ (in millions), where x is the number of years after 1977. (Source: *The World Almanac and Book of Facts 2006*)

7. The percentage of eighth-graders who reported using alcohol is given by $y = -0.7x + 27$, where x is the number of years after 1993. (Source: *The World Almanac and Book of Facts 2006*)

8. The average cost, in dollars, of higher education at 2-yr institutions can be approximated by the equation $y = 334x + 5252$, where x is the number of years after 1999. (Source: www.infoplease.com)

 Practice Makes Perfect!

Determine the slope described in each situation.
(*See Objective 2.*)

9. The Cooper River Bridge in Charleston, South Caroline, is North America's longest cable-stayed bridge. The steepest part of the bridge  occurs on a section rises 88.7 ft for a 1584-ft run. What is the grade of the bridge on this section? (Source: www.cooperriverbridge.org, http://www.cooperriverbridge.org)

10. Filbert Street in San Francisco, California, is one of the steepest streets in the city and in America. The street rises 63 ft for a 200-ft run. What is the grade of the street? Express the grade as a fraction and a percentage. (Source: http://wikitravel.org/en/San_Francisco)

11. A roof rises vertically 12 in. for each 12 in. it extends horizontally. What is the pitch of the roof?

12. What is the pitch of a roof that rises 6 in. for each 12 in. it extends horizontally?

13. The American National Standards Institute facilitates the development of national standards for the steepness of wheelchair ramps. The national standard for constructing wheelchair ramps specifies that a slope of $\frac{1}{12}$ is preferable, but a slope of $\frac{1}{8}$ is acceptable when the former is not possible. If a builder constructs a wheelchair ramp with a rise of 3 in. for every run of 24 in., does the slope of the ramp meet the guidelines?

14. If a door is 15 in. above the ground, how long must the wheelchair ramp be for its slope to be $\frac{1}{12}$? Express answer in inches and feet.

Determine if the two lines are parallel, perpendicular, or neither. (*See Objective 3.*)

15. $y = \frac{4}{3}x - 5$ and $y = \frac{4}{3}x + 5$

16. $y = -6x + 1$ and $y = -6x$

17. $y = 9x + 3$ and $y = -9x - 7$

18. $y = 7x + 2$ and $7x - y = 14$

19. $x + 3y = -6$ and $y = 3x + 4$

20. $2x - y = 8$ and $4x - 2y = -4$

21. $x + 5y = 10$ and $3x - 15y = 0$

22. $y = 3$ and $y = 3x$

23. $x = 2$ and $5x + 2 = 0$

24. $y - 5 = 0$ and $y = -2$

25. $y = 2x - 4$ and $y = -2x + 3$

26. $y = -2x - 4$ and $y = \frac{1}{2}x + 5$

27. $y = 5x - 4$ and $y = -\frac{1}{5}x$

28. $3x + y = 6$ and $y = 3x - 4$

29. $3x + y = 6$ and $y = -\frac{1}{3}x - 6$

30. $4x + 3y = 6$ and $y = \frac{3}{4}x - 5$

31. $x + 7y = 21$ and $7x - y = -14$

32. $2x + 5y = -10$ and $5x + 2y = 20$

33. $x = 5$ and $y = 2$

34. $y = -4$ and $y = \frac{1}{4}x$

Graph the given line and the line that is parallel to it through the given point. (*See Objective 4.*)

35. $y = 2x - 4$; $(-2, 2)$ 36. $y = -3x + 3$; $(0, -2)$

37. $y = \frac{1}{5}x - 1$; $(5, 5)$ 38. $y = -\frac{2}{3}x + 4$; $(0, 0)$

39. $x + y = 4$; $(0, -2)$ 40. $x - y = 3$; $(-3, -1)$

41. $x = 2$; $(-1, 4)$ 42. $x = -3$; $(4, 2)$

43. $y = -4$; $(0, 3)$ 44. $y = 1$; $(-4, -2)$

Graph the given line and the line that is perpendicular to it through the given point. (*See Objective 4.*)

45. $y = 2x - 4$; $(-2, 2)$ 46. $y = -3x + 3$; $(0, -2)$

47. $y = \frac{1}{5}x - 1$; $(5, 5)$ 48. $y = -\frac{2}{3}x + 4$; $(0, 0)$

49. $x + y = 4$; $(0, -2)$ 50. $x - y = 3$; $(-3, -1)$

51. $x = 2$; $(-1, 4)$ 52. $x = -3$; $(4, 2)$

53. $y = -4$; $(0, 3)$ 54. $y = 1$; $(-4, -2)$

 Mix 'Em Up!

Solve each problem.

55. What is the pitch of a roof that rises vertically 12 in. for each 5 in. it extends horizontally?

56. What is the pitch of a roof that rises vertically 10 in. for each 15 in. it extends horizontally?

57. Determine if the lines are parallel, perpendicular, or neither: $y = -\dfrac{1}{7}x + 14$ and $y = 7x + 7$.

58. Determine if the lines are parallel, perpendicular, or neither: $y = \dfrac{5}{6}x - 3$ and $y = \dfrac{6}{5}x + \dfrac{1}{3}$.

59. Graph the line that is parallel to $y = 2x + 1$ and passes through the point $(-3, 2)$.

60. Graph the line that is parallel to $y = \dfrac{2}{3}x - 4$ and passes through the point $(3, 1)$.

61. Graph the line that is perpendicular to $y = -5x + 1$ and passes through the point $(1, 1)$.

62. Graph the line that is perpendicular to $y = -6x + 5$ and passes through the point $(-1, 4)$.

63. Determine if the lines are parallel, perpendicular, or neither: $y = 3$ and $y = -\dfrac{1}{3}$.

64. Determine if the lines are parallel, perpendicular, or neither: $x = \dfrac{2}{5}$ and $x = -\dfrac{5}{2}$.

Interpret the meaning of the slope and y-intercept in the context of each situation.

65. Suzy's car is in the shop, so she is using a rental car. The total cost, in dollars, for Suzy to rent a compact car is given by $y = 149x$, where x is the number of weeks the car is rented.

66. Mindy has a wedding cake business. Mindy's total weekly cost, in dollars, for making wedding cakes is given by $y = 20x + 50$, where x is the number of cakes Mindy makes each week.

67. The average cost, in dollars, of higher education at 4-yr institutions can be approximated by the equation $y = 892x + 12{,}034$, where x is the number of years after 1999. (Source: www.infoplease.com)

68. The median annual income, in dollars, for men who complete high school can be modeled by the equation $y = 627.5x + 30{,}818$, where x is the number of years after 1996. (Source: www.infoplease.com)

69. The median annual income, in dollars, for women who complete high school can be modeled by the equation $y = 641x + 21{,}506$, where x is the number of years after 1996. (Source: www.infoplease.com)

70. The percent of America's high school seniors who abuse alcohol can be modeled by the equation $y = -1.13x + 77.1$, where x is the number of years after 2003.

 You Be the Teacher!

Correct each student's errors, if any.

71. Explain the steps to graph the line that is parallel to $6x - 2y = 4$ and passes through the point $(1, 2)$.

Sandee's work:
The slope of the line from the given equation is 6. From the point $(1, 2)$, go up 6 units and right 1 unit.

72. Graph the line that is perpendicular to $-3x + y = 4$ and passes through the point $(0, 5)$.

Aisha's work:

$$-3x + y = 4$$
$$\underline{+3x \quad\;\; +3x}$$

$y = 3x + 4$. So, the slope of the perpendicular line is $-\dfrac{1}{3}$. The y-intercept of this line is $(0, 4)$. So, to get another point on the graph, I must go down 1 unit and right 3 units. So another point on this line is $(3, 3)$.

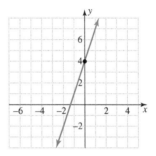

73. Determine if the lines are parallel, perpendicular, or neither: $y = 2x$ and $y = -\dfrac{1}{2}$.

Warren's work: perpendicular

74. Determine if the lines are parallel, perpendicular, or neither: $x = -2$ and $y = -2$.

Crystal's work: parallel

 Calculate It!

Graph each pair of equations in the ZSquare format on a graphing calculator to determine if the lines are parallel, perpendicular, or neither.

75. $y = 5x - 3$ and $y = \dfrac{1}{5}x + 3$

76. $y = -\dfrac{4}{3}x + 5$ and $y = \dfrac{3}{4}x + 1$

77. $y = \dfrac{1}{6}x + 2$ and $y = -6x - 4$

78. $y = 2x + 6$ and $y = 2x + 3$

79. $0.4x + 1.2y = 3.5$ and $1.8x - 0.6y = 4.9$

80. $5.6x - 3.5y = 8.9$ and $2.4x - 1.5y = 2.7$

 Think About It!

Write equations of lines whose graphs are parallel and perpendicular to the graph of the given equation.

81. $y = 5x - 3$

82. $y = -\dfrac{2}{3}x$

83. $4x + 3y = 12$

84. $7x - 2y = 4$

| SECTION 3.5 | **Writing Equations of Lines** |

▶ **OBJECTIVES**

As a result of completing this section, you will be able to

1. Write the equation of a line given its slope and *y*-intercept.
2. Write the equation of a line given a point and a slope.
3. Write the equation of a line given two points.
4. Write the equation of a line given a point and a relationship to another line.
5. Solve application problems.
6. Troubleshoot common errors.

The average price of a movie ticket in the United States in 1996 was \$4.42. The average price of a movie ticket in the United States in 2010 was \$7.89. Write a linear equation that models the average price of a movie ticket. To answer this question, we need to know how to use two points to write the equation of a line. (Source: http://www.natoonline.org/statisticstickets.htm)

In Sections 3.1 to 3.4, we were given the equation of a line and we had to find certain information about it. In this section, we will be given specific information about a line and will write the equation of the line that satisfies the given conditions. The answers to these types of problems will be linear equations of the form $y = mx + b$ or $Ax + By = C$.

Use the Slope and *y*-Intercept to Write an Equation of a Line

| **Objective 1** ▶ |

Write the equation of a line given its slope and *y*-intercept.

The first case we consider when writing equations of lines is when we are given a line's slope and *y*-intercept. Recall that the slope-intercept form of a line is $y = mx + b$, where m is the slope and b is the *y*-intercept.

So, if we know the slope and *y*-intercept, we will substitute these values into the slope-intercept form to obtain the equation.

> **Procedure: Writing the Equation of a Line Given Its Slope and *y*-Intercept**
>
> The line's slope is the value of m and the *y*-coordinate of the *y*-intercept is the value b.
>
>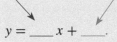
>
> So, the equation is $\qquad y = \underline{\ \ } x + \underline{\ \ \ }.$

| **Objective 1 Examples** | **Write the equation of the line that satisfies the given information.** |

1a. $m = 4$ and the *y*-intercept is $(0, 5)$ **1b.** $m = -\dfrac{1}{2}$ and the *y*-intercept is $(0, -4)$

1c. $m = 0$ and the *y*-intercept is $(0, 1)$

1d.

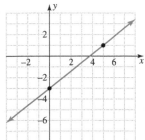

Solutions   **1a.** The slope is $m = 4$ and the value of $b = 5$. So, the equation is

$$y = 4x + 5$$

1b. The slope is $m = -\dfrac{1}{2}$ and the value of $b = -4$. So, the equation is

$$y = -\dfrac{1}{2}x - 4$$

1c. The slope is $m = 0$ and the value of $b = 1$. So, the equation is

$$y = 0x + 1 \quad \text{or} \quad y = 1$$

1d. The graph crosses the y-axis at $(0, -3)$, so $b = -3$.

To determine the slope, we calculate the *rise* and *run* between the two points. The slope is $m = \dfrac{\text{rise}}{\text{run}} = \dfrac{4}{5}$.

So, the equation is $y = \dfrac{4}{5}x - 3$.

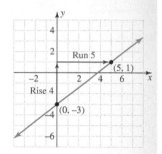

✔ **Student Check 1**   Write the equation of the line that satisfies the given information.

a. $m = -2$ and the y-intercept is $(0, 9)$ **b.** $m = \dfrac{7}{3}$ and the y-intercept is $(0, -1)$

c. $m = 0$ and the y-intercept is $\left(0, \dfrac{2}{3}\right)$

d.

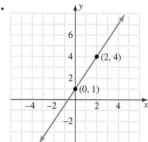

Use a Point and a Slope to Write an Equation of a Line

Objective 2 ▶

Write the equation of a line given a point and a slope.

When writing the equation of a line when the given point is not the y-intercept, we have to perform some work to determine the y-intercept and the value of b. There are two methods that we can use to determine the value of b.

- One method uses the slope-intercept form of a line $y = mx + b$.
- The other method uses another form of a line called the *point-slope form*.

The point-slope form is derived from the slope formula using (x_1, y_1) as the given point and (x, y) as any other point on the line.

$$m = \frac{y - y_1}{x - x_1}$$

Multiply both sides of the slope formula by the LCD $= x - x_1$ to get the point-slope form of a line.

$$y - y_1 = m(x - x_1)$$

> **Definition:** The **point-slope form** of a line is $y - y_1 = m(x - x_1)$, where (x_1, y_1) is a point on the line and m is the slope of the line.

To write the equation of a line given a point and a slope, we will use one of the two methods described next.

Procedure: Writing the Equation of a Line Given a Point and a Slope

Method 1: Use the slope-intercept form of a line	**Method 2:** Use the point-slope form of a line
Step 1: In the equation $y = mx + b$, substitute the given point (x, y) for the x-value and the y-value in the equation. Substitute the slope for m.	**Step 1:** In the equation $y - y_1 = m(x - x_1)$, substitute the given point (x_1, y_1) and the slope for m.
Step 2: Solve the resulting equation for b.	**Step 2:** Simplify both sides of the equation.
Step 3: Write the equation in the form $y = mx + b$ by substituting the given slope and the value of b.	**Step 3:** Write the equation in either standard form or slope-intercept form.
Step 4: Check the equation.	**Step 4:** Check the equation.

Objective 2 Examples Write the equation of the line that has the given slope and passes through the given point. Write the answer in both slope-intercept form and standard form.

2a. $m = 3$; $(5, -2)$ **2b.** $m = -\dfrac{3}{4}$; $(1, 5)$

2c. $m = 0$; $(-2, 1)$ **2d.** undefined slope; $(-3, 9)$

Solutions **2a.**

Method 1	**Method 2**
$m = 3$ and $(x, y) = (5, -2)$	$m = 3$ and $(x_1, y_1) = (5, -2)$
$y = mx + b$	$y - y_1 = m(x - x_1)$
$-2 = 3(5) + b$	$y - (-2) = 3(x - 5)$
$-2 = 15 + b$	$y + 2 = 3x - 15$
$-2 - 15 = 15 + b - 15$	$y + 2 - 2 = 3x - 15 - 2$
$-17 = b$	$y = 3x - 17$
Equation in slope-intercept form:	Equation in slope-intercept form:
$y = 3x - 17$	$y = 3x - 17$

Equation in standard form:

$$y = 3x - 17 \qquad \text{Begin with the slope-intercept form.}$$
$$y - 3x = 3x - 17 - 3x \qquad \text{Subtract } 3x \text{ from each side.}$$
$$-3x + y = -17 \qquad \text{Simplify.}$$
$$-1(-3x + y) = -1(-17) \qquad \text{Multiply each side by } -1.$$
$$3x - y = 17 \qquad \text{Simplify.}$$

Check: Show that $(5, -2)$ is a solution of $y = 3x - 17$ or $3x - y = 17$.

$$y = 3x - 17 \qquad \text{Begin with the slope-intercept form.}$$
$$-2 = 3(5) - 17 \qquad \text{Let } x = 5 \text{ and } y = -2.$$
$$-2 = 15 - 17 \qquad \text{Simplify.}$$
$$-2 = -2 \qquad \text{Simplify.}$$

Since the resulting equation is true, our work is correct.

2b.

Method 1	**Method 2**
$m = -\dfrac{3}{4}$ and $(x, y) = (1, 5)$	$m = -\dfrac{3}{4}$ and $(x_1, y_1) = (1, 5)$
$y = mx + b$	$y - y_1 = m(x - x_1)$
$5 = -\dfrac{3}{4}(1) + b$	$y - 5 = -\dfrac{3}{4}(x - 1)$
$5 = -\dfrac{3}{4} + b$	$y - 5 = -\dfrac{3}{4}x + \dfrac{3}{4}$
$4(5) = 4\left(-\dfrac{3}{4} + b\right)$	$4(y - 5) = 4\left(-\dfrac{3}{4}x + \dfrac{3}{4}\right)$
$20 = -3 + 4b$	$4y - 20 = -3x + 3$
$20 + 3 = -3 + 4b + 3$	$4y - 20 + 20 = -3x + 3 + 20$
$23 = 4b$	$4y = -3x + 23$
$\dfrac{23}{4} = b$	$y = -\dfrac{3}{4}x + \dfrac{23}{4}$
Equation in slope-intercept form:	Equation in slope-intercept form:
$y = -\dfrac{3}{4}x + \dfrac{23}{4}$	$y = -\dfrac{3}{4}x + \dfrac{23}{4}$

Equation in standard form:

$$y = -\frac{3}{4}x + \frac{23}{4}$$ 　Begin with the slope-intercept form.

$$4(y) = 4\left(-\frac{3}{4}x + \frac{23}{4}\right)$$ 　Multiply each side by 4.

$$4y = -3x + 23$$ 　Simplify.

$$4y + 3x = -3x + 23 + 3x$$ 　Add $3x$ to each side.

$$3x + 4y = 23$$ 　Simplify.

Check: Show that $(1, 5)$ is a solution of $y = -\dfrac{3}{4}x + \dfrac{23}{4}$ or $3x + 4y = 23$.

$$3x + 4y = 23$$ 　Begin with the standard form.

$$3(1) + 4(5) = 23$$ 　Let $x = 1$ and $x = 5$.

$$3 + 20 = 23$$ 　Simplify.

$$23 = 23$$ 　Simplify.

Since the resulting equation is true, our work is correct.

2c.

Method 1	**Method 2**
$m = 0$ and $(x, y) = (-2, 1)$	$m = 0$ and $(x_1, y_1) = (-2, 1)$
$y = mx + b$	$y - y_1 = m(x - x_1)$
$1 = 0(-2) + b$	$y - 1 = 0[x - (-2)]$
$1 = 0 + b$	$y - 1 = 0$
$1 = b$	$y - 1 + 1 = 0 + 1$
	$y = 1$
Equation: $y = 0x + 1$ or $y = 1$	Equation: $y = 1$

Recall that a line with a slope of zero is horizontal and that horizontal lines are of the form $y = k$, where k is the y-value of a point on the line.

2d. A line with undefined slope is a vertical line. A vertical line is written as $x = h$, where h is the x-value of a point on the line. Since the given point is $(-3, 9)$, $h = -3$. So, the equation of the line is $x = -3$.

✔ **Student Check 2** Write the equation of the line that has the given slope and passes through the given point. Write the answer in both slope-intercept form and standard form.

a. $m = -5; (3, 1)$
b. $m = -\dfrac{2}{7}; (4, 5)$

c. $m = 0; (-6, 9)$
d. undefined slope; $(8, 1)$

Note: *Using method 1 or 2 produces the same linear equation. When using the slope-intercept form, we must substitute the values of m and b into $y = mx + b$ to obtain the equation. When using the point-slope form, the equation is obtained through the process.*

Use Two Points to Write the Equation of a Line

Objective 3 ▶

Write the equation of a line given two points.

In Objectives 1 and 2, we were given the slope and a point on the line. If we know only two points on the line, we can still apply the two methods discussed earlier to write the equation of the line through these points. The only difference is that now we will have to calculate the slope of the line using the slope formula. Once the slope is known, we can use method 1 or method 2 to write the equation of the line.

Procedure: Writing the Equation of a Line Given Two Points

Step 1: Find the slope of the line using the slope formula, $m = \dfrac{y_2 - y_1}{x_2 - x_1}$.

Step 2: Use the slope and one of the given points to find the equation of the line using either the slope-intercept form of a line or the point-slope form of a line.

Objective 3 Examples Write the equation of the line, in slope-intercept form, that passes through the given points.

3a. $(4, -3)$ and $(-2, 9)$ **3b.** $(-4, 0)$ and $(4, 12)$ **3c.** $(-2, 4)$ and $(-2, 11)$

Solutions **3a.** Find the slope of the line.

$$m = \frac{y_2 - y_1}{x_2 - x_1} \qquad \text{State the slope formula.}$$

$$m = \frac{9 - (-3)}{-2 - 4} \qquad \text{Let } (x_1, y_1) = (4, -3) \text{ and } (x_2, y_2) = (-2, 9).$$

$$m = \frac{12}{-6} \qquad \text{Simplify numerator and denominator.}$$

$$m = -2 \qquad \text{Simplify result.}$$

Use the point-slope form of a line to write the equation. Either of the given points can be used in the point-slope form as shown.

$m = -2, (x_1, y_1) = (4, -3)$	$m = -2, (x_1, y_1) = (-2, 9)$
$y - y_1 = m(x - x_1)$	$y - y_1 = m(x - x_1)$
$y - (-3) = -2(x - 4)$	$y - (9) = -2[x - (-2)]$
$y + 3 = -2x + 8$	$y - 9 = -2(x + 2)$
$y + 3 - 3 = -2x + 8 - 3$	$y - 9 = -2x - 4$
$y = -2x + 5$	$y - 9 + 9 = -2x - 4 + 9$
	$y = -2x + 5$

Check: Show that $(4, -3)$ and $(-2, 9)$ are solutions of the equation $y = -2x + 5$.

(x, y)	$y = -2x + 5$	Solution?
$(4, -3)$	$-3 = -2(4) + 5$ $-3 = -8 + 5$ $-3 = -3$	Yes
$(-2, 9)$	$9 = -2(-2) + 5$ $9 = 4 + 5$ $9 = 9$	Yes

3b. Find the slope of the line.

$$m = \frac{y_2 - y_1}{x_2 - x_1}$$
Begin with the slope formula.

$$m = \frac{12 - 0}{4 - (-4)}$$
Let $(x_1, y_1) = (-4, 0)$ and $(x_2, y_2) = (4, 12)$.

$$m = \frac{12}{8}$$
Simplify the numerator and denominator.

$$m = \frac{3}{2}$$
Simplify the result.

Use the slope-intercept form of a line to write the equation. Either of the given points can be used in the slope-intercept form to find b.

$m = \frac{3}{2}; (x, y) = (-4, 0)$

$y = mx + b$

$0 = \frac{3}{2}(-4) + b$

$0 = -6 + b$

$0 + 6 = -6 + b + 6$

$6 = b$

Equation: $y = \frac{3}{2}x + 6$

$m = \frac{3}{2}; (x, y) = (4, 12)$

$y = mx + b$

$12 = \frac{3}{2}(4) + b$

$12 = 6 + b$

$12 - 6 = 6 + b - 6$

$6 = b$

Equation: $y = \frac{3}{2}x + 6$

Check: Show that $(-4, 0)$ and $(4, 12)$ are solutions of the equation $y = \frac{3}{2}x + 6$.

(x, y)	$y = \frac{3}{2}x + 6$	Solution?
$(-4, 0)$	$0 = \frac{3}{2}(-4) + 6$ $0 = -6 + 6$ $0 = 0$	Yes
$(4, 12)$	$12 = \frac{3}{2}(4) + 6$ $12 = 6 + 6$ $12 = 12$	Yes

3c. Find the slope of the line.

$$m = \frac{y_2 - y_1}{x_2 - x_1}$$
Begin with the slope formula.

$$m = \frac{11 - 4}{-2 - (-2)}$$
Let $(x_1, y_1) = (-2, 4)$ and $(x_2, y_2) = (-2, 11)$.

$$m = \frac{7}{0}$$
Simplify the numerator and denominator.

$$m = \text{undefined}$$
Simplify the result.

Since the slope is undefined, the line is vertical. Recall that a vertical line is represented by an equation of the form $x = h$, where h is the x-coordinate of a point on the line. Since the given points are $(-2, 4)$ and $(-2, 11)$, $h = -2$. So, the equation of the line is $x = -2$.

✓ **Student Check 3** Write the equation of the line, in slope-intercept form, that passes through the given points.

a. $(3, -7)$ and $(1, 3)$ **b.** $(-1, 2)$ and $(5, -6)$ **c.** $(4, -1)$ and $(6, -1)$

Use a Point and a Line to Write the Equation of a Line

Objective 4 ▶

Write the equation of a line given a point and a relationship to another line.

The last case we will discuss is writing the equation of a line given a point and a relationship to another line. We will be given the equation of a parallel or perpendicular line. Knowing this information enables us to find the slope of the unknown line. If the lines are parallel, the slopes are the same. If the lines are perpendicular, the slopes are negative reciprocals. Once the slope is determined, the methods to find the equation of a line, as previously stated, apply.

> **Procedure: Writing the Equation of a Line Given a Point and a Relationship to Another Line**
>
> **Step 1:** Determine the slope of the given line by writing the equation in slope-intercept form.
> **Step 2:** Determine the slope of the unknown line.
>   **a.** The slope will be the same as the given line if the lines are parallel.
>   **b.** The slope will be the negative reciprocal of the given line if the lines are perpendicular.
> **Step 3:** Use the slope from step 2 and the given point to write the equation of the unknown line using either method 1 or method 2.

Objective 4 Examples   Write the equation of the line that passes through the given point and is either parallel or perpendicular to the given line.

4a. $(1, -5)$ and parallel to $4x - y = 8$; Write answer in slope-intercept form.
4b. $(-2, 4)$ and perpendicular to $y = 5x - 6$; Write answer in standard form.
4c. $(-2, 3)$ and parallel to $x = 5$
4d. $(4, -7)$ and perpendicular to $x = 5$

Solutions   **4a.** We first find the slope of $4x - y = 8$ by writing it in slope-intercept form.

$$4x - y = 8$$
$$4x - y - 4x = 8 - 4x \qquad \text{Subtract } 4x \text{ from each side.}$$
$$-y = -4x + 8 \qquad \text{Simplify.}$$
$$-1(-y) = -1(-4x + 8) \qquad \text{Multiply each side by } -1.$$
$$y = 4x - 8 \qquad \text{Simplify.}$$

The slope of the line $4x - y = 8$ is $m = 4$. Since the lines are parallel, the slope of the unknown line is also $m = 4$.

Now we find the equation of the line with slope $m = 4$ that passes through $(1, -5)$.

$$y - y_1 = m(x - x_1) \qquad \text{State the point-slope form.}$$
$$y - (-5) = 4(x - 1) \qquad \text{Let } m = 4, (x_1, y_1) = (1, -5).$$
$$y + 5 = 4x - 4 \qquad \text{Simplify and distribute.}$$
$$y + 5 - 5 = 4x - 4 - 5 \qquad \text{Subtract 5 from each side.}$$
$$y = 4x - 9 \qquad \text{Simplify.}$$

So, the equation of the line through $(1, -5)$ and parallel to $4x - y = 8$ is $y = 4x - 9$.

4b. We first find the slope of $y = 5x - 6$. Since the equation is in slope-intercept form, the slope of the line is $m = 5$. The lines are perpendicular, so the slope of the unknown line is the negative reciprocal of 5. Hence, the slope of the unknown line is $m = -\dfrac{1}{5}$.

Now we find the equation of the line with $m = -\dfrac{1}{5}$ that passes through $(-2, 4)$.

$$y - y_1 = m(x - x_1)$$

$$y - (4) = -\frac{1}{5}[x - (-2)] \qquad \text{Let } m = -\tfrac{1}{5}, (x_1, y_1) = (-2, 4).$$

$$y - 4 = -\frac{1}{5}(x + 2) \qquad \text{Simplify.}$$

$$y - 4 = -\frac{1}{5}x - \frac{2}{5} \qquad \text{Apply the distributive property.}$$

$$5(y - 4) = 5\left(-\frac{1}{5}x - \frac{2}{5}\right) \qquad \text{Multiply each side by 5.}$$

$$5y - 20 = -x - 2 \qquad \text{Apply the distributive property.}$$

$$5y - 20 + x = -x - 2 + x \qquad \text{Add } x \text{ to each side.}$$

$$x + 5y - 20 = -2 \qquad \text{Simplify.}$$

$$x + 5y - 20 + 20 = -2 + 20 \qquad \text{Add 20 to each side.}$$

$$x + 5y = 18 \qquad \text{Simplify.}$$

So, the equation of the line through $(-2, 4)$ and perpendicular to $y = 5x - 6$ is $x + 5y = 18$.

4c. We first find the slope of $x = 5$. The equation $x = 5$ represents a vertical line and, therefore, has undefined slope. Since the lines are parallel, the slope of the unknown line is also undefined.

The unknown line is a vertical line that passes through the point $(2, -3)$. So, its equation is $x = -2$.

4d. From part c, the slope of $x = 5$ is undefined.

Since the lines are perpendicular, the slope of the unknown line is the negative reciprocal of the slope of the given line. Because the slope of the given line is undefined, we can't find its negative reciprocal. We do know, however, that a line perpendicular to a vertical line is a horizontal line with slope $m = 0$.

So, the equation of the horizontal line through $(4, -7)$ is $y = -7$.

✓ **Student Check 4**   Write the equation of the line that goes through $(2, -3)$ that is

a. parallel to $y = \dfrac{1}{2}x + 6$ **b.** perpendicular to $y = \dfrac{1}{2}x + 6$

c. parallel to $x = -7$ **d.** perpendicular to $x = -7$

Applications

Objective 5 ▶

Solve application problems.

There are two types of application problems that we will examine. One type involves knowing an **initial value** (or beginning value, that is, the y-value that corresponds to $x = 0$) and information about how that value changes. The other type involves knowing two data points that describe a particular situation.

> **Procedure: Solving Problems Given an Initial Value and Rate of Change**
>
> **Step 1:** Identify the initial (or beginning) value of the given quantity. This value represents b in the slope-intercept form of a line.
> **Step 2:** Identify how the initial value changes. This is the rate of change, or slope m.
> **Step 3:** Write the equation in slope intercept form $y = mx + b$ by substituting the values of m and b into the equation.

> **Procedure: Solving Problems Given Two Data Points**
>
> **Step 1:** Find the slope of the line between the two points.
> **Step 2:** Use one data point and the slope to determine the equation by using either the slope-intercept form or the point-slope form of a line.

Objective 5 Examples   **Find the equation that represents each situation. Then use the equation to answer the questions.**

5a. Juanita has just been hired as a public school teacher. Her starting salary is $35,000. She will get a raise of $1500 each year she works at the school.

   i. Write a linear equation that represents her salary, where x is the number of years worked.

   ii. Use the equation to determine Juanita's salary after she has worked at the school for 10 yr.

   iii. Use the equation to determine how long she will have to work to have a salary of $80,000.

Solution   **5a.**   **i.** The starting salary is the initial value, so $b = 35,000$. The raise she gets each year is the rate of change or slope.

$$m = \frac{\text{change in } y}{\text{change in } x} = \frac{\text{change in salary}}{\text{change in years}} = \frac{1500}{1} = 1500$$

The equation that models Juanita's salary is $y = 1500x + 35,000$, where x is the number of years worked.

   ii. Juanita's salary after working 10 years is found by setting $x = 10$.

$y = 1500x + 35,000$	Begin with the model.
$y = 1500(10) + 35,000$	Replace x with 10.
$y = 15,000 + 35,000$	Multiply.
$y = 50,000$	Add.

So, Juanita's salary after 10 yr is $50,000.

   iii. To determine how long she needs to work to have a salary of $80,000, we set $y = 80,000$.

$y = 1500x + 35,000$	Begin with the model.
$80,000 = 1500x + 35,000$	Replace y with 80,000.
$80,000 - 35,000 = 1500x + 35,000 - 35,000$	Subtract 35,000 from each side.
$45,000 = 1500x$	Simplify.
$\dfrac{45,000}{1500} = \dfrac{1500x}{1500}$	Divide each side by 1500.
$30 = x$	Simplify.

Juanita's salary will be $80,000 after she works 30 yr at the school.

5b. The average price of a movie ticket in 1996 was $4.42. The average price of a movie ticket in 2010 was $7.89. (Source: http://www.natoonline.org/statisticstickets.htm)

   i. Write a linear equation that represents the average price of a movie ticket, where x is the years after 1996.

   ii. Use the equation to predict the average price of a movie ticket in 2015.

   iii. In what year will the average price of a movie ticket be $10.67?

Solution   **5b.** In this problem, x represents the years after 1996 and y represents the average price of a movie ticket. The year 1996 corresponds to the x-value of 0 and the year 2010 corresponds to the x-value of 14 since $2010 - 1996 = 14$. So, the two ordered pairs given in the problem are $(0, 4.42)$ and $(14, 7.89)$.

   i. To write the equation, we first find the slope.

$$m = \frac{y_2 - y_1}{x_2 - x_1}$$ Begin with the slope formula.

$$m = \frac{7.89 - 4.42}{14 - 0}$$ Let $(x_1, y_1) = (0, 4.42)$ and $(x_2, y_2) = (14, 7.89)$.

$$m = \frac{3.47}{14}$$ Simplify the numerator and denominator.

$$m \approx 0.25$$ Simplify the result.

One of the given points is $(0, 4.42)$. This is the y-intercept; therefore, $b = 4.42$. Knowing the slope $m = 0.25$ and $b = 4.42$, the equation that models the average price of a movie ticket is $y = 0.25x + 4.42$.

   ii. The average price in 2015 is found by setting $x = 19$.

$$y = 0.25x + 4.42$$ Begin with the model.

$$y = 0.25(19) + 4.42$$ Replace x with 19.

$$y = 4.75 + 4.42$$ Multiply.

$$y = 9.17$$ Add.

The average price of a movie ticket in 2015 will be $9.17.

   iii. To find when the average price is $10.67, we set $y = 10.67$ and solve for x.

$$y = 0.25x + 4.42$$ Begin with the model.

$$10.67 = 0.25x + 4.42$$ Replace y with 10.67.

$$10.67 - 4.42 = 0.25x + 4.42 - 4.42$$ Subtract 4.42 from each side.

$$6.25 = 0.25x$$ Simplify.

$$\frac{6.25}{0.25} = \frac{0.25x}{0.25}$$ Divide each side by 0.25.

$$25 = x$$ Simplify.

The average price of a movie ticket will be $10.67 in 25 years or in 2021.

✔ Student Check 5   Find the equation that represents each situation. Then use the equation to answer the questions.

   a. Ryan bought an SUV for $30,000. The value of the SUV decreases by $3000 each year.

     i. Write a linear equation that represents the value of the SUV, where x is the age in years.

     ii. Use the equation to find the value of the SUV after 4 yr.

     iii. When will the value of the SUV be $9000?

   b. In Fall 2000, there were approximately 15.3 million students enrolled in post-secondary degree granting institutions. In Fall 2009, there were approximately 20.2 million students enrolled in post-secondary degree granting institutions. (Source: National Center for Education Statistics)

    i. Assuming that enrollment is growing linearly, write an equation that represents the number of students (in millions) enrolled in post-secondary degree granting institutions x years after 2000.

    ii. Use the equation to estimate the enrollment in 2012.

    iii. When will the enrollment reach 28.26 million?

Objective 6 ▶

Troubleshoot common errors.

Troubleshooting Common Errors

Some common errors for writing equations of lines are shown next.

Objective 6 Examples **A problem and an incorrect solution are given. Provide the correct solution and an explanation of the error.**

6a. Write the equation of the line through $(3, 0)$ with slope $m = \dfrac{2}{3}$.

Incorrect Solution	Correct Solution and Explanation
Since the point $(3, 0)$ is given, the value of $b = 3$. So, the equation is $$y = \frac{2}{3}x + 3$$	The point $(3, 0)$ is the x-intercept, *not* the y-intercept. So, we use the slope-intercept form to find the value of b. $$y = mx + b$$ $$0 = \frac{2}{3}(3) + b$$ $$0 = 2 + b$$ $$-2 = b$$ So, the equation is $y = \frac{2}{3}x - 2$.

6b. Write the equation of the line through $(2, -1)$ and $(5, -2)$.

Incorrect Solution	Correct Solution and Explanation
The slope is $$m = \frac{5 - 2}{-2 - (-1)} = \frac{3}{-1} = -3.$$ So, the equation is $$y - y_1 = m(x - x_1)$$ $$y - (-2) = -3(x - 5)$$ $$y + 2 = -3x + 15$$ $$y = -3x + 13$$	The error was made in calculating the slope of the line. The slope should be $$m = \frac{\text{change in } y}{\text{change in } x} = \frac{-2 - (-1)}{5 - 2} = -\frac{1}{3}$$ We can use the point-slope form to find the equation. $$y - y_1 = m(x - x_1)$$ $$y - (-2) = -\frac{1}{3}(x - 5)$$ $$y + 2 = -\frac{1}{3}x + \frac{5}{3}$$ $$y + 2 - 2 = -\frac{1}{3}x + \frac{5}{3} - 2$$ $$y = -\frac{1}{3}x + \frac{5}{3} - \frac{6}{3}$$ $$y = -\frac{1}{3}x - \frac{1}{3}$$

6c. Write the equation of the line through $(4, -1)$ that is perpendicular to $y = 2x + 4$.

Incorrect Solution	Correct Solution and Explanation
The slope of the given line is $m = 2$. The slope of a perpendicular line is $m = -\dfrac{1}{2}$. The value of $b = 4$. So, the equation is $$y = -\frac{1}{2}x + 4$$	The error was made in calculating the value of b. The y-intercept of the given line is $(0, 4)$, but this is not necessarily the y-intercept of the perpendicular line. We must use the slope-intercept form or point-slope form to find the equation. Use $m = -\dfrac{1}{2}$ and $(4, -1)$ to get $$y = mx + b$$ $$-1 = -\frac{1}{2}(4) + b$$ $$-1 = -2 + b$$ $$1 = b$$ So, the equation is $y = -\dfrac{1}{2}x + 1$.

ANSWERS TO STUDENT CHECKS

Student Check 1 **a.** $y = -2x + 9$ **b.** $y = \dfrac{7}{3}x - 1$
 c. $y = \dfrac{2}{3}$ **d.** $y = \dfrac{3}{2}x + 1$

Student Check 2 **a.** $y = -5x + 16$ **b.** $y = -\dfrac{2}{7}x + \dfrac{43}{7}$
 c. $y = 9$ **d.** $x = 8$

Student Check 3 **a.** $y = -5x + 8$ **b.** $y = -\dfrac{4}{3}x + \dfrac{2}{3}$
 c. $y = -1$ **d.** $x = -8$

Student Check 4 **a.** $y = \dfrac{1}{2}x - 4$ **b.** $y = -2x + 1$
 c. $x = 2$ **d.** $y = -3$

Student Check 5 **a. i.** $y = -3000x + 30{,}000$
  ii. \$18,000 **iii.** 7 yr
  b. i. $y = 0.54x + 15.3$ **ii.** 21.78 million **iii.** 2024

SUMMARY OF KEY CONCEPTS

1. If the slope and y-intercept of an equation are known, writing the equation that satisfies this information is immediate. The value of m and b are substituted into the slope-intercept form $y = mx + b$.

2. There are three other situations that provide enough information for an equation of a line to be written. They are:
   - a point and a slope
   - two points
   - a point and a line parallel or perpendicular

   In each of these situations, the slope must be determined. If it is not given, use the slope formula or the relationship to a given line to find it. After the slope is found, use it with one of the points in either the point-slope form or the slope-intercept form to write the equation of the line.

3. If a line is described as vertical or with undefined slope, the equation of the line will be of the form $x = h$, where h is the x-coordinate of the given point.

4. If a line is described as horizontal or with zero slope, the equation of the line will be of the form $y = k$, where k is the y-coordinate of the given point.

5. In application problems, we will either know the slope (how the values change) and y-intercept (initial value) from the problem or we will be given two points that enable us to write the equation.

GRAPHING CALCULATOR SKILLS

The graphing calculator has the ability to calculate the equation of the line if provided enough information. At this point, it is more beneficial to use the calculator to check our work instead of allowing it to do the work for us.

Example: Use the calculator to verify that $y = \dfrac{3}{2}x + 6$ is the equation of the line that goes through the points $(-4, 0)$ and $(4, 12)$.

Enter the equation in the equation editor.

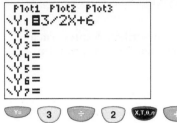

Graph the equation and use the TRACE feature to determine if the points $(-4, 0)$ and $(4, 12)$ lie on the line. Press TRACE, enter the x-value of the first point and press enter. Notice that this takes us to the point $(-4, 0)$ on the graph.

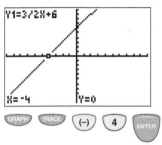

Now enter the x-value of the second point. Notice that the display is $x = 4$ and $y = 12$. The point is not shown on the graph since it is outside of the standard viewing window.

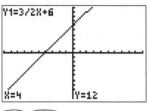

Another way to check to see if the equation contains the points as solutions is to use the TABLE feature. Press 2nd GRAPH and verify that the points $(4, 12)$ and $(-4, 0)$ are in the table.

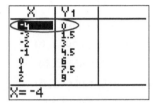

SECTION 3.5 / EXERCISE SET

Write About It!

Use complete sentences in your answer to each question.

1. How can you determine the equation of a line if you know its slope and y-intercept?

2. How can you determine the equation of a line if you know two points on the line?

3. Which method do you prefer to find the equation of a line—using the slope-intercept form or using the point-slope form? Why?

4. Explain how to determine the equation of the line that passes through the point $(-1, 3)$ and is parallel to $y = 4$.

5. Explain how to determine the equation of the line that passes through the point $(3, -2)$ and is perpendicular to $2x - 1 = 0$.

Determine if each statement is true or false. If a statement is false, explain why it is false.

6. The equation of the line with slope 7 that passes through the point $(5, 0)$ is $y = 7x + 5$.

7. The equation of the line that passes through the point $(2, 6)$ and is perpendicular to $y = 3x + 4$ is
$$y = -\frac{1}{3}x + 6.$$

8. The equation of the line that passes through the point $(3, -4)$ and is parallel to $y = 3 - 2x$ is $y = 3x - 13$.

Practice Makes Perfect!

Write the equation of the line that satisfies the given information. (See Objective 1.)

9. $m = 4, b = 5$

10. $m = -2, b = 1$

11. $m = 3$, passes through $(0, -4)$

12. $m = \frac{1}{2}$, passes through $(0, 7)$

13. $m = -\frac{2}{3}$ with y-intercept $(0, -10)$

14. $m = \frac{4}{5}$ with y-intercept $(0, 20)$

15. $m = 0$, passes through $(0, 7)$

16. $m = 0$, passes through $(0, -14)$

17. $m = -\frac{1}{9}$, passes through $\left(0, -\frac{4}{9}\right)$

18. $m = \frac{3}{7}$, passes through $\left(0, \frac{2}{7}\right)$

Write the equation of the line that has the given slope and passes through the given point. Express each answer in slope-intercept form when possible. (See Objective 2.)

19. $m = 4; (1, 5)$ 20. $m = 2; (4, 3)$

21. $m = -5; (-2, -1)$ 22. $m = 6; (3, 0)$

23. $m = -1; (-4, 0)$ 24. $m = \frac{1}{2}; (4, 0)$

25. $m = -\frac{5}{2}; (-2, 0)$ 26. $m = -7; (1, 6)$

27. $m = 8$; $(-3, -9)$ **28.** $m = -12$; $(-4, 50)$

29. $m = 6$; $(0, 0)$ **30.** $m = -1$; $(0, 0)$

31. $m = -8$; $(0, 0)$ **32.** $m = 4$; $(0, 0)$

33. $m = 0$; $(2, -5)$ **34.** $m = 0$; $(2, 8)$

35. $m = \dfrac{2}{3}$; $(6, -8)$ **36.** $m = \dfrac{1}{9}$; $(-9, 2)$

37. $m = -\dfrac{3}{5}$; $(2, -1)$ **38.** $m = -\dfrac{1}{4}$; $(6, -1)$

39. $m =$ undefined; $(9, 3)$ **40.** $m =$ undefined; $(-5, 4)$

Write the equation of the line that passes through the two points. Express each answer in slope-intercept form when possible. (*See Objective 3.*)

41. $(-4, 1)$ and $(2, 7)$ **42.** $(-2, -5)$ and $(2, -1)$

43. $(-4, -8)$ and $(2, 4)$ **44.** $(-5, 15)$ and $(1, -3)$

45. $(-1, -6)$ and $(5, 6)$ **46.** $(-3, -14)$ and $(2, 11)$

47. $(4, -4)$ and $(12, 0)$ **48.** $(8, -3)$ and $(-8, -15)$

49. $(-9, 8)$ and $(-3, 6)$ **50.** $(-11, 8)$ and $(11, -4)$

51. $(-2, 5)$ and $(4, 5)$ **52.** $(1, -7)$ and $(-4, -7)$

53. $(2, -1)$ and $(2, 6)$ **54.** $(-3, 5)$ and $(-3, 2)$

Write the equation of the line that passes through the given point and is either parallel or perpendicular to the given line. Express each answer in slope-intercept form when possible. (*See Objective 4.*)

55. $(0, -4)$, parallel to $y = -3x + 6$

56. $(0, 3)$, parallel to $y = 7x - 3$

57. $(-7, 8)$, parallel to $2x + y = 4$

58. $(5, -2)$, parallel to $3x - y = 9$

59. $(3, -5)$, parallel to $4x + 3y = -12$

60. $(-6, -1)$, parallel to $5x - 6y = -30$

61. $(-1, 4)$, parallel to $y = 3$

62. $(7, -3)$, parallel to $y = -4$

63. $(-1, 4)$, parallel to $x = 3$

64. $(2, -7)$, parallel to $x = -9$

65. $(0, -4)$, perpendicular to $y = -3x + 6$

66. $(0, 3)$, perpendicular to $y = 7x - 3$

67. $(-6, 8)$, perpendicular to $2x + y = 4$

68. $(3, -5)$, perpendicular to $3x - y = 9$

69. $(-8, 1)$, perpendicular to $4x + 3y = -12$

70. $(-5, 9)$, perpendicular to $5x - 6y = -30$

71. $(-1, 4)$, perpendicular to $y = 3$

72. $(11, -2)$, perpendicular to $y = -7$

73. $(-4, 8)$, perpendicular to $x = \dfrac{1}{2}$

74. $(1, -6)$, perpendicular to $x = -4$

 Mix 'Em Up!

Write the equation of the line described. Express each answer in slope-intercept form and in standard form.

75. $m = -\dfrac{1}{2}$, $(-2, 4)$

76. $m = -\dfrac{2}{3}$, $(3, -3)$

77. $(3.5, -4.7)$ and $(-6.3, -3.3)$

78. $(5.3, 0)$ and $(6.2, -1.8)$

79. $m =$ undefined, $(-10, 12)$

80. $m = 0$, $(-15, 20)$

81. $(4, -5)$, parallel to $y = -6x + 3$

82. $(-1, 3)$, perpendicular to $y = -\dfrac{1}{10}x + 15$

83. $(5, 4)$ and $(5, -6)$

84. $(-1, -3)$ and $(2, -3)$

85. $m = 5$, y-intercept: $(0, -5)$

86. $m = -6$, x-intercept: $(12, 0)$

87. $m = 2.5$, $(-4, 7)$

88. $m = -1.6$, $(5, -3)$

89. x-intercept $(-4, 0)$ and y-intercept $(0, 5)$

90. x-intercept $(3, 0)$ and y-intercept $(0, 2)$

91. $(0, 5)$, parallel to $3x - 5y = 4$

92. $\left(0, \dfrac{3}{5}\right)$, perpendicular to $6x - 3y = 12$

93. $(-2, 4)$, perpendicular to $x = -5$

94. $(-4, 0)$, parallel to $x = 10$

Write a linear equation to model each situation and use the model to answer each question.

95. Pedro has a new job as a computer programmer. His starting salary is $45,000. He will receive a raise of $2000 each year he works with the company.

    a. Write a linear equation that represents Pedro's salary, where x is the years he has worked with the company.

    b. What is Pedro's salary after 5 yr of working for the company?

    c. How long does he have to work for the company to earn a salary of $71,000?

96. Sue has a new job as a pharmaceutical sales representative. Her starting salary is $50,000. She earns a bonus based on her sales. Her bonus is 25% of her total sales for the year.

    a. Write a linear equation that represents Sue's yearly income, where x is the total sales for the year.

    b. If Sue's sales total $30,000 for the year, what is her income?

    c. How much does Sue need to sell to have an income of $65,000?

97. Shanika registers for her first semester in college. Her tuition includes fees of $400 plus $150 per credit hour.

    a. Write a linear equation that represents Shanika's total tuition, where x is the number of credit hours Shanika will take.

    b. What is Shanika's tuition if she registers for 12 credit hours?

    c. How many hours does Shanika take if her tuition is $1300?

98. James joins a fitness club. There is a one-time membership fee of $300 plus a monthly charge of $35.

    a. Write a linear equation that represents the total cost of joining the fitness club, where x is the number of months James is a member.

    b. What is the total cost of joining the fitness club for 1 yr?

    c. How many months has James been a member if this total cost is $1560?

99. Abdul purchases a new car for $20,000. The car's value decreases by $1500 each year.

    a. Write a linear equation that represents the value of the car, where x is the age of the car in years.

    b. What is the car's value after 6 yr?

    c. What will be the age of the car when its value is $5000?

100. One of the fastest growing counties in the United States is Flagler County, Florida. In 2004, its population was approximately 69,000. In 2010, the population was approximately 96,000. (Source: U.S. Census Bureau)

    a. Assuming this growth is linear, write an equation that approximates the population of Flagler County where x is the number of years after 2004.

    b. What is the estimated population of Flagler County in 2015?

    c. When will the population reach 150,000?

101. The number of single-family, existing homes sold in Florida in January 2010 was approximately 10,700. In January 2011, the number of homes sold was approximately 12,200. (Source: http://media.living.net)

    a. Assuming linear growth, write an equation that approximates the number of homes sold in Florida where x is the number of years after 2010.

    b. If this trend continues, how many homes will be sold in January 2015?

102. There were approximately 40 million Social Security beneficiaries in 1990 and approximately 61.4 million beneficiaries in 2012. (Source: www.ssa.gov)

    a. Write a linear equation that approximates the number of Social Security beneficiaries (in millions), where x is the number of years after 1990.

    b. If this growth continues, how many beneficiaries will there be in 2015?

103. The following graph shows the billions of dollars spent on spectator sports in the United States for the years 1990–2003. (Source: www.infoplease.com)

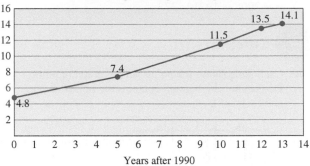

Billions Spent on Spectator Sports

    a. Use the points (0, 4.8) and (13, 14.1) to write a linear equation that models the money spent on spectator sports (in billions of dollars) x years after 1990.

    b. If this growth continues, how much will be spent on spectator sports in 2015?

104. The median age of women at their first wedding was 20.3 in 1950. In 2010, the median age of women at their first wedding was 26.1. (Source: www.infoplease.com)

    a. Assuming linear growth, write an equation that represents the median age of women at their first wedding, where x is the number of years after 1950.

    b. Use the equation to predict the median age of women on the date of their first wedding in 2015.

You Be the Teacher!

Correct each student's errors, if any.

105. Find the equation of the line that is parallel to $4x - 3y = 6$ and passes through the point (8, 3).

Vivian's work:

$$4x - 3y = 6$$
$$\underline{-4x \qquad\qquad -4x}$$
$$\frac{-3y}{-3} = \frac{-4x}{-3} + \frac{6}{-3}$$

$y = -\dfrac{4}{3}x - 2$. So, the slope is $y = -\dfrac{4}{3}$. Since it passes through the point (8, 3), the equation of the line is

$$y = -\frac{4}{3}x + 3.$$

106. Find the equation of the line that passes through the points $(-3, 4)$ and $\left(0, \dfrac{1}{2}\right)$.

William's work: $m = -\dfrac{7}{6}$ and the equation is

$$y = -\frac{7}{6}x + 4$$

107. Find the equation of the line with slope -2 that passes through the point $(5, -1)$.

Desi's work: $y = -2x - 1$

108. Suppose x = number of years after 1971. What does $x = 0$ represent? How do I figure out what x is for the year 2008?

Marla's work: $x = 0$ represents 1971. For 2008, $x = 2008 - 1971 = 37$.

 Calculate It!

109. Each table shows the points on the graph of a line. Determine which graph contains the points $(0, 2)$ and $(-1, -3)$.

a.

X	Y1	
-3	17	
-2	12	
-1	7	
0	2	
1	-3	
2	-8	
3	-13	

X=3

b.

X	Y1	
-3	-13	
-2	-8	
-1	-3	
0	2	
1	7	
2	12	
3	17	

X=3

c.

X	Y1	
-3	-5	
-2	-4	
-1	-3	
0	-2	
1	-1	
2	0	
3	1	

X=3

110. Each table shows the points on the graph of a line. Determine which graph contains the points $(0, -5)$ and $(-2, 4)$.

a.

X	Y1	
-4	3	
-3	3.5	
-2	4	
-1	4.5	
0	5	
1	5.5	
2	6	

X=-4

b.

X	Y1	
-3	-18.5	
-2	-14	
-1	-9.5	
0	-5	
1	-.5	
2	4	
3	8.5	

X=3

c.

X	Y1	
-4	13	
-3	8.5	
-2	4	
-1	-.5	
0	-5	
1	-9.5	
2	-14	

X=-4

SECTION 3.6   **Functions**

Suppose Greg registers for classes online. What would happen if Greg input his student ID number and another student's name came up? Besides being annoyed, Greg would have to contact the Registrar's office and inform them that his ID is not functioning correctly; it corresponds to another student.

In this section, we will learn the definition of a function and how functions apply to our everyday lives.

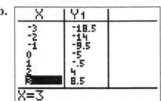

Relations

If we look up the definition of a relation, we will find that most every description involves the words connection and/or association. This idea extends to the definition of a mathematical *relation*, as well. A relation, in mathematics, describes the connection or association between two sets of information. This description is denoted by a set of ordered pairs.

We use relations every day. When we check out at the grocery store, each item scanned corresponds to a price. When a topic in Google is entered, many search results are displayed. If we register for classes online and enter the course title, a list of class options will be shown. In each of these cases, we are pairing together a piece of information from one set to a piece of information from another set.

Objective 1 ▶

Identify a relation and its domain and range.

The set of pairs is called a relation. The first piece of information is the *input value*, or *x*-value. The second piece of information is the *output value*, or *y*-value. The set of all *x*-coordinates of a relation is called the *domain* and the set of all *y*-coordinates of a relation is called the *range*.

Definition: A **relation** is a set of ordered pairs in which the first coordinate of the ordered pairs comes from a set called the **domain** and the second coordinate of the ordered pairs comes from a set called the **range**.

Relations can be expressed in various forms: a graph, a table, a mapping, a set of ordered pairs, or an equation. Some examples are shown.

Graph	Table	Mapping
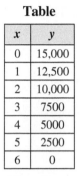		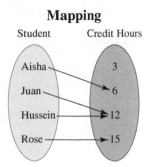

Table

x	y
0	15,000
1	12,500
2	10,000
3	7500
4	5000
5	2500
6	0

Set of ordered pairs

$\{(1, 3), (2, 4), (3, 5), (4, 6)\}$

Equation

$y = 0.24x + 4.42$

When we state the domain or range of a relation, it is not necessary to list values more than once.

Objective 1 Examples | **Express each described relation in the requested form. State the domain and range of each relation.**

1a. A class was surveyed to determine which students were enrolled in a specific number of hours. The results are shown in the following mapping. Write this relation as a set of ordered pairs.

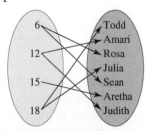

1b. According to T-Mobile.com, each text message sent or received in the United States costs $0.20 if you do not have a messaging plan. Express the relation between the number of text messages sent or received and the cost associated with them as an equation.

1c. The network TV ad revenue (in millions) for the NCAA Division I Men's Basketball Championship is given in the table. Express this relation in a graph.
(Source: http://www.internetadsales.com/march-madness-advertising-trends-report)

Year	2000	2001	2002	2003	2004	2005	2006	2007	2008	2009
Revenue (in millions)	319	318	358	380	451	475	500	520	643	589

Solutions

1a. The x-value of the ordered pairs represents the number of credit hours and the y-value represents the student. So, the set of ordered pairs for this relation is

{(6, Rosa), (6, Sean), (12, Amari), (12, Judith), (15, Aretha), (18, Todd), (18, Julia)}.

The domain of the relation is {6, 12, 15, 18} and the range of the relation is {Todd, Amari, Rosa, Julia, Sean, Aretha, Judith}.

1b. The x-value of the ordered pairs represents the number of text messages and the y-value represents the cost. The equation for the relation is $y = 0.20x$. Since the number of text messages sent or received can be 0, 1, 2, and so on, the domain of the relation is {0, 1, 2, 3, 4, 5, 6, . . .} and the range is {0, 0.20, 0.40, 0.60, 0.80, 1, 1.20, . . .}.

1c. The x-value of the ordered pairs represents the year and the y-value represents the revenue (in millions). The graph is shown.

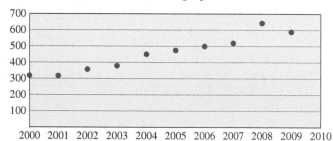

The domain is {2000, 2001, 2002, 2003, 2004, 2005, 2006, 2007, 2008, 2009}, and the range is {318, 319, 358, 380, 451, 475, 500, 520, 589, 643}.

✓ Student Check 1

Express each described relation in the requested form. State the domain and range of each relation. For b, use only whole numbers for the input.

a. A class was surveyed to find out each student's declared major. The results are shown in the mapping. Write this relation as a set of ordered pairs.

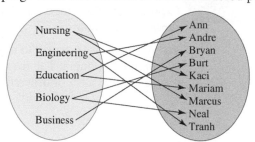

b. Skype allows calls from computer-to-computer or computer-to-land line or mobile phone. Calls from computer-to-computer are free but calls made from a computer to phones have an associated cost. Calls from a computer in the United States to a phone in Iraq cost approximately $0.39 per minute. Express the relation between the number of minutes and the associated cost of using Skype to call Iraq from the United States as an equation. (Source: http://www.skype.com/intl/en-us/prices/payg-rates#cc=IQ)

c. The percentage of Americans who are obese (BMI is greater than or equal to 30) is given in the table. Express this relation as a graph. (Source: http://apps.nccd.cdc.gov/brfss/display.asp and http://www.data360.org/index.aspx)

Year	Percentage	Year	Percentage
1995	15.9	2003	22.8
1996	16.8	2004	23.2
1997	16.6	2005	24.4
1998	18.3	2006	25.1
1999	19.8	2007	26.3
2000	20.1	2008	26.6
2001	21.1	2009	27.2
2002	22.2		

Functions

Objective 2 ▶

Determine if a relation is a function.

We defined a relation as a set of ordered pairs—a correspondence between a set of inputs and outputs. A *function* is a special type of relation.

> **Definition:** A **function** is a relation in which each member of the domain (or each input) corresponds to exactly one member of the range (or an output). In other words, each input, or x-value, can have only one output, or y-value.

In the opening of this section, we mentioned a student ID that corresponds to more than one output. It corresponds to the student who was trying to register as well as to the other student whose name appeared on the screen. This was certainly due to entry error. It doesn't make sense for a student ID to correspond to more than one student. Except in the case of an error, this relation is a function. Each input (student ID) can only correspond to one output (student).

> **Procedure: Determining if a Relation Is a Function**
>
> **Step 1:** Determine how many y-values correspond to each x-value.
>   **a.** If there is only one corresponding y-value for each x-value, then the relation is a function.
>   **b.** If at least one x-value corresponds to more than one y-value, then the relation is *not* a function.

Objective 2 Examples

Determine if each relation is a function. If not, explain why.

2a. $\{(-4, 0), (-3, 2), (-3, -2), (0, 4), (0, -4)\}$

2b. $\{(-2, -3), (-1, -3), (0, -3), (1, -3), (2, -3), (3, -3)\}$

2c. Let the relation be defined by the mapping, where x is the math grade and y is the student earning the grade.

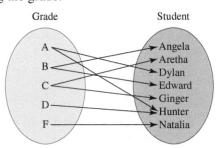

2d. $y = |x|$

Solutions

2a. This is *not a function* since the x-value of -3 corresponds to two y-values, 2 and -2. Also, the x-value of 0 corresponds to two y-values, -4 and 4.

2b. This *is a function* since each x-value corresponds to only one y-value.

2c. This is *not a function* since the grade of A, B, and C correspond to more than one student. For example, the grade of A corresponds to Dylan and Hunter. The grade of B corresponds to Angela and Edward. The grade of C corresponds to Aretha and Ginger.

2d. This *is a function* since each input, or x-value, has only one possible output, or y-value. When a value is substituted for x, there is only one possible y-value. Some examples are

$x = -3; y = |-3| = 3 \rightarrow (-3, 3)$

$x = 2; y = |2| = 2 \rightarrow (2, 2)$

✓ **Student Check 2**   Determine if each relation is a function. If not, explain why.

   a. $\{(5, 1), (5, -2), (5, 6), (5, 3), (5, 0)\}$

   b. $\{(1, 5), (-2, 5), (6, 5), (3, 5), (0, 5)\}$

   c. Let the relation be defined by the mapping, where x is the student and y is the grade earned in math class.

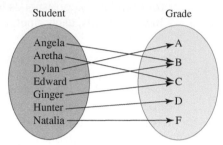

   d. $y = x^2$

Vertical Line Test

Objective 3 ▶

Use the vertical line test.

We will now examine graphs of relations and see how we can use them to determine if a graph represents a function. Graphing some of the preceding examples will show us an important characteristic of the graph of a function.

Shown next are some examples of functions and their corresponding graphs.

$$\{(-2, -3), (-1, -3), (0, -3),$$
$$(1, -3), (2, -3), (3, -3)\}$$

$$y = |x|$$

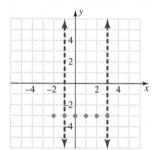

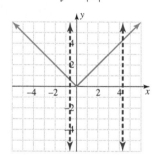

If each x-value corresponds to only one y-value, we know the relation is a function. Notice that when we draw vertical lines through the graph of the relations, each vertical line touches the graph only once.

Shown next are examples of relations that are *not* functions.

$$\{(-4, 0), (-3, 2), (-3, -2),$$
$$(0, 4), (0, -4)\}$$

$$x = y^2$$

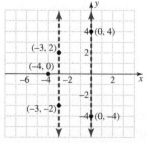

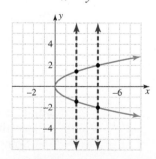

Notice that when an x-coordinate is paired with more than one y-coordinate, a vertical line can be drawn that will intersect the graph at more than one point.

These illustrations provide us a method for determining if a relation is a function by examining its graph. This method involves performing the *vertical line test*.

> **Property: Vertical Line Test**
>
> **1.** If all vertical lines drawn on the graph of a relation intersect the graph in at most one point, then the graph is a function.
> **2.** If at least one vertical line intersects the graph in more than one point, then the graph is *not* a function.

Objective 3 Examples   **Use the vertical line test to determine if each relation is a function.**

3a.
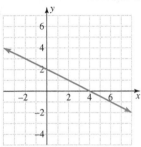

3b.
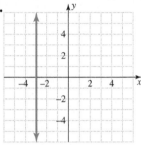

3c.
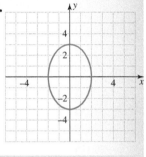

Solutions   **3a.** The relation graphed *is a function* since every vertical line intersects the graph in at most one point.

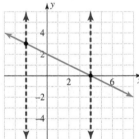

3b. The graph *is not a function* since a vertical line through $x = -3$ intersects infinitely many points of the graph.

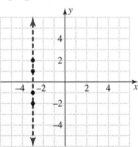

3c. The graph *is not a function* since any vertical line drawn between the *x*-values of -2 and 2 intersects the graph in two points.

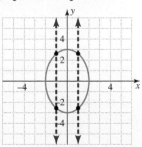

✓ **Student Check 3**   Use the vertical line test to determine if each relation is a function.

a.

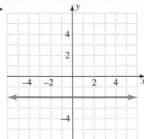

b.

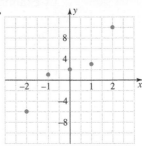

c.

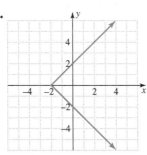

d.

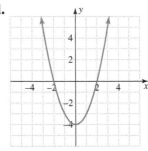

 Note: *All lines, except vertical lines, pass the vertical line test. Therefore, every linear equation of the form y = mx + b is a function.*

Function Notation

Objective 4 ▶

Use function notation.

An example of a linear equation that is a function is $y = 4x - 5$. This equation states a rule for writing ordered pairs where each x-coordinate is paired with exactly one y-coordinate. Function notation is a convenient way to determine this pairing of coordinates.

Suppose we want to find the y-value that corresponds to $x = 3$ for $y = 4x - 5$. We determine the y-value by replacing the variable x with 3.

$$y = 4x - 5$$
$$y = 4(3) - 5$$
$$y = 12 - 5 = 7$$

So, the output value of $4x - 5$ is 7 when $x = 3$. In other words, the point $(3, 7)$ lies on the graph of $y = 4x - 5$.

In mathematics, we use letters such as f, g, and h to name functions. The symbol $f(x)$ is read "f of x" and means "the function of x." This notation is called **function notation**.

Definition: Function Notation

The notation $f(x)$ is used to denote that y is a function of x.

$$y = f(x) \qquad \text{Name of function}$$

Output value   Input value

In this notation, x is the input value and $f(x)$ is the output value.

We can use function notation to write an equivalent equation to $y = 4x - 5$, which is $f(x) = 4x - 5$. The equations are equivalent because $y = f(x)$.

The notation $f(3)$ means to replace x with 3 and find the corresponding y-value. For example,

$$f(x) = 4x - 5$$
$$f(3) = 4(3) - 5 = 7$$

So, when $x = 3$, $y = 7$ or $[f(3) = 7]$ and an ordered pair solution of the equation is $(3, 7)$. Note that this is exactly what we obtain when we evaluate $y = 4x - 5$ at $x = 3$. This process is called *evaluating a function*.

> **Procedure: Evaluating a Function $f(x)$ at $x = k$**
>
> **Step 1:** If the function is given in terms of an equation, replace the variable with the number k that is given in parentheses and simplify.
>
> **Step 2:** If the function is given in terms of a set, a mapping, or a table, find the y-value that corresponds to the x-value of k.
>
> **Step 3:** If the function is given in terms of a graph, find the ordered pair on the graph whose x-value is k. The corresponding y-value is the result of $f(k)$.

Objective 4 Examples / **Evaluate each function at the given value.**

4a. Find $f(0)$ if $f(x) = 3x - 12$.

4b. Find $g(-3)$ if $g(x) = x^2 + 1$.

4c. Find $c(4)$ if the function $c(x)$ is given by Y_1.

4d. Find $f(2)$ if $f(x)$ is given by the following graph.

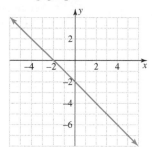

Solutions **4a.** $f(0) = 3(0) - 12 = 0 - 12 = -12$

4b. $g(-3) = (-3)^2 + 1 = 9 + 1 = 10$

4c. The point $(4, -8)$ is one of the ordered pairs in the table. So, $c(4) = -8$

4d. The point $(2, -4)$ lies on the graph of the function. So, $f(2) = -4$.

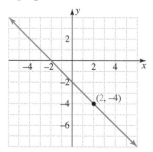

Student Check 4 Evaluate each function at the given value.

a. Find $f(0)$ if $f(x) = 9x + 4$.

b. Find $g(-5)$ if $g(x) = -2x^2 - 4x + 1$.

c. Find $c(2)$ if the function $c(x)$ is given by Y_1.

d. Find $f(-2)$ if $f(x)$ is given by the following graph.

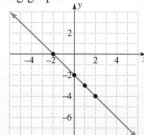

Use Function Notation to Represent Linear Equations in Two Variables

Objective 5 ▶

Write a linear equation in two variables in function notation.

We will now focus on a particular type of function, a *linear function*. Recall that the graph of $Ax + By = C$ or $y = mx + b$ is a line. All of these lines are also functions, except for vertical lines since they do not pass the vertical line test. Therefore, we can use function notation to write equations of non-vertical lines. When lines are written in the form $y = mx + b$, we can replace y with $f(x)$ and put it in function notation.

> **Definition:** A function of the form $f(x) = mx + b$, for m and b real numbers is a **linear function**.

> **Procedure: Writing a Linear Equation in Two Variables in Function Notation**
>
> **Step 1:** Solve the equation for y, if needed.
> **Step 2:** Replace y with $f(x)$.

Objective 5 Examples   Write each linear equation in two variables in function notation. Then, find $f(0)$, $f(2)$, and $f(-1)$ and write the corresponding ordered pairs.

5a. $x + 2y = 4$ **5b.** $y = 7$

Solutions **5a.**

$$x + 2y = 4$$
$$x + 2y - x = 4 - x \qquad \text{Subtract } x \text{ from each side.}$$
$$2y = -x + 4 \qquad \text{Simplify.}$$
$$\frac{2y}{2} = \frac{-x}{2} + \frac{4}{2} \qquad \text{Divide each side by 2.}$$
$$y = -\frac{1}{2}x + 2 \qquad \text{Simplify.}$$
$$f(x) = -\frac{1}{2}x + 2 \qquad \text{Replace } y \text{ with } f(x).$$

Now we must evaluate $f(0), f(2),$ and $f(-1)$.

$$f(0) = -\frac{1}{2}(0) + 2 = 0 + 2 = 2$$
$$f(2) = -\frac{1}{2}(2) + 2 = -1 + 2 = 1$$
$$f(-1) = -\frac{1}{2}(-1) + 2 = \frac{1}{2} + 2 = \frac{1}{2} + \frac{4}{2} = \frac{5}{2}$$

The corresponding ordered pairs are $(0, 2)$, $(2, 1)$, and $\left(-1, \frac{5}{2}\right)$.

5b. The equation is already solved for y. So, in function notation, it is $f(x) = 7$. Recall that $y = 7$ is equivalent to $y = 0x + 7$. So, $f(x) = 0x + 7$.

$$f(0) = 0(0) + 7 = 0 + 7 = 7$$
$$f(2) = 0(2) + 7 = 0 + 7 = 7$$
$$f(-1) = 0(-1) + 7 = 0 + 7 = 7$$

The corresponding ordered pairs are $(0, 7)$, $(2, 7)$, and $(-1, 7)$.

✓ **Student Check 5**   Write each linear equation in two variables in function notation. Then find $f(0)$, $f(3)$, and $f(-2)$ and write the corresponding ordered pairs.

a. $2x + 3y = 6$ **b.** $y = 1$

Objective 6 ▶

Apply functions to real-life applications.

Applications

Without functions, many things in life would be very chaotic. Functions enable our mail to be delivered properly since each address can only correspond to one location. Functions enable our login to different programs to work properly since each login name must correspond to exactly one person. Every social security number must correspond to exactly one person. The list of these types of situations is infinite.

In Example 6, we will see a function that relates to a real-world situation.

Objective 6 Examples

According to the U.S. Department of Labor, the fastest-growing occupation for the years 2006–2016 is a network system and data communications analyst. The function $f(x) = 1343x + 45{,}657$ approximates the median salary in the United States for x years of experience in this field. (Sources: www.bls.gov and www.payscale.com)

6a. Find the median salary with 5 yr of experience.

6b. How many years of experience is required for the median salary to be $60,430?

Solutions

6a. The input x represents the number of years of experience and $f(x)$, or y, represents the median salary. To find the salary with 5 yr of experience, we let $x = 5$.

$$f(x) = 1343x + 45{,}657$$
$$f(5) = 1343(5) + 45{,}657 = 6715 + 45{,}657 = 52{,}372$$

The median salary is $52,372 with 5 yr of experience.

6b. To find how many years of experience is required for the salary to be $60,430, we set $f(x) = 60{,}430$ and solve for x.

$$1343x + 45{,}657 = 60{,}430$$
$$1343x + 45{,}657 - 45{,}657 = 60{,}430 - 45{,}657 \quad \text{Subtract 45,657 from each side.}$$
$$1343x = 14{,}773 \quad \text{Divide each side by 1343.}$$
$$x = 11 \quad \text{Simplify.}$$

The median salary is $60,430 with 11 yr of experience.

✓ Student Check 6

The number of cell-phone subscribers in the United States, in millions, between 1985 and 2009 can be approximated by $f(x) = 0.7082x^2 - 4.0318x + 4.8661$, where x is the number of years after 1985. Use the function to determine the number of cell phone subscribers in the United States in 2015. (Source: www.infoplease.com)

Objective 7 ▶

Troubleshoot common errors.

Troubleshooting Common Errors

Some common errors associated with the concept of functions are shown next.

Objective 7 Examples

A problem and an incorrect solution are given. Provide the correct solution and an explanation of the error.

7a. Is the relation $\{(-2, 3), (-1, 0), (0, -1), (1, 0), (2, 3)\}$ a function?

Incorrect Solution	Correct Solution and Explanation
The relation is not a function since two x-values correspond to the same y-value. Both -2 and 2 have a y-value of 3.	The relation *is* a function. The definition of a function is that each x-value have only one output. It is OK if the output values repeat. Since every value of x has only one output, this relation is a function.

7b. Write the linear equation $x = 5$ in function notation.

Incorrect Solution	Correct Solution and Explanation
The equation $x = 5$ in function notation is $f(x) = 5$.	The equation $x = 5$ is a vertical line and is *not* a function. It cannot be written in function notation.

7c. If $f(x) = 7x + 9$, find $f(2)$.

Incorrect Solution	Correct Solution and Explanation
$f(2) = (7x + 9)(2) = 14x + 18$	The notation $f(2)$ means to evaluate the function at $x = 2$. It does not mean multiplication. $$f(2) = 7(2) + 9 = 14 + 9 = 23$$

ANSWERS TO STUDENT CHECKS

Student Check 1   a. {(Nursing, Kaci), (Nursing, Marcus), (Engineering, Andre), (Engineering, Tranh), (Education, Ann), (Education, Mariam), (Biology, Burt), (Biology, Neal), (Business, Bryan)}

Domain = {nursing, engineering, education, biology, business}

Range = {Ann, Andre, Bryan, Burt, Kaci, Mariam, Marcus, Neal, Tranh}

b. $y = 0.39$; Domain = {0, 1, 2, 3, ...}, Range = {0, 0.39, 0.78, 1.17, ...}

Domain = {1995, 1996, 1997, 1998, 1999, 2000, 2001, 2002, 2003, 2004, 2005, 2006, 2007, 2008, 2009}

Range = {15.9, 16.6, 16.8, 18.3, 19.8, 20.1, 21.1, 22.2, 22.8, 23.2, 24.4, 25.1, 26.3, 26.6, 27.2}

Student Check 2   a. no, the x-value of 5 corresponds to more than one y-value.   **b.** yes   **c.** yes   **d.** yes

Student Check 3   a. yes   **b.** yes   **c.** no   **d.** yes

Student Check 4   a. $f(0) = 4$ **b.** $g(-5) = -29$ **c.** $c(2) = -6$ **d.** $f(-2) = 0$

Student Check 5

a. $f(x) = -\dfrac{2}{3}x + 2$; $f(0) = 2$; $f(3) = 0$; $f(-2) = \dfrac{10}{3}$

b. $f(x) = 1$; $f(0) = 1$; $f(3) = 1$; $f(-2) = 1$

Student Check 6   521.29 million cell phone subscribers.

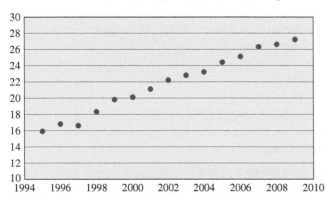

SUMMARY OF KEY CONCEPTS

1. A relation is a set of ordered pairs. The set of x-values of the ordered pairs is the domain. The set of y-values of the ordered pairs is the range.

2. A function is a relation in which each input has only one output, that is, for every x in the domain of the function, there is exactly one y-value. The x-values in the list of ordered pairs cannot repeat for the relation to be a function.

3. The vertical line test can be used to determine if a graph represents a function. If any vertical line intersects a graph in more than one point, the graph does *not* represent a function.

4. Function notation, $f(x)$, is another name for the output value y. It shows that y is dependent on the value of x. When a specific value of x is inside the function notation, we evaluate the function at that value.

a. In an equation, we do this by substituting the variable with the given number.

b. In a table or graph, we find the ordered pair in the table or on the graph with the given x value.

The y-value of this ordered pair is the value we need.

5. Linear equations in two variables of the form $y = mx + b$ can be written in function notation as $f(x) = mx + b$. Equations of

the form $Ax + By = C$ must be solved for y before they can be written in function notation. Vertical lines are not functions.

6. Functions occur in many real-world situations. It is important to understand from the problem what the input and output of the function represent.

GRAPHING CALCULATOR SKILLS

The graphing calculator can assist us in evaluating functions when the function is given in terms of an equation. Input the function into the calculator to find the requested information.

Example: Find $f(0)$ if $f(x) = |x - 4|$.

Solution: Input the equation into the calculator and use the Table or TRACE feature to find the y-value.

Use the Table to find the y-value when $x = 0$.

Graph the equation and use the TRACE feature to find the point that has an x-value of 0.

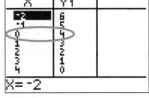

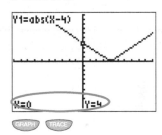

So, $f(0) = 4$.

SECTION 3.6 / EXERCISE SET

✍ Write About It!

Use complete sentences in your answer to each exercise.

1. Define a relation and its domain and range.

2. How do we determine if a relation is a function?

3. What is the vertical line test?

4. Define a function.

5. What is function notation?

6. What is the difference between an equation of a line and a linear equation in two variables in function notation?

Determine if each statement is true or false. If a statement is false, explain why it is false.

7. If $f(3) = 5$, then $(3, 5)$ is a point on the graph of the function f.

8. If $f(0) = -7$, then the x-intercept is $(-7, 0)$.

9. If $f(8) = 0$, then the y-intercept is $(0, 8)$.

10. A relation is always a function.

11. If any vertical line intersects a graph in more than one point, the graph represents a function.

12. A relation always has a domain and range.

🎹 Practice Makes Perfect!

Express each relation in Exercises 13–16 as a set of ordered pairs. State the domain and range of each relation. (*See Objective 1.*)

13. The following table shows the 2009 total health expenditure per capita in the selected countries. (Source: http://www.who.int/en)

Countries	Per Capita
Canada	$4400
France	$4800
Italy	$3300
Japan	$3300
Republic of Korea	$1100
United Kingdom	$3300
United States	$7400

14. The following table shows the number of days that selected U.S. metropolitan areas failed to meet acceptable air-quality standards in 2008. (Source: http://worldalmanac.com)

Metropolitan Statistical Areas	Number of Days
Atlanta	4
Bakersfield	26
Baltimore	4
Detroit	1
Houston	2
Los Angeles	28
Memphis	1
New Orleans	0
New York	1
Orange County	1
Phoenix	84
Sacramento	20
Washington, DC	3

15. The mapping below shows the number of major professional championships earned by each professional golfer. (Source: http://worldalmanac.com)

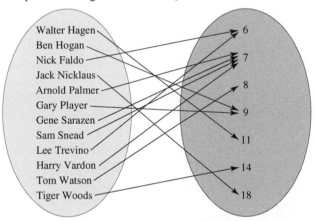

16. The results of the Indy Racing League (IRL) winners from 1996 to 2008 are shown in the mapping. (Source: http://worldalmanac.com)

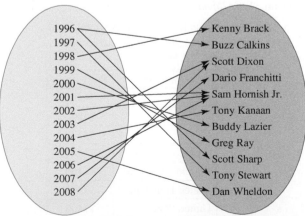

17. The residential average price of electricity in New Jersey was 14.52 cents per kilowatt-hour in 2009. Express the relation between the number of kilowatt-hours (in hundreds) and the cost (in dollars) associated with them as an equation. (Source: www.think-energy.net)

18. The residential average price of electricity in Alaska was 15.09 cents per kilowatt-hour in 2009. Express the relation between the number of kilowatt-hours (in hundreds) and the cost (in dollars) associated with them as an equation. (Source: www.think-energy.net)

Determine if each relation is a function. If not, explain why. (*See Objective 2.*)

19. $\{(2, -5), (2, -1), (2, 8), (2, 14), (2, 18)\}$

20. $\{(-14, -3), (-7, -3), (-5, -3), (6, -3), (12, -3)\}$

21. $\{(-20, 6), (-9, 6), (0, 6), (1, 6), (4, 6)\}$

22. $\{(4, -8), (4, -1), (4, 0), (4, 7), (4, 11)\}$

23. $\{(-2, 2), (-2, 4), (1, 8), (5, 12), (0, 9)\}$

24. $\{(-5, 8), (-5, -7), (1, 15), (-1, 0), (2, 10)\}$

25. Assign to each student the grade earned in their physics class.

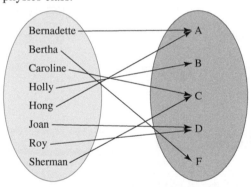

26. Assign to each student the grade earned in their math class.

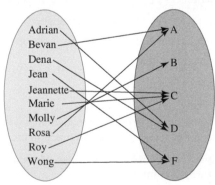

27. $y = 2x + 5$

28. $y = -4x + 3$

29. $y = |x - 1|$

30. $y = 2 - |x|$

31. $x = |y - 2|$

32. $x = |y + 3|$

Use the vertical line test to determine if each relation is a function. (*See Objective 3.*)

33.

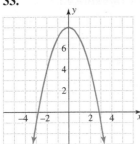

34.

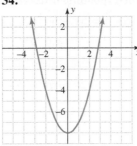

35.

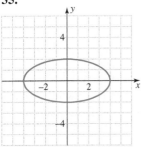

36.

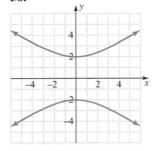

37.

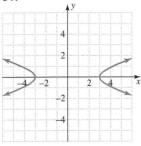

38.

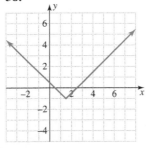

Evaluate each function at the given value. (*See Objective 4.*)

39. $f(-4)$ if $f(x) = -6x + 1$ **40.** $f(0)$ if $f(x) = 2x - 7$

41. $g(5)$ if $g(x) = 2x^2 - 1$ **42.** $g(-3)$ if $g(x) = -x^2 + 2$

43. $h(-2)$ if $h(x) = |5x - 4|$ **44.** $h(-5)$ if $h(x) = |3x - 1|$

45. $f(10)$ if $f(x) = -x^2 + 7x - 6$

46. $f(-1)$ if $f(x) = 3x^2 - 6x + 12$

47. Find $c(3)$ if the function $c(x)$ is given by Y_1.

X	Y1	
-3	9	
-2	7	
-1	5	
0	3	
1	1	
2	-1	
3	-3	
X=3		

48. Find $c(-2)$ if the function $c(x)$ is given by Y_1.

X	Y1	
-3	-13	
-2	-10	
-1	-7	
0	-4	
1	-1	
2	2	
3	5	
X=3		

49. Find $f(-3)$ if the function $f(x)$ is given by Y_1.

X	Y1	
-3	-9.6	
-2	-7	
-1	-4.4	
0	-1.8	
1	.8	
2	3.4	
3	6	
X=3		

50. Find $f(-2)$ if the function $f(x)$ is given by Y_1.

X	Y1	
-4	17	
-3	13.4	
-2	9.8	
-1	6.2	
0	2.6	
1	-1	
2	-4.6	
X= -4		

51. Find x if $c(x) = -4$ and the function $c(x)$ is given by Y_1.

X	Y1	
-3	-13	
-2	-10	
-1	-7	
0	-4	
1	-1	
2	2	
3	5	
X=3		

52. Find x if $f(x) = 3.4$ and the function $f(x)$ is given by Y_1.

X	Y1	
-3	-9.6	
-2	-7	
-1	-4.4	
0	-1.8	
1	.8	
2	3.4	
3	6	
X=3		

53. Find $f(-4)$ if $f(x)$ is given by the graph.

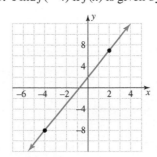

54. Find $f(-2)$ if $f(x)$ is given by the graph.

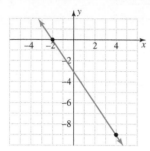

55. Find $f(2)$ if $f(x)$ is given by the graph.

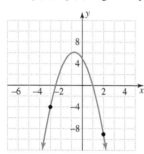

56. Find $f(-1)$ if $f(x)$ is given by the graph.

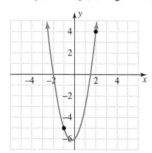

57. Find x if $f(x) = -6$ and $f(x)$ is given by the graph.

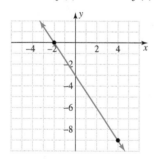

58. Find x if $f(x) = 7$ and $f(x)$ is given by the graph.

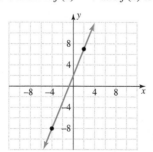

Write each linear equation in two variables in function notation. Then find $f(0)$, $f(3)$, and $f(-2)$ and write the corresponding ordered pairs. (*See Objective 4.*)

59. $y = 5x - 6$ **60.** $y = -7x + 3$ **61.** $4x - y = 9$

62. $3x + 5y = 6$ **63.** $2y - 7 = 0$ **64.** $2y + 3 = 0$

Solve each problem. (*See Objective 5.*)

65. Assume that the S&P 500 index historical prices (in dollars) can be modeled by the function $f(x) = 1.56x^4 - 35x^3 + 247x^2 - 567x + 1381$, where x is the number of years after 2000.

    a. Use the function to find the S&P 500 index price in 2009. How close is the answer if the actual estimate was 1106?

    b. Use the function to predict the S&P 500 index for year 2012.

66. Assume that the NASDAQ Composite historical prices (in dollars) can be modeled by the function $f(x) = 8.271x^4 - 106.302x^3 + 346.835x^2 + 34.085x + 1462.356$, where x is the number of years after 2002.

    a. Use the function to find the NASDAQ composite price in 2006. How close is the answer if the actual estimate was $2457?

    b. Use the function to predict the NASDAQ composite price for year 2010.

67. An open-top box is constructed by cutting four squares of width x in. from the corners of a rectangular piece of cardboard with dimensions 12 in. by 18 in. and then folding up the sides. The volume of the box is given by the function $V(x) = 216x - 60x^2 + 4x^3$, where $0 \le x \le 6$. Evaluate $V(2)$ and interpret the answer.

68. An open-top box is constructed by cutting four squares of width x in. from the corners of a rectangular piece of cardboard with dimensions 24 in. by 30 in. and then folding up the sides. The volume of the box is given by the function $V(x) = 720x - 108x^2 + 4x^3$, where $0 \le x \le 12$. Evaluate $V(5)$ and interpret the answer.

 Mix 'Em Up!

Determine if each relation is a function. If not, explain why.

69. The following table shows the 2009 total health expenditure per capita in the selected countries. (Source: http://www.who.int/en)

Countries	Per Capita
Canada	$4400
France	$4800
Italy	$3300
Japan	$3300
Republic of Korea	$1100
United Kingdom	$3300
United States	$7400

70. The following table shows the number of days that selected U.S. metropolitan areas failed to meet acceptable air-quality standards in 2008. (Source: http://worldalmanac.com)

Metropolitan Statistical Areas	Number of Days
Atlanta	4
Bakersfield	26
Baltimore	4
Detroit	1
Houston	2
Los Angeles	28
Memphis	1
New Orleans	0
New York	1
Orange County	1
Phoenix	84
Sacramento	20
Washington, DC	3

71. The mapping below shows the number of major professional championships earned by each professional golfer. (Source: http://worldalmanac.com)

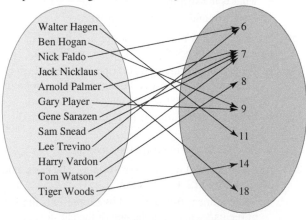

72. The results of the Indy Racing League (IRL) winners from 1996 to 2008 are shown in the mapping. (Source: http://worldalmanac.com)

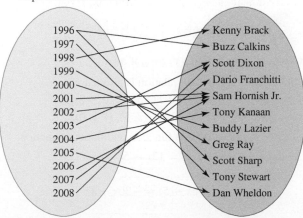

73. {(3, 9), (3, −14), (3, 4), (3, −2), (3, −8)}

74. {(9, 3), (−14, 3), (4, 3), (−2, 3), (−8, 3)}

75. {(1, 3), (−2, 7), (−2, 1), (0, 5), (4, 0)}

76. {(6, −3), (1, −3), (0, 11), (−5, 0), (2, 0)}

77. Assign to each student the grade earned in their chemistry class.

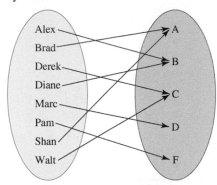

78. Assign to each grade the student who earned the grade in a computer programming class.

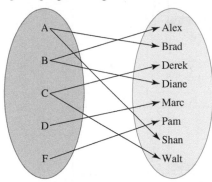

79. $y = 7$ **80.** $x = 4$

81. $y = \dfrac{1}{2}x + 4$ **82.** $y = -\dfrac{5}{2}x + 1$

83.

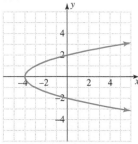

84.

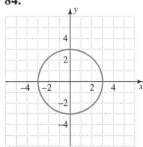

85.

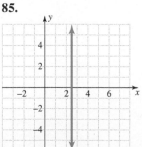

86.

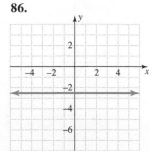

87.

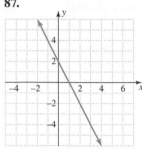

88.

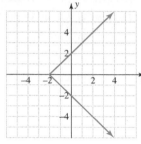

Find the requested information.

89. $f(-2)$ if $f(x) = -5x^2 + 7x + 1$

90. $f(-1)$ if $f(x) = -x^2 + 6x + 18$

91. The function $f(x)$ is given by Y_1. Find **a.** $f(4)$ and **b.** x if $f(x) = -10$.

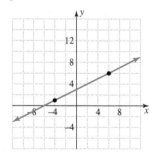

92. The function $c(x)$ is given by Y_1. Find **a.** $c(1)$ and **b.** x if $c(x) = 3.5$.

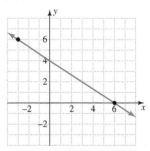

93. The graph of $f(x)$ is given. Find **a.** $f(6)$ and **b.** x if $f(x) = 0$.

94. The graph of $f(x)$ is given. Find **a.** $f(0)$ and **b.** x if $f(x) = 0$.

95. Find $f(4.2)$ if $f(x) = 1.5x - 3.2$.

96. Find $f(-0.8)$ if $f(x) = -2.4x + 8.4$.

Write each linear equation in two variables in function notation. Then find $f(0), f(3)$, and $f(-2)$ and write the corresponding ordered pairs.

97. $9x - 2y = 10$ **98.** $7x + 14y = 4$

Solve each problem.

99. Brasil has a new job as a computer information analyst. His starting salary is $52,000. He will receive a raise of $1600 each year he works with the company.

    a. Write a linear function $f(x)$ that represents Brasil's salary, where x is the number of years he has worked with the company.

    b. Find $f(5)$ and interpret the answer.

100. Lois has a new job as a math instructor at a local community college. Her starting salary is $41,500 with a regular teaching load of 30 credit hours. If she teaches beyond the regular load, she earns $950 per credit hour.

    a. Write a linear function $f(x)$ that represents Lois' yearly income, where x is the number of extra credit hours she teaches.

    b. If Lois teaches two additional 4-credit-hour courses, what is her income?

    c. Find $f(6)$ and interpret the answer.

101. Lee registers for his first semester in college. His tuition includes fees of $565 plus $135 per credit hour.

    a. Write a linear function $f(x)$ that represents Lee's total tuition, where x is the number of credit hours Lee takes during the first semester.

    b. What is Lee's tuition if he registers for 12 credit hours?

    c. Find $f(15)$ and interpret the answer.

102. Aziz purchased a new car for $18,500. The car's value decreases by $1250 each year.

    a. Write a linear function $f(x)$ that represents the value of the car, where x is the age of the car in years.

    b. What is the car's value after 4 yr?

    c. Find $f(8)$ and interpret the answer.

You Be the Teacher!

Correct each student's errors, if any.

103. Determine if each relation is a function.

    a. $\{(3, 1), (7, 5), (1, 2), (5, 2), (0, -2)\}$

    b. $\{(1, 3), (5, 7), (2, 1), (2, 5), (-2, 0)\}$

    Vivian's work: **a.** a function and **b.** not a function

104. Find $f(-3)$ if $f(x) = 12x - 5$.

    William's work:
    $f(-3) = -3(12x - 5) = -36x + 15$

105. Find $f(0)$ and x if $f(x) = 0$ from the given graph.

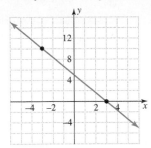

Brian's work: $f(0) = 3$ and $f(5) = 0$

106. Find $f(0)$ and x if $f(x) = 0$ from the given table.

X	Y₁
-2	-6
-1	-4.5
0	-3
1	-1.5
2	0
3	1.5
4	3

X= -2

Marla's work: $f(-3) = 0$ and $f(0) = 2$

 Calculate It!

Input the function in a graphing calculator. Use both the TRACE and TABLE features to find $f(0)$ and x if $f(x) = 0$.

107. $f(x) = 0.25x - 1.75$

108. $f(x) = -0.06x^2 + 0.5x + 0.36$

109. $f(x) = |2x - 1| - 3$

110. The table shows the points on the graph of a function. Determine which table shows a line that contains the points $(0, -5)$ and $(-2, 4)$.

a.

X	Y₁
-4	3
-3	3.5
-2	4
-1	4.5
0	5
1	5.5
2	6

X= -4

b.

X	Y₁
-3	-18.5
-2	-14
-1	-9.5
0	-5
1	-.5
2	4
3	8.5

X=3

c.

X	Y₁
-4	13
-3	8.5
-2	4
-1	-.5
0	-5
1	-9.5
2	-14

X= -4

GROUP ACTIVITY / Use a Linear Equation to Model Population Growth

1. Go to http://www.google.com/publicdata?ds=uspopulation and select a state or expand the state to select a specific county.

2. After you make your selection, a graph will appear in the main window. Use your cursor to trace the graph to see the population for each year. Write two ordered pairs of the form (year, population) for the state or county that you have selected.

3. Use the two ordered pairs from step 2 to write a linear equation that models the population of the selected state/county.

4. What is the slope of the model? What does it mean in the context of the problem?

5. What is the y-intercept of the model? What does it mean in the context of the problem?

6. Use the model to predict the population of your selected state/county in the year 2020.

Linear Equations in Two Variables

💡 **What's the big idea?** Now that you have completed Chapter 3, you should be able to see a connection between a linear equation in two variables, solutions of the equation, and the graph of the equation. You should also be able to identify important characteristics of the linear equation from any of its different forms.

The Tools

Listed below are the key terms, skills, formulas, and properties you should know for this chapter.

The page reference is provided if you need additional help with the given topic. The Study Tips will assist in your preparation for an exam.

Study Tips

1. Learn all of the terms, formulas, and properties. Make flash cards and have someone quiz you.
2. Rework problems from the exercises and also the ones you worked in class. Work additional problems from the review exercises.
3. Review the summaries of key concepts.
4. Work the chapter test.
5. Be sure to review the online resources for additional study materials.

Terms

Domain 265	Paired data 200	Scatter plot 200
Function notation 270	Parallel lines 242	Slope 227
Function 267	Perpendicular lines 242	Slope-intercept form 230
Horizontal line 216	Point-slope form 250	Solution of an equation 194
Initial value 239, 256	Quadrants 196	Vertical line 216
Linear equation in two variables 209	Range 265	x-axis 196
Linear function 272	Rate of change 239	x-intercept 213
Ordered pair 194	Rectangular coordinate system 196	y-axis 196
Origin 196	Relation 265	y-intercept 213

Formulas and Properties

- Equation of a horizontal line   216
- Equation of a vertical line   216
- Linear equation in two variables   209
- Point-slope form of a line   250
- Slope formula   227
- Slope-intercept form of a line   230
- Standard form of a line   209
- Vertical line test   269

CHAPTER 3 / SUMMARY

How well do you know this chapter? Complete the following questions to find out. Take a look back at the section if you need help.

SECTION 3.1 Equations and the Rectangular Coordinate System

1. A solution of an equation in two variables is a(n) _____ ____ that makes the equation a(n) ____ statement.

2. In an ordered pair, the first coordinate tells you how to move ___ or ____ from the origin. The second coordinate tells you how far to move ___ or ____ from the ____.

3. To identify an ordered pair of a point on the coordinate system, determine how far ____ or ____ the point is from the origin and how far ___ or ____ the point is from the ____.

4. _____ are the four regions formed by the two axes of the rectangular coordinate system. They are labeled with _____ numerals moving in the _____ direction.

5. The graph of an equation in two variables is composed of all _____ ____ that are _____ of the equation. An equation in two variables has _____ many solutions.

6. If a point lies on the graph of an equation, it is a(n) _____ of the equation.

7. We used _____ and _____ to solve application problems. From these representations, _____ _____ must be identified to answer questions.

SECTION 3.2 Graphing Linear Equations

8. The standard form of a linear equation in two variables is _____.

9. To graph a line by plotting points, ___ points are required. However, a total of ____ points is recommended so that one of the points serves as a check.

10. The point where a graph crosses the x-axis is called the _____. To find this point, set _ = 0 and solve the resulting equation. The point where a graph crosses the y-axis is called the _____. To find this point, set _ = 0 and solve the resulting equation.

11. If the value of the constant in a linear equation equals zero, then the graph of the equation will pass through the _____. To graph such an equation, _____.

12. The equation of a horizontal line has the form _____.

13. The equation of a vertical line has the form _____.

SECTION 3.3 The Slope of a Line

14. The slope of a line measures the _____ of the line.

15. The slope is defined as the ratio of the _____ change to the _____ change. It is also called ___ over ___.

16. The slope-intercept form of a line is _____. The m represents the _____ and b represents the _____.

17. To graph a line using the slope and y-intercept, plot the _____ first. From this point, use the _____ to move

to another point on the line. For instance, a slope of $\frac{4}{3}$ would mean to _____ and _____.

18. A horizontal line has a slope of ____ and a vertical line has a slope that is _____.

19. The slope formula is _____.

20. A positive slope corresponds to a line that is _____. A negative slope corresponds to a line that is _____. A slope of zero corresponds to a _____ line. An undefined slope corresponds to a _____ line.

SECTION 3.4 More About Slope

21. In application problems, the value of b represents the _____ value. The value m represents the _____.

22. Two lines are _____ if they have the same ____.

23. Two lines are _____ if their slopes are _____ reciprocals.

SECTION 3.5 Writing Equations of Lines

24. To write the equation of a line, we must know the _____ and at least one _____. The _____ form of a line or the _____ form of a line can be used to find the equation.

SECTION 3.6 Functions

25. A(n) _____ is a set of ordered pairs. The set of x-values is the _____, and the set of y-values is the ____.

26. A _____ is a relation in which each input value corresponds to _____ output value.

27. The _____ line test can be used to determine if a graph represents a function. If every vertical line touches at most ___ ____ on the graph, then the graph is a function.

28. In function notation, we use the symbol ___ for y.

29. To write a linear equation in two variables in function notation, we must first solve the equation for __.

30. If $f(a) = b$, then the point _____ is on the graph of $f(x)$.

CHAPTER 3 / REVIEW EXERCISES

SECTION 3.1

Determine algebraically if the given ordered pair is a solution of the equation. (See Objective 1.)

1. $(1, -1)$; $3x - 2y = 5$ 2. $(-3, -0.75)$; $x + 6y = -7.5$

3. $(-2, 1)$; $y = |3 - x|$ 4. $(4, 1)$; $y = -x^2 + 8x - 15$

5. $(-2, 0)$; $y = \dfrac{x - 2}{x + 2}$ 6. $(0, 3)$; $y = \dfrac{2x - 3}{x + 1}$

Plot each ordered pair on the Cartesian coordinate system and state the quadrant or axis where each point is located. (See Objective 2.)

7. $(-5, 1)$

8. $(2, 0)$

9. $(-3, -2.5)$

10. $\left(\dfrac{7}{2}, 4\right)$

11. $(0, -4)$

12. $(1, -6)$

Identify each point plotted on the coordinate system and state the quadrant or axis where each point is located. (See Objective 2.)

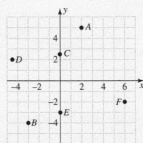

13. A

14. B

15. C

16. D

17. E

18. F

Identify the quadrant or axis where each point is located. (*See Objective 2.*)

19. $(-1, 0)$ **20.** $(2, -7)$

21. $(-3.2, -1.6)$ **22.** $(5.2, 0.8)$

Graph each equation using a table of solutions. (*See Objective 3.*)

23. $y = 5 - 3x$ **24.** $y = |5 - 2x|$

25. $y = 4 - x^2$ **26.** $y = x^3$

Use the given graph of an equation to determine if each ordered pair is a solution of the equation. (*See Objective 4.*)

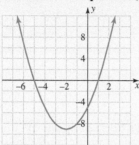

27. $(-5, 0)$ **28.** $(0, 4)$

29. $(2, 4)$ **30.** $(1, 0)$

Use the given graph to complete each ordered pair that is a solution. (*See Objective 4.*)

31. $(\_\_\_, 0)$ **32.** $(0, \_\_\_)$

33. $(1, \_\_\_)$ **34.** $(\_\_\_, 5)$

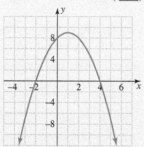

Solve each problem. (*See Objective 5.*)

35. The following graph shows the expenditures for national health in the United States between 1995 and 2006. (Source: http://worldalmanac.com)

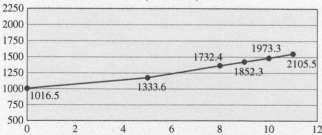

National Health Expenditures in United States, in Billions (1995–2006)

    a. Write an ordered pair for each point labeled on the graph.

    b. Between what consecutive years did the expenditures increase the least? By how much?

    c. Based on the graph, how would you describe expenditures for national health?

SECTION 3.2

Graph each equation. Identify at least two points on the graph. (*See Objectives 2–4.*)

36. $8x + y = 16$ **37.** $3y + 6 = 0$

38. $y + 3x = 0$ **39.** $5x - 10 = 0$

40. $y = 0.2x + 1.5$ **41.** $y = -\dfrac{2}{3}x - 1$

42. $0.6x - 0.1y = 1.2$

Use the equation to answer each question. (*See Objective 5.*)

43. The emissions of carbon monoxide, the principal air pollutants in the United States during 1970–2008, are on the decline. The emissions of carbon monoxide in millions of tons can be modeled by the equation $y = -3.36x + 213$, where x is the number of years after 1970. (Source: U.S. Environmental Protection Agency, Office of Air Quality Planning and Standards)

    a. Find the y-intercept and interpret its meaning.

    b. Find the x-intercept and interpret its meaning.

    c. What was the emission of carbon monoxide in 2005?

    d. What was the emission of carbon monoxide in 2009?

    e. Use the intercepts to graph the equation.

SECTION 3.3

Find the slope of the line that passes through the given points or that is defined by the given equation. (*See Objectives 1, 3, and 4.*)

44. $(3, -2)$ and $(3, 1)$ **45.** $y = 4$

46. $\left(-2, -\dfrac{2}{5}\right)$ and $\left(4, \dfrac{3}{5}\right)$ **47.** $5x - 6y = 1$

48. $0.7x + 0.14y = 2.1$

49.

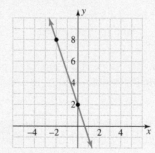

50.

x	y
-2	-3
-1	-1
0	1
1	3
2	5

Graph each line using the given slope and point. Label two points on the line. (*See Objective 3.*)

51. $m = \dfrac{2}{7}, (0, -2)$ **52.** $m = -1, (0, 5)$

53. $m = -0.5, (2, -6)$ **54.** $m = $ undefined, $(-3, 1)$

55. $m = 0, (1, 2)$ **56.** $m = -4, (0, 0)$

SECTION 3.4

Interpret the meaning of the slope and *y*-intercept in the context of each situation. (*See Objective 1.*)

57. Monica's car is in the shop, so she is using a rental car. The total cost in dollars for Monica to rent a compact car is given by $y = 135x$, where x is the number of weeks the car is rented.

58. The per capita income in dollars can be modeled by the equation $y = 1155.6x + 23{,}507$, where x is the number of years after 1995. (Source: www.infoplease.com)

59. What is the pitch of a roof that rises 10 in. for every horizontal move of 6 in.? (*See Objective 2.*)

60. Determine if the lines are parallel, perpendicular, or neither: (*See Objective 3.*) $y = -\dfrac{2}{3}x + 1$ and $y = \dfrac{3}{2}x - 2$.

61. Graph the line that is parallel to $y = -x + 3$ and passes through the point $(2, -6)$. (*See Objective 4.*)

62. Graph the line that is perpendicular to $y = -2x + 3$ and passes through the point $(-4, -1.5)$. (*See Objective 4.*)

SECTION 3.5

Write the equation of each line described. Express each answer in slope-intercept form if possible. (*See Objectives 1–4.*)

63. $m = -\dfrac{1}{4}, (-8, 1)$

64. $(5.2, -2.3)$ and $(7.2, -4.2)$

65. $m = $ undefined, $(-1, 2)$

66. $(1, -2)$, parallel to $y = 4x - 1$

67. $(2, -4)$ and $(2, 6)$

68. x-intercept $(6, 0)$ and y-intercept $(0, -1)$

69. $(4, 1)$, perpendicular to $2x + 3y = 1$

70. $(-1, 0)$, perpendicular to $x = -5$

Solve each problem. (*See Objective 5.*)

71. Amy has a new job as a graphic designer. Her starting salary is $43,500. She will receive a raise of $2160 each year she works with the company.

    a. Write a linear equation that represents Amy's salary where x is the years she has worked with the company.

    b. What is Amy's salary after 4 yr of working for the company?

    c. How long does she have to work for the company to earn a salary of $69,420?

72. Adam registers for his first semester in college. His tuition, includes fees of $525 plus $145 per credit hour.

    a. Write a linear equation that represents Adam's total tuition, where x is the number of credit hours Adam will take.

    b. What is Adam's tuition if he registers for 12 credit hours?

    c. How many hours is Adam taking if his tuition is $1830?

SECTION 3.6

Determine if each relation is a function. If not, explain why. (*See Objective 2.*)

73. The following table shows the richest Americans in 2009 by their net worth. (Source: http://www.infoplease.com/us/statistics/richest-americans.html)

Name	Net Worth, in Billions
Bill Gates	$50
Warren Buffett	$40
Larry Ellison	$27
Christy Walton	$21.5
Jim Walton	$19.6
Alice Walton	$19.3
S. Robson Walter	$19
Michael Bloomberg	$27.5
Charles Koch	$16
David Koch	$16

74. The following table shows the gas mileage of the "meanest" (i.e., most polluting) vehicles of 2010.

Meanest Vehicles	Miles per Gallon
1. Lamborghini Mucielago	18
2. Bugatti Veyron	8
3. Bentley Azure/Brooklands	9
4. Maybach 57S	10
5. Dodge Ram 2500 Mega Cab	12
6. Ford F-250	12
7. Bentley Continental	10
8. Ferrari 612 Scaglietti	9
9. Mercedes-Benz ML 63 AMG	11
10. Chevrolet Suburban K2500	13

(Source: http://www.infoplease.com)

75. $\{(2, 8), (2, -1), (1, 5), (-2, 1), (-4, 0)\}$

76. Assign to each student the grade earned in their beginning algebra class.

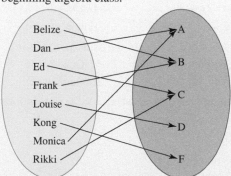

77. $x = 1$

78. $y = -\dfrac{1}{5}x$

Use the vertical line test to determine if each relation is a function. (*See Objective 3.*)

79.
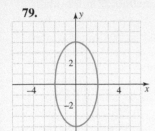

80.
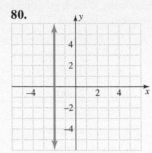

81.
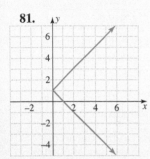

82.
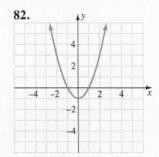

Find the requested information. (*See Objective 4.*)

83. $f(-3)$ if $f(x) = -2x^2 + x + 3$

84. The function $f(x)$ is given by Y_1. Find **a.** $f(-2)$ and **b.** x if $f(x) = -14$.

X	Y1
-2	-19
-1	-10
0	-5
1	-4
2	-7
3	-14
4	-25

X=4

85. The graph of $f(x)$ is given below. Find **a.** $f(0)$ and **b.** x if $f(x) = 4$.

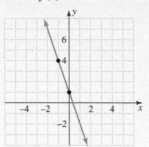

86. Find $f(1.5)$ if $f(x) = 3.4x - 0.5$.

Write each linear equation in two variables in function notation. Then find $f(0), f(3)$, and $f(-2)$ and write the corresponding ordered pairs. (*See Objective 5.*)

87. $2x - 7y = 21$ **88.** $y = |4x + 1|$

Solve each problem. (*See Objective 5.*)

89. Rikki has a new job as a math instructor at a local community college. Her starting salary is $43,200 with a regular annual teaching load of 30 credit hours. If she teaches beyond the regular load, she earns $1020 per credit hour.

   a. Write a linear function $f(x)$ that represents Rikki's yearly income, where x is the number of extra credit hours Rikki teaches.

   b. If Rikki teaches two 3-credit-hour courses in addition to her regular load, what is her income?

   c. Find $f(9)$ and interpret the answer.

90. Wong purchases a new car for $16,400. The car's value decreases by $1325 each year.

   a. Write a linear function that represents the value of the car, where x is the age of the car in years.

   b. What is the car's value after 3 yr?

   c. Find $f(5)$ and interpret the answer.

CHAPTER 3 TEST / LINEAR EQUATIONS IN TWO VARIABLES

1. The point $(2, -4)$ is a solution of the equation
   a. $y = 2x + 8$ **b.** $x = -4$
   c. $2x - y = 8$ **d.** $y = |x - 6|$

2. If $a > 0$ and $b < 0$, then the point (a, b) lies in
   a. Quadrant I   **b.** Quadrant II
   c. Quadrant III   **d.** Quadrant IV

Graph each equation by creating a table of solutions.

3. $y = x^2 + 1$

4. $y = |x + 2|$

5. Is the point $(-1, 1)$ a solution of the equation shown in the graph?

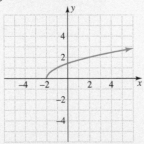

6. The average price (rounded to the nearest dollar) of brand name prescription drugs for certain years is shown in the table.

Year	1995	2000	2005	2006	2007
Average price of brand name drugs	40	65	98	107	120

   a. Write an ordered pair for each year, where x is the number of years after 1995 and y is the average price.

   b. Interpret the meaning of the first and last ordered pairs.

   c. Plot the ordered pairs to form a scatter plot of the given data.

7. Graph each line by finding the x- and y-intercepts.

   a. $2x - y = 6$ **b.** $y = \dfrac{1}{5}x + 1$

   c. $x + 3y = 0$

8. The equation $x = 5$ is a _____ line.

9. The equation $y = -1$ is a _____ line.

10. Suppose an airplane descends at a rate of 400 ft/min from an altitude of 8000 ft above the ground.

   a. Write a linear equation that represents the plane's altitude, y, after x min.

   b. Find the x-intercept and state its meaning.

   c. Find the y-intercept and state its meaning.

11. State the slope of each line.

   a. $y = \dfrac{4}{5}x + 3$ **b.** $3x + y = 9$

   c. $x = 4$ **d.** $y = 7$

   e. Through $(4, -1)$ and $(6, 5)$

   f. Parallel to $y = 7x + 4$

   g. Perpendicular to $y = -3x$

   h.

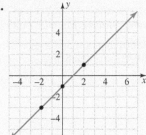

12. Graph the line $y = -\dfrac{3}{2}x + 4$ using the slope and y-intercept. Label the y-intercept and at least one other point.

13. Determine if the lines $5x + 2y = 10$ and $4x - 10y = -4$ are parallel, perpendicular, or neither.

14. The equation $y = 138.41x + 2961.8$ models the average undergraduate cost of attending a public 2-yr college x years after 1986. (Source: http://nces.ed.gov/fastfacts)

   a. What are the slope and y-intercept?

   b. What do they mean in the context of this problem?

   c. Use this model to find the cost of attending a 2-yr college in 2016.

15. Write the equation of the line that satisfies the given conditions.

   a. $m = \dfrac{3}{4}$ and passes through $(0, -6)$

   b. $m = -2$ and passes through $(-7, 4)$

   c. passes through $(4, -1)$ and $(6, 5)$

   d. passes through $(2, -8)$ and parallel to $y = 4x - 1$

   e. passes through $(3, 0)$ and perpendicular to $y = -3x + 9$

16. Suppose the enrollment of Einstein Community College was 16,000 in 1998 and 22,000 in 2004. If x is the number of years after 1998 and y is the college's enrollment, write a linear equation that represents Einstein's enrollment x years after 1998.

17. Which of the following relations is not a function?

   a. $\{(-2, 4), (-1, 1), (0, 0), (1, 1), (2, 4)\}$

   b. $\{(x, y) \mid x = \text{person and } y = \text{person's Social Security number.}\}$

   c. 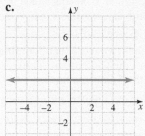   **d.**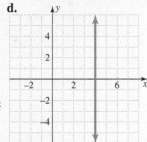

18. Find $f(0)$ and $f(-2)$ for $f(x) = x^2 - 3x + 2$.

19. Write the linear equation $4x - 7y = -14$ in function notation. Then find $f(-14)$. Write the corresponding ordered pair.

20. Explain what it means for a relation to be a function. Provide a real-life example of a function.

CUMULATIVE REVIEW EXERCISES / CHAPTERS 1–3

1. Evaluate each absolute value expression. (*Section 1.1, Objective 5*)

   a. $|6|$ **b.** $\left|-\dfrac{3}{4}\right|$ **c.** $-|-24|$

2. Perform the each operation and simplify each result. (*Section 1.2, Objectives 2–7*)

   a. $\dfrac{15}{16} + \dfrac{7}{12} - \dfrac{17}{18}$

   b. $6\dfrac{2}{5} \cdot 3\dfrac{3}{4} \div 12$

   c. $4\dfrac{1}{5} - 1\dfrac{2}{3} + \dfrac{4}{15}$

3. Use the order of operations to simplify each numerical expression. (*Section 1.3, Objectives 1 and 2*)

   a. $-\left(-\dfrac{3}{2}\right)^3$ **b.** $\dfrac{1}{6}(10 - 8)^3 - \dfrac{5}{3}$

   c. $2(7)^2 - 5(12) + 1$ **d.** $5[20 - 2|13 - 2(4 - 1)|]$

   e. $\dfrac{13 + \sqrt{13^2 + 5(13)(6)}}{2 \cdot 6}$

4. Evaluate each expression for the given values of the variables. (*Section 1.3, Objective 3*)

   a. $\dfrac{2x + 3}{x - 1}$ for $x = 0, 1, 2, 3$

   b. $b^2 - 4ac$ for $a = 5, b = 2, c = 15$

5. The height of a baseball hit upward with an initial velocity of 112 ft/sec from an initial height of 8 ft is represented by the expression $-16t^2 + 112t + 8$, where t is the number of seconds after the ball has been hit. What is the height of the ball after 4 sec? (*Section 1.3, Objective 6*)

6. Perform the indicated operation and simplify. (*Sections 1.4 and 1.5, Objectives 1 and 2*)

   a. $5.4 - 9.2 + (-8.4) - (-11.5)$

   b. $\dfrac{5}{6} - \dfrac{3}{4} - \left(-\dfrac{1}{4}\right)$

   c. $30 - \{7 - [4 - (2 - \sqrt{20 + 5})]\}$

Write the mathematical expression needed to solve each problem and then answer any questions.
(*Sections 1.4 and 1.5, Objective 3*)

7. Sherry has $235.65 in her checking account. She deposits her paycheck of $567.45. Sherry writes a check for phone bill for $47.82, for groceries for $135.93, and then for rent for $475. What is Sherry's checking account balance?

8. Selena wants to mix a 4% salt solution with a 30% solution to get a 25% solution. Let x oz represent the volume of the 4% salt solution. If Selena wants 50 oz of 25% salt solution, write an expression that represents the volume of the 30% salt solution.

9. In a rectangle, the length is two more than four times the width. If w represents the width, write an expression that represents the length of the rectangle.

10. Find the complement and the supplement of an angle whose measure is 63°.

11. Perform the indicated operation and simplify. (*Section 1.6, Objectives 1–3*)

   a. $(-5.4)(3.6)$

   b. $(-36)\left(\dfrac{5}{16}\right)\left(-\dfrac{8}{15}\right)$

   c. $-(-5)^3$

   d. $\dfrac{-7}{0}$

   e. $6(-15)(-4)$

   f. $\left(\dfrac{15}{14}\right) \div \left(-\dfrac{10}{21}\right)$

12. Evaluate each expression for the given values of the variables. (*Section 1.6, Objective 4*)

   a. $b^2 - 4ac$ for $a = -2, b = 1, c = 5$

   b. $\dfrac{|3 + x|}{x - 1}$ for $x = -1, 0, 1$

13. Apply the commutative, associative, and/or distributive properties to simplify each expression. (*Section 1.7, Objectives 2 and 3*)

   a. $13 + c - (-25)$

   b. $-4(x + 3y - 8)$

   c. $\left(x - \dfrac{11}{6}\right) + 2$

   d. $8\left(\dfrac{3}{4}x - \dfrac{1}{2}\right)$

14. Simplify each algebraic expression. (*Section 1.8, Objectives 3 and 4*)

   a. $3(4x - 5) - 16x$

   b. $-(2x - 7) - 3(-2x + 5)$

   c. $4.8 - 0.3(-0.5x + 1.4)$

   d. $15\left(\dfrac{2}{5}x - \dfrac{1}{3}\right) - 10\left(\dfrac{1}{2}x + \dfrac{4}{5}\right)$

   e. $5x^2 - 7(x + x^2) - 8x$

15. Determine whether each of the following is an expression or an equation. (*Section 2.1, Objective 1*)

   a. $5x - 28 = 2x - 12$

   b. $x - 10 + 3x + 5$

16. Determine if the given numbers are solutions of the equation. (*Section 2.1, Objective 2*)

   a. Is $x = -2, x = 0$, or $x = 1$ a solution of $7x - 3 = -x - 19$?

   b. Is $x = -2, x = 0$, or $x = 3$ a solution of $x - 2 = -4x + 13$?

17. For each exercise, define the variable, write an equation, and solve the problem. (*Section 2.1, Objectives 3 and 4*)

   a. The sum of a number and 32 is the same as three times the number. Find the number.

   b. The difference of three times a number and -5 equals 7 less than the number. Find the number.

   c. One angle is 14° more than another angle. Their sum is 90°. Find the measures of each angle.

18. Solve each equation. (*Section 2.2, Objectives 2 and 3; Sections 2.3 and 2.4, Objectives 1 and 2*)

   a. $3(x - 2) - 28 = 5(x - 1) + 9$

   b. $5(1 - x) - 7x = 13 - (2 - 4x)$

   c. $-\dfrac{2}{5}(5x + 10) + 3 = x - 5$

   d. $-0.2y + 14.1 = 2.3y + 4.1$

For each exercise: (1) assign a variable to the unknown, (2) write an equation that represents the situation, (3) solve the equation, and (4) answer the question using complete sentences. (*Section 2.2, Objective 4; Section 2.3, Objectives 3 and 4*)

19. The cost y to rent a midsize car for one day is $68 plus $0.36 per mile. This daily cost is represented by the equation $y = 68 + 0.36x$, where x is the number of miles driven. How many miles has the car been driven if the cost of the rental car is $199.40?

20. The sum of three consecutive odd integers is 75. Find the integers.

21. The sum of the two smaller of three consecutive integers is the same as 30 less than three times the largest integer. Find the integers.

22. Solve each equation. If the equation is a contradiction, write the solution as ∅. If the equation is an identity, write the solution as ℝ. (*Section 2.4, Objectives 1–4*)

   a. $7(2y - 1) - (y - 5) = 3(y + 2) + 8$

   b. $0.7x + 1.1x = 7.2$

   c. $0.015x + 0.024(480 - x) = 8.82$

   d. $-\dfrac{1}{2}(x - 3) = \dfrac{1}{3}(2x + 9) + \dfrac{1}{6}$

23. The formula $A = P(1 + r)^t$ is used to calculate the amount of money in an account at the end of t years if P is invested at an annual interest rate, r. (*Section 2.5, Objective 1*)

   a. Find the value of A if $960 is invested for 3 yr at 1.4% annual interest.

   b. Find the value of P if $2,930.88 is expected to be accumulated at the end of 2 yr at 1.6% annual interest.

24. The revenue R, unit price p, and sale level x of a product are related by the formula $R = xp$. (*Section 2.5, Objective 1*)

   a. Find R if $p = 86.50 and $x = 40$.

   b. Find p if $R = 3212.50 and $x = 125$.

25. Solve each formula for the specified variable. (*Section 2.5, Objective 3*)

   a. $V = lwh$ for w **b.** $C = 8l + 3w$ for w

   c. $5x - 9y = 18$ for y **d.** $0.04x + 0.03y = 7.2$ for x

26. Find the measure of each unknown angle. (*Section 2.5, Objective 5*)

   a. Find the measure of an angle whose complement is 40° less than the measure of the angle.

   b. The supplement of an angle is 12° more than three times its complement. Find the measure of the angle.

27. Find the measure of each angle labeled in the figure. (*Section 2.5, Objectives 4 and 5*)

   a.

$(x + 11)°$ $(3x - 31)°$

   b.

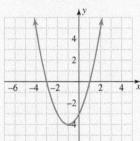

$(5x - 12)°$ $(2x + 48)°$

28. In triangle ABC, the measure of angle B is 36° less than the measure of angle A. The measure of angle C is 21° more than the measure of angle A. Find the measure of each angle in the triangle. (*Section 2.5, Objective 5*)

Write an equation that represents each situation, solve the equation, and write the answer in complete sentences. (*Section 2.6, Objectives 1–5*)

29. A 46-in. LED-LCD HDTV is on sale for $1202.50. This is 35% off the original price. What is the original selling price of the TV?

30. A suit is on sale for $138.75. If this price is 25% off the original price, what is the original price of the suit?

31. Sue invested $3000 in two accounts. She invested $1800 in a mutual fund that pays 2.1% annual interest and the remaining account in a money market fund that pays 1.2% annual interest. How much interest did she earn from the two accounts in 1 yr?

32. Solve each inequality. Graph the solution set and write the solution set in interval notation and in set-builder notation. (*Section 2.7, Objectives 1–4*)

   a. $0.3a - 1.2 \geq -3.6$ **b.** $-7 < 2b - 15 < 19$

   c. $2x + (x - 9) \geq 8x + 16$

33. Determine algebraically if the given ordered pair is a solution of the equation. (*Section 3.1, Objective 1*)

   a. $(2, -5)$; $2x - y = 15$

   b. $(3, -1)$; $y = -x^2 - 3x + 1$

   c. $(1, 5)$; $y = |4x + 1|$

34. Plot each ordered pair on the Cartesian coordinate system and state the quadrant or the axis where each point is located. (*Section 3.1, Objective 2*)

   a. $(2, -7)$ **b.** $(-3, -9)$

   c. $(-6, 0)$ **d.** $(5, -3)$

   e. $(0, 5)$

35. Graph each equation using a table of solutions. (*Section 3.1, Objective 3*)

   a. $y = 2x - 6$ **b.** $y = 3 - x^2$

   c. $y = |x| + 3$ **d.** $y = 4 - |x|$

36. Use the given graph to complete each ordered pair that is a solution. (*Section 3.1, Objective 4*)

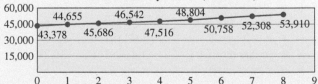

   a. $(\_\_\_\_, 0)$ **b.** $(0, \_\_\_\_)$

   c. $(-1, \_\_\_\_)$

37. The graph shows the average annual salary of full-time teachers in public elementary and secondary schools, 2000 to 2008. (*Section 3.1, Objective 5*) (Source: http://nces.ed.gov)

Average Annual Salary of Teachers in Public Elementary and Secondary Schools (2002–2008)

60,000	44,655		46,542		48,804				
45,000	43,378	45,686		47,516		50,758	52,308	53,910	
30,000									
15,000									
0	1	2	3	4	5	6	7	8	9

   a. Write an ordered pair for each point labeled on the graph.

   b. Between what consecutive years did the average annual salary increase most? By how much?

38. The table lists the annual profit (in billions of dollars) for Delta Airlines between 2000 and 2006. (*Section 3.1, Objective 5*) (Sources: *The Atlanta Journal-Constitution*, Sept. 15, 2005 and www.delta.com)

Years After 2000	0	1	2	3	4	5	6
Annual profit (in billions)	1.23	−1.21	−1.27	−0.773	−5.2	−1.46	0.058

   a. Write an ordered pair (x, y) that corresponds to each year between 2000 and 2006, where x is the number of years after 2000 and y is the profit of Delta Airlines (in billions of dollars).

   b. Interpret the meaning of the first and last ordered pairs from part (a) in the context of the problem.

   c. In what year was the profit the highest? The lowest?

   d. Make a scatter plot of the data.

39. Graph each equation. Identify at least two points on the graph. (*Section 3.2, Objectives 2–4*)

   a. $2x + 7y = 14$ **b.** $y + 5 = 0$ **c.** $x − 6 = 0$
   d. $3x + y = 0$ **e.** $x − 0.5y = 1$ **f.** $y = −2x + 1$
   g. $0.4x − 0.5y = 2$

40. The number of full-time classroom teachers (in thousands) in elementary and secondary schools in the United States can be modeled by the equation $y = 32.5x + 3649.1$, where x is the number of years after 2008. (*Section 3.2, Objective 5*) (Source: http://www.census.gov/compendia/statab/2010/tables/10s0216.pdf)

   a. Find the y-intercept and interpret its meaning. Is this realistic?

   b. How many full-time classroom teachers in elementary and secondary schools does the model predict in 2010?

   c. How many full-time classroom teachers in elementary and secondary schools does the model predict in 2014?

   d. Use the intercepts to graph the equation.

41. Find the slope of each line described. (*Section 3.3, Objectives 1, 3, 4*)

   a. $y = −\dfrac{2}{3}x + 4$

   b. passes through $(2, −4)$ and $(4, 0)$

   c. $y = 6$ **d.** $x = −5$

   e. passes through $(−3.2, −1.4)$ and $(1.2, −10.2)$

   f. $5x − 3y = 2$

   g.

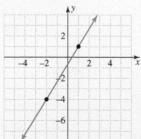

42. Graph each line described. Label two points on the line. (*Section 3.4, Objective 2*)

   a. $m = 2, (0, 5)$ **b.** $m = 5, (0, 3)$
   c. $m = −3, (0, 1)$ **d.** $m = −4, b = 3$
   e. $m = 0.5, b = −2$ **f.** $m =$ undefined, $(−2, 1)$
   g. $m = 0, (3, −4)$

43. The number of associate degrees (in thousands) conferred in all higher education institutions can be modeled by the equation $y = 10.93x + 73{,}764$, where x is the number of years after 2008. Interpret the meaning of the slope and y-intercept in the context of the problem. Use the slope and y-intercept to create a table of three ordered pairs that satisfy the given equation. (*Section 3.4, Objective 1*)

44. Determine if each pair of lines is parallel, perpendicular, or neither. (*Section 3.4, Objective 3*)

   a. $y = \dfrac{3}{2}x + 1$ and $y = \dfrac{3}{2}x − 4$

   b. $2x + 5y = 1$ and $2x − 5y = 2$

   c. $y = 1$ and $x = 2$

   d. $y = 0.5x$ and $y = −2x$

45. Graph the line that is parallel to $y = −3x + 2$ and passes through the point $(−4, 1)$. (*Section 3.4, Objective 4*)

46. Graph the line that is perpendicular to $y = 2x − 3$ and passes through the point $(−2, 3)$. (*Section 3.4, Objective 4*)

47. Write the equation of the line described. Express your answer in slope-intercept form and in standard form if possible. (*Section 3.5, Objectives 1–4*)

   a. $(−5, 6)$ and $(2, −1)$

   b. $m =$ undefined, $(3, 6)$

   c. $m = 0, (0, 12)$

   d. $(−2, 4)$, parallel to $2x + y = 5$

   e. $(3, −5)$, perpendicular to $x + 3y = 1$

   f. x-intercept $(−4, 0)$ and y-intercept $(0, 3)$

   g. $m = −4$, x-intercept: $(2, 0)$

48. Michael enrolls in a fitness club. There is a one-time membership fee of $250 plus a monthly charge of $40. (*Section 3.5, Objective 5*)

   a. Write a linear equation that represents the total cost of joining the fitness club, where x is the number of months Michael is a member.

   b. How much money has Michael spent for his fitness club membership if he has been a member for 1 year?

   c. How long has Michael been a member of the club if he pays the fitness club a total of $1450?

49. State the domain and range of each relation. Determine if each relation is a function. (*Section 3.6, Objectives 1–3*)

a. The following table shows the average number of paid vacation days per year employees receive in nine countries. (Source: http://www.infoplease.com)

Countries	Average Number of Paid Vacation Days per Year
Italy	42
France	37
Germany	35
Brazil	34
United Kingdom	28
Canada	26
Korea	25
Japan	25
United States	13

b. The following table shows the green score for the "greenest" vehicles in 2010. (Source: http://www.infoplease.com)

Greenest Vehicles	Green Score
1. Honda Civic CX	57
2. Toyota Prius	52
3. Honda Civic Hybrid	51
4. Smart For Two Convertible/ Coupe	50
5. Honda Insight	50
6. Ford Fusion Hybrid/ Mercury Milan Hybrid	47
7. Toyota Yaris	46
8. Nissan Altima Hybrid	46
9. Mini Cooper	45
10. Chevrolet Cobalt XFE/ Pontiac G5 XFE	45

c. $\{(-3, 4), (-3, 1), (0, 9), (-5, 1), (2, 4)\}$

d. $x = 6$

e. $y = 3x + 4$

50. Find the requested information. (*Section 3.6, Objective 4*)

a. Find $f(0)$ if $f(x) = 5x - 2$.

b. Find $g(-5)$ if $g(x) = -3x^2 - x + 7$.

c. Find $f(2)$ and $f(5)$ if $f(x)$ is given by the following graph.

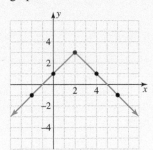

d. Find $g(-1)$ and find x such that $g(x) = 3.5$ if the function $g(x)$ is given by Y_1.

X	Y₁	
-3	-9	
-2	-6.5	
-1	-4	
0	-1.5	
1	1	
2	3.5	
3	6	

X= -3

51. Write each linear equation in two variables in function notation. Then find $f(0)$, $f(3)$, and $f(-2)$ and write the corresponding ordered pairs. (*Section 3.6, Objective 4*)

a. $4x + y = 3$ **b.** $y = 0.3x - 0.6$

c. $y = |3x - 2|$

52. Beva has a new job as a chemistry instructor at a local community college. Her starting salary is $41,500 with a regular teaching load of 30 credit hours. If she teaches beyond the regular load, she earns $950 per additional credit hour. (*Section 3.6, Objective 5*)

a. Write a linear function $f(x)$ that represents Beva's yearly income, where x is the number of extra credit hours she teaches in an academic year.

b. If Beva teaches two additional 4-credit-hour courses in an academic year, what is her income?

c. Find $f(6)$ and interpret the answer.

Systems of Linear Equations in Two and Three Variables

Study Skills

Studying math is very much like studying a foreign language. The way to learn a foreign language is to learn the alphabet, learn words, learn the rules of grammar, and so on. Learning mathematics takes a similar approach. You must learn the notation and symbols, definitions, mathematical rules and processes, and so on. If you do not have a firm foundation with the basics of mathematics, then it will be very difficult to build upon this foundation.

To learn mathematics, you must employ certain study skills. Some of these include

- Taking notes effectively.
- Reviewing notes on a regular basis.
- Studying new material as soon after class as possible.
- Seeking help on topics you do not understand before the next class meeting.
- Thinking about your performance and not comparing yourself with classmates since everyone comes from a different mathematical background.

Question For Thought: What study techniques have been successful for you in other courses? Can they work for you in this class? What is your biggest obstacle in studying?

Chapter Outline

Coming Up...

In Section 4.4, we will learn how to use systems of equations to solve this problem. As of 2010, Nintendo DS and Nintendo Wii were the two top-selling gaming units in the United States. A total of 81 million gaming units were sold. If there were 13 million more Nintendo DS units sold than Nintendo Wii units, how many of each gaming unit was sold? (Source: http://www.vgchartz.com/home.php)

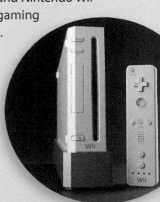

"Keep your mind on your objective, and persist until you succeed. Study, think, and plan."

—W. Clement Stone (Philanthropist, Author)

SECTION 4.1   **Solving Systems of Linear Equations Graphically**

In this chapter, we will focus on solving systems of linear equations in two and three variables. We will use our graphing skills from Chapter 3 to introduce one method of solving systems of linear equations. In addition, two other methods for solving systems of linear equations will be presented. Applications of linear systems will also be discussed.

▶ **OBJECTIVES**

As a result of completing this section, you will be able to

1. Determine if an ordered pair is a solution of a system of linear equations.

2. Determine the solution of a system of linear equations from a graph.

3. Solve a system of linear equations graphically.

4. Solve special cases of systems of linear equations graphically.

5. Determine how the lines in a system of linear equations relate, the number of solutions, and the type of system without graphing.

6. Solve applications using systems of linear equations.

7. Troubleshoot common errors.

Simone is considering two different cell phone plans but is not sure which plan is best for her. CellOne Company offers a plan that charges $60 a month for 1000 min and $0.40 for each additional minute. Phone-a-Friend Company offers a plan that charges $70 a month for 1000 min and $0.25 for each additional minute. Which plan should Simone choose?

To solve this problem, we will use a system of linear equations. We will learn how to solve the system graphically in this section.

Solutions of Systems of Linear Equations

A **system of linear equations** is a set of two or more linear equations that must be solved together. The focus in this chapter will be on solving linear systems that contain two linear equations in two variables. Some examples of systems of linear equations in two variables are shown.

$$\begin{cases} x + 6y = -11 \\ 2x - y = 4 \end{cases} \qquad \begin{cases} y = \dfrac{1}{2}x + 1 \\ x - y = 3 \end{cases} \qquad \begin{cases} x + y = 20 \\ 0.05x + 0.10(20 - x) = 1.75 \end{cases}$$

System equations are often grouped together with a brace to indicate that they belong to the system. A **solution of a system of linear equations in two variables** is an ordered pair that satisfies both equations in the system.

Objective 1 ▶

Determine if an ordered pair is a solution of a system of linear equations.

Procedure: Determining if an Ordered Pair Is a Solution of a Linear System

Step 1: Substitute the given ordered pair into each equation in the system.

Step 2: Simplify the resulting equations.

Step 3: Determine if the resulting equations are true or false.

    a. If both equations are true, then the ordered pair is a solution of the system.

    b. If one or both equations are false, then the ordered pair is not a solution of the system.

Objective 1 Examples   Determine if the given ordered pair is a solution of the system.

1a. $(1, -2)$, $\begin{cases} x + 6y = -11 \\ 2x - y = 4 \end{cases}$ **1b.** $(-5, -4)$, $\begin{cases} 3x + 2y = -23 \\ 4x - y = -24 \end{cases}$

Solutions **1a.** Let $(x, y) = (1, -2)$.

$$x + 6y = -11 \qquad\qquad 2x - y = 4$$
$$1 + 6(-2) = -11 \qquad 2(1) - (-2) = 4$$
$$1 - 12 = -11 \qquad\qquad 2 + 2 = 4$$
$$-11 = -11 \qquad\qquad 4 = 4$$
$$\text{True} \qquad\qquad\qquad \text{True}$$

Since the ordered pair makes both equations true, $(1, -2)$ is a solution of the system.

1b. Let $(x, y) = (-5, -4)$.

$$3x + 2y = -23 \qquad\qquad 4x - y = -24$$
$$3(-5) + 2(-4) = -23 \qquad 4(-5) - (-4) = -24$$
$$-15 - 8 = -23 \qquad\qquad -20 + 4 = -24$$
$$-23 = -23 \qquad\qquad -16 = -24$$
$$\text{True} \qquad\qquad\qquad \text{False}$$

Since the ordered pair makes one of the equations false, $(-5, -4)$ is not a solution of the system.

✔ Student Check 1 Determine if the given ordered pair is a solution of the system.

a. $(-6, 3)$, $\begin{cases} 2x + y = 15 \\ x - 4y = -18 \end{cases}$ **b.** $\left(\dfrac{1}{2}, 3\right)$, $\begin{cases} y = 8x - 1 \\ 6x - y = 0 \end{cases}$

Use a Graph to Determine Solutions of a System

Objective 2 ▶

Determine the solution of a system of linear equations from a graph.

As we know from Objective 1, a solution of a system of linear equations in two variables is an ordered pair that satisfies both equations. The system containing the equations $x - 3y = 6$ and $x + y = 2$ is shown in the following graph. Each of these linear equations has infinitely many solutions. When we graph the two lines on the same coordinate system, we find that they share a common point. This common point is the solution of the system.

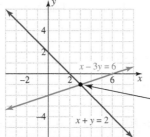

$(3, -1)$ is a point on both lines and is the solution of the system containing these two linear equations.

Graphically, a solution of a system of two linear equations is the ordered pair that the two lines have in common; that is, the point where the two lines intersect. This is called the **point of intersection**.

> **Procedure: Determining the Solution of a System of Linear Equations from a Graph**
>
> **Step 1:** Identify the point where the two lines intersect.
> **Step 2:** Check that this point is the solution of the system by substituting the coordinates in each of the equations.

Objective 2 Examples   Use the graph of each linear system to determine its solution.

2a. $\begin{cases} x - y = 4 \\ 2x + y = 2 \end{cases}$

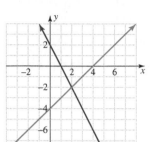

2b. $\begin{cases} y = 3x - 6 \\ y = -\dfrac{5}{2}x + 5 \end{cases}$

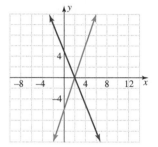

Solutions   **2a.** The lines intersect at the point $(2, -2)$.

Check:

$x - y = 4$	$2x + y = 2$
$2 - (-2) = 4$	$2(2) + (-2) = 2$
$2 + 2 = 4$	$4 - 2 = 2$
$4 = 4$	$2 = 2$
True	True

Both equations are true, so $(2, -2)$ is the solution of the system.

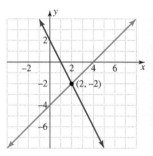

2b. The lines intersect at the point $(2, 0)$.

Check:

$y = 3x - 6$	$y = -\dfrac{5}{2}x + 5$
$0 = 3(2) - 6$	$0 = -\dfrac{5}{2}(2) + 5$
$0 = 6 - 6$	$0 = -5 + 5$
$0 = 0$	$0 = 0$
True	True

Both equations are true, so $(2, 0)$ is the solution of the system.

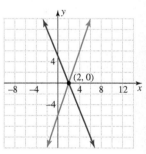

✓ Student Check 2   Use the graph of the linear system to determine its solution.

a. $\begin{cases} x + y = 5 \\ x - 4y = -10 \end{cases}$

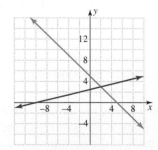

b. $\begin{cases} y = \dfrac{2}{3}x - 2 \\ y = -\dfrac{1}{3}x - 5 \end{cases}$

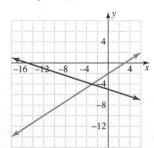

Use Graphing to Solve Systems of Linear Equations

Objective 3 ▶

Solve a system of linear equations graphically.

To solve a system graphically we need to find the point of intersection of the graphs of the equations. Since the systems will contain linear equations in two variables, we can use the methods presented in Chapter 3 to graph the equations in the systems. Recall that we can graph linear equations by plotting points, finding the x- and y-intercepts, or by using the slope and y-intercept.

Precision is very important when solving a system graphically. Be sure to use graph paper and a straight edge to construct the graphs. Solving a system graphically is not as precise as other methods that we will study later in this chapter but it is a method that enables us to visualize the solution.

Procedure: Solving a System of Linear Equations Graphically

Step 1: Graph each equation in the system on the same set of axes.

Step 2: Identify the ordered pair of the intersection point. This is the solution of the system.

Step 3: Check that the ordered pair is the solution by substituting the point in each equation.

Objective 3 Examples Solve each system of linear equations graphically.

3a. $\begin{cases} x + y = -3 \\ x - y = 1 \end{cases}$　　　　**3b.** $\begin{cases} 2x - y = 8 \\ y = -2 \end{cases}$

Solutions　**3a.** We graph the equations in the system by finding the intercepts.

$x + y = -3$		
x	y	(x, y)
$x + 0 = -3$ $x = -3$	0	$(-3, 0)$
0	$0 + y = -3$ $y = -3$	$(0, -3)$

$x - y = 1$		
x	y	(x, y)
$x - 0 = 1$ $x = 1$	0	$(1, 0)$
0	$0 - y = 1$ $-y = 1$ $y = -1$	$(0, -1)$

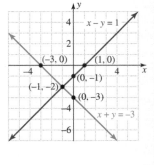

The solution of the system is the ordered pair $(-1, -2)$ since the two lines intersect at this point.

Check:

$$
\begin{array}{c}
x + y = -3 \\
-1 + (-2) = -3 \\
-3 = -3
\end{array}
\qquad\qquad
\begin{array}{c}
x - y = 1 \\
-1 - (-2) = 1 \\
-1 + 2 = 1 \\
1 = 1
\end{array}
$$

　　　　　　　　True　　　　　　　　　　　　　　True

Since $(-1, -2)$ makes each equation in the system true, the solution set is $\{(-1, -2)\}$.

3b. Graph each equation in the system using its slope and y-intercept by solving each equation for y.

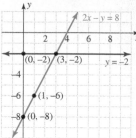

$$2x - y = 8$$
$$-y = -2x + 8$$
$$y = 2x - 8$$

So, the slope is $m = 2 = \dfrac{2}{1}$ and the y-intercept

is $(0, -8)$. Plot $(0, -8)$ and move up
2 units and right 1 unit.

The line $y = -2$ is in slope-intercept form. It has slope $m = 0$ and y-intercept $(0, -2)$. It is a horizontal line through $(0, -2)$.

The solution of the system is the ordered pair $(3, -2)$ since the lines intersect at this point. The ordered pair $(3, -2)$ makes each equation true, so the solution set is $\{(3, -2)\}$.

✔ **Student Check 3**   Solve each system of linear equations graphically.

a. $\begin{cases} x - 3y = -6 \\ 4x - 3y = 12 \end{cases}$ **b.** $\begin{cases} y = -\dfrac{2}{5}x + 4 \\ x = -5 \end{cases}$

Special Cases of Systems of Linear Equations

Objective 4 ▶

Solve special cases of systems of linear equations graphically.

In Examples 1–3, the graphs of the lines in the system of linear equations intersected once, so there was one solution of the system of linear equations. There are two other possibilities for the graphs of the lines in a system of linear equations. The lines in the system can be parallel or they can be the same line, as shown.

Parallel Lines	Same Lines
The two lines do not share any common points. So, there is *no solution* of the system.	Every point on the line is a point of intersection. So, there are *infinitely many solutions* of the system.

- A system that contains parallel lines and has no solution is called an **inconsistent system**.
- A system that contains the same lines with infinitely many solutions is called a **consistent system with dependent equations**.
- A system that contains intersecting lines with one solution is called a **consistent system with independent equations**.

> **Procedure: Determining Solutions of Systems That Are Special Cases**
>
> **Step 1:** Graph each equation in the system on the same set of axes.
> > **a.** If the lines are parallel, then the system has no solution. (The lines will have the same slope and different y-intercepts.)
> > **b.** If the lines are the same, then the system has infinitely many solutions. (The lines will have the same slope and same y-intercept.)
>
> **Step 2:** Write the solution set appropriately.

Objective 4 Examples | Solve each system of equations graphically.

4a. $\begin{cases} 2x - y = 6 \\ 4x - 2y = -8 \end{cases}$ **4b.** $\begin{cases} x + 3y = 9 \\ -2x - 6y = -18 \end{cases}$

Solutions   **4a.** Graph each equation using its slope and y-intercept by first solving each equation for y.

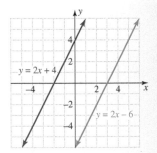

$$
\begin{array}{l|l}
2x - y = 6 & 4x - 2y = -8 \\
-y = -2x + 6 & -2y = -4x - 8 \\
y = 2x - 6 & y = 2x + 4 \\
m = 2 = \dfrac{2}{1} & m = 2 = \dfrac{2}{1} \\
\text{y-intercept: } (0, -6) & \text{y-intercept: } (0, 4)
\end{array}
$$

From the slope-intercept form of the lines, we see that the lines have slope $m = 2$ and different y-intercepts. So, the two lines in this system are parallel, which means there is no solution of the system. The system is inconsistent. We state the solution set as the empty set, $\varnothing$.

4b. Graph each equation using its slope and y-intercept.

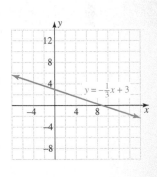

$$
\begin{array}{l|l}
x + 3y = 9 & -2x - 6y = -18 \\
3y = -x + 9 & -6y = 2x - 18 \\
y = -\dfrac{1}{3}x + 3 & y = -\dfrac{2}{6}x + 3 \\
m = -\dfrac{1}{3} & y = -\dfrac{1}{3}x + 3 \\
\text{y-intercept: } (0, 3) & m = -\dfrac{1}{3} \\
 & \text{y-intercept: } (0, 3)
\end{array}
$$

From the slope-intercept form of the lines, we see that the lines have slope $m = -\dfrac{1}{3}$ and the same y-intercept. Lines with the same slope and same y-intercept are the same line. The equations are dependent and the system is consistent.

    The solution set is written as $\{(x, y) \mid x + 3y = 9\}$. This set states that every ordered pair on the graph of $x + 3y = 9$ is a solution of the system.

> **Note:** *The equations in this system are multiples of one another. If we multiply the first equation by −2, we get the second equation. This type of relationship always exists when the system of linear equations contains dependent equations.*
>
> $$-2(x + 3y = 9) \text{ gives us } -2x - 6y = -18$$

☑ **Student Check 4** Solve each system of equations graphically.

a. $\begin{cases} x + 5y = -5 \\ \dfrac{1}{15}x + \dfrac{1}{3}y = -\dfrac{1}{3} \end{cases}$ b. $\begin{cases} 6x - 8y = 24 \\ y = \dfrac{3}{4}x + 3 \end{cases}$

Objective 5 ▶

Determine how the lines in a system of linear equations relate, the number of solutions, and the type of system without graphing.

Determine the Relationship of Lines in a System of Linear Equations Without Graphing

As we have seen, there are three possibilities for the solutions of systems of linear equations in two variables. The system can have one solution, no solution, or infinitely many solutions. The number of solutions is determined by the slopes and y-intercepts of the lines in the system. The following table summarizes the different possibilities.

Intersecting Lines	Parallel Lines	Same Lines
One solution	No solution	Infinitely many solutions
Different slopes; Same or different y-intercepts	Same slopes; Different y-intercepts	Same slopes; Same y-intercepts
Consistent system with independent equations	Inconsistent system	Consistent system with dependent equations

Objective 5 Examples Determine how the lines in each system of linear equations relate, the number of solutions of the system, and the type of system without graphing.

5a. $\begin{cases} 7x + 3y = -6 \\ 2x - 5y = 10 \end{cases}$ 5b. $\begin{cases} \dfrac{3}{2}x + \dfrac{5}{3}y = 1 \\ \dfrac{9}{5}x + 2y = \dfrac{6}{5} \end{cases}$

Solutions 5a.

	$7x + 3y = -6$	$2x - 5y = 10$
Slope-intercept forms	$3y = -7x - 6$ $\dfrac{3y}{3} = -\dfrac{7x}{3} - \dfrac{6}{3}$ $y = -\dfrac{7}{3}x - 2$	$-5y = -2x + 10$ $\dfrac{-5y}{-5} = \dfrac{-2x}{-5} + \dfrac{10}{-5}$ $y = \dfrac{2}{5}x - 2$
Slopes	$m = -\dfrac{7}{3}$	$m = \dfrac{2}{5}$
y-intercepts	$(0, -2)$	$(0, -2)$
How do lines relate?	Because the slopes are different, the two lines intersect.	
Number of solutions	One solution	
Type of system	The system is consistent with independent equations.	

5b.

	$\frac{3}{2}x + \frac{5}{3}y = 1$	$\frac{9}{5}x + 2y = \frac{6}{5}$
Slope-intercept forms	$6\left(\frac{3}{2}x + \frac{5}{3}y\right) = 6(1)$ $9x + 10y = 6$ $10y = -9x + 6$ $y = -\frac{9}{10}x + \frac{6}{10}$ $y = -\frac{9}{10}x + \frac{3}{5}$	$5\left(\frac{9}{5}x + 2y\right) = 5\left(\frac{6}{5}\right)$ $9x + 10y = 6$ $10y = -9x + 6$ $y = -\frac{9}{10}x + \frac{6}{10}$ $y = -\frac{9}{10}x + \frac{3}{5}$
Slopes	$m = -\frac{9}{10}$	$m = -\frac{9}{10}$
y-intercepts	$\left(0, \frac{3}{5}\right)$	$\left(0, \frac{3}{5}\right)$
How do lines relate?	Because the slopes and the y-intercepts are the same, the two lines are the same.	
Number of solutions	Infinitely many solutions	
Type of system	The system is consistent with dependent equations.	

✔ **Student Check 5**   Determine how the lines in each system of linear equations relate, the number of solutions of the system, and the type of system without graphing.

a. $\begin{cases} 6x + 2y = 10 \\ 15x + 5y = 25 \end{cases}$ b. $\begin{cases} \frac{3}{4}x - 2y = 8 \\ x + 6y = 12 \end{cases}$

Applications

Objective 6 ▶

Solve applications using systems of linear equations.

The skills learned in this section apply to many real-life situations. We can use graphs and their equations to solve systems that represent these situations.

Objective 6 Example

The world's rural and urban populations are modeled by the following graph. Use the graph to approximate the year when the rural and urban populations are projected to be equal. Approximate the population at this time. (Source: http://esa.un.org/unup/)

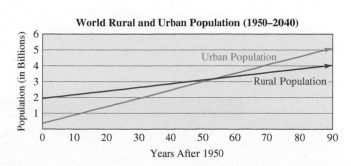

World Rural and Urban Population (1950–2040)

Solution   The point of intersection of the graphs is approximately (54, 3.25). So, in 54 yr after 1950, or 2004, the urban and rural populations of the world were equal and were approximately 3.25 billion each.

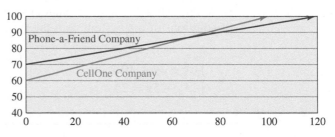

World Rural and Urban Population (1950–2040)

✓ **Student Check 6**   Simone is considering two different cell phone plans but is not sure which plan is best for her. CellOne Company offers a plan that charges \$60 a month for 1000 min and \$0.40 for each additional minute. Phone-a-Friend Company offers a plan that charges \$70 a month for 1000 min and \$0.25 for each additional minute. The system that is used to solve this problem is $\begin{cases} y = 0.40x + 60 \\ y = 0.25x + 70 \end{cases}$, where x is the number of additional minutes. The graph of the system is given. After approximately how many minutes do the two plans have the same cost? What is this approximate cost of the plans? Based on this information, which plan should Simone choose?

Objective 7 ▶

Troubleshoot common errors.

Troubleshooting Common Errors

Some common errors associated with solving systems by graphing are shown next.

Objective 7 Examples   **A problem and an incorrect solution are given. Provide the correct solution and an explanation of the error.**

7a. The graph of a system is shown. What is the solution of the system?

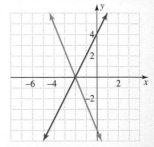

Incorrect Solution	Correct Solution and Explanation
The solution of the system is (0, −2).	The order of the coordinates of the solution is incorrect. The x-value comes first and then the y-value. So, the solution of the system is (−2, 0).

7b. Solve the system $\begin{cases} y = 3x + 7 \\ y = 4x + 6 \end{cases}$ by graphing.

Incorrect Solution	Correct Solution and Explanation
The graph of the lines from my calculator is shown. 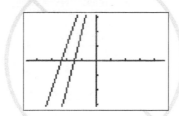 Since the lines do not intersect, there is no solution.	The lines have different slopes, so they will intersect. We must adjust the viewing window to see their intersection. 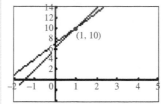 The lines intersect at the point (1, 10). So, this is the solution of the system.

ANSWERS TO STUDENT CHECKS

Student Check 1   **a.** no   **b.** yes

Student Check 2   **a.** $\{(2, 3)\}$ **b.** $\{(-3, -4)\}$

Student Check 3   **a.** $\{(6, 4)\}$ **b.** $\{(-5, 6)\}$

Student Check 4   **a.** $\{(x, y) \mid x + 5y = -5\}$
   b. $\varnothing$

Student Check 5   **a.** same line, infinitely many solutions, consistent system with dependent equations

   b. intersecting lines, one solution, consistent system with independent equations

Student Check 6   65 min, $85; If Simone generally talks less than 65 additional minutes, she should choose CellOne Company. Otherwise, she should select Phone-a-Friend Company.

SUMMARY OF KEY CONCEPTS

1. A solution of a system of linear equations in two variables is an ordered pair that makes both equations in the system true. The solution is the point where the graphs intersect.
2. There are three possible solutions of a system of linear equations in two variables.
   a. One ordered pair—two lines intersect in one point (consistent system, independent equations)
   b. No solution—two lines are parallel and never intersect (inconsistent system)

   c. Infinitely many solutions—two lines are coinciding and intersect at every point on the line (consistent system, dependent equations)
3. The slopes and y-intercepts of the lines in the system determine the type of solution the system will have.
   a. If the slopes of the lines are different, then the lines intersect.
   b. If the slopes are the same and the y-intercepts are different, the lines are parallel.
   c. If the slopes and the y-intercepts are the same, the lines are coinciding.

GRAPHING CALCULATOR SKILLS

To solve a system on a calculator, rewrite each equation in slope-intercept form ($y = mx + b$), enter both equations in the equation editor, graph the equations, and execute the Intersect command.

Example: Solve the system $\begin{cases} x + y = -3 \\ x - y = 1 \end{cases}$.

Solution: Solve each equation for y.

$$x + y = -3 \qquad x - y = 1$$
$$y = -x - 3 \qquad -y = -x + 1$$
$$\qquad\qquad\qquad y = x - 1$$

Enter the equations in the equation editor.

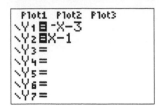

Graph the equations.

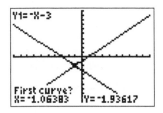

Execute the Intersect command.

(Press Enter to accept First curve. Note the first equation is displayed in the upper left corner.)

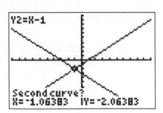

(Press Enter to accept Second curve. Note the second equation is displayed in the upper left corner.)

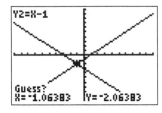

To input Guess, move the cursor left or right to the point of intersection and press Enter.

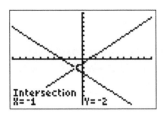

The point of intersection is displayed at the bottom of the screen. So, the solution of the system is $(-1, -2)$. It is always important to verify the answer algebraically as well.

SECTION 4.1 / EXERCISE SET

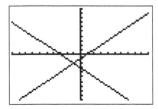

Write About It!

Use complete sentences in your answer to each exercise.

1. What is a system of linear equations?

2. When you graph the equations in a linear system, what part of the graph represents the solution of the system?

3. What are the steps for solving a system of linear equations by graphing?

4. **a.** What are the three possible solutions that a system of linear equations can have?

   b. How can you determine which type of solution a system of linear equations has without graphing?

Determine if each statement is true or false. If the statement is false, explain why.

5. The system $\begin{cases} y = 4x + 5 \\ y = -4x - 3 \end{cases}$ has no solution.

6. The ordered pair $(4, -5)$ is a solution of the system $\begin{cases} 2x - y = 3 \\ x + y = -1 \end{cases}$.

7. The system $\begin{cases} 2x + y = 6 \\ 4x + 2y = 12 \end{cases}$ has no solution.

8. The system $\begin{cases} x - 2y = 4 \\ 3x - 6y = 6 \end{cases}$ has infinitely many solutions.

Practice Makes Perfect!

Determine if the given ordered pair is a solution of the system. (*See Objective 1.*)

9. $(1, 3)$, $\begin{cases} x - 4y = -11 \\ 2x + y = 5 \end{cases}$

10. $(4, 2)$, $\begin{cases} x - y = 2 \\ 3x + y = 14 \end{cases}$

11. $(-3, 5)$, $\begin{cases} 5x - 3y = 0 \\ x - y = -8 \end{cases}$

12. $(4, -7)$, $\begin{cases} 4x + 3y = 5 \\ 9x - 2y = 22 \end{cases}$

13. $(0, -8)$, $\begin{cases} y = 2x - 8 \\ y + 8 = 0 \end{cases}$

14. $(3, 5)$, $\begin{cases} y = -x + 8 \\ x = 3 \end{cases}$

15. $\left(\dfrac{1}{2}, -\dfrac{1}{3}\right)$, $\begin{cases} 6x - 9y = 6 \\ 4x + 6y = 4 \end{cases}$

16. $\left(-\dfrac{2}{5}, \dfrac{4}{3}\right)$, $\begin{cases} 5x + 3y = 2 \\ 10x - 6y = -12 \end{cases}$

Use the graph of each system of linear equations to determine its solution set. (*See Objective 2.*)

17. $\begin{cases} 2x + y = 0 \\ x - y = -3 \end{cases}$

18. $\begin{cases} 2x - y = -3 \\ x - y = -3 \end{cases}$

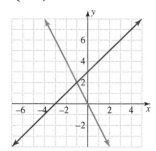
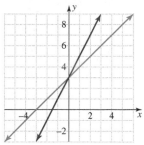

19. $\begin{cases} x + 2y = 4 \\ 3x - 2y = 4 \end{cases}$

20. $\begin{cases} x - 4y = 4 \\ 3x - 4y = -4 \end{cases}$

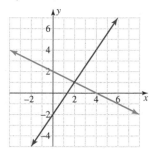
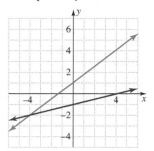

21. $\begin{cases} 2x - y = 0 \\ 4x - 2y = 8 \end{cases}$

22. $\begin{cases} x = -1 \\ x + y = 3 \end{cases}$

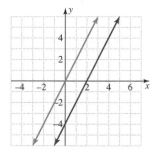
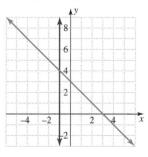

23. $\begin{cases} 2x + y = 4 \\ y = -2 \end{cases}$

24. $\begin{cases} x + y = 2 \\ 3x + 3y = -9 \end{cases}$

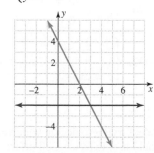
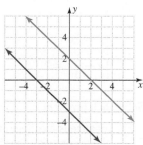

25. $\begin{cases} 2x - y = 0 \\ 4x - 2y = 0 \end{cases}$

26. $\begin{cases} x + y = 2 \\ 3x + 3y = 6 \end{cases}$

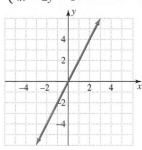
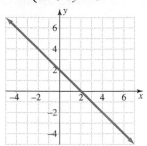

Solve each system of linear equations by graphing.
(*See Objective 3.*)

27. $\begin{cases} y = x + 1 \\ y = 2x - 2 \end{cases}$

28. $\begin{cases} y = -3x + 4 \\ y = x \end{cases}$

29. $\begin{cases} y = -\dfrac{2}{3}x + 2 \\ y = \dfrac{1}{6}x - 3 \end{cases}$

30. $\begin{cases} y = 4x - 1 \\ y = \dfrac{1}{2}x + 6 \end{cases}$

31. $\begin{cases} 3x + y = 1 \\ 2x - y = 9 \end{cases}$

32. $\begin{cases} x - y = 3 \\ x + 2y = -6 \end{cases}$

33. $\begin{cases} 5x - y = -4 \\ 3x - y = 0 \end{cases}$

34. $\begin{cases} 3x + 2y = -12 \\ 4x - 3y = 1 \end{cases}$

35. $\begin{cases} x + 5y = -10 \\ y = -3 \end{cases}$

36. $\begin{cases} 6x + 2y = -14 \\ y = 5 \end{cases}$

37. $\begin{cases} x - 4y = 8 \\ x = -4 \end{cases}$

38. $\begin{cases} 7x + y = 0 \\ x = 1 \end{cases}$

39. $\begin{cases} x = 5 \\ y = -1 \end{cases}$

40. $\begin{cases} x = -2 \\ y = 3 \end{cases}$

Solve each system of linear equations by graphing.
(*See Objective 4.*)

41. $\begin{cases} 2x + y = 7 \\ \dfrac{3}{2}x + \dfrac{3}{4}y = \dfrac{15}{4} \end{cases}$

42. $\begin{cases} -3x + y = 6 \\ \dfrac{2}{3}x - \dfrac{2}{9}y = \dfrac{8}{9} \end{cases}$

43. $\begin{cases} x + 4y = -8 \\ y = -\dfrac{1}{4}x + 2 \end{cases}$

44. $\begin{cases} x - 6y = 3 \\ y = \dfrac{1}{6}x - \dfrac{1}{2} \end{cases}$

45. $\begin{cases} 4x + 2y = 5 \\ y = -2x + \dfrac{5}{2} \end{cases}$

46. $\begin{cases} 3x - 5y = 2 \\ -15x + 25y = -10 \end{cases}$

Determine how the lines in each system of linear equations relate, the number of solutions of the system, and the type of system without graphing. (*See Objective 5.*)

47. $\begin{cases} y = x + 2 \\ y = x - 3 \end{cases}$

48. $\begin{cases} y = \dfrac{1}{4}x - 1 \\ y = \dfrac{1}{4}x + 4 \end{cases}$

49. $\begin{cases} 3x + y = -3 \\ 9x + 3y = -9 \end{cases}$

50. $\begin{cases} y = \dfrac{1}{2}x + 1 \\ x - 2y = -2 \end{cases}$

51. $\begin{cases} x - \dfrac{1}{5}y = 4 \\ x - \dfrac{2}{10}y = -2 \end{cases}$

52. $\begin{cases} 2x - y = 6 \\ 4x - 2y = 12 \end{cases}$

53. $\begin{cases} 6x + 2y = 4 \\ 15x + 5y = 10 \end{cases}$

54. $\begin{cases} x + 2y = 4 \\ 2x + 4y = 0 \end{cases}$

Solve each problem. (*See Objectives 2 and 6.*)

In business, a company breaks even when the revenue generated from selling a particular number of items is equal to the total cost of making the same number of items. In other words, a company breaks even when its revenue equals its cost.

55. A shoe factory can make sneakers for $25 per pair and the factory has fixed monthly costs of $15,000. The shoe store sells each pair of sneakers for $50. If x is the number of pairs of sneakers sold in a month, the monthly cost of making sneakers is given by $C = 25x + 15,000$. The monthly revenue for selling x pairs of sneakers is given by $R = 50x$. Use the system of equations

$$\begin{cases} C = 25x + 15,000 \\ R = 50x \end{cases}$$ and its graph to answer the questions that follow.

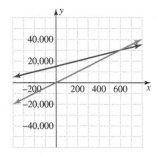

a. What is the point of intersection of the graphs?

b. What does the point of intersection mean in the context of the problem?

56. It costs a local donut shop $0.50 to make each donut and the shop has fixed monthly costs of $2000. The shop sells each donut for $1.50. If x represents the number of donuts made in a month, the total monthly cost to make x donuts is $C = 0.50x + 2000$. The revenue from

selling x donuts is $R = 1.50x$. Use the system of equations $\begin{cases} C = 0.50x + 2000 \\ R = 1.50x \end{cases}$ and its graph to answer the questions that follow.

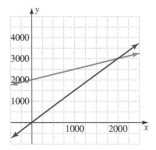

a. What is the point of intersection of the graphs?

b. What does the point of intersection mean in the context of the problem?

 Mix 'Em Up!

Solve each system of linear equations graphically. Then state if the system is inconsistent, consistent with dependent equations, or consistent with independent equations.

57. $\begin{cases} y = 8x - 4 \\ y = 4x - 6 \end{cases}$

58. $\begin{cases} y = -3x + 4 \\ y = -3x + 2 \end{cases}$

59. $\begin{cases} x - y = 4 \\ 2x = 2y + 8 \end{cases}$

60. $\begin{cases} 4x - y = -3 \\ -3x + y = 6 \end{cases}$

61. $\begin{cases} x + 5y = -9 \\ y = -2x \end{cases}$

62. $\begin{cases} 3x - y = -4 \\ y = 2x \end{cases}$

63. $\begin{cases} x + 2y = -7 \\ y = -3 \end{cases}$

64. $\begin{cases} 5x - y = -4 \\ x = -2 \end{cases}$

65. $\begin{cases} 0.25x + y = 0.25 \\ y = -0.8x + 3 \end{cases}$

66. $\begin{cases} 0.3x - 0.2y = 2.1 \\ y = 0.2x + 2.5 \end{cases}$

67. $\begin{cases} 2x - 3y = 1 \\ y = \dfrac{2}{3}x - \dfrac{1}{3} \end{cases}$

68. $\begin{cases} y = -2x \\ 2x + y = 3 \end{cases}$

69. $\begin{cases} y = -1.5x \\ y = 2.5x + 12 \end{cases}$

70. $\begin{cases} y = 1.2x - 1.6 \\ 0.2x - 0.1y = 0.72 \end{cases}$

71. $\begin{cases} y = 3x + 1 \\ x - \dfrac{1}{3}y = 2 \end{cases}$

72. $\begin{cases} x - \dfrac{1}{4}y = 2 \\ y = 2x - 2 \end{cases}$

Solve each problem.

73. Johanne is a college student trying to decide which cell phone company she should use. InTouch Cell offers a monthly rate of $30 plus $0.40 for each minute over 200. FriendsConnect offers a monthly rate of $60 per month plus $0.20 for each minute over 200. The monthly cost of InTouch Cell can be expressed by $y = 0.40x + 30$ and the monthly cost of FriendsConnect

can be expressed by $y = 0.20x + 60$, where x is the number of minutes over 200. Use the graph of the system formed by these two equations to answer the questions that follow.

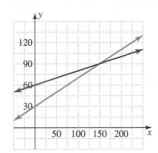

a. What is the point of intersection of this system?

b. What is the meaning of the intersection point in the context of the problem?

c. If Johanne expects to use 300 min each month, which plan should she choose? Why?

74. Janis is trying to decide which shipping company she should use to mail a package. You Gotta Be Shipping Me charges a flat fee of $25 plus $2 per pound. ShipItQuick charges a flat fee of $50 plus $1 per pound. For a package that weighs x lb, the cost charged by You Gotta Be Shipping Me is $y = 2x + 25$. The cost charged by ShipItQuick is $y = x + 50$. Use the graph of the system formed by these two equations to answer the questions that follow.

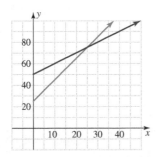

a. What is the point of intersection of this system?

b. What is the meaning of the intersection point in the context of this problem?

c. If Janis needs to mail a package that weighs 40 lb, which company should she use? Why?

 You Be the Teacher!

Correct each student's errors, if any.

75. Determine if $(1, 3)$ is a solution of the system $\begin{cases} 3x - y = 0 \\ x + y = 3 \end{cases}$.

Janeane's work:

$3(1) - 3 = 0$

$3 - 3 = 0$.

This is true, so $(1, 3)$ is a solution of the system.

76. Solve the system $\begin{cases} 4x - 2y = 4 \\ 6x + 2y = 11 \end{cases}$ graphically.

Jocelyn's work:

Equation 1: $-2y = -4x + 4$

$y = \dfrac{-4x + 4}{-2} = 2x - 2$. So, $y = 2x - 2$.

Equation 2: $2y = -6x + 11$

$y = \dfrac{-6x + 11}{2} = -3x + 5.5$. So, $y = -3x + 5.5$.

77. Explain how to identify the solution of the system from the graph.

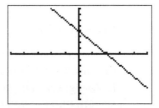

Emelle's work:

Since there is only one line, there is no solution and the system is inconsistent.

 Calculate It!

Use a graphing calculator to solve each system. Round each answer to the nearest hundredth when needed.

78. $\begin{cases} 3x + y = 7 \\ y = -2x + 2 \end{cases}$ **79.** $\begin{cases} 4x - y = 3 \\ y = 2x + 5 \end{cases}$

80. $\begin{cases} 2x + 3y = 6 \\ 2x - y = 4 \end{cases}$ **81.** $\begin{cases} 4x + 2y = 3 \\ 5x - y = -2 \end{cases}$

 Think About It!

Write a system of linear equations in two variables that has the given ordered pair as its solution.

82. $(3, -2)$ **83.** $(-4, 1)$ **84.** $(0, 0)$ **85.** $\left(\dfrac{1}{2}, 3\right)$

A system of linear equations in two variables contains the given equation. Write the other equation in the system that would make the system satisfy the stated condition.

86. $y = \dfrac{1}{3}x - 8$, inconsistent

87. $y = -\dfrac{5}{2}x + 2$, inconsistent

88. $x - 4y = 8$, consistent with dependent equations

89. $3x + 5y = 2$, consistent with dependent equations

Solving Systems of Linear Equations by Substitution

In Section 4.1 we solved systems by graphing. Graphing by hand, however, is not always the most precise method to solve a system. Consider the system

$$\begin{cases} 2x - 3y = -3 \\ 4x + 9y = 14 \end{cases}$$

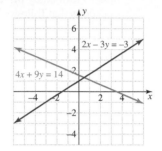

The graph of this system is shown. The point of intersection does not occur at integer values, so it is very difficult to know the exact solution based on this graph. In this section, we will learn an algebraic method to solve systems precisely.

Substitution

Objective 1 ▶

Solve a system of linear equations by substitution.

The method of substitution is a precise way to solve a system of linear equations because we will obtain an exact answer, rather than having to approximate our answer from a graph. The goal of the **substitution method** is to substitute one equation of the system into the other equation of the system so the new equation has just one variable.

Consider the system $\begin{cases} x + 2y = 4 \\ y = x - 1 \end{cases}$. The second equation has the variable y expressed in terms of x. When we replace the variable y in the first equation with its equivalent expression, $x - 1$, from the second equation, we obtain an equation with just one variable, which we can solve for x.

$$
\begin{array}{ll}
x + 2y = 4 & \\
x + 2(x - 1) = 4 & \text{Replace } y \text{ with } x - 1. \\
x + 2x - 2 = 4 & \text{Apply the distributive property.} \\
3x - 2 = 4 & \text{Combine like terms.} \\
3x - 2 + 2 = 4 + 2 & \text{Add 2 to each side.} \\
3x = 6 & \text{Simplify.} \\
x = \dfrac{6}{3} & \text{Divide each side by 3.} \\
x = 2 & \text{Simplify.}
\end{array}
$$

Now that we know the value of x, we can find the corresponding y-value by replacing x with 2 in either of the given equations. This yields

$$
\begin{array}{ll}
y = x - 1 & \text{Begin with the second equation.} \\
y = 2 - 1 & \text{Replace } x \text{ with 2.} \\
y = 1 & \text{Simplify.}
\end{array}
$$

So, the solution set of the system $\begin{cases} x + 2y = 4 \\ y = x - 1 \end{cases}$ is $\{(2, 1)\}$. The steps used to solve this system are generalized next.

Procedure: Solving a System of Linear Equations by Substitution

Step 1: Solve one of the equations in the system for one of the variables.
Step 2: Substitute the expression found in step 1 into the other equation in the system.
Step 3: Solve the resulting equation.
Step 4: Substitute the value found in step 3 back into one of the equations to find the value of the other variable.
Step 5: The solution of the system is the ordered pair (x, y).
Step 6: Check the solution in the system.

| Objective 1 Examples | Solve each system of linear equations by substitution. |

1a. $\begin{cases} y = x - 4 \\ 2x + y = 5 \end{cases}$ **1b.** $\begin{cases} 3x + 2y = -2 \\ x - 6y = -14 \end{cases}$ **1c.** $\begin{cases} 2x - 3y = -3 \\ 4x + 9y = 14 \end{cases}$

1d. $\begin{cases} x + y = 42 \\ 0.05x + 0.10y = 3.50 \end{cases}$

Solutions   **1a.** The first equation is already solved for y, $y = x - 4$.
Substitute and solve.

$2x + y = 5$	Begin with the second equation.
$2x + (x - 4) = 5$	Substitute $x - 4$ for y.
$3x - 4 = 5$	Combine like terms.
$3x - 4 + 4 = 5 + 4$	Add 4 to each side.
$3x = 9$	Simplify.
$\dfrac{3x}{3} = \dfrac{9}{3}$	Divide each side by 3.
$x = 3$	Simplify.

Now find the value of y. (For reference, both equations are shown but only one is necessary.)

$y = x - 4$		$2x + y = 5$	
$y = 3 - 4$	Substitute 3 for x.	$2(3) + y = 5$	Substitute 3 for x.
$y = -1$	Simplify.	$6 + y = 5$	Simplify.
		$6 + y - 6 = 5 - 6$	Subtract 6 from each
		$y = -1$	side and simplify.

So, the solution is the ordered pair $(3, -1)$. Check the solution in each of the equations in the system.

$y = x - 4$	$2x + y = 5$
$-1 = 3 - 4$	$2(3) + (-1) = 5$
$-1 = -1$	$6 - 1 = 5$
	$5 = 5$
True	True

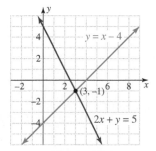

Figure 4.1

Since $(3, -1)$ makes both equations true, the solution set is $\{(3, -1)\}$. Because there is one solution, this system contains lines that intersect at this point. The system is consistent with independent equations. Graphing the system confirms the solution, as well as shown in Figure 4.1.

1b. Solve the second equation for x since its coefficient is 1.

$x - 6y = -14$	
$x - 6y + 6y = -14 + 6y$	Add $6y$ to each side.
$x = 6y - 14$	

Substitute and solve.

$3x + 2y = -2$	Begin with the first equation.
$3(6y - 14) + 2y = -2$	Substitute $6y - 14$ for x.
$18y - 42 + 2y = -2$	Apply the distributive property.
$20y - 42 = -2$	Combine like terms.
$20y - 42 + 42 = -2 + 42$	Add 42 to each side.
$20y = 40$	Simplify.
$y = \dfrac{40}{20}$	Divide each side by 20.
$y = 2$	Simplify.

Now find the value of x.

$$x - 6y = -14 \qquad \text{Begin with the second equation.}$$
$$x - 6(2) = -14 \qquad \text{Substitute 2 for } y.$$
$$x - 12 = -14 \qquad \text{Simplify.}$$
$$x - 12 + 12 = -14 + 12 \qquad \text{Add 12 to each side.}$$
$$x = -2 \qquad \text{Simplify.}$$

So, the solution is the ordered pair $(-2, 2)$. It makes each equation in the system true, so the solution set of the system is $\{(-2, 2)\}$.

1c. None of the coefficients of the variables is 1 or -1. If we solve the first equation for x, we will get a fraction. The fraction, however, will be cleared when we substitute into the other equation since the coefficient of x is a multiple of 2.

$$2x - 3y = -3$$
$$2x - 3y + 3y = -3 + 3y \qquad \text{Add } 3y \text{ to each side.}$$
$$2x = 3y - 3 \qquad \text{Simplify.}$$
$$\frac{2x}{2} = \frac{3y - 3}{2} \qquad \text{Divide each side by 2.}$$
$$x = \frac{3y - 3}{2} \qquad \text{Simplify.}$$

Substitute and solve.

$$4x + 9y = 14 \qquad \text{Begin with the second equation.}$$
$$4\left(\frac{3y - 3}{2}\right) + 9y = 14 \qquad \text{Substitute } \frac{3y - 3}{2} \text{ for } x.$$
$$2(3y - 3) + 9y = 14 \qquad \text{Simplify. Note that } \frac{4}{2} = 2.$$
$$6y - 6 + 9y = 14 \qquad \text{Apply the distributive property.}$$
$$15y - 6 = 14 \qquad \text{Combine like terms.}$$
$$15y - 6 + 6 = 14 + 6 \qquad \text{Add 6 to each side.}$$
$$15y = 20 \qquad \text{Simplify.}$$
$$\frac{15y}{15} = \frac{20}{15} \qquad \text{Divide each side by 15.}$$
$$y = \frac{4}{3} \qquad \text{Simplify.}$$

Now find the value of x.

$$2x - 3y = -3$$
$$2x - 3\left(\frac{4}{3}\right) = -3 \qquad \text{Substitute } \frac{4}{3} \text{ for } y.$$
$$2x - 4 = -3 \qquad \text{Simplify.}$$
$$2x - 4 + 4 = -3 + 4 \qquad \text{Add 4 to each side.}$$
$$2x = 1 \qquad \text{Simplify.}$$
$$x = \frac{1}{2} \qquad \text{Divide each side by 2.}$$

So, the solution is the ordered pair $\left(\frac{1}{2}, \frac{4}{3}\right)$. It makes each equation in the system true, so the solution set is $\left\{\left(\frac{1}{2}, \frac{4}{3}\right)\right\}$.

1d. Since both variables in the first equation have a coefficient of 1, we can solve for either x or y. We choose to solve for y.

$$x + y = 42$$
$$x + y - x = 42 - x \qquad \text{Subtract } x \text{ from each side.}$$
$$y = -x + 42$$

Substitute and solve.

$$0.05x + 0.10y = 3.50 \qquad \text{Begin with the second equation.}$$
$$0.05x + 0.10(-x + 42) = 3.50 \qquad \text{Substitute } -x + 42 \text{ for } y.$$
$$0.05x - 0.10x + 4.2 = 3.50 \qquad \text{Apply the distributive property.}$$
$$-0.05x + 4.2 = 3.5 \qquad \text{Combine like terms.}$$
$$-0.05x + 4.2 - 4.2 = 3.5 - 4.2 \qquad \text{Subtract 4.2 from each side.}$$
$$-0.05x = -0.7 \qquad \text{Simplify.}$$
$$\frac{-0.05x}{-0.05} = \frac{-0.7}{-0.05} \qquad \text{Divide each side by } -0.05.$$
$$x = 14 \qquad \text{Simplify.}$$

Now find the value of x.

$$x + y = 42$$
$$14 + y = 42 \qquad \text{Substitute 14 for } x.$$
$$14 + y - 14 = 42 - 14 \qquad \text{Subtract 14 from each side.}$$
$$y = 28 \qquad \text{Simplify.}$$

So, the solution is the ordered pair $(14, 28)$. It makes each equation in the system true, so the solution set is $\{(14, 28)\}$.

✓ **Student Check 1**   Solve each system of linear equations by substitution.

a. $\begin{cases} 4x + y = -2 \\ y = 7x + 9 \end{cases}$ 　　　**b.** $\begin{cases} 5x - y = 10 \\ 3x + 4y = 6 \end{cases}$

c. $\begin{cases} 9x - 6y = 11 \\ 3x + 6y = -7 \end{cases}$ 　　　**d.** $\begin{cases} x + y = 50 \\ 0.25x + 0.05y = 8.50 \end{cases}$

Special Cases of Systems of Linear Equations

Objective 2 ▶

Solve special cases of systems of linear equations using substitution.

From Section 4.1, we know that there are two special cases of systems of linear equations—systems with no solution and systems with infinitely many solutions. Recall that a system with no solution consists of lines whose graphs are parallel; a system with infinitely many solutions consists of two lines that are the same.

> **Procedure: Solving a Special Case of a System of Linear Equations by Substitution**
>
> **Step 1:** Solve one of the equations in the system for one of the variables.
> **Step 2:** Substitute the expression found in step 1 into the other equation in the system.
> **Step 3:** Solve the resulting equation.
> **Step 4:** The resulting equation will be either a contradiction or an identity.
>
>   **a.** If the resulting equation is a false statement (i.e., a contradiction), then there is no solution of the system.
>   **b.** If the resulting equation is a true statement (i.e., an identity), then there are infinitely many solutions.

Objective 2 Examples	**Solve each system of linear equations by substitution.**

2a. $\begin{cases} x + 5y = -1 \\ 2x + 10y = 5 \end{cases}$ **2b.** $\begin{cases} 9x - 3y = 27 \\ 3x - y = 9 \end{cases}$

Solutions   **2a.** Solve the first equation for x since its coefficient is 1.

$$x + 5y = -1$$
$$x + 5y - 5y = -1 - 5y \qquad \text{Subtract } 5y \text{ from each side.}$$
$$x = -5y - 1$$

Substitute and solve.

$$2x + 10y = 5 \qquad \text{Begin with the second equation.}$$
$$2(-5y - 1) + 10y = 5 \qquad \text{Substitute } -5y - 1 \text{ for } x.$$
$$-10y - 2 + 10y = 5 \qquad \text{Apply the distributive property.}$$
$$-2 = 5 \qquad \text{Combine like terms.}$$

Because we obtained a contradiction, $-2 = 5$, in the substitution process, this system has no solution. This means that the two lines are parallel and that the system is inconsistent. So, the solution set is the empty set, or $\varnothing$.

2b. Solve the second equation for y since its coefficient is -1.

$$3x - y = 9$$
$$3x - y - 3x = 9 - 3x \qquad \text{Subtract } 3x \text{ from each side.}$$
$$-y = -3x + 9 \qquad \text{Simplify.}$$
$$-1(-y) = -1(-3x + 9) \qquad \text{Multiply each side by } -1.$$
$$y = 3x - 9 \qquad \text{Simplify.}$$

Substitute and solve.

$$9x - 3y = 27 \qquad \text{Begin with the first equation.}$$
$$9x - 3(3x - 9) = 27 \qquad \text{Substitute } 3x - 9 \text{ for } y.$$
$$9x - 9x + 27 = 27 \qquad \text{Apply the distributive property.}$$
$$27 = 27 \qquad \text{Combine like terms.}$$

Because we obtained an identity, $27 = 27$, in the substitution process, this system has infinitely many solutions. The lines are the same and the system is consistent with dependent equations. The solution set consists of all ordered pairs on the line, which is denoted as $\{(x, y) \mid 3x - y = 9\}$.

✓ **Student Check 2**   Solve each system of linear equations by substitution.

a. $\begin{cases} 8x - 12y = -8 \\ 2x - 3y = -2 \end{cases}$ **b.** $\begin{cases} 10x + 5y = -20 \\ 8x + 4y = 10 \end{cases}$

Determine the Relationship of Lines in a System of Linear Equations

Objective 3 ▶

Determine how the lines in a system of linear equations relate, the number of solutions, and the type of system.

It is helpful to understand the equations that result in the substitution process and what they mean as far as how the lines in the system relate to one another, the number of solutions of the system, and the type of system. The following table summarizes the three possibilities.

Intersecting Lines	Parallel Lines	Same Lines
After substituting and simplifying each side of the equation, the variable remains and we get something of the form $x = k$ or $y = k$, where k is a constant.	After substituting and simplifying each side of the equation, a *false* statement or contradiction results; for example, $3 = 8$.	After substituting and simplifying each side of the equation, a *true* statement or identity results; for example, $3 = 3$.
One solution	No solution	Infinitely many solutions
Consistent system with independent equations	Inconsistent system	Consistent system with dependent equations

---

Objective 3 Examples

Determine how the lines in each system of linear equations relate, the number of solutions of the system, and the type of system from the given information. State the solution set of the system.

3a. $\begin{cases} \dfrac{2}{3}x - 4y = 5 \\ 4x - 24y = 30 \end{cases}$; Substitution process yields $30 = 30$.

3b. $\begin{cases} 5x - 3y = -15 \\ x + y = -3 \end{cases}$; Substitution process yields $y = 0$.

Solutions

3a. Since the substitution process yields a true statement, or identity, this system consists of lines that are the same. There are infinitely many solutions. The system is consistent with dependent equations. The solution set is $\{(x, y) \mid 4x - 24y = 30\}$.

3b. Since one of the variables remains in the resulting equation, this system has intersecting lines. There is one solution. The system is consistent with independent equations. We need to find the value of x to write the solution set.

$$y = 0: x + 0 = -3$$
$$x = -3$$

So, the solution set is $\{(-3, 0)\}$.

✓ Student Check 3

Determine how the lines in each system of linear equations relate, the number of solutions of the system, and the type of system from the given information. State the solution set of the system.

a. $\begin{cases} x + 5y = 20 \\ 9x - y = -4 \end{cases}$; Substitution process yields $x = 0$.

b. $\begin{cases} \dfrac{1}{4}x + 7y = 3 \\ 2x + 56y = 8 \end{cases}$; Substitution process yields $24 = 8$.

Objective 4 ▶

Troubleshoot common errors.

Troubleshooting Common Errors

Some common errors associated with substitution are illustrated next.

Objective 4 Examples / **A problem and an incorrect solution are given. Provide the correct solution and an explanation of the error.**

4a. Solve the system $\begin{cases} y = 4x - 9 \\ 3x - y = 6 \end{cases}$.

Incorrect Solution	Correct Solution and Explanation
$y = 4x - 9$ $3x - 4x - 9 = 6$ $-x - 9 = 6$ $-x = 15$ $x = -15$ $y = 4(15) - 9$ $y = 60 - 9$ $y = 51$ The solution set is $\{(-15, 51)\}$.	The expression $4x - 9$ should be put in parentheses because the negative sign must be distributed. $3x - (4x - 9) = 6$ $3x - 4x + 9 = 6$ $-x + 9 = 6$ $-x = -3$ $x = 3$ $y = 4(3) - 9$ $y = 12 - 9$ $y = 3$ So, the solution set is $\{(3, 3)\}$.

4b. Solve the system $\begin{cases} x + 2y = 5 \\ 4x + y = 6 \end{cases}$.

Incorrect Solution	Correct Solution and Explanation
$x = 5 - 2y$ $5 - 2y + 2y = 5$ $5 = 5$ Since this is true, there are infinitely many solutions. The solution set is $\{(x, y) \mid x + 2y = 5\}$.	Both equations must be used in the substitution process. The expression $5 - 2y$ should be substituted for x in the second equation. $4x + y = 6$ $4(5 - 2y) + y = 6$ $20 - 8y + y = 6$ $20 - 7y = 6$ $-7y = -14$ $y = 2$ $x = 5 - 2y = 5 - 2(2) = 5 - 4 = 1$ The solution set is $\{(1, 2)\}$.

ANSWERS TO STUDENT CHECKS

Student Check 1   **a.** $\{(-1, 2)\}$ **b.** $\{(2, 0)\}$
   c. $\left\{\left(\dfrac{1}{3}, -\dfrac{4}{3}\right)\right\}$ **d.** $\{(30, 20)\}$
Student Check 2   **a.** $\{(x, y) \mid 2x - 3y = -2\}$ **b.** $\varnothing$

Student Check 3   **a.** intersecting lines, one solution, consistent system with independent equations, $\{(0, 4)\}$
   b. parallel lines, no solution, inconsistent system, $\varnothing$

SUMMARY OF KEY CONCEPTS

1. Substitution is an algebraic method that enables us to find the exact solution of a system of linear equations.

2. To solve a linear system by substitution,

   - Choose one of the equations and solve it for either x or y.
   - Once we solve for x or y, we substitute this value into the other equation. We will then have a simpler linear equation with one variable to solve.

   - Once we have found the value of one variable, we substitute this value back into one of the equations to find the other variable's value.

3. If, after substitution, the resulting equation is a contradiction or identity, then the system has no solution or infinitely many solutions, respectively.

4. Answers can be checked by substituting the ordered pair into both equations or by graphing.

GRAPHING CALCULATOR SKILLS

A graphing calculator can be used to check our work. In Section 4.1, the Intersect command was shown. We can also check by substituting the x- and y-coordinates of the solution into each equation.

Example: Check that $\left(\dfrac{1}{2}, \dfrac{4}{3}\right)$ is the solution of $\begin{cases} 2x - 3y = -3 \\ 4x + 9y = 14 \end{cases}$.

Solution: Substitute the coordinates of the ordered pair into the two equations in the system.

```
2(1/2)-3(4/3)
                -3
4(1/2)+9(4/3)
                14
```

Since the ordered pair makes each equation true, the solution is correct.

SECTION 4.2 / EXERCISE SET

 Write About It!

Use complete sentences in your answer to each exercise.

1. Why is substitution more precise than graphing to find the solutions of a system of linear equations?

2. Explain how to solve a system of linear equations by substitution.

3. When using substitution to solve a system of linear equations, what will happen if there are infinitely many solutions?

4. When using substitution to solve a system of linear equation, what will happen if there is no solution?

5. When using substitution to solve a system of linear equations, what will happen if there is only one solution?

6. Suppose that after using substitution to solve a system of linear equations, you derive $x = 0$. Does this mean that there is no solution of the system? Explain.

7. When expressing the solution of a system of linear equations with infinitely many solutions, does it matter which equation is used in the set notation? Explain.

8. After using substitution to solve for one of the two variables in a system of linear equations, explain how to solve for the remaining variable.

 Practice Makes Perfect!

Solve each system of linear equations by substitution. (See Objective 1.)

9. $\begin{cases} x = 2y + 1 \\ x + 3y = 6 \end{cases}$

10. $\begin{cases} x = 3y + 2 \\ x + 5y = 10 \end{cases}$

11. $\begin{cases} y = 5x - 3 \\ 2x + 2y = 18 \end{cases}$

12. $\begin{cases} y = 4x - 1 \\ 4x + 3y = -19 \end{cases}$

13. $\begin{cases} y = \dfrac{2}{3}x + 9 \\ x + 6y = 24 \end{cases}$

14. $\begin{cases} y = -\dfrac{4}{7}x + \dfrac{2}{7} \\ -2x + 5y = 16 \end{cases}$

15. $\begin{cases} x - 2y = 7 \\ 3x - y = 6 \end{cases}$

16. $\begin{cases} 2x - y = 5 \\ x - 5y = 4 \end{cases}$

17. $\begin{cases} 2x = 4y - 6 \\ 6x - 3y = 0 \end{cases}$

18. $\begin{cases} 3y = 3x + 9 \\ 4x - 5y = 10 \end{cases}$

19. $\begin{cases} 5x = 2y + 3 \\ 5x + 4y = 7 \end{cases}$

20. $\begin{cases} 7x + 2y = 4 \\ 2y = 3x + 1 \end{cases}$

21. $\begin{cases} -4x + 3y = 5 \\ 2x + 6y = -1 \end{cases}$

22. $\begin{cases} 3x - 5y = 2 \\ -6x + 3y = 10 \end{cases}$

Solve each system of linear equations by substitution. (*See Objective 2.*)

23. $\begin{cases} 4x - y = 6 \\ y = 4x - 3 \end{cases}$

24. $\begin{cases} x = -5y + 10 \\ 3x + 15y = 30 \end{cases}$

25. $\begin{cases} 2x - 4y = 6 \\ -x + 2y = -3 \end{cases}$

26. $\begin{cases} 2x - y = 3 \\ 14x = 7y + 28 \end{cases}$

27. $\begin{cases} 2x + y = 3 \\ 4x + 2y = 0 \end{cases}$

28. $\begin{cases} 3x + 6y = 3 \\ 5x = -10y + 5 \end{cases}$

29. $\begin{cases} x - 3y = 4 \\ 2x - 6y = 8 \end{cases}$

30. $\begin{cases} 6x = 2y + 2 \\ 9x - 3y = 6 \end{cases}$

Determine how the lines in each system of linear equations relate, the number of solutions of the system, and the type of system from the given information. State the solution set of each system. (*See Objective 3.*)

31. $\begin{cases} x + 4y = 5 \\ 2x + 8y = 1 \end{cases}$; Substitution process yields $0 = 9$.

32. $\begin{cases} -2x + 4y = 1 \\ 5x - 10y = 5 \end{cases}$; Substitution process yields $-2 = 1$.

33. $\begin{cases} 6x - 3y = 3 \\ 2x = y + 1 \end{cases}$; Substitution process yields $3 = 3$.

34. $\begin{cases} 2x = 3y + 4 \\ 6x - 9y = 12 \end{cases}$; Substitution process yields $12 = 12$.

35. $\begin{cases} y = \dfrac{1}{3}x - 5 \\ 2x - y = 15 \end{cases}$; Substitution process yields $x = 6$.

36. $\begin{cases} y = -\dfrac{5}{3}x - 9 \\ 6x - y = -37 \end{cases}$; Substitution process yields $x = -6$.

 Mix 'Em Up!

Solve each system of linear equations by substitution.

37. $\begin{cases} 3x - 7y = -4 \\ x = 3 - 2y \end{cases}$

38. $\begin{cases} 4x - 5y = 23 \\ y = 1 - 2x \end{cases}$

39. $\begin{cases} y = -\dfrac{3}{4}x + 1 \\ 3x - 2y = 34 \end{cases}$

40. $\begin{cases} y = -\dfrac{5}{2}x + \dfrac{1}{2} \\ -7x = 2y + 9 \end{cases}$

41. $\begin{cases} \dfrac{x}{2} + 3y = 1 \\ x = -6y + 2 \end{cases}$

42. $\begin{cases} \dfrac{x}{3} + y = 2 \\ 2x + 6y = 12 \end{cases}$

43. $\begin{cases} 5x + 2y = 12 \\ 2x - y = 3 \end{cases}$

44. $\begin{cases} 8x = 12 - 2y \\ 4x + y = 3 \end{cases}$

45. $\begin{cases} 3x - 2y = 6 \\ -x + 4y = 3 \end{cases}$

46. $\begin{cases} \dfrac{x}{2} + \dfrac{y}{3} = -2 \\ 3x + 2y = 6 \end{cases}$

47. $\begin{cases} \dfrac{x}{5} + \dfrac{y}{10} = \dfrac{3}{5} \\ \dfrac{x}{4} - \dfrac{y}{2} = \dfrac{1}{4} \end{cases}$

48. $\begin{cases} 10y = 1 - 5x \\ 5x - 10y = -1 \end{cases}$

49. $\begin{cases} y = -1.8x + 1.7 \\ 0.6x - 3.5y = 25.1 \end{cases}$

50. $\begin{cases} y = 6.2x - 14.5 \\ 4.8x - 2.5y = -1.2 \end{cases}$

You Be the Teacher!

Answer each student's questions.

51. Clarice: When I solve the system $\begin{cases} 4x + 2y = 8 \\ x + \dfrac{1}{2}y = 1 \end{cases}$ using substitution, all of the variables cancel out. Am I doing something wrong?

52. Janki: When I solve the system $\begin{cases} 9x - 15y = -15 \\ y = \dfrac{3}{5}x + 1 \end{cases}$ using substitution, all of the variables and constants cancel out. Am I doing something wrong?

Correct each student's errors, if any.

53. Solve the system by substitution: $\begin{cases} 4x - 2y = 10 \\ 3x + 2y = 4 \end{cases}$.

Carrel's work: I used the first equation to solve for y:

$4x - 2y = 10 \Rightarrow -2y = -4x + 10 \Rightarrow y = \dfrac{-4x + 10}{2} = -2x + 10$

Substituting this into the second equation for y:

$3x + 2(-2x + 10) = 4$

$3x - 4x + 20 = 4$

$-x + 20 = 4 \Rightarrow -x = -16 \Rightarrow x = 16$. Substituting this value into the equation I solved for y gives me:

$y = -2(16) + 10 = -22$. So, my answer is $(16, -22)$.

54. Solve the system $\begin{cases} 2x - y = 3 \\ 3x + 2y = 1 \end{cases}$ by substitution.

Sarah's work:

I used the first equation to solve for y:

$2x - y = 3 \Rightarrow -y = -2x + 3 \Rightarrow y = 2x + 3$

Substituting this into the second equation for y:

$3x + 2(2x + 3) = 1$

$3x + 4x + 3 = 1$

$7x + 3 = 1 \Rightarrow 7x = -2 \Rightarrow x = -\dfrac{2}{7}$. Substituting this value into the equation I solved for y gives me:

$y = 2\left(-\dfrac{2}{7}\right) + 3 = -\dfrac{4}{7} + 3 = \dfrac{17}{7}$.

So, my answer is $\left(-\dfrac{3}{7}, \dfrac{22}{7}\right)$.

 Calculate It!

Use a graphing calculator to determine if the ordered pair is a solution of the given system.

55. $\begin{cases} 3x + y = 7 \\ y = -2x + 2 \end{cases}$; $(5, -8)$ **56.** $\begin{cases} 5x - y = 2 \\ y = 3x - 1 \end{cases}$; $\left(\dfrac{1}{2}, \dfrac{1}{2}\right)$

Solve each system by substitution. Then use a graphing calculator and the Intersect command to check answers.

57. $\begin{cases} 10x + 9y = -23 \\ 15x - 12y = 144 \end{cases}$ **58.** $\begin{cases} 9x + 5y = 2 \\ 3x - 4y = 12 \end{cases}$

 Think About It!

59. The expression $4x - 2$ is substituted for y in the second equation of a system. If the solution of the system is $(3, 10)$, what is the equation of the other line in the system?

60. The expression $\dfrac{1}{3}y + 1$ is substituted for x in the second equation of a system. If the solution of the system is $(-1, -6)$, what is the equation of the other line in the system?

61. The expression $-5x + 3$ is substituted for y in the second equation of a system. If the solution set of the system is $\varnothing$, what is the equation of the other line in the system?

62. The expression $7y - 1$ is substituted for x in the second equation of a system. If the solution set of the system is $\varnothing$, what is the equation of the other line in the system?

63. If one equation in a system is $2x + y = 6$ and the equation that results after substitution is $0 = 5$, what is the equation of the other line in the system?

64. If one equation in a system is $x - 3y = 4$ and the equation that results after substitution is $0 = -1$, what is the equation of the other line in the system?

65. If one equation in a system is $2x + y = 6$ and the equation that results after substitution is $0 = 0$, what is the equation of the other line in the system?

66. If one equation in a system is $x - 3y = 4$ and the equation that results after substitution is $0 = 0$, what is the equation of the other line in the system?

SECTION 4.3   **Solving Systems of Linear Equations by Elimination**

▶ **OBJECTIVES**

As a result of completing this section, you will be able to

1. Solve a system of linear equations by elimination.
2. Solve special cases of systems of linear equations by elimination.
3. Determine how the lines in a system of linear equations relate, the number of solutions, and the type of system.
4. Solve application problems.
5. Troubleshoot common errors.

As of 2010, Nintendo DS and Nintendo Wii were the two top-selling gaming units in the United States. A total of 81 million gaming units have been sold. If there have been 13 million more Nintendo DS units sold than Nintendo Wii units, how many of each gaming unit has been sold in the United States? (Source: http://www.vgchartz.com/home.php)

In this section, we will learn another method for solving systems of linear equations in two variables and how to apply these methods to solving a problem such as the one just stated.

The Elimination Method

Objective 1 ▶

Solve a system of linear equations by elimination.

While substitution is a precise algebraic method to solve a system of linear equations, it can be tedious if the coefficients of the variables in the system are not 1. In this section, we will explore another method to solve systems, the **elimination method** or **addition method**.

The goal of the elimination method is to add the two equations in the system together so that one of the variables is eliminated—that is, the terms containing that variable add to zero. This process creates an equation in just one variable that enables us to solve for one of the variables in the system.

Consider the system $\begin{cases} 3x + y = 5 \\ -3x + 3y = 7 \end{cases}$.

$$\begin{cases} 3x + y = 5 \\ -3x + 3y = 7 \end{cases}$$ Note the coefficients of x, 3 and -3, are opposites.

$0x + 4y = 12$ Add the equations in the system to eliminate x.

$4y = 12$ Simplify.

$\dfrac{4y}{4} = \dfrac{12}{4}$ Divide each side by 4.

$y = 3$ Simplify.

Now we solve for x by substituting 3 for y in one of the equations.

$3x + y = 5$ Choose one of the original equations.

$3x + 3 = 5$ Substitute 3 for y.

$3x + 3 - 3 = 5 - 3$ Subtract 3 from each side.

$3x = 2$ Simplify.

$x = \dfrac{2}{3}$ Divide each side by 3.

So, the solution set of the system is $\left\{ \left(\dfrac{2}{3}, 3 \right) \right\}$.

In the system, $\begin{cases} 4x - y = 3 \\ 5x + 2y = 7 \end{cases}$, neither the coefficients of x nor y are opposites. To eliminate the variable y, we can multiply the first equation by 2.

$$\begin{cases} 2(4x - y) = 2(3) \\ 5x + 2y = 7 \end{cases}$$ Multiply the first equation by 2.

$$\begin{cases} 8x - 2y = 6 \\ 5x + 2y = 7 \end{cases}$$ The coefficients of the y-terms in each equation are now opposites.

$13x + 0y = 13$ Add the equations to eliminate y.

$13x = 13$ Simplify.

$x = 1$ Divide each side by 13.

Now we solve for y by substituting 1 for x in one of the two equations.

$5x + 2y = 7$ Choose one of the original equations.

$5(1) + 2y = 7$ Substitute 1 for x.

$5 + 2y = 7$ Simplify.

$5 + 2y - 5 = 7 - 5$ Subtract 5 from each side.

$2y = 2$ Simplify.

$\dfrac{2y}{2} = \dfrac{2}{2}$ Divide each side by 2.

$y = 1$ Simplify.

So, the solution set of the system is $\{(1, 1)\}$.

The elimination method is based on the addition property of equality. This property enables us to add equivalent expressions to each side of an equation and not change the solution.

> **Property: Addition Property of Equality**
> $$\text{If } A = B, \text{ then } A + C = B + C.$$

The steps that were used to solve the preceding systems of linear equations are generalized as follows.

> **Procedure: Solving a System of Linear Equations by Elimination**
>
> **Step 1:** Write each equation in the system in standard form $(Ax + By = C)$.
> **Step 2:** Create a new system with opposites as coefficients of one of the variables, if necessary.
>   **a.** Choose the variable to be eliminated.
>   **b.** Find a nonzero number, a, and multiply one or both equations by the number that will produce coefficients of a and $-a$ in the new system.
> **Step 3:** Add the equations together.
> **Step 4:** Solve the resulting equation for the variable.
> **Step 5:** Substitute this value into one of the original equations to find the value of the other variable.
> **Step 6:** Check by substituting the ordered pair into the equations in the original system.

Objective 1 Examples   Solve each system of linear equations by elimination.

1a. $\begin{cases} 2x - y = 4 \\ 3x + y = 1 \end{cases}$ **1b.** $\begin{cases} 4x + 3y = 12 \\ 2x - 5y = 6 \end{cases}$ **1c.** $\begin{cases} 7x - 2y = 4 \\ 3x + 9y = -10 \end{cases}$

Solutions   **1a.** The coefficients of y, -1 and 1, are opposites, so we add the equations in this form.

$$\begin{cases} 2x - y = 4 \\ 3x + y = 1 \end{cases}$$

$$5x = 5 \qquad \text{Add the equations.}$$

$$\frac{5x}{5} = \frac{5}{5} \qquad \text{Divide each side by 5.}$$

$$x = 1 \qquad \text{Simplify.}$$

Now solve for y, by substituting 1 in either of the original equations for x.

$$2x - y = 4 \qquad \text{Begin with the first equation.}$$

$$2(1) - y = 4 \qquad \text{Substitute 1 for } x.$$

$$2 - y = 4 \qquad \text{Simplify.}$$

$$2 - y - 2 = 4 - 2 \qquad \text{Subtract 2 from each side.}$$

$$-y = 2 \qquad \text{Simplify.}$$

$$-1(-y) = -1(2) \qquad \text{Multiply each side by } -1.$$

$$y = -2 \qquad \text{Simplify.}$$

The solution of the system is the ordered pair $(1, -2)$.

Check:

$$
\begin{array}{c|c}
2x - y = 4 & 3x + y = 1 \\
2(1) - (-2) = 4 & 3(1) + (-2) = 1 \\
2 + 2 = 4 & 3 - 2 = 1 \\
4 = 4 & 1 = 1 \\
\text{True} & \text{True}
\end{array}
$$

Since $(1, -2)$ makes each equation true, the solution set is $\{(1, -2)\}$.

1b. Neither the coefficients of x nor y are opposites. We choose to eliminate the variable x. The coefficients of x, 4 and 2, divide evenly into 4. We must create an equivalent system with equations that contain $4x$ and $-4x$. The first equation contains $4x$ so nothing needs to be done to this equation. To create $-4x$ in the second equation, multiply the equation by $\dfrac{-4}{2} = -2$.

$$
\begin{cases} 4x + 3y = 12 \\ 2x - 5y = 6 \end{cases} \longrightarrow \begin{cases} 4x + 3y = 12 \\ -2(2x - 5y) = -2(6) \end{cases} \longrightarrow \begin{cases} 4x + 3y = 12 \\ -4x + 10y = -12 \end{cases}
$$

Now add the equations to solve for y.

$$
\begin{cases} 4x + 3y = 12 \\ -4x + 10y = -12 \end{cases}
$$

$$
\begin{aligned}
13y &= 0 && \text{Add the equations.} \\
\frac{13y}{13} &= \frac{0}{13} && \text{Divide each side by 13.} \\
y &= 0 && \text{Simplify.}
\end{aligned}
$$

Now solve for x, by substituting 0 for y in either of the original equations.

$$
\begin{aligned}
2x - 5y &= 6 && \text{Begin with the second equation.} \\
2x - 5(0) &= 6 && \text{Substitute 0 for } y. \\
2x &= 6 && \text{Simplify.} \\
\frac{2x}{2} &= \frac{6}{2} && \text{Divide each side by 2.} \\
x &= 3 && \text{Simplify.}
\end{aligned}
$$

The solution of the system is the ordered pair $(3, 0)$. It makes each equation true, so the solution set is $\{(3, 0)\}$.

1c. Neither the coefficients of x nor y are opposites. We choose to eliminate the variable y. The coefficients of y, -2 and 9, divide evenly into 18. We must create an equivalent system with equations that contain $18y$ and $-18y$. To create $-18y$ in the first equation, multiply the equation by $\dfrac{-18}{-2} = 9$. To create $18y$ in the second equation, multiply by $\dfrac{18}{9} = 2$.

$$
\begin{cases} 7x - 2y = 4 \\ 3x + 9y = -10 \end{cases} \longrightarrow \begin{cases} 9(7x - 2y) = 9(4) \\ 2(3x + 9y) = 2(-10) \end{cases} \longrightarrow \begin{cases} 63x - 18y = 36 \\ 6x + 18y = -20 \end{cases}
$$

Now add the equations to solve for x.

$$
\begin{cases} 63x - 18y = 36 \\ 6x + 18y = -20 \end{cases}
$$

$$
69x = 16 \qquad \text{Add the equations.}
$$

$$\frac{69x}{69} = \frac{16}{69} \qquad \text{Divide each side by 69.}$$

$$x = \frac{16}{69} \qquad \text{Simplify.}$$

Now we solve for y by substituting $\dfrac{16}{69}$ for x in either of the original equations.

$$3x + 9y = -10 \qquad \text{Begin with the second equation.}$$

$$3\left(\frac{16}{69}\right) + 9y = -10 \qquad \text{Substitute } \frac{16}{69} \text{ for } x.$$

$$\frac{16}{23} + 9y = -10 \qquad \text{Simplify.}$$

$$23\left(\frac{16}{23} + 9y\right) = 23(-10) \qquad \text{Multiply each side by the LCD, 23.}$$

$$16 + 207y = -230 \qquad \text{Simplify.}$$

$$16 + 207y - 16 = -230 - 16 \qquad \text{Subtract 16 from each side.}$$

$$207y = -246 \qquad \text{Simplify.}$$

$$y = -\frac{246}{207} \qquad \text{Divide each side by 207.}$$

$$y = -\frac{82}{69} \qquad \text{Simplify the fraction.}$$

The solution of the system is the ordered pair $\left(\dfrac{16}{69}, -\dfrac{82}{69}\right)$. It makes each equation true, so the solution set is $\left\{\left(\dfrac{16}{69}, -\dfrac{82}{69}\right)\right\}$.

✓ **Student Check 1**   Solve each system of linear equations by elimination.

a. $\begin{cases} x + 4y = 8 \\ 3x - 4y = 4 \end{cases}$
b. $\begin{cases} 6x + 9y = 15 \\ 2x + 5y = 11 \end{cases}$
c. $\begin{cases} 7x - 3y = 6 \\ 5x + 4y = 8 \end{cases}$

Special Cases of Systems of Linear Equations

Objective 2 ▶

Solve special cases of systems of linear equations by elimination.

We now see how elimination can be used to obtain solutions of special systems of linear equations.

> **Procedure: Solving a Special System of Linear Equations by Elimination**
>
> **Step 1:** Apply the steps shown in Objective 1.
> **Step 2:** The equation that results from adding the two equations in the system together will be a contradiction or an identity.
>   **a.** If the resulting equation is a false statement (i.e., a contradiction), then there is no solution of the system and the lines are parallel.
>   **b.** If the resulting equation is a true statement (i.e., an identity), then there are infinitely many solutions and the lines are the same.

Objective 2 Examples   Solve each system of linear equations by elimination.

2a. $\begin{cases} x + 5y = -5 \\ 2x + 10y = 5 \end{cases}$
2b. $\begin{cases} 3x - y = 6 \\ 12x - 4y = 24 \end{cases}$

Solutions   **2a.** We choose to eliminate the variable x. The number that the coefficients of x, 1 and 2, divide into evenly is 2. We must create an equivalent system that contains $2x$ and $-2x$. The second equation contains $2x$. To obtain $-2x$ in the first equation, we multiply it by $\dfrac{-2}{1} = -2$.

$$\begin{cases} x + 5y = -5 \\ 2x + 10y = 5 \end{cases} \longrightarrow \begin{cases} -2(x + 5y) = -2(-5) \\ 2x + 10y = 5 \end{cases} \longrightarrow \begin{cases} -2x - 10y = 10 \\ 2x + 10y = 5 \end{cases}$$

Now add the equations to solve for y.

$$\begin{cases} -2x - 10y = 10 \\ \underline{2x + 10y = 5} \\ 0 = 15 \end{cases}$$ Add the equations.

The resulting equation, $0 = 15$, is a contradiction, or a false statement. Therefore, the system has no solution. The lines are parallel and the solution set is the empty set, $\varnothing$.

2b. We choose to eliminate the variable x. The number that the coefficients of x, 3 and 12, divide into evenly is 12. We must create an equivalent system that contains $12x$ and $-12x$. The second equation contains $12x$. We multiply the first equation by $\dfrac{-12}{3} = -4$ to obtain $-12x$.

$$\begin{cases} 3x - y = 6 \\ 12x - 4y = 24 \end{cases} \longrightarrow \begin{cases} -4(3x - y) = -4(6) \\ 12x - 4y = 24 \end{cases} \longrightarrow \begin{cases} -12x + 4y = -24 \\ 12x - 4y = 24 \end{cases}$$

Now add the equations to solve for y.

$$\begin{cases} -12x + 4y = -24 \\ \underline{12x - 4y = 24} \\ 0 = 0 \end{cases}$$ Add the equations.

The resulting equation, $0 = 0$, is an identity, or a true statement. Therefore, the system has infinitely many solutions and the lines are the same. The solution set of the system is $\{(x, y) \mid 3x - y = 6\}$.

✔ **Student Check 2**   Solve each system of linear equations by elimination.

a. $\begin{cases} 8x - 10y = 4 \\ 12x - 15y = 6 \end{cases}$ **b.** $\begin{cases} 6x - 2y = -8 \\ 3x - y = -2 \end{cases}$

Determine the Relationship of the Lines in a System of Linear Equations

Objective 3 ▶

Determine how the lines in a system of linear equations relate, the number of solutions, and the type of system.

Just as we have seen with graphing and substitution, the elimination method for solving systems of linear equations produces one of three types of solutions. It is helpful to understand the equations that result after adding the equations in the system and what they mean as far as how the lines in the system relate to one another, the number of solutions of the system, and the type of system. The following table summarizes the three possibilities.

Intersecting Lines	**Parallel Lines**	**Same Lines**
After adding the two equations, we will get an equation of the form $x = k$ or $y = k$, where k is a constant.	After adding the equations, a *false* statement or contradiction results; for example, $0 = 8$.	After adding the equations, a *true* statement or identity results; for example, $0 = 0$.
One solution	No solution	Infinitely many solutions
Consistent system with independent equations	Inconsistent system	Consistent system with dependent equations

Objective 3 Examples Determine how the lines in each system of linear equations relate, the number of solutions of the system, and the type of system from the given information. State the solution set of the system.

3a. $\begin{cases} 4x + y = 12 \\ x - 5y = 3 \end{cases}$; The elimination process yields $21y = 0$.

3b. $\begin{cases} \dfrac{3}{2}x - \dfrac{7}{3}y = \dfrac{1}{6} \\ \dfrac{9}{5}x - \dfrac{14}{5}y = \dfrac{1}{5} \end{cases}$; The elimination process yields $0 = 0$.

Solutions **3a.** The equation $21y = 0$ can be solved for y. We get $y = 0$. Therefore, this system contains intersecting lines with one solution. The system is consistent with independent equations. The solution is found by replacing y with 0 in one of the given equations.

$$4x + y = 12$$
$$4x + 0 = 12$$
$$4x = 12$$
$$x = 3$$

The solution set is $\{(3, 0)\}$.

3b. The equation $0 = 0$ is a true statement, or an identity. This means that the equations in the system are the same line and that there are infinitely many solutions. The system is consistent with dependent equations. The solution set is $\{(x, y) \mid 9x - 14y = 1\}$.

 Note: *Note that if we multiply the first equation by 6, we get*

$$6\left(\frac{3}{2}x - \frac{7}{3}y = \frac{1}{6}\right) \longrightarrow 9x - 14y = 1$$

If we multiply the second equation by 5, we get

$$5\left(\frac{9}{5}x - \frac{14}{5}y = \frac{1}{5}\right) \longrightarrow 9x - 14y = 1$$

☑ **Student Check 3**   Determine how the lines in each system of linear equations relate, the number of solutions of the system, and the type of system from the given information. State the solution set of the system.

a. $\begin{cases} y = \dfrac{3}{2}x - 6 \\ 9x - 6y = 24 \end{cases}$; The elimination process yields $0 = -12$.

b. $\begin{cases} 10x + \dfrac{1}{3}y = 8 \\ 7x - y = 13 \end{cases}$; The elimination process yields $37x = 37$.

Applications

Objective 4 ▶

Solve application problems.

Many word problems that we solve involve more than one unknown. When this is the case, systems are the best method to employ to solve the problem. A system enables us to assign a different variable for each of the unknown values.

We assign two different variables to the two unknown values, and then use a system of linear equations in two variables to solve the problem. There are four general steps we will follow to solve such problems.

Procedure: General Problem Solving Strategy

Step 1: Read the problem and determine the two unknown values. Assign a different variable to represent each of these values.

Step 2: Read the problem and write two equations that define a relationship between the two variables.

Step 3: Solve the system of equations using elimination.

Step 4: Write the answer to the problem.

Objective 4 Example   As of 2010, Nintendo DS and Nintendo Wii were the two top-selling gaming units in the United States. A total of 81 million gaming units have been sold. If there have been 13 million more Nintendo DS units sold than Nintendo Wii units, how many of each gaming unit has been sold in the United States? (Source: http://www.vgchartz.com/home.php)

Solution   What are the unknowns? The unknowns are the number of Nintendo DS units and the number of Nintendo Wii units sold.

Let x = number of Nintendo DS units sold.
Let y = number of Nintendo Wii games sold.

What is known? The total gaming units sold was 81 million. There were 13 million more DS units sold than Wii units. The system of equations that represents this information is

$$\begin{cases} x + y = 81 \\ x = y + 13 \end{cases}$$

Solve the system using elimination.

$\begin{cases} x + y = 81 \\ x - y = 13 \end{cases}$ Rewrite the second equation in standard form by subtracting y from each side.

$\overline{\qquad 2x = 94 \qquad}$ Add the equations.

$\dfrac{2x}{2} = \dfrac{94}{2}$ Divide each side by 2.

$x = 47$ Simplify.

Now solve for y by replacing x with 47 in either equation.

$x + y = 81$	Begin with the first equation.
$47 + y = 81$	Substitute 47 for x.
$47 + y - 47 = 81 - 47$	Subtract 47 from each side.
$y = 34$	Simplify.

As of 2010, there have been 47 million Nintendo DS units sold and 34 million Nintendo Wii gaming units sold in the United States.

✓ Student Check 4 In November 2009, the Xbox 360 and the PlayStation 3 (PS3) were the third and fourth top-selling gaming units. A total of 1,529,900 gaming units were sold. If there were 109,100 more Xbox 360 units sold than PS3 units, how many of each gaming unit was sold? (Source: http://www.1up.com/do/newsStory?cId=3177273)

Objective 5 ▶

Troubleshoot common errors.

Troubleshooting Common Errors

A common error associated with elimination is shown next.

Objective 5 Example A problem and an incorrect solution are given. Provide the correct solution and an explanation of the error.

Solve the system $\begin{cases} 6x - 5y = 27 \\ 3x + y = 3 \end{cases}$ by elimination.

Incorrect Solution	Correct Solution and Explanation
$\begin{cases} 6x - 5y = 27 \\ 3x + y = 3 \end{cases}$	When multiplying the second equation by 5, we must distribute 5 to each side of the equation.
$\begin{cases} 6x - 5y = 27 \\ 15x + 5y = 3 \end{cases}$ $21x = 30$ $x = \dfrac{30}{21}$ $x = \dfrac{10}{7}$ $3\left(\dfrac{10}{7}\right) + y = 3$ $\dfrac{30}{7} + y = 3$ $30 + 7y = 21$ $7y = -9$ $y = -\dfrac{9}{7}$ The solution set is $\left\{\left(\dfrac{10}{7}, -\dfrac{9}{7}\right)\right\}$.	$\begin{cases} 6x - 5y = 27 \\ 15x + 5y = 15 \end{cases}$ $21x = 42$ $x = 2$ $3(2) + y = 3$ $6 + y = 3$ $y = -3$ The solution set is $\{(2, -3)\}$.

ANSWERS TO STUDENT CHECKS

Student Check 1   **a.** $\left\{\left(3, \dfrac{5}{4}\right)\right\}$ **b.** $\{(-2, 3)\}$

c. $\left\{\left(\dfrac{48}{43}, \dfrac{26}{43}\right)\right\}$

Student Check 2   **a.** $\{(x, y)\,|\,8x - 10y = 4\}$
    b. $\varnothing$

Student Check 3   **a.** parallel lines, no solution, inconsistent system, $\varnothing$ **b.** intersecting lines, one solution, consistent system with independent equations, $\{(1, -6)\}$

Student Check 4   There were 819,500 Xbox 360 units sold and 710,400 PS3 units sold.

SUMMARY OF KEY CONCEPTS

1. Elimination is an algebraic method that enables us to find the exact solution of a system of linear equations.
2. To solve a linear system by elimination, the equations in the system should be written in standard form. The goal is to add the equations together so that one of the variables is eliminated. Multiply one or both equations by some nonzero number so that the coefficients of one of the variables are opposites, if needed.
3. If the resulting equation is a contradiction or an identity, then the system has no solution or infinitely many solutions, respectively.
4. Answers can be checked by substituting the ordered pair into both equations or by graphing.
5. Systems can be used to solve problems involving two unknowns. From the information given, we need to determine the two unknowns and the two equations that represent the situation.

SECTION 4.3 / EXERCISE SET

 ### Write About It!

Use complete sentences in your answer to each exercise.

1. Explain how to solve a system of linear equations using elimination.
2. How do you determine the numbers by which to multiply the equations in a system in order to apply elimination?
3. When solving a system of linear equations by elimination, what will happen if there are infinitely many solutions?
4. When solving a system of linear equations by elimination, what will happen if there is no solution?

 ### Practice Makes Perfect!

Solve each system of linear equations by elimination. (See Objective 1.)

5. $\begin{cases} 7x + 4y = 5 \\ -7x + 2y = 1 \end{cases}$ 6. $\begin{cases} 6x - y = 9 \\ 6x + y = 0 \end{cases}$

7. $\begin{cases} x - 4y = -2 \\ 3x + 2y = -13 \end{cases}$ 8. $\begin{cases} -12x + y = 12 \\ 8x + y = 9 \end{cases}$

9. $\begin{cases} 4x - 5y = 17 \\ 8x + 2y = -2 \end{cases}$ 10. $\begin{cases} 10x - 3y = 1 \\ -2x + 6y = 16 \end{cases}$

11. $\begin{cases} 7x + 6y = 10 \\ 3x + 9y = 12 \end{cases}$ 12. $\begin{cases} 10x - 3y = -69 \\ x + \dfrac{3y}{2} = -\dfrac{3}{2} \end{cases}$

13. $\begin{cases} \dfrac{x}{6} + \dfrac{y}{8} = -1 \\ -3x - 2y = 4 \end{cases}$ 14. $\begin{cases} \dfrac{x}{4} - \dfrac{3y}{4} = -1 \\ 4x - 7y = 44 \end{cases}$

Solve each system of linear equations by elimination. (See Objective 2.)

15. $\begin{cases} 4x - 6y = 5 \\ -2x + 3y = 4 \end{cases}$ 16. $\begin{cases} -3x + 7y = -4 \\ 24x - 56y = 32 \end{cases}$

17. $\begin{cases} 5x - 8y = 10 \\ 10x - 16y = 20 \end{cases}$ 18. $\begin{cases} 9x - 3y = 12 \\ 3x - y = -10 \end{cases}$

19. $\begin{cases} 6x + 18y = 20 \\ \dfrac{x}{6} + \dfrac{y}{2} = \dfrac{5}{9} \end{cases}$ 20. $\begin{cases} 14x - 21y = 28 \\ \dfrac{x}{3} - \dfrac{y}{2} = -\dfrac{2}{3} \end{cases}$

Determine how the lines in each system of linear equations relate, the number of solutions of the system, and the type of system from the given information. State the solution set of the system. (See Objective 3.)

21. $\begin{cases} 7x - 2y = 14 \\ y = \dfrac{7}{2}x - 7 \end{cases}$; The elimination process yields $0 = 0$.

22. $\begin{cases} 2x + 5y = 12 \\ y = -0.4x + 2.4 \end{cases}$; The elimination process yields $0 = 0$.

23. $\begin{cases} x - 11y = 22 \\ -3x + 33y = -60 \end{cases}$; The elimination process yields $0 = 6$.

24. $\begin{cases} y = 1.5x - 7.5 \\ 3x - 2y = 18 \end{cases}$; The elimination process yields $0 = 3$.

25. $\begin{cases} 2x + 9y = 5 \\ -x + 4y = 6 \end{cases}$; The elimination process yields $17y = 17$.

26. $\begin{cases} 2x - 5y = 15 \\ y = -\dfrac{3}{5}x + 1 \end{cases}$; The elimination process yields $5x = 20$.

Solve each problem. (*See Objective 4.*)

27. In 2008, the two largest toy manufacturers, Mattel and Hasbro, reported a total net sales of $9.939 billion. If the reported net sales of Mattel was $2.124 billion less than twice that of Hasbro, find the net sales of each company. (Source: http://www.ita.doc.gov/td/ocg/toyoutlook_09.pdf)

28. In 2010, there were 25.8 million people, diagnosed and undiagnosed, in the United States affected by diabetes. If the number of people diagnosed with diabetes is 4.8 million more than twice the number of people undiagnosed, find the number of diagnosed and undiagnosed cases of diabetes in the United States (Source: http://diabetes.niddk.nih.gov/dm/pubs/statistics/#fast)

29. In 2010, there were a total of 6.43 million homes sold or foreclosed in the United States. If the number of homes sold was 7.92 million less than four times the number of foreclosures, find the number of homes sold and foreclosed. (Sources: http://www.census.gov/const/newressales.pdf and http://www.tampabay.com/news/business/banking/florida-ranks-second-in-number-of-foreclosures-for-2010/1145229)

30. In May 2011, the top two automobile corporations, General Motors Corporation and Ford Motor Company, sold a total of approximately 413,000 vehicles in the United States. If the Ford Motor Company sold 250,000 less than twice the sales of General Motors, find the number of vehicles sold by each company. (Source: http://www.motorintelligence.com/m_frameset.html)

 Mix 'Em Up!

Solve each system of linear equations by elimination.

31. $\begin{cases} 18x + 8y = 11 \\ 9x + 4y = 5 \end{cases}$

32. $\begin{cases} 4x - 15y = -6 \\ 12x - 3y = 42 \end{cases}$

33. $\begin{cases} 2x = 5y + 8 \\ 3x - y = 1 \end{cases}$

34. $\begin{cases} 12x - 6 = 10y \\ -6x + 5y = -3 \end{cases}$

35. $\begin{cases} 3x - 2y = 6 \\ 15x = 10y - 12 \end{cases}$

36. $\begin{cases} -6y = -5x + 4 \\ 20x - 3y = 19 \end{cases}$

37. $\begin{cases} 2x + 5y = 4 \\ \dfrac{x}{10} + \dfrac{y}{4} = \dfrac{1}{5} \end{cases}$

38. $\begin{cases} 7x - 3y = 2 \\ -\dfrac{x}{4} + \dfrac{y}{7} = -\dfrac{1}{14} \end{cases}$

39. $\begin{cases} \dfrac{x}{10} - y = \dfrac{5}{2} \\ 4x + 5y = 1 \end{cases}$

40. $\begin{cases} 2x - 8y = 13 \\ \dfrac{x}{8} - y = 0 \end{cases}$

41. $\begin{cases} 0.4x - 1.4y = 5.88 \\ -1.6x + 2.5y = -13.6 \end{cases}$

42. $\begin{cases} 4.5x + 1.2y = 5.1 \\ 1.5x = 6.4y + 18.7 \end{cases}$

43. $\begin{cases} 3.5x = 1.2y - 11.81 \\ -0.6x + 2.4y = -3.9 \end{cases}$

44. $\begin{cases} 3.6x - 0.5y = 8.01 \\ 0.4x = 2.6y + 3.18 \end{cases}$

 You Be the Teacher!

Answer each student's question or correct the errors, if any.

45. Solve the system $\begin{cases} 3x + 4y = 8 \\ 5x - 3y = 1 \end{cases}$.

Cory: I'm not sure which variable I should eliminate. Which variable should I eliminate in this system and why?

46. Solve the system $\begin{cases} 2x - 5y = 6 \\ 6x - 2y = 1 \end{cases}$.

Andy's work:

$\begin{cases} -3(2x - 5y) = 6 \\ 6x - 2y = 1 \end{cases}$

$\begin{cases} -6x + 15y = 6 \\ \underline{6x - 2y = 1} \end{cases}$
$\qquad\qquad 13y = 7$
$\qquad\qquad\quad y = \dfrac{7}{13}$

$2x - 5y = 6$

$2x - 5\left(\dfrac{7}{13}\right) = 6$

$2x - \dfrac{35}{13} = 6$

$2x = 6 + \dfrac{35}{13}$

$2x = \dfrac{113}{13}$

$x = \dfrac{113}{26}$

So, my answer is $\left(\dfrac{113}{26}, \dfrac{7}{13}\right)$.

47. Solve the system $\begin{cases} -12x + 15y = 10 \\ 14x + 6y = 11 \end{cases}$.

Jalen: I have decided to eliminate the x-variable. How do I determine the numbers to multiply each equation by to eliminate x?

48. Solve the system $\begin{cases} 6x + 28y = 25 \\ 8x - 70y = 15 \end{cases}$.

Edna: I know that 6 and 8 will each divide evenly into 48, so I write an equivalent system that contains $48x$ and $-48x$ in it. A friend told me that I should have used the least common multiple of 6 and 8, which is 24. Will I get the same solution if I use 48 instead of 24? Please explain.

Calculate It!

Use elimination to solve each system algebraically. Then use a graphing calculator and the Intersect command to check the answer.

49. $\begin{cases} 3x - 5y = 21 \\ 4x + y = 5 \end{cases}$ **50.** $\begin{cases} 5x + 6y = 8 \\ 2x + 3y = 4 \end{cases}$

51. $\begin{cases} 6x = 4y + 5 \\ x - y = 2 \end{cases}$ **52.** $\begin{cases} -9y = 10 - 10x \\ 15x - 12y = -3 \end{cases}$

PIECE IT TOGETHER   SECTIONS 4.1–4.3

Determine if the given ordered pair is a solution of the system. (*Section 4.1, Objective 1*)

1. Is $\left(\dfrac{8}{7}, 0\right)$ a solution of $\begin{cases} 7x - 8y = 8 \\ 14x - 16y = 16 \end{cases}$?

2. Is $\left(0, -\dfrac{10}{3}\right)$ a solution of $\begin{cases} x - 3y = 10 \\ 2x - 6y = -5 \end{cases}$?

Solve each system by graphing. (*Section 4.1, Objective 3*)

3. $\begin{cases} x + y = -1 \\ 2x - y = -8 \end{cases}$ **4.** $\begin{cases} x - 4y = 8 \\ x + y = -2 \end{cases}$

5. $\begin{cases} x + 2y = 1 \\ 3x + 6y = -5 \end{cases}$ **6.** $\begin{cases} 3x - y = 1 \\ 3y = 9x - 3 \end{cases}$

Solve each system by substitution or elimination. (*Sections 4.2 and 4.3, Objectives 1 and 2*)

7. $\begin{cases} 4x + 3y = 14 \\ x + 2y = 1 \end{cases}$ **8.** $\begin{cases} 5x - 2y = 14 \\ 3x + y = 4 \end{cases}$

9. $\begin{cases} \dfrac{x}{2} - \dfrac{y}{4} = 1 \\ x = 4y \end{cases}$ **10.** $\begin{cases} 8x - 2y = 26 \\ 2x = 3y - 1 \end{cases}$

11. $\begin{cases} 2x + y = 3 \\ 4x + 2y = 0 \end{cases}$ **12.** $\begin{cases} 3x + 6y = 3 \\ 5x = -10y + 5 \end{cases}$

13. $\begin{cases} -3x + 4y = 2 \\ -2x + 5y = 6 \end{cases}$ **14.** $\begin{cases} 2x - 5y = 2 \\ 5x - 2y = 5 \end{cases}$

15. $\begin{cases} -5x + 4y = 6 \\ 2x + 3y = 1 \end{cases}$ **16.** $\begin{cases} 2x - 7y = -26 \\ 3x + 2y = 136 \end{cases}$

17. $\begin{cases} x - 6y = 2 \\ 3x = 18y + 6 \end{cases}$ **18.** $\begin{cases} 5x - y = 2 \\ 10x = 2y - 4 \end{cases}$

Determine how the lines in each system of linear equations relate, the number of solutions of the system, and the type of system from the given information. State the solution set of the system. (*Sections 4.2 and 4.3, Objective 3*)

19. $\begin{cases} y = -4x \\ 4x + y = 2 \end{cases}$; The elimination process yields $0 = -2$.

20. $\begin{cases} y = 2.5x - 7.1 \\ 6x - y = 19.7 \end{cases}$; The substitution process yields $3.5x = 12.6$.

Applications of Systems of Linear Equations in Two Variables

Sam and Lori flew from Atlanta, Georgia, to San Francisco, California, a distance of approximately 2135 mi. The flight from Atlanta to San Francisco took 5 hr since the plane was flying against the wind. The return trip took 4.5 hr since the plane was flying with the wind. What was the speed of the plane in still air and what was the speed of the wind?

In this section, we will learn how to use a system of linear equations in two variables to solve this problem and other applications. The applications in this section are similar to the applications covered in Section 2.6. In Section 2.6, there were two unknowns in the problem and we expressed both unknowns in terms of one variable. We then used one linear equation to solve the problem.

Sometimes, it is easier to assign a different variable to each unknown. There are four general steps we will follow to solve such problems.

Procedure: General Problem Solving Strategy

Step 1: Read the problem and determine the two unknown values. Assign a different variable to represent each of these values.

Step 2: Read the problem and write two equations that define a relationship between the two variables.

Step 3: Solve the system of equations using substitution or elimination.

Step 4: Write the answer to the problem.

Money Applications

Objective 1 ▶

Solve money applications.

Some of the applications covered in this objective require us to know how to find the value of a collection of objects. The following table illustrates how to find the total worth of a collection of books, for example, if each book in the collection has the same value.

Number of Books	Value of Each Book	Total Worth of Collection
3	$10	$3(10) = \$30$
7	$15	$7(15) = \$105$
10	$20	$10(20) = \$200$
x	$30	$x(30) = 30x$

So, if we know the value of an individual item, we can determine the worth of a collection of these items by multiplying the value of the individual item by the total number of items in the collection. This concept can be extended to money as we have seen before when dealing with applications of linear equations in one variable. Suppose we have a total of 30 coins consisting of dimes and quarters. The total value of the collection of coins is provided in the following table for several different cases.

Number of Dimes	Number of Quarters	Value of Dimes + Value of Quarters =	Total Worth of Collection
6	$30 - 6 = 24$	$6(0.10) + 24(0.25) = \$0.60 + \$6.00 =$	$\$6.60$
20	$30 - 20 = 10$	$20(0.10) + 10(0.25) = \$2.00 + \$2.50 =$	$\$4.50$
12	$30 - 12 = 18$	$12(0.10) + 18(0.25) = \$1.20 + \$4.50 =$	$\$5.70$
x	$30 - x = y$	$x(0.10) + y(0.25) = 0.10x + 0.25y$	

The objects in the collection can extend beyond coins. They might be a collection of tickets or a collection of hours at work, as we will see in Example 1.

| Objective 1 Examples | Solve each problem using a system of linear equations. |

1a. A movie theater sold 2500 tickets. Adult tickets cost $8.25 each and student tickets cost $6.00 each. If the movie theater collected a total of $17,475 for the tickets, how many adult tickets and student tickets were sold?

Solution

1a. What is unknown? The number of adult tickets and the number of student tickets sold are unknown.

Let a = the number of adult tickets sold.

Let s = the number of student tickets sold.

What is known? The total number of tickets sold is 2500. The total amount of money collected is $17,475. We can organize this information in a table.

Unknowns	Number of tickets sold × Price per ticket = Total money collected		
Adult tickets sold	a	$8.25	$8.25a$
Student tickets sold	s	$6.00	$6.00s$
Totals	2500		$17,475

The system that represents these facts is

$$\begin{cases} a + s = 2500 \\ 8.25a + 6.00s = 17{,}475 \end{cases}$$

A movie theater sold 2500 tickets.

The movie theater collected $17,475.

We solve the system using elimination since both equations are in standard form.

$$\begin{cases} -6(a + s = 2500) \\ 8.25a + 6.00s = 17{,}475 \end{cases} \longrightarrow \begin{cases} -6.00a - 6.00s = -15{,}000 \\ 8.25a + 6.00s = 17{,}475 \end{cases}$$

Multiply the first equation by −6.

$$2.25a = 2475$$
$$a = 1100$$

Add the equations.

Divide each side by 2.25.

Now solve for s.

$$a + s = 2500$$
$$1100 + s = 2500$$
$$1100 + s - 1100 = 2500 - 1100$$
$$s = 1400$$

Begin with the first equation.

Substitute 1100 for a.

Subtract 1100 from each side.

Simplify.

The movie theater sold 1100 adult tickets and 1400 student tickets.

1b. Monte shows his children a jar of coins that consists of only quarters and nickels. He tells them there are 300 coins totaling $51. How many quarters and nickels are in the jar?

Solution

1b. What is unknown? The number of quarters and the number of nickels in the jar are unknown.

Let q = number of quarters.
Let n = number of nickels.

What is known? There are 300 coins in the jar. The value of the coins is $51. We can organize the information in a table.

Unknowns	Number of coins × Value per coin = Total value of coins		
Quarters	q	$0.25	$0.25q$
Nickels	n	$0.05	$0.05n$
Totals	300		$51

The system that represents these facts is

$$\begin{cases} q + n = 300 \\ 0.25q + 0.05n = 51 \end{cases}$$ There is a total of 300 coins in the jar.
The total value of the collection is $51.

We solve the system by substitution and solve the first equation for q.

$$q + n = 300$$ Begin with the first equation.
$$q + n - n = 300 - n$$ Subtract n from each side.
$$q = 300 - n$$ Simplify.

Now we substitute this expression in place of q in the other equation.

$$0.25q + 0.05n = 51$$ Begin with the second equation.
$$0.25(300 - n) + 0.05n = 51$$ Substitute $300 - n$ for q.
$$75 - 0.25n + 0.05n = 51$$ Apply the distributive property.
$$75 - 0.20n = 51$$ Combine like terms.
$$75 - 0.20n - 75 = 51 - 75$$ Subtract 75 from each side.
$$-0.20n = -24$$ Simplify.
$$n = \frac{-24}{-0.20}$$ Divide each side by -0.20.
$$n = 120$$ Simplify.

Finally we solve for q.

$$q + n = 300$$ Begin with the first equation.
$$q + 120 = 300$$ Replace n with 120.
$$q + 120 - 120 = 300 - 120$$ Subtract 120 from each side.
$$q = 180$$ Simplify.

Monte's jar contains 120 nickels and 180 quarters.

1c. Janna is working two jobs to help pay her way through college. She works as an office assistant and a store clerk. One week she earns $440 by working 25 hr as an office assistant and 20 hr as a store clerk. Another week she earns $390 by working 15 hr as an office assistant and 30 hr as a store clerk. What is Janna's hourly wage as an office assistant and a store clerk?

Solution **1c.** What is unknown? The hourly wage Janna makes as an office assistant and a store clerk is unknown.

Let $x =$ Janna's hourly wage as an office assistant.
Let $y =$ Janna's hourly wage as a store clerk.

What is known? Janna's salary for two different weeks is given and shown in the following table.

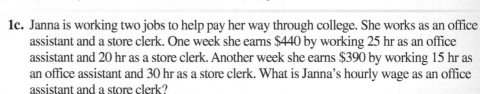

Unknowns	Hourly wage	Hours worked for Week 1	Total paid	Hours worked for Week 2	Total paid
Office assistant wage	x	25	$25x$	15	$15x$
Store clerk wage	y	20	$20y$	30	$30y$
Totals			440		390

The system that represents these facts is

$$\begin{cases} 25x + 20y = 440 \\ 15x + 30y = 390 \end{cases}$$

Week 1 earnings were \$440.

Week 2 earnings were \$390.

Solve the system using elimination.

$$\begin{cases} 3(25x + 20y = 440) \\ -2(15x + 30y = 390) \end{cases} \longrightarrow \begin{cases} 75x + 60y = 1320 \\ -30x - 60y = -780 \end{cases}$$

Multiply the first equation by 3.

Multiply the second equation by −2.

$$45x = 540$$ Add the equations.

$$\frac{45x}{45} = \frac{540}{45}$$ Divide each side by 45.

$$x = 12$$ Simplify.

Now we solve for y.

$$25x + 20y = 440$$ Begin with the first equation.

$$25(12) + 20y = 440$$ Substitute 12 for x.

$$300 + 20y = 440$$ Simplify.

$$300 + 20y - 300 = 440 - 300$$ Subtract 300 from each side.

$$20y = 140$$ Simplify.

$$\frac{20y}{20} = \frac{140}{20}$$ Divide each side by 20.

$$y = 7$$ Simplify.

So, Janna earns \$12 an hour as an office assistant and \$7 an hour as a store clerk.

✔ **Student Check 1**   Solve each problem using a system of linear equations.

a. Admission to a football game is \$10 for students and \$14 for general admission. If 300 tickets are sold and \$3500 is collected, how many student tickets and general admission tickets are sold?

b. Thomas has a collection of coins consisting of nickels and dimes. If he has 500 coins totaling \$40, how many nickels and dimes are in his collection?

c. Two families go to the movies together. One family purchases four drinks and two large popcorns for \$26. The other family purchases two drinks and three large popcorns for \$24. How much does each drink and popcorn cost?

Investment Applications

Objective 2 ▶

Solve investment applications.

Investment problems in this course deal primarily with *simple interest*. *Interest* is a fee paid for the use of money. Interest is calculated as a percentage of the amount of money invested. Simple interest is the most basic type of interest that can be earned.

> **Property: Simple Interest Formula**
>
> Let P be the principal or initial money invested, r the annual interest rate (as a decimal), and t the length of the investment in years, then the earned interest I is
>
> $$I = Prt$$

Suppose we invest \$5000 in two different savings accounts. One account earns 5% annual interest and the other account earns 4% annual interest. The following table shows how much annual interest is earned from the two accounts if the money is invested as shown.

	Principal ×	Rate ×	Time =	Interest
Account 1	$3000	0.05	1	(3000)(0.05)(1) = $150
Account 2	$2000	0.04	1	(2000)(0.04)(1) = $80
Totals	$5000			$230

If the amount invested in each account is unknown, we can assign a variable to represent those amounts. The amount of interest earned is shown in the table.

	Principal ×	Rate ×	Time =	Interest
Account 1	x	0.05	1	$(x)(0.05)(1) = 0.05x$
Account 2	y	0.04	1	$(y)(0.04)(1) = 0.04y$
Totals	$5000			$0.05x + 0.04y$

We can use these ideas to solve word problems involving investments.

Objective 2 Example   Sonja invests $3000 in two different accounts. She invests part of her money in a savings account that earns 4% simple interest and the rest of her money in a money market account that earns 8% simple interest. If she earns a total of $200 interest in 1 yr, how much did she invest in each account? Use a system of linear equations to solve this problem.

Solution   What is unknown? The amount of money invested at 4% and at 8% is unknown.

Let x = amount invested at 4%.
Let y = amount invested at 8%.

What is known? The total amount invested is $3000. The total interest earned is $200. We can organize the information in a table.

	Principal ×	Rate ×	Time =	Interest
Savings account	x	0.04	1	$(x)(0.04)(1) = 0.04x$
Money market account	y	0.08	1	$(y)(0.08)(1) = 0.08y$
Totals	$3000			$200

The system that represents these facts is

$$\begin{cases} x + y = 3000 & \text{Total invested is \$3000.} \\ 0.04x + 0.08y = 200 & \text{Total interest earned is \$200.} \end{cases}$$

We solve the system using elimination but first clear the decimals from the system by multiplying the second equation by 100.

$$\begin{cases} x + y = 3000 \\ 100(0.04x + 0.08y = 200) \end{cases} \longrightarrow \begin{cases} x + y = 3000 \\ 4x + 8y = 20{,}000 \end{cases}$$

Now we multiply the first equation by -8 to eliminate the y-terms from the equations.

$$\begin{cases} -8(x + y = 3000) \\ 4x + 8y = 20{,}000 \end{cases} \longrightarrow \begin{cases} -8x - 8y = -24{,}000 \\ \underline{4x + 8y = 20{,}000} \end{cases}$$

Multiply the first equation by -8.

$$-4x = -4000 \qquad \text{Add the equations.}$$

$$\frac{-4x}{-4} = \frac{-4000}{-4} \qquad \text{Divide each side by } -4.$$

$$x = 1000 \qquad \text{Simplify.}$$

Next we solve for y.

$$x + y = 3000 \qquad \text{Begin with the first equation.}$$
$$1000 + y = 3000 \qquad \text{Replace } x \text{ with 1000.}$$
$$1000 + y - 1000 = 3000 - 1000 \qquad \text{Subtract 1000 from each side.}$$
$$y = 2000 \qquad \text{Simplify.}$$

Sonja invested $1000 into the 4% account and $2000 into the 8% account.

✔ Student Check 2 Georgia receives an inheritance of $100,000. She invests part of the money in a certificate of deposit (CD) that earns 8% simple interest and the rest of the money in an international mutual fund that earns 12% simple interest. If Georgia earns $11,000 in interest for 1 yr, how much did she invest in each account? Use a system of linear equations to solve this problem.

Mixture Applications

Objective 3 ▶

Solve mixture applications.

In Section 2.6, we learned about mixture problems and discovered that to find the amount of substance in a solution, we multiply the amount of the given substance by the strength of the solution. To solve these types of applications with a system, we will assign a variable for each of the unknowns and write two equations that represent the facts in the problem.

Objective 3 Example A dairy farmer produces two types of milk. How many gallons of a 3.25% milk fat solution and how many gallons of a 1% milk fat solution should be mixed together to obtain 750 gal of a 2% milk fat solution?

Solution What is unknown? The gallons of the 3.25% milk fat solution and the gallons of the 1% milk fat solution are unknown.

Let x = number of gallons of 3.25% milk fat solution.
Let y = number of gallons of 1% milk fat solution.

What is known? The total mixture is 750 gal. The final mixture is 2% milk fat. We can organize the information in a table.

Unknowns	Amount of Solution	Strength of Solution	Amount of Milk Fat
Gallons of 3.25% milk fat	x	$3.25\% = 0.0325$	$0.0325x$
Gallons of 1% milk fat	y	$1\% = 0.01$	$0.01y$
Final mixture	750	$2\% = 0.02$	$0.02(750) = 15$

The system that represents the situation is

$$\begin{cases} x + y = 750 & \text{Total obtained is 750 gal.} \\ 0.0325x + 0.01y = 15 & \text{Mixture needs to be 2\% milk fat.} \end{cases}$$

We solve the system using substitution and solve the first equation for y.

$$x + y = 750$$
$$x + y - x = 750 - x \qquad \text{Subtract } x \text{ from each side.}$$
$$y = 750 - x$$

Now we solve for x.

$0.0325x + 0.01y = 15$	Begin with the second equation.
$0.0325x + 0.01(750 - x) = 15$	Substitute $750 - x$ for y.
$0.0325x + 7.5 - 0.01x = 15$	Apply the distributive property.
$0.0225x + 7.5 = 15$	Combine like terms.
$0.0225x + 7.5 - 7.5 = 15 - 7.5$	Subtract 7.5 from each side.
$0.0225x = 7.5$	Simplify.
$x = \dfrac{7.5}{0.0225}$	Divide each side by 0.0225.
$x \approx 333.33$	Approximate.

Now we solve for y.

$x + y = 750$	Begin with the first equation.
$333.33 + y \approx 750$	Substitute 333.33 for x.
$333.33 + y - 333.33 \approx 750 - 333.33$	Subtract 333.33 from each side.
$y \approx 416.67$	Approximate.

The farmer needs to mix 333.33 gal of the 3.25% milk fat solution with 416.67 gal of the 1% milk fat solution to obtain 750 gal of a 2% milk fat solution.

 **Student Check 3**   A chemist has a 50% acid solution. She needs 30 L of a 15% acid solution. How many liters of water and how many liters of the 50% acid solution must she mix together to obtain 30 L of a 15% acid solution? Use a system of linear equations to solve this problem.

Distance Applications

Objective 4 ▶

Solve distance applications.

In Chapter 2, we learned that the total *distance* traveled equals the rate (or speed) times the time elapsed. Algebraically, we write the **distance formula** as $d = rt$, where r is the rate or speed and t is the time traveled. We will apply this concept to flying.

If a plane is flying *directly with the wind*, the plane will arrive at its destination sooner than it would if flying in still air since the speed of the wind increases the plane's speed. If, however, a plane flies *directly against the wind*, the plane will take longer to reach its destination since the speed of the wind decreases the speed of the plane. The following table provides some illustrations of how the speed of the wind can affect the speed of the plane.

Speed of the Plane in Still Air	Speed of the Wind	Speed of Plane Flying with the Wind	Speed of Plane Flying against the Wind
500 mph	20 mph	$500 + 20 = 520$ mph	$500 - 20 = 480$ mph
600 mph	x mph	$600 + x$ mph	$600 - x$ mph

Note the following:

- Flying directly *with* the wind increases the speed of the plane by the speed of the wind.
- Flying directly *against* the wind decreases the speed of the plane by the speed of the wind.

Objective 4 Examples   Solve each problem using a system of linear equations.

4a. Sam and Lori flew from Atlanta, Georgia (ATL), to San Francisco, California (SFO), a distance of approximately 2135 mi. The flight from Atlanta to San Francisco took 5 hr since the plane was flying against the wind. The return trip took 4.5 hr since the plane was flying with the wind. What was the speed of the plane in still air and what was the speed of the wind?

Solution   **4a.** What is unknown? The speed of the plane and the speed of the wind are unknown.

Let x = speed of the plane.
Let y = speed of the wind.

What is known? The distance between the cities is 2135 mi. The trip against the wind was 5 hr. The trip with the wind was 4.5 hr. We can organize the information in a table as shown.

	Rate $\times$	Time $=$	Distance
ATL to SFO	$x - y$	5	2135
SFO to ATL	$x + y$	4.5	2135

The system that represents the situation is

$$\begin{cases} 5(x - y) = 2135 \\ 4.5(x + y) = 2135 \end{cases}$$

We solve the system using elimination after first applying the distributive property. Then we clear decimals by multiplying the second equation by 10.

$$\begin{cases} 5x - 5y = 2135 \\ 4.5x + 4.5y = 2135 \end{cases} \longrightarrow \begin{cases} 5x - 5y = 2135 \\ 10(4.5x + 4.5y = 2135) \end{cases} \longrightarrow \begin{cases} 5x - 5y = 2135 \\ 45x + 45y = 21{,}350 \end{cases}$$

Now multiply the first equation by 9 to eliminate the y-terms from the system.

$$\begin{cases} 9(5x - 5y = 2135) \\ 45x + 45y = 21{,}350 \end{cases} \longrightarrow \begin{cases} 45x - 45y = 19{,}215 \\ 45x + 45y = 21{,}350 \end{cases}$$ Multiply the first equation by 9.

$$90x = 40{,}565$$ Add the equations.

$$\frac{90x}{90} = \frac{40{,}565}{90}$$ Divide each side by 90.

$$x \approx 450.7$$ Approximate.

Next solve for y.

$$5(x - y) = 2135$$ Begin with the first equation.

$$5(450.7 - y) \approx 2135$$ Replace x with 450.7.

$$\frac{5(450.7 - y)}{5} \approx \frac{2135}{5}$$ Divide each side by 5.

$$450.7 - y \approx 427$$ Simplify.

$$450.7 - y - 450.7 \approx 427 - 450.7$$ Subtract 450.7 from each side.

$$-y \approx -23.7$$ Simplify.

$$\frac{-y}{-1} \approx \frac{-23.7}{-1}$$ Divide each side by -1.

$$y \approx 23.7$$ Approximate.

So, the speed of the plane is approximately 450.7 mph and the speed of the wind is approximately 23.7 mph.

4b. At 9 A.M., David leaves his home traveling south on his bike at an average speed of 10 mph. At 11 A.M., his wife Judy leaves their home to find David to give him an urgent message, since he left his cell phone at home. She travels in her car at an average speed of 65 mph. How long will it take for Judy to reach David?

Solution **4b.** What is unknown? The time that David and Judy each travel is unknown.

Let x = time David travels.

Let y = time Judy travels.

What is known? David's speed is 10 mph and Judy's speed is 65 mph. The distance traveled by both David and Judy is the same. The time Judy travels is 2 hr less than David's since she left 2 hr after David. We can organize this information in a table.

	Rate	×	Time	=	Distance
David	10		x		$10x$
Judy	65		y		$65y$

The system that represents the situation is

$$\begin{cases} 10x = 65y \\ y = x - 2 \end{cases}$$

Distances are the same.

Judy's time is 2 hr less than David's time.

Because the second equation is solved for y, we solve the system using substitution.

$$10x = 65y$$ Begin with the first equation.

$$10x = 65(x - 2)$$ Replace y with $x - 2$.

$$10x = 65x - 130$$ Apply the distributive property.

$$10x - 65x = 65x - 130 - 65x$$ Subtract $65x$ from each side.

$$-55x = -130$$ Simplify.

$$\frac{-55x}{-55} = \frac{-130}{-55}$$ Divide each side by −55.

$$x \approx 2.4$$ Approximate.

Finally, we solve for y.

$$y = x - 2$$ Begin with the second equation.

$$y \approx 2.4 - 2$$ Substitute 2.4 for x.

$$y \approx 0.4$$ Approximate.

It will take Judy about 0.4 hr or 24 min to reach David.

✓ **Student Check 4** Solve each problem using a system of linear equations.

a. A plane can travel 1800 mi with the wind in 4 hr. The return trip against the wind takes 6 hr. What is the speed of the plane and the speed of the wind?

b. Cherie and her mom live 330 mi apart. They want to meet each other so that Cherie's kids can go home with her mom for a few days. Cherie leaves at 8 A.M. and travels toward her mom at an average speed of 50 mph. Her mom leaves at 8:30 A.M. and travels toward Cherie at an average speed of 65 mph. What time will they meet? (Round solutions to the nearest minute.)

Geometry Applications

Objective 5 ▶

Solve geometry applications.

We will use some topics from Chapter 2 to solve applications involving geometry problems. Some of the geometry problems we will solve involve perimeter of a rectangle and the relationships between complementary and supplementary angles. Recall these important facts.

- The **perimeter of a rectangle** is $2l + 2w$, where l is the length and w is the width.
- **Supplementary angles** are two angles whose measures add to $180°$.
- **Complementary angles** are two angles whose measures add to $90°$.

Objective 5 Examples   **Solve each problem using a system of linear equations.**

5a. The perimeter of a high school basketball court is 268 ft. The length of the court is 16 ft less than two times the width of the court. Find the dimensions of the court.

Solution   **5a.** What is unknown? The length and the width of the basketball court are unknown.

Let $l =$ the length of the court.
Let $w =$ the width of the court.

What is known? The perimeter of the court is 268 ft. The length is 16 ft less than two times the width. The system that represents the problem is

$$\begin{cases} 2l + 2w = 268 \\ l = 2w - 16 \end{cases}$$

Because the second equation is solved for l, we solve the system using substitution.

$2l + 2w = 268$	Begin with the first equation.
$2(2w - 16) + 2w = 268$	Substitute $(2w - 16)$ for l.
$4w - 32 + 2w = 268$	Apply the distributive property.
$6w - 32 = 268$	Combine like terms.
$6w - 32 + 32 = 268 + 32$	Add 32 to each side.
$6w = 300$	Simplify.
$\dfrac{6w}{6} = \dfrac{300}{6}$	Divide each side by 6.
$w = 50$	Simplify.

Next we solve for the length.

$l = 2w - 16$	Begin with the second equation.
$l = 2(50) - 16$	Substitute 50 for w.
$l = 100 - 16$	Multiply.
$l = 84$	Subtract.

So, the dimensions of the basketball court are 84 ft by 50 ft.

5b. Two angles are supplementary. The measure of one angle is 6° less than twice the measure of the other angle. Find the measure of each angle.

Solution   **5b.** What is unknown? The measure of each angle is unknown.

Let $x =$ the measure of one angle.
Let $y =$ the measure of its supplement.

What is known? The angles are supplementary, which means their sum is 180°. The measure of one angle is 6° less than twice the measure of the other. The system that represents the problem is

$$\begin{cases} x + y = 180 \\ x = 2y - 6 \end{cases}$$

Because the second equation is solved for x, we solve the system by substitution.

$x + y = 180$	Begin with the first equation.
$2y - 6 + y = 180$	Substitute $2y - 6$ for x.
$3y - 6 = 180$	Combine like terms.
$3y - 6 + 6 = 180 + 6$	Add 6 to each side.
$3y = 186$	Simplify.
$\dfrac{3y}{3} = \dfrac{186}{3}$	Divide each side by 3.
$y = 62$	Simplify.

Now we solve for x.

$x = 2y - 6$	Begin with the second equation.
$x = 2(62) - 6$	Substitute 62 for y.
$x = 124 - 6$	Multiply.
$x = 118$	Subtract.

So, the measure of one angle is 118°, and the measure of its supplement is 62°.

 **Student Check 5**   Solve each problem using a system of linear equations.

   a. A regulation size tennis court has a perimeter of 228 ft. The length of the court is 6 ft more than twice the width. Find the dimensions of the court.

   b. Two angles are complementary. The measure of one angle is 10° more than the measure of the other. Find the measure of each angle.

Objective 6 ▶

Troubleshoot common errors.

Troubleshooting Common Errors

Some common errors associated with applications of systems are shown next. Most of the errors are the result of setting up the equations in the system incorrectly.

Objective 6 Examples   **A problem and an incorrect solution are given. Provide the correct solution and an explanation of the error.**

6a. Micah invests money in two different accounts, a CD that earns 5% annual interest and a money market account that earns 8% annual interest. In the money market account, he invests $2000 less than 3 times the amount he invests in the CD account. If he earns a total of $3320 in yearly interest, how much does he invest in each account? Write the system that represents this situation.

Incorrect Solution	Correct Solution and Explanation
Let $x =$ amount invested in the CD account. Let $y =$ amount invested in the money market account.	The error was made in the first equation. To represent 2000 less than 3 times an amount is $3x - 2000$.
$\begin{cases} y = 2000 - 3x \\ 0.05x + 0.08y = 3320 \end{cases}$	$\begin{cases} y = 3x - 2000 \\ 0.5x + 0.08y = 3320 \end{cases}$

6b. A chemist needs to make a 20% iodine solution. He has a 50% iodine solution and a 10% iodine solution. How much of each should he mix together to obtain 60 mL of a 20% iodine solution? Write the system that represents this situation.

Incorrect Solution	Correct Solution and Explanation
Let x = amount of 50% iodine solution. Let y = amount of 10% iodine solution.	The error was made in the second equation. The left side of the equation represents the amount of pure iodine in the two solutions, so the right side should also represent the amount of pure iodine.
$$\begin{cases} x + y = 60 \\ 0.5x + 0.1y = 60 \end{cases}$$	$$\begin{cases} x + y = 60 \\ 0.5x + 0.1y = 0.2(60) \end{cases}$$

ANSWERS TO STUDENT CHECKS

Student Check 1   **a.** 175 student tickets and 125 general tickets   **b.** 200 nickels and 300 dimes

  c. $3.75 per drink and $5.50 per popcorn

Student Check 2 $75,000 invested in the international mutual fund and $25,000 in the CD

Student Check 3   21 L of water and 9 L of the 50% acid solution

Student Check 4   **a.** The plane's speed is 375 mph and the wind speed is 75 mph.   **b.** They will meet at approximately 11:09 A.M.

Student Check 5   **a.** The tennis court is 78 ft long by 36 ft wide.   **b.** The angles are 50° and 40°.

SUMMARY OF KEY CONCEPTS

While many different types of applications were solved in this section, there is really only one method that we use to solve these problems. The steps to solve an application with a system of two linear equations in two variables are

1. Determine the two unknowns in the problem. Use two different variables to represent these unknowns.

2. Write two equations that relate these unknowns to each other.

3. Once the system is set up, use substitution or elimination to solve the system.

SECTION 4.4 / EXERCISE SET

 Write About It!

Use complete sentences in your answer to each exercise.

1. Explain the general strategy you should follow to solve application problems.

2. This section contains applications similar to those found in Section 2.6. How does the approach to solving these problems differ in this section from the approach given in Section 2.6?

3. Give an example of a mixture application.

4. In Objective 3, an explanation was given about the relationship between the wind and the net speed of a plane. Suppose that the speed of a plane in still air is 450 mph. If this plane is flying against a wind current of x mph, will the net speed of the plane be $450 - x$ mph or $x - 450$ mph? Explain.

 Practice Makes Perfect!

Solve each problem using a system of linear equations. Round answers to two decimal places when applicable. (*See Objective 1.*)

5. Jamie's middle school put on a production of *The Wizard of Oz*. They charged students and school employees $5 per ticket, and everyone else was charged $10 per ticket. Five hundred people attended the show, and the receipts totaled $3000. How many students and school employees attended the production?

6. Angie's youth group raised money by having a car wash. The group charged $3 to wash a car and $5 to wash an SUV. At the end of the day, they had

washed 46 vehicles and raised $180. How many SUVs were washed by the youth group?

7. Janeane takes a bag of dimes and quarters to a coin counting machine and opts to have the value of her money converted to a gift card. If 113 coins are counted and the value of the gift card is $19.55, how many dimes did she deposit in the machine?

8. Elise collects half-dollars and silver dollars. After several years, Elise takes her coins to the bank. She has 130 coins that total $90. How many half-dollar coins does she have?

9. Rachel buy a 2-lb bag of candied pecans and candied peanuts for $17. The candied pecans sell for $9.50 per pound, and the candied peanuts sell for $5.50 per pound. How many pounds of each type of nut are included in the bag?

10. Lori makes her own trail mix at a natural-foods grocery store. She mixes Spanish almonds costing $7.50 per pound with dried dates costing $9 per pound. If a 3-lb bag cost Lori $24, how many pounds of Spanish almonds and how many pounds of dried dates are included in the mixture?

11. Lisa works for a community food bank. She buys 100 cans of green beans and 70 cans of tuna for $92.50. The next day, she purchases 60 cans of green beans and 80 cans of tuna for $84. How much is she charged for each can of green beans and each can of tuna?

12. A video store has a month-long sale on movies. Pauline, who loves movies, purchases two new DVDs and three used DVDs for $46. Pauline returns the following week and purchases three new DVDs and five used DVDs for $71.50. How much does the video store charge for each new DVD and each used DVD?

Solve each problem using a system of linear equations. (*See Objective 2.*)

13. Behnaz invests $500 in two different accounts. She invests part of her money in a savings account that earns 5% simple interest and the rest of her money in a high-risk junk bond that earns 15% simple interest. If Behnaz's accounts earn $37.50 interest in 1 yr, how much does she invest in each account?

14. Haazim has $1500 that he invests in two different accounts. He invests part of his money in a savings account that earns 4.5% simple interest and he invests the rest of his money in a CD that earns 9% simple interest. If Haazim's accounts earn $90 interest in 1 yr, how much does he invest in each account?

Solve each problem using a system of linear equations. (*See Objective 3.*)

15. A dairy farmer produces two types of milk: milk with 1% milk fat and milk with 4% milk fat. How many gallons of each type of milk should he mix to obtain 300 gal of milk with 2% milk fat?

16. A dairy farmer produces two types of milk: milk with 1.5% milk fat and milk with 3.25% milk fat. How many gallons of each type of milk should he mix to obtain 420 gal of milk with 2% milk fat?

17. Joe's Diner purchases whole milk (3.25% milk fat) and skim milk (0% milk fat). Some of Joe's customers order glasses of 2% lowfat milk. How many liters of whole milk and skim milk must Joe combine to obtain 26 L of 2% lowfat milk?

18. Asa likes the taste of a 1% milk fat solution. How much whole milk and skim milk must Asa mix together to obtain 13 oz of milk with 1% milk fat? (Recall that whole milk is 3.25% milk fat and skim milk is 0% milk fat.)

19. A chemist needs 5 mL of a solution that is 20% acid. In her lab, she has a container of solution that is 50% acid and another container of solution that is 10% acid. How much of each acid solution must the chemist combine to get the desired 5 mL of 20% acid?

20. A chemist needs 12 L of a solution that is 15% acid. In his lab, he has a supply of a 30% acid solution and a 10% acid solution. How many liters of each acid solution must he combine to make 12 L of a 15% acid solution?

21. A chemist needs 10 L of a solution that is 25% acid. In his lab, he has a 40% acid solution. How much water and how much 40% acid solution must the chemist combine to make 10 L of a 25% acid solution?

22. A chemist needs 20 mL of a solution that is 10% acid. In her lab, she has a supply of pure acid. How much water and how much pure acid must the chemist combine to make 20 mL of a 10% acid solution?

Solve each problem using a system of linear equations. Round answers to the nearest hundredth when applicable. (*See Objective 4.*)

23. The distance between Newark Liberty International Airport in New Jersey and Denver International Airport in Colorado is approximately 1600 mi. Ming's flight from Newark to Denver takes 3.5 hr flying against the jet stream, but her return flight takes only 3 hr flying with the jet stream. What was the plane's speed in still air, and what was the speed of the jet stream?

24. Roland, who lives in Phoenix, is planning to fly 2900 mi to Honolulu during his winter break. By checking online flight times, Roland finds a flight from Phoenix to Honolulu that takes 6.5 hr traveling against the wind. He finds a return flight from Honolulu to Phoenix that takes only 5.75 hr traveling with the wind in the same type of plane. What is the speed of the plane in still air, and what is the speed of the wind?

Similar to wind currents, water currents affect the rates boats travel. Suppose that a boat in still water travels 20 mph and that the speed of the water current is *x* mph.

If a boat is traveling *upstream* (against the current), the net speed of the boat is *slower: speed of boat − speed of current* = 20 − *x* **mph**

If a boat is traveling *downstream* (with the current), the net speed of the boat is *faster: speed of boat + speed of current* = 20 + *x* **mph**

25. Mitchell and his friends rent a fishing boat. They travel 36 mi upstream in 1 hr. Their return trip takes only 0.75 hr. What was the speed of the current? What was the speed of Mitchell's boat in still water?

26. Ira leaves his dock and sails 60 mi upstream in 1.5 hr. It takes him only 1 hr to return to his dock. What was the speed of the current? What was the speed of Ira's boat in still water?

27. Drivers A and B enter the pit row at the same time during a professional car race. Driver A stops for gas and a tire change and returns to the race immediately at an average speed of 160 mph. Driver B has mechanical problems that take 10 min to resolve. He enters the race at an average speed of 180 mph. Assuming the drivers do not make any more stops, how long will it take for Driver B to catch up to Driver A?

28. A fugitive leaves the scene of an accident traveling by car at an average speed of 80 mph. The police arrive at the scene of the accident 15 min later. They travel in the same direction as the fugitive at an average speed of 90 mph. How long will it take before the police catch the fugitive?

Solve each problem using a system of linear equations. (*See Objective 5.*)

29. The perimeter of a rectangle is 160 ft. If the length of the rectangle is 4 ft less than the width, find the dimensions of the rectangle.

30. The perimeter of a rectangle is 70 in. If the width of the rectangle is 5 in. more than 3 times the length, find the dimensions of the rectangle.

31. Two angles are complementary. One angle is 3° less than twice the other angle. Find the measure of the larger angle.

32. Two angles are supplementary. One angle is 6° more than 5 times the other angle. Find the measure of the smaller angle.

33. Sandeep fences off part of his yard for his new Labrador retriever. He encloses a rectangular region with 50 yd of chain link fence. If Sandeep wants the length of the dog yard to be 5 yd longer than its width, find the dimensions of the dog yard he can build.

34. Chang makes wooden frames for paintings with elaborate hand-carved details. He has enough wood pieces to make a rectangular frame with a perimeter of 120 in. If the width of the frame is 10 in. less than its length, find the dimensions of the frame Chang made.

35. Becky and Elaine are aspiring ballerinas. They are each working toward doing a perfect six-o'clock position, but neither can do it just yet. Elaine's back leg can rise 20° higher than Becky's back leg. The sum of the angles formed by the legs of both ballerinas is 280°. What angle is each ballerina able to form with her legs? Who is closer to her six-o'clock position goal?

36. Sisters Chelsea and Cheyanne just started taking ballet classes. They are working toward executing the arabesque position in which their legs form a right angle. Chelsea, the older of the sisters, is able to raise her back leg 25° higher than Cheyanne and still keep her balance. The sum of the angles formed by the legs of the sisters when attempting their best arabesque positions is 135°. What angle is each sister able to make?

🌀 Mix 'Em Up!

Solve each problem using a system of linear equations.

37. John Carpenter's *Halloween* (1978) gave Jamie Lee Curtis her big break as an actor. *Halloween* and *Halloween H20: 20 Years Later* (1998) grossed approximately $102 million at the box office in the United States. *Halloween* grossed $8 million less than *Halloween H20*. How much did each movie gross at the box office? (Source: Wikipedia)

38. Two angles are supplementary. The larger angle is 5° more than four times the smaller angle. What is the measure of each angle?

39. Two angles are supplementary. One angle is 6° less than five times the other angle. What is the measure of each angle?

40. Two angles are complementary. The larger angle is 21° less than twice the measure of the smaller one. What is the measure of each angle?

41. Two angles are complementary. One angle is 28° more than three times the measure of the other one. What is the measure of each angle?

42. Leigh and Ryan are experienced cyclists who like challenging cycling routes. Ryan gives Leigh a 15-min head start on a 50-mi route. On this route, Leigh cycles an average of 10 mph. On the same route, Ryan averages 15 mph. How long will it take Ryan to catch up to Leigh?

43. Susan leaves her job and heads east at 55.5 mph. Vera leaves the same workplace 36 min later and heads east at 74 mph. How long will it take Vera to catch up to Susan?

44. Paint thinner is often mixed with paint to lessen the appearance of brush strokes. A rule of thumb is to use a mixture that is 60% paint and 40% paint thinner. Jennifer has a large supply of pure paint, and a large supply of a mixture that is 50% paint (and 50% paint thinner). How much pure paint and how much paint mixture must Jennifer mix together so that she will have 3 gal of a 60% paint mixture?

45. An angle measures 6° more than twice its complement. What is the measure of each angle?

46. An angle measures 36° more than four times its supplement. What is the measure of each angle?

47. A plane travels 3159.5 mi in 7.1 hr traveling with the wind. It takes 8.9 hr for the plane to travel the same distance against the wind. What is the speed of the plane in still air and what is the speed of the wind?

48. A plane travels 1505 mi in 3.5 hr traveling with the wind. It takes 4.3 hr for the plane to travel the same distance against the wind. What is the speed of the plane in still air and what is the speed of the wind?

49. Steven has $10,000 to invest. He invests part of the money in a savings account that pays 1.4% annual interest and the rest of the money in a CD that pays 3.1% annual interest. If he earns $268.52 interest in 1 yr, how much does he invest in each account?

50. Michelle has $6650 to invest. She invests part of the money in a stock that pays a 3.8% annual dividend and the rest of the money in a mutual fund that pays 1.6% annual interest. If she earns $228.50 in investment income in 1 yr, how much does she invest in each account?

51. One week, Danielle buys 7 cans of tuna and 6 cans of soup for $13.69. The next week, she buys 3 cans of tuna and 2 cans of soup for $5.13. What is the price of each can?

52. One week, Lucia buys 6 cans of soup and 1 can of mushrooms for $15.24. The next week, she buys 9 cans of soup and 3 cans of mushrooms for $25.11. What is the price of each can?

 You Be the Teacher!

Correct each student's errors, if any.

53. Jim has $10,650 to invest. He invests part of the money in a mutual fund that pays 1.1% annual interest and the rest of the money in a money market fund that pays 4.5% annual interest. If he earns $248.05 interest in 1 yr, how much did he invest in each account?

Steve's work:

Let x be the amount invested at 1.1% and y at 4.5%.

$$\begin{cases} 1.1x + 4.5y = 10{,}650 \\ x + y = 248.05 \end{cases}$$

I solve for x and substitute into the first equation.

$$x = 248.05 - y$$
$$1.1(248.05 - y) + 4.5y = 10{,}650$$
$$2728.55 - 1.1y + 4.5y = 10{,}650$$
$$3.5y = 7921.45$$
$$y = 2263.27$$

So, $x = 248.05 - 2263.27 = -2015.22$.

54. Todd needs to make 266 L of a 46% antifreeze solution. How much pure antifreeze and how much of a 16% antifreeze solution should he mix together to get 266 L of a 46% antifreeze solution?

Andrew's work:

Let x be the volume of the pure antifreeze solution and y be the volume of a 16% antifreeze solution.

$$\begin{cases} x + y = 266(0.46) \\ 0x + 0.16y = 266 \end{cases}$$

I solve for y using the second equation.

$$y = 266/0.16 = 1662.5$$

SECTION 4.5   **Solving Systems of Linear Equations in Three Variables and Their Applications**

► OBJECTIVES

As a result of completing this section, you will be able to

1. Solve a system of linear equations in three variables by elimination.

2. Solve applications of linear systems in three variables.

3. Troubleshoot common errors.

An international flight from Atlanta, Georgia, to Dubai, United Arab Emirates, consists of first-class, business-class, and coach travelers. A first-class ticket costs $4700. A business-class ticket costs $2300 and a coach ticket costs $750. The airline made $342,000 in airfare revenue for one trip from Atlanta to Dubai. The flight had a total of 300 paying passengers. If there were four times as many coach travelers as first-class and business-class travelers combined, how many of each type of traveler were on this flight?

Notice that there are three unknowns in this problem. Solving this problem requires us to use a system that consists of three equations and three variables. In this section, we will learn how to set up and solve a system involving three linear equations and three unknowns.

Solving a System of Linear Equations in Three Variables

Objective 1 ►

Solve a system of linear equations in three variables by elimination.

In business, engineering, and other fields, systems that contain more than two equations and more than two unknowns are needed to represent real-life situations. These systems are solved using computers or calculators. Systems involving more than two unknowns can be difficult to solve by hand, so we will only solve systems of three linear equations in three variables algebraically. We will use elimination, which was covered in Section 4.2. Substitution can also be used to solve these systems but can be tedious to use.

> **Definition:** A **system of linear equations in three variables** can be represented by
> $$\begin{cases} Ax + By + Cz = D \\ Ex + Fy + Gz = H \\ Ix + Jy + Kz = L \end{cases}$$

Notice that each variable in a linear system in three variables is raised to an exponent of 1. A **solution of a linear system in three variables** is an *ordered triple* (x, y, z) that makes each equation in the system a true statement.

Just as a graph of linear equations in two variables is a line in two-dimensional space, the graph of a linear equation in three variables is a plane in three-dimensional space provided the coefficients of all the variables are nonzero. Linear systems in three variables have solution sets similar to linear systems in two variables.

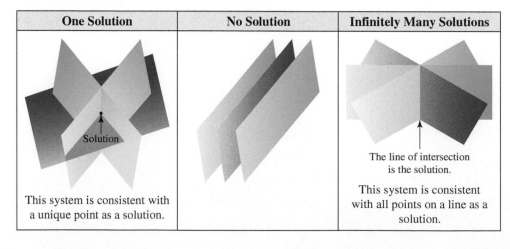

One Solution	**No Solution**	**Infinitely Many Solutions**
This system is consistent with a unique point as a solution.		The line of intersection is the solution. This system is consistent with all points on a line as a solution.

One Solution	No Solution	Infinitely Many Solutions
	These systems are inconsistent with no common point of the three planes.	

Visualizing solutions of linear systems in three variables can be difficult. It is sufficient to understand that the solution of such a system consists of the points where the three planes intersect. A plane is a flat surface—think of a wall or ceiling that extends indefinitely on all four sides.

To solve a system of three linear equations in three variables, we will create an equivalent system of two linear equations with two variables. To do this, we work with pairs of equations from the original system and eliminate the same variable from each pair of equations obtaining the new system we need.

Procedure: Solving a System of Linear Equations in Three Variables Using Elimination

Step 1: Write all equations in standard form.
Step 2: Pair together two of the equations and eliminate one variable from this set of equations.
Step 3: Pair together two other equations and eliminate the variable that was eliminated in step 2.
Step 4: Form a system of the two equations obtained from steps 2 and 3. Solve this system by elimination.
Step 5: Substitute the solution found in step 4 into one of the equations in the system from step 4 to find the value of another variable.
Step 6: Substitute the two known values into one of the original equations to find the remaining value.
Step 7: Check the solution by substituting into all three equations in the given system.

Note: *If in the process of solving the system, we obtain a*

- *False statement (for example, 0 = 12), then the system has no solution.*
- *True statement (for example, 12 = 12), then the system has infinitely many solutions.*

Objective 1 Examples   **Solve each system by elimination. The equations in the system have been numbered for reference.**

1a. $\begin{cases} x + 4y + 2z = -1 & (1) \\ 2x - 4y - z = 6 & (2) \\ 5x + 4y - z = 23 & (3) \end{cases}$ **1b.** $\begin{cases} x - 2y + z = 6 & (1) \\ 2x + y - z = 0 & (2) \\ 4x - 8y + 4z = -12 & (3) \end{cases}$

1c. $\begin{cases} x + 4y - 5z = 8 & (1) \\ y - 3z = 5 & (2) \\ x + 2y + z = -2 & (3) \end{cases}$

Solutions   **1a.** We form pairs of equations and eliminate the variable y from each pair.

Pair together Equations (1) and (2) and add.

$$
\begin{array}{rl}
x + 4y + 2z = -1 & (1) \\
\underline{2x - 4y - z = 6} & (2) \\
3x + z = 5 & (4)
\end{array}
$$

Pair together Equations (2) and (3) and add.

$$
\begin{array}{rl}
2x - 4y - z = 6 & (2) \\
\underline{5x + 4y - z = 23} & (3) \\
7x - 2z = 29 & (5)
\end{array}
$$

Now form a new system using Equations (4) and (5) and solve by elimination.

$$
\begin{cases}
3x + z = 5 & (4) \\
7x - 2z = 29 & (5)
\end{cases}
$$

We choose to eliminate z from this system, so we multiply the first equation by 2.

$$
\begin{cases}
3x + z = 5 \\
7x - 2z = 29
\end{cases}
\rightarrow
\begin{cases}
2(3x + z = 5) \\
7x - 2z = 29
\end{cases}
\rightarrow
\begin{cases}
6x + 2z = 10 \\
7x - 2z = 29
\end{cases}
$$

Now add the equations in the new system to solve for x.

$$
\begin{array}{rl}
6x + 2z = 10 & \text{Equation (4)} \\
\underline{7x - 2z = 29} & \text{Equation (5)} \\
13x = 39 & \text{Add.} \\
x = 3 & \text{Divide each side by 13.}
\end{array}
$$

Now we can solve for z by replacing x with 3 into one of the new equations, Equation (4) or Equation (5).

$$
\begin{array}{rl}
3x + z = 5 & \text{Equation (4)} \\
3(3) + z = 5 & \text{Substitute 3 for } x. \\
9 + z = 5 & \text{Simplify.} \\
z = -4 & \text{Subtract 9 from each side.}
\end{array}
$$

Solve for y using one of the original equations.

$$
\begin{array}{rl}
x + 4y + 2z = -1 & \text{Equation (1)} \\
3 + 4y + 2(-4) = -1 & \text{Substitute 3 for } x \text{ and } -4 \text{ for } z. \\
3 + 4y - 8 = -1 & \text{Simplify.} \\
4y - 5 = -1 & \text{Combine like terms.} \\
4y = 4 & \text{Add 5 to each side.} \\
y = 1 & \text{Divide each side by 4.}
\end{array}
$$

The solution for the system is $x = 3$, $y = 1$, and $z = -4$, or we write the solution set as $\{(3, 1, -4)\}$.

We can check the solution by substituting these values in the three original equations.

$$
\begin{array}{c|c|c}
x + 4y + 2z = -1 & 2x - 4y - z = 6 & 5x + 4y - z = 23 \\
3 + 4(1) + 2(-4) = -1 & 2(3) - 4(1) - (-4) = 6 & 5(3) + 4(1) - (-4) = 23 \\
3 + 4 - 8 = -1 & 6 - 4 + 4 = 6 & 15 + 4 + 4 = 23 \\
-1 = -1 & 6 = 6 & 23 = 23 \\
\text{True} & \text{True} & \text{True}
\end{array}
$$

1b. We form pairs of equations and eliminate the variable z from each pair.

Pair together Equations (1) and (2) and add.

$$\begin{cases} x - 2y + z = 6 & (1) \\ \underline{2x + y - z = 0} & (2) \\ 3x - y = 6 & (4) \end{cases}$$

Pair together Equations (1) and (3). To eliminate z, we must multiply Equation (1) by -4.

$$\begin{cases} -4(x - 2y + z = 6) & (1) \\ 4x - 8y + 4z = -12 & (3) \end{cases}$$

$$\begin{cases} -4x + 8y - 4z = -24 \\ \underline{4x - 8y + 4z = -12} \\ 0 = -36 \quad (5) \end{cases}$$

Because we obtained a false statement, $0 = -36$, this system is an inconsistent system and has no solution.

1c. Since Equation (2) is missing the variable x, we can use it as one of the equations in the system of two linear equations.

$$y - 3z = 5$$

Now we pair together Equations (1) and (3) and eliminate x. We multiply Equation (3) by -1 and add.

$$\begin{cases} x + 4y - 5z = 8 & (1) \\ -1(x + 2y + z = -2) & (3) \end{cases} \rightarrow \begin{cases} x + 4y - 5z = 8 \\ \underline{-x - 2y - z = 2} \\ 2y - 6z = 10 \quad (4) \end{cases}$$

Now pair Equations (2) and (4).

$$\begin{cases} y - 3z = 5 & (2) \\ 2y - 6z = 10 & (4) \end{cases}$$

Solve the system by multiplying Equation (2) by -2 to eliminate y and then add.

$$\begin{cases} -2(y - 3z = 5) & (2) \\ 2y - 6z = 10 & (4) \end{cases} \rightarrow \begin{cases} -2y + 6z = -10 \\ \underline{2y - 6z = 10} \\ 0 = 0 \end{cases}$$

Because we obtained a true statement, $0 = 0$, this system has infinitely many solutions. We can represent those solutions in terms of the variable z, a real number. Equation (2) expresses y in terms of z, $y = 3z + 5$. We substitute $y = 3z + 5$ in Equation (1) to solve for x in terms of z.

$$x + 4y - 5z = 8$$
$$x + 4(3z + 5) - 5z = 8 \qquad \text{Replace } y \text{ with } 3z + 5.$$
$$x + 12z + 20 - 5z = 8 \qquad \text{Apply the distributive property.}$$
$$x + 7z + 20 = 8 \qquad \text{Combine like terms.}$$
$$x = -7z - 12 \qquad \text{Subtract } 7z \text{ and } 20 \text{ from each side.}$$

The solution set is $\{(-7z - 12, 3z + 5, z) | z \text{ is a real number}\}$.

✓ Student Check 1   Solve each system by elimination.

a. $\begin{cases} x + 6y - 5z = -14 \\ 3x - 6y + z = -10 \\ 7x + 6y + z = -26 \end{cases}$ **b.** $\begin{cases} x + 4y - z = -10 \\ 6x + 2y + z = 8 \\ 4x - 6y + 3z = 16 \end{cases}$ **c.** $\begin{cases} x - 3y + 2z = -1 \\ -5x + 16y - 14z = 13 \\ 4x - 10y = 12 \end{cases}$

Applications of Linear Systems in Three Variables

Many of the word problems we solved that involved systems of two linear equations can be extended to situations with three unknowns—mixture problems, investment problems, geometry problems, and the like. If a problem contains three unknown quantities, then this is our clue that we must use a system of three linear equations and three variables to solve it.

> **Procedure: Solving an Application of Linear Systems in Three Variables**
>
> **Step 1:** Read the problem carefully to determine the unknown quantities and assign a variable to represent each of them.
> **Step 2:** From the problem determine three equations that define the relationships between the unknowns.
> **Step 3:** Solve the system using elimination.
> **Step 4:** Check the answer by substituting into the three equations in the system.

Objective 2 Examples   Use a linear system in three variables to solve each problem.

2a. An international flight from Atlanta, Georgia to Dubai, United Arab Emirates, consists of first-class, business-class, and coach travelers. A first-class ticket costs $4700. A business-class ticket costs $2300 and a coach ticket costs $750. The airline made $342,000 in airfare revenue for one trip from Atlanta to Dubai. The flight had a total of 300 paying passengers. If there were four times as many coach travelers as first-class and business-class travelers combined, how many of each type of traveler were on this flight?

Solution   **2a.** What is unknown? The number of first-class passengers, the number of business-class passengers, and the number of coach passengers are unknown.

> Let f = number of first-class passengers.
> Let b = number of business-class passengers.
> Let c = number of coach passengers.

We use the given facts to write a system that represents the problem.

$$\begin{cases} f + b + c = 300 \\ 4700f + 2300b + 750c = 342{,}000 \\ c = 4(f + b) \end{cases}$$

There are a total of 300 passengers.

Total revenue is $342,000.

There are 4 times as many coach travelers as the other two combined.

We rewrite the system so that each equation is in standard form.

$$\begin{cases} f + b + c = 300 & (1) \\ 4700f + 2300b + 750c = 342{,}000 & (2) \\ 4f + 4b - c = 0 & (3) \end{cases}$$

To solve the system using elimination, we pair together Equations (1) and (2) and Equations (1) and (3) and eliminate c from each pair.

To eliminate c from Equations (1) and (2), multiply the first equation by -750 and add.

$$\begin{cases} -750(f + b + c = 300) & (1) \\ 4700f + 2300b + 750c = 342{,}000 & (2) \end{cases} \rightarrow \begin{cases} -750f - 750b - 750c = -225{,}000 & (1) \\ \underline{4700f + 2300b + 750c = 342{,}000} & (2) \\ 3950f + 1550b = 117{,}000 & (4) \end{cases}$$

To eliminate c from Equations (1) and (3), add the equations.

$$\begin{cases} f + b + c = 300 & (1) \\ 4f + 4b - c = 0 & (3) \\ \hline 5f + 5b = 300 & (5) \end{cases}$$

Now solve the system $\begin{cases} 3950f + 1550b = 117{,}000 & (4) \\ 5f + 5b = 300 & (5) \end{cases}$.

We can multiply Equation (5) by -310 to eliminate the variable b. Then add.

$$\begin{cases} 3950f + 1550b = 117{,}000 & (4) \\ -310(5f + 5b = 300) & (5) \end{cases} \rightarrow \begin{cases} 3950f + 1550b = 117{,}000 \\ -1550f - 1550b = -93{,}000 \\ \hline 2400f = 24{,}000 \\ f = 10 \end{cases}$$

Now we solve for b:

$5f + 5b = 300$	Equation (5)
$5(10) + 5b = 300$	Substitute 10 for f.
$50 + 5b = 300$	Simplify.
$5b = 250$	Subtract 50 from each side.
$b = 50$	Divide each side by 5.

Now solve for c:

$f + b + c = 300$	Equation (1)
$10 + 50 + c = 300$	Substitute 10 for f and 50 for b.
$60 + c = 300$	Simplify.
$c = 240$	Subtract 60 from each side.

So, the flight contained 10 first-class passengers, 50 business-class passengers, and 240 coach passengers. We can check our answer by substituting the three values into each equation in the original system.

$f + b + c = 300$	$4700f + 2300b + 750c$	$4f + 4b - c = 0$
$10 + 50 + 240 = 300$	$= 342{,}000$	$4(10) + 4(50) - 240 = 0$
$300 = 300$	$4700(10) + 2300(50) + 750(240)$	$40 + 200 - 240 = 0$
	$= 342{,}000$	$0 = 0$
	$47{,}000 + 115{,}000 + 180{,}000$	
	$= 342{,}000$	
	$342{,}000 = 342{,}000$	

2b. A retired couple made a profit of $300,000 on the sale of their family business. They invest this money in three areas: a high-risk stock fund with an expected annual rate of return of 15%, a low-risk stock fund with an expected annual rate of return of 10%, and government bonds at an annual return of 8%. To protect their investment, they invest twice as much money in the low-risk stock fund as the high-risk stock fund and use the remainder to buy bonds. How much did the couple allocate per investment if they earned $32,800 annually on their investment?

Solution   **2b.** What is unknown? The amounts invested in the high-risk stock fund, the low-risk stock fund, and government bonds are the unknowns.

Let $x =$ amount invested in the high-risk stock fund.
Let $y =$ amount invested in the low-risk stock fund.
Let $z =$ amount invested in government bonds.

We use the given facts to write a system that represents the problem.

$$\begin{cases} x + y + z = 300{,}000 \\ 0.15x + 0.10y + 0.08z = 32{,}800 \\ y = 2x \end{cases}$$

Total invested is $300,000.
Total earned is $32,800.
Twice as much is invested in the low-risk stock fund as the high-risk stock fund.

We rewrite each equation in standard form and clear the decimals by multiplying the second equation by 100.

$$\begin{cases} x + y + z = 300{,}000 \\ 100(0.15x + 0.10y + 0.08z = 32{,}800) \\ 2x - y = 0 \end{cases} \rightarrow \begin{cases} x + y + z = 300{,}000 & (1) \\ 15x + 10y + 8z = 3{,}280{,}000 & (2) \\ 2x - y = 0 & (3) \end{cases}$$

Solve the system using elimination. We will use Equation (3) as one of the equations in the new system since it only contains two variables. Then we pair together Equations (1) and (2) and eliminate the variable z.

$$\begin{cases} x + y + z = 300{,}000 & (1) \\ 15x + 10y + 8z = 3{,}280{,}000 & (2) \end{cases}$$

Multiply Equation (1) by -8 to eliminate z and add the equations.

$$\begin{cases} -8(x + y + z = 300{,}000) & (1) \\ 15x + 10y + 8z = 3{,}280{,}000 & (2) \end{cases} \rightarrow \begin{cases} -8x - 8y - 8z = -2{,}400{,}000 & (1) \\ \underline{15x + 10y + 8z = 3{,}280{,}000} & (2) \\ 7x + 2y = 880{,}000 & (4) \end{cases}$$

Now solve the system $\begin{cases} 2x - y = 0 & (3) \\ 7x + 2y = 880{,}000 & (4) \end{cases}$. Multiply Equation (3) by 2 and add the equations.

$$\begin{cases} 2(2x - y = 0) & (3) \\ 7x + 2y = 880{,}000 & (4) \end{cases} \rightarrow \begin{cases} 4x - 2y = 0 & (3) \\ \underline{7x + 2y = 880{,}000} & (4) \\ 11x = 880{,}000 \\ x = 80{,}000 \end{cases}$$

Solve for y:

$$y = 2x \qquad \text{Equation (3)}$$
$$y = 2(80{,}000) \quad \text{Substitute 80,000 for } x.$$
$$y = 160{,}000 \quad \text{Simplify.}$$

Solve for z:

$$x + y + z = 300{,}000 \qquad \text{Equation (1)}$$
$$80{,}000 + 160{,}000 + z = 300{,}000 \quad \text{Substitute 80,000 for } x \text{ and } 160{,}000 \text{ for } y.$$
$$240{,}000 + z = 300{,}000 \quad \text{Simplify.}$$
$$z = 60{,}000 \quad \text{Subtract 240,000 from each side.}$$

The couple invested $80,000 in the high-risk stock fund, $160,000 in the low-risk stock fund, and $60,000 in government bonds.

✓ Student Check 2

Solve each problem using a linear system with three equations and three variables.

a. A music hall hosted an awards ceremony. There were three types of tickets available to the public—first mezzanine ($539 each), second mezzanine ($392 each), and third mezzanine ($294 each). A total of 2300 tickets were sold totaling $1,004,500. The number of 1st mezzanine tickets sold was 500 less than the other two combined. Find the number of each type of ticket that was sold.

b. A young couple has saved six months of living expenses, which is $30,000. They invest this money in three different accounts—a 1-yr CD that pays 5% annual interest, a high-interest savings account that pays 4.5% annual interest, and a money market account that pays 10% annual interest. To keep their money as liquid as possible, they invest three times as much money in the high-interest savings account as the amount invested in the CD. Find the amount of money invested in each account if they earn $1925 interest in 1 yr.

Objective 3 ▶

Troubleshoot common errors.

Troubleshooting Common Errors

Some common errors associated with systems of linear equations in three variables are shown.

Objective 3 Example

A problem and an incorrect solution are given. Provide the correct solution and an explanation of the error.

Solve the system $\begin{cases} 4x - 5y + 3z = 3 \\ 2x + 5y - z = 9 \\ x + 2y + z = 8 \end{cases}$.

Incorrect Solution	Correct Solution and Explanation
Add Equations (1) and (2) to get $$6x + 2z = 12$$ Add Equations (2) and (3) to get $$3x + 7y = 17$$ Now solve $\begin{cases} 6x + 2z = 12 \\ 3x + 7y = 17 \end{cases}$. There is no solution since the system doesn't have the same variables.	The mistake is that different variables were eliminated in the pairs of equations. We must eliminate the same variable. So, we can eliminate the variable y from Equations (2) and (3) by multiplying Equation (2) by -2 and Equation (3) by 5. This yields $$\begin{cases} -4x - 10y + 2z = -18 \\ 5x + 10y + 5z = 40 \end{cases}$$ Adding the equations gives us $$x + 7z = 22$$ Now we solve the system $\begin{cases} 6x + 2z = 12 \\ x + 7z = 22 \end{cases}$.

$$\begin{cases} 6x + 2z = 12 \\ -6(x + 7z = 22) \end{cases} \rightarrow \begin{cases} 6x + 2z = 12 \\ \underline{-6x - 42z = -132} \\ -40z = -120 \\ z = 3 \end{cases}$$

Now solve for x:

$$x + 7z = 22$$
$$x + 7(3) = 22$$
$$x + 21 = 22$$
$$x = 1$$

Now solve for y:

$$x + 2y + z = 8$$
$$1 + 2y + 3 = 8$$
$$4 + 2y = 8$$
$$2y = 4$$
$$y = 2$$

The solution set is $\{(1, 2, 3)\}$.

ANSWERS TO STUDENT CHECKS

Student Check 1   a. $\{(-4, 0, 2)\}$ **b.** no solution

c. $\left\{\left(x, \dfrac{2x-6}{5}, \dfrac{x-23}{10}\right)\middle| x \text{ is a real number}\right\}$

Student Check 2   a. There were 900 first mezzanine, 1100 second mezzanine, and 300 third mezzanine tickets sold.   **b.** \$5000 in CD, \$15,000 in savings account, and \$10,000 in money market account

SUMMARY OF KEY CONCEPTS

1. The first step in solving a linear system in three variables is to use elimination to reduce it to a system involving two variables.
   - Pair together two different sets of equations from the linear system in three variables and eliminate the same variable from each pair.
   - The resulting equations form a linear system in two variables. Solve this system using elimination.

- The solution for a linear system in three variables may be an ordered triple, no solution, or infinitely many solutions.

2. To solve an application with a system in three variables, determine three different equations that represent the given information and solve it.

SECTION 4.5 / EXERCISE SET

 Write About It!

Use complete sentences in your answer to each exercise.

1. Describe the solution of a system of three linear equations.

2. Explain how to reduce a system of three linear equations to a system of two linear equations.

3. When you obtain a false statement during the process of solving a system of three linear equations, what is the solution of the system?

4. When you obtain a true statement during the process of solving a system of three linear equations, what is the solution of the system?

5. How do you know that you must use a system of three linear equations to solve a word problem?

6. Let A be the total amount you have to invest, and let r, s, and t be the annual rate of return in percents for three investments. Describe how to set up the first two equations in solving this investment problem.

 Practice Makes Perfect!

Solve each system of linear equations in three variables by elimination. (*See Objective 1.*)

7. $\begin{cases} x + 6y + 2z = 5 \\ 2x + 9y + z = -2 \\ 3x - y - 4z = -16 \end{cases}$

8. $\begin{cases} -2x + 5y + 7z = -11 \\ 2x - 6y + 3z = -10 \\ 3x + y + 3z = -2 \end{cases}$

9. $\begin{cases} x + 2y + 4z = -15 \\ 6x - 5y - 2z = -3 \\ -x - 4y - 3z = 9 \end{cases}$

10. $\begin{cases} -2x + 5y + 7z = -11 \\ 2x - 6y + 3z = -10 \\ 3x + y + 3z = -2 \end{cases}$

11. $\begin{cases} 2x + y - 3z = 0 \\ -5x + 2y + 3z = -9 \\ 4x + 4y - 5z = -1 \end{cases}$

12. $\begin{cases} 2x - 3y - 6z = -4 \\ -4x + y + z = -4 \\ 2x + 7y + z = -10 \end{cases}$

13. $\begin{cases} 3x - y + 6z = 2 \\ -2x + 3y + z = 6 \\ 2x + 5y - 4z = -1 \end{cases}$

14. $\begin{cases} x + 3y - 2z = 15 \\ x + y - z = 8 \\ 3x + 2y = 8 \end{cases}$

15. $\begin{cases} x + 5y - 3z = -4 \\ x - y - z = 10 \\ x + 3y = -4 \end{cases}$

16. $\begin{cases} -x + y + 2z = -3 \\ 4x + 3y - 5z = 14 \\ 5x - 4y = -23 \end{cases}$

17. $\begin{cases} x + 5y + 7z = 30 \\ -2x + z = 5 \\ 6x - 6y + 5z = -3 \end{cases}$

18. $\begin{cases} 2x + 3y + 4z = 4 \\ 5x + 7z = 22 \\ 9x + y - 2z = 23 \end{cases}$

19. $\begin{cases} -3x + 4y - 7z = 2 \\ x + 3z = 0 \\ 6x + 8y - 5z = 31 \end{cases}$

20. $\begin{cases} 2x - 3y - 2z = -3 \\ x - 2z = 2 \\ 5x + 2y - 4z = 0 \end{cases}$

21. $\begin{cases} 5x + 6y + 6z = 35 \\ y + z = 5 \\ -x + y + z = 4 \end{cases}$

22. $\begin{cases} 2x + y - 3z = 4 \\ y - z = 0 \\ 3x - y - 2z = 6 \end{cases}$

23. $\begin{cases} -4x + y + 5z = 0 \\ y - 27z = -48 \\ -3x + y - 3z = -12 \end{cases}$

24. $\begin{cases} -x + 5y + 5z = -19 \\ 2y + 3z = -8 \\ x + y + 4z = -5 \end{cases}$

25. $\begin{cases} x - y - 2z = 3 \\ -2y + z = -7 \\ -2x + 4y + 3z = 2 \end{cases}$

26. $\begin{cases} 5x + 2y + 6z = 0 \\ 2y + 26z = 11 \\ -4x - y + 3z = 15 \end{cases}$

27. $\begin{cases} 1.5x + y - 2.5z = 4 \\ 0.1x + 0.1y - 0.3z = 0.7 \\ 1.5x - 1.5y + 0.5z = -0.5 \end{cases}$

28. $\begin{cases} 0.3x + 0.2y + 0.6z = -1.1 \\ 3x - 2y - 1.5z = 3.5 \\ -x + 0.8y + 0.2z = 0 \end{cases}$

29. $\begin{cases} 2x - 5y - 3z = -39 \\ 17x - 23z = -7 \\ -3x - y + 4z = -10 \end{cases}$ **30.** $\begin{cases} 2x + 5y + 3z = 0 \\ -7x - 13z = 8 \\ x - y + 2z = 5 \end{cases}$

31. $\begin{cases} x - y - 4z = 11 \\ x - \dfrac{1}{2}y + \dfrac{3}{2}z = -6 \\ -\dfrac{5}{3}x - \dfrac{2}{3}y + z = 1 \end{cases}$ **32.** $\begin{cases} \dfrac{3}{2}x - \dfrac{1}{2}y + 3z = 1 \\ 2x - y - z = -4 \\ x + \dfrac{5}{2}y - 2z = -\dfrac{1}{2} \end{cases}$

Solve each problem using a linear system. (*See Objective 2.*)

33. An international flight from London to Singapore consists of first-class, business-class, and coach travelers. A first-class ticket costs $18,000. A business-class ticket costs $7400 and a coach ticket costs $1600. The airline made $858,400 in airfare revenue for one trip from London to Singapore. The flight had a total of 336 paying passengers. If there were six times as many coach travelers as first-class and business-class travelers combined, how many of each type of traveler were on this flight?

34. An international flight from Paris to Singapore consists of first-class, business-class, and coach travelers. A first-class ticket costs $11,200. A business-class ticket costs $5300 and a coach ticket costs $1100. The airline made $431,300 in airfare revenue for one trip from Paris to Singapore. The flight had a total of 240 paying passengers. If there were seven times as many coach travelers as first-class and business-class travelers combined, how many of each type of traveler were on this flight?

35. An international flight from Los Angeles to Hong Kong consists of first-class, business-class, and coach travelers. A first-class ticket costs $15,200. A business-class ticket costs $5700 and a coach ticket costs $1160. The airline made $746,980 in airfare revenue for one trip from Los Angeles to Hong Kong. The flight had a total of 371 paying passengers. If there were six times as many coach travelers as first-class and business-class travelers combined, how many of each type of traveler were on this flight?

36. An international flight from Honolulu to Hong Kong consists of first-class, business-class, and coach travelers. A first-class ticket costs $11,200. A business-class ticket costs $5300 and a coach ticket costs $1100. The airline made $636,300 in airfare revenue for one trip from Honolulu to Hong Kong. The flight had a total of 350 paying passengers. If there were six times as many coach travelers as first-class and business-class travelers combined, how many of each type of traveler were on this flight?

37. The Winters have saved $15,000. They invest this money in three different accounts—1-yr CD that pays 4.2% annual interest, a high-interest savings account that pays 5.6% annual interest, and a money market

account that pays 4.1% annual interest. They invest twice as much money in the high-interest savings account as the amount invested in the CD. Find the amount invested in each account if they earn $735.90 interest in 1 yr.

38. The Browns invest $29,500 in three different accounts—1-yr CD that pays 4.1% annual interest, a high-interest savings account that pays 7.9% annual interest, and a money market account that pays 3.3% annual interest. They invest twice as much money in the high-interest savings account as the amount invested in the CD. Find the amount invested in each account if they earn $1681.50 interest in 1 yr.

39. A concert is held at a music hall. There are three types of tickets available to the public—first mezzanine ($120 each), second mezzanine ($105 each), and third mezzanine ($65 each). A total of 10,300 tickets are sold totaling $1,091,500. The number of first mezzanine tickets sold is 200 less than twice the other two combined. Find the number of tickets sold in each mezzanine.

40. A concert is held at a music hall. There are three types of tickets available to the public—first mezzanine ($185 each), second mezzanine ($120 each), and third mezzanine ($70 each). A total of 11,450 tickets are sold totaling $1,701,500. The number of first mezzanine tickets sold is 400 less than twice the other two combined. Find the number of tickets sold in each mezzanine.

41. There are three types of tickets sold for a college football game—main ($115 each), terrace ($65 each), and grandstand ($50 each). A total of 3040 tickets are sold for a revenue of $199,050. The number of the grandstand tickets sold is 100 less than the other two combined. Find the number of tickets sold in each section.

42. There are three types of tickets sold for a college baseball championship game—main ($110 each), terrace ($70 each), and grandstand ($35 each). A total of 3000 tickets are sold for a revenue of $147,500. The number of the grandstand tickets sold is 300 less than twice the other two combined. Find the number of tickets sold in each section.

 Mix 'Em Up!

Solve each system of linear equations.

43. $\begin{cases} x + 3y - 6z = -11 \\ -5x + 4y + 3z = -10 \\ 4x + y - 5z = -3 \end{cases}$ **44.** $\begin{cases} 4x + y - 6z = -19 \\ -5x + 4y + z = -18 \\ x + 5y - 3z = -25 \end{cases}$

45. $\begin{cases} x + 5y - 2z = 28 \\ 3y - z = 16 \\ x + 2y - z = 12 \end{cases}$ **46.** $\begin{cases} x - 2y + 4z = 3 \\ -5y + 14z = -4 \\ 3x - y - 2z = 13 \end{cases}$

47. $\begin{cases} 0.1x - 0.2y - 0.4z = 1.3 \\ 0.7x + 0.3y + 0.2z = 1.8 \\ 1.5x + 4y + z = 5.5 \end{cases}$

48. $\begin{cases} 0.5x - y - z = 0 \\ 0.5x + y - z = -2 \\ x + y + 0.4z = 4.2 \end{cases}$

49. $\begin{cases} 5x + 6y + z = 43 \\ 9x + 11z = -8 \\ -2x + 3y - 5z = 6 \end{cases}$

50. $\begin{cases} 4x - 5y - 6z = 31 \\ 8x - 7y = 2 \\ -2x + 3y + 5z = -8 \end{cases}$

51. $\begin{cases} 4x + \dfrac{1}{2}y + z = -23 \\ 2y + 3z = 31 \\ \dfrac{1}{4}x + \dfrac{1}{4}y = 0 \end{cases}$

52. $\begin{cases} x + \dfrac{2}{3}y + \dfrac{5}{3}z = -1 \\ 3y - z = 17 \\ \dfrac{8}{5}x + y = \dfrac{17}{5} \end{cases}$

Solve each problem using a linear system.

53. The Shutters invest $42,500 in three different accounts—a 1-yr CD that pays 4.9% annual interest, a savings account that pays 8% annual interest, and a money market account that pays 3.3% annual interest. They invest three times as much money in the savings account as the amount invested in the CD. Find the amount invested in each account if they earn $2870.45 interest in 1 yr.

54. The Chans invest $35,000 in three different accounts— a 1-yr CD that pays 4.2% annual interest, a savings account that pays 6.8% annual interest, and a money market account that pays 2.7% annual interest. They invest three times as much money in the savings account as the amount invested in the CD. Find the amount invested in each account if they earn $1911 interest in 1 yr.

55. There are three types of tickets sold for a Las Vegas show—category 1 ($130 each), category 2 ($115 each), category 3 ($95 each). A total of 1500 tickets are sold for a revenue of $168,500. The number of category 3 tickets sold is 500 less than the other two combined. Find the number of tickets sold in each category.

56. There are three types of tickets sold for a concert— floor ($300 each), lower level ($160 each), and upper level ($110 each). A total of 5000 tickets are sold for a revenue of $770,000. The number of upper level tickets sold is 1000 less than the other two combined. Find the number of tickets sold in each section.

57. An international flight from Los Angeles to Shanghai consists of first-class, business-class, and coach travelers. A first-class ticket costs $5662. A business-class ticket costs $2400 and a coach ticket costs $1000. The airline made $444,234 in airfare revenue for one trip from Los Angeles to Shanghai. The flight had a total of 357 paying passengers. If the number of coach travelers was 35 more than six times the first-class and business-class travelers combined, how many of each type of traveler were on this flight?

58. An international flight from Los Angeles to Seoul consists of first-class, business-class, and coach travelers. A first-class ticket costs $7690. A business-class ticket costs $3900 and a coach ticket costs $700. The airline made $402,240 in airfare revenue for one trip from Los Angeles to Seoul. The flight had a total of 341 paying passengers.

If the number of coach travelers was 11 less than seven times the first-class and business-class travelers combined, how many of each type of traveler were on this flight?

59. There are three types of tickets sold for a college football game—main ($120 each), terrace ($65 each), and grandstand ($45 each). A total of 3990 tickets are sold for a revenue of $251,050. The number of the grandstand tickets sold is 300 less than twice the other two combined. Find the number of tickets sold in each.

60. There are three types of tickets sold for a college baseball game—main ($125 each), terrace ($70 each), and grandstand ($30 each). A total of 6970 tickets are sold for a revenue of $341,150. The number of the grandstand tickets sold is 500 less than twice the other two combined. Find the number of tickets sold in each.

 You Be the Teacher!

Correct each student's errors, if any.

61. Solve $\begin{cases} 3x - 4y + 6z = 3 \\ x - y - 2z = 8 \\ -2x + 7y - 3z = -17 \end{cases}$.

Marian's work:

$3x - 4y + 6z = 3 \rightarrow 3x - 4y + 6z = 3$
$x - y - 2z = 8 \rightarrow -3(x - y - 2z = 8)$

$\quad\ 3x - 4y + 6z = 3$
$\underline{-3x - 3y - 6z = 24}$
$\qquad\quad -7y = 27$

$$y = -\frac{27}{7}$$

62. Solve $\begin{cases} x + 2y + 3z = -1 \\ 4y + 4z = -5 \\ 3x - 2y + z = 7 \end{cases}$.

Josh's work:

$\begin{cases} x + 2y + 3z = -1 \\ \underline{3x - 2y + z = 7} \\ \quad 4x + 4z = 6 \end{cases}$

$\begin{cases} 4y + 4z = -5 \\ 4x + 4z = 6 \end{cases} \rightarrow \begin{array}{r} 4y + 4z = -5 \\ \underline{-4x - 4z = -6} \\ -4x + 4y = -11 \end{array}$

No solution

 Think About It!

63. Write a system of linear equations in three variables that has $(5, -1, 4)$ as its solution.

64. Write a system of linear equations in three variables that has $(-2, 3, 7)$ as its solution.

65. The equation $y = ax^2 + bx + c$ has solutions of $(1, 6)$, $(-1, 8)$, and $(2, 11)$. Use a system of three linear equations to find the values of a, b, and c. [Hint: Substitute each ordered pair, (x, y), in $y = ax^2 + bx + c$ to obtain three equations with unknowns of a, b, and c.

66. The equation $y = ax^2 + bx + c$ has solutions of $(1, 2)$, $(2, 2)$, and $(3, 4)$. Use a system of three linear equations to find the values of a, b, and c. [Hint: Substitute each ordered pair, (x, y), in $y = ax^2 + bx + c$ to obtain three equations with unknowns of a, b, and c.

GROUP ACTIVITY   The Mathematics of Choosing a Design

Mike and Amy purchased a lakefront lot for their dream home. The lot is rectangular and borders the lake on one side. Mike and Amy want to fence the other three sides of the property so that the perimeter of the fenced area is 2000 ft. They have $25,000 budgeted for this job. The cost of several types of fencing materials is shown in the table.

Type	Cost per Linear Foot
4-ft-high chain link	$6
$4\frac{1}{2}$ ft wood fence	$15
Vinyl fence	$30
Picket fence	$20
6-ft-high cedar privacy fence	$11

1. Draw a diagram of the lake property and label the sides of the property with appropriate variables.

2. Choose two different fencing options, one to be used for the front of the property and the other to be used for the remaining two sides.

3. Write a system of linear equations that can be used to determine the amount of each type of fencing needed to enclose the property.

4. Solve the system using an appropriate method. Round to the nearest hundredth, if necessary.

5. Explain what the solution of the system means in the context of the problem.

6. Find the area that is enclosed by the selected fencing design.

7. Repeat the process two additional times.

8. Which design yields the largest enclosed area of the property?

Systems of Linear Equations in Two and Three Variables

What's the big idea? Systems can be used to solve applications involving two and three unknowns. Now that we have completed chapter 4, we should be able to solve these types of systems with two unknowns in three different ways and to identify the number of solutions of the system. We should also be able to solve systems in three variables and their applications using elimination.

The Tools

Listed below are the key terms, skills, formulas, and properties you should know for this chapter.

The page reference is provided if you need additional help with the given topic. The Study Tips will assist in your preparation for an exam.

Study Tips

1. Learn all of the terms, formulas, and properties. Make flash cards and have someone quiz you.
2. Rework problems from the exercises and also the ones you worked in class. Work additional problems from the review exercises.
3. Review the summaries of key concepts.
4. Work the chapter test.
5. Be sure to review the online resources for additional study materials.

Terms

Addition method   317
Consistent system with dependent equations   298
Consistent system with independent equations   298
Elimination method   317
Inconsistent system   298

Solution of a linear system in three variables   344
Solution of a system of linear equations in two variables   294
Substitution method   308
System of linear equations   294
System of linear equations in three variables   344

Formulas and Properties

- Addition property of equality   319
- Complementary angles   338
- Distance formula   335

- Perimeter of a rectangle   338
- Simple interest formula   332
- Supplementary angles   338

CHAPTER 4 / SUMMARY

How well do you know this chapter? Complete the following questions to find out. Take a look back at the section if you need help.

SECTION 4.1 Solving Systems of Linear Equations Graphically

1. A system of linear equations consists or two or more equations that must be _____ _____ .
2. A solution of a system of linear equations in two variables is an _____ ___ that satisfies both equations.

3. To solve a system of linear equations in two variables by graphing, we graph each _____ on the same coordinate system and find their _____ ___.
4. A system of linear equations in two variables has three possible types of solutions: ___ solution, __ solution, or _____ ____ solutions.
5. The lines in a system of linear equations will either _____ one another, be _____ to each other, or be the ___ line.

6. A system of linear equations that has at least one solution is called a _____ system.

7. A system of linear equations that has no solution is called an _____ system.

8. If the two equations in a system of linear equations are multiples of one another, then the equations are _____. Otherwise, the equations are _____.

SECTION 4.2 Solving Systems of Linear Equations by Substitution

9. To solve a system of linear equations by substitution, we must first ____ one of the equations for one of the _____ and _____ this expression into the other equation.

10. If after substitution, the resulting equation is a contradiction, then the system of linear equations is _____ and has __ solution.

11. If after substitution, the resulting equation is an identity, then the system of linear equations is _____ with _____ equations and has _____ solutions.

SECTION 4.3 Solving Systems of Linear Equations by Elimination

12. The goal of elimination is to __ the equations in a system of linear equations so that one variable is _____.

13. For a variable to be eliminated from a system of linear equations, its coefficients must be _____ of one another. If they are not, we must _____ one or both equations by some nonzero number.

14. If after adding the two equations in a system of linear equations together, the resulting equation is an identity, this means that the system contains two lines that are the ____ and there are _____ ____ solutions.

15. If after adding the two equations in a system of linear equations together, the resulting equation is a contradiction, this means that the system contains two lines that are _____ and there is __ solution of the system.

SECTION 4.4 Applications of Systems of Linear Equations in Two Variables

16. Applications that involve two unknowns can be solved using a _____ of equations.

17. We assign two _____ to the unknowns and write two different _____ that relate the two unknowns and then solve using _____ or _____.

SECTION 4.5 Solving Systems of Linear Equations in Three Variables and Their Applications

18. To solve a linear system in three variables, we use _____ to reduce it to a system involving ___ variables.

19. To solve an application with a system of three variables, determine the three _____ and the three different _____ that represent the given information.

CHAPTER 4 / REVIEW EXERCISES

SECTION 4.1

Solve each system of linear equations graphically. Then state if the system is consistent with dependent equations, consistent with independent equations, or inconsistent. (See Objectives 3–5.)

1. $\begin{cases} y = 2x + 1 \\ y = 4x - 3 \end{cases}$ **2.** $\begin{cases} 3x - y = 2 \\ y = 3x - 2 \end{cases}$

3. $\begin{cases} x - 2y = 0 \\ 3y + x = 10 \end{cases}$ **4.** $\begin{cases} 2x - y = 7 \\ x = 3 \end{cases}$

5. $\begin{cases} x + y = -5 \\ y = \dfrac{1}{7}x + 3 \end{cases}$ **6.** $\begin{cases} y = 4x - 1 \\ x - \dfrac{1}{4}y = 2 \end{cases}$

Solve each problem. (See Objectives 2 and 6.)

7. George is trying to decide which Qwest home phone service he should use. The basic service starts at $15 per month with unlimited local calling and $0.15 per minute for long distance calls. The Home Phone Plus package starts at $30 per month with unlimited local calling and $0.05 per minute for long distance calls per month, the cost per month for the

basic service is $y = 15 + 0.15x$. The cost per month for the Home Phone Plus package is $y = 30 + 0.05x$. Use the graph of the system formed by these two equations to answer the questions that follow.

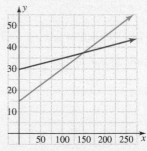

a. What is the point of intersection of this system?

b. What is the meaning of the point of intersection in the context of this problem?

c. If George makes an average of 200 min of long distance calls per month, which plan should he use? Why?

SECTION 4.2

Solve each system of linear equations by substitution. (See Objectives 1 and 2.)

8. $\begin{cases} x + 2y = 1 \\ 7x - 5y = 26 \end{cases}$ **9.** $\begin{cases} y = -\dfrac{1}{4}x + \dfrac{7}{2} \\ x + y = 2 \end{cases}$

10. $\begin{cases} y = -2x - 15 \\ 4x + 2y = -30 \end{cases}$ 11. $\begin{cases} -2x + 5y = 27 \\ y = 7 \end{cases}$

12. $\begin{cases} x = -0.75y - 3.5 \\ 0.25x + 0.25y = -1.25 \end{cases}$

13. $\begin{cases} y = \dfrac{5}{3}x + 2 \\ \dfrac{1}{3}x - \dfrac{1}{5}y = -\dfrac{6}{5} \end{cases}$ 14. $\begin{cases} y = 2x \\ 0.3x - 0.4y = 1.5 \end{cases}$

SECTION 4.3

Solve each system by elimination. (*See Objectives 1–3.*)

15. $\begin{cases} 3x - 7y = 4 \\ 5x - 6y = 18 \end{cases}$ 16. $\begin{cases} 4x + y = -2 \\ x + 5y = 9 \end{cases}$

17. $\begin{cases} 4x + y = 16 \\ y = -4x + 12 \end{cases}$ 18. $\begin{cases} 3x - y = -1 \\ 2x - \dfrac{2}{3}y = -\dfrac{2}{3} \end{cases}$

19. $\begin{cases} 0.7x + 0.1y = 1.7 \\ -0.9x + y = 1.2 \end{cases}$ 20. $\begin{cases} \dfrac{1}{2}x - \dfrac{2}{3}y = -2 \\ \dfrac{1}{4}x + \dfrac{5}{6}y = 6 \end{cases}$

SECTION 4.4

Solve each problem. (*See Objectives 1–5.*)

21. Two angles are supplementary. The larger angle is 28° less than three times the smaller angle. What is the measure of each angle?

22. Two angles are complementary. The larger angle is 6° less than twice the measure of the smaller one. What is the measure of each angle?

23. Matt and Dimitri are experienced cyclists who like challenging cycling routes. Dimitri gives Matt a 20-min head start on the route. On this route, Matt cycles an average of 8 mph. On the same route, Dimitri averages 10 mph. How long will it take Dimitri to catch up to Matt? $\left(\text{Hint: 20 min} = \dfrac{1}{3} \text{ hr} \right)$

24. Joseph has a 40% alcohol solution and a 20% alcohol solution. How many liters of the 40% solution and how many liters of the 20% solution must he mix together to obtain 100 L of the 36% alcohol solution?

25. A plane travels 2305 mi in 4.15 hr when it is traveling with the wind. It takes 4.56 hr for the plane to travel the same distance against the wind. What is the speed of the plane in still air and what is the speed of the wind?

26. Sandy has $8600 to invest. She invests part of the money in a savings account that pays 1.2% annual interest and the rest of the money in a CD that pays 2.8% annual interest. If she earns $202.40 interest in 1 yr, how much did she invest in each account?

27. Two families visit an amusement park. One family pays $272 for four adults and six children to be admitted. Another family pays $154 for two adults and four children to be admitted. How much does it cost for each adult and each child to be admitted to the park?

28. Two families eat at a buffet restaurant. One family pays $190 for six adults and eight children. Another family pays $104.50 for two adults and seven children. How much does it cost for each adult and each child?

SECTION 4.5

Solve each system of linear equations. (*See Objective 1.*)

29. $\begin{cases} 2x + y - 2z = -1 \\ -4x - 3y - 3z = 6 \\ -x + 8y - 2z = 29 \end{cases}$ 30. $\begin{cases} 3x + 3y + 5z = 25 \\ x + y + z = 5 \\ 3x - 3y + 4z = 14 \end{cases}$

31. $\begin{cases} 7x + 6y + z = -34 \\ 9x + 2z = -10 \\ x + 9y + z = -34 \end{cases}$ 32. $\begin{cases} 6x + y = 28 \\ 5x + 4y - 5z = 7 \\ -x - 3y + 3z = 7 \end{cases}$

33. $\begin{cases} 5x + y + z = 10 \\ 3y + 13z = -45 \\ 2x + y + 3z = -5 \end{cases}$ 34. $\begin{cases} -5x + 6y + 3z = -10 \\ y + 8z = -15 \\ -x + y - z = 1 \end{cases}$

Solve each problem using a system of linear equations in three variables. (*See Objective 2.*)

35. Ted and Kristi invest $35,750 in three different accounts—a 1-yr CD that pays 3.5% annual interest, a savings account that pays 4.5% annual interest, and a money market account that pays 2.5% annual interest. They invest twice as much money in the savings account as the amount invested in the CD. Find the amount invested in each account if they earn $1443.75 interest in a year.

36. A baseball game offers three types of admission tickets. Dugout-level tickets are $120 each, club-level tickets are $75 each, and general admission tickets are $65 each. A total of 7050 tickets are sold totaling $647,500. The number of dugout-level tickets sold was 350 less than the other two combined. How many of each type of ticket were sold?

37. An international flight from Los Angeles to Beijing consists of first-class, business-class, and coach travelers. A first-class ticket costs $9500. A business-class ticket costs $4600 and a coach ticket costs $650. The airline made $306,300 in airfare revenue for the trip from Los Angeles to Beijing. The flight had a total of 224 paying passengers. If the number of coach travelers was 6 times as many as first-class and business-class combined, how many of each type of traveler were on the flight?

CHAPTER 4 TEST / SYSTEMS OF LINEAR EQUATIONS IN TWO AND THREE VARIABLES

1. The ordered pair that is a solution of the system
$$\begin{cases} 4x - y = 3 \\ y = 6x - 4 \end{cases} \text{ is}$$

   a. $(2, 8)$

   b. $\left(\dfrac{1}{2}, -1\right)$

   c. $\left(-\dfrac{1}{2}, -7\right)$

   d. $(-2, -11)$

2. Use graphing to solve the system $\begin{cases} x - y = -2 \\ y = -\dfrac{2}{3}x - 3 \end{cases}$.

3. Without graphing each system of linear equations, complete the table.

   a. $\begin{cases} 7x - 4y = 8 \\ \dfrac{14}{8}x - y = 2 \end{cases}$

Slope-intercept form		
Slopes		
y-intercepts		
How do lines relate?		
Number of solutions		
Type of system		
Solution set		

   b. $\begin{cases} y = \dfrac{1}{5}x + 3 \\ x - 5y = 10 \end{cases}$

Slope-intercept form		
Slopes		
y-intercepts		
How do lines relate?		
Number of solutions		
Type of system		
Solution set		

4. The graph shows the percentage of Americans who use the newspaper or Internet as their source for news, where x is the years after 2003. (Source: http://pewresearch.org/pubs/1066/internet-overtakes-newspapers-as-news-outlet)

National and International News Sources for Americans

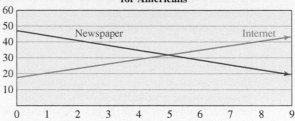

   a. Approximate the point of intersection.

   b. Interpret what the point of intersection means in the context of this problem.

5. What are the three methods to solve a system of linear equations in two variables by hand? Which method is the least precise? Why? How do you decide when to use the other two methods?

6. Solve each system by substitution.

   a. $\begin{cases} x - 8y = 13 \\ 5y - 9x = -50 \end{cases}$

   b. $\begin{cases} -4x - y = 15 \\ 3x + 5y = -41 \end{cases}$

7. Solve each system by elimination.

   a. $\begin{cases} 7x + 3y = -22 \\ 6x - 3y = -30 \end{cases}$

   b. $\begin{cases} 3x - 2y = 19 \\ 2x + 6y = -46 \end{cases}$

8. Solve each system using any method.

   a. $\begin{cases} 24x - 4y = 24 \\ 6x = y - 6 \end{cases}$

   b. $\begin{cases} \dfrac{x}{5} + y = \dfrac{11}{5} \\ \dfrac{3x}{5} + \dfrac{y}{2} = -\dfrac{17}{5} \end{cases}$

   c. $\begin{cases} x + \dfrac{1}{4}y = y - 4 \\ \dfrac{1}{3}x + y = x + y \end{cases}$

Use a system of equations to solve each problem.

9. Ms. Jones invests $18,000 in two accounts, one yielding 8% interest and the other yielding 9%. She earns a total of $1490 in interest at the end of the year. How much was invested in each account?

10. Diane is the manager of a 10,000-seat ballpark and plans to host a special event to raise money for the Cystic Fibrosis Foundation. She plans to sell VIP seats for $100 each and regular admission seats for $37.50. If Diane wants to raise $500,000 by selling every seat, how many of seats should she designate as VIP seats?

11. Dominick is on a flight from Chicago O'Hare Airport to London Heathrow Airport, a distance of approximately 3950 mi. He notices on his ticket that the travel time from Chicago to London is 7.5 hr and the return trip is 9 hr. The trip to London is with the wind and the trip from London is against the wind. What is the average speed of the plane in still air and the average speed of the wind? Round the answers to the nearest integer.

12. The Lincoln Memorial Reflecting Pool is the largest reflecting pool in Washington, DC. It is in the shape of a rectangle whose perimeter is 1338 m. Its length is 6 m more than 12 times its width. Find the length and width of the reflecting pool.

13. Solve each system of linear equations in three variables.

a. $\begin{cases} x + y + z = 3 \\ x + 2y + 3z = 9 \\ 2x - y - z = 0 \end{cases}$ **b.** $\begin{cases} 2x + 3y - z = 4 \\ x - 2y = 5 \\ x - 9y + z = 1 \end{cases}$

c. $\begin{cases} x - y + 5z = -2 \\ 2x + y + 4z = 2 \\ 2x + 4y - 2z = 8 \end{cases}$

14. Juan receives an inheritance of $30,000. He invests his money in three different investments—a money-market account, a CD, and stocks. The money-market account will provide a return of 7%, the CD will provide a return of 5%, and the stocks will provide a return of 10%. Juan is feeling risky, so he invests twice as much money in the stocks as he does in the CD. How much does he invest in each investment if he earns a total of $2420 during his first year?

CUMULATIVE REVIEW EXERCISES / CHAPTERS 1–4

1. Perform each operation and simplify the result. (*Section 1.2, Objectives 2–7*)

a. $\dfrac{9}{10} + \dfrac{1}{2} - \dfrac{3}{5}$ **b.** $5\dfrac{1}{3} \cdot 2\dfrac{1}{4} \div 6$

c. $3\dfrac{1}{6} - 1\dfrac{3}{4} + \dfrac{7}{12}$

2. Use the order of operations to simplify each expression. (*Section 1.3, Objectives 1 and 2*)

a. $-(-2.5)^4$ **b.** $\dfrac{2}{3}(10 - 8)^2 - \dfrac{5}{6}$

c. $3(4)^2 - 6(7) + 11$ **d.** $4[7 - |9 - (5 - 3)|]$

e. $\dfrac{5 + \sqrt{5^2 - 4(2)(3)}}{2 \cdot 2}$

3. The expression $27.94x + 41.91y$ represents the weekly earnings of a mining worker, where x is the number of regular hours worked in a week and y is the number of overtime hours worked in a week. Find the weekly salary of a mining worker who worked 40 regular hours and 10 overtime hours. (Source: http://www.bls.gov) (*Section 1.3, Objective 6*)

4. Perform the operation and simplify. (*Sections 1.4 and 1.5, Objectives 1 and 2*)

a. $4.2 - 2.9 + (-1.6) - (-7.5)$

b. $\dfrac{5}{7} - \dfrac{1}{2} - \left(-\dfrac{13}{14}\right)$

c. $10 - \{6 - [3 - (1 - \sqrt{17 - 1})]\}$

5. Perry has $115.25 in his checking account. He deposits his paycheck of $627.85. Perry writes a check for the phone bill for $56.29, for groceries for $156.23, and then for rent for $525. What is Perry's checking account balance? (*Section 1.4, Objective 3*)

6. Nate wants to mix a 10% salt solution with a 40% salt solution to get a 28% salt solution. Let x represent the volume of the 10% salt solution. If Nate wants to make 80 oz of a 28% salt solution, write an expression that represents the volume of the 40% salt solution. (*Section 1.5, Objective 3*)

7. In a rectangle, the length is six less than twice times the width. If w represents the width, write an expression that represents the length of the rectangle. (*Section 1.5, Objective 3*)

8. Perform the indicated operation and simplify. (*Section 1.6, Objectives 1–3*)

a. $(-12.4)(-7.5)$ **b.** $(-76)\left(\dfrac{14}{19}\right)\left(-\dfrac{10}{21}\right)$

c. $(-4)^4$ **d.** $0 \div 16$

e. $\left(4\dfrac{1}{6}\right) \div \left(3\dfrac{3}{4}\right)$

9. Simplify each algebraic expression. (*Section 1.8, Objectives 3 and 4*)

a. $-(3a - 1) - 4(-5a + 2)$

b. $5.8 - 0.2(-6b + 36)$

c. $12\left(\dfrac{1}{2}x - \dfrac{5}{6}\right) - 6\left(\dfrac{1}{3}x + 1\right)$

d. $12x^2 - 8(x + 2x^2) + x$

10. Determine if the given numbers are solutions of the equation. (*Section 2.1, Objective 2*)

    a. Is $x = -2$, $x = 0$, or $x = 2$ a solution of $4x - 1 = x + 5$?

    b. Is $x = -3$, $x = 0$, or $x = 3$ a solution of $3x + 11 = -2x - 4$?

11. For each problem, define the variable, write an equation, and solve the problem. (*Section 2.1, Objectives 3 and 4*)

    a. The difference of 40 and a number is the same as three times the number. Find the number.

    b. The sum of three times a number and -5 equals 7 less than the number. Find the number.

    c. One angle is 34° more than another angle. Their sum is 90°. Find the measure of each angle.

12. Solve each equation. (*Section 2.2, Objectives 2 and 3; Sections 2.3 and 2.4, Objectives 1 and 2*)

    a. $5(x - 1) - 14 = 2(x - 3) + 11$

    b. $-\dfrac{2}{7}(7x + 21) + 9 = x - 6$

    c. $-1.2y + 24.1 = 2.3y + 3.1$

13. Solve each equation. If the equation is a contradiction, write the solution as Ø. If the equation is an identity, write the solution as ℝ. (*Section 2.4, Objectives 1–4*)

    a. $4(z - 2) - (z + 6) = 3(z + 2) - 20$

    b. $0.012x + 0.019(890 - x) = 14.46$

    c. $-\dfrac{1}{4}(x - 3) = \dfrac{1}{2}(2x + 9) + 1$

14. The revenue R, unit price p, and sales level x of a product are related by the formula $R = xp$. (*Section 2.5, Objective 1*)

    a. Find R if $p = \$23.50$ and $x = 50$.

    b. Find p if $R = \$9480$ and $x = 395$.

15. Solve each formula for the specified variable. (*Section 2.5, Objective 3*)

    a. $V = \pi r^2 h$ for h

    b. $C = 2l + 5w$ for w

    c. $3x - 2y = 60$ for y

    d. $0.035x + 0.028y = 2.1$ for x

16. Solve each problem. (*Section 2.5, Objective 5*)

    a. Find the measure of an angle whose complement is 21° less than the measure of the angle.

    b. The supplement of an angle is 32° more than three times its complement. Find the measure of the angle.

17. In triangle ABC, the measure of angle B is 20° less than the measure of angle A. The measure of angle C is 11° more than the measure of angle A. Find the measure of each angle in the triangle. (*Section 2.5, Objective 5*)

Solve each problem. Write an appropriate equation, solve the equation, and write the answer in complete sentences. (*Section 2.6, Objectives 1–5*)

18. A suit is on sale for \$110.55. If this price is 33% off the original price, what is the original price of the suit?

19. Candice invests \$3400 in two accounts. She invests \$2600 in a mutual fund that pays 2.1% annual interest and the remaining amount in a money market fund that pays 1.0% annual interest. How much interest will she earn from the two accounts in 1 yr?

20. Solve each inequality. For each inequality, graph the solution set, and write the solution set in interval notation, and in set-builder notation. (*Section 2.7, Objectives 1–4*)

    a. $0.2b - 3.8 \geq -4.2$ **b.** $-1 < 4a + 3 < 15$

21. Complete a table of values, plot the resulting points, and use the points to draw the graph of the equation. (*Section 3.1, Objective 3*)

    a. $y = -2x + 6$ **b.** $y = x^2 - 5$

    c. $y = 2 - |x - 1|$

22. Use the given graph to complete each ordered pair that is a solution. (*Section 3.1, Objective 4*)

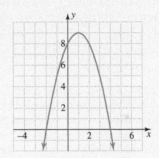

    a. (_____ , 0)   **b.** (0, _____)

    c. (1, _____)   **d.** (2, _____)

23. The graph shows the seasonally adjusted unemployment rate in the month of September from 2000 to 2010. (Source: http://data.bls.gov) (*Section 3.1, Objective 5*)

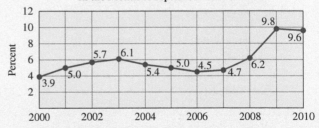

Unemployment Rate (Seasonally Adjusted) in the Month of September 2000 to 2010

    a. Write an ordered pair for each point labeled on the graph.

    b. Between what consecutive years did the unemployment rate increase most? By how much?

    c. Between what consecutive years did the unemployment rate decrease most? By how much?

24. The table lists the net income as reported (in millions of dollars) for Southwest Airlines between 2003 and 2009. (Source: http://phx.corporate-ir.net) (*Section 3.1, Objective 5*)

Years After 2003	0	1	2	3	4	5	6
Net income as reported (in millions)	$372	$217	$484	$499	$645	$178	$99

   a. Write an ordered pair (x, y) that corresponds to each year between 2003 and 2009, where x is the number of years after 2003 and y is the net income as reported for Southwest Airlines (in millions of dollars).

   b. Interpret the meaning of first and last ordered pairs from part (a) in the context of the problem.

   c. In what year was the net income the highest? The lowest?

   d. Make a scatter plot of the data.

25. Graph each equation. Identify at least two points on the graph. (*Section 3.2, Objectives 2–4*)

   a. $5x - 3y = 15$ **b.** $2y - x = 0$

   c. $2x - 3 = 0$ **d.** $1 - 0.5y = 0$

   e. $0.6x + 0.1y = 12$

26. The yearly cost of jet fuel and oil (in millions of dollars) for Southwest Airlines can be modeled by the equation $y = 457.7x + 1724.8$, where x is the number of years after 2005. (Source: http://www.southwest.com/htm/cs/investor_relations/ar2010.html) (*Section 3.2, Objective 5*)

   a. Find the y-intercept and interpret its meaning.

   b. What was the yearly cost of jet fuel and oil for Southwest Airlines in 2009?

   c. What will the yearly cost of jet fuel and oil be for Southwest Airlines in 2013?

   d. Graph the equation.

27. Find the slope of each line (*Section 3.3, Objectives 1, 3, and 4*)

   a. $y = -\dfrac{1}{2}x + 1$

   b. Passes through $(-1, -3)$ and $(-2, 0)$

   c. $y = -2$

   d. Passes through $(-3.2, -0.4)$ and $(-3.2, -10.2)$

   e. $x - 4y = 1$

28. Use the given information to graph the line. Label two points on the line. (*Section 3.4, Objective 2*)

   a. $m = -3$, passes through $(0, 0)$

   b. $y = 3x - 5$

   c. $m =$ undefined, passes through $(-1, 3)$

   d. $m = 0$, passes through $(-3, 2)$

29. The average price, in dollars per barrel, of jet fuel in the United States can be modeled by $y = 8.38x + 24.02$, where x is the years after 2000. Interpret the meaning of the slope and y-intercept in the context of the problem. (Source: http://www.airlines.org/Economics/DataAnalysis/Pages/AnnualCrudeOilandJetFuelPrices.aspx) (*Section 3.4, Objective 3*)

30. Determine if the lines are parallel, perpendicular, or neither. (*Section 3.4, Objective 3*)

   a. $3x + 7y = 1$ and $0.14y = -0.06x + 1.2$

   b. $y = 0.6x$ and $y = -0.6x$

31. Write the equation of each line. Express the answer in slope-intercept form and in standard form. (*Section 3.5, Objectives 1–4*)

   a. passes through $(-1, 4)$ and $(-2, 0)$

   b. $m = 0$, passes through $(0, 12)$

   c. passes through $(2, -2)$, perpendicular to $x + 3y = 1$

32. Janis plans to enroll in a fitness club. There is a one-time membership fee of $195 plus a monthly charge of $35. (*Section 3.5, Objective 5*)

   a. Write a linear equation that represents the cost of joining the fitness club, where x is the number of months Janis is a member.

   b. How much money has Janis spent for her fitness club membership if she has been a member for 1 yr?

   c. How long has Janis been a member of the club if she has paid a total of $825?

33. State the domain and range of each relation. Determine if the relation is a function. (Source: http://www.eia.gov) (*Section 3.6, Objectives 1–3*)

   a. The average residential natural gas prices (in dollars per thousand cubic feet) from 2004 to 2010 is shown.

Years	Average Residential Natural Gas Prices
2004	10.75
2005	12.70
2006	13.73
2007	13.08
2008	13.89
2009	12.14
2010	11.20

   b. $y = 2$ **c.** $y = -2x$

34. Find the requested information. (*Section 3.6, Objective 4*)

   a. Find $f(0)$ if $f(x) = 12x - 9$.

   b. Find $g(-2)$ if $g(x) = 3x^2 - 2x$.

   c. Find $f(-1)$ and $f(4)$ if $f(x)$ is given by the graph.

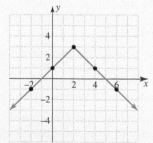

d. Find $g(-3)$ and find x for which $g(x) = -4$ if the function $g(x)$ is given by Y_1.

X	Y₁
-3	-9
-2	-6.5
-1	-4
0	-1.5
1	1
2	3.5
3	6

X=-3

35. Write each equation in two variables in function notation. Then find $f(0)$, $f(3)$, and $f(-2)$ and write the corresponding ordered pairs. (*Section 3.6, Objectives 4 and 5*)

a. $x + 2y = 1$

b. $y = 0.2x^2 - 0.1x$

36. Shawn has a new job as an English instructor at a local community college. His annual salary is $42,400 with a regular teaching load of 30 credit hours. If he teaches beyond the regular load in a year, he earns $1050 per credit hour. (*Section 3.6, Objective 6*)

a. Write a linear function $f(x)$ that represents Shawn's yearly income, where x is the number of additional credit hours he teaches in a year.

b. If Shawn teaches two extra 3-credit-hour courses in a year, what is his income?

c. Find $f(2)$ and interpret the answer.

37. Solve each system of equations graphically. Then state if the system is consistent with dependent equations, consistent with independent equations, or inconsistent. (*Section 4.1, Objectives 3–5*)

a. $\begin{cases} y = 3x - 5 \\ y = x - 1 \end{cases}$

b. $\begin{cases} y = 5x + 1 \\ y = 5x - 1 \end{cases}$

c. $\begin{cases} x - y = 3 \\ x = 3 \end{cases}$

d. $\begin{cases} x - 2y = -5 \\ -x + 3y = 7 \end{cases}$

e. $\begin{cases} 3x - y = 12 \\ x = \dfrac{1}{3}y + 4 \end{cases}$

38. Rikki needs to rent a refrigerator for her apartment. Household Rentals offers a monthly rental rate of $42 plus a one-time installation fee of $60. Rent and Go offers a monthly rental rate of $50 per month. The monthly cost of Household Rentals can be expressed by $y = 42x + 60$ and the monthly cost of Rent and Go can be expressed by $y = 50x$, where x is the number of months. Use the graph of the system formed by these two equations to answer each question. (*Section 4.1, Objectives 2 and 6*)

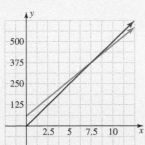

a. Approximate the point of intersection of this system.

b. What is the meaning of the intersection point in the context of the problem?

c. If Rikki expects to sign a 12-month rental of the refrigerator, which company should she choose? Why?

Solve each system of linear equations using substitution. (*Section 4.2, Objectives 1 and 2*)

39. $\begin{cases} 7x - 5y = 16 \\ x + 3y = 6 \end{cases}$

40. $\begin{cases} x - 6y = 4 \\ 4x + 2y = 3 \end{cases}$

41. $\begin{cases} \dfrac{1}{3}x - y = 2 \\ x = 3y + 5 \end{cases}$

Solve each system using elimination. (*Section 4.3, Objectives 1–3*)

42. $\begin{cases} 3x - 10y = -56 \\ 4x + 8y = 32 \end{cases}$

43. $\begin{cases} 2x + y = 5 \\ 3y = -6x + 15 \end{cases}$

44. $\begin{cases} 6x - 8y = 13 \\ \dfrac{5}{6}x + \dfrac{1}{2}y = 1 \end{cases}$

45. $\begin{cases} 1.5x - 0.4y = 5.8 \\ 0.5x + 2.4y = -4.4 \end{cases}$

Solve each problem using a system of linear equations. (*Section 4.4, Objectives 1–5*)

46. Two angles are supplementary. The larger angle is 12° more than three times the smaller angle. What is the measure of each angle?

47. Two angles are complementary. The larger angle is 24° less than twice the measure of the smaller one. What is the measure of each angle?

48. Wendy leaves her job and heads south at 45 mph. Jay leaves the same workplace 20 min later and heads south at 60 mph. How long will it take Jay to catch up to Wendy?

49. Alison needs to make 250 L of a 60% alcohol solution. How much pure alcohol and how much 50% alcohol should be mixed together to make 250 L of a 60% alcohol solution?

50. A plane travels 1260 mi with the wind in 2.8 hr. It takes 3.6 hr for the plane to travel the same distance against the wind. What is the speed of the plane in still air and what is the speed of the wind?

Solve each system of linear equations in three variables. (*Section 4.5, Objective 1*)

51. $\begin{cases} 2x - 3y + 5z = 13 \\ -x + 5y - 3z = 2 \\ 5x - 6y + z = -9 \end{cases}$

52. $\begin{cases} 3x + y - 3z = -4 \\ -5x + 5y + 4z = 1 \\ 6x - 7y + z = -6 \end{cases}$

53. $\begin{cases} -3x + y + 2z = -9 \\ 9x + 2y + 2z = 10 \\ 6x + 7y = -23 \end{cases}$

54. $\begin{cases} 2x + y + z = 5 \\ -5x + 6y + 4z = -53 \\ 3x + y = 11 \end{cases}$

55. $\begin{cases} -x + 6y + 6z = 22 \\ 2y + z = 5 \\ x + 2y - 2z = -2 \end{cases}$

56. $\begin{cases} x + y + 3z = -1 \\ 3y + 5z = -4 \\ 3x - 3y - z = 5 \end{cases}$

57. $\begin{cases} 2x - y - 3z = 9 \\ x + y = 12 \\ -3x + y + 4z = 13 \end{cases}$ **58.** $\begin{cases} -3x - 2y + 4z = 30 \\ 13x + 4y = 15 \\ 5x + y + 2z = 28 \end{cases}$

Solve each problem using a system of linear equations.
(*Section 4.5, Objective 2*)

59. A concert is held at a music hall. There are three types of tickets available to the public—first mezzanine ($190 each), second mezzanine ($85 each), and third mezzanine ($70 each). A total of 5050 tickets are sold totaling $666,250. The number of first mezzanine tickets sold is 50 less than the other two types combined. Find the number of tickets sold in each mezzanine.

60. The Gasaways have saved 6 months of living expenses, which is $32,700. They invest this money in three different accounts—a 1-yr CD that pays 3.4% annual interest, a high-interest savings account that pays 4.6% annual interest, and a money-market account that pays 3.2% annual interest. To keep their money as liquid as possible, they plan to invest four times as much money in the high-interest savings account as the amount they invest in the CD. Find the amount they should invest in each account if they earned $1330.89 in interest in a year.

Exponents, Polynomials, and Polynomial Functions

Goal Setting

Having goals is essential to your success in life, in college, and in the classroom. Without goals, it is difficult to know what you are working toward. Take the time to write down the goals you hope to accomplish by being in this class and by being in college. Start with specific and realistic goals. Here are some examples.

- I will review my class notes each day.
- I will spend 1 hour per day working homework problems from this class.
- I will ask questions if I do not understand.
- I will attend class each day.

Be sure that your goals are attainable. Review these goals often so that you do not lose focus. Defining goals may require you to make changes in other areas of your life. Be willing to do what is necessary to accomplish your dreams.

? *Question For Thought:* What do you hope to accomplish by attending college? How does this class help you reach that goal? What do you hope to accomplish in this class?

Chapter Outline

Coming Up...

In Section 5.1, we will learn how to work with a formula that involves negative exponents that calculates the monthly payment for repaying a loan.

> **"** Where there is no vision, there is no hope. **"**
>
> —George Washington Carver (American Scientist and Inventor)

Rules of Exponents, Zero and Negative Exponents

In **Chapter 1,** the concept of an exponent was defined and introduced briefly. In this chapter, we will look more closely at exponents and their properties. We will also examine a new class of functions, polynomial functions. Operations of polynomials and polynomial equations will also be studied.

▶ OBJECTIVES

As a result of completing this section, you will be able to

1. Apply the product of like bases rule for exponents.
2. Apply the quotient of like bases rule for exponents.
3. Simplify expressions with zero as an exponent.
4. Simplify expressions with negative exponents.
5. Apply a combination of properties and definitions.
6. Solve application problems.
7. Troubleshoot common errors.

Negative exponents arise in many formulas. One such formula is the amortization formula, the formula to determine the monthly payment needed to repay a loan. The expression for the monthly payment needed to repay a home loan of $200,000 at an annual interest rate of 5% in 360 months is

$$\frac{200{,}000\left(\dfrac{0.05}{12}\right)}{1 - \left(1 + \dfrac{0.05}{12}\right)^{-360}}$$

The graphing calculator skills will illustrate how to simplify this expression.

In this section, we will learn some of the basic rules for exponents. These rules are used quite extensively in algebra, so it is very important that you become familiar with them.

The Product of Like Bases

Recall from Chapter 1 that exponents are used to write repeated factors in a more compact form.

Objective 1 ▶

Apply the product of like bases rule for exponents.

> **Definition: Exponent**
>
> For a real number b and a natural number n,
>
> $$\underset{\text{Base}}{\uparrow}\ b^{\overset{\text{Exponent}}{n}} = \underbrace{b \cdot b \cdot b \cdots b}_{b \text{ is a factor } n \text{ times}}$$

The exponent n tells us how many times b is repeated as a factor. Expressions that contain exponents, such as 2^5 and y^4, are called **exponential expressions** or *powers*. Operations such as addition, subtraction, multiplication, division, and raising to an exponent can be performed on exponential expressions. The definition of an exponent can be used to develop rules for simplifying these types of exponential expressions.

The first rule relates to products of exponential expressions with the same base, what we call like bases. In the examples shown in the table, we obtain each product by first using the definition of the exponent. Then we observe a rule that relates the final result to the original problem.

Based on Definition	Observed Rule
$b \cdot b = b^1 \cdot b^1 = b^2$	$b \cdot b = b^{1+1} = b^2$
$b^2 \cdot b = b^2 \cdot b^1 = (b \cdot b) \cdot b = b^3$	$b^2 \cdot b = b^{2+1} = b^3$
$b^2 \cdot b^3 = (b \cdot b) \cdot (b \cdot b \cdot b) = b^5$	$b^2 \cdot b^3 = b^{2+3} = b^5$

So, we observe that when exponential expressions with like bases are multiplied, the exponent of the product is the *sum* of the exponents of the like bases.

> **Property: The Product of Like Bases Rule for Exponents**
> For a real number b and natural numbers m and n,
> $$b^m \cdot b^n = b^{m+n}$$

The product rule can be applied *only* if the bases of each factor are the same.

| **Objective 1 Examples** | Simplify each expression by applying the product of like bases rule. |

1a. $t^7 \cdot t^4 \cdot t$ **1b.** $(-5)^3 \cdot (-5)^2$ **1c.** $(-2a^7)(-3a^2)$ **1d.** $(-4xy^2)(-5x^2y^4)$

Solutions   **1a.**

$$t^7 \cdot t^4 \cdot t = t^7 \cdot t^4 \cdot t^1 \qquad \text{Recall } t = t^1.$$
$$= t^{7+4+1} \qquad \text{Apply the product of like bases rule.}$$
$$= t^{12} \qquad \text{Add the exponents.}$$

1b.

$$(-5)^3 \cdot (-5)^2 = (-5)^{3+2} \qquad \text{Apply the product of like bases rule.}$$
$$= (-5)^5 \qquad \text{Add the exponents.}$$
$$= -3125 \qquad \text{Simplify the exponent.}$$

1c.

$$(-2a^7)(-3a^2) = (-2)(-3)a^7 \cdot a^2 \qquad \text{Apply the commutative property.}$$
$$= 6a^{7+2} \qquad \text{Multiply and apply the product of like bases rule.}$$
$$= 6a^9 \qquad \text{Simplify.}$$

1d. $(-4x^1y^2)(-5x^2y^4) = (-4)(-5)x^1x^2y^2y^4 \qquad \text{Apply the commutative property.}$
$$= 20x^{1+2}y^{2+4} \qquad \text{Multiply and apply the product of like bases rule.}$$
$$= 20x^3y^6 \qquad \text{Simplify.}$$

| ✔ **Student Check 1** | Simplify each expression by applying the product of like bases rule. |

a. $y^6 \cdot y^2 \cdot y^{12}$ **b.** $(-3)^6 \cdot (-3)^{11}$ **c.** $(7b^5)(-4b^7)$ **d.** $(-2x^4y^5)(-8x^3y^7)$

The Quotient of Like Bases

| **Objective 2** ▶ | Now we find a rule for simplifying quotients of exponential expressions with like bases. The following example illustrates this concept. Assume that the denominator is not zero. |

Apply the quotient of like bases rule for exponents.

Based on Definition

$$\frac{a^6}{a^4} = \frac{a \cdot a \cdot a \cdot a \cdot a \cdot a}{a \cdot a \cdot a \cdot a} \qquad \begin{array}{l}\text{Six factors of } a \text{ in the} \\ \text{numerator and four factors} \\ \text{of } a \text{ in the denominator}\end{array}$$

$$= \frac{\overset{1}{\cancel{a}} \cdot \overset{1}{\cancel{a}} \cdot \overset{1}{\cancel{a}} \cdot \overset{1}{\cancel{a}} \cdot a \cdot a}{\underset{1}{\cancel{a}} \cdot \underset{1}{\cancel{a}} \cdot \underset{1}{\cancel{a}} \cdot \underset{1}{\cancel{a}}} \qquad \begin{array}{l}\text{Divide out the common} \\ \text{factors.}\end{array}$$

$$= \frac{a \cdot a}{1} \qquad \text{Write the remaining factors.}$$

$$= a^2 \qquad \begin{array}{l}\text{Express the answer with} \\ \text{exponents.}\end{array}$$

Observed Rule

The same result is obtained by subtracting the exponents.

$$\frac{a^6}{a^4} = a^{6-4}$$

$$= a^2$$

So, we observe that when exponential expressions with like bases are divided, the exponent of the quotient is the *difference* of the exponents of the exponential expressions in the numerator and denominator.

Property: The Quotient of Like Bases Rule for Exponents

For a real number b ($b \neq 0$) and natural numbers m and n,

$$\frac{b^m}{b^n} = b^{m-n}$$

The quotient of like bases rule can be applied *only* if the bases of the exponential expression in the numerator and denominator are the same.

Objective 2 Examples | Simplify each expression by applying the quotient of like bases rule.

2a. $\dfrac{a^6}{a^2}$ **2b.** $\dfrac{4^5}{4^3}$ **2c.** $\dfrac{-10x^4}{2x^3}$ **2d.** $\dfrac{-18a^8b^{10}}{-9a^4b^2}$

Solutions

2a.
$$\frac{a^6}{a^2} = a^{6-2} \qquad \text{Apply the quotient of like bases rule.}$$
$$= a^4 \qquad \text{Subtract the exponents.}$$

2b.
$$\frac{4^5}{4^3} = 4^{5-3} \qquad \text{Apply the quotient of like bases rule.}$$
$$= 4^2 \qquad \text{Subtract the exponents.}$$
$$= 16 \qquad \text{Simplify the exponent.}$$

2c.
$$\frac{-10x^4}{2x^3} = \frac{-10}{2} \cdot \frac{x^4}{x^3} \qquad \text{Divide the coefficients and the like bases.}$$
$$= -5x^{4-3} \qquad \text{Apply the quotient of like bases rule.}$$
$$= -5x^1 \qquad \text{Subtract the exponents.}$$
$$= -5x \qquad \text{Rewrite } x^1 \text{ as } x.$$

2d.
$$\frac{-18a^8b^{10}}{-9a^4b^2} = \frac{-18}{-9} \cdot \frac{a^8}{a^4} \cdot \frac{b^{10}}{b^2} \qquad \text{Divide the coefficients and the like bases.}$$
$$= 2a^{8-4}b^{10-2} \qquad \text{Apply the quotient of like bases rule.}$$
$$= 2a^4b^8 \qquad \text{Subtract the exponents.}$$

☑ Student Check 2 | Simplify each expression by applying the quotient of like bases rule.

a. $\dfrac{y^{10}}{y}$ **b.** $\dfrac{(-1)^{11}}{(-1)^5}$ **c.** $\dfrac{-21r^8}{-3r^6}$ **d.** $\dfrac{-16x^9y^{12}}{4x^6y^3}$

The Zero Exponent

Objective 3 ▶

Simplify expressions with zero as an exponent.

Examples 1 and 2 used only natural number (that is, positive integer) exponents. We now investigate how to define an exponent of zero. Consider the expression, $\dfrac{x^5}{x^5}$. We know that to divide exponential expressions with like bases we subtract the exponents. So,

$$\frac{x^5}{x^5} = x^{5-5} = x^0$$

On the other hand, we also know that any nonzero number divided by itself is 1. So, it is also true that

$$\frac{x^5}{x^5} = 1$$

We have that $\frac{x^5}{x^5} = x^0$ and $\frac{x^5}{x^5} = 1$, so it must follow that $x^0 = 1$. This leads to the following definition.

Definition: The Zero Exponent

For a real number b ($b \neq 0$),

$$b^0 = 1$$

So, any nonzero number raised to an exponent of 0 is 1.

Objective 3 Examples Use the definition of the zero exponent to simplify each expression. Assume all bases are nonzero real numbers.

3a. 3^0 **3b.** $(-6)^0$ **3c.** $-(-4)^0$ **3d.** $8b^0$ **3e.** $(5x + 4)^0$

Solutions **3a.** $3^0 = 1$

3b. $(-6)^0 = 1$

3c. The opposite of -4 to the 0 is $-(-4)^0 = -(1) = -1$.

3d. Eight times b to the 0 is $8b^0 = 8(1) = 8$.

3e. $(5x + 4)^0 = 1$

✓ Student Check 3 Use the definition of the zero exponent to simplify each expression. Assume all bases are nonzero real numbers.

   a. 17^0 **b.** $(-9)^0$ **c.** $-(-10)^0$ **d.** $3d^0$ **e.** $(7a - 1)^0$

Negative Exponents

Objective 4 ▶

Simplify expressions with negative exponents.

We now know how to work with exponents that are whole numbers $\{0, 1, 2, \ldots\}$. It is important to understand how to work with exponents that are negative integers as well, such as 2^{-3}. It is impossible to repeat the base 2 a "-3" times, so we need another way to simplify this exponential expression. The following example will help us understand the concept of a negative exponent.

Based on Definition	**Based on the Quotient of Like Bases Rule**

$$\frac{a^3}{a^6} = \frac{a \cdot a \cdot a}{a \cdot a \cdot a \cdot a \cdot a \cdot a}$$ Three factors of a in the numerator and six factors of a in the denominator

$$= \frac{\overset{1}{\cancel{a}} \cdot \overset{1}{\cancel{a}} \cdot \overset{1}{\cancel{a}}}{\underset{1}{\cancel{a}} \cdot \underset{1}{\cancel{a}} \cdot \underset{1}{\cancel{a}} \cdot a \cdot a \cdot a}$$ Divide out the common factors.

$$= \frac{1}{a^3}$$ Simplify.

$$\frac{a^3}{a^6} = a^{3-6}$$

$$= a^{-3}$$

So, we see that $\dfrac{a^3}{a^6} = \dfrac{1}{a^3}$ and $\dfrac{a^3}{a^6} = a^{-3}$. Therefore,

$$a^{-3} = \dfrac{1}{a^3}$$

An exponential expression with a *negative exponent* is equivalent to the reciprocal of the base of the expression raised to the positive exponent.

Definition: Negative Exponent

For a real number b ($b \neq 0$) and a positive integer n,

$$b^{-n} = \dfrac{1}{b^n}$$

Procedure: Evaluating an Expression Raised to a Negative Exponent

Step 1: Identify the base of the negative exponent.
Step 2: Rewrite the expression as 1 divided by the base to its positive exponent.
Step 3: Simplify the expression.

Objective 4 Examples   Simplify each expression. Write each answer with positive exponents. Assume all bases are nonzero real numbers.

4a. 6^{-2} **4b.** $(-4)^{-3}$ **4c.** -2^{-4} **4d.** $\left(\dfrac{5}{4}\right)^{-2}$ **4e.** $6x^{-2}$ **4f.** $\dfrac{x^{-3}}{y^{-4}}$

Solutions   **4a.** $6^{-2} = \dfrac{1}{6^2}$ Apply the definition of a negative exponent.

$\phantom{6^{-2}} = \dfrac{1}{36}$ Simplify the exponent.

4b. $(-4)^{-3} = \dfrac{1}{(-4)^3}$ Apply the definition of a negative exponent.

$\phantom{(-4)^{-3}} = \dfrac{1}{-64}$ Simplify the exponent.

$\phantom{(-4)^{-3}} = -\dfrac{1}{64}$ Simplify.

4c. Without parentheses, the negative exponent only applies to the base 2.

$-2^{-4} = -\dfrac{1}{2^4}$ Apply the definition of a negative exponent.

$\phantom{-2^{-4}} = -\dfrac{1}{16}$ Simplify the exponent.

4d. $\left(\dfrac{5}{4}\right)^{-2} = \dfrac{1}{\left(\dfrac{5}{4}\right)^2}$ Apply the definition of a negative exponent.

$\phantom{\left(\dfrac{5}{4}\right)^{-2}} = \dfrac{1}{\dfrac{25}{16}}$ Simplify the exponent.

$\phantom{\left(\dfrac{5}{4}\right)^{-2}} = 1 \cdot \dfrac{16}{25}$ Multiply by the reciprocal of $\dfrac{25}{16}$, which is $\dfrac{16}{25}$.

$\phantom{\left(\dfrac{5}{4}\right)^{-2}} = \dfrac{16}{25}$ Simplify.

4e. The base of the negative exponent is x. Rewrite x^{-2} with a positive exponent and multiply by 6.

$$6x^{-2} = 6 \cdot \frac{1}{x^2} \qquad \text{Apply the definition of a negative exponent to } x^{-2}.$$

$$= \frac{6}{x^2} \qquad \text{Multiply.}$$

4f. We must apply the definition of a negative exponent twice.

$$\frac{x^{-3}}{y^{-4}} = \frac{\dfrac{1}{x^3}}{\dfrac{1}{y^4}} \qquad \text{Apply the definition of a negative exponent to the numerator and denominator.}$$

$$= \frac{1}{x^3} \div \frac{1}{y^4} \qquad \text{Write the numerator divided by the denominator.}$$

$$= \frac{1}{x^3} \cdot \frac{y^4}{1} \qquad \text{Multiply by the reciprocal of } \frac{1}{y^4}, \text{ or } \frac{y^4}{1}.$$

$$= \frac{y^4}{x^3} \qquad \text{Simplify.}$$

So, note that $\dfrac{x^{-3}}{y^{-4}} = \dfrac{y^4}{x^3}$.

 Student Check 4 Simplify each expression. Write each answer with positive exponents. Assume all bases are nonzero real numbers.

a. 7^{-3} **b.** $(-2)^{-4}$ **c.** -5^{-2} **d.** $\left(\dfrac{1}{4}\right)^{-1}$ **e.** $3y^{-3}$ **f.** $\dfrac{a^{-5}}{b^{-2}}$

Example 4 illustrates two important facts about negative exponents.

Fact 1: When a factor crosses the fraction bar, the sign of its exponent changes.

$$\frac{1}{b^{-n}} = b^n \quad \text{and} \quad b^{-n} = \frac{1}{b^n}, \quad \text{for } n \text{ an integer and } b \neq 0$$

Fact 2: A fraction raised to an exponent is equivalent to its reciprocal raised to the opposite of the exponent.

$$\left(\frac{a}{b}\right)^{-n} = \left(\frac{b}{a}\right)^n$$

Combining Properties and Definitions

Objective 5 ▶

Apply a combination of properties and definitions.

Now that we know how to work with all integers as exponents, the product and quotient of like bases rules apply for integer exponents, not just natural numbers. We now simplify expressions that require us to apply both rules.

Objective 5 Examples Simplify each expression. Write each answer with positive exponents. Assume all bases are nonzero real numbers.

5a. $a^{-4} \cdot a^{-6} \cdot a^5$ **5b.** $(-4x^2y^{-3})(2x^{-4}y^6)$ **5c.** $\dfrac{6b^3}{2b^{-2}}$ **5d.** $\dfrac{8a^3b^{-5}}{16a^7b^{-2}}$

Solutions   **5a.** $a^{-4} \cdot a^{-6} \cdot a^5 = a^{-4+(-6)+5}$ Apply the product of like bases rule.

$= a^{-5}$ Add the exponents.

$= \dfrac{1}{a^5}$ Rewrite with positive exponents.

5b. $(-4x^2y^{-3})(2x^{-4}y^6) = (-4)(2)x^2x^{-4}y^{-3}y^6$ Apply the commutative property.

$= -8x^{2+(-4)}y^{-3+6}$ Multiply coefficients and apply the product of like bases rule.

$= -8x^{-2}y^3$ Add exponents.

$= -8 \cdot \dfrac{1}{x^2} \cdot y^3$ Rewrite with positive exponents.

$= -\dfrac{8y^3}{x^2}$ Simplify.

5c. $\dfrac{6b^3}{2b^{-2}} = \dfrac{6}{2} \cdot b^{3-(-2)}$ Divide the coefficients and apply the quotient of like bases rule.

$= 3b^5$ Simplify.

5d. $\dfrac{8a^3b^{-5}}{16a^7b^{-2}} = \dfrac{8}{16} \cdot a^{3-7}b^{-5-(-2)}$ Divide the coefficients and apply the quotient of like bases rule.

$= \dfrac{1}{2}a^{-4}b^{-3}$ Simplify the coefficient and subtract the exponents.

$= \dfrac{1}{2} \cdot \dfrac{1}{a^4} \cdot \dfrac{1}{b^3}$ Rewrite with positive exponents.

$= \dfrac{1}{2a^4b^3}$ Simplify.

✓ **Student Check 5**   Simplify each expression. Write each answer with positive exponents. Assume all bases are nonzero real numbers.

a. $b^{-6} \cdot b^2 \cdot b^5$ **b.** $(-5x^{-3}y)(4xy^{-2})$ **c.** $\dfrac{4a^5}{10a^{-4}}$ **d.** $\dfrac{9x^{-2}y^4}{27x^{-5}y^6}$

Applications

Objective 6 ▶

Solve application problems.

There are many applications in real life that use exponents. Exponents appear in formulas for area, volume, surface area of geometric figures, population growth, investments, the growth of bacteria, and the half-life of chemical elements. In Example 6 we simplify some exponential expressions as well as construct expressions that represent a specific situation.

Objective 6 Examples   Solve each problem by evaluating an expression or by constructing an expression that represents the situation. Round each answer to two decimal places, when needed.

6a. A soup can has a diameter of 6.7 cm and a height of 10.2 cm. How much steel is needed to produce one soup can? The formula for surface area is $S = 2\pi r^2 + 2\pi rh$. (Use $\pi \approx 3.14$.)

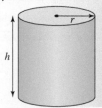

Solution   **6a.** $S = 2\pi r^2 + 2\pi rh$ Begin with the surface area formula.

           $S = 2\pi(3.35)^2 + 2\pi(3.35)(10.2)$ Replace h with 10.2 and r with $\dfrac{6.7}{2} = 3.35$.

           $S = 2\pi(11.2225) + 2\pi(34.17)$ Simplify the exponent and multiply.

           $S \approx 70.4773 + 214.5876$ Multiply and approximate.

           $S \approx 285.0649$ Add.

So, approximately 285.06 cm² of steel are needed to produce one soup can.

6b. Write an expression in terms of the height for the surface area of a can if its radius is one-half its height. Recall that the formula for surface area is $S = 2\pi r^2 + 2\pi rh$.

Solution   **6b.** Let h represent the height of the can. Then $r = \dfrac{h}{2}$.

           $S = 2\pi r^2 + 2\pi rh$ Begin with the surface area formula.

           $S = 2\pi\left(\dfrac{h}{2}\right)^2 + 2\pi\left(\dfrac{h}{2}\right)(h)$ Replace r with $\dfrac{h}{2}$.

           $S = 2\pi\left(\dfrac{h^2}{4}\right) + \dfrac{2\pi h^2}{2}$ Simplify the exponent and apply the product of like bases rule.

           $S = \dfrac{\pi h^2}{2} + \dfrac{2\pi h^2}{2}$ Simplify the first term.

           $S = \dfrac{3\pi h^2}{2}$ Add the fractions.

6c. The area of a rectangle is given by $A = 32x^3y$ and the length is $l = 8x$. What expression represents the width of the rectangle?

Solution   **6c.** Since $A = lw$, the width of the rectangle can be found by dividing both sides by l to obtain $w = \dfrac{A}{l}$.

           $w = \dfrac{A}{l}$ State the formula to find w.

           $w = \dfrac{32x^3y}{8x}$ Replace A with $32x^3y$ and l with $8x$.

           $w = \dfrac{32}{8} \cdot \dfrac{x^3}{x} \cdot y$ Divide coefficients and the like bases.

           $w = 4x^2y$ Simplify the coefficient and subtract exponents.

6d. The formula to determine the monthly payment to repay a loan of P dollars, at an annual interest rate of r (as a decimal), in n months is given by

$$A = \dfrac{P\left(\dfrac{r}{12}\right)}{1 - \left(1 + \dfrac{r}{12}\right)^{-n}}$$

Write the expression to find the monthly payment to repay a home loan of \$200,000 at an annual interest rate of 5% in 360 months and then simplify the expression.

Solution   **6d.** $A = \dfrac{P\left(\dfrac{r}{12}\right)}{1 - \left(1 + \dfrac{r}{12}\right)^{-n}}$ State the formula.

$A = \dfrac{200{,}000\left(\dfrac{0.05}{12}\right)}{1 - \left(1 + \dfrac{0.05}{12}\right)^{-360}}$ Replace P with 200,000, r with 0.05, and n with 360.

$A = \$1073.64$ Approximate using a calculator.

So, the monthly payment to repay the $200,000 loan is $1073.64.

 **Student Check 6**   Solve each problem by evaluating an expression or by constructing an expression that represents the situation. Round each answer to two decimal places, when needed.

a. A sugar cone has a radius of 2.5 cm and a height of 11.2 cm. Find how much ice cream the cone will hold. The volume of a cone is $V = \dfrac{1}{3}\pi r^2 h$. (Use $\pi \approx 3.14$.)

b. Suppose the height of a cone is 6 times its radius. Write an expression in terms of the radius that represents the volume of the cone.

c. The area of a rectangle is $A = 60x^3y^4$. The length of the rectangle is $l = 15x^2y^3$. What expression represents the width of the rectangle?

d. Use the formula in Example 6d to find the monthly payment to repay a loan of $30,000 at 6% interest in 60 months.

Objective 7 ▶

Troubleshoot common errors.

Troubleshooting Common Errors

Some common errors associated with the product and quotient of like bases rules and negative exponents are shown.

Objective 7 Examples   A problem and an incorrect solution are given. Provide the correct solution and an explanation of the error.

7a. Simplify $y^2 \cdot y^6$.

Incorrect Solution	Correct Solution and Explanation
$y^2 \cdot y^6 = y^{12}$	The product of like bases rule states that when we multiply exponential expressions with like bases, we *add* exponents not multiply them. So, the correct answer is $y^2 \cdot y^6 = y^{2+6} = y^8$.

7b. Simplify $5^2 \cdot 5^3$.

Incorrect Solution	Correct Solution and Explanation
$5^2 \cdot 5^3 = 25^5$	The product of like bases rule states that when we multiply exponential expressions with like bases, we must keep the base the same. So, the correct answer is $5^2 \cdot 5^3 = 5^5 = 3125$.

7c. Rewrite 2^{-3} with positive exponents.

Incorrect Solution	Correct Solution and Explanation
$2^{-3} = -8$	A negative exponent doesn't make the sign of the expression negative; it means to take the reciprocal of the base. $$2^{-3} = \frac{1}{2^3} = \frac{1}{8}$$

7d. Simplify $\dfrac{x^{-5}}{x^{-6}}$.

Incorrect Solution	Correct Solution and Explanation
$\dfrac{x^{-5}}{x^{-6}} = x^{-5-6} = x^{-11} = \dfrac{1}{x^{11}}$	The denominator's exponent is negative, so when we subtract it, we get a positive number. $$\frac{x^{-5}}{x^{-6}} = x^{-5-(-6)} = x^{-5+6} = x^1 = x$$

ANSWERS TO STUDENT CHECKS

Student Check 1   **a.** y^{20} **b.** -3^{17} **c.** $-28b^{12}$
d. $16x^7y^{12}$

Student Check 2   **a.** y^9 **b.** 1   **c.** $7r^2$ **d.** $-4x^3y^9$

Student Check 3   **a.** 1   **b.** 1   **c.** -1 **d.** 3   **e.** 1

Student Check 4   **a.** $\dfrac{1}{343}$ **b.** $\dfrac{1}{16}$ **c.** $-\dfrac{1}{25}$ **d.** 4
e. $\dfrac{3}{y^3}$ **f.** $\dfrac{b^2}{a^5}$

Student Check 5   **a.** b **b.** $-\dfrac{20}{x^2y}$ **c.** $\dfrac{2a^9}{5}$ **d.** $\dfrac{x^3}{3y^2}$

Student Check 6   **a.** $73.27\ \text{cm}^3$ **b.** $V = 2\pi r^3$
c. $w = 4xy$ **d.** $579.98

SUMMARY OF KEY CONCEPTS

1. The product of like bases rule only applies when the bases are the same. It states that the product of exponential expressions with like bases is the like base raised to the sum of its exponents.

2. The quotient of like bases rule only applies when the bases are the same. The quotient of exponential expressions with like bases is the like base raised to the difference of its exponents. If the resulting exponent is negative, rewrite the expression with a positive exponent.

3. Any nonzero base raised to the zero exponent is 1. Be careful with negative signs. Note the difference in the following expressions.

$$(-b)^0 = 1 \quad \text{but} \quad -b^0 = -1$$

4. Negative exponents involve taking reciprocals. The base with the negative exponent always crosses the fraction bar for the exponent to become positive. The expression b^{-n} is the reciprocal of b^n. That is, $b^{-n} = \dfrac{1}{b^n}$.

GRAPHING CALCULATOR SKILLS

The calculator skills for this section involve working with negative exponents and evaluating formulas.

Example 1: Simplify 7^{-3} and express the result as a fraction.

Solution:

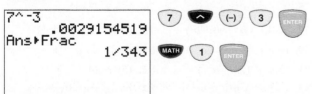

Example 2: Find the monthly payment to repay a home loan of $200,000 at an annual interest rate of 5% in 360 months. (Refer to Example 6d for the monthly payment formula.)

Solution:

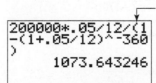

Note that the denominator must be entered in parentheses

SECTION 5.1 / EXERCISE SET

 Write About It!

Use complete sentences in your answer to each exercise.

1. Explain why a nonzero real number raised to the zero exponent is 1.

2. Explain the difference between $(-5)^4$ and -5^4.

3. Explain why $3^{-2} \neq -3^2$.

4. Use an example to show how to simplify a nonzero real number raised to a negative exponent.

5. Can we apply the product of like bases rule of exponents to $(-7)^m (-7)^n$? Explain.

6. Can we apply the product of like bases rule of exponents to $a^m b^n$, $a \neq b$? Explain.

7. Can we apply the quotient of like bases rule of exponents to $\dfrac{x^{-4}}{x^4}$? Explain.

8. Can we apply the quotient of like bases rule of exponents to $\dfrac{a^m}{b^n}$, $a \neq b$? Explain.

 Practice Makes Perfect!

Simplify each expression by applying the product of like bases rule. (*See Objective 1.*)

9. $r^3 \cdot r^4$

10. $s^8 \cdot s^2$

11. $m \cdot m^4$

12. $x \cdot x^7$

13. $x^2 \cdot x^3 \cdot x$

14. $y^3 \cdot y^5 \cdot y$

15. $m^3 \cdot m^5 \cdot m^4$

16. $n^4 \cdot n^5 \cdot n^6$

17. $3^{11} \cdot 3^4$

18. $4^2 \cdot 4^{10}$

19. $(-2)^5 \cdot (-2)^9$

20. $(-5)^6 \cdot (-5)^8$

21. $(5x^3) \cdot (-4x^2)$

22. $(-3y^8) \cdot (6y^3)$

23. $(-12a^4) \cdot (-5a^6)$

24. $(15y^7) \cdot (-5y^{12})$

25. $(-2x^2y^3)(-3x^4y^4)$

26. $(6x^5y)(-12x^7y^3)$

27. $(-3a^4b^2)(5a^9b^7)$

28. $(-8a^2b^4)(-10a^6b^8)$

Simplify each expression by applying the quotient of like bases rule. (*See Objective 2.*)

29. $\dfrac{x^{13}}{x^7}$

30. $\dfrac{y^{19}}{y^{11}}$

31. $\dfrac{(-1)^9}{(-1)^2}$

32. $\dfrac{(-1)^{12}}{(-1)^4}$

33. $\dfrac{2^9}{2^5}$

34. $\dfrac{3^{12}}{3^7}$

35. $\dfrac{45p^9}{5p}$

36. $\dfrac{-18q^7}{-6q^3}$

37. $\dfrac{-72r^{15}}{12r^4}$

38. $\dfrac{68s^9}{-4s^3}$

39. $\dfrac{-12x^9y^{13}}{6xy^9}$

40. $\dfrac{-36a^3b^8}{-9ab^3}$

41. $\dfrac{112p^{19}q^5}{7p^{13}q}$

42. $\dfrac{-120r^{10}s^7}{-15r^2s^6}$

Simplify each expression. Assume all bases are nonzero real numbers. (*See Objective 3.*)

43. 29^0

44. 13^0

45. $(-20)^0$

46. $(-15)^0$

47. $-(-18)^0$

48. $-(-25)^0$

49. $(6c)^0$

50. $(9a)^0$

51. $(2b + 3)^0$

52. $(3x - 4)^0$

53. $2(4z - 3)^0$

54. $-2(b - 1)^0$

Simplify each expression. Express each answer with positive exponents. Assume all bases are nonzero real numbers. (*See Objective 4.*)

55. 2^{-5}

56. 3^{-4}

57. $(-4)^2$

58. $(-5)^{-2}$

59. $\left(\dfrac{2}{3}\right)^{-3}$

60. $\left(\dfrac{2}{5}\right)^{-1}$

61. $\left(-\dfrac{1}{5}\right)^{-4}$

62. $\left(-\dfrac{7}{12}\right)^{-2}$

63. $-(-3)^{-4}$

64. $-(-4)^{-3}$

65. $4v^{-5}$

66. $-15w^{-8}$

67. $\dfrac{y^{-6}}{y^{12}}$

68. $\dfrac{x^{-4}}{x^5}$

69. $\dfrac{a^{-4}}{b^{-3}}$

70. $\dfrac{x^{-6}}{y^{-2}}$

71. $\dfrac{p^2}{q^{-4}}$

72. $\dfrac{a^{-1}}{b^3}$

Simplify each expression. Write each answer with positive exponents. Assume all bases are nonzero real numbers. (*See Objective 5.*)

73. $q^{-7} \cdot q^{-12}$

74. $p^{-4} \cdot p^{-9}$

75. $x^5 \cdot x^3 \cdot x^{-14}$

76. $x^3 \cdot x^{-10} \cdot x^4$

77. $(-4a^7 \cdot b^{-12})(a^{10}b^2)$

78. $(-s^3t^{-7})(-3s^{-4}t^{10})$

79. $(-r^4t^{-9})(-6r^{-7}t^{-1})$

80. $(-3p^{-5} \cdot q^{-11})(-p^{-12}q^6)$

81. $(5m^5n^8)(2m^{-2}n^{-7})$

82. $(5y^6z^2)(4y^{-5}z^7)$

83. $(-5p^{-7} \cdot h^2)(3p^3h^{-3})$

84. $(-3x^5y^4)(-9x^{-2}y^{-8})$

85. $(2m^{-9} \cdot n^{-10})(-m^9n^{-8})$

86. $(4m^{-8} \cdot n^{-6})(-m^2n^6)$

87. $\dfrac{-5u^{-3}v^{10}}{-15u^{-6}v^8}$

88. $\dfrac{10cd^{-4}}{40c^{-3}d^{-6}}$

89. $\dfrac{-18u^{-3}v^{-1}}{5u^{11}v^2}$

90. $\dfrac{-14m^3n^{-12}}{-24m^{12}n^7}$

91. $\dfrac{-6e^{-5}t^{10}}{-18e^3t^{-5}}$

92. $\dfrac{-16f^{-5}g^{12}}{-20f^{-3}g^{-6}}$

Solve each problem. Use $\pi \approx 3.14$ and round each answer to two decimal places where applicable. (*See Objective 6.*)

93. A soup can has a diameter of 6.8 cm and a height of 12.4 cm. How much steel is needed to produce one can? Use the formula for surface area $S = 2\pi r^2 + 2\pi rh$.

94. A soup can has a diameter of 8.1 cm and a height of 11.1 cm. How much steel is needed to produce one can? Use the formula for surface area $S = 2\pi r^2 + 2\pi rh$.

95. Write an expression in terms of the radius for the surface area of a can if the height is three-fourths of its radius.

96. Write an expression in terms of the radius for the surface area of a can if the height is one-third of its radius.

97. The area of a rectangle is given by $A = 42x^3y$ and the length is $l = 3x$. What expression represents the width of the rectangle?

98. The area of a rectangle is given by $A = 18x^4y^5$ and the length is $l = 2x^2y^2$. What expression represents the width of the rectangle?

99. The area of a triangle is given by $A = \dfrac{1}{2}bh$. If the base of a triangle is $b = 2x^4y$ and the height is $h = 6xy^7$, what expression represents the area of the triangle?

100. The area of a triangle is given by $A = \dfrac{1}{2}bh$. If the base of a triangle is $b = 4x^4y^2$ and the height is $h = 9x^4y^5$, what expression represents the area of the triangle?

The formula to determine the monthly payment A to repay a loan of P dollars in n months, at an annual interest rate of r (as a decimal), is given by

$$A = \dfrac{P\left(\dfrac{r}{12}\right)}{1 - \left(1 + \dfrac{r}{12}\right)^{-n}}$$

Find the monthly payment needed to repay each loan. Round each answer to the nearest cent.

101. Home loan of $150,000 at an annual interest rate of 4.3% in 360 months

102. Home loan of $258,000 at an annual interest rate of 5.2% in 360 months

The formula to determine the accumulated amount A for an investment of P dollars at an annual interest rate of r (as a decimal) compounded quarterly for t years is given by

$$A = P\left(1 + \dfrac{r}{4}\right)^{4t}$$

Find the accumulated amount for each investment. Round each answer to the nearest cent.

103. $5600 at an annual interest rate of 2.5% compounded quarterly for 2 yr

104. $12,600 at an annual interest rate of 3.2% compounded quarterly for 4 yr

Mix 'Em Up!

Simplify each expression and write the result with positive exponents. Assume all bases are nonzero real numbers.

105. $-7(6m)^0$

106. $12(34a)^0$

107. $-19d^{-5}$

108. $5y^{-9}$

109. $\dfrac{11^9}{11^4}$

110. $\dfrac{5^4}{5^6}$

111. $\dfrac{90p^4}{5p}$

112. $\dfrac{-80q^{11}}{-4q^5}$

113. $\dfrac{s^{-4}}{b^{-12}}$

114. $\dfrac{c^{-3}}{d^2}$

115. $(-2)^5 \cdot (-2)^{-8}$

116. $(-5)^9 \cdot (-5)^{12}$

117. $\dfrac{3u^{-1}v^{-7}}{12u^{-9}v^7}$

118. $\dfrac{7x^7y^{-9}}{63x^{-3}y^5}$

119. $(-4d^6r^9)(2d^3r^{-7})$

120. $(-2a^7b^{-2})(5a^{-6}b^4)$

121. $-c^0 + 3$

122. $-2c^0 + 5$

123. $\dfrac{-20a^{-8}n}{-a^{13}n^3}$

124. $\dfrac{12u^{-8}v}{-10u^6v^2}$

125. $(2b + 3)^0$

126. $-2(b - 1)^0$

127. $7^{-3} \cdot 7^3$

128. $9^{-4} \cdot 9^4$

129. $\dfrac{8t^{-7}z^6}{10t^{-14}z^4}$

130. $\dfrac{9a^4b^{12}}{-18a^{-10}b^4}$

131. $-(-3)^{-6}$

132. $\left(-\dfrac{1}{5}\right)^{-4}$

133. $(-c^{-3}p^{-1})(5c^4p^{-5})$

134. $(-3m^{-1}k^{-6})(m^{-2}k)$

Find the monthly payment needed to repay each loan. Round each answer to the nearest cent.

135. Home loan of $125,000 at an annual interest rate of 3.9% in 360 months

136. Car loan of $6450 at an annual interest rate of 2.4% in 60 months

The formula to determine the amount P of radioactive material present at time t, where P_0 is the original amount of the material and m is the half-life of the material, is given by

$$P = P_0 \cdot 2^{-t/m}$$

The half-life of a radioactive substance is the time it takes for half of a certain amount of radioactive substance to decay.

137. The radioactive isotope sodium-24 is used in medical and nonmedical applications. It can be used to detect circulatory problems in patients as well as used to detect leaks in oil pipe lines. The half-life of sodium-24 is 15 hr. If a hospital has 20 g of sodium-24, how many grams will be present after 15 hr? 60 hr? (Sources: http://www.chemistryexplained.com/elements/P-T/Sodium.html and http://www.3rd1000.com/nuclear/halflife.htm)

138. The radioactive isotope gallium-67 can be used to locate sites of infection in a patient, tumor imaging, and chemotherapy for pediatric patients. Its half-life is approximately 3 days. If the Imaging Center has 400 mg of gallium-67, how much will be present after 3 days? after 15 days? (Source: http://www.radiochemistry.org/nuclearmedicine/radioisotopes/ex_iso_medicine.htm)

 You Be the Teacher!

Correct each student's errors, if any.

139. Simplify $-2^4 \cdot x^3 \cdot x^2$.

Isabella's work:
$-2^4 \cdot x^3 \cdot x^2 = 16x^{3+2} = 16x^5$

140. Simplify $(2^{-3}x^4 y)(x^{-5}y^2)$.

Jason's work:

$(2^{-3}x^4 y)(x^{-5}y^2) = -6x^{4-5}y^{1+2} = -6x^{-1}y^3 = \dfrac{-6y^3}{x}$

141. Simplify $\dfrac{c^4 d^7}{c^{-3} d^2}$.

Kyle's work:

$$\frac{c^4 d^7}{c^{-3} d^2} = c^{4-3} d^{7-2} = cd^5$$

142. Simplify $3500\left(1 + \dfrac{0.0225}{12}\right)^{12(4)}$.

Fora's work:

```
3500(1+.0225/12)
^12*4
          14318.26883
```

 Calculate It!

Explain and correct errors in each screen shot, if any.

143. Simplify $2^5 \cdot 2^8$.

```
4^13
          67108864
```

144. Simplify $(-3)^5 \cdot (-3)^4$.

```
9^9
          387420489
```

For Exercises 145 and 146, use the formula

$$A = \frac{P\left(\dfrac{r}{12}\right)}{1 - \left(1 + \dfrac{r}{12}\right)^{-n}}$$

to determine the monthly payment A needed to repay each loan of P dollars in n months at an annual interest rate r (as a decimal).

145. Home loan of \$165,000 at an annual interest rate of 4.5% in 360 months

```
165000(.045/12)/
1-(1+.045/12)^(-
360)
          618.4901043
```

146. Car loan of \$3200 at an annual interest rate of 2.9% in 60 months

```
3200(.29/12)/(1-
(1+.29/12)^(-60)
)
          101.5739843
```

 Think About It!

Use the properties of exponents to simplify each expression.

147. $3^{4x} \cdot 3^{x+1}$

148. $2^{3(a+2)} \cdot 2$

149. $\dfrac{5^{2a}}{5^{a-1}}$

150. $\dfrac{4^{x+3}}{4^{x+4}}$

More Rules of Exponents and Scientific Notation

OBJECTIVES

As a result of completing this section, you will be able to

1. Apply the power rules for exponents.

2. Simplify exponential expressions using a combination of exponent properties.

3. Convert between standard notation and scientific notation.

4. Perform operations with numbers in scientific notation.

5. Troubleshoot common errors.

Objective 1 ▶

Apply the power rule for exponents.

Blood platelets are very small disk-shaped particles that are necessary for clotting blood. The diameter of a single blood platelet is approximately 0.003 mm. A healthy person produces approximately 150 billion platelets each day. If the platelets could be placed side by side, how many miles would they extend? (Source: http://medical-dictionary.thefreedictionary.com/ Blood+platelets)

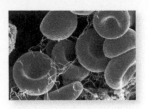

To answer this question, we will represent each number in a special form called scientific notation and then use properties of exponents to find their product.

Power Rules for Exponents

We will now investigate how to find powers of exponential expressions. The following illustrations show how to use the definition of an exponent to obtain the result as well as a rule that can be applied to the original problem to obtain the result.

Based on Definition	**Observed Rule**
$(x^2)^3 = x^2 \cdot x^2 \cdot x^2$ Three factors of x^2	The same result is obtained by multiplying the exponents.
$= x^{2+2+2}$ Apply the product of like bases rule.	$(x^2)^3 = x^{2\cdot3}$
$= x^6$ Add the exponents.	$= x^6$

So, we observe that when an exponential expression is raised to an exponent, the original exponents are multiplied to obtain the exponent in the final result. This brings us to the power of a power rule.

> **Property: The Power of a Power Rule for Exponents**
>
> For a real number b and integers m and n,
>
> $$(b^m)^n = b^{m \cdot n}$$

The base of an exponential expression can be a product, a quotient, a sum, or a difference of expressions. When the base of an exponential expression contains an operation, the base must be written in parentheses when it is expressed in factored form. The entire expression in parentheses repeats when the exponent is applied to it. The following example illustrates how to simplify a power of a product.

Based on Definition	**Observed Rule**
$(3x^4)^2 = (3x^4)(3x^4)$ Two factors of $3x^4$	The same result is the obtained by squaring 3 and squaring x^4.
$= 3 \cdot 3 \cdot x^4 \cdot x^4$ Group coefficients and like bases.	$(3x^4)^2 = (3)^2(x^4)^2$
$= 9x^8$ Multiply the coefficients and add the exponents.	$= 9x^8$

So, we observe that when a product is raised to an exponent, the result is the same as applying the exponent to each factor in the base. This brings us to another property of exponents.

> **Property: The Power of a Product Rule for Exponents**
>
> For real numbers a and b and an integer n,
>
> $$(ab)^n = a^n b^n$$

A similar result occurs when an exponent is applied to a quotient.

Based on Definition	**Observed Rule**

Based on Definition

$$\left(\frac{y}{3}\right)^2 = \left(\frac{y}{3}\right)\left(\frac{y}{3}\right)$$ Two factors of $\frac{y}{3}$

$$= \frac{y \cdot y}{3 \cdot 3}$$ Group coefficients and like bases.

$$= \frac{y^2}{9}$$ Multiply the coefficients and add the exponents.

Observed Rule

The same result is the obtained by squaring y and squaring 3.

$$\left(\frac{y}{3}\right)^2 = \frac{(y)^2}{(3)^2}$$

$$= \frac{y^2}{9}$$

So, we observe that when a quotient is raised to an exponent, the exponent is applied to both the numerator and denominator. This leads to another property.

Property: The Power of a Quotient Rule for Exponents

For real numbers a and b $(b \neq 0)$ and n an integer,

$$\left(\frac{a}{b}\right)^n = \frac{a^n}{b^n}$$

Procedure: Applying the Power Rules

Step 1: If the base is a product, apply the power of a power rule.
Step 2: If the base is a quotient, apply the power of a quotient rule.
Step 3: Apply the power of a power rule to simplify any exponential expressions.
Step 4: Rewrite all exponents as positive exponents.

Objective 1 Examples Simplify each expression by applying the power rules. Write each result using positive exponents. Assume all bases are nonzero real numbers.

1a. $(x^{-5})^2$ **1b.** $[(-3)^4]^5$ **1c.** $(-6x^{-3})^{-1}$

1d. $(-4a^5b^{-2})^3$ **1e.** $\left(\frac{2a}{5b^2}\right)^3$ **1f.** $\left(\frac{4x^2}{3z^{-1}}\right)^{-2}$

Solutions **1a.** $(x^{-5})^2 = x^{-5 \cdot 2}$ Apply the power of a power rule.

$\qquad\qquad = x^{-10}$ Multiply the exponents.

$\qquad\qquad = \dfrac{1}{x^{10}}$ Rewrite with a positive exponent.

1b. $[(-3)^4]^5 = (-3)^{4 \cdot 5}$ Apply the power of a power rule.

$\qquad\qquad\quad = (-3)^{20}$ Multiply the exponents.

$\qquad\qquad\quad = 3^{20}$ Recall a negative base to an even exponent is positive.

1c. $(-6x^{-3})^{-1} = (-6)^{-1}(x^{-3})^{-1}$ Apply the power of a product rule.

$\qquad\qquad\qquad = \dfrac{1}{(-6)^1} \cdot x^3$ Apply the definition of a negative exponent and the power of a power rule.

$\qquad\qquad\qquad = \dfrac{1}{-6} \cdot x^3$ Simplify the exponent.

$\qquad\qquad\qquad = -\dfrac{x^3}{6}$ Multiply.

1d. $(-4a^5b^{-2})^3 = (-4)^3(a^5)^3(b^{-2})^3$ 　　Apply the power of a product rule.

$$= -64a^{15}b^{-6}$$ 　　Simplify the exponent and apply the power of a power rule.

$$= -64a^{15} \cdot \frac{1}{b^6}$$ 　　Rewrite with positive exponents.

$$= -\frac{64a^{15}}{b^6}$$ 　　Simplify.

1e. $\left(\dfrac{2a}{5b^2}\right)^3 = \dfrac{(2a)^3}{(5b^2)^3}$ 　　Apply the power of a quotient rule.

$$= \frac{(2)^3(a)^3}{(5)^3(b^2)^3}$$ 　　Apply the power of a product rule in the numerator and denominator.

$$= \frac{8a^3}{125b^6}$$ 　　Simplify the exponents and apply the power of a power rule.

1f. $\left(\dfrac{4x^2}{3z^{-1}}\right)^{-2} = \dfrac{(4x^2)^{-2}}{(3z^{-1})^{-2}}$ 　　Apply the power of a quotient rule.

$$= \frac{4^{-2}x^{-4}}{3^{-2}z^2}$$ 　　Apply the power of a product rule in the numerator and denominator.

$$= \frac{3^2}{4^2x^4z^2}$$ 　　Rewrite with positive exponents.

$$= \frac{9}{16x^4z^2}$$ 　　Simplify.

✓ **Student Check 1** Simplify each expression by applying the power rules. Write each result using positive exponents. Assume all bases are nonzero real numbers.

a. $(y^{-3})^3$ 　　　**b.** $[(-2)^7]^3$ 　　　**c.** $(-11x^{-4})^2$

d. $(-2m^{-6}n^4)^5$ 　　**e.** $\left(\dfrac{6x^2}{7y}\right)^3$ 　　**f.** $\left(\dfrac{-2a^{-1}}{5b^4}\right)^{-3}$

Note: *Once we become comfortable with the rules, we can omit some of the steps. For instance, in Example 1f, we can directly apply the exponent of −2 to each factor in the numerator and denominator.*

Combine Properties of Exponents

Objective 2 ▶

Simplify exponential expressions using a combination of exponent properties.

We now review all of the exponent properties presented in Sections 5.1 and 5.2 so that we can combine several properties in one problem.

Property: Summary of Exponent Rules

If a and b are real numbers and m and n are integers, then

Product of like bases: 　　　　$b^m \cdot b^n = b^{m+n}$

Quotient of like bases: 　　　　$\dfrac{b^m}{b^n} = b^{m-n}, b \neq 0$

Zero exponent: 　　　　　　　　$b^0 = 1, b \neq 0$

Negative exponent: 　　　　　　$b^{-n} = \dfrac{1}{b^n}, b \neq 0$

Negative exponent of a fraction: 　$\left(\dfrac{a}{b}\right)^{-n} = \left(\dfrac{b}{a}\right)^n, a \neq 0, b \neq 0$

Power of a power:	$(b^m)^n = b^{mn}$
Power of a product:	$(ab)^n = a^n b^n$
Power of a quotient:	$\left(\dfrac{a}{b}\right)^n = \dfrac{a^n}{b^n},\ b \neq 0$

Procedure: **Simplifying Exponential Expressions**

When problems require us to use multiple properties to simplify, it is generally best to

Step 1: Clear parentheses by applying the power rule for products or quotients.
Step 2: If like bases occur, apply the product or quotient rule.
Step 3: Finally, rewrite any negative exponents and simplify any exponential expressions.

Objective 2 Examples   **Simplify each expression and write each result using positive exponents. Assume all bases are nonzero real numbers.**

2a. $(-3x^5 y^{-2})^3 \cdot x^{-7} y^5$ **2b.** $\left(\dfrac{5a^{-2}}{4a^3}\right)^{-4}$ **2c.** $\left(\dfrac{6b^4}{a^2}\right)^{-1}\left(\dfrac{6a^{-4}}{b^{-2}}\right)^2$

Solutions

2a.

$(-3x^5 y^{-2})^3 \cdot x^{-7} y^5 = (-3)^3 (x^5)^3 (y^{-2})^3 \cdot x^{-7} y^5$ — Apply the power of a product rule.

$= -27x^{15} y^{-6} \cdot x^{-7} y^5$ — Simplify the exponent and apply the power of a power rule.

$= -27x^{15} x^{-7} y^{-6} y^5$ — Apply the commutative property.

$= -27x^{15+(-7)} y^{-6+5}$ — Apply the product of like bases rule.

$= -27x^8 y^{-1}$ — Simplify.

$= -\dfrac{27x^8}{y}$ — Rewrite with positive exponents.

2b.

$\left(\dfrac{5a^{-2}}{4a^3}\right)^{-4} = \dfrac{(5a^{-2})^{-4}}{(4a^3)^{-4}}$ — Apply the power of a quotient rule.

$= \dfrac{5^{-4} a^8}{4^{-4} a^{-12}}$ — Apply the power of a product rule and the power of a power rule.

$= \dfrac{4^4 a^8 a^{12}}{5^4}$ — Rewrite with positive exponents.

$= \dfrac{256 a^{20}}{625}$ — Simplify the exponents and apply the product of like bases rule.

2c.

$\left(\dfrac{6b^4}{a^2}\right)^{-1}\left(\dfrac{6a^{-4}}{b^{-2}}\right)^2 = \dfrac{6^{-1} b^{-4}}{a^{-2}} \cdot \dfrac{6^2 a^{-8}}{b^{-4}}$ — Apply the power of a quotient rule for each fraction. Apply the power of a power rule.

$= \dfrac{6^{-1} 6^2 a^{-8} b^{-4}}{a^{-2} b^{-4}}$ — Multiply the fractions.

$= 6^{-1+2} a^{-8-(-2)} b^{-4-(-4)}$ — Apply the product of like bases rule and the quotient of like bases rule.

$= 6^1 a^{-6} b^0$ — Simplify.

$= \dfrac{6(1)}{a^6}$ — Rewrite with positive exponents and simplify the exponents.

$= \dfrac{6}{a^6}$ — Simplify.

☑ **Student Check 2**   Simplify each expression and write each result using positive exponents. Assume all bases are nonzero real numbers.

   a. $(-2m^{-4}n^2)^4 \cdot m^{-3}n^5$ **b.** $\left(\dfrac{3x^{-3}}{2x^5}\right)^{-2}$ **c.** $\left(\dfrac{8h^3}{r^4}\right)^{-2}\left(\dfrac{8r^{-3}}{h^{-1}}\right)^3$

Scientific Notation

Objective 3 ▶

Convert between standard notation and scientific notation.

In real life, we deal with numbers that are very large and very small. For example, extremely large numbers are needed to represent the amount of information stored on a computer. Information stored on computers is measured in bytes.

- It takes 1 byte (B) to digitally store a single character.
- A typewritten page takes 2000 bytes or 2 kilobytes (kB) to store.
- A movie takes 1 billion bytes or 1 gigabyte (GB) to store.
- An academic research library would take 2 trillion bytes or 2 terabytes (TB) to store.
- It would take 5 quintillion bytes or 5 exabytes (EB) to store all words ever spoken by humans. (Source: http://www2.sims.berkeley.edu/research/projects/how-much -info/datapowers.html)

We can also find some very small numbers when dealing with the body and things in nature.

- The thickness of human skin is 0.07 in.
- The diameter of a skin cell is 0.001181 in.
- The diameter of a red blood cell is 0.000315 in.
- The diameter of an influenza virus is 0.00000472 in.
- The diameter of a water molecule is 0.0000000108 in.
  (Source: http://learn.genetics.utah.edu/content/begin/cells/scale/)

We now use exponents to write very large or very small numbers in *scientific notation*. Because our numbering system is based on the number 10, every digit in the number represents a power of 10. Recall the place value of the digits in a number.

Ten-thousands	Thousands	Hundreds	Tens	Ones	●	Tenths	Hundredths
$10{,}000 = 10^4$	$1000 = 10^3$	$100 = 10^2$	$10 = 10^1$	$1 = 10^0$	Decimal point	$\dfrac{1}{10} = 10^{-1}$	$\dfrac{1}{100} = 10^{-2}$

So, every number in standard notation can be written as a product of a number, called the *coefficient*, and a power of 10. For example,

$$50{,}000 = 5 \times 10{,}000 = 5 \times 10^4$$
$$5000 = 5 \times 1000 = 5 \times 10^3$$
$$500 = 5 \times 100 = 5 \times 10^2$$
$$50 = 5 \times 10 = 5 \times 10^1$$
$$5 = 5 \times 1 = 5 \times 10^0$$
$$0.5 = 5 \times 0.1 = 5 \times \frac{1}{10} = 5 \times 10^{-1}$$
$$0.05 = 5 \times 0.01 = 5 \times \frac{1}{100} = 5 \times 10^{-2}$$

When the exponent of 10 is positive, the number (e.g., 50,000, 5000, 500, and 50) is greater than or equal to 10. When the exponent of 10 is negative, the number (e.g., 0.5 and 0.05) is between 0 and 1. When the exponent of 10 is zero, the number (e.g., 5) is greater than or equal to 1 but less than 10. The power of 10 represents the place value of the coefficient, 5.

Definition: A number written in the form $a \times 10^n$ is a number written in **scientific notation** as long as $1 \le a < 10$ and n is an integer. The number a is called the coefficient.

Procedure: Writing a Number in Scientific Notation

Step 1: Move the original decimal point to the left or right until you reach a number between 1 and 10. This number is the coefficient.

Step 2: The exponent of 10 represents the number of decimal places that you moved in step 1. If the decimal is moved left, the exponent of 10 is positive. If the decimal is moved right, the exponent of 10 is negative.

Procedure: Converting a Number from Scientific Notation to Standard Notation

Step 1: Drop " $\times 10^n$".

Step 2: If $n > 0$, move the decimal point to the right n places.

Step 3: If $n < 0$, move the decimal point to the left $|n|$ places.

Note: *Helpful Method to Remember How to Convert to the Standard Form*

Think of a number line. If a number is positive, we move to the right. If the number is negative, we move to the left. This is the same movement needed to convert numbers from scientific notation to standard form.

Objective 3 Examples   **Write the given number in either scientific or standard notation.**

3a. The storage needed for all words spoken by humans is 5 exabytes or 5 quintillion bytes. Write this number in scientific notation. (Source: http://www2 .sims.berkeley.edu/research/projects/how-much-info/datapowers.html)

3b. The diameter of a red blood cell is about 0.000315 in. Write this number in scientific notation. (Source: http://www.wadsworth.org/chemheme/heme/microscope/rbc.htm)

3c. The memory capacity of the brain is 1×10^{12} bytes. Write this number in standard form. (Source: http://www.moah.org/exhibits/archives/brains/technology.html)

3d. The radius of a carbon atom is 1.34×10^{-8} in. Write this number in standard form. (Source: http://learn.genetics.utah.edu/content/begin/cells/scale/)

Solutions   **3a.** 5 quintillion $= 5,000,000,000,000,000,000.0 = 5 \times 10^{18}$

Move the decimal point left 18 places

3b. $0.000315 = 3.15 \times 10^{-4}$

Move the decimal point right 4 places

3c. $1 \times 10^{12} = 1,000,000,000,000.$

Move the decimal point right 12 places

3d. $1.34 \times 10^{-8} = 0.0000000134$

Move the decimal point left 8 places

✓ Student Check 3   Write the given number in either scientific or standard notation.

a. The world population in 2011 was approximately 6,900,000,000. Write this number in scientific notation. (Source: http://www.census.gov/main/www/popclock.html)

    b. A hydrogen atom is about 0.00000005 mm in diameter. Write this number in scientific notation. (Source: http://web.jjay.cuny.edu/~acarpi/NSC/3-atoms.htm)

    c. The population of the United States in 2011 was approximately 3.1×10^8. Write this number in standard form. (Source: http://www.census.gov/main/www/popclock.html)

    d. A carbon atom has a mass of 2×10^{-23} g. Write this number in standard form. (Source: http://www.cavendishscience.org/phys/mole/mole.htm)

Operations with Scientific Notation

Objective 4 ▶

Perform operations with numbers in scientific notation.

It is nearly impossible to perform calculations with numbers that are very large or very small in standard notation on a calculator. It is much easier to handle calculations with these types of numbers if they are written in scientific notation. We will use properties of exponents to perform the computations.

Objective 4 Examples   Use scientific notation to solve each problem. Convert each answer to standard notation.

4a. In 2011, the national debt of the United States was approximately $14,096,000,000,000 and the population of the United States was approximately 310,000,000. If the debt was evenly distributed among all people in the country, how much would each person owe? (Sources: http://www.usdebtclock.org/ and http://www.census.gov/main/www/poplock.html)

Solution   **4a.** We must divide the national debt by the population.

$$\frac{14,096,000,000,000}{310,000,000} = \frac{1.4096 \times 10^{13}}{3.1 \times 10^8}$$ Convert each number to scientific notation.

$$= \frac{1.4096}{3.1} \times \frac{10^{13}}{10^8}$$ Divide the coefficients and divide the powers of 10.

$$\approx 0.4547 \times 10^5$$ Apply the quotient of like bases rule.

$$\approx \$45,470$$ Convert to standard notation.

So, each person would owe approximately $45,470 to pay off the national debt.

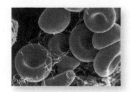

4b. The diameter of a blood platelet is approximately 0.003 mm. A healthy person produces approximately 150 billion platelets each day. If the platelets could be placed side by side, how many miles would they extend? (Note: There are 1,609,344 mm in 1 mi.) (Source: http://medical-dictionary.thefreedictionary.com/Blood+platelets)

Solution   **4b.** We must multiply the number of platelets produced by their diameter.

150,000,000,000 × 0.003

$$= (1.5 \times 10^{11})(3 \times 10^{-3})$$ Write each number in scientific notation.

$$= (1.5 \times 3) \times (10^{11} \times 10^{-3})$$ Multiply the coefficients and the powers of 10.

$$= 4.5 \times 10^8 \text{ mm}$$ Apply the product of like bases rule.

$$= 450,000,000 \text{ mm}$$ Convert to standard notation.

Dividing 450,000,000 mm by 1,609,344 gives us the number of miles the platelets would extend. So, the blood platelets would extend approximately 280 mi if placed side by side.

✓ **Student Check 4** Use scientific notation to solve each problem. Convert each answer to standard notation.

    a. As of June 2010, there were approximately 1,970,000,000 worldwide users of the Internet. Europe had approximately 475,000,000 users of the Internet. What percent of the worldwide users were in Europe? (Source: http://www.internetworldstats.com/)

    b. The Milky Way Galaxy is approximately 100,000 light-years in diameter. One light-year is the distance light can travel in a year, or approximately 5.9 trillion mi. How many miles wide is the Milky Way Galaxy? (Source: http://www.hartrao.ac.za/other/howfar/howfar.html)

Objective 5 ▶

Troubleshoot common errors.

Troubleshooting Common Errors

Some common errors associated with properties of exponents are shown.

Objective 5 Examples **A problem and an incorrect solution are given. Provide the correct solution and an explanation of the error.**

5a. Simplify $(-4x^3)^2$.

Incorrect Solution	Correct Solution and Explanation
$(-4x^3)^2 = -4x^9$	The exponent was not applied to -4 and the exponents were not multiplied. $$(-4x^3)^2 = (-4)^2(x^3)^2 = 16x^6$$

5b. Simplify $(2y^{-5})^{-3}$.

Incorrect Solution	Correct Solution and Explanation
$(2y^{-5})^{-3} = -8y^{15}$	The exponent was not applied to the coefficient of 2 correctly. $$(2y^{-5})^{-3} = (2)^{-3}(y^{-5})^{-3}$$ $$= \frac{1}{8}y^{15}$$ $$= \frac{y^{15}}{8}$$

ANSWERS TO STUDENT CHECKS

Student Check 1 **a.** $\dfrac{1}{y^9}$ **b.** -2^{21} **c.** $\dfrac{121}{x^8}$ **d.** $-\dfrac{32n^{20}}{m^{30}}$

  e. $\dfrac{216x^6}{343y^3}$ **f.** $-\dfrac{125a^3b^{12}}{8}$

Student Check 2 **a.** $\dfrac{16n^{13}}{m^{19}}$ **b.** $\dfrac{4x^{16}}{9}$ **c.** $\dfrac{8}{rh^3}$

Student Check 3 **a.** 6.9×10^9 **b.** 5×10^{-8}
  c. 310,000,000 **d.** 0.000 000 000 000 000 000 000 02

Student Check 4 **a.** 24% **b.** 590 quadrillion mi

SUMMARY OF KEY CONCEPTS

1. The power of a power rule applies when an exponential expression is raised to an exponent. In this case, the exponents are multiplied to simplify the expression. It is important that we know when to add and when to multiply the exponents. Note the difference in the following expressions.

$$x^3 \cdot x^2 = x^5 \quad \text{but} \quad (x^3)^2 = x^6$$

2. The power of a product rule applies when the base is a product. We apply the exponent to each factor and simplify. Note the difference in the expressions $3x^4$ and $(3x)^4$.

$$3x^4 = 3 \cdot x \cdot x \cdot x \cdot x \qquad \text{(The base is } x.)$$
$$(3x)^4 = (3x)(3x)(3x)(3x) = 3^4x^4 = 81x^4. \quad \text{(The base is } 3x.)$$

3. The power of a quotient rule applies when the base is a quotient or a fraction. We apply the exponent to the numerator and denominator. To simplify the resulting expression, we may have to use the power of a power rule and/or the power of a product rule.

4. Scientific notation is a useful way to express very large or very small numbers. A number of the form $a \times 10^n$

is a number written in scientific notation as long as $1 \le a < 10$ and n is an integer.

- If the exponent in scientific notation is negative, the number has a value between 0 and 1.
- If the exponent in scientific notation is positive, the number has a value greater than or equal to 10.
- A number is converted from scientific notation to standard notation by moving the decimal point as indicated by the exponent of 10. A positive exponent tells us to move the decimal to the right. A negative exponent tells us to move the decimal point to the left.

5. Operations with numbers written in scientific notation can be performed by using the properties of exponents.

GRAPHING CALCULATOR SKILLS

The calculator skills for this section include entering exponential expressions and working with scientific notation.

Example 1: Simplify $(-6)^4 \cdot (-6)^8$.

Solution:

```
(-6)^4*(-6)^8
              2176782336
(-6)^12
              2176782336
```

So, $(-6)^4 \cdot (-6)^8 = (-6)^{12} = 6^{12}$.

Example 3: Enter the number 10,000,000,000,000 on a calculator and interpret the display.

Solution:

```
10000000000000
              1E13
```

Notice that the display is 1 E 13. This means that the number entered is equivalent to 1×10^{13}. "E" represents "times 10 to the exponent of."

Example 2: Simplify $[(-3)^4]^5$.

Solution:

```
((-3)^4)^5
              3486784401
(-3)^20
              3486784401
```

So, $[(-3)^4]^5 = (-3)^{20} = 3^{20}$.

Example 4: Enter the number 7.53×10^{-10} on the calculator. To enter "E" on the calculator, access the EE function (2nd ,) on the calculator.

Solution:

```
7.53E-10
              7.53E-10
7.53*10^-10
              7.53E-10
```

SECTION 5.2 / EXERCISE SET

 Write About It!

Use complete sentences in your answer to each exercise.

1. Explain how to apply the power of a product rule to $(ab)^m$.

2. Explain how to apply the power of a quotient rule to $\left(\dfrac{a}{b}\right)^m$.

3. Explain why $(-2x^m)^n \ne 2^n x^{mn}$.

4. Define the scientific notation of a number.

5. Use an example to explain how to convert a number less than one from standard form to scientific notation.

6. Use an example to explain how to convert a number in scientific notation to standard form.

7. Use an example to explain how to convert a number greater than 10 from standard form to scientific notation.

8. Name two advantages of using scientific notation in the calculation of very large or very small numbers.

Practice Makes Perfect!

Simplify each expression by applying the power rules. Assume all bases are nonzero real numbers. Express each answer with positive exponents. (See Objective 1.)

9. $(b^3)^5$

10. $(y^7)^2$

11. $(5^{-3})^2$

12. $(6^9)^{-2}$

13. $[(-2)^3]^{-5}$

14. $[(-3)^2]^{-4}$

15. $[(-6)^{-7}]^5$

16. $[(-5)^{-3}]^6$

17. $(2a^9b)^5$

18. $(5u^7v^2)^3$

19. $(-4r^9s^2)^4$

20. $(-5u^3v^{11})^2$

21. $(-6x^{-3})^{-2}$

22. $(-3v^{-6})^{-1}$

23. $\left(\dfrac{2x^3}{3y}\right)^4$

24. $\left(-\dfrac{7r^2}{10s^4}\right)^2$

25. $\left(\dfrac{3u^{-1}}{2v^3}\right)^{-4}$

26. $\left(\dfrac{2a^{-3}}{3b}\right)^{-3}$

Simplify each expression using the rules of exponents. Write each answer with positive exponents. Assume all bases are nonzero real numbers (See Objective 2.)

27. $(a^7b^{-5})^3 \cdot a^{-17}b^{-3}$

28. $(p^5q^{-7})^6 \cdot p^{-16}q^{-3}$

29. $(-2r^3s^{-2})^4 \cdot r^{-10}s^{-5}$

30. $(-4u^9v^{-10})^3 \cdot u^{16}v^{12}$

31. $\left(-\dfrac{9x^{-4}}{4x^4}\right)^{-2}$

32. $\left(-\dfrac{2x^{-1}}{3x^2}\right)^{-5}$

33. $\left(\dfrac{7q^8}{6q^{-10}}\right)^2$

34. $\left(\dfrac{13u^2}{2u^{-11}}\right)^2$

35. $\left(\dfrac{9a^{-4}}{b^{-3}}\right)^4\left(\dfrac{9b^{-1}}{a^{-6}}\right)^{-4}$

36. $\left(\dfrac{3x^{-5}}{y^{-1}}\right)^3\left(\dfrac{3y^{-4}}{x^{-7}}\right)^{-3}$

37. $\left(\dfrac{12h^{-7}}{k^{-4}}\right)^3\left(\dfrac{12k^9}{h^6}\right)^{-2}$

38. $\left(\dfrac{10p^{-4}}{q^{-2}}\right)^{-2}\left(\dfrac{10q^{-5}}{p^{-3}}\right)^5$

39. $\dfrac{(y^{-1}z^8)^4}{2^5y^3z^{15}}$

40. $\dfrac{(3^{-1}a^2b^8)^3}{5a^{10}b^{14}}$

41. $\dfrac{(2^{-3}s^{-2}t^{12})^{-1}}{s^3t^{12}}$

42. $\dfrac{(r^{-5}s^{-8})^4}{2^4r^{-8}s^{10}}$

Write each number in an alternate form, either in scientific or standard notation. (See Objective 3.)

43. 0.000000603

44. 0.000000563

45. 945,000

46. 812,000,000

47. 4.65×10^4

48. 6.68×10^3

49. 4.52×10^{-3}

50. 1.89×10^{-5}

Use scientific notation to perform the indicated operation. Express each answer in scientific notation. (See Objective 4.)

51. $(7.6 \times 10^{-20})(4.8 \times 10^{-5})$

52. $(1.8 \times 10^3)(5 \times 10^{-8})$

53. $(4 \times 10^{13})(5.3 \times 10^{-8})$

54. $(4 \times 10^{-18})(1.4 \times 10^{-3})$

55. $(2 \times 10^{-5})^5$

56. $(5 \times 10^{-6})^3$

57. $(0.071)(120,000)$

58. $(0.15)(900,000)$

59. $(0.00096)(970,000,000,000)$

60. $(0.000000054)(0.0000008)$

61. $\dfrac{1.62 \times 10^{-11}}{4.5 \times 10^3}$

62. $\dfrac{2.496 \times 10^8}{3.9 \times 10^2}$

63. $\dfrac{804,000}{0.00012}$

64. $\dfrac{18,200}{0.0000013}$

65. $\dfrac{(1.29 \times 10^3)(4 \times 10^{12})}{3 \times 10^{-4}}$

66. $\dfrac{(6.27 \times 10^5)(9 \times 10^8)}{5.7 \times 10^{-1}}$

67. $\dfrac{(6.052 \times 10^7)(7 \times 10^{-12})}{8.9 \times 10^6}$

68. $\dfrac{(6.02 \times 10^{-7})(5 \times 10^{-2})}{3.5 \times 10^{-11}}$

Use scientific notation to solve each problem. Convert each result to standard notation. (See Objective 4.)

69. As of June 2010, there were approximately 1,970,000,000 worldwide users of the Internet. Japan had approximately 99,000,000 users of the Internet. What percent of the worldwide users were in Japan? (Source: http://www.internetworldstats.com)

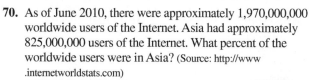

70. As of June 2010, there were approximately 1,970,000,000 worldwide users of the Internet. Asia had approximately 825,000,000 users of the Internet. What percent of the worldwide users were in Asia? (Source: http://www.internetworldstats.com)

71. As of June 2010, the world population was approximately 6,800,000,000. Asia had a population of approximately 3,800,000,000. What percent of the world population lived in Asia? (Source: http://www.internetworldstats.com)

72. As of June 2010, the world population was approximately 6,800,000,000. The United States had a population of approximately 310,000,000. What percent of the world population lived in the United States? (Source: http://www.internetworldstats.com)

73. As of April 2011, the public sector net debt of the United Kingdom was £910,000,000,000 and the population of the UK was approximately 60,000,000. If the debt was evenly distributed among all people in the UK, how much debt would each person owe? (Source: http://www.statistics.gov.uk/default.asp)

74. As of April 2011, the national debt of Canada was approximately $576,000,000,000 USD and the population of Canada was approximately 34,000,000. If the debt was evenly distributed among all people in Canada, how much debt would each person owe? (Sources: http://www.cia.gov/library/publicatons/the-world-factbook/index.html and http://taxpayer.com/federal/national-debt-clock-tour-2011)

75. One hydrogen atom has a diameter of approximately 5×10^{-8} mm. How many hydrogen atoms would it take to form a line 1 in. long if they were put side by side? Note: 1 in. is equivalent to 25.4 mm or 2.54×10^1 mm. (Source: http://web.jjay.cuny.edu/~acarpi/NSC/3-atoms.htm)

76. The average red blood cell is 3×10^{-4} in. in diameter. The diameter of the head of a pin is approximately 7.87×10^{-2} in. How many red blood cells would it take to stretch the length of the pin's diameter if they were put side by side? (Source: http://www.wadsworth.org/chemheme/heme/microscope/rbc.htm)

 Mix 'Em Up!

Simplify each expression using the rules of exponents. Express each answer with positive exponents. Assume all bases are nonzero real numbers.

77. $(-5x^6y^{-2})^{-3}$

78. $(-3u^{-3}v^2)^{-2}$

79. $(-2a^7b^{-1})^5$

80. $(-4p^{-5}q^8)^4$

81. $(-3v^{-10})^{-4}$

82. $(-2y^5)^{-3}$

83. $(7s^{-5})^{-2}$

84. $(5r^8)^{-4}$

85. $\left(-\dfrac{x^{-2}}{3}\right)^{-3}$

86. $\left(-\dfrac{y^{-6}}{5}\right)^{-2}$

87. $\left(\dfrac{q^{-4}}{6}\right)^2$

88. $\left(\dfrac{p^{-9}}{2}\right)^5$

89. $\left(\dfrac{6u^{-5}}{11v^3}\right)^{-2}$

90. $\left(-\dfrac{5p^{-1}}{2q^2}\right)^{-3}$

91. $\left(-\dfrac{9x^{-5}}{10y^3}\right)^{-3}$

92. $\left(\dfrac{2a^6}{7b^{-1}}\right)^2$

93. $(2a^3b^{-11})^4 \cdot a^{-14}b^3$

94. $(3r^8s^{-14})^2 \cdot r^{-7}s^{12}$

95. $(-x^2y^{-13})^3 \cdot x^9y^7$

96. $(-a^3b^{-7})^5 \cdot a^6b^4$

97. $\dfrac{(5^{-3}s^{-8}t^{-10})^{-1}}{4^{-1}s^{-3}t^9}$

98. $\dfrac{(r^{-9}s^{-3})^2}{2^4r^{12}s^{-7}}$

99. $\dfrac{(y^{-3}z^{-9})^2}{2y^6z^{-7}}$

100. $\dfrac{(2^4r^7s^{10})^{-1}}{5^{-3}r^{-4}s^{-20}}$

101. $\left(\dfrac{3u^{-7}}{5u^{-3}}\right)^{-2}$

102. $\left(\dfrac{10y^{-1}}{y^{-8}}\right)^2$

103. $\left(-\dfrac{y^8}{4y^{11}}\right)^4$

104. $\left(-\dfrac{2q^{-8}}{3q^{-7}}\right)^{-5}$

105. $\left(\dfrac{12p^{-3}}{q^6}\right)^{-3}\left(\dfrac{12q^{-1}}{p^{-7}}\right)^5$

106. $\left(\dfrac{5h^{-6}}{k^7}\right)^{-1}\left(\dfrac{5k^{-7}}{h^6}\right)^4$

107. $\left(\dfrac{15r}{h^8}\right)\left(\dfrac{15h^{-9}}{r^{10}}\right)^{-1}$

108. $\left(\dfrac{8u^2}{v^{-6}}\right)^{-2}\left(\dfrac{8v^{-8}}{u^{-7}}\right)^2$

Use scientific notation to perform the indicated operation. Express each answer in scientific notation.

109. $(5 \times 10^{-12})(7.7 \times 10^5)$

110. $(5.2 \times 10^{-9})(8.9 \times 10^{23})$

111. $\dfrac{(5.4464 \times 10^{11})(5.5 \times 10^5)}{4.6 \times 10^{12}}$

112. $\dfrac{(2.4948 \times 10^5)(2.5 \times 10^{-9})}{2.7 \times 10^9}$

113. $(0.000032)(280,000,000)$

114. $(0.0000042)(12,500)$

115. $\dfrac{7.885 \times 10^7}{9.5 \times 10^{-4}}$

116. $\dfrac{(9.672 \times 10^5)(5.5 \times 10^{-4})}{1.3 \times 10^{-9}}$

117. $(1.7 \times 10^6)^2$

118. $(3.1 \times 10^{-5})^3$

Use scientific notation to solve each problem. Convert each answer to standard notation.

119. In 2009 the average number of monthly visits to the top 500 retail websites was 2,580,000,000. The number of Internet users in the United States was approximately 228,000,000. Find the average number of monthly visits per Internet user in the United States. Round to two decimal places. (Source: http://www.internetretailer.com)

120. Amazon was the largest Web retailer in 2010. It had sales of $34,200,000,000. Their international sales were approximately $15,000,000,000 in 2010. Find the percent of Amazon's total sales that were international. Round to the nearest hundredth of a percent. (Sources: http://www.internetretailer.com and http://www.amazon.com)

 You Be the Teacher!

Correct each student's errors, if any.

121. Simplify $(-4a^3b^2)^3$.

Adrian's work:

$(-4a^3b^2)^3 = -12a^9b^6$

122. Simplify $\left(\dfrac{5x^{-2}}{2x^5}\right)^{-3}$.

Joshua's work:

$$\left(\dfrac{5x^{-2}}{2x^5}\right)^{-3} = \dfrac{-125x^6}{-8x^{-15}}$$

$$= \dfrac{125x^{-9}}{8}$$

$$= \dfrac{125}{8x^9}$$

123. Simplify $\dfrac{6.3 \times 10^{-21}}{1.4 \times 10^{-6}}$.

Atesa's work:

$$\dfrac{6.3 \times 10^{-21}}{1.4 \times 10^{-6}} = 4.9 \times 10^{-27}$$

124. Simplify $(4.1 \times 10^8)(2.9 \times 10^{-5})$.

Mark's work:

$$(4.1 \times 10^{-8})(2.8 \times 10^{-5}) = 6.9 \times 10^{-3}$$

 Calculate It!

Correct the error in each screen shot.

125. $\dfrac{1.512 \times 10^{-12}}{7.2 \times 10^{-4}}$

```
1.512*10^-12/7.2
*10^-4
          2.1E-17
```

126. $\dfrac{2.47 \times 10^2}{1.3 \times 10^{-7}}$

```
2.47E2-1.3E-7
    246.9999999
```

 Think About It!

127. What expression needs to be squared to obtain $49a^2b^{10}$? Check your result.

128. What expression needs to be squared to obtain $144x^4y^6$? Check your result.

129. What expression needs to be cubed to obtain $-\dfrac{27r^3}{s^6}$? Check your result.

130. What expression needs to be cubed to obtain $\dfrac{125x^{12}}{y^9}$? Check your result.

131. Use the properties of exponents to simplify $\dfrac{(x^{3a})^2}{x^{a-6}}$.

132. Use the properties of exponents to simplify $\dfrac{(2^{4x})^3 \cdot 2^{x-1}}{2^{x+3}}$.

| **SECTION 5.3** | **Polynomial Functions, Addition and Subtraction of Polynomials** |

▶ **OBJECTIVES**

As a result of completing this section, you will be able to

1. Identify the coefficient and degree of a term.

2. Classify a polynomial and identify its terms, leading coefficient, and degree.

3. Evaluate polynomial functions.

4. Add and subtract polynomials and polynomial functions.

5. Use polynomials to write expressions for real-world situations.

6. Troubleshoot common errors.

Objective 1 ▶

Identify the coefficient and degree of a term.

In 2010 the Burj Khalifa, in Dubai, became the tallest skyscraper in the world, reaching 2717 ft. If a penny is dropped from the top of this building, its height above the ground h, in feet, t seconds after it is dropped is given by

$$h(t) = -16t^2 + 2717$$

This function is an example of a *polynomial function*. We can evaluate it at different values to determine the height of the penny using the concepts we will learn from this chapter.

Terms

A *term* is a number or the product of a number and powers of variables. If a term only contains a number, it is called a **constant term**.

> **Definition:** A **term** is an expression of the form ax^n, where a is a real number and n is a whole number. The real number a is the **coefficient** of the term. The exponent n is the *degree* of the term.

Terms can have one variable or multiple variables. Each term has a degree.

> **Definition:** The **degree of a term** is the sum of all the exponents of the variables in the term.

Some terms and their degrees are shown in the table.

Term	Degree
$4x = 4x^1$	1
$-3y^2$	2
$7ab^3 = 7a^1b^3$	$1 + 3 = 4$
$5 = 5x^0$	0

Objective 1 Examples   Determine the coefficient and the degree of each term.

Problems	Solutions		
		Coefficient	**Degree**
1a. $-3x$	$-3x = -3x^1$	-3	1
1b. $5.3a^2$	$5.3a^2$	5.3	2
1c. $-\dfrac{y}{2}$	$-\dfrac{y}{2} = -\dfrac{1}{2}y^1$	$-\dfrac{1}{2}$	1
1d. $\dfrac{b^3}{6}$	$\dfrac{b^3}{6} = \dfrac{1}{6}b^3$	$\dfrac{1}{6}$	3
1e. -10	$-10 = -10x^0$	-10	0
1f. $7x^4y^2$	$7x^4y^2$	7	$4 + 2 = 6$

✓ Student Check 1   Determine the coefficient and the degree of each term.

   a. $-y$ **b.** $6.7h^5$ **c.** $-\dfrac{r}{4}$ **d.** $\dfrac{x^2}{5}$ **e.** -7 **f.** $4a^3b^8$

Polynomials

Objective 2 ▶

Classify a polynomial and identify its terms, leading coefficient, and degree.

In Section 1.3, we discussed the concept of algebraic expressions. *Polynomials* are special types of algebraic expressions in which the variables have whole numbers as exponents. Polynomials are very important to the study of algebra.

> **Definition:** A **polynomial** is an algebraic expression that consists of a finite sum of terms of the form ax^n, where a is a real number and n is a whole number. The **standard form** is to write the polynomials so that the degrees of the terms are in descending order.

An example of a polynomial is $-3x^4 + 2x^2 - 5x + 7$. This polynomial is written in standard form because the exponents of x are in descending order. The expressions $\dfrac{2}{x}$, $6x^{-2} + 4x^{-1}$, and $\sqrt{y^2 + 3}$ are not polynomials since a variable occurs in the denominator, has a negative exponent, or occurs inside a square root, respectively. Polynomials can have one, two, three, or more terms. The following table shows examples of each of these cases and provides the special name given to some of these polynomials.

Number of Terms	Name	Examples
One	**Monomial**	$5x, -3y, 7t, \frac{2}{3}ab$
Two	**Binomial**	$x - 4, \frac{1}{2}t - 1, y + 2, -3a - 7$
Three	**Trinomial**	$x^2 - xy + 6y^2, 4y^2 + 3y - 1$
Four or more	**Polynomial**	$-7y^3 + 9y^2 + 6y - 5$

Polynomials, like terms, have a degree as well.

> **Definition:** The **degree of a polynomial** is the largest degree of the terms in the polynomial.

A polynomial can be classified by not only its number of terms but also by its degree.

Degree	Type of Polynomial
0	Constant
1	Linear or first-degree
2	Quadratic or second-degree
3	Cubic or third-degree
4	Quartic or fourth-degree
5	Quintic or fifth-degree

> **Definition:** The **leading coefficient** of a single-variable polynomial is the coefficient of the term with the largest degree.

An example of a polynomial with its degree and leading coefficient is shown.

$$\overset{\text{Degree}}{\underset{\text{Leading coefficient}}{-5y^2 + 4y + 3}}$$

The degree and leading coefficient of a polynomial are very important when we graph polynomial functions.

The leading coefficient for a multiple-variable polynomial is not defined, so we use not applicable (n/a).

Objective 2 Examples	Classify each polynomial and identify its terms, degree, type, and leading coefficient, if applicable.

Problems	Solutions		
	Terms and Classification	Degree and Type	Leading Coefficient
2a. $-x^2 + 7x - 6$	$-1x^2, 7x, -6$ Trinomial	2 Quadratic	-1
2b. $4.2x^3$	$4.2x^3$ Monomial	3 Cubic	4.2

Problems	Solutions		
	Terms and Classification	**Degree and Type**	**Leading Coefficient**
2c. $\frac{3}{5}x + 2$	$\frac{3}{5}x^1, 2$ Binomial	1 Linear	$\frac{3}{5}$
2d. $3x^3y - 4x^2y^3 + 2xy^2$	$3x^3y^1, -4x^2y^3, 2x^1y^2$ Trinomial	5 Quintic	n/a

✓ **Student Check 2** Classify each polynomial and identify its terms, degree, type, and leading coefficient, if applicable.

a. $-4y^4 + y^3 + 3y$ **b.** $x^2 + 16x$ **c.** $9\pi r^3$ **d.** $8r^2s^5 - 7rs^8 + 9rs$

Polynomial Functions

Objective 3 ▶

Evaluate polynomial functions.

In Chapter 3, we discussed linear functions. Recall a linear function is a function of the form $f(x) = ax + b$, where a and b are real numbers. Linear functions are *polynomial functions* of degree 1, because the expression $ax + b$ is a first-degree polynomial. Some other examples of polynomial functions are $f(x) = x^2 - 4x + 3$ and $P(x) = x^3 + 5x^2 - 7x + 1$.

> **Definition:** A **polynomial function** is a function of the form
>
> $$P(x) = a_nx^n + a_{n-1}x^{n-1} + \cdots + a_1x + a_0$$
>
> where $a_n, a_{n-1}, \ldots, a_1$, and a_0 are real numbers and n is a whole number.

Function notation provides a convenient way for us to evaluate polynomials. Recall that to evaluate a function is to find the output value that corresponds to a given input value.

> **Procedure: Evaluating a Polynomial Function**
>
> **Step 1:** Replace the independent variable, x, with the assigned value.
> **Step 2:** Simplify the result to find the output value.

Objective 3 Examples   Evaluate each function for the given value.

3a. Find $f(0)$ if $f(x) = 3x^2 - 5x - 8$.

Solution   **3a.**

$f(x) = 3x^2 - 5x - 8$

$f(0) = 3(0)^2 - 5(0) - 8$ Replace x with 0.

$f(0) = 3(0) - 0 - 8$ Simplify.

$f(0) = -8$ Add the resulting values.

3b. Find $P(-3)$ if $P(x) = -4x^3 - x^2 + 5x + 2$.

Solution   **3b.** $P(x) = -4x^3 - x^2 + 5x + 2$

$P(-3) = -4(-3)^3 - (-3)^2 + 5(-3) + 2$ Replace x with -3.

$P(-3) = -4(-27) - (9) - 15 + 2$ Simplify the exponents.

$P(-3) = 108 - 9 - 15 + 2$ Multiply.

$P(-3) = 86$ Add the resulting values.

3c. In 2010 the Burj Khalifa in Dubai became the tallest skyscraper in the world reaching 2717 ft. If a penny is dropped from the top of this building, its height above ground h, in feet, t seconds after it is dropped is given by $h(t) = -16t^2 + 2717$. Find $h(5)$ and $h(12)$ and interpret the results. (Source: http://www .emporis.com)

Solution   **3c.** We evaluate the function at $t = 5$ and $t = 12$.

$t = 5$	$t = 12$
$h(t) = -16t^2 + 2717$	$h(t) = -16t^2 + 2717$
$h(5) = -16(5)^2 + 2717$	$h(12) = -16(12)^2 + 2717$
$h(5) = -16(25) + 2717$	$h(12) = -16(144) + 2717$
$h(5) = -400 + 2717$	$h(12) = -2304 + 2717$
$h(5) = 2317$	$h(12) = 413$

So, the penny will be 2317 ft above the ground 5 sec after it is dropped and 413 ft above the ground 12 sec after it is dropped.

3d. The number of persons from Mexico obtaining legal permanent resident status in the United States from 1999–2008 can be modeled by the function

$$f(x) = 1028.12x^3 - 15{,}178.23x^2 + 58{,}211.6x + 141{,}625.65$$

where x is the number of years after 1999. Find $f(5)$ and interpret the result. (Source: www.dhs.gov)

Solution   **3d.** We evaluate the function at $x = 5$.

$$f(x) = 1028.12x^3 - 15{,}178.23x^2 + 58{,}211.6x + 141{,}625.65$$

$$f(5) = 1028.12(5)^3 - 15{,}178.23(5)^2 + 58{,}211.6(5) + 141{,}625.65$$

$$f(5) \approx 181{,}743$$

So, in 2004 (5 yr after 1999), approximately 181,743 people from Mexico obtained legal permanent resident status in the United States.

 Student Check 3   Evaluate each function for the given value.

a. Find $f(0)$ if $f(x) = 9x^2 + 3x - 1$.

b. Find $P(-2)$ if $P(x) = -x^3 + 7x^2 - x + 4$.

c. The Eiffel Tower is 1063 ft tall. If a penny is dropped from the top of this tower, its height above ground h, in feet, t seconds after it is dropped is given by $h(t) = -16t^2 + 1063$. Find $h(3)$ and $h(8)$ and interpret the results. (Source: http://www.emporis.com)

d. The average price of a gallon of gasoline in the United States in the first week of January between 2006 and 2011 can be modeled by $p(x) = -0.1251x^5 + 1.5647x^4 - 6.7497x^3 + 11.539x^2 - 6.1691x + 2.236$, where x is the years after 2006. Find $p(5)$ and interpret its meaning. (Source: http://www.eia.gov)

Objective 4 ▶
Add and subtract polynomials and polynomial functions.

Adding and Subtracting Polynomials

Before we add polynomials, we review some facts about simplifying algebraic expressions. Recall that only algebraic expressions with *like terms* can be combined. **Like terms** are terms that contain the same variables raised to the same exponents.

Like Terms	Unlike Terms
$-a, \dfrac{1}{2}a$	$3y^2, 3y$
$\dfrac{2x^2}{3}, -x^2$	$4x, 7$

To add like terms, we add their coefficients and the variable component remains the same.

$$6x^2 + 3x^2 = (6 + 3)x^2 = 9x^2$$

Since polynomials consist of terms, we add polynomials by combining the like terms of the polynomials. When polynomials are added or subtracted, it is most often the case that each polynomial will be written in a set of parentheses.

> **Procedure: Adding Polynomials**
>
> **Step 1:** Remove the parentheses.
> **Step 2:** Group like terms together.
> **Step 3:** Combine like terms.
> **Step 4:** Write the answer in standard form.

The ability to add polynomials enables us to subtract them as well, since subtraction is defined as adding the opposite. The rule $a - b = a + (-b)$ also extends to polynomials.

> **Property: Subtracting Polynomials**
>
> If P and Q are polynomials, then
>
> $$P - Q = P + (-Q)$$

To find the **opposite of a polynomial**, we multiply the polynomial by -1.

Opposite of $(x - 4) = -(x - 4) = -1(x - 4) = -x + 4$

Opposite of $(2y^2 - 3y + 5) = -(2y^2 - 3y + 5) = -1(2y^2 - 3y + 5) = -2y^2 + 3y - 5$

Notice that each sign of the polynomial is *changed* when the opposite is found.

> **Procedure: Subtracting Polynomials**
>
> **Step 1:** Find the opposite of the polynomial that is being subtracted.
> **Step 2:** Combine like terms.
> **Step 3:** Write the answer in standard form.

> **Note:** *We can add or subtract polynomial functions to obtain another polynomial function by simply adding or subtracting the two polynomials.*

Objective 4 Examples   **Perform each indicated operation.**

4a. $(0.3x - 5.2) + (x + 9.4)$ **4b.** $(9y^2 - 3y + 7) + (-y^2 - 6y + 1)$

4c. $\left(\dfrac{3}{2}a^2 - \dfrac{1}{4}ab + 2b^2\right) + \left(\dfrac{5}{2}a^2 - \dfrac{3}{4}ab + b^2\right)$

4d. $(0.3x - 5.2) - (x + 9.4)$ **4e.** $(9y^2 - 3y + 7) - (-y^2 - 6y + 1)$

4f. $\left(\dfrac{3}{2}a^2 - \dfrac{1}{4}ab + 2b^2\right) - \left(\dfrac{5}{2}a^2 - \dfrac{3}{4}ab + b^2\right)$

4g. Let $f(x) = 3x^2 - 4x + 1$ and $g(x) = 2x - 3$. Find $f(x) + g(x)$ and $f(x) - g(x)$.

Solutions   **4a.** $(0.3x - 5.2) + (x + 9.4)$

$\qquad\qquad = 0.3x - 5.2 + x + 9.4$ Remove parentheses.

$\qquad\qquad = 0.3x + x - 5.2 + 9.4$ Group like terms.

$\qquad\qquad = 1.3x + 4.2$ Combine like terms. Recall $x = 1x$.

4b. $(9y^2 - 3y + 7) + (-y^2 - 6y + 1)$

$\qquad\qquad = 9y^2 - 3y + 7 - y^2 - 6y + 1$ Remove parentheses.

$\qquad\qquad = 9y^2 - y^2 - 3y - 6y + 7 + 1$ Group like terms.

$\qquad\qquad = 8y^2 - 9y + 8$ Combine like terms.

4c. $\left(\dfrac{3}{2}a^2 - \dfrac{1}{4}ab + 2b^2\right) + \left(\dfrac{5}{2}a^2 - \dfrac{3}{4}ab + b^2\right)$

$\qquad = \dfrac{3}{2}a^2 - \dfrac{1}{4}ab + 2b^2 + \dfrac{5}{2}a^2 - \dfrac{3}{4}ab + b^2$ Remove parentheses.

$\qquad = \dfrac{3}{2}a^2 + \dfrac{5}{2}a^2 - \dfrac{1}{4}ab - \dfrac{3}{4}ab + 2b^2 + b^2$ Group like terms.

$\qquad = \dfrac{8}{2}a^2 - \dfrac{4}{4}ab + 3b^2$ Combine like terms.

$\qquad = 4a^2 - ab + 3b^2$ Simplify each fraction.

4d. $(0.3x - 5.2) - (x + 9.4)$

$\qquad = (0.3x - 5.2) - 1(x + 9.4)$ Find the opposite by multiplying by -1.

$\qquad = 0.3x - 5.2 - x - 9.4$ Apply the distributive property.

$\qquad = 0.3x - x - 5.2 - 9.4$ Group like terms.

$\qquad = -0.7x - 14.6$ Combine like terms.

4e. $(9y^2 - 3y + 7) - (-y^2 - 6y + 1)$

$\qquad = (9y^2 - 3y + 7) - 1(-y^2 - 6y + 1)$ Find the opposite by multiplying by -1.

$\qquad = 9y^2 - 3y + 7 + y^2 + 6y - 1$ Apply the distributive property.

$\qquad = 9y^2 + y^2 - 3y + 6y + 7 - 1$ Group like terms.

$\qquad = 10y^2 + 3y + 6$ Combine like terms.

4f. $\left(\dfrac{3}{2}a^2 - \dfrac{1}{4}ab + 2b^2\right) - \left(\dfrac{5}{2}a^2 - \dfrac{3}{4}ab + b^2\right)$

$\qquad = \left(\dfrac{3}{2}a^2 - \dfrac{1}{4}ab + 2b^2\right) - 1\left(\dfrac{5}{2}a^2 - \dfrac{3}{4}ab + b^2\right)$ Find the opposite by multiplying by -1.

$\qquad = \dfrac{3}{2}a^2 - \dfrac{1}{4}ab + 2b^2 - \dfrac{5}{2}a^2 + \dfrac{3}{4}ab - b^2$ Apply the distributive property.

$\qquad = \dfrac{3}{2}a^2 - \dfrac{5}{2}a^2 - \dfrac{1}{4}ab + \dfrac{3}{4}ab + 2b^2 - b^2$ Group like terms.

$\qquad = -\dfrac{2}{2}a^2 + \dfrac{2}{4}ab + b^2$ Combine like terms.

$\qquad = -a^2 + \dfrac{1}{2}ab + b^2$ Simplify fractions.

4g. Since both $f(x)$ and $g(x)$ are polynomial functions, we add and subtract them by adding and subtracting the polynomials.

$f(x) + g(x)$
$= (3x^2 - 4x + 1) + (2x - 3)$
$= 3x^2 - 4x + 1 + 2x - 3$
$= 3x^2 - 4x + 2x + 1 - 3$
$= 3x^2 - 2x - 2$

$f(x) - g(x)$
$= (3x^2 - 4x + 1) - (2x - 3)$
$= (3x^2 - 4x + 1) - 1(2x - 3)$
$= 3x^2 - 4x + 1 - 2x + 3$
$= 3x^2 - 4x - 2x + 1 + 3$
$= 3x^2 - 6x + 4$

☑ **Student Check 4**   Perform each indicated operation.

a. $(2.1p - 3.9) + (7.5p - 5.1)$ **b.** $(y^2 - 9y + 10) + (y^2 - y + 1)$

c. $\left(\dfrac{6}{7}x^2 + \dfrac{3}{4}xy + 4y^2\right) + \left(\dfrac{1}{7}x^2 - \dfrac{7}{4}xy + 2y^2\right)$

d. $(2.1p - 3.9) - (7.5p - 5.1)$ **e.** $(y^2 - 9y + 10) - (y^2 - y + 1)$

f. $\left(\dfrac{6}{7}x^2 + \dfrac{3}{4}xy + 4y^2\right) - \left(\dfrac{1}{7}x^2 - \dfrac{7}{4}xy + 2y^2\right)$

g. Let $f(x) = -2x^2 - 6x - 5$ and $g(x) = -5x + 6$. Find $f(x) + g(x)$ and $f(x) - g(x)$.

Note: *We can also add and subtract polynomials vertically by aligning like terms.*

$(9y^2 - 3y + 7) + (-y^2 - 6y + 1)$ $(9y^2 - 3y + 7) - (-y^2 - 6y + 1)$

$\begin{array}{r} 9y^2 - 3y + 7 \\ -y^2 - 6y + 1 \\ \hline 8y^2 - 9y + 8 \end{array}$ $\begin{array}{r} 9y^2 - 3y + 7 \\ -(-y^2 - 6y + 1) \\ \hline 10y^2 + 3y + 6 \end{array}$

Objective 5 ▶

Use polynomials to write expressions for real-world situations.

Applications

Polynomials are used to represent real-life situations. Polynomial models occur in geometry, physics, engineering, and business. In Example 5, we write a polynomial expression that represents each situation.

Objective 5 Examples   **Write a polynomial that represents each situation and use it to find the requested information.**

5a. One side of a triangle is 2 ft shorter than the cube of the shortest side. The other side is 3 ft longer than the square of the shortest side. Find a polynomial that represents the perimeter of the triangle. What is the perimeter if the shortest side is 2 ft?

Solution   **5a.** Let x represent the length of the shortest side of the triangle. We translate the statements to obtain an expression for the other two sides of the triangle.

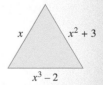

2 ft shorter than the cube of the shortest side, x, is $x^3 - 2$.

3 ft longer than the square of the shortest side, x, is $x^2 + 3$.

The perimeter is the sum of the length of the sides.

$P = x + (x^3 - 2) + (x^2 + 3)$ Find the sum of the three sides.

$P = x + x^3 - 2 + x^2 + 3$ Add the polynomials.

$P = x^3 + x^2 + x + 1$ Combine like terms.

If the shortest side is 2 ft, then the perimeter is

$P = x^3 + x^2 + x + 1$ State the expression for perimeter.

$P = (2)^3 + (2)^2 + (2) + 1$ Substitute 2 for x.

$P = 8 + 4 + 2 + 1$ Simplify the exponents.

$P = 15$ ft   Add.

So, the perimeter of the triangle whose shortest side is 2 ft is 15 ft.

5b. The operator of a hot dog stand can make hot dogs for $0.25 each. The cost to operate his hot dog stand is $500 per month. So, the monthly cost of operating the hot dog stand is $C(h) = 0.25h + 500$, where h is the number of hot dogs made in a month. Each hot dog sells for $0.99. The revenue (in dollars) made from selling h hot dogs in a month is $R(h) = 0.99h$. Write a polynomial function that represents the profit (in dollars) from selling h hot dogs in a month. What is the profit from selling 800 hot dogs in a month? [Note: Profit, $P(h)$, is defined as revenue minus cost, $R(h) - C(h)$.]

Solution   **5b.** $P(h) = R(h) - C(h)$ Write the profit function.

$P(h) = 0.99h - (0.25h + 500)$ Replace $R(h)$ and $C(h)$ with the appropriate functions.

$P(h) = 0.99h - 1(0.25h + 500)$ Subtract by multiplying the second polynomial by -1.

$P(h) = 0.99h - 0.25h - 500$ Apply the distributive property.

$P(h) = 0.74h - 500$ Combine like terms.

The profit from selling 800 hot dogs in a month is represented by $P(800)$.

$P(h) = 0.74h - 500$ State the polynomial function.

$P(800) = 0.74(800) - 500$ Replace h with 800.

$P(800) = 92$ Simplify.

So, the profit from selling 800 hot dogs in a month is $92.

☑ **Student Check 5**   Write a polynomial that represents each situation and use it to find the requested information.

a. Write a polynomial function that represents the perimeter of a trapezoid with sides of lengths $y + 2$, $4y^2 + y + 5$, $2y - 3$, and $3y^2 - 2y + 1$ units.

b. A Kiwanis club is organizing a pancake breakfast as a fund-raiser. The cost of preparing the breakfast is $1.25 per person and the cost to rent the location for the breakfast is $200. So, the total cost (in dollars) is $C(x) = 1.25x + 200$, where x is the number of people who attend the breakfast. A ticket to attend the breakfast is $6 per person. So, the revenue (in dollars) is $R(x) = 6x$, where x is the number of tickets sold. Write a polynomial function that represents the profit for the fund-raiser. What is the profit from selling 100 tickets?

Objective 6 ▶
Troubleshoot common errors.

Troubleshooting Common Errors

Some common errors associated with evaluating, adding, and subtracting polynomials are shown next.

Objective 6 Examples / **A problem and an incorrect solution are given. Provide the correct solution and an explanation of the error.**

6a. Find $f(-4)$ if $f(x) = -2x^2 + 3x - 5$.

Incorrect Solution	Correct Solution and Explanation
$f(-4) = -2(-4)^2 + 3(-4) - 5$ $f(-4) = 64 - 12 - 5$ $f(-4) = 47$	The mistake was made in evaluating the first term $-2(-4)^2$. Based on the order of operations, we square -4 first and then multiply by -2. $$-2(-4)^2 = -2(16) = -32$$ $$f(-4) = -2(-4)^2 + 3(-4) - 5$$ $$f(-4) = -32 - 12 - 5$$ $$f(-4) = -49$$

6b. Simplify $(x^2 + 3x - 4) + (x^2 - 7x + 1)$.

Incorrect Solution	Correct Solution and Explanation
$(x^2 + 3x - 4) + (x^2 - 7x + 1)$ $= x^2 + 3x - 4 + x^2 - 7x + 1$ $= x^2 + x^2 + 3x - 7x - 4 + 1$ $= x^4 - 4x - 3$	The sum of x^2 and x^2 is not x^4 These are like terms so their coefficients should be added to get $1x^2 + 1x^2 = 2x^2$. $(x^2 + 3x - 4) + (x^2 - 7x + 1)$ $= x^2 + 3x - 4 + x^2 - 7x + 1$ $= x^2 + x^2 + 3x - 7x - 4 + 1$ $= 2x^2 - 4x - 3$

6c. Simplify $(x^2 + 3x - 4) - (x^2 - 7x + 1)$.

Incorrect Solution	Correct Solution and Explanation
$(x^2 + 3x - 4) - (x^2 - 7x + 1)$ $= x^2 + 3x - 4 - x^2 - 7x + 1$ $= x^2 - x^2 + 3x - 7x - 4 + 1$ $= -4x - 3$	Each sign of the polynomial being subtracted will change. $(x^2 + 3x - 4) - (x^2 - 7x + 1)$ $= x^2 + 3x - 4 - x^2 + 7x - 1$ $= x^2 - x^2 + 3x + 7x - 4 - 1$ $= 10x - 5$

ANSWERS TO STUDENT CHECKS

Student Check 1   a. $-1, 1$ **b.** $6.7, 5$ **c.** $-\frac{1}{4}, 1$
d. $\frac{1}{5}, 2$ **e.** $-7, 0$ **f.** $4, 11$

Student Check 2   a. $-4y^4, y^3, 3y$; trinomial; degree $= 4$, quartic; leading coefficient $= -4$ **b.** $x^2, 16x$; binomial; degree $= 2$, quadratic; leading coefficient $= 1$ **c.** $9\pi r^3$; monomial; degree $= 3$, cubic; leading coefficient $= 9\pi$
d. $8r^2s^5, -7rs^8, 9rs$; trinomial; degree $= 9$, ninth-degree; leading coefficient: n/a

Student Check 3   a. -1 **b.** 42 **c.** $h(3) = 919$, $h(8) = 39$; After 3 sec the penny will be 919 ft above the ground and, after 8 sec, it will be 39 ft above the ground.

d. $p(5) \approx 3.15$; The average price of gas was about \$3.15 in the first week of January 2011.

Student Check 4   a. $9.6p - 9$
b. $2y^2 - 10y + 11$ **c.** $x^2 - xy + 6y^2$
d. $-5.4p + 1.2$ **e.** $-8y + 9$
f. $\frac{5}{7}x^2 + \frac{5}{2}xy + 2y^2$
g. $-2x^2 - 11x + 1$; $-2x^2 - x - 11$

Student Check 5   a. $P = 7y^2 + 2y + 5$
b. $P(x) = 4.75x - 200$; $P(100) = \$275$

SUMMARY OF KEY CONCEPTS

1. An expression of the form ax^n is a term with coefficient a and degree n.

2. Polynomials consist of a finite sum of terms of the form ax^n.

- If the polynomial consists of one term, it is a monomial.
- If it consists of two terms, it is a binomial.
- If it consists of three terms, it is a trinomial.
- The degree of a term is the sum of the exponents of the variables.
- The degree of a polynomial is the largest degree of the terms in the polynomial.

- The leading coefficient of a single-variable polynomial is the coefficient of the term with the largest degree.

3. To evaluate a polynomial function at a given value, replace the variable with the given value and use the order of operations to simplify the expression.

4. Polynomials are added by combining like terms. When we add, the exponent stays the same and the coefficients are added.

5. To subtract two polynomials, we add the first polynomial to the opposite of the second polynomial. The opposite is found by multiplying the polynomial by -1.

GRAPHING CALCULATOR SKILLS

The graphing calculator can be used to evaluate polynomial functions and to verify polynomial arithmetic.

Example 1: Find $f(1)$ if $f(t) = -16t^2 + 32t + 100$.

Method 1: Enter the resulting numerical expression on the calculator.

```
-16(1)²+32(1)+10
0
              116
```

Method 2: Enter the function in the equation editor and use the table to evaluate it.

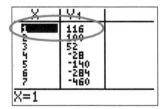

So, $f(1) = 116$.

Example 2: Verify that $(x^2 - x + 6) - (x^2 + 4x - 3) = -5x + 9$.

Solution: Enter the given problem in the equation editor for Y_1 and the result in Y_2. Then compare the table entries. If the columns agree, then the polynomial operation has been performed correctly.

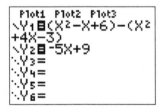

Since the Y_1 and Y_2 columns agree for each value of x, we can conclude that the subtraction is correct.

SECTION 5.3 / EXERCISE SET

 Write About It!

Use complete sentences in your answers to the following exercises.

1. Define a second-degree polynomial.

2. Explain the difference between the degree of a term and the degree of the polynomial.

3. Explain the difference between the coefficient of a term and the leading coefficient.

4. Explain how to evaluate a polynomial function for a given x-value, that is $x = c$.

5. Use an example to explain how to add two polynomials.

6. Use an example to explain how to add two polynomials vertically.

7. Use an example to explain how to subtract two polynomials vertically.

8. Give an example of a common mistake made by students when subtracting two polynomials.

Practice Makes Perfect!

Determine the coefficient and degree of each term.
(*See Objective 1.*)

9. $2b^3$

10. $4.7z^2$

11. $-c^5$

12. $-w^4$

13. $-\dfrac{x^2}{3}$

14. $\dfrac{y^3}{4}$

15. 12

16. $\dfrac{1}{2}$

17. $3a^5b^3$

18. $-2cd^4$

19. $-7.5u^2v^4w$

20. $12.3xy^6z^3$

Classify each polynomial and identify its terms. State its degree, type, and leading coefficient, if applicable.
(*See Objective 2.*)

	Terms/Classification	Degree/Type	Leading Coefficient
21. $3a^3 + 2a^2 - 5a + 7$			
22. $-2p^3 - 5p^2 + 4p + 11$			
23. $-11r^4 + 13r + 2$			
24. $\dfrac{3}{2}s^2 - 7s + 10$			
25. $32abc^2$			
26. $49rs^2t^2$			
27. $7 - 16s$			
28. $2t - 8$			

Evaluate each function. (*See Objective 3.*)

If $f(x) = 4x^2 - 5x + 1$ and $g(x) = -x^4 + 5x^2 + 3x - 10$, find the following.

29. $f(0)$

30. $g(0)$

31. $f(3)$

32. $f(-2)$

33. $g(-1)$

34. $g(3)$

35. $f(2) + g(1)$

36. $f(-3) + g(-2)$

37. $f(-1) - g(2)$

38. $f(5) - g(-3)$

If $P(x) = 12x^2 + 6x - 3$ and $Q(x) = -12x + 1$, find the following.

39. $P\left(\dfrac{1}{2}\right)$

40. $P\left(\dfrac{1}{3}\right)$

41. $Q\left(-\dfrac{1}{2}\right)$

42. $Q\left(-\dfrac{1}{6}\right)$

43. $P\left(\dfrac{1}{4}\right) + Q\left(\dfrac{1}{3}\right)$

44. $P\left(\dfrac{1}{3}\right) + Q\left(-\dfrac{1}{4}\right)$

45. $P\left(\dfrac{3}{4}\right) - Q\left(\dfrac{2}{3}\right)$

46. $P\left(\dfrac{2}{3}\right) - Q\left(-\dfrac{5}{6}\right)$

47. The Shanghai World Financial Center in China is 1614 ft high. If a penny is dropped from the top of this building, its height h, in feet, t sec after it is dropped is given by $h(t) = -16t^2 + 1614$. Find $h(4)$ and $h(9)$ and interpret the results.

48. The Taipei 101 Tower in Taipei, Taiwan, is 509 m high. If a dime is dropped from the top of this building, its height h, in meters, t sec after it is dropped is given by $h(t) = -9.8t^2 + 509$. Find $h(5)$ and $h(7)$ and interpret the results.

49. Video game revenues, in billions of dollars, in the United States can be modeled by the function $R(x) = 0.004167x^3 + 0.5482x^2 - 1.6917x + 11.9286$, where x is the number of years after 2002. Find $R(5)$ and $R(7)$ and interpret the results. Round to one decimal place. (Source: http://www.bls.gov/data/)

50. The unemployment rate in percent for men over 20 yr old can be modeled by the function $f(x) = 0.04227x^3 - 0.5832x^2 + 2.2242x + 2.6042$, where x is the number of years after 1999. Find $f(8)$ and $f(11)$ and interpret the results. Round to one decimal place. (Source: http://www.bls.gov/data/)

Perform the indicated operation. (*See Objective 4.*)

51. $(11y - 13) + (-20y - 3)$

52. $(-10x - 4) + (23x + 13)$

53. $(-2y + 7) - (-15y - 8)$

54. $(-3x + 4) - (5x - 15)$

55. $(7.1a - 3.8) + (-4.3a + 1.4)$

56. $(-9.4b - 7.5) + (6.2b - 12.1)$

57. $(0.6a + 10.3) - (2.1a + 6.1)$

58. $(9.7b - 0.1) - (-3.1b + 2)$

59. $(3y^2 - 12y + 25) + (-11y^2 + 2y - 5)$

60. $(11x^2 - 20x - 14) + (18x^2 + 7x - 17)$

61. $(9x^2 + 14x + 25) - (-15x^2 + 28x - 15)$

62. $(-9y^2 - 4y + 4) - (25y^2 - 9y + 14)$

63. $(8.2q^2 + 7.5q + 7.8) + (-8.8q^2 - 14.6q + 1.8)$

64. $(0.6p^2 - 6.4p - 7.2) + (-2.3p^2 + 5.2p + 15.5)$

65. $\left(\dfrac{10}{7}x^2 + \dfrac{5}{12}xy + 12y^2\right) - \left(-\dfrac{4}{7}x^2 - \dfrac{7}{12}xy + 16y^2\right)$

66. $\left(\dfrac{1}{2}a^2 - \dfrac{5}{14}ab - 5b^2\right) + \left(\dfrac{9}{2}a^2 - \dfrac{23}{14}ab + 18b^2\right)$

67. Let $f(x) = 5x^2 + 7x - 21$ and $g(x) = -3x^2 + 3x + 13$. Find $f(x) + g(x)$ and $f(x) - g(x)$.

68. Let $f(x) = 8x^2 + 3x - 10$ and $g(x) = -13x + 2$. Find $f(x) + g(x)$ and $f(x) - g(x)$.

Solve each problem. (*See Objective 5*.)

69. One side of a triangle is 2 in. shorter than the cube of the shortest side. The other side is 4 in. longer than the cube of the shortest side. Find a polynomial that represents the perimeter of the triangle. What is the perimeter if the shortest side is 8 in.?

70. One side of a triangle is 1 in. shorter than the square of the shortest side. The other side is 2 in. longer than the shortest side. Find a polynomial that represents the perimeter of the triangle. What is the perimeter if the shortest side is 3 in.?

71. Write a polynomial function that represents the perimeter of a rectangle with sides of lengths $2y^2 + 11y - 2$ units and $y^2 + 6y - 5$ units.

72. Write a polynomial function that represents the perimeter of a triangle with sides of lengths $6x - 15$ ft, $4x^2 - 4x - 2$ ft and $5x^2 + 4x - 2$ ft.

73. Write a polynomial function that represents the perimeter of a trapezoid with sides of lengths $4y - 7$ in., $2y + 5$ in., $5y^2 - 11y - 5$ in. and $9y^2 - 5y + 6$ in.

74. Write a polynomial function that represents the perimeter of a trapezoid with sides of lengths $6x - 8$ cm, $4x - 1$ cm, $9x^2 + 5x + 6$ cm and $8x^2 - 4x - 14$ cm.

75. A Boy Scout troop is organizing a barbeque chicken dinner as a fund-raiser. The cost of preparing the dinner is \$1.45 per person and the cost to rent the location for the dinner is \$310. So, the total cost (in dollars) is $C(x) = 1.45x + 310$, where x is the number of people who attend the dinner. A ticket to attend the dinner is \$7.25. So, the revenue (in dollars) is $R(x) = 7.25x$, where x is the number of tickets sold.

   a. Write a polynomial function that represents the profit for the fund-raiser.

   b. What is the profit from selling 185 tickets?

76. A Kiwanis club is organizing a spaghetti dinner as a fund-raiser. The cost of preparing the dinner is \$1.25 per person and the cost to rent the location for the dinner is \$280. So, the total cost (in dollars) is $C(x) = 1.25x + 280$, where x is the number of people who attend the dinner. A ticket to attend the dinner is \$7.50. So, the revenue (in dollars) is $R(x) = 7.5x$, where x is the number of tickets sold.

   a. Write a polynomial function that represents the profit for the fund-raiser.

   b. What is the profit from selling 150 tickets?

Mix 'Em Up!

Classify each polynomial and identify its terms. State its degree, type, and leading coefficient, if applicable.

	Terms/Classification	Degree/Type	Leading Coefficient
77. $6.25b^2 - 0.64$			
78. $\dfrac{1}{27}z^3 - 8$			
79. $5a^3b^2c$			
80. $-6p^2q^4r$			
81. $0.1x^3 + 1.6x^2 - 2.9x + 1.8$			
82. $-0.1y^3 - 0.9y^2 + 1.2y - 2.1$			
83. $-\dfrac{x^2}{5}$			
84. $\dfrac{v}{7}$			

Evaluate each function. If $f(x) = 2.4x^2 - 1.5x + 1.2$ and $g(x) = -3.6x + 4.8$, find the following.

85. $f(5)$ **86.** $g(-5)$ **87.** $f(-2)$

88. $g(2)$ **89.** $f(0) + g(-1)$ **90.** $f(2) + g(0)$

91. $f(-1) - g(1)$ **92.** $f(6) - g(-6)$

Perform each indicated operation.

93. $(3x - 10) + (18x + 16)$

94. $(-5y + 12) + (-14y - 4)$

95. $(-9r - 11) - (-10r + 3)$

96. $(4s + 5) - (8s - 13)$

97. $(-0.2a + 10.4) + (-9.8a + 5.1)$

98. $(-2.7b - 8.3) + (-6.9b - 5)$

99. $(20a^2 - 8a - 10) - (8a^2 + 31a - 16)$

100. $(11b^2 + 23b - 18) - (-15b^2 + 17b + 6)$

101. $\left(\frac{4}{3}p^2 - \frac{1}{4}pq - 11q^2\right) + \left(\frac{8}{3}p^2 - \frac{15}{4}pq + 2q^2\right)$

102. $\left(\frac{5}{2}x^2 + \frac{3}{5}xy - 7y^2\right) - \left(\frac{1}{2}x^2 - \frac{2}{5}xy + y^2\right)$

103. $(1.8x^2 + 5x - 0.6) - (-2.6x^2 - 13.5x - 8.6)$

104. $(-1.5y^2 - 1.7y + 6.3) - (-7.4y^2 - 14.6x + 21.5)$

Solve each problem.

105. The average cost of electricity per 500 kWh in any U.S. city can be modeled by the function $C(x) = -0.0168x^4 + 0.3194x^3 - 1.6895x^2 + 3.7336x + 45.5973$, where x is the number of years after 1999. Find C(7) and C(10) and interpret the results. Round to the nearest cent. (Source: http://www.bls.gov/data/)

106. The average quarterly domestic airline fuel consumption, in millions of gallons, can be modeled by the function $Q(x) = -0.6x^4 + 1.51x^3 + 60.94x^2 - 329.1x + 4561.2$, where x is the number of years after 2000. Find Q(6) and Q(9) and interpret the results. Round to the nearest million gallons. (Source: http://www.transtats.bts.gov)

107. A street vendor sells lemonade. She can make each drink for $0.45 Her fixed cost to operate her stand is $320 per month. So, the monthly cost (in dollars) of operating the stand is $C(x) = 0.45x + 320$, where x is the number of lemonades sold in a month. If she sells lemonades for $1.60 each, the revenue (in dollars) made from selling x lemonades in a month is $R(x) = 1.6x$.

   a. Write a polynomial function that represents the profit of the lemonade stand.
   b. What is the profit from selling 1200 lemonades in a month?

108. A Girl Scout troop is organizing a pancake breakfast as a fund-raiser. The cost of preparing the breakfast is $1.05 per person and the cost to rent the location for the breakfast is $330. So, the total cost (in dollars) is $C(x) = 1.05x + 330$, where x is the number of people who attend the breakfast. A ticket to attend the breakfast is $5.10. So, the revenue (in dollars) is $R(x) = 5.10x$, where x is the number of tickets sold.
   a. Write a polynomial function that represents the profit for the fund-raiser.
   b. What is the profit from 205 tickets?

 You Be the Teacher!

Correct each student's errors, if any.

109. Evaluate $g(-4)$ when $g(x) = -x^2 - 7x + 6$.

Chase's work:

$g(-4) = -(-4)^2 - 7(-4) + 6$
$= 16 + 28 + 6$
$= 50$

110. Evaluate $P(2)$ when $P(x) = -4x^2 + 12x - 5$.

Josh's work:

$P(2) = -4(2)^2 + 12(2) - 5$
$= 16 + 24 - 5$
$= 35$

111. Add the polynomials $(a^2 + 23a - 18) + (a^2 + 4a - 33)$.

Andrea's work:

$(a^2 + 23a - 18) + (a^2 + 4a - 33) = a^4 + 27a^2 - 51$

112. Subtract the polynomials $(9b^2 - 15b + 2) - (-2b^2 - 6b + 7)$.

Matt's work:

$(9b^2 - 15b + 2) - (-2b^2 - 6b + 7)$
$= 9b^2 - 15b + 2 + 2b^2 - 6b + 7$
$= 11b^2 - 21b + 9$

 Calculate It!

For Exercises 113 and 114, identify the mistake in the calculator entry and determine the correct answer.

113. Evaluate $f(-5)$ for $f(x) = -3x^2 + 10x - 7$.

```
-3*-5²+10*-5-7
            18
```

114. Evaluate $g(-1)$ for $g(x) = -x^2 + 6x - 4$.

```
--1²+6*-1-4
          -9
```

115. Verify $(-2x^2 + 5x - 4) + (5x^2 - 7x + 3) = 3x^2 - 2x - 1$.

 Think About It!

Write an example of a polynomial that satisfies the given conditions.

116. A trinomial with degree 3 and leading coefficient of -2.

117. A trinomial with degree 4 and leading coefficient of 7 .

118. A monomial with degree 0.

119. A monomial with degree 1.

PIECE IT TOGETHER / SECTIONS 5.1–5.3

Simplify each expression by applying the properties of exponents. Assume all variables are nonzero. (*Sections 5.1 and 5.2, Objectives 1–4*)

1. $(-15a^7) \cdot (-4a^{11})$

2. $(-3m^6)^5$

3. $-\left(-\dfrac{2x}{3y}\right)^5$

4. $\dfrac{-18y^5}{6y^2}$

5. $(-15)^0$

6. $\left(-\dfrac{5}{4}\right)^{-3}$

7. $\dfrac{x^5}{y^{-6}}$

8. $(-2s^{-4})^{-3}$

Write each number in scientific or standard notation. (*Section 5.2, Objectives 3 and 4*)

9. 0.0000009547

10. 2.38×10^5

Use exponent rules to simplify each expression. Write each answer in scientific notation. (*Section 5.2, Objectives 3 and 4*)

11. $(3.0 \times 10^{-4})(6.0 \times 10^{12})$

12. $\dfrac{18.0 \times 10^{15}}{6.0 \times 10^{-6}}$

Identify the leading coefficient and the degree of each polynomial. (*Section 5.3, Objective 2*)

13. $12x^2 - 45x + 3$

14. $54p^3 - 20p^2 + p - 16p^4 + 35$

Evaluate each polynomial function for the given value. (*Section 5.3, Objective 3*)

15. $P(h) = -3h^2 + 17; \quad P(-2)$

16. $f(y) = 5y^2 - 9; \quad f = \left(\dfrac{1}{2}\right)$

Perform the indicated operation. Write each answer in standard form. (*Section 5.3, Objective 4*)

17. $(9h^2 + 2hk + 5k^2) + (-2h^2 - 4hk - k^2)$

18. $(y^2 - 7y + 16) - (5y^2 - 4y - 23)$

19. $(3y^2 + y - 11) - (y^2 + 4y + 5)$

20. $(8a^4 - 2a^2 + 4a - 7) - (2a^4 + 5a^2 - a + 6)$

SECTION 5.4 / Multiplication of Polynomials and Polynomial Functions

▶ OBJECTIVES

As a result of completing this section, you will be able to

1. Multiply a monomial by a polynomial using the distributive property.

2. Multiply polynomials and polynomial functions.

3. Write expressions that represent real-life situations.

4. Troubleshoot common errors.

When an airline company charges $100 per seat, they can sell 120 tickets. For each $10 increase in price, they sell one less ticket. If x represents the number of increases in price, then $100 + 10x$ is the expression for the ticket price and $120 - x$ is the number of tickets sold. In this section, we will learn how to write an expression for the amount of money the airline company makes for this trip.

Multiplying Monomials and Polynomials

In Section 5.4, we learned that we can add and subtract polynomials by combining like terms. To multiply polynomials, we use some skills that we have learned in previous sections. The tools we need are the product of like bases rule for exponents and the distributive property.

Recall the product of like bases rule for exponents enables us to multiply exponential expressions with the same base.

Objective 1 ▶

Multiply a monomial by a polynomial using the distributive property.

Property	Example
$a^m a^n = a^{m+n}$	$2x(5x) = 2 \cdot 5 \cdot x^1 \cdot x^1$ $= 10x^2$

Recall the distributive property enables us to multiply an expression by a sum or difference.

Property	Example
$a(b + c) = ab + ac$	$5(x + 4) = 5(x) + 5(4)$ $= 5x + 20$

> **Procedure: Multiplying a Monomial by a Polynomial**
>
> **Step 1:** Distribute the monomial to each term of the polynomial.
> **Step 2:** Multiply the coefficients and multiply any like bases by adding the exponents.

Objective 1 Examples Multiply the expressions using the distributive property.

1a. $5x(x + 4)$ **1b.** $3y^2(6y^2 - 2y + 9)$

1c. $-2a^3(a^3 - 8)$ **1d.** $4x^2y(3x^2y^2 - 2xy + 1)$

Solutions **1a.** $5x(x + 4)$

$\quad\quad = 5x(x) + 5x(4)$ Apply the distributive property.

$\quad\quad = 5x^2 + 20x$ Simplify each product.

1b. $3y^2(6y^2 - 2y + 9)$

$\quad\quad = 3y^2(6y^2) - 3y^2(2y) + 3y^2(9)$ Apply the distributive property.

$\quad\quad = 3 \cdot 6 \cdot y^2 \cdot y^2 - 3 \cdot 2 \cdot y^2 \cdot y^1 + 3 \cdot 9 \cdot y^2$ Apply the commutative property.

$\quad\quad = 18y^4 - 6y^3 + 27y^2$ Simplify each product.

1c. $-2a^3(a^3 - 8)$

$\quad\quad = -2a^3(a^3) - (-2a^3)(8)$ Apply the distributive property.

$\quad\quad = -2a^6 + 16a^3$ Simplify each product.

1d. $4x^2y(3x^2y^2 - 2xy + 1)$

$\quad\quad = 4x^2y(3x^2y^2) - 4x^2y(2xy) + 4x^2y(1)$

$\quad\quad = 4 \cdot 3 \cdot x^2 \cdot x^2 \cdot y \cdot y^2 - 4 \cdot 2 \cdot x^2 \cdot x \cdot y \cdot y + 4 \cdot 1x^2y$

$\quad\quad = 12x^4y^3 - 8x^3y^2 + 4x^2y$

✓ Student Check 1 Multiply the expressions using the distributive property.

    a. $2y(y + 3)$ **b.** $8x^2(2x^2 - 5x + 1)$ **c.** $-7b^5(b^3 - 2)$ **d.** $3ab^3(5a^2b + ab - 1)$

Multiplying Polynomials and Polynomial Functions

Objective 2 ▶

Multiply polynomials and polynomial functions.

The distributive property can be used when we multiply a binomial by a polynomial. We will, in fact, have to use the distributive property twice. For instance, to multiply $(x + 3)$ by the polynomial $(x^2 + x + 6)$, we distribute the trinomial to each term in the binomial, x and 3. Then we apply the distributive property again to distribute x and 3 to each term in the trinomial, $(x^2 + x + 6)$.

$\quad (x + 3)(x^2 + x + 6)$

$\quad\quad = x(x^2 + x + 6) + 3(x^2 + x + 6)$ Apply the distributive property.

$\quad\quad = x(x^2) + x(x) + x(6) + 3(x^2) + 3(x) + 3(6)$ Apply the distributive property.

$\quad\quad = x^3 + x^2 + 6x + 3x^2 + 3x + 18$ Multiply.

$\quad\quad = x^3 + x^2 + 3x^2 + 6x + 3x + 18$ Group like terms.

$\quad\quad = x^3 + 4x^2 + 9x + 18$ Add.

This multiplication results in each term of the binomial being multiplied by each term of the trinomial. So, we can perform the multiplication by distributing the first term of the binomial to each term of the trinomial and then distributing the second term of the binomial to each term of the trinomial. (This is shown in the second step above.) This concept can be extended in order to multiply any two polynomials.

Procedure: Multiplying Polynomials and Polynomial Functions

Step 1: Distribute each term of the first polynomial to each term of the second polynomial.
Step 2: Simplify the resulting products.
Step 3: Combine like terms.
Step 4: Write the polynomial in standard form.

Objective 2 Examples | **Find each product.**

2a. $(x + 4)(x + 3)$ **2b.** $(2y + 4)(2y + 4)$ **2c.** $(x - 2)(x^2 + 2x + 4)$
2d. $(2m + 3n)(m - 5n)$ **2e.** Find the product of $(x^2 + 2x - 4)$ and $(x^2 - 8x + 3)$.
 2f. Let $f(x) = x^2 + 2x - 4$ and $g(x) = 8x + 3$. Find $f(x)g(x)$.

Solutions | **2a.** $(x + 4)(x + 3)$

    $= x(x) + x(3) + 4(x) + 4(3)$ Apply the distributive property.

    $= x^2 + 3x + 4x + 12$ Simplify each product.

    $= x^2 + 7x + 12$ Combine like terms.

2b. $(2y + 4)(2y + 4)$

    $= 2y(2y) + 2y(4) + 4(2y) + 4(4)$ Apply the distributive property.

    $= 4y^2 + 8y + 8y + 16$ Simplify each product.

    $= 4y^2 + 16y + 16$ Combine like terms.

 Note: *Another way to represent* $(2y + 4)(2y + 4)$ *is* $(2y + 4)^2$. *Note the base is* $(2y + 4)$ *and it is repeated as a factor two times.*

2c. Apply the distributive property and combine like terms.

    $(x - 2)(x^2 + 2x + 4)$

    $= x(x^2) + x(2x) + x(4) - 2(x^2) - 2(2x) - 2(4)$

    $= x^3 + 2x^2 + 4x - 2x^2 - 4x - 8$

    $= x^3 + 0x^2 + 0x - 8$

    $= x^3 - 8$

2d. Apply the distributive property and combine like terms.

    $(2m + 3n)(m - 5n)$

    $= 2m(m) - 2m(5n) + 3n(m) - 3n(5n)$

    $= 2m^2 - 10mn + 3mn - 15n^2$

    $= 2m^2 - 7mn - 15n^2$

2e. $(x^2 + 2x - 4)(x^2 - 8x + 3)$

    $= x^2(x^2) + x^2(-8x) + x^2(3) + 2x(x^2) + 2x(-8x) + 2x(3) - 4(x^2)$

       $-4(-8x) - 4(3)$

    $= x^4 - 8x^3 + 3x^2 + 2x^3 - 16x^2 + 6x - 4x^2 + 32x - 12$

    $= x^4 - 6x^3 - 17x^2 + 38x - 12$

2f. Since $f(x)$ and $g(x)$ are polynomials, we can find their product as we have done previously.

$$\begin{aligned}
f(x)g(x) &= (x^2 + 2x - 4)(8x + 3) \\
&= x^2(8x) + x^2(3) + 2x(8x) & \text{Apply the distributive property.} \\
&\quad + 2x(3) - 4(8x) - 4(3) \\
&= 8x^3 + 3x^2 + 16x^2 + 6x - 32x - 12 & \text{Multiply resulting terms.} \\
&= 8x^3 + 19x^2 - 26x - 12 & \text{Combine like terms.}
\end{aligned}$$

 Student Check 2   Find each product.

a. $(x + 5)(x + 6)$ **b.** $(8y + 3)(8y + 3)$ **c.** $(a - 4)(a^2 + 4x + 16)$
d. $(4c - d)(6c + 7d)$ **e.** Find the product of $(y^2 - 5y + 7)$ and $(y^2 + 2y - 9)$.
f. Let $f(x) = (x^2 - 3x + 2)$ and $g(x) = (2x - 5)$. Find $f(x)g(x)$.

Note: *We can also multiply polynomials vertically.*

$$\begin{array}{r}
x^2 + 2x + 4 \\
\times \qquad x - 2 \\
\hline
-2x^2 - 4x - 8 \\
x^3 + 2x^2 + 4x \\
\hline
x^3 + 0x^2 + 0x - 8
\end{array}$$

Multiply -2 by each term in the first row.
Multiply x by each term in the first row.
Line up like terms and combine them.

Use Polynomials to Represent Real-Life Situations

Objective 3 ▶

Write expressions that represent real-life situations.

In Chapter 6, we will learn how to solve real-life problems that involve multiplying polynomials. Example 3 illustrates how the expressions involved in the equations arise.

Objective 3 Examples   **Use polynomial multiplication to solve each problem.**

3a. When an airline company charges $100 per seat, they can sell 120 tickets. For each $10 increase in price, they sell one less ticket. If x represents the number of increases in price, then $100 + 10x$ is the expression for the ticket price and $120 - x$ is the number of tickets sold. What simplified polynomial represents the amount of revenue the airline company earns for the trip?

3b. A homeowner plans to remodel his master bathroom. He wants to use an expensive tile to make a uniform border around the floor of the shower stall. If the shower stall measures 3 ft by 4 ft, what simplified polynomial represents the area of the shower floor that is not covered by the expensive tile?

Solutions   **3a.** Revenue is calculated by multiplying the price per ticket by the number of tickets sold. To find the revenue, we must multiply the price, $(100 + 10x)$, by the number of tickets sold, $(120 - x)$.

$$\begin{aligned}
\text{Revenue} &= (100 + 10x)(120 - x) \\
&= 100(120) + 100(-x) + 10x(120) + 10x(-x) \\
&= 12{,}000 - 100x + 1200x - 10x^2 \\
&= -10x^2 + 1100x + 12{,}000
\end{aligned}$$

3b. Begin by drawing a diagram to determine the dimensions of the floor not covered by the expensive tile.

Let x represent the width of the border with the expensive tile. Then the dimensions of the floor not covered by the expensive tile are $3 - 2x$ and $4 - 2x$.

The area of a rectangle is length times width. The expression that represents the area of the shower floor not covered by the expensive tile is

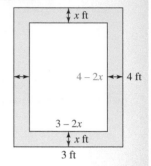

$$A = (3 - 2x)(4 - 2x)$$
$$= 3(4) + 3(-2x) - 2x(4) - 2x(-2x)$$
$$= 12 - 6x - 8x + 4x^2$$
$$= 4x^2 - 14x + 12$$

✔ **Student Check 3**   Use polynomial multiplication to solve each problem.

a. An apartment manager can rent all 150 units if he charges $400 per month rent. If he increases the price by $50, he rents two fewer apartments. If x represents the number of increases in price, then the rent for an apartment is given by $400 + 50x$ and the number of apartments rented is $150 - 2x$. What simplified polynomial represents the total rent collected each month?

b. A gardener has a rectangular garden that measures 5 ft by 7 ft. She wants to put a uniform border of gravel around each side of the garden. If x represents the width of the border, what simplified polynomial represents the area of the garden and the border?

Objective 4 ▶

Troubleshoot common errors.

Troubleshooting Common Errors

Some common errors for multiplying polynomials are shown next.

Objective 4 Examples   **A problem and an incorrect solution are given. Provide the correct solution as well as an explanation of the error.**

4a. Simplify $-5x(3x^2 + 4x - 1)$.

Incorrect Solution	Correct Solution and Explanation
$-5x(3x^2 + 4x - 1)$ $= -15x^3 - 20x - 5x$ $= -15x^3 - 25x$	When $-5x$ and $4x$ are multiplied, we get $-20x^2$. We must add the exponents of x together. Also, when $-5x$ is distributed to the last term, -1, we get $5x$. $-5x(3x^2 + 4x - 1) = -15x^3 - 20x^2 + 5x$

4b. Simplify $(x - 2)(x + 7)$.

Incorrect Solution	Correct Solution and Explanation
$(x - 2)(x + 7)$ $= x + 7x - 2x - 14$ $= 6x - 14$	The error was made in multiplying the first two terms. We should get $x \cdot x = x^2$. $(x - 2)(x + 7) = x^2 + 7x - 2x - 14$ $\qquad\qquad\qquad = x^2 + 5x - 14$

ANSWERS TO STUDENT CHECKS

Student Check 1   **a.** $2y^2 + 6y$ **b.** $16x^4 - 40x^3 + 8x^2$
  c. $-7b^8 + 14b^5$ **d.** $15a^3b^4 + 3a^2b^4 - 3ab^3$
Student Check 2   **a.** $x^2 + 11x + 30$ **b.** $64y^2 + 48y + 9$
  c. $a^3 - 64$ **d.** $24c^2 + 22cd - 7d^2$

  e. $y^4 - 3y^3 - 12y^2 + 59y - 63$
  f. $2x^3 - 11x^2 + 19x - 10$
Student Check 3   **a.** $-100x^2 + 6700x + 60{,}000$
  b. $4x^2 + 24x + 35$

SUMMARY OF KEY CONCEPTS

1. Polynomial multiplication is based on the distributive property and also uses the product of like bases rule for exponents, $x^m \cdot x^n = x^{m+n}$.

2. To multiply polynomials, distribute each term of the first polynomial to each term of the second polynomial.

Combine like terms to simplify the resulting expression. Polynomials can also be multiplied using a vertical format.

3. Polynomials can be used to represent real-life situations. These problems will be visited later in the chapter when we learn how to solve equations containing polynomials.

GRAPHING CALCULATOR SKILLS

The graphing calculator can be used to verify polynomial multiplication.

Example: Verify that $(x + 4)(x + 3) = x^2 + 7x + 12$.

Solution: Enter the product in Y_1 and the result in Y_2. If our multiplication is correct, the y-values will agree for each value of x.

```
Plot1 Plot2 Plot3
\Y1◼(X+4)(X+3)
\Y2◼X²+7X+12
\Y3=
\Y4=
\Y5=
\Y6=
\Y7=
```

X	Y1	Y2
-2	2	2
-1	6	6
0	12	12
1	20	20
2	30	30
3	42	42
4	56	56

X = -2

Since the Y_1 and Y_2 columns agree, the product is correct.

SECTION 5.4 / EXERCISE SET

Write About It!

Use complete sentences in your answer to each exercise.

1. Explain the procedure for multiplying a monomial and a polynomial.

2. Explain the procedure for multiplying a two polynomials.

Practice Makes Perfect!

Multiply the expressions using the distributive property.
(See Objective 1.)

3. $6x(2x^2)$ 4. $5x^2(-7x^3)$

5. $2x(x + 5)$ 6. $3x(4x - 6)$

7. $-x(x^2 + 4x - 1)$ 8. $-2x(3x^2 + 5x - 2)$

9. $4a^2b(2ab + 3)$ 10. $2x^2y(3xy - 1)$

11. $3a^2(a^3 + 4a^2 + 4)$ 12. $5a^2(4a^3 - a^2 + 2a)$

13. $-5b^3(b^5 - 7b^4)$ 14. $-6b^3(3b^6 - 2b^3)$

15. $-10r^2s(3r^2s^2 - rs + 5)$

16. $-11ab^2(9a^2b^2 - 2ab + 4)$

Find each product. *(See Objective 2.)*

17. $(x + 1)(x + 2)$ 18. $(x + 5)(x + 2)$

19. $(x - 3)(x + 5)$ 20. $(x - 2)(x + 6)$

21. $(3x + 5)(2x - 1)$ 22. $(4x - 3)(5x + 2)$

23. $(5a + 2)(5a + 2)$ 24. $(6y + 7)(6y + 7)$

25. $(8r - 3)(8r - 3)$ 26. $(11a - 1)(11a - 1)$

27. $(2x - y)(x + 3y)$ 28. $(3x + y)(x - 4y)$

29. $(3r + 2s)(r - s)$ 30. $(10a - 7b)(a - 2b)$

31. $(a - 1)(a^2 - 3a + 2)$ 32. $(a + 2)(a^2 + 4a - 6)$

33. $(6y - 1)(2y^2 + 3y + 1)$ 34. $(5y + 2)(3y^2 - 4y + 2)$

35. $(3x - y)(9x^2 + 3xy + y^2)$

36. $(2a + 3b)(4a^2 - 6ab + 9b^2)$

37. Find the product of $b^2 - 6b + 5$ and $b^2 + 4b - 2$.

38. Find the product of $3b^2 - 4b + 2$ and $b^2 - 7b + 7$.

39. Find the product of $5y^2 + y - 1$ and $2y^2 + 3y - 2$.

40. Find the product of $4y^2 + 5y - 3$ and $y^2 + 2y + 4$.

41. Let $f(x) = 8x^2 - 7x - 5$ and $g(x) = x - 3$. Find $f(x)g(x)$.

42. Let $f(x) = 3x^2 - 5x + 2$ and $g(x) = 2x + 7$. Find $f(x)g(x)$.

43. Let $f(x) = 10x^2 + 3x - 6$ and $g(x) = 3x - 2$. Find $f(x)g(x)$.

44. Let $f(x) = 6x^2 - 2x - 5$ and $g(x) = 3x + 10$. Find $f(x)g(x)$.

Use polynomial multiplication to solve each problem. (See Objective 3.)

45. When a hotel charges $140 per room, 80 rooms are rented. For each $5 increase in price, they rent one less room. If x represents the number of price increases,  then $140 + 5x$ represents the room charge and $80 - x$ represents the number of rooms rented. What simplified polynomial represents the revenue the hotel earns?

46. When a car rental company charges a daily rate of $25 for a compact car, 40 compact cars are rented. For each $3 increase in the daily rate, two less cars are rented. If x represents the number of price increases, then $25 + 3x$ represents the daily rate and $(40 - 2x)$ represents the number of cars rented. What simplified polynomial represents the revenue the car rental company earns?

47. A rectangular garden measures 12 ft by 15 ft. A gravel path of equal width is to be laid around the garden. If x is the width of the gravel path, what simplified polynomial represents the area of both the garden and the path?

48. A picture measures 10 in. by 12 in. If the width of the frame is x in., what simplified polynomial represents the area of the framed picture?

 Mix 'Em Up!

Find each product.

49. $(4 - x)(x + 4)$

50. $(1 - x)(2x + 3)$

51. $pq(3pq - 4)$

52. $ab(1 - 2ab)$

53. $(a + 3)(a^3 - 2a + 8)$

54. $(a + 8)(a^2 - 6a + 9)$

55. $-4n^2(3n^3 - 2n^2 + 1)$

56. $-6m^3(m^3 + 6m^2 + 5)$

57. $\left(\frac{2}{3}x - 2y\right)\left(\frac{1}{3}x + 5y\right)$

58. $\left(\frac{5}{4}a - 3b\right)\left(\frac{1}{4}a - b\right)$

59. $\left(5x + \frac{2}{3}\right)\left(2x - \frac{1}{3}\right)$

60. $\left(4x - \frac{2}{7}\right)\left(x + \frac{5}{7}\right)$

61. $(r + 2s)(r^2 - 2rs + 4s^2)$

62. $(5x - y)(25x^2 + 5xy + y^2)$

63. $(x^2 + 3x - 4)(2x^2 - 5x + 1)$

64. $(5x^2 - x + 6)(4x^2 - 3x + 2)$

Solve each problem. (See Objective 4.)

65. An apartment manager can rent all 100 units if he charges $600 per month rent. If he increases the price by $40, he rents two less apartments. If x represents the number of increases in price, then the rent for an apartment is given by $600 + 40x$ dollars and the number of apartments rented is $100 - 2x$.

   a. What simplified polynomial represents the total rent collected each month?

   b. If the manager decides to increase the rent to $720 per month, how much total rent can he collect each month?

66. An apartment manager can rent all 120 units if he charges $700 per month rent. If he increases the price by $60, he rents one less apartment. If x represents the number of increases in price, then the rent for an apartment is given by $700 + 60x$ and the number of apartments rented is $120 - x$.

   a. What simplified polynomial represents the total rent collected each month?

   b. If the manager decides to increase the rent to $820 per month, how much total rent can he collect each month?

Find the area of each figure.

67.

$2y + 1$ in.

$2y + 1$ in.

68.

$4x$ cm

$3x - 2$ cm

69.

$6a$ cm

$3a - 4$ cm

70.

$2b$ mm

$5b - 2$ mm

You Be the Teacher!

Correct each student's errors, if any.

71. Simplify $(x - 4)(x + 5)$.

   Corina's work: $(x - 4)(x + 5) = x^2 - 20$.

72. Simplify $(3x - 2)(x^2 + 5x - 1)$.

   Dean's work: $(3x - 2)(x^2 + 5x - 1)$

   $= 3x^3 + 15x - 3x - 2x^2 - 10x - 2$

   $= 3x^3 - 2x^2 + 2x - 2$

73. A rectangular swimming pool bordered by a concrete sidewalk of uniform width measures 50 m by 20 m. If x represents the width of the sidewalk, what simplified polynomial represents the area of the pool?

Christen's work: $(50 + 2x)(20 + 2x)$

74. A picture measures 8 in. by 12 in. A rectangular frame is placed around the picture. If the width of the frame is x in., what simplified polynomial represents the area of the picture combined with its frame.

Jonathan's work: $(8 + x)(12 + x)$

 **Calculate It!**

Multiply the polynomials and use a calculator to verify each product.

75. $(6 - x)(2x + 3)$ **76.** $(4 - x)(x + 3)$

77. $(x^2 - 5x)(2x^2 + x - 4)$

78. $(3a^2 - a + 2)(5a^2 + 2a - 1)$

SECTION 5.5 Special Products

The area of a square in which each side is $x + 5$ ft is

$$(x + 5)(x + 5) \quad \text{or} \quad (x + 5)^2 \text{ ft}^2$$

In this section, we will learn how to simplify this special product, the square of a binomial.

$x + 5$ ft

The Product of Binomials

Objective 1 ▶

Simplify products of two binomials.

In Section 5.4, we learned that polynomials can be multiplied using the distributive property. All of the problems in this section can be multiplied in the same manner. However, there are some products that repeatedly come up in algebra and it is helpful if we develop patterns to simplify these products. The first of these special products is the product of two binomials. Recall the following product.

$$(x + 3)(x + 4) = x(x) + x(4) + 3(x) + 3(4)$$
$$= x^2 + 4x + 3x + 12$$
$$= x^2 + 7x + 12$$

This product was formed by distributing x to $(x + 4)$ and then 3 to $(x + 4)$. To ensure that we get all of the terms in the product, we can remember the word FOIL.

$$\textbf{FOIL} = \textbf{First} + \textbf{Outer} + \textbf{Inner} + \textbf{Last}$$

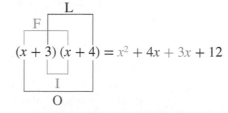

$$(x + 3)(x + 4) = x^2 + 4x + 3x + 12$$

Procedure: Multiplying Two Binomials Using the FOIL Method

Step 1: Find the product of the two first (F) terms of the binomials.
Step 2: Find the product of the two outer (O) terms of the binomials.
Step 3: Find the product of the two inner (I) terms of the binomials.
Step 4: Find the product of the two last (L) terms of the binomials.
Step 5: Add the products and combine like terms.

 Note: *It is very important to understand that this method applies only to the product of two binomials. This method does not extend to other types of products. Also note that this method is exactly what was covered in Section 5.4; it is simply that we are giving the process a name, the FOIL method.*

Objective 1 Examples / **Find the product of the binomials using the FOIL method.**

1a. $(y - 5)(y - 7)$ **1b.** $(2x - 3)(5x + 2)$ **1c.** $(a^2 + 9)(a^2 - 4)$

Solutions   **1a.**

$$\begin{array}{cccc} & F & O & I \quad\;\; L \\ (y - 5)(y - 7) = (y)(y) & + (y)(-7) & - 5(y) & - 5(-7) \end{array}$$
$$= y^2 - 7y - 5y + 35$$
$$= y^2 - 12y + 35$$

1b.

$$\begin{array}{cccc} & F & O & I \quad\;\; L \\ (2x - 3)(5x + 2) = (2x)(5x) & + (2x)(2) & - 3(5x) & - 3(2) \end{array}$$
$$= 10x^2 + 4x - 15x - 6$$
$$= 10x^2 - 11x - 6$$

1c.

$$\begin{array}{cccc} & F & O & I \quad\;\; L \\ (a^2 + 9)(a^2 - 4) = (a^2)(a^2) & + (a^2)(-4) & + 9(a^2) & + 9(-4) \end{array}$$
$$= a^4 - 4a^2 + 9a^2 - 36$$
$$= a^4 + 5a^2 - 36$$

✔ **Student Check 1**   Find the product of the binomials using the FOIL method.
  a. $(x + 6)(x + 2)$ **b.** $(4y - 7)(3y + 8)$ **c.** $(a^2 - 2)(a^2 + 10)$

The Square of a Binomial

Objective 2 ▶

Simplify the square of a binomial.

We now examine how to simplify a product that involves the same binomial base, like $(x + 4)^2$, by applying the definition of the exponent. The exponent of 2 tells us to multiply the base together two times.

$$(x + 4)^2 = (x + 4)(x + 4)$$

This gives us the product of two binomials, so we can apply the FOIL method in order to multiply.

$$\begin{array}{cccc} & F & O \quad\;\; I & L \\ (x + 4)^2 = (x + 4)(x + 4) = (x)(x) & + 4(x) & + 4(x) & + 4(4) \end{array}$$
$$= x^2 + 4x + 4x + 16$$
$$= x^2 + 8x + 16$$

So, $(x + 4)^2 = x^2 + 8x + 16$.

The FOIL method or the distributive property can always be used to square a binomial, but there is also a pattern that is worth noting. From the example, we can make the following observations.

1. The *square of a binomial* is a trinomial. $(x + 4)^2 = x^2 + 8x + 16$

2. The first term of the trinomial is the square of the first term of the binomial.
   x^2 is the first term of the trinomial and is the first term of the binomial squared,
   $(x)^2 = x^2$

3. The middle term of the trinomial is twice the product of the terms of the binomial.
   $8x$ is the middle term and is twice the product of the terms in the binomial,
   $2(x)(4) = 8x$

4. The last term of the trinomial is the square of the last term of the binomial.
   16 is the last term and is the second term of the binomial squared,
   $(4)^2 = 16$

In mathematical notation, we write this relationship as follows.

> **Property: The Square of a Binomial**
>
> For a and b real numbers,
> $$(a + b)^2 = a^2 + 2ab + b^2$$
> $$(a - b)^2 = a^2 - 2ab + b^2$$

We can verify this rule using geometry by creating a square whose length is $a + b$. The area of the large square is

$$A = (a + b)(a + b) = (a + b)^2$$

The area of the large square is also equal to the sum of the area of the four rectangles; that is,

$$A = a^2 + ab + ab + b^2 = a^2 + 2ab + b^2$$

So, we conclude that

$$(a + b)^2 = a^2 + 2ab + b^2$$

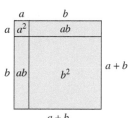

The trinomials that result from squaring a binomial are called **perfect square trinomials**.

> **Procedure: Squaring a Binomial**
>
> **Step 1:** Square the first term.
> **Step 2:** Find twice the product of the terms in the binomial.
> **Step 3:** Square the last term.
> **Step 4:** Add the products found in steps 1 through 3.
> **Step 5:** Verify the result using the FOIL method.

Objective 2 Examples Find each product.

2a. $(x + 6)^2$ **2b.** $(3a - 7b)^2$ **2c.** $(b^3 + 5)^2$

Solutions **2a.** $(x + 6)^2 = (x)^2 + 2(x)(6) + (6)^2$ Apply the squaring property.
$\qquad\qquad\quad = x^2 + 12x + 36$ Simplify.

2b. $(3a - 7b)^2 = (3a)^2 - 2(3a)(7b) + (7b)^2$ Apply the squaring property.
$\qquad\qquad\quad\;\; = 9a^2 - 42ab + 49b^2$ Simplify.

2c. $(b^3 + 5)^2 = (b^3)^2 + 2(b^3)(5) + (5)^2$ Apply the squaring property.
$\qquad\qquad\quad = b^6 + 10b^3 + 25$ Simplify.

✓ Student Check 2 Find each product.

    a. $(y - 8)^2$ **b.** $(4x + 3y)^2$ **c.** $(a^4 - 3)^2$

The Product of Conjugates

Objective 3 ▶

Simplify the product of conjugates.

Conjugates are binomials with the same terms connected by opposite signs. For example,

$$x + 3 \text{ and } x - 3 \text{ are conjugates.}$$
$$4b + 5 \text{ and } 4b - 5 \text{ are conjugates.}$$

To find the product of conjugates, we can apply the FOIL method or the distributive property.

$$(x + 3)(x - 3) = x^2 - 3x + 3x - 9 = x^2 - 9$$

$$(4b + 5)(4b - 5) = 16b^2 - 20b + 20b - 25 = 16b^2 - 25$$

When we multiply conjugates, note that the outer and inner products are opposites of one another and, therefore, add to zero. The product of the first terms minus the product of the last terms remains. Since the first and last terms are the same in conjugates, this is equivalent to the *difference of the squares* of the terms. That is,

$$\begin{array}{ccccccc} & & \text{Square of} & & \text{Square of} & & \\ & & \text{the first} & - & \text{the last} & & \\ & & \text{term} & & \text{term} & & \\ (x + 3)(x - 3) & = & (x)^2 & - & (3)^2 & = & x^2 - 9 \\ (4b + 5)(4b - 5) & = & (4b)^2 & - & (5)^2 & = & 16b^2 - 25 \end{array}$$

Property: The Product of Conjugates

For a and b real numbers,

$$(a + b)(a - b) = a^2 - b^2$$

Procedure: Finding the Product of Conjugates

Step 1: Find the square of the first term.
Step 2: Find the square of the last term.
Step 3: Subtract the results from step 1 and step 2.
Step 4: Verify the result using the FOIL method.

 Note: *The product of conjugates always results in the difference of two squares.*

Objective 3 Examples   Find each product.

3a. $(x + 8)(x - 8)$ **3b.** $(6a - 7b)(6a + 7b)$ **3c.** $(a^2 - 5)(a^2 + 5)$

Solutions   **3a.** $(x + 8)(x - 8) = (x)^2 - (8)^2$ Apply the product of conjugates property.
$= x^2 - 64$ Simplify.

3b. $(6a - 7b)(6a + 7b) = (6a)^2 - (7b)^2$ Apply the product of conjugates property.
$= 36a^2 - 49b^2$ Simplify.

3c. $(a^2 - 5)(a^2 + 5) = (a^2)^2 - (5)^2$ Apply the product of conjugates property.
$= a^4 - 25$ Simplify.

✓ **Student Check 3**   Find each product.
a. $(x + 1)(x - 1)$ **b.** $(9c - 2d)(9c + 2d)$ **c.** $(a^3 - 6)(a^3 + 6)$

Higher Powers of Binomials

We now examine how to raise a binomial to a larger exponent, such as 3 or 4. We begin by applying the definition of the exponent. That is, we repeat the base as a factor as indicated by the exponent.

$$(x + 2)^3 = (x + 2)(x + 2)(x + 2)$$ Repeat the base three times.

$$= (x + 2)(x + 2)^2$$ Rewrite as a product of one factor and the square of the factor.

$$= (x + 2)(x^2 + 4x + 4)$$ Apply the rule for squaring a binomial.

$$= x^3 + 4x^2 + 4x + 2x^2 + 8x + 8$$ Apply the distributive property.

$$= x^3 + 6x^2 + 12x + 8$$ Combine like terms.

$$(x + 2)^4 = (x + 2)(x + 2)(x + 2)(x + 2)$$ Repeat the base four times.

$$= (x + 2)^2(x + 2)^2$$ Rewrite as two factors which are both squares of $(x + 2)$.

$$= (x^2 + 4x + 4)(x^2 + 4x + 4)$$ Apply the rule for squaring a binomial.

$$= x^4 + 4x^3 + 4x^2 + 4x^3 + 16x^2$$
$$+ 16x + 4x^2 + 16x + 16$$ Apply the distributive property.

$$= x^4 + 8x^3 + 24x^2 + 32x + 16$$ Combine like terms.

There are patterns that we could memorize to compute higher powers of binomials, but these formulas are beyond the scope of this course. For now, we can apply the rules for polynomial multiplication and special products.

Procedure: Finding Higher Powers of a Binomial

Step 1: Rewrite as a product of the base repeated the number of times indicated by the exponent.

Step 2: Rewrite pairs of factors as the square of the binomial and use the rule for square binomials to simplify these expressions.

Step 3: Distribute the remaining polynomials to complete the multiplication.

Step 4: Combine like terms and write the answer in standard form.

Simplify the expression $(x - 5)^4$.

Solution

$$(x - 5)^4 = (x - 5)(x - 5)(x - 5)(x - 5)$$ Repeat the base four times.

$$= (x - 5)^2(x - 5)^2$$ Rewrite as two factors of $(x - 5)^2$.

$$= (x^2 - 10x + 25)(x^2 - 10x + 25)$$ Square each binomial.

$$= x^4 - 10x^3 + 25x^2 - 10x^3 + 100x^2$$
$$- 250x + 25x^2 - 250x + 625$$ Apply the distributive property.

$$= x^4 - 20x^3 + 150x^2 - 500x + 625$$ Combine like terms.

 **Student Check 4**   Simplify the expression $(y + 8)^3$.

Troubleshooting Common Errors

Some common errors related to special products are shown next.

Objective 5 Examples   A problem and an incorrect solution are given. Provide the correct solution and an explanation of the error.

5a. Simplify $(x + 9)^2$.

Incorrect Solution	Correct Solution and Explanation
$(x + 9)^2 = x^2 + 81$	Since the base is not a product, we must use the definition of the exponent or use the formula for squaring a binomial. We cannot just square each term. $$(x + 9)^2 = x^2 + 18x + 81$$

5b. Simplify $(6x + 5y)(6x - 5y)$.

Incorrect Solution	Correct Solution and Explanation
$(6x + 5y)(6x - 5y)$ $= 36x^2 - 60xy - 25y^2$	The error was made in combining the middle terms. Since the outer and inner products are opposites of one another, their sum is zero. $$(6x + 5y)(6x - 5y) = 36x^2 - 25y^2$$

ANSWERS TO STUDENT CHECKS

Student Check 1   a. $x^2 + 8x + 12$ **b.** $12y^2 + 11y - 56$
   c. $a^4 + 8a^2 - 20$

Student Check 2   a. $y^2 - 16y + 64$ **b.** $16x^2 + 24xy + 9y^2$
   c. $a^8 - 6a^4 + 9$

Student Check 3   a. $x^2 - 1$ **b.** $81c^2 - 4d^2$ **c.** $a^6 - 36$
Student Check 4 $y^3 + 24y^2 + 192y + 512$

SUMMARY OF KEY CONCEPTS

1. All of the products discussed in this section can be found by applying the distributive property. It is helpful to be able to apply the patterns shown in this section so that we can develop speed to simplify expressions.

2. Two binomials can be multiplied using the FOIL method. FOIL stands for *first, outer, inner*, and *last*.

3. The square of a binomial is *always* a trinomial. The middle term results from twice the product of the terms

in the binomial. The first and last terms are the squares of the first and last terms of the binomial.

4. Conjugates are expressions of the form $a + b$ and $a - b$. Their product is the difference of two squares.

5. Higher powers of binomials are found by repeating the base and then using the rules developed within this section and the last section to simplify them.

SECTION 5.5   EXERCISE SET

 Write About It!

Use complete sentences in your answer to each exercise.

1. Explain the FOIL method.

2. Is FOIL different than the distributive property? Explain.

3. Explain why a trinomial results when a binomial is squared.

4. Give examples of conjugates and explain what happens when they are multiplied.

 Practice Makes Perfect!

Find the product of the binomials using the FOIL method. Write each answer in standard form. (*See Objective 1.*)

5. $(x - 6)(x - 1)$ 6. $(x - 2)(x - 4)$

7. $(x + 5)(x - 7)$ 8. $(x + 6)(x - 9)$

9. $(3x - 2y)(5x + y)$ 10. $(2x + 3y)(4x - y)$

11. $(a^2 + 1)(a^2 - 4)$ 12. $(a^2 + 9)(a^2 - 3)$

13. $(4y^2 + 5)(3y^2 - 2)$ 14. $(2y^2 - 1)(3y^2 + 4)$

15. $(2r^2 - 6s)(6r + 8s)$ 16. $(3a^2 + 7b)(a - 3b)$

17. $\left(2x - \dfrac{3}{5}\right)\left(3x + \dfrac{2}{5}\right)$ **18.** $\left(5x - \dfrac{3}{4}\right)\left(x - \dfrac{1}{2}\right)$

19. $(1.5x + 0.6y)(2.4x + 1.8y)$

20. $(0.8a - 1.7b)(0.6a + 2.8b)$

21. $(1 - 3x)(4x + 5)$ **22.** $(6x - 5)(2 - 3x)$

Use a special product to multiply. Write each answer in standard form. (*See Objective 2.*)

23. $(x + 4)^2$ **24.** $(x + 9)^2$

25. $(2y - 1)^2$ **26.** $(3y - 2)^2$

27. $(a^2 - 2)^2$ **28.** $(a^2 - 3)^2$

29. $(4c + 5d)^2$ **30.** $(5a - 6b)^2$

31. $\left(3a - \dfrac{2}{5}\right)^2$ **32.** $\left(2b - \dfrac{1}{3}\right)^2$

33. $(1.5p - 0.6)^2$ **34.** $(0.9p + 2.5)^2$

35. $(3x^2 - 7y)^2$ **36.** $(5a^2 + 8b)^2$

Use a special product to multiply. Write each answer in standard form. (*See Objective 3.*)

37. $(x - 7)(x + 7)$ **38.** $(x + 4)(x - 4)$

39. $(x + 2y)(x - 2y)$ **40.** $(a + 3b)(a - 3b)$

41. $(2y + 4)(2y - 4)$ **42.** $(3y + 5)(3y - 5)$

43. $(7r - 3s)(7r + 3s)$ **44.** $(9p + 2q)(9p - 2q)$

45. $(1.3x - 2.5)(1.3x + 2.5)$ **46.** $(0.8y + 1.5)(0.8y - 1.5)$

47. $(a^2 + 1)(a^2 - 1)$ **48.** $(a^2 + 4b^2)(a^2 - 4b^2)$

49. $(c^3 - 8d^3)(c^3 + 8d^3)$ **50.** $(b^3 - 9)(b^3 + 9)$

51. $\left(x^2 - \dfrac{1}{7}\right)\left(x^2 + \dfrac{1}{7}\right)$ **52.** $\left(y^2 - \dfrac{4}{5}\right)\left(y^2 + \dfrac{4}{5}\right)$

53. $\left(\dfrac{3}{4}c^3 + d\right)\left(\dfrac{3}{4}c^3 - d\right)$ **54.** $\left(\dfrac{2}{9}a^3 + b\right)\left(\dfrac{2}{9}a^3 - b\right)$

Perform the indicated operation. Write each answer in standard form. (*See Objective 4.*)

55. $(x + 1)^3$ **56.** $(x + 4)^3$

57. $(y - 2)^3$ **58.** $(y - 3)^3$

59. $(2y - 3)^3$ **60.** $(3y + 1)^3$

61. $\left(b + \dfrac{1}{2}\right)^3$ **62.** $\left(c - \dfrac{1}{5}\right)^3$

63. $(a + 1)^4$ **64.** $(a + 2)^4$

 Mix 'Em Up!

Perform the indicated operation. Write each answer in standard form.

65. $(x - 12y)(x + 12y)$ **66.** $(a + 13b)(a - 13b)$

67. $(3x + 9)(2x - 5)$ **68.** $(8x + 3)(9x - 4)$

69. $(3a + 2)^2$ **70.** $(5b - 7)^2$

71. $(y + 11)^3$ **72.** $(a + 6)^3$

73. $\left(a^2 - \dfrac{3}{7}\right)\left(a^2 + \dfrac{3}{7}\right)$ **74.** $\left(b - \dfrac{4}{3}\right)\left(b + \dfrac{4}{3}\right)$

75. $\left(\dfrac{1}{2}c^3 + 5d\right)\left(\dfrac{1}{2}c^3 - 5d\right)$ **76.** $\left(2a - \dfrac{1}{5}b^3\right)\left(2a + \dfrac{1}{5}b^3\right)$

77. $(2p - 1)^3$ **78.** $(2p - 1)^4$

79. $(10x - 7y)(10x + 7y)$ **80.** $(15a - 2b)(15a + 2b)$

81. $\left(\dfrac{1}{3}x - 1\right)^2$ **82.** $\left(\dfrac{1}{5}x + 2\right)^2$

83. $(1.6x - 4.5)(0.2x + 1.2)$ **84.** $(0.6x - 3.1)(0.7x - 1.9)$

85. $(x^2 - 3y)(x^2 + 3y)$ **86.** $(x^3 + 8y^2)(x^3 - 8y^2)$

 You Be the Teacher!

Correct each student's errors, if any.

87. Simplify $(3x - 8)^2$.

Mark's work: $(3x - 8)^2 = 9x^2 + 64$

88. Simplify $(5y + 2)^3$.

Joan's work: $(5y + 2)^3 = 125y^3 + 8$

89. Simplify $(4x + 3)(4x - 3)$.

Antoine's work: $(4x + 3)(4x - 3)$
$= 16x^2 - 12x + 12x - 9$
$= 16x^2 - 24x - 9$

Calculate It!

Use a graphing calculator to determine if each product is correct. If it is not correct, state the correct product.

90. $(5x - 7)^2 = 25x^2 - 70x + 49$

91. $(11x - 8)^2 = 121x^2 + 64$

92. $(4x - 5)^3 = 64x^3 - 125$

93. $(3x + 2)^3 = 27x^3 + 8$

94. $(3x + 7)(3x - 7) = 9x^2 + 49$

95. $(2x + 5)(2x - 5) = 4x^2 - 25$

Think About It!

96. Show how to derive the formula for $(a + b)^2$ using the FOIL method.

97. Show how to derive the formula for $(a - b)^2$ using the FOIL method.

98. Write an example of a difference of two squares. What product of conjugate produces it?

▶ **SECTION 5.6**	**Division of Polynomials**

The area of this pool table is $4x^3y + 2x^2y$ ft^2 and its length is $2x^2y$ ft. How can we express the width of the table?

In this section, we will learn how to divide polynomials, which will enable us to answer this question.

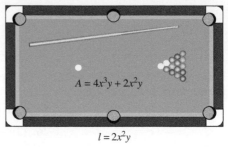

$A = 4x^3y + 2x^2y$

$l = 2x^2y$

Dividing Polynomials by Monomials

Objective 1 ▶

Divide a polynomial by a monomial.

We conclude this chapter with a discussion of the last operation of polynomials, division. To divide a polynomial by a monomial, we use a basic property of fractions. The following example reviews this property.

$$\frac{14}{7} + \frac{21}{7} = \frac{14 + 21}{7} = \frac{35}{7} = 5$$

This property reminds us that to add fractions, they must have the same denominator. To use it for a property relating to division, we need to view it in reverse order.

$$\frac{14 + 21}{7} = \frac{14}{7} + \frac{21}{7} = 2 + 3 = 5$$

In this order, we see that if a sum is divided by a single term, each term in the numerator must be divided by the denominator.

This leads to the following property.

> **Property: Property of Division for Fractions**
>
> If A, B, and C are monomials with $C \neq 0$, then
>
> $$\frac{A + B}{C} = \frac{A}{C} + \frac{B}{C}$$

Throughout this section, we will assume that all denominators are nonzero.

> **Procedure: Dividing a Polynomial by a Monomial**
>
> **Step 1:** Rewrite the problem so that each term in the numerator is divided by the monomial in the denominator.
>
> **Step 2:** Simplify each expression using properties of exponents. Recall $\frac{x^m}{x^n} = x^{m-n}$.
>
> **Step 3:** Check by multiplication. If $\frac{a}{b} = c$, then $bc = a$. That is, if we multiply the quotient by the divisor, we should obtain the dividend.

Objective 1 Examples	**Find each quotient and check by multiplying.**

1a. Divide $5x^2 - 10x$ by $5x$.

1b. $\dfrac{12y^4 - 18y^3 + 6y^2}{-6y^2}$

1c. Divide $9a^3 + 3a^2 - 6a$ by $9a^2$.

Solutions   **1a.** $\dfrac{5x^2 - 10x}{5x} = \dfrac{5x^2}{5x} - \dfrac{10x}{5x}$ *Apply the property of division for fractions.*

$= x - 2$ *Simplify each quotient.*

Check: $5x(x - 2) = 5x(x) - 5x(2)$
$$= 5x^2 - 10x$$

Since we obtained the dividend, our work is correct.

1b. Rewrite the problem by applying the property of division for fractions. Be careful with signs since the divisor is negative.

$$\frac{12y^4 - 18y^3 + 6y^2}{-6y^2} = \frac{12y^4}{-6y^2} - \frac{18y^3}{-6y^2} + \frac{6y^2}{-6y^2}$$
$$= -2y^2 + 3y - 1$$

Check: $-6y^2(-2y^2 + 3y - 1) = -6y^2(-2y^2) - 6y^2(3y) - 6y^2(-1)$
$$= 12y^4 - 18y^3 + 6y^2$$

Since we obtained the dividend, our quotient is correct.

1c. $\dfrac{9a^3 + 3a^2 - 6a}{9a^2} = \dfrac{9a^3}{9a^2} + \dfrac{3a^2}{9a^2} - \dfrac{6a}{9a^2}$ Apply the property of division for fractions.

$$= a + \frac{1}{3} - \frac{2}{3}a^{-1}$$ Simplify each quotient.

$$= a + \frac{1}{3} - \frac{2}{3} \cdot \frac{1}{a}$$ Apply the definition of a negative exponent.

$$= a + \frac{1}{3} - \frac{2}{3a}$$ Simplify.

Check: $9a^2\left(a + \dfrac{1}{3} - \dfrac{2}{3a}\right) = 9a^2(a) + 9a^2\left(\dfrac{1}{3}\right) - 9a^2\left(\dfrac{2}{3a}\right)$
$$= 9a^3 + 3a^2 - 6a$$

Since we obtained the dividend, our work is correct.

✔ **Student Check 1**   Find each quotient and check by multiplying.

a. Divide $8y^2 - 24y$ by $8y$.

b. $\dfrac{14x^4 - 21x^3 + 7x^2}{-7x^2}$

c. Divide $6b^5 + 15b^3 - 24b^2$ by $3b^4$.

Objective 2 ▶

Divide a polynomial by a binomial.

Dividing Polynomials by Binomials

When the denominator of a quotient is not a monomial, we cannot apply the division property for fractions or the rules of exponents directly. We must use a long division process to perform the division. The long division process is exactly the process used to divide real numbers. Review the following example for the steps used in long division.

$$\begin{array}{r} 21 \\ 15\overline{)325} \\ -30 \\ \hline 25 \\ -15 \\ \hline 10 \end{array}$$

Divide: $\dfrac{32}{15} = 2$.

Multiply: $2(15) = 30$.

Subtract and bring down the next digit.

Divide: $\dfrac{25}{15} = 1$.

Multiply: $1(15) = 15$.

Subtract.

The divisor, 15, doesn't divide into 10, so we are done. We say that $\frac{325}{15} = 21 + \frac{10}{15}$.

We can check our answer: Quotient • Divisor + Remainder = Dividend. That is,

$$21(15) + 10 = 325$$

$$315 + 10 = 325 \quad \text{True}$$

This brings us to the four basic steps of long division:

divide, multiply, subtract, and *bring down*

The same basic steps apply to **long division of polynomials**. When applying long division to polynomials, the dividend (numerator) must be written in standard form. If a term is missing, we write it with a coefficient of 0.

The following example illustrates how long division applies to polynomials.

$$
\begin{array}{r}
2x + 3 \\
x + 4 \overline{\smash{)}\, 2x^2 + 11x + 12} \\
\underline{-(2x^2 + 8x)} \\
3x + 12 \\
\underline{-(3x + 12)} \\
0
\end{array}
$$

Divide: $\frac{2x^2}{x} = 2x$.

Multiply: $2x(x + 4) = 2x^2 + 8x$.

Subtract and bring down the next term.

Divide: $\frac{3x}{x} = 3$.

Multiply: $3(x + 4) = 3x + 12$.

Subtract and bring down.

Procedure: Dividing a Polynomial by a Binomial Using Long Division

Step 1: Write the numerator (dividend) in standard form writing any missing terms with a coefficient of zero.

Step 2: Divide the first term of the binomial in the denominator (divisor) into the first term of the polynomial in the numerator.

Step 3: Multiply this result by the binomial and line up like terms.

Step 4: Subtract the polynomials.

Step 5: Repeat this process until the degree of the polynomial that results from subtraction is less than the degree of the binomial.

Step 6: Check by multiplying: Quotient × Divisor + Remainder = Dividend.

Objective 2 Examples Find each quotient and check by multiplying.

2a. Divide $x^2 + 2x - 35$ by $x - 5$.

2b. $\dfrac{x^3 + 8}{x + 2}$

2c. $\dfrac{7y - 6 + 3y^2}{3y + 4}$

Solutions **2a.**

$$
\begin{array}{r}
x + 7 \\
x - 5 \overline{\smash{)}\, x^2 + 2x - 35} \\
\underline{-(x^2 - 5x)} \\
7x - 35 \\
\underline{-(7x - 35)} \\
0
\end{array}
$$

Divide x^2 by x, which is x.

Multiply x by $x - 5$, which is $x^2 - 5x$.

Subtract and bring down -35.

Divide $7x$ by x to get 7.

Multiply 7 by $x - 5$, which is $7x - 35$.

Subtract.

Check: $(x + 7)(x - 5) = x^2 + 2x - 35$

Because the quotient times the divisor is the dividend, our result is correct.

So, $\dfrac{x^2 + 2x - 35}{x - 5} = x + 7$.

2b. Write the dividend as $x^3 + 0x^2 + 0x + 8$, so that like terms can be aligned.

$$
\begin{array}{r}
x^2 - 2x + 4 \\
x + 2 \overline{)\, x^3 + 0x^2 + 0x + 8} \\
\underline{-(x^3 + 2x^2)} \\
-2x^2 + 0x \\
\underline{-(-2x^2 - 4x)} \\
4x + 8 \\
\underline{-(4x + 8)} \\
0
\end{array}
$$

Divide x^3 by x, which is x^2.
Multiply x^2 by $x + 2$, which is $x^3 + 2x^2$.
Subtract and bring down $0x$.
Divide $-2x^2$ by x to get $-2x$.
Multiply $-2x$ by $x + 2$, which is $-2x^2 - 4x$.
Subtract and bring down 8.
Divide $4x$ by x, which is 4.
Multiply 4 by $x + 2$, which is $4x + 8$.
Subtract.

Check:

$$(x^2 - 2x + 4)(x + 2) = x^3 + 2x^2 - 2x^2 - 4x + 4x + 8$$
$$= x^3 + 8$$

Because the quotient times the divisor is the dividend, our result is correct.

So, $\dfrac{x^3 + 8}{x + 2} = x^2 - 2x + 4$.

2c. Write the numerator in standard form, $3y^2 + 7y - 6$.

$$
\begin{array}{r}
y + 1 \\
3y + 4 \overline{)\, 3y^2 + 7y - 6} \\
\underline{-(3y^2 + 4y)} \\
3y - 6 \\
\underline{-(3y + 4)} \\
-10
\end{array}
$$

Divide $3y^2$ by $3y$, which is y.
Multiply y by $3y + 4$, which is $3y^2 + 4y$.
Subtract and bring down -6.
Divide $3y$ by $3y$ to get 1.
Multiply 1 by $3y + 4$, which is $3y + 4$.
Subtract and get -10.

Check:

$$(y + 1)(3y + 4) - 10 = 3y^2 + 4y + 3y + 4 - 10$$
$$= 3y^2 + 7y - 6$$

Because the quotient times the divisor plus the remainder is the dividend, our result is correct. So, $\dfrac{3y^2 + 7y - 6}{3y + 4} = y + 1 - \dfrac{10}{3y + 4}$.

✓ **Student Check 2** Find each quotient and check by multiplying.

a. Divide $x^2 - x - 72$ by $x - 9$.

b. $\dfrac{a^3 + 125}{a + 5}$

c. $\dfrac{4y^2 - 4 + 10y}{4y - 2}$

Applications

Objective 3 ▶
Solve application problems.

The application problems we will solve involve finding an expression that represents the length of a side of a geometric figure.

Procedure: Solving Applications of Division of Polynomials

Step 1: Write the formula for the area, volume, or perimeter given.
Step 2: Write a quotient that represents the expression we need to find.
Step 3: Divide using the methods from Objective 1 or 2.

Objective 3 Example The area of this pool table is $4x^3y + 2x^2y$ ft^2 and its length is $2x^2y$ ft. Write an expression for the width of the pool table.

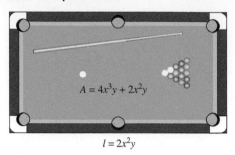

$A = 4x^3y + 2x^2y$

$l = 2x^2y$

Solution We know the area of a rectangle is $A = lw$. So, to solve for w, we must divide the area by the length of the rectangle. That is, $w = \dfrac{A}{l}$.

$$w = \frac{A}{l}$$

$$w = \frac{4x^3y + 2x^2y}{2x^2y}$$

$$w = \frac{4x^3y}{2x^2y} + \frac{2x^2y}{2x^2y}$$

$$w = 2x + 1$$

Check: $(2x + 1)(2x^2y) = 4x^3y + 2x^2y$

So, the width can be represented by $2x + 1$ ft.

✓ Student Check 3 The area of a parallelogram is $9a^3b^3 - 12a^2b^2$ square units. Its height is $3a^2b$ units. Find an expression for the base of the parallelogram.

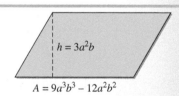

$h = 3a^2b$

$A = 9a^3b^3 - 12a^2b^2$

Objective 4 ▶

Troubleshoot common errors.

Troubleshooting Common Errors

Some common errors related to dividing polynomials are shown next.

Objective 4 Examples A problem and an incorrect solution are given. Provide the correct solution and an explanation of the error.

4a. Simplify $\dfrac{6x^2 + 14x}{2x}$.

Incorrect Solution	Correct Solution and Explanation
$$\frac{6x^3 + 14x}{2x} = \frac{6x^3}{2x} + 14x$$ $$= 3x^2 + 14x$$	We must divide $2x$ into each term in the numerator of the fraction. $$\frac{6x^3 + 14x}{2x} = \frac{6x^3}{2x} + \frac{14x}{2x}$$ $$= 3x^2 + 7$$

4b. Simplify $\dfrac{x^2 - 5x + 6}{x - 2}$.

Incorrect Solution	Correct Solution and Explanation
$$\begin{array}{r} x - 7 \\ x-2\overline{)x^2 - 5x + 6} \\ \underline{x^2 - 2x} \\ -7x + 6 \\ \underline{-7x + 14} \\ 20 \end{array}$$ $$\dfrac{x^2 - 5x + 6}{x - 2} = x - 7 + \dfrac{20}{x - 2}$$	We must change the signs when subtracting. $$\begin{array}{r} x - 3 \\ x-2\overline{)x^2 - 5x + 6} \\ \underline{-(x^2 - 2x)} \\ -3x + 6 \\ \underline{-(-3x + 6)} \\ 0 \end{array}$$ So, $\dfrac{x^2 - 5x + 6}{x - 2} = x - 3$.

ANSWERS TO STUDENT CHECKS

Student Check 1   **a.** $y - 3$ **b.** $-2x^2 + 3x - 1$
   c. $2b + \dfrac{5}{b} - \dfrac{8}{b^2}$

Student Check 2   **a.** $x + 8$ **b.** $a^2 - 5a + 25$
   c. $y + 3 + \dfrac{2}{4y - 2}$
Student Check 3 $3ab^2 - 4b$

SUMMARY OF KEY CONCEPTS

1. To divide a polynomial by a monomial, divide the monomial into each term of the polynomial. Use rules of exponents to simplify each term. Check by multiplying.
2. Dividing by a binomial requires long division. Repeat the steps involved in long division of real numbers—*divide, multiply, subtract,* and *bring down.* Continue these steps until the remainder has a degree less than the degree

of the divisor. The quotient times the divisor plus the remainder should equal the dividend.
3. Division of polynomials can be used to find expressions that represent the length of a side of a geometric figure if an area or perimeter is given and one of the dimensions is known.

GRAPHING CALCULATOR SKILLS

A graphing calculator can be used to check quotients for problems that involve a single variable.

Example: Verify that $\dfrac{x^3 + 8}{x + 2} = x^2 - 2x + 4$.

Method 1: Input the original division problem and the answer into Y_1 and Y_2, respectively. Then compare the table of values. They should agree everywhere except the value of x that makes the denominator zero.

```
Plot1 Plot2 Plot3
\Y1 (X^3+8)/(X+2
)
\Y2 X2-2X+4
\Y3=
\Y4=
\Y5=
\Y6=
```

X	Y1	Y2
-3	19	19
-2	ERR:	12
-1	7	7
0	4	4
1	3	3
2	4	4
3	7	7
X=-3		

Method 2: Use the calculator to check the multiplication. The quotient times the divisor plus the remainder should equal the dividend. Input the quotient times the divisor plus the remainder in Y_1 and the dividend in Y_2. Compare the table of values. If they agree, then the answer is correct.

```
Plot1 Plot2 Plot3
\Y1 (X+2)(X2-2X+
4)
\Y2 X^3+8
\Y3=
\Y4=
\Y5=
\Y6=
```

X	Y1	Y2
-3	-19	-19
-2	0	0
-1	7	7
0	8	8
1	9	9
2	16	16
3	35	35
X=-3		

Since the tables agree for Y_1 and Y_2, the answer is correct.

SECTION 5.6 / EXERCISE SET

Write About It!

Use complete sentences in your answer to each exercise.

1. Explain how to divide a polynomial by a monomial.

2. How can you determine if you divided a polynomial by a monomial correctly?

3. Explain how to divide a polynomial by a binomial.

4. Explain how to check your answer if you divided a polynomial by a binomial and ended up with a remainder.

5. How do you set up your division process if the dividend has a missing term? For example, show how you would set up $\dfrac{x^3 - 7x + 2}{x - 2}$.

6. Explain how the degrees of the dividend, divisor, and remainder are related.

Practice Makes Perfect!

Find each quotient. Check each answer by multiplying. (*See Objective 1.*)

7. Divide $6x^2 - 12x$ by $6x$.

8. Divide $4x^2 - 16x$ by $2x$.

9. Divide $10x^3 - 2x^2 + 4x$ by $2x$.

10. Divide $7x^3 - 14x^2 + 21x$ by $7x$.

11. Divide $3a^4 - 12a^3 - 6a^2$ by $3a^2$.

12. Divide $8a^5 - 4a^4 + 16a^3$ by $4a^3$.

13. $\dfrac{12y^5 - 16y^3 + 8y^2}{-4y^2}$

14. $\dfrac{21y^6 - 18y^5 + 9y^2}{-3y^2}$

15. $\dfrac{20x^4 - 10x^3 - 5x}{5x}$

16. $\dfrac{45x^7 - 30x^6 + 60x^5}{15x^5}$

17. $\dfrac{18y^3 - 24y^2 + 12y}{-6y}$

18. $\dfrac{24y^6 - 8y^4 + 16y^3}{8y^3}$

19. Divide $14a^3 - 7a^2 + 28$ by $14a$.

20. Divide $12a^4 - 18a^3 - 3a^2$ by $6a^3$.

21. Divide $5x^4 - 15x^3 + 45x^2$ by $15x^2$.

22. Divide $10x^6 - 20x^3 - 15x$ by $10x$.

23. $\dfrac{3a^2 - 2a + 6}{3a^2}$

24. $\dfrac{5a^2 - 6a}{5a^2}$

25. $\dfrac{12x^3 - 24x^2}{12x^3}$

26. $\dfrac{4x^5 - 16x^4 + 8}{4x^5}$

27. $\dfrac{13y^7 - 26y^6 + 26y^5}{-26y^5}$

28. $\dfrac{11y^8 - 22y^7 + 11y^6}{-22y^6}$

29. $\dfrac{6y^3 - 48y^2 - 12}{24y^2}$

30. $\dfrac{7y^4 - 63y^3 + 14}{49y^3}$

Find each quotient. Check each answer by multiplying. (*See Objective 2.*)

31. Divide $x^2 - x - 12$ by $x + 3$.

32. Divide $x^2 - 3x - 10$ by $x + 2$.

33. Divide $x^2 - 16$ by $x - 4$.

34. Divide $x^2 - 25$ by $x + 5$.

35. $\dfrac{x^3 + 1}{x - 1}$

36. $\dfrac{x^3 + 64}{x + 4}$

37. $\dfrac{4a^2 - 8a}{a - 2}$

38. $\dfrac{6a^2 + 30a}{a + 5}$

39. $\dfrac{2x^2 + 5x - 3}{x + 3}$

40. $\dfrac{3x^2 - 7x - 20}{x - 4}$

41. Divide $x^2 - 6x - 5$ by $x + 1$.

42. Divide $x^2 + 3x - 4$ by $x + 2$.

43. Divide $3y^2 - 5y + 7$ by $y - 6$.

44. Divide $5y^2 - 2y + 3$ by $y - 1$.

45. $\dfrac{2x^2 - 5x + 7}{x + 7}$

46. $\dfrac{3x^2 + 18x - 115}{x + 10}$

47. $\dfrac{6y^3 - 5y^2 - 16y + 22}{2y - 3}$

48. $\dfrac{8y^3 - 18y^2 + 19y - 2}{4y + 1}$

Solve each problem. (*See Objective 3.*)

49. The area of a rectangle is $3x^2y - 6xy^2$ square units and the length of the rectangle is $3xy$ units. Find an expression for the width of the rectangle.

50. The area of a rectangle is $6xy^2 + 14x^2y$ square units and the length of the rectangle is $2xy$ units. Find an expression for the width of the rectangle.

51. The area of a parallelogram is $144x^2y + 96x^3y^3$ square units. If the height of the parallelogram is $12x^2y$ units, then find the length of its base.

52. The area of a parallelogram is $52x^2y^3 - 39xy^2$ square units. If the height of the parallelogram is $13xy^2$ units, then find the length of its base.

Mix 'Em Up!

Find each quotient.

53. Divide $6a^3 + 3a^2 - 18$ by $3a$.

54. Divide $4a^5 - 5a^4 + 6a^3$ by a^2.

55. Divide $3x^2 - 11x - 4$ by $x - 4$.

56. Divide $6x^2 - 3x - 30$ by $2x - 5$.

57. $\dfrac{6x^5 - 4x^4 + 2x^3}{-2x^3}$

58. $\dfrac{20x^4 - 28x^3 + 12x^2}{-4x^2}$

59. $\dfrac{7x^3 - x^2 + 14}{x + 2}$

60. $\dfrac{3x^3 - 20x - 8}{x - 3}$

61. $\dfrac{6x^3 - x^2 - 5x + 17}{2x + 3}$

62. $\dfrac{5x^4 + x^3 - 10x^2 - 2x + 1}{5x + 1}$

63. If the area of a parallelogram is $12xz^2 - 6x^2z$ square units and its height is $6xz$ units, find the length of its base.

64. If the area of a rectangle is $4x^3y^3 - 14x^2y$ square units and its length is $2x^2y$ units, find the width of the rectangle.

 You Be the Teacher!

Correct each student's errors, if any.

65. Divide: $\dfrac{5x^2 + 2x - 3}{5x^2}$

Britney's work:
$$\frac{5x^2 + 2x - 3}{5x^2} = \frac{\cancel{5x^2} + 2x - 3}{\cancel{5x^2}} = 2x - 3$$

66. Divide: $\dfrac{2x^3 - 4x^2 - 1}{2x^3}$

Susan's work:
$$\frac{2x^3 - 4x^2 - 1}{2x^3} = \frac{\cancel{2x^3} - 4x^2 - 1}{\cancel{2x^3}} = -4x^2 - 1$$

67. Divide: $x^2 - 2x - 3$ by $x + 3$

Damien's work:
$$
\begin{array}{r}
x + 1 \quad \longleftarrow \text{Answer} \\
x + 3 \overline{)\, x^2 - 2x - 3} \\
\underline{-x^2 + 3x} \\
x - 3 \\
\underline{-x + 3} \\
0
\end{array}
$$

68. Divide: $x^3 + 5x^2 - 8x - 12$ by $x - 6$

Kit's work:
$$
\begin{array}{r}
x^2 - x - 2 \quad \longleftarrow \text{Answer} \\
x - 6 \overline{)\, x^3 + 5x^2 - 8x - 12} \\
\underline{-x^3 - 6x^2} \\
-x^2 - 8x \\
\underline{+x^2 + 6x} \\
-2x - 12 \\
\underline{+2x + 12} \\
0
\end{array}
$$

 Calculate It!

Use a graphing calculator to determine if each quotient is correct. If the answer is incorrect, find the correct answer.

69. $\dfrac{x^4 - 3x^2}{x^4} = -3x^2$

70. $\dfrac{5x^2 + 3x - 1}{5x^2} = 3x - 1$

71. $\dfrac{x^3 + 125}{x + 5} = x^2 - 5x + 25$

72. $\dfrac{x^3 + 1}{x + 1} = x^2 - 1$

| SECTION 5.7 | Synthetic Division and the Remainder Theorem |

▶ **OBJECTIVES**

As a result of completing this section, you will be able to

1. **Use synthetic division to divide a polynomial by a binomial of the form $x - c$, $c \neq 0$.**

2. **Use the remainder theorem to evaluate polynomials.**

3. **Troubleshoot common errors.**

Objective 1 ▶

Use synthetic division to divide a polynomial by a binomial of the form $x - c$, $c \neq 0$.

Synthetic Division

Synthetic division is a process that can be used as a shortcut for long division as long as the divisor is a binomial of the form $x - c$. The expression on the left below is an example of long division. The expression on the right is the same long division problem with the variables removed.

$$
\begin{array}{r}
2x + 3 \\
x + 4 \overline{)\, 2x^2 + 11x + 12} \\
\underline{-(2x^2 + 8x)} \\
3x + 12 \\
\underline{-(3x + 12)}
\end{array}
\qquad
\begin{array}{r}
2 + 3 \\
1 \quad 4 \overline{)\, 2 \ + 11 \ + 12} \\
\underline{2 \ + 8} \\
3 \ + 12 \\
3 \ + 12
\end{array}
$$

Notice that after each multiplication step, the first terms subtract to zero ($2x^2 - 2x^2 = 0$ and $3x - 3x = 0$), leaving the term that determines the quotient. Also remember that when we subtract, the signs change.

This leads us to the process of synthetic division. In order to use synthetic division, the divisor must be of the form, $x - c$. Since the divisor in the previous example is

$x + 4$, or $x - (-4)$, $c = -4$. We use a symbol that looks like an upside down division symbol to perform our work.

$$
\begin{array}{r|rrr}
-4 & 2 & 11 & 12 \\
\hline
& 2 & &
\end{array}
$$
Bring down the first term.

$$
\begin{array}{r|rrr}
-4 & 2 & 11 & 12 \\
& & -8 & \\
\hline
& 2 & &
\end{array}
$$
Multiply -4 and 2 and write result in the second column.

$$
\begin{array}{r|rrr}
-4 & 2 & 11 & 12 \\
& & -8 & \\
\hline
& 2 & 3 &
\end{array}
$$
Add the values in the second column.

$$
\begin{array}{r|rrr}
-4 & 2 & 11 & 12 \\
& & -8 & -12 \\
\hline
& 2 & 3 &
\end{array}
$$
Multiply -4 and 3 and write result in the third column.

$$
\begin{array}{r|rrr}
-4 & 2 & 11 & 12 \\
& & -8 & -12 \\
\hline
& 2 & 3 & 0
\end{array}
$$
Add the values in the third column.

So, $2x^2 + 11x + 12$ divide by $x + 4$ is $2x + 3$ with remainder 0.

Procedure: Dividing a Polynomial by a Binomial of the Form $x - c$ Using Synthetic Division

Step 1: Write down the coefficients of the polynomial in the numerator, inserting zeros for any missing terms.

Step 2: Determine the value of c from the divisor and write it to the left of the inverted division symbol.

Step 3: Bring down the coefficient of the first term and multiply it by the number c. Write the result under the number in the second column.

Step 4: Add the numbers in the second column.

Step 5: Multiply the result from step 4 by the value of c and write this result under the number in the third column.

Step 6: Add the numbers in the third column.

Step 7: Continue this process until there are no more columns.

Step 8: The quotient is on the bottom row. The numbers are the coefficients of the polynomial whose degree is one less than the degree of the dividend. The last number in the row is the remainder.

Objective 1 Examples   Use synthetic division to divide the polynomials.

1a. Divide $5x^2 - 7x + 3$ by $x - 2$.   **1b.** Divide $4x^3 - 5x^2 + 2x - 9$ by $x + 1$.

Solutions   **1a.** The coefficients of the dividend are $5, -7,$ and 3 and the value of c is 2 since the divisor is $x - 2$.

$$
\begin{array}{r|rrr}
2 & 5 & -7 & 3 \\
\hline
& 5 & &
\end{array}
$$
Write the coefficients of the dividend and bring down 5 in the first column.

$$
\begin{array}{c|ccc}
2 & 5 & -7 & 3 \\
  &   & 10 &   \\
\hline
  & 5 &   &
\end{array}
$$

Multiply 2 and 5 and record the result, 10, in the second column.

$$
\begin{array}{c|ccc}
2 & 5 & -7 & 3 \\
  &   & 10 &   \\
\hline
  & 5 & 3 &
\end{array}
$$

Add the values in the second column, -7 and 10, and write the result, 3, below the line.

$$
\begin{array}{c|ccc}
2 & 5 & -7 & 3 \\
  &   & 10 & 6 \\
\hline
  & 5 & 3 &
\end{array}
$$

Multiply 2 and 3 and record the result, 6, in the third column.

$$
\begin{array}{c|ccc}
2 & 5 & -7 & 3 \\
  &   & 10 & 6 \\
\hline
  & 5 & 3 & 9
\end{array}
$$

Add the values in the third column, 3 and 6, and write the result, 9, below the line.

The numbers 5 and 3 are the coefficients of the quotient polynomial and 9 is the remainder. The degree of the quotient polynomial is one less than the degree of the dividend.

So, $\dfrac{5x^2 - 7x + 3}{x - 2} = 5x + 3 + \dfrac{9}{x - 2}$.

1b. The coefficients of the dividend are $4, -5, 2,$ and -9. The value of c is -1, since the divisor is $x + 1 = x - (-1)$.

$$
\begin{array}{c|cccc}
-1 & 4 & -5 & 2 & -9 \\
   &   &   &   &   \\
\hline
   & 4 &   &   &
\end{array}
$$

Write the coefficients of the dividend and bring down 4 in the first column.

$$
\begin{array}{c|cccc}
-1 & 4 & -5 & 2 & -9 \\
   &   & -4 &   &   \\
\hline
   & 4 &   &   &
\end{array}
$$

Multiply -1 and 4 and write the result, -4, in the second column.

$$
\begin{array}{c|cccc}
-1 & 4 & -5 & 2 & -9 \\
   &   & -4 &   &   \\
\hline
   & 4 & -9 &   &
\end{array}
$$

Add the values in the second column, -5 and -4, and record the result, -9, below the line.

$$
\begin{array}{c|cccc}
-1 & 4 & -5 & 2 & -9 \\
   &   & -4 & 9 &   \\
\hline
   & 4 & -9 &   &
\end{array}
$$

Multiply -1 and -9 and write the result, 9, in the third column.

$$
\begin{array}{c|cccc}
-1 & 4 & -5 & 2 & -9 \\
   &   & -4 & 9 &   \\
\hline
   & 4 & -9 & 11 &
\end{array}
$$

Add the values in the third column, 2 and 9, and write the result, 11, below the line.

$$
\begin{array}{c|cccc}
-1 & 4 & -5 & 2 & -9 \\
   &   & -4 & 9 & -11 \\
\hline
   & 4 & -9 & 11 &
\end{array}
$$

Multiply -1 and 11 and write the result, -11, in the fourth column.

$$
\begin{array}{c|cccc}
-1 & 4 & -5 & 2 & -9 \\
   &   & -4 & 9 & -11 \\
\hline
   & 4 & -9 & 11 & -20
\end{array}
$$

Add the values in the fourth column, -9 and -11, and write the result, -20, below the line.

The numbers 4, -9, and 11 are the coefficients of the quotient polynomial and -20 is the remainder. The degree of the quotient polynomial is one less than the degree of the dividend.

So, $\dfrac{4x^3 - 5x^2 + 2x - 9}{x + 1} = 4x^2 - 9x + 11 - \dfrac{20}{x + 1}$.

✔ **Student Check 1**   Use synthetic division to divide the polynomials.

   a. Divide $6x^2 + 8x - 3$ by $x - 4$.   **b.** Divide $2x^3 + 3x^2 - x + 5$ by $x + 2$.

The Remainder Theorem

Objective 2 ▶

Use the remainder theorem to evaluate polynomials.

There is an interesting relationship between the remainder of dividing a polynomial $P(x)$ by the quantity $x - c$ and the value of $P(c)$. In Example 1a, we found that the remainder after dividing $P(x) = 5x^2 - 7x + 3$ by $x - 2$ is 9. When we evaluate $P(x)$ for $x = 2$, we get

$$
\begin{aligned}
P(2) &= 5(2)^2 - 7(2) + 3 \\
&= 5(4) - 14 + 3 \\
&= 20 - 14 + 3 \\
&= 6 + 3 \\
&= 9
\end{aligned}
$$

So, $P(2) = 9$, which is the remainder obtained after dividing $P(x)$ by $x - 2$. This result can be stated as the *remainder theorem*.

> **Property: The Remainder Theorem**
> The remainder when dividing $P(x)$ by $x - c$ is $P(c)$.

Objective 2 Example **Use the remainder theorem and synthetic division to find P(3) if**
$$P(x) = 9x^4 - 3x^3 + 7x^2 - 5x + 6$$

Solution   The coefficients of the polynomial are 9, -3, 7, -5, and 6. The value of c is 3.

$$
\begin{array}{r|rrrrr}
3 & 9 & -3 & 7 & -5 & 6 \\
  &   & 27 & 72 & 237 & 696 \\
\hline
  & 9 & 24 & 79 & 232 & 702
\end{array}
$$

The remainder after dividing $P(x)$ by $x - 3$ is 702. So, $P(3) = 702$.

✔ **Student Check 2**   Use the remainder theorem and synthetic division to find $P(-5)$ if
$$P(x) = 3x^4 + 2x^3 - 5x^2 + 4x + 1$$

Objective 3 ▶

Troubleshoot common errors.

Troubleshooting Common Errors

A common error associated with synthetic division is illustrated.

Objective 3 Example	A problem and an incorrect solution are given. Provide the correct solution and an explanation of the error.

Use synthetic division to divide $7x^2 + 8x - 9$ by $x + 1$.

Incorrect Solution	Correct Solution and Explanation
$\begin{array}{r|rrr} 1 & 7 & 8 & -9 \\ & & 7 & 15 \\ \hline & 7 & 15 & 6 \end{array}$ So, $\dfrac{7x^2 + 8x - 9}{x + 1}$ $= 7x + 15 + \dfrac{6}{x + 1}$.	Synthetic division works with divisors of the form $x - c$. Since $x + 1 = x - (-1)$, $c = -1$. $\begin{array}{r|rrr} -1 & 7 & 8 & -9 \\ & & -7 & -1 \\ \hline & 7 & 1 & -10 \end{array}$ So, $\dfrac{7x^2 + 8x - 9}{x + 1} = 7x + 1 - \dfrac{10}{x + 1}$.

ANSWERS TO STUDENT CHECKS

Student Check 1   **a.** $6x + 32 + \dfrac{125}{x - 4}$

 b. $2x^2 - x + 1 + \dfrac{3}{x + 2}$

Student Check 2 $P(-5) = 1481$

SUMMARY OF KEY CONCEPTS

1. Synthetic division is a shortcut for dividing a polynomial by a binomial of the form $x - c$. Only the coefficients are used in this process.

2. The remainder theorem tells us that the remainder after dividing a polynomial $P(x)$ by a binomial of the form $x - c$ is equivalent to $P(c)$.

SECTION 5.7 EXERCISE SET

Write About It!

Use complete sentences in your answer to each exercise.

1. Can you apply the synthetic division to any form of divisor? Explain.

2. In synthetic division, what is the first step in writing the dividend?

3. In your own words, explain how to write the quotient and remainder after completing synthetic division.

4. Explain how to use synthetic division to evaluate a polynomial function at a given x-value.

5. Show why $\dfrac{x^3 + 64}{x + 4} \neq x^2 + 16$ using synthetic division.

Practice Makes Perfect!

Find each quotient using synthetic division. (See Objective 1.)

6. $\dfrac{3x^2 + 5x - 28}{x + 4}$

7. $\dfrac{2x^2 + 3x - 2}{x + 2}$

8. $\dfrac{2x^2 - x - 6}{x - 2}$

9. $\dfrac{3x^2 - 15x - 42}{x - 7}$

10. $(x^2 + 14x + 27) \div (x + 10)$

11. $(2x^2 - x - 11) \div (x + 3)$

12. $\dfrac{5x^3 + 24x^2 - 28x + 48}{x + 6}$

13. $\dfrac{7x^3 - 2x^2 - 21x - 12}{x + 1}$

14. $\dfrac{x^3 - 13x - 12}{x - 4}$

15. $\dfrac{x^3 - 2x^2 + 45}{x + 3}$

16. $\dfrac{2x^3 - 15x^2 + 25x + 8}{x - 3}$

17. $\dfrac{6x^3 + 47x^2 - 6x - 7}{x + 8}$

18. $\dfrac{x^3 + 27}{x + 3}$

19. $\dfrac{x^3 - 64}{x - 4}$

20. $\dfrac{x^3 + 8}{x - 2}$

21. $\dfrac{x^3 - 216}{x + 6}$

22. $\dfrac{3x^2 + x - 2}{x + 1}$

23. $\dfrac{8x^2 - 30x + 18}{x - 3}$

24. $\dfrac{3x^3 - 17x^2 + 29x - 36}{x - 4}$ **25.** $\dfrac{x^3 - x^2 - 14x + 8}{x - 4}$

26. $\dfrac{4x^3 + 6x^2 + 13}{x + 2}$ **27.** $\dfrac{2x^3 - 35x + 24}{x - 3}$

Use the remainder theorem and synthetic division to evaluate $P(x)$ at the specified x-value. (*See Objective 2.*)

28. $P(x) = 2x^3 + 3x^2 + 8x - 17,\ P(1)$

29. $P(x) = -8x^3 + 2x^2 + x + 10,\ P(2)$

30. $P(x) = -8x^3 + 4x^2 + 1,\ P(2)$

31. $P(x) = 5x^3 + 12x^2 + 8x - 5,\ P(-1)$

32. $P(x) = 4x^4 + 10x^3 - 3x^2 + 4x - 2,\ P(-3)$

33. $P(x) = 2x^4 + 8x^3 - 9x^2 + 2x - 12,\ P(-5)$

34. $P(x) = x^4 - 5x^3 + 8x^2 + 6x - 13,\ P(-1)$

35. $P(x) = x^4 + 7x^2 - 10x - 12,\ P(-2)$

36. $P(x) = 4x^3 + x^2 + 10,\ P(-2)$

37. $P(x) = 2x^4 - 13x^2 + x - 39,\ P(3)$

38. $P(x) = 3x^3 + 4x^2 - 5x - 18,\ P(2)$

39. $P(x) = -3x^3 + 6x^2 - 5x - 5,\ P(-1)$

40. $P(x) = 5x^4 + 8x^3 - 6x + 4,\ P(-1)$

41. $P(x) = 5x^4 - x^2 - 25x - 3,\ P(2)$

 You Be the Teacher!

Correct each student's errors, if any.

42. Simplify $(5x^3 - 2x - 16) \div (x - 2)$ using synthetic division.

Bert's work:

$$
\begin{array}{r|rrr}
2 & 5 & -2 & -16 \\
  &   & 10 & 16 \\
\hline
  & 5 & 8 & 0
\end{array}
$$

So, $5x^2 + 8$.

 Think About It!

43. If $x = 5$ is a zero of the polynomial $P(x)$, what is the remainder after dividing $P(x)$ by $(x - 5)$? Why?

 GROUP ACTIVITY   **The Mathematics of Finance**

The amortization formula gives the monthly payment P for repaying a loan of A dollars at an annual interest rate of r (as a decimal) for t yr.

$$
P = \dfrac{A \cdot \dfrac{r}{12}}{1 - \left(1 + \dfrac{r}{12}\right)^{-12t}}
$$

1. Show why the amortization formula can be written as

$$
P = \dfrac{\dfrac{r}{12}}{1 - \left(1 + \dfrac{r}{12}\right)^{-12t}} \cdot A.
$$

2. Use the Internet, newspapers, a local bank, or credit union to determine the current rate for a 30-yr fixed rate mortgage.

3. Use the interest rate you found in part 2 along with $t = 30$ yr to determine the coefficient of A in the amortization formula in part 1. Round the value to five decimal places.

4. Now the amortization formula in part 2 should look like $P = $ _____ A. This is a linear equation that we can write as $y = $ _____ x, where y represents P and x represents A. Use this equation to complete the following table. Round answers to two decimal places.

x	y
100,000	
150,000	
200,000	
250,000	
300,000	

5. What do the y-values represent in the table in part 4?

6. Based on your financial situation, choose a loan amount you could afford. If you paid the loan off in 30 yr, what is the total amount that you would pay for the loan? How much would you pay in interest for the loan?

7. Repeat parts 2–6 with a 15-yr fixed rate loan.

Note: The payment that results from the amortization formula only represents principal plus interest. A monthly mortgage payment typically includes an amount to be paid into escrow. Escrow funds cover such items as property taxes and home insurance payments. Taxes and insurance can add several hundred dollars to the payment determined by the formula.

Exponents, Polynomials, and Polynomial Functions

What's the big idea? Polynomials are very important to the study of math, chemistry, science, economics, and other fields. Polynomials and polynomial functions are used to model real-life phenomena in these fields. Since polynomials are constructed of terms of the form ax^n, the properties of exponents allow us to perform operations on the polynomials.

The Tools

Listed below are the key terms, skills, formulas, and properties you should know for this chapter.

The page reference is provided if you need additional help with the given topic. The Study Tips will assist in your preparation for an exam.

Study Tips

1. Learn all of the terms, formulas, and properties. Make flash cards and have someone quiz you.
2. Rework problems from the exercises and also the ones you worked in class. Work additional problems from the review exercises.
3. Review the summaries of key concepts.
4. Work the chapter test.
5. Be sure to review the online resources for additional study materials.

Terms

Binomial 392	Leading coefficient 392	Polynomial function 393
Coefficient 390	Like terms 394	Scientific notation 384
Conjugates 413	Monomial 392	Synthetic division 425
Degree of a polynomial 392	Perfect square trinomial 413	Trinomial 392
Degree of a term 391	Polynomial 392	

Formulas and Properties

- Difference of two squares 414
- FOIL 411
- Long division of polynomials 420
- Negative exponent 370
- Negative power of a fraction 000
- Power of a power rule 379
- Power of a product rule 379
- Power of a quotient rule 380
- Product of like bases rule 367
- Property of division for fractions 418
- Quotient of like bases rule 368
- Remainder theorem 428
- Square of a binomial 413
- Zero exponent 369

CHAPTER 5 / SUMMARY

How well do you know this chapter? Complete the following questions to find out. Take a look back at the section if you need help.

SECTION 5.1 Rules of Exponents, Zero and Negative Exponents

1. When exponential expressions with like bases are multiplied, the resulting exponent is the ____ of the exponents in the original expression. The base stays the ____. This rule can be written as $x^a \cdot x^b = $ ____.

2. When exponential expressions with like bases are divided, the resulting exponent is the _____ of the exponents in the original expression. The base stays the ____. This rule can be written as $\dfrac{x^a}{x^b} = $ ____.

3. When a nonzero number is raised to the exponent of zero, the result is ____.

4. A negative exponent is equivalent to the _____ of the base raised to the _____ exponent. This can be stated as $b^{-n} = $ __ for $b \neq 0$.

SECTION 5.2 More Rules of Exponents and Scientific Notation

5. When an exponential expression is raised to an exponent, the original exponents are _____ to obtain the exponent in the final result. This rule can be written as $(x^a)^b = $ ___.

6. When a product is raised to an exponent, the exponent must be applied to each _____ in the base. This rule can be written as $(xy)^b = $ ____.

7. When a quotient is raised to an exponent, the exponent must be applied to the _____ and _____ of the base. This rule can be written as $\left(\dfrac{x}{y}\right)^b = $ ___.

8. A number written in the form $a \times 10^n$ is a number written in _____ _____ as long as $1 \leq a < 10$ and n is an integer.

9. When a number is greater than 10 and the number is written in scientific notation, the exponent is _____.

10. When a number is greater than 0 and less than 1 and the number is written in scientific notation, the exponent is _____.

SECTION 5.3 Polynomial Functions, Addition and Subtraction of Polynomials

11. _____ consist of a finite number of terms of the form ax^n, where a is a real number and n is a whole number.

12. _____ with one term are called _____.

13. _____ with two terms are called _____.

14. _____ with three terms are called _____.

15. When the terms are written with their exponents in descending order, the polynomial is in _____ ____.

16. The degree of a polynomial in a single variable is the _____ exponent on the variable.

17. The _____ _____ is the coefficient of the term with the largest degree.

18. The degree of a term with more than one variable is the ____ of the exponents of the variables.

19. The degree of a polynomial in more than one variable is the _____ degree of the terms in the polynomial.

20. To evaluate a polynomial function, _____ the variable with the given number and simplify.

21. To add polynomials, combine ____ _____.

22. To subtract polynomials, add the _____ of the polynomial. When we find the _____ of a polynomial, all of the signs are changed.

SECTION 5.4 Multiplication of Polynomials and Polynomial Functions

23. When we multiply a monomial by a polynomial, we apply the _____ property and the product of like bases rule for exponents.

24. When polynomials are multiplied, each term in the first polynomial must be _____ to each term in the second polynomial.

SECTION 5.5 Special Products

25. Two binomials can be multiplied using the _____ method. This only works for two _____.

26. When a binomial is squared, the result is always a(n) _____. The rule for squaring a binomial may be written as $(a + b)^2 = $ _____. The result of squaring a binomial is called a(n) _____ _____ _____.

27. _____ are binomials of the form $a + b$ and $a - b$.

28. When $(a + b)$ and $(a - b)$ are multiplied, the result is _____. This is called the _____ of two _____.

29. When we raise a binomial to a larger exponent, we must apply the definition of the _____. We cannot apply the exponent to each ____ in the binomial.

SECTION 5.6 Division of Polynomials

30. To divide a polynomial by a monomial, we must divide ____ term of the _____ by the monomial. This can be written as $\dfrac{A + B}{C} = $ _____.

31. To divide a polynomial by a binomial, we must use ___ _____. Repeat the steps involved in long division of real numbers: _____, _____, _____ and _____ _____.

32. To check division of polynomials, we can use _____. The dividend should equal the _____ times the _____ plus the _____.

SECTION 5.7 Synthetic Division and the Remainder Theorem

33. Synthetic division is a shortcut for dividing a polynomial by a binomial of the form _____. Only the _____ are used in the process.

34. The remainder theorem tells us that the remainder after dividing a polynomial $P(x)$ by a binomial of the form $x - c$ is equivalent to ____.

CHAPTER 5 / REVIEW EXERCISES

SECTION 5.1

Simplify each expression using the rules of exponents. Write each answer with positive exponents. Assume all variables are nonzero. (*See Objectives 1–5.*)

1. $-4a^{-4}$

2. $-(-2)^{-5}$

3. $-(-3)^{-3}$

4. $(-p^{-2}q^{-4})(2p^5q^{-7})$

5. $(-5m^{-7}n^{-2})(m^{-3}n)$

6. $\dfrac{-12a^{-5}n}{6a^9n^4}$

7. $\dfrac{16s^{-2}t^6}{10s^{-4}t^4}$

8. $\dfrac{5u^{-9}v^{-4}}{20u^{-1}v^{-4}}$

9. $\dfrac{9x^6y^{-5}}{63x^{-2}y^9}$

10. $(2a+1)^0$

11. $-2(b-3)^0$

12. $-9(2x)^0$

13. $10(36y)^0$

14. $-a^0 + 6$

15. $-2b^0 - 3$

Solve each problem. (*See Objective 6.*)

16. The formula to find the amount of substance, P, with a half-life of m units that remains after time t units is $P = P_0 \cdot 2^{-t/m}$, where P_0 is the original amount of the substance. Suppose the half-life of a substance is 8 days. If there is originally 16 g of the substance, how much will remain after

　a. 16 days?　　**b.** 32 days?　　**c.** 64 days?

17. The formula to find the monthly payment P to repay a loan is $P = \dfrac{A\left(\dfrac{r}{12}\right)}{1 - \left(1 + \dfrac{r}{12}\right)^{-n}}$, where A is the amount of the loan, r is the interest rate (in decimal form), and n is the number of monthly payments. Suppose that Samuel takes out a student loan for \$12,000 at an interest rate of 4.2% for 6 yr. What is Samuel's monthly payment?

SECTION 5.2

Simplify each expression using the rules of exponents. Write each answer with positive exponents. Assume all variables are nonzero. (*See Objectives 1 and 2.*)

18. $(2x^4)^{-3}$

19. $(3y^{-5})^{-2}$

20. $[(5a)^3]^{-1}$

21. $[(2b)^{-4}]^{-2}$

22. $\left(\dfrac{5p}{q^2}\right)^{-2}$

23. $\left(\dfrac{9a}{b^0}\right)^{-3}$

24. $\left(\dfrac{3c^0}{2d^{-4}}\right)^{-2}$

25. $\left(-\dfrac{x^{-4}}{7y^3}\right)^{-3}$

Convert each number in standard notation to scientific notation and each number in scientific notation to standard notation. (*See Objectives 3 and 4.*)

26. 3,500,000

27. 0.0000891

28. 4.32×10^9

29. 2.78×10^{-5}

30. Population of Ethiopia in 2010: 88,010,000 (Source: http://www.factmonster.com)

31. Exxon Mobil 2010 Profit: 3.05×10^{10} U.S. dollars (Source: http://money.cnn.com)

Simplify each expression using the rules of exponents. Write answers in scientific notation. (*See Objective 4.*)

32. $(6.9 \times 10^5)(2.8 \times 10^{-12})$

33. $\dfrac{2.834 \times 10^{-9}}{8.72 \times 10^{-3}}$

34. The population of Japan in 2011 was approximately 126,500,000 and the land area is approximately 141,000 mi^2. If the land was evenly distributed among all of the people in Japan, how much land would each person have? Write answer in scientific notation. (Source: The World Factbook at http://www.cia.gov)

SECTION 5.3

Classify each polynomial, write it in standard form, and identify its terms, leading coefficient, and degree. (*See Objectives 1 and 2.*)

35. $11 - 5x + 9x^2$

36. $4 - x^4$

Evaluate each polynomial function for the given value. (*See Objective 3.*)

37. $f(x) = -x^2 + 3x + 7, f(-2)$

38. $f(x) = 2y^3 - 4y - 18, f(3)$

39. A company's total cost, in dollars, to make x items is given by $C(x) = 12x^3 - 5x^2 + 100$. Find $C(20)$ and interpret the results.

40. The height h, in feet, of a ball thrown upward from the top of a 100-ft building with an initial velocity of 80 ft/sec is given by $h(t) = -16t^2 + 80t + 100$, where t is the number of seconds after the ball is thrown. Find $h(2.5)$ and $h(3)$ and interpret the results.

Perform the indicated operation. Write each answer in standard form. (*See Objective 4.*)

41. $(12 - 9x^2 - 5x) - (-7 + 4x^2 + 8x)$

42. $(2x^4 + 17x^3 - 11) + (3x^4 - 12x^3 + 19)$

43. $\left(-\dfrac{6}{5}x^2 + \dfrac{7}{5}x - \dfrac{4}{5}\right) + \left(\dfrac{1}{5}x^2 + \dfrac{3}{5}x - \dfrac{1}{5}\right)$

44. $(1.5x^2 + 2.1x - 3.1) - (-0.8x^2 + 7.2x - 8.4)$

45. Add $(-5a^2 + 2ab - 16b^2)$ and $(-2a^2 - 4ab + 11b^2)$.

46. Subtract $(7x^2 - xy + 6y^2)$ from $(-3x^2 + 4xy - y^2)$.

47. $(6x^2 + 4xy - 3y^2) - (-5x^2 + 2xy + y^2) + (-7x^2 - 2xy + 4y^2)$

48. $f(x) = 3x^2 - 5x + 6$ and $g(x) = -2x^2 - 11x + 13$. Find $f(x) + g(x)$ and $f(x) - g(x)$.

Solve each problem. (*See Objective 5.*)

49. Write a simplified polynomial that represents the perimeter of the trapezoid.

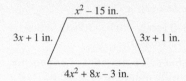

$x^2 - 15$ in.

$3x + 1$ in. $3x + 1$ in.

$4x^2 + 8x - 3$ in.

50. Suppose the profit (in dollars) generated by selling x items is $P(x) = -0.2x^2 + 124x + 75$ and the total cost (in dollars) is $C(x) = 12x + 320$. Find a function that represents the company's revenue.

SECTION 5.4

Use the distributive property to find each product.
(*See Objectives 1–3.*)

51. $p(-3p + 5)$

52. $6x^2y(5xy - 2)$

53. $(7 - x)(x + 7)$

54. $\left(\frac{2}{3}x - 2\right)\left(\frac{1}{3}x + 5\right)$

55. $(x^2 + 2x - 5)(3x^2 - 7x + 4)$

56. $(a + 3b)(a^2 - 3ab + 9b^2)$

57. $(2r - 3s)(4r^2 + 6rs + 9s^2)$

58. An apartment manager can rent all 100 units if he charges $720 per month rent. If he increases the price by $50, he rents two less apartments. If x represents the number of increases in price, then the rent for an apartment is given by $720 + 50x$ dollars and the number of apartments rented is $100 - 2x$. (*See Objective 4.*)

    a. What simplified polynomial represents the total rent collected each month?

    b. If the manager decides to increase the rent to $820 per month, how much total rent can he collect?

SECTION 5.5

Perform the indicated operation. Write each answer in standard form. (*See Objectives 1–4.*)

59. $(4x + 5)(3x - 1)$

60. $(1.4x - 2.5)(0.1x + 1.5)$

61. $(3a + 2b)^2$

62. $\left(\frac{3}{5}a - b\right)^2$

63. $(11x - 7y)(11x + 7y)$

64. $(x - 12y)(x + 12y)$

65. $\left(\frac{2}{3}a^3 + 5b\right)\left(\frac{2}{3}a^3 - 5b\right)$

66. $(a + 12)^3$

SECTION 5.6

Perform the indicated operation. (*See Objectives 1–3.*)

67. Divide $8a^3 + 4a^2 - 24$ by $4a$.

68. $\dfrac{12x^4 - 6x^3 + 14x^2}{-2x^2}$

69. $\dfrac{5x^3 + 31x^2 + 29x - 3}{x + 5}$

70. $\dfrac{8x^3 - 14x^2 + 7x - 13}{2x - 3}$

71. If the area of a parallelogram is $28xy^2 - 21x^2y$ square units and its height is $7xy$ units, find an expression that represents the length of its base.

72. If the area of a triangle is $32xy^2 - 24x^2y$ square units and its height is $4xy$ units, find an expression that represents the length of its base.

SECTION 5.7

Use synthetic division to find each quotient. (*See Objective 1.*)

73. $\dfrac{5x^2 - 12x - 32}{x - 4}$

74. $\dfrac{7x^3 + 43x^2 + 34x - 30}{x + 5}$

75. $\dfrac{x^3 - 8x^2 - 20}{x - 6}$

76. $\dfrac{x^3 + 4x - 12}{x + 3}$

Use the remainder theorem and synthetic division to evaluate $P(x)$ at the specified value. (*See Objective 2.*)

77. $P(x) = 5x^3 - 8x^2 + 10x - 4$, $P(-2)$

78. $P(x) = -2x^3 + 3x^2 + 2x - 18$, $P(-4)$

CHAPTER 5 TEST / EXPONENTS, POLYNOMIALS, AND POLYNOMIAL FUNCTIONS

1. When simplified $(4x^5)(-2x^3)^2$ is
   a. $16x^{11}$ b. $-8x^{11}$
   c. $-8x^{10}$ d. $16x^{10}$

2. -8^0 is equivalent to
   a. -8 b. -1
   c. 1 d. 0

3. When simplified $\dfrac{a^6}{a^{-2}}$ is equivalent to
   a. a^4 b. a^8
   c. $\dfrac{1}{a^3}$ d. $\dfrac{1}{a^{12}}$

4. When 0.0000037 is written in scientific notation, it is
   a. 37×10^5 b. 37×10^{-5}
   c. 3.7×10^6 d. 3.7×10^{-6}

5. When simplified $(-4x^2 - 6x + 2) - (3x^2 + 2x - 7)$ is
   a. $-7x^2 - 4x + 5$ b. $-7x^2 - 8x + 9$
   c. $x^2 - 4x - 5$ d. $x^2 - 8x + 9$

6. When simplified $-3x^2(x^2 - 3x - 1)$ is
   a. $-3x^4 + 9x^3 + 3x^2$ b. $-3x^4 - 9x^3 - 3x^2$
   c. $-3x^4 - 12x^2$ d. $-3x^4 + 6x^2$

7. When simplified $(3x + 1)(3x - 1)$ is
   a. $9x^2 + 1$ b. $9x^2 - 6x - 1$
   c. $9x^2 - 1$ d. $9x^2 + 6x - 1$

8. When simplified $(x - 5)^2$ is
   a. $x^2 + 25$ b. $x^2 - 25$
   c. $x^2 + 10x + 25$ d. $x^2 - 10x + 25$

9. $\dfrac{8x^3 - 6x^2 - x + 5}{2x}$ is equivalent to

 a. $4x^2 - 3x - 2 + \dfrac{5x}{2}$ b. $4x^2 - 3x - \dfrac{1}{2} + \dfrac{5}{2x}$

 c. $-2x^2 - x + 5$ d. $4x^2 - 3x + 5$

10. If the final row when using synthetic division to dividing $P(x)$ by $x + 4$ is 1 -6 27 -112, then
 a. $P(4) = -112$ b. $P(4) = 112$
 c. $P(-4) = -112$ d. $P(-4) = 112$

Simplify each expression using the rules of exponents. Write each answer with positive exponents. Assume all variables are nonzero.

11. $(-6a^3)(2a^{-4})^2$ 12. $\left(-\dfrac{3x^5 y}{2x^{-1}y^3}\right)^{-2}$

13. $-4x^0 + (-4)^0$

Simplify each expression.

14. $(-6)^2$ 15. -6^2 16. 6^{-2}

17. -6^{-2} 18. $(-6)^{-2}$

Complete the table for each polynomial.

	Polynomial	Standard Form	Type	Degree	Leading Coefficient
19.	$-4y - y^2 + 5$				
20.	$1 - 27x^3$				
21.	$9a^6$				

22. Simplify: $(6x^4y^3 - 2x^3y^2 + x^2y) + (2x^4y^3 + 3x^3y^2 - 7x^2y)$.

23. What is the degree of the polynomial that results in Exercise 22?

24. Evaluate the polynomial in Exercise 22 when $x = -2$ and $y = 3$.

Perform the indicated operation.

25. $(2x + 5) - (3x - 7)$ 26. $(2x + 5)(3x - 7)$

27. $(5a^3 + 2b)^2$ 28. $(5a^3 + 2b)(5a^3 - 2b)$

29. $(3x - 5)(9x^2 + 15x + 25)$ 30. $(3x - 5)^3$

31. $\dfrac{24a^2b^3 - 12ab^2 + 4ab}{4ab}$

32. $\dfrac{x^3 - 27}{x + 3}$ (use long division)

33. $\dfrac{2x^2 - 5x + 3}{x - 1}$ (use synthetic division)

Solve each problem.

34. The distance traveled in 1 light-year is approximately 6 trillion mi. The North Star, Polaris, is approximately 300 light-years away from Earth. Convert each of these numbers to scientific notation and then perform an operation using these numbers that will determine how many miles away the North Star is from Earth.

35. Write the polynomials that represent the perimeter *and* area of a rectangle whose length is $5x^2 + 3x + 1$ units and whose width is $6x + 7$ units.

36. Amy owns a scarf making business. The cost (in dollars) for making x scarves in a month can be represented by $C(x) = 2x + 50$ and the revenue (in dollars) for selling x scarves in a month is $R(x) = 12x$.
 a. Find a polynomial function that represents the profit for Amy's business. Recall profit is revenue minus cost.
 b. Find the profit for selling 30 scarves in a month.

CUMULATIVE REVIEW EXERCISES / CHAPTERS 1–5

1. Perform each operation and simplify the results. (*Section 1.2, Objectives 2–7*)

 a. $2\dfrac{1}{2} \cdot 3\dfrac{1}{5} \div 6\dfrac{1}{4}$ b. $5\dfrac{1}{2} + 2\dfrac{3}{7} - 4\dfrac{2}{7}$

2. According to the Bureau of Labor Statistics, the average hourly rate of utility workers was $29.64 in September 2010. The expression $29.64x + 44.46y$ represents the average weekly earnings of a utility worker, where x is the number of regular hours worked in a week and y is the number of overtime hours worked in a week. Find the weekly earnings of a utility worker if she worked 40 regular hours and 12 overtime hours in September 2010. (*Section 1.3, Objective 6*)

3. Simplify each expression. (*Sections 1.3–1.5, Objectives 1 and 2*)
 a. $3(5)^2 - (-4)^3 - 11 \cdot 5$
 b. $13.2 - 8.7 + (-6.4) - (-12.5)$

4. Warren wants to mix a 10% antifreeze solution with a 30% antifreeze solution to get an 18% antifreeze solution. Let x oz represent the volume of the 10% antifreeze solution. If Warren wants to get 120 oz of 18% antifreeze solution, write an expression that represents the volume of the 30% antifreeze solution. (*Sections 1.4 and 1.5, Objective 3*)

5. Perform the indicated operation and simplify. (*Section 1.6, Objectives 1–3*)

 a. $(-5.3)(-12.2)$ b. $(-54)\left(-\dfrac{25}{18}\right)(0)$

 c. $26 \div 0$

6. Simplify each algebraic expression. (*Section 1.8, Objectives 3 and 4*)
 a. $-5(2a - 9) + 3(-a + 8)$
 b. $15.8 - 1.2(9b - 2)$
 c. $2x^2 - (4x - 3x^2) + 9x$

7. Determine whether the following is an expression or an equation. (*Section 2.1, Objective 1*)

   a. $16x - 29 = 2x - 10$ **b.** $27x - 5 - 13x + 40$

8. For each problem, define the variable, write an equation, and solve the problem. (*Section 2.1, Objective 3 and 4*)

   a. The product of a number and 4 is the same as the sum of twice the number and 14.

   b. The cost to rent a mid-size car for 1 day is $62 plus $0.42 per mile. This daily cost is represented by the expression $62 + 0.42x$, where x is the number of miles driven. How many miles had the car been driven if the rental cost was $314?

9. Solve each equation. (*Section 2.2, Objective 2 and 3; Sections 2.3 and 2.4, Objectives 1–4*)

   a. $4(x - 3) - 20 = 7(x + 5) - 19$

   b. $-\dfrac{1}{4}(2x + 1) + \dfrac{9}{4} = x$

   c. $-14.5x + 20.5 = 5(-2.9x + 4.3)$

10. Find the measure of an angle whose complement is 28° more than the measure of the angle. (*Section 2.5, Objective 5*)

11. Bianca invests $9600 in two accounts. She invests $6500 in a certificate of deposit that pays 2.3% annual interest and the remaining amount in a money market fund that pays 1.6% annual interest. How much interest did she earn from the two accounts in 1 yr? (*Section 2.6, Objectives 1–5*)

12. Solve each inequality. Graph the solution set, and write the solution set in interval notation and set-builder notation. (*Section 2.7, Objectives 1–4*)

   a. $6x + 2(x + 22) < 11x - 16$

   b. $-1.6 < 0.4a + 4.3 < 1.5$

13. The table gives the percentage of reported flight operations that arrived on time as defined by the Department of Transportation (DOT) for US Airways between 2003 and 2009. (*Section 3.1, Objective 5*)

Years after 2003	0	1	2	3	4	5	6
Percentage of reported flight operations arriving on time	82.0	75.7	77.8	76.9	68.7	80.1	80.9

   a. Write an ordered pair (x, y) that corresponds to each year between 2003 and 2009, where x is the number of years after 2003 and y is percentage of reported flight operations that arrived on time as defined by DOT for US Airways.

   b. Interpret the meaning of first and last ordered pairs from Part (a) in the context of the problem.

   c. In what year was the percentage of reported flight operations that arrived on time the highest for US Airways? The lowest?

   d. Create a scatter plot of the data.

14. Graph each equation. Identify at least two points on the graph. (*Section 3.2, Objectives 2–4*)

   a. $5x - 2y = 10$ **b.** $x - 0.5y = 0$ **c.** $0.6x + 0.1 = 1.3$

15. The gallons of fuel consumed (in millions) by a major airline can be modeled by the equation $y = 38x + 330.4$, where x is the number of years after 2005. (*Section 3.2, Objective 5*)

   a. Find the y-intercept of the equation and interpret its meaning.

   b. How many gallons of fuel were consumed by the airline in 2010?

   c. How many gallons of fuel were consumed by the airline in 2013?

   d. Use the information to graph the equation.

16. Find the slope of each line. (*Section 3.3, Objectives 1, 3, and 4*)

   a. The line that passes through the point $(-4, -1)$ and $(3, 5)$

   b. $y = 6$ **c.** $x + 4 = 0$

   d. $x - 2y = 6$

17. The total operating revenue (in billions of dollars) for a major department store can be modeled by the equation $y = -0.281x + 12.01$, where x is the number of years after 2006. Interpret the meaning of the slope and y-intercept in the context of the problem. (*Section 3.4, Objectives 1 and 2*)

18. Graph each line. Label two points on the line. (*Section 3.4, Objective 2*)

   a. $m = \dfrac{2}{3}$, passes through $(-2, 3)$

   b. $y = 2x - 4$

19. Determine if the lines are parallel, perpendicular, or neither. (*Section 3.4, Objective 3*)

   a. $x + 2y = 5$ and $2y - x = 6$

   b. $y = 0.6x$ and $6x - 10y = 5$

20. Write the equation of the line, in slope-intercept form and standard form, that satisfies the given information. (*Section 3.5, Objectives 1–4*)

   a. passes through $(3, -1)$ perpendicular to $x + 3y = 1$

   b. passes through $(6, -5)$ parallel to $x - 6y = 7$

   c. passes through $(3, -2)$ with undefined slope

21. The table shows the average residential natural gas prices (in dollars per thousand cubic feet) from 2006 to 2010. (Source: http://www.eia.doe.gov/naturalgas) (*Section 3.5, Objective 5*)

Year	2006	2007	2008	2009	2010
Average Residential Natural Gas Prices	$13.73	$13.08	$13.89	$12.14	$11.19

   a. Use the points $(1, 13.08)$ and $(4, 11.19)$ to write a linear equation that represents the average residential natural gas prices (in dollars per thousand cubic feet), where x is the number of years after 2006.

b. Find the average residential natural gas price in 2011.

c. Find the average residential natural gas price in 2013.

22. State the domain and range of each relation. Determine if the relation is a function. (*Section 3.6, Objectives 1–3*)

a. $\{(-4, 1), (1, -4), (0, 5), (-2, 3), (3, 1)\}$

b. $x = 3$ **c.** $y = -2x + 12$

23. Find the requested information. (*Section 3.6, Objective 4*)

a. Find $f(0)$ if $f(x) = 15x - 6$.

b. Find $g(-2)$ if $g(x) = -x^2 + 3x + 2$.

c. Find $f(-2)$ and solve $f(x) = 3$ if $f(x)$ is given by the following graph.

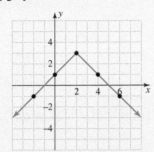

24. Write the linear equation $4x + 5y = 24$ in function notation. Then find $(-1), f(0)$, and $f(3)$ and write the corresponding ordered pairs. (*Section 3.6, Objective 4*)

25. Solve each system using substitution. (*Section 4.2, Objectives 1 and 2*)

a. $\begin{cases} 7x - y = -25 \\ x + 3y = 9 \end{cases}$ **b.** $\begin{cases} x - 4y = 9 \\ 5x + 2y = 23 \end{cases}$

26. Solve each system using elimination. (*Section 4.3, Objectives 1–3*)

a. $\begin{cases} 4x - 7y = -59 \\ 5x + 3y = -15 \end{cases}$ **b.** $\begin{cases} 2x + y = -1 \\ 3y = -6x - 3 \end{cases}$

c. $\begin{cases} 1.1x - 0.2y = 5.8 \\ 0.4x + 0.1y = 4.7 \end{cases}$

27. Solve each problem using a system of equations. (*Section 4.4 Objectives 1–5*)

a. Two angles are supplementary. The larger angle is 20° more than three times the smaller angle. What is the measure of each angle?

b. Wayne leaves his job and heads south at 50 mph. Joy leaves the same workplace 10 min later and heads south at 60 mph. How long will it take Joy to catch up to Wayne?

28. Solve each system using elimination. (*Section 4.5, Objective 1*)

a. $\begin{cases} x - 2y - 6z = -14 \\ -x + 4y + 8z = 22 \\ 3x - y + 7z = -2 \end{cases}$ **b.** $\begin{cases} x + 2y = -1 \\ 3y - z = -10 \\ 2x + 5z = 26 \end{cases}$

c. $\begin{cases} x + 3y - 6z = 7 \\ y - 4z = 5 \\ -x - 2y + 2z = -2 \end{cases}$ **d.** $\begin{cases} 3x - 2y + 4z = 35 \\ 7x + 4y = -9 \\ 5x + y + 2z = 18 \end{cases}$

29. Simplify each expression. Write each answer with positive exponents. Assume all variables are nonzero. (*Section 5.1, Objectives 1–5*)

a. $-8(10b)^0$ **b.** $-2c^0 + 5$

c. $-\left(-\dfrac{10}{3}\right)^{-3}$ **d.** $(3c^7q^2)(-6c^{-12}q)$

e. $(2x^5y^4)(-4x^{-5}y^{-8})$ **f.** $\dfrac{-12pq^{-8}}{3p^{-5}q^{-6}}$

30. Write an expression in terms of the radius for the surface area of a can if the height is three times its radius. (*Section 5.1, Objective 6*)

31. Simplify each expression using the rules of exponents. Write each answer with positive exponents. Assume all variables are nonzero. (*Section 5.2, Objectives 1 and 2*)

a. $(-3x^4y^{10})^3$ **b.** $(-2a^{-5})^{-3}$

c. $\left(\dfrac{8p^2}{7q^{-3}}\right)^{-2}$ **d.** $\dfrac{(4y^{10}z^{-9})^{-3}}{y^{-6}z^{-17}}$

32. Use scientific notation to perform the indicated operations. Express each answer in scientific notation. (*Section 5.2, Objectives 3 and 4*)

a. $(1.2 \times 10^4)^3$ **b.** $\dfrac{0.297}{5,500,000}$

c. $\dfrac{(9.36 \times 10^{-3})(7.5 \times 10^{-1})}{4.5 \times 10^7}$

33. Solve each problem. (*Section 5.2, Objective 5*)

a. The formula $P = P_0 \cdot 2^{-t/m}$ gives the amount of radioactive material present at time t, where P_0 is the original amount of the material and m is the half-life of the material. Suppose the half-life of a substance is 12 weeks. If the original amount of the substance is 10 g, how much substance remains after a. 12 weeks? b. 36 weeks?

b. The formula to find the monthly payment P to repay a loan is $P = \dfrac{A\left(\dfrac{r}{12}\right)}{1 - \left(1 + \dfrac{r}{12}\right)^{-n}}$, where A is the amount of the loan, r is the interest rate (in decimal form), and n is the number of monthly payments. What is the monthly payment for an 8-yr student loan of \$15,800 at an interest rate of 5.1%?

34. Complete the table for each polynomial, if applicable. (*Section 5.3, Objectives 1 and 2*)

	Terms/ Classification	Degree/ Type	Leading Coefficient
a. $2x^3 - 5x + 17$			
b. $2y^2 - 10y - 12$			
c. $-ab^2c$			
d. $9 - 5t$			

35. If $f(x) = x^2 - 6x + 2$ and $g(x) = -x^3 + 8x - 1$, find the following. (*Section 5.3, Objective 3*)

   a. $f(0) - g(-2)$ **b.** $f(2) + g(-1)$

36. Simplify each expression. (*Section 5.3, Objective 4*)

   a. $(-19q^2 - 15q - 4) + (23q^2 - 24q - 15)$

   b. $(-7b^2 + 9b - 17) - (-22b^2 - 8b + 15)$

   c. $\left(\dfrac{2}{3}r^2 + \dfrac{4}{3}rs - 8s^2\right) + \left(\dfrac{4}{3}r^2 - \dfrac{7}{3}rs + s^2\right)$

   d. $\left(\dfrac{19}{10}s^2 + \dfrac{5}{3}st - 9t^2\right) - \left(\dfrac{9}{10}s^2 + \dfrac{8}{3}st - 6t^2\right)$

37 Write a polynomial function that represents each situation. (*Section 5.3, Objectives 5*)

   a. Find the perimeter of the trapezoid.

$5x^2 - 1$ m

$x + 3$ m $x + 3$ m

$8x^2 + 3x - 4$ m

   b. Suppose the profit (in dollars) generated by selling x items is $P(x) = -0.3x^2 + 143x + 105$ and the total cost (in dollars) is $C(x) = 18x + 360$. Find the company's revenue function.

38. Simplify each product. (*Section 5.4, Objectives 1–3*)

   a. $5x^3y^2(3xy - 1)$ **b.** $(5 - 2x)(2x + 5)$

   c. $(a + 4b)(a^2 - 4ab + 16b^2)$ **d.** $\left(\dfrac{1}{4}x - 1\right)\left(\dfrac{3}{4}x + 6\right)$

   e. $(3x^2 + 5x - 1)(2x^2 - 6x + 1)$

39. An apartment manager can rent all 100 units if he charges $750 per month rent. If he increases the price by $60, he rents three fewer apartments. If x represents the number of increases in price, then the rent for

an apartment is given by $750 + 60x$ dollars and the number of apartments rented is $100 - 3x$. (*Section 5.4, Objective 4*)

   a. What simplified polynomial represents the total rent collected each month?

   b. If the manager decides to increase the rent to $870 per month, how much total rent can he collect each month?

40. Perform the indicated operation. Write each answer in standard form. (*Section 5.5, Objectives 1–4*)

   a. $(4x + 1)(3x - 5)$ **b.** $\left(\dfrac{2}{3}a^3 + 5b\right)\left(\dfrac{2}{3}a^3 - 5b\right)$

   c. $(2p - 1)^3$ **d.** $(11x - 5y)(11x + 5y)$

   e. $\left(\dfrac{2}{3}a + b\right)^2$ **f.** $(1.4x - 1.5)(0.1x + 2.5)$

41. Perform the indicated operation. (*Section 5.6, Objectives 1–3*)

   a. Divide $12a^3 + 4a^2 - 24$ by $4a$.

   b. $\dfrac{12a^4 - 48a^3 + 8a - 16}{16a^2}$

   c. $\dfrac{2x^3 + 20x^2 + 18x - 15}{x + 3}$

   d. If the area of a parallelogram is $32xy^3 - 16x^2y$ units^2 and its height is $8xy$ units, find the length of its base.

42. Use synthetic division to find each quotient. (*Section 5.7, Objective 1*)

   a. $\dfrac{3x^2 - 4x + 5}{x - 6}$ **b.** $\dfrac{x^3 - x + 2}{x + 3}$

43. Use synthetic division and the remainder theorem to find $P(-5)$ for $P(x) = 4x^2 - x + 2$. (*Section 5.7, Objective 2*)

Factoring Polynomials and Polynomial Equations

Motivation

Motivation is a key element in helping us accomplish our goals like completing this class and ultimately a college degree. Something motivated you to enroll in college, so keep that vision in mind as you complete each and every course. Some courses are going to require more energy and time to complete but stay motivated through the process. Obstacles or circumstances that challenge us have a tendency to make us less motivated to finish what we started. So find what motivates you and continue to dwell on that until you achieve your goal.

? *Question For Thought:* What is your motivation to complete this course? to complete your college education? How does this impact you in your daily life?

Chapter Outline

Coming Up...

In Section 6.6, we will solve a quadratic equation that enables us to determine how long a B.A.S.E. jumper has to open his parachute before reaching the ground after he falls off of a 1600-ft structure. (B.A.S.E. stands for buildings, antennas, structures, and earth formations.)

> "Get a clear vision of your goals. Convince yourself that you can achieve them. Purposefully do something each day that advances you toward your goal."
>
> —Andrea Hendricks

SECTION 6.1 Greatest Common Factor and Grouping

In Chapter 5 we learned how to multiply polynomials. The focus of this chapter is on the reverse process—that is, given a polynomial, find the factors that must be multiplied to get the polynomial. We will learn how to factor binomials, trinomials, and other polynomials. We will then show how factoring can be used to solve quadratic equations and their applications.

▶ **OBJECTIVES**

As a result of completing this section, you will be able to

1. Find the greatest common factor of a set of integers.
2. Find the greatest common factor of a set of terms.
3. Use the greatest common factor to factor a polynomial.
4. Use grouping to factor polynomials.
5. Troubleshoot common errors.

A company's profit can be represented by $-5q^2 + 200q$, where q is the number of units sold in hundreds. In Section 6.6, we will solve an application involving this expression that requires us to factor it.

In Chapter 5, we began with two polynomials and multiplied them to obtain another polynomial. For instance,

$$2x(x^2 + 3) = 2x(x^2) + 2x(3) = 2x^3 + 6x$$

So, the polynomials $2x$ and $x^2 + 3$ are called **factors** of $2x^3 + 6x$ and the product $2x(x^2 + 3)$ is called the *factored form*.

In this chapter, we will be given a polynomial and must be able to find the polynomials whose product is the given polynomial. This process, called **factoring**, is the reverse of multiplication.

One of the first steps when factoring a polynomial is to determine if the terms in the polynomial have a common factor. So, we will begin our factoring techniques with finding the greatest common factor and then extend this to a method called grouping.

The Greatest Common Factor of a Set of Integers

Objective 1 ▶

Find the greatest common factor of a set of integers.

In the product $3 \cdot 4 = 12$, the numbers 3 and 4 are called the *factors* of 12. To *factor a number* means to write the number as a product. Often, we are interested in determining the factors that numbers have in common, specifically the largest of these factors. This is called the *greatest common factor*.

> **Definition:** The **greatest common factor** (GCF) of a set of integers is the largest integer that is a factor of each integer in the set.

Suppose we want to find the GCF of 42 and 60. Then we need to find the largest integer that is a factor of both of these numbers. There are two approaches that we can take.

First, we can list all of the factors of each number and find the largest factor that is in both lists.

Factors of 42: 1, 2, 3, 6, 7, 14, 21, 42

Factors of 60: 1, 2, 3, 4, 5, 6, 10, 12, 15, 20, 30, 60

> The common factors are 1, 2, 3, and 6 and the greatest common factor is 6.

This method works great for small integers but may not be the best method for larger integers.

Another way we can find the GCF is to use prime factorizations. Recall from Chapter 1, we used factor trees to find the prime factorization of numbers. The prime factorizations of 42 and 60 are shown.

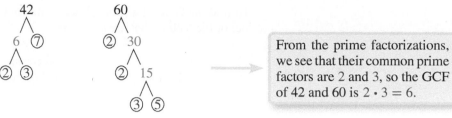

So, $42 = 2 \cdot 3 \cdot 7.$ So, $60 = 2 \cdot 2 \cdot 3 \cdot 5.$

From the prime factorizations, we see that their common prime factors are 2 and 3, so the GCF of 42 and 60 is $2 \cdot 3 = 6.$

Note: *We can also write the factorizations horizontally as shown.*

$$42 = 6 \cdot 7 = 2 \cdot 3 \cdot 7$$
$$60 = 2 \cdot 30 = 2 \cdot 2 \cdot 15 = 2 \cdot 2 \cdot 3 \cdot 5$$

Procedure: Finding the GCF of a Set of Integers

Step 1: Write each integer as a product of prime factors.
Step 2: Determine the prime factors that are common to each integer.
Step 3: The GCF is the product of these common factors. If there is no common factor, then the GCF is 1.

Objective 1 Examples Find the GCF of each set of integers.

1a. 56 and 42   **1b.** 18, 54, and 81   **1c.** 48 and 35

Solutions **1a.** $56 = 8 \cdot 7 = 2 \cdot 2 \cdot 2 \cdot 7$
$42 = 6 \cdot 7 = 2 \cdot 3 \cdot 7$

Since the common factors are 2 and 7, the GCF of 56 and 42 is $2 \cdot 7 = 14.$

1b. $18 = 2 \cdot 9 = 2 \cdot 3 \cdot 3$
$54 = 6 \cdot 9 = 2 \cdot 3 \cdot 3 \cdot 3$
$81 = 9 \cdot 9 = 3 \cdot 3 \cdot 3 \cdot 3$

Since the common factors are 3 and 3, the GCF of 18, 54, and 81 is $3 \cdot 3 = 9.$

1c. $48 = 4 \cdot 12 = 2 \cdot 2 \cdot 2 \cdot 2 \cdot 3$
$35 = 5 \cdot 7$

The numbers do not share any common factors. Therefore, the GCF of 48 and 35 is 1.

✓ Student Check 1 Find the GCF of each set of integers.
a. 60 and 36   **b.** 75, 51, and 99   **c.** 32 and 45

The Greatest Common Factor of a Set of Terms

Objective 2 ▶

Find the greatest common factor of a set of terms.

We can extend the idea of the GCF of integers to finding the greatest common factor of terms. Because terms involve integer coefficients and variables raised to exponents, we first need to determine how to find the greatest common factor of variables. Since we already know how to find the GCF of the integer coefficients, we will combine these results to find the GCF of a set of terms.

To find the GCF of x^3 and x^5, we write each term in expanded form and find the product of the factors that are common.

$$x^3 = \textcircled{x} \cdot \textcircled{x} \cdot \textcircled{x}$$
$$x^5 = \textcircled{x} \cdot \textcircled{x} \cdot \textcircled{x} \cdot x \cdot x$$

The common factors are three factors of x. So, the GCF is $x \cdot x \cdot x = x^3$.

 Note: *The GCF of the variables in the terms is the common variable raised to the smallest exponent of the terms.*

Procedure: Finding the GCF of a Set of Terms

Step 1: Find the GCF of the coefficients of the terms.
Step 2: Find the GCF of the variables of the terms.
Step 3: The GCF of the terms is the product of the GCF of the coefficients and the GCF of the variable factors.

Objective 2 Examples   Find the GCF of each set of terms.

2a. $4x^2$ and $10x$ \qquad\qquad **2b.** $-12a^3$ and $42a^4$

2c. $6x^2y$, $15xy^3$, and $20xy$ \qquad **2d.** $4(x - 2)$ and $y(x - 2)$

Solutions   **2a.** $4x^2 = 2 \cdot 2 \cdot x^2$
$10x = 2 \cdot 5 \cdot x$

The greatest common factor of 4 and 10 is 2. The greatest common factor of x^2 and x is x since 1 is the smallest exponent of the variable. So, the GCF of $4x^2$ and $10x$ is $2x$.

2b. $-12a^3 = -1 \cdot 2 \cdot 2 \cdot 3 \cdot a^3$
$42a^4 = 2 \cdot 3 \cdot 7 \cdot a^4$

The greatest common factor of -12 and 42 is $2 \cdot 3 = 6$. The greatest common factor of a^3 and a^4 is a^3 since 3 is the smallest exponent of the variable. So, the GCF of $-12a^3$ and $42a^4$ is $6a^3$.

2c. $6x^2y = 2 \cdot 3 \cdot x^2 \cdot y$
$15xy^3 = 3 \cdot 5 \cdot x \cdot y^3$
$20xy = 2 \cdot 2 \cdot 5 \cdot x \cdot y$

The coefficients do not have a common factor, so their GCF is 1. The greatest common factor of x^2y, xy^3, and xy is xy since 1 is the smallest exponent of both x and y. So, the GCF of $6x^2y$, $15xy^3$, and $20xy$ is xy.

2d. $4(x - 2) = 2 \cdot 2(x - 2)$
$y(x - 2)$

The greatest common factor of the coefficients is 1. The common variable factor, $(x - 2)$, has a smallest exponent of 1, so the GCF of the variable factors is $(x - 2)$. So, the GCF of $4(x - 2)$ and $y(x - 2)$ is $1(x - 2)$ or $(x - 2)$.

✓ Student Check 2   Find the GCF of each set of terms.

a. $6y$ and $9y^2$ \qquad\qquad **b.** $24b^5$ and $32b^4$

c. $8x^3y^2$, $20x^2y$, and $25x^4y^3$ \qquad **d.** $6(y + 4)$ and $z(y + 4)$

Factor out the Greatest Common Factor from a Polynomial

To use the greatest common factor when factoring, we apply the distributive property in reverse by "factoring out" the GCF. Recall:

$$2x(x^2 + 3) = 2x(x^2) + 2x(3) = 2x^3 + 6x$$

When this statement is written in reverse, we have

$$2x^3 + 6x = 2x(x^2) + 2x(3) = \underbrace{2x}(x^2 + 3)$$

$2x$ is the GCF   Factored form
of $2x^3$ and $6x$ of $2x^3 + 6x$

So, the polynomial $2x^3 + 6x$ is written as a product, or in factored form, where one of its factors is the GCF of its terms.

Property: The Distributive Property

For a, b, and c real numbers,

$$ab + ac = a(b + c)$$

Note: *a represents the greatest common factor of each term of the polynomial.*

Procedure: Factoring a Polynomial Using the Greatest Common Factor

Step 1: Find the GCF of the terms of the polynomial.
Step 2: Rewrite each term of the polynomial as a product of the GCF and the remaining factor.
Step 3: Apply the distributive property to factor out the GCF. The GCF "a" is written on the outside of parentheses and the remaining factors $(b + c)$ are written on the inside of the parentheses.
Step 4: Check the factorization by multiplying.

Objective 3 Examples   Use the GCF to factor each polynomial.

3a. $6x^2 - 18x$ **3b.** $14xy^3 + 49x^2y^2 + 7xy^2$

3c. $-2a^3 + 8a$ **3d.** $x(x + 5) + 3(x + 5)$

3e. A company's profit can by represented by $-5q^2 + 200q$, where q is the number of units sold in hundreds. Factor this polynomial.

Solutions   **3a.** The GCF of $6x^2$ and $18x$ is $6x$.

$$6x^2 - 18x = 6x(x) - 6x(3)$$ Rewrite each term as a product involving $6x$.
$$= 6x(x - 3)$$ Apply the distributive property.

Check: $6x(x - 3) = 6x(x) - 6x(3) = 6x^2 - 18x$

3b. The GCF of $14xy^3$, $49x^2y^2$, $7xy^2$ is $7xy^2$.

$14xy^3 + 49x^2y^2 + 7xy^2$
$= 7xy^2(2y) + 7xy^2(7x) + 7xy^2(1)$ Rewrite each term as a product involving $7xy^2$.
$= 7xy^2(2y + 7x + 1)$ Apply the distributive property.

Check: $7xy^2(2y + 7x + 1) = 7xy^2(2y) + 7xy^2(7x) + 7xy^2(1)$
$= 14xy^3 + 49x^2y^2 + 7xy^2$

3c. The GCF of $-2a^3$ and $8a$ is $2a$. When the first term of a polynomial is negative, it is common to make the GCF negative. Therefore, we will factor out $-2a$.

$-2a^3 + 8a = -2a(a^2) + (-2a)(-4)$ Rewrite each term as a product involving $-2a$.
$= -2a(a^2 - 4)$ Apply the distributive property.

Check: $-2a(a^2 - 4) = -2a(a^2) + (-2a)(-4) = -2a^3 + 8a$

We could have also factored out $2a$.

$-2a^3 + 8a = 2a(-a^2) + (2a)(4)$ Rewrite each term as a product involving $2a$.
$= 2a(-a^2 + 4)$ Apply the distributive property.

Check: $2a(-a^2 + 4) = 2a(-a^2) + (2a)(4) = -2a^3 + 8a$

 Note: *The difference between the two factorizations is the signs.*

3d. The GCF of $x(x + 5)$ and $3(x + 5)$ is $x + 5$. Each term is already expressed as a product of $(x + 5)$ and another factor.

$x(x + 5) + 3(x + 5) = (x + 5)(x + 3)$ Apply the distributive property.

Check: $(x + 5)(x + 3) = (x + 5)x + (x + 5)(3) = x(x + 5) + 3(x + 5)$.

3e. The GCF of $-5q^2$ and $200q$ is $5q$. Since the first term is negative, we will factor out $-5q$.

$-5q^2 + 200q$
$= -5q(q) + (-5q)(-40)$ Rewrite each term as a product involving $-5q$.
$= -5q(q - 40)$ Apply the distributive property.

Check: $-5q(q - 40) = -5q(q) - 5q(-40) = -5q^2 + 200q$

☑ Student Check 3   Use the GCF to factor each polynomial.

a. $5x^2 - 20x$ **b.** $21x^2y^4 + 63x^3y^2$

c. $-8a^3 + 72a^2 + 16a$ **d.** $x(x - 9) + 5(x - 9)$

e. The expression $8x^2 + 20x$ represents the area of a rectangle. Factor the expression to determine an expression for the length and width of the rectangle.

Grouping

Objective 4 ▶

Use grouping to factor polynomials.

The grouping method of factoring applies to polynomials with four or more terms in which pairs of terms have a common factor. If, after factoring out the GCF from pairs of terms, the remaining factors are the same, the polynomial can be factored. Consider the following polynomial.

$x^3 + x^2 + 2x + 2 = (x^3 + x^2) + (2x + 2)$ Group the first two terms and the last two terms.
$= x^2(x + 1) + 2(x + 1)$ Factor out the GCF from each pair of terms.
$= (x + 1)(x^2 + 2)$ Factor out the common binomial of $(x + 1)$.

In this illustration, note that after the GCF was factored out of each pair of terms, the binomial that remained was the same, namely $(x + 1)$. If, however, the binomial is not the same after removing the GCF from the two pairs of terms, then different terms may need to be grouped together. If no grouping produces a common binomial, then we can conclude that the polynomial cannot be factored and is a **prime polynomial**.

Procedure: Using Grouping to Factor a Polynomial

Step 1: Group together pairs of terms.
Step 2: Factor out the GCF from each pair of terms.
Step 3: Factor out the common binomial from both terms, if possible.
     a. If the binomial is not the same, try rearranging the grouping of terms.
     b. Repeat the preceding steps.
     c. If this doesn't result in a common binomial factor, then the polynomial cannot be factored.
Step 4: Check the answer by multiplying.

Objective 4 Examples   Use grouping to factor each polynomial. Check answer by multiplying.

4a. $12x^2 + 18x + 10x + 15$ **4b.** $5x^3 - 35x^2 + x - 7$

4c. $3x^3 - 7x^2 - 6x + 14$ **4d.** $6x^3 + 3x^2 + 6x + 3$

Solutions   **4a.** $12x^2 + 18x + 10x + 15$

$= (12x^2 + 18x) + (10x + 15)$ Group together the first two and last two terms.

$= 6x(2x + 3) + 5(2x + 3)$ Factor out the GCF from each pair of terms.

$= (2x + 3)(6x + 5)$ Factor out the common binomial, $2x + 3$.

Check: $(2x + 3)(6x + 5) = 12x^2 + 10x + 18x + 15$

Because of the commutative property of addition, this is equivalent to the given polynomial.

Note: *If one grouping works, then another grouping may also work. For example, grouping the first and third terms and the second and fourth terms gives us the following.*

$12x^2 + 18x + 10x + 15$

$= (12x^2 + 10x) + (18x + 15)$ Group together the first two and last two terms.

$= 2x(6x + 5) + 3(6x + 5)$ Factor out the GCF from each pair of terms.

$= (6x + 5)(2x + 3)$ Factor out the common binomial, $6x + 5$.

4b. $5x^3 - 35x^2 + x - 7$

$= (5x^3 - 35x^2) + (x - 7)$ Group together the first two and last two terms.

$= 5x^2(x - 7) + 1(x - 7)$ Factor out the GCF from the first pair of terms. Write $(x - 7)$ as $1(x - 7)$.

$= (x - 7)(5x^2 + 1)$ Factor out the common binomial, $x - 7$.

Check: $(x - 7)(5x^2 + 1) = 5x^3 + x - 35x^2 - 7$

4c. Group together the first two and the last two terms. Be sure to keep the negative sign on $6x$ in the grouping. Because the negative sign is on the first term of the second pair, we need to factor out a negative GCF.

$3x^3 - 7x^2 - 6x + 14$

$\quad = (3x^3 - 7x^2) + (-6x + 14)$ Group together the first two and last two terms.

$\quad = x^2(3x - 7) - 2(3x - 7)$ Factor out the GCF from each pair of terms.

$\quad = (3x - 7)(x^2 - 2)$ Factor out the common binomial, $3x - 7$.

Check: $(3x - 7)(x^2 - 2) = 3x^3 - 6x - 7x^2 + 14$

4d. Notice that each term in this polynomial has a common factor of 3. We factor this out first.

$$6x^3 + 3x^2 + 6x + 3 = 3(2x^3 + x^2 + 2x + 1)$$

Now we will factor $2x^3 + x^2 + 2x + 1$ by grouping.

$2x^3 + x^2 + 2x + 1$

$\quad = (2x^3 + x^2) + (2x + 1)$ Group together the first two and last two terms.

$\quad = x^2(2x + 1) + 1(2x + 1)$ Factor out the GCF from each pair of terms.

$\quad = (2x + 1)(x^2 + 1)$ Factor out the common binomial, $2x + 1$.

So, the complete factorization is

$$6x^3 + 3x^2 + 6x + 3 = 3(2x^3 + x^2 + 2x + 1)$$
$$= 3(2x + 1)(x^2 + 1)$$

Check: $3(2x + 1)(x^2 + 1) = (6x + 3)(x^2 + 1) = 6x^3 + 6x + 3x^2 + 3$

 **Student Check 4**   Use grouping to factor each polynomial. Check answer by multiplying.

    a. $8y^2 - 2y + 20y - 5$ **b.** $6x^4 - 42x^3 + x - 7$

    c. $2x^3 - 3x^2 - 12x + 18$ **d.** $12y^3 + 8y^2 + 48y + 32$

Objective 5 ▶

Troubleshoot common errors.

Troubleshooting Common Errors

Some common errors related to factoring using the GCF and grouping are shown next.

Objective 5 Examples   **A problem and an incorrect solution are given. Provide the correct solution and an explanation of the error.**

5a. Find the GCF of $120x^3y^2$ and $50x^4y$.

Incorrect Solution	Correct Solution and Explanation
$120 = 6 \cdot 20 = 2 \cdot 3 \cdot 2 \cdot 2 \cdot 5$ $50 = 5 \cdot 10 = 5 \cdot 2 \cdot 5$ The GCF of the coefficients is $2 \cdot 5 = 10$. The GCF of the variables is x^4y^2. So, the GCF is $10x^4y^2$.	An error was made in finding the GCF of the variables. The GCF of the variables is the smallest exponent, not the largest exponent. Therefore, the GCF of the variables is x^3y. So, the GCF is $10x^3y$.

5b. Factor $4x^3 - 12x$.

Incorrect Solution	Correct Solution and Explanation
$4x^3 - 12x^2$ $2x(2x^2 - 6x)$	A common factor was factored out but not the greatest common factor. The terms in the binomial still have something in common. The GCF of the terms in the binomial is $4x^2$. $4x^3 - 12x^2 = 4x^2(x - 3)$

5c. Factor $m^2 + 5m - 2mn - 10n$.

Incorrect Solution	Correct Solution and Explanation
$m^2 + 5m - 2mn - 10n$ $(m^2 + 5m) - (2mn - 10n)$ $m(m + 5) - 2n(m - 5)$ Since the binomials are not the same, this is prime.	Because the sign of $2mn$ is negative, we need to include the sign in the grouping. Then, factor out a negative GCF from the last two terms. $m^2 + 5m - 2mn - 10n$ $(m^2 + 5m) + (-2mn - 10n)$ $m(m + 5) - 2n(m + 5)$ $(m + 5)(m - 2n)$

5d. Factor $x^2 + 4x + 3x + 12$ by grouping.

Incorrect Solution	Correct Solution and Explanation
$x^2 + 4x + 3x + 12$ $(x^2 + 4x) + (3x + 12)$ $x(x + 4) + 3(x + 4)$ $(x + 3)(x + 4)(x + 4)$ $(x + 3)(x + 4)^2$ Since the binomials are not the same, this is prime.	The error was made in not factoring out the common binomial $(x + 4)$ correctly. $x^2 + 4x + 3x + 12$ $(x^2 + 4x) + (3x + 12)$ $x(x + 4) + 3(x + 4)$ $(x + 4)(x + 3)$

ANSWERS TO STUDENT CHECKS

Student Check 1 **a.** 12 **b.** 3 **c.** 1

Student Check 2 **a.** $3y$ **b.** $8b^4$ **c.** x^2y **d.** $(y + 4)$

Student Check 3 **a.** $5x(x - 4)$ **b.** $21x^2y^2(y^2 + 3x)$
 c. $-8a(a^2 - 9a - 2)$ **d.** $(x - 9)(x + 5)$
 e. $4x$ and $2x + 5$ are the length and width of the rectangle.

Student Check 4 **a.** $(4y - 1)(2y + 5)$ **b.** $(x - 7)(6x^3 + 1)$
 c. $(2x - 3)(x^2 - 6)$ **d.** $4(3y + 2)(y^2 + 4)$

SUMMARY OF KEY CONCEPTS

1. The greatest common factor (GCF) of a set of integers is the largest integer that is a factor of each integer. We can find this by comparing the list of factors or by writing the integers as a product of prime numbers.

2. For common variables, the GCF is the variable raised to the smallest exponent on the variables. The GCF of a group of terms is the product of the GCF of the coefficients and the GCF of the common variables.

3. To factor means to express as a product. Factoring is the reverse of multiplication. The distributive property enables us to factor a polynomial by factoring out the GCF. We identify the GCF of the terms in the polynomial and rewrite each term as a product of the GCF and another factor. The factored form is the GCF times the sum of the remaining factors. Factoring can always be checked by multiplying.

4. Grouping is used to factor a polynomial with four or more terms. Pair together terms. Factor out the GCF from each pair of terms. The remaining binomial must be the same to factor further. If the binomial is the same, factor it out and multiply by the remaining factors. If it is not the same, try regrouping terms. A polynomial that cannot be factored is a prime polynomial.

GRAPHING CALCULATOR SKILLS

The graphing calculator can be used in to find factors of integers as well as to find the GCF of a set of integers.

Example: Find factors of 42.

Solution: To find factors of 42 by hand, we divide it by 1, 2, 3, ... To perform this task using a calculator, we divide 42 by x and then view the table. If the corresponding value in the Y_1 column is an integer, then x is a factor of 42.

From the table, we see that factors of 42 are 1, 2, 3, 4, 6, 7, 14, 21, and 42. Some factorizations of 42 are 1 • 42, 2 • 21, 3 • 14, and 6 • 7.

Example: Find the GCF of 60 and 84.

Solution: The calculator has a built-in command called gcd which stands for the greatest common divisor, or greatest common factor. To access the command, press MATH and use the NUM menu, option 9.

So, the GCF of 60 and 84 is 12.

SECTION 6.1 / EXERCISE SET

Write About It!

Use complete sentences in your answer to each exercise.

1. Define the greatest common factor.

2. Explain how to find the GCF of a set of terms.

3. What does it mean to factor a polynomial?

4. Explain how to check if your factoring is correct.

5. Explain how to factor by grouping.

6. Explain when the factoring is completed.

7. Most math operations have an inverse, which "undoes" the operation (subtraction undoes addition, etc.). What is the inverse operation for factoring? Illustrate your answer with an example.

8. What is the inverse operation for factoring out the GCF? Illustrate your answer with an example.

Practice Makes Perfect!

Find the GCF of each set of integers. (*See Objective 1.*)

9. 42 and 30

10. 45 and 27

11. 54 and 24

12. 80, 32, and 48

13. 60, 45, and 90

14. 57, 84, and 18

15. 44, 66, and 110

16. 26, 65, and 91

Find the GCF of each set of terms. (*See Objective 2.*)

17. $6x^2$ and $3x^3$

18. $4x^2$ and $8x^3$

19. $-15a^2$ and $10a$

20. $-18a^3$ and $6a$

21. $-24b^3$ and $16b^2$

22. $-36b^4$ and $24b^5$

23. $12a^2b^3$ and $16a^3b^4$

24. $20a^5b^3$ and $12a^4b^2$

25. $-8ab^2$ and $14a^2b$

26. $-6a^3b^2$ and $15a^2b^3$

27. $4y, 16y^2$, and $8y^3$

28. $8y, 12y^2$, and $24y^4$

29. $6y^3, 18y^5$, and 24

30. $8y^4, 12y^3$, and 16

31. $5(x + 1)$ and $3(x + 1)$

32. $6(x + 3)$ and $7(x + 3)$

33. $8(y - 12)$ and $x(y - 12)$

34. $10(y + 9)$ and $x^2(y + 9)$

35. $6(a + 2)$ and $3(a + 2)$

36. $12(a + 4)$ and $14(a + 4)$

37. $3x(b - 3)$ and $5(b + 3)$

38. $5x(b + 7)$ and $4(b + 5)$

Use the GCF to factor each polynomial. (*See Objective 3.*)

39. $2x^2 - 4x$

40. $3x^2 - 9x$

41. $4a^4 - 16a^3$

42. $5a^5 - 15a^3$

43. $-3y^2 + 6y$

44. $-10y^3 + 12y^2$

45. $-36b^5 - 24b^3 - 12b^2$

46. $-48b^6 - 36b^5 + 12b^3$

47. $45x^3 - 30x^2 + 15x$

48. $18y^4 - 27y^3 - 9y^2$

49. $2a^2 + 2a - 2$

50. $5a^2 + 25a - 10$

51. $-8x^3 + 20x^2 + 4x$

52. $-6y^3 - 12y^2 + 6y$

53. $x(x + 3) + 2(x + 3)$

54. $x(x - 11) + 6(x - 11)$

55. $6(x - 2) + x(x - 2)$

56. $5(x - 5) + x(x - 5)$

57. $4(x + 3) - x(x + 3)$

58. $3(x + 9) - x(x + 9)$

59. $x^2(x + 1) - 3x(x + 1)$

60. $2x^2(x - 7) - x(x - 7)$

61. The expression $6y^3 + 16y^2$ represents the area of a rectangle. Factor the expression to determine expressions that could represent the length and width of the rectangle.

62. The expression $8y^3 + 12y^2$ represents the area of a rectangle. Factor the expression to determine expressions that could represent the length and width of the rectangle.

63. A company's profit can be represented by $-20q^2 + 500q$, where q is the number of units sold in thousands. Factor this polynomial.

64. A company's profit can be represented by $-15p^2 + 450p$, where p is the number of units sold in thousands. Factor this polynomial.

Use grouping to factor each polynomial. (*See Objective 4.*)

65. $6x^2 - 2x + 3x - 1$

66. $5x^2 + 2x + 20x + 8$

67. $4b^3 + 4b^2 + 3b + 3$

68. $2y^3 - 6y^2 + 5y - 15$

69. $5a - 20 - a^2 + 4a$

70. $6x - 12 - 5x^2 + 10x$

71. $r^3 + 5r^2 - 8r - 40$

72. $15s^3 - 10s^2 - 3s + 2$

73. $4a^3 - 20a^2 - a - 5$

74. $3c^3 + 6c^2 - c + 2$

75. $32x^2 - 56x + 12x - 21$

76. $14t^2 + 21t + 10t + 15$

77. $3x^3 + 12x^2 - 18x - 72$

78. $15y^3 + 15y - 35y^2 - 35$

79. $8x^3 - 40x^2 - 4x + 20$

80. $15d^3 - 15d^2 - 5d + 5$

 Mix 'Em Up!

Factor each polynomial.

81. $9y^2 - 81y$

82. $6r^2 - 21r$

83. $-16a^3b + 48a^2b - 8ab$

84. $30c^3d - 30c^2d + 6cd$

85. $3x^2 - 6x + 2x - 4$

86. $21z^2 - 35z + 3z - 5$

87. $3x^3 + 5x^2 - 6x + 10$

88. $9s^3 + 21s^2 - 3s + 7$

89. $36x^2 + 12x - 6x - 2$

90. $15t^2 - 45t + 12t - 36$

91. $-16a^4 + 48a^3 + 16a^2$

92. $-15b^4 - 18b^3 + 3b^2$

93. $4x^3(2x - 1) - 3x(2x - 1)$

94. $6y(y + 3) - 42(y + 3)$

95. $4a^2 + 6a - 6$

96. $3b^3 + 5b^2 - b$

97. $6x^2 + 30x - 5x - 25$

98. $4z^2 - 10z - 6z + 15$

99. The expression $6z^4 + 24z^2$ represents the area of a parallelogram. Factor the expression to determine expressions that could represent the base and height of the parallelogram.

100. The expression $5z^4 + 30z^2$ represents the area of a parallelogram. Factor the expression to determine expressions that could represent the base and height of the parallelogram.

101. The surface area of a circular cone is the sum of the area of the base and the area of the lateral surface, $\pi r^2 + \pi rl$, where r is the radius of the base and l is the slant height of the cone. Factor the expression.

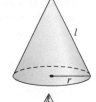

102. The surface area of a square pyramid is given by $x^2 + 2xh$, where x is the length of the square base and h is the slant height of the lateral face of the pyramid. Factor the expression.

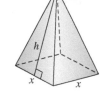

 You Be the Teacher!

Correct each student's errors, if any.

103. Find the GCF of x^2 and x^3.

Stephanie's work:

The GCF of x^2 and x^3 is x^3.

104. Find the GCF of $12a^3$ and $6a^4$.

Jessica's work:

The GCF of $12a^3$ and $6a^4$ is $12a^4$.

105. Factor $2x^2 - 6x - x + 3$.

Tammy's work:

$2x(x + 3) - 1(x + 3) = (x + 3)(2x - 1)$

106. Factor $6x^3 + 15x^2 - 18x - 45$.

Danny's work:

$3x^2(2x + 5) - 9(2x + 5) = (2x + 5)(3x^2 - 9)$

 Calculate It!

Find the GCF of each group of terms. Use a calculator to find the GCF of the coefficients of the given terms.

107. $60x^5$ and $72x^6$

108. $112x^8$ and $42x^3$

109. $512x^3$ and $48x^2$

110. $216x^6$ and $144x^5$

Think About It!

Write a binomial, in factored form, whose greatest common factor is given.

111. $6x$

112. $8a$

113. $2y^2$

114. $7b^3$

115. $-2ab$

116. $-4x^2y$

SECTION 6.2 **Factoring Trinomials**

▶ **OBJECTIVES**

As a result of completing this section, you will be able to

1. Factor trinomials of the form $x^2 + bx + c$.
2. Factor trinomials that have a common factor.
3. Troubleshoot common errors.

A ball is thrown upward with an initial velocity of 16 ft/sec from a height of 32 ft above the ground. The height, in feet, of the ball t sec after it is thrown can be represented by $-16t^2 + 16t + 32$. In Section 6.6, we will solve an equation that enables us to determine when the ball reaches the ground. This process requires us to factor the trinomial, $-16t^2 + 16t + 32$.

As discussed in Section 6.1, factoring is the reverse of multiplication. Factoring trinomials is like solving a puzzle. There are clues to the answer provided in the given problem. This section will show us how to use the given clues to factor a trinomial.

Factoring $x^2 + bx + c$

Objective 1 ▶

Factor trinomials of the form $x^2 + bx + c$.

Our goal is to express a trinomial of the form $x^2 + bx + c$ as a product. This trinomial has a leading coefficient of 1. Some examples of this type of trinomial are

$$x^2 + 9x + 14 \qquad y^2 - 8y + 15$$
$$x^2 - 2x - 35 \qquad y^2 + 4y - 12$$

The products that produce these trinomials are shown.

$$(x + 7)(x + 2) = x^2 + 7x + 2x + 14 = x^2 + 9x + 14$$
$$(y - 3)(y - 5) = y^2 - 3y - 5y + 15 = y^2 - 8y + 15$$
$$(x - 7)(x + 5) = x^2 + 5x - 7x - 35 = x^2 - 2x - 35$$
$$(y + 6)(y - 2) = y^2 - 2y + 6y - 12 = y^2 + 4y - 12$$

If we reverse the preceding products, we begin with a trinomial and end with its factored form. The following table shows some important relationships between the binomial factors and the trinomial.

Trinomial	Factored Form	Important Observations
$x^2 + 9x + 14$	$= (x + 7)(x + 2)$	7 and 2 are factors of 14 whose sum is 9.
$y^2 - 8y + 15$	$= (y - 3)(y - 5)$	-3 and -5 are factors of 15 whose sum is -8.
$x^2 - 2x - 35$	$= (x - 7)(x + 5)$	-7 and 5 are factors of -35 whose sum is -2.
$y^2 + 4y - 12$	$= (y + 6)(y - 2)$	6 and -2 are factors of -12 whose sum is 4.

Observe that when the sign of the last term is *positive*, the binomial factors have the *same* sign. For example,

$$x^2 + 9x + 14 = (x + 7)(x + 2)$$
$$y^2 - 8y + 15 = (y - 3)(y - 5)$$

When the sign of the last term is *negative*, the binomial factors have *opposite* signs.

$$x^2 - 2x - 35 = (x - 7)(x + 5)$$
$$y^2 + 4y - 12 = (y + 6)(y - 2)$$

Note also that the first two terms of the binomial factors have a product equal to the first term of the trinomial.

Property: Factoring $x^2 + bx + c$

For b and c real numbers, the factored form of $x^2 + bx + c$ is

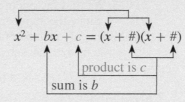

$$x^2 + bx + c = (x + \#)(x + \#)$$

product is c

sum is b

Procedure: Factoring a Trinomial of the Form $x^2 + bx + c$

Step 1: List pairs of factors of the last term of the trinomial, c.

Step 2: Determine the signs of the factors that produce the correct product.
    a. If the last term is positive, both factors are of the same sign.
    b. If the last term is negative, the factors are opposite signs.

Step 3: Choose the pair of factors whose sum is the middle coefficient, b. If no such factors exist, then the trinomial is prime and cannot be factored.

Step 4: Arrange the pair of factors in the binomials.

Step 5: Check answer by multiplying.

Objective 1 Examples

Factor each trinomial. Check answer by multiplying.

1a. $a^2 + 20a + 75$ **1b.** $h^2 - 13hk + 36k^2$ **1c.** $y^2 + 3y + 4$

1d. $a^4 - 9a^2 + 18$ **1e.** $x^2 - 6x - 16$ **1f.** $x^2 + 4x - 10$

1g. $a^2 + 7a - 30$ **1h.** $x^2 - 3xy - 54y^2$

Solutions

1a. We need factors of 75 whose sum is 20. Since the last term is positive, the signs of the factors are the same. Since the sum is positive, we consider only positive factors. The factors that satisfy the conditions are 5 and 15. So, the factorization is

Factors of 75	Sum of the Factors
1, 75	$1 + 75 = 76$
3, 25	$3 + 25 = 28$
5, 15	$5 + 15 = 20$

$$a^2 + 20a + 75 = (a + 5)(a + 15)$$

Check: $(a + 5)(a + 15) = a^2 + 15a + 5a + 75 = a^2 + 20a + 75$

1b. We find factors of 36 whose sum is -13. Since the last term is positive, the signs of the factors are the same. Since the sum is negative, we consider only negative factors.
    The factors that satisfy the conditions are -4 and -9. For the last term to have a product of $36k^2$, we use $-4k$ and $-9k$ in the factorization.

Factors of 36	Sum of the Factors
$-1, -36$	$-1 + (-36) = -37$
$-2, -18$	$-2 + (-18) = -20$
$-3, -12$	$-3 + (-12) = -15$
$-4, -9$	$-4 + (-9) = -13$
$-6, -6$	$-6 + (-6) = -12$

$$h^2 - 13hk + 36k^2 = (h - 4k)(h - 9k)$$

Check: $(h - 4k)(h - 9k) = h^2 - 9hk - 4hk + 36k^2 = h^2 - 13hk + 36k^2$

1c. We find factors of 4 whose sum is 3. Since the sum is positive, we consider only positive factors.
    There are not any factors of 4 whose sum is 3. So, the trinomial $y^2 + 3y + 4$ is *prime*.

Factors of 4	Sum of the Factors
1, 4	$1 + 4 = 5$
2, 2	$2 + 2 = 4$

1d. The first term is a^4. The factors that produce a^4 are a^2 and a^2. The last term is 18. We find factors of 18 whose sum is −9. Since the sum is negative, we consider only negative factors.

Factors of 18	Sum of the Factors
−1, −18	−1 + (−18) = −19
−2, −9	−2 + (−9) = −11
−3, −6	−3 + (−6) = −9

The factors that satisfy the conditions are −3 and −6. So, the factorization is

$$a^4 - 9a^2 + 18 = (a^2 - 3)(a^2 - 6)$$

Check: $(a^2 - 3)(a^2 - 6) = a^4 - 6a^2 - 3a^2 + 18 = a^4 - 9a^2 + 18$

1e. We find factors of −16 whose sum is −6. Since the last term is negative, the factors have opposite signs.
The factors that satisfy the conditions are 2 and −8. So, the factorization is

Factors of −16	Sum of the Factors
1, −16	1 + (−16) = −15
2, −8	2 + (−8) = −6
4, −4	4 + (−4) = 0
−1, 16	−1 + 16 = 15
−2, 8	−2 + 8 = 6

$$x^2 - 6x - 16 = (x + 2)(x - 8)$$

Check: $(x + 2)(x - 8) = x^2 - 8x + 2x - 16 = x^2 - 6x - 16$

1f. We find factors of −10 whose sum is 4. Since the last term is negative, the factors have opposite signs.
None of the factors satisfy the required condition, so $x^2 + 4x - 10$ is *prime*.

Factors of −10	Sum of the Factors
−1, 10	−1 + 10 = 9
−2, 5	−2 + 5 = 3
1, −10	1 + (−10) = −9
2, −5	2 + (−5) = −3

1g. We find factors of −30 whose sum is 7. Since the last term is negative, the factors have opposite signs.
The factors that satisfy the conditions are −3 and 10. So, the factorization is

Factors of −30	Sum of the Factors
−1, 30	−1 + 30 = 29
−2, 15	−2 + 15 = 13
−3, 10	−3 + 10 = 7
−5, 6	−5 + 6 = 1
1, −30	1 + (−30) = −29
2, −15	2 + (−15) = −13
3, −10	3 + (−10) = −7
5, −6	5 + (−6) = −1

$$a^2 + 7a - 30 = (a - 3)(a + 10)$$

Check: $(a - 3)(a + 10) = a^2 + 10a - 3a - 30$
$$= a^2 + 7a - 30$$

1h. We find factors of −54 whose sum is −3. Since the last term is negative, the factors have opposite signs.
The factors 6 and −9 satisfy the conditions. For the last term to have a product of $-54y^2$, we use $6y$ and $-9y$ in the factorization.

Factors of −54	Sum of the Factors
1, −54	1 + (−54) = −53
2, −27	2 + (−27) = −25
3, −18	3 + (−18) = −15
6, −9	6 + (−9) = −3
−1, 54	−1 + 54 = 53
−2, 27	−2 + 27 = 25
−3, 18	−3 + 18 = 15
−6, 9	−6 + 9 = 3

$$x^2 - 3yx - 54y^2 = (x + 6y)(x - 9y)$$

Check: $(x + 6y)(x - 9y) = x^2 - 9xy + 6xy - 54y^2$
$$= x^2 - 3xy - 54y^2$$

✓ Student Check 1 Factor each trinomial and check answer by multiplying.

a. $x^2 + 12x + 32$ **b.** $a^2 - 38ab + 72b^2$ **c.** $y^2 + 13y + 24$

d. $r^4 - 11r^2 + 30$ **e.** $x^2 - x - 42$ **f.** $a^2 + 12a - 45$

g. $x^2 + 11x - 18$ **h.** $x^2 - xy - 12y^2$

Factoring Trinomials with Common Factors

Objective 2 ▶

Factor trinomials that have a common factor.

One of the key steps in factoring a polynomial is to factor out any common factors first. After the common factors are factored out, the remaining polynomial may need to be factored further.

Procedure: Factoring Trinomials with a Common Factor

Step 1: Factor out the GCF.

Step 2: If the remaining trinomial has a coefficient of 1, determine if it can be factored further.

Step 3: Check by multiplying.

Objective 2 Examples

Factor each polynomial completely. Check the answer by multiplying.

2a. $2x^2 - 12x + 16$ **2b.** $-5x^3 - 15x^2 + 20x$

2c. A ball is thrown upward with an initial velocity of 16 ft/sec from a height of 32 ft above the ground. The height, in feet, of the ball t seconds after it is thrown can be represented by $-16t^2 + 16t + 32$.

Solutions

2a.

$$2x^2 - 12x + 16 = 2(x^2 - 6x + 8)$$ Factor out the GCF, 2.

$$= 2(x - 4)(x - 2)$$ The factors of 8 whose sum is -6 are -4 and -2.

Check: $2(x - 4)(x - 2) = (2x - 8)(x - 2)$

$$= 2x^2 - 4x - 8x + 16$$

$$= 2x^2 - 12x + 16$$

2b. The terms of the trinomial have a common factor of $5x$. Since the leading term is negative, we factor out $-5x$.

$$-5x^3 - 15x^2 + 20x = -5x(x^2 + 3x - 4)$$ Factor out the GCF, $-5x$.

$$= -5x(x - 1)(x + 4)$$ The factors of -4 whose sum is 3 are 4 and -1.

Check: $-5x(x - 1)(x + 4) = (-5x^2 + 5x)(x + 4)$

$$= -5x^3 - 20x^2 + 5x^2 + 20x$$

$$= -5x^3 - 15x^2 + 20x$$

2c. The terms of the trinomial have a GCF of 16. Since the leading term is negative, we factor out -16.

$$-16t^2 + 16t + 32 = -16(t^2 - t - 2)$$ Factor out the GCF, -16.

$$= -16(t - 2)(t + 1)$$ The factors of -2 whose sum is -1 are -2 and 1.

Check: $-16(t - 2)(t + 1) = (-16t + 32)(t + 1)$

$$= -16t^2 - 16t + 32t + 32$$

$$= -16t^2 + 16t + 32$$

☑ Student Check 2

Factor each polynomial completely. Check the answer by multiplying.

a. $3y^2 + 33y + 84$ **b.** $-6x^3 + 6x^2 + 12x$

c. The height of a ball thrown upward with an initial velocity of 32 ft/sec from a height of 128 ft above the ground can be represented by $-16t^2 + 32t + 128$, where t is the number of seconds after the ball is thrown.

Objective 3 ▶	Troubleshooting Common Errors
Troubleshoot common errors.	Some common errors associated with factoring trinomials of the form $x^2 + bx + c$ are shown next.

Objective 3 Examples	A problem and an incorrect solution are given. Provide the correct solution and an explanation of the error.

3a. Factor $x^2 - x - 6$.

Incorrect Solution	Correct Solution and Explanation
The factors that produce -6 are -2 and 3. So, the factorization is $$x^2 - x - 6 = (x - 2)(x + 3)$$	If we multiply the two binomials, we get $x^2 + x - 6$. The middle sign is wrong since $-2 + 3 = 1$, not -1. We need factors of -6 whose sum is -1, the middle coefficient. The factors are -3 and 2. $$x^2 - x - 6 = (x - 3)(x + 2)$$

3b. Factor $y^2 + 2y + 35$.

Incorrect Solution	Correct Solution and Explanation
The factors that produce 35 that can give us 2 are 7 and 5. So, the factorization is $$y^2 + 2y + 35 = (y + 5)(y + 7)$$	The difference of 7 and 5 is 2 but the sum of the factors is 12. There are not any factors of 35 whose sum is 2. Therefore, $$y^2 + 2y + 35 \text{ is prime.}$$

3c. Factor $4x^3 + 8x^2 - 32x$.

Incorrect Solution	Correct Solution and Explanation
The GCF is $4x$. When we factor $4x$ out, we get $x^2 + 2x - 8$. This factors as $$(x + 4)(x - 2)$$	The error made was not keeping the GCF with the remaining factors. $$4x^3 + 8x^2 - 32x = 4x(x^2 + 2x - 8)$$ $$= 4x(x + 4)(x - 2)$$

ANSWERS TO STUDENT CHECKS

Student Check 1   **a.** $(x + 4)(x + 8)$
   b. $(a - 36b)(a - 2b)$ **c.** prime
   d. $(r^2 - 5)(r^2 - 6)$ **e.** $(x - 7)(x + 6)$

f. $(a + 15)(a - 3)$ **g.** prime   **h.** $(x - 4y)(x + 3y)$
Student Check 2   **a.** $3(y + 7)(y + 4)$ **b.** $-6x(x - 2)(x + 1)$
   c. $-16(t + 2)(t - 4)$

SUMMARY OF KEY CONCEPTS

Here is a summary of the "clues" given in a trinomial of the form $x^2 + bx + c$ that will enable us to factor it.

1. The last term of the trinomial tells us the number to factor. So, begin by making a list of all the factors of the number c.

2. The sign of the last term tells us if the factors are going to have the same sign or different signs.

   a. If $c > 0$, the factors will have the same sign.

   b. If $c < 0$, the factors will have different signs.

3. The sign of the middle coefficient, b, tells us the specific signs.

   a. If the signs of the factors are the same, they are the same as the sign of the middle coefficient.

   b. If the signs of the factors are different, the sign of the middle coefficient is the sign of the larger factor.

4. The coefficient of the middle term, b, tells us the sum of the factors of c.

5. From the list of factors, we select the pair whose sum is b. If there are no such factors, then the trinomial is prime.

6. Always check the answer by multiplying.

7. If the leading coefficient is not 1, then look for a common factor to factor out first.

GRAPHING CALCULATOR SKILLS

The graphing calculator can assist us in finding factors of a number and their sum. The calculator can also be used to verify our factored form.

Example 1: Find the factors of -75 whose sum is -10.

Recall, to factor numbers on the calculator, we can divide the number by x and view the table for different values. So, to find the factors of -75, we enter $-75/X$ as Y_1. To find the sum of the factors, we enter $X + Y_1$.

Now we view the table.

The first two columns are the factors of -75. The last column shows the sum of the factors. So, -5 and 15 are the factors of -75 whose sum is 10.

Example 2: Verify that $x^2 - 6x - 16 = (x - 8)(x + 2)$.

Enter the trinomial in Y_1 and the factored form in Y_2. If the factored form is correct, these two expressions will have the same value when evaluated for different values of x. We can confirm this using the table feature.

Since the values of Y_1 and Y_2 are the same for every value of x, our factoring is correct.

SECTION 6.2 / EXERCISE SET

Write About It!

Use complete sentences in your answer to each exercise.

1. Use an example to explain how to factor a trinomial in the form $x^2 + bx + c = 0$, where c is a positive number.

2. Use an example to explain how to factor a trinomial in the form $x^2 + bx + c = 0$, where c is a negative number.

3. Explain the conditions the factors of the constant term must satisfy in Exercises 7 and 8.

4. Explain the conditions the factors of the constant term must satisfy in Exercises 33 and 34.

Practice Makes Perfect!

Factor each trinomial completely. Check by multiplying. (*See Objective 1.*)

5. $a^2 + 6a + 5$ 6. $a^2 + 7a + 6$

7. $x^2 + 5x + 6$ 8. $x^2 + 9x + 20$

9. $x^2 - 17x + 72$ 10. $x^2 - 15x + 26$

11. $a^2 - 10a + 12$ 12. $a^2 + 12a + 6$

13. $b^2 - 18b + 80$ 14. $b^2 - 10b + 24$

15. $x^2 + 21xy + 90y^2$ 16. $a^2 + 20ab + 64b^2$

17. $h^2 - 17hk + 70k^2$ 18. $c^2 - 16cd + 48d^2$

19. $r^2 + 13r + 40$ 20. $s^2 + 20s + 99$

21. $a^2 + a - 20$ 22. $c^2 + 4c - 21$

23. $x^2 + 12x - 45$ 24. $y^2 + 8y - 9$

25. $x^2 + 9xy - 10y^2$
26. $a^2 - ab - 72b^2$
27. $h^2 + 4h - 4$
28. $k^2 + 10k - 14$
29. $b^2 - 7b - 60$
30. $r^2 + 5r - 66$
31. $a^2 - 4ab - 12b^2$
32. $c^2 - 6cd - 27d^2$
33. $h^2 - 11h - 42$
34. $k^2 - 6k - 40$
35. $x^2 - 4xy - 45y^2$
36. $s^2 + st - 56t^2$

Factor each trinomial completely. Check by multiplying. (See Objective 2.)

37. $4x^2 - 12x - 72$
38. $3x^2 + 3x - 270$
39. $5y^2 - 15y - 20$
40. $6y^2 - 30y - 84$
41. $-4a^2 - 40a - 36$
42. $-8b^2 + 40b + 48$
43. $3r^3 - 21r^2 - 24r$
44. $5s^3 - 30s^2 - 80s$
45. $-6x^3 + 6x^2 + 72x$
46. $-4y^3 - 16y^2 + 84y$
47. $9p^4 - 9p^3 - 270p^2$
48. $3q^4 - 6q^3 + 3q^2$
49. $5x^3y + 5x^2y - 30xy$
50. $6x^3y - 36x^2y + 30xy$
51. $-3a^3b + 39a^2b - 126ab$
52. $-2a^3b - 6a^2b + 80ab$

53. The height of a ball thrown upward with an initial velocity of 96 ft/sec from a height of 640 ft above the ground can be represented by $-16t^2 + 96t + 640$ where t is the number of seconds after the ball is thrown.

54. The height of a ball thrown upward with an initial velocity of 128 ft/sec from a height of 768 ft above the ground can be represented by $-16t^2 + 128t + 768$, where t is the number of seconds after the ball is thrown.

55. The volume of an open box with height x in. is given by the expression $x^3 + 4x^2 - 12x$ in.^{3}.

56. The volume of an open box with height x in. is given by the expression $x^3 + 7x^2 - 8x$ in.^{3}.

 Mix 'Em Up!

Factor each trinomial completely.

57. $x^2 - 15x + 56$
58. $x^2 + 2x - 63$
59. $3y^3 - 12y^2 + 12y$
60. $-2x^3 - 4x^2 - 2x$
61. $p^2 - 15pq + 54q^2$
62. $c^2 + 19cd + 60d^2$
63. $x^2 + 4x + 7$
64. $t^2 - 5t - 10$
65. $-6x^4 + 24x^3 + 126x^2$
66. $4a^3 + 12a^2 - 160a$
67. $r^2 + 16rs + 39s^2$
68. $h^2 - 8hk - 48k^2$
69. $3x^4 + 33x^3 + 54x^2$
70. $-10x^3 + 60x^2 + 270x$
71. $-2s^2 - 16s + 40$
72. $a^2 + 10ab - 9b^2$
73. $x^2 + 4xy + 5y^2$

74. The volume of an open box with height x in. is given by the expression $x^3 + 5x^2 - 36x$ in.^{3}.

75. The volume of an open box with height x in. is given by the expression $x^3 + 9x^2 - 36x$ in.^{3}.

76. A company's profit can be represented by $-20q^2 + 700q - 5000$, where q is the number of units sold, in thousands.

77. A company's profit can be represented by $-16q^2 + 160q + 9600$, where q is the number of units sold, in thousands.

 You Be the Teacher!

Explain each situation to a student.

78. Explain how to factor $a^2 - 6a + 8$.

79. Explain how to factor $b^2 + 2b - 24$.

Correct each student's errors, if any.

80. Factor $-3x^3 + 12x^2 - 36$.

Johannes's work:
$$-3x^3 + 12x - 36 = -3x(x^2 + 4x - 12)$$
$$= -3x(x - 6)(x + 2)$$

81. Factor $2x^2 - 18xy + 36y^2$.

Sarah's work:
$$2x^2 - 18xy + 36y^2 = 2(x^2 - 9x + 18)$$
$$= 2(x - 6)(x - 3)$$

 Calculate It!

Use the tables provided to write the factorization of each trinomial.

82. $x^2 - 4x - 60$

Plot1 Plot2 Plot3
\Y1目-60/X
\Y2目Y1+X
\Y3=
\Y4=
\Y5=
\Y6=
\Y7=

X	Y1	Y2
1	-60	-59
2	-30	-28
3	-20	-17
4	-15	-11
5	-12	-7
6	-10	-4
7	-8.571	-1.571
X=6		

83. $x^2 - 17x + 42$

Plot1 Plot2 Plot3
\Y1目42/X
\Y2目Y1+X
\Y3=
\Y4=
\Y5=
\Y6=
\Y7=

X	Y1	Y2
-18	-2.333	-20.33
-17	-2.471	-19.47
-16	-2.625	-18.63
-15	-2.8	-17.8
-14	-3	-17
-13	-3.231	-16.23
-12	-3.5	-15.5
X=-18		

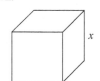

 **Think About It!**

Write a trinomial of the form $x^2 + bx + c$ that is factorable and has the given last term. Explain how you determined the middle term of your trinomial.

84. 15   **85.** 28   **86.** −12   **87.** −30

Write a trinomial of the form $x^2 + bx + c$ that is prime with the given last term. Explain how you determined the middle term of your trinomial.

88. 14   **89.** −8

SECTION 6.3 More on Factoring Trinomials

▶ OBJECTIVES

As a result of completing this section, you will be able to

1. **Factor trinomials using trial and error.**
2. **Factor trinomials using grouping.**
3. **Factor perfect square trinomials.**
4. **Troubleshoot common errors.**

The annual profit, in thousands of dollars, for a mobile phone company can be represented by $-4x^2 + 102x - 50$, where x is the total number of mobile phones sold for the year, in thousands. In Section 6.6, we will solve an equation that enables us to determine how many phones need to be sold for the company to break even. This process requires us to factor the trinomial $-4x^2 + 102x - 50$.

This section will present two different methods for factoring trinomials of the form $ax^2 + bx + c$. Both methods will produce the same results. Your instructor may choose to present only one of these methods.

Factoring by Trial and Error

Objective 1 ▶

Factor trinomials using trial and error.

In Objective 1 of Section 6.2, we dealt primarily with trinomials whose leading coefficient was one. Now we will deal with trinomials whose leading coefficient is not one. Some examples are

$$6x^2 + 19x + 10 \qquad 4y^2 - 29y + 7 \qquad 12c^2 + 4c - 5$$

The products that produce these trinomials are shown. Pay particular attention to the relationship between the two binomials and the trinomial.

$$(2x + 5)(3x + 2) = 6x^2 + 4x + 15x + 10 = 6x^2 + 19x + 10$$
$$(4y - 1)(y - 7) = 4y^2 - 28y - y + 7 = 4y^2 - 29y + 7$$
$$(6c + 5)(2c - 1) = 12c^2 - 6c + 10c - 5 = 12c^2 + 4c - 5$$

By reversing the preceding statements, we get the factored form of the trinomials.

Trinomial Factored Form	Important Observations
$6x^2 + 19x + 10 = (2x + 5)(3x + 2)$	$2x$ and $3x$ are factors of $6x^2$. 5 and 2 are factors of 10. The sum of the outer and inner products, $4x$ and $15x$, is $19x$.
$4y^2 - 29y + 7 = (4y - 1)(y - 7)$	$4y$ and y are factors of $4y^2$. -1 and -7 are factors of 7. The sum of the outer and inner products, $-28y$ and $-y$, is $-29y$.
$12c^2 + 4c - 5 = (6c + 5)(2c - 1)$	$6c$ and $2c$ are factors of $12c^2$. 5 and -1 are factors of -5. The sum of the outer and inner products, $-6c$ and $10c$, is $4c$.

So we conclude the following:

1. The first terms of the binomials are factors of the first term of the trinomial.
2. The last terms of the binomials are factors of the last term of the trinomial.
3. The sum of the outer and inner products is the middle term of the trinomial.

We can use this as a guideline for factoring trinomials using the **trial-and-error method**. The process of trial and error involves arranging all possible factors of the first term of the trinomial and factors of the last term of the trinomial until we obtain the correct outer and inner products.

Property: **Factoring $ax^2 + bx + c$ by Trial and Error**

For a, b, and c real numbers and $a \neq 0$, the factored form of $ax^2 + bx + c$ is

$$\overset{\text{F}\quad \text{O}+\text{I}\quad \text{L}}{ax^2 + bx + c} = (\underline{}\,x + \underline{})(\underline{}\,x + \underline{})$$

Procedure: **Factoring a Trinomial of the Form $ax^2 + bx + c$ Using Trial and Error**

Step 1: Factor out any common factors, if possible.
Step 2: List the factors of the first term of the trinomial.
Step 3: List the factors of the last term of the trinomial.
Step 4: Determine the appropriate signs of the factors in Steps 2 and 3.
Step 5: Arrange these factors in two binomials until the product produces the given trinomial.
Step 6: Check by multiplying.

When trying different factors, be mindful of the following.

• If the first arrangement of factors doesn't work, obtain another combination by reversing the position of the factors of the last term.
• If the trinomial does not have a common factor, the binomial factors cannot have a common factor. So, we can eliminate any combinations in which one or both of the binomials has a common factor in this case.
• If none of the combinations work, then the trinomial is *prime*.

Objective 1 Examples Factor each trinomial using trial and error.

1a. $3x^2 + 10x + 8$ **1b.** $2a^2 - 5a - 12$ **1c.** $15x^2y - 21xy + 6y$

1d. The annual profit, in thousands of dollars, for a mobile phone company can be represented by $-4x^2 + 102x - 50$, where x is the total number of mobile phones sold for the year, in thousands.

Solutions **1a.** We find the factors of $3x^2$ and the factors of 8. Since the last term of the trinomial is positive and the middle term is positive, we need to use only pairs of positive factors.

Factors of $3x^2$ and 8		Possible Factors	
$3x, x$	1, 8	$(3x+1)(x+8) = 3x^2 + 24x + x + 8$ $= 3x^2 + 25x + 8$	Wrong middle term
		$(3x+8)(x+1) = 3x^2 + 3x + 8x + 8$ $= 3x^2 + 11x + 8$	Wrong middle term
	2, 4	$(3x+2)(x+4) = 3x^2 + 12x + 2x + 8$ $= 3x^2 + 14x + 8$	Wrong middle term
		$(3x+4)(x+2) = 3x^2 + 6x + 4x + 8$ $= 3x^2 + 10x + 8$	Correct middle term

So, the factorization is $3x^2 + 10x + 8 = (3x+4)(x+2)$.

1b. We find the factors of $2a^2$ and the factors of -12. Since the last term of the trinomial is negative, the factors of -12 must have opposite signs.

Factors of $2a^2$ and -12		Possible Factors	
$2a, a$	$1, -12$	$(2a + 1)(a - 12) = 2a^2 - 12a + a - 12$ $= 2a^2 - 11a - 12$	Wrong middle term
		$(2a - 12)(a + 1)$ no need to check	Common factor
	$-1, 12$	$(2a - 1)(a + 12) = 2a^2 + 24a - a - 12$ $= 2a^2 + 23a - 12$	Wrong middle term
		$(2a + 12)(a - 1)$ no need to check	Common factor
	$2, -6$	$(2a + 2)(a - 6)$ no need to check	Common factor
		$(2a - 6)(a + 2)$ no need to check	Common factor
	$-2, 6$	$(2a - 2)(a + 6)$ no need to check	Common factor
		$(2a + 6)(a - 2)$ no need to check	Common factor
	$3, -4$	$(2a + 3)(a - 4) = 2a^2 - 8a + 3a - 12$ $= 2a^2 - 5a - 12$	Correct middle term
		$(2a - 4)(a + 3)$ no need to check	Common factor
	$-3, 4$	$(2a - 3)(a + 4) = 2a^2 + 8a - 3a - 12$ $= 2a^2 + 5a - 12$	Wrong middle term
		$(2a + 4)(a - 3)$ no need to check	Common factor

So, the factorization is $2a^2 - 5a - 12 = (2a + 3)(a - 4)$.

1c. The terms in this trinomial have a GCF of $3y$. We factor this out first.

$$15x^2y - 21xy + 6y = 3y(5x^2 - 7x + 2)$$

Now we use trial and error to factor $5x^2 - 7x + 2$. Since the last term is positive, the signs of its factors are the same. To obtain a negative sum, the factors of the last term are both negative.

Factors of $5x^2$ and 2		Possible Factors	
$5x, x$	$-1, -2$	$(5x - 1)(x - 2) = 5x^2 - 10x - x + 2$ $= 5x^2 - 11x + 2$	Wrong middle term
		$(5x - 2)(x - 1) = 5x^2 - 5x - 2x + 2$ $= 5x^2 - 7x + 2$	Correct middle term

So, the factorization is

$$15x^2y - 21xy + 6y = 3y(5x^2 - 7x + 2)$$
$$= 3y(5x - 2)(x - 1)$$

1d. The terms in this trinomial have a GCF of 2. Since the first term is negative, we factor out -2.

$$-4x^2 + 102x - 50 = -2(2x^2 - 51x + 25)$$

Now we use trial and error to factor $2x^2 - 51x + 25$.

Factors of $2x^2$ and 25		Possible Factors	
$2x, x$	$-1, -25$	$\begin{aligned}(2x - 1)(x - 25) &= 2x^2 - 50x - x + 25 \\ &= 2x^2 - 51x + 25\end{aligned}$	Correct middle term
		$\begin{aligned}(2x - 25)(x - 1) &= 2x^2 - 2x - 25x + 25 \\ &= 2x^2 - 27x + 25\end{aligned}$	Wrong middle term
	$-5, -5$	$\begin{aligned}(2x - 5)(x - 5) &= 2x^2 - 10x - 5x + 25 \\ &= 2x^2 - 15x + 25\end{aligned}$	Wrong middle term

So, the factorization is

$$\begin{aligned}-4x^2 + 102x - 50 &= -2(2x^2 - 51x + 25) \\ &= -2(2x - 1)(x - 25)\end{aligned}$$

☑ **Student Check 1**   Factor each polynomial using trial and error.

a. $5x^2 + 21x + 4$ **b.** $7x^2 - 11x - 6$ **c.** $12xy^3 + 2xy^2 - 30xy$

d. The profit, in thousands, of a toy manufacturer is $-9x^2 + 129x + 90$, where x is in hundreds.

Factoring by Grouping

Objective 2 ▶

Factor trinomials using grouping.

Factoring by trial and error is a method that works but it can be a lengthy and tedious process, especially when the first term of the trinomial and/or the last term have several factors. The method illustrated in this objective provides us with a very systematic approach to factoring trinomials.

This process expands the middle term of the trinomial to two terms, specifically the outer and inner products that would result from multiplying the two binomial factors together. This expanded polynomial now has four terms, which we can factor by grouping. This method is basically FOIL in reverse.

We will use the products that we examined at the beginning of this section to illustrate this concept. It is important for us to determine how the outer and inner products relate to the terms in the trinomial.

Trinomial and Its Factored Form	Observations
$6x^2 + 19x + 10$ $6x^2 + 4x + 15x + 10$ $(2x + 5)(3x + 2)$	The coefficients of the outer and inner terms are 4 and 15. Note that $4(15) = 60$ and $4 + 15 = 19$. Also note that $ac = 6(10) = 60$.
$12c^2 + 4c - 5$ $12c^2 - 6c + 10c - 5$ $(6c + 5)(2c - 1)$	The coefficients of the outer and inner terms are -6 and 10. Note that $-6(10) = -60$ and $-6 + 10 = 4$. Also note that $ac = 12(-5) = -60$.
$4y^2 - 29y + 7$ $4y^2 - 28y - y + 7$ $(4y - 1)(y - 7)$	The coefficients of the outer and inner terms are -28 and -1. Note that $-28(-1) = 28$ and $-28 + (-1) = -29$. Also note that $ac = 4(7) = 28$.

From the observations, we can conclude that the coefficients of the outer and inner terms have a product equal to ac and have a sum equal to the middle coefficient of the trinomial. We summarize this as follows.

Property: **Factoring $ax^2 + bx + c$ by Grouping**

The coefficients of the outer and inner terms are factors of the number $a \cdot c$, whose sum is b.

$$ax^2 + bx + c = ax^2 + \underbrace{b_1 x + b_2 x}_{} + c = (\underline{}x + \underline{})(\underline{}x + \underline{})$$

$$b_1 \cdot b_2 = a \cdot c$$
$$b_1 + b_2 = b \qquad \text{Obtained by grouping}$$

Procedure: **Factoring a Trinomial of the Form $ax^2 + bx + c$ Using Grouping**

Step 1: Factor out any common factors.

Step 2: Find the product of the leading coefficient and the constant term; that is, $a \cdot c$.

Step 3: List the factors of this number to find the pair of factors whose sum is b, the middle coefficient of the trinomial. If there is no such pair, the trinomial is prime.

Step 4: Replace the middle term of the trinomial with a sum that uses the factors from Step 3.

Step 5: Factor by grouping.

Step 6: Check by multiplying.

Objective 2 Examples Factor each trinomial by grouping. Check by multiplying.

2a. $15x^2 - 16x + 4$ **2b.** $7y^2 + 41y - 6$ **2c.** $-16x^3 + 26x^2 + 12x$

Solutions **2a.** The terms of this trinomial contain no common factors other than 1. This trinomial is in the form $ax^2 + bx + c$, where $a = 15$ and $c = 4$. The product $a \cdot c = 15(4) = 60$.

We find the factors of 60 whose sum is $b = -16$, the middle coefficient. Since the product is positive and the sum is negative, both factors of 60 must be negative. The factors whose sum produces the correct middle coefficient are -6 and -10.

$a \cdot c = 60$	Sum of Factors
$-1, -60$	$-1 + (-60) = -61$
$-2, -30$	$-2 + (-30) = -32$
$-3, -20$	$-3 + (-20) = -23$
$-4, -15$	$-4 + (-15) = -19$
$-6, -10$	$-6 + (-10) = -16$

$$
\begin{aligned}
15x^2 - 16x + 4 &= 15x^2 - 6x - 10x + 4 &&\text{Replace } -16x \text{ with } -6x - 10x.\\
&= (15x^2 - 6x) + (-10x + 4) &&\text{Group together first two and last two terms.}\\
&= 3x(5x - 2) - 2(5x - 2) &&\text{Factor out the GCF from each pair of terms.}\\
&= (5x - 2)(3x - 2) &&\text{Factor out the common binomial, } (5x - 2).
\end{aligned}
$$

So, the factorization is $15x^2 - 16x + 4 = (5x - 2)(3x - 2)$.

Check: $(5x - 2)(3x - 2) = 15x^2 - 10x - 6x + 4$
$$= 15x^2 - 16x + 4$$

2b. The terms of this trinomial contain no common factors other than 1. This trinomial is in the form $ax^2 + bx + c$, where $a = 7$ and $c = -6$. The product $a \cdot c = 7(-6) = -42$.

We find the factors of -42 whose sum is the middle coefficient, $b = 41$. The signs of the factors of -42 are opposite since it is negative. The factors whose sum produces the correct middle coefficient are -1 and 42.

$a \cdot c = -42$	Sum of Factors
$-1, 42$	$-1 + 42 = 41$
$1, -42$	$1 + (-42) = -41$
$-2, 21$	$-2 + 21 = 19$
$2, -21$	$2 + (-21) = -19$
$-3, 14$	$-3 + 14 = 11$
$3, -14$	$3 + (-14) = -11$
$-6, 7$	$-6 + 7 = 1$
$6, -7$	$6 + (-7) = -1$

$$
\begin{aligned}
7y^2 + 41y - 6 &= 7y^2 - y + 42y - 6 && \text{Replace } 41y \text{ with } -y + 42y. \\
&= (7y^2 - y) + (42y - 6) && \text{Group together first two and last two terms.} \\
&= y(7y - 1) + 6(7y - 1) && \text{Factor out the GCF from each pair of terms.} \\
&= (7y - 1)(y + 6) && \text{Factor out the common binomial, } (7y - 1).
\end{aligned}
$$

So, the factorization is $7y^2 + 41x - 6 = (7y - 1)(y + 6)$.

Check: $(7y - 1)(y + 6) = 7y^2 + 42y - y - 6$
$\qquad\qquad\qquad\qquad\; = 7y^2 + 41y - 6$

> **Note:** *The order in which we insert the two terms will not affect the factorization. For instance, we could replace* $41y$ *with* $42y + (-1y)$ *or* $42y - y$.
>
> $$
> \begin{aligned}
> 7y^2 + 41y - 6 &= 7y^2 + 42y - y - 6 && \text{Replace } 41y \text{ with } -y + 42y. \\
> &= (7y^2 + 42y) + (-y - 6) && \text{Group together first two and last two terms.} \\
> &= 7y(y + 6) - 1(y + 6) && \text{Factor out the GCF from each pair of terms.} \\
> &= (y + 6)(7y - 1) && \text{Factor out the common binomial, } (7y - 1).
> \end{aligned}
> $$

2c. The terms in the trinomial have a common factor of $-2x$. We factor this out.

$$-16x^3 + 26x^2 + 12x = -2x(8x^2 - 13x - 6)$$

Now, we factor $8x^2 - 13x - 6$, which is of the form $ax^2 + bx + c$, where $a = 8$ and $c = -6$. The product $a \cdot c = 8(-6) = -48$.

We find the factors of -48 whose sum is $b = -13$. The signs of the factors of -48 are opposite since it is negative. The factors whose sum produces the correct middle coefficient are 3 and -16.

$a \cdot c = -48$	Sum of Factors
$-1, 48$	$-1 + 48 = 47$
$1, -48$	$1 + (-48) = -47$
$-2, 24$	$-2 + 24 = 22$
$2, -24$	$2 + (-24) = -22$
$-3, 16$	$-3 + 16 = 13$
$3, -16$	$3 + (-16) = -13$
$-4, 12$	$-4 + 12 = 8$
$4, -12$	$4 + (-12) = -8$
$-6, 8$	$-6 + 8 = 2$
$6, -8$	$6 + (-8) = -2$

$$
\begin{aligned}
8x^2 - 13x - 6 &= 8x^2 + 3x - 16x - 6 && \text{Replace } -13x \text{ with } 3x - 16x. \\
&= (8x^2 + 3x) + (-16x - 6) && \text{Group together first two and last two terms.} \\
&= x(8x + 3) - 2(8x + 3) && \text{Factor out the GCF from each pair of terms.} \\
&= (8x + 3)(x - 2) && \text{Factor out the common binomial, } (8x + 3).
\end{aligned}
$$

So, the factorization of the original trinomial is

$$-16x^3 + 26x^2 + 12x = -2x(8x^2 - 13x - 6)$$
$$= -2x(8x + 3)(x - 2)$$

Check: $-2x(8x + 3)(x - 2) = (-16x^2 - 6x)(x - 2)$
$$= -16x^3 + 32x^2 - 6x^2 + 12$$
$$= -16x^3 + 26x^2 + 12x$$

✔ **Student Check 2** Factor each trinomial using grouping. Check by multiplying.
a. $4x^2 + 13x + 3$ **b.** $6x^2 - x - 12$ **c.** $-40y^3 - 44y^2 + 32y$

Perfect Square Trinomials

Objective 3 ▶

Factor perfect square trinomials.

Recall that **perfect square trinomials** are trinomials whose factored form is obtained from squaring a binomial. In Section 5.5, we learned the rules for squaring binomials.

$(a + b)^2 = a^2 + 2ab + b^2$ Example: $(3x + 4)^2 = 9x^2 + 24x + 16$

$(a - b)^2 = a^2 - 2ab + b^2$ Example: $(x - 5)^2 = x^2 - 10x + 25$

So, in factored form, we have

$a^2 + 2ab + b^2 = (a + b)^2$ Example: $9x^2 + 24x + 16 = (3x + 4)^2$

$a^2 - 2ab + b^2 = (a - b)^2$ Example: $x^2 - 10x + 25 = (x - 5)^2$

Recall that some examples of perfect squares are $1, 4, 9, 16, 25, \ldots$, and $x^2, x^4, x^6, \ldots$.

> **Procedure: Identifying and Factoring a Perfect Square Trinomial**
>
> **Step 1:** Is the first term of the trinomial a perfect square? If so, write it as a^2.
> **Step 2:** Is the last term of the trinomial a perfect square and positive? If so, write it as b^2.
> **Step 3:** Is the middle term twice the product of a and b or $2ab$? If so, the trinomial is a perfect square trinomial. If not, the trinomial is not a perfect square trinomial.
> **Step 4:** Factor the trinomial as $(a + b)^2$ or $(a - b)^2$, as appropriate.

Objective 3 Examples

Determine if each trinomial is a perfect square trinomial and factor, if possible.
3a. $y^2 - 8y + 16$ **3b.** $4x^2 + 20x + 25$ **3c.** $4a^2 + 8a + 1$

Solutions **3a.** The first term is a perfect square: $y^2 = (y)^2$
The last term is a perfect square: $16 = (4)^2$
The middle term is twice the product of $a \cdot b$: $8y = 2(y)(4)$
So, $y^2 - 8y + 16$ is a perfect square trinomial. It is factored as

$$y^2 - 8y + 16 = (y - 4)^2$$

> **Note:** We can also factor $y^2 - 8y + 16$ using the techniques of Section 6.1. We find the factors of 16 whose sum is −8. The factors are −4 and −4.
> So, $y^2 - 8y + 16 = (y - 4)(y - 4)$ or $(y - 4)^2$.

3b. The first term is a perfect square: $4x^2 = (2x)^2$
The last term is a perfect square: $25 = (5)^2$
The middle term is twice the product of $a \cdot b$: $20x = 2(2x)(5)$
So, $4x^2 + 20x + 25$ is a perfect square trinomial. It is factored as

$$4x^2 + 20x + 25 = (2x + 5)^2$$

> **Note:** We can also factor $4x^2 + 20x + 25$ using the techniques presented in this section.
>
> $$4x^2 + 20x + 25 \qquad a \cdot c = 4 \cdot 25 = 100$$
>
> We find the factors of 100 whose sum is 20. The factors are 10 and 10.
>
> | | |
> |---|---|
> | $4x^2 + 20x + 25 = 4x^2 + 10x + 10x + 25$ | Replace $20x$ with $10x + 10x$. |
> | $= (4x^2 + 10x) + (10x + 25)$ | Group together first two and last two terms. |
> | $= 2x(2x + 5) + 5(2x + 5)$ | Factor out the GCF from each pair of terms. |
> | $= (2x + 5)(2x + 5)$ | Factor out the common binomial, $(2x + 5)$. |
> | $= (2x + 5)^2$ | Write in exponential form. |

3c. The first term is a perfect square: $4a^2 = (2a)^2$

The last term is a perfect square: $1 = (1)^2$

The middle term is *not* twice the product of $a \cdot b$: $8a \neq 2(2a)(1)$.

Since the trinomial is not a perfect square trinomial, we use trial and error or grouping to factor it. If we use grouping, we need to find factors of $a \cdot c = 4(1) = 4$ whose sum is the middle coefficient, 8. Since the factors of 4 are 1 and 4 or 2 and 2, this polynomial can't be factored since the sum of the factors is not 8. So, $4a^2 + 8a + 1$ is prime.

✔ **Student Check 3**   Determine if each trinomial is a perfect square trinomial and factor, if possible.

   a. $x^2 + 6x + 9$ **b.** $16x^2 - 24x + 9$ **c.** $y^2 + 4y + 16$

Objective 4 ▶

Troubleshoot common errors.

Troubleshooting Common Errors

Some common errors related to factoring trinomials are shown next.

Objective 4 Examples / **A problem and an incorrect solution are given. Provide the correct solution and an explanation of the error.**

4a. Factor $10r^2 + 47r - 15$ by trial and error.

Incorrect Solution	Correct Solution and Explanation
The first terms are factors of $10r^2$ and the last terms are factors of -15. So, factors of $10r^2$ are $5r$ and $2r$ and factors of -15 are -5 and 3. $10r^2 + 47r - 15$ $(5r + 3)(2r - 5)$ 	While the first term and last term of the product are $10r^2$ and -15, the middle term is not $47r$. If we multiply the binomials together, we get $10r^2 - 19r - 15$. The factors of $10r^2$ also include $10r$ and r. The factors of -15 also include $-3, 5; -1, 15;$ and $1, -15$. If we try all possible arrangements, the one that works is $$10r^2 + 47r - 15$$ $$(10r - 3)(r + 5)$$ Note that the outer and inner products are $50r - 3r = 47r$.

4b. Factor $18y^2 - 17y + 4$ by grouping.

Incorrect Solution	Correct Solution and Explanation
To use grouping, we need to find the factors of $18 \cdot 4 = 72$ which add to 17. This is 8 and 9. $$18y^2 - 17y + 4$$ $$18y^2 - 8y + 9y + 4$$ $$2y(9y - 4) + 1(9y + 4)$$ $$(9y - 4)(2y + 1)$$	The factors of 72 need to add to -17. Also, the binomial after grouping in the incorrect solution is not the same. The factors that we need are -8 and -9. $$18y^2 - 17y + 4$$ $$18y^2 - 8y - 9y + 4$$ $$2y(9y - 4) - 1(9y - 4)$$ $$(9y - 4)(2y - 1)$$

ANSWERS TO STUDENT CHECKS

Student Check 1 **a.** $(5x + 1)(x + 4)$ **b.** $(7x + 3)(x - 2)$
 c. $2xy(2y - 3)(3y + 5)$ **d.** $-3(3x + 2)(x - 15)$

Student Check 2 **a.** $(4x + 1)(x + 3)$ **b.** $(3x + 4)(2x - 3)$
 c. $-4y(5y + 8)(2y - 1)$

Student Check 3 **a.** $(x + 3)^2$ **b.** $(4x - 3)^2$ **c.** prime

SUMMARY OF KEY CONCEPTS

1. Two methods can be used to factor a trinomial. We can use the trial-and-error method or the grouping method.

2. The trial-and-error method requires us to try every combination of factors of the first term with factors of the last term until the correct middle term is obtained.

3. Grouping is a methodical process that requires us to find factors of the product of the leading coefficient and the last term that adds to the middle coefficient.

4. Perfect square trinomials are special trinomials that can be factored as a binomial squared. Memorizing the pattern is helpful but not absolutely necessary. These trinomials can be factored using the methods of factoring trinomials.

5. Every factoring problem can be checked by multiplying.

SECTION 6.3 / EXERCISE SET

 Write About It!

Use complete sentences in your answer to each exercise.

1. Explain the steps to factor a trinomial of the form $ax^2 + bx + c$, $a \neq 1$, by trial and error.

2. Explain the steps to factor a trinomial of the form $ax^2 + bx + c$, $a \neq 1$, by grouping.

3. When factoring a trinomial, how do you determine the signs of the last terms in the binomial factors?

4. What is a perfect square trinomial?

5. Find the missing factor of $6x^2 + 7x - 3 = (3x - 1)$ $(\,?\,)$ and explain how you arrived at your answer.

6. Use an example to show how to factor a trinomial which can be written as a perfect square trinomial.

7. Explain what types of numbers you need to find to factor Exercises 17 and 18 by trial and error.

8. Explain what types of numbers you need to find to factor Exercises 33 and 34 by grouping.

 Practice Makes Perfect!

Factor each trinomial completely using trial and error. Check by multiplying. (*See Objective 1.*)

9. $3x^2 + 13x + 4$

10. $2y^2 + 5y + 2$

11. $4s^2 - 29s + 7$

12. $5t^2 - 17t + 6$

13. $7a^2 - 41a - 6$

14. $7h^2 - 2h - 5$

15. $3x^2 + 20x - 7$

16. $5m^2 + m - 4$

17. $4k^2 - 16k + 15$

18. $9x^2 + 15x + 4$

19. $6y^2 - 17y - 14$

20. $15a^2 - a - 6$

21. $10t^2 + 11t - 8$

22. $21b^2 + 20b - 9$

23. $2y^2 + 3y + 5$

24. $4s^2 + 8s + 5$

25. $3m^2 - 10m - 7$

26. $6a^2 + 7a - 2$

27. $16k^2 - 4k - 6$

28. $12x^2 - 48x + 45$

29. $-7b^2 + b + 6$

30. $-4y^2 + 16y - 15$

31. $8h^3 - 12h^2 - 8h$

32. $15x^3 - 70x^2 + 40x$

Factor each trinomial completely using grouping. Check by multiplying. (*See Objective 2.*)

33. $3y^2 + 7y + 4$

34. $2a^2 + 13a + 6$

35. $3m^2 - 8m + 5$

36. $2t^2 - 11t + 12$

37. $5k^2 - 28k - 12$

38. $7m^2 - 4m - 3$

39. $11h^2 + 19h - 6$

40. $6b^2 + 7b - 10$

41. $4x^2 + 33x + 8$

42. $10y^2 + 13y + 4$

43. $9r^2 - 29r + 6$

44. $8m^2 - 38m + 9$

45. $15u^2 - 14u - 8$

46. $12s^2 - 11s - 15$

47. $4a^2 + 20a - 11$

48. $18s^2 + 17s - 15$

49. $8x^2 + 13x + 6$

50. $6y^2 + 13y - 12$

51. $-5a^2 - 13a - 6$

52. $24x^3 - 6x^2 - 9x$

53. $36h^2k - 6hk - 20k$

54. $60p^2q - 27pq - 60q$

55. The annual operating income, in millions of dollars, of an airline company from 2007 to 2009 can be modeled by the expression $2120x^2 - 4505x + 530$, where x is the number of years after 2007.

56. The net nonoperating income, in millions of dollars, of an airline company from 2007 to 2009 be modeled by the expression $-256x^2 + 704x - 288$, where x is the number of years after 2007.

Factor each perfect square trinomial. (*See Objective 3.*)

57. $x^2 + 4x + 4$

58. $m^2 + 10m + 25$

59. $b^2 - 6b + 9$

60. $y^2 - 14y + 49$

61. $25r^2 + 20r + 4$

62. $48h^2 + 24h + 3$

63. $18u^2 - 60u + 50$

64. $36a^2 + 84a + 49$

 **Mix 'Em Up!**

Use either trial and error or grouping to factor each trinomial completely.

65. $2x^2 + 15x + 18$

66. $2y^2 + 25y + 50$

67. $12t^2 - 5t - 2$

68. $16s^2 + 34s - 15$

69. $81b^2 - 36b + 4$

70. $25c^2 + 40c + 16$

71. $40u^2 - 2u - 21$

72. $33x^2 - 49x - 10$

73. $-6a^2 + 29a + 5$

74. $-5y^2 + 7y + 6$

75. $4x^2 + 6x - 28$

76. $18t^2 + 12t - 48$

77. $5u^2 - 2u + 3$

78. $6r^2 - 5r + 4$

79. $45p^2 + 120p + 80$

80. $12q^2 - 84q + 147$

81. $10a^2b - 29ab - 21b$

82. $32c^2 - 52cd + 6d^2$

83. $52x^2 - 34xy + 4y^2$

84. $96p^2q - 156pq + 45q$

85. $18a^2c - 84abc + 98b^2c$

86. $150x^2z + 120xyz + 24y^2z$

87. The profit, in millions, of a video game company is given by the expression $-60x^2 + 555x + 450$, where x is the number of units sold, in thousands.

88. The profit, in millions, of a software company is given by the expression $-150x^2 + 1425x + 4500$, where x is the number of units sold, in thousands.

 You Be the Teacher!

Correct each student's errors, if any.

89. Factor $3y^2 + 11y - 4$.

Tearra's work:

$3y^2 + 11y - 4 = (3y + 1)(y - 4)$

90. Factor $-6x^2 + 31x - 35$.

Charlene's work:

$6x^2 - 31x + 35 = (3x - 5)(2x - 7)$

91. Factor $5x^2 + 13x - 6$ by grouping.

Darlene's work:

$5x^2 - 15x + 2x - 6 = 5x(x - 3) + 2(x - 3)$
$= (x - 3)(5x + 2)$

92. Factor $3r^2 + 12r + 12$ by grouping.

Coppola's work:

$3r^2 + 6r + 6r + 12 = 3r(r + 2) + 6(r + 2)$
$= 3(r + 2)^2(r + 2)$
$= 3(r + 2)^3$

 Calculate It!

93. To factor $33x^2 - 49x - 10$ by grouping, what number do we need to factor? Use a calculator to list the factors of this number. Find the pair of factors that produce the correct middle term. Complete your factoring by hand.

94. Use a calculator to verify which, if any, of the following factorizations of $40b^2 - 11b - 21$ is correct.

  a. $(10b - 7)(4b + 3)$ **b.** $(10b + 7)(4b - 3)$

 Think About It!

Write a trinomial, $a \neq 1$, that satisfies the given conditions and factor it.

95. $a \cdot c = -36$ and $b = -5$

96. $a \cdot c = -20$ and $b = 19$

97. $a \cdot c = 144$ and $b = 24$

98. $a \cdot c = 324$ and $b = 36$

For what values of b is the trinomial factorable?

99. $2x^2 + bx + 5$

100. $6x^2 + bx + 2$

101. $5x^2 + bx - 4$

102. $3x^2 + bx - 7$

SECTION 6.4

Factoring Binomials

▶ OBJECTIVES

As a result of completing this section, you will be able to

1. Factor the difference of two squares.
2. Factor the sum and difference of two cubes.
3. Factor binomials with higher exponents.
4. Troubleshoot common errors.

Objective 1 ▶

Factor the difference of two squares.

The west face of Mount Thor in Canada is home to the Earth's greatest vertical drop. The cliff is approximately 4096 ft high. The expression $-16t^2 + 4096$ represents the height, in feet, of an object dropped off the top of Mount Thor t sec after it was dropped. In Section 6.6, we will solve an equation that determines how long it takes for a falling object to reach the ground. This process requires us to factor $-16t^2 + 4096$. (Source: http://en.wikipedia.org/wiki/Extremes_on_Earth#Greatest_vertical_drop)

So far in this chapter, we have learned how to factor trinomials. We will now finish our journey into factoring by learning how to factor three types of binomials.

The Difference of Two Squares

In Section 5.5, we learned that the product of two conjugates results in the *difference of two squares*. That is,

$$(a - b)(a + b) = a^2 - b^2$$

An example is

$$(x - 3)(x + 3) = (x)^2 - (3)^2 = x^2 - 9$$

When we reverse this statement, we have a method that enables us to factor the difference of two squares. The difference of two squares can be factored into the product of two conjugates. The factored form is

$$x^2 - 9 = (x)^2 - (3)^2 = (x - 3)(x + 3)$$

Property: The Difference of Two Squares

$$a^2 - b^2 = (a - b)(a + b)$$

It is very helpful to be able to recognize the difference of two squares. To do this, we must be able to identify perfect squares. Some perfect squares are

$$1, 4, 9, 16, 25, \ldots, \quad \text{and} \quad x^2, x^4, x^6, \ldots$$

Some examples of the difference of two squares are

$$x^2 - 4 \qquad 16a^2 - 25 \qquad 49r^2 - 81s^2$$

Notice that each term in the binomial is a perfect square and the terms are connected by subtraction.

Procedure: Factoring the Difference of Two Squares

Step 1: Factor out any common factor.
Step 2: Rewrite the first term as a^2, where a is what must be squared to obtain the first term.
Step 3: Rewrite the second term as b^2, where b is what must be squared to obtain the second term.
Step 4: Factor as $(a - b)(a + b)$.
Step 5: Check by multiplying.

Objective 1 Examples Use the difference of two squares property to factor each binomial.

1a. $y^2 - 49$ **1b.** $100 - x^2$ **1c.** $2x^3 - 72x$ **1d.** $49r^2 - \dfrac{1}{9}s^2$ **1e.** $x^2 + 4$

1f. The west face of Mount Thor in Canada is home to the Earth's greatest vertical drop. The cliff is approximately 4096 ft high. The expression $-16t^2 + 4096$ represents the height, in feet, of an object dropped off the top of Mount Thor t sec after it was dropped. Factor $-16t^2 + 4096$. (Source: http://en.wikipedia.org/wiki/Extremes_on_Earth#Greatest_vertical_drop)

Solutions   **1a.** $y^2 - 49 = (y)^2 - (7)^2$ Write as $a^2 - b^2$.

$= (y - 7)(y + 7)$ Apply the difference of two squares property.

1b. $100 - x^2 = (10)^2 - (x)^2$ Write as $a^2 - b^2$.

$= (10 - x)(10 + x)$ Apply the difference of two squares property.

1c. The terms are not perfect squares but they have a common factor, $2x$.

$2x^3 - 72x = 2x(x^2 - 36)$ Factor out the common factor, $2x$.

$= 2x(x^2 - 6^2)$ Write the binomial as $a^2 - b^2$.

$= 2x(x - 6)(x + 6)$ Apply the difference of two squares property.

1d. $49r^2 - \dfrac{1}{9}s^2 = (7r)^2 - \left(\dfrac{1}{3}s\right)^2$ Write as $a^2 - b^2$.

$= \left(7r - \dfrac{1}{3}s\right)\left(7r + \dfrac{1}{3}s\right)$ Apply the difference of two squares property.

1e. Each term is a perfect square, but this binomial is the **sum of squares**, not the difference of squares. This binomial cannot be factored as the product of conjugates.

A common answer for the factorization of this binomial is $(x + 2)^2$. But when we square this binomial, we get

$$(x + 2)^2 = (x + 2)(x + 2) = x^2 + 4x + 4$$

This is not the same as the given binomial, $x^2 + 4$. So, $(x + 2)^2$ is not the factorization.

We can think of $x^2 + 4$ as $x^2 + 0x + 4$. Using the rules from Section 6.2, we find the factors of 4 whose sum is 0. No such factors exist. So, $x^2 + 4$ is a *prime* polynomial.

> **Note:** *The sum of squares $a^2 + b^2$ cannot be factored unless there is a GCF.*

1f. The terms in the binomial, $-16t^2 + 4096$, have a common factor, -16.

$-16t^2 + 4096 = -16(t^2 - 256)$ Factor out the GCF, -16.

$= -16(t^2 - 16^2)$ Write the binomial as $a^2 - b^2$.

$= -16(t - 16)(t + 16)$ Apply the difference of two squares property.

✔ **Student Check 1**   Use the difference of squares property to factor each binomial.

a. $x^2 - 64$ **b.** $25 - y^2$ **c.** $3x^3 - 3x$ **d.** $9a^2 - 4b^2$ **e.** $4y^2 + 25$

f. B.A.S.E. jumping is an activity whereby people fall off structures with only a brief period before they must open a parachute to survive. If a person free-falls off a 1600-ft structure, his height can be represented by $-16t^2 + 1600$, where t is the number of seconds after the fall.

The Sum and Difference of Two Cubes

Objective 2 ▶

Factor the sum and difference of two cubes.

We will now factor binomials that are the sum and difference of two cubes; that is, binomials of the form $a^3 + b^3$ and $a^3 - b^3$. Again, it will be helpful to identify perfect cubes. Perfect cubes come from cubing integers and powers of x. A list of some perfect cubes is shown.

Note: *Perfect Cubes:*

$$1, 8, 27, 64, 125, 216 \ldots, \quad \text{and} \quad x^3, x^6, x^9, x^{12}, x^{15} \ldots$$

In Section 5.4, we found the following product.

$$(x + 2)(x^2 - 2x + 4) = x(x^2) - x(2x) + x(4) + 2(x^2) - 2(2x) + 2(4)$$
$$= x^3 - 2x^2 + 4x + 2x^2 - 4x + 8$$
$$= x^3 + 8$$

The result of the product of this binomial and trinomial is the *sum of two cubes*,

$$x^3 + 8 = x^3 + 2^3$$

If we reverse the statement, we obtain the factorization of $x^3 + 8$.

$$x^3 + 8 = x^3 + 2^3 = (x + 2)(x^2 - 2x + 4)$$

Notice the binomial in the factorization, $(x + 2)$, consists of the terms that we cube to get x^3 and 8. The trinomial results from a special pattern, which is shown. The pattern also applies to the difference of two cubes.

Property: Sum and Difference of Two Cubes

$$a^3 + b^3 = (a + b)(a^2 - ab + b^2)$$
$$a^3 - b^3 = (a - b)(a^2 + ab + b^2)$$

Some observations:

1. The sign of the binomial factor is the same as the given binomial.
2. The first term of the trinomial factor, a^2, is the first term of the binomial factor squared.
3. The middle term of the trinomial factor, ab, is the product of the terms in the binomial. Its sign is always the opposite of the sign in the binomial factor.
4. The last term of the trinomial factor, b^2, is the last term of the binomial factor squared. Its sign is always positive.

Note: *Helpful hint to remember the signs of the factors is the word SOAP.*

S = same sign

O = opposite sign

AP = always positive

$$\overset{\text{Same}}{} \qquad \overset{\text{Always positive}}{}$$
$$x^3 + 8 = (x + 2)(x^2 - 2x + 4)$$
$$\underset{\text{Opposite}}{}$$

Procedure: Factoring the Sum or Difference of Two Cubes

Step 1: Factor out any common factor.
Step 2: Rewrite the first term as a^3, where a is what must be cubed to obtain the first term.
Step 3: Rewrite the second term as b^3, where b is what must be cubed to obtain the second term.
Step 4: Apply the sum and difference of two cubes property.
    a. If the binomial is a sum of cubes, the first factor is $(a + b)$ and the second factor is $(a^2 - ab + b^2)$.
    b. If the binomial is a difference of cubes, the first factor is $(a - b)$ and the second factor is $(a^2 + ab + b^2)$.
Step 5: Check by multiplying.

Objective 2 Examples   **Factor each binomial.**

2a. $x^3 - 125$ **2b.** $8r^3 - s^3$ **2c.** $m^3 + 216n^3$ **2d.** $54y^3 + 128$

Solutions   **2a.** $\begin{aligned} x^3 - 125 &= (x)^3 - (5)^3 \\ &= (x - 5)[x^2 + (x)(5) + 5^2] \\ \\ &= (x - 5)(x^2 + 5x + 25) \end{aligned}$

Write as $a^3 - b^3$.
Apply the property with $a = x$ and $b = 5$.
Simplify.

Check: $\begin{aligned}(x - 5)(x^2 + 5x + 25) &= x^3 + 5x^2 + 25x - 5x^2 - 25x - 125 \\ &= x^3 - 125\end{aligned}$

2b. $\begin{aligned} 8r^3 - s^3 &= (2r)^3 - (s)^3 \\ &= (2r - s)[(2r)^2 + (2r)(s) + s^2] \\ \\ &= (2r - s)(4r^2 + 2rs + s^2) \end{aligned}$

Write as $a^3 - b^3$.
Apply the property with $a = 2r$ and $b = s$.
Simplify.

2c. $\begin{aligned} m^3 + 216n^3 &= m^3 + (6n)^3 \\ &= (m + 6n)[m^2 - (m)(6n) + (6n)^2] \\ \\ &= (m + 6n)(m^2 - 6mn + 36n^2) \end{aligned}$

Write as $a^3 + b^3$.
Apply the property with $a = m$ and $b = 6n$.
Simplify.

2d. The terms in the binomial are not perfect cubes. They do, however, have a common factor, 2.

$\begin{aligned} 54y^3 + 128 &= 2(27y^3 + 64) \\ &= 2[(3y)^3 + (4)^3] \\ &= 2(3y + 4)[(3y)^2 - (3y)(4) + 4^2] \\ \\ &= 2(3y + 4)(9y^2 - 12y + 16) \end{aligned}$

Factor out the GCF, 2.
Write the binomial as $a^3 + b^3$.
Apply the property with $a = 3y$ and $b = 4$.
Simplify.

✓ **Student Check 2**   Factor each binomial.

a. $y^3 - 64$ **b.** $27a^3 - b^3$ **c.** $x^3 + 8y^3$ **d.** $-24m^3 - 3$

Factoring Binomials with Exponents Greater Than 3

Objective 3 ▶

Factor binomials with higher exponents.

Binomials containing exponents greater than 3 may not initially appear to satisfy the patterns we have discussed thus far. But if we can write the binomial as the difference of two squares, sum of two cubes, or difference of two cubes, then we can factor the polynomial.

> **Procedure: Factoring Binomials with Higher Exponents**
>
> **Step 1:** Factor out common factors, if applicable.
> **Step 2:** Determine if the binomial can be expressed as $a^2 - b^2$, $a^3 - b^3$, or $a^3 + b^3$.
>   **a.** If all of the exponents are even, try applying the difference of squares property.
>   **b.** If all of the exponents are multiples of 3, try applying the sum or difference of cubes property.
> **Step 3:** Make sure that any resulting factors cannot be factored any further.
> **Step 4:** Check by multiplying.

Objective 3 Examples   **Factor each binomial completely.**

3a. $x^4 - 81$ **3b.** $y^6 + 64$ **3c.** $125r^3s^9 - 216t^{15}$

Solutions **3a.** Both terms are perfect squares since $x^4 = (x^2)^2$ and $81 = 9^2$, so we can apply the difference of two squares property.

$$x^4 - 81 = (x^2)^2 - 9^2 \qquad \text{Write as } a^2 - b^2.$$

$$= (x^2 - 9)(x^2 + 9) \qquad \text{Apply the difference of two squares property.}$$

$$= (x - 3)(x + 3)(x^2 + 9) \qquad \text{Apply the difference of two squares property on } x^2 - 9.$$

3b. The terms in this binomial are both perfect squares and perfect cubes.

$$y^6 + 64 = (y^3)^2 + 8^2 \quad \text{or} \quad y^6 + 64 = (y^2)^3 + 4^3$$

We cannot factor the sum of squares over the real numbers, so we must factor the binomial as the sum of cubes.

$$y^6 + 64 = (y^2)^3 + 4^3 \qquad \text{Write as } a^3 + b^3.$$

$$= (y^2 + 4)[(y^2)^2 - (y^2)(4) + 4^2] \qquad \text{Apply the property with } a = y^2 \text{ and } b = 4.$$

$$= (y^2 + 4)(y^4 - 4y^2 + 16) \qquad \text{Simplify.}$$

3c. All the powers in this binomial are multiples of 3 and the coefficients are perfect cubes, so this is the difference of two cubes.

$$125r^3s^9 - 216t^{15}$$

$$= (5rs^3)^3 - (6t^5)^3 \qquad \text{Write as } a^3 - b^3.$$

$$= (5rs^3 + 6t^5)[(5rs^3)^2 - (5rs^3)(6t^5) + (6t^5)^2] \qquad \text{Apply the property with } a = 5rs^3 \text{ and } b = 6t^5.$$

$$= (5rs^3 + 6t^5)(25r^2s^6 - 30rs^3t^5 + 36t^{10}) \qquad \text{Simplify by applying the properties of exponents.}$$

✓ **Student Check 3** Factor each binomial completely.
a. $x^4 - 16$ **b.** $y^6 + 1$ **c.** $16m^4n^{10} - 25p^6$

Objective 4 ▶

Troubleshoot common errors.

Troubleshooting Common Errors

Some common errors with applying the factoring rules for binomials are shown next.

Objective 4 Examples **A problem and an incorrect solution are given. Provide the correct solution and an explanation of the error.**

4a. Factor $m^2 - 36$.

Incorrect Solution	Correct Solution and Explanation
$m^2 - 36$ $(m - 6)(m - 6)$ $(m - 6)^2$	If we square $(m - 6)$, we get $m^2 - 12m + 36$. This is not the given binomial. The binomial is a difference of two squares and factors as the product of conjugates. $$m^2 - 36 = (m - 6)(m + 6)$$

4b. Factor $4y^2 + 49$.

Incorrect Solution	Correct Solution and Explanation
$4y^2 + 49$ $(2y + 7)(2y + 7)$ $(2y + 7)^2$	If we square $(2y + 7)$, we get $4y^2 + 28y + 49$. This is not the given binomial. The binomial is a sum of two squares, not the difference of two squares. It cannot be factored over the real numbers. $4y^2 + 49$ is prime.

4c. Factor $a^3 - 27$.

Incorrect Solution	Correct Solution and Explanation
$a^3 - 27$ $(a - 3)^3$	When we cube $(a - 3)$, we get $a^3 - 9a^2 + 27a - 27$. This does not yield $a^3 - 27$. The binomial is a difference of two cubes and is factored in the form $(a - b)(a^2 + ab + b^2)$. $a^3 - 27 = (a - 3)(a^2 + 3a + 9)$

ANSWERS TO STUDENT CHECKS

Student Check 1 **a.** $(x - 8)(x + 8)$ **b.** $(5 - y)(5 + y)$
  c. $3x(x - 1)(x + 1)$ **d.** $(3a - 2b)(3a + 2b)$ **e.** prime
  f. $-16(t - 10)(t + 10)$

Student Check 2 **a.** $(y - 4)(y^2 + 4y + 16)$
  b. $(3a - b)(9a^2 + 3ab + b^2)$ **c.** $(x + 2y)(x^2 - 2xy + 4y^2)$
  d. $-3(2m + 1)(4m^2 - 2m + 1)$

Student Check 3 **a.** $(x - 2)(x + 2)(x^2 + 4)$
  b. $(y^2 + 1)(y^4 - y^2 + 1)$
  c. $(4m^2n^5 - 5p^3)(4m^2n^5 + 5p^3)$

SUMMARY OF KEY CONCEPTS

1. The difference of two squares can be factored as a product of conjugates: $a^2 - b^2 = (a - b)(a + b)$. The sum of two squares is prime.

2. The sum and difference of two cubes can be factored as
$$a^3 + b^3 = (a + b)(a^2 - ab + b^2)$$
$$a^3 - b^3 = (a - b)(a^2 + ab + b^2)$$

3. Binomials with larger exponents may be factored into the difference of two squares, difference of two cubes, or the sum of two cubes. The key is to rewrite each term of the binomial as a quantity that is squared or cubed.

GRAPHING CALCULATOR SKILLS

The graphing calculator can be used to verify our factorization as described in previous sections. It can also be used to generate a list of perfect squares and perfect cubes that might help in identifying the values of a and b in the factoring properties.

Example 1: List the perfect squares.

Solution: Enter x squared in the equation editor. Then view the table. The values in Y_1 are the perfect squares. The corresponding x-value is the number that must be squared to obtain the value in Y_1.

Example 2: List the perfect cubes.

Solution: Enter x cubed in the equation editor. Then view the table. The values in Y_1 are the perfect cubes. The corresponding x-value is the number that must be cubed to obtain the value in Y_1.

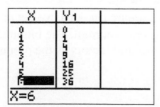

SECTION 6.4 / EXERCISE SET

Write About It!

Use complete sentences in your answer to each exercise.

1. Explain how to factor a binomial that is the difference of two squares.

2. Explain how to factor a binomial that is the difference of two cubes.

3. Explain how to factor a binomial that is the sum of two cubes.

4. Use an example to show how to factor a binomial that can be written as a difference of two squares.

5. Use the example, $125a^3 + 27b^3$, to explain how to factor a sum of two cubes.

6. Use the example, $64a^3 - 27b^3$, to explain how to factor a difference of two cubes.

7. Explain why $x^2 + 9$ is prime.

8. Explain how to factor the binomial: $a^4 - b^4$.

Practice Makes Perfect!

Use the difference of two squares property to factor each binomial, if possible. (*See Objective 1.*)

9. $x^2 - 1$
10. $x^2 - 4$
11. $y^2 - 36$
12. $y^2 - 9$
13. $2x^3 - 8x$
14. $5x^3 - 125x$
15. $9a^2 - 4$
16. $16a^2 - 1$
17. $25b^2 - 49$
18. $64b^2 - 25$
19. $121 - 4x^2$
20. $81 - 49x^2$
21. $8t^2 - 72$
22. $7t^2 - 112$
23. $9m^2 - 256$
24. $121m^2 - 4$
25. $3x^5 - 48x^3$
26. $2x^7 - 18x^5$
27. $9a^2 + 4$
28. $25 + 49b^2$
29. $121y^2 - 144$
30. $81y^2 - 64$
31. $64a^2 - 9b^2$
32. $25a^2 - 16b^2$
33. $m^2n^2 - 121$
34. $m^2n^2 - 256$
35. $9x^2 - \dfrac{1}{4}$
36. $4y^2 - \dfrac{9}{25}$
37. $\dfrac{1}{9}a^2 - \dfrac{4}{49}$
38. $\dfrac{1}{4}b^2 - \dfrac{16}{25}$

Use the difference of two cubes property to factor each binomial, if possible. (*See Objective 2.*)

39. $x^3 - 8$
40. $x^3 - 1$
41. $a^3 - 64$
42. $a^3 - 125$
43. $2t^4 - 54t$
44. $3t^4 - 24t$
45. $125y^3 - 1$
46. $64y^3 - 27$
47. $1 - 8t^3$
48. $8 - 27t^3$
49. $m^3 - 216n^3$
50. $8m^3 - 1$
51. $\dfrac{1}{27}s^3 - 1$
52. $\dfrac{1}{8}t^3 - 125$

Use the sum of two cubes property to factor each binomial, if possible. (*See Objective 2.*)

53. $y^3 + 27$
54. $y^3 + 1$
55. $a^3 + 64$
56. $a^3 + 125$
57. $5t^4 + 320t$
58. $-3t^4 - 24t$
59. $1000 + 125b^2$
60. $8 + 216b^2$
61. $8m^3 + n^3$
62. $m^3 + 27n^3$
63. $a^3 + \dfrac{1}{27}$
64. $b^3 + \dfrac{8}{125}$

Factor each binomial completely. (*See Objective 3.*)

65. $x^4 - 16$
66. $x^4 - 625$
67. $2y^6 + 250$
68. $3p^6 + 24$
69. $256 - y^6$
70. $1 - 16y^6$
71. $r^4 - \dfrac{81}{16}$
72. $s^4 - \dfrac{1}{625}$
73. $16a^4 - 81$
74. $1 - 256a^4$

Mix 'Em Up!

Factor each binomial completely.

75. $9y^3 - 36y$
76. $36m^2 - 81n^2$
77. $256 - a^2$
78. $4m^2 - 121$
79. $169 - b^2$
80. $625 - c^2$
81. $8a^3 - 125$
82. $4y^3 - 324$
83. $r^2 + 100s^2$
84. $s^4 + 16t^4$
85. $a^3 + 1000$
86. $8y^3 + 64$
87. $16x^4 - 81$
88. $256x^4 - 1$
89. $2y^6 - 54$
90. $5y^6 + 40$
91. $5x^4y^4 - 405z^4$
92. $7p^4q^4 - 7r^4$
93. $64r^2 - 49s^2$
94. $100m^2 - 49n^2$
95. $0.16a^2 - 0.25b^2$
96. $0.01c^2 - 0.09d^2$
97. $8x^3 - \dfrac{1}{125}y^3$
98. $27x^3 + \dfrac{1}{8}y^3$
99. $\dfrac{1}{16}x^4 - \dfrac{1}{81}$
100. $\dfrac{1}{625}y^4 - \dfrac{1}{16}$

You Be the Teacher!

Correct each student's errors, if any.

101. Factor $x^2 - 64x$.

Sam's work: $x^2 - 64x = (x + 8)(x - 8)$

102. Factor $y^2 + 25$.

Cameron's work: $y^2 + 25 = (y + 5)^2$

103. Factor $x^3 + 64$.

Hyunh's work: $x^3 + 64 = (x + 4)(x^2 + 4x + 16)$

104. Factor $y^3 - 8$.

Dawn's work: $y^3 - 8 = (y - 2)^3$

 Calculate It!

Factor each binomial and use a calculator to verify it.

105. $x^3 - 8$

106. $x^3 + 27$

107. $81x^4 - 16$

108. $10,000x^4 - 1$

 PIECE IT TOGETHER / **SECTIONS 6.1–6.4**

Factor each polynomial completely.

1. $3x^3 + 15x^2 - 18x - 90$

2. $16s^3 - 48s^2 - s + 3$

3. $x^2 + 13x + 36$

4. $y^2 - y - 30$

5. $a^2 + 14a + 7$

6. $b^2 - 19b + 60$

7. $5y^2 + 15y - 20$

8. $6y^2 + 30y - 84$

9. $p^3q^2 + 2p^2q - 35p$

10. $5r^3s^2 - 14r^2s - 24r$

11. $75x^2 - 65x + 10$

12. $10y^2 - 13y + 4$

13. $4r^2 - 20r + 25$

14. $50h^2 + 20h + 2$

15. $4y^2 - \dfrac{49}{25}$

16. $\dfrac{1}{9}a^2 + \dfrac{4}{15}$

17. $x^2 - 9y^2$

18. $1000a^3 + b^3$

19. $8x^3 - 125y^3$

20. $81a^4 - 1$

SECTION 6.5 | **Solving Quadratic Equations and Other Polynomial Equations by Factoring**

▶ **OBJECTIVES**

As a result of completing this section, you will be able to

1. Solve quadratic equations by factoring.

2. Solve equations of degree 3 or higher by factoring.

3. Troubleshoot common errors.

B.A.S.E. jumping is an activity whereby people leap off structures with only a brief period before they must open a parachute to survive. If a person free-falls from a 1600-ft structure, his height can be represented by $-16t^2 + 1600$, where t is the number of seconds after the fall. The jumper must open the parachute well before reaching the ground, to give it time to open fully and slow the descent. If the jumper does not open his parachute, when does he reach the ground?

To answer this question, we must solve the equation $-16t^2 + 1600 = 0$. This is an example of a *quadratic equation*. We will learn the steps needed to solve this type of equation in this section.

Objective 1 ▶

Solve quadratic equations by factoring.

Solve Quadratic Equations by Factoring

So far in this chapter, we have learned how to factor different types of polynomials. Now we are going to use our knowledge of factoring to solve a new type of equation called a quadratic equation.

> **Definition:** A **quadratic equation** is an equation that can be written in the form $ax^2 + bx + c = 0$, where a, b, and c are real numbers and $a \neq 0$. This form is called the **standard form** of a quadratic equation.

Some examples of quadratic equations in standard form are

$$2x^2 + 7x - 4 = 0 \qquad y^2 + 5y + 6 = 0 \qquad h^2 - 4 = 0$$

Some examples of quadratic equations that are not in standard form are

$$a^2 - a = 12 \qquad t^2 = 5t \qquad (x - 3)(x + 1) = 8$$

There are numerous ways to solve quadratic equations, but in this section we are going to focus on just one method—solving quadratic equations by factoring. Other methods will be discussed in Chapter 11.

Suppose two numbers are multiplied and the product is zero. Then at least one of the numbers has to be zero. For example,

$$-4 \cdot 0 = 0 \quad \text{and} \quad 0 \cdot \frac{1}{2} = 0$$

The only way for a product to equal zero is if one of the factors is zero. This concept is called the *zero products property*.

Property: Zero Products Property

If a and b are real numbers and $a \cdot b = 0$, then

$$a = 0 \quad \text{or} \quad b = 0$$

This property provides a method for solving quadratic equations. It states that if one side of the equation is expressed in factored form and the other side of the equation is zero, the solutions are found by setting each factor equal to zero.

Procedure: Solving Quadratic Equations by Factoring

Step 1: Write the quadratic equation in standard form, if necessary.
Step 2: Factor the resulting polynomial.
Step 3: Use the zero products property to solve the equation.
Step 4: Check the solutions in the original equation.

Objective 1 Examples Solve each quadratic equation by factoring.

1a. $(x + 1)(x - 3) = 0$ **1b.** $4y(y - 5) = 0$ **1c.** $x^2 - 4x - 5 = 0$
1d. $x^2 = 49$ **1e.** $4x^2 + 20x + 25 = 0$ **1f.** $3y^2 + 4 = 7y$
1g. $12 - 5x = 2x^2$ **1h.** $(x - 1)(x + 2) = -2$

Solutions **1a.** We can apply the zero products property to solve the equation since the left side of the equation is in factored form and the right side is zero.

$$(x + 1)(x - 3) = 0$$

$x + 1 = 0$	or $\quad x - 3 = 0$	Apply the zero products property.
$x + 1 - 1 = 0 - 1$	$x - 3 + 3 = 0 + 3$	Solve each equation.
$x = -1$	$x = 3$	

The solutions are -1 and 3, and the solution set is $\{-1, 3\}$. Check solution by replacing the variable with its corresponding value.

$x = -1$:

$$(x + 1)(x - 3) = 0$$
$$(-1 + 1)(-1 - 3) = 0$$
$$(0)(-4) = 0$$
$$0 = 0 \quad \text{True}$$

$x = 3$:

$$(x + 1)(x - 3) = 0$$
$$(3 + 1)(3 - 3) = 0$$
$$(4)(0) = 0$$
$$0 = 0 \quad \text{True}$$

1b.
$$4y(y - 5) = 0$$
The equation is factored and equal to zero.

$$4y = 0 \qquad \text{or} \qquad y - 5 = 0$$
Apply the zero products property.

$$\frac{4y}{4} = \frac{0}{4} \qquad\qquad y - 5 + 5 = 0 + 5$$
Solve each equation.

$$y = 0 \qquad\qquad\qquad y = 5$$

The solutions are 0 and 5, and the solution set is $\{0, 5\}$.

Check:

$y = 0$: $\qquad\qquad\qquad\qquad\qquad\qquad\qquad$ $y = 5$:

$$4y(y - 5) = 0 \qquad\qquad\qquad\qquad 4y(y - 5) = 0$$
$$4(0)(0 - 5) = 0 \qquad\qquad\qquad\qquad 4(5)(5 - 5) = 0$$
$$(0)(-5) = 0 \qquad\qquad\qquad\qquad 20(0) = 0$$
$$0 = 0 \quad \text{True} \qquad\qquad\qquad\qquad 0 = 0 \quad \text{True}$$

1c.
$$x^2 - 4x - 5 = 0$$
The equation is in standard form.

$$(x - 5)(x + 1) = 0$$
Factor the trinomial on the left side.

$$x - 5 = 0 \qquad \text{or} \qquad x + 1 = 0$$
Apply the zero products property.

$$x - 5 + 5 = 0 + 5 \qquad x + 1 - 1 = 0 - 1$$
Solve each equation.

$$x = 5 \qquad\qquad\qquad x = -1$$

The solutions are 5 and -1, and the solution set is $\{-1, 5\}$.

Check:

$x = 5$: $\qquad\qquad\qquad\qquad\qquad\qquad\qquad$ $x = -1$:

$$x^2 - 4x - 5 = 0 \qquad\qquad\qquad\qquad x^2 - 4x - 5 = 0$$
$$(5)^2 - 4(5) - 5 = 0 \qquad\qquad\qquad (-1)^2 - 4(-1) - 5 = 0$$
$$25 - 20 - 5 = 0 \qquad\qquad\qquad\qquad 1 + 4 - 5 = 0$$
$$5 - 5 = 0 \qquad\qquad\qquad\qquad 5 - 5 = 0$$
$$0 = 0 \quad \text{True} \qquad\qquad\qquad\qquad 0 = 0 \quad \text{True}$$

1d.
$$x^2 = 49$$
The equation is not in standard form.

$$x^2 - 49 = 49 - 49$$
Subtract 49 from each side.

$$x^2 - 49 = 0$$
Simplify.

$$(x - 7)(x + 7) = 0$$
Factor the binomial on the left side.

$$x - 7 = 0 \qquad \text{or} \qquad x + 7 = 0$$
Apply the zero products property.

$$x - 7 + 7 = 0 + 7 \qquad x + 7 - 7 = 0 - 7$$
Solve each equation.

$$x = 7 \qquad\qquad\qquad x = -7$$

The solutions are 7 and -7, and the solution set is $\{-7, 7\}$. We can write the solution set as $\{\pm 7\}$ to represent both the positive and negative value of 7.

Check:

$x = 7$: $\qquad\qquad\qquad\qquad\qquad\qquad\qquad$ $x = -7$:

$$x^2 = 49 \qquad\qquad\qquad\qquad\qquad x^2 = 49$$
$$(7)^2 = 49 \qquad\qquad\qquad\qquad\qquad (-7)^2 = 49$$
$$49 = 49 \quad \text{True} \qquad\qquad\qquad\qquad 49 = 49 \quad \text{True}$$

1e.

$4x^2 + 20x + 25 = 0$	The equation is in standard form.
$(2x + 5)(2x + 5) = 0$	Factor the trinomial on the left side.
$2x + 5 = 0$	Apply the zero products property.
$2x + 5 - 5 = 0 - 5$	Subtract 5 from each side.
$2x = -5$	Simplify.
$x = -\dfrac{5}{2}$	

The solution is $-\dfrac{5}{2}$, and the solution set is $\left\{-\dfrac{5}{2}\right\}$.

Check:

$x = -\dfrac{5}{2}$:

$$4x^2 + 20x + 25 = 0$$

$$4\left(-\frac{5}{2}\right)^2 + 20\left(-\frac{5}{2}\right) + 25 = 0$$

$$4\left(\frac{25}{4}\right) - 50 + 25 = 0$$

$$25 - 50 + 25 = 0$$

$$0 = 0 \quad \text{True}$$

1f.

$3y^2 + 4 = 7y$	The equation is not in standard form.
$3y^2 + 4 - 7y = 7y - 7y$	Subtract $7y$ from each side.
$3y^2 - 7y + 4 = 0$	Simplify and write in standard form.
$3y^2 - 3y - 4y + 4 = 0$	Replace $-7y$ with $-3y - 4y$.
$3y(y - 1) - 4(y - 1) = 0$	Factor by grouping.
$(y - 1)(3y - 4) = 0$	

$y - 1 = 0$	or	$3y - 4 = 0$	Apply the zero products property.
$y - 1 + 1 = 0 + 1$		$3y - 4 + 4 = 0 + 4$	Solve each equation.
$y = 1$		$3y = 4$	
		$\dfrac{3y}{3} = \dfrac{4}{3}$	
		$y = \dfrac{4}{3}$	

The solutions are 1 and $\dfrac{4}{3}$, and the solution set is $\left\{1, \dfrac{4}{3}\right\}$.

Check:

$y = 1$:

$$3y^2 + 4 = 7y$$
$$3(1)^2 + 4 = 7(1)$$
$$3(1) + 4 = 7$$
$$3 + 4 = 7$$
$$7 = 7 \quad \text{True}$$

$y = \dfrac{4}{3}$:

$$3y^2 + 4 = 7y$$

$$3\left(\frac{4}{3}\right)^2 + 4 = 7\left(\frac{4}{3}\right)$$

$$3\left(\frac{16}{9}\right) + 4 = \frac{28}{3}$$

$$\frac{48}{9} + \frac{36}{9} = \frac{28}{3}$$

$$\frac{84}{9} = \frac{28}{3}$$

$$\frac{28}{3} = \frac{28}{3} \quad \text{True}$$

1g. We need to perform operations on each side of the equation to make one side equal to zero. We can either make the left side of the equation zero or the right side of the equation zero.

Method 1	**Method 2**
Make the right side of equation zero.	Make the left side of equation zero.

Method 1

$$12 - 5x = 2x^2$$
$$12 - 5x - 2x^2 = 2x^2 - 2x^2$$
$$-2x^2 - 5x + 12 = 0$$
$$-1(-2x^2 - 5x + 12) = -1(0)$$
$$2x^2 + 5x - 12 = 0$$
$$2x^2 - 3x + 8x - 12 = 0$$
$$x(2x - 3) + 4(2x - 3) = 0$$
$$(2x - 3)(x + 4) = 0$$
$$2x - 3 = 0 \quad \text{or} \quad x + 4 = 0$$
$$x = \frac{3}{2} \qquad\qquad x = -4$$

Method 2

$$12 - 5x = 2x^2$$
$$12 - 5x - 12 + 5x = 2x^2 - 12 + 5x$$
$$0 = 2x^2 + 5x - 12$$
$$0 = 2x^2 - 3x + 8x - 12$$
$$0 = x(2x - 3) + 4(2x - 3)$$
$$0 = (2x - 3)(x + 4)$$
$$2x - 3 = 0 \quad \text{or} \quad x + 4 = 0$$
$$x = \frac{3}{2} \qquad\qquad x = -4$$

The solutions are $\frac{3}{2}$ and -4, and the solution set is $\left\{-4, \frac{3}{2}\right\}$.

Check:

$x = \frac{3}{2}$:

$$12 - 5x = 2x^2$$
$$12 - 5\left(\frac{3}{2}\right) = 2\left(\frac{3}{2}\right)^2$$
$$12 - \frac{15}{2} = 2\left(\frac{9}{4}\right)$$
$$\frac{24}{2} - \frac{15}{2} = \frac{18}{4}$$
$$\frac{9}{2} = \frac{9}{2} \quad \text{True}$$

$x = -4$:

$$12 - 5x = 2x^2$$
$$12 - 5(-4) = 2(-4)^2$$
$$12 + 20 = 2(16)$$
$$32 = 32 \quad \text{True}$$

Note: *Both methods provide the same solutions, but the second method provides a positive coefficient on the squared term in fewer steps. In general, we want to write the equation in standard form in a way that makes the coefficient of the squared term positive.*

1h. The left side of the equation is in factored form, but we cannot apply the zero products property because the right side of the equation is *not* zero. So, we first multiply the binomials on the left.

$(x - 1)(x + 2) = -2$	The equation is not in standard form.
$x^2 + 2x - x - 2 = -2$	Multiply the two binomials on the left side.
$x^2 + x - 2 = -2$	Combine like terms on the left side.

Now that the left side of the equation has been simplified, we can write the equation in standard form.

$$x^2 + x - 2 + 2 = -2 + 2 \qquad \text{Add 2 to each side.}$$
$$x^2 + x = 0 \qquad \text{Simplify.}$$
$$x(x + 1) = 0 \qquad \text{Factor.}$$
$$x = 0 \quad \text{or} \quad x + 1 = 0 \qquad \text{Apply the zero products property.}$$
$$x + 1 - 1 = 1 - 1 \qquad \text{Solve the resulting equations.}$$
$$x = -1$$

The solutions are 0 or -1 and the solution set is $\{-1, 0\}$.

Check:

$x = 0$:

$$(x - 1)(x + 2) = -2$$
$$(0 - 1)(0 + 2) = -2$$
$$(-1)(2) = -2$$
$$-2 = -2 \quad \text{True}$$

$x = -1$:

$$(x - 1)(x + 2) = -2$$
$$(-1 - 1)(-1 + 2) = -2$$
$$(-2)(1) = -2$$
$$-2 = -2 \quad \text{True}$$

✓ **Student Check 1**   Solve each quadratic equation by factoring.

a. $(x - 10)(x + 8) = 0$ **b.** $2b(4b + 3) = 0$ **c.** $y^2 - 15y + 36 = 0$

d. $r^2 = 36$ **e.** $9a^2 - 12a + 4 = 0$ **f.** $4a^2 - 5 = -8a$

g. $7x = 3x^2 - 20$ **h.** $(x - 4)(x - 2) = 8$

Recall that the degree of a polynomial is the largest degree of the terms of the polynomial. So, the degree of each of the equations in Example 1 is 2. Note that each of the equations in Example 1 has two solutions except for 1e. Example 1e has 1 solution that repeats. The degree of an equation determines the maximum number of solutions of that equation.

Note: *The degree of a polynomial equation determines the maximum number of solutions of the equation.*

Solving Polynomial Equations

Objective 2 ▶

Solve equations of degree 3 or higher by factoring.

Polynomial equations are equations that involve polynomials. A quadratic equation is a special case of a polynomial equation whose degree is 2. The following table shows examples of polynomial equations, the type of equation, its degree, and its maximum number of solutions.

Example	Type of Equation	Degree	Maximum Number of Solutions
$x^2 - x - 6 = 0$	Quadratic	2	2
$2y^3 - 18y = 0$	Cubic	3	3
$a^4 - 5a^2 + 4 = 0$	Quartic	4	4

We now learn how to solve polynomial equations that have a degree of 3 or higher by factoring. The process of solving these equations is similar to the process we used to solve quadratic equations. We must write the polynomial equation in standard form, factor the polynomial, and apply the zero products property.

Procedure: Solving Polynomial Equations of Degree 3 or Higher

Step 1: Write the equation in standard form, that is "polynomial $= 0$."
Step 2: Factor the polynomial.
Step 3: Apply the zero products property and set each factor equal to zero.
Step 4: Solve the resulting equations.
Step 5: Check each solution in the original equation.

Solve each equation by factoring.

2a. $(x - 2)(4x - 1)(x + 6) = 0$ **2b.** $2y^3 - 18y = 0$

2c. $(3x - 5)(5x^2 + 9x - 2) = 0$ **2d.** $a^4 - 5a^2 + 4 = 0$

Solutions   **2a.** We set each factor equal to zero and solve.

$$(x - 2)(4x - 1)(x + 6) = 0$$

$x - 2 = 0$	$4x - 1 = 0$	$x + 6 = 0$
$x - 2 + 2 = 0 + 2$	$4x - 1 + 1 = 0 + 1$	$x + 6 - 6 = 0 - 6$
$x = 2$	$4x = 1$	$x = -6$
	$\dfrac{4x}{4} = \dfrac{1}{4}$	
	$x = \dfrac{1}{4}$	

The solution set is $\left\{ -6, \dfrac{1}{4}, 2 \right\}$. Check each solution.

$x = 2$:

$$(x - 2)(4x - 1)(x + 6) = 0$$
$$(2 - 2)[4(2) - 1](2 + 6) = 0$$
$$(0)(7)(8) = 0$$
$$0 = 0$$

$x = \dfrac{1}{4}$:

$$(x - 2)(4x - 1)(x + 6) = 0$$
$$\left(\dfrac{1}{4} - 2 \right)\left[4\left(\dfrac{1}{4} \right) - 1 \right]\left(\dfrac{1}{4} + 6 \right) = 0$$
$$\left(-\dfrac{7}{4} \right)(0)\left(\dfrac{25}{4} \right) = 0$$
$$0 = 0$$

$x = -6$:

$$(x - 2)(4x - 1)(x + 6) = 0$$
$$(-6 - 2)[4(-6) - 1](-6 + 6) = 0$$
$$(-8)(-25)(0) = 0$$
$$0 = 0$$

Since each solution makes the equation true, our solution set is correct.

2b.

$$2y^3 - 18y = 0$$
$$2y(y^2 - 9) = 0$$ Factor out the common factor.

$$2y(y - 3)(y + 3) = 0$$ Apply the difference of two squares property.

$2y = 0$	$y - 3 = 0$	$y + 3 = 0$	Apply the zero products property.
$\dfrac{2y}{2} = \dfrac{0}{2}$	$y - 3 + 3 = 0 + 3$	$y + 3 - 3 = 0 - 3$	Solve the resulting equations.
$y = 0$	$y = 3$	$y = -3$	

The solution set is $\{-3, 0, 3\}$. Check each solution.

$y = 0$:
$$2y^3 - 18y = 0$$
$$2(0)^3 - 18(0) = 0$$
$$2(0) - 0 = 0$$
$$0 - 0 = 0$$
$$0 = 0$$

$y = 3$:
$$2y^3 - 18y = 0$$
$$2(3)^3 - 18(3) = 0$$
$$2(27) - 54 = 0$$
$$54 - 54 = 0$$
$$0 = 0$$

$y = -3$:
$$2y^3 - 18y = 0$$
$$2(-3)^3 - 18(-3) = 0$$
$$2(-27) + 54 = 0$$
$$-54 + 54 = 0$$
$$0 = 0$$

Since each solution makes the equation true, our solution set is correct.

2c.
$$(3x - 5)(5x^2 + 9x - 2) = 0$$
$$(3x - 5)(5x - 1)(x + 2) = 0 \qquad \text{Factor the trinomial.}$$
$$3x - 5 = 0 \quad \text{or} \quad 5x - 1 = 0 \quad \text{or} \quad x + 2 = 0 \qquad \begin{array}{l}\text{Apply the zero}\\\text{products property.}\end{array}$$

$$\begin{array}{ccc} 3x = 5 & 5x = 1 & x = -2 \\ x = \dfrac{5}{3} & x = \dfrac{1}{5} & \end{array} \qquad \text{Solve each equation.}$$

The solution set is $\left\{-2, \dfrac{1}{5}, \dfrac{5}{3}\right\}$. Each solution checks.

2d.
$$a^4 - 5a^2 + 4 = 0$$
$$(a^2 - 4)(a^2 - 1) = 0 \qquad \text{Factor the trinomial.}$$
$$(a - 2)(a + 2)(a - 1)(a + 1) = 0 \qquad \begin{array}{l}\text{Factor the difference}\\\text{of squares.}\end{array}$$
$$a - 2 = 0 \quad \text{or} \quad a + 2 = 0 \quad \text{or} \quad a - 1 = 0 \quad \text{or} \quad a + 1 = 0 \qquad \begin{array}{l}\text{Apply the zero}\\\text{products property.}\end{array}$$

$$\begin{array}{cccc} a = 2 & a = -2 & a = 1 & a = -1 \end{array} \qquad \text{Solve each equation.}$$

The solution set is $\{-2, -1, 1, 2\}$. We can also write the four solutions as $\{\pm 1, \pm 2\}$. Each solution checks.

✔ **Student Check 2**   Solve each equation by factoring.

a. $(a - 1)(3a + 6)(a - 5) = 0$ **b.** $16x - 4x^3 = 0$

c. $(y + 5)(y^2 - 3y + 2) = 0$ **d.** $b^4 - 13b^2 + 36 = 0$

Objective 3 ▶

Troubleshoot common errors.

Troubleshooting Common Errors

Some common errors related to solving quadratic equations and polynomial equations are shown next.

Objective 3 Examples   **A problem and an incorrect solution are given. Provide the correct solution and an explanation of the error.**

3a. Solve $3x(x - 4) = 0$.

Incorrect Solution	Correct Solution and Explanation
$3x(x - 4) = 0$ $3x = 0 \quad \text{or} \quad x - 4 = 0$ $x = -3 \qquad\qquad x = 4$ The solution set is $\{-3, 4\}$.	The error was made in solving the equation $3x = 0$. To solve this equation, we must divide both sides by 3 to get $x = \dfrac{0}{3} = 0$. So, the solution set is $\{0, 4\}$.

3b. Solve $x^2 = 9x$.

Incorrect Solution	Correct Solution and Explanation
$$x^2 = 9x$$ $$x^2 - 9x = 0$$ $$(x - 3)(x + 3) = 0$$ $$x - 3 = 0 \quad \text{or} \quad x + 3 = 0$$ $$x = 3 \qquad\qquad x = -3$$ The solution set is $\{-3, 3\}$.	The binomial $x^2 - 9$ factors as $(x - 3)(x + 3)$ but $x^2 - 9x$ is not the difference of two squares. We must factor out a common factor of x first. $$x^2 = 9x$$ $$x^2 - 9x = 0$$ $$x(x - 9) = 0$$ $$x = 0 \quad \text{or} \quad x - 9 = 0$$ $$x = 9$$ So, the solution set is $\{0, 9\}$.

3c. Solve $(x - 6)(x + 3) = -8$.

Incorrect Solution	Correct Solution and Explanation
$$(x - 6)(x + 3) = -8$$ $$x - 6 = -8 \quad \text{or} \quad x + 3 = -8$$ $$x = -2 \qquad\qquad x = -11$$ The solution set is $\{-11, -2\}$.	We cannot apply the zero products property since the right side is not equal to zero. So, we must first write the equation in standard form. $$(x - 6)(x + 3) = -8$$ $$x^2 + 3x - 6x - 18 = -8$$ $$x^2 - 3x - 10 = 0$$ $$(x - 5)(x + 2) = 0$$ $$x - 5 = 0 \quad \text{or} \quad x + 2 = 0$$ $$x = 5 \qquad\qquad x = -2$$ So, the solution set is $\{-2, 5\}$.

3d. Solve $4x^2 - 5x - 6 = 0$.

Incorrect Solution	Correct Solution and Explanation
$$4x^2 - 5x - 6 = 0$$ $$(4x - 3)(x + 2) = 0$$ $$4x - 3 = 0 \quad \text{or} \quad x + 2 = 0$$ $$x = \frac{3}{4} \qquad\qquad x = -2$$ The solution set is $\left\{-2, \frac{3}{4}\right\}$.	The error was made in assigning the signs of the binomial factors. Because the middle term of the trinomial is negative, the larger of the outer and inner product must be negative, so the negative sign should be on 2. $$4x^2 - 5x - 6 = 0$$ $$(4x + 3)(x - 2) = 0$$ $$4x + 3 = 0 \quad \text{or} \quad x - 2 = 0$$ $$x = -\frac{3}{4} \qquad\qquad x = 2$$ The solution set is $\left\{-\frac{3}{4}, 2\right\}$.

3e. Solve $2x^3 - 18x = 0$.

Incorrect Solution	Correct Solution and Explanation
$2x^3 - 18x = 0$ $$\frac{2x^3 - 18x}{2x} = \frac{0}{2x}$$ $x^2 - 9 = 0$ $(x-3)(x+3) = 0$ $x - 3 = 0 \quad \text{or} \quad x + 3 = 0$ $x = 3 \qquad\qquad x = -3$ The solution set is $\{-3, 3\}$.	The degree of the equation is 3, so we should have 3 solutions. The error was made in dividing both sides by a variable. We cannot divide by a variable because we might be dividing by zero. This also loses one of the solutions. We should begin by factoring out the common factor. $2x^3 - 18x = 0$ $2x(x^2 - 9) = 0$ $2x(x-3)(x+3) = 0$ $2x = 0 \quad \text{or} \quad x - 3 = 0 \quad \text{or} \quad x + 3 = 0$ $x = 0 \qquad\qquad x = 3 \qquad\qquad x = -3$ The solution set is $\{-3, 0, 3\}$.

ANSWERS TO STUDENT CHECKS

Student Check 1   **a.** $\{-8, 10\}$ **b.** $\left\{-\dfrac{3}{4}, 0\right\}$ **c.** $\{3, 12\}$

d. $\{-6, 6\}$ **e.** $\dfrac{2}{3}$ **f.** $\left\{-\dfrac{5}{2}, \dfrac{1}{2}\right\}$ **g.** $\left\{-\dfrac{5}{3}, 4\right\}$

h. $\{0, 6\}$

Student Check 2   **a.** $\{-2, 1, 5\}$ **b.** $\{-2, 0, 2\}$

c. $\{-5, 1, 2\}$ **d.** $\{-3, -2, 2, 3\}$

SUMMARY OF KEY CONCEPTS

1. The zero products property provides the basis for solving polynomial equations. It states that if the product of factors is zero, then one of those factors must equal zero.

2. To solve a quadratic equation, write the equation in standard form; that is, make one side of the equation zero and factor the other side. Lastly, set each factor equal to 0 and solve the resulting equations.

3. To solve an equation with degree of 3 or higher, write the equation in the form "polynomial = 0." Factor the resulting polynomial. Lastly, set each factor equal to 0 and solve the resulting equations.

GRAPHING CALCULATOR SKILLS

The graphing calculator can be used to verify solutions of quadratic equations.

Example: Verify that the solutions of $x^2 - 8x - 20 = 0$ are $x = 10$ or $x = 2$.

Solution: To verify the solutions, the equation must be in standard form. Then enter the polynomial from the equation into Y_1.

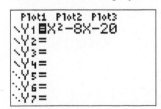

Then view the table. For the solutions to be correct, the Y_1 column should equal zero for the given x-values. Press 2nd Graph and view the x-values of 2 and 10.

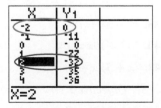 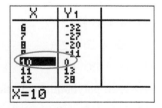

Notice that the Y_1 value for $x = 2$ is -32. This means that 2 is *not* a solution of the equation.

From the table, we see that the Y_1 value for $x = -2$ is 0. So, -2 is a solution. The Y_1 value for $x = 10$ is 0, so 10 is a solution of the equation as well.

SECTION 6.5 / EXERCISE SET

 Write About It!

Use complete sentences in your answer to each exercise.

1. Explain the zero products property.

2. What is a quadratic equation?

3. Explain how to solve a quadratic equation by using the zero products property.

4. Explain how the zero products property can be used to solve polynomial equations of degree 3 or higher.

5. Explain the difference in solving the equations $(x - 5)(x - 1) = 12$ and $(x - 5)(x - 1) = 0$.

6. Explain the difference in solving the equations $4x(x - 5)(x - 1) = 0$ and $4(x - 5)(x - 1) = 0$.

 Practice Makes Perfect!

Solve each equation. (*See Objective 1.*)

7. $(x + 2)(x - 5) = 0$
8. $(x - 1)(x + 4) = 0$
9. $x(x - 3) = 0$
10. $x(x + 1) = 0$
11. $(10 - x)(x + 7) = 0$
12. $(4 - x)(x + 2) = 0$
13. $(2x - 1)(3x + 6) = 0$
14. $(3x + 4)(4x - 1) = 0$
15. $5(6 - x) = 0$
16. $3(1 - x) = 0$
17. $-4(x + 12)(2x - 3) = 0$
18. $-2(4x - 5)(x + 5) = 0$
19. $x(3x - 6)(x + 6) = 0$
20. $x(5x - 10)(x + 3) = 0$
21. $2x(x - 7) = 0$
22. $3x(x - 4) = 0$
23. $(x + 7)(x - 1)(x + 3) = 0$
24. $(x + 8)(x - 9)(x + 1) = 0$
25. $x^2 - x - 6 = 0$
26. $x^2 - 2x - 8 = 0$
27. $x^2 - 2x + 1 = 0$
28. $x^2 - 6x + 9 = 0$
29. $x^2 + 7x + 10 = 0$
30. $x^2 + 10x + 24 = 0$
31. $2x^2 - x - 3 = 0$
32. $3x^2 + 5x - 2 = 0$
33. $x^2 = 25$
34. $x^2 = 16$
35. $(x - 6)(x + 7) = -42$
36. $(x + 3)(x - 8) = -24$
37. $x^2 + 4x = 32$
38. $x^2 - 3x = 18$
39. $6x^2 + 20 = 23x$
40. $3x^2 - 19x = 14$
41. $(x - 5)(x + 3) = -7$
42. $(x + 7)(x + 2) = 24$

Solve each polynomial equation. (*See Objective 2.*)

43. $2x^3 - 18x = 0$
44. $3x^3 - 12x = 0$
45. $(2x - 1)(x^2 - 6x - 16) = 0$
46. $(4x + 3)(x^2 + 5x - 14) = 0$
47. $x^4 - 5x^2 + 4 = 0$
48. $x^4 - 20x^2 + 64 = 0$

49. $5x^4 - 15x^3 - 90x^2 = 0$
50. $4x^4 - 8x^3 + 4x^2 = 0$
51. $5x^3 = 5x$
52. $2x^3 = 32x$
53. $(49x^2 - 1)(4x^2 - 9) = 0$
54. $(16x^2 - 25)(9x^2 - 4) = 0$

 Mix 'Em Up!

Solve each equation.

55. $4x^2 - 5x - 6 = 0$
56. $5x^2 + 3x - 2 = 0$
57. $(x + 9)^2 = 0$
58. $(2x - 1)^2 = 0$
59. $(x + 2)(x + 11) = 22$
60. $(2x - 3)(x - 4) = 12$
61. $(3 - x)(10 + 3x) = 0$
62. $(12 - 5x)(6 - x) = 0$
63. $-4x^2 + x + 5 = 0$
64. $(x^2 - 9)(x^2 - 36) = 0$
65. $(x^2 - 100)(x^2 - 1) = 0$
66. $4x^2 + 10x - 24 = 0$
67. $4x^2 = 16$
68. $5x^2 = 245$
69. $3x^3 - 363x = 0$
70. $4x^4 - 324x^2 = 0$
71. $4x^4 - 37x^2 + 9 = 0$
72. $9x^4 - 13x^2 + 4 = 0$
73. $10x^2 + 12x = 16$
74. $12x^2 + 14 = 34x$
75. $3x(x + 2)(x - 1) = 0$
76. $2x(x - 5)(x - 4) = 0$
77. $25x^2 + 9 = 0$
78. $16x^2 + 1 = 0$
79. $7x^3 - 63x = 0$
80. $5x^3 - 125x = 0$
81. $6x^2 + 38x + 40 = 0$
82. $20x^2 + 94x + 18 = 0$
83. $(2x - 3)(x^2 + x - 20) = 0$
84. $(7x + 2)(x^2 + 3x - 18) = 0$

 You Be the Teacher!

Correct each student's errors, if any.

85. Solve: $(x - 3)(x - 1) = 3$.

Bruce's work:

$$(x - 3)(x - 1) = 3$$
$$x - 3 = 3 \quad \text{or} \quad x - 1 = 3$$
$$x = 6 \qquad\qquad x = 4$$

86. Solve: $5x(x - 4) = 0$.

Jordan's work:

$$5x(x - 4) = 0$$
$$5x = 0 \quad \text{or} \quad 2x - 4 = 0$$
$$x = 5 \qquad\qquad x = 2$$

87. Solve: $2(x + 1) = 0$.

Chris's work:

$$2(x + 1) = 0$$
$$x = 2 \quad \text{or} \quad x + 1 = 0$$
$$x = -1$$

88. Give an example of a quadratic equation with 1 solution.

Taylor's answer: $x^2 - 9 = 0$

 Calculate It!

Use a graphing calculator to determine if the solutions are correct. If they are not, state the correct solutions.

89. $x^2 + 3x - 54 = 0, x = -6, x = -9$

90. $4x^2 + 7x - 2 = 0, x = \frac{1}{4}, x = 2$

 Think About It!

Write a polynomial equation that has the given solutions.

91. $x = 2$ and $x = 5$

92. $x = -1$ and $x = 4$

93. $x = 0, x = -3$

94. $x = 0, x = \frac{1}{2}$

95. $x = 4$ (occurs twice as a solution)

96. $x = -2$ (occurs twice as a solution)

97. $x = 1, x = 3,$ and $x = -4$ **98.** $x = 0, x = \frac{2}{5}, x = -6$

SECTION 6.6 | **Applications of Quadratic Equations**

▶ **OBJECTIVES**

As a result of completing this section, you will be able to use quadratic equations to

1. Solve applications involving area.

2. Solve applications involving consecutive integers.

3. Solve applications involving revenue and profit.

4. Solve applications using the Pythagorean theorem.

5. Solve applications given a model.

6. Troubleshoot common errors.

Suppose that when an airline charges $150 per seat, they can sell 100 tickets. For each $5 decrease in price, they sell 10 additional tickets. How much should the airline charge for each ticket if they want to generate $15,000 in revenue?

This and other real-world situations can be modeled by quadratic equations. In physics, quadratic equations can be used to determine the exact speed or position of moving objects. In business, quadratic models can be used to determine information about a company's revenue and profit. We will study these and other applications in this section.

Applications Involving Area

Some area formulas involve the multiplication of two measurements. For example,

Objective 1 ▶

Solve applications involving area.

$$\text{Area of a rectangle} = \text{length} \times \text{width} \qquad \text{Area of a triangle} = \frac{1}{2} \times \text{base} \times \text{height}$$

$w \qquad A = l \cdot w$

l

$h \qquad A = \frac{1}{2} \cdot b \cdot h$

b

When we multiply the two measurements in these area formulas, we get square units. For example, if a rectangle has length 3 in. and width 4 in., its area is

$$A = lw$$
$$A = (3 \text{ in.})(4 \text{ in.})$$
$$A = 12 \text{ in.}^2$$

So, it is not surprising that quadratic equations can often be used to model applications involving area.

We can use the following general problem solving strategy to guide us in solving the applications in this section.

Procedure: Using the General Problem Solving Strategy

Step 1: Read the question carefully to determine what is known and what is unknown. Assign a variable to the unknown, as needed. Make note of any formulas that apply and draw a picture of the situation, if possible.

Step 2: Translate the words into a mathematical equation.

Step 3: Solve the equation.

Step 4: Check the answer.

Step 5: Answer the original question and make sure the answer is reasonable in the context of the problem.

Objective 1 Examples   Solve each problem.

1a. Find the dimensions of a rectangle that has a length of $2x + 1$ units, a width of $x - 3$ units, and an area of 22 units².

Solution   **1a.** What is unknown? The length is $2x + 1$.

The width is $x - 3$.

What is known? The area is 22 units² and $A = lw$.

Translate and solve: Use the area formula to write the equation.

$(2x + 1)(x - 3) = 22$	Apply the area formula.
$2x^2 - 6x + x - 3 = 22$	Multiply the binomials.
$2x^2 - 5x - 3 = 22$	Combine like terms.
$2x^2 - 5x - 3 - 22 = 22 - 22$	Subtract 22 from each side.
$2x^2 - 5x - 25 = 0$	Simplify.
$(2x + 5)(x - 5) = 0$	Factor the resulting trinomial.
$2x + 5 = 0$ or $x - 5 = 0$	Apply the zero products property.
$2x = -5$ $x = 5$	Solve each equation.
$x = -\dfrac{5}{2}$	

Now we need to find the value of the length and width. Replace the variable in the expressions for the length and width with the possible solutions.

$x = -\dfrac{5}{2}$		$x = 5$	
$l = 2x + 1$	$w = x - 3$	$l = 2x + 1$	$w = x - 3$
$= 2\left(-\dfrac{5}{2}\right) + 1$	$= -\dfrac{5}{2} - \dfrac{6}{2}$	$= 2(5) + 1$	$= 5 - 3$
$= -5 + 1$	$= -\dfrac{11}{2}$	$= 10 + 1$	$= 2$
$= -4$		$= 11$	

The value $-\dfrac{5}{2}$ doesn't make sense in the context of this situation since it makes the length and width of the rectangle negative. So, the length of the rectangle is 11 units and the width is 2 units.

1b. Jentezen plans to pour concrete in a uniform width around his rectangular pool to form a walkway. If his pool measures 20 ft by 50 ft and he has enough concrete to cover 800 ft², how wide can the walkway be?

Solution

1b. What is unknown? The width of the walkway is unknown. Let x represent the width of the walkway.

What is known? The pool measures 20 ft × 50 ft. The area of the walkway is 800 ft². The formula that applies is $A = lw$.

Translate and solve: We need an expression that represents the area of the walkway. The area of the walkway is the area of the pool and walkway combined minus the area of the pool.

area of the pool — area of pool = area of the
and walkway   walkway

$(50 + 2x)(20 + 2x) - (50)(20) = 800$	Write the equation.
$1000 + 100x + 40x + 4x^2 - 1000 = 800$	Find each product.
$4x^2 + 140x = 800$	Combine like terms.
$4x^2 + 140x - 800 = 800 - 800$	Subtract 800 from each side.
$4x^2 + 140x - 800 = 0$	Simplify.
$4(x^2 + 35x - 200) = 0$	Factor out the GCF, 4.
$4(x + 40)(x - 5) = 0$	Factor the trinomial.
$x + 40 = 0$ or $x - 5 = 0$	Apply the zero products property.
$x = -40$ $x = 5$	Solve each equation.

Since x represents the width of the walkway, the negative value doesn't make sense in this context. Therefore, the walkway should be 5 ft wide.

✓ Student Check 1

a. Find the length of the base and height of a triangle if the base is $4x$ units, the height is $x + 2$ units, and the area is 30 units².

b. Kathleen built a rectangular koi pond in her backyard that measures 5 ft by 4 ft. She purchased gravel to spread uniformly around the pond for decoration. If Kathleen has enough gravel to cover 52 ft², how wide can she make the gravel border?

Applications Involving Consecutive Integers

Objective 2 ▶

Solve applications involving consecutive integers.

In Section 2.3, we learned that **consecutive integers** are integers that are successive, or follow one another without interruption. We also learned how to express consecutive integers algebraically as shown.

Property: Consecutive Integer Representation

Consecutive integers: $x, x + 1, x + 2, \ldots$, where x is the first integer.
Consecutive odd integers: $x, x + 2, x + 4, \ldots$, where x is the first odd integer.
Consecutive even integers: $x, x + 2, x + 4, \ldots$, where x is the first even integer.

| **Objective 2 Examples** | Solve each problem. |

2a. The product of two consecutive integers is 90. Find the integers.

Solution   **2a.** What is unknown? The two consecutive integers are unknown.

Let x, $x + 1$ represent the two integers.

What is known? The product of the two integers is 90.

Translate and solve: The product of the integers is $x(x + 1)$ and is equal to 90.

$x(x + 1) = 90$	Write the equation.
$x^2 + x = 90$	Multiply the expressions on the left.
$x^2 + x - 90 = 90 - 90$	Subtract 90 from each side.
$x^2 + x - 90 = 0$	Write in standard form.
$(x + 10)(x - 9) = 0$	Factor.
$x + 10 = 0 \quad$ or $\quad x - 9 = 0$	Apply the zero products property.
$x = -10 \qquad\qquad x = 9$	Solve each equation.

We have two possible values for x.

If $x = -10$, then $x + 1 = -10 + 1 = -9$.	If $x = 9$, then $x + 1 = 9 + 1 = 10$.
-10 and -9 are consecutive integers with a product of 90.	9 and 10 are consecutive integers with a product of 90.

So, there are two pairs of integers that solve this problem. The integers are either -10 and -9 or 9 and 10.

2b. The product of two consecutive positive even integers is 14 more than their sum. Find the integers.

Solution   **2b.** What is unknown? The two consecutive positive even integers are unknown.

Let x and $x + 2$ represent the consecutive even integers.

What is known? The product of two integers is 14 more than their sum.

Translate and solve: The product of the integers is $x(x + 2)$. The sum of the integers is $x + x + 2$.

product of two integers is 14 more than their sum

$x(x + 2) = x + x + 2 + 14$	Write the equation.
$x^2 + 2x = 2x + 16$	Simplify each side.
$x^2 + 2x - 2x - 16 = 2x + 16 - 2x - 16$	Subtract 2x and 16 from each side.
$x^2 - 16 = 0$	Write the equation in standard form.
$(x - 4)(x + 4) = 0$	Factor.
$x - 4 = 0 \quad$ or $\quad x + 4 = 0$	Apply the zero products property.
$x = 4 \qquad\qquad x = -4$	Solve each equation.

Because the integers are positive, the value -4 must be discarded. So, if 4 is one integer, the next even integer is $x + 2 = 4 + 2 = 6$. So, the integers are 4 and 6.

| ✔ **Student Check 2** | Solve each problem. |

a. The product of two consecutive integers is 72. Find the integers.

b. The product of two consecutive odd integers is 23 more than their sum. Find the integers.

Applications Involving Revenue and Profit

Objective 3 ▶

Solve applications involving revenue and profit.

In Chapter 5, we learned that **revenue** is calculated by multiplying the price of an item by the number of items sold and that **profit** is the difference of revenue and cost. We can use our knowledge of revenue, profit, and quadratic equations to solve applications involving these quantities.

Objective 3 Examples **Solve each problem.**

3a. The annual profit, in thousands of dollars, for a mobile phone company can be represented by $p = -4x^2 + 102x - 50$, where x is the total number of mobile phones sold for the year, in thousands. Find the number of phones the company needs to sell to break even; that is, find when profit is equal to zero dollars.

Solution   **3a.**

$-4x^2 + 102x - 50 = p$	Begin with the profit model.
$-4x^2 + 102x - 50 = 0$	Replace p with 0.
$-2(2x^2 - 51x + 25) = 0$	Factor out the GCF.
$-2(2x - 1)(x - 25) = 0$	Factor the trinomial.
$2x - 1 = 0$ or $x - 25 = 0$	Apply the zero products property.
$x = \dfrac{1}{2}$ $x = 25$	Solve each equation.
$x = 0.5$	

So, the company will break even when they sell 0.5 thousand (or 500) phones or when they sell 25 thousand (or 25,000) phones.

3b. Suppose that when an airline charges $150 per seat, they can sell 100 tickets. For each $5 decrease in price, they sell 10 additional tickets. If x represents the number of price decreases, then $150 - 5x$ is the expression for the ticket price and $100 + 10x$ is the number of tickets sold. How much should the airline charge for each ticket if they want to generate $18,200 in revenue?

Solution   **3b.** What is unknown? The price the airline should charge for each ticket is unknown. This is represented by the quantity $150 - 5x$.

What is known? The revenue generated is $18,200.

Translate and solve: Since revenue = price times items sold, we have the following.

price $\times$ items sold = revenue	
$(150 - 5x)(100 + 10x) = 18{,}200$	Write the equation.
$15{,}000 + 1500x - 500x - 50x^2 = 18{,}200$	Simplify the product.
$-50x^2 + 1000x + 15{,}000 = 18{,}200$	Combine like terms.
$-50x^2 + 1000x + 15{,}000 - 18{,}200 = 18{,}200 - 18{,}200$	Subtract 18,200 from each side.
$-50x^2 + 1000x - 3200 = 0$	Simplify.
$-50(x^2 - 20x + 64) = 0$	Factor out the GCF, −50.
$-50(x - 16)(x - 4) = 0$	Factor the trinomial.
$x - 16 = 0$ or $x - 4 = 0$	Apply the zero products property.
$x = 16$ $x = 4$	Solve each equation.

We have two possible solutions, $x = 16$ or $x = 4$. To find the corresponding price the airline should charge for each of these values, substitute the values into the expression $150 - 5x$.

If $x = 16$, then the price is $150 - 5(16) = 70$ or $70.

If $x = 4$, then the price is $150 - 5(4) = 130$ or $130.

So, the airline could charge $70 per ticket or $130 per ticket in order to generate $18,200 in revenue.

☑ **Student Check 3**   Solve each problem.

a. The profit, in hundreds of dollars, for a company is given by $p = -x^2 + 23x - 60$, where x is the number of units sold. Find how many units need to be sold for the company to break even.

b. When a boutique charges \$30 for one of its designer shirts, they can sell 25 shirts. For each \$10 increase in price, they sell five fewer shirts. If x represents the number of price increases, then $30 + 10x$ represents the price of the shirt and $25 - 5x$ represents the number of shirts sold. What price should the boutique charge for a designer shirt if they want to generate \$800 in revenue?

Applications Involving the Pythagorean Theorem

Objective 4 ▶

Solve applications using the Pythagorean theorem.

The Greek mathematician Pythagoras is credited with discovering a special relationship that exists between the lengths of the sides of a *right triangle* (i.e., a triangle with a 90° angle). This special relationship is called the *Pythagorean theorem*.

Property: Pythagorean Theorem

For any right triangle, where a and b are the lengths of the legs and c is the length of the hypotenuse (the side opposite the right angle),

$$a^2 + b^2 = c^2$$

$$(\text{leg})^2 + (\text{other leg})^2 = (\text{hypotenuse})^2$$

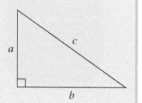

Objective 4 Examples   Solve each problem.

4a. A 13-ft ladder leans against the side of a house. If the distance from the ground to the bottom of the ladder is 5 ft, how far up the house does the ladder reach?

Solution   **4a.** What is unknown? The distance from the ground to where the ladder touches the house is unknown. Let x represent this distance.

What is known? The length of the ladder is 13 ft. The distance from the house to the ladder is 5 ft.

Translate and solve: Since we are dealing with a right triangle, we can use the Pythagorean theorem.

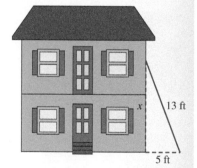

$a^2 + b^2 = c^2$	State the Pythagorean theorem.
$x^2 + 5^2 = 13^2$	Replace a, b, and c with x, 5, and 13, respectively.
$x^2 + 25 = 169$	Simplify.
$x^2 + 25 - 169 = 169 - 169$	Subtract 169 from each side.
$x^2 - 144 = 0$	Simplify.
$(x - 12)(x + 12) = 0$	Factor.
$x - 12 = 0$ or $x + 12 = 0$	Apply the zero products property.
$x = 12$ $x = -12$	Solve each equation.

Since x represents a length, the only possible answer is 12. So, the ladder touches the house at a height of 12 ft.

4b. Donna and Jerry leave their home at the same time one morning. Jerry travels north, and Donna travels east. When they are 10 mi apart, Donna has traveled 2 mi more than Jerry. How many miles have Donna and Jerry traveled?

Solution   **4b.** What is unknown? The distance that Jerry and Donna traveled is unknown. Let x represent the distance Jerry traveled. Then $x + 2$ represents the distance Donna traveled.

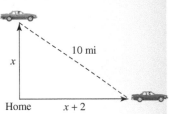

What is known? The distance between the two is 10 mi.

Translate and solve: We are dealing with a right triangle, so we use the Pythagorean theorem to get an equation.

$a^2 + b^2 = c^2$	State the Pythagorean theorem.
$x^2 + (x + 2)^2 = 10^2$	Replace a, b, and c, with x, $x + 2$, and 10, respectively.
$x^2 + x^2 + 4x + 4 = 100$	Square the binomial.
$2x^2 + 4x + 4 = 100$	Combine like terms.
$2x^2 + 4x + 4 - 100 = 100 - 100$	Subtract 100 from each side.
$2x^2 + 4x - 96 = 0$	Simplify.
$2(x^2 + 2x - 48) = 0$	Factor out the GCF.
$2(x + 8)(x - 6) = 0$	Factor the polynomial.
$x + 8 = 0$ or $x - 6 = 0$	Apply the zero products property.
$x = -8 \qquad\qquad x = 6$	Solve each equation.

Since x represents a distance, the only possible answer is 6. So, Jerry traveled 6 mi and Donna traveled $x + 2 = 6 + 2 = 8$ mi.

 **Student Check 4**   Solve each problem.

a. A 20-ft ladder leans against the side of a house. If distance from the house to the ladder is 12 ft, what is the distance the ladder reaches on the house?

b. One leg of a triangle is 1 unit longer than the other leg. If the hypotenuse measures 5 units, then what is the length of each leg?

Objective 5 ▶
Solve applications given a model.

Applications Involving a Model

Now we will use a quadratic model to solve some application problems.

Objective 5 Examples   **Solve each problem.**

5a. A B.A.S.E. jumper leaps from a 1600-ft structure. His height, in feet, can be represented by $s = -16t^2 + 1600$, where t is the number of seconds after he has jumped. The jumper must open his parachute before he reaches the ground. If the jumper does not open his parachute, how long will it take before he reaches the ground?

Solution   **5a.** If the jumper reaches the ground, his height above the ground will be 0 ft. So, we replace s with 0 and solve the resulting equation.

$s = -16t^2 + 1600$	Begin with the given model.
$0 = -16t^2 + 1600$	Replace s with 0.
$0 = -16(t^2 - 100)$	Factor out the common factor, -16.
$0 = -16(t - 10)(t + 10)$	Factor the binomial.
$t - 10 = 0$ or $t + 10 = 0$	Apply the zero products property.
$t = 10 \qquad\qquad t = -10$	Solve.

Since t represents the number of seconds after the jumper leaps off the building, only the positive answer makes sense for this problem. So, the jumper would reach the ground in 10 sec. He has less than 10 sec to open his parachute after jumping.

5b. A ball is thrown upward with an initial velocity of 16 ft/sec from a height of 32 ft above the ground. The height, in feet, of the ball t sec after it is thrown can be represented by $s = -16t^2 + 16t + 32$. How long will it take for the ball to reach the ground?

Solution   **5b.** When the ball reaches the ground, its height above the ground is 0 ft.

$s = -16t^2 + 16t + 32$	Begin with the given model.
$0 = -16t^2 + 16t + 32$	Replace s with 0.
$0 = -16(t^2 - t - 2)$	Factor out the common factor, -16.
$0 = -16(t - 2)(t + 1)$	Factor the trinomial.
$t - 2 = 0$ or $t + 1 = 0$	Apply the zero products property.
$t = 2$ $t = -1$	Solve.

The negative value doesn't make sense in this context. So, the ball reaches the ground 2 sec after it was thrown.

✓ Student Check 5   Solve each problem.

a. Suppose a penny is dropped from the top of the Empire State Building, which is approximately 1296 ft high. The quadratic equation $s = -16t^2 + 1296$ gives the distance from the ground (in feet) that the penny will be after t sec. How long will it take the penny to reach the ground?

b. The height of a ball thrown upward with an initial velocity of 32 ft/sec from a height of 128 ft above the ground can be represented by $-16t^2 + 32t + 128$, where t is the number of seconds after the ball is thrown. How long will it take the ball to reach the ground?

Objective 6 ▶

Troubleshoot common errors.

Troubleshooting Common Errors

Some of the common errors associated with applications of quadratic equations are shown. Most of the problems occur from setting up the equation incorrectly. Others come from simplifying the equation incorrectly.

Objective 6 Examples   **A problem and an incorrect solution are given. Provide the correct solution and provide an explanation of the error.**

6a. The product of two consecutive odd integers is 63. Write an equation that can be used to find the integers.

Incorrect Solution	Correct Solution and Explanation
$x(x + 1) = 63$	Because the integers are consecutive odd integers, the equation is $x(x + 2) = 63$

6b. The sides of a right triangle are consecutive integers. Find the length of each side.

Incorrect Solution	Correct Solution and Explanation
The sides are x, $x + 1$, and $x + 2$. So, the equation is $$x^2 + (x + 1)^2 = (x + 2)^2$$ $$x^2 + x^2 + 1 = x^2 + 4$$ $$2x^2 + 1 = x^2 + 4$$ $$2x^2 + 1 - x^2 - 4 = x^2 + 4 - x^2 - 4$$ $$x^2 - 3 = 0$$ Can't factor so there are no solutions.	The binomials were not squared correctly. Remember that $$(a + b)^2 = a^2 + 2ab + b^2$$ So, we get $$x^2 + (x + 1)^2 = (x + 2)^2$$ $$x^2 + x^2 + 2x + 1 = x^2 + 4x + 4$$ $$2x^2 + 2x + 1 = x^2 + 4x + 4$$ $$x^2 - 2x - 3 = 0$$ $$(x - 3)(x + 1) = 0$$ $$x = 3 \quad \text{or} \quad x = -1$$ So, the sides of the right triangle are 3, 4, and 5 units.

ANSWERS TO STUDENT CHECKS

Student Check 1   a. The base is 12 units and the height is 5 units.   **b.** The width of the gravel border is 2 ft.

Student Check 2   a. The integers are 8 and 9 or −9 and −8.   **b.** The integers are 5 and 7 or −5 and −3.

Student Check 3   a. The company will break even after selling 3 units or 20 units.   **b.** The shirts should be sold for $40 each.

Student Check 4   a. The ladder reaches 16 ft up the house.   **b.** The legs of the triangle are 3 units and 4 units.

Student Check 5   a. The penny will reach the ground in 9 sec.   **b.** The ball will reach the ground in 4 sec.

SUMMARY OF KEY CONCEPTS

1. When solving applications involving area, identify the shape of the object whose area is being examined. Next, write down the area formula for that shape. Draw a picture to visualize the problem.

2. Consecutive integers are integers that follow one another without interruption.

    a. Consecutive integers can be expressed as x, $x + 1$, ...

    b. Consecutive even integers can be expressed as x, $x + 2$, . . . , where x is an even integer.

    c. Consecutive odd integers can be expressed as x, $x + 2$, . . . , where x is an odd integer.

3. When solving applications involving revenue, remember the formula: revenue = price × quantity. When solving applications involving profit, remember the formula: profit = revenue − cost.

4. The Pythagorean theorem defines a relationship between the sides of a right triangle. The Pythagorean theorem is $(\text{leg})^2 + (\text{other leg})^2 = (\text{hypotenuse})^2$.

5. If a model is given, replace the appropriate variable with the known value.

GRAPHING CALCULATOR SKILLS

The graphing calculator can help us solve equations. One method is to use the Table feature. If we input the two sides of the equation into Y_1 and Y_2, respectively, we can use the Table feature to determine the value of x for which the two sides are equal.

Example: Solve $(150 - 5x)(100 + 10x) = 18,200$ using the Table feature.

Solution: Enter each side of the equation into the equation editor. Examine the table for the value of x that makes $Y_1 = Y_2$.

Plot1 Plot2 Plot3
\Y1◘(150−5X)(100 +10X)
\Y2◘18200
\Y3=
\Y4=
\Y5=
\Y6=

X	Y1	Y2
0	15000	18200
1	15950	18200
2	16800	18200
3	17550	18200
4	18200	18200
5	18750	18200
6	19200	18200

X=4

X	Y1	Y2
12	19800	18200
13	19550	18200
14	19200	18200
15	18750	18200
16	18200	18200
17	17550	18200
18	16800	18200

X=16

From the table, we see the solutions are 4 and 16.

SECTION 6.6 / EXERCISE SET

Write About It!

Use complete sentences in your answer to each exercise.

1. Explain the first two steps in solving an application problem.

2. Give examples of consecutive integers.

3. Explain what expressions represent consecutive odd and consecutive even integers.

4. What is revenue? What is a profit?

Practice Makes Perfect!

Solve each problem involving area. (*See Objective 1.*)

5. Allison wants to section off a rectangular portion of her yard for a flowerbed. She wants the flowerbed to be 8 ft longer than it is wide. If the area of the flowerbed will be 48 ft², find the dimensions of the flowerbed.

6. A county park builds rectangular pool. The length of the pool is 10 ft more than twice its width. If the area of the pool is 1000 ft², find the dimensions of the pool.

7. Janna wants to mat her wedding picture before framing it. Her picture measures 14 in. by 20 in. and she wants the mat in uniform width around the picture. How wide can the mat be if the area of the picture and mat is 1360 in.²?

8. Mona wants to professionally frame her son's first school photo. The photo is 8 in. by 10 in. She wants to mat the picture with an even width around the picture. How wide should the mat be if the area of the photo and the frame is 360 in.²?

9. Jensen wants to pour concrete in a uniform width around his pool that measures 30 ft by 40 ft. If he has enough concrete to cover an area of 624 ft², then what should be the width of the concrete walkway that is created?

10. Persia wants to put mulch in a uniform width around a rectangular reflection pond that she built in her backyard. If the pond measures 8 ft by 12 ft and she has enough mulch to cover 300 ft², then what should be the width of the mulch border?

Solve each problem involving consecutive integers. (*See Objective 2.*)

11. The product of two positive consecutive integers is 132. Find the integers.

12. The product of two positive consecutive integers is 20. Find the integers.

13. The product of two consecutive even integers is 34 more than their sum. Find the integers.

14. The product of two consecutive even integers is 2 more than their sum. Find the integers.

15. The product of two consecutive odd integers is equal to 3 times the larger integer. Find the integers.

16. The product of two consecutive odd integers equals 9 more than 6 times the larger integer. Find the integers.

Solve each problem involving revenue and profit. (*See Objective 3.*)

17. A luxury kennel provides accommodations for cats. When the kennel charges a daily rate of $45, they can board 10 cats. For each $2 decrease in price, 1 more cat is boarded. If x represents the number of price decreases, then $45 - 2x$ represents the daily boarding rate and $10 + x$ represents the number of cats boarded. How many cats do they need to board to make $375 in revenue?

18. Ashley monograms T-shirts. When she charges $5 per name, she monograms 50 T-shirts. For every $1 increase in price, five fewer T-shirts are monogrammed. If x represents the number of price increases, then $5 + x$ represents the price to monogram a name and $50 - 5x$ represents the number of T-shirts monogrammed. How much should Ashley charge to monogram a name to earn $250 in revenue?

19. When an airline charges $200 per seat, they can sell 50 tickets. For every $2 decrease in price, five more tickets are sold. If x represents the number of price decreases, then $200 - 2x$ represents the ticket price and $50 + 5x$ represents the number of passengers. How many passengers are needed for the airline to earn $18,000 in revenue?

20. When an online shoe seller charges $100 for a pair of wedge sandals, 50 pairs are sold. For each $5 decrease in price, 10 more pairs are sold. If x represents the number of price decreases, then $100 - 5x$ represents the price of each pair of sandals and $50 + 10x$ represents the number of pairs of sandals sold. How much should the online shoe seller charge for each pair of sandals to earn $5000 in revenue?

21. The profit, in millions, of a video game company is given by $p = -60x^2 + 555x + 450$, where x is the number of units sold, in thousands. How many units need to be sold for the company to break even?

22. The profit, in millions, of a software company is given by $p = -150x^2 + 1425x + 4500$, where x is the number of units sold, in thousands. How many units need to be sold for the company to break even?

Find the lengths of the sides of each right triangle. (*See Objective 4.*)

23.

$x - 1$ $x + 1$

x

24.

x $x + 4$

$x + 2$

25.

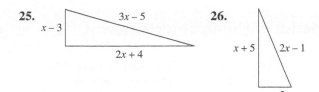

26.

Solve each problem. (*See Objective 4.*)

27. A 15-ft ladder leans against the side of a house. If the distance from the bottom of the ladder to the house is 3 ft shorter than the distance from the ground to the top of the ladder, then what is the distance from the ground to the top of the ladder?

28. The length of a ladder leaning against a house is 3 ft more than twice the distance from the bottom of the ladder to the house. If the distance from the bottom of the house to the top of the ladder is 12 ft, what is the distance from the bottom of the ladder to the house?

29. Heidi and Spencer leave a restaurant in two separate cars at the same time. Heidi travels south and Spencer travels west. When they are 20 mi apart, Heidi has traveled 4 mi less than Spencer. How many miles have Heidi and Spencer each traveled?

30. Theodore and Wallace leave their dorm room at the same time. Theodore travels north and Wallace travels west. When they are 30 mi apart, Theodore has traveled 6 mi more than the distance Wallace has traveled. How far did Wallace travel?

Solve each problem. (*See Objective 5.*)

31. A penny is dropped from a 144-ft bridge. The distance between the penny and the ground can be approximated by $s = -16t^2 + 144$ (in feet), where t is the time (in seconds) after the penny is dropped. How long will it take the penny to hit the ground?

32. How long will it take the penny from Exercise 31 to be 80 ft from the ground?

33. The distance of a ball thrown directly upwards from the ground is $s = -16t^2 + 120t$ (in feet), where t is the time (in seconds) after the ball is thrown. When will the ball be 104 ft in the air?

34. A toy rocket is projected upward from the ground with an initial velocity of 96 ft/sec. The distance of the rocket from the ground is $s = -16t^2 + 96t$ (in feet), where t is the time (in seconds) after the rocket is launched. When will the rocket reach 80 ft?

35. A sky diver is dropped from a plane at 3600 ft. His distance from the ground can be approximated by $s = -16t^2 + 3600$ (in feet), where t is the time (in seconds) after the diver is dropped. How long will it take the diver to reach the ground if his parachute doesn't open?

36. How long will it take the diver from Exercise 35 to be 2000 ft from the ground?

37. Young performs at birthday parties for children. The revenue generated from performing at x birthday parties is $R = x^2 + 250x$ (in dollars). How many birthday parties must Young perform at to earn $2600 in revenue?

38. The total cost of producing x items is given by $C = 3x^2 - 22x + 10$ (in dollars). How many items can be produced at a total cost of $90?

39. Joyce's Bakery makes creative birthday cakes. The profit generated from the sale of x cakes is $P = 100x - x^2$ (in dollars). How many cakes must be sold for Joyce's Bakery to have a profit of $2500?

40. Leroy is a popular deejay. He determined that his revenue from hosting x parties is approximately $R = x^2 + 244x$ (in dollars). How many parties does Leroy have to host to earn $1500 in revenue?

41. The height of a ball thrown upward with an initial velocity of 96 ft/sec from a height of 640 ft above the ground can be represented by $S = -16t^2 + 96t + 640$, where t is the number of seconds after the ball is thrown. How long will it take the ball to reach the ground?

42. The height of a ball thrown upward with an initial velocity of 128 ft/sec from a height of 768 ft above the ground can be represented by $S = -16t^2 + 128t + 768$, where t is the number of seconds after the ball is thrown. How long will it take the ball to reach the ground?

 You Be the Teacher!

Correct each student's errors, if any.

43. The product of two positive odd consecutive integers is 1 less than twice their sum. Find the integers.

Chris's work:
$$x = \text{1st odd integer}$$
$$x + 1 = \text{2nd odd integer}$$
$$x(x + 1) = 2(2x + 1) - 1$$
$$x^2 + x = 4x + 1 - 1$$
$$x^2 - 3x = 0$$
$$x(x - 3) = 0$$
$$x = 0 \quad \text{or} \quad x = 3$$
$$x + 1 = 1 \quad \text{or} \quad x = 4$$

44. Set up the equation that models the situation: "When a local theater charges $6 per seat, they can sell 200 tickets. For every $1 increase in price, 5 less tickets are sold. How many tickets are needed to earn $1800?"

Jordan's work:
$$x = \text{number of tickets}$$
$$(6 + x)(5x - 200) = 1800$$

 Calculate It!

Use the table feature in a graphing calculator to solve each equation.

45. $(10 + x)(120 - 3x) = 1443$

46. $(15 + x)(200 - 5x) = 3570$

GROUP ACTIVITY The Mathematics of Home Entertainment

The table shows consumer spending on video home systems (VHS) and universal media disks (UMD), digital video discs (DVD), blue-ray discs (BD) and high definition TVs (HDTV), and digital movies. (Source: http://www.degonline.org)

U.S. Consumer Spending for Home Entertainment Rentals and Purchases (in billions)

Year	VHS/UMD	DVD	BD/HDTV	Digital	Total
1999	$12.2	$ 1.1	$0.0	$0.6	$13.9
2000	$11.4	$ 2.4	$0.0	$0.7	$14.5
2001	$10.9	$ 5.3	$0.0	$0.7	$16.9
2002	$ 9.6	$ 8.6	$0.0	$0.7	$19.0
2003	$ 6.9	$13.1	$0.0	$0.7	$20.7
2004	$ 4.4	$16.7	$0.0	$0.7	$21.8
2005	$ 2.1	$18.9	$0.0	$0.8	$21.7
2006	$ 0.4	$20.2	$0.0	$1.0	$21.6
2007	$ 0.1	$19.7	$0.3	$1.3	$21.4
2008	$ 0.1	$18.4	$0.9	$1.6	$21.0
2009	$ 0.0	$15.8	$1.5	$2.1	$19.4
2010	$ 0.0	$14.0	$2.3	$2.5	$18.8

The following graph shows a scatter plot of the data presented in the table for three of the categories.

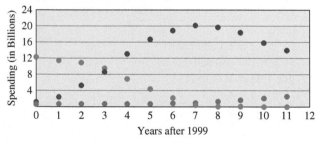

1. Determine which data each scatter plot represents. Describe the trend shown by each scatter plot.

2. Connect the points in a smooth curve for each scatter plot to form three graphs. Approximate the three points of intersection for the graphs and interpret the results.

3. A model for the red graph is

   $y = -0.0411x^3 + 0.3327x^2 + 2.4591x + 0.2187$.

   A model for the blue graph is

   $y = 0.0305x^3 - 0.4126x^2 - 0.212x + 12.334$

   and a model for the green graph is

   $y = 0.0267x^2 - 0.1408x + 0.7643$.

   Write equations in standard form that can be used to determine the point of intersection for the following graphs. Do not solve the equations.

   a. the blue and green graphs

   b. the red and green graphs

   c. the blue and red graphs

4. The equations in step 3 are examples of polynomial equations that cannot be solved by factoring. So, use a graphing calculator to approximate the point of intersection for the green and red graphs and interpret the results.

Factoring Polynomials and Polynomial Equations

What's the big idea? The ability to factor polynomials provides us with the foundation for solving polynomial equations. This skill is utilized with a basic property of real numbers; that is, if a product is equal to zero, then one of the factors must be zero. Once we know how to solve polynomial equations, we can then solve applications that are modeled by this type of equation.

The Tools

Listed below are the key terms, skills, formulas, and properties you should know for this chapter.

The page reference is provided if you need additional help with the given topic. The Study Tips will assist in your preparation for an exam.

Study Tips

1. Learn all of the terms, formulas, and properties. Make flash cards and have someone quiz you.
2. Rework problems from the exercises and also the ones you worked in class. Work additional problems from the review exercises.
3. Review the summaries of key concepts.
4. Work the chapter test.
5. Be sure to review the online resources for additional study materials.

Terms

Consecutive integers 487	Perfect square trinomial 463	Revenue and profit 489
Factor 440	Polynomial equation 479	Standard form 474
Factoring 440	Prime polynomial 445	Sum of squares 468
Greatest common factor 440	Quadratic equation 474	Trial-and-error method 458

Formulas and Properties

- Area formulas 485
- Difference of two cubes property 469
- Difference of two squares property 467
- Distributive property 443
- Pythagorean theorem 490
- Sum of two cubes property 469
- Zero products property 475

CHAPTER 6 / SUMMARY

How well do you know this chapter? Complete the following questions to find out. Take a look back at the section if you need help.

SECTION 6.1 Greatest Common Factor and Grouping

1. To factor a polynomial means to write the polynomial as a(n) _____.

2. The _____ _____ _____ of a set of integers is the largest factor that is common to each integer.

3. The GCF of a set of variable terms is the common variable raised to the _____ exponent of the terms.

4. The GCF of a set of terms is the _____ of the GCF of the _____ and the GCF of the _____ _____.

5. To factor using the GCF, we apply the _____ _____ in reverse. That is, $ab + ac =$ _____.

6. The _____ method applies to factoring polynomials with four or more terms.

7. All factoring can be checked by _____.

SECTION 6.2 Factoring Trinomials

8. To factor a trinomial of the form $x^2 + bx + c$, we must find the factors of __ whose sum is __.

9. If the last term of the trinomial is positive, the signs in the binomial factors are the _____.

10. If the last term of the trinomial is negative, the signs in the binomial factors are _____.

11. Before factoring a trinomial, we should factor out any _____ _____, if necessary.

SECTION 6.3 More on Factoring Trinomials

12. There are two methods for factoring trinomials of the form $ax^2 + bx + c$, $a \neq 1$. We can use ____ and ____ or the method of ____.

13. A ____ ____ trinomial factors as a binomial squared.

SECTION 6.4 Factoring Binomials

14. A binomial of the form $a^2 - b^2$ can be factored as ____. An example is ____.

15. A binomial of the form $a^3 - b^3$ can be factored as ____. An example is ____.

16. A binomial of the form $a^3 + b^3$ can be factored as ____. An example is ____.

17. The sum of squares ____ ____ ____ over the real numbers. An example is ____.

SECTION 6.5 Solving Quadratic Equations and Other Polynomial Equations by Factoring

18. The zero products property states that if a product is zero, then one of the ____ must be ____.

19. A quadratic equation is an equation of the form ____, where $a \neq 0$.

20. To solve a quadratic equation by factoring, write the equation in ____ ____, ____ the polynomial, set each ____ equal to ____, and ____.

21. Polynomial equations are equations that involve ____.

22. A polynomial equation with degree 3 is called a(n) ____ equation. A polynomial equation with degree 4 is called a(n) ____ equation.

23. The degree of an equation determines the maximum number of ____ of the equation.

SECTION 6.6 Applications of Quadratic Equations

24. Quadratic equations can be used to solve problems involving ____ of rectangles and triangles.

25. Consecutive integers can be represented as ____. Consecutive odd integers can be represented as ____. Consecutive even integers can be represented as ____.

26. ____ is calculated by multiplying the price of an item by the number of items sold and ____ is the difference of revenue and cost.

27. The Pythagorean theorem is ____, where a and b represent the ____ and c represents the ____.

CHAPTER 6 / REVIEW EXERCISES

SECTION 6.1

Find the GCF of the terms in each set. (*See Objectives 1 and 2.*)

1. $12x^3$ and $28x$
2. $6a^2$ and 15
3. $-20y^2$, $36y^4$, and $28y$
4. $-44ab^3$ and $33a^2b^2$
5. $112r^4s^2$ and $84r^3$
6. $10xy$, $16xy^2$, and $18x^3y$

Factor each polynomial. (*See Objectives 3 and 4.*)

7. $2x^2 - 6x + 5x - 15$
8. $3x^2 + 10x + 3x + 10$
9. $7x^2 + 28x - 3x - 12$
10. $12b^3 - 60b^2$
11. $4c^2(3c - 2d) - 5d(3c - 2d)$
12. $7ac - 21ad + 2bc - 6bd$
13. $-10p^2 - 35p$
14. $-3a^2 + 15a - 6$
15. $3x^2(x + 1) - (x - 1)$
16. $6x^2 + 21xy - 10xy - 35y^2$

17. The surface area of a cylindrical can is given as $2\pi r^2 + 2\pi rh$, where r is the radius of the base and h is the height of the cylinder.

18. The volume of a box is given by $2x^2 + 24x$ in.^{3}.

SECTION 6.2

Factor each trinomial. (*See Objectives 1 and 2.*)

19. $x^2 + 3x - 28$
20. $6x^3 + 12x^2 - 90x$
21. $4y^2 + y + 3$
22. $4a^3b + 44a^2b^2 + 96ab^3$
23. $-x^4 + 9x^3 - 20x^2$
24. $x^2 + 16xy + 60y^2$
25. $-5x^2y + 10xy - 30y$
26. $a^3 + 5a^2b - 84ab^2$

SECTION 6.3

Use either trial and error or grouping to factor each trinomial completely. (*See Objectives 1–3.*)

27. $4a^2 - 28a + 49$
28. $3x^2 - 13x - 10$
29. $-3y^2 - 2y + 21$
30. $8t^2 + 2t - 45$
31. $16x^2 - 12x + 9$
32. $9a^2 + 30ab + 25b^2$
33. $-14h^2 + 28h - 7$
34. $4x^2 + x + 1$
35. $26x^2 - 6xy - 20y^2$
36. $18x^2y^4 - 2z^6$
37. $-108a^2 - 9ab + 42b^2$
38. $p^3q^6 + 8r^9$
39. $-12h^4 + 12h^2 - 3$
40. $25u^2 - 10uv + v^2$

41. The profit, in thousands of dollars, of a video game company is given by the expression $-60x^2 + 456x + 192$, where x is the number of units sold, in hundreds.

42. The profit, in millions of dollars, of a footwear company is given by the expression $-150x^2 + 1150x + 3500$, where x is the number of units sold, in thousands.

SECTION 6.4

Factor each binomial completely. (*See Objectives 1–3.*)

43. $9y^4 - 72y$ **44.** $8a^2 - 50b^2$ **45.** $64x^3y + 125y$

46. $32r^4s^4 - 2t^4$ **47.** $x^6y^6 + z^6$ **48.** $a^4 + 100b^4$

49. $a^3 - 1000b^3c^3$ **50.** $343x^3 - 8$ **51.** $625y^4 - 16z^4$

52. $a^{15}b^{12} - 216c^6$ **53.** $\frac{1}{8}a^3 + b^3c^3$ **54.** $x^4y^4 - \frac{1}{16}z^4$

55. $x^3 + 27y^5$ **56.** $0.36r^6 - 0.49s^8$

SECTION 6.5

Solve each equation. (*See Objectives 1 and 2.*)

57. $5x^2 - 8x = 4$ **58.** $(3x + 4)^2 = 0$

59. $(x + 6)(x + 5) = 30$ **60.** $-2x^2 + 11x - 14 = 0$

61. $6x^3 + 15x^2 - 9x = 0$ **62.** $9x^2 = 25$

63. $7(x + 4)(x - 12) = 0$

64. $(4x - 1)(x^2 + 6x - 16) = 0$

65. $75x^3 - 3x = 0$ **66.** $x^4 - 5x^2 + 4 = 0$

SECTION 6.6

Solve each problem. (*See Objectives 1–5.*)

67. George wants to section off a rectangular portion of his yard for a vegetable garden. He wants the garden to be 7 ft longer than it is wide. If the area of the garden will be 144 ft², find the dimensions of the garden.

68. Shutter wants to pour concrete in a uniform width around his backyard pool that measures 35 ft by 40 ft. If he has enough concrete to cover an area of 850 ft², what should be the width of the concrete walkway that is created?

69. The product of two positive consecutive integers is 240. Find the integers.

70. The product of two positive consecutive even integers is equal to 6 more than 3 times the sum of the two integers. Find the integers.

Find the lengths of the sides of each right triangle.

71.
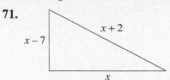

72.
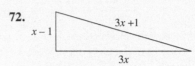

73. A 25-ft ladder leans against the side of a house. If the distance from the bottom of the ladder to the house is 5 ft shorter than the distance from the ground to the top of the ladder, then what is the distance from the ground to the top of the ladder?

74. Anna and Bob leave a restaurant in two separate cars at the same time. Anna travels north and Bob travels east. When they are 26 mi apart, Anna has traveled 14 mi less than Bob. How many miles have Anna and Bob each traveled?

75. A sky diver is dropped from a plane at 2500 ft. His distance from the ground can be approximated by $s = -16t^2 + 2500$ (in feet), where t is the time (in seconds) after the diver is dropped. How long will it take the diver to land on the ground if his parachute doesn't open?

76. Wing plays the piano for weddings. The revenue generated from performing at x weddings is $R = x^2 + 160x$ (in dollars). At how many weddings must Wing perform to receive $2625 in revenue?

CHAPTER 6 TEST / FACTORING POLYNOMIALS AND POLYNOMIAL EQUATIONS

1. The greatest common factor of $4x^5 - 8x^4 + 12x^3$ is
 a. 4
 b. $4x$
 c. $4x^3$
 d. $4x^5$

2. The complete factorization of $x^2 + 2x - 3$ is
 a. $(x - 1)(x - 3)$ **b.** $(x - 1)(x + 3)$
 c. $(x + 1)(x - 3)$ **d.** $(x + 1)(x + 3)$

3. One of the prime factors of $6t^2 - 19t - 20$ is
 a. $t + 5$ **b.** $2t + 5$
 c. $6t + 5$ **d.** $t + 1$

4. The prime factorization of $16x^2 + 48x + 36$ is
 a. $(8x + 12)(2x + 3)$ **b.** $4(2x + 3)^2$
 c. $4(2x - 3)^2$ **d.** $(4x + 9)^2$

5. The prime factorization of $16x^2 - 25$ is
 a. $(8x - 5)(2x + 5)$ **b.** $(8x - 5)(8x + 5)$
 c. $(16x - 25)(16x + 25)$ **d.** $(4x - 5)(4x + 5)$

6. One of the prime factors of $8x^3 + 27$ is
 a. $4x^2 - 6x + 9$ **b.** $4x^2 - 6x - 9$
 c. $4x^2 + 6x + 9$ **d.** $2x^2 - 6x + 3$

7. The prime factorization of $ab^2 + ac + b^2d + cd$ is
 a. $(a + c)(d + b^2)$ **b.** $(a + d)(b^2 + c)$
 c. $a(b^2 + c) + (b^2 + c)$ **d.** $(a + d)^2(b^2 + c)$

8. The solution set for $x^2 = 16$ is
 a. $\{4\}$ **b.** $\{-4\}$
 c. $\{16\}$ **d.** $\{-4, 4\}$

9. The solution set for $6x^2 + x = 2$ is
 a. $\left\{-\frac{3}{2}, 2\right\}$ **b.** $\left\{-\frac{1}{2}, \frac{2}{3}\right\}$
 c. $\left\{-\frac{3}{2}, \frac{1}{2}\right\}$ **d.** $\left\{-\frac{2}{3}, \frac{1}{2}\right\}$

Factor each polynomial completely.

10. $5x^2 + 10x$

11. $20 - 5y - 4p + yp$

12. $r^3 + r^2 - 4r - 4$

13. $3a^2 - 3a - 18$

14. $9a^2 + 6a - 8$

15. $10a^2 - 35a - 20$

16. $9x^2 + 30x + 25$

17. $81p^2 - 49$

18. $4a^2 + 25$

19. $125x^3 - 27y^3$

20. $81t^4 - 16$

21. $64y^6 + 1$

Solve each equation.

22. $(x - 5)(x + 6) = 0$

23. $4x^2 = 36$

24. $a^2 + a = 56$

25. $6z^2 = 5z + 6$

26. $(y + 1)(y - 3) = 5$

27. $(x - 4)(x + 3)(x + 5) = 0$

28. $x^4 - 29x^2 + 100 = 0$

Solve each problem.

29. Find the lengths of the sides of a right triangle if the legs are consecutive even integers and the hypotenuse is 10 units.

30. The Grand Canyon Skywalk is a glass walkway that stands about 3600 ft above the floor of the Grand Canyon. The distance an object would be above the floor of the canyon if dropped from the skywalk is given by $s = -16t^2 + 3600$ ft after t sec. How long will it take for a rock dropped from the skywalk to reach the floor of the canyon? (Source: http://www.grandcanyonskywalk.com /skywalk.html)

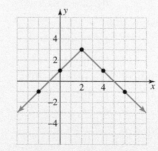

CUMULATIVE REVIEW EXERCISES / CHAPTERS 1–6

1. Perform the indicated operation and simplify. (*Sections 1.3–1.6, Objectives 1 and 2*)

  a. $-(-3)^4 + (-2)^5 - 8 \cdot 9$

  b. $15.1 - 8.6 + (-4.3) - (-7.8)$

  c. $(-7.3)(0)(-11.2)$

2. Simplify each algebraic expression. (*Section 1.8, Objectives 3 and 4*)

  a. $-12(3a - 4) + 5(-a + 9)$

  b. $14.2 - 1.2(2b - 2.5)$

  c. $6x^2 - (-5x^2 + 16x - 5) + 12x - 11$

3. Solve each equation. If the equation is a contradiction, write the solution as ∅. If the equation is an identity, write the solution as ℝ. (*Section 2.2, Objectives 2 and 3; Section 2.3, Objectives 1 and 2; Section 2.4, Objectives 1–4*)

  a. $-(2x - 9) - 21 = 5(x - 1)$

  b. $\frac{2}{3}(2x - 5) + \frac{5}{6} = x$

  c. $-4.5x + 20.5 = 3(-1.5x + 4.3)$

4. Ellen invests $5400 in two accounts. She invests $4100 in a certificate of deposit that pays 1.6% annual interest and the remaining amount in a money market fund that pays 1.9% annual interest. How much interest does she earn from the two accounts in 1 yr? (*Section 2.6, Objective 2*)

5. Solve each inequality. For each inequality, graph the solution set, write the solution set in interval notation and in set-builder notation. (*Section 2.7, Objectives 1–4*)

  a. $9x + 3(x - 10) < 8x - 10$

  b. $-2.4 < 0.6a + 1.8 < 3.9$

6. Graph each line. Label two points on the line. (*Section 3.2, Objective 2*)

  a. $m = \frac{2}{5}$, passes through $(-10, 6)$

  b. $y = 2x + 7$

7. The U.S. Bureau of Labor Statistics conducted a Consumer Expenditure Survey to find out how many U.S. households own desktops and laptops. The percent of computer ownership from 2005 to 2008 can be modeled by the equation $y = 2.45x + 68.6$, where x is the number of years after 2005. (Source: http://www.bls.gov) (*Section 3.2, Objective 5*)

  a. Find the y-intercept and interpret its meaning in the context of the problem.

  b. Based on the model, what is the percent of computer ownership in 2011?

  c. Based on the model, what is the percent of computer ownership in 2013?

  d. Graph the equation.

8. Write the equation of each line. Express each answer in slope-intercept form and in standard form. (*Section 3.5, Objectives 1–4*)

  a. $(10, -5)$, perpendicular to $x - 2y = 9$

  b. $(2, -1)$, parallel to $6x - y = 2$

  c. $m = 0$, passes through $(3, -4)$

9. Find the requested information. (*Section 3.6, Objective 4*)

  a. Find $f(0)$ if $f(x) = 5x^2 - x + 3$.

  b. Find $g(-3)$ if $g(x) = -x^2 + 2x$.

  c. Find $f(3)$ and find all x such that $f(x) = -1$ if $f(x)$ is given by the graph.

10. Solve each system of equations graphically. Then state if the system is consistent with dependent equations, consistent with independent equations, or inconsistent. (*Section 4.1, Objectives 3 and 4*)

   a. $\begin{cases} y = x - 4 \\ y = -3x \end{cases}$ **b.** $\begin{cases} x - y = 2 \\ y = -3 \end{cases}$

11. Solve each system of linear equations using substitution. (*Section 4.2, Objective 1*)

   a. $\begin{cases} 3x - 2y = -8 \\ x + 5y = 3 \end{cases}$ **b.** $\begin{cases} 4x - y = 15 \\ 2x + 7y = -45 \end{cases}$

12. Solve the systems using elimination. (*Section 4.3, Objective 1*)

   a. $\begin{cases} 3x - 5y = -13 \\ -2x + 9y = 20 \end{cases}$ **b.** $\begin{cases} \dfrac{2}{3}x - \dfrac{1}{4}y = 2 \\ \dfrac{5}{6}x + \dfrac{1}{2}y = \dfrac{5}{2} \end{cases}$

13. Colleen is working two jobs to help pay her way through college. She works as a peer tutor in the college math lab and a part-time receptionist at a doctor's office. One week she earns \$390 by working 15 hr in the doctor's office and 25 hr in the math lab. Another week she earns \$240 by working 15 hr in the doctor's office and 5 hr in the math lab. What is Colleen's hourly wage as a part-time receptionist and as a peer tutor? (*Section 4.4, Objective 1*)

14. A chemist has a 40% alcohol solution. He needs 40 gal of a 25% alcohol solution. How many gallons of water and how many gallons of the 40% alcohol solution must he mix together to obtain 40 gal of a 25% alcohol solution? (*Section 4.4, Objective 3*)

15. Use the exponent rules to simplify each expression. Write each answer with positive exponents. (*Section 5.1, Objectives 1–4*)

   a. $(-p^{-2}q^{-4})(2p^5q^{-7})$ **b.** $\dfrac{-12a^{-5}n}{6a^9n^{-4}}$

   c. $-2b^0 - 3$

16. Use the exponent rules to simplify each expression. Write each answer with positive exponents. (*Section 5.2, Objectives 1 and 2*)

   a. $(3^{-1}x^5)^{-3}$ **b.** $\left(\dfrac{5x^2}{y^{-4}}\right)^{-3}$

   c. $\left(-\dfrac{2c^{-3}}{d^4}\right)^{-3}$

17. The formula $P = P_0 \cdot 2^{-t/m}$ gives the amount of substance, P, present at time t, where P_0 is the original amount of the substance and m is its half-life. Suppose the half-life of a substance is 12 weeks. If the original amount of the substance is 10 g, how much substance will be left after (*Section 5.1, Objective 6*)

   a. 12 weeks?   **b.** 36 weeks?

18. Use exponent rules to simplify each expression. Write answer in scientific notation. (*Section 5.2, Objective 4*)

   a. $(4.2 \times 10^6)(3.5 \times 10^{-20})$ **b.** $\dfrac{7.925 \times 10^{-14}}{6.34 \times 10^{-5}}$

19. The Consumer Assistance to Recycle and Save Act (CARS), also known as the "Cash for Clunkers" Program, was signed into law June 24, 2009, to provide incentives to motorists to trade in less fuel-efficient vehicles for more fuel-efficient ones. When the program ended, there were about 700,000 car sales with rebate applications totaling \$2.877 billion. If the rebate was evenly distributed among the car buyers, how much would each buyer receive? Write answer in scientific notation. (1 billion = 10^9) (Source: http://www.cars.gov/carsreport/index.html) (*Section 5.2, Objective 4*)

20. A company's total cost to make x items is $24x^3 - 15x^2 + 1900$ dollars. How much does it cost the company to make 18 items? (*Section 5.3, Objective 3*)

21. The height, in feet, of a ball projected upward from the top of a 240-ft building with an initial velocity of 128 ft/sec is given by $-16t^2 + 128t + 240$, where t is the number of seconds after the ball is thrown. Find the height of the ball. (*Section 5.3, Objective 3*)

   a. 3 sec after it is thrown   **b.** 4 sec after it is thrown

22. Perform each operation. Write answers in standard form. (*Section 5.3, Objective 4*)

   a. $(21 - 15x^2 - 9x) + (-16 + 10x^2 + 28x)$

   b. $(32x^3 + 19x^2 - 16x) - (20x^3 + 7x^2 - 22x)$

   c. Subtract $(4x^2 - 6xy + 9y^2)$ from $(-3x^2 + 4xy - 13y^2)$.

   d. $(5x^2 + 11xy - 3y^2) - (-6x^2 + 14xy + y^2) + (-12x^2 - 18xy - 6y^2)$

23. Suppose the profit generated by selling x items is $-0.6x^2 + 98x + 213$ dollars and the total cost is $28x + 460$ dollars. Find an expression that represents the company's revenue. Write answer in standard form. (*Section 5.3, Objectives 4 and 5*)

24. Simplify each product. (*Section 5.4, Objectives 1 and 2*)

   a. $6x^2y(2xy^2 - 8)$ **b.** $(13 - 4x)(3x + 1)$

   c. $(3a - 2b)(9a^2 + 6ab + 4b^2)$

   d. $\left(\dfrac{1}{5}x - 3\right)\left(\dfrac{3}{5}x + 2\right)$

   e. $(x^2 + 6x - 5)(3x^2 - 4x + 1)$

25. Perform the indicated operation. Write answers in standard form. (*Section 5.5, Objectives 1–4*)

   a. $(5x + 3)(4x - 1)$ **b.** $(5b + 12)^3$

   c. $(p - 3)^3$ **d.** $(15x - y)(15x + y)$

   e. $(2r + 3s)(4r^2 - 6rs + 9s^2)$

26. Perform the indicated operation. (*Section 5.6, Objectives 1 and 2*)

   a. Divide $15x^3 + 6x^2 - 24x + 7$ by $3x$.

   b. $\dfrac{5x^3 - 18x^2 + 23x - 12}{x - 4}$

27. Factor each trinomial completely. (*Section 6.2, Objectives 1 and 2; Section 6.3, Objectives 1 and 2*)

   a. $x^3 + 3x^2 - 108x$ **b.** $10y^3 - 25y^2 - 60y$

   c. $3x^2 + 4x + 8$ **d.** $-7x^4 + 91x^2 - 252$

   e. $-9x^3 + 36x^2 - 36x$

28. The volume of an open box with height x inches is given by the expression $x^3 - 7x^2 - 60x$ in.^{3}. Factor this trinomial. (*Section 6.2, Objectives 1 and 2*)

29. Factor each trinomial completely. (*Section 6.2, Objectives 1 and 2; Section 6.3, Objectives 1 and 2*)

    a. $x^2 - 15x + 54$ **b.** $3y^3 - 18y^2 + 27y$

    c. $-12x^3 - 54x^2 + 30x$ **d.** $a^2 - 16ab + 48b^2$

    e. $2x^2y + 42xy + 216y$ **f.** $x^2 + 6xy + 7y^2$

30. A company's profit can be represented by $-14q^2 + 350q + 8400$, where q is the number of units sold in thousands. Factor this polynomial. (*Section 6.3, Objectives 1 and 2*)

31. Use either trial and error or grouping to factor each trinomial completely. (*Section 6.3, Objectives 1 and 2*)

    a. $5p^2 + 24p + 27$ **b.** $5pq^2 - 6pq - 27p$

    c. $30x^2 - x - 7$ **d.** $16x^2 + 56xy + 49y^2$

    e. $117a^2b^2 + 3ab - 12$ **f.** $3p^2 + 4pq + 7q^2$

32. The profit, in millions of dollars, of a video game company is given by $p = -60x^2 + 555x + 450$, where x is the number of units sold in thousands. How many units need to be sold for the company to break even? (*See Section 6.6, Objective 3*)

33. Factor completely. (*Section 6.4, Objectives 1–3*)

    a. $5x^3 - 45x$ **b.** $50m^2 - 32n^2$

    c. $54x^3 - 128y^3$ **d.** $2y^3 + 250z^3$

    e. $r^2 + 64s^2$ **f.** $81a^4 - 16$

    g. $3c^6 - 24$ **h.** $x^4y^4 - 16z^4$

    i. $\dfrac{1}{4}a^2 - \dfrac{9}{25}b^2$

34. Solve each equation. (*Section 6.5, Objectives 1 and 2*)

    a. $4x^2 - 5x = 6$ **b.** $5x^2 - 2 = 3x$

    c. $4x^2 + 4x + 1 = 0$ **d.** $(x + 2)(x + 11) = 22$

    e. $(2 - x)(3 + 5x) = 0$ **f.** $-3x^2 + 17x - 20 = 0$

    g. $x^4 - 10x^2 + 9 = 0$ **h.** $3x(x + 2)(x - 5) = 0$

    i. $4x^2 + 25 = 0$ **j.** $48x^2 + 116x + 70 = 0$

35. Solve each problem. (*Section 6.6, Objectives 1–5*)

    a. Patty wants to section off a rectangular portion of her yard for a flowerbed. She wants the flowerbed to be 3 ft longer than it is wide. If the area of the flowerbed will be 54 ft^2, find the dimensions of the flowerbed.

    b. The product of two positive consecutive integers is 156. Find the integers.

    c. The product of two consecutive odd integers equals 12 more than 9 times the larger integer. Find the integers.

    d. When an express bus company charges $40 per seat from New York City to Philadelphia, they can sell 90 tickets. For every $2 decrease in price, 5 more tickets are sold. If x represents the number of price decreases, then $40 - 2x$ represents the ticket price and $90 + 5x$ represents the number of passengers. How many passengers are needed for the express bus company to earn $2400?

    e. Find the lengths of the sides of the right triangle.

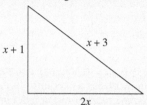

    f. The profit, in millions, of a furniture company is given by $p = -40x^2 + 464x + 192$, where x is the number of units sold in thousands. How many units need to be sold for the company to break even?

    g. Kelly and Sue leave their dorm room at the same time. Kelly travels north and Sue travels west. When they are 30 mi apart, Kelly has traveled 6 mi more than the distance Sue has traveled. How far did Sue travel?

    h. The total cost of producing x items is given by $C = 12x^2 - 540x + 6000$ dollars, where x is the number of units sold. How many items can be produced at a total cost of $1800?

    i. The height of a ball thrown upward with an initial velocity of 128 ft/sec from a height of 144 ft above the ground can be represented by $s = -16t^2 + 128t + 144$, where t is the number of seconds after the ball is thrown. How long will it take the ball to reach the ground?

Rational Expressions, Functions, and Equations

Learning Strategies

Albert Einstein is probably the epitome of an incredible learner. Could you ever be like him? What gave him the capacity to acquire all of his knowledge? While you and I are not Albert Einstein, there are things that we can do to improve our learning.

- Find out what kind of learner you are—visual, auditory, kinesthetic, and the like. The Internet offers many different learning styles inventories that you can complete to determine this information. One is also provided in Chapter S.
- As soon as class ends, take a few moments to write down a summary of the main points that were presented in class.
- It has been proven that most people forget up to 40% of new material within a few hours of being exposed to it. Therefore, you should find the time to review class notes within a couple hours of attending class.
- Review your notes frequently over the next few days so that the concepts get transferred to your long-term memory.
- Distribute your studying over several sessions rather than one long cram session. Studies have shown that students who distribute their practice over several sessions remember 67% more than those who learn by cramming. (Source: John J. Donovan and David J. Radosevich, "A Meta-Analytic Review of the Distribution of Practice Effect: Now You See It, Now You Don't," *Journal of Applied Psychology*, 84 [1999]: 795–805)

Question For Thought: What type of learner are you? How long after new material is presented do you review it?

Chapter Outline

Coming Up...

In Section 7.5, we will evaluate the function $C(t) = \dfrac{300t}{36t^2 + 3.3}$ to calculate the concentration, in micrograms per milliliter, of a drug in the bloodstream t hr after it is administered.

"Imagination is more important than knowledge. For while knowledge defines all we currently know and understand, imagination points to all we might yet discover and create."

— Albert Einstein (Mathematician and scientist)

SECTION 7.1 Rational Functions and Simplifying Rational Expressions

Chapters 5 and 6 introduced us to polynomials, operations on polynomials, and factoring polynomials. In this chapter, our focus will be on quotients of polynomials, or rational expressions and equations involving these types of expressions. It is critical that we have a good understanding of the prior chapters as the skills we learned will be used quite extensively.

▶ **OBJECTIVES**

As a result of completing this section, you will be able to

1. Evaluate rational expressions and functions.
2. Find the domain of a rational function.
3. Simplify rational expressions.
4. Write equivalent forms of rational expressions that include negative signs.
5. Solve applications of rational expressions.
6. Troubleshoot common errors.

A high school class is organizing its 25-yr reunion. The venue, DJ, decorations, and other expenses total $6000 and the dinner costs $35 per person. The total cost for x attendees is $C(x) = 35x + 6000$. The cost per person is the expression, $\dfrac{C(x)}{x} = \dfrac{35x + 6000}{x}$. What is the cost per person if 100 people attend? 200 attend? 300 attend?

The expression in the preceding example is a rational expression. In this section, we learn how to evaluate and simplify these expressions.

Evaluating Rational Expressions and Functions

At this point, we have studied only rational numbers. Rational numbers, as you recall, are numbers that can be expressed as a quotient of integers, such as $-\dfrac{5}{3}, \dfrac{15}{2}, 0.25 = \dfrac{1}{4}, -2 = \dfrac{-2}{1}$. A *rational expression* is also a quotient. It is a quotient of two polynomials.

Objective 1 ▶

Evaluate rational expressions and functions.

Definition: The expression $\dfrac{P}{Q}, Q \neq 0$, is a **rational expression** if P and Q are both polynomials.

Some examples of rational expressions are

$$\dfrac{x^2 - x - 6}{x} \qquad -\dfrac{5}{y + 1} \qquad b^2 - 4 = \dfrac{b^2 - 4}{1}$$

The fraction $\dfrac{\sqrt{x + 2}}{x - 1}$, for example, is not a rational expression since the numerator is not a polynomial.

When we use function notation to represent quotients of polynomials, we have a *rational function*.

Definition: A **rational function** is a function of the form $f(x) = \dfrac{P(x)}{Q(x)}$, where $P(x)$ and $Q(x)$ are polynomial functions and $Q(x) \neq 0$.

The definitions of a rational expression and a rational function have a restriction that the denominator cannot equal zero. Recall the following facts about fractions.

- When the *numerator* of a fraction is zero, the value of the fraction is zero. For instance, $\dfrac{0}{5} = 0$.

504

- When the *denominator* of a fraction is zero, the value of the fraction is undefined. For example, $\dfrac{5}{0}$ is undefined.

So, the requirement that the denominator cannot equal zero is stated to ensure that the expression is defined.

When working with rational expressions and negative signs, recall that when the numerator or denominator of a fraction is negative, we can make the entire fraction negative. For instance,

$$\frac{-2}{3} = -\frac{2}{3} \quad \text{and} \quad \frac{2}{-3} = -\frac{2}{3}$$

We state this in the following property.

Property: Negative Rational Expressions

If P and Q are polynomials, and $Q \neq 0$, then

$$\frac{-P}{Q} = \frac{P}{-Q} = -\frac{P}{Q}$$

Throughout the course, we have evaluated algebraic expressions and functions. Evaluating a rational expression or function is no different.

Procedure: Evaluating a Rational Expression or Rational Function

Step 1: Replace the variable with the given number.
Step 2: Apply the order of operations to simplify the resulting expression.

Objective 1 Examples

Evaluate each expression or function for the given values. Round each answer to two decimal places when necessary.

1a. $\dfrac{3x^2 - 5x}{x - 2}$; $x = -50, 0, 1.9, 2, 2.001,$ and 100

1b. $f(x) = \dfrac{4x + 5}{x + 1}$. Find $f(-25), f(-1.05), f(-1), f(0), f\left(\dfrac{1}{4}\right),$ and $f(100)$.

Solutions

1a.

x	$\dfrac{3x^2 - 5x}{x - 2}$
-50	$\dfrac{3(-50)^2 - 5(-50)}{-50 - 2} = \dfrac{3(2500) + 250}{-52} = \dfrac{7500 + 250}{-52} = \dfrac{7750}{-52} \approx -149.04$
0	$\dfrac{3(0)^2 - 5(0)}{0 - 2} = \dfrac{0}{-2} = 0$
1.9	$\dfrac{3(1.9)^2 - 5(1.9)}{1.9 - 2} = \dfrac{3(3.61) - 9.5}{-0.1} = \dfrac{10.83 - 9.5}{-0.1} = \dfrac{1.33}{-0.1} = -13.3$
2	$\dfrac{3(2)^2 - 5(2)}{2 - 2} = \dfrac{3(4) - 10}{0} = \dfrac{12 - 10}{0} = \dfrac{2}{0}$ undefined
2.001	$\dfrac{3(2.001)^2 - 5(2.001)}{2.001 - 2} = \dfrac{3(4.004001) - 10.005}{0.001} = \dfrac{2.007003}{0.001} \approx 2007.00$
100	$\dfrac{3(100)^2 - 5(100)}{100 - 2} = \dfrac{3(10,000) - 500}{98} = \dfrac{30,000 - 500}{98} = \dfrac{29,500}{98} \approx 301.02$

1b.

x	$f(x) = \dfrac{4x + 5}{x + 1}$
-25	$f(-25) = \dfrac{4(-25) + 5}{-25 + 1} = \dfrac{-100 + 5}{-24} = \dfrac{-95}{-24} \approx 3.96$
-1.05	$f(-1.05) = \dfrac{4(-1.05) + 5}{-1.05 + 1} = \dfrac{-4.2 + 5}{-0.05} = \dfrac{0.8}{-0.05} = -16$
-1	$f(-1) = \dfrac{4(-1) + 5}{-1 + 1} = \dfrac{1}{0}$ undefined
0	$f(0) = \dfrac{4(0) + 5}{0 + 1} = \dfrac{5}{1} = 5$
$\dfrac{1}{4}$	$f\left(\dfrac{1}{4}\right) = \dfrac{4\left(\dfrac{1}{4}\right) + 5}{\dfrac{1}{4} + 1} = \dfrac{1 + 5}{\dfrac{1}{4} + \dfrac{4}{4}} = \dfrac{6}{\dfrac{5}{4}} = 6 \cdot \dfrac{4}{5} = \dfrac{24}{5} = 4.8$
100	$f(100) = \dfrac{4(100) + 5}{100 + 1} = \dfrac{405}{101} \approx 4.01$

✓ **Student Check 1**   Evaluate each expression or function for the given values. Round each answer to two decimal places when necessary.

a. $\dfrac{x^2 + 5x}{x - 4}$; $x = -100, 0, 3.98, 4, 4.001$, and 200

b. $f(x) = \dfrac{6x - 2}{x + 5}$. Find $f(-25)$, $f(-5)$, $f(-4.99)$, $f(0)$, $f\left(\dfrac{1}{2}\right)$, and $f(100)$.

The Domain of a Rational Function

Objective 2 ▶

Find the domain of a rational function.

In Examples 1a and 1b, $x = 2$ makes $\dfrac{3x^2 - 5x}{x - 2}$ undefined and the value $x = -1$ makes $\dfrac{4x + 5}{x + 1}$ undefined. The expressions are undefined at these values because they make the denominator of the rational expression equal to zero. The values that make the denominator of a rational function zero are the values that must be excluded from the *domain of a rational function*.

> **Definition:** The **domain of a rational function** consists of all real numbers except for the values that make the denominator equal to zero. In other words, the domain is the set of numbers for which the rational expression is defined.

> **Procedure: Finding the Domain of a Rational Function**
>
> **Step 1:** Set the denominator equal to zero and solve the resulting equation.
> **Step 2:** The domain is all real numbers, excluding the solutions of the equation obtained in step 1.
> **Step 3:** Write the domain in either set-builder notation or interval notation.

Objective 2 Examples	**Find the domain of each rational function.**

2a. $f(x) = \dfrac{2x + 3}{4x - 8}$ **2b.** $f(x) = \dfrac{x^2 + 4}{x^2 - 5x - 14}$ **2c.** $f(x) = \dfrac{x - 1}{5}$

Solutions   **2a.**

$$4x - 8 = 0 \qquad\qquad\qquad \text{Set the denominator equal to zero.}$$
$$4x - 8 + 8 = 0 + 8 \qquad\qquad \text{Add 8 to each side.}$$
$$4x = 8 \qquad\qquad\qquad\qquad \text{Simplify.}$$
$$x = 2 \qquad\qquad\qquad\qquad \text{Divide each side by 4.}$$

So, the domain of the function is all real numbers, except 2. We write this as $\{x \mid x$ is a real number, $x \neq 2\}$ or $(-\infty, 2) \cup (2, \infty)$.

2b. Since the denominator is a quadratic expression, we use factoring to solve the resulting equation.

$$x^2 - 5x - 14 = 0 \qquad\qquad \text{Set the denominator equal to zero.}$$
$$(x - 7)(x + 2) = 0 \qquad\qquad \text{Factor the trinomial.}$$
$$x - 7 = 0 \quad \text{or} \quad x + 2 = 0 \qquad \text{Apply the zero products property.}$$
$$x = 7 \qquad\qquad x = -2 \qquad \text{Solve each equation.}$$

The domain of the function is all real numbers, except 7 and -2. We write this as $\{x \mid x$ is a real number and $x \neq 7, x \neq -2\}$ or $(-\infty, -2) \cup (-2, 7) \cup (7, \infty)$.

2c. $5 = 0 \qquad\qquad\qquad\qquad\qquad\qquad\quad$ Set the denominator equal to zero.

Since this statement is false, there is no solution of the equation, which indicates that there are no values that need to be restricted from the domain. The domain of the function is all real numbers, or $(-\infty, \infty)$.

✓ **Student Check 2**   Find the domain of each rational function.

a. $f(x) = \dfrac{3x + 10}{5x + 5}$ **b.** $f(x) = \dfrac{x + 4}{x^2 - 9}$ **c.** $f(x) = \dfrac{7x - 4}{3}$

Objective 3 ▶
Simplify rational expressions.

The Fundamental Property of Rational Expressions

A fraction is in **lowest terms** if 1 is the only common factor of its numerator and denominator. The fraction $\dfrac{26}{39}$, for example, is not in lowest terms because the common factor of 26 and 39 is 13. To simplify the fraction $\dfrac{26}{39}$, we must factor the numerator and denominator and divide out the common factor, 13.

$$\frac{26}{39} = \frac{2 \cdot 13}{3 \cdot 13} = \frac{2}{3} \cdot \frac{13}{13} = \frac{2}{3} \cdot 1 = \frac{2}{3}$$

Likewise, a rational expression is in lowest terms if 1 is the only common factor of its numerator and denominator. Consider the expression, $\dfrac{a^2 - 9}{a^2 - 4a + 3}$. To simplify this expression, we must factor the numerator and denominator and divide out the common factors.

$$\frac{a^2 - 9}{a^2 - 4a + 3} = \frac{(a - 3)(a + 3)}{(a - 1)(a - 3)} = \frac{a + 3}{a - 1} \cdot \frac{a - 3}{a - 3} = \frac{a + 3}{a - 1} \cdot 1 = \frac{a + 3}{a - 1}$$

Note that the original rational expression, $\dfrac{a^2 - 9}{a^2 - 4a + 3}$, is equivalent to the simplified rational expression, $\dfrac{a + 3}{a - 1}$, except for the values that make either denominator zero.

The process of writing a rational expression in lowest terms is called *simplifying the expression*. To simplify a rational expression, we use the *fundamental property of rational expressions*.

Property: Fundamental Property of Rational Expressions

For a rational expression $\dfrac{P}{Q}$ and a polynomial R ($Q, R \neq 0$),

$$\frac{PR}{QR} = \frac{P}{Q} \cdot \frac{R}{R} = \frac{P}{Q} \cdot 1 = \frac{P}{Q}$$

The key in simplifying rational expressions is recognizing quotients that are equivalent to 1 or −1.

- A quotient has a value of 1 when an expression is divided by itself. For example, assuming the denominator is not zero,

$$\frac{3x}{3x} = 1 \qquad \frac{y + 4}{y + 4} = 1 \qquad \frac{2a - 5}{2a - 5} = 1$$

- A quotient has a value of −1 when an expression and its opposite are divided. For example, assuming the denominator is not zero,

$$\frac{5}{-5} = -1 \qquad \frac{-2y}{2y} = -1 \qquad \frac{t - 2}{2 - t} = -1$$

$$2 - t = -t + 2$$
$$= -1(t - 2)$$

Property: Dividing Opposites

The expressions $x - y$ and $y - x$ are opposites of one another. As long as $x \neq y$,

$$\frac{x - y}{y - x} = -1$$

Procedure: Simplifying a Rational Expression

Step 1: Rewrite the numerator and denominator in their factored forms.
Step 2: Divide the numerator and denominator by their common factors.

Note: *We will assume that the variables in a rational expression do not represent the values that make the denominator zero.*

Objective 3 Examples   **Simplify each rational expression.**

3a. $\dfrac{4y - 2}{2 - 4y}$

3b. $\dfrac{5x^2 - 10x}{5x^2 - 20}$

3c. $\dfrac{x^2 - x - 12}{27 + x^3}$

3d. $\dfrac{6x^2 + 11x - 10}{3x^3 - 2x^2 + 6x - 4}$

Solutions   **3a.** $\dfrac{4y-2}{2-4y} = \dfrac{4y-2}{-1(4y-2)}$ Rewrite the denominator as the opposite of $4y-2$.

$= \dfrac{1}{-1} \cdot \dfrac{4y-2}{4y-2}$ The common factor of the numerator and denominator is $4y-2$.

$= \dfrac{1}{-1} \cdot 1$ Note that $\dfrac{4y-2}{4y-2} = 1$.

$= -1$ Simplify.

We can also apply the property of dividing opposites to note that

$$\dfrac{4y-2}{2-4y} = -1$$

3b. $\dfrac{5x^2-10x}{5x^2-20} = \dfrac{5x(x-2)}{5(x-2)(x+2)}$ Factor the numerator and denominator completely.

$= \dfrac{x}{x+2} \cdot \dfrac{5(x-2)}{5(x-2)}$ The common factor of the numerator and denominator is $5(x-2)$.

$= \dfrac{x}{x+2} \cdot 1$ Note that $\dfrac{5(x-2)}{5(x-2)} = 1$.

$= \dfrac{x}{x+2}$ Simplify.

We can also show the simplification by dividing the numerator and denominator by their common factors.

$\dfrac{5x^2-10x}{5x^2-20} = \dfrac{5x(x-2)}{5(x-2)(x+2)}$

$= \dfrac{\overset{1}{\cancel{5}}x\overset{1}{\cancel{(x-2)}}}{\underset{1}{\cancel{5}}\underset{1}{\cancel{(x-2)}}(x+2)}$

$= \dfrac{x}{x+2}$

> Divide the numerator and denominator by their common factors, 5 and $x-2$. Write a "1" above and below the common factors to show that
> $$\dfrac{5}{5} = 1 \text{ and } \dfrac{x-2}{x-2} = 1.$$

3c.

Method 1: Factor out the common factor from the numerator and denominator and show that the quotient is 1.

$\dfrac{x^2-x-12}{27+x^3} = \dfrac{(x-4)(x+3)}{(3+x)(9+3x+x^2)}$

$= \dfrac{x-4}{9+3x+x^2} \cdot \dfrac{x+3}{3+x}$

$= \dfrac{x-4}{9+3x+x^2} \cdot 1$

$= \dfrac{x-4}{x^2+3x+9}$

Method 2: Divide the numerator and denominator by their common factor.

$\dfrac{x^2-x-12}{27+x^3} = \dfrac{(x-4)\overset{1}{\cancel{(x+3)}}}{\underset{1}{\cancel{(3+x)}}(9+3x+x^2)}$

$= \dfrac{x-4}{x^2+3x+9}$

3d. $\dfrac{6x^2+11x-10}{3x^3-2x^2+6x-4} = \dfrac{\overset{1}{\cancel{(3x-2)}}(2x+5)}{\underset{1}{\cancel{(3x-2)}}(x^2+2)}$ Factor the numerator and denominator.

$= \dfrac{2x+5}{x^2+2}$ Divide the numerator and denominator by their common factor, $3x-2$.

✔ **Student Check 3**   Simplify each rational expression.

a. $\dfrac{7p - 5}{5 - 7p}$

b. $\dfrac{3y^2 - 12y}{3y^2 - 48}$

c. $\dfrac{x^2 + 6x + 5}{1 + x^3}$

d. $\dfrac{9a^2 - 20a - 21}{a^3 - 3a^2 + 7a - 21}$

Negative Rational Expressions

Objective 4 ▶

Write equivalent forms of rational expressions that include negative signs.

When working with rational expressions, it is helpful to recognize equivalent forms especially when negative signs are involved. When a rational expression is negative, we can write it in two equivalent forms.

> **Property: Rule of Negative Rational Expressions**
>
> If N and D are polynomials, and $D \neq 0$, then
>
> $$-\frac{N}{D} = \frac{-N}{D} = \frac{N}{-D}$$

Objective 4 Examples   Use the rule of negative rational expressions to write two equivalent forms of each rational expression.

4a. $-\dfrac{4x + 1}{x - 2}$

4b. $\dfrac{-3x + 5}{x - 1}$

Solutions **4a.** $-\dfrac{4x + 1}{x - 2} = \dfrac{-(4x + 1)}{x - 2}$ Apply $-\dfrac{N}{D} = \dfrac{-N}{D}$. $\left|\right.$ $-\dfrac{4x + 1}{x - 2} = \dfrac{4x + 1}{-(x - 2)}$ Apply $-\dfrac{N}{D} = \dfrac{N}{-D}$.

$\qquad\qquad\qquad = \dfrac{-4x - 1}{x - 2}$ Distribute. $\qquad\qquad = \dfrac{4x + 1}{-x + 2}$ Distribute.

$\qquad\qquad\qquad\qquad\qquad\qquad\qquad\qquad\qquad\qquad\qquad\qquad\;\; = \dfrac{4x + 1}{2 - x}$ Apply the commutative property.

So,

$$-\frac{4x + 1}{x - 2} = \frac{-4x - 1}{x - 2} = \frac{4x + 1}{2 - x}$$

4b. Because the first term in the numerator is negative, we can factor out -1 from the numerator.

$$\frac{-3x + 5}{x - 1} = \frac{-1(3x - 5)}{x - 1} = \frac{-(3x - 5)}{x - 1}$$

Now we can write the two equivalent expressions.

$\dfrac{-(3x - 5)}{x - 1} = -\dfrac{3x - 5}{x - 1}$ Apply $\dfrac{-N}{D} = -\dfrac{N}{D}$. $\left|\right.$ $\dfrac{-(3x - 5)}{x - 1} = \dfrac{3x - 5}{-(x - 1)}$ Apply $\dfrac{-N}{D} = \dfrac{N}{-D}$.

$\qquad\qquad\qquad\qquad\qquad\qquad\qquad\qquad\qquad\qquad\qquad = \dfrac{3x - 5}{-x + 1}$ Distribute.

$\qquad\qquad\qquad\qquad\qquad\qquad\qquad\qquad\qquad\qquad\qquad = \dfrac{3x - 5}{1 - x}$ Apply the commutative property.

So,

$$\frac{-3x + 5}{x - 1} = -\frac{3x - 5}{x - 1} = \frac{3x - 5}{1 - x}$$

✓ **Student Check 4**   Use the rule of negative rational expressions to write two equivalent forms of each rational expression.

a. $-\dfrac{3x-1}{6x+1}$

b. $\dfrac{-x-12}{x+7}$

Applications

Objective 5 ▶

Solve applications of rational expressions.

We now look at some rational expressions that describe real-life events and evaluate them to find some specific information.

Objective 5 Examples   Solve each problem.

5a. A high school class is organizing its 25-yr reunion. The venue, DJ, decorations, and other expenses total $6000 and the dinner costs $35 per person. The total cost for x attendees is $C(x) = 35x + 6000$. The cost per person is the expression, $\dfrac{C(x)}{x} = \dfrac{35x + 6000}{x}$. What is the cost per person if 100 people attend? 200 attend? 300 attend?

Solution   **5a.** We evaluate $\dfrac{C(x)}{x} = \dfrac{35x + 6000}{x}$ for $x = 100$, $x = 200$, and $x = 300$.

x	$\dfrac{C(x)}{x} = \dfrac{35x + 6000}{x}$	Interpretation
100	$\dfrac{C(100)}{100} = \dfrac{35(100) + 6000}{100}$ $= \dfrac{3500 + 6000}{100}$ $= \dfrac{9500}{100}$ $= 95$	The reunion will cost $95 per person if 100 people attend.
200	$\dfrac{C(200)}{200} = \dfrac{35(200) + 6000}{200}$ $= \dfrac{7000 + 6000}{200}$ $= \dfrac{13,000}{200}$ $= 65$	The reunion will cost $65 per person if 200 people attend.
300	$\dfrac{C(300)}{300} = \dfrac{35(300) + 6000}{300}$ $= \dfrac{10,500 + 6000}{300}$ $= \dfrac{16,500}{300}$ $= 55$	The reunion will cost $55 per person if 300 people attend.

5b. A utility company burns coal to generate electricity. The cost C (in dollars) of removing p percent of the smokestack pollutants is given by $C = \dfrac{70,000p}{100 - p}$. What is the cost of removing 75% of the pollutants? What is the cost of removing 95% of the pollutants? What happens to the cost as more and more pollutants are removed?

Solution   **5b.** We first evaluate $C = \dfrac{70{,}000p}{100 - p}$ for $p = 75$ and $p = 95$.

p	$C = \dfrac{70{,}000p}{100 - p}$	Interpretation
75	$C = \dfrac{70{,}000(75)}{100 - 75} = \dfrac{5{,}250{,}000}{25} = 210{,}000$	The cost to remove 75% of pollutants is $210,000.
95	$C = \dfrac{70{,}000(95)}{100 - 95} = \dfrac{6{,}650{,}000}{5} = 1{,}330{,}000$	The cost to remove 95% of pollutants is $1,330,000.

As the percent of pollutants to be removed increases, the cost of removing the pollutants increases.

✓ Student Check 5   Solve each problem.

a. A fitness club charges a one-time membership fee of $120 when a new member signs a contract. A new member is charged a monthly fee of $48, and the one-time fee is divided equally among the monthly payments. If x represents the number of months a new member commits to in the contract, the expression $\dfrac{120 + 48x}{x}$ represents the member's monthly payment. What is the monthly payment if a member signs a contract for 6 months? 12 months? 24 months?

b. The body mass index provides an indication of whether a person is underweight, at a normal weight, overweight, or obese. It is calculated using the formula $\text{BMI} = \dfrac{703w}{h^2}$, where w is a person's weight in pounds and h is a person's height in inches. Find the BMI of a person who is 5 ft 8 in. tall and weighs 175 lb. Based on the chart, what can you conclude about the person's weight?

BMI	Weight Status
Below 18.5	Underweight
18.5–24.9	Normal
25.0–29.9	Overweight
30.0 and Above	Obese

Objective 6 ▶

Troubleshoot common errors.

Troubleshooting Common Errors

Some of the common mistakes made with rational expressions are shown next.

Objective 6 Examples   **A problem and an incorrect solution are given. Provide the correct solution and an explanation of the error.**

6a. Find the value of $\dfrac{3x + 2}{x - 1}$ for $x = 1$.

Incorrect Solution	Correct Solution and Explanation
$\dfrac{3x + 2}{x - 1} = \dfrac{3(1) + 2}{1 - 1}$ $= \dfrac{3 + 2}{0} = \dfrac{5}{0} = 0$	The process is correct but division by zero is undefined. So, $\dfrac{5}{0}$ is undefined. Therefore, $\dfrac{3x + 2}{x - 1}$ is undefined for $x = 1$.

6b. Find the values that make $\dfrac{x+1}{x^2-4}$ undefined.

Incorrect Solution	Correct Solution and Explanation
The expression $\dfrac{x+1}{x^2-4}$ is undefined for $x=-1$ and $x=2$ since $x=-1$ makes the numerator zero and $x=2$ makes the denominator zero.	Rational expressions are undefined when the denominator is equal to zero, not the numerator. We must set the denominator equal to zero and solve. $$x^2-4=0$$ $$(x-2)(x+2)=0$$ $$x-2=0 \quad \text{or} \quad x+2=0$$ $$x=2 \qquad\qquad x=-2$$ So, it is undefined for $x=2$ and $x=-2$.

6c. Find the domain of $f(x)=\dfrac{4x-5}{3x}$.

Incorrect Solution	Correct Solution and Explanation
$$3x=0$$ $$x=-3$$ The domain is all real numbers except -3 or $(-\infty,-3)\cup(-3,\infty)$.	It is correct to set the denominator equal to zero. However, the equation was solved incorrectly. Instead of subtracting 3 from each side, we must divide each side by 3. $$3x=0$$ $$\dfrac{3x}{3}=\dfrac{0}{3}$$ $$x=0$$ So, the domain is all real numbers except 0 or $(-\infty,0)\cup(0,\infty)$.

6d. Simplify the expression $\dfrac{6x+2}{6}$.

Incorrect Solution	Correct Solution and Explanation
$$\dfrac{\cancel{6}x+2}{\cancel{6}}=x+2$$	We cannot divide 6 out of the first term and not the second term of the numerator. To simplify the expression, we must factor the numerator. $$\dfrac{6x+2}{6}=\dfrac{2(3x+1)}{2\cdot 3}=\dfrac{3x+1}{3}$$ The expression is simplified as far as possible since 3 is not a factor of the numerator. We can rewrite the expression by dividing each term in the numerator by 3. $$\dfrac{3x+1}{3}=\dfrac{3x}{3}+\dfrac{1}{3}=x+\dfrac{1}{3}$$

ANSWERS TO STUDENT CHECKS

Student Check 1   **a.** $-91.35, 0, -1787.02,$ undefined,

$36{,}013.001,$ and 209.18 **b.** $\dfrac{38}{5} = 7.6,$ undefined, $-3194,$

$-\dfrac{2}{5} = -0.4, \dfrac{2}{11} \approx 0.18,$ and $\dfrac{598}{105} \approx 5.70$

Student Check 2   **a.** $(-\infty, -1) \cup (-1, \infty)$
b. $(-\infty, -3) \cup (-3, 3) \cup (3, \infty)$ **c.** $(-\infty, \infty)$

Student Check 3   **a.** -1 **b.** $\dfrac{y}{y+4}$ **c.** $\dfrac{x+5}{x^2 - x + 1}$
d. $\dfrac{9a+7}{a^2+7}$

Student Check 4   **a.** $\dfrac{1-3x}{6x+1}$ or $\dfrac{3x-1}{-6x-1}$

b. $-\dfrac{x+12}{x+7}$ or $\dfrac{x+12}{-x-7}$

Student Check 5   **a.** The payment for a 6-month contract would be \$68 per month, for a 12-month contract would be \$58 per month, and for a 24-month contract would be \$53 per month.   **b.** $26.6,$ overweight

SUMMARY OF KEY CONCEPTS

1. A rational expression or a rational function is a quotient of two polynomials, in which the denominator cannot be zero. Rational expressions can be evaluated by substituting the given value in place of the variable.

2. Rational functions are undefined for the values that make the denominator equal to zero. So, the domain of a rational function is all real numbers excluding the values that make the denominator equal to 0.

3. To simplify a rational expression, factor the numerator and denominator completely and divide out their common factors.

4. If a rational expression has a negative sign in front of it, -1 can be distributed through the numerator or the denominator, but not both.

GRAPHING CALCULATOR SKILLS

We can use the Store command in the calculator to evaluate rational expressions. When entering rational expressions in the calculator, parentheses are important to preserve the order of operations.

Example: Use the Store command to evaluate the expression $\dfrac{x+1}{4x-3}$ for $x = 3$.

Solution: Enter the function in the equation editor.

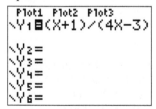

Go to the home screen and store the value 3 for the variable x.

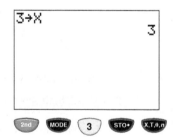

Evaluate the function Y_1. Use the Fraction command to put the answer in Fraction form.

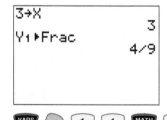

So, the answer is $\dfrac{4}{9}$.

SECTION 7.1   EXERCISE SET

 Write About It!

Use complete sentences in your answer to each exercise.

1. Describe the difference between a rational expression and a rational function.

2. Describe how to find the domain of a rational function.

3. Explain how to simplify a rational expression.

4. Is every polynomial a rational expression? Explain.

5. Is the expression $\dfrac{4x+6}{x}$ in simplified form? Why or why not?

6. Is the expression $\dfrac{4x^2+6x}{x}$ in simplified form? Why or why not?

Practice Makes Perfect!

Evaluate each expression or function for the given values. Round each answer to two decimal places when necessary. (See Objective 1.)

7. $\dfrac{3x + 4}{x - 3}$, $x = -2, 3.001, 2.99, 200, -100$

8. $\dfrac{5x - 2}{x - 2}$, $x = 3, 2.001, 1.99, 100, -200$

9. $\dfrac{3x^2 + x + 1}{x + 4}$, $x = 1, -3.99, -4.01, 200, -200$

10. $\dfrac{x^2 + 2x + 3}{x - 2}$, $x = -4, 2.001, 1.9, 100, -100$

11. $f(x) = \dfrac{4x + 3}{x + 2}$, $f(0), f(-1.99), f(-2.01), f(100), f(-100)$

12. $f(x) = \dfrac{3x + 8}{x - 5}$, $f(0), f(5.01), f(4.9), f(100), f(-100)$

Find the domain of each rational function. Write each answer in set-builder notation. (See Objective 2.)

13. $f(x) = \dfrac{4x + 11}{x - 3}$

14. $f(x) = \dfrac{5x - 4}{x + 9}$

15. $f(x) = \dfrac{3x - 1}{x^2 - 49}$

16. $f(x) = \dfrac{x - 5}{x^2 - 36}$

17. $f(x) = \dfrac{x + 3}{x^2 - 2x}$

18. $f(x) = \dfrac{2x + 1}{x^2 + 3x}$

19. $f(x) = \dfrac{x^2 - 3x - 4}{2x + 7}$

20. $f(x) = \dfrac{x^2 + 3x + 2}{5x - 2}$

Find the domain of each rational function. Write each answer in interval notation. (See Objective 2.)

21. $f(x) = \dfrac{2x + 3}{x - 5}$

22. $f(x) = \dfrac{6x - 1}{x + 7}$

23. $f(x) = \dfrac{3x}{4x^2 - 1}$

24. $f(x) = \dfrac{4x}{25x^2 - 4}$

25. $f(x) = \dfrac{x + 1}{3x^2 - 2x}$

26. $f(x) = \dfrac{4x - 3}{2x^2 + 5x}$

27. $f(x) = \dfrac{2x^2 - x - 3}{2x + 1}$

28. $f(x) = \dfrac{x^2 - 4x - 5}{x}$

29. $f(x) = \dfrac{3x^2 + 1}{2x - 5}$

30. $f(x) = \dfrac{x^2 + 4}{3x - 2}$

Simplify each rational expression. (See Objective 3.)

31. $\dfrac{12x^2 - 10x}{18x^2 - 3x - 10}$

32. $\dfrac{3x^2 + 6x}{2x^2 - 3x - 14}$

33. $\dfrac{6x^2 - 5x}{18x^2 - 3x - 10}$

34. $\dfrac{2x^2 + 5x}{2x^2 - x - 15}$

35. $\dfrac{3x^2 + 2x}{9x^2 - 4}$

36. $\dfrac{30x^2 + 75x}{12x^2 - 75}$

37. $\dfrac{3x + 2}{2 + 3x}$

38. $\dfrac{2x + 5}{5 + 2x}$

39. $\dfrac{x - 2}{2 - x}$

40. $\dfrac{3 - x}{x - 3}$

41. $\dfrac{3 + x}{x - 3}$

42. $\dfrac{4 + 3x}{3x - 4}$

43. $\dfrac{30x^2 + 12x}{50x^2 - 8}$

44. $\dfrac{6x^2 - 8x}{18x^2 - 32}$

45. $\dfrac{2x^2 + 17x + 35}{x^3 + 125}$

46. $\dfrac{5x^2 + 11x - 12}{x^3 + 27}$

47. $\dfrac{6x^2 - 7x - 10}{x^3 - 8}$

48. $\dfrac{9x^2 + 15x - 14}{27x^3 - 8}$

49. $\dfrac{6x^2 + 11x + 5}{24x^3 + 20x^2 + 18x + 15}$

50. $\dfrac{21x^2 - 2x - 8}{6x^3 - 4x^2 + 21x - 14}$

51. $\dfrac{4x^2 + 18x + 8}{8x^3 + 4x^2 + 26x + 13}$

52. $\dfrac{15x^2 - 36x - 27}{5x^3 + 3x^2 + 20x + 12}$

Use the rule of negative rational expressions to write two equivalent forms of each rational expression. (See Objective 4.)

53. $-\dfrac{x - 3}{2x + 4}$

54. $-\dfrac{2x - 3}{4}$

55. $-\dfrac{-3x + 4}{x - 1}$

56. $-\dfrac{-x + 8}{2x - 1}$

57. $\dfrac{-3x + 1}{x - 4}$

58. $\dfrac{-4x + 8}{x - 2}$

59. $\dfrac{-4x - 9}{2x + 1}$

60. $\dfrac{-2x + 15}{3x - 12}$

Solve each problem. (See Objective 6.)

61. A high school class is organizing its 40-yr reunion. The venue, DJ, decorations, and other expenses total $4500 and the dinner costs $15 per person. The total cost for x attendees is $C(x) = 15x + 4500$. The cost per person is the expression $\dfrac{C(x)}{x} = \dfrac{15x + 4500}{x}$.
    a. What is the cost per person if 50 people attend?
    b. 75 attend?
    c. 100 attend?

62. A high school class is organizing its 15-yr reunion. The venue, DJ, decorations, and other expenses total $3000 and the dinner costs $50 per person. The total cost for x attendees is $C(x) = 50x + 3000$. The cost per person is the expression $\dfrac{C(x)}{x} = \dfrac{50x + 3000}{x}$.
    a. What is the cost per person if 100 people attend?
    b. 125 attend?
    c. 150 attend?

63. A utility company burns coal to generate electricity. The cost C (in dollars) of removing p amount (percent) of the smokestack pollutants is given by $C = \dfrac{82{,}500p}{100 - p}$.

   a. What is the cost of removing 80% of the pollutants?
   b. 95% of the pollutants?

64. A utility company burns coal to generate electricity. The cost C (in dollars) of removing p amount (percent) of the smokestack pollutants is given by $C = \dfrac{87{,}500p}{100 - p}$.

   a. What is the cost of removing 75% of the pollutants?
   b. 98% of the pollutants?

65. A fitness club charges a one-time membership fee of $180 when a new member signs a contract. The new member is charged a monthly fee of $54, and the one-time fee is divided equally among the monthly payments. If x represents the number of months a new member commits to in the contract, the expression $\dfrac{180 + 54x}{x}$ represents the member's monthly payment.

   a. What is the monthly payment if a member signs a contract for 6 months?
   b. 12 months?
   c. 24 months?

66. A fitness club charges a one-time membership fee of $90 when a new member signs a contract. The new member is charged a monthly fee of $48, and the one-time fee is divided equally among the monthly payments. If x represents the number of months a new member commits to in the contract, the expression $\dfrac{90 + 48x}{x}$ represent the member's monthly payment.

   a. What is the monthly payment if a member signs a contract for 3 months?
   b. 6 months?
   c. 12 months?

 Mix 'Em Up!

Evaluate each expression or function for the given values. Round each answer to two decimal places when necessary.

67. $\dfrac{3x^2 - 4x + 5}{x + 1}$, $x = 4, -0.99, -1.1, 100, -100$

68. $\dfrac{4x^2 + x - 5}{x + 3}$, $x = 0, -2.999, -3.01, 100, -100$

69. $f(x) = \dfrac{3x - 8}{x + 6}$, $f(0), f(-5.99), f(-6.01), f(100), f(-100)$

70. $f(x) = \dfrac{3x + 4}{x - 4}$, $f(0), f(4.01), f(3.9), f(100), f(-100)$

Find the domain of each rational function. Write each answer in both set-builder and interval notation.

71. $f(x) = \dfrac{6x - 5}{3x + 4}$

72. $f(x) = \dfrac{x + 9}{5x - 12}$

73. $f(x) = \dfrac{2x + 7}{4x^2 - 3x}$

74. $f(x) = \dfrac{x + 6}{x^2 + 7x}$

75. $f(x) = \dfrac{2x}{4x^2 - 9}$

76. $f(x) = \dfrac{x^2 - x - 6}{x + 4}$

77. $f(x) = \dfrac{x^2 - 1}{x - 3}$

78. $f(x) = \dfrac{x + 3}{9x^2 - 25}$

Simplify each rational expression.

79. $\dfrac{72^2 + 27x}{48x^2 - 102x - 45}$

80. $\dfrac{18x^2 + 6x}{9x^2 - 42x - 15}$

81. $\dfrac{12x^2 - 14x}{36x^2 + 42x - 98}$

82. $\dfrac{4x^2 + 14x}{24x^2 + 64x - 70}$

83. $\dfrac{8x^2 - 32x}{2x^2 - 32}$

84. $\dfrac{20x^2 + 8x}{25x^2 - 4}$

85. $\dfrac{6x^2 - 4x}{18x^2 - 8}$

86. $\dfrac{30x^2 + 45x}{12x^2 - 27}$

87. $\dfrac{16x^2 - 28x - 30}{64x^3 + 27}$

88. $\dfrac{14x^2 + 3x - 2}{8x^3 + 1}$

89. $\dfrac{12x^2 + 22x - 20}{27x^3 - 8}$

90. $\dfrac{6x^2 - 38x - 80}{x^3 - 512}$

91. $\dfrac{5x^2 + 14x - 24}{x^3 + 4x^2 - 10x - 40}$

92. $\dfrac{9x^2 - 9x + 2}{15x^3 - 10x^2 - 33x + 22}$

93. $\dfrac{7x^2 - 24x + 9}{2x^3 - 6x^2 - 5x + 15}$

94. $\dfrac{30x^2 - 84x + 48}{5x^3 - 4x^2 + 30x - 24}$

Write two equivalent forms of each rational expression.

95. $\dfrac{-9x + 18}{x - 10}$

96. $\dfrac{-2x + 7}{x + 1}$

97. $-\dfrac{x^2 + 3x}{4x - 1}$

98. $-\dfrac{x^2 - x}{-3x + 2}$

Solve each problem.

99. A high school class is organizing its 10-yr reunion. The venue, DJ, decorations, and other expenses total $4500 and the dinner costs $25 per person. The total cost for x attendees is $C(x) = 25x + 4500$. The cost per person is the expression, $\dfrac{C(x)}{x} = \dfrac{25x + 4500}{x}$.

   a. What is the cost per person if 100 people attend?
   b. 125 attend?
   c. 150 attend?

100. A high school class is organizing its 5-yr reunion. The venue, DJ, decorations, and other expenses total $7500 and the dinner costs $40 per person. The total cost for x attendees is $C(x) = 40x + 7500$. The cost per person is the expression $\dfrac{C(x)}{x} = \dfrac{40x + 7500}{x}$.

   a. What is the cost per person if 75 people attend?

   b. 100 attend?

   c. 125 attend?

101. A utility company burns coal to generate electricity. The cost C (in dollars) of removing p amount (percent) of the smokestack pollutants is given by $C = \dfrac{67{,}500p}{100 - p}$

   a. What is the cost of removing 85% of the pollutants?

   b. 97% of the pollutants?

102. A utility company burns coal to generate electricity. The cost C (in dollars) of removing p amount (percent) of the smokestack pollutants is given by $C = \dfrac{72{,}000p}{100 - p}$

   a. What is the cost of removing 82% of the pollutants?

   b. 96% of the pollutants?

103. A fitness club charges a one-time membership fee of $90 when a new member signs a contract. The new member is charged a monthly fee of $51, and the one-time fee is divided equally among the monthly payments. If x represents the number of months a new member commits to in the contract, the expression $\dfrac{90 + 51x}{x}$ represents the member's monthly payment.

   a. What is the monthly payment if a member signs a contract for 18 months?

   b. 24 months?

   c. 30 months?

104. A fitness club charges a one-time membership fee of $105 when a new member signs a contract. The new member is charged a monthly fee of $27, and the one-time fee is divided equally among the monthly payments. If x represents the number of months a new member commits to in the contract, the expression $\dfrac{105 + 27x}{x}$ represents the member's monthly payment.

   a. What is the monthly payment if a member signs a contract for 3 months?

   b. 6 months?

   c. 12 months?

 You Be the Teacher!

Correct each student's errors, if any.

105. Find the domain of the function $f(x) = \dfrac{3x - 1}{2x}$.

   Michael's work:

   $2x \neq 0$

   $x \neq -2$

   The domain of the function is $(-\infty, -2) \cup (-2, \infty)$.

106. Find the domain of the function $f(x) = \dfrac{x - 3}{x + 1}$.

   Romona's work:

   $x - 3 \neq 0$

   $x \neq 3$

   The domain of the function is $(-\infty, 3) \cup (3, \infty)$.

107. Simplify the rational expression $\dfrac{9x^2 - 4}{3x^2 + x - 2}$.

   Tom's work:

   $\dfrac{9x^2 - 4}{3x^2 + x - 2} = \dfrac{\cancel{9}x^{\cancel{2}\,3} - \cancel{4}^2}{\cancel{3}x^{\cancel{2}} + x - \cancel{2}} = \dfrac{1}{x}$

108. Simplify the rational expression $\dfrac{4x^2 - x - 3}{2x^2 + x - 3}$.

   Taylor's work:

   $\dfrac{4x^2 - x - 3}{2x^2 + x - 3} = \dfrac{\cancel{4}x^{\cancel{2}\,2} - \cancel{x} - \cancel{3}^{1}}{\cancel{2}x^{\cancel{2}} + \cancel{x} - \cancel{3}} = 2$

 Calculate It!

109. Use the Store command to evaluate $f(x) = \dfrac{5x + 4}{2x - 1}$ for $x = 2$.

110. Use the Store command to evaluate $f(x) = \dfrac{3x - 1}{x + 2}$ for $x = -6$.

 Think About It!

Write a rational function that has the given domain.

111. $(-\infty, 3) \cup (3, \infty)$

112. $(-\infty, -2) \cup (-2, \infty)$

113. $(-\infty, 1) \cup (1, 5) \cup (5, \infty)$

114. $(-\infty, -2) \cup (-2, 2) \cup (2, \infty)$

Multiplication and Division of Rational Expressions

▶ **OBJECTIVES**

As a result of completing this section, you will be able to

1. Multiply rational expressions.
2. Divide rational expressions.
3. Solve problems requiring conversion of units.
4. Troubleshoot common errors.

The distance for most races and marathons is given in kilometers (K or km) rather than miles. How many miles are in a 5K race? Once we know the conversion rate between kilometers and miles, we can use multiplication of rational expressions to answer this question.

Multiplying Rational Expressions

Multiplying rational expressions is very similar to multiplying numeric fractions. Recall

$$\frac{5}{3} \cdot \frac{2}{7} = \frac{5 \cdot 2}{3 \cdot 7} = \frac{10}{21}$$

Objective 1 ▶

Multiply rational expressions.

We multiply the numerators together, multiply the denominators together, and then express in the lowest terms.

Property: Multiplying Rational Expressions

If $\dfrac{A}{B}$ and $\dfrac{C}{D}$ are rational expressions, where B and D are not zero, then

$$\frac{A}{B} \cdot \frac{C}{D} = \frac{A \cdot C}{B \cdot D}$$

Procedure: Multiplying Rational Expressions

Step 1: Factor each numerator and denominator, if possible.
Step 2: Divide out common factors.
Step 3: The result is the quotient of the product of the remaining factors in the numerators and the product of the remaining factors in the denominators.

Objective 1 Examples   Multiply the rational expressions. Write answers in lowest terms.

1a. $\dfrac{12a^2}{5} \cdot \dfrac{10}{3a^3}$ **1b.** $\dfrac{x^2 + 5x + 6}{x + 3} \cdot \dfrac{3}{x + 2}$ **1c.** $-\dfrac{2y}{y^2 - 16} \cdot \dfrac{3y^2 - 10y - 8}{12y^2 + 8y}$

Solutions   **1a.** $\dfrac{12a^2}{5} \cdot \dfrac{10}{3a^3} = \dfrac{2 \cdot 2 \cdot 3 \cdot a^2}{5} \cdot \dfrac{2 \cdot 5}{3 \cdot a^2 \cdot a}$ Factor. The common factors are $3a^2$ and 5.

$= \dfrac{2 \cdot 2 \cdot 2 \cdot 1}{1 \cdot a}$ Divide out the common factors and multiply the rational expressions.

$= \dfrac{8}{a}$ Simplify.

We can also multiply first and then simplify.

$\dfrac{12a^2}{5} \cdot \dfrac{10}{3a^3} = \dfrac{(12a^2)(10)}{(5)(3a^3)}$ Multiply the rational expressions.

$= \dfrac{120a^2}{15a^3}$ Simplify the products.

$= \dfrac{15a^2(8)}{15a^2(a)}$ Factor out the common factor of the numerator and denominator.

$= \dfrac{8}{a}$ Apply the fundamental property of rational expressions.

1b. $\dfrac{x^2 + 5x + 6}{x + 3} \cdot \dfrac{3}{x + 2}$

$= \dfrac{(x + 3)(x + 2)}{x + 3} \cdot \dfrac{3}{x + 2}$ Factor. The common factors are $x + 3$ and $x + 2$.

$= \dfrac{1 \cdot 3}{1}$ Divide out the common factors and multiply the remaining factors.

$= \dfrac{3}{1}$ Simplify the products.

$= 3$ Divide.

1c. $-\dfrac{2y}{y^2 - 16} \cdot \dfrac{3y^2 - 10y - 8}{12y^2 + 8y}$

$= -\dfrac{2y}{(y - 4)(y + 4)} \cdot \dfrac{(3y + 2)(y - 4)}{2 \cdot 2y(3y + 2)}$ Factor. The common factors are $2y$, $3y + 2$, and $y - 4$.

$= -\dfrac{1}{1 \cdot (y + 4) \cdot 2}$ Divide out the common factors and multiply the rational expressions.

$= -\dfrac{1}{2(y + 4)}$ Simplify.

✔ **Student Check 1**   Multiply the rational expressions. Write answers in lowest terms.

a. $-\dfrac{12x^3}{5y^2} \cdot \dfrac{y^3}{6x^4}$ **b.** $\dfrac{x^2 - 6x + 5}{4x + 4} \cdot \dfrac{4}{x - 5}$ **c.** $-\dfrac{10h}{h^2 + 6h + 9} \cdot \dfrac{2h^2 + 11h + 15}{4h^2 + 10h}$

Dividing Rational Expressions

Objective 2 ▶

Divide rational expressions.

In Chapter 1, we learned that dividing by a number is the same as multiplying by its reciprocal. Recall that the *reciprocal* of $\dfrac{a}{b}$ is $\dfrac{b}{a}$, where $a \neq 0$ and $b \neq 0$. For example, we know that

$$36 \div 9 = 4, \text{ but we also know that } 36 \times \dfrac{1}{9} = \dfrac{36}{1} \times \dfrac{1}{9} = \dfrac{36}{9} = 4.$$

So, division by 9 is the same as multiplication by $\dfrac{1}{9}$. Notice the reciprocal of 9 or $\dfrac{9}{1}$ is $\dfrac{1}{9}$.

The same rule holds true when dividing rational expressions: dividing by a rational expression is the same as multiplying by the reciprocal of the expression.

Property: Dividing Rational Expressions

If $\dfrac{A}{B}$ and $\dfrac{C}{D}$ are rational expressions, where B, C, and D are not zero, then

$$\dfrac{A}{B} \div \dfrac{C}{D} = \dfrac{A}{B} \cdot \dfrac{D}{C} = \dfrac{A \cdot D}{B \cdot C}$$

Procedure: Dividing Rational Expressions

Step 1: Convert the division to multiplication by the reciprocal of the second fraction.

Step 2: Perform the steps to multiply the expressions.

Step 3: Express the answer in lowest terms.

Objective 2 Examples   Divide the rational expressions. Write answers in lowest terms.

2a. $\dfrac{4a^7}{3} \div \dfrac{8a^8}{21b}$

2b. $\dfrac{-3y^2 + 3y}{5y^2 + 10y} \div \dfrac{y^2 + 8y - 9}{15}$

2c. $\dfrac{4a^2 - a - 5}{a^3 + 1} \div \dfrac{16a^2 - 25}{-2a^3 + 2a^2 - 2a}$

Solutions   **2a.** $\dfrac{4a^7}{3} \div \dfrac{8a^8}{21b} = \dfrac{4a^7}{3} \cdot \dfrac{21b}{8a^8}$ Multiply by the reciprocal of $\dfrac{8a^8}{21b}$.

$= \dfrac{2 \cdot 2 \cdot a^7}{3} \cdot \dfrac{7 \cdot 3b}{2 \cdot 2 \cdot 2 \cdot a^7 \cdot a}$ Factor and identify the common factors.

$= \dfrac{1 \cdot 7b}{1 \cdot 2a}$ Divide out the common factors and multiply the rational expressions.

$= \dfrac{7b}{2a}$ Simplify.

2b. $\dfrac{-3y^2 + 3y}{5y^2 + 10y} \div \dfrac{y^2 + 8y - 9}{15}$

$= \dfrac{-3y^2 + 3y}{5y^2 + 10y} \cdot \dfrac{15}{y^2 + 8y - 9}$ Multiply by the reciprocal of $\dfrac{y^2 + 8y - 9}{15}$.

$= \dfrac{-3y(y - 1)}{5y(y + 2)} \cdot \dfrac{3 \cdot 5}{(y + 9)(y - 1)}$ Factor and identify the common factors.

$= \dfrac{-3 \cdot 3 \cdot 1}{1(y + 2)(y + 9)}$ Divide out the common factors and multiply the rational expressions.

$= \dfrac{-9}{(y + 2)(y + 9)}$ Simplify.

$= -\dfrac{9}{(y + 2)(y + 9)}$ Apply the rule of negative rational expressions.

2c. $\dfrac{4a^2 - a - 5}{a^3 + 1} \div \dfrac{16a^2 - 25}{-2a^3 + 2a^2 - 2a}$

$= \dfrac{4a^2 - a - 5}{a^3 + 1} \cdot \dfrac{-2a^3 + 2a^2 - 2a}{16a^2 - 25}$ Multiply by the reciprocal of the second fraction.

$= \dfrac{(4a - 5)(a + 1)}{(a + 1)(a^2 - a + 1)} \cdot \dfrac{-2a(a^2 - a + 1)}{(4a - 5)(4a + 5)}$ Factor and identify the common factors.

$= \dfrac{1(-2a)}{1(4a + 5)}$ Divide out the common factors and multiply the expressions.

$= \dfrac{-2a}{4a + 5}$ or $-\dfrac{2a}{4a + 5}$ Simplify.

✔ **Student Check 2**   Divide the rational expressions. Write answers in lowest terms.

a. $\dfrac{3a^3}{14b^2} \div \dfrac{9a^2}{12b}$

b. $\dfrac{3x^2 + 10x - 8}{6x^3 - 4x^2} \div \dfrac{x^2 + 7x + 12}{2x^2}$

c. $\dfrac{8y^3 - 27}{2y^2 + 5y + 3} \div \dfrac{12y^2 + 18y + 27}{4y^2 - 9}$

Unit Conversion

Objective 3 ▶

Solve problems requiring
conversion of units.

There are many real-life situations that require us to convert between different units of a quantity. When cooking, we may need to convert a measurement from liters to cups. In the medical field, we may need to convert a quantity from pounds to kilograms. In the section opener, we stated an example of the need to convert a distance from kilometers to miles.

To convert from one unit to another, we will use the fact that when equal quantities are divided, the result is 1. For example, we know that 60 min is equal to 1 hr, so

$$\frac{60 \text{ min}}{1 \text{ hr}} = 1$$

We use this fact to convert 5 hr to minutes, for example.

$$5 \text{ hr} = 5 \text{ hr} \times 1 = 5 \; \cancel{\text{hr}} \times \frac{60 \text{ min}}{1 \; \cancel{\text{hr}}} = 5 \times 60 \text{ min} = 300 \text{ min}$$

Property: Rule for Converting Units

If a and b are equal quantities, then $\dfrac{a}{b} = 1$.

Procedure: Converting a Unit to Another Unit

Step 1: Write the given unit as a product of itself and 1.
Step 2: Replace 1 with a quotient that involves equivalent units. The numerator represents the new units and the denominator represents the given units.
Step 3: Divide out the common factor and multiply the remaining fractions.

Objective 3 Examples   **Answer each question.**

3a. The dosage of medicine that a patient should receive is often based on the patient's weight in kilograms (kg). If a patient weights 165 pounds (lb), what is his weight in kilograms? (Note: 1 lb ≈ 0.45 kg.)

Solution   **3a.** $165 \text{ lb} = 165 \text{ lb} \cdot 1$

$$\approx \frac{165 \; \cancel{\text{lb}}}{1} \cdot \frac{0.45 \text{ kg}}{1 \; \cancel{\text{lb}}}$$

$$\approx \frac{165(0.45 \text{ kg})}{1 \cdot 1}$$

$$\approx 74.25 \text{ kg}$$

3b. The Willis Tower (formerly known as the Sears Tower) is the tallest building in the United States with a height of 527 m. What is the height of the Willis Tower in feet? (Note: 1 m ≈ 3.28 ft.) (Source: Wikipedia)

Solution   **3b.** $527 \text{ m} = 527 \text{ m} \cdot 1$

$$\approx \frac{527 \; \cancel{\text{m}}}{1} \cdot \frac{3.28 \text{ ft}}{1 \; \cancel{\text{m}}}$$

$$\approx \frac{527(3.28 \text{ ft})}{1 \cdot 1}$$

$$\approx 1728.56 \text{ ft}$$

3c. The distance for most races and marathons is given in kilometers (K or km) rather than miles. How many miles are in a 5K race? (Note: 1 km ≈ 0.62 mi.)

Solution   **3c.** $5\text{ K} = 5\text{ km} \cdot 1$

$$\approx \frac{5\ \cancel{\text{km}}}{1} \cdot \frac{0.62\text{ mi}}{1\ \cancel{\text{km}}}$$

$$\approx \frac{5(0.62\text{ mi})}{1 \cdot 1}$$

$$\approx 3.1\text{ mi}$$

✓ **Student Check 3**   Use the conversion equivalents given in Example 3 to answer each question.

a. An average-size cat weighs 3.3 kg. How many pounds does an average cat weigh?

b. The length of an Olympic-size pool is 50 m. What is this length in feet?

c. The New York Marathon is 42.2 km. What is this distance in miles?

Objective 4 ▶   ## Troubleshooting Common Errors

Troubleshoot common errors.   A common error related to dividing rational expressions is shown.

Objective 4 Example   **A problem and an incorrect solution are given. Provide the correct solution and an explanation of the error.**

Perform the operation $\dfrac{x^2 + 4x + 3}{x - 3} \div (x^2 - 9)$.

Incorrect Solution	**Correct Solution and Explanation**
$\dfrac{x^2 + 4x + 3}{x - 3} \div (x^2 - 9)$	We must multiply by the reciprocal of $x^2 - 9$, which is $\dfrac{1}{x^2 - 9}$.
$= \dfrac{x^2 + 4x + 3}{x - 3} \cdot \dfrac{(x^2 - 9)}{1}$	$\dfrac{x^2 + 4x + 3}{x - 3} \div (x^2 - 9)$
$= \dfrac{(x + 3)(x + 1)}{\cancel{x - 3}} \cdot \dfrac{(\cancel{x - 3})(x + 3)}{1}$	$= \dfrac{x^2 + 4x + 3}{x - 3} \cdot \dfrac{1}{(x^2 - 9)}$
$= (x + 3)^2(x + 1)$	$= \dfrac{(x + 3)(x + 1)}{x - 3} \cdot \dfrac{1}{(x - 3)(\cancel{x + 3})}$
	$= \dfrac{x + 1}{(x - 3)^2}$

ANSWERS TO STUDENT CHECKS

Student Check 1   **a.** $-\dfrac{2y}{5x}$ **b.** $\dfrac{x - 1}{x + 1}$ **c.** $-\dfrac{5}{h + 3}$

Student Check 2   **a.** $\dfrac{2a}{7b}$ **b.** $\dfrac{1}{x + 3}$ **c.** $\dfrac{(2y - 3)^2}{3(y + 1)}$

Student Check 3   **a.** 7.33 lb   **b.** 164 ft   **c.** 26.164 mi

SUMMARY OF KEY CONCEPTS

1. The key to multiplying rational expressions is factoring. We begin by factoring each of the numerators and denominators. Then we divide out any common factors and multiply the remaining factors.

2. Division of rational expressions is just like dividing numerical fractions. We first convert the division to

multiplication by the reciprocal of the second rational expression and follow the steps for multiplying fractions.

3. The key to unit conversions is multiplying the expression to be converted by a form of 1. The form of 1 involves a ratio of two equivalent quantities.

GRAPHING CALCULATOR SKILLS

A graphing calculator can be used to verify products and quotients.

Example: Verify that $\dfrac{x^2 - 36}{x + 1} \div \dfrac{2x + 12}{x - 1} = \dfrac{(x - 6)(x - 1)}{2(x + 1)}$.

Solution: Enter the original problem in Y_1 and the result in Y_2. Be very careful about the use of parentheses to enter these expressions. If the numerator or denominator is a sum or difference, it must be entered in parentheses.

View the Table to compare the y-values. If the y-values agree for each x-value for which both Y_1 and Y_2 are defined, then the answer is correct.

Note: The original problem and the final answer may have different values for which the expressions are undefined. The y-value has an ERROR message for the values at which it is undefined.

SECTION 7.2 / EXERCISE SET

Write About It!

Use complete sentences in your answer to each exercise.

1. Explain the process of multiplying rational expressions.

2. Explain how to convert division by a polynomial to multiplication.

3. Explain the process of dividing rational expressions.

4. Explain the two general rules that are used when converting units.

Practice Makes Perfect!

Multiply the rational expressions. Write each answer in lowest terms. (See Objective 1.)

5. $\dfrac{4a^3}{3} \cdot \dfrac{9}{12a^5}$

6. $\dfrac{16a^7}{9} \cdot \dfrac{18}{a^2}$

7. $\dfrac{6a}{b^4} \cdot \dfrac{3b^5}{14a^2}$

8. $\dfrac{4a^4}{3b} \cdot \dfrac{9b^2}{a^3}$

9. $\dfrac{3xy^2}{16y} \cdot \dfrac{4x}{6x^2y}$

10. $\dfrac{12x^3y}{5x} \cdot \dfrac{6y^2}{12xy^4}$

11. $\dfrac{x^2 - 8x + 15}{x + 1} \cdot \dfrac{2}{x - 3}$

12. $\dfrac{x - 2}{5} \cdot \dfrac{x + 7}{x^2 - 3x + 2}$

13. $\dfrac{x^2 - 1}{x + 3} \cdot \dfrac{4x + 12}{x + 1}$

14. $\dfrac{x^2 - 4}{x - 4} \cdot \dfrac{3x - 12}{x + 2}$

15. $\dfrac{-5x}{(x - 5)^2} \cdot \dfrac{2x - 10}{15x^3}$

16. $\dfrac{-2x^3}{(x + 6)^3} \cdot \dfrac{4x + 24}{6x^2}$

17. $\dfrac{2x + 1}{x - 3} \cdot \dfrac{x^2 - 9}{2x^2 + 7x + 3}$

18. $\dfrac{3x - 2}{x + 1} \cdot \dfrac{x^2 - 1}{3x^2 - 5x + 2}$

19. $\dfrac{x^3 - 8}{x + 3} \cdot \dfrac{4}{x - 2}$

20. $\dfrac{2}{x^3 + 1} \cdot \dfrac{x + 1}{x + 2}$

21. $\dfrac{x - 3}{2x^2 + 4x} \cdot 4x$

22. $\dfrac{x - 1}{3x^3 - 6x^2} \cdot 6x^2$

23. $5x^3 \cdot \dfrac{2x + 3}{15x^2 - 5x}$

24. $10x^2 \cdot \dfrac{3x + 5}{4x^5 - 8x^3}$

Divide the rational expressions. Write each answer in lowest/terms. (*See Objective 2.*)

25. $\dfrac{2a^4}{5} \div \dfrac{10a^3}{3}$

26. $\dfrac{3a^6}{14} \div \dfrac{21a^7}{9}$

27. $\dfrac{12a^2}{5b} \div \dfrac{3ab}{20}$

28. $\dfrac{6a^3}{25b^2} \div \dfrac{12a^2b}{5}$

29. $\dfrac{2xy^4}{x^3} \div \dfrac{6x^2y}{5x}$

30. $\dfrac{8x^2y^3}{5y} \div \dfrac{16xy^4}{15y}$

31. $\dfrac{x+3}{2x} \div \dfrac{x^2-9}{8}$

32. $\dfrac{x-6}{3x} \div \dfrac{x^2-36}{x^2}$

33. $\dfrac{x^2}{x^2-1} \div \dfrac{4x}{x-1}$

34. $\dfrac{x^2-49}{7x^3} \div \dfrac{x+7}{14x}$

35. $\dfrac{x^2+2x-15}{x-2} \div (x-3)$

36. $\dfrac{x^2+5x+6}{x+6} \div (x+2)$

37. $\dfrac{x^2-6x+1}{x^2-2x} \div \dfrac{x-1}{x}$

38. $\dfrac{x^2-8x+10}{x^3-4x^2} \div \dfrac{x+1}{x^2}$

39. $\dfrac{x^2-5x-14}{2x-2} \div \dfrac{x^2-6x-7}{x-1}$

40. $\dfrac{3x+12}{x^2-2x-3} \div \dfrac{x+4}{x^2-x-2}$

41. $\dfrac{x^2-3x}{x^2-1} \div \dfrac{2x^2-6x}{x^2+2x-3}$

42. $\dfrac{3x^2-3x}{x^2-4} \div \dfrac{x^2+6x-7}{x^2-x-6}$

Use the given conversion equivalents to answer each question. (*See Objective 3.*)

43. A dosage of medicine is based on a patient's weight in kilograms. If a patient weighs 180 lb, what is his weight in kilograms? (1 lb ≈ 0.45 kg)

44. A dosage of medicine is based on a patient's weight in kilograms. If a patient weighs 210 lb, what is his weight in kilograms? (1 lb ≈ 0.45 kg)

45. Joanie watched a British television show on which a woman claimed to have lost 3 stone (st) in weight in 5 months. How many pounds did the woman lose? (1 st ≈ 14 lb)

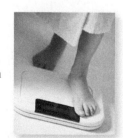

46. Danielle has an interview with a London modeling agency. Her weight is 135 lb, but she wants to know her weight in stones for the modeling agency. How many stones does Danielle weigh? (1 st ≈ 14 lb)

47. The height of the Eiffel Tower in Paris is 324 m. What is its height in feet? (1 m ≈ 3.28 ft) (Source: http://www.tour-eiffel.com/)

48. The height of the Ferris wheel at Hershey Park in Pennsylvania is approximately 100 ft. What is its height in meters? (1 m ≈ 3.28 ft)
(Source: http://www.hersheypark.com/rides/detail.php?q=yes&id=34)

49. Rose saves $200 to spend during her trip to Barcelona. How many Euros will Rose be able to spend? (1 U.S. dollar ≈ 0.694 Euros)

50. When Eliza returns from her trip to Russia, she has 624 rubles. How many U.S. dollars does Eliza have? (1 U.S. dollar ≈ 24.71 Russian rubles)

 Mix 'Em Up!

Perform each operation and simplify each answer.

51. $\dfrac{6x^2y^4}{5} \cdot \dfrac{21x}{20y^6}$

52. $\dfrac{12a^2}{35b} \div \dfrac{4a^{12}}{7b^5}$

53. $\dfrac{x^2+3x}{4} \cdot \dfrac{2x-10}{x^3}$

54. $\dfrac{2x-1}{2x^2-3x+1} \cdot \dfrac{4x-4}{x+3}$

55. $\dfrac{x^2-4x-12}{x^2} \div \dfrac{x^2+4x+4}{3x}$

56. $\dfrac{x^2-1}{x+3} \div \dfrac{x^2-5x+4}{x^2-9}$

57. $\dfrac{x^3+16}{x-2} \div (x+2)$

58. $\dfrac{x+4}{3x^2-10x-8} \cdot (6x+4)$

59. $\dfrac{x^3+8}{x^2-2x-8} \div \dfrac{2x^2-4x+8}{x^2-x-12}$

60. $\dfrac{x^3-1}{x^2-6x+5} \div \dfrac{3x^2+3x+3}{x^2-25}$

61. $\dfrac{x^3+3x^2+5x+15}{x^2-9} \cdot \dfrac{x^2-2x-3}{x^3+5x}$

62. $\dfrac{x^3-4x^2+x-4}{x^2-3x-4} \cdot \dfrac{x^2-6x-7}{2x^3+2x}$

63. $\dfrac{8x^2-20x}{2x^2+x-15} \div \dfrac{6x^3+2x^2}{3x^2+10x+3}$

64. $\dfrac{35x^2+41x+12}{7x^2+18x+8} \div \dfrac{25x^2-1}{5x^2+9x-2}$

65. $\dfrac{x^2+8x+15}{x^2+x-20} \cdot \dfrac{2x^2-x-28}{2x^2+x-21}$

66. $\dfrac{2x^2 + 11x + 9}{3x^2 + 4x - 4} \cdot \dfrac{3x^2 - 8x + 4}{2x^2 + 5x - 18}$

67. $\dfrac{x^4 - 16}{5x^2 - 10x} \div \dfrac{x^3 - 3x^2 + 4x - 12}{x^2 - 3x}$

68. $\dfrac{3x^3 - 2x^2 + 3x - 2}{18x^2 - 12x} \div \dfrac{x^4 - 1}{5x^2 + 5x}$

Use the given conversion equivalents to answer each question.

69. The distance between Frankfurt and Würzburg is 217 km. What is this distance in miles? (1 km ≈ 0.62 mi)

70. The distance between London and Paris is 340 km. What is this distance in miles? (1 km ≈ 0.62 mi)

71. How many minutes are in 0.7 hr? (1 hr = 60 min)

72. How many minutes are in 1.2 hr? (1 hr = 60 min)

73. Convert 45 kg to pounds. (1 lb ≈ 0.45 kg)

74. Convert 98 kg to pounds. (1 lb ≈ 0.45 kg)

 You Be the Teacher!

Correct each student's errors, if any.

75. Convert 6 ft into meters. (1 m ≈ 3.28 ft)

Marissa's work:

$$\dfrac{x\ m}{3.28\ ft} = \dfrac{6\ ft}{1\ m}$$

$$x = 6 \cdot 3.28 = 19.68\ m$$

76. Convert 0.88 hr into minutes. (1 hr = 60 min)

Rozalynn's work:

$$\dfrac{x\ min}{1\ hr} = \dfrac{60\ min}{0.88\ hr}$$

77. Simplify $\dfrac{x+3}{2x^3} \div 4x^2$.

Rupert's work:

$$\dfrac{x+3}{2x^3} \div 4x^2 = \dfrac{x+3}{2x^3} \cdot \dfrac{4x^2}{1}$$

$$= \dfrac{x+3}{2x^{\overset{2}{\cancel{6}}}_{x}} \cdot \dfrac{\cancel{4x^2}}{1} = \dfrac{(x+3) \cdot 2}{x \cdot 1} = \dfrac{2x+6}{x}$$

78. Simplify $\dfrac{x^2 - 9}{2x + 8} \div \dfrac{x+3}{4x}$.

Norman's work:

$$\dfrac{x^2 - 9}{2x + 8} \div \dfrac{x+3}{4x} = \dfrac{(x+3)(x-3)}{2\cancel{(x+4)}} \div \dfrac{\cancel{x+4}}{4x}$$

$$= \dfrac{(x+3)(x-3)}{2} \div \dfrac{1}{4x}$$

$$= \dfrac{(x+3)(x-3)}{\underset{1}{\cancel{2}}} \cdot \dfrac{\overset{2}{\cancel{4}}x}{1}$$

$$= \dfrac{(x^2 - 9) \cdot 2x}{1} = 2x^3 - 18x$$

 Calculate It!

Use a graphing calculator to verify each product and quotient.

79. $\dfrac{25x^2 - 49}{2x + 6} \div \dfrac{5x^2 + 3x - 14}{x^2 + 5x + 6} = \dfrac{5x + 7}{2}$

80. $\dfrac{3x^2 - 12}{2x^2 + x - 10} \div \dfrac{x^3 + 8}{4x + 10} = \dfrac{6}{x^2 - 2x + 4}$

81. $\dfrac{x^2 - 9}{x + 2} \cdot \dfrac{x^2 + 2x}{3x + 9} = \dfrac{x^2 - 3x}{3}$

82. $\dfrac{x^2 - 16}{x^2 - 4x} \cdot \dfrac{4x - 8}{5x + 20} = \dfrac{4(x - 2)}{5x}$

SECTION 7.3   **Addition and Subtraction of Rational Expressions with Like Denominators and the Least Common Denominator**

▶ **OBJECTIVES**

As a result of completing this section, you will be able to

1. Add rational expressions with like denominators.

2. Subtract rational expressions with like denominators.

3. Find the least common denominator of rational expressions.

4. Write equivalent rational expressions.

5. Troubleshoot common errors.

Have you ever heard the expression that you cannot compare apples to oranges? This saying lends itself perfectly to fractions. In order to compare two fractions, they must be the same kind of "fruit." Think of the denominator as the type of fruit. We learned in Chapter 1 that fractions must have a common denominator before we can compare them, or add them together. The same is true for rational expressions.

Adding Rational Expressions with Like Denominators

Objective 1 ▶

Add rational expressions with like denominators.

As we know, we can add fractions when their denominators are the same. For example,

$$\frac{2}{7} + \frac{3}{7} = \frac{2+3}{7} = \frac{5}{7}$$

This is also true with rational expressions.

> **Property:** **Adding Rational Expressions**
>
> If $\dfrac{A}{C}$ and $\dfrac{B}{C}$ are rational expressions with $C \neq 0$, then
>
> $$\frac{A}{C} + \frac{B}{C} = \frac{A+B}{C}$$

> **Procedure:** **Adding Rational Expressions with Like Denominators**
>
> **Step 1:** Add the numerators and place the sum over the like denominator.
> **Step 2:** Simplify, if possible.

Objective 1 Examples / **Add the rational expressions. Write each answer in lowest terms.**

1a. $\dfrac{3}{x} + \dfrac{5}{x}$ **1b.** $\dfrac{2}{5a} + \dfrac{8}{5a}$ **1c.** $\dfrac{2x}{x^2 + 6x + 5} + \dfrac{2}{x^2 + 6x + 5}$

Solutions **1a.**

$$\frac{3}{x} + \frac{5}{x} = \frac{3+5}{x}$$ Add the numerators and place over the denominator.

$$= \frac{8}{x}$$ Simplify the numerator.

1b.

$$\frac{2}{5a} + \frac{8}{5a} = \frac{2+8}{5a}$$ Add the numerators and place over the denominator.

$$= \frac{10}{5a}$$ Simplify the numerator.

$$= \frac{5 \cdot 2}{5 \cdot a}$$ Factor.

$$= \frac{2}{a}$$ Simplify.

1c. $\dfrac{2x}{x^2 + 6x + 5} + \dfrac{2}{x^2 + 6x + 5} = \dfrac{2x+2}{x^2 + 6x + 5}$ Add the numerators and place over the denominator.

$$= \frac{2(x+1)}{(x+5)(x+1)}$$ Factor the numerator and the denominator.

$$= \frac{2}{x+5}$$ Simplify.

✔ **Student Check 1**   Add the rational expressions. Write each answer in lowest terms.

a. $\dfrac{11}{2y} + \dfrac{3}{2y}$ **b.** $\dfrac{1}{x+1} + \dfrac{5}{x+1}$ **c.** $\dfrac{2x+5}{6x^2 + 13x - 5} + \dfrac{4x-7}{6x^2 + 13x - 5}$

Subtracting Rational Expressions with Like Denominators

Objective 2 ▶

Subtract rational expressions with like denominators.

Subtracting fractions is similar to adding fractions in that we must have like denominators. If we have like denominators, we subtract the numerators and put the result over the common denominator. For example,

$$\frac{2}{5} - \frac{3}{5} = \frac{2 - 3}{5}$$

Property: Subtracting Rational Expressions

If $\dfrac{A}{C}$ and $\dfrac{B}{C}$ are rational expressions with $C \neq 0$, then

$$\frac{A}{C} - \frac{B}{C} = \frac{A - B}{C}$$

Note: *When we subtract rational expressions we must apply the negative sign to the entire expression in the second numerator. This may require us to apply the distributive property.*

Procedure: Subtracting Rational Expressions with Like Denominators

Step 1: Subtract the numerators and place the difference over the like denominator.
Step 2: Simplify, if possible.

Objective 2 Examples   Subtract the rational expressions. Write each answer in lowest terms.

2a. $\dfrac{5t}{t - 6} - \dfrac{6t}{t - 6}$ **2b.** $\dfrac{6x^2 - 4x}{5x + 10} - \dfrac{5x^2 - 6x}{5x + 10}$ **2c.** $\dfrac{4x}{x^2 + x - 12} - \dfrac{3x - 4}{x^2 + x - 12}$

Solutions   **2a.**

$$\frac{5t}{t - 6} - \frac{6t}{t - 6} = \frac{5t - 6t}{t - 6}$$

Subtract the numerators and place over the denominator.

$$= \frac{-t}{t - 6}$$

Simplify the numerator.

$$= -\frac{t}{t - 6}$$

Apply the rule of negative rational expressions.

2b. $\dfrac{6x^2 - 4x}{5x + 10} - \dfrac{5x^2 - 6x}{5x + 10} = \dfrac{6x^2 - 4x - (5x^2 - 6x)}{5x + 10}$

Subtract the numerators and place over the denominator.

$$= \frac{6x^2 - 4x - 5x^2 + 6x}{5x + 10}$$

Distribute -1 to the numerator of the second fraction.

$$= \frac{x^2 + 2x}{5x + 10}$$

Simplify the numerator.

$$= \frac{x(x + 2)}{5(x + 2)}$$

Factor the numerator and denominator.

$$= \frac{x}{5}$$

Simplify.

2c. $\dfrac{4x}{x^2 + x - 12} - \dfrac{3x - 4}{x^2 + x - 12} = \dfrac{4x - (3x - 4)}{x^2 + x - 12}$ Subtract the numerators and place over the denominator.

$\qquad\qquad\qquad\qquad\qquad\quad = \dfrac{4x - 3x + 4}{x^2 + x - 12}$ Distribute -1 to the numerator of the second fraction.

$\qquad\qquad\qquad\qquad\qquad\quad = \dfrac{x + 4}{(x + 4)(x - 3)}$ Simplify the numerator and factor the denominator.

$\qquad\qquad\qquad\qquad\qquad\quad = \dfrac{1}{x - 3}$ Simplify.

✓ Student Check 2   Subtract the rational expressions. Write each answer in lowest terms.

a. $\dfrac{2a}{a - 4} - \dfrac{7a}{a - 4}$ **b.** $\dfrac{3b}{b^2 - 5b} - \dfrac{2b + 5}{b^2 - 5b}$ **c.** $\dfrac{5m^2 + m}{m^2 - 2m - 3} - \dfrac{4m^2 - 4m}{m^2 - 2m - 3}$

The Least Common Denominator

Objective 3 ▶

Find the least common denominator of rational expressions.

Recall that if fractions have unlike denominators, we can find their least common denominator (LCD) and convert each fraction to an equivalent fraction with the LCD in order to add or subtract them.

Consider the rational numbers $\dfrac{5}{12}$ and $\dfrac{7}{30}$. To find the **least common denominator** of these fractions, we begin with the prime factorization of each denominator.

$12 = 4 \cdot 3 = 2 \cdot 2 \cdot 3$
$30 = 6 \cdot 5 = 2 \cdot 3 \cdot 5$

> The LCD is a number that 12 and 30 divide into, that is, it is a number that contains the factors from each number. So, the LCD of 12 and 30 is
>
> Factors of 30
> $$\text{LCD} = 2 \cdot 2 \cdot 3 \cdot 5 = 60$$
> Factors of 12

The LCD is the product of the common factors of the denominators and the other factors from each of the denominators. That is, the LCD contains all of the factors from each denominator.

We can also find the LCD using the exponents of the factors in the denominators. Rewrite the prime factorization using exponents.

$12 = 4 \cdot 3 = 2 \cdot 2 \cdot 3 = 2^2 \cdot 3$
$30 = 6 \cdot 5 = 2 \cdot 3 \cdot 5$ $\longrightarrow$ $\boxed{\text{LCD} = 2^2 \cdot 3 \cdot 5 = 60}$

So, the LCD is the product of all the different factors raised to the largest exponent that the factor occurs in any one denominator. So, if a factor occurs three times in one denominator and two times in another denominator, it must occur three times in the LCD.

Procedure: Finding the Least Common Denominator

Step 1: Write each denominator in factored form.
Step 2: "Build" the LCD by forming the product of all factors from the first denominator.
Step 3: Include any additional factors from the other denominators that have not been included in Step 2.
Step 4: The product of the factors from Steps 2 and 3 is the LCD. Leave the LCD in factored form.

Note: *The least common denominator is the same as the least common multiple.*

Objective 3 Examples Determine the least common denominator of the given rational expressions.

3a. $\dfrac{1}{3ab}, \dfrac{2}{15a^2}$ **3b.** $\dfrac{x}{x+1}, \dfrac{4-x}{x^2+2x+1}$ **3c.** $\dfrac{5}{t}, \dfrac{t}{t-1}$

3d. $\dfrac{7}{2y+18}, \dfrac{y}{y^2-81}, \dfrac{y+2}{y^2-10y+9}$ **3e.** $\dfrac{x+1}{x-6}, \dfrac{2}{6-x}$

Solutions **3a.** $3ab = 3 \cdot a \cdot b$
        $15a^2 = 3 \cdot 5 \cdot a \cdot a$ $\longrightarrow$ $\text{LCD} = 3 \cdot a \cdot b \cdot 5 \cdot a = 15a^2b$

When we include the factors of $15a^2$ in the LCD, we only need to include the factors of 5 and a since the factors of 3 and a are already included from the first denominator.

3b. $x + 1 = (x+1)$
        $x^2 + 2x + 1 = (x+1)(x+1)$ $\longrightarrow$ $\text{LCD} = (x+1)(x+1) = (x+1)^2$

The exponent of $(x+1)$ is 1 in the first denominator and 2 in the second denominator, so its exponent is 2 in the LCD.

3c. $t = (t)$
        $t + 1 = (t+1)$ $\longrightarrow$ $\text{LCD} = (t)(t+1)$

Since the denominators do not share any common factors, the LCD is the product of the denominators.

3d. $2y + 18 = 2(y+9)$
        $y^2 - 81 = (y+9)(y-9)$ $\longrightarrow$ $\text{LCD} = 2(y+9)(y-9)(y-1)$
        $y^2 - 10y + 9 = (y-9)(y-1)$

3e. The denominators $x - 6$ and $6 - x$ are opposites of one another. When the denominators are opposite, the LCD can be either one of the denominators. That is, it can be $x - 6$ or $6 - x$, since we can go from one expression to the other by multiplying by -1.

        $x - 6 = (x-6)$ $\text{LCD} = (x-6)$
                                or
        $6 - x = -x + 6 = -1(x-6)$ $\longrightarrow$ $\text{LCD} = -1(x-6) = 6 - x$

✓ Student Check 3 Determine the least common denominator of the given rational expressions.

a. $\dfrac{5}{18xy^2}, \dfrac{1}{4xy}$ **b.** $\dfrac{x}{x+4}, \dfrac{-2}{x^2+8x+16}$ **c.** $\dfrac{a}{a+6}, \dfrac{7}{a-2}$

d. $\dfrac{-1}{3y-21}, \dfrac{y}{y^2-49}, \dfrac{y+2}{2y^2+9y-35}$ **e.** $\dfrac{3}{x-1}, \dfrac{5}{1-x}$

Objective 4 ▶

Write equivalent rational expressions.

Equivalent Rational Expressions

Now that we know how to find the LCD of rational expressions, we will write equivalent forms of rational expressions. This is required before we can add or subtract rational expressions with unlike denominators.

Converting a fraction to an equivalent fraction is a skill that was presented in Chapter 1. Recall that we can write the fraction $\frac{3}{8}$ as a fraction with a denominator of 16 by multiplying the fraction by $\frac{2}{2}$, a form of 1.

$$\frac{3}{8} = \frac{3}{8} \cdot \frac{2}{2} = \frac{6}{16}$$

This skill is based on the *fundamental property of rational expressions*, which states that for $N, C \neq 0$,

$$\frac{N \cdot C}{D \cdot C} = \frac{N}{D}$$

We use this property to write equivalent forms of rational expressions as well. The identity property of multiplication states that any number multiplied by 1 is unchanged in value. Therefore, we can multiply a rational expression $\frac{N}{D}$ by a form of 1 to obtain an equivalent expression.

Property: Equivalent Rational Expressions

For $D, C \neq 0$,

$$\frac{N}{D} = \frac{N}{D} \cdot 1 = \frac{N}{D} \cdot \frac{C}{C} = \frac{N \cdot C}{D \cdot C}$$

Some examples of **equivalent rational expressions** are as follows.

Equivalent Fractions		
$\dfrac{2}{3}$	$\dfrac{2}{3} \cdot \dfrac{5}{5} = \dfrac{10}{15}$	$\dfrac{2}{3} \cdot \dfrac{x}{x} = \dfrac{2x}{3x}$
$\dfrac{7}{x+2}$	$\dfrac{7}{x+2} \cdot \dfrac{2}{2} = \dfrac{14}{2x+4}$	$\dfrac{7}{x+2} \cdot \dfrac{2x-3}{2x-3} = \dfrac{14x-21}{2x^2+x-6}$

We see that $\frac{2}{3}, \frac{10}{15}$, and $\frac{2x}{3x}$ are equivalent rational expressions since we obtain the last two from the first by multiplying by 1 in the form of $\frac{5}{5}$ and $\frac{x}{x}$, respectively.

Also, $\frac{7}{x+2}, \frac{14}{2x+4}$, and $\frac{14x-21}{2x^2+x-6}$ are equivalent rational expressions since we obtain the last two from the first by multiplying by 1 in the form of $\frac{2}{2}$ and $\frac{2x-3}{2x-3}$, respectively.

Procedure: Finding an Equivalent Form of a Rational Expression

Step 1: Multiply the numerator and denominator of a rational expression by the same nonzero value.

Step 2: Simplify each product.

Objective 4 Examples   **Find the numerator that will make the two rational expressions equivalent.**

4a. $\dfrac{7}{3x} = \dfrac{?}{6x^2}$

4b. $\dfrac{3a}{4b} = \dfrac{?}{24ab^2}$

4c. $\dfrac{x+2}{x-1} = \dfrac{?}{x^2+3x-4}$

4d. $\dfrac{5x}{3x+6} = \dfrac{?}{3x^2-12}$

Solutions **4a.** Write the prime factorization of the two denominators. Determine the factors that are not in the original denominator. These factors are used in the form of 1.

Original denominator	$3x = 3 \cdot x$
Denominator in the equivalent fraction	$6x^2 = 2 \cdot 3 \cdot x \cdot x$
What do we multiply the original denominator by to obtain the second denominator?	$2x$

We must multiply by 1 in the form of $\dfrac{2x}{2x}$. So, the equivalent fraction with the given denominator is

$$\frac{7}{3x} = \frac{7}{3x} \cdot 1 = \frac{7}{3x} \cdot \frac{2x}{2x} = \frac{7(2x)}{(3x)(2x)} = \frac{14x}{6x^2}$$

4b.

Original denominator	$4b = 2 \cdot 2 \cdot b$
Denominator in the equivalent fraction	$24ab^2 = 2 \cdot 2 \cdot 2 \cdot 3 \cdot a \cdot b \cdot b$
What do we multiply the original denominator by to obtain the second denominator?	$2 \cdot 3 \cdot a \cdot b = 6ab$

We must multiply by 1 in the form of $\dfrac{6ab}{6ab}$. So, the equivalent fraction with the given denominator is

$$\frac{3a}{4b} = \frac{3a}{4b} \cdot 1 = \frac{3a}{4b} \cdot \frac{6ab}{6ab} = \frac{3a(6ab)}{(4b)(6ab)} = \frac{18a^2b}{24ab^2}$$

4c.

Original denominator	$x - 1$
Denominator in the equivalent fraction	$x^2 + 3x - 4 = (x - 1)(x + 4)$
What do we multiply the original denominator by to obtain the second denominator?	$(x + 4)$

We must multiply by 1 in the form of $\dfrac{x + 4}{x + 4}$. So, the equivalent fraction with the given denominator is

$$\frac{x + 2}{x - 1} = \frac{x + 2}{x - 1} \cdot 1 = \frac{x + 2}{x - 1} \cdot \frac{(x + 4)}{(x + 4)} = \frac{(x + 2)(x + 4)}{(x - 1)(x + 4)} = \frac{x^2 + 6x + 8}{x^2 + 3x - 4}$$

4d.

Original denominator	$3x + 6 = 3(x + 2)$
Denominator in the equivalent fraction	$3x^2 - 12 = 3(x^2 - 4) = 3(x - 2)(x + 2)$
What do we multiply the original denominator by to obtain the second denominator?	$(x - 2)$

We must multiply by 1 in the form of $\dfrac{x - 2}{x - 2}$. So, the equivalent fraction with the given denominator is

$$\frac{5x}{3x + 6} = \frac{5x}{3(x + 2)} \cdot 1 = \frac{5x}{3(x + 2)} \cdot \frac{(x - 2)}{(x - 2)} = \frac{(5x)(x - 2)}{3(x + 2)(x - 2)} = \frac{5x^2 - 10x}{3x^2 - 12}$$

✓ **Student Check 4** Find the numerator that will make the two rational expressions equivalent.

a. $\dfrac{6}{7y} = \dfrac{?}{35y^2}$ **b.** $\dfrac{3a}{4b} = \dfrac{?}{36ab^3}$

c. $\dfrac{x - 3}{x - 5} = \dfrac{?}{x^2 - 9x + 20}$ **d.** $\dfrac{2b + 1}{4b + 32} = \dfrac{?}{4b^2 - 256}$

Objective 5

Troubleshoot common errors.

Troubleshooting Common Errors

Some common errors for adding and subtracting rational expressions, finding the LCD, and writing equivalent rational expressions are shown next.

Objective 5 Examples   **A problem and an incorrect solution are given. Provide the correct solution and an explanation of the error.**

5a. $\dfrac{6}{x} + \dfrac{1}{x}$

Incorrect Solution	Correct Solution and Explanation
$$\dfrac{6}{x} + \dfrac{1}{x} = \dfrac{7}{2x}$$	When we add rational expressions, we do not add their denominators. We add the numerators and place their sum over the common denominator. $$\dfrac{6}{x} + \dfrac{1}{x} = \dfrac{6+1}{x} = \dfrac{7}{x}$$

5b. $\dfrac{y+5}{y+4} - \dfrac{y+3}{y+4}$

Incorrect Solution	Correct Solution and Explanation
$$\dfrac{y+5}{y+4} - \dfrac{y+3}{y+4} = \dfrac{y+5-y+3}{y+4}$$ $$= \dfrac{8}{y+4}$$	When we subtract two rational expressions, the negative sign must be applied to the entire second numerator, not just the first term of the numerator. $$\dfrac{y+5}{y+4} - \dfrac{y+3}{y+4} = \dfrac{y+5-(y+3)}{y+4}$$ $$= \dfrac{y+5-y-3}{y+4}$$ $$= \dfrac{2}{y+4}$$

5c. Determine the numerator that will make the expressions equivalent: $\dfrac{4x}{x-1} = \dfrac{?}{x^2-1}$

Incorrect Solution	Correct Solution and Explanation
$$\dfrac{4x}{x-1}\left(\dfrac{x}{x}\right) = \dfrac{4x^2}{x^2-1}$$	If we multiply x by $x-1$, we get $x^2 - x$, not $x^2 - 1$. The second denominator $x^2 - 1$ is equal to $(x-1)(x+1)$, so we must multiply the initial fraction by 1 in the form of $\dfrac{x+1}{x+1}$. $$\dfrac{4x}{x-1}\left(\dfrac{x+1}{x+1}\right) = \dfrac{4x^2 + 4x}{x^2 - 1}$$

5d. Find the LCD of $\dfrac{4}{x}$ and $\dfrac{3}{x+5}$.

Incorrect Solution	Correct Solution and Explanation
The LCD is $x + 5$ since x is included in the expression $x + 5$.	The LCD consists of factors of each denominator. In the second denominator, $x + 5$, the x is a term and not a factor. So, the LCD is $x(x + 5)$.

ANSWERS TO STUDENT CHECKS

Student Check 1   **a.** $\dfrac{7}{y}$ **b.** $\dfrac{6}{x+1}$ **c.** $\dfrac{2}{2x+5}$

Student Check 2   **a.** $-\dfrac{5a}{a-4}$ or $\dfrac{5a}{4-a}$ **b.** $\dfrac{1}{b}$

c. $\dfrac{m^2+5m}{(m-3)(m+1)}$ or $\dfrac{m(m+5)}{(m-3)(m+1)}$

Student Check 3   **a.** $36xy^2$ **b.** $(x+4)(x+4)$ or $(x+4)^2$
   c. $(a+6)(a-2)$ **d.** $3(y-7)(y+7)(2y-5)$
   e. $x-1$ or $1-x$

Student Check 4   **a.** $30y$ **b.** $27a^2b^2$ **c.** $x^2-7x+12$
   d. $2b^2-15b-8$

SUMMARY OF KEY CONCEPTS

1. If the rational expressions being added have like denominators, then add the numerators and place the sum over the common denominator. Simplify the result, if possible.

2. If the rational expressions being subtracted have like denominators, then subtract the numerators and place the difference over the common denominator. Remember to distribute -1 to all terms in the numerator of the second fraction. Simplify the result, if possible.

3. To find the LCD of rational expressions, factor all of the denominators. Build the LCD by forming the product of all the factors from the first denominator. Multiply this by

any factors from the other denominators that have not yet been included.

4. To convert a rational expression to an equivalent fraction in which the denominator of the equivalent fraction is known, we have to determine what factor was multiplied by the original denominator to obtain the new denominator. Rewrite the original denominator and the new denominator in terms of their prime factors. Identify the factors that are missing from the original denominator and multiply the original numerator by this factor to obtain an equivalent fraction.

GRAPHING CALCULATOR SKILLS

We can use a graphing calculator to find the least common multiple of two numbers. This can assist us in finding the least common denominator for some rational expressions.

Example: Find the coefficient of the least common denominator of the two rational expressions: $\dfrac{5}{14x^4y^5}$, $\dfrac{x}{104y^4}$

Solution: We need to find the least common multiple (lcm) of 14 and 104. The lcm is found by pressing Math and accessing the Num menu and selecting option 8.

So, the LCD of the rational expressions is $728x^4y^5$.

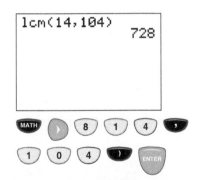

SECTION 7.3   EXERCISE SET

 **Write About It!**

Use complete sentences in your answer to each exercise.

1. Explain how to add or subtract rational expressions with like denominators.

2. Explain the fundamental property of rational expressions.

3. What does it mean for two rational expressions to be equivalent?

4. Explain how to create a rational expression that is equivalent to a given rational expression.

5. Explain the process for finding the least common denominator of rational expressions.

6. Explain why $\dfrac{1}{-x-1}$ and $-\dfrac{1}{x-1}$ are not equivalent.

Practice Makes Perfect!

Add or subtract the rational expressions. Write each answer in lowest terms. (*See Objectives 1 and 2.*)

7. $\dfrac{4}{3x} + \dfrac{2}{3x}$

8. $\dfrac{4}{5x} + \dfrac{1}{5x}$

9. $\dfrac{3x}{x-1} - \dfrac{2x}{x-1}$

10. $\dfrac{6x}{x+2} - \dfrac{x}{x+2}$

11. $\dfrac{y}{y-1} + \dfrac{2y+3}{y-1}$

12. $\dfrac{3y}{4y+1} + \dfrac{y+1}{4y+1}$

13. $\dfrac{2x}{(2x-1)(x-4)} - \dfrac{1}{(2x-1)(x-4)}$

14. $\dfrac{4x}{(4x-5)(x+2)} - \dfrac{5}{(4x-5)(x+2)}$

15. $\dfrac{y}{y^2-1} + \dfrac{1}{y^2-1}$

16. $\dfrac{y}{y^2-25} + \dfrac{5}{y^2-25}$

17. $\dfrac{3x-2}{x+1} - \dfrac{x-3}{x+1}$

18. $\dfrac{6x-7}{x+7} - \dfrac{3x-6}{x+7}$

19. $\dfrac{4x}{x^2+x-6} - \dfrac{3x+2}{x^2+x-6}$

20. $\dfrac{5x-8}{(x-2)(x-7)} - \dfrac{6x-10}{(x-2)(x-7)}$

21. $\dfrac{3x-5}{(x-3)(x+1)} - \dfrac{4x-8}{(x-3)(x+1)}$

22. $\dfrac{6x+7}{x^2-3x-4} - \dfrac{5x+11}{x^2-3x-4}$

23. $\dfrac{x+1}{x-3} + \dfrac{2x-7}{x-3}$

24. $\dfrac{4x-3}{2x+1} + \dfrac{8x+9}{2x+1}$

Find the least common denominator of the given rational expressions. (*See Objective 3.*)

25. $\dfrac{1}{5a}, \dfrac{3}{15a}$

26. $\dfrac{3}{10a}, \dfrac{5}{20a}$

27. $\dfrac{7}{2x}, \dfrac{3}{6x^2}$

28. $\dfrac{2}{5x}, \dfrac{-6}{20x^3}$

29. $\dfrac{3}{4x}, \dfrac{1}{3x^2}$

30. $\dfrac{10}{7x^2}, \dfrac{11}{6x}$

31. $\dfrac{3}{xy^2}, \dfrac{3}{x^3y}$

32. $\dfrac{a+1}{2a^3b}, \dfrac{b}{7a}$

33. $\dfrac{10}{3ab^3}, \dfrac{6}{5a^5b^4}$

34. $\dfrac{9}{12x^3y^2}, \dfrac{4}{8xy^2}$

35. $\dfrac{x}{x(x-5)}, \dfrac{2}{x}$

36. $\dfrac{2x}{x(3x-4)}, \dfrac{x^2}{3x-4}$

37. $\dfrac{4}{x+2}, \dfrac{5}{x^2+2x}$

38. $\dfrac{1}{5-x}, \dfrac{x+1}{5x-x^2}$

39. $\dfrac{x+1}{x-3}, \dfrac{x}{x-1}$

40. $\dfrac{2x-1}{x+5}, \dfrac{4x}{x-6}$

41. $\dfrac{3}{x^2+7x+10}, \dfrac{5}{x+2}$

42. $\dfrac{8}{x^2+3x-18}, \dfrac{6}{x-3}$

43. $\dfrac{6x}{x^2+2x-35}, \dfrac{4x}{x+7}$

44. $\dfrac{2x}{x^2-6x+5}, \dfrac{x^2}{x-1}$

45. $\dfrac{3x}{x^2-9}, \dfrac{4x}{x^2-6x+9}$

46. $\dfrac{2x}{x^2-25}, \dfrac{x}{x^2-x-20}$

47. $\dfrac{2x}{4x^2-9}, \dfrac{5x}{6x^2-5x-6}$

48. $\dfrac{-9}{6x^2-13x-5}, \dfrac{8}{2x^2+3x-20}$

Find the numerator that makes each equation true. (*See Objective 4.*)

49. $\dfrac{1}{2x} = \dfrac{?}{6x}$

50. $\dfrac{2}{5x} = \dfrac{?}{15x}$

51. $\dfrac{2}{x^2} = \dfrac{?}{4x^2}$

52. $\dfrac{3}{y} = \dfrac{?}{4y}$

53. $\dfrac{4x}{y} = \dfrac{?}{xy}$

54. $\dfrac{3x^2}{y^2} = \dfrac{?}{2y^2}$

55. $\dfrac{x-1}{y} = \dfrac{?}{3y}$

56. $\dfrac{x+4}{y} = \dfrac{?}{5y}$

57. $\dfrac{2a}{b} = \dfrac{?}{3ab}$

58. $\dfrac{3}{4b} = \dfrac{?}{8ab}$

59. $\dfrac{2x}{x-1} = \dfrac{?}{(x-1)(x+2)}$

60. $\dfrac{x^2}{x+3} = \dfrac{?}{(x+3)(x-5)}$

61. $\dfrac{x+3}{x-1} = \dfrac{?}{3x-3}$

62. $\dfrac{x-6}{x+7} = \dfrac{?}{2x+14}$

63. $\dfrac{x}{x+1} = \dfrac{?}{3x+3}$

64. $\dfrac{y}{4-y} = \dfrac{?}{8-2y}$

65. $\dfrac{1-2x}{x} = \dfrac{?}{x^2}$

66. $\dfrac{3+5x}{x} = \dfrac{?}{2x}$

67. $\dfrac{3x}{1-2x} = \dfrac{?}{4x^2-2x}$

68. $\dfrac{5x}{3-4x} = \dfrac{?}{12x^2-9x}$

69. $\dfrac{x}{x+3} = \dfrac{?}{x^2-9}$

70. $\dfrac{2}{x+5} = \dfrac{?}{x^2-25}$

71. $\dfrac{t^3+t}{t-2} = \dfrac{?}{2t-4}$

72. $\dfrac{t^2-7t}{3t+5} = \dfrac{?}{12t+20}$

73. $\dfrac{4a-5}{3a+12} = \dfrac{?}{6ab+24b}$

74. $\dfrac{2a-1}{5a-6} = \dfrac{?}{20ab-24b}$

Mix 'Em Up!

Add or subtract the rational expressions. Write each answer in lowest terms.

75. $\dfrac{-3x}{3x+8}+\dfrac{19x}{3x+8}$

76. $\dfrac{8x}{x+7}-\dfrac{9x}{x+7}$

77. $\dfrac{6a+5}{9a-1}+\dfrac{3a-2}{9a-1}$

78. $\dfrac{5b-6}{2b-15}+\dfrac{b-4}{2b-15}$

79. $\dfrac{5x+3}{2x+7}-\dfrac{3x-4}{2x+7}$

80. $\dfrac{22x-9}{12x-5}-\dfrac{10x-4}{12x-5}$

81. $\dfrac{3x}{9x^2-4}-\dfrac{2}{9x^2-4}$

82. $\dfrac{x}{x^2-36}+\dfrac{6}{x^2-36}$

83. $\dfrac{2x}{4x^2-1}+\dfrac{1}{4x^2-1}$

84. $\dfrac{x}{x^2-100}-\dfrac{10}{x^2-100}$

85. $\dfrac{3x+7}{(x-2)(x+5)}-\dfrac{6x+22}{(x-2)(x+5)}$

86. $\dfrac{4y-7}{y^2+7y-18}-\dfrac{2y-3}{y^2+7y-18}$

Find the least common denominator of the rational expressions.

87. $\dfrac{x}{x-9},\dfrac{3}{9x-x^2}$

88. $\dfrac{1}{x-12},\dfrac{5x}{12x-x^2}$

89. $\dfrac{x}{4x+3},\dfrac{3}{x+1}$

90. $\dfrac{5}{x+3},\dfrac{x}{x^2-2x-15}$

91. $\dfrac{6x}{25x^2-4},\dfrac{-x}{5x^2+13x-6}$

92. $\dfrac{5x}{4x+6},\dfrac{3}{4x^2+12x+9}$

93. $\dfrac{x}{15x-6},\dfrac{7}{25x^2-20x+4}$

94. $\dfrac{7}{6x^2-5x-4},\dfrac{-5}{2x^2-9x-5}$

Find the numerator that makes the two rational expressions equivalent.

95. $\dfrac{x+1}{x-6}=\dfrac{?}{x^2-6x}$

96. $\dfrac{2x}{x-4}=\dfrac{?}{2x^2-7x-4}$

97. $\dfrac{x-1}{3x+6}=\dfrac{?}{12x^2+24x}$

98. $\dfrac{3x}{x+1}=\dfrac{?}{x^2+4x+3}$

99. $\dfrac{3}{4a}=\dfrac{?}{12a^2}$

100. $\dfrac{2}{xy}=\dfrac{?}{3x^2y}$

101. $\dfrac{5x}{7-2x}=\dfrac{?}{8x^2-28x}$

102. $\dfrac{x}{1-3x}=\dfrac{?}{15x^2-5x}$

 You Be the Teacher!

Correct each student's errors, if any.

103. Add $\dfrac{2}{5x}+\dfrac{3}{5x}$.

Ricky's work:

$$\dfrac{2}{5x}+\dfrac{3}{5x}=\dfrac{5}{5x}=\dfrac{\cancel{5}}{\cancel{5}x}=x$$

104. Subtract $\dfrac{4x+3}{2x-5}-\dfrac{6x+1}{2x-5}$.

Rodney's work:

$$\dfrac{4x+3}{2x-5}-\dfrac{6x+1}{2x-5}=\dfrac{4x+3-6x+1}{2x-5}=\dfrac{-2x+4}{2x-5}$$

105. Find the least common denominator of $\dfrac{3}{x+2}$ and $\dfrac{1}{x}$.

Caridee's work:

The least common denominator of $\dfrac{3}{x+2}$ and $\dfrac{1}{x}$ is $x+2$.

106. Find the least common denominator of $\dfrac{3}{x+1}$ and $\dfrac{x}{x^2-1}$.

Meg's work:

The least common denominator of $\dfrac{3}{x+1}$ and $\dfrac{x}{x^2-1}$ is x^2-1.

| SECTION 7.4 | **Addition and Subtraction of Rational Expressions with Unlike Denominators** |

▶ **OBJECTIVES**

As a result of completing this section, you will be able to

1. Add rational expressions with unlike denominators.
2. Subtract rational expressions with unlike denominators.
3. Troubleshoot common errors.

Two rectangles are placed side by side. The length of one rectangle is $\dfrac{4}{x+1}$ and the length of the other rectangle is $\dfrac{2}{x-4}$. What is the total length formed by the two rectangles? To answer this question, we must simplify $\dfrac{4}{x+1} + \dfrac{2}{x-4}$. These are rational expressions with unlike denominators.

Adding Rational Expressions with Unlike Denominators

Objective 1 ▶

Add rational expressions with unlike denominators.

To add fractions with unlike denominators, we can use their LCD. For example, the LCD of $\dfrac{7}{6}$ and $\dfrac{5}{22}$ is 66. So, we multiply $\dfrac{7}{6}$ by 1 in the form of $\dfrac{11}{11}$ and we multiply $\dfrac{5}{22}$ by 1 in the form of $\dfrac{3}{3}$.

$$\begin{aligned}
\frac{7}{6} + \frac{5}{22} &= \frac{7}{6} \cdot \frac{11}{11} + \frac{5}{22} \cdot \frac{3}{3} \\
&= \frac{77}{66} + \frac{15}{66} \\
&= \frac{92}{66} \\
&= \frac{46}{33}
\end{aligned}$$

We add rational expressions with unlike denominators in this same way. We will apply the skills we learned in Section 7.3 to determine the LCD, rewrite equivalent rational expressions, and convert rational expressions to an equivalent fraction using the LCD. Then we will be able to add the expressions as we did in Section 7.3.

> **Procedure: Adding Rational Expressions with Unlike Denominators**
>
> **Step 1:** Determine the LCD of the rational expressions.
> **Step 2:** Convert each rational expression to an equivalent fraction with the LCD as its denominator.
> **Step 3:** Add the rational expressions.
> **Step 4:** Simplify, if possible.

Objective 1 Examples   **Add the rational expressions. Write each answer in lowest terms.**

1a. $\dfrac{3}{x} + \dfrac{5}{x^2}$ 　　　　　　　**1b.** $\dfrac{4}{x+1} + \dfrac{2}{x-4}$

1c. $\dfrac{5b}{b^2 - 9} + \dfrac{6}{2b^2 - b - 15}$ 　　**1d.** $\dfrac{3}{x-1} + \dfrac{2}{1-x}$

Solutions　**1a.** $\dfrac{3}{x} + \dfrac{5}{x^2} = \dfrac{3}{x} \cdot \dfrac{x}{x} + \dfrac{5}{x^2}$　　Write equivalent fractions with the LCD, x^2.

$= \dfrac{3x}{x^2} + \dfrac{5}{x^2}$　　Simplify the product.

$= \dfrac{3x + 5}{x^2}$　　Add the numerators and place over the LCD.

1b. $\dfrac{4}{x+1} + \dfrac{2}{x-4} = \dfrac{4}{x+1} \cdot \dfrac{x-4}{x-4} + \dfrac{2}{x-4} \cdot \dfrac{x+1}{x+1}$ Write equivalent fractions with the LCD, $(x+1)(x-4)$.

$\qquad = \dfrac{4x-16}{(x+1)(x-4)} + \dfrac{2x+2}{(x+1)(x-4)}$ Simplify each product.

$\qquad = \dfrac{4x-16+2x+2}{(x+1)(x-4)}$ Add the numerators and place over the LCD.

$\qquad = \dfrac{6x-14}{(x+1)(x-4)}$ Simplify.

1c. Factor the denominators to determine the LCD.

$$b^2 - 9 = (b-3)(b+3)$$
$$2b^2 - b - 15 = (2b+5)(b-3) \quad \longrightarrow \quad \text{LCD} = (b-3)(b+3)(2b+5)$$

Multiply each expression by a form of 1 that contains the factor needed for the original denominator to become the LCD. Then add the numerators of the like denominators.

$\dfrac{5b}{b^2-9} + \dfrac{6}{2b^2-b-15} = \dfrac{5b}{(b-3)(b+3)} \cdot \dfrac{2b+5}{2b+5} + \dfrac{6}{(2b+5)(b-3)} \cdot \dfrac{b+3}{b+3}$

$\qquad = \dfrac{5b(2b+5)}{(b-3)(b+3)(2b+5)} + \dfrac{6(b+3)}{(b-3)(b+3)(2b+5)}$

$\qquad = \dfrac{10b^2+25b}{(b-3)(b+3)(2b+5)} + \dfrac{6b+18}{(b-3)(b+3)(2b+5)}$

$\qquad = \dfrac{10b^2+25b+6b+18}{(b-3)(b+3)(2b+5)}$

$\qquad = \dfrac{10b^2+31b+18}{(b-3)(b+3)(2b+5)}$

1d. $\dfrac{3}{x-1} + \dfrac{2}{1-x} = \dfrac{3}{x-1} + \dfrac{2}{-(x-1)}$ Rewrite $1-x$ as $-(x-1)$.

$\qquad = \dfrac{3}{x-1} + \dfrac{-2}{x-1}$ Apply the rule of negative rational expressions.

$\qquad = \dfrac{3-2}{x-1}$ Add the numerators and place over the LCD.

$\qquad = \dfrac{1}{x-1}$ Simplify.

✓ Student Check 1   Add the rational expressions. Write each answer in lowest terms.

a. $\dfrac{x^2}{x^2-7x} + \dfrac{2x}{x-7}$ **b.** $\dfrac{5}{x+6} + \dfrac{3}{x+2}$

c. $\dfrac{6y}{y^2-4} + \dfrac{4}{3y^2+7y+2}$ **d.** $\dfrac{6}{a-5} + \dfrac{9}{5-a}$

Subtracting Rational Expressions with Unlike Denominators

Objective 2 ▶

Subtract rational expressions with unlike denominators.

Subtracting rational expressions with unlike denominators is similar to adding rational expressions with unlike denominators. After the LCD is determined and each rational expression has been converted to an equivalent rational expression, we subtract the numerators. So, the key is that we subtract the entire expression in the numerator of the rational expression being subtracted. To subtract this numerator is to add the opposite of it.

> **Procedure: Subtracting Rational Expressions with Unlike Denominators**
> **Step 1:** Determine the LCD of the rational expressions.
> **Step 2:** Convert each rational expression to an equivalent fraction with the LCD as its denominator.
> **Step 3:** Subtract the rational expressions.
> **Step 4:** Simplify, if possible.

Objective 2 Examples   **Subtract the rational expressions. Write each answer in lowest terms.**

2a. $\dfrac{2a}{a^2 - 3a} - \dfrac{1}{2a - 6}$ **2b.** $\dfrac{4}{y - 1} - \dfrac{7}{4y}$

2c. $2 - \dfrac{3x}{x + 4}$ **2d.** $\dfrac{x}{3x^2 + 19x - 14} - \dfrac{2}{x^2 + 15x + 56}$

Solutions   **2a.** Factor the denominators to determine the LCD.

$$a^2 - 3a = a(a - 3)$$
$$2a - 6 = 2(a - 3)$$ $\longrightarrow$ $\text{LCD} = 2a(a - 3)$

$\dfrac{2a}{a^2 - 3a} - \dfrac{1}{2a - 6} = \dfrac{2a}{a(a - 3)} \cdot \dfrac{2}{2} - \dfrac{1}{2(a - 3)} \cdot \dfrac{a}{a}$ Write equivalent fractions with the LCD, $2a(a - 3)$.

$= \dfrac{4a}{2a(a - 3)} - \dfrac{a}{2a(a - 3)}$ Simplify each product.

$= \dfrac{4a - a}{2a(a - 3)}$ Subtract the numerators and place over the LCD.

$= \dfrac{3a}{2a(a - 3)}$ Simplify.

$= \dfrac{3}{2(a - 3)}$ Divide out the common factor, a.

2b. $\dfrac{4}{y - 1} - \dfrac{7}{4y} = \dfrac{4}{y - 1} \cdot \dfrac{4y}{4y} - \dfrac{7}{4y} \cdot \dfrac{y - 1}{y - 1}$ Write equivalent fractions with the LCD, $4y(y - 1)$.

$= \dfrac{4(4y)}{4y(y - 1)} - \dfrac{7(y - 1)}{4y(y - 1)}$ Simplify each product.

$= \dfrac{16y - 7(y - 1)}{4y(y - 1)}$ Subtract the numerators and place over the LCD.

$= \dfrac{16y - 7y + 7}{4y(y - 1)}$ Apply the distributive property in the numerator.

$= \dfrac{9y + 7}{4y(y - 1)}$ Simplify.

2c. $2 - \dfrac{3x}{x + 4} = \dfrac{2}{1} \cdot \dfrac{x + 4}{x + 4} - \dfrac{3x}{x + 4}$ Write equivalent fractions with the LCD, $x + 4$.

$= \dfrac{2(x + 4)}{x + 4} - \dfrac{3x}{x + 4}$ Simplify the product.

$= \dfrac{2x + 8 - 3x}{x + 4}$ Subtract the numerators and place over the LCD.

$= \dfrac{-x + 8}{x + 4}$ Simplify.

$= -\dfrac{x - 8}{x + 4}$ Recall $\dfrac{-N}{D} = -\dfrac{N}{D}$ and that $-x + 8 = -(x - 8)$.

2d. Factor the denominators to determine the LCD.

$$3x^2 + 19x - 14 = (3x - 2)(x + 7)$$
$$x^2 + 15x + 56 = (x + 8)(x + 7)$$

$\longrightarrow$ $\text{LCD} = (3x - 2)(x + 8)(x + 7)$

$$\frac{x}{3x^2 + 19x - 14} - \frac{2}{x^2 + 15x + 56}$$

$$= \frac{x}{(3x - 2)(x + 7)} \cdot \frac{x + 8}{x + 8} - \frac{2}{(x + 8)(x + 7)} \cdot \frac{3x - 2}{3x - 2}$$

$$= \frac{x(x + 8)}{(3x - 2)(x + 8)(x + 7)} - \frac{2(3x - 2)}{(3x - 2)(x + 8)(x + 7)}$$

$$= \frac{x^2 + 8x - 2(3x - 2)}{(3x - 2)(x - 8)(x + 7)}$$

$$= \frac{x^2 + 8x - 6x + 4}{(3x - 2)(x - 8)(x + 7)}$$

$$= \frac{x^2 + 2x + 4}{(3x - 2)(x - 8)(x + 7)}$$

✔ **Student Check 2** Subtract the rational expressions. Write each answer in lowest terms.

a. $\dfrac{2}{a} - \dfrac{4}{3a^2}$

b. $\dfrac{3y}{2y + 1} - \dfrac{2}{5y}$

c. $6 - \dfrac{2}{x + 3}$

d. $\dfrac{3x}{x^2 + 5x + 6} - \dfrac{4}{2x^2 + 3x - 2}$

Objective 3 ▶

Troubleshoot common errors.

Troubleshooting Common Errors

A common error associated with subtracting rational expressions with unlike denominators is shown.

Objective 3 Example **A problem and an incorrect solution are given. Provide the correct solution and an explanation of the error.**

Simplify $\dfrac{3}{x + 5} - \dfrac{6}{x - 1}$.

Incorrect Solution	Correct Solution and Explanation
$\dfrac{3}{x + 5} - \dfrac{6}{x - 1}$	The negative sign was not distributed to the numerator of the second expression.
$= \dfrac{3}{x + 5} \cdot \dfrac{x - 1}{x - 1} - \dfrac{6}{x - 1} \cdot \dfrac{x + 5}{x + 5}$	$\dfrac{3}{x + 5} - \dfrac{6}{x - 1}$
$= \dfrac{3x - 3 - 6x + 30}{(x - 1)(x + 5)}$	$= \dfrac{3}{x + 5} \cdot \dfrac{x - 1}{x - 1} - \dfrac{6}{x - 1} \cdot \dfrac{x + 5}{x + 5}$
$= \dfrac{-3x + 27}{(x - 1)(x + 5)}$	$= \dfrac{3x - 3 - 6x - 30}{(x - 1)(x + 5)}$
	$= \dfrac{-3x - 33}{(x - 1)(x + 5)}$

ANSWERS TO STUDENT CHECKS

Student Check 1   **a.** $\dfrac{3x}{x-7}$ **b.** $\dfrac{8x+28}{(x+2)(x+6)}$

c. $\dfrac{18y^2+10y-8}{(y-2)(y+2)(3y+1)}$ **d.** $-\dfrac{3}{a-5}$

Student Check 2   **a.** $\dfrac{6a-4}{3a^2}$ **b.** $\dfrac{15y^2-4y-2}{5y(2y+1)}$

c. $\dfrac{6x+16}{x+3}$ **d.** $\dfrac{6x^2-7x-12}{(x+2)(x+3)(2x-1)}$

SUMMARY OF KEY CONCEPTS

1. If the rational expressions being added or subtracted do not have the same denominator, then convert each expression to an equivalent fraction with the LCD as its denominator. Then add or subtract the numerators and put the result over the LCD. Simplify, if possible.

2. When subtracting rational expressions, be sure to apply the negative sign to the entire numerator of the expression being subtracted.

GRAPHING CALCULATOR SKILLS

A graphing calculator can help us determine if we worked a problem correctly.

Example: Verify that $\dfrac{2x}{x^2-3x} - \dfrac{1}{x-3} = \dfrac{1}{x-3}$.

Solution: Enter the original problem in Y_1 and the simplified expression in Y_2. Compare the y-values of these expressions. If they agree, then we performed the operation correctly. The only values for which the two quantities may not agree are the values where the expressions are undefined.

Since the two columns agree, except for the values where the expressions are undefined, our answer is correct.

SECTION 7.4 / EXERCISE SET

 Write About It!

Use complete sentences in your answer to each exercise.

1. Explain how to add or subtract rational expressions with different denominators.

2. What is a key step in subtracting rational expressions?

 Practice Makes Perfect!

Add or subtract the rational expressions. Write each answer in lowest terms. (*See Objectives 1 and 2.*)

3. $\dfrac{5}{x} + \dfrac{1}{x^2}$

4. $\dfrac{3}{x^3} + \dfrac{2}{x^2}$

5. $\dfrac{1}{3x} - \dfrac{1}{6x^2}$

6. $\dfrac{4}{5x} - \dfrac{2}{15x^3}$

7. $\dfrac{1}{x} - \dfrac{2}{x(x+2)}$

8. $\dfrac{1}{x} - \dfrac{9}{x(x+9)}$

9. $\dfrac{100}{r-5} + \dfrac{50}{r}$

10. $\dfrac{20}{r+1} + \dfrac{40}{r}$

11. $\dfrac{4}{x-1} + \dfrac{3x}{(x-1)(x+2)}$

12. $\dfrac{2}{x+4} + \dfrac{x}{(x+4)(x-2)}$

13. $\dfrac{4x}{x+1} + \dfrac{x^2}{x^2+6x+5}$

14. $\dfrac{x}{x-6} + \dfrac{x-1}{x^2-5x-6}$

15. $\dfrac{2x}{x^2-9} - \dfrac{1}{x+3}$

16. $\dfrac{5}{x^2-1} - \dfrac{3}{x+1}$

17. $\dfrac{x}{x+3} + \dfrac{4}{x+1}$

18. $\dfrac{2x}{x-1} + \dfrac{x}{x-4}$

19. $\dfrac{3x}{x-2} - \dfrac{x}{x-3}$

20. $\dfrac{6x}{x+9} - \dfrac{x}{x-4}$

21. $\dfrac{3x}{x-2} - \dfrac{2-x}{x}$

22. $\dfrac{2x}{x-3} - \dfrac{3-x}{x}$

23. $\dfrac{2x}{x+3} - \dfrac{5}{4}$

24. $\dfrac{x}{3x-1} - \dfrac{2}{7}$

25. $\dfrac{3}{x} + \dfrac{2}{x+5}$

26. $\dfrac{x+2}{x-2} + \dfrac{3}{x}$

27. $5 - \dfrac{2}{x-2}$

28. $2 - \dfrac{5}{x+4}$

29. $4 + \dfrac{6}{x-5}$

30. $8 + \dfrac{6}{x+1}$

31. $\dfrac{4}{x^2-1} - \dfrac{5}{2x^2-x-1}$

32. $\dfrac{6}{9x^2-1} - \dfrac{3}{6x^2+13x-5}$

33. $\dfrac{5}{2x^2-5x-3} + \dfrac{2}{4x^2-9x-9}$

34. $\dfrac{2}{x^2-x-20} + \dfrac{3}{3x^2-14x-5}$

 ## Mix 'Em Up!

Add or subtract the rational expressions. Write each answer in lowest terms.

35. $\dfrac{3x}{x-2} + \dfrac{4x}{2-x}$

36. $\dfrac{x+1}{x-3} + \dfrac{2x}{x+3}$

37. $\dfrac{5}{x-10} + \dfrac{2}{x+2}$

38. $\dfrac{4x-1}{x+1} - \dfrac{3x}{x-1}$

39. $\dfrac{3x}{x^2-8x-9} + \dfrac{2x-1}{x-9}$

40. $\dfrac{3x}{x^2-3x-18} - \dfrac{3x^2}{x-6}$

41. $5 + \dfrac{1}{x+1}$

42. $4 + \dfrac{2x}{x-3}$

43. $\dfrac{4x}{x-1} + \dfrac{3x-6}{1-x}$

44. $\dfrac{2x^2}{x^2-6x-40} + \dfrac{5x+3}{x+4}$

45. $\dfrac{8}{x^2-9} - \dfrac{5}{x^2+6x+9}$

46. $\dfrac{3}{x^2-8x+16} + \dfrac{1}{x^2-16}$

47. $\dfrac{6}{x^2-3x+2} - \dfrac{4}{x^2-5x+6}$

48. $\dfrac{1}{x^2+x-42} + \dfrac{2}{x^2-5x-6}$

49. $\dfrac{450}{r-10} + \dfrac{300}{r+3}$

50. $\dfrac{150}{r+40} + \dfrac{60}{r+20}$

 ## You Be the Teacher!

Correct each student's errors, if any.

51. Add $\dfrac{7a}{3} + \dfrac{7a}{4}$.

Barry's work:

$$\dfrac{7a}{3} + \dfrac{7a}{4} = \dfrac{7a}{7} = \dfrac{\cancel{7}a}{\cancel{7}} = a$$

52. Subtract $\dfrac{x+3}{4x+1} - \dfrac{2x-1}{4x-1}$.

Darren's work:

$$\dfrac{2x}{4x+1} - \dfrac{x}{4x-1} = \dfrac{2x}{4x+1} + \dfrac{x}{4x+1} = \dfrac{2x+x}{4x+1} = \dfrac{3x}{4x+1}$$

Calculate It!

Use a graphing calculator to verify that each operation has been performed correctly.

53. $\dfrac{3x-1}{x+3} + \dfrac{-4x+2}{x+3} = \dfrac{-x+1}{x+3}$

54. $\dfrac{6x-4}{2x-5} - \dfrac{3x}{2x-5} = \dfrac{3x-4}{2x-5}$

55. $\dfrac{2x+4}{x^2+6x+9} - \dfrac{x}{x+3} = -\dfrac{x^2+x-4}{(x+3)^2}$

56. $\dfrac{-5x}{x^2+2x-8} + \dfrac{3x}{x-2} = \dfrac{x(3x+7)}{(x+4)(x-2)}$

PIECE IT TOGETHER   SECTION 7.1–7.4

Find the value(s) that make each rational expression undefined. (*Section 7.1, Objective 2*)

1. $\dfrac{x-1}{2x-10}$

2. $\dfrac{2x^2+5x-3}{x^2+3x-10}$

Simplify each rational expression. (*Section 7.1, Objective 3*)

3. $\dfrac{10x}{5x^2+15}$

4. $\dfrac{x^2-5x+6}{x^2+4x-12}$

5. $\dfrac{4x^2-25}{6x^2+13x-5}$

6. $\dfrac{x^3-125}{x^2-25}$

Use the rule of negative rational expressions to write two equivalent forms of the rational expression. (*Section 7.1, Objective 4*)

7. $\dfrac{-5x-1}{x+1}$

Perform each operation with the rational expressions. Write each answer in lowest terms. (*Section 7.2, Objectives 1 and 2; Section 7.3, Objectives 1 and 2; Section 7.4, Objectives 1 and 2*)

8. $\dfrac{x^2-9}{x-9} \cdot \dfrac{5x-45}{x+3}$

9. $\dfrac{-6x}{(x+1)^2} \cdot \dfrac{3x+3}{18x^2}$

10. $\dfrac{x^3 + 8}{4x - 20} \cdot \dfrac{x - 5}{x + 2}$

11. $\dfrac{3x + 2}{x + 2} \cdot \dfrac{x^2 - 4}{3x^2 + 5x + 2}$

17. $\dfrac{y + 10}{y + 5} + \dfrac{2y + 5}{y + 5}$

12. $\dfrac{25a^2}{6b^3} \div \dfrac{5ab^2}{18}$

13. $\dfrac{x^2 - 36}{5x^3} \div \dfrac{x - 6}{10x}$

18. $\dfrac{4x}{x^2 + x - 12} - \dfrac{3x + 3}{x^2 + x - 12}$

14. $\dfrac{x^2 - 2x - 8}{x - 4} \div (x + 2)$

15. $\dfrac{x^2 - 3x - 10}{x^3 - 5x^2} \div \dfrac{x + 2}{x^2}$

19. $\dfrac{5x}{x^2 - 49} - \dfrac{2}{x + 7}$

16. $\dfrac{x^2 + 4x - 12}{3x + 6} \div \dfrac{x^2 + x - 30}{2x + 4}$

20. $\dfrac{4}{x^2 - 1} - \dfrac{5}{2x^2 - x - 1}$

| **SECTION 7.5** | **Simplifying Complex Fractions** |

The function $C(t) = \dfrac{300t}{36t^2 + 3.3}$ describes the concentration, in micrograms per milliliter, of a drug in the bloodstream t hr after it is administered. Find the concentration of the drug 24 min after it is administered. To determine this information, we must evaluate the function for $t = \dfrac{24}{60} = \dfrac{2}{5}$ hr. This gives us

$$C\left(\dfrac{2}{5}\right) = \dfrac{300\left(\dfrac{2}{5}\right)}{36\left(\dfrac{2}{5}\right)^2 + 3.3}$$

This expression is a *complex fraction* because the numerator and denominator of the fraction contain a fraction. We will learn two different methods to simplify complex fractions.

Simplifying Complex Fractions—Method 1

Objective 1 ▶

Simplify complex fractions using division (method 1).

> **Definition:** A **complex fraction** is a fraction in which the numerator and/or denominator of a fraction contains a fraction or rational expression.

Some examples of complex fractions are

$$\dfrac{\dfrac{1}{3}}{\dfrac{2}{5}} \qquad \dfrac{2 - \dfrac{1}{4}}{\dfrac{1}{8}} \qquad \dfrac{\dfrac{6}{x}}{x + \dfrac{1}{x}}$$

  Numerator of the complex fraction

  Denominator of the complex fraction

Our goal is to simplify a complex fraction by writing it as a rational expression, $\dfrac{P}{Q}$, where P and Q ($Q \neq 0$) are polynomials with no common factors. In other words, our goal is to remove fractions from the numerator and denominator of the complex fraction.

The first method to simplify a complex fraction requires us to combine the terms in the numerator and the terms in the denominator and write them as a single fraction. Then we can apply the rules for dividing fractions.

Procedure: Simplifying Complex Fractions Using Division (Method 1)

Step 1: Add or subtract the fractions in the numerator and/or denominator of the complex fraction, if necessary. This will result in a fraction divided by a fraction.

Step 2: Divide the resulting fractions by multiplying the fraction in the numerator by the reciprocal of the denominator.

$$\frac{\dfrac{a}{b}}{\dfrac{c}{d}} = \frac{a}{b} \div \frac{c}{d} = \frac{a}{b} \cdot \frac{d}{c}$$

Step 3: Simplify the answer by dividing out any common factors.

Objective 1 Examples | Simplify each complex fraction using method 1.

1a. $\dfrac{\dfrac{4a^2}{5b}}{\dfrac{10ab}{6b^2}}$

1b. $\dfrac{\dfrac{7x-14}{6x+6}}{\dfrac{x^2-4}{x+1}}$

1c. $\dfrac{x+y}{\dfrac{1}{x}+\dfrac{1}{y}}$

Solutions **1a.** $\dfrac{\dfrac{4a^2}{5b}}{\dfrac{10ab}{6b^2}} = \dfrac{4a^2}{5b} \div \dfrac{10ab}{6b^2}$

Rewrite the complex fraction as a division problem.

$= \dfrac{4a^2}{5b} \cdot \dfrac{6b^2}{10ab}$

Multiply by the reciprocal of $\dfrac{10ab}{6b^2}$, which is $\dfrac{6b^2}{10ab}$.

$= \dfrac{4a^2 \cdot 6b^2}{5b \cdot 10ab}$

Multiply the rational expressions.

$= \dfrac{24a^2b^2}{50ab^2}$

Simplify.

$= \dfrac{12a}{25}$

Divide out the common factor, $2ab^2$.

1b. $\dfrac{\dfrac{7x-14}{6x+6}}{\dfrac{x^2-4}{x+1}} = \dfrac{7x-14}{6x+6} \div \dfrac{x^2-4}{x+1}$

Rewrite the complex fraction as a division problem.

$= \dfrac{7x-14}{6x+6} \cdot \dfrac{x+1}{x^2-4}$

Multiply by the reciprocal of $\dfrac{x^2-4}{x+1}$, which is $\dfrac{x+1}{x^2-4}$.

$= \dfrac{7(x-2)}{6(x+1)} \cdot \dfrac{x+1}{(x+2)(x-2)}$

Factor each numerator and denominator.

$= \dfrac{7\cancel{(x-2)}}{6\cancel{(x+1)}} \cdot \dfrac{\cancel{(x+1)}}{(x+2)\cancel{(x-2)}}$

Divide out the common factors.

$= \dfrac{7}{6(x+2)}$

Multiply the remaining factors.

1c.

$$\frac{x+y}{\dfrac{1}{x}+\dfrac{1}{y}} = \frac{x+y}{\dfrac{1}{x}\cdot\dfrac{y}{y}+\dfrac{1}{y}\cdot\dfrac{x}{x}}$$

Write the fractions in the denominator as equivalent fractions with an LCD of xy.

$$= \frac{x+y}{\dfrac{y}{xy}+\dfrac{x}{xy}}$$

Simplify the products in the denominator.

$$= \frac{x+y}{\dfrac{y+x}{xy}}$$

Add the fractions in the denominator.

$$= \frac{x+y}{1}\div\frac{y+x}{xy}$$

Rewrite the complex fraction as a division problem.

$$= \frac{x+y}{1}\cdot\frac{xy}{y+x}$$

Multiply by the reciprocal of $\dfrac{y+x}{xy}$, which is $\dfrac{xy}{y+x}$.

$$= \frac{\cancel{x+y}}{1}\cdot\frac{xy}{\cancel{y+x}}$$

Divide out the common factor, $x+y$.

$$= \frac{xy}{1}$$

Multiply the remaining factors.

$$= xy$$

Simplify.

☑ **Student Check 1**   Simplify each complex fraction using method 1.

a. $\dfrac{\dfrac{3}{8y^2}}{\dfrac{15x}{14y}}$

b. $\dfrac{\dfrac{9x-12}{6x+9}}{\dfrac{9x^2-16}{2x+3}}$

c. $\dfrac{\dfrac{1}{a}+\dfrac{2}{b}}{2a+b}$

Simplifying Complex Fractions—Method 2

Objective 2 ▶

Simplify complex fractions using multiplication by the LCD (method 2).

Another method to simplify complex fractions is to eliminate the fractions from the numerator and denominator rather than combine the fractions in the numerator and denominator. We do this by multiplying the numerator and denominator of the complex fraction by the LCD of all the fractions in the complex fraction.

> **Procedure: Simplifying a Complex Fraction Using Multiplication (Method 2)**
>
> **Step 1:** Determine the LCD of all the fractions in the complex fraction.
> **Step 2:** Multiply the numerator and denominator of the complex fraction by the LCD.
> **Step 3:** Simplify the resulting fraction.

Objective 2 Examples   Simplify each complex fraction using method 2.

2a. $\dfrac{\dfrac{4a^2}{5b}}{\dfrac{10ab}{6b^2}}$

2b. $\dfrac{x+y}{\dfrac{1}{x}+\dfrac{1}{y}}$

2c. $\dfrac{\dfrac{1}{y}-\dfrac{1}{xy}}{\dfrac{1}{x}-\dfrac{1}{x^2}}$

Solutions   **2a.** The LCD of the fractions is $30b^2$.

$$\frac{\dfrac{4a^2}{5b}}{\dfrac{10ab}{6b^2}} = \frac{\left(\dfrac{4a^2}{5b}\right)\cdot(30b^2)}{\left(\dfrac{10ab}{6b^2}\right)\cdot(30b^2)}$$ Multiply the numerator and denominator by the LCD.

$$= \frac{4a^2\cdot 6b}{10ab\cdot 5}$$ Simplify the products $\left(\dfrac{30b^2}{5b}=6b \text{ and } \dfrac{30b^2}{6b^2}=5\right)$.

$$= \frac{24a^2b}{50ab}$$ Multiply the remaining factors.

$$= \frac{12a}{25}$$ Divide out the common factor, $2ab$.

2b. The LCD of the fractions is xy.

$$\frac{x+y}{\dfrac{1}{x}+\dfrac{1}{y}} = \frac{(x+y)\cdot xy}{\left(\dfrac{1}{x}+\dfrac{1}{y}\right)\cdot xy}$$ Multiply the numerator and denominator by the LCD.

$$= \frac{xy(x+y)}{\dfrac{1}{x}\cdot xy+\dfrac{1}{y}\cdot xy}$$ Apply the distributive property in the denominator.

$$= \frac{xy(x+y)}{y+x}$$ Simplify the products in the denominator.

$$= \frac{xy\cancel{(x+y)}}{\cancel{y+x}}$$ Divide out the common factor, $x+y$.

$$= xy$$ Simplify.

2c. The LCD of the fractions is x^2y.

$$\frac{\dfrac{1}{y}-\dfrac{1}{xy}}{\dfrac{1}{x}-\dfrac{1}{x^2}} = \frac{\left(\dfrac{1}{y}-\dfrac{1}{xy}\right)\cdot x^2y}{\left(\dfrac{1}{x}-\dfrac{1}{x^2}\right)\cdot x^2y}$$ Multiply the numerator and denominator by the LCD.

$$= \frac{\left(\dfrac{1}{y}\right)x^2y-\left(\dfrac{1}{xy}\right)x^2y}{\left(\dfrac{1}{x}\right)x^2y-\left(\dfrac{1}{x^2}\right)x^2y}$$ Apply the distributive property in the numerator and denominator.

$$= \frac{\left(\dfrac{1}{\cancel{y}}\right)\dfrac{x^2\cancel{y}}{1}-\left(\dfrac{1}{\cancel{xy}}\right)\dfrac{\overset{x}{\cancel{x^2}y}}{1}}{\left(\dfrac{1}{\cancel{x}}\right)\dfrac{\overset{x}{x^2}y}{1}-\left(\dfrac{1}{\cancel{x^2}}\right)\dfrac{x^2y}{1}}$$ Divide out the common factors of each product.

$$= \frac{x^2-x}{xy-y}$$ Simplify each product.

$$= \frac{x(x-1)}{y(x-1)}$$ Factor the numerator and denominator.

$$= \frac{x}{y}$$ Divide out the common factor, $x-1$.

✓ Student Check 2 Simplify each complex fraction using method 2.

a. $\dfrac{\dfrac{3}{8y^2}}{\dfrac{15x}{14y}}$

b. $\dfrac{\dfrac{1}{a} + \dfrac{2}{b}}{2a + b}$

c. $\dfrac{\dfrac{1}{ab} + \dfrac{2}{a^2b}}{\dfrac{1}{b^2} + \dfrac{2}{ab^2}}$

Problems Involving Complex Fractions

Objective 3 ▶

Solve problems involving complex fractions.

Complex fractions can occur when we use different formulas, such as slope or midpoint, or when we evaluate functions. Complex fractions can also result from problems involving negative exponents. Recall that $x^{-n} = \dfrac{1}{x^n}$, provided $x \neq 0$.

Objective 3 Examples   **Solve each problem.**

3a. The function $C(t) = \dfrac{300t}{36t^2 + 3.3}$ describes the concentration, in micrograms per milliliter (µg/mL), of a drug in the bloodstream t hr after it is administered. Find the concentration of the drug 24 min after it is administered.

Solution   **3a.** We evaluate the function for $t = \dfrac{24}{60} = \dfrac{2}{5}$ hr.

$$C\left(\frac{2}{5}\right) = \frac{300\left(\dfrac{2}{5}\right)}{36\left(\dfrac{2}{5}\right)^2 + 3.3}$$ Replace t with $\dfrac{2}{5}$.

$$= \frac{120}{36\left(\dfrac{4}{25}\right) + 3.3}$$ Simplify the numerator and denominator.

$$= \frac{120}{\dfrac{144}{25} + \dfrac{33}{10}}$$ Simplify the product in the denominator and write 3.3 as $\dfrac{33}{10}$.

$$= \frac{(120) \cdot 50}{\left(\dfrac{144}{25}\right) \cdot 50 + \left(\dfrac{33}{10}\right) \cdot 50}$$ Multiply the numerator and denominator by the LCD, 50.

$$= \frac{6000}{288 + 165}$$ Simplify the products.

$$= \frac{6000}{453}$$ Add the denominators.

$$\approx 13.25$$ Approximate.

So, in 24 min, there will be approximately 13.25 µg/mL of the drug in the patient's bloodstream.

3b. Find the slope of the line that passes through the points $\left(\frac{1}{4}, -\frac{2}{3}\right)$ and $\left(-\frac{5}{6}, \frac{1}{12}\right)$.

Solution   **3b.**

$$m = \frac{y_2 - y_1}{x_2 - x_1}$$

State the slope formula.

$$m = \frac{\dfrac{1}{12} - \left(-\dfrac{2}{3}\right)}{-\dfrac{5}{6} - \dfrac{1}{4}}$$

Substitute the appropriate values:

$(x_1, y_1) = \left(\dfrac{1}{4}, -\dfrac{2}{3}\right)$ and $(x_2, y_2) = \left(-\dfrac{5}{6}, \dfrac{1}{12}\right)$

$$m = \frac{\left(\dfrac{1}{12} + \dfrac{2}{3}\right) \cdot 12}{\left(-\dfrac{5}{6} - \dfrac{1}{4}\right) \cdot 12}$$

Multiply the numerator and denominator by the LCD, 12.

$$m = \frac{\dfrac{1}{12} \cdot 12 + \dfrac{2}{3} \cdot 12}{-\dfrac{5}{6} \cdot 12 - \dfrac{1}{4} \cdot 12}$$

Apply the distributive property in the numerator and denominator.

$$m = \frac{1 + 8}{-10 - 3}$$

Simplify each product.

$$m = \frac{9}{-13}$$

Add.

$$m = -\frac{9}{13}$$

Simplify.

3c. Rewrite with positive exponents and then simplify: $\dfrac{2^{-2} + 3^{-1}}{6^{-1}}$.

Solution   **3c.** Recall $2^{-2} = \dfrac{1}{2^2} = \dfrac{1}{4}$, $3^{-1} = \dfrac{1}{3^1} = \dfrac{1}{3}$, and $6^{-1} = \dfrac{1}{6^1} = \dfrac{1}{6}$.

$$\frac{2^{-2} + 3^{-1}}{6^{-1}} = \frac{\dfrac{1}{4} + \dfrac{1}{3}}{\dfrac{1}{6}}$$

Rewrite each negative exponent with positive exponents.

$$= \frac{\dfrac{3}{12} + \dfrac{4}{12}}{\dfrac{1}{6}}$$

Convert the fractions in the numerator to equivalent fractions with the LCD, 12.

$$= \frac{\dfrac{7}{12}}{\dfrac{1}{6}}$$

Add the fractions in the numerator.

Divide the fractions by multiplying by the reciprocal of $\dfrac{1}{6}$.

$$= \frac{7}{12} \cdot \frac{6}{1}$$

$$= \frac{7}{2}$$

Simplify.

✓ Student Check 3 Solve each problem.

a. Let $f(x) = \dfrac{x + 3}{2x}$. Find $f\left(\dfrac{1}{3}\right)$.

b. Find the slope of the line that passes through the points $\left(\dfrac{3}{2}, \dfrac{1}{5}\right)$ and $\left(\dfrac{3}{10}, \dfrac{5}{4}\right)$.

c. Rewrite with positive exponents and then simplify: $\dfrac{4^{-2} + 2^{-3}}{2^{-1}}$.

Objective 4 ▶

Troubleshoot common errors.

Troubleshooting Common Errors

Some common errors associated with complex fractions are shown.

Objective 4 Examples A problem and an incorrect solution are given. Provide the correct solution and an explanation of the error.

4a. Simplify $\dfrac{x + \dfrac{1}{x}}{\dfrac{2}{x^2}}$.

Incorrect Solution	Correct Solution and Explanation
$\dfrac{x + \dfrac{1}{x}}{\dfrac{2}{x^2}} = x + \dfrac{1}{x} \cdot \dfrac{x^2}{2} = x + \dfrac{x}{2}$	We must combine the fractions in the numerator before multiplying by the reciprocal of the fraction in the denominator. $$\dfrac{x + \dfrac{1}{x}}{\dfrac{2}{x^2}} = \dfrac{\dfrac{x^2}{x} + \dfrac{1}{x}}{\dfrac{2}{x^2}} = \dfrac{\dfrac{x^2 + 1}{x}}{\dfrac{2}{x^2}}$$ $$= \dfrac{x^2 + 1}{x} \cdot \dfrac{x^2}{2}$$ $$= \dfrac{x(x^2 + 1)}{2}$$

4b. Simplify $\dfrac{2^{-2} + 3^{-1}}{6^{-1}}$.

Incorrect Solution	Correct Solution and Explanation
$\dfrac{2^{-2} + 3^{-1}}{6^{-1}} = \dfrac{6^1}{2^2 + 3^1}$ $= \dfrac{6}{4 + 3}$ $= \dfrac{6}{7}$	Since the numerator is a sum, we must rewrite the terms as fractions and then simplify the resulting complex fraction. $$\dfrac{2^{-2} + 3^{-1}}{6^{-1}} = \dfrac{\dfrac{1}{4} + \dfrac{1}{3}}{\dfrac{1}{6}} = \dfrac{\dfrac{1}{4} \cdot 12 + \dfrac{1}{3} \cdot 12}{\dfrac{1}{6} \cdot 12}$$ $$= \dfrac{3 + 4}{2}$$ $$= \dfrac{7}{2}$$

ANSWERS TO STUDENT CHECKS

Student Check 1   **a.** $\dfrac{7}{20xy}$ **b.** $\dfrac{1}{3x+4}$ **c.** $\dfrac{1}{ab}$

Student Check 2   **a.** $\dfrac{7}{20xy}$ **b.** $\dfrac{1}{ab}$ **c.** $\dfrac{b}{a}$

Student Check 3   **a.** 5   **b.** $-\dfrac{7}{8}$ **c.** $\dfrac{3}{8}$

SUMMARY OF KEY CONCEPTS

1. There are two methods to simplify a complex fraction. Method 1 requires us to combine the numerators and/or denominators of the complex fractions into a single fraction so that we have one fraction divided by another fraction. Then we multiply by the reciprocal of the denominator to simplify the result.

2. Method 2 requires us to eliminate the fractions from the numerator and/or denominator of the complex fraction initially. This is done by multiplying the numerator and the denominator of the complex fraction by the LCD of all the fractions within the complex fraction. Combine any like terms and simplify the resulting fraction.

SECTION 7.5 / EXERCISE SET

 Write About It!

Use complete sentences in your answer to each exercise.

1. Explain how to simplify the complex fraction $\dfrac{\dfrac{6x^3}{20y^3}}{\dfrac{36x^4}{15y}}$ using method 1.

2. Explain how to simplify the complex fraction $\dfrac{\dfrac{6x^3}{20y^3}}{\dfrac{36x^4}{15y}}$ using method 2.

3. Explain how to simplify the complex fraction $\dfrac{\dfrac{1}{a}+\dfrac{5}{b}}{10a+2b}$.

4. Find the mistakes in the work of simplifying the complex fraction $\dfrac{\dfrac{49}{x^2}-4}{\dfrac{7}{x}-2}$.

$$\dfrac{\dfrac{49}{x^2}-4}{\dfrac{7}{x}-2}=\dfrac{\dfrac{\overset{7}{\cancel{x}}\cancel{49}}{\cancel{x^2}}-4^2}{\dfrac{\cancel{7}}{\cancel{x}}-\cancel{2}}=\dfrac{7}{x}-4$$

5. Explain why the solution is incorrect.
$$\dfrac{4x^{-2}-y^{-2}}{2x^{-1}-y^{-1}}\neq\dfrac{2x-y}{4x^2-y^2}=\dfrac{1}{4x-y}$$

6. Explain why the solution is incorrect.
$$\dfrac{\dfrac{3}{5x-4}+\dfrac{3}{4}}{x}=\dfrac{\dfrac{3}{5x}-\dfrac{3}{4}+\dfrac{3}{4}}{x}=\dfrac{3}{5x^2}$$

7. Explain how to simplify the complex fraction in Exercise 6.

8. Explain how to simplify the expression $\dfrac{3^{-1}+4^{-1}}{2^{-3}}$.

 Practice Makes Perfect!

Simplify each complex fraction using method 1. (See Objective 1.)

9. $\dfrac{\dfrac{7x}{10y^3}}{\dfrac{21x^2}{9y}}$

10. $\dfrac{\dfrac{3r}{7s^2}}{\dfrac{28r^3}{8s}}$

11. $\dfrac{\dfrac{3a}{10b^4}}{\dfrac{6a^2}{36b^3}}$

12. $\dfrac{\dfrac{1}{4y^3}}{\dfrac{16x}{24y^2}}$

13. $\dfrac{\dfrac{15a+12}{a-7}}{\dfrac{25a^2-16}{5a-35}}$

14. $\dfrac{\dfrac{24b+16}{6b+8}}{\dfrac{9b^2-4}{9b+12}}$

15. $\dfrac{\dfrac{8a}{4a+10}}{\dfrac{a^2}{6a+15}}$

16. $\dfrac{\dfrac{7b}{6b+12}}{\dfrac{2b^2}{3b+6}}$

17. $\dfrac{\dfrac{8}{r}+\dfrac{5}{s}}{5r+8s}$

18. $\dfrac{\dfrac{3}{a}+\dfrac{2}{b}}{2a+3b}$

19. $\dfrac{\dfrac{5}{x}-\dfrac{3}{y}}{9x-15y}$

20. $\dfrac{\dfrac{3}{r}-\dfrac{4}{s}}{20r-15s}$

Simplify each complex fraction using method 2.
(*See Objective 2.*)

21. $\dfrac{\dfrac{3x^2}{21y^3}}{\dfrac{35x^3}{28}}$

22. $\dfrac{\dfrac{12r^2}{2s}}{\dfrac{25r^4}{20}}$

23. $\dfrac{\dfrac{10a^3}{18b^3}}{\dfrac{20a^4}{21b^5}}$

24. $\dfrac{\dfrac{4r}{15s^2}}{\dfrac{36r^4}{40s^3}}$

25. $\dfrac{\dfrac{2}{x}-\dfrac{3}{y}}{9x-6y}$

26. $\dfrac{\dfrac{1}{a}+\dfrac{7}{b}}{14a+2b}$

27. $\dfrac{\dfrac{8}{x}+\dfrac{1}{y}}{5x+40y}$

28. $\dfrac{\dfrac{3}{r}-\dfrac{7}{s}}{28r-12s}$

29. $\dfrac{\dfrac{6}{rs}-\dfrac{3}{r^2s}}{\dfrac{12}{s^2}-\dfrac{6}{rs^2}}$

30. $\dfrac{\dfrac{10}{ab}+\dfrac{8}{a^2b}}{\dfrac{15}{b^2}+\dfrac{12}{ab^2}}$

31. $\dfrac{\dfrac{7}{xy}-\dfrac{8}{x^2y}}{\dfrac{35}{y^2}-\dfrac{40}{xy^2}}$

32. $\dfrac{\dfrac{24}{ab}+\dfrac{6}{a^2b}}{\dfrac{4}{b^2}+\dfrac{1}{ab^2}}$

33. $\dfrac{\dfrac{21}{xy}+\dfrac{12}{x^2}}{\dfrac{7}{y^2}+\dfrac{4}{xy}}$

34. $\dfrac{\dfrac{14}{ab}-\dfrac{7}{a^2}}{\dfrac{16}{b^2}-\dfrac{8}{ab}}$

Evaluate each rational function for the given *x*-value.
(*See Objective 3.*)

35. $f(x)=\dfrac{3x+4}{7x-2}, f\left(\dfrac{4}{3}\right)$

36. $f(x)=\dfrac{7x-9}{3x+2}, f\left(\dfrac{6}{7}\right)$

37. $f(x)=\dfrac{2x+1}{x+6}, f\left(-\dfrac{2}{7}\right)$

38. $f(x)=\dfrac{x-5}{3x-2}, f\left(-\dfrac{3}{4}\right)$

Find the slope of the line that passes through the given
points. (*See Objective 3.*)

39. $\left(\dfrac{1}{3},\dfrac{2}{9}\right)$ and $\left(-\dfrac{1}{5},\dfrac{4}{9}\right)$

40. $\left(\dfrac{3}{2},-\dfrac{6}{7}\right)$ and $\left(\dfrac{3}{7},\dfrac{5}{2}\right)$

41. $\left(\dfrac{1}{10},-\dfrac{2}{3}\right)$ and $\left(-\dfrac{2}{5},-\dfrac{1}{10}\right)$

42. $\left(\dfrac{3}{2},-\dfrac{1}{7}\right)$ and $\left(-\dfrac{3}{10},-\dfrac{1}{5}\right)$

Rewrite with positive exponents and then simplify each
expression. (*See Objective 3.*)

43. $\dfrac{3^{-3}-6^{-2}}{4^{-2}}$

44. $\dfrac{2^{-1}-8^{-2}}{2^{-3}}$

45. $\dfrac{3^{-3}+3^{-2}}{2^{-1}}$

46. $\dfrac{6^{-2}-7^{-1}}{6^{-1}}$

47. $\dfrac{4^{-2}-2^{-2}}{5^{-1}}$

48. $\dfrac{4^{-2}+6^{-1}}{2^{-1}}$

 Mix 'Em Up!

Simplify each complex fraction using method 1 or method 2.

49. $\dfrac{\dfrac{3x}{9y^4}}{\dfrac{24x^2}{18y^5}}$

50. $\dfrac{\dfrac{4a}{12b^2}}{\dfrac{18a^3}{32b^4}}$

51. $\dfrac{\dfrac{3a}{4a+8}}{\dfrac{a^2}{8a+16}}$

52. $\dfrac{\dfrac{7x}{18x+30}}{\dfrac{x^2}{6x+10}}$

53. $\dfrac{\dfrac{8x+4}{20x-35}}{\dfrac{4x^2-1}{28x-49}}$

54. $\dfrac{\dfrac{5a-15}{6a-12}}{\dfrac{a^2-9}{8a-16}}$

55. $\dfrac{\dfrac{6}{x}+\dfrac{5}{y}}{20x+24y}$

56. $\dfrac{\dfrac{3}{r}-\dfrac{1}{s}}{4r-12s}$

57. $\dfrac{\dfrac{5}{a}-\dfrac{7}{b}}{21a-15b}$

58. $\dfrac{\dfrac{6}{a}+\dfrac{1}{b}}{2a+12b}$

59. $\dfrac{\dfrac{8}{rs}+\dfrac{1}{r^2s}}{\dfrac{32}{s^2}+\dfrac{4}{rs^2}}$

60. $\dfrac{\dfrac{14}{xy}+\dfrac{49}{x^2y}}{\dfrac{2}{y^2}+\dfrac{7}{xy^2}}$

61. $\dfrac{\dfrac{2}{xy}-\dfrac{1}{x^2}}{\dfrac{12}{y^2}-\dfrac{6}{xy}}$

62. $\dfrac{\dfrac{42}{ab}-\dfrac{49}{a^2}}{\dfrac{6}{b^2}-\dfrac{7}{ab}}$

63. $\dfrac{\dfrac{9}{x^2}-\dfrac{25}{y^2}}{\dfrac{3}{x}+\dfrac{5}{y}}$

64. $\dfrac{\dfrac{25}{r^2}-\dfrac{4}{s^2}}{\dfrac{5}{r}+\dfrac{2}{s}}$

65. $\dfrac{\dfrac{16}{a^2}-\dfrac{1}{b^2}}{\dfrac{12}{a}-\dfrac{3}{b}}$

66. $\dfrac{\dfrac{1}{x^2}-\dfrac{49}{y^2}}{\dfrac{5}{x}-\dfrac{35}{y}}$

67. $\dfrac{r^{-2}-81s^{-2}}{2r^{-1}+18s^{-1}}$

68. $\dfrac{100a^{-2}-b^{-2}}{30a^{-1}+3b^{-1}}$

69. $\dfrac{x^{-2}-9y^{-2}}{5x^{-1}-15y^{-1}}$

70. $\dfrac{a^{-2}-36b^{-2}}{4a^{-1}-24b^{-1}}$

71. $\dfrac{\dfrac{2}{6x-1}+2}{x}$

72. $\dfrac{\dfrac{1}{7y+6}-\dfrac{1}{6}}{y}$

73. $\dfrac{\dfrac{4}{2a-7}+\dfrac{4}{7}}{a}$

74. $\dfrac{\dfrac{3}{b+2}-\dfrac{3}{2}}{b}$

Evaluate each rational function for the given x-value.

75. $f(x)=\dfrac{7x+5}{x-2},\ f\!\left(\dfrac{1}{4}\right)$

76. $f(x)=\dfrac{3x-5}{9x-4},\ f\!\left(\dfrac{5}{2}\right)$

77. $f(x)=\dfrac{x}{4x+1},\ f\!\left(-\dfrac{3}{7}\right)$

78. $f(x)=\dfrac{x}{8x+3},\ f\!\left(-\dfrac{6}{7}\right)$

Find the slope of the line that passes through the given points.

79. $\left(\dfrac{1}{6},-\dfrac{5}{3}\right)$ and $\left(-4,\dfrac{2}{3}\right)$

80. $\left(-\dfrac{5}{4},-6\right)$ and $\left(5,\dfrac{2}{7}\right)$

81. $\left(-\dfrac{1}{6},-\dfrac{3}{4}\right)$ and $\left(5,\dfrac{3}{2}\right)$

82. $\left(2,-\dfrac{7}{6}\right)$ and $\left(-\dfrac{4}{9},-\dfrac{3}{2}\right)$

Simplify each rational expression.

83. $\dfrac{8^{-1}-3^{-3}}{6^{-3}}$

84. $\dfrac{6^{-3}-3^{-3}}{6^{-2}}$

85. $\dfrac{8^{-2}-6^{-1}}{4^{-3}}$

86. $\dfrac{9^{-2}+3^{-1}}{3^{-3}}$

 You Be the Teacher!

Correct each student's errors, if any.

87. Simplify: $\dfrac{\dfrac{2}{x}+\dfrac{5}{y}}{10x+4y}$.

Jen's work:

$$\dfrac{\dfrac{2}{x}+\dfrac{5}{y}}{10x+4y}=\dfrac{\dfrac{2y+5x}{xy}}{2(5x+2y)}=\dfrac{2y+5x}{xy}\cdot\dfrac{2(2y+5x)}{1}$$

$$=\dfrac{(5x+2y)^2}{2xy}$$

88. Simplify: $\dfrac{\dfrac{3}{x}-\dfrac{2}{y}}{\dfrac{1}{x}+\dfrac{1}{y}}$.

Marie's work:

$$\dfrac{\dfrac{3}{x}-\dfrac{2}{y}}{\dfrac{1}{x}+\dfrac{1}{y}}=\left(\dfrac{3}{x}-\dfrac{2}{y}\right)(x+y)=\dfrac{3y-2x}{xy}(x+y)$$

$$=\dfrac{(x+y)(3y-2x)}{xy}$$

89. Simplify: $\dfrac{4x^{-2}-y^{-2}}{2x^{-1}-y^{-1}}$.

Al's work:

$$\dfrac{4x^{-2}-y^{-2}}{2x^{-1}-y^{-1}}=\dfrac{2x-y}{4x^2-y^2}=\dfrac{2x-y}{(2x-y)(2y+x)}=\dfrac{1}{2y+x}$$

90. Simplify: $\dfrac{7^{-2}+8^{-1}}{2^{-3}}$.

Sally's work:

$$\dfrac{7^{-2}+8^{-1}}{2^{-3}}=\dfrac{2^3}{7^2+8}=\dfrac{8}{49+8}=\dfrac{8}{57}$$

 Calculate It!

Use a calculator for each problem.

91. Let $f(x)=\dfrac{x-2}{4x-3}$. Find $f\!\left(-\dfrac{5}{8}\right)$. Find the error in the screen shot and explain why $-\dfrac{53}{16}$ is incorrect.

```
-5/8-2/4(-5/8)-3
            -3.3125
Ans▶Frac
            -53/16
```

92. Let $f(x)=\dfrac{6x+1}{x-4}$. Find $f\!\left(\dfrac{2}{13}\right)$. Find the error in the screen shot and explain why $-\dfrac{79}{26}$ is incorrect.

```
6*2/13+1/2/13-4
          -3.038461538
Ans▶Frac
            -79/26
```

93. Find the slope of the line that passes through $\left(-\dfrac{4}{3},-\dfrac{1}{2}\right)$ and $\left(-\dfrac{5}{3},4\right)$.

94. Find the slope of the line that passes through $\left(\dfrac{1}{3},\dfrac{5}{9}\right)$ and $\left(-7,-\dfrac{5}{2}\right)$.

95. Simplify $\dfrac{4^{-2}+8^{-2}}{2^{-3}}$.

96. Simplify $\dfrac{9^{-3}-4^{-1}}{6^{-2}}$.

Solving Rational Equations

▶ **OBJECTIVES**

As a result of completing this section, you will be able to

1. Solve rational equations.
2. Solve rational equations for a specific variable.
3. Troubleshoot common errors.

Suppose Dionne travels 165 mi to visit a friend. If she wants her trip to take 2 hr, how fast would she have to drive? To answer this question, we need to solve the equation $2 = \dfrac{165}{s}$, where s is in mph.

This is an example of a *rational equation*. We will learn how to solve this type of equation in this section.

Rational Equations

Objective 1 ▶

Solve rational equations.

A **rational equation** is an equation that contains a rational expression. Some examples are

$$\frac{4}{x} = \frac{3}{x+1} \qquad \frac{1}{x-2} + \frac{1}{x+2} = \frac{3}{x^2-4} \qquad \frac{2}{m} + 4 = \frac{3}{2}$$

Recall that in Chapter 2, we solved linear equations that contained fractions. To clear fractions, we applied the multiplication property of equality by multiplying each side of the equation by the LCD. An example is shown.

$$\frac{1}{3}x + \frac{2}{9} = \frac{1}{4}x$$

$$36\left(\frac{1}{3}x + \frac{2}{9}\right) = 36\left(\frac{1}{4}x\right) \qquad \text{Multiply each side by the LCD, 36.}$$

$$36\left(\frac{1}{3}x\right) + 36\left(\frac{2}{9}\right) = 9x \qquad \text{Apply the distributive property and simplify.}$$

$$12x + 8 = 9x \qquad \text{Simplify each product.}$$

$$3x = -8 \qquad \text{Subtract 9x and 8 from each side.}$$

$$x = -\frac{8}{3} \qquad \text{Divide each side by 3.}$$

When we solve rational equations, the LCD will involve algebraic expressions. The multiplication property of equality states that we must multiply both sides of an equation by a *nonzero number*. So, we must determine the values of the variable that make the LCD equal to zero. These values must be excluded from the solution set since they will make the denominator of the rational expression zero and the rational expression undefined.

> **Procedure: Solving Rational Equations**
>
> **Step 1:** Determine the LCD of the rational expressions in the equation.
> **Step 2:** Multiply each side of the equation by the LCD.
> **Step 3:** Solve the resulting equation.
> **Step 4:** Check the solutions in the original equation. Write the solution set.

> **Note:** *If one of the solutions is a value that makes the LCD zero, then it is an extraneous solution and must be discarded from the solution set.*

| **Objective 1 Examples** | **Solve each rational equation.** |

1a. $\dfrac{3}{x} + 5 = 8$ **1b.** $\dfrac{1}{m-3} + 2 = \dfrac{1}{m-3}$ **1c.** $\dfrac{4}{x^2} + \dfrac{5}{x} = -1$

1d. $\dfrac{r+2}{r+8} = \dfrac{2}{r}$ **1e.** $\dfrac{-4x}{3x^2+5x+2} + \dfrac{x}{3x+2} = \dfrac{-4}{x+1}$

Solutions   **1a.**

$$\frac{3}{x} + 5 = 8$$

$$x\left(\frac{3}{x} + 5\right) = x(8)$$ Multiply each side by the LCD, x.

$$(x)\frac{3}{x} + (x)(5) = 8x$$ Apply the distributive property.

$$3 + 5x = 8x$$ Simplify.

$$3 + 5x - 5x = 8x - 5x$$ Subtract $5x$ from each side.

$$3 = 3x$$ Simplify.

$$\frac{3}{3} = \frac{3x}{3}$$ Divide each side by 3.

$$1 = x$$ Simplify.

Check:

$$\frac{3}{x} + 5 = 8$$ Original equation.

$$\frac{3}{1} + 5 = 8$$ Replace x with 1.

$$3 + 5 = 8$$ Simplify.

$$8 = 8$$ True

So the solution set is $\{1\}$.

1b.

$$\frac{1}{m-3} + 2 = \frac{1}{m-3}$$

$$(m-3)\left(\frac{1}{m-3} + 2\right) = (m-3)\left(\frac{1}{m-3}\right)$$ Multiply each side by the LCD, $m-3$.

$$(m-3)\left(\frac{1}{m-3}\right) + 2(m-3) = 1$$ Apply the distributive property and simplify the right side.

$$1 + 2m - 6 = 1$$ Simplify each product.

$$2m - 5 = 1$$ Combine like terms.

$$2m - 5 + 5 = 1 + 5$$ Add 5 to each side.

$$2m = 6$$ Simplify.

$$\frac{2m}{2} = \frac{6}{2}$$ Divide each side by 2.

$$m = 3$$ Simplify.

Check:

$$\frac{1}{m-3} + 2 = \frac{1}{m-3}$$ Original equation

$$\frac{1}{3-3} + 2 = \frac{1}{3-3}$$ Replace m with 3.

$$\frac{1}{0} + 2 = \frac{1}{0}$$ Simplify.

The proposed solution, 3, makes the LCD equal to zero and the rational expression undefined. So, 3 is an *extraneous* solution and must be excluded from the solution set. So, the solution set is the empty set, or Ø.

1c.
$$\frac{4}{x^2} + \frac{5}{x} = -1$$

$$x^2\left(\frac{4}{x^2} + \frac{5}{x}\right) = x^2(-1) \qquad \text{Multiply each side by the LCD, } x^2.$$

$$x^2\left(\frac{4}{x^2}\right) + x^2\left(\frac{5}{x}\right) = -x^2 \qquad \text{Apply the distributive property.}$$

$$4 + 5x = -x^2 \qquad \text{Simplify each product.}$$

The resulting equation is quadratic, so we must write the equation in standard form and solve it by factoring.

$$4 + 5x + x^2 = -x^2 + x^2 \qquad \text{Add } x^2 \text{ to each side.}$$
$$x^2 + 5x + 4 = 0 \qquad \text{Simplify and write in standard form.}$$
$$(x + 1)(x + 4) = 0 \qquad \text{Factor.}$$
$$x + 1 = 0 \quad \text{or} \quad x + 4 = 0 \qquad \text{Apply the zero products property.}$$
$$x = -1 \qquad\qquad x = -4 \qquad \text{Solve each equation.}$$

Both -4 and -1 make the equation true and are, therefore, solutions of the equation. So, the solution set is $\{-4, -1\}$.

> **Note:** *Neither -4 nor -1 makes the denominator in the original equation equal to zero. The value of 0 makes the rational expression undefined and could not be a solution.*

1d.
$$\frac{r + 2}{r + 8} = \frac{2}{r}$$

$$r(r + 8)\frac{r + 2}{r + 8} = r(r + 8)\frac{2}{r} \qquad \text{Multiply each side by the LCD, } r(r + 8).$$

$$r(r + 2) = 2(r + 8) \qquad \text{Simplify each product.}$$

$$r^2 + 2r = 2r + 16 \qquad \text{Apply the distributive property.}$$

Note that the resulting equation is quadratic, so we write the equation in standard form and solve it by factoring.

$$r^2 + 2r - 2r - 16 = 2r + 16 - 2r - 16 \qquad \text{Subtract } 2r \text{ and 16 from each side.}$$

$$r^2 - 16 = 0 \qquad \text{Simplify.}$$

$$(r - 4)(r + 4) = 0 \qquad \text{Factor.}$$

$$r - 4 = 0 \quad \text{or} \quad r + 4 = 0 \qquad \text{Apply the zero products property.}$$

$$r = 4 \qquad\qquad r = -4 \qquad \text{Solve each equation.}$$

The values 4 and -4 make the equation true, so the solution set is $\{-4, 4\}$.

> **Note:** *Neither 4 nor -4 makes the denominator in the original equation equal to zero. The values 0 or -8 make the rational expression undefined and could not be solutions.*

1e. The denominators are $3x^2 + 5x + 2 = (3x + 2)(x + 1)$, $(3x + 2)$, and $(x + 1)$. So, the LCD is $(3x + 2)(x + 1)$.

$$\frac{-4x}{3x^2 + 5x + 2} + \frac{x}{3x + 2} = \frac{-4}{x + 1}$$

$$(3x + 2)(x + 1)\left(\frac{-4x}{(3x + 2)(x + 1)} + \frac{x}{3x + 2}\right) = (3x + 2)(x + 1)\left(\frac{-4}{x + 1}\right)$$

$$(3x + 2)(x + 1)\frac{-4x}{(3x + 2)(x + 1)} + (3x + 2)(x + 1)\frac{x}{3x + 2} = -4(3x + 2)$$

$$-4x + x(x + 1) = -12x - 8$$

$$-4x + x^2 + x = -12x - 8$$

$$x^2 - 3x = -12x - 8$$

$$x^2 - 3x + 12x + 8 = -12x - 8 + 12x + 8$$

$$x^2 + 9x + 8 = 0$$

$$(x + 8)(x + 1) = 0$$

$$x + 8 = 0 \quad \text{or} \quad x + 1 = 0$$

$$x = -8 \qquad\qquad x = -1$$

The value -1 makes the expression undefined and must be excluded from the solution set. The value -8 makes the equation true, and is, therefore, a solution of the equation. So, the solution set is $\{-8\}$.

✔ **Student Check 1**   Solve each rational equation.

a. $\dfrac{4}{x} = 1 + \dfrac{2}{x}$ **b.** $\dfrac{2}{y + 7} - 1 = \dfrac{2}{y + 7}$ **c.** $1 + \dfrac{2}{x} = \dfrac{8}{x^2}$

d. $\dfrac{a - 7}{a + 3} = \dfrac{4}{a}$ **e.** $\dfrac{x}{2x + 1} - \dfrac{13 + x}{2x^2 - 3x - 2} = \dfrac{-3}{x - 2}$

Solving Rational Equations for a Specific Variable

Objective 2 ▶

Solve rational equations for a specific variable.

The skills used for solving rational equations provide the framework for rewriting formulas for specific variables. When the formula contains a rational expression, we must multiply each side of the equation by the LCD to enable us to solve for a different variable.

Objective 2 Examples   Solve each rational equation for the specified variable.

2a. $a = \dfrac{GM}{r^2}$ for M (acceleration of gravity)   **2b.** $a = \dfrac{f - v}{t}$ for v

2c. $\dfrac{1}{a} + \dfrac{1}{b} = \dfrac{1}{t}$ for b

Solutions   **2a.** $a = \dfrac{GM}{r^2}$ Identify the variable to be isolated.

$r^2(a) = r^2\left(\dfrac{GM}{r^2}\right)$ Multiply each side by the LCD, r^2.

$ar^2 = GM$ Simplify each side.

$\dfrac{ar^2}{G} = \dfrac{GM}{G}$ Divide each side by G.

$ar^2 = M$ Simplify.

2b.

$$a = \frac{f - v}{t}$$ Identify the variable to be isolated.

$$t(a) = t\left(\frac{f - v}{t}\right)$$ Multiply each side by the LCD, t.

$$at = f - v$$ Simplify.

$$at - f = f - v - f$$ Subtract f from each side.

$$at - f = -v$$ Simplify.

$$-1(at - f) = -1(-v)$$ Multiply each side by -1.

$$-at + f = v$$ Simplify.

2c.

$$\frac{1}{a} + \frac{1}{b} = \frac{1}{t}$$ Identify the variable to be isolated.

$$abt\left(\frac{1}{a} + \frac{1}{b}\right) = abt\left(\frac{1}{t}\right)$$ Multiply each side by the LCD, abt.

$$abt\left(\frac{1}{a}\right) + abt\left(\frac{1}{b}\right) = ab$$ Apply the distributive property.

$$bt + at = ab$$ Simplify each product.

$$bt + at - bt = ab - bt$$ Subtract bt from each side.

$$at = b(a - t)$$ Factor out b from the terms on the right side.

$$\frac{at}{a - t} = \frac{b(a - t)}{a - t}$$ Divide each side by $a - t$.

$$\frac{at}{a - t} = b$$ Simplify.

✔ **Student Check 2**   Solve each rational equation for the specified variable.

a. $d = \dfrac{m}{v}$ for m (density)   **b.** $r^2 = \dfrac{V}{ph}$ for p **c.** $\dfrac{1}{a} + \dfrac{1}{b} = \dfrac{1}{t}$ for t

---

Objective 3 ▶

Troubleshoot common errors.

Troubleshooting Common Errors

Some common errors related to rational equations are shown next.

Objective 3 Examples   **A problem and an incorrect solution are given. Provide the correct solution and an explanation of the error.**

3a. Solve $\dfrac{x}{x - 2} + 3 = \dfrac{2}{x - 2}$.

Incorrect Solution	Correct Solution and Explanation
$$\dfrac{x}{x - 2} + 3 = \dfrac{2}{x - 2}$$ $$(x - 2)\left(\dfrac{x}{x - 2}\right) + 3(x - 2) = \left(\dfrac{2}{x - 2}\right)(x - 2)$$ $$x + 3x - 6 = 2$$ $$4x - 6 = 2$$ $$4x - 6 + 6 = 2 + 6$$ $$4x = 8$$ $$x = 2$$ So, the solution set is $\{2\}$.	The error was made in not excluding the solution 2 from the solution set. This value makes the denominator in the original equation zero and, therefore, the rational expression undefined. So, the solution set is the empty set, $\varnothing$.

3b. Solve $\dfrac{4}{x+5} - \dfrac{3}{x-2} = \dfrac{1}{x^2 + 3x - 10}$.

Incorrect Solution	Correct Solution and Explanation
$\dfrac{4}{x+5} - \dfrac{3}{x-2} = \dfrac{1}{x^2 + 3x - 10}$ $\dfrac{4}{x+5} - \dfrac{3}{x-2} = \dfrac{1}{(x+5)(x-2)}$ $(x+5)(x-2)\left(\dfrac{4}{x+5} - \dfrac{3}{x-2}\right)$ $= (x+5)(x-2)\left[\dfrac{1}{(x+5)(x-2)}\right]$ $4x - 8 - 3x + 15 = 1$ $x + 7 = 1$ $x = -6$	The error was made in writing the equation that results after the fractions are cleared. The -3, rather than 3, should be distributed to $(x+5)$. $4x - 8 - 3x - 15 = 1$ $x - 23 = 1$ $x = 24$ So, the solution set is $\{24\}$.

ANSWERS TO STUDENT CHECKS

Student Check 1 **a.** $\{2\}$ **b.** $\varnothing$ **c.** $\{-4, 2\}$
d. $\{-1, 12\}$ **e.** $\{-5\}$

Student Check 2 **a.** $m = vd$ **b.** $p = \dfrac{V}{r^2 h}$ **c.** $t = \dfrac{ab}{b+a}$

SUMMARY OF KEY CONCEPTS

1. The method to solve a rational equation involves clearing fractions from the equation by multiplying by the LCD. Once the fractions are cleared, solve the resulting equation. It is essential to check the solutions in the original equation to make sure that the proposed solution is not extraneous.

2. We can solve formulas for a specific variable as well. If the formula has a fraction, clear the fractions by multiplying by the LCD and then isolate the specified variable.

GRAPHING CALCULATOR SKILLS

We can use the Table feature of a graphing calculator to check our possible answers when solving rational equations.

Example: Solve $x + \dfrac{6}{x-2} = \dfrac{3x}{x-2} + 1$.

Solution: We first solve the equation by hand.

$$(x-2)\left(x + \dfrac{6}{x-2}\right) = \left(\dfrac{3x}{x-2} + 1\right)(x-2)$$

$$(x-2)x + (x-2)\dfrac{6}{x-2} = \dfrac{3x}{x-2}(x-2) + 1(x-2)$$

$$x^2 - 2x + 6 = 3x + x - 2$$

$$x^2 - 2x + 6 = 4x - 2$$

$$x^2 - 6x + 8 = 0$$

$$(x-4)(x-2) = 0$$

$$x - 4 = 0 \qquad x - 2 = 0$$

$$x = 4 \qquad x = 2$$

We now need to check our answers in the original equation.

Enter two functions in the calculator. The first function Y_1 is the left side of the equation and the second function Y_2 is the right side of the equation.

Next we examine the table of values.

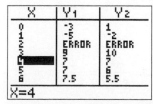

Since the two y-values are equal at $x = 4$, this confirms 4 is a solution. At $x = 2$, the y-values have the ERROR message. This means that 2 makes the rational expression undefined. Therefore, 2 cannot be a solution.

SECTION 7.6 / EXERCISE SET

Write About It!

Use complete sentences in your answer to each exercise.

1. What is a rational equation?

2. Explain the steps to solve a rational equation.

3. Why must we check our answers when solving rational equations?

4. Explain the difference between a rational expression and a rational equation.

5. Explain the difference in the use of the LCD in $\dfrac{4}{x+5} + \dfrac{5}{(x-3)(x+5)}$ and $\dfrac{4}{x+5} + \dfrac{5}{(x-3)(x+5)} = 1$.

6. Explain how to solve $\dfrac{1}{a} + \dfrac{2}{b} = \dfrac{3}{c}$ for a.

Practice Makes Perfect!

Solve each rational equation. (See Objective 1.)

7. $\dfrac{4}{x} = 8$

8. $\dfrac{5}{x} = 10$

9. $\dfrac{16}{2x+3} = 4$

10. $\dfrac{8}{3x+2} = 1$

11. $\dfrac{x}{3} = \dfrac{x+1}{4}$

12. $\dfrac{y}{5} = \dfrac{y-2}{6}$

13. $\dfrac{y-4}{3} + \dfrac{1}{2} = \dfrac{y}{3}$

14. $\dfrac{a+3}{4} - \dfrac{2}{5} = \dfrac{a}{4}$

15. $\dfrac{4}{3a+5} = -2$

16. $\dfrac{4}{2y-8} = 5$

17. $\dfrac{9}{2x} + 2 = 12$

18. $\dfrac{5}{3x} + 3 = 7$

19. $1 - \dfrac{5}{x} = -\dfrac{4}{x^2}$

20. $2 - \dfrac{1}{x} = \dfrac{6}{x^2}$

21. $5 + \dfrac{3}{t+1} = -\dfrac{2}{t+1}$

22. $\dfrac{3}{t+2} = \dfrac{6}{t+2} - 2$

23. $\dfrac{3}{x+3} = \dfrac{4}{x-2}$

24. $\dfrac{1}{x-5} = \dfrac{2}{x+4}$

25. $\dfrac{3}{x+1} - \dfrac{x}{x^2-1} = \dfrac{2}{x-1}$

26. $\dfrac{4}{x+2} + \dfrac{3x}{x^2-4} = \dfrac{7}{x-2}$

27. $\dfrac{x^2}{x+9} = \dfrac{49}{x+9}$

28. $\dfrac{3x^2}{x+10} = \dfrac{12}{x+10}$

29. $\dfrac{x^2}{x^2+7x+6} = \dfrac{1}{x+1}$

30. $\dfrac{x^2}{2x^2+5x-12} = \dfrac{-x}{2x-3}$

31. $\dfrac{5}{2x-1} + \dfrac{3x}{2x^2+7x-4} = \dfrac{4}{x+4}$

32. $\dfrac{4x}{3x^2-13x-10} - \dfrac{1}{x-5} = \dfrac{2}{3x+2}$

33. $\dfrac{x^2-10}{2x^2+5x-7} = \dfrac{x}{2x+1}$

34. $\dfrac{x^2-1}{3x^2-13x-6} = \dfrac{2x}{6x+1}$

Solve each rational equation for the specified variable. (See Objective 2.)

35. $A = \dfrac{bh}{2}$ for h (area of triangle)

36. $A = \dfrac{bh}{2}$ for b (area of triangle)

37. $P = \dfrac{f}{a}$ for a (pressure)   38. $P = \dfrac{f}{a}$ for f (pressure)

39. $a = \dfrac{v-v_0}{t}$ for t (acceleration)

40. $a = \dfrac{v-v_0}{t}$ for v (acceleration)

41. $F = \dfrac{mv^2}{r}$ for r (force)   42. $F = \dfrac{mv^2}{r}$ for m (force)

43. $u = \dfrac{P^2L}{2AE}$ for A (mechanics, strain energy)

44. $u = \dfrac{P^2L}{2AE}$ for E (mechanics, strain energy)

45. $G = \dfrac{E}{2(I+v)}$ for I (engineering, shear stress)

46. $G = \dfrac{E}{2(I+v)}$ for v (engineering, shear stress)

47. $\dfrac{1}{a} + \dfrac{2}{b} = \dfrac{3}{c}$ for b 48. $\dfrac{4}{r} - \dfrac{5}{s} = \dfrac{1}{t}$ for r

Mix 'Em Up!

Solve each rational equation.

49. $\dfrac{a}{3} = \dfrac{3a-1}{7}$

50. $y = \dfrac{2y-5}{6}$

51. $\dfrac{x-5}{2} - \dfrac{1}{6} = \dfrac{4x}{3}$

52. $\dfrac{4y+1}{5} + \dfrac{7}{10} = \dfrac{3y}{2}$

53. $\dfrac{4a}{10a+3} = \dfrac{2}{5}$

54. $\dfrac{2a}{6a-5} = \dfrac{1}{3}$

55. $\dfrac{5}{7x} - 1 = 14$

56. $\dfrac{3}{4x} + 2 = 11$

57. $\dfrac{2x^2 - 3x - 2}{x^2 - 6} = \dfrac{2x + 1}{x}$

58. $\dfrac{x^2 + 2x}{2x^2 - 5x - 3} = \dfrac{2x}{4x + 3}$

59. $3 + \dfrac{13}{x} = \dfrac{10}{x^2}$

60. $2 + \dfrac{7}{x} = -\dfrac{3}{x^2}$

61. $3 + \dfrac{4}{t - 2} = -\dfrac{1}{t - 2}$

62. $\dfrac{2}{t + 5} = \dfrac{1}{t + 5} - 4$

63. $\dfrac{1}{3x + 2} = \dfrac{2}{5x - 1}$

64. $\dfrac{5}{2x - 7} = \dfrac{3}{x + 3}$

65. $\dfrac{3}{2x + 3} - \dfrac{2x}{4x^2 - 9} = \dfrac{2}{2x - 3}$

66. $\dfrac{4}{x - 5} + \dfrac{2x}{x^2 - 25} = \dfrac{6}{x + 5}$

Solve each rational equation for the specified variable.

67. $A = \dfrac{1}{2}h(a + b)$ for b (area of trapezoid)

68. $A = \dfrac{1}{2}h(a + b)$ for h (area of trapezoid)

69. $\dfrac{PV}{T} = k$ for V (combined gas law)

70. $\dfrac{PV}{T} = k$ for T (combined gas law)

71. $\dfrac{1}{R} = \dfrac{1}{R_1} + \dfrac{1}{R_2}$ for R (resistors in parallel)

72. $\dfrac{1}{R} = \dfrac{1}{R_1} + \dfrac{1}{R_2}$ for R_1 (resistors in parallel)

73. $m = \dfrac{y_2 - y_1}{x_2 - x_1}$ for y_2 (slope formula)

74. $m = \dfrac{y_2 - y_1}{x_2 - x_1}$ for x_2 (slope formula)

 You Be the Teacher!

Correct each student's errors, if any.

75. Solve $\dfrac{3}{x + 1} = \dfrac{3}{x + 4}$.

Baxter's work:

$$(\cancel{x + 1})(x + 4)\dfrac{3}{\cancel{x + 1}} = \dfrac{3}{\cancel{x + 4}}(x + 1)(\cancel{x + 4})$$

$$3(x + 4) = 3(x + 1)$$

$$3x + 12 = 3x + 3$$

$$0 = -9$$

So, $x = 0$, $x = -9$.

76. Solve $\dfrac{x^2}{x - 1} = \dfrac{1}{x - 1}$.

Lydia's work:

$$(\cancel{x - 1})\dfrac{x^2}{\cancel{x - 1}} = \dfrac{1}{\cancel{x - 1}}(\cancel{x - 1})$$

$$x^2 = 1$$

$$x^2 - 1 = 0$$

$$(x - 1)(x + 1) = 0$$

$$x - 1 = 0 \quad x + 1 = 0$$

$$x = 1 \qquad x = -1$$

So, the solutions are 1 and -1.

77. Solve $\dfrac{2}{x + 2} - \dfrac{5}{3x - 1} = \dfrac{4x}{3x^2 + 5x - 2}$.

Holly's work:

$$\dfrac{2}{x + 2} - \dfrac{5}{3x - 1} = \dfrac{4x}{(3x - 1)(x + 2)}$$

$$(3x - 1)(\cancel{x + 2})\dfrac{2}{\cancel{x + 2}} - \dfrac{5}{\cancel{3x - 1}}(\cancel{3x - 1})(x + 2)$$

$$= \dfrac{4x}{(\cancel{3x - 1})(\cancel{x + 2})}(\cancel{3x - 1})(\cancel{x + 2})$$

$$2(3x - 1) - 5(x + 2) = 4x$$

$$6x - 2 - 5x + 10 = 4x$$

$$x + 8 = 4x$$

$$8 = 3x$$

$$x = \dfrac{8}{3}$$

So, the solution is $\dfrac{8}{3}$.

78. Solve $\dfrac{2}{x + 1} - \dfrac{1}{2x - 1} = \dfrac{4x}{2x^2 + x - 1}$.

Eddie's work:

$$\dfrac{2}{x + 1} - \dfrac{1}{2x - 1} = \dfrac{4x}{2x^2 + x - 1}$$

$$(2x - 1)(\cancel{x + 1})\dfrac{2}{\cancel{x + 1}} - \dfrac{1}{\cancel{2x - 1}}(\cancel{2x - 1})(x + 1)$$

$$= \dfrac{4x}{(\cancel{2x - 1})(\cancel{x + 1})}(\cancel{2x - 1})(\cancel{x + 1})$$

$$2(2x - 1) - (x + 1) = 4x$$

$$4x - 2 - x + 1 = 4x$$

$$3x - 1 = 4x$$

$$x = -1$$

So, the solution is -1.

 Calculate It!

Solve each rational equation. Use a graphing calculator to check the answers.

79. $\dfrac{3}{x} + \dfrac{5}{x^2} = 2$

80. $\dfrac{2}{x} + \dfrac{1}{x^2} = 3$

81. $\dfrac{x}{x + 5} - \dfrac{3}{x - 5} = 1$

82. $\dfrac{2x}{x + 6} - \dfrac{2}{x - 6} = 1$

| SECTION 7.7 | **Proportions and Applications of Rational Equations** |

On April 20, 2010, the worst oil spill in U.S. history occurred when the oil spilled from BP Oil's Deepwater Horizon oil-drilling rig in the Gulf of Mexico. It is estimated that 5000 barrels of oil were spilled in the ocean each day until July 15. On July 5, 2010, BP announced that its one-day oil recovery effort accounted for 24,980 barrels of oil, and the burning off of 57.1 million ft³ of natural gas. If the total oil collected for the spill was estimated at 657,300 barrels, how many cubic feet of natural gas was released? (Source: Wikipedia)

In this section, we use ratios and proportions to answer this and other real-life questions.

Proportions

| **Objective 1** ▶ | A **ratio** relates the measurement of two quantities. Ratios can be expressed in two ways: (1) with a colon, or (2) as a fraction. For example, the fact that 5000 barrels of |
| Solve proportions. | oil were released into the Gulf of Mexico each day can be written as a ratio. The ratio of oil spilled to days can be expressed as 5000:1 (read "5000 to 1") or the ratio can be |

expressed as the fraction $\frac{5000}{1}$. We will express ratios as fractions in this section.

A **proportion** is an equation with two ratios set equal to each other. A proportion is an equation of the form $\frac{a}{b} = \frac{c}{d}$, where b and $d \neq 0$. When a proportion includes variables, it is a rational equation. Therefore, we can apply the methods for solving rational equations to solve proportions. When we multiply a proportion by its LCD, it is equivalent to the equation that results from cross multiplication.

$$\frac{a}{b} = \frac{c}{d}$$

$$bd\left(\frac{a}{b}\right) = bd\left(\frac{c}{d}\right) \qquad \text{Multiply by the LCD, } bd.$$

$$ad = bc \qquad \text{Simplify.}$$

$$\frac{a}{b} \diagdown \frac{c}{d}$$

$$ad = bc$$

cross products

Definition: Cross Products

If b and $d \neq 0$, and $\frac{a}{b} = \frac{c}{d}$, then the **cross products** are ad and bc. The cross products are equal. So,

$$ad = bc$$

For example, since $\frac{2}{3} = \frac{4}{6}$, then

$$2 \cdot 6 = 3 \cdot 4$$
$$12 = 12$$

Procedure: Solving a Proportion

Step 1: Multiply each side of the equation by the LCD or cross multiply.
Step 2: Solve for the indicated variable.
Step 3: Check the answer in the original equation.

| **Objective 1 Examples** | **Solve each proportion.** |

1a. $\dfrac{4}{5} = \dfrac{x}{10}$ **1b.** $\dfrac{x+2}{4} = \dfrac{2}{3}$ **1c.** $\dfrac{1}{2x} = \dfrac{5}{9}$

Solutions   **1a.** **Method 1:** Multiply by the LCD.   **Method 2:** Cross multiply.

$$\frac{4}{5} = \frac{x}{10}$$ $$\frac{4}{5} = \frac{x}{10}$$

$$10\left(\frac{4}{5}\right) = 10\left(\frac{x}{10}\right)$$ Multiply each side by 10. $4(10) = 5x$ Cross multiply.

$$2 \cdot 4 = 1 \cdot x$$ Divide out the common factors. $40 = 5x$ Simplify.

$$8 = x$$ Simplify. $$\frac{40}{5} = \frac{5x}{5}$$ Divide each side by 5.

$$8 = x$$ Simplify.

So, the solution set is $\{8\}$. We can check by substituting 8 in the original equation.

1b. **Method 1:** Multiply by the LCD.   **Method 2:** Cross multiply.

$$\frac{x+2}{4} = \frac{2}{3}$$ $$\frac{x+2}{4} = \frac{2}{3}$$

$$12\left(\frac{x+2}{4}\right) = 12\left(\frac{2}{3}\right)$$ Multiply each side by 12. $3(x+2) = 2(4)$ Cross multiply.

$$3(x+2) = 4 \cdot 2$$ Divide out the common factors. $3x + 6 = 8$ Simplify.

$$3x + 6 = 8$$ Distribute. $3x + 6 - 6 = 8 - 6$ Subtract 6 from each side.

$$3x + 6 - 6 = 8 - 6$$ Subtract 6 from each side. $3x = 2$ Simplify.

$$3x = 2$$ Simplify. $$x = \frac{2}{3}$$ Divide each side by 3.

$$x = \frac{2}{3}$$ Divide each side by 3.

So, the solution set is $\left\{\frac{2}{3}\right\}$. We can check by substituting $\frac{2}{3}$ in the original equation.

1c. **Method 1:** Multiply by the LCD.   **Method 2:** Cross multiply.

$$\frac{1}{2x} = \frac{5}{9}$$ $$\frac{1}{2x} = \frac{5}{9}$$

$$18x\left(\frac{1}{2x}\right) = 18x\left(\frac{5}{9}\right)$$ Multiply each side by 18x. $1(9) = 2x(5)$ Cross multiply.

$$9(1) = 2x(5)$$ Divide out common factors. $9 = 10x$ Simplify.

$$9 = 10x$$ Simplify. $$\frac{9}{10} = x$$ Divide each side by 10.

$$\frac{9}{10} = x$$ Divide each side by 10.

So, the solution set is $\left\{\frac{9}{10}\right\}$. We can check by substituting $\frac{9}{10}$ in the original equation.

☑ **Student Check 1**   Solve each proportion.

a. $\dfrac{3}{4} = \dfrac{x}{6}$ **b.** $\dfrac{x-3}{5} = \dfrac{x}{2}$ **c.** $\dfrac{3}{x} = \dfrac{7}{10}$

Applications

Objective 2 ▶

Use proportions in various real-world applications.

Proportions can be used to solve problems from the real world. Recall the oil spill ratio, $\frac{5000}{1}$. Suppose we want to determine the barrels of oil spilled after 87 days, the length of time before the oil was contained. We can write the following proportion to determine this information.

$$\text{Number of barrels} \rightarrow \frac{5000}{1} = \frac{x}{87} \leftarrow \text{Number of barrels}$$
$$\text{Number of days} \rightarrow \qquad\qquad \leftarrow \text{Number of days}$$

To solve the proportion, we cross multiply to get

$$\frac{5000}{1} = \frac{x}{87}$$
$$1(x) = 5000(87)$$
$$x = 435{,}000$$

So, 435,000 barrels of oil were spilled in the Gulf after 87 days.

Procedure: Writing a Proportion

Step 1: Define the ratio so that the numerator represents one quantity and the denominator represents the other quantity.

Step 2: Write the proportion, setting the known ratio equal to the unknown ratio. Use a variable to represent the unknown quantity.

Step 3: Solve the proportion.

Objective 2 Examples   Use a proportion to solve each problem.

2a. Jennifer's favorite chocolate chip cookie recipe calls for 1 cup of brown sugar and it yields 36 cookies. If Jennifer wants to make only 24 cookies, how much brown sugar should she use?

Solution   **2a.** Let the ratio be cups of brown sugar to the number of servings in the recipe.

$$\text{Cups of brown sugar} \rightarrow \frac{1}{36} = \frac{x}{24} \qquad \text{Write the proportion, where } x \text{ is the number}$$
$$\text{Number of servings} \rightarrow \qquad\qquad \text{of cups needed to make 24 cookies.}$$

$$1(24) = x(36) \qquad \text{Cross multiply.}$$

$$24 = 36x \qquad \text{Simplify each product.}$$

$$\frac{24}{36} = x \qquad \text{Divide each side by 36.}$$

$$\frac{2}{3} = x \qquad \text{Simplify.}$$

So, Jennifer should use $\frac{2}{3}$ cup of brown sugar to make 24 cookies.

2b. When Iason returns to the United States from Greece, he has 245 Euros. If the conversion rate is 1 Euro = $1.50, how much money in U.S. dollars does Iason have?

Solution **2b.** Let the ratio be Euros to U.S. dollars.

$$\frac{\text{Euros} \rightarrow 245}{\text{U.S. dollars} \rightarrow x} = \frac{1}{1.50}$$

Write the proportion, where x is the number of U.S. dollars for 245 Euros.

$$245(1.50) = x(1)$$ Cross multiply.

$$367.5 = x$$ Simplify.

So, Iason has the equivalent of $367.50.

2c. On July 5, 2010, BP announced that its one-day oil recovery effort accounted for 24,980 barrels of oil, and the burning off of 57.1 million ft³ of natural gas. If the total oil collected for the spill was estimated at 657,300 barrels, how many cubic feet of natural gas was released? (Source: Wikipedia)

Solution **2c.** Let the ratio be barrels of oil to cubic feet of natural gas in millions. Let x represent the cubic feet of natural gas released for 657,300 barrels of oil.

$$\frac{\text{Barrels}}{\text{Cubic feet in millions}} \rightarrow \frac{24{,}980}{57.1} = \frac{657{,}300}{x}$$

Write the appropriate proportion.

$$57.1x\left(\frac{24{,}980}{57.1}\right) = 57.1x\left(\frac{657{,}300}{x}\right)$$

Multiply each side by the LCD, 57.1x.

$$24{,}980x = 37{,}531{,}830$$ Simplify.

$$x = \frac{37{,}531{,}830}{24{,}980}$$ Divide each side by 24,980.

$$x \approx 1502.5$$ Simplify.

So, if the total oil collected was 657,300 barrels, then about 1,502.5 million ft³ of natural gas was burned off.

☑ **Student Check 2** Use a proportion to solve each problem.

a. A punch recipe that serves 40 calls for $1\frac{3}{4}$ cups lemon juice. How many cups of lemon juice is needed to serve 100 people?

b. How many U.S. dollars are in 3000 Pakistani rupee if the conversion rate is 1 U.S. dollar = 61.2 Pakistan rupee?

c. If 1500 people were surveyed in a small town and 850 of them were Republicans, how many in the town of 10,000 are Republicans? Round to the nearest integer.

Similar Triangles

Objective 3 ▶

Use proportions to solve similar triangle problems.

Similar triangles have the same shape, but different sizes. Corresponding angles of similar triangles are equal, and the lengths of corresponding sides of similar triangles are proportional. Triangles *ABC* and *LMN* are similar triangles.

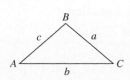

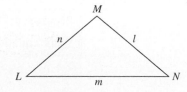

The measure of angle A is equal to the measure of angle L, the measure of angle B is equal to the measure of angle M, and the measure of angle C is equal to the measure of angle N. Since the sides of the similar triangles are proportional, we can write

$$\frac{a}{l} = \frac{b}{m} = \frac{c}{n}$$

Objective 3 Example ⟋ **Triangles *DEF* and *JKL* are similar. Find the missing length, *l*. Note the figures are not drawn to scale.**

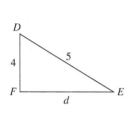
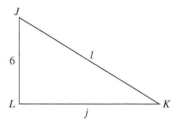

Solution

$$\frac{4}{6} = \frac{5}{l}$$ Write a proportion with ratios of corresponding sides of the triangle.

$$4l = 30$$ Cross multiply.

$$l = \frac{30}{4}$$ Divide each side by 4.

$$l = 7.5$$ Simplify.

✓ **Student Check 3**   If $d = 7$, use the result of Example 3 to find the missing length j in triangle *JKL*.

Work Problems

Objective 4 ▶

Solve work problems.

Now we will solve problems from other applications of rational equations. The first of these involves work applications. Work applications involve finding the amount of time it takes two individuals to complete a task when working together.

For example, if Jodi can do a job in 6 hr, then she completes $\frac{1}{6}$ of the job in 1 hr. If Kyle can do the same job in 4 hr, then he completes $\frac{1}{4}$ of the job in 1 hr. To determine the time it will take for Jodi and Kyle to complete the job if they work together, we add up the portion of the job that can be completed in 1 hr by each of them working alone.

$$\frac{1}{4} + \frac{1}{6} = \frac{3}{12} + \frac{2}{12}$$

$$= \frac{5}{12}$$

So, $\frac{5}{12}$ represents the portion of the job that can be completed in 1 hr. Since $\frac{5}{12} = \frac{1}{\frac{12}{5}}$, it will take Jodi and Kyle $\frac{12}{5} = 2.4$ hr to complete the job if they work together. This leads to the following property.

Property: Work Equation

Work equations are of the form

$$\frac{1}{a} + \frac{1}{b} = \frac{1}{t}$$

where a is the time for one person to complete the job working alone, b is the time for another person to complete the job working alone, and t is the time for the two people working together to complete the job.

Objective 4 Example

Suppose it takes Atesa 5 hr to rake and bag the leaves in her yard when she works alone. Her son, Hunter, can complete the job by himself in 3 hr. How long would it take Atesa and Hunter to complete the job if they worked together?

Solution

What is unknown? The time it takes for Atesa and Hunter to complete the job working together. Let x represent this time in hours.

What is known? Atesa can complete the job in 5 hr. Hunter can complete the job in 3 hr.

So, we can apply the property to write the equation.

$$\frac{1}{a} + \frac{1}{b} = \frac{1}{t}$$ State the work equation.

$$\frac{1}{5} + \frac{1}{3} = \frac{1}{x}$$ Replace the times working alone, 5 hr and 3 hr, and the time working together, x.

$$15x\left(\frac{1}{5} + \frac{1}{3}\right) = 15x\left(\frac{1}{x}\right)$$ Multiply each side by the LCD, $15x$.

$$15x\left(\frac{1}{5}\right) + 15x\left(\frac{1}{3}\right) = 15$$ Apply the distributive property.

$$3x + 5x = 15$$ Simplify each product.

$$8x = 15$$ Combine like terms.

$$x = \frac{15}{8}$$ Divide each side by 8.

$$x \approx 1.89$$ Approximate the solution.

So, together they can complete the job in approximately 1.89 hr.

✓ Student Check 4

Helena cleans newly constructed homes for a real estate company. Alone, Helena can clean a 2400 ft² home in 4 hr. Her partner Kenneth can clean the same size home in 6 hr. How long would it take them to clean the home if they worked together?

Motion Problems

Objective 5 ▶

Solve motion problems.

A *rate* is a ratio of distance to time. A rate that we are familiar with is the speed of a car, usually given as $\frac{miles}{hour}$ or $\frac{kilometers}{hour}$, depending on the country where one lives. We know that rate (speed) is equal to distance (miles) divided by time (hours), that is, $r = \frac{d}{t}$. A more familiar form is $d = rt$, that is distance is equal to rate times time. This is true no matter what unit is being used for distance or time.

> **Property: Distance-Rate-Time Formula**
>
> If an object travels a distance d at rate r for time t, then $d = rt$. Solving for r, we obtain the formula $r = \frac{d}{t}$. Solving for t, we obtain the formula $t = \frac{d}{r}$.

Objective 5 Example

An airplane flies 1573 mi from Denver to Philadelphia in the same amount of time another airplane flies 1090 mi from New York to Orlando. If the speed of the Denver flight is 220 mph faster than the New York flight, find the speed of the Denver flight.

Solution

What is unknown? The speed of the each flight is unknown. Let x represent the speed of the New York flight. Then $x + 220$ is the speed of the Denver flight.

What is known? The total distance from Denver to Philadelphia is 1573 mi. The total distance from New York to Orlando is 1090 mi.

We can use a table to summarize the information. We complete the distance and rate columns with the known and unknown information. We then use the distance-rate-time formula to complete the time column. So, time is distance divided by rate.

	Distance (mi) =	Rate (mph)	×	Time (hr)
Denver flight	1573	$x + 220$		$\dfrac{1573}{x + 220}$
New York flight	1090	x		$\dfrac{1090}{x}$

Since both flights take the same amount of time, we write an equation by setting the expressions for time equal.

$$\frac{1573}{x + 220} = \frac{1090}{x}$$ Begin with the model.

$$1573x = 1090(x + 220)$$ Cross multiply.

$$1573x = 1090x + 239{,}800$$ Apply the distributive property.

$$1573x - 1090x = 1090x + 239{,}800 - 1090x$$ Subtract 1090x from each side.

$$483x = 239{,}800$$ Simplify.

$$x \approx 496.48$$ Divide each side by 483.

The New York flight traveled at approximately 496.48 mph, so the speed of the Denver flight was approximately $496.48 + 220$ or 716.48 mph.

✓ **Student Check 5**   Courtney ran 3.5 mi in the same amount of time it took Tammy to run 6 mi. If Courtney ran 2 mph slower than Tammy, how fast did each woman run?

Objective 6 ▶

Troubleshoot common errors.

Troubleshooting Common Errors

Some common errors associated with applications of rational equations are shown next.

Objective 6 Examples  **A problem and an incorrect solution are given. Provide the correct solution and an explanation of the error.**

6a. Marice can complete a job in 4 hr working alone and Gwen can complete the job in 3 hr working alone. How long can they complete the job working together?

Incorrect Solution	Correct Solution and Explanation
$$\dfrac{4 + 3}{2} = 3.5 \text{ hr}$$ (crossed out)	We must find the portions that they each can do in an hour and set the sum equal to the portion they can do together in an hour. Let x represent the time working together to complete the job. $$\frac{1}{4} + \frac{1}{3} = \frac{1}{x}$$ $$12x\left(\frac{1}{4} + \frac{1}{3}\right) = 12x\left(\frac{1}{x}\right)$$ $$3x + 4x = 12$$ $$7x = 12$$ $$x = \frac{12}{7} \approx 1.7 \text{ hr}$$

6b. The triangles are similar. Find the value of x. Note the figures are not drawn to scale.

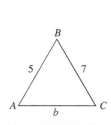 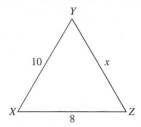

Incorrect Solution	Correct Solution and Explanation
$$\frac{b}{8} = \frac{7}{x}$$ $$bx = 56$$ $$x = \frac{56}{b}$$	We must use a known ratio in our equation. $$\frac{5}{10} = \frac{7}{x}$$ $$5x = 70$$ $$x = 14$$

ANSWERS TO STUDENT CHECKS

Student Check 1   a. 4.5   **b.** -2 **c.** $\dfrac{30}{7}$

Student Check 2   a. 4.38 cups   **b.** 49.02 U.S. dollars
   c. 5667 Republicans

Student Check 3   10.5

Student Check 4   It would take them 2.4 hr to clean the home.

Student Check 5   Courtney ran 2.8 mph and Tammy ran 4.8 mph.

SUMMARY OF KEY CONCEPTS

1. A ratio relates the measures of two quantities and can be expressed as a fraction. A proportion is an equation in which two ratios are equal.

2. Proportions are solved by multiplying both sides of the equation by the LCD or by cross multiplying.

3. Proportions can be used to solve many real-world problems. To solve a problem, set up a proportion with the known ratio equal to the unknown ratio. In these ratios, let the numerators represent one quantity and the denominators represent the other quantity.

4. Similar triangles are triangles in which the lengths of corresponding sides are proportional. To find a missing side, set up a proportion.

5. In work problems, we either find how long it will take one person working alone or two people working together to complete a job. We must represent the portion of the job that each person can do in one hour. We then add the individual portions and set equal to $\dfrac{1}{x}$, the portion of the job they can complete together.

6. To solve motion problems, use the formula $d = rt$. Make a chart and complete the columns. One of the columns is the given information. Variables are used to represent one of the other columns. The remaining column is obtained by applying the $d = rt$ formula. Write an appropriate equation and solve.

SECTION 7.7   EXERCISE SET

 Write About It!

Use complete sentences in your answer to each exercise.

1. Explain the meaning of a ratio.

2. Explain the meaning of a proportion.

3. What is the difference between a ratio and a proportion?

4. Explain how to solve a proportion.

5. What does it mean for two triangles to be similar?

6. Explain how to set up a work problem.

 Practice Makes Perfect!

Solve each proportion. (*See Objective 1.*)

7. $\dfrac{2}{3} = \dfrac{x}{6}$

8. $\dfrac{1}{4} = \dfrac{x}{12}$

9. $\dfrac{1}{6} = \dfrac{x}{20}$

10. $\dfrac{1}{5} = \dfrac{x}{12}$ **11.** $\dfrac{x+1}{4} = \dfrac{3}{5}$ **12.** $\dfrac{x-4}{3} = \dfrac{2}{5}$

13. $\dfrac{\frac{1}{2}}{12} = \dfrac{3}{x}$ **14.** $\dfrac{\frac{2}{3}}{8} = \dfrac{2}{x}$ **15.** $\dfrac{3.65}{2} = \dfrac{x}{10}$

16. $\dfrac{12}{2.88} = \dfrac{100}{x}$ **17.** $\dfrac{100}{x+3} = \dfrac{250}{x+5}$

18. $\dfrac{300}{x+1} = \dfrac{400}{x-4}$

Use proportions to solve each problem. (*See Objective 2.*)

Use the Country Biscuits recipe for Exercises 19–22.

> **Country Biscuits**
>
> 2 cups all-purpose flour
> $\dfrac{3}{4}$ cup shortening
> 1 cup milk
> 1 tsp salt
> 3 tsp baking powder
>
> Directions: Combine the ingredients in a bowl. On a floured surface, roll out the dough and cut out 12 biscuits. Bake on a cookie sheet for 375° F until lightly browned. (Makes 12 biscuits.)

19. How much flour should you use to increase the recipe to 20 biscuits?

20. How much shortening should you use if you want to make 20 biscuits?

21. If you only have $\dfrac{1}{2}$ cup of flour, approximately how many biscuits can you make?

22. If you only have $\dfrac{1}{3}$ cup of milk, approximately how many biscuits can you make?

23. How many U.S. dollars are in 15 Saudi riyals? (Note: 1 U.S. dollar = 3.75 Saudi riyals.)

24. How many U.S. dollars are in 1500 South Korean won? (Note: 1 U.S. dollar = 1056.15 South Korean won.)

25. How many Swiss francs are in 10 U.S. dollars? (Note: 1 U.S. dollar = 0.8238 Swiss francs.)

26. How many Japanese yen are in 15,000 U.S. dollars? (Note: 1 U.S. dollar = 79.19 Japanese yen.)

27. If 1 in. represents 120 mi on a map, how many inches represent the distance between Los Angeles, California, and Denver, Colorado, which is 1020 mi?

28. If 1 in. represents 40 mi on a map, how many inches represent the distance between Boston, Massachusetts, and Baltimore, Maryland, which is 408 mi?

29. If 1 cm represents 210 km on a map, how many centimeters represent the distance between Philadelphia, Pennsylvania, and Chicago, Illinois, which is 1224.3 km?

30. If 1 cm represents 210 km on a map, how many centimeters represent the distance between New Orleans, Louisiana, and San Diego, California, which is 2927.4 km?

31. An object weighing 1 lb on Earth weighs 0.166 lb on the moon. Find Cynthia's weight on the moon if she weighs 125 lb on Earth.

32. An object weighing 1 lb on Earth weighs 0.377 lb on Mars. Find Andrew's weight on Mars if he weighs 180 lb on Earth.

33. An object weighing 1 lb on Earth weighs 0.067 lb on Pluto. Find Louise's weight on Earth if she weighs 6.03 lb on Pluto.

34. An object weighing 1 lb on Earth weighs 2.364 lb on Jupiter. Find Mark's weight on Earth if he weighs 531.9 lb on Jupiter.

Use similar triangles *ABC* and *XYZ* for Exercises 35 and 36. Note the figures are not drawn to scale. (*See Objective 3.*)

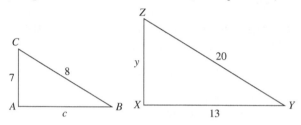

35. Find *y*. **36.** Find *c*.

Use similar triangles *ABC* and *XYZ* for Exercises 37 and 38. Note the figures are not drawn to scale (*See Objective 3.*)

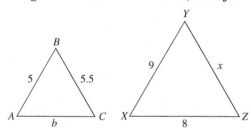

37. Find *b*. **38.** Find *x*.

Solve each work problem. (*See Objective 4.*)

39. Lynn and Paul enjoy working on jigsaw puzzles. On average, it takes Lynn 6 hr to complete a 1000-piece puzzle. It takes Paul 10 hr to complete a 1000-piece puzzle on average. How long will it take Lynn and Paul to complete a 1000-piece puzzle working together?

40. Lindsey distributes a town newspaper by foot. It takes her 3 hr to deliver the paper to the customers in her assigned area. Jamie can deliver newspapers in the same area in 2 hr. How long will it take Lindsey and Jamie to distribute the newspapers if they work together?

Solve each motion problem. (*See Objective 5.*)

41. It takes Margie 4 hr to drive to a conference. If she drives at a constant rate of 50 mph, how far away is the conference?

42. How long will it take Howie to drive 264 mi if he is traveling a constant rate of 60 mph?

43. Cassie runs 3 mi in the time that it takes Tran to walk 1 mi. Cassie runs 2 mph faster than Tran walks. At what speed does each person travel?

44. Esther runs 5 mi in the same time that it takes Janice to run 6 mi. If Esther runs 1 mph slower than Janice, find the speed that each woman runs.

45. An airplane flies 1960 mi in the same amount of time it takes a second plane to fly 1610 mi. If the speed of the first airplane is 100 mph faster than the speed of the second airplane, find the speed of each airplane.

46. An airplane flies 1350 mi in the same amount of time it takes a second plane to fly 1200 mi. If the speed of the first airplane is 60 mph faster than the speed of the second airplane, find the speed of each airplane.

 Mix 'Em Up!

Solve each problem.

47. A supermarket has a special on chunk white albacore tuna—three cans for $2.85. How much will it cost to purchase seven cans of tuna?

48. Mandy and Bridgette together can decorate a wedding venue in 1 hr. Mandy decorates a similar size venue alone in 3 hr. How long will it take Bridgette to decorate the wedding venue alone?

Triangles *ABC* and *STU* are similar. Use the information in the diagram for Exercises 49 and 50. Note that figures are not drawn to scale.

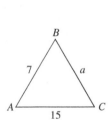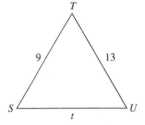

49. Find *a*.   **50.** Find *t*.

51. An order of six chicken nuggets from a popular fast food restaurant has 250 cal. Jillian is trying to cut back on her calories and only eats four of the nuggets. How many calories does she consume?

52. If 1 in. represents 120 mi on a map, how many inches represent the distance between Danbury, Connecticut, and Miami, Florida, which is 1338 mi?

53. An airplane flies 1320 mi in the same amount of time it takes a second plane to fly 1100 mi. If the speed of the first airplane is 80 mph more than the speed of the second airplane, find the speed of each airplane.

54. If 1 cm represents 210 km on a map, how many centimeters represent the distance between Buffalo, New York, and Des Moines, Iowa, which is 1375.5 km?

55. An object weighing 1 lb on Earth weighs 0.378 lb on Mercury. Find Frank's weight on Mercury if he weighs 195 lb on Earth.

56. An object weighing 1 lb on Earth weighs 1.125 lb on Neptune. Find Lucia's weight on Earth if she weighs 117 lb on Neptune.

57. Olivia walks 5 mi in the same time that Carol jogs 8 mi. If Carol jogs 2 mph faster than Olivia walks, then how fast does Carol jog?

58. A pump can fill a tank in 2 hr. A second pump can fill the same tank in 4 hr. How long will it take to fill the tank if both pumps are used?

 You Be the Teacher!

Determine if the student provides the correct equation to answer the given question. If not, provide the correct equation. Then solve the equation to answer the question.

59. A recipe that serves 20 people calls for 3 cups of cheese. How many cups of cheese are needed if the recipe is being increased to serve 25 people?

Jacqueline's work:

$$\frac{20}{3} = \frac{25}{x}$$

60. How many Hong Kong dollars are in 30 U.S. dollars? (Note: 1 U.S. dollar = 7.8125 Hong Kong dollars.)

Helen's work:

$$\frac{x}{30} = \frac{1}{7.8125}$$

61. Annette can clean an office building in 5 hr. It takes her work partner 9 hr to clean the same building. How long would it take for Annette and her partner to clean the building together?

Scott's work:

Let x = number of hours worked, $\frac{x}{5} + \frac{x}{9} = 14$

62. Use the similar triangles *ABC* and *STU* to find *a*. Note the figures are not drawn to scale.

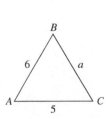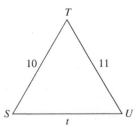

Suzanne's work:

$$\frac{5}{t} = \frac{a}{11}$$

GROUP ACTIVITY   /   Mathematics of Operating a Vehicle

Rational expressions can be used to determine the cost efficiency of our cars. Because the issue of oil and the economy are so important these days, we will examine the cost of gas per year for a particular car.

1. Select a traditional gas-powered vehicle to examine from http://www.fueleconomy.gov. Determine the average mpg for the selected vehicle.

2. Select a vehicle to examine from the list of the most fuel efficient cars at http://www.thesupercars.org/ford/most-fuel-efficient-cars/. Determine the average mpg for this vehicle.

3. Calculate the average number of miles that you drive in a week. Use this information to set up a proportion to calculate the number of miles that you drive in a year.

4. What is the cost per gallon of gas where you live?

5. Convert the mpg of each vehicle to gallons per mile. Write these values as ratios.

6. Calculate the cost/year of each vehicle using the following equation.

$$\frac{\text{cost}}{\text{year}} = \frac{\text{miles}}{\text{year}} \cdot \frac{\text{cost}}{\text{gallon}} \cdot \frac{\text{gallons}}{\text{mile}}$$

7. Use the cost/year that was found in step 6 to determine the cost/month and the cost/week.

8. How do the efficiencies of the two vehicles compare to one another?

9. What percent of your weekly or monthly expenses does the cost of gas represent?

Rational Expressions, Functions, and Equations

What's the big idea? Now that you have completed Chapter 7, you should understand that the quotient of polynomials is a rational expression. You should be able to simplify rational expressions and to perform operations on rational expressions. Factoring is a key skill that is used to reduce rational expressions, multiply and divide rational expressions, and to find the LCD. The LCD is used in adding and subtracting rational expressions, simplifying complex fractions, and in solving rational equations.

The Tools

Listed below are the key terms, skills, formulas, and properties you should know for this chapter.

The page reference is provided if you need additional help with the given topic. The Study Tips will assist in your preparation for an exam.

Study Tips

1. Learn all of the terms, formulas, and properties. Make flash cards and have someone quiz you.
2. Rework problems from the exercises and also the ones you worked in class. Work additional problems from the review exercises.
3. Review the summaries of key concepts.
4. Work the chapter test.
5. Be sure to review the online resources for additional study materials.

Terms

Complex fraction   542
Cross products   560
Domain of a rational function   506
Equivalent rational
  expressions   530

Least common denominator   528
Lowest terms   507
Opposites   508
Proportion   560
Rate   565

Ratio   560
Rational equation   552
Rational expression   504
Rational function   504
Similar triangles   563

Formulas and Properties

- Adding rational expressions   526
- Distance-rate-time formula   565
- Dividing rational
  expressions   519
- Dividing opposites   508

- Fundamental property of rational
  expressions   508
- Multiplying rational
  expressions   518

- Negative rational
  expressions   505
- Subtracting rational
  expressions   527
- Work equation   564

CHAPTER 7 / SUMMARY

How well do you know this chapter? Complete the following questions to find out. Take a look back at the section if you need help.

SECTION 7.1 Rational Functions and Simplifying Rational Expressions

1. A rational expression is a(n) _____ of two _____.

2. A rational function is a function of the form _____, where $P(x)$ and $Q(x)$ are _____ and $Q(x)$ is not zero.

3. To evaluate a rational expression or function, substitute the given _____ in place of the _____ and simplify.

4. When the numerator of a rational expression is zero, the value of the expression is _____. When the denominator of a rational expression is zero, the expression is _____.

5. To find the values that make a rational expression undefined, set the _____ equal to _____ and solve.

6. The domain of a rational function is the set of all real numbers excluding the values that make the expression _____.

7. To simplify a rational expression, _____ the numerator and denominator and divide out any common _____.

8. The expressions $x - y$ and _____ are opposites. The quotient of opposites is _____.

9. The expression $-\dfrac{A}{B}$ equivalent to _____ and _____.

SECTION 7.2 Multiplication and Division of Rational Expressions

10. To multiply rational expressions, _____ each numerator and denominator, _____ out common factors, and _____ the remaining factors in the numerator and denominator.

11. To divide rational expressions, convert the division to a product of the first expression and the _____ of the second expression. That is, $\dfrac{A}{B} \div \dfrac{C}{D} =$ _____.

12. To convert units, we use the fact that if a and b are equal quantities, then $\dfrac{a}{b} =$ __, and the fact that multiplying a quantity by 1 _____ change the quantity.

SECTION 7.3 Addition and Subtraction of Rational Expressions with Like Denominators and the Least Common Denominator

13. To add or subtract rational expressions, they must have the _____ denominators.

14. The _____ is the product of all different factors raised to the _____ power that the factor occurs in any one denominator.

15. When we multiply a rational expression by a form of 1, we obtain a(n) _____ rational expression.

SECTION 7.4 Addition and Subtraction of Rational Expressions with Unlike Denominators

16. To add or subtract rational expressions with different denominators, we must find the _____ and convert each fraction to a(n) _____ fraction with the _____ as its denominator. Then we can add or subtract the expressions.

SECTION 7.5 Simplifying Complex Fractions

17. A(n) _____ fraction is a fraction that contains fractions in its numerator and/or denominator.

18. There are two methods to simplify a complex fraction. We may either _____ _____ or multiply the numerator and denominator of the complex fraction by the _____.

SECTION 7.6 Solving Rational Equations

19. A(n) _____ equation is an equation that contains a rational expression.

20. To solve rational equation, multiply both sides by the _____ and then solve the resulting equation.

21. A solution that makes the LCD equal to _____ or the rational expression _____ is a(n) _____ solution and must be excluded from the solution set.

SECTION 7.7 Proportions and Applications of Rational Equations

22. A(n) _____ relates the measurement of two quantities and can be expressed with a(n) _____ or as a(n) _____.

23. A(n) _____ is an equation with two _____ set equal to each other.

24. _____ are triangles with the same _____ but different _____. Corresponding angles are _____ and the lengths of corresponding sides are _____.

25. Work applications involve finding the time it takes for two individuals to complete a task when _____ _____.

26. To solve motion problems, use the formula _____.

CHAPTER 7 / REVIEW EXERCISES

SECTION 7.1

Perform each operation. Round each answer to two decimal places when necessary. (See Objective 1.)

1. $\dfrac{2x^2 + x + 3}{x - 4}$, $x = 1, 4.01, 3.99, 100$

2. $\dfrac{x^2 - 5x + 1}{x - 2}$, $x = 0, 1.99, 2.01, 100$

3. $f(x) = \dfrac{5x + 2}{x + 2}$, $f(0), f(-1.99), f(-2.01), f(100)$

4. $f(x) = \dfrac{2x - 1}{x + 3}$, $f(0), f(-3.01), f(-2.99), f(100)$

Find the domain of each rational function. Write each answer in both set-builder and interval notation. (See Objective 2.)

5. $f(x) = \dfrac{2x - 11}{x - 10}$

6. $f(x) = \dfrac{5x + 12}{3x - 2}$

7. $f(x) = \dfrac{2x - 1}{x^2 - 36}$

8. $f(x) = \dfrac{2x - 3}{x^2 - 49}$

9. $f(x) = \dfrac{x - 6}{2x^2 + 4x}$

10. $f(x) = \dfrac{x + 5}{2x^2 - 6x}$

Simplify each rational expression. (See Objective 3.)

11. $\dfrac{6x^2 - 2x}{6x^2 + 70x - 24}$

12. $\dfrac{20x^2 + 25x}{12x^2 - 5x - 25}$

13. $\dfrac{3x^2 - 5x}{9x^2 - 25}$

14. $\dfrac{x^2 + 2x}{x^2 - 100}$

15. $\dfrac{6x^2 - 4x - 2}{27x^3 + 1}$

16. $\dfrac{7x^2 - 17x + 6}{x^3 - 8}$

17. $\dfrac{5x^2 + 37x + 14}{2x^3 + 14x^2 + 15x + 105}$

18. $\dfrac{3x^2 - x - 2}{3x^3 + 2x^2 + 9x + 6}$

19. $\dfrac{x^3 - 6x^2 - 4x + 24}{x^2 - 8x + 12}$

20. $\dfrac{x^3 + 8x^2 - 9x - 72}{x^2 + 5x - 24}$

Write two forms equivalent of each rational expression. (*See Objective 4.*)

21. $\dfrac{2x - 9}{-x - 10}$

22. $\dfrac{-3x - 1}{x - 4}$

23. $-\dfrac{-x^2 + 7x}{x^2 - 5}$

24. $-\dfrac{2x^2 - x}{-3x^2 + 2}$

Solve each problem. (*See Objective 6.*)

25. A high school class is organizing its 10-yr reunion. The venue, DJ, decorations, and other expenses total $3600 and the dinner will cost $30 per person. The total cost for x attendees is $C(x) = 30x + 3600$. The cost per person is the expression, $\dfrac{C(x)}{x} = \dfrac{30x + 3600}{x}$. What is the cost per person if 80 people attend? 100 attend? 120 attend?

26. A fitness club charges a one-time membership fee of $90 when a new member signs a contract. The new member is charged a monthly fee of $54, and the one-time fee is divided equally among the monthly payments. If x represents the number of months a new member commits to in a contract, the expression $\dfrac{90 + 54x}{x}$ represents the member's monthly payment. What is the member's monthly payment if the member signs a contract for 12 months? 18 months? 24 months?

SECTION 7.2

Perform each operation and simplify the answer. (*See Objectives 1 and 2.*)

27. $\dfrac{14x^3y^2}{5} \cdot \dfrac{20x}{21y^5}$

28. $\dfrac{24a^3}{25b} \div \dfrac{6a^{15}}{5b^8}$

29. $\dfrac{2x^2 + 6x}{6} \cdot \dfrac{3x + 3}{x^3}$

30. $\dfrac{6x + 4}{3x^2 - 13x - 10} \cdot \dfrac{3x - 15}{12x - 12}$

31. $\dfrac{x^2 - 3x - 10}{4x} \div \dfrac{x^2 - 25}{x^2}$

32. $\dfrac{2x^2 - 8}{x + 7} \div \dfrac{x^2 - 5x - 14}{x^2 - 49}$

33. $\dfrac{x^3 + 1}{x - 1} \div (x + 1)$

34. $\dfrac{2x - 3}{2x^2 + 3x - 9} \cdot (2x + 2)$

35. $\dfrac{x^2 + x - 12}{x^3 - 27} \div \dfrac{x^2 + 8x + 16}{2x^2 + 6x + 18}$

36. $\dfrac{x^2 - 1}{x^2 - 2x - 3} \div \dfrac{x^3 - 1}{2x - 6}$

37. $\dfrac{x^3 + 125}{x^2 + 3x - 10} \div \dfrac{x^3 - 5x^2 + 25x}{x^2 + 2x}$

38. $\dfrac{x^3 - 2x^2 - 7x + 14}{x^2 - 4} \cdot \dfrac{x^2 + x - 2}{x^3 - 7x}$

39. $\dfrac{x^4 - 81}{6x^2 - 18x} \div \dfrac{x^3 + 4x^2 + 9x + 36}{2x^2 + 8x}$

40. $\dfrac{x^3 - 1}{x^4 - 1} \div \dfrac{x^2 - 2x - 3}{x^3 - 3x^2 + x - 3}$

Use the given conversion equivalents to solve each problem. (*See Objective 3.*)

41. The distance between London and Manchester is 321.7 km. What is this distance in miles? (Note: 1 km ≈ 0.62 mi.)

42. The distance between Paris and Rome is 892.7 mi. What is this distance in kilometers? (Note: 1 km ≈ 0.62 mi.)

43. How many minutes are in 0.35 hr? (Note: 1 hr = 60 min.)

44. How many minutes are in 1.5 hr? (Note: 1 hr = 60 min.)

45. Convert 28.8 kg to pounds. (Note: 1 lb ≈ 0.45 kg.)

46. Convert 150 lb to kilograms. (Note: 1 lb ≈ 0.45 kg.)

SECTION 7.3

Find the least common denominator of each pair of rational expressions. (*See Objective 3.*)

47. $\dfrac{1}{x + 3}, \dfrac{5}{-6x - 2x^2}$

48. $\dfrac{x}{4x + 1}, \dfrac{2}{x - 3}$

49. $\dfrac{5}{x + 7}, \dfrac{x}{3x^2 + 19x - 14}$

50. $\dfrac{5}{6x + 4}, \dfrac{6}{9x^2 + 12x + 4}$

Find the numerator that makes each equation true. (*See Objective 4.*)

51. $\dfrac{7}{2x} = \dfrac{?}{10x^2}$

52. $\dfrac{x - 9}{x + 1} = \dfrac{?}{x^2 + x}$

53. $\dfrac{x}{2x + 5} = \dfrac{?}{2x^2 + 9x + 10}$

54. $\dfrac{x}{1 - 6x} = \dfrac{?}{18x^2 - 3x}$

SECTION 7.4

Add or subtract the rational expressions. Write each answer in lowest terms. (*See Objectives 1 and 2.*)

55. $\dfrac{5x}{x - 6} - \dfrac{9x}{6 - x}$

56. $\dfrac{x - 4}{x - 2} + \dfrac{x}{x + 2}$

57. $\dfrac{x}{5x + 2} + \dfrac{2x}{2x + 5}$

58. $\dfrac{7}{12x^3} - \dfrac{5}{18x^2}$

59. $\dfrac{3}{x - 12} + \dfrac{2}{x + 8}$

60. $\dfrac{2x - 5}{x + 1} - \dfrac{3x + 1}{x + 1}$

61. $\dfrac{5x-4}{x^2-6x-7}+\dfrac{3x}{x-7}$

62. $\dfrac{3x}{2x^2-5x-3}-\dfrac{x^2}{2x+1}$

63. $2+\dfrac{2x}{5x+3}$

64. $\dfrac{5}{x^2-6x+9}-\dfrac{4}{x^2-9}$

SECTION 7.5

Simplify each complex fraction using method 1 or method 2. (*See Objectives 1 and 2.*)

65. $\dfrac{\dfrac{18}{3x^2}}{\dfrac{42}{8x^4}}$

66. $\dfrac{\dfrac{7}{35y^6}}{\dfrac{32}{4y^3}}$

67. $\dfrac{6-\dfrac{a}{b}}{5+\dfrac{b}{a}}$

68. $\dfrac{\dfrac{15x}{3x+2}}{\dfrac{5x^2}{9x^2-4}}$

69. $\dfrac{\dfrac{5}{2x}+\dfrac{1}{2}}{35+7x}$

70. $\dfrac{\dfrac{24ab}{9a^2-6ab+b^2}}{\dfrac{18ab}{3a-b}}$

71. $\dfrac{\dfrac{7}{x}-\dfrac{3}{y}}{\dfrac{49}{x^2}-\dfrac{9}{y^2}}$

72. $\dfrac{\dfrac{3}{r}+\dfrac{1}{s}}{\dfrac{9}{r^2}-\dfrac{1}{s^2}}$

Evaluate each rational function for the given *x*-value. (*See Objective 3.*)

73. $f(x)=\dfrac{3x-2}{x-7},\ f\left(-\dfrac{2}{3}\right)$

74. $f(x)=\dfrac{5x-4}{10x+9},\ f\left(\dfrac{1}{2}\right)$

75. $f(x)=\dfrac{5x+1}{x},\ f\left(\dfrac{2}{5}\right)$

76. $f(x)=\dfrac{2x-7}{x},\ f\left(\dfrac{3}{4}\right)$

Find the slope of the line that passes through the given points. (*See Objective 3.*)

77. $\left(\dfrac{3}{10},\dfrac{1}{8}\right)$ and $\left(-\dfrac{5}{2},\dfrac{3}{10}\right)$

78. $\left(\dfrac{1}{10},-\dfrac{7}{5}\right)$ and $\left(-\dfrac{1}{10},\dfrac{3}{5}\right)$

79. $\left(-\dfrac{3}{2},-\dfrac{1}{4}\right)$ and $\left(\dfrac{7}{6},-\dfrac{1}{4}\right)$

80. $\left(\dfrac{5}{6},-\dfrac{7}{4}\right)$ and $\left(\dfrac{5}{6},\dfrac{5}{9}\right)$

SECTION 7.6

Solve each rational equation. (*See Objective 1.*)

81. $\dfrac{b}{5}=\dfrac{2b-3}{6}$

82. $\dfrac{y-7}{5}-\dfrac{14}{15}=\dfrac{2y}{3}$

83. $\dfrac{2x-5}{3}-1=\dfrac{2x}{3}$

84. $\dfrac{6x}{8x-1}=\dfrac{3}{4}$

85. $\dfrac{2}{3x}+5=9$

86. $\dfrac{3x^2-20}{x^2-8}=\dfrac{3x+2}{x}$

87. $5+\dfrac{18}{x}=\dfrac{8}{x^2}$

88. $\dfrac{2}{t-6}=\dfrac{3}{t-6}-5$

89. $\dfrac{1}{4x-3}=\dfrac{3}{6x-7}$

90. $\dfrac{3}{3x+1}-\dfrac{2x}{9x^2-1}=\dfrac{2}{3x-1}$

Solve each rational equation for the specified variable. (*See Objective 2.*)

91. $A=\dfrac{1}{2}h(y_0+2y_1+y_2)$ for y_1 (trapezoidal rule)

92. $A=\dfrac{1}{2}h(y_0+2y_1+y_2)$ for h (trapezoidal rule)

93. $\dfrac{1}{R}=\dfrac{1}{R_1}+\dfrac{1}{R_2}+\dfrac{1}{R_3}$ for R_2 (resistors in parallel)

94. $t=\dfrac{d}{x+y}$ for x

SECTION 7.7

Solve each problem. (*See Objectives 1–5.*)

95. The Weis supermarket has a special on Campbell's chunky roasted beef tips and vegetables soup—three cans for $5.97. How much will it cost to purchase eight cans of soup?

96. Working together, Warren and Diane together can mow a lawn in $\dfrac{3}{4}$ hr. Warren can mow a lawn of similar size alone in 1 hr. How long will it take Diane to mow a lawn of similar size alone?

Triangles *ABC* and *STU* are similar. Use the information in the diagrams for Exercises 83 and 84. Note the figures are not drawn to scale.

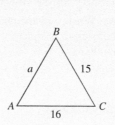
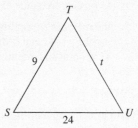

97. Find a.

98. Find t.

99. If 1 in. represents 120 mi on a map, how many inches represent the distance between Harrisburg, Pennsylvania, and Atlanta, Georgia, which is 720 mi?

100. If 1 cm represents 210 km on a map, how many centimeters represent the distance between Topeka, Kansas, and Sacramento, California, which is 2730 km?

101. An object weighing 1 lb on Earth weighs 0.378 lb on Mercury. Find Edmond's weight on Mercury if he weighs 165 lb on Earth.

102. An airplane flies 1122 mi in the same amount of time it takes a second plane to fly 946 mi. If the speed of the first airplane is 80 mph faster than the speed of the second airplane, find the speed of each airplane.

103. Renee walks 4 mi in the same time that Jane jogs 7.2 mi. If Jane jogs 2 mph faster than Renee walks, then how fast does Jane jog?

104. A pump fills a tank in 3 hr. A second pump fills the same tank in 2 hr. How long will it take to fill the tank if both pumps are used?

CHAPTER 7 TEST / RATIONAL EXPRESSIONS, FUNCTIONS, AND EQUATIONS

1. When simplified completely, $\dfrac{2t-1}{t+2} \cdot \dfrac{t^2+2t}{2t^2+t-1}$ is equivalent to

   a. $\dfrac{t}{t+1}$

   b. $\dfrac{t(2t-1)}{2t^2+t+1}$

   c. $\dfrac{t^2+2t}{(t+2)(t+1)}$

   d. $\dfrac{t}{t-1}$

2. When simplified completely, $\dfrac{x}{x^2-1} \div \dfrac{x-1}{x^2-3x-4}$ is equivalent to

   a. $\dfrac{x-1}{x}$

   b. $\dfrac{x(x-4)}{(x-1)^2}$

   c. $\dfrac{x}{x-1}$

   d. $\dfrac{x}{(x+1)^2(x-4)}$

3. When simplified completely, $\dfrac{x^2}{x+4} + \dfrac{4x}{x+4}$ is equivalent to

   a. x

   b. $\dfrac{4x^3}{x+4}$

   c. $\dfrac{5x}{x+4}$

   d. $\dfrac{x^2+4x}{x+4}$

4. When simplified completely, $\dfrac{5-2x}{x+7} - \dfrac{x+3}{x+7}$ is equivalent to

   a. $\dfrac{-3x+2}{0}$

   b. $\dfrac{-3x+8}{x+7}$

   c. $\dfrac{-x+8}{x+7}$

   d. $\dfrac{-3x+2}{x+7}$

5. The solution set for $\dfrac{6}{3x} - \dfrac{2}{x} = 4$ is

   a. $\{0\}$

   b. $\left\{\dfrac{1}{12}\right\}$

   c. $\{12\}$

   d. $\varnothing$

6. The solution set for $\dfrac{x}{x-1} = \dfrac{2x}{x+1}$ is

   a. $\{0, 3\}$

   b. $\{3\}$

   c. $\{0\}$

   d. $\varnothing$

7. $\dfrac{x-7}{(x-1)(x+3)}$ is undefined for the x value(s)

   a. 0

   b. $7, -3,$ and 1

   c. -1 and 3

   d. 1 and -3

8. Find the value of $\dfrac{2x^2-4x+5}{3x+1}$ for $x = -1$.

9. Write each expression in lowest terms.

   a. $\dfrac{5-3x}{9x^2-25}$ **b.** $\dfrac{2x^2-5x-7}{x^2+5x+4}$ **c.** $\dfrac{8a^3-27}{4a^2-9}$

10. What numerator makes the statement $\dfrac{6y}{5y+10} = \dfrac{?}{5y^2-20}$ true?

11. Find the LCD of each pair of rational expressions.

   a. $\dfrac{3}{8x^2}, \dfrac{5}{10x^3}$

   b. $\dfrac{4}{b}, -\dfrac{2}{b+5}$

   c. $\dfrac{x+5}{x^2+6x-7}, \dfrac{3x-1}{2x^2+3x-5}$

12. Perform each indicated operation.

   a. $\dfrac{4}{x-2} - \dfrac{6}{2-x}$

   b. $\dfrac{2a}{a+1} - \dfrac{5}{3a}$

   c. $\dfrac{3}{2x+8} - \dfrac{x}{x^2-16}$

13. Simplify each complex fraction.

   a. $\dfrac{\frac{12a^3}{5b^2}}{\frac{21a^4}{10b}}$

   b. $\dfrac{\frac{1}{a}-\frac{1}{b}}{\frac{1}{a}+\frac{1}{b}}$

14. Solve each equation.

   a. $\dfrac{5}{x} + \dfrac{3}{2} = \dfrac{x+1}{x}$

   b. $1 - \dfrac{4}{x} - \dfrac{21}{x^2} = 0$

   c. $\dfrac{x}{x+3} + \dfrac{6}{x+1} = \dfrac{6}{x^2+4x+3}$

15. Solve each problem.

   a. The average weight of a newborn baby is 3.2 kg. Given that 1 lb ≈ 0.45 kg, convert the average weight of a newborn baby to pounds.

   b. The triangles are similar. Use the information in the diagram to find the length of a. Note the figures are not drawn to scale.

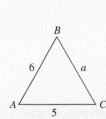 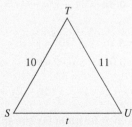

   c. Mary and Sue have a maid service. Mary cleans an average-size house in 3 hr when she works alone. Sue cleans an average-size house in 4 hr when she works alone. How long will it take the two of them to clean an average-size house if they work together?

   d. Todd runs 10 mi in the same time that Andrea walks 8 mi. Todd runs 1 mph faster than Andrea walks. How fast does Todd run?

16. Find the domain of $f(x) = \dfrac{x^2+4}{x^2-x}$.

CUMULATIVE REVIEW EXERCISES / CHAPTERS 1–7

Perform each operation and simplify each result. (*Section 1.2, Objectives 2–7*)

1. $\dfrac{8}{15} + \dfrac{7}{12}$ **2.** $3\dfrac{1}{4} - 1\dfrac{5}{6}$ **3.** $12\dfrac{3}{5} \cdot 4\dfrac{2}{7}$

Perform the indicated operation and simplify. (*Sections 1.4 and 1.5, Objectives 1 and 2*)

4. $2 - 12 + (-23) - (-10)$

5. $3.1 - (-4.5) + (-6.7) + 7.2$

6. $17 - \left\{ 8 - \left[4 - \left(1 - \sqrt{25 - 16} \right) \right] \right\}$

7. $-(-3)^3 + (-14) - |5 - 9|$

8. The lowest recorded temperature in Montana was $-70°F$ at Rogers Pass on January 20, 1954. The highest recorded temperature in Montana was $117°F$ at Medicine Lake on July 5, 1937. What is the difference between the highest and lowest temperatures? (*Section 1.5, Objective 3*)

Evaluate each expression for the given values of the variables. (*Section 1.6, Objective 4*)

9. $\sqrt{(x_1 - x_2)^2 + (y_1 - y_2)^2}$ for $x_1 = -5$, $x_2 = 7$, $y_1 = 13$, $y_2 = 8$

10. $\dfrac{|x + 1|}{2x - 4}$ for $x = -1, 0, 2$

Solve each equation. (*Section 2.2, Objectives 2 and 3; Sections 2.3 and 2.4, Objectives 1 and 2*)

11. $-3(x - 5) + 8 = 4(x - 1) - 22$

12. $5.2x - 6.3 + 1.7x = 9.4x - 12.3$

13. $\dfrac{1}{7}(28x - 14) + 12 = x + 5$

Solve each equation. (*Section 2.4, Objectives 1–4*)

14. $6(2a + 1) - 2(a + 4) = 5(2a + 1) - 7$

15. $\dfrac{2}{5}x + 3 = \dfrac{3}{10}x$

16. $0.015x + 0.024(950 - x) = 20.28$

17. $\dfrac{3}{2}(x + 1) = \dfrac{1}{3}(5x - 2) + \dfrac{5}{6}$

Solve each formula for the specified variable. (*Section 2.5, Objective 3*)

18. $S = 2\pi r^2 + 2\pi rh$ for h **19.** $C = 4l + 8w$ for l

20. $x + 5y = 10$ for y **21.** $0.06x + 0.03y = 4.2$ for x

22. A coat is on sale for $144. If this price is 60% off the original price, what is the original price of the coat? (*Section 2.6, Objective 1*)

Determine if the lines are parallel, perpendicular, or neither. (*Section 3.4, Objective 3*)

23. $y = \dfrac{3}{2}x + 5$ and $y = \dfrac{3}{2}x - 1$.

24. $y = 2$ and $x = 5$ **25.** $y = 0.25x$ and $y = -4x$

Write the equation of each line described. Express each answer in slope-intercept form and in standard form. (*Section 3.5, Objectives 2 and 4*)

26. $(-8, 5)$ perpendicular to $4x - 3y = 1$

27. $(-1, 4)$ parallel to $6x + y = 5$

28. $m =$ undefined, passes through $(3, -4)$

29. Ming has a new job as a History instructor at a college. Her starting salary is $43,500 with a yearly teaching load of 30 credit-hours. If she teaches beyond the regular load, she earns $1050 per credit-hour. (*Section 3.6, Objectives 4–6*)

   a. Write a linear function $f(x)$ that represents Ming's yearly income, where x is the number of extra credit-hours taught in a year.

   b. If Ming has a 39 credit-hour teaching load in one school year, what is her income?

   c. Find $f(8)$ and interpret the answer.

Solve each system of linear equations using substitution. (*Section 4.2, Objectives 1 and 2*)

30. $\begin{cases} 2x - 3y = 16 \\ x + 7y = -9 \end{cases}$ **31.** $\begin{cases} \dfrac{1}{2}x - y = -4 \\ \dfrac{1}{4}x + 3y = 5 \end{cases}$

Solve each system of linear equations using elimination. (*Section 4.3, Objectives 1–3*)

32. $\begin{cases} 4x - 3y = 78 \\ -5x + 2y = 40 \end{cases}$ **33.** $\begin{cases} \dfrac{2}{3}x - \dfrac{1}{4}y = -6 \\ \dfrac{5}{6}x + \dfrac{1}{2}y = -1 \end{cases}$

Use the appropriate exponent rules to simplify each expression. Write each answer with positive exponents. (*Section 5.1, Objectives 1–4; Section 5.2, Objectives 1 and 2*)

34. $(5^{-1}x^{-3}y^4)^2$ **35.** $\left(-\dfrac{4x^{-1}}{y^3} \right)^3$

36. $(4^{-1}x^6)^{-2}$ **37.** $\left(\dfrac{3x^2}{y^{-7}} \right)^{-3}$

Perform each operation. (*Section 5.3, Objectives 4 and 5*)

38. $(18 - 5x^2 - 29x) + (-12 + 25x^2 + 37x)$

39. Subtract $(4x^2 - 9xy + 6y^2)$ from $(-3x^2 + 12xy - 15y^2)$.

40. $(9x^2 + xy - 3y^2) - (-6x^2 + 11xy + 3y^2) + (-2x^2 - 18xy - 6y^2)$

41. Suppose the profit generated by selling x items is $-0.4x^2 + 86x + 198$ dollars and the total cost to manufacture x items is $23x + 420$ dollars. Find an expression that represents the company's revenue. Write the answer in standard form.

Perform the indicated operation. (*Section 5.4, Objectives 1 and 2; Section 5.5, Objectives 1–4*)

42. $3ab(a^2 - 6ab + 5b^2)$ **43.** $(7x - 3)(4x + 5)$

44. $(a + 2)^4$

45. $(2b - 1)^3$

46. $(12x - 7y)(12x + 7y)$

47. $(3r - 2s)(9r^2 + 6rs + 4s^2)$

Perform the indicated operation. (*Section 5.6, Objectives 1 and 2*)

48. Divide $27x^3 - 12x^2 + 24x + 6$ by $6x$

49. $\dfrac{6x^3 - 18x^2 + 21x - 10}{x + 3}$

Factor each trinomial completely. (*Section 6.1, Objectives 1 and 2*)

50. $x^3 + 3x^2 - 10x$

51. $10y^3 + 25y^2 - 60y$

52. $3x^2 - 4x + 8$

53. $-7a^4 - 91a^2 - 252$

54. $-9b^3 + 36b^2 - 36b$

Factor each trinomial completely. (*Section 6.2, Objectives 1 and 2*)

55. $3y^3 - 24y^2 + 48y$

56. $-12x^3 + 54x^2 + 30x$

57. $a^2 - 15ab + 36b^2$

58. $2x^2y + 34xy + 144y$

59. $x^2 + 7xy + 8y^2$

Use either trial and error or grouping to factor the trinomials completely. (*Section 6.3, Objectives 1–3*)

60. $35a^2 + 31a + 6$

61. $40ab^2 - 35ab - 90a$

62. $12x^2 - 2xy - 30y^2$

63. $9x^2 + 12xy + 4y^2$

64. $90a^2b^2c + 3abc - 3c$

65. $3p^2 + 4pq - 7q^2$

Factor completely. (*Section 6.4, Objectives 1–4*)

66. $98a^2 - 18b^2$

67. $5x^3 + 40y^3$

68. $y^3 - 1000z^3$

69. $a^2 + 16b^2$

70. $a^6 - 64$

71. $x^4 - 81y^4$

Solve each equation. (*Section 6.5, Objectives 1 and 2*)

72. $(x + 2)(x - 8) = 24$

73. $(5 - x)(4 + x) = 0$

74. $x^4 - 5x^2 + 4 = 0$

75. $x^3 - 3x^2 - 10x = 0$

Solve each problem. (*Section 6.6, Objectives 1–5*)

76. The product of two consecutive odd integers is 143. Find the integers.

77. When an express bus company charges $40 per seat from Boston to Princeton, they can sell 90 tickets. For every $1 decrease in price, 2 more tickets are sold. If x represents the number of price decreases, then $40 - x$ represents the ticket price and $90 + 2x$ represents the number of passengers. How many passengers are needed for the express bus company to collect $3300?

78. The height of a ball thrown upward with an initial velocity of 48 ft/sec from a height of 160 ft above the ground can be represented by $s = -16t^2 + 48t + 160$, where t is the time in seconds after the ball is thrown. How long will it take the ball to reach the ground?

Evaluate each expression for the given. (*Section 7.1, Objective 1*)

79. $\dfrac{x^2 - 3x}{x + 2}$ for $x = -2, 0, 3$ **80.** $\dfrac{4x}{x - 1}$ for $x = -1, 0, 1$

81. A company's revenue in dollars can be represented by $40q^2 + 600q$, where q is the number of items sold. If the average revenue per item is given by $\dfrac{40q^2 + 600q}{q}$, what is the average revenue per item when 10 items are sold? 20 items are sold? (*Section 7.1, Objective 1*)

82. The volume in cubic inches of an open box with height x in. is given by the expression $x^3 + 5x^2 - 24x$. If the area of the base of the box is given by $\dfrac{x^3 + 5x^2 - 24x}{x}$, what is the area of the base when the height is 5 in.? 6 in.? (*Section 7.1, Objective 1*)

Find the value(s) that make each rational function undefined. Then simplify each rational function. (*Section 7.1, Objectives 2 and 3*)

83. $f(x) = \dfrac{2x}{x^2 + 5x}$

84. $f(x) = \dfrac{x^2 - 1}{x^2 + 5x - 6}$

85. $f(x) = \dfrac{x^3 - 8}{x^2 + 3x - 10}$

86. $f(x) = \dfrac{3x^2 - 6x}{x^3 - 4x}$

Use the rule of negative rational expressions to write two equivalent forms of each rational expression. (*Section 7.1, Objective 4*)

87. $\dfrac{-3x + 6}{x + 4}$

88. $-\dfrac{1 - x}{2x - 3}$

Perform each operation and simplify the answer. (*Section 7.2, Objectives 1 and 2*)

89. $\dfrac{x^2 - 3x}{4} \cdot \dfrac{2x - 10}{x^3 - 9x}$

90. $\dfrac{x^2 + 4x - 12}{6x^2} \div \dfrac{x^2 - 4x + 4}{3x}$

91. $\dfrac{x^3 - 1}{x + 1} \div (x - 1)$

92. $\dfrac{x^3 + 4x - 7x^2 - 28}{x^2 - 49} \cdot \dfrac{x^2 + 10x + 21}{x^3 + 4x}$

93. $\dfrac{8x^3 + 4x^2}{2x^2 + 7x + 3} \div \dfrac{2x^2 + 10x}{x^2 + 8x + 15}$

94. $\dfrac{x^4 - 1}{x^2 + 2x + 1} \cdot \dfrac{x^2 - 4x - 5}{x^3 + x - 5x^2 - 5}$

Use the given conversion equivalents to answer each question. (*Section 7.2, Objective 3*)

95. The distance between London and Rylstone in the United Kingdom is 224 mi. What is this distance in inches on the map? (Note: 1 in. = 18 mi.)

96. How many minutes are in 0.8 hr? (Note: 1 hr = 60 min.)

Find the numerator that makes each equation true. (*Section 7.3, Objective 4*)

97. $\dfrac{1}{3a} = \dfrac{?}{18a^2b}$

98. $\dfrac{7}{xy} = \dfrac{?}{6x^2y}$

99. $\dfrac{x + 2}{x - 3} = \dfrac{?}{2x^2 - 6x}$

100. $\dfrac{x}{1 - 2x} = \dfrac{?}{6x^2 - 3x}$

Find the least common denominator of the rational expressions. (*Section 7.3, Objective 3*)

101. $\dfrac{x}{6x + 15}, \dfrac{1}{x}$

102. $\dfrac{5x}{2x + 3}, \dfrac{3}{x - 4}$

103. $\dfrac{5}{x - 3}, \dfrac{x}{x^2 + 4x - 21}$

104. $\dfrac{5x}{x^2 - 16}, \dfrac{7x}{5x^2 + 19x - 4}$

Add or subtract the rational expressions. Write each answer in lowest terms. (*Section 7.4, Objectives 1 and 2*)

105. $\dfrac{x}{x - 4} - \dfrac{2x}{x - 4}$

106. $\dfrac{4}{x - 6} + \dfrac{5}{x + 3}$

107. $\dfrac{4x}{x - 2} + \dfrac{3x - 6}{2 - x}$

108. $\dfrac{2x^2}{x^2 - 25} - \dfrac{2x}{x - 5}$

109. $4 + \dfrac{3x}{x - 1}$

110. $\dfrac{4x^2}{x^2 + 7x - 30} - \dfrac{2x + 1}{x + 10}$

111. $\dfrac{6}{x^2 - 9} - \dfrac{7}{x^2 - 6x + 9}$

112. $\dfrac{250}{x - 8} + \dfrac{100}{x + 3}$

Simplify each complex fraction using method 1 or method 2. (*Section 7.5, Objectives 1 and 2*)

113. $\dfrac{\dfrac{18}{3x^6}}{\dfrac{24}{5x^4}}$

114. $\dfrac{4 - \dfrac{a}{b}}{2 + \dfrac{a}{b}}$

115. $\dfrac{\dfrac{4y^2 - 9}{20y^2}}{\dfrac{2y - 3}{5y}}$

116. $\dfrac{\dfrac{1}{a} - \dfrac{2}{3}}{4a - 6}$

117. $\dfrac{\dfrac{8xy}{2x - y}}{\dfrac{6xy}{2x^2 + 5xy - 3y^2}}$

118. $\dfrac{\dfrac{4}{x} - \dfrac{3}{y}}{\dfrac{16}{x^2} - \dfrac{9}{y^2}}$

Evaluate each rational function for the given x-value. (*Section 7.4, Objective 3*)

119. $f(x) = \dfrac{3x - 2}{x + 6}, f\left(-\dfrac{8}{3}\right)$ **120.** $f(x) = \dfrac{3x + 10}{9x + 5}, f\left(-\dfrac{5}{3}\right)$

Find the slope of the line that passes through the given points. (*Section 7.4, Objective 3*)

121. $\left(\dfrac{4}{5}, \dfrac{3}{4}\right)$ and $\left(\dfrac{2}{3}, \dfrac{1}{2}\right)$

122. $\left(\dfrac{5}{2}, -\dfrac{5}{6}\right)$ and $\left(\dfrac{2}{5}, \dfrac{1}{3}\right)$

Simplify each rational expression. (*Section 7.4, Objective 3*)

123. $\dfrac{2^{-3} + 3^{-1}}{2^{-1}}$

124. $\dfrac{5^{-1} + 4^{-3}}{2^{-3}}$

Solve each rational equation. (*Section 7.6, Objective 1*)

125. $\dfrac{a}{3} = \dfrac{a - 5}{4}$

126. $\dfrac{3y - 2}{4} + \dfrac{5}{6} = \dfrac{2y}{3}$

127. $\dfrac{2a}{8a - 5} = \dfrac{1}{4}$

128. $\dfrac{x^2 + 3x + 2}{x^2 - 4} = \dfrac{x + 2}{x}$

129. $1 = \dfrac{2}{x} + \dfrac{35}{x^2}$

130. $6 - \dfrac{1}{t - 2} = \dfrac{5}{t - 2}$

131. $\dfrac{3}{7x - 5} = \dfrac{2}{4x - 3}$

132. $\dfrac{3}{2x - 3} - \dfrac{2x}{4x^2 - 9} = \dfrac{2}{2x + 3}$

Solve each rational equation for the specified variable. (*Section 7.6, Objective 2*)

133. $\dfrac{1}{R} = \dfrac{1}{R_1} + \dfrac{1}{R_2} + \dfrac{1}{R_3}$ for R_1 (resistors in parallel)

134. $m = \dfrac{y_2 - y_1}{x_2 - x_1}$ for x_1 (slope formula)

Solve each problem. (*Section 7.7, Objectives 1–5*)

135. A supermarket has a special on chunky soup—three cans for $3.87. How much will it cost to purchase seven cans of chunky soup?

136. Triangles *ABC* and *STU* are similar. Use the information in the diagram to find *a* and *t*. Note the figures are not drawn to scale.

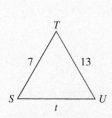 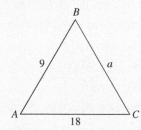

137. If 1 in. represents 120 mi on a map, how many inches represent the distance between Las Vegas, Nevada, and Denver, Colorado, which is 750 mi?

138. An airplane flies 1500 mi in the same amount of time it takes a second plane to fly 1350 mi. If the speed of the first airplane is 50 mph faster than the speed of the second airplane, find the speed of each airplane.

More on Functions and Graphs; Variation

Purpose

What drives and motivates you on this journey called life? Why are you pursuing a college education? Why do you make certain choices? The answers to these questions provide insight into your *purpose* in life. Your purpose is what determines how you set your goals. Without purpose, we go through life without a "map." So, it is important to take a few moments to consider why you do what you do. Purpose is not something to be met by a certain timeline; it is the underlying reason why we make the choices we make. It is easy to lose sight of that purpose when things get hectic and uncertain.

We not only have a purpose in life, but also in the courses we complete. Once you have this purpose, then everything you do in the class should be done to promote that purpose. I call this *studying with a purpose.*

? *Question For Thought:* What is your purpose in life? What is your purpose in this course? In what activities can you engage to promote your purpose?

Chapter Outline

Coming Up...

In Section 8.3, we will learn how to use piecewise-defined functions to represent federal income tax rates.

"What is success? I think it is a mixture of having a flair for the thing that you are doing; knowing that it is not enough, that you have got to have hard work and a certain sense of purpose."

—Margaret Thatcher (Former Prime Minister of the United Kingdom)

Copyright Lady Thatcher. Reprinted with permission from www.margaretthatcher.org, the website of the Margaret Thatcher Foundation.

SECTION 8.1 / The Domain and Range of Functions

In this chapter, we will study functions in more depth. We will learn how to find the domain and range of a function given various forms and will focus on linear, nonlinear, and piecewise-defined functions. Lastly, the concept of variation will be presented.

▶ **OBJECTIVES**

As a result of completing this section, you will be able to

1. Find the domain and range of a function given a set, a mapping, or a table.
2. Find the domain and range of a function given a graph.
3. Find the domain of a function given an equation.
4. Apply the concept of domain to real-world situations.
5. Troubleshoot common errors.

Many websites require us to enter a login name and a password to access their site. The correspondence of a login name and password form a function since each login name can have only one password. Have you ever entered your login name incorrectly and received the message that your login name is invalid? This is because the incorrect login name is not in the *domain* of this function.

In this section, we will define and determine the domain and range of a function.

Domain and Range of a Function

In Section 3.6, we learned how to find the domain and range of a relation. Because every function is a relation, the steps for finding the domain and range of a function are the same.

Objective 1 ▶

Find the domain and range of a function given a set, a mapping, or a table.

> **Definition: Domain and Range of a Function**
>
> The **domain** is the set of *x*-values in the given set of points.
> The **range** is the set of *y*-values in the given set of points.

Objective 1 Examples / **Find the domain and range of each function.**

1a. $\{(-2, 4), (-1, 1), (0, 0), (1, 1), (2, 4)\}$

1b. The table shows the five most populated states along with the party of their Governor in 2011. (Source: en.wikipedia.org)

State	California	Texas	New York	Florida	Illinois
Party	Democrat	Republican	Democrat	Republican	Democrat

Solutions

1a. The domain is the set of *x*-values and is $\{-2, -1, 0, 1, 2\}$.
The range is the set of *y*-values and is $\{0, 1, 4\}$.

1b. The domain is the set of *x*-values and is $\{$California, Texas, New York, Florida, Illinois$\}$.
The range is the set of *y*-values and is $\{$Democrat, Republican$\}$.

☑ **Student Check 1** Find the domain and range of each function.

a. $\{(-3, -27), (-2, -8), (-1, -1), (0, 0), (1, 1)\}$

b. The table shows the total number of degree-granting institutions in the United States in the fall of each year. (Source: National Center for Education Statistics)

Year	1869	1919	1969	1989	1999	2005	2006	2007
Number of institutions	563	1041	2525	3535	4084	4276	4314	4352

Reading a Graph to Find the Domain and Range

Objective 2 ▶

Find the domain and range of a function given a graph.

Recall that functions can also be represented by graphs. Graphs that pass the vertical line test represent functions. We can use a graph to determine the domain and range of a function. We use interval notation to represent the domain and range of a function when a graph is a smooth, continuous curve. If a graph consists of a finite set of points, we list the domain and range explicitly.

Between what values of *x* is the graph contained? ⇢

Procedure: Finding the Domain of a Function from Its Graph

Step 1: Determine the *x*-value of the leftmost point on the graph.
Step 2: Determine the *x*-value of the rightmost point on the graph.
Step 3: The domain is the interval between the *x*-values of the points found in steps 1 and 2.

Note: *If the graph extends indefinitely from the left to the right, the domain is all real numbers, denoted by* $(-\infty, \infty)$.

Between what values of *y* is the graph contained? ⇢

Procedure: Finding the Range of a Function from Its Graph

Step 1: Determine the *y*-value of the lowest point on the graph.
Step 2: Determine the *y*-value of the highest point on the graph.
Step 3: The range is the interval between the *y*-values of the points found in steps 1 and 2

Note: *If the graph extends indefinitely from the bottom to the top, the range is all real numbers, denoted by* $(-\infty, \infty)$.

Objective 2 Examples Find the domain and range of each function.

2a.

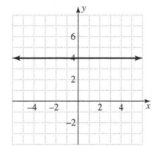

2b.

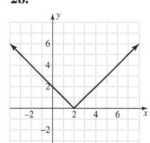

2c.

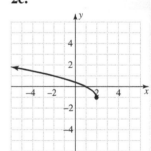

Solutions **2a.** The graph extends indefinitely to the left and right, so the domain is $(-\infty, \infty)$. The lowest and highest points on the graph are both on the line $y = 4$ since every point on the line has a *y*-value of 4. So, the range is $\{4\}$. (See Figure 8.1.1.)

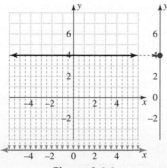

Figure 8.1.1

2b. The graph extends indefinitely to the left and right, so the domain is $(-\infty, \infty)$. The lowest point on the graph is $(2, 0)$. The graph doesn't have a highest point since it extends indefinitely upward. So, the range is $[0, \infty)$. (See Figure 8.1.2.)

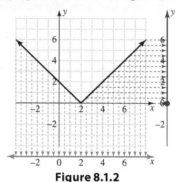

Figure 8.1.2

2c. The graph extends indefinitely to the left. The rightmost point is $(2, -1)$. So, the domain is $(-\infty, 2]$. The lowest point on the graph is $(2, -1)$. The graph extends indefinitely upward. So, the range is $[-1, \infty)$. (See Figure 8.1.3.)

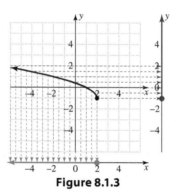

Figure 8.1.3

✔ **Student Check 2**   Find the domain and range of each function.

a.   **b.**   **c.**

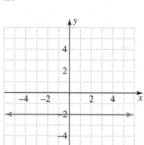   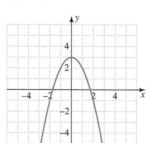   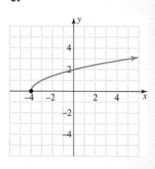

Using an Equation to Find Its Domain

Objective 3 ▶

Find the domain of a function given an equation.

When examining an equation to find its domain, the goal is to determine what values of x make the function *defined*. A function is defined for a value of x if the corresponding y-value is a real number. If there are any values of x that make the function *undefined*, they must be excluded from the domain of the function.

Definition: The **domain of a function** is the set of real numbers that makes the function defined.

Recall that linear functions and polynomial functions have a domain of $(-\infty, \infty)$, since there are no values of x for which these types of function are undefined. Recall from Section 7.1 that a rational function is undefined for the value(s) that make the denominator of the function zero. Another type of function that is undefined for some values is one that involves the square root of an algebraic expression. From Section 1.1, we know that $\sqrt{9} = 3$ since $3 \cdot 3 = 9$, but $\sqrt{-4}$ is not a real number since no number squared is -4.

So, if a function is a rational function or a function involving the square root of an algebraic expression, the function can have values that make it undefined. Examples of these types of functions are

$$f(x) = \frac{1}{x + 5} \qquad g(x) = \sqrt{x + 3}$$

The rational function $f(x)$ is undefined when the denominator is equal to zero, or for

$$x + 5 = 0$$
$$x = -5$$

So, the domain of $f(x)$ is all real numbers except $x = -5$. We write this as $(-\infty, -5) \cup (-5, \infty)$.

The function $g(x)$ is undefined when the expression inside the square root is negative since the square root of a negative number is not a real number. So, as long as the expression inside the square root is positive or zero, the function is defined. We must solve the following inequality to determine the domain of $g(x)$.

$$x + 3 \geq 0$$
$$x \geq -3$$

So, the domain of $g(x)$ is $[-3, \infty)$.

Procedure: Determining the Domain of a Function from its Equation

Step 1: If the function does not contain either a variable in the denominator or the square root of an expression, the domain is all real numbers.

Step 2: If the function contains a fraction with a variable in the denominator, we must set the denominator equal to zero to find the value(s) that make(s) the function undefined. The real numbers excluding these values is the domain of the function.

Step 3: If the function involves the square root of an expression, we must set the expression inside the square root greater than or equal to zero to find its domain.

Objective 3 Examples   **Find the domain of each function.**

3a. $f(x) = 3x - 5$ **3b.** $f(x) = \dfrac{2}{x - 3}$ **3c.** $f(x) = \sqrt{x - 4}$

Solutions   **3a.** The expression $3x - 5$ is defined for all real numbers. So, the domain is $(-\infty, \infty)$.

3b. The expression $\dfrac{2}{x - 3}$ is undefined for $x - 3 = 0$ or $x = 3$. The domain is all real numbers except $x = 3$. We write this set as $(-\infty, 3) \cup (3, \infty)$.

3c. The expression $\sqrt{x - 4}$ is defined for $x - 4 \geq 0$ or $x \geq 4$. We write this set as $[4, \infty)$.

✓ **Student Check 3**   Find the domain of each function.

    a. $f(x) = 2x + 1$ **b.** $f(x) = \dfrac{4}{x + 7}$ **c.** $f(x) = \sqrt{x - 2}$

Applications

Objective 4 ▶

Apply the concept of domain to real-world situations.

We have already established that functions occur in many real-world situations. When a student uses a student ID number to register for classes, the number is paired with the student name so that the student gets registered for the appropriate classes. The domain of this function is the set of student ID numbers and the range is the set of students assigned to an ID.

    Another example of a function involves e-mail. When a person e-mails someone, an e-mail address is paired with a particular recipient. The domain in this function is the e-mail address and the range is the recipient. The list could go on with these types of examples.

    In each of these situations, the domain and range of these functions has to be defined. Someone has to set up restrictions on what is going to be a reasonable value for the domain (input) and the range (output) of the function.

    In Example 4, we are in charge of determining an appropriate domain of a particular function. These examples come from problems that will be discussed later in the book.

Objective 4 Examples   **Determine an appropriate domain for each situation.**

4a. The revenue R for selling d donuts a month for a bakery is given by

$$R(d) = 0.65d - 4500$$

4b. The area of the rectangle can be represented by the function $A(x) = x(10 - 2x)$.

x ▭

    $10 - 2x$

Solutions   **4a.** It doesn't make sense to sell a negative number of donuts or a partial donut. Only whole donuts can be sold. So, the reasonable domain for this situation is $d = 0, 1, 2, 3, \ldots$, or the domain is the set of whole numbers, denoted by $\mathbb{W} = \{0, 1, 2, 3, \ldots\}$.

> **Note:** *It is not practical for a bakery to make infinitely many donuts in a month. There is an upper limit to the domain but we need more information to determine what that is.*

4b. Since the expressions x and $10 - 2x$ represent the lengths of the sides of a rectangle, it is appropriate that these values be positive. So, to find the domain we have to solve the compound inequality, $x > 0$ and $10 - 2x > 0$.

$x > 0$ and	$10 - 2x > 0$	Set each side greater than zero.
	$10 - 2x - 10 > 0 - 10$	Subtract 10 from each side.
	$-2x > -10$	Simplify.
	$\dfrac{-2x}{-2} < \dfrac{-10}{-2}$	Divide each side by -2 and reverse the inequality symbol.
	$x < 5$	Simplify.

The intersection of the sets $x > 0$ and $x < 5$ is the interval $(0, 5)$. So, the domain of $A(x)$ is $(0, 5)$.

✓ **Student Check 4**   Determine an appropriate domain for each situation.

   a. The monthly cost of a cell phone plan is $C(x) = 69.99 + 0.45x$, where x is the number of minutes over 4000.

   b. The area of the rectangle is $A(x) = x(18 - 2x)$.

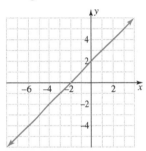

x

$18 - 2x$

Objective 5 ▶
Troubleshoot common errors.

Troubleshooting Common Errors

A common error associated with the domain and range of functions is shown.

Objective 5 Example | **A problem and an incorrect solution are given. Provide the correct solution and an explanation of the error.**

Find the domain and range of the given function.

Incorrect Solution	**Correct Solution and Explanation**
The domain is $[-2, \infty)$ and the range is $[2, \infty)$.	The graph of the function extends indefinitely to the left of $x = -2$. So, the domain is $(-\infty, \infty)$. The graph also extends indefinitely below $y = 2$. So, the range is $(-\infty, \infty)$.

ANSWERS TO STUDENT CHECKS

Student Check 1   **a.** Domain = {−3, −2, −1, 0, 1}
   Range = {−27, −8, −1, 0, 1}
   b. Domain = {1869, 1919, 1969, 1989, 1999, 2005, 2006, 2007}
   Range = {563, 1041, 2525, 3535, 4084, 4276, 4314, 4352}

Student Check 2   **a.** Domain = $(-\infty, \infty)$ and Range = {−2}
   b. Domain = $(-\infty, \infty)$ and Range = $(-\infty, 3]$
   c. Domain = $[-4, \infty)$ and Range = $[0, \infty)$

Student Check 3   **a.** Domain = $(-\infty, \infty)$ **b.** Domain = $(-\infty, -7) \cup (-7, \infty)$ **c.** Domain = $[2, \infty)$

Student Check 4   **a.** $\mathbb{W} = \{0, 1, 2, 3, \ldots\}$ **b.** $(0, 9)$

SUMMARY OF KEY CONCEPTS

1. The domain of a function is the set of x-values for which the function is defined. The range is the set of corresponding y-values. In a set, table, or mapping, the x- and y-values are listed explicitly. To write the domain and range, list the appropriate values in a set.

2. To find the domain and range from a graph, state the intervals of x and y, respectively, that contain the graph. The domain is found by reading the graph horizontally from left to right and the range is found by reading the graph vertically from the bottom up.

3. The domain of a function that is represented algebraically is the set of real numbers that make the function defined.

   a. If a function contains a fraction with variables in the denominator, exclude the values that make the denominator zero.

   b. If a function contains a square root, exclude the values that make the expression inside the square root negative.

4. When dealing with real-world applications, the domain consists not only of the values that make the function defined, but also the values that are reasonable. For example, a negative number may make the function defined but it may not make sense in the context of the application.

GRAPHING CALCULATOR SKILLS

A graphing calculator can assist us in determining values for the domain of a function that is represented algebraically. We input the function into the calculator and then either graph it or examine the table of ordered pairs to determine input values that make the function defined or undefined.

Example 1: Find the domain of $f(x) = \dfrac{2}{x - 3}$.

Solution:

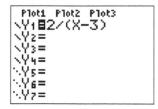

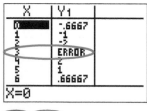

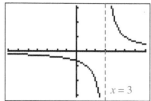

Note $x = 3$ makes the function undefined as evidenced by the error message in the Y_1-column of the table and also the gap in the graph. So, the graph and table show us that the domain is $(-\infty, 3) \cup (3, \infty)$.

Example 2: Find the domain of $f(x) = \sqrt{x - 4}$.

Solution:

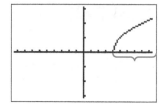

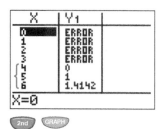

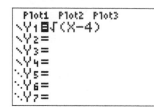

Note that the graph shows the leftmost x-value is 4 and it extends indefinitely to the right. The table shows an error message for all values less than 4, which means these numbers are not included in the domain. So, the table and graph show us that the domain is $[4, \infty)$.

SECTION 8.1   EXERCISE SET

Write About It!

Use complete sentences in your answer to each exercise.

1. Explain how to find the domain of a function that is given as a set of ordered pairs.

2. Explain how to find the domain of a function from its graph.

3. Explain how to find the range of a function that is given as a set of ordered pairs.

4. Explain how to find the range of a function from its graph.

5. Explain how to find the domain of an algebraic function with a fraction.

6. Explain how to find the domain of an algebraic function with a square root.

Practice Makes Perfect!

Find the domain and range of each function.
(See Objective 1.)

7. $\{(-3, 1), (-2, -4), (-1, 0), (0, 5)\}$

8. $\{(-1, 6), (0, 4), (1, -10), (2, -7)\}$

9. $\{(-10, 4), (0, 3), (2, 4), (5, -7)\}$

10. $\{(-9, -11), (-8, 11), (6, 11), (7, -9)\}$

11. The number of highway miles per gallon is a function of the vehicle model. (Source: http:www.fueleconomy.gov)

Vehicle Model	Highway mpg
2011 Ford Edge AWD	26
2011 Honda CR-V 4WD	27
2011 Subaru Forester AWD	27
2011 Toyota RAV4 4WD	27

12. The number of city miles per gallon is a function of the vehicle model. (Source: http:www.fueleconomy.gov)

Vehicle Model	City mpg
2011 Ford Edge AWD	18
2011 Honda CR-V 4WD	21
2011 Subaru Forester AWD	21
2011 Toyota RAV4 4WD	21

13. The total fall enrollment in U.S. degree-granting institutions is a function of the year. (Source: http://nces.ed.gov/)

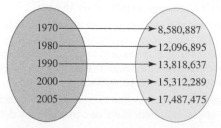

14. The percentage of male high school seniors who felt that having lots of money was "very important" is a function of the year. (Source: http://nces.ed.gov/)

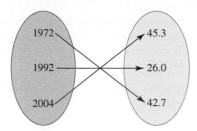

15. The percentage of female high school seniors who felt that having lots of money was "very important" is a function of the year. (Source: http://nces.ed.gov/)

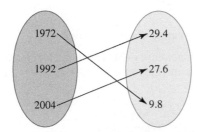

16. Percentage of male high school seniors who felt that having strong friendships was "very important" is a function of the year. (Source: http://nces.ed.gov/)

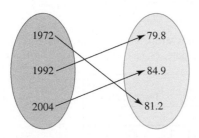

Find the domain and range of each function.
(*See Objective 2.*)

17.

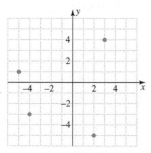

18.

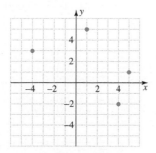

19.

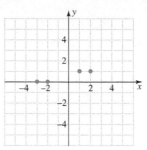

20.

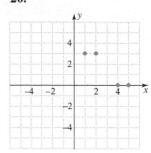

21.

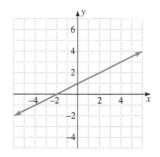

22.

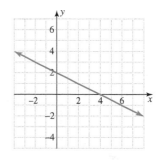

23.

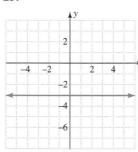

24.

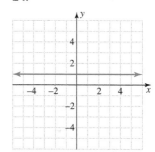

25.

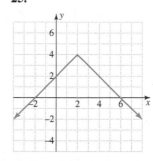

26.

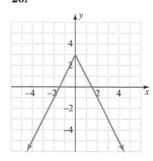

27.

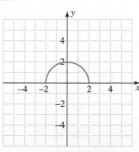

28.

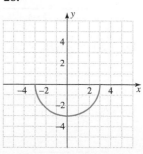

29.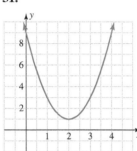

30.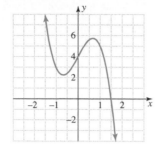

31.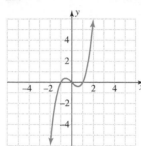

32.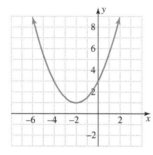

Find the domain of each function. (*See Objective 3.*)

33. $f(x) = 4x - 7$

34. $f(x) = 2x + 1$

35. $f(x) = \dfrac{3}{5}x + 2$

36. $f(x) = -\dfrac{9}{2}x - 5$

37. $g(x) = \dfrac{1}{x}$

38. $g(x) = \dfrac{2}{x}$

39. $h(x) = \dfrac{x}{x + 5}$

40. $h(x) = \dfrac{x}{x - 7}$

41. $f(x) = \sqrt{x + 6}$

42. $f(x) = \sqrt{5x + 10}$

43. $f(x) = \sqrt{6 - 8x}$

44. $f(x) = \sqrt{5 - 3x}$

Find the domain of each function described. (*See Objective 4.*)

45. $A(x) = x(8 - x)$

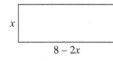

46. $A(x) = x(2 - x)$

47. $A(x) = (4 - 2x)(5 - x)$

48. $A(x) = (9 - 3x)(7 - x)$

49. $A(x) = \dfrac{1}{2}x(18 - 2x)$

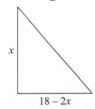

50. $A(x) = \dfrac{1}{2}x(10 - 2x)$

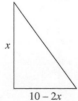

51. Vivienne owns a wedding cake business. The price she charges per wedding cake when x cakes are scheduled to be made in a month is given by the equation $p = 700 - 50x$.

52. Janelle sells gourmet praline cookies through the Internet. The price she charges per order depends on the dozens of cookies ordered. If x dozen cookies are ordered, the price of the order is $p = 150 - 6x$.

53. The total cost of joining a gym for x months can be modeled by the equation $C = 75 + 40x$.

54. The total cost of joining a professional organization for x years is $C = 75x$.

 Mix 'Em Up!

Find the domain and range of each function.

55. The average temperature in Stone Mountain, Georgia, is a function of the month. (Source: http://www.accuweather.com)

Month	Average Temperature
January 2011	39°F
February 2011	49°F
March 2011	55°F
April 2011	63°F

56. The average high temperature for Amarillo, Texas, is a function of the date. (Source: http://www.accuweather.com)

Date	Average High Temperature
September 4	85°F
September 11	83°F
September 18	81°F
September 25	79°F

57. $\{(0, -2), (4, -3), (13, 44), (20, -3)\}$.

58. $\{(-4, 21), (-2, 13), (0, -14), (4, 21)\}$.

59.

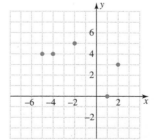

60.

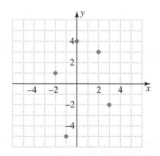

61.

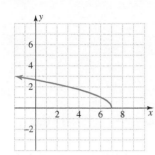

62.

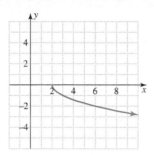

Find the domain of each function.

63. $h(x) = \sqrt{3x - 7}$.

64. $h(x) = \sqrt{7x + 8}$.

65. $g(x) = x + 4$

66. $g(x) = \dfrac{1}{2}x - 3$

67. $f(x) = \dfrac{x + 10}{x - 19}$

68. $f(x) = \dfrac{x - 8}{x + 12}$

69. $V(x) = x(8 - 2x)(14 - 2x)$

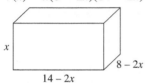

70. $V(x) = 3(24 - 3x)(6 - x)$

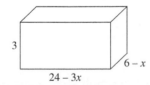

71. The cost that Boris pays for x months of lawn service is modeled by the equation $C = 125x$.

72. Olivia regularly pays for manicures. When she makes x visits to the nail salon in a month, Olivia's monthly manicure cost can be modeled by the equation $C = 15x$.

 You Be the Teacher!

Correct each student's errors, if any.

73. Find the domain of $f(x) = 4x - 1$.
Petra's work:
$$4x - 1 \geq 0$$
$$4x \geq 1$$
$$x \geq \dfrac{1}{4}$$
The domain is $\left[\dfrac{1}{4}, \infty\right)$.

74. Find the domain of $f(x) = \dfrac{x + 2}{x - 1}$.
Rob's work: The domain is $(-\infty, -2) \cup (-2, \infty)$.

75. Find the domain of the graph.

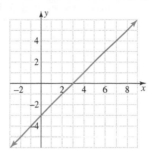

Stephanie's work: The domain is $[3, \infty)$.

 Calculate It!

Determine which of the given values are in the domain of the function and sketch a graph of each function.

76. $f(x) = \sqrt{9 - x^2}$; $x = -4$, $x = -3$, $x = 0$, $x = 2$, $x = 5$

77. $f(x) = \sqrt{x^2 - 9}$; $x = -4$, $x = -3$, $x = 0$, $x = 2$, $x = 5$

78. $g(x) = \sqrt{x^2 - 16}$; $x = -4$, $x = -3$, $x = 0$, $x = 2$, $x = 5$

79. $g(x) = \sqrt{16 - x^2}$; $x = -4$, $x = -3$, $x = 0$, $x = 2$, $x = 5$

Think About It!

Draw an example of a graph of a function that has the given domain and range. Answers vary.

80. Domain $(-\infty, \infty)$ and range $(-\infty, \infty)$

81. Domain $(-\infty, \infty)$ and range $[0, \infty)$

82. Domain $(-\infty, \infty)$ and range $(-\infty, 2]$

83. Domain $(-\infty, \infty)$ and range $\{3\}$

84. Domain $[0, \infty)$ and range $[0, \infty)$

85. Domain $[4, \infty)$ and range $[3, \infty)$

86. Domain $[-3, 3]$ and range $\{-2\}$

87. Domain $[-2, 2]$ and range $[-4, 4]$

| SECTION 8.2 | **Graphing and Writing Linear Functions** |

The status dropout rate represents the percentage of 16- to 24-yr-olds who are not enrolled in school and have not earned a high school diploma or GED. In 1980, the status dropout rate for Hispanics was 35.2%. In 2009, the status dropout rate for Hispanics was 17.6%. Write a linear function that represents the status dropout rate for Hispanics in terms of the years after 1980. (Source: http://nces.ed.gov/fastfacts/display.asp?id=16)

To address this problem, we will use the skills learned in Chapter 3 to write the equation of the line that satisfies the given information and then we will write the equation using function notation.

Graphing Linear Functions

Objective 1 ▶

Graph a linear function and determine its domain and range.

In Chapter 3, a linear function was defined as a function of the form $f(x) = mx + b$, where m and b are real numbers. Recall also that the notation $f(x)$ is another way to represent the variable y. So, we can think of the function $f(x) = mx + b$ as the equation $y = mx + b$. This is nothing more than the slope-intercept form of the line, where m is the slope and $(0, b)$ is the y-intercept.

To graph a linear function, we can use the skills from Chapter 3. We find two ordered pairs that satisfy the given function.

Procedure: Graphing a Linear Function of the Form $f(x) = mx + b$

Step 1: Replace $f(x)$ with y.
Step 2: Find at least two ordered pairs that satisfy the equation.
    a. Substitute two values of x in the function to find two points.
    b. Find the intercepts. The value $f(0)$ corresponds to the y-intercept and the solution of $f(x) = 0$ corresponds to the x-intercept.
    c. Use the slope and y-intercept to find two points.
Step 3: Plot the points and draw the line that contains them.

Objective 1 Examples   Graph each linear function. Label at least two points on the graph. Determine the domain and range of each function.

    1a. $f(x) = -\dfrac{1}{2}x + 4$ **1b.** $f(x) = 1$

Solutions   **1a.** $f(x) = -\dfrac{1}{2}x + 4$

         $y = -\dfrac{1}{2}x + 4$

x	$y = -\dfrac{1}{2}x + 4$	(x, y)
0	$y = -\dfrac{1}{2}(0) + 4 = 4$	$(0, 4)$
2	$y = -\dfrac{1}{2}(2) + 4 = 3$	$(2, 3)$
4	$y = -\dfrac{1}{2}(4) + 4 = 2$	$(4, 2)$

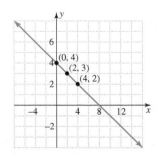

The domain of the function is $(-\infty, \infty)$ since the graph extends indefinitely to the left and right. The range of the function is $(-\infty, \infty)$ since the graph extends indefinitely from the bottom to the top.

1b. $f(x) = 1$

$y = 1$

The graph of this function is a horizontal line since it is of the form $y = k$, where k is a real number. So, the graph is a horizontal line through the point $(0, 1)$.

The domain of the function is $(-\infty, \infty)$ since the graph extends indefinitely to the left and right. The range is $\{1\}$ since the only y-value represented on the graph is $y = 1$.

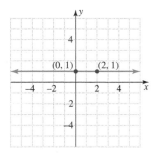

✔ Student Check 1   Graph each linear function. Label at least two points on the graph. Determine the domain and range of each function.

  a. $f(x) = 3x - 1$ **b.** $f(x) = -5$

Applications of Linear Functions

Objective 2 ▶

Use linear functions to solve application problems.

In the application problems in this section, a linear function that models a real-world situation is given. We will obtain information from the function and also construct a graph of the function.

Objective 2   Examples

The function $f(x) = -422x + 6050$ **approximates the annual crude oil production in thousands of barrels in the Florida field in** x **after 1997. (Source: http://www.eia .gov/dnav/pet/pet_crd_crpdn_adc_mbbl_a.htm)**

2a. Find $f(0)$ and write the corresponding ordered pair and interpret its meaning.

2b. Find the solution of $f(x) = 0$ and write the corresponding ordered pair and interpret its meaning.

2c. Use the information from parts 1 and 2 to graph the function.

Solutions   **2a.** $f(x) = -422x + 6050$

$f(0) = -422(0) + 6050$ Replace x with 0.

$f(0) = 6050$ Simplify.

The corresponding ordered pair is $(0, 6050)$, which is the y-intercept. It means that the annual crude oil production in the Florida field in 0 yr after 1997, or 1997, was 6050 thousand barrels.

2b. $f(x) = -422x + 6050$

$0 = -422x + 6050$ Replace $f(x)$ with 0.

$0 + 422x = -422x + 6050 + 422x$ Add 422x to each side.

$422x = 6050$ Simplify.

$\dfrac{422x}{422} = \dfrac{6050}{422}$ Divide each side by 422.

$x = 14.34$ Simplify.

The corresponding ordered pair is $(14.34, 0)$, which is the x-intercept. It means that approximately 14 yr after 1997, or 2011, the annual crude oil production in the Florida field was 0 thousand barrels.

2c. The graph of the function is found by plotting (0, 6050)and (14.34, 0). We need to graph in Quadrant I only since it doesn't make sense to have a negative crude oil production or negative years. Since the largest y-value is 6050, we will let each tic mark on the y-axis represent 1000 units.

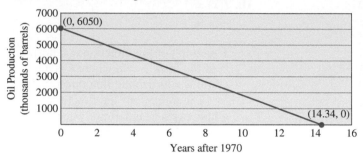

✓ **Student Check 2** The median annual income for men with a Master's degree is given by the function $f(x) = 1826x + 47,609$, where x is the number of years after 1990. (Source: www .infoplease.com)

    a. Find $f(0)$ and interpret its meaning.

    b. Find the median annual income for the years 2000 and 2015.

Writing Linear Functions

Objective 3 ▶

Write a linear function that satisfies given conditions.

Since a linear function is an equation of the form $y = mx + b$, we can apply the skills from Chapter 3 to write a linear function. Recall that to write the equation of a line, we must know the slope and a point on the line. This information can be given to us in several ways. We might be given

- the slope and y-intercept.
- the slope and a point other than the y-intercept.
- two points on the line.
- a point and a function whose graph is parallel or perpendicular to the graph of the unknown function.

Recall that if $f(x) = y$, then (x, y) is a point that lies on the graph of the function $f(x)$.

In each of the described situations, we will be able to use the point-slope form of a line to determine the linear function.

> **Property: Point-Slope Form**
> The line with slope m that contains the point (x_1, y_1) is given by
> $$y - y_1 = m(x - x_1)$$

If the slope is not given explicitly, we will have to use the slope formula if we know two points on the line.

> **Property: Slope Formula**
> The slope of the line through (x_1, y_1) and (x_2, y_2) is given by
> $$m = \frac{y_2 - y_1}{x_2 - x_1}$$

Also recall that parallel lines have the same slope and perpendicular lines have slopes that are negative reciprocals.

Procedure: Writing a Linear Function

Step 1: Determine the slope of the line.

Step 2: Use the slope and one point in the point-slope form. Note: If the given point is the y-intercept, then substitute the value of m and b into the slope-intercept form.

Step 3: Solve the equation for y.

Step 4: Replace y with $f(x)$.

Objective 3 Examples   **Write the linear function that satisfies the given information.**

3a. $m = \dfrac{2}{5}$ and $f(0) = -7$ **3b.** $m = \dfrac{2}{3}$ and $f(-5) = 1$ **3c.** $f(1) = -4$ and $f(3) = 6$

3d. $f(6) = -4$ and the graph of f is parallel and then perpendicular to $g(x) = \dfrac{1}{2}x + 5$.

Solutions   **3a.** The given information represents one point on the graph of the line, $(0, -7)$. This is the y-intercept. Since the slope and y-intercept are given, we know that $m = \dfrac{2}{5}$ and $b = -7$.

$$y = mx + b \qquad \text{Begin with the slope-intercept form.}$$

$$y = \frac{2}{5}x - 7 \qquad \text{Replace } m \text{ with } \frac{2}{5} \text{ and } b \text{ with } -7.$$

$$f(x) = \frac{2}{5}x - 7 \qquad \text{Replace } y \text{ with } f(x).$$

3b. The given information represents one point on the graph of the line, $(-5, 1)$.

$$y - y_1 = m(x - x_1) \qquad \text{Begin with the point-slope form.}$$

$$y - 1 = \frac{2}{3}[x - (-5)] \qquad \text{Replace } m \text{ with } \frac{2}{3}, x_1 \text{ with } -5, \text{ and } y_1 \text{ with } 1.$$

$$y - 1 = \frac{2}{3}(x + 5) \qquad \text{Simplify inside brackets.}$$

$$3(y - 1) = 3\left[\frac{2}{3}(x + 5)\right] \qquad \text{Multiply each side by 3.}$$

$$3y - 3 = 2(x + 5) \qquad \text{Apply the distributive property and simplify.}$$

$$3y - 3 = 2x + 10 \qquad \text{Apply the distributive property.}$$

$$3y - 3 + 3 = 2x + 10 + 3 \qquad \text{Add 3 to each side.}$$

$$3y = 2x + 13 \qquad \text{Simplify}$$

$$\frac{3y}{3} = \frac{2x + 13}{3} \qquad \text{Divide each side by 3.}$$

$$y = \frac{2}{3}x + \frac{13}{3} \qquad \text{Simplify.}$$

$$f(x) = \frac{2}{3}x + \frac{13}{3} \qquad \text{Replace } y \text{ with } f(x).$$

3c. The given information represents two points on the graph of the line, $(1, -4)$ and $(3, 6)$. We first find the slope of the line passing through these points.

$$m = \frac{y_2 - y_1}{x_2 - x_1} \qquad \text{Begin with the slope formula.}$$

$$m = \frac{6 - (-4)}{3 - 1} \qquad \text{Let } (x_1, y_1) = (1, -4) \text{ and } (x_2, y_2) = (3, 6).$$

$$m = \frac{10}{2} \qquad \text{Simplify the numerator and denominator.}$$

$$m = 5 \qquad \text{Simplify the quotient.}$$

Now we use the slope $m = 5$ with one of the given points, $(3, 6)$ in the point-slope form.

$$
\begin{array}{ll}
y - y_1 = m(x - x_1) & \text{Begin with the point-slope form.} \\
y - 6 = 5(x - 3) & \text{Replace } m \text{ with 5, } x_1 \text{ with 3, and } y_1 \text{ with 6.} \\
y - 6 = 5x - 15 & \text{Apply the distributive property on the right.} \\
y - 6 + 6 = 5x - 15 + 6 & \text{Add 6 to each side.} \\
y = 5x - 9 & \text{Simplify.} \\
f(x) = 5x - 9 & \text{Replace } y \text{ with } f(x).
\end{array}
$$

3d. The given information represents one point on the graph of the line, $(6, -4)$.

The slope of the given linear function, $g(x)$, is $m = \dfrac{1}{2}$.

Equation of the Parallel Line	Equation of the Perpendicular Line
Since parallel lines have the same slope, the slope of $f(x)$ is also $m = \dfrac{1}{2}$.	Since perpendicular lines have slopes that are negative reciprocals, the slope of $f(x)$ is $m = -2$.
We use the slope, $m = \dfrac{1}{2}$, with the given point $(6, -4)$ in the point-slope form.	We use the slope, $m = -2$, with the given point $(6, -4)$ in the point-slope form.
$$\begin{array}{c} y - y_1 = m(x - x_1) \\ y - (-4) = \dfrac{1}{2}(x - 6) \\ y + 4 = \dfrac{1}{2}x - 3 \\ y + 4 - 4 = \dfrac{1}{2}x - 3 - 4 \\ y = \dfrac{1}{2}x - 7 \\ f(x) = \dfrac{1}{2}x - 7 \end{array}$$	$$\begin{array}{c} y - y_1 = m(x - x_1) \\ y - (-4) = -2(x - 6) \\ y + 4 = -2x + 12 \\ y + 4 - 4 = -2x + 12 - 4 \\ y = -2x + 8 \\ f(x) = -2x + 8 \end{array}$$

✓ **Student Check 3**   Write the linear function that satisfies the given information.

a. $m = -6$ and $f(0) = -2$ **b.** $m = \dfrac{1}{5}; f(-9) = 2$ **c.** $f(-1) = 4$ and $f(1) = -8$

d. $f(1) = -6$ and is parallel and then perpendicular to $g(x) = -\dfrac{1}{3}x + 3$

Objective 4 ▶

Write a linear function that models a real-life situation.

Real-life Models of Linear Functions

We examine two types of application problems in this section. One type involves knowing an initial value and information about how that value changes. The other type involves knowing two data points that describe a particular situation.

Procedure: Writing Linear Functions That Model Real-Life Situations

Step 1: Identify the initial (or beginning) value of the given quantity. This value is b in the slope-intercept form of the line.

Step 2: Identify how the initial value changes. This information is used to find the slope, m.

Step 3: Write the linear function, $f(x) = mx + b$, by substituting the values of m and b in the function.

Procedure: Writing Linear Functions Given Two Data Points

Step 1: Identify two points given in the problem.

Step 2: Find the slope of the linear function containing these two points.

Step 3: Use one data point and the slope in the point-slope form to determine the function.

Objective 4 Examples | **Solve each problem.**

4a. David recently took a job as a sales representative for a pharmaceutical company. He gets a base salary of $30,000 a year plus 5% commission on his total annual sales.

 i. Write a linear function that represents his income as a function of x, his annual sales.

 ii. Use the function to determine David's income if his annual sales are $150,000.

 iii. Use the function to determine the sales he needs to earn an income of $70,000.

Solution **4a.** **i.** The base salary, $30,000, is the initial value b. This is David's income if no sales are made. The commission rate, $5\% = 0.05$, is the value m. The commission rate tells us that David gets an additional $0.05 for each dollar in sales. So, the linear function that represents his salary is $f(x) = 0.05x + 30,000$.

 ii. To find David's income if his sales are $150,000, replace x with 150,000.

$$f(x) = 0.05x + 30,000$$
$$f(150,000) = 0.05(150,000) + 30,000 \quad \text{Replace } x \text{ with 150,000.}$$
$$f(150,000) = 7500 + 30,000 \quad \text{Multiply 0.05 and 150,000.}$$
$$f(150,000) = \$37,500 \quad \text{Add.}$$

David's income will be $37,500 if his annual sales are $150,000.

 iii. If David's income is $50,000, his annual sales can be found by solving the equation $f(x) = 50,000$.

$$f(x) = 0.05x + 30,000$$
$$50,000 = 0.05x + 30,000 \quad \text{Replace } f(x) \text{ with 50,000.}$$
$$50,000 - 30,000 = 0.05x + 30,000 - 30,000 \quad \text{Subtract 30,000 from each side.}$$
$$20,000 = 0.05x \quad \text{Simplify.}$$
$$\frac{20,000}{0.05} = \frac{0.05x}{0.05} \quad \text{Divide each side by 0.05.}$$
$$400,000 = x \quad \text{Simplify.}$$

For David to have an income of $50,000, his annual sales must total $400,000.

4b. The *status dropout rate* represents the percentage of 16- to 24-yr-olds who are not enrolled in school and have not earned a high school diploma or GED. In 1980, the status dropout rate for Hispanics was 35.2%. In 2009, the status dropout rate for Hispanics was 17.6%. (Source: http://nces.ed.gov/fastfacts/display.asp?id=16)

    i. Write a linear function that represents the status dropout rate for Hispanics in terms of the number of years after 1980.

    ii. Use the function to estimate the status dropout rate for Hispanics in 2015.

    iii. When will the status dropout rate for Hispanics reach 10%? Round to the nearest year.

Solution   **4b.**   **i.** Since x is the number of years after 1980, the year 1980 corresponds to $x = 1980 - 1980 = 0$. The year 2009 corresponds to $x = 2009 - 1980 = 29$. The given information corresponds to the ordered pairs $(0, 35.2)$ and $(29, 17.6)$. To find the linear function, we first find the slope.

$$m = \frac{y_2 - y_1}{x_2 - x_1} = \frac{17.6 - 35.2}{29 - 0} = \frac{-17.6}{29} \approx -0.61$$

The status dropout rate in 1980 represents the initial value, so $b = 35.2$. So, the linear equation that represents the status dropout rate for Hispanics x years after 1980 is $y = -0.61x + 35.2$. In function notation, we write this as $f(x) = -0.61x + 35.2$.

    ii.

$f(x) = -0.61x + 35.2$	Begin with the model.
$f(35) = -0.61(35) + 35.2$	Replace x with 35 (2015−1980 = 35).
$f(35) = 13.85$	Simplify.

So, the status dropout rate for Hispanics in 2015 will be about 13.85%.

    iii. A status dropout rate of 10% corresponds to $f(x) = 10$.

$f(x) = -0.61x + 35.2$	Begin with the model.
$10 = -0.61x + 35.2$	Replace $f(x)$ with 10.
$10 - 35.2 = -0.61x + 35.2 - 35.2$	Subtract 35.2 from each side.
$-25.2 = -0.61x$	Simplify.
$\dfrac{-25.2}{-0.61} = \dfrac{-0.61x}{-0.61}$	Divide each side by −0.61.
$41 \approx x$	Approximate.

So, the status dropout rate for Hispanics will reach 10% in about 41 yr after 1980, or 2021.

✓ **Student Check 4**   Solve each problem.

  a. Joe bought a truck for $30,000. The value of the truck decreases by $3000 each year.

    i. Write a linear function that represents the value of the truck, where x is its age in years.

    ii. Use the function to find the value of the truck after 4 yr.

    iii. When will the value of the truck be $9000?

  b. In 1980, the average price of a movie ticket in the United States was $2.69. The average price of a movie ticket in 2010 was $7.89. (Source: http://www.natoonline.org/statistics.htm)

    i. Write a linear function that represents the average price of a movie ticket as a function of the years after 1980.

    ii. Use the function to predict the average price of a movie ticket in 2015.

    iii. In what year will the average price of a movie ticket be $10?

Objective 5 ▶

Troubleshoot common errors.

Troubleshooting Common Errors

Some common errors associated with graphing and writing linear functions are shown next.

Objective 5 Examples

Provide the correct solution and an explanation of the error.

5a. Graph $f(x) = \frac{1}{2}x$.

Incorrect Solution	Correct Solution and Explanation
The y-intercept of the line is $(0, 0)$ and the slope is $\frac{1}{2}$. So, the graph is 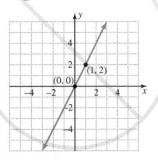	The slope is $m = \frac{1}{2}$, which means to move up 1 and right 2 from the origin. 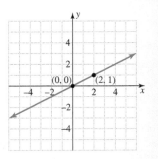

5b. Write the linear function that satisfies $f(3) = -2$ and $f(-1) = 6$.

Incorrect Solution	Correct Solution and Explanation
The points are $(3, -2)$ and $(-1, 6)$. So, the slope is $$m = \frac{y_2 - y_1}{x_2 - x_1}$$ $$m = \frac{-1 - 3}{6 - (-2)}$$ $$m = \frac{-4}{8}$$ $$m = -\frac{1}{2}$$ $$y - y_1 = m(x - x_1)$$ $$y - 6 = -\frac{1}{2}[x - (-1)]$$ $$y - 6 = -\frac{1}{2}(x + 1)$$ $$2y - 12 = -x + 1$$ $$2y - 12 + 12 = -x + 1 + 12$$ $$2y = -x + 13$$ $$y = -\frac{1}{2}x + \frac{13}{2}$$ $$f(x) = -\frac{1}{2}x + \frac{13}{2}$$	The error was made in substituting the values into the slope formula. The y-values are in the numerator and the x-values are in the denominator. $$m = \frac{y_2 - y_1}{x_2 - x_1}$$ $$m = \frac{6 - (-2)}{-1 - 3}$$ $$m = \frac{8}{-4}$$ $$m = -2$$ $$y - y_1 = m(x - x_1)$$ $$y - 6 = -2[x - (-1)]$$ $$y - 6 = -2(x + 1)$$ $$y - 6 = -2x - 2$$ $$y - 6 + 6 = -2x - 2 + 6$$ $$y = -2x + 4$$ $$f(x) = -2x + 4$$

ANSWERS TO STUDENT CHECKS

Student Check 1   a.

Domain = $(-\infty, \infty)$,
Range = $(-\infty, \infty)$

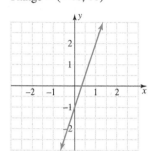

b.

Domain = $(-\infty, \infty)$,
Range = $\{-5\}$

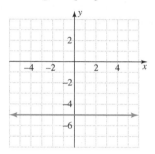

Student Check 2   a. $f(0) = 47,609$. In 0 years after 1990, or 1990, the median annual income for men with a Master's degree is $47,609.   **b.** The median annual income in 2000 was $65,869. The median annual income in 2015 will be $93,259.

Student Check 3   a. $f(x) = -6x - 2$

b. $f(x) = \frac{1}{5}x + \frac{19}{5}$ **c.** $f(x) = -6x - 2$

d. Parallel: $y = -\frac{1}{3}x - \frac{17}{3}$, Perpendicular: $y = 3x - 9$

Student Check 4   a. i. $f(x) = -3000x + 30,000$
ii. $18,000   **iii.** in 7 yr
b. i. $f(x) = 0.17x + 2.69$ **ii.** 8.64   **iii.** 2023

SUMMARY OF KEY CONCEPTS

1. A linear function, $f(x) = mx + b$, can be graphed by rewriting it in the form $y = mx + b$. We can plot points, find the x and y-intercepts, or use the slope and y-intercept to obtain the graph. The domain of all linear functions is all real numbers, $(-\infty, \infty)$. The range of all linear functions that are not horizontal lines is also all real numbers, $(-\infty, \infty)$. The range of a linear function of the form $f(x) = c$ is $\{c\}$.

2. If a linear function represents a real-life situation, we can use it to obtain specific information. It is important to understand what the input and output values represent.

3. To write a linear function, one of the following criteria must be given.

   a. The slope and y-intercept: substitute the given information into the slope-intercept form, $f(x) = mx + b$.

   b. A function value and a slope: substitute the given information into the point-slope form, solve for y, and replace y with $f(x)$.

   c. Two function values: write the function values as ordered pairs, find the slope using the slope formula, then substitute one ordered pair and the slope into the point-slope form, solve for y, and replace y with $f(x)$.

   d. A function value and a function whose graph is parallel or perpendicular: find the slope by using the relationship between parallel and perpendicular lines and follow the steps in part (c).

4. In application problems, we will either know the slope (how the values change) and y-intercept (initial value) from the problem or we will be given two points that enable us to write the function that represents the problem.

GRAPHING CALCULATOR SKILLS

As we have seen previously, we can use the graphing calculator to graph linear functions and to verify that a linear function satisfies given conditions.

Example: Verify that $f(x) = \frac{3}{2}x + 6$ is the function that satisfies $f(-4) = 0$ and $f(4) = 12$.

Solution: Enter the function in the equation editor.

Method 1: Use the TRACE feature to determine if the points $(-4, 0)$ and $(4, 12)$ lie on the line. Press TRACE, enter the x-value of the first point and press enter. Notice how this takes us to the point $(-4, 0)$ on the graph.

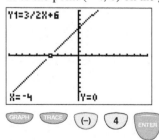

Now enter the *x*-value of the second point. Notice how the display is *x* = 4 and *y* = 12. The point is not shown since it is outside of the standard viewing window.

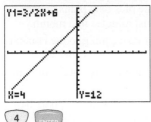

Method 2: Use the TABLE feature. Press 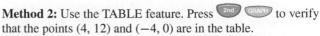 to verify that the points (4, 12) and (−4, 0) are in the table.

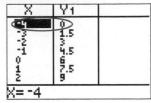

SECTION 8.2 / EXERCISE SET

Write About It!

Use complete sentences in your answers to the following exercises.

1. Describe how you know that a function is a linear function. Provide an example.

2. How can you graph a linear function?

3. How do you find the intercepts of linear function?

4. Do all linear functions have a domain of all real numbers? Why or why not? Do all linear functions have a range of all real numbers? Why or why not?

Determine if each statement is true or false. If a statement is false, explain why.

5. The linear function with slope 7 that passes through the point (5, 0) is $f(x) = 7x + 5$.

6. The linear function that passes through the point (2, 6) and perpendicular to $f(x) = 3x + 4$ is $f(x) = -\frac{1}{3}x + 6$.

7. If $f(0) = -7$, then the *x*-intercept is (−7, 0).

8. If $f(8) = 0$, then the *y*-intercept is (0, 8).

Practice Makes Perfect!

Graph each linear function. Label at least two points on the graph. State the domain and range of each function. (See Objective 1.)

9. $f(x) = 3x - 4$

10. $f(x) = 2x - 7$

11. $f(x) = \frac{4}{5}x + 3$

12. $f(x) = \frac{2}{3}x + 1$

13. $f(x) = -\frac{1}{3}x + 3$

14. $f(x) = -\frac{1}{4}x + 2$

15. $f(x) = \frac{3}{2}x - 4$

16. $f(x) = \frac{5}{3}x - 2$

17. $f(x) = -4$

18. $f(x) = -1.5$

19. $f(x) = 7$

20. $f(x) = 0$

21. $f(x) = 2x$

22. $f(x) = -6x$

Solve each problem. (See Objective 2.)

23. The percent of high school students who smoke cigarettes is on the decline. The percent of students who smoke can be modeled by the linear function $f(x) = -2.5x + 38$, where *x* is the number of years after 1997. (Source: National Cancer Institute)

    a. Find $f(0)$ and interpret its meaning in the context of the problem.

    b. Find $f(3)$ and $f(8)$ and interpret their meaning in the context of the problem.

    c. Solve $f(x) = 3$ and interpret its meaning in the context of the problem.

24. The percent of adults aged 20 − 74 who are at a healthy weight is modeled by the linear function $f(x) = -0.6x + 52$, where *x* is the number of years after 1972. (Source: National Center for Health Statistics)

    a. Find $f(0)$ and interpret its meaning in the context of the problem.

    b. Find $f(33)$ and $f(38)$ and interpret their meaning in the context of the problem.

    c. Solve $f(x) = 28$ and interpret its meaning in the context of the problem.

25. The owner of a limousine rental company purchases a stretch limousine for $110,000. For tax purposes, the owner uses straight-line depreciation for reporting the value of the limo. The depreciated value of the limo is given by the linear function $f(x) = -22,000x + 110,000$, where *x* is the age of the limo in years.

    a. Find $f(0)$ and interpret its meaning in the context of the problem.

    b. Find $f(3)$ and $f(4)$ and interpret their meaning in the context of the problem.

    c. Solve $f(x) = 0$ and interpret its meaning in the context of the problem.

26. The number of associate degrees conferred in all higher education institutions in the United States can be modeled by the linear function $f(x) = 22,740x + 559,900$, where *x*

is the number of years after 1998. (Source: http://nces.ed.gov/programs/coe/indicator_fsu.asp)

a. Find $f(0)$ and interpret its meaning in the context of the problem.

b. Find $f(10)$ and $f(20)$ and interpret their meaning in the context of the problem.

c. Solve $f(x) = 901{,}000$ and interpret its meaning in the context of the problem.

Write the linear function $f(x)$ that satisfies the given information. (*See Objective 3.*)

27. $m = 4; f(0) = 5$

28. $m = -2; f(0) = 1$

29. $m = \dfrac{1}{2}; f(0) = 7$

30. $m = -\dfrac{2}{3}; f(0) = -10$

31. $m = 8; f(9) = -3$

32. $m = -12; f(-5) = -1$

33. $m = \dfrac{4}{5}; f(-10) = 3$

34. $m = -\dfrac{1}{3}; f(6) = -5$

35. $f(-4) = 1$ and $f(2) = 7$

36. $f(-2) = -5$ and $f(2) = -1$

37. $f(-6) = -8$ and $f(3) = 4$

38. $f(3) = 10$ and $f(-4) = 6$

39. $f(-7) = 8$ and parallel to $g(x) = -2x + 4$

40. $f(5) = -2$ and parallel to $g(x) = 3x - 9$

41. $f(3) = -5$ and parallel to $g(x) = -\dfrac{4}{3}x - 4$

42. $f(-6) = -1$ and parallel to $g(x) = \dfrac{5}{6}x - 5$

43. $f(0) = 3$ and perpendicular to $g(x) = 7x - 3$

44. $f(0) = -4$ and perpendicular to $g(x) = -3x + 6$

45. $f(-5) = 2$ and perpendicular to $g(x) = 5x + 1$

46. $f(1) = -3$ and perpendicular to $g(x) = 6x + 5$

Solve each problem. (*See Objective 4.*)

47. Laura is a sales representative for an apparel company. She earns a base salary of \$40,000 plus a 5% commission of her annual sales.

a. Write a linear function that models Laura's income as a function of x, her annual sales.

b. Use the function to determine Laura's income if her annual sales are \$100,000.

c. Use the function to determine the sales she needs to earn an income of \$65,000.

48. Derek has been recently hired as a sales representative for a sports company. He earns a base salary of \$25,000 plus a 10% commission.

a. Write a linear function that models Derek's income as a function of x, his total sales.

b. Use the function to determine Derek's income if his total sales are \$600,000.

c. Use the function to determine the sales he needs to earn an income of \$100,000.

49. CellOne offers a voice plan for a cell phone with 1000 min for \$64.99 per month. A data plan can be

added for an additional \$35 each month. The phone service charges \$0.25 for each minute over 1000 min.

a. Write a linear function $f(x)$ that represents the total monthly cost for the voice plus data plans, where x is the number of additional minutes in a month.

b. Find $f(300)$ and interpret its meaning in the context of the problem.

c. Solve $f(x) = 124.99$ and interpret its meaning in the context of the problem.

50. The Verizon Wireless Nationwide Talk & Text 900 plan costs \$79.99 per month for 900 anytime minutes. Additional anytime minutes cost \$0.40 each. There is also a one-time \$35 activation fee. (Source: www.letstalk.com)

a. Write a linear function $f(x)$ that represents the first monthly cost of this cell phone plan as a function of x, the number of additional anytime minutes.

b. Find $f(100)$ and interpret its meaning in the context of the problem.

c. Solve $f(x) = 214.99$ and interpret its meaning in the context of the problem.

51. The graph shows the number of people (aged 7 and older) who played golf at least once during the year from 1999 to 2009. (Source: http://www.nsga.org)

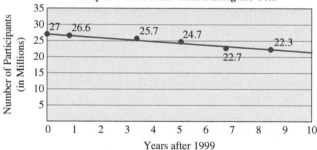

a. Use the points $(0, 27)$ and $(10, 22.3)$ to write a linear function that models the number of people who have played golf at least once during the year, x yr after 1999.

b. If this trend continues, how many people will play golf at least once during 2014?

52. The annual expenditures, in dollars, for raising a child born in 2010 is shown in the table. (Source: http://247wallst.com/2011/06/24/the-fifty-year-soaring-cost-to-raise-a-child/2/)

Age (in years)	Annual Expenditures
0	11,950
1	12,260
2	12,580
3	12,940
4	13,280
5	13,620
6	13,860

Age (in years)	Annual Expenditures
7	14,220
8	14,590
9	15,950
10	16,360
11	17,790
12	18,150
13	18,620

a. Use the first and last data points to write a linear function that models the annual expenditures for raising a child born in 2010 who is x yr old. Round slope to two decimal places.

b. Use the function to estimate the annual expenditures for raising a 14-yr-old who was born in 2010.

c. At what age are the annual expenditures $20,159.28?

53. Postsecondary teacher is one of the occupations with the largest job growth. In 2008, there were 1699.2 thousand postsecondary teachers. It is projected that the United States will need 1956.1 thousand postsecondary teachers in 2018. (Source: http://www.bls.gov/emp/ep_table_104.htm)

a. Write a linear function that represents the number, in thousands, of postsecondary teachers needed in the United States as a function of x, the number of years after 2008.

b. Use the function to estimate the number of postsecondary teachers needed in 2015.

54. One of the industries with the largest decline in employment in the United States is the postal service. In 2008, there were 747.5 thousand jobs in the postal service industry. It is projected that there will only be 650 thousand jobs in the postal service industry in 2018. (Source: http://www.bls.gov/opub/mlr/2009/11/art4full.pdf)

a. Write a linear function that models the number of jobs, in thousands, of postal service workers in the United States as a function of x yr after 2008.

b. Use the function to estimate when there will be 494 thousand jobs in the postal service industry in the United States.

 Mix 'Em Up!

Graph each linear function. Label at least two points on the graph. State the domain and range of each function.

55. $f(x) = -6x + 9$

56. $f(x) = 2x + 7$

57. $f(x) = \dfrac{4}{9}x$

58. $f(x) = -\dfrac{2}{3}x$

59. $f(x) = -5$

60. $f(x) = 4$

61. $f(x) = \dfrac{8}{3}x - 1$

62. $f(x) = -\dfrac{5}{2}x + 3$

Write the linear function $f(x)$ that satisfies the given information.

63. $m = 5; f(0) = -5$

64. $m = -1; f(0) = \dfrac{3}{4}$

65. $m = -\dfrac{1}{2}; f(-2) = 4$

66. $m = -\dfrac{2}{3}; f(3) = -3$

67. $f(4) = 1$ and $f(-4) = -1$

68. $f(-5) = 0$ and $f(6) = -10$

69. $f(-1) = 3$ and parallel to $g(x) = -6x + 3$

70. $f(-10) = 1$ and parallel to $g(x) = \dfrac{1}{4}x$

71. $f(0) = 5$ and perpendicular to $g(x) = \dfrac{3}{5}x - \dfrac{4}{5}$

72. $f(-5) = 4$ and perpendicular to $g(x) = -\dfrac{1}{2}x + 1$

73. $f(3) = 7$ and parallel to $g(x) = -1$

74. $f(-2) = 8$ and parallel to $g(x) = \dfrac{1}{2}$

Solve each problem.

75. The cost to rent a mid-size car for one day is $72 plus $0.42 per mile.

a. Write a linear function that models the cost to rent the car, where x is the number of miles driven.

b. Find $f(0)$ and interpret its meaning in the context of the problem.

c. Find $f(350)$ and interpret their meaning in the context of the problem.

d. Solve $f(x) = 101.4$ and interpret its meaning in the context of the problem.

76. A general guideline for taxi cab rates in Seattle, Washington, is an initial fee of $2.50 plus $2.00 per mile. (Source: http://www.taxigrab.com/Washington/seattle-taxi-service.html)

a. Write a linear function $f(x)$ that models the cost of a cab ride, where x is the number of miles traveled.

b. Find $f(5)$ and interpret its meaning in the context of the problem.

c. Solve $f(x) = 6.50$ and interpret its meaning in the context of the problem.

77. When a 160-lb person plays basketball for 1 hr, he will burn 584 Cal. When a 240-lb person plays basketball for 1 hr, he will burn 872 Cal. (Source: http://www.mayoclinic.com/health/exercise/SM00109)

a. Write a linear function $f(x)$ that models the calories burned by playing basketball for 1 hr, where x is the person's weight in pounds.

b. Find $f(200)$ and interpret its meaning in the context of the problem.

c. Solve $f(x) = 656$ and interpret its meaning in the context of the problem.

78. When a 160-lb person practices Tae Kwon Do for 1 hr, he will burn 730 Cal. When a 200-lb person practices Tae Kwon Do for 1 hr, he will burn 910 Cal. (Source: http://www.mayoclinic.com/health/exercise/SM00109)

a. Write a linear function $f(x)$ that models the calories burned by practicing Tae Kwon Do for 1 hr, where x is the person's weight in pounds.

b. Find $f(150)$ and interpret its meaning in the context of the problem.

c. Solve $f(x) = 1000$ and interpret its meaning in the context of the problem.

 You Be the Teacher!

Correct each student's errors, if any.

79. Find the linear function such that $f(8) = 3$ and the graph is parallel to $g(x) = \frac{4}{3}x - 2$.

Vivian's work:

The slope is $\frac{4}{3}$. Since it passes through the point $(8, 3)$, the linear function is $f(x) = \frac{4}{3}x + 3$.

80. Find the linear function such that $f(-3) = 4$ and $f(0) = \frac{1}{2}$.

William's work:

$$m = \frac{0 - (-3)}{\frac{1}{2} - 4} = \frac{3}{\frac{1}{2} - \frac{8}{2}} = \frac{3}{-\frac{7}{2}} = 3 \cdot -\frac{2}{7} = -\frac{6}{7}$$

So, $f(x) = -\frac{6}{7}x + \frac{1}{2}$.

 Calculate It!

Write the linear function that satisfies the given conditions and then use a graphing calculator to verify the solution.

81. $f(-3) = 1$ and $f(0) = 7$

82. $f(4) = 0$ and $f(-2) = 6$

83. $f(-6) = 2$ and $f(5) = -3$

84. $f(10) = 3$ and $f(-8) = 9$

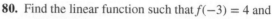

PIECE IT TOGETHER   **SECTIONS 8.1 AND 8.2**

Find the domain and range of each function. (*Section 8.1, Objectives 1 and 2*)

1.

x	y
−7	3
0	4
2	4
5	5

2.

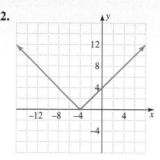

Find the domain of each function. (*Section 8.1, Objective 3*)

3. $f(x) = \dfrac{3x + 2}{x}$

4. $f(x) = \sqrt{4 - 9x}$

Graph each function and state its domain and range. (*Section 8.2, Objective 2*)

5. $f(x) = \frac{4}{5}x + 3$

6. $f(x) = 8$

Write a linear function that satisfies the conditions. (*Section 8.2, Objective 3*)

7. $m = \frac{9}{2}$; $f(0) = -5$

8. $m = 0$; $f(-1) = 4$

9. $f(-2) = 7$ and $f(3) = -2$

10. $f(-4) = 1$ and parallel to $g(x) = \frac{1}{2}x - 7$

11. $f(8) = -6$ and perpendicular to $g(x) = -\frac{1}{3}x + 2$

Solve each problem. (*Section 8.2, Objectives 2 and 4*)

12. An electrician charges \$50 for a house call plus \$45 per hour for his services.

a. Write a linear function $f(x)$ for the cost of a house call as a function of x hr.

b. Find $f(3)$ and interpret its meaning in the context of the problem.

c. Solve $f(x) = 117.50$ and interpret its meaning in the context of the problem.

13. The 2009 federal income tax for a person filing single with an income over \$33,950 but not over \$82,250 was \$4675 plus 25% of the amount over \$33,950.

a. Write a linear function $f(x)$ that represents the tax owed, where x is the amount of income over \$33,950.

b. Find $f(41,050)$ and interpret its meaning in the context of the problem.

c. Solve $f(x) = 8687.50$ and interpret its meaning in the context of the problem.

| SECTION 8.3 | Graphing Nonlinear and Piecewise-Defined Functions; Shifting and Reflecting Functions |

OBJECTIVES

As a result of completing this section, you will be able to

1. **Graph a nonlinear function and find its domain and range.**
2. **Shift graphs of functions vertically and/or horizontally.**
3. **Reflect graphs of functions.**
4. **Graph a piecewise-defined function.**
5. **Solve application problems.**
6. **Troubleshoot common errors.**

The parking fees for a hospital parking garage are shown in the table. Write a function that represents the fee as a function of the time parked.

To answer this question, we must know how to write and use *piecewise-defined functions.*

Time	Fee
0–30 min	Free
30 min–60 min	$3
60 min–90 min	$6
90 min–120 min	$9
120 min–24 hr	$12

Graphing Nonlinear Functions

In Section 3.1, we graphed equations in two variables that were not linear. To graph a nonlinear function encompasses the same skills; the only additional step is to replace $f(x)$ with y. Then we make a table of solutions, plot the solutions, and connect the points in the pattern formed. In Example 1, we graph three basic nonlinear functions—the squaring function, absolute value function, and the square root function. After we obtain their graph, we use them to determine the domain and range of each function.

Objective 1 ▶

Graph a nonlinear function and find its domain and range.

> **Note:** *A nonlinear function is a function that is not linear.*

Objective 1 Examples   **Graph each function and state its domain and range.**

1a. $f(x) = x^2$ **1b.** $f(x) = \sqrt{x}$

Solutions   **1a.**

x	$y = x^2$	(x, y)
-2	$y = (-2)^2 = 4$	$(-2, 4)$
-1	$y = (-1)^2 = 1$	$(-1, 1)$
0	$y = (0)^2 = 0$	$(0, 0)$
1	$y = (1)^2 = 1$	$(1, 1)$
2	$y = (2)^2 = 4$	$(2, 4)$

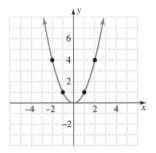

The graph of the squaring function, $f(x) = x^2$, is a *parabola*. The domain of the function is all real numbers, or $(-\infty, \infty)$, since the graph extends to the left and right indefinitely. The range of the graph is $[0, \infty)$, since the bottom-most point is $(0, 0)$ and the graph extends indefinitely upward.

1b. Recall from Section 8.1 that the domain of $f(x) = \sqrt{x}$ is $\{x \,|\, x \geq 0\}$. So, we can input only nonnegative real numbers for x.

x	$y = \sqrt{x}$	(x, y)
0	$y = \sqrt{0} = 0$	$(0, 0)$
1	$y = \sqrt{1} = 1$	$(1, 1)$
2	$y = \sqrt{2} \approx 1.41$	$(2, 1.41)$
3	$y = \sqrt{3} \approx 1.73$	$(3, 1.73)$
4	$y = \sqrt{4} = 2$	$(4, 2)$

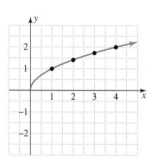

The graph of the square root function, $f(x) = \sqrt{x}$, is a *half parabola* that opens to the right. The domain of the function is $[0, \infty)$, since the leftmost point is $(0, 0)$ and the graph extends to the right indefinitely. The range of the graph is $[0, \infty)$, since the bottommost point is $(0, 0)$ and the graph extends indefinitely upward.

✔ **Student Check 1** Graph each function and state its domain and range.

a. $f(x) = x^3$ **b.** $f(x) = |x|$

Vertical and/or Horizontal Shifts of Graphs of Functions

Objective 2 ▶

Shift graphs of functions vertically and/or horizontally.

We now take the graphs of the basic functions from Objective 1 and perform operations on them, shifting their graphs vertically and/or horizontally. These types of changes are called **translations**. Example 2 will illustrate the following properties.

Property: Functions of the Form $y = f(x) + k$, for $k > 0$

- The graph of $y = f(x) + k$ is the graph of $f(x)$ shifted up k units.
- The graph of $y = f(x) - k$ is the graph of $f(x)$ shifted down k units.

Property: Functions of the Form $y = f(x - h)$, for $h > 0$

- The graph of $y = f(x - h)$ is the graph of $f(x)$ shifted right h units.
- The graph of $y = f(x + h)$ is the graph of $f(x)$ shifted left h units.

Objective 2 Examples	**Graph each function and state its domain and range.**

2a. $f(x) = x^2 + 3$ **2b.** $f(x) = \sqrt{x} - 2$

2c. $f(x) = |x + 3|$ **2d.** $f(x) = |x + 1| - 2$

Solutions   **2a.**

x	$y = x^2 + 3$	(x, y)
-2	$y = (-2)^2 + 3 = 7$	$(-2, 7)$
-1	$y = (-1)^2 + 3 = 4$	$(-1, 4)$
0	$y = (0)^2 + 3 = 3$	$(0, 3)$
1	$y = (1)^2 + 3 = 4$	$(1, 4)$
2	$y = (2)^2 + 3 = 7$	$(2, 7)$

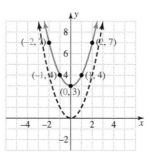

The domain of $f(x)$ is $(-\infty, \infty)$, and its range is $[3, \infty)$. The graph of $f(x) = x^2 + 3$ is the graph of $y = x^2$ shifted up 3 units.

2b.

x	$y = \sqrt{x} - 2$	(x, y)
0	$y = \sqrt{0} - 2 = -2$	$(0, -2)$
1	$y = \sqrt{1} - 2 = -1$	$(1, -1)$
2	$y = \sqrt{2} - 2 \approx -0.59$	$(2, -0.59)$
3	$y = \sqrt{3} - 2 \approx -0.27$	$(3, -0.27)$
4	$y = \sqrt{4} - 2 = 0$	$(4, 0)$

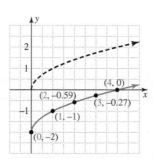

The domain of $f(x)$ is $[0, \infty)$, and its range is $[-2, \infty)$. The graph of $f(x) = \sqrt{x} - 2$ is the graph of $y = \sqrt{x}$ shifted down 2 units.

2c.

x	$y = \lvert x + 3 \rvert$	(x, y)
-4	$y = \lvert -4 + 3 \rvert = 1$	$(-4, 1)$
-3	$y = \lvert -3 + 3 \rvert = 0$	$(-3, 0)$
-2	$y = \lvert -2 + 3 \rvert = 1$	$(-2, 1)$
-1	$y = \lvert -1 + 3 \rvert = 2$	$(-1, 2)$
0	$y = \lvert 0 + 3 \rvert = 3$	$(0, 3)$
1	$y = \lvert 1 + 3 \rvert = 4$	$(1, 4)$
2	$y = \lvert 2 + 3 \rvert = 5$	$(2, 5)$

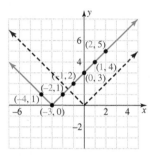

The domain of $f(x)$ is $(-\infty, \infty)$, and its range is $[0, \infty)$. The graph of $f(x) = \lvert x + 3 \rvert$ is the graph of $y = \lvert x \rvert$ shifted left 3 units.

2d.

x	$y = \lvert x + 1 \rvert - 2$	(x, y)
-3	$y = \lvert -3 + 1 \rvert - 2 = 0$	$(-3, 0)$
-2	$y = \lvert -2 + 1 \rvert - 2 = -1$	$(-2, -1)$
-1	$y = \lvert -1 + 1 \rvert - 2 = -2$	$(-1, -2)$
0	$y = \lvert 0 + 1 \rvert - 2 = -1$	$(0, -1)$
1	$y = \lvert 1 + 1 \rvert - 2 = 0$	$(1, 0)$
2	$y = \lvert 2 + 1 \rvert - 2 = 1$	$(2, 1)$

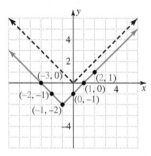

The domain of $f(x)$ is $(-\infty, \infty)$, and its range is $[-2, \infty)$. The graph of $f(x) = \lvert x + 1 \rvert - 2$ is the graph of $y = \lvert x \rvert$ shifted left 1 unit and down 2 units.

✔ **Student Check 2** Graph each function and state its domain and range.

  a. $f(x) = \lvert x \rvert - 4$ **b.** $f(x) = x^2 + 2$

  c. $f(x) = \sqrt{x - 5}$ **d.** $f(x) = (x - 1)^2 - 4$

Reflections

Objective 3 ▶

Reflect graphs of functions.

Another way that graphs of functions can be transformed is by reflecting them. When we see the reflection of a group of trees in a body of water, we see the mirror image of the trees in the water. This is the case with reflections of graphs. A **reflection** is the mirror image of the graph across one of its axes. We will focus only on reflections over the x-axis. Example 3 will illustrate the following property.

> **Property: Functions of the Form $y = -f(x)$**
>
> The graph of $y = -f(x)$ is the graph of $f(x)$ reflected over the x-axis.

Objective 3 Example Graph $f(x) = -\sqrt{x}$ and state its domain and range.

Solution

x	$y = -\sqrt{x}$	(x, y)
0	$y = -\sqrt{0} = 0$	$(0, 0)$
1	$y = -\sqrt{1} = -1$	$(1, -1)$
2	$y = -\sqrt{2} \approx -1.41$	$(2, -1.41)$
3	$y = -\sqrt{3} \approx -1.73$	$(3, -1.73)$
4	$y = -\sqrt{4} = 2$	$(4, -2)$

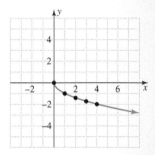

The domain of $f(x)$ is $[0, \infty)$, and its range is $(-\infty, 0]$. The graph of $f(x) = -\sqrt{x}$ is the graph of $y = \sqrt{x}$ reflected over the x-axis.

✓ Student Check 3 Graph $f(x) = -|x|$ and state its domain and range.

Piecewise-Defined Functions

Objective 4 ▶

Graph a piecewise-defined function.

Now that we have an understanding of functions and their graphs, we will turn our attention to a new kind of function, called a **piecewise-defined function**. These are functions that are comprised of two or more functions. Each function in the piecewise-defined function is defined on a specific part, or piece, of the domain. The function rule used to obtain the function value depends on the value of x.

Suppose that a function has the rule $y = x + 2$ for $x \le 1$ and $y = 2x$ for $x > 1$. We can write this as one function, which is a piecewise-defined function. It becomes

$$f(x) = \begin{cases} x + 2, & x \le 1 \\ 2x, & x > 1 \end{cases}$$

When we evaluate a function of this form, we must determine the piece we use when we input the assigned value. For instance,

$$f(0) = 0 + 2 = 0 \quad \text{since} \quad 0 \le 1.$$
$$f(3) = 2(3) = 6 \quad \text{since} \quad 3 > 1.$$
$$f(1) = 1 + 2 = 3 \quad \text{since} \quad 1 \le 1.$$

> **Procedure: Graphing a Piecewise-Defined Function**
>
> **Step 1:** Write each function in the piecewise-defined function in terms of y.
> **Step 2:** Create a table of solutions for each function using the stated domain. For the endpoint of the domain, determine if an open circle or closed circle is needed.
> > **a.** If the inequality symbol is $<$ or $>$, use an open circle.
> > **b.** If the inequality symbol is $\le$ or $\ge$, use a closed circle.
> **Step 3:** Graph each piece of the function on the domain for which it is defined.

Objective 4 Examples / **Graph each function.**

4a. $f(x) = \begin{cases} x - 1, \text{ if } x \le 1 \\ 2x, \quad\text{ if } x > 1 \end{cases}$ **4b.** $f(x) = \begin{cases} -3x + 3, \text{ if } x < 0 \\ \dfrac{1}{2}x - 2, \text{ if } x \ge 0 \end{cases}$

Solutions   **4a.** The piecewise-defined function consists of the pieces $y = x - 1$ and $y = 2x$. For the first piece, we must choose values of x that are less than or equal to 1. For the second piece, we must choose values of x that are greater than 1. Because this includes numbers, such as, 1.01, 1.005, and 1.00003, we must determine what happens at $x = 1$ for this piece as well, even though it will not be included on the graph.

$y = x - 1$ (for $x \le 1$)

x	$y = x - 1$	(x, y)
1	$y = 1 - 1 = 0$	• $(1, 0)$
0	$y = 0 - 1 = -1$	$(0, -1)$
-1	$y = -1 - 1 = -2$	$(-1, -2)$
-2	$y = -2 - 1 = -3$	$(-2, -3)$

$y = 2x$ (for $x > 1$)

x	$y = 2x$	(x, y)
1	$y = 2(1) = 2$	◦ $(1, 2)$
2	$y = 2(2) = 4$	$(2, 4)$
3	$y = 2(3) = 6$	$(3, 6)$

Each piece of the graph is a **ray**. A ray is a half line that extends indefinitely in one direction only. Notice that the point $(1, 0)$ is marked with a closed circle since this piece of the function, $y = x - 1$, is defined at $x = 1$. The point $(1, 2)$ is marked with an open circle since this piece of the function, $y = 2x$, is not defined for $x = 1$.

    The domain of the function is all real numbers, or $(-\infty, \infty)$, and its range is $(-\infty, 0] \cup (2, \infty)$.

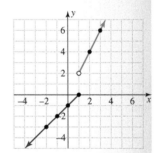

4b. This piecewise-defined function consists of the pieces $y = -3x + 3$ and $y = \dfrac{1}{2}x - 2$. For the first piece, we must choose values of x that are less than 0. Because this includes values such as $x = -0.5, -0.01$, and -0.0003, we must determine what happens at $x = 0$ for this piece, but it will not be included in the graph. For the second piece, we must include values for x that are greater than or equal to 0.

$y = -3x + 3$ (for $x < 0$)

x	$y = -3x + 3$	(x, y)
0	$y = -3(0) + 3 = 3$	◦ $(0, 3)$
-1	$y = -3(-1) + 3 = 6$	$(-1, 6)$
-2	$y = -3(-2) + 3 = 9$	$(-2, 9)$

$y = \dfrac{1}{2}x - 2$ (for $x \ge 0$)

x	$y = \dfrac{1}{2}x - 2$	(x, y)
0	$y = \dfrac{1}{2}(0) - 2 = -2$	• $(0, -2)$
1	$y = \dfrac{1}{2}(1) - 2 = -\dfrac{3}{2}$	$\left(1, -\dfrac{3}{2}\right)$
2	$y = \dfrac{1}{2}(2) - 2 = -1$	$(2, -1)$

Notice that the point $(0, 3)$ is marked with an open circle since this piece of the function, $y = -3x + 3$, is not defined at $x = 0$. The point $(0, -2)$ is marked with a closed circle since this piece of the function, $y = \dfrac{1}{2}x - 2$, is defined for $x = 0$. The domain of the function $(-\infty, \infty)$, and its range is $[-2, \infty)$.

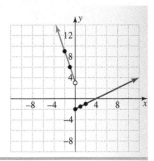

✓ **Student Check 4** Graph each function.

a. $f(x) = \begin{cases} -x + 4, & x < 2 \\ 2x - 4, & x \geq 2 \end{cases}$

b. $f(x) = \begin{cases} -\dfrac{2}{3}x + 2, & x \leq 0 \\ \dfrac{1}{4}x - 4, & x > 0 \end{cases}$

Applications

Objective 5 ▶

Solve application problems.

Piecewise-defined functions are used to model many real-life situations. For instance, the federal income tax brackets, fees owed for a parking garage, and the cost of a cell phone plan can each be modeled by a piecewise-defined function. Example 5 illustrates how we can write a piecewise-defined function to model a real-life situation and use it to answer questions.

Objective 5 Examples Solve each problem.

5a. The parking fees for a hospital parking garage are shown in the table. Write a function that represents the fee as a function of the time parked. Then graph the function.

Time	Fee
0–30 min	Free
30 min–60 min	$3
60 min–90 min	$6
90 min–120 min	$9
120 min–24 hr	$12

Solution

5a. Since the fee paid is dependent on the amount of time parked, we let x represent the time and y represent the fee in dollars. There are five different fees, so the piecewise-defined function will consist of five pieces. We write the function for each piece and then the corresponding values of x for which the piece is defined.

$$f(x) = \begin{cases} 0, & 0 \leq x \leq 30 \\ 3, & 30 < x \leq 60 \\ 6, & 60 < x \leq 90 \\ 9, & 90 < x \leq 120 \\ 12, & 120 < x \leq 1440 \end{cases}$$

The graph is as follows.

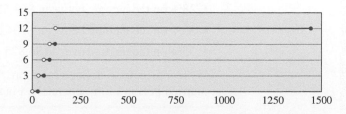

5b. Federal income tax rates are based on salary and filing status. The tax owed for a person filing single is as shown.

10% on income between $0 and $8375

15% on the income between $8375 and $34,000 plus $837.50

25% on the income between $34,000 and $82,400 plus $4,681.25

28% on the income between $82,400 and $171,850 plus $16,781.25

33% on the income between $171,850 and $373,650 plus $41,827.25

35% on the income over $373,650 plus $108,421.25

Write a piecewise-defined function, $f(x)$, that represents the tax owed for a person filing single. Find $f(100,000)$ and interpret its meaning.

Solution   **5b.** Since the tax owed is dependent on income, we let x represent the income and y represent the tax owed in dollars. There are six different categories of taxes owed, so the piecewise-defined function will have six pieces. We write the function for each piece and its corresponding values for which it is defined.

$$f(x) = \begin{cases} 0.10x, & 0 \le x \le 8375 \\ 0.15(x - 8375) + 837.50, & 8375 < x \le 34,000 \\ 0.25(x - 34,000) + 4681.25, & 34,000 < x \le 82,400 \\ 0.28(x - 82,400) + 16781.25, & 82,400 < x \le 171,850 \\ 0.33(x - 171,850) + 41827.25, & 171,850 < x \le 373,650 \\ 0.35(x - 373,650) + 108421.25, & x > 373,650 \end{cases}$$

To find $f(100,000)$, we use the fourth piece of the function.

$$f(100,000) = 0.28(100,000 - 82,400) + 16,781.25$$

$$f(100,000) = 4928 + 16,781.25$$

$$f(100,000) = 21,709.25$$

A single person who makes $100,000 owes $21,709.25 in federal income taxes for the year.

✓ Student Check 5   Solve each problem.

a. A cell phone company offers a monthly talk plan that costs $60 for 900 min and $0.40 for each additional minute over 900 min. Write a piecewise-defined function that represents the situation and graph the function.

b. The tax owed for a person that is married but filing separately is as shown.

- 10% on the income between $0 and $8,375

- 15% on the income between $8,375 and $34,000 plus $837.50

- 25% on the income between $34,000 and $68,650 plus $4,681.25

- 28% on the income between $68,650 and $104,625 plus $13,343.75

- 33% on the income between $104,625 and $186,825 plus $23,416.75

- 35% on the income over $186,825 plus $50,542.75

Write a piecewise-defined function, $f(x)$, that represents the tax owed for a married person filing separately. Find $f(75,000)$ and interpret its meaning.

Objective 6 ▶
Troubleshoot common errors.

Troubleshooting Common Errors

Some common errors associated with piecewise-defined functions are shown next.

Objective 6 Examples / **A problem and an incorrect solution are given. Provide the correct solution and an explanation of the error.**

6a. Graph $f(x) = \begin{cases} x - 1, & \text{if } x < 4 \\ -\dfrac{1}{4}x + 4, & \text{if } x \geq 4 \end{cases}$.

Incorrect Solution	Correct Solution and Explanation
	The error was made in not applying the stated restrictions on each piece of the function.

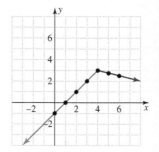

6b. For $f(x) = \begin{cases} 3x + 2, & \text{if } x < 0 \\ -\dfrac{2}{5}x + 6, & \text{if } x \geq 0 \end{cases}$, find $f(0)$

Incorrect Solution	Correct Solution and Explanation
$f(0) = 3(0) + 2 = 2$ and $f(0) = -\dfrac{2}{5}(0) + 6 = 6$ So, $f(0) = 2$ and 6.	Recall that a function has only one output value for each input value. The first piece is defined only for values less than zero. The second piece is defined at $x = 0$. $$f(0) = -\dfrac{2}{5}(0) + 6$$ $$f(0) = 6$$

ANSWERS TO STUDENT CHECKS

Student Check 1

a. Domain $= (-\infty, \infty)$,
  Range $= (-\infty, \infty)$

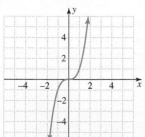

b. Domain $= (-\infty, \infty)$,
  Range $= [0, \infty)$

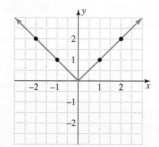

Student Check 2

a. Domain $= (-\infty, \infty)$,
  Range $= [-4, \infty)$

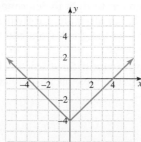

Student Check 2

b. Domain $= (-\infty, \infty)$,
Range $= [2, \infty)$

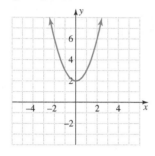

c. Domain $= [5, \infty)$,
Range $= [0, \infty)$

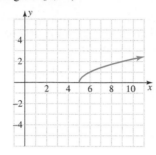

d. Domain $= (-\infty, \infty)$,
Range $= [-4, \infty)$

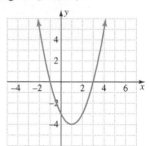

Student Check 3

Domain $= (-\infty, \infty)$,
Range $= (-\infty, 0]$

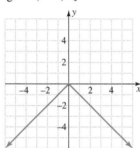

Student Check 4

a.

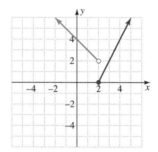

b.

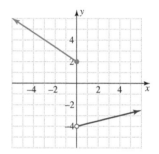

Student Check 5

a. $f(x) = \begin{cases} 60, & 0 \le x \le 900 \\ 60 + 0.40(x - 900), & x > 900 \end{cases}$

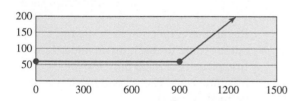

b.

$$f(x) = \begin{cases} 0.10x, & 0 \le x \le 8375 \\ 0.15(x - 8375) + 837.50, & 8375 < x \le 34{,}000 \\ 0.25(x - 34{,}000) + 4681.25, & 34{,}000 < x \le 68{,}650 \\ 0.28(x - 68{,}650) + 13{,}343.75, & 68{,}650 < x \le 104{,}625 \\ 0.33(x - 104{,}625) + 23{,}416.75, & 104{,}625 < x \le 186{,}825 \\ 0.35(x - 186{,}825) + 50{,}542.75, & x > 186{,}825 \end{cases}$$

$f(75{,}000) = 15{,}121.75$ A married person filing separately who earns $75,000 owes $15,121.75 in federal income tax.

SUMMARY OF KEY CONCEPTS

1. Nonlinear functions can be graphed by plotting points. It is important to graph enough points to determine the shape of the graph. The domain is the set of x-values for which the graph is contained. We read the graph from left to right to determine the domain. The range is the set of y-values for which the graph is contained. We read the graph from bottom to top to determine the range.

2. A piecewise-defined function is comprised of two or more functions that are defined on a specific piece of the

domain of the function. We can graph by plotting points on each piece. The points where the pieces change will have either an open or closed circle. If the inequality is $\le$ or $\ge$, the point will be closed. If the inequality is $<$ or $>$, the point will be open.

3. Piecewise-defined functions are used to model many real-life situations. To write a piecewise-defined function, we must determine how many different functions are involved and on which values these functions are defined.

GRAPHING CALCULATOR SKILLS

The graphing calculator can graph nonlinear functions and piecewise-defined functions.

Example 1: Graph $f(x) = |x + 2| - 3$.

Solution: To graph a nonlinear function, enter the function into the equation editor and graph.

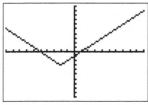

Example 2: Graph

$$f(x) = \begin{cases} 1 - 2x, & x < 3 \\ x + 1, & x \geq 3 \end{cases}.$$

Solution: To graph piecewise-defined functions, we must enter each piece of the function as a separate equation. We surround the function with parentheses and then divide it by the restriction on x,

also written in parentheses. So, $Y_1 = (1 - 2x)/(x < 3)$ and $Y_2 = (x + 1)/(x \geq 3)$. Recall that we use $\boxed{\text{2nd}}$ $\boxed{\text{MATH}}$ to select an inequality symbol.

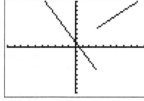

Note: The calculator will not show the open or closed circle. We can determine the appropriate endpoint using the trace key or the table.

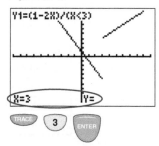

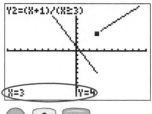

The first function has no corresponding y-value when $x = 3$, so we put an open circle on this endpoint. The second function has a y-value of 4 when $x = 3$, so we put a closed circle on this point.

When we examine the table, by pressing $\boxed{\text{2nd}}$ $\boxed{\text{GRAPH}}$, we find that there is an error in the Y_1 column when $x = 3$ but $Y_2 = 4$ when $x = 3$. So, the error message tells us to make the endpoint an open circle.

X	Y₁	Y₂
0	1	ERR:
1	-1	ERR:
2	-3	ERR:
3	ERR:	4
4	ERR:	5
5	ERR:	6
6	ERR:	7

X=0

SECTION 8.3   /   EXERCISE SET

✍ Write About It!

Use complete sentences in your answer to each exercise.

1. Explain the meaning of a piecewise-defined function.

2. Explain how to graph a piecewise-defined function.

3. Explain how the graphs of $f(x), f(x) + k$, and $f(x) - k$, for $k > 0$, relate to each other.

4. Explain how the graphs of $f(x), f(x - h)$, and $f(x + h)$, for $h > 0$, relate to each other.

5. Explain how the graphs of $f(x)$ and $-f(x)$ relate to each other.

6. Explain how the graph of $g(x) = -|x - 2| - 3$ relates to the graph of $f(x) = |x|$.

🎹 Practice Makes Perfect!

Graph each function and state its domain and range. (See Objectives 1–3.)

7. $f(x) = x^2 - 4$

8. $f(x) = -x^2 + 1$

9. $f(x) = (x - 3)^2$

10. $f(x) = (x + 1)^2$

11. $f(x) = |x - 2| + 1$

12. $f(x) = |x + 3| - 4$

13. $f(x) = \sqrt{x} + 2$

14. $f(x) = \sqrt{x} - 4$

15. $f(x) = \sqrt{x - 1}$

16. $f(x) = \sqrt{x + 2}$

Graph each piecewise-defined function. (See Objective 4.)

17. $f(x) = \begin{cases} -3x - 2, & x < 0 \\ x + 1, & x \geq 0 \end{cases}$

18. $f(x) = \begin{cases} \frac{1}{2}x + 1, & x < 0 \\ -\frac{1}{2}x + 2, & x \geq 0 \end{cases}$

19. $f(x) = \begin{cases} 1, & x \leq 0 \\ 2x - 1, & x > 0 \end{cases}$

20. $f(x) = \begin{cases} -2, & x \leq 0 \\ x + 1, & x > 0 \end{cases}$

21. $f(x) = \begin{cases} 3 - x, & x < 1 \\ 2, & x > 1 \end{cases}$

22. $f(x) = \begin{cases} 1 - 2x, & x < 1 \\ -1, & x > 1 \end{cases}$

23. $f(x) = \begin{cases} \frac{1}{2}x - 2, & x \leq -1 \\ -\frac{1}{2}x + 2, & x > -1 \end{cases}$

24. $f(x) = \begin{cases} -\frac{2}{3}x, & x \leq 1 \\ \frac{2}{3}x + 1, & x > 1 \end{cases}$

Solve each problem. (*See Objective 5.*)

25. The table shows the cost of mailing a first-class letter with the United States Postal Service in 2011. Write a function that represents the cost of mailing a letter as a function of x, the letter's weight. Then graph the function. (Source: http://www.usps.com/prices/first-class-mail-prices.htm)

Weight Not Over	Price
1 oz	$0.44
2 oz	$0.64
3 oz	$0.84
3.5 oz	$1.04

26. Parking rates for an economy lot for an international airport are shown in the table. Write a function that represents the parking rate as a function of x, the hours parked. Then graph the function.

Hours	Rate
1 or less	$2.00
2 or less	$4.00
3 or less	$5.00
5 or less	$6.00
5–24	$9.00

27. The recommended dosage (in milligrams) for a child's medicine is based on the weight of the child as shown in the table. Write a function that represents the dosage as a function of x, the child's weight in pounds. Then graph the function. (Source: http://assets.babycenter.com/ims/Content/ibuprofen_chart_pdf.pdf)

Weight	Dosage
Up to 17 lb	50 mg
Up to 23 lb	75 mg
Up to 35 lb	100 mg
Up to 47 lb	150 mg
Up to 59 lb	200 mg
Up to 71 lb	250 mg
Up to 95 lb	300 mg
Greater than 95 lb	400 mg

28. The NYU School of Business undergraduate tuition and fees for full-time students for the 2011 academic year is shown in the table. Write a function that represents tuition as a function of x, the number of units taken. Then graph the function. (Source: http://www.nyu.edu/bursar/tuition.fees/rates11/ugstern.html)

Units	Tuition and Fees
12–18	$19,903
19 or more	Additional $1231 per unit

29. The monthly charges for one county's residential water bill with a $\frac{3}{4}$-in. meter are calculated by adding a base charge, sewer charge, and a rate based on the number of thousands of gallons used as shown in the table. Write a function that represents the monthly charges as a function of x, the thousands of gallons used.

Thousands of Gallons	Base Charge	Sewer Charge	Residential Rate (per thousand gal)
Less than 8	$7.50	$5.89	$4.38
Less than 12	$7.50	$5.89	$6.57
12 or more	$7.50	$5.89	$8.76

30. A wireless family talk and text plan charges $90 per month for up to 700 min. Each additional minute costs $0.45 per minute. Write a function that represents the monthly cost as a function of x, the number of minutes. Then find $f(1000)$ and interpret its meaning in the context of the problem.

 Mix 'Em Up!

Graph each function. State the domain and range.

31. $f(x) = -\sqrt{x} + 2$

32. $f(x) = \sqrt{x} - 3$

33. $f(x) = -|x + 2| + 1$

34. $f(x) = |x - 1| - 3$

35. $f(x) = x^3 - 1$

36. $f(x) = -x^3 + 1$

37. $f(x) = \sqrt{x} + 3$

38. $f(x) = \sqrt{x - 3}$

39. $f(x) = (x - 1)^2 - 1$

40. $f(x) = -(x + 2)^2 + 4$

Graph each piecewise-defined function.

41. $f(x) = \begin{cases} -2, & x < 1 \\ x - 1, & x \geq 1 \end{cases}$

42. $f(x) = \begin{cases} -x + 2, & x \leq 1 \\ -1, & x > 1 \end{cases}$

43. $f(x) = \begin{cases} -x - 3, & x < -1 \\ -2, & x > -1 \end{cases}$

44. $f(x) = \begin{cases} -2, & x < -2 \\ x, & x > -2 \end{cases}$

45. $f(x) = \begin{cases} -2, & x \leq 0 \\ 3, & x > 0 \end{cases}$ **46.** $f(x) = \begin{cases} 2, & x < 0 \\ -1, & x \geq 0 \end{cases}$

47. $f(x) = \begin{cases} x, & x < 1 \\ -x, & x \geq 1 \end{cases}$ **48.** $f(x) = \begin{cases} -x, & x \leq -1 \\ x, & x > -1 \end{cases}$

Solve each problem.

49. Taxi fare in New York City is shown in the table. Write a function that represents the taxi fare as a function of x, the number of miles driven between 6 A.M. and 4 P.M. Then graph the function.

Miles	Rate
Initial fare	$2.50
Each $\frac{1}{5}$ mi	$0.40
Peak surcharge (after 4 P.M. until 8 P.M. Monday–Friday)	$1.00
Night surcharge (after 8 P.M. until 6 A.M.)	$0.50

50. Taxi fare in Los Angeles, California, is shown in the table. Write a function that represents the taxi fare as a function of x, the number of miles driven. Then graph the function.

Miles	Rate
Initial fare	$2.85
Each $\frac{1}{9}$ mi	$0.30

 You Be the Teacher!

Correct each student's errors, if any.

51. Graph $f(x) = \begin{cases} 2x, & x < 1 \\ -2x + 4, & x > 1 \end{cases}$.

Danielle's work:

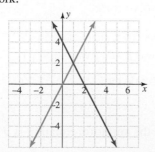

52. $f(x) = -|x + 3|$

Paul's work:

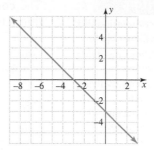

 Calculate It!

Use a graphing calculator to graph each function.

53. $f(x) = |x^2 - 4|$ **54.** $f(x) = |8 - x^3|$

55. $f(x) = \begin{cases} (x + 1)^2, & x < 1 \\ 5, & x \geq 1 \end{cases}$

56. $f(x) = \begin{cases} 3, & x \leq 1 \\ (x - 2)^2, & x > 1 \end{cases}$

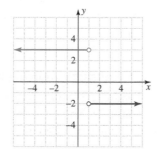 **Think About It!**

Given the graph of the piecewise-defined function, $y = f(x)$, state the domain and range. Write the equation for the function.

57. **58.**

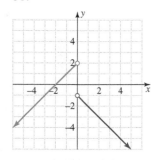

SECTION 8.4   Variation and Applications

The cost of a house in Florida is directly proportional to the size of the house. If a 2600-ft² house costs $195,000, then what is the cost of a 3500-ft² house? We will learn a formula for direct variation to solve this problem.

Direct Variation

Direct variation may be new terminology, but it is a relationship that we have used many times before. Here are some examples of direct variation and their mathematical models.

Objective 1 ▶

Solve problems involving direct variation.

Distance traveled is directly proportional to the time traveled.	$d = 60t$
Salary varies directly as hours worked.	$s = 15h$
The circumference of a circle varies directly as the diameter of the circle.	$C = \pi d$
The sales tax on a car is directly proportional to the price of the car.	$t = 0.07p$

Each of the preceding equations is written in a similar form: one quantity is equal to a constant times another quantity.

> **Definition: Direct Variation**
>
> We say *y varies directly as x*, or *y is directly proportional to x*, if
> $$y = kx$$
> for some nonzero constant k, called the **constant of variation** or **constant of proportionality**.

Another way to think of direct variation is that y is directly proportional to x if $\dfrac{y}{x} = k$, where k is a nonzero constant. In other words, the quotient of y and x is a constant.

> **Procedure: Solving Problems Involving Direct Variation**
>
> **Step 1:** If y varies directly as x, then $y = kx$.
> **Step 2:** Solve for the constant of variation by substituting the known values of x and y.

Objective 1 Examples   Use direct variation to solve each problem.

1a. Suppose y varies directly as x. If y is 10 when x is 5, find the constant of variation and the direct variation equation.

Solution   **1a.** Since y varies directly as x, we can write $y = kx$. To find the constant of variation, let $x = 5$ and $y = 10$ and solve for k.

$$y = kx \qquad \text{State the direct variation equation.}$$
$$10 = k(5) \qquad \text{Replace } x \text{ with 5 and } y \text{ with 10.}$$
$$\frac{10}{5} = k \qquad \text{Divide each side by 5.}$$
$$2 = k \qquad \text{Simplify.}$$

So, the constant of variation is 2 and the variation equation is $y = 2x$.

1b. The cost of a house in Florida is directly proportional to the size of the house. If a 2600-ft² house costs \$195,000, then what is the cost of a 3500-ft² house?

Solution   **1b.** Let c represent the cost of a house and s represent its size. Since c is directly proportional to s, we know that $c = ks$.

We use the fact that a 2600-ft² house costs \$195,000 to find the constant of variation.

$c = ks$	Write the variation equation.
$195{,}000 = k(2600)$	Replace c with 195,000 and k with 2600.
$\dfrac{195{,}000}{2600} = k$	Divide each side by 2600.
$75 = k$	Simplify.

So, the variation equation is $c = 75s$.

We use this equation to find the cost of a 3500-ft² house.

$c = 75s$	Write the variation equation.
$c = 75(3500)$	Replace s with 3500.
$c = 262{,}500$	Simplify.

So, the cost of a 3500-ft² house is \$262,500.

1c. Determine if each table shows that y varies directly as x. If the table represents direct variation, find the variation equation.

Table 1

x	y
3	2
6	4
9	6
12	8

Table 2

x	y
1	5
2	10
4	15
8	20

Solution   **1c.** For y to be directly proportional to x, $\dfrac{y}{x}$ must be constant.

Table 1			Table 2		
x	y	$\dfrac{y}{x}$	x	y	$\dfrac{y}{x}$
3	2	$\dfrac{2}{3}$	1	5	$\dfrac{5}{1} = 5$
6	4	$\dfrac{4}{6} = \dfrac{2}{3}$	2	10	$\dfrac{10}{2} = 5$
9	6	$\dfrac{6}{9} = \dfrac{2}{3}$	4	15	$\dfrac{15}{4}$
12	8	$\dfrac{8}{12} = \dfrac{2}{3}$	8	20	$\dfrac{20}{8} = \dfrac{5}{2}$

In Table 1, we see that for each ordered pair, $\dfrac{y}{x} = \dfrac{2}{3}$, so y varies directly as x. So, the variation equation is $y = \dfrac{2}{3}x$.

In Table 2, the first two entries have $\dfrac{y}{x} = 5$ but the last two entries do not satisfy this relationship. Therefore, Table 2 is not an example of direct variation.

1d. Which graph shows y varies directly as x? If the graph represents direct variation, find the variation equation.

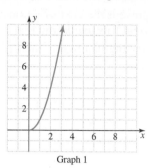

Graph 1

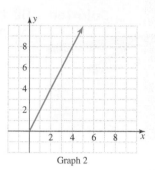

Graph 2

Solution **1d.** Direct variation is an equation of the form $y = kx$. The graph of this type of equation is a straight line with slope $m = k$ and y-intercept $(0, 0)$. In other words, it is a straight line through the origin.

Graph 2 is a line through the origin and so y is directly proportional to x.

To find the constant of proportionality, we need to find $\dfrac{y}{x}$ for points on the line. The points $(1, 2)$, $(2, 4)$, $(3, 6)$, ..., are on the line, so the constant of proportionality is $\dfrac{6}{3} = \dfrac{4}{2} = \dfrac{2}{1} = 2$. Therefore, the variation equation is $y = 2x$.

✓ **Student Check 1** Use direct variation to solve each problem.

a. Suppose y varies directly as x. If $y = -4$ when $x = 1$, find the constant of variation and the variation equation.

b. Hooke's law of elasticity states that the distance a spring stretches vertically is directly proportional to the weight on the end of the spring. If a 10-lb weight attached to the spring stretches the spring 2 in., find the distance that a 35-lb weight stretches the spring.

c. Determine if each table shows that y varies directly as x. If the table represents direct variation, find the variation equation.

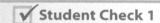

Table 1

x	y
1	2
2	4
3	8
4	16

Table 2

x	y
1	-2
2	-4
4	-8
8	-16

d. Which graph shows y varies directly as x? If the graph represents direct variation, find the variation equation.

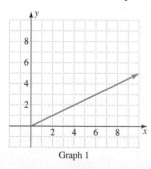

Graph 1

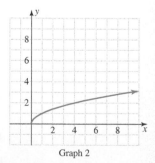

Graph 2

Inverse Variation

When the values of two quantities change in an opposite manner, they are inversely related. So, if one quantity increases, the other will decrease. Some examples of this relationship are shown in the following table.

For a fixed distance, the time at which one travels is inversely proportional to the average rate.	$t = \dfrac{100}{r}$
For a fixed area, the length of a rectangle is inversely proportional to its width.	$l = \dfrac{50}{w}$
For a salaried employee, the hourly wage, w, is inversely proportional to the hours worked, h.	$w = \dfrac{1500}{h}$

Two quantities x and y are said to be *inversely proportional* to each other if y is proportional to the *reciprocal* of x.

Definition: Inverse Variation

We say *y varies inversely as x*, or *y is indirectly proportional to x*, if

$$y = \frac{k}{x}$$

for some nonzero constant k, called the *constant of variation* or *constant of proportionality*.

Another way to think of inverse variation is that y is inversely proportional to x if $xy = k$, where k is a nonzero constant. In other words, the product and x and y is a constant.

Procedure: Solving Problems Involving Inverse Variation

Step 1: If y varies inversely as x, then $y = \dfrac{k}{x}$.

Step 2: Solve for the constant of variation by substituting the known values of x and y.

Use indirect variation to solve each problem.

2a. Suppose y varies inversely as x. If $y = 5$ when $x = 30$, find the constant of proportionality and the variation equation.

Solution   **2a.** Since y varies inversely as x, $y = \dfrac{k}{x}$. To find the constant of proportionality, let $y = 5$ and $x = 30$.

$$y = \frac{k}{x} \qquad \text{Write the variation equation.}$$

$$5 = \frac{k}{30} \qquad \text{Replace } y \text{ with 5 and } x \text{ with 30.}$$

$$30(5) = 30\left(\frac{k}{30}\right) \qquad \text{Multiply each side by 30.}$$

$$150 = k \qquad \text{Simplify.}$$

So, the constant of proportionality is 150 and the variation equation is $y = \dfrac{150}{x}$.

2b. Ohm's law states that the current, I, in an electrical conductor varies inversely as the resistance, R, of the conductor. If the current is 6 amperes (amp) when the resistance is 170 ohms (Ω), what is the current when the resistance is 34 Ω?

Solution   2b. I varies inversely as R, so $I = \dfrac{k}{R}$.

$$I = \frac{k}{R}$$ Write the variation equation.

$$6 = \frac{k}{170}$$ Replace I with 6 and R with 170.

$$170(6) = 170\left(\frac{k}{170}\right)$$ Multiply each side by 170.

$$1020 = k$$ Simplify.

So, the constant of variation is 1020 and the variation equation is $I = \dfrac{1020}{R}$.

We use this equation to find the current when the resistance is 34 Ω.

$$I = \frac{1020}{R}$$ State the variation equation.

$$I = \frac{1020}{34}$$ Replace R with 34.

$$I = 30$$ Simplify.

So, the current is 30 amp when the resistance is 34 Ω.

2c. Determine if each table shows that y is inversely proportional to x. Find the variation equation for a table that has this relationship.

Table 1

x	y
1	8
2	4
4	2
16	$\dfrac{1}{2}$

Table 2

x	y
1	16
2	32
4	64
8	128

Solution   2c. For a set of points to be inversely related, their product must be a constant.

Table 1

x	y	xy
1	8	$1(8) = 8$
2	4	$2(4) = 8$
4	2	$4(2) = 8$
16	$\dfrac{1}{2}$	$16\left(\dfrac{1}{2}\right) = 8$

Table 2

x	y	xy
1	16	$1(16) = 16$
2	32	$2(32) = 64$
4	64	$4(64) = 256$
8	128	$8(128) = 1024$

In Table 1, the product $xy = 8$ for each pair of points. So, y is inversely proportional to x and the variation equation is $y = \dfrac{8}{x}$.

In Table 2, the product of x and y is not the same for the ordered pairs. This table does not represent inverse variation.

✔ **Student Check 2**   Use inverse variation to solve each problem.

a. Suppose y is inversely proportional to x. If $y = 8$ when $x = 3$, find the constant of proportionality and the variation equation.

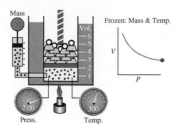

Mass

Frozen: Mass & Temp.

Press. Temp.

b. Boyle's law states that if the temperature remains the same, the volume V of a gas is inversely proportional to the pressure P applied to it. A given mass of a gas occupies 240 mL at a pressure of 800 mm of mercury (mm Hg). If the temperature remains constant, what volume will the gas occupy if the pressure is increased to 1200 mm Hg?

c. Determine if each table shows that y is inversely proportional to x. Find the variation equation for table that has this relationship.

Table 1

x	y
2	30
3	20
4	15
5	12

Table 2

x	y
2	120
3	180
4	240
5	300

Objective 3 ▶

Solve problems involving joint and combined variation.

Joint and Combined Variation

Joint variation occurs when a quantity can be expressed as the product of a constant and two or more variables.

> **Definition: Joint Variation**
>
> We say *z varies jointly as x and y* or *z is jointly proportional to x and y* if
>
> $$z = kxy \quad \text{or} \quad \frac{z}{xy} = k$$
>
> where k is the *constant of proportionality* or the *constant of variation*.

An example of joint variation is the area of a rectangle, $A = lw$. Area is equal to the product of the constant 1 and length and width.

> **Definition: Combined variation** involves combinations of direct, inverse, or joint variation.

Objective 3 Examples Use joint or combined variation to solve each problem.

3a. Suppose z varies jointly as x and y. Find an equation that relates the variables if $z = 2$ when $x = 4$ and $y = -3$.

Solution **3a.** Since z varies jointly as x and y, we know $z = kxy$.

$$z = kxy \qquad \text{Write the variation equation.}$$
$$2 = k(4)(-3) \qquad \text{Let } z = 2, x = 4, \text{ and } y = -3.$$
$$2 = -12k \qquad \text{Simplify.}$$
$$\frac{2}{-12} = k \qquad \text{Divide each side by } -12.$$
$$-\frac{1}{6} = k \qquad \text{Simplify.}$$

So, the constant of variation is $-\frac{1}{6}$ and the variation equation is $z = -\frac{1}{6}xy$.

3b. When an object is in motion, it is said to have kinetic energy. Kinetic energy, K, varies jointly as the mass, m, of an object and the square of the object's velocity, v. Write a variation formula that represents this relationship.

Solution   **3b.** K varies jointly as m and the square of v. So,

$$K = kmv^2, \text{ where } k \text{ is the constant of proportionality.}$$

3c. The area A of a trapezoid varies jointly as the height h and the sum of its bases b_1 and b_2. Find the equation of joint variation if $A = 48$ cm^2 when $h = 8$ cm, $b_1 = 5$ cm, and $b_2 = 7$ cm.

Solution   **3c.** Since A varies jointly as h and $b_1 + b_2$, we write

$$A = kh(b_1 + b_2)$$

We find k by substituting the given values.

$$A = kh(b_1 + b_2) \qquad \text{State the variation equation.}$$
$$48 = k(8)(5 + 7) \qquad \text{Let } A = 48, h = 8, b_1 = 5, b_2 = 7.$$
$$48 = k(8)(12) \qquad \text{Simplify in parentheses.}$$
$$48 = 96k \qquad \text{Multiply 8 and 12.}$$
$$\frac{48}{96} = k \qquad \text{Divide each side by 96.}$$
$$\frac{1}{2} = k \qquad \text{Simplify.}$$

So, $A = \dfrac{1}{2}h(b_1 + b_2)$.

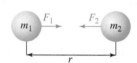

3d. Newton's law of universal gravitation states that every object in the universe attracts every other object with a force, F, directed along the line through the centers of the two objects that is jointly proportional to the objects' masses, m_1 and m_2, and inversely proportional to the square of the distance, r, between the two masses. Write a variation formula that represents this relationship.

Solution   **3d.** F varies jointly as m_1 and m_2 and is inversely proportional to the square of the distance, r. So,

$$F = \frac{km_1m_2}{r^2}, \text{ where } k \text{ is the constant of proportionality.}$$

3e. The time T, in hours, required for a satellite to complete a circular orbit around the Earth varies directly as the radius r of the orbit measured from the center of the Earth and inversely as the orbital velocity v in miles per hour. If a satellite traveling at 17,000 mph completes an orbit 4500 miles from the center of the Earth in 100 min, find the constant of variation. Then determine how long it will take a satellite to complete one orbit if it is circling the earth at 4800 mi from the center of the Earth at 16,000 mph.

Solution   **3e.** T varies directly as r and inversely as v, so $T = \dfrac{kr}{v}$. To find k, substitute the given values.

$$T = \frac{kr}{v} \qquad \text{Write the variation equation.}$$

$$\frac{5}{3} = \frac{k(4500)}{17{,}000} \qquad \text{Let } T = 100 \text{ min} = \frac{100}{60} \text{ hr} = \frac{5}{3} \text{ hr,}$$

$$\frac{5}{3} \cdot \frac{17{,}000}{4500} = \frac{k(4500)}{17{,}000} \cdot \frac{17{,}000}{4500} \qquad r = 4500, \text{ and } v = 17{,}000.$$
$$\text{Multiply each side by } \frac{17{,}000}{4500}.$$

$$\frac{85{,}000}{13{,}500} = k \qquad \text{Simplify.}$$

$$6.296 = k \qquad \text{Divide.}$$

So, the constant of variation is 6.296 and the variation equation is $T = \dfrac{6.296r}{v}$.

We use the variation equation to find how long it will take a satellite to complete one orbit if it is circling the Earth at 4800 mi from the center of the Earth at 16,000 mph.

$$T = \frac{6.296r}{v} \qquad \text{Write the variation equation.}$$

$$T = \frac{6.296(4800)}{16,000} \qquad \text{Let } r = 4800 \text{ and } v = 16,000.$$

$$T = 1.89 \text{ hours} \qquad \text{Simplify.}$$

So, it will take approximately 1 hr and 53 min for the satellite to complete one orbit.

✔ **Student Check 3**   Use joint or combined variation to solve each problem.

a. Suppose z varies jointly as x and y. Find an equation that relates the variables if $z = 6$ when $x = 3$ and $y = -4$.

b. The volume of a cylinder varies jointly as the square of its radius, r, and its height, h. Write a variation formula that represents this relationship.

c. The surface area, A, of a cylinder varies jointly as the radius, r, and the sum of the radius and the height, h. A cylinder with height 8 cm and radius 4 cm has a surface area of 96π cm^2. Find the surface area of a cylinder with radius 3 cm and height 10 cm.

d. Newton's second law states that the acceleration, a, of an object is directly proportional to the force, F, acting on it and inversely proportional to the mass, m, of the object. Write a variation formula that represents this relationship.

e. The maximum load, l, that a cylindrical column with a circular cross section can hold varies directly as the fourth power of the diameter, d, and inversely as the square of the height, h. A 9-m column that is 2 m in diameter can support 64 metric tons. Find the constant of variation and then determine how many metric tons can be supported by a 9-m column that is 3 m in diameter.

Objective 4 ▶

Troubleshoot common errors.

Troubleshoot Common Errors

Some common errors associated with variation are shown.

Objective 4 Examples   A problem and an incorrect solution are given. Provide the correct solution and an explanation of the error.

4a. Suppose y varies directly as the cube of x. Find the variation equation if $y = 16$ when $x = 2$.

Incorrect Solution	Correct Solution and Explanation
$y = k(3x)$	Since y varies directly as the cube of x, we should have
$16 = k(3 \cdot 2)$	$y = kx^3$
$16 = 6k$	$16 = k(2)^3$
$\dfrac{16}{6} = k$	$16 = 8k$
$\dfrac{8}{3} = k$	$2 = k$
$y = \dfrac{8}{3}(3k) = 8k$	$y = 2x^3$

4b. Suppose y varies inversely as x. Find the constant of variation if $y = 12$ when $x = 3$.

Incorrect Solution	Correct Solution and Explanation
$y = \dfrac{x}{k}$ $12 = \dfrac{3}{k}$ $12k = 3$ $k = \dfrac{3}{12}$ $k = \dfrac{1}{4}$	The constant of variation goes in the numerator and the variable x is in the denominator. $y = \dfrac{k}{x}$ $12 = \dfrac{k}{3}$ $36 = k$

ANSWERS TO STUDENT CHECKS

Student Check 1 **a.** $k = -4$, $y = -4x$ **b.** $d = 7$ in.

c. Table 2, $y = -2x$ **d.** Graph 1, $y = \dfrac{1}{2}x$

Student Check 2 **a.** $k = 24$, $y = \dfrac{24}{x}$ **b.** $V = 160$ mL

c. Table 1, $y = \dfrac{60}{x}$

Student Check 3 **a.** $z = -\dfrac{1}{2}xy$ **b.** $V = kr^2h$

c. $A = 78\pi$ cm^2 **d.** $a = \dfrac{kF}{m}$ **e.** 324 metric tons

SUMMARY OF KEY CONCEPTS

1. When two quantities x and y are related by $y = kx$ for some constant k, then we say y varies directly as x.

2. When two quantities x and y are related by $y = \dfrac{k}{x}$ for some constant k, then we say y varies indirectly or inversely as x.

3. Quantities can be jointly proportional. If $z = kxy$, then z varies jointly as x and y. Combined variation involves combinations of direct, inverse, or joint variation.

SECTION 8.4 / EXERCISE SET

Write About It!

Use complete sentences in your answer to each exercise.

1. Define direct variation. **2.** Define inverse variation.

3. Define joint variation.

For Exercises 4–8, translate each equation into words using variation.

4. $A = \pi r^2$ **5.** $V = \dfrac{1}{3}\pi r^2 h$ **6.** $V = \dfrac{4}{3}\pi r^3$

7. $G = \dfrac{km_1 m_2}{r^2}$ **8.** $y = \dfrac{k}{\sqrt[3]{x}}$

Practice Makes Perfect!

Suppose y varies directly as x. Find the constant of variation and the equation of variation for each case. (See Objective 1.)

9. $y = 35$ when $x = 7$ **10.** $y = 24$ when $x = 3$

11. $y = -4$ when $x = -6$ **12.** $y = -12$ when $x = -10$

13. $y = 1.2$ when $x = 2$ **14.** $y = 36$ when $x = 15$

Solve each problem. (See Objective 1.)

15. The cost of a house in California is directly proportional to the size of the house. If a 1200-ft^2 house costs \$180,000, then what is the cost of a 2200-ft^2 house?

16. The cost of a house in Pennsylvania is directly proportional to the size of the house. If a 1800-ft² house costs $210,000, then what is the cost of a 2700-ft² house?

Hooke's law of elasticity states that the distance a spring stretches vertically is directly proportional to the weight on the end of the spring. (*See Objective 1.*)

17. If a 5-lb weight attached to the spring stretches the spring 6 in., find the distance that a 17-lb weight stretches the spring.

18. If an 8-lb weight attached to the spring stretches the spring 3 in., find the distance that a 48-lb weight stretches the spring.

Determine if each table shows that *y* varies directly as *x*. If the table represents direct variation, $y = kx$, find the variation equation. (*See Objective 1.*)

19.

Table 1			Table 2	
x	y		x	y
2	1		2	−1
4	2		4	2
6	3		6	−3
8	4		8	4

20.

Table 1			Table 2	
x	y		x	y
2	2		7	3
4	8		14	6
6	18		21	9
8	32		28	12

Determine if each graph shows that *y* varies directly as *x*. If the graph represents direct variation, $y = kx$, find the variation equation. (*See Objective 1.*)

21.

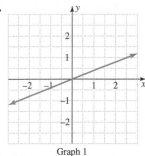

Graph 1   Graph 2

22.

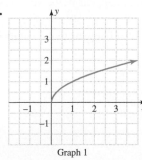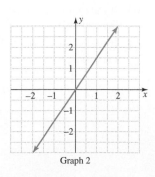

Graph 1   Graph 2

Suppose *y* varies inversely as *x*. Find the constant of variation and the equation of variation for each case. (*See Objective 2.*)

23. $y = \dfrac{1}{2}$ when $x = 6$ **24.** $y = 2$ when $x = 2$

25. $y = -3$ when $x = -\dfrac{1}{2}$ **26.** $y = -0.3$ when $x = -2$

27. $y = 0.32$ when $x = 5$ **28.** $y = 3$ when $x = 0.7$

Boyle's law states that if the temperature remains the same, the volume *V* of a gas is inversely proportional to the pressure *P* applied to it. (*See Objective 2.*)

29. A given mass of a gas occupies 930 mL at a pressure of 450 mm Hg. If the temperature remains constant, what volume will the gas occupy if the pressure is increased to 1350 mm Hg?

30. A given mass of a gas occupies 360 mL at a pressure of 430 mm Hg. If the temperature remains constant, what volume will the gas occupy if the pressure is increased to 1200 mm Hg?

Ohm's law states that if the voltage remains the same, the current, *I*, in an electrical conductor varies inversely as the resistance, *R*, of the conductor. (*See Objective 2.*)

31. If the current is 41 amp when the resistance is 270 Ω, what is the current when the resistance is 300 Ω?

32. If the current is 57 amp when the resistance is 320 Ω, what is the current when the resistance is 480 Ω?

Determine if each table shows that *y* varies inversely as *x*. If the table represents inverse variation, $y = \dfrac{k}{x}$, find the variation equation. (*See Objective 2.*)

33.

Table 1			Table 2	
x	y		x	y
2	1		0.5	5
4	2		2	1.25
6	3		4	0.625
8	4		8	0.3125

34.

Table 1			Table 2	
x	y		x	y
0.5	7		1	3
1	3.5		2	6
2	1.75		3	9
3	0.875		4	12

Suppose z varies jointly as x and y. Find the constant of variation and the equation of variation for each case. (See Objective 3.)

35. $z = 10$ when $x = 2$ and $y = 4$

36. $z = 40$ when $x = 3$ and $y = 10$

37. $z = \dfrac{1}{3}$ when $x = 2$ and $y = 3$

38. $z = \dfrac{1}{4}$ when $x = 3$ and $y = \dfrac{5}{4}$

39. $z = 0.9$ when $x = 3$ and $y = 12$

40. $z = 0.36$ when $x = 0.5$ and $y = 9$

Find the constant of variation and variation equation for each situation. (See Objective 3.)

41. y varies jointly as the square of p and cube of q; $y = 8$ when $p = 3$ and $q = 2$.

42. z varies jointly as x and square of y; $z = 16$ when $x = 5$ and $y = 2$.

43. c varies jointly as the square root of a and b; $c = 3.6$ when $a = 4$ and $b = 1.2$.

44. F varies jointly as the cube root of G and square of H; $F = 4.5$ when $G = 27$ and $H = 0.5$.

45. w varies directly as the square of u and inversely as the cube of v; $w = 1.8$ when $u = 1.5$ and $v = 2$.

46. z varies directly as the square root of x and inversely as the fourth power of y; $z = 2.7$ when $x = 2.25$ and $y = 0.2$.

The maximum load that a cylindrical column with a circular cross section can hold varies directly as the fourth power of the diameter and inversely as the square of the height. (See Objective 3.)

47. A 12-m column that is 2.4 m in diameter can support 72 metric tons. Find the constant of variation and then determine how many metric tons can be supported by a 10-m column that is 3.2 m in diameter.

48. A 10-m column that is 2 m in diameter can support 72 metric tons. Find the constant of variation and then determine how many metric tons can be supported by a 9-m column that is 3 m in diameter.

 Mix 'Em Up!

Find the constant of variation and the variation equation for each situation.

49. y varies directly as the cube root of x; $y = 0.5$ when $x = 125$.

50. b varies directly as square root of a; $b = 5$ when $a = 4$.

51. y varies inversely as the square of x; $y = 2.5$ when $x = 1.2$.

52. F varies inversely as the cube of G; $F = 4.8$ when $G = 1.5$.

53. w varies directly as the cube of u and inversely as the square root of v; $w = 24$ when $u = 2$ and $v = 25$.

54. c varies directly as a and inversely as the cube root of b; $c = 0.9$ when $a = 1.8$ and $b = 0.064$.

Boyle's law states that if the temperature remains the same, the volume V of a gas is inversely proportional to the pressure P applied to it.

55. A given mass of a gas occupies 520 mL at a pressure of 880 mm Hg. If the temperature remains constant, what volume will the gas occupy if the pressure is increased to 1040 mm Hg?

56. A given mass of a gas occupies 880 mL at a pressure of 680 mm Hg. If the temperature remains constant, what volume will the gas occupy if the pressure is increased to 1700 mm Hg?

Hooke's law of elasticity states that the distance a spring stretches vertically is directly proportional to the weight on the end of the spring.

57. If a 10-lb weight attached to the spring stretches the spring 7 in., find the distance that a 43-lb weight stretches the spring.

58. If a 24-lb weight attached to the spring stretches the spring 16 in., find the distance that a 27-lb weight stretches the spring.

Ohm's law states that if the voltage remains the same, the current, I, in an electrical conductor varies inversely as the resistance, R, of the conductor.

59. If the current is 59 amp when the resistance is 480 Ω, what is the current when the resistance is 600 Ω?

60. If the current is 54 amp when the resistance is 320 Ω, what is the current when the resistance is 450 Ω?

The volume of a cone varies jointly as the height and the square of the radius of the base.

61. If the height is doubled and the radius is halved, what happens to the volume of the cone?

62. If the height is halved and the radius is doubled, what happens to the volume of the cone?

The volume of a sphere varies as the cube of the radius.

63. If the radius is halved, what happens to the volume of the sphere?

64. If the radius is doubled, what happens to the volume of the sphere?

You Be the Teacher!

Correct each student's errors, if any.

65. Find the equation of variation if y varies directly as the square root of x and $y = 48$ when $x = 4$.

Joseph's work:

$y = kx^2$
$48 = k(4)^2$
$48 = 16k$
$k = 3$
$y = 3x^2$

66. Find the equation of variation if z varies inversely as the square of w and $z = 6$ when $w = 9$.

Mike's work:

$z = \dfrac{k}{\sqrt{w}}$

$6 = \dfrac{k}{\sqrt{9}}$

$18 = k$

$z = \dfrac{18}{\sqrt{w}}$

GROUP ACTIVITY / The Mathematics of Income Taxes

The U.S. federal income tax is based on a graduated tax system, which means that taxpayers with higher incomes are taxed at higher rates that those with lower incomes. To find the tax tables for the current year, go to http://www.irs.gov/> Forms and Publications > Tax Tables. Scroll to the end of the web page to find the tax rate schedules.

1. Write a piecewise-defined function for each category (single, married filing jointly, married filing separately, and head of household.)

2. Use the Internet to research the average income of the career you are pursuing.

3. Use the piecewise-defined functions to determine the amount of federal income tax owed for the average income for each of the four categories.

4. Pauline and Michael are contemplating marriage and are curious about the exact amount of the marriage penalty. The marriage penalty is a term used to describe the situation in which a couple filing their taxes as married filing jointly pays more in tax than two people filing single.

   a. Calculate Pauline's tax, as a single person, on her income of $80,000.

   b. Calculate Michael's tax, as a single person, on his income of $85,000.

   c. Sum the taxes to find the total amount that Pauline and Michael owe when they file as two single people.

   d. As a married couple, their combined income would be $165,000. Calculate the tax they would owe on their combined income if they file as married filing jointly.

   e. Compare the values parts c and d. Is there a marriage penalty for this income level and if so, what is it?

More on Functions and Graphs; Variation

What's the big idea? Now that we have completed Chapter 8, we should be able to graph linear, nonlinear, and piecewise-defined functions and determine their domains and ranges. We should also be able to solve applications modeled by these types of functions. Lastly, we should be able to work with quantities that are directly proportional and inversely proportional by defining an appropriate variation equation.

The Tools

Listed below are the key terms, skills, formulas, and properties you should know for this chapter.

The page reference is provided if you need additional help with the given topic. The Study Tips will assist in your preparation for an exam.

Study Tips

1. Learn all of the terms, formulas, and properties. Make flash cards and have someone quiz you.
2. Rework problems from the exercises and also the ones you worked in class. Work additional problems from the review exercises.
3. Review the summaries of key concepts.
4. Work the chapter test.
5. Be sure to review the online resources for additional study materials.

Terms

Combined variation 621	Domain of a function 582	Piecewise-defined function 607
Constant of proportionality 616	Function 580	Range 580
Constant of variation 616	Inverse variation 619	Ray 608
Direct variation 616	Joint variation 621	Reflection 606
Domain 580	Nonlinear function 603	Translation 604

Formulas and Properties

- Functions of the form
  $y = f(x) + k$ 604
- Functions of the form
  $y = f(x - h)$ 604

- Functions of the form
  $y = -f(x)$ 607

- Point-slope form   592
- Slope formula   592

CHAPTER 8 / SUMMARY

How well do you know this chapter? Complete the following questions to find out. Take a look back at the section if you need help.

SECTION 8.1 The Domain and Range of Functions

1. The domain of a function is the set of _____.
2. The range of a function is the set of _____.
3. To find the domain from a graph, read it from ____ to ____.
4. To find the range from a graph, read it from _____ to ___.

5. To determine the domain of an algebraic function, we must determine the values of x for which the function is _____.
6. If the function involves a fraction with a variable in the denominator, we must exclude, from the domain, values that make the _____ zero since division by zero is _____.
7. If the function involves the square root of an algebraic expression, we must find the values that make the expression inside the square root _____ ____ or ____ __ zero to find its domain.

SECTION 8.2 Graphing and Writing Linear Functions

8. A linear function is a function of the form _____. It can be graphed by rewriting it in the form _____. Its graph is a(n) _____. The domain of all linear functions is _____. The range of all linear functions, $m \neq 0$, is _____.

9. The function $f(x) = c$ has the graph of a(n) _____ line. Its domain is _____ and its range is ___.

10. To write a linear function, we must know a(n) ____ and a(n) _____. If two function values are given, we must write them as _____ ____ to find the ____ of the line.

11. If two linear functions have graphs that are parallel lines, then their slopes are the _____. If two linear functions have graphs that are perpendicular lines, then their slopes are _____ _____.

12. In application problems, the slope tells us the ____ of _____ and the y-intercept tells us the _____ value.

SECTION 8.3 Graphing Nonlinear and Piecewise-Defined Functions; Shifting and Reflecting Functions

13. To graph a nonlinear function, we can ____ _____. We need enough points to determine the _____ of the graph. The domain can be determined by reading the graph from ___ to ____ and its range can be determine by reading the graph from _____ to ___.

14. A function of the form $f(x) + k$, for $k > 0$ is the graph of $f(x)$ shifted ___ k units. A function of the form $f(x) - k$, for $k > 0$, is the graph of $f(x)$ shifted ____ k units.

15. A function of the form $f(x - h)$, for $h > 0$ is the graph of $f(x)$ shifted _____ h units. A function of the form $f(x + h)$, for $h > 0$ is the graph of $f(x)$ shifted ___ h units.

16. A function of the form $-f(x)$ is the graph of $f(x)$ reflected across the _____.

17. A(n) _____ function is a function comprised of two or more functions which are defined on a piece of the domain of the function. The ordered pairs where the pieces change will have either a(n) ____ or _____ circle. If the inequality is $\leq$ or $\geq$, the point will be _____. If the inequality is $<$ or $>$, the point will be ____.

SECTION 8.4 Variation and Applications

18. When y varies directly as x, we write $y =$ ___. The quotient of y and x is _____.

19. When y varies indirectly as x, we write $y =$ ___. The product of y and x is _____.

20. When z varies jointly as x and y, we write $z =$ ___.

21. _____ variation involves combinations of direct, inverse, or joint variation.

CHAPTER 8 / REVIEW EXERCISES

SECTION 8.1

Find the domain and range of each function. (*See Objective 1.*)

1. $\{(-5, -4), (-1, -2), (0, 11), (2, 13)\}$

2. $\{(-4, 1), (-3, 3), (-2, 0), (4, 8)\}$

3. The temperature (in degrees Fahrenheit) in New York on June 26, 2010, is a function of the time of day. (Source: www.accuweather.com)

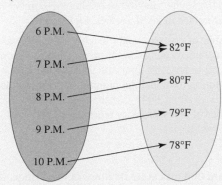

4. The average high temperature (in degrees Fahrenheit) for Chicago, Illinois, is a function of the date. (Source: www.weather.com)

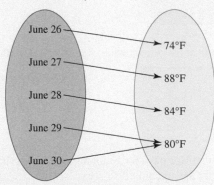

Find the domain and range of each function represented by the graph. (*See Objective 2.*)

5.

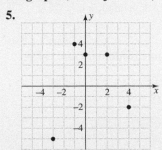

6.

7.

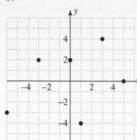

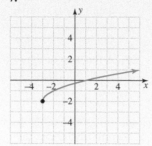

8.

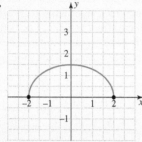

Find the domain of each function. (*See Objective 3.*)

9. $f(x) = \dfrac{x+6}{x-1}$ **10.** $f(x) = \dfrac{x}{x+2}$

11. $h(x) = \sqrt{2x-8}$ **12.** $h(x) = \sqrt{3x+6}$

Determine an appropriate domain for each situation.
(*See Objective 4.*)

13. The volume of the box shown is $V = x(12 - 2x)(24 - 2x)$.

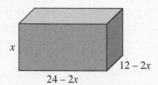

14. The volume of the box shown is $V = 4(18 - 3x)(10 - x)$.

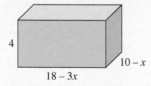

15. The cost that Kos pays for x months of lawn service is modeled by the equation $C = 40x$.

16. Bea regularly pays for manicures. If she makes x visits to the nail salon in a month, Bea's monthly manicure cost can be modeled by the equation $C = 18x$.

SECTION 8.2

Graph each function and state its domain and range.
(*See Objective 1.*)

17. $f(x) = -2x + 6$ **18.** $f(x) = -\dfrac{1}{2}x$

19. $f(x) = 6.5$ **20.** $f(x) = 2$

Solve each problem. (*See Objective 2.*)

21. The average yearly cost of attending a 2-yr public institution is modeled by the function $f(x) = 335.6x + 4889.5$, where x is the number of years after 2000. (Source: www.infoplease.com)

   a. Find the y-intercept and interpret its meaning in the context of the problem.

   b. Find $f(5)$ and $f(11)$ and interpret their meaning in the context of the problem.

   c. Solve $f(x) = 9923.5$ and interpret its meaning in the context of the problem.

22. A car's value is given by $f(x) = -1850x + 18{,}500$, where x is the age of the car in years.

   a. Find the y-intercept and interpret its meaning in the context of the problem.

   b. Find $f(4)$ and interpret its meaning in the context of the problem.

   c. Solve $f(x) = 0$ and interpret its meaning in the context of the problem.

Write a linear function that satisfies the given conditions.
(*See Objective 3.*)

23. $m = -\dfrac{3}{5}, f(0) = 4$ **24.** $m = \dfrac{1}{7}, f(0) = -6$

25. $m = -6, f(-5) = 0$ **26.** $m = 3, f(1.4) = 0$

27. $f(-1.6) = 5.2$ and $f(-4.6) = 8.2$

28. $f\left(\dfrac{1}{2}\right) = \dfrac{2}{3}$ and $f\left(\dfrac{5}{4}\right) = \dfrac{1}{6}$

29. $f(4.8) = 0$ and perpendicular to $f(x) = \dfrac{1}{5}x - 20$

30. $f(-3) = 4$ and parallel to $f(x) = -\dfrac{5}{6}x + 1$

Solve each problem. (*See Objective 4.*)

31. In 2004, the number of registered nurses was 2394 thousand. The number of registered nurses is predicted to be 3096 thousand in 2014. Write a linear function to model the number of registered nurses, x yr after 2004. Use the function to predict the number of registered nurses in 2020. (Source: www.bls.gov)

32. The number of postsecondary teachers in 2004 was 1628 thousand. The number of postsecondary teachers is predicted to be 2153 thousand in 2014. Write a linear function to model the number of postsecondary teachers, x yr after 2004. Use the function to predict the number of postsecondary teachers in 2018. (Source: www.bls.gov)

SECTION 8.3

Graph each function and state its domain and range.
(*See Objectives 1–3.*)

33. $f(x) = x^2 - 1$ **34.** $f(x) = (x - 1)^2$

35. $f(x) = |x + 3| - 4$ **36.** $f(x) = -|x - 2| + 2$

37. $f(x) = \sqrt{x - 1}$ **38.** $f(x) = \sqrt{x + 5}$

Graph each function and state its domain and range. (*See Objective 4.*)

39. $f(x) = \begin{cases} -2x, & x < 0 \\ x + 1, & x \geq 0 \end{cases}$ 40. $\begin{cases} \frac{1}{2}x - 4, & x < -2 \\ 3x, & x \geq -2 \end{cases}$

Solve each problem. (*See Objective 5.*)

41. The parking rates for one of the short-term parking lots at Hartsfield-Jackson Atlanta International Airport are as follows.

    $2 per hour (first 2 hr)
    $3 per hour (next 4 hr)
    $32 (6–24 hr)
    $36 per day (each additional day)

    Write a piecewise-defined function that models the cost of parking for x hr of parking, where $x \leq 72$. (Source: http://www.atlanta-airport.com/parking/Parking_Rates.aspx)

42. A telecommunications provider advertises a individual cell phone plan with a monthly cost of $59.99 for 900 min. Each additional minute costs $0.40 per minute. Write a piecewise-defined function that represents the monthly cost of the plan for x min. Then find the cost of 1500 min.

SECTION 8.4

Find the constant of variation and the variation equation for each situation. (*See Objectives 1–3.*)

43. r varies directly as the square of s; $r = 8$ when $s = 2$.

44. b varies directly as cube of a; $b = 54$ when $a = 3$.

45. y varies inversely as the cube root of x; $y = 2$ when $x = 64$.

46. M varies inversely as the square root of N; $M = 4.8$ when $N = 2.25$.

47. w varies directly as the cube root of u and inversely as the square of v; $w = 6$ when $u = 27$ and $v = 5$.

48. z varies directly as x and inversely as the cube of y; $z = 6$ when $x = 24$ and $y = 2$.

Solve each problem. (*See Objectives 1–3.*)

49. Boyle's law states that if the temperature remains the same, the volume V of a gas is inversely proportional to the pressure P applied to it. A given mass of a gas occupies 320 mL at a pressure of 600 mm Hg. If the temperature remains constant, what volume will the gas occupy if the pressure is increased to 1200 mm Hg?

50. The area A of a circle varies directly as the square of the radius. A circle with radius 6 cm has an area of 36π cm². Find the area of a circle with radius 21 cm.

51. The volume of a cylinder varies jointly as the square of the radius and the height. If the radius is halved and the height is doubled, what happens to the volume of the cylinder?

52. The cost of a house in Iowa is directly proportional to the size of the house. If a 2000-ft² house costs $210,000, then what is the cost of a 3350-ft² house?

CHAPTER 8 TEST / MORE ON FUNCTIONS AND GRAPHS; VARIATION

1. The function that doesn't have a domain of all real numbers is
   a. $f(x) = \dfrac{x}{3}$ **b.** $f(x) = -3$ **c.** $f(x) = x^2 - 2$ **d.** $\dfrac{4}{x}$

2. The function that has a range of $(-\infty, 2]$ is
   a. $f(x) = x^2 - 2$ **b.** $f(x) = -x^2 + 2$
   c. $f(x) = -x^2 - 2$ **d.** $f(x) = x^2 - 2$

3. The range of the function shown is
   a. $[-3, \infty)$ **b.** $(-\infty, \infty)$ **c.** $[-4, \infty)$ **d.** $[-2, \infty)$

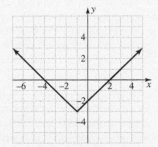

4. The graph of the linear function $f(x) = \dfrac{4}{3}x - 2$ is found by
   a. plotting $(0, 2)$ and moving up 4 units and right 3 units
   b. plotting $(0, 2)$ and moving up 3 units and right 4 units
   c. plotting $(0, -2)$ and moving up 4 units and right 3 units
   d. plotting $(0, -2)$ and moving up 3 units and right 4 units

5. The linear function for which $f(2) = -1$ and $f(-3) = 5$ is
   a. $f(x) = \dfrac{6}{5}x - \dfrac{17}{5}$ **b.** $f(x) = -\dfrac{6}{5}x + \dfrac{7}{5}$
   c. $f(x) = -\dfrac{6}{5}x - \dfrac{7}{5}$ **d.** $f(x) = \dfrac{6}{5}x + \dfrac{17}{5}$

6. For $f(x) = \begin{cases} x + 3, & x \leq 1 \\ -x + 2, & x > 1 \end{cases}$, the points where the pieces change are
   a. $(1, 4)$ open and $(1, 1)$ closed
   b. $(1, 4)$ closed and $(1, 1)$ open
   c. $(1, 4)$ open and $(1, 3)$ closed
   d. $(1, 4)$ closed and $(1, 3)$ open

7. The equation that shows that a is directly proportional to b is

a. $b = ka$ **b.** $a = \dfrac{k}{b}$

c. $\dfrac{a}{b} = k$ **d.** $ab = k$

Graph each function and state its domain and range.

8. $f(x) = -\dfrac{3}{4}x + 4$ **9.** $f(x) = -2$

10. $f(x) = x^2 - 4$ **11.** $f(x) = |x - 3|$

12. $f(x) = \sqrt{4 - x}$ **13.** $f(x) = \begin{cases} 2x - 1, & x < 1 \\ 3, & x \geq 1 \end{cases}$

Write a linear function that satisfies the given conditions.

14. $m = -\dfrac{4}{5}$ and $f(0) = 7$ **15.** $f(2) = -3$ and $f(-1) = 2$

16. $f(4) = 0$ and the graph is parallel to $f(x) = \dfrac{2}{3}x + 1$

17. $f(-5) = 6$ and the graph is perpendicular to $f(x) = 5x - 4$

Solve each problem.

18. If the area of a rectangle is represented by $A(x) = x(3x - 2)$, what are the possible values of x?

19. In fall 2010, Arizona State University was the largest public university campus with an enrollment of 58,371. In fall 2008, its enrollment was 52,734. Write a linear function, $f(x)$, that models the enrollment of ASU x yr after 2008. Then use the function to find $f(6)$ and interpret its meaning in the context of the problem. (Source: Wikipedia)

20. The 2007 *US News and World Report* published an evaluation of the best Nursing Schools in the United States. The report stated that the University of Washington was the best nursing school. The tuition at UW is given in the following table. Use the following information to write a piecewise-defined formula for tuition based on the number of credit hours x. (Source: http://www.washington.edu/admin/pb/home/pdf/tuition/2010-11/tf-sea-aut%201.pdf)

Credit Hours	Tuition
2–9	$270.71 per credit hour plus $0.93
10–18	$2708
19–30	$2708 plus $271 per hour over 18 hours

21. The amount of money, y, Constance earns is directly proportional to the hours, x, she works. If she makes $172 for working 8 hr, how much money does she earn for working 20 hr? Find the direct variation equation and the constant of proportionality.

22. For people who practice martial arts, the force, f, needed to break a board, varies inversely with the length, l, of the board. If it takes 5 lb of pressure to break a board 2 ft long, how many pounds of pressure will it take to break a board that is 6 ft long?

CUMULATIVE REVIEW EXERCISES / CHAPTERS 1–8

Use the order of operations to simplify each expression. (*Section 1.3, Objectives 1 and 2*)

1. $(-2)^5$ **2.** $-\left(-\dfrac{3}{5}\right)^3$ **3.** $\dfrac{1}{6}(12 - 8)^2 - \dfrac{5}{3}$

4. $2(9)^2 - 6(8) + 12$ **5.** $\dfrac{-16 + \sqrt{16^2 - 4(12)(-3)}}{2(12)}$

Evaluate each expression for the given values. (*Section 1.3, Objective 3*)

6. $1.4x + 2.6y$ for $x = 5$ and $y = 2$

7. $b^2 - 4ac$ for $a = 6, b = 9, c = 2$

Simplify each expression. (*Sections 1.4 and 1.5, Objectives 1 and 2*)

8. $2 - 12 + (-23) - (-10)$

9. $3.1 - (-4.5) + (-6.7) + 7.2$

10. $-(-3)^3 + (-14) - |5 - 9|$

Simplify each expression. (*Section 1.6, Objectives 1–3*)

11. $(-4.5)(1.8)$ **12.** $(-12)\left(\dfrac{-11}{20}\right)$ **13.** $-(-5)^3$

Evaluate each expression for the given values. (*Section 1.6, Objective 4*)

14. $\sqrt{(x_1 - x_2)^2 + (y_1 - y_2)^2}$ for $x_1 = -1, x_2 = 3, y_1 = 13, y_2 = 16$

15. $\dfrac{|x - 1|}{2x - 4}$ for $x = -1, 0, 1, 2$

Solve each equation. (*Section 2.4, Objectives 1–4*)

16. $5(3a + 4) - 6(a + 1) = 3(3a + 1) - 7$

17. $0.014x + 0.021(870 - x) = 14.63$

Use the table to answer the questions. (*Section 3.1, Objective 5*)

18. The table shows the average fare per revenue passenger flying AirTran Airways between 2005 and 2009. The average fare was calculated as total passenger revenue divided by total number of passengers. (Source: http://investor.airtran.com/phoenix.zhtml?c=64267&p=irol-reportsAnnual)

Years after 2005	0	1	2	3	4
Average fare	83.93	90.51	92.47	98.04	87.05

a. Write an ordered pair (x, y) that corresponds to each year between 2005 and 2009, where x is the number of years after 2005 and y is the average fare, excluding transportation taxes, per revenue passenger flying AirTran Airways.

b. Interpret the meaning of first and last ordered pairs from part (a) in the context of the problem.

c. In what year was the average fare per revenue passenger the highest? The lowest?

d. Make a scatter plot of the data.

Write the equation of each line that satisfies the given criteria. Express each answer in slope-intercept form and in standard form. (*Section 3.5, Objectives 1–4*)

19. $(-6, 1)$ perpendicular to $3x + 5y = 15$

20. $(-1, -4)$ parallel to $2x + y = 0$

Solve each system of equations graphically. Then state if the system is consistent with independent equations, consistent with dependent equations, or inconsistent. (*Section 4.2, Objectives 1 and 2*)

21. $\begin{cases} 2x - y = -8 \\ x + 5y = 7 \end{cases}$ **22.** $\begin{cases} y = 5x + 2 \\ -5x + y = -5 \end{cases}$

Simplify each expression using the rules of exponents. Write each answer with positive exponents. (*Section 5.1, Objectives 1–4; Section 5.2, Objectives 1 and 2*)

23. $(-2^{-1}a^{-5}b^2)^3$ **24.** $\left(\dfrac{3x^2}{y^{-3}}\right)^2$

Perform each operation. (*Section 5.3, Objective 4; Section 5.4, Objectives 1 and 2; Section 5.5, Objectives 1 and 3; Section 5.6, Objectives 1 and 2*)

25. $(5p^2 + 2pq - 8q^2) - (-2q^2 + 14pq + q^2) + (3p^2 - 6pq - 4q^2)$

26. $7xy(2x^2 - xy + 3y^2)$ **27.** $(4x - 9y)(4x + 9y)$

28. $(a - 2b)(a + 2b)(a^2 + 4b^2)$

29. Divide $28x^3 - 12x^2 + 24x + 16$ by $4x$.

30. $\dfrac{4x^3 - 10x^2 + 21x - 18}{x - 2}$

Factor each polynomial completely. (*Sections 6.1 and 6.2, Objectives 1 and 2; Section 6.3, Objectives 1–3; Section 6.4, Objectives 1–4*)

31. $50x^2y - 40xy^2 + 8y^3$ **32.** $a^4 - 26a^2 + 25$

33. $36x^2 - 3x - 5$ **34.** $27a^3 + 8b^3$

35. $25a^2 + b^2$ **36.** $1000p^3 - q^3$

37. $c^6 - 1$

Solve each equation. (*Section 6.5, Objectives 1 and 2*)

38. $2x^2 = 9x + 35$ **39.** $(x^2 - 1)(x^2 - 12) = -24$

40. $x^3 - 3x^2 - 10x = 0$

Solve each problem. (*Section 6.6, Objectives 1–5*)

41. The profit, in millions of dollars, of a furniture company is given by $p = -6x^2 + 138x + 2520$, where x is the number of units sold in thousands. How many units need to be sold for the company to break even?

42. The product of two consecutive positive odd integers is 99. Find the integers.

Evaluate each expression for the given values. (*Section 7.1, Objectives 2 and 3*)

43. $\dfrac{x^2 + x}{x - 2}$ for $x = -1, 0, 2$ **44.** $\dfrac{4y}{2y - 3}$ for $y = -1, 0, 1.5$

45. A company's revenue can be represented by $R(q) = -20q^2 + 800q$, where q is the number of items sold and $R(q)$ is in dollars. If the unit price of each item is given by $\dfrac{R(q)}{q}$, what is the unit price of each item when 10 items are sold? 20 items are sold?

Find the domain of each rational function. Then simplify each function. (*Section 7.1, Objectives 2 and 3*)

46. $f(x) = \dfrac{x + 4}{3x^2 + 10x - 8}$ **47.** $f(x) = \dfrac{2x^2 + 6x}{x^3 - 9x}$

Write two equivalent forms of each rational expression. (*Section 7.1, Objective 4*)

48. $\dfrac{x + 5}{-x + 7}$ **49.** $-\dfrac{2 - x}{7x + 3}$

Perform each operation and simplify the answer. (*Section 7.2, Objectives 1 and 2*)

50. $\dfrac{2x^2 + 4x}{x^2 + x - 12} \cdot \dfrac{3 - x}{x^2 + 2x}$ **51.** $\dfrac{x^2 + 6x - 27}{x^2 + 2x} \div \dfrac{x^2 - 81}{x^2 - 9x}$

52. $\dfrac{x^2 - 9}{x^3 + 3x^2 + 9x + 27} \div \dfrac{x^2 - 5x + 6}{x^4 + 5x^2 - 36}$

53. $\dfrac{x^4 - 16}{x^2 - 4} \cdot \dfrac{x^2 - 3x + 2}{x^4 + 3x^2 - 4}$

Use the given conversion equivalents to solve each problem. (*Section 7.2, Objective 3*)

54. The distance between San Francisco, California, and Pittsburgh, Pennsylvania, is 2592 mi. What is this distance in inches on the map? (Note: 1 in. = 360 mi)

55. Convert 13.5 kg to pounds. (Note: 1 lb = 0.45 kg.)

Add or subtract the rational expressions. Write each answer in lowest terms. (*Section 7.4, Objectives 1 and 2*)

56. $\dfrac{4x - 7}{x - 3} - \dfrac{x - 8}{3 - x}$ **57.** $\dfrac{3x^2}{x^2 - 2x - 24} - \dfrac{x + 1}{x - 6}$

58. $\dfrac{4}{x - 3} + \dfrac{2x}{x^2 + 4x - 21}$ **59.** $\dfrac{6}{x^2 - 1} - \dfrac{7}{x^2 + 2x + 1}$

60. $\dfrac{3}{x^3 + 8} - \dfrac{4}{2x^2 - 4x + 8}$

Simplify each complex fraction using method 1 or method 2. (*Section 7.5, Objectives 1 and 2*)

61. $\dfrac{\dfrac{8a^2b}{28x^4}}{\dfrac{6ab^2}{14x^6}}$ **62.** $\dfrac{\dfrac{1}{a} + \dfrac{3}{4}}{6a + 8}$ **63.** $\dfrac{\dfrac{4}{x} + \dfrac{3}{y}}{\dfrac{16}{x^2} - \dfrac{9}{y^2}}$

Use the methods for simplifying complex numbers to solve each problem. (*Section 7.5, Objective 3*)

64. Find the slope of the line passing through $\left(-\dfrac{2}{3}, \dfrac{4}{5}\right)$ and $\left(\dfrac{1}{6}, -\dfrac{3}{10}\right)$.

65. Simplify the expression: $\dfrac{4^{-2} - 5^{-1}}{2^{-3}}$.

Solve each rational equation. (*Section 7.6, Objective 1*)

66. $2 + \dfrac{7}{x} = \dfrac{9}{x^2}$

67. $\dfrac{2}{5x - 7} = \dfrac{3}{3x - 2}$

68. $\dfrac{6}{2x^2 - x - 3} - \dfrac{5}{4x^2 - 9} = \dfrac{4}{2x^2 + 5x + 3}$

Solve each rational equation for the specified variable. (*Section 7.6, Objective 2*)

69. $\dfrac{1}{a} = \dfrac{2}{b} + \dfrac{3}{c} + \dfrac{4}{d}$ for b

70. $x_m = \dfrac{x_1 + x_2}{2}$ for x_1 (midpoint formula)

71. An airplane can fly 1350 mi in the same amount of time it takes a second plane to fly 1500 mi. If the speed of the first airplane is 50 mph slower than the speed of the second airplane, find the speed of each airplane. (*Section 7.7, Objectives 1–5*)

Determine the domain of each function. (*Section 8.1, Objective 3*)

72. $f(x) = \dfrac{4x - 5}{x^2 - 4x}$

73. $f(x) = \sqrt{3 - x}$

Graph each function and state its domain and range. (*Section 8.2, Objective 2*)

74. $f(x) = \dfrac{3}{4}x - 2$

75. $f(x) = 2$

76. $f(x) = -|x - 1| + 2$

Solve each problem. (*Section 8.2, Objectives 1–4*)

77. The expenditure per pupil in public elementary and secondary schools is modeled by the function $f(x) = 370.4x + 5969.8$, where x is the number of years after 1994. (Source: www.infoplease.com)

   a. Find the y-intercept and interpret its meaning in the context of the problem.

   b. Find $f(7)$ and interpret its meaning in the context of the problem.

   c. Solve $f(x) = 12{,}637$ and interpret its meaning in the context of the problem.

78. The number of deaths caused by cancer in the United States was approximately 208.7 per 100,000 people in 1996 and 180.7 per 100,000 people in 2006. (Source: http://www.cdc.gov)

   a. Assuming this trend continues, write a linear function that models the number of deaths caused by cancer (per 100,000 people), where x is the number of years after 1996.

   b. Use the function to predict the number of deaths caused by cancer in 2016.

Graph each piecewise-defined function. (*Section 8.3, Objective 1*)

79. $f(x) = \begin{cases} -x + 4, & x < 0 \\ \dfrac{5}{2}x - 2, & x \geq 0 \end{cases}$

80. $f(x) = \begin{cases} 4x - 3, & x \leq 2 \\ -1, & x > 2 \end{cases}$

Solve each problem. (*Section 8.4, Objectives 1–3*)

81. Find the variation equation if z varies jointly as the square of x and the sum of u and v given that $z = 50$ when $x = 2.5$, $u = 1.2$ and $v = 0.8$.

82. Boyle's law states that if the temperature remains the same, the volume V of a gas is inversely proportional to the pressure P applied to it. A given mass of a gas occupies 340 mL at a pressure of 800 mm Hg. If the temperature remains constant, what volume will the gas occupy if the pressure is increased to 1600 mm Hg?

Inequalities and Absolute Value

Creativity

Success in a math class depends largely on the ability to be creative. The willingness to think "outside of the box" is necessary for critical thinking and problem solving to take place. It is not enough to learn the rules and definitions; you must be able to apply them correctly. While you are not going to create new mathematics in this class, you will create new associations with the concepts. So, keep an open mind and allow yourself the time to truly think about what you are learning and how it applies to the things you know mathematically as well as in your everyday life.

Question For Thought: What are some things that stifle your creativity when it comes to learning math? Do you feel you are more creative in other areas of your life? The site http://www.virtualsalt.com/crebook1.htm has a great exposition on creative thinking. Review this and find one thing that you can apply to your life.

Chapter Outline

Coming Up...

In Section 9.1, we will learn how to use compound inequalities to determine the weight necessary for a person to have a normal BMI. Body mass index (BMI) is a measure of how much body fat a person has. A normal BMI is between 18.5 and 24.9. The formula to calculate BMI is

$BMI = \dfrac{703w}{h^2}$, where w is weight (in pounds) and h is height (in inches).

"There is no doubt that creativity is the most important human resource of all. Without creativity, there would be no progress, and we would be forever repeating the same patterns."

—Edward de Bono (Authority in the Field of Creative Thinking)

SECTION 9.1 / Compound Inequalities

In Chapter 1, we introduced the concept of the union and intersection of sets. This concept will be expanded to apply to solving special types of inequalities, compound inequalities. We will also solve linear inequalities in two variables and systems of linear inequalities. We will use the tools from Chapter 2 on solving linear inequalities in one variable to assist us here. Furthermore, a new type of equation and inequality related to absolute value will be presented.

▶ **OBJECTIVES**

As a result of completing this section, you will be able to

1. Determine the intersection of two sets.
2. Determine the union of two sets.
3. Solve compound inequalities involving "and."
4. Solve compound inequalities involving "or."
5. Solve applications of compound inequalities.
6. Troubleshoot common errors.

Body mass index (BMI) is a measure of how much body fat a person has. A normal BMI is between 18.5 and 24.9. The formula to calculate BMI is $\text{BMI} = \dfrac{703w}{h^2}$, where w is weight (in pounds) and h is height (in inches). Find the weight necessary for a person with a height of 5 ft 4 in. to have a normal BMI. To solve this problem, we need to find the solution of the inequality

$$18.5 \le \frac{703w}{64^2} \le 24.9$$

This is an example of a *compound inequality*, which we will discuss in this section.

The Intersection of Sets

Objective 1 ▶

Determine the intersection of two sets.

In Chapter 1, we discussed the concept of sets and their intersection and union. We will explore this concept further as we deal with sets which are intervals of the real numbers. Knowing how to find the intersection and union of sets will enable us to solve special types of inequalities, namely, compound inequalities.

We first investigate the *intersection* of two sets. To do this, think about the intersection of two streets. The intersection is where the streets cross; that is, the part of the road that the two streets have in common. This idea extends to sets, as discussed in Section 1.1.

Recall, the intersection of two sets is the set of elements that the two sets have in common.

> **Definition: The Intersection of Two Sets**
> Let A and B be sets. The **intersection** of A and B is denoted by $A \cap B$ and is the set of elements that are members of both sets A **and** B.
>
>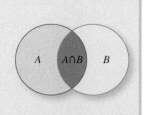
>
> In the figure, the intersection is represented by the purple piece.

> **Procedure: Finding the Intersection of Two Intervals A and B**
>
> **Step 1:** Draw the graph of each interval.
> **Step 2:** Find the interval of the real number line that contains members from both sets, that is, where the two sets overlap.

Objective 1 Examples	Find the intersection of the sets. Draw the graph of the intersection and write each solution set in interval notation and set-builder notation.

1a. $A = (-\infty, 7]$ and $B = [1, \infty)$ **1b.** $A = (3, \infty)$ and $B = (0, \infty)$

1c. $A = [4, \infty)$ and $B = (-\infty, 2]$

Solutions   **1a.**

Set	Graph	Interval
A		$(-\infty, 7]$ $\{x \mid x \le 7\}$
B		$[1, \infty)$ $\{x \mid x \ge 1\}$
$A \cap B$	Overlaps here 	$[1, 7]$ $\{x \mid 1 \le x \le 7\}$

So, $A \cap B = [1, 7]$.

1b.

Set	Graph	Interval
A		$(3, \infty)$ $\{x \mid x > 3\}$
B		$(0, \infty)$ $\{x \mid x > 0\}$
$A \cap B$	Overlaps here 	$(3, \infty)$ $\{x \mid x > 3\}$

So, $A \cap B = (3, \infty)$.

1c.

Set	Graph	Interval
A		$[4, \infty)$ $\{x \mid x \ge 4\}$
B		$(-\infty, 2]$ $\{x \mid x \le 2\}$
$A \cap B$	No overlap 	$\{\ \}$ $\varnothing$

Notice that the two sets do not overlap; therefore, there is no intersection. So, $A \cap B$ is the empty set, or $\varnothing$.

✓ **Student Check 1**   Find the intersection of the sets. Draw the graph of the intersection and write each solution set in interval notation and set-builder notation.

a. $A = [5, \infty)$ and $B = (-\infty, 6)$ **b.** $A = (-\infty, -1)$ and $B = (-\infty, -4)$

c. $A = (-\infty, -3)$ and $B = (5, \infty)$

The Union of Sets

Objective 2 ▶

Determine the union of two sets.

We will now examine another operation on sets called the *union*. The union of sets is the joining together of the elements from each set. The word "union" conjures up images of joining together people or things for a common purpose. For instance, groups of players come together to form a team, and workers often join unions to have someone look out for their interests.

Definition: The Union of Two Sets

Let A and B be sets. The **union** of A and B, denoted $A \cup B$, is the set of elements that belong to either set A **or** B.

In the diagram, $A \cup B$ consists of everything in set A and everything in set B.

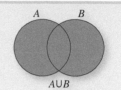

Procedure: Finding the Union of Two Intervals A and B

Step 1: Draw the graph of each interval.
Step 2: Combine each interval on one number line to form the union of the two sets.

Objective 2 Examples

Find the union of the sets. Draw the graph of each union and write the solution set in interval notation and set-builder notation.

2a. $A = (-\infty, 3)$ and $B = (6, \infty)$
2b. $A = (0, \infty)$ and $B = (3, \infty)$
2c. $A = (-\infty, 5]$ and $B = [-2, \infty)$

Solutions

2a.

Set	Graph	Interval
A	◄———————) + + + + + + + ► −1 0 1 2 3 4 5 6 7 8 9	$(-\infty, 3)$ $\{x \mid x < 3\}$
B	◄ + + + + + + (+ + + ► −1 0 1 2 3 4 5 6 7 8 9	$(6, \infty)$ $\{x \mid x > 6\}$
$A \cup B$	◄———————) + + (+ + + ► −1 0 1 2 3 4 5 6 7 8 9 Include all members of set A and all members of set B	$(-\infty, 3) \cup (6, \infty)$ $\{x \mid x < 3 \text{ or } x > 6\}$

So, $A \cup B = (-\infty, 3) \cup (6, \infty)$.

2b.

Set	Graph	Interval
A	◄ + + + + + (———————► −5 −4 −3 −2 −1 0 1 2 3 4 5	$(0, \infty)$ $\{x \mid x > 0\}$
B	◄ + + + + + + + + (+ + ► −5 −4 −3 −2 −1 0 1 2 3 4 5	$(3, \infty)$ $\{x \mid x > 3\}$
$A \cup B$	◄ + + + + + (———————► −5 −4 −3 −2 −1 0 1 2 3 4 5 Include all members of set A and all members of set B. Note all members of set B are included in set A, so we do not need to list these again.	$(0, \infty)$ $\{x \mid x > 0\}$

So, $A \cup B = (0, \infty)$.

2c.

Set	Graph	Interval
A	−5 −4 −3 −2 −1 0 1 2 3 4 5	$(-\infty, 5]$ $\{x \mid x \le 5\}$
B	−5 −4 −3 −2 −1 0 1 2 3 4 5	$[-2, \infty)$ $\{x \mid x \ge -2\}$
$A \cup B$	−5 −4 −3 −2 −1 0 1 2 3 4 5 Include all members of set A and all members of set B. The two sets together cover the entire number line.	$(-\infty, \infty)$ $\{x \mid x \in \mathbb{R}\}$

Notice that the union of the two sets covers the entire set of real numbers. So, $A \cup B = (-\infty, \infty)$.

✓ **Student Check 2** Find the union of the sets. Draw the graph of each union and write the solution set in interval notation and set-builder notation.

  a. $A = [4, \infty)$ and $B = (-\infty, 0]$ **b.** $A = (-\infty, -2)$ and $B = (-\infty, 3)$
  c. $A = [-1, \infty)$ and $B = (-\infty, 5]$

Compound Inequalities Joined by "And"

Objective 3 ▶

Solve compound inequalities involving "and."

Compound inequalities are mathematical compound sentences. Recall that a compound sentence is formed by joining two simple sentences by a coordinating conjunction, such as "and" or "or." A compound inequality is defined similarly.

> **Definition:** A **compound inequality** consists of two inequalities joined by the terms "and" or "or."

We first focus on compound inequalities joined by the term "and." Recall that a compound sentence joined by the term "and" is a statement in which two conditions must be met. Suppose an advertisement for a job states that "Candidates must have a Bachelor's degree *and* candidates must have 5 years of experience." Both conditions must be met for a person to be eligible to apply for the job.

Suppose we want to solve the inequality: $x \ge -1$ and $x \le 5$. Solutions of this compound inequality are numbers that satisfy both parts of the inequality. Consider the possible solutions.

Value	$x \ge -1$ and $x \le 5$	Is the value a solution?
$x = 3$	$3 \ge -1$ and $3 \le 5$	Because 3 makes both inequalities true, the compound inequality is *true*. So, 3 is a solution of the compound inequality.
$x = -2$	$-2 \ge -1$ and $-2 \le 5$	Because −2 makes only the second inequality true, the compound inequality is *false*. So, −2 is not a solution of the compound inequality.

So, solutions of the compound inequality, $x \ge -1$ and $x \le 5$, are values that are both greater than or equal to −1 and also less than or equal to 5. The solutions lie between and include −1 and 5. This can be written in a more compact way as $-1 \le x \le 5$. Thus, the solution set is $[-1, 5]$ or $\{x \mid -1 \le x \le 5\}$.

The values of the variable that make both inequalities true are solutions of a compound inequality joined by "and." So, solutions must lie in the *intersection* of the solution sets of each individual inequality in the compound inequality.

Procedure: Solving a Compound Inequality Involving "And"

Step 1: Find the solution set of inequality 1.
Step 2: Find the solution set of inequality 2.
Step 3: Find the intersection of the solution sets of inequalities 1 and 2.
Step 4: Write the final solution set in interval notation and provide its graph.

Objective 3 Examples   Solve each compound inequality. Write each solution set in interval notation and graph the solution set.

3a. $x + 5 \geq 8$ and $-2x \geq -10$ **3b.** $3y + 1 < 5$ and $2(y - 3) < 2$

3c. $4 - 2x > 0$ and $5x \geq 15$ **3d.** $2x - 1 \leq 5$ and $2x - 1 \geq -5$

3e. $-7 \leq 4x + 3 \leq 10$

Solutions   **3a.** $x + 5 \geq 8$ and $-2x \geq -10$

$$x + 5 - 5 \geq 8 - 5 \qquad \frac{-2x}{-2} \leq \frac{-10}{-2}$$

$$x \geq 3 \qquad\qquad x \leq 5$$

Now find the intersection of the solution sets of the two inequalities.

Inequality 1	$x \geq 3$ $[3, \infty)$	← +−+−+−+−[+−+−+−+−+−+−→ −1 0 1 2 3 4 5 6 7 8 9
Inequality 2	$x \leq 5$ $(-\infty, 5]$	←+−+−+−+−+−+−]−+−+−+−+−→ −1 0 1 2 3 4 5 6 7 8 9
Intersection of the inequalities	$[3, 5]$	←+−+−+−+[−+−+−]−+−+−+−→ −1 0 1 2 3 4 5 6 7 8 9

Check: Numbers in the intersection of the two inequalities will make the compound inequality true. Numbers not in the intersection of the two inequalities will make the compound inequality false.

$x = 4$: $x + 5 \geq 8$ and $-2x \geq -10$ | $x = 0$: $x + 5 \geq 8$ and $-2x \geq -10$

$4 + 5 \geq 8$ and $-2(4) \geq -10$ | $0 + 5 \geq 8$ and $-2(0) \geq -10$

$9 \geq 8$ and $-8 \geq -10$ | $5 \geq 8$ and $0 \geq -10$

Note that 4 makes both inequalities true and so is a solution of the compound inequality. The value 0 makes one of the inequalities false and is, therefore, not a solution of the compound inequality. So, the solution set is [3, 5].

3b. $3y + 1 < 5$ and $2(y - 3) < 2$

$3y + 1 - 1 < 5 - 1$ $2y - 6 < 2$

$\qquad 3y < 4$ $2y - 6 + 6 < 2 + 6$

$\qquad y < \dfrac{4}{3}$ $2y < 8$

$\qquad\qquad\qquad\qquad\qquad y < 4$

Now find the intersection of the solution sets of the two inequalities.

Inequality 1	$y < \dfrac{4}{3}$ $\left(-\infty, \dfrac{4}{3}\right)$	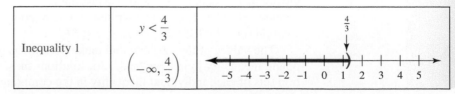

Inequality 2	$y < 4$ $(-\infty, 4)$	
Intersection of the inequalities	$\left(-\infty, \dfrac{4}{3}\right)$	

So, the solution set is $\left(-\infty, \dfrac{4}{3}\right)$.

3c.
$$4 - 2x > 0 \qquad \text{and} \qquad 5x \geq 15$$

$$4 - 2x - 4 > 0 - 4 \qquad\qquad \frac{5x}{5} \geq \frac{15}{5}$$

$$-2x > -4 \qquad\qquad\qquad x \geq 3$$

$$\frac{-2x}{-2} < \frac{-4}{-2}$$

$$x < 2$$

Now find the intersection of the solution sets of the two inequalities.

Inequality 1	$x < 2$ $(-\infty, 2)$	
Inequality 2	$x \geq 3$ $[3, \infty)$	
Intersection of the inequalities	$\varnothing$	

So, the solution set is the empty set, or $\varnothing$.

3d.
$$2x - 1 \leq 5 \qquad \text{and} \qquad 2x - 1 \geq -5$$

$$2x - 1 + 1 \leq 5 + 1 \qquad\qquad 2x - 1 + 1 \geq -5 + 1$$

$$2x \leq 6 \qquad\qquad\qquad 2x \geq -4$$

$$x \leq 3 \qquad\qquad\qquad x \geq -2$$

Now find the intersection of the solution sets of the two inequalities.

Inequality 1	$x \leq 3$ $(-\infty, 3]$	
Inequality 2	$x \geq -2$ $[-2, \infty)$	
Intersection of the inequalities	$[-2, 3]$	

So, the solution set is $[-2, 3]$.

Note: *This problem can also be worked using a compact form. The inequality* $2x - 1 \geq -5$ *is the same as* $-5 \leq 2x - 1$*. So, we have that*

$$-5 \leq 2x - 1 \quad \text{and} \quad 2x - 1 \leq 5$$

This is equivalent to $-5 \le 2x - 1 \le 5$.
This three-part inequality can be solved by applying the same operation to each part. The goal is to isolate the variable in the middle.

$$
\begin{array}{ccccl}
-5 & \le & 2x - 1 & \le & 5 \\
-5 + 1 & \le & 2x - 1 + 1 & \le & 5 + 1 \qquad \text{Add 1 to each part.} \\
-4 & \le & 2x & \le & 6 \qquad \text{Simplify.} \\
\dfrac{-4}{2} & \le & \dfrac{2x}{2} & \le & \dfrac{6}{2} \qquad \text{Divide each part by 2.} \\
-2 & \le & x & \le & 3 \qquad \text{Simplify.}
\end{array}
$$

3e.
$$
\begin{array}{ccccl}
-7 & \le & 4x + 3 & \le & 10 \\
-7 - 3 & \le & 4x + 3 - 3 & \le & 10 - 3 \qquad \text{Subtract 3 from each part.} \\
-10 & \le & 4x & \le & 7 \qquad \text{Simplify.} \\
\dfrac{-10}{4} & \le & \dfrac{4x}{4} & \le & \dfrac{7}{4} \qquad \text{Divide each part by 4.} \\
-\dfrac{5}{2} & \le & x & \le & \dfrac{7}{4} \qquad \text{Simplify.}
\end{array}
$$

So, the solution set is $\left[-\dfrac{5}{2}, \dfrac{7}{4} \right]$ and the graph is

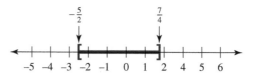

✓ **Student Check 3**   Solve each compound inequality. Write each solution set in interval notation and graph the solution set. Solve (d) also using the compact form.

   a. $x + 4 \ge 1$ and $-3x \ge -6$ **b.** $2y - 3 > 7$ and $4(y + 2) > 4$
   c. $8 + 5x > -2$ and $7x \le -21$ **d.** $1 - 9x < 8$ and $1 - 9x > -8$
   e. $-6 \le 7x + 5 \le 19$

Compound Inequalities Joined by "Or"

Objective 4 ▶

Solve compound inequalities involving "or."

We now focus on compound inequalities joined by the term "or." Recall that a compound sentence joined by the term "or" is a statement in which one of two conditions must be met. Suppose the advertisement for a job states that "Candidates must have a Bachelor's degree *or* candidates must have 5 years of experience." In this case, only one of the conditions is necessary for a person to apply for the job. Note a person may also meet both conditions and be eligible to apply.

Suppose we want to solve the inequality: $x < -1$ or $x > 5$. Solutions of this compound inequality are numbers that satisfy at least one part of the inequality. Consider the possible solutions.

Value	$x < -1$ or $x > 5$	Is the value a solution?
$x = 3$	$3 < -1$ or $3 > 5$	Because 3 makes both inequalities false, the compound inequality is *false*. So, 3 is a not a solution of the compound inequality.
$x = -2$	$-2 < -1$ or $-2 > 5$	Because -2 makes the first inequality true, the compound inequality is *true*. So, -2 is a solution of the compound inequality.
$x = 7$	$7 < -1$ or $7 > 5$	Because 7 makes the second inequality true, the compound inequality is *true*. So, 7 is a solution of the compound inequality.

The values of the variable that make at least one inequality true are solutions of a compound inequality joined by "or." Thus, values that lie in the *union* of the solution sets of each inequality are solutions of the compound inequality.

> **Procedure: Solving a Compound Inequality Involving "Or"**
>
> **Step 1:** Find the solution set of inequality 1.
> **Step 2:** Find the solution set of inequality 2.
> **Step 3:** Find the union of the solution sets of inequalities 1 and 2.
> **Step 4:** Write the final solution set in interval notation and provide its graph.

Objective 4 Examples Solve each compound inequality. Write each solution set in interval notation and graph the solution set.

4a. $x + 3 < -1$ or $2x > -4$ **4b.** $\frac{2}{3}x - 4 \geq 2$ or $-3(x + 1) \geq 5$

4c. $2x - 4 > -8$ or $7x - 5 < 2$

Solutions **4a.**

$$x + 3 < -1 \qquad \text{or} \qquad 2x > -4$$

$$x + 3 - 3 < -1 - 3 \qquad \frac{2x}{2} > \frac{-4}{2}$$

$$x < -4 \qquad\qquad\qquad x > -2$$

Now find the union of the two solution sets.

Inequality 1	$x < -4$ $(-\infty, -4)$	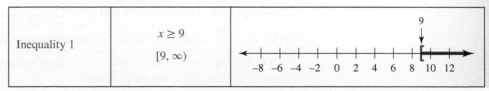
Inequality 2	$x > -2$ $(-2, \infty)$	
Union of two sets	$(-\infty, -4) \cup (-2, \infty)$	

So, the solution set is $(-\infty, -4) \cup (-2, \infty)$.

4b.

$$\frac{2}{3}x - 4 \geq 2 \qquad \text{or} \qquad -3(x + 1) \geq 5$$

$$\qquad\qquad\qquad\qquad -3x - 3 \geq 5$$

$$3\left(\frac{2}{3}x - 4\right) \geq 3(2) \qquad -3x - 3 + 3 \geq 5 + 3$$

$$2x - 12 \geq 6 \qquad\qquad -3x \geq 8$$

$$2x - 12 + 12 \geq 6 + 12 \qquad \frac{-3x}{-3} \leq \frac{8}{-3}$$

$$2x \geq 18 \qquad\qquad\qquad x \leq -\frac{8}{3}$$

$$x \geq 9$$

Now find the union of the two solution sets.

Inequality 1	$x \geq 9$ $[9, \infty)$	

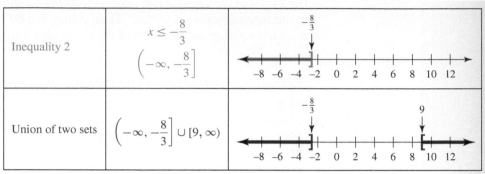

| Inequality 2 | $x \leq -\frac{8}{3}$ $\left(-\infty, -\frac{8}{3}\right]$ | |
| Union of two sets | $\left(-\infty, -\frac{8}{3}\right] \cup [9, \infty)$ | |

So, the solution set is $\left(-\infty, -\frac{8}{3}\right] \cup [9, \infty)$.

4c.

$$\begin{array}{ll}
2x - 4 > -8 & \text{or} \qquad 7x - 5 < 2 \\
2x - 4 + 4 > -8 + 4 & \qquad 7x - 5 + 5 < 2 + 5 \\
2x > -4 & \qquad 7x < 7 \\
x > -2 & \qquad x < 1
\end{array}$$

Now find the union of the two solution sets.

Inequality 1	$x > -2$ $(-2, \infty)$	
Inequality 2	$x < 1$ $(-\infty, 1)$	
Union of two sets	$(-\infty, \infty)$	

So, the solution set is all real numbers, $\mathbb{R}$, or $(-\infty, \infty)$.

✔ **Student Check 4**   Solve each compound inequality. Write each solution set in interval notation and graph the solution set.

a. $x - 1 > 4$ or $-2x > -10$ **b.** $\frac{9}{4}x - 5 \geq -3$ or $6(x + 2) < 7$

c. $-4x - 9 > -1$ or $2 - 3x < 11$

Applications

Objective 5 ▶

Solve applications of compound inequalities.

Applications of compound inequalities arise when there are two conditions that define a situation. When both conditions must be satisfied, an "and" statement is used. When only one of the conditions must be satisfied, an "or" statement is used.

Note: *When the word "between" is used in the following examples and in the exercises, we will assume the endpoints are included.*

Objective 5 Examples   Write a compound inequality that represents each situation and solve the inequality. Use a complete sentence to answer each question.

5a. Body mass index (BMI) is a measure of how much body fat a person has. A normal BMI is between 18.5 and 24.9. The formula to calculate BMI is

$$\text{BMI} = \frac{703w}{h^2},$$ where w is weight (in pounds) and h is height (in inches). If Sondra is 5 ft 4 in. tall, how much should she weigh to have a normal BMI? Round answer to the nearest whole number.

Solution **5a.** Sondra's height in inches is 64. Her BMI is represented by $\dfrac{703w}{64^2} = \dfrac{703w}{4096}$. So, we need to find the solution of the following inequality.

$$18.5 \le \frac{703w}{4096} \le 24.9$$

$$4096(18.5) \le 4096\left(\frac{703w}{4096}\right) \le 4096(24.9) \qquad \text{Multiply each part by 4096.}$$

$$75{,}776 \le 703w \le 101{,}990.4 \qquad \text{Simplify.}$$

$$\frac{75{,}776}{703} \le \frac{703w}{703} \le \frac{101{,}990.4}{703} \qquad \text{Divide each part by 703.}$$

$$108 \le w \le 145 \qquad \text{Approximate.}$$

So, Sondra should weigh between 108 and 145 lb to have a normal BMI.

5b. Neal needs to earn a B in his math class to keep his 3.0 GPA. His final grade is calculated using the following percentages: tests (45%), quizzes (20%), projects (10%), and final (25%). Neal has a 73 test average, 85 quiz average, and 82 project average. What grades could he earn on the final exam for his average to be between 80 and 89? (The maximum grade on the final exam is 100.)

Solution **5b.** The unknown is Neal's final exam grade. So, we let f represent the final exam grade. Neal's final average is computed as shown.

$$\text{Final average} = 0.45(73) + 0.20(85) + 0.10(82) + 0.25f$$
$$= 32.85 + 17 + 8.2 + 0.25f$$
$$= 58.05 + 0.25f$$

For Neal's final average to be between 80 and 89, we solve the following inequality.

$$80 \le 58.05 + 0.25f \le 89$$

$$80 - 58.05 \le 58.05 + 0.25f - 58.05 \le 89 - 58.05 \qquad \text{Subtract 58.05 from each part.}$$

$$21.95 \le 0.25f \le 30.95 \qquad \text{Simplify.}$$

$$\frac{21.95}{0.25} \le \frac{0.25f}{0.25} \le \frac{30.95}{0.25} \qquad \text{Divide each part by 0.25.}$$

$$87.5 \le f \le 123.8 \qquad \text{Simplify.}$$

To get a B in the course, Neal needs to score between 88 and 100 since 100 is the highest grade possible. This also tells us that Neal cannot earn a grade higher than a B, since a grade of 123.8 would be needed to earn an 89 average.

✓ Student Check 5 Write a compound inequality that represents each situation and solve the inequality. Use a complete sentence to answer each question.

a. Bryan is 6 ft tall. How much should he weigh to have a normal BMI? Use the formula in Example 5(a).

b. Ameena is attempting to earn a C in her chemistry class. Her final grade is computed as follows: 60% (test average), 15% (quizzes), and 25% (final exam). Ameena has a test average of 62 and a quiz average of 78. What grades could she earn on her final to have a final grade between 70 and 79?

Objective 6 ▶	Troubleshooting Common Errors
Troubleshoot common errors.	Some common errors related to solving compound inequalities are shown.

Objective 6 Examples	A problem and an incorrect solution are given. Provide the correct solution and an explanation of the error.

6a. Solve $5x - 5 < -5$ or $-2x < 0$.

Incorrect Solution	Correct Solution and Explanation
$5x - 5 < -5$ or $-2x < 0$ $5x < 0$ or $x > 0$ $x < 0$ The solution set is all real numbers, or $\mathbb{R}$.	Since neither inequality includes zero as a solution, the union can't include zero. So, the solution set is $(-\infty, 0) \cup (0, \infty)$.

6b. Solve $5x - 5 \leq -5$ or $-2x \leq 0$.

Incorrect Solution	Correct Solution and Explanation
$5x - 5 \leq -5$ or $-2x \leq 0$ $5x \leq 0$ or $x \geq 0$ $x \leq 0$ The solution set is the empty set, or $\varnothing$.	Since both inequalities include zero as a solution, the intersection will include zero. So, the solution set is $\{0\}$.

ANSWERS TO STUDENT CHECKS

Student Check 1

a.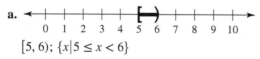

$[5, 6)$; $\{x \mid 5 \leq x < 6\}$

b.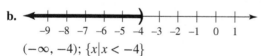

$(-\infty, -4)$; $\{x \mid x < -4\}$

c.

$\varnothing$

Student Check 2   **a.** $(-\infty, 0] \cup (4, \infty)$ **b.** $(-\infty, 3)$
c. $(-\infty, \infty)$

Student Check 3   **a.** $[-3, 2]$

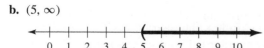

b. $(5, \infty)$

c. $\varnothing$

d. $\left(-\dfrac{7}{9}, 1\right)$

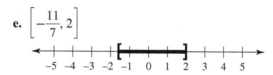

e. $\left[-\dfrac{11}{7}, 2\right]$

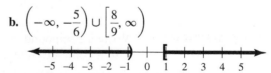

Student Check 4   a. $(-\infty, 5) \cup (5, \infty)$

b. $\left(-\infty, -\dfrac{5}{6}\right) \cup \left[\dfrac{8}{9}, \infty\right)$

c. $(-\infty, \infty)$

Student Check 5   a. Bryan should weigh between 136 and 184 lb.   **b.** Ameena must score between 85 and 100 on her final to make a 70–79 in chemistry.

SUMMARY OF KEY CONCEPTS

1. The intersection of two sets is the set of elements the sets have in common.

2. The union of two sets is the set of all elements from each set.

3. To solve a compound inequality involving "and" means to find the intersection of the solution sets of each inequality.

4. To solve a compound inequality involving "or" means to find the union of the solution sets of each inequality.

GRAPHING CALCULATOR SKILLS

We can use a graphing calculator to graph the solution set of a compound inequality. It is best to solve the inequality by hand and then use the calculator to confirm the solution. Enter the compound inequality into Y_1. Insert parentheses around each inequality. Use the Test menu (2nd Math) to access the inequality symbols and the coordinating conjunction (2nd Math, Logic). Then graph the solution. The interval that contains the solutions will be highlighted above the x-axis.

Example 1: Solve $x + 6 < -1$ or $2x > -4$.

Solution:

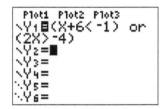

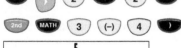

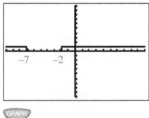

So, the solution set is $(-\infty, -7) \cup (-2, \infty)$.

Example 2: $x + 5 \geq 8$ and $-2x \geq -10$.

Solution:

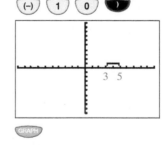

So, the solution set is $[3, 5]$.

SECTION 9.1 / EXERCISE SET

✏️ Write About It!

Use complete sentences in your answer to each exercise.

1. Explain the meaning of the intersection of sets.

2. Explain the meaning of the union of sets.

3. Is it possible for the intersection of two sets to be the empty set? Explain.

4. Is it possible for the union of two sets to be the empty set? Explain.

5. What is a compound inequality?

6. Explain the difference of the graphs of compound inequalities $-a < x < a$ and $x < -a$ or $x > a$, where $a > 0$.

7. How do you find the solutions of a compound inequality involving "and"?

8. How do you find the solutions of a compound inequality involving "or"?

Practice Makes Perfect!

Find the intersection and union of the following sets. Write each solution set in interval notation. (*See Objectives 1 and 2.*)

9. $A = (-10, 6)$, $B = [-4, 2]$

10. $A = (-8, 3)$, $B = [-7, 1]$

11. $A = [-10, -3]$, $B = [-9, -3]$

12. $A = [-13, -5]$, $B = [-12, -5]$

13. $A = [6, 20]$, $B = (6, 20)$

14. $A = [-7, 5]$, $B = (-7, 5)$

15. $A = (-\infty, 7)$, $B = [-1, 7]$

16. $A = (-\infty, -2]$, $B = [-4, -2)$

17. $A = (-11, \infty)$, $B = (-\infty, 11)$

18. $A = (-\infty, 53)$, $B = (-53, \infty)$

19. $A = [-12, 5]$, $B = (-10, 8)$

20. $A = [-2, 10]$, $B = (-4, 6)$

Solve each compound inequality. Write each solution set in interval notation. (*See Objective 3.*)

21. $x \leq 5$ and $x \geq -1$

22. $x \leq 1$ and $x \geq -10$

23. $x > 3$ and $x < 2$

24. $x \leq 8$ and $x > 10$

25. $2x + 4 > 8$ and $3x < 9$

26. $9x + 3 > 12$ and $10x \leq 40$

27. $3x + 6 \geq 3$ and $7x - 13 < -20$

28. $6x - 1 > 17$ and $4x - 5 \leq 29$

29. $6 - 4x > 2$ and $1 - 3x < 5$

30. $7 - 8x > -1$ and $2 - 5x < 2$

31. $9x + 11 \geq 2$ and $4x - 3 \leq -7$

32. $12x + 1 \geq 13$ and $5x - 6 \leq -1$

33. $5(3x - 2) \leq 15$ and $2(4x + 1) \geq 10$

34. $4(2x - 3) \leq 6$ and $3(5x + 10) \geq 0$

35. $x > 7$ and $x > 5$

36. $x \leq -3$ and $x \leq -7$

37. $6x - 5 > 1$ and $7x + 6 > 20$

38. $4x - 3 > 5$ and $5x - 1 \geq 9$

39. $3x - 2 \geq 10$ and $1 - 3x < 40$

40. $2x - 7 > 0$ and $1 - 2x < 15$

41. $6 - 3x > 4$ and $2 - 2x > 10$

42. $3 - 10x \leq 23$ and $10 - 11x \leq 43$

43. $3x + 9 \geq 4$ and $9x > -15$

44. $4x - 1 \leq 11$ and $3x < 9$

45. $2(4x - 3) < 4$ and $6(x - 5) < 12$

46. $4(2x + 1) > 5$ and $8(x - 2) > 3$

Solve each compound inequality. Write each solution set in interval notation. (*See Objective 4.*)

47. $x < -3$ or $x > 1$

48. $x < -4$ or $x > 0$

49. $x \geq -7$ or $x < 3$

50. $x > 2$ or $x \leq 10$

51. $x \leq 0$ or $x \leq 6$

52. $x \geq 10$ or $x \geq 15$

53. $11x - 3 \geq 52$ or $6x \leq 42$

54. $13x < -52$ or $3 - 5x > 17$

55. $12x \leq -60$ or $7 - 9x > 3$

56. $13x \geq 65$ or $11 - 5x > 6$

57. $5x + 3 > 8$ or $7x - 4 \leq -3$

58. $2x + 5 > 6$ or $3x - 1 \leq -4$

59. $5x + 1 \leq 7$ or $6x - 2 < 1$

60. $4x - 7 < 9$ or $2x + 3 < 4$

61. $10(2 - 9x) \geq 20$ or $2(1 + 3x) > 6$

62. $10(3 - x) \leq 15$ or $3(2 - 5x) > 13$

Write a compound inequality that represents each situation and solve the inequality. Use a complete sentence to answer each question. (*See Objective 5.*)

63. If Lawrence's quiz scores are 90, 82, 75, and 78, what must he earn on his fifth quiz to have an average between 80 and 89?

64. If Emma's quiz scores are 99, 89, 80, and 70, what must she earn on her fifth quiz to have an average between 80 and 89?

65. Sakinah's course grade is based on a weighted average. The weights are as follows: homework (15%), quizzes (25%), tests (40%), and final exam (20%). If Sakinah's homework average is 100, quiz average is 90, and test average is 86, what must she score on the final exam to have an average between 90 and 100?

66. Su Ming's course grade is based on a weighted average. The weights are as follows: homework (15%), quizzes (25%), tests (40%), and final exam (20%). If Su Ming's homework average is 75, quiz average is 70, test average is 83, what must she score on the final exam for her course average to be between 80 and 89?

67. A cell phone company charges a monthly fee of $29.99 for 300 anytime cell phone minutes and $0.40 per additional minute. How many additional minutes can a subscriber use if he budgets between $40 and $50 for his phone bill?

68. A cell phone company charges a monthly fee of $49.99 for 800 anytime cell phone minutes and $0.25 per additional minute. If Callie budgets a range of $70 to $100 each month for her phone bill, how many additional minutes can she afford to use?

69. Miguel pays $3000 to rent a hall for his fiftieth birthday party. The buffet costs $18 per person. If Miguel has between $4000 and $5000 to spend on his party, how many people can he invite to his birthday party?

70. Marianne is planning a graduation party for her daughter. She finds a banquet hall that charges $850 and a caterer who charges $20 per person. If Marianne can spend between $2000 and $3000 on the party, how many people can attend the party?

 Mix 'Em Up!

Solve each problem. Write each answer in interval notation when applicable.

71. $[-7, 8] \cup (-3, 12)$ **72.** $(-20, 5) \cup [-5, 9]$

73. $[-20, 9] \cap (-20, 9]$ **74.** $[-1, 2] \cap (-1, 2]$

75. $(-\infty, 6] \cup (6, \infty)$ **76.** $(-\infty, 3) \cup [3, \infty)$

77. $(-\infty, 3) \cap (6, \infty)$ **78.** $(-\infty, 5] \cap [10, \infty)$

79. $(-\infty, 6.5) \cup [-3.5, 12.2)$

80. $(-\infty, 6.5) \cap [-3.5, 12.2)$

81. $\varnothing \cap [2, 7]$ **82.** $\varnothing \cap (-9, 7)$

83. $\varnothing \cup (-10, \infty)$ **84.** $\varnothing \cup [-1, \infty)$

85. $7x + 6 < 11$ or $5x - 5 > 12$

86. $10x - 5 \geq 2$ and $4x + 3 < 10$

87. $2(1 - 4x) > 10$ and $4x - 5 < 11$

88. $3(2 - x) > -6$ and $6x + 1 < 7$

89. $\frac{2}{3}x - \frac{7}{6} > \frac{5}{6}$ or $\frac{3}{5}x + \frac{1}{2} < \frac{7}{10}$

90. $\frac{3}{4}x - \frac{1}{8} > 1$ or $\frac{5}{6}x + 2 < \frac{2}{3}$

91. $1 - \frac{3}{7}x < \frac{1}{2}$ and $1 + \frac{1}{2}x < \frac{3}{5}$

92. $2 - \frac{5}{2}x > \frac{13}{2}$ and $\frac{7}{3}x - 1 > \frac{2}{5}$

93. $3 - 4x \leq 0$ or $6x + 1 > 3$

94. $2 - 5x \leq 1$ or $4x - 3 > 5$

95. $8x + 9 > 3$ and $7x - 2 > -3$

96. $7x - 1 \geq 5$ and $3x + 4 > -2$

97. $3x - 5 > 7$ or $2x + 6 < 1$

98. $12x - 4 \leq 8$ and $3x + 2 > 2$

99. $0.3x + 0.12 < 3.27$ and $0.4x - 1.36 > 0.54$

100. $1.56x - 3.2 \geq 2.26$ and $0.8x + 5.6 \leq 2.4$

101. $2x + 6 < 8$ or $4x - 9 < 2$

102. $8x - 3 \leq 3$ or $5x - 1 < 7$

Write a compound inequality that represents each situation and solve the inequality. Use a complete sentence to answer each question.

103. Frederick needs the work of an electrician. He finds an electrician that charges $50 to make a service call plus $30 per hour. If Frederick plans to spend between $110 and $500, how many hours can he afford to hire the electrician?

104. Roberto hires a professional organizer to organize his home office. The professional organizer charges a flat fee of $75 plus $25 per hour. If Roberto budgets between $600 and $800 to get his office organized, how many hours can he afford to hire the professional organizer?

105. Maddy wants to maintain a test average between an 80 and 89. If Maddy's test scores are 85, 96, 89, and 78, what must she score on her fifth test?

106. Isaac wants to maintain a quiz average between 70 and 79. If Isaac's quiz scores are 84, 69, 73, 81, and 70, what must he score on his sixth quiz?

 You Be the Teacher!

Correct each student's errors, if any.

107. Solve $10 - 3x < 4$ and $5x + 23 > 3$.

Josh's work:

$10 - 3x < 4$ and $5x + 23 > 3$

$-3x < -6$ and $5x > -20$

$x < 2$ and $x > -4$

The solution set is $(-4, 2)$.

108. Solve $28 - 4x > 4$ or $28 - 4x < -4$.

Mark's work:

$28 - 4x > 4$ or $28 - 4x < -4$

$-4x > -24$ or $-4x < -32$

$x > 6$ or $x < 8$

The solution set is $(-\infty, \infty)$.

 Calculate It!

Solve each compound inequality and use a graphing calculator to confirm each solution.

109. $6x + 7 > 1$ and $5x - 2 < 13$

110. $3x - 2 > 10$ and $4x - 5 < 23$

111. $5x - 7 \leq 33$ or $7x > 14$

112. $2x \geq 18$ or $3x - 4 < 23$

| SECTION 9.2 | **Absolute Value Equations** |

▶ **OBJECTIVES**

As a result of completing this section, you will be able to

1. Solve absolute value equations of the form $|X| = k$, where k is a real number.

2. Solve absolute value equations of the form $|X| = |Y|$.

3. Solve applications of absolute value equations.

4. Troubleshoot common errors.

A memory card is measured with a ruler with centimeter gradations. The width is reported to be 2.5 cm. Because of the inaccuracy of measuring, the actual width is 2.5 ± 0.1 cm. This states that the distance between the exact width and the measured width is 0.1 cm. What is the actual width of the memory card? To solve this problem, we must solve the equation $|w - 2.5| = 0.1$, where w represents the exact width of the memory card.

In this section, we will learn how to solve equations containing absolute values.

Absolute Value Equations

In Section 1.1, the absolute value of a number was presented. The absolute value of a number is the number's distance from zero on a real number line. We can visualize this on a number line as shown. Note that both 5 and -5 are 5 units from zero. So, their absolute values are the same.

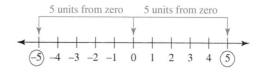

In fact, every real number and its opposite have the same absolute value, since they are the same distance from zero on a number line. Some examples are shown.

$$|-5| = 5 \text{ and } |5| = 5 \qquad |-6| = 6 \text{ and } |6| = 6 \qquad |-7| = 7 \text{ and } |7| = 7$$

Objective 1 ▶

Solve absolute value equations of the form $|X| = k$, where k is a real number.

Important Facts About Absolute Value

- The absolute value of any real number is nonnegative.
- There are two numbers (a number and its opposite) that have the same absolute value. For instance, both -5 and 5 have the same absolute value, which is 5.
- The number zero has an absolute value of zero.

In this section, we will solve **absolute value equations**. Some examples are

$$|x| = 5 \qquad |y - 4| = 7 \qquad |2x - 1| - 3 = 2$$

We will use the following property to solve absolute value equations.

Property: Property 1 for Absolute Value Equations

Let k be a real number.

1. If $|X| = k$ and $k > 0$, then $X = k$ or $X = -k$.
2. If $|X| = 0$, then $X = 0$.
3. If $|X| = k$ and $k < 0$, then there are no solutions.

Note: *The expression inside the absolute value can be a single variable or a variable expression.*

Procedure: Solving an Absolute Value Equation

Step 1: Isolate the absolute value on one side of the equation and the constant on the other side.

Step 2: Set the expression inside the absolute value equal to the number(s) whose absolute value is the constant.

a. If the constant is positive, there are two solutions.
b. If the constant is zero, there is one solution.
c. If the constant is negative, there are no solutions.
Step 3: Check the solutions by substituting them in the original equation.

Objective 1 Examples	**Solve each equation.**

1a. $|x| = 5$ **1b.** $|a| = -3$ **1c.** $|y - 4| = 7$ **1d.** $|2x - 1| - 3 = 2$

1e. $\left|\dfrac{4x + 6}{5}\right| + 1 = 1$

Solutions

1a. We set the variable x equal to the numbers whose absolute value is 5, namely 5 and -5.

$$|x| = 5$$
$$x = 5 \quad \text{or} \quad x = -5$$

So, the solution set is $\{-5, 5\}$ or $\{\pm 5\}$. We can check by replacing x with 5 and -5 in the original equation.

1b. The equation asks us to find all real numbers a whose absolute value, or distance from zero, is -3. There is no real number whose absolute value is a negative number. Therefore, the solution set of this equation is the empty set, or $\varnothing$.

1c. We set the expression $y - 4$ equal to the numbers whose absolute value is 7, namely 7 and -7.

$$|y - 4| = 7$$

$y - 4 = 7$	or $\quad y - 4 = -7$	Apply property 1.
$y - 4 + 4 = 7 + 4$	$y - 4 + 4 = -7 + 4$	Add 4 to each side.
$y = 11$	$y = -3$	Simplify.

Check: $y = 11$: $\qquad\qquad\qquad\qquad\qquad y = -3$:

$|y - 4| = 7$	$|y - 4| = 7$
$|11 - 4| = 7$	$|-3 - 4| = 7$
$|7| = 7$	$|-7| = 7$
$7 = 7 \quad$ True	$7 = 7 \quad$ True

So, the solution set is $\{-3, 11\}$.

1d. We must first isolate the absolute value expression.

$$|2x - 1| - 3 = 2$$

$|2x - 1| - 3 + 3 = 2 + 3$	Add 3 to each side.
$|2x - 1| = 5$	Simplify.

$2x - 1 = 5$	or $\quad 2x - 1 = -5$	Apply property 1.
$2x - 1 + 1 = 5 + 1$	$2x - 1 + 1 = -5 + 1$	Add 1 to each side.
$2x = 6$	$2x = -4$	Simplify.
$\dfrac{2x}{2} = \dfrac{6}{2}$	$\dfrac{2x}{2} = \dfrac{-4}{2}$	Divide each side by 2.
$x = 3$	$x = -2$	Simplify.

So, the solution set is $\{-2, 3\}$. We can check by replacing x with -2 and 3 in the original equation.

1e. We must first isolate the absolute value expression.

$$\left|\frac{4x+6}{5}\right| + 1 = 1$$

$$\left|\frac{4x+6}{5}\right| + 1 - 1 = 1 - 1 \qquad \text{Subtract 1 from each side.}$$

$$\left|\frac{4x+6}{5}\right| = 0 \qquad \text{Simplify.}$$

$$\frac{4x+6}{5} = 0 \qquad \text{Apply property 1.}$$

$$5\left(\frac{4x+6}{5}\right) = 5(0) \qquad \text{Multiply each side by 5.}$$

$$4x + 6 = 0 \qquad \text{Simplify.}$$

$$4x + 6 - 6 = 0 - 6 \qquad \text{Subtract 6 from each side.}$$

$$4x = -6 \qquad \text{Simplify.}$$

$$\frac{4x}{4} = \frac{-6}{4} \qquad \text{Divide each side by 4.}$$

$$x = -\frac{3}{2} \qquad \text{Simplify.}$$

So, the solution set is $\left\{-\frac{3}{2}\right\}$. We can check by replacing x with 0 in the original equation.

✓ **Student Check 1**   Solve each equation.

    a. $|x| = 7$ **b.** $|a| = -10$ **c.** $|y+1| = 6$ **d.** $|3x+2| - 4 = 8$

    e. $\left|\dfrac{5x-9}{4}\right| + 3 = 3$

Absolute Value Equations, $|X| = |Y|$

Objective 2 ▶

Solve absolute value equations of the form $|X| = |Y|$.

To solve equations containing two absolute values equal to one another, we need to understand when the absolute value of two expressions is equal. This occurs when the expressions have the same distance from zero. So, the expressions must either be the *same* or *opposite*. The following are examples of absolute value expressions that are equal.

$|5| = |5|$ $|-5| = |-5|$ The expressions inside the absolute values are the same.

$|5| = |-5|$ The expressions inside the absolute values are opposites.

We use the following property to solve equations containing two absolute values.

> **Property: Property 2 for Absolute Value Equations**
> If $|X| = |Y|$, then $X = Y$ or $X = -Y$.

> **Procedure: Solving an Equation Containing Two Absolute Value Expressions**
>
> **Step 1:** Set the expressions inside the absolute values equal to one another.
> **Step 2:** Set the expression inside the first absolute value equal to the opposite of the expression inside the second absolute value.
> **Step 3:** Solve the resulting equations.
> **Step 4:** Check the solutions by substituting them in the original equation.

Objective 2 Examples   Solve each equation.

2a. $|x - 2| = |3x - 4|$ **2b.** $\left|\dfrac{y-1}{5}\right| = \left|\dfrac{y}{2}\right|$

Solutions   **2a.**
$$|x - 2| = |3x - 4|$$

$$x - 2 = 3x - 4 \qquad \text{or} \qquad x - 2 = -(3x - 4)$$
$$x - 2 - x = 3x - 4 - x \qquad\qquad x - 2 = -3x + 4$$
$$-2 = 2x - 4 \qquad\qquad x - 2 + 3x = -3x + 4 + 3x$$
$$-2 + 4 = 2x - 4 + 4 \qquad\qquad 4x - 2 = 4$$
$$2 = 2x \qquad\qquad 4x - 2 + 2 = 4 + 2$$
$$\frac{2}{2} = \frac{2x}{2} \qquad\qquad 4x = 6$$
$$1 = x \qquad\qquad \frac{4x}{4} = \frac{6}{4}$$
$$x = \frac{3}{2}$$

So, the solution set is $\left\{1, \dfrac{3}{2}\right\}$. We can check by replacing x with 1 and $\dfrac{3}{2}$.

2b.
$$\left|\frac{y-1}{5}\right| = \left|\frac{y}{2}\right|$$

$$\frac{y-1}{5} = \frac{y}{2} \qquad \text{or} \qquad \frac{y-1}{5} = -\frac{y}{2}$$
$$10\left(\frac{y-1}{5}\right) = 10\left(\frac{y}{2}\right) \qquad\qquad 10\left(\frac{y-1}{5}\right) = 10\left(-\frac{y}{2}\right)$$
$$2(y - 1) = 5y \qquad\qquad 2(y - 1) = -5y$$
$$2y - 2 = 5y \qquad\qquad 2y - 2 = -5y$$
$$2y - 2 - 2y = 5y - 2y \qquad\qquad 2y - 2 - 2y = -5y - 2y$$
$$-2 = 3y \qquad\qquad -2 = -7y$$
$$\frac{-2}{3} = \frac{3y}{3} \qquad\qquad \frac{-2}{-7} = \frac{-7y}{-7}$$
$$-\frac{2}{3} = y \qquad\qquad \frac{2}{7} = y$$

So, the solution set is $\left\{-\dfrac{2}{3}, \dfrac{2}{7}\right\}$. We can check by replacing y with $-\dfrac{2}{3}$ and $\dfrac{2}{7}$.

✓ Student Check 2   Solve each equation.

a. $|5z - 7| = |z + 1|$ **b.** $\left|\dfrac{x+3}{2}\right| = \left|\dfrac{2x}{5}\right|$

Applications

Objective 3 ▶

Solve applications of absolute value equations.

Since the absolute value of a number represents that number's distance from zero on a number line, distance between objects is a key application of absolute value equations. In measurements, this distance is often referred to as absolute error. In science-related fields, accuracy in measurement is vital. However, there will almost always be error in measurement due to the precision of the measuring instruments and human error.

> **Definition:** **Absolute error** is the absolute value (sometimes called magnitude) of the difference between the measured value and the exact value. We can use the following formula to represent the absolute error.
>
> $E_{abs} = |x - a|$, where a is the approximated value and x is the exact measure.

Objective 3 Examples　　Solve each problem.

3a. A memory card is measured with a ruler with centimeter gradations. The width is reported to be 2.5 cm. The absolute error is 0.1 cm. Find the possible values for the exact width of the memory card.

Solution　**3a.** Let w represent the exact width of the memory card. The approximated value is 2.5 cm and the absolute error is 0.1 cm. To find the exact width, we must solve the following equation.

$$|w - 2.5| = 0.1$$

$w - 2.5 = 0.1$　　or　　$w - 2.5 = -0.1$	Apply property 1.
$w - 2.5 + 2.5 = 0.1 + 2.5$　　$w - 2.5 + 2.5 = -0.1 + 2.5$	Add 2.5 to each side.
$w = 2.6$　　　　　$w = 2.4$	Simplify.

So, the exact width of the memory card is at most 2.6 cm and at least 2.4 cm.

3b. Find the absolute error when the value of 3.14 is used for π. Round the answer to four decimal places.

Solution　**3b.** We can find the absolute error using the given formula since we know the approximated value is 3.14 and the exact value is π. So, $a = 3.14$ and $x = \pi$.

$$E_{abs} = |x - a|$$
$$E_{abs} = |\pi - 3.14|$$
$$E_{abs} = 0.0016$$

So, the absolute error is 0.0016.

✔ Student Check 3　　Solve each problem.

a. Suppose a dog measures 80 cm long with a 3 cm absolute error. Find the possible values for the exact length of the dog.

b. Find the absolute error when the value of 2.7 is used for the number e, where $e = 2.71828 \ldots$. Round the answer to four decimal places.

Objective 4 ▶

Troubleshoot common errors.

Troubleshooting Common Errors

Some common errors associated with solving absolute value equations are shown.

Objective 4 Examples	A problem and an incorrect solution are given. Provide the correct solution and an explanation of the error.

4a. Solve: $|2x - 3| = 7$.

Incorrect Solution	Correct Solution and Explanation
$$|2x - 3| = 7$$ $$2x - 3 = 7$$ $$2x = 10$$ $$x = 5$$ The solution set is $\{5\}$.	There are two solutions of this equation since there are two numbers whose absolute value is 7. $$|2x - 3| = 7$$ $$2x - 3 = 7 \quad \text{or} \quad 2x - 3 = -7$$ $$2x = 10 \qquad\qquad 2x = -4$$ $$x = 5 \qquad\qquad x = -2$$ The solution set is $\{-2, 5\}$.

4b. Solve: $|y + 2| - 1 = 4$.

Incorrect Solution	Correct Solution and Explanation
$$|y + 2| - 1 = 4$$ $$y + 2 - 1 = 4 \quad \text{or} \quad y + 2 - 1 = -4$$ $$y + 1 = 4 \qquad\qquad y + 1 = -4$$ $$y = 3 \qquad\qquad y = -5$$ The solution set is $\{-5, 3\}$.	The absolute value expression must be isolated prior to solving. We must first add 1 each side. $$|y + 2| - 1 = 4$$ $$|y + 2| = 5$$ $$y + 2 = 5 \quad \text{or} \quad y + 2 = -5$$ $$y = 3 \qquad\qquad y = -7$$ The solution set is $\{-7, 3\}$.

4c. Solve: $|3a + 2| = -7$.

Incorrect Solution	Correct Solution and Explanation
$$|3a + 2| = -7$$ $$3a + 2 = -7 \quad \text{or} \quad 3a + 2 = 4$$ $$3a = -9 \qquad\qquad 3a = 2$$ $$a = -3 \qquad\qquad a = \frac{2}{3}$$ The solution set is $\left\{-3, \frac{2}{3}\right\}$.	Because the absolute value expression is equal to a negative number, this equation has no solution. The absolute value of any real number is always greater than or equal to zero. So, the solution set is the empty set, or $\varnothing$.

ANSWERS TO STUDENT CHECKS

Student Check 1   **a.** $\{\pm 7\}$ **b.** $\varnothing$ **c.** $\{-7, 5\}$
     d. $\left\{-\frac{14}{3}, \frac{10}{3}\right\}$ **e.** $\left\{\frac{9}{5}\right\}$

Student Check 2   **a.** $\{1, 2\}$ **b.** $\left\{-15, -\frac{5}{3}\right\}$

Student Check 3   **a.** The length of the dog is at most 83 cm and at least 77 cm.
     b. The absolute error is 0.0183.

SUMMARY OF KEY CONCEPTS

1. To solve an absolute value equation, we must write the equation so that it is in the form $|X| = c$, where c is a real number. If the constant c is positive, there are two solutions. If the constant c is zero, there is one solution. If the constant c is negative, there are no solutions.

2. To solve an equation in which the absolute value of one expression is equal to the absolute value of another expression, we must form two equations. One equation is formed by setting the two expressions equal to one another. The other equation is formed by setting one expression equal to the opposite of the other expression.

3. A key application of absolute value involves the error in measurement. The absolute error is calculated as the absolute value of the difference of the exact value and the approximated value.

GRAPHING CALCULATOR SKILLS

We can use the graphing calculator to verify solutions of absolute value equations.

Example: Verify that the solutions of the equation $|2x - 1| - 3 = 2$ are -2 and 3.

Solution:

Method 1: Enter the solutions as the stored value for the variable x. Then compute the value of the left side of the equation. The result should be the right side of the equation.

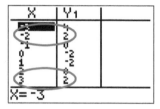

Method 2: Enter the left side of the equation in the equation editor. Verify that the table shows that the corresponding y-value of $x = -2$ and $x = 3$ is $y = 2$.

Since $Y_1 = 2$ for both $x = -2$ and $x = 3$, we know they are solutions of the equation.

SECTION 9.2 / EXERCISE SET

Write About It!

Use complete sentences in your answer to each exercise.

1. Can the absolute value of an expression equal zero? Explain.

2. Can the absolute value of an expression equal a negative number? Explain.

3. Explain the steps for solving an equation with one absolute value expression.

4. Explain the steps for solving an equation with two absolute value expressions.

Practice Makes Perfect!

Solve each equation. (*See Objective 1.*)

5. $|x| = 10$
6. $|x| = 1$
7. $|2x + 3| = 2$
8. $|3x - 1| = 5$

9. $|4 - y| = 6$
10. $|6 - y| = 3$
11. $|a| = -9$
12. $|a| = -4$
13. $|4x + 3| = -6$
14. $|5x - 1| = -2$
15. $|4x + 3| - 6 = 8$
16. $|6x - 5| + 1 = 3$
17. $|4 - 7x| + 8 = 2$
18. $|1 - 3x| + 5 = 4$
19. $\left|\dfrac{1 - 4x}{5}\right| + 5 = 5$
20. $\left|\dfrac{2 - 3x}{6}\right| + 1 = 1$
21. $\left|\dfrac{5x + 3}{2}\right| + 4 = 9$
22. $\left|\dfrac{6x + 5}{3}\right| + 2 = 3$

Solve each equation. (*See Objective 2.*)

23. $|4x + 3| = |5x - 1|$
24. $|6x - 3| = |7x + 4|$
25. $|5 - x| = |3x + 2|$
26. $|5 - 2x| = |x + 5|$
27. $|x - 1| = |6x - 6|$
28. $|4x + 8| = |3x - 9|$
29. $|3x + 7| = |2x - 8|$
30. $|7x - 10| = |10x - 5|$
31. $\left|\dfrac{y - 3}{2}\right| = |y + 2|$
32. $\left|\dfrac{y + 1}{3}\right| = |2y - 1|$

33. $\left|\dfrac{a+2}{5}\right| = |3a+3|$ **34.** $\left|\dfrac{a+4}{6}\right| = |a-4|$

35. $\left|\dfrac{x-6}{3}\right| = \left|\dfrac{3x}{4}\right|$ **36.** $\left|\dfrac{x-4}{4}\right| = \left|\dfrac{2x}{7}\right|$

Solve each problem. (*See Objective 3.*)

37. Estimate the absolute error when the value $\dfrac{22}{7}$ is used for π. Round to five decimal places.

38. Estimate the absolute error when the value 3.141593 is used for π. Round to 10 decimal places.

39. Estimate the absolute error when the value 2.72 is used for e, where $e = 2.718281828.\ldots$ Round to five decimal places.

40. Estimate the absolute error when the value 2.71828 is used for e, where $e = 2.71828182846.\ldots$ Round to 10 decimal places.

41. Suppose a digital scale reflects a weight (in pounds) with an absolute error of 0.4 lb. If someone weighs 184.3 lb according to the scale, find the possible values for the exact weight of the person.

42. Suppose a countertop food scale reflects a weight (in ounces) with an absolute error of 0.1 oz. If a piece of chicken weighs 3 oz according to the scale, find the possible values for the exact weight of the piece of chicken.

43. Jana Lipsey and Linda Easley are running for the same congressional seat. Currently, 51% of those polled will reportedly vote for Lipsey and 45% of those polled will reportedly vote for Easley. If the pollsters report an absolute error of 3 points, find the possible values for the actual percentage of those polled who will reportedly vote for Lipsey and Easley.

44. Alexis Johnson and Steve Hunter are running for the same office. Currently, 42% of those polled will vote for Johnson and 48% of those polled will vote for Hunter. If the pollsters report an absolute error of 2 points, find the possible values for the actual percentage of those polled who will reportedly vote for Johnson and Hunter.

 Mix 'Em Up!

Solve each problem.

45. $|10x-3| = 15$ **46.** $|8x-11| = 12$

47. $|4-5x| = 4$ **48.** $|6-2x| = 3$

49. $\left|\dfrac{7x+5}{3}\right| = \left|\dfrac{2x}{5}\right|$ **50.** $\left|\dfrac{4x-9}{6}\right| = \left|\dfrac{3x}{5}\right|$

51. $\left|\dfrac{5x-6}{2}\right| - 1 = 12$ **52.** $\left|\dfrac{3x+10}{4}\right| + 2 = 5$

53. $|10-3x| + 8 = 3$ **54.** $|2-5x| + 3 = 1$

55. $|5x+2| = |6x-7|$ **56.** $|8x-9| = |2-3x|$

57. $\left|\dfrac{9x}{5}\right| = 3$ **58.** $\left|\dfrac{10x}{3}\right| = 2$

59. $|2x+1| = 0$ **60.** $|7x-8| = 0$

61. $|0.4x-1.5| = 6.3$ **62.** $|0.3x+4.2| = 12.6$

63. $\left|\dfrac{9x+10}{5}\right| - 3 = 12$ **64.** $\left|\dfrac{8x-3}{4}\right| + 4 = 9$

65. $|2.6-0.2x| = 14.6$ **66.** $|5.4-0.6x| = 13.8$

67. Suppose a machine with an absolute error reading of 1.3 units reports the measurement of an object as 10.98 units. What are the possible values for the actual measurement of the object?

68. Suppose a machine with an absolute error reading of 5.65 units reports the measurement of an object as 2564.42 units. What are the possible values for the actual measurement of the object?

69. Suppose the actual length of an object is 4.569 m. If this length is estimated by 4.6 m, what is the absolute error?

70. Suppose the actual length of an object is 10.32 in. If this length is estimated by 9 in., what is the absolute error?

71. A poll shows that candidate A will receive 30% of the vote and candidate B will receive 39% of the vote. If this poll has an absolute error of 5%, what are the possible values for how much of the vote each candidate will receive?

72. A poll shows that candidate C will receive 60% of the vote and candidate D will receive 29% of the vote. If this poll has an absolute error of 7%, what are the possible values for how much of the vote each candidate will receive?

You Be the Teacher!

Correct each student's errors, if any.

73. Solve $|4x-5| = 11$.
Kelly's work:
$$|4x-5| = 11$$
$$4x-5 = 11$$
$$4x = 16$$
$$x = 4$$

So, the solution set is $\{4\}$.

74. Solve $|9x+1| - 4 = -2$.
Kyle's work:
The solution set is $\varnothing$ since the right side of the equation is negative.

75. Solve $|x + 5| = -3$.

Mark's work:

$|x + 5| = -3$

$x + 5 = 3$ or $x + 5 = -3$

$x = -2$ $x = -8$

So, the solution set is $\{-8, -2\}$.

76. Solve $|2x - 3| + 5 = 8$.

Ashley's work:

$|2x - 3| + 5 = 8$

$2x - 3 + 5 = 8$ or $2x - 3 + 5 = -8$

$2x + 2 = 8$ $2x + 2 = -8$

$2x = 6$ $2x = -10$

$x = 3$ $x = -5$

So, the solution set is $\{-5, 3\}$.

 Calculate It!

Solve each absolute value equation. Use a graphing calculator to check each answer.

77. $|17x - 5| = 33$ **78.** $|13x + 15| = 21$

79. $\left| \dfrac{6x + 11}{4} \right| - 2 = 5$ **80.** $\left| \dfrac{5x - 12}{4} \right| + 5 = 8$

 Think About It!

81. For what values of x and a does the absolute value equation $|x - a| = x$ have solutions?

82. For what values of x and a does the absolute value equation $|x - a| = -x$ have solutions?

83. Write an example of an absolute value equation that has one solution.

84. Write an example of an absolute value equation that has no solution.

| SECTION 9.3 | Absolute Value Inequalities |

▶ **OBJECTIVES**

As a result of completing this section, you will be able to

1. Solve absolute value inequalities involving $<$ or $\leq$.
2. Solve absolute value inequalities involving $>$ or $\geq$.
3. Solve special cases of absolute value inequalities.
4. Solve applications of absolute value inequalities.
5. Solve absolute value inequalities using test points.
6. Troubleshoot common errors.

During the 2008 presidential election, one poll showed Obama leading McCain by 47% to 44%. This poll had a margin of error of 3%. According to the poll, what percent of votes could Obama actually get? To answer this question, we must solve the inequality $|p - 47| \leq 3$, where p is equal to the percent of people that actually voted for Obama. (Source: www.foxnews.com)

In this section, we will learn how to solve inequalities involving absolute values.

Absolute Value Inequalities with $<$ or $\leq$

When we solve an absolute value equation, we are looking for values that have a specific distance from 0 on the number line. But when we solve absolute value inequalities, we are looking for numbers whose distance from 0 may be less than or greater than a given number.

Suppose we want to solve $|x| \leq 2$. This means that we need to find all real numbers whose distance from zero is less than or equal to 2. Consider the following number line.

Objective 1 ▶

Solve absolute value inequalities involving $<$ or $\leq$.

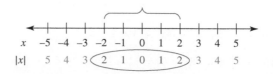

So, we see from the number line that numbers between, and including, -2 and 2 have an absolute value less than or equal to 2. So, the solution set is $[-2, 2]$. This leads to a property that will enable us to solve absolute value inequalities of the form $|X| < k$ or $|X| \leq k$.

Property: Property 1 for Absolute Value Inequalities

Let k be a positive real number.

If $|X| < k$, then $X < k$ and $X > -k$.
If $|X| \leq k$, then $X \leq k$ and $X \geq -k$.

Note: *These compound inequalities can also be written as*

$$-k < X < k \quad and \quad -k \leq X \leq k$$

Procedure: Solving an Inequality of the Form $|X| < k$ or $|X| \leq k$

Step 1: Isolate the absolute value expression on one side of the inequality.
Step 2: Apply property 1 to remove the absolute value sign.
Step 3: Solve the resulting compound inequality.
Step 4: Graph the solution set and write the answer in interval notation.

Objective 1 Examples   **Solve each absolute value inequality.**

1a. $|x| < 5$ **1b.** $|y + 3| \leq 4$ **1c.** $|2x - 1| - 5 \leq 4$

Solutions   **1a.** We must find all numbers whose distance from zero is less than 5. We can do this in two ways. We can write the two appropriate inequalities joined by "and" or we can write the compound inequality using the compact form.

	Method 1	**Method 2**	
	$\lvert x \rvert < 5$	$\lvert x \rvert < 5$	
	$x < 5$ and $x > -5$	$-5 < x < 5$	Apply property 1.

Graph:

$$\xleftarrow{\quad} \; +\!\!+\!\!+\!\!(\!+\!\!+\!\!+\!\!+\!\!+\!\!)\!\!+\!\!+\!\!+ \; \xrightarrow{\quad}$$
$$-10 \;\; -8 \;\; -6 \;\; -4 \;\; -2 \quad 0 \quad 2 \quad 4 \quad 6 \quad 8 \quad 10$$

Interval: $(-5, 5)$

1b. We solve the inequality using two separate inequalities.

$$|y + 3| \leq 4$$

$y + 3 \leq 4$	and	$y + 3 \geq -4$	Apply property 1.
$y + 3 - 3 \leq 4 - 3$		$y + 3 - 3 \geq -4 - 3$	Subtract 3 from each side.
$y \leq 1$		$y \geq -7$	Simplify.

Graph:

$$\xleftarrow{\quad} \; +\![\!+\!\!+\!\!+\!\!+\!\!+\!\!+\!\!+\!\!+\!]\!+ \; \xrightarrow{\quad}$$
$$-8 \;\; -7 \;\; -6 \;\; -5 \;\; -4 \;\; -3 \;\; -2 \;\; -1 \quad 0 \quad 1 \quad 2$$

Interval: $[-7, 1]$

1c. We first isolate the absolute value expression and then apply property 1. We use the compact form to solve the inequality.

$$|2x - 1| - 5 \le 4$$

$$|2x - 1| - 5 + 5 \le 4 + 5 \qquad \text{Add 5 to each side.}$$

$$|2x - 1| \le 9 \qquad \text{Simplify.}$$

$$-9 \le 2x - 1 \le 9 \qquad \text{Apply property 1.}$$

$$-9 + 1 \le 2x - 1 + 1 \le 9 + 1 \qquad \text{Add 1 to each part.}$$

$$-8 \le 2x \le 10 \qquad \text{Simplify.}$$

$$\frac{-8}{2} \le \frac{2x}{2} \le \frac{10}{2} \qquad \text{Divide each part by 2.}$$

$$-4 \le x \le 5 \qquad \text{Simplify.}$$

Graph:

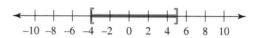

Interval: $[-4, 5]$

✓ **Student Check 1**   Solve each absolute value inequality.

    a. $|x| < 10$ **b.** $|y - 4| \le 5$ **c.** $|4x + 3| - 1 \le 6$

Absolute Value Inequalities with > or ≥

Objective 2 ▶

Solve absolute value
inequalities involving > or ≥.

The graph from Objective 1 will help us solve the equation $|x| \ge 2$.

The numbers larger than and including 2 or less than and including −2 have an absolute value greater than or equal to 2. So, the solution set is $(-\infty, -2] \cup [2, \infty)$. This leads to a property that enables us to solve absolute value inequalities of the form $|X| > k$ or $|X| \ge k$.

Property: Property 2 for Absolute Value Inequalities

Let k be a positive real number.

If $|X| > k$, then $X < -k$ or $X > k$.

If $|X| \ge k$, then $X \le -k$ or $X \ge k$.

Procedure: Solving an Inequality of the Form $|X| > k$ or $|X| \ge k$

Step 1: Isolate the absolute value expression on one side of the inequality.
Step 2: Apply property 2 to remove the absolute value sign.
Step 3: Solve the resulting compound inequality.
Step 4: Graph the solution set and write the answer in interval notation.

| **Objective 2 Examples** | **Solve each absolute value inequality.** |

2a. $|x| > 5$ **2b.** $|y + 3| \geq 4$ **2c.** $|2x - 1| - 5 > 4$

Solutions   **2a.**

$$|x| > 5$$
$$x > 5 \quad \text{or} \quad x < -5 \qquad \text{Apply property 2.}$$

Graph:
$$-10 \ -8 \ -6 \ -4 \ -2 \quad 0 \quad 2 \quad 4 \quad 6 \quad 8 \quad 10$$

Interval: $(-\infty, -5) \cup (5, \infty)$

2b.

$$|y + 3| \geq 4$$

$y + 3 \geq 4$	or	$y + 3 \leq -4$	Apply property 2.
$y + 3 - 3 \geq 4 - 3$		$y + 3 - 3 \leq -4 - 3$	Subtract 3 from each side.
$y \geq 1$		$y \leq -7$	Simplify.

Graph: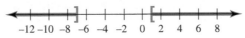
$$-12 \ -10 \ -8 \ -6 \ -4 \ -2 \quad 0 \quad 2 \quad 4 \quad 6 \quad 8$$

Interval: $(-\infty, -7] \cup [1, \infty)$

2c. We first isolate the absolute value expression and then apply property 2.

$$|2x - 1| - 5 > 4$$
$$|2x - 1| - 5 + 5 > 4 + 5 \qquad \text{Add 5 to each side.}$$
$$|2x - 1| > 9 \qquad \text{Simplify.}$$

$2x - 1 > 9$	or	$2x - 1 < -9$	Apply property 2.
$2x - 1 + 1 > 9 + 1$		$2x - 1 + 1 < -9 + 1$	Add 1 to each side.
$2x > 10$		$2x < -8$	Simplify.
$\dfrac{2x}{2} > \dfrac{10}{2}$		$\dfrac{2x}{2} < \dfrac{-8}{2}$	Divide each side by 2.
$x > 5$		$x < -4$	Simplify.

Graph:
$$-10 \ -8 \ -6 \ -4 \ -2 \quad 0 \quad 2 \quad 4 \quad 6 \quad 8 \quad 10$$

Interval: $(-\infty, -4) \cup (5, \infty)$

| ✓ **Student Check 2** | **Solve each absolute value inequality.** |

a. $|x| > 10$ **b.** $|y - 4| \geq 5$ **c.** $|4x + 3| - 1 > 6$

Special Cases of Absolute Value Inequalities

| **Objective 3** ▶ | We will now investigate how to solve absolute value inequalities in which the number opposite the absolute value expression is negative. We will use the number line to solve $|x| < -3$ and $|x| > -3$. |
| Solve special cases of absolute value inequalities. | |

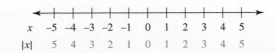

| x | -5 | -4 | -3 | -2 | -1 | 0 | 1 | 2 | 3 | 4 | 5 |
| $|x|$ | 5 | 4 | 3 | 2 | 1 | 0 | 1 | 2 | 3 | 4 | 5 |

$\lvert x \rvert < -3$	There are no real numbers whose absolute value is less than -3. So, the solution set of this inequality is the empty set, $\varnothing$.
$\lvert x \rvert > -3$	Every real number has an absolute value that is greater than -3. So, the solution set of this inequality is all real numbers, $\mathbb{R}$, or $(-\infty, \infty)$.

Note that the absolute value of any real number is nonnegative. That is, the absolute value of any real number is always greater than or equal to zero. This tells us two important facts:

- The absolute value of a number will never be less than a negative number.
- The absolute value of a number will always be greater than a negative number.

These facts yield the following property.

> **Property: Property 3 for Absolute Value Inequalities**
>
> Let k be a negative real number.
>
> If $\lvert X \rvert < k$ or $\lvert X \rvert \le k$, then the solution set is $\varnothing$.
> If $\lvert X \rvert > k$ or $\lvert X \rvert \ge k$, then the solution set is $(-\infty, \infty)$ or all real numbers, $\mathbb{R}$.

So, inequalities of the form $\lvert X \rvert < k$, $\lvert X \rvert \le k$, $\lvert X \rvert > k$, or $\lvert X \rvert \ge k$, where $k < 0$, will either have no solution or all real numbers as solutions.

The last property deals with absolute value inequalities in which the number opposite the absolute value expression is zero.

> **Property: Property 4 for Absolute Value Inequalities**
>
> **a.** $\lvert X \rvert < 0$ This inequality has no solution since the absolute value of any real number is nonnegative.
>
> **b.** $\lvert X \rvert \le 0$ The only solution that satisfies this inequality is $X = 0$.
>
> **c.** $\lvert X \rvert > 0$ This inequality is solved using property 2.
> $X > 0$ or $X < 0$. The only point not included is $X = 0$.
>
> **d.** $\lvert X \rvert \ge 0$ This inequality has all real numbers as its solution since the absolute value of any real number is always positive or zero.

Objective 3 Examples   **Solve each absolute value inequality.**

3a. $\lvert x \rvert < -5$ **3b.** $\lvert y + 1 \rvert \le -7$ **3c.** $\lvert a \rvert > -5$

3d. $\lvert 3b + 5 \rvert + 4 \ge 1$ **3e.** $\left\lvert \dfrac{4x + 1}{3} \right\rvert > 0$

Solutions   **3a.** The solution set is the empty set, or $\varnothing$, since the absolute value of a number is never less than -5.

3b. Since the absolute value of a number is never less than or equal to -7, the solution set is the empty set, or $\varnothing$.

3c. Since the absolute value of a number is always greater than -5, the solution set is all real numbers, or $(-\infty, \infty)$. The graph of the solution set is

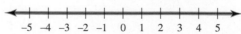

3d. We must first isolate the absolute value by subtracting 4 from each side.

$$|3b + 5| + 4 \geq 1$$

$$|3b + 5| + 4 - 4 \geq 1 - 4 \qquad \text{Subtract 4 from each side.}$$

$$|3b + 5| \geq -3 \qquad \text{Simplify.}$$

Since the absolute value of a number is always greater than or equal to -3, the solution set is all real numbers, or $(-\infty, \infty)$. The graph is

3e.
$$\left|\frac{4x + 1}{3}\right| > 0$$

$\dfrac{4x + 1}{3} > 0$ or	$\dfrac{4x + 1}{3} < 0$	Apply property 4.
$3\left(\dfrac{4x + 1}{3}\right) > 3(0)$	$3\left(\dfrac{4x + 1}{3}\right) < 3(0)$	Multiply each side by 3.
$4x + 1 > 0$	$4x + 1 < 0$	Simplify.
$4x + 1 - 1 > 0 - 1$	$4x + 1 - 1 < 0 - 1$	Subtract 1 from each side.
$4x > -1$	$4x < -1$	Simplify.
$\dfrac{4x}{4} > \dfrac{-1}{4}$	$\dfrac{4x}{4} < \dfrac{-1}{4}$	Divide each side by 4.
$x > -\dfrac{1}{4}$	$x < -\dfrac{1}{4}$	Simplify.

So, the solution set is $\left(-\infty, -\dfrac{1}{4}\right) \cup \left(-\dfrac{1}{4}, \infty\right)$. The graph is

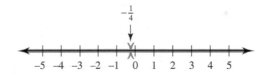

✔ **Student Check 3** Solve each absolute value inequality.

   a. $|x| < -4$ **b.** $|y - 8| \leq -1$ **c.** $|a| > -3$

   d. $|2b - 7| + 6 \geq 3$ **e.** $\left|\dfrac{3x - 5}{4}\right| > 0$

Objective 4 ▶

Solve applications of absolute value inequalities.

Applications

Applications of absolute value inequalities arise when a certain quantity must have a certain distance from another quantity. Some examples are shown.

Objective 4 Examples   **Solve each problem using an absolute value inequality.**

   4a. During the 2008 presidential election, one poll showed Obama leading McCain by 47% to 44%. This poll had a margin of error of 3%. According to the poll, what percent of votes would Obama actually get? (Source: www .foxnews.com)

Solution   **4a.** Let p represent the percent of people who said they would vote for Obama. Since the margin of error is 3%, the absolute value of the difference between the percent surveyed and the actual percentage that voted for Obama is less than or equal to 3. So, we solve the following inequality.

$$|p - 47| \leq 3$$
$$-3 \leq p - 47 \leq 3 \qquad \text{Apply property 1.}$$
$$-3 + 47 \leq p - 47 + 47 \leq 3 + 47 \qquad \text{Add 47 to each part.}$$
$$44 \leq p \leq 50 \qquad \text{Simplify.}$$

Based on the poll, Obama would have received between 44% and 50% of the votes.

4b. A hazardous chemical spill of titanium tetrachloride occurred at mile marker 331 on Interstate 40 in Tennessee. Emergency management personnel stated that motorists should be a minimum of 12 mi from the site of the spill. What are safe locations on Interstate 40 for motorists?

Solution   **4b.** Let x represent the mile marker of a motorist. The distance from the site of the spill to the motorist should be at least 12 mi. Since motorists can be on either side of the spill, we must solve the following inequality.

$$|x - 331| \geq 12$$

$x - 331 \geq 12$ $\qquad$ or	$x - 331 \leq -12$	Apply property 2.
$x - 331 + 331 \geq 12 + 331$	$x - 331 + 331 \leq -12 + 331$	Add 331 to each side.
$x \geq 343$	$x \leq 319$	Simplify.

Motorists should be located at or below mile marker 319 or located at or above mile marker 343 to be a safe distance from the spill.

✓ Student Check 4   Solve each word problem with an absolute value inequality.

a. A Gallup Poll, taken in August 2008, revealed that 48% of U.S. workers are completely satisfied with their job. The poll had a margin of error of 5%. What is the range of workers who are completely satisfied with their job? (Source: www.gallup.com)

b. A tractor trailer accident at mile marker 51 on I-68 in West Virginia caused a hazardous chemical spill. Emergency management personnel advised that motorists should be at least 8 mi from the site of the spill. What are safe locations on I-68 for motorists?

Using Test Points to Solve Absolute Value Inequalities

Objective 5 ▶

Solve absolute value inequalities using test points.

In the first two objectives, compound inequalities were used to solve absolute value inequalities. Absolute value inequalities can be solved using an alternate method involving test points. In this method, the graph of the solution set is first constructed and, from this, the solution set is obtained.

We examine the graph of the solution set of the inequality we solved in Example 2 part (b). The inequality $|y + 3| \geq 4$ was shown to have the following solution.

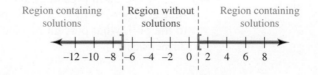

Notice that the number line is separated into regions separated by the numbers -7 and 1. The significance of the numbers -7 and 1 is that they are the solutions of the equation $|y + 3| = 4$.

$$|y + 3| = 4$$

$$y + 3 = -4 \qquad \text{or} \qquad y + 3 = 4$$
$$y + 3 - 3 = -4 - 3 \qquad\qquad y + 3 - 3 = 4 - 3$$
$$y = -7 \qquad\qquad\qquad y = 1$$

The parts of the number line that are shaded include solutions of the inequality. If numbers from these regions are substituted into the inequality, a true statement will result. The values -10 and 3 are included in the shaded regions. When we substitute these numbers in for y, we get

$$|-10 + 3| \geq 4 \rightarrow |-7| \geq 4 \rightarrow 7 \geq 4 \quad \text{True}$$

$$|3 + 3| \geq 4 \rightarrow |6| \geq 4 \rightarrow 6 \geq 4 \quad \text{True}$$

The part of the number line that is *not* shaded includes numbers that are *not* solutions of the inequality. If numbers from this region are substituted into the inequality, a false statement will result. The value -4 is in the region that is not shaded. When we substitute this value for y, we get

$$|-4 + 3| \geq 4 \rightarrow |-1| \geq 4 \rightarrow 1 \geq 4 \quad \text{False}$$

So, this example illustrates that the solutions of the associated absolute value equation separate the number line into intervals of numbers that are or are not solutions of the absolute value inequality.

Procedure: Solving an Absolute Value Inequality Using Test Points

Step 1: Solve the associated absolute value equation. This equation comes from replacing the inequality symbol with an equals sign.

Step 2: Place the solutions of the equation on a number line.

Step 3: Select a number (a **test point**) from each of the resulting intervals and substitute it into the original inequality to determine if it is a solution.

Step 4: If the test point is a solution of the inequality, shade the interval of the number line that includes this point. If the test point is not a solution of the inequality, do not shade this interval of the number line.

Step 5: Put the appropriate symbol on the endpoints of the intervals.

    a. If the absolute value inequality contains $\leq$ or $\geq$, the endpoints are solutions of the inequality. Brackets are used to represent this.

    b. If the absolute value inequality contains $<$ or $>$, the endpoints are *not* solutions of the inequality. Parentheses are used to represent this.

Step 6: Write the solution set in interval notation.

Objective 5 Examples   Use test points to solve each absolute value inequality.

5a. $|x - 2| < 5$ **5b.** $|2y - 7| \geq -1$

Solutions   **5a.** Solve the equation $|x - 2| = 5$.

$$x - 2 = -5 \qquad \text{or} \qquad x - 2 = 5$$
$$x - 2 + 2 = -5 + 2 \quad \text{or} \quad x - 2 + 2 = 5 + 2$$
$$x = -3 \qquad \text{or} \qquad x = 7$$

Place the solutions of the equation on a number line to form intervals.

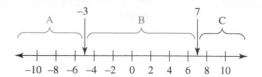

Test a point from intervals A, B, and C.

Interval	Test Point	$|x - 2| < 5$	True/False
A: $(-\infty, -3)$	-6	$|-6 - 2| < 5$ $|-8| < 5$ $8 < 5$	False
B: $(-3, 7)$	0	$|0 - 2| < 5$ $|-2| < 5$ $2 < 5$	True
C: $(7, \infty)$	9	$|9 - 2| < 5$ $|7| < 5$ $7 < 5$	False

We shade interval B since its test point is a solution of the inequality. The endpoints are not included since the original inequality is $<$.

So, the interval notation for the solution set is $(-3, 7)$.

5b. Solve the equation $|2y - 7| = -1$. This equation has no solution since the constant on the right side is negative.

Since there are no solutions of the equation, the number line is not separated into any additional intervals. The entire number line is the interval we need to test.

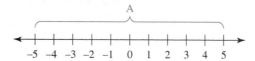

Test a point from interval A.

Interval	Test Point	$|2y - 7| \geq -1$	True/False
A: $(-\infty, \infty)$	0	$|2(0) - 7| \geq -1$ $|-7| \geq -1$ $7 \geq -1$	True

Since the test point is a solution of the inequality, we shade interval A.

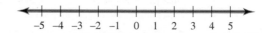

The solution set is $(-\infty, \infty)$ or all real numbers.

✓ **Student Check 5**   Use test points to solve each absolute value inequality.

a. $|x + 5| \geq 4$ **b.** $|6a - 1| < -2$

Objective 6 ▶

Troubleshoot common errors.

Troubleshooting Common Errors

Some common errors associated with absolute value inequalities are shown.

Objective 6 Examples

A problem and an incorrect solution are given. Provide the correct solution and an explanation of the error.

6a. Solve $|x + 2| < 5$.

Incorrect Solution	Correct Solution and Explanation
$|x + 2| < 5$ $|x + 2| < 5$ and $|x + 2| > -5$ $|x| < 3$ and $|x| > -7$ So, the solution set is $(-7, 3)$.	When we write the compound inequality that satisfies the absolute value inequality, we no longer include the absolute value bars. $|x + 2| < 5$ $x + 2 < 5$ and $x + 2 > -5$ $x < 3$ and $x > -7$

6b. Solve $|y - 5| + 3 < 2$.

Incorrect Solution	Correct Solution and Explanation
$|y - 5| + 3 < 2$ $y - 5 + 3 < 2$ or $y - 5 + 3 > -2$ $y - 2 < 2$ or $y - 2 > -2$ $y < 4$ or $y > 0$ Solution: $(-\infty, \infty)$	We must first isolate the absolute value expression. We should subtract 3 from each side. $|y - 5| + 3 < 2$ $|y - 5| + 3 - 3 < 2 - 3$ $|y - 5| < -1$ The absolute value of a real number is never less than a negative number. So, the solution set is $\varnothing$.

ANSWERS TO STUDENT CHECKS

Student Check 1   **a.** $(-10, 10)$ **b.** $[-1, 9]$

   c. $\left[-\dfrac{5}{2}, 1\right]$

Student Check 2   **a.** $(-\infty, -10) \cup (10, \infty)$

   b. $(-\infty, -1] \cup [9, \infty)$ **c.** $\left(-\infty, -\dfrac{5}{2}\right) \cup (1, \infty)$

Student Check 3   **a.** $\varnothing$ **b.** $\varnothing$ **c.** $(-\infty, \infty)$

   d. $(-\infty, \infty)$ **e.** $\left(-\infty, \dfrac{5}{3}\right) \cup \left(\dfrac{5}{3}, \infty\right)$

Student Check 4   **a.** According to the poll, 43% to 53% of U.S. workers are completely satisfied with their job.

   b. Motorists should be at or below mile marker 43 or at or above mile marker 59.

Student Check 5   **a.** $(-\infty, -9] \cup [-1, \infty)$ **b.** $\varnothing$

SUMMARY OF KEY CONCEPTS

1. Solving an absolute value inequality involves solving a compound inequality. If the absolute value of an expression is less than a positive number, the compound inequality involves *and*. $|X| < k$, is equivalent to $X < k$ and $X > -k$, or we can write this as $-k < X < k$, where $k > 0$.

2. If the absolute value of an expression is greater than a positive number, the compound inequality involves *or*. $|X| > k$ is equivalent to $X < -k$ or $X > k$, where $k > 0$.

3. Special cases of absolute value inequalities arise when the constant $k < 0$ or when $k = 0$.

- If the absolute value of an expression is less than a negative number, the inequality has no solution since the absolute value will never be less than a negative number.

- If the absolute value of an expression is greater than a negative number, then the solution is all real numbers. This is because the absolute value of any number is positive and will always be greater than a negative number.

- The inequality $|X| < 0$ has no solution. The inequality $|X| \leq 0$ has only one solution, $X = 0$. The inequality $|X| \geq 0$ has all real numbers as solutions. To solve $|X| > 0$, we must solve it by solving $X > 0$ or $X < 0$.

4. To solve applications of absolute values, the expression inside the absolute value represents the distance between two quantities. The number that the absolute value is greater than or less than is the given distance.

5. Test points can be used to solve absolute value inequalities. The solutions of the associated equation must be found first. The solutions partition the number line into regions that must be tested. A test point from each region is substituted into the original inequality to determine whether it is a solution. The regions that test true are shaded. Brackets or parentheses should be placed on the end points of the regions.

GRAPHING CALCULATOR SKILLS

A graphing calculator has the capacity to graph the solution set of an absolute value inequality. At this point, make sure that you know how to solve these by hand and use the calculator only to check your work.

Example: Solve $|y + 3| \leq 4$.

Solution: Enter the inequality in the equation editor and graph.

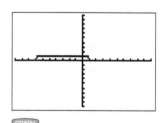

We TRACE along the graph to the endpoints to determine whether they should be included or not. If $y = 1$, then the

endpoint is included. If $y = 0$, the endpoint is not included. Notice both $x = -7$ and $x = 1$ correspond to a y-value of 1, so they are included in the solution set. So, the solution set is $[-7, 1]$.

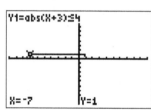 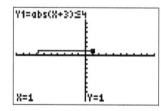

The TABLE also shows a picture of the solution set. The x-values whose y-value is 1 are part of the solution of the inequality. The x-values whose y-value is 0 are *not* solutions of the inequality. The table confirms the solution $[-7, 1]$.

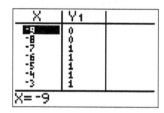 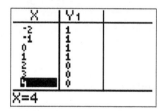

SECTION 9.3   EXERCISE SET

Write About It!

Use complete sentences in your answer to each exercise.

1. Explain how to solve an absolute value inequality involving a less than symbol.

2. Explain how to solve an absolute value inequality involving a greater than symbol.

3. Describe the special cases that may arise when solving an absolute value inequality.

4. Use an example to explain how to use test points to solve an absolute value inequality, $|x - a| < b$, $b > 0$.

5. Use an example to explain how to use test points to solve an absolute value inequality, $|x - a| > b$, $b > 0$.

6. Use an example to explain how to solve an absolute value inequality, $|x - a| \leq 0$.

 Practice Makes Perfect!

Solve each inequality. Write each answer in interval notation. (See Objective 1.)

7. $|x| \leq 6$

8. $|x| \leq 3$

9. $|y - 4| < 19$

10. $|y + 6| < 20$

11. $|2x + 3| < 4$

12. $|4x - 3| < 5$

13. $|3x - 1| \leq 6$

14. $|6x - 7| \leq 12$

15. $|2 - 5x| < 1$

16. $|6 - 3x| < 7$

17. $|1 - 4x| \leq 5$

18. $|3 - 9x| \leq 15$

19. $|3x - 1| + 5 < 12$

20. $|5x + 3| + 8 < 10$

21. $|7 + 4x| - 1 < 4$

22. $|6 - 2x| - 8 < 3$

23. $6 + |8x - 5| \leq 16$

24. $5 + |3x + 2| < 15$

Solve each inequality. Write each answer in interval notation. (See Objective 2.)

25. $|y| > 4$

26. $|y| > 9$

27. $|x - 3| \geq 2$

28. $|x + 7| \geq 12$

29. $|5x + 5| > 17$

30. $|6x + 7| > 8$

31. $|10x - 4| > 16$

32. $|8x - 3| > 13$

33. $|8 - 4x| > 6$

34. $|7 - 3x| > 5$

35. $|2x + 7| - 3 \geq 6$

36. $|3x - 5| - 9 \geq 1$

37. $|6 - 2x| + 3 \geq 11$

38. $|7 - 4x| + 6 \geq 13$

39. $4 + |6x - 1| > 9$

40. $7 + |2x + 9| > 15$

Solve each inequality. Write each answer in interval notation. (See Objective 3.)

41. $|a| \leq -1$

42. $|a| \leq -5$

43. $|x + 5| < 0$

44. $|3x - 6| < 0$

45. $\left|\dfrac{x + 1}{2}\right| < 0$

46. $\left|\dfrac{2x - 3}{5}\right| < 0$

47. $|2x + 5| + 6 \leq 3$

48. $|5x - 1| + 8 \leq 2$

49. $|x| > -8$

50. $|x| > -2$

51. $|4x + 10| > 0$

52. $|5x - 6| > 0$

53. $|3x - 18| \leq 0$

54. $|15 - 5x| \leq 0$

55. $\left|\dfrac{4x - 3}{2}\right| > 0$

56. $\left|\dfrac{x + 3}{4}\right| > 0$

57. $|4 - 3x| + 6 \leq 3$

58. $|2 - 7x| + 9 \leq 6$

59. $|7 - x| \leq 0$

60. $|6 - 2x| \leq 0$

61. $5 + |4x + 6| > 5$

62. $1 + |2x + 5| > 1$

63. $\left|\dfrac{3x - 1}{5}\right| + 4 > 6$

64. $\left|\dfrac{2x + 5}{3}\right| - 1 \geq 3$

65. $\left|\dfrac{5x + 3}{2}\right| - 3 \leq 1$

66. $\left|\dfrac{x - 4}{6}\right| + 3 < 8$

Use an absolute value inequality to solve each problem. (See Objective 4.)

67. A 2011 poll revealed that 74% of Americans think education and training in science, technology, engineering, and mathematics (STEM) is very important to U.S. competitiveness and our future economic prosperity. If the margin of error of the poll was 3.1%, what is the possible range of percentages of Americans who think that STEM education and training is very important to U.S. competitiveness and prosperity? (Source: www.researchamerica.org)

68. A 2010 poll revealed that 88% of Americans with private health insurance thought that the quality of their health care was excellent or good. If the margin of error of the poll was 4%, what is the possible range of percentages of Americans who were satisfied with the quality of their health care? (Source: www.gallup.com)

69. Rita read in a magazine that the best way to maintain one's weight loss is making sure one's weight stays within 5 lb of his or her goal weight. If Rita's goal weight is 145 lb, how much can she weigh while maintaining her weight loss?

70. Randal wants to stay within 5 lb of his ideal weight of 185 lb. How much can Randal weigh while meeting his goal?

Use test points to solve each absolute value inequality. (See Objective 5.)

71. $|x| > 16$

72. $|x| > 25$

73. $|2x - 5| < 11$

74. $|6x - 3| < 21$

75. $|1 - 4x| \geq 5$

76. $|3 - 6x| \geq 3$

77. $|y - 4| \leq 0$

78. $|y + 3| \leq 0$

79. $|5x + 4| - 6 \geq 10$

80. $|7x + 3| - 8 \geq 2$

81. $|7 - 7x| < -3$

82. $|5 - 3x| < -4$

83. $|x| > -1$

84. $|x| > -7$

85. $|x - 6| \geq 0$

86. $|x + 3| \geq 0$

87. $|6x - 3| > -3$

88. $|3x - 2| > -5$

89. $|2x - 10| > 0$

90. $|0.4x - 1.2| > 0$

91. $|2x - 5| + 9 \leq 21$

92. $|3x - 1| + 7 \leq 18$

93. $|2 - x| - 5 < 16$

94. $|3 - 2x| - 7 < 23$

Mix 'Em Up!

Solve each inequality. Write each answer in interval notation.

95. $|6x + 9| \leq 17$ **96.** $|4x + 10| \leq 2$

97. $|a| > -9$ **98.** $|a| \geq -\dfrac{1}{2}$

99. $|2y - 1| - 8 < -2$ **100.** $|4y - 3| - 9 < -1$

101. $-|x| < 6$ **102.** $-|x| < -10$

103. $|5x + 6| - 2 \geq 7$ **104.** $|8x - 3| - 4 \geq 5$

105. $|2x - 12| > 0$ **106.** $|5x + 15| > 0$

107. $|8 - 4x| \leq 0$ **108.** $|24 - 6x| \leq 0$

109. $\left|\dfrac{2x - 6}{5}\right| + 1 \leq 2$ **110.** $\left|\dfrac{2 - 3x}{4}\right| - 2 \geq 1$

111. $3 - |x + 4| < 5$ **112.** $1 - |x + 3| < 6$

113. $3.2 - |x - 2.1| \geq 8.6$

114. $12.5 + |x + 1.3| \geq 3.5$

115. $\left|\dfrac{1 - x}{2}\right| + 3 < 3$

116. $\left|\dfrac{2 + x}{3}\right| - 2 \geq -2$

117. $2.8 + |x - 0.9| \leq 6.3$

118. $4.5 + |x - 2.4| \leq 7.2$

Use test points to solve each inequality.

119. $|x - 3| \leq -3$ **120.** $|x + 9| \leq 0$

121. $|x - 8| < 9$ **122.** $|x + 3| < 11$

123. $|6x + 3| - 1 \geq 4$ **124.** $|5x + 8| + 3 \geq 6$

125. A 2008 *USA Today*/Gallup Poll found that 53% of Americans described themselves as angry about the country's financial crisis. If the margin of error was 3%, what are the possible percentages of Americans polled who described themselves as angry about the country's financial crisis? (Source: www.gallup.com)

126. A 2007/2008 Gallup Poll found that approximately 28% of people polled in Latin America with a college degree desired to move to another country. If the margin of error was 4%, what are the possible percentages of people in Latin America with a college degree who desired to move to another country. (Source: www.gallup.com)

127. A truck overturn results in a hazardous chemical spill. Emergency management personnel advise that motorists should be restricted from being within a 10-mi radius of the spill. If the spill happens at mile

marker 35 on Interstate 20 in South Carolina, what are the safe locations on Interstate 20 for motorists?

128. A truck overturn on the New Jersey Turnpike results in a hazardous chemical spill. Emergency management personnel advise that motorists should be restricted from being within a 25-mi radius of the spill. If the spill happens at mile marker 40 on the New Jersey Turnpike, what are the safe locations on the turnpike for motorists?

You Be the Teacher!

Answer each student's questions.

129. Lee: Will there always be two disjoint intervals in the solution set when I solve an absolute value inequality of the form $|X| > k$? Please explain.

130. Brielle: When solving an absolute value inequality of the form $|X| < k$, will the answer always be a single interval? Please explain.

Correct each student's errors, if any.

131. Solve $|2x - 5| < 13$.

Ed's work:

$$|2x - 5| < 13$$
$$2x - 5 < 13 \quad \text{or} \quad 2x - 5 > -13$$
$$2x < 18 \quad \text{or} \quad 2x > -8$$
$$x < 9 \quad \text{or} \quad x > -4$$

132. Solve $|4x + 1| + 2 \geq 15$.

Monica's work:

$$|4x + 1| + 2 \geq 15$$
$$4x + 1 + 2 \geq 15 \quad \text{or} \quad 4x + 1 + 2 \leq -15$$
$$4x \geq 12 \quad \text{or} \quad 4x \leq -18$$
$$x \geq 3 \quad \text{or} \quad x \leq -4.5$$

Calculate It!

Find the test points of the inequalities algebraically and then use a graphing calculator to solve the inequality.

133. $|8x + 4| < 12$ **134.** $|7x + 3| < 10$

135. $|5x - 14| \geq 14$ **136.** $|4x - 9| \geq 9$

Think About It!

137. Solve the inequality: $|x + 5| > |x - 7|$

138. Solve the inequality: $|2x - 6| \le |x + 4|$

139. Use test points to solve: $|x - 3| > |x + 2|$

140. Use test points to solve: $|2x - 3| \ge |x + 1|$

PIECE IT TOGETHER　　SECTIONS 9.1–9.3

Solve each compound inequality. Write each solution set in interval notation and graph the solution set. (*Section 9.1, Objectives 3 and 4*)

1. $4x - 5 \le -13$ and $4x - 5 \ge 13$

2. $2x - 1 > 7$ or $-3(x + 1) > 15$

Solve the problem (*Section 9.1, Objective 5*)

3. Body mass index (BMI) is a measure of how much body fat a person has. A normal BMI is between 18.5 and 24.9. The formula to calculate BMI is $\text{BMI} = \dfrac{703w}{h^2}$, where w is weight (in pounds) and h is height (in inches). If Conner is 6 ft 2 in. tall, how much should he weigh to have a normal BMI? Round the answer to the nearest whole number.

Solve each equation. (*Section 9.2, Objectives 1 and 2*)

4. $|2x - 11| = 3$

5. $\left|\dfrac{3x - 1}{7}\right| + 1 = 3$

6. $|x + 2| + 5 = 4$

7. $|x - 3| = 0$

8. $|2x + 5| = |4x - 17|$

9. $\left|\dfrac{y - 3}{2}\right| = \left|\dfrac{y}{6}\right|$

Solve each problem. (*Section 9.2, Objective 3*)

10. Suppose the length of an object is 5.35 in. The absolute error is 0.0125 in. Find the possible values for the exact length of the object.

11. Estimate the absolute error when the value 2.717 is used for e, where $e = 2.7182818$. Round to four decimal places.

Solve each absolute value inequality. Write each answer in interval notation. (*Section 9.3, Objectives 1–3*)

12. $|x - 10| + 4 \le 18$

13. $|2y + 9| > 9$

14. $|5x - 3| + 7 < 4$

15. $|3x + 2| - 1 > -5$

16. $|x - 5| + 10 \le 10$

17. $|y + 3| - 8 > -8$

Solve the problem. (*Section 9.3, Objective 4*)

18. Allison wants to stay within 4 lb of her ideal weight of 120 lb. How much can Allison weigh while meeting her goal?

Use test points to solve each absolute value inequality. (*Section 9.3, Objective 5*)

19. $|2x - 6| < 8$

20. $|5x + 6| > 16$

SECTION 9.4　　Linear Inequalities in Two Variables

▶ OBJECTIVES

As a result of completing this section, you will be able to

1. Determine if an ordered pair is a solution of a linear inequality in two variables.

2. Graph the solution set of a linear inequality in two variables.

3. Solve applications of linear inequalities in two variables.

4. Troubleshoot common errors.

The final grade in Tyrone's math class is based on his test average and final exam grade. The test average counts for 70% of his final grade and the final exam counts for 30% of his final grade. If Tyrone wants to have at least a 70 average in the course, what are some possible combinations of test average and final exam grades to produce the desired result?

To solve this problem, we must know how to solve linear inequalities in two variables.

Solutions of Linear Inequalities in Two Variables

At this point, we have discussed solving systems of linear equations in two variables both graphically and algebraically. In the next section, we will learn how to identify solutions of systems of linear inequalities in two variables based on their graphs. To do this, we first need to learn how to graph a single *linear inequality in two variables*.

OBJECTIVE 1 ▶

Determine if an ordered pair is a solution of a linear inequality in two variables.

> **Definition:** A **linear inequality in two variables** is an inequality that can be written in one of the following ways, where A, B, C, m, and b are real numbers with A and B not both zero.
>
> $Ax + By > C$ $\qquad$ $Ax + By \geq C$ $\qquad$ $Ax + By < C$ $\qquad$ $Ax + By \leq C$
>
> $y > mx + b$ $\qquad$ $y \geq mx + b$ $\qquad$ $y < mx + b$ $\qquad$ $y \leq mx + b$

Some examples of linear inequalities in two variables are $x + y > 3$ and $y \leq \frac{1}{4}x + 5$.

Like a solution of linear equations in two variables, a **solution of a linear inequality in two variables** is an ordered pair (x, y) that makes the inequality true.

> **Procedure: Determining if an Ordered Pair Is a Solution of a Linear Inequality in Two Variables**
>
> **Step 1:** Replace the variables with the corresponding values of x and y.
> **Step 2:** Simplify the resulting inequality.
> **Step 3:** If the ordered pair makes the inequality true, then the ordered pair is a solution. If the ordered pair makes the inequality false, then the ordered pair is not a solution.

Objective 1 Examples Determine which ordered pairs are solutions of the given inequality.

1a. $2x + y > 6$; $(2, 8)$, $(3, 0)$, $(-1, 3)$, $(-2, -1)$, $(4, 0)$, $(6, -2)$

1b. $y \leq \frac{1}{2}x - 4$; $(0, 1)$, $(3, -5)$, $(8, 0)$, and $(0.1, 0.3)$

Solutions **1a.**

(x, y)	$2x + y > 6$	True or False?	Solution?
$(2, 8)$	$2(2) + 8 > 6$ $12 > 6$	True	Yes
$(3, 0)$	$2(3) + 0 > 6$ $6 > 6$	False	No
$(-1, 3)$	$2(-1) + 3 > 6$ $1 > 6$	False	No
$(-2, -1)$	$2(-2) + (-1) > 6$ $-5 > 6$	False	No
$(4, 0)$	$2(4) + 0 > 6$ $8 > 6$	True	Yes
$(6, -2)$	$2(6) + (-2) > 6$ $10 > 6$	True	Yes

So, the ordered pairs $(2, 8)$, $(4, 0)$, and $(6, -2)$ are solutions of $2x + y > 6$.

1b.

(x, y)	$y \leq \frac{1}{2}x - 4$	True or False?	Solution?
$(0, 1)$	$1 \leq \frac{1}{2}(0) - 4$ $1 \leq -4$	False	No
$(3, -5)$	$-5 \leq \frac{1}{2}(3) - 4$ $-5 \leq -\frac{5}{2}$	True	Yes

(x, y)	$y \le \dfrac{1}{2}x - 4$	True or False?	Solution?
$(8, 0)$	$0 \le \dfrac{1}{2}(8) - 4$ $0 \le 0$	True	Yes
$(0.1, 0.3)$	$0.3 \le \dfrac{1}{2}(0.1) - 4$ $0.3 \le -3.95$	False	No

So, $(3, -5)$ and $(8, 0)$ are solutions of $y \le \dfrac{1}{2}x - 4$.

✓ Student Check 1   Determine if the ordered pair is a solution of the given inequality.

a. $2x - 5y > -10;\ (0, 0)$ **b.** $y \ge \dfrac{1}{3}x + 4;\ (-6, 1)$ **c.** $x < -1;\ (-2, 3)$

Graphing Linear Inequalities in Two Variables

Objective 2 ▶

Graph the solution set of a linear inequality in two variables.

Now that we know what it means for an ordered pair to be a solution of a linear inequality in two variables, we will graph the solution set of a linear inequality in two variables. Just as there are infinitely many solutions of linear equations in one variable, there are infinitely many solutions of linear inequalities in two variables.

To graph a linear inequality in two variables, we begin by graphing the linear equation in two variables that is formed by replacing the inequality with an equals sign. This line is called the **boundary line** and divides the plane into two regions called **half-planes**. The region above or to the right of the line is the **upper half-plane** and the region below or to the left of the line is the **lower half-plane**. Either the upper half-plane or the lower half-plane contains solutions of the inequality. The boundary line may or may not be included in the solution set.

To determine the graph of the solution set of the inequality $2x + y > 6$, we use the boundary line, $2x + y = 6$, and the information we found in Example 1a. First, we graph the linear equation associated with this inequality: $2x + y = 6$. This equation has intercepts of $(3, 0)$ and $(0, 6)$. The graph is shown next along with the ordered pairs from Example 1a.

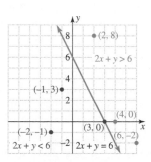

(x, y)	$2x + y > 6$	Solution?
$(2, 8)$	$2(2) + 8 > 6$ $12 > 6$	Yes
$(4, 0)$	$2(4) + 0 > 6$ $8 > 6$	Yes
$(6, -2)$	$2(6) + (-2) > 6$ $10 > 6$	Yes
$(3, 0)$	$2(3) + 0 > 6$ $6 > 6$	No
$(-1, 3)$	$2(-1) + 3 > 6$ $1 > 6$	No
$(-2, -1)$	$2(-2) + (-1) > 6$ $-5 > 6$	No

The points $(2, 8)$, $(4, 0)$, and $(6, -2)$ are solutions of $2x + y > 6$. Notice these points are all located above the line $2x + y = 6$, or in the upper half-plane. The points $(-1, 3)$ and $(-2, -1)$ are not solutions of $2x + y > 6$. Notice these points are located below the line $2x + y = 6$, or in the lower half-plane. We also found that the point $(3, 0)$ is not a solution of $2x + y > 6$. However, this point is located on the line $2x + y = 6$.

From the graph we observe the following:
- Solutions of the inequality are all located on the same side of the graph of the associated equation.
- The point $(3, 0)$ lies on the line, $2x + y = 6$, but is not a solution of the inequality.
- None of the points that lie below the line are solutions of the inequality.

> **Property: The Graph of a Linear Inequality in Two Variables**
>
> The graph of the solution set of a linear inequality in two variables has three parts:
> - The region, or half-plane, that contains solutions of the inequality.
> - The region, or half-plane, that contains no solutions of the inequality.
> - The boundary line that divides the plane into these two regions.

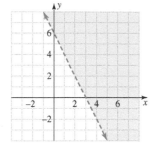

When we shade the half-plane that contains solutions of the inequality, $2x + y > 6$, we get the graph shown to the left. Notice that the half-plane that contains no solutions of $2x + y > 6$ is not shaded. The line $2x + y = 6$ is dashed since the line does not contain solutions.

> **Note:**
>
> - *If one point in a half-plane is a solution of an inequality, then all points in that half-plane are also solutions.*
> - *If one point in a half-plane is not a solution of an inequality, then none of the points in that half-plane are solutions.*

These facts enable us to use a **test-point method** to graph linear inequalities in two variables. This method is described next.

> **Procedure: Graphing a Linear Inequality in Two Variables Using the Test-Point Method**
>
> **Step 1:** Graph the associated linear equation in two variables by replacing the inequality symbol with an equals sign. This line forms the boundary between the region that contains solutions and the region that does not contain solutions.
>   **a.** The boundary line is solid if the inequality symbol is $\leq$ or $\geq$.
>   (This means that the points on the line are solutions of the inequality.)
>   **b.** The boundary line is dashed if the inequality symbol is $<$ or $>$.
>   (This means that the points on the line are not solutions of the inequality.)
>
> **Step 2:** Identify a point in either one of the half-planes formed by the boundary line, but not on the boundary line, and test it in the original inequality.
>   **a.** If the point makes the original inequality *true*, then shade the half-plane that contains the test point.
>   **b.** If the point makes the original inequality *false*, then shade the half-plane that does not contain the test point.

> **Note:** *The point $(0, 0)$ is the easiest point to test. Any other point in the plane will work as long as the point is not on the boundary line. If the boundary line goes through the origin, then $(0, 0)$ cannot be used as the test point.*

Objective 2 Examples   Graph the solution set of each linear inequality in two variables.

    2a. $x - 2y > 4$ **2b.** $y \leq -\dfrac{2}{3}x + 3$ **2c.** $x \geq -2$ **2d.** $y < 5$ **2e.** $y + 3x \geq 0$

Solutions   **2a.** We first graph the associated equation $x - 2y = 4$.

To graph $x - 2y = 4$, we find the x- and y-intercepts as shown in the table. The line is dashed since the original inequality is $<$.

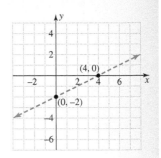

x	y	(x, y)
0	-2	$(0, -2)$
4	0	$(4, 0)$

Next, we use $(0, 0)$ as a test point since the graph of $x - 2y = 4$ does not go through the origin.

$$x - 2y > 4 \qquad \text{Begin with the original inequality.}$$

$$0 - 2(0) > 4 \qquad \text{Replace } x \text{ and } y \text{ with 0.}$$

$$0 > 4 \qquad \text{Simplify.}$$

Since the test point makes the original inequality false, solutions of the inequality are located in the half-plane that does *not* contain $(0, 0)$. The solutions of the inequality are located in the half-plane below the boundary line, or the lower half-plane. Recall that none of the points on the boundary line are solutions since the inequality symbol is $>$. We shade the region below the boundary line to denote the graph of $x - 2y > 4$.

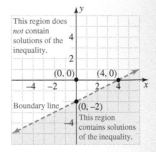

2b. We first graph the associated equation $y = -\dfrac{2}{3}x + 3$.

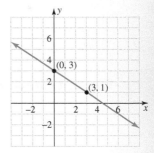

The y-intercept of $y = -\dfrac{2}{3}x + 3$ is $(0, 3)$ and

the slope $m = -\dfrac{2}{3}$. Plot $(0, 3)$ and then move

down 2 units and right 3 to get to another point on the line. The boundary line is solid since the inequality is $\leq$.

Next, we use $(0, 0)$ as a test point since the graph

of $y = -\dfrac{2}{3}x + 3$ does not go through the origin.

$$y \leq -\frac{2}{3}x + 3 \qquad \text{Begin with the original inequality.}$$

$$0 \leq -\frac{2}{3}(0) + 3 \qquad \text{Replace } x \text{ and } y \text{ with 0.}$$

$$0 \leq 0 + 3 \qquad \text{Simplify.}$$

$$0 \leq 3 \qquad \text{Simplify.}$$

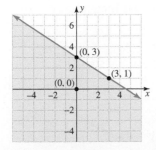

Since $(0, 0)$ makes the original inequality true, we shade the half-plane that contains $(0, 0)$. All points on the boundary line and in the lower half-plane are

solutions of the inequality, $y \leq -\dfrac{2}{3}x + 3$.

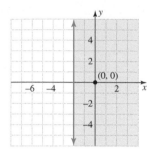

2c. We first graph the associated equation $x = -2$. The line $x = -2$ is a vertical line through $(-2, 0)$. The boundary line is solid since the inequality symbol is $\geq$.

Next, we test the point $(0, 0)$ since it is not on the line, $x = -2$.

$x \geq -2$ Begin with the original inequality.

$0 \geq -2$ Replace x with 0.

Since $(0, 0)$ makes the inequality true, we shade the half-plane that contains $(0, 0)$. This is the upper, or right, half-plane. All points on the boundary line and in the right half-plane are solutions of the inequality, $x \geq -2$.

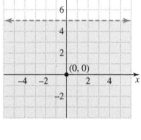

2d. We first graph the associated equation $y = 5$. The line $y = 5$ is a horizontal line through $(0, 5)$. The line is dashed since the inequality symbol is $<$.

Next, we test the point $(0, 0)$ since it is not on the line, $y = 5$.

$y < 5$ Begin with the original inequality.

$0 < 5$ Replace y with 0.

Since $(0, 0)$ makes the original inequality true, we shade the half-plane that contains $(0, 0)$. Only points in the lower half-plane are solutions of the inequality; none of the points on the boundary line are solutions.

2e. We graph the associated equation $y + 3x = 0$. This line is equivalent to $y = -3x$, which has y-intercept of $(0, 0)$ and slope $m = -3$. Plot $(0, 0)$ and move down 3 units and right 1 to get to another point on the line. The boundary line is solid since the inequality is $\geq$.

Since $(0, 0)$ is on the boundary line, $y = -3x$, we must test a point other than $(0, 0)$, say, $(1, 1)$.

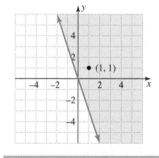

$y + 3x \geq 0$ Begin with the original inequality.

$(1) + 3(1) \geq 0$ Replace x and y with 1.

$4 \geq 0$ Simplify.

Since $(1, 1)$ makes the original inequality true, we shade the half-plane that contains $(1, 1)$. All points on the boundary line and in the upper half-plane are solutions of the inequality, $y + 3x \geq 0$.

✓ Student Check 2

Graph the solution set of each linear inequality in two variables.

a. $3x + y > 6$ **b.** $y \leq 2x + 1$ **c.** $x < -2$

d. $y \geq 3$ **e.** $y + 2x \geq 0$

Applications

Objective 3 ▶

Solve applications of linear inequalities in two variables.

Applications of linear inequalities in two variables are very important for a branch of mathematics called *linear programming*. This field of mathematics is used to determine how to maximize or minimize different values. While this topic is beyond the scope of this class, we will explore other applications of linear inequalities.

Objective 3 Example

The final grade in Tyrone's math class is based on his test average and his final exam grade. The test average counts for 70% of his final grade and the final exam counts for 30% of his final grade. If Tyrone wants to have at least a 70 in the course, what are some possible combinations of test average and final exam grades that produce the desired result? (Note the final exam grade and test average cannot exceed 100.)

Solution What is unknown? The test average and the final exam grade are unknown.

Let x = test average and let y = final exam grade.

Recall that 70% and 30% written as decimals are 0.70 and 0.30, respectively. Since the test average counts for 70% and the final counts for 30%, the final grade is computed as

$$0.70x + 0.30y$$

For the final grade to be at least 70, Tyrone must obtain a grade of 70 or higher. Therefore, we must solve the linear inequality

$$0.70x + 0.30y \geq 70$$

The graph of the boundary line $0.70x + 0.30y = 70$ is solid ($\geq$) and goes through the points (100, 0) and (70, 70). We graph only in the first quadrant since negative grades are not possible. We test (0, 0) since it does not lie on the boundary line, $0.70x + 0.30y = 70$.

$0.70x + 0.30y \geq 70$	Begin with the original inequality.
$0.70(0) + 0.30(0) \geq 70$	Replace x and y with 0.
$0 + 0 \geq 70$	Simplify.
$0 \geq 70$	Simplify.

Since the point (0, 0) makes the inequality false, we shade the portion of the plane that does not contain the origin, that is, the upper half-plane.

Any point contained in the shaded region satisfies the inequality. Points in this region give possible values for Tyrone's test average and final exam grade, provided both values are less than or equal to 100. Some possible solutions to this problem are shown on the graph in green. The final average each point yields is shown in the following table.

Test Average, x	Final Exam Grade, y	Final Average $= 0.70x + 0.30y$
70	80	$0.70(70) + 0.30(80) = 73$
70	70	$0.70(70) + 0.30(70) = 70$
85	60	$0.70(85) + 0.30(60) = 77.5$
90	30	$0.70(90) + 0.30(30) = 72$
100	0	$0.70(100) + 0.30(0) = 70$

There are many other combinations of test averages and final exam grades that provide the desired course grade.

The points labeled in magenta (50, 90), (70, 60), and (80, 30) are ordered pairs that do not satisfy the inequality and would, therefore, not make the desired final grade.

✓ **Student Check 3** The final grade in Lorna's history class is based on her midterm exam grade and her final exam grade. The midterm exam counts for 40% of her grade and her final exam counts for 60% of her grade. Find three possible combinations of grades she can earn on the midterm and final exams to have at least an 80 in the class.

Objective 4 ▶
Troubleshoot common errors.

Troubleshooting Common Errors

Some common errors associated with linear inequalities in two variables are shown next. Be careful to draw the boundary line correctly.

Objective 4 Examples A problem and an incorrect solution are given. Provide the correct solution and an explanation of the error.

4a. Graph the solution set of $y > -x + 3$.

Incorrect Solution	Correct Solution and Explanation
The boundary line is $y = -x + 3$. The point $(0, 0)$ makes the inequality false.	Everything is correct except for the boundary line. It should be dashed since the inequality is $>$.
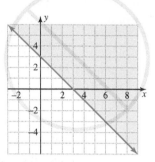	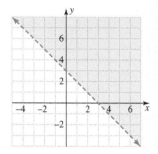

4b. Graph the solution set of $x - 5y < 10$.

Incorrect Solution	Correct Solution and Explanation
The boundary line is $x - 5y = 10$. The point $(0, 0)$ makes the inequality true. The graph is	Since $(0, 0)$ makes the inequality true, the graph must include this point. We shade the half-plane that contains the origin.
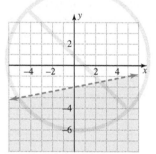	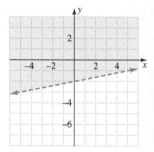

ANSWERS TO STUDENT CHECKS

Student Check 1 **a.** yes **b.** no **c.** yes

Student Check 2

a.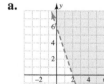

b.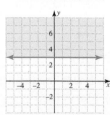

e.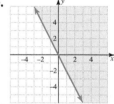

c.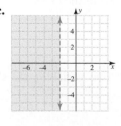

d.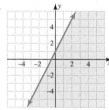

Student Check 3 Answers vary; $(80, 80)$, $(60, 100)$, and $(70, 90)$.

SUMMARY OF KEY CONCEPTS

1. Solutions of linear inequalities in two variables are ordered pairs that satisfy the linear inequality. There are infinitely many solutions of linear inequalities in two variables.

2. Solutions of linear inequalities can be illustrated by a graph. The solutions of a linear inequality in two variables lie in a half-plane that is formed by the graph of the associated linear equation. A single test point can be used to determine which half-plane contains solutions.

3. To use a linear inequality in two variables to solve an applied problem, graph the inequality as shown in this section. Any ordered pair in the shaded region is a solution of the problem.

GRAPHING CALCULATOR SKILLS

To graph a linear inequality in two variables using the graphing calculator, we must first solve the inequality for y. If the resulting inequality is of the form $y < mx + b$, shade below the boundary line. If the resulting inequality is of the form $y > mx + b$, shade above the boundary line. If the inequality is $<$ or $>$, the boundary line will be dashed and if the inequality is $\leq$ or $\geq$, then the boundary line is solid.

Example: Solve $x - 2y > 4$.

Solution: We first solve the inequality for y.

$$-2y > -x + 4$$

$$y < \frac{1}{2}x - 2$$

The inequality reverses because we divide both sides by -2. Enter the associated equation. For the calculator to graph the appropriate region, we need to "tell it" which region to shade—the region above the boundary line ($>$ or $\geq$) or the region below ($<$ or $\leq$). Move the cursor to the left of Y_1. Press ENTER until we reach the symbol ▜ (shade above) or ◣ (shade below).

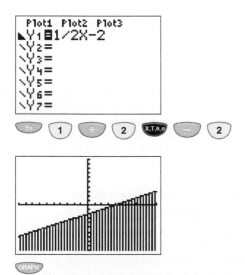

Alternate Method for TI-84 Plus Users: Use the APPS feature to graph inequalities.

Example: Solve $y > -2x - 4$.

Press ⬜ and ⬜. Scroll down in the menu and select Inequalz.

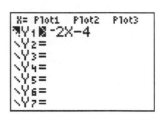

Press any key to continue. Press ⬜ ⬜ to select the $>$ symbol and then enter the remaining part of the inequality. Press ⬜

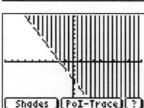

SECTION 9.4 / EXERCISE SET

Write About It!

Use complete sentences in your answer to each exercise.

1. How do you determine if an ordered pair (x, y) is a solution of a linear inequality?

2. How do you determine the boundary line of a linear inequality?

3. How will you know whether the boundary line of the graph of a linear inequality should be a dotted line or a solid line?

4. How many points in a half-plane must be tested to determine if the half-plane includes solutions of a linear inequality?

Practice Makes Perfect!

Determine if the ordered pair is a solution of the given inequality. (*See Objective 1.*)

5. $3x - 4y < 12$; $(1, 1)$

6. $2x + y < -3$; $(0, 1)$

7. $y \geq 5x - 1$; $(4, -1)$

8. $y \leq -2x + 3$; $(0, 0)$

9. $x + y < 9$; $(10, 3)$

10. $x - y > -5$; $(-2, -4)$

11. $y > \frac{1}{2}x + 8$; $(6, -3)$

12. $y < -\frac{1}{3}x - 11$; $(9, 0)$

13. $x \geq 2y - 4$; $(0, 2)$

14. $y + 2 < 0$; $(8, -1)$

15. $x - 3 < 0$; $(-4, 11)$

16. $x + 15 > 1$; $(13, -2)$

17. $4x - 2y > 1$; $\left(\frac{1}{2}, \frac{3}{4}\right)$

18. $x - 2y < 3$; $\left(\frac{2}{3}, \frac{1}{6}\right)$

Graph the solution set of each linear inequality in two variables. (*See Objective 2.*)

19. $2x + y > 4$

20. $3x + y < 5$

21. $y \leq 4x - 8$

22. $y \geq 3x - 6$

23. $y < -2x + 12$

24. $y > -3x - 9$

25. $y \leq 5x - 15$

26. $y \geq 10x - 20$

27. $x - y < 2$

28. $y - 7x \leq 0$

29. $y < 3x$

30. $y > 2x$

31. $3x - 3y \geq 0$

32. $4x - y > -3$

33. $x > 8$

34. $x < -1$

35. $y < -4$

36. $y > 3$

Use a linear inequality in two variables to solve each problem. (*See Objective 3.*)

For Exercises 37 and 38: Dawn's algebra course grade is based on her test average and her final exam grade. Her test average counts for 60% of her course grade and the final exam counts for 40% of her course grade.

37. If Dawn wants to have at least an 80 in the course, (a) write a linear inequality in two variables that represents this situation, (b) graph the linear inequality, and (c) give three possible combinations of test averages and final exam grades that produce a course grade of at least 80.

38. If Dawn wants to have at least a 70 in the course, (a) write a linear inequality in two variables that represents this situation, (b) graph the linear inequality, and (c) give three possible combinations of test averages and final exam grades that produce a course grade of at least 70.

For Exercises 39 and 40: Jenna works at a bookstore for $8 per hour and at the school library for $10.50 per hour.

39. If Jenna wants to make at least $300 per week, (a) write a linear inequality in two variables that represents this situation; (b) graph the linear inequality; and (c) give three possible combinations of hours that Jenna could work at each job to make at least $300 per week.

40. If Jenna wants to make at least $450 per week, (a) write a linear inequality in two variables that represents this situation; (b) graph the linear inequality; and (c) give three possible combinations of hours that Jenna could work at each job to make at least $450 per week.

For Exercises 41 and 42: Shenita charges $15 to hem a pair of pants and $25 to tailor a suit jacket.

41. If Shenita wants to make at least $500 in one week from tailoring, (a) write a linear inequality in two variables that represents this situation; (b) graph the linear inequality; and (c) if Shenita wants to make at least $500 in one week, give three possible combinations of the number of pants she must hem and the number of suit jackets she must tailor.

42. If Shenita wants to make at least $650 in one week from tailoring, (a) write a linear inequality in two variables that represents this situation; (b) graph the linear inequality; and (c) if Shenita wants to make at least $650 in one week, give three possible combinations of the number of pants she must hem and the number of suit jackets she must tailor.

 Mix 'Em Up!

Determine if the ordered pair is a solution of the given inequality.

43. $x + 3y < 6$; $(0, 0)$

44. $x - 7y < 1$; $(10, 3)$

45. $y > \frac{1}{2}x + 8$; $(8, -5)$

46. $y < -\frac{1}{3}x - 11$; $(-9, 0)$

47. $x > 12y$; $(1, 0)$

48. $y > 6x$; $(0, -1)$

49. $y - 11 < 4$; $(-3, 25)$

50. $x + 5 \geq 1$; $(-2, 4)$

51. $6x - y \leq 1$; $(0.2, -0.5)$

52. $x + 8y > 3$; $(-1.4, 0.5)$

Graph the solution set of each linear inequality in two variables.

53. $x + 3y < 2$

54. $2x - 4y \leq 3$

55. $5x - 2y > 0$

56. $3x + 2y < 0$

57. $y + 3 \geq 0$

58. $2x - 3 < 0$

59. $0.2x + 0.3y < 1.5$

60. $-0.3x + 0.4y \geq 1.2$

For Exercises 61 and 62: Ted's biology course grade is based on his test average and his final exam grade. His test average counts for 75% of his course grade and his final exam counts for 25% of his course grade.

61. If Ted wants to have at least a 90 in the course, (a) write a linear inequality in two variables that represents this situation, (b) graph the linear inequality, and (c) give three possible combinations of test averages and final exam grades that produce a course grade of at least 90.

62. If Ted wants to have at least an 80 in the course, (a) write a linear inequality in two variables that represents this situation, (b) graph the linear inequality, and (c) give three possible combinations of test averages and final exam grades that produce a course grade of at least 80.

For Exercises 63 and 64: Brasil works at a grocery store for $10.50 per hour and as a student worker for $8.50 per hour.

63. If Brasil wants to make at least $350 per week, (a) write a linear inequality in two variables that represents this situation, (b) graph the linear inequality, and (c) give three possible combinations of hours that Brasil could work at each job to make at least $350 per week.

64. If Brasil wants to make at least $450 per week, (a) write a linear inequality in two variables that represents this situation, (b) graph the linear inequality, and (c) give three possible combinations of hours that Brasil could work at each job to make at least $450 per week.

 You Be the Teacher!

Correct each student's error, if any.

65. Graph $4x + y < 8$.
Jamie's work: The boundary line is $4x + y = 8$ and my test point is $(0, 0)$.
$4(0) + 0 < 8$
$0 < 8$
True

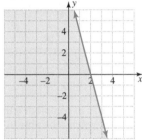

66. Graph $x + 3y \geq 6$.
Vince's work: The boundary line is $x + 3y > 6$ and use $(0, 0)$ as a test point.
$0 + 3(0) > 6$
$0 > 6$
False

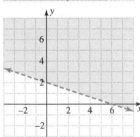

67. Graph $x - 4y > 0$.
Carolyn's work: The boundary line is $x - 4y = 0$.
I shade the upper half-plane since the inequality symbol is $>$.

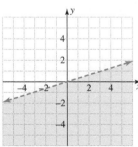

68. Graph $2x \leq y$.
Brent's work:
I graph $2x = y$ and shade the lower half-plane since the symbol is $\leq$.

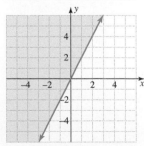

Calculate It!

Use a graphing calculator to graph each linear inequality in two variables.

69. $3x + y \geq 1$

70. $6x + y \leq 2$

71. $x + 3y < 2$

72. $x - 5y < 4$

Think About It!

73. Give an example of a linear inequality for which the point (0, 0) cannot be used as the test point to determine which half-plane to shade.

74. Is it possible for a linear inequality in two variables to have one point as a solution? Explain.

75. Is it possible for a linear inequality in two variables to have only its boundary line as a solution? Explain.

76. Graph the inequalities in parts a–d and use your results to answer parts e–g.

a. $y > -2x + 4$ **b.** $y > \dfrac{1}{3}x - 1$

c. $y > 4x$ **d.** $y > 1$

e. Which half-plane was shaded in each of these inequalities?

f. What can you conclude about the graph of the solution of $y > mx + b$?

g. How does the graph of $y \geq mx + b$ differ from the graph of $y > mx + b$?

77. Graph the inequalities in parts a–d and use your results to answer parts e–g.

a. $y < -2x + 4$ **b.** $y < \dfrac{1}{3}x - 1$

c. $y < 4x$ **d.** $y < 1$

e. Which half-plane was shaded in each of these inequalities?

f. What can you conclude about the graph of the solution of $y < mx + b$?

g. How does the graph of $y \leq mx + b$ differ from the graph of $y < mx + b$?

SECTION 9.5   Systems of Linear Inequalities in Two Variables

▶ OBJECTIVES

As a result of completing this section, you will be able to

1. Solve a system of linear inequalities in two variables.
2. Solve application problems.
3. Troubleshoot common errors.

The Parent Teacher Association (PTA) of a local school is selling rolls of wrapping paper and boxed chocolates for a fund-raiser. The PTA can order at most 300 items. Each roll of wrapping paper costs $2 and each box of chocolates costs $3. The PTA can spend no more than $1200 on these items. From past experience, the PTA knows that they sell at least twice as many boxes of chocolates as they do rolls of wrapping paper.

How many rolls of wrapping paper and boxes of chocolates should the PTA order for their conditions to be satisfied?

Solving this problem requires us to satisfy more than one linear inequality in two variables. In this section, we will learn how to solve a system of linear inequalities.

Solving Systems of Linear Inequalities in Two Variables

Objective 1 ▶

Solve a system of linear inequalities in two variables.

A **system of linear inequalities** is a set of two or more linear inequalities that must be solved together. The **solution set of a system of linear inequalities in two variables** is the set of all ordered pairs that satisfies each inequality in the system.

In Section 4.1, we learned that the solution set of a system of linear equations in two variables consists of the point where the two lines intersect. Similarly, the solution set of a system of linear inequalities in two variables consists of the points where the two inequalities intersect. Graphically, this is the set of all ordered pairs where the two half-planes in the system intersect.

> **Procedure: Solving a System of Linear Inequalities in Two Variables**
>
> **Step 1:** Graph each linear inequality on the same coordinate system using the method presented in Section 9.4.
>
> **Step 2:** Identify the intersection of the shaded regions. This is the solution of the system.

| **Objective 1 Examples** | **Solve each system of linear inequalities.** |

$$\textbf{1a. } \begin{cases} y > -x + 3 \\ y \le 2x - 4 \end{cases} \qquad \textbf{1b. } \begin{cases} 2x + y < 6 \\ x - y \le 4 \end{cases} \qquad \textbf{1c. } \begin{cases} y > -2 \\ x < 3 \end{cases}$$

Solutions

1a. We graph each inequality and find the intersection.

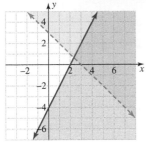

Figure 9.5.1

$y > -x + 3$: The boundary line is $y = -x + 3$ and is dashed. The line $y = -x + 3$ goes through $(0, 3)$ and $(3, 0)$. We test $(0, 0)$.

$$y > -x + 3$$
$$0 > -(0) + 3$$
$$0 > 3 \qquad \text{False}$$

Since $(0, 0)$ makes the original inequality false, we shade the upper half-plane that doesn't contain $(0, 0)$. The graph of the solution set of $y > -x + 3$ is shown in blue in Figure 9.5.1.

$y \le 2x - 4$: The boundary line is $y = 2x - 4$ and is solid. The line $y = 2x - 4$ goes through $(0, -4)$ and $(2, 0)$. We test $(0, 0)$.

$$y \le 2x - 4$$
$$0 \le 2(0) - 4$$
$$0 \le -4 \qquad \text{False}$$

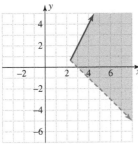

Figure 9.5.2

Since $(0, 0)$ makes the original inequality false, we shade the lower half-plane that doesn't contain $(0, 0)$. The graph of the solution set of $y \le 2x - 4$ is shown in magenta in Figure 9.5.1.

The solution set of the system is the region where the two half-planes intersect as shown in Figure 9.5.2.

1b. We graph each inequality and find the intersection.

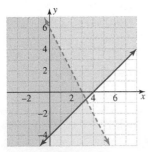

Figure 9.5.3

$2x + y < 6$: The boundary line is $2x + y = 6$ and is dashed. The line goes through $(0, 6)$ and $(3, 0)$. We test $(0, 0)$.

$$2x + y < 6$$
$$2(0) + 0 < 6$$
$$0 < 6 \qquad \text{True}$$

Since $(0, 0)$ makes the original inequality true, we shade the lower half-plane, which contains $(0, 0)$. The solution set of $2x + y < 6$ is shown in blue in Figure 9.5.3.

$x - y \le 4$: The boundary line is $x - y = 4$ and is solid. The line goes through the $(0, -4)$ and $(4, 0)$. We test $(0, 0)$.

$$x - y \le 4$$
$$0 - 0 \le 4$$
$$0 \le 4 \qquad \text{True}$$

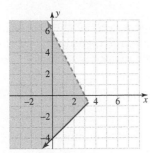

Figure 9.5.4

Since $(0, 0)$ makes the original inequality true, we shade the upper half-plane, which contains $(0, 0)$. The solution set of $x - y \le 4$ is shown in magenta as shown in Figure 9.5.3.

The solution set of the system is the intersection of the two regions as shown in Figure 9.5.4.

1c. We graph each inequality and find the intersection.

$y > -2$: The boundary line is $y = -2$ and is dashed. The line is a horizontal line through $(0, -2)$. We test $(0, 0)$.

$$y > -2$$
$$0 > -2 \qquad \text{True}$$

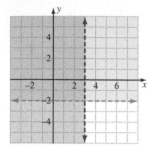

Figure 9.5.5

Since (0, 0) makes the inequality true, we shade the upper half-plane which contains (0, 0). The graph of $y > -2$ is shown in blue in Figure 9.5.5.

$x < 3$: The boundary line is $x = 3$ and is dashed. The line is a vertical line through (3, 0). We test (0, 0).

$$x < 3$$
$$0 < 3 \quad \text{True}$$

Since (0, 0) makes the inequality true, we shade the half-plane that contains (0, 0). The graph of $x < 3$ is shown in magenta in Figure 9.5.5.

The solution set of the system is the intersection of the shaded regions as shown in Figure 9.5.6.

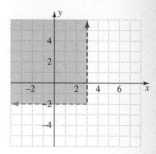

Figure 9.5.6

✓ **Student Check 1**   Solve each system of linear inequalities.

a. $\begin{cases} y \geq x + 5 \\ y \geq -x + 2 \end{cases}$ **b.** $\begin{cases} 3x - y < 3 \\ x + 2y > 4 \end{cases}$ **c.** $\begin{cases} x \geq -2 \\ y \leq 4 \end{cases}$

Applications

Objective 2 ▶

Solve application problems.

Systems of linear inequalities can be used to solve problems in business as well as other areas. If the variables in the problem satisfy several constraints that can be represented by inequalities, then a system is used. Key phrases such as "at most," "at least," "not more than," and "not less than" indicate that an inequality is needed.

> **Procedure: Solving an Application Involving a System of Linear Inequalities**
>
> **Step 1:** Determine the unknowns and define variables for them.
> **Step 2:** Write a system of linear inequalities to represent the constraints given in the problem.
> **Step 3:** Solve the system graphically.
> **Step 4:** Ordered pairs within the solution set satisfy all of the constraints in the problem.

Objective 2 Example   The Parent Teacher Association (PTA) of a local school is selling rolls of wrapping paper and boxed chocolates for a fund-raiser. The PTA can order at most 300 items. Each roll of wrapping paper costs $2 and each box of chocolates costs $3. The PTA can spend no more than $1200 on these items. From past experience, the PTA knows that they sell at least twice as many boxes  of chocolates as they do rolls of wrapping paper. How many rolls of wrapping paper and boxes of chocolates should the PTA order for their conditions to be satisfied? Provide three specific examples that are in the solution set.

Solution   What is unknown? The number of rolls of wrapping paper and the number of boxes of chocolates that should be ordered are unknown.

Let w = number of rolls of wrapping paper (first coordinate).

Let c = number of boxes of chocolate (second coordinate).

What is known? At most 300 items can be ordered. Each roll of wrapping paper costs $2 and each box of chocolates costs $3. The PTA can't spend more than $1200. The boxes of chocolates sold are at least twice as many as the rolls of

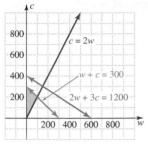

Figure 9.5.7

wrapping paper sold. The total number of each item sold is nonnegative. We can represent these facts in the system as shown.

$$\begin{cases} 2w + 3c \le 1200 \\ w + c \le 300 \\ c \ge 2w \\ c, w \ge 0 \end{cases}$$

The solution set of this system is shown in Figure 9.5.7.

Any ordered pair in the shaded region or on the boundary lines satisfies the system. Some possible combinations that will satisfy the PTA's requirements are stated next.

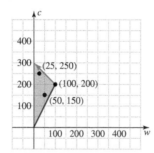

50 rolls of wrapping paper and 150 boxes of chocolates

100 rolls of wrapping paper and 200 boxes of chocolates

25 rolls of wrapping paper and 250 boxes of chocolates

✓ Student Check 2   A motel chain plans to open a new motel. The motel will have a combination of double rooms and king-size rooms. The motel will have at most 200 rooms. Based on consumer trends, the motel knows that they should have at least 125 double rooms. They also know that they need at least twice as many double rooms as king-size rooms. Use a system of linear inequalities to find at least three combinations of double rooms and king-size rooms that will satisfy these conditions.

Objective 3 ▶

Troubleshoot common errors.

Troubleshooting Common Errors

A common error associated with systems of linear inequalities is shown next.

Objective 3 Example   A problem and an incorrect solution are given. Provide the correct solution and an explanation of the error.

Solve the system $\begin{cases} y > x + 2 \\ y < \dfrac{1}{4}x - 4 \end{cases}$.

Incorrect Solution	Correct Solution and Explanation
Each inequality is graphed here	The error was made in not extending the graphs to the left. If we adjust our scale, we get the following graph.

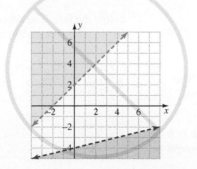

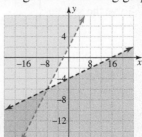

Since the graphs do not overlap, there is no solution for this system.

So, the solution set is

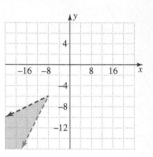

ANSWERS TO STUDENT CHECKS

Student Check 1 **a.**

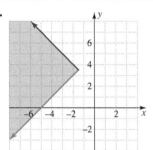

c.

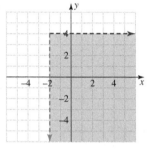

b.

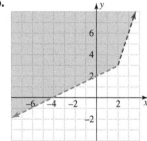

Student Check 2

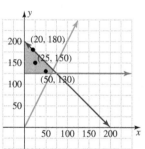

SUMMARY OF KEY CONCEPTS

Solving a system of linear inequalities requires us to graph each linear inequality on the same coordinate system. The solution set for the system is the intersection of the half-planes formed by each inequality.

Be sure to draw the boundary lines correctly for each inequality. If the inequality symbol is $<$ or $>$, then the boundary line is dashed. If the inequality symbol is $\leq$ or $\geq$, then the boundary line is solid.

GRAPHING CALCULATOR SKILLS

Example: Solve the system $\begin{cases} 2x + y < 6 \\ x - y \leq 4 \end{cases}$ using a graphing calculator.

Solution: Solve each inequality in the system for y. This gives us $\begin{cases} y < -2x + 6 \\ y \geq x - 4 \end{cases}$.

Enter each inequality in the equation editor. Since the first inequality involves a $<$ symbol, select the option to graph

below the line. The second inequality involves a $\geq$ symbol, so we select the option to graph above the line.

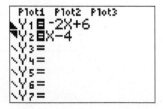

Press GRAPH to view the intersection of the two inequalities.

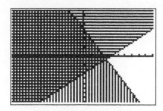

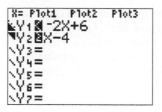

Press GRAPH to view the solution set of the system.

Alternative Method for TI-84 Plus Users: Use the APPS menu to graph the inequalities.

Press Y=, APPS, arrow down to Inequalz, and press any key to enter the program.

Select the appropriate inequality symbol by using the Alpha key and the appropriate function key.

SECTION 9.5   EXERCISE SET

 Write About It!

Use complete sentences in your answer to each exercise.

1. Describe the steps to solve a system of linear inequalities.

2. Explain how the solution of a system of linear inequalities could be a straight line.

3. Explain how to determine two points (x, y) that are solutions of the system $\begin{cases} y \geq 3 \\ x < 4 \end{cases}$

4. Explain how to determine two points (x, y) that are solutions of the system $\begin{cases} x \geq 10 \\ y < 7 \end{cases}$.

 Practice Makes Perfect!

Solve each system of linear inequalities. (*See Objective 1.*)

5. $\begin{cases} y > x + 1 \\ y \leq 2x - 1 \end{cases}$

6. $\begin{cases} y > 3x + 3 \\ y \leq -x + 5 \end{cases}$

7. $\begin{cases} y \geq 4x - 8 \\ y > -2x + 4 \end{cases}$

8. $\begin{cases} y \geq 6x + 6 \\ y > -3x + 6 \end{cases}$

9. $\begin{cases} 5x - y > 5 \\ x - y < 7 \end{cases}$

10. $\begin{cases} 4x + y < 4 \\ -7x + y > -3 \end{cases}$

11. $\begin{cases} x \geq 5 \\ y < -4 \end{cases}$

12. $\begin{cases} x \geq -3 \\ y < 9 \end{cases}$

13. $\begin{cases} 3x + y \geq 6 \\ y \geq 2 \end{cases}$

14. $\begin{cases} 2x + y \geq 5 \\ x \geq -3 \end{cases}$

15. $\begin{cases} x \leq 2y \\ x - 4y \geq 4 \end{cases}$

16. $\begin{cases} x \leq 3y \\ x - 2y \geq -1 \end{cases}$

17. $\begin{cases} x > 2 \\ y \leq -1 \end{cases}$

18. $\begin{cases} x > -4 \\ y \leq -7 \end{cases}$

19. $\begin{cases} y \geq 10 \\ x < -8 \end{cases}$

20. $\begin{cases} y \geq -6 \\ x \leq 5 \end{cases}$

21. $\begin{cases} 3x - 4y \geq 12 \\ x < -5 \end{cases}$

22. $\begin{cases} 5x + 2y \geq 6 \\ x \geq 1 \end{cases}$

23. $\begin{cases} 2x + 3y \leq 9 \\ y \leq 5 \end{cases}$

24. $\begin{cases} 3x - 5y \leq 10 \\ y \geq -2 \end{cases}$

25. $\begin{cases} 4x + y \geq 8 \\ 4x + y \leq -1 \end{cases}$

26. $\begin{cases} 3x - 2y \geq 6 \\ -3x + 2y \geq 4 \end{cases}$

For Exercises 27 and 28, (a) write a system of linear inequalities that represents the situation, (b) solve it by graphing, and (c) find at least three combinations of rolls of wrapping paper and boxes of chocolates that satisfy the given conditions. (*See Objective 2.*)

27. The Parent Teacher Association (PTA) of a local school is selling rolls of wrapping paper and boxed chocolates for a fund-raiser. The PTA can order at most 1200 items. Each roll of wrapping paper costs $1.50 and each box of chocolates costs $4. The PTA can spend no more than $4200 on these items. From past experience, the PTA knows that they sell at least twice as many boxes of chocolates as they do rolls of wrapping paper.

28. The Parent Teacher Association (PTA) of a local school is selling rolls of wrapping paper and boxed chocolates for a fund-raiser. The PTA can order at most 750 items. Each roll of wrapping paper costs $2 and each box of chocolates costs $6.50. The PTA can spend no more than $5100 on these items. From past experience, the PTA knows that they sell at least twice as many boxes of chocolates as they do rolls of wrapping paper.

For Exercises 29 and 30, use the following information. In the business world, the price per item, in dollars, is determined by the demand equation $y = D(x)$, where x is the number of items sold. The consumer's surplus is the area of the graph of the solution set of the system of linear inequalities

$$\begin{cases} y \le D(x) \\ y \ge k \\ x \ge 0 \end{cases} \text{, where } k \text{ is a fixed price}$$

29. The demand equation for an item is $y = D(x) = 20 - 0.05x$ and $k = 10$. (a) Write a system of linear inequalities and solve it by graphing and (b) find the consumer's surplus.

30. The demand equation for an item is $y = D(x) = 30 - 0.45x$ and $k = 7.5$. (a) Write a system of linear inequalities and solve it by graphing and (b) find the consumer's surplus.

 Mix 'Em Up!

Solve each system of linear inequalities.

31. $\begin{cases} y \ge -\dfrac{1}{2}x \\ y \le 2x \end{cases}$

32. $\begin{cases} y > 3x \\ y < -\dfrac{1}{3}x + 2 \end{cases}$

33. $\begin{cases} y \ge 2x - 1 \\ y \le 2x + 3 \end{cases}$

34. $\begin{cases} y < -x \\ y > -x + 3 \end{cases}$

35. $\begin{cases} 3x - y \le 2 \\ x \ge -2 \end{cases}$

36. $\begin{cases} y \le 2 \\ x + 2y \ge -6 \end{cases}$

37. $\begin{cases} y \ge \dfrac{2}{3}x + 4 \\ y \ge \dfrac{1}{4}x - 1 \end{cases}$

38. $\begin{cases} y < 4x + 2 \\ y > -2x + 5 \end{cases}$

39. $\begin{cases} y \ge -2 \\ x \le 2 \end{cases}$

40. $\begin{cases} x + 2y \ge -6 \\ x + 2y \le 4 \end{cases}$

The price per item, in dollars, is determined by the supply equation $y = S(x)$, where x is the number of items produced. The producer's surplus is the area of the graph of the solution set of the system of linear inequalities

$$\begin{cases} y \ge S(x) \\ y \le k \\ x \ge 0 \end{cases} \text{, where } k \text{ is a fixed price}$$

41. The supply equation for an item is $y = S(x) = 25 + 0.10x$ and $k = 50$. (a) Write a system of linear inequalities and solve it by graphing and (b) find the producer's surplus.

42. The supply equation for an item is $y = S(x) = 8 + 0.004x$ and $k = 20$. (a) Write a system of linear inequalities and solve it by graphing and (b) find the producer's surplus.

 You Be the Teacher!

Correct each student's error, if any.

43. Solve $\begin{cases} x - 2y \le 2 \\ -x + y \ge 1 \end{cases}$.

Bernie's work: For the first inequality, I graphed the line $x - 2y = 2$ and shaded the lower half-plane since the inequality is $\le$. For the second inequality, I graphed the line $-x + y = 1$ and shaded the upper half-plane since the inequality is $\ge$.

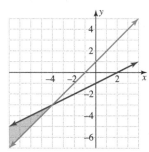

44. Solve $\begin{cases} y > 2x \\ x - 2y > 4 \end{cases}$.

Yvonne's work: For the first inequality, I graphed the line $y = 2$ as a dotted line and shaded the upper half-plane since the inequality is $>$. For the second inequality, I graphed the line $x - 2y = 4$ as a dotted line and shaded the upper half-plane since the inequality is $>$.

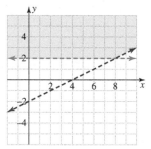

 Calculate It!

Use a graphing calculator to solve each system of linear inequalities.

45. $\begin{cases} y < 4x - 1 \\ y \ge -2x + 3 \end{cases}$

46. $\begin{cases} y \ge 3x - 6 \\ y > x \end{cases}$

47. $\begin{cases} 4x - 2y > 5 \\ -3x + y < 1 \end{cases}$

48. $\begin{cases} x - 3y \le 6 \\ 2x + y \ge -1 \end{cases}$

 Think About It!

49. Give an example of a system of linear inequalities that has no solution.

50. Give an example of a system of linear inequalities that has a straight line as a solution.

GROUP ACTIVITY   The Mathematics of Controlling Waste

Your employer has assigned you with the task of controlling waste in the cutting room for a wrapping paper manufacturer. The company wants to lose less than 1% of wrapping paper for

each run of 2000 sheets of paper. The reported width of each piece of square wrapping paper is 1 yd. The absolute error of the cutting machines is 0.25 in. in all directions. Determine if the current machines are within the desired standards. (Recall from Section 9.2 that $E_{abs} = |x - a|$, where x is the exact measure and a is the approximated value.)

1. Let x be the actual length of the cut wrapping paper. Write an absolute value inequality that shows that the absolute error of one side of the paper is within 0.25 in.

2. Solve the inequality from part 1. Interpret the meaning in the context of the problem.

3. What are the maximum and minimum values for the area of a cut sheet of wrapping paper?

4. What is the maximum error of the area of the cut paper?

5. If the cutter wastes the maximum amount of paper (as found in part 4) for each piece of paper in the run, how much total waste would there be?

6. How many sheets of wrapping paper does the cutter waste for a run of 2000 sheets of paper?

7. What percentage of waste does this represent?

8. What would you report to your employer?

Inequalities and Absolute Value

What's the big idea? Chapter 9 provides us with the skills to solve a new class of equations — absolute value equations. We also use the methods of solving compound inequalities to solve absolute value inequalities. The methods for solving linear equations in two variables were extended to solving linear inequalities in two variables. The methods for solving systems of linear equations in two variables were extended to solving systems of linear inequalities in two variables.

The Tools

Listed below are the key terms, skills, formulas, and properties you should know for this chapter.

The page reference is provided if you need additional help with the given topic. The Study Tips will assist in your preparation for an exam.

Study Tips

1. Learn all of the terms, formulas, and properties. Make flash cards and have someone quiz you.
2. Rework problems from the exercises and also the ones you worked in class. Work additional problems from the review exercises.
3. Review the summaries of key concepts.
4. Work the chapter test.
5. Be sure to review the online resources for additional study materials.

Terms

Absolute error   654
Absolute value equation   650
Boundary line   673
Compound inequality   639
Half-plane   673
Intersection   636

Linear inequality in two
   variables   672
Lower half-plane   673
Solution of a linear inequality
   in two variables   672

Solution set of a system of linear
   inequalities   682
System of linear inequalities   682
Test point method   674
Union   638
Upper half-plane   673

Formulas and Properties

- Absolute value equations
  (Property 1)   650
- Absolute value equations
  (Property 2)   652

- Absolute value inequalities
  (Property 1)   659
- Absolute value inequalities
  (Property 2)   660

- Absolute value inequalities
  (Property 3)   662
- Absolute value inequalities
  (Property 4)   662

CHAPTER 9 / SUMMARY

How well do you know this chapter? Complete the following questions to find out. Take a look back at the section if you need help.

SECTION 9.1 Compound Inequalities

1. The _____ of two sets consists of the elements the sets have in common.

2. The _____ of two sets consists of the elements that are in either set.

3. A(n) _____ _____ consists of two inequalities joined by "and" or "or."

4. To solve a compound inequality using "and" requires us to find the _____ of the solution sets.

5. To solve a compound inequality using "or" requires us to find the _____ of the solution sets.

SECTION 9.2 Absolute Value Equations

6. The absolute value of a number is its _____ from zero on the number line.

7. The absolute value of any real number is _____.

8. If $|X| = k$, $k > 0$, then _____ or _____.

9. If $|X| = 0$, then _____.

10. If $|X| = k$, $k < 0$, then there is ___ _____.

11. To solve an absolute value equation, we must first _____ the absolute value on one side of the equation.

12. If $|X| = |Y|$, then _____ or _____. That is, the expressions inside the absolute value must be the ____ or _____.

13. The _____ _____ is the absolute value of the difference between the measured value and the exact value.

SECTION 9.3   Absolute Value Inequalities

14. To solve $|X| < k$, $k > 0$, is equivalent to solving the compound inequality _____ ___ _____. This can be written in a compact form, _____.

15. To solve $|X| > k$, $k > 0$, is equivalent to solving the compound inequality _____ __ _____.

16. If $|X| < k$, $k < 0$, then there is ___ _____.

17. If $|X| > k$, $k < 0$, then the solution set is ___ ___ _____.

18. Another method to solving absolute value inequalities involves using ___ _____.

SECTION 9.4   Linear Inequalities in Two Variables

19. A linear inequality in two variables is an inequality of the form _____.

20. Solutions of linear inequalities in two variables are _____ _____ that make the inequality ____.

21. To graph a linear inequality in two variables, we must first graph the _____ ____, which is formed by replacing the inequality symbol with a(n) _____ ____. If the inequality symbol is ___ or ___, the boundary line is solid. If the inequality symbol is ___ or ___, the boundary line is dashed.

22. The boundary line divides the plane into two _____. One _____ contains _____ of the inequality and the other does not.

23. After the boundary line is drawn, we must use a(n) ___ ____ in the original inequality to determine which half-plane contains solutions of the inequality. The ideal ___ ____ is _____ as long as the boundary line does not go through the origin.

SECTION 9.5   Systems of Linear Inequalities in Two Variables

24. The solution of a system of linear inequalities in two variables consists of the _____ of the half-planes formed from each linear inequality.

CHAPTER 9 / REVIEW EXERCISES

SECTION 9.1

Find the intersection or union of the two sets. Write each answer in interval notation when applicable. (See Objectives 1 and 2.)

1. $(-\infty, 0) \cap (-4, \infty)$

2. $[-10, 3] \cup (-4, 6)$

3. $(-\infty, 4) \cup [4, \infty)$

4. $(-\infty, 10]) \cap (10, \infty)$

5. $(-\infty, 8.6) \cup [-2.9, 13.8)$

6. $(-\infty, 3.2) \cap [-1.8, 16.4)$

7. $\varnothing \cap [4, 12]$

8. $\varnothing \cap (-2, 1)$

9. $\varnothing \cup (-4, \infty)$

10. $\varnothing \cup [8, \infty)$

Solve each compound inequality. Write each answer in interval notation when applicable. (See Objectives 3 and 4.)

11. $8x + 3 < 19$ or $3x - 1 > 14$

12. $6x - 7 \geq 5$ and $4x - 3 < 9$

13. $2(1 - 5x) > 12$ and $3x - 4 < 14$

14. $6(2 - x) < -4$ and $4x + 3 < 10$

15. $4 - 3x \leq 0$ or $8x + 3 > -5$

16. $2 - 5x \leq 1$ or $4x - 3 > 5$

17. $3x - 5 > 7$ and $2x + 6 < 1$

18. $12x - 4 \leq 8$ and $3x + 2 > 8$

19. $0.7x + 0.34 < 4.274$ and $0.4x - 2.56 > -1.7$

20. $1.24x - 4.35 \geq -2.49$ and $0.8x + 2.6 \leq 2.4$

Solve each problem using compound inequalities. (See Objective 5.)

21. Margie wants to maintain a test average between an 80 and 89. If Margie's test scores thus far are 81, 70, 85, and 76, what must she score on her fifth test to make the desired grade? (Assume the highest grade possible is 100.)

22. Grace wants to make a quiz average between 70 and 79. If Grace's quiz scores are 65, 72, 65, 71, and 68, then what must she score on her sixth quiz to get the desired grade? (Assume the highest grade possible is 100.)

23. Michael needs someone to trim two-thirds of his bamboo forest. He finds a company that charges $35 per hour plus $90 to haul the trees away. If Michael plans to spend between $200 and $500, for how many hours can he afford to hire the tree-trimmer?

24. Patty hires a professional organizer to organize her home office. The professional organizer charges a flat fee of $80 plus $32 per hour. If Patty budgets $750 to $1000 to get her office organized, for how many hours can she afford to hire the professional organizer?

SECTION 9.2

Solve each equation. (See Objective 1.)

25. $|12x - 5| = 25$

26. $|5x - 16| = 14$

27. $|7 - 2x| = 18$

28. $|9 - 4x| = 3$

29. $\left|\dfrac{8x - 4}{2}\right| - 1 = 6$

30. $\left|\dfrac{2x + 9}{5}\right| + 1 = 4$

31. $\left|\dfrac{2x}{3}\right| = -12$

32. $\left|\dfrac{15x}{4}\right| = 0$

33. $|3x + 2| = 0$

34. $|6x - 9| = 0$

35. $|0.6x - 2.5| = 7.1$

36. $|0.4x + 4.1| = 11.5$

Solve each equation. (*See Objective 2.*)

37. $\left|\dfrac{5x+1}{3}\right| = \left|\dfrac{x}{6}\right|$ **38.** $\left|\dfrac{3x-1}{4}\right| = \left|\dfrac{x}{4}\right|$

39. $|2x+10| = |4x-8|$ **40.** $|6x-11| = |5-4x|$

Solve each problem using an absolute value equation. (*See Objective 3.*)

41. Suppose the actual length of an object is 6.245 m. If the length is approximated to be 6.512 m, what is the absolute error?

42. Suppose the actual length of an object is 20.35 in. If the length is approximated to be 22.83 in., what is the absolute error?

43. Suppose a machine with an absolute error reading of 1.15 units reports the measurement of an object as 12.98 units. What is the actual measurement of the object?

44. Suppose a machine with an absolute error reading of 8.25 units reports the measurement of an object as 3163.78 units. What is the actual measurement of the object?

45. A poll shows that candidate A will receive 32% of the votes and candidate B will receive 36% of the votes. If the poll has an absolute error of 4%, how much of the vote will each candidate receive?

46. A poll shows that candidate C will receive 58% of the votes and candidate D will receive 32% of the votes. If the poll has an absolute error of 5%, how much of the vote will each candidate receive?

SECTION 9.3

Solve each absolute value inequality. Write each answer in interval notation. (*See Objectives 1–3.*)

47. $|3x+2| \le 16$ **48.** $|12-4x| \ge 8$

49. $-|x| \le -2$ **50.** $14 - |x+3| > 8$

51. $10 - |5-2x| < 7$ **52.** $3.2 + |x-0.6| \ge 5.4$

53. $-|x| > 7$ **54.** $|4-x| > 0$

55. $|2x+10| \le 0$ **56.** $|2-9x| < 0$

57. $|8-3x| \ge 0$ **58.** $4.6 - |x-2.3| \le 9.6$

59. $|b| > -12$ **60.** $|b| \ge -\dfrac{4}{5}$

61. $|5y-9| - 2 < -12$ **62.** $|3y-2| - 1 < -14$

Use an absolute value inequality to solve each problem. (*See Objective 4.*)

63. A November 2009 Gallup Poll found that 38% of Americans rated healthcare coverage in this country as excellent or good. If the margin of error was 4 percentage points, what are the possible percentages of Americans rating healthcare coverage as excellent or good? (Source: www.gallup.com)

64. A June 2010 Gallup Poll found that approximately 54% of Americans said they were energized Americans who were well-rested and did not experience ailments such as physical pain, headache, cold or flu. If the margin

of error was 3 percentage points, what are the possible percentages of Americans who were energized? (Source: www.gallup.com)

65. A truck derailment results in a hazardous chemical spill. Emergency management personnel advise that motorists should be restricted from being within a 12-mi radius of the spill. If the spill happens at mile marker 226 on Pennsylvania Turnpike 76 in Pennsylvania, what are the safe locations on Interstate 76 for motorists?

66. A truck derailment results in a hazardous chemical spill. Emergency management personnel advise that motorists should be restricted from being within a 20-mi radius of the spill. If the spill happens at mile marker 74 on a turnpike, what are the safe locations on the turnpike for motorists?

Use test points to solve each inequality. (*See Objective 5.*)

67. $|x-12| < 19$ **68.** $|x+7| > 21$

69. $|1-10x| - 3 \ge 18$ **70.** $|3-2x| + 5 \ge 20$

SECTION 9.4

Determine if each ordered pair is a solution of the given inequality. (*See Objective 1.*)

71. $3x - y < 5$; $(0, 1)$ **72.** $y > -\dfrac{1}{2}x$; $(8, -5)$

73. $3x > 2y$; $(0, 1)$ **74.** $y + 5 < 7$; $(-3, 1)$

75. $0.5x - y \le 1.2$; $(0.3, -0.1)$ **76.** $7y - x < 0$; $(2, -3)$

Graph each linear inequality in two variables. (*See Objective 2.*)

77. $x - 2y \ge 3$ **78.** $6 - 2y > 0$

79. $x + 3 \le 1$ **80.** $0.1x + 0.4y > 1.6$

Use a linear inequality in two variables to solve each problem. (*See Objective 3.*)

81. Suppose your geography course grade is based on your test average and your final exam average. Your test average counts for 80% of your course grade and your final exam counts for 20% of your course grade. You want to have at least a 90% average in the course. Write a linear inequality in two variables that represents this situation. Graph the linear inequality.

82. Gail works two jobs to put herself through college. She works at a grocery store for $11.50 per hour, and she works at the school as a student worker for $8.25 per hour. Gail wants to make at least $450 per week. Write a linear inequality in two variables that represents this situation. Graph the linear inequality.

SECTION 9.5

Solve each system of linear inequalities. (*See Objective 1.*)

83. $\begin{cases} y \ge -3x - 6 \\ y \le 4x \end{cases}$ **84.** $\begin{cases} y \ge \dfrac{1}{2}x - 5 \\ 2y \le x + 2 \end{cases}$

85. $\begin{cases} x + 2y > -4 \\ x < 2 \end{cases}$ **86.** $\begin{cases} x \ge -1 \\ y \le 3 \end{cases}$

Solve the problem using a system of linear inequalities. (*See Objective 2.*)

87. The price per item is determined by the demand equation $y = D(x)$, where x is the number of items sold.

The consumer's surplus is the area of the solution to the following system of linear inequalities.

$$\begin{cases} y \le D(x) \\ y \ge k \\ x \ge 0 \end{cases} \text{, where } k \text{ is a fixed price}$$

The demand equation for an item is $D(x) = 25 - 0.10x$ and $k = 10$. Set up a system of linear inequalities and find the consumer's surplus.

CHAPTER 9 TEST / INEQUALITIES AND ABSOLUTE VALUE

1. The interval that represents $(2, \infty) \cup (4, \infty)$ is
   a. $\varnothing$ b. $(2, 4)$ c. $(4, \infty)$ d. $(2, \infty)$

2. The solution set of $|3x - 5| - 4 = -1$ is

   a. $\left\{\dfrac{2}{3}, \dfrac{8}{3}\right\}$ b. $\left\{\dfrac{8}{3}, \dfrac{10}{3}\right\}$ c. $\left\{\dfrac{8}{3}\right\}$ d. $\varnothing$

3. The solution set for $|x + 3| > -2$ is
   a. $(-5, \infty)$ b. $\varnothing$ c. $(-\infty, \infty)$ d. $(-5, -1)$

Solve each equation. Write each answer in a solution set.

4. $|7a + 4| - 5 = 8$ 5. $\left|\dfrac{1}{2}b - \dfrac{2}{3}\right| + 5 = 1$

6. $|3 - 4x| = \left|\dfrac{1}{5}x + 2\right|$

Solve each inequality. Graph the solution set and write the solution set in interval and set-builder notation.

7. $-0.25x + 4 > -1$ and $\dfrac{1}{2}x - \dfrac{2}{3} < x$

8. $7(4 - x) + 3 < -2(x + 5)$ or $6(x + 1) - 4(2x - 3) > 7x$

9. $|8x + 3| \le 5$ 10. $|3x - 2| - 1 \ge 6$

Use an appropriate equation or inequality to solve each problem. Use complete sentences to state the answer.

11. Juan's final grade in his math class is based on a weighted average as follows.

Test average	50%
Homework	10%
Quizzes	15%
Final exam	25%

If Juan has a 79 test average, 95 homework average, and 83 quiz average, what scores can he make on his final exam to have a final grade of a C (70 to 79)?

12. A utility company installed a power line in a subdivision at 100 ft from the entrance of the subdivision. The utility company has an easement that is 40 ft on either side of the power line. Write an absolute value equation that will determine where the easement begins and ends, where x represents the location of the easement, and solve it.

13. In a portion of an interstate highway, the speed limit is 65 mph. Highway patrol officers stop vehicles traveling at speeds that are more than 15 mph from the speed limit. Write an absolute value inequality that models this situation, where s is the speed of the vehicle, and solve the inequality.

14. The demand equation for an item is $y = D(x) = 25 - 0.10x$ and $k = 10$.
    a. Set up a system of linear inequalities and
    b. find the consumer's surplus.

15. Using complete sentences, explain the steps used to graph the solution of a linear inequality in two variables.

Graph each linear inequality.

16. $x - 4y \ge -8$ 17. $y > -\dfrac{3}{5}x$

18. $x < 2$ 19. $y \ge 1$

20. Solve the system of linear inequalities $\begin{cases} x + y < 6 \\ y > \dfrac{1}{2}x - 4 \end{cases}$.

CUMULATIVE REVIEW EXERCISES / CHAPTERS 1–9

1. Perform each operation and simplify the result. (*Section 1.2, Objectives 2–7*)

   a. $\dfrac{8}{15} + \dfrac{7}{12}$ b. $3\dfrac{1}{4} - 1\dfrac{5}{6}$ c. $12\dfrac{3}{5} \cdot 4\dfrac{2}{7}$

2. Perform each operation and simplify. (*Sections 1.4 and 1.5, Objectives 1 and 2*)
   a. $2 - 12 + (-23) - (-10)$
   b. $3.1 - (-4.5) + (-6.7) + 7.2$
   c. $17 - \{8 - [4 - (1 - \sqrt{25 - 16})]\}$
   d. $-(-3)^3 + (-14) - |5 - 9|$

3. The lowest recorded temperature in Arizona was $-40°F$ at the Hawley Lake on January 7, 1971. The highest recorded temperature in Arizona was $128°F$ at Lake Havasu City on June 29, 1994. What is the difference between the highest and lowest temperatures? (Source: http://www.netstate.com/states/geography/az_geography.htm) (*Section 1.5, Objective 3*)

4. Evaluate each expression for the given values of the variables. (*Section 1.6, Objective 4*)

   a. $\sqrt{(x_1 - x_2)^2 + (y_1 - y_2)^2}$ for
   $x_1 = -5, x_2 = 7, y_1 = 13, y_2 = 8$

b. $\dfrac{|x+1|}{2x-4}$ for $x = -1, 0, 2$

5. Solve each equation. (*Section 2.2, Objectives 2 and 3; Sections 2.3 and 2.4, Objectives 1 and 2*)

   a. $-3(x-5) + 8 = 4(x-1) - 22$

   b. $5.2x - 6.3 + 1.7x = 9.4x - 12.3$

   c. $\dfrac{1}{7}(28x - 14) + 12 = x + 5$

6. Solve each equation. (*Section 2.4, Objectives 1–4*)

   a. $6(2a+1) - 2(a+4) = 5(2a+1) - 7$

   b. $\dfrac{2}{5}x + 3 = \dfrac{3}{10}x$

   c. $0.015x + 0.024(950 - x) = 20.28$

   d. $\dfrac{3}{2}(x+1) = \dfrac{1}{3}(5x - 2) + \dfrac{5}{6}$

7. Solve each formula for the specified variable. (*Section 2.5, Objective 3*)

   a. $S = 2\pi r^2 + 2\pi rh$ for h **b.** $C = 4l + 8w$ for l

   c. $x + 5y = 10$ for y **d.** $0.06x + 0.03y = 4.2$ for x

8. A coat is on sale for $144. If this price is 60% off the original price, what is the original price of the coat? (*Section 2.6, Objectives 1–5*)

9. Determine if the lines are parallel, perpendicular, or neither. (*Section 3.4, Objective 3*)

   a. $y = \dfrac{3}{2}x + 5$ and $y = \dfrac{3}{2}x - 1$

   b. $y = 2$ and $x = 5$ **c.** $y = 0.25x$ and $y = -4x$

10. Write the equation of the line described. Write each answer in slope-intercept form and in standard form. (*Section 3.5, Objectives 2 and 4*)

   a. through $(-8, 5)$ perpendicular to $4x - 3y = 1$

   b. through $(-1, 4)$ parallel to $6x + y = 5$

   c. $m =$ undefined, through $(3, -4)$

11. Amanda has a new job as a chemistry instructor at a local community college. Her starting salary is $43,500 with a regular teaching load of 30 credit-hours. If she teaches beyond the regular load in an academic year, she earns $1050 per credit hour. (*Section 3.6, Objective 5*)

   a. Write a linear function $f(x)$ that represents Amanda's yearly income, where x is the number of extra credit-hours she teaches in an academic year.

   b. If Amanda teaches 39 credit-hours in an academic year, what is her income?

   c. Find $f(8)$ and interpret the answer.

12. Solve each system using substitution. (*Section 4.2, Objectives 1 and 2*)

   a. $\begin{cases} 2x - 3y = 16 \\ x + 7y = -9 \end{cases}$ **b.** $\begin{cases} \dfrac{1}{2}x - y = -4 \\ \dfrac{1}{4}x + 3y = 5 \end{cases}$

13. Solve each system using elimination. (*Section 4.3, Objectives 1 and 2*)

   a. $\begin{cases} 4x - 3y = 78 \\ -5x + 2y = 40 \end{cases}$

   b. $\begin{cases} \dfrac{2}{3}x - \dfrac{1}{4}y = -6 \\ \dfrac{5}{6}x + \dfrac{1}{2}y = -1 \end{cases}$

14. Solve each system of linear equations in three variables using elimination. (*Section 4.5, Objective 1*)

   a. $\begin{cases} -4x + 2y + z = 14 \\ -5x + 3z = 3 \\ 5x - 3y - 8z = 8 \end{cases}$ **b.** $\begin{cases} 3x - 7y - 6z = -25 \\ x - z = 1 \\ 2x - y - 7z = -2 \end{cases}$

15. Evaluate and simplify each expression. Write each answer with positive exponents. (*Section 5.1, Objectives 1–4*)

   a. $(2x^5y^4)(-4x^{-5}y^{-8})$ **b.** $(5x^3y)(-3x^{-3}y^{-4})$

   c. $\dfrac{-16x^3y^7}{2xy^{10}}$ **d.** $\dfrac{-6r^6t^{-10}}{20r^{-11}t^{-5}}$

16. Simplify each expression using the rules of exponents. Write each answer with positive exponents. (*Section 5.2, Objectives 1 and 2*)

   a. $\dfrac{(4y^{10}z^{-9})^{-3}}{y^{-6}z^{-17}}$ **b.** $\dfrac{(x^{-5}y^{-1})^{-1}}{6^2xy^{-4}}$

17. Use scientific notation to perform the indicated operation. Express each answer in scientific notation. (*Section 5.2, Objectives 3 and 4*)

   a. $(1.2 \times 10^4)^3$ **b.** $\dfrac{1.476 \times 10^{29}}{4.1 \times 10^{18}}$

   c. $\dfrac{0.297}{5,500,000}$

   d. $\dfrac{(9.36 \times 10^{-3})(7.5 \times 10^{-1})}{4.5 \times 10^7}$

18. Simplify each expression. (*Section 5.3, Objective 4; Section 5.4, Objectives 1 and 2*)

   a. $(-9a + 6) + (-12a - 13)$

   b. $(5p + 10) - (-17p + 9)$

   c. $(-19q^2 - 15q - 4) + (23q^2 - 24q - 15)$

   d. $(-7b^2 + 9b - 17) - (-22b^2 - 8b + 15)$

   e. $3ab(a^2 - 6ab + 5b^2)$

   f. $(3r - 2s)(9r^2 + 6rs + 4s^2)$

   g. $(5r - 3)(4r - 1)$

19. Perform the indicated operation. Write each answer in standard form. (*Section 5.5, Objectives 1–4*)

   a. $(7x - 3)(4x + 5)$ **b.** $(6y - 5)(y + 2)$

   c. $(2a - b)^2$ **d.** $(12x - 7y)(12x + 7y)$

   e. $(a + 2)^4$ **f.** $(2b - 1)^3$

20. Perform the indicated operation. (*Section 5.6, Objectives 1 and 2*)

   a. Divide $27x^3 - 12x^2 + 24x + 6$ by $6x$.

   b. $\dfrac{6x^3 - 18x^2 + 21x - 10}{x + 3}$

21. Factor each trinomial completely. (*Section 6.2, Objectives 1 and 2*)

  a. $3y^3 - 24y^2 + 48y$ **b.** $-12x^3 + 54x^2 + 30x$

  c. $a^2 - 15ab + 36b^2$ **d.** $2x^2y + 34xy + 144y$

  e. $x^2 + 7xy + 8y^2$

22. Use either trial and error or grouping to factor each trinomial completely. (*Section 6.3, Objectives 1–3*)

  a. $35a^2 + 31a + 6$ **b.** $40ab^2 - 35ab - 90a$

  c. $12x^2 - 2xy - 30y^2$ **d.** $9x^2 + 12xy + 4y^2$

  e. $90a^2b^2c + 3abc - 3c$ **f.** $3p^2 + 4pq - 7q^2$

23. Factor each binomial completely. (*Section 6.4, Objectives 1–3*)

  a. $98a^2 - 18b^2$ **b.** $5x^3 + 40y^3$

  c. $y^3 - 1000z^3$ **d.** $a^2 + 16b^2$

  e. $a^6 - 64$ **f.** $x^4 - 81y^4$

24. Solve each equation. (*Section 6.5, Objectives 1 and 2*)

  a. $(x + 2)(x - 8) = 24$ **b.** $(5 - x)(4 + x) = 0$

  c. $x^4 - 5x^2 + 4 = 0$ **d.** $x^3 - 3x^2 - 10x = 0$

25. Solve each problem. (*Section 6.6, Objectives 1–5*)

  a. The product of two consecutive odd integers is 143. Find the integers.

  b. When Philly Express Bus Company charges $40 per seat from New York City to Philadelphia, they can sell 90 tickets. For every $1 decrease in price, 2 more tickets are sold. If x represents the number of price decreases, then $40 - x$ represents the ticket price and $90 + 2x$ represents the number of passengers. How many tickets must be sold for the Philly Express Bus Company to earn $3300?

  c. The height of a ball thrown upward with an initial velocity of 48 ft/sec from a height of 160 ft above the ground can be represented by $s = -16t^2 + 48t + 160$, where t is the number of seconds after the ball is thrown. How long will it take the ball to reach the ground?

  d. The profit, in millions of dollars, of a toy company is given by $p = -12x^2 + 180x + 3000$, where x is the number of units sold in thousands. How many units need to be sold for the company to break even?

26. Evaluate each rational expression or function for the given values. (*Section 7.1, Objective 1*)

  a. $f(x) = \dfrac{x^2 - 3x}{x + 2}$ for $x = -2, 0, 3$

  b. $\dfrac{4y}{y - 1}$ for $y = -1, 0, 1$

27. Find the domain of each rational function. (*Section 7.1, Objective 2*)

  a. $f(x) = \dfrac{x^2 - 1}{x - 6}$ **b.** $f(x) = \dfrac{2x}{x^2 + 5x}$

28. Simplify each rational expression. (*Section 7.1, Objective 3*)

  a. $\dfrac{x^3 - 8}{x^2 + 3x - 10}$ **b.** $\dfrac{3x^2 - 6x}{x^3 - 4x}$

29. Write two equivalent forms of each rational expression. (*Section 7.1, Objective 4*)

  a. $\dfrac{-3x + 6}{x + 4}$ **b.** $-\dfrac{1 - x}{2x - 3}$

30. Solve each problem. (*Section 7.1, Objective 5*)

  a. A company's revenue can be represented by $40q^2 + 600q$, where q is the number of items sold. If the unit price of each item is given by $\dfrac{40q^2 + 600q}{q}$, what is the unit price of each item when 10 items are sold? 20 items are sold?

  b. The volume of an open box with height x in. is given by the expression $x^3 + 5x^2 - 24x$ in.^{3}. If the area of the base of the box is given by $\dfrac{x^3 + 5x^2 - 24x}{x}$, what is the area of the base when the height is 5 in.? 6 in.? (*Section 7.1, Objective 1*)

31. Perform each operation and simplify each answer. (*Section 7.2, Objectives 1 and 2*)

  a. $\dfrac{x^2 - 3x}{4} \cdot \dfrac{2x - 10}{x^3 - 9x}$

  b. $\dfrac{x^2 + 4x - 12}{6x^2} \div \dfrac{x^2 - 4x + 4}{3x}$

  c. $\dfrac{x^3 - 1}{x + 1} \div (x - 1)$

  d. $\dfrac{x^3 + 4x - 7x^2 - 28}{x^2 - 49} \cdot \dfrac{x^2 + 10x + 21}{x^3 + 4x}$

  e. $\dfrac{8x^3 + 4x^2}{2x^2 + 7x + 3} \div \dfrac{2x^2 + 10x}{x^2 + 8x + 15}$

  f. $\dfrac{x^4 - 1}{x^2 + 2x + 1} \cdot \dfrac{x^2 - 4x - 5}{x^3 + x - 5x^2 - 5}$

32. Use the given conversion equivalents to solve each problem. (*Section 7.2, Objective 3*)

  a. The distance between London and Rylstone in the United Kingdom is 224 mi. What is this distance in inches on the map? (1 in. = 18 mi)

  b. How many minutes are in 0.8 hr? (1 hr = 60 min)

  c. Convert 2.25 kg to pounds. (1 lb = 0.45 kg)

33. Find the least common denominator of each rational expression. (*Section 7.3, Objective 3*)

  a. $\dfrac{x}{6x + 15}, \dfrac{1}{5x + 2x^2}$ **b.** $\dfrac{5x}{2x + 3}, \dfrac{3}{x - 4}$

  c. $\dfrac{5}{x - 3}, \dfrac{x}{x^2 + 4x - 21}$ **d.** $\dfrac{5x}{x^2 - 16}, \dfrac{7x}{5x^2 + 19x - 4}$

  e. $\dfrac{7}{x^3 - 8}, \dfrac{-5}{3x^2 + 6x + 12}$

34. Find the numerator that makes each rational expression equivalent. (*Section 7.3, Objective 4*)

  a. $\dfrac{5}{3a} = \dfrac{?}{18a^2b}$ **b.** $\dfrac{7}{xy} = \dfrac{?}{6x^2y}$

  c. $\dfrac{x + 2}{x - 3} = \dfrac{?}{2x^2 - 6x}$ **d.** $\dfrac{x}{1 - 2x} = \dfrac{?}{6x^2 - 3x}$

35. Add or subtract the rational expressions. Write each answer in lowest terms. (*Section 7.3, Objectives 1 and 2; Section 7.4, Objectives 1 and 2*)

a. $\dfrac{x}{x-4} - \dfrac{2x}{x-4}$ **b.** $\dfrac{4}{x-6} + \dfrac{5}{x+3}$

c. $\dfrac{4x}{x-2} + \dfrac{3x-6}{2-x}$ **d.** $\dfrac{2x^2}{x^2-25} - \dfrac{2x}{x-5}$

e. $4 + \dfrac{3x}{x-1}$ **f.** $\dfrac{4x^2}{x^2+7x-30} - \dfrac{2x+1}{x+10}$

g. $\dfrac{6}{x^2-9} - \dfrac{7}{x^2-6x+9}$

h. $\dfrac{3}{x^2+3x+2} + \dfrac{4}{x^2+7x+6}$

i. $\dfrac{250}{x-8} + \dfrac{100}{x+3}$

36. Simplify each complex fraction using method 1, and then method 2. (*Section 7.5, Objectives 1 and 2*)

a. $\dfrac{\frac{18}{3x^6}}{\frac{24}{5x^4}}$ **b.** $\dfrac{4-\frac{a}{b}}{2+\frac{a}{b}}$ **c.** $\dfrac{\frac{4y^2-9}{20y^2}}{\frac{2y-3}{5y}}$ **d.** $\dfrac{\frac{1}{a}-\frac{2}{3}}{4a-6}$

e. $\dfrac{\frac{8xy}{2x-y}}{\frac{6xy}{2x^2+5xy-3y^2}}$ **f.** $\dfrac{\frac{4}{x}-\frac{3}{y}}{\frac{16}{x^2}-\frac{9}{y^2}}$

g. $\dfrac{\frac{x^2-16}{x^2+3x+9}}{\frac{x^2+x-12}{x^3-27}}$ **h.** $\dfrac{1-\frac{2x}{5x-2}}{2+\frac{1}{5x-2}}$

37. Write a complex fraction that represents each situation, and then simplify the complex fraction. (*Section 7.5, Objective 3*)

a. Find the slope of the line passing through the points $\left(\dfrac{1}{2},\dfrac{2}{3}\right)$ and $\left(-\dfrac{1}{4},6\right)$

b. Rewrite with positive exponents and then simplify $\dfrac{5^{-1}+2^{-3}}{4^{-2}}$.

38. Solve each rational equation. (*Section 7.6, Objective 1*)

a. $\dfrac{a}{3} = \dfrac{a-5}{4}$ **b.** $\dfrac{3y-2}{4} + \dfrac{5}{6} = \dfrac{2y}{3}$

c. $\dfrac{2a}{8a-5} = \dfrac{1}{4}$ **d.** $1 = \dfrac{2}{x} + \dfrac{35}{x^2}$

e. $6 - \dfrac{1}{t-2} = \dfrac{5}{t-2}$ **f.** $\dfrac{3}{7x-5} = \dfrac{2}{4x-3}$

g. $\dfrac{3}{2x-3} - \dfrac{2x}{4x^2-9} = \dfrac{2}{2x+3}$

39. Solve each rational equation for the specified variable. (*Section 7.6, Objective 2*)

a. $\dfrac{1}{R} = \dfrac{1}{R_1} + \dfrac{1}{R_2} + \dfrac{1}{R_3}$ for $R1$ (resistors in parallel)

b. $m = \dfrac{y_2-y_1}{x_2-x_1}$ for x_1 (slope formula)

40. Solve each problem. (*Section 7.7, Objectives 1–5*)

a. A supermarket has a special on tuna—3 cans for $3.87. How much will it cost to purchase 7 cans of tuna?

b. The total calories for six chicken nuggets from a popular fast food restaurant is 250 Cal. Jill is trying to cut back on her calories and eats only three of the nuggets. How many calories does she consume?

c. If 1 in. represents 120 mi on a map, how many inches represent the distance between Las Vegas, Nevada, and Denver, Colorado, which is 750 mi.

d. An airplane flies 1500 mi in the same amount of time it takes a second plane to fly 1350 mi. If the speed of the first airplane is 50 mph faster than the speed of the second airplane, find the speed of each airplane.

e. If 1 lb on Earth weighs the same as 1.125 lb on Neptune, find Lucia's weight on Earth if she weighs 144 lb on Neptune.

41. Find the domain of each function. (*Section 8.1, Objectives 2 and 3*)

a.

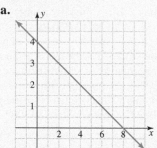

b. $f(x) = \sqrt{x+3}$

42. Graph each linear function and state its domain and range. (*Section 8.2, Objective 1*)

a. $f(x) = -\dfrac{1}{4}x + 8$ **b.** $f(x) = -4$

c. $f(x) = -x$ **d.** $f(x) = 2x - 18$

43. Write the linear function that satisfies the given information. (*Section 8.2, Objective 3*)

a. $m = 0; f(-1) = 7$ **b.** $m = \dfrac{1}{9}; f(-9) = 2$

c. $f(2) = -4$, parallel to $g(x) = \dfrac{4}{5}x + 6$

d. $f(7) = -9$, perpendicular to $g(x) = -\dfrac{1}{6}x + 5$

44. When a 160-lb person swims laps for 1 hr, he burns 511 Cal. When a 240-lb person swims laps for 1 hr, he burns 637 Cal. (*Section 8.2, Objective 4*)

a. Write a linear function $f(x)$ that models the calories burned by swimming laps for 1 hr, where x is the person's weight in pounds.

b. Find $f(130)$ and interpret its meaning in the context of the problem.

c. Solve $f(x) = 700$ and interpret its meaning in the context of the problem.

45. Graph each function and state its domain and range. (*Section 8.3, Objectives 1 and 2*)

a. $f(x) = x^2 - 9$ **b.** $f(x) = -|x + 2|$

c. $f(x) = \begin{cases} -2x + 1, & x < -1 \\ x, & x \geq -1 \end{cases}$ **d.** $f(x) = \begin{cases} -2, & x \leq 2 \\ \dfrac{1}{2}x, & x > 2 \end{cases}$

46. Find the constant of variation and the equation of variation for each case. (*Section 8.4, Objectives 1–3*)

a. F varies jointly as the cube of G and square root of H; $F = 12$ when $G = 2$ and $H = 0.25$

b. w varies directly as the square of u and inversely as the cube root of v; $w = 7.5$ when $u = 5$ and $v = 64$

47. Find the intersection and union of the sets. Write each solution set in interval notation. (*Section 9.1, Objectives 1 and 2*)

a. $A = [-13, 1]$, $B = [-12, 4]$

b. $A = (4, 20]$, $B = [3, 18)$

48. Solve each compound inequality. Write each solution set in interval notation. (*Section 9.1, Objective 3*)

a. $x > -7$ and $x > -5$

b. $x \leq 3$ and $x \leq 7$

49. Solve each compound inequality. Write each solution set in interval notation. (*Section 9.1, Objective 4*)

a. $x \geq 8$ or $x \geq 15$

b. $5x + 13 > 18$ or $7x - 14 \leq -13$

50. Solve each equation. (*Section 9.2, Objectives 1 and 2*)

a. $|11 - x| = 23$ **b.** $|7 - 2x| = |3x + 16|$

c. $|2x + 10| = 0$ **d.** $|26 - 2x| = 0$

51. Solve each problem. (*Section 9.2, Objective 3*)

a. Suppose a digital scale reflects a weight (in pounds) with an absolute error of 0.2 lb. If someone weighs 110.1 lb according to the scale, find the exact weight of the person.

b. Suppose a countertop food scale reflects a weight (in ounces) with an absolute error of 0.1 oz. If a piece of chicken weighs 3 oz according to the scale, find the exact weight of the piece of chicken.

52. Solve each inequality. Write each solution set in interval notation. (*Section 9.3, Objectives 1–3 and 5*)

a. $|2x + 3| \geq 7$ **b.** $|3x - 1| - 6 < -2$

c. $|5x + 6| + 1 \geq -7$

53. Determine if the ordered pair is a solution of the given inequality. (*Section 9.4, Objective 1*)

a. $3x - 9y < 5$; $(0, 2)$ **b.** $y \leq -2x + 8$; $(-2, 1)$

54. Graph each linear inequality in two variables. (*Section 9.4, Objective 2*)

a. $x + 5y \leq 10$ **b.** $4x - y \leq 16$

c. $x + y < 0$ **d.** $x - 4 \leq 0$

55. Brenda works two jobs to finance her college education. She works as an office assistant earning \$8.75 an hour and as a tutor earning \$11.50 per hour. She wants to make at least \$280 per week.

a. Write a linear inequality in two variables that represents this situation.

b. Graph the linear inequality.

c. Give three possible combinations of hours that Brenda can work at each job to make at least \$280 per week. (*Section 9.4, Objective 3*)

56. Solve each system of linear inequalities. (*Section 9.5, Objective 1*)

a. $\begin{cases} y \geq 5x - 5 \\ y \leq -4x + 8 \end{cases}$ **b.** $\begin{cases} x \geq 2 \\ x - 3y \leq 0 \end{cases}$

c. $\begin{cases} y \geq -3x - 4 \\ y \leq -3x + 1 \end{cases}$ **d.** $\begin{cases} x \geq 1 \\ y \leq 4 \end{cases}$

57. The price per item is determined by the supply equation $y = S(x)$, where x is the number of items produced. The producer's surplus is the area of the solution to the following system of linear inequalities, where k is a fixed price. (*Section 9.5, Objective 2*)

$$\begin{cases} y \geq S(x) \\ y \leq k \\ x \geq 0 \end{cases}$$

a. If the supply equation for an item is $y = S(x) = 21 + 0.40x$ and $k = 53$, set up a system of linear inequalities and solve it.

b. Find the producer's surplus.

Rational Exponents, Radicals, and Complex Numbers

Refocus

Do you need to refocus your efforts to be successful? Maybe it has been a long term for you and it is a struggle to stay motivated and attentive to your studies. Don't give up. There is light at the end of the tunnel. Take a moment to remember why you enrolled in college and in this class. Keep sight of your short- and long-term goals. We know it will be worth it!

"What comes first, the compass or the clock? Before one can truly manage time (the clock), it is important to know where you are going, what your priorities and goals are, in which direction you are headed (the compass). Where you are headed is more important than how fast you are going. Rather than always *focusing* on what's urgent, learn to focus on what is really important." (Author Unknown)

Question For Thought: What are the things that distract your from accomplishing tasks or drain you of your time and energy? (Examples: Facebook, Web surfing, gaming, talking to co-workers, watching TV, etc.) How can you refocus your time and energy toward your studies?

Chapter Outline

Coming Up...

In Section 10.1, we will use the function $f(d) = 3.13\sqrt{d}$, which determines the wave speed, in meters per second, of surface waves in water that is d meters deep to find the wave speed of a tsunami in the Pacific Ocean. (Source: http://oceanservice.noaa.gov/facts/oceandepth.html)

"The best advice I ever came across on the subject of concentration is: Wherever you are, be there. When you work, work. When you play, play. Don't mix the two."

—Jim Rohn (Entrepreneur, Author, Motivational Speaker)

SECTION 10.1 / Radicals and Radical Functions

Chapter 10 deals with the concept of roots and radicals. We will review square roots and introduce cube roots and higher roots. This chapter will also explore operations with radical expressions and solve equations containing them. We will learn how roots are connected to exponents and how to express some roots as complex numbers.

▶ OBJECTIVES

As a result of completing this section, you will be able to

1. Simplify square roots.
2. Simplify cube roots.
3. Simplify nth roots.
4. Approximate roots.
5. Simplify expressions of the forms $(\sqrt[n]{a})^n$ and $\sqrt[n]{a^n}$.
6. Find the domain of radical functions and evaluate radical functions.
7. Graph square root and cube root functions.
8. Troubleshoot common errors.

The function $f(d) = 3.13\sqrt{d}$ determines the wave speed, in meters per second, of surface waves in water that is d meters deep. The fastest surface waves are those caused by tsunamis. The average depth of the Pacific Ocean is approximately 4300 m. Estimate the wave speed of a tsunami in the Pacific Ocean.

To determine the wave speed of a tsunami in the Pacific Ocean, we have to simplify $3.13\sqrt{4300}$. In this section, we will learn how to simplify roots, simplify radical expressions, and graph radical functions.

Square Roots

Recall that the *square root* of a number, b, is the number that must be squared to obtain b. Since $(-3)^2 = 9$ and $3^2 = 9$, we say that the square roots of 9 are -3 and 3. We write this mathematically as

$$\sqrt{9} = 3 \quad \text{and} \quad -\sqrt{9} = -3$$

The notation $\sqrt{9}$ denotes the *principal square root* of 9, which is the nonnegative square root, or 3. The notation $-\sqrt{9}$ denotes the *negative square root* of 9, which is -3.

Objective 1 ▶

Simplify square roots.

> **Definition:** A number a is a **square root** of b if $a^2 = b$, for $b \geq 0$. The notation $\sqrt{b}$ denotes the **principal square root** of b and $-\sqrt{b}$ denotes the **negative square root** of b.

A nonnegative number whose square root is an integer value is a **perfect square**. For example, $\sqrt{9} = 3$, so 9 is a perfect square. It is very helpful to know perfect squares when simplifying square roots. Some of the perfect squares are shown.

$$
\begin{array}{llll}
1^2 = 1 & 6^2 = 36 & 11^2 = 121 & (b)^2 = b^2 \\
2^2 = 4 & 7^2 = 49 & 12^2 = 144 & (b^2)^2 = b^4 \\
3^2 = 9 & 8^2 = 64 & 13^2 = 169 & (b^3)^2 = b^6 \\
4^2 = 16 & 9^2 = 81 & 14^2 = 196 & (b^4)^2 = b^8 \\
5^2 = 25 & 10^2 = 100 & 15^2 = 225 & (b^5)^2 = b^{10}
\end{array}
$$

The square root of a negative number, however, is *not a real number*. For instance, the notation $\sqrt{-9}$ represents the number whose square is -9. Recall that $(3)^2 = 9$ and $(-3)^2 = 9$, so there is not a real number whose square is -9. This concept will be revisited in Section 10.7.

The expression $\sqrt{9}$ is an example of a **radical**, which is the root of a number. The notation used to denote a radical is defined as follows.

> **Definition: Radical**
>
> The expression $\sqrt{b}$ is called a **radical**, or **radical expression**. The symbol $\sqrt{}$ is called a **radical sign** and b is called the **radicand**.
>
> $$\sqrt{b} \quad \longleftarrow \text{ Radical Sign}$$
> $$\uparrow$$
> $$\text{Radicand}$$

> **Procedure: Simplifying Square Roots**
>
> **Step 1:** Find the expression that must be squared to obtain the radicand. If the radicand is negative, the square root is not a real number.
> **Step 2:** Check by squaring the result. The square should yield the radicand.

Objective 1 Examples Simplify each expression. Assume all variables represent positive real numbers.

1a. $\sqrt{49}$ **1b.** $\sqrt{\dfrac{4}{9}}$ **1c.** $-\sqrt{4}$ **1d.** $\sqrt{-4}$ **1e.** $\sqrt{0.36}$ **1f.** $\sqrt{y^8}$ **1g.** $\sqrt{16b^6}$

Solutions **1a.** $\sqrt{49} = 7$ because $7^2 = 49$. Note that $\sqrt{49} = \sqrt{7^2} = 7$.

1b. $\sqrt{\dfrac{4}{9}} = \dfrac{2}{3}$ because $\left(\dfrac{2}{3}\right)^2 = \dfrac{4}{9}$. Note that $\sqrt{\dfrac{4}{9}} = \sqrt{\left(\dfrac{2}{3}\right)^2} = \dfrac{2}{3}$.

1c. $-\sqrt{4} = -2$

1d. $\sqrt{-4}$ is not a real number.

1e. $\sqrt{0.36} = 0.6$ because $0.6^2 = 0.36$. Note that $\sqrt{0.36} = \sqrt{(0.6)^2} = 0.6$.

1f. We must find the expression that we square to get y^8: $(?)^2 = y^8$. Recall that when we raise a power to an exponent, we multiply exponents. For instance, $(y^4)^2 = y^8$. So,

$$\sqrt{y^8} = y^4$$

Note that $\sqrt{y^8} = \sqrt{(y^4)^2} = y^4$.

1g. We must find the expression that we square to get $16b^6$: $(?)^2 = 16b^6$. The power of a product rule for exponents tells us that $(4b^3)^2 = (4)^2(b^3)^2 = 16b^6$, so

$$\sqrt{16b^6} = 4b^3$$

Note that $\sqrt{16b^6} = \sqrt{(4b^3)^2} = 4b^3$.

✓ Student Check 1 Simplify each expression. Assume all variables represent positive real numbers.

a. $\sqrt{64}$ **b.** $\sqrt{\dfrac{25}{36}}$ **c.** $-\sqrt{49}$ **d.** $\sqrt{-49}$ **e.** $\sqrt{0.04}$ **f.** $\sqrt{h^{10}}$ **g.** $\sqrt{9a^{12}}$

Cube Roots

Objective 2 ▶

Simplify cube roots.

The idea of square roots can be extended to other roots. The *cube root* of a number b, denoted $\sqrt[3]{b}$, is the number that must be cubed to obtain b.

> **Definition:** The **cube root** of a real number b, denoted $\sqrt[3]{b}$, is the number that must be cubed to obtain b. That is,
>
> $$\sqrt[3]{b} = a \text{ if and only if } a^3 = b$$

For example,

$$\sqrt[3]{64} = 4 \text{ because } 4^3 = 64$$

$$\sqrt[3]{-64} = -4 \text{ because } (-4)^3 = -64$$

This example illustrates that the cube root of a negative number is a real number, unlike the square root of a negative number.

A real number whose cube root is an integer is a **perfect cube**. For example, the number 64 is a perfect cube since its cube root is 4. It is very helpful to know perfect cubes when simplifying cube roots. Some of the perfect cubes are shown.

$1^3 = 1$	$6^3 = 216$	$(b)^3 = b^3$
$2^3 = 8$	$7^3 = 343$	$(b^2)^3 = b^6$
$3^3 = 27$	$8^3 = 512$	$(b^3)^3 = b^9$
$4^3 = 64$	$9^3 = 729$	$(b^4)^3 = b^{12}$
$5^3 = 125$	$10^3 = 1000$	$(b^5)^3 = b^{15}$

Procedure: Simplifying Cube Roots

Step 1: Find the expression that must be cubed to obtain the radicand.
Step 2: Check by cubing the result. The cube should yield the radicand.

Objective 2 Examples   Simplify each expression.

2a. $\sqrt[3]{27}$ **2b.** $\sqrt[3]{-125}$ **2c.** $\sqrt[3]{\dfrac{1}{8}}$ **2d.** $\sqrt[3]{x^{12}}$ **2e.** $\sqrt[3]{-64y^6}$

Solutions   **2a.** $\sqrt[3]{27} = 3$ because $3^3 = 27$. Note that $\sqrt[3]{27} = \sqrt[3]{3^3} = 3$.

2b. $\sqrt[3]{-125} = -5$ because $(-5)^3 = -125$. Note that $\sqrt[3]{-125} = \sqrt[3]{(-5)^3} = -5$.

2c. $\sqrt[3]{\dfrac{1}{8}} = \dfrac{1}{2}$ because $\left(\dfrac{1}{2}\right)^3 = \dfrac{1}{8}$. Note that $\sqrt[3]{\dfrac{1}{8}} = \sqrt[3]{\left(\dfrac{1}{2}\right)^3} = \dfrac{1}{2}$.

2d. We must find the expression we cube to get x^{12}: $(?)^3 = x^{12}$. Recall that $(x^4)^3 = x^{12}$. So,

$$\sqrt[3]{x^{12}} = x^4$$

Note that $\sqrt[3]{x^{12}} = \sqrt[3]{(x^4)^3} = x^4$.

2e. We must find the expression we cube to get $-64y^6$: $(?)^3 = -64y^6$. Recall that $(-4y^2)^3 = (-4)^3(y^2)^3 = -64y^6$. So,

$$\sqrt[3]{-64y^6} = -4y^2$$

Note that $\sqrt[3]{-64y^6} = \sqrt[3]{(-4y^2)^3} = -4y^2$.

✔ Student Check 2   Simplify each expression.

a. $\sqrt[3]{343}$ **b.** $\sqrt[3]{-8}$ **c.** $\sqrt[3]{\dfrac{27}{64}}$ **d.** $\sqrt[3]{a^3}$ **e.** $\sqrt[3]{-125y^9}$

Objective 3 ▶

Simplify *n*th roots.

The *n*th Root

We can generalize this concept of roots and define an *nth root*. The square root is the expression that is squared to obtain the radicand. The cube root is the expression that is cubed to obtain the radicand. The fourth root is the expression that is raised to the fourth to obtain the radicand and so on.

> **Definition:** For a natural number $n \geq 2$, the ***n*th root** of a number b, denoted $\sqrt[n]{b}$, is the value that must be raised to the exponent of n to obtain b.
>
> $$\sqrt[n]{b} = a \text{ if and only if } a^n = b$$
>
> The number n is called the **index**. For square roots, the value of $n = 2$ but this is generally omitted.

The *n*th root is not a real number for all indices. It depends on whether n is even or odd.

- When the value of n is *even* $(2, 4, 6, \ldots)$, the radicand must be a *nonnegative number* for the *n*th root to be a real number.

$$\sqrt[4]{16} = 2 \text{ because } 2^4 = 16 \text{ but } \sqrt[4]{-16} \text{ is not a real number}$$

Recall that when we raise a nonzero number to an *even* exponent, the result is positive.

- When the value of n is *odd* $(3, 5, 7, \ldots)$, the radicand can be any real number for the *n*th root to be a real number.

$$\sqrt[5]{32} = 2 \text{ because } 2^5 = 32$$
$$\sqrt[5]{-32} = -2 \text{ because } (-2)^5 = -32$$

It is helpful to know the following powers when simplifying *n*th roots.

$1^4 = 1$	$1^5 = 1$	$(b)^4 = b^4$	$(b)^5 = b^5$
$2^4 = 16$	$2^5 = 32$	$(b^2)^4 = b^8$	$(b^2)^5 = b^{10}$
$3^4 = 81$	$3^5 = 243$	$(b^3)^4 = b^{12}$	$(b^3)^5 = b^{15}$
$4^4 = 256$	$4^5 = 1024$	$(b^4)^4 = b^{16}$	$(b^4)^5 = b^{20}$

> **Procedure: Simplifying *n*th Roots**
> **Step 1:** Determine the expression that must be raised to the *n*th to obtain the radicand.
> **Step 2:** Check by raising the result to the exponent of n, $(\text{result})^n = \text{radicand}$.

Objective 3 Examples   Simplify each expression. Assume all variables represent positive real numbers.

3a. $\sqrt[4]{256}$ **3b.** $\sqrt[4]{-625}$ **3c.** $\sqrt[5]{1024}$ **3d.** $\sqrt[5]{-243x^{10}}$ **3e.** $\sqrt[6]{x^6 y^{12}}$

Solutions   **3a.** $\sqrt[4]{256} = 4$ because $4^4 = 256$. Note that $\sqrt[4]{256} = \sqrt[4]{4^4} = 4$.

3b. $\sqrt[4]{-625}$ is not a real number because there is no real number that we can raise to the fourth to get -625.

3c. $\sqrt[5]{1024} = 4$ because $4^5 = 1024$. Note that $\sqrt[5]{1024} = \sqrt[5]{4^5} = 4$.

3d. We must find the expression that, when raised to the fifth, is $-243x^{10}$. By the properties of exponents, we know that $(-3x^2)^5 = (-3)^5(x^2)^5 = -243x^{10}$. So,
$$\sqrt[5]{-243x^{10}} = -3x^2$$
Note that $\sqrt[5]{-243x^{10}} = \sqrt[5]{(-3x^2)^5} = -3x^2$.

3e. We must find the expression that, when raised to the sixth, is $x^6 y^{12}$. By the properties of exponents, we know that $(xy^2)^6 = (x)^6(y^2)^6 = x^6 y^{12}$. So,
$$\sqrt[6]{x^6 y^{12}} = xy^2$$
Note that $\sqrt[6]{x^6 y^{12}} = \sqrt[6]{(xy^2)^6} = xy^2$.

✔ **Student Check 3**   Simplify each expression. Assume all variables represent positive real numbers.

a. $\sqrt[4]{625}$ **b.** $\sqrt[4]{-81}$ **c.** $\sqrt[5]{1}$ **d.** $\sqrt[5]{-32x^5}$ **e.** $\sqrt[7]{x^{14}y^7}$

Approximating Roots

Objective 4 ▶

Approximate roots.

In Objectives 1–3, we simplified some roots that were rational numbers. For instance,

$$\sqrt{49} = \sqrt{7^2} = 7$$
$$\sqrt[3]{27} = \sqrt[3]{3^3} = 3$$
$$\sqrt[4]{256} = \sqrt[4]{4^4} = 4$$

Each of the radicands can be written as a factor raised to an exponent equal to the index. When the radicand is in this form, the radical simplifies to a *rational number*. If the radicand cannot be written as a factor raised to an exponent equal to the index, the radical is an *irrational number*. For the purpose of this discussion, we will assume that the radicals are real numbers.

Consider $\sqrt{5}$. To simplify this expression, we must find the number whose square is 5. There is no integer whose square is 5, but this does not mean that the square root does not exist. Because $2^2 = 4$ and $3^2 = 9$, $\sqrt{5}$ has a value between 2 and 3.

$$4 < 5 < 9$$
$$\sqrt{4} < \sqrt{5} < \sqrt{9}$$
$$2 < \sqrt{5} < 3$$

Using a calculator, we find that $\sqrt{5} \approx 2.24$. This number is irrational since its decimal form never terminates or repeats. Note that the symbol "$\approx$" means "approximately equal to." Depending on the instructions, we will be asked to provide either the exact value of a radical or an approximation of it.

$\sqrt{5}$ is the *exact value*, but 2.24 is the *approximated value*.

Procedure: Approximating Roots

Step 1: Use a calculator to find the decimal approximation. (Calculator instructions are provided at the end of the section.)

Step 2: Determine if the answer is reasonable. Find the two perfect squares (if the index is 2), perfect cubes (if the index is 3), and so on that the radicand lies between. Take the root of these numbers. The root should lie between these two values.

Step 3: Check the solution by raising the decimal approximation to the index. This number should be very close to the radicand. It will not be exact since the value has been approximated.

Objective 4 Examples   Use a calculator to approximate each root. Round each answer to two decimal places. Verify the reasonableness of the result.

4a. $\sqrt{50}$ **4b.** $\sqrt[3]{15}$

Solutions **4a.** $\sqrt{50} \approx 7.07$

Check: The perfect squares that 50 lies between are 49 and 64.

$$\sqrt{49} < \sqrt{50} < \sqrt{64}$$
$$7 < \sqrt{50} < 8$$
$$7 < 7.07 < 8$$

Also, note that $7.07^2 = 49.9849$. So, the approximation is reasonable.

4b. $\sqrt[3]{15} \approx 2.47$

Check: The perfect cubes that 15 lies between are 8 and 27.

$$\sqrt[3]{8} < \sqrt[3]{15} < \sqrt[3]{27}$$
$$2 < \sqrt[3]{15} < 3$$
$$2 < 2.47 < 3$$

Also, note that $2.47^3 = 15.069223$. So, the approximation is reasonable.

✓ **Student Check 4** Use a calculator to approximate each root. Round each answer to two decimal places. Verify the reasonableness of the result.

a. $\sqrt{72}$ **b.** $\sqrt[3]{60}$

Expressions of the Forms $\left(\sqrt[n]{a}\right)^n$ and $\sqrt[n]{a^n}$

Objective 5 ▶

Simplify expressions of the forms $\left(\sqrt[n]{a}\right)^n$ and $\sqrt[n]{a^n}$.

We defined the nth root of a number to be the number that must be raised to the nth to obtain the radicand. In other words, when we raise the nth root to the exponent of n, we get the radicand. This is true as long as the nth root is a real number. For instance,

$$\left(\sqrt{25}\right)^2 = 5^2 = 25$$
$$\left(\sqrt[3]{-8}\right)^3 = (-2)^3 = -8$$

Note: $\sqrt{-16}$ is not a real number, so $\left(\sqrt{-16}\right)^2$ can't be simplified with what we know now. We will learn how to simplify this expression in Section 10.7.

Property: Simplifying $\left(\sqrt[n]{a}\right)^n$

$\left(\sqrt[n]{a}\right)^n = a$, provided $\sqrt[n]{a}$ is a real number.

The following table illustrates simplified forms of $\sqrt[n]{a^n}$.

n even (2, 4, 6, . . .)	n odd (3, 5, 7, . . .)
$\sqrt[4]{2^4} = \sqrt[4]{16} = 2$	$\sqrt[3]{4^3} = \sqrt[3]{64} = 4$
$\sqrt{3^2} = \sqrt{9} = 3$	$\sqrt[5]{2^5} = \sqrt[5]{32} = 2$
$\sqrt[4]{(-2)^4} = \sqrt[4]{16} = 2$ Note: $2 = -(-2)$	$\sqrt[5]{(-2)^5} = \sqrt[5]{-32} = -2$
$\sqrt{(-3)^2} = \sqrt{9} = 3$ Note: $3 = -(-3)$	$\sqrt[3]{(-4)^3} = \sqrt[3]{-64} = -4$

These examples illustrate that when we simplify $\sqrt[n]{a^n}$, the result is a when the index is odd. But when the index is even, the result is a if $a \geq 0$ or $-a$ if $a < 0$.

Instead of stating the two different cases when the index is even, we can just state that $\sqrt[n]{a^n} = |a|$.

> **Property: Simplifying $\sqrt[n]{a^n}$**
>
> For a real number a and
>
> $$n \text{ an even positive integer, } \sqrt[n]{a^n} = |a|.$$
> $$n \text{ an odd positive integer, } \sqrt[n]{a^n} = a.$$

Objective 5 Examples   Simplify each expression. Assume the radical expressions with variables are real numbers.

5a. $\left(\sqrt{17}\right)^2$ **5b.** $\left(\sqrt{x+1}\right)^2$ **5c.** $\left(\sqrt[3]{-5}\right)^3$ **5d.** $\sqrt{(-6)^2}$

5e. $\sqrt{(3x-1)^2}$ **5f.** $\sqrt[3]{(-7)^3}$ **5g.** $\sqrt[5]{(x+4)^5}$

Solutions   **5a.** Since $\sqrt{17}$ is a real number, we apply the property $\left(\sqrt[n]{a}\right)^n = a$.

$$\left(\sqrt{17}\right)^2 = 17$$

5b. Since $\sqrt{x+1}$ is a real number, we apply the property $\left(\sqrt[n]{a}\right)^n = a$.

$$\left(\sqrt{x+1}\right)^2 = x+1$$

5c. Since $\sqrt[3]{-5}$ is a real number, we apply the property $\left(\sqrt[n]{a}\right)^n = a$.

$$\left(\sqrt[3]{-5}\right)^3 = -5$$

5d. The index is even, so we apply the property $\sqrt[n]{a^n} = |a|$.

$$\sqrt{(-6)^2} = |-6| = 6$$

5e. The index is even, so we apply the property $\sqrt[n]{a^n} = |a|$.

$$\sqrt{(3x-1)^2} = |3x-1|$$

5f. The index is odd, so we apply the property $\sqrt[n]{a^n} = a$.

$$\sqrt[3]{(-7)^3} = -7$$

5g. The index is odd, so we apply the property $\sqrt[n]{a^n} = a$.

$$\sqrt[5]{(x+4)^5} = x+4$$

✓ Student Check 5   Simplify each expression. Assume the radical expressions with variables are real numbers.

a. $\left(\sqrt{40}\right)^2$ **b.** $\left(\sqrt{x-2}\right)^2$ **c.** $\left(\sqrt[3]{-9}\right)^3$ **d.** $\sqrt{(-4)^2}$

e. $\sqrt{(x+5)^2}$ **f.** $\sqrt[3]{(-11)^3}$ **g.** $\sqrt[5]{(3x)^5}$

Radical Functions

Objective 6 ▶

Find the domain of radical functions and evaluate radical functions.

A **radical function** is a function of the form $f(x) = \sqrt[n]{x}$. Some examples of radical functions are

$$f(x) = \sqrt{x-2} \qquad f(x) = \sqrt{x} + 3 \qquad f(x) = \sqrt{2x+1} \qquad f(x) = \sqrt[3]{x-5}$$

We know these are functions because there is only one output for each value of x. Recall the square root function denotes only the nonnegative root, so there is only one output for each square root.

Recall the *domain of a function* is the set of *x*-values for which the function is defined, or is a real number. As shown earlier in this section, we know that there are values for which the square root, or any even root, is not a real number. When we simplify the *even root of a negative number*, the result is not a real number, such as $\sqrt{-16}$. However, when we simplify the *odd root of a negative number*, the result is a real number, such as $\sqrt[3]{-8} = -2$. So, finding the domain of a radical function depends on the index.

Procedure: Finding the Domain of a Radical Function

Step 1: If the index is *odd*, the function $f(x) = \sqrt[n]{x}$ is defined for all real numbers. So, the domain is $(-\infty, \infty)$, or $\mathbb{R}$.

Step 2: If the index is *even*, the function $f(x) = \sqrt[n]{x}$ is defined for all nonnegative real numbers. So, the domain is $[0, \infty)$ or $\{x \mid x \geq 0\}$. The key is that the radicand must be greater than or equal to 0.

We can evaluate radical functions just as we have done throughout the course.

Procedure: Evaluating a Radical Function

Step 1: Replace the variable with the given value.
Step 2: Simplify the resulting expression using the order of operations.

Objective 6 Examples Find the domain of each function and then evaluate the function for the values.

6a. $f(x) = \sqrt{x + 3}$; $x = 0, x = -3, x = -5$

Solution **6a.** The domain of the function $f(x) = \sqrt{x + 3}$ is the set of values for which the radicand is greater than or equal to zero.

$$x + 3 \geq 0$$
$$x + 3 - 3 \geq 0 - 3$$
$$x \geq -3$$

The domain is $[-3, \infty)$ or $\{x \mid x \geq -3\}$.

$f(x) = \sqrt{x + 3}$	$f(x) = \sqrt{x + 3}$	$f(x) = \sqrt{x + 3}$
$f(0) = \sqrt{0 + 3}$	$f(-3) = \sqrt{-3 + 3}$	$f(-5) = \sqrt{-5 + 3}$
$f(0) = \sqrt{3}$	$f(-3) = \sqrt{0}$	$f(-5) = \sqrt{-2}$
	$f(-3) = 0$	not a real number

Note: −5 *is not greater than or equal to* −3, *so it is not in the domain of the function. So,* $f(-5)$ *is undefined. We can also determine this by substituting* −5 *in the function in place of* x *as shown.*

6b. $f(x) = \sqrt{5x - 4}$; $x = \dfrac{4}{5}, x = 4, x = 0$

Solution **6b.** The domain of the function $f(x) = \sqrt{5x - 4}$ is the set of values for which the radicand is greater than or equal to zero.

$$5x - 4 \geq 0$$
$$5x - 4 + 4 \geq 0 + 4$$
$$5x \geq 4$$
$$x \geq \frac{4}{5}$$

The domain is $\left[\dfrac{4}{5}, \infty\right)$ or $\left\{x \mid x \geq \dfrac{4}{5}\right\}$.

$$f(x) = \sqrt{5x - 4} \qquad\qquad f(x) = \sqrt{5x - 4} \qquad\qquad f(x) = \sqrt{5x - 4}$$

$$f\left(\dfrac{4}{5}\right) = \sqrt{5\left(\dfrac{4}{5}\right) - 4} \qquad f(4) = \sqrt{5(4) - 4} \qquad f(0) = \sqrt{5(0) - 4}$$

$$f\left(\dfrac{4}{5}\right) = \sqrt{4 - 4} \qquad\qquad f(4) = \sqrt{20 - 4} \qquad\qquad f(0) = \sqrt{0 - 4}$$

$$f\left(\dfrac{4}{5}\right) = \sqrt{0} \qquad\qquad f(4) = \sqrt{16} \qquad\qquad f(0) = \sqrt{-4}$$

$$f\left(\dfrac{4}{5}\right) = 0 \qquad\qquad f(4) = 4 \qquad\qquad \text{not a real number}$$

> **Note:** 0 is not greater than or equal to $\dfrac{4}{5}$, so it is not in the domain of the function. So, $f(0)$ is undefined. We can also determine this by substituting 0 in the function in place of x as shown.

6c. $f(x) = \sqrt[3]{2x + 5}$; $x = -0, x = 0, x = \dfrac{3}{2}$

Solution   **6c.** Because the index is odd, the domain of the function $f(x) = \sqrt[3]{2x + 5}$ is the set of all real numbers or $(-\infty, \infty)$.

$$f(x) = \sqrt[3]{2x + 5} \qquad\qquad f(x) = \sqrt[3]{2x + 5} \qquad\qquad f(x) = \sqrt[3]{2x + 5}$$

$$f(-2) = \sqrt[3]{2(-2) + 5} \qquad f(0) = \sqrt[3]{2(0) + 5} \qquad f\left(\dfrac{3}{2}\right) = \sqrt[3]{2\left(\dfrac{3}{2}\right) + 5}$$

$$f(-2) = \sqrt[3]{-4 + 5} \qquad\qquad f(0) = \sqrt[3]{0 + 5} \qquad\qquad f\left(\dfrac{3}{2}\right) = \sqrt[3]{3 + 5}$$

$$f(-2) = \sqrt[3]{1} \qquad\qquad f(0) = \sqrt[3]{5} \qquad\qquad f\left(\dfrac{3}{2}\right) = \sqrt[3]{8}$$

$$f(-2) = 1 \qquad\qquad\qquad\qquad\qquad\qquad\qquad\quad f\left(\dfrac{3}{2}\right) = 2$$

6d. The function $f(d) = 3.13\sqrt{d}$ determines the wave speed, in meters per second, of surface waves in water that is d meters deep. The fastest surface waves are those caused by tsunamis. The average depth of the Pacific Ocean is approximately 4300 m (about 2.7 mi). Estimate the wave speed of a tsunami in the Pacific Ocean. (Sources: http://oceanservice.noaa.gov/facts/oceandepth.html and http://www.smccd.net/accounts/bramalln/documents/waterwaves.pdf)

Solution   **6d.** The domain of $f(x) = 3.13\sqrt{d}$ is $[0, \infty)$. To find the wave speed of a tsunami in the Pacific Ocean, we must evaluate the function for $d = 4300$.

$$f(x) = 3.13\sqrt{d}$$

$$f(4300) = 3.13\sqrt{4300}$$

$$f(4300) \approx 205.25$$

So, a tsunami in the Pacific Ocean has a speed of approximately 205.25 m/s, or 459.13 mph.

✓ **Student Check 6**   Find the domain of each function and then evaluate the function for the values.

    a. $f(x) = \sqrt{x + 10}$; $x = -15$, $x = 0$, $x = 6$

    b. $f(x) = \sqrt{7x - 2}$; $x = 0$, $x = 4$, $x = \dfrac{2}{7}$

    c. $f(x) = \sqrt[3]{4x - 3}$; $x = -6$, $x = 1$, $x = \dfrac{3}{4}$

    d. The pirate boat is an amusement park ride that swings back and forth giving its riders a sense of weightlessness. The time, in seconds, that it takes to make a complete swing is given by the function $f(h) = 2\pi\sqrt{\dfrac{h}{32}}$, where h is the height in feet of the swinging bar. Find the time it takes to make a complete swing if the height of the bar is 60 ft.

Objective 7 ▶

Graph square root and cube root functions.

The Graphs of Square Root and Cube Root Functions

Recall that we graph a function by finding points that satisfy the function. We plot these points and then connect the points with a smooth curve to draw the graph.

Objective 7 Examples   **Graph each function.**

    7a. $f(x) = \sqrt{x}$ **7b.** $f(x) = \sqrt[3]{x}$

Solutions   **7a.** Since the domain of the function is $[0, \infty)$, we use only values that are greater than or equal to 0 for x.

x	$y = \sqrt{x}$	(x, y)
0	$\sqrt{0} = 0$	$(0, 0)$
1	$\sqrt{1} = 1$	$(1, 1)$
2	$\sqrt{2} \approx 1.4$	$(2, 1.4)$
4	$\sqrt{4} = 2$	$(4, 2)$
9	$\sqrt{9} = 3$	$(9, 3)$

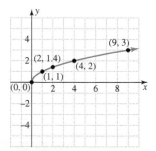

7b. Since the function is defined for all real numbers, we can use any values for x. It is beneficial to use both positive and negative values when selecting the points, so we can see the shape of the entire graph.

x	$y = \sqrt[3]{x}$	(x, y)
-8	$\sqrt[3]{-8} = -2$	$(-8, -2)$
-1	$\sqrt[3]{-1} = -1$	$(-1, -1)$
0	$\sqrt[3]{0} = 0$	$(0, 0)$
1	$\sqrt[3]{1} = 1$	$(1, 1)$
8	$\sqrt[3]{8} = 2$	$(8, 2)$

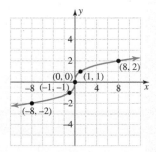

✓ **Student Check 7**   Graph each function.

    a. $f(x) = \sqrt{x - 2}$ **b.** $f(x) = \sqrt[3]{x + 1}$

Objective 8 ▶

Troubleshoot common errors.

Troubleshooting Common Errors

Some of the common errors associated with radicals are shown.

Objective 8 Examples

A problem and an incorrect solution are given. Provide the correct solution and an explanation of the error.

8a. Simplify $\sqrt{-25}$.

Incorrect Solution	Correct Solution and Explanation
$\sqrt{-25} = -5$	Note that $(-5)^2 = 25$ not -25. So, $\sqrt{-25}$ is not a real number.

8b. Simplify $\sqrt[3]{-8}$.

Incorrect Solution	Correct Solution and Explanation
$\sqrt[3]{-8}$ is not a real number.	The cube root of -8 is the number whose cube is -8. Because $(-2)^3 = -8$, $\sqrt[3]{-8} = -2$.

8c. Simplify $\sqrt{(-4)^2}$.

Incorrect Solution	Correct Solution and Explanation
$\sqrt{(-4)^2} = -4$	Because the index is even, we apply the property $\sqrt[n]{a^n} = |a|$. $\sqrt{(-4)^2} = |-4| = 4$ or $\sqrt{(-4)^2} = \sqrt{16} = 4$.

8d. Simplify $\sqrt{(2x - 5)^2}$.

Incorrect Solution	Correct Solution and Explanation
$\sqrt{(2x - 5)^2} = 2x - 5$	The square root symbol always denotes the positive root. Since the expression $2x - 5$ is negative for some values, the answer is the absolute value of $2x - 5$. Because the index is even, we apply the property $\sqrt[n]{a^n} = |a|$. $\sqrt{(2x - 5)^2} = |2x - 5|$

ANSWERS TO STUDENT CHECKS

Student Check 1   **a.** 8   **b.** $\dfrac{5}{6}$ **c.** -7
   d. not a real number   **e.** 0.2   **f.** h^5 **g.** $3a^6$

Student Check 2   **a.** 7   **b.** -2 **c.** $\dfrac{3}{4}$ **d.** a **e.** $-5y^3$

Student Check 3   **a.** 5   **b.** not a real number   **c.** 1
   d. $-2x$ **e.** x^2y

Student Check 4   **a.** 8.49   **b.** 3.91

Student Check 5   **a.** 40   **b.** $x - 2$ **c.** -9 **d.** 4
   e. $|x + 5|$ **f.** -11 **g.** $3x$

Student Check 6   a. Domain $= [-10, \infty)$; $f(-15)$ is undefined, $f(0) = \sqrt{10}$, $f(6) = 4$

b. Domain $= \left[\dfrac{2}{7}, \infty\right)$, $f(0)$ is undefined, $f(4) = \sqrt{26}$, $f\left(\dfrac{2}{7}\right) = 0$

c. Domain $= (-\infty, \infty)$; $f(-6) = -3$, $f(1) = 1$, $f\left(\dfrac{3}{4}\right) = 0$

d. Domain $= [0, \infty)$; $f(60) \approx 8.60$ sec

Student Check 7

a.

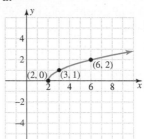

b.

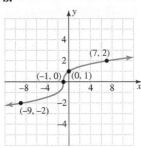

SUMMARY OF KEY CONCEPTS

1. The square root of a real number $a \geq 0$ is the number b such that $b^2 = a$. In symbols, $\sqrt{a} = b$, where $a, b \geq 0$.
2. The cube root of a real number a is the number b such that $b^3 = a$. In symbols, $\sqrt[3]{a} = b$.
3. The nth root of a real number a is the number b such that $b^n = a$. In symbols, $\sqrt[n]{a} = b$. If n is even, $a \geq 0$

for $\sqrt[n]{a}$ to be a real number. The number a is called the radicand and n is called the index. Because of the relationship between exponents and roots, it is important to know the following powers. Each column continues indefinitely.

x	Perfect Squares x^2	Perfect Cubes x^3	Perfect Fourths x^4	Perfect Fifths x^5
1	1	1	1	1
2	4	8	16	32
3	9	27	81	243
4	16	64	256	1024
5	25	125	625	
6	36	216		
7	49	343		
8	64			
9	81			
10	100			
11	121			
12	144			
13	169			
14	196			
15	225			

	Perfect Squares	Perfect Cubes	Perfect Fourths	Perfect Fifths
b	b^2	b^3	b^4	b^5
b^2	b^4	b^6	b^8	b^{10}
b^3	b^6	b^9	b^{12}	b^{15}
b^4	b^8	b^{12}	b^{16}	b^{20}
b^5	b^{10}	b^{15}	b^{20}	b^{25}

4. The roots of numbers that cannot be written as a factor raised to the index are irrational numbers. The calculator must be used to approximate these roots.
5. When the exponent and index are both n, we have $\left(\sqrt[n]{a}\right)^n = a$, provided $\sqrt[n]{a}$ is a real number. Also, $\sqrt[n]{a^n} = |a|$ for n even and $\sqrt[n]{a^n} = a$ for n odd.
6. A radical function is a function that involves a radical expression. The domain of a radical expression with an

even index is found by setting the radicand ≥ 0. The domain of a radical expression with an odd index is all real numbers as long as the radicand is defined for all real numbers.

7. Graph a radical function by plotting a few points whose x-values are in the domain of the function. Then connect the points with a smooth curve.

GRAPHING CALCULATOR SKILLS

We can use a graphing calculator to simplify square roots, cube roots, and radicals with index greater than 3. We can also graph radical functions on the calculator.

Example 1: Simplify $\sqrt{2704}$.

Solution: The calculator gives us the first parenthesis after the radical sign. So, we need to close the parenthesis after the radicand is entered.

Example 2: Simplify $\sqrt[3]{1728}$.

Solution: To access the cube root command, we use the MATH menu. The calculator provides the first parenthesis after the radical sign. We need to close the parenthesis after the radicand is entered.

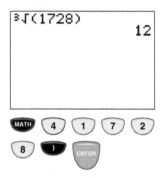

Example 3: Simplify $\sqrt[4]{1296}$.

Solution: To simplify radicals with indices greater than 3, we must use the xth-root function. We first enter the index and

then go to the MATH menu and choose $\sqrt[x]{}$. Last, we enter the radicand in parentheses.

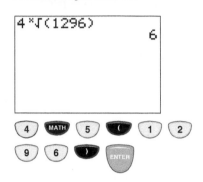

Example 4: Graph $f(x) = \sqrt[3]{x+1}$ and $f(x) = \sqrt{x-5}$.

Solution:

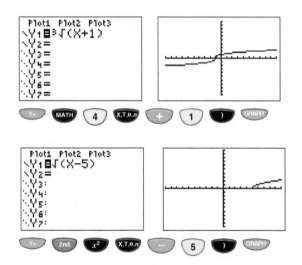

SECTION 10.1 / EXERCISE SET

 Write About It!

Use complete sentences in your answer to each exercise.

1. Explain the difference between a radical expression and a radicand.

2. Explain the difference between $\sqrt{64}$ and $\sqrt[3]{64}$.

3. Explain the difference between $\sqrt{(x-3)^2}$ and $\left(\sqrt{x-3}\right)^2$. Assume the radicals are real numbers.

4. Explain why $\sqrt[4]{-256}$ is not a real number.

5. Use an example to explain how to find the cube root of a number.

6. Explain how to find the domain of the function $f(x) = \sqrt{2x+9}$.

7. Explain how to find the domain of the function $f(x) = \sqrt[3]{2x - 1}$.

8. Explain how to simplify $\sqrt[5]{x^{10}}$.

 Practice Makes Perfect!

Simplify each expression. Assume all variables represent positive real numbers. (*See Objective 1.*)

9. $\sqrt{81}$

10. $\sqrt{100}$

11. $-\sqrt{25}$

12. $-\sqrt{49}$

13. $\sqrt{-64}$

14. $\sqrt{-9}$

15. $\sqrt{\dfrac{16}{49}}$

16. $\sqrt{\dfrac{25}{64}}$

17. $-\sqrt{\dfrac{9}{100}}$

18. $-\sqrt{\dfrac{25}{4}}$

19. $-\sqrt{-16}$

20. $-\sqrt{-81}$

21. $\sqrt{1.44}$

22. $\sqrt{0.81}$

23. $\sqrt{s^4}$

24. $\sqrt{t^8}$

25. $\sqrt{81a^{10}}$

26. $\sqrt{121b^{12}}$

27. $-\sqrt{16h^6}$

28. $-\sqrt{36k^{10}}$

29. $\sqrt{\dfrac{4x^{16}}{9}}$

30. $\sqrt{\dfrac{64y^{14}}{25}}$

Find each cube root. (*See Objective 2.*)

31. $\sqrt[3]{64}$

32. $\sqrt[3]{125}$

33. $\sqrt[3]{-1000}$

34. $\sqrt[3]{-512}$

35. $\sqrt[3]{\dfrac{64}{125}}$

36. $\sqrt[3]{\dfrac{27}{8}}$

37. $\sqrt[3]{a^6}$

38. $\sqrt[3]{b^9}$

39. $\sqrt[3]{64x^6}$

40. $\sqrt[3]{27y^{12}}$

41. $\sqrt[3]{-1000c^3}$

42. $\sqrt[3]{-125d^{15}}$

43. $\sqrt[3]{0.064a^6}$

44. $\sqrt[3]{0.216b^{12}}$

45. $\sqrt[3]{-\dfrac{729x^3}{125}}$

46. $\sqrt[3]{-\dfrac{27y^6}{64}}$

47. $-\sqrt[3]{-\dfrac{343a^{12}}{125}}$

48. $-\sqrt[3]{-\dfrac{27y^9}{1000}}$

Simplify each expression. Assume all variables represent positive real numbers. (*See Objective 3.*)

49. $\sqrt[4]{16}$

50. $\sqrt[5]{243}$

51. $\sqrt[5]{-32}$

52. $\sqrt[4]{-16}$

53. $\sqrt[4]{256x^4}$

54. $\sqrt[4]{81y^8}$

55. $\sqrt[5]{-32x^{15}y^{10}}$

56. $\sqrt[5]{-243a^5b^{30}}$

57. $\sqrt[7]{128a^{14}b^7}$

58. $\sqrt[7]{c^{28}d^{21}}$

59. $\sqrt[6]{-x^6y^{12}}$

60. $\sqrt[6]{-x^{18}y^{30}}$

61. $\sqrt[6]{64a^{12}b^{24}}$

62. $\sqrt[6]{x^{18}y^6}$

Use a calculator to approximate each root. Round the answer to two decimal places. Verify that the answer is reasonable. (*See Objective 4.*)

63. $\sqrt{42}$

64. $\sqrt{95}$

65. $\sqrt[3]{145}$

66. $\sqrt[3]{387}$

Simplify each expression. Assume all radical expressions with variables are real numbers. (*See Objective 5.*)

67. $(\sqrt{65})^2$

68. $(\sqrt{48})^2$

69. $(\sqrt{(-10)})^2$

70. $(\sqrt{(-24)})^2$

71. $\sqrt[3]{(-7)^3}$

72. $\sqrt[3]{(-12)^3}$

73. $(\sqrt{x - 5})^2$

74. $(\sqrt{x + 3})^2$

75. $(\sqrt{2x - 3})^2$

76. $(\sqrt{x + 6})^2$

77. $\sqrt{(2x - 5)^2}$

78. $\sqrt{(7x + 10)^2}$

79. $\sqrt[5]{(6x)^5}$

80. $\sqrt[5]{(15y)^5}$

81. $\sqrt[7]{(13a)^7}$

82. $\sqrt[7]{(3b)^7}$

Find the domain of each radical function and write it in interval notation. Evaluate each function for the indicated values. (*See Objective 6.*)

83. $f(x) = \sqrt{x + 5}$; $x = -6$, $x = 0$, $x = 4$

84. $f(x) = \sqrt{x - 2}$; $x = 0$, $x = 3$, $x = 7$

85. $f(x) = \sqrt{6x + 1}$; $x = -1$, $x = -\dfrac{1}{6}$, $x = 4$

86. $f(x) = \sqrt{5x - 3}$; $x = 0$, $x = \dfrac{3}{5}$, $x = 2$

87. $f(x) = \sqrt[3]{x - 3}$; $x = -5$, $x = 0$, $x = 3$

88. $f(x) = \sqrt[3]{x + 7}$; $x = -7$, $x = 0$, $x = 1$

89. The function $f(S) = \sqrt{\dfrac{S}{4\pi}}$ gives the radius of a sphere with surface area S. Find the radius of a sphere whose surface area is 400 ft². Round the answer to two decimal places.

90. Use the function given in Exercise 89 to find the radius of a sphere whose surface area is 650 m². Round the answer to two decimal places.

91. The function $f(L) = 2\pi\sqrt{\dfrac{L}{32}}$ gives the period in seconds of a simple pendulum L feet long. Find the exact value of the period of a simple pendulum if its length is 64 ft.

92. Use the function in Exercise 91 to find the exact value of the period of a simple pendulum if its length is 128 ft.

Graph each radical function using the accompanying tables. (*See Objective 7.*)

93. $f(x) = \sqrt{x - 3}$

x	3	4	7	8
$f(x)$				

94. $f(x) = \sqrt{x + 2}$

x	-2	-1	2	6
$f(x)$				

95. $f(x) = \sqrt{2x + 5}$

x	–2.5	–1	0	2
$f(x)$				

96. $f(x) = \sqrt{2x - 3}$

x	1.5	2	4	6
$f(x)$				

97. $f(x) = \sqrt[3]{x} - 2$

x	–6	0	1	2	3	10
$f(x)$						

98. $f(x) = \sqrt[3]{x} + 2$

x	–10	–2	–1	0	6
$f(x)$					

99. $f(x) = \sqrt[3]{x - 2}$

x	–8	–1	0	1	8	12
$f(x)$						

100. $f(x) = \sqrt[3]{x + 2}$

x	–8	–1	0	1	8
$f(x)$					

 Mix 'Em Up!

Simplify each radical. Assume all radical expressions with variables are real numbers.

101. $-\sqrt{400x^6}$ **102.** $\sqrt{196y^2}$

103. $\sqrt{225h^2k^8}$ **104.** $\sqrt{81m^6n^{10}}$

105. $\sqrt{(5x - 9)^2}$ **106.** $\sqrt{(8x + 3)^2}$

107. $\sqrt{0.49a^4}$ **108.** $\sqrt{1.21b^6}$

109. $\left(\sqrt{5x - 6}\right)^2$ for $x \geq \dfrac{6}{5}$

110. $\left(\sqrt{7x + 2}\right)^2$ for $x \geq -\dfrac{2}{7}$

111. $\sqrt[3]{64a^3}$ **112.** $\sqrt[3]{27b^6}$

113. $\sqrt[5]{-32r^{10}t^5}$ **114.** $\sqrt[4]{81x^4y^{12}}$

115. $\sqrt{\dfrac{25r^{12}}{49}}$ **116.** $\sqrt{\dfrac{9t^{10}}{121}}$

117. $-\sqrt[4]{625r^8s^4}$ **118.** $\sqrt[5]{-243a^{15}b^{20}}$

119. $\sqrt[6]{a^6b^{18}}$ **120.** $\sqrt[6]{c^{12}d^6}$

121. $\sqrt[7]{x^{21}y^{14}}$ **122.** $\sqrt[7]{x^{14}y^7}$

Find the domain of each radical function and write it in interval notation. Evaluate each function for the indicated values.

123. $f(x) = \sqrt{3x + 4}$; $x = -\dfrac{4}{3}$, $x = 0$, $x = 2$

124. $f(x) = \sqrt{2x - 7}$; $x = 0$, $x = \dfrac{7}{2}$, $x = 5$

125. $f(x) = \sqrt{3 - x}$; $x = -6$, $x = 0$, $x = 4$

126. $f(x) = \sqrt{5 - x}$; $x = -4$, $x = 0$, $x = 5$

127. $f(x) = \sqrt[3]{x + 4}$; $x = -4$, $x = 0$, $x = 4$

128. $f(x) = \sqrt[3]{x - 5}$; $x = -3$, $x = 0$, $x = 6$

For Exercises 129 and 130, the initial speed s, in miles per hour, of a car that skids to a stop can be calculated by measuring the distance d, in feet, of the skid marks. The formula is $s = \sqrt{30kd}$, where k is the coefficient of friction, which depends on the road surface and condition.

Road Surface and Condition	k
Dry tar	1.0
Wet tar	0.5
Dry concrete	0.8
Wet concrete	0.4

129. Find the initial speed of a car if the distance of the skid marks is 145 ft on a wet tar road.

130. Find the initial speed of a car if the distance of the skid marks is 280 ft on a dry concrete road.

Graph each radical function using the accompanying tables.

131. $f(x) = \sqrt{3 - 2x}$

x	–3	0	1	1.5
$f(x)$				

132. $f(x) = \sqrt{1 - 2x}$

x	–3	–1	0	0.5
$f(x)$				

133. $f(x) = 1 - \sqrt[3]{x}$

x	–8	–1	0	1	8
$f(x)$					

134. $f(x) = 1 + \sqrt[3]{x}$

x	–8	–1	0	1	8
$f(x)$					

You Be the Teacher!

Correct each student's errors, if any.

135. Find the domain of the function $f(x) = \sqrt{x - 1}$.

Vincent's work:

$x \neq 1$

$(-\infty, 1) \cup (1, \infty)$

136. Find the domain of the function $f(x) = \sqrt[3]{2x - 5}$.

Tom's work:

$2x - 5 \geq 0$

$2x \geq 5$

$x \geq \dfrac{5}{2}$

$\left[\dfrac{5}{2}, \infty\right)$

137. Simplify: $-\sqrt{-9x^6y^{12}}$.

Ashley's work:

$-\sqrt{-9x^6y^{12}} = 3x^3y^6$

138. Simplify: $-\sqrt[4]{-81x^4}$.

Kimberly's work:

$-\sqrt[4]{-81x^4} = 3x$

 **Calculate It!**

Use a calculator to find each root and round each answer to two decimal places.

139. $\sqrt{1834}$ **140.** $\sqrt[3]{432}$

141. $\sqrt[5]{1492}$ **142.** $\sqrt[7]{654}$

SECTION 10.2 / Rational Exponents

▶ **OBJECTIVES**

As a result of completing this section, you will be able to

1. Simplify expressions of the form $b^{1/n}$.

2. Simplify expressions of the form $b^{m/n}$.

3. Simplify expressions of the form $b^{-m/n}$.

4. Use exponent rules to simplify expressions with rational exponents.

5. Use rational exponents to simplify radical expressions.

6. Solve applications with rational exponents.

7. Troubleshoot common errors.

The compound interest formula is given by $A = P\left(1 + \dfrac{r}{n}\right)^{nt}$,

where A is the amount of money in an account in t years if P dollars is invested at an annual interest rate r, compounded n times a year. How much money will be in an account in 6 months $\left(\dfrac{1}{2} \text{ of a year}\right)$ if \$3000 is invested at an annual interest rate of 5% compounded annually?

To answer this question, we must simplify the expression

$$A = 3000\left(1 + \frac{0.05}{1}\right)^{1(1/2)} = 3000(1.05)^{1/2}$$

Notice that this expression has an exponent that is a fraction. This is called a *fractional exponent*, or a *rational exponent*. In this section, we will learn how to deal with this type of exponent. We will also discover how rational exponents are related to radicals.

Rational Exponents of the Form $b^{1/n}$

Until this point, we have dealt only with integer exponents. We know the following.

- A positive integer exponent indicates how many times a base is repeated as a factor. For example, $4^3 = 4 \cdot 4 \cdot 4$.
- A nonzero number raised to the exponent of 0 is 1. For example, $4^0 = 1$.
- A negative integer exponent indicates that we take the reciprocal of the base to the positive exponent. For example, $4^{-3} = \dfrac{1}{4^3}$.

We will now learn how to deal with rational exponents of the form $\dfrac{1}{2}, \dfrac{1}{3}, \dfrac{1}{4}, \dfrac{1}{5}$, and so on.

Using a calculator, we can determine the value of the exponential expressions in the left column of the following table. The value of each exponential expression is also the same as a value of a radical expression, as shown.

 **Objective 1** ▶

Simplify expressions of the form $b^{1/n}$.

Value Obtained by a Calculator	Radical Expression with the Same Value
$4^{1/2} = 2$	$\sqrt{4} = 2$
$9^{1/2} = 3$	$\sqrt{9} = 3$
$16^{1/2} = 4$	$\sqrt{16} = 4$
$8^{1/3} = 2$	$\sqrt[3]{8} = 2$
$27^{1/3} = 3$	$\sqrt[3]{27} = 3$

This example shows equivalent expressions that will help us define an exponent of the form $\frac{1}{n}$. Note the following.

$$4^{1/2} = \sqrt{4}$$
$$9^{1/2} = \sqrt{9}$$
$$16^{1/2} = \sqrt{16}$$
$$8^{1/3} = \sqrt[3]{8}$$
$$27^{1/3} = \sqrt[3]{27}$$

So, **rational exponents** are simply another way to write a radical expression. Notice that the denominator of a rational exponent is the index of the equivalent radical expression.

Definition: $b^{1/n}$

For n a positive integer greater than 1 and $\sqrt[n]{b}$ a real number,
$$b^{1/n} = \sqrt[n]{b}$$

Recall that the nth root of a number b is defined as $\sqrt[n]{b} = a$ if and only if $a^n = b$. This definition also holds true for rational exponents.

$$(b^{1/n})^n = b^{(1/n)(n)} = b^1 = b$$

Procedure: Evaluating Expressions of the Form $b^{1/n}$

Step 1: Convert the expression to a radical of the form $\sqrt[n]{b}$.
    a. The denominator of the fractional exponent becomes the index of the radical.
    b. The base of the fractional exponent becomes the radicand of the radical.
Step 2: Simplify the resulting radical expression, if possible.

Objective 1 Examples   Rewrite each expression as a radical expression and simplify, if possible.

    1a. $81^{1/2}$ **1b.** $(-8)^{1/3}$ **1c.** $y^{1/4}$ **1d.** $-144^{1/2}$ **1e.** $(-144)^{1/2}$ **1f.** $(32x^{10})^{1/5}$

Solutions   **1a.** $81^{1/2} = \sqrt{81} = 9$

    1b. $(-8)^{1/3} = \sqrt[3]{-8} = -2$

    1c. $y^{1/4} = \sqrt[4]{y}$

    1d. The base of the exponent is 144. So, we find the opposite of the square root of 144.

$$-144^{1/2} = -\sqrt{144} = -12$$

    1e. The base of the exponent is -144. So, we find the square root of -144.

$$(-144)^{1/2} = \sqrt{-144} \text{ not a real number}$$

    1f. This problem is equivalent to the fifth root of $32x^{10}$. So, we must find the expression that when raised to the exponent of 5 gives us $32x^{10}$.

$$(30x^{10})^{1/5} = \sqrt[5]{32x^{10}} = 2x^2$$

✓ Student Check 1   Rewrite each expression as a radical expression and simplify, if possible.

    a. $25^{1/2}$ **b.** $(-64)^{1/3}$ **c.** $x^{1/5}$ **d.** $-36^{1/2}$ **e.** $(-36)^{1/2}$ **f.** $(81x^{12})^{1/4}$

Rational Exponents of the Form $b^{m/n}$

Objective 2 ▶

Simplify expressions of the form $b^{m/n}$.

To be thorough in our discussion of rational exponents, we need to consider rational exponents whose numerators are different from one. In other words, we will define what it means to raise an expression to exponents such as $\frac{2}{3}, \frac{3}{2},$ and $\frac{3}{4}$. The power of a power rule for exponents helps us define this type of rational exponent. Recall that

$$(b^m)^n = b^{mn}$$

To simplify $16^{3/4}$, we think of it as follows.

$$16^{3/4} = 16^{(1/4)(3)} = (16^{1/4})^3 = \left(\sqrt[4]{16}\right)^3 = 2^3 = 8$$

Definition: $b^{m/n}$

If m and n are positive integers greater than 1 and $\sqrt[n]{b}$ a real number, then

$$b^{m/n} = \left(\sqrt[n]{b}\right)^m \quad \text{or} \quad \sqrt[n]{b^m}$$

Note: *The exponent m can be applied after the root is simplified or before the root is taken.*

Procedure: Evaluating an Expression of the Form $b^{m/n}$

Step 1: Convert the expression to radical form $\left(\sqrt[n]{b}\right)^m$.
    a. The base of the exponent is the radicand of the radical.
    b. The denominator of the fractional exponent is the index of the radical.
    c. The numerator of the fractional exponent is the exponent of the radical expression.
Step 2: Simplify the radical expression.
Step 3: Simplify the exponential expression.

Objective 2 Examples Rewrite each expression as a radical expression and simplify, if possible. Assume all variable bases represent positive real numbers.

    2a. $16^{5/4}$ **2b.** $-25^{3/2}$ **2c.** $(-64)^{4/3}$ **2d.** $\left(\frac{1}{8}\right)^{2/3}$ **2e.** $(x-3)^{5/6}$

Solutions **2a.**

$$\begin{aligned} 16^{5/4} &= \left(\sqrt[4]{16}\right)^5 && \text{Apply the definition of } b^{m/n}. \\ &= 2^5 && \text{Simplify the radical expression.} \\ &= 32 && \text{Simplify the exponential expression.} \end{aligned}$$

2b.

$$\begin{aligned} -25^{3/2} &= -\left(\sqrt{25}\right)^3 && \text{Apply the definition of } b^{m/n}. \\ &= -5^3 && \text{Simplify the radical expression.} \\ &= -125 && \text{Simplify the exponential expression.} \end{aligned}$$

Notice the base is 25, not -25, because there are no parentheses in the given problem.

2c.

$$\begin{aligned} (-64)^{4/3} &= \left(\sqrt[3]{-64}\right)^4 && \text{Apply the definition of } b^{m/n}. \\ &= (-4)^4 && \text{Simplify the radical expression.} \\ &= 256 && \text{Simplify the exponential expression.} \end{aligned}$$

Notice the base is -64 since it is contained in parentheses in the given problem.

2d.

$$\left(\frac{1}{8}\right)^{2/3} = \left(\sqrt[3]{\frac{1}{8}}\right)^{2}$$ Apply the definition of $b^{m/n}$.

$$= \left(\frac{1}{2}\right)^{2}$$ Simplify the radical expression.

$$= \frac{1}{4}$$ Simplify the exponential expression.

2e.

$$(x - 3)^{5/6} = \left(\sqrt[6]{x - 3}\right)^{5}$$ Apply the definition of $b^{m/n}$.

$$= \sqrt[6]{(x - 3)^{5}}$$ Apply the exponent of 5.

✔ **Student Check 2**   Rewrite each expression as a radical expression and simplify, if possible. Assume all variable bases represent positive real numbers.

a. $9^{3/2}$ **b.** $-81^{3/4}$ **c.** $(-27)^{2/3}$ **d.** $\left(\frac{1}{32}\right)^{3/5}$ **e.** $(x + 2)^{6/7}$

Rational Exponents of the Form $b^{-m/n}$

Objective 3 ▶

Simplify expressions of the form $b^{-m/n}$.

Rational exponents can also be negative. In this case, we have to use the definition of the negative exponent to help us rewrite the expression. Recall that for $n > 0$, $x^{-n} = \frac{1}{x^{n}}$. So, using this definition together with the definition of $b^{m/n}$, we get the following.

> **Definition: $b^{-m/n}$**
> For $b \neq 0$,
> $$b^{-m/n} = \frac{1}{b^{m/n}}$$

> **Procedure: Evaluating an Expression of the Form $b^{-m/n}$**
> **Step 1:** Rewrite the expression with a positive exponent.
> **Step 2:** Convert the expression to its equivalent radical form $\left(\sqrt[n]{b}\right)^{m}$.
> **Step 3:** Simplify the radical expression.
> **Step 4:** Simplify the exponential expression, if needed.

Objective 3 Examples   Rewrite each expression with a positive exponent and simplify.

3a. $16^{-1/2}$ **3b.** $(-8)^{-4/3}$ **3c.** $-81^{-3/4}$

Solutions   **3a.**

$$16^{-1/2} = \frac{1}{16^{1/2}}$$ Apply the definition of $b^{-m/n}$.

$$= \frac{1}{\sqrt{16}}$$ Apply the definition of $b^{1/n}$.

$$= \frac{1}{4}$$ Simplify the radical expression.

3b.

$$(-8)^{-4/3} = \frac{1}{(-8)^{4/3}}$$ Apply the definition of $b^{-m/n}$.

$$= \frac{1}{\left(\sqrt[3]{-8}\right)^{4}}$$ Apply the definition of $b^{m/n}$.

$$= \frac{1}{(-2)^4}$$ Simplify the radical expression.

$$= \frac{1}{16}$$ Simplify the exponential expression.

3c. $$-81^{-3/4} = -\frac{1}{81^{3/4}}$$ Apply the definition of $b^{-m/n}$.

$$= -\frac{1}{\left(\sqrt[4]{81}\right)^3}$$ Apply the definition of $b^{m/n}$.

$$= -\frac{1}{(3)^3}$$ Simplify the radical expression.

$$= -\frac{1}{27}$$ Simplify the exponential expression.

✔ **Student Check 3**   Rewrite each expression with a positive exponent and simplify.

     a. $32^{-1/5}$ **b.** $(-27)^{-2/3}$ **c.** $-25^{-3/2}$

Simplify Exponential Expressions Involving Rational Exponents

Objective 4 ▶

Use exponent rules to simplify expressions with rational exponents.

The exponent rules that we learned previously work not only for integer exponents, but also for rational exponents. A review of the exponent rules follows.

Property: Summary of Exponent Rules

If a and b are real numbers and m and n are integers, then

Product of like bases: $b^m \cdot b^n = b^{m+n}$

Quotient of like bases: $\dfrac{b^m}{b^n} = b^{m-n}, b \neq 0$

Power of a power: $(b^m)^n = b^{mn}$
Power of a product: $(ab)^n = a^n b^n$

Power of a quotient: $\left(\dfrac{a}{b}\right)^n = \dfrac{a^n}{b^n}, b \neq 0$

Zero exponent: $b^0 = 1$

Negative exponent: $b^{-n} = \dfrac{1}{b^n}, b \neq 0$

Objective 4 Examples   Simplify each expression and write each result with positive exponents. Assume all variables represent positive real numbers.

     4a. $4^{4/3} \cdot 4^{2/3}$ **4b.** $x^{3/5} \cdot x^{-4/5}$ **4c.** $y^{1/2} \cdot y^{1/4}$ **4d.** $\dfrac{8^{1/5}}{8^{6/5}}$

     4e. $(a^4)^{3/2}$ **4f.** $(8x^6)^{2/3}$ **4g.** $\left(\dfrac{3b^{1/2}c^{-1/3}}{b^2}\right)^2$ **4h.** $x^{1/3}(x^{1/3} + 2x^{2/3})$

Solutions   **4a.** $4^{4/3} \cdot 4^{2/3} = 4^{4/3+2/3}$ Apply the product of like bases rule.

$$= 4^{6/3}$$ Add the exponents.

$$= 4^2$$ Simplify the exponent.

$$= 16$$ Simplify.

     4b. $x^{3/5} \cdot x^{-4/5} = x^{3/5+(-4/5)}$ Apply the product of like bases rule.

$$= x^{-1/5}$$ Add the exponents.

$$= \frac{1}{x^{1/5}}$$ Apply the negative exponent rule.

4c.
$$y^{1/2} \cdot y^{1/4} = y^{1/2+1/4}$$
Apply the product of like bases rule.
$$= y^{2/4+1/4}$$
Find a common denominator.
$$= y^{3/4}$$
Add the exponents.

4d.
$$\frac{8^{1/5}}{8^{6/5}} = 8^{1/5-6/5}$$
Apply the quotient of like bases rule.
$$= 8^{-5/5}$$
Subtract the exponents.
$$= 8^{-1}$$
Simplify the exponent.
$$= \frac{1}{8^1}$$
Apply the negative exponent rule.
$$= \frac{1}{8}$$
Simplify.

4e.
$$(a^4)^{3/2} = a^{4(3/2)}$$
Apply the power of a power rule.
$$= a^6$$
Multiply exponents.

4f.
$$(8x^6)^{2/3} = (8)^{2/3}(x^6)^{2/3}$$
Apply the power of a product rule.
$$= \left(\sqrt[3]{8}\right)^2 x^{6(2/3)}$$
Apply the definition of $b^{m/n}$ and the power of a power rule.
$$= 4x^4$$
Simplify the radical expression and multiply exponents.

4g.
$$\left(\frac{3b^{1/2}c^{-1/3}}{b^2}\right)^2 = \frac{(3b^{1/2}c^{-1/3})^2}{(b^2)^2}$$
Apply the power of a quotient rule.
$$= \frac{3^2 b^{(1/2)(2)}c^{(-1/3)(2)}}{b^4}$$
Apply the power of a product rule in the numerator and the power of a power rule in the denominator.
$$= \frac{9b^1 c^{-2/3}}{b^4}$$
Simplify 3^2 and multiply exponents.
$$= 9b^{1-4}c^{-2/3}$$
Apply the quotient of like bases rule.
$$= 9b^{-3}c^{-2/3}$$
Subtract the exponents.
$$= \frac{9}{b^3 c^{2/3}}$$
Apply the negative exponent rule.

4h.
$$x^{1/3}(x^{1/3} + 2x^{2/3}) = x^{1/3}(x^{1/3}) + x^{1/3}(2x^{2/3})$$
Apply the distributive property.
$$= x^{1/3+1/3} + 2x^{1/3+2/3}$$
Apply the product of like bases rule.
$$= x^{2/3} + 2x^{3/3}$$
Add the exponents.
$$= x^{2/3} + 2x^1$$
Simplify the exponent.
$$= x^{2/3} + 2x$$
Recall $x^1 = x$.

✓ **Student Check 4**　Simplify each expression and write each result with positive exponents. Assume all variables represent positive real numbers.

a. $7^{1/6} \cdot 7^{5/6}$　　**b.** $y^{3/7} \cdot y^{-6/7}$　　**c.** $x^{1/3} \cdot x^{1/6}$　　**d.** $\dfrac{3^{2/3}}{3^{5/3}}$

e. $(a^4)^{5/4}$　　**f.** $(9x^6)^{3/2}$　　**g.** $\left(\dfrac{2x^{2/3}y^{-1/2}}{x^2}\right)^3$　　**h.** $x^{1/4}(x^{1/4} - 3x^{3/4})$

Objective 5 ▶

Use rational exponents to simplify radical expressions.

Simplifying Radical Expressions

Radical expressions can often be simplified if we write the radical as an exponential expression. Rational exponents also enable us to multiply radicals that have the same radicand but difference indices.

Procedure: Simplifying Radical Expressions

Step 1: Convert the radical expression to an exponential expression.
Step 2: Apply the properties of exponents.
Step 3: Write the result in radical form.

Objective 5 Examples

Use rational exponents to simplify each expression. Assume all variables represent positive real numbers.

5a. $\sqrt[3]{x^6}$ **5b.** $\sqrt[6]{x^3}$ **5c.** $\sqrt[4]{9}$ **5d.** $\sqrt[3]{2} \cdot \sqrt{2}$

Solutions

5a.

$$\sqrt[3]{x^6} = x^{6/3}$$ Convert the radical to a rational exponent.

$$= x^2$$ Simplify the exponent.

5b.

$$\sqrt[6]{x^3} = x^{3/6}$$ Convert the radical to a rational exponent.

$$= x^{1/2}$$ Simplify the exponent.

$$= \sqrt{x}$$ Convert the rational exponent to a radical.

5c.

$$\sqrt[4]{9} = \sqrt[4]{3^2}$$ Rewrite 9 as an exponent, $9 = 3^2$.

$$= 3^{2/4}$$ Convert the radical to a rational exponent.

$$= 3^{1/2}$$ Simplify the exponent.

$$= \sqrt{3}$$ Convert the rational exponent to a radical.

5d.

$$\sqrt[3]{2} \cdot \sqrt{2} = 2^{1/3} \cdot 2^{1/2}$$ Convert each radical to a rational exponent.

$$= 2^{1/3 + 1/2}$$ Apply the product of like bases rule.

$$= 2^{2/6 + 3/6}$$ Find a common denominator.

$$= 2^{5/6}$$ Add the exponents.

$$= \sqrt[6]{2^5}$$ Convert the rational exponent to radical form.

$$= \sqrt[6]{32}$$ Simplify the radicand.

✓ Student Check 5

Use rational exponents to simplify each expression. Assume all variables represent positive real numbers.

a. $\sqrt{x^8}$ **b.** $\sqrt[8]{x^2}$ **c.** $\sqrt[4]{4}$ **d.** $\sqrt[3]{5} \cdot \sqrt[4]{5}$

Applications

Objective 6 ▶

Solve applications with rational exponents.

Rational exponents are used in many formulas, such as those for compound interest and wind chill. Any formula that contains a radical can be written with rational exponents. Formulas that involve exponents are applications of rational exponents when those exponents are rational numbers. Example 6 illustrates a few of these applications.

Objective 6 Examples

Solve each problem.

6a. The compound interest formula is $A = P\left(1 + \dfrac{r}{n}\right)^{nt}$, where A is the amount of money in an account in t years if P dollars is invested at an annual interest rate r, compounded n times a year. How much money will be in an account in 6 months if $3000 is invested at an annual interest rate of 5% compounded annually?

Solution   **6a.** $P = 3000$, $r = 0.05$, $n = 1$, and $t = \dfrac{1}{2}$.

$$A = P\left(1 + \frac{r}{n}\right)^{nt}$$ State the compound interest formula.

$$A = 3000\left(1 + \frac{0.05}{1}\right)^{1(1/2)}$$ Let $P = 3000$, $r = 0.05$, $n = 1$, and $t = \dfrac{1}{2}$.

$$A = 3000(1.05)^{1/2}$$ Simplify the exponent and in parentheses.

$$A \approx 3074.09$$ Approximate the value.

So, $3074.09 will be in the account after 6 months.

6b. The wind chill temperature, WC, in °F, can be determined by the given formula, where T is the air temperature in °F and V is the wind speed in mph.

$$WC = 35.74 + 0.6215T - 35.75V^{0.16} + 0.4275TV^{0.16}$$

Use the formula to determine the wind chill temperature if the air temperature is 45°F and the wind speed is 15 mph. Round answer to the nearest hundredth. (Source: http://www.weather.gov/om/windchill/)

Solution   **6b.** Replace T with 45 and V with 15.

$$WC = 35.74 + 0.6215T - 35.75V^{0.16} + 0.4275TV^{0.16}$$

$$WC = 35.74 + 0.6215(45) - 35.75(15)^{0.16} + 0.4275(45)(15)^{0.16}$$

$$WC \approx 35.74 + 27.9675 - 35.75(1.54232) + 0.4275(45)(1.54232)$$

$$WC \approx 35.74 + 27.9675 - 55.13794 + 29.67038$$

$$WC \approx 38.24$$

So, the wind chill temperature in these conditions is approximately 38.24°F.

✓ **Student Check 6**   Solve each problem.

a. How much money will be in an account after 3 months if $5000 is invested at an annual interest rate of 4% compounded quarterly ($n = 4$)?

b. Use the wind chill formula to calculate the wind chill temperature when the air temperature is 20°F and the wind speed is 30 mph. Round the answer to the nearest hundredth of a degree.

Objective 7 ▶

Troubleshoot common errors.

Troubleshooting Common Errors

Some common errors associated with rational exponents are shown next.

Objective 7 Examples   **A problem and an incorrect solution are given. Provide the correct solution and an explanation of the error.**

7a. Simplify $(-27)^{2/3}$.

Incorrect Solution	Correct Solution and Explanation
$(-27)^{2/3} = \left(\sqrt[3]{-27}\right)^2 = -3^2 = -9$	The error was made in not applying the exponent of 2 to -3. $(-27)^{2/3} = \left(\sqrt[3]{-27}\right)^2 = (-3)^2 = 9$

7b. Simplify $5^{1/4} \cdot 5^{3/4}$.

Incorrect Solution	Correct Solution and Explanation
$5^{1/4} \cdot 5^{3/4} = 25^{4/4} = 25^1 = 25$	When we multiply like bases, the bases stay the same. $$5^{1/4} \cdot 5^{3/4} = 5^{4/4} = 5^1 = 5$$

ANSWERS TO STUDENT CHECKS

Student Check 1   **a.** 5   **b.** -4 **c.** $\sqrt[5]{x}$
d. -6 **e.** not a real number   **f.** $3x^3$

Student Check 2   **a.** 27   **b.** -27 **c.** 9   **d.** $\dfrac{1}{8}$
e. $\sqrt[7]{(x+2)^6}$

Student Check 3   **a.** $\dfrac{1}{2}$ **b.** $\dfrac{1}{9}$ **c.** $-\dfrac{1}{125}$

Student Check 4   **a.** 7   **b.** $\dfrac{1}{y^{3/7}}$ **c.** $x^{1/2}$
d. $\dfrac{1}{3}$ **e.** a^5 **f.** $27x^9$ **g.** $\dfrac{8}{x^4 y^{3/2}}$ **h.** $x^{1/2} - 3x$

Student Check 5   **a.** x^4 **b.** $\sqrt[4]{x}$ **c.** $\sqrt{2}$
d. $\sqrt[12]{5^7} = \sqrt[12]{78,125}$

Student Check 6   **a.** $\$5050.00$ **b.** $1.30\ °F$

SUMMARY OF KEY CONCEPTS

1. Radical expressions can be converted to exponential form, or vice versa, by using these formulas. If $\sqrt[n]{b}$ is a real number, then
$$b^{1/n} = \sqrt[n]{b}$$
$$b^{m/n} = \left(\sqrt[n]{b}\right)^m \quad \text{or} \quad \sqrt[n]{b^m}$$
$$b^{-m/n} = \left(\sqrt[n]{b}\right)^{-m} = \frac{1}{\left(\sqrt[n]{b}\right)^m}$$

2. All of the exponent rules are defined for rational exponents. If a and b are real numbers and m and n are rational numbers, then

   Product of like bases: $b^m \cdot b^n = b^{m+n}$

   Quotient of like bases: $\dfrac{b^m}{b^n} = b^{m-n},\ b \neq 0$

Power of a power: $(b^m)^n = b^{mn}$
Power of a product: $(ab)^n = a^n b^n$

Power of a quotient: $\left(\dfrac{a}{b}\right)^n = \dfrac{a^n}{b^n},\ b \neq 0$

Zero exponent: $b^0 = 1$

Negative exponent: $b^{-n} = \dfrac{1}{b^n},\ b \neq 0$

3. A radical expression can often be simplified if we convert it to exponential form. This allows us to simplify the exponents and, thereby, simplify the expression. After the exponents are simplified, the expression can be converted back to radical form.

GRAPHING CALCULATOR SKILLS

We can use the graphing calculator to simplify expressions with rational exponents. The key is to put parentheses around the entire exponent.

Example: Simplify $(-32)^{6/5}$.

Solution:

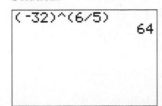

SECTION 10.2 / EXERCISE SET

 Write About It!

Use complete sentences in your answer to each exercise.

1. Define a rational exponent and give an example.

2. Explain how an expression with a rational exponent is related to a radical.

3. Explain how to apply the rules of exponents to simplify $x^{2/5} \cdot x^{1/3}$.

4. Explain how to apply the rules of exponents to simplify $\dfrac{x^{1/2}}{x^{1/3}}$.

5. Explain how to simplify $32^{-3/5}$ by converting it to a radical.

6. Explain why $2^{1/3} \cdot 3^{1/2} \neq 6^{5/6}$.

 Practice Makes Perfect!

Rewrite each expression as a radical expression and simplify, if possible. Assume all variables represent positive real numbers. (*See Objective 1.*)

7. $49^{1/2}$

8. $25^{1/2}$

9. $121^{1/2}$

10. $169^{1/2}$

11. $(-64)^{1/3}$

12. $(-125)^{1/3}$

13. $-225^{1/2}$

14. $-9^{1/2}$

15. $(-64)^{1/2}$

16. $(-225)^{1/2}$

17. $x^{1/3}$

18. $y^{1/4}$

19. $5^{1/4}$

20. $20^{1/5}$

21. $64^{1/6}$

22. $625^{1/4}$

23. $(8x^6)^{1/3}$

24. $(125y^9)^{1/3}$

25. $(243a^{15})^{1/5}$

26. $(625b^{12})^{1/4}$

Rewrite each expression as a radical and simplify, if possible. Assume all variables represent positive real numbers. (*See Objective 2.*)

27. $16^{3/4}$

28. $27^{4/3}$

29. $25^{3/2}$

30. $49^{3/2}$

31. $-16^{5/4}$

32. $-625^{3/4}$

33. $-27^{2/3}$

34. $-64^{2/3}$

35. $(-125)^{2/3}$

36. $(-8)^{4/3}$

37. $\left(\dfrac{1}{27}\right)^{2/3}$

38. $\left(\dfrac{1}{1000}\right)^{2/3}$

39. $\left(\dfrac{16}{625}\right)^{3/4}$

40. $\left(\dfrac{256}{81}\right)^{3/4}$

41. $(64x^9y^6)^{2/3}$

42. $(27x^6y^{12})^{4/3}$

43. $(32x^{15}y^{10})^{4/5}$

44. $(81x^{12}y^8)^{5/4}$

45. $\left(\dfrac{a^6}{64b^3}\right)^{2/3}$

46. $\left(\dfrac{x^{10}}{243y^5}\right)^{3/5}$

47. $(x+8)^{4/5}$

48. $(2x+1)^{2/5}$

49. $(x+3)^{2/3}$

50. $(3x-1)^{2/3}$

Rewrite each expression with positive exponents and simplify. Assume all variables represent positive real numbers. (*See Objective 3.*)

51. $64^{-1/3}$

52. $125^{-1/3}$

53. $-32^{-2/5}$

54. $-243^{-3/5}$

55. $(-125)^{-2/3}$

56. $(-216)^{-2/3}$

57. $(32x^{10})^{-2/5}$

58. $(125y^9)^{-2/3}$

59. $(64a^6b^{12})^{-5/6}$

60. $(c^{10}d^{25})^{-4/5}$

61. $(x^{12}y^{-15})^{-2/3}$

62. $(a^{-8}b^{12})^{-5/4}$

Simplify each expression. Assume all variables represent positive real numbers. Write each answer with positive exponents. (*See Objective 4.*)

63. $5^{1/6} \cdot 5^{5/6}$

64. $3^{2/5} \cdot 3^{3/5}$

65. $2^{4/3} \cdot 2^{2/3}$

66. $6^{4/5} \cdot 6^{6/5}$

67. $x^{3/5} \cdot x^{1/2}$

68. $y^{4/7} \cdot y^{1/3}$

69. $a^{1/6} \cdot a^{-5/6}$

70. $b^{1/2} \cdot b^{-5/6}$

71. $\dfrac{5^{1/5}}{5^{6/5}}$

72. $\dfrac{7^{1/3}}{7^{4/3}}$

73. $\dfrac{x^{1/2}}{x^{2/5}}$

74. $\dfrac{y^{3/7}}{y^{1/3}}$

75. $(x^3y^6)^{4/3}$

76. $(a^3b^{12})^{5/3}$

77. $(32x^5y^{15})^{6/5}$

78. $(729a^{18}b^6)^{5/6}$

79. $(2x^{2/3}y^{-1/2})^3$

80. $(2a^{-3/5}b^{1/3})^5$

81. $\left(\dfrac{x^{3/5}y^{1/3}}{x^{1/2}}\right)^5$

82. $\left(\dfrac{a^{3/2}b^{2/3}}{a^{1/3}}\right)^3$

83. $\dfrac{(x^{5/6}y^{1/5})^3}{y}$

84. $\dfrac{(a^{2/7}b^{1/3})^7}{b^3}$

85. $x^{1/2}(x^{1/2}+x^{1/3})$

86. $y^{2/5}(y^{3/5}-y^{1/5})$

87. $(2a^{1/2}-3b^{1/3})(a^{1/2}+4b^{1/3})$

88. $(7x^{2/5}+2y^{1/2})(x^{2/5}+y^{1/2})$

89. $(4r^{1/2}-s^{2/3})(4r^{1/2}+s^{2/3})$

90. $(a^{3/5}+3b^{2/3})(a^{3/5}-3b^{2/3})$

Use rational exponents to simplify each expression. Assume all variables represent positive real numbers. (*See Objective 5.*)

91. $\sqrt[3]{a^{12}}$

92. $\sqrt[5]{b^{15}}$

93. $\sqrt[6]{x^2}$

94. $\sqrt[15]{y^3}$

95. $\sqrt[6]{a^3}$

96. $\sqrt[4]{b^2}$

97. $\sqrt[6]{8}$

98. $\sqrt[6]{125}$

99. $\sqrt[4]{49}$

100. $\sqrt[4]{25}$

101. $\sqrt[6]{27a^3}$

102. $\sqrt[6]{8b^3}$

103. $\sqrt[4]{121x^2}$

104. $\sqrt[4]{49y^2}$

105. $\sqrt[5]{2} \cdot \sqrt{2}$

106. $\sqrt[3]{3} \cdot \sqrt{3}$

107. $\sqrt[5]{x^3} \cdot \sqrt{x}$

108. $\sqrt[3]{y} \cdot \sqrt{y^3}$

109. $\sqrt[6]{a} \cdot \sqrt[3]{a^2}$

110. $\sqrt{b} \cdot \sqrt[4]{b^3}$

For Exercises 111–114, use the compound interest formula $A = P\left(1 + \dfrac{r}{n}\right)^{nt}$, where A is the amount of money in an account in t years if P dollars is invested at an annual interest rate r, compounded n times a year. Round each answer to the nearest cent. (*See Objective 6.*)

111. How much money will be in an account after 2 months if you invest \$1000 at an annual interest rate of 2% compounded annually ($n = 1$)?

112. How much money will be in an account after 4 months if you invest \$1500 at an annual interest rate of 2.5% compounded annually ($n = 1$)?

113. How much money will be in an account after 2 months if you invest \$800 at an annual interest rate of 2% compounded quarterly ($n = 4$)?

114. How much money will be in an account after 4 months if you invest \$2400 at an annual interest rate of 1.5% compounded quarterly ($n = 4$)?

For Exercises 115 and 116, use the wind chill formula, $WC = 35.74 + 0.6215T - 35.75V^{0.16} + 0.4275TV^{0.16}$, where T is the air temperature in °F and V is the wind speed in mph. Round each answer to the nearest hundredth of a degree. (*See Objective 6.*)

115. What is the wind chill temperature when the air temperature is 32°F and the wind speed is 20 mph?

116. What is the wind chill temperature when the air temperature is 0°F and the wind speed is 25 mph?

 Mix 'Em Up!

Rewrite each expression as a radical expression and simplify, if possible. Assume all variables represent positive real numbers.

117. $(-8x^9)^{1/3}$

118. $(-64y^6)^{1/3}$

119. $\left(\dfrac{x^3}{8y^6}\right)^{5/3}$

120. $\left(\dfrac{a^8}{81b^4}\right)^{1/4}$

121. $(125a^3b^6)^{2/3}$

122. $(8c^3d^6)^{5/3}$

123. $(243x^{-10}y^5)^{3/5}$

124. $(625a^8b^{-4})^{3/4}$

125. $(y + 5)^{4/3}$

126. $(2x + 3)^{3/2}$

Simplify each expression and write each answer with positive exponents. Assume all variables represent positive real numbers.

127. $(-64x^6)^{-1/3}$

128. $(125y^{-3})^{-2/3}$

129. $2^{5/6} \cdot 2^{7/6}$

130. $3^{4/5} \cdot 3^{6/5}$

131. $(16x^8)^{-3/4}$

132. $(-27y^6)^{-2/3}$

133. $(a^{-6}b^{12})^{-5/3}$

134. $(-c^5d^{-15})^{-3/5}$

135. $r^{1/4} \cdot r^{5/3}$

136. $s^{4/7} \cdot s^{-1/3}$

137. $(-x^5y^{10})^{2/5}$

138. $(-a^3b^9)^{2/3}$

139. $\dfrac{x^{2/7}}{x^{1/2}}$

140. $\dfrac{y^{1/5}}{y^{2/3}}$

141. $(3x^{2/5}y^{-2/3})^5$

142. $(4a^{-1/2}b^{5/6})^3$

143. $\dfrac{(x^{-3/7}y^{1/2})^4}{x^3}$

144. $\dfrac{(a^{-5/2}b^{1/2})^2}{a^{4/3}}$

145. $(3x^{1/2} - y^{1/3})(3x^{1/2} + y^{1/3})$

146. $(2a^{2/5} + 3b^{1/2})(2a^{2/5} - 3b^{1/2})$

147. $r^{1/2}(5r^{1/2} + r^{2/3})$

148. $s^{2/3}(s^{2/3} - 4s^{1/2})$

149. $(5x^{1/2} - 3y^{1/3})(2x^{1/2} + 4y^{1/3})$

150. $(a^{2/5} - 3b^{1/2})(a^{2/5} - 5b^{1/2})$

Use rational exponents to simplify each expression. Assume all variables represent positive real numbers.

151. $\sqrt[3]{64x^6}$

152. $\sqrt[5]{32y^{20}}$

153. $\sqrt[6]{343a^3}$

154. $\sqrt[6]{216b^3}$

155. $\sqrt[4]{\dfrac{49}{x^4}}$

156. $\sqrt[4]{\dfrac{9}{y^4}}$

157. $\sqrt{x} \cdot \sqrt[3]{x^5}$

158. $\sqrt[4]{y^3} \cdot \sqrt[4]{y}$

Solve each problem.

159. How much money will be in an account after 8 months if you invest \$500 at an annual interest rate of 1.8% compounded quarterly ($n = 4$)? Round to the nearest cent.

160. How much money will be in an account after 9 months if you invest \$600 at an annual interest rate of 2.4% compounded quarterly ($n = 4$)? Round to the nearest cent.

161. On top of Mount Everest, the wind speed can reach 115 mph and the temperature is $-15°F$. Use the wind chill formula to determine the wind chill temperature in these conditions. (Source: http://www.mounteverest2008.com/)

162. The formula $v = 5.289(T + 4)^{3/2}$ gives the wind speed in miles per hour of a tornado with a Torro intensity value, T. A violent tornado has an intensity of T-10. Estimate the wind speed of this tornado. Round the answer to the nearest hundredth. (Source: http://www.torro.org.uk/TORRO/severeweather/Tscaleorigin.php)

 You Be the Teacher!

Correct each student's errors, if any.

163. Simplify: $5^{1/3} \cdot 5^{2/3}$.

Celine's work: $5^{1/3} \cdot 5^{2/3} = 25^{1/3+2/3} = 25^{3/3} = 25^1 = 25$

164. Simplify: $(-16x^8)^{3/4}$. Assume $x \geq 0$.

Christine's work: $(-16x^8)^{3/4} = -16^{3/4}(x^8)^{3/4} = -8x^6$

165. Simplify: $-(-x^3y^6)^{2/3}$. Assume $x, y \geq 0$.

Tony's work:
$-(-x^3y^6)^{2/3} = (x^3y^6)^{2/3} = (x^3)^{2/3}(y^6)^{2/3} = x^2y^4$

166. Simplify: $-\sqrt[4]{9x^2}$. Assume $x \geq 0$.

Kim's work: $-\sqrt[4]{9x^2} = -3x$

 Calculate It!

Use a calculator to simplify each expression. Write the answer to two decimal places. If possible, convert the result to fractional form.

167. $(36)^{5/3}$

168. $(7)^{-2/5}$

169. $(32)^{-4/5}$

170. $(125)^{-4/3}$

Simplifying Radical Expressions and the Distance Formula

▶ **OBJECTIVES**

As a result of completing this section, you will be able to

1. Use the product rule to simplify radicals.
2. Use the quotient rule to simplify radicals.
3. Use the distance formula.
4. Troubleshoot common errors.

Suppose we want to find the distance between Key West, Florida, and Pensacola, Florida, as shown on the map (Source: www.siteatlas.com). We can impose a grid on the map, label the coordinates of the two cities and then use a distance formula to find the distance between them. In this section, we will learn the distance formula and two basic rules of radicals that enable us to simplify radical expressions.

Simplifying Radical Expressions Using the Product Rule

Objective 1 ▶

Use the product rule to simplify radicals.

We know that $\sqrt{25}$ and $\sqrt[3]{8}$ are radical expressions that are not in their *simplest form* since $\sqrt{25} = 5$ and $\sqrt[3]{8} = 2$. When the radicand can be written as a number raised to the index, the radical can be simplified fairly easily. If a radicand contains a factor raised to the index, it can also be simplified. Before this is illustrated, we need to learn the *product rule for radicals*. The following illustration shows some radical expressions that are equivalent.

$$\sqrt{4 \cdot 4} = \sqrt{16} = 4 \quad \middle| \quad \sqrt{4} \cdot \sqrt{4} = 2 \cdot 2 = 4 \longrightarrow \sqrt{4 \cdot 4} = \sqrt{4} \cdot \sqrt{4} = 4$$
$$\sqrt[3]{8 \cdot 27} = \sqrt[3]{216} = 6 \quad \middle| \quad \sqrt[3]{8} \cdot \sqrt[3]{27} = 2 \cdot 3 = 6 \longrightarrow \sqrt[3]{8 \cdot 27} = \sqrt[3]{8} \cdot \sqrt[3]{27} = 6$$

These examples show that the nth root of a product is the same as the product of the nth roots.

Property: Product Rule for Radicals

If $\sqrt[n]{a}$ and $\sqrt[n]{b}$ are real numbers, and n is a positive integer, then
$$\sqrt[n]{a \cdot b} = \sqrt[n]{a} \cdot \sqrt[n]{b}$$

For a radical to be in its simplest form, the radicand cannot contain any factor raised to the index. Some examples of radicals that are not in their simplest form are $\sqrt{12}$, $\sqrt[3]{16}$, $\sqrt{y^7}$, and $\sqrt[3]{x^5}$.

- $\sqrt{12}$ is not in simplest form since $12 = 4 \cdot 3$ and 4 is a perfect square factor of 12.
- $\sqrt[3]{16}$ is not in simplest form since $16 = 8 \cdot 2$ and 8 is a perfect cube factor of 16.
- $\sqrt{y^7}$ is not in simplest form since $y^7 = y^6 \cdot y$ and y^6 is a perfect square factor of y^7.
- $\sqrt[3]{x^5}$ is not in simplest form since $x^5 = x^3 \cdot x^2$ and x^3 is a perfect cube factor of x^5.

> **Note:** *The key to simplifying higher roots lies in our ability to recognize perfect powers. Refer to the chart in Section 10.1 for more perfect powers.*
>
> *Perfect squares: $1, 4, 9, 16, 25, 36, \ldots, x^2, x^4, x^6, \ldots$*
> *Perfect cubes: $1, 8, 27, 64, 125, 216, \ldots, x^3, x^6, x^9, \ldots$*
> *Perfect fourths: $1, 16, 81, 256, \ldots, x^4, x^8, x^{12}, \ldots$*

To simplify radicals, we use properties that were presented in Sections 8.1 and 8.2. We will assume all variables represent positive real numbers,

Property	Examples
$\sqrt[n]{a^n} = a$	$\sqrt{x^2} = x$, $\sqrt[3]{x^3} = x$, and $\sqrt[4]{(x^3)^4} = x^3$
$\sqrt[n]{a^m} = a^{m/n}$	$\sqrt[3]{x^6} = x^{6/3} = x^2$ and $\sqrt{x^4} = x^{4/2} = x^2$

There are two methods that we can use to simplify radical expressions. One method relies on listing the factors for the radicand and the other method relies on expressing the radicand in its prime factorization.

> **Procedure: Simplifying Radical Expressions**
>
> **Method 1**
>
> **Step 1:** List the pairs of factors of the radicand.
> **Step 2:** Select the pair of factors for which one factor is a *perfect power whose exponent is the index.*
> **Step 3:** Rewrite the radicand as the product of numbers found in Step 2. Write the perfect power first.
> **Step 4:** Apply the product rule and extract any roots.
>
> **Method 2**
>
> **Step 1:** Rewrite the radicand as a product of its prime factors.
> **Step 2:** Apply the product rule and extract any roots.

> **Note:** *For a radical expression to be completely simplified, the radicand cannot contain any exponents greater than the index.*

Objective 1 Examples Simplify each radical completely. Assume all variables represent positive real numbers.

1a. $\sqrt{80}$ **1b.** $2\sqrt{72}$ **1c.** $\sqrt{y^5}$ **1d.** $\sqrt{24x^3}$ **1e.** $\sqrt[3]{54}$ **1f.** $\sqrt[4]{80a^7b^8}$

Solutions **1a.**

Method 1: Factors of 80	**Method 2:** Prime factorization of 80
1, 80	$80 = 4 \cdot 20$
2, 40	$= 2 \cdot 2 \cdot 4 \cdot 5$
4, 20 4 and 16 are both perfect square factors.	$= 2 \cdot 2 \cdot 2 \cdot 2 \cdot 5$
5, 16 We will use 16 since it is larger.	$= 2^4 \cdot 5$
8, 10	

$\sqrt{80} = \sqrt{16 \cdot 5}$	Rewrite 80 as $16 \cdot 5$.	$\sqrt{80} = \sqrt{2^4 \cdot 5}$	Rewrite 80 as $2^4 \cdot 5$.
$= \sqrt{16}\sqrt{5}$	Apply the product rule.	$= \sqrt{2^4}\sqrt{5}$	Apply the product rule.
$= 4\sqrt{5}$	Write $\sqrt{16}$ as 4.	$= 2^2\sqrt{5}$	Write $\sqrt{2^4}$ as 2^2.
		$= 4\sqrt{5}$	Simplify 2^2.

1b.

Method 1: Factors of 72	**Method 2:** Prime factorization of 72
1, 72	$72 = 2 \cdot 36$
2, 36	$= 2 \cdot 6 \cdot 6$
3, 24 36, 4, and 9 are perfect square factors.	$= 6^2 \cdot 2$
4, 18 We will use 36 since it is the largest.	
6, 12	
8, 9	

$2\sqrt{72} = 2\sqrt{36 \cdot 2}$	Rewrite 72 as $36 \cdot 2$.	$2\sqrt{72} = 2\sqrt{6^2 \cdot 2}$	Rewrite 72 as $6^2 \cdot 2$.
$= 2\sqrt{36}\sqrt{2}$	Apply the product rule.	$= 2\sqrt{6^2}\sqrt{2}$	Apply the product rule.
$= 2(6)\sqrt{2}$	Write $\sqrt{36}$ as 6.	$= 2(6)\sqrt{2}$	Write $\sqrt{6^2}$ as 6.
$= 12\sqrt{2}$	Multiply the coefficients.	$= 12\sqrt{2}$	Multiply the coefficients.

1c. We apply method 1 to simplify this radical. The factors of y^5 are

$$y^4, y$$
$$y^3, y^2$$

Both y^4 and y^2 are perfect squares. We use y^4 since it is the largest perfect square factor.

$\sqrt{y^5} = \sqrt{y^4 \cdot y}$	Rewrite y^5 as $y^4 \cdot y$.
$= \sqrt{y^4}\sqrt{y}$	Apply the product rule for radicals.
$= y^2\sqrt{y}$	Write $\sqrt{y^4}$ as y^2.

1d. We apply method 1 to simplify this radical. The factors of 24 and x^3 are

1, 24		
2, 12	x^2, x	4 and x^2 are perfect square factors.
3, 8		
4, 6		

$\sqrt{24x^3} = \sqrt{4x^2 \cdot 6x}$	Rewrite $24x^3$ as $4x^2 \cdot 6x$.
$= \sqrt{4x^2}\sqrt{6x}$	Apply the product rule.
$= 2x\sqrt{6x}$	Write $\sqrt{4x^2}$ as $2x$.

1e. We need to find perfect cube factors of 54 since the index is 3.

Method 1: Factors of 54	**Method 2:** Prime factorization of 54

Method 1: Factors of 54

1, 54

2, 27   27 is a perfect cube.

3, 18

6, 9

$\sqrt[3]{54} = \sqrt[3]{27 \cdot 2}$ Rewrite 54 as 27 · 2.

$\quad\ = \sqrt[3]{27}\sqrt[3]{2}$ Apply the product rule.

$\quad\ = 3\sqrt[3]{2}$ Write $\sqrt[3]{27}$ as 3.

Method 2: Prime factorization of 54

$54 = 6 \cdot 9$

$\quad = 2 \cdot 3 \cdot 3 \cdot 3$

$\quad = 2 \cdot 3^3$

$\sqrt[3]{54} = \sqrt[3]{3^3 \cdot 2}$ Rewrite 54 as 3^3 · 2.

$\quad\ = \sqrt[3]{3^3}\sqrt[3]{2}$ Apply the product rule.

$\quad\ = 3\sqrt[3]{2}$ Write $\sqrt[3]{3^3}$ as 3.

1f. We apply method 2 to simplify this radical. The prime factorization of 80 is

$$80 = 8 \cdot 10$$
$$= 2 \cdot 2 \cdot 2 \cdot 2 \cdot 5$$
$$= 2^4 \cdot 5$$

A perfect fourth factor of a^7 is a^4, so $a^7 = a^4 \cdot a^3$. Note that b^8 is a perfect fourth since $b^8 = (b^2)^4$.

$$\sqrt[4]{80a^7b^8} = \sqrt[4]{2^4 \cdot 5a^4a^3(b^2)^4}$$
$$= \sqrt[4]{2^4a^4(b^2)^4}\sqrt[4]{5a^3}$$
$$= 2ab^2\sqrt[4]{5a^3}$$

 ✓ **Student Check 1**   Simplify each radical completely. Assume all variables represent positive real numbers.

   a. $\sqrt{18}$ **b.** $-3\sqrt{76}$ **c.** $\sqrt{a^9}$ **d.** $\sqrt{32y^{10}}$ **e.** $\sqrt[3]{128}$ **f.** $\sqrt[4]{162x^6y^4}$

Note: *In Example 1b, we would have obtained the same result if we used a different pair of factors. We would just have more steps of simplification. For instance,*

$$2\sqrt{72} = 2\sqrt{9 \cdot 8}$$ Rewrite 72 as 9 · 8.
$$= 2\sqrt{9}\sqrt{8}$$ Apply the product rule.
$$= 2(3)\sqrt{8}$$ Write $\sqrt{9}$ as 3.
$$= 6\sqrt{8}$$ Multiply the coefficients.
$$= 6\sqrt{4 \cdot 2}$$ Rewrite 8 as 4 · 2.
$$= 6\sqrt{4}\sqrt{2}$$ Apply the product rule.
$$= 6(2)\sqrt{2}$$ Write $\sqrt{4}$ as 2.
$$= 12\sqrt{2}$$ Multiply the coefficients.

If we do not use the largest perfect square factor, the radicand will contain another perfect square factor. Therefore, the steps to simplify have to be repeated.

Objective 2 ▶

Use the quotient rule to simplify radicals.

Simplifying Radical Expressions Using the Quotient Rule

When the radicand is a quotient, we can simplify it by applying the root to the numerator and denominator. Consider the following example.

$$\sqrt{\frac{36}{4}} = \sqrt{9} = 3 \qquad \frac{\sqrt{36}}{\sqrt{4}} = \frac{6}{2} = 3 \longrightarrow \sqrt{\frac{36}{4}} = \frac{\sqrt{36}}{\sqrt{4}} = 3$$

This example illustrates another important property of radicals. The nth root of a quotient is the same as the quotient of the nth roots.

Property: Quotient Rule for Radicals

If $\sqrt[n]{a}$ and $\sqrt[n]{b}$ are real numbers with $b \neq 0$ and n is a positive integer, then

$$\sqrt[n]{\frac{a}{b}} = \frac{\sqrt[n]{a}}{\sqrt[n]{b}}$$

For a radical to be in its simplest form,

- There cannot be any fractions in the radical.
- There cannot be any radicals in the denominator of a fraction. In Section 10.5, we will learn how to remove a radical from a denominator.

Procedure: Simplifying the nth Root of a Quotient

Step 1: If the radicand is a quotient, apply the root to the numerator and denominator.

Step 2: Rewrite each radicand as a product of a perfect power whose exponent is the index and another factor.

Step 3: Apply the product rule.

Objective 2 Examples   Simplify each radical completely. Assume the variables represent positive real numbers.

2a. $\sqrt{\dfrac{4}{9}}$ **2b.** $\sqrt[3]{\dfrac{x}{8}}$ **2c.** $\sqrt{\dfrac{20}{9y^4}}$ **2d.** $\sqrt[3]{\dfrac{192x^5}{27}}$

Solutions   **2a.** $\sqrt{\dfrac{4}{9}} = \dfrac{\sqrt{4}}{\sqrt{9}} = \dfrac{2}{3}$

2b. $\sqrt[3]{\dfrac{x}{8}} = \dfrac{\sqrt[3]{x}}{\sqrt[3]{8}} = \dfrac{\sqrt[3]{x}}{2}$ or $\dfrac{1}{2}\sqrt[3]{x}$

2c. $\sqrt{\dfrac{20}{9y^4}} = \dfrac{\sqrt{20}}{\sqrt{9y^4}}$ Apply the quotient rule for radicals.

$= \dfrac{\sqrt{4 \cdot 5}}{3y^2}$ Write the numerator with a perfect square factor and simplify the denominator.

$= \dfrac{\sqrt{4}\sqrt{5}}{3y^2}$ Apply the product rule for radicals.

$= \dfrac{2\sqrt{5}}{3y^2}$ Simplify.

2d. We apply the quotient rule for radicals and then find the factors of the radicand that are perfect cubes. The factors of 192 and x^5 are

$$1, 192 \qquad\qquad x^4, x$$
$$2, 96 \qquad\qquad x^3, x^2$$
$$3, 64$$
$$4, 48$$
$$6, 32$$
$$8, 24$$
$$12, 16$$

$$\sqrt[3]{\frac{192x^5}{27}} = \frac{\sqrt[3]{192x^5}}{\sqrt[3]{27}} \qquad \text{Apply the quotient rule for radicals.}$$

$$= \frac{\sqrt[3]{64 \cdot 3x^3x^2}}{\sqrt[3]{27}} \qquad \text{Write the appropriate factors of the radicand.}$$

$$= \frac{\sqrt[3]{64x^3}\sqrt[3]{3x^2}}{3} \qquad \text{Apply the product rule for radicals and simplify the denominator.}$$

$$= \frac{4x\sqrt[3]{3x^2}}{3} \qquad \text{Simplify.}$$

✔ **Student Check 2**   Simplify each radical completely. Assume the variables represent positive real numbers.

a. $\sqrt{\dfrac{81}{49}}$ **b.** $\sqrt[3]{\dfrac{a^2}{64}}$ **c.** $\sqrt{\dfrac{44}{9x^2}}$ **d.** $\sqrt[3]{\dfrac{40y^8}{343}}$

The Distance Formula

Objective 3 ▶

Use the distance formula.

The distance formula provides a way for us to measure the distance between two ordered pairs on a coordinate system. The distance formula is based on the Pythagorean theorem, $a^2 + b^2 = c^2$.

If we let the distance between (x_1, y_1) and (x_2, y_2) be represented by d, then we have by the Pythagorean theorem

$$(x_2 - x_1)^2 + (y_2 - y_1)^2 = d^2$$

So, $d = \sqrt{(x_2 - x_1)^2 + (y_2 - y_1)^2}$.

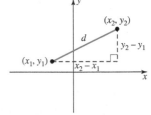

Property: Distance Formula

If (x_1, y_1) and (x_2, y_2) are two points, then the distance between them is
$$d = \sqrt{(x_2 - x_1)^2 + (y_2 - y_1)^2}$$

Procedure: Finding the Distance Between Two Points

Step 1: Label one point (x_1, y_1) and the other point (x_2, y_2).
Step 2: Substitute the values in the formula.
Step 3: Simplify the radicand.
Step 4: Simplify the radical by applying the product rule.

Objective 3 Examples Find the distance between each pair of points. State the exact distance by expressing answers with simplified radicals and approximate them to two decimal places.

3a. $(-3, 5)$ and $(3, 8)$

Solution **3a.** $d = \sqrt{(x_2 - x_1)^2 + (y_2 - y_1)^2}$ State the distance formula.

$= \sqrt{[3 - (-3)]^2 + (8 - 5)^2}$ Let $(x_1, y_1) = (-3, 5)$ and $(x_2, y_2) = (3, 8)$.

$= \sqrt{(6)^2 + (3)^2}$ Simplify inside each set of parentheses.

$= \sqrt{36 + 9}$ Simplify the exponents.

$= \sqrt{45}$ Add.

$= \sqrt{9 \cdot 5}$ Write 45 with a perfect square factor.

$= 3\sqrt{5}$ Apply the product rule for radicals and simplify.

The exact distance between $(-3, 5)$ and $(3, 8)$ is $3\sqrt{5}$ units, or approximately 6.71 units.

3b. $\left(4, \sqrt{7}\right)$ and $(-7, 0)$

Solution **3b.** Let $(x_1, y_1) = \left(4, \sqrt{7}\right)$ and $(x_2, y_2) = (-7, 0)$.

$d = \sqrt{(x_2 - x_1)^2 + (y_2 - y_1)^2}$ State the distance formula.

$= \sqrt{(-7 - 4)^2 + \left(0 - \sqrt{7}\right)^2}$ Let $(x_1, y_1) = (4, \sqrt{7})$ and $(x_2, y_2) = (-7, 0)$.

$= \sqrt{(-11)^2 + \left(-\sqrt{7}\right)^2}$ Simplify inside each set of parentheses.

$= \sqrt{121 + 7}$ Simplify the exponents.

$= \sqrt{128}$ Add.

$= \sqrt{64 \cdot 2}$ Write 128 with a perfect square factor.

$= 8\sqrt{2}$ Apply the product rule for radicals and simplify.

The exact distance between $\left(4, \sqrt{7}\right)$ and $(-7, 0)$ is $8\sqrt{2}$ units, or approximately 11.31 units.

3c. Use the map to approximate the distance between Key West, Florida, and Pensacola, Florida. Each tick mark represents 60 mi. (Source: http://www.sitesatlas .com/Maps/index.htm)

Solution   **3c.** We first identify coordinates of the two cities. Key West is approximately located at $(60, -180)$ and Pensacola is approximately located at $(-240, 180)$.

$$d = \sqrt{(x_2 - x_1)^2 + (y_2 - y_1)^2}$$ State the distance formula.

$$= \sqrt{[180 - (-180)]^2 + (-240 - 60)^2}$$ Let $(x_1, y_1) = (60, -180)$ and $(x_2, y_2) = (-240, 180)$.

$$= \sqrt{(360)^2 + (-300)^2}$$ Simplify inside both sets of parentheses.

$$= \sqrt{129,600 + 90,000}$$ Simplify the exponents.

$$= \sqrt{219,600}$$ Add.

$$= \sqrt{3600 \cdot 61}$$ Write 219,600 with a perfect square factor.

$$= 60\sqrt{61}$$ Apply the product rule for radicals and simplify.

So, the distance from Key West to Pensacola is $60\sqrt{61}$ mi or approximately 468.61 mi.

✔ **Student Check 3**   Find the distance between each pair of points. State the exact distances by expressing answers with simplified radicals and approximate them to two decimal places.

a. $(-4, 2)$ and $(4, 6)$

b. $\left(5, \sqrt{3}\right)$ and $(-2, 0)$

c. Use the map to approximate the distance between Orlando $(90, 90)$ and Boca Raton $(180, -60)$. Each tick mark represents 60 mi. (Source: http://www .sitesatlas.com/Maps/index.htm)

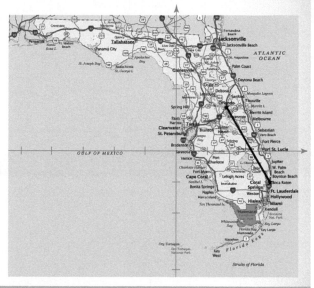

Objective 4 ▶

Troubleshoot common errors.

Troubleshooting Common Errors

Some common errors associated with simplifying radical expressions are shown.

Objective 4 Examples   A problem and an incorrect solution are given. Provide the correct solution and an explanation of the error.

4a. Simplify $\sqrt[3]{24}$.

Incorrect Solution	Correct Solution and Explanation
$\sqrt[3]{24} = \sqrt[3]{4 \cdot 6} = 2\sqrt[3]{6}$	The index is 3 so we should remove perfect cube factors not perfect square factors. $$\sqrt[3]{24} = \sqrt[3]{8 \cdot 3} = 2\sqrt[3]{3}$$

4b. Simplify $\sqrt{162x^4}$.

Incorrect Solution	Correct Solution and Explanation
$\sqrt{162x^4} = \sqrt{9 \cdot 2x^2} = 3x\sqrt{2}$	The factors inside the radical should be factors of the original radicand. $$\sqrt{162x^4} = \sqrt{81 \cdot 2x^4}$$ $$= \sqrt{81x^4}\sqrt{2}$$ $$= 9x^2\sqrt{2}$$

4c. Find the distance between $(6, -1)$ and $(4, 3)$.

Incorrect Solution	Correct Solution and Explanation
$$d = \sqrt{(4 - 6) + [3 - (-1)]}$$ $$= \sqrt{-2 + 4}$$ $$= \sqrt{2}$$	The distance formula has a square on each of the differences inside the radical. $$d = \sqrt{(4 - 6)^2 + [3 - (-1)]^2}$$ $$= \sqrt{(-2)^2 + (4)^2}$$ $$= \sqrt{4 + 16}$$ $$= \sqrt{20}$$ $$= \sqrt{4 \cdot 5}$$ $$= 2\sqrt{5}$$

ANSWERS TO STUDENT CHECKS

Student Check 1　**a.** $3\sqrt{2}$　**b.** $-6\sqrt{19}$　**c.** $a^4\sqrt{a}$
d. $4y^5\sqrt{2}$　**e.** $4\sqrt[3]{2}$　**f.** $3xy\sqrt[4]{2x^2}$

Student Check 2　**a.** $\dfrac{9}{7}$　**b.** $\dfrac{1}{4}\sqrt[3]{a^2}$　**c.** $\dfrac{2\sqrt{11}}{3x}$

d. $\dfrac{2y^2\sqrt[3]{5y^2}}{7}$

Student Check 3　**a.** $4\sqrt{5}$ units or 8.94 units
b. $2\sqrt{13}$ units or 7.21 units
c. $30\sqrt{34}$ or 174.93 mi

SUMMARY OF KEY CONCEPTS

1. The product rule for radicals tells us that the nth root of a product is the product of nth roots. This enables us to simplify radicals by writing the radicand as a product involving a power whose exponent is the index.

2. The quotient rule for radicals tells us that the nth root of a quotient is the quotient of nth roots.

3. The distance between two points (x_1, y_1) and (x_2, y_2) is $d = \sqrt{(x_2 - x_1)^2 + (y_2 - y_1)^2}$.

GRAPHING CALCULATOR SKILLS

The graphing calculator can help us learn powers, find factors of numbers, and check answers.

Example 1: Find perfect squares, cubes, and fourths.

Solution: Enter x raised to the appropriate exponent and view the table. The values in the Y_1, Y_2, and Y_3 column are perfect squares, cubes, and fourths, respectively.

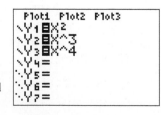

Example 2: Find factors of 54.

Solution: In the equation editor, , enter 54/x for Y_1. View the TABLE.

X	Y₁
1	54
2	27
3	18
4	13.5
5	10.8
6	9
7	7.7143

X=1

From the table, we see that the factors of 54 are 1 and 54, 2 and 27, 3 and 18, 6 and 9.

Example 3: Show that $\sqrt{54} = 3\sqrt{6}$ and that $\sqrt[3]{54} = 3\sqrt[3]{2}$.

Solution:

```
√(54)
          7.348469228
3√(6)
          7.348469228
```

Since the decimal values are the same, $\sqrt{54} = 3\sqrt{6}$.

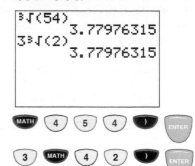

Since the decimal values are the same, $\sqrt[3]{54} = 3\sqrt[3]{2}$.

SECTION 10.3 / EXERCISE SET

 Write About It!

Use complete sentences in your answer to each exercise.

1. What does it mean for a square root expression to be written in simplified form?

2. What does it mean for a cube root expression to be written in simplified form?

3. How can the product rule for radicals be used to simplify a radical expression?

4. How can the quotient rule for radicals be used to simplify a radical expression?

5. Explain how to simplify $\sqrt{72x^2y^4}$.

6. Explain how to simplify $\sqrt[3]{\dfrac{54x^5}{125y^{15}}}$.

Practice Makes Perfect!

Simplify each radical completely. Assume all variables represent positive real numbers. (*See Objective 1.*)

7. $\sqrt{72}$

8. $3\sqrt{50}$

9. $-2\sqrt{90}$

10. $-3\sqrt{75}$

11. $\sqrt{-68}$

12. $\sqrt{-18}$

13. $5\sqrt{288}$

14. $3\sqrt{500}$

15. $\sqrt{48a^8}$

16. $\sqrt{80b^4}$

17. $\sqrt{300x^5}$

18. $\sqrt{98y^3}$

19. $\sqrt{54a^7}$

20. $\sqrt{112b^{13}}$

21. $\sqrt{125r^9t^{11}}$

22. $\sqrt{76x^3y^5}$

23. $\sqrt{144x^5y^6}$

24. $\sqrt{49r^4t^{13}}$

25. $\sqrt[3]{250}$

26. $\sqrt[3]{72}$

27. $\sqrt[3]{-2000}$

28. $\sqrt[3]{-40}$

29. $\sqrt[3]{192a^6}$

30. $\sqrt[3]{80b^9}$

31. $\sqrt[3]{56x^8y^3}$

32. $\sqrt[4]{80x^7y^{10}}$

33. $\sqrt[5]{64x^7y^{12}}$

34. $\sqrt[5]{160a^{17}b^9}$

Simplify each radical completely. Assume all variables represent positive real numbers. (*See Objective 2.*)

35. $\sqrt{\dfrac{45}{49}}$

36. $\sqrt{\dfrac{48}{121}}$

37. $\sqrt{\dfrac{32x^5}{9}}$

38. $\sqrt{\dfrac{125y^9}{4}}$

39. $\sqrt{\dfrac{375x^7}{27}}$

40. $\sqrt{\dfrac{196y^5}{18}}$

41. $\sqrt[3]{\dfrac{16r^8}{125}}$

42. $\sqrt[3]{\dfrac{375t^5}{8}}$

Find the distance between the points. Express each answer with simplified radicals. (*See Objective 3.*)

43. $(4, -1)$ and $(-5, -13)$

44. $(10, 0)$ and $(30, -15)$

45. $(3, -11)$ and $(-12, -3)$

46. $(-5, 13)$ and $(-17, 29)$

47. $(6, -4\sqrt{5})$ and $(-1, -5\sqrt{5})$

48. $(9, -3\sqrt{2})$ and $(12, -6\sqrt{2})$

49. $(-9, 5\sqrt{3})$ and $(-13, 3\sqrt{3})$

50. $(-1, 9\sqrt{15})$ and $(4, 10\sqrt{15})$

51. $\left(\dfrac{4}{5}, -\dfrac{1}{5}\right)$ and $\left(\dfrac{12}{5}, \dfrac{14}{5}\right)$

52. $(0, 2)$ and $\left(\dfrac{1}{2}, \dfrac{8}{3}\right)$

53. $(-0.5, 0.1)$ and $(-1.7, 0.6)$

54. $(0.7, -0.6)$ and $(6.7, 0.5)$

55. Use the map to approximate the distance between Missoula, Montana $(-180, 180)$, and Billings, Montana $(90, 90)$. Each tick mark represents 90 mi. Round the answer to two decimal places. (Source: http://www .sitesatlas.com/Maps/index.htm)

56. Use the map to approximate the distance between Ciudad Guayana, Venezuela $(400, 400)$, and Oranjestad, Aruba $(-400, 800)$. Each tick mark represents 400 km. Round the answer to two decimal places. (Source: http:// www.sitesatlas.com/Maps/index.htm)

 Mix 'Em Up!

Simplify each radical. Assume all variables represent positive real numbers.

57. $-\sqrt{48a^6b^4}$

58. $\sqrt{128x^8y^2}$

59. $\sqrt{450h^7k^{11}}$

60. $\sqrt{162m^3n^5}$

61. $\sqrt{\dfrac{75r^5}{16s^4}}$

62. $\sqrt{\dfrac{98a^3}{25b^2}}$

63. $\sqrt[3]{160a^5b^7}$

64. $\sqrt[3]{108c^4d^8}$

65. $\sqrt[5]{-486r^7t^{13}}$

66. $\sqrt[5]{-96x^6y^{11}}$

67. $\sqrt[3]{\dfrac{128x^8}{27y^6}}$

68. $\sqrt[3]{-\dfrac{375r^{11}}{216t^9}}$

69. $\dfrac{\sqrt{192a^7b^2}}{\sqrt{75}}$

70. $\dfrac{\sqrt{98y^5}}{\sqrt{18x^4}}$

71. $\dfrac{\sqrt{24d^7}}{\sqrt{75c^4}}$

72. $\dfrac{\sqrt{56y^7}}{\sqrt{63x^6}}$

Find the distance between the points. Express each answer with simplified radicals.

73. $(-10, 0)$ and $(14, -10)$

74. $(6, -2)$ and $(46, 7)$

75. $(5, -\sqrt{6})$ and $(13, -3\sqrt{6})$

76. $(8, -8\sqrt{5})$ and $(1, -7\sqrt{5})$

77. $\left(1, \dfrac{11}{6}\right)$ and $\left(\dfrac{1}{2}, \dfrac{5}{2}\right)$ **78.** $\left(\dfrac{1}{4}, \dfrac{1}{12}\right)$ and $\left(-\dfrac{37}{12}, -\dfrac{2}{3}\right)$

79. $(3, -10\sqrt{2})$ and $(7, -8\sqrt{2})$

80. $(9, 9\sqrt{3})$ and $(18, 10\sqrt{3})$

81. $(-0.04, 0.02)$ and $(0.05, -0.38)$

82. $(3.5, -7)$ and $(5.5, -8.5)$

83. $\left(\dfrac{1}{2}, \dfrac{2}{3}\right)$ and $\left(-\dfrac{1}{4}, \dfrac{5}{3}\right)$ **84.** $\left(\dfrac{1}{3}, \dfrac{7}{6}\right)$ and $\left(-\dfrac{1}{3}, \dfrac{29}{12}\right)$

 You Be the Teacher!

Correct each student's errors, if any.

85. Simplify the expression $\sqrt{63x^7}$.

Vincent's work:

$$\sqrt{63x^7} = \sqrt{9 \cdot 7x^2x^5}$$
$$= 3x\sqrt{7x^5}$$

86. Simplify the expression $\sqrt[3]{64a^{10}b^4}$.

Emily's work:

$$\sqrt[3]{64a^{10}b^4} = \sqrt{8^2a^{10}b^4} = 8a^5b^2$$

87. Find the distance between the points $(3, -1)$ and $(11, -16)$.

Doris's work:

$$\sqrt{(11 - 3)^2 + (-16 + 1)^2} = \sqrt{(8)^2 + (-15)^2}$$
$$= \sqrt{64 + 225}$$
$$= 8 + 15$$
$$= 23$$

88. Find the distance between the points $(-5, 3)$ and $(-9, 6)$.

Nancy's work:

$$\sqrt{(-9 - 5)^2 + (6 + 3)^2} = \sqrt{(-14)^2 + (9)^2}$$
$$= \sqrt{196 + 81}$$
$$= \sqrt{277}$$

 Calculate It!

Use a calculator to verify the simplification of each radical expression.

89. $\sqrt{54x^5} = 3x^2\sqrt{6x}$ **90.** $\sqrt{\dfrac{98}{25x^4}} = \dfrac{7\sqrt{2}}{5x^2}$

91. $\sqrt[5]{96x^{12}} = 2x^2\sqrt[5]{3x^2}$ **92.** $\sqrt[3]{250x^7} = 5x^2\sqrt[3]{3x}$

| **Adding, Subtracting, and Multiplying Radical Expressions**

▶ OBJECTIVES

As a result of completing this section, you will be able to

1. Add or subtract radicals.
2. Multiply radical expressions.
3. Solve applications.
4. Troubleshoot common errors.

A homeowner wants to fence in her rectangular garden to keep animals out. The dimensions of the fenced area are $5\sqrt{2}$ ft by $10\sqrt{2}$ ft. How many linear feet of fencing will the homeowner need? To answer this question, we need to find the perimeter of the fenced area, which is

$$P = 5\sqrt{2} + 5\sqrt{2} + 10\sqrt{2} + 10\sqrt{2}$$

To simplify this expression, we must add radical expressions. In this section we will learn this and other operations of radicals.

Adding and Subtracting Radical Expressions

Objective 1 ▶

Add or subtract radicals.

To add radical expressions together, they must be of the same form. Recall we can only add variable expressions that have like terms. For instance,

$$2x + 4x = (2 + 4)x = 6x$$

This same rule applies to radicals. We can only add them if they are *like radicals*.

$$2\sqrt{5} + 4\sqrt{5} = (2 + 4)\sqrt{5} = 6\sqrt{5}$$

> **Definition:** **Like radicals** are radicals with the same index and the same radicand.

Some examples of like and unlike radicals are shown.

Like Radicals	**Unlike Radicals**
$2\sqrt{5}, 4\sqrt{5}$	$2\sqrt{3}, 3\sqrt{5}$
$-\sqrt{6}, \sqrt{6}$	$\sqrt{6}, \sqrt[3]{6}$

Just because radicals are unlike does not mean that we cannot combine them. Consider, for example, $\sqrt{8}$ and $\sqrt{50}$. These radicals are not simplified since the radicands contain perfect square factors. When simplified, $\sqrt{8} = \sqrt{4 \cdot 2} = 2\sqrt{2}$ and $\sqrt{50} = \sqrt{25 \cdot 2} = 5\sqrt{2}$. So, after simplification, the radicals are alike and can therefore be added or subtracted.

> **Procedure:** **Adding or Subtracting Radical Expressions**
>
> **Step 1:** Use the product rule or quotient rule for radicals to simplify each radical, if necessary.
> **Step 2:** Add or subtract like radicals by adding their coefficients and keeping the radical the same.

Objective 1 Examples | Perform the indicated operation.

1a. $\sqrt{6} + \sqrt{6}$ **1b.** $\sqrt{15} - 3\sqrt{15}$ **1c.** $4\sqrt[3]{5} + 3\sqrt{5}$

1d. $\sqrt{12} + 2\sqrt{48} - \sqrt{75}$ **1e.** $\sqrt[3]{16} + 5\sqrt[3]{54}$ **1f.** $-3x\sqrt{4x} + 2\sqrt{x^3}$

1g. $\sqrt[3]{\dfrac{16a^4}{27}} - 4a\sqrt[3]{54a}$

Solutions | **1a.**

$$\sqrt{6} + \sqrt{6} = 1\sqrt{6} + 1\sqrt{6}$$ Recall $\sqrt{6} = 1\sqrt{6}$.

$$= (1 + 1)\sqrt{6}$$ Add the coefficients.

$$= 2\sqrt{6}$$ Simplify.

1b. $\sqrt{15} - 3\sqrt{15} = 1\sqrt{15} - 3\sqrt{15}$ Recall $\sqrt{15} = 1\sqrt{15}$.

$\qquad\qquad\qquad = (1-3)\sqrt{15}$ Subtract the coefficients.

$\qquad\qquad\qquad = -2\sqrt{15}$ Simplify.

1c. The radicals, $\sqrt[3]{5}$ and $\sqrt{5}$, are not like radicals since their indices are different. So, this expression cannot be simplified any further.

1d. $\sqrt{12} + 2\sqrt{48} - \sqrt{75}$

$\qquad = \sqrt{4 \cdot 3} + 2\sqrt{16 \cdot 3} - \sqrt{25 \cdot 3}$ Find perfect square factors.

$\qquad = \sqrt{4}\sqrt{3} + 2\sqrt{16}\sqrt{3} - \sqrt{25}\sqrt{3}$ Apply the product rule.

$\qquad = 2\sqrt{3} + 2(4)\sqrt{3} - 5\sqrt{3}$ Simplify each square root.

$\qquad = 2\sqrt{3} + 8\sqrt{3} - 5\sqrt{3}$ Multiply the coefficients.

$\qquad = (2 + 8 - 5)\sqrt{3}$ Add like radicals.

$\qquad = 5\sqrt{3}$ Simplify.

1e. $\sqrt[3]{16} + 5\sqrt[3]{54} = \sqrt[3]{8 \cdot 2} + 5\sqrt[3]{27 \cdot 2}$ Find perfect cube factors.

$\qquad\qquad\qquad = \sqrt[3]{8}\sqrt[3]{2} + 5\sqrt[3]{27}\sqrt[3]{2}$ Apply the product rule.

$\qquad\qquad\qquad = 2\sqrt[3]{2} + 5(3)\sqrt[3]{2}$ Simplify each cube root.

$\qquad\qquad\qquad = 2\sqrt[3]{2} + 15\sqrt[3]{2}$ Multiply the coefficients.

$\qquad\qquad\qquad = (2 + 15)\sqrt[3]{2}$ Add like radicals.

$\qquad\qquad\qquad = 17\sqrt[3]{2}$ Simplify.

1f. $-3x\sqrt{4x} + 2\sqrt{x^3} = -3x\sqrt{4 \cdot x} + 2\sqrt{x^2 \cdot x}$ Find perfect square factors.

$\qquad\qquad\qquad = -3x\sqrt{4}\sqrt{x} + 2\sqrt{x^2}\sqrt{x}$ Apply the product rule.

$\qquad\qquad\qquad = -3x(2)\sqrt{x} + 2x\sqrt{x}$ Simplify each square root.

$\qquad\qquad\qquad = -6x\sqrt{x} + 2x\sqrt{x}$ Multiply the coefficients.

$\qquad\qquad\qquad = (-6x + 2x)\sqrt{x}$ Add like radicals.

$\qquad\qquad\qquad = -4x\sqrt{x}$ Simplify.

1g. $\sqrt[3]{\dfrac{16a^4}{27}} - 4a\sqrt[3]{54a}$

$\quad = \dfrac{\sqrt[3]{16a^4}}{\sqrt[3]{27}} - 4a\sqrt[3]{27 \cdot 2a}$ Apply the quotient rule for radicals and find a perfect cube factor of $54a$.

$\quad = \dfrac{\sqrt[3]{8a^3 \cdot 2a}}{3} - 4a\sqrt[3]{27}\sqrt[3]{2a}$ Find a perfect cube factor of $16a^4$ and apply the product rule for radicals.

$\quad = \dfrac{\sqrt[3]{8a^3}\sqrt[3]{2a}}{3} - 4a(3)\sqrt[3]{2a}$ Apply the product rule for radicals and simplify $\sqrt[3]{27}$.

$\quad = \dfrac{2a\sqrt[3]{2a}}{3} - 12a\sqrt[3]{2a}$ Simplify $\sqrt[3]{8a^3}$ and multiply the coefficients.

$\quad = \dfrac{2a\sqrt[3]{2a}}{3} - \dfrac{36a\sqrt[3]{2a}}{3}$ Convert the second expression to a fraction with the common denominator, 3.

$\quad = \dfrac{2a\sqrt[3]{2a} - 36a\sqrt[3]{2a}}{3}$ Combine the fractions.

$$= \frac{(2a - 36a)\sqrt[3]{2a}}{3}$$ Subtract the like radicals.

$$= -\frac{34a\sqrt[3]{2a}}{3}$$ Simplify.

✔ **Student Check 1**   Perform the indicated operation.

a. $\sqrt{10} + \sqrt{10}$ **b.** $\sqrt{7} - 5\sqrt{7}$ **c.** $6\sqrt[3]{2} + 4\sqrt{2}$

d. $\sqrt{24} + 2\sqrt{54} - \sqrt{6}$ **e.** $\sqrt[3]{24} + 9\sqrt[3]{81}$ **f.** $-7y\sqrt{16y} + 4\sqrt{y^3}$

g. $\sqrt{\dfrac{20b^5}{49}} - 2b^2\sqrt{45b}$

Multiplying Radical Expressions

Objective 2 ▶

Multiply radical expressions.

In Section 10.3, the product rule for radicals was introduced to enable us to simplify radical expressions. The rule stated that, for $\sqrt[n]{a}$ and $\sqrt[n]{b}$ real numbers,

$$\sqrt[n]{a \cdot b} = \sqrt[n]{a} \cdot \sqrt[n]{b}$$

Rewriting this property provides us with a method to multiply radicals.

> **Property: Product Rule for Radicals**
>
> For $\sqrt[n]{a}$ and $\sqrt[n]{b}$ real numbers,
>
> $$\sqrt[n]{a} \cdot \sqrt[n]{b} = \sqrt[n]{a \cdot b}$$

The product rule states that the product of nth roots is the nth root of the product of the radicands. Note that the radicals must have the same index to apply this rule. This property, along with the distributive property, enables us to multiply radical expressions.

> **Procedure: Multiplying Radical Expressions**
>
> **Step 1:** Use the distributive property to eliminate parentheses, if needed.
> **Step 2:** Multiply two radicals with the same index by forming the root of the product of the radicands.
> **Step 3:** Simplify the product.
> **Step 4:** Simplify the radical, if necessary.

Objective 2 Examples   **Simplify each product and write each answer in simplest radical form.**

2a. $\sqrt{2} \cdot \sqrt{6}$ **2b.** $(3\sqrt{7})(2\sqrt{8})$ **2c.** $\sqrt[3]{4x^2} \cdot \sqrt[3]{2x}$ **2d.** $\sqrt{6}(2 + \sqrt{3})$

2e. $(2\sqrt{3} + \sqrt{5})(\sqrt{3} - 2\sqrt{5})$ **2f.** $(2 + \sqrt{7})^2$ **2g.** $(2 + \sqrt{7})(2 - \sqrt{7})$

Solutions   **2a.** $\sqrt{2} \cdot \sqrt{6} = \sqrt{2 \cdot 6}$ Apply the product rule.

$= \sqrt{12}$ Simplify the radicand.

$= \sqrt{4 \cdot 3}$ Find a perfect square factor of 12.

$= \sqrt{4} \cdot \sqrt{3}$ Apply the product rule.

$= 2\sqrt{3}$ Simplify.

2b. $(3\sqrt{7})(2\sqrt{8}) = 3 \cdot 2 \cdot \sqrt{7}\sqrt{8}$ Apply the commutative property.

$= 6\sqrt{7 \cdot 8}$ Multiply the coefficients and apply the product rule.

$= 6\sqrt{56}$ Simplify the radicand.

$= 6\sqrt{4 \cdot 14}$ Find a perfect square factor of 56.

$$= 6\sqrt{4}\sqrt{14} \qquad \text{Apply the product rule.}$$
$$= 6(2)\sqrt{14} \qquad \text{Simplify the square root.}$$
$$= 12\sqrt{14} \qquad \text{Multiply the coefficients.}$$

2c. $\sqrt[3]{4x^2} \cdot \sqrt[3]{2x} = \sqrt[3]{4x^2 \cdot 2x}$ 　　　 Apply the product rule.

$$= \sqrt[3]{8x^3} \qquad \text{Simplify the radicand.}$$
$$= 2x \qquad \text{Simplify.}$$

2d. $\sqrt{6}(2 + \sqrt{3}) = \sqrt{6}(2) + \sqrt{6}(\sqrt{3})$ 　　 Apply the distributive property.

$$= 2\sqrt{6} + \sqrt{6 \cdot 3} \qquad \text{Apply the product rule.}$$
$$= 2\sqrt{6} + \sqrt{18} \qquad \text{Simplify the radicand.}$$
$$= 2\sqrt{6} + \sqrt{9 \cdot 2} \qquad \text{Find a perfect square factor of 18.}$$
$$= 2\sqrt{6} + \sqrt{9}\sqrt{2} \qquad \text{Apply the product rule.}$$
$$= 2\sqrt{6} + 3\sqrt{2} \qquad \text{Simplify the square root.}$$

2e. Apply the FOIL method to multiply the two radical expressions.

$$(2\sqrt{3} + \sqrt{5})(\sqrt{3} - 2\sqrt{5}) = 2\sqrt{3}\sqrt{3} - 2\sqrt{3}(2\sqrt{5}) + \sqrt{5}\sqrt{3} - \sqrt{5}(2\sqrt{5})$$
$$= 2\sqrt{9} - 4\sqrt{15} + \sqrt{15} - 2\sqrt{25}$$
$$= 2(3) + (-4 + 1)\sqrt{15} - 2(5)$$
$$= 6 - 3\sqrt{15} - 10$$
$$= -4 - 3\sqrt{15}$$

2f. A binomial can be squared by applying the FOIL method or by applying the rule for squaring a binomial.

Method 1	**Method 2**
$(2 + \sqrt{7})^2 = (2 + \sqrt{7})(2 + \sqrt{7})$	Recall $(a + b)^2 = a^2 + 2ab + b^2$.
$= 2 \cdot 2 + 2\sqrt{7} + 2\sqrt{7} + \sqrt{7}\sqrt{7}$	$(2 + \sqrt{7})^2 = 2^2 + 2 \cdot 2\sqrt{7} + (\sqrt{7})^2$
$= 4 + 4\sqrt{7} + \sqrt{49}$	$= 4 + 4\sqrt{7} + 7$
$= 4 + 4\sqrt{7} + 7$	$= 11 + 4\sqrt{7}$
$= 11 + 4\sqrt{7}$	

2g. Conjugates can be multiplied by applying the FOIL method or by applying the product of conjugates property.

Method 1	**Method 2**
$(2 + \sqrt{7})(2 - \sqrt{7}) = 2 \cdot 2 - 2\sqrt{7} + 2\sqrt{7} - \sqrt{7}\sqrt{7}$	Recall $(a + b)(a - b) = a^2 - b^2$.
$= 4 + 0\sqrt{7} - \sqrt{49}$	$(2 + \sqrt{7})(2 - \sqrt{7}) = 2^2 - (\sqrt{7})^2$
$= 4 - 7$	$= 4 - 7$
$= -3$	$= -3$

✓ Student Check 2 　 Simplify each product and write each answer in simplest radical form.

　　a. $\sqrt{8} \cdot \sqrt{3}$ 　　**b.** $(2\sqrt{6})(5\sqrt{15})$ 　　**c.** $\sqrt[4]{9x^3} \cdot \sqrt[4]{9x}$ 　　**d.** $\sqrt{3}(4 - \sqrt{15})$

　　e. $(7\sqrt{2} + \sqrt{6})(2\sqrt{2} - 3\sqrt{6})$ 　　**f.** $(3 + \sqrt{5})^2$ 　　**g.** $(3 + \sqrt{5})(3 - \sqrt{5})$

Applications

Objective 3 ▶

Solve applications.

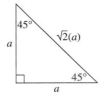

The applications for adding, subtracting, and multiplying radicals that we will solve involve perimeter and area formulas and finding sides of special right triangles. We will examine two special triangles. One is a 45°-45°-90° triangle and the other is a 30°-60°-90° triangle.

The **45°-45°-90° triangle** is a triangle with angles of 45°, 45°, and 90°. The legs of this type of triangle are equal and the length of the hypotenuse of the triangle is $\sqrt{2}$ times the length of a leg.

The **30°-60°-90° triangle** is a triangle with angles of 30°, 60°, and 90°. If a is the length of the side opposite the 30° angle, then the length of the other leg is $\sqrt{3}$ times the value of a, and the length of the hypotenuse is twice the value of a.

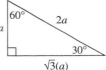

Objective 3 Examples

Solve each problem. Write each answer in simplest radical form and approximate to two decimal places.

3a. Find the perimeter and area of the triangle in Figure 10.1.

Solution **3a.**

$P = 3\sqrt{2} + 3\sqrt{2} + 6$ Add the lengths of each side.

$P = (3 + 3)\sqrt{2} + 6$ Add like radicals.

$P = 6\sqrt{2} + 6$ Simplify.

So, the perimeter of the triangle is $6\sqrt{2} + 6$ in. or approximately 14.49 in.

Figure 10.1

$3\sqrt{2}$ in.   6 in. $3\sqrt{2}$ in.

$A = \dfrac{1}{2}bh$ State the area formula.

$A = \dfrac{1}{2}(3\sqrt{2})(3\sqrt{2})$ Substitute $3\sqrt{2}$ for b and h.

$A = \dfrac{1}{2}(3)(3)\sqrt{2}\sqrt{2}$ Apply the commutative property.

$A = \dfrac{1}{2}(9)\sqrt{2 \cdot 2}$ Multiply the coefficients and apply the product rule for radicals.

$A = \dfrac{1}{2}(9)\sqrt{4}$ Simplify the radicand.

$A = \dfrac{1}{2}(9)(2)$ Simplify the square root.

$A = \dfrac{1}{2}(18)$ Multiply.

$A = 9$ Simplify.

So, the area of the triangle is 9 in².

3b. Find the length of a room's diagonal if the room measures (i) 12 ft by 12 ft and (ii) $12\sqrt{3}$ ft by $12\sqrt{3}$ ft.

Solution **3b.** (i) A square cut by a diagonal produces two 45°-45°-90° right triangles. So, a square room cut in half by the diagonal is an example of a 45°-45°-90° triangle.

The hypotenuse of the right triangle is $\sqrt{2}$ times the length of the leg. Therefore, the length of the diagonal is

$$h = \sqrt{2}(12) = 12\sqrt{2}$$

So, the room's diagonal is $12\sqrt{2}$ ft or approximately 16.97 ft.

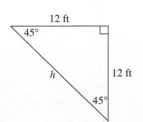

(ii) If the room is $12\sqrt{3}$ ft by $12\sqrt{3}$ ft, the diagonal of the room is $\sqrt{2}$ times the length of the leg. Therefore, the length of the diagonal is

$$h = \sqrt{2}(12\sqrt{3}) = 12\sqrt{2}\sqrt{3} = 12\sqrt{6}$$

So, the room's diagonal is $12\sqrt{6}$ ft or approximately 29.39 ft.

Note: *Since we are working with right triangles, we can check our work by using the Pythagorean theorem, $a^2 + b^2 = c^2$.*

$$a^2 + b^2 = c^2 \rightarrow (12\sqrt{3})^2 + (12\sqrt{3})^2 = (12\sqrt{6})^2$$

$$144(3) + 144(3) = 144(6)$$

$$432 + 432 = 864$$

$$864 = 864$$

Since this is a true statement, our work is correct.

3c. The triangle shown in Figure 10.2 is a 30°-60°-90° triangle. Find the length of the other leg and the hypotenuse.

Solution   **3c.** The side opposite the 30° angle is given, $a = 5\sqrt{15}$. The hypotenuse is twice the value of a. So,

$$c = 2a$$
$$c = 2(5\sqrt{15})$$
$$c = 10\sqrt{15}$$

60°
c
$5\sqrt{15}$ cm
30°
b

Figure 10.2

The leg opposite the 60° angle is $b = \sqrt{3}(a)$. So,

$b = \sqrt{3}(a)$	State the relationship between a and b.
$b = \sqrt{3}(5\sqrt{15})$	Replace a with $5\sqrt{15}$.
$b = 5\sqrt{45}$	Multiply the radicals.
$b = 5\sqrt{9 \cdot 5}$	Find a perfect square factor of 45.
$b = 5(3)\sqrt{5}$	Apply the product rule.
$b = 15\sqrt{5}$	Simplify.

Check: $a^2 + b^2 = c^2 \rightarrow (5\sqrt{15})^2 + (15\sqrt{5})^2 = (10\sqrt{15})^2$

$$25(15) + 225(5) = 100(15)$$
$$375 + 1125 = 1500$$
$$1500 = 1500$$

So, the other leg is $15\sqrt{5}$ cm, or approximately 33.54 cm, and the hypotenuse is $10\sqrt{15}$ cm, or approximately 38.73 cm.

✓ **Student Check 3**   Solve each problem. Write each radical in simplest radical form and approximate to two decimal places.

   a. Find the perimeter and area of the triangle shown

   b. Find the length of the hypotenuse in a 45°-45°-90° triangle if the length of each leg is $2\sqrt{10}$ in.

   c. In a 30°-60°-90° triangle, the leg opposite the 30° angle measures $2\sqrt{6}$ ft. Find the length of the other leg and the hypotenuse.

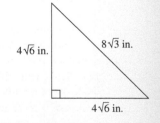

$4\sqrt{6}$ in.

$8\sqrt{3}$ in.

$4\sqrt{6}$ in.

Objective 4 ▶	**Troubleshooting Common Errors**
Troubleshoot common errors.	Some common errors associated with adding, subtracting, and multiplying radicals are shown.

Objective 4 Examples	A problem and an incorrect solution are given. Provide the correct solution and an explanation of the error.

4a. Add: $\sqrt{3} + \sqrt{3}$.

Incorrect Solution	**Correct Solution and Explanation**
$\sqrt{3} + \sqrt{3} = \sqrt{6}$	To add like radicals, we add the coefficients of the radicals, not the radicands. $$\sqrt{3} + \sqrt{3} = 1\sqrt{3} + 1\sqrt{3} = 2\sqrt{3}$$

4b. Simplify: $\left(4\sqrt{7}\right)^2$.

Incorrect Solution	**Correct Solution and Explanation**
$\left(4\sqrt{7}\right)^2 = 4(7) = 28$	Because we have a product squared, we must square each factor. $$\left(4\sqrt{7}\right)^2 = 4^2\left(\sqrt{7}\right)^2 = 16(7) = 112$$

4c. Simplify: $\left(\sqrt{3} + \sqrt{5}\right)^2$.

Incorrect Solution	**Correct Solution and Explanation**
$\left(\sqrt{3} + \sqrt{5}\right)^2 = 3 + 5 = 8$	We cannot square each term in the binomial. We must use the FOIL method or the squaring pattern. $$\left(\sqrt{3} + \sqrt{5}\right)^2 = \left(\sqrt{3}\right)^2 + 2\left(\sqrt{3}\right)\left(\sqrt{5}\right) + \left(\sqrt{5}\right)^2$$ $$= 3 + 2\sqrt{15} + 5$$ $$= 8 + 2\sqrt{15}$$

ANSWERS TO STUDENT CHECKS

Student Check 1   a. $2\sqrt{10}$ **b.** $-4\sqrt{7}$ **c.** $3\sqrt[3]{5}$
  d. can't be combined **e.** $7\sqrt{6}$ **f.** $29\sqrt[3]{3}$
  g. $-24y\sqrt{y}$ **h.** $-\dfrac{40b^2\sqrt{5b}}{7}$

Student Check 2   a. $2\sqrt{6}$ **b.** $30\sqrt{10}$ **c.** $3x$
  d. $4\sqrt{3} - 3\sqrt{5}$ **e.** $10 - 38\sqrt{3}$ **f.** $14 + 6\sqrt{5}$ **g.** 4

Student Check 3   a. $8\sqrt{6} + 8\sqrt{3}$ in. or 33.45 in.; 48 in.^{2}
  b. $4\sqrt{5}$ in.   **c.** $6\sqrt{2}$ ft; $4\sqrt{6}$ ft

SUMMARY OF KEY CONCEPTS

1. Like radicals are radicals with the same index and same radicand. Like radicals can be added by adding their coefficients and keeping the radical the same. If the radicals are unlike, try simplifying them according to the methods shown in Section 10.3. After simplifying the radicals, combine them, if possible.

2. Multiply radicals using the product rule for radicals. If the indices are the same, multiply the radicands.

3. Special cases of right triangles can be solved using the relationships to the right.

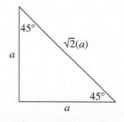

45°-45°-90° triangle

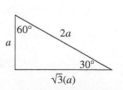

30°-60°-90° triangle

GRAPHING CALCULATOR SKILLS

We can use the graphing calculator to verify that our answers are correct.

Example 1: Verify that $\sqrt{12} + 2\sqrt{48} - \sqrt{75} = 5\sqrt{3}$.

Solution:

```
√(12)+2√(48)-√(7
5)
          8.660254038
5√(3)
          8.660254038
```

The two radical expressions have the same decimal value and are, therefore, equivalent.

Example 2: Verify that $(2 + \sqrt{7})^2 = 11 + 4\sqrt{7}$.

Solution:

```
(2+√(7))²
          21.58300524
11+4√(7)
          21.58300524
```

The two radical expressions have the same decimal value and are, therefore, equivalent.

SECTION 10.4 / EXERCISE SET

 Write About It!

Use complete sentences in your answer to each exercise.

1. Define like radicals.
2. Explain how to add and subtract radical expressions.
3. Explain how to multiply radical expressions.
4. Can you multiply like radicals? Give an example.
5. Explain why $\sqrt{18x} + \sqrt{12x} \neq \sqrt{30x^2} = x\sqrt{30}$.
6. Explain why $\sqrt[3]{10x^2} - \sqrt[3]{2x^2} \neq \sqrt[3]{8x^2}$.
7. Explain how to simplify $(\sqrt{x} + \sqrt{y})^2$.
8. Explain why $\sqrt{a} + \sqrt{a} \neq \sqrt{2a}$.

 Practice Makes Perfect!

Add or subtract. (*See Objective 1.*)

9. $3\sqrt{6} + 2\sqrt{6}$
10. $11\sqrt{5} + \sqrt{5}$
11. $-2\sqrt{7} - 12\sqrt{7}$
12. $3\sqrt{5} - 15\sqrt{5}$
13. $2\sqrt{54} + 7\sqrt{216}$
14. $3\sqrt{125} + 5\sqrt{20}$
15. $10\sqrt{8x^3} - 4\sqrt{18x^3}$
16. $7y^2\sqrt{27y} + 2\sqrt{75y^5}$
17. $-7\sqrt[3]{5} + 6\sqrt[3]{5} - 8\sqrt[3]{5}$
18. $16\sqrt[3]{11} - 9\sqrt[3]{11} - 2\sqrt[3]{11}$
19. $8\sqrt[3]{16} - \sqrt[3]{54} + 18\sqrt[3]{2}$
20. $6\sqrt[3]{40} - 2\sqrt[3]{135} + 8\sqrt[3]{5}$
21. $-15\sqrt[3]{24a^3b} + 9\sqrt[3]{81a^3b} + 2\sqrt[3]{6a^3b}$
22. $4\sqrt[3]{250xy^3} + 8\sqrt[3]{54xy^3} - 7\sqrt[3]{2y^3}$
23. $\dfrac{3\sqrt{5}}{4} + \sqrt{\dfrac{20}{9}}$
24. $\dfrac{3\sqrt{7}}{2} - \sqrt{\dfrac{28}{25}}$
25. $\dfrac{7\sqrt{20}}{3} - \sqrt{\dfrac{125}{9}}$
26. $\dfrac{\sqrt{45}}{4} - \sqrt{\dfrac{80}{9}}$

27. $\dfrac{7\sqrt[3]{24}}{4} - \sqrt[3]{\dfrac{375}{64}}$
28. $\dfrac{\sqrt[3]{80}}{5} - \sqrt[3]{\dfrac{270}{343}}$
29. $\dfrac{\sqrt{50r^3}}{9} + \sqrt{\dfrac{98r^3}{9}}$
30. $\dfrac{s^2\sqrt{300s}}{7} - \sqrt{\dfrac{48s^5}{49}}$

Simplify each product and write each answer in simplest radical form. (*See Objective 2.*)

31. $\sqrt{32} \cdot \sqrt{7}$
32. $\sqrt{18} \cdot \sqrt{3}$
33. $(6\sqrt{20})(2\sqrt{8})$
34. $(5\sqrt{12})(6\sqrt{28})$
35. $(2\sqrt{12x^2})(\sqrt{8x^3})$
36. $(\sqrt{98y})(2\sqrt{27y^2})$
37. $\sqrt[3]{32x}\sqrt[3]{18x^4}$
38. $\sqrt[3]{243y^2}\sqrt[3]{15y^8}$
39. $\sqrt{7}(5 + \sqrt{21})$
40. $\sqrt{5}(16 + \sqrt{15})$
41. $(\sqrt{5} - 2\sqrt{30})(3\sqrt{5} + \sqrt{30})$
42. $(4\sqrt{7} - \sqrt{35})(2\sqrt{7} + \sqrt{35})$
43. $(3\sqrt{2} - \sqrt{6})^2$
44. $(2\sqrt{5} + \sqrt{10})^2$
45. $(1 - 2\sqrt{10})^2$
46. $(5 + 3\sqrt{7})^2$
47. $(2 - \sqrt{6x})(2 + \sqrt{6x})$
48. $(5 - \sqrt{2y})(5 + \sqrt{2y})$
49. $(2\sqrt{3x} - 5\sqrt{2y})(\sqrt{3x} + \sqrt{2y})$
50. $(3\sqrt{7a} + 2\sqrt{5b})(\sqrt{7a} - 6\sqrt{5b})$

Solve each problem. (*See Objective 3.*)

In Exercises 51–54, find the perimeter and area of the triangle shown. Write each answer in simplest radical form and approximate answers to two decimal places.

51. $a = 18\sqrt{2}$ cm and $c = 36$ cm
52. $a = 11\sqrt{2}$ cm and $c = 22$ cm
53. $a = 10\sqrt{14}$ ft and $c = 20\sqrt{7}$ ft
54. $a = 3\sqrt{6}$ ft and $c = 6\sqrt{3}$ ft

55. In a 30°-60°-90° triangle, the leg opposite the 30° angle measures $4\sqrt{21}$ cm. Find the length of the other leg and the hypotenuse.

56. In a 30°-60°-90° triangle, the leg opposite the 30° angle measures $14\sqrt{6}$ m. Find the length of the other leg and the hypotenuse.

 Mix 'Em Up!

Perform each operation and write each answer in simplest radical form.

57. $11\sqrt{6} - 3\sqrt{6} + \sqrt{6}$ **58.** $9\sqrt{5} + 4\sqrt{5} - 12\sqrt{5}$

59. $21\sqrt[3]{24r^3t} - 5\sqrt[3]{81r^3t}$ **60.** $12\sqrt[3]{40xy^6} - 6\sqrt[3]{135xy^6}$

61. $-8\sqrt[3]{16a^3b} + 4\sqrt[3]{54a^3b} + 10\sqrt[3]{3a^3b}$

62. $7\sqrt{108a^2b} - 6\sqrt{75a^2b}$ **63.** $10\sqrt{98cd^3} - 16\sqrt{50cd^3}$

64. $6\sqrt[3]{80xy^3} + 11\sqrt[3]{270xy^3} - 9\sqrt[3]{40xy^3}$

65. $5\sqrt[4]{16x^4y^7} - x\sqrt[4]{81y^7}$ **66.** $12\sqrt[4]{5a^4b^9} - 2a\sqrt[4]{48b^9}$

67. $(5 - \sqrt{3a})(5 + \sqrt{3a})$ **68.** $(3 + \sqrt{5b})(3 - \sqrt{5b})$

69. $-5\sqrt{2x^2} + 8x\sqrt{2} - \sqrt{18x^2}$

70. $7\sqrt{5y^3} - 8\sqrt{20y^3} + 3y\sqrt{45y}$

71. $\dfrac{3\sqrt{75r^2t}}{2} - \sqrt{\dfrac{363r^2t}{49}}$ **72.** $\dfrac{\sqrt{45ab^3}}{2} - \sqrt{\dfrac{80ab^3}{9}}$

73. $\sqrt[3]{\dfrac{375a}{64}} - \dfrac{\sqrt[3]{24a}}{5}$ **74.** $\sqrt[3]{\dfrac{135}{8b^3}} + \dfrac{\sqrt[3]{40}}{3b}$

75. $\sqrt[5]{16x^2y}\,\sqrt[5]{10x^6y^4}$ **76.** $\sqrt[5]{81a^3b^2}\,\sqrt[5]{6a^8b^4}$

77. $(8\sqrt{2} - \sqrt{6})(\sqrt{2} + 3\sqrt{6})$

78. $(\sqrt{20} - \sqrt{3})(2\sqrt{20} + 9\sqrt{3})$

79. $(4\sqrt{3} - \sqrt{15})^2$ **80.** $(3\sqrt{6} + \sqrt{2})^2$

81. $\sqrt{6}(2 - \sqrt{12})$ **82.** $\sqrt{5}(3 + \sqrt{30})$

83. $(\sqrt{5x} - 2\sqrt{3y})(\sqrt{5x} + 4\sqrt{3y})$

84. $(7\sqrt{6a} + 4\sqrt{2b})(\sqrt{6a} - \sqrt{2b})$

In Exercises 85–88, find the perimeter and area of the triangle shown. Write each answer in simplest radical form and approximate answers to two decimal places.

85. $a = 21\sqrt{2}$ m and $c = 42$ m

86. $a = 16\sqrt{2}$ m and $c = 32$ m

87. $a = 18\sqrt{6}$ in. and $c = 36\sqrt{3}$ in.

88. $a = 7\sqrt{10}$ in. and $c = 14\sqrt{5}$ in.

89. In a 30°-60°-90° triangle, the leg opposite the 30° angle measures $5\sqrt{21}$ cm. Find the length of the other leg and the hypotenuse.

90. In a 30°-60°-90° triangle, the leg opposite the 30° angle measures $18\sqrt{6}$ m. Find the length of the other leg and the hypotenuse.

 You Be the Teacher!

Correct each student's errors, if any.

91. Simplify $\sqrt{28x^2y} + \sqrt{63x^2y}$.

Vincent's work:

$$\sqrt{28x^2y} + \sqrt{63x^2y} = \sqrt{4 \cdot 7x^2y} + \sqrt{9 \cdot 7x^2y}$$
$$= 2x\sqrt{7y} + 3x\sqrt{7y}$$
$$= 5x^2\sqrt{49y^2}$$
$$= 35x^2y$$

92. Simplify $3\sqrt{75a^4b} - a^2\sqrt{98b}$.

Daron's work:

$$3\sqrt{75a^4b} - a^2\sqrt{98b} = 3\sqrt{25 \cdot 3a^4b} - a^2\sqrt{49 \cdot 2b}$$
$$= 15a^2\sqrt{3b} - 7a^2\sqrt{2b}$$
$$= 8a^2\sqrt{b}$$

93. Simplify $(2\sqrt{x} - y)^2$.

Doris's work:

$$(2\sqrt{x} - y)^2 = (2\sqrt{x})^2 + (-y)^2$$
$$= 4x + y^2$$

94. Simplify $(\sqrt{7} - \sqrt{3})^2$.

Nancy's work:

$$(\sqrt{7} - \sqrt{3})^2 = (\sqrt{7})^2 - (\sqrt{3})^2$$
$$= 7 - 3$$
$$= 4$$

 Calculate It!

Use a calculator to verify the simplification of each radical expression.

95. $\sqrt{72} + 3\sqrt{50} - 6\sqrt{18} = 3\sqrt{2}$

96. $10\sqrt{48} - 3\sqrt{192} + 6\sqrt{12} = 28\sqrt{3}$

97. $7\sqrt[3]{40} - 2\sqrt[3]{135} = 8\sqrt[3]{5}$

98. $(3\sqrt{12} - 2\sqrt{5})^2 = 128 - 24\sqrt{15}$

PIECE IT TOGETHER   SECTIONS 10.1–10.4

Simplify each expression. Assume all variables represent positive real numbers. (*Section 10.1, Objectives 1–3*)

1. $\sqrt{49a^{12}}$ **2.** $\sqrt[3]{\dfrac{125x^9}{8y^6}}$

3. $\sqrt[5]{-243a^{10}b^{25}}$ **4.** $\sqrt[7]{128a^7b^{21}}$

Simplify each expression. (*Section 10.1, Objective 5*)

5. $\sqrt{(2x + 1)^2}$ **6.** $\sqrt[7]{(5b)^7}$

Graph the radical function using the accompanying table. (*Section 10.1, Objective 7*)

7. $f(x) = \sqrt[3]{x} - 2$

x	–8	–1	0	1	8	12
$f(x)$						

Simplify each expression and write each answer with positive exponents. Assume all variables represent positive real numbers. (*Section 10.2, Objectives 1–4*)

8. $(-16)^{1/2}$

9. $\left(\dfrac{1}{27}\right)^{2/3}$

10. $(64x^9y^6)^{2/3}$

11. $(32x^{10})^{-2/5}$

12. $(a^{-8}b^{12})^{-3/4}$

13. $\dfrac{(x^{5/6}y^{1/5})^3}{y}$

14. $\dfrac{(a^{2/7}b^{1/3})^7}{b^3}$

15. $(12x^{2/5} - 5y^{1/2})(x^{2/5} + y^{1/2})$

16. $(7r^{1/2} - s^{2/3})(7r^{1/2} + s^{2/3})$

Simplify each radical. Assume all variables represent positive real numbers. (*Section 10.3, Objectives 1 and 2*)

17. $3\sqrt{75}$

18. $\sqrt{144wx^5y^6}$

19. $\sqrt[3]{40a^7b^{12}}$

20. $\sqrt[4]{48x^5y^{11}}$

21. $\sqrt{\dfrac{80x^5}{9y^2}}$

Find the distance between the points. Write the answer with simplified radicals. (*Section 10.3, Objective 3*)

22. $\left(-9, 5\sqrt{3}\right)$ and $\left(-13, 3\sqrt{3}\right)$

Perform each operation and write each answer in simplest radical form. (*Section 10.4, Objectives 1 and 2*)

23. $14\sqrt{98} - 6\sqrt{8}$

24. $8\sqrt[3]{16} - \sqrt[3]{54} + 19\sqrt[3]{2}$

25. $\left(2\sqrt{12x^4}\right)\left(\sqrt{8x^3}\right)$

26. $\left(\sqrt{5} - 2\sqrt{30}\right)\left(3\sqrt{5} + \sqrt{30}\right)$

27. $\left(3\sqrt{2} - \sqrt{6}\right)^2$

SECTION 10.5   Dividing Radical Expressions and Rationalizing

▶ OBJECTIVES

As a result of completing this section, you will be able to

1. Divide radical expressions.
2. Rationalize the denominator.
3. Rationalize the denominator using conjugates.
4. Troubleshoot common errors.

Objective 1 ▶

Divide radical expressions.

In Section 10.4, we learned how to add, subtract, and multiply radicals. We will now learn how to divide radicals using the quotient rule for radicals. We will also learn how to rewrite radical expressions by removing the radical from the denominator.

Dividing Radical Expressions

In Section 10.3, the quotient rule for radicals was introduced. The rule stated that if $\sqrt[n]{a}$ and $\sqrt[n]{b}$ are real numbers and $b \neq 0$,

$$\sqrt[n]{\dfrac{a}{b}} = \dfrac{\sqrt[n]{a}}{\sqrt[n]{b}}$$

This property enables us to divide radical expressions.

> **Property: Quotient Rule for Radicals**
>
> For $\sqrt[n]{a}$ and $\sqrt[n]{b}$ real numbers with $b \neq 0$,
>
> $$\dfrac{\sqrt[n]{a}}{\sqrt[n]{b}} = \sqrt[n]{\dfrac{a}{b}}$$

In this section, we will continue to assume that all variables represent positive real numbers.

> **Procedure: Dividing Radical Expressions**
>
> **Step 1:** Rewrite the problem as the nth root of the quotient of the radicands in the numerator and denominator.
>
> **Step 2:** Simplify the quotient.
>
> **Step 3:** Simplify the radical, if necessary.

| **Objective 1 Examples** | **Simplify each quotient and write each answer in simplest radical form.** |

1a. $\dfrac{\sqrt{48}}{\sqrt{3}}$ **1b.** $\dfrac{\sqrt{21x^3}}{\sqrt{3x}}$ **1c.** $\dfrac{\sqrt{40}}{5\sqrt{2}}$ **1d.** $\dfrac{7\sqrt[3]{16x^5}}{\sqrt[3]{2x}}$ **1e.** $\dfrac{\sqrt[4]{64a^5b}}{\sqrt[4]{ab^{-3}}}$

Solutions

1a.
$$\dfrac{\sqrt{48}}{\sqrt{3}} = \sqrt{\dfrac{48}{3}}$$ Apply the quotient rule for radicals.
$$= \sqrt{16}$$ Simplify the radicand.
$$= 4$$ Simplify the square root.

1b.
$$\dfrac{\sqrt{21x^3}}{\sqrt{3x}} = \sqrt{\dfrac{21x^3}{3x}}$$ Apply the quotient rule for radicals.
$$= \sqrt{7x^2}$$ Simplify the radicand.
$$= x\sqrt{7}$$ Simplify: $\sqrt{x^2} = x$.

1c.
$$\dfrac{1\sqrt{40}}{5\sqrt{2}} = \dfrac{1}{5}\sqrt{\dfrac{40}{2}}$$ Apply the quotient rule for radicals.
$$= \dfrac{1}{5}\sqrt{20}$$ Simplify the radicand.
$$= \dfrac{1}{5}\sqrt{4 \cdot 5}$$ Find a perfect square factor of 20.
$$= \dfrac{1}{5}(2)\sqrt{5}$$ Apply the product rule for radicals and simplify.
$$= \dfrac{2}{5}\sqrt{5}$$ Multiply the coefficients.

1d.
$$\dfrac{7\sqrt[3]{16x^5}}{\sqrt[3]{2x}} = 7\sqrt[3]{\dfrac{16x^5}{2x}}$$ Apply the quotient rule for radicals.
$$= 7\sqrt[3]{8x^4}$$ Simplify the radicand.
$$= 7\sqrt[3]{8x^3}\sqrt[3]{x}$$ Find a perfect square factor of $8x^4$.
$$= 7(2x)\sqrt[3]{x}$$ Simplify.
$$= 14x\sqrt[3]{x}$$ Multiply the coefficients.

1e.
$$\dfrac{\sqrt[4]{64a^5b}}{\sqrt[4]{ab^{-3}}} = \sqrt[4]{\dfrac{64a^5b}{ab^{-3}}}$$ Apply the quotient rule for radicals.
$$= \sqrt[4]{64a^{5-1}b^{1-(-3)}}$$ Apply the quotient of like bases rule.
$$= \sqrt[4]{64a^4b^4}$$ Simplify.
$$= \sqrt[4]{16a^4b^4}\sqrt[4]{4}$$ Find a perfect fourth factor of $64a^4b^4$.
$$= 2ab\sqrt[4]{4}$$ Simplify.

| ✔ **Student Check 1** | **Simplify each quotient and write each answer in simplest radical form.** |

a. $\dfrac{\sqrt{50}}{\sqrt{2}}$ **b.** $\dfrac{\sqrt{30y^5}}{\sqrt{10y}}$ **c.** $\dfrac{\sqrt{24}}{4\sqrt{3}}$ **d.** $\dfrac{5\sqrt[3]{54y^7}}{\sqrt[3]{2y}}$ **e.** $\dfrac{\sqrt[4]{48a^9b}}{\sqrt[4]{ab^{-3}}}$

Rationalizing the Denominator

| **Objective 2** ▶ |
| **Rationalize the denominator.** |

Now we examine quotients, such as $\dfrac{\sqrt{2}}{\sqrt{3}}$, in which the radicand in the denominator does not divide evenly into the radicand in the numerator. Such a quotient is not in simplest radical form since there is either a radical in the denominator or a fraction in the radical. Our goal is to rewrite the quotient as an equivalent expression that doesn't contain a radical in the denominator.

The process of removing the radical from the denominator of a fraction is called **rationalizing the denominator**. It is called rationalizing because the original denominator is an irrational number but after we go through this process, the denominator is a rational number. We will use the following facts.

- Multiplying a fraction by a form of 1 produces an equivalent fraction.

$$\frac{a}{b} \cdot \frac{c}{c} = \frac{a \cdot c}{b \cdot c}$$

- An nth root raised to an exponent of n removes the radical.

$$\left(\sqrt[n]{x}\right)^n = x, \text{ provided } \sqrt[n]{x} \text{ is a real number}$$

Procedure: Rationalizing the Denominator

Step 1: Multiply the fraction by a form of 1.
  a. If the radical in the denominator is a square root, multiply the numerator and denominator by the square root expression that makes the radicand in the denominator a perfect square.
  b. If the radical in the denominator is a cube root, multiply the numerator and denominator by the cube root expression that makes the radicand in the denominator a perfect cube.
  c. If the radical in the denominator is an nth root, multiply the numerator and denominator by the nth root expression that makes the radicand in the denominator a perfect nth.

Step 2: Simplify the products in the numerator and denominator.

Objective 2 Examples   Simplify each expression by rationalizing the denominator.

2a. $\dfrac{\sqrt{3}}{\sqrt{7}}$ **2b.** $\dfrac{1}{\sqrt{2x}}$ **2c.** $\dfrac{2}{\sqrt[3]{9}}$ **2d.** $\dfrac{6\sqrt[4]{5}}{\sqrt[4]{2}}$

Solutions   **2a.** $\dfrac{\sqrt{3}}{\sqrt{7}} = \dfrac{\sqrt{3}}{\sqrt{7}} \cdot \dfrac{\sqrt{7}}{\sqrt{7}}$ Multiply by a form of 1, $\dfrac{\sqrt{7}}{\sqrt{7}}$.

$\quad = \dfrac{\sqrt{3} \cdot \sqrt{7}}{\sqrt{7} \cdot \sqrt{7}}$ Multiply the numerators and denominators.

$\quad = \dfrac{\sqrt{21}}{\sqrt{49}}$ Apply the product rule for radicals.

$\quad = \dfrac{\sqrt{21}}{7}$ Simplify the denominator.

2b. $\dfrac{1}{\sqrt{2x}} = \dfrac{1}{\sqrt{2x}} \cdot \dfrac{\sqrt{2x}}{\sqrt{2x}}$ Multiply by a form of 1, $\dfrac{\sqrt{2x}}{\sqrt{2x}}$.

$\quad = \dfrac{1 \cdot \sqrt{2x}}{\sqrt{2x} \cdot \sqrt{2x}}$ Multiply the numerators and denominators.

$\quad = \dfrac{\sqrt{2x}}{\sqrt{4x^2}}$ Apply the product rule for radicals.

$\quad = \dfrac{\sqrt{2x}}{2x}$ Simplify the denominator.

2c. We must make the radicand in the denominator, 9, a perfect cube. Since 27 is a perfect cube, we use the factor $\sqrt[3]{3}$.

$$\frac{2}{\sqrt[3]{9}} = \frac{2}{\sqrt[3]{9}} \cdot \frac{\sqrt[3]{3}}{\sqrt[3]{3}}$$ Multiply by a form of 1, $\frac{\sqrt[3]{3}}{\sqrt[3]{3}}$.

$$= \frac{2 \cdot \sqrt[3]{3}}{\sqrt[3]{9} \cdot \sqrt[3]{3}}$$ Multiply the numerators and denominators.

$$= \frac{2\sqrt[3]{3}}{\sqrt[3]{27}}$$ Apply the product rule for radicals.

$$= \frac{2\sqrt[3]{3}}{3}$$ Simplify the denominator.

2d. We must make the radicand in the denominator, 2, a perfect fourth. Since 16 is a perfect fourth, we use the factor $\sqrt[4]{8}$.

$$\frac{6\sqrt[4]{5}}{\sqrt[4]{2}} = \frac{6\sqrt[4]{5}}{\sqrt[4]{2}} \cdot \frac{\sqrt[4]{8}}{\sqrt[4]{8}}$$ Multiply by a form of 1, $\frac{\sqrt[4]{8}}{\sqrt[4]{8}}$.

$$= \frac{6\sqrt[4]{5} \cdot \sqrt[4]{8}}{\sqrt[4]{2} \cdot \sqrt[4]{8}}$$ Multiply the numerators and denominators.

$$= \frac{6\sqrt[4]{40}}{\sqrt[4]{16}}$$ Apply the product rule for radicals.

$$= \frac{6\sqrt[4]{40}}{2}$$ Simplify the denominator.

$$= 3\sqrt[4]{40}$$ Simplify the fraction.

✔ Student Check 2 Simplify each expression by rationalizing the denominator.

a. $\frac{\sqrt{5}}{\sqrt{6}}$ **b.** $\frac{\sqrt{10}}{\sqrt{3x}}$ **c.** $\frac{5}{\sqrt[3]{2}}$ **d.** $\frac{7\sqrt[5]{3}}{\sqrt[5]{2}}$

Rationalizing the Denominator Using Conjugates

Objective 3 ▶

Rationalize the denominator using conjugates.

If the denominator of a fraction contains a sum or difference of square roots, then the form of 1 that must be used to rationalize the denominator involves the *conjugate* of the denominator. Recall that **conjugates** are binomial expressions that only differ in the sign of a term. Two important facts are

- $a + b$ and $a - b$ are conjugates.
- $(a + b)(a - b) = a^2 - b^2$

In general, the product of conjugates will not contain any radical terms.

Procedure: Rationalizing the Denominator Using Conjugates

Step 1: Multiply the fraction by a form of 1. The form of 1 is the conjugate of the denominator divided by itself.
Step 2: Multiply the numerators and denominators.
Step 3: Simplify the expressions in the numerators and denominators.

Objective 3 Examples Simplify each expression. Rationalize the denominators using conjugates.

3a. $\frac{6}{\sqrt{2} + \sqrt{3}}$ **3b.** $\frac{4 + \sqrt{5}}{3 - \sqrt{2}}$

Solutions **3a.** The conjugate of the denominator, $\sqrt{2} + \sqrt{3}$, is $\sqrt{2} - \sqrt{3}$.

$$\frac{6}{\sqrt{2} + \sqrt{3}} = \frac{6}{\sqrt{2} + \sqrt{3}} \cdot \frac{\sqrt{2} - \sqrt{3}}{\sqrt{2} - \sqrt{3}}$$ Multiply by a form of 1, $\frac{\sqrt{2} - \sqrt{3}}{\sqrt{2} - \sqrt{3}}$.

$$= \frac{6(\sqrt{2} - \sqrt{3})}{(\sqrt{2} + \sqrt{3})(\sqrt{2} - \sqrt{3})}$$ Multiply the numerators and denominators.

$$= \frac{6\sqrt{2} - 6\sqrt{3}}{(\sqrt{2})^2 - (\sqrt{3})^2}$$ Apply the distributive property and the product of conjugates property.

$$= \frac{6\sqrt{2} - 6\sqrt{3}}{2 - 3}$$ Simplify.

$$= \frac{6\sqrt{2} - 6\sqrt{3}}{-1}$$ Combine the terms in the denominator.

$$= -6\sqrt{2} + 6\sqrt{3}$$ Simplify the fraction.

3b. The conjugate of $3 - \sqrt{2}$ is $3 + \sqrt{2}$.

$$\frac{4 + \sqrt{5}}{3 - \sqrt{2}} = \frac{4 + \sqrt{5}}{3 - \sqrt{2}} \cdot \frac{3 + \sqrt{2}}{3 + \sqrt{2}}$$ Multiply by a form of 1, $\frac{3 + \sqrt{2}}{3 + \sqrt{2}}$.

$$= \frac{(4 + \sqrt{5})(3 + \sqrt{2})}{(3 - \sqrt{2})(3 + \sqrt{2})}$$ Multiply the numerators and denominators.

$$= \frac{12 + 4\sqrt{2} + 3\sqrt{5} + \sqrt{10}}{(3)^2 - (\sqrt{2})^2}$$ Apply the FOIL method and the product of conjugates property.

$$= \frac{12 + 4\sqrt{2} + 3\sqrt{5} + \sqrt{10}}{9 - 2}$$ Simplify.

$$= \frac{12 + 4\sqrt{2} + 3\sqrt{5} + \sqrt{10}}{7}$$ Combine the terms in the denominator.

✔**Student Check 3**   Simplify each expression. Rationalize the denominators using conjugates.

a. $\dfrac{5}{\sqrt{6} - \sqrt{2}}$ **b.** $\dfrac{7 + \sqrt{3}}{5 - \sqrt{7}}$

Objective 4 ▶

Troubleshoot common errors.

Troubleshooting Common Errors

Some common errors with dividing radicals and rationalizing denominators are shown.

Objective 4 Examples   **A problem and an incorrect solution are given. Provide the correct solution and an explanation of the error.**

4a. Simplify $\dfrac{\sqrt{6}}{3}$.

Incorrect Solution	Correct Solution and Explanation
$\dfrac{\sqrt{6}}{3} = \sqrt{2}$	The denominator does not have a radical, so we cannot simplify the quotient. $\dfrac{\sqrt{6}}{3} = \dfrac{1}{3}\sqrt{6}$

4b. Simplify by rationalizing the denominator. $\dfrac{\sqrt[3]{5}}{\sqrt[3]{2}}$.

Incorrect Solution	Correct Solution and Explanation
$\dfrac{\sqrt[3]{5}}{\sqrt[3]{2}} = \dfrac{\sqrt[3]{5}}{\sqrt[3]{2}} \cdot \dfrac{\sqrt[3]{2}}{\sqrt[3]{2}} = \dfrac{\sqrt[3]{10}}{2}$	We need to multiply by an expression that will make the denominator a perfect cube. The radicand of the denominator has a single factor of 2. So, we need to get two more factors of 2, or $2^2 = 4$. $$\dfrac{\sqrt[3]{5}}{\sqrt[3]{2}} = \dfrac{\sqrt[3]{5}}{\sqrt[3]{2}} \cdot \dfrac{\sqrt[3]{4}}{\sqrt[3]{4}} = \dfrac{\sqrt[3]{20}}{\sqrt[3]{8}} = \dfrac{\sqrt[3]{20}}{2}$$

ANSWERS TO STUDENT CHECKS

Student Check 1 **a.** 5 **b.** $y^2\sqrt{3}$ **c.** $\dfrac{\sqrt{2}}{2}$

d. $15y^2$ **e.** $2a^2b\sqrt[4]{3}$

Student Check 2 **a.** $\dfrac{\sqrt{30}}{6}$ **b.** $\dfrac{\sqrt{30x}}{3x}$

c. $\dfrac{5\sqrt[3]{4}}{2}$ **d.** $\dfrac{7\sqrt[5]{48}}{2}$

Student Check 3 **a.** $\dfrac{5\sqrt{6} + 5\sqrt{2}}{4}$

b. $\dfrac{35 + 7\sqrt{7} + 5\sqrt{3} + \sqrt{21}}{18}$

SUMMARY OF KEY CONCEPTS

1. We can divide radical expressions with the same index by dividing the radicands and then taking the root of the quotient.
2. Rationalizing the denominator is the process of removing the radical from the denominator.
   a. If there is a single term in the denominator, multiply the fraction by a form of 1 that makes
the radicand in the denominator a perfect square if the index is 2, a perfect cube if the index is 3, and so on.
   b. If the denominator is a binomial, multiply the numerator and denominator of the fraction by the conjugate of the denominator.

GRAPHING CALCULATOR SKILLS

The calculator can be used to check our answers when rationalizing expressions. Enter the given expression and the result to determine if their values are the same. If they are, then the expressions are equivalent.

Example: Verify that $\dfrac{1}{\sqrt{2}} = \dfrac{\sqrt{2}}{2}$.

Solution:

```
1/√(2)
          .7071067812
√(2)/2
          .7071067812
```

Since the decimal values of the two expressions are equal, the expressions are equivalent.

SECTION 10.5 / EXERCISE SET

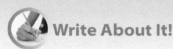 **Write About It!**

Use complete sentences in your answer to each exercise.

1. Explain how to divide radical expressions.

2. Explain why we need to rationalize denominators.

3. Define the conjugate of $a + \sqrt{b}$, where a and b are real numbers and $b > 0$.

4. Explain how we multiply $a + \sqrt{b}$ and its conjugate, where a and b are real numbers and $b > 0$.

5. Explain how to rationalize the denominator of the expression $\dfrac{1}{\sqrt[5]{6}}$.

6. Explain why $\dfrac{5}{\sqrt{11} - \sqrt{10}} \neq 5$.

 Practice Makes Perfect!

Simplify each quotient and write each answer in simplest radical form. (*See Objective 1.*)

7. $\dfrac{\sqrt{48}}{\sqrt{3}}$

8. $\dfrac{\sqrt{180}}{\sqrt{5}}$

9. $\dfrac{\sqrt{300s^{15}}}{\sqrt{3s^7}}$

10. $\dfrac{\sqrt{216t^{11}}}{\sqrt{6t^5}}$

11. $\dfrac{\sqrt{150a^3}}{\sqrt{6a^{13}}}$

12. $\dfrac{\sqrt{128b^3}}{\sqrt{2b^7}}$

13. $\dfrac{\sqrt{24c^{13}}}{\sqrt{2c^{10}}}$

14. $\dfrac{\sqrt{96d^{15}}}{\sqrt{2d^4}}$

15. $\dfrac{\sqrt{294}}{21\sqrt{3}}$

16. $\dfrac{\sqrt{224}}{14\sqrt{7}}$

17. $\dfrac{\sqrt{54}}{12\sqrt{3}}$

18. $\dfrac{\sqrt{90}}{21\sqrt{5}}$

19. $\dfrac{\sqrt{24a^6}}{18\sqrt{2a}}$

20. $\dfrac{\sqrt{126b^9}}{6\sqrt{2b^4}}$

21. $\dfrac{6\sqrt[3]{112t^{11}}}{\sqrt[3]{7t^2}}$

22. $\dfrac{4\sqrt[3]{120s^{18}}}{\sqrt[3]{3s^9}}$

23. $\dfrac{9\sqrt[3]{810x^{18}}}{\sqrt[3]{10x^3}}$

24. $\dfrac{3\sqrt[3]{2500y^{11}}}{\sqrt[3]{5y^5}}$

25. $\dfrac{7\sqrt[5]{192r^{13}}}{\sqrt[5]{3r^2}}$

26. $\dfrac{5\sqrt[5]{486y^{13}}}{\sqrt[5]{2y^7}}$

27. $\dfrac{\sqrt{192a^5b^3}}{\sqrt{75a^{-2}b^9}}$

28. $\dfrac{\sqrt{98x^{-1}y^7}}{\sqrt{18x^3y^2}}$

29. $\dfrac{9\sqrt[3]{810x^{18}y^6}}{\sqrt[3]{10x^3y^2}}$

30. $\dfrac{6\sqrt[4]{144u^{14}v^{-1}}}{\sqrt[4]{3u^6v^{-10}}}$

31. $\dfrac{10\sqrt[3]{189a^{20}b^{12}}}{\sqrt[3]{7ab^{-4}}}$

32. $\dfrac{4\sqrt[4]{1215s^{17}t^8}}{\sqrt[4]{3s^2t^{-9}}}$

Simplify each expression by rationalizing the denominators. (*See Objective 2.*)

33. $\dfrac{\sqrt{44}}{\sqrt{6}}$

34. $\dfrac{\sqrt{21}}{\sqrt{15}}$

35. $\dfrac{\sqrt{18}}{\sqrt{3a}}$

36. $\dfrac{\sqrt{2}}{\sqrt{6x}}$

37. $\dfrac{\sqrt{15}}{\sqrt{12y}}$

38. $\dfrac{\sqrt{49}}{\sqrt{21x}}$

39. $\dfrac{2}{\sqrt[3]{25}}$

40. $\dfrac{7}{\sqrt[3]{2}}$

41. $\dfrac{6}{\sqrt[5]{9a^2}}$

42. $\dfrac{12}{\sqrt[5]{4b}}$

43. $\dfrac{6}{\sqrt[6]{8x}}$

44. $\dfrac{15}{\sqrt[6]{25y}}$

45. $\dfrac{18}{\sqrt[4]{3x^3}}$

46. $\dfrac{24}{\sqrt[4]{2y}}$

Simplify each expression. Rationalize the denominator using its conjugate. (*See Objective 3.*)

47. $\dfrac{12}{\sqrt{6} + \sqrt{3}}$

48. $\dfrac{8}{\sqrt{11} - \sqrt{7}}$

49. $\dfrac{10}{\sqrt{11} - \sqrt{3}}$

50. $\dfrac{6}{\sqrt{10} + \sqrt{6}}$

51. $\dfrac{4 - \sqrt{6}}{2 - \sqrt{5}}$

52. $\dfrac{2 - \sqrt{2}}{5 + \sqrt{3}}$

53. $\dfrac{2 - \sqrt{3}}{2 + \sqrt{3}}$

54. $\dfrac{10 - \sqrt{7}}{4 + \sqrt{7}}$

55. $\dfrac{6 + \sqrt{15}}{5 - \sqrt{15}}$

56. $\dfrac{10 - \sqrt{10}}{2 - \sqrt{10}}$

57. $\dfrac{h}{\sqrt{h + 100} - 10}$

58. $\dfrac{k}{\sqrt{k + 16} - 4}$

 Mix 'Em Up!

Simplify each expression. Assume all variables represent positive real numbers.

59. $\dfrac{\sqrt{640x^3}}{\sqrt{10x}}$

60. $\dfrac{\sqrt{300y^5}}{\sqrt{3y}}$

61. $\dfrac{\sqrt{245a^4}}{\sqrt{5a^8}}$

62. $\dfrac{\sqrt{68b^{10}}}{\sqrt{17b^2}}$

63. $\dfrac{10\sqrt[3]{540x^{13}}}{\sqrt[3]{10x^2}}$

64. $\dfrac{8\sqrt[3]{112y^{14}}}{\sqrt[3]{7y^7}}$

65. $\dfrac{4\sqrt[3]{405a^{12}}}{\sqrt[3]{3a^3}}$

66. $\dfrac{3\sqrt[3]{1500b^{15}}}{\sqrt[3]{6b^9}}$

67. $\dfrac{\sqrt{600s^{21}}}{\sqrt{2s^4}}$

68. $\dfrac{\sqrt{96d^{15}}}{\sqrt{2d^4}}$

69. $\dfrac{\sqrt{150x^4}}{20\sqrt{3x}}$

70. $\dfrac{3\sqrt[4]{112a^9}}{\sqrt[4]{7a^2}}$

71. $\dfrac{2\sqrt[4]{486b^{13}}}{\sqrt[4]{6b^2}}$

72. $\dfrac{\sqrt{96y^6}}{18\sqrt{2y^3}}$

73. $\dfrac{7\sqrt[5]{320r^8}}{\sqrt[5]{2r^2}}$

74. $\dfrac{6\sqrt[5]{3402t^{14}}}{\sqrt[5]{2t^3}}$

75. $\dfrac{\sqrt{96a^7b^{-3}}}{2\sqrt{2a^{-2}b}}$

76. $\dfrac{\sqrt{1080x^{-1}y^7}}{8\sqrt{6x^3y^2}}$

77. $\dfrac{5\sqrt[3]{54x^{12}y}}{\sqrt[3]{2x^5y^{-5}}}$

78. $\dfrac{3\sqrt[3]{48x^{10}y^{-11}}}{\sqrt[3]{2x^3y^{-2}}}$

79. $\dfrac{\sqrt{216a^7}}{15\sqrt{2b^4}}$

80. $\dfrac{\sqrt{120x^5}}{8\sqrt{6y^2}}$

81. $\dfrac{6\sqrt[5]{800r^9s^{-16}}}{\sqrt[5]{5r^2s^{-6}}}$

82. $\dfrac{3\sqrt[5]{576r^{16}s^{-12}}}{\sqrt[5]{2r^4s^{-2}}}$

Simplify each expression by rationalizing the denominator.

83. $\dfrac{\sqrt{28}}{\sqrt{6}}$

84. $\dfrac{\sqrt{40}}{\sqrt{14}}$

85. $\dfrac{3}{\sqrt[4]{18a^3}}$

86. $\dfrac{15}{\sqrt[4]{125b}}$

87. $\dfrac{8}{\sqrt{21}+\sqrt{5}}$

88. $\dfrac{6}{\sqrt{11}-\sqrt{3}}$

89. $\dfrac{\sqrt{42}}{\sqrt{7x}}$

90. $\dfrac{\sqrt{75}}{\sqrt{15y}}$

91. $\dfrac{9}{\sqrt[3]{12}}$

92. $\dfrac{12}{\sqrt[3]{45}}$

93. $\dfrac{1+\sqrt{2}}{-3+\sqrt{5}}$

94. $\dfrac{7-\sqrt{5}}{3-\sqrt{3}}$

95. $\dfrac{-3-\sqrt{21}}{5-\sqrt{21}}$

96. $\dfrac{-9+\sqrt{14}}{4+\sqrt{14}}$

97. $\dfrac{5+\sqrt{13}}{4-\sqrt{13}}$

98. $\dfrac{-4-\sqrt{6}}{2+\sqrt{6}}$

99. $\dfrac{6}{\sqrt[5]{4x^2}}$

100. $\dfrac{8}{\sqrt[5]{48y^3}}$

101. $\dfrac{2-\sqrt{6}}{1+\sqrt{3}}$

102. $\dfrac{1-\sqrt{15}}{3-\sqrt{5}}$

103. $\dfrac{a}{\sqrt{a+36}+6}$

104. $\dfrac{b}{\sqrt{b+49}-7}$

You Be the Teacher!

Correct each student's errors, if any.

105. Simplify $\dfrac{\sqrt{28}}{14}$.

Andrew's work:

$$\dfrac{\sqrt{28}}{14} = \dfrac{\sqrt{28}^2}{14_1} = \sqrt{2}$$

106. Simplify by rationalizing the denominator: $\dfrac{6}{\sqrt[5]{4x^2}}$.

Rebecca's work:

$$\dfrac{6}{\sqrt[5]{4x^2}} = \dfrac{6\sqrt[5]{4^4x^3}}{\sqrt[5]{4x^2}\sqrt[5]{4^4x^3}}$$

$$= \dfrac{6\sqrt[5]{256x^3}}{4x}$$

$$= \dfrac{3\sqrt[5]{256x^3}}{2x}$$

107. Simplify by rationalizing the denominator: $\dfrac{\sqrt{7}-\sqrt{3}}{\sqrt{7}+\sqrt{3}}$.

Roy's work:

$$\dfrac{\sqrt{7}-\sqrt{3}}{\sqrt{7}+\sqrt{3}} = \dfrac{(\sqrt{7}-\sqrt{3})}{(\sqrt{7}+\sqrt{3})} \cdot \dfrac{(\sqrt{7}+\sqrt{3})}{(\sqrt{7}+\sqrt{3})}$$

$$= \dfrac{7-3}{7+2\sqrt{21}+3}$$

$$= \dfrac{4}{10+2\sqrt{21}}$$

$$= \dfrac{\overset{2}{4}}{\underset{5}{10}+\overset{1}{2}\sqrt{21}}$$

$$= \dfrac{2}{5+\sqrt{21}}$$

108. Simplify by rationalizing the denominator: $\dfrac{\sqrt{6}}{\sqrt{12}-\sqrt{3}}$.

Rita's work:

$$\dfrac{\sqrt{6}}{\sqrt{12}-\sqrt{3}} = \dfrac{\sqrt{6}}{\sqrt{12-3}}$$

$$= \dfrac{\sqrt{6}}{\sqrt{9}}$$

$$= \dfrac{\sqrt{6}}{3}$$

$$= \dfrac{\sqrt{6}^2}{3^1}$$

$$= \sqrt{2}$$

Calculate It!

Use a calculator to verify the simplification of each radical expression.

109. $\dfrac{12}{\sqrt{11}-\sqrt{2}} = \dfrac{4(\sqrt{11}+\sqrt{2})}{3}$

110. $\dfrac{-6+\sqrt{3}}{-11-\sqrt{5}} = \dfrac{66-6\sqrt{5}-11\sqrt{3}+\sqrt{15}}{116}$

111. $\dfrac{12}{\sqrt[6]{3}} = 4\sqrt[6]{243}$

112. $\dfrac{16}{\sqrt[5]{20}} = \dfrac{8\sqrt[5]{5000}}{5}$

Radical Equations and Their Applications

Mosteller's formula is used to calculate a person's body surface area (BSA). This value is often used in calculating drug dosages. A person's BSA in square meters is given by $\text{BSA} = \sqrt{\dfrac{hw}{3131}}$, where h is a person's height in inches and w is a person's weight in pounds. The average BSA for women is 1.6 m^2. If a woman is 5 ft tall, how much should she weigh to have an average BSA?

In this section, we will learn how square roots can be used to solve this problem.

Radical Equations

Now that we know how to work with radical expressions, we will solve equations that contain these expressions. This type of equation is called a **radical equation**. Some examples of radical equations are

$$\sqrt{x - 1} = 3 \qquad \sqrt{4x - 1} = \sqrt{3x} \qquad \sqrt[3]{2x + 3} = 6$$

To solve these equations, our goal is to produce an equivalent equation that doesn't contain radicals. We use the property that we learned in Section 10.1 to accomplish this goal. Recall that for $a \geq 0$,

$$\left(\sqrt[n]{a}\right)^n = a$$

Here are some illustrations of this property. We will assume the radicands are nonnegative.

Radical Expression	Value of n	$\left(\sqrt[n]{a}\right)^n$
$\sqrt{x - 1}$	$n = 2$	$\left(\sqrt{x - 1}\right)^2 = x - 1$
$\sqrt[3]{2x + 3}$	$n = 3$	$\left(\sqrt[3]{2x + 3}\right)^3 = 2x + 3$

In addition to this property, we must apply a fundamental principle of solving equations; what we do to one side of an equation, we must do to the other side. This means that if we have to square one side of the equation to remove a radical, we must square the other side. If we have to cube one side of an equation to remove a radical, we must cube the other side. This leads to the following property.

> **Property: Power Property**
>
> If $a = b$, then $a^n = b^n$.

This property tells us that if two quantities are equal, then the quantities raised to the nth are equal as well. The equation that results from raising each side to the nth does not necessarily have the same solutions as the original equation. Consider the following illustration.

Original equation	$x = -4$	Solution set: $\{-4\}$
Squared equation	$(x)^2 = (-4)^2$ $x^2 = 16$ $x^2 - 16 = 0$ $(x - 4)(x + 4) = 0$ $x = 4 \text{ or } x = -4$	Solution set: $\{-4, 4\}$

Notice that the solution set of the squared equation contains an extra solution, 4, that is *not* a solution of the original equation. In this case, 4 is an **extraneous solution** that results from squaring the original equation.

When we apply the power property, we find solutions of the original equation, but we can also get solutions that do not satisfy the original equation. That is, we can obtain extraneous solutions when we apply the power property. For this reason, we must *always* check the proposed solutions in the original equation.

> **Procedure: Solving Radical Equations**
>
> **Step 1:** Isolate the radical expression on one side of the equation.
> **Step 2:** Raise each side of the equation to the exponent equal to the index and simplify each side.
> **Step 3:** If the equation still contains a radical expression, repeat Steps 1 and 2.
> **Step 4:** Solve the resulting equation.
> **Step 5:** Check the proposed solutions in the original equation. Discard any extraneous solutions.
> **Step 6:** Write the solution set of the equation.

Objective 1 Examples   **Solve each radical equation.**

1a. $\sqrt{x-1} = 5$ **1b.** $\sqrt[3]{2x+3} = 6$ **1c.** $\sqrt{y-3} + 4 = 2$

1d. $\sqrt{5x-2} = \sqrt{3x}$ **1e.** $\sqrt{y-4} = y - 10$ **1f.** $\sqrt{x+1} - \sqrt{x} = 1$

Solutions   **1a.**

$$\sqrt{x-1} = 5$$

$(\sqrt{x-1})^2 = (5)^2$	Apply the power property.
$x - 1 = 25$	Simplify each side.
$x - 1 + 1 = 25 + 1$	Add 1 to each side.
$x = 26$	Simplify.

Check:

$\sqrt{x-1} = 5$	Original equation
$\sqrt{26-1} = 5$	Replace x with 26.
$\sqrt{25} = 5$	Simplify.
$5 = 5$	Simplify.

The solution 26 makes the original equation true. So, the solution set is $\{26\}$.

1b.

$$\sqrt[3]{2x+3} = 6$$

$(\sqrt[3]{2x+3})^3 = (6)^3$	Apply the power property.
$2x + 3 = 216$	Simplify each side.
$2x + 3 - 3 = 216 - 3$	Subtract 3 from each side.
$2x = 213$	Simplify.
$\dfrac{2x}{2} = \dfrac{213}{2}$	Divide each side by 2.
$x = \dfrac{213}{2}$	Simplify.

Check:

$\sqrt[3]{2x+3} = 6$	Original equation
$\sqrt[3]{2\left(\dfrac{213}{2}\right) + 3} = 6$	Replace x with $\dfrac{213}{2}$.
$\sqrt[3]{213 + 3} = 6$	Multiply 2 and $\dfrac{213}{2}$.
$\sqrt[3]{216} = 6$	Add.
$6 = 6$	Simplify.

The solution $\dfrac{213}{2}$ makes the original equation true, so the solution set is $\left\{\dfrac{213}{2}\right\}$.

1c. The square root expression is not isolated. Before squaring, we must subtract 4 from each side of the equation.

$\sqrt{y-3} + 4 = 2$	
$\sqrt{y-3} + 4 - 4 = 2 - 4$	Subtract 4 from each side.

$$\sqrt{y-3} = -2 \qquad \text{Simplify.}$$
$$\left(\sqrt{y-3}\right)^2 = (-2)^2 \qquad \text{Apply the power property.}$$
$$y - 3 = 4 \qquad \text{Simplify each side.}$$
$$y - 3 + 3 = 4 + 3 \qquad \text{Add 3 to each side.}$$
$$y = 7 \qquad \text{Simplify.}$$

Check:
$$\sqrt{y-3} + 4 = 2 \qquad \text{Original equation}$$
$$\sqrt{7-3} + 4 = 2 \qquad \text{Replace } y \text{ with 7.}$$
$$\sqrt{4} + 4 = 2 \qquad \text{Simplify the radicand.}$$
$$2 + 4 = 2 \qquad \text{Simplify the square root.}$$
$$6 = 2 \qquad \text{Simplify.}$$

The solution 7 makes the original equation false and, therefore, must be excluded from the solution set. It is an extraneous solution. So, the solution set is the empty set, or ∅.

> **Note:** We can conclude that the equation has no solution without going through all of the steps to solve and check. After we isolate the radical, we obtain the equation, $\sqrt{y-3} = -2$. This statement is always false. Because $\sqrt{y-3}$ represents the principal, or positive, square root of $y - 3$, it will never equal a negative number.

1d. Each square root expression is isolated on one side of the equation. So, we can apply the power property and square each side.

$$\sqrt{5x-2} = \sqrt{3x}$$
$$\left(\sqrt{5x-2}\right)^2 = \left(\sqrt{3x}\right)^2 \qquad \text{Square each side.}$$
$$5x - 2 = 3x \qquad \text{Simplify each side.}$$
$$5x - 2 + 2 = 3x + 2 \qquad \text{Add 2 to each side.}$$
$$5x = 3x + 2 \qquad \text{Simplify.}$$
$$5x - 3x = 3x + 2 - 3x \qquad \text{Subtract } 3x \text{ from each side.}$$
$$2x = 2 \qquad \text{Simplify.}$$
$$\frac{2x}{2} = \frac{2}{2} \qquad \text{Divide each side by 2.}$$
$$x = 1 \qquad \text{Simplify.}$$

Check:
$$\sqrt{5x-2} = \sqrt{3x} \qquad \text{Original equation}$$
$$\sqrt{5(1)-2} = \sqrt{3(1)} \qquad \text{Replace } x \text{ with 1.}$$
$$\sqrt{5-2} = \sqrt{3} \qquad \text{Multiply.}$$
$$\sqrt{3} = \sqrt{3} \qquad \text{Simplify.}$$

The solution 1 makes the original equation true. So, the solution set is $\{1\}$.

1e.
$$\sqrt{y-4} = y - 10$$
$$\left(\sqrt{y-4}\right)^2 = (y-10)^2 \qquad \text{Apply the power property.}$$
$$y - 4 = y^2 - 20y + 100 \qquad \text{Simplify. Recall } (a-b)^2 = a^2 - 2ab + b^2.$$

To solve the resulting quadratic equation, we must write the equation in standard form, that is, get one side of the equation equal to 0. Then we solve the equation by factoring.

$$y - 4 - y = y^2 - 20y + 100 - y \qquad \text{Subtract } y \text{ from each side.}$$
$$-4 = y^2 - 21y + 100 \qquad \text{Simplify.}$$
$$-4 + 4 = y^2 - 21y + 100 + 4 \qquad \text{Add 4 to each side.}$$

$$0 = y^2 - 21y + 104 \qquad \text{Simplify.}$$
$$0 = (y - 13)(y - 8) \qquad \text{Factor.}$$
$$y - 13 = 0 \ \text{ or } \ y - 8 = 0 \qquad \text{Set each factor equal to 0.}$$
$$y = 13 \qquad y = 8 \qquad \text{Solve the resulting equations.}$$

Check: Replace y with 13. | Replace y with 8.

$$\sqrt{y - 4} = y - 10 \qquad\qquad \sqrt{y - 4} = y - 10$$
$$\sqrt{13 - 4} = 13 - 10 \qquad\qquad \sqrt{8 - 4} = 8 - 10$$
$$\sqrt{9} = 3 \qquad\qquad\qquad \sqrt{4} = -2$$
$$3 = 3 \quad \text{True} \qquad\qquad 2 = -2 \quad \text{False}$$

The only solution that makes the original equation true is 13. So, the solution set is $\{13\}$.

1f. We must isolate one of the radicals on one side of the equation before squaring.

$$\sqrt{x + 1} - \sqrt{x} = 1$$
$$\sqrt{x + 1} - \sqrt{x} + \sqrt{x} = 1 + \sqrt{x} \qquad \text{Add } \sqrt{x} \text{ to each side.}$$
$$\sqrt{x + 1} = 1 + \sqrt{x} \qquad \text{Simplify.}$$
$$\left(\sqrt{x + 1}\right)^2 = \left(1 + \sqrt{x}\right)^2 \qquad \text{Apply the power property.}$$
$$x + 1 = 1 + 2\sqrt{x} + x \qquad \text{Simplify each side.}$$

Since the equation still has a radical expression, we must isolate the radical $2\sqrt{x}$ on one side of the equation.

$$x + 1 - 1 = 1 + 2\sqrt{x} + x - 1 \quad \text{Subtract 1 from each side.}$$
$$x = 2\sqrt{x} + x \qquad \text{Simplify.}$$
$$x - x = 2\sqrt{x} + x - x \qquad \text{Subtract } x \text{ from each side.}$$
$$0 = 2\sqrt{x} \qquad \text{Simplify.}$$
$$(0)^2 = \left(2\sqrt{x}\right)^2 \qquad \text{Apply the power property.}$$
$$0 = 4x \qquad \text{Simplify: } (2\sqrt{x})^2 = (2)^2(\sqrt{x})^2 = 4x.$$
$$\frac{0}{4} = \frac{4x}{4} \qquad \text{Divide each side by 4.}$$
$$0 = x \qquad \text{Simplify.}$$

The solution 0 makes the original equation true. So, the solution set is $\{0\}$.

✓ Student Check 1   Solve each radical equation.

a. $\sqrt{x + 3} = 4$ **b.** $\sqrt[3]{4x - 3} = 5$ **c.** $\sqrt{3a - 2} + 5 = 1$

d. $\sqrt{3x - 1} = \sqrt{2x}$ **e.** $\sqrt{z + 2} + 4 = z$ **f.** $\sqrt{3x - 2} - \sqrt{x} = 2$

Objective 2 ▶

Solve problems involving the Pythagorean theorem and other formulas.

Applications

Some applications of radical equations deal with right triangles. Recall the *Pythagorean theorem* shows the relationship between the sides of a right triangle.

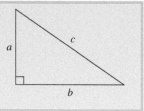

Property: Pythagorean Theorem

For a right triangle with legs a and b and hypotenuse c,

$$a^2 + b^2 = c^2$$

We will also solve some other applications that are based on formulas.

Objective 2 Examples   **Solve each problem.**

2a. Find the exact length of the unknown side of the right triangle in Figure 10.4.

Solution   **2a.**

$a^2 + b^2 = c^2$	State the Pythagorean theorem.
$5^2 + b^2 = 15^2$	Replace a with 5 and c with 15.
$25 + b^2 = 225$	Simplify: $5^2 = 25$ and $15^2 = 225$.
$25 + b^2 - 25 = 225 - 25$	Subtract 25 from each side.
$b^2 = 200$	Simplify.
$b = \sqrt{200}$	Solve for b.
$b = \sqrt{100 \cdot 2}$	Factor 200 as $100 \cdot 2$.
$b = 10\sqrt{2}$	Apply the product rule.

The value of the unknown leg of the right triangle is $10\sqrt{2}$ ft, or approximately 14.14 ft.

2b. Mosteller's formula is used to calculate a person's body surface area (BSA). This value is often used in calculating drug dosages. A person's BSA in square meters is given by $\text{BSA} = \sqrt{\dfrac{hw}{3131}}$, where h is a person's height in inches and w is a person's weight in pounds. (i) Find the BSA for a female who is 5 ft 6 in. tall and weighs 165 lb. (ii) The average BSA for women is 1.6 m². If a woman is 5 ft tall, how much should she weigh to have an average BSA?

Solution   **2b. i.**

$\text{BSA} = \sqrt{\dfrac{hw}{3131}}$	State the BSA formula.
$\text{BSA} = \sqrt{\dfrac{(66)(165)}{3131}}$	Replace h with 66 and w with 165.
$\text{BSA} = \sqrt{\dfrac{10{,}890}{3131}}$	Multiply 66(165).
$\text{BSA} \approx 1.86$	Approximate the value.

So, the BSA for a female who is 5 ft 6 in. tall and weighs approximately 165 lb is approximately 1.86 m².

5 ft · **15 ft** · b

Figure 10.4

ii. Let BSA = 1.6, $h = 60$ in., and solve for w.

$$BSA = \sqrt{\frac{hw}{3131}} \qquad \text{State the BSA formula.}$$

$$1.6 = \sqrt{\frac{60w}{3131}} \qquad \text{Replace BSA with 1.6 and } h \text{ with 60.}$$

$$(1.6)^2 = \left(\sqrt{\frac{60w}{3131}}\right)^2 \qquad \text{Square each side.}$$

$$2.56 = \frac{60w}{3131} \qquad \text{Simplify each side.}$$

$$2.56(3131) = \left(\frac{60w}{3131}\right)(3131) \qquad \text{Multiply each side by 3131.}$$

$$8015.36 = 60w \qquad \text{Simplify.}$$

$$\frac{8015.36}{60} = \frac{60w}{60} \qquad \text{Divide each side by 60.}$$

$$133.6 \approx w \qquad \text{Approximate.}$$

A woman 5 ft tall should weigh approximately 134 lb to have an average BSA.

✓ **Student Check 2** Solve each problem.

a. One of the ropes that is used to tie down a tent needs to be replaced. Determine the distance from the tent to the tie-down stake if the tent is 6 ft tall and the tie-down stake is placed 4 ft from the tent.

b. i. Find the BSA of a person who is 5 ft 4 in. and weighs 140 lb.
ii. The average BSA of a man is 1.9 m². If a man is 5 ft 8 in. tall, how much should he weigh to have an average BSA?

Objective 3 ▶
Troubleshoot common errors.

Troubleshooting Common Errors

Some common errors associated with radical equations are shown.

Objective 3 Examples A problem and an incorrect solution are given. Provide the correct solution and an explanation of the error.

3a. Solve $\sqrt{3x - 5} + 2 = 7$.

Incorrect Solution	Correct Solution and Explanation
$\sqrt{3x-5}+2=7$ $(\sqrt{3x-5}+2)^2 = 7^2$ $3x-5+4=49$ $3x-1=49$ $3x=50$ $x=\frac{50}{3}$ $\left\{\frac{50}{3}\right\}$	We must isolate the radical before squaring. $\sqrt{3x-5}+2=7$ $\sqrt{3x-5}+2-2=7-2$ $\sqrt{3x-5}=5$ $(\sqrt{3x-5})^2=5^2$ $3x-5=25$ $3x=30$ $x=10$ The solution set is $\{10\}$.

3b. Solve $\sqrt{z+2}+4=z$.

Incorrect Solution	Correct Solution and Explanation
$\sqrt{z+2}+4=z$ $\sqrt{z+2}=z-4$ $(\sqrt{z+2})^2=(z-4)^2$ $z+2=z^2-8z+16$ $0=z^2-9z+14$ $0=(z-2)(z-7)$ $z=2$ or $z=7$ $\{2,7\}$	The equation was solved correctly but the solutions were not checked to determine if the solutions are extraneous. **Check:** $z=2$: $\sqrt{z+2}+4=z$ $\sqrt{2+2}+4=2$ $\sqrt{4}+4=2$ $2+4=2$ $6=2$ False **Check:** $z=7$: $\sqrt{z+2}+4=z$ $\sqrt{7+2}+4=7$ $\sqrt{9}+4=7$ $3+4=7$ $7=7$ True So, the solution set is $\{7\}$.

ANSWERS TO STUDENT CHECKS

Student Check 1 **a.** $\{13\}$ **b.** $\{32\}$ **c.** $\varnothing$ **d.** $\{1\}$ **e.** $\{7\}$ **f.** $\{9\}$

Student Check 2 **a.** $2\sqrt{13}$ ft **b. i.** 1.69 m² **ii.** 166.2 lb

SUMMARY OF KEY CONCEPTS

1. There are five main steps that we must follow to solve equations with radicals.
  • First, isolate the radical on one side of the equation.
  • Second, raise both sides of the equation to the exponent equal to the index and simplify. Use the FOIL method when appropriate.
  • Third, if the equation contains a radical, repeat the first two steps.
  • Fourth, solve the resulting equation.

  • Fifth, check the proposed answers in the original equation. It is possible that extraneous solutions may result, so do not skip the checking step.

2. When solving applications involving the Pythagorean theorem, it is helpful to draw a diagram of the right triangle. Be sure to identify which sides are legs and which side is the hypotenuse. Then substitute the known values into $a^2+b^2=c^2$ and solve.

GRAPHING CALCULATOR SKILLS

As we have seen in previous sections, we can use a graphing calculator to check the answers found when solving square root equations.

Example: In Example 1 part (c), we found that $\sqrt{y-3}+4=2$ has no solution. Use a graphing calculator to verify this result.

Solution: We can determine where the graphs of the expressions on the left and right side of the equation intersect. Enter the left side of the original equation as Y_1 and the right side of the original equation as Y_2. Note: We use the variable x in the calculator, rather than y like we have in our equation.

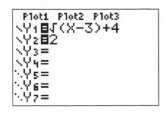

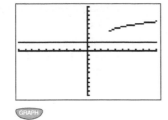

GRAPH

Since the two graphs do not intersect, the equation has no solution.

We can also use the table to check the solution. The proposed solution for this problem is 7. When we examine the table, we notice that when x is 7, the values of Y_1 and Y_2 are not equal. Therefore, 7 is not a solution of the equation.

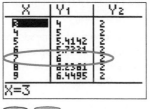

2nd GRAPH

SECTION 10.6 / EXERCISE SET

Write About It!

Use complete sentences in your answer to each exercise.

1. Explain how to apply the power property to solve a radical equation of the form $\sqrt[n]{x} = a$, a is a real number.

2. Explain why it is important to check the solutions when solving radical equations.

3. Explain how to solve an equation with two radicals, such as $\sqrt{x-5} + 1 = \sqrt{x}$.

4. Explain how to solve an equation with one radical, such as $\sqrt{x+5} + 1 = x$.

5. Explain how to determine whether 655 is a solution of the equation $\sqrt[3]{2x+21} - 9 = 2$.

6. Explain why 3 and -2 are not solutions of the equation $\sqrt{70 - 15x} = x - 8$.

Practice Makes Perfect!

Solve each radical equation. (See Objective 1.)

7. $\sqrt{2x - 19} = 1$

8. $\sqrt{3y - 11} = 8$

9. $\sqrt{4a + 1} = 9$

10. $\sqrt{9b + 22} = 7$

11. $\sqrt{7r + 15} - 6 = 2$

12. $\sqrt{5r - 6} + 7 = 19$

13. $\sqrt{5r - 1} + 7 = 3$

14. $\sqrt{6s + 5} - 1 = -10$

15. $\sqrt[3]{2x - 9} = 1$

16. $\sqrt[3]{3y + 11} = 5$

17. $\sqrt[3]{3x + 1} = -2$

18. $\sqrt[3]{6x + 2} = -4$

19. $\sqrt{6x + 37} = x + 5$

20. $\sqrt{2y - 5} = y - 4$

21. $\sqrt{32 - 2a} = a - 4$

22. $\sqrt{-4b - 7} = b + 3$

23. $\sqrt{2x + 27} = x + 6$

24. $\sqrt{-25y + 19} = y - 7$

25. $\sqrt{7y - 3} - 1 = y$

26. $\sqrt{14r - 62} + 1 = r$

27. $\sqrt{2x + 7} = \sqrt{x} + 2$

28. $\sqrt{3y + 97} = \sqrt{y} + 9$

29. $\sqrt{3z + 61} = \sqrt{z} - 9$

30. $\sqrt{3w + 121} = \sqrt{w} - 9$

31. $\sqrt{6x - 5} = 1 + \sqrt{4x - 4}$

32. $\sqrt{25x + 19} = 4 + \sqrt{17x - 21}$

33. $\sqrt{4r + 29} = 1 + \sqrt{2r + 26}$

34. $\sqrt{6x + 31} = 1 + \sqrt{4x + 24}$

35. $\sqrt{31 - 5x} = -4 + \sqrt{3x - 9}$

36. $\sqrt{-13x + 29} = -1 + \sqrt{-11x + 20}$

Solve each problem. (See Objective 2.)

37. Find the length of the missing side of a right triangle if one of the legs is 12 ft and the hypotenuse is 37 ft.

38. Find the length of the missing side of a right triangle if one of the legs is 30 cm and the hypotenuse is 34 cm.

39. Find the length of the hypotenuse of a right triangle if the lengths of the two legs are 11 cm and 60 cm.

40. Find the length of the hypotenuse of a right triangle if the lengths of the two legs are 15 m and 8 m.

41. Find the length of the hypotenuse of a right triangle if the lengths of the two legs are 12 ft and 16 ft.

42. Find the length of the hypotenuse of a right triangle if the lengths of the two legs are 15 in. and 20 in.

43. One of the ropes that is used to tie down a tent needs to be replaced. Determine the distance from the top of the tent to the tie-down stake if the tent is 10 ft tall and the tie-down stake is placed 5 ft from the tent.

44. One of the ropes that is used to tie down a tent needs to be replaced. Determine the distance from the top of the tent to the tie-down stake if the tent is 9 ft tall and the tie-down stake is placed 3 ft from the tent.

For Exercises 45–48, a person's body surface area (BSA) in square meters is given by $BSA = \sqrt{\dfrac{hw}{3131}}$, where h is a person's height in inches and w is a person's weight in pounds. Round each answer to two decimal places.

45. (a) Find the BSA of a person who is 5 ft 11 in. and weighs 233 lb. (b) The average BSA of a man is 1.9 m². If a man is 6 ft 5 in. tall, how much should he weigh to have an average BSA?

46. (a) Find the BSA of a person who is 6 ft 5 in. and weighs 180 lb. (b) The average BSA of a man is 1.9 m². If a man is 5 ft 5 in. tall, how much should he weigh to have an average BSA?

47. (a) Find the BSA of a person who is 5 ft 4 in. and weighs 136 lb. (b) The average BSA of a woman is 1.6 m². If a woman is 4 ft 6 in. tall, how much should she weigh to have an average BSA?

48. (a) Find the BSA of a person who is 4 ft 8 in. and weighs 140 lb. (b) The average BSA of a woman is 1.6 m². If a woman is 5 ft 7 in. tall, how much should she weigh to have an average BSA?

 Mix 'Em Up!

Solve each radical equation.

49. $\sqrt{12x + 21} - 7 = 2$ 50. $\sqrt{11y - 13} + 2 = 5$

51. $\sqrt{6a + 7} - 1 = -10$ 52. $\sqrt{3b - 5} + 7 = 1$

53. $\sqrt[4]{5x - 7} + 1 = 3$ 54. $\sqrt[4]{6y - 1} - 8 = -5$

55. $\sqrt[3]{7a - 13} = 2$ 56. $\sqrt[3]{4y + 19} - 6 = -1$

57. $\sqrt{16x + 20} - 5 = x$ 58. $\sqrt{21y + 15} = y + 5$

59. $\sqrt{7x + 39} = x + 3$ 60. $\sqrt{5b - 24} + 4 = b$

61. $\sqrt{3x - 11} + 7 = \sqrt{x}$ 62. $\sqrt{2y + 7} + 2 = \sqrt{y}$

63. $\sqrt{3r - 23} - 1 = \sqrt{r}$ 64. $\sqrt{2s - 63} + 3 = \sqrt{s}$

65. $\sqrt{2x - 3} = -1 + \sqrt{4x - 8}$

66. $\sqrt{3x + 6} = 1 + \sqrt{x + 3}$

67. $\sqrt{26 - 5x} = -4 + \sqrt{3x - 6}$

68. $\sqrt{11 - 7y} = -4 + \sqrt{y + 3}$

69. $\sqrt{13x + 3} = 1 + \sqrt{11x - 2}$

70. $\sqrt{5y - 1} = -3 + \sqrt{11y + 14}$

Solve each problem. Round each answer to two decimal places.

71. Find the length of the missing side of a right triangle if one of the legs is 24 ft and the hypotenuse is 25 ft.

72. Find the length of the missing side of a right triangle if one of the legs is 16 cm and the hypotenuse is 34 cm.

For Exercises 73 and 74, a person's body surface area (BSA) in square meters is given by $BSA = \sqrt{\dfrac{hw}{3131}}$, where h is a person's height in inches and w is a person's weight in pounds.

73. (a) Find the BSA of a person who is 6 ft 4 in. tall and weighs 236 lb. (b) The average BSA of a man is 1.9 m². If a man is 5 ft 9 in. tall, how much should he weigh to have an average BSA?

74. (a) Find the BSA of a person who is 4 ft 4 in. and weighs 103 lb. (b) The average BSA of a woman is 1.6 m². If a woman is 5 ft 9 in. tall, how much should she weigh to have an average BSA?

For Exercises 75 and 76, the period T, in seconds, of a simple pendulum is the time it takes for it to make a complete swing and can be approximated by

$$T = 2\pi\sqrt{\dfrac{L}{9.8}}$$

where L is the length of the pendulum in meters.

75. Find the exact value of the period of a simple pendulum if its length is 2.8 m.

76. Find the exact value of the period of a simple pendulum if its length is 12.25 m.

For Exercises 77 and 78, the speed s, in miles per hour, of a car that skids to a stop can be calculated by measuring the distance d, in feet, of the skid marks. The formula is

$$s = \sqrt{30kd}$$

where k is the coefficient of friction which depends on the road surface and condition. Round answers to one decimal place.

Road Surface and Condition	k
Dry tar	1.0
Wet tar	0.5
Dry concrete	0.8
Wet concrete	0.4

77. Find the speed of a car if it leaves skid marks 190 ft long on a dry concrete road.

78. Find the speed of a car if it leaves skid marks 90 ft long on a wet tar road.

 You Be the Teacher!

Correct each student's errors, if any.

79. Solve the equation $\sqrt{4x + 25} = 9$.

Mary Beth's work:

$$\sqrt{4x + 25} = 9$$
$$2x + 5 = 9$$
$$2x = 4$$
$$x = 2$$

80. Solve the equation $\sqrt[3]{2x - 5} + 2 = 3$.

Rebecca's work:

$$\sqrt[3]{2x - 5} + 2 = 3$$
$$2x - 5 + 8 = 27$$
$$2x + 3 = 27$$
$$2x = 24$$
$$x = 12$$

81. Solve the equation $\sqrt{13 - 3x} = x - 3$.

Darius's work:

$$\sqrt{13 - 3x} = x - 3$$
$$13 - 3x = x^2 + 9$$
$$0 = x^2 + 3x - 4$$
$$0 = (x + 4)(x - 1)$$
$$x = -4, 1$$

82. Solve the equation $\sqrt{12x - 11} = 3 + \sqrt{6x - 14}$.

Carol's work:

$$\sqrt{12x - 11} = 3 + \sqrt{6x - 14}$$
$$12x - 11 = 9 + 6x - 14$$
$$6x = 6$$
$$x = 1$$

83. Solve the equation $\sqrt{2x - 23} = \sqrt{x} - 1$.

James's work:

$$\sqrt{2x - 23} = \sqrt{x} - 1$$
$$2x - 23 = x + 1$$
$$x = 24$$

84. Solve the equation $\sqrt{x + 7} = -2 + \sqrt{5x + 15}$.

Randy's work:

$$\sqrt{x + 7} = -2 + \sqrt{5x + 15}$$
$$x + 7 = 4 + 5x + 15$$
$$-4x = 12$$
$$x = -3$$

 **Calculate It!**

Solve each equation and then use a calculator to check the solutions.

85. $\sqrt{4x + 57} = x + 3$
86. $\sqrt[3]{4x + 19} - 10 = -7$
87. $\sqrt{8x + 49} = 3 + \sqrt{2x + 10}$
88. $\sqrt{5x + 15} = 4 + \sqrt{-3x + 7}$

Complex Numbers

▶ **OBJECTIVES**

As a result of completing this section, you will be able to

1. Write square roots of negative numbers as complex numbers.
2. Add and subtract complex numbers.
3. Multiply complex numbers.
4. Divide complex numbers.
5. Find powers of i.
6. Troubleshoot common errors.

In Section 10.1, we discussed square roots such as $\sqrt{-16}$, $\sqrt{-8}$, and $\sqrt{-7}$. We stated that these are not real numbers since there is no real number that, when squared, equals a negative number. While it is true that they are not real numbers, we now introduce a new set of numbers, called *complex numbers*, that are used to represent these expressions. Complex numbers are used in many aspects of science, such as engineering. They are used in particular in electrical engineering.

Complex Numbers

Centuries ago, mathematicians realized that the square root of negative numbers arose when solving certain types of equations. They actually referred to these types of solutions as "imaginary" or "ghostly." Mathematicians devised a new number system to include numbers like $\sqrt{-16}$ and $\sqrt{-8}$. This new number system is called the *complex number* system. The *imaginary unit*, represented by the symbol i, is included in this system. The imaginary unit is a number that can be squared to obtain a negative number.

Objective 1 ▶

Write square roots of negative numbers as complex numbers.

Definition: Imaginary Unit i

The number i is the number whose square is -1.

$$i = \sqrt{-1} \quad \text{and} \quad i^2 = -1$$

We use i to write numbers like $\sqrt{-16}$ as the product of a real number and i. Since $i = \sqrt{-1}$,

$$\sqrt{-16} = \sqrt{-1 \cdot 16} = \sqrt{-1} \cdot \sqrt{16} = i(4) \text{ or } 4i.$$

Definition: Any number of the form $a + bi$, where a and b are real numbers, is a **complex number**. This form is the standard form of a complex number, where a is the real part and b is the imaginary part of the complex number.

If $a = 0$, then $a + bi = 0 + bi = bi$. This number is purely imaginary.
If $b = 0$, then $a + bi = a + 0(i) = a$. This number is purely real.

Some examples of complex numbers are $3 + 2i$, $-1 - 5i$, $0 + i\sqrt{7}$, and $-2 + 0i$.

Note: *The number $3 + 2i$ represents $3 + \sqrt{-4}$ since $\sqrt{-4} = \sqrt{-1 \cdot 4} = \sqrt{-1} \cdot \sqrt{4} = 2i$.*

The set of complex numbers includes all of the real numbers since any real number a can be written as $a + 0i$. We can visualize the complex number system as shown.

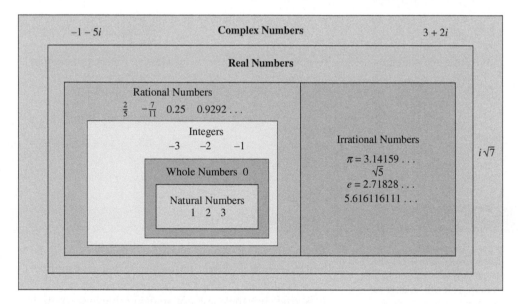

Property: Square Root of a Negative Number

For c a real number and $c > 0$,

$$\sqrt{-c} = \sqrt{-1 \cdot c} = \sqrt{-1}\sqrt{c} = i\sqrt{c}$$

Objective 1 Examples Write each expression in terms of i.

Problems	Solutions
1a. $\sqrt{-16}$	$\sqrt{-16} = \sqrt{16 \cdot -1} = \sqrt{16}\sqrt{-1} = 4i$
1b. $\sqrt{-7}$	$\sqrt{-7} = \sqrt{-1 \cdot 7} = \sqrt{-1}\sqrt{7} = i\sqrt{7}$
1c. $\sqrt{-8}$	$\sqrt{-8} = \sqrt{-1 \cdot 4 \cdot 2} = \sqrt{4 \cdot -1}\sqrt{2} = 2i\sqrt{2}$

Problems	Solutions
1d. $4 + \sqrt{-25}$	$4 + \sqrt{-25} = 4 + \sqrt{25 \cdot -1} = 4 + 5i$
1e. $\dfrac{8 + \sqrt{-36}}{2}$	$\dfrac{8 + \sqrt{-36}}{2} = \dfrac{8 + \sqrt{-1 \cdot 36}}{2}$
	$\qquad = \dfrac{8 + \sqrt{-1}\sqrt{36}}{2}$
	$\qquad = \dfrac{8 + 6i}{2}$
	$\qquad = \dfrac{8}{2} + \dfrac{6i}{2}$
	$\qquad = 4 + 3i$
1f. $\dfrac{3 + \sqrt{-12}}{4}$	$\dfrac{3 + \sqrt{-12}}{4} = \dfrac{3 + \sqrt{-1 \cdot 4 \cdot 3}}{4}$
	$\qquad = \dfrac{3 + \sqrt{4 \cdot -1}\sqrt{3}}{4}$
	$\qquad = \dfrac{3 + 2i\sqrt{3}}{4}$
	$\qquad = \dfrac{3}{4} + \dfrac{2\sqrt{3}}{4}i$
	$\qquad = \dfrac{3}{4} + \dfrac{\sqrt{3}}{2}i$

✓ **Student Check 1**   Write each expression in terms of i.

    a. $\sqrt{-49}$ **b.** $\sqrt{-5}$ **c.** $\sqrt{-20}$

    d. $6 + \sqrt{-4}$ **e.** $\dfrac{10 + \sqrt{-25}}{5}$ **f.** $\dfrac{2 + \sqrt{-18}}{6}$

Adding and Subtracting Complex Numbers

Objective 2 ▶

Add and subtract complex numbers.

Any operation that can be performed on real numbers can also be performed on complex numbers. That is, we can add, subtract, multiply, and divide complex numbers.

Operations with complex numbers are done the same way we perform operations on polynomials. To add or subtract complex numbers, combine like terms. Like terms are the real parts and the imaginary parts.

> **Procedure: Adding or Subtracting Complex Numbers**
>
> **Step 1:** If adding, remove parentheses and combine the real parts and the imaginary parts.
> **Step 2:** If subtracting, remove the parentheses by distributing -1 to the number being subtracted. Combine the real parts and imaginary parts.

Objective 2 Examples   Perform the indicated operation on the complex numbers.

    2a. $(4 - 5i) + (7 - 9i)$ **2b.** $(4 - 5i) - (7 - 9i)$ **2c.** $2i + (7 - i) - (3 + i)$

Solutions   **2a.** $(4 - 5i) + (7 - 9i) = 4 - 5i + 7 - 9i$ Clear parentheses.

                          $= 4 + 7 - 5i - 9i$ Group real parts and imaginary parts.

                          $= 11 - 14i$ Add.

2b. We first find the opposite of $7 - 9i$.

$$(4 - 5i) - 1(\overset{\frown}{7 - 9i}) = 4 - 5i - 7 + 9i \qquad \text{Clear parentheses.}$$
$$= 4 - 7 - 5i + 9i \qquad \text{Group real parts and imaginary parts.}$$
$$= -3 + 4i \qquad \text{Add.}$$

2c. We first find the opposite of $3 + i$.

$$2i + (7 - i) - 1(\overset{\frown}{3 + i}) = 2i + 7 - i - 3 - i \qquad \text{Clear parentheses.}$$
$$= 7 - 3 + 2i - i - i \qquad \text{Group real parts and imaginary parts.}$$
$$= 4 + 0i \qquad \text{Add.}$$
$$= 4$$

✓ **Student Check 2**   Perform the indicated operation on the complex numbers.
   a. $(2 + 3i) + (8 - 5i)$ **b.** $(2 + 3i) - (8 - 5i)$ **c.** $4 + (8 + 7i) - (4 + 6i)$

Multiplying Complex Numbers

Objective 3 ▶

Multiply complex numbers.

We multiply complex numbers just like we multiply polynomials. We will use the distributive property and the FOIL method. The key is to express the answers in the form $a + bi$. After multiplying, we must rewrite all instances of i^2 as -1.

> **Procedure: Multiplying Complex Numbers**
>
> **Step 1:** Rewrite any square roots of negative numbers as complex numbers.
> **Step 2:** If a single term is multiplied by a binomial, then distribute.
> **Step 3:** If the complex numbers being multiplied are binomials, then use the FOIL method.
> **Step 4:** If necessary, rewrite i^2 as -1 and combine like terms.
> **Step 5:** Write the answer in the form $a + bi$.

Objective 3 Examples   Simplify each product of complex numbers and write each answer in standard form.

3a. $\sqrt{-25} \cdot \sqrt{-4}$ **3b.** $3i(2 + 6i)$ **3c.** $(4 - 5i)(7 - 9i)$
3d. $(4 - 5i)(4 + 5i)$ **3e.** $(1 - 3i)^2$

Solutions   **3a.** $\sqrt{-25} \cdot \sqrt{-4} = (5i)(2i)$ Write each radical as a complex number.
$$= 10i^2 \qquad \text{Multiply.}$$
$$= 10(-1) \qquad \text{Rewrite } i^2 \text{ as } -1.$$
$$= -10 \qquad \text{Multiply.}$$

3b. $3i(2 + 6i) = (3i)(2) + (3i)(6i)$ Apply the distributive property.
$$= 6i + 18i^2 \qquad \text{Multiply.}$$
$$= 6i + 18(-1) \qquad \text{Rewrite } i^2 \text{ as } -1.$$
$$= 6i - 18 \qquad \text{Simplify.}$$
$$= -18 + 6i \qquad \text{Write in standard form.}$$

3c. $(4 - 5i)(7 - 9i)$
$$= 4(7) - 4(9i) - 5i(7) - 5i(-9i) \qquad \text{Apply the FOIL method.}$$
$$= 28 - 36i - 35i + 45i^2 \qquad \text{Multiply.}$$
$$= 28 - 71i + 45(-1) \qquad \text{Rewrite } i^2 \text{ as } -1 \text{ and combine like terms.}$$
$$= 28 - 71i - 45 \qquad \text{Simplify.}$$
$$= -17 - 71i \qquad \text{Combine like terms.}$$

3d. $(4 - 5i)(4 + 5i)$

$$
\begin{aligned}
&= 4(4) + 4(5i) - 5i(4) - 5i(5i) &&\text{Apply the FOIL method.}\\
&= 16 + 20i - 20i - 25i^2 &&\text{Simplify.}\\
&= 16 - 25(-1) &&\text{Rewrite } i^2 \text{ as } -1 \text{ and combine like terms.}\\
&= 16 + 25 &&\text{Simplify.}\\
&= 41 &&\text{Combine like terms.}
\end{aligned}
$$

3e. We can square the complex number by either repeating the base or applying the property to square a binomial.

Method 1	**Method 2**
$(1 - 3i)^2 = (1 - 3i)(1 - 3i)$	$(1 - 3i)^2 = (1)^2 - 2(1)(3i) + (3i)^2$
$\quad = 1 - 3i - 3i + 9i^2$	$\quad = 1 - 6i + 9i^2$
$\quad = 1 - 6i + 9(-1)$	$\quad = 1 - 6i + 9(-1)$
$\quad = 1 - 6i - 9$	$\quad = 1 - 6i - 9$
$\quad = -8 - 6i$	$\quad = -8 - 6i$

 **Student Check 3**   Simplify each product of complex numbers and write each answer in standard form.

a. $\sqrt{-25} \cdot \sqrt{-16}$ **b.** $7i(3 + 2i)$ **c.** $(2 - 3i)(4 - i)$
d. $(2 - 3i)(2 + 3i)$ **e.** $(4 - 5i)^2$

Dividing Complex Numbers

Objective 4 ▶

Divide complex numbers.

The goal in dividing complex numbers is to write the result as a complex number in standard form $a + bi$. That is, we must remove the complex number from the denominator. This will involve the product of conjugates. Complex numbers also have conjugates. For example, the numbers $4 + 5i$ and $4 - 5i$ are *complex conjugates*.

> **Definition:** The **complex conjugate** of $a + bi$ is $a - bi$.

Some other examples of complex conjugates are shown in the table.

Complex Number	Complex Conjugate
$-2 - 7i$	$-2 + 7i$
$5i = 0 + 5i$	$-5i = 0 - 5i$
$1 + \sqrt{11}i$	$1 - \sqrt{11}i$
$\dfrac{3}{5} - \dfrac{\sqrt{2}}{5}i$	$\dfrac{3}{5} + \dfrac{\sqrt{2}}{5}i$
$4 = 4 + 0i$	$4 = 4 - 0i$

Note from Example 3 part (d) that the product of complex conjugates is a real number, that is,

$$(4 - 5i)(4 + 5i) = 16 + 25 = 41$$

We generalize this relationship as follows.

> **Property: Product of Complex Conjugates**
> $$(a + bi)(a - bi) = a^2 + b^2$$

We use this fact to divide complex numbers.

> **Procedure: Dividing Complex Numbers**
> **Step 1:** If necessary, write the square root of a negative number as an imaginary number.

Step 2: Multiply the numerator and the denominator by the complex conjugate of the denominator.
Step 3: Simplify the result.

Objective 4 Examples | **Simplify each quotient of complex numbers and write answers in standard form.**

4a. $\dfrac{\sqrt{-36}}{\sqrt{-4}}$ **4b.** $\dfrac{5-3i}{4i}$ **4c.** $\dfrac{9}{1+i}$ **4d.** $\dfrac{-2+i}{4+3i}$

Solutions **4a.** $\dfrac{\sqrt{-36}}{\sqrt{-4}} = \dfrac{6i}{2i}$ Rewrite the numerator and denominator as complex numbers.

$= \dfrac{6i(-2i)}{2i(-2i)}$ Multiply the numerator and denominator by the complex conjugate of $2i$, $-2i$.

$= \dfrac{-12i^2}{-4i^2}$ Simplify each product.

$= \dfrac{-12(-1)}{-4(-1)}$ Rewrite i^2 as -1.

$= \dfrac{12}{4}$ Simplify the numerator and denominator.

$= 3$ Divide.

Note that the standard form of 3 is $3 + 0i$. When the number is purely real, there is no need to write it in this form.

4b. $\dfrac{5-3i}{4i} = \dfrac{(5-3i)(-4i)}{(4i)(-4i)}$ Multiply the numerator and denominator by $-4i$.

$= \dfrac{-20i + 12i^2}{-16i^2}$ Simplify each product.

$= \dfrac{-20i + 12(-1)}{-16(-1)}$ Rewrite i^2 as -1.

$= \dfrac{-20i - 12}{16}$ Simplify.

$= \dfrac{-20i}{16} - \dfrac{12}{16}$ Divide each term in the numerator by 16.

$= -\dfrac{12}{16} - \dfrac{20}{16}i$ Write in the standard form of a complex number.

$= -\dfrac{3}{4} - \dfrac{5}{4}i$ Simplify the real and imaginary parts.

4c. $\dfrac{9}{1+i} = \dfrac{(9)(1-i)}{(1+i)(1-i)}$ Multiply the numerator and denominator by $1-i$.

$= \dfrac{9-9i}{1-i^2}$ Simplify each product.

$= \dfrac{9-9i}{1-(-1)}$ Rewrite i^2 as -1.

$= \dfrac{9-9i}{2}$ Simplify.

$= \dfrac{9}{2} - \dfrac{9}{2}i$ Divide each term in the numerator by 2.

4d. $\dfrac{-2+i}{4+3i} = \dfrac{(-2+i)(4-3i)}{(4+3i)(4-3i)}$ Multiply the numerator and denominator by $4-3i$.

$= \dfrac{-8+6i+4i-3i^2}{16-9i^2}$ Simplify each product.

$= \dfrac{-8+10i-3(-1)}{16-9(-1)}$ Rewrite i^2 as -1.

$= \dfrac{-8+10i+3}{16+9}$ Simplify.

$= \dfrac{-5+10i}{25}$ Combine like terms.

$= \dfrac{-5}{25} + \dfrac{10}{25}i$ Write in the standard form of a complex number.

$= -\dfrac{1}{5} + \dfrac{2}{5}i$ Simplify the real and imaginary parts.

✔ **Student Check 4**   Simplify each quotient of complex numbers. Write answers in standard form.

a. $\dfrac{\sqrt{-64}}{\sqrt{-16}}$ **b.** $\dfrac{4+3i}{5i}$ **c.** $\dfrac{8}{3-i}$ **d.** $\dfrac{1+2i}{4+i}$

Powers of i

Objective 5 ▶

Find powers of i.

We can raise complex numbers to exponents just as we can raise real numbers to exponents. So, we will examine what happens when we find powers of i. Note from the following examples that all powers of i simplify to one of four quantities: $1, -1, i,$ or $-i$. We find the powers of i by using the fact that $i^2 = -1$.

$i^1 = i$ $\qquad\qquad i^5 = i^4 \cdot i = 1 \cdot i = i$

$i^2 = -1$ $\qquad\qquad i^6 = i^4 \cdot i^2 = (1)(-1) = -1$

$i^3 = i^2(i) = -1i = -i$ $\qquad i^7 = i^4 \cdot i^3 = (1)(-i) = -i$

$i^4 = i^2 \cdot i^2 = (-1)(-1) = 1$ $\qquad i^8 = i^4 \cdot i^4 = (i^4)^2 = (1)^2 = 1$

We could continue finding the powers of i in this manner, but it is not necessary since there is a pattern to the powers of i. Observe the following.

- i^5 has the same value as i^1 and 5 divided by 4 has remainder 1.
- i^6 has the same value as i^2 and 6 divided by 4 has remainder 2.
- i^7 has the same value as i^3 and 7 divided by 4 has remainder 3.
- i^8 has the same value as i^4 and 8 divided by 4 has remainder 0. (Recall $i^0 = 1$.)

So, if we know the remainder of the exponent of i divided by 4, then we can find the value of i raised to that exponent.

Property: The Power of i

Let n be an integer. Then

$$i^n = i^r$$

where r is the remainder of n divided by 4.

> **Procedure:** **Finding a Power of i**
>
> **Step 1:** Divide the exponent of i by 4 and find its remainder (0, 1, 2, or 3).
> **Step 2:** The result of i to this exponent is as follows.
> **a.** 1 if the remainder is 0.
> **b.** i if the remainder is 1.
> **c.** -1 if the remainder is 2.
> **d.** $-i$ if the remainder is 3.

Objective 5 Examples Find each power of i.

5a. i^{14} **5b.** i^{49} **5c.** i^{23} **5d.** i^{-16}

Solutions **5a.** $\dfrac{14}{4} = 7\text{R}2$, so $i^{14} = i^2 = -1$. **5b.** $\dfrac{49}{4} = 12\text{R}1$, so $i^{49} = i^1 = i$.

5c. $\dfrac{23}{4} = 5\text{R}3$, so $i^{23} = i^3 = -i$. **5d.** $-\dfrac{16}{4} = -4\text{R}0$, so $i^{-16} = i^0 = 1$.

☑ Student Check 5 Find each power of i.

a. i^{13} **b.** i^{35} **c.** i^{30} **d.** i^{-24}

Objective 6 ▶

Troubleshoot common errors.

Troubleshooting Common Errors

Some common errors related to complex numbers are shown.

Objective 6 Examples **A problem and an incorrect solution are given. Provide the correct solution and an explanation of the error.**

6a. Simplify $\sqrt[3]{-8}$.

Incorrect Solution	Correct Solution and Explanation
$\sqrt[3]{-8} = 2i$	The cube root of a negative number is a negative number. The square root of a negative number is imaginary. So, $\sqrt[3]{-8} = -2$ but $\sqrt{-8} = 2i\sqrt{2}$.

6b. Perform the operation: $(2 + 3i) + (-7 + 6i)$.

Incorrect Solution	Correct Solution and Explanation
$(2 + 3i) + (-7 + 6i)$ $= 2 - 7 + 3i + 6i$ $= -5 + 9i$ $= -5 + 9(-1)$ $= -5 - 9$ $= -14$	The value of i is not -1. Recall $i^2 = -1$. $(2 + 3i) + (-7 + 6i) = 2 - 7 + 3i + 6i$ $ = -5 + 9i$

ANSWERS TO STUDENT CHECKS

Student Check 1   **a.** $7i$ **b.** $i\sqrt{5}$ **c.** $2i\sqrt{5}$
     d. $6 + 2i$ **e.** $2 + i$ **f.** $\dfrac{1}{3} + \dfrac{\sqrt{2}}{2}i$

Student Check 2   **a.** $10 - 2i$ **b.** $-6 + 8i$ **c.** $8 + i$

Student Check 3   **a.** -20 **b.** $-14 + 21i$
     c. $5 - 14i$ **d.** 13 **e.** $-9 - 40i$

Student Check 4   **a.** 2 **b.** $\dfrac{3}{5} - \dfrac{4}{5}i$ **c.** $\dfrac{12}{5} + \dfrac{4}{5}i$
     d. $\dfrac{6}{17} + \dfrac{7}{17}i$

Student Check 5   **a.** i **b.** $-i$ **c.** -1 **d.** 1

SUMMARY OF KEY CONCEPTS

1. A complex number is a number that can be written in the form $a + bi$, where $i^2 = -1$ and $i = \sqrt{-1}$.
2. Add and subtract complex numbers by combining like terms.
3. Multiply complex numbers by applying the distributive property and the FOIL method. Rewrite i^2 as -1 and write the answer in the form $a + bi$.
4. Divide complex numbers by multiplying the numerator and denominator by the complex conjugate of the denominator. Write the answer in the form $a + bi$.
5. Any power of i is 1, -1, i, or $-i$. The result of i to an exponent is the same as i raised to the remainder of dividing the given exponent by 4.

GRAPHING CALCULATOR SKILLS

The calculator can express square roots of negative numbers in terms of i as well as perform operations with complex numbers.

Example 1: Simplify $4 + \sqrt{-25}$.

Solution: Change the calculator MODE from REAL to $a + bi$.

```
NORMAL  SCI  ENG
FLOAT  0123456789
RADIAN  DEGREE
FUNC  PAR  POL  SEQ
CONNECTED  DOT
SEQUENTIAL  SIMUL
REAL  a+bi  re^θi
FULL  HORIZ  G-T
SET CLOCK 01/01/02 20:55
```

```
4+√(-25)
            4+5i
```

Example 2: Perform the operations (a) $(4 - 5i) - (7 - 9i)$ and (b) $(4 - 5i)(7 - 9i)$.

Solution: To enter the imaginary unit i, press 2nd .

```
(4-5i)-(7-9i)
            -3+4i
(4-5i)(7-9i)
           -17-71i
```

SECTION 10.7 / EXERCISE SET

Write About It!

Use complete sentences in your answer to each exercise.

1. Define an imaginary number.
2. Define a complex number.
3. Define the conjugate of $a + bi$, where a and b are real numbers.
4. Explain how to add and subtract complex numbers.
5. Explain how to multiply complex numbers.
6. Explain the role of the conjugate in dividing complex numbers.

7. Show why $(5 - i)^2 \neq 24$.
8. Show why $\sqrt{-4} \cdot \sqrt{-4} \neq 4$.

Practice Makes Perfect!

Write each expression in terms of i. (See Objective 1.)

9. $\sqrt{-36}$
10. $\sqrt{-25}$
11. $\sqrt{-75}$
12. $\sqrt{-60}$
13. $1 - \sqrt{-4}$
14. $8 - \sqrt{-225}$
15. $10 + \sqrt{-16}$
16. $5 + \sqrt{-400}$

17. $\dfrac{2 + \sqrt{-81}}{2}$

18. $\dfrac{8 + \sqrt{-196}}{6}$

19. $\dfrac{7 - \sqrt{-9}}{15}$

20. $\dfrac{12 - \sqrt{-36}}{10}$

21. $\dfrac{10 - \sqrt{-363}}{14}$

22. $\dfrac{9 - \sqrt{-52}}{17}$

23. $\dfrac{5 + \sqrt{-32}}{11}$

24. $\dfrac{6 + \sqrt{-405}}{15}$

Perform the indicated operation on the complex numbers. (*See Objective 2.*)

25. $(3 - 6i) + (7 + 10i)$ 26. $(2 + 5i) + (5 - 7i)$

27. $(-3 - 4i) - (1 - 9i)$ 28. $(-5 + 9i) - (-2 - 8i)$

29. $(-1 + 7i) + (2 + 12i) + (-11 + 5i)$

30. $(-2 + 8i) - (5 - 2i) + (-9 - 6i)$

31. $(5 - 8i) - (1 + 9i) - (-3 + 2i)$

32. $(6 - i) + (6 + 3i) - (-6 + 5i)$

Simplify each product of complex numbers and write each answer in standard form. (*See Objective 3.*)

33. $\sqrt{-4} \cdot \sqrt{-49}$ 34. $\sqrt{-64} \cdot \sqrt{-9}$

35. $\sqrt{9} \cdot \sqrt{-25}$ 36. $\sqrt{16} \cdot \sqrt{-225}$

37. $\sqrt{-9}(2 + \sqrt{-100})$ 38. $\sqrt{-25}(-3 - \sqrt{-16})$

39. $3i(7 + 4i)$ 40. $2i(9 - 8i)$

41. $(1 + \sqrt{-25})(-3 - \sqrt{-16})$

42. $(2 - \sqrt{-49})(1 + \sqrt{-4})$

43. $(9 + 2i)(-2 + 3i)$ 44. $(3 - 7i)(11 + 2i)$

45. $(3 - 4i)(6 + 7i)$ 46. $(1 - 4i)(9 + 3i)$

47. $(\sqrt{5} + 3i)(2\sqrt{5} - i)$ 48. $(3\sqrt{6} - 5i)(\sqrt{6} + 2i)$

49. $(3 - \sqrt{-49})^2$ 50. $(1 + \sqrt{-25})^2$

51. $(6 - 7i)^2$ 52. $(5 - 4i)^2$

53. $(8 + i)^2$ 54. $(2 + 3i)^2$

55. $(\sqrt{7} - 2i)^2$ 56. $(\sqrt{10} + 3i)^2$

57. $(1 - \sqrt{-25})(1 + \sqrt{-25})$

58. $(3 - \sqrt{-1})(3 + \sqrt{-1})$

59. $(2 + 5i)(2 - 5i)$ 60. $(3 - 5i)(3 + 5i)$

Simplify each quotient of complex numbers and write each answer in standard form. (*See Objective 4.*)

61. $\dfrac{\sqrt{-4}}{\sqrt{-25}}$ 62. $\dfrac{\sqrt{-81}}{\sqrt{-49}}$

63. $\dfrac{3}{5i}$ 64. $\dfrac{7}{3i}$

65. $-\dfrac{12}{8i}$ 66. $-\dfrac{25}{15i}$

67. $\dfrac{-3 + 5i}{2i}$ 68. $\dfrac{1 - 3i}{6i}$

69. $\dfrac{-3 + i}{-6 + 7i}$ 70. $\dfrac{1 + 10i}{-2 + 5i}$

71. $\dfrac{7 + 9i}{1 - 3i}$ 72. $\dfrac{11 - 3i}{8 - i}$

73. $\dfrac{-2 - 4i}{7 - i}$ 74. $\dfrac{13 + 4i}{7 + 9i}$

Find each power of *i*. (*See Objective 5.*)

75. i^{30} 76. i^{52} 77. i^{21} 78. i^{51}

79. i^{82} 80. i^{47} 81. i^{-12} 82. i^{-7}

 **Mix 'Em Up!**

Simplify each complex number and write each answer in standard form.

83. $\dfrac{4 - \sqrt{-196}}{8}$ 84. $\dfrac{10 - \sqrt{-25}}{25}$

85. $\dfrac{8 + \sqrt{-225}}{10}$ 86. $\dfrac{1 + \sqrt{-400}}{4}$

87. $(7 - 8i) + (-4 + 11i)$ 88. $(-2 + 10i) + (3 + 2i)$

89. $(9 + 2i)^2$ 90. $(-8 + 2i)^2$

91. $(3 - i) + (-1 - 6i)$ 92. $(-5 - 7i) - (8 - 2i)$

93. $(-1 - 3i) - (6 + 4i) - (13 + 4i)$

94. $(-3 + 9i) - (5 + 5i) - (-12 + 3i)$

95. $\dfrac{\sqrt{32}}{\sqrt{-36}}$ 96. $\dfrac{\sqrt{48}}{\sqrt{-64}}$

97. $\sqrt{-9}(\sqrt{49} + \sqrt{-25})$ 98. $\sqrt{-4}(\sqrt{64} - \sqrt{-9})$

99. $\dfrac{-6 + 15i}{3i}$ 100. $\dfrac{12 - 3i}{6i}$

101. $(7 - \sqrt{-16})^2$ 102. $(1 + \sqrt{-36})^2$

103. $9i(8 - 3i)$ 104. $4i(-6 + 4i)$

105. $(5 + \sqrt{-1})(-4 - \sqrt{-9})$

106. $(1 - \sqrt{-81})(3 + \sqrt{-49})$

107. $(-5 + i)(15 - 3i)$ 108. $(-6 + 4i)(-2 + 2i)$

109. $(2\sqrt{7} + 4i)(\sqrt{7} - 5i)$ 110. $(3 - 2i\sqrt{5})(1 + i\sqrt{5})$

111. $\dfrac{7 + 9i}{1 - i}$ 112. $\dfrac{-13 + 4i}{1 - 6i}$

113. $(\sqrt{7} - 6i)(\sqrt{7} + 6i)$ 114. $(5 - i\sqrt{3})(5 + i\sqrt{3})$

115. $(8 - 6i)(8 + 6i)$ 116. $(12 + 5i)(12 - 5i)$

117. $\dfrac{12}{18i}$ 118. $-\dfrac{15}{36i}$

119. $\dfrac{13 + 5i}{7 + 3i}$ 120. $\dfrac{5 + i}{5 + 3i}$

Find each power of i.

121. i^{41}

122. i^{71}

123. i^{192}

124. i^{130}

125. $(2i)^7$

126. $(3i)^5$

127. $(-2i)^{-3}$

128. $(-5i)^{-1}$

 You Be the Teacher!

Correct each student's errors, if any.

129. Multiply: $\sqrt{-9} \cdot \sqrt{-16}$.

Andrew's work:
$$\sqrt{-9} \cdot \sqrt{-16} = \sqrt{-9 \cdot -16} = \sqrt{144} = 12$$

130. Subtract: $\left(6 - \sqrt{-4}\right) - \left(11 + \sqrt{-9}\right)$.

Monica's work:

$$\left(6 - \sqrt{-4}\right) - \left(11 + \sqrt{-9}\right) = \left(6 + \sqrt{4}\right) - (11 + 3i)$$
$$= 6 + 2 - 11 + 3i$$
$$= -3 + 3i$$

131. Simplify $\left(4 - i\sqrt{3}\right)^2$.

Jeannette's work:
$$\left(4 - i\sqrt{3}\right)^2 = 16 - 9i^2 = 16 + 9 = 25$$

132. Simplify $\dfrac{3 - 4i}{6 + 2i}$.

Mark's work:

$$\frac{3 - 4i}{6 + 2i} = \frac{3 - 4i}{6 + 2i} \cdot \frac{6 + 2i}{6 + 2i}$$
$$= \frac{18 - 24i + 6i - 8i^2}{36 + 4i^2}$$
$$= \frac{26 - 18i}{32} = \frac{13}{16} - \frac{9}{16}i$$

 Calculate It!

Use a calculator to verify each simplification.

133. $(3 - 5i)(7 + 2i) = 31 - 29i$

134. $\left(6 - \sqrt{-48}\right)^2 = -12 - 48i\sqrt{3}$

135. $\dfrac{-1 - 3i}{3 + 4i} = -\dfrac{3}{5} - \dfrac{1}{5}i$　　**136.** $(-5i)^{-3} = -\dfrac{1}{125}i$

 GROUP ACTIVITY ／ **Solutions of Quadratic Equations**

In Chapter 11, we will solve quadratic equations that do not factor. The solutions of these equations will involve radical expressions. We must use radical operations to verify that the number is a solution of the given equation. Radical operations are also used to simplify the solutions.

1. Write each radical expression in simplified radical form.
   a. $-2 \pm \sqrt{50}$
   b. $\dfrac{6 \pm \sqrt{20}}{2}$
   c. $-1 \pm \sqrt{-4}$

2. Verify that the values in part 1 are solutions of the corresponding equations. (1a are solutions of 2a, 1b are solutions of 2b, and 1c are solutions of 2c.)
   a. $(x + 2)^2 = 50$
   b. $x^2 - 6x + 4 = 0$
   c. $3(x + 1)^2 + 12 = 0$

3. Find the decimal approximation of the radicals in 1a and 1b. Round to the nearest hundredth.

4. Use the decimal approximations to show that the values are solutions of the corresponding equations given in 2a and 2b.

Chapter 10 / REVIEW

Rational Exponents, Radicals, and Complex Numbers

What's the big idea? Now that you have completed Chapter 10, you should be able to see a connection between radical expressions and exponents. You should also be able to simplify radicals, perform operations with radicals, and solve equations containing radical expressions. Lastly, you should be able to express square roots of negative numbers as complex numbers.

The Tools

Listed below are the key terms, skills, formulas, and properties you should know for this chapter.

The page reference is provided if you need additional help with the given topic. The Study Tips will assist in your preparation for an exam.

Study Tips

1. Learn all of the terms, formulas, and properties. Make flash cards and have someone quiz you.
2. Rework problems from the exercises and also the ones you worked in class. Work additional problems from the review exercises.
3. Review the summaries of key concepts.
4. Work the chapter test.
5. Be sure to review the online resources for additional study materials.

Terms

Conjugates   749
Complex conjugates   767
Complex number   764
Cube root   701
Extraneous solution   754
Imaginary unit i 764
Index   703
Like radicals   737

Negative square root   700
nth root   703
Perfect cube   702
Perfect square   700
Principal square root   700
Radical   700
Radical equation   754
Radical expression   701

Radical function   706
Radical sign   701
Radicand   701
Rational exponent   716
Rationalizing the denominator   748
Square root   700

Formulas and Properties

- $b^{1/n}$ 716
- $b^{m/n}$ 717
- $b^{-m/n}$ 718
- $\sqrt[n]{a^n}$ 706
- $(\sqrt[n]{a})^n$ 705
- Cube root   701
- Distance formula   731

- nth root   703
- Power of i 769
- Power property   754
- Product of complex conjugates   767
- Product rule for radicals   726 & 739
- Pythagorean theorem   758

- Quotient rule for radicals   730 & 746
- Rules of exponents   719
- Special triangles, 30°-60°-90° and 45°-45°-90°   741
- Square root of a negative number   764

CHAPTER 10 / SUMMARY

How well do you know this chapter? Complete the following questions to find out. Take a look back at the section if you need help.

SECTION 10.1 Radicals and Radical Functions

1. The square root of a real number A is a real number b if _____.

2. The symbol ____ denotes the principal or _____ square root.

3. The number inside the radical is called the _____ .

4. If $A < 0$, then the $\sqrt{A}$ is __ a(n) _____.

5. The cube root of a real number A is a real number b if _____.

6. The nth root of a real number A is a real number b if _____.

7. If n is an even number, _____ for the nth root of A to be a real number.

8. If n is an odd number, _____ for the nth root of A to be a real number.

9. If n is even, then $\sqrt[n]{a^n} = $ __ for $a \geq 0$ and $\sqrt[n]{a^n} = $ __ for $a < 0$.

10. If n is odd, then $\sqrt[n]{a^n} = $ __ for all values of a.

11. As long as $\sqrt[n]{a}$ is a real number, $(\sqrt[n]{a})^n = $ __.

12. If n is even, the domain of $f(x) = \sqrt[n]{x}$ is _____ .

13. If n is odd, the domain of $f(x) = \sqrt[n]{x}$ is _____.

SECTION 10.2 Rational Exponents

14. An exponent with a fraction is called a(n) _____ exponent.

15. For n a positive integer greater than 1 and $\sqrt[n]{b}$ a real number, $b^{1/n} = $ ___.

16. If m and n are positive integers greater than 1 and $\sqrt[n]{b}$ is a real number, then $b^{m/n} = $ ____.

17. If m and n are positive integers greater than 1 and $\sqrt[n]{b}$ is a real number, then for $b \neq 0$, $b^{-m/n} = $ ___.

18. The exponent rules work for integer exponents but also for _____ exponents.

SECTION 10.3 Simplifying Radical Expressions and the Distance Formula

19. The product rule for radicals states that for a and b real numbers, $\sqrt[n]{ab} = $ _____.

20. For a radical to be in its simplest form, the radicand cannot contain any _____ factors or any exponents greater than or equal to the ____.

21. The quotient rule for radicals states that for a and b real numbers and $b \neq 0$, $\sqrt[n]{\dfrac{a}{b}} = $ ___.

22. The distance between two points (x_1, y_1) and (x_2, y_2) is _____.

SECTION 10.4 Adding, Subtracting, and Multiplying Radical Expressions

23. Like radicals are radicals with the same _____ and the same ____.

24. To add like radicals, combine the _____ of the like radical expressions then keep the _____ the same.

25. Radicals should be _____ before adding.

26. We can multiply radicals with the same ____. In this case, we multiply the _____ and simplify the product.

27. We can multiply radical expressions by applying the _____ property and the ____ method.

28. The product of _____ doesn't contain a radical.

SECTION 10.5 Dividing Radical Expressions and Rationalizing

29. We can divide radicals with the same ____. In this case, divide the _____ and simplify the quotient.

30. The process of removing a radical from the denominator of a fraction is called _____.

31. To remove a radical from the denominator, we must multiply the fraction by a form of ___ that makes the denominator a perfect power whose exponent is the ____.

32. To remove a radical from the denominator if the denominator is a binomial, we must multiply by a form of one that involves the _____ of the denominator.

SECTION 10.6 Radical Equations and Their Applications

33. The power property states that if $a = b$, then _____.

34. When we apply the power property, we obtain the solutions of the original equation but we can also obtain _____ solutions. Therefore, we must ____ the proposed solutions in the original equation.

35. To apply the power property, the radical expression must be _____ to one side of the equation.

36. If an equation contains more than one _____ expression, we may have to apply the power property ____ .

37. The Pythagorean theorem states that if a and b are legs of a right triangle and c is a hypotenuse, then _____.

SECTION 10.7 Complex Numbers

38. The _____ can be defined as the number whose square is equal to -1. We write it as ... $= \sqrt{-1}$ and __ $= -1$.

39. A complex number is a number of the form ____, where _ and _ are real numbers.

40. We add and subtract complex numbers by _____ like terms.

41. We multiply complex numbers by applying the _____ property and the ____ method.

42. To divide complex numbers, we multiply the numerator and denominator by the _____ of the denominator. If the denominator is $a + bi$, then we will multiply by ____ .

43. To find i^n find the _____ of n divided by _. Then $i^n = $ _.

CHAPTER 10 / REVIEW EXERCISES

SECTION 10.1

Simplify each radical. Assume all radicands with even indices are positive real numbers. (See Objectives 1–3.)

1. $-\sqrt{196t^{20}}$ **2.** $\sqrt{4y^{12}}$ **3.** $\sqrt{100h^8k^2}$

4. $\sqrt{625m^{10}n^6}$ **5.** $\sqrt{\dfrac{81s^{16}}{25}}$ **6.** $\sqrt{\dfrac{144t^6}{49}}$

7. $\sqrt{3.61a^6}$ **8.** $\sqrt{2.25c^{14}}$ **9.** $\sqrt{(3x-1)^2}$

10. $(\sqrt{5x+3})^2$ for $x \geq -\dfrac{3}{5}$ **11.** $\sqrt[3]{-1000b^3}$

12. $\sqrt[5]{-32a^{10}b^5}$ **13.** $\sqrt[4]{64x^8y^4}$

14. $-\sqrt[4]{81a^4b^{16}}$ **15.** $\sqrt[6]{4096a^{12}b^6}$

16. $\sqrt[7]{x^7y^{28}}$

Find the domain of each radical function. Write answers in interval notation. Evaluate the radical functions for the indicated values. Round answers to two decimal places as needed. (See Objective 6.)

17. $f(x) = \sqrt{8x-5}$; $x = -2$, $x = 0$, $x = 2$

18. $f(x) = \sqrt{7x-9}$; $x = -4$, $x = 0$, $x = 6$

19. $f(x) = \sqrt[3]{3-8x}$; $x = -3$, $x = 0$, $x = 2$

20. $f(x) = \sqrt[3]{4-x}$; $x = -4$, $x = 0$, $x = 4$

Graph each radical function using the accompanying tables. (See Objective 7.)

21. $f(x) = \sqrt{16-3x}$

x	$f(x)$
-3	
0	
1	
4	

22. $f(x) = 2 - \sqrt[3]{x}$

x	$f(x)$
-8	
-1	
0	
1	
8	

SECTION 10.2

Rewrite each expression as a radical expression and simplify, if possible. Assume all variables represent positive real numbers. (See Objectives 1 and 2.)

23. $(-64x^{12})^{1/3}$ **24.** $(-27y^9)^{2/3}$

25. $(1000a^6b^3)^{2/3}$ **26.** $(125c^3d^6)^{4/3}$

27. $(32x^{-15}y^5)^{3/5}$ **28.** $(y+5)^{2/3}$

29. $(2x-3)^{2/5}$ **30.** $\left(\dfrac{a^{12}}{81b^8}\right)^{3/4}$

Simplify each expression and write each answer with positive exponents. Assume all variables represent positive real numbers. (See Objectives 3 and 4.)

31. $(-125a^3)^{-1/3}$ **32.** $(27y^{-6})^{-2/3}$

33. $(x^{-3}y^{12})^{-4/3}$ **34.** $(-r^{10}s^{-5})^{-2/5}$

35. $2^{5/4} \cdot 2^{3/4}$ **36.** $3^{4/7} \cdot 3^{3/7}$

37. $\dfrac{x^{4/7}}{x^{2/3}}$ **38.** $\dfrac{y^{3/5}}{y^{1/2}}$

39. $(-32x^{15}y^{20})^{3/5}$ **40.** $(-64a^6b^3)^{2/3}$

41. $(5x^{1/2} - 3y^{1/3})(5x^{1/2} + 3y^{1/3})$

42. $(a^{4/5} - 2b^{1/2})(a^{4/5} + 6b^{1/2})$

43. $x^{1/2}(2x^{1/2} + 3x^{2/3})$ **44.** $2s^{2/3}(3s^{2/3} - 4s^{1/2})$

Use rational exponents to simplify each expression. Assume all variables represent positive real numbers. (See Objective 5.)

45. $\sqrt[3]{125x^9}$ **46.** $\sqrt[5]{32y^{25}}$

47. $\sqrt[6]{u^3v^6}$ **48.** $\sqrt[6]{x^6y^9}$

49. $\sqrt[4]{\dfrac{36}{c^6}}$ **50.** $\sqrt[4]{\dfrac{9}{y^2}}$

51. $\sqrt{8} \cdot \sqrt[3]{4}$ **52.** $\sqrt[3]{9} \cdot \sqrt[6]{3}$

Solve each problem. (See Objective 6.)

53. How much money will be in an account in 6 months if you invest $1200 at an annual interest rate of 1.2% compounded quarterly ($n = 4$)?

54. What expression would you need to enter on a scientific calculator to evaluate $\sqrt[5]{650}$? Use the calculator to approximate the value and round your answer to two decimal places.

SECTION 10.3

Simplify each radical. Assume all variables represent positive real numbers. (See Objectives 1 and 2.)

55. $\sqrt{360a^5b^{12}}$ **56.** $\sqrt{\dfrac{50r^9}{49s^2}}$

57. $\sqrt[3]{250x^8y^{16}}$ **58.** $\sqrt[5]{-64r^9t^{18}}$

59. $\sqrt[3]{\dfrac{2000x^5}{27y^{12}}}$ **60.** $\sqrt[4]{\dfrac{32r^{15}}{81t^{12}}}$

61. $\dfrac{\sqrt{54d^{11}}}{\sqrt{18c^6}}$ **62.** $\dfrac{\sqrt[3]{14y^7}}{\sqrt[3]{56x^6}}$

Find the distance between each pair of points. Write each answer in simplest radical form. (*See Objective 3.*)

63. $\left(3\sqrt{2}, -\sqrt{6}\right)$ and $\left(5\sqrt{2}, -3\sqrt{6}\right)$

64. $\left(-2, 10\sqrt{5}\right)$ and $\left(1, 7\sqrt{5}\right)$

65. $\left(\dfrac{\sqrt{3}}{2}, -4\sqrt{2}\right)$ and $\left(\dfrac{5\sqrt{3}}{2}, -2\sqrt{2}\right)$

66. $\left(\dfrac{\sqrt{5}}{3}, \dfrac{3}{4}\right)$ and $\left(-\dfrac{\sqrt{5}}{6}, \dfrac{1}{4}\right)$

SECTION 10.4

Perform each operation and write each answer in simplest radical form. (*See Objectives 1 and 2.*)

67. $-9\sqrt{6} + 14\sqrt{6} + \sqrt{6}$

68. $4\sqrt{14x^3} - 8\sqrt{14x^3} - 9x\sqrt{14x}$

69. $-7a\sqrt{216ab^2} + 15b\sqrt{294a^3}$

70. $13\sqrt{45cd^5} - 2d^2\sqrt{245cd}$

71. $9\sqrt[3]{3xy^3} - 5y\sqrt[3]{27x} + \sqrt[3]{24xy^3}$

72. $14t\sqrt[5]{729r^5t} - 10r\sqrt[5]{96t^6}$

73. $\dfrac{7\sqrt{9a^2b}}{2} + \sqrt{\dfrac{16a^2b}{25}}$

74. $\dfrac{\sqrt{80ab^3}}{3} - \sqrt{\dfrac{45ab^3}{4}}$

75. $\sqrt[3]{\dfrac{24x^2y^3}{125}} - \dfrac{\sqrt[3]{375x^2y^3}}{5}$

76. $\sqrt[3]{\dfrac{162a}{125b^3}} + \dfrac{\sqrt[3]{48a}}{3b}$

77. $\sqrt[5]{10x^4y^{-10}}\sqrt[5]{16x^8y^{15}}$ 78. $\sqrt[3]{96a^5b^6}\sqrt[3]{6a^7b^{-4}}$

79. $\sqrt{7}\left(5 - \sqrt{14}\right)$ 80. $\sqrt{5}\left(6 + \sqrt{15}\right)$

81. $\left(\sqrt{3} - 2\sqrt{6}\right)\left(\sqrt{3} + 5\sqrt{6}\right)$

82. $\left(3\sqrt{2} - \sqrt{10}\right)^2$ 83. $\left(7 - \sqrt{2a}\right)\left(7 + \sqrt{2a}\right)$

84. $\left(2\sqrt{12a} + 5\sqrt{b}\right)\left(\sqrt{3a} - \sqrt{16b}\right)$

In Exercises 85 and 86, find the perimeter and area of the triangle. Approximate each answer to one decimal place. (*See Objective 3.*)

85. $a = 13\sqrt{2}$ and $c = 26$ 86. $a = 6\sqrt{2}$ and $c = 12$

87. In a 30°-60°-90° triangle, the leg opposite the 30° angle measures $12\sqrt{6}$ cm. Find the length of the other leg and the hypotenuse.

88. In a 30°-60°-90° triangle, the leg opposite the 30° angle measures $8\sqrt{2}$ m. Find the length of the other leg and the hypotenuse.

SECTION 10.5

Perform each operation and write each answer in simplest radical form. Assume all variables represent positive real numbers. (*See Objective 1.*)

89. $\dfrac{\sqrt{150x^{15}}}{\sqrt{2x^6}}$ 90. $\dfrac{\sqrt[3]{48t^9}}{\sqrt[3]{3t^8}}$

91. $\dfrac{\sqrt{126y^7}}{2\sqrt{7y^2}}$ 92. $\dfrac{5\sqrt[3]{72a^{10}}}{\sqrt[3]{3a^2}}$

93. $\dfrac{6\sqrt[5]{640c^{12}}}{\sqrt[5]{5c^4}}$ 94. $\dfrac{7\sqrt[4]{320a^{15}}}{\sqrt[4]{5a^5}}$

95. $\dfrac{\sqrt[4]{64a^{11}b^{11}}}{2\sqrt[4]{2a^6b^{-3}}}$ 96. $\dfrac{3\sqrt[3]{945a^8b^{-4}}}{\sqrt[3]{7a^2b^{-9}}}$

Simplify by rationalizing each denominator. (*See Objectives 2 and 3.*)

97. $\dfrac{\sqrt{40}}{\sqrt{15x}}$ 98. $\dfrac{6}{\sqrt[5]{16x^3}}$

99. $\dfrac{12}{\sqrt[4]{27a}}$ 100. $\dfrac{18}{\sqrt[5]{4b^3}}$

101. $\dfrac{2}{\sqrt{7} + \sqrt{5}}$ 102. $\dfrac{2 + \sqrt{7}}{5 - \sqrt{7}}$

103. $\dfrac{5 - \sqrt{6}}{\sqrt{3} - 2}$ 104. $\dfrac{\sqrt{2}}{\sqrt{6} - \sqrt{2}}$

105. $\dfrac{a}{\sqrt{a + 4} - 2}$ 106. $\dfrac{b}{\sqrt{b + 16} - 4}$

SECTION 10.6

Solve each radical equation. (*See Objective 1.*)

107. $\sqrt{2x - 5} + 2 = 1$ 108. $\sqrt{12b - 23} + 2 = 7$

109. $\sqrt[4]{3x - 5} - 6 = -4$ 110. $\sqrt[3]{7y + 6} + 9 = 12$

111. $\sqrt{2x - 17} + 8 = x$ 112. $\sqrt{72 - 4y} + 3 = y$

113. $\sqrt{2y + 124} - 8 = \sqrt{y}$ 114. $\sqrt{2z + 41} + 5 = \sqrt{z}$

115. $\sqrt{6x + 31} = 2 + \sqrt{2x + 11}$

116. $\sqrt{3x + 10} + 2 = \sqrt{7x + 22}$

Solve each problem. (*See Objective 2.*)

117. Find the length of the missing side of a right triangle if one of the legs is 35 ft and the hypotenuse is 37 ft.

118. Find the length of the hypotenuse of a right triangle if the lengths of the two legs are 40 in. and 9 in.

119. One of the ropes that is used to tie down a tent needs to be replaced. Determine the distance from the top of the tent to the tie-down stake if the tent is 9 ft tall and the tie-down stake is placed 3 ft from the tent.

120. A person's body surface area (BSA) in square meters is given by $BSA = \sqrt{\dfrac{hw}{3131}}$, where h is a person's height in inches and w is a person's weight in pounds.

  a. Find the BSA of a person who is 4 ft 9 in. and weighs 116 lb.

  b. The average BSA of a woman is 1.9 m². If a woman is 5 ft 11 in. tall, how much should she weigh to have an average BSA?

SECTION 10.7

Perform each operation and write each answer in standard form. (*See Objectives 1–5.*)

121. $(3 - 5i) + (-11 + 7i)$ **122.** $(-7 + 9i) - (13 - 2i)$

123. $(-4 - 2i) - (9 + 5i) + (13 + 14i)$

124. $(-9 + 4i) + (1 + 3i) - (-15 + 7i)$

125. $\sqrt{-9} \cdot \sqrt{-25}$ **126.** $6i(-8 - 5i)$

127. $(-2 + 3i)(5 - 4i)$ **128.** $(6 - 3i)(-4 + 8i)$

129. $(1 - \sqrt{6}i)(4 + 3\sqrt{6}i)$ **130.** $(3 + \sqrt{5}i)^2$

131. $(\sqrt{11} - 2i)(\sqrt{11} + 2i)$ **132.** $\dfrac{-8 + 12i}{4i}$

133. $\dfrac{9 + i}{1 + 3i}$ **134.** $\dfrac{1 + 5i}{4 - 3i}$

135. $(2i)^6$ **136.** $(-3i)^5$

137. i^{-7} **138.** $(-i)^{-9}$

CHAPTER 10 TEST / RATIONAL EXPONENTS, RADICALS, AND COMPLEX NUMBERS

1. $-\sqrt{25a^{16}} =$
  a. $5a^4$ **b.** $-5a^4$ **c.** $5a^8$ **d.** $-5a^8$

2. $\sqrt{-256} =$
  a. -16 **b.** 16 **c.** -128 **d.** not a real number

3. $\sqrt[3]{-729} =$
  a. 9 **b.** -9 **c.** 243 **d.** not a real number

4. $-\sqrt[4]{16x^{12}} =$
  a. $-2x^3$ **b.** $2x^3$ **c.** $-4x^3$ **d.** $4x^3$

5. The best approximation for $\sqrt[7]{12{,}516}$ is
  a. 3.849 **b.** 1788 **c.** 111.875 **d.** 783.125

6. $\sqrt{75} - \sqrt{12} + \sqrt{27} =$
  a. $\sqrt{60}$ **b.** $6\sqrt{3}$ **c.** $2\sqrt{15}$ **d.** $10\sqrt{3}$

7. $(\sqrt{5} - 1)(2\sqrt{5} + 7) =$
  a. $-7 + 7\sqrt{5}$ **b.** $3 - 5\sqrt{5}$ **c.** $3 + 5\sqrt{5}$
  d. $-7 - 5\sqrt{5}$

8. $(4\sqrt{3} - 6\sqrt{2})^2 =$
  a. $120 - 48\sqrt{6}$ **b.** $-24 - 48\sqrt{6}$ **c.** 120
  d. -24

9. $\dfrac{5\sqrt{12}}{10\sqrt{2}} =$
  a. $\sqrt{3}$ **b.** $\dfrac{\sqrt{6}}{2}$ **c.** 3 **d.** 12

10. The solution set of $\sqrt{m^2 + 5m - 8} = m + 1$ is
  a. $\{3\}$ **b.** $\left\{\dfrac{9}{5}\right\}$ **c.** $\{-3\}$ **d.** $\{2\}$

11. The solution set of $\sqrt{x + 6} = x$ is
  a. $\{-2, 3\}$ **b.** $\{-2\}$ **c.** $\{3\}$ **d.** $\varnothing$

12. $625^{3/4} =$
  a. 15 **b.** 25 **c.** 125 **d.** 5

13. $4^{3/5} \cdot 4^{7/5} =$
  a. 8 **b.** 16 **c.** 256 **d.** 4

14. $\dfrac{16^{-5/4}}{16^{-3/4}}$
  a. 8 **b.** 4 **c.** $\dfrac{1}{8}$ **d.** $\dfrac{1}{4}$

15. $(32k^{10})^{1/5} =$
  a. $2k^2$ **b.** $32k^2$ **c.** $32k^{50}$ **d.** $2k^{50}$

16. When simplified, $3i(5 + 2i) - 4i$ is
  a. $-6 + 11i$ **b.** $-6 + 4i$ **c.** $5i$ **d.** $17i$

Simplify each radical. Assume all variables represent positive real numbers.

17. $\sqrt{50x^4y^3}$ **18.** $\sqrt{\dfrac{49a^6}{16}}$

19. $\sqrt[3]{40}$ **20.** $\sqrt[3]{\dfrac{8}{27b^6}}$

21. $\sqrt[4]{162a^7}$

Perform each operation and write each answer in simplest radical form. Rationalize the denominators when necessary.

22. $\sqrt{6x} \cdot \sqrt{2x}$ **23.** $\sqrt[3]{4y^2} \cdot \sqrt[3]{6y^2}$

24. $(\sqrt{4x - 9})^2$ **25.** $\dfrac{\sqrt{18b^3}}{\sqrt{2b}}$

26. $\dfrac{1}{\sqrt{3}}$ **27.** $\dfrac{6\sqrt[3]{2}}{\sqrt[3]{5}}$

28. $\dfrac{\sqrt{3}}{\sqrt{5} + \sqrt{2}}$ **29.** $\sqrt{48} - 5\sqrt{300}$

30. $2\sqrt[3]{16} - 4\sqrt[3]{54}$ **31.** $(\sqrt{5} + 6)^2$

32. $(\sqrt{5} + 6)(\sqrt{5} - 6)$

Solve each radical equation.

33. $\sqrt{x + 2} = 3$

34. $\sqrt{5x + 2} + 4 = 3$

35. $\sqrt{2x + 1} = x - 1$

36. A right triangle has legs that are each $4\sqrt{5}$ units. Find the simplest radical expression that represents the length of the hypotenuse. Then find the perimeter and area of the triangle.

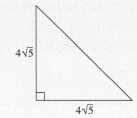

Write each complex number in terms of *i*.

37. $\sqrt{-64}$

38. $4 + \sqrt{-9}$

Perform each operation with the complex numbers.

39. $(-2 + 10i) + (-5 - 5i) + (12 + 2i)$

40. $3i(-7 + 4i)$

41. $(9 + 4i)(9 - 4i)$

42. $(9 + 4i)^2$

43. $\dfrac{4}{3i}$

44. $\dfrac{13}{1 + 5i}$

CUMULATIVE REVIEW EXERCISES / CHAPTERS 1–10

Translate each problem into a linear equation and solve it to find the information. (*Section 2.3, Objectives 3 and 4*)

1. The sum of two consecutive even integers is 66. Find the numbers.

2. The sum of two consecutive integers is 43. Find the numbers.

3. In 2011, the annual salary of the Secretary of State, Hillary Clinton, was $199,700. The annual salary of the Speaker of the House of Representatives, John Boehner, was $23,800 more than the salary of Secretary Clinton. What was the annual salary of Speaker Boehner? (Source: http://usgovinfo.about.com)

4. In 2011, the annual salary of the rank-and-file members of the House and Senate was $174,000. The annual salary of the Majority Leader was $19,400 more than the salary of the rank-and-file members of the House and Senate. What was the annual salary for the Majority Leader? (Source: http://usgovinfo.about.com)

Solve each formula for the specified variable. (*Section 2.5, Objective 3*)

5. $S = 2\pi rh + \pi r^2$ for h

6. $A = P(1 + rt)^n$ for P

Solve each compound inequality. Write the solution in interval notation. (*Section 9.1, Objectives 3 and 4*)

7. $4(1 - 3x) < -5$ and $8x - 2 < 14$

8. $6 - 2x \le 0$ or $5x - 2 \le 3$

Solve each absolute value inequality. (*Section 9.3, Objectives 1 and 2*)

9. $|4 - 3x| + 5 \le 12$

10. $|5x + 1| - 3 > 16$

Find the slope of each line formed by the ordered pairs. (*Section 3.4, Objective 1*)

11. $(3, 4)$ and $(-9, 12)$

12. $(-5, 1)$ and $(-14, 10)$

Find $f(a + 2)$ for each function. (*Section 3.6, Objective 4*)

13. $f(x) = x^2 - 3x - 5$

14. $f(x) = -3x + 8$

Rewrite the equation using function notation, if possible. Identify m and b. (*Section 3.6, Objective 5*)

15. $6x - 5y = 10$

16. $7y + 21 = 0$

Write the equation of the line that satisfies the given information. Write each answer in slope-intercept form. (*Section 3.5, Objectives 3 and 4*)

17. $(-3, -14)$ and $(2, 11)$

18. passes through $(3, -8)$ and parallel to $y = 5$

Solve each system of linear equations graphically. (*Section 4.1, Objectives 2 and 3*)

19. $\begin{cases} -5x + y = 24 \\ 10x - 2y = 16 \end{cases}$

20. $\begin{cases} x + 3y = 9 \\ x + 2y = 0 \end{cases}$

Solve each system of linear equations using substitution or elimination. (*Section 4.2, Objective 1; Section 4.3, Objective 1*)

21. $\begin{cases} x + 7y = 16 \\ -2x + 4y = 22 \end{cases}$

22. $\begin{cases} 0.6x - 0.1y = 0.5 \\ -0.5x + 0.3y = 1.1 \end{cases}$

23. $\begin{cases} \dfrac{1}{3}x - y = 1 \\ -\dfrac{1}{2}x + y = \dfrac{3}{4} \end{cases}$

24. $\begin{cases} \dfrac{4}{9}x - \dfrac{1}{3}y = 4 \\ \dfrac{1}{6}x + \dfrac{3}{2}y = -5 \end{cases}$

Solve each problem. (*Section 4.4, Objectives 1–6*)

25. A chemist has a 48% acid solution. She needs 80 L of an 18% acid solution. How many liters of water and how many liters of the 48% acid solution must she mix together to obtain 80 L of an 18% acid solution?

26. A chemist has a 64% alcohol solution. He needs 30 gal of a 24% alcohol solution. How many gallons of water and how many gallons of the 64% alcohol solution must he mix together to obtain 30 gal of a 24% alcohol solution?

27. Jennifer invests $8000 in two different accounts. She invests part of her money in a savings account that earns 1.25% simple interest and the rest of her money in a money market fund that earns 0.9% simple interest. If she earns a total of $94.75 interest in 1 yr, how much did she invest in each account?

28. Nancy invests $15,000 in two different accounts. She invests part of her money in a savings account that earns 1.5% simple interest and the rest of her money in a CD that earns 4% simple interest. If she earns a total of $512.50 interest in 1 yr, how much did she invest in each account?

Simplify each expression. Write each answer with positive exponents. (*Section 5.1, Objectives 1 and 2*)

29. $-9(7a)^0$

30. $-2(-3x)^0$

31. $\left(\dfrac{3}{5}\right)^{-2}$

32. $\left(-\dfrac{2}{7}\right)^{-3}$

Solve each problem. (*Section 5.1, Objective 6*)

33. Write an expression for the surface area of a can if the height is twice its radius.

34. Write an expression for the surface area of a can if the radius is twice its height.

Simplify each expression using the rules of exponents. Write each answer with positive exponents. (*Section 5.1, Objectives 2–4; Section 5.2, Objective 1*)

35. $(-2x^2y^8)^4$

36. $(-3a^{-6})^{-2}$

37. $\left(\dfrac{3p^3}{5q^{-4}}\right)^{-3}$

38. $\left(\dfrac{2r^3}{7s^{-4}}\right)^{-2}$

Simplify each expression. (*Section 5.3, Objective 3; Section 5.4, Objective 3, and Section 5.5, Objective 2*)

39. $(2x^2 + 6x + 16) - (-3x^2 + 8x - 10)$

40. $(-7y^2 + 8y - 4) - (8y^2 - 6y + 13)$

41. $(a + 5)(a^2 - 5a + 25)$ **42.** $(2b - 1)(4b^2 + 2b + 1)$

43. $(a + 2b)^2$ **44.** $(7c - d)^2$

Use grouping to factor each polynomial completely, if possible. (*Section 6.1, Objective 3*)

45. $6x^3 + 21x^2 - 10x - 35$ **46.** $15y^3 - 3y^2 + 10y + 2$

Factor each trinomial. (*Section 6.2, Objectives 1 and 2*)

47. $a^6 - 2a^3 - 35$ **48.** $d^8 + 9d^4 - 36$

Factor completely. (*Section 6.4, Objectives 1 and 2*)

49. $x^4 - 81$ **50.** $625y^4 - 1$

51. $a^3 + 1000$ **52.** $64b^3 - 1$

Solve each equation by factoring. (*Section 6.5, Objective 2*)

53. $y(4y + 1) = 3$ **54.** $z(2z + 3) = 14$

Simplify each rational expression. (*Section 7.1, Objective 3*)

55. $\dfrac{x^2 - 49}{5x^2 - 29x - 42}$ **56.** $\dfrac{16x^2 - 1}{4x^2 - 11x - 3}$

Multiply or divide the rational expressions. Write each answer in simplest form. (*Section 7.2, Objectives 1 and 2*)

57. $\dfrac{x^2 - 5x}{x^2 - 2x - 15} \cdot \dfrac{2x^2 - x - 21}{4x^2}$

58. $\dfrac{x^3 - 8}{x^2 + 3x - 10} \div \dfrac{2x^2 + 4x + 8}{x^2 + 4x - 5}$

Perform each operation using long or synthetic division. (*Section 5.7, Objectives 2 and 3*)

59. $\dfrac{2x^3 - 7x^2 - 16}{x - 4}$ **60.** $\dfrac{3x^3 - 79x - 20}{x + 5}$

Add or subtract the rational expressions. Write each answer in lowest terms. (*Section 7.4, Objective 2*)

61. $\dfrac{2}{x^2 - 4} + \dfrac{3}{x^2 - 3x - 10}$

62. $\dfrac{x + 1}{9x^2 - 1} - \dfrac{1}{12x - 4}$

Simplify each complex fraction. (*Section 7.5, Objectives 1 and 2*)

63. $\dfrac{\dfrac{3}{a} + \dfrac{5}{b}}{5a + 3b}$ **64.** $\dfrac{\dfrac{7}{a} - \dfrac{3}{b}}{3a - 7b}$

Solve each rational equation. (*Section 7.6, Objective 1*)

65. $\dfrac{5}{x - 5} - \dfrac{4}{x + 1} = \dfrac{7}{x^2 - 4x - 5}$

66. $\dfrac{2}{3x - 2} - \dfrac{3}{x - 4} = \dfrac{12}{3x^2 - 14x + 8}$

Boyle's law states that if the temperature remains the same, the volume *V* of a gas is inversely proportional to the pressure *P* applied to it. (*Section 8.4, Objective 2*)

67. A given mass of a gas occupies 252 mL at 640 mm Hg. If the temperature remains constant, what volume will the gas occupy if the pressure is increased to 720 mm Hg?

68. A given mass of a gas occupies 320 mL at 540 mm Hg. If the temperature remains constant, what volume will the gas occupy if the pressure is increased to 900 mm Hg?

Find the constant of variation and the equation of variation for each problem. (*Section 8.4, Objectives 2 and 3*)

69. *x* varies jointly as the cube of *y* and the sum of *u* and *v*; $x = 125$ when $y = 2.5$, $u = 2.3$, and $v = 1.7$

70. *F* varies inversely as the square root of *G*; $F = 1.2$ when $G = 6.25$.

Simplify each expression. Assume all variables represent positive real numbers. (*Section 10.1, Objectives 1–3*)

71. $\sqrt{\dfrac{49x^6}{4}}$ **72.** $\sqrt{\dfrac{64y^{12}}{25}}$

73. $\sqrt[3]{\dfrac{125x^6}{64y^3}}$ **74.** $\sqrt[3]{\dfrac{a^9}{8b^{12}}}$

75. $\sqrt[7]{128x^{14}y^7}$ **76.** $\sqrt[7]{-a^{14}b^{21}}$

Find the domain of each radical function and write in interval notation. Evaluate each function for the indicated values. (*Section 10.1, Objective 6*)

77. $f(x) = \sqrt{5 - x}$; $x = -4, x = 1, x = 6$

78. $f(x) = \sqrt{x - 4}$; $x = 5, x = 4, x = 3$

Rewrite each expression as a radical expression and simplify, if possible. Assume all variables represent positive real numbers. (*Section 10.2, Objectives 1–3*)

79. $(64x^9)^{1/3}$ **80.** $(27y^{15})^{1/3}$

81. $(-64)^{2/3}$ **82.** $(-36)^{3/2}$

83. $(64a^6b^{12})^{-5/6}$ **84.** $(a^{-8}b^{12})^{-5/4}$

Simplify each expression. Assume all variables represent positive real numbers. Write each answer with positive exponents. (*Section 10.2, Objective 4*)

85. $\dfrac{x^{3/4}}{x^{1/3}}$ **86.** $\dfrac{y^{1/3}}{y^{-1/6}}$

87. $(6x^{2/5} - y^{1/2})(x^{2/5} - 3y^{1/2})$ **88.** $(r^{1/2} - 2s^{2/3})(r^{1/2} - 2s^{2/3})$

Use rational exponents to simplify each expression. (*Section 10.2, Objective 5*)

89. $\sqrt[12]{a^3}$ **90.** $\sqrt[6]{b^3}$

91. $\sqrt[10]{32}$ **92.** $\sqrt[6]{27}$

Simplify each radical expression. Assume all variables represent positive real numbers. (*Section 10.3, Objectives 1 and 2*)

93. $\sqrt{72x^4}$ **94.** $\sqrt{50y^6}$

95. $\sqrt{98a^5b^4}$ **96.** $\sqrt{48r^2t^7}$

97. $\sqrt[3]{\dfrac{8x^4}{y^3}}$ **98.** $\sqrt[3]{\dfrac{125a^7}{b^6}}$

Find the distance between each pair of points. Write answers with simplified radicals. (*Section 10.3, Objective 3*)

99. $(3, -6)$ and $(1, 2)$

100. $(-1, 9)$ and $(6, 10)$

Perform the indicated operation and write answers in simplest radical form. (*Section 10.4, Objectives 1 and 2*)

101. $2\sqrt{50x^2} - \sqrt{18x^2} + 3x\sqrt{98}$

102. $5\sqrt[3]{40} + 3\sqrt[3]{135} - 4\sqrt[3]{5}$

103. $\sqrt{3}(2 + \sqrt{6})$

104. $(\sqrt{5} + \sqrt{10})(3\sqrt{5} - \sqrt{40})$

Simplify each quotient of radical expressions and write answers in simplest radical form. (*Section 10.5, Objective 1*)

105. $\dfrac{\sqrt{150x^{14}}}{\sqrt{3x^{11}}}$ **106.** $\dfrac{\sqrt{54y^{11}}}{\sqrt{2y^8}}$

107. $\dfrac{\sqrt{75c^4d^9}}{\sqrt{3cd^3}}$ **108.** $\dfrac{\sqrt{20x^8y^3}}{\sqrt{5x^5y^{-1}}}$

Simplify each expression by rationalizing the denominator. (*Section 10.5, Objectives 2 and 3*)

109. $\dfrac{10}{\sqrt{5a}}$ **110.** $\dfrac{6}{\sqrt{2b}}$

111. $\dfrac{10}{\sqrt{7} + \sqrt{3}}$

112. $\dfrac{6}{\sqrt{11} - \sqrt{2}}$

Solve each radical equation. (*Section 10.6, Objective 1*)

113. $\sqrt{3r + 4} - 1 = 6$ **114.** $\sqrt{2t - 1} + 5 = 8$

115. $\sqrt{14 - 2x} + 3 = x$ **116.** $\sqrt{2x + 12} - 2 = x$

Perform each operation on the complex numbers and write each answer in standard form. (*Section 10.7, Objectives 2–4*)

117. $(1 - 4i) + (-3 + 11i)$ **118.** $(-8 + 9i) - (10 + 4i)$

119. $(3 - 2i)^2$ **120.** $(1 - 2i)(1 + 2i)$

121. $\dfrac{2 + 12i}{4i}$ **122.** $\dfrac{15 + 2i}{-3i}$

123. $\dfrac{3 + 5i}{2 - i}$ **124.** $\dfrac{8 + 5i}{1 + 3i}$

Find each power of i. (*Section 10.7, Objective 5*)

125. i^{16} **126.** i^6 **127.** i^{-3} **128.** i^{-5}

Quadratic Equations and Functions and Nonlinear Inequalities

Reflection

As you approach the end of this book and perhaps the end of the course, take a few moments to reflect on the past weeks. Chances are this is not the last math course required for your program of study (talk to your instructor or advisor to be sure). It is our hope that you will take not only math skills from this course but also skills that can help you succeed in other classes. It is important to understand the methods that helped you successfully understand and retain the material as well as those things that did not work as well for you. You should also leave the course with information about the concepts you understand well and the ones that you do not.

Question For Thought: What is working in this course that you want to continue doing as you take other classes? What are things you did not do in this course that you want to start doing as you take other classes? What are some things that did not work that you need to stop doing as you take other classes?

Chapter Outline

Coming Up...

In Section 11.5, we will learn how to find the maximum height of a ball that is tossed upward with an initial velocity of 30 ft/sec from a height of 6 ft, and whose height is modeled by the function $f(t) = -16t^2 + 30t + 6$.

"Everyone thinks of changing the world, but no one thinks of changing himself."

—Leo Tolstoy (Novelist)

SECTION 11.1 / **Quadratic Functions and Their Graphs**

In Chapter 6, we learned how to solve quadratic equations by factoring. Factoring, however, does not solve every quadratic equation. In this chapter, we will present several additional methods for solving quadratic equations. We will also explore the graphs of quadratic functions and solve polynomial and rational inequalities.

▶ **OBJECTIVES**

As a result of completing this section, you will be able to

1. Graph a function of the form $f(x) = x^2 + k$.
2. Graph a function of the form $f(x) = (x - h)^2$.
3. Graph a function of the form $f(x) = ax^2$.
4. Graph a function of the form $f(x) = a(x - h)^2 + k$.
5. Solve equations of the form $a(x - h)^2 + k = 0$ graphically.
6. Troubleshoot common errors.

In Chapter 3, we learned how to graph an equation in two variables by plotting points. In fact, we graphed a basic quadratic equation of the form $y = x^2$ and found its shape to be a parabola. Many structures in the real world are parabolic-shaped. In this section, we will examine the graphs of quadratic functions more closely.

Graphs of Functions of the Form $f(x) = x^2 + k$

Objective 1 ▶

Graph a function of the form $f(x) = x^2 + k$.

> **Definition:** A **quadratic function** is a function of the form $f(x) = ax^2 + bx + c$, where a, b, and c are real numbers with $a \neq 0$. The graph of a quadratic function is a **parabola**.

Before we examine graphs of the form $f(x) = x^2 + k$, we will review the graph of $y = x^2$. A table of values is shown for reference.

x	$y = x^2$	(x, y)
-3	$(-3)^2 = 9$	$(-3, 9)$
-2	$(-2)^2 = 4$	$(-2, 4)$
-1	$(-1)^2 = 1$	$(-1, 1)$
$-\dfrac{1}{2}$	$\left(-\dfrac{1}{2}\right)^2 = \dfrac{1}{4}$	$\left(-\dfrac{1}{2}, \dfrac{1}{4}\right)$
0	$(0)^2 = 0$	$(0, 0)$
$\dfrac{1}{2}$	$\left(\dfrac{1}{2}\right)^2 = \dfrac{1}{4}$	$\left(\dfrac{1}{2}, \dfrac{1}{4}\right)$
1	$(1)^2 = 1$	$(1, 1)$
2	$(2)^2 = 4$	$(2, 4)$
3	$(3)^2 = 9$	$(3, 9)$

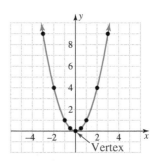

This graph represents a function because it passes the vertical line test. The domain of the function is $(-\infty, \infty)$ because the graph extends indefinitely in the left and right directions. The range of the function is $[0, \infty)$ because the graph begins from $(0, 0)$ in the bottom-most direction and extends upward indefinitely. The point $(0, 0)$ is the lowest point on the graph. The x-intercept and y-intercept of the graph are also $(0, 0)$.

Parabolas have some important characteristics.

1. Parabolas are smooth curves.
2. Parabolas have a highest or lowest point, called the **vertex**.
3. Parabolas are symmetric about their vertex. If a parabola is folded vertically in half through its vertex, the left side and the right side of the parabola are mirror images of one another. This means that x-values with the same distance from the vertex have the same y-values.
4. The vertical line through the vertex is called the **axis of symmetry**. The axis of symmetry is of the form $x = h$, where h is the x-value of the vertex.
5. The direction that a parabola opens is determined by a, the coefficient of the x^2 term. If $a > 0$, the parabola opens up and if $a < 0$, the parabola opens down.

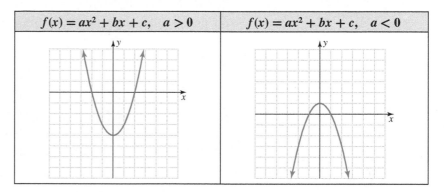

Key points on the graph of a parabola are

1. the x-intercepts, points where the graph crosses the x-axis and of the form $(x, 0)$,
2. the y-intercept, point where the graph crosses the y-axis and of the form $(0, y)$,
3. and the vertex.

The following graph is a parabola with its key points identified.

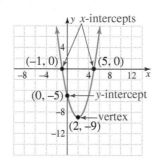

Now that we have an understanding of the basic shape and characteristics of a parabola, we will graph transformations of the basic parabola formed by adding or subtracting a number from the term x^2; that is, we will graph equations of the form $y = x^2 + k$.

Procedure: Graphing a Function of the Form $f(x) = x^2 + k$

Step 1: Make a table of values.
Step 2: Plot the points.
Step 3: Connect the points with a smooth curve.

Objective 1 Examples   **Create a table of values to graph each function. Identify the x-intercepts, y-intercept, vertex, the axis of symmetry, domain, and range. Explain how the graph of the function relates to the graph of $f(x) = x^2$.**

1a. $f(x) = x^2 - 4$ **1b.** $f(x) = x^2 + 1$

Solutions   **1a.** A table of values and the graph of $f(x) = x^2 - 4$ are shown. The graph of $f(x) = x^2$ is shown for comparison.

x	$f(x) = x^2 - 4$	(x, y)
-2	$(-2)^2 - 4 = 4 - 4 = 0$	$(-2, 0)$
-1	$(-1)^2 - 4 = 1 - 4 = -3$	$(-1, -3)$
0	$(0)^2 - 4 = 0 - 4 = -4$	$(0, -4)$
1	$(1)^2 - 4 = 1 - 4 = -3$	$(1, -3)$
2	$(2)^2 - 4 = 4 - 4 = 0$	$(2, 0)$

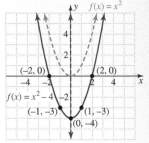

We obtain the following information from the graph.

x-intercepts	$(-2, 0)$ and $(2, 0)$
y-intercept	$(0, -4)$
Vertex	$(0, -4)$
Axis of symmetry	$x = 0$
Domain	$(-\infty, \infty)$
Range	$[-4, \infty)$

The graph of $f(x) = x^2 - 4$ is the graph of $f(x) = x^2$ shifted *down* 4 units.

1b. A table of values and the graph of $f(x) = x^2 + 1$ are shown. The graph of $f(x) = x^2$ is shown for comparison.

x	$f(x) = x^2 + 1$	(x, y)
-2	$(-2)^2 + 1 = 4 + 1 = 5$	$(-2, 5)$
-1	$(-1)^2 + 1 = 1 + 1 = 2$	$(-1, 2)$
0	$(0)^2 + 1 = 0 + 1 = 1$	$(0, 1)$
1	$(1)^2 + 1 = 1 + 1 = 2$	$(1, 2)$
2	$(2)^2 + 1 = 4 + 1 = 5$	$(2, 5)$

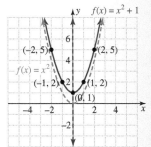

We obtain the following information from the graph.

x-intercepts	None
y-intercept	$(0, 1)$
Vertex	$(0, 1)$
Axis of symmetry	$x = 0$
Domain	$(-\infty, \infty)$
Range	$[1, \infty)$

The graph of $f(x) = x^2 + 1$ is the graph of $f(x) = x^2$ shifted *up* 1 unit.

✓ **Student Check 1**   Create a table of values to graph each function. Identify the x-intercepts, y-intercept, vertex, the axis of symmetry, domain, and range. Explain how the graph of the function relates to the graph of $f(x) = x^2$.

   a. $f(x) = x^2 + 2$ **b.** $f(x) = x^2 - 1$

> **Fact:** Important Features of a Parabola of the Form $f(x) = x^2 + k$
>
> The graph of $f(x) = x^2 + k$ is the graph of $f(x) = x^2$ shifted up or down.
>
> - If $k > 0$, the graph of $f(x) = x^2 + k$ is the graph of $f(x) = x^2$ shifted up k units.
> - If $k < 0$, the graph of $f(x) = x^2 + k$ is the graph of $f(x) = x^2$ shifted down $|k|$ units.
> - The vertex is $(0, k)$ and the axis of symmetry is the vertical line $x = 0$, or the y-axis.

Graphs of Functions of the Form $f(x) = (x - h)^2$

Objective 2 ▶

Graph a function of the form $f(x) = (x - h)^2$.

Now we examine what happens to the graph of $f(x) = x^2$ when we add or subtract a number from x before squaring.

> **Procedure: Graphing a Function of the Form $f(x) = (x - h)^2$**
>
> **Step 1:** Make a table of values.
> **Step 2:** Plot the points.
> **Step 3:** Connect the points with a smooth curve.

Objective 2 Examples Create a table of values to graph each function. Identify the x-intercepts, y-intercept, vertex, the axis of symmetry, domain, and range. Explain how the graph of the function relates to the graph of $f(x) = x^2$.

2a. $f(x) = (x + 1)^2$ **2b.** $f(x) = (x - 2)^2$

Solutions **2a.** A table of values and the graph of $f(x) = (x + 1)^2$ are shown. The graph of $f(x) = x^2$ is shown for comparison.

x	$f(x) = (x + 1)^2$	(x, y)
-3	$(-3 + 1)^2 = (-2)^2 = 4$	$(-3, 4)$
-2	$(-2 + 1)^2 = (-1)^2 = 1$	$(-2, 1)$
-1	$(-1 + 1)^2 = (0)^2 = 0$	$(-1, 0)$
0	$(0 + 1)^2 = (1)^2 = 1$	$(0, 1)$
1	$(1 + 1)^2 = (2)^2 = 4$	$(1, 4)$

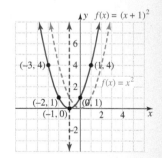

We obtain the following information from the graph.

x-intercept	$(-1, 0)$
y-intercept	$(0, 1)$
Vertex	$(-1, 0)$
Axis of symmetry	$x = -1$
Domain	$(-\infty, \infty)$
Range	$[0, \infty)$

The graph of $f(x) = (x + 1)^2$ is the graph of $f(x) = x^2$ shifted *left* 1 unit.

2b. A table of values and the graph of $f(x) = (x - 2)^2$ are shown. The graph of $f(x) = x^2$ is shown for comparison.

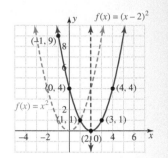

x	$f(x) = (x - 2)^2$	(x, y)
-1	$(-1 - 2)^2 = (-3)^2 = 9$	$(-1, 9)$
0	$(0 - 2)^2 = (-2)^2 = 4$	$(0, 4)$
1	$(1 - 2)^2 = (-1)^2 = 1$	$(1, 1)$
2	$(2 - 2)^2 = (0)^2 = 0$	$(2, 0)$
3	$(3 - 2)^2 = (1)^2 = 1$	$(3, 1)$
4	$(4 - 2)^2 = (2)^2 = 4$	$(4, 4)$

We obtain the following information from the graph.

x-intercept	$(2, 0)$
y-intercept	$(0, 4)$
Vertex	$(2, 0)$
Axis of symmetry	$x = 2$
Domain	$(-\infty, \infty)$
Range	$[0, \infty)$

The graph of $f(x) = (x - 2)^2$ is the graph of $f(x) = x^2$ shifted *right* 2 units.

 Student Check 2   Create a table of values to graph each function. Identify the x-intercepts, y-intercept, vertex, the axis of symmetry, domain, and range. Explain how the graph of the function relates to the graph of $f(x) = x^2$.

a. $f(x) = (x + 2)^2$ **b.** $f(x) = (x - 1)^2$

Fact: Important Features of Parabolas of the Form $f(x) = (x - h)^2$

The graph of $f(x) = (x - h)^2$ is the graph of $f(x) = x^2$ shifted left or right.

- *If $h > 0$, the graph of $f(x) = (x - h)^2$ is the graph of $f(x) = x^2$ shifted right h units.*
- *If $h < 0$, the graph of $f(x) = (x - h)^2$ is the graph of $f(x) = x^2$ shifted left $|h|$ units.*
- *The vertex is $(h, 0)$ and the axis of symmetry is the vertical line $x = h$.*

Graphs of Functions of the Form $f(x) = ax^2$

Objective 3 ▶

Graph a function of the form $f(x) = ax^2$.

The previous two objectives illustrate that the values h and k, from the functions $f(x) = (x - h)^2$ and $f(x) = x^2 + k$, shift the graph of $f(x) = x^2$ left or right h units and up or down k units. We now investigate the effects of multiplying the function, $f(x) = x^2$, by a constant.

Objective 3 Examples   Create a table of values to graph each function. Identify the x-intercepts, y-intercept, vertex, the axis of symmetry, domain, and range. Explain how the graph of the function relates to the graph of $f(x) = x^2$.

3a. $f(x) = 2x^2$ **3b.** $f(x) = \dfrac{1}{2}x^2$ **3c.** $f(x) = -x^2$

Solutions   **3a.** A table of values and the graph of $f(x) = 2x^2$ are shown. The graph of $f(x) = x^2$ is shown for comparison.

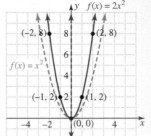

x	$f(x) = 2x^2$	(x, y)
-2	$2(-2)^2 = 2(4) = 8$	$(-2, 8)$
-1	$2(-1)^2 = 2(1) = 2$	$(-1, 2)$
0	$2(0)^2 = 2(0) = 0$	$(0, 0)$
1	$2(1)^2 = 2(1) = 2$	$(1, 2)$
2	$2(2)^2 = 2(4) = 8$	$(2, 8)$

We obtain the following information from the graph.

x-intercept	$(0, 0)$
y-intercept	$(0, 0)$
Vertex	$(0, 0)$
Axis of symmetry	$x = 0$
Domain	$(-\infty, \infty)$
Range	$[0, \infty)$

The graph of $f(x) = 2x^2$ is more *narrow* than the graph of $f(x) = x^2$. We say that $f(x) = 2x^2$ is the graph of $f(x) = x^2$ *vertically stretched* by a factor of 2. The y-values of $f(x) = 2x^2$ are twice as large as the y-values of $f(x) = x^2$, which has the effect of pulling the graph of $f(x) = x^2$ upward; that is, stretching it vertically.

3b. A table of values and the graph of $f(x) = \dfrac{1}{2}x^2$ are shown. The graph of $f(x) = x^2$ is shown for comparison.

x	$f(x) = \dfrac{1}{2}x^2$	(x, y)
-2	$\dfrac{1}{2}(-2)^2 = \dfrac{1}{2}(4) = 2$	$(-2, 2)$
-1	$\dfrac{1}{2}(-1)^2 = \dfrac{1}{2}(1) = \dfrac{1}{2}$	$\left(-1, \dfrac{1}{2}\right)$
0	$\dfrac{1}{2}(0)^2 = \dfrac{1}{2}(0) = 0$	$(0, 0)$
1	$\dfrac{1}{2}(1)^2 = \dfrac{1}{2}(1) = \dfrac{1}{2}$	$\left(1, \dfrac{1}{2}\right)$
2	$\dfrac{1}{2}(2)^2 = \dfrac{1}{2}(4) = 2$	$(2, 2)$

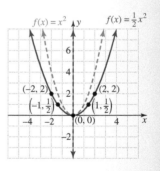

We obtain the following information from the graph.

x-intercept	$(0, 0)$
y-intercept	$(0, 0)$
Vertex	$(0, 0)$
Axis of symmetry	$x = 0$
Domain	$(-\infty, \infty)$
Range	$[0, \infty)$

The graph of $f(x) = \frac{1}{2}x^2$ is *wider* than the graph of $f(x) = x^2$. The graph of $f(x) = \frac{1}{2}x^2$ is the graph of $f(x) = x^2$ *vertically shrunk* by a factor of $\frac{1}{2}$. The y-values of $f(x) = \frac{1}{2}x^2$ are half as large as the y-values of $f(x) = x^2$. The effect is that the graph is being pushed downward, that is, vertically shrunk.

3c. A table of values and the graph of $f(x) = -x^2$ are shown. The graph of $f(x) = x^2$ is shown for comparison.

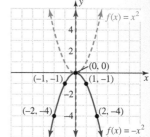

x	$f(x) = -x^2$	(x, y)
-2	$-(-2)^2 = -(4) = -4$	$(-2, -4)$
-1	$-(-1)^2 = -(1) = -1$	$(-1, -1)$
0	$-(0)^2 = -(0) = 0$	$(0, 0)$
1	$-(1)^2 = -(1) = -1$	$(1, -1)$
2	$-(2)^2 = -(4) = -4$	$(2, -4)$

We obtain the following information from the graph.

x-intercept	$(0, 0)$
y-intercept	$(0, 0)$
Vertex	$(0, 0)$
Axis of symmetry	$x = 0$
Domain	$(-\infty, \infty)$
Range	$(-\infty, 0]$

The graph of $f(x) = -x^2$ is the graph of $y = x^2$ *reflected over the x-axis*.

 Student Check 3   Create a table of values to graph each function. Identify the x-intercepts, y-intercept, vertex, the axis of symmetry, domain, and range. Explain how the graph of the function relates to the graph of $f(x) = x^2$.

   a. $y = 3x^2$ **b.** $y = \frac{1}{3}x^2$ **c.** $y = -2x^2$

Fact: Important Features of a Parabola of the Form $f(x) = ax^2$

The graph of $f(x) = ax^2$ is the graph of $f(x) = x^2$ stretched, shrunk, and/or reflected over the x-axis.

- If $a > 1$, the graph of $f(x) = ax^2$ is the graph of $f(x) = x^2$ stretched vertically by a factor of a.
- If $0 < a < 1$, the graph of $f(x) = ax^2$ is the graph of $f(x) = x^2$ shrunk vertically by a factor of a.
- If $a < 0$, the graph of $y = ax^2$ is also reflected over the x-axis.
- The vertex is $(0, 0)$ and the axis of symmetry is the vertical line $x = 0$.
- If $a > 0$, the parabola opens up. If $a < 0$, the parabola opens down.

Graphs of Functions of the Form $f(x) = a(x - h)^2 + k$

Objective 4 ▶

Graph a function of the form $f(x) = a(x - h)^2 + k$.

We now incorporate what we know about the values of a, h, and k and the graphs of functions of the form $f(x) = x^2 + k$, $f(x) = (x - h)^2$, and $f(x) = ax^2$ to graph functions of the form $f(x) = a(x - h)^2 + k$. The following table summarizes functions from Examples 1–3 and the effects of a, h, and k.

Function	Vertex	Direction	Function in the form $f(x) = a(x - h)^2 + k$	Values of a, h, and k
$f(x) = x^2 - 4$	$(0, -4)$	opens up	$f(x) = 1(x - 0)^2 - 4$	$a = 1$, $(h, k) = (0, -4)$
$f(x) = x^2 + 1$	$(0, 1)$	opens up	$f(x) = 1(x - 0)^2 + 1$	$a = 1$, $(h, k) = (0, 1)$
$f(x) = (x + 1)^2$	$(-1, 0)$	opens up	$f(x) = 1[x - (-1)]^2 + 0$	$a = 1$, $(h, k) = (-1, 0)$
$f(x) = (x - 2)^2$	$(2, 0)$	opens up	$f(x) = 1(x - 2)^2 + 0$	$a = 1$, $(h, k) = (2, 0)$
$f(x) = 2x^2$	$(0, 0)$	opens up	$f(x) = 2(x - 0)^2 + 0$	$a = 2$, $(h, k) = (0, 0)$
$f(x) = \dfrac{1}{2}x^2$	$(0, 0)$	opens up	$f(x) = \dfrac{1}{2}(x - 0)^2 + 0$	$a = \dfrac{1}{2}$, $(h, k) = (0, 0)$
$f(x) = -x^2$	$(0, 0)$	opens down	$f(x) = -1(x - 0)^2 + 0$	$a = -1$, $(h, k) = (0, 0)$

Property: Vertex Form of a Quadratic Function $f(x) = a(x - h)^2 + k$

The graph of $f(x) = a(x - h)^2 + k$ is the graph of $f(x) = x^2$ where

- The vertex of the graph is the point (h, k).
- The axis of symmetry is the vertical line through the vertex, $x = h$.
- The parabola opens up if $a > 0$ and opens down if $a < 0$.

$$f(x) = a(x - h)^2 + k \quad (a > 0) \qquad f(x) = a(x - h)^2 + k \quad (a < 0)$$

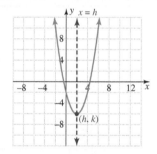
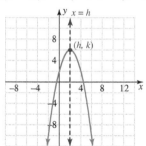

Procedure: Graphing a Function of the Form $f(x) = a(x - h)^2 + k$

Step 1: Identify the vertex (h, k).

Step 2: Plot two additional points, one on each side of the vertex. Find the y-intercept if one of the two additional points does not provide this information.

Step 3: Connect the points with a parabola.

Objective 4 Examples

Graph each function by plotting the vertex and two additional points. Identify the x-intercepts, y-intercept, vertex, the axis of symmetry, domain, and range. Explain how the graph of the function relates to the graph of $y = x^2$.

4a. $f(x) = (x - 1)^2 - 4$ **4b.** $f(x) = (x + 3)^2 + 1$ **4c.** $f(x) = -2(x - 2)^2 + 1$

Solutions

4a. The vertex of $f(x) = 1(x - 1)^2 - 4$ is $(h, k) = (1, -4)$. The graph opens up since $a = 1 > 0$.

	x	$y = (x - 1)^2 - 4$	(x, y)
x-value to the left of the vertex	-1	$(-1 - 1)^2 - 4 = 0$	$(-1, 0)$
vertex	1	-4	$(1, -4)$
x-value to the right of the vertex	3	$(3 - 1)^2 - 4 = 0$	$(3, 0)$
y-intercept	0	$(0 - 1)^2 - 4 = -3$	$(0, -3)$

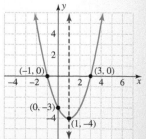

We obtain the following information from the graph.

x-intercepts	$(-1, 0)$ and $(3, 0)$
y-intercept	$(0, -3)$
Vertex	$(1, -4)$
Axis of symmetry	$x = 1$
Domain	$(-\infty, \infty)$
Range	$[-4, \infty)$

The graph of $f(x) = (x - 1)^2 - 4$ is the graph of $f(x) = x^2$ shifted right 1 unit and down 4 units.

4b. The vertex of $f(x) = 1[x - (-3)]^2 + 1$ is $(h, k) = (-3, 1)$. The graph opens up since $a = 1 > 0$.

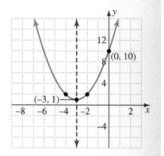

	x	$y = (x + 3)^2 + 1$	(x, y)
x-value to the left of the vertex	-4	$(-4 + 3)^2 + 1 = 2$	$(-4, 2)$
vertex	-3	1	$(-3, 1)$
x-value to the right of the vertex	-2	$(-2 + 3)^2 + 1 = 2$	$(-2, 2)$
y-intercept	0	$(0 + 3)^2 + 1 = 10$	$(0, 10)$

We obtain the following information from the graph.

x-intercepts	none
y-intercept	$(0, 10)$
Vertex	$(-3, 1)$
Axis of symmetry	$x = -3$
Domain	$(-\infty, \infty)$
Range	$[1, \infty)$

The graph of $f(x) = (x + 3)^2 + 1$ is the graph of $f(x) = x^2$ shifted left 3 units and up 1 unit.

4c. The vertex of $f(x) = -2(x - 2)^2 + 1$ is $(h, k) = (2, 1)$. The graph opens down since $a = -2 < 0$.

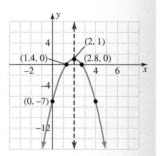

	x	$y = -2(x - 2)^2 + 1$	(x, y)
x-value to the left of the vertex	0	$-2(0 - 2)^2 + 1 = -7$	$(0, -7)$
vertex	2	1	$(2, 1)$
x-value to the right of the vertex	4	$-2(4 - 2)^2 + 1 = -7$	$(4, -7)$

We find the following information from the graph. The x-intercepts are approximated from the graph. We will learn how to find these precisely later in this chapter.

x-intercepts	Approximately $(1.4, 0)$ and $(2.8, 0)$
y-intercept	$(0, -7)$
Vertex	$(2, 1)$
Axis of symmetry	$x = 2$
Domain	$(-\infty, \infty)$
Range	$(-\infty, 1]$

The graph of $f(x) = -2(x - 2)^2 + 1$ is the graph of $f(x) = x^2$ shifted right 2 units, reflected over the x-axis and stretched by a factor of 2, and shifted up 1 unit.

✓ **Student Check 4**   Graph each function by plotting the vertex and two additional points. Identify the x-intercepts, y-intercept, vertex, the axis of symmetry, domain, and range. Explain how the graph of the function relates to the graph of $f(x) = x^2$.

   a. $f(x) = (x - 2)^2 - 9$ **b.** $f(x) = (x + 4)^2 + 3$ **c.** $f(x) = -(x - 3)^2 + 4$

Note: *While we obtain the graphs using a table of values, it is important to note that there is an order of the transformations. We must first translate horizontally, reflect over the x-axis, stretch or shrink vertically, and finally translate vertically.*

Solve Quadratic Equations Using a Graph

Objective 5 ▶

Solve equations of the form $a(x - h)^2 + k = 0$ graphically.

In Sections 11.2 and 11.3, we will learn several methods to solve quadratic equations algebraically. We can also solve quadratic equations graphically. Being able to visualize the solutions will assist us in later sections.

Property: The Graph of $f(x) = a(x - h)^2 + k$ and Solutions of $a(x - h)^2 + k = 0$

If $(r, 0)$ and/or $(s, 0)$ are points on the graph of $y = a(x - h)^2 + k$, then r and s are solutions of the equation $a(x - h)^2 + k = 0$.

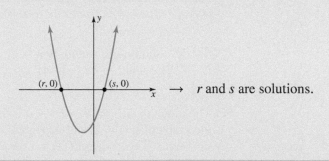

r and s are solutions.

So, the solutions of the equation $a(x - h)^2 + k = 0$ correspond to the x-values whose y-values are zero. Graphically, these are the x-intercepts of the function $f(x) = a(x - h)^2 + k$. There are three possible cases.

- If the graph of the parabola crosses the x-axis two times, $a(x - h)^2 + k = 0$ has two real solutions.
- If the graph of the parabola crosses the x-axis one time, $a(x - h)^2 + k = 0$ has one real solution.

- If the graph of the parabola doesn't cross the x-axis, $a(x - h)^2 + k = 0$ has no real solutions.

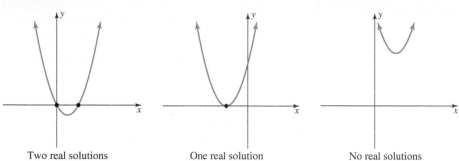

Two real solutions   One real solution   No real solutions

Procedure: Solving an Equation of the Form $a(x - h)^2 + k = 0$ Graphically

Step 1: Graph the function $f(x) = a(x - h)^2 + k$.
Step 2: State the x-intercepts of the function.
Step 3: The x-value(s) of the x-intercepts are the solutions of the equation.

Objective 5 Examples   Graph each function and then use the graph to solve the equation $f(x) = 0$.

5a. $f(x) = x^2 - 4$ **5b.** $f(x) = -3(x + 4)^2$ **5c.** $f(x) = (x - 2)^2 + 4$

Solutions   **5a.** The graph of $f(x) = x^2 - 4$ is the same as the graph of $f(x) = x^2$ shifted down 4 units. The graph of $f(x) = x^2 - 4$ crosses the x-axis twice at the x-intercepts of $(-2, 0)$ and $(2, 0)$.

So, the solutions of the equation $x^2 - 4 = 0$ are -2 and 2. The solution set is $\{\pm 2\}$.

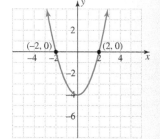

5b. The graph of $f(x) = -3(x + 4)^2$ is the same as the graph of $f(x) = x^2$ shifted left 4 units but opens down since $a = -3$. It is also more narrow than the graph of $f(x) = x^2$. The graph of $f(x) = -3(x + 4)^2$ crosses the x-axis once at the x-intercept of $(-4, 0)$.

So, the solution of the equation $-3(x + 4)^2 = 0$ is -4. The solution set is $\{-4\}$.

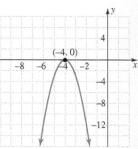

5c. The graph of $f(x) = (x - 2)^2 + 4$ is the same as the graph of $f(x) = x^2$ shifted right 2 units and up 4 units. The graph of $f(x) = (x - 2)^2 + 4$ does not cross the x-axis.

There are no values of x for which $y = 0$. So, there are no real solutions of this equation.

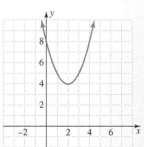

✓ Student Check 5   Graph each function and then use the graph to solve the equation $f(x) = 0$.

a. $f(x) = (x - 3)^2 - 4$ **b.** $f(x) = 2(x + 1)^2$ **c.** $f(x) = -x^2 - 2$

Objective 6 ▶

Troubleshoot common errors.

Troubleshooting Common Errors

A common error associated with graphing quadratic functions is shown.

Objective 6 Example ▶ **A problem and an incorrect solution are given. Provide the correct solution and an explanation of the error.**

Find the vertex of $y = (x + 3)^2$.

Incorrect Solution	Correct Solution and Explanation
The vertex is (3, 0).	The vertex is the point (h, k) when the function is written in the form $y = a(x - h)^2 + k$. $$y = (x + 3)^2 = [x - (-3)]^2 + 0$$ So, the vertex is $(-3, 0)$.

ANSWERS TO STUDENT CHECKS

Student Check 1   a. none; (0, 2); (0, 2); $x = 0$; $(-\infty, \infty)$; $[2, \infty)$; It is the graph of $f(x) = x^2$ shifted up 2 units.

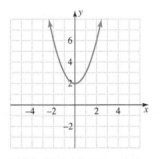

b. $(-1, 0)$ and $(1, 0)$; $(0, -1)$; $(0, -1)$; $x = 0$; $(-\infty, \infty)$; $[-1, \infty)$; It is the graph of $f(x) = x^2$ shifted down 1 unit.

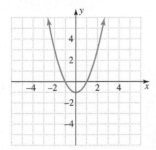

Student Check 2   a. $(-2, 0)$; (0, 4); $(-2, 0)$; $x = -2$; $(-\infty, \infty)$; $[0, \infty)$; It is the graph of $f(x) = x^2$ shifted left 2 units.

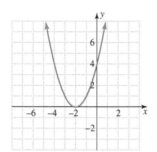

b. (1, 0); (0, 1); (1, 0); $x = 1$; $(-\infty, \infty)$; $[0, \infty)$; It is the graph of $f(x) = x^2$ shifted right 1 unit.

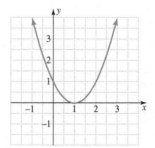

Student Check 3   a. (0, 0); (0, 0); (0, 0); $x = 0$; $(-\infty, \infty)$; $[0, \infty)$; It is the graph of $f(x) = x^2$ vertically stretched by a factor of 3.

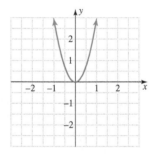

b. (0, 0); (0, 0); (0, 0); $x = 0$; $(-\infty, \infty)$; $[0, \infty)$; It is the graph of $f(x) = x^2$ vertically shrunk by a factor of $\dfrac{1}{3}$.

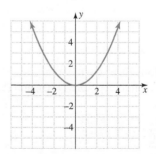

c. (0, 0); (0, 0); (0, 0), $x = 0$;
$(-\infty, \infty)$; $(-\infty, 0]$; It is the
graph of $f(x) = x^2$ vertically
stretched by a factor of 2 and
reflected over the x-axis.

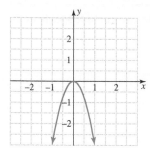

Student Check 4   a. $(-1, 0)$ and
$(5, 0)$; $(0, -5)$; $(2, -9)$; $x = 2$;
$(-\infty, \infty)$; $[-9, \infty)$; It is the graph
of $f(x) = x^2$ shifted right 2 units and
down 9 units.

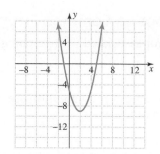

b. None; (0, 19); $(-4, 3)$; $x = -4$;
$(-\infty, \infty)$; $[-4, \infty)$; It is the
graph of $f(x) = x^2$ shifted left
4 units and up 3 units.

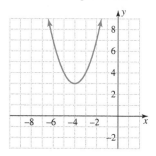

c. (1, 0) and (5, 0); (0, −5); (3, 4);
$x = 3$; $(-\infty, \infty)$; $(-\infty, 4]$;
It is the graph of $y = x^2$ shifted
right 3 units up 4 units, and
reflected over the x-axis.

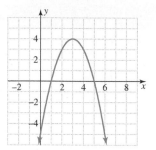

Student Check 5   a. {1, 5}
b. {−1}   **c.** no real solutions

SUMMARY OF KEY CONCEPTS

1. The graph of a function of the form $f(x) = a(x - h)^2 + k$
is a parabola.

  a. The vertex is the ordered pair (h, k).

  b. If $a > 0$, the parabola opens upward.

  c. If $a < 0$, the parabola opens downward.

  d. The domain of the function is $(-\infty, \infty)$ and the range
  is $[k, \infty)$ if $a > 0$ and $(-\infty, k]$ if $a < 0$.

  e. The axis of symmetry is the vertical line through the
  vertex, $x = h$.

2. Solutions of the equation $a(x - h)^2 + k = 0$ are the
x-intercepts of the graph of $f(x) = a(x - h)^2 + k$.

  a. If the graph of $f(x) = a(x - h)^2 + k$ crosses the x-axis
  two times, there are two real solutions of the equation
  $a(x - h)^2 + k = 0$.

  b. If the graph of $f(x) = a(x - h)^2 + k$ crosses the x-axis
  one time, there is one real solution of the equation
  $a(x - h)^2 + k = 0$.

  c. If the graph of $f(x) = a(x - h)^2 + k$ does not cross
  the x-axis, there are no real solutions of the equation
  $a(x - h)^2 + k = 0$.

GRAPHING CALCULATOR SKILLS

We can use the graphing calculator to find the x- and y-intercepts and the vertex of a parabola.

Example: Graph $f(x) = (x - 3)^2 - 4$. Find the x- and y-intercepts and the vertex.

Solution: Enter the function and graph. From the table, we
see the x-intercepts are (1, 0), and (5, 0) and the y-intercept
is (0, 5).

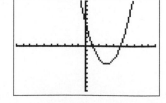

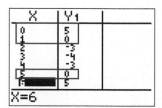

To find the vertex, press ⟨2nd⟩ ⟨TRACE⟩ and choose either
minimum (if the parabola opens up) or maximum (if the
parabola opens down). Once the selection is made, we are

prompted to enter a left bound, a right bound, and a guess. The left bound is a point on the left side of the vertex. The right bound is a point on the right side of the vertex. So, move to the appropriate point and press ENTER. To enter a guess, move the cursor between the markings made from the left and right bound and press ENTER.

The x-intercepts are found using the zero function. Press  and select zero. We are prompted to enter a left and right bound of the zero as well. The left bound is a point on the left side of the x-intercept and the right bound is a point on the right side of the x-intercept. To enter a guess, move between the markings placed by the bounds and press ENTER.

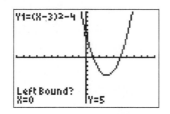 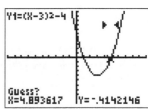

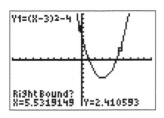 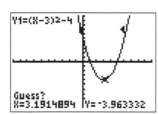

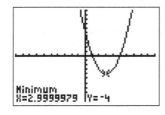 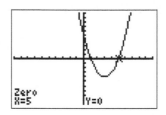

SECTION 11.1 / EXERCISE SET

Write About It!

Use complete sentences in your answer to each exercise.

1. Describe the graph of the function $f(x) = x^2 + k$, $k > 0$.
2. Describe the graph of the function $f(x) = (x - h)^2$.
3. Describe the graph of the function $f(x) = -(x - h)^2 + k$.
4. Describe the graph of the function $f(x) = a(x - h)^2 + k$.

Practice Makes Perfect!

Create a table of values to graph each function. Identify the x-intercepts, y-intercept, vertex, the axis of symmetry, domain, and range. Explain how the graph of the function relates to the graph of $f(x) = x^2$. (*See Objective 1.*)

5. $f(x) = x^2 + 5$
6. $f(x) = x^2 + 7$
7. $f(x) = x^2 - 4$
8. $f(x) = x^2 - 25$
9. $f(x) = x^2 - 2.25$
10. $f(x) = x^2 - 12.96$
11. $f(x) = x^2 + 5.76$
12. $f(x) = x^2 + 6.25$

(*See Objective 2.*)

13. $f(x) = (x + 3)^2$
14. $f(x) = (x + 5)^2$
15. $f(x) = (x - 4)^2$
16. $f(x) = (x - 6)^2$
17. $f(x) = \left(x + \dfrac{3}{2}\right)^2$
18. $f(x) = \left(x + \dfrac{3}{4}\right)^2$
19. $f(x) = \left(x - \dfrac{5}{8}\right)^2$
20. $f(x) = \left(x - \dfrac{5}{2}\right)^2$

(*See Objective 3.*)

21. $f(x) = 4x^2$
22. $f(x) = 5x^2$
23. $f(x) = \dfrac{1}{4}x^2$
24. $f(x) = \dfrac{1}{5}x^2$
25. $f(x) = -3x^2$
26. $f(x) = -6x^2$
27. $f(x) = -\dfrac{1}{5}x^2$
28. $f(x) = -\dfrac{1}{10}x^2$
29. $f(x) = -2.5x^2$
30. $f(x) = -1.6x^2$

Graph each function by plotting the vertex and two additional points. Identify the x-intercepts, y-intercept, the axis of symmetry, domain, and range. Explain how the graph of the function relates to the graph of $y = x^2$. (See Objective 4.)

31. $f(x) = (x - 1)^2 - 9$ **32.** $f(x) = (x - 3)^2 - 4$

33 $f(x) = (x + 3)^2 + 4$ **34.** $f(x) = (x + 2)^2 + 5$

35. $f(x) = -(x - 5)^2 - 2$ **36.** $f(x) = -(x + 4)^2 - 2$

37. $f(x) = 2(x + 4)^2 - 18$ **38.** $f(x) = 3(x - 2)^2 - 12$

39. $f(x) = 4(x - 5)^2 + 1$ **40.** $f(x) = 4(x + 1)^2 + 9$

41. $f(x) = -\dfrac{1}{2}(x + 4)^2 + 18$ **42.** $f(x) = -\dfrac{2}{3}(x + 3)^2 + 6$

43. $f(x) = \dfrac{2}{3}(x - 2)^2 - 6$ **44.** $f(x) = \dfrac{3}{4}(x + 2)^2 - 12$

Graph each function and use the graph to solve the equation $f(x) = 0$. (See Objective 5.)

45. $f(x) = x^2 - 16$ **46.** $f(x) = x^2 - 12.25$

47. $f(x) = 12(x + 2)^2 - 3$ **48.** $f(x) = 4(x - 1)^2 - 9$

49. $f(x) = -2(x + 6)^2 - 3$ **50.** $f(x) = -3(x - 1)^2 - 6$

51. $f(x) = \dfrac{1}{2}(x - 1)^2 - 2$ **52.** $f(x) = \dfrac{1}{3}(x - 2)^2 + 3$

53. $f(x) = -\dfrac{1}{2}(x - 1)^2 + 2$ **54.** $f(x) = -\dfrac{1}{3}(x + 2)^2 + 3$

 Mix 'Em Up!

Graph each function by plotting the vertex and two additional points. Identify the x-intercepts, y-intercept, axis of symmetry, domain, and range. Explain how the graph of the function relates to the graph of $f(x) = x^2$.

55. $f(x) = x^2 - \dfrac{25}{9}$ **56.** $f(x) = x^2 - \dfrac{9}{4}$

57. $f(x) = \left(x - \dfrac{7}{2}\right)^2$ **58.** $f(x) = \left(x + \dfrac{5}{2}\right)^2$

59. $f(x) = -x^2 + 1$ **60.** $f(x) = -x^2 + 4$

61. $f(x) = 2(x + 1)^2 - 8$ **62.** $f(x) = 3(x - 2)^2 - 12$

63. $f(x) = -8(x - 5)^2 + 2$ **64.** $f(x) = -18(x + 1)^2 - 2$

65. $f(x) = (x - 2.5)^2 - 12.96$ **66.** $f(x) = (x - 1.5)^2 - 1.96$

67. $f(x) = -\dfrac{1}{2}(x + 1)^2 + 2$ **68.** $f(x) = -\dfrac{3}{2}(x - 3)^2 + 6$

69. $f(x) = x^2 + \dfrac{1}{4}$ **70.** $f(x) = x^2 + \dfrac{1}{9}$

71. $f(x) = \dfrac{1}{4}(x - 1)^2 + 1$ **72.** $f(x) = \dfrac{1}{2}(x - 1)^2 + 1$

73. $f(x) = -(x + 1.8)^2 + 20.25$

74. $f(x) = -(x + 3.2)^2 + 6.25$

75. $f(x) = 2(x - 1.2)^2 - 6.48$

76. $f(x) = 3(x + 1.5)^2 - 21.87$

Graph each function and use the graph to solve the equation $f(x) = 0$.

77. $f(x) = (x - 2)^2 - 2.25$

78. $f(x) = (x + 2)^2 - 12.25$

79. $f(x) = 2(x + 2)^2 + 4.5$

80. $f(x) = 3(x - 1.5)^2 + 0.75$

81. $f(x) = 2(x - 2)^2 - 4.5$

82. $f(x) = 3(x + 1.5)^2 - 0.75$

83. $f(x) = -\dfrac{1}{2}(x - 3)^2 + 8$

84. $f(x) = -\dfrac{4}{3}(x + 1)^2 + 3$

85. $f(x) = -2(x - 1)^2 - 5$

86. $f(x) = -6(x + 1)^2 - 2$

 You Be the Teacher!

Correct each student's errors, if any.

87. Describe how the graph of $f(x) = (x - 3)^2 + 25$ relates to the graph of $f(x) = x^2$.

Patel's work: The graph of $y = x^2$ is shifted left 3 units and down 25 units.

88. Describe how the graph of $f(x) = -(x + 9)^2 - 4$ relates to the graph of $f(x) = x^2$.

Blake's work: The graph of $y = x^2$ is shifted left 9 units, down 4 units, and reflected over the x-axis.

89. Describe how the graph of $f(x) = -\dfrac{1}{2}(x - 5)^2 + 4$ relates to the graph of $f(x) = x^2$.

Mary's work: The graph of $y = x^2$ is shifted right 5 units, up 4 units, reflected over the x-axis, and shrunk by a factor of 2.

90. Describe how the graph of $f(x) = 2(x + 7)^2 + 3$ relates to the graph of $f(x) = x^2$.

Jeannette's work: The graph of $y = x^2$ is shifted left 7 units, up 3 units, and stretched by a factor of 2.

Calculate It!

Use a graphing calculator to find the x-intercepts and the vertex of each quadratic function. Round each answer to two decimal places.

91. $f(x) = 2.7x^2 - 5.2x - 4.8$

92. $f(x) = -1.6x^2 + 7.3x + 8.1$

| **SECTION 11.2** | **Solving Quadratic Equations Using the Square Root Property and Completing the Square** |

One of the world's highest commercial bungee jumps is off of the Bloukrans River Bridge in South Africa. The jump takes place from a platform below the roadway of the bridge and is 216 m (710 ft) above the ground. A jumper is in free fall until the elasticity of the cord affects the rate of fall. While the jumper is free-falling, his height, in feet, above the valley floor is $h = -16t^2 + 710$, where t is in seconds. How many seconds will it take for the jumper to be 410 ft above the valley floor? (Source: http://www.faceadrenalin.com/)

To answer this question, we must solve the equation $-16t^2 + 710 = 410$. In this section, we will learn how to solve problems of this type.

Using the Square Root Property to Solve Quadratic Equations

In Chapter 6, we learned that a **quadratic equation** is an equation of the form $ax^2 + bx + c = 0$, where a, b, and c are real numbers with $a \neq 0$. We also learned how to solve these equations by applying the zero products property. We will review this property for reference.

To solve the following equations using the zero products property, we first write the equation in standard form, $ax^2 + bx + c = 0$. Then we factor and set each factor equal to zero.

Objective 1 ▶
Solve quadratic equations by applying the square root property.

$$
\begin{array}{ccc}
x^2 = 36 & a^2 = 25 & y^2 = 49 \\
x^2 - 36 = 0 & a^2 - 25 = 0 & y^2 - 49 = 0 \\
(x - 6)(x + 6) = 0 & (a - 5)(a + 5) = 0 & (y - 7)(y + 7) = 0 \\
x - 6 = 0 \ \text{or} \ x + 6 = 0 & a - 5 = 0 \ \text{or} \ a + 5 = 0 & y - 7 = 0 \ \text{or} \ y + 7 = 0 \\
x = 6 \ \text{or} \ x = -6 & a = 5 \ \text{or} \ a = -5 & y = 7 \ \text{or} \ y = -7 \\
\{-6, 6\} & \{-5, 5\} & \{-7, 7\}
\end{array}
$$

These equations illustrate that when an equation is of the form $x^2 = k$, where k is a perfect square, there are two solutions of the equation, the positive and negative square root of the value k. Thus, we can also solve the equation $x^2 = 36$ as shown.

$$
\begin{aligned}
x^2 &= 36 \\
x &= \pm\sqrt{36} \\
x &= \pm 6
\end{aligned}
$$

This leads to another important property for solving quadratic equations.

> **Property: The Square Root Property**
>
> If $a^2 = k$, where k is a real number, then $a = \sqrt{k}$ or $a = -\sqrt{k}$, or $a = \pm\sqrt{k}$.

> **Note:** *If the radicand is a negative number, then the equation has two complex, nonreal solutions since the square root of a negative number is imaginary. Recall $\sqrt{-k} = i\sqrt{k}$, for $k > 0$.*

> **Procedure: Solving a Quadratic Equation Using the Square Root Property**
>
> **Step 1:** Isolate the squared term to one side of the equation.
> **Step 2:** Remove the square on the variable by applying the square root property.
> **Step 3:** Solve the resulting equations.
> **Step 4:** Check solutions by substituting the values into the original equation.

Objective 1 Examples	Solve each equation by applying the square root property. Write each radical in simplest form.

1a. $y^2 = 10$ **1b.** $a^2 + 3 = 7$ **1c.** $2y^2 - 5 = 9$

1d. $(x + 3)^2 = 25$ **1e.** $(3a - 4)^2 - 1 = 7$ **1f.** $(5y + 2)^2 = -49$

Solutions

1a. $y^2 = 10$

$\quad y = \pm\sqrt{10}$ Apply the square root property.

The solutions are $-\sqrt{10}$ or $\sqrt{10}$. The solution set is $\{-\sqrt{10}, \sqrt{10}\}$ or $\{\pm\sqrt{10}\}$.

Check: Let $y = \sqrt{10}$.   Let $y = -\sqrt{10}$.

$y^2 = 10$	$y^2 = 10$
$(\sqrt{10})^2 = 10$	$(-\sqrt{10})^2 = 10$
$10 = 10$	$10 = 10$

Original equation.
Replace y with proposed solution.
Apply $(\sqrt[n]{x})^n = x$.

1b. $\quad\quad a^2 + 3 = 7$

$a^2 + 3 - 3 = 7 - 3$ Subtract 3 from each side.

$\quad\quad\quad a^2 = 4$ Simplify.

$\quad\quad\quad a = \pm\sqrt{4}$ Apply the square root property.

$\quad\quad\quad a = \pm 2$ Simplify.

The solutions are -2 or 2, and the solution set is $\{-2, 2\}$ or $\{\pm 2\}$. The check is left for the reader.

1c. $\quad\quad 2y^2 - 5 = 9$

$2y^2 - 5 + 5 = 9 + 5$ Add 5 to each side.

$\quad\quad\quad 2y^2 = 14$ Simplify.

$\quad\quad\quad \dfrac{2y^2}{2} = \dfrac{14}{2}$ Divide each side by 2.

$\quad\quad\quad y^2 = 7$ Simplify.

$\quad\quad\quad y = \pm\sqrt{7}$ Apply the square root property.

The solutions are $-\sqrt{7}$ or $\sqrt{7}$, and the solution set is $\{-\sqrt{7}, \sqrt{7}\}$ or $\{\pm\sqrt{7}\}$. The check is left for the reader.

1d. $\quad\quad\quad (x + 3)^2 = 25$

$\quad\quad\quad x + 3 = \pm\sqrt{25}$ Apply the square root property.

$\quad\quad\quad x + 3 = \pm 5$ Simplify.

$\quad\quad x + 3 - 3 = \pm 5 - 3$ Subtract 3 from each side.

$\quad\quad\quad\quad x = -3 \pm 5$ Rewrite using the commutative property.

$x = -3 + 5 \quad$ or $\quad x = -3 - 5$ Separate the two solutions.

$x = 2 \quad\quad\quad\quad\quad x = -8$ Solve the resulting equations.

The solutions are -8 or 2, and the solution set is $\{-8, 2\}$. The check is left for the reader.

1e.
$$(3a - 4)^2 - 1 = 7$$

$(3a - 4)^2 - 1 + 1 = 7 + 1$ Add 1 to each side.

$(3a - 4)^2 = 8$ Simplify.

$3a - 4 = \pm\sqrt{8}$ Apply the square root property.

$3a - 4 = \pm 2\sqrt{2}$ Simplify: $\sqrt{8} = \sqrt{4 \cdot 2} = 2\sqrt{2}$.

$3a - 4 + 4 = \pm 2\sqrt{2} + 4$ Add 4 to each side.

$3a = 4 \pm 2\sqrt{2}$ Rewrite using the commutative property.

$a = \dfrac{4 \pm 2\sqrt{2}}{3}$ Divide each side by 3.

The exact solutions are $\dfrac{4 + 2\sqrt{2}}{3}$ or $\dfrac{4 - 2\sqrt{2}}{3}$, and the solution set is $\left\{ \dfrac{4 + 2\sqrt{2}}{3}, \dfrac{4 - 2\sqrt{2}}{3} \right\}$ or $\left\{ \dfrac{4 \pm 2\sqrt{2}}{3} \right\}$. The approximate solutions are $\dfrac{4 + 2\sqrt{2}}{3} \approx 2.28$ and $\dfrac{4 - 2\sqrt{2}}{3} \approx 0.39$.

1f.
$$(5y + 2)^2 = -49$$

$5y + 2 = \pm\sqrt{-49}$ Apply the square root property.

$5y + 2 = \pm 7i$ Simplify: $\sqrt{-49} = 7i$.

$5y + 2 - 2 = \pm 7i - 2$ Subtract 2 from each side.

$5y = -2 \pm 7i$ Rewrite using the commutative property.

$\dfrac{5y}{5} = \dfrac{-2 \pm 7i}{5}$ Divide each side by 5.

$y = \dfrac{-2 \pm 7i}{5}$ Simplify.

The solutions are $\dfrac{-2 + 7i}{5}$ or $\dfrac{-2 - 7i}{5}$, and the solution set is $\left\{ \dfrac{-2 + 7i}{5}, \dfrac{-2 - 7i}{5} \right\}$ or $\left\{ \dfrac{-2 \pm 7i}{5} \right\}$.

✓ **Student Check 1**   Solve each equation by applying the square root property. Write each radical in simplest form.

a. $y^2 = 21$ **b.** $a^2 + 7 = 43$ **c.** $3y^2 - 8 = 22$

d. $(x + 7)^2 = 16$ **e.** $(5a - 7)^2 = 12$ **f.** $(7y + 6)^2 = -1$

Perfect Square Trinomials

Objective 2 ▶

Create a perfect square trinomial and express it in factored form.

We can use the square root property to solve a quadratic equation as long as one side of the equation can be expressed as a binomial squared. Squared binomials result from factoring perfect square trinomials. We will learn how to create perfect square trinomials from binomials and then use this in Objective 3 to solve a quadratic equation.

Recall that a **perfect square trinomial** is a trinomial that results from squaring a binomial. Some examples of these trinomials are shown.

$$(x + 3)^2 = x^2 + 6x + 9$$
$$(x + 5)^2 = x^2 + 10x + 25$$
$$(x - 6)^2 = x^2 - 12x + 36$$
$$(x - 10)^2 = x^2 - 20x + 100$$

In a perfect square trinomial with a leading coefficient of 1, the constant term relates to the coefficient of x in an interesting way.

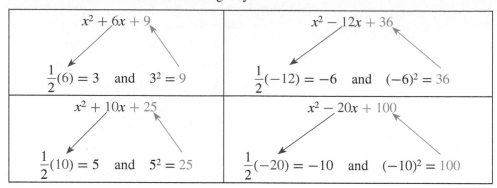

$x^2 + 6x + 9$	$x^2 - 12x + 36$
$\frac{1}{2}(6) = 3$ and $3^2 = 9$	$\frac{1}{2}(-12) = -6$ and $(-6)^2 = 36$
$x^2 + 10x + 25$	$x^2 - 20x + 100$
$\frac{1}{2}(10) = 5$ and $5^2 = 25$	$\frac{1}{2}(-20) = -10$ and $(-10)^2 = 100$

So, in a perfect square trinomial whose leading coefficient is 1, the constant term is half of the coefficient of x, squared. For $x^2 + bx + c$ to be a perfect square trinomial, $c = \left(\frac{b}{2}\right)^2$. We can use this fact to create a perfect square trinomial from a binomial of the form, $x^2 + bx$. This process is called **completing the square**.

Procedure: Creating a Perfect Square Trinomial from $x^2 + bx$

Step 1: Find half of the value of b and square it.
Step 2: Add this number to the expression $x^2 + bx$.
Step 3: Factor the trinomial as $\left(x + \frac{b}{2}\right)^2$.

Objective 2 Examples   Create a perfect square trinomial from each binomial by completing the square and then write it in factored form.

2a. $x^2 + 4x$ **2b.** $y^2 - 8y$ **2c.** $x^2 - 3x$

Solutions   **2a.** In $x^2 + 4x$, the value of $b = 4$, so

$$c = \left(\frac{4}{2}\right)^2 = (2)^2 = 4$$

Therefore, $x^2 + 4x + 4$ is a perfect square trinomial, and it factors as
$$x^2 + 4x + 4 = (x + 2)(x + 2) = (x + 2)^2$$

2b. In $y^2 - 8y$, the value of $b = -8$, so

$$c = \left(\frac{-8}{2}\right)^2 = (-4)^2 = 16$$

Therefore, $y^2 - 8y + 16$ is a perfect square trinomial, and it factors as
$$y^2 - 8y + 16 = (y - 4)(y - 4) = (y - 4)^2$$

2c. In $x^2 - 3x$, the value of $b = -3$, so

$$c = \left(\frac{-3}{2}\right)^2 = \frac{9}{4}$$

Therefore, $x^2 - 3x + \frac{9}{4}$ is a perfect square trinomial, and it factors as

$$x^2 - 3x + \frac{9}{4} = \left(x - \frac{3}{2}\right)\left(x - \frac{3}{2}\right) = \left(x - \frac{3}{2}\right)^2$$

✓ **Student Check 2** Create a perfect square trinomial from each binomial by completing the square and then write it in factored form.

a. $x^2 + 12x$ **b.** $y^2 - 2y$ **c.** $x^2 - 5x$

Completing the Square to Solve $x^2 + bx + c = 0$

Objective 3 ▶

Solve a quadratic equation of the form $x^2 + bx + c = 0$ by completing the square.

We now apply the process of *completing the square* to solving a quadratic equation. Our goal is to make one side of the equation a perfect square trinomial, which we will factor and then apply the square root property to solve the equation.

Consider the equation $x^2 - 6x + 2 = 0$. This equation cannot be solved by factoring. So, we want to isolate the binomial $x^2 - 6x$ on one side of the equation and then complete the square.

$x^2 - 6x = -2$	Move the constant term to the right side of the equation.
$x^2 - 6x + 9 = -2 + 9$	Make $x^2 - 6x$ a perfect square trinomial by completing the square. Take half of -6 and square it. Add this number to each side of the equation.
$(x - 3)^2 = 7$	Factor the perfect square trinomial and simplify the right side of the equation.
$x - 3 = \pm\sqrt{7}$	Apply the square root property.
$x = 3 \pm \sqrt{7}$	Solve for the variable by adding 3 to each side.

> **Procedure: Solving an Equation of the Form $x^2 + bx + c = 0$ by Completing the Square**
>
> **Step 1:** Move the constant term to the right side.
> **Step 2:** Complete the square by finding half of the coefficient of x and squaring it. Add this number to each side of the equation.
> **Step 3:** Factor the perfect square trinomial and simplify the right side.
> **Step 4:** Apply the square root property to solve the equation.

Objective 3 Examples   Solve each equation by completing the square.

3a. $x^2 + 8x - 5 = 0$ **3b.** $y^2 - y - 4 = 0$

Solutions   **3a.**

$x^2 + 8x - 5 = 0$	
$x^2 + 8x - 5 + 5 = 0 + 5$	Add 5 to each side.
$x^2 + 8x = 5$	Simplify.
$x^2 + 8x + 16 = 5 + 16$	Add $\left(\dfrac{8}{2}\right)^2 = 4^2 = 16$ to each side to complete the square.
$(x + 4)^2 = 21$	Factor the trinomial and simplify the right side.
$x + 4 = \pm\sqrt{21}$	Apply the square root property.
$x = -4 \pm \sqrt{21}$	Subtract 4 from each side.

The solution set is $\{-4 - \sqrt{21}, -4 + \sqrt{21}\}$ or $\{-4 \pm \sqrt{21}\}$.

3b. $y^2 - y - 4 = 0$

$$y^2 - y - 4 + 4 = 0 + 4 \qquad \text{Add 4 to each side.}$$

$$y^2 - y = 4 \qquad \text{Simplify.}$$

$$y^2 - 1y + \frac{1}{4} = 4 + \frac{1}{4} \qquad \text{Add } \left(\frac{-1}{2}\right)^2 = \frac{1}{4} \text{ to each side to complete the square.}$$

$$\left(y - \frac{1}{2}\right)^2 = \frac{17}{4} \qquad \text{Factor the trinomial and simplify the right side.}$$

$$y - \frac{1}{2} = \pm\sqrt{\frac{17}{4}} \qquad \text{Apply the square root property.}$$

$$y - \frac{1}{2} = \pm\frac{\sqrt{17}}{2} \qquad \text{Apply the quotient rule for radicals.}$$

$$y = \frac{1}{2} \pm \frac{\sqrt{17}}{2} \qquad \text{Add } \frac{1}{2} \text{ to each side.}$$

The solution set is $\left\{\dfrac{1}{2} + \dfrac{\sqrt{17}}{2}, \dfrac{1}{2} - \dfrac{\sqrt{17}}{2}\right\}$ or $\left\{\dfrac{1}{2} \pm \dfrac{\sqrt{17}}{2}\right\}$.

 Student Check 3   Solve each equation by completing the square.

   a. $a^2 + 10a + 20 = 0$ **b.** $x^2 - 3x - 11 = 0$

Completing the Square to Solve $ax^2 + bx + c = 0$, $a \neq 1$

Objective 4 ▶

Solve a quadratic equation of the form $ax^2 + bx + c = 0$ ($a \neq 1$) by completing the square.

In the equations in Example 3, the coefficient of the squared term is 1. This must be the case in order to complete the square. Consider the following trinomial.

$$(2x - 3)^2 = 4x^2 - 12x + 9$$

This is a perfect square trinomial, but the relationship between the constant term and the middle term that we discussed earlier does not exist; that is,

$$\frac{1}{2}(-12) = -6 \quad \text{but} \quad (-6)^2 = 36 \neq 9$$

So, to solve a quadratic equation with $a \neq 1$ by completing the square, we must write an equivalent equation with $a = 1$. We do this by dividing each side of the equation by a.

Procedure: Solving $ax^2 + bx + c = 0$, $a \neq 1$, by Completing the Square

Step 1: Divide each side of the equation by a, the coefficient of the squared term.
Step 2: Move the constant term to the right side of the equation.
Step 3: Complete the square to make one side a perfect square trinomial. Add this number to each side of the equation.
Step 4: Factor the trinomial and simplify the other side.
Step 5: Apply the square root property and solve the equation.

Objective 4 Examples   Solve each equation by completing the square.

   4a. $2x^2 - 24x + 30 = 0$ **4b.** $3x^2 - 2x - 5 = 0$

Solutions **4a.** $2x^2 - 12x + 30 = 0$

$\dfrac{2x^2 - 12x + 30}{2} = \dfrac{0}{2}$ Divide each side by 2.

$x^2 - 6x + 15 = 0$ Simplify.

$x^2 - 6x + 15 - 15 = 0 - 15$ Subtract 15 from each side.

$x^2 - 6x = -15$ Simplify.

$x^2 - 6x + 9 = -15 + 9$ Add $\left(\dfrac{-6}{2}\right)^2 = (-3)^2 = 9$ to each side.

$(x - 3)^2 = -6$ Factor and simplify.

$x - 3 = \pm\sqrt{-6}$ Apply the square root property.

$x = 3 \pm i\sqrt{6}$ Add 3 to each side and simplify $\sqrt{-6}$.

The solution set is $\{3 + i\sqrt{6}, 3 - i\sqrt{6}\}$ or $\{3 \pm i\sqrt{6}\}$.

4b. $3x^2 - 2x - 5 = 0$

$\dfrac{3x^2 - 2x - 5}{3} = \dfrac{0}{3}$ Divide each side by 3.

$x^2 - \dfrac{2}{3}x - \dfrac{5}{3} = 0$ Simplify.

$x^2 - \dfrac{2}{3}x - \dfrac{5}{3} + \dfrac{5}{3} = 0 + \dfrac{5}{3}$ Add $\dfrac{5}{3}$ to each side.

$x^2 - \dfrac{2}{3}x = \dfrac{5}{3}$ Simplify.

$x^2 - \dfrac{2}{3}x + \dfrac{1}{9} = \dfrac{5}{3} + \dfrac{1}{9}$ Add $\left(\dfrac{1}{2} \cdot \dfrac{-2}{3}\right)^2 = \left(\dfrac{-1}{3}\right)^2 = \dfrac{1}{9}$ to each side to complete the square.

$\left(x - \dfrac{1}{3}\right)^2 = \dfrac{16}{9}$ Factor and simplify, $\dfrac{5}{3} + \dfrac{1}{9} = \dfrac{15}{9} + \dfrac{1}{9} = \dfrac{16}{9}$.

$x - \dfrac{1}{3} = \pm\sqrt{\dfrac{16}{9}}$ Apply the square root property.

$x - \dfrac{1}{3} = \pm\dfrac{4}{3}$ Apply the quotient rule for radicals.

$x = \dfrac{1}{3} \pm \dfrac{4}{3}$ Add $\dfrac{1}{3}$ to each side.

So, $x = \dfrac{1}{3} + \dfrac{4}{3} = \dfrac{5}{3}$ or $x = \dfrac{1}{3} - \dfrac{4}{3} = -\dfrac{3}{3} = -1$. The solution set is $\left\{-1, \dfrac{5}{3}\right\}$.

✓ Student Check 4 Solve each equation by completing the square.
  a. $4x^2 - 8x + 20 = 0$ **b.** $2y^2 - y - 3 = 0$

Applications

Objective 5 ▶

Solve applications of quadratic equations.

Quadratic equations model many real-life situations. Some of these include the area of geometric figures, the height of an object, and investment-related problems. For these types of problems, there are formulas we use to write equations that solve them.

Objective 5 Examples Solve each problem.

  5a. The base of the Great Pyramid of Giza, Egypt, is a square. If the area of the base of the pyramid is 570,780.25 ft², how long is each side of the base of the pyramid? Recall the area of a square is $A = s^2$. (Source: http://www.plim.org/greatpyramid.html)

Solution   5a.

$$s^2 = A$$ State the area formula.

$$s^2 = 570,785.25$$ Replace A with 570,785.25.

$$s = \pm\sqrt{570,785.25}$$ Apply the square root property.

$$s = \pm 755.5$$ Simplify.

Because the variable s represents the length of the side of a square, the negative value must be discarded. Therefore, the length of each side of the base of the pyramid is 755.5 ft.

5b. One of the world's highest commercial bungee jumps is off of the Bloukrans River Bridge in South Africa. The jump takes place from a platform below the roadway of the bridge and is 710 ft above the valley floor. A jumper is in free fall until the elasticity of the cord affects the rate of fall. While the jumper is free-falling, his height, in feet, above the valley floor is $h = -16t^2 + 710$, where t is in seconds. How many seconds will it take for the jumper to be 410 ft above the valley floor? (Source: http://www.faceadrenalin.com/)

Solution   5b.

$$h = -16t^2 + 710$$ State the given model.

$$-16t^2 + 710 = 410$$ Replace h with 410.

$$-16t^2 + 710 - 710 = 410 - 710$$ Subtract 710 from each side.

$$-16t^2 = -300$$ Simplify.

$$\frac{-16t^2}{-16} = \frac{-300}{-16}$$ Divide each side by -16.

$$t^2 = \frac{300}{16}$$ Simplify.

$$t = \pm\sqrt{\frac{300}{16}}$$ Apply the square root property.

$$t = \pm\frac{\sqrt{300}}{\sqrt{16}}$$ Apply the quotient rule for radicals.

$$t = \pm\frac{10\sqrt{3}}{4}$$ Simplify.

Approximating the solutions gives us $t \approx 4.33$ sec or $t \approx -4.33$ sec. We must discard the negative value because time can't be negative. So, the jumper reaches 410 ft above the valley floor in approximately 4.33 sec.

5c. At what rate would you have to invest $1000 into a 2-yr CD to have $1060.90 once the CD matures? Use the formula $A = P(1 + r)^2$, where P is the amount invested and r is the annual interest rate.

Solution   5c.

$$P(1 + r)^2 = A$$ State the given model.

$$1000(1 + r)^2 = 1060.90$$ Replace P with 1000 and A with 1060.90.

$$\frac{1000(1 + r)^2}{1000} = \frac{1060.90}{1000}$$ Divide each side by 1000.

$$(1 + r)^2 = 1.0609$$ Simplify.

$$1 + r = \pm\sqrt{1.0609}$$ Apply the square root property.

$$1 + r = \pm 1.03$$ Simplify the radical.

$$r = -1 \pm 1.03$$ Subtract 1 from each side.

$$r = -1 + 1.03 \quad \text{or} \quad r = -1 - 1.03$$ Write two equivalent expressions for r.

$$r = 0.03 \qquad\qquad r = -2.03$$ Simplify.

The negative value doesn't make sense, so the money needs to be invested at a 3% annual interest rate.

☑ **Student Check 5** Solve each problem.

a. A square city block covers an area of 435,600 ft². What is the length of the side of the block?

b. In 2010, the Burj Khalifa in Dubai officially opened as the tallest building in the world at 2717 ft. If a penny is dropped from the top of this building, its height, in feet, t sec after it is dropped is given by $h = -16t^2 + 2717$. How many seconds will it take for the penny to reach the ground? (Source: http://www.burjkhalifa.ae/language/en-us/home.aspx)

c. At what rate would you have to invest $25,000 in a 2-yr CD to have $28,090 once the CD matures? Use the formula $A = P(1 + r)^2$, where P is the amount invested and r is the annual interest rate.

Zeros of Quadratic Functions

Objective 6 ▶

Find zeros of quadratic functions of the form $f(x) = a(x - h)^2 + k$.

In Section 11.1, we learned how to graph functions of the form $f(x) = a(x - h)^2 + k$ by plotting points and shifting and reflecting (if necessary) the graph of $f(x) = x^2$. We will apply the square root property and complete the square to find zeros of quadratic functions more precisely. The **zeros of a quadratic function** are the values of x for which $y = f(x) = 0$. The zeros correspond to the x-intercepts of the graph.

> **Procedure: Finding Zeros of Quadratic Functions**
>
> **Step 1:** Set $f(x) = 0$.
> **Step 2:** Complete the square or apply the square root property to solve the equation.
> **Step 3:** Confirm by graphing the function.

Objective 6 Examples Find the zeros of each function. State the x-intercepts of the graph of the function. Confirm the solutions by graphing.

6a. $f(x) = (x - 1)^2 - 4$ **6b.** $f(x) = (x + 2)^2 - 3$ **6c.** $f(x) = 2(x - 3)^2 + 8$

Solutions **6a.**

$$(x - 1)^2 - 4 = 0 \qquad \text{Set } f(x) = 0.$$
$$(x - 1)^2 = 4 \qquad \text{Add 4 to each side.}$$
$$x - 1 = \pm\sqrt{4} \qquad \text{Apply the square root property.}$$
$$x = 1 \pm 2 \qquad \text{Add 1 to each side and simplify the square root.}$$
$$x = 1 + 2 \quad \text{or} \quad x = 1 - 2 \qquad \text{Find the two values for } x.$$
$$x = 3 \qquad\qquad\quad x = -1 \qquad \text{Simplify.}$$

The zeros of the function are $x = 3$ and $x = -1$. So, the x-intercepts are $(3, 0)$ and $(-1, 0)$. The graph of this function confirms our work is correct.

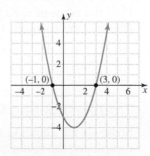

6b.

$$(x + 2)^2 - 3 = 0$$ Set $f(x) = 0$.

$$(x + 2)^2 = 3$$ Add 3 to each side.

$$x + 2 = \pm\sqrt{3}$$ Apply the square root property.

$$x = -2 \pm \sqrt{3}$$ Subtract 2 from each side.

$$x = -2 + \sqrt{3} \quad \text{or} \quad x = -2 - \sqrt{3}$$ Find the two values for x.

$$x \approx -0.27 \qquad\qquad x \approx -3.73$$ Approximate the solutions.

The zeros of the function are approximately $x \approx -3.73$ and $x \approx -0.27$. So, the x-intercepts are $(-3.73, 0)$ and $(-0.27, 0)$. The graph of this function confirms our work is correct.

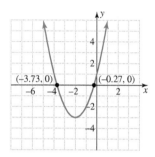

6c.

$$2(x - 3)^2 + 8 = 0$$ Set $f(x) = 0$.

$$2(x - 3)^2 = -8$$ Subtract 8 from each side.

$$\frac{2(x - 3)^2}{2} = \frac{-8}{2}$$ Divide each side by 2.

$$(x - 3)^2 = -4$$ Simplify.

$$x - 3 = \pm\sqrt{-4}$$ Apply the square root property.

$$x = 3 \pm 2i$$ Add 3 to each side and simplify the square root.

$$x = 3 + 2i \quad \text{or} \quad x = 3 - 2i$$ Find the two values for x.

Because the solutions are not real numbers, there are no zeros of this function. The graph confirms this since it does not cross the x-axis.

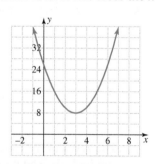

✓ **Student Check 6**   Find the zeros of each function. State the x-intercepts of the graph of the function. Confirm the solutions by graphing.

a. $f(x) = (x + 4)^2 - 1$ **b.** $f(x) = (x - 5)^2 - 3$ **c.** $f(x) = -3(x + 1)^2 - 3$

Objective 7 ▶

Troubleshoot common errors.

Troubleshooting Common Errors

Some of the common errors associated with applying the square root property and completing the square are shown.

Objective 7 Examples	A problem and an incorrect solution are given. Provide the correct solution and an explanation of the error.

7a. Solve $(x - 1)^2 = 9$.

Incorrect Solution	**Correct Solution and Explanation**
$(x - 1)^2 = 9$ $x - 1 = 3$ $x = 4$ The solution set is $\{4\}$.	When we apply the square root property, we take both the positive and negative square root. $(x - 1)^2 = 9$ $x - 1 = \pm 3$ $x = 1 \pm 3$ $x = 1 + 3 \quad \text{or} \quad x = 1 - 3$ $x = 4 \qquad \text{or} \quad x = -2$ The solution set is $\{-2, 4\}$.

7b. Solve $x^2 - 8x - 1 = 0$ by completing the square.

Incorrect Solution	**Correct Solution and Explanation**
$x^2 - 8x - 1 = 0$ $x^2 - 8x = 1$ $x^2 - 8x + 16 = 1$ $(x - 4)^2 = 1$ $x - 4 = \pm 1$ $x = 4 \pm 1$ $x = 5 \quad \text{or} \quad x = -3$ The solution set is $\{-3, 5\}$.	When we complete the square, we add the number to both sides of the equation. $x^2 - 8x - 1 = 0$ $x^2 - 8x = 1$ $x^2 - 8x + 16 = 1 + 16$ $(x - 4)^2 = 17$ $x - 4 = \pm\sqrt{17}$ $x = 4 \pm \sqrt{17}$ $x = 4 + \sqrt{17} \quad \text{or} \quad x = 4 - \sqrt{17}$ The solution set is $\{4 \pm \sqrt{17}\}$.

ANSWERS TO STUDENT CHECKS

Student Check 1   **a.** $\{\pm\sqrt{21}\}$ **b.** $\{\pm 6\}$ **c.** $\{\pm\sqrt{10}\}$

     d. $\{-11, -3\}$ **e.** $\left\{\dfrac{7 \pm 2\sqrt{3}}{5}\right\}$ **f.** $\left\{\dfrac{-6 \pm i}{7}\right\}$

Student Check 2   **a.** $x^2 + 12x + 36 = (x + 6)^2$

     b. $y^2 - 2y + 1 = (y - 1)^2$

     c. $x^2 - 5x + \dfrac{25}{4} = \left(x - \dfrac{5}{2}\right)^2$

Student Check 3   **a.** $\{-5 \pm \sqrt{5}\}$ **b.** $\left\{\dfrac{3}{2} \pm \dfrac{\sqrt{53}}{2}\right\}$

Student Check 4   **a.** $\{1 \pm 2i\}$ **b.** $\left\{-1, \dfrac{3}{2}\right\}$

Student Check 5   **a.** 660 ft   **b.** approximately 13.03 sec

     c. 6%

Student Check 6   **a.** The zeros are $x = -5, -3$. The x-intercepts are $(-5, 0)$ and $(-3, 0)$.   **b.** The zeros are $x = 5 \pm \sqrt{3}$. The x-intercepts are approximately $(6.73, 0)$ and $(3.27, 0)$.   **c.** The zeros are $x = -1 \pm i$, so there are no x-intercepts of the graph.

SUMMARY OF KEY CONCEPTS

1. When an equation is of the form $a^2 = k$, where k is a real number, the square root property can be applied. The property enables us to remove the square from the expression on the left side of the equation and set that expression equal to the positive and negative square root of the number on the right side of the equation. If a squared expression is equal to a negative number, the equation has complex, nonreal solutions.

2. A perfect square trinomial results from squaring a binomial. If given a binomial of the form $x^2 + bx$, we can

make it a perfect square trinomial by adding the number $c = \left(\dfrac{b}{2}\right)^2$. That is, adding half of the coefficient of x, squared, makes the binomial a perfect square trinomial.

3. Completing the square can be used to solve any quadratic equation. To apply this method, we must
   a. Make the leading coefficient 1. If it is not 1, divide each side by the given coefficient.
   b. Isolate the constant term on one side of the equation.
   c. Find the number that completes the square and add it to both sides of the equation.

d. Factor the trinomial as a binomial squared and simplify the constant side.
   e. Apply the square root property and solve.

4. To solve applications of quadratic equations use the appropriate mathematical formula. Substitute the known values and apply the square root property to solve the resulting equation.

5. To find zeros of a quadratic function of the form $f(x) = a(x - h)^2 + k$, set $f(x) = 0$ and apply the square root property to solve the equation. The zeros correspond to the x-intercepts on the graph.

GRAPHING CALCULATOR SKILLS

The calculator can be used to approximate the solutions of quadratic equations and to check solutions. The solutions of a quadratic equation correspond to the x-intercepts of the graph.

Example 1: Show that $\dfrac{4 + 2\sqrt{2}}{3} \approx 2.28$ and $\dfrac{4 - 2\sqrt{2}}{3} \approx 0.39$.

Solution: Enter parentheses around the numerator of the fraction. Close the parentheses after the radicand.

```
(4+2√(2))/3
        2.276142375
(4-2√(2))/3
        .3905242918
```

Example 2: Show that $\dfrac{4 + 2\sqrt{2}}{3}$ is a solution of $(3a - 4)^2 = 8$.

Method 1 Store the solution as the value of x. Enter the expression on the left side of the equation. If the result agrees with the right side of the equation, then the value entered is a solution of the equation.

```
(4+2√(2))/3→X
        2.276142375
(3X-4)²
                8
```

Method 2 Graph the function $f(x) = (3a - 4)^2 - 8$ and determine if the given value corresponds to an x-intercept of the graph. After the graph is displayed, press TRACE and enter the given value. If it is a solution, the y-value is 0.

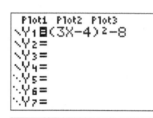

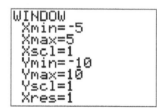

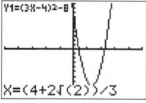

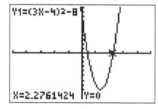

SECTION 11.2 / EXERCISE SET

 Write About It!

Use complete sentences in your answer to each exercise.

1. Explain the difference in solving the quadratic equation $x^2 - a^2 = 0$ using factoring and applying the square root property.

2. Use an example to explain how to complete the square for an expression of the form $x^2 + bx$, $b > 0$.

3. Explain how to complete the square for the expression $2x^2 - 3x$.

4. What is the connection between the solutions of a quadratic equation and the zeros of the corresponding quadratic function?

5. If the solutions of a quadratic equation are complex, what does it mean about the graph of the corresponding quadratic function?

6. If the solutions of a quadratic equation are two distinct real numbers, what does it mean about the graph of the corresponding quadratic function?

Practice Makes Perfect!

Solve each equation by applying the square root property. (See Objective 1.)

7. $x^2 + 5 = 54$ **8.** $y^2 + 10 = 46$

9. $r^2 - 3 = 61$ **10.** $s^2 - 8 = 17$

11. $49a^2 - 9 = -8$ **12.** $25b^2 - 7 = 74$

13. $81x^2 + 9 = 13$ **14.** $9y^2 + 5 = 69$

15. $z^2 - 2100 = 0$ **16.** $z^2 - 175 = 0$

17. $a^2 + 180 = 0$ **18.** $b^2 + 24 = 0$

19. $(4x - 1)^2 - 5 = 20$ **20.** $(5y + 7)^2 - 8 = -4$

21. $(7z - 10)^2 + 9 = 45$ **22.** $(4w + 8)^2 - 7 = 18$

23. $(3a - 1)^2 = 48$ **24.** $(4b - 7)^2 - 6 = 0$

25. $(6z + 5)^2 = -150$ **26.** $(9w - 2)^2 + 3 = 0$

Create a perfect square trinomial from each binomial by completing the square and then express it in factored form. (See Objective 2.)

27. $x^2 + 18x$ **28.** $y^2 + 16y$

29. $a^2 - 14a$ **30.** $b^2 - 24b$

31. $r^2 + r$ **32.** $s^2 + 7s$

33. $a^2 - 9a$ **34.** $b^2 - 15b$

Solve each equation by completing the square. (See Objective 3.)

35. $x^2 - 6x - 7 = 0$ **36.** $y^2 - 8y + 16 = 0$

37. $r^2 + 5r - 50 = 0$ **38.** $s^2 - 3s - 10 = 0$

39. $x^2 - 10x + 21 = 0$ **40.** $y^2 + 6y + 5 = 0$

Solve each equation by completing the square. (See Objective 4.)

41. $8x^2 - 2x - 3 = 0$ **42.** $3y^2 - 11y - 42 = 0$

43. $2a^2 + 17a + 21 = 0$ **44.** $4b^2 + 27b + 18 = 0$

45. $4x^2 + 16x + 5 = 0$ **46.** $3y^2 + 12y + 7 = 0$

47. $2r^2 + 8r + 13 = 0$ **48.** $2r^2 + 3r + 3 = 0$

Solve each problem. (See Objective 5.)

49. A downtown square city block in Portland, Oregon, is 68,000 ft². What is the length of each side of the city block? Round the answer to two decimal places. (Source: http://www.land4ever.com/block.htm)

50. A downtown square city block in Chicago, Illinois, is 217,000 ft². What is the length of each side of the city block? Round the answer to two decimal places. (Source: http://www.land4ever.com/block.htm)

51. If a penny is dropped from the top of the Central Plaza in Hong Kong, its height, in meters, t sec after it is dropped is given by $s = -9.8t^2 + 374$. How many seconds will it take for the penny to reach the ground? Round the answer to two decimal places. (Source: http://www.centralplaza.com.hk)

52. If a penny is dropped from the top of the Willis Tower in Chicago, its height, in meters, t sec after it is dropped is given by $s = -9.8t^2 + 442$. How many seconds will it take for the penny to reach the ground? Round the answer to two decimal places. (Source: http://www.willistower.com)

53. If a penny is dropped from the top of the Chrysler Building in New York, its height, in feet, t sec after it is dropped is given by $s = -16t^2 + 1046$. How many seconds will it take for the penny to reach the ground? Round the answer to two decimal places. (Source: http://www.nyc.com)

54. If a penny is dropped from the top of the Taipei Tower in Taiwan, its height, in feet, t sec after it is dropped is given by $s = -16t^2 + 1667$. How many seconds will it take for the penny to reach the ground? Round the answer to two decimal places. (Source: http://www.taipei-101.com.tw)

55. At what rate would you have to invest $5100 in a 2-yr CD to have $5421.10 once the CD matures? Use the formula $A = P(1 + r)^2$, where P is the amount invested and r is the annual interest rate.

56. At what rate would you have to invest $14,200 in a 2-yr CD to have $14,773.68 once the CD matures? Use the formula $A = P(1 + r)^2$, where P is the amount invested and r is the annual interest rate.

57. At what rate would you have to invest $25,600 in a 2-yr CD to have $27,370.39 once the CD matures? Use the formula $A = P(1 + r)^2$, where P is the amount invested and r is the annual interest rate.

58. At what rate would you have to invest $4800 in a 2-yr CD to have $5052.84 once the CD matures? Use the formula $A = P(1 + r)^2$, where P is the amount invested and r is the annual interest rate.

Find the zeros of each function. State the x-intercepts of the graph of the function. (See Objective 6.)

59. $f(x) = (x + 7)^2 - 36$ **60.** $f(x) = (x - 8)^2 - 81$

61. $f(x) = (3x + 2)^2 - 1$ **62.** $f(x) = (3x - 1)^2 - 4$

63. $f(x) = (x + 1)^2 + 24$ **64.** $f(x) = (x - 11)^2 + 20$

65. $f(x) = (x - 3)^2 - 50$ **66.** $f(x) = (x + 11)^2 - 48$

67. $f(x) = (10x + 1)^2 - 3$ **68.** $f(x) = (5x - 1)^2 - 50$

69. $f(x) = (3x + 5)^2 + 84$ **70.** $f(x) = (x - 3)^2 + 45$

 ## Mix 'Em Up!

Solve each equation by applying the square root property or by completing the square.

71. $64x^2 - 1 = 24$ **72.** $49y^2 - 6 = 3$

73. $r^2 + 99 = -1$ **74.** $49s^2 + 41 = 5$

75. $z^2 = 252$ **76.** $w^2 = -243$

77. $(2x + 10)^2 + 8 = 57$ **78.** $(4y - 8)^2 + 5 = 14$

79. $(9z + 2)^2 + 75 = 0$ **80.** $(7w + 3)^2 + 50 = 5$

81. $(4a - 3)^2 - 21 = 0$ **82.** $(3b - 6)^2 - 7 = 0$

83. $4(4x - 3)^2 + 8 = 0$

84. $2(3w + 8)^2 + 20 = 0$

85. $3(6x - 1)^2 - 900 = 0$

86. $2(5w - 4)^2 - 270 = 0$

87. $x^2 + 5x - 36 = 0$

88. $y^2 + 2y - 63 = 0$

89. $2r^2 - r - 6 = 0$

90. $8s^2 - 2s - 3 = 0$

91. $2x^2 + 7x = 0$

92. $5x^2 - 3x = 0$

93. $2r^2 - r + 2 = 0$

94. $2s^2 + s + 1 = 0$

95. $4x(x - 3) + 3 = 0$

96. $8y(2y + 1) = 4$

97. $3a(3a - 8) + 19 = 0$

98. $4y(y + 5) + 39 = 0$

Create a perfect square trinomial from each binomial by completing the square and then express it in factored form.

99. $x^2 + 8x$

100. $y^2 - 26y$

101. $a^2 - 1.8a$

102. $b^2 - 0.6b$

103. $r^2 + 0.9a$

104. $s^2 + 0.5s$

105. $r^2 + \dfrac{1}{2}r$

106. $s^2 + \dfrac{5}{3}s$

107. $a^2 - \dfrac{3}{2}a$

108. $b^2 - \dfrac{7}{4}b$

Solve each problem.

109. The largest church clock face in Europe is on the steeple of St. Peter's Church in Zurich, Switzerland. It has an area of approximately 639.73 ft². What is the diameter of the clock face? Round the answer to two decimal places. (Source: http://en.wikipedia.org)

110. The Busch Stadium in St. Louis, Missouri, is almost a perfect circle. It has an area of approximately 502,654.82 ft². What is the diameter of the stadium? (Source: http://www.baseball-statistics.com/)

111. BASE jumping is an extreme sport in which people parachute off of fixed structures such as buildings, antennas, spans (bridges), or earth (cliffs). The Perrine Bridge in Twin Falls, Idaho, is 486 ft above the Snake River. The height of a base jumper t sec after the start of his fall can be represented by $s = -16t^2 + 486$. How many seconds does the jumper have before he reaches the Snake River without opening his parachute? Round the answer to two decimal places. (Source: http://en.wikipedia.org/wiki/BASE_jumping)

112. The mountain Kjerag in Lysefjorden, Norway, is a BASE jumping site with a drop of 984 m. The height of a base jumper t sec after the start of his fall can be represented by $h = -9.8t^2 + 984$. How many seconds does the jumper have before reaching the ground without opening his parachute? Round the answer to two decimal places. (Source: http://en.wikipedia.org/wiki/Kjerag)

113. At what rate would you have to invest $12,600 in a 2-yr CD to have $13,263.72 once the CD matures? Use the formula $A = P(1 + r)^2$, where P is the amount invested and r is the annual interest rate.

114. At what rate would you have to invest $21,200 in a 2-yr CD to have $21,711.85 once the CD matures? Use the formula $A = P(1 + r)^2$, where P is the amount invested and r is the annual interest rate.

 You Be the Teacher!

Correct each student's errors, if any.

115. Solve $4x^2 - 9 = 0$ using the square root property.

Josh's work:
$$4x^2 - 9 = 0$$
$$2x + 3i = 0$$
$$2x = -3i$$
$$x = -\dfrac{3}{2}i$$

116. Solve $(3x + 1)^2 + 9 = 25$ using the square root property.

Matt's work:
$$(3x + 1)^2 + 9 = 25$$
$$3x + 1 + 3 = 5$$
$$3x = 1$$
$$x = \dfrac{1}{3}$$

117. Complete the square for $x^2 + 14x$ and factor the trinomial.

Allison's work:
$$x^2 + 14x + 196 = (x + 14)^2$$

118. Solve $3x^2 + 8x + 6 = 0$ by completing the square.

Michelle's work:
$$3x^2 + 8x + 6 = 0$$
$$x^2 + \dfrac{8}{3}x = -6$$
$$x^2 + \dfrac{8}{3}x + \dfrac{16}{9} = -6$$
$$\left(x + \dfrac{4}{3}\right)^2 = -6$$
$$x + \dfrac{4}{3} = \pm i\sqrt{6}$$
$$x = -\dfrac{4}{3} \pm i\sqrt{6}$$

 Calculate It!

Use a graphing calculator to determine if the two expressions are the same after completing the square.

119. $x^2 - 15x + \dfrac{225}{4}$ and $\left(x - \dfrac{15}{2}\right)^2$

120. $x^2 - \dfrac{7}{10}x + \dfrac{49}{400}$ and $\left(x - \dfrac{7}{20}\right)^2$

121. $x^2 - 2.3x + 1.3225$ and $(x - 1.15)^2$

122. $x^2 - \dfrac{5}{7}x + \dfrac{25}{196}$ and $\left(x - \dfrac{5}{14}\right)^2$

Solving Quadratic Equations Using the Quadratic Formula

▶ **OBJECTIVES**

As a result of completing this section, you will be able to

1. Solve quadratic equations using the quadratic formula.

2. Use the discriminant to determine the number and types of solutions of a quadratic equation.

3. Solve applications of quadratic equations.

4. Troubleshoot common errors.

The height of a basketball, in feet, thrown upward by a player at a free throw line is given by $h = -16t^2 + 22t + 7$. When will the ball reach the height of the basketball hoop, which is 10 ft from the floor? To answer this question, we must solve the equation $-16t^2 + 22t + 7 = 10$.

In this section, we will learn another method for solving quadratic equations that enables us to solve this problem.

The Quadratic Formula

Objective 1 ▶

Solve quadratic equations using the quadratic formula.

Thus far, we have learned three methods to solve quadratic equations. These include applying the zero products property (factoring), applying the square root property, and completing the square.

- Factoring is an efficient method but it does not solve every equation.
- The square root property only applies to equations of the form $a^2 = k$, where k is a real number.
- Completing the square does solve every quadratic equation but can be somewhat tedious, especially if the leading coefficient is not 1 and the coefficient of the x term is not even.

Another method for solving quadratic equations involves using a formula. This formula, the *quadratic formula*, is derived from completing the square on the standard form of the quadratic equation, $ax^2 + bx + c = 0$, where $a \neq 0$.

Derivation of the Quadratic Formula

Begin with $ax^2 + bx + c = 0$ and complete the square.

$$\frac{ax^2 + bx + c}{a} = \frac{0}{a}$$

Divide each side by a to get a coefficient of 1 on x^2.

$$x^2 + \frac{b}{a}x + \frac{c}{a} = 0$$

Simplify.

$$x^2 + \frac{b}{a}x = -\frac{c}{a}$$

Subtract $\frac{c}{a}$ from each side.

$$x^2 + \frac{b}{a}x + \frac{b^2}{4a^2} = -\frac{c}{a} + \frac{b^2}{4a^2}$$

Complete the square $\left(\frac{1}{2} \cdot \frac{b}{a}\right)^2 = \frac{b^2}{4a^2}$ and add to each side.

$$\left(x + \frac{b}{2a}\right)^2 = \frac{b^2 - 4ac}{4a^2}$$

Factor the left and simplify the right.
$$-\frac{c}{a} + \frac{b^2}{4a^2} = -\frac{4ac}{4a \cdot a} + \frac{b^2}{4a^2} = \frac{b^2 - 4ac}{4a^2}$$

$$x + \frac{b}{2a} = \pm\sqrt{\frac{b^2 - 4ac}{4a^2}}$$

Apply the square root property.

$$x + \frac{b}{2a} = \pm\frac{\sqrt{b^2 - 4ac}}{\sqrt{4a^2}}$$

Apply the quotient rule for radicals.

$$x = -\frac{b}{2a} \pm \frac{\sqrt{b^2 - 4ac}}{2a}$$

Simplify the radical and subtract $\frac{b}{2a}$ from each side.

$$x = \frac{-b \pm \sqrt{b^2 - 4ac}}{2a}$$

Add the fractions.

So, this formula depends on the values of a, b, and c from the standard form of the quadratic equation.

Property: The Quadratic Formula

If $ax^2 + bx + c = 0$ $(a \neq 0)$, then

$$x = \frac{-b \pm \sqrt{b^2 - 4ac}}{2a}$$

Procedure: Solving an Equation Using the Quadratic Formula

Step 1: Write the equation in standard form, $ax^2 + bx + c = 0$.
Step 2: Identify the values of a, b, and c from the standard form.
Step 3: Substitute the values of a, b, and c in the quadratic formula.
Step 4: Use the order of operations to simplify the resulting expression.
Step 5: Simplify the quotient as much as possible.
Step 6: Check solutions by substituting the values into the original equation.

Objective 1 Examples **Solve each equation using the quadratic formula.**

1a. $x^2 - 6x + 9 = 0$ **1b.** $y^2 + 4y = 7$ **1c.** $6x^2 + x = 0$ **1d.** $\dfrac{2y^2}{3} + \dfrac{1}{6}y + 4 = 0$

Solutions **1a.** The equation is in standard form, so

$$1x^2 - 6x + 9 = 0 \rightarrow a = 1, b = -6, c = 9$$

$x = \dfrac{-b \pm \sqrt{b^2 - 4ac}}{2a}$ 　　　　　State the quadratic formula.

$x = \dfrac{-(-6) \pm \sqrt{(-6)^2 - 4(1)(9)}}{2(1)}$ 　　Replace a, b, and c with the appropriate values.

$x = \dfrac{6 \pm \sqrt{36 - 36}}{2}$ 　　　　　Simplify the expression.

$x = \dfrac{6 \pm \sqrt{0}}{2}$ 　　　　　　Simplify the radicand.

$x = \dfrac{6 \pm 0}{2}$ 　　　　　　Simplify the radical expression.

$x = \dfrac{6 + 0}{2} = \dfrac{6}{2} = 3$ 　or　 $x = \dfrac{6 - 0}{2} = \dfrac{6}{2} = 3$　Write the two values for x.

So, the solution set is {3}. This quadratic equation has *one repeating rational* solution.

1b. We first write the equation in standard form and identify a, b, and c.

$y^2 + 4y = 7$

$1y^2 + 4y - 7 = 0$ 　　　　　Subtract 7 from each side.

$a = 1, b = 4, c = -7$ 　　　　Identify a, b, and c.

$y = \dfrac{-b \pm \sqrt{b^2 - 4ac}}{2a}$ 　　　　State the quadratic formula.

$y = \dfrac{-(4) \pm \sqrt{(4)^2 - 4(1)(-7)}}{2(1)}$ 　　Replace a, b, and c with the appropriate values.

$y = \dfrac{-4 \pm \sqrt{16 + 28}}{2}$ 　　　　Simplify the expression.

$y = \dfrac{-4 \pm \sqrt{44}}{2}$ 　　　　　Simplify the radicand.

$$y = \frac{-4 \pm \sqrt{4 \cdot 11}}{2}$$ Find a perfect square factor of 44.

$$y = \frac{-4 \pm 2\sqrt{11}}{2}$$ Apply the product rule for radicals.

$$y = \frac{2(-2 \pm \sqrt{11})}{2}$$ Factor the numerator.

$$y = -2 \pm \sqrt{11}$$ Divide out the common factor, 2.

$$y = -2 + \sqrt{11} \quad \text{or} \quad y = -2 - \sqrt{11}$$ Write the two values for x.

The solution set is $\{-2 + \sqrt{11}, -2 - \sqrt{11}\}$ or $\{-2 \pm \sqrt{11}\}$. This quadratic equation has *two irrational* solutions.

1c. The equation is in standard form, so

$$6x^2 + 1x + 0 = 0 \rightarrow a = 6, b = 1, c = 0$$

$$x = \frac{-b \pm \sqrt{b^2 - 4ac}}{2a}$$ State the quadratic formula.

$$x = \frac{-(1) \pm \sqrt{(1)^2 - 4(6)(0)}}{2(6)}$$ Replace a, b, and c with the appropriate values.

$$x = \frac{-1 \pm \sqrt{1 - 0}}{12}$$ Simplify the expression.

$$x = \frac{-1 \pm \sqrt{1}}{12}$$ Simplify the radicand.

$$x = \frac{-1 \pm 1}{12}$$ Simplify the radical expression.

$$x = \frac{-1 + 1}{12} = \frac{0}{12} = 0 \quad \text{or}$$ Write the two values for x.

$$x = \frac{-1 - 1}{12} = \frac{-2}{12} = -\frac{1}{6}$$

The solution set is $\left\{-\frac{1}{6}, 0\right\}$. This quadratic equation has *two rational* solutions.

1d. We first clear the fractions before identifying a, b, and c.

$$6\left(\frac{2y^2}{3} + \frac{1}{6}y + 4\right) = 6(0)$$ Multiply each side of the equation by the LCD, 6.

$$4y^2 + 1y + 24 = 0$$ Simplify.

$$a = 4, b = 1, c = 24$$ Identify a, b, and c.

$$y = \frac{-b \pm \sqrt{b^2 - 4ac}}{2a}$$ State the quadratic formula.

$$y = \frac{-(1) \pm \sqrt{(1)^2 - 4(4)(24)}}{2(4)}$$ Replace a, b, and c with the appropriate values.

$$y = \frac{-1 \pm \sqrt{1 - 384}}{8}$$ Simplify the expression.

$$y = \frac{-1 \pm \sqrt{-383}}{8}$$ Simplify the radicand.

$$y = \frac{-1 \pm i\sqrt{383}}{8}$$ Simplify the radical expression.

The solution set is $\left\{\frac{-1 \pm i\sqrt{383}}{8}\right\}$. This quadratic equation has *no real* solutions but two *complex, nonreal* solutions.

✓ **Student Check 1**   Solve each equation using the quadratic formula.

a. $x^2 - 12x + 36 = 0$ **b.** $5h^2 = 4h - 1$

c. $7x^2 + 2x = 0$ **d.** $\frac{3}{2}x^2 + \frac{1}{4}x - 2 = 0$

The Discriminant of a Quadratic Equation

Objective 2 ▶

Use the discriminant to determine the number and types of solutions of a quadratic equation.

Example 1 illustrates that a quadratic equation will have either two real solutions, one real solution, or no real solutions. In Section 11.1, we discovered these same possibilities by graphing. Recall that the solutions of a quadratic equation $f(x) = 0$ correspond to the x-intercepts of the graph of $y = f(x)$.

   We can also determine the number and types of solutions of a quadratic equation by the number inside the radical of the quadratic formula, $b^2 - 4ac$. This number is called the *discriminant*.

Definition: Discriminant

For $ax^2 + bx + c = 0$ $(a \neq 0)$, the discriminant is $b^2 - 4ac$.

Description of $b^2 - 4ac$	Types of Solutions
Positive and a perfect square	Two rational solutions
Positive and not a perfect square	Two irrational solutions
Zero	One repeating rational solution
Negative	Two complex, nonreal solutions

The discriminant can also tell us whether or not a quadratic equation can be solved by factoring. If the discriminant is a positive perfect square, then the equation can also be solved by factoring. In Example 1 part (c), the discriminant is $b^2 - 4ac = 1$. So, the equation can also be solved by factoring.

$$6x^2 + x = 0$$
$$x(6x + 1) = 0$$
$$x = 0 \quad \text{or} \quad 6x + 1 = 0$$
$$x = 0 \qquad\qquad x = -\frac{1}{6}$$

So, there may be more than one method that can be used to solve a quadratic equation.

Procedure: Using the Discriminant to Determine the Number and Types of Solutions

Step 1: Write the equation in standard form.
Step 2: Identify the values of a, b, and c.
Step 3: Substitute the values into the expression $b^2 - 4ac$.
Step 4: Use the chart in the definition box to determine the number and types of solutions.

Objective 2 Examples   Find the discriminant of each equation and use it to determine the number and types of solutions.

2a. $3x^2 - 5x + 6 = 0$ **2b.** $7x^2 + x = 6$ **2c.** $4x(x + 2) = 7$

Solutions   **2a.** The equation is in standard form, so $a = 3$, $b = -5$, $c = 6$. The discriminant is

$$b^2 - 4ac = (-5)^2 - 4(3)(6)$$
$$= 25 - 72$$
$$= -47$$

A negative discriminant means that the equation has two complex, nonreal solutions.

2b. First write the equation in standard form.

$$7x^2 + x = 6$$
$$7x^2 + 1x - 6 = 0 \qquad \text{Subtract 6 from each side.}$$

So, $a = 7$, $b = 1$, $c = -6$.

$$b^2 - 4ac = (1)^2 - 4(7)(-6)$$
$$= 1 + 168$$
$$= 169$$

The discriminant is a positive perfect square, so the equation $7x^2 + x - 6 = 0$ has two rational solutions. This equation is also factorable.

2c. First write the equation in standard form.

$$4x(x + 2) = 7$$
$$4x^2 + 8x = 7 \qquad \text{Apply the distributive property.}$$
$$4x^2 + 8x - 7 = 0 \qquad \text{Subtract 7 from each side.}$$

So, $a = 4$, $b = 8$, and $c = -7$.

$$b^2 - 4ac = (8)^2 - 4(4)(-7)$$
$$= 64 + 112$$
$$= 176$$

The discriminant is positive but not a perfect square, so the equation $4x(x + 2) = 7$ has two irrational solutions.

✓ **Student Check 2**   Find the discriminant of each equation and use it to determine the number and types of solutions.

a. $x^2 + x + 1 = 0$ **b.** $2x^2 - 3x = 5$ **c.** $5x(x - 2) = 6$

Objective 3 ▶

Solve applications of quadratic equations.

Applications

In Chapter 6, we solved applications of quadratic equations by factoring. We will solve similar types of applications but will use the quadratic formula to solve the problems.

Objective 3 Examples   Solve each problem.

3a. The height, in feet, of a basketball thrown upward by a player at a free throw line is $h = -16t^2 + 22t + 7$ after t seconds. To the nearest hundredth of a second, when will the ball reach the height of the basketball hoop, which is 10 ft from the floor?

Solution   **3a.**

$$-16t^2 + 22t + 7 = h \qquad \text{State the given model.}$$
$$-16t^2 + 22t + 7 = 10 \qquad \text{Replace the value of } h \text{ with 10.}$$
$$-16t^2 + 22t + 7 - 10 = 10 - 10 \qquad \text{Subtract 10 from each side.}$$
$$-16t^2 + 22t - 3 = 0 \qquad \text{Simplify.}$$
$$16t^2 - 22t + 3 = 0 \qquad \text{Multiply each side by } -1.$$

The equation is in standard form, so $a = 16$, $b = -22$, and $c = 3$.

$$t = \frac{-b \pm \sqrt{b^2 - 4ac}}{2a}$$

State the quadratic formula.

$$t = \frac{-(-22) \pm \sqrt{(-22)^2 - 4(16)(3)}}{2(16)}$$

Replace a, b, and c with the appropriate values.

$$t = \frac{22 \pm \sqrt{484 - 192}}{32}$$

Simplify the expression.

$$t = \frac{22 \pm \sqrt{292}}{32}$$

Simplify the radicand.

$$t = \frac{22 + \sqrt{292}}{32} \approx 1.22 \quad \text{or}$$

Write the two solutions and approximate their value.

$$t = \frac{22 - \sqrt{292}}{32} \approx 0.15$$

So, the ball reaches a height of 10 ft in approximately 0.15 sec after the ball is thrown and in approximately 1.22 sec after the ball is thrown.

3b. A company knows that the cost, in dollars, to produce x items is given by the cost function $C(x) = 5x^2 + 800x$. It also knows that the revenue from selling x items is given by the revenue function $R(x) = 1000x + 200$. How many items does the company need to produce to break even, that is, when does revenue = cost?

Solution   3b.

$$C(x) = R(x)$$

$$5x^2 + 800x = 1000x + 200$$

Write the model.

$$5x^2 + 800x - 1000x - 200 = 0$$

Subtract $1000x$ and 200 from each side.

$$5x^2 - 200x - 200 = 0$$

Simplify.

$$\frac{5x^2 - 200x - 200}{5} = \frac{0}{5}$$

Divide each side by 5.

$$x^2 - 40x - 40 = 0$$

Simplify.

The equation is in standard form, so $a = 1$, $b = -40$, and $c = -40$.

$$x = \frac{-b \pm \sqrt{b^2 - 4ac}}{2a}$$

State the quadratic formula.

$$x = \frac{-(-40) \pm \sqrt{(-40)^2 - 4(1)(-40)}}{2(1)}$$

Replace a, b, and c with the appropriate values.

$$x = \frac{40 \pm \sqrt{1600 + 160}}{2}$$

Simplify the expression.

$$x = \frac{40 \pm \sqrt{1760}}{2}$$

Simplify the radicand.

$$x = \frac{40 + \sqrt{1760}}{2} \approx 41 \quad \text{or}$$

Write the two solutions and find their values.

$$x = \frac{40 - \sqrt{1760}}{2} \approx -1$$

Because it doesn't make sense to produce -1 items, the company needs to produce 41 items to break even.

3c. The owners of a child care center want to construct a playground at the back of their building, using the building as one of its borders. They can afford 200 ft of fencing. What dimensions should they construct the playground to have an enclosed area of 5000 ft²?

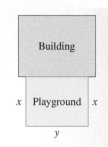

Solution **3c.** Construct a diagram of the situation. Let x represent the length of the playground and let y represent the width.

The total length of the fence is 200 ft, so $2x + y = 200$. The area of the playground is 5000 ft², so $xy = 5000$. Since we have two variables and two equations, we must solve one of the equations for one of the variables and substitute the expression in the other equation.

When we solve $2x + y = 200$ for y, we get $y = 200 - 2x$. Now we substitute $200 - 2x$ in $xy = 5000$ in place of y.

$xy = 5000$	Begin with the second equation.
$x(200 - 2x) = 5000$	Replace y with $200 - 2x$.
$200x - 2x^2 = 5000$	Apply the distributive property.
$-2x^2 + 200x - 5000 = 0$	Subtract 5000 from each side.
$x^2 - 100x + 2500 = 0$	Divide each side by -2.

The equation is in standard form, so $a = 1$, $b = -100$, and $c = 2500$.

$x = \dfrac{-b \pm \sqrt{b^2 - 4ac}}{2a}$	State the quadratic formula.
$x = \dfrac{-(-100) \pm \sqrt{(-100)^2 - 4(1)(2500)}}{2(1)}$	Replace a, b, and c with the appropriate values.
$x = \dfrac{100 \pm \sqrt{10{,}000 - 10{,}000}}{2}$	Simplify the expression.
$x = \dfrac{100 \pm \sqrt{0}}{2}$	Simplify the radicand.
$x = \dfrac{100 \pm 0}{2}$	Simplify the radical.
$x = 50$	Simplify.

So, the child care center should construct the playground with a length of 50 ft. The width of the playground is

$$y = 200 - 2x$$
$$= 200 - 2(50)$$
$$= 200 - 100$$
$$= 100 \text{ ft}$$

3d. At noon, Nguyen left on his bike and traveled due north with an average speed of 5 mph. One hour later, Tranh left from the same point and traveled due east with an average speed of 7 mph. At what time will the boys be 50 mi apart?

Solution **3d.** Construct a diagram and label the distances traveled. Let x represent the time that Nguyen has traveled. Since Nguyen travels at 5 mph, his total distance traveled is $5x$ mi. Tranh left an hour later, so his time traveled is $x - 1$. Tranh's total distance traveled is $7(x - 1)$ mi. Since Nguyen traveled due north and Tranh traveled due east, a right triangle is formed.

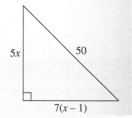

So, we use the Pythagorean theorem to solve the problem.

$$a^2 + b^2 = c^2$$ State the Pythagorean theorem.

$$(5x)^2 + [7(x - 1)]^2 = (50)^2$$ Replace a, b, and c with the legs and hypotenuse.

$$25x^2 + 49(x^2 - 2x + 1) = 2500$$ Square each expression.

$$25x^2 + 49x^2 - 98x + 49 = 2500$$ Apply the distributive property.

$$74x^2 - 98x + 49 - 2500 = 2500 - 2500$$ Combine like terms. Subtract 2500 from each side.

$$74x^2 - 98x - 2451 = 0$$ Simplify.

The equation is in standard form, so $a = 74$, $b = -98$, and $c = -2451$.

$$x = \frac{-b \pm \sqrt{b^2 - 4ac}}{2a}$$ State the quadratic formula.

$$x = \frac{-(-98) \pm \sqrt{(-98)^2 - 4(74)(-2451)}}{2(74)}$$ Replace a, b, and c with the appropriate values.

$$x = \frac{98 \pm \sqrt{735,100}}{148}$$ Simplify the expression.

$$x = \frac{98 + \sqrt{735,100}}{148} \approx 6.46$$ Write the two solutions and approximate their values.

$$x = \frac{98 - \sqrt{735,100}}{148} \approx -5.13$$

Because time is not negative, we can omit -5.13 as a solution. So, Nguyen and Tranh will be 50 mi apart in 6.46 hr after noon, or at approximately 6:28 P.M.

✔ **Student Check 3**   Solve each problem.

a. The height, in feet, of a ball thrown upward from a height of 10 ft is $h = -16t^2 + 32t + 10$ after t seconds. How many seconds will it take for the ball to reach a height of 20 ft? Round the answer to the nearest hundredth.

b. A company knows that the cost, in thousands of dollars, to produce x hundred items is given by the cost function $C(x) = -0.1x^2 + 2x + 5$. It also knows that the revenue, in thousands of dollars, from selling x hundred items is given by the revenue function $R(x) = 0.35x$. How many items does the company need to produce to break even, that is, when will revenue = cost? Round the answer to the nearest integer.

c. A farmer wants to enclose a pen at the back of his barn, using the barn as one of its borders. The farmer has 400 ft of fencing. What dimensions should the pen have to enclose an area of 16,000 ft²?

d. Susan and her husband, Jim, both have to travel out of town. Susan leaves at 7 A.M. and travels west at 60 mph. Jim leaves 2 hr later and travels south at 75 mph. At what time will Susan and Jim be 200 mi apart?

Objective 4 ▶

Troubleshoot common errors.

Troubleshooting Common Errors

Some common errors associated with the quadratic formula are shown.

Objective 4 Examples	A problem and an incorrect solution are given. Provide the correct solution and an explanation of the error.

4a. Solve $x^2 - 7x = 9$.

Incorrect Solution	Correct Solution and Explanation
$x^2 - 7x = 9$ So, $a = 1$, $b = -7$, and $c = 9$. $x = \dfrac{-b \pm \sqrt{b^2 - 4ac}}{2a}$ $x = \dfrac{-(-7) \pm \sqrt{(-7)^2 - 4(1)(9)}}{2(1)}$ $x = \dfrac{7 \pm \sqrt{13}}{2}$ The solution set is $\left\{ \dfrac{7 \pm \sqrt{13}}{2} \right\}$.	The equation must be written in standard form before using the quadratic formula. $x^2 - 7x - 9 = 0$ So, $a = 1$, $b = -7$, and $c = -9$. $x = \dfrac{-b \pm \sqrt{b^2 - 4ac}}{2a}$ $x = \dfrac{-(-7) \pm \sqrt{(-7)^2 - 4(1)(-9)}}{2(1)}$ $x = \dfrac{7 \pm \sqrt{85}}{2}$ The solution set is $\left\{ \dfrac{7 \pm \sqrt{85}}{2} \right\}$.

4b. Solve $x^2 + 4x + 2 = 0$.

Incorrect Solution	Correct Solution and Explanation
$x^2 + 4x + 2 = 0$ So, $a = 1$, $b = 4$, and $c = 2$. $x = \dfrac{-b \pm \sqrt{b^2 - 4ac}}{2a}$ $x = \dfrac{-(4) \pm \sqrt{(4)^2 - 4(1)(2)}}{2(1)}$ $x = \dfrac{-4 \pm \sqrt{8}}{2}$ $x = -2 \pm \sqrt{8}$ $x = -2 \pm 2\sqrt{2}$ The solution set is $\{-2 \pm 2\sqrt{2}\}$.	The denominator of 2 must divide into *each* term in the numerator. Simplify the radical first and then divide out the common factor. $x = \dfrac{-4 \pm \sqrt{8}}{2}$ $x = \dfrac{-4 \pm 2\sqrt{2}}{2}$ $x = \dfrac{2(-2 \pm \sqrt{2})}{2}$ $x = -2 \pm \sqrt{2}$ The solution set is $\{-2 \pm \sqrt{2}\}$.

ANSWERS TO STUDENT CHECKS

Student Check 1 **a.** $\{6\}$ **b.** $\left\{ \dfrac{2 \pm i}{5} \right\}$ **c.** $\left\{ -\dfrac{2}{7}, 0 \right\}$

d. $\left\{ \dfrac{-1 \pm \sqrt{193}}{12} \right\}$

Student Check 2 **a.** -3; two complex, nonreal solutions **b.** 49; two rational solutions **c.** 220; two irrational solutions

Student Check 3 **a.** The ball will reach 10 ft in 0.39 sec and 1.61 sec. **b.** The company needs to produce 1912 items to break even. **c.** The pen can either be 55.3 ft by 289.4 ft or 144.7 ft by 110.6 ft. **d.** They will be 200 mi apart in approximately 3 hr or at 10 A.M.

SUMMARY OF KEY CONCEPTS

1. We now have four different methods for solving quadratic equations.

   a. The zero products property (factoring)—doesn't solve all equations.

   b. The square root property—solves equations of the form $a^2 = k$.

   c. Completing the square—solves all equations but can be very tedious.

   d. The quadratic formula—solves all equations that are in standard form.

2. Quadratic equations have at most two real solutions. The discriminant, $b^2 - 4ac$, can be used to determine the number and type of solutions of a quadratic equation. If the discriminant is positive, there are two real solutions. If it is zero, there is one real solution. If it is negative, there are no real solutions.

SECTION 11.3 / EXERCISE SET

Write About It!

Use complete sentences in your answer to each exercise. For Exercises 1–4, explain the method that you would use to solve the given equation.

1. $(x - 2)(x - 3) = 0$
2. $(x + 4)^2 = 24$
3. $x^2 - 4x + 2 = 0$
4. $x^2 - 3x + 4 = 0$

5. Explain the advantage of using the quadratic formula to solve a quadratic equation.

6. Explain what you must do to an equation before using the quadratic formula to solve it.

7. How can you use the discriminant to determine the type and number of solutions of a quadratic equation?

8. How can you use the discriminant to determine if the equation can be solved by factoring?

Practice Makes Perfect!

Solve each equation using the quadratic formula. (See Objective 1.)

9. $x^2 - 7x + 12 = 0$
10. $y^2 + 4y - 12 = 0$
11. $x^2 - 9x + 13 = 0$
12. $9y^2 - 6y = 1$
13. $x(x + 8) + 18 = 0$
14. $y(y + 6) + 12 = 0$
15. $2a^2 + 11a = 0$
16. $3b^2 - 2b = 0$
17. $81z^2 = 18z - 11$
18. $6w^2 = 24w - 25$
19. $a^2 = -20a - 77$
20. $4b^2 = 8b + 9$
21. $x^2 - \frac{1}{3}x - 8 = 0$
22. $b^2 + \frac{8}{15}b + \frac{1}{15} = 0$
23. $z^2 - \frac{5}{3}z + \frac{1}{18} = 0$
24. $w^2 - \frac{1}{4}w - \frac{5}{16} = 0$

Find the discriminant and use it to determine the number and types of solutions of each equation. (See Objective 2.)

25. $8x^2 - 5x - 2 = 0$
26. $10y^2 - 4y - 3 = 0$
27. $3a^2 - 4a + 7 = 0$
28. $8b^2 + 4b + 1 = 0$

29. $8r^2 - 7r - 1 = 0$
30. $2s^2 - s - 3 = 0$
31. $4a^2 - 12a + 9 = 0$
32. $25b^2 + 20b + 4 = 0$

Solve each problem. (See Objective 3.)

33. The height, in feet, of a ball thrown upward from a height of 9 ft is $h = -16t^2 + 45t + 9$ after t seconds. How many seconds will it take for the ball to reach a height of 24 ft? Round the answer to two decimal places.

34. The height, in feet, of a ball thrown upward from a height of 11 ft is $h = -16t^2 + 50t + 11$ after t seconds. How many seconds will it take for the ball to reach a height of 20 ft? Round the answer to two decimal places.

35. A company knows that the cost, in thousands of dollars, to produce x hundred items is given by the cost function $C(x) = -0.2x^2 + 10x + 5$. It also knows that the revenue, in thousands of dollars, from selling x hundred items is given by the revenue function $R(x) = 0.1x$. How many items does the company need to produce to break even, that is, when does revenue = cost?

36. A company knows that the cost, in thousands of dollars, to produce x hundred items is given by the cost function $C(x) = -0.3x^2 + 5x + 10$. It also knows that the revenue, in thousands of dollars, from selling x hundred items is given by the revenue function $R(x) = 3.85x$. How many items does the company need to produce to break even, that is, when does revenue = cost?

37. A farmer wants to enclose a pen at the back of his barn, using the barn as one of its borders. The farmer has 320 ft of fencing. What dimensions should the pen have to enclose an area of 12,800 ft²?

38. A farmer wants to enclose a pen at the back of his barn, using the barn as one of its borders. The farmer has 460 ft of fencing. What dimensions should the pen have to enclose an area of 26,450 ft²?

39. Josh and his wife, Christy, each have to travel out of town for meetings. The towns where they are meeting are 250 mi apart and are due north and due east from their home. Josh leaves at 9 A.M. and travels at 50 mph, and Christy leaves 2 hr later and travels at 75 mph. When will they reach their destinations and be 250 mi apart?

40. Mike and Brian each have to travel out of town for meetings. The towns where they are meeting are 230 mi apart and are due east and due south from their home. Mike leaves at 9 A.M. and travels at 60 mph, and Brian leaves 1 hr later and travels at 70 mph. When will they reach their destinations and be 230 mi apart?

 Mix 'Em Up!

Solve each equation using the quadratic formula.

41. $x^2 + 6x - 16 = 0$

42. $y^2 - 4y - 32 = 0$

43. $3a^2 + 4a - 20 = 0$

44. $2b^2 + 13b + 20 = 0$

45. $2x^2 - 16x + 27 = 0$

46. $4y^2 + 8y - 11 = 0$

47. $x(x - 3) + 6 = 0$

48. $2y(y + 4) + 15 = 0$

49. $5a^2 + 9a = 0$

50. $3b^2 + b = 0$

51. $2z^2 = 10z - 5$

52. $w^2 = 2w + 21$

53. $a^2 = \dfrac{1}{2}a + \dfrac{9}{16}$

54. $\dfrac{1}{7}b^2 = b - \dfrac{23}{28}$

55. $x^2 - \dfrac{7}{6}x - 4 = 0$

56. $b^2 + \dfrac{3}{4}b + \dfrac{1}{8} = 0$

57. $z^2 - \dfrac{2}{3}z + \dfrac{1}{3} = 0$

58. $w^2 + \dfrac{1}{2}w + \dfrac{7}{4} = 0$

Find the discriminant and use it to determine the number and types of solutions of each equation.

59. $x^2 + 2x - 6 = 0$

60. $3y^2 - 5y - 10 = 0$

61. $9a^2 + a + 2 = 0$

62. $2b^2 + 3b + 11 = 0$

63. $2r^2 - 9r - 18 = 0$

64. $3s^2 - 11s - 42 = 0$

65. $9a^2 + 30a + 25 = 0$

66. $4b^2 - 28b + 49 = 0$

Solve each problem.

67. A company knows that the cost, in thousands of dollars, to produce x hundred items is given by the cost function $C(x) = -0.2x^2 + x + 5$. It also knows that the revenue, in thousands of dollars, from selling x hundred items is given by the revenue function $R(x) = 1.45x$. How many items does the company need to produce to break even, that is, when does revenue = cost?

68. A company knows that the cost, in thousands of dollars, to produce x hundred items is given by the cost function $C(x) = -0.5x^2 + 2x + 3$. It also knows that the revenue, in thousands of dollars, from selling x hundred items is given by the revenue function $R(x) = 3.25x$. How many items does the company need to produce to break even, that is, when does revenue = cost?

69. A farmer wants to enclose a pen at the back of his barn, using the barn as one of its borders. The farmer has 280 ft of fencing. What dimensions should the pen have to enclose an area of 9800 ft²?

70. A farmer wants to enclose a pen at the back of his barn, using the barn as one of its borders. The farmer has 640 ft of fencing. What dimensions should the pen have to enclose an area of 51,200 ft²?

71. Mark and Nina each have to travel out of town for meetings. The towns where they are meeting are  220 mi apart and are due west and due north from their home. Mark leaves at 7 A.M. and travels at 60 mph, and Nina leaves 2 hr later and travels at 77 mph. When will they reach their destinations and be 220 mi apart?

72. Pat and Nancy each have to travel out of town for meetings. The towns where they are meeting are 170 mi apart and are due east and due south from their home. Pat leaves at 6 A.M. and travels at 65 mph, and Nancy leaves 3 hr later and travels at 77 mph. When will they reach their destinations and be 170 mi apart?

 You Be the Teacher!

Correct each student's errors, if any.

73. Solve the equation $x^2 - 3x - 4 = 0$ using the quadratic formula.

Josh's work:

$$x^2 - 3x - 4 = 0$$

$$x = \frac{3 \pm \sqrt{-3^2 - 4(-4)}}{2}$$

$$x = \frac{3 \pm \sqrt{-9 + 16}}{2}$$

$$x = \frac{3 \pm \sqrt{7}}{2}$$

74. Solve the equation $4x^2 + 16x + 13 = 0$ using the quadratic formula.

Matt's work:

$$4x^2 + 16x + 13 = 0$$

$$x = \frac{-16 \pm \sqrt{16^2 - 4(4)(13)}}{4}$$

$$x = \frac{-16 \pm \sqrt{256 - 208}}{4}$$

$$x = \frac{-16 \pm \sqrt{48}}{4}$$

$$x = -4 \pm \sqrt{12}$$

$$x = -4 \pm 2\sqrt{3}$$

75. Solve the equation $2x^2 - 5x - 8 = 0$ using the quadratic formula and approximate the solutions to two decimal places.

Bruce's work:

$2x^2 - 5x - 8 = 0$

$$x = \frac{5 \pm \sqrt{5^2 - 4(2)(-8)}}{4}$$

$$x = \frac{5 \pm \sqrt{89}}{4}$$

```
5+√(89)/4
          7.358495283
5-√(89)/4
          2.641504717
```

The approximate solutions are 7.36 and 2.64.

76. Solve the equation $3x^2 + 5x - 6 = 0$ using the quadratic formula and approximate the solutions to two decimal places.

Beth's work:

$3x^2 + 5x - 6 = 0$

$$x = \frac{-5 \pm \sqrt{-5^2 - 4(3)(-6)}}{6}$$

```
-5+√(-5²-4*3*-6)
/6
          -3.8573909
-5-√(-5²-4*3*-6)
/6
          -6.1426091
```

So, the approximate solutions are -3.86 and -6.14.

PIECE IT TOGETHER   SECTIONS 11.1–11.3

Graph each function by plotting the vertex and two additional points. Identify the *x*-intercepts, *y*-intercept, the vertex, the axis of symmetry, domain, and range. Explain how the graph of the function relates to the graph of $y = x^2$. (*Section 11.1, Objective 4*)

1. $f(x) = (x - 2)^2 + 6$ **2.** $f(x) = (x + 4)^2 - 9$

Graph each function and use the graph to solve the equation $f(x) = 0$. State the zeros of $f(x)$ and the *x*-intercepts of the graph of $f(x)$. (*Section 11.1, Objective 5; Section 11.2, Objective 6*)

3. $f(x) = (x - 5)^2 - 9$ **4.** $f(x) = -(x + 3)^2 + 1$

Solve each equation by applying the square root property, completing the square, or using the quadratic formula. (*Section 11.2, Objectives 1 and 4; Section 11.3, Objective 1*)

5. $(7a + 1)^2 + 2 = 6$ **6.** $18x^2 - 8 = 90$

7. $3y^2 - y - 14 = 0$ **8.** $2a^2 + 10a + 13 = 0$

Create a perfect square trinomial from each binomial by completing the square and then write it in factored form. (*Section 11.2, Objective 2*)

9. $y^2 - 22y$ **10.** $b^2 - 7b$

Find the discriminant and use it to determine the number and types of solutions of each equation. (*Section 11.3, Objective 2*)

11. $2x^2 + 2x - 3 = 0$ **12.** $x^2 - 6x + 16 = 0$

13. $3y^2 - 14y - 5 = 0$ **14.** $25y^2 - 20y + 4 = 0$

Solve each problem. (*Section 11.2, Objective 5; Section 11.3, Objective 3*)

15. If a penny is dropped from the top of the John Hancock Center in Chicago, its height, in meters, t sec after it is dropped is given by $s = -9.8t^2 + 344$. How many seconds will it take for the penny to reach the ground? Round the answer to two decimal places. (Source: http://www.emporis.com)

16. A company knows that the cost, in thousands of dollars, to produce x hundred items is given by the cost function $C(x) = -0.3x^2 + 2x + 6$. It also knows that the revenue, in thousands of dollars, from selling x hundred items is given by the revenue function $R(x) = 2.3x$. How many items does the company need to produce to break even, that is, when does revenue = cost?

SECTION 11.4   **Solving Equations Using Quadratic Methods**

▶ **OBJECTIVES**

As a result of completing this section, you will be able to

1. Solve rational equations.
2. Solve radical equations.
3. Solve higher degree polynomial equations.
4. Solve equations by substitution.
5. Solve applications of equations that use quadratic methods.
6. Troubleshoot common errors.

A local YMCA is holding a special biathlon event that consists of 1 mi of running and $\frac{1}{2}$ mi of swimming. Taylor finished the event in 25 min. He swam an average of 4 mph less than he ran. What was Taylor's speed running and swimming?

To solve this problem, we must solve a rational equation that can be written as a quadratic equation. We then will use one of the methods we have learned for solving quadratic equations: factoring, applying the square root property, completing the square, or using the quadratic formula. In this section, we will use these methods to solve rational, radical, and higher degree polynomial equations.

Rational Equations

Objective 1 ▶

Solve rational equations.

Recall that a **rational equation** is an equation that contains a rational expression. Generally speaking, these are equations that contain variables in the denominator. When we solve a rational equation, we clear the fractions and then solve the resulting equation. The key to working with these equations is that we must check our solutions to make sure we do not have an extraneous solution; that is, one that makes one of the rational expressions undefined.

> **Procedure: Solving a Rational Equation**
>
> **Step 1:** Multiply each side of the equation by the LCD to clear fractions.
> **Step 2:** Solve the resulting equation. (In this section, the resulting equation will be quadratic.)
> **Step 3:** Exclude from the solution set any possible solutions that make any of the denominators zero.
> **Step 4:** Check by substituting the possible solutions in the original equation.

Objective 1 Examples   Solve each equation.

1a. $\dfrac{14}{x} = x - 5$ **1b.** $\dfrac{7y}{y-3} - \dfrac{y+2}{y+3} = \dfrac{4}{y^2 - 9}$

Solutions   **1a.** $\dfrac{14}{x} = x - 5$

$x\left(\dfrac{14}{x}\right) = x(x - 5)$ Multiply each side by the LCD, x.

$14 = x^2 - 5x$ Simplify.

$0 = x^2 - 5x - 14$ Subtract 14 from each side.

$0 = (x - 7)(x + 2)$ Factor.

$x - 7 = 0$ or $x + 2 = 0$ Set each equation equal to zero.

$x = 7$ $x = -2$ Solve each equation.

We can check the solutions in the original equation.

Let $x = 7$:

$$\frac{14}{x} = x - 5$$

$$\frac{14}{7} = 7 - 5$$

$$2 = 2$$

True

Let $x = -2$:

$$\frac{14}{x} = x - 5$$

$$\frac{14}{-2} = -2 - 5$$

$$-7 = -7$$

True

Since both solutions make the original equation true, the solution set is $\{-2, 7\}$.

1b. Factoring the last denominator gives us $y^2 - 9 = (y - 3)(y + 3)$. So, the LCD is $(y - 3)(y + 3)$. Multiply each side by the LCD and simplify.

$$\frac{7y}{y - 3} - \frac{y + 2}{y + 3} = \frac{4}{y^2 - 9}$$

$$(y - 3)(y + 3)\frac{7y}{y - 3} - (y - 3)(y + 3)\frac{y + 2}{y + 3} = (y - 3)(y + 3)\frac{4}{(y - 3)(y + 3)}$$

$$7y(y + 3) - (y + 2)(y - 3) = 4$$

$$7y^2 + 21y - (y^2 - y - 6) = 4$$

$$7y^2 + 21y - y^2 + y + 6 = 4$$

$$6y^2 + 22y + 2 = 0$$

$$3y^2 + 11y + 1 = 0$$

This equation does not factor, so we will use the quadratic formula to solve it with $a = 3, b = 11, c = 1$.

$$y = \frac{-b \pm \sqrt{b^2 - 4ac}}{2a}$$

State the quadratic formula.

$$y = \frac{-(11) \pm \sqrt{(11)^2 - 4(3)(1)}}{2(3)}$$

Replace a, b, and c with the appropriate values.

$$y = \frac{-11 \pm \sqrt{121 - 12}}{6}$$

Simplify the expression.

$$y = \frac{-11 \pm \sqrt{109}}{6}$$

Simplify the radicand.

Since neither solution makes the denominators zero, the solution set is

$$\left\{\frac{-11 + \sqrt{109}}{6}, \frac{-11 - \sqrt{109}}{6}\right\} \text{ or } \left\{\frac{-11 \pm \sqrt{109}}{6}\right\}.$$

✓ **Student Check 1**   Solve each equation.

a. $\dfrac{20}{x} = x + 8$

b. $\dfrac{3a}{a - 1} + \dfrac{a + 2}{2a} = \dfrac{1}{2a^2 - 2a}$

Radical Equations

Objective 2 ▶

Solve radical equations.

Radical equations are equations containing radicals. In this section, we will use only square roots, not higher roots, in the equations. To solve these equations, recall that we first isolate the square root on one side of the equation and then square both sides to eliminate the radical. These equations are like rational equations in that we must check the possible solutions to ensure they are not extraneous solutions.

Procedure: Solving a Radical Equation

Step 1: Isolate the radical expression to one side of the equation.
Step 2: Square each side of the equation to eliminate the radical.
Step 3: Solve the resulting quadratic equation.
Step 4: Check the possible solutions and exclude any extraneous solutions.

Objective 2 Examples Solve each equation.

2a. $2x = \sqrt{7x + 2}$ **2b.** $p - 2\sqrt{p} = 8$

Solutions **2a.**

$$2x = \sqrt{7x + 2}$$

$$(2x)^2 = \left(\sqrt{7x + 2}\right)^2 \qquad \text{Square each side.}$$

$$4x^2 = 7x + 2 \qquad \text{Simplify.}$$

$$4x^2 - 7x - 2 = 0 \qquad \text{Subtract } 7x \text{ and } 2 \text{ from each side.}$$

$$(4x + 1)(x - 2) = 0 \qquad \text{Factor the resulting equation.}$$

$$4x + 1 = 0 \quad \text{or} \quad x - 2 = 0 \qquad \text{Set each factor equal to 0.}$$

$$x = -\frac{1}{4} \qquad x = 2 \qquad \text{Solve each equation.}$$

We check the solutions in the original equation.

$x = -\frac{1}{4}$:

$$2x = \sqrt{7x + 2}$$

$$2\left(-\frac{1}{4}\right) = \sqrt{7\left(-\frac{1}{4}\right) + 2}$$

$$-\frac{1}{2} = \sqrt{\frac{1}{4}}$$

$$-\frac{1}{2} = \frac{1}{2}$$

False

$x = 2$:

$$2x = \sqrt{7x + 2}$$

$$2(2) = \sqrt{7(2) + 2}$$

$$4 = \sqrt{16}$$

$$4 = 4$$

True

The only solution that checks is 2. So, the solution set is $\{2\}$.

2b.

$$p - 2\sqrt{p} = 8$$

$$p - 8 = 2\sqrt{p} \qquad \text{Subtract 8 and add } 2\sqrt{p} \text{ to each side.}$$

$$(p - 8)^2 = \left(2\sqrt{p}\right)^2 \qquad \text{Square each side.}$$

$$p^2 - 16p + 64 = 4p \qquad \text{Simplify.}$$

$$p^2 - 20p + 64 = 0 \qquad \text{Subtract } 4p \text{ from each side.}$$

$$(p - 16)(p - 4) = 0 \qquad \text{Factor the trinomial.}$$

$$p - 16 = 0 \quad \text{or} \quad p - 4 = 0 \qquad \text{Set each factor equal to zero.}$$

$$p = 16 \qquad p = 4 \qquad \text{Solve each equation.}$$

We check the solutions in the original equation.

$p = 16$:

$$p - 2\sqrt{p} = 8$$
$$16 - 2\sqrt{16} = 8$$
$$16 - 2(4) = 8$$
$$16 - 8 = 8$$
$$8 = 8$$

True

$p = 4$:

$$p - 2\sqrt{p} = 8$$
$$4 - 2\sqrt{4} = 8$$
$$4 - 2(2) = 8$$
$$4 - 4 = 8$$
$$0 = 8$$

False

The only solution that checks is 16. So, the solution set is $\{16\}$.

✓ **Student Check 2**   Solve each equation.

    a. $3x = \sqrt{5x + 4}$ **b.** $a - \sqrt{a} = 12$

Higher Degree Polynomial Equations

Objective 3 ▶

Solve higher degree
polynomial equations.

Higher degree polynomial equations are polynomial equations with a degree greater than 2. We will solve these types of equations by initially applying the zero products property. That is, we set the equation equal to zero and factor. In Section 6.8, we solved these types of equations. An example is shown.

$$x^3 = 64x$$

$x^3 - 64x = 0$	Write the equation in standard form.
$x(x^2 - 64) = 0$	Factor out the GCF.
$x(x - 8)(x + 8) = 0$	Factor the difference of squares.
$x = 0$ or $x - 8 = 0$ or $x + 8 = 0$	Set each factor equal to zero.
$x = 8$ $x = -8$	Solve.

The solution set is $\{-8, 0, 8\}$. Note that the degree of the polynomial is 3 and there are three solutions. Notice that we were able to factor the polynomial until it consisted of three linear factors.

In this section, we will solve polynomial equations that we can initially factor. But we will find that some of the factors can't be factored any further. So, we may have to apply the square root property, complete the square, or use the quadratic formula to solve the equation. Consider the following example.

$$x^3 = -64x$$

$x^3 + 64x = 0$	Write the equation in standard form.
$x(x^2 + 64) = 0$	Factor out the GCF.
$x = 0$ or $x^2 + 64 = 0$	Set each factor equal to zero.
$x^2 = -64$	Write the equation in the form $a^2 = k$.
$x = \pm\sqrt{-64}$	Apply the square root property.
$x = \pm 8i$	Simplify the radical.

So, the solution set is $\{0, -8i, 8i\}$. Note that we were able to factor after we wrote the equation in standard form, but then we couldn't factor $x^2 + 64$ since it is the sum of squares. So, we had to set that factor equal to zero and apply another technique of solving quadratic equations, the square root property. Also note the degree of the equation is 3 and there are three solutions (one real solution and two imaginary solutions).

Procedure: Solving a Higher Degree Polynomial Equation

Step 1: Write the equation in standard form (one side is equal to zero).
Step 2: Factor the polynomial.
Step 3: Set each factor equal to zero and solve.

Objective 3 Examples Solve each equation.

3a. $x^4 - 5x^2 - 36 = 0$ **3b.** $4y^4 - 5y^2 = -1$ **3c.** $x^3 = 8$

Solutions **3a.**

$$x^4 - 5x^2 - 36 = 0$$

$$(x^2 - 9)(x^2 + 4) = 0 \qquad \text{Factor the polynomial.}$$

$$x^2 - 9 = 0 \quad \text{or} \quad x^2 + 4 = 0 \qquad \text{Set each factor equal to zero.}$$

$$x^2 = 9 \qquad\qquad x^2 = -4 \qquad \text{Write each equation in the form } a^2 = k.$$

$$x = \pm\sqrt{9} \qquad\quad x = \pm\sqrt{-4} \qquad \text{Apply the square root property.}$$

$$x = \pm 3 \qquad\qquad x = \pm 2i \qquad \text{Simplify each radical.}$$

So, the solution set is $\{-3, 3, -2i, 2i\}$ or $\{\pm 3, \pm 2i\}$. We can check the solutions.

Let $x = -3$:

$$x^4 - 5x^2 - 36 = 0$$
$$(-3)^4 - 5(-3)^2 - 36 = 0$$
$$81 - 5(9) - 36 = 0$$
$$81 - 45 - 36 = 0$$
$$36 - 36 = 0$$
$$0 = 0$$

Let $x = 3$:

$$x^4 - 5x^2 - 36 = 0$$
$$(3)^4 - 5(3)^2 - 36 = 0$$
$$81 - 5(9) - 36 = 0$$
$$81 - 45 - 36 = 0$$
$$36 - 36 = 0$$
$$0 = 0$$

Let $x = -2i$:

$$x^4 - 5x^2 - 36 = 0$$
$$(-2i)^4 - 5(-2i)^2 - 36 = 0$$
$$16i^4 - 5(4i^2) - 36 = 0$$
$$16 - 5(-4) - 36 = 0$$
$$16 + 20 - 36 = 0$$
$$0 = 0$$

Let $x = 2i$:

$$x^4 - 5x^2 - 36 = 0$$
$$(2i)^4 - 5(2i)^2 - 36 = 0$$
$$16i^4 - 5(4i^2) - 36 = 0$$
$$16 - 5(-4) - 36 = 0$$
$$16 + 20 - 36 = 0$$
$$0 = 0$$

3b.

$$4y^4 - 5y^2 = -1$$

$$4y^4 - 5y^2 + 1 = 0 \qquad \text{Add 1 to each side.}$$

$$(4y^2 - 1)(y^2 - 1) = 0 \qquad \text{Factor.}$$

$$4y^2 - 1 = 0 \quad \text{or} \quad y^2 - 1 = 0 \qquad \text{Set each factor equal to zero.}$$

$$y^2 = \frac{1}{4} \qquad\qquad y^2 = 1 \qquad \text{Write each equation in the form } a^2 = k.$$

$$y = \pm\sqrt{\frac{1}{4}} \qquad\quad y = \pm\sqrt{1} \qquad \text{Apply the square root property.}$$

$$y = \pm\frac{1}{2} \qquad\qquad y = \pm 1 \qquad \text{Simplify each radical.}$$

The solution set is $\left\{-\dfrac{1}{2}, -1, \dfrac{1}{2}, 1\right\}$ or $\left\{\pm\dfrac{1}{2}, \pm 1\right\}$.

3c.

$$x^3 = 8$$

$$x^3 - 8 = 0 \qquad \text{Write the equation in standard form.}$$

$$(x - 2)(x^2 + 2x + 4) = 0 \qquad \text{Factor.}$$

$$x - 2 = 0 \quad \text{or} \quad x^2 + 2x + 4 = 0 \qquad \text{Set each factor equal to zero.}$$

$$x = 2 \quad x = \frac{-b \pm \sqrt{b^2 - 4ac}}{2a} \qquad \text{Solve. Apply the quadratic formula.}$$

$$x = \frac{-(2) \pm \sqrt{(2)^2 - 4(1)(4)}}{2(1)}$$ Let $a = 1$, $b = 2$, and $c = 4$.

$$x = \frac{-2 \pm \sqrt{-12}}{2}$$ Simplify the radicand.

$$x = \frac{-2 \pm \sqrt{-4 \cdot 3}}{2}$$ Find a perfect square factor of -12.

$$x = \frac{-2 \pm 2i\sqrt{3}}{2}$$ Apply the product rule for radicals.

$$x = -1 \pm i\sqrt{3}$$ Simplify.

The solution set is $\left\{2, -1 + i\sqrt{3}, -1 - i\sqrt{3}\right\}$ or $\left\{2, -1 \pm i\sqrt{3}\right\}$.

 **Student Check 3** Solve each equation.

a. $x^4 - 15x^2 - 16 = 0$ **b.** $9x^4 - 10x^2 = -1$ **c.** $x^3 = 27$

Note: *Example 3 illustrates that the degree of the equation relates to the number of solutions of the equation. In parts 3a and 3b, the degree is 4 and there are 4 solutions. In part 3c, the degree is 3 and there are 3 solutions. It turns out that the degree determines the maximum number of solutions since solutions can repeat.*

Solving Equations Using Substitution

Objective 4 ▶

Solve equations by substitution.

To solve an equation by substitution requires us to replace an expression by a variable so that a quadratic equation results. The following table shows some examples of how this is done.

Original Equation	Expression to be Replaced	New Equation
$x^4 - 9x^2 - 10 = 0$ $(x^2)^2 - 9x^2 - 10 = 0$	$u = x^2$	$(u)^2 - 9u - 10 = 0$
$(x - 1)^2 + 5(x - 1) + 4 = 0$	$u = (x - 1)$	$(u)^2 + 5(u) + 4 = 0$
$x^{2/3} + 8x^{1/3} + 16 = 0$ $(x^{1/3})^2 + 8x^{1/3} + 16 = 0$	$u = x^{1/3}$	$(u)^2 + 8u + 16 = 0$

Note that the expression that we replaced with u

- is the base of the squared term.
- is also found in the middle term of the original equation.

After we replace the expression with u, the resulting equation is a quadratic equation that we can solve using the methods of this chapter. When we solve this new equation, we find the values of u that satisfy it. Our goal is to solve for the variable in the original equation. So, after we solve the new equation, we must use these solutions to solve for the original variable.

Procedure: Solving an Equation Using Substitution

Step 1: Let "u" equal the base of the squared term.
Step 2: Rewrite the equation using u.
Step 3: Solve the resulting quadratic equation.
Step 4: Back-substitute to solve for the original variable.

Objective 4 Examples Solve each equation by substitution.

4a. $(x + 4)^2 - 5(x + 4) + 6 = 0$ **4b.** $x^{2/5} - 3x^{1/5} - 4 = 0$

Solutions **4a.** The expression $(x + 4)$ is the base of the squared term and it also repeats in the middle term of the equation. So, we let $u = x + 4$.

$$(x + 4)^2 - 5(x + 4) + 6 = 0$$

$$u^2 - 5u + 6 = 0 \qquad \text{Substitute } u \text{ for } x + 4.$$

$$(u - 3)(u - 2) = 0 \qquad \text{Factor.}$$

$$u - 3 = 0 \quad \text{or} \quad u - 2 = 0 \qquad \text{Apply the zero products property.}$$

$$u = 3 \qquad\qquad u = 2 \qquad \text{Solve.}$$

Now we solve for x using the equation $u = x + 4$.

$$u = 3 \qquad\qquad u = 2 \qquad \text{State the solutions for } u.$$

$$x + 4 = 3 \qquad x + 4 = 2 \qquad \text{Substitute } x + 4 \text{ for } u.$$

$$x = -1 \qquad\quad x = -2 \qquad \text{Subtract 4 from each side of each equation.}$$

Both solutions check, so the solution set is $\{-2, -1\}$.

4b. The expression $x^{2/5} = (x^{1/5})^2$, so let $u = x^{1/5}$.

$$x^{2/5} - 3x^{1/5} - 4 = 0$$

$$(x^{1/5})^2 - 3x^{1/5} - 4 = 0 \qquad \text{Write } x^{2/5} \text{ as } (x^{1/5})^2.$$

$$u^2 - 3u - 4 = 0 \qquad \text{Substitute } u \text{ for } x^{1/5}.$$

$$(u - 4)(u + 1) = 0 \qquad \text{Factor.}$$

$$u - 4 = 0 \quad \text{or} \quad u + 1 = 0 \qquad \text{Apply the zero products property.}$$

$$u = 4 \qquad\qquad u = -1 \qquad \text{Solve.}$$

Now we solve for x using the equation $u = x^{1/5}$.

$$u = 4 \qquad\qquad u = -1 \qquad \text{State the solutions for } u.$$

$$x^{1/5} = 4 \qquad\quad x^{1/5} = -1 \qquad \text{Substitute } x^{1/5} \text{ for } u.$$

$$(x^{1/5})^5 = 4^5 \qquad (x^{1/5})^5 = (-1)^5 \qquad \text{Apply the power property.}$$

$$x = 1024 \qquad\quad x = -1 \qquad \text{Simplify.}$$

Both solutions check, so the solution set is $\{-1, 1024\}$.

☑ **Student Check 4** Solve each equation by substitution.

a. $(x - 2)^2 - 4(x - 2) + 3 = 0$ **b.** $x^{2/3} + 2x^{1/3} - 8 = 0$

Applications

Objective 5 ▶

Solve applications of equations that use quadratic methods.

Distance and work problems are often modeled by equations that require us to use quadratic methods to solve them. Example 5 illustrates these concepts.

Objective 5 Examples Solve each problem.

5a. A local YMCA is holding a special biathlon event that consists of 1 mi of running and $\frac{1}{2}$ mi of swimming. Taylor finished the event in 25 min. He swam an average of 4 mph less than he ran. What was his speed running and swimming?

Solution   **5a.** We let x represent the speed running and $x - 4$ represent the speed swimming. We organize the information in a table. Recall $d = rt$ or $t = \dfrac{d}{r}$.

	Distance	Rate	Time
Running	1 mi	x	$\dfrac{1}{x}$
Swimming	$\dfrac{1}{2}$ mi	$x - 4$	$\dfrac{\frac{1}{2}}{x-4} = \dfrac{1}{2(x-4)}$

The total time running and swimming is 25 min, which is $\dfrac{25}{60} = \dfrac{5}{12}$ hr.

$$\frac{1}{x} + \frac{1}{2(x-4)} = \frac{5}{12} \qquad \text{Sum of times is 25 min.}$$

$$12x(x-4)\left(\frac{1}{x}\right) + 12x(x-4)\left[\frac{1}{2(x-4)}\right]$$

$$= 12x(x-4)\left(\frac{5}{12}\right) \qquad \text{Multiply each side by the LCD.}$$

$$12(x-4) + 6x = 5x(x-4) \qquad \text{Simplify each product.}$$

$$12x - 48 + 6x = 5x^2 - 20x \qquad \text{Apply the distributive property.}$$

$$0 = 5x^2 - 38x + 48 \qquad \text{Write equation in standard form.}$$

$$0 = (5x - 8)(x - 6) \qquad \text{Factor.}$$

$$5x - 8 = 0 \quad \text{or} \quad x - 6 = 0 \qquad \text{Apply the zero products property.}$$

$$x = \frac{8}{5} \qquad\qquad x = 6 \qquad \text{Solve each equation.}$$

The only solution that makes sense is 6 since $\dfrac{8}{5}$ makes the swimming speed negative. So, Taylor ran at a speed of 6 mph and swam at a speed of 2 mph.

5b. Peter and Jason own a lawn-mowing service. Together, they can mow a lawn in 24 min. Peter mows 20 min faster than John. How long will it take each man working alone to mow the lawn?

Solution   **5b.** The unknowns are the times it takes each man to mow the lawn alone. Since Peter's time is 20 min faster than John's time, we let

$$x = \text{time for Jason to mow the lawn}$$

$$x - 20 = \text{time for Peter to mow the lawn}$$

Recall that we need to find the portion of the job completed in 1 min.

	Time to Complete the Job	Portion Completed in 1 Min
Peter	$x - 20$	$\dfrac{1}{x-20}$
Jason	x	$\dfrac{1}{x}$
Working together	24	$\dfrac{1}{24}$

So, the sum of the portions of the job completed in 1 min working alone is equal to the portion of the job completed in 1 min working together.

$$\frac{1}{x-20} + \frac{1}{x} = \frac{1}{24} \qquad \text{Write the model.}$$

$$24x(x-20)\frac{1}{x-20} + 24x(x-20)\frac{1}{x} = 24x(x-20)\frac{1}{24}$$ Multiply each side by the LCD.

$$24x + 24(x-20) = x(x-20)$$ Simplify each product.

$$24x + 24x - 480 = x^2 - 20x$$ Apply the distributive property.

$$48x - 480 = x^2 - 20x$$ Combine like terms.

$$0 = x^2 - 68x + 480$$ Write the equation in standard form.

$$0 = (x-60)(x-8)$$ Factor.

$$x - 60 = 0 \quad \text{or} \quad x - 8 = 0$$ Apply the zero products property.

$$x = 60 \qquad\qquad x = 8$$ Solve.

The only solution that makes sense is 60 since 8 makes Peter's time negative. Therefore, it takes Jason 60 min to mow the lawn alone and Peter $60 - 20 = 40$ min to mow the lawn alone.

☑ **Student Check 5**

Solve each problem.

a. Maria and Juan travel out of town for work. Juan travels 245 mi and Maria travels 420 mi. Maria's speed is 14 mph less than Juan's. If their total time traveling was 11 hr, what was each of their speeds?

b. Tom and Barbara can clean their house in 3 hr working together. Working alone, Tom takes 2 hr longer than Barbara to clean their house. How long will it take each of them working alone to clean their house?

Objective 6 ▶

Troubleshoot common errors.

Troubleshooting Common Errors

Some common errors associated with solving equations using quadratic methods are shown.

Objective 6 Examples

A problem and an incorrect solution are given. Provide the correct solution and an explanation of the error.

6a. Solve $3x = \sqrt{2 - 7x}$.

Incorrect Solution	Correct Solution and Explanation
$$3x = \sqrt{2-7x}$$ $$(3x)^2 = (\sqrt{2-7x})^2$$ $$9x^2 = 2 - 7x$$ $$9x^2 + 7x - 2 = 0$$ $$(9x-2)(x+1) = 0$$ $$9x - 2 = 0 \quad \text{or} \quad x + 1 = 0$$ $$x = \frac{2}{9} \qquad\qquad x = -1$$ The solution set is $\left\{-1, \frac{2}{9}\right\}$.	When solving a radical equation, we must check for extraneous solutions. $$x = \frac{2}{9}: \quad 3\left(\frac{2}{9}\right) = \sqrt{2 - 7\left(\frac{2}{9}\right)}$$ $$\frac{2}{3} = \sqrt{2 - \frac{14}{9}}$$ $$\frac{2}{3} = \sqrt{\frac{4}{9}}$$ $$\frac{2}{3} = \frac{2}{3} \quad \text{True}$$ $$x = -1: \quad 3(-1) = \sqrt{2 - 7(-1)}$$ $$-3 = \sqrt{2 + 7}$$ $$-3 = \sqrt{9}$$ $$-3 = 3 \quad \text{False}$$ So, the solution set is $\left\{\frac{2}{9}\right\}$.

6b. $x^4 - 13x^2 + 36 = 0$.

Incorrect Solution	Correct Solution and Explanation
$$x^4 - 13x^2 + 36 = 0$$ $$(x^2 - 4)(x^2 - 9) = 0$$ $$x^2 = 4 \quad \text{or} \quad x^2 = 9$$ $$x = 2 \qquad\qquad x = 3$$ The solution set is $\{2, 3\}$.	The degree of the equation is 4, which means we should have four solutions. When we solve the two quadratic equations, we get $$x = \pm 2 \quad \text{or} \quad x = \pm 3$$ The solution set is $\{-3, -2, 2, 3\}$.

6c. $(y + 3)^2 - 4(y + 3) - 32 = 0$.

Incorrect Solution	Correct Solution and Explanation
Let $u = (y + 3)$: $$(y + 3)^2 - 4(y + 3) - 32 = 0$$ $$u^2 - 4u - 32 = 0$$ $$(u - 8)(u + 4) = 0$$ $$u = 8 \quad \text{or} \quad u = -4$$ The solution set is $\{-4, 8\}$.	We need to back-substitute and find the solutions for y. $$u = 8 \quad \text{or} \quad u = -4$$ $$y + 3 = 8 \qquad y + 3 = -4$$ $$y = 5 \qquad\qquad y = -7$$ The solution set is $\{-7, 5\}$.

ANSWERS TO STUDENT CHECKS

Student Check 1 **a.** $\{-10, 2\}$ **b.** $\left\{ \dfrac{-1 \pm \sqrt{85}}{14} \right\}$

Student Check 2 **a.** $\{1\}$ **b.** $\{16\}$

Student Check 3 **a.** $\{-4, 4, -i, i\}$ **b.** $\left\{ -1, 1, -\dfrac{1}{3}, \dfrac{1}{3} \right\}$

c. $\left\{ 3, \dfrac{-3 \pm 3i\sqrt{3}}{2} \right\}$

Student Check 4 **a.** $\{3, 5\}$ **b.** $\{-64, 8\}$

Student Check 5 **a.** Juan's speed was 70 mph and Maria's speed was 56 mph. **b.** It will take Barbara 5.16 hr and Tom 7.16 hr to clean their house working alone.

SUMMARY OF KEY CONCEPTS

1. We must apply quadratic techniques to solve rational equations if, after clearing the fractions, a quadratic equation results. The solution set includes only those solutions that do not make any of the rational expressions undefined.

2. After eliminating the radicals from a square root equation, a quadratic equation may result. We solve the resulting equation using any of the methods we have learned in the previous sections. The possible solutions must be checked in the original equation. Exclude any extraneous solutions from the solution set.

3. We solve higher degree polynomial equations by initially factoring the polynomial. Once we obtain either linear or quadratic factors, we set the factors equal to zero and solve. The degree indicates the number of solutions including repetitions.

4. Other equations can be solved by substituting a variable for an appropriate expression that creates a quadratic equation. The key is to back-substitute to find the solutions of the original equation.

5. Work and distance problems can result in rational equations. We solve these by clearing the fractions by multiplying by the LCD.

GRAPHING CALCULATOR SKILLS

The graphing calculator can be used to verify solutions or to determine which solutions might be extraneous.

Example 1: Determine if $-\dfrac{1}{4}$ and 2 are solutions of $2x = \sqrt{7x + 2}$.

Solution: Enter the left side and right side of the equation into Y_1 and Y_2, respectively.

 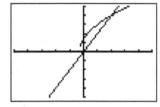

Notice that the graphs of the two functions intersect at only one point. This means there is only one solution of the equation. From the graph we see that the two functions intersect at $x = 2$.

Example 2: Find the solutions of $x^4 - 13x^2 + 36 = 0$ graphically.

Solution: Let $Y_1 = x^4 - 13x^2 + 36$. The solutions of $Y_1 = 0$ correspond to the x-intercepts of the graph.

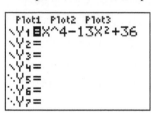 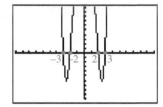

The graph crosses the x-axis at $x = -3, -2, 2, 3$. So, these values are in the solution set. Recall that if an equation has complex solutions, the graph will not intersect the x-axis at these values.

SECTION 11.4 / EXERCISE SET

Write About It!

Use complete sentences in your answer to each exercise. For Exercises 1 and 2, explain how to change the equation to a quadratic form. Do not solve the equation.

1. $21x = 8 + \dfrac{4}{x}$

2. $3x = \sqrt{8 - 6x}$

3. Explain how to use the substitution method to solve the equation $(2x - 1)^2 - 2(2x - 1) - 8 = 0$.

4. If you solve the equation in Exercise 3 by simplifying the equation and writing it in standard form, $ax^2 + bx + c = 0$, do we get the same answer?

5. After solving the radical equation using substitution, explain the next step required to finish the problem.

$$x^{1/2} - 2x^{1/4} - 15 = 0$$
$$w = x^{1/4}$$
$$w^2 - 2w - 15 = 0$$
$$(w - 5)(w + 3) = 0$$
$$w = 5 \quad \text{or} \quad w = -3$$

6. Can we use the substitution method to solve the following equation? Explain.
$$3(x + 2)^2 + 10(x - 2) - 8 = 0$$

Practice Makes Perfect!

Solve each rational equation. (*See Objective 1.*)

7. $6x = 1 + \dfrac{15}{x}$

8. $7x = 31 - \dfrac{12}{x}$

9. $\dfrac{2x}{x - 1} + \dfrac{14x + 52}{x + 1} = -\dfrac{76}{x^2 - 1}$

10. $\dfrac{3x}{x + 4} + \dfrac{6x + 12}{x - 4} = \dfrac{33}{x^2 - 16}$

11. $\dfrac{5x}{x - 2} - \dfrac{4x + 28}{x + 2} = \dfrac{23}{x^2 - 4}$

12. $\dfrac{3x}{x - 3} - \dfrac{2x + 21}{x + 3} = \dfrac{41}{x^2 - 9}$

13. $\dfrac{7x}{2x - 3} + \dfrac{x - 6}{x} = \dfrac{14}{2x^2 - 3x}$

14. $\dfrac{3x}{3x + 5} + \dfrac{x - 11}{2x} = -\dfrac{29}{3x^2 + 5x}$

15. $\dfrac{10x}{2x + 1} + \dfrac{x - 3}{x} = -\dfrac{3}{2x^2 + x}$

16. $\dfrac{x}{3x - 2} + \dfrac{x - 2}{x} = \dfrac{4}{3x^2 - 2x}$

17. $\dfrac{6x}{x - 2} + \dfrac{3x - 12}{x} = \dfrac{21}{x^2 - 2x}$

18. $\dfrac{9x}{x - 2} - \dfrac{9x + 6}{5x} = \dfrac{26}{5x^2 - 10x}$

19. $\dfrac{x}{x+6} + \dfrac{x-12}{x} = -\dfrac{78}{x^2+6x}$

20. $\dfrac{x}{x-5} - \dfrac{4x+24}{5x} = \dfrac{96}{5x^2-25x}$

Solve each radical equation. (*See Objective 2.*)

21. $x = \sqrt{15x-56}$

22. $x = \sqrt{-5x+66}$

23. $x+2 = \sqrt{8x+16}$

24. $x+6 = \sqrt{17x+60}$

25. $3x = \sqrt{14x+8}$

26. $2x = \sqrt{13x+12}$

27. $14x + 3\sqrt{x} = 27$

28. $15x - 4\sqrt{x} = 4$

29. $x + 14\sqrt{x} = -45$

30. $4x + 7\sqrt{x} = -3$

31. $4x - 19\sqrt{x} = -12$

32. $3x - 14\sqrt{x} = -15$

Solve each polynomial equation. (*See Objective 3.*)

33. $x^4 - 24x^2 - 25 = 0$

34. $9x^4 + 80x^2 - 9 = 0$

35. $9x^4 - 13x^2 + 4 = 0$

36. $4x^4 - 17x^2 + 4 = 0$

37. $x^4 + 10x^2 = -9$

38. $4x^4 + 37x^2 = -9$

39. $x^4 - 16 = 0$

40. $81x^4 - 625 = 0$

41. $x^3 = 64$

42. $x^3 = -125$

Solve each equation using substitution. (*See Objective 4.*)

43. $5(x-3)^2 + 23(x-3) - 10 = 0$

44. $15(3x-1)^2 - 31(3x-1) + 14 = 0$

45. $9(x+1)^2 + 12(x+1) + 4 = 0$

46. $4(x-2)^2 - 20(x-2) + 25 = 0$

47. $3x^{2/3} - 8x^{1/3} + 4 = 0$

48. $2x^{2/3} + x^{1/3} - 3 = 0$

49. $2x^{2/5} + 3x^{1/5} + 1 = 0$

50. $x^{2/5} + x^{1/5} - 6 = 0$

Solve each problem. Round each answer to two decimal places. (*See Objective 5.*)

51. Carolina and Marilyn own a lawn service. Together they can mow the lawns in one development in 4 hr. Carolina does all the mowing 2 hr faster than Marilyn. How long will it take each working alone to do the job?

52. Melanie and Jim own a house cleaning company. Together they can clean the houses in one development in 5 hr. Melanie does all the house cleaning 4 hr slower than Jim. How long will it take each working alone to do the job?

53. Tony drove a distance of 500 mi to visit his friend Johnson in Palmyra, Pennsylvania. During the return trip, he was caught in a rainstorm and had to decrease his speed by 10 mph. If the return trip took 1 hr longer than the trip to Palmyra, find his original speed.

54. Allison drove a distance of 240 mi to visit her friend Ricki in Northridge, California. During the

return trip, she tried a new toll road and was able to increase her speed by 14 mph. If her return trip took 1 hr less time than the trip to Northridge, find her original speed.

 Mix 'Em Up!

Solve each equation.

55. $x = -3 + \dfrac{18}{x}$

56. $5x = -36 + \dfrac{32}{x}$

57 $x = \sqrt{16x-55}$

58. $x = \sqrt{11x-28}$

59. $\dfrac{5x}{x-3} + \dfrac{20x+5}{x+3} = -\dfrac{7}{x^2-9}$

60. $\dfrac{3x}{x-1} + \dfrac{x-14}{x+1} = \dfrac{10}{x^2-1}$

61. $8x - 15\sqrt{x} = 27$

62. $2x + 17\sqrt{x} = 30$

63. $\dfrac{2x}{x+2} + \dfrac{2x+12}{x-2} = \dfrac{1}{x^2-4}$

64. $\dfrac{2x}{x+4} + \dfrac{2x+4}{x-4} = \dfrac{12}{x^2-16}$

65. $2x - 11\sqrt{x} = -15$

66. $7x - 37\sqrt{x} = -10$

67. $4x = -25 - \dfrac{25}{x}$

68. $3x = 29 - \dfrac{56}{x}$

69. $\dfrac{5x}{x+3} + \dfrac{20x+15}{x-3} = \dfrac{13}{x^2-9}$

70. $\dfrac{5x}{x+4} + \dfrac{11x+16}{x-4} = \dfrac{48}{x^2-16}$

71. $4x = \sqrt{-8x+15}$

72. $3x = \sqrt{20x-4}$

73. $12(5x+6)^2 + 7(5x+6) - 10 = 0$

74. $4(x-1)^2 + 4(x-1) + 1 = 0$

75. $\dfrac{x}{3x+5} - \dfrac{x-4}{x} = \dfrac{20}{3x^2+5x}$

76. $\dfrac{4x}{x-1} - \dfrac{9x+5}{5x} = \dfrac{1}{x^2-x}$

77. $4x^4 - 29x^2 = -25$

78. $9x^4 - 16x^2 = 25$

79. $x^{2/3} - 3x^{1/3} - 4 = 0$

80. $x^{2/5} + x^{1/5} - 2 = 0$

81. $x + 5 = \sqrt{15x+49}$

82. $x - 1 = \sqrt{-3x+13}$

83. $(x-2)^2 - 6(x-2) + 9 = 0$

84. $(x+4)^2 - 10(x+4) + 25 = 0$

85. $25x + 35\sqrt{x} = -6$

86. $7x + 9\sqrt{x} = -2$

87. $x^4 - 625 = 0$

88. $16x^4 - 81 = 0$

89. $3(4x-3)^2 + 16(4x-3) + 5 = 0$

90. $2(2x+1)^2 - 9(2x+1) + 10 = 0$

91. $8x^3 - 125 = 0$

92. $x^3 + 1 = 0$

Solve each problem. Round each answer to two decimal places.

93. Ron and Ted own a printing service. Together they can complete all print jobs in 3 hr. Ron does all the print jobs 1 hr faster than Ted. How long will it take each working alone to do the job?

94. Albert and Alex own a lawn service. Together they can mow the lawns in one development in 3 hr. Albert does all the mowing 2 hr faster than Alex. How long will it take each working alone to do the job?

95. Ruth drove a distance of 480 mi to visit her friend, Brian, in Irvine, California. During her trip home, she had to decrease her speed by 12 mph because of a traffic accident. If her trip back home took 2 hr longer than the trip to Irvine, find her original speed.

96. Seung drove a distance of 280 mi to visit her parents in Atlanta, Georgia. During her trip home, she was caught in a rainstorm and had to decrease her speed by 15 mph. If her trip back home took 1.5 hr longer than her trip to Atlanta, find her original speed.

 You Be the Teacher!

Correct each student's errors, if any.

97. Solve the equation $\dfrac{2x}{x+4} - \dfrac{x-6}{x-4} = \dfrac{32}{x^2-16}$.

Jeung's work:

$$\frac{2x}{x+4} - \frac{x-6}{x-4} = \frac{32}{x^2-16}$$

$$\frac{2x}{x+4} - \frac{x-6}{x-4} = \frac{32}{(x+4)(x-4)}$$

$$2x(x-4) - (x-6)(x+4) = 32$$

$$2x^2 - 8x - x^2 - 2x - 24 = 32$$

$$x^2 - 10x - 56 = 0$$

$$(x-14)(x+4) = 0$$

$$x = 14 \quad \text{or} \quad x = -4$$

98. Solve the equation $x = \sqrt{5x+6}$.

Mark's work:

$$x = \sqrt{5x+6}$$

$$x^2 = 5x + 6$$

$$x^2 - 5x - 6 = 0$$

$$(x-6)(x+1) = 0$$

$$x = 6 \quad \text{or} \quad x = -1$$

99. Solve the equation $4x^4 + 7x^2 - 36 = 0$.

Ken's work:

$$4x^4 + 7x^2 - 36 = 0$$

$$(4x^2 - 9)(x^2 + 4) = 0$$

$$4x^2 - 9 = 0 \quad \text{or} \quad x^2 + 4 = 0$$

$$x^2 = \frac{9}{4} \qquad\qquad x^2 = -4$$

$$x = \frac{3}{2} \qquad\qquad x = 2i$$

100. Solve the equation $x^{2/3} - 3x^{1/3} - 4 = 0$.

Beth's work: Let $u = x^{1/3}$.

$$x^{2/3} - 3x^{1/3} - 4 = 0$$

$$u^2 - 3u - 4 = 0$$

$$(u-4)(u+1) = 0$$

$$u - 4 = 0 \quad \text{or} \quad u + 1 = 0$$

$$u = 4 \qquad\qquad u = -1$$

The solution set is $\{-1, 4\}$.

 Calculate It!

Use a graphing calculator to verify or find solutions of each equation.

101. Determine if 2 and 8 are solutions of $x + 3 = \sqrt{16x - 7}$.

102. Determine if 6 and −4 are solutions of $x - 2 = \sqrt{28 - 2x}$.

103. Find the solutions of $3x = \dfrac{8}{x} - 10$ graphically.

104. Find the solutions of $25x^4 - 29x^2 + 4 = 0$ graphically.

| SECTION 11.5 | More on Graphing Quadratic Functions |

▶ OBJECTIVES

As a result of completing this section, you will be able to

1. Graph a quadratic function of the form $f(x) = ax^2 + bx + c$ by converting it to the form $f(x) = a(x - h)^2 + k$.

2. Graph a quadratic function of the form $f(x) = ax^2 + bx + c$ using the vertex formula.

3. Solve applications of finding maximum or minimum values.

4. Troubleshoot common errors.

Objective 1 ▶

Graph a quadratic function of the form $f(x) = ax^2 + bx + c$ by converting it to the form $f(x) = a(x - h)^2 + k$.

A ball is tossed upward with an initial velocity of 30 ft/sec from a height of 6 ft. The height of the ball t sec after it is tossed is given by $f(t) = -16t^2 + 30t + 6$. What is the maximum height the ball will reach and when will it reach this height?

Since the height of the ball is modeled by a quadratic function, its graph is a parabola. We know from Section 11.1 that the parabola has its highest or lowest point at its vertex. To answer this question, we must find the vertex of this function. We will learn methods that enable us to do that in this section.

Graphing Quadratic Functions

Recall from Section 11.1 that the vertex of the parabola given by $y = a(x - h)^2 + k$ is the point (h, k). The vertex, along with a few other key points, is used to determine the shape of a parabola. In this section, the quadratic functions have the form $f(x) = ax^2 + bx + c$. If we can convert this form to the vertex form, $f(x) = a(x - h)^2 + k$, we will know the vertex of the given function. We complete the square to make the conversion.

Recall, we can make $x^2 + 10x$ a perfect square by adding $\left(\dfrac{10}{2}\right)^2 = (5)^2 = 25$; that is, $x^2 + 10x + 25 = (x + 5)^2$.

Procedure: Converting $f(x) = ax^2 + bx + c$ to the Form $f(x) = a(x - h)^2 + k$

Step 1: Write $f(x) = ax^2 + bx + c$ as $y = ax^2 + bx + c$.
Step 2: Move the constant term c to the opposite side of the equation.
Step 3: Divide each side by a if $a \neq 1$.
Step 4: Add the number that completes the square to each side of the equation.
Step 5: Factor the perfect square trinomial as (binomial)^2 and simplify the other side.
Step 6: Isolate y by moving the constant back to the other side.
Step 7: Replace y with $f(x)$.

Objective 1 Examples

Convert $f(x) = ax^2 + bx + c$ to the vertex form, $f(x) = a(x - h)^2 + k$. Identify the vertex and intercepts, and graph each function.

1a. $f(x) = x^2 - 4x + 3$ **1b.** $f(x) = -x^2 + 6x - 5$ **1c.** $f(x) = 2x^2 + 4x - 1$

Solutions   **1a.**

$$f(x) = x^2 - 4x + 3$$

$$y = x^2 - 4x + 3 \qquad \text{Let } f(x) = y.$$

$$y - 3 = x^2 - 4x \qquad \text{Subtract 3 from each side.}$$

$$y - 3 + 4 = x^2 - 4x + 4 \qquad \text{Add } \left(\dfrac{-4}{2}\right)^2 = (-2)^2 = 4 \text{ to each side.}$$

$$y + 1 = (x - 2)^2 \qquad \text{Simplify and factor } x^2 - 4x + 4 = (x - 2)^2.$$

$$y = (x - 2)^2 - 1 \qquad \text{Subtract 1 from each side.}$$

$$f(x) = (x - 2)^2 - 1 \qquad \text{Replace } y \text{ with } f(x).$$

So, the vertex of $f(x) = (x - 2)^2 - 1$ is $(2, -1)$.

x-intercepts: Let $y = 0$ or $f(x) = 0$.	y-intercept: Let $x = 0$ or find $f(0)$.
$y = x^2 - 4x + 3$	$f(x) = x^2 - 4x + 3$
$0 = x^2 - 4x + 3$	$f(0) = (0)^2 - 4(0) + 3$
$0 = (x - 1)(x - 3)$	$f(0) = 3$
$x - 1 = 0$ or $x - 3 = 0$	The y-intercept is $(0, 3)$.
$x = 1$ $x = 3$	
The x-intercepts are $(3, 0)$ and $(1, 0)$.	

Note that $a = 1$. Since $a > 0$, the parabola opens up. So, the graph is

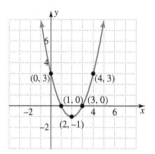

1b.

$$f(x) = -x^2 + 6x - 5$$

$y = -x^2 + 6x - 5$	Let $f(x) = y$.
$y + 5 = -x^2 + 6x$	Add 5 to each side.
$\dfrac{y + 5}{-1} = \dfrac{-x^2 + 6x}{-1}$	Divide each side by the coefficient of x^2, -1.
$-y - 5 = x^2 - 6x$	Simplify.
$-y - 5 + 9 = x^2 - 6x + 9$	Add $\left(\dfrac{-6}{2}\right)^2 = (-3)^2 = 9$ to each side.
$-y + 4 = (x - 3)^2$	Simplify and factor $x^2 - 6x + 9 = (x - 3)^2$.
$-y = (x - 3)^2 - 4$	Subtract 4 from each side.
$y = -(x - 3)^2 + 4$	Multiply each side by -1.
$f(x) = -(x - 3)^2 + 4$	Replace y with $f(x)$.

So, the vertex of $f(x) = -(x - 3)^2 + 4$ is $(3, 4)$.

x- intercepts: Let $y = 0$ or $f(x) = 0$.	y-intercept: Let $x = 0$ or find $f(0)$.
$y = -x^2 + 6x - 5$	$f(x) = -x^2 + 6x - 5$
$0 = -x^2 + 6x - 5$	$f(0) = -(0)^2 + 6(0) - 5$
$-1(0) = -1(-x^2 + 6x - 5)$	$f(0) = -5$
$0 = x^2 - 6x + 5$	The y-intercept is $(0, -5)$.
$0 = (x - 1)(x - 5)$	
$x - 1 = 0$ or $x - 5 = 0$	
$x = 1$ $x = 5$	
The x-intercepts are $(1, 0)$ and $(5, 0)$.	

Note that $a = -1$. Since $a < 0$, the parabola opens down. So, the graph is

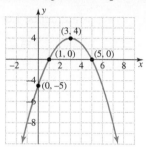

1c.

$$f(x) = 2x^2 + 4x - 1$$

$y = 2x^2 + 4x - 1$	Let $f(x) = y$.
$y + 1 = 2x^2 + 4x$	Add 1 to each side.
$\dfrac{y + 1}{2} = \dfrac{2x^2 + 4x}{2}$	Divide each side by the coefficient of x^2, 2.
$\dfrac{y}{2} + \dfrac{1}{2} = x^2 + 2x$	Simplify.
$\dfrac{y}{2} + \dfrac{1}{2} + 1 = x^2 + 2x + 1$	Add $\left(\dfrac{2}{2}\right)^2 = (1)^2 = 1$ to each side.
$\dfrac{y}{2} + \dfrac{3}{2} = (x + 1)^2$	Simplify and factor $x^2 + 2x + 1 = (x + 1)^2$.
$\dfrac{y}{2} = (x + 1)^2 - \dfrac{3}{2}$	Subtract $\dfrac{3}{2}$ from each side.
$2\left(\dfrac{y}{2}\right) = 2\left[(x + 1)^2 - \dfrac{3}{2}\right]$	Multiply each side by 2.
$y = 2(x + 1)^2 - 3$	Simplify.
$f(x) = 2(x + 1)^2 - 3$	Replace y with $f(x)$.

So, the vertex of $f(x) = 2[x - (-1)]^2 - 3$ is $(-1, -3)$.

x-intercepts: Let $y = 0$ or $f(x) = 0$.

$$y = 2x^2 + 4x - 1$$
$$0 = 2x^2 + 4x - 1$$
$$x = \frac{-(4) \pm \sqrt{(4)^2 - 4(2)(-1)}}{2(2)}$$
$$x = \frac{-4 \pm \sqrt{24}}{4}$$
$$x = \frac{-4 \pm 2\sqrt{6}}{4}$$
$$x = \frac{2(-2 \pm \sqrt{6})}{2(2)}$$
$$x = \frac{-2 \pm \sqrt{6}}{2}$$
$$x = \frac{-2 + \sqrt{6}}{2} \quad \text{or} \quad x = \frac{-2 - \sqrt{6}}{2}$$
$$x \approx 0.22 \qquad\qquad x \approx -2.22$$

The x-intercepts are $(0.22, 0)$ and $(-2.22, 0)$.

y-intercept: Let $x = 0$ or find $f(0)$.

$$f(x) = 2x^2 + 4x - 1$$
$$f(0) = 2(0)^2 + 4(0) - 1$$
$$f(0) = -1$$

The y-intercept is $(0, -1)$.

Note that $a = 2$. Since $a > 0$, the parabola opens up. So, the graph is

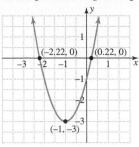

☑ **Student Check 1**   Convert $f(x) = ax^2 + bx + c$ to the vertex form, $f(x) = a(x - h)^2 + k$. Identify the vertex and intercepts, and graph each function.

   a. $f(x) = x^2 - 10x + 9$ **b.** $f(x) = -x^2 - 8x + 3$ **c.** $f(x) = 3x^2 + 12x - 2$

The Vertex Formula

Objective 2 ▶

Graph a quadratic function of the form $f(x) = ax^2 + bx + c$ using the vertex formula.

Completing the square to obtain the form $f(x) = a(x - h)^2 + k$ can often be tedious. There is a formula that we can use to find the vertex of a parabola. In Example 1, notice that the vertex is always located halfway between the two x-intercepts. The following table shows this result.

Example	x-Intercepts	x-Value of Vertex	Average of the x-Values of the x-Intercepts
1a	$(1, 0)$ and $(3, 0)$	$x = 2$	$\dfrac{1 + 3}{2} = \dfrac{4}{2} = 2$
1b	$(1, 0)$ and $(5, 0)$	$x = 3$	$\dfrac{1 + 5}{2} = \dfrac{6}{2} = 3$
1c	$(-2.22, 0)$ and $(0.22, 0)$	$x = -1$	$\dfrac{-2.22 + 0.22}{2} = \dfrac{-2}{2} = -1$

So, if the x-intercepts are $x = \dfrac{-b + \sqrt{b^2 - 4ac}}{2a}$ and $x = \dfrac{-b - \sqrt{b^2 - 4ac}}{2a}$, then the x-value of the vertex can be found by averaging these x-values. This gives us

$$x = \frac{\dfrac{-b + \sqrt{b^2 - 4ac}}{2a} + \dfrac{-b - \sqrt{b^2 - 4ac}}{2a}}{2}$$

$$x = \frac{\dfrac{-b + \sqrt{b^2 - 4ac} - b - \sqrt{b^2 - 4ac}}{2a}}{2}$$

$$x = \frac{-2b}{2a} \cdot \frac{1}{2}$$

$$x = -\frac{b}{2a}$$

Note that the y-value of the vertex is found by substituting the x-value in the function.

Property: The Vertex Formula

The x-value of the vertex of the graph of $f(x) = ax^2 + bx + c$ is $x = -\dfrac{b}{2a}$.

The y-value is found by substituting this value of x in the function, or $f\left(-\dfrac{b}{2a}\right)$.

	The Key Points on a Parabola
x-intercepts	Recall x-*intercepts* are solutions of the equation $f(x) = 0$ or $$ax^2 + bx + c = 0$$ Since this type of equation has either 2, 1, or 0 real solutions, there will be 2 x-intercepts, 1 x-intercept, or no x-intercepts.
y-intercept	The y-*intercept* is found by replacing x with 0, that is, $f(0)$. $$y = a(0)^2 + b(0) + c = c$$ This shows that the y-intercept of $f(x) = ax^2 + bx + c$ is the point $(0, c)$. A quadratic function has only one y-intercept.
Vertex	The *vertex* is of the form $\left[-\dfrac{b}{2a}, f\left(-\dfrac{b}{2a} \right) \right]$.

Objective 2 Examples Graph each function by finding the intercepts, the vertex, and additional points, if needed.

2a. $f(x) = -x^2 + 1$ **2b.** $f(x) = x^2 + 4x - 5$

Solutions **2a.** Note that $f(x) = -1x^2 + 0x + 1$, so $a = -1$, $b = 0$, and $c = 1$.

x-intercepts	$-x^2 + 1 = 0$ $-x^2 = -1$ $x^2 = 1$ $x = \pm\sqrt{1}$ $x = \pm 1$ $(1, 0)$ and $(-1, 0)$
y-intercept	$f(0) = -(0)^2 + 1$ $f(0) = 1$ $(0, 1)$
Vertex	$x = -\dfrac{b}{2a} = -\dfrac{0}{2(-1)} = -\dfrac{0}{-2} = 0$ $f(0) = -(0)^2 + 1 = 0 + 1 = 1$ $(0, 1)$

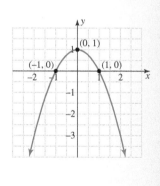

2b. Note that $a = 1$, $b = 4$, and $c = -5$.

x-intercepts	$x^2 + 4x - 5 = 0$ $(x + 5)(x - 1) = 0$ $x + 5 = 0$ or $x - 1 = 0$ $x = -5$ or $x = 1$ $(-5, 0)$ and $(1, 0)$
y-intercept	$f(0) = (0)^2 + 4(0) - 5$ $f(0) = -5$ $(0, -5)$
Vertex	$x = -\dfrac{b}{2a} = -\dfrac{4}{2(1)} = -\dfrac{4}{2} = -2$ $f(-2) = (-2)^2 + 4(-2) - 5$ $f(-2) = 4 - 8 - 5$ $f(-2) = -9$ $(-2, -9)$

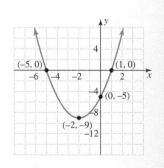

✔ **Student Check 2**   Graph each function by finding the intercepts, the vertex, and additional points, if needed.

   a. $y = -x^2 + 4$ **b.** $y = x^2 + 6x$

Maximum and Minimum Values of Quadratic Functions

Objective 3 ▶

Solve applications of finding the maximum and minimum values.

We have learned that the vertex of a parabola corresponds to either the highest or lowest point on the graph of a quadratic function. When a parabola opens up, the vertex is the lowest point on the graph. When a parabola opens down, the vertex is the highest point on the graph.

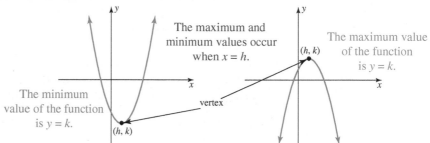

We can use this information to solve applications that ask us to determine where the maximum or minimum values of a function occur. If the given model is a quadratic function, we find the vertex of the parabola to solve the problem. For example, the graph of $f(x) = -x^2 + 6x - 5$ is shown.

Since the vertex of the parabola is (3, 4), we say that the function has a maximum value of 4 when $x = 3$. Note that the parabola extends indefinitely downward, so there is no minimum value.

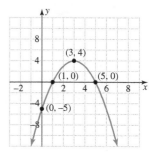

> **Procedure: Solving Applications by Finding Maximum or Minimum Values**
>
> **Step 1:** Identify the value of a and determine if the parabola opens up ($a > 0$) or down ($a < 0$).
> **Step 2:** Determine if the vertex is the maximum or minimum point of the graph of the function.
>   **a.** If the parabola opens up, the vertex is the minimum point.
>   **b.** If the parabola opens down, the vertex is the maximum point.
> **Step 3:** Find the vertex of the quadratic function.
> **Step 4:** The x-value of the vertex is where the maximum or minimum occurs.
> **Step 5:** The y-value of the vertex is the maximum or minimum value of the function.

Objective 3 Example   A ball is tossed upward with an initial velocity of 30 ft/sec from a height of 6 ft. The height of the ball t sec after it is tossed is given by $f(t) = -16t^2 + 30t + 6$. What is the maximum height the ball will reach and when will it reach this height?

Solution   The quadratic function opens down since the value of $a = -16 < 0$. Therefore, the maximum value of the function occurs at the vertex. We find the vertex by using the vertex formula.

$$f(t) = -16t^2 + 30t + 6 \rightarrow a = -16, b = 30, c = 6$$

We substitute these values into the vertex formula, $t = -\dfrac{b}{2a}$.

$$t = -\frac{b}{2a} = -\frac{30}{2(-16)} = -\frac{30}{-32} = \frac{15}{16} \approx 0.94$$

$$f\left(\frac{15}{16}\right) = -16\left(\frac{15}{16}\right)^2 + 30\left(\frac{15}{16}\right) + 6$$

$$= -16\left(\frac{225}{256}\right) + \frac{225}{8} + 6$$

$$= -\frac{225}{16} + \frac{450}{16} + \frac{96}{16}$$

$$= \frac{321}{16} \approx 20.06$$

The vertex of the function $f(t) = -16t^2 + 30t + 6$ is approximately $(0.94, 20.06)$. So, the maximum height of the ball is 20.06 ft. It will reach this height about 0.94 sec after the ball has been thrown.

✓ **Student Check 3**   The monthly profit, in thousands of dollars, from selling x hundred computers is given by $f(x) = -10x^2 + 56x + 37$. How many computers must be sold in a month to maximize profit? What is the maximum monthly profit?

Objective 4 ▶

Troubleshoot common errors.

Troubleshooting Common Errors

A common error associated with converting a quadratic function from standard form to vertex form is shown.

Objective 4 Example   A problem and an incorrect solution are given. Provide the correct solution and an explanation of the error.

Write $y = x^2 - 6x + 5$ in the form $y = a(x - h)^2 + k$.

Incorrect Solution	Correct Solution and Explanation
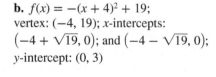	When we complete the square, we must add that number to both sides of the equation.

$$\begin{aligned} y &= x^2 - 6x + 5 \\ y - 5 &= x^2 - 6x + 9 \\ y - 5 &= (x - 3)^2 \\ y &= (x - 3)^2 + 5 \end{aligned}$$

$$\begin{aligned} y - 5 + 9 &= x^2 - 6x + 9 \\ y + 4 &= (x - 3)^2 \\ y &= (x - 3)^2 - 4 \end{aligned}$$

ANSWERS TO STUDENT CHECKS

Student Check 1   a. $f(x) = (x - 5)^2 - 16$; vertex: $(5, -16)$; x-intercepts: $(1, 0)$ and $(9, 0)$; y-intercept: $(0, 9)$ **b.** $f(x) = -(x + 4)^2 + 19$; vertex: $(-4, 19)$; x-intercepts: $(-4 + \sqrt{19}, 0)$; and $(-4 - \sqrt{19}, 0)$; y-intercept: $(0, 3)$ **c.** $f(x) = 3(x + 2)^2 - 14$; vertex: $(-2, -14)$; x-intercepts: $\left(\dfrac{-6 + \sqrt{42}}{3}, 0\right)$ and $\left(\dfrac{-6 - \sqrt{42}}{3}, 0\right)$; y-intercept: $(0, -2)$

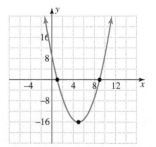

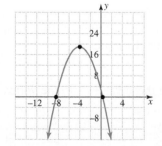

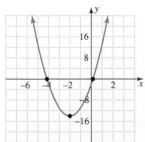

Student Check 2 **a.** vertex: $(0, 4)$; x-intercepts: $(-2, 0)$ and $(2, 0)$; y-intercept: $(0, 4)$

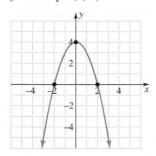

b. vertex: $(-3, -9)$; x-intercepts: $(-6, 0)$ and $(0, 0)$; y-intercept: $(0, 0)$

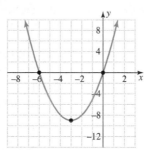

Student Check 3 The company must sell 2.8 hundred or 280 computers in a month to maximize profit. The maximum monthly profit is 115.4 thousand or $115,400.

SUMMARY OF KEY CONCEPTS

1. Complete the square to convert a quadratic function from standard form, $f(x) = ax^2 + bx + c$, to vertex form, $f(x) = a(x - h)^2 + k$. The steps that are used to solve an equation by completing the square are also used in this process. We can then readily identify the vertex of the parabola when written in the vertex form.

2. The vertex formula can be used to find the vertex of

$$f(x) = ax^2 + bx + c. \text{ The } x\text{-value of the vertex is } x = -\frac{b}{2a}$$

and the corresponding y-value is found by substituting this x-value into the function and solving for y.

3. For quadratic models, the maximum or minimum value will occur at its vertex. The y-value of the vertex is the maximum or minimum value of the function. The x-value of the vertex is where this maximum or minimum occurs.

GRAPHING CALCULATOR SKILLS

The graphing calculator can be used to graph quadratic functions.

Example: Verify that $(2, -1)$ is the vertex of $y = x^2 - 4x + 3$. Also verify that $(1, 0)$ and $(3, 0)$ are the x-intercepts and $(0, 3)$ is the y-intercept.

Solution: Enter the function into the calculator. View the table and graph.

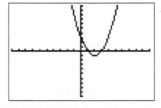

From the table, we see the points $(0, 3)$, $(1, 0)$, and $(3, 0)$ are solutions and the graph confirms these are intercepts. The point $(2, -1)$ is also the lowest point on the graph and is the vertex of the parabola.

SECTION 11.5 / EXERCISE SET

 Write About It!

Use complete sentences in your answer to each exercise.

1. Explain how to convert a quadratic function of the form $y = x^2 + bx + c$ to the vertex form $y = (x - h)^2 + k$.

2. Explain how to find the x-intercepts of a quadratic function in vertex form, $y = (x - h)^2 + k$.

3. Explain which form, standard or vertex, makes it easier to find the y-intercept.

4. Explain which form, standard or vertex, makes it easier to find the x-intercepts.

5. What can you say about the location of the vertex and the two x-intercepts?

6. Explain how the solutions of the quadratic equation, $ax^2 + bx + c = 0$, are related to the x-intercepts of the quadratic function, $f(x) = ax^2 + bx + c$.

7. Explain how the quadratic formula is related to the vertex formula.

$$ax^2 + bx + c = 0$$

$$x = \frac{-b \pm \sqrt{b^2 - 4ac}}{2a}$$

8. Explain how to apply the vertex formula to find the maximum or minimum value of a quadratic function.

 Practice Makes Perfect!

Convert $f(x) = ax^2 + bx + c$ to the vertex form, $f(x) = a(x - h)^2 + k$. Identify the vertex and intercepts, and graph each function. *(See Objective 1.)*

9. $f(x) = x^2 + 2x - 15$
10. $f(x) = x^2 - 4x - 5$
11. $f(x) = x^2 + 4x - 60$
12. $f(x) = x^2 + 12x + 35$
13. $f(x) = x^2 + 3x - 18$
14. $f(x) = x^2 - 7x + 12$
15. $f(x) = -x^2 - 2x + 8$
16. $f(x) = -x^2 - 6x + 7$
17. $f(x) = -x^2 - 5x + 36$
18. $f(x) = -x^2 + x + 2$
19. $f(x) = 2x^2 + 12x + 2$
20. $f(x) = 3x^2 + 6x - 3$
21. $f(x) = -5x^2 - 10x + 7$
22. $f(x) = -3x^2 + 12x - 5$
23. $f(x) = -2x^2 + 12x - 27$
24. $f(x) = -2x^2 + 8x - 25$
25. $f(x) = 2x^2 + 8x + 11$
26. $f(x) = 3x^2 - 6x + 9$

Use the vertex formula to determine the vertex of each function. *(See Objective 2.)*

27. $f(x) = x^2 - 9x + 2$
28. $f(x) = x^2 + 5x + 7$
29. $f(x) = -2x^2 + 6x + 1$
30. $f(x) = -3x^2 - 5x + 6$
31. $f(x) = 5x^2 + 3x + 1$
32. $f(x) = 5x^2 + 2x - 7$

Graph each quadratic function by finding the intercepts, the vertex, and additional points, if needed. *(See Objective 2.)*

33. $f(x) = x^2 - 10$
34. $f(x) = x^2 - 7$
35. $f(x) = -x^2 - 5$
36. $f(x) = x^2 + 3$
37. $f(x) = x^2 + 6x$
38. $f(x) = x^2 - 5x$
39. $f(x) = -x^2 + 7x$
40. $f(x) = -x^2 - 3x$
41. $f(x) = (x - 2)^2 - 1$
42. $f(x) = (x + 4)^2 - 9$
43. $f(x) = -(x + 3)^2 + 1$
44. $f(x) = -(x - 2)^2 + 9$
45. $f(x) = 2x^2 - 9x - 5$
46. $f(x) = 3x^2 + 2x - 5$

Solve each problem. *(See Objective 3.)*

47. The monthly revenue, in thousands of dollars, from selling x hundred computers is given by $f(x) = -2x^2 + 40x + 235$. How many computers must be sold in a month to maximize revenue? What is the maximum monthly revenue?

48. The monthly revenue, in thousands of dollars, from selling x hundred computers is given by

$f(x) = -17x^2 + 85x - 35$. How many computers must be sold in a month to maximize revenue? What is the maximum monthly revenue?

49. The monthly profit, in thousands of dollars, from selling x thousand graphing calculators is given by $f(x) = -3x^2 + 63x + 17$. How many graphing calculators must be sold in a month to maximize profit? What is the maximum monthly profit?

50. The monthly revenue, in thousands of dollars, from selling x hundred laptops is given by $f(x) = -6x^2 + 75x - 23$. How many laptops must be sold in a month to maximize revenue? What is the maximum monthly revenue?

51. A ball is tossed upward with an initial velocity of 48 ft/sec from a height of 118 ft. The height of the ball t sec after it is tossed is given by $f(t) = -16t^2 + 48t + 118$. What is the maximum height the ball will reach and when will it reach this height?

52. A ball is tossed upward with an initial velocity of 67.2 ft/sec from a height of 21 ft. The height of the ball t sec after it is tossed is given by $f(t) = -16t^2 + 67.2t + 21$. What is the maximum height the ball will reach and when will it reach this height?

53. A ball is tossed upward with an initial velocity of 102.4 ft/sec from a height of 15 ft. The height of the ball t sec after it is tossed is given by $f(t) = -16t^2 + 102.4t + 15$. What is the maximum height the ball will reach and when will it reach this height?

54. A ball is tossed upward with an initial velocity of 54.4 ft/sec from a height of 99 ft. The height of the ball t sec after it is tossed is given by $f(t) = -16t^2 + 54.4t + 99$. What is the maximum height the ball will reach and when will it reach this height?

Mix 'Em Up!

Graph each quadratic function by finding the intercepts, the vertex, and additional points, if needed.

55. $f(x) = x^2 + 11x + 24$
56. $f(x) = x^2 - 3x + 2$
57. $f(x) = -x^2 - 3x + 10$
58. $f(x) = -x^2 - 3x - 2$
59. $f(x) = -(x - 4)^2 + 20$
60. $f(x) = 4x^2 - 8x - 1$
61. $f(x) = 2x^2 - 5x$
62. $f(x) = -3x^2 + 12x$
63. $f(x) = 4\left(x + \frac{1}{2}\right)^2 + 3$
64. $f(x) = -3(x - 2)^2 + 12$
65. $f(x) = 4(x + 1.4)^2 - 6.25$
66. $f(x) = -2(x - 2.1)^2 + 11.52$
67. $f(x) = x^2 + 5.6x - 21.32$
68. $f(x) = -x^2 + 6.4x - 7.68$

Solve each problem.

69. The unemployment rate among individuals aged 25 and older who are not high school graduates can be modeled by $f(x) = 0.0051x^2 - 0.0177x + 0.0542$, where x is the number of years after 2004. In what year did the unemployment rate reach its minimum value? What was the minimum unemployment rate? (Source: http://www.trends.collegeboard.org)

70. The unemployment rate among individuals aged 25 and older who have a bachelor's degree or higher can be modeled by $f(x) = 0.0028x^2 - 0.011x + 0.0289$, where x is the number of years after 2004. In what year did the unemployment rate reach its minimum value? What was the minimum unemployment rate? (Source: http://www.trends.collegeboard.org)

71. The mean maximum temperature for Topeka, Kansas can be modeled by $f(x) = -2.1585x^2 + 30.564x - 20.28$, where $x = 1$ corresponds to January, $x = 2$ corresponds to February and so on. In what month does this temperature reach a maximum? What is the maximum temperature? (Source: http://www.esrl.noaa.gov/psd/data/usstations/)

72. The mean maximum temperature for Bismark, North Dakota can be modeled by $f(x) = -2.5503x^2 + 36.542x - 49.625$, where $x = 1$ corresponds to January, $x = 2$ corresponds to February and so on. In what month does this temperature reach a maximum? What is the maximum temperature? (Source: http://www.esrl.noaa.gov/psd/data/usstations/)

 You Be the Teacher!

Correct each student's errors, if any.

73. Convert $y = -x^2 + 6x - 1$ to vertex form.

Elaine's work:
$$y = -x^2 + 6x - 1$$
$$y + 1 = -x^2 + 6x$$
$$-y + 1 = x^2 + 6x$$
$$-y + 1 + 9 = x^2 + 6x + 9$$
$$-y + 10 = (x + 3)^2$$
$$-y = (x + 3)^2 - 10$$
$$y = -(x + 3)^2 + 10$$

74. Use the vertex formula to find the coordinates of the vertex of the function, $f(x) = -3x^2 + 12x + 8$.

Mark's work:
$$x = -\frac{12}{3(2)} = -2$$
$$f(-2) = -3(-2)^2 + 12(-2) + 8$$
$$= 12 - 14 + 8$$
$$= 6$$

75. Find the x-intercepts of $f(x) = -x^2 + 5x + 6$.

Loretta's work:
$$-x^2 + 5x + 6 = 0$$
$$-(x^2 - 5x + 6) = 0$$
$$-(x - 2)(x - 3) = 0$$
$$x = -2, x = 3$$

76. Graph $f(x) = -2(x - 1)^2 + 8$.

Allison's work:
$$x\text{-intercepts: } -2(x - 1)^2 + 8 = 0$$
$$(-2x + 2)^2 = -8$$
$$-2x + 2 = 2i\sqrt{2}$$

no x-intercepts

y-intercept: $f(0) = -2(0 - 1)^2 + 8 = 4 + 8 = 12$

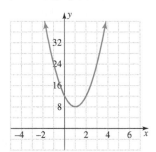

 Calculate It!

Use a graphing calculator to verify the vertex form of each quadratic function.

77. $f(x) = x^2 - 8x + 17 = (x - 4)^2 + 1$

78. $f(x) = -x^2 - 4x = -(x + 2)^2 + 4$

Solving Polynomial and Rational Inequalities in One Variable

▶ OBJECTIVES

As a result of completing this section, you will be able to

1. Solve quadratic inequalities using the boundary number method.
2. Solve higher degree polynomial inequalities using the boundary number method.
3. Solve rational inequalities using the boundary number method.
4. Find solutions of nonlinear inequalities from a graph.
5. Solve applications of nonlinear inequalities.
6. Troubleshoot common errors.

Objective 1 ▶

Solve quadratic inequalities using the boundary number method.

The monthly profit, in thousands of dollars, from selling x thousand digital cameras is given by $f(x) = -4x^2 + 80x + 14$. How many cameras must be sold to have a profit greater than \$350,000? To solve this problem, we can solve the inequality $-4x^2 + 80x + 14 > 350$.

In this section, we use our knowledge of solving polynomial and rational equations to solve polynomial and rational inequalities, or *nonlinear inequalities*.

The Boundary Number Method

In Sections 11.2 and 11.3, we learned how to solve quadratic equations of the form $ax^2 + bx + c = 0$. We now learn how to solve inequalities where one side is a quadratic expression and the other side is zero. These inequalities are called **quadratic inequalities**. Some examples of quadratic inequalities are

$$x^2 + x - 6 < 0 \qquad x^2 - 8x > 9$$

Consider the inequality $x^2 + x - 6 < 0$. Solving this inequality requires us to find the values of x that make the expression $x^2 + x - 6$ negative. It is impossible to substitute every value of x in $x^2 + x - 6$ to determine which ones make the expression negative. So, we use a method, called the *boundary number method*, to solve this problem. **Boundary numbers** are solutions of the equation formed by replacing the inequality with an equals sign. These values form boundaries of regions on the number line.

To use this method, we first find the solutions of the equation $x^2 + x - 6 = 0$.

$$x^2 + x - 6 = 0$$
$$(x + 3)(x - 2) = 0$$
$$x + 3 = 0 \quad \text{or} \quad x - 2 = 0$$
$$x = -3 \qquad\qquad x = 2$$

The solutions tell us that $x^2 + x - 6$ is zero for only $x = -3$ and $x = 2$. So, all other values of x will make the expression $x^2 + x - 6$ either positive or negative. To determine which x-values make the expression positive or negative, we test one point from each of the regions formed by the solutions of the equation. Marking the solutions on a number line enables us to select our test points. It is sufficient to test only one point from each region.

Region	Intervals of x	Test Value	Test	True/False
A	Values less than -3	-4	$(-4)^2 + (-4) - 6 < 0$ $64 - 4 - 6 < 0$ $54 < 0$	False
B	Values between -3 and 2	0	$(0)^2 + (0) - 6 < 0$ $0 + 0 - 6 < 0$ $-6 < 0$	True
C	Values greater than 2	3	$(3)^2 + (3) - 6 < 0$ $9 + 3 - 6 < 0$ $6 < 0$	False
	Boundary number	-3	$(-3)^2 + (-3) - 6 < 0$ $9 - 3 - 6 < 0$ $0 < 0$	False
	Boundary number	2	$(2)^2 + (2) - 6 < 0$ $4 + 2 - 6 < 0$ $0 < 0$	False

The test point in region B satisfies the inequality, so all of the solutions of the inequality lie in region B. The boundary numbers -3 and 2 are not included in the solution set since they do not satisfy the inequality. Therefore, we will use a parenthesis on these numbers. Recall that the inequality symbol $<$ tells us that the boundary numbers are not included in the solution set.

We shade region B and place the appropriate symbols on the boundary numbers. The graph and solution set are

Solution set: $(-3, 2)$

Procedure: Solving a Quadratic Inequality Using the Boundary Number Method

Step 1: Find the boundary numbers by solving the associated equation.
Step 2: Mark the boundary numbers on a number line.
Step 3: Test one point from each region formed by the boundary numbers in the original inequality.
Step 4: Shade the region that contains the test point that satisfies the inequality.
Step 5: Determine the appropriate symbols to use on the boundary numbers.
　a. If the inequality is $<$ or $>$, then use parentheses on the boundary numbers.
　b. If the inequality is $\leq$ or $\geq$, use brackets on the boundary numbers.
Step 6: Write the solution in interval notation.

Objective 1 Examples　Solve each inequality using the boundary number method. Write each solution in interval notation.

1a. $(x - 2)(x + 4) \leq 0$　**1b.** $x^2 - 8x > 9$　**1c.** $(x - 1)^2 \leq 0$　**1d.** $(x + 3)^2 < -25$

Solutions　**1a.**

$$(x - 2)(x + 4) = 0 \qquad \text{Write the associated equation.}$$
$$x - 2 = 0 \quad \text{or} \quad x + 4 = 0 \qquad \text{Set each factor equal to zero.}$$
$$x = 2 \qquad\qquad x = -4 \qquad \text{Solve.}$$

Mark the boundary numbers, -4 and 2, on the number line.

Region	Interval	Test Value	Test	True/False
A	Values less than -4	-5	$(-5 - 2)(-5 + 4) \leq 0$ $(-7)(-1) \leq 0$ $7 \leq 0$	False
B	Values between -4 and 2	0	$(0 - 2)(0 + 4) \leq 0$ $(-2)(4) \leq 0$ $-8 \leq 0$	True
C	Values greater than 2	3	$(3 - 2)(3 + 4) \leq 0$ $(1)(7) \leq 0$ $7 \leq 0$	False

We shade region B and place brackets on the boundary numbers since they satisfy inequality. The graph and solution set are

Solution set: $[-4, 2]$

1b.

$$x^2 - 8x = 9 \qquad \text{Write the associated equation.}$$
$$x^2 - 8x - 9 = 0 \qquad \text{Write the equation in standard form.}$$
$$(x - 9)(x + 1) = 0 \qquad \text{Factor.}$$
$$x - 9 = 0 \quad \text{or} \quad x + 1 = 0 \qquad \text{Set each factor equal to zero.}$$
$$x = 9 \qquad\qquad x = -1 \qquad \text{Solve.}$$

Mark the boundary numbers -1 and 9 on the number line.

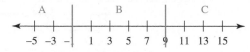

Region	Intervals of x	Test Value	Test	True/False
A	Values less than -1	-2	$(-2)^2 - 8(-2) > 9$ $4 + 16 > 9$ $20 > 9$	True
B	Values between -1 and 9	0	$(0)^2 - 8(0) > 9$ $0 - 0 > 9$ $0 > 9$	False
C	Values greater than 9	10	$(10)^2 - 8(10) > 9$ $100 - 80 > 9$ $20 > 9$	True

We shade regions A and C and place parentheses on the boundary numbers since the inequality symbol is $>$. The graph and solution set are

Solution set: $(-\infty, -1) \cup (9, \infty)$

1c. $(x - 1)^2 = 0 \qquad \text{Write the associated equation.}$
$$x - 1 = \pm 0 \qquad \text{Apply the square root property.}$$
$$x = 1 \qquad \text{Add 1 to each side.}$$

Mark the boundary number 1 on the number line.

Region	Intervals of x	Test Value	Test	True/False
A	Values less than 1	0	$(0 - 1)^2 \leq 0$ $(-1)^2 \leq 0$ $1 \leq 0$	False
B	Values greater than 1	2	$(2 - 1)^2 \leq 0$ $(1)^2 \leq 0$ $1 \leq 0$	False

The two regions do not satisfy the inequality but the boundary number does satisfy it. So, the solution set consists of a single value, 1. The graph and solution set are

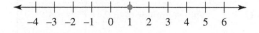

Solution set: $\{1\}$

1d. $(x + 3)^2 = -25 \qquad \text{Write the associated equation.}$
$$x + 3 = \pm\sqrt{-25} \qquad \text{Apply the square root property.}$$
$$x + 3 = \pm 5i \qquad \text{Simplify the radical.}$$
$$x = -3 \pm 5i \qquad \text{Subtract 3 from each side.}$$

Because the solutions are not real numbers, there are no boundary numbers to mark on the number line. So, the entire number line is a region itself.

Region	Intervals of x	Test Value	Test	True/False
A	Any real number	0	$(0 + 3)^2 < -25$ $(3)^2 < -25$ $9 < -25$	False

Since this region doesn't contain values that satisfy the inequality, there are no solutions of the inequality. So, the solution set is the empty set, $\varnothing$.

 Note: *The fact that there is no solution could have also been obtained from the inequality. The expression on the left is always positive. A positive number will never be less than a negative number. So, this is a contradiction and has no solution.*

 **Student Check 1**   Solve each inequality using the boundary number method. Write each solution in interval notation.

     a. $(x - 6)(x - 3) > 0$ **b.** $x^2 + 4x \geq 21$ **c.** $(x - 5)^2 \leq 0$ **d.** $(x + 8)^2 < -16$

Higher Degree Polynomial Inequalities

Objective 2 ▶

Solve higher degree polynomial inequalities using the boundary number method.

We solve higher degree polynomial inequalities in exactly the same way that we solve quadratic inequalities. The only difference is that we solve a higher degree equation instead of a quadratic equation to find the boundary numbers. We mark the numbers on a number line and test points from each of the corresponding regions.

Objective 2 Examples   Solve each inequality using the boundary number method. Write each solution set in interval notation.

     2a. $x(x - 3)(x + 2) > 0$ **2b.** $x^4 - 5x^2 + 4 \leq 0$

Solutions   **2a.**

$$x(x - 3)(x + 2) = 0 \qquad \text{Write the associated equation.}$$

$$x = 0 \quad x - 3 = 0 \quad x + 2 = 0 \qquad \text{Set each factor equal to zero.}$$

$$x = 3 \qquad x = -2 \qquad \text{Solve.}$$

Mark the boundary numbers -2, 0, and 3 on a number line.

Region	Intervals of x	Test Value	Test	True/False
A	Values less than -2	-4	$-4(-4 - 3)(-4 + 2) > 0$ $-4(-7)(-2) > 0$ $-56 > 0$	False
B	Values between -2 and 0	-1	$-1(-1 - 3)(-1 + 2) > 0$ $-1(-4)(1) > 0$ $4 > 0$	True

Region	Intervals of x	Test Value	Test	True/False
C	Values between 0 and 3	1	$1(1-3)(1+2) > 0$ $1(-2)(3) > 0$ $-6 > 0$	False
D	Values greater than 3	4	$4(4-3)(4+2) > 0$ $4(1)(6) > 0$ $24 > 0$	True

Regions B and D contain solutions of the inequality. The boundary numbers are not part of the solution set. The graph and solution set are

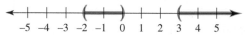

Solution set: $(-2, 0) \cup (3, \infty)$

2b.

$$x^4 - 5x^2 + 4 = 0 \qquad \text{Write the associated equation.}$$
$$(x^2 - 4)(x^2 - 1) = 0 \qquad \text{Factor.}$$
$$(x + 2)(x - 2)(x + 1)(x - 1) = 0 \qquad \text{Apply the difference of squares.}$$
$$x + 2 = 0, x - 2 = 0, x + 1 = 0, x - 1 = 0 \qquad \text{Set each factor equal to zero.}$$
$$x = -2, \quad x = 2, \quad x = -1, \quad x = 1 \qquad \text{Solve.}$$

Mark the boundary numbers $-2, -1, 1,$ and 2 on a number line.

Region	Intervals of x	Test Value	Test	True/False
A	Values less than -2	-3	$x^4 - 5x^2 + 4 \le 0$ $(-3)^4 - 5(-3)^2 + 4 \le 0$ $81 - 5(9) + 4 \le 0$ $40 \le 0$	False
B	Values between -2 and -1	-1.5	$x^4 - 5x^2 + 4 \le 0$ $(-1.5)^4 - 5(-1.5)^2 + 4 \le 0$ $5.0625 - 5(2.25) + 4 \le 0$ $-2.1875 \le 0$	True
C	Values between -1 and 1	0	$x^4 - 5x^2 + 4 \le 0$ $(0)^4 - 5(0)^2 + 4 \le 0$ $0 - 0 + 4 \le 0$ $4 \le 0$	False
D	Values between 1 and 2	1.5	$x^4 - 5x^2 + 4 \le 0$ $(1.5)^4 - 5(1.5)^2 + 4 \le 0$ $5.0625 - 5(2.25) + 4 \le 0$ $-2.1875 \le 0$	True
E	Values greater than 2	3	$x^4 - 5x^2 + 4 \le 0$ $(3)^4 - 5(3)^2 + 4 \le 0$ $81 - 5(9) + 4 \le 0$ $40 \le 0$	False

Regions B and D contain solutions of the inequality. The boundary numbers are part of the solution set since the inequality is $\le$. The graph and solution set are

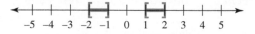

Solution set: $[-2, -1] \cup [1, 2]$

✓ **Student Check 2**   Solve each inequality using the boundary number method. Write each solution set in interval notation.

a. $(x - 4)(x - 7)(x + 3) < 0$ **b.** $x^4 - 10x^2 + 9 \ge 0$

Rational Inequalities

Objective 3 ▶

Solve rational inequalities using the boundary number method.

Rational inequalities are inequalities that involve rational expressions. They are solved in much the same way as the previous types of inequalities. Boundary numbers, however, come not only from the solutions of the associated equation, but also from the values that make the denominators zero; that is, that make the rational expression undefined.

Procedure: Solving a Rational Inequality

Step 1: Find solutions of the associated equation.
Step 2: Find the values that make each rational expression undefined.
Step 3: Mark the numbers from steps 1 and 2 on a number line.
Step 4: Test points from each of the regions formed by the boundary numbers.
Step 5: Shade the regions that tested true.
Step 6: Determine if the boundary numbers are included. Boundary numbers that come from the solutions of the equation are only included if the inequality symbol is $\leq$ or $\geq$. Boundary numbers that make the denominator zero are *never* included in the solution set.
Step 7: Write the solution in interval notation.

Objective 3 Examples Solve each inequality using the boundary number method. Write each solution set in interval notation.

3a. $\dfrac{x-1}{x+2} \geq 0$ **3b.** $\dfrac{7}{x-3} < -1$

Solutions   **3a.** We find the boundary numbers.

Associated Equation	**Values That Make the Expression Undefined**
$\dfrac{x-1}{x+2} = 0$	
$(x+2)\dfrac{x-1}{x+2} = 0(x+2)$	$x+2 = 0$
$x - 1 = 0$	$x = -2$
$x = 1$	

Mark the boundary numbers -2 and 1 on a number line.

Region	Intervals of x	Test Value	Test	True/False
A	Values less than -2	-5	$\dfrac{-5-1}{-5+2} \geq 0$ $\dfrac{-6}{-3} \geq 0$ $2 \geq 0$	True
B	Values between -2 and 1	0	$\dfrac{0-1}{0+2} \geq 0$ $\dfrac{-1}{2} \geq 0$ $-\dfrac{1}{2} \geq 0$	False
C	Values greater than 1	3	$\dfrac{3-1}{3+2} \geq 0$ $\dfrac{2}{5} \geq 0$	True

Regions A and C contain solutions of the inequality. The boundary number 1 is included in the solution set but -2 is *not* included in the solution. The graph and solution set are

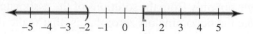

Solution set: $(-\infty, -2) \cup [1, \infty)$

3b. We find the boundary numbers.

Associated Equation	Values That Make the Expression Undefined
$\dfrac{7}{x-3} = -1$	
$(x-3)\dfrac{7}{x-3} = -1(x-3)$	$x - 3 = 0$
$7 = -x + 3$	$x = 3$
$4 = -x$	
$-4 = x$	

Mark the boundary numbers -4 and 3 on a number line.

Region	Intervals of x	Test Value	Test	True/False
A	Values less than -4	-5	$\dfrac{7}{-5-3} < -1$ $\dfrac{7}{-8} < -1$ $-0.875 < -1$	False
B	Values between -4 and 3	0	$\dfrac{7}{0-3} < -1$ $\dfrac{7}{-3} < -1$ $-2.33 < -1$	True
C	Values greater than 3	4	$\dfrac{7}{4-3} < -1$ $\dfrac{7}{1} < -1$ $7 < -1$	False

Region B contains solutions of the inequality. Neither of the boundary numbers is included in the solution. The graph and solution set are

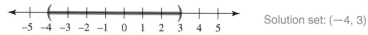

Solution set: $(-4, 3)$

✔ **Student Check 3**   Solve each inequality using the boundary number method. Write each solution set in interval notation.

a. $\dfrac{x-5}{x+1} \le 0$ b. $\dfrac{4}{x-8} \ge -2$

Solving Nonlinear Inequalities Graphically

Objective 4 ▶

Find solutions of nonlinear inequalities from a graph.

We can also find solutions of inequalities from a graph. When we solve an inequality of the form $f(x) < 0$, we find the values of x whose corresponding y-values are negative. Points with negative y-values lie in Quadrants III and IV. So, our solution will consist of the intervals of x whose points lie *below* the x-axis.

When we solve an inequality of the form $f(x) > 0$, we find the values of x whose corresponding y-values are positive. Points with positive y-values lie in Quadrants I and II. So, our solution will consist of the intervals of x whose points lie *above* the x-axis.

To solve an inequality graphically, we must find the solutions of the associated equation. These solutions correspond to the x-intercepts and are our boundary numbers. The solutions divide the graph into regions where the graph of $f(x)$ is either above or below the x-axis. Consider the graph of $f(x) = x^2 - 4$. The boundary numbers and x-intercepts are $x = -2$ and $x = 2$.

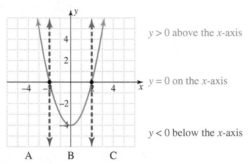

Note that we have three regions.

Region A consists of x values less than -2.

Region B consists of x values between -2 and 2.

Region C consists of x values greater than 2.

In regions A and C, the graph lies *above* the x-axis which means the y-values are *positive*. In region B, the graph lies *below* the x-axis which means the y-values are *negative*.

To solve the inequality $f(x) > 0$ or $x^2 - 4 > 0$ using the graph of $f(x)$, we make a table as shown.

Region	Intervals of x	Graph Above/Below x-axis	Is $f(x) > 0$?	Solution?
A	Values less than -2	Above	Yes	Yes
B	Values between -2 and 2	Below	No	No
C	Values greater than 2	Above	Yes	Yes
Boundary numbers: $-2, 2$		On x-axis	No	No

So, the solution set of $f(x) > 0$ or $x^2 - 4 > 0$ consists of regions A and C. We write the solution set as $(-\infty, -2) \cup (2, \infty)$.

Procedure: Solving a Nonlinear Inequality Graphically

Step 1: Rewrite the inequality so that one side of it is zero, that is, put it in the form $f(x) > 0$, $f(x) \geq 0$, $f(x) < 0$, or $f(x) \leq 0$.

Step 2: Draw the graph of the function $f(x)$. Find the x-intercepts, or boundary numbers.

Step 3: Draw a dashed vertical line through the x-intercepts to determine the regions in the coordinate plane.

Step 4: Complete a chart to determine the solution of the inequality.
  a. If the graph is below the x-axis, $f(x) < 0$.
  b. If the graph is above the x-axis, $f(x) < 0$.

Step 5: Write the solution set in interval notation.
  a. The boundary numbers will be included in the solution set if the inequality is $\leq$ or $\geq$.
  b. The boundary values will not be included in the solution set if the inequality is $<$ or $>$.
  c. If the problem involves a rational inequality, do not include a boundary number in the solution set that makes the expression undefined.

Objective 4 Examples Use a graph to solve each associated inequality. Write each solution set in interval notation.

4a. The graph of $f(x) = \frac{1}{4}x^2 - \frac{1}{4}x - 3$ is given. Use this graph to solve the inequality $\frac{1}{4}x^2 - \frac{1}{4}x - 3 \leq 0$.

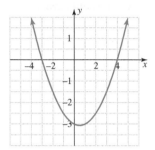

Solution **4a.** The boundary numbers are -3 and 4 since these are the x-intercepts of the graph. We can show the regions by drawing a vertical line through the boundary numbers.

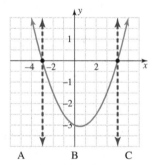

Region	Intervals of x	Graph Above/Below x-axis	Is $f(x) \leq 0$?	Solution?
A	Values less than -3	Above	No	No
B	Values between -3 and 4	Below	Yes	Yes
C	Values greater than 4	Above	No	No
Boundary numbers: $-3, 4$		On the x-axis	Yes	Yes

So, the solution set consists of region B and the boundary numbers. The solution set is $[-3, 4]$.

4b. The graph of $f(x) = x(x - 2)(x + 2)$ is shown. Use this to solve $x(x - 2)(x + 2) > 0$.

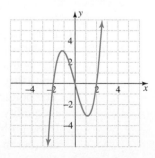

Solution **4b.** The boundary numbers are $-2, 0$, and 2 since these are the x-intercepts of the graph. We can show the regions on the graph by drawing a vertical line through these boundary numbers.

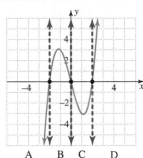

Region	Intervals of x	Graph Above/Below x-axis	Is $f(x) > 0$?	Solution?
A	Values less than -2	Below	No	No
B	Values between -2 and 0	Above	Yes	Yes
C	Values between 0 and 2	Below	No	No
D	Values greater than 2	Above	Yes	Yes
Boundary numbers: $-2, 0, 2$		On the x-axis	No	No

The solution set consists of regions B and D. The solution set is $(-2, 0) \cup (2, \infty)$.

✓ **Student Check 4**   Use the graph to solve each associated inequality. Write each solution set in interval notation.

a. The graph of $f(x) = -x^2 + 1$ is given below. Use this graph to solve $-x^2 + 1 > 0$.

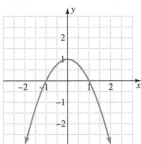

b. The graph of $f(x) = (x - 1)(x + 2)(x - 3)$ is given. Use the graph to solve $(x - 1)(x + 2)(x - 3) \le 0$.

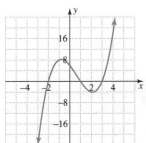

Objective 5 ▶

Solve applications of nonlinear inequalities.

Applications

Nonlinear inequalities are used to represent situations in which a quantity modeled by a nonlinear function is less than or greater than another quantity. Key phrases to denote less than or more than will determine the appropriate symbol to use.

Objective 5 Example   The monthly profit, in thousands of dollars, from selling x thousand digital cameras is given by $P(x) = -4x^2 + 76x + 14$. How many cameras must be sold to have a profit greater than \$350,000?

Solution

$$\text{Profit} > 350 \qquad \text{Profit is greater than \$350,000.}$$
$$-4x^2 + 76x + 14 > 350 \qquad \text{Replace profit with the given function.}$$

Now we solve the quadratic inequality using the boundary number method.

$-4x^2 + 76x + 14 = 350$	Write the associated equation.
$-4x^2 + 76x + 14 - 350 = 350 - 350$	Subtract 350 from each side.
$-4x^2 + 76x - 336 = 0$	Simplify.
$-4(x^2 - 19x + 84) = 0$	Factor out the GCF, -4.
$-4(x - 12)(x - 7) = 0$	Factor the trinomial.
$x - 12 = 0 \quad \text{or} \quad x - 7 = 0$	Set each factor equal to 0.
$x = 12 \qquad\qquad x = 7$	Solve the resulting equations.

Mark the boundary numbers 7 and 12 on the number line.

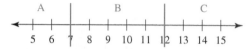

Region	Intervals of x	Test Value	Test	True/False
A	Values less than 7	0	$-4x^2 + 76x + 14 > 350$ $-4(0)^2 + 76(0) + 14 > 350$ $14 > 350$	False
B	Values between 7 and 12	8	$-4x^2 + 76x + 14 > 350$ $-4(8)^2 + 76(8) + 14 > 350$ $366 > 350$	True
C	Values greater than 12	13	$-4x^2 + 76x + 14 > 350$ $-4(13)^2 + 76(13) + 14 > 350$ $326 > 350$	False

The solution set of the inequality is the interval $(7, 12)$, which means that the profit will be greater than \$350,000 if the company sells between 7000 and 12,000 cameras.

✓ Student Check 5   A rocket is launched upward from the ground at an initial velocity of 64 ft/sec. The height of the rocket t sec after it is launched is $h(t) = -16t^2 + 64t$. When will the rocket be greater than 48 ft above the ground?

Objective 6 ▶

Troubleshoot common errors.

Troubleshooting Common Errors

Some common errors associated with nonlinear inequalities are shown.

Objective 6 Examples   **A problem and an incorrect solution are given. Provide the correct solution and an explanation of the error.**

6a. Solve $x^2 - x - 2 > 0$.

Incorrect Solution	Correct Solution and Explanation
$x^2 - x - 2 = 0$ $(x - 2)(x + 1) = 0$ $x = 2, x = -1$ Regions A and C test true. The solution set is $(-\infty, -1] \cup [2, \infty)$.	The boundary numbers are correct and regions A and C do test true. The error is that the symbol on the boundary numbers should be ( ) and not [ ]. The solution set should be $(-\infty, -1) \cup (2, \infty)$.

6b. Solve $(x + 3)^2 > -4$.

Incorrect Solution	Correct Solution and Explanation
$(x + 3)^2 = -4$ $x + 3 = \pm\sqrt{-4}$ $x + 3 = \pm 2i$ $x = -3 \pm 2i$ There are no boundary numbers, so the solution of the inequality is the empty set, $\varnothing$.	While there are no boundary numbers, we cannot automatically conclude that there is no solution until we test a point in the region. The only region to test is the entire number line. We test 0: $$(0 + 3)^2 > -4$$ $$(3)^2 > -4$$ $$9 > -4 \quad \text{True}$$ Since the test point makes the inequality true, the entire region contains solutions. Therefore, the solution set is $(-\infty, \infty)$. Note: We could have noticed this from the inequality. Any number squared is always greater than a negative number.

ANSWERS TO STUDENT CHECKS

Student Check 1 **a.** $(-\infty, 3) \cup (6, \infty)$ **b.** $(-\infty, -7] \cup [3, \infty)$
  c. $\{5\}$ **d.** $\varnothing$
Student Check 2 **a.** $(-\infty, -3) \cup (4, 7)$
  b. $(-\infty, -3] \cup [-1, 1] \cup [3, \infty)$

Student Check 3 **a.** $(-1, 5]$ **b.** $(-\infty, 6] \cup (8, \infty)$
Student Check 4 **a.** $(-1, 1)$ **b.** $(-\infty, -2] \cup [1, 3]$
Student Check 5 between 1 and 3 sec

SUMMARY OF KEY CONCEPTS

1. The boundary number method can be used to solve any inequality. We use it to solve quadratic inequalities, higher degree polynomial inequalities, and rational inequalities.

   a. Find the boundary numbers by finding solutions of the associated equation and any values of x that make the function undefined.

   b. Mark the boundary numbers on the number line.

   c. Test a point from each region formed by the boundary numbers in the *original* inequality.

   d. Shade the regions that contain solutions. Determine the appropriate symbol on the endpoints.

   e. Write the solution set in interval notation.

2. We can also use a graph to solve an inequality. Draw the graph of $f(x)$. Intervals of x whose y-values lie above the x-axis are solutions of the inequality $f(x) > 0$. Intervals of x whose y-values lie below the x-axis are solutions of the inequality $f(x) < 0$.

GRAPHING CALCULATOR SKILLS

A graphing calculator can help us test points in an inequality. The calculator can also graph the solution set of an inequality.

Example 1: Use the calculator to test points to solve $x^2 - 8x > 9$.

Solution: After solving the associated equation, we know the boundary numbers are -1 and 9. Enter the left and right side of the inequality in Y_1 and Y_2, respectively. Now we find the values of x for which $Y_1 > Y_2$.

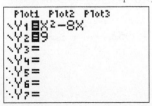

To test points in the regions formed by the boundary numbers of -1 and 9, we can view the table. We choose test points of -2, 2, and 10.

For $x = -2$, we have $20 > 9$ which is true.
For $x = 2$, we have $-12 > 9$ which is false
For $x = 10$, we have $20 > 9$ which is true.

So, regions A and C are included in the solution set.

Example 2: Graph the solution set of $x^2 - 8x > 9$.

Solution: Enter the complete inequality in Y_1. Use
MATH to select the inequality symbol and then graph.

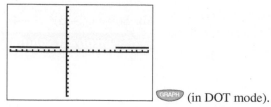

GRAPH (in DOT mode).

The solution of the inequality is shaded slightly above the x-axis. We must determine whether or not the endpoints are included, but the graph shows us that regions A and C contain solutions of the inequality.

SECTION 11.6 / EXERCISE SET

 Write About It!

Use complete sentences in your answer to each exercise.

1. Explain how to find the boundary numbers of a polynomial or rational inequality.

2. Explain how to determine the appropriate symbols, parentheses or brackets, on the boundary numbers when solving polynomial inequalities.

3. When solving the rational inequality, what can you say about the boundary numbers that make the rational expression undefined?

4. If there is no solution of a quadratic inequality, $ax^2 + bx + c \leq 0$, what can you say about the graph of the corresponding quadratic function $f(x) = ax^2 + bx + c$?

5. If there is no solution of a quadratic inequality, $ax^2 + bx + c \geq 0$, what can you say about the graph of the corresponding quadratic function $f(x) = ax^2 + bx + c$?

6. Consider a quadratic inequality of the form $(x + b)^2 \leq c$ and explain how to determine the solution if $c < 0$ and $c = 0$.

7. Consider a quadratic inequality of the form $(x + b)^2 \geq c$ and explain how to determine the solution if $c > 0$ and $c = 0$.

Practice Makes Perfect!

Solve each inequality using the boundary number method. Write each solution in interval notation. (See Objective 1.)

8. $(x + 2)(x - 5) > 0$
9. $(x - 1)(x + 12) > 0$
10. $(x - 3)(x - 8) \geq 0$
11. $(x + 5)(x + 2) \geq 0$
12. $(x + 4)(x - 6) < 0$
13. $(x + 10)(x + 2) < 0$
14. $x^2 + 6x \leq 40$
15. $x^2 - 2x \leq 35$

16. $x^2 - 5x > 24$
17. $x^2 + 5x > 36$
18. $(x + 4)^2 > 0$
19. $(x - 2)^2 > 0$
20. $(x + 5)^2 \geq 0$
21. $(x + 6)^2 \geq 0$
22. $3x^2 + 5x + 2 \leq 0$
23. $2x^2 - 13x + 20 \leq 0$
24. $5x^2 + 7x - 6 > 0$
25. $6x^2 - x - 1 > 0$
26. $(2x - 1)^2 \leq 0$
27. $(3x + 5)^2 \leq 0$
28. $(3x + 4)^2 < 0$
29. $(4x - 1)^2 < 0$
30. $(x + 7)^2 + 9 < 0$
31. $(x - 1)^2 + 4 < 0$

Solve each inequality using the boundary number method. Write each solution in interval notation. (See Objective 2.)

32. $x(x + 3)(x - 6) \geq 0$
33. $x(x - 3)(x + 4) \geq 0$
34. $(x + 8)(x + 1)(x - 5) < 0$
35. $(x - 9)(x + 2)(x - 2) < 0$
36. $(3x - 5)(x + 2)(2x - 9) > 0$
37. $(7x + 2)(x - 1)(7x - 3) > 0$
38. $x^4 - 13x^2 + 36 \geq 0$
39. $9x^4 - 10x^2 + 1 > 0$
40. $x^4 - 26x^2 + 25 \leq 0$
41. $x^4 - 3x^2 - 4 \leq 0$
42. $x^4 + 5x^2 + 6 < 0$
43. $x^4 + 8x^2 + 7 > 0$

Solve each rational inequality using the boundary number method. Write each solution in interval notation. (See Objective 3.)

44. $\dfrac{x + 3}{x - 4} \geq 0$
45. $\dfrac{x - 5}{x + 1} \geq 0$
46. $\dfrac{2x + 5}{x - 3} \leq 0$
47. $\dfrac{3x - 8}{x + 6} \leq 0$
48. $\dfrac{x + 3}{2x - 5} > 0$
49. $\dfrac{x + 4}{x - 9} > 0$
50. $\dfrac{3}{x - 5} \geq 2$
51. $\dfrac{7}{x + 3} \geq 1$
52. $\dfrac{5}{x - 6} \leq -2$
53. $\dfrac{4}{x + 8} \leq -2$

Use the graph of $y = f(x)$ to solve each inequality. Write each solution in interval notation. (*See Objective 4.*)

54. $(x + 6)(x - 2) \leq 0$

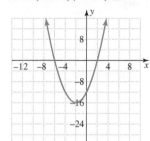

55. $(2x + 5)(x - 3) \leq 0$

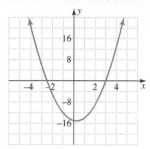

56. $(x + 4)(x - 1) > 0$

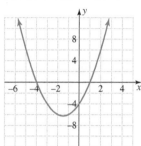

57. $(x - 7)(x - 2) > 0$

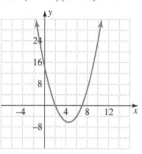

58. $x(x + 5)(x - 4) \geq 0$

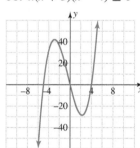

59. $x(x - 6)(x - 3) \geq 0$

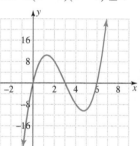

60. $2x^2 + 5x - 3 \leq 0$

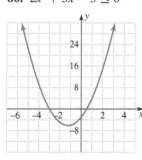

61. $2x^2 + 3x - 20 \leq 0$

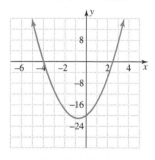

62. $5x^2 - 13x + 6 > 0$

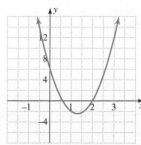

63. $2x^2 + x - 3 > 0$

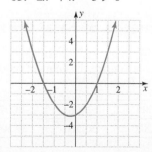

Solve each inequality using its graph. Write each solution in interval notation. (*See Objective 4.*)

64. $x^2 - 2x - 15 \geq 0$

65. $x^2 - 13x + 40 > 0$

66. $x^2 - 12x + 20 < 0$

67. $x^2 + 6x - 40 \leq 0$

68. $3x^2 + 4x - 4 \geq 0$

69. $9x^2 + 12x + 4 \geq 0$

70. $x^2 + 6x + 9 < 0$

71. $2x^2 + x - 15 \leq 0$

Solve each problem. (*See Objective 5.*)

72. The weekly profit, in hundreds of dollars, from selling x hundred shirts is given by $P(x) = -x^2 + 8x - 12$. How many shirts must be sold for the company to make a profit?

73. The profit, in dollars, for selling x units is given by $P(x) = -x^2 + 20x$. How many units must be sold to have a profit greater than $75?

74. A projectile is shot upward from ground level at an initial velocity of 256 ft/sec. Its height in t sec is given by $h(t) = -16t^2 + 256t$. When will the height of the projectile be greater than 768 ft above the ground?

75. A cannon ball is shot into the air from a cannon at an initial velocity of 320 ft/sec from a height of 10 ft. Its height in t sec is given by $h(t) = -16t^2 + 320t + 10$. When will the height of the cannon ball be more than 586 ft above the ground?

 Mix 'Em Up!

Solve each inequality using the boundary number method or graphing. Write each solution in interval notation.

76. $(x - 3)(x + 9) \leq 0$

77. $(x - 5)(x - 2) \leq 0$

78. $(x - 4)(x + 6) > 0$

79. $(x - 8)(x + 3) > 0$

80. $x^2 - 3x \leq 28$

81. $x^2 - 7x > 60$

82. $(x - 11)^2 > 0$

83. $(x - 1)^2 \leq 0$

84. $4x^2 + 11x + 6 > 0$

85. $2x^2 - x - 1 < 0$

86. $\dfrac{1}{x - 3} > -1$

87. $\dfrac{4}{x + 5} > 1$

88. $(x + 7)(x + 2)(x - 1) \geq 0$

89. $(x - 10)(x + 5)(x - 4) \geq 0$

90. $\dfrac{x - 4}{5x - 2} \geq 0$

91. $\dfrac{2x + 5}{x - 3} \geq 0$

92. $x(5x - 2)(2x + 7) < 0$

93. $x(3x - 2)(5x - 6) < 0$

94. $\dfrac{x + 7}{x - 1} \leq 0$

95. $\dfrac{x - 11}{x + 6} \leq 0$

96. $x^4 + 5x^2 - 36 < 0$

97. $x^4 - 26x^2 + 25 > 0$

98. $x^4 + 7x^2 + 10 > 0$

99. $x^4 + 4x^2 + 3 < 0$

100. $\dfrac{2}{x - 5} \leq -1$

101. $\dfrac{3}{x - 6} \leq 1$

Use the graph of $y = f(x)$ to solve each associated inequality. Write each solution in interval notation.

102. $(x + 5)(x - 4) \le 0$

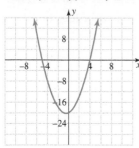

103. $(2x - 5)(x + 1) \le 0$

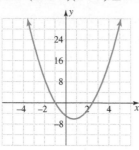

104. $(x + 7)(x - 2) > 0$

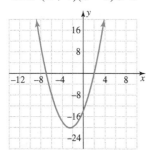

105. $(2x + 7)(x - 3) > 0$

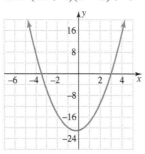

106. $x(2x + 7)(x - 3) \ge 0$

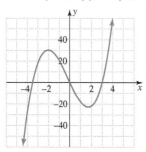

107. $(x - 3)(x - 6)(x + 3) \ge 0$

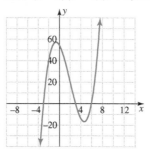

108. $x^4 - 9x^2 < 0$

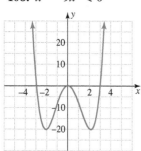

109. $x^4 - 4x^2 > 0$

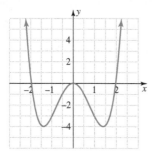

 You Be the Teacher!

Correct each student's errors, if any.

110. Solve the inequality $x(x + 5)(2x - 1) \le 0$.

Elaine's work:

The solutions of the associated equation are -5 and $\frac{1}{2}$.

The inequality is true when $x \ge -5$ and $x \le \frac{1}{2}$. The

solution is $\left[-5, \frac{1}{2}\right]$

111. Solve the inequality $\dfrac{5}{x + 3} \ge 2$.

Mark's work:

$$\frac{5}{x + 3} \ge 2$$
$$5 \ge 2x + 6$$
$$2x + 1 \le 0$$
$$x \le -\frac{1}{2}$$
$$\left(-\infty, -\frac{1}{2}\right]$$

 Calculate It!

Use a calculator to graph the solution set of each inequality. Write each solution in interval notation.

112. $(x + 5)(x - 7) \le 0$

113. $(x + 4)(x - 1) > 0$

114. $\dfrac{2x - 1}{x + 3} \ge 0$

115. $\dfrac{x + 1}{x - 3} \le 0$

GROUP ACTIVITY   **The Mathematics of Fatal Crashes**

The following table shows the number of drivers involved in fatal crashes during the hours of 6:00 A.M. and 5:59 P.M. for the years 2002 to 2009. (Source: http://www-nrd.nhtsa.dot.gov/Pubs/811402EE.pdf)

Year	2002	2003	2004	2005	2006	2007	2008	2009
Number of Drivers	31,135	31,863	31,686	31,820	30,566	29,307	26,377	23,625

1. Write the ordered pairs that correspond to the information given in the table, where x is the years after 2002.

2. Plot the ordered pairs by hand, on a graphing calculator, or in a spreadsheet.

3. Do the data look quadratic? If so, what can you say about the value of a for the quadratic function that models the data? Will the function that models the data have a maximum or minimum value?

4. Determine the quadratic function, $f(x) = ax^2 + bx + c$, that represents the data.

   a. What is the y-intercept of the data? This is the value of c in the function.

   b. Choose two other points from the table. Write two equations by substituting these ordered pairs into the function $f(x) = ax^2 + bx + c$.

   c. You should now have two equations with two unknowns, a and b. Use either substitution or elimination to solve for a and b. Round answers to two decimal places.

   d. Write the quadratic function that models the data, using the values of a, b, and c.

5. Find the vertex of your model and interpret its meaning in the context of the problem.

6. Use your model to approximate when the number of drivers involved in fatal crashes will be 10,000.

7. Now use a graphing calculator or spreadsheet to determine the quadratic function that models these data.

Quadratic Equations and Functions and Nonlinear Inequalities

-☼- **What's the big idea?** Now that we have completed Chapter 11, we should be able to solve a quadratic equation using various methods and see how the solutions of the equation relate to the graph of the corresponding function. We should also be able to use the properties of parabolas to solve problems that relate to maximizing and minimizing functions. We should be able to apply our skills for solving polynomial and rational equations to solving polynomial and rational inequalities.

The Tools

Listed below are the key terms, skills, formulas, and properties you should know for this chapter.

The page reference is provided if you need additional help with the given topic. The Study Tips will assist in your preparation for an exam.

Study Tips

1. Learn all of the terms, formulas, and properties. Make flash cards and have someone quiz you.
2. Rework problems from the exercises and also the ones you worked in class. Work additional problems from the review exercises.
3. Review the summaries of key concepts.
4. Work the chapter test.
5. Be sure to review the online resources for additional study materials.

Terms

Axis of symmetry   785
Boundary number   848
Completing the square   802
Discriminant   816
Higher degree polynomial
    equation   828

Parabola   784
Perfect square trinomial   801
Quadratic equation   799
Quadratic function   784
Quadratic inequalities   848

Radical equation   826
Rational equation   825
Rational inequalities   853
Vertex   785
Zeros of a quadratic function   807

Formulas and Properties

- Discriminant   816
- Quadratic formula   814
- Quadratic function   807

- Square root property   799
- Vertex formula   841

- Vertex form of a quadratic
    function   791

CHAPTER 11 / SUMMARY

How well do you know this chapter? Complete the following questions to find out. Take a look back at the section if you need help.

SECTION 11.1 Quadratic Functions and Their Graphs

1. The graph of the basic quadratic function $f(x) = x^2$ is a(n) _____.

2. The highest or lowest point on a parabola is called the _____.

3. Parabolas are _____ about their vertex. The vertical line through the vertex is called the ___ __ _____.

4. The graph of $f(x) = x^2 + k$ is the graph of $f(x) = x^2$ shifted ___ or ____. If $k > 0$, the graph goes ___ k units. If $k < 0$, the graph goes ____ $|k|$ units. Its vertex is ____.

5. The graph of $f(x) = (x - h)^2$ is the graph of $f(x) = x^2$ shifted ___ or ____. If $h > 0$, the graph goes ____ h units. If $h < 0$, the graph goes ___ $|h|$ units. Its vertex is ____.

6. The graph of $f(x) = ax^2$ is the graph of $f(x) = x^2$ _____, _____, or _____ over the x-axis. If $a > 1$, the parabola opens ___ and is more _____ than the graph of $f(x) = x^2$. If $0 < a < 1$, the parabola opens ___ and is _____ than the graph of $f(x) = x^2$. If $a < 0$, the parabola opens _____.

7. The vertex form of a quadratic function is _____. The vertex of the parabola is _____. The x-intercepts are found by solving the equation _____ and the y-intercept is ____.

8. If the point $(r, 0)$ is on the graph of a quadratic function $f(x) = a(x - h)^2 + k$, then _____ is a solution of the equation $a(x - h)^2 + k = 0$. The solution is also called a(n) ____ of the function.

SECTION 11.2 Solving Quadratic Equations Using the Square Root Property and Completing the Square

9. The square root property states that if $x^2 = k$, where k is a real number, then $x =$ ____ or $x =$ _____.

   a. If $k > 0$, the equation has _____ real solution(s).

   b. If $k = 0$, the equation has _____ real solution(s).

   c. If $k < 0$, the equation has _____ complex, nonreal solution(s).

10. To apply the square root property, the squared expression must be _____ to one side and equal to a(n) _____.

11. A(n) _____ _____ _____ results from squaring a binomial.

12. In a perfect square trinomial with a leading coefficient of 1, the _____ term is half of the _____ of

    x, _____. In symbols, $c =$ _____.

13. When we add the number that makes a binomial a perfect square trinomial, we are _____ ___ _____.

14. To solve an equation by completing the square,

    • The coefficient of the squared term must be ___.

    • If it is not, we _____ both sides of the equation by the coefficient.

    • Move the _____ term to the side opposite the variable terms.

    • Find the number that _____ the _____ and ___ it to both sides of the equation.

    • _____ the perfect square trinomial and simplify the other side.

    • Apply the _____ ___ _____ and solve the equation.

SECTION 11.3 Solving Quadratic Equations Using the Quadratic Formula

15. The _____ _____ is derived by completing the

    square. If $ax^2 + bx + c = 0$, then $x =$ _____.

16. The _____ of a quadratic equation determines the number and types of solutions. Its value is _____.

    a. If this value is a positive perfect square, there is/are ___ _____ solution(s).

b. If this value is positive but not a perfect square, there is/are ____ _____ solution(s).

c. If this value is zero, there is/are ____ _____ solution(s).

d. If this value is negative, there is/are ____ _____, _____ solution(s).

SECTION 11.4 Solving Equations Using Quadratic Methods

17. A(n) _____ equation is an equation that contains a rational expression. Multiplying both sides by the ____ clears fractions. We must exclude any solutions that make the expression _____.

18. A(n) _____ equation is an equation that contains radicals. We must isolate the _____ and _____ both sides. We must check for _____ solutions.

19. We can solve some other equations by _____. We replace the base of the squared term with __ to create a(n) _____ equation. We must back _____ to solve for the original variable.

SECTION 11.5 More on Graphing Quadratic Functions

20. We can convert $f(x) = ax^2 + bx + c$ to the vertex form $f(x) = a(x - h)^2 + k$ by _____ the _____.

21. The vertex of $f(x) = ax^2 + bx + c$ is (____, _____).

22. The key points on the graph of a parabola are the _____, the _____, and the _____.

23. If a parabola opens up, the vertex is the _____ point. If a parabola opens down, the vertex is the _____ point. If (h, k) is the vertex, then the maximum or minimum value of a quadratic function is __ and occurs at _____.

SECTION 11.6 Solving Polynomial and Rational Inequalities in One Variable

24. One method for solving polynomial inequalities is the _____ _____ _____. It requires us to solve the _____ equation which is found by replacing the _____ symbol with a(n) _____ sign.

25. The _____ _____ are marked on a number line and one point from each region is _____ in the original inequality. It is sufficient to test only ___ point from each region.

26. If the inequality is $<$ or $>$, the boundary numbers are ___ _____ in the solution set. We use _____ for them in the interval notation.

27. If the inequality is $\leq$ or $\geq$, the boundary numbers are _____ in the solution set. We use _____ for them in the interval notation.

28. The boundary numbers for a rational inequality come from the solutions of the _____ equation as well as values that make the expression _____.

29. The boundary numbers of $f(x) = 0$ are also the _____ of the graph of $f(x)$. To solve $f(x) > 0$, we find the regions where the graph of $f(x)$ is _____ the x-axis. To solve $f(x) < 0$, we find the regions where the graph of $f(x)$ is _____ the x-axis.

CHAPTER 11 / REVIEW EXERCISES

SECTION 11.1

Graph each function. Identify the x-intercepts, y-intercept, vertex, axis of symmetry, domain, and range. Explain how the graph of the function relates to the graph of $y = x^2$. (*See Objectives 1–4.*)

1. $f(x) = x^2 + 9$

2. $f(x) = \left(x - \dfrac{3}{2}\right)^2$

3. $f(x) = -\dfrac{1}{2}(x - 5)^2 + 2$ **4.** $f(x) = 4(x + 1)^2 - 16$

Graph each function and use its graph to solve $f(x) = 0$. (*See Objective 5.*)

5. $f(x) = -3(x + 8)^2 + 12$ **6.** $f(x) = -\dfrac{5}{4}(x - 1)^2 + 5$

7. $f(x) = 4.5(x - 2)^2 - 18$ **8.** $f(x) = 3(x + 1)^2 + 10$

SECTION 11.2

Solve each equation by applying the square root property. (*See Objective 1.*)

9. $z^2 - 15 = 3$

10. $25x^2 + 10 = 131$

11. $9(y - 5)^2 - 2 = 47$

12. $25(2b + 3)^2 + 49 = 13$

13. $25(y - 3)^2 - 11 = 38$

14. $9(z - 6)^2 + 2 = 22$

15. $(3b - 2)^2 + 73 = 9$

16. $(3w - 1)^2 + 33 = 8$

Create a perfect square trinomial from each binomial by completing the square and then write it in factored form. (*See Objective 2.*)

17. $x^2 + 12x$ **18.** $s^2 + 15s$ **19.** $r^2 - \dfrac{3}{2}r$ **20.** $s^2 + \dfrac{5}{2}s$

Solve each equation by completing the square. (*See Objectives 3 and 4.*)

21. $x^2 + 6x - 9 = 0$

22. $y^2 - 10y + 5 = 0$

23. $9r^2 - 18r - 7 = 0$

24. $4r^2 + 24r + 45 = 0$

25. $2x(x - 1) - 7 = 0$

26. $2x(x + 1) = 1$

Solve each problem. (*See Objective 5.*)

27. A square city block covers an area of 67,600 ft². What is the length of the side of the block?

28. If a ball is dropped from the top of the Q1 Tower in Gold Coast, its height, in feet, t sec after it is dropped is given by $s = -16t^2 + 1058$. How many seconds will it take for the ball to reach the ground? Round the answer to two decimal places. (Source: http://www.skypoint.com.au)

29. At what rate would you have to invest $25,500 in a 2-yr CD to have $26,270.74 once the CD matures? Use the formula $A = P(1 + r)^2$, where P is the amount invested and r is the annual interest rate.

SECTION 11.3

Solve each equation using the quadratic formula. (*See Objective 1.*)

30. $3x^2 - 14x + 16 = 0$ **31.** $7y^2 - 34y - 5 = 0$

32. $3a^2 + 17a + 22 = 0$ **33.** $20b^2 - b - 12 = 0$

34. $5x(5x + 18) + 76 = 0$ **35.** $3x(3x + 4) = 13$

36. $x^2 = 10x - 31$ **37.** $z^2 - 6z = -16$

Use the discriminant to determine the number and types of solutions of each equation. (*See Objective 2.*)

38. $5x^2 - 9x - 2 = 0$ **39.** $6y^2 + 4y + 5 = 0$

40. $11a^2 - 7a + 7 = 0$ **41.** $3b^2 + 4b - 8 = 0$

42. $4r^2 - 20r + 25 = 0$ **43.** $9s^2 - 24s - 16 = 0$

Solve each problem. (*See Objective 3.*)

44. A company knows that the cost, in thousands of dollars, to produce x hundred items is given by the cost function $C(x) = -0.35x^2 + 2x + 4$. It also knows that the revenue, in thousands of dollars, from selling x hundred items is given by the revenue function $R(x) = 0.55x$. How many items does the company need to produce to break even, that is, when does revenue = cost? Round the answer to the nearest integer.

45. A farmer wants to enclose a pen at the back of his barn, using the barn as one of its borders. The farmer has 320 ft of fencing. What dimensions should the pen have to enclose an area of 12,800 ft²?

46. The height, in feet, of a ball thrown upward from a height of 6 ft is $h = -16t^2 + 56t + 6$, where t is in seconds. How many seconds will it take for the ball to reach a height of 30 ft?

47. Josh and his wife, Christy, each have to travel out of town for meetings. The towns where they are meeting are 210 mi apart and are due east and due south from their home. Josh leaves at 9 A.M. and travels at 54 mph, and Christy leaves 1 hr later and travels at 72 mph. When will they reach their destinations and be 210 mi apart?

SECTION 11.4

Solve each equation. (*See Objectives 1 and 2.*)

48. $3x = 22 + \dfrac{80}{x}$ **49.** $8x = -16 - \dfrac{11}{x}$

50. $\dfrac{4x}{x + 1} + \dfrac{5x - 31}{x - 1} = \dfrac{52}{x^2 - 1}$

51. $\dfrac{4x}{x - 2} + \dfrac{12x + 8}{x + 2} = \dfrac{7}{x^2 - 4}$

52. $\sqrt{18 - 2x} = x - 5$ **53.** $\sqrt{10x + 24} = x + 4$

54. $x + 5 = 4\sqrt{x + 1}$ **55.** $x - 3 = \sqrt{5 - 2x}$

56. $x + 5\sqrt{x} + 4 = 0$ **57.** $20x + \sqrt{x} = 12$

Solve each equation using substitution, if needed. (*See Objectives 3 and 4.*)

58. $100x^4 - 21x^2 - 1 = 0$ **59.** $x^4 - 20x^2 + 64 = 0$

60. $16x^4 = 32x^2 + 9$ **61.** $4x^4 = 9x^2 + 100$

62. $25(x - 4)^2 + 20(x - 4) + 3 = 0$

63. $4(2x + 3)^2 + 5(2x + 3) - 6 = 0$

64. $(5x + 2)^2 + 6(5x + 2) - 16 = 0$

65. $\dfrac{3}{(x+2)^2} + \dfrac{10}{x+2} - 8 = 0$

66. $\dfrac{3}{(4x+1)^2} - \dfrac{4}{4x+1} - 4 = 0$

67. $x^{2/3} + x^{1/3} - 20 = 0$ **68.** $3x^{1/2} - 23x^{1/4} + 40 = 0$

69. $2x^{2/5} + 3x^{1/5} - 2 = 0$ **70.** $x^{4/3} - 13x^{2/3} + 36 = 0$

Solve each problem. (*See Objective 5.*)

71. Mel and Jay own a lawn service. Together they can mow lawns in one development in 4 hr. Mel does all the mowing 3 hr faster than Jay. How long will it take each working alone to do the job? Round answers to two decimal places.

72. Monica drove a distance of 250 mi to visit her parents in Hershey, Pennsylvania. During the return trip she was caught in a rainstorm and had to decrease her speed by 12 mph. If her return trip took 1 hr longer than her trip to Hershey, find her original speed. Round the answer to two decimal places.

SECTION 11.5

Graph each quadratic function by finding the intercepts, the vertex, and additional points, if needed. (*See Objectives 1 and 2.*)

73. $f(x) = x^2 + 2x - 24$ **74.** $f(x) = -x^2 + 4x + 5$

Solve each problem. (*See Objective 3.*)

75. The monthly profit, in thousands of dollars, from selling x thousand digital cameras is given by $f(x) = -3x^2 + 60x + 50$. How many digital cameras must be sold in a month to maximize profit? What is the maximum monthly profit?

76. A ball is tossed upward with an initial velocity of 89.6 ft/sec from a height of 60 ft. The height of the ball t sec after it is tossed is given by $f(t) = -16t^2 + 89.6t + 60$. What is the maximum height the ball will reach and when will it reach this height?

SECTION 11.6

Solve each inequality using the boundary number method. Write each solution set in interval notation. (*See Objectives 1–3.*)

77. $(x+9)(x-4) \le 0$ **78.** $(3x+8)(x-3) \ge 0$

79. $x^2 - 4x > 32$ **80.** $3x^2 + 15 < 14x$

81. $x^2 + 12x + 36 > 0$ **82.** $25x^2 + 10x + 1 \le 0$

83. $(3x-5)^2 + 1 < 0$ **84.** $(2x-3)^2 - 16 > 0$

85. $(2x+3)(x-4)(x+5) \le 0$

86. $(2x-5)(x+2)(x-6) \ge 0$

87. $x(7x-3)(5x+12) > 0$

88. $x(x+2)(4x+15) < 0$ **89.** $25x^4 - 29x^2 + 4 > 0$

90. $x^4 - 16 \le 0$ **91.** $\dfrac{6}{2x-5} \ge -3$

92. $\dfrac{6}{x+5} \le 2$ **93.** $\dfrac{2x-7}{x+4} \le 0$

94. $\dfrac{5x+12}{x-2} \ge 0$

Use the graph of $y = f(x)$ to solve each associated inequality. Write each solution set in interval notation. (*See Objective 4.*)

95. $(x+5)(2x-3) \ge 0$

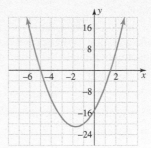

96. $(2x+7)^2 \le 0$

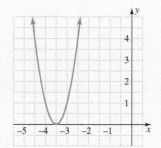

97. $(x-3)(x+6)(x+1) \le 0$

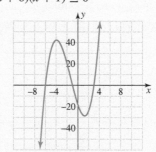

98. $x(2x-7)(x+4) > 0$

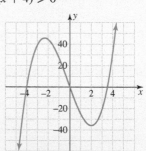

Solve each problem. (*See Objective 5.*)

99. A toy rocket is launched from the ground with an initial velocity of 160 ft/sec. Its height in t sec is given by $h(t) = -16t^2 + 160t$. When will the rocket be more than 384 ft above the ground?

100. The profit, in thousands of dollars, of selling x hundred units of an item is given by $P(x) = -2x^2 + 16x - 14$. How many items must be sold for the company to have a profit?

CHAPTER 11 TEST / QUADRATIC EQUATIONS AND FUNCTIONS AND NONLINEAR INEQUALITIES

1. What are the four methods for solving a quadratic equation? Which method(s) can be used to solve any quadratic equation?

2. What determines the number and types of solutions of a quadratic equation? What expression is used to calculate this number? State the different cases.

3. The vertex of $f(x) = 2(x + 5)^2 - 3$ is
   a. $(5, 3)$
   b. $(5, -3)$
   c. $(-5, 3)$
   d. $(-5, -3)$

4. The solution set of $(x + 5)^2 + 2 = 9$ is
   a. $\{-4\}$
   b. $\{-10, -4\}$
   c. $\{-5 + \sqrt{7}\}$
   d. $\{-5 - \sqrt{7}, -5 + \sqrt{7}\}$

5. The number that makes $x^2 - 20x$ a perfect square trinomial is
   a. 10
   b. 20
   c. 100
   d. 400

6. If the discriminant of a quadratic equation is 33, the equation has
   a. one rational solution
   b. two rational solutions
   c. two irrational solutions
   d. two complex nonreal solutions

7. Using the graph of $y = ax^2 + bx + c$, the solution set of $ax^2 + bx + c = 0$ is

   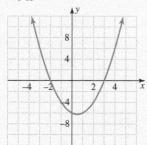

   a. $\{-3, 2\}$
   b. $\{-2, 3\}$
   c. $\{6\}$
   d. $\{-6\}$

8. Using the graph in Exercise 7, the solution set of $ax^2 + bx + c < 0$ is
   a. $(-\infty, -2) \cup (3, \infty)$
   b. $(-\infty, -2] \cup [3, \infty)$
   c. $(-2, 3)$
   d. $[-2, 3]$

9. The vertex of $y = 2x^2 - 8x + 3$ is
   a. $(2, -5)$
   b. $(2, 11)$
   c. $(-2, 27)$
   d. $(-2, 11)$

10. The boundary number(s) used to solve $\dfrac{x + 5}{x - 2} \geq 0$ is (are)
    a. $x = -5$
    b. $x = 2$
    c. $x = -5, x = 2$
    d. no boundary numbers

Solve each equation.

11. $y^2 + 16 = 8y$

12. $2a^2 - 3a = 4$

13. $b^2 + b + 5 = 0$

14. $5x = 19 - \dfrac{12}{x}$

15. $8 - x = \sqrt{1 - 32x}$

16. $\dfrac{5x}{x + 1} - \dfrac{4x - 19}{x - 1} = \dfrac{4}{x^2 - 1}$

17. $x^4 - 8x^2 + 12 = 0$

18. $(x - 1)^{2/3} + 5(x - 1)^{1/3} + 4 = 0$

Graph each quadratic function by finding the intercepts, vertex, and additional points, if necessary. State the domain and range.

19. $f(x) = x^2 - 4x - 12$

20. $f(x) = -(x + 3)^2 + 4$

21. $f(x) = 2x^2 - 6x + 5$

22. If $5000 is invested in a savings account for 2 yr, the amount in the account may be represented by $A = 5000(1 + r)^2$, where r is the annual interest rate. Find the interest rate that is required for an investment of $5000 to grow to $6050 in 2 yr.

23. The Devil's Pool is a swimming area located atop the Victoria Falls along the Zambezi River, which is located on the border of Zambia and Zimbabwe. Swimmers can literally look over the edge of the falls. One swimmer was trying to take a picture over the edge and dropped his camera. The height, in meters, of his camera t sec after he dropped it is represented by $h = -9.8t^2 + 108$. How many seconds (to the nearest tenth) will it take for the camera to reach the bottom of the falls? (Source: http://goafrica.about.com/od/zambia/p/devilspool.htm)

24. The national unemployment rate (not seasonally adjusted) in January for the years 2004–2010 is shown in the table. The unemployment rate is the percent of the U.S. labor force who is unemployed and can be modeled by the equation $y = 0.39x^2 - 1.69x + 6.61$, where x is the years after 2004. Assuming the unemployment rate continues to grow in this manner, in what year will the unemployment rate be 25? (Source: http://www.google.com/publicdata/home)

Year	Unemployment Rate
2004	6.3
2005	5.7
2006	5.1
2007	5
2008	5.4
2009	8.5
2010	10.6

Solve each inequality using the boundary number method or graphing. Write each solution set in interval notation.

25. $x^2 - 3x - 10 \geq 0$

26. $(x + 5)(x - 3)(x - 6) < 0$

27. $\dfrac{2x}{x - 2} \geq 0$

28. $(2x - 5)^2 < 0$

CUMULATIVE REVIEW EXERCISES / CHAPTERS 1–11

Evaluate each expression for the given values. (*Section 1.3, Objective 3*)

1. $b^2 - 4ac$ for $a = -2$, $b = 5$, $c = 3$

2. $b^2 - 4ac$ for $a = 4$, $b = -7$, $c = -1$

Translate each sentence into an algebraic inequality. Use the variable x to represent the unknown number. (*Section 2.1, Objective 3*)

3. Four less twice a number is greater than the number plus six.

4. Two less than a number is less than two more than three times the number.

Solve each linear equation and check the answer. (*Section 2.3, Objective 3*)

5. $15 + 3x = 6(x - 4)$ 6. $26 - 6x = -4(x - 9)$

Translate each problem into a linear equation and solve the problem. (*Section 2.3, Objectives 4 and 5*)

7. Twice the sum of the two consecutive odd integers is 96. Find the two numbers.

8. Six less than three times a number yields 27. Find the number.

Solve each formula for the specified variable. (*Section 2.5, Objective 3*)

9. $p = \dfrac{3}{4}q - rs$ for r 10. $A = P(1 + rt)$ for t

Solve each compound inequality. Write each solution set in interval notation. (*Section 9.1, Objectives 3 and 4*)

11. $3x - 6 \le 0$ and $2 - 5x \le 12$

12. $2x + 7 > 1$ or $4 + x < -1$

Find $f(1)$ for each function. (*Section 3.6, Objective 4*)

13. $f(x) = 4 - x^2$ 14. $f(x) = 8x + 5$

Rewrite each linear equation using function notation, if possible. Identify m and b. (*Section 3.6, Objective 5*)

15. $6x + 7 = 3$ 16. $2x + 3y = 0$

Find the slope and y-intercept of each line from its equation. Write the y-intercept as an ordered pair. (*Section 3.3, Objective 2*)

17. $f(x) = 4x - 1$ 18. $5x + 2y = 3$

Write the equation of each line that passes through the given point and is either parallel or perpendicular to the given line. Write each answer in slope-intercept form. (*Section 3.4, Objective 4*)

19. $(2, -5)$, parallel to $2x + y = 15$

20. $(-8, 3)$, perpendicular to $x - 3y = 1$

Solve each system of linear equations using elimination. (*Section 4.3, Objectives 1 and 2*)

21. $\begin{cases} 12x - y = 43 \\ -5x + 2y = -29 \end{cases}$ 22. $\begin{cases} 6x + 7y = 50 \\ 8x - 3y = 42 \end{cases}$

Solve each problem. (*Section 4.4, Objective 4*)

23. A plane can travel 1560 mi with the wind in 4 hr. The return trip against the wind takes 5 hr. What are the speed of the plane in still air and the speed of the wind?

24. A plane can travel 3240 mi with the wind in 6 hr. The return trip against the wind takes 7.5 hr. What are the speed of the plane in still air and the speed of the wind?

Simplify each expression. Write each answer with positive exponents. (*Section 5.1, Objectives 2–4; Section 5.2, Objectives 1 and 2*)

25. $(-5x^2y^3)(13x^5y^2)$ 26. $(7a^4b)(-4a^2b^3)$

27. $\dfrac{28p^{12}q^5}{21p^3q}$ 28. $\dfrac{36r^{16}s^5}{60r^8s^2}$

29. $(a^4b^{-5})^3 \cdot a^{-10}b^8$ 30. $(x^{-6}y^3)^3 \cdot x^{20}y^{-20}$

31. $\left(\dfrac{2x^{-2}}{y^{-3}}\right)^3 \left(\dfrac{2y^{-2}}{x^{-1}}\right)^{-5}$ 32. $\left(\dfrac{5a^{-1}}{b^{-4}}\right)^2 \left(\dfrac{5b^{-3}}{a^{-2}}\right)^{-3}$

Simplify each expression. (*Section 5.5, Objectives 1 and 3*)

33. $(2x^2 - 5)(3x^2 - 10)$ 34. $(7y^2 - 4)(2y^2 + 3)$

35. $(a + 5)(a - 5)(a + 2)$ 36. $(b - 3)(b + 3)(b + 1)$

Factor each polynomial completely. (*Section 6.1, Objective 4*)

37. $4x^4 - 12x^3 + 8x^2 - 24x$ 38. $6y^4 + 6y^3 + 15y^2 + 15y$

Factor completely. (*Section 6.2, Objective 1; Section 6.3, Objectives 1 and 2*)

39. $4x^2 + 2x - 12$

40. $a^2 + 13a + 12$

Factor completely. (*Section 6.4, Objectives 1–3*)

41. $8x^3 + 125$ 42. $a^4 - 16b^4$

Solve each equation by factoring. (*Section 6.5, Objective 2*)

43. $(x + 1)^2 + (x + 2)^2 = (x + 3)^2$

44. $(y + 2)^2 + (y + 4)^2 = (y + 6)^2$

Multiply or divide the rational expressions. Write each answer in simplest form. (*Section 7.2, Objectives 1 and 2*)

45. $\dfrac{x^2 - 4}{x^2 - x - 6} \cdot \dfrac{2x^2 - x - 15}{4x^2 + 10x}$

46. $\dfrac{x^2 + 8x + 16}{x^3 + 64} \div \dfrac{5x^2 + 20x}{5x}$

Use the remainder theorem and synthetic division to evaluate $P(x)$ at the specified x-value. (*Section 5.7, Objective 4*)

47. $P(x) = 3x^2 - 7x + 4$, $P(3)$

48. $P(x) = x^3 - 3x + 9$, $P(1)$

Add or subtract the rational expressions. Write each answer in lowest terms. (*Section 7.4, Objectives 1 and 2*)

49. $\dfrac{4}{x^2 - 4x + 4} + \dfrac{3}{x^2 + 3x - 10}$

50. $\dfrac{x}{3x^2 - 5x - 2} - \dfrac{2x}{9x^2 + 6x + 1}$

Solve each problem. (*Section 7.6, Objective 2*)

51. A number minus eight times its reciprocal is 2. Find the number.

52. A number plus four times its reciprocal is 4. Find the number.

Solve each problem. (*Section 8.4, Objective 3*)

53. The cost of gas for a trip is directly proportional to the number of miles traveled and inversely proportional to the gas mileage of the car. If the cost of gas for a 150-mi trip in a car with gas mileage of 20 mpg is $26.25, what is the cost of gas for a 240-mi trip in a car with gas mileage of 16 mpg?

54. The cost of gas for a trip is directly proportional to the number of miles traveled and inversely proportional to the gas mileage of the car. If the cost of gas for a 390-mi trip in a car with gas mileage of 19.5 mpg is $83.00, what is the cost of gas for a 240-mi trip in a car with gas mileage of 48 mpg?

Simplify each expression. Assume that all variables represent positive real numbers. (*Section 10.1, Objectives 1 and 2*)

55. $\sqrt{\dfrac{75x^8}{3}}$ **56.** $\sqrt{\dfrac{98y^{10}}{2}}$ **57.** $\sqrt[3]{\dfrac{64a^9}{b^6}}$ **58.** $\sqrt[3]{\dfrac{c^3}{27d^6}}$

Rewrite each expression as a radical expression and simplify, if possible. Assume all variables represent positive real numbers. (*Section 10.2, Objective 3*)

59. $(a^{-15}b^6)^{-2/3}$ **60.** $(x^8y^{-12})^{-3/4}$

Simplify each expression. Assume all variables represent positive real numbers. Write each answer with positive exponents. (*Section 10.2, Objective 4*)

61. $(x^{1/3} - y^{1/4})(x^{1/3} + y^{1/4})$ **62.** $(r^{1/2} - s^{1/3})^2$

Simplify each radical. Assume all variables represent positive real numbers. (*Section 10.3, Objectives 1 and 2*)

63. $\sqrt{80x^3y^6}$ **64.** $\sqrt{18r^4t}$ **65.** $\sqrt[3]{\dfrac{a^5}{b^3}}$ **66.** $\sqrt[3]{\dfrac{c^4}{b^6}}$

Perform the indicated operation and write each answer in simplest radical form. (*Section 10.4, Objectives 1 and 2*)

67. $(3 + \sqrt{5})^2$ **68.** $(\sqrt{10} - 4)^2$

Divide the radical expressions and write each answer in simplest radical form. (*Section 10.5, Objective 1*)

69. $\dfrac{\sqrt{80c^7d^{14}}}{\sqrt{2c^3d^3}}$ **70.** $\dfrac{\sqrt{36x^8y^3}}{\sqrt{3x^5y^{-2}}}$

Solve each radical equation. (*Section 10.6, Objective 1*)

71. $\sqrt{3x + 4} - x = 2$ **72.** $\sqrt{2t - 1} + t = 8$

Perform the operation on the complex numbers and write each answer in standard form. (*Section 10.7, Objectives 2–4*)

73. $\dfrac{5i}{3 - i}$ **74.** $\dfrac{10i}{1 + 7i}$

Create a table of values to graph each function. Identify the *x*-intercepts, *y*-intercept, vertex, and the axis of symmetry. Explain how the graph of the function relates to the graph of $y = x^2$. State the domain and range of the function. (*Section 11.1, Objective 4*)

75. $f(x) = (x - 1)^2 + 3$ **76.** $f(x) = (x + 2)^2 - 1$

77. $f(x) = -(x - 2)^2 + 1$ **78.** $f(x) = -(x + 3)^2 + 1$

Graph each function and use the graph to solve the equation $f(x) = 0$. (*Section 11.1, Objective 5*)

79. $f(x) = \dfrac{1}{2}(x + 2)^2 - 8$ **80.** $f(x) = -(x - 5)^2 + 1$

Solve each equation by completing the square. (*Section 11.2, Objectives 3 and 4*)

81. $x^2 + 8x + 9 = 0$ **82.** $y^2 + 12y + 26 = 0$

83. $4r^2 - 8r + 29 = 0$ **84.** $4x^2 + 12x + 13 = 0$

Find the zeros of each function. State the *x*-intercepts of the graph of the function. (*Section 11.2, Objective 6*)

85. $f(x) = 3(x - 1)^2 - 12$ **86.** $f(x) = \dfrac{1}{2}(x + 4)^2 - 18$

Solve each equation using the quadratic formula. (*Section 11.3, Objective 1*)

87. $b^2 - 2b + 82 = 0$ **88.** $a^2 + 4a + 53 = 0$

89. $x^2 + 6x - 8 = 0$ **90.** $y^2 - 4y - 10 = 0$

Use the discriminant to determine the number and types of solutions of each equation. (*Section 11.3, Objective 2*)

91. $4x^2 - 5x + 6 = 0$ **92.** $3y^2 - 7y - 5 = 0$

93. $x^2 - 6x + 9 = 0$ **94.** $4y^2 + 7y + 3 = 0$

Solve each problem. (*Section 11.3, Objective 3*)

95. John wants to enclose a pen at the back of his barn, using the barn as one of its borders. John has 440 ft of fencing. What dimensions should the pen have to enclose an area of 24,000 ft²?

96. Ivan wants to enclose a pen at the back of his barn, using the barn as one of its borders. Ivan has 380 ft of fencing. What dimensions should the pen have to enclose an area of 12,000 ft²?

Solve each equation. (*Section 11.4, Objectives 1–4*)

97. $\dfrac{x}{2x + 1} + \dfrac{2x}{2x - 1} = \dfrac{5}{4x^2 - 1}$

98. $\dfrac{5x}{x - 6} - \dfrac{2x}{x + 5} = \dfrac{26}{x^2 - x - 30}$

99. $\sqrt{5x + 6} = x$ **100.** $\sqrt{5x + 14} = x$

101. $x^{2/3} - 4x^{1/3} - 5 = 0$ **102.** $x^{1/2} - 6x^{1/4} + 9 = 0$

Convert $y = ax^2 + bx + c$ to the vertex form, $y = a(x - h)^2 + k$. Identify the vertex and intercepts, and graph the function. (*Section 11.5, Objective 1*)

103. $f(x) = 2x^2 - 12x + 19$

104. $f(x) = x^2 + x - 2$

Solve each inequality using the boundary number method. Write each solution set in interval notation. (*Section 11.6, Objective 1*)

105. $3x^2 - 7x \geq 20$ **106.** $2x^2 + x \leq 28$

Use the given graph of $y = f(x)$ to solve each associated inequality. Write each solution set in interval notation. (*Section 11.6, Objective 4*)

107. $x(2x + 5)(x - 3) > 0$

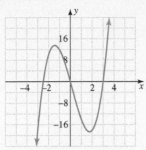

108. $x^4 - 10x^2 + 9 < 0$

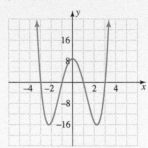

Exponential and Logarithmic Functions

Greatness

When we think of greatness, perhaps we think of star athletes such as Peyton Manning or Cal Ripken. *Fortune Magazine* published an article entitled "Secrets of Greatness." The findings of this article are very pertinent to college students. In response to the question, what does it take to be great? the article stated

> Well, folks, it's not so simple. For one thing, you do not possess a natural gift for a certain job, because targeted natural gifts don't exist . . . You are not a born CEO or investor or chess grandmaster. You will achieve greatness only through an enormous amount of hard work over many years. . . . The good news is that your lack of a natural gift is irrelevant—talent has little or nothing to do with greatness. You can make yourself into any number of things, and you can even make yourself great. The best people in any field are those who devote the most hours to what the researchers call "deliberate practice." It is activity that's explicitly intended to improve performance, that reaches for objectives just beyond one's level of competence, provides feedback on results and involves high levels of repetition.

The researchers questioned why some people are more motivated than others to devote their lives to this deliberate practice to become great. Do you have the motivation to become great? You do not have to have some special gift or talent. You are the only thing you need to make yourself great! (Source: http://money.cnn.com/magazines/fortune/fortune_archive/2006/10/30/8391794/index.htm)

Question For Thought: What do you desire to be great in? What will it take for you to achieve this? What can you do today that will enable you to accomplish greatness?

Chapter Outline

Coming Up. . .

In Section 12.6, we will learn how to use logarithms to determine how proficient we are at a task after we no longer practice it by using the model $P = 90 - 12.5 \ln d$, where P is the proficiency rate and d is the number of days after practicing.

"I long to accomplish a great and noble task, but it is my chief duty to accomplish small tasks as if they were great and noble."

—Helen Keller (Author, Teacher)

Operations and Composition of Functions

We have covered a few concepts related to functions throughout this text. In this chapter, we will study how to combine functions in different ways and how to find the inverse of a function, if it exists. Lastly, we will cover two other families of functions, exponential and logarithmic.

▶ **OBJECTIVES**

As a result of completing this section, you will be able to

1. Add, subtract, multiply, and divide functions.
2. Evaluate combined functions.
3. Find the composition of two functions.
4. Solve applications.
5. Troubleshoot common errors.

Josh makes $25 an hour, so his income can be represented by the function $f(x) = 25x$, where x is the number of hours he works. He pays 15% of his income in taxes. The amount of tax he owes can be represented by the function $g(x) = 0.15x$, where x is his income. Write a function that represents the total taxes Josh owes as a function of the number of hours he works.

To complete this exercise, we need to know how to combine two functions.

Function Operations

Recall a function represents a real number. So, we can perform operations on them just as we do with real numbers. We can add, subtract, multiply, and divide functions. In fact, we have already seen some examples of this in earlier sections. When we perform these types of operations on functions, we get another function as its result. For example, suppose $f(x) = 2x + 1$ and $g(x) = 3x$, we can add these functions together to obtain

$$f(x) + g(x) = 2x + 1 + 3x = 5x + 1$$

The *domain* of this combined function is the intersection of the domains of the individual functions. Adding, subtracting, multiplying, and dividing functions to generate a new function is called the *algebra of functions*.

Objective 1 ▶

Add, subtract, multiply, and divide functions.

Property: The Algebra of Functions

If we let $f(x)$ and $g(x)$ be two functions, we have

Sum	$(f + g)(x) = f(x) + g(x)$
Difference	$(f - g)(x) = f(x) - g(x)$
Product	$(f \cdot g)(x) = f(x) \cdot g(x)$
Quotient	$\left(\dfrac{f}{g}\right)(x) = \dfrac{f(x)}{g(x)}, g(x) \neq 0$

Property: The Domain of the Combined Functions

Let the domain of $f(x)$ be set A and the domain of $g(x)$ be set B. Then the domain of $(f + g)(x)$, $(f - g)(x)$, $(f \cdot g)(x)$ is $A \cap B$. The domain of $\left(\dfrac{f}{g}\right)(x)$ is $A \cap B$, excluding the values that make the denominator zero.

Objective 1 Examples | Find the sum, difference, product, and quotient of the functions. State the domain of each combined function.

1a. $f(x) = 3x + 2, g(x) = x - 5$ **1b.** $f(x) = \sqrt{x}, g(x) = x^2 + x$

Solutions

1a. Both $f(x)$ and $g(x)$ are linear functions since they are of the form $y = mx + b$. The domain of each function is all real numbers or $(-\infty, \infty)$, and the intersection of their domains is $(-\infty, \infty)$.

$(f + g)(x) = f(x) + g(x) = (3x + 2) + (x - 5)$ Domain of $f + g$ is $(-\infty, \infty)$.

$= 3x + 2 + x - 5$

$= 4x - 3$

$(f - g)(x) = f(x) - g(x) = (3x + 2) - (x - 5)$ Domain of $f - g$ is $(-\infty, \infty)$.

$= 3x + 2 - x + 5$

$= 2x + 7$

$(f \cdot g)(x) = f(x) \cdot g(x) = (3x + 2)(x - 5)$ Domain of $f \cdot g$ is $(-\infty, \infty)$.

$= 3x^2 - 15x + 2x - 10$

$= 3x^2 - 13x - 10$

$\left(\dfrac{f}{g}\right)(x) = \dfrac{f(x)}{g(x)} = \dfrac{3x + 2}{x - 5}$ $g(x) \neq 0$

                                $x - 5 \neq 0$

                                $x \neq 5$

$= \dfrac{3x + 2}{x - 5}$ Domain of $\dfrac{f}{g}$ is $(-\infty, 5) \cup (5, \infty)$.

1b. The function $f(x)$ is a square root function and is defined for $x \geq 0$ or $[0, \infty)$. The function $g(x)$ is a quadratic function and has domain $(-\infty, \infty)$. The domain of f intersected with the domain of g is $[0, \infty)$.

$(f + g)(x) = f(x) + g(x) = \left(\sqrt{x}\right) + (x^2 + x)$ Domain of $f + g$ is $[0, \infty)$.

$= \sqrt{x} + x^2 + x$

$(f - g)(x) = f(x) - g(x) = \left(\sqrt{x}\right) - (x^2 + x)$ Domain of $f - g$ is $[0, \infty)$.

$= \sqrt{x} - x^2 - x$

$(f \cdot g)(x) = f(x) \cdot g(x) = \left(\sqrt{x}\right)(x^2 + x)$ Domain of $f \cdot g$ is $[0, \infty)$.

$= x^2\sqrt{x} + x\sqrt{x}$

$\left(\dfrac{f}{g}\right)(x) = \dfrac{f(x)}{g(x)} = \dfrac{\sqrt{x}}{x^2 + x}$ $g(x) \neq 0$

                                $x^2 + x \neq 0$

                                $x(x + 1) \neq 0$

                                $x \neq 0$ or $x \neq -1$

$= \dfrac{\sqrt{x}}{x^2 + x}$ Domain of $\dfrac{f}{g}$ is $(0, \infty)$.

✓ Student Check 1   Find the sum, difference, product, and quotient of the functions. State the domain of each combined function.

a. $f(x) = x^2 - x + 3$, $g(x) = 2x + 1$ **b.** $f(x) = \sqrt{2x}$, $g(x) = x^2 - 4$

Evaluating Combined Functions

Objective 2 ▶

Evaluate combined functions.

Recall we have evaluated functions in several sections. When we *evaluate a function*, we find the y-value that corresponds to the given x-value. We can also evaluate the sum, difference, product, or quotient of functions. When we evaluate a combined function, we add, subtract, multiply, or divide the y-values of the two functions. For example, let $f(x) = 3x + 2$ and $g(x) = x - 5$, to find $(f - g)(6)$, we can either find $f(6) - g(6)$ or we can find $(f - g)(x)$ and evaluate it when $x = 6$.

Method 1	**Method 2**

$$f(6) = 3(6) + 2 = 18 + 2 = 20$$
$$g(6) = 6 - 5 = 1$$
$$f(6) - g(6) = 20 - 1 = 19$$

$$f(x) - g(x) = 3x + 2 - (x - 5)$$
$$= 3x + 2 - x + 5$$
$$= 2x + 7$$

So,

$$(f - g)(6) = 2(6) + 7$$
$$= 12 + 7$$
$$= 19$$

Procedure: Evaluating Combined Functions

Method 1 Evaluate each function at the given x-value. Then find the sum, difference, product, or quotient of the results, as required.

Method 2 Find the combined function first and then evaluate this function at the given x-value.

  Note: *If a function is given graphically or numerically, then we need to apply method 1.*

Objective 2 Examples   **Evaluate each function.**

2a. Let $f(x) = x^2 - x + 3$ and $g(x) = 2x + 1$. Find $(f + g)(4)$.

Solution   2a.

Method 1	**Method 2**

$$(f + g)(4) = f(4) + g(4)$$
$$f(4) = (4)^2 - (4) + 3 = 16 - 4 + 3 = 15$$
$$g(4) = 2(4) + 1 = 8 + 1 = 9$$
$$(f + g)(4) = f(4) + g(4)$$
$$= 15 + 9$$
$$= 24$$

$$(f + g)(x) = f(x) + g(x)$$
$$= x^2 - x + 3 + (2x + 1)$$
$$= x^2 + x + 4$$
$$(f + g)(4) = (4)^2 + (4) + 4$$
$$= 16 + 4 + 4$$
$$= 24$$

2b. Use the graphs to find $(f + g)(3)$, $(f - g)(0)$, $(f \cdot g)(-2)$, and $\left(\dfrac{f}{g}\right)(1)$.

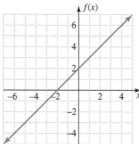   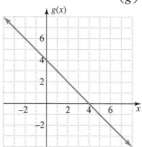

Solution   2b. We first identify the points on each graph that correspond to the x-values of 3, 0, -2, and 1.

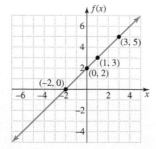   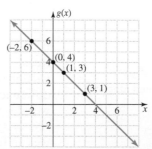

$(3, 5)$ is on the graph of $f(x)$, so $f(3) = 5$. → $(f + g)(3) = f(3) + g(3)$

$(3, 1)$ is on the graph of $g(x)$, so $g(3) = 1$. $= 5 + 1$

                                                                                 $= 6$

$(0, 2)$ is on the graph of $f(x)$, so $f(0) = 2$. → $(f - g)(0) = f(0) - g(0)$

$(0, 4)$ is on the graph of $g(x)$, so $g(0) = 4$. $= 2 - 4$

                                                                                 $= -2$

$(-2, 0)$ is on the graph of $f(x)$, so $f(-2) = 0$. → $(f \cdot g)(-2) = f(-2) \cdot g(-2)$

$(-2, 6)$ is on the graph of $g(x)$, so $g(-2) = 6$. $= (0)(6)$

                                                                                 $= 0$

$(1, 3)$ is on the graph of $f(x)$, so $f(1) = 3$. → $\left(\dfrac{f}{g}\right)(1) = \dfrac{f(1)}{g(1)}$

$(1, 3)$ is on the graph of $g(x)$, so $g(1) = 3$.

                                                                                 $= \dfrac{3}{3}$

                                                                                 $= 1$

2c. Use the table to find $(f + g)(-3)$, $(f - g)(-2)$, $(f \cdot g)(0)$, and $\left(\dfrac{f}{g}\right)(3)$.

x	$f(x)$	$g(x)$
-3	8.5	8
-2	7	3
-1	5.5	0
0	4	-1
1	2.5	0
2	1	3
3	-0.5	8

Solution   **2c.** $(f + g)(-3) = f(-3) + g(-3) = 8.5 + 8 = 16.5$

$(f - g)(-2) = f(-2) - g(-2) = 7 - 3 = 4$

$(f \cdot g)(0) = f(0) \cdot g(0) = (4)(-1) = -4$

$\left(\dfrac{f}{g}\right)(3) = \dfrac{f(3)}{g(3)} = \dfrac{-0.5}{8} = -\dfrac{1}{2} \cdot \dfrac{1}{8} = -\dfrac{1}{16}$

✓ Student Check 2   Evaluate each function.

a. Let $f(x) = 3x^2 - 2x + 1$ and $g(x) = 5x + 4$. Find $(f + g)(2)$.

b. Use the graphs to find $(f + g)(0)$, $(f - g)(-1)$, $(f \cdot g)(3)$, and $\left(\dfrac{f}{g}\right)(4)$.

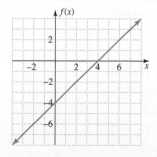
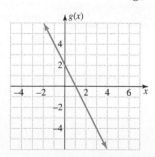

c. Use the table to find $(f + g)(1)$, $(f - g)(2)$, $(f \cdot g)(-3)$, and $\left(\dfrac{f}{g}\right)(0)$.

x	$f(x)$	$g(x)$
-4	7	-13
-3	6	-11
-2	5	-9
-1	4	-7
0	3	-5
1	2	-3
2	1	-1

Function Composition

Objective 3 ▶

Find the composition
of two functions.

Functions can also be combined in another way. The *composition* of functions takes one function's output and makes it the input of another function. So, we are inputting a function into a function. In more mathematical terms, we are evaluating a function at a function.

Consider the two functions, $f(x) = x^2$ and $g(x) = x + 1$. If we begin with $x = 3$ and find $f(3)$, we get $f(3) = 3^2 = 9$. Now if we evaluate $g(x)$ at $x = 9$, we get $g(9) = 9 + 1 = 10$. This can be represented by the following diagram.

$$x = 3 \quad \Large\rangle\, f(x) = x^2 \,\rangle \quad f(3) = 9 \quad \Large\rangle\, g(x) = x + 1 \,\rangle \quad g(9) = 10$$

So, $g(9) = 10$, but since $f(3) = 9$, it follows that $g(9) = g\big(f(3)\big)$. The notation $g\big(f(3)\big)$ means "g composed with f." It is more commonly denoted as $(g \circ f)(3)$ which is read as "g of f of 3." We generalize this as follows.

> **Definition: The Composition of Functions**
>
> If f and g are functions, then the function f composed with the function g can be represented as $(f \circ g)(x)$ and
> $$(f \circ g)(x) = f\big(g(x)\big)$$
> The function g composed with the function f can be represented as $(g \circ f)(x)$ and
> $$(g \circ f)(x) = g\big(f(x)\big)$$

The domain of a composed function is the domain of the final function, excluding any values that make the inner function undefined.

> **Procedure: Finding f Composed with g**
>
> **Step 1:** Substitute the function g for the variable in the function f.
> **Step 2:** Simplify.

Objective 3 Examples   Find each composition.

3a. Let $f(x) = 2x + 5$ and $g(x) = 3x - 1$. Find $(f \circ g)(x)$ and $(g \circ f)(x)$ and state their domains.

Solution   **3a.** $(f \circ g)(x) = f\big(g(x)\big)$ Apply the definition of composition.

$\qquad\qquad = f(3x - 1)$ Replace $g(x)$ with $3x - 1$.

$\qquad\qquad = 2(3x - 1) + 5$ Substitute $3x - 1$ for x in the function f.

$\qquad\qquad = 6x - 2 + 5$ Apply the distributive property.

$\qquad\qquad = 6x + 3$ Simplify.

$$(g \circ f)(x) = g\bigl(f(x)\bigr) \qquad \text{Apply the definition of composition.}$$
$$= g(2x + 5) \qquad \text{Replace } f(x) \text{ with } 2x + 5.$$
$$= 3(2x + 5) - 1 \qquad \text{Substitute } 2x + 5 \text{ for } x \text{ in the function } g.$$
$$= 6x + 15 - 1 \qquad \text{Apply the distributive property.}$$
$$= 6x + 14 \qquad \text{Simplify.}$$

The functions $f \circ g$ and $g \circ f$ are both linear functions. The functions f and g are linear as well. Therefore, the domain of $f \circ g$ and $g \circ f$ is $(-\infty, \infty)$.

3b. Let $f(x) = x^2 - 2x + 3$ and $g(x) = \sqrt{x + 1}$. Find $(f \circ g)(3)$.

Solution **3b.** $(f \circ g)(3) = f\bigl(g(3)\bigr)$ Apply the definition of composition.

$$= f(2) \qquad \text{Evaluate } g(3) = \sqrt{3 + 1} = \sqrt{4} = 2.$$
$$= (2)^2 - 2(2) + 3 \qquad \text{Evaluate } f(2).$$
$$= 4 - 4 + 3 \qquad \text{Simplify.}$$
$$= 3 \qquad \text{Simplify.}$$

We can also find $(f \circ g)(x)$ and then evaluate this function at $x = 3$.

$$(f \circ g)(x) = f\bigl(g(x)\bigr) \qquad \text{Apply the definition of composition.}$$
$$= f\bigl(\sqrt{x + 1}\bigr) \qquad \text{Replace } g(x) \text{ with } \sqrt{x + 1}.$$
$$= \bigl(\sqrt{x + 1}\bigr)^2 - 2\bigl(\sqrt{x + 1}\bigr) + 3 \qquad \text{Substitute } \sqrt{x + 1} \text{ for } x \text{ in } f(x).$$
$$= x + 1 - 2\sqrt{x + 1} + 3 \qquad \text{Apply } \bigl(\sqrt[n]{x}\bigr)^n = x.$$
$$(f \circ g)(3) = 3 + 1 - 2\sqrt{3 + 1} + 3 \qquad \text{Replace } x \text{ with } 3.$$
$$= 7 - 2\sqrt{4} \qquad \text{Combine like terms and simplify the radicand.}$$
$$= 7 - 2(2) \qquad \text{Simplify the radical.}$$
$$= 7 - 4 \qquad \text{Multiply.}$$
$$= 3 \qquad \text{Subtract.}$$

3c. Use the graphs of f and g to find $(g \circ f)(4)$.

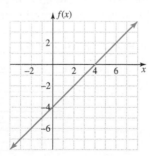 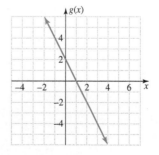

Solution **3c.** $(g \circ f)(4) = g\bigl(f(4)\bigr)$ Apply the definition of composition.

$$= g(0) \qquad \text{Since } (4, 0) \text{ is on the graph of } f(x), f(4) = 0.$$
$$= 2 \qquad \text{Since } (0, 2) \text{ is on the graph of } g(x), g(0) = 2.$$

3d. Use the table to find $(g \circ f)(0)$ and $(f \circ g)(3)$.

x	$f(x)$	$g(x)$
0	1	−1
1	0	1
2	−1	3
3	−2	5
4	−3	7
5	−4	9
6	−5	11

Solution   **3d.**

$(g \circ f)(0) = g\big(f(0)\big)$	$(f \circ g)(3) = f\big(g(3)\big)$	Apply the definition of composition.
$= g(1)$	$= f(5)$	The table shows that $f(0) = 1$ and $g(3) = 5$.
$= 1$	$= -4$	The table shows that $g(1) = 1$ and $f(5) = -4$.

✓ **Student Check 3**   Find each composition.

a. Let $f(x) = 5x + 7$ and $g(x) = 4x - 8$. Find $(f \circ g)(x)$ and $(g \circ f)(x)$ and state their domains.

b. Let $f(x) = x^2 - 4x + 1$ and $g(x) = |x + 3|$. Find $(g \circ f)(3)$.

c. Use the graphs of f and g to find $(f \circ g)(5)$.

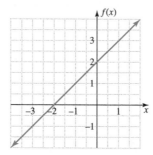   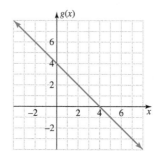

d. Use the table to find $(g \circ f)(1)$ and $(f \circ g)(2)$.

x	$f(x)$	$g(x)$
0	1	−1
1	0	1
2	−1	3
3	−2	5
4	−3	7
5	−4	9
6	−5	11

Applications

Objective 4 ▶

Solve applications.

Applications of combining functions occur in many areas of our lives though we may not think of them as such. There are applications in business, physics, biology, and so on. Examples 4a and 4b illustrate some of these applications.

| **Objective 4 Examples** | **Solve each problem using an appropriate combination of functions.** |

4a. A bookstore buys math textbooks from a publishing company for $80 each and has other fixed costs of $200. So, the bookstore's total cost of the math books can be represented by the function $C(x) = 80x + 200$, where x is the number of math books purchased. The bookstore sells the math books at a 40% markup, or at $112 each. So, the bookstore's total revenue from selling x math books can be represented by the function $R(x) = 112x$. Write a function that represents the profit from selling x math books. (Recall: Profit = Revenue − Cost.)

Solution   **4a.**
$$P(x) = R(x) - C(x) \qquad \text{Write the definition of profit.}$$
$$= 112x - (80x + 200) \qquad \text{Replace } R(x) \text{ and } C(x) \text{ with the given functions.}$$
$$= 112x - 80x - 200 \qquad \text{Apply the distributive property.}$$
$$= 32x - 200 \qquad \text{Combine like terms.}$$

So, the function $P(x) = 32x - 200$ represents the profit from selling x math books.

4b. Josh makes $25 an hour, so his income can be represented by the function $f(x) = 25x$, where x is the number of hours he works. He pays 15% of his income in taxes. The amount of tax he owes can be represented by the function $g(x) = 0.15x$, where x is his income. Write a function that represents the total taxes Josh owes as a function of the number of hours he works.

Solution   **4b.** To find the total taxes Josh owes, we have to determine his total income and then find 15% of that amount. This is the composition of functions because the output of the income function becomes the input of the tax function. We will let $t(x)$ represent the total taxes as a function of the number of hours Josh works.

$$t(x) = (g \circ f)(x) \qquad \text{Write } t(x) \text{ as a composition of } g \text{ and } f.$$
$$= g\big(f(x)\big) \qquad \text{Apply the definition of the composition.}$$
$$= g(25x) \qquad \text{Replace } f(x) \text{ with 25x.}$$
$$= 0.15(25x) \qquad \text{Substitute 25x for } x \text{ in } g(x).$$
$$= 3.75x \qquad \text{Simplify.}$$

The function $t(x) = 3.75x$ represents the amount of taxes Josh owes as a function of the number of hours he works. So, Josh pays $3.75 in taxes for each hour he works.

| ☑ **Student Check 4** | Solve each problem using an appropriate combination of functions. |

a. An online company sells only one product each day. They pay $60 for each item they sell and have other fixed costs of $500 a day. So, the company's total daily cost can be represented by the function $C(x) = 60x + 500$, where x is the number of items sold. The company sells the items for $85 each. So, the total revenue from selling x items can be represented by the function $R(x) = 85x$. Write a function that represents the daily profit from selling x items.

b. Sarah makes $20 an hour, so her income can be represented by the function $f(x) = 20x$, where x is the number of hours she works. She pays 12% of her income in taxes. The amount of tax she owes can be represented by the function $g(x) = 0.12x$, where x is her income. Write a function that represents the total taxes as a function of the number of hours Sarah works.

| **Objective 5** ▶ | **Troubleshooting Common Errors** |
| Troubleshoot common errors. | Some common errors associated with combining functions are shown. |

Objective 5 Examples / A problem and an incorrect solution are given. Provide the correct solution and an explanation of the error.

5a. Let $f(x) = 4x - 1$ and $g(x) = 7x + 9$. Find $(f - g)(x)$.

Incorrect Solution	Correct Solution and Explanation
$\begin{aligned}(f - g)(x) &= f(x) - g(x) \\ &= 4x - 1 - 7x + 9 \\ &= -3x + 8\end{aligned}$	We must distribute -1 to each term of $g(x)$. $\begin{aligned}(f - g)(x) &= f(x) - g(x) \\ &= 4x - 1 - (7x + 9) \\ &= 4x - 1 - 7x - 9 \\ &= -3x - 10\end{aligned}$

5b. Let $f(x) = 4x - 1$ and $g(x) = 7x + 9$. Find $f(g(x))$.

Incorrect Solution	Correct Solution and Explanation
$\begin{aligned}f(g(x)) &= (4x - 1)(7x + 9) \\ &= 28x^2 + 36x - 7x - 9 \\ &= 28x^2 + 29x - 9\end{aligned}$	The product of the functions is written as $(f \cdot g)(x)$ or $f(x) \cdot g(x)$ but $f(g(x))$ represents the composition. $\begin{aligned}f(g(x)) &= f(7x + 9) \\ &= 4(7x + 9) - 1 \\ &= 28x + 36 - 1 \\ &= 28x + 35\end{aligned}$

ANSWERS TO STUDENT CHECKS

Student Check 1   a. $(f + g)(x) = x^2 + x + 4$, domain $= (-\infty, \infty)$; $(f - g)(x) = x^2 - 3x + 2$, domain $= (-\infty, \infty)$; $(f \cdot g)(x) = 2x^3 - x^2 + 5x + 3$; domain $= (-\infty, \infty)$; $\left(\dfrac{f}{g}\right)(x) = \dfrac{x^2 - x + 3}{2x + 1}$; domain $= \left(-\infty, -\dfrac{1}{2}\right) \cup \left(-\dfrac{1}{2}, \infty\right)$

b. $(f + g)(x) = x^2 + \sqrt{2x} - 4$, domain $= [0, \infty)$; $(f - g)(x) = -x^2 + \sqrt{2x} + 4$, domain $= [0, \infty)$; $(f \cdot g)(x) = x^2\sqrt{2x} - 4\sqrt{2x}$; domain $= [0, \infty)$; $\left(\dfrac{f}{g}\right)(x) = \dfrac{\sqrt{2x}}{x^2 - 4}$; domain $= [0, 2) \cup (2, \infty)$

Student Check 2   a. 23   **b.** $-2, -9, 4, 0$
c. $-1, 2, -66, -\dfrac{3}{5}$

Student Check 3   a. $f \circ g = 20x - 33$, $g \circ f = 20x + 20$; domain of both is $(-\infty, \infty)$ **b.** 1   **c.** 1   **d.** $-1, -2$

Student Check 4   a. $R(x) = 25x - 500$ **b.** $t(x) = 2.4x$

SUMMARY OF KEY CONCEPTS

1. We can add, subtract, multiply, and divide functions by performing these operations on the expressions that represent the functions.

$$(f + g)(x) = f(x) + g(x)$$
$$(f - g)(x) = f(x) - g(x)$$
$$(f \cdot g)(x) = f(x) \cdot g(x)$$
$$\left(\frac{f}{g}\right)(x) = \frac{f(x)}{g(x)}$$

2. When a function's output value is input into another function, we are evaluating the composition of functions.

3. To compose two functions together, replace the variable x in the outer function with the expression that represents the inner function.

$$(f \circ g)(x) = f(g(x))$$
$$(g \circ f)(x) = g(f(x))$$

GRAPHING CALCULATOR SKILLS

A graphing calculator can perform operations on functions and evaluate combined functions.

Example: Let $f(x) = 4x - 1$ and $g(x) = 7x + 9$. Find the following:

a. $(f + g)(-2)$ **b.** $(f \cdot g)(-2)$ **c.** $(f \circ g)(-2)$

Solution: Enter the functions f and g into Y_1 and Y_2. Then let Y_3 equal the operation to be performed. We can then view the table. The Y_3 value for $x = -2$ is the value of the combined function.

a. $(f + g)(-2)$

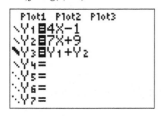

So, $(f + g)(-2) = -14$.

b. $(f \cdot g)(-2)$

So, $(f \cdot g)(-2) = 45$.

c. $(f \circ g)(-2) = f(g(-2))$

So, $(f \circ g)(-2) = f(g(-2)) = -21$.

Note: To access Y_1 and Y_2, press VARS, select Y-VARS, 1, 1 (for Y_1) or 2 (for Y_2).

SECTION 12.1 / EXERCISE SET

Write About It!

Use complete sentences in your answer to each exercise.

1. Explain how to add and subtract functions, $f(x)$ and $g(x)$. Give an example of each.

2. Explain how to multiply binomial functions, $f(x)$ and $g(x)$. Give an example.

3. Explain how to determine the restriction on the domain when finding the quotient of two functions, $f(x)$ and $g(x)$.

4. Explain how to evaluate the sum, $(f + g)(2)$.

5. Explain how to evaluate the difference, $(f - g)(2)$.

6. Explain how to find the composite function, $f(g(x))$, if $f(x) = 2x^2 - x$ and $g(x) = 5x - 1$.

7. Explain how to find the composite function, $g(f(x))$, if $f(x) = 2x^2 - x$ and $g(x) = 5x - 1$.

8. If $f(x) = |7x - 2|$ and $g(x) = 2x - 6$, explain how to evaluate $f(g(2))$ and $g(f(2))$.

Practice Makes Perfect!

Find the sum, difference, product, and quotient of the functions. State the domain of each combined function. (*See Objective 1.*)

9. $f(x) = 4x + 11$ and $g(x) = x + 8$

10. $f(x) = 7x - 3$ and $g(x) = 2x + 1$

11. $f(x) = \sqrt{x}$ and $g(x) = 11x - 2$

12. $f(x) = 5\sqrt{2x}$ and $g(x) = 3x - 12$

13. $f(x) = \sqrt{5x}$ and $g(x) = x^2 + 9$

14. $f(x) = 2\sqrt{x}$ and $g(x) = 4x^2 + 1$

15. $f(x) = 4x - 3$ and $g(x) = -x^2 + 5x$

16. $f(x) = 6x + 1$ and $g(x) = x^2 - 3x$

Let $f(x) = x^2 - 5$, $g(x) = -3x^2 + x - 4$, $h(x) = 2x + 3$. Evaluate each function. (*See Objective 2.*)

17. $(f + g)(-3)$ 18. $(f + h)(-2)$

19. $(f - g)(5)$ 20. $(h - g)(2)$

21. $(f \cdot g)(-2)$ 22. $(f \cdot h)(3)$

23. $\left(\dfrac{f}{g}\right)(4)$

24. $\left(\dfrac{g}{h}\right)(-1)$

25. $(g + h)\left(\dfrac{1}{2}\right)$

26. $(g - h)\left(-\dfrac{1}{3}\right)$

Use the graphs of $f(x)$, $g(x)$, and $h(x)$ to find the value of each combined function. (See Objective 2.)

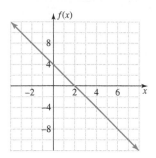
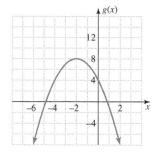

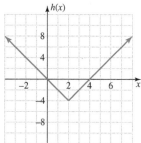

27. $(f + g)(0)$ **28.** $(g + h)(-2)$ **29.** $(f - h)(4)$

30. $(g - h)(-4)$ **31.** $(f \cdot g)(0)$ **32.** $(g \cdot h)(-2)$

33. $(f \circ h)(5)$ **34.** $\left(\dfrac{f}{g}\right)(2)$ **35.** $\left(\dfrac{f}{h}\right)(4)$

36. $\left(\dfrac{g}{f}\right)(-4)$

Use the table to find the value of each combined function. (See Objective 2.)

37. $(g + h)(0)$

38. $(f + g)(-3)$

39. $(f - g)(2)$

40. $(g - h)(-2)$

41. $(f \cdot g)(-1)$

42. $(g \cdot h)(-2)$

43. $\left(\dfrac{f}{g}\right)(1)$

44. $\left(\dfrac{g}{h}\right)(-1)$

x	f(x)	g(x)	h(x)
−3	15	−23	29
−2	3	−17	13
−1	−3	−11	3
0	−3	−5	−1
1	3	1	1
2	15	7	9
3	33	13	23

For the given functions $f(x)$ and $g(x)$, find the composition functions $(f \circ g)(x)$ and $(g \circ f)(x)$ and state their domain. (See Objective 3.)

45. $f(x) = x - 2$ and $g(x) = 12x - 11$

46. $f(x) = 5x + 7$ and $g(x) = 6x + 1$

47. $f(x) = 5x + 9$ and $g(x) = x - 4$

48. $f(x) = 3x - 15$ and $g(x) = 2x + 1$

For the given functions $f(x)$ and $g(x)$, evaluate each composition function. (See Objective 3.)

49. Let $f(x) = 2x^2 + 7x - 5$ and $g(x) = 3x - 4$. Find $(f \circ g)(2)$ and $(g \circ f)(-3)$.

50. Let $f(x) = x^2 - 3x$ and $g(x) = 4 - 2x$. Find $(f \circ g)(4)$ and $(g \circ f)(-1)$.

51. Let $f(x) = 3x^2 - 4x + 1$ and $g(x) = \sqrt{x + 18}$. Find $(f \circ g)(-2)$ and $(g \circ f)(3)$.

52. Let $f(x) = 2x^2 - 3x - 9$ and $g(x) = \sqrt{x - 3}$. Find $(f \circ g)(7)$ and $(g \circ f)(4)$.

53. Let $f(x) = x^2 - 5x + 9$ and $g(x) = |x + 7|$. Find $(f \circ g)(-6)$ and $(g \circ f)(-1)$.

54. Let $f(x) = 3x^2 - 3x + 10$ and $g(x) = |x + 6|$. Find $(f \circ g)(-5)$ and $(g \circ f)(3)$.

Use the graphs to find each value. (See Objective 3.)

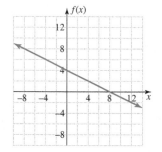
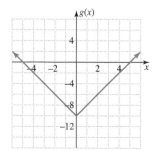

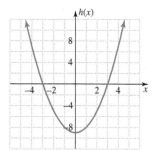

55. $(g \circ f)(0)$ **56.** $(h \circ g)(5)$ **57.** $(g \circ f)(2)$

58. $(f \circ g)(-3)$ **59.** $(g \circ h)(-2)$ **60.** $(h \circ f)(2)$

Use the table to find each value. (See Objective 3.)

61. Find $(f \circ g)(-2)$ and $(g \circ f)(-2)$.

x	f(x)	g(x)
−2	7	−19
7	16	17
−19	−10	−87

62. Find $(f \circ g)(0)$ and $(g \circ f)(0)$.

x	$f(x)$	$g(x)$
0	−5	3
−5	−10	13
3	−2	−3

63. Find $(f \circ g)(-3)$ and $(g \circ f)(-3)$.

x	$f(x)$	$g(x)$
−3	−11	1
−11	−19	−31
1	−7	17

64. Find $(f \circ g)(2)$ and $(g \circ f)(2)$.

x	$f(x)$	$g(x)$
2	−5	9
−5	−12	−5
9	2	23

Solve each problem. (*See Objective 4.*)

65. Amiee makes $18 an hour, so her income can be represented by the function $f(x) = 18x$, where x is the number of hours she works. She pays 15% of her income in taxes. The amount of tax she owes can be represented by the function $g(x) = 0.15x$, where x is her income. Write a function that represents the total taxes Amiee owes as a function of the number of hours she works.

66. Blake makes $36 an hour, so his income can be represented by the function $f(x) = 36x$, where x is the number of hours he works. He pays 30% of his income in taxes. The amount of tax he owes can be represented by the function $g(x) = 0.30x$, where x is his income. Write a function that represents the total taxes Blake owes as a function of the number of hours he works.

67. A bookstore buys math textbooks from a publishing company for $105 each and has other fixed costs of $421. So, the bookstore's total cost of the math books can be represented by the function $C(x) = 105x + 421$, where x is the number of math books purchased. The bookstore sells the math books at a 36% markup or at $142.80 each. So, the bookstore's total revenue from selling x math books can be represented by the function $R(x) = 142.80x$. Write a function that represents the profit from selling x math books.

68. A bookstore buys biology textbooks from a publishing company for $60 each and has other fixed costs of $474. So, the bookstore's total cost of the biology books can be represented by the function $C(x) = 60x + 474$, where x is the number of biology books purchased. The bookstore sells the biology books at a 41% markup or at $84.60 each. So, the bookstore's total revenue from selling x biology books can be represented by the function $R(x) = 84.60x$. Write a function that represents the profit from selling x biology books.

69. An organization is planning a bingo fund-raiser. The cost of using the hall is $27 per person. The cost of food is $20 per person and the cost of prizes is $16 for every four players. Let x represent the number of players.

a. Write a function that represents the cost of the hall, the cost of food, and the cost of prizes.

b. Write a function that represents the total cost of the bingo fund-raiser as a function of the number of players.

70. An organization is planning a golf outing. The cost of using the golf course is $39 per person. The cost of food is $23 per person and the cost of prizes is $60 for every six golfers. Let x represent the number of golfers.

a. Write a function that represents the cost of the golf course, the cost of food, and the cost of prizes.

b. Write a function that represents the total cost of the golf outing as a function of the number of golfers.

 Mix 'Em Up!

Find each function. State the restriction for each quotient function.

For Exercises 71–76, let $f(x) = 2x + 1$, $g(x) = 3x^2 - 2x$, and $h(x) = 4 - x$.

71. $(f + g)(x)$ and $(f \cdot g)(x)$ **72.** $(f + h)(x)$ and $(f \cdot h)(x)$

73. $(g \cdot h)(x)$ and $\left(\dfrac{h}{g}\right)(x)$ **74.** $(g - h)(x)$ and $\left(\dfrac{g}{h}\right)(x)$

75. $(f \circ g)(x)$ and $(g \circ f)(x)$ **76.** $(f \circ h)(x)$ and $(h \circ f)(x)$

For Exercises 77–82, let $f(x) = 2x^2 + 3$, $g(x) = x^2 - 7x + 6$, and $h(x) = 5 - 3x$.

77. $(f + g)(x)$ and $(f - g)(x)$

78. $(g + h)(x)$ and $(g - h)(x)$

79. $(f \cdot g)(x)$ and $\left(\dfrac{f}{g}\right)(x)$ **80.** $(g \cdot h)(x)$ and $\left(\dfrac{g}{h}\right)(x)$

81. $(f \circ h)(x)$ and $(h \circ f)(x)$ **82.** $(g \circ h)(x)$ and $(h \circ g)(x)$

Let $f(x) = x^2 + 2$, $g(x) = x^2 - 4x - 5$, $h(x) = 2x - 1$. Evaluate each combined function.

83. $(f + g)(4)$ **84.** $(f - h)(-3)$

85. $(f \cdot g)(1)$ **86.** $(h \cdot g)(-2)$

87. $\left(\dfrac{f}{g}\right)(2)$ **88.** $\left(\dfrac{g}{h}\right)(3)$

89. $(f \circ g)(2)$ **90.** $(g \circ f)(-1)$

91. $(g \circ h)\left(\dfrac{1}{2}\right)$ **92.** $(h \circ f)\left(-\dfrac{1}{2}\right)$

Use the graphs of $f(x)$, $g(x)$, and $h(x)$ to find the value of each combined function.

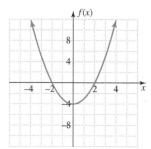

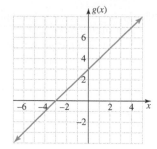

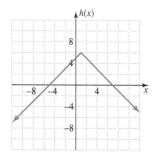

93. $(f + g)(0)$ 94. $(g + h)(-3)$ 95. $(f - h)(3)$

96. $(g - h)(-5)$ 97. $(f \cdot g)(1)$ 98. $(g \cdot h)(-2)$

99. $\left(\dfrac{f}{g}\right)(2)$ 100. $\left(\dfrac{f}{h}\right)(7)$ 101. $(f \circ g)(-2)$

102. $(h \circ g)(-2)$ 103. $(h \circ f)(3)$ 104. $(f \circ h)(-5)$

Use the table to find the value of each combined function.

105. $(g + h)(0)$

106. $(f + g)(5)$

107. $(f - g)(-1)$

108. $(h - g)(1)$

109. $(f \cdot g)(2)$

110. $(g \cdot h)(4)$

x	$f(x)$	$g(x)$	$h(x)$
-3	23	-8	2
-2	16	-6	1
-1	11	-4	0
0	8	-2	1
1	5	0	2
2	8	2	3
3	11	4	4
4	16	6	5
5	23	8	6

111. $\left(\dfrac{f}{g}\right)(3)$ 112. $\left(\dfrac{g}{h}\right)(5)$ 113. $(f \circ g)(1)$

114. $(g \circ h)(-3)$ 115. $(h \circ h)(2)$ 116. $(g \circ g)(0)$

Solve each problem.

117. Vicki makes \$10.25 an hour, so her income can be represented by the function $f(x) = 10.25x$, where x is the number of hours she works. She pays 15% of her income in taxes. The amount of tax she owes can be represented by the function $g(x) = 0.15x$, where x is her income. Write a function that represents the total taxes Vicki owes as a function of the number of hours she works.

118. Emilie makes \$38 an hour, so her income can be represented the function $f(x) = 38x$, where x is the number of hours she works. She pays 30% of her income in taxes. The amount of tax she owes can be represented by the function $g(x) = 0.30x$, where x is her income. Write a function that represents the total taxes Emilie owes as a function of the number of hours she works.

119. A bookstore buys math textbooks from a publishing company for \$110 each and has other fixed costs of \$214. So, the bookstore's total cost of the math books can be represented by the function $C(x) = 110x + 214$, where x is the number of math books purchased. The bookstore sells the math books at a 24% markup or at \$136.40 each. So, the bookstore's total revenue from selling x math books can be represented by the function $R(x) = 136.40x$. Write a function that represents the profit from selling x math books.

120. A bookstore buys physics textbooks from a publishing company for \$90 each and has other fixed costs of \$167. So, the bookstore's total cost of the physics books can be represented by the function $C(x) = 90x + 167$, where x is the number of physics books purchased. The bookstore sells the physics books at a 27% markup or at \$114.30 each. So, the bookstore's total revenue from selling x physics books can be represented by the function $R(x) = 114.30x$. Write a function that represents the profit from selling x physics books.

121. An organization is planning a bingo fund-raiser. The cost of using the hall is \$29 per person. The cost of food is \$15 per person and the cost of prizes is \$48 for every four players. Let x represent the number of players.

    a. Write a function that represents the cost of the hall, the cost of food, and the cost of prizes.

    b. Write a function that represents the total cost of the bingo fund-raiser as a function of the number of players.

122. An organization is planning a golf outing. The cost of using the golf course is \$45 per person. The cost of food is \$26 per person and the cost of prizes is \$40 for every eight players. Let x represent the number of golfers.

    a. Write a function that represents the cost of the golf course, the cost of food, and the cost of prizes.

    b. Write a function that represents the total cost of the golf outing as a function of the number of golfers.

You Be the Teacher!

Correct each student's errors, if any.

123. Find the difference function $(f - g)(x)$ for $f(x) = 2x^2 + 4x - 10$ and $g(x) = 3x^2 - 2x + 9$.

Melanie's work:

$$(f - g)(x) = (2x^2 + 4x - 10) - (3x^2 - 2x + 9)$$
$$= 2x^2 + 4x - 10 - 3x^2 - 2x + 9$$
$$= -x^2 + 2x - 1$$

124. Find the quotient function $\left(\dfrac{f}{g}\right)(x)$ for $f(x) = 9x - 4$ and $g(x) = x^2 - 4x$.

Don's work:

$$\left(\frac{f}{g}\right)(x) = \frac{9x - 4}{x^2 - 4x}, x \neq -2, 2$$

125. Find the composition function $(f \circ g)(x)$ for $f(x) = 4x - 9$ and $g(x) = 5 - 2x$.

Fred's work:

$$(f \circ g)(x) = 5 - 2(4x - 9)$$
$$= 5 - 8x + 18$$
$$= -8x + 23$$

126. Find the composition function $(g \circ f)(x)$ for $f(x) = x^2 - 3x$ and $g(x) = 2x + 1$.

Deb's work:

$$(g \circ f)(x) = 2(x^2 - 3x)^2 + 1$$
$$= (2x^2 - 6x)^2 + 1$$
$$= 4x^4 + 36x^2 + 1$$

Calculate It!

Use a calculator to find each value.

127. Let $f(x) = 2x + 5$ and $g(x) = 3 - 7x$. Find $(f \circ g)(-4)$, $(f \circ g)(-2)$, and $(f \circ g)(2)$.

128. Let $f(x) = x^2 + 1$ and $g(x) = 2 - x$. Find $(g \circ f)(-2)$, $(g \circ f)(1)$, and $(g \circ f)(3)$.

Think About It!

Let $f(x) = x^2$, $g(x) = 2x + 3$, and $h(x) = \sqrt{x}$. Write each function as a composition of two of these functions.

129. $c(x) = 2x^2 + 3$

130. $c(x) = (2x + 3)^2$

131. $c(x) = \sqrt{2x + 3}$

132. $c(x) = 2\sqrt{x} + 3$

SECTION 12.2 — Inverse Functions

▶ OBJECTIVES

As a result of completing this section, you will be able to

1. Determine if a function is one-to-one.
2. Use the horizontal line test.
3. Find the inverse of a function.
4. Find the equation of the inverse of a function.
5. Evaluate inverse functions.
6. Graph a function and its inverse.
7. Verify two functions are inverses.
8. Troubleshoot common errors.

We have studied many functions throughout this course, linear, quadratic, polynomial, radical, rational, and so on. Our goal in this section is to find a function that enables us to "undo" the effects of that function; that is, to find the *inverse* of a function.

One-to-One Functions

Often, it is necessary for us to represent a formula, or a function, for "undoing" the operations that have been performed on a number. This formula or function is called an inverse function. The *inverse function* is a function that takes us back to the place where we started before the original function was applied. The following are some examples of inverses in real life.

Original "function"	Inverse "function"
Sit down	Stand up
Open a door	Close a door
Turn a light on	Turn a light off

Objective 1 ▶

Determine if a function is one-to-one.

We see from these examples that the inverse function is what must be performed to get us back to where we started. Can every function be inverted? That is, are there operations that cannot be undone? Consider the following examples.

	Original "function"	**Inverse "function"**
	Crack an egg	Cannot be undone
	Cut your hair	Cannot be undone
	Spill a drink	Cannot be undone

These examples are operations that cannot be undone. We cannot get back to the egg before it was cracked. We cannot get back to our hairstyle after our hair has been cut. We cannot get back to the original drink after the drink has spilled.

We will use this idea to consider the inverse of a mathematical function. The following table shows examples of performing an operation on a given number, and then performing another operation that takes us back to that original number.

Original Number	Function	Inverse Function	Original Number?
3	Add 2 $3 + 2 = 5$	Subtract 2 $5 - 2 = 3$	Yes
5	Multiply by 4 $5 \cdot 4 = 20$	Divide by 4 $\dfrac{20}{4} = 5$	Yes
-3	Square $(-3)^2 = 9$	Square root $\sqrt{9} = 3$	No

We see from the table that squaring is a mathematical function that does *not* have an inverse function. This is because if we square a negative number, we get a positive number. When we take the square root of a positive number, the result is positive. We will never get back to the original negative number. Only special types of functions have inverse functions. These functions are called *one-to-one functions*.

Definition: A function is a **one-to-one function** if each y-value corresponds to only one x-value. If $f(a) = f(b)$ implies that $a = b$, then f is one-to-one.

Procedure: Determining if a Function Is One-to-One

Step 1: Examine the y-values in the given points.
Step 2: If any y-value corresponds to more than one x-value, then the function is *not* one-to-one.

Objective 1 Examples   Determine if each function is one-to-one.

1a. $f = \{(1, 2), (3, 4), (5, 6), (0, 1)\}$
1b. $f = \{(-2, 4), (-1, 1), (0, 0), (1, 1), (2, 4)\}$

1c.

x	y
-4	2
2	2
4	2

1d.

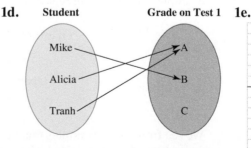

1e.

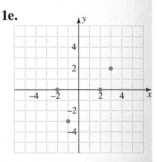

Solutions   **1a.** This function is one-to-one because every value of y corresponds to only one value of x.

1b. This function is *not* one-to-one because the *y*-values of 1 and 4 correspond to more than one *x*-value.

1c. This function is *not* one-to-one because the *y*-value of 2 corresponds to more than one *x*-value.

1d. This function is *not* one-to-one since the *y*-value of "A" corresponds to more than one *x*-value, Alicia and Tranh.

1e. The points plotted on the graph are $(-1, -3)$, $(-2, 0)$, $(2, 0)$, and $(3, 2)$. Since the *y*-value of 0 corresponds to more than one *x*-value, -2 and 2, this function is *not* one-to-one.

✔ **Student Check 1**   Determine if each function is one-to-one.

    a. $f = \{(-3, 8), (-1, 4), (0, 2), (1, 0), (2, -2)\}$

    b. $f = \{(-4, 0), (4, 0), (0, 4)\}$

c.

x	y
−1	7
0	7
3	7

d.

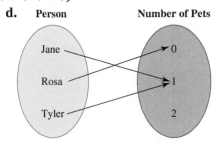

e.

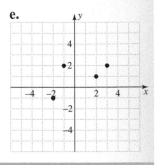

The Horizontal Line Test

Recall that we use a vertical line test to determine if a graph represents a function. If every vertical line drawn through a graph intersects the graph at no more than one point, then the graph represents a function.

    We use a *horizontal line test* to determine if a function is one-to-one. Consider the graph in Figure 12.1. If we draw a horizontal line through $y = 3$, we see that it touches two points on the graph of the function, $(-1, 3)$ and $(3, 3)$. This means that the *y*-value of 3 corresponds to two *x*-values, -1 and 3; therefore, this function is *not* one-to-one.

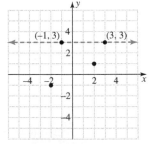

Figure 12.1

> **Property: Horizontal Line Test**
>
> If every horizontal line intersects the graph of a function at no more than one point, the function is one-to-one.

Objective 2 Examples   **Use the horizontal line test to determine if each graph represents a one-to-one function.**

2a.

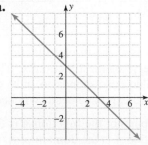

2b.

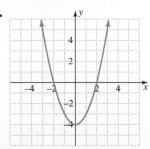

Solutions **2a.** 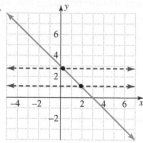   **2b.**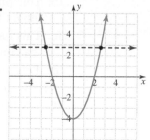

| The function is one-to-one since every horizontal line intersects the graph at no more than one point. | The function is *not* one-to-one since there is a horizontal line that intersects the graph at two points. |

 Student Check 2   Use the horizontal line test to determine if each graph represents a one-to-one function.

a. 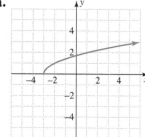   **b.**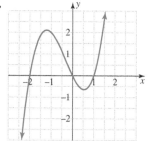

Note: *All linear equations are one-to-one functions except for vertical and horizontal lines. Recall that vertical lines are not functions since they fail the vertical line test. Horizontal lines are functions, but they are not one-to-one functions since they do not pass the horizontal line test.*

The Inverse of a Function

Objective 3 ▶

Find the inverse of a function.

Before we defined a one-to-one function, we briefly introduced the idea of the inverse of a function. As we know, one-to-one functions pass both the vertical and horizontal line tests. One-to-one functions are also special because for each one-to-one function, we can find its *inverse function*.

To find the inverse of a function, we interchange the x- and y-values. In other words, if a function $f(x)$ takes a value x to a value y, then the inverse of the function $f(x)$, denoted $f^{-1}(x)$ (read "f inverse of x"), takes the value y back to the value x. Consider the two following functions denoted by A and B.

	Function	**1-1**	**Inverse—interchange x and y**	**Is the inverse a function?**
A	$\{(1, 3), (2, 5), (3, 7), (4, 9)\}$	Yes	$\{(3, 1), (5, 2), (7, 3), (9, 4)\}$	Yes
B	$\{(-2, 4), (-1, 1), (0, 0), (1, 1), (2, 4)\}$	No	$\{(4, -2), (1, -1), (0, 0), (1, 1), (4, 2)\}$	No, the x-value of 4 corresponds to 2 y-values.

In the preceding table, A is a one-to-one function. When we interchange the x- and y-values, we get the inverse of the function A. The inverse of A, denoted A^{-1}, is a function since each x-value corresponds to only one y-value. Notice that since the coordinates of each ordered pair have been switched,

- the domain (set of inputs) of the function A is the range (set of outputs) of A^{-1}.
- the range of the function A is the domain of A^{-1}.

We also note that the function B is not one-to-one. When we interchange the x- and y-values, we get an inverse relation but *not* an inverse function. This is because $x = 4$ corresponds to the y-values of -1 and 1.

Definition: If a function f is one-to-one, then the inverse of f, denoted f^{-1} exists. The **inverse function** f^{-1} consists of all ordered pairs (y, x) such that (x, y) belongs to the function f. The domain of f is the range of f^{-1}. The range of f is the domain of f^{-1}.

Procedure: Finding the Inverse of a Function

Step 1: Determine if the function is one-to-one. If the function is not one-to-one, then it doesn't have an inverse function.

Step 2: Interchange the x- and y-values of the ordered pairs.

Objective 3 Examples Determine if each function is one-to-one and if so, state its inverse.

3a. $f = \{(0, 1), (1, 2), (2, 3), (3, 4)\}$

3b.

State (x)	Governor in 2011 (y)
Alabama	Robert Bentley
Florida	Rick Scott
Georgia	Nathan Deal
North Carolina	Beverly Perdue
South Carolina	Nikki Haley
Tennessee	Bill Haslam

(Source: http://www.multistate.com/site .nsf/G_L2011?OpenPage)

3c.

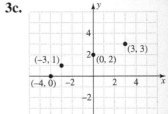

Solutions **3a.** The function f is one-to-one since each y-value corresponds to only one x-value. So, we interchange the x- and y-values to find the inverse function.

$$f^{-1} = \{(1, 0), (2, 1), (3, 2), (4, 3)\}$$

3b. The table represents a one-to-one function since each y-value corresponds to only one x-value. The inverse function interchanges the x- and y-values. So, the table is the inverse function.

Governor in 2011 (x)	State (y)
Robert Bentley	Alabama
Rick Scott	Florida
Nathan Deal	Georgia
Beverly Perdue	North Carolina
Nikki Haley	South Carolina
Bill Haslam	Tennessee

3c. The function is one-to-one since it passes the horizontal line test. We find the inverse function by graphing the ordered pairs that result from interchanging the x- and y-coordinates. So, the graph is the inverse function.

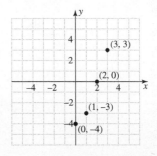

✓ **Student Check 3** Determine if each function is one-to-one and if so, state its inverse.

a. $f = \{(-3, -27), (-2, -8), (-1, -1), (0, 0), (1, 1), (2, 8), (3, 27)\}$

b.

Year Term Began (input)	President of the United States (output)
2009	Barack Hussein Obama
2001	George Walker Bush
1993	William Jefferson Clinton
1989	George Herbert Walker Bush
1981	Ronald Wilson Reagan
1977	James Earl Carter, Jr.
1974	Gerald Rudolph Ford
1969	Richard Milhous Nixon

c.

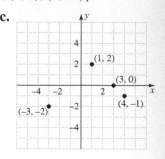

Finding the Equation of the Inverse of a Function

Objective 4 ▶

Find the equation of the inverse of a function.

We can apply the concept from the last objective to find the equation of the inverse of a function. The functions we work with in this objective are one-to-one functions, so that we know the inverse exists.

> **Procedure: Finding the Equation of the Inverse of a One-to-One Function**
>
> **Step 1:** Replace $f(x)$ with y.
> **Step 2:** Interchange x and y.
> **Step 3:** Solve the equation for y.
> **Step 4:** Replace y with the notation $f^{-1}(x)$.

Objective 4 Examples Find the equation of the inverse of each function.

4a. $f(x) = 2x - 6$ **4b.** $f(x) = (x + 1)^3$ **4c.** $f(x) = \dfrac{4}{x - 3}$

Solutions **4a.**

$f(x) = 2x - 6$

$y = 2x - 6$ Replace $f(x)$ with y.

$x = 2y - 6$ Interchange x and y.

$x + 6 = 2y$ Add 6 to each side.

$\dfrac{x + 6}{2} = y$ Divide each side by 2.

$f^{-1}(x) = \dfrac{x + 6}{2}$ Replace y with $f^{-1}(x)$.

$f^{-1}(x) = \dfrac{1}{2}x + 3$ Simplify.

4b.

$f(x) = (x + 1)^3$

$y = (x + 1)^3$ Replace $f(x)$ with y.

$x = (y + 1)^3$ Interchange x and y.

$\sqrt[3]{x} = y + 1$ Take the cube root of each side.

$\sqrt[3]{x} - 1 = y$ Subtract 1 from each side.

$f^{-1}(x) = \sqrt[3]{x} - 1$ Replace y with $f^{-1}(x)$.

4c. $f(x) = \dfrac{4}{x-3}$

$y = \dfrac{4}{x-3}$ Replace $f(x)$ with y.

$x = \dfrac{4}{y-3}$ Interchange x and y.

$x(y-3) = 4$ Multiply each side by $(y-3)$.

$xy - 3x = 4$ Apply the distributive property.

$xy = 4 + 3x$ Add $3x$ to each side.

$y = \dfrac{4 + 3x}{x}$ Divide each side by x.

$f^{-1}(x) = \dfrac{4 + 3x}{x}$ Replace y with $f^{-1}(x)$.

☑ **Student Check 4**   Find the equation of the inverse of each function.

   a. $f(x) = 4x - 8$ **b.** $f(x) = (x-2)^3$ **c.** $f(x) = \dfrac{1}{x+6}$

Evaluating Inverse Functions

Objective 5 ▶

Evaluate inverse functions.

We will now learn how to use function notation with inverse functions. Recall that if the point (x, y) is on the graph of f, then we write $f(x) = y$. For instance, if the point $(2, -3)$ is on the graph of f, then $f(2) = -3$.

   Knowing one point on the graph of f provides us with information about a point on the graph of f^{-1}. Recall that if the point (x, y) is on the graph of f, then the point (y, x) is on the graph of f^{-1}. Therefore, $f^{-1}(y) = x$. So, knowing that $(2, -3)$ is on the graph of f tells us that $(-3, 2)$ is on the graph of f^{-1}. Therefore, $f^{-1}(-3) = 2$.

Objective 5 Examples   **Evaluate each function using the given information.**

   5a. If $f(6) = 10$, what is $f^{-1}(10)$?   **5b.** If $f^{-1}(3) = -4$, what is $f(-4)$?

Solutions   **5a.** $f(6) = 10$ means that $(6, 10)$ is on the graph of f. Therefore, $(10, 6)$ is on the graph of f^{-1}. So, $f^{-1}(10) = 6$.

   5b. $f^{-1}(3) = -4$ means that $(3, -4)$ is on the graph of f^{-1}. Therefore, $(-4, 3)$ is on the graph of f. So, $f(-4) = 3$.

☑ **Student Check 5**   Evaluate each function using the given information.

   a. If $f(-1) = 5$, what is $f^{-1}(5)$?   **b.** If $f^{-1}\left(\dfrac{1}{2}\right) = 7$, what is $f(7)$?

The Graph of a Function and the Graph of Its Inverse

Objective 6 ▶

Graph a function and its inverse.

The graph of a function and the graph of its inverse relate to each other in a special way. Consider the graphs of the functions from Example 4a, $f(x) = 2x - 6$ and $f^{-1}(x) = \dfrac{1}{2}x + 3$. The points on the graph of the function and the graph of its inverse are mirror images of each other across the line $y = x$. We say that the graph of $f(x)$ and the graph of $f^{-1}(x)$ are *symmetric* about the line $y = x$.

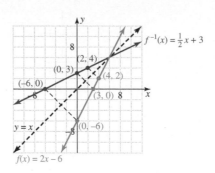

Procedure: Graphing the Inverse of a One-to-One Function

Step 1: Draw the line of symmetry, $y = x$.
Step 2: Interchange the coordinates of ordered pairs on the graph of the function f.
Step 3: Use the line of symmetry to assist in drawing the graph of the inverse function.

Objective 6 Examples   Graph the inverse of each function.

6a.

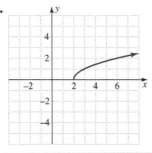

6b.

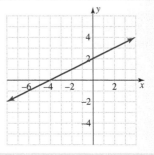

Solutions   **6a.** The inverse is shown in blue.

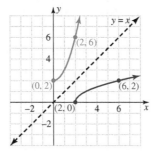

6b. The inverse is shown in blue.

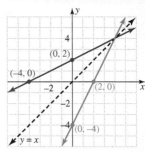

✔ **Student Check 6**   Graph the inverse of each function.

a.

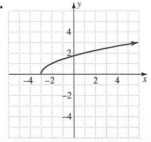

b.

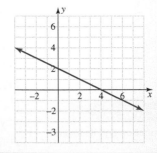

Verifying Functions are Inverses

Objective 7 ▶

Verify two functions are inverses.

Two functions are inverses of one another if, when we apply a function and then apply its inverse, we end up with the original input value. For instance, we know that if $f(x) = 2x - 6$, then its inverse function is $f^{-1}(x) = \frac{1}{2}x + 3$. We evaluate the function f at 0 and then apply the inverse function to that output value.

$$f(x) = 2x - 6$$
$$f(0) = 2(0) - 6$$
$$f(0) = -6$$

Now, we evaluate the inverse function, f^{-1}, at $x = -6$.

$$f^{-1}(x) = \frac{1}{2}x + 3$$
$$f^{-1}(-6) = \frac{1}{2}(-6) + 3$$
$$f^{-1}(-6) = -3 + 3$$
$$f^{-1}(-6) = 0$$

So, the function f takes the value 0 to -6 and the function f^{-1} takes the value -6 back to 0. When we compose a function with its inverse, we always get back to the starting value.

> **Property: The Composition of Inverse Functions**
> The inverse of a one-to-one function f is the function f^{-1} such that
> $$(f \circ f^{-1})(x) = x \text{ and } (f^{-1} \circ f)(x) = x$$

> **Procedure: Verifying Two One-to-One Functions, f and f^{-1}, Are Inverses**
> **Step 1:** Find $(f \circ f^{-1})(x)$ and $(f^{-1} \circ f)(x)$.
> **Step 2:** If both of the compositions produce x, then the functions f and f^{-1} are inverses.

Objective 7 Examples Verify that the functions are inverses of each other.

7a. $f(x) = 4x + 12$ and $f^{-1}(x) = \frac{1}{4}x - 3$ **7b.** $f(x) = x^3 - 1$ and $f^{-1}(x) = \sqrt[3]{x + 1}$

Solutions **7a.**

$$(f \circ f^{-1})(x) = f\left(f^{-1}(x)\right)$$
$$= f\left(\frac{1}{4}x - 3\right)$$
$$= 4\left(\frac{1}{4}x - 3\right) + 12$$
$$= x - 12 + 12$$
$$= x$$

$$(f^{-1} \circ f)(x) = f^{-1}\left(f(x)\right)$$
$$= f^{-1}(4x + 12)$$
$$= \frac{1}{4}(4x + 12) - 3$$
$$= x + 3 - 3$$
$$= x$$

Since both compositions produce x, the two functions are inverses.

7b.

$$(f \circ f^{-1})(x) = f\left(f^{-1}(x)\right)$$
$$= f\left(\sqrt[3]{x + 1}\right)$$
$$= \left(\sqrt[3]{x + 1}\right)^3 - 1$$
$$= x + 1 - 1$$
$$= x$$

$$(f^{-1} \circ f)(x) = f^{-1}\left(f(x)\right)$$
$$= f^{-1}(x^3 - 1)$$
$$= \sqrt[3]{x^3 - 1 + 1}$$
$$= \sqrt[3]{x^3}$$
$$= x$$

Since both compositions produce x, the two functions are inverses.

✓ **Student Check 7**   Verify that the functions are inverses of each other.

$\quad$ **a.** $f(x) = 3x - 6$ and $f^{-1}(x) = \dfrac{1}{3}x + 2$ $\quad$ **b.** $f(x) = (x + 5)^3$ and $f^{-1}(x) = \sqrt[3]{x} - 5$

Objective 8 ▶

Troubleshoot common errors.

Troubleshooting Common Errors

Some common errors associated with inverse functions are shown.

Objective 8 Examples   **A problem and an incorrect solution are given. Provide the correct solution and an explanation of the error.**

8a. Determine if the function $f(x) = x^2 + 4$ is one-to-one and if so, find the equation of its inverse.

Incorrect Solution	Correct Solution and Explanation
$f(x) = x^2 + 4$ $y = x^2 + 4$ $x = y^2 + 4$ $x - 4 = y^2$ $\sqrt{x - 4} = y$	The function $f(x) = x^2 + 4$ is a parabola and is not one-to-one. Therefore, it does not have an inverse function.

8b. Find $f^{-1}(x)$ if $f(x) = 4x - 5$.

Incorrect Solution	Correct Solution and Explanation
$f^{-1}(x) = \dfrac{1}{4x - 5}$	The notation $f^{-1}(x)$ represents the inverse function. It is not the same as a negative exponent. So, to find $f^{-1}(x)$, $\quad y = 4x - 5$ $\quad x = 4y - 5$ $\quad x + 5 = 4y$ $\quad \dfrac{x + 5}{4} = y$ So, $f^{-1}(x) = \dfrac{1}{4}x + \dfrac{5}{4}$.

8c. Show that if $f(x) = \dfrac{1 + x}{x}$, then $\dfrac{1}{x - 1}$ is its inverse.

Incorrect Solution	Correct Solution and Explanation
$(f \circ f^{-1})(x) = f\left(f^{-1}(x)\right)$ $= f\left(\dfrac{1}{x - 1}\right)$ $= 1 + \dfrac{1}{x - 1}$ $= \dfrac{x - 1 + 1}{x - 1}$ $= \dfrac{x}{x - 1}$ Since $f \circ f^{-1} \neq x$, the functions are not inverses.	The expression $\dfrac{1}{x - 1}$ should replace both occurrences of x in the function f. $f\left(\dfrac{1}{x - 1}\right) = \dfrac{1 + \dfrac{1}{x - 1}}{\dfrac{1}{x - 1}}$ $= \dfrac{\left(1 + \dfrac{1}{x - 1}\right)(x - 1)}{\left(\dfrac{1}{x - 1}\right)(x - 1)}$ $= \dfrac{x - 1 + 1}{1}$ $= x$ So, $f \circ f^{-1} = x$. It can be shown that $f^{-1} \circ f$ is also x.

ANSWERS TO STUDENT CHECKS

Student Check 1 **a.** yes **b.** no **c.** no **d.** no **e.** no

Student Check 2 **a.** one-to-one **b.** not one-to-one

Student Check 3 **a.** one-to-one, $\{(-27, -3), (-8, -2),$
$(-1, -1), (0, 0), (1, 1), (8, 2), (27, 3)\}$

b. one-to-one,

President of the United States (input)	Year Term Began (output)
Barack Hussein Obama	2009
George Walker Bush	2001
William Jefferson Clinton	1993
George Herbert Walker Bush	1989
Ronald Wilson Reagan	1981
James Earl Carter, Jr.	1977
Gerald Rudolph Ford	1974
Richard Milhous Nixon	1969

c. one-to-one,

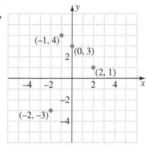

Student Check 4 **a.** $f^{-1}(x) = \frac{1}{4}x + 2$

b. $f^{-1}(x) = \sqrt[3]{x} + 2$ **c.** $f^{-1}(x) = \frac{1 - 6x}{x}$

Student Check 5 **a.** -1 **b.** $\frac{1}{2}$

Student Check 6 **a.** **b.**

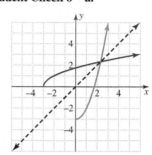 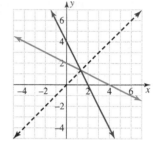

Student Check 7 **a.** f and f^{-1} are inverses. **b.** f and f^{-1} are inverses.

SUMMARY OF KEY CONCEPTS

1. A function is one-to-one if every y-value corresponds to only one x-value.

2. The horizontal line test is used to determine if the graph of a function is one-to-one. If every horizontal line crosses the graph at no more than one point, then the graph is one-to-one.

3. To find the inverse of a one-to-one function, interchange the ordered pairs of the function f.

4. To find the equation of the inverse of a one-to-one function, replace $f(x)$ with y and interchange the x- and y-variables. Then solve the new equation for y. This result is $f^{-1}(x)$.

5. If (x, y) is on the graph of a one-to-one function f, then $f^{-1}(y) = x$ since the point (y, x) is on the graph of f^{-1}.

6. The graphs of a one-to-one function and its inverse are symmetric about the line $y = x$.

7. We can verify two functions f and f^{-1} are inverses by showing that $(f \circ f^{-1})(x) = x$ and $(f^{-1} \circ f)(x) = x$.

GRAPHING CALCULATOR SKILLS

A graphing calculator can draw the graph of an inverse function. It can also assist us in verifying that two functions are inverses.

Example 1: Draw the inverse of the function $f(x) = 2x - 6$.

Solution: Input the function into the equation editor and graph the function.

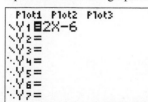

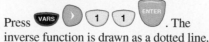

To draw the inverse of the function, press

 PRGM **8** . Then enter the function for which you want the inverse drawn.

Press **VARS** **▷** **1** **1** **ENTER** . The inverse function is drawn as a dotted line.

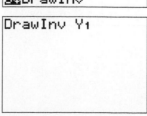

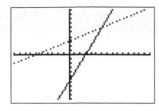

Note that the graphs might not look symmetric about the line $y = x$ unless we use the square viewing window.

Example 2: Verify that $f(x) = 2x - 6$ and $f^{-1}(x) = \dfrac{1}{2}x + 3$ are inverses.

Solution: Enter the two functions into Y_1 and Y_2. In Y_3 and Y_4, enter the expressions for the compositions of the two functions.

```
Plot1 Plot2 Plot3
\Y1∎2X-6
\Y2∎1/2X+3
\Y3∎Y1(Y2)
\Y4∎Y2(Y1)
\Y5=
\Y6=
\Y7=
```

View the TABLE and examine the columns for Y_3 and Y_4.

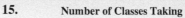

As we can see from the TABLE, the values in Y_3 and Y_4 equal the corresponding x-value; therefore, the two functions are inverses.

SECTION 12.2 / EXERCISE SET

✍ Write About It!

Use complete sentences in your answer to each exercise.

1. Define a one-to-one function.
2. Explain how to determine if a function is one-to-one.
3. Define the inverse of a one-to-one function f.
4. Explain how to find the equation of the inverse of a one-to-one function.
5. Explain how to graph the inverse of a one-to-one function from the graph of the function.
6. Explain how to determine if two functions are inverses of each other.

🎹 Practice Makes Perfect!

Determine if each function is one-to-one. (*See Objective 1.*)

7. $f = \{(1, 4), (2, 7), (3, 10), (4, 13), (5, 16)\}$
8. $f = \{(-3, 9), (-2, 7), (-1, 5), (0, 3), (1, 1)\}$
9. $f = \{(-2, 7), (-1, 4), (0, 3), (1, 4), (2, 7)\}$
10. $f = \{(-1, 0), (0, 3), (1, 4), (2, 3), (3, 0)\}$
11. $f = \{(-2, 1), (0, 1), (2, 1), (3, 1), (5, 1)\}$
12. $f = \{(-2, -8), (0, 0), (1, 1), (2, 8), (3, 27)\}$

13.

x	-2	0	2	4	6
y	-3	-3	-3	-3	-3

14.

x	-1	0	1	2	3
y	-2	-1	2	9	28

15.
Number of Classes Taking

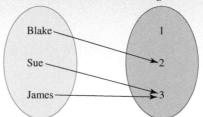

16.
Number of TV Sets

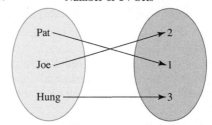

17.   18.

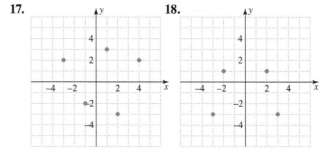

Use the horizontal line test to determine if each graph is a one-to-one function. (*See Objective 2.*)

19.

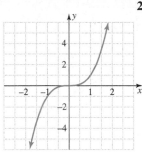

20.

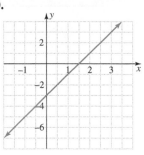

21.

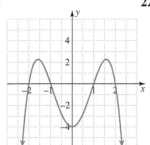

22.

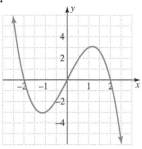

23.

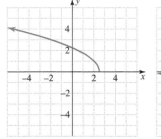

24.

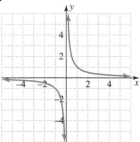

Determine if each function is one-to-one and if so, state its inverse. (*See Objective 3.*)

25. $f(x) = \{(-2, 3), (0, 4), (1, 6), (2, -1), (3, 1)\}$

26. $f(x) = \{(-3, 1), (1, 2), (3, 3), (4, 5), (5, 7)\}$

27. $f(x) = \{(-5, 4), (1, 3), (4, 2), (6, 3), (7, 1)\}$

28. $f(x) = \{(-1, 5), (0, 9), (2, 6), (5, -4), (7, 9)\}$

29.

City (x)	Mayor in 2011 (y)
New York City	Michael Bloomberg
Los Angeles	Antonio Villaraigosa
Chicago	Rahm Emanuel
Houston	Annise Parker
Phoenix	Phil Gordon

(Source: http://www.citymayors.com)

30.

City(x)	Number of States with the same city name (y)
Riverside	46
Fairview	43
Franklin	42
Midway	40
Fairfield	39
Pleasant Valley	39

(Source: http://geography.about.com/od/lists/a/placename50.htm)

31.

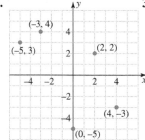

32.

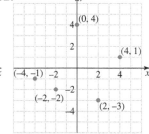

Find the equation of the inverse of each function. (*See Objective 4.*)

33. $f(x) = x - 8$ **34.** $f(x) = x + 5$

35. $f(x) = 11x - 2$ **36.** $f(x) = 5x - 9$

37. $f(x) = \dfrac{3x - 5}{2}$ **38.** $f(x) = \dfrac{3x + 7}{6}$

39. $f(x) = (2x + 9)^3$ **40.** $f(x) = (4x - 7)^3$

41. $f(x) = 8x^3 + 1$ **42.** $f(x) = 27x^3 + 5$

43. $f(x) = \dfrac{15}{x + 3}$ **44.** $f(x) = \dfrac{10}{x - 1}$

Evaluate each function using the given information. (*See Objective 5.*)

45. If $f(-10) = 6$, what is $f^{-1}(6)$?

46. If $f(4) = 9$, what is $f^{-1}(9)$?

47. If $f\left(\dfrac{3}{2}\right) = \dfrac{1}{5}$, what is $f^{-1}\left(\dfrac{1}{5}\right)$?

48. If $f\left(-\dfrac{7}{4}\right) = \dfrac{1}{6}$, what is $f^{-1}\left(\dfrac{1}{6}\right)$?

49. If $f^{-1}(3) = -5$, what is $f(-5)$?

50. If $f^{-1}(-4) = 12$, what is $f(12)$?

51. If $f^{-1}\left(-\dfrac{1}{6}\right) = \dfrac{5}{4}$, what is $f\left(\dfrac{5}{4}\right)$?

52. If $f^{-1}\left(\dfrac{7}{10}\right) = -\dfrac{1}{4}$, what is $f\left(-\dfrac{1}{4}\right)$?

Graph the inverse of the given function on the same set of axes. (*See Objective 6.*)

53.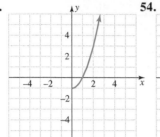

54.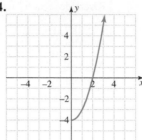

55.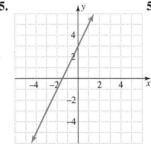

56.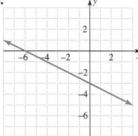

Determine if the functions are inverses of one other. (*See Objective 7.*)

57. $f(x) = 3x - 4$, $g(x) = \dfrac{1}{3}x + \dfrac{4}{3}$

58. $f(x) = 4x - 1$, $g(x) = \dfrac{1}{4}x + \dfrac{1}{4}$

59. $f(x) = (2x + 1)^3$, $g(x) = \dfrac{\sqrt[3]{x} - 1}{2}$

60. $f(x) = (5x - 9)^3$, $g(x) = \dfrac{\sqrt[3]{x} + 9}{5}$

61. $f(x) = \dfrac{2x + 1}{4}$, $g(x) = 2x + \dfrac{1}{2}$

62. $f(x) = \dfrac{3x - 5}{2}$, $g(x) = \dfrac{2}{3}x - \dfrac{5}{3}$

63. $f(x) = x^3 - 2$, $g(x) = \sqrt[3]{x + 2}$

64. $f(x) = 8x^3 + 9$, $g(x) = \dfrac{\sqrt[3]{x} - 9}{2}$

65. $f(x) = x^3 + 8$, $g(x) = \sqrt[3]{x} - 2$

66. $f(x) = x^3 - 27$, $g(x) = \sqrt[3]{x} + 3$

 Mix 'Em Up!

Determine if each function is one-to-one.

67. $f = \{(2, 4), (1, 7), (-2, 4), (-1, 7), (0, 5)\}$

68. $f = \{(-1, -1.5), (0, -1), (3, 0.5), (5, 1.5), (6, 2)\}$

69.

x	-5	-3	-1	1	3
y	-3	-1	1	5	7

70.

x	-4	-2	0	2	4
y	1	4	9	4	25

71.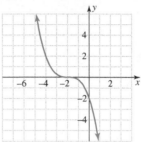

72.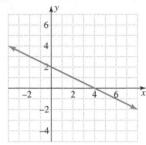

73.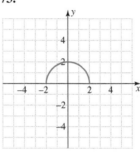

74.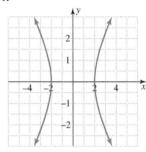

75.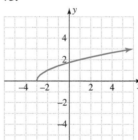

76.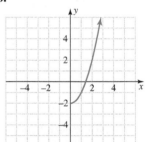

Determine if each function is one-to-one and if so, state its inverse.

77. $f(x) = \{(-2, -8), (0, 2), (1, 7), (2, 12), (4, 22)\}$

78. $f(x) = \{(-3, -2.3), (-1, -1.1), (2, 0.7), (5, 2.5), (9, 4.9)\}$

79. $f(x) = \{(-2.5, 11.5), (0, -1), (-1.5, 3.5), (2.5, 11.5), (1.5, 3.5)\}$

80.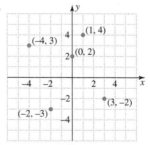

81.

State (x)	Senator in 2011 (y)
Connecticut	Joseph Lieberman
South Carolina	Lindsey Graham
Hawaii	Daniel Inouye
Georgia	Johnny Isakson
Louisiana	Mary Landrieu

(Source: http://www.senate.gov/index.htm)

82.

Car Model (x)	Average Highway MPG (y)
Volkswagen Jetta	42
Toyota Yaris	36
Honda Fit	35
Hyundai Accent	36
Chevrolet Aveo	35

(Source: http://www.fueleconomy.gov)

Find the equation of the inverse of each function.

83. $f(x) = 7x + 3$

84. $f(x) = 8x + 12$

85. $f(x) = \dfrac{5}{8x - 3}$

86. $f(x) = \dfrac{13}{6x + 1}$

87. $f(x) = 4x^3 + 2$

88. $f(x) = 3x^3 - 5$

89. $f(x) = \dfrac{6x - 2}{5}$

90. $f(x) = \dfrac{3x + 10}{6}$

91. $f(x) = (5x + 4)^3$

92. $f(x) = (3x + 1)^3$

Evaluate each function given the information.

93. If $f(-7) = -2$, what is $f^{-1}(-2)$?

94. If $f(4.5) = -1.5$, what is $f^{-1}(-1.5)$?

95. If $f\left(-\dfrac{4}{5}\right) = \dfrac{3}{8}$, what is $f^{-1}\left(\dfrac{3}{8}\right)$?

96. If $f^{-1}\left(\dfrac{2}{7}\right) = -\dfrac{1}{4}$, what is $f\left(-\dfrac{1}{4}\right)$?

Graph the inverse of each function.

97.

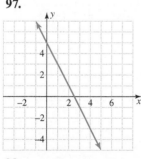

98.

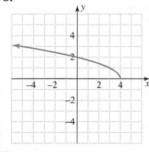

99.

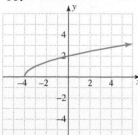

100.

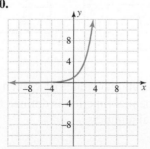

You Be the Teacher!

Correct each student's errors, if any.

101. Find the equation of the inverse of the function
$f(x) = \dfrac{3x - 7}{8}$.

Eric's work:

$f(x) = \dfrac{3x - 7}{8}$

$f^{-1}(x) = \dfrac{1}{f(x)} = \dfrac{8}{3x - 7}$

102. Find the equation of the inverse of the function
$f(x) = 8x^3 + 27$.

Mona's work:

$y = 8x^3 + 27$

$x = 8y^3 + 27$

$\sqrt[3]{x} = 2y + 3$

$\sqrt[3]{x} - 3 = 2y$

$y = f^{-1}(x) = \dfrac{\sqrt[3]{x} - 3}{2}$

Calculate It!

Use both the TABLE and graphing features of a graphing calculator to determine if the functions are inverses of one another.

103. $f(x) = 2x + 4$ and $g(x) = \dfrac{1}{2}x - 2$

104. $f(x) = (x - 1)^3$ and $g(x) = \sqrt[3]{x} + 1$

Think About It!

105. Find the inverse function, $f^{-1}(x)$, if $f(x)$ is a linear function and $f(3) = -6$ and $f(0) = 2$.

106. Find the inverse function, $f^{-1}(x)$, if $f(x)$ is a linear function and $f(-2) = 5$ and $f(6) = 13$.

107. Find the inverse function, $f^{-1}(x)$, if $f(x)$ is a linear function and $f(4) = -1$ and $f(3) = 0$.

108. Find the inverse function, $f^{-1}(x)$, if $f(x)$ is a linear function and $f(7) = -2$ and $f(-1) = 0$.

Exponential Functions

▶ **OBJECTIVES**

As a result of completing this section, you will be able to

1. Graph exponential functions.
2. Solve exponential equations of the form $b^x = b^y$.
3. Solve applications of exponential functions.
4. Troubleshoot common errors.

The number of Facebook accounts (in millions) can be modeled by $f(x) = 108.8449(1.0658)^x$, where x is the months after August 2008. Use this model to estimate the number of Facebook accounts in December 2013. (Source: http://www.benphoster.com/facebook-user-growth-chart-2004-2010/)

To solve this problem, we must evaluate the function for $x = 64$. This type of function is called an exponential function. In this section, we will learn to work with functions and equations of this type.

Exponential Functions

Objective 1 ▶

Graph exponential functions.

Recall from previous chapters that expressions that contain exponents, such as 3^4 and x^3, are called exponential expressions. In this section, we will look at functions that involve exponential expressions. In these functions, the base is constant and the exponent changes. This type of function is called an *exponential function*.

> **Definition:** An **exponential function** is a function of the form $f(x) = b^x$, where $b > 0$ and $b \neq 1$, and x is a real number.

Note that in the definition of an exponential function, $b > 0$ and $b \neq 1$.

- If the base of an exponential function is 1, then $1^x = 1$ for all x, which is a horizontal line and not an exponential function.
- If the base is a negative value, then the function would not be defined for all values. For instance, $(-4)^{1/2}$ is not a real number.

Consider the function $f(x) = 4^x$. Recall $x^{m/n} = \left(\sqrt[n]{x}\right)^m$ and $x^{-n} = \dfrac{1}{x^n}$.

x	$y = 4^x$	(x, y)
-2	$4^{-2} = \dfrac{1}{16}$	$\left(-2, \dfrac{1}{16}\right)$
$-\dfrac{3}{2}$	$4^{-3/2} = \dfrac{1}{8}$	$\left(-\dfrac{3}{2}, \dfrac{1}{8}\right)$
-1	$4^{-1} = \dfrac{1}{4}$	$\left(-1, \dfrac{1}{4}\right)$
$-\dfrac{1}{2}$	$4^{-1/2} = \dfrac{1}{2}$	$\left(-\dfrac{1}{2}, \dfrac{1}{2}\right)$

x	$y = 4^x$	(x, y)
0	$4^0 = 1$	$(0, 1)$
$\dfrac{1}{2}$	$4^{1/2} = 2$	$\left(\dfrac{1}{2}, 2\right)$
1	$4^1 = 4$	$(1, 4)$
$\dfrac{3}{2}$	$4^{3/2} = 8$	$\left(\dfrac{3}{2}, 8\right)$
2	$4^2 = 16$	$(2, 16)$

To evaluate the function for irrational values of x, such as $\sqrt{2}$, we must use a calculator; $4^{\sqrt{2}} \approx 7.10$. So, an exponential function is defined for all real numbers. That is, the domain of $f(x) = b^x$ is all real numbers.

Using the points obtained in the previous tables, we can graph the function $f(x) = 4^x$. (See Figure 12.2.)

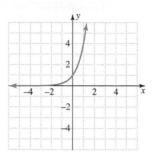

Figure 12.2

Property: Characteristics of the Graph of $y = b^x, b > 0, b \neq 1$

- The domain of the function is $(-\infty, \infty)$.
- The range is $(0, \infty)$.
- The y-intercept is $(0, 1)$.
- There is no x-intercept. The graph gets very close to the x-axis but never touches it.
- The point $(1, b)$ lies on the graph.
- The function is one-to-one.
- If $b > 1$, the function is increasing.
- If $0 < b < 1$, the function is decreasing.

$f(x) = b^x$ for $b > 1$

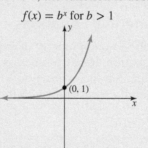

$f(x) = b^x$ for $0 < b < 1$

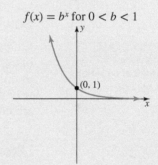

Objective 1 Examples Graph each exponential function. Plot and label at least three points on the graph. State the domain and range of each function.

1a. $f(x) = 2^x$ **1b.** $f(x) = \left(\dfrac{1}{2}\right)^x$

Solutions **1a.**

x	$y = 2^x$	(x, y)
-2	$2^{-2} = \dfrac{1}{4}$	$\left(-2, \dfrac{1}{4}\right)$
-1	$2^{-1} = \dfrac{1}{2}$	$\left(-1, \dfrac{1}{2}\right)$
0	$2^0 = 1$	$(0, 1)$
1	$2^1 = 2$	$(1, 2)$
2	$2^2 = 4$	$(2, 4)$

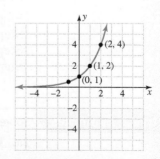

The domain is $(-\infty, \infty)$, and the range is $(0, \infty)$.

1b.

x	$y = \left(\dfrac{1}{2}\right)^x$	(x, y)
-2	$\left(\dfrac{1}{2}\right)^{-2} = 4$	$(-2, 4)$
-1	$\left(\dfrac{1}{2}\right)^{-1} = 2$	$(-1, 2)$
0	$\left(\dfrac{1}{2}\right)^{0} = 1$	$(0, 1)$
1	$\left(\dfrac{1}{2}\right)^{1} = \dfrac{1}{2}$	$\left(1, \dfrac{1}{2}\right)$
2	$\left(\dfrac{1}{2}\right)^{2} = \dfrac{1}{4}$	$\left(2, \dfrac{1}{4}\right)$

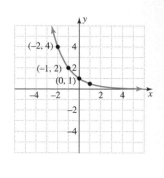

The domain is $(-\infty, \infty)$, and the range is $(0, \infty)$.

✔ **Student Check 1**   Graph each exponential function. Plot and label at least three points on the graph. State the domain and range of each function.

   a. $f(x) = 3^x$ **b.** $f(x) = \left(\dfrac{1}{3}\right)^x$

Solving Exponential Equations

Objective 2 ▶

Solve exponential equations of the form $b^x = b^y$.

We can use the one-to-one property of exponential functions to solve a special type of exponential equation. Recall a function is one-to-one if every y-value corresponds to only one x-value. So, the expression b^x is different for every value of x. The only way that exponential expressions with the same base are equal is if their exponents are equal.

> **Property: One-to-One Property of b^x**
>
> If $b > 0$ and $b \neq 1$, then $b^x = b^y$ if and only if $x = y$.

This property indicates that if each side of an equation can be expressed with the same base, then the exponents of the bases must be equal. Later in the chapter, we will learn how to solve exponential equations in which the sides of the equation cannot be expressed with the same base.

> **Procedure: Solving Exponential Equations of the Form $b^x = b^y$**
>
> **Step 1:** Rewrite each side of the equation as a power of the same base.
> **Step 2:** Set the exponents equal.
> **Step 3:** Solve the resulting equation.
> **Step 4:** Check the solution by substitution.

Objective 2 Examples   Solve each equation.

   2a. $3^x = 27$ **2b.** $4^x = 32$ **2c.** $16^{x-1} = 64^x$

Solutions **2a.** $3^x = 27$

$3^x = 3^3$ Rewrite each side with the same base.

$x = 3$ Apply the one-to-one property of b^x.

The solution set is $\{3\}$. To check the solution, replace x with 3 in the original equation.

2b. $4^x = 32$

$(2^2)^x = 2^5$ Rewrite each side with the same base.

$2^{2x} = 2^5$ Apply the power of a power rule, $(b^m)^n = b^{mn}$.

$2x = 5$ Apply the one-to-one property of b^x.

$x = \dfrac{5}{2}$ Divide each side by 2.

The solution set is $\left\{\dfrac{5}{2}\right\}$. The check is left for the reader.

2c. $16^{x-1} = 64^x$

$(4^2)^{x-1} = (4^3)^x$ Rewrite each side with the same base.

$4^{2x-2} = 4^{3x}$ Apply the power of a power rule, $(b^m)^n = b^{mn}$.

$2x - 2 = 3x$ Apply the one-to-one property of b^x.

$-2 = x$ Subtract 2x from each side.

The solution set is $\{-2\}$. The check is left for the reader.

 **Student Check 2** Solve each equation.

 a. $10^x = 100$ **b.** $8^x = 4$ **c.** $9^{x+2} = 27^x$

Applications

Objective 3 ▶

Solve applications of exponential functions.

Exponential functions are used to model quantities that grow or decline very rapidly, such as the population of a country, the growth of bacteria, the amount of a chemical element, or the amount of money in an account.

> **Property: Compound Interest Formula**
>
> The compound interest formula determines the amount of money A that will be in an account if P dollars is invested for t years at an annual interest rate r, compounded n times a year.
>
> $$A = P\left(1 + \frac{r}{n}\right)^{nt}$$

Objective 3 Examples Solve each problem.

3a. Sandra invests $5000 in a savings account that pays 5% annual interest compounded quarterly. Use the compound interest formula to determine the amount of money she will have in the account at the end of 3 yr.

Solution   **3a.** $A = P\left(1 + \dfrac{r}{n}\right)^{nt}$ State the compound interest formula.

$A = 5000\left(1 + \dfrac{0.05}{4}\right)^{4(3)}$ Substitute the values: $P = 5000$, $r = 0.05$, $n = 4$, and $t = 3$.

$A = 5000(1.0125)^{12}$ Simplify the expression in parentheses.

$A \approx \$5803.77$ Approximate.

So, Sandra will have $5803.77 at the end of 3 yr.

3b. The number of Facebook accounts is shown in the following table. The number of accounts (in millions) can be modeled by $f(x) = 108.8449(1.0658)^x$, where x is the months after August 2008. Assuming this growth continues, use this model to estimate the number of Facebook accounts in December 2013. (Source: http://www.benphoster.com/facebook-user-growth-chart-2004-2010/)

Month	Total Accounts (in millions)
August 2008 ($x = 0$)	100
January 2009 ($x = 5$)	150
December 2009 ($x = 16$)	350
July 2010 ($x = 23$)	500
February 2011 ($x = 30$)	650

Solution   **3b.** December 2013 is 64 months after August 2008.

$f(x) = 108.8449(1.0658)^x$ State the model.

$f(64) = 108.8449(1.0658)^{64}$ Replace x with 64.

$f(64) \approx 6427.67$ Approximate.

According to the model, there will be about 6427.67 million Facebook accounts, or 6,427,670,000 accounts in December 2013.

✓ Student Check 3   Solve each problem.

a. Use the compound interest formula to determine the amount Fred will have in a savings account at the end of 10 yr if he invests $2000 at 4.5% annual interest compounded monthly.

b. The population of Russia (in millions) can be modeled by the function $f(x) = 146.25(0.996)^x$, where x is the number of years after 2000. What is an estimate for the population in 2000? 2015? 2050? (Source: http://www.google.com/publicdata)

Objective 4 ▶
Troubleshoot common errors.

Troubleshooting Common Errors

Some common errors associated with exponential functions are shown.

| Objective 4 Examples | A problem and an incorrect solution are given. Provide the correct solution and an explanation of the error. |

4a. Graph $y = 2^x$.

Incorrect Solution	Correct Solution and Explanation
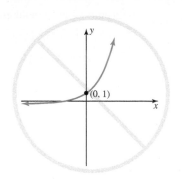	The graph is incorrect because the output of $2^x > 0$. The graph should not cross the x-axis or go below the x-axis. The correct graph is 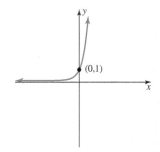

4b. Find the amount Sung will have in an account after 7 yr if he invests $3000 at 5% annual interest compounded monthly.

Incorrect Solution	Correct Solution and Explanation
$A = P\left(1 + \dfrac{r}{n}\right)^{nt}$ $A = 3000\left(1 + \dfrac{0.05}{12}\right)^{12(7)}$ $A = 22{,}074.40$	The error occurred in entering the expression on a calculator. Be sure to put the entire exponent in parentheses. This will give us the correct amount of $A = \$4254.11$

ANSWERS TO STUDENT CHECKS

Student Check 1

a.
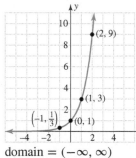
domain $= (-\infty, \infty)$
range $= (0, \infty)$

b.
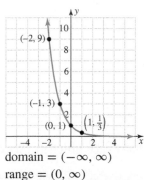
domain $= (-\infty, \infty)$
range $= (0, \infty)$

Student Check 2   a. $\{2\}$ **b.** $\left\{\dfrac{2}{3}\right\}$ **c.** $\left\{-\dfrac{1}{2}\right\}$

Student Check 3   a. \$3133.99   **b.** about 142.3 million, about 135.61 million, about 109.85 million

SUMMARY OF KEY CONCEPTS

1. An exponential function is a function of the form $y = b^x$, $b > 0$, $b \neq 1$.
  - The domain of the function is $(-\infty, \infty)$ and the range is $(0, \infty)$.
  - The y-intercept is $(0, 1)$ and there is no x-intercept.
  - If $b > 1$, the function is increasing. If $0 < b < 1$, the function is decreasing.

2. If an equation can be written in the form $b^x = b^y$, $b > 0$, then we can use the one-to-one property of exponential functions to conclude that $x = y$.

3. Exponential functions model many real-life situations. The compound interest formula is such a model. It is given by $A = P\left(1 + \dfrac{r}{n}\right)^{nt}$.

GRAPHING CALCULATOR SKILLS

A graphing calculator can be used to graph exponential functions as well as solve application problems.

Example 1: Graph $y = 2^x$.

Solution: Enter the function in the equation editor. Graph and review the TABLE.

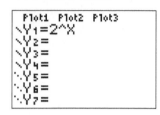

Example 2: Simplify $A = 3000\left(1 + \dfrac{0.05}{12}\right)^{12(7)}$.

Solution: Insert parentheses around the entire exponent.

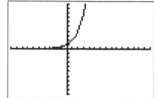

SECTION 12.3 / EXERCISE SET

Write About It!

Use complete sentences in your answer to each exercise.

1. Describe the behavior of the graph of $f(x) = 6^x$ at each end of the x-axis.

2. Describe the behavior of the graph of $f(x) = \left(\dfrac{1}{8}\right)^x$ at each end of the x-axis.

3. Explain how to solve the equation $5^{2x-1} = 625$.

4. Explain how to solve the equation $25^{2x-1} = 625$.

5. Explain how to solve the equation $4^{x+1} = \dfrac{1}{64}$ by using the base 4.

6. Explain how to solve the equation, $4^{x+1} = \dfrac{1}{64}$ by using the base 2.

Practice Makes Perfect!

Graph each exponential function. Plot and label at least three points on the graph. State the domain and range of each function. (See Objective 1.)

7. $y = 5^x$

8. $y = 6^x$

9. $y = \left(\dfrac{1}{5}\right)^x$

10. $y = \left(\dfrac{2}{3}\right)^x$

11. $y = \left(\dfrac{2}{5}\right)^x$

12. $y = \left(\dfrac{1}{6}\right)^x$

Solve each exponential equation. (See Objective 2.)

13. $5^x = 125$

14. $6^x = 36$

15. $\left(\dfrac{1}{2}\right)^x = 8$

16. $\left(\dfrac{1}{3}\right)^x = 81$

17. $\left(\dfrac{3}{2}\right)^x = \dfrac{16}{81}$

18. $\left(\dfrac{5}{3}\right)^x = \dfrac{27}{125}$

19. $27^x = 9$

20. $216^x = 36$

21. $125^x = \dfrac{1}{25}$

22. $4^{x-2} = 64^x$

23. $2^{3x-1} = 128$

24. $25^{2x+3} = 125$

25. $\left(\dfrac{1}{2}\right)^{3x+2} = 32^x$

26. $\left(\dfrac{1}{3}\right)^{2x-1} = 27^x$

Solve each problem. (See Objective 3.)

For Exercises 27 and 28, use the formula $A = P\left(1 + \dfrac{r}{n}\right)^{nt}$.

27. Hixon invests $6000 in a savings account that pays 4% annual interest compounded quarterly. Determine the amount of money he will have in the account at the end of 3 yr.

28. Ruth invests $7000 in a savings account that pays 1.3% annual interest compounded quarterly. Determine the amount of money she will have in the account at the end of 6 yr.

29. The population of the
United Kingdom (in millions)
can be modeled by the
function $f(x) = 62.06(1.005)^x$,
where x is the number of years
after 2009. What is an estimate
for the population in 2015?

2025? 2050? Round answers to two decimal places.
(Source: http://www.census.gov/population/international/data/idb/
informationGateway.php)

30. The population of the United States (in millions) can be
modeled by the function $f(x) = 281.4(1.0093)^x$, where
x is the number of years after 2000. What is an estimate
for the population in 2010? 2015? 2025? Round
answers to two decimal places. (Source: http://www.census
.gov/prod/cen2010/briefs/c2010br-01.pdf)

31. The radioactive isotope sodium-24 is used to detect
circulatory problems in patients. The function
$A(t) = 20(0.9548)^t$ determines the amount of
sodium-24, in grams, that a hospital has in storage after
t hr. How many grams does the hospital have initially?
How many grams will remain after 24 hr? after 48 hr?
(Sources: http://www.chemistryexplained.com/elements/P-T/
Sodium.html and http://www.3rd1000.com/nuclear/halflife.htm)

32. Rachel has been on a medication for several months
and has reached a steady level of 150 mg in her body. It
is time for Rachel to discontinue taking the medication.
The function $A(t) = 150(0.7071)^t$ determines the
amount of drug, in milligrams, that remains in Rachel's
body t days after she discontinues the drug. How much
is in her body 1 week after she stops taking the drug?
2 weeks after?

 Mix 'Em Up!

**Graph each exponential function. State the domain and
range of the function.**

33. $f(x) = 8^x$ **34.** $f(x) = 9^x$ **35.** $f(x) = 4.5^x$

36. $f(x) = 1.6^x$ **37.** $f(x) = \left(\dfrac{1}{10}\right)^x$ **38.** $f(x) = \left(\dfrac{3}{5}\right)^x$

Solve each exponential equation.

39. $\left(\dfrac{2}{5}\right)^x = \dfrac{625}{16}$ **40.** $\left(\dfrac{2}{3}\right)^x = \dfrac{243}{32}$

41. $9^{x+2} = 243$ **42.** $3^{4x+7} = 3$

43. $16^{5x+2} = 64$ **44.** $2^{3x+8} = 4$

45. $27^{5x+1} = \dfrac{1}{9}$ **46.** $64^{x-1} = \dfrac{1}{256}$

47. $3^{2x+3} = 27$ **48.** $2^{x+5} = 32$

49. $10^{3x+2} = 1000$ **50.** $10^{2x-1} = 100$

51. $3^{2x-5} = 27^x$ **52.** $2^{5x-5} = 4^x$

53. $3^{3x-6} = 81^x$ **54.** $8^{4x+8} = 16^x$

55. $27^{2x-2} = \dfrac{1}{3}$ **56.** $2^{2x+4} = \dfrac{1}{8}$

Solve each problem.

57. Seyfried invests $9500 in a savings account that pays
2.6% annual interest compounded monthly. Use the
compound interest formula to determine the amount of
money she will have in the account at the end of 5 yr.

58. Morgan invests $7200 in a savings account that pays
2.1% annual interest compounded monthly. Use the
compound interest formula to determine the amount of
money he will have in the account at the end of 6 yr.

59. The population of China (in billions) can be modeled
by the function $f(x) = 1.33(1.003)^x$, where x is the
number of years after 2009. What is an estimate for the
population in 2012? 2020? 2030? Round answers to two
decimal places. (Source: http://www.census.gov/population/
international/data/idb/informationGateway.php)

60. The population of France (in millions)
can be modeled by the function
$f(x) = 64.57(1.004)^x$, where x is the
number of years after 2009. What is
an estimate for the population in
2010? 2025? 2040? Round answers
to two decimal places. (Source: http://
www.census.gov/population/international/
data/idb/informationGateway.php)

61. Iodine-131 is a radioactive isotope that is used in
treating thyroid cancer and in imaging the thyroid
and can also be used in nuclear reactors. The
function $A(t) = 30(0.917)^t$ determines the amount, in
milligrams, of the isotope that a hospital has in storage
after t days. How much does the hospital have initially?
How much of the substance will remain after 8 days?
after 30 days? (Source: http://www.ead.anl.gov)

62. Gallium-67 is a radioactive isotope that is used
in imaging various inflammatory conditions and
tumors such as Hodgkin's disease. The function
$A(t) = 10(0.8085)^t$ determines the amount, in
milliliters, of the substance that an imaging center has
in storage after t days. How much of the substance will
remain after 2 days? after 1 week? (Source: http://www
.nordion.com)

You Be the Teacher!

Correct each student's errors, if any.

63. Solve the equation, $16^{3x-2} = 8$.

Ellen's work:

$$16^{3x-2} = 8$$
$$8^{2(3x-2)} = 8$$
$$2(3x - 2) = 0$$
$$6x - 4 = 0$$
$$6x = 4$$
$$x = \dfrac{4}{6} = \dfrac{2}{3}$$

64. Solve the equation, $125^{x-7} = \dfrac{1}{25}$.

Fourlas's work:

$$125^{x-7} = \frac{1}{25}$$

$$5^{3x-7} = 5^{-2}$$

$$3x - 7 = -2$$

$$3x = 5$$

$$x = \frac{5}{3}$$

 **Calculate It!**

Use a graphing calculator to graph or evaluate each function. Round each answer to two decimal places if appropriate.

65. $y = \left(\dfrac{1}{5}\right)^x$

66. $y = 7^x$

67. Evaluate $114.5(5.14)^{-0.72x}$ at $x = 5$.

68. Evaluate $2500\left(1 + \dfrac{0.023}{12}\right)^{12t}$ at $t = 8$.

SECTION 12.4 Logarithmic Functions

► OBJECTIVES

As a result of completing this section, you will be able to

1. Use logarithmic notation to write an exponential equation and vice versa.

2. Simplify logarithmic expressions.

3. Solve logarithmic equations using exponential notation.

4. Graph basic logarithmic functions.

5. Troubleshoot common errors.

The graph shows the annual carbon dioxide (CO_2) emissions for a car with the given gas mileage. The annual tons of CO_2 emitted can be modeled by $f(x) = 13.272(0.8369)^x$, where x is the mileage of a car in miles per gallon. Determine the gas mileage of a car that emits 1 ton of CO_2 annually. To solve this problem, we must solve

$$13.272(0.8369)^x = 1$$

Solving this equation requires us to apply a property that enables us to move the variable from the exponent. In this section, we will introduce the concept of logarithms, which we will apply in Section 12.7 to solve this problem.

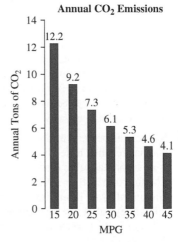

Annual CO₂ Emissions

(Source: http://www.fueleconomy.gov/)

Logarithms

Objective 1 ►

Use logarithmic notation to write an exponential equation and vice versa.

Consider the exponential function $f(x) = 2^x$. Because $f(x) = 2^x$ is a one-to-one function, we know it has an inverse. To find the inverse of $f(x) = 2^x$ algebraically, we must interchange the variables and solve for y. This gives us

$$f(x) = 2^x$$
$$y = 2^x$$
$$x = 2^y$$

We have not yet learned a technique that enables us to solve for the variable y. So, we need a new symbol, called a *logarithm*, to denote an equivalent form of y. The symbol $\log_b x$ is read "log base b of x" and means the exponent of b that produces x. So, in this example, we say that

$$y = \log_2 x$$

This notation states that y is the exponent of 2 that produces x. So, a logarithm is simply an exponent. As we can see, logarithmic functions arise as we find the inverse of an exponential function. So, the inverse of $f(x) = 2^x$ is $f^{-1}(x) = \log_2 x$.

Definition: Logarithm

For $b > 0$, $b \ne 1$,

Logarithmic form
↓
$$y = \log_b x \text{ is equivalent to } b^y = x$$
↑
Exponential form

for every $x > 0$ and every real number y. The value x is called the **argument** of the logarithmic expression.

Some examples of this definition are as follows.

$$2^3 = 8 \text{ is equivalent to } \log_2 8 = 3.$$
$$5^2 = 25 \text{ is equivalent to } \log_5 25 = 2.$$
$$3^{-1} = \frac{1}{3} \text{ is equivalent to } \log_3\left(\frac{1}{3}\right) = -1.$$

Objective 1 Examples **Rewrite each logarithmic equation in exponential form and each exponential equation in logarithmic form.**

Problems	Solutions
1a. $\log_2 16 = 4$	$\log_2 16 = 4$ is equivalent to $2^4 = 16$.
1b. $\log_5 1 = 0$	$\log_5 1 = 0$ is equivalent to $5^0 = 1$.
1c. $\log_3 3 = 1$	$\log_3 3 = 1$ is equivalent to $3^1 = 3$.
1d. $\log_4\dfrac{1}{16} = -2$	$\log_4\dfrac{1}{16} = -2$ is equivalent to $4^{-2} = \dfrac{1}{16}$.
1e. $5^3 = 125$	$5^3 = 125$ is equivalent to $\log_5 125 = 3$.
1f. $3^{-1} = \dfrac{1}{3}$	$3^{-1} = \dfrac{1}{3}$ is equivalent to $\log_3\dfrac{1}{3} = -1$.
1g. $8^{2/3} = 4$	$8^{2/3} = 4$ is equivalent to $\log_8 4 = \dfrac{2}{3}$.
1h. $\sqrt{4} = 2$	$\sqrt{4} = 4^{1/2} = 2$ is equivalent to $\log_4 2 = \dfrac{1}{2}$.

✔ Student Check 1 Rewrite each logarithmic equation in exponential form and each exponential equation in logarithmic form.

a. $\log_3 81 = 4$ **b.** $\log_2 1 = 0$ **c.** $\log_6 6 = 1$ **d.** $\log_5\dfrac{1}{125} = -3$

e. $6^2 = 36$ **f.** $4^{-1} = \dfrac{1}{4}$ **g.** $9^{3/2} = 27$ **h.** $\sqrt{9} = 3$

Evaluating Logarithmic Expressions

Objective 2 ▶

Simplify logarithmic expressions.

Now that we know that $\log_b x$ means to find the exponent of b that produces x, we can simplify logarithmic expressions. It is helpful to be able to express in words what the logarithm denotes.

$\log_4 16$ means the exponent of 4 that produces 16. Since $4^2 = 16$, $\log_4 16 = 2$.
$\log_{10} 10$ means the exponent of 10 that produces 10. Since $10^1 = 10$, $\log_{10} 10 = 1$.
$\log_2 1$ means the exponent of 2 produces that 1. Since $2^0 = 1$, $\log_2 1 = 0$.

It is also helpful to recall the following properties of exponents.

$$b^0 = 1$$
$$b^1 = b$$
$$b^{-n} = \frac{1}{b^n}$$
$$b^{1/n} = \sqrt[n]{b}$$
$$b^{m/n} = \sqrt[n]{b^m}$$

Recall also that when we raise a base b, $b > 0$ and $b \neq 1$, to any number, the result is always positive. So, the logarithm of a negative number is undefined. For instance,

$$\log_2(-4) \text{ means the exponent of 2 that produces } -4.$$

There is no exponent of 2 that produces a negative value, so $\log_2(-4)$ is undefined.

Procedure: Simplifying a Logarithmic Expression of the Form $\log_b x$

Step 1: Find the exponent of b that produces x by using the exponential form.

Step 2: The value of the logarithmic expression is this exponent.

Objective 2 Example | Simplify each logarithmic expression.

Problems	Solutions
2a. $\log_6 1$	$6^0 = 1 \Rightarrow \log_6 1 = 0$
2b. $\log_4 4$	$4^1 = 4 \Rightarrow \log_4 4 = 1$
2c. $\log_3 \dfrac{1}{27}$	$3^{-3} = \dfrac{1}{27} \Rightarrow \log_3 \dfrac{1}{27} = -3$
2d. $\log_2 \sqrt{2}$	$2^{1/2} = \sqrt{2} \Rightarrow \log_2 \sqrt{2} = \dfrac{1}{2}$
2e. $\log_{1/2} 4$	$\left(\dfrac{1}{2}\right)^{-2} = 4 \Rightarrow \log_{1/2} 4 = -2$
2f. $\log_5 (-25)$	$\log_5(-25)$ is undefined since $5^y \neq -25$ for any real number y.

✓ **Student Check 2** | Simplify each logarithmic expression.

a. $\log_9 1$ **b.** $\log_5 5$ **c.** $\log_2 \dfrac{1}{4}$ **d.** $\log_6 \sqrt[3]{6}$

e. $\log_{1/3} 3$ **f.** $\log_7 (-49)$

Note: *Examples 2a and 2b show us two important properties of logarithms. For $b > 0$, $b \neq 1$, we have*

$$\log_b 1 = 0$$
$$\log_b b = 1$$

Solving Logarithmic Equations

Objective 3 ▶

Solve logarithmic equations using exponential notation.

We can use the exponential form of a logarithmic equation to solve logarithmic equations.

Procedure: Solving a Logarithmic Equation

Step 1: Rewrite the equation in exponential form.

Step 2: Solve the resulting equation.

Objective 3 Examples | Solve each equation.

3a. $\log_2 32 = x$ **3b.** $\log_6 x = 3$ **3c.** $\log_x 49 = 2$ **3d.** $\log_b b = x$

Solutions | **3a.** $\log_2 32 = x$

$\quad\quad 2^x = 32$ Rewrite as an exponential equation.

$\quad\quad 2^x = 2^5$ Rewrite 32 with a base of 2.

$\quad\quad\quad x = 5$ Apply the one-to-one property of b^x.

The solution set is $\{5\}$. To check, note that $\log_2 32 = 5$ since $2^5 = 32$.

3b. $\log_6 x = 3$

$\qquad 6^3 = x$ \qquad Rewrite as an exponential equation.

$\qquad 216 = x$ \qquad Simplify the exponent.

The solution set is $\{216\}$.

3c. $\log_x 49 = 2$

$\qquad x^2 = 49$ \qquad Rewrite as an exponential equation.

$\qquad x = \pm\sqrt{49}$ \qquad Apply the square root property.

$\qquad x = \pm 7$ \qquad Simplify the square root.

Because the base of a logarithm must be greater than zero, the only solution is $x = 7$. So, the solution set is $\{7\}$.

3d. $\log_b b = x$

$\qquad b^x = b$ \qquad Rewrite as an exponential equation.

$\qquad b^x = b^1$ \qquad Write b as b^1.

$\qquad x = 1$ \qquad Apply the one-to-one property of b^x.

The solution set is $\{1\}$.

✓ **Student Check 3** Solve each equation.

a. $\log_8 64 = x$ \qquad **b.** $\log_{10} x = 2$ \qquad **c.** $\log_x 8 = 3$ \qquad **d.** $\log_b b^2 = x$

Graphing Basic Logarithmic Functions

Objective 4 ▶

Graph basic logarithmic functions.

Recall in Objective 1, we found that the inverse of $f(x) = 2^x$ is $f^{-1}(x) = \log_2 x$. The inverse function, $f^{-1}(x) = \log_2 x$, is called a *logarithmic function*.

Definition: For $x > 0$ and $b > 0$ $(b \neq 1)$, a function of the form $f(x) = \log_b x$ is called a **logarithmic function**. The domain of $f(x)$ is $(0, \infty)$ and the range is $(-\infty, \infty)$.

Tables of values for $f(x) = 2^x$ and $f^{-1}(x) = \log_2 x$ are shown. Recall we interchange the x- and y-values to obtain the inverse function.

x	$f(x) = 2^x$
-2	$\dfrac{1}{4}$
-1	$\dfrac{1}{2}$
0	1
1	2
2	4

x	$f^{-1}(x) = \log_2 x$
$\dfrac{1}{4}$	-2
$\dfrac{1}{2}$	-1
1	0
2	1
4	2

When we plot the points for f and f^{-1}, we see that their graphs are symmetric about the line $y = x$.

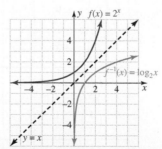

Because $y = b^x$ and $y = \log_b x$ are inverse functions, we know some characteristics about the graph of the logarithmic function.

Characteristics of the Graph of $f(x) = b^x, b > 0, b \neq 1$	Characteristics of the Graph of $f(x) = \log_b x, b > 0, b \neq 1$
The domain of the function is $(-\infty, \infty)$.	The domain of the function is $(0, \infty)$.
The range is $(0, \infty)$.	The range is $(-\infty, \infty)$.
The y-intercept is $(0, 1)$.	The x-intercept is $(1, 0)$.
There is no x-intercept. The graph gets very close to the x-axis but never touches it.	There is no y-intercept. The graph gets very close to the y-axis but never touches it.
The point $(1, b)$ lies on the graph.	The point $(b, 1)$ lies on the graph.
The function is one-to-one.	The function is one-to-one.
If $b > 1$, the function is increasing.	If $b > 1$, the function is increasing.
If $0 < b < 1$, the function is decreasing.	If $0 < b < 1$, the function is decreasing.

Procedure: Graphing a Function of the Form $y = \log_b x$

Step 1: Convert the equation to its exponential form $b^y = x$.
Step 2: Substitute values for y and solve for x.
Step 3: Plot the ordered pairs (x, y) and connect the points with a smooth curve.

Objective 4 Examples Graph each logarithmic function. State the domain and range of each function.

4a. $f(x) = \log_3 x$ **4b.** $f(x) = \log_{1/3} x$

Solutions **4a.** The function $f(x) = \log_3 x$ is equivalent to $3^y = x$. So, we make a table, assign values to y and solve for x.

$x = 3^y$	y	(x, y)
$3^{-2} = \dfrac{1}{9}$	-2	$\left(\dfrac{1}{9}, -2\right)$
$3^{-1} = \dfrac{1}{3}$	-1	$\left(\dfrac{1}{3}, -1\right)$
$3^0 = 1$	0	$(1, 0)$
$3^1 = 3$	1	$(3, 1)$
$3^2 = 9$	2	$(9, 2)$

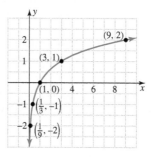

The domain is $(0, \infty)$ and the range is $(-\infty, \infty)$. Notice that the x-intercept is $(1, 0)$ and there is no y-intercept.

4b. The equation $y = \log_{1/3} x$ is equivalent to $\left(\dfrac{1}{3}\right)^y = x$. So, we make a table, assigning values to y and solving for x. We get

$x = \left(\dfrac{1}{3}\right)^y$	y	(x, y)
$\left(\dfrac{1}{3}\right)^{-2} = 9$	-2	$(9, -2)$
$\left(\dfrac{1}{3}\right)^{-1} = 3$	-1	$(3, -1)$
$\left(\dfrac{1}{3}\right)^0 = 1$	0	$(1, 0)$
$\left(\dfrac{1}{3}\right)^1 = \dfrac{1}{3}$	1	$\left(\dfrac{1}{3}, 1\right)$
$\left(\dfrac{1}{3}\right)^2 = \dfrac{1}{9}$	2	$\left(\dfrac{1}{9}, 2\right)$

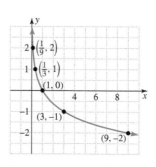

The domain is $(0, \infty)$ and the range is $(-\infty, \infty)$. Notice that the x-intercept is $(1, 0)$ and there is no y-intercept.

✔ **Student Check 4**   Graph each logarithmic function. State the domain and range of each function.

     a. $y = \log_4 x$ **b.** $y = \log_{1/4} x$

Objective 5 ▶

Troubleshoot common errors.

Troubleshooting Common Errors

Some common errors associated with logarithms are shown.

Objective 5 Examples   **A problem and an incorrect solution are given. Provide the correct solution and an explanation of the error.**

5a. Write $3^{-2} = \dfrac{1}{9}$ in logarithmic form.

Incorrect Solution	Correct Solution and Explanation
$3^{-2} = \dfrac{1}{9}$ is equivalent to $\log_3(-2) = \dfrac{1}{9}$.	The logarithm is always equal to the exponent of the base. So, $3^{-2} = \dfrac{1}{9}$ is equivalent to $\log_3 \dfrac{1}{9} = -2$.

5b. Simplify $\log_2(-4)$.

Incorrect Solution	Correct Solution and Explanation
$\log_2(-4) = -2$	The argument of a logarithm must be greater than zero. So, $\log_2(-4)$ is not defined. There is no exponent of 2 that will produce -4. Note that $2^{-2} = \dfrac{1}{4}$ not -4.

ANSWERS TO STUDENT CHECKS

Student Check 1   **a.** $3^4 = 81$ **b.** $2^0 = 1$ **c.** $6^1 = 6$

     d. $5^{-3} = \dfrac{1}{125}$ **e.** $\log_6 36 = 2$ **f.** $\log_4 \dfrac{1}{4} = -1$

     g. $\log_9 27 = \dfrac{3}{2}$ **h.** $\log_9 3 = \dfrac{1}{2}$

Student Check 2   **a.** 0 **b.** 1 **c.** -2 **d.** $\dfrac{1}{3}$ **e.** -1
     f. undefined

Student Check 3   **a.** $\{2\}$ **b.** $\{100\}$ **c.** $\{2\}$ **d.** $\{2\}$

Student Check 4

a.

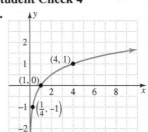

b.

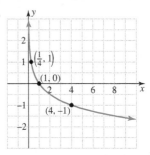

SUMMARY OF KEY CONCEPTS

1. For $b > 0$ and $b \neq 1$, $y = \log_b x$ is equivalent to $b^y = x$. The log base b of x means to find the exponent of b that produces x.

2. We can simplify logarithmic expressions by finding the exponent of the base that produces the argument. Some key logarithms for $b > 0$, $b \neq 1$ are as follows.

     a. $\log_b 1 = 0$ since $b^0 = 1$.
     b. $\log_b b = 1$ since $b^1 = b$.
     c. $\log_b b^x = x$ since $b^x = b^x$.

3. We can solve logarithmic equations by converting the equation to its equivalent exponential form.

4. We can graph logarithmic functions by converting them to their equivalent exponential form. Some of the key properties are

     • The domain of the function is $(0, \infty)$ and the range is $(-\infty, \infty)$.

     • The x-intercept is $(1, 0)$. There is no y-intercept. The graph gets very close to the y-axis but never touches it.

     • The point $(b, 1)$ lies on the graph.

     • The function is one-to-one.

     • If $b > 1$, the function is increasing and if $0 < b < 1$, the function is decreasing.

SECTION 12.4 / EXERCISE SET

 Write About It!

Use complete sentences in your answer to each exercise.

1. Define a logarithm.
2. Explain how to write the exponential equation, $x^a = b$, in logarithmic form.
3. Explain how to write the logarithmic equation, $\log_w y = z$, in exponential form.
4. Explain why $\log_a 1 = 0$, $a > 0$.
5. Explain why $\log_a a = 1$, $a > 0$.
6. Explain how to set up a table to graph a logarithmic function.
7. Explain how to simplify the expression $\log_b b^5$.
8. Explain how to simplify the expression $a^{\log_a 2}$.

 Practice Makes Perfect!

Rewrite each logarithmic equation in exponential form and each exponential equation in logarithmic form. (*See Objective 1.*)

9. $\log_2 32 = 5$
10. $\log_{10} 100 = 2$
11. $\log_8 8 = 1$
12. $\log_5 5 = 1$
13. $\log_7 1 = 0$
14. $\log_{12} 1 = 0$
15. $\log_2 \dfrac{1}{16} = -4$
16. $\log_4 \dfrac{1}{64} = -3$
17. $6^3 = 216$
18. $12^2 = 144$
19. $11^1 = 11$
20. $10^1 = 10$
21. $15^0 = 1$
22. $2^0 = 1$
23. $7^{-2} = \dfrac{1}{49}$
24. $10^{-3} = \dfrac{1}{1000}$
25. $4^{3/2} = 8$
26. $125^{2/3} = 25$
27. $\sqrt{144} = 12$
28. $\sqrt[3]{64} = 4$
29. $\sqrt[4]{81} = 3$
30. $\sqrt[3]{1000} = 10$

Simplify each expression. (*See Objective 2.*)

31. $\log_5 1$
32. $\log_{11} 1$
33. $\log_{13} 13$
34. $\log_7 7$
35. $\log_{1/2} \dfrac{1}{2}$
36. $\log_{1/5} \dfrac{1}{5}$
37. $\log_2 \dfrac{1}{32}$
38. $\log_3 \dfrac{1}{243}$
39. $\log_{1/2} 4$
40. $\log_{1/3} 81$
41. $\log_3 (-9)$
42. $\log_{10} (-10)$

Solve each logarithmic equation. (*See Objective 3.*)

43. $\log_5 625 = x$
44. $\log_3 243 = x$
45. $\log_6 216 = x$
46. $\log_{10} 1000 = x$
47. $\log_8 x = 3$
48. $\log_3 x = 4$
49. $\log_x 343 = 3$
50. $\log_x 100 = 2$
51. $\log_{10} 1 = x$
52. $\log_7 1 = x$
53. $\log_a a^3 = x$
54. $\log_a a^5 = x$
55. $\log_b b^{-2} = x$
56. $\log_b b^{-6} = x$

Graph each logarithmic function. Plot and label at least three points. State the domain and range of each function. (*See Objective 4.*)

57. $y = \log_5 x$
58. $y = \log_{10} x$
59. $y = \log_{1.5} x$
60. $y = \log_{2.5} x$
61. $y = \log_{1/2} x$
62. $y = \log_{1/10} x$

 Mix 'Em Up!

Solve each logarithmic equation.

63. $\log_5 125 = x$
64. $\log_2 128 = x$
65. $\log_c c^4 = x$
66. $\log_c c^8 = x$
67. $\log_5 x = 4$
68. $\log_3 x = 5$
69. $\log_x \dfrac{1}{125} = 3$
70. $\log_x \dfrac{1}{100} = 2$
71. $\log_x 3 = \dfrac{1}{3}$
72. $\log_x 2 = \dfrac{1}{4}$
73. $\log_{1.2} 1 = x$
74. $\log_{5.4} 1 = x$
75. $\log_{1/4} 4 = x$
76. $\log_{1/5} 25 = x$
77. $\log_d d^{-3} = x$
78. $\log_d d^{-7} = x$
79. $\log_6 \sqrt{6} = x$
80. $\log_{10} \sqrt{10} = x$

Graph each logarithmic function. Plot and label at least three points. State the domain and range of each function.

81. $y = \log_8 x$
82. $y = \log_{1.6} x$
83. $y = \log_{0.4} x$
84. $y = \log_{0.8} x$

 You Be the Teacher!

Correct each student's errors, if any.

85. Solve the equation $\log_{1/4} 16 = x$.

Ahmed's work:

$\log_{1/4} 16 = x$

$\dfrac{1}{4} x = 16$

$x = 64$

86. Solve the equation $\log_x (-25) = 2$.

Celina's work:

$\log_x (-25) = 2$

$x^2 = \dfrac{1}{25}$

$x = \dfrac{1}{5}$

PIECE IT TOGETHER / **SECTIONS 12.1–12.4**

1. Let $f(x) = 4x + 15$ and $g(x) = 3 - 2x$. Find $f + g, f - g,$ $f \cdot g, \dfrac{f}{g}, f \circ g,$ and $g \circ f$. State the domain of each combined function. (*Section 12.1, Objectives 1 and 3*)

Let $f(x) = 5x^2 + 1$ and $g(x) = 6 - x^2$. **Evaluate each function.** (*Section 12.1, Objective 2*)

2. $(f + g)(-3)$

3. $(f - g)(4)$

4. $(f \cdot g)(3)$

5. $\left(\dfrac{f}{g}\right)(-2)$

Determine if each function is one-to-one. If so, find its inverse. Verify the functions are inverses. (*Section 12.2, Objectives 1–4*)

6. $f(x) = \{(-4, 1), (-2, 5), (0, 5), (2, 6)\}$

7. $f(x) = \dfrac{x}{x - 4}$

Graph each function. Plot and label at least three points on the graph. State the domain and range of each function. (*Section 12.3, Objective 1, and Section 12.4, Objective 4*)

8. $f(x) = 1.5^x$

9. $f(x) = \log_{1/5} x$

Solve each exponential equation. (*Section 12.3, Objective 2*)

10. $3^{2x-5} = 27^x$

11. $2^{5x-5} = 4^x$

12. The population of Mexico (in millions), can be modeled by the function $f(x) = 111.21(1.0112)^x$, where x is the number of years after 2009. What is an estimate for the population in 2012? 2020? 2030? Round answers to the nearest million.

Simplify each expression. (*Section 12.4, Objective 2*)

13. $\log_8 \dfrac{1}{512}$

14. $\log_{1/5} 125$

15. $\log_{3.5} 3.5$

16. $\log_4(-16)$

Solve each logarithmic equation. (*Section 12.4, Objective 3*)

17. $\log_6 x = 4$

18. $\log_x 32 = 5$

SECTION 12.5 / Properties of Logarithms

▶ OBJECTIVES

As a result of completing this section, you will be able to

1. Apply the product rule for logarithms.
2. Apply the quotient rule for logarithms.
3. Apply the power rule for logarithms.
4. Rewrite logarithmic expressions using two or more rules.
5. Apply inverse properties of logarithms.
6. Troubleshoot common errors.

Objective 1 ▶

Apply the product rule for logarithms.

In Section 12.4, we introduced the concept of a logarithm and discovered that a logarithm is actually an exponent. In this section, we will apply properties of exponents and inverse properties to develop properties of logarithms.

The Product Rule for Logarithms

The first rule that we will develop is the product rule for logarithms, or logs for short. That is, we want to discover a rule for simplifying the log of a product. Consider the following expressions.

Based on Definition	**Observed Rule**
$\log_2(4 \cdot 8) = \log_2 32 = 5$ since $2^5 = 32$	The log of a product can be obtained by adding the log of each factor.
$\log_2 4 = 2$ and $\log_2 8 = 3$ since $2^2 = 4$ and $2^3 = 8$	$\log_2(4 \cdot 8) = \log_2 4 + \log_2 8$ $5 = 2 + 3$

> **Property: Product Rule for Logarithms**
>
> For $x, y,$ and $b > 0$ with $b \neq 1$,
> $$\log_b(xy) = \log_b x + \log_b y$$

This property states that the log of a product is the sum of the log of the factors. To verify this property, we use the product of like bases rule for exponents, $b^m \cdot b^n = b^{m+n}$. We assign variables to each exponential expression and write them in their logarithmic form.

$$x = b^m \text{ is equivalent to } \log_b x = m$$
$$y = b^n \text{ is equivalent to } \log_b y = n$$

Next multiply the left sides and right sides of the exponential equations to obtain $x \cdot y = b^m \cdot b^n$. So, $x \cdot y = b^{m+n}$. If we write this equation in its logarithmic form, we have

$$\log_b(x \cdot y) = m + n$$

Since $m = \log_b x$ and $n = \log_b y$, we have

$$\log_b(xy) = \log_b x + \log_b y$$

> **Procedure: Applying the Product Rule for Logarithms**
>
> **Step 1:** If the expression involves the sum of two logarithms with the same base, then the expression can be combined into one logarithm. It is equal to the logarithm of the product of the arguments of the two logarithms.
>
> **Step 2:** If the expression involves the logarithm of an expression with factors, then the expression can be written as the sum of the logarithms of each factor.

Objective 1 Examples Use the product rule for logarithms to write each expression as a single logarithm. Simplify, if possible.

1a. $\log_2 3 + \log_2 5$ **1b.** $\log_4 \dfrac{1}{3} + \log_4 12$ **1c.** $\log_3 x + \log_3(x + 1)$

Solutions **1a.**
$$\log_2 3 + \log_2 5 = \log_2(3 \cdot 5)$$
$$= \log_2 15$$

1b.
$$\log_4 \frac{1}{3} + \log_4 12 = \log_4\left(\frac{1}{3} \cdot 12\right)$$
$$= \log_4 4$$
$$= 1$$

1c.
$$\log_3 x + \log_3(x + 1) = \log_3[x(x + 1)]$$
$$= \log_3(x^2 + x)$$

✓ Student Check 1 Use the product rule for logarithms to write each expression as a single logarithm. Simplify, if possible.

1a. $\log_6 2 + \log_6 7$ **1b.** $\log_5 \dfrac{1}{2} + \log_5 10$ **1c.** $\log_2 x + \log_2(x - 3)$

The Quotient Rule for Logarithms

Objective 2 ▶

Apply the quotient rule for logarithms.

Now we investigate what happens when we simplify the log of a quotient.

Based on Definition	**Observed Rule**
$\log_2 \dfrac{4}{8} = \log_2 \dfrac{1}{2} = -1$ since $2^{-1} = \dfrac{1}{2}$	The log of a quotient can be obtained by subtracting the log of the numerator and the log of the denominator. $\log_2 \dfrac{4}{8} = \log_2 4 - \log_2 8$ $-1 = 2 - 3$

> **Property: Quotient Rule for Logarithms**
>
> For x, y, and $b > 0$ and $b \neq 1$,
>
> $$\log_b \frac{x}{y} = \log_b x - \log_b y$$

This property states that the log of a quotient is the difference of the log of the numerator and denominator. To verify this property, we use the quotient of like bases rule for exponents, $\dfrac{b^m}{b^n} = b^{m-n}$. We assign variables to each exponential expression and write them in their logarithmic form.

$$x = b^m \text{ is equivalent to } \log_b x = m$$
$$y = b^n \text{ is equivalent to } \log_b y = n$$

Next divide the left sides and right sides of the exponential equations to obtain $\dfrac{x}{y} = \dfrac{b^m}{b^n}$. So, $\dfrac{x}{y} = b^{m-n}$. If we write this equation in its logarithmic form, we have

$$\log_b \frac{x}{y} = m - n$$

Since $m = \log_b x$ and $n = \log_b y$, we have

$$\log_b \frac{x}{y} = \log_b x - \log_b y$$

> **Procedure: Applying the Quotient Rule for Logarithms**
>
> **Step 1:** If the expression involves the difference of two logarithms with the same base, then the expression can be combined into one logarithm. It is equal to the logarithm of the quotient of the arguments of the two logarithms.
>
> **Step 2:** If the expression involves the logarithm of an expression with a quotient, then the expression can be written as the difference of the logarithm of the numerator and the logarithm of the denominator.

Objective 2 Examples Use the quotient rule for logarithms to write each expression as a single logarithm. Simplify, if possible.

2a. $\log_2 20 - \log_2 5$ **2b.** $\log_3 x - \log_3 7$ **2c.** $\log_{10}(x + 2) - \log_{10}(x - 3)$

Solutions **2a.** $\log_2 20 - \log_2 5 = \log_2 \dfrac{20}{5}$

$$= \log_2 4$$
$$= 2$$

2b. $\log_3 x - \log_3 7 = \log_3 \dfrac{x}{7}$

2c. $\log_{10}(x + 2) - \log_{10}(x - 3) = \log_{10}\left(\dfrac{x + 2}{x - 3}\right)$

✓ Student Check 2 Use the quotient rule for logarithms to write each expression as a single logarithm. Simplify, if possible.

a. $\log_3 54 - \log_3 2$ **b.** $\log_2 y - \log_2 5$ **c.** $\log_5(x + 1) - \log_5(x - 2)$

Power Rule for Logarithms

Objective 3 ▶

Apply the power rule for logarithms.

The last rule of logarithms we develop is the power rule.

Based on Definition	**Observed Rule**
$\log_2 4^3 = \log_2 64 = 6$ since $2^6 = 64$	The log of a power can be obtained by multiplying the exponent by the log of the base. $\log_2 4^3 = 3\log_2 4$ $6 = 3(2)$

> **Property: Power Rule for Logarithms**
>
> For x and $b > 0$ with $b \neq 1$,
> $$\log_b x^n = n \log_b x$$

This property states that the log of an exponential expression is the exponent times the log of the base of the exponential expression. To verify this property, we must use the power of a power rule for exponents, $(b^m)^n = b^{mn}$. So, we assign a variable to b^m and convert it to its logarithmic form.

$$x = b^m \text{ is equivalent to } \log_b x = m$$

So, if we raise each side of the exponential equation to an exponent of n, we get

$$x^n = (b^m)^n$$
$$x^n = b^{mn}$$

If we write this in logarithmic form, we have

$$\log_b x^n = mn$$

Since $\log_b x = m$, we have

$$\log_b x^n = [\log_b x]n$$
$$\log_b x^n = n \log_b x$$

This rule enables us to solve exponential equations in Section 12.6.

> **Procedure: Applying the Power Rule for Logarithms**
>
> **Step 1:** If the expression involves a coefficient of a logarithm, then the expression can be written with a coefficient of one. The coefficient of the logarithm becomes the exponent of the argument of the logarithm.
>
> **Step 2:** If the expression involves the logarithm of an exponential expression, then the exponent of the exponential expression becomes the coefficient of the logarithm.

Objective 3 Examples **Use the power rule for logarithms to rewrite each expression.**

3a. $\log_2 3^5$ **3b.** $\log_4 x^2$ **3c.** $\log_3 \sqrt{6}$ **3d.** $\log_{10} 10^{x-1}$

Solutions   **3a.** $\log_2 3^5 = 5\log_2 3$

3b. $\log_4 x^2 = 2\log_4 x$

3c. $\log_3 \sqrt{6} = \log_3 6^{1/2}$

$$= \frac{1}{2}\log_3 6$$

3d. $\log_{10}10^{x-1} = (x-1)\log_{10}10$

$= (x-1)(1)$

$= x-1$

✓ **Student Check 3** Use the power rule for logarithms to rewrite each expression.

a. $\log_4 2^5$ **b.** $\log_3 x^4$ **c.** $\log_2\sqrt{7}$ **d.** $\log_{10}10^{x+2}$

Applying Two or More Logarithmic Properties

Objective 4 ▶

Rewrite logarithmic expressions using two or more rules.

To rewrite some expressions, we must apply more than one rule of logarithms. A summary of the rules is provided.

Property: Summary of Logarithmic Properties

Product rule: $\log_b xy = \log_b x + \log_b y$

Quotient rule: $\log_b \dfrac{x}{y} = \log_b x - \log_b y$

Power rule: $\log_b x^n = n\log_b x$

Procedure: Rewriting Logarithmic Expressions as a Single Logarithm

Step 1: Apply the power rule by making the coefficients of the logarithm the exponent of the argument of the logarithm.

Step 2: Work from left to right, applying the product or quotient rules as necessary.

Procedure: Expanding a Single Logarithm as a Combination of Logarithmic Expressions

1. If the argument of the logarithm is a quotient, apply the quotient rule.
   a. If either the numerator or denominator is a product, apply the product rule.
   b. If any of the arguments of the logarithms have a power, apply the power rule.
2. If the argument of the logarithm is a product, apply the product rule. If any of the factors has an exponent, apply the power rule.

Objective 4 Examples

Rewrite each logarithmic expression. For parts (a)–(c), write the expression as a single logarithm. For parts (d) and (e), expand the logarithmic expression.

4a. $3\log_9 2 + 4\log_9 3$ **4b.** $2\log_3 x + \dfrac{1}{2}\log_3(x+1)$

4c. $\log_4 x - 2\log_4 y$ **4d.** $\log_2\left(\dfrac{3x}{y}\right)$ **4e.** $\log_5\left(\dfrac{x^3}{y^4}\right)$

Solutions **4a.** $3\log_9 2 + 4\log_9 3 = \log_9 2^3 + \log_9 3^4$ Apply the power rule for logarithms.

$= \log_9 8 + \log_9 81$ Simplify the exponents.

$= \log_9(8 \cdot 81)$ Apply the product rule for logarithms.

$= \log_9 648$ Simplify.

4b. $2\log_3 x + \dfrac{1}{2}\log_3(x+1)$

$= \log_3 x^2 + \log_3(x+1)^{1/2}$ Apply the power rule for logarithms.

$= \log_3 x^2 + \log_3 \sqrt{x+1}$ Apply $b^{\frac{1}{n}} = \sqrt[n]{b}$.

$= \log_3\left(x^2\sqrt{x+1}\right)$ Apply the product rule for logarithms.

4c. $\log_4 x - 2\log_4 y = \log_4 x - \log_4 y^2$ Apply the power rule for logarithms.

$= \log_4\left(\dfrac{x}{y^2}\right)$ Apply the quotient rule for logarithms.

4d. $\log_2\left(\dfrac{3x}{y}\right) = \log_2 3x - \log_2 y$ Apply the quotient rule for logarithms.

$= \log_2 3 + \log_2 x - \log_2 y$ Apply the product rule for logarithms.

4e. $\log_5\left(\dfrac{x^3}{y^4}\right) = \log_5 x^3 - \log_5 y^4$ Apply the quotient rule for logarithms.

$= 3\log_5 x - 4\log_5 y$ Apply the power rule for logarithms.

✔ **Student Check 4**   Rewrite each logarithmic expression. For parts (a)–(c), write the expression as a single logarithm. For parts (d) and (e), expand the logarithmic expression.

a. $2\log_6 3 + 3\log_6 2$ **b.** $3\log_2 y + \dfrac{1}{3}\log_2(y+2)$

c. $\log_3 x - 4\log_3 y$ **d.** $\log_4\left(\dfrac{3a}{b}\right)$ **e.** $\log_7\left(\dfrac{x^2}{y^3}\right)$

Inverse Properties of Logarithms

Objective 5 ▶

Apply inverse properties of logarithms.

In Section 12.4 we learned that the functions $f(x) = b^x$ and $f(x) = \log_b x$ $(b > 0, b \neq 1)$ are inverse functions, and in Section 12.2 we learned that when we compose a function and its inverse, we get the value x. That is, $(f \circ f^{-1})(x) = x$ and $(f^{-1} \circ f)(x) = x$. Applying these facts to $f(x) = b^x$ and $f^{-1}(x) = \log_b x$, we get

$$(f \circ f^{-1})(x) = f(\log_b x) = b^{\log_b x} = x$$

$$(f^{-1} \circ f)(x) = f^{-1}(b^x) = \log_b b^x = x$$

> **Property: Inverse Properties of Logarithms**
> For $b > 0, b \neq 1$,
> $$\log_b b^x = x$$
> $$b^{\log_b x} = x, \ x > 0$$

Objective 5 Examples   Simplify each expression.

Problems	Solutions
5a. $\log_{10} 10^5$	$\log_{10} 10^5 = 5$
5b. $\log_2 2^{x+1}$	$\log_2 2^{x+1} = x + 1$
5c. $3^{\log_3 9}$	$3^{\log_3 9} = 9$
5d. $6^{\log_6(x+2)}$	$6^{\log_6(x+2)} = x + 2$

Student Check 5 Simplify each expression.

    a. $\log_5 5^4$ **b.** $\log_3 3^{x-5}$ **c.** $2^{\log_2 8}$ **d.** $10^{\log_{10}(x-1)}$

> **Note:** *The problems in Example 5 can be simplified in another way. For instance, we can apply the power rule to simplify* $\log_{10} 10^5$. *We get* $\log_{10} 10^5 = 5\log_{10} 10 = 5(1) = 5$. *We can also simplify* $3^{\log_3 9}$ *by first evaluating* $\log_3 9$. *Since* $\log_3 9 = 2$, *we have* $3^{\log_3 9} = 3^2 = 9$.

Objective 6 ▶

Troubleshoot common errors.

Troubleshooting Common Errors

Some common errors associated with logarithmic properties are shown.

Objective 6 Examples A problem and an incorrect solution are given. Provide the correct solution and an explanation of the error.

6a. Rewrite $\log_5(x^2 + 3)$, if possible.

Incorrect Solution	Correct Solution and Explanation
$\log_5(x^2 + 3) = \log_5 x^2 + \log_5 3$ $= 2\log_5 x + \log_5 3$	The sum of logarithms comes from the logarithm of a product, not the logarithm of a sum. The original expression cannot be rewritten. The expression $\log_5(3x^2)$ is written as $\log_5 3 + 2\log_5 x$.

6b. Rewrite $\dfrac{\log_5 3}{\log_5 7}$, if possible.

Incorrect Solution	Correct Solution and Explanation
$\dfrac{\log_5 3}{\log_5 7} = \log_5 3 - \log_5 7$	The difference of logarithms comes from the logarithm of a quotient, not the quotient of logarithms. The original expression cannot be rewritten. The expression $\log_5 \dfrac{3}{7}$ is written as $\log_5 3 - \log_5 7$.

6c. Rewrite $(\log_2 x)^4$, if possible.

Incorrect Solution	Correct Solution and Explanation
$(\log_2 x)^4 = 4\log_2 x$	The power rule can only be applied if the exponent is on the expression inside the logarithm, not when the exponent is on the entire logarithmic expression. The original expression cannot be rewritten. The expression $\log_2 x^4$ is written as $4\log_2 x$.

ANSWERS TO STUDENT CHECKS

Student Check 1   **a.** $\log_6 14$ **b.** 1   **c.** $\log_2(x^2 - 3x)$

Student Check 2   **a.** 3   **b.** $\log_2 \dfrac{y}{5}$ **c.** $\log_5\left(\dfrac{x+1}{x-2}\right)$

Student Check 3   **a.** $5\log_4 2 = \dfrac{5}{2}$ **b.** $4\log_3 x$

     c. $\dfrac{1}{2}\log_2 7$ **d.** $x + 2$

Student Check 4   **a.** $\log_6 72$ **b.** $\log_2\left(y^3 \sqrt[3]{y+2}\right)$

     c. $\log_3 \dfrac{x}{y^4}$ **d.** $\log_4 3 + \log_4 a - \log_4 b$

     e. $2\log_7 x - 3\log_7 y$

Student Check 5   **a.** 4   **b.** $x - 5$ **c.** 8   **d.** $x - 1$

SUMMARY OF KEY CONCEPTS

1. The product, quotient, and power rules for logarithms come from the properties of exponents. These rules enable us to rewrite a logarithmic expression as a single logarithm or as a combination of logarithms. The rules are, for x, y, and $b > 0$ with $b \neq 1$,

   Product rule: $\log_b xy = \log_b x + \log_b y$

   Quotient rule: $\log_b \dfrac{x}{y} = \log_b x - \log_b y$

   Power rule: $\log_b x^n = n\log_b x$

2. The inverse properties of logarithms are, for $b > 0$, $b \neq 1$,

   $$\log_b b^x = x$$
   $$b^{\log_b x} = x, \; x > 0$$

SECTION 12.5 / EXERCISE SET

Write About It!

Use complete sentences in your answer to each exercise.

1. Explain how to apply the product rule for logarithms.
2. Explain how to apply the quotient rule for logarithms.
3. Explain how to apply the power rule for logarithms.
4. Explain the difference between $\log_a x^m$ and $(\log_a x)^m$.
5. Explain why $\log_a(x + y) \neq \log_a x + \log_a y$. Give an example.
6. Explain why $\log_a(x - y) \neq \log_a x - \log_a y$. Give an example.

Practice Makes Perfect!

Use the product rule for logarithms to write each expression as a single logarithm. Simplify, if possible. Assume all logarithmic arguments are positive numbers. (*See Objective 1.*)

7. $\log_2 5 + \log_2 3$
8. $\log_5 4 + \log_5 2$
9. $\log_3 5 + \log_3 6$
10. $\log_9 7 + \log_9 6$
11. $\log_3 \dfrac{2}{5} + \log_3 20$
12. $\log_7 \dfrac{1}{3} + \log_7 12$
13. $\log_2 \dfrac{1}{5} + \log_2 10$
14. $\log_6 \dfrac{2}{7} + \log_6 21$
15. $\log_4 7 + \log_4 x$
16. $\log_{10} 6 + \log_{10} y$
17. $\log_2 x + \log_2 (x - 4)$
18. $\log_3 x + \log_3 (x + 5)$
19. $\log_a(x + 2) + \log_a(x - 3)$
20. $\log_b(x + 3) + \log_b(x + 4)$

Use the quotient rule for logarithms to write each expression as a single logarithm. Simplify, if possible. Assume all logarithmic arguments are positive numbers. (*See Objective 2.*)

21. $\log_2 15 - \log_2 3$
22. $\log_4 36 - \log_4 2$
23. $\log_5 45 - \log_5 9$
24. $\log_3 12 - \log_3 4$
25. $\log_3 x - \log_3 8$
26. $\log_2 x - \log_2 5$

27. $\log_6 24x - \log_6 2x$
28. $\log_7 42y - \log_7 3y$
29. $\log_2(x - 2) - \log_2(x + 3)$
30. $\log_3 x - \log_3(x + 6)$
31. $\log_9(x^2 - 25) - \log_9(x - 5)$
32. $\log_{10}(x^2 - 4) - \log_{10}(x + 2)$

Use the power rule for logarithms to rewrite each expression. Assume all logarithmic arguments are positive numbers. (*See Objective 3.*)

33. $\log_3 5^2$
34. $\log_2 7^3$
35. $\log_5 x^4$
36. $\log_{11} y^6$
37. $\log_3(x - 2)^4$
38. $\log_5(x + 3)^6$
39. $\log_3 \sqrt[5]{x}$
40. $\log_6 \sqrt[3]{y}$
41. $\log_{10} \sqrt{10}$
42. $\log_{12} \sqrt[5]{12}$
43. $\log_{10} 10^{2x-3}$
44. $\log_3 3^{4x+5}$

Write each expression as a single logarithm. Assume all logarithmic arguments are positive numbers. (*See Objective 4.*)

45. $2\log_3 5 + 3\log_3 2$
46. $5\log_9 2 + \log_9 6$
47. $3\log_6 3 + 2\log_6 5$
48. $3\log_{10} 5 + 2\log_{10} 4$
49. $4\log_5 x + \dfrac{1}{5}\log_5(x + 6)$
50. $2\log_4 y + \dfrac{1}{3}\log_4(y - 2)$
51. $\log_{10} x - 3\log_{10} y$
52. $3\log_2 x - 5\log_2 y$
53. $2\log_5 a - 4\log_5 b$
54. $8\log_{10} a - 2\log_{10} b$

Expand each logarithm. Assume all logarithmic arguments are positive numbers. (*See Objective 4.*)

55. $\log_{10} a^5 \sqrt{a - 3}$
56. $\log_6 b^7 \sqrt[3]{b + 5}$
57. $\log_2 \dfrac{x^{15}}{y^3}$
58. $\log_7 \dfrac{a^6}{b^8}$
59. $\log_4 \dfrac{3x}{y^2}$
60. $\log_6 \dfrac{10a^2}{b}$
61. $\log_3 \dfrac{x^2}{27y}$
62. $\log_7 \dfrac{z^3}{49w^2}$

Simplify each expression. Assume all logarithmic arguments are positive numbers. (*See Objective 5.*)

63. $\log_{10}10^6$

64. $\log_9 9^5$

65. $\log_3 3^{4x-5}$

66. $\log_{10}10^{3x+7}$

67. $3^{\log_3 6}$

68. $2^{\log_2 x}$

69. $4^{\log_4(x+5)}$

70. $5^{\log_5(2x+3)}$

71. $2^{\log_2(2x)}$

72. $3^{\log_3(4y+5)}$

 Mix 'Em Up!

Use the rules for logarithms to write each expression as a single logarithm. Simplify, if possible. Assume all logarithmic arguments are positive numbers.

73. $\log_3 6 + \log_3 14 - \log_3 7$

74. $\log_6 4 - \log_6 2 + \log_6 5$

75. $\log_5 x - \log_5 y + \log_5 z$

76. $\log_9 a + \log_9 b - \log_9 c$

77. $\log_{10}\dfrac{4}{3} + \log_{10}18 - \log_{10}8$

78. $\log_7\dfrac{3}{2} - \log_7 21 + \log_7 28$

79. $\log_3 20 - \log_3\dfrac{15}{4} + \log_3 6$

80. $\log_5 8 - \log_5\dfrac{6}{7} + \log_5 3$

81. $2\log_9 6 + \log_9 5 - \dfrac{1}{3}\log_9 64$

82. $2\log_2 3 + \log_2 10 - \dfrac{1}{3}\log_2 125$

83. $\log_3 x + \log_3(x-1) - \log_3(x^2+4)$

84. $\log_4(y-3) - \log_4(y+1) + \log_4(y+3)$

85. $2\log_{10}x - \dfrac{1}{2}\log_{10}x - 3\log_{10}(x+1)$

86. $2\log_5 y - \dfrac{3}{2}\log_5 y - 2\log_5(y+4)$

87. $3\log_6 a + 2\log_6 b - 5\log_6 c$

88. $2\log_7 x - 4\log_7 y + \log_7 z$

89. $\log_{10}a - 2\log_{10}b - \dfrac{1}{3}\log_{10}c$

90. $\log_9 r - \dfrac{1}{2}\log_9 s + 3\log_9 t$

91. $\log_6(x^3-8) - \log_6(x-2)$

92. $\log_2(x^3+8) - \log_2(x+2)$

Write each expression as a combination of logarithms. Assume all logarithmic arguments are positive numbers.

93. $\log_3\dfrac{5x^2}{y}$

94. $\log_6\dfrac{7a}{b^2}$

95. $\log_7\sqrt{\dfrac{6x}{y}}$

96. $\log_4\sqrt[3]{\dfrac{a}{5b}}$

97. $\log_2\left(x^5 y^3\sqrt{x+3}\right)$

98. $\log_4\left(a^6 b^3\sqrt[5]{b+1}\right)$

99. $\log_{10}\left(\dfrac{100a^3}{b^5}\right)$

100. $\log_5\left(\dfrac{25x^2}{y^6}\right)$

101. $\log_4\dfrac{r^2(r+3)}{s^3}$

102. $\log_{10}\dfrac{x(x^2+1)}{y^4}$

Simplify each expression. Assume all logarithmic arguments are positive numbers.

103. $\log_{10}10^{6x+5}$

104. $\log_3 3^{x+5}$

105. $\log_a a^x$

106. $\log_b b^y$

107. $5^{\log_5 16x}$

108. $2^{\log_2 7y}$

109. $a^{\log_a(2x+1)}$

110. $3^{4\log_3 y}$

111. $6^{2\log_6(3a)}$

112. $5^{4\log_5 2y}$

You Be the Teacher!

Correct each student's errors, if any.

113. Write the expression $\log_3 4 + \log_3 x$ as a single logarithm.

Karen's work:

$\log_3(4+x)$

114. Write the expression $\log_{10}15 - \log_{10}5$.

Mona's work: $\log_{10}15 - \log_{10}5 = \log_{10}(15-5) = \log_{10}10 = 1$

115. Simplify the expression $(\log_2 8)^2$.

Myra's work:

$(\log_2 8)^2 = (\log_2 2^3)^2 = \log_2 2^6 = 6$

116. Simplify the expression $3^{2\log_3 5}$.

Cara's work:

$3^{2\log_3 5} = 3^{\log_3 10} = 10$

The Common Log, Natural Log, and Change-of-Base Formula

Suppose you learn a new task over a one-week period and you become about 90% proficient at it. However, as time goes by, you don't practice and you begin forgetting how to do the task though some of your ability to complete the task is retained. This situation can be modeled with the equation: $P = 90 - 12.5 \ln d$, where P is the percent proficiency and d is the number of days without practicing. What is your proficiency level if 5 days go by without practicing? 10 days? 30 days?

The expression $\ln d$ represents a special type of logarithm. In this section, we will learn about this logarithm along with another special logarithm. These are logarithms with the bases of 10 and the number e.

Common Logarithms

Objective 1 ▶

Simplify common logarithms.

On a calculator, we find a special command, the $\boxed{\text{LOG}}$ command. This represents a logarithm with an understood base. If we use this command to simplify some logarithms, we can determine what it means. Using a calculator, we find the following.

```
log(10)
                    1
log(100)
                    2
log(1000)
                    3
```

   Note: *On a calculator, we must enter the argument of a log in parentheses.*

To determine the base of the LOG function, we let x represent the base. So, we have the following logarithmic statements and their equivalent exponential form.

$$\log_x 10 = 1 \rightarrow x^1 = 10$$
$$\log_x 100 = 2 \rightarrow x^2 = 100$$
$$\log_x 1000 = 3 \rightarrow x^3 = 1000$$

Solving each of these equations for x gives us $x = 10$. Therefore, the $\boxed{\text{LOG}}$ command is used to simplify logarithms with a base of 10. So, the $\boxed{\text{LOG}}$ function tells us the exponent of 10 that will produce the argument of the logarithm. A logarithm with a base of 10 is called a common logarithm and is written as $\log x$.

Definition: For $x > 0$,

$$y = \log x \text{ is equivalent to } y = \log_{10} x$$

The equivalent exponential form of the **common logarithm** is $10^x = y$.

So, the common logarithm means the exponent of 10 that produces the argument value.

Procedure: Simplifying Common Logarithms

Step 1: If possible, rewrite the argument of the logarithm as a power of 10; that is, as 10^x.

Step 2: Apply the inverse property to simplify the logarithm. Recall $\log 10^x = x$ since the base of the logarithm and the base of the exponent are the same.

Objective 1 Examples Simplify each common logarithm.

Problems	Solutions
1a. $\log 100$	$\log 100 = \log 10^2 = 2$
1b. $\log \dfrac{1}{100}$	$\log \dfrac{1}{100} = \log \dfrac{1}{10^2} = \log 10^{-2} = -2$
1c. $\log 1$	$\log 1 = \log 10^0 = 0$
1d. $\log \sqrt{10}$	$\log \sqrt{10} = \log 10^{1/2} = \dfrac{1}{2}$

✓ Student Check 1   Simplify each common logarithm.

    a. $\log 10$ **b.** $\log \dfrac{1}{10}$ **c.** $\log 1000$ **d.** $\log \sqrt[3]{10}$

Natural Logarithms

Objective 2 ▶

Simplify natural logarithms.

Natural logarithms are another type of logarithm that has a special notation. Natural logarithms are denoted by $\log_e x$ but are usually written as $\ln x$. We read this as "el en of x." On a calculator, the function $\boxed{\text{LN}}$ denotes the natural logarithm. The natural logarithm is a logarithm with a base of e. Recall that e is an irrational number and is approximately equal to 2.71828. This number arises in various real-life situations. It is used in financing, exponential growth and decay, and other applications.

> **Definition: Natural Logarithm**
>
> For $x > 0$,
>
> $$y = \ln x \text{ is equivalent to } y = \log_e x$$
>
> The equivalent exponential form of $y = \ln x$ is $e^y = x$.

So, the natural logarithm means to find the exponent of e that produces the argument value.

> **Procedure: Simplifying Natural Logarithms**
>
> **Step 1:** If possible, rewrite the argument as a power of e; that is, as e^x.
> **Step 2:** Apply the inverse property to simplify the logarithm. $\ln(e^x) = x$ since the base of the logarithm and the base of the exponent are the same.

Objective 2 Examples Simplify each natural logarithm.

Problems	Solutions
2a. $\ln e$	$\ln e = \ln e^1 = 1$
2b. $\ln e^3$	$\ln e^3 = 3$
2c. $\ln 1$	$\ln 1 = \ln e^0 = 0$
2d. $\ln \dfrac{1}{e^2}$	$\ln \dfrac{1}{e^2} = \ln e^{-2} = -2$

✓ Student Check 2   Simplify each natural logarithm.

    a. $\ln e^2$ **b.** $\ln \sqrt{e}$ **c.** $\ln \dfrac{1}{e}$

Simplifying Common and Natural Logarithms on a Calculator

Objective 3 ▶

Use a calculator to approximate common and natural logarithms.

We are not restricted to using only powers of 10 and powers of e as arguments of common logarithms and natural logarithms, respectively. We can use a calculator to simplify common and natural logarithms for any positive real number.

Objective 3 Examples   Use a calculator to approximate each logarithm to two decimal places.

3a. $\log 5$ **3b.** $\ln 10$

Solutions   **3a.** Press  to get $\log 5 \approx 0.70$.

Check: $10^{0.70} \approx 5.01$. Because of rounding, the value will not be exactly 5 but it is very close, so our work is correct.

3b. Press ⬤LN⬤ ⬤1⬤ ⬤0⬤ ⬤⬤ to get $\ln 10 \approx 2.30$.

Check: $e^{2.30} \approx 9.97$.

✔ **Student Check 3**   Use a calculator to approximate each logarithm to two decimal places.

   a. $\log 80$ **b.** $\ln 17$

Solving Logarithmic Equations

Objective 4 ▶

Solve simple logarithmic equations.

We can solve simple logarithmic equations by converting them to their exponential form.

> **Procedure: Solving a Logarithmic Equation Using LOG and LN**
>
> **Step 1:** Write the equation in exponential form.
>   **a.** If the equation is of the form $y = \log x$, rewrite it as $10^y = x$.
>   **b.** If the equation is of the form $y = \ln x$, rewrite it as $e^y = x$.
> **Step 2:** Solve the resulting equation.

Objective 4 Examples   Solve each equation for x. Give an exact solution and approximate the solution to two decimal places, if necessary.

4a. $\log x = 4$ **4b.** $\log(x + 1) = 2$ **4c.** $\log(2x - 3) = -0.5$
4d. $\ln x = 2$ **4e.** $\ln(2x) = 7$

Solutions   **4a.** $\log x = 4$

$10^4 = x$ Write in exponential form.

$x = 10,000$ Simplify.

So, the solution set is $\{10,000\}$.

4b. $\log(x + 1) = 2$

$10^2 = x + 1$ Write in exponential form.

$100 = x + 1$ Simplify.

$99 = x$ Subtract 1 from each side.

So, the solution set is $\{99\}$.

4c. $\log(2x - 3) = -0.5$

$\quad\quad 10^{-0.5} = 2x - 3$ Write in exponential form.

$\quad\quad 10^{-0.5} + 3 = 2x$ Add 3 to each side.

$\quad\quad \dfrac{10^{-0.5} + 3}{2} = x$ Divide each side by 2.

$\quad\quad\quad\quad x \approx 1.66$ Simplify.

So, the solution set is $\left\{ \dfrac{10^{-0.5} + 3}{2} \right\}$ or $\{1.66\}$.

4d. $\quad\quad\quad \ln x = 2$

$\quad\quad\quad\quad e^2 = x$ Write in exponential form.

$\quad\quad\quad\quad x \approx 7.39$ Simplify.

So, the solution set is $\{e^2\}$ or $\{7.39\}$.

4e. $\quad\quad \ln(2x) = 7$

$\quad\quad\quad\quad e^7 = 2x$ Write in exponential form.

$\quad\quad\quad\quad \dfrac{e^7}{2} = x$ Divide each side by 2.

$\quad\quad\quad\quad x \approx 548.32$ Simplify.

So, the solution set is $\left\{ \dfrac{e^7}{2} \right\}$ or $\{548.32\}$.

✓ **Student Check 4**   Solve each equation for x. Give an exact solution and approximate the solution to two decimal places, if necessary.

    a. $\log x = 3$ **b.** $\log(x - 4) = 1$ **c.** $\log(5x + 2) = -0.25$

    d. $\ln x = 5$ **e.** $\ln(6x) = 3$

The Change-of-Base Formula

Objective 5 ▶

Use the change-of-base formula.

A calculator has built-in functions for only the common logarithm and the natural logarithm; that is, logs with bases of 10 and e. If we need to simplify $\log_3 7$, we need a way to express it in terms of a common logarithm or natural logarithm. So, let $y = \log_3 7$, which is equivalent to $3^y = 7$. This is an exponential equation in which the left side and right side cannot be expressed with the same base. So, we have to use a property of logarithms to solve for y. Recall the power rule states $\log x^n = n \log x$. It is also true that $\ln x^n = n \ln x$. So, we apply the common log to each side, which we can do since logarithms are one-to-one functions.

$\quad\quad\quad\quad 3^y = 7$ Write the exponential form of $y = \log_3 7$.

$\quad\quad\quad \log 3^y = \log 7$ Apply the common log to each side.

$\quad\quad y \log 3 = \log 7$ Apply the power rule for logarithms.

$\quad\quad\quad\quad y = \dfrac{\log 7}{\log 3}$ Divide each side by $\log 3$.

So, $\log_3 7 = \dfrac{\log 7}{\log 3}$. Now we can use our calculator to approximate the value of this logarithm. We could have also taken the natural log of both sides to get $\log_3 7 = \dfrac{\ln 7}{\ln 3}$.

Definition: **The Change-of-Base Formula**

For $a, b > 0$, $b \neq 1$,

$$\log_b a = \frac{\log a}{\log b} \quad \text{or} \quad \frac{\ln a}{\ln b}$$

Procedure: **Using the Change-of-Base Formula**

Step 1: Rewrite the logarithm using either the common log or the natural log.
  a. The log of the argument goes in the numerator.
  b. The log of the base goes in the denominator. (Think "basement.")
Step 2: Use a calculator to simplify the resulting expression.

Objective 5 Examples   Use the change-of-base formula to rewrite each logarithm. Approximate its value to two decimal places.

5a. $\log_2 20$ **5b.** $\log_3 0.25$ **5c.** $f(x) = \log_4 x$

Solutions   **5a.** $\log_2 20 = \dfrac{\log 20}{\log 2}$ $\log_2 20 = \dfrac{\ln 20}{\ln 2}$

≈ 4.32 ≈ 4.32

Check: $2^{4.32} \approx 19.97$

5b. $\log_3 0.25 = \dfrac{\log 0.25}{\log 3}$ $\log_3 0.25 = \dfrac{\ln 0.25}{\ln 3}$

≈ -1.26 ≈ -1.26

Check: $3^{-1.26} \approx 0.25$

5c. $f(x) = \log_4 x = \dfrac{\log x}{\log 4}$ $f(x) = \log_4 x = \dfrac{\ln x}{\ln 4}$

✓ Student Check 5   Use the change-of-base formula to rewrite each logarithm. Approximate its value to two decimal places.

a. $\log_4 20$ **b.** $\log_5 0.1$ **c.** $f(x) = \log_6 x$

Applications

Objective 6 ▶

Solve applications of logarithms.

Logarithms are used to find the magnitude of an earthquake, the acidity of a solution, the brightness of a star, the color ratio between two stars, and the intensity level of a sound. Logarithms are also used to solve equations that have variable exponents. These types of equations come up when dealing with compound interest or exponential growth and decay. Section 12.7 will develop some of these ideas further. Example 6 provides examples of applications that can be modeled using logarithms.

Objective 6 Examples	**Solve each problem.**

6a. Suppose you learn a new task over a 1-week period and you become about 90% proficient at it. However, as time goes by, you don't practice and begin to forget how to do the task, though some of your ability to complete the task is retained. This situation can be modeled by the equation $P = 90 - 12.5 \ln d$, where P is the percent proficiency and d is the number of days without practicing. What is the proficiency level if 5 days go by without practicing? 10 days? 30 days? Round answers to two decimal places.

Solution **6a.** Let $d = 5$:

$$P = 90 - 12.5 \ln d$$
$$P = 90 - 12.5 \ln 5$$
$$P \approx 69.88$$

Let $d = 10$:

$$P = 90 - 12.5 \ln d$$
$$P = 90 - 12.5 \ln 10$$
$$P \approx 61.22$$

Let $d = 30$:

$$P = 90 - 12.5 \ln d$$
$$P = 90 - 12.5 \ln 30$$
$$P \approx 47.49$$

The proficiency level will be 69.88% after 5 days without practicing, 61.22% after 10 days without practicing, and 47.49% after 30 days without practicing.

6b. The moment magnitude scale, denoted M_W, is a scale that measures the energy released by an earthquake and is now preferred over the Richter scale for medium to large earthquakes. The formula is $M_W = \frac{2}{3} \log_{10} M_0 - 10.7$, where M_0 is the magnitude of the seismic moment, or the energy released, in dyne centimeters. As of July 2011, the largest reported moment is 2.5×10^{30} dyn · cm for the 1960 Chile earthquake. Compute the moment magnitude for this earthquake. (Source: http://earthquake.usgs.gov/learn/topics/measure.php)

Solution **6b.** $M_W = \frac{2}{3} \log_{10} M_0 - 10.7$ State the moment magnitude formula.

$M_W = \frac{2}{3} \log_{10}(2.5 \times 10^{30}) - 10.7$ Replace M_0 with 2.5×10^{30}.

$M_W \approx \frac{2}{3}(30.39794) - 10.7$ Simplify the logarithm.

$M_W \approx 9.6$ Simplify.

✓ Student Check 6 Solve each problem.

a. Suppose you learn a new task over a 1-week period and you become about 85% proficient at it. However, as time goes by, you don't practice and begin to forget how to do the task. Some of your ability to complete the task is retained. This situation can be modeled with the equation: $P = 85 - 12.5 \ln d$, where P is the percent proficiency and d is the number of days without practicing. What is your proficiency level if 5 days go by without practicing? 10 days? 30 days? Round answers to two decimal places.

b. As of July 2011, the second largest reported moment is 7.5×10^{29} dyn·cm for the 1964 Alaska earthquake. Use the moment magnitude scale to compute the moment magnitude for this earthquake. (Source: http://earthquake.usgs.gov/learn/topics/measure.php)

Objective 7 ▶	**Troubleshooting Common Errors**
Troubleshoot common errors.	Some common errors associated with common and natural logs are shown.

Objective 7 Examples   **A problem and an incorrect solution are given. Provide the correct solution and an explanation of the error.**

7a. Use the change-of-base formula to approximate $\log_2 3$ to two decimal places.

Incorrect Solution	Correct Solution and Explanation
$\log_2 3 = \dfrac{\log 2}{\log 3} \approx 0.63$	The log of the base goes in the denominator, not the numerator. $$\log_2 3 = \dfrac{\log 3}{\log 2} \approx 1.58$$

7b. Solve $\log 4x = 2$.

Incorrect Solution	Correct Solution and Explanation
$\log 4x = 2$ $e^2 = 4x$ $\dfrac{e^2}{4} = x$	The notation log represents the common logarithm, which has base 10, not e. $$\log 4x = 2$$ $$10^2 = 4x$$ $$100 = 4x$$ $$\dfrac{100}{4} = x$$ $$25 = x$$

ANSWERS TO STUDENT CHECKS

Student Check 1 **a.** 1   **b.** -1 **c.** 3   **d.** $\dfrac{1}{3}$

Student Check 2 **a.** 2   **b.** $\dfrac{1}{2}$ **c.** -1

Student Check 3 **a.** 1.90   **b.** 2.83

Student Check 4 **a.** $\{1000\}$ **b.** $\{14\}$

   c. $\left\{\dfrac{10^{-0.25} - 2}{5}\right\}$ or $\{-0.29\}$

   d. $\{e^5\}$ or $\{148.41\}$ **e.** $\left\{\dfrac{e^3}{6}\right\}$ or $\{3.35\}$

Student Check 5 **a.** 2.16   **b.** -1.43

   c. $f(x) = \dfrac{\log x}{\log 6}$ or $\dfrac{\ln x}{\ln 6}$

Student Check 6 **a.** 64.88%, 56.22%, 42.49%   **b.** 9.22

SUMMARY OF KEY CONCEPTS

1. A common logarithm is $y = \log x$, which is equivalent to $y = \log_{10} x$. It represents the exponent of 10 that produces x; that is, $10^y = x$. If the argument of the logarithm is a power of 10, we can use the inverse property of logarithms, $\log 10^x = x$, to simplify the expression.

2. A natural logarithm is $y = \ln x$, which is equivalent to $y = \log_e x$. It represents the exponent of e that produces x; that is, $e^y = x$. If the argument of the logarithm is a power of e, we can use the inverse property of logarithms, $\ln e^x = x$, to simplify the expression.

3. A calculator can be used to simplify common and natural logarithms when the argument is not a power of the base. Use the LOG and LN keys.

4. A simple logarithmic equation can be solved by converting the equation to its exponential form.

5. Any logarithm can be converted to a common log or natural log using the change-of-base formula. For $a, b > 0, b \neq 1$,

$$\log_b a = \dfrac{\log a}{\log b} \text{ or } \dfrac{\ln a}{\ln b}$$

GRAPHING CALCULATOR SKILLS

A graphing calculator can be used to simplify common and natural logarithms. We can also graph basic logarithmic functions.

Example 1: Find $\log 50$ and $\ln 50$.

Solution:

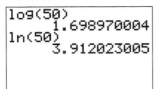

Example 2: Simplify $\log_4 9$.

Solution: $\log_4 9 = \dfrac{\log 9}{\log 4}$

Example 3: Graph $y = \log x$.

Solution:

Example 4: Graph $y = \log_2 x$.

Solution: By the change-of-base formula, $y = \log_2 x = \dfrac{\log x}{\log 2}$.

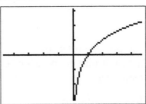

SECTION 12.6 / EXERCISE SET

Write About It!

Use complete sentences in your answer to each exercise.

1. What is the base of $y = \log x$? Write the equivalent exponential form.

2. What is the base of $y = \ln x$? Write the equivalent exponential form.

3. Explain how to apply the change-of-base formula. Give an example.

4. Explain why $\dfrac{\log 5}{\log 3}$, $\dfrac{\ln 5}{\ln 3}$, and $\log_3 5$ are the same.

5. Explain why $\ln 1$ and $\log 1$ are both equal to zero.

6. Explain how to evaluate each expression exactly.
   a. $\log_2 64$ b. $\log_4 64$
   c. $\log_8 64$ d. $\log_{1/2} 64$

7. Explain how to apply the change-of-base formula to evaluate each expression.
   a. $\log_3 64$ b. $\log_5 64$ c. $\log_{3.75} 64$

8. Explain the connection between the parts of Exercises 7 and 8.

Practice Makes Perfect!

Simplify each common logarithm without a calculator. (See Objective 1.)

9. $\log 10^5$ 10. $\log 10^7$ 11. $\log 0.001$

12. $\log 0.0001$ 13. $\log \dfrac{1}{1000}$ 14. $\log \dfrac{1}{10,000}$

15. $\log \sqrt[3]{10}$ 16. $\log \sqrt[3]{100}$

17. $\log \sqrt[5]{100}$ 18. $\log \sqrt[5]{1000}$

Simplify each natural logarithm without a calculator. (See Objective 2.)

19. $\ln e^4$ 20. $\ln e^{-3}$ 21. $\ln e^8$

22. $\ln e^{-1}$ 23. $\ln \dfrac{1}{e^4}$ 24. $\ln \dfrac{1}{e^6}$

25. $\ln \sqrt[3]{e}$ 26. $\ln \sqrt[3]{e^2}$ 27. $\ln \sqrt[5]{e^3}$

28. $\ln \sqrt[4]{e^3}$

Approximate each logarithm to two decimal places. (See Objective 3.)

29. $\log 20$ 30. $\log 45$ 31. $\log 0.085$

32. $\log 0.0054$ **33.** $\log e$ **34.** $\log \sqrt{e}$

35. $\ln 30$ **36.** $\ln 100$ **37.** $\ln \sqrt{10}$

38. $\ln \sqrt[3]{100}$

Solve each equation for x. Give an exact solution and approximate the solution to two decimal places, if necessary. (See Objective 4.)

39. $\log x = 6$ **40.** $\log x = -1.5$

41. $\ln x = 3$ **42.** $\ln x = -4$

43. $\log(x - 4) = 0.6$ **44.** $\log(x + 10) = 1.5$

45. $\ln(x - 4) = -2$ **46.** $\ln(x - 8) = -2.2$

47. $\log(5x + 12) = -4.4$ **48.** $\log(9x + 3) = 3.1$

49. $\ln(2x - 10) = 1.9$ **50.** $\log(4x - 7) = -5.9$

51. $\ln(3x + 5) = -0.9$ **52.** $\ln(12x - 3) = -3.4$

53. $\log(2x) = -1$ **54.** $\log(7x) = 2$

Use the change-of-base formula to rewrite each logarithm. Approximate each value to two decimal places. (See Objective 5.)

55. $\log_{12} 39.5$ **56.** $\log_7 90.1$ **57.** $\log_4 2.4$

58. $\log_2 0.00518$ **59.** $\log_5 0.395$ **60.** $\log_6 0.75$

Solve each problem. (See Objective 6.)

61. Samuel becomes about 95% proficient at a new task over a 1-week period. The equation $P = 95 - 28.782 \log d$ models his percent proficiency P after d days of not practicing the new task. What is Samuel's proficiency level after 2 days without practicing? 7 days? 14 days? Round answers to two decimal places.

62. Bethany becomes about 90% proficient at a new task over a 1-week period. The equation $P = 90 - 28.782 \log d$ models her percent proficiency P after d days of not practicing the new task. What is Bethany's proficiency level after 4 days without practicing? 8 days? 20 days? Round answers to two decimal places.

For Exercises 63 and 64, compute the moment magnitude given by $M_w = \dfrac{2}{3} \log_{10} M_0 - 10.7$, where M_0 is the magnitude of the seismic moment, or the energy released, in dyne centimeters for each earthquake. Round each answer to two decimal places.

63. The reported magnitude of the seismic moment for the earthquake in Tohoku off the coast of Honshu, Japan in March 2011 was 4.04×10^{29} dyn·cm. (Sources: http://earthquake.usgs.gov/learn/topics/measure.php and http://warships1discussionboards.yuku.com/reply/212609/Re-USGS-Press-Release-9-0-Earthquake-no-reactors-string-)

64. The reported magnitude of the seismic moment for the earthquake in Haiti in January 2010 was 4.4×10^{26} dyn·cm. (Source: http://earthquake.usgs.gov/learn/topics/measure.php)

Mix 'Em Up!

Simplify each logarithm without a calculator.

65. $\log 10^{5/2}$ **66.** $\log \dfrac{1}{10^7}$ **67.** $\ln e^{5/3}$

68. $\ln \sqrt[4]{e}$ **69.** $\log \sqrt[5]{10}$ **70.** $\log \dfrac{1}{\sqrt[4]{10}}$

71. $\ln \sqrt[7]{e^4}$ **72.** $\ln \dfrac{1}{e^6}$

Approximate each logarithm to two decimal places.

73. $\log 54$ **74.** $\log 0.0279$ **75.** $\ln 67$

76. $\ln 238$ **77.** $\ln \sqrt[3]{10}$ **78.** $\ln \sqrt{120}$

79. $\ln 0.00812$ **80.** $\ln 0.0572$

Solve each equation for x. Give an exact solution and approximate the solution to two decimal places, if necessary.

81. $\ln(x - 8) = 5.5$ **82.** $\ln(x + 13) = 7.1$

83. $5 \log(6x - 1) = 8.5$ **84.** $6 \ln(2x - 3) = 9.6$

85. $\log(10x - 9) = 1.2$ **86.** $\log(5x - 6) = 1.9$

87. $2 \ln(4x + 9) = 3$ **88.** $3 \ln(7x - 11) = -6.9$

89. $\log(5x) = 1$ **90.** $\ln(10x) = 2.7$

91. $\log(4x + 3) = 0$ **92.** $\ln(4x - 5) = 0$

Use the change-of-base formula to rewrite each logarithm. Approximate each value to two decimal places.

93. $\log_{2.3} 638.7$ **94.** $\log_{8.5} 50.9$

95. $\log_{0.24} 3.1$ **96.** $\log_{1/3} 0.85$

Solve each problem.

97. Harry has become about 88% proficient at a new task over a 1-week period. The equation $P = 88 - 12.5 \ln d$ models his percent proficiency P after d days of not practicing the new task. What is Harold's proficiency level after 6 days without practicing? 12 days? 18 days? Round answers to two decimal places.

98. Alyse has become about 92% proficient at a new task over a 1-week period. The equation $P = 92 - 12.5 \ln d$ models her percent proficiency P after d days of not practicing the new task. What is Alyse's proficiency level after 5 days without practicing? 10 days? 15 days? Round answers to two decimal places.

For Exercises 99 and 100, the formula $D(x) = 10 \log(10^{12} x)$, represents the intensity of sound, in decibels (dB) where x is the intensity of sound in watts per square meter (W/m²). Compute the intensity of each sound in decibels.

99. The intensity of a rocket launching is 10^6 W/m².

100. The intensity of a soft whisper is 10^{-9} W/m².

You Be the Teacher!

Correct each student's errors, if any.

101. Solve $\log(3x - 5) = 2$.

Brad's work:

$$\log(3x - 5) = 2$$
$$3x - 5 = 10$$
$$3x = 15$$
$$x = 5$$

102. Solve $2\ln(x - 1) = 4.8$.

Mona's work:

$$2\ln(x - 3) = 4.8$$
$$\ln(2x - 6) = 4.8$$
$$2x - 6 = e^{4.8}$$
$$2x = e^{4.8} + 6$$
$$x = \frac{e^{4.8} + \cancel{6}^3}{\cancel{2}}$$
$$x = e^{4.8} + 3$$
$$x \approx 124.510$$

103. Solve $\log(4x - 15) = 1.7$.

Tracy's work:

$$\log(4x - 15) = 1.7$$
$$4x - 15 = 10^{1.7}$$
$$4x = 10^{1.7} + 15$$
$$x = \frac{10^{1.7} + 15}{4}$$

The screen shot of Tracy's calculator is shown.

```
10^1.7+15/4
        53.86872336
log(4*Ans-15)
        2.302059991
```

104. Solve $\ln(5x + 3) = -1.5$.

Brian's work:

$$\ln(5x + 3) = -1.5$$
$$5x + 3 = e^{-1.5}$$
$$5x = e^{-1.5} - 3$$
$$x = \frac{e^{-1.5} - 3}{5}$$

The screen shot of Brian's calculator is shown.

```
e^(-1.5)-3/5
        -.3768698399
ln(5*-Ans+3)
        .1094379122
```

Calculate It!

Apply the change-of-base formula and simplify each logarithm exactly.

105. $\log_2 32$ **106.** $\log_5 625$

SECTION 12.7   **Exponential and Logarithmic Equations and Applications**

▶ **OBJECTIVES**

As a result of completing this section, you will be able to

1. Solve exponential equations.
2. Solve logarithmic equations.
3. Solve applications that involve exponential and logarithmic models.
4. Troubleshoot common errors.

Maurice invests $10,000 in a money market account that earns 4% annual interest compounded quarterly. How long will it take the account to grow to $20,000? To answer this question, we need to solve the equation

$$10,000\left(1 + \frac{0.04}{4}\right)^{4t} = 20,000$$

In this section, we will learn how to solve exponential and logarithmic equations.

Exponential Equations

Objective 1 ▶

Solve exponential equations.

In Section 12.3, we learned how to solve exponential equations using the one-to-one property of exponential functions, which stated that for $b > 0$, $b \neq 1$, $b^x = b^y$ if and only if $x = y$. For example,

$$2^x = 8$$
$$2^x = 2^3$$
$$x = 3$$

If the left and right sides of an equation cannot be expressed as a power of the same base, we must use a property of logarithms to solve the exponential equation. In the equation $2^x = 10$, we cannot express 10 as a power of 2, so we must apply a property that enables us to remove the variable from the exponent. The power rule of logarithms provides us with this ability. Recall the power rule states that $\log_b a^n = n \log_b a$.

To apply this property, we must first take the logarithm of each side of the equation. Remember that what we do to one side of an equation, we must do to the other. Because common logarithms and natural logarithms can be approximated on a calculator, we will use them in solving the equations.

$$2^x = 10$$

$\log(2^x) = \log(10)$ Take the common log of each side.

$x \log 2 = 1$ Apply the power rule for logarithms.

$x = \dfrac{1}{\log 2}$ Divide each side by log 2.

$x \approx 3.32$ Approximate.

This leads us to an important property for solving equations.

Property: One-to-One Property of Logarithms

For $x, y, b > 0$, $b \neq 1$, if $x = y$, then
$$\log_b x = \log_b y$$

Procedure: Solving an Exponential Equation Using Logarithms

Step 1: Isolate the exponential expression to one side of the equation, if necessary.
Step 2: Apply the one-to-one property of logs using common logs or natural logs.
Step 3: Apply the power rule for logarithms to make the exponent the coefficient of the logarithm.
Step 4: Solve the resulting equation.
Step 5: Check the proposed solution by substituting it into the original equation.

Objective 1 Examples   **Solve each equation. Give an exact solution and an approximation of the solution to two decimal places.**

1a. $5^x = 30$ **1b.** $e^x = 12$ **1c.** $2^{x+1} = 6$ **1d.** $1.05^{12x} = 2$ **1e.** $4(0.6^x) = 12$

Solutions   **1a.** $5^x = 30$

$\log 5^x = \log 30$ Apply the one-to-one property of logs.

$x \log 5 = \log 30$ Apply the power rule for logarithms.

$x = \dfrac{\log 30}{\log 5}$ Divide each side by log 5.

$x \approx 2.11$ Approximate.

The solution set is $\left\{ \dfrac{\log 30}{\log 5} \right\}$ or $\{2.11\}$. The solution can be checked in the original equation.

1b. Since the base is e, we take the natural logarithm of each side.

$e^x = 12$

$\ln e^x = \ln 12$ Apply the one-to-one property of logs.

$x = \ln 12$ Apply the inverse property, $\ln e^x = x$.

$x \approx 2.48$ Approximate.

The solution set is $\{\ln 12\}$ or $\{2.48\}$. The solution can be checked in the original equation.

1c.
$$2^{x+1} = 6$$

$$\log 2^{x+1} = \log 6 \qquad \text{Apply the one-to-one property of logs.}$$

$$(x+1)\log 2 = \log 6 \qquad \text{Apply the power rule for logs.}$$

$$x+1 = \frac{\log 6}{\log 2} \qquad \text{Divide each side by log 2.}$$

$$x = \frac{\log 6}{\log 2} - 1 \qquad \text{Subtract 1 from each side.}$$

$$x \approx 1.58 \qquad \text{Approximate.}$$

The solution set is $\left\{ \dfrac{\log 6}{\log 2} - 1 \right\}$ or $\{1.58\}$. The solution can be checked in the original equation.

1d.
$$1.05^{12x} = 2$$

$$\ln 1.05^{12x} = \ln 2 \qquad \text{Apply the one-to-one property of logs.}$$

$$12x \ln 1.05 = \ln 2 \qquad \text{Apply the power rule for logs.}$$

$$x = \frac{\ln 2}{12 \ln 1.05} \qquad \text{Divide each side by } 12 \ln 1.05.$$

$$x \approx 1.18 \qquad \text{Approximate.}$$

The solution set is $\left\{ \dfrac{\ln 2}{12 \ln 1.05} \right\}$ or $\{1.18\}$. The solution can be checked in the original equation.

1e.
$$4(0.6^x) = 12$$

$$0.6^x = \frac{12}{4} \qquad \text{Divide each side by 4.}$$

$$0.6^x = 3 \qquad \text{Simplify.}$$

$$\log 0.6^x = \log 3 \qquad \text{Apply the one-to-one property of logs.}$$

$$x \log 0.6 = \log 3 \qquad \text{Apply the power rule for logs.}$$

$$x = \frac{\log 3}{\log 0.6} \qquad \text{Divide each side by log 0.6.}$$

$$x \approx -2.15 \qquad \text{Approximate.}$$

The solution set is $\left\{ \dfrac{\log 3}{\log 0.6} \right\}$ or $\{-2.15\}$. The solution can be checked in the original equation.

✔ **Student Check 1** Solve each equation. Give an exact solution and an approximation of the solution to two decimal places.

a. $7^x = 53$ **b.** $e^x = 25$ **c.** $3^{x-2} = 80$ **d.** $1.02^{4x} = 3$ **e.** $5(0.2)^x = 2$

Logarithmic Equations

Objective 2 ▶

Solve logarithmic equations.

To solve a logarithmic equation, we may have to apply some of the properties we learned in Section 12.5 that enable us to combine two or more logarithms into a single logarithm. Then we can convert the logarithmic equation to its exponential form to solve it.

> **Procedure: Solving a Logarithmic Equation**
>
> **Step 1:** Write each side of the equation as a single logarithm, if necessary.
>   **a.** $\log_b x + \log_b y = \log_b(xy)$
>   **b.** $\log_b x - \log_b y = \log_b \dfrac{x}{y}$
> **Step 2:** If the equation is in the form $\log_b x = y$, convert it to $b^y = x$ to solve the equation.
> **Step 3:** Exclude any values that make the arguments of the logarithms negative or zero.

Objective 2 Examples   Solve each equation.

2a. $\log_2(x + 1) = 4$ **2b.** $\log_3 x + \log_3(x - 8) = 2$

2c. $\log_4(x + 3) - \log_4(2x) = 1$ **2d.** $2 \log x - \log(x + 6) = 0$

Solutions   **2a.** Note that $x + 1 > 0$ for the log to be defined, so $x > -1$.

$$\log_2(x + 1) = 4$$

$2^4 = x + 1$	Rewrite in exponential form.
$16 = x + 1$	Simplify the exponent.
$15 = x$	Subtract 1 from each side.

Check:

$\log_2(15 + 1) = 4$	Replace x with 15.
$\log_2 16 = 4$	Simplify.
$4 = 4$	Simplify the logarithm.

Since 15 makes the equation true, the solution set is $\{15\}$.

2b. Note that $x > 0$ and $x - 8 > 0$, or $x > 8$, for the log to be defined. So, $x > 8$ satisfies both of these conditions.

$$\log_3 x + \log_3(x - 8) = 2$$

$\log_3 x(x - 8) = 2$	Apply the product rule.
$3^2 = x(x - 8)$	Rewrite in exponential form.
$9 = x^2 - 8x$	Simplify and apply the distributive property.
$0 = x^2 - 8x - 9$	Subtract 9 from each side.
$0 = (x - 9)(x + 1)$	Factor.
$x - 9 = 0$ or $x + 1 = 0$	Apply the zero products property.
$x = 9$ $x = -1$	Solve each equation.

Because $x > 8$, we must exclude -1 from the solution set since it makes the argument of the logarithm negative. So, the solution set is $\{9\}$.

2c. Note that $x + 3 > 0$, or $x > -3$, and $x > 0$ for the log to be defined. So, $x > 0$ satisfies both of these conditions.

$$\log_4(x + 3) - \log_4(2x) = 1$$

$\log_4 \dfrac{x + 3}{2x} = 1$	Apply the quotient rule for logs.
$4^1 = \dfrac{x + 3}{2x}$	Rewrite in exponential form.

$$4 = \frac{x+3}{2x} \qquad \text{Simplify the exponent.}$$

$$8x = x + 3 \qquad \text{Multiply each side by } 2x \text{ and simplify.}$$

$$7x = 3 \qquad \text{Subtract } x \text{ from each side and simplify.}$$

$$x = \frac{3}{7} \qquad \text{Divide each side by 7.}$$

The solution set is $\left\{\dfrac{3}{7}\right\}$. We can check the solution in the original equation.

2d. Note that $x > 0$ and $x + 6 > 0$, or $x > -6$, for the log to be defined. So, $x > 0$ satisfies both of these conditions.

$$2 \log x - \log(x + 6) = 0$$

$$\log x^2 - \log(x + 6) = 0 \qquad \text{Apply the power rule for logs.}$$

$$\log \frac{x^2}{x+6} = 0 \qquad \text{Apply the quotient rule for logs.}$$

$$10^0 = \frac{x^2}{x+6} \qquad \text{Rewrite in exponential form.}$$

$$(x+6)1 = \frac{x^2}{x+6}(x+6) \qquad \text{Simplify and multiply by the LCD, } x + 6.$$

$$x + 6 = x^2 \qquad \text{Simplify.}$$

$$0 = x^2 - x - 6 \qquad \text{Subtract } x \text{ and 6 from each side.}$$

$$0 = (x - 3)(x + 2) \qquad \text{Factor.}$$

$$x - 3 = 0 \quad \text{or} \quad x + 2 = 0 \qquad \text{Apply the zero products property.}$$

$$x = 3 \qquad \qquad x = -2 \qquad \text{Solve each equation.}$$

We must discard -2 from the solution set since it makes the argument of the logarithm negative. So, the solution set is $\{3\}$.

✔ **Student Check 2** Solve each equation.

a. $\log_3(x - 1) = 1$ 　　　　　　　**b.** $\log_2(x) + \log_2(x - 2) = 3$

c. $\log_5(2x - 5) - \log_5(x) = 2$ 　　**d.** $2\log(x) - \log(2x + 15) = 0$

Applications

Objective 3 ▶

Solve applications that involve exponential and logarithmic models.

Exponential and logarithmic models arise in many different areas. We have already seen a few of these in this chapter. In Example 6, we will use the properties of exponents to solve for the variable. One of the formulas that we will use is the compound interest formula. We will also use a formula for continuous compound interest.

$$\text{Compound interest formula: } A = P\left(1 + \frac{r}{n}\right)^{nt}$$

Property: Continuous Compound Interest Formula

If P dollars are invested in an account that pays $r\%$ compounded continuously, then the amount in the account after t years is

$$A = Pe^{rt}$$

Objective 3 Examples   Solve each problem.

3a. David invests $200 in an account that pays 5% compounded continuously. How much will he have in the account after 2 yr? How long will it take his account to grow to $1000?

Solution   **3a.** To find the amount David will have after two years, we use the compound interest formula with $P = 200$, $r = 0.05$, and $t = 2$.

$A = Pe^{rt}$	State the compound interest formula.
$A = 200e^{(0.05)(2)}$	Let $P = 200$, $r = 0.05$, and $t = 2$.
$A = 200e^{0.1}$	Simplify the exponent.
$A \approx 221.03$	Approximate.

After 2 yr, David will have $221.03 in his account.

To determine how long it will take for his account to grow to $1000, we must solve the equation $A = Pe^{rt}$ for t.

$A = Pe^{rt}$	State the compound interest formula.
$1000 = 200\,e^{0.05t}$	Let $A = 1000$, $P = 200$, and $r = 0.05$.
$\dfrac{1000}{200} = e^{0.05t}$	Divide each side by 200.
$5 = e^{0.05t}$	Simplify.
$\ln 5 = \ln e^{0.05t}$	Apply the one-to-one property of logs.
$\ln 5 = 0.05t$	Apply the inverse properties of logs.
$\dfrac{\ln 5}{0.05} = t$	Divide each side by 0.05.
$32.19 \approx t$	Approximate.

It will take about 32 yr for David's account to grow to $1000.

3b. Maurice invests $10,000 in a money market account that earns 4% annual interest compounded quarterly. How long will it take his account to grow to $20,000?

Solution   **3b.** We need to use the compound interest formula and solve for t.

$P\left(1 + \dfrac{r}{n}\right)^{nt} = A$	State the compound interest formula.
$10{,}000\left(1 + \dfrac{0.04}{4}\right)^{4t} = 20{,}000$	Let $P = 10{,}000$, $r = 0.04$, $n = 4$, and $A = 20{,}000$.
$(1.01)^{4t} = \dfrac{20{,}000}{10{,}000}$	Divide each side by 10,000 and simplify in parentheses.
$(1.01)^{4t} = 2$	Simplify.
$\log(1.01)^{4t} = \log 2$	Apply the one-to-one property of logs.
$4t \log 1.01 = \log 2$	Apply the power rule for logs.
$t = \dfrac{\log 2}{4 \log 1.01}$	Divide each side by $4 \log 1.01$.
$t \approx 17.42$	Approximate.

So, it will take about 17 yr for Maurice's account to grow to $20,000.

3c. In 2009, the country of India, the second most populous country in the world, had a population of approximately 1.166 billion. Its population is growing at a rate of 1.548% per year. Its population (in billions) can be modeled by $f(t) = 1.166e^{0.01548t}$, where t is the number of years after 2009. What is an estimate for the population of India in 2015 if it continues to grow at this rate? How long will it take the population to reach 1.5 billion? (Source: https://www.cia.gov)

Solution **3c.** The population in 2015 is found by replacing t with $2015 - 2009 = 6$.

$f(t) = 1.166e^{0.01548t}$	State the given model.
$f(6) = 1.166e^{0.01548(6)}$	Replace t with 6.
$f(6) = 1.166e^{0.09288}$	Simplify the exponent.
$f(6) \approx 1.28$	Approximate.

In 2015, the population of India will be approximately 1.28 billion.

To determine when the population will reach 1.5 billion, we replace y with 1.5 and solve for t.

$y = 1.166e^{0.01548t}$	
$1.5 = 1.166e^{0.01548t}$	Replace y with 1.5.
$\dfrac{1.5}{1.166} = e^{0.01548t}$	Divide each side by 1.166.
$1.28645 \approx e^{0.01548t}$	Approximate.
$\ln 1.28645 \approx \ln e^{0.01548t}$	Apply the one-to-one property of logs.
$\ln 1.28645 \approx 0.01548t$	Apply the inverse property of logs.
$\dfrac{\ln 1.28645}{0.01548} \approx t$	Divide each side by 0.01548.
$16.27 \approx t$	Approximate.

In approximately 16 yr after 2009, or 2025, the population of India will reach 1.5 billion.

✓ Student Check 3 Solve each problem.

a. Jeanette invests $500 in an account that pays 6% compounded continuously. How much will she have in the account after 4 yr? How long will it take her account to grow to $2000?

b. Suppose Cantrell invests $30,000 in a money market account that earns 5% annual interest compounded monthly. How long will it take his account to grow to $60,000?

c. In 2009, the country of China, the most populous country in the world, had a population of approximately 1.339 billion. Its population is growing at a rate of 0.655% per year. Its population (in billions) can be modeled by $y = 1.339e^{0.00655t}$, where t is the number of years after 2009. What is an estimate for the population of China in 2015 if it continues to grow at this rate? How long will it take the population to reach 1.5 billion? (Source: https://www.cia.gov)

Objective 4 ▶

Troubleshoot common errors.

Troubleshooting Common Errors

Some common errors associated with solving exponential and logarithmic equations are shown.

Objective 4 Examples   **A problem and an incorrect solution are given. Provide the correct solution and an explanation of the error.**

4a. Solve $4^x = 12$.

Incorrect Solution	Correct Solution and Explanation
$$4^x = 12$$ $$\log 4^x = \log 12$$ $$x \log 4 = \log 12$$ $$x = \frac{\log 12}{\log 4}$$ $$x = 3$$	The logarithm is a function; we cannot divide out logs from the numerator and denominator of a fraction. We must divide the log values. $$x = \frac{\log 12}{\log 4} \approx 1.792$$

4b. Solve $\log x + \log(x - 3) = 1$.

Incorrect Solution	Correct Solution and Explanation
$$\log x + \log(x - 3) = 1$$ $$\log x(x - 3) = 1$$ $$10^1 = x(x - 3)$$ $$10 = x^2 - 3x$$ $$x^2 - 3x - 10 = 0$$ $$(x - 5)(x + 2) = 0$$ $$x = 5, x = -2$$ The solution set is $\{-2, 5\}$.	The logarithm of a negative number is undefined. Therefore, we must exclude -2 from the solution set. So, the solution set is $\{5\}$.

ANSWERS TO STUDENT CHECKS

Student Check 1   **a.** $\left\{ \dfrac{\ln 53}{\ln 7} \right\}$ or $\{2.04\}$

b. $\{\ln 25\}$ or $\{3.22\}$ **c.** $\left\{ \dfrac{\log 80 + 2\log 3}{\log 3} \right\}$ or $\{5.99\}$

d. $\left\{ \dfrac{\log 3}{4\log 1.02} \right\}$ or $\{13.87\}$ **e.** $\left\{ \dfrac{\log 0.4}{\log 0.2} \right\}$ or $\{0.57\}$

Student Check 2   **a.** $\{4\}$ **b.** $\{4\}$ **c.** $\varnothing$ **d.** $\{5\}$
Student Check 3   **a.** \$635.62; 23.10 yr   **b.** 13.89 yr
   c. 1.39 billion; approximately 17 yr or in 2026

SUMMARY OF KEY CONCEPTS

1. Exponential equations are solved by applying the one-to-one property of logarithms, which enables us to take the logarithm of both sides. Any logarithm will work but we generally use the common logarithm or the natural logarithm since we can use our calculator to simplify those expressions. The exponential expression should be isolated before taking the log of each side.

2. Logarithmic equations are solved by converting the equation to exponential form. Before doing this, we must combine any logarithms. We apply the product rule and quotient rule for logarithms to combine the logarithmic expressions. After solving the equation, we must exclude any values that make the argument of the logarithm negative or zero.

GRAPHING CALCULATOR SKILLS

Most of the skills in this section have been presented in earlier sections. The one new skill is using the number e.

Example: Simplify $1.166e^{0.01548(6)}$.

Solution: To enter e raised to an exponent, use or .

```
1.166e^(.01548*6
)
        1.279486836
```

SECTION 12.7 / EXERCISE SET

Write About It!

Use complete sentences in your answer to each exercise.

1. Explain why you can take either the common or natural log of each side when solving the exponential equation $7^x = 15$.

2. Explain how to solve the exponential equation $3e^{x-1} = 48$ using an appropriate logarithm.

3. Explain how to solve the exponential equation $6(10^{x+2}) = 102$ using an appropriate logarithm.

4. Explain how to solve the logarithmic equation $\log_4(2x - 1) = 3$.

5. Explain how to solve the equation $\log_6 x + \log_6(x - 16) = 2$.

6. Explain how to solve the equation $\log_2(6x) - \log_2(2x - 10) = 3$.

Practice Makes Perfect!

Solve each equation. Give an exact solution and an approximation of the solution to two decimal places. (See Objective 1.)

7. $10^x = 29$ 8. $10^x = 40$ 9. $e^x = 6$

10. $10^x = 9$ 11. $10^{7x} = 90$ 12. $10^{5x} = 88$

13. $e^{4x} = 13$ 14. $e^{6x} = 31$ 15. $7^{x-2} = 38$

16. $4^{x+9} = 12$ 17. $(5.75)^{4x} = 19$ 18. $(6.71)^{2x} = 11$

19. $9(0.53)^{5x} = 94$ 20. $3(1.26)^{4x} = 92$

Solve each logarithmic equation. (See Objective 2.)

21. $\log_5(x - 18) = 1$ 22. $\log_2(x - 3) = 5$

23. $\log_2(2x - 9) = 2$ 24. $\log_3(9x - 7) = 3$

25. $\log_2 x + \log_2(x + 2) = 3$

26. $\log_6 x + \log_6(x + 35) = 2$

27. $\log_2 x + \log_2(x - 3) = 2$

28. $\log_4 x + \log_4(x - 6) = 2$

29. $\log_4 4x - \log_4(4x - 5) = 2$

30. $\log_4 6x - \log_4(3x - 12) = 1$

31. $\log 2x - \log(x - 8) = 1$

32. $\log_5 5x - \log_5(x - 11) = 2$

33. $\log_2 6x - \log_2(5x + 11) = 2$

34. $\log 2x - \log(x + 6) = 2$

35. $2\log_3 x - \log_3(x + 6) = 1$

36. $2\log_4 x - \log_4(x + 3) = 1$

37. $2\log x - \log(7x - 12) = 0$

38. $2\log_7 x - \log_7(6x + 16) = 0$

Solve each problem. Round each answer to two decimal places. (See Objective 3.)

39. Hunter invests $990 in an account that pays 2.8% compounded continuously. How much will he have in the account after 4 yr? How long will it take his account to grow to $1500?

40. Camille invests $1320 in an account that pays 2.5% compounded continuously. How much will she have in the account after 2 yr? How long will it take her account to grow to $2500?

41. In 2011, Mexico had a population of approximately 113.724 million. Its population is growing at a rate of 1.102% per year. Its population (in millions) can be modeled by $y = 113.724e^{0.01102t}$, where t is the number of years after 2011. What is an estimate for the population of Mexico in 2021 if it continues to grow at this rate? How long will it take the population to reach 120 million? (Source: http://www.cia.gov)

42. The annual tons of Carbon Dioxide (CO_2) emitted can be modeled by $f(x) = 13.272(0.8369)^x$, where x is the mileage of a car in miles per gallon. What is an estimate for the annual tons of CO_2 emitted for a car that gets 30 mpg? Determine the gas mileage of a car that emits 1 ton of CO_2 annually.

43. The amount of the blood thinner Warfarin (in milligrams) in a person's body t hr after ingestion is $y = a_0 e^{-0.01733t}$, where a_0 is the initial amount taken. If Wing takes 10 mg of Warfarin, how much will be in her body after 4 hr? (Source: http://www.aafp.org/afp/990201ap/635.html)

44. The amount of ibuprofen (in milligrams) in a person's body t hr after ingestion is $y = a_0 e^{-0.347t}$, where a_0 is the initial amount taken. If Lisa takes 600 mg of ibuprofen, how much ibuprofen will be in her body after 5 hr? (Source: http://www.rxlist.com/ibuprofen-drug.htm)

 Mix 'Em Up!

Solve each equation. For solutions of the exponential equations, give an exact solution and an approximation of the solution to two decimal places.

45. $\log_4(5x - 4) = 2$

46. $\log_2(4x - 12) = 3$

47. $e^{4.6x} = 37.5$

48. $e^{1.7x} = 109$

49. $\log_6 x + \log_6(x - 9) = 2$

50. $\log_5 x + \log_5(x - 24) = 2$

51. $12^{6x-8} = 2$

52. $6^{5x-2} = 30$

53. $\log_6 4x - \log_6(x - 2) = 2$

54. $\log_2 3x - \log_2(x - 6) = 2$

55. $2\ln x - \ln(x + 42) = 0$

56. $2\log x - \log(7x - 6) = 0$

57. $7(4.08)^{4x} = 1$

58. $5(3.49)^{4x} = 1$

59. $\log_3(2x + 1) = 4$

60. $\log_5(12x + 5) = 3$

61. $\log_4 x + \log_4(x + 6) = 2$

62. $\log_6 x + \log_6(x - 5) = 2$

63. $5^{7x+1} = 21$

64. $7^{2x+8} = 5$

65. $\log_2 6x - \log_2(x - 5) = 1$

66. $\log 5x - \log(x + 7) = 1$

67. $\log_2 6x - \log_2(5x + 11) = 2$

68. $\log 2x - \log(x + 6) = 2$

69. $2(6.25)^x = 4$

70. $9(3.67)^{6x} = 57$

71. $2\log_5 x - \log_5(x - 4) = 2$

72. $2\log_2 x - \log_2(2x - 6) = 3$

73. $10^{4.7x} = 17.5$

74. $10^{2.1x} = 29.5$

75. $2\ln x - \ln(5x + 36) = 0$

76. $2\log_7 x - \log_7(13x - 40) = 0$

Solve each problem. Round each answer to two decimal places.

77. Kristen invests $1870 in an account that pays 4.6% compounded continuously. How much will she have

in the account after 4 yr? How long will it take her account to grow to $2400?

78. Michael invests $1650 in an account that pays 3.8% compounded continuously. How much will he have in the account after 6 yr? How long will it take his account to grow to $2800?

79. Lance invests $10,500 into a money market account that earns 5% annual interest compounded quarterly. How long will it take his account to grow to $23,000?

80. Lawton invests $12,000 in a money market account that earns 1.2% annual interest compounded semiannually. How long will it take his account to grow to $14,500?

81. In 2011, Japan had a population of approximately 126.476 million. Its population is decreasing at a rate of 0.278% per year. Its population (in millions) can be modeled by $y = 126.476 e^{-0.00278t}$, where t is the number of years after 2011. What is an estimate of the population of Japan in 2029 if it continues to grow at this rate? How long will it take the population to reach 115 million? (Source: http://www.cia.gov)

82. In 2011, the United States had a population of approximately 313.232 million. Its population is growing at a rate of 0.963% per year. Its population (in millions) can be modeled by $y = 313.232 e^{0.00963t}$, where t is the number of years after 2011. What is an estimate of the population of the United States in 2026 if it continues to grow at this rate? How long will it take the population to reach 340 million? (Source: http://www.cia.gov)

83. The amount of acetaminophen (in milligrams) in a person's body t hr after ingestion is $y = a_0 e^{-0.231t}$, where a_0 is the initial amount taken. If Isabel takes 600 mg of acetaminophen, how much will be in her body after 3 hr? (Source: http://www.rxlist.com/tylenol-drug.htm)

84. The amount of blood pressure medication Lisinopril in a person's body t hr after ingestion is $y = a_0 e^{-0.0578t}$, where a_0 is the initial amount taken. If Jason takes 10 mg of Lisinopril, how much Lisinopril will be in his body after 12 hr? (Source: http://www.drugs.com/pro/lisinopril.html)

 You Be the Teacher!

Correct each student's errors, if any.

85. Solve $8(2)^{3x} = 48$.

Lois's work:

$$8(2)^{3x} = 48$$
$$16^{3x} = 48$$
$$3x \ln 16 = \ln 48$$
$$x = \frac{\ln 48}{3 \ln 16}$$
$$x \approx 0.465$$

86. Solve $8^{5x-6} = 13$.

Chuck's work:

$$8^{5x-6} = 13$$
$$5x - 6\ln 8 = \ln 13$$
$$5x = 6\ln 8 + \ln 13$$
$$x = \frac{6\ln 8 + \ln 13}{5}$$
$$x \approx 3.008$$

87. Solve $\log_6 x + \log_6(x - 35) = 2$.

Ulrey's work:

$$\log_6 x + \log_6(x - 35) = 2$$
$$\log_6(2x - 35) = 2$$
$$2x - 35 = 6^2$$
$$2x = 71$$
$$x = \frac{71}{2}$$

88. Solve $2\log_2 x - \log_2(8x - 12) = 0$.

Soner's work:

$$2\log_2 x - \log_2(8x - 12) = 0$$
$$\log_2(2x - 8x + 12) = 0$$
$$-6x + 12 = 0$$
$$x = 2$$

 Calculate It!

Solve each equation and use a calculator to check each solution.

89. $4.02^{8x} = 64$

90. $4^{7x-2} = 18$

91. $\log_6 x + \log_6(x + 5) = 2$

92. $\log_3 3x - \log_3(3x - 10) = 2$

 GROUP ACTIVITY / **The Mathematics of Financing a College Education**

The average in-state cost (tuition and fees) for attending a public 4-yr institution averaged $7605 in the 2010–2011 academic year. On average, college costs tend to increase about 8% each year. The  function $f(x) = 7605(1.08)^x$ models the average cost of a college education x yr after 2010. (Sources: http://trends.collegeboard.org/college_pricing/ and http://www.finaid.org/savings/tuition-inflation.phtml).

1. Planning for the future is important to you, so you want to determine the amount you need to save to pay for your child's college education. Use the given function to determine the average cost of college each year your child attends college. Assume your child was born in 2010 and that he or she will enter college at age 18 for 4 yr.

2. What is the total cost of your child's 4-yr college education?

3. The amount in step 2 is the amount that you need to save. The formula

$$FV = PMT\frac{\left(1 + \dfrac{i}{12}\right)^n - 1}{\dfrac{i}{12}}$$

gives the future value (FV) of an ordinary annuity, where i is the annual interest rate of a savings account, n is the number of payments, and PMT is the payment amount. Use the formula to determine the payment necessary to have the funds for your child's college education in 18 yr if you pay this amount each month into an account that earns 6% annual interest. (Note: An annuity is any sequence of equal periodic payments. If payments are made at the end of each time interval, then the annuity is called an ordinary annuity.)

Exponential and Logarithmic Functions

-⚬- **What's the big idea?** Functions can be combined in several ways to produce other functions. Functions that are one-to-one can be inverted to find the function that "un-does" the effect of the function. Exponential and logarithmic functions are inverses of one another. They are very important in many fields. The properties of exponents enable us to develop properties of logarithms.

The Tools

Listed below are the key terms, skills, formulas, and properties you should know for this chapter.

The page reference is provided if you need additional help with the given topic. The Study Tips will assist in your preparation for an exam.

Study Tips

1. Learn all of the terms, formulas, and properties. Make flash cards and have someone quiz you.
2. Rework problems from the exercises and also the ones you worked in class. Work additional problems from the review exercises.
3. Review the summaries of key concepts.
4. Work the chapter test.
5. Be sure to review the online resources for additional study materials.

Terms

Argument 910	Inverse function 891	Natural logarithm 927
Common logarithm 926	Logarithm 910	One-to-one function 888
Exponential function 902	Logarithmic function 913	

Formulas and Properties

- Change-of-base formula   930
- Composition of functions   878
- Compound interest formula   905
- Continuous compound interest formula   939
- Difference of functions   874
- Horizontal line test   889

- Inverse properties of logarithms   922
- One-to-one property of b^x 904
- One-to-one property of logarithms   936
- Product of functions   884

- Product rule for logarithms   917
- Power rule for logarithms   920
- Quotient of functions   874
- Quotient rule for logarithms   919
- Sum of functions   874

CHAPTER 12 / SUMMARY

How well do you know this chapter? Complete the following questions to find out. Take a look back at the section if you need help.

SECTION 12.1 Operations and Composition of Functions

1. We can perform _____ on functions just like we do real numbers. We can ___ them, _____ them, _____ them, and _____ them. The domain of the combined function is the _____ of the

domains of the individual functions. For the quotient function, we must also exclude values that make the denominator ____.

2. To evaluate a combined function, we can either evaluate the functions _____ and then perform the operations on the results or we can find the _____ function first and then evaluate this function.

3. The composition of functions takes one function's _____ and makes it the other function's _____. The notation $(f \circ g)(x) = $ _____ and $(g \circ f)(x) = $ _____.

SECTION 12.2 Inverse Functions

4. The _____ function is a function that takes us back to the value we started with before the original function was applied.

5. Only _____ functions have inverse functions. These are functions such that every input value corresponds to a(n) _____ output value.

6. The _____ line test is used to determine if a function is one-to-one. If each _____ line touches the graph at no more than one point, the function is one-to-one.

7. To find the inverse of a one-to-one function, we _____ the x- and y-variables. The inverse of $f(x)$ is denoted by _____. If (x, y) belongs to $f(x)$, then _____ belongs to _____.

8. To find the equation of the inverse function, we replace $f(x)$ with __, interchange _____, solve the equation for __, and replace y with _____.

9. If $f(x) = y$, then _____.

10. The graph of a function and its inverse are _____ about the line _____.

11. To verify that $f(x)$ and $f^{-1}(x)$ are inverses, _____ and _____.

SECTION 12.3 Exponential Functions

12. Exponential functions are functions of the form _____, where _____ and _____. The domain is _____ and the range is _____. The points _____ and _____ are on the graph of this function. If $b > 1$, the function is _____ and if $0 < b < 1$, the function is _____. The graph of a basic exponential function gets very close to the _____ but never _____ it.

13. We can use the one-to-one property of b^x to solve exponential equations. It states that if _____, then _____.

14. The compound interest formula is an application of exponential functions and is _____.

SECTION 12.4 Logarithmic Functions

15. The expression $\log_b x$ represents the exponent of __ that produces __. Two of the basic logarithms are $\log_b(1) =$ __, $\log_b b =$ __

16. $y = \log_b x$ written in exponential form is _____.

17. A logarithmic function is a function of the form _____, where $b > 0$, $b \neq 1$. The domain of the function is _____ and the range is _____. The x-intercept is _____ and there is ___ y-intercept.

18. To graph an equation of the form $f(x) = \log_b x$, we should convert the equation to its _____ form and substitute values for __ and solve for __. The graph of a basic logarithmic functions gets very close to the _____ but never _____ it.

SECTION 12.5 Properties of Logarithms

19. The log of a product is the _____ of the logs of the _____. The product rule for logs is _____, for $x, y, b > 0$, and $b \neq 1$.

20. The log of a quotient is the _____ of the logs of the _____ and _____. The quotient rule for logs is _____, for $x, y, b > 0$, and $b \neq 1$.

21. The log of an exponential expression is the _____ times the log of the _____ of the exponential expression. The power rule for logs is _____, for $x, y, b > 0$ and $b \neq 1$.

22. The inverse properties of logs state that _____ and _____ both equal x.

SECTION 12.6 The Common Log, Natural Log, and Change-of-Base-Formula

23. The common log is written as _____ and it represents the exponent of __ that produces x.

24. The natural log is written as ___ and it represents the exponent of __ that produces x.

25. The change-of-base formula is $\log_b a =$ _____ or ___ and it enables us to calculate logs with bases other than __ or __.

SECTION 12.7 Exponential and Logarithmic Equations and Applications

26. To solve an exponential equation, we must isolate the _____ and then take the ___ of each side.

27. To solve a logarithmic equation, we must combine the logs into a _____ log and then convert the equation to its _____ form.

CHAPTER 12 / REVIEW EXERCISES

SECTION 12.1

Find each combined function and state its domain. (*See Objectives 1–3.*)

Let $f(x) = 2x - 7$, $g(x) = x^2 - x - 2$, and $h(x) = 4 - 2x$.

1. $(f + g)(x)$ and $(f - g)(x)$ 2. $(f \cdot h)(x)$ and $\left(\dfrac{f}{h}\right)(x)$

3. $(f \circ h)(x)$ and $(h \circ f)(x)$ 4. $(f \cdot g)(x)$ and $\left(\dfrac{f}{g}\right)(x)$

Let $f(x) = 2x^2 - x$, $g(x) = x^2 + 3x - 7$, $h(x) = 5x + 3$. **Evaluate each function.** (*See Objectives 1–3.*)

5. $(f + g)(-2)$ 6. $(f - h)(3)$ 7. $(g \cdot h)(0)$

8. $\left(\dfrac{g}{h}\right)(-1)$ 9. $(f \circ g)(2)$ 10. $(h \circ f)\left(-\dfrac{1}{2}\right)$

Use the graphs of $f(x)$, $g(x)$, and $h(x)$ to find each value. (*See Objectives 1–3.*)

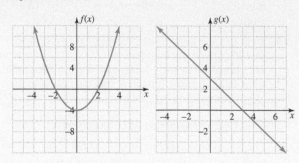

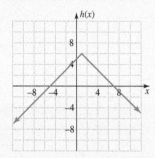

11. $(f + g)(0)$ 12. $(f - h)(3)$ 13. $(f \cdot g)(0)$

14. $(g \cdot h)(-2)$ 15. $\left(\dfrac{f}{g}\right)(2)$ 16. $(f \circ g)(5)$

17. $(h \circ g)(2)$ 18. $(h \circ f)(0)$

Use the table to find each value. (*See Objectives 1–3.*)

x	$f(x)$	$g(x)$	$h(x)$
-3	15	13	0
-2	9	10	-1
-1	5	7	-2
0	3	4	-3
1	3	1	-4
2	5	-2	-5
3	9	-5	-4
4	15	-8	-3
5	23	-11	-2

19. $(g + h)(0)$ 20. $(f - g)(-3)$ 21. $(f \cdot g)(2)$

22. $\left(\dfrac{f}{h}\right)(4)$ 23. $(f \circ g)(2)$ 24. $(g \circ g)(2)$

Solve each problem. (*See Objective 4.*)

25. Loretta makes $12.50 an hour, so her income can be represented by the function $f(x) = 12.50x$, where x is the number of hours she works. She pays 18% of her income in taxes. The amount of tax she owes can be represented by the function $g(x) = 0.18x$, where x is her income. Write a function that represents the total taxes Loretta owes as a function of the number of hours she works?

26. An organization is planning a bingo fund-raiser. There are several costs involved for this event. The cost of using the hall is $25 per person. The cost of food is $18 per person and the cost of prizes is $52 for every four players. Let x represent the number of players. Write a function that represents the cost of the hall, the cost of food, and the cost of prizes. Then write a function that represents the total cost of the fund-raiser as a function of players.

SECTION 12.2

Determine if each function is one-to-one. (*See Objectives 1 and 2.*)

27. $f = \{(-2, 1), (1, 5), (-1, 2), (1, 0), (0, 4)\}$

28. $f = \left\{(0, -5), (0, 2), \left(-\sqrt{3}, 4\right), \left(\sqrt{3}, 4\right)\right\}$

29.

x	-2	-1	0	1	2
y	-4	-3	0	3	4

30.

x	-4	-2	0	2	4
y	5	1	3	4	6

31.   32.

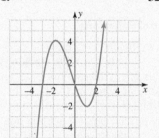
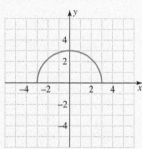

33.   34.

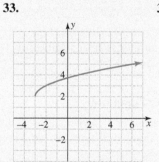
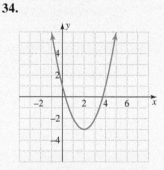

Find the inverse of each function. (*See Objectives 3 and 4.*)

35. $f(x) = \{(-2, -5), (0, 1), (3, 5), (2, 3), (4, 7)\}$

36. $f(x) = \{(-3, -3.2), (-1, -2.5), (2, 0.9), (5, 2.1), (9, 6.9)\}$

37.

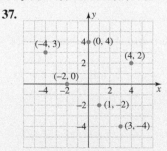

38.

State (x)	Congressmen in 2009 (y)
Arizona	Harry Mitchell
Kansas	Lynn Jenkins
Florida	Connie Mack
Illinois	Debbie Halvorson
Ohio	Marcy Kaptur

39. $f(x) = 5x - 4$

40. $f(x) = \dfrac{2x - 11}{5}$

41. $f(x) = 2x^3 + 9$

42. $f(x) = \dfrac{6}{3 - 5x}$

Evaluate each function using the given information. (*See Objective 5.*)

43. If $f(-5) = -1$, what is $f^{-1}(-1)$?

44. If $f(7.2) = -2.4$, what is $f^{-1}(-2.4)$?

45. If $f\left(-\dfrac{1}{6}\right) = \dfrac{9}{5}$, what is $f^{-1}\left(\dfrac{9}{5}\right)$?

46. If $f(0) = 8.4$, what is $f^{-1}(8.4)$?

Graph the inverse of each function. (*See Objective 6.*)

47.

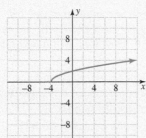

48.

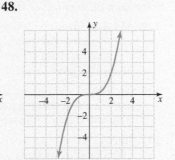

Determine if the functions are inverses of one other. (*See Objective 7.*)

49. $f(x) = 5x - 12,\ g(x) = \dfrac{1}{5}x + \dfrac{12}{5}$

50. $f(x) = (4x - 11)^3,\ g(x) = \dfrac{\sqrt[3]{x} + 11}{4}$

51. $f(x) = 6x^5 + 9,\ g(x) = \dfrac{\sqrt[5]{x - 9}}{6}$

52. $f(x) = \dfrac{2}{1 - 3x},\ g(x) = \dfrac{1}{2} - \dfrac{3x}{2}$

SECTION 12.3

Graph each exponential function. Plot and label at least three points on the graph. State the domain and range of each function. (*See Objective 1.*)

53. $y = 5^x$

54. $y = 0.3^x$

Solve each exponential equation. (*See Objective 2.*)

55. $\left(\dfrac{3}{7}\right)^x = \dfrac{49}{9}$

56. $5^{3x-2} = 625$

57. $2^{x+3} = 1$

58. $\left(\dfrac{1}{2}\right)^{3x-1} = 8$

59. $25^{3-2x} = \dfrac{1}{5}$

60. $3^{5x-7} = 9^x$

Solve each problem. Round each answer to two decimal places. (*See Objective 3.*)

61. Elaine invests $3500 in a saving account that pays 2.2% annual interest compounded quarterly. Use the compound interest formula to determine the amount of money she will have in the account at the end of 2 yr.

62. The population of China (in billions) can be modeled by $y = 1.34(1.005)^x$, where x is the number of years after 2011. What is an estimate for the population in 2015? 2025? 2035? (Source: http://www.cia.gov)

SECTION 12.4

Rewrite each logarithmic equation in exponential form and each exponential equation in logarithmic form. (*See Objective 1.*)

63. $\log_2 128 = 7$

64. $\log_{1/10} \dfrac{1}{100} = 2$

65. $9^{3/2} = 27$

66. $\left(\dfrac{1}{81}\right)^{-3/4} = 27$

Simplify each expression. (*See Objective 2.*)

67. $\log_{1/5} 1$

68. $\log_{0.25} 0.25$

69. $\log_3 \dfrac{1}{27}$

70. $\log 0.01$

71. $\log_{1/2} 32$

72. $\log_2(-4)$

Solve each equation for x. (*See Objective 3.*)

73. $\log_6 216 = x$

74. $\log_a \sqrt{a} = x,\ a > 0$

75. $\log_4 x = 3$

76. $\log_x \dfrac{1}{64} = 3$

77. $\log_{3.6} 1 = x$

78. $\log_c c^{-4} = x,\ c > 0$

Graph each logarithmic function. Plot and label at least three points. State the domain and range of each function. (*See Objective 4.*)

79. $y = \log_5 x$

80. $y = \log_{0.6} x$

SECTION 12.5

Use the rules for logarithms to write each expression as a single logarithm. Simplify, if possible. Assume all logarithmic arguments are positive numbers. (*See Objectives 1–4.*)

81. $\log_3 8 + \log_3 7 - \log_3 28$

82. $\log_4 24 - \log_4 3 + \log_4 8$

83. $2\log_5 x - 3\log_5 y + 5\log_5 z$

84. $\dfrac{1}{2}\log_7 a + 3\log_7 b - \dfrac{1}{3}\log_7 c$

Write each expression as a combination of logarithms. (*See Objectives 1–4.*)

85. $\log_4 \dfrac{7x^3}{y^2}$ **86.** $\log_2 \sqrt[3]{\dfrac{a^2}{2b}}$

87. $\log_{10}\left(x^2 y^5 \sqrt{x-1}\right)$ **88.** $\log_{10} \dfrac{100(x^2+4)}{y^4}$

Simplify each expression. Assume all logarithmic arguments are positive numbers. (*See Objective 5.*)

89. $\log_{10} 10^{5x+7}$ **90.** $\log_2 2^{x-7}$

91. $\log_b b^x$ **92.** $6^{\log_6 15x}$

93. $a^{\log_a(3x-5)}$ **94.** $e^{2\log_e(x)}$

SECTION 12.6

Simplify each logarithm without a calculator. (*See Objectives 1 and 2.*)

95. $\log 10^{5/3}$ **96.** $\ln \dfrac{1}{e^4}$

97. $\log \sqrt[3]{100}$ **98.** $\ln \sqrt[3]{e^4}$

Approximate each logarithm to two decimal places. (*See Objective 3.*)

99. $\log 6.589$ **100.** $\log 0.0158$

101. $\ln 58.3$ **102.** $\ln \sqrt[3]{7.5}$

Solve each equation for x. Give an exact solution and approximate the solution to two decimal places, if necessary. (*See Objective 4.*)

103. $\ln(2x-5)=4$ **104.** $2\log(7x+3)=6$

105. $\ln(x-2.5)=0$ **106.** $3\ln(2x+17)=4.5$

107. $\ln(5x)=1$ **108.** $\log(0.25x)=1$

Use the change-of-base formula to rewrite each logarithm. Approximate its value to two decimal places. (*See Objective 5.*)

109. $\log_{3.24} 8.56$ **110.** $\log_{0.075} 0.469$

111. $\log_{3/2} 16.43$ **112.** $\log_{1/3} 0.986$

Solve each problem. Round each answer to two decimal places. (*See Objective 6.*)

113. Carrie becomes about 92% proficient at a new task after a 1-week period. The equation $P = 92 - 28.782\log d$ models her percent proficiency P after d days without practicing the new skill. What is Carrie's proficiency level after 5 days without practicing? 10 days?

114. Mark becomes about 96% proficient at a new task after a 1-week period. The equation $P = 96 - 12.5\ln d$ models his percent proficiency P after d days without practicing the new skill. What is Mark's proficiency level after 7 days without practicing? 12 days?

SECTION 12.7

Solve each equation. For solutions of exponential equations, give an exact solution and an approximation of the solution to two decimal places. (*See Objectives 1 and 2.*)

115. $10^{5.2x} = 32.6$

116. $e^{1.2x} = 45$

117. $7^{3x-2} = 28$

118. $10^{5x-2} = 3.5$

119. $\ln x + \ln(2x+1) = \ln 15$

120. $\log 2x - \log(x-4) = 1$

121. $\log_3(5x-2) = 4$

122. $\log_5(1-3x) = 2$

123. $\log_2 x + \log_2(3x-22) = 4$

124. $6(2.17)^{3x} = 84$

125. $12(0.54)^{8x} = 3$

126. $\log_3(5x+2) - \log_3(x-1) = 2$

127. $2\log_2 x - \log_2(3x+8) = 1$

128. $2\ln x - \ln(56-x) = 0$

Solve each problem. (*See Objective 3.*)

129. Madison invests $1270 in an account that pays 2.6% compounded continuously. (a) How much will she have in the account after 5 yr? (b) How long will it take her account to grow to $2750?

130. Troy invests $2500 in a money market account that earns 2.2% annual interest compounded monthly. How long will it take his money to grow to $4200?

131. In 2011, Brazil had a population of approximately 203.43 million. Its population is growing at a rate of 1.134%. The population (in millions) can be modeled by $y = 203.43e^{0.01134t}$, where t is the number of years after 2011. (a) What is an estimate for the population of Brazil in 2023 if it continues to grow at this rate? (b) How long will it take the population to reach 210 million? (Source: https://www.cia.gov/library/publications/the-world-factbook/geos/br.html)

132. The amount of ibuprofen (in milligrams) in a person's body t hour after ingestion is $y = a_0 e^{-0.347t}$, where a_0 is the initial amount taken. If Lynn takes 1000 mg of ibuprofen, how much ibuprofen will be in his body after 4 hr? Round to the nearest integer. (Source: http://www.rxlist.com/ibuprofen-drug.htm)

CHAPTER 12 TEST / EXPONENTIAL AND LOGARITHMIC FUNCTIONS

1. If $f(x) = 5x + 3$ and $g(x) = x^2 - 1$, then $(f \circ g)(x)$ is
  a. $25x^2 + 2$　　　　　　**b.** $25x^2 + 30x + 8$
  c. $5x^2 - 2$　　　　　　　**d.** $5x^3 + 3x^2 - 5x - 3$

2. The function that is one-to-one is
  a.　　　　　　　　　　　**b.**

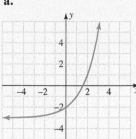　　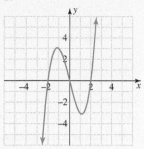

  c.　　　　　　　　　　　**d.**

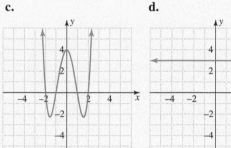

3. The statement that is false about the function $f(x) = 5^x$ is
  a. The graph goes through the point $(1, 0)$.
  b. The domain of the function is $(-\infty, \infty)$.
  c. The range of the function is $(0, \infty)$.
  d. The graph is increasing.

4. The equation $4^3 = 64$ is equivalent to
  a. $\log_3 64 = 4$　　　　　**b.** $\log_4 64 = 3$
  c. $\log_3 4 = 64$　　　　　**d.** $\log_4 3 = 64$

5. The expression $2\log_b 5 + \log_b 3$ is equivalent to
  a. $\log_b 13$　　　　　　　**b.** $\log_b 28$
  c. $\log_b 30$　　　　　　　**d.** $\log_b 75$

6. When simplified, $\ln\dfrac{1}{e} + \log\dfrac{1}{1000} + e^{\ln 5} - 10^{\log 7}$ is
  a. 2　　　　　　　　　　　**b.** 5
  c. -6　　　　　　　　　　**d.** -2

7. The solution set of $6^x = 12$ is
  a. $\{2\}$　　　　　　　**b.** $\left\{\dfrac{\ln 12}{\ln 6}\right\}$
  c. $\{\ln 6\}$　　　　　　**d.** $\{\ln 2\}$

8. A conference committee is planning a reception. The cost of using a banquet hall is \$10 per person. The cost of food is \$25 per person and the cost of prizes is \$40 for every three attendees. Let x represent the number of attendees. Write a function that represents the cost of the reception hall, the cost of food, and the cost of the prizes. Finally write a function that represents the total cost of the reception as a function of attendees.

9. Let $f(x) = \dfrac{9}{x - 2}$ and $g(x) = 4x + 1$. Find the following.
  a. $(f + g)(1)$
  b. $(f \cdot g)(x)$ and state the domain.
  c. $(f \circ g)(7)$　　　　　**d.** $f^{-1}(x)$

10. Graph each function. State the domain and range. Then sketch the graph of the inverse of the function on the same set of axes.
  a. $f(x) = 3^x$　　　　　　**b.** $f(x) = \log_4 x$

11. Solve each equation.
  a. $2^{x+1} = \dfrac{1}{4}$　　　　　**b.** $9^x = 27$
  c. $5^{x-2} = 25^x$　　　　　**d.** $\log_4 \dfrac{1}{16} = x$
  e. $\log x = 0$　　　　　　**f.** $\log_x 81 = 2$
  g. $\ln(x + 1) = 3$　　　　**h.** $2\log(5x) = 4$
  i. $3e^{x+1} = 12$

12. Use the properties of logarithms to rewrite the expression $4\log_2 x - \log_2 y + \dfrac{1}{3}\log_2 z$ as a single logarithms.

13. Juan invests \$2000 in an account that earns 4% annual interest compounded quarterly.
  a. How much will Juan have in the account at the end of 5 yr.
  b. How long will it take the account to grow to \$10,000?

14. The human memory model given by $f(t) = 75 - 6\ln(t + 1)$ determines the percentage of information the average person retains t months after the information was presented.
  a. What percentage of information is retained 6 months after its presentation?
  b. For how long does the average person retain 60% of presented information?

CUMULATIVE REVIEW EXERCISES / CHAPTERS 1–12

Use the order of operations to simplify each expression.
(Section 1.3, Objective 3)

  1. $25 - [(12 - 7) - (6 - 10)] - 28 \div 7 \cdot 4$

  2. $\dfrac{17 - 3\sqrt{13 - 4}}{|4 - 5|^2}$

Solve each problem. (*Section 1.3, Objective 6*)

3. Jonathan has collected 165 dimes and nickels. If x represents the number of dimes he collected, write an expression for the number of nickels.

4. Referring to Exercise 3, write an algebraic expression that represents the total value of Jonathan's coins.

Solve each linear equation. (*Section 2.3, Objective 3*)

5. $6(x - 4) - x = 5(x + 2)$ 6. $3(2 - x) - 2x = x - 12$

Solve each equation. (*Section 9.2, Objectives 1 and 2*)

7. $|4x + 1| = |2x - 7|$ 8. $|3x - 5| = 4$

Find the slope of the line formed by the ordered pairs. (*Section 3.3, Objective 1*)

9. $(10, 9)$ and $(-4, -7)$ 10. $(-2, 3)$ and $(6, -5)$

Find the domain of each function. (*Section 8.1, Objective 3*)

11. $f(x) = 4x^3 - 5$ 12. $f(x) = \sqrt{12 - 4x}$

Rewrite each equation using function notation, if possible. Identify m and b. (*Section 8.2, Objective 2*)

13. $4x - 7y = 28$ 14. $x = -2$

Determine if the two lines are parallel, perpendicular, or neither. (*Section 3.4, Objective 3*)

15. $y = -\dfrac{2}{5}x + 2$ and $y = \dfrac{2}{5}x$

16. $y = 2x + 1$ and $4x - 2y = 3$

Solve each system of linear equations graphically. (*Section 4.1, Objectives 2 and 3*)

17. $\begin{cases} 2x - y = 4 \\ y = x \end{cases}$ 18. $\begin{cases} 2y + 3x = 12 \\ x + y = 2 \end{cases}$

Solve each system of linear equations using substitution. (*Section 4.2, Objectives 1 and 2*)

19. $\begin{cases} y = 5x + 11 \\ x + 4y = 2 \end{cases}$ 20. $\begin{cases} 7x - 15y = 51 \\ 3x + 4y = 1 \end{cases}$

21. $\begin{cases} -0.1x + 0.6y = 7.8 \\ 0.4x + 0.5y = 12.3 \end{cases}$ 22. $\begin{cases} x + 6y = -3 \\ y = \dfrac{2}{3}x - 3 \end{cases}$

Solve each problem. (*Section 4.4, Objective 5*)

23. Two angles are complementary. The measure of one angle is 15° less than twice the measure of the other angle. Find the measure of each angle.

24. Two angles are supplementary. The measure of one angle is 12° more than three times the measure of the other angle. Find the measure of each angle.

Solve each system of linear equations in three variables using elimination. (*Section 4.5, Objective 1*)

25. $\begin{cases} 3x - y - 2z = -1 \\ -x + y + 2z = 3 \\ 4x - 3y + z = 13 \end{cases}$ 26. $\begin{cases} x + y - z = 9 \\ x - y - 2z = 3 \\ 2x - 3y + z = -8 \end{cases}$

Solve each system of linear inequalities in two variables. (*Section 9.4, Objective 1*)

27. $\begin{cases} 2x - y \ge -4 \\ y \le x \end{cases}$ 28. $\begin{cases} x - 3 < 0 \\ y > 2x \end{cases}$

Simplify each expression. Write each answer with positive exponents. (*Section 5.1, Objectives 2–4; Section 5.2, Objectives 1 and 2*)

29. $(-15x^2) \cdot (2x^{13})$ 30. $(4rs^3)(-16r^7s^9)$

31. $\dfrac{a^{12}b}{a^5b^7}$ 32. $\dfrac{p^2q^9}{p^{13}q^4}$

33. $\dfrac{(5x^3y^{-1})^{-3}}{x^{-7}y^{-10}}$ 34. $\dfrac{(r^{-4}s^3)^{-2}}{(3rs)^{-3}}$

Simplify each expression. (*Section 5.3, Objective 4; Section 5.5, Objectives 1–4*)

35. $(13a - 23) + (-15a + 17)$

36. $(12p - 10) - (-8p + 1)$

37. $(7x - 4)(2x + 5)$ 38. $(6y - 1)^2$

39. $(3x - 2)^3$ 40. $(a + 4)^3$

41. Write a polynomial function that represents the perimeter of a triangle with sides of lengths: $8y - 5$, $3y^2 - 12y - 7$, and $2y^2 + 16y + 9$. (*Section 5.4, Objective 4*)

42. Write a polynomial function that represents the perimeter of a trapezoid with sides of lengths: $4x - 1$, $10x$, $8x^2 + 3x$, and $2x^2 - 5x + 9$. (*See Section 5.4, Objective 4*)

Use grouping to factor each polynomial completely. (*Section 6.1, Objective 3*)

43. $w^3 + 3w - 2w^2 - 6$ 44. $4x^3 - 7x^2 + 4x - 7$

Factor each trinomial. (*Section 6.3, Objectives 1–3*)

45. $25x^2 + 40xy + 16y^2$ 46. $12x^2 - 4xy - 5y^2$

Factor completely. (*Section 6.4, Objectives 1–3*)

47. $8x^3 + 125y^3$ 48. $a^3 - b^3$

49. $16x^2 - 8x + 1 - 9y^2$ 50. $y^2 + 10y + 25 - 49z^2$

Solving each equation by factoring. (*Section 6.5, Objective 2*)

51. $10x^2 + 33x - 7 = 0$ 52. $3y^2 - 10y - 8 = 0$

53. $y(2y - 5) = 52$ 54. $a(2a - 15) = 27$

Simplify each rational expression. (*Section 7.1, Objective 3*)

55. $\dfrac{x^2 + 2x - 8}{x^3 - 8}$ 56. $\dfrac{6x^2 + 2x}{9x^2 - 1}$

Multiply or divide the rational expressions. Write each answer in simplest form. (*Section 7.2, Objectives 1 and 2*)

57. $\dfrac{4x^2 - 16}{15x + 10} \cdot \dfrac{3x^2 - x - 2}{x^2 + x - 2}$

58. $\dfrac{x^3 + 125}{(x + 5)^2} \div \dfrac{2x^2 - 10x + 50}{3x^2 + 6x - 45}$

Perform each operation. (*Section 5.7, Objectives 1–3*)

59. $\dfrac{-2x^5 + 3x^4 - 4x^3}{2x^3}$

60. $\dfrac{3x^3 - 6x^2 - 4x}{6x}$

61. $\dfrac{7x^3 + 10x^2 - 14}{x + 1}$

62. $\dfrac{12x^3 + 5x + 2}{2x - 1}$

Add or subtract the rational expressions. (*Section 7.4, Objectives 1 and 2*)

63. $\dfrac{3}{x^2 - 16} + \dfrac{2}{x^2 - 3x - 4}$

64. $\dfrac{4}{x^2 - 9} - \dfrac{5}{x^2 + x - 12}$

65. $\dfrac{3}{x^2 + 2x + 1} - \dfrac{2}{x^2 - 1}$

66. $\dfrac{2}{4x^2 + 12x + 9} + \dfrac{1}{4x^2 - 9}$

Simplify each complex fraction using method 1 or method 2. (*Section 7.5, Objectives 1 and 2*)

67. $\dfrac{\dfrac{6c}{3ab^2}}{\dfrac{4c^2}{2a^2b}}$

68. $\dfrac{\dfrac{2z^2}{5x^2y}}{\dfrac{z}{10xy^2}}$

69. $\dfrac{\dfrac{1}{x} - \dfrac{2}{y}}{2x - y}$

70. $\dfrac{\dfrac{3}{r} + \dfrac{1}{s}}{r + 3s}$

Find the slope of the line that goes through the two given points. (*Section 7.5, Objective 3*)

71. $\left(-\dfrac{5}{6}, \dfrac{1}{4}\right)$ and $\left(\dfrac{1}{3}, -\dfrac{11}{8}\right)$

72. $\left(\dfrac{1}{2}, -\dfrac{1}{3}\right)$ and $\left(\dfrac{1}{10}, -\dfrac{1}{3}\right)$

Simplify each rational expression. (*Section 7.5, Objective 3*)

73. $\dfrac{3^{-2} + 2^{-4}}{2^{-2}}$

74. $\dfrac{1 + 5^{-2}}{2^{-2}}$

Solve each rational equation. (*Section 7.6, Objective 1*)

75. $\dfrac{5}{y} + \dfrac{2}{y - 5} = \dfrac{3}{y^2 - 5y}$

76. $\dfrac{2}{x} - \dfrac{1}{x + 3} = \dfrac{3}{x^2 + 3x}$

Solve each rational equation for the specified variable. (*Section 7.6, Objective 2*)

77. $\dfrac{1}{a} = \dfrac{2}{b} - \dfrac{3}{c}$ for b

78. $\dfrac{4}{x} + \dfrac{1}{y} = \dfrac{5}{z}$ for z

Simplify each expression. Assume all variables represent positive real numbers. Write each answer with positive exponents. (*Section 10.2, Objectives 1–4*)

79. $(x^{-12}y^8)^{3/4}$

80. $(x^{-6}y^9)^{-2/3}$

81. $(a^{1/2} - b^{1/3})(a^{1/2} + 2b^{1/3})$

82. $(2x^{2/5} - y^{1/4})^2$

Simplify each radical expression. Assume all variables represent positive real numbers. (*Section 10.3, Objectives 1 and 2*)

83. $\sqrt{120x^5y^3}$

84. $\sqrt{90rt^7}$

85. $\sqrt[3]{\dfrac{x^4}{8y^3}}$

86. $\sqrt[5]{\dfrac{32a^7}{b^5}}$

Perform the indicated operation and write each answer in simplest radical form. (*Section 10.4, Objectives 1 and 2*)

87. $(2 - \sqrt{3})^2$

88. $(\sqrt{7} + 1)^2$

Solve each radical equation. (*Section 10.6, Objective 1*)

89. $\sqrt{6x + 28} + 2 = x$

90. $\sqrt{4t + 60} = t$

Graph each function $y = f(x)$ by plotting points. Identify the x-intercepts, y-intercept, vertex, and the axis of symmetry. Explain how the graph of the function relates to the graph of $y = x^2$. State the domain and range of the function. (*Section 11.1, Objectives 1–4*)

91. $f(x) = (x + 3)^2 - 1$

92. $f(x) = -(x + 2)^2 + 4$

Solve each equation by completing the square. (*Section 11.2, Objectives 1–4*)

93. $x^2 - 4x + 1 = 0$

94. $x^2 + 6x + 14 = 0$

Solve each equation using the quadratic formula. (*Section 11.3, Objective 1*)

95. $x^2 - 2x + 50 = 0$

96. $x^2 + 6x + 4 = 0$

Solve each equation. (*Section 11.4, Objectives 1–4*)

97. $6x^{2/3} - 7x^{1/3} - 3 = 0$

98. $x - 5x^{1/2} + 6 = 0$

Use the graph of $y = f(x)$ to solve each associated inequality. Write each solution set in interval notation. (*Section 11.6, Objective 4*)

99. $x(2x + 5)(x - 3) < 0$

100. $x^4 - 10x^2 + 9 > 0$

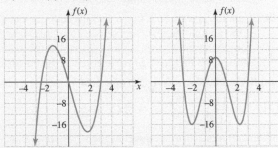

Find the sum, difference, product, and quotients of the functions. State the domain of the combined functions. Also, state the restriction for each quotient function. (*Section 12.1, Objective 1*)

101. $f(x) = 5x + 4$ and $g(x) = 3x - 2$

102. $f(x) = \sqrt{x + 2}$ and $g(x) = 3x - 6$

Let $f(x) = x^2 - 3x + 5$ and $g(x) = 4x + 3$. Evaluate each function. (*Section 12.1, Objectives 2 and 3*)

103. $(f + g)(-2)$

104. $(f \cdot g)(2)$

105. $(f \circ g)(3)$

106. $\left(\dfrac{f}{g}\right)(-1)$

For the given functions, $f(x)$ and $g(x)$, find the composition functions, $(f \circ g)(x)$ and $(g \circ f)(x)$. State the domain. (*Section 12.1, Objective 3*)

107. $f(x) = 3x - 5$ and $g(x) = 4 - x$

108. $f(x) = 7x - 10$ and $g(x) = 2x$

Determine if each function is one-to-one. (*Section 12.2, Objectives 1 and 2*)

109. $f = \{(-9, 3), (-7, 2), (-5, 1), (3, 0), (2, -1)\}$

110. $f = \{(-3, 1), (-2, 4), (0, 5), (3, 2), (2, 2)\}$

Find the equation of the inverse of each function. (*Section 12.2, Objective 4*)

111. $f(x) = (7x - 2)^3$ **112.** $f(x) = x^3 + 125$

Evaluate each function given the information. (*Section 12.2, Objective 5*)

113. If $f^{-1}(6.9) = 1.5$, what is $f(1.5)$?

114. If $f(0) = 2.1$, what is $f^{-1}(2.1)$?

Graph each exponential function. Plot and label at least three points on the graph. State the domain and range of the function. (*Section 12.3, Objective 1*)

115. $y = 0.2^x$ **116.** $y = 3^x$

Solve each exponential equation. (*Section 12.3, Objective 2*)

117. $4^{2x-5} = 8^x$ **118.** $27^{3x-4} = 9^{4x}$

For Exercises 121–123, use the formula $A = P\left(1 + \dfrac{r}{n}\right)^{nt}$.

119. Joshua invests $1850 in a savings account that pays 2.5% annual interest compounded quarterly. Use the compound interest formula to determine the amount of money he will have in the account at the end of 2 yr.

120. Candice invests $5200 in a savings account that pays 1.8% annual interest compounded monthly. Use the compound interest formula to determine the amount of money she will have in the account at the end of 3 yr.

Simplify each expression. (*Section 12.4, Objectives 1 and 2*)

121. $\log_3 \dfrac{1}{81}$ **122.** $\log_{1/2} 1$

Solve each logarithmic equation for x. (*Section 12.4, Objective 3*)

123. $\log_3 x = 5$ **124.** $\log_x 144 = 2$

Graph each logarithmic function. Plot and label at least three points. State the domain and range of the function. (*Section 12.4, Objective 4*)

125. $y = \log_{1/6} x$ **126.** $y = \log_4 x$

Use the properties of logarithms to write each expression as a single logarithm. Simplify, if possible. (*Section 12.5, Objectives 1–3*)

127. $\log_3(x + 5) + \log_3(x - 4)$

128. $\log_6 x + \log_6(2x + 1)$ **129.** $\log_4 x - \log_4(x + 6)$

130. $\log_2 x - \log_2 10$ **131.** $\log(x + 2)^5$

132. $\ln(x - 3)^7$

Rewrite each logarithmic expression. If the expression is a single logarithm, write the expression as a combination of logarithms. If the expression is a combination of logarithms, write the expression as a single logarithm. (*Section 12.5, Objective 4*)

133. $\ln x^3 \sqrt{x - 1}$ **134.** $\log_6 \dfrac{y^3}{\sqrt{y + 12}}$

135. $5\log_{10} x - 2\log_{10} y$ **136.** $\dfrac{1}{2}\log_2 x + 2\log_2 y$

Simplify each the expression. (*Section 12.5, Objective 5*)

137. $\log_6 6^{3x+2}$ **138.** $\log_{10} 10^{5x-4}$

139. $5^{\log_5(2x+7)}$ **140.** $3^{\log_3(2x)}$

Simplify each common or natural logarithm without a calculator. (*Section 12.6, Objectives 1 and 2*)

141. $\log 10^9$ **142.** $\log \dfrac{1}{1000}$ **143.** $\ln \dfrac{1}{e^8}$ **144.** $\ln \sqrt[5]{e^3}$

Solve each equation for x. Give an exact solution and approximate the solution to two decimal places, if necessary. (*Section 12.6, Objective 4*)

145. $\log(5x - 4) = 1.5$ **146.** $\ln(2x + 21) = 4.5$

Solve each equation. Give an exact solution and an approximation of the solution to two decimal places. (*Section 12.7, Objectives 1 and 2*)

147. $(0.75)^{8x} = 72$ **148.** $(0.44)^{5x} = 91$

149. $\log_5(2x - 7) = 1$ **150.** $\log(3x + 46) = 2$

Conic Sections and Nonlinear Systems

Responsibility

A responsible person is someone who is both trustworthy and dependable. It's a trait that people put their confidence in and it may make the difference in you beating out the competition for a job or completing a college degree.

As a student, you must be responsible for your education. Remember, college is a privilege and a magnificent way for you to grow yourself while at the same time making yourself extremely marketable and competitive in today's workplace. Some ways to demonstrate responsibility as a student are to

- make sure you've done the work your instructor has assigned.
- be an active learner by reading ahead in the book, asking questions, and seeking help.
- own up to mistakes, fix them, and move on toward your goal.

If you model yourself around the attributes given in this book, you're well on your way to success in both college and career.

Question For Thought: Would your current employer or teacher describe you as a responsible person? How do you define responsibility?

Chapter Outline

Coming Up...

In Section 13.2, we will learn how to graph an equation that can be used to model the orbit of the planets around the sun.

> "The willingness to accept responsibility for one's own life is the source from which self-respect springs."
>
> —Joan Didion (Author)

The Parabola and the Circle

In **Chapter 11,** we studied quadratic functions that are transformations of the function $f(x) = x^2$. The graphs of these functions are vertical parabolas. In this chapter, we study other types of equations that produce horizontal parabolas, circles, ellipses, and hyperbolas. To conclude the chapter, we will solve nonlinear systems of equations and inequalities.

▶ **OBJECTIVES**

As a result of completing this section, you will be able to

1. Graph parabolas of the form $x = a(y - k)^2 + h$.
2. Graph circles of the form $(x - h)^2 + (y - k)^2 = r^2$.
3. Write the equation of a circle given its center and radius.
4. Complete the square to write the equation of a circle in standard form.
5. Troubleshoot common errors.

Centuries ago, mathematicians studied the graphs of parabolas and circles as simply mathematical objects. But these graphs have been found to have significant value in the study of science and in real-life architectural structures. The parabola, for example, is found in the design of satellite dishes and is also used to model projectile motion.

The parabola is an example of a conic section. **Conic sections** are curves that are obtained by the intersection of a cone and a plane. In this chapter, we will study the four types of curves that result from this intersection: the parabola, the circle, the ellipse, and the hyperbola.

  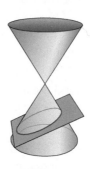 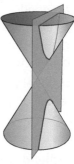

| Parabola | Circle | Ellipse | Hyperbola |

Parabolas of the Form $x = a(y - k)^2 + h$

Objective 1 ▶

Graph parabolas of the form $x = a(y - k)^2 + h$.

Recall in Chapter 11, we studied quadratic functions of the form $f(x) = a(x - h)^2 + k$. This function is the graph of a *vertical* parabola whose vertex is (h, k) and opens upward if $a > 0$ and opens downward if $a < 0$. Now we will turn our attention to equations of the form $x = a(y - k)^2 + h$.

We will first examine the equations $x = y^2$ and $x = -y^2$. To determine their graphs, we plot ordered pairs that satisfy the equations. Since the variable x is isolated, we assign values to y and solve for x.

Graph of $x = y^2$

$x = y^2$	y	(x, y)
$(-3)^2 = 9$	-3	$(9, -3)$
$(-2)^2 = 4$	-2	$(4, -2)$
$(-1)^2 = 1$	-1	$(1, -1)$
$(0)^2 = 0$	0	$(0, 0)$
$(1)^2 = 1$	1	$(1, 1)$
$(2)^2 = 4$	2	$(4, 2)$
$(3)^2 = 9$	3	$(9, 3)$

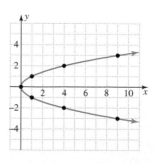

Plotting these points gives us the graph of a *horizontal* parabola. This equation is a relation but not a function since it fails the vertical line test. The vertex of the parabola is $(0, 0)$ and the axis of symmetry is the horizontal line $y = 0$. The parabola opens to the right.

Graph of $x = -y^2$

$x = -y^2$	y	(x, y)
$-(-3)^2 = -9$	-3	$(-9, -3)$
$-(-2)^2 = -4$	-2	$(-4, -2)$
$(-1)^2 = -1$	-1	$(-1, -1)$
$-(0)^2 = 0$	0	$(0, 0)$
$-(1)^2 = -1$	1	$(-1, 1)$
$-(2)^2 = -4$	2	$(-4, 2)$
$-(3)^2 = -9$	3	$(-9, 3)$

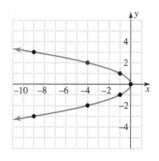

These two graphs illustrate that $x = a(y - k)^2 + h$ is the equation of a horizontal parabola that opens right if $a > 0$ and opens left if $a < 0$. The vertex of the parabola is (h, k) and the axis of symmetry is $y = k$. Just as the value of a stretches or shrinks a vertical parabola, it does the same to a horizontal parabola. If $|a| > 1$, then the parabola is horizontally stretched (made more narrow). If $|a| < 1$, then the parabola is horizontally shrunk (made wider).

Equation	$x = a(y - k)^2 + h$	Vertex	Direction
$x = y^2 - 4$	$x = 1(y - 0)^2 - 4$	$(-4, 0)$	opens right
$x = (y + 1)^2$	$x = 1[y - (-1)]^2 + 0$	$(0, -1)$	opens right
$x = -(y - 2)^2 + 3$	$x = -1(y - 2)^2 + 3$	$(3, 2)$	opens left

---

Property: Standard Form of a Parabola

The graph of $x = a(y - k)^2 + h$ is the graph of $x = y^2$ where

- The vertex of the graph is the point (h, k).
- The axis of symmetry is the horizontal line through the vertex, $y = k$.
- The parabola opens right if $a > 0$ and opens left if $a < 0$.

It is helpful to note the differences between the function $f(x) = a(x - h)^2 + k$ and the equation $x = a(y - k)^2 + h$, as shown. These two forms are the *standard forms* of a **parabola**.

$f(x) = a(x - h)^2 + k,$
$a > 0$

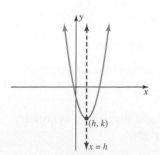

$f(x) = a(x - h)^2 + k,$
$a < 0$

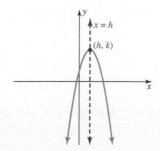

$x = a(y - k)^2 + h,$
$a > 0$

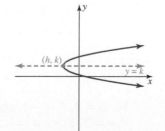

$x = a(y - k)^2 + h,$
$a < 0$

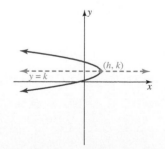

Procedure: Graphing a Parabola of the Form $x = a(y - k)^2 + h$

Step 1: If the equation is not in standard form, complete the square to write it in this form.

Step 2: Identify and plot the vertex, (h, k).

Step 3: Determine the axis of symmetry, $y = h$.

Step 4: Determine the direction the parabola opens.

    a. If $a > 0$, the parabola opens to the right.

    b. If $a < 0$, the parabola opens to the left.

Step 5: Plot additional points by finding the x- and y-intercepts or by using symmetry.

Step 6: Connect the points with a smooth curve to form a parabola.

Objective 1 Examples **Graph each equation by plotting the vertex and at least two additional points. State the axis of symmetry.**

1a. $x = (y - 1)^2 - 4$ **1b.** $x = -2y^2 + 2$ **1c.** $x = y^2 - 4y + 3$

Solutions **1a.** The equation is in standard form with $a = 1$, $h = -4$, and $k = 1$, so the vertex is $(-4, 1)$ and the axis of symmetry is the horizontal line $y = 1$. Since $a = 1 > 0$, the parabola opens to the right.

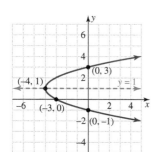

x-intercept ($y = 0$):

$$x = (y - 1)^2 - 4$$
$$x = (0 - 1)^2 - 4$$
$$x = (-1)^2 - 4$$
$$x = 1 - 4$$
$$x = -3$$

The x-intercept is $(-3, 0)$.

y-intercept ($x = 0$):

$$x = (y - 1)^2 - 4$$
$$0 = (y - 1)^2 - 4$$
$$4 = (y - 1)^2$$
$$\pm\sqrt{4} = y - 1$$
$$\pm 2 = y - 1$$
$$1 \pm 2 = y$$

The y-intercepts are $(0, 3)$ and $(0, -1)$.

1b. The equation, written in standard form, is $x = -2(y - 0)^2 + 2$, so the vertex is $(2, 0)$ and the axis of symmetry is $y = 0$. Since $a = -2 < 0$, the parabola opens to the left.

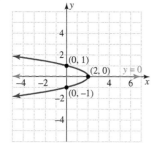

x-intercept ($y = 0$):

$$x = -2y^2 + 2$$
$$x = -2(0)^2 + 2$$
$$x = 2$$

The x-intercept is $(2, 0)$.

y-intercept ($x = 0$):

$$x = -2y^2 + 2$$
$$0 = -2y^2 + 2$$
$$2y^2 = 2$$
$$y^2 = 1$$
$$y = \pm 1$$

The y-intercepts are $(0, 1)$ and $(0, -1)$.

1c. We complete the square to write the equation in the form $x = a(y - k)^2 + h$.

$$x = y^2 - 4y + 3$$
$$x - 3 = y^2 - 4y \qquad \text{Subtract 3 from each side.}$$
$$x - 3 + 4 = y^2 - 4y + 4 \qquad \text{Add } \left(-\frac{4}{2}\right)^2 = 4 \text{ to each side.}$$
$$x + 1 = (y - 2)^2 \qquad \text{Simplify the left side and factor the trinomial.}$$
$$x = (y - 2)^2 - 1 \qquad \text{Subtract 1 from each side.}$$

So, $a = 1$, $h = -1$, and $k = 2$. The vertex is $(-1, 2)$ and the axis of symmetry is $y = 2$. Since $a = 1 > 0$, the parabola opens to the right.

Note that we can use the vertex formula, $-\dfrac{b}{2a}$, to find the vertex as well. Since the equation is of the form $x = ay^2 + by + c$, the vertex is

$$y = -\frac{b}{2a} = -\frac{-4}{2(1)} = \frac{4}{2} = 2$$

$$x = (2)^2 - 4(2) + 3 = 4 - 8 + 3 = -1$$

So, the vertex formula also gives us the vertex of $(-1, 2)$.

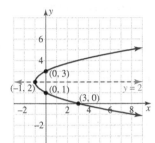

x-intercept $(y = 0)$:

$$x = y^2 - 4y + 3$$
$$x = (0)^2 - 4(0) + 3$$
$$x = 3$$

The *x*-intercept is $(3, 0)$.

y-intercept $(x = 0)$:

$$x = y^2 - 4y + 3$$
$$0 = y^2 - 4y + 3$$
$$0 = (y - 3)(y - 1)$$
$$y - 3 = 0 \quad \text{or} \quad y - 1 = 0$$
$$y = 3 \qquad\qquad y = 1$$

The *y*-intercepts are $(0, 3)$ and $(0, 1)$.

✅ **Student Check 1** Graph each equation by plotting the vertex and at least two additional points. State the axis of symmetry.

a. $x = (y - 3)^2 - 1$ **b.** $x = -y^2 + 4$ **c.** $x = y^2 - 6y + 5$

Circles

Objective 2 ▶

Graph circles of the form $(x - h)^2 + (y - k)^2 = r^2$.

A **circle** is defined to be the set of all points (x, y) that have the same distance, called the **radius**, from a fixed point, called the **center**. The following figure shows the graph of a circle whose center is at the origin, $(0, 0)$, and whose radius is r. The Pythagorean theorem provides the equation of the circle, which is $x^2 + y^2 = r^2$.

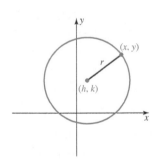

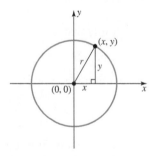

Figure 13.1

Figure 13.1 shows a circle whose center is not at the origin but at a point (h, k). The distance formula gives us $r = \sqrt{(x - h)^2 + (y - k)^2}$, or after squaring each side,

$$r^2 = (x - h)^2 + (y - k)^2$$

Property: Standard Form of a Circle

A circle with center (h, k) and radius r is given by the equation

$$(x - h)^2 + (y - k)^2 = r^2$$

If the center is $(0, 0)$, then the equation is $x^2 + y^2 = r^2$.

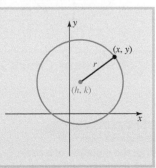

Note that in the equation of a circle, the coefficients of x^2 and y^2 are the same and the terms are connected by addition. Some examples are shown in the table.

Equation	Center	Radius
$x^2 + y^2 = 4$	$(0, 0)$	$r = \sqrt{4} = 2$
$x^2 + (y - 5)^2 = 9$	$(0, 5)$	$r = \sqrt{9} = 3$
$(x + 1)^2 + y^2 = 25$	$(-1, 0)$	$r = \sqrt{25} = 5$
$(x - 2)^2 + (y + 3)^2 = 2$	$(2, -3)$	$r = \sqrt{2}$

> **Procedure: Graphing a Circle with the Equation $(x - h)^2 + (y - k)^2 = r^2$**
> **Step 1:** Identify the center (h, k).
> **Step 2:** Identify the radius r.
> **Step 3:** Plot the center and then plot four key points that are r units away from the center. These four points are $(h + r, k)$, $(h - r, k)$, $(h, k + r)$, and $(h, k - r)$.
> **Step 4:** Draw a circle through these four points.

Objective 2 Examples   **Graph each circle. Identify the center, the radius, and four key points.**

2a. $x^2 + y^2 = 4$ **2b.** $(x - 1)^2 + (y + 3)^2 = 9$ **2c.** $(x + 2)^2 + y^2 = 5$

Solutions   **2a.** The equation is equivalent to $(x - 0)^2 + (y - 0)^2 = 2^2$. So, the center is $(0, 0)$ and the radius is 2. The four key points that are 2 units from the center are

$$(0 + 2, 0) = (2, 0)$$
$$(0 - 2, 0) = (-2, 0)$$
$$(0, 0 + 2) = (0, 2)$$
$$(0, 0 - 2) = (0, -2)$$

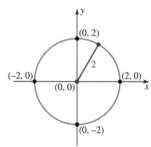

2b. The equation is equivalent to $(x - 1)^2 + [y - (-3)]^2 = 3^2$. So, the center is $(1, -3)$ and the radius is 3. The four key points that are 3 units from the center are

$$(1 + 3, -3) = (4, -3)$$
$$(1 - 3, -3) = (-2, -3)$$
$$(1, -3 + 3) = (1, 0)$$
$$(1, -3 - 3) = (1, -6)$$

2c. The equation is equivalent to $[x - (-2)]^2 + (y - 0)^2 = \left(\sqrt{5}\right)^2$. So, the center is $(-2, 0)$ and the radius is $\sqrt{5}$. The four key points that are $\sqrt{5}$ units from the center are

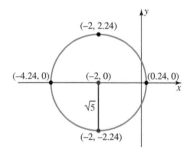

$$\left(-2 + \sqrt{5}, 0\right) \approx (0.24, 0)$$
$$\left(-2 - \sqrt{5}, 0\right) \approx (-4.24, 0)$$
$$\left(-2, 0 + \sqrt{5}\right) \approx (-2, 2.24)$$
$$\left(-2, 0 - \sqrt{5}\right) \approx (-2, -2.24)$$

✓ Student Check 2   Graph each circle. Identify the center, the radius, and four key points.

a. $x^2 + y^2 = 16$ **b.** $(x + 4)^2 + (y - 2)^2 = 4$ **c.** $x^2 + (y - 2)^2 = 15$

Objective 3 ▶

Write the equation of a circle given its center and radius.

Write the Standard Form of a Circle

If we know the center and radius of a circle, we can write its equation by substituting these values into the standard form of a circle.

Procedure: Writing the Equation of a Circle

Step 1: Let the center be (h, k) and the radius be r.

Step 2: Substitute the values into the standard form $(x - h)^2 + (y - k)^2 = r^2$ and simplify.

Objective 3 Examples   Write the equation of each circle given its center and radius.

3a. center $= (-5, 4)$ and radius $= 7$ **3b.** center $= (0, 0)$ and radius $= 5$

Solutions   **3a.** $(x - h)^2 + (y - k)^2 = r^2$ State the standard form.

$[x - (-5)]^2 + (y - 4)^2 = 7^2$ Let $(h, k) = (-5, 4)$, $r = 7$.

$(x + 5)^2 + (y - 4)^2 = 49$ Simplify.

3b. $(x - h)^2 + (y - k)^2 = r^2$ State the standard form.

$(x - 0)^2 + (y - 0)^2 = 5^2$ Let $(h, k) = (0, 0)$, $r = 5$.

$x^2 + y^2 = 25$ Simplify.

✓ **Student Check 3**   Write the equation of each circle given its center and radius.

a. center $= (5, -2)$ and radius $= 10$ **b.** center $= (0, 0)$ and radius $= 6$

Complete the Square to Write the Equation of a Circle

Objective 4 ▶

Complete the square to write the equation of a circle in standard form.

Just as with parabolas, we sometimes have to complete the square to write the equation of a circle in standard form. For a circle, we have to complete the square twice, once with the x-terms and once with the y-terms.

Procedure: Completing the Square to Write the Equation of a Circle in Standard Form

Step 1: Isolate the constant on one side of the equation.

Step 2: Group the x-terms and y-terms together.

Step 3: If the coefficient of x^2 and/or y^2 is not 1, factor out the coefficient from the x-terms and y-terms, as necessary.

Step 4: Find the number that makes the x-terms a perfect square trinomial and add this to each side of the equation. Find the number that makes the y-terms a perfect square trinomial and add this to each side of the equation.

Step 5: Factor the two trinomials and simplify the right side.

Step 6: Identify the center and radius.

Objective 4 Example   Write the equation $x^2 + y^2 - 4x + 6y = 3$ as a circle in standard form and identify its center and radius.

Solution $x^2 + y^2 - 4x + 6y = 3$ The constant is already isolated on one side.

$(x^2 - 4x) + (y^2 + 6y) = 3$ Group the x-terms and y-terms together.

$(x^2 - 4x + 4) + (y^2 + 6y + 9) = 3 + 4 + 9$ Complete the squares and add to each side.

$(x - 2)^2 + (y + 3)^2 = 16$ Factor the left side and simplify the right side.

So, the center of the circle is $(2, -3)$ and the radius is $\sqrt{16} = 4$.

✓ **Student Check 4**   Write the equation $x^2 + y^2 - 2x + 10y = 10$ as a circle in standard form and identify its center and radius.

Objective 5 ▶ Troubleshooting Common Errors

Troubleshoot common errors.

Some common errors associated with the parabola and circle are shown.

Objective 5 Examples A problem and an incorrect solution are given. Provide the correct solution and an explanation of the error.

5a. Find the vertex of $x = (y - 4)^2 + 3$.

Incorrect Solution	Correct Solution and Explanation
The vertex of $x = (y - 4)^2 + 3$ is (4, 3).	The vertex of a horizontal parabola $x = a(y - k)^2 + h$ is (h, k). So, the vertex of $x = (y - 4)^2 + 3$ is (3, 4).

5b. Find the center and radius of $(x + 3)^2 + (y - 4)^2 = 3$.

Incorrect Solution	Correct Solution and Explanation
The center is $(3, -4)$ and the radius is 3.	The equation in standard form is $$(x + 3)^2 + (y - 4)^2 = 3$$ $$[x - (-3)]^2 + (y - 4)^2 = (\sqrt{3})^2$$ So, the center is $(-3, 4)$ and the radius is $\sqrt{3}$.

ANSWERS TO STUDENT CHECKS

Student Check 1 **a.**

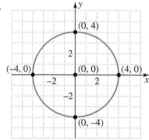

b.

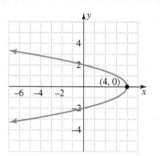

c.

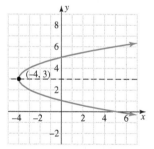

Student Check 2 **a.**

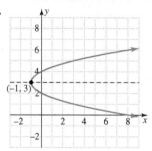

b.

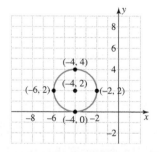

c.

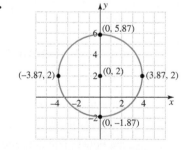

Student Check 3 **a.** $(x - 5)^2 + (y + 2)^2 = 100$
b. $x^2 + y^2 = 36$

Student Check 4 $(x - 1)^2 + (y + 5)^2 = 36$, center = $(1, -5)$ and radius = 6

SUMMARY OF KEY CONCEPTS

1. The equation $x = a(y - k)^2 + h$ is a horizontal parabola with vertex (h, k). The parabola opens to the right if $a > 0$ and to the left if $a < 0$.

2. The equation $(x - h)^2 + (y - k)^2 = r^2$ represents a circle with center (h, k) and radius r.

  a. We can identify the center and radius from the standard form of a circle.

  b. If we know the center and radius, we can write the standard form of a circle.

  c. We may have to complete the square to write the standard form of a circle.

GRAPHING CALCULATOR SKILLS

The graphing calculator can graph horizontal parabolas and circles; however, we must solve the equations for y. Because the y-variable is squared, we apply the square root property to solve for y and will, therefore, have two expressions to enter in the calculator to graph the equation.

Example 1: Graph $x = (y - 1)^2 - 4$.

Solution: Solve the equation for y.

$$x = (y - 1)^2 - 4$$
$$x + 4 = (y - 1)^2 \qquad \text{Add 4 to each side.}$$
$$\pm\sqrt{x + 4} = y - 1 \qquad \text{Apply the square root property.}$$
$$1 \pm \sqrt{x + 4} = y \qquad \text{Add 1 to each side.}$$

Enter the two expressions, $1 + \sqrt{x + 4}$ and $1 - \sqrt{x + 4}$, to graph the parabola. Y_1 is the top half of the parabola and Y_2 is the bottom half of the parabola.

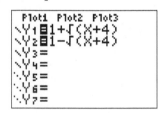 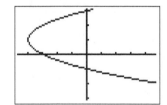

Example 2: Graph $x^2 + y^2 = 4$.

Solution: Solve the equation for y.

$$x^2 + y^2 = 4$$
$$y^2 = 4 - x^2 \qquad \text{Subtract } x^2 \text{ from each side.}$$
$$y = \pm\sqrt{4 - x^2} \qquad \text{Apply the square root property.}$$

Enter the two expressions, $\sqrt{4 - x^2}$ and $-\sqrt{4 - x^2}$, to graph the circle. It is best to graph a circle in the ZSquare window so that the graph will not be distorted. Y_1 is the top half of the circle and Y_2 is the bottom half of the circle.

 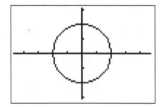

SECTION 13.1 / EXERCISE SET

 Write About It!

Use complete sentences in your answer to each exercise.

1. Describe the graph of $x = \dfrac{1}{2}(y - 3)^2 + 1$.

2. Describe the graph of $x = -2(y + 1)^2 - 3$.

3. Describe how to graph the equation $x = -y^2 + 25$.

4. Explain how to write the equation, $x = y^2 + 8y - 12$ in standard form.

5. State the advantages of writing the equation of a circle in standard form.

6. Explain how to write the equation of the circle when the center (h, k) and radius r are given.

7. Explain how to write the equation of a circle, $x^2 + y^2 - 6x + 10y = 15$, in standard form.

8. Describe the graph of the equation $x^2 + y^2 + 4x + 5 = 0$.

 Practice Makes Perfect!

Graph each equation by plotting the vertex and at least two additional points. State the axis of symmetry. (*See Objective 1.*)

9. $x = (y + 2)^2 + 3$ **10.** $x = (y + 1)^2 + 5$

11. $x = (y - 2)^2 + 1$ **12.** $x = (y - 4)^2 + 2$

13. $x = (y - 3)^2 - 4$ **14.** $x = (y - 2)^2 - 6$

15. $x = -(y + 2)^2 + 4$ **16.** $x = -(y - 4)^2 + 5$

17. $x = -(y - 2)^2 - 1$ **18.** $x = -(y + 2)^2 - 2$

19. $x = -y^2 + 9$ **20.** $x = -y^2 - 4$

21. $x = y^2 - 9$ **22.** $x = y^2 - 6$

23. $x = y^2 + 2y - 8$ **24.** $x = -y^2 + 6y - 8$

25. $x = y^2 + 3y + 2$ **26.** $x = y^2 + y - 2$

Graph each circle. Identify the center, the radius, and four key points. (*See Objective 2.*)

27. $x^2 + y^2 = 25$ **28.** $x^2 + y^2 = 9$

29. $(x - 2)^2 + (y - 1)^2 = 9$ **30.** $(x + 1)^2 + (y - 3)^2 = 4$

31. $x^2 + (y + 3)^2 = 25$ **32.** $(x - 1)^2 + y^2 = 4$

Write the equation of each circle given its center and radius. (*See Objective 3.*)

33. center $= (-4, 5)$ and radius $= 1$

34. center $= (-2, 3)$ and radius $= 3$

35. center $= (4, -2)$ and radius $= \sqrt{3}$

36. center $= (1, -6)$ and radius $= \sqrt{7}$

37. center $= (-2, -7)$ and radius $= \sqrt{6}$

38. center $= (-3, -1)$ and radius $= \sqrt{2}$

Write the equation of each circle in standard form and identify the center and radius. (*See Objective 4.*)

39. $x^2 + y^2 - 12x - 8y = -51$

40. $x^2 + y^2 - 4x - 16y = -52$

41. $x^2 + y^2 + 2x + 2y = 2$

42. $x^2 + y^2 + 10x + 4y = -13$

43. $x^2 + y^2 - 6x = 16$

44. $x^2 + y^2 + 12x = 64$

45. $x^2 + y^2 - x + y = \dfrac{1}{2}$ **46.** $x^2 + y^2 + 3x - y = \dfrac{3}{2}$

 Mix 'Em Up!

Graph each equation. If it is a parabola, plot the vertex and at least two additional points, and state the axis of symmetry. If it is a circle, identify the center, the radius, and the four key points.

47. $x = (y + 1)^2 - 5$ **48.** $x = (y - 1)^2 - 4$

49. $(x + 1)^2 + (y - 1)^2 = 1$ **50.** $(x - 2)^2 + (y + 3)^2 = 9$

51. $x = 2(y - 3)^2 + 4$ **52.** $x = 2(y + 2)^2 - 3$

53. $x = -3(y - 1)^2 + 1$ **54.** $x = -4(y + 2)^2 + 5$

55. $(x - 1)^2 + y^2 = 16$ **56.** $(x + 2)^2 + y^2 = 4$

57. $x = \dfrac{1}{2}(y - 1)^2 + 3$ **58.** $x = \dfrac{1}{3}(y + 1)^2 - 2$

59. $x = -2y^2 + 4$ **60.** $x = -3y^2 + 1$

61. $x = 2y^2 + 3$ **62.** $x = 2y^2 - 6$

63. $x = y^2 + 2y - 3$ **64.** $x = y^2 + 6y + 8$

65. $x = -y^2 + 2y + 24$ **66.** $x = -y^2 + 6y - 4$

67. $x = y^2 - 5y + 5$ **68.** $x = y^2 - 7y + 13$

Write the equation of each circle given its center and radius.

69. center $= (-1, -2)$ and radius $= 4$

70. center $= (-5, -6)$ and radius $= 12$

71. center $= (-4, 6)$ and radius $= \sqrt{6}$

72. center $= (-1, 4)$ and radius $= \sqrt{15}$

73. center $= (-7, 0)$ and radius $= \sqrt{5}$

74. center $= (-4, 0)$ and radius $= \sqrt{3}$

75. center $= \left(0, -\dfrac{3}{4}\right)$ and radius $= \dfrac{1}{2}$

76. center $= \left(0, \dfrac{2}{3}\right)$ and radius $= \dfrac{1}{3}$

77. center $= (1.2, -2.5)$ and radius $= 4.5$

78. center $= (-1.6, 3.2)$ and radius $= 1.5$

79.

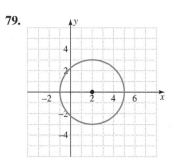

80.

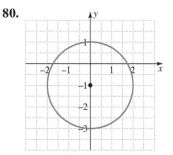

Write the equation of each circle in standard form and identify its center and radius.

81. $x^2 + y^2 + 14x + 6y = -22$

82. $x^2 + y^2 + 2x - 6y = -9$

83. $x^2 + y^2 + 6x - 12y = 0$

84. $x^2 + y^2 - 4x + 2y = 0$

85. $x^2 + y^2 + 8x - 6y = 0$

86. $x^2 + y^2 - 10x - 24y = 0$

87. $x^2 + y^2 - 2y = 8$

88. $x^2 + y^2 - 12y = 45$

89. $x^2 + y^2 - 3x - 5y = \dfrac{1}{2}$

90. $x^2 + y^2 + 7x - 3y = \dfrac{3}{2}$

91. $x^2 + y^2 - 1.6x + 4.2y = 0.71$

92. $x^2 + y^2 + 3.6x - 5.6y = 29.88$

 You Be the Teacher!

Correct each student's errors, if any.

93. Find the vertex from the equation $x = (y + 1)^2 - 2$.

Eric's work:

$(-1, -2)$

94. Write the equation of the circle in standard form:
$x^2 + y^2 - 10x + 8y = 9$.

Monica's work:

$(x - 5)^2 + (y + 4)^2 = 9$

 Calculate It!

Use a graphing calculator to graph each function and find the vertex.

95. $x = (y - 1.2)^2 - 1.5$

96. $x = -2(y + 2.4)^2 + 3.2$

Use a graphing calculator to graph each circle.

97. $(x - 2)^2 + (y + 3)^2 = 9$

98. $(x + 3)^2 + (y - 1)^2 = 25$

| SECTION 13.2 | The Ellipse and the Hyperbola |

▶ **OBJECTIVES**

As a result of completing this section, you will be able to

1. Identify the standard form of an ellipse and graph an ellipse.

2. Identify the standard form of a hyperbola and graph a hyperbola.

3. Troubleshoot common errors.

The other two conic sections that we will discuss are the ellipse and the hyperbola. Both of these shapes occur in real-life situations. The orbits of the planets, moons, and satellites are elliptical. Ellipses are also used in microscopes, telescopes, and many architectural designs. The hyperbola is used to model the path of a comet, telescope lenses, and gears for some machines.

Ellipses

An ellipse basically looks like an oval. It is a circle that has been flattened or stretched. Mathematically, we define an **ellipse** as the set of all points such that the sum of their distances from two fixed points is constant. The two fixed points are called **foci**. The midpoint of the foci is the **center** of the ellipse.

Objective 1 ▶

Identify the standard form of an ellipse and graph an ellipse.

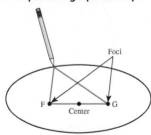

Figure 13.2

To draw an ellipse by hand, we can tie the ends of a piece of string to two pins that are stuck in a sheet of paper. Then, keeping the string tight with the point of a pencil, allow the pencil to trace a path around the pins. The resulting curve is an ellipse, with the two pins, or fixed points, representing its foci. (See Figure 13.2.)

Property: Standard Form of an Ellipse

The equation

$$\frac{x^2}{a^2} + \frac{y^2}{b^2} = 1$$

is an ellipse with center $(0, 0)$. The x-intercepts are $(a, 0)$ and $(-a, 0)$, and the y-intercepts are $(0, b)$ and $(0, -b)$.

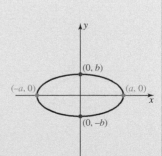

The equation

$$\frac{(x-h)^2}{a^2} + \frac{(y-k)^2}{b^2} = 1$$

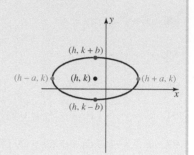

is an ellipse with center (h, k).

Procedure: Graphing an Ellipse

Step 1: Identify the center and the values of a and b.

Step 2: Plot the center.

Step 3: From the center, move a units right and left to find two key points on the ellipse.

Step 4: From the center, move b units up and down to find two more key points on the ellipse.

Step 5: Connect the points with a smooth curve.

Objective 1 Examples   **Graph each ellipse. Label four key points and the center on the graph.**

1a. $\dfrac{x^2}{16} + \dfrac{y^2}{9} = 1$ **1b.** $25x^2 + 4y^2 = 100$ **1c.** $\dfrac{(x-2)^2}{36} + \dfrac{(y+4)^2}{9} = 1$

Solutions   **1a.** The equation represents an ellipse whose center is at $(0, 0)$ and can be written as

$$\frac{x^2}{4^2} + \frac{y^2}{3^2} = 1$$

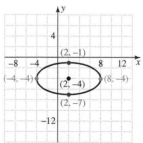

Figure 13.3

Plot $(0, 0)$ and move 4 units to the right and left, which takes us to the x-intercepts of $(-4, 0)$ and $(4, 0)$. Then move 3 units up and down, which takes us to the y-intercepts of $(0, 3)$ and $(0, -3)$. (See Figure 13.3.)

1b. The equation is not in standard form. To be in standard form, the constant on one side must be 1.

$$25x^2 + 4y^2 = 100$$

$$\frac{25x^2}{100} + \frac{4y^2}{100} = \frac{100}{100} \qquad \text{Divide each side by 100.}$$

$$\frac{x^2}{4} + \frac{y^2}{25} = 1 \qquad \text{Simplify.}$$

$$\frac{x^2}{2^2} + \frac{y^2}{5^2} = 1 \qquad \text{Identify } a \text{ and } b.$$

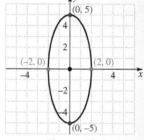

Figure 13.4

So, this is an ellipse with center $(0, 0)$. The x-intercepts are $(2, 0)$ and $(-2, 0)$, and the y-intercepts are $(0, 5)$ and $(0, -5)$. (See Figure 13.4.)

1c. The equation represents an ellipse whose center is $(2, -4)$ and can be written as

$$\frac{(x-2)^2}{6^2} + \frac{[y-(-4)]^2}{3^2} = 1$$

Figure 13.5

Since $a = 6$, we move 6 units right and left from the center $(2, -4)$ to $(2 + 6, -4) = (8, -4)$ and $(2 - 6, -4) = (-4, -4)$. Since $b = 3$, we move 3 units up and down from the center $(2, -4)$ to $(2, -4 + 3) = (2, -1)$ and $(2, -4 - 3) = (2, -7)$. (See Figure 13.5.)

Student Check 1   Graph each ellipse. Label four key points and the center on the graph.

    a. $\dfrac{x^2}{36} + \dfrac{y^2}{25} = 1$ **b.** $9x^2 + 4y^2 = 36$ **c.** $\dfrac{(x+1)^2}{4} + \dfrac{(y-3)^2}{16} = 1$

Note: *From Example 1, we observe the following.*
1. In the equation of an ellipse, the coefficients of x^2 and y^2 are not the same. If they were the same, the equation would represent a circle.
2. If $a > b$, then the ellipse is horizontal.
3. If $a < b$, then the ellipse is vertical.

Hyperbolas

Objective 2 ▶

Identify the standard form of a hyperbola and graph a hyperbola.

The last conic section we will discuss is the hyperbola. The graph of a **hyperbola** consists of two branches and is defined as the set of points such that the absolute value of the difference of their distances from two fixed points is constant. The fixed points are the **foci** and the midpoint of the foci is the **center** of the hyperbola.

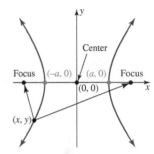

Property: Standard Form of a Hyperbola

The equation

$$\frac{x^2}{a^2} - \frac{y^2}{b^2} = 1$$

is a hyperbola with center $(0, 0)$. The x-intercepts are $(a, 0)$ and $(-a, 0)$.

The equation

$$\frac{y^2}{b^2} - \frac{x^2}{a^2} = 1$$

is a hyperbola with center $(0, 0)$. The y-intercepts are $(0, b)$ and $(0, -b)$.

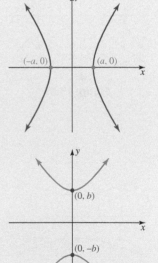

We will construct graphs of hyperbolas whose center is only at the origin. To draw the graph of a hyperbola $\dfrac{x^2}{a^2} - \dfrac{y^2}{b^2} = 1$, we will make use of a rectangle that has vertices at

$$(a, b), (-a, b), (a, -b), \text{ and } (-a, -b)$$

The diagonals of this rectangle form lines, called **asymptotes**, that the branches of the hyperbola approach. These lines are not part of the graph but enable us to draw the graph with more accuracy as shown in the figure to the left.

Procedure: Graphing a Hyperbola

Step 1: Plot the intercepts of the graph.

   a. The graph of $\dfrac{x^2}{a^2} - \dfrac{y^2}{b^2} = 1$ is a horizontal hyperbola and has x-intercepts at $(a, 0)$ and $(-a, 0)$.

   b. The graph of $\dfrac{y^2}{b^2} - \dfrac{x^2}{a^2} = 1$ is a vertical hyperbola and has y-intercepts at $(0, b)$ and $(0, -b)$.

Step 2: Plot the four vertices of the rectangle: (a, b), $(-a, b)$, $(a, -b)$, and $(-a, -b)$.

Step 3: Draw the two diagonals of the rectangle with dashed lines, extending beyond the rectangle. These are the asymptotes.

Step 4: Draw the two branches of the hyperbola going through the intercepts and approaching the asymptotes.

Objective 2 Examples   Graph each hyperbola.

   2a. $\dfrac{x^2}{16} - \dfrac{y^2}{9} = 1$ **2b.** $16y^2 - 4x^2 = 64$

Solutions   **2a.** The equation represents a horizontal hyperbola with the center at the origin and is equivalent to

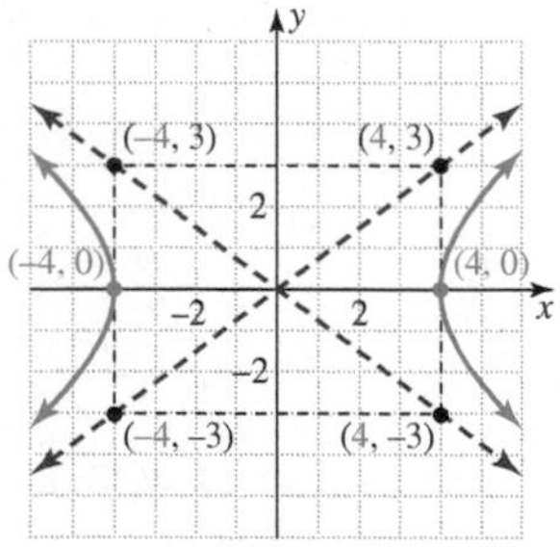

$$\frac{x^2}{4^2} - \frac{y^2}{3^2} = 1$$

Because $a = 4$, the x-intercepts are $(4, 0)$ and $(-4, 0)$. Since $b = 3$, the four vertices of the rectangle are $(4, 3)$, $(-4, 3)$, $(4, -3)$, and $(-4, -3)$. The diagonals of the rectangle are drawn and extended. (See Figure 13.6.)

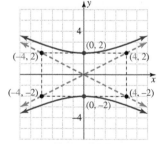

Figure 13.6

2b. The equation is not in the standard form of a hyperbola. The constant on one side of the equation must be 1.

$$16y^2 - 4x^2 = 64$$

$$\frac{16y^2}{64} - \frac{4x^2}{64} = \frac{64}{64} \qquad \text{Divide each side by 64.}$$

$$\frac{y^2}{4} - \frac{x^2}{16} = 1 \qquad \text{Simplify.}$$

$$\frac{y^2}{2^2} - \frac{x^2}{4^2} = 1 \qquad \text{Identify } a \text{ and } b.$$

So, this equation represents a vertical hyperbola with center at the origin. Since $b = 2$, the y-intercepts are $(0, 2)$ and $(0, -2)$. Since $a = 4$, the four vertices of the rectangle are $(4, 2)$, $(-4, 2)$, $(4, -2)$, and $(-4, -2)$. The diagonals of the rectangle are drawn and extended. (See Figure 13.7.)

Figure 13.7

✓ Student Check 2   Graph each hyperbola.

   a. $\dfrac{x^2}{25} - \dfrac{y^2}{16} = 1$ **b.** $49y^2 - 4x^2 = 196$

Objective 3 ▶

Troubleshoot common errors.

Troubleshooting Common Errors

Some common errors associated with graphing ellipses and hyperbolas are shown.

Objective 3 Examples — A problem and an incorrect solution are given. Provide the correct solution and an explanation of the error.

3a. Graph $\dfrac{x^2}{9} + \dfrac{y^2}{25} = 1$.

Incorrect Solution	Correct Solution and Explanation
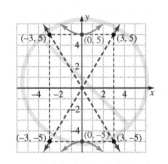	Since $a = 3$ and $b = 5$, the ellipse is vertical. The intercepts are $(0, 5)$, $(0, -5)$, $(3, 0)$, and $(-3, 0)$. 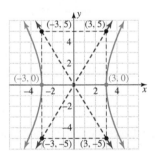

3b. Graph $\dfrac{x^2}{9} - \dfrac{y^2}{25} = 1$.

Incorrect Solution	Correct Solution and Explanation
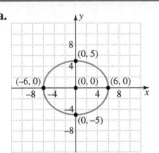	The hyperbola is horizontal since the x^2-term is positive. The x-intercepts are $(3, 0)$ and $(-3, 0)$. 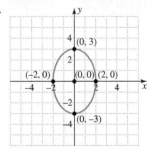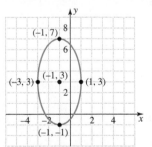

ANSWERS TO STUDENT CHECKS

Student Check 1
a. 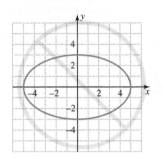 b. 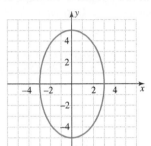 c.

Student Check 2
a. 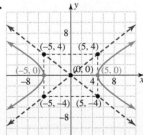 b.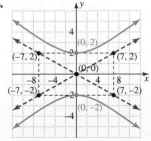

SUMMARY OF KEY CONCEPTS

1. An ellipse is defined by the equation
$$\frac{(x-h)^2}{a^2} + \frac{(y-k)^2}{b^2} = 1.$$

   a. The center is (h, k).

   b. If $a > b$, the ellipse is horizontal. If $a < b$, the ellipse is vertical.

   c. The four key points on the graph of an ellipse are $(h + a, k)$, $(h - a, k)$, $(h, k + b)$, $(h, k - b)$.

2. A hyperbola whose center is $(0, 0)$ is given by the equation
$$\frac{x^2}{a^2} - \frac{y^2}{b^2} = 1 \quad \text{or} \quad \frac{y^2}{b^2} - \frac{x^2}{a^2} = 1$$

   a. If the x^2-term is positive, the hyperbola opens horizontally.

   b. If the y^2-term is positive, the hyperbola opens vertically.

   c. Use the intercepts and the diagonals of the rectangle formed by (a, b), $(-a, b)$, $(a, -b)$, and $(-a, -b)$ to construct the graph.

GRAPHING CALCULATOR SKILLS

The graphing calculator can be used to graph ellipses and hyperbolas if the equation is solved for y. When graphing ellipses and hyperbolas, it is best to use the ZDecimal Setting. If the window is not large enough to view the graph, then use multiples of 4.7 for the Xmin and Xmax and multiples of 3.1 for the Ymin and Ymax.

Example 1: Graph $\dfrac{x^2}{9} + \dfrac{y^2}{25} = 1$.

Solution:
$$\frac{x^2}{9} + \frac{y^2}{25} = 1$$

$$225\left(\frac{x^2}{9} + \frac{y^2}{25}\right) = 225(1) \qquad \text{Multiply each side by the LCD, 225.}$$

$$25x^2 + 9y^2 = 225 \qquad \text{Simplify.}$$

$$9y^2 = 225 - 25x^2 \qquad \text{Subtract } 25x^2 \text{ from each side.}$$

$$y^2 = \frac{225 - 25x^2}{9} \qquad \text{Divide each side by 9.}$$

$$y = \pm\sqrt{\frac{225 - 25x^2}{9}} \qquad \text{Apply the square root property.}$$

Now enter the two expressions in the equation editor. Change the window to view the intercepts and graph.

Example 2: Graph $\dfrac{x^2}{9} - \dfrac{y^2}{25} = 1$.

Solution:
$$\frac{x^2}{9} - \frac{y^2}{25} = 1$$

$$225\left(\frac{x^2}{9} - \frac{y^2}{25}\right) = 225(1) \qquad \text{Multiply each side by the LCD, 225.}$$

$$25x^2 - 9y^2 = 225 \qquad \text{Simplify.}$$

$$-9y^2 = 225 - 25x^2 \qquad \text{Subtract } 25x^2 \text{ from each side.}$$

$$y^2 = \frac{225 - 25x^2}{-9} \qquad \text{Divide each side by } -9.$$

$$y = \pm\sqrt{\frac{25x^2 - 225}{9}} \qquad \text{Apply the square root property and simplify the radicand.}$$

Now enter the two expressions in the equation editor. Change the window to view the intercepts and graph.

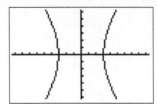

SECTION 13.2 / EXERCISE SET

 Write About It!

Use complete sentences in your answer to each exercise.

1. Describe how to find the four key points of the graph of the ellipse $\frac{x^2}{9} + \frac{y^2}{4} = 1$.

2. Describe how to graph the ellipse $\frac{x^2}{9} + \frac{y^2}{4} = 1$.

3. Describe how to find the x-intercepts of the graph of the hyperbola $\frac{x^2}{25} - \frac{y^2}{4} = 1$.

4. Describe how to find the two asymptotes for the graph of the hyperbola $\frac{x^2}{25} - \frac{y^2}{4} = 1$.

5. Describe how to graph the hyperbola $\frac{x^2}{25} - \frac{y^2}{4} = 1$.

6. Describe the difference between the two hyperbolas $\frac{x^2}{9} - \frac{y^2}{4} = 1$ and $\frac{y^2}{9} - \frac{x^2}{4} = 1$.

7. Explain how to use a graphing calculator to graph the ellipse $\frac{x^2}{144} + \frac{y^2}{225} = 1$. Describe what will happen if the viewing window is set to $[-10, 10]$ by $[-10, 10]$.

8. Explain how to use graphing calculator to graph the hyperbola $\frac{y^2}{144} - \frac{x^2}{225} = 1$. Describe what will happen if the viewing window is set to $[-10, 10]$ by $[-10, 10]$.

 Practice Makes Perfect!

Graph each ellipse. Label four key points and the center on the graph. (*See Objective 1.*)

9. $\frac{x^2}{4} + \frac{y^2}{36} = 1$ 10. $\frac{x^2}{49} + \frac{y^2}{9} = 1$

11. $\frac{x^2}{25} + y^2 = 1$ 12. $x^2 + \frac{y^2}{9} = 1$

13. $49x^2 + 4y^2 = 196$ 14. $x^2 + 64y^2 = 64$

15. $4x^2 + 25y^2 = 1$ 16. $9x^2 + 64y^2 = 1$

17. $\frac{(x-1)^2}{36} + \frac{(y+3)^2}{25} = 1$ 18. $\frac{(x+5)^2}{4} + \frac{(y-1)^2}{64} = 1$

19. $\frac{(x+3)^2}{16} + \frac{(y-2)^2}{81} = 1$ 20. $\frac{(x-2)^2}{25} + \frac{(y-1)^2}{4} = 1$

Graph each hyperbola and its asymptotes. Label the intercepts. (*See Objective 2.*)

21. $\frac{y^2}{9} - \frac{x^2}{25} = 1$ 22. $\frac{y^2}{16} - \frac{x^2}{4} = 1$

23. $\frac{x^2}{25} - \frac{y^2}{64} = 1$ 24. $\frac{x^2}{49} - \frac{y^2}{4} = 1$

25. $x^2 - 4y^2 = 4$ 26. $25y^2 - x^2 = 25$

 Mix 'Em Up!

Identify each equation as a parabola, circle, ellipse, or hyperbola. Sketch the graph of each equation.

27. $x^2 + 36y^2 = 36$ 28. $x^2 + \frac{y^2}{64} = 1$

29. $x = y^2 - 6y + 8$ 30. $x = -y^2 - 3y + 4$

31. $2x^2 = 8 - 2y^2$ 32. $x^2 + y^2 = 6y + 16$

33. $x^2 = 9y^2 - 36$ 34. $y^2 = 4x^2 - 16$

35. $(x-1)^2 + \frac{(y+3)^2}{49} = 1$ 36. $\frac{(x+2)^2}{4} + (y+1)^2 = 1$

37. $\frac{x^2}{16} - y^2 = 4$ 38. $y^2 = 25x^2 + 1$

39. $\frac{x^2}{2.25} - y^2 = 1$ 40. $\frac{y^2}{6.25} - x^2 = 1$

41. $x = y^2 + 7y - 8$ 42. $x = -y^2 + 5y - 3.69$

 You Be the Teacher!

Correct each student's errors, if any.

43. Graph the ellipse $\frac{x^2}{25} + \frac{y^2}{4} = 1$.

     Randy's work:

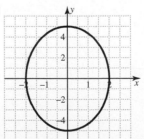

44. Graph the hyperbola $\dfrac{x^2}{49} - \dfrac{y^2}{4} = 1$.

Monica's work:

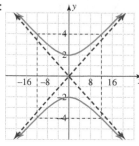

 **Calculate It!**

Use a graphing calculator to graph each equation.

45. $\dfrac{(x - 0.6)^2}{2.25} + (y + 1.6)^2 = 1$

46. $4y^2 - 25x^2 = 49$

PIECE IT TOGETHER   SECTIONS 13.1–13.2

Graph each equation. For parabolas, plot the vertex and two additional points, and state the axis of symmetry. For circles, identify the center, the radius, and four key points. For ellipses, label four key points and the center on the graph. For hyperbolas, use a rectangle and asymptotes to graph. (*Sections 13.1 and 11.2, Objectives 1 and 2*)

1. $x = y^2 + 6y - 27$

2. $x = -y^2 - 4y + 21$

3. $x^2 + y^2 = 100$

4. $x^2 + y^2 = 36$

5. $x^2 + (y + 1)^2 = 4$

6. $(x + 2)^2 + y^2 = 1$

7. $25x^2 + 9y^2 = 225$

8. $81x^2 + 4y^2 = 324$

9. $9x^2 - y^2 = 9$

10. $y^2 - 16x^2 = 16$

Write the equation of each circle in standard form and identify the center and radius. (*Section 13.1, Objective 4*)

11. $x^2 + y^2 + 10x - 14y + 58 = 0$

12. $x^2 + y^2 - 2x - 12y + 36 = 0$

SECTION 13.3   Solving Nonlinear Systems of Equations

▶ OBJECTIVES

As a result of completing this section, you will be able to

1. Solve a nonlinear system by substitution.

2. Solve a nonlinear system by elimination.

3. Troubleshoot common errors.

Objective 1 ▶

Solve a nonlinear system by substitution.

In Chapter 4, we learned how to solve systems of linear equations by graphing, substitution, and elimination. While graphing is a good method to confirm answers and to determine the number of solutions of a system, it is not very precise. So, we will focus on the methods of substitution and elimination.

Use Substitution to Solve Nonlinear Systems of Equations

A **nonlinear system of equations** is a system of equations in which at least one of the equations is not linear. A **solution of a nonlinear system** of equations in two variables is an ordered pair that satisfies each equation in the system. Because at least one of the equations is not linear, the number of solutions will vary. A nonlinear system can have no solution, one solution, two solutions, three solutions, four solutions, and so on. Here are some illustrations of the different possibilities.

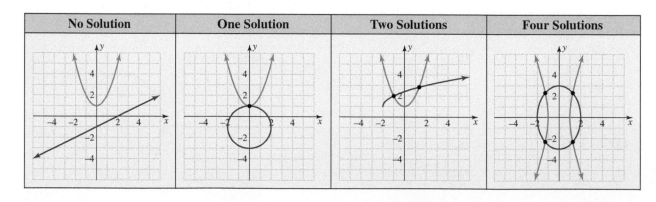

> **Procedure: Solving a Nonlinear System by Substitution**
>
> **Step 1:** Solve one of the equations for one of the variables.
> **Step 2:** Substitute this expression into the other equation and solve the equation.
> **Step 3:** Back substitute to find the other coordinate of each solution.
> **Step 4:** Check by substituting the solutions into the original equations
>   in the system.

Objective 1 Examples   **Solve each system by substitution.**

1a. $\begin{cases} y = (x - 1)^2 \\ x - 3y = -1 \end{cases}$ **1b.** $\begin{cases} x = \dfrac{1}{2}y^2 + 4 \\ y = \sqrt{x - 1} \end{cases}$ **1c.** $\begin{cases} x^2 + y^2 = 1 \\ x - y = 2 \end{cases}$

Solutions   **1a.** The first equation is solved for y so we substitute it into the second equation and solve for x.

$$x - 3y = -1 \qquad \text{Begin with the second equation.}$$
$$x - 3(x - 1)^2 = -1 \qquad \text{Replace } y \text{ with } (x - 1)^2.$$
$$x - 3(x^2 - 2x + 1) = -1 \qquad \text{Square the binomial.}$$
$$x - 3x^2 + 6x - 3 = -1 \qquad \text{Apply the distributive property.}$$
$$-3x^2 + 7x - 2 = 0 \qquad \text{Add 1 to each side and combine like terms.}$$
$$3x^2 - 7x + 2 = 0 \qquad \text{Multiply each side by } -1.$$
$$(3x - 1)(x - 2) = 0 \qquad \text{Factor.}$$
$$3x - 1 = 0 \quad \text{or} \quad x - 2 = 0 \qquad \text{Apply the zero products property.}$$
$$x = \frac{1}{3} \qquad\qquad x = 2 \qquad \text{Solve each equation.}$$

Now solve for the other variable by replacing x with the values $\dfrac{1}{3}$ and 2 in the first equation.

Let $x = \dfrac{1}{3}$. | Let $x = 2$.

$$y = (x - 1)^2$$
$$y = \left(\frac{1}{3} - 1\right)^2$$
$$y = \left(-\frac{2}{3}\right)^2$$
$$y = \frac{4}{9}$$

$$y = (x - 1)^2$$
$$y = (2 - 1)^2$$
$$y = (1)^2$$
$$y = 1$$

Now check the proposed solutions of $\left(\dfrac{1}{3}, \dfrac{4}{9}\right)$ and $(2,\ 1)$ in the original system.

Let $(x, y) = \left(\dfrac{1}{3}, \dfrac{4}{9}\right)$. | Let $(x, y) = (2,\ 1)$.

$$\begin{cases} y = (x - 1)^2 \\ x - 3y = -1 \end{cases} \qquad\qquad \begin{cases} y = (x - 1)^2 \\ x - 3y = -1 \end{cases}$$

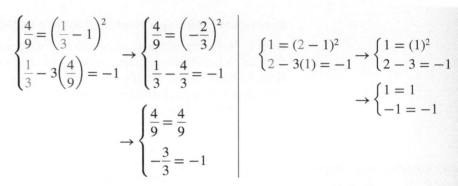

$$\begin{cases} \dfrac{4}{9} = \left(\dfrac{1}{3} - 1\right)^2 \\ \dfrac{1}{3} - 3\left(\dfrac{4}{9}\right) = -1 \end{cases} \rightarrow \begin{cases} \dfrac{4}{9} = \left(-\dfrac{2}{3}\right)^2 \\ \dfrac{1}{3} - \dfrac{4}{3} = -1 \end{cases}$$

$$\rightarrow \begin{cases} \dfrac{4}{9} = \dfrac{4}{9} \\ -\dfrac{3}{3} = -1 \end{cases}$$

$$\begin{cases} 1 = (2 - 1)^2 \\ 2 - 3(1) = -1 \end{cases} \rightarrow \begin{cases} 1 = (1)^2 \\ 2 - 3 = -1 \end{cases}$$

$$\rightarrow \begin{cases} 1 = 1 \\ -1 = -1 \end{cases}$$

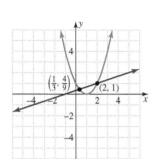

Figure 13.8

Since both ordered pairs make each equation in the system true, the solutions are $\left(\dfrac{1}{3}, \dfrac{4}{9}\right)$ and $(2, 1)$. So, the solution set is written as $\left\{\left(\dfrac{1}{3}, \dfrac{4}{9}\right), (2, 1)\right\}$. The graph of the system is shown in Figure 13.8. Notice that the graphs intersect at the points $\left(\dfrac{1}{3}, \dfrac{4}{9}\right)$ and $(2, 1)$.

1b. The second equation is solved for y so we substitute it into the first equation and solve for x.

$$x = \frac{1}{2}y^2 + 4 \qquad \text{Begin with the first equation.}$$

$$x = \frac{1}{2}\left(\sqrt{x - 1}\right)^2 + 4 \qquad \text{Replace } y \text{ with } \sqrt{x - 1}.$$

$$x = \frac{1}{2}(x - 1) + 4 \qquad \text{Simplify.}$$

$$x = \frac{1}{2}x - \frac{1}{2} + 4 \qquad \text{Apply the distributive property.}$$

$$2(x) = 2\left(\frac{1}{2}x - \frac{1}{2} + 4\right) \qquad \text{Multiply each side by the LCD, 2.}$$

$$2x = x - 1 + 8 \qquad \text{Simplify.}$$

$$2x = x + 7 \qquad \text{Combine like terms.}$$

$$x = 7 \qquad \text{Subtract } x \text{ from each side.}$$

Now solve for the other variable by replacing x with 7 in the second equation.

$$y = \sqrt{x - 1}$$

$$y = \sqrt{7 - 1}$$

$$y = \sqrt{6}$$

Now check the proposed solution $\left(7, \sqrt{6}\right)$ in the original system.

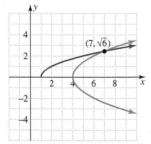

Figure 13.9

$$\begin{cases} x = \dfrac{1}{2}y^2 + 4 \\ y = \sqrt{x - 1} \end{cases} \rightarrow \begin{cases} 7 = \dfrac{1}{2}(\sqrt{6})^2 + 4 \\ \sqrt{6} = \sqrt{7 - 1} \end{cases} \rightarrow \begin{cases} 7 = \dfrac{1}{2}(6) + 4 \\ \sqrt{6} = \sqrt{6} \end{cases} \rightarrow \begin{cases} 7 = 3 + 4 \\ \sqrt{6} = \sqrt{6} \end{cases}$$

Since the ordered pair makes each equation in the system true, the solution set is $\left\{\left(7, \sqrt{6}\right)\right\}$. The graph of the system is shown in Figure 13.9. Notice that the graphs intersect at the point $\left(7, \sqrt{6}\right)$.

Note: After we found that $x = 7$, we could have solved for y using the other equation.

$$x = \frac{1}{2}y^2 + 4 \qquad \text{Begin with the first equation.}$$

$$7 = \frac{1}{2}y^2 + 4 \qquad \text{Substitute 7 for } x.$$

$$7 - 4 = \frac{1}{2}y^2 + 4 - 4 \qquad \text{Subtract 4 from each side.}$$

$$3 = \frac{1}{2}y^2 \qquad \text{Simplify.}$$

$$2(3) = 2\left(\frac{1}{2}y^2\right) \qquad \text{Multiply each side by the LCD, 2.}$$

$$6 = y^2 \qquad \text{Simplify.}$$

$$\pm\sqrt{6} = y \qquad \text{Apply the square root property.}$$

So, this would give us two proposed solutions $(7, \sqrt{6})$ and $(7, -\sqrt{6})$. When we check $(7, -\sqrt{6})$, we find that it doesn't check in the second equation.

$$y = \sqrt{x - 1}$$

$$-\sqrt{6} = \sqrt{7 - 1} \qquad \text{Replace } x \text{ with 7 and } y \text{ with } -\sqrt{6}.$$

$$-\sqrt{6} = \sqrt{6} \qquad \text{Simplify.}$$

Since the resulting equation is false, we discard the solution $(7, -\sqrt{6})$.

1c. Solve the second equation for x. $x - y = 2$

$$x = y + 2$$

Now substitute this expression in the first equation and solve for y.

$$x^2 + y^2 = 1 \qquad \text{Begin with the first equation.}$$

$$(y + 2)^2 + y^2 = 1 \qquad \text{Replace } x \text{ with } y + 2.$$

$$y^2 + 4y + 4 + y^2 = 1 \qquad \text{Square the binomial.}$$

$$2y^2 + 4y + 3 = 0 \qquad \text{Subtract 1 from each side and combine like terms.}$$

To solve the equation, we apply the quadratic formula with $a = 2$, $b = 4$, and $c = 3$.

$$y = \frac{-b \pm \sqrt{b^2 - 4ac}}{2a} \qquad \text{State the quadratic formula.}$$

$$y = \frac{-4 \pm \sqrt{4^2 - 4(2)(3)}}{2(2)} \qquad \text{Replace } a \text{ with 2, } b \text{ with 4, and } c \text{ with 3.}$$

$$y = \frac{-4 \pm \sqrt{16 - 24}}{4} \qquad \text{Simplify each term in the radical.}$$

$$y = \frac{-4 \pm \sqrt{-8}}{4} \qquad \text{Simplify the radical.}$$

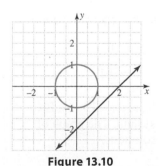

Figure 13.10

Since $\sqrt{-8}$ is not a real number, the system of equations has no solution. We confirm this by graphing as shown in Figure 13.10. So, the solution set is $\varnothing$.

✓ **Student Check 1** Solve each system by substitution.

a. $\begin{cases} y = (x + 2)^2 \\ x + y = 4 \end{cases}$ **b.** $\begin{cases} x = \dfrac{1}{2}y^2 + 2 \\ y = \sqrt{x + 4} \end{cases}$ **c.** $\begin{cases} x^2 + y^2 = 4 \\ x - y = 3 \end{cases}$

Use Elimination to Solve Nonlinear Systems of Equations

Objective 2 ▶

Solve a nonlinear system by elimination.

Recall that elimination is a method that requires us to add the equations in the system so that one variable is eliminated. The coefficients of the eliminated variable must be opposites. If they are not opposites, we have to multiply one or both of the equations by a nonzero number.

> **Procedure: Solving a Nonlinear System by Elimination**
>
> **Step 1:** Write the equations in the same form.
> **Step 2:** Determine the variable to be isolated. If its coefficients are not opposites, multiply one or both of the equations by the number that will make them opposites.
> **Step 3:** Add the equations.
> **Step 4:** Solve the resulting equation.
> **Step 5:** Back substitute to determine the other coordinate of each solution.

Objective 2 Example

Solve the system $\begin{cases} x^2 + y^2 = 10 \\ 2x^2 - y^2 = 17 \end{cases}$ by elimination.

Solution   We can add the equations in the system since the coefficients of the y^2-terms are opposites and the equations are already in the same form.

$$
\begin{cases} x^2 + y^2 = 10 \\ \underline{2x^2 - y^2 = 17} \end{cases}
$$

$3x^2 = 27$	Add the equations.
$x^2 = 9$	Divide each side by 3.
$x = \pm\sqrt{9}$	Apply the square root property.
$x = \pm 3$	Simplify.

Now we find the y-coordinates for each of the solutions.

Let $x = 3$.

$$x^2 + y^2 = 10$$
$$(3)^2 + y^2 = 10$$
$$9 + y^2 = 10$$
$$y^2 = 1$$
$$y = \pm 1$$

Let $x = -3$.

$$x^2 + y^2 = 10$$
$$(-3)^2 + y^2 = 10$$
$$9 + y^2 = 10$$
$$y^2 = 1$$
$$y = \pm 1$$

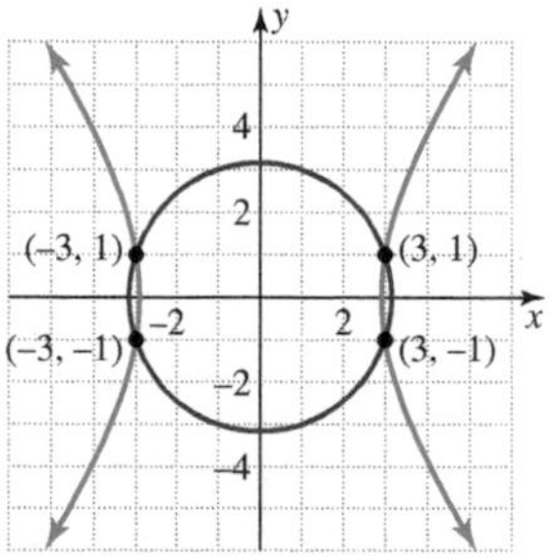

Figure 13.11

So, the four solutions of the system are $(3, 1)$, $(3, -1)$, $(-3, 1)$, and $(-3, -1)$. Be sure to check each solution in the original equations in the system. The graph of the system is shown in Figure 13.11 and confirms the four solutions.

✓ **Student Check 2**

Solve the system $\begin{cases} x^2 + y^2 = 16 \\ 4x^2 - y^2 = 4 \end{cases}$ by elimination.

Objective 3 ▶

Troubleshoot common errors.

Troubleshooting Common Errors

Some common errors associated with solving nonlinear systems of equations are shown.

Objective 3 Example A problem and an incorrect solution are given. Provide the correct solution and an explanation of the error.

Solve the system $\begin{cases} x^2 + y^2 = 13 \\ 4x^2 - y^2 = 7 \end{cases}$ by elimination.

Incorrect Solution	Correct Solution and Explanation
Adding the equations, we get	When the square root property is applied, there are two solutions. So, $x = \pm 2$. When we solve for y, we get
$5x^2 = 20$	
$x^2 = 4$	$x = 2$:
$x = 2$	$(2)^2 + y^2 = 13$
Solving for y:	$4 + y^2 = 13$
$(2)^2 + y^2 = 13$	$y^2 = 9$
$4 + y^2 = 13$	$y = \pm 3$
$y^2 = 9$	$x = -2$:
$y = \pm 3$	$(-2)^2 + y^2 = 13$
The solutions are $(2, -3)$ and $(2, 3)$.	$4 + y^2 = 13$
	$y^2 = 9$
	$y = \pm 3$
	The solutions are $(2, -3)$, $(2, 3)$, $(-2, -3)$, and $(-2, 3)$.

ANSWERS TO STUDENT CHECKS

Student Check 1 **a.** $\{(-5, 9), (0, 4)\}$ **b.** $\{(8, 2\sqrt{3})\}$ **Student Check 2** **a.** $\{(2, 2\sqrt{3}), (2, -2\sqrt{3}), (-2, 2\sqrt{3}),$
c. $\varnothing$ $(-2, -2\sqrt{3})\}$

SUMMARY OF KEY CONCEPTS

1. Nonlinear systems can be solved by substitution. We solve one equation for one variable and substitute this expression into the other equation in the system. We then back substitute to solve for the other variable. Solutions can be checked algebraically or graphically.

2. Nonlinear systems can also be solved by elimination. We transform the equations so that the coefficients of one of the variables are opposites, and then add the equations. We solve the resulting equation and back substitute to solve for the other variable.

SECTION 13.3 / EXERCISE SET

 Write About It!

Use complete sentences in your answer to each exercise.

1. What is the number of possible solutions of a system of equations consisting of a parabola and a line?

2. What is the number of possible solutions of a system of equations consisting of a circle and either a circle or an ellipse?

3. Explain how to apply the first two steps of the substitution method to solve the following system. Do not solve.

$$\begin{cases} x^2 + y^2 = 100 \\ 2x - y = 20 \end{cases}$$

4. Explain how to apply the first two steps of the substitution method to solve the following system. Do not solve.

$$\begin{cases} y = x^2 + 8 \\ 2x - y = 23 \end{cases}$$

5. Explain how to apply the first two steps of the elimination method to solve the following system. Do not solve.
$$\begin{cases} 2x^2 + 4y^2 = 36 \\ 2y^2 - 5x^2 = -78 \end{cases}$$

6. Explain how to apply the first two steps of the elimination method to solve the following system. Do not solve.
$$\begin{cases} 2x^2 + 2y^2 = 40 \\ y^2 + 3x^2 = 52 \end{cases}$$

7. Explain how to choose the method, elimination or substitution, to solve the following system. Do not solve.
$$\begin{cases} x^2 + y^2 = 25 \\ 4x + 5y = 8 \end{cases}$$

8. Explain how to choose the method, elimination or substitution, to solve the following system. Do not solve.
$$\begin{cases} 4x^2 + 3y^2 = 79 \\ x^2 + 3y^2 = 76 \end{cases}$$

 Practice Makes Perfect!

Solve each system by substitution. (*See Objective 1.*)

9. $\begin{cases} y = (x - 5)^2 \\ 7x + y = 25 \end{cases}$

10. $\begin{cases} y = (x - 3)^2 \\ -x + y = 9 \end{cases}$

11. $\begin{cases} y = (x + 1)^2 \\ -5x + y = 1 \end{cases}$

12. $\begin{cases} y = (x + 6)^2 \\ -8x + y = 36 \end{cases}$

13. $\begin{cases} x = \dfrac{1}{4}y^2 + 3 \\ y = \sqrt{3x - 8} \end{cases}$

14. $\begin{cases} x = \dfrac{3}{8}y^2 - 4 \\ y = \sqrt{10x - 4} \end{cases}$

15. $\begin{cases} x = \dfrac{2}{3}y^2 - 4 \\ y = \sqrt{3x + 9} \end{cases}$

16. $\begin{cases} x = \dfrac{1}{4}y^2 + 8 \\ y = \sqrt{2x - 7} \end{cases}$

17. $\begin{cases} y = x^2 - 5 \\ -x + y = -3 \end{cases}$

18. $\begin{cases} y = x^2 + 2 \\ 4x - y = 1 \end{cases}$

19. $\begin{cases} x^2 + y^2 = 100 \\ x - y = 2 \end{cases}$

20. $\begin{cases} x^2 + y^2 = 25 \\ x + y = 1 \end{cases}$

Solve each system by elimination. (*See Objective 2.*)

21. $\begin{cases} x^2 + y^2 = 10 \\ 5x^2 + 8y^2 = 53 \end{cases}$

22. $\begin{cases} x^2 + y^2 = 25 \\ x^2 + 2y^2 = 41 \end{cases}$

23. $\begin{cases} x^2 + y^2 = 29 \\ x^2 - 4y^2 = 9 \end{cases}$

24. $\begin{cases} x^2 + y^2 = 40 \\ x^2 - y^2 = 32 \end{cases}$

25. $\begin{cases} 3x^2 + y^2 = 28 \\ y^2 - x^2 = 12 \end{cases}$

26. $\begin{cases} x^2 + 4y^2 = 52 \\ x^2 - 2y^2 = 28 \end{cases}$

27. $\begin{cases} x^2 + 2y^2 = 54 \\ 6x^2 + y^2 = 49 \end{cases}$

28. $\begin{cases} x^2 + y^2 = 20 \\ x^2 + 2y^2 = 24 \end{cases}$

29. $\begin{cases} x^2 + 4y^2 = 65 \\ x^2 + y^2 = 17 \end{cases}$

30. $\begin{cases} 5x^2 + 2y^2 = 38 \\ y^2 - x^2 = 5 \end{cases}$

31. $\begin{cases} y = 2x^2 - 4 \\ y = x^2 + 3x \end{cases}$

32. $\begin{cases} y = 4x^2 + 2 \\ y = 3x^2 + 3x \end{cases}$

 Mix 'Em Up!

Solve each system by substitution or elimination.

33. $\begin{cases} y = (x - 8)^2 \\ x + y = 64 \end{cases}$

34. $\begin{cases} y = (x + 4)^2 \\ -3x + y = 16 \end{cases}$

35. $\begin{cases} x = \dfrac{3}{8}y^2 - 4 \\ y = \sqrt{10x - 4} \end{cases}$

36. $\begin{cases} x = \dfrac{1}{7}y^2 + 11 \\ y = \sqrt{3x - 5} \end{cases}$

37. $\begin{cases} x = \dfrac{2}{3}y^2 - 4 \\ y = \sqrt{x + 5} \end{cases}$

38. $\begin{cases} x = \dfrac{3}{4}y^2 - 3 \\ y = \sqrt{2x - 4} \end{cases}$

39. $\begin{cases} x^2 + y^2 = 25 \\ 2x + y = 5 \end{cases}$

40. $\begin{cases} x^2 + y^2 = 100 \\ 3x - y = 10 \end{cases}$

41. $\begin{cases} y = x^2 - 9 \\ x + y = -3 \end{cases}$

42. $\begin{cases} y = x^2 + 2 \\ 5x + y = -4 \end{cases}$

43. $\begin{cases} y = x^2 - 2 \\ -x + y = 4 \end{cases}$

44. $\begin{cases} y = x^2 + 4 \\ 7x + y = -8 \end{cases}$

45. $\begin{cases} x^2 + y^2 = 100 \\ y^2 - x^2 = 28 \end{cases}$

46. $\begin{cases} x^2 + y^2 = 10 \\ 5x^2 - 2y^2 = 43 \end{cases}$

47. $\begin{cases} 5x^2 + 4y^2 = 21 \\ 7y^2 - 2x^2 = 26 \end{cases}$

48. $\begin{cases} 4x^2 + y^2 = 32 \\ x^2 + y^2 = 20 \end{cases}$

49. $\begin{cases} y = 9x^2 - 25 \\ y = 3x^2 - 5x \end{cases}$

50. $\begin{cases} y = 2x^2 - 8 \\ y = x^2 - 2x \end{cases}$

You Be the Teacher!

Correct each student's errors, if any.

51. Solve $\begin{cases} y = (x - 1)^2 \\ 8x + y = 1 \end{cases}$ by substitution.

Rosa's work:

Solve the second equation for y and substitute into the first equation.

$$y = 1 - 8x$$
$$1 - 8x = (x - 1)^2$$
$$1 - 8x = x^2 + 1$$
$$0 = x^2 + 8x$$
$$0 = x(x + 8)$$
$$x = 0 \quad \text{or} \quad x = -8$$

Substitute the values into the first equation to solve for y.

$x = 0$: $y = (0 - 1)^2 = 1$

$x = -8$: $y = (-8 - 1)^2 = 81$

$\{(0, 1), (-8, 81)\}$

52. Solve $\begin{cases} x = \frac{1}{4}y^2 - 2 \\ y = \sqrt{6x + 4} \end{cases}$ by substitution.

Ricki's work:

Substitute the second equation into the first equation.

$x = \frac{1}{4}(6x + 4)^2 - 2$

$4x = (6x + 4)^2 - 8$

$4x = 36x^2 + 48x + 16 - 8$

$0 = 36x^2 + 44x + 8$

$0 = 4(9x^2 + 11x + 2)$

$0 = 4(9x + 2)(x + 1)$

$x = -1 \quad \text{or} \quad x = -\frac{2}{9}$

Substitute the values into the second equation to solve for y.

$x = -1: y = \sqrt{-6 + 4} = \sqrt{-2}$

$x = -\frac{2}{9}: y = \sqrt{-\frac{4}{3} + 4} = \sqrt{\frac{8}{3}}$

$\left\{ \left(-\frac{2}{9}, \sqrt{\frac{8}{3}} \right) \right\}$

53. Solve $\begin{cases} 3x^2 + 2y^2 = 12 \\ 3x^2 - y^2 = -3 \end{cases}$ by elimination.

Allison's work:

Multiply the second equation by -1 and add to the first equation.

$\begin{array}{rcl} 3x^2 + 2y^2 &=& 12 \\ -3x^2 - y^2 &=& -3 \\ \hline y^2 &=& 9 \\ y &=& 3 \end{array}$

Substitute the value into the second equation to solve for y.

$y = 3: 3x^2 - 9 = -3$

$3x^2 = 6$

$x^2 = 2$

$x = \sqrt{2}$

$\{(\sqrt{2}, 3)\}$

54. Solve $\begin{cases} 4x^2 + y^2 = 28 \\ x^2 + y^2 = 19 \end{cases}$ by elimination.

Sally's work:

Multiply the second equation by -1 and add to the first equation.

$\begin{array}{rcl} 4x^2 + y^2 &=& 28 \\ -x^2 - y^2 &=& -19 \\ \hline 4x^2 &=& 9 \end{array}$

$x^2 = \frac{9}{4}$

$x = \frac{3}{2}$

Substitute the value into the first equation to solve for y.

$x = \frac{3}{2}: \frac{9}{4} + y^2 = 19$

$y^2 = \frac{67}{4}$

$y = \frac{\sqrt{67}}{2} \quad \left\{ \left(\frac{3}{2}, \frac{\sqrt{67}}{2} \right) \right\}$

 Calculate It!

Use a graphing calculator to solve each nonlinear system of equations.

55. $\begin{cases} y = x^2 - 2 \\ x + y = 10 \end{cases}$ **56.** $\begin{cases} 2x^2 + y^2 = 11 \\ y^2 - 3x^2 = 6 \end{cases}$

SECTION 13.4   Solving Nonlinear Inequalities and Systems of Inequalities

▶ OBJECTIVES

As a result of completing this section, you will be able to

1. Graph a nonlinear inequality.
2. Solve a nonlinear system of inequalities.
3. Troubleshoot common errors.

Objective 1 ▶

Graph a nonlinear inequality.

In Section 9.4, we solved linear inequalities in two variables and then applied this to solving systems of linear inequalities in Section 9.5. The method that we learned to solve linear inequalities in two variables will be used to solve nonlinear inequalities. This will be the foundation for solving nonlinear systems of inequalities, as well.

Nonlinear Inequalities

A **nonlinear inequality** is an inequality that contains an expression that is not linear. Some examples of nonlinear inequalities are

$$x^2 + y^2 > 9 \qquad x \leq y^2 + 1$$

When we graph a nonlinear inequality in two variables, our goal is to find the region of the coordinate plane that contains solutions of the inequality. We begin by finding the boundary curve of the region of solutions and then test points to determine where the solutions are located. This is exactly the process that we used with linear inequalities in two variables.

Procedure: Graphing a Nonlinear Inequality in Two Variables

Step 1: Graph the associated nonlinear equation.
    a. If the inequality symbol is $\leq$ or $\geq$, draw the graph as a solid curve.
    b. If the inequality symbol is $<$ or $>$, draw the graph as a dashed curve.

Step 2: Test a point from each region to determine if it makes the inequality true or false.
    a. If the test point makes the inequality true, shade the region that contains the point.
    b. If the test point makes the inequality false, do not shade the region containing the test point.

Step 3: The solution set consists of the shaded regions.

Objective 1 Examples   **Graph each inequality.**

1a. $y > (x - 1)^2 + 3$ **1b.** $\dfrac{x^2}{16} + \dfrac{y^2}{9} < 1$ **1c.** $\dfrac{x^2}{4} - \dfrac{y^2}{9} \leq 1$

Solutions   **1a.** Begin by graphing the parabola $y = (x - 1)^2 + 3$ with a dashed curve since the inequality is $>$. The parabola opens up because $a > 0$ and it has a vertex of $(1, 3)$.

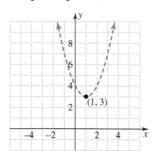

The boundary curve divides the plane into two regions, the region inside the parabola and the region outside of the parabola. We test a point from each region.

Test point	$y > (x - 1)^2 + 3$	True/False
$(0, 0)$	$0 > (0 - 1)^2 + 3$ $0 > 4$	False
$(1, 4)$	$4 > (1 - 1)^2 + 3$ $4 > 3$	True

Since the point $(1, 4)$ satisfies the inequality, the region containing this point is the solution set of the inequality. Only the points inside the parabola are solutions of $y > (x - 1)^2 + 3$. This is shown on the graph.

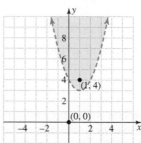

1b. We graph the ellipse $\frac{x^2}{16} + \frac{y^2}{9} = 1$ with a dashed curve since the inequality is $<$.

The ellipse is centered at the origin and contains the key points $(4, 0)$, $(-4, 0)$, $(0, 3)$, and $(0, -3)$.

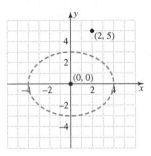

The boundary curve divides the plane into two regions, the region inside the ellipse and the region outside the ellipse. We test a point from each region.

Test point	$\frac{x^2}{16} + \frac{y^2}{9} < 1$	True/False
$(0, 0)$	$\frac{0^2}{16} + \frac{0^2}{9} < 1$ $0 < 1$	True
$(2, 5)$	$\frac{2^2}{16} + \frac{5^2}{9} < 1$ $\frac{1}{4} + \frac{25}{9} < 1$	False

Because the point $(0, 0)$ makes the inequality true, the solution set consists of the points inside the ellipse.

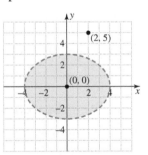

1c. Begin by graphing the hyperbola $\frac{x^2}{4} - \frac{y^2}{9} = 1$ with a solid curve since the inequality symbol is $\leq$. The hyperbola is horizontal and has x-intercepts $(2, 0)$ and $(-2, 0)$.

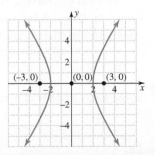

The boundary curves divide the plane into three regions, the region to the left of the left branch of the hyperbola, the region between the branches of the hyperbola, and the region to the right of the right branch of the hyperbola. We test a point in each region.

Test point	$\dfrac{x^2}{4} - \dfrac{y^2}{9} \le 1$	True/False
$(0, 0)$	$\dfrac{0^2}{4} - \dfrac{0^2}{9} \le 1$ $0 \le 1$	True
$(3, 0)$	$\dfrac{3^2}{4} - \dfrac{0^2}{9} \le 1$ $\dfrac{9}{4} - 0 \le 1$	False
$(-3, 0)$	$\dfrac{(-3)^2}{4} - \dfrac{0^2}{9} \le 1$ $\dfrac{9}{4} - 0 \le 1$	False

Since the point $(0, 0)$ satisfies the inequality, the solution set consists of the graph of the hyperbola and all points between the branches.

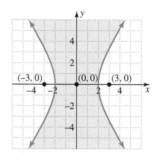

✔ **Student Check 1**   Graph each inequality.

   a. $y > (x - 2)^2 - 4$ **b.** $\dfrac{x^2}{4} + \dfrac{y^2}{16} < 1$ **c.** $\dfrac{y^2}{16} - \dfrac{x^2}{9} \le 1$

Nonlinear Systems of Inequalities

Objective 2 ▶

Solve a nonlinear system of inequalities.

A **nonlinear system of inequalities** is a system of inequalities in which at least one of the inequalities is not linear. When we studied systems of linear inequalities, we learned that the solution set is the intersection of the solution sets of each inequality in the system. This is exactly the same for a nonlinear system of inequalities.

> **Procedure: Solving a Nonlinear System of Inequalities**
>
> **Step 1:** Graph each inequality in the system.
> **Step 2:** Find the intersection of the graphs of these inequalities. The intersection is
>   the solution set of the system.

Objective 2 Examples   Solve each system of inequalities.

   2a. $\begin{cases} x - y \ge 1 \\ x^2 + y^2 \le 4 \end{cases}$ **2b.** $\begin{cases} \dfrac{x^2}{9} + \dfrac{y^2}{25} < 1 \\ \dfrac{y^2}{16} - \dfrac{x^2}{4} < 1 \end{cases}$

Solutions   **2a.** Graph the inequalities and find their intersection.

$$x - y \geq 1$$

The boundary line is $x - y = 1$, which goes through the points $(1, 0)$ and $(0, -1)$. The points on and below the line satisfy the inequality.

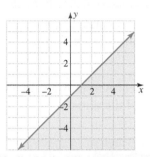

$$x^2 + y^2 \leq 4$$

The boundary curve is $x^2 + y^2 = 4$, which is a circle centered at the origin with radius 2. The points on and inside the circle satisfy the inequality.

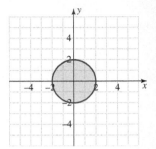

Graphing the two inequalities on the same coordinate system enables us to find the solution. The region where the two graphs intersect is the solution set of the system as shown in purple.

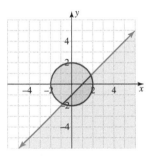

2b. Graph the inequalities and find their intersection.

$$\frac{x^2}{9} + \frac{y^2}{25} < 1$$

The boundary curve is the ellipse $\frac{x^2}{9} + \frac{y^2}{25} = 1$, which goes through the key points $(3, 0)$, $(-3, 0)$, $(0, 5)$, and $(0, -5)$. Only the points inside the ellipse satisfy the inequality.

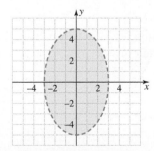

$$\frac{y^2}{16} - \frac{x^2}{4} < 1$$

The boundary curve is the hyperbola $\frac{y^2}{16} - \frac{x^2}{4} < 1$, which has y-intercepts at $(0, 4)$ and $(0, -4)$. The points between the branches satisfy the inequality.

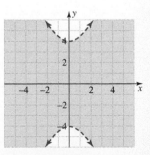

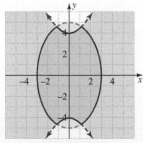

Figure 13.12

Graphing the two inequalities on the same coordinate system enables us to find the solution. The region where the two graphs intersect is the solution set of the system as shown in purple in Figure 13.12.

✓ **Student Check 2**   Solve each system of inequalities.

a. $\begin{cases} x + y \geq 2 \\ x^2 + y^2 \leq 9 \end{cases}$ b. $\begin{cases} \dfrac{x^2}{4} + \dfrac{y^2}{16} < 1 \\ \dfrac{y^2}{9} - \dfrac{x^2}{25} < 1 \end{cases}$

Objective 3 ▶

Troubleshoot common errors.

Troubleshooting Common Errors

Some common errors associated with graphing nonlinear inequalities and solving nonlinear systems of inequalities are shown.

Objective 3 Examples   **A problem and an incorrect solution are given. Provide the correct solution and an explanation of the error.**

3a. Graph $x > (y - 3)^2 - 4$.

Incorrect Solution	Correct Solution and Explanation
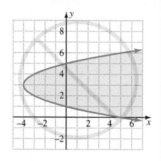	Because the inequality symbol is $>$, the boundary curve is not included in the solution set. 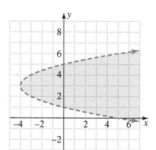

3c. Solve $\begin{cases} y \leq 3x \\ \dfrac{x^2}{4} - \dfrac{y^2}{9} \geq 1 \end{cases}$.

Incorrect Solution	Correct Solution and Explanation
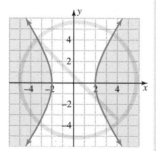	The solution of the linear inequality includes all points on and below the line $y = 3x$. Since the line $y = 3x$ does not intersect the left branch of the hyperbola, none of those points can be included in the solution of the system. 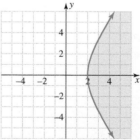

ANSWERS TO STUDENT CHECKS

Student Check 1   **a.**

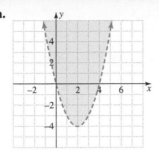

b.

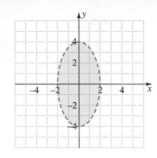

c.

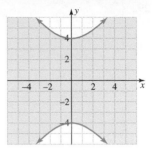

Student Check 2   **a.**

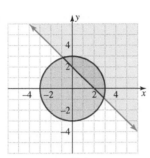

b.

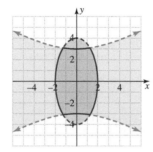

SUMMARY OF KEY CONCEPTS

1. Graphing a nonlinear inequality is similar to graphing a quadratic inequality. We graph the boundary curve, dashed or solid according to the inequality symbol. We test a point in each of the regions, and then shade the region containing the test point that satisfies the inequality. The most difficult part of this process is graphing the boundary curve. Be sure that you are familiar with the basic shapes of all the conic sections.

2. Solving a nonlinear system of inequalities is similar to solving a system of linear inequalities. We graph each inequality and then find the intersection of the graphs to determine the solution set of the system.

SECTION 13.4 / EXERCISE SET

 Write About It!

Use complete sentences in your answer to each exercise.

1. Explain how to graph the boundary curve when graphing a nonlinear inequality.

2. Explain how to use test points to determine the graph of a nonlinear inequality.

3. Explain how to draw the boundary curve for the inequality $x^2 - 4y^2 \le 16$.

4. Explain how to use test points to determine the graph of the solution set of $x^2 - 4y^2 \le 16$.

5. Explain how to draw the boundary curves for the inequalities in the system $\begin{cases} y > x^2 - 4 \\ \dfrac{x^2}{9} + \dfrac{y^2}{16} < 1 \end{cases}$.

6. Explain how to solve the system $\begin{cases} y > x^2 - 4 \\ \dfrac{x^2}{9} + \dfrac{y^2}{16} < 1 \end{cases}$.

 Practice Makes Perfect!

Graph each inequality. (*See Objective 1.*)

7. $y \le (x - 2)^2 + 6$

8. $y \le (x - 1)^2 - 4$

9. $y \ge (x + 3)^2 - 4$

10. $y \ge (x - 1)^2 + 4$

11. $y < -2(x - 4)^2 + 1$

12. $y < -(x + 2)^2 + 9$

13. $y > -2(x + 3)^2 + 8$

14. $y > -2(x - 1)^2 + 10$

15. $x^2 + y^2 \le 36$

16. $x^2 + y^2 \le 100$

17. $x^2 + y^2 \ge 16$

18. $x^2 + y^2 \ge 81$

19. $x^2 + y^2 < 100$

20. $x^2 + y^2 < 49$

21. $x^2 + y^2 > 1$

22. $x^2 + y^2 > 144$

23. $\dfrac{x^2}{4} + y^2 \le 1$

24. $x^2 + \dfrac{y^2}{25} \le 1$

25. $\dfrac{x^2}{4} - \dfrac{y^2}{9} < 1$

26. $\dfrac{y^2}{49} - \dfrac{x^2}{4} < 1$

27. $4x^2 - y^2 \ge 64$

28. $y^2 - 4x^2 \ge 100$

Solve each system. (*See Objective 2.*)

29. $\begin{cases} x^2 + y^2 < 16 \\ x + 6y > 12 \end{cases}$

30. $\begin{cases} x^2 + y^2 < 25 \\ 2x + 3y < 6 \end{cases}$

31. $\begin{cases} x^2 + y^2 < 4 \\ x - 2y > 6 \end{cases}$

32. $\begin{cases} x^2 + y^2 < 1 \\ 2x - 5y > 10 \end{cases}$

33. $\begin{cases} x^2 + y^2 \le 9 \\ x + 5y \le 8 \end{cases}$

34. $\begin{cases} x^2 + y^2 \le 9 \\ 2x - y \ge 4 \end{cases}$

35. $\begin{cases} x^2 + y^2 \le 4 \\ x - y \le -3 \end{cases}$

36. $\begin{cases} x^2 + y^2 \le 1 \\ 2x + y \le -4 \end{cases}$

37. $\begin{cases} x^2 + y^2 \ge 100 \\ 3x - 4y \le 12 \end{cases}$

38. $\begin{cases} x^2 + y^2 \ge 4 \\ x + 2y \le 2 \end{cases}$

39. $\begin{cases} \dfrac{x^2}{4} + \dfrac{y^2}{49} \le 1 \\ \dfrac{y^2}{4} - \dfrac{x^2}{64} \le 1 \end{cases}$

40. $\begin{cases} \dfrac{x^2}{4} + \dfrac{y^2}{49} \le 1 \\ \dfrac{y^2}{4} - \dfrac{x^2}{64} \ge 1 \end{cases}$

41. $\begin{cases} x^2 + \dfrac{y^2}{81} < 1 \\ \dfrac{x^2}{4} - \dfrac{y^2}{144} > 1 \end{cases}$

42. $\begin{cases} x^2 + \dfrac{y^2}{81} \le 1 \\ \dfrac{x^2}{4} - \dfrac{y^2}{64} \ge 1 \end{cases}$

43. $\begin{cases} \dfrac{x^2}{16} - \dfrac{y^2}{9} \le 1 \\ \dfrac{x^2}{4} + \dfrac{y^2}{64} \ge 1 \end{cases}$

44. $\begin{cases} \dfrac{y^2}{16} - \dfrac{x^2}{4} \le 1 \\ \dfrac{x^2}{25} + \dfrac{y^2}{81} \ge 1 \end{cases}$

45. $\begin{cases} \dfrac{x^2}{4} - \dfrac{y^2}{25} \le 1 \\ \dfrac{y^2}{25} - \dfrac{x^2}{4} \le 1 \end{cases}$

46. $\begin{cases} \dfrac{x^2}{16} - \dfrac{y^2}{9} \le 1 \\ \dfrac{y^2}{9} - \dfrac{x^2}{16} \le 1 \end{cases}$

 Mix 'Em Up!

Graph each inequality or solve each system.

47. $y \ge x^2 - 9$

48. $y \le \dfrac{1}{2}x^2 - 8$

49. $\dfrac{x^2}{5} + \dfrac{y^2}{5} \ge 1$

50. $\dfrac{x^2}{4} + \dfrac{y^2}{2.25} \le 1$

51. $\dfrac{x^2}{2.25} - \dfrac{y^2}{6.25} \ge 1$

52. $\dfrac{y^2}{12.25} - \dfrac{x^2}{1.44} \ge 1$

53. $\begin{cases} y \le (x-3)^2 + 1 \\ x^2 + y^2 \le 36 \end{cases}$

54. $\begin{cases} y \ge (x-1)^2 + 4 \\ x^2 + y^2 \le 10 \end{cases}$

55. $\begin{cases} y < -(x+2)^2 + 5 \\ x^2 + y^2 > 9 \end{cases}$

56. $\begin{cases} y < -(x-1)^2 - 1 \\ x^2 + y^2 > 4 \end{cases}$

57. $\begin{cases} \dfrac{x^2}{4} + y^2 \ge 1 \\ x^2 + \dfrac{y^2}{25} \ge 1 \end{cases}$

58. $\begin{cases} \dfrac{y^2}{49} + \dfrac{x^2}{4} < 1 \\ \dfrac{x^2}{25} + \dfrac{y^2}{4} > 1 \end{cases}$

59. $\begin{cases} 4x^2 - y^2 \le 64 \\ 9x^2 + y^2 \ge 9 \end{cases}$

60. $\begin{cases} y^2 - 4x^2 \ge 16 \\ \dfrac{x^2}{25} + \dfrac{y^2}{81} \le 1 \end{cases}$

61. $\begin{cases} x^2 + y^2 < 16 \\ x + y \le 3 \end{cases}$

62. $\begin{cases} x^2 + y^2 \le 9 \\ 3x - 2y \ge 6 \end{cases}$

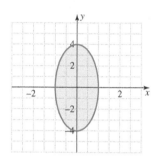 **You Be the Teacher!**

Correct each student's errors, if any.

63. Graph the inequality $\dfrac{x^2}{16} + y^2 \le 1$.

Josh's work:

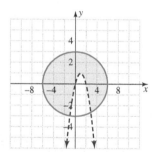

64. Solve the system $\begin{cases} y < -(x-1)^2 + 1 \\ x^2 + 4y^2 \le 36 \end{cases}$.

Deb's work:

Calculate It!

Use a graphing calculator to solve each inequality.

65. $y > (x+1)^2 - 4$

66. $\begin{cases} x^2 + 4y^2 \le 4 \\ 2x - 3y \ge 6 \end{cases}$

Think About It!

Give an example of a system of nonlinear inequalities that satisfies the given conditions.

67. The solution set is inside two ellipses, including the boundaries.

68. The solution set is outside a circle and inside an ellipse, including the boundaries.

69. The solution set is outside an ellipse and between two branches of a vertical hyperbola, excluding the boundaries.

GROUP ACTIVITY / **The Mathematics of Orbits**

In the early 1600s, Johannes Kepler discovered that planets orbit the sun in a path that resembles an ellipse with the sun being one of the foci. This is Kepler's first law of planetary motion and is sometimes called the law of ellipses.

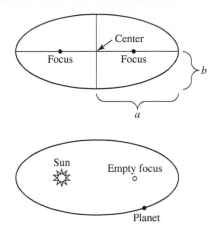

We know that the standard form of an ellipse is $\frac{x^2}{a^2} + \frac{y^2}{b^2} = 1$, where a is the horizontal distance from the origin, or center, to the ellipse, and b is the vertical distance from the origin, or center, to the ellipse. We can also describe ellipses by their eccentricity. The eccentricity, e, is a measure of how circular the ellipse is. If $e = 0$, then the ellipse is actually a circle. As e gets closer to 1, the ellipse is long and narrow. Eccentricity is defined as the ratio of the distance from the origin to a foci and the value of a or b, whichever is larger. We will assume $a > b$ for this activity. So, we define e with the formula

$$e = \frac{\sqrt{a^2 - b^2}}{a}$$

The eccentricity and value of a, in astronomical units (AU), for each planet and the dwarf planet, Pluto, are given in the table. (Source: http://www.windows2universe.org/our_solar_system/planets_table.html)

Planet or Dwarf Planet	Value of a (AU)	Orbital Eccentricity (e)
Mercury	0.3871	0.2056
Venus	0.7233	0.0068
Earth	1.000	0.0167
Mars	1.5273	0.0934
Jupiter	5.2028	0.0483
Saturn	9.5388	0.0560
Uranus	19.1914	0.0461
Neptune	30.0611	0.0097
Pluto	39.5294	0.2482

1. Which orbit is least eccentric, that is, most circular?

2. Which orbit is most eccentric, that is, longest and narrowest?

3. Solve the eccentricity formula for b. Then use the information in the table to find the value of b for each planet and dwarf planet. Round to four decimal places.

4. Write the equation that models the path of each orbit. Leave the equation in exponential form.

5. Perihelion and aphelion are the nearest and farthest points on the orbit, as shown in the following figure. The perihelion distance is $a(1 - e)$ and the aphelion distance is $a(1 + e)$. Find the perihelion and aphelion distances, in astronomical units and in miles, for three orbits. (Note: 1 AU ≈ 92,955,807 mi.)

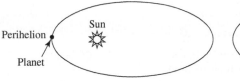

Chapter 13 / REVIEW

Conic Sections and Nonlinear Systems

What's the big idea? Now that we have completed Chapter 13, we should be able to recognize the equations and graphs of the conic sections, which are parabolas, circles, ellipses, and hyperbolas. We should also be able to solve inequalities involving the equations represented by conic sections and systems that involve these nonlinear equations.

The Tools

Listed below are the key terms, skills, formulas, and properties you should know for this chapter.

The page reference is provided if you need additional help with the given topic. The Study Tips will assist in your preparation for an exam.

Study Tips

1. Learn all of the terms, formulas, and properties. Make flash cards and have someone quiz you.
2. Rework problems from the exercises and also the ones you worked in class. Work additional problems from the review exercises.
3. Review the summaries of key concepts.
4. Work the chapter test.
5. Be sure to review the online resources for additional study materials.

Terms

Asymptotes 967
Center of a circle 959
Center of a hyperbola 967
Center of an ellipse 965
Circle 959
Conic sections 956

Ellipse 965
Foci of a hyperbola 967
Foci of an ellipse 965
Hyperbola 967
Nonlinear inequality 979
Nonlinear system of equations 972

Nonlinear system of
  inequalities 982
Parabola 957
Radius 959
Solution of a nonlinear
  system 972

Formulas and Properties

- Standard form of a circle 959
- Standard form of a hyperbola 967

- Standard form of a parabola 957
- Standard form of an ellipse 965

CHAPTER 13 / SUMMARY

How well do you know this chapter? Complete the following questions to find out. Take a look back at the section if you need help.

SECTION 13.1 The Parabola and the Circle

1. The equation $x = a(y - k)^2 + h$ is a(n) _____ parabola with vertex ____. The parabola opens ____ if $a > 0$ and opens ___ if $a < 0$.

2. The equation $(x - h)^2 + (y - k)^2 = r^2$ is the standard form of a(n) ____ with center ____ and radius _.

SECTION 13.2 The Ellipse and the Hyperbola

3. The equation $\dfrac{(x - h)^2}{a^2} + \dfrac{(y - k)^2}{b^2} = 1$ is the standard form of a(n) _____ with center ____. If $a > b$, the _____ is _____. If $a < b$, the _____ is _____.

4. The x- and y-intercepts of an ellipse with center $(0, 0)$ are ____, _____, ____, and _____.

5. A hyperbola centered at the origin is represented by _____ or _____. If the x^2-term is positive, the

hyperbola opens _____. If the y^2-term is positive, the hyperbola opens _____.

6. A(n) _____ is constructed to graph the hyperbola. The vertices are ____, ____, ____, and _____. The diagonals through this rectangle are called _____.

SECTION 13.3 Solving Nonlinear Systems of Equations

7. A nonlinear system can be solved by _____ or _____.

8. Solutions of nonlinear systems can be checked _____ or _____.

9. The solution set of a nonlinear system may consist of _ points, _ point, _ points, or more points.

SECTION 13.4 Solving Nonlinear Inequalities and Systems of Inequalities

10. A nonlinear inequality can be graphed using ___ ____. We first graph the _____ ____ and then test a point in each ____.

11. To solve a nonlinear system of inequalities, we graph each _____ and then find the _____ of the graphs of the inequalities in the system.

CHAPTER 13 / REVIEW EXERCISES

SECTION 13.1

Graph each parabola by plotting the vertex and two additional points. State the axis of symmetry. (See Objective 1.)

1. $x = (y + 3)^2 - 4$
2. $x = -(y - 3)^2 + 9$
3. $x = 2(y - 1)^2 - 10$
4. $x = -3(y + 4)^2 + 1$
5. $x = \frac{1}{2}(y - 2)^2$
6. $x = -\frac{1}{5}(y + 3)^2$
7. $x = 2y^2 + 7$
8. $x = 2y^2 - 10$
9. $x = y^2 - 4y - 5$
10. $x = -y^2 + y + 1$
11. $x = -y^2 + 6y + 7$
12. $x = y^2 - 2y - 24$

Graph each circle. Identify the center, the radius, and four key points. (See Objective 2.)

13. $x^2 + y^2 = 49$
14. $(x + 2)^2 + (y - 8)^2 = 25$
15. $(x - 4)^2 + (y + 5)^2 = 9$
16. $(x + 3)^2 + y^2 = 16$

Write the equation of each circle given its center and radius. (See Objective 3.)

17. center = $(-1, -3)$ and radius = 1
18. center = $(5, -2)$ and radius = 4
19. center = $(-6, 7)$ and radius = $\sqrt{10}$
20. center = $\left(0, \frac{3}{5}\right)$ and radius = $\frac{3}{2}$
21. center = $(3.5, 1.5)$ and radius = 4.5
22. center = $(-2.5, 0.5)$ and radius = 1.5

Write the equation of each circle in standard form and identify its center and radius. (See Objective 4.)

23. $x^2 + y^2 - 2x - 14y = 150$
24. $x^2 + y^2 - 2x - 4y = 54$
25. $x^2 + y^2 - 20y = 116$
26. $x^2 + y^2 - 2x = 26$
27. $x^2 + y^2 - x + 3y = \frac{1}{2}$
28. $x^2 + y^2 - 0.8x + 1.2y = 0.29$

SECTION 13.2

Identify each equation as a parabola, circle, ellipse, or hyperbola. Sketch the graph of each equation. (See Objectives 1 and 2.)

29. $x^2 + 49y^2 = 49$
30. $x = -y^2 + 5y + 6$
31. $x^2 + y^2 = 10y + 11$
32. $x^2 = 4y^2 + 25$
33. $x^2 - \frac{y^2}{9} = 1$
34. $x^2 = 4y^2 - 25$
35. $y^2 = 4x^2 + 1$
36. $\frac{x^2}{3.24} - y^2 = 1$

SECTION 13.3

Solve each system by substitution or elimination. (See Objectives 1 and 2.)

37. $\begin{cases} y = (x - 5)^2 \\ 3x + y = 25 \end{cases}$
38. $\begin{cases} x = \frac{2}{3}y^2 + 2 \\ y = \sqrt{2x - 5} \end{cases}$
39. $\begin{cases} x^2 + y^2 = 25 \\ 4x - 5y = 31 \end{cases}$
40. $\begin{cases} y = x^2 + 6 \\ 5x + y = 2 \end{cases}$
41. $\begin{cases} y = x^2 - 12 \\ 2x + y = -9 \end{cases}$
42. $\begin{cases} 3x^2 + 2y^2 = 315 \\ 3y^2 - x^2 = 27 \end{cases}$
43. $\begin{cases} x^2 + y^2 = 29 \\ 5x^2 + 4y^2 = 120 \end{cases}$
44. $\begin{cases} 2x^2 + 3y^2 = 59 \\ x^2 - y^2 = 7 \end{cases}$
45. $\begin{cases} y = 11x^2 - 3 \\ y = 3x^2 + 2x \end{cases}$
46. $\begin{cases} y = 6x^2 - 3 \\ y = x^2 + 2x \end{cases}$
47. $\begin{cases} xy = 1 \\ 4x - 6y = 5 \end{cases}$
48. $\begin{cases} xy = 1 \\ 3x - 14y = 1 \end{cases}$

SECTION 13.4

Graph each inequality. (See Objective 1.)

49. $y \geq (x + 3)^2 - 7$
50. $y \leq (x - 6)^2 + 2$
51. $\frac{x^2}{25} + \frac{y^2}{25} > 1$
52. $\frac{x^2}{16} + \frac{y^2}{49} \leq 1$
53. $\frac{x^2}{9} - \frac{y^2}{16} > 1$
54. $\frac{y^2}{9} - x^2 < 1$

Solve each system. (*See Objective 2.*)

55. $\begin{cases} y \geq x^2 \\ x^2 + y^2 \leq 2 \end{cases}$ **56.** $\begin{cases} y < x^2 - 4 \\ x^2 + y^2 > 10 \end{cases}$ **57.** $\begin{cases} \dfrac{x^2}{16} + \dfrac{y^2}{9} \geq 1 \\ \dfrac{x^2}{9} + \dfrac{y^2}{4} \leq 1 \end{cases}$ **58.** $\begin{cases} 4x^2 - y^2 \leq 8 \\ x^2 + 3y^2 \geq 28 \end{cases}$

CHAPTER 13 TEST / CONIC SECTIONS AND NONLINEAR SYSTEMS

1. The vertex of $x = -2(y + 4)^2 - 5$ is
   a. $(-4, -5)$ **b.** $(4, -5)$
   c. $(-5, 4)$ **d.** $(-5, -4)$

2. The radius of $x^2 + 2x + y^2 - 4y = 8$ is
   a. 13   **b.** $\sqrt{13}$
   c. 8   **d.** $\sqrt{8}$

3. The equation $x^2 + 4y^2 = 4$ represents the graph of a(n)
   a. parabola   **b.** circle
   c. ellipse   **d.** hyperbola

4. The graph of $\dfrac{y^2}{9} - \dfrac{x^2}{4} = 1$ has
   a. x-intercepts of $(-2, 0)$ and $(2, 0)$.
   b. x-intercepts of $(-3, 0)$ and $(3, 0)$.
   c. y-intercepts of $(0, -2)$ and $(0, 2)$.
   d. y-intercepts of $(0, -3)$ and $(0, 3)$.

Graph each equation and label any key points of the graph.

5. $x = 2(y + 1)^2 - 8$ **6.** $x^2 + y^2 = 25$
7. $x^2 + y^2 - 4x + 6y = 3$

8. $9x^2 + y^2 = 9$ **9.** $\dfrac{(x + 2)^2}{25} + \dfrac{(y - 1)^2}{9} = 1$

10. $9x^2 - y^2 = 9$ **11.** $\dfrac{y^2}{4} - \dfrac{x^2}{9} = 1$

Graph each inequality.

12. $x > y^2 - 9$ **13.** $x^2 - 4y^2 \leq 16$

Solve each system of nonlinear equations.

14. $\begin{cases} x^2 + y^2 = 5 \\ x + y = 3 \end{cases}$ **15.** $\begin{cases} y = \sqrt{x + 6} \\ x + 2y^2 = 6 \end{cases}$

16. $\begin{cases} x^2 + 3y^2 = 19 \\ x^2 - 3y^2 = 13 \end{cases}$

Solve each system of nonlinear inequalities.

17. $\begin{cases} x + 3y \leq 6 \\ x^2 + \dfrac{y^2}{9} \leq 1 \end{cases}$ **18.** $\begin{cases} x^2 + y^2 < 16 \\ \dfrac{x^2}{4} - \dfrac{y^2}{9} > 1 \end{cases}$

CUMULATIVE REVIEW EXERCISES / CHAPTERS 1–13

Evaluate each expression for the given values. (*Section 1.3, Objective 3*)

1. $-x^3 + 2x^2 - x + 4$ for $x = 3$
2. $12x - 21y$ for $x = 2, y = -1$

Simply each expression. (*Section 1.8, Objective 4*)

3. $-2x(8 - 3x) - 9x - 2x(x - 6)$
4. $\dfrac{3}{5}x + \dfrac{1}{4} - \dfrac{1}{2}x + \dfrac{1}{6}$

Find the measure of the angles in each figure. (*Section 2.5, Objective 5*)

5.   **6.**

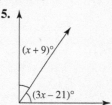

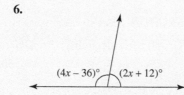

Solve each problem. (*Section 9.2, Objective 3*)

7. Suppose the digital scale reflects a weight (in pounds) with an absolute error of 0.3 lb. If someone weighs 104.6 lb according to the scale, find the exact weight of the person.

8. Suppose a countertop food scale reflects a weight (in ounces) with an absolute error of 0.1 oz. If a piece of cheese weighs 4.5 oz according to the scale, find the exact weight of the piece of cheese.

Find the slope of each line formed by the ordered pairs. (*Section 3.3, Objective 1*)

9. $(-12, 7)$ and $(-8, -7)$
10. $(-3, 2)$ and $(0, 2)$

Express each relation as a graph. (*Section 3.6, Objective 1*)

11. The price of Apple Inc. stock at the end of March of the specified year is given in the table. (Source: http://www.dailyfinance.com/)

Year	2007	2008	2009	2010	2011
Price per share (in dollars)	92.91	143.50	105.12	235.00	348.51

12. The price of Exxon Mobil Corp. stock at the end of March of the specified year is given in the table.
(Source: http://www.dailyfinance.com)

Year	2007	2008	2009	2010	2011
Price per share (in dollars)	75.45	84.58	68.10	67.77	84.13

Find the requested information. (*Section 3.6, Objective 4*)

13. Find $f(-3)$ if $f(x) = 2x^2 - x$.

14. Find $f(1)$ if $f(x) = 4 - x^2$.

Find the slope and y-intercept of each line from its equation. Write the y-intercept as an ordered pair. Graph the line using its slope and y-intercept. Label at least two points on the graph. (*Section 3.3, Objectives 2 and 3*)

15. $2x + 5y = 20$

16. $y + 3 = 0$

Write the equation of each line that satisfies the given conditions. Express your answer in slope-intercept form. (*Section 4.4, Objective 4*)

17. passes through $(2, -6)$, parallel to $y = 2$

18. passes through $(-3, 1)$, perpendicular to $x + 2y = 1$

Graph each linear inequality in two variables. (*Section 9.4, Objective 2*)

19. $x + 4y \le 12$

20. $3x - y \le 6$

Determine how the lines relate, the number of solutions of the system, and the type of system without graphing. (*Section 4.1, Objective 4*)

21. $\begin{cases} y = 5x + 11 \\ x + 4y = 2 \end{cases}$

22. $\begin{cases} 6x + 8y = 5 \\ 3x + 4y = 1 \end{cases}$

Solve each problem. (*Section 4.4, Objective 1*)

23. 550 tickets were sold at a benefit concert at a college. Adult tickets cost \$12 and student tickets cost \$5. If a total of \$4150 was collected for the tickets, how many adult tickets and student tickets were sold?

24. 1500 tickets were sold at a movie theater. Adult tickets cost \$9.50 and student tickets cost \$5.50. If a total of \$12,170 was collected for the tickets, how many adult tickets and student tickets were sold?

Solve each system of linear equations in three variables using elimination. (*Section 4.5, Objective 1*)

25. $\begin{cases} x - y - 3z = -15 \\ -x + 2y + z = 8 \\ 2x + y + 3z = 9 \end{cases}$

26. $\begin{cases} 2x + 5y - z = 17 \\ x - y + 8z = -7 \\ x - 3y + 6z = -9 \end{cases}$

Evaluate and simplify each expression. Express all answers with positive exponents. (*Section 5.1, Objectives 1–4*)

27. $4x^0 - 5$

28. $-7(5x)^0$

29. $(3a^4b^{-7})(-5a^{-3}b^7)$

30. $(12x^8y^5)(2x^{-6}y^{-5})$

Simplify each expression. (*Section 5.3, Objective 4; Section 5.4, Objective 3; Section 5.5, Objectives 1 and 2*)

31. $(2x^2 - 3x + 5) - (4x^2 + 7x + 1)$

32. $2(5a - 9)(2a + 3)$

33. $3(5c - d)^2$

34. $(x - 2y)(x^2 + 2xy + 4y^2)$

Factor each trinomial completely. (*Section 6.1, Objective 4; Section 6.2, Objectives 1–3; Section 6.3, Objectives 1 and 2; Section 6.4, Objectives 1–3*)

35. $(x - 2)^2 - 5(x - 2) - 14$

36. $3(y + 4)^2 + 4(y + 4) - 4$

37. $3x^4 + 2x^3 - 3x - 2$

38. $4y^4 - 5y^3 + 4y - 5$

Solve each polynomial equation. (*Section 6.5, Objective 3*)

39. $18x^3 - 8x = 0$

40. $12x^3 - 12x^2 + 3x = 0$

Simplify each rational expression. (*Section 7.4, Objective 3*)

41. $\dfrac{2x^2 + 8x + 32}{x^3 - 64}$

42. $\dfrac{6x^2 - 18x + 54}{x^3 + 27}$

Add or subtract the rational expressions. (*Section 7.4, Objectives 1 and 2*)

43. $\dfrac{5x}{9x^2 - 6x + 1} - \dfrac{x}{3x^2 + 5x - 2}$

44. $\dfrac{6}{x^2 + 4x - 5} - \dfrac{2}{x^2 - 25}$

Evaluate each rational function at the given value. (*Section 7.1, Objective 1*)

45. $f(x) = \dfrac{x + 1}{x - 2}, \ f\left(\dfrac{7}{5}\right)$

46. $f(x) = \dfrac{2x + 3}{x + 1}, \ f\left(-\dfrac{5}{3}\right)$

Solve each rational equation. (*Section 7.6, Objective 1*)

47. $\dfrac{4}{x^2 - 2x} - \dfrac{7}{2 - x} = \dfrac{5}{x}$

48. $\dfrac{6}{y^2 - 3y} + \dfrac{5}{3 - y} = \dfrac{4}{y}$

Simplify each radical expression. Assume all variables represent positive real numbers. (*Section 10.3, Objectives 1 and 2*)

49. $\dfrac{\sqrt{56x^3y^9}}{\sqrt{2x^4y^{-3}}}$

50. $\dfrac{\sqrt{100x^{-3}y^6}}{\sqrt{5x^2y^{-2}}}$

Perform the indicated operation and write each answer in simplest radical form. (*Section 10.4, Objectives 1 and 2*)

51. $(7 - 2\sqrt{3})(5 + 6\sqrt{3})$ **52.** $(5 - 3\sqrt{2})^2$

Perform the indicated operation and write each answer in standard form. (*Section 10.7, Objectives 2–4*)

53. $\dfrac{2 + i}{3 - i}$ **54.** $\dfrac{1 + 2i}{4 + 3i}$

Graph each function and then use the graph to solve the equation $f(x) = 0$. (*Section 11.1, Objective 5*)

55. $f(x) = \dfrac{1}{3}(x - 1)^2 - 3$ **56.** $f(x) = -\dfrac{1}{2}(x + 1)^2 + 2$

Solve each equation by completing the square. (*Section 11.2, Objectives 1–4*)

57. $2x^2 + 2x - 1 = 0$ **58.** $3y^2 - 2y + 1 = 0$

Use the discriminant to determine the number and type of solutions of each equation. (*Section 11.3, Objective 2*)

59. $x^2 - 3x + 6 = 0$ **60.** $2y^2 - 7y - 5 = 0$

61. $x^2 - 10x + 25 = 0$ **62.** $4y^2 + 9y + 2 = 0$

Solve each equation. (*Section 11.4, Objectives 1–4*)

63. $x^4 + 5x^2 - 36 = 0$ **64.** $x - 7\sqrt{x} + 10 = 0$

65. $x^{2/5} - 2x^{1/5} - 3 = 0$ **66.** $x^{1/2} + 3x^{1/4} - 4 = 0$

Solve each inequality using the boundary number method. Write the solution set in interval notation. (*Section 11.6, Objectives 1–3*)

67. $x^2 - 4 < 0$ **68.** $x^2 + x > 20$

Find the sum, difference, product, and quotient of the functions. State the domain of each combined function. Also, state the restriction for the quotient function. (*Section 12.1, Objective 1*)

69. $f(x) = x^2 + 4$ and $g(x) = x^2 - 2x - 3$

70. $f(x) = \sqrt{x + 1}$ and $g(x) = \sqrt{x - 3}$

Use the graphs of $f(x)$ and $g(x)$ to find the values of each combined function. (*Section 12.1, Objectives 1–3*)

$f(x)$ $g(x)$

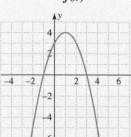   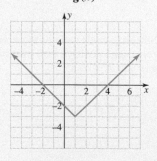

71. $(f + g)(1)$ **72.** $(f \cdot g)(2)$

73. $\left(\dfrac{f}{g}\right)(-1)$ **74.** $(f \circ g)(3)$

For the given functions $f(x)$ and $g(x)$, find the composition functions $(f \circ g)(x)$ and $(g \circ f)(x)$. State the domain of each function. (*Section 12.1, Objective 3*)

75. $f(x) = 5x - 1$ and $g(x) = 2 - 3x$

76. $f(x) = 4x - 10$ and $g(x) = 6x$

Find the equation of the inverse of each function. (*Section 12.2, Objectives 3 and 4*)

77. $f(x) = 8x^3 - 5$ **78.** $f(x) = x^5 + 32$

Evaluate each function given the information. (*Section 12.2, Objective 5*)

79. If $f^{-1}(-1) = 5$, what is $f(5)$?

80. If $f(2.5) = 0$, what is $f^{-1}(0)$?

Solve each exponential equation. (*Section 12.3, Objective 2*)

81. $\left(\dfrac{1}{2}\right)^{2x-5} = 8^x$ **82.** $\left(\dfrac{1}{3}\right)^{2x-5} = 9^{4x}$

Simplify each expression. (*Section 12.4, Objectives 1 and 2*)

83. $\log_5 \dfrac{1}{625}$ **84.** $\log_{1/2} 4$

Solve each logarithmic equation for x. (*Section 12.4, Objective 3*)

85. $\log_{0.5} x = 2$ **86.** $\log_x 121 = 2$

Rewrite each logarithmic expression. If the expression is a single logarithm, write the expression as a combination of logarithms. If the expression is a combination of logarithms, write the expression as a single logarithm. (*Section 12.5, Objectives 1–4*)

87. $\ln\left[\sqrt{x}(x + 2)^3\right]$ **88.** $\log_7\left(\dfrac{y^2}{\sqrt{y - 3}}\right)$

89. $2\log_{10} x + 3\log_{10} y - \log_{10} z$

90. $\log_2 x - 5\log_2 y - \log_2 z$

Evaluate each expression. (*Section 12.5, Objective 5*)

91. $\log_5 5^{2x+6}$ **92.** $\log_{10} 10^{x-3}$

93. $e^{\ln(x+1)}$ **94.** $1.5^{\log_{1.5}(4x)}$

Solve each equation for x. Give an exact solution and approximate the solution to two decimal places, if necessary. (*Section 12.6, Objectives 3 and 4*)

95. $\log(2x + 9) = 1.8$ **96.** $\ln(3x - 5) = 4.2$

Solve each equation. Give an exact solution and an approximation of the solution to two decimal places. (*Section 12.7, Objectives 1 and 2*)

97. $6(1.035)^{4x} = 42$ **98.** $12(2.44)^{3x} = 96$

99. $\log_4(3x - 11) = 3$ **100.** $\log(3x + 52) = 2$

101. $\log x + \log(3x + 2) = 0$

102. $\log x - \log(3x + 2) = 0$

Graph each parabola by plotting the vertex and two additional points. State the axis of symmetry. (*Section 13.1, Objective 1*)

103. $x = y^2 + 6y - 27$ **104.** $x = -y^2 - 4y + 21$

Graph each circle. Identify the center, the radius, and four key points. (*Section 13.1, Objective 2*)

105. $x^2 + y^2 = 100$ **106.** $x^2 + y^2 = 36$

107. $x^2 + (y + 1)^2 = 4$ **108.** $(x + 2)^2 + y^2 = 1$

Write the equation of each circle given its center and radius. (*Section 13.1, Objective 3*)

109. center $= (5, -6)$ and radius $= 9$

110. center $= (3, -1)$ and radius $= \sqrt{7}$

Write the equation of each circle in standard form and identify the center and radius. (*Section 13.1, Objective 4*)

111. $x^2 + y^2 - 4x + 2y = 4$

112. $x^2 + y^2 + 10x - 4y + 28 = 0$

Graph each ellipse. Label four key points and the center on the graph. (*Section 13.2, Objective 1*)

113. $25x^2 + 9y^2 = 225$ **114.** $81x^2 + 4y^2 = 324$

Graph each hyperbola. (*Section 13.2, Objective 2*)

115. $9x^2 - y^2 = 9$ **116.** $y^2 - 16x^2 = 16$

Solve each system by substitution or elimination. (*Section 13.3, Objectives 1 and 2*)

117. $\begin{cases} y = \dfrac{2}{3}x^2 - 4 \\ x = \sqrt{3y + 9} \end{cases}$ **118.** $\begin{cases} y = x^2 - 8 \\ x = \sqrt{2y + 7} \end{cases}$

119. $\begin{cases} y = (x - 1)^2 \\ -4x + y = 1 \end{cases}$ **120.** $\begin{cases} y = (x - 6)^2 \\ 8x + y = 36 \end{cases}$

121. $\begin{cases} x^2 + 2y^2 = 32 \\ x^2 - 2y^2 = 8 \end{cases}$ **122.** $\begin{cases} x^2 + y^2 = 12 \\ 4x^2 + y^2 = 36 \end{cases}$

123. $\begin{cases} y = 4x^2 - x \\ y = 2x^2 + 3 \end{cases}$ **124.** $\begin{cases} y = 2x^2 + x - 7 \\ y = x^2 - x + 8 \end{cases}$

Graph each inequality. (*Section 13.4, Objective 1*)

125. $\dfrac{x^2}{64} + \dfrac{y^2}{9} \geq 1$ **126.** $\dfrac{x^2}{4} + \dfrac{y^2}{81} \geq 1$

127. $\dfrac{x^2}{81} - \dfrac{y^2}{4} > 1$ **128.** $\dfrac{y^2}{9} - x^2 > 1$

Solve each system. (*Section 13.4, Objective 2*)

129. $\begin{cases} x^2 + y^2 < 64 \\ x - 4y < 12 \end{cases}$ **130.** $\begin{cases} x^2 + y^2 > 81 \\ 3x + 2y > 18 \end{cases}$

131. $\begin{cases} x^2 + \dfrac{y^2}{2.25} \leq 1 \\ \dfrac{y^2}{0.25} - x^2 \leq 1 \end{cases}$ **132.** $\begin{cases} x^2 - \dfrac{y^2}{0.81} \geq 1 \\ \dfrac{x^2}{6.25} + \dfrac{y^2}{1.96} \geq 1 \end{cases}$

Student Answer Appendix

CHAPTER 1

Section 1.1

1. A rational number is a number that can be written as a ratio of two integers, p and q, with q not equal to zero.
3. The opposite of a number is the number that has the same distance from 0 but lies on the opposite side of 0.
5. False, for example $\frac{5}{2}$ is a rational number but is not an integer.
7. False, 0 is a whole number but is not a natural number.
9. False, for example $\frac{1}{3} = 0.333333\ldots$ and is rational.
11. integer, $-\frac{7}{1}$ rational, real
13. natural, whole, integer, rational, real
15. $\frac{5}{2}$ rational, real   17. $\frac{15}{8}$ rational, real   19. $\frac{1}{9}$ rational, real
21. natural, whole, integer, $\frac{9}{1}$ rational, real
23. irrational; 3.87, real   25. irrational; -3.14, real
27. natural, whole, integer, rational, real
29. integer, rational, real
31. natural, whole, integer, rational, real   33. Answers vary.
35. Answers vary.   37. Answers vary.   39. 2

41.

43.

45.

47.

49.

51.

53.

55.

57. $>$ 59. $=$ 61. $<$ 63. $=$ 65. $>$ 67. $>$
69. -6 71. 1   73. $-\frac{1}{4}$ 75. 5.1   77. π 79. $1.\overline{3}$
81. 5   83. 7.5   85. $\frac{1}{3}$ 87. $-\frac{9}{11}$ 89. -7
91. rational, real; $\frac{25}{2}$ 93. rational, real
95. natural, whole, integer, rational, real, $\frac{96}{1}$
97. rational, real; $\frac{56}{9}$ 99. irrational, real; 3.46
101. rational, real; $\frac{137}{10000}$ 103. irrational, real; 4.12
105. integer, rational, real; $-\frac{1293}{1}$

107.

109.

111. $\frac{99}{70} > \sqrt{2}$ 113. $-|-9| < \sqrt{81}$ 115. $-\left|-\frac{10}{3}\right| < \frac{\pi}{2}$
117. $-\frac{14}{3}$ 119. $-4.\overline{29}$ 121. -6 123. 9.5
125. No. The opposite of a positive number is negative but the opposite of a negative number is positive.
127. Since $\frac{5}{7} \approx 0.71$ and $\frac{7}{10} = 0.7, \frac{5}{7} > \frac{7}{10}$.
129. 0.52, irrational   131. 3.73, irrational   133. 4.22, rational
135. 7, rational   137. -23, rational

Section 1.2

1. Answers vary.   3. Answers vary.   5. Answers vary.
7. a. Answers vary.   b. Answers vary.
9. Answers vary.   11. False. It simplifies to $\frac{5}{8}$.   13. True.
15. False. The reciprocal of 0 would be $\frac{1}{0}$ which is undefined.
17. False. The LCD of 2 and 5 is 10. We don't add the denominators.
19. $2 \cdot 2 \cdot 3$ 21. $2 \cdot 2 \cdot 2 \cdot 5$ 23. $2 \cdot 2 \cdot 2 \cdot 2 \cdot 3 \cdot 3$
25. $\frac{165}{767}$ 27. $\frac{34}{53}$ 29. Google: $\frac{15}{23}$; Yahoo! $\frac{121}{920}$ 31. $\frac{1}{3}$
33. 8   35. $\frac{3}{4}$ 37. $\frac{3}{10}$ 39. $\frac{3}{14}$ 41. 1   43. 25
45. 10   47. $\frac{5}{6}$ 49. $\frac{56}{27}$ 51. $\frac{1}{9}$ 53. 16   55. 6
57. 4   59. $\frac{4}{7}$ 61. $\frac{4}{5}$ 63. $9\frac{1}{2}$ 65. 12   67. 8
69. 54   71. 2   73. $\frac{9}{14}$ 75. $\frac{35}{8}$ 77. $\frac{7}{6}$ 79. $\frac{9}{2}$
81. $\frac{11}{84}$ 83. $5\frac{5}{6}$ 85. $\frac{3}{5}$ 87. $\frac{7}{8}$ 89. $\frac{7}{10}$ 91. $\frac{36}{5}$

93. $\frac{22}{15}$ **95.** $\frac{5}{2}$ **97.** $2\frac{7}{12}$ **99.** $\frac{1}{27}$ **101.** $\frac{8}{5}$

103. $\frac{3}{8} = \frac{3 \cdot 2}{8 \cdot 2} = \frac{6}{16}$

105. $2\frac{1}{3} \times 3\frac{3}{4} = \frac{7}{3} \times \frac{15}{4} = \frac{7 \times 15}{3 \times 4} = \frac{7 \times 5}{4} = \frac{35}{4} = 8\frac{3}{4}$

107. $\frac{117}{77}$ **109.** $\frac{5}{3}$ **111.** $\frac{9}{14}$

Section 1.3

1. Answers vary. **3.** Answers vary. **5.** Answers vary.
7. Answers vary. **9.** False, the translation is $3x - 4$.
11. 256 **13.** -27 **15.** 15.625 **17.** -0.216 **19.** $-\frac{9}{25}$
21. $\frac{32}{243}$ **23.** 17 **25.** 22 **27.** 2 **29.** $\frac{26}{35}$ **31.** $\frac{5}{4}$
33. 19 **35.** 22 **37.** 36 **39.** 1 **41.** 28 **43.** $\frac{4}{9}$

45.

x	$3x + 2$
0	2
$\frac{1}{3}$	3
2	8
4	14

47.

x	$\dfrac{3x - 2}{x}$
1	1
2	2
3	$\frac{7}{3}$
4	$\frac{5}{2}$

49.

x	$\dfrac{x + 3}{x - 1}$
2	5
3	3
4	$\frac{7}{3}$
5	2

51. 6 **53.** $\frac{2}{3}$ **55.** 10 **57.** 36 **59.** 5 **61.** 1
63. $x = 2$ is not a solution. **65.** $a = 5$ is a solution.
67. $x = 3$ is not a solution. **69.** $y = 1$ is a solution.
71. $x + 4$ **73.** $3x + 2$ **75.** $x - 9$ **77.** $4x - 3$ **79.** $10x$
81. $3(x + 15)$ **83.** $4x + 3$ **85.** $2(x - 7)$ **87.** $\frac{3x}{4}$
89. $5000 - x$ **91.** $2w - 3$ **93.** $x + 215$ **95.** 4392 ft
97. \$1276.28 **99.** \$42,200 **101.** \$1275 **103.** 146.5 ft
105. 216 **107.** 52 **109.** 2.744 **111.** 20 **113.** $\frac{16}{9}$
115. $1\frac{8}{15}$ **117.** $\frac{8}{5}$ **119.** 42 **121.** $\frac{1}{2}$

123.

x	$5x - 3$
1	2
2	7
3	12
4	17

125. 9 **127.** 24 **129.** 13 **131.** $x = -3$ is not a solution.
133. $5(x + 11)$ **135.** $6x - 18$ **137.** $2(15x)$ **139.** $14(x + 9)$
141. $50 - x$ oz **143.** $3w + 2$ **145.** \$1656.12 **147.** 206 ft
149. \$2743
151. $3 \cdot 2^2 - 2 \cdot 2 + 1 = 3 \cdot 4 - 4 + 1 = 12 - 4 + 1 = 9$
153. $30 - 2(3)^2 = 30 - 2(9) = 30 - 18 = 12$
155. $-3 \cdot 2^2 + 28 \div 7 \cdot 4 + 36 = -3 \cdot 4 + 4 \cdot 4 + 36$
$= -12 + 16 + 36 = 40$
157. 1, 11.625, 27.368 **159.** 0, 4, 8.49

Section 1.4

1. Answers vary. **3.** 12.21 **5.** -9 **7.** -8.45 **9.** $\frac{3}{2}$
11. $-\frac{13}{5}$ **13.** $-\frac{17}{20}$ **15.** $-4\frac{1}{2}$ **17.** -25 **19.** -11

21. -2 **23.** 4 **25.** -4 **27.** 2.1 **29.** -2.6 **31.** $-\frac{2}{15}$
33. $-1\frac{1}{3}$ **35.** 4 **37.** -15 **39.** 302 million euros
41. -5.1 **43.** -2.46 **45.** $-\frac{19}{6}$ **47.** $\frac{11}{7}$ **49.** $-2\frac{1}{4}$
51. -1 **53.** -0.6 **55.** $-5 + 2 = -3$ **57.** $-3 + (-4) = -7$
59. $12 + (-4) = 8$ **61.** $-2 + 9 = 7$ **63.** \$290 million
65. \$2.67 **67.** \$279.54 **69.** $-$\$5488 million
71. Answers vary. **73.** -14.01 **75.** $-3\frac{9}{10}$ **77.** $-10\frac{9}{20}$

Piece It Together Sections 1.1–1.4

1. $-\frac{15}{2}, 2.5, 6\frac{1}{2}$: rational and real; $-\sqrt{3}$: irrational and real

2. $<$ **3.** $=$ **4.** $\frac{2}{5}$ **5.** $\frac{1}{6}$ **6.** $\frac{1}{9}$ **7.** 49 **8.** $\frac{4}{5}$
9. $\frac{1}{36}$ **10.** $\frac{5}{12}$ **11.** $\frac{9}{2}$ **12.** 4.41 **13.** -81
14. 42 **15.** 1 **16.** 18
17.

x	$\dfrac{2}{5}x + 3$
0	3
1	$\frac{17}{5}$
5	5
10	7

18. 5 **19.** $x \div 6$ **20.** $2(x + 4)$ **21.** $\frac{1}{5}x + 3$
22. $2x - 8$ **23.** -10 **24.** -14 **25.** $\frac{2}{3}$

Section 1.5

1. Answers vary. **3.** False, $3 - (-3) = 3 + 3 = 6$ **5.** -7
7. 6.8 **9.** -8 **11.** -78 **13.** 8 **15.** 4 **17.** $-\frac{5}{3}$
19. -3 **21.** 3 **23.** $\frac{5}{3}$ **25.** $-5\frac{3}{5}$ **27.** -8 **29.** 8
31. $5\frac{3}{5}$ **33.** 2.6 **35.** -4 **37.** -18 **39.** 27
41. -4.1 **43.** $\frac{5}{8}$ **45.** -12 **47.** 6 **49.** -18
51. 87°C **53.** 20,976.2 ft
55.

Year	Net Income	Change in Net Income
2005	\$ 9,320	n/a
2006	\$ 14,494	\$5174
2007	\$ 50,545	\$36,051
2008	$-$\$266,334	$-$\$316,879
2009	\$134,662	\$400,996

57. $c = 68°$ **59.** $c = 60°$ **61.** 38° and 128° **63.** 17° and 107°
65. -8.7 **67.** $-\frac{23}{26}$ **69.** 8 **71.** $-\frac{11}{4}$ **73.** 3 **75.** 9.4
77. 10 **79.** -4.1 **81.** $\frac{1}{3}$ **83.** 2.75 **85.** -41
87. 88°F **89.** $c = 40°$ **91.** 21° and 111°
93. $\frac{1}{5} - \left(-1\frac{3}{4}\right) = \frac{1}{5} - \left(-\frac{7}{4}\right) = \frac{1}{5} + \frac{7}{4}$

$= \frac{1 \times 4}{5 \times 4} + \frac{7 \times 5}{4 \times 5} = \frac{4}{20} + \frac{35}{20} = \frac{39}{20}$

95. 7.93 **97.** $-5\frac{5}{6}$

Section 1.6

1. Answers vary. **3.** Answers vary. **5.** Answers vary.
7. Answers vary. **9.** Answers vary. **11.** -48 **13.** -3

15. 3   **17.** $-\dfrac{8}{35}$ **19.** $\dfrac{8}{35}$ **21.** -11.5 **23.** -3240

25. 1   **27.** 6   **29.** -168 **31.** 0   **33.** -82 **35.** 64

37. 125   **39.** 81   **41.** -64 **43.** 16   **45.** -169

47. -625 **49.** $\dfrac{1}{16}$ **51.** $\dfrac{27}{8}$ **53.** 0.64   **55.** 1.728

57. -4 **59.** -7 **61.** 8   **63.** -25 **65.** -12 **67.** 1

69. 0   **71.** undefined   **73.** $\dfrac{5}{2}$ **75.** $-\dfrac{7}{2}$

77.

x	$2x^2 - 5x + 1$
-2	19
0	1
$\dfrac{1}{2}$	-1
2	-1

79.

x	$-2x^2 + 4x + 3$
-2	-13
-1	-3
$\dfrac{1}{2}$	$\dfrac{9}{2}$
2	3

81. 5

83.

x	$\dfrac{\lvert x - 2 \rvert}{x + 1}$
-2	-4
-1	Undefined
0	2
1	$\dfrac{1}{2}$
2	0

85. $-\dfrac{1}{5}$ **87.** 97   **89.** $V = \dfrac{1}{3}\pi r^2 h$; 45,239 in.^{3}

91. a.

Date	Closing Price	Daily Change in Price	Percent Change in Price
10/8/2010	$48.75	$0.87	0.0182 or 1.82%
10/7/2010	$47.88	$-$0.29	-0.0060 or -0.60%
10/6/2010	$48.17	$0.07	0.0015 or 0.15%
10/5/2010	$48.10	$0.47	0.0099 or 0.99%
10/4/2010	$47.63	n/a	n/a

b. $0.28

93. -35 **95.** -30.3 **97.** 35   **99.** 81   **101.** $-\dfrac{3}{28}$

103. 9   **105.** -625 **107.** $\dfrac{8}{125}$ **109.** 256   **111.** -0.0016

113. -14.7 **115.** 1   **117.** -196 **119.** 0   **121.** undefined

123. $\dfrac{2}{45}$ **125.** $\dfrac{5}{2}$ **127.** $-\dfrac{125}{216}$ **129.** -60 **131.** 25

133. 52   **135.** -95 **137.** 10   **139.** $-\dfrac{3}{5}$ **141.** $-\dfrac{8}{17}$

143.

x	$\dfrac{\lvert x - 2 \rvert}{x + 3}$
-3	Undefined
-2	4
0	$\dfrac{2}{3}$
1	$\dfrac{1}{4}$
2	0

145. $V = lwh$; 55,589 in.^{3} 241 gal   **147.** $\dfrac{0}{a} = 0$ and $\dfrac{a}{0}$ is undefined.

149. $\dfrac{12}{15} \div \left(-\dfrac{10}{9}\right) = -\dfrac{12}{15} \cdot \dfrac{9}{10} = -\dfrac{\overset{6}{\cancel{12}}}{\underset{5}{\cancel{15}}} \cdot \dfrac{\overset{3}{\cancel{9}}}{\underset{5}{\cancel{10}}} = -\dfrac{18}{25}$ **151.** 1.44

153. $\dfrac{9}{16}$ **155.** 18

Section 1.7

1. Answers vary.   **3.** Answers vary.   **5.** Answers vary.

7. Answers vary.   **9.** Answers vary.   **11.** $7, -\dfrac{1}{7}$

13. $-25, \dfrac{1}{25}$ **15.** $-\dfrac{3}{5}, \dfrac{5}{3}$ **17.** $\dfrac{4}{9}, -\dfrac{9}{4}$ **19.** $-3x, \dfrac{1}{3x}$

21. $6a, -\dfrac{1}{6a}$ **23.** $-\dfrac{x}{6}, \dfrac{6}{x}$ **25.** $\dfrac{7a}{2}, -\dfrac{2}{7a}$ **27.** $x + 11$

29. $a - 13$ **31.** $x - 16$ **33.** $27x$ **35.** $16a$ **37.** $-70x$

39. $c - 9$ **41.** $x - 10$ **43.** $-6x$ **45.** $21a$ **47.** $x + \dfrac{1}{4}$

49. $a + \dfrac{19}{7}$ **51.** $2x$ **53.** $-\dfrac{3x}{2}$ **55.** $4x - 28$

57. $3x + 15$ **59.** $33x - 44$ **61.** $12x + 18y - 30$

63. $5x - 2y + 9$ **65.** $-16x + 3$ **67.** $-6x + 4$

69. $-18x + 3$ **71.** $-15x, \dfrac{1}{15x}$ **73.** $-\dfrac{6x}{7}, \dfrac{7}{6x}$ **75.** $11, -\dfrac{1}{11}$

77. $-\dfrac{9}{5}, \dfrac{5}{9}$ **79.** $x + 20$ **81.** $28x$ **83.** $x - 36$ **85.** $18a$

87. $x - 14$ **89.** $a - \dfrac{1}{10}$ **91.** $8.4a$ **93.** $a + \dfrac{1}{8}$

95. $x + 12y - 5$ **97.** $-14x$ **99.** $-1.5x + 1$

101. $-10x - 15$ **103.** $-3x$ **105.** $-x + 2$

107. $0.56x - 0.96$ **109.** $3(2a) = 6a$

111. $-(2 - x + 3y) = -2 + x - 3y$

Section 1.8

1. Answers vary.   **3.** Answers vary.

5. Answers vary.   **7.** Answers vary.

Expression	Total Number of Terms	Variable Term(s)	Constant Term(s)	Coefficient of Each Variable Term
9.	1	$-5x$	None	-5
11.	2	$2y$	-5	2
13.	3	$x^2, -5x$	3	$1, -5$
15.	4	$-3x^4, -1x^2, 4x$	-1	$-3, -1, 4$
17.	1	$\dfrac{1}{2}x$	None	$\dfrac{1}{2}$

19. like terms   **21.** like terms   **23.** unlike terms

25. unlike terms   **27.** $13x$ **29.** $11a$ **31.** $-12x$ **33.** $19x^2$

35. $-18x^2$ **37.** not like terms   **39.** not like terms   **41.** $\dfrac{9x}{4}$

43. $\dfrac{7a}{5}$ **45.** $-2x^2 - 3x$ **47.** $-12x^2 - 25x$ **49.** $2x - 7$

51. $15x + 10$ **53.** $2x - 23$ **55.** $-2x - 22$ **57.** $-x - 16$

59. $15x + 6$ **61.** $-12y - 4$

Expression	Total Number of Terms	Variable Term(s)	Constant Term(s)	Coefficient of Each Variable Term
63.	2	$-1x$	12	-1
65.	3	$-3x^2, 7x$	-2	$-3, 7$
67.	4	$-x^4, 3x^2, -5x$	6	$-1, 3, -5$
69.	2	$\dfrac{3}{2}x$	$-\dfrac{5}{4}$	$\dfrac{3}{2}$
71.	3	$0.5x^2, -1.8x$	-2.1	$0.5, -1.8$

73. $-5x - 18$ **75.** $-x - 8$ **77.** $0.6x - 3$ **79.** $7x + 2$

81. $-8y + 12$ **83.** $1.4x - 2.9$ **85.** $5x - 16$

87. $5 - 3(2 + 4) = 5 - 3(6) = 5 - 18 = -13$

89. $x^2 + x^2 = 1x^2 + 1x^2 = 2x^2$

91. $4 + 2(5 + x) = 4 + 10 + 2x = 14 + 2x$

93. $-6b - 4b = -10b$

CHAPTER 1 SUMMARY

Section 1.1

1. set   **2.** natural (or counting)   **3.** whole   **4.** integers
5. rational   **6.** irrational   **7.** is less than; is greater than
8. opposite   **9.** absolute value

Section 1.2

10. prime   **11.** quotient; real; numerator; denominator
12. simplest form
13. fundamental property; fractions; Answers vary.
14. reciprocals   **15.** multiply; reciprocal   **16.** common
17. least common denominator

Section 1.3

18. exponent   **19.** base; exponent; power
20. order; operations; grouping; exponents; multiply;
divide; add; subtract
21. variable; letter   **22.** algebraic expression
23. variable; given value   **24.** equation   **25.** solution

Section 1.4

26. absolute values; sign   **27.** absolute values; larger absolute value

Section 1.5

28. adding; opposite; $a + (-b)$

Section 1.6

29. positive; negative   **30.** positive   **31.** negative
32. multiplying; reciprocal; $a \cdot \dfrac{1}{b}$ **33.** zero; undefined

Section 1.7

34. commutative; $b + a$ **35.** associative; $(a + b) + c$
36. zero   **37.** identity; zero; 0   **38.** inverse; opposite; $-a$
39. commutative; ba **40.** associative; $(ab)c$ **41.** one
42. identity; one; 1   **43.** inverse; reciprocal; $\dfrac{1}{a}$
44. distributive; $ab + ac$ **45.** negative; change

Section 1.8

46. term; terms   **47.** variable term   **48.** constant term
49. coefficient   **50.** like; unlike   **51.** coefficients; variable
52. parentheses; like terms

CHAPTER 1 REVIEW EXERCISES

Section 1.1

1. rational, real   **3.** irrational, real, 31.42   **5.** Answers vary, {0}
7. Answers vary, {1.8}
9.

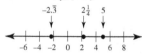

11. >   **13.** $-1.\overline{12}$ **15.** $3\dfrac{1}{5}$ **17.** 5.8   **19.** $\dfrac{6}{13}$

Section 1.2

21. $2 \cdot 3 \cdot 3 \cdot 5$ **23.** $3 \cdot 3 \cdot 3 \cdot 5 \cdot 7$ **25.** $\dfrac{2}{3}$ **27.** $\dfrac{8}{9}$
29. $\dfrac{41}{42}$ **31.** 4   **33.** $\dfrac{3}{8}$

Section 1.3

35. 160   **37.** 0   **39.** $-\dfrac{4}{15}$ **41.** 29.6   **43.** 70
45. $-18 - 7 = -25$ **47.** $-11 - 6 = -17$
49. $(-13 + 7) - 9 = -15$ **51.** $5 + (-2) = 3$
53. $15 + (-3) = 12$ **55.** $-18 - 7 = -25$ **57.** $2(-35) = -70$

59. $0.3(12) = -3.6$ **61.** $\dfrac{28}{42} = \dfrac{2}{3}$ **63.** $\dfrac{-32}{-48} = \dfrac{2}{3}$
65. $7x - 12$ **67.** $2(x)(45)$ **69.** $6\left(\dfrac{x}{-9}\right)$ **71.** 256 ft
73. $4w + 2$

Section 1.4

75. 0.27   **77.** $-\dfrac{9}{2}$ **79.** $-2\dfrac{7}{12}$

Section 1.5

81. 4.41   **83.** $5\dfrac{2}{5}$ **85.** 7   **87.** -24 **89.** -2.6
91. $510 million   **93.** $55.10

Section 1.6

95. -2 **97.** $\dfrac{4}{15}$ **99.** -1296 **101.** $\dfrac{8}{27}$
103. undefined   **105.** $-\dfrac{49}{12}$ **107.** -90

109.

x	y
-2	$\dfrac{3}{5}$
-1	$\dfrac{1}{4}$
0	$-\dfrac{1}{3}$
1	$\dfrac{3}{2}$
2	-5

111.

x	y
-2	2
-1	-1
0	-2
1	-1
2	2

113.

x	y
-4	-21
-2	-11
0	1
2	9
4	19

115. 2   **117.** 35   **119.** 16

Section 1.7

121. commutative property of addition
123. inverse property of addition
125. associative property of multiplication
127. identity property of multiplication
129. associative property of multiplication
131. $-15x + 20y - 10$ **133.** $-3x + \dfrac{3}{2}$

Section 1.8

Expression	Total Number of Terms	Variable Term(s)	Constant Term(s)	Coefficient of each Variable Term
135.	3	$-20x^4$, $13x^2$	5	-20, 13
137.	3	$-0.2x^2$, $3.2x$	3.6	-0.2, 3.2

139. unlike terms   **141.** unlike terms   **143.** $-28b$
145. $16x - 3$ **147.** $0.06x + 2.3$ **149.** $-8y - 1$ **151.** $13x^2$
153. $1.2x + 2.5$ **155.** $2.1x - 12$

CHAPTER 1 TEST

1. d   **3.** b   **5. a.** $\dfrac{7}{2}$ **b.** 3   **c.** 41   **d.** 64
7. a. $\dfrac{31}{50}$ **b.** 4

9. a. commutative property of multiplication
  b. inverse property of multiplication
  c. associative property of addition
  d. inverse property of addition
  e. identity property of multiplication
  f. identity property of addition
  g. commutative property of addition

11. a. $12x + 28$ **b.** $48x - 40$ **c.** $-2x - 3$ **d.** $6x - 2$
  e. $3x + 3$ **f.** $3x^2 - 9x - 2$ **g.** $\frac{5}{2}x$ **h.** $12x + 13$

13. a. $x + 2$ **b.** $x - 5$ **c.** $3x + 4$ **d.** $2(x + 7)$
  e. $\frac{4x}{3}$ **f.** $\frac{1}{4}x - 6$

15. a.

Date	Closing Price of Google Stock	Daily Change in Price
11/2/2009	$533.99	-2.13
10/30/2009	$536.12	-14.93
10/29/2009	$551.05	10.75
10/28/2009	$540.30	-7.99
10/27/2009	$548.29	-5.92
10/26/2009	$554.21	n/a

  b. -4.044 **17.** 434.89 in.^{3}

CHAPTER 2

Section 2.1

1. Answers vary.   **3.** Answers vary.   **5.** Answers vary.
7. expression   **9.** equation   **11.** expression   **13.** equation
15. equation

17.

$x = -2$	Is a solution
$x = -1$	Is not a solution
$x = 0$	Is not a solution
$x = 1$	Is not a solution
$x = 2$	Is not a solution

19.

$x = -2$	Is not a solution
$x = -1$	Is not a solution
$x = 0$	Is a solution
$x = 1$	Is not a solution
$x = 2$	Is not a solution

21.

$x = -2$	Is not a solution
$x = -1$	Is a solution
$x = 0$	Is not a solution
$x = 1$	Is not a solution
$x = 2$	Is a solution

23.

$x = -2$	Is not a solution
$x = -1$	Is a solution
$x = 0$	Is not a solution
$x = 1$	Is a solution
$x = 2$	Is not a solution

25.

$x = -2$	Is not a solution
$x = -1$	Is not a solution
$x = 0$	Is not a solution
$x = 1$	Is not a solution
$x = 2$	Is not a solution

27. $x + 4 = -3$ **29.** $x - 7 = 4$ **31.** $2x = 18$ **33.** $\frac{x}{4} = 8$

35. $2x + 7 = x - 1$ **37.** $\frac{3}{x} = \frac{1}{3}x$ **39.** $3x - 4 = 6 + x$

41. $3x + 4x = 5x - 7$ **43.** $4[x + (-2)] = 2x$
45. $4x - (-12) = 50 + x$ **47.** $2x + 18 = x - 27$
49. $x + x - 13° = 90°$ **51.** $x + x - 4.8 = 25.8$
53. a. $2l + 2w = 32$ **b.** $l = w + 4$ or $w = l - 4$
55. a. $l + h = 1.31$ **b.** $l = h - 0.254$
57. a. $e + m = 3.1$ **b.** $e + m = \frac{1}{2}t$

59. equation   **61.** expression   **63.** equation

65.

$x = -4$	Is not a solution
$x = -1$	Is not a solution
$x = 0$	Is not a solution
$x = 4$	Is not a solution
$x = \frac{5}{14}$	Is a solution

67.

$x = -4$	Is a solution
$x = -2$	Is not a solution
$x = 0$	Is not a solution
$x = 5$	Is a solution
$x = \frac{1}{2}$	Is not a solution

69. $2(x + 11) = 5x$ **71.** $4x + (-10) = 20 + x$
73. $2x + 4 = 3x - 5$ **75.** $4 - 5x = 2x - 6$
77. $x + x + 25° = 180°$ **79.** $x + x - 19° = 90°$
81. $x + x - 7.7 = 46.3$
83. a. $x + y = 69.1$ **b.** $x = y - 1.5$
85. Answers vary. Three less than the square of a number is 10.
87. Answers vary. Two more than half a number is 6.

89.

$x = -1$	Is not a solution
$x = 0$	Is a solution
$x = 1$	Is a solution

91.

$x = 0$	Is not a solution
$x = \frac{1}{2}$	Is a solution
$x = 1$	Is not a solution

93. Twice the difference of a number and 5 is 6 more than the number.
95. Answers vary. $2x - 1 = 8 - x$

Section 2.2

1. Answers vary.   **3.** Answers vary.
5. False, we must subtract 7 from both sides.   **7.** True
9. linear equation
11. not a linear equation because the largest exponent is 2
13. linear equation   **15.** $\{-8\}$ **17.** $\{4\}$ **19.** $\{6\}$
21. $\{1\}$ **23.** $\{-1\}$ **25.** $\{-2\}$ **27.** $\{-4.1\}$ **29.** $\{6\}$
31. $\{13\}$ **33.** $\{-11\}$ **35.** $\{7\}$ **37.** $\{7\}$ **39.** $\{-2\}$
41. $\{-28\}$ **43.** $\{18\}$ **45.** $\{13\}$ **47.** The number is 15.
49. The number is -45.   **51.** The markup is $3750.
53. The taxes and fees are $78.89.
55. The original price of the mixer is $349.99.
57. Zuckerberg's net worth was $6.9 billion and Moskovitz's net worth was $1.4 billion.
59. $\{0\}$ **61.** $\{8\}$ **63.** $\{5\}$ **65.** $\{-6\}$ **67.** $\{13\}$
69. $\{2.86\}$ **71.** $\{-0.9\}$ **73.** $\{0\}$ **75.** The number is -7.
77. The number is -3.   **79.** The number is -7.   **81.** $2505
83. Williams earned $20 million and Sharapova earned $25 million.
85. $2(x - 1) + 3 = x - 4$
  $2x - 2 + 3 = x - 4$
  $2x + 1 = x - 4$
  $x + 1 = -4$
  $x = -5$

87. correct   **89.** $\{-6\}$ **91.** $\{3\}$ **93.** $\{-31.36\}$
95. Answers vary.

Section 2.3

1. Answers vary.   **3.** Answers vary.
5. False, we must divide both sides by -1 to solve for p.   **7.** True
9. False, we use $x, x + 2, x + 4$, etc.   **11.** $\{5\}$ **13.** $\{-3\}$
15. $\{-40\}$ **17.** $\{-11\}$ **19.** $\{6\}$ **21.** $\{10\}$
23. $\{9\}$ **25.** $\{12\}$ **27.** $\{32\}$ **29.** $\{2\}$ **31.** $\{6\}$
33. $\{12\}$ **35.** $\{-1.2\}$ **37.** $\{-8\}$ **39.** $\{-3\}$ **41.** $\left\{\frac{11}{9}\right\}$
43. $\{1\}$ **45.** $\{-0.4\}$ **47.** $\{1.5\}$
49. The car was be driven 150 mi   **51.** 175 nickels
53. 78 quarters   **55.** -40 and -38 **57.** 20, 21, 22
59. $-1, 1, 3$ **61.** $\{2\}$ **63.** $\{4\}$ **65.** $\{-1\}$ **67.** $\{-8\}$
69. $\{1\}$ **71.** $\left\{-\frac{1}{4}\right\}$ **73.** $\{0\}$ **75.** $\{-15\}$ **77.** $\{-2.4\}$
79. $\{0.4\}$ **81.** $\{-3\}$ **83.** $\{2\}$ **85.** -9 **87.** 25
89. -147 **91.** 30   **93.** -1 **95.** 150

97. The car was driven 210 mi.   **99.** $-40, -39, -38,$ and -37

101. $-8, -6,$ and -4

103. $\dfrac{x}{2} + 3 - 3 = 7 - 3$

$$2 \cdot \dfrac{x}{2} = 2 \cdot 4$$
$$x = 8$$

105. $2x - 5 = 4x + 10$

107. Let x, $x + 2$, and $x + 4$ be the three consecutive odd integers.

$$3(x + 4) = 87 + x + (x + 2)$$
$$3x + 12 = 89 + 2x$$
$$x = 77$$

The three odd integers are 77, 79, and 81.

109. $\{-7\}$ **111.** $\{6\}$

Section 2.4

1. Answers vary.   **3.** Answers vary.

5. Answers vary.   **7.** Answers vary.   **9.** True

11. False, we must multiply 3 by 8 to get $2y + 24 = y$.

13. $\{-2\}$ **15.** $\{30\}$ **17.** $\left\{-\dfrac{5}{7}\right\}$ **19.** $\{4\}$ **21.** $\{-11\}$

23. $\left\{-\dfrac{15}{2}\right\}$ **25.** $\left\{-\dfrac{18}{7}\right\}$ **27.** $\left\{\dfrac{17}{2}\right\}$ **29.** $\{50\}$

31. $\{65\}$ **33.** $\{3500\}$ **35.** $\{25\}$ **37.** $\varnothing$ **39.** $\varnothing$

41. $\varnothing$ **43.** $\mathbb{R}$

45. $\mathbb{R}$ **47.** $\varnothing$ **49.** $\{0\}$ **51.** $\{-1\}$ **53.** $\left\{\dfrac{30}{7}\right\}$

55. $\{500\}$ **57.** $\left\{-\dfrac{19}{7}\right\}$ **59.** $\{700\}$ **61.** $\{-5\}$

63. $\{1400\}$ **65.** $\mathbb{R}$ **67.** $\{5\}$ **69.** $\varnothing$ **71.** $\{25\}$

73. $\{120\}$ **75.** $\left\{\dfrac{37}{3}\right\}$

77. $\dfrac{x}{5} = \dfrac{1}{2} - \dfrac{x}{3}$, LCD = 30   **79.** $3x - 5 = 3(x - 5)$

$$\dfrac{30x}{5} = \dfrac{30}{2} - \dfrac{30x}{3}$$ $3x - 5 = 3x - 15$
$$6x = 15 - 10x$$ $3x - 5 - 3x = 3x - 15 - 3x$
$$16x = 15$$ $-5 \neq -15$
$$x = \dfrac{15}{16}$$ Therefore, the solution is $\varnothing$.

81. $\mathbb{R}$ **83.** $\{-21\}$ **85.** $\{22\}$

87. Answers vary.   **89.** Answers vary.

Piece It Together   Sections 2.1–2.4

1. equation   **2.** expression

3. not a linear equation because the highest power is 2

4. linear equation   **5.** $\{10\}$ **6.** $\{-7.2\}$ **7.** $\{-6\}$

8. $\{75\}$ **9.** $\{27\}$ **10.** $\{7\}$ **11.** $\{-0.6\}$ **12.** $\{19\}$

13. $\{-6\}$ **14.** $\{-26\}$ **15.** $\{-36\}$ **16.** $\left\{\dfrac{4}{3}\right\}$ **17.** $\{10\}$

18. $\mathbb{R}$ **19.** $-\dfrac{15}{2}$ **20.** 15, 17, 19

Section 2.5

1. Answers vary.   **3.** Answers vary.   **5.** Answers vary.

7. Answers vary.

9. False, we can't combine $16 - 8x$. We should get $y = 8x - 16$.

11. True   **13.** $8830.50   **15.** $12,484.80   **17.** 300 mi

19. 55 mph   **21.** 2 hr   **23.** 100 in.³   **25.** 2 yd   **27.** 25 in.²

29. 8 in.   **31.** 28 ft   **33.** 5 m   **35.** 24 ft

37. 50.24 cm; 200.96 cm²   **39.** 25 ft; 1962.5 ft²   **41.** $l = \dfrac{A}{w}$

43. $c = P - a - b$ **45.** $x = -2y + 4$ **47.** $x = \dfrac{1}{4}y + 2$

49. $x = 4y - 20$ **51.** $x = \dfrac{3}{7}y - 3$ **53.** $r = \dfrac{A - P}{Pt}$

55. $30°$ **57.** $20°$ **59.** $20°$ **61.** $28°$ **63.** $35°$

65. $53°, 127°$ **67.** $62°, 62°$ **69.** $35°, 30°, 115°$

71. $45°, 53°, 82°$ **73.** $5408   **75.** 146 m   **77.** $3120

79. 56.52 in.³   **81.** 144 mi   **83.** $x = -\dfrac{2}{3}y + 2$

85. 25 m   **87.** $15,905.73   **89.** $x = \dfrac{R}{p}$ **91.** $x = 2.5y + 250$

93. $51°$ **95.** $w = \dfrac{C - 6l}{4}$ **97.** $52°$ and $128°$

99. $51°, 51°,$ and $78°$ **101.** 43 m   **103.** $y = -0.4x + 1600$

105. 19 cm; 1133.54 cm²   **107.** $29°$ and $29°$ **109.** $x = \dfrac{1}{5}y - 2$

111. $68°$ **113.** $107°, 50°,$ and $23°$ **115.** $72°, 63°,$ and $45°$

117. 8.01 in.; 59.03 in.; 201.46 in.²

119. $7a - 20 = 5a + 8$ **121.** $A = 2\pi r^2 + 2\pi rh$

$$2a - 20 = 8$$ $A - 2\pi r^2 = 2\pi rh$
$$2a = 28$$ $\dfrac{A - 2\pi r^2}{2\pi r} = \dfrac{2\pi rh}{2\pi r}$
$$a = 14$$
$$7a - 20 = 7(14) - 20 = 78$$ $h = \dfrac{A - 2\pi r^2}{2\pi r}$
$$5a + 8 = 5(14) + 8 = 78$$

The two angles are both $78°$.

123. $\{0.2\}$ **125.** $\{25\}$

Section 2.6

1. Answers vary.   **3.** Answers vary.

5. False, the amounts are x and $6000 - x$.   **7.** $874.99

9. $280.49   **11.** $80   **13.** $42,600   **15.** $30,000

17. $1750   **19.** $2500   **21.** 21.82%   **23.** 4.79%

25. $5000 in the high risk account and $15,000 in the less risky account

27. $10,000 in the money market account and $7000 in a CD

29. 30 nickels and 90 quarters   **31.** 20 $10 bills and 35 $20 bills

33. 5 gal of 50% alcohol solution

35. 250 gal of 3% milk fat solution and 500 gal of 4.5% milk fat solution

37. 3.5 hr

39. 5.25 hr to the meeting; 3.75 hr to home; 525 mi total distance

41. 2.75 hr   **43.** $2219.98   **45.** 7.8%

47. 44 $1 bills and 37 $5 bills   **49.** $450   **51.** $283.25

53. 5.3 hr

55. 320 gal of 2% milk fat solution and 180 gal of 4.5% milk fat solution

57. 2.5 hr

59. $4900 in the high risk account and $4200 in the less risky account

61. 288 dimes and 278 quarters   **63.** $37

65. 17 L of 8% hydrochloric and 51 L of 72% hydrochloric acid solution

67. 14.13% increase   **69.** $72   **71.** 1.2 hr

73. Let x be the price of the car without tax.

$$x + 0.08x = 52,629.48$$
$$1.08x = 52,629.48$$
$$x = \dfrac{52,629.48}{1.08}$$
$$x = 48,731$$

The price of the car without tax is $48,731.

75. Average percent $= \dfrac{34.06 - 30.17}{30.17}$

$$= \dfrac{3.89}{30.17}$$
$$= 0.128936$$
$$\approx 12.9\%$$

The average percent increase is about 12.9%.

77. Let x be the number of liters of pure acid and $210 - x$ be the number of liters of 40% solution.

$$1.00x + 0.40(210 - x) = 0.60(210)$$
$$0.60x + 84 = 126$$
$$0.60x = 42$$
$$x = 70$$
$$210 - x = 140$$

Hong needs 70 L of pure acid and 140 L of 40% solution.

Section 2.7

1. Answers vary.   **3.** Answers vary.   **5.** Answers vary.

7. False, $2 - 3 = -1$ which is not greater than -1.

9. True   **11.** False, reverse the inequality symbols.

13. $x > -2$

$-7 \; -6 \; -5 \; -4 \; -3 \; -2 \; -1 \;\; 0 \;\; 1 \;\; 2 \;\; 3$

$(-2, \infty)$ $\{x | x > -2\}$

15. $a < 1$

$-4 \; -3 \; -2 \; -1 \;\; 0 \;\; 1 \;\; 2 \;\; 3 \;\; 4 \;\; 5 \;\; 6$

$(-\infty, 1)$ $\{a | a < 1\}$

17. $x < -6$

$-11 \,-10 \,-9 \; -8 \; -7 \; -6 \; -5 \; -4 \; -3 \; -2 \; -1$

$(-\infty, -6)$ $\{x | x < -6\}$

19. $x > 7$

$2 \;\; 3 \;\; 4 \;\; 5 \;\; 6 \;\; 7 \;\; 8 \;\; 9 \;\; 10 \;\; 11 \;\; 12$

$(7, \infty)$ $\{x | x > 7\}$

21. $x \leq -3$

$-8 \; -7 \; -6 \; -5 \; -4 \; -3 \; -2 \; -1 \;\; 0 \;\; 1 \;\; 2$

$(-\infty, -3]$ $\{x | x \leq -3\}$

23. $x \leq \dfrac{1}{4}$

$-5 \; -4 \; -3 \; -2 \; -1 \;\; 1 \;\; 2 \;\; 3 \;\; 4 \;\; 5$
$\frac{1}{4}$

$\left(-\infty, \dfrac{1}{4}\right]$ $\left\{x \middle| x \leq \dfrac{1}{4}\right\}$

25. $x < 10$

$5 \;\; 6 \;\; 7 \;\; 8 \;\; 9 \;\; 10 \;\; 11 \;\; 12 \;\; 13 \;\; 14 \;\; 15$

$(-\infty, 10)$ $\{x | x < 10\}$

27. $x \leq -\dfrac{5}{4}$

$-5 \; -4 \; -3 \; -2 \;\; 0 \;\; 1 \;\; 2 \;\; 3 \;\; 4 \;\; 5$
$-\frac{5}{4}$

$\left(-\infty, -\dfrac{5}{4}\right]$ $\left\{x \middle| x \leq -\dfrac{5}{4}\right\}$

29. $-2 \leq x \leq 5$

$-4 \; -3 \; -2 \; -1 \;\; 0 \;\; 1 \;\; 2 \;\; 3 \;\; 4 \;\; 5 \;\; 6$

$[-2, 5]$ $\{x | -2 \leq x \leq 5\}$

31. $3 \leq x < 10$

$2 \;\; 3 \;\; 4 \;\; 5 \;\; 6 \;\; 7 \;\; 8 \;\; 9 \;\; 10 \;\; 11 \;\; 12$

$[3, 10)$ $\{x | 3 \leq x < 10\}$

33. $-7 \leq x < 3$

$-7 \; -6 \; -5 \; -4 \; -3 \; -2 \; -1 \;\; 0 \;\; 1 \;\; 2 \;\; 3$

$[-7, 3)$ $\{x | -7 \leq x < 3\}$

35. $-\dfrac{7}{3} \leq x \leq 8$

$-6 \; -4 \;\; 0 \;\; 2 \;\; 4 \;\; 6 \;\; 8 \;\; 10 \;\; 12 \;\; 14$
$-\frac{7}{3}$

$\left[-\dfrac{7}{3}, 8\right]$ $\left\{x \middle| -\dfrac{7}{3} \leq x \leq 8\right\}$

37. $x > -4$

$-9 \; -8 \; -7 \; -6 \; -5 \; -4 \; -3 \; -2 \; -1 \;\; 0 \;\; 1$

$(-4, \infty)$ $\{x | x > -4\}$

39. $x \geq 7$

$2 \;\; 3 \;\; 4 \;\; 5 \;\; 6 \;\; 7 \;\; 8 \;\; 9 \;\; 10 \;\; 11 \;\; 12$

$[7, \infty)$ $\{x | x \geq 7\}$

41. $y < 0$

$-5 \; -4 \; -3 \; -2 \; -1 \;\; 0 \;\; 1 \;\; 2 \;\; 3 \;\; 4 \;\; 5$

$(-\infty, 0)$ $\{y | y < 0\}$

43. $x \leq -4$

$-9 \; -8 \; -7 \; -6 \; -5 \; -4 \; -3 \; -2 \; -1 \;\; 0 \;\; 1$

$(-\infty, -4]$ $\{x | x \leq -4\}$

45. $y > -1$

$-6 \; -5 \; -4 \; -3 \; -2 \; -1 \;\; 0 \;\; 1 \;\; 2 \;\; 3 \;\; 4$

$(-1, \infty)$ $\{y | y > -1\}$

47. $x > -3$

$-8 \; -7 \; -6 \; -5 \; -4 \; -3 \; -2 \; -1 \;\; 0 \;\; 1 \;\; 2$

$(-3, \infty)$ $\{x | x > -3\}$

49. $y > 4$

$-1 \;\; 0 \;\; 1 \;\; 2 \;\; 3 \;\; 4 \;\; 5 \;\; 6 \;\; 7 \;\; 8 \;\; 9$

$(4, \infty)$ $\{y | y > 4\}$

51. $a < -8$

$-13 \,-12 \,-11 \,-10 \,-9 \; -8 \; -7 \; -6 \; -5 \; -4 \; -3$

$(-\infty, -8)$ $\{a | a < -8\}$

53. $a \leq 2$

$-3 \; -2 \; -1 \;\; 0 \;\; 1 \;\; 2 \;\; 3 \;\; 4 \;\; 5 \;\; 6 \;\; 7$

$(-\infty, 2]$ $\{a | a \leq 2\}$

55. $x < -27$

$-32 \,-31 \,-30 \,-29 \,-28 \,-27 \,-26 \,-25 \,-24 \,-23 \,-22$

$(-\infty, -27)$ $\{x | x < -27\}$

57. $y < -2$

$-7 \; -6 \; -5 \; -4 \; -3 \; -2 \; -1 \;\; 0 \;\; 1 \;\; 2 \;\; 3$

$(-\infty, -2)$ $\{y | y < -2\}$

59. $a \leq 2$

$-3 \; -2 \; -1 \;\; 0 \;\; 1 \;\; 2 \;\; 3 \;\; 4 \;\; 5 \;\; 6 \;\; 7$

$(-\infty, 2]$ $\{a | a \leq 2\}$

61. $x \leq \dfrac{13}{4}$

$-2 \; -1 \;\; 0 \;\; 1 \;\; 2 \;\; 3 \;\; 4 \;\; 5 \;\; 6 \;\; 7 \;\; 8$
$\frac{13}{4}$

$\left(-\infty, \dfrac{13}{4}\right]$ $\left\{x \middle| x \leq \dfrac{13}{4}\right\}$

63. $x < 18$

$13 \;\; 14 \;\; 15 \;\; 16 \;\; 17 \;\; 18 \;\; 19 \;\; 20 \;\; 21 \;\; 22 \;\; 23$

$(-\infty, 18)$ $\{x | x < 18\}$

65. $y \leq \dfrac{19}{3}$

$1 \;\; 2 \;\; 3 \;\; 4 \;\; 5 \;\; 6 \;\; 7 \;\; 8 \;\; 9 \;\; 10 \;\; 12$
$\frac{19}{3}$

$\left(-\infty, \dfrac{19}{3}\right]$ $\left\{y \middle| y \leq \dfrac{19}{3}\right\}$

67. $x < 100$

$95 \;\; 96 \;\; 97 \;\; 98 \;\; 99 \;\; 100 \; 101 \; 102 \; 103 \; 104 \; 105$

$(-\infty, 100)$ $\{x | x < 100\}$

69. $-13 < a < 8$

$-20 \,-16 \,-12 \; -8 \; -4 \;\; 0 \;\; 4 \;\; 8 \;\; 12 \;\; 16 \;\; 20$

$(-13, 8)$ $\{a | -13 < a < 8\}$

71. $-2 \le x \le 5$

$[-2, 5]$ $\{x | -2 \le x \le 5\}$

73. $-7 \le x \le -3$

$[-7, -3]$ $\{x | -7 \le x \le -3\}$

75. $-3 \le a \le -1$

$[-3, -1]$ $\{a | -3 \le a \le -1\}$

77. $-\dfrac{7}{6} < a < -\dfrac{1}{2}$

$\left(-\dfrac{7}{6}, -\dfrac{1}{2}\right)$ $\left\{a \middle| -\dfrac{7}{6} < a < -\dfrac{1}{2}\right\}$

79. $-2 < x < 7$

$(-2, 7)$ $\{x | -2 < x < 7\}$

81. $x \ge 3$

$[3, \infty)$ $\{x | x \ge 3\}$

83. $x > -6$

$(-6, \infty)$ $\{x | x > -6\}$

85. $y > -\dfrac{1}{21}$

$\left(-\dfrac{1}{21}, \infty\right)$ $\left\{y \middle| y > -\dfrac{1}{21}\right\}$

87. $x < -19$

$(-\infty, -19)$ $\{x | x < -19\}$

89. at least 84 on the fifth test   **91.** 89.8 on the final exam
93. at most 2602 min

95. $x > 24$

$(24, \infty)$ $\{x | x > 24\}$

97. $x > 2$

$(2, \infty)$ $\{x | x > 2\}$

99. $a \le 27.5$

$(-\infty, 27.5]$ $\{a | a \le 27.5\}$

101. $y \le 64$

$(-\infty, 64]$ $\{y | y \le 64\}$

103. $x \ge -\dfrac{27}{5}$

$\left[-\dfrac{27}{5}, \infty\right)$ $\left\{x \middle| x \ge -\dfrac{27}{5}\right\}$

105. $-5 < a < 9$

$(-5, 9)$ $\{a | -5 < a < 9\}$

107. $-2 \le x \le 4$

$[-2, 4]$ $\{x | -2 \le x \le 4\}$

109. $0 < a < 7$

$(0, 7)$ $\{a | 0 < a < 7\}$

111. $-\infty < x < \infty$

$(-\infty, \infty)$ $\{x | -\infty < x < \infty\}$

113. $\varnothing$

$\varnothing$ $\varnothing$

115. $x \ge -8$

$[-8, \infty)$ $\{x | x \ge -8\}$

117. $4 \le x \le 6$

$[4, 6]$ $\{x | 4 \le x \le 6\}$

119. $-\dfrac{15}{2} < x < \dfrac{19}{2}$

$\left(-\dfrac{15}{2}, \dfrac{19}{2}\right)$ $\left\{x \middle| -\dfrac{15}{2} < x < \dfrac{19}{2}\right\}$

121. $-4 < x < 64$

$(-4, 64)$ $\{x | -4 < x < 64\}$

123. at least 81 on the fourth test   **125.** 79 on the final exam
127. 231 attendees   **129.** at most 2187 mi
131. $-2x + 5 - 5 < 11 - 5$

$$-2x < 6$$
$$\dfrac{-2x}{-2} > \dfrac{6}{-2}$$
$$x > -3$$

The solution is $(-3, \infty)$
133. $3(2x - 5) + 9 > 2(3x + 1)$

$$6x - 15 + 9 > 6x + 2$$
$$6x - 6 > 6x + 2$$
$$-6 > 2$$

The solution is $\varnothing$.
135. $x < -7$
137. $x < 6$

CHAPTER 2 SUMMARY

Section 2.1

1. equation   **2.** solution
3. is equal to; the result is; is the same as; equals; is

Section 2.2

4. $Ax + B = C$; $2x + 3 = 7$ (Answers vary.)
5. $x = c$; $c = x$ **6.** Equivalent   **7.** add; subtract
8. simplify; variable; constants

Section 2.3

9. multiply; divide   **10.** reciprocal
11. Consecutive; 1, 2, 3, . . .; 1, 3, 5, . . .; 2, 4, 6, . . .; $x + 1$; $x + 2$

Section 2.4

12. least common denominator   **13.** power of 10
14. conditional   **15.** contradiction; the empty set or $\varnothing$
16. identity; all real numbers or $\mathbb{R}$

Section 2.5

17. formula   **18.** substitute   **19.** distance   **20.** circumference
21. area   **22.** 90°; 180°   **23.** 90°; straight
24. $90 - x$; $180 - x$ **25.** opposite; equal   **26.** 180°

Section 2.6

27. per hundred; $\dfrac{30}{100}$; 0.30   **28.** Interest; simple interest formula
29. pure; percentage (or concentration)   **30.** value
31. rate; time; rt

Section 2.7

32. $ax + b < c$ **33.** solution   **34.** graph
35. bracket; parenthesis   **36.** interval notation
37. ∞; $-\infty$; parenthesis   **38.** set-builder
39. add; subtract   **40.** positive   **41.** negative; reverse
42. $\leq$; $\geq$

CHAPTER 2 REVIEW EXERCISES

Section 2.1

1. equation   **3.** expression   **5.**

$x = -4$	Is not a solution
$x = -1$	Is not a solution
$x = 0$	Is not a solution
$x = 4$	Is not a solution
$x = -\dfrac{4}{5}$	Is a solution

7. $4[x + (-17)] = 2x$ **9.** $4x - (-3) = x - 45$
11. $x + x - 17° = 180°$

Section 2.2

13. {4}   **15.** {−8}   **17.** {−9}   **19.** {5.6}
21. The number is $-\dfrac{9}{2}$.   **23.** The number is 5.
25. The number is 28.
27. The net worth of Christy Walton was $21.5 billion and the net worth of Philip Knight was $9.5 billion.

Section 2.3

29. {6}   **31.** {−16}   **33.** $\left\{\dfrac{7}{18}\right\}$ **35.** {−20}   **37.** {0.5}
39. −18   **41.** −36   **43.** 12
45. The integers are −49 and −48.

Section 2.4

47. {5}   **49.** {40}   **51.** {−4}   **53.** $\mathbb{R}$ **55.** $\varnothing$

Section 2.5

57. $1236.27   **59.** 155 mi   **61.** $7000   **63.** 1695.6 in.³
65. 14 m   **67.** 340 m²   **69.** $p = \dfrac{R}{x}$ **71.** $y = -\dfrac{7}{3}x + 7$
73. $x = -0.4y + 700$ **75.** 57.5º   **77.** 50º
79. 49° and 131°   **81.** 118°, 40°, and 22°

Section 2.6

83. $1800   **85.** $81   **87.** 4.5% increase
89. 80 half-dollars and 40 quarters
91. 9 L of 79% hydrochloric and 63 L of 27% hydrochloric acid solution
93. 7.6 hr   **95.** 38, 40, and 42

Section 2.7

97. $a \geq -14$	(graph: -19 to -9, bracket at -14)
	$[-14, \infty)$ $\{a \mid a \geq -14\}$
99. $a \leq 16$	(graph: 11 to 21, bracket at 16)
	$(-\infty, 16]$ $\{a \mid a \leq 16\}$
101. $-1 < a < 7$	(graph: -6 to 14, parentheses at -1 and 7)
	$(-1, 7)$ $\{a \mid -1 < a < 7\}$
103. $x < -6$	(graph: -11 to -1, parenthesis at -6)
	$(-\infty, -6)$ $\{x \mid x < -6\}$
105. $y \geq -\dfrac{1}{4}$	(graph: -5 to 5, bracket at $-\dfrac{1}{4}$)
	$\left[-\dfrac{1}{4}, \infty\right)$ $\left\{y \mid y \geq -\dfrac{1}{4}\right\}$
107. $-9.8 < y < -0.4$	(graph: -9 to 0, parentheses at -9.8 and -0.4)
	$(-9.8, -0.4)$ $\{y \mid -9.8 < y < -0.4\}$

109. at least 88 on the fourth test   **111.** 185 attendees

CHAPTER 2 TEST

1. b   **3.** c
5. a. $x + (x - 4007) = 112{,}499$
   b. There were approximately 58,253 deaths in the Vietnam War and approximately 54,246 deaths in the Korean War.
7. $\left\{-\dfrac{2}{9}\right\}$ **9.** {6}   **11.** {4}   **13.** {7000}   **15.** $\varnothing$
17. 15°C   **19.** d   **21.** $(-\infty, 13]$
23. a. I could travel at most 29.9 mi.   **b.** No
25. Joann needs a 60.4 or higher on the final to have at least an 80 in the class.
27. Measures of the angles are 25°, 25°, and 130°.
29. David invested $800 in the 4% savings account and $2200 in the CD.
31. Dr. Cleves needs to mix 23.3 mL of water with the 10 mL of 50% glycerol solution to obtain a 15% glycerol solution.
33. They jog for approximately 0.62 hr or about 37 minutes.

CHAPTERS 1 AND 2 CUMULATIVE REVIEW EXERCISES

1. a. rational number, real number
 b. irrational number, real number, 3.61
 c. integer, rational number, real number; $-\dfrac{47}{1}$

3. a.   **3. b.**

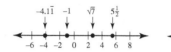

5. a. $-1.3\overline{3}$ **b.** -7.5 **c.** $1\dfrac{2}{3}$

7. a. $2^2 \cdot 3^2 \cdot 5$ **b.** $2 \cdot 7^2$ **c.** $2^2 \cdot 5^3$

9. a. $\dfrac{2}{5}$ **b.** $\dfrac{125}{258}$

11. a. 12.25   **b.** $\dfrac{8}{343}$ **c.** $\dfrac{9}{4}$ **d.** 64   **e.** 49   **f.** $\dfrac{8}{3}$

13. a. $x = 9$ is a solution.   **b.** $y = 4$ is not a solution.

15. 150 ft   **17.** \$1436.75   **19.** The difference is 7815.2 ft.

21. Dave's checking account balance is \$164.79.   **23.** 12°; 102°

25. a. 5   **b.** $-\dfrac{1}{4}, -\dfrac{1}{3}, -\dfrac{3}{2}, -5$, undefined

27. a. $c - 3$ **b.** $-2x + 20y + 14$ **c.** $b + \dfrac{17}{10}$
 d. $-4x + 15$

29. a. $-70x$ **b.** $-6x - 15$
 c. $-11x - 2$ **d.** $0.08x + 1.7$ **e.** $2a + 2$ **f.** $14x^2 - 5x$

31. a. $x = -2$ is a solution; $x = 0$ is not a solution; $x = 2$ is not a solution
 b. $x = -2$ is not a solution; $x = 0$ is not a solution; $x = 2$ is a solution

33. a. $\{-7\}$ **b.** $\{1.5\}$ **c.** $\{1.8\}$ **d.** $\mathbb{R}$ **e.** $\{-3\}$
 f. $\left\{-\dfrac{3}{5}\right\}$ **g.** $\{2.8\}$ **h.** $\mathbb{R}$ **35.** 550 mi

37. 30, 32, and 34.   **39. a.** \$4945.08   **b.** \$5000.00

41. a. \$6750   **b.** \$24.99

43. a. $l = \dfrac{A}{w}$ **b.** $w = \dfrac{C - 5l}{6}$ **c.** $y = \dfrac{3x - 24}{8} = \dfrac{3}{8}x - 3$
 d. $x = -\dfrac{5}{2}y + 225$

45. a. 47° and 133°   **b.** 96°   **c.** 21°, 46°, and 113°

47. \$1500   **49.** \$52

51. a. $a \le 14; (-\infty, 14];$
 $\{a \mid a \le 14\}$

 b. $-3 < a < 5; (-3, 5);$
 $\{a \mid -3 < a < 5\}$

 c. $x < -10; (-\infty, -10)$
 $\{x \mid x < -10\}$

 d. $-4.2 \le y \le 5.2; [-4.2, 5.2]$
 $\{y \mid -4.2 \le y \le 5.2\}$

53. at most 1200 mi

CHAPTER 3

Section 3.1

1. Answers vary.   **3.** Answers vary.
5. False, the ordered pair does not satisfy the equation.
7. False, the y-value is zero.
9. False, the point is in Quadrant IV.
11. yes   **13.** no   **15.** no   **17.** yes   **19.** yes   **21.** yes
23. yes   **25.** no   **27.** yes   **29.** I   **31.** II   **33.** IV
35. IV   **37.** III   **39.** y-axis   **41.** x-axis   **43.** IV

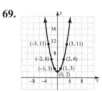

45. $(4, -5)$, IV   **47.** $(1, 3)$, I   **49.** $(-7, 0)$, x-axis
51. $(-3, -5)$, III   **53.** $(-4, -1)$, III   **55.** IV   **57.** I
59. y-axis   **61.** I   **63.** x-axis

65.   **67.**   **69.**

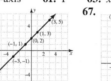

71.   **73.**

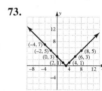

75. no   **77.** yes   **79.** yes   **81.** yes   **83.** no   **85.** yes
87. $(0, 4)$ **89.** $(2, 0)$ **91.** $(3, -2)$
93. a. $(0, 4.7), (1, 5.8), (2, 6.0), (3, 5.5), (4, 5.1), (5, 4.6), (6, 4.6),$
 $(7, 5.8), (8, 9.3), (9, 9.6)$
 b. The point $(0, 4.7)$ means that in 0 yr after 2001, or in 2001,
 the unemployment rate was 4.7%. The point $(9, 9.6)$ means
 that 9 yr after 2001, or in 2010, the unemployment rate
 was 9.6%.
 c. The unemployment rate was the highest in 2010 and the lowest
 in 2006 and 2007.
 d.

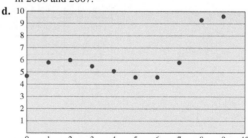

95. a. $(0, 20.71), (1, 21.38), (2, 22.31), (3, 23.19), (4, 23.96),$
 $(5, 25.39), (6, 26.77), (7, 27.35)$
 b. The national average hourly salary for nurses in 1998
 was \$20.71.
 c. The national average hourly salary for nurses in 2002 was
 \$23.96.
 d. The average hourly salary has increased \$6.64 from 1998
 to 2005.
 e. The hourly salaries of nurses increased from 1998 to 2005.
97. yes   **99.** yes   **101.** no   **103.** yes   **105.** yes

107. no **109.** III **111.** *x*-axis **113.** II **115.** I

117. (5, 4), I **119.** (3, 0), *x*-axis **121.** (−3, 3), II
123. (3, −6), IV
125. 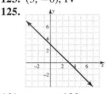 **127.** **129.**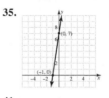

131. yes **133.** no **135.** (0, −3) and (4, −3) **137.** (2, 1)
139. (1, 0)
141. a. (0, 31.8), (1, 33.3), (2, 36.7), (3, 31.4), (4, 28.3), (5, 31.2),
(6, 33.3), (7, 34.4), (8, 35.9), (9, 36.7)
   b. The expenditures were the greatest 2 yr after 1998, or in 2000
   and in 9 yr after 1998, or 2007.
   c. The expenditures decreased by $5.3 billion between 2000 and
   2001 and $3.3 billion between 2001 and 2002.
   d. The expenditures for U.S. airline transportation increased from
   1998 to 2000, decreased from 2000 to 2002, and increased
   from 2002 to 2007.
143. Always begin in the upper right corner and then rotate
counterclockwise.
145. Valerie did not apply the negative sign after she squared the number.

x	$y = -x^2 + 1$	(x, y)
−2	$-(-2)^2 + 1 = -4 + 1 = -3$	(−2, −3)
−1	$-1(-1)^2 + 1 = -1 + 1 = 0$	(−1, 0)
0	$-(0)^2 + 1 = 0 + 1 = 1$	(0, 1)
1	$-(1)^2 + 1 = -1 + 1 = 0$	(1, 0)
2	$-(2)^2 + 1 = -4 + 1 = -3$	(2, −1)

147. 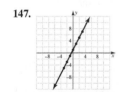 **149.**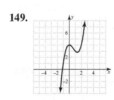

Section 3.2
   1. Answers vary. **3.** Answers vary.
   5. False, the graph goes through (2, 0) and (0, −14).
   7. False, the graph has no *y*-intercept. **9.** True **11.** yes
13. no **15.** yes **17.** yes **19.** yes
21.  **23.** 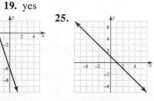 **25.**

27. **29.** **31.**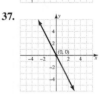

33. 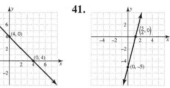 **35.** 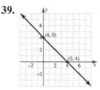 **37.**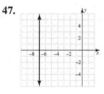

39. **41.** 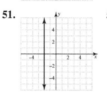 **43.**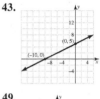

45. **47.** 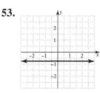 **49.**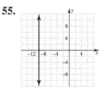

51. 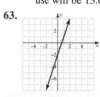 **53.** 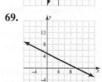 **55.**

57.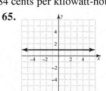

59. a. *y*-intercept is (0, 49,217). The median annual income for men
   with a master's degree is $49,217 in 1991.
   b. The median annual income for men in 2001 is $67,947.
   c. The median annual income for men in 2012 is $88,550.
61. a. *y*-intercept is (0, 8.154). The average retail price of electricity
   for residential use was 8.154 cents per kilowatt-hour in 2001.
   b. The average retail price of electricity for residential use was
   11.709 cents per kilowatt-hour in 2010.
   c. The expected average retail price of electricity for residential
   use will be 13.684 cents per kilowatt-hour in 2015.

63. 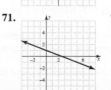 **65.** 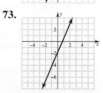 **67.**

69. **71.** **73.**

75. *y*-axis

77.

79.

81.

83.

85.

87.

89.

91. a. The *y*-intercept is (0, 38). It means 38% of high school students were users of cigarettes in 1997.
   b. 30.5% of high school students used cigarettes in 2000.
   c. 5.5% of high school students used in cigarettes in 2010.
   d.

93. a. The *x*-intercept is (77.7, 0). It means about 77.7 yr after 1998, or in yr 2076, 0% of men aged 18 yr and older will be cigarette smokers. No, it is not realistic because there will most likely always be smokers.
   b. The *y*-intercept is (0, 25.64). It means 25.64% of the men aged 18 yr and older were cigarette smokers in 1998.
   c. 23.33% in 2005 and is very close to the actual estimate.
   d. 21.02% in 2012
   e.

   f. About 41 yr after 1998 or in year 2039.

95. a. The *y*-intercept is (0, 19.83). It means 19.83% of the U.S. adults aged 20 yr and over were obese in 1997.
   b. 27.64% in 2008. It is very close to the actual estimate.
   c. 30.48% in 2012
   d.

   e. The percent of U.S. adults aged 20 yr and over who are obese is increasing.

97. Two points because this allows us to determine the direction of the line

99. Write the equation as $x + 0y = 7$, replace *y* with 0, and solve for *x* to get (7, 0).

101. Choose any nonzero number for *x* and substitute into the equation to solve for *y*.

103. Answers vary.   **105.** (2, 0) and (0, 6)

107. a. Answers vary.   **b.** (10, 0)   **c.** (0, 20) and (10, 0)
   d. Answers vary.   **e.** Answers vary.

109. The *y*-values increase by 2 units for each unit increase in *x*. The coefficient of *x* is also 2.

x	*y*
−2	0
−1	2
0	4
1	6
2	8

Section 3.3

1. Answers vary.   **3.** Answers vary.

5. False, the slope is the coefficient of *x* only if the equation is in slope-intercept form.

7. False, $x = 5$ is a vertical line and its slope is undefined.

9. False, the slope is the change in *y* over the change in *x*, so the slope is $-\dfrac{6}{1}$ or -6.

11. $m = \dfrac{1}{3}$ **13.** $m = 1$ **15.** $m = $ undefined   **17.** $m = 0$

19. $m = -3$ **21.** $m = \dfrac{2}{3}$ **23.** $m = -2$ **25.** $m = 0$

27. $m = \dfrac{1}{3}$

29. $m = 4$; (0, −3) The *x*-values increase by 1 unit, the *y*-values increase by 4 units or the *x*-values decrease by 1 unit, the *y*-values decrease by 4 units.

31. $m = -2$; (0, −1) The *x*-values increase by 1 unit, the *y*-values decrease by 2 units or the *x*-values decrease by 1 unit, the *y*-values increase by 2 units.

33. $m = \dfrac{1}{2}$; $\left(0, \dfrac{3}{2}\right)$ The *x*-values increase by 2 units, the *y*-values increase by 1 unit or the *x*-values decrease by 2 units, the *y*-values decrease by 1 unit.

35. $m = 5$; (0, 0) The *x*-values increase by 1 unit, the *y*-values increase by 5 units or the *x*-values decrease by 1 unit, the *y*-values decrease by 5 units.

37. $y = -4x + 8$; $m = -4$; (0, 8) The *x*-values increase by 1 unit, the *y*-values decrease by 4 units or the *x*-values decrease by 1 unit, the *y*-values increase by 4 units.

39. $y = -\dfrac{7}{4}x$; $m = -\dfrac{7}{4}$; (0, 0) The *x*-values increase by 4 units, the *y*-values decrease by 7 units or the *x*-values decrease by 4 units, the *y*-values increase by 7 units.

41. vertical line, cannot be written in slope-intercept form, slope is undefined

43. $y = 0x - 3$, $m = 0$, no change in the *y*-value as the *x*-value increases by 1 unit

45. $y = \dfrac{4}{3}$, $m = 0$, no change in the *y*-value as the *x*-value increases by 1 unit

47. vertical line, cannot be written in slope-intercept form, slope is undefined

49.

51.

53.

55.

57.

59.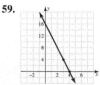

61.   **63.**   **65.**

67.

69. $m = \dfrac{1}{2}$ **71.** $m = \dfrac{8}{3}$ **73.** $m = 0$ **75.** $m = -\dfrac{2}{3}$

77. $m = -\dfrac{3}{2}$ **79.** $m = \dfrac{3}{5}$ **81.** $m = $ undefined

83. $m = -0.4$ **85.** $m = -1.4$ **87.** $m = \dfrac{2}{7}$

89. $m = -\dfrac{5}{3}$

91.   **93.**   **95.**

97. 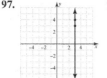   **99.**

101. correct   **103.** correct   **105.** $m = -\dfrac{1}{2}$

107. Answers vary.   **109.** Answers vary.

111. Answers vary.   **113.** Answers vary.

   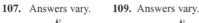

Piece It Together Sections 3.1–3.3

1. yes   **2.** no   **3.** III   **4.** IV   **5.** III   **6.** I
7. no   **8.** yes
9.   **10.**   **11.**

12.   **13.**   **14.**

15. $m = \dfrac{2}{3}$ **16.** $m = 0$

17. $y = \dfrac{3}{2}x - 3;\ m = \dfrac{3}{2};\ (0, -3)$; As the x-values increase by 2 units, the y-values increase by 3 units.

18. $y = -\dfrac{5}{3}x - 5;\ m = -\dfrac{5}{3};\ (0, -5)$; As the x-values increase by 3 units, the y-values decrease by 5 units.

19.   **20.**

Section 3.4

1. The y-intercept denotes the expenditure per student in fall enrollment for public elementary and secondary education was $7672.73 in 2001. The slope means that expenditures are increasing at a rate of $374.15 per year.
   The ordered pairs: $(4, 9169.33)$, $(9, 11{,}040.08)$, $(11, 11{,}788.38)$

3. The y-intercept denotes that 38% of high school students smoked in 1997. The slope means that the percent of students who are smokers is decreasing at a rate of 2.5% annually.
   The ordered pairs: $(8, 18)$, $(13, 5.5)$, $(15, 0.5)$

5. The initial value of the limo was $110,000 and depreciates at a rate of $22,000 annually.
   The ordered pairs: $(2, 66{,}000)$, $(4, 22{,}000)$, $(5, 0)$

7. The percent of eighth-graders who reported using alcohol in 1993 was 27% and decreases at a rate of 0.7% annually.
   The ordered pairs: $(12, 18.6)$, $(17, 15.1)$, $(19, 13.7)$

9. $\dfrac{887}{15840};\ 5.6\%$ **11.** 1

13. $\dfrac{1}{8}$; Yes, meets the guidelines.   **15.** parallel   **17.** neither
19. perpendicular   **21.** neither   **23.** parallel   **25.** neither
27. perpendicular   **29.** neither   **31.** perpendicular
33. perpendicular

35.   **37.**   **39.**

41.   **43.**   **45.**

47.   **49.**   **51.**

53.   **55.** $\dfrac{12}{5}$ **57.** perpendicular

59.   **61.**

63. parallel
65. The y-intercept is (0, 0), which means the initial cost is $0. The total cost increases at a rate of $149 per week.
67. The y-intercept is (0, 12,034) which means the cost of higher education at 4-yr institutions was $12,034 in 1999. The cost increases at a rate of $892 per year.
69. The y-intercept is (0, 21,506) which means the median annual income for women who completed high school was $21,506 in 1996. The income increases at a rate of $641 per year.
71. We must write the equation in slope-intercept form first.

$$6x - 2y = 4$$
$$-2y = -6x + 4$$
$$y = \frac{-6}{-2}x + \frac{4}{-2}$$
$$y = 3x - 2$$
$$\text{slope} = 3$$

From the point (1, 2), go up 3 units and right 1 unit.

73. $y = -\frac{1}{2}$ is a horizontal line and $m = 0$. $y = 2x$ is not a horizontal line and $m = 2$. The lines are neither parallel nor perpendicular.

75. neither   **77.** perpendicular   **79.** perpendicular
81. Answers vary; A parallel line is $y = 5x + 1$ and a perpendicular line is $y = -\frac{1}{5}x + 7$.
83. Answers vary; A parallel line is $y = -\frac{4}{3}x - 1$ and a perpendicular line is $y = \frac{3}{4}x + 5$.

Section 3.5

1. Answers vary.   **3.** Answers vary.   **5.** Answers vary.
7. False, (2, 6) is not the y-intercept of the line, so we must find b.

The equation is $y = -\frac{1}{3}x + \frac{20}{3}$.

9. $y = 4x + 5$ **11.** $y = 3x - 4$ **13.** $y = -\frac{2}{3}x - 10$
15. $y = 7$ **17.** $y = -\frac{1}{9}x - \frac{4}{9}$ **19.** $y = 4x + 1$
21. $y = -5x - 11$ **23.** $y = -x - 4$ **25.** $y = -\frac{5}{2}x - 5$
27. $y = 8x + 15$ **29.** $y = 6x$ **31.** $y = -8x$ **33.** $y = -5$
35. $y = \frac{2}{3}x - 12$ **37.** $y = -\frac{3}{5}x + \frac{1}{5}$ **39.** $x = 9$
41. $y = x + 5$ **43.** $y = 2x$ **45.** $y = 2x - 4$
47. $y = \frac{1}{2}x - 6$ **49.** $y = -\frac{1}{3}x + 5$
51. $y = 5$ **53.** $x = 2$ **55.** $y = -3x - 4$ **57.** $y = -2x - 6$
59. $y = -\frac{4}{3}x - 1$ **61.** $y = 4$ **63.** $x = -1$
65. $y = \frac{1}{3}x - 4$ **67.** $y = \frac{1}{2}x + 11$ **69.** $y = \frac{3}{4}x + 7$
71. $x = -1$ **73.** $y = 8$ **75.** $y = -\frac{1}{2}x + 3$; $x + 2y = 6$
77. $y = -\frac{1}{7}x - 4.2$; $x + 7y = -29.4$ **79.** $x = -10$
81. $y = -6x + 19$; $6x + y = 19$ **83.** $x = 5$
85. $y = 5x - 5$; $5x - y = 5$ **87.** $y = 2.5x + 17$; $5x - 2y = -34$
89. $y = \frac{5}{4}x + 5$; $-5x + 4y = 20$ **91.** $y = \frac{3}{5}x + 5$; $3x - 5y = -25$
93. $y = 4$ **95. a.** $y = 2000x + 45,000$ **b.** $55,000   **c.** 13 yr
97. a. $y = 150x + 400$ **b.** $2200   **c.** 6 credits
99. a. $y = -1500x + 20,000$ **b.** $11,000   **c.** 10 yr

101. a. $y = 1500x + 10,700$ **b.** 18,200 homes
103. a. $y = 0.72x + 4.8$ **b.** approximately $22.8 billion
105. $y = \frac{4}{3}x - 2$. So, the slope is $\frac{4}{3}$. Since it passes through the point (8, 3), the equation of the line is $y = \frac{4}{3}x - \frac{23}{3}$.
107. $y - (-1) = -2(x - 5)$ **109.** b
$$y + 1 = -2x + 10$$
$$y = -2x + 9$$

Section 3.6

1. Answers vary.   **3.** Answers vary.   **5.** Answers vary.
7. True
9. False, $f(8) = 0$ corresponds to (8, 0) which is the x-intercept.
11. False, the vertical line can only touch one point on the graph for it to be a function.
13. Relation = {(Canada, 4400), (France, 4800), (Italy, 3300), (Japan, 3300), (Republic of Korea, 1100), (United Kingdom, 3300), (United States, 7400)}; Domain = {Canada, France, Italy, Japan, Republic of Korea, United Kingdom, United States}; Range = {1100, 3300, 4400, 4800, 7400}
15. Relation = {(Walter Hagen, 11), (Ben Hogan, 9), (Nick Faldo, 6), (Jack Nicklaus, 18), (Arnold Palmer, 7), (Gary Player, 9), (Gene Sarazen, 7), (Sam Snead, 7), (Lee Trevino, 6), (Harry Vardon, 7), (Tom Watson, 8), (Tiger Woods, 14)}; Domain = {Walter Hagen, Ben Hogan, Nick Faldo, Jack Nicklaus, Arnold Palmer, Gary Player, Gene Sarazen, Sam Snead, Lee Trevino, Harry Vardon, Tom Watson, Tiger Woods}; Range = {6, 7, 8, 9, 11, 14, 18}
17. The cost $y = 14.52x$, where x is the number of kilowatt-hours in hundreds; Domain = {0, 1, 2, 3, ...}; Range = {0, 14.52, 29.04, 43.56, ...}
19. not a function, The x-value of 2 corresponds to more than one y-value.   **21.** a function
23. not a function; The x-value of –2 corresponds to more than one y-value.   **25.** a function   **27.** a function   **29.** a function
31. not a function, because $x = 1$ corresponds to $y = 3$ and $y = 1$
33. a function   **35.** not a function   **37.** not a function
39. 25   **41.** 49   **43.** 14   **45.** −36   **47.** −3   **49.** −9.6
51. 0   **53.** −8   **55.** −9   **57.** 2
59. $f(x) = 5x - 6$;
$f(0) = -6, f(3) = 9, f(-2) = -16$;
$(0, -6), (3, 9), (-2, -16)$
61. $f(x) = 4x - 9$;
$f(0) = -9, f(3) = 3, f(-2) = -17$;
$(0, -9), (3, 3), (-2, -17)$
63. $f(x) = \frac{7}{2}$;
$f(0) = \frac{7}{2}, f(3) = \frac{7}{2}, f(-2) = \frac{7}{2}$;
$\left(0, \frac{7}{2}\right), \left(3, \frac{7}{2}\right), \left(-2, \frac{7}{2}\right)$
65. a. $1005.16, about $100.84 off   **b.** $2013.16
67. 224; The volume of the open-top box is 224 in.^{3}.
69. Relation = {(Canada, 4400), (France, 4800), (Italy, 3300), (Japan, 3300), (Republic of Korea, 1100), (United Kingdom, 3300), (United States, 7400)}; Domain = {Canada, France, Italy, Japan, Republic of Korea, United Kingdom, United States}; Range = {1100, 3300, 4400, 4800, 7400}
71. a function   **73.** not a function   **75.** not a function
77. a function   **79.** a function   **81.** a function
83. not a function   **85.** not a function   **87.** a function
89. −33   **91. a.** −25   **b.** −1   **93. a.** 6   **b.** −6   **95.** 3.1
97. $f(x) = \frac{9}{2}x - 5$;
$f(0) = -5, f(3) = \frac{17}{2}, f(-2) = -14$;
$(0, -5), \left(3, \frac{17}{2}\right), (-2, -14)$

99. a. $y = 1600x + 52,000$
   b. \$60,000; If Brasil has worked with the company for 5 yr, his salary would be \$60,000.
101. a. $f(x) = 135x + 565$ **b.** \$2185
   c. \$2590; It means Lee will take 15 credits.
103. Correct! **105.** $f(0) = 5$ and $f(3) = 0$
107. $f(0) = -1.75; x = 7$ **109.** $f(0) = -2; x = -1, 2$

CHAPTER 3 SUMMARY

Section 3.1

1. ordered pair; true **2.** left; right; up; down; x-axis
3. left; right; up; down; x-axis
4. Quadrants; Roman; counterclockwise
5. ordered pairs; solutions; infinitely **6.** solution
7. graphs; tables; ordered pairs

Section 3.2

8. $Ax + By = C$ **9.** two; three **10.** x-intercept; y; y-intercept; x
11. origin; substitute a nonzero value for x and solve for y to get another point
12. $y = k$, where k is a real number
13. $x = h$, where h is a real number

Section 3.3

14. steepness **15.** vertical; horizontal; rise; run
16. $y = mx + b$; slope; y-coordinate of the y-intercept
17. y-intercept; slope; move up 4 units; move right 3 units
18. zero; undefined **19.** $m = \dfrac{y_2 - y_1}{x_2 - x_1}$
20. rising; falling; horizontal; vertical

Section 3.4

21. initial; rate of change **22.** parallel; slope
23. perpendicular; negative

Section 3.5

24. slope; point; slope-intercept; point-slope

Section 3.6

25. relation; domain; range **26.** function; exactly one
27. vertical; one point **28.** $f(x)$ **29.** y **30.** (a, b)

CHAPTER 3 REVIEW EXERCISES

Section 3.1

1. yes **3.** no **5.** no **7.** II **9.** III **11.** y-axis

13. $(2, 5)$, I **15.** $(0, 2.5)$, y-axis **17.** $(0, -3)$, y-axis
19. x-axis **21.** III
23. **25.**

27. yes **29.** no **31.** $(-2, 0)$ and $(4, 0)$ **33.** $(1, 9)$

35. a. $(0, 1016.5), (5, 1333.6), (8, 1732.4), (9, 1852.3), (10, 1973.3),$ $(11, 2105.5)$
   b. The expenditures increased the least between 1995 and 2000, by \$337.1 billon in 5 yr.
   c. The expenditures for national health are increasing.

Section 3.2

37. **39.** **41.**

43. a. The y-intercept is $(0, 213)$. The emission of carbon monoxide was about 213 metric tons in 1970.
   b. The x-intercept is $(63, 0)$. In 2033, the emission of carbon monoxide is 0 metric tons. No, it is not realistic.
   c. The emission of carbon monoxide was about 95.4 million tons in 2005.
   d. The emission of carbon monoxide was about 82 million tons in 2009.
   e.

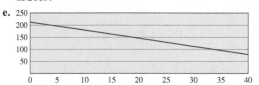

Section 3.3

45. $m = 0$ **47.** $m = \dfrac{5}{6}$ **49.** $m = -3$

51. **53.** **55.**

Section 3.4

57. The y-intercept is $(0, 0)$, which means the initial cost is \$0. The total cost increases at a rate of \$135 per week.

59. $\dfrac{5}{3}$ **61.**

Section 3.5

63. $y = -\dfrac{1}{4}x - 1$ **65.** $x = -1$ **67.** $x = 2$ **69.** $y = \dfrac{3}{2}x - 5$
71. a. $y = 2160x + 43,500$ **b.** \$52,140 **c.** 12 yr

Section 3.6

73. a function
75. not a function since $x = 2$ corresponds to two different y-values.
77. not a function since $(1, 0)$ and $(1, 2)$, to name two points, both satisfy the equation. So, $x = 1$ corresponds to more than one y-value.
79. not a function **81.** not a function **83.** -18
85. a. 1 **b.** -1
87. $f(x) = \dfrac{2}{7}x - 3$;
   $f(0) = -3, f(3) = -\dfrac{15}{7}, f(-2) = -\dfrac{25}{7}$;
   $(0, -3), \left(3, -\dfrac{15}{7}\right), \left(-2, -\dfrac{25}{7}\right)$

89. a. $f(x) = 1020x + 43,200$ **b.** $49,320

c. $52,380; It means Rikki taught an additional 9 credit hours.

CHAPTER 3 TEST

1. c **3.** **5.** yes

7. a. $(0, -6); (3, 0)$ **b.** $(0, 1); (-5, 0)$ **c.** $(0, 0)$

9. horizontal

11. a. $m = \dfrac{4}{5}$ **b.** $m = -3$ **c.** $m =$ undefined **d.** $m = 0$

e. $m = 3$ **f.** $m = 7$ **g.** $m = \dfrac{1}{3}$ **h.** $m = 1$

13. perpendicular

15. a. $y = \dfrac{3}{4}x - 6$ **b.** $y = -2x - 10$ **c.** $y = 3x - 13$

d. $y = 4x - 16$ **e.** $y = \dfrac{1}{3}x - 1$

17. d **19.** $f(x) = \dfrac{4}{7}x + 2; f(-14) = -6; (-14, -6)$

CHAPTERS 1–3 CUMULATIVE REVIEW EXERCISES

1. a. 6 **b.** $\dfrac{3}{4}$ **c.** -24

3. a. $\dfrac{27}{8}$ **b.** $-\dfrac{1}{3}$ **c.** 39 **d.** 30 **e.** $\dfrac{13 + \sqrt{559}}{12}$

5. 200 ft **7.** $144.35 **9.** $2 + 4w$

11. a. -19.44 **b.** 6 **c.** 125 **d.** undefined

e. 360 **f.** $-\dfrac{9}{4}$

13. a. $c + 38$ **b.** $-4x - 12y + 32$ **c.** $x + \dfrac{1}{6}$ **d.** $6x - 4$

15. a. equation **b.** expression

17. a. 16 **b.** -6 **c.** 38° and 52°

19. 365 mi **21.** 25, 26, 27 **23. a.** $1000.89 **b.** $2839.30

25. a. $w = \dfrac{V}{lh}$ **b.** $w = \dfrac{C - 8l}{3}$ **c.** $y = \dfrac{5}{9}x - 2$

d. $x = -\dfrac{3}{4}y + 180$

27. a. 61°, 119° **b.** 88°, 88° **29.** $1850 **31.** $52.20

33. a. $(2, -5)$ is not a solution **b.** $(3, -1)$ is not a solution

c. $(1, 5)$ is a solution

35. a. **b.**

c. **d.**

37. a. $(0, 43,378), (1, 44,655), (2, 45,686), (3, 46,542), (4, 47,516),$
$(5, 48,804), (6, 50,758), (7, 52,308), (8, 53,910)$

b. Between 2007 and 2008; $1954

39. a. 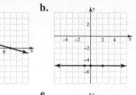 **b.** **c.**

d. **e.** **f.**

g.

41. a. $-\dfrac{2}{3}$ **b.** 2 **c.** 0 **d.** undefined **e.** -2

f. $\dfrac{5}{3}$ **g.** $\dfrac{5}{3}$

43. $m = 10.93, (0, 73,764);$ The number of associate degrees conferred in all higher education institutions is increasing at a rate of 10,930 per year. In 2008, the number of associate degrees conferred was approximately 73,764,000.

x	y (in thousands)
0	73,764
4	73,807.72
8	73,851.44

45.

47. a. $y = -x + 1; x + y = 1$ **b.** $x = 3$ **c.** $y = 12$

d. $y = -2x; 2x + y = 0$ **e.** $y = 3x - 14; 3x - y = 14$

f. $y = \dfrac{3}{4}x + 3; 3x - 4y = -12$ **g.** $y = -4x + 8; 4x + y = 8$

49. a. D = {Italy, France, Germany, Brazil, United Kingdom, Canada, Korea, Japan, U.S.}, R = {13, 25, 26, 28, 34, 35, 37, 42}, the relation is a function

b. D = {Honda Civic CX, Toyota Prius, Honda Civic Hybrid, Smart For Two Convertible/Coupe, Honda Insight, Ford Fusion Hybrid/Mercury Milan Hybrid, Toyota Yaris, Nissan Altima Hybrid, Mini Cooper, Chevrolet Cobalt XFE/Pontiac G5 XFE}, R = {45, 46, 47, 50, 51, 52, 57}, The relation is a function.

c. D = {$-5, -3, 0, 2$}, R = {1, 4, 9}, The relation is not a function.

d. D = {6}, R = $(-\infty, \infty)$, $x = 6$ is not a function

e. D = $(-\infty, \infty)$, R = $(-\infty, \infty)$, $y = 3x + 4$ is a function

51. a. $f(x) = -4x + 3;$
$f(0) = 3, f(3) = -9,$
$f(-2) = 11; (0, 3),$
$(3, -9), (-2, 11)$

b. $f(x) = 0.3x - 0.6;$
$f(0) = -0.6, f(3) = 0.3,$
$f(-2) = -1.2; (0, -0.6),$
$(3, 0.3), (-2, -1.2)$

c. $f(x) = |3x - 2|;$
$f(0) = 2, f(3) = 7,$
$f(-2) = 8; (0, 2), (3, 7),$
$(-2, 8)$

CHAPTER 4

Section 4.1

1. Answers vary.
3. Answers vary.
5. False, the slopes are different so the system has a solution.
7. False, the system has infinitely many solutions.
9. yes   11. no   13. yes   15. no   17. $\{(-1, 2)\}$
19. $\{(2, 1)\}$ 21. $\varnothing$ 23. $\{(3, -2)\}$
25. $\{(x, y)\,|\,2x - y = 0\}$

27.   29.   31.

33.   35.   37.

39.

41. $\varnothing$ 43. $\varnothing$ 45. $\{(x, y)\,|\,4x + 2y = 5\}$
47. parallel lines, no solution, inconsistent system
49. same line, infinitely many solutions, consistent system with dependent equations
51. parallel lines, no solution, inconsistent system
53. same line, infinitely many solutions, consistent system with dependent equations
55. **a.** (600, 30,000)
    b. The company will break even when 600 sneakers are sold. Both cost and revenue equal $30,000.

57. $\left\{\left(-\dfrac{1}{2}, -8\right)\right\}$, consistent system with independent equations

59. $\{(x, y)\,|\,x - y = 4\}$, consistent system with dependent equations

61. $\{(1, -2)\}$, consistent system with independent equations

63. $\{(-1, -3)\}$, consistent system with independent equations

65. $\{(5, -1)\}$, consistent system with independent equations

67. $\{(x, y)\,|\,2x - 3y = 1\}$, consistent system with dependent equations

69. $\left\{\left(-3, \dfrac{9}{2}\right)\right\}$, consistent system with independent equations

71. $\varnothing$, inconsistent system

73. **a.** (150, 90)
    b. Both plans will have the same cost of $90 if Johanne expects to use 350 min
    c. She should use the InTouch Cell plan because it will cost her $70.
75. The ordered pair must be checked in both equations. Substituting (1, 3) into the second gives us $1 + 3 = 4 \neq 3$. This is false, so (1, 3) is not a solution of the system.
77. Since there is only one line, the lines are the same. There is an infinite number of solutions and the system is consistent with dependent equations.
79. $\{(4, 13)\}$ 81. $\{(-0.07, 1.64)\}$
83. Answers vary, $\begin{cases} x + y = -3 \\ x - y = -5 \end{cases}$.
85. Answers vary, $\begin{cases} 2x + y = 4 \\ 4x - 3y = -7 \end{cases}$.
87. Answers vary, $y = -\dfrac{5}{2}x - 1$.
89. Answers vary, $-9x - 15y = -6$.

Section 4.2

1. Answers vary.   3. Answers vary.   5. Answers vary.
7. Answers vary.   9. $\{(3, 1)\}$ 11. $\{(2, 7)\}$ 13. $\{(-6, 5)\}$
15. $\{(1, -3)\}$ 17. $\{(1, 2)\}$ 19. $\left\{\left(\dfrac{13}{15}, \dfrac{2}{3}\right)\right\}$
21. $\left\{\left(-\dfrac{11}{10}, \dfrac{1}{5}\right)\right\}$ 23. $\varnothing$ 25. $\{(x, y)\,|\,-x + 2y = -3\}$
27. $\varnothing$ 29. $\{(x, y)\,|\,x - 3y = 4\}$
31. The system has no solution and the lines are parallel. The system is inconsistent. The solution set is the empty set, $\varnothing$
33. The equations give the same line and there are infinitely many solutions. The system is consistent with dependent solutions. The solution set is $\{(x, y)\,|\,2x = y + 1\}$
35. The system contains intersecting lines with one solution. The system is consistent with independent equations. The solution set is $\{(6, -3)\}$.
37. $\{(1, 1)\}$ 39. $\{(8, -5)\}$ 41. $\{(x, y)\,|\,x = -6y + 2\}$
43. $\{(2, 1)\}$ 45. $\left\{\left(3, \dfrac{3}{2}\right)\right\}$ 47. $\left\{\left(\dfrac{13}{5}, \dfrac{4}{5}\right)\right\}$
49. $\{(4.5, -6.4)\}$
51. When the variables cancel out, there is either no solution or infinitely many solutions. In this case, the remaining statement is false, so there is no solution.
53. $4x - 2y = 10 \Rightarrow -2y = -4x + 10 \Rightarrow y = \dfrac{-4x + 10}{-2} = 2x - 5$
    $3x + 2(2x - 5) = 4 \Rightarrow 3x + 4x - 10 = 4 \Rightarrow 7x = 14 \Rightarrow x = 2$
    $y = 2(2) - 5 = 4 - 5 = -1$
    So, the answer is $\{(2, -1)\}$.

55. yes   **57.** $\{(4, -7)\}$ **59.** Answers vary, $2x - y = -4$.
61. Answers vary though $m = -5$ and $b \ne 3$; $y = -5x - 6$.
63. Answers vary. $2x + y = 1$.   **65.** Answers vary, $4x + 2y = 12$.

Section 4.3

1. Answers vary.   **3.** Answers vary.

5. $\left\{\left(\dfrac{1}{7}, 1\right)\right\}$ **7.** $\left\{\left(-4, -\dfrac{1}{2}\right)\right\}$ **9.** $\left\{\left(\dfrac{1}{2}, -3\right)\right\}$

11. $\left\{\left(\dfrac{2}{5}, \dfrac{6}{5}\right)\right\}$ **13.** $\{(36, -56)\}$ **15.** $\varnothing$

17. $\{(x, y) \mid 5x - 8y = 10\}$ **19.** $\{(x, y) \mid 6x + 18y = 20\}$
21. The equations give the same line and there are infinitely many solutions. The system is consistent with dependent solutions. The solution set is $\{(x, y) \mid 7x - 2y = 14\}$.
23. The system has no solution and the lines are parallel. The system is inconsistent. The solution set is the empty set, $\varnothing$.
25. The system contains intersecting lines with one solution. The system is consistent with independent equations. The solution set is $\{(-2, 1)\}$.
27. Mattel reported net sales of $5.918 billion and Hasbro reported net sales of $4.021 billion.
29. Foreclosures: 2.87 million, Homes sold: 3.56 million.

31. $\varnothing$ **33.** $\left\{\left(-\dfrac{3}{13}, -\dfrac{22}{13}\right)\right\}$ **35.** $\varnothing$

37. $\{(x, y) \mid 2x + 5y = 4\}$ **39.** $\left\{\left(3, -\dfrac{11}{5}\right)\right\}$ **41.** $\{(3.5, -3.2)\}$

43. $\{(-4.3, -2.7)\}$
45. It doesn't matter which variable you choose to eliminate. If you choose to eliminate x, you must find the number that both 3 and 5 divide into evenly. If you choose to eliminate y, you must find the number that both 4 and 3 divide into. In either case, we will multiply both equations by a nonzero number.
47. We must find the number that both 12 and 14 divide into evenly. We can write the prime factorization of each number to get $12 = 2 \cdot 2 \cdot 3$ and $14 = 2 \cdot 7$. So, the number that both 12 and 14 divide into is $2 \cdot 2 \cdot 3 \cdot 7 = 84$. We multiply the first equation by $\dfrac{-84}{-12} = 7$ and the second equation by $\dfrac{84}{14} = 6$.

49. $\{(2, -3)\}$ **51.** $\left\{\left(-\dfrac{3}{2}, -\dfrac{7}{2}\right)\right\}$

Piece It Together Sections 4.1–4.3

1. yes   **2.** no
3.   **4.**

5. $\varnothing$ **6.** $\{(x, y) \mid 3x - y = 1\}$

7. $\{(5, -2)\}$ **8.** $\{(2, -2)\}$ **9.** $\left\{\left(\dfrac{16}{7}, \dfrac{4}{7}\right)\right\}$

10. $\{(4, 3)\}$ **11.** $\varnothing$ **12.** $\{(x, y) \mid 3x + 6y = 3\}$ **13.** $\{(2, 2)\}$

14. $\{(1, 0)\}$ **15.** $\left\{\left(-\dfrac{14}{23}, \dfrac{17}{23}\right)\right\}$ **16.** $\{(36, 14)\}$

17. $\{(x, y) \mid x - 6y = 2\}$ **18.** $\varnothing$

19. The system has no solution and the lines are parallel. The system is inconsistent. The solution set is the empty set, $\varnothing$.
20. The system contains intersecting lines with one solution. The system is consistent with independent equations. The solution set is $\{(3.6, 1.9)\}$.

Section 4.4

1. Answers vary.   **3.** Answers vary.
5. 400 students and school employees   **7.** 58 dimes
9. 1.5 lb candied pecans, 0.5 lb candied peanuts
11. green beans: $0.40, tuna: $0.75
13. 5%: $375, 15%: $125
15. 1%: 200 gal, 8%: 100 gal
17. whole milk: 16 L, skim milk: 10 L
19. 50%: 1.25 mL, 10%: 3.75 mL
21. water: 3.75 L, 40%: 6.25 L
23. plane: 495.24 mph, jet stream: 38.10 mph
25. current: 6 mph, boat: 42 mph   **27.** 1 hr 20 min
29. length: 38 ft, width: 42 ft   **31.** 59°
33. length: 15 yd, width: 10 yd
35. Becky (130°); Elaine is closer (150°).
37. *Halloween:* $47 million, *Halloween H20:* $55 million
39. 31°, 149°   **41.** 15.5°, 74.5°
43. 1.8 hr   **45.** 28°, 62°
47. 400 mph, 45 mph   **49.** $7560 at 3.1% and $2440 at 1.4%
51. $0.85 for a can of tuna, $1.29 for a can of soup
53.
$$x + y = 10{,}650$$
$$0.011x + 0.045y = 248.05$$
$$x = 10{,}650 - y \Rightarrow 0.011(10{,}650 - y) + 0.045y = 248.05$$
$$117.15 - 0.011y + 0.045y = 248.05$$
$$0.034y = 130.9$$
$$y = 3850$$
$$x = 10{,}650 - 3850 = 6800$$
$6800 at 1.1% and $3850 at 4.5%

Section 4.5

1. Answers vary. A solution of a system of three variables will be an ordered triple (x, y, z) that satisfies all three equations in the system.
3. Answers vary. A false statement means that there is no solution of the system of equations.
5. Answers vary. We know we must use a system of three linear equations to solve a word problem if there are three unknowns.
7. $\{(1, -1, 5)\}$ **9.** $\{(-1, 1, -4)\}$ **11.** $\{(2, -1, 1)\}$
13. $\{(-1, 1, 1)\}$ **15.** $\{(5, -3, -2)\}$ **17.** $\{(-1, 2, 3)\}$
19. $\{(3, 1, -1)\}$ **21.** $\{(1, -z + 5, z) \mid z$ is a real number$\}$
23. $\{(8z - 12, 27z - 48, z) \mid z$ is a real number$\}$
25. no solution   **27.** $\{(-2, -3, -4)\}$
29. no solution   **31.** $\{(-2, -1, -3)\}$
33. 4 in first-class, 44 in business-class, and 288 in coach
35. 8 in first-class, 45 in business-class, and 318 in coach
37. $3900 in CD, $7800 in high-interest savings account, $3300 in money market account
39. 6800 in first mezzanine, 1200 in second mezzanine, and 2300 in third mezzanine
41. 470 in main, 1100 in terrace, 1470 in grandstand   **43.** $\{(1, -2, 1)\}$
45. $\left\{\left(\dfrac{1}{3}z + \dfrac{4}{3}, \dfrac{1}{3}z + \dfrac{16}{3}, z\right) \mid z$ is a real number$\right\}$ **47.** $\{(3, 1, -3)\}$
49. no solution   **51.** $\{(-8, 8, 5)\}$
53. $9350 in CD, $28,050 in high-interest savings account, $5100 in money market account
55. 400 in category 1600 in category 2500 in category 3
57. 7 in first-class, 39 in business-class, and 311 in coach
59. 780 in main, 650 in terrace, 2560 in grandstand

61.
$$\begin{cases} 3x - 4y + 6z = 3 \\ -3(x - y - 2z = 8) \end{cases} \rightarrow \begin{cases} 3x - 4y + 6z = 3 \\ \underline{-3x + 3y + 6z = -24} \end{cases}$$
$$-y + 12z = -21$$

$$\begin{cases} 2(x - y - 2z = 8) \\ -2x + 7y - 3z = -17 \end{cases} \rightarrow \begin{cases} 2x - 2y - 4z = 16 \\ \underline{-2x + 7y - 3z = -17} \end{cases}$$
$$5y - 7z = -1$$

$$\begin{cases} 5(-y + 12z = -21) \\ 5y - 7z = -1 \end{cases} \rightarrow \begin{cases} -5y + 60z = -105 \\ \underline{5y - 7z = -1} \end{cases}$$
$$53z = -106$$
$$z = -2$$

$5y - 7z = -1$	$x - y - 2z = 8$
$5y - 7(-2) = -1$	$x - (-3) - 2(-2) = 8$
$5y + 14 = -1$	$x + 3 + 4 = 8$
$5y = -15$	$x + 7 = 8$
$y = -3$	$x = 1$

The solution set is $\{(1, -3, -2)\}$.

63. Answers vary. One example is
$$x + y + z = 8$$
$$x - y - z = 2$$
$$2x + y + z = 13$$

65. $a = 2, b = -1, c = 5$

CHAPTER 4 SUMMARY

Section 4.1
1. solved simultaneously
2. ordered pair
3. equation; intersection point
4. one; no; infinitely many
5. intersect; parallel; same
6. consistent
7. inconsistent
8. dependent; independent

Section 4.2
9. solve; variables; substitute
10. inconsistent; no   **11.** consistent; dependent; infinitely many

Section 4.3
12. add; eliminated   **13.** opposites; multiply
14. same; infinitely many   **15.** parallel; no

Section 4.4
16. system
17. variables; equations; substitution; elimination

Section 4.5
18. elimination, two   **19.** unknowns, equations

CHAPTER 4 REVIEW EXERCISES

Section 4.1
1. consistent system with independent equations
3. consistent system with independent equations

5. consistent system with independent equations

7. a. $(150, 37.5)$
 b. The cost of making 150 min of long distance calls will be the same for both plans at $37.50.
 c. The x-value 200 is to the right of the intersection point. The Home Phone Plus graph (purple) is lower than the Qwest (blue) for this x-value. So the cost for Home Phone Plus is less than the cost for Qwest. He should use Qwest.

Section 4.2
9. $\{(-2, 4)\}$ **11.** $\{(4, 7)\}$ **13.** $\varnothing$

Section 4.3
15. $\{(6, 2)\}$ **17.** $\varnothing$ **19.** $\{(2, 3)\}$

Section 4.4
21. $52°, 128°$ **23.** $\dfrac{4}{3}$ hr   **25.** 530 mph, 25 mph
27. $41 for an adult and $18 for a child

Section 4.5
29. $\{(-3, 3, -1)\}$ **31.** $\{(-2, -4, 4)\}$
33. $\left\{ \left(5 + \dfrac{2}{3}z, -15 - \dfrac{13}{3}z, z \right) \middle| z \text{ is a real number} \right\}$
35. $11,000 in CD, $22,000 in high-interest savings account, $2750 in money market account
37. 7 first-class travelers, 25 business-class travelers, and 192 coach travelers

CHAPTER 4 TEST

1. b
3. a.

Slope-intercept form	$y = \dfrac{7}{4}x - 2$	$y = \dfrac{7}{4}x - 2$
Slopes	$m = \dfrac{7}{4}$	$m = \dfrac{7}{4}$
y-intercepts	$(0, -2)$	$(0, -2)$
How do lines relate?	The lines are the same.	
Number of solutions	Infinitely many solutions	
Type of system	Consistent system with dependent equations	
Solution set	$\{(x, y) \mid 7x - 4y = 8\}$	

 b.

Slope-intercept form	$y = \dfrac{1}{5}x + 3$	$y = \dfrac{1}{5}x - 2$
Slopes	$m = \dfrac{1}{5}$	$m = \dfrac{1}{5}$
y-intercepts	$(0, 3)$	$(0, -2)$
How do lines relate?	The lines are parallel.	
Number of solutions	No solution	
Type of system	Inconsistent system	
Solution set	$\varnothing$	

5. Answers vary. The three methods are graphing, substitution, and elimination. Graphing is the least precise since it is impossible to determine exact fractional solutions when I graph by hand. I would use substitution if one of the equations in the system is already solved for one of the variables. I would use elimination for almost every other system as long as I first put the equations in standard form.
7. a. $\{(-4, 2)\}$ **b.** $\{(1, -8)\}$
9. Ms. Jones invested $13,000 in the 8% account and $5000 in the 9% account.
11. The average speed of the plane is approximately 483 mph and the average speed of the wind is approximately 44 mph.
13. a. $\{(1, -2, 4)\}$ **b.** $\varnothing$ **c.** $\{(-3z, 2z + 2, z) \mid z \text{ is a real number}\}$

CHAPTERS 1–4 CUMULATIVE REVIEW EXERCISES

1. a. $\dfrac{4}{5}$ **b.** 2   **c.** 2   **3.** \$1536.70   **5.** \$5.58

7. $2w - 6$ **9. a.** $17a - 7$ **b.** $1.2b - 1.4$ **c.** $4x - 16$
d. $-4x^2 - 7x$

11. a. 10   **b.** -1 **c.** 28°, 62°

13. a. $\mathbb{R}$ **b.** $\{350\}$ **c.** $\left\{-\dfrac{19}{5}\right\}$

15. a. $h = \dfrac{V}{\pi r^2}$ **b.** $w = \dfrac{C - 2l}{5}$ **c.** $y = \dfrac{3}{2}x - 30$

d. $x = 60 - \dfrac{4}{5}y$ **17.** 63°, 43°, 74°   **19.** \$62.60

21. a.   **b.**   **c.**

23. a. (2000, 3.9), (2001, 5.0),(2002, 5.7), (2003, 6.1), (2004, 5.4), (2005, 5.0), (2006, 4.5), (2007, 4.7), (2008, 6.2), (2009, 9.8),(2010, 9.6)
b. The rate increased most between 2008 and 2009. It increased by 3.6%.
c. The rate decreased most between 2003 and 2004. It decreased by 0.7%.

25. a.   **b.**   **c.**

d.   **e.**

27. a. $-\dfrac{1}{2}$ **b.** -3 **c.** 0   **d.** undefined   **e.** $\dfrac{1}{4}$

29. The y-intercept (0, 24.02) and means that in 2000, the average price of jet fuel was about \$24.02 per barrel. The slope is 8.38 and means that the average price of jet fuel increased by about \$8.38 per barrel per year.

31. a. $y = 4x + 8$ or $-4x + y = 8$ **b.** $y = 12$
c. $y = 3x - 8$ or $3x - y = 8$

33. a. Domain = {2004, 2005, 2006, 2007, 2008, 2009, 2010}; Range = {10.75, 12.70, 13.73, 13.08, 13.89, 12.14, 11.20}; The relation is a function.
b. Domain = $(-\infty, \infty)$; Range = {2}; The relation is a function.   **c.** Domain = $(-\infty, \infty)$; Range = $(-\infty, \infty)$; The relation is a function.

35. a. $f(x) = -\dfrac{1}{2}x + \dfrac{1}{2}; \left(0, \dfrac{1}{2}\right), \left(-2, \dfrac{3}{2}\right), (3, -1)$
b. $f(x) = 0.2x^2 - 0.1x$; (0, 0),(-2, 1), (3, 1.5)

37. a. {(2, 1)}, one solution, consistent system with independent equations

b. parallel lines, no solution, inconsistent system

c. {(3, 0)}, one solution, consistent system with independent equations

d. {(-1, 2)}, one solution, consistent system with independent equations
e. infinite number of solutions, consistent system with dependent equations

39. {(3, 1)}   **41.** ∅   **43.** $\{(x, y) \mid 2x + y = 5\}$
45. {(3.2, -2.5)}   **47.** 38°, 52°
49. 50 L of pure alcohol and 200 L of 50% alcohol
51. {(1, 3, 4)}   **53.** {(2, -5, 1)}
55. The system has infinitely many solutions.
The solutions are $\left\{\left(3 + z, \dfrac{1}{2}z - \dfrac{5}{2}, z\right) \middle| z \text{ is a real number}\right\}$.
57. ∅
59. 2500 in first mezzanine, 850 in second mezzanine, and 1700 in third mezzanine

CHAPTER 5

Section 5.1

1. Consider the expression $\dfrac{x^3}{x^3}$. It simplifies to 1. Applying the quotient of like bases rule, we have $\dfrac{x^3}{x^3} = x^{3-3} = x^0$. So a nonzero real number raised to the zero exponent is 1.

3. In 3^{-2}, the exponent -2 applies only to the base 3. So, $3^{-2} = \dfrac{1}{3^2} = \dfrac{1}{9}$. In -3^2, the expression is the opposite of 3^2, which is 9. So, $-3^2 = -9$. Therefore, $3^{-2} \neq -3^2$.

5. Yes, we can apply the product of like bases rule of exponents because the bases are the same.
7. Yes, we can apply the quotient of like bases rule of exponents because the bases are the same.

9. r^7 **11.** m^5 **13.** x^6 **15.** m^{12} **17.** 3^{15} **19.** 2^{14}
21. $-20x^5$ **23.** $60a^{10}$ **25.** $6x^2y^7$ **27.** $-15a^{13}b^7$ **29.** x^6
31. -1 **33.** 16   **35.** $9p^8$ **37.** $-6r^{11}$ **39.** $-2x^8y^4$
41. $16p^6q^4$ **43.** 1   **45.** 1   **47.** -1 **49.** 1   **51.** 1
53. 2   **55.** $\dfrac{1}{32}$ **57.** $\dfrac{1}{16}$ **59.** $\dfrac{27}{8}$ **61.** 625   **63.** $-\dfrac{1}{81}$
65. $\dfrac{4}{v^5}$ **67.** $\dfrac{1}{y^{18}}$ **69.** $\dfrac{b^3}{a^4}$ **71.** p^2q^4 **73.** $\dfrac{1}{q^{19}}$ **75.** $\dfrac{1}{x^6}$
77. $-\dfrac{4a^{17}}{b^{10}}$ **79.** $\dfrac{6}{r^3t^{10}}$ **81.** $10m^3n$ **83.** $-\dfrac{15}{hp^4}$
85. $-\dfrac{2}{n^{18}}$ **87.** $\dfrac{u^3v^2}{3}$ **89.** $-\dfrac{18}{5u^{14}v^3}$ **91.** $\dfrac{t^{15}}{3e^8}$
93. 337.36 cm²   **95.** $\dfrac{7}{2}\pi r^2$ **97.** $14x^2y$ **99.** $6x^5y^8$
101. \$742.31   **103.** \$5886.20   **105.** -7

107. $-\dfrac{19}{d^5}$ 109. 11^5 or 161,051   111. $18p^3$ 113. $\dfrac{b^{12}}{s^4}$

115. $-\dfrac{1}{8}$ 117. $\dfrac{u^8}{4v^{14}}$ 119. $-8d^9r^2$ 121. 2   123. $\dfrac{20}{a^{21}n^2}$

125. 1   127. 1   129. $\dfrac{4t^7z^2}{5}$ 131. $-\dfrac{1}{729}$ 133. $-\dfrac{5c}{p^6}$

135. \$589.59   137. 10 g, 1.25 g

139. $-2^4 \cdot x^3 \cdot x^2 = -2 \cdot 2 \cdot 2 \cdot 2x^{3+2} = -16x^5$

141. $\dfrac{c^4d^7}{c^{-3}d^2} = c^{4-(-3)}d^{7-2} = c^7d^5$

143. When applying the product of like bases rule, we add the exponents but keep the base the same.

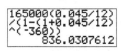

The correct answer is 8192.

145. Enter the expression in the denominator with a pair of parentheses.

```
165000(0.045/12)
/(1-(1+0.045/12)
^(-360))
            836.0307612
```

The correct answer is \$836.03.

147. 35^{x+1} 149. 5^{a+1}

Section 5.2

1. We apply the exponent m to each factor in the base that is a product. So, $(ab)^m = a^mb^m$.
3. We apply the exponent n to each factor in the base. The factors in the base are -2 and x^m. So, $(-2x^m)^n = (-2)^n(x^m)^n = (-2)^nx^{mn} \neq -2^nx^{mn}$.
5. Answers vary; For example, to convert 0.000234 to a number in scientific notation, we move the decimal point right 4 places. So, $0.000234 = 2.34 \times 10^{-4}$.
7. Answers vary; For example, to convert 9800 to a number in scientific notation, we move the decimal point left 3 places. So, $9800 = 9.8 \times 10^3$.

9. b^{15} 11. $\dfrac{1}{15{,}625}$ 13. $-\dfrac{1}{32{,}768}$ 15. $-\dfrac{1}{6^{35}}$

17. $32a^{45}b^5$ 19. $256r^{36}s^8$ 21. $\dfrac{x^6}{36}$ 23. $\dfrac{16x^{12}}{81y^4}$

25. $\dfrac{16u^4v^{12}}{81}$ 27. $\dfrac{a^4}{b^{18}}$ 29. $\dfrac{16r^2}{s^{13}}$ 31. $\dfrac{16x^{16}}{81}$ 33. $\dfrac{49q^{36}}{36}$

35. $\dfrac{b^{16}}{a^{40}}$ 37. $\dfrac{12}{h^9k^6}$ 39. $\dfrac{z^{17}}{32y^7}$ 41. $\dfrac{8}{st^{24}}$ 43. 6.03×10^{-7}

45. 9.45×10^5 47. 46,500   49. 0.00452   51. 3.648×10^{-24}

53. 2.12×10^6 55. 3.2×10^{-24} 57. 8.52×10^3

59. 9.312×10^8 61. 3.6×10^{-15} 63. 6.7×10^9

65. 1.72×10^{19} 67. 4.76×10^{-11} 69. 5.03%

71. 55.88%   73. £15,166.67   75. 508,000,000 hydrogen atoms

77. $-\dfrac{y^6}{125x^{18}}$ 79. $-\dfrac{32a^{35}}{b^5}$ 81. $\dfrac{v^{40}}{81}$ 83. $\dfrac{s^{10}}{49}$ 85. $-27x^6$

87. $\dfrac{1}{36q^8}$ 89. $\dfrac{121u^{10}v^6}{36}$ 91. $-\dfrac{1000x^{15}y^9}{729}$ 93. $\dfrac{16}{a^2b^{41}}$

95. $\dfrac{x^{15}}{y^{32}}$ 97. $500s^{11}t$ 99. $\dfrac{1}{2y^{12}z^{11}}$ 101. $\dfrac{25u^8}{9}$

103. $\dfrac{1}{256y^{12}}$ 105. $144p^{44}q^{13}$ 107. hr^{11} 109. 3.85×10^{-6}

111. 6.512×10^4 113. 8.96×10^3 115. 8.3×10^{10}

117. 2.89×10^{12} 119. 11.32 visits per month

121. $(-4a^3b^2)^3 = (-4)^3(a^3)^3(b^2)^3$
$= (-4)(-4)(-4)a^9b^6$
$= -64a^9b^6$

123. $\dfrac{6.3 \times 10^{-21}}{1.4 \times 10^{-6}} = \dfrac{6.3}{1.4} \times 10^{-21-(-6)}$
$= 4.5 \times 10^{-15}$

125.
```
1.512*10^-12/(7.
2*10^-4)
          2.1E-9
1.512E-12/7.2E-4
          2.1E-9
```
The correct answer is 2.1×10^{-9}.

127. $7ab^5$; $(7ab^5)^2 = 49a^2b^{10}$ ($-7ab^5$ is also acceptable.)

129. $-\dfrac{3r}{s^2}$; $\left(-\dfrac{3r}{s^2}\right)^3 = -\dfrac{27r^3}{s^6}$ 131. x^{5a+6}

Section 5.3

1. Answers vary.   3. Answers vary.   5. Answers vary.
7. Answers vary.   9. 2, 3   11. $-1, 5$ 13. $-\dfrac{1}{3}, 2$
15. 12, 0   17. 3, 8   19. $-7.5, 7$

	Terms/Classification	Degree/Type	Leading Coefficient
21.	$3a^3, 2a^2, -5a, 7$; Polynomial	3; Cubic	3
23.	$-11r^4, 13r, 2$; Trinomial	4; Quartic	-11
25.	$32abc^2$; Monomial	4; Quartic	n/a
27.	$7, -16s$; Binomial	1; Linear	-16

29. 1   31. 22   33. -9 35. 4   37. 10   39. 3
41. 7   43. $-\dfrac{15}{4}$ 45. $\dfrac{61}{4}$
47. The penny will be 1358 ft above the ground 4 sec after it is dropped. The penny will be 318 ft above the ground 9 sec after it is dropped.
49. Video game revenues were approximately \$17.7 billion in 2007 and approximately \$28.4 billion in 2009.
51. $-9y - 16$ 53. $13y + 15$ 55. $2.8a - 2.4$
57. $-1.5a + 4.2$ 59. $-8y^2 - 10y + 20$
61. $24x^2 - 14x + 40$ 63. $-0.6q^2 - 7.1q + 9.6$
65. $2x^2 + xy - 4y^2$ 67. $f(x) + g(x) = 2x^2 + 10x - 8$ and $f(x) - g(x) = 8x^2 + 4x - 34$ 69. $P(x) = 2x^3 + x + 2$; 1034 in.
71. $P(y) = 6y^2 + 34y - 14$ units   73. $P(y) = 14y^2 - 10y - 1$ in.
75. a. $P(x) = 5.8x - 310$; b. $P(185) = \$763$

	Terms/Classification	Degree/Type	Leading Coefficient
77.	$6.25b^2, -0.64$; Binomial	2; Quadratic	6.25
79.	$5a^3b^2c$; Monomial	6; Sixth degree	n/a
81.	$0.1x^3, 1.6x^2, -2.9x, 1.8$; Polynomial	3; Cubic	0.1
83.	$-\dfrac{x^2}{5}$; Monomial	2; Quadratic	$-\dfrac{1}{5}$

85. 53.7   87. 13.8   89. 9.6   91. 3.9   93. $21x + 6$
95. $r - 14$ 97. $-10a + 15.5$ 99. $12a^2 - 39a + 6$
101. $4p^2 - 4pq - 9q^2$ 103. $4.4x^2 + 18.5x + 8$
105. The average cost of electricity was approximately \$58.16 per 500 kWh in 2006 and \$65.38 per 500 kWh in 2009.
107. a. $p(x) = 1.15x - 320$ b. $p(1200) = \$1060$
109. $g(-4) = -(-4)^2 - 7(-4) + 6$
$= -16 + 28 + 6$
$= 18$
111. $(a^2 + 23a - 18) + (a^2 + 4a - 33) = a^2 + 23a - 18 + a^2 + 4a - 33$
$= 2a^2 + 27a - 51$
113. The mistake is that -5 was not entered in parentheses. The correct answer is $f(-5) = -132$.

115.
```
Plot1 Plot2 Plot3
\Y1■(-2X²+5X-4)+
(5X²-7X+3)
\Y2■3X²-2X-1
\Y3=
\Y4=
\Y5=
\Y6=
```
```
 X   Y1   Y2
-1   -11  -11
 0    0    0
 1    0    0
 2    7    7
 3   20   20
 4   39   39
 5   64   64
Y2■3X²-2X-1
```

117. Answers vary, but an example is $7x^4 - 6x^3 + 3$.
119. Answers vary, but an example is $-4x$.

Piece It Together Sections 5.1–5.3

1. $60a^{18}$ **2.** $-243m^{30}$ **3.** $\dfrac{32x^5}{243y^5}$ **4.** $-3y^3$ **5.** 1

6. $-\dfrac{64}{125}$ **7.** x^5y^6 **8.** $-\dfrac{s^{12}}{8}$ **9.** 9.547×10^{-7}

10. $238{,}000$ **11.** 1.8×10^9 **12.** 3.0×10^{21}

13. leading coefficient 12, degree 2

14. leading coefficient -16, degree 4

15. 5 **16.** $-\dfrac{31}{4}$ **17.** $7h^2 - 2hk + 4k^2$ **18.** $-4y^2 - 3y + 39$

19. $2y^2 - 3y - 16$ **20.** $6a^4 - 7a^2 + 5a - 13$

Section 5.4

1. Answers vary. **3.** $12x^3$ **5.** $2x^2 + 10x$ **7.** $-x^3 - 4x^2 + x$

9. $8a^3b^2 + 12a^2b$ **11.** $3a^5 + 12a^4 + 12a^2$ **13.** $-5b^8 + 35b^7$

15. $-30r^4s^3 + 10r^3s^2 - 50r^2s$ **17.** $x^2 + 3x + 2$ **19.** $x^2 + 2x - 15$

21. $6x^2 + 7x - 5$ **23.** $25a^2 + 20a + 4$ **25.** $64r^2 - 48r + 9$

27. $2x^2 + 5xy - 3y^2$ **29.** $3r^2 - rs - 2s^2$

31. $a^3 - 4a^2 + 5a - 2$ **33.** $12y^3 + 16y^2 + 3y - 1$

35. $27x^3 - y^3$ **37.** $b^4 - 2b^3 - 21b^2 + 32b - 10$

39. $10y^4 + 17y^3 - 9y^2 - 5y + 2$ **41.** $8x^3 - 31x^2 + 16x + 15$

43. $30x^3 - 11x^2 - 24x + 12$ **45.** $-5x^2 + 260x + 11{,}200$

47. $4x^2 + 54x + 180$ **49.** $-x^2 + 16$ **51.** $3p^2q^2 - 4pq$

53. $a^4 + 3a^3 - 2a^2 + 2a + 24$ **55.** $-12n^5 + 8n^4 - 4n^2$

57. $\dfrac{2}{9}x^2 + \dfrac{8}{3}xy - 10y^2$ **59.** $10x^2 - \dfrac{1}{3}x - \dfrac{2}{9}$ **61.** $r^3 + 8s^3$

63. $2x^4 + x^3 - 22x^2 + 23x - 4$

65. a. $-80x^2 + 2800x + 60{,}000$ **b.** $\$67{,}680$ **67.** $4y^2 + 4y + 1$ in.^{2}

69. $9a^2 - 12a$ cm^2

71. We must multiply each term in the first binomial by each term in the second binomial. This gives us $(x - 4)(x + 5) = x^2 + 5x - 4x - 20 = x^2 + x - 20$.

73. The dimensions of the pool are $50 - 2x$ and $20 - 2x$, so the area is $(50 - 2x)(20 - 2x) = 4x^2 - 140x + 1000$ m^2.

75. $-2x^2 + 9x + 18$ **77.** $2x^4 - 9x^3 - 9x^2 + 20x$

Section 5.5

1. Answers vary. **3.** Answers vary. **5.** $x^2 - 7x + 6$

7. $x^2 - 2x - 35$ **9.** $15x^2 - 7xy - 2y^2$ **11.** $a^4 - 3a^2 - 4$

13. $12y^4 + 7y^2 - 10$ **15.** $12r^3 + 16r^2s - 36rs - 48s^2$

17. $6x^2 - x - \dfrac{6}{25}$ **19.** $3.6x^2 + 4.14xy + 1.08y^2$

21. $-12x^2 - 11x + 5$ **23.** $x^2 + 8x + 16$ **25.** $4y^2 - 4y + 1$

27. $a^4 - 4a^2 + 4$ **29.** $16c^2 + 40cd + 25d^2$

31. $9a^2 - \dfrac{12}{5}a + \dfrac{4}{25}$ **33.** $2.25p^2 - 1.8p + 0.36$

35. $9x^4 - 42x^2y + 49y^2$ **37.** $x^2 - 49$ **39.** $x^2 - 4y^2$

41. $4y^2 - 16$ **43.** $49r^2 - 9s^2$ **45.** $1.69x^2 - 6.25$

47. $a^4 - 1$ **49.** $c^6 - 64d^6$ **51.** $x^4 - \dfrac{1}{49}$ **53.** $\dfrac{9}{16}c^6 - d^2$

55. $x^3 + 3x^2 + 3x + 1$ **57.** $y^3 - 6y^2 + 12y - 8$

59. $8y^3 - 36y^2 + 54y - 27$ **61.** $b^3 + \dfrac{3}{2}b^2 + \dfrac{3}{4}b + \dfrac{1}{8}$

63. $a^4 + 4a^3 + 6a^2 + 4a + 1$ **65.** $x^2 - 144y^2$ **67.** $6x^2 + 3x - 45$

69. $9a^2 + 12a + 4$ **71.** $y^3 + 33y^2 + 363y + 1331$

73. $a^4 - \dfrac{9}{49}$ **75.** $\dfrac{1}{4}c^6 - 25d^2$ **77.** $8p^3 - 12p^2 + 6p - 1$

79. $100x^2 - 49y^2$ **81.** $\dfrac{1}{9}x^2 - \dfrac{2}{3}x + 1$ **83.** $0.32x^2 + 1.02x - 5.4$

85. $x^4 - 9y^2$

87. We ct just square each term. We must apply the rule for squaring a binomial. $(3x - 8)^2 = 9x^2 - 48x + 64$

89. Antoine didn't combine like terms correctly. The middle terms add to zero. The result should be $16x^2 - 9$.

91. incorrect, $121x^2 - 176x + 64$

93. incorrect, $27x^3 + 54x^2 + 36x + 8$ **95.** correct

97. $(a - b)^2 = (a - b)(a - b) = a^2 - ab - ab + b^2 = a^2 - 2ab + b^2$

Section 5.6

1. Answers vary. **3.** Answers vary. **5.** Answers vary.

7. $x - 2$ **9.** $5x^2 - x + 2$ **11.** $a^2 - 4a - 2$

13. $-3y^3 + 4y - 2$ **15.** $4x^3 - 2x^2 - 1$ **17.** $-3y^2 + 4y - 2$

19. $a^2 - \dfrac{a}{2} + \dfrac{2}{a}$ **21.** $\dfrac{x^2}{3} - x + 3$ **23.** $1 - \dfrac{2}{3a} + \dfrac{2}{a^2}$

25. $1 - \dfrac{2}{x}$ **27.** $-\dfrac{y^2}{2} + y - 1$ **29.** $\dfrac{y}{4} - 2 - \dfrac{1}{2y^2}$ **31.** $x - 4$

33. $x + 4$ **35.** $x^2 + x + 1 + \dfrac{2}{x - 1}$ **37.** $4a$ **39.** $2x - 1$

41. $x - 7 + \dfrac{2}{x + 1}$ **43.** $3y + 13 + \dfrac{85}{y - 6}$ **45.** $2x - 19 + \dfrac{140}{x + 7}$

47. $3y^2 + 2y - 5 + \dfrac{7}{2y - 3}$ **49.** $x - 2y$ units

51. $12 + 8xy^2$ units **53.** $2a^2 + a - \dfrac{6}{a}$ **55.** $3x + 1$

57. $-3x^2 + 2x - 1$ **59.** $7x^2 - 15x + 30 - \dfrac{46}{x + 2}$

61. $3x^2 - 5x + 5 + \dfrac{2}{2x + 3}$ **63.** $2z - x$ units

65. $\dfrac{5x^2 + 2x - 3}{5x^2} = \dfrac{5x^2}{5x^2} + \dfrac{2x}{5x^2} - \dfrac{3}{5x^2} = 1 + \dfrac{2}{5x} - \dfrac{3}{5x^2}$

67. $x - 5 + \dfrac{12}{x + 3}$ **69.** incorrect; The correct answer is $1 - \dfrac{3}{x^2}$.

71. correct

Section 5.7

1. No, synthetic division can only be used when the divisor is a binomial of the form $x - c$.

3. The last row of numbers represent the coefficients of the quotient polynomial and the remainder. The degree of the quotient is one less than the degree of the dividend.

5. Answers vary. **7.** $2x - 1$ **9.** $3x + 6$

11. $2x - 7 + \dfrac{10}{x + 3}$ **13.** $7x^2 - 9x - 12$ **15.** $x^2 - 5x + 15$

17. $6x^2 - x + 2 - \dfrac{23}{x + 8}$ **19.** $x^2 + 4x + 16$

21. $x^2 - 6x + 36 - \dfrac{432}{x + 6}$ **23.** $8x - 6$ **25.** $x^2 + 3x - 2$

27. $2x^2 + 6x - 17 - \dfrac{27}{x - 3}$ **29.** $P(2) = -44$ **31.** $P(-1) = -6$

33. $P(-5) = 3$ **35.** $P(-2) = 52$ **37.** $P(3) = 9$

39. $P(-1) = 9$ **41.** $P(2) = 23$

43. If $x = 5$ is a zero of $P(x)$, then we know that $P(5) = 0$. By the remainder theorem, this is equivalent to the remainder after dividing $P(x)$ by $x - 5$. So, the remainder is 0.

CHAPTER 5 SUMMARY

Section 5.1

1. sum; same; x^{a+b} **2.** difference; same; x^{a-b} **3.** one

4. reciprocal; positive; $\dfrac{1}{b^n}$

Section 5.2

5. multiplied; x^{ab} **6.** factor; $x^b y^b$

7. numerator; denominator; $\dfrac{x^b}{y^b}$ **8.** scientific; notation

9. positive **10.** negative

Section 5.3

11. Polynomials **12.** Polynomials; monomials

13. Polynomials; binomials **14.** Polynomials; trinomials

15. standard; form **16.** largest **17.** leading; coefficient

18. sum **19.** largest **20.** replace **21.** like; terms

22. opposite; opposite

Section 5.4

23. distributive   **24.** distributed

Section 5.5

25. FOIL; binomials
26. trinomial; $a^2 + 2ab + b^2$; perfect; square; trinomial
27. Conjugates   **28.** $a^2 - b^2$; difference; squares
29. exponent; term

Section 5.6

30. each; polynomial; $\dfrac{A}{C} + \dfrac{B}{C}$

31. long; division; divide; multiply; subtract; bring; down
32. multiplication; quotient; divisor; remainder

Section 5.7

33. $x - c$; coefficients   **34.** $P(c)$

CHAPTER 5 REVIEW EXERCISES

Section 5.1

1. $-\dfrac{4}{a^4}$ **3.** $\dfrac{1}{27}$ **5.** $-\dfrac{5}{m^{10}n}$ **7.** $\dfrac{8}{5}s^2t^2$ **9.** $\dfrac{x^8}{7y^{14}}$
11. -2 **13.** 10 **15.** -5 **17.** $188.84

Section 5.2

19. $\dfrac{y^{10}}{9}$ **21.** $256b^8$ **23.** $\dfrac{1}{729a^3}$ **25.** $-343x^{12}y^9$
27. 8.91×10^{-5} **29.** 0.0000278 **31.** $30,500,000,000
33. 3.25×10^{-7}

Section 5.3

35. trinomial; $9x^2 - 5x + 11$; terms: $9x^2$, $-5x$, and 11; leading coefficient: 9, degree: 2
37. $f(-2) = -3$
39. $C(20) = 94{,}100$; It costs \$94,100 to make 20 items.
41. $-13x^2 - 13x + 19$ **43** $-x^2 + 2x - 1$
45. $-7a^2 - 2ab - 5b^2$ **47.** $4x^2$ **49.** $5x^2 + 14x - 16$ in.

Section 5.4

51. $-3p^2 + 5p$ **53.** $-x^2 + 49$ **55.** $3x^4 - x^3 - 25x^2 + 43x - 20$
57. $8r^3 - 27s^3$

Section 5.5

59. $12x^2 + 11x - 5$ **61.** $9a^2 + 12ab + 4b^2$ **63.** $121x^2 - 49y^2$
65. $\dfrac{4}{9}a^6 - 25b^2$

Section 5.6

67. $2a^2 + a - \dfrac{6}{a}$ **69.** $5x^2 + 6x - 1 + \dfrac{2}{x+5}$

71. $4y - 3x$ units

Section 5.7

73. $5x + 8$ **75.** $x^2 - 2x - 12 - \dfrac{92}{x-6}$ **77.** -96

CHAPTER 5 TEST

1. a.   **3.** b.   **5.** b.   **7.** c.   **9.** b.   **11.** $-\dfrac{24}{a^5}$
13. -3 **15.** -36 **17.** $-\dfrac{1}{36}$

	Standard Form	Type	Degree	Leading Coefficient
19.	$-y^2 - 4y + 5$	trinomial	2	-1
21.	$9a^6$	monomial	6	9

23. 7   **25.** $-x + 12$ **27.** $25a^6 + 20a^3b + 4b^2$
29. $27x^3 - 125$ **31.** $6ab^2 - 3b + 1$ **33.** $2x - 3$
35. Perimeter $10x^2 + 18x + 16$ units; Area $30x^3 + 53x^2 + 27x + 7$ units^2

CHAPTERS 1–5 CUMULATIVE REVIEW EXERCISES

1. a. $1\dfrac{7}{25}$ **b.** $3\dfrac{9}{14}$ **3. a.** 84   **b.** 10.6
5. a. 64.66   **b.** 0   **c.** not defined
7. a. equation   **b.** expression
9. a. $\{-16\}$ **b.** $\left\{\dfrac{4}{3}\right\}$ **c.** $\varnothing$

11. Bianca earned \$199.10.
13. a. $(0, 82.0), (1, 75.7), (2, 77.8), (3, 76.9), (4, 68.7), (5, 80.1), (6, 80.9)$
   b. In 2003, 82.0% of US Airways flight operations arrived on time. In 2009, 80.9% of US Airways flight operations arrived on time.
   c. In 2003, the percentage of US Airways flight operations that arrived on time was the highest. In 2007, the percentage of US Airways flight operations that arrived on time was the lowest.
   d.

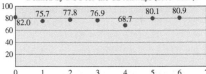

Percent of Flight Operations Arriving on Time as Defined by DOT for the US Airways (2003–2009)

15. a. $(0, 330.4)$; In 2005, the airline consumed 330.4 million gal of fuel.
   b. In 2010, the airline consumed 520.4 million gal of fuel.
   c. In 2013, the airline consumed 634.4 million gal of fuel.
   d.

17. $m = -0.281$; The operating revenue is decreasing by \$0.281 billion per year. $(0, 12.01)$; In 2006, the total operating revenue was \$12.01 billion.
19. a. neither   **b.** parallel
21. a. $y = -0.63x + 13.71$
   b. The average residential natural gas price in 2011 is \$10.56.
   c. The average residential natural gas price in 2013 is \$9.30.
23. a. $f(0) = -6$ **b.** $g(-2) = -8$ **c.** $f(-2) = -1; x = 2$
25. a. $\{(-3, 4)\}$ **b.** $\{(5, -1)\}$

27. a. $40°, 140°$ **b.** $\dfrac{5}{6}$ hr or 50 min

29. a. -8 **b.** 3   **c.** $\dfrac{27}{1000}$ **d.** $-\dfrac{18q^3}{c^5}$

e. $-\dfrac{8}{y^4}$ **f.** $-\dfrac{4p^6}{q^2}$

31. a. $-27x^{12}y^{30}$ **b.** $-\dfrac{a^{15}}{8}$ **c.** $\dfrac{49}{64p^4q^6}$ **d.** $\dfrac{z^{44}}{64y^{24}}$

33. a. 5 g; 1.25 g   **b.** $200.78
35. a. 11   **b.** -14 **37. a.** $P(x) = 13x^2 + 5x + 1$
    b. $R(x) = -0.3x^2 + 161x + 465$
39. a. $-180x^2 + 3750x + 75{,}000$ **b.** $81,780
41. a. $3a^2 + a - \dfrac{6}{a}$ **b.** $\dfrac{3}{4}a^2 - 3a + \dfrac{1}{2a} - \dfrac{1}{a^2}$

   c. $2x^2 + 14x - 24 + \dfrac{57}{x + 3}$ **d.** $4y^2 - 2x$ units
43. $P(-5) = 107$

CHAPTER 6

Section 6.1

1. The greatest common factor of a set of integers is the largest integer that is a factor of each integer.
3. To factor a polynomial is to write it as a product.
5. For polynomials with four or more terms, group pairs of terms by factoring out the GCF. Factor out the resulting binomial as the GCF.
7. The reverse operation of factoring is multiplication. For example, $(3x + y)(a + 2b) = 3ax + ay + 6bx + 2by$.
9. 6   **11.** 6   **13.** 15   **15.** 22   **17.** $3x^2$ **19.** $5a$
21. $8b^2$ **23.** $4a^2b^3$ **25.** $2ab$ **27.** $4y$ **29.** 6   **31.** $x + 1$
33. $y - 12$ **35.** $3(a + 2)$ **37.** 1   **39.** $2x(x - 2)$
41. $4a^3(a - 4)$ **43.** $-3y(y - 2)$ **45.** $-12b^2(3b^3 + 2b + 1)$
47. $15x(3x^2 - 2x + 1)$ **49.** $2(a^2 + a - 1)$
51. $-4x(2x^2 - 5x - 1)$ **53.** $(x + 3)(x + 2)$ **55.** $(x - 2)(x + 6)$
57. $(x + 3)(4 - x)$ **59.** $x(x + 1)(x - 3)$
61. $2y^2$ and $3y + 8$ are the length and width of the rectangle.
63. $-20q(q - 25)$ **65.** $(3x - 1)(2x + 1)$ **67.** $(4b^2 + 3)(b + 1)$
69. $(5 - a)(a - 4)$ **71.** $(r + 5)(r^2 - 8)$
73. Cannot be factored using grouping   **75.** $(8x + 3)(4x - 7)$
77. $3(x^2 - 6)(x + 4)$ **79.** $4(2x^2 - 1)(x - 5)$ **81.** $9y(y - 9)$
83. $-8ab(2a^2 - 6a + 1)$ **85.** $(3x + 2)(x - 2)$
87. Cannot be factored using grouping   **89.** $2(6x - 1)(3x + 1)$
91. $-16a^2(a^2 - 3a - 1)$ **93.** $x(4x^2 - 3)(2x - 1)$
95. $2(2a^2 + 3a - 3)$ **97.** $(6x - 5)(x + 5)$
99. Answers vary; $6z^2$ and $z^2 + 4$ are the base and height of the parallelogram.
101. $\pi r(r + 1)$ **103.** x^2
105. $2x(x - 3) - 1(x - 3) = (2x - 1)(x - 3)$ **107.** $12x^5$ **109.** $16x^2$
111. Answers vary; $12x^2 - 6x = 6x(2x - 1)$
113. Answers vary; $4y^3 - 6y^2 = 2y^2(2y - 3)$
115. Answers vary; $-10a^2b + 12ab^2 = -2ab(5a - 6b)$

Section 6.2

1. Answers vary. For example, factor $x^2 + 7x + 10$. We need to find the factors of 10 whose sum is 7. Since 10 is positive and the sum is positive, we will consider only positive factors of 10: 1 and 10, 2 and 5. Since $2 + 5 = 7$, the expression is factored as $(x + 2)(x + 5)$.
3. Answers vary. In Exercise 7, we need to find the factors of 6 whose sum is 5. Since the last term 6 is positive and the sum is positive, we will consider only positive factors of 6. In Exercise 8, we need to find the factors of 20 whose sum is 9. We will consider only positive factors of 20.
5. $(a + 1)(a + 5)$ **7.** $(x + 2)(x + 3)$ **9.** $(x - 8)(x - 9)$
11. prime   **13.** $(b - 8)(b - 10)$ **15.** $(x + 6y)(x + 15y)$
17. $(h - 7k)(h - 10k)$ **19.** $(r + 5)(r + 8)$ **21.** $(a - 4)(a + 5)$
23. $(x - 3)(x + 15)$ **25.** $(x - y)(x + 10y)$ **27.** prime
29. $(b + 5)(b - 12)$ **31.** $(a + 2b)(a - 6b)$ **33.** $(h + 3)(h - 14)$
35. $(x + 5y)(x - 9y)$ **37.** $4(x + 3)(x - 6)$ **39.** $5(y + 1)(y - 4)$
41. $-4(a + 1)(a + 9)$ **43.** $3r(r - 8)(r + 1)$
45. $-6x(x + 3)(x - 4)$ **47.** $9p^2(p + 5)(p - 6)$
49. $5xy(x - 2)(x + 3)$ **51.** $-3ab(a - 6)(a - 7)$
53. $-16(t + 4)(t - 10)$ **55.** $x(x + 6)(x - 2)$
57. $(x - 7)(x - 8)$ **59.** $3y(y - 2)(y - 2)$
61. $(p - 6q)(p - 9q)$ **63.** prime   **65.** $-6x^2(x + 3)(x - 7)$

67. $(r + 3s)(r + 13s)$ **69.** $3x^2(x + 2)(x + 9)$
71. $-2(s + 10)(s - 2)$ **73.** prime   **75.** $x(x + 12)(x - 3)$
77. $-16(q - 30)(q + 20)$
79. Find the factors of -24 whose sum is 2. They are -4 and 6, $(b + 6)(b - 4)$.
81. $2x^2 - 18xy + 36y^2 = 2(x^2 - 9xy + 18y^2) = 2(x - 3y)(x - 6y)$
83. $(x - 3)(x - 14)$
85. The factors of 28 are 1 and 28, 2 and 14, 4 and 7, -7 and -4, -14 and -2, and -28 and -1. So, for the last term to be positive, both factors have to have the same sign. Since b is the sum of the factors, the value of b could be 29, 16, 11, -11, -16, or -29. So, an example is $x^2 - 16x + 28 = (x - 2)(x - 14)$.
87. The factors of -30 are 1 and -30, 2 and -15, 3 and -10, 5 and -6, 6 and -5, 10 and -3, 15 and -2, and 30 and -1. Since b is the sum of the factors, the value of b could be -29, -13, -7, -1, 1, 7, 13, or 29. An example is $x^2 + 13x - 30 = (x + 15)(x - 2)$.
89. The factors of -8 are 1 and -8, 2 and -4, 4 and -2, and 8 and -1. For the trinomial to be factorable with a last term of -8, the value of b would have to be -7, -2, 2, or 7. So, as long as b is different from these values, the trinomial is prime. An example is $x^2 + 5x - 8$.

Section 6.3

1. Trial and error requires us to try every combination of factors of the first term, a, with factors of the last term, c, until the correct middle term, b, is obtained.
3. If the last term in the trinomial is positive, then the signs of the last terms of the binomial factors are both positive or both negative. If the last term in the trinomial is negative, then the signs of the last terms of the binomial factors have opposite signs.
5. We write $6x^2 = (3x)(2x)$ and $-3 = (-1)(3)$. Therefore, $6x^2 + 7x - 3 = (3x - 1)(2x + 3)$. We check the factoring by multiplication.
7. In Exercise 17, we need to find every combination of factors of 4 with factors of 15 and add to -16. In Exercise 18, we need to find every combination of factors of 9 with factors of 4 and add to 15.
9. $(3x + 1)(x + 4)$ **11.** $(4s - 1)(s - 7)$ **13.** $(7a + 1)(a - 6)$
15. $(3x - 1)(x + 7)$ **17.** $(2k - 3)(2k - 5)$ **19.** $(2y - 7)(3y + 2)$
21. $(2t - 1)(5t + 8)$ **23.** prime   **25.** prime
27. $2(4k - 3)(2k + 1)$ **29.** $-(7b + 6)(b - 1)$
31. $4h(2h + 1)(h - 2)$ **33.** $(3y + 4)(y + 1)$ **35.** $(3m - 5)(m - 1)$
37. $(5k + 2)(k - 6)$ **39.** $(11h - 3)(h + 2)$ **41.** $(4x + 1)(x + 8)$
43. $(9r - 2)(r - 3)$ **45.** $(5u + 2)(3u - 4)$ **47.** $(2a + 11)(2a - 1)$
49. prime   **51.** $-(5a + 3)(a + 2)$ **53.** $2k(6h - 5)(3h + 2)$
55. $265(8x - 1)(x - 2)$ **57.** $(x + 2)^2$ **59.** $(b - 3)^2$
61. $(5r + 2)^2$ **63.** $2(3u - 5)^2$ **65.** $(2x + 3)(x + 6)$
67. $(4t + 1)(3t - 2)$ **69.** $(9b - 2)^2$ **71.** $(4u - 3)(10u + 7)$
73. $-(6a + 1)(a - 5)$ **75.** $2(2x + 7)(x - 2)$ **77.** prime
79. $5(3p + 4)^2$ **81.** $b(5a + 3)(2a - 7)$ **83.** $2(13x - 2y)(2x - y)$
85. $2c(3a - 7b)^2$ **87.** $-15(4x + 3)(x - 10)$ **89.** $(3y - 1)(y + 4)$
91. $5x^2 + 15x - 2x - 6 = 5x(x + 3) - 2(x + 3) = (5x - 2)(x + 3)$
93. $(3x - 5)(11x + 2)$ **95.** $6x^2 - 5x - 6 = (2x - 3)(3x + 2)$
97. $9x^2 + 24x + 16 = (3x + 4)^2$ **99.** $b = -11, -7, 7, 11$
101. $b = -19, -8, -1, 1, 8, 19$

Section 6.4

1. Remove any GCF. Write the first term as a^2 and the second term as b^2. Factor the binomial as a product of conjugates, $(a + b)(a - b)$.
3. Write the first term as a^3 and the second term as b^3. The sum of two cubes is a product of a binomial and a trinomial. The sign of the binomial is the same as the sign of the given binomial, $a + b$. The trinomial is $a^2 - ab + b^2$.
5. Answers vary; $125a^3 + 27b^3 = (5a)^3 + (3b)^3 = (5a + 3b)$ $(25a^2 - 15ab + 9b^2)$
7. We cannot write $x^2 + 9$ as a product of two binomials.
9. $(x + 1)(x - 1)$ **11.** $(y + 6)(y - 6)$ **13.** $2x(x + 2)(x - 2)$
15. $(3a + 2)(3a - 2)$ **17.** $(5b + 7)(5b - 7)$
19. $(11 + 2x)(11 - 2x)$ **21.** $8(t + 3)(t - 3)$

23. $(3m + 16)(3m - 16)$ **25.** $3x^3(x + 4)(x - 4)$ **27.** prime

29. $(11y + 12)(11y - 12)$ **31.** $(8a + 3b)(8a - 3b)$

33. $(mn + 11)(mn - 11)$ **35.** $\left(3x + \dfrac{1}{2}\right)\left(3x - \dfrac{1}{2}\right)$

37. $\left(\dfrac{1}{3}a + \dfrac{2}{7}\right)\left(\dfrac{1}{3}a - \dfrac{2}{7}\right)$ **39.** $(x - 2)(x^2 + 2x + 4)$

41. $(a - 4)(a^2 + 4a + 16)$ **43.** $2t(t - 3)(t^2 + 3t + 9)$

45. $(5y - 1)(25y^2 + 5y + 1)$ **47.** $(1 - 2t)(1 + 2t + 4t^2)$

49. $(m - 6n)(m^2 + 6mn + 36n^2)$ **51.** $\left(\dfrac{1}{3}s - 1\right)\left(\dfrac{1}{9}s^2 + \dfrac{1}{3}s + 1\right)$

53. $(y + 3)(y^2 - 3y + 9)$ **55.** $(a + 4)(a^2 - 4a + 16)$

57. $5t(t + 4)(t^2 - 4t + 16)$ **59.** $125(8 + b^2)$

61. $(2m + n)(4m^2 - 2mn + n^2)$ **63.** $\left(a + \dfrac{1}{3}\right)\left(a^2 - \dfrac{1}{3}a + \dfrac{1}{9}\right)$

65. $(x^2 + 4)(x + 2)(x - 2)$ **67.** $2(y^2 + 5)(y^4 - 5y^2 + 25)$

69. $(16 + y^3)(16 - y^3)$ **71.** $\left(r^2 + \dfrac{9}{4}\right)\left(r + \dfrac{3}{2}\right)\left(r - \dfrac{3}{2}\right)$

73. $(4a^2 + 9)(2a + 3)(2a - 3)$ **75.** $9y(y + 2)(y - 2)$

77. $(16 + a)(16 - a)$ **79.** $(13 + b)(13 - b)$

81. $(2a - 5)(4a^2 + 10a + 25)$ **83.** prime

85. $(a + 10)(a^2 - 10a + 100)$ **87.** $(4x^2 + 9)(2x + 3)(2x - 3)$

89. $2(y^2 - 3)(y^4 + 3y^2 + 9)$ **91.** $5(x^2y^2 + 9z^2)(xy + 3z)(xy - 3z)$

93. $(8r + 7s)(8r - 7s)$ **95.** $(0.4a - 0.5b)(0.4a + 0.5b)$

97. $\left(2x - \dfrac{1}{5}y\right)\left(4x^2 + \dfrac{2}{5}xy + \dfrac{1}{25}y^2\right)$

99. $\left(\dfrac{1}{4}x^2 + \dfrac{1}{9}\right)\left(\dfrac{1}{2}x - \dfrac{1}{3}\right)\left(\dfrac{1}{2}x + \dfrac{1}{3}\right)$

101. $x^2 - 64x = x(x - 64)$ **103.** $x^3 + 64 = (x + 4)(x^2 - 4x + 16)$

105. $(x - 2)(x^2 + 2x + 4)$ **107.** $(9x^2 + 4)(3x - 2)(3x + 2)$

Piece It Together Sections 6.1–6.4

1. $3(x^2 - 6)(x + 5)$ **2.** $(4s + 1)(4s - 1)(s - 3)$

3. $(x + 4)(x + 9)$ **4.** $(y - 6)(y + 5)$ **5.** prime

6. $(b - 4)(b - 15)$ **7.** $5(y + 4)(y - 1)$ **8.** $6(y - 2)(y + 7)$

9. $p(pq - 5)(pq + 7)$ **10.** $r(rs - 4)(5rs + 6)$

11. $5(3x - 2)(5x - 1)$ **12.** $(2y - 1)(5y - 4)$ **13.** $(2r - 5)^2$

14. $2(5h + 1)^2$ **15.** $\left(2y + \dfrac{7}{5}\right)\left(2y - \dfrac{7}{5}\right)$ **16.** prime

17. $(x + 3y)(x - 3y)$ **18.** $(10a + b)(100a^2 - 10ab + b^2)$

19. $(2x - 5y)(4x^2 + 10xy + 25y^2)$ **20.** $(9a^2 + 1)(3a + 1)(3a - 1)$

Section 6.5

1. If the product of two numbers is zero, then at least one of the numbers has to be zero.

3. Write the quadratic equation in standard form. Factor the resulting polynomial. Use the zero products property to solve.

5. In $(x - 5)(x - 1) = 12$, you need to write the equation in standard form by multiplying the binomials on the left and subtracting 12 from both sides. Now factor the left side and apply the zero products property to solve. In $(x - 5)(x - 1) = 0$, you can apply the zero products property to solve.

7. $\{-2, 5\}$ **9.** $\{0, 3\}$ **11.** $\{-7, 10\}$ **13.** $\left\{-2, \dfrac{1}{2}\right\}$

15. $\{6\}$ **17.** $\left\{-12, \dfrac{3}{2}\right\}$ **19.** $\{-6, 0, 2\}$ **21.** $\{0, 7\}$

23. $\{-7, -3, 1\}$ **25.** $\{-2, 3\}$ **27.** $\{1\}$ **29.** $\{-5, -2\}$

31. $\left\{-1, \dfrac{3}{2}\right\}$ **33.** $\{-5, 5\}$ **35.** $\{-1, 0\}$ **37.** $\{-8, 4\}$

39. $\left\{\dfrac{4}{3}, \dfrac{5}{2}\right\}$ **41.** $\{-2, 4\}$ **43.** $\{-3, 0, 3\}$ **45.** $\left\{-2, \dfrac{1}{2}, 8\right\}$

47. $\{-2, -1, 1, 2\}$ **49.** $\{-3, 0, 6\}$ **51.** $\{-1, 0, 1\}$

53. $\left\{-\dfrac{3}{2}, -\dfrac{1}{7}, \dfrac{1}{7}, \dfrac{3}{2}\right\}$ **55.** $\left\{-\dfrac{3}{4}, 2\right\}$ **57.** $\{-9\}$

59. $\{-13, 0\}$ **61.** $\left\{-\dfrac{10}{3}, 3\right\}$ **63.** $\left\{-1, \dfrac{5}{4}\right\}$

65. $\{-10, -1, 1, 10\}$ **67.** $\{-2, 2\}$ **69.** $\{-11, 0, 11\}$

71. $\left\{-3, -\dfrac{1}{2}, \dfrac{1}{2}, 3\right\}$ **73.** $\left\{-2, \dfrac{4}{5}\right\}$ **75.** $\{-2, 0, 1\}$

77. $\varnothing$ **79.** $\{-3, 0, 3\}$ **81.** $\left\{-5, -\dfrac{4}{3}\right\}$ **83.** $\left\{-5, \dfrac{3}{2}, 4\right\}$

85. $(x - 3)(x - 1) = 3$
$$x^2 - 4x + 3 = 3$$
$$x^2 - 4x = 0$$
$$x(x - 4) = 0$$
$$x = 0 \quad \text{or} \quad x = 4$$

87. $2(x + 1) = 0$
$$x + 1 = 0$$
$$x = -1$$

89. $\{-9, 6\}$ **91.** $x^2 - 7x + 10 = 0$ **93.** $x^2 + 3x = 0$

95. $x^2 - 8x + 16 = 0$ **97.** $(x - 1)(x - 3)(x + 4) = 0$

Section 6.6

1. First, read the problem and identify the known and unknown. Assign a variable to the unknown and set up any formula. Second, translate the words into a mathematical equation.

3. If x is an odd integer, then $x, x + 2, x + 4, \ldots$ represent consecutive odd integers. If x is an even integer, then $x, x + 2, x + 4, \ldots$ represent consecutive even integers.

5. 4 ft by 12 ft   **7.** 10 in.   **9.** 4 ft   **11.** 11, 12

13. 6 and 8 or -6 and -4 **15.** 3 and 5 or -2 and 0   **17.** 25 cats

19. 100 or 450   **21.** 10,000 units   **23.** 4, 3, and 5

25. 7, 24, and 25   **27.** 12 ft

29. Heidi traveled 12 mi and Spencer traveled 16 mi.

31. 3 sec   **33.** 1 and 6.5 sec   **35.** 15 sec

37. 10 parties   **39.** 50 cakes   **41.** 10 sec

43.
$$x(x + 2) = 2(2x + 2) - 1$$
$$x^2 + 2x = 4x + 4 - 1$$
$$x^2 - 2x - 3 = 0$$
$$(x - 3)(x + 1) = 0$$
$$x = 3 \quad \text{or} \quad x = -1$$
The two consecutive positive odd integers are 3 and 5.

45. 3, 27

CHAPTER 6 SUMMARY

Section 6.1

1. product   **2.** greatest common factor   **3.** smallest

4. product; coefficients; variable expressions

5. distributive property; $a(b + c)$ **6.** grouping

7. multiplying

Section 6.2

8. $c; b$ **9.** same   **10.** different   **11.** common factor

Section 6.3

12. trial; error; grouping   **13.** perfect square

Section 6.4

14. $(a - b)(a + b)$; Answers vary; $x^2 - 16$

15. $(a - b)(a^2 + ab + b^2)$; Answers vary; $x^3 - 8$

16. $(a + b)(a^2 - ab + b^2)$; Answers vary; $x^3 + 27$

17. does not factor; Answers vary; $x^2 + 4$

Section 6.5

18. factors; zero   **19.** $ax^2 + bx + c = 0$

20. standard form; factor; factor; zero; solve

21. polynomials   **22.** cubic; quartic   **23.** solutions

Section 6.6

24. area

25. $x, x + 1, x + 2, \ldots; x, x + 2, x + 4, \ldots; x, x + 2, x + 4, \ldots$

26. Revenue; profit   **27.** $a^2 + b^2 = c^2$; legs; hypotenuse

CHAPTER 6 REVIEW EXERCISES

Section 6.1

1. $4x$ **3.** $4y$ **5.** $28r^3$ **7.** $(2x + 5)(x - 3)$
9. $(7x - 3)(x + 4)$ **11.** $(4c^2 - 5d)(3c - 2d)$ **13.** $-5p(2p + 7)$
15. prime   **17.** $2\pi r(r + h)$

Section 6.2

19. $(x - 4)(x + 7)$ **21.** prime   **23.** $-x^2(x - 4)(x - 5)$
25. $-5y(x^2 - 2x + 6)$

Section 6.3

27. $(2a - 7)^2$ **29.** $-(3y - 7)(y + 3)$ **31.** prime
33. $-7(2h^2 - 4h + 1)$ **35.** $2(13x + 10y)(x - y)$
37. $-3(12a - 7b)(3a + 2b)$ **39.** $-3(2h^2 - 1)^2$
41. $-12(5x + 2)(x - 8)$

Section 6.4

43. $9y(y - 2)(y^2 + 2y + 4)$ **45.** $y(4x + 5)(16x^2 - 20x + 25)$
47. $(x^2y^2 + z^2)(x^4y^4 - x^2y^2z^2 + z^4)$
49. $(a - 10bc)(a^2 + 10abc + 100b^2c^2)$
51. $(25y^2 + 4z^2)(5y + 2z)(5y - 2z)$
53. $\left(\frac{1}{2}a + bc\right)\left(\frac{1}{4}a^2 - \frac{1}{2}abc + b^2c^2\right)$ **55.** prime

Section 6.5

57. $\left\{-\frac{2}{5}, 2\right\}$ **59.** $\{-11, 0\}$ **61.** $\left\{-3, 0, \frac{1}{2}\right\}$
63. $\{-4, 12\}$ **65.** $\left\{-\frac{1}{5}, 0, \frac{1}{5}\right\}$

Section 6.6

67. 9 ft by 16 ft   **69.** 15, 16   **71.** 8, 15, and 17   **73.** 20 ft
75. 12.5 sec

CHAPTER 6 TEST

1. c   **3.** c   **5.** d   **7.** b   **9.** d   **11.** $(4 - y)(5 - p)$
13. $3(a - 3)(a + 2)$ **15.** $5(2a + 1)(a - 4)$ **17.** $(9p - 7)(9p + 7)$
19. $(5x - 3y)(25x^2 + 15xy + 9y^2)$ **21.** $(4y^2 + 1)(16y^4 - 4y^2 + 1)$
23. $\{-3, 3\}$ **25.** $\left\{-\frac{2}{3}, \frac{3}{2}\right\}$ **27.** $\{-5, -3, 4\}$
29. The legs are 6 and 8 units.

CHAPTERS 1–6 CUMULATIVE REVIEW EXERCISES

1. a. -185 **b.** 10   **c.** 0
3. a. $\{-1\}$ **b.** $\left\{\frac{15}{2}\right\}$ **c.** $\varnothing$
5. a.
![number line from -15 to -5 with arrow left from -10]
$(-\infty, -10)$
$\{x \mid x < -10\}$
b.
![number line from -12 to 8 segment between -7 and 3.5]
$(-7, 3.5)$
$\{a \mid -7 < a < 3.5\}$
7. a. (0, 68.6), In 2005, 68.6% of the U.S. households owned desktops and laptops.
b. The percent of computer ownership is predicted to be 83.3 in 2011.

c. The percent of computer ownership is predicted to be 88.2 in 2013.
d.

9. a. $f(0) = 3$ **b.** $g(-3) = -15$
c. $f(3) = 2; x = -2, x = 6$
11. a. $\{(-2, 1)\}$ **b.** $\{(2, -7)\}$
13. Her hourly wage as a part-time receptionist is \$13.50 and her hourly wage as a peer tutor is \$7.50.
15. a. $-\dfrac{2p^3}{q^{11}}$ **b.** $-\dfrac{2n^5}{a^{14}}$ **c.** -5
17. a. 5 g   **b.** 1.25 g   **19.** \$4.11 × 10^3
21. a. 480 ft   **b.** 496 ft   **23.** $-0.6x^2 + 126x + 673$ dollars
25. a. $20x^2 + 7x - 3$ **b.** $125b^3 + 900b^2 + 2160b + 1728$
c. $p^3 - 9p^2 + 27p - 27$ **d.** $225x^2 - y^2$ **e.** $8r^3 + 27s^3$
27. a. $x(x + 12)(x - 9)$ **b.** $5y(2y + 3)(y - 4)$ **c.** prime
d. $-7(x + 3)(x - 3)(x + 2)(x - 2)$ **e.** $-9x(x - 2)^2$
29. a. $(x - 9)(x - 6)$ **b.** $3y(y - 3)^2$ **c.** $-6x(2x - 1)(x + 5)$
d. $(a - 12b)(a - 4b)$ **e.** $2y(x + 12)(x + 9)$ **f.** prime
31. a. $(5p + 9)(p + 3)$ **b.** $p(5q + 9)(q - 3)$
c. $(2x - 1)(15x + 7)$ **d.** $(4x + 7y)^2$
e. $3(13ab - 4)(3ab + 1)$ **f.** prime
33. a. $5x(x - 3)(x + 3)$ **b.** $2(5m + 4n)(5m - 4n)$
c. $2(3x - 4y)(9x^2 + 12xy + 16y^2)$
d. $2(y + 5z)(y^2 - 5yz + 25z^2)$
e. prime   **f.** $(9a^2 + 4)(3a + 2)(3a - 2)$
g. $3(c^2 - 2)(c^4 + 2c^2 + 4)$ **h.** $(x^2y^2 + 4z^2)(xy + 2z)(xy - 2z)$
i. $\left(\frac{1}{2}a - \frac{3}{5}b\right)\left(\frac{1}{2}a + \frac{3}{5}b\right)$
35. a. 9 ft by 6 ft   **b.** 12 and 13   **c.** -1 and -3, 10 and 12
d. 150 passengers   **e.** 3, 4, and 5   **f.** 12,000 units
g. 18 mi   **h.** 10 items or 35 items   **i.** 9 sec

CHAPTER 7

Section 7.1

1. The difference between a rational expression and a rational function is that only a rational function uses function notation to represent the quotient of polynomials. Both have the restriction that the denominator cannot equal zero.
3. To simplify a rational expression, factor the numerator and denominator completely and divide out their common factors.
5. It is in simplest form because x is not a common factor in the numerator.
7. 0.4, 13,003, -1297, 3.07, 2.87
9. 1, 4477.03, -4523.03, 589.22, -611.23
11. 1.5, -496, 504, 3.95, 4.05
13. $\{x \mid x \text{ is a real number and } x \neq 3\}$
15. $\{x \mid x \text{ is a real number and } x \neq -7, x \neq 7\}$
17. $\{x \mid x \text{ is a real number and } x \neq 0, x \neq 2\}$
19. $\left\{x \mid x \text{ is a real number and } x \neq -\frac{7}{2}\right\}$ **21.** $(-\infty, 5) \cup (5, \infty)$
23. $\left(-\infty, -\frac{1}{2}\right) \cup \left(-\frac{1}{2}, \frac{1}{2}\right) \cup \left(\frac{1}{2}, \infty\right)$
25. $(-\infty, 0) \cup \left(0, \frac{2}{3}\right) \cup \left(\frac{2}{3}, \infty\right)$ **27.** $\left(-\infty, -\frac{1}{2}\right) \cup \left(-\frac{1}{2}, \infty\right)$

29. $\left(-\infty, \frac{5}{2}\right) \cup \left(\frac{5}{2}, \infty\right)$ **31.** $\dfrac{2x}{3x+2}$ **33.** $\dfrac{x}{3x+2}$

35. $\dfrac{x}{3x-2}$ **37.** 1   **39.** -1 **41.** can't be simplified

43. $\dfrac{3x}{5x-2}$ **45.** $\dfrac{2x+7}{x^2-5x+25}$ **47.** $\dfrac{6x+5}{x^2+2x+4}$

49. $\dfrac{x+1}{4x^2+3}$ **51.** $\dfrac{2(x+4)}{4x^2+13}$ **53.** $\dfrac{x-3}{-2x-4}$ or $\dfrac{-x+3}{2x+4}$

55. $\dfrac{-3x+4}{-x+1}$ or $\dfrac{3x-4}{x-1}$ **57.** $\dfrac{3x-1}{-x+4}$ or $\dfrac{-3x-1}{x-4}$

59. $\dfrac{4x+9}{-2x-1}$ or $\dfrac{-4x+9}{2x+1}$

61. a. The reunion will cost $105 per person if 50 people attend
   b. $75 per person if 75 attend,
   c. $60 per person if 100 attend
63. a. The cost to remove 80% of the pollutants is $330,000.
   b. The cost to remove 95% of the pollutants is $1,567,500.
65. a. The monthly payment is $84 for 6 months
   b. $69 for 1 month,   **c.** $61.50 for 24 months
67. 7.4, 1190.03, -130.3, 293.12, -307.12
69. $-\dfrac{4}{3}$, -2597, 2603, 2.75, 3.28

71. $\left\{x\middle|x \text{ is a real number and } x \neq -\dfrac{4}{3}\right\}; \left(-\infty, -\dfrac{4}{3}\right) \cup \left(-\dfrac{4}{3}, \infty\right)$

73. $\left\{x\middle|x \text{ is a real number and } x \neq 0, x \neq \dfrac{3}{4}\right\};$

   $(-\infty, 0) \cup \left(0, \dfrac{3}{4}\right) \cup \left(\dfrac{3}{4}, \infty\right)$

75. $\left\{x\middle|x \text{ is a real number and } x \neq -\dfrac{3}{2}, x \neq \dfrac{3}{2}\right\};$

   $\left(-\infty, -\dfrac{3}{2}\right) \cup \left(-\dfrac{3}{2}, \dfrac{3}{2}\right) \cup \left(\dfrac{3}{2}, \infty\right)$

77. $\{x|x \text{ is a real number and } x \neq 3\}; (-\infty, 3) \cup (3, \infty)$

79. $\dfrac{3x}{2x-5}$ **81.** $\dfrac{x}{3x+7}$ **83.** $\dfrac{4x}{x+4}$ **85.** $\dfrac{x}{3x+2}$

87. $\dfrac{2(2x-5)}{16x^2-12x+9}$ **89.** $\dfrac{2(2x+5)}{9x^2+6x+4}$ **91.** $\dfrac{5x-6}{x^2-10}$

93. $\dfrac{7x-3}{2x^2-5}$ **95.** $-\dfrac{9x-18}{x-10}$ or $\dfrac{9x-18}{-x+10}$

97. $\dfrac{-x^2-3x}{4x-1}$ or $\dfrac{x^2+3x}{-4x+1}$

99. a. The reunion will cost $70 per person if 100 people attend
   b. $61 per person if 125 attend,
   c. $55 per person if 150 attend
101. a. The cost to remove 85% of the pollutants is $382,500.
   b. The cost to remove 97% of the pollutants is $2,182,500.
103. a. The monthly payment is $56 for 18 months
   b. $54.75 for 24 months,
   c. $54 for 30 months
105. $2x \neq 0$
   $x \neq 0$
   The domain of the function is $(-\infty, 0) \cup (0, \infty)$.
107. $\dfrac{9x^2-4}{3x^2+x-2} = \dfrac{(3x-2)(3x+2)}{(x+1)(3x-2)}$

   $= \dfrac{(3x+2)\cancel{(3x-2)}^{\,1}}{(x+1)\cancel{(3x-2)}^{\,1}}$

   $= \dfrac{3x+2}{x+1}$

109.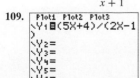

111. Answers vary; $f(x) = \dfrac{x+5}{x-3}$

113. Answers vary; $f(x) = \dfrac{2x}{(x-1)(x-5)}$

Section 7.2

1. Factor the numerator and denominator completely. Divide out the common factors. Multiply the remaining factors.
3. Change the division to multiplication of the reciprocal, factor all numerators and denominators completely, divide out the common factors, and multiply the remaining factors.

5. $\dfrac{1}{a^2}$ **7.** $\dfrac{9b}{7a}$ **9.** $\dfrac{1}{8}$ **11.** $\dfrac{2(x-5)}{x+1}$ **13.** $4(x-1)$

15. $-\dfrac{2}{3x^2(x-5)}$ **17.** 1   **19.** $\dfrac{4(x^2+2x+4)}{x+3}$

21. $\dfrac{2(x-3)}{x+2}$ **23.** $\dfrac{x^2(2x+3)}{3x-1}$ **25.** $\dfrac{3a}{25}$ **27.** $\dfrac{16a}{b^2}$

29. $\dfrac{5y^3}{3x^3}$ **31.** $\dfrac{4}{x(x-3)}$ **33.** $\dfrac{x}{4(x+1)}$ **35.** $\dfrac{x+5}{x-2}$

37. $\dfrac{x^2-6x+1}{(x-2)(x+1)}$ **39.** $\dfrac{x+2}{2(x+1)}$ **41.** $\dfrac{x+3}{2(x+1)}$

43. about 81 kg   **45.** about 42 lb   **47.** about 1062.7 ft

49. about 138.8 Euros   **51.** $\dfrac{63x^3}{50y^2}$ **53.** $\dfrac{(x+3)(x-5)}{2x^2}$

55. $\dfrac{3(x-6)}{x(x+2)}$ **57.** $\dfrac{x^3+16}{(x-2)(x+2)}$ **59.** $\dfrac{x+3}{2}$ **61.** $\dfrac{x+1}{x}$

63. $\dfrac{2}{x}$ **65.** $\dfrac{x+3}{x-3}$ **67.** $\dfrac{x+2}{5}$ **69.** about 134.54 mi

71. 42 min   **73.** about 100 lb   **75.** 6 ft = 6 ft · 1

$\approx \dfrac{6 \text{ ft} \cdot 1 \text{ m}}{3.28 \text{ ft}}$

$\approx \dfrac{6}{3.28 \text{ m}}$

$\approx 1.829 \text{ m}$

77. $\dfrac{x+3}{2x^3} \div 4x^2 = \dfrac{x+3}{2x^3} \cdot \dfrac{1}{4x^2} = \dfrac{x+3}{8x^5}$

79.

Since the y-values are the same, the quotient is correct.

81.

Since the y-values are the same, the product is correct.

Section 7.3

1. You add or subtract the numerators and place the result over the like denominator.
3. Two rational expressions are equivalent if one rational expression can be obtained from the other rational expression by multiplying a form of 1.

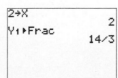

5. Factor each denominator completely. The LCD will include all factors in the first denominator and any additional ones from other denominators that are not in the first denominator.

7. $\dfrac{2}{x}$ **9.** $\dfrac{x}{x-1}$ **11.** $\dfrac{3y+3}{y-1}$ **13.** $\dfrac{1}{x-4}$ **15.** $\dfrac{1}{y-1}$

17. $\dfrac{2x+1}{x+1}$ **19.** $\dfrac{1}{x+3}$ **21.** $-\dfrac{1}{x+1}$ **23.** $\dfrac{3(x-2)}{x-3}$

25. $15a$ **27.** $6x^2$ **29.** $12x^2$ **31.** x^3y^2 **33.** $15a^5b^4$

35. $x(x-5)$ **37.** $x(x+2)$ **39.** $(x-3)(x-1)$

41. $(x+2)(x+5)$ **43.** $(x-5)(x+7)$ **45.** $(x+3)(x-3)(x-3)$

47. $(2x+3)(2x-3)(3x+2)$ **49.** 3   **51.** 8   **53.** $4x^2$

55. $3(x-1)$ **57.** $6a^2$ **59.** $2x^2+4$ **61.** $3x+9$

63. $3x$ **65.** $x-2x^2$ **67.** $-6x^2$ **69.** x^2-3x

71. $2t^3+2t$ **73.** $8ab-10b$ **75.** $\dfrac{16x}{3x+8}$ **77.** $\dfrac{9a+3}{9a-1}$

79. 1   **81.** $\dfrac{1}{3x+2}$ **83.** $\dfrac{1}{2x-1}$ **85.** $-\dfrac{3}{x-2}$

87. $x(x-9)$ **89.** $(4x+3)(x+1)$ **91.** $(5x+2)(5x-2)(x+3)$

93. $3(5x-2)^2$ **95.** $x(x+1)$ **97.** $4x(x-1)$ **99.** $9a$

101. $-20x^2$ **103.** $\dfrac{2}{5x}+\dfrac{3}{5x}=\dfrac{5}{5x}=\dfrac{\not{5}}{\not{5}x}=\dfrac{1}{x}$

105. $x(x+2)$

Section 7.4

1. To add or subtract rational expressions with different denominators, you need to find the least common denominator, convert each fraction to an equivalent fraction with the LCD as its denominator, add or subtract the numerators, and simplify, if possible.

3. $\dfrac{5x+1}{x^2}$ **5.** $\dfrac{2x-1}{6x^2}$ **7.** $\dfrac{1}{x+2}$ **9.** $\dfrac{50(3r-5)}{r(r-5)}$

11. $\dfrac{7x+8}{(x-1)(x+2)}$ **13.** $\dfrac{5x(x+4)}{(x+1)(x+5)}$ **15.** $\dfrac{1}{x-3}$

17. $\dfrac{x^2+5x+12}{(x+3)(x+1)}$ **19.** $\dfrac{x(2x-7)}{(x-2)(x-3)}$ **21.** $\dfrac{4(x^2-x+1)}{x(x-2)}$

23. $\dfrac{3(x-5)}{4(x+3)}$ **25.** $\dfrac{5(x+3)}{x(x+5)}$ **27.** $\dfrac{5x-12}{x-2}$ **29.** $\dfrac{4x-14}{x-5}$

31. $\dfrac{3x-1}{(x+1)(x-1)(2x+1)}$ **33.** $\dfrac{24x+17}{(2x+1)(x-3)(4x+3)}$

35. $-\dfrac{x}{x-2}$ **37.** $\dfrac{7x-10}{(x-10)(x+2)}$ **39.** $\dfrac{2x^2+4x-1}{(x-9)(x+1)}$

41. $\dfrac{5x+6}{x+1}$ **43.** $\dfrac{x+6}{x-1}$ **45.** $\dfrac{3(x+13)}{(x+3)^2(x-3)}$

47. $\dfrac{2(x-7)}{(x-1)(x-2)(x-3)}$ **49.** $\dfrac{150(5r-11)}{(r-10)(r+3)}$

51. $\dfrac{7a}{3}+\dfrac{7a}{4}=\dfrac{28a+21a}{12}=\dfrac{49a}{12}$

53.

Since the y-values are the same, the addition has been performed correctly.

55.
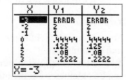
Since the y-values are the same, the subtraction has been performed correctly.

Piece It Together Sections 7.1–7.4

1. $x=5$ **2.** $x=-5$ and $x=2$ **3.** $\dfrac{2x}{x^2+3}$ **4.** $\dfrac{x-3}{x+6}$

5. $\dfrac{2x-5}{3x-1}$ **6.** $\dfrac{x^2+5x+25}{x+5}$ **7.** $\dfrac{5x+1}{x+1}$ or $\dfrac{-5x-1}{-x-1}$

8. $5(x-3)$ **9.** $-\dfrac{1}{x(x+1)}$ **10.** $\dfrac{x^2-2x+4}{4}$ **11.** $\dfrac{x-2}{x+1}$

12. $\dfrac{15a}{b^5}$ **13.** $\dfrac{2(x+6)}{x^2}$ **14.** 1   **15.** 1   **16.** $\dfrac{2(x-2)}{3(x-5)}$

17. 3   **18.** $\dfrac{1}{x+4}$ **19.** $\dfrac{3x+14}{(x-7)(x+7)}$

20. $\dfrac{3x-1}{(2x+1)(x+1)(x-1)}$

Section 7.5

1. Rewrite the complex fraction by multiplying the fraction in the numerator by the reciprocal of the fraction in the denominator. Simplify by dividing out common factors.

3. Find the LCD of the fractions in the numerator of the complex fraction. Multiply the numerator and denominator of the complex fraction by the LCD and simplify. Factor the denominator completely. Divide out the common factors.

5. You need to rewrite each negative exponent as a reciprocal and simplify the resulting complex fraction.

7. You need to multiply the numerator and denominator by the LCD and simplify.

9. $\dfrac{3}{10xy^2}$ **11.** $\dfrac{9}{5ab}$ **13.** $\dfrac{15}{5a-4}$ **15.** $\dfrac{12}{a}$ **17.** $\dfrac{1}{rs}$

19. $-\dfrac{1}{3xy}$ **21.** $\dfrac{4}{35xy^3}$ **23.** $\dfrac{7b^2}{12a}$ **25.** $-\dfrac{1}{3xy}$ **27.** $\dfrac{1}{5xy}$

29. $\dfrac{s}{2r}$ **31.** $\dfrac{y}{5x}$ **33.** $\dfrac{3y}{x}$ **35.** $\dfrac{12}{11}$ **37.** $\dfrac{3}{40}$ **39.** $-\dfrac{5}{12}$

41. $-\dfrac{17}{15}$ **43.** $\dfrac{4}{27}$ **45.** $\dfrac{8}{27}$ **47.** $-\dfrac{15}{16}$ **49.** $\dfrac{y}{4x}$

51. $\dfrac{6}{a}$ **53.** $\dfrac{28}{10x-5}$ **55.** $\dfrac{1}{4xy}$ **57.** $-\dfrac{1}{3ab}$ **59.** $\dfrac{s}{4r}$

61. $\dfrac{y}{6x}$ **63.** $\dfrac{3y-5x}{xy}$ **65.** $\dfrac{4b+a}{3ab}$ **67.** $\dfrac{s-9r}{2rs}$

69. $\dfrac{3x+y}{5xy}$ **71.** $\dfrac{12}{6x-1}$ **73.** $\dfrac{8}{14a-49}$ **75.** $-\dfrac{27}{7}$

77. $\dfrac{3}{5}$ **79.** $-\dfrac{14}{25}$ **81.** $\dfrac{27}{62}$ **83.** 19   **85.** $-\dfrac{29}{3}$

87. $\dfrac{\dfrac{2}{x}+\dfrac{5}{y}}{10x+4y}=\dfrac{\dfrac{2y+5x}{xy}}{2(5x+2y)}$

$=\dfrac{2y+5x}{xy}\div\dfrac{2(2y+5x)}{1}$

$=\dfrac{2y+5x}{xy}\cdot\dfrac{1}{2(2y+5x)}$

$=\dfrac{\overset{1}{\cancel{2y+5y}}}{xy}\cdot\dfrac{1}{2\underset{1}{\cancel{(2y+5x)}}}=\dfrac{1}{2xy}$

89. $\dfrac{4x^{-2} - y^{-2}}{2x^{-1} - y^{-1}} = \dfrac{\dfrac{4}{x^2} - \dfrac{1}{y^2}}{\dfrac{2}{x} - \dfrac{1}{y}}$

$= \dfrac{\left(\dfrac{4}{x^2} - \dfrac{1}{y^2}\right)x^2 y^2}{\left(\dfrac{2}{x} - \dfrac{1}{y}\right)x^2 y^2}$

$= \dfrac{4y^2 - x^2}{2xy^2 - x^2 y}$

$= \dfrac{(2y + x)\overset{1}{\cancel{(2y - x)}}}{xy\underset{1}{\cancel{(2y - x)}}}$

$= \dfrac{2y + x}{xy}$

91. The numerator and denominator of the complex fraction must be entered in parentheses.

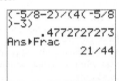

93.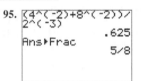

95.

Section 7.6

1. A rational equation is an equation that contains a rational expression.

3. We must check the answers because if one of the solutions is a value that makes the LCD zero, then it is an extraneous solution and must be discarded from the solution set.

5. The LCD for both problems is the same, $(x - 3)(x + 5)$. To add fractions, you convert the fraction $\dfrac{4}{(x + 5)}$ to an equivalent fraction with the LCD as the denominator. To solve the equation, you multiply both sides by the LCD and solve the resulting equation.

7. $\left\{\dfrac{1}{2}\right\}$ **9.** $\left\{\dfrac{1}{2}\right\}$ **11.** $\{3\}$ **13.** $\varnothing$ **15.** $\left\{-\dfrac{7}{3}\right\}$

17. $\left\{\dfrac{9}{20}\right\}$ **19.** $\{4, 1\}$ **21.** $\{-2\}$ **23.** $\{-18\}$ **25.** $\varnothing$

27. $\{-7, 7\}$ **29.** $\{-2, 3\}$ **31.** $\varnothing$ **33.** $\left\{-\dfrac{5}{4}, -2\right\}$

35. $h = \dfrac{2A}{b}$ **37.** $a = \dfrac{f}{P}$ **39.** $t = \dfrac{v - v_0}{a}$ **41.** $r = \dfrac{mv^2}{F}$

43. $A = \dfrac{P^2 L}{2uE}$ **45.** $I = \dfrac{E}{2G} - v$ **47.** $b = \dfrac{2ac}{3a - c}$

49. $\left\{\dfrac{3}{2}\right\}$ **51.** $\left\{-\dfrac{16}{5}\right\}$ **53.** $\varnothing$ **55.** $\left\{\dfrac{1}{21}\right\}$

57. $\left\{-\dfrac{1}{2}, 3\right\}$ **59.** $\left\{-5, \dfrac{2}{3}\right\}$ **61.** $\left\{\dfrac{1}{3}\right\}$ **63.** $\{-5\}$

65. $\varnothing$ **67.** $b = \dfrac{2A - ha}{h}$ **69.** $V = \dfrac{kT}{P}$ **71.** $R = \dfrac{R_1 R_2}{R_1 + R_2}$

73. $y_2 = mx_2 - mx_1 + y_1$

75. $\cancel{(x + 1)}(x + 4)\dfrac{3}{\cancel{x + 1}} = \dfrac{3}{\cancel{x + 4}}(x + 1)\cancel{(x + 4)}$

$3(x + 4) = 3(x + 1)$
$3x + 12 = 3x + 3$
$0 = -9$

So, there is no solution.

77. $\dfrac{2}{x + 2} - \dfrac{5}{3x - 1} = \dfrac{4x}{(3x - 1)(x + 2)}$

$(3x - 1)\cancel{(x + 2)}\dfrac{2}{\cancel{x + 2}} - \dfrac{5}{\cancel{3x - 1}}\cancel{(3x - 1)}(x + 2)$

$= \dfrac{4x}{\cancel{(3x - 1)}\cancel{(x + 2)}}\cancel{(3x - 1)}\cancel{(x + 2)}$

$2(3x - 1) - 5(x + 2) = 4x$
$6x - 2 - 5x - 10 = 4x$
$x - 12 = 4x$
$-12 = 3x$
$x = -4$

So, the solution is $x = -4$.

79. $\left\{-1, \dfrac{5}{2}\right\}$ **81.** $\left\{\dfrac{5}{4}\right\}$

Section 7.7

1. A ratio relates the measurement of two quantities.

3. A ratio is an expression and a proportion is an equation.

5. Two triangles that are similar have the same shape but different sizes. Their corresponding angles are equal and the lengths of their corresponding sides are proportional.

7. $\{4\}$ **9.** $\left\{\dfrac{10}{3}\right\}$ **11.** $\left\{\dfrac{7}{5}\right\}$ **13.** $\{72\}$ **15.** $\{18.25\}$

17. $\left\{-\dfrac{5}{3}\right\}$ **19.** $\dfrac{10}{3}$ cups   **21.** 3 biscuits   **23.** 4 U.S. dollars

25. 8.24 Swiss francs   **27.** 8.5 in.   **29.** 5.83 cm   **31.** 20.75 lb

33. 90 lb   **35.** $\dfrac{35}{2}$ **37.** $\dfrac{40}{9}$ **39.** $\dfrac{15}{4}$ hr   **41.** 200 mi

43. Cassie runs at 3 mph, and Tran walks at 1 mph.

45. The first airplane flies at 560 mph, and the second airplane flies at 460 mph.

47. $6.65   **49.** $\dfrac{91}{9}$ **51.** $\dfrac{500}{3}$ Cal

53. The first airplane flies at 480 mph, and the second airplane flies at 400 mph.

55. 73.71 lb   **57.** $\dfrac{16}{3}$ mph   **59.** correct; $x = \dfrac{15}{4}$ cups

61. $\dfrac{1}{5} + \dfrac{1}{9} = \dfrac{1}{x}$

$x = \dfrac{45}{14}$ hr

CHAPTER 7 SUMMARY

Section 7.1

1. quotient; polynomials   **2.** $\dfrac{P(x)}{Q(x)}$; polynomials

3. value; variable   **4.** zero; undefined   **5.** denominator; zero

6 undefined   **7.** factor; factors   **8.** $y - x$; -1

9. $\dfrac{-A}{B}$; $\dfrac{A}{-B}$

Section 7.2

10. factor; divide; multiply   **11.** reciprocal; $\dfrac{A}{B} \cdot \dfrac{D}{C}$

12. 1; does not

Section 7.3

13. same   **14.** least common denominator; largest   **15.** equivalent

Section 7.4

16. LCD; equivalent; LCD

Section 7.5

17. complex
18. combine the fractions in the numerators and denominators and multiply the reciprocal; LCD of all the fractions

Section 7.6

19. rational   **20.** LCD   **21.** zero; undefined; extraneous

Section 7.7

22. ratio; colon; fraction   **23.** proportion; ratios
24. Similar triangles; shape; sizes; equal; proportional
25. working together   **26.** $d = rt$

CHAPTER 7 REVIEW EXERCISES

Section 7.1

1. $-2, 3917.02, -3883.02, 209.41$ **3.** $1, -795, 805, 4.92$
5. $\{x \mid x \text{ is a real number and } x \neq 10\}$ or $(-\infty, 10) \cup (10, \infty)$
7. $\{x \mid x \text{ is a real number and } x \neq 6, x \neq -6\}$ or $(-\infty, -6) \cup (-6, 6) \cup (6, \infty)$
9. $\{x \mid x \text{ is a real number and } x \neq -2, x \neq 0\}$ or $(-\infty, -2) \cup (-2, 0) \cup (0, \infty)$
11. $\dfrac{x}{x + 12}$ **13.** $\dfrac{x}{3x + 5}$ **15.** $\dfrac{2x - 2}{9x^2 - 3x + 1}$
17. $\dfrac{5x + 2}{2x^2 + 15}$ **19.** $x + 2$ **21.** $\dfrac{-2x + 9}{x + 10}$ or $-\dfrac{2x - 9}{x + 10}$
23. $\dfrac{x^2 - 7x}{x^2 - 5}$ or $\dfrac{-x^2 + 7x}{-x^2 + 5}$ **25.** $75, \$66, \$60$

Section 7.2

27. $\dfrac{8x^4}{3y^3}$ **29.** $\dfrac{(x + 3)(x + 1)}{x^2}$ **31.** $\dfrac{x(x + 2)}{4(x + 5)}$ **33.** $\dfrac{x^2 - x + 1}{x - 1}$
35. $\dfrac{2}{x + 4}$ **37.** $\dfrac{x + 2}{x - 2}$ **39.** $\dfrac{x + 3}{3}$ **41.** about 199.5 mi
43. 21 min   **45.** about 64 lb

Section 7.3

47. $2x(x + 3)$ **49.** $(x + 7)(3x - 2)$ **51.** $35x$ **53.** $x(x + 2)$

Section 7.4

55. $\dfrac{14x}{x - 6}$ **57.** $\dfrac{12x^2 + 9x}{(5x + 2)(2x + 5)}$ **59.** $\dfrac{5x}{(x - 12)(x + 8)}$
61. $\dfrac{3x^2 + 8x - 4}{(x - 7)(x + 1)}$ **63.** $\dfrac{12x + 6}{5x + 3}$

Section 7.5

65. $\dfrac{8x^2}{7}$ **67.** $\dfrac{a(6b - a)}{b(5a + b)}$ **69.** $\dfrac{1}{14x}$ **71.** $\dfrac{xy}{7y + 3x}$
73. $\dfrac{12}{23}$ **75.** $\dfrac{15}{2}$ **77.** $-\dfrac{1}{16}$ **79.** 0

Section 7.6

81. $\left\{\dfrac{15}{4}\right\}$ **83.** $\varnothing$ **85.** $\left\{\dfrac{1}{6}\right\}$ **87.** $\left\{\dfrac{2}{5}, -4\right\}$ **89.** $\left\{\dfrac{1}{3}\right\}$
91. $y_1 = \dfrac{2A - hy_0 - hy_2}{2h}$ **93.** $R_2 = \dfrac{RR_1R_3}{R_1R_3 - RR_3 - RR_1}$

Section 7.7

95. $15.92   **97.** 6   **99.** 6 in.   **101.** 62.37 lb   **103.** 4.5 mph

CHAPTER 7 TEST

1. a   **3.** a   **5.** d   **7.** d
9. a. $-\dfrac{1}{3x + 5}$ **b.** $\dfrac{2x - 7}{x + 4}$ **c.** $\dfrac{4a^2 + 6a + 9}{2a + 3}$
11. a. $40x^3$ **b.** $b(b + 5)$ **c.** $(x + 7)(x - 1)(2x + 5)$
13. a. $\dfrac{8}{7ab}$ **b.** $\dfrac{b - a}{a + b}$
15. a. about 7.1 lb   **b.** $a = \dfrac{33}{5}$ **c.** 1.7 hr   **d.** 5 mph

CHAPTERS 1–7 CUMULATIVE REVIEW EXERCISES

1. $\dfrac{67}{60}$ **3.** 54   **5.** 8.1   **7.** 9   **9.** 13   **11.** $\{7\}$
13. $\left\{-\dfrac{5}{3}\right\}$ **15.** $\{-30\}$ **17.** $\{8\}$ **19.** $l = \dfrac{C - 8w}{4}$
21. $x = 70 - \dfrac{1}{2}y$ **23.** parallel   **25.** perpendicular
27. $y = -6x - 2; 6x + y = -2$
29. a. $f(x) = 1050x + 43,500$ **b.** \$52,950
   c. $f(8) = \$51,900$; Ming's income for teaching 8 additional credit-hour is \$51,900.
31. $\{(-4, 2)\}$ **33.** $\{(-6, 8)\}$ **35.** $-\dfrac{64}{x^3y^9}$ **37.** $\dfrac{1}{27x^6y^{21}}$
39. $-7x^2 + 21xy - 21y^2$ **41.** $-0.4x^2 + 109x + 618$
43. $28x^2 + 23x - 15$ **45.** $8b^3 - 12b^2 + 6b - 1$
47. $27r^3 - 8s^3$ **49.** $6x^2 - 36x + 129 - \dfrac{397}{x + 3}$
51. $5y(2y - 3)(y + 4)$ **53.** $-7(a^2 + 4)(a^2 + 9)$ **55.** $3y(y - 4)^2$
57. $(a - 3b)(a - 12b)$ **59.** prime   **61.** $5a(8b + 9)(b - 2)$
63. $(3x + 2y)^2$ **65.** $(3p + 7q)(p - q)$
67. $5(x + 2y)(x^2 - 2xy + 4y^2)$ **69.** prime
71. $(x^2 + 9y^2)(x - 3y)(x + 3y)$ **73.** $\{-4, 5\}$ **75.** $\{-2, 0, 5\}$
77. 110 passengers
79.

x	y
-2	Undefined
0	0
3	0

81. \$1000; \$1400   **83.** $x = 0, -5; \dfrac{2}{x + 5}$
85. $x = -5, 2; \dfrac{x^2 + 2x + 4}{x + 5}$ **87.** $-\dfrac{3x - 6}{x + 4}$ or $\dfrac{3x - 6}{-x - 4}$
89. $\dfrac{x - 5}{2(x + 3)}$ **91.** $\dfrac{x^2 + x + 1}{x + 1}$ **93.** $2x$
95. $\dfrac{112}{9}$ or $12\dfrac{4}{9}$ in.   **97.** $6ab$ **99.** $2x^2 + 4x$
101. $x(6x + 15)$ **103.** $(x + 7)(x - 3)$ **105.** $-\dfrac{x}{x - 4}$
107. $\dfrac{x + 6}{x - 2}$ **109.** $\dfrac{7x - 4}{x - 1}$ **111.** $\dfrac{-x - 39}{(x + 3)(x - 3)^2}$
113. $\dfrac{5}{4x^2}$ **115.** $\dfrac{2y + 3}{4y}$ **117.** $\dfrac{4(x + 3y)}{3}$ **119.** -3
121. $\dfrac{15}{8}$ **123.** $\dfrac{11}{12}$ **125.** $\{-15\}$ **127.** $\varnothing$ **129.** $\{-5, 7\}$
131. $\left\{\dfrac{1}{2}\right\}$ **133.** $R_1 = \dfrac{R_2R_3 - RR_3 - RR_2}{RR_2R_3}$ **135.** \$9.03
137. 6.25 in.

CHAPTER 8

Section 8.1

1. The domain of a function is the set of x-values in the given set of ordered pairs.
3. The range of a function is the set of y-values in the given set of ordered pairs.
5. To find the domain of an algebraic function with a fraction, exclude any number that makes the denominator of the fraction equal to zero or the function undefined.
7. Domain $= \{-3, -2, -1, 0\}$, Range $= \{-4, 0, 1, 5\}$
9. Domain $= \{-10, 0, 2, 5\}$, Range $= \{-7, 3, 4\}$
11. Domain $= \{2011$ Ford Edge AWD, 2011 Honda CR-V 4WD, 2011 Subaru Forester AWD, 2011 Toyota RAV4 4WD$\}$, Range $= \{26, 27\}$
13. Domain $= \{1970, 1980, 1990, 2000, 2005\}$, Range $= \{8{,}580{,}887, 12{,}096{,}895, 13{,}818{,}637, 15{,}312{,}289, 17{,}487{,}475\}$
15. Domain $= \{1972, 1992, 2004\}$, Range $= \{9.8, 27.6, 29.4\}$
17. Domain $= \{-5, -4, 2, 3\}$, Range $= \{-5, -3, 1, 4\}$
19. Domain $= \{-3, -2, 1, 2\}$, Range $= \{0, 1\}$
21. Domain $= (-\infty, \infty)$, Range $= (-\infty, \infty)$
23. Domain $= (-\infty, \infty)$, Range $= \{-3\}$
25. Domain $= (-\infty, \infty)$, Range $= (-\infty, 4]$
27. Domain $= [-2, 2]$, Range $= [0, 2]$
29. Domain $= (-\infty, \infty)$, Range $= (-\infty, \infty)$
31. Domain $= (-\infty, \infty)$, Range $= [1, \infty)$
33. $(-\infty, \infty)$ 35. $(-\infty, \infty)$ 37. $(-\infty, 0) \cup (0, \infty)$
39. $(-\infty, -5) \cup (-5, \infty)$ 41. $[-6, \infty)$ 43. $\left(-\infty, \dfrac{3}{4}\right]$
45. $(0, 8)$ 47. $(0, 2)$ 49. $(0, 9)$ 51. $\{1, 2, \ldots, 13\}$
53. $\{0, 1, 2, \ldots\}$
55. Domain $= \{$January 2011, February 2011, March 2011, April 2011$\}$, Range $= \{39°F, 49°F, 55°F, 63°F\}$
57. Domain $= \{0, 4, 13, 20\}$, Range $= \{-3, -2, 44\}$
59. Domain $= \{-5, -4, -2, 1, 2\}$, Range $= \{0, 3, 4, 5\}$
61. Domain $= (-\infty, 7]$, Range $= [0, \infty)$ 63. $\left[\dfrac{7}{3}, \infty\right)$
65. Domain $= (-\infty, \infty)$ 67. Domain $= (-\infty, 19) \cup (19, \infty)$
69. Domain $= (0, 4)$ 71. Domain $= \{0, 1, 2, 3, \ldots\}$
73. If the function was $f(x) = \sqrt{4x - 1}$, then Petra's work would be correct. The function $f(x) = 4x - 1$ is defined for all real numbers. So, the domain is $(-\infty, \infty)$.
75. The graph extends left and right indefinitely, so the domain is $(-\infty, \infty)$.

77.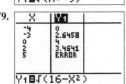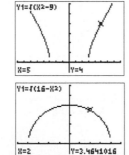

79.

81. Answers vary.   83. Answers vary.

85. Answers vary.   87. Answers vary.

Section 8.2

1. A function is a linear function if it is in the form $f(x) = mx + b$, where m and b are real numbers. An example is $f(x) = \dfrac{3}{5}x + 1$.
3. The x-intercept is found by replacing y, or $f(x)$, with 0 and solving for x. In other words, we must solve the equation $f(x) = 0$. The y-intercept is found by replacing x with 0 and solving for y. In other words, we find $f(0)$.
5. False. The given point is the x-intercept, not the y-intercept. The function is $f(x) = 7x - 35$.
7. False. $f(0) = -7$ corresponds to $(0, -7)$, which is the y-intercept.
9. Domain $= (-\infty, \infty)$,   11. Domain $= (-\infty, \infty)$,
Range $= (-\infty, \infty)$ Range $= (-\infty, \infty)$

13. Domain $= (-\infty, \infty)$,   15. Domain $= (-\infty, \infty)$,
Range $= (-\infty, \infty)$ Range $= (-\infty, \infty)$

17. Domain $= (-\infty, \infty)$,   19. Domain $= (-\infty, \infty)$,
Range $= \{-4\}$ Range $= \{7\}$

  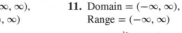

21. Domain $= (-\infty, \infty)$,
Range $= (-\infty, \infty)$

23. **a.** $f(0) = 38$; So, 38% of high school students smoked cigarettes in 1997.
    b. $f(3) = 30.5$ and $f(8) = 18$; In 3 yr after 1997, or 2000, 30.5% of high school students were smokers and in 8 yr after 1997, or 2005, 18% of high school students smoked cigarettes.
    c. $x = 14$; So, in 14 yr after 1997, or 2011, 3% of high school students smoked cigarettes.
25. **a.** $f(0) = 110{,}000$; The initial value of the limo is \$110,000.
    b. $f(3) = 44{,}000$ and $f(4) = 22{,}000$; The value of the limo when it is 3 yr old \$44,000 and its value is \$22,000 when it is 4 yr old.
    c. $x = 5$; In 5 yr, the value of the limo is \$0.
27. $f(x) = 4x + 5$ 29. $f(x) = \dfrac{1}{2}x + 7$ 31. $f(x) = 8x - 75$

33. $f(x) = \dfrac{4}{5}x + 11$ **35.** $f(x) = x + 5$ **37.** $f(x) = \dfrac{4}{3}x$

39. $f(x) = -2x - 6$ **41.** $f(x) = -\dfrac{4}{3}x - 1$

43. $f(x) = -\dfrac{1}{7}x + 3$ **45.** $f(x) = -\dfrac{1}{5}x + 1$

47. a. $f(x) = 0.05x + 40,000$ **b.** $45,000$ **c.** $500,000$

49. a. $f(x) = 99.99 + 0.25x$

    b. $f(300) = 174.99$; The total monthly charge for an additional 300 min is $174.99.

    c. $x = 100$; The total monthly charge for an additional 100 min is $124.99.

51. a. $f(x) = -0.47x + 27$ **b.** 19.95 million

53. a. $f(x) = 25.69x + 1699.2$ **b.** 1979.0 thousand

55. Domain $= (-\infty, \infty)$,   **57.** Domain $= (-\infty, \infty)$,
    Range $= (-\infty, \infty)$ Range $= (-\infty, \infty)$

59. Domain $= (-\infty, \infty)$,   **61.** Domain $= (-\infty, \infty)$,
    Range $= \{-5\}$ Range $= (-\infty, \infty)$

63. $f(x) = 5x - 5$ **65.** $f(x) = -\dfrac{1}{2}x + 3$ **67.** $f(x) = \dfrac{1}{4}x$

69. $f(x) = -6x - 3$ **71.** $f(x) = -\dfrac{5}{3}x + 5$ **73.** $f(x) = 7$

75. a. $f(x) = 0.42x + 72$

    b. $f(0) = 72$; The cost of renting the car for 1-day without driving it is $72.

    c. $f(350) = 219$; The cost of the 1-day rental is $219 if the car is driven 350 mi.

    d. $x = 70$; The cost of the 1-day rental is $101.40 if the car is driven 70 mi.

77. a. $f(x) = 3.6x + 8$

    b. $f(200) = 728$; A 200-lb person will burn 728 Cal by playing basketball for 1 hr.

    c. $x = 180$; A 180-lb person will burn 656 Cal by playing basketball for 1 hr.

79. The given point is not the y-intercept, so we have to use the point-slope form to write the equation of the function.

$$y - 3 = \frac{4}{3}(x - 8)$$
$$y - 3 = \frac{4}{3}x - \frac{32}{3}$$
$$y = \frac{4}{3}x - \frac{32}{3} + 3$$
$$y = \frac{4}{3}x - \frac{32}{3} + \frac{9}{3}$$
$$y = \frac{4}{3}x - \frac{23}{3}$$

    So, $f(x) = \dfrac{4}{3}x - \dfrac{23}{3}$.

81. $f(x) = 2x + 7$ **83.** $f(x) = -\dfrac{5}{11}x - \dfrac{8}{11}$

Piece it Together Sections 8.1 AND 8.2

1. Domain $= \{-7, 0, 2, 5\}$, Range $= \{3, 4, 5\}$

2. Domain $= (-\infty, \infty)$, Range $= [0, \infty)$

3. $(-\infty, 0) \cup (0, \infty)$ **4.** $\left(-\infty, \dfrac{4}{9}\right]$

5. Domain $= (-\infty, \infty)$,   **6.** Domain $= (-\infty, \infty)$,
    Range $= (-\infty, \infty)$ Range $= \{8\}$

7. $f(x) = \dfrac{9}{2}x - 5$ **8.** $f(x) = 4$ **9.** $f(x) = -\dfrac{9}{5}x + \dfrac{17}{5}$

10. $f(x) = \dfrac{1}{2}x + 3$ **11.** $f(x) = 3x - 30$

12. a. $f(x) = 45x + 50$

    b. $f(3) = 185$; The cost of a 3-hr house call is $185.

    c. $x = 1.5$; The cost of a 1.5-hr house call is $117.50.

13. a. $f(x) = 0.25x + 4675$

    b. $f(41,050) = 14,937.50$; The tax owed on an income of 75,000 is $41,937.50.

    c. $x = 16,050$; The tax owed on an income of $50,000 is $8687.50.

Section 8.3

1. A piecewise-defined function is a function that is comprised of two or more functions defined for a piece of the domain.

3. The graph of $f(x) + k$ is the graph of $f(x)$ shifted up k units. The graph of $f(x) - k$ is the graph of $f(x)$ shifted down k units.

5. The graph of $-f(x)$ is the graph of $f(x)$ reflected across the x-axis.

7. Domain $= (-\infty, \infty)$,   **9.** Domain $= (-\infty, \infty)$,
    Range $= [-4, \infty)$ Range $= [0, \infty)$

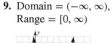

11. Domain $= (-\infty, \infty)$,   **13.** Domain $= [0, \infty)$,
    Range $= [1, \infty)$ Range $= [2, \infty)$

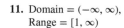 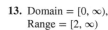

15. Domain $= (1, \infty]$,   **17.**
    Range $= [0, \infty)$

 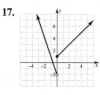
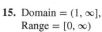

19.   **21.**   **23.**

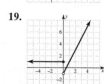 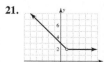 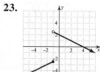

25. $f(x) = \begin{cases} 0.44, & 0 < x < 1 \\ 0.64, & 1 \le x < 2 \\ 0.84, & 2 \le x < 3 \\ 1.04, & 3 \le x < 3.5 \end{cases}$

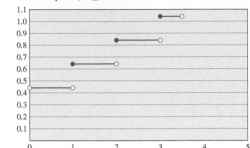

27. $f(x) = \begin{cases} 50, & 0 < x \le 17 \\ 75, & 17 < x \le 23 \\ 100, & 23 < x \le 35 \\ 150, & 35 < x \le 47 \\ 200, & 47 < x \le 59 \\ 250, & 59 < x \le 71 \\ 300, & 71 < x \le 95 \\ 400, & x > 95 \end{cases}$

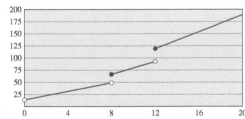

29. $f(x) = \begin{cases} 13.39 + 4.38x, & 0 < x < 8 \\ 13.39 + 6.57x, & 8 \le x < 12 \\ 13.39 + 8.76x, & x \ge 12 \end{cases}$

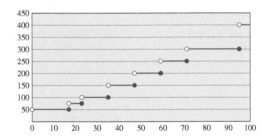

31. Domain $= [0, \infty)$, Range $= (-\infty, 2]$ **33.** Domain $= (-\infty, \infty)$, Range $= (-\infty, 1]$

35. Domain $= (-\infty, \infty)$, Range $= (-\infty, \infty)$ **37.** Domain $= [0, \infty)$, Range $= [3, \infty)$

39. Domain $= (-\infty, \infty)$, Range $= [-1, \infty)$ **41.**

43.   **45.**   **47.**

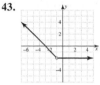

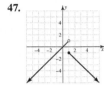

49. $f(x) = \begin{cases} 2.50 & x \le \dfrac{1}{5} \\ 2x + 2.50 & x > \dfrac{1}{5} \end{cases}$

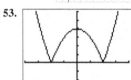

51.   **53.**

55.   **57.** Domain $= (-\infty, 1) \cup (1, \infty)$, Range $= \{-2, 3\}$

$f(x) = \begin{cases} 3 & x < 1 \\ -2 & x > 1 \end{cases}$

Section 8.4

1. Direct variation describes the situation where y varies directly as x or y is directly proportional to x. The equation is $y = kx$, where k is the constant of variation.

3. Joint variation describes the situation where z varies jointly as x and y or z is jointly proportional to x and y. The equation is $z = kxy$, where k is the constant of variation.

5. V varies jointly as the square of r and h.

7. G varies jointly as m_1 and m_2 and inversely as the square of r.

9. $k = 5; y = 5x$ **11.** $k = \dfrac{2}{3}; y = \dfrac{2}{3}x$ **13.** $k = 0.6; y = 0.6x$

15. \$330,000   **17.** 20.4 in.   **19.** Table 1; $y = \dfrac{1}{2}x$

21. Graph 1; $y = 0.4x$ **23.** $k = 3; y = \dfrac{3}{x}$ **25.** $k = \dfrac{3}{2}; y = \dfrac{3}{2x}$

27. $k = 1.6; y = \dfrac{1.6}{x}$ **29.** 310 mL   **31.** 36.9 amp

33. Table 2; $y = \dfrac{2.5}{x}$ **35.** $k = \dfrac{5}{4}; z = \dfrac{5}{4}xy$

37. $k = \dfrac{1}{18}; z = \dfrac{1}{18}xy$ **39.** $k = 0.025; z = 0.025xy$

41. $k = \dfrac{1}{9}; y = \dfrac{1}{9}p^2q^3$ **43.** $k = 1.5; c = 1.5b\sqrt{a}$

45. $k = 6.4; w = \dfrac{6.4u^2}{v^3}$ **47.** $k = 312.5; 327.68$ metric tons

49. $y = \dfrac{1}{10}\sqrt[3]{x}$ **51.** $y = \dfrac{3.6}{x^2}$ **53.** $w = \dfrac{15u^3}{\sqrt{v}}$ **55.** 440 mL

57. 30.1 in.   **59.** 47.2 amp   **61.** Original volume is halved.

63. New volume is one-eighth of the original volume.

65. $y = k\sqrt{x}$
$48 = k\sqrt{4}$
$48 = 2k$
$k = 24$
$y = 24\sqrt{x}$

CHAPTER 8 SUMMARY

Section 8.1

1. x-values   **2.** y-values   **3.** left; right   **4.** bottom; top
5. defined   **6.** denominator; undefined
7. greater; than; equal; to

Section 8.2

8. $f(x) = mx + b;\ y = mx + b;$ line; $(-\infty, \infty);\ (-\infty, \infty)$
9. horizontal; $(-\infty, \infty);\ \{c\}$ **10.** point; slope; ordered; pairs; slope
11. same; negative; reciprocals   **12.** rate; change; initial

Section 8.3

13. plot; points; shape; left; right; bottom; top   **14.** up; down
15. right; left   **16.** x-axis
17. piecewise-defined; open; closed; closed; open

Section 8.4

18. kx; constant   **19.** $\dfrac{k}{x}$; constant   **20.** kxy **21.** Combined

CHAPTER 8 REVIEW EXERCISES

Section 8.1

1. Domain = $\{-5, -1, 0, 2\}$, Range = $\{-4, -2, 11, 13\}$
3. Domain = $\{6$ P.M., 7 P.M., 8 P.M., 9 P.M., 10 P.M.$\}$,
Range = $\{78°F, 79°F, 80°F, 82°F\}$
5. Domain = $\{-3, -1, 0, 2, 4\}$, Range = $\{-5, -2, 3, 4\}$
7. Domain = $[-3, \infty)$, Range = $[-2, \infty)$ **9.** $(-\infty, 1) \cup (1, \infty)$
11. $[4, \infty)$ **13.** $(0, 6)$ **15.** $\{0, 1, 2, 3, \dots\}$

Section 8.2

17. Domain = $(-\infty, \infty)$, Range = $(-\infty, \infty)$

19. Domain = $(-\infty, \infty)$, Range = $\{6.5\}$

21. a. $f(0) = 4889.5$; The average yearly cost of attending a 2-yr public institution was $4889.50 in 2000.
b. $f(5) = 6567.5$ and $f(11) = 8581.1$; The average yearly cost of attending a 2-yr public institution was $6567.50 in 2005 and $8581.10 in 2011.
c. $x = 15$; In 15 yr after 2000, or 2015, the average yearly cost of attending a 2-yr public institution will be $9923.50.
23. $f(x) = -\dfrac{3}{5}x + 4$ **25.** $f(x) = -6x - 30$
27. $f(x) = -x + 3.6$ **29.** $f(x) = -5x + 24$
31. $f(x) = 70.2x + 2394$;
$f(16) = 3517.2$ thousand or 3,517,200

Section 8.3

33. Domain = $(-\infty, \infty)$, Range = $[-1, \infty)$

35. Domain = $(-\infty, \infty)$, Range = $[-4, \infty)$

37. Domain = $[1, \infty)$, Range = $[0, \infty)$

39. Domain = $(-\infty, \infty)$, Range = $(0, \infty)$

41. $f(x) = \begin{cases} 2x, & 0 < x \le 2 \\ 3x, & 2 < x \le 6 \\ 32, & 6 < x \le 24 \\ 68, & 24 < x \le 48 \\ 104, & 48 < x \le 72 \end{cases}$

Section 8.4

43. $k = 2;\ r = 2s^2$ **45.** $k = 8;\ y = \dfrac{8}{\sqrt[3]{x}}$
47. $k = 50;\ w = \dfrac{50\sqrt[3]{u}}{v^2}$ **49.** 160 mL
51. The new volume is half of the original volume.

CHAPTER 8 TEST

1. d   **3.** a   **5.** b   **7.** c

9. Domain = $(-\infty, \infty)$, Range = $\{-2\}$

11. Domain = $(-\infty, \infty)$, Range = $[0, \infty)$

13. Domain = $(-\infty, \infty)$, Range = $(-\infty, 1) \cup \{3\}$

15. $f(x) = -\dfrac{5}{3}x + \dfrac{1}{3}$

17. $f(x) = -\dfrac{1}{5}x + 5$
19. $f(x) = 2818.5x + 52,734;\ f(6) = 69,645$; In 2014, the enrollment of ASU will be 69,645.
21. $k = 21.5;\ y = 21.5x$; Constance earns $430 for working 20 hr.

CHAPTERS 1–8 CUMULATIVE REVIEW EXERCISES

1. -32 **3.** 1   **5.** $\dfrac{1}{6}$ **7.** 33   **9.** 8.1
11. -8.1 **13.** 125
15. $-\dfrac{1}{3}, -\dfrac{1}{4}, 0,$ undefined   **17.** $\{520\}$
19. $y = \dfrac{5}{3}x + 11;\ 5x - 3y = -33$

21. $\{(-3, 2)\}$, consistent with independent equations

23. $-\dfrac{b^6}{8a^{15}}$

25. $10p^2 - 18pq - 13q^2$　**27.** $16x^2 - 81y^2$

29. $7x^2 - 3x + 6 + \dfrac{4}{x}$　**31.** $2y(5x - 2y)^2$　**33.** $(3x + 1)(12x - 5)$

35. prime　**37.** $(c - 1)(c^2 + c + 1)(c + 1)(c^2 - c + 1)$

39. $\{-2, 2, -3, 3\}$　**41.** 35 thousand units

43. 0, 0, undefined　**45.** \$600; \$400

47. $(-\infty, -3) \cup (-3, 0) \cup (0, 3) \cup (3, \infty); \dfrac{2}{x - 3}$

49. $\dfrac{x - 2}{7x + 3}$ or $\dfrac{2 - x}{-7x - 3}$　**51.** $\dfrac{x - 3}{x + 2}$　**53.** $\dfrac{x - 2}{x + 1}$

55. 30 lb　**57.** $\dfrac{2x^2 - 5x - 4}{(x - 6)(x + 4)}$　**59.** $\dfrac{-x + 13}{(x - 1)(x + 1)^2}$

61. $\dfrac{2ax^2}{3b}$　**63.** $\dfrac{xy}{4y - 3x}$　**65.** $-\dfrac{11}{10}$　**67.** $\left\{\dfrac{17}{9}\right\}$

69. $b = \dfrac{2acd}{cd - 3ad - 4ac}$　**71.** 450 mph, 500 mph

73. $(-\infty, 3]$

75. Domain $= (-\infty, \infty)$, Range $= \{2\}$

77. **a.** $f(0) = 5969.8$; The expenditure per pupil was \$5969.80 in 1994.

b. $f(7) = 8562.6$; In 2001, the expenditure per pupil was \$8562.60.

c. $x = 18$; In 2012, the expenditure per pupil was \$12,637.

79.　　　　　　　　　　　　　　**81.** $z = 4x^2(u + v)$

CHAPTER 9

Section 9.1

1. The intersection of two sets is the set of elements that the two sets have in common.

3. The intersection of sets A and B can be the empty set if the two sets do not have any elements in common.

5. A compound inequality consists of two inequalities joined by the terms "and" or "or."

7. Find the solution set of inequality 1 and the solution set of inequality 2. Find the intersection of the solution sets of inequalities 1 and 2 and write the final solution in interval notation and provide its graph.

9. $A \cap B = [-4, 2], A \cup B = (-10, 6)$

11. $A \cap B = [-9, -3], A \cup B = [-10, -3]$

13. $A \cap B = (6, 20), A \cup B = [6, 20]$

15. $A \cap B = [-1, 7), A \cup B = (-\infty, 7]$

17. $A \cap B = (-11, 11), A \cup B = (-\infty, \infty)$

19. $A \cap B = (-10, 5], A \cup B = [-12, 8)$

21. $[-1, 5]$　**23.** $\varnothing$　**25.** $(2, 3)$　**27.** $\varnothing$

29. $\left(-\dfrac{4}{3}, 1\right)$　**31.** $\{-1\}$　**33.** $\left[1, \dfrac{5}{3}\right]$　**35.** $(7, \infty)$

37. $(2, \infty)$　**39.** $[4, \infty)$　**41.** $(-\infty, -4)$

43. $\left(-\dfrac{5}{3}, \infty\right)$　**45.** $\left(-\infty, \dfrac{5}{4}\right)$　**47.** $(-\infty, -3) \cup (1, \infty)$

49. $(-\infty, \infty)$　**51.** $(-\infty, 6]$　**53.** $(-\infty, \infty)$

55. $\left(-\infty, \dfrac{4}{9}\right)$　**57.** $\left(-\infty, \dfrac{1}{7}\right] \cup (1, \infty)$

59. $\left(-\infty, \dfrac{6}{5}\right]$　**61.** $(-\infty, 0] \cup \left(\dfrac{2}{3}, \infty\right)$

63. Lawrence must earn between 75 and 100 on his fifth quiz.

65. Sakinah must score between 90.5 and 100 on the final exam.

67. The subscriber can use between 25 and 50 additional minutes.

69. Miguel can invite between 55 and 111 people to his birthday party.

71. $[-7, 12)$　**73.** $(-20, 9]$　**75.** $(-\infty, \infty)$　**77.** $\varnothing$

79. $(-\infty, 12.2)$　**81.** $\varnothing$　**83.** $(-10, \infty)$

85. $\left(-\infty, \dfrac{5}{7}\right) \cup \left(\dfrac{17}{5}, \infty\right)$　**87.** $(-\infty, -1)$

89. $\left(-\infty, \dfrac{1}{3}\right) \cup (3, \infty)$　**91.** $\varnothing$　**93.** $\left(\dfrac{1}{3}, \infty\right)$

95. $\left(-\dfrac{1}{7}, \infty\right)$　**97.** $\left(-\infty, -\dfrac{5}{2}\right) \cup (4, \infty)$　**99.** $(4.75, 10.5)$

101. $\left(-\infty, \dfrac{11}{4}\right)$

103. Frederick can afford to hire the electrician for 2 to 15 hr.

105. Maddy can score between 52 and 97 on her fifth test.

107. $10 - 3x < 4$　　and　　$5x + 23 > 3$
$\quad\quad -3x < -6$　　and　　$5x > -20$
$\quad\quad\quad x > 2$　　and　　$\quad x > -4$
The intersection is $x > 2$. The solution set is $(2, \infty)$.

109. $(-1, 3)$　　**111.** $(-\infty, \infty)$

Section 9.2

1. The number zero has an absolute value of zero.

3. Isolate the absolute value on one side of the equation and the constant on the other side. Set the expression inside the absolute value equal to the number(s) whose absolute value is the constant. (i) If the constant is positive, there are two solutions. (ii) If the constant is zero, there is one solution. (iii) If the constant is negative, there are no solutions.

5. $\{-10, 10\}$　**7.** $\left\{-\dfrac{5}{2}, -\dfrac{1}{2}\right\}$　**9.** $\{-2, 10\}$

11. $\varnothing$　**13.** $\varnothing$　**15.** $\left\{-\dfrac{17}{4}, \dfrac{11}{4}\right\}$　**17.** $\varnothing$

19. $\left\{\dfrac{1}{4}\right\}$　**21.** $\left\{\dfrac{7}{5}, -\dfrac{13}{5}\right\}$　**23.** $\left\{-\dfrac{2}{9}, 4\right\}$　**25.** $\left\{-\dfrac{7}{2}, \dfrac{3}{4}\right\}$

27. $\{1\}$　**29.** $\left\{\dfrac{1}{5}, -15\right\}$　**31.** $\left\{-\dfrac{1}{3}, -7\right\}$

29. $\left\{\dfrac{1}{5}, -15\right\}$　**31.** $\left\{-\dfrac{1}{3}, -7\right\}$　**33.** $\left\{-\dfrac{13}{14}, -\dfrac{17}{16}\right\}$

35. $\left\{-\dfrac{24}{5}, \dfrac{24}{13}\right\}$　**37.** 0.00126　**39.** 0.00172

41. The exact weight of the person can be a minimum of 183.9 lb and a maximum of 184.7 lb.

43. The actual percentage of those polled who will reportedly vote for Lipsey is between 48% and 54% and for Easley is between 42% and 48%.

45. $\left\{-\dfrac{6}{5}, \dfrac{9}{5}\right\}$ **47.** $\left\{0, \dfrac{8}{5}\right\}$ **49.** $\left\{-\dfrac{25}{29}, -\dfrac{25}{41}\right\}$

51. $\left\{\dfrac{32}{5}, -4\right\}$ **53.** $\varnothing$ **55.** $\left\{\dfrac{5}{11}, 9\right\}$

57. $\left\{-\dfrac{5}{3}, \dfrac{5}{3}\right\}$ **59.** $\left\{-\dfrac{1}{2}\right\}$ **61.** $\{-12, 19.5\}$

63. $\left\{-\dfrac{85}{9}, \dfrac{65}{9}\right\}$ **65.** $\{-60, 86\}$

67. The actual measurement of the object is between 9.68 units and 12.28 units.

69. The absolute error is 0.031 m.

71. Candidate A will receive between 25% and 35% of the vote and candidate B will receive between 34% and 44% of the vote.

73. $|4x - 5| = 11$

$4x - 5 = 11$ or $4x - 5 = -11$

$4x = 16$ $4x = -6$

$x = 4$ $x = -\dfrac{3}{2}$

So, the solution set is $\left\{-\dfrac{3}{2}, 4\right\}$.

75. The solution set is $\varnothing$ since the absolute value expression is equal to a negative number.

77. $\left\{-\dfrac{28}{17}, \dfrac{38}{17}\right\}$ **79.** $\left\{-\dfrac{13}{2}, \dfrac{17}{6}\right\}$

81. $x \geq 0$ and $a = 0$ or $x \geq 0$ and $a = 2x$

83. Answers vary. $|2x - 6| = 0$

Section 9.3

1. Isolate the absolute value expression to one side of the inequality. Remove the absolute value sign by writing $X < k$ and $X > -k$. Solve the resulting compound inequality.

3. After isolating the absolute value expression to one side of the inequality, you have a negative number on the other side. If the inequality sign is $<$ or $\leq$, then there is no solution. Otherwise, all real numbers.

5. Answers will vary.   **7.** $[-6, 6]$ **9.** $(-15, 23)$

11. $\left(-\dfrac{7}{2}, \dfrac{1}{2}\right)$ **13.** $\left[-\dfrac{5}{3}, \dfrac{7}{3}\right]$ **15.** $\left(\dfrac{1}{5}, \dfrac{3}{5}\right)$

17. $\left[-1, \dfrac{3}{2}\right]$ **19.** $\left(-2, \dfrac{8}{3}\right)$ **21.** $\left(-3, -\dfrac{1}{2}\right)$

23. $\left\{-\dfrac{5}{8}, \dfrac{15}{8}\right\}$ **25.** $(-\infty, -4) \cup (4, \infty)$

27. $(-\infty, 1] \cup [5, \infty)$ **29.** $\left(-\infty, -\dfrac{22}{5}\right) \cup \left(\dfrac{12}{5}, \infty\right)$

31. $\left(-\infty, -\dfrac{6}{5}\right) \cup (2, \infty)$ **33.** $\left(-\infty, \dfrac{1}{2}\right) \cup \left(\dfrac{7}{2}, \infty\right)$

35. $(-\infty, -8] \cup [1, \infty)$ **37.** $(-\infty, -1] \cup [7, \infty)$

39. $\left(-\infty, -\dfrac{2}{3}\right) \cup (1, \infty)$ **41.** $\varnothing$

43. $\varnothing$ **45.** $\varnothing$ **47.** $\varnothing$ **49.** $\mathbb{R}$

51. $\left(-\infty, -\dfrac{10}{4}\right) \cup \left(-\dfrac{10}{4}, \infty\right)$ **53.** $\{6\}$

55. $\left(-\infty, \dfrac{3}{4}\right) \cup \left(\dfrac{3}{4}, \infty\right)$ **57.** $\varnothing$ **59.** $\{7\}$

61. $\left(-\infty, -\dfrac{3}{2}\right) \cup \left(-\dfrac{3}{2}, \infty\right)$ **63.** $(-\infty, -3) \cup \left(\dfrac{11}{3}, \infty\right)$

65. $\left[-\dfrac{11}{5}, 1\right]$

67. According to the poll, 70.9% to 77.1% of Americans think that STEM education and training is important to U.S. prosperity and competitiveness.

69. Rita can weigh between 140 and 150 lb.

71. $(-\infty, -16) \cup (16, \infty)$ **73.** $(-3, 8)$

75. $(-\infty, -1] \cup \left[\dfrac{3}{2}, \infty\right)$ **77.** $\{4\}$

79. $(-\infty, -4] \cup \left[\dfrac{12}{5}, \infty\right)$ **81.** $\varnothing$

83. $\mathbb{R}$ **85.** $\mathbb{R}$ **87.** $\mathbb{R}$ **89.** $(-\infty, 5) \cup (5, \infty)$

91. $\left[-\dfrac{7}{2}, \dfrac{17}{2}\right]$ **93.** $(-19, 23)$ **95.** $\left[-\dfrac{13}{3}, \dfrac{4}{3}\right]$

97. $\mathbb{R}$ **99.** $\left(-\dfrac{5}{2}, \dfrac{7}{2}\right)$ **101.** $\mathbb{R}$

103. $(-\infty, -3] \cup \left[\dfrac{3}{5}, \infty\right)$ **105.** $(-\infty, 6) \cup (6, \infty)$

107. $\{2\}$ **109.** $\left[\dfrac{1}{2}, \dfrac{11}{2}\right]$ **111.** $\mathbb{R}$ **113.** $\varnothing$

115. $\varnothing$ **117.** $[-2.6, 4.4]$ **119.** $\varnothing$ **121.** $(-1, 17)$

123. $\left(-\infty, -\dfrac{4}{3}\right] \cup \left[\dfrac{1}{3}, \infty\right)$

125. According to the poll, 50% to 56% of Americans described themselves as angry about the country's financial crisis.

127. The safe locations on Interstate 20 for motorists will be outside markers 25 and 45.

129. There will always be two disjoint intervals in the solution set if $k > 0$.

131. $|2x - 5| < 13$

$-13 < 2x - 5 < 13$

$-8 < 2x < 18$

$-4 < x < 9$

The solution set is $(-4, 9)$.

133. $(-2, 1)$ **135.** $(-\infty, 0] \cup \left[\dfrac{28}{5}, \infty\right)$

137. $(1, \infty)$ **139.** $\left(-\infty, \dfrac{1}{2}\right)$

Piece It Together Sections 9.1–9.3

1. $\varnothing$

2. $(-\infty, -6) \cup (4, \infty)$

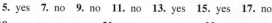

3. Conner should weigh between 144 lb and 194 lb.

4. $\{4, 7\}$ **5.** $\left\{-\dfrac{13}{3}, 5\right\}$ **6.** $\varnothing$ **7.** $\{3\}$ **8.** $(2, 11)$

9. $\left\{-\dfrac{9}{4}, \dfrac{9}{2}\right\}$

10. The exact length of the object is at most 5.3625 in. and at least 5.3375 in.

11. 0.0012828   **12.** $[-4, 24]$ **13.** $(-\infty, -9) \cup (0, \infty)$

14. $\varnothing$ **15.** $(-\infty, \infty)$ **16.** $\{5\}$ **17.** $(-\infty, -3) \cup (-3, \infty)$

18. Allison can weigh between 116 lb and 124 lb.

19. $(-1, 7)$ **20.** $\left(-\infty, -\dfrac{22}{5}\right) \cup (2, \infty)$

Section 9.4

1. Answers vary. Substitute the coordinates for x and y. If a true inequality results, then the ordered pair is a solution. If a false inequality results, then the ordered pair is not a solution.

3. Answers vary. If the inequality is strictly greater than or less than, then the boundary line will be dotted. If the inequality is greater than or equal to or less than or equal to, then the boundary line will be solid.

5. yes   **7.** no   **9.** no   **11.** no   **13.** yes   **15.** yes   **17.** no

19.   **21.**   **23.**

25.
27.
29.

31.
33.
35.

37. a. $0.6x + 0.4y \geq 80$
b.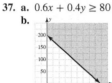
c. Answers vary. Three possible combinations are (80, 90), (90, 70), and (80, 80).

39. a. $8x + 10.50y \geq 300$
b.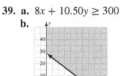
c. Answers vary. bookstore: 10 hr & library: 25 hr, bookstore: 12 hr & library: 20 hr, bookstore: 20 hr & library: 15 hr

41. a. $15x + 25y \geq 500$
b.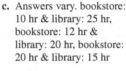
c. Answers vary. 20 pants & 15 jackets, 5 pants & 20 jackets, 15 pants & 15 jackets.

43. yes   **45.** no   **47.** yes   **49.** no   **51.** no

53.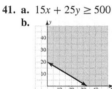
55.
57.

59.

61. a. $0.75x + 0.25y \geq 90$
b.
c. Answers vary. Three possible combinations are (95, 80), (88, 98), and (90, 90).

63. a. $10.50x + 8.5y \geq 350$
b.
c. Answers vary. grocery store: 24 hr & school: 12 hr, grocery store: 22 hr & school: 14 hr, grocery store: 20 hr & school: 17 hr

65. Jamie shaded the correct half-plane and has the correct boundary line. The mistake was made in drawing the boundary line solid. It should be dashed since the inequality is <.

67. Since y is not isolated on the left side of the inequality, Carolyn cannot shade based on the inequality symbol. In this form, we must use a test point. We use (4, 2) since the boundary line goes through (0, 0).
$$4 - 4(2) > 0$$
$$4 - 8 > 0$$
$$-4 > 0$$
False
We shade the half-plane that doesn't contain (4, 2).

69.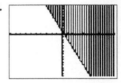
71.

73. Answers vary. $x + y > 0$ or any inequality of the form $Ax + By > 0$, $Ax + By \geq 0$, $Ax + By < 0$, and $Ax + By \leq 0$.

75. Answers vary. No, if the boundary line is part of the solution, it will be in addition to a half-plane.

77. a.
b.
c.
d.
e. lower half-plane
f. The solution set will always be the lower half-plane.
g. Both will have the lower half-plane as solutions but $y \leq mx + b$ will have a solid boundary line and $y < mx + b$ will have a dashed boundary line.

Section 9.5

1. Answers vary. First graph each linear inequality. The overlapping portions of the graphs represent the solution of the system of linear inequalities.
3. Answers vary. Graph the system and then choose any two points that lie in the solution space. Two points that are in the solution set of this system are (3, 4) and (0, 5).

5.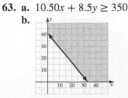
7.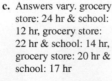
9.

11. 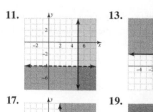   **13.** 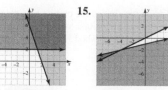   **15.**

17. 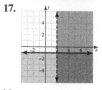   **19.**   **21.**

23. 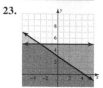   **25.**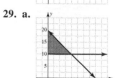

27. a. $x + y \leq 1200$; $1.5x + 4y \leq 4200$; $y \geq 2x$

b.   **c.** Answers vary. $(0, 1050)$, $(400, 800)$, $(240, 960)$

29. a. 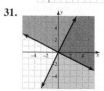   **b.** 1000

31.   **33.** 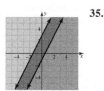   **35.**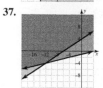

37. 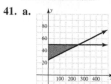   **39.**

41. a. 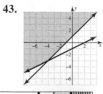   **b.** 3125   **43.**

45. 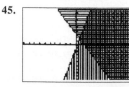   **47.**

49. Answers vary. $x > 3$ and $x < 3$.

CHAPTER 9 SUMMARY

Section 9.1

1. intersection;   **2.** union   **3.** compound; inequality
4. intersection   **5.** union

Section 9.2

6. distance   **7.** nonnegative   **8.** $X = k$; $X = -k$
9. $X = 0$ **10.** no; solution   **11.** isolate;
12. $X = Y$; $X = -Y$; same; opposite   **13.** absolute; error

Section 9.3

14. $X < k$; and; $X > -k$; $-k < X < k$
15. $X > k$; or; $X < -k$ **16.** no; solution
17. all; real; numbers   **18.** test; points

Section 9.4

19. $y < mx + b$ or $Ax + By < C$ **20.** ordered; pairs; true
21. boundary; line; equals; sign; $\leq$; $\geq$; $<$; $>$
22. half-planes; half-plane; solutions
23. test; point; test; point; $(0, 0)$

Section 9.5

24. intersection

CHAPTER 9 REVIEW EXERCISES

Section 9.1

1. $(-4, 0)$ **3.** $(-\infty, \infty)$ **5.** $(-\infty, 13.8)$ **7.** $\varnothing$
9. $(-4, \infty)$ **11.** $(-\infty, 2) \cup (5, \infty)$ **13.** $(-\infty, -1)$
15. $(-1, \infty)$ **17.** $\varnothing$ **19.** $(2.15, 5.62)$
21. She must score between 88 and 100.
23. Between 3.14 hr and 11.71 hr

Section 9.2

25. $\left\{-\dfrac{5}{3}, \dfrac{5}{2}\right\}$ **27.** $\left\{-\dfrac{11}{2}, \dfrac{25}{2}\right\}$ **29.** $\left\{-\dfrac{5}{4}, \dfrac{9}{4}\right\}$

31. $\varnothing$ **33.** $\left\{-\dfrac{2}{3}\right\}$ **35.** $\left\{-\dfrac{23}{3}, 16\right\}$ **37.** $\left\{-\dfrac{2}{9}, -\dfrac{2}{11}\right\}$

39. $\left\{-\dfrac{1}{3}, 9\right\}$ **41.** 0.267 m   **43.** 11.83 units or 14.13 units

45. Candidate A will receive 28% or 36% and candidate B will receive 32% or 40%.

Section 9.3

47. $\left[-6, \dfrac{14}{3}\right]$ **49.** $(-\infty, -2] \cup [2, \infty)$ **51.** $(-\infty, 1) \cup (4, \infty)$
53. $\varnothing$ **55.** $\{-5\}$ **57.** $(-\infty, \infty)$ **59.** $(-\infty, \infty)$
61. $\varnothing$ **63.** 34% or 42%
65. Motorists should be at or beyond mile marker 238 or at or before mile marker 214.

67. $(-7, 31)$ **69.** $(-\infty, -2] \cup \left[\dfrac{11}{5}, \infty\right)$

Section 9.4

71. Yes   **73.** No   **75.** Yes

77.   **79.**   **81.** $0.8x + 0.2y \geq 90$

Section 9.5

83.   **85.**

87.

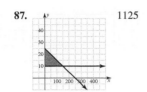

1125

CHAPTER 9 TEST

1. d   **3.** c   **5.** ∅

7.

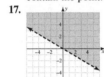

$$\left(-\frac{4}{3}, 20\right) \quad \left\{x \,\middle|\, -\frac{4}{3} < x < 20\right\}$$

9.

$$\left[-1, \frac{1}{4}\right] \quad \left\{x \,\middle|\, -1 \le x < \frac{1}{4}\right\}$$

11. Juan can make between, and including, 34.2 and 70.2 to have a final grade of a *C*.

13. $|s - 65| > 15$; Officers stop vehicles traveling at speeds greater than 80 mph or less than 50 mph.

15. Answers vary. The first step is to graph the boundary line, which is the equation formed by replacing the inequality symbol with an equals sign. Draw the line solid if the inequality is ≤ or ≥ and dashed if the inequality is < or >. Test a point not on the boundary line in the original inequality. If the point makes the inequality true, shade the half-plane that contains the point. If the point makes the inequality false, shade the half-plane that doesn't contain the point.

17.

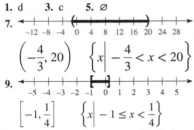

19.

CHAPTERS 1–9 CUMULATIVE REVIEW EXERCISES

1. a. $\dfrac{67}{60}$ **b.** $1\dfrac{5}{12}$ **c.** 54   **3.** 168°F

5. a. {7}   **b.** {2.4}   **c.** $\left\{-\dfrac{5}{3}\right\}$

7. a. $h = \dfrac{S - 2\pi r^2}{2\pi r}$ **b.** $l = \dfrac{C - 8w}{4}$

   c. $y = -\dfrac{1}{5}x + 2$ **d.** $x = 70 - \dfrac{1}{2}y$

9. a. parallel   **b.** perpendicular   **c.** perpendicular

11. a. $f(x) = 1050x + 43{,}500$ **b.** \$52,950   **c.** $f(8) = \$51{,}900$; Amanda will earn \$51,900 if she teaches 38 credit-hours in an academic year.

13. a. $\left\{\left(-\dfrac{276}{7}, -\dfrac{550}{7}\right)\right\}$ **b.** {(−6, 8)}

15. a. $-\dfrac{8}{y^4}$ **b.** $-\dfrac{15}{y^3}$ **c.** $-\dfrac{8x^2}{y^3}$ **d.** $\dfrac{3r^{17}}{10t^5}$

17. a. 1.728×10^{12} **b.** 3.6×10^{10} **c.** 5.4×10^{-8} **d.** 1.56×10^{-10}

19. a. $28x^2 + 23x - 15$ **b.** $6y^2 + 7y - 10$ **c.** $4a^2 - 4ab + b^2$ **d.** $144x^2 - 49y^2$ **e.** $a^4 + 8a^3 + 24a^2 + 32a + 16$ **f.** $8b^3 - 12b^2 + 6b - 1$

21. a. $3y(y - 4)^2$ **b.** $-6x(2x + 1)(x - 5)$ **c.** $(a - 3b)(a - 12b)$ **d.** $2y(x + 8)(x + 9)$ **e.** prime

23. a. $2(7a - 3b)(7a + 3b)$ **b.** $5(x + 2y)(x^2 - 2xy + 4y^2)$ **c.** $(y - 10z)(y^2 + 10yz + 100z^2)$ **d.** prime

   e. $(a + 2)(a^2 - 2a + 4)(a - 2)(a^2 + 2a + 4)$ **f.** $(x^2 + 9y^2)(x - 3y)(x + 3y)$

25. a. The integers are 11 and 13 or −11 and −13.   **b.** The bus company must sell 110 tickets.   **c.** It will take 5 sec for the ball to reach the ground.   **d.** 25,000 units

27. a. $(-\infty, -6) \cup (-6, \infty)$ **b.** $(-\infty, -5) \cup (-5, 0) \cup (0, \infty)$

29. a. $-\dfrac{3x - 6}{x + 4}$ or $\dfrac{3x - 6}{-x - 4}$ **b.** $\dfrac{-1 + x}{2x - 3}$ or $\dfrac{1 - x}{-2x + 3}$

31. a. $\dfrac{x - 5}{2(x + 3)}$ **b.** $\dfrac{x + 6}{2x(x - 2)}$ **c.** $\dfrac{x^2 + x + 1}{x + 1}$ **d.** $\dfrac{x + 3}{x}$ **e.** $2x$ **f.** $x - 1$

33. a. $6x^2 + 15x$ **b.** $(2x + 3)(x - 4)$ **c.** $(x + 7)(x - 3)$ **d.** $(x + 4)(x - 4)(5x - 1)$ **e.** $3(x - 2)(x^2 + 2x + 4)$

35. a. $-\dfrac{x}{x - 4}$ **b.** $\dfrac{9(x - 2)}{(x - 6)(x + 3)}$ **c.** $\dfrac{x + 6}{x - 2}$ **d.** $\dfrac{-10x}{(x + 5)(x - 5)}$ **e.** $\dfrac{7x - 4}{x - 1}$ **f.** $\dfrac{(2x + 3)(x + 1)}{(x + 10)(x - 3)}$ **g.** $\dfrac{-x - 39}{(x + 3)(x - 3)^2}$ **h.** $\dfrac{7x + 26}{(x + 2)(x + 1)(x + 6)}$ **i.** $\dfrac{350x - 50}{(x - 8)(x + 3)}$

37. a. $-\dfrac{64}{9}$ **b.** $\dfrac{26}{5}$ **39. a.** $R_1 = \dfrac{RR_2 R_3}{R_2 R_3 - RR_3 - RR_2}$ **b.** $x_1 = \dfrac{mx_2 - y_2 + y_1}{m}$

41. a. Domain = $(-\infty, \infty)$, Range = $(-\infty, \infty)$ **b.** $[-3, \infty)$

43. a. $f(x) = 7$ **b.** $f(x) = \dfrac{1}{9}x + 3$ **c.** $f(x) = \dfrac{4}{5}x - \dfrac{28}{5}$ **d.** $f(x) = 6x - 51$

45. a.

b.

   c.

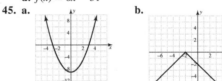

   d.

47. a. $A \cap B = [-12, 1]$, $A \cup B = [-13, 4]$ **b.** $A \cap B = (4, 18)$, $A \cup B = [3, 20]$

49. a. $[8, \infty)$ **b.** $\left(-\infty, \dfrac{1}{7}\right] \cup (1, \infty)$

51. a. The exact weight of the person is 109.9 lb or 110.3 lb.   **b.** The exact weight of the piece of chicken is 2.9 oz or 3.1 oz.

53. a. yes   **b.** yes

55. a. $8.75x + 11.50y \ge 280$

   b.

   c. Answers vary. Brenda can work 10 hr as an office assistant and 17 hr as a tutor, 15 hr as an office assistant and 13 hr as a tutor, or 9 hr as an office assistant and 18 hr as a tutor to earn at least \$280 each week.

57. a.

   b. 1280

CHAPTER 10

Section 10.1

1. The radical expression is the nth root of a, $\sqrt[n]{a}$, and a is the radicand.

3. $\sqrt{(x-3)^2} = |x-3|$ and $(\sqrt{x-3})^2 = x-3, x \geq 3$

5. Answers vary. To find the cube root of 125, find b such that $b^3 = 125$. So, $b = 5$.

7. Since it is a cube root function, the domain is $(-\infty, \infty)$.

9. 9   **11.** -5 **13.** not a real number   **15.** $\dfrac{4}{7}$ **17.** $-\dfrac{3}{10}$

19. not a real number   **21.** 1.2   **23.** s^2 **25.** $9a^5$ **27.** $-4h^3$

29. $\dfrac{2x^8}{3}$ **31.** 4   **33.** -10 **35.** $\dfrac{4}{5}$ **37.** a^2 **39.** $4x^2$

41. $-10c$ **43.** $0.4a^2$ **45.** $-\dfrac{9x}{5}$ **47.** $\dfrac{7a^4}{5}$ **49.** 2   **51.** -2

53. $4x$ **55.** $-2x^3y^2$ **57.** $2a^2b$ **59.** not a real number

61. $2a^2b^4$ **63.** 6.48   **65.** 5.25   **67.** 65   **69.** not a real number

71. -7 **73.** $x-5$ **75.** $2x-3$ **77.** $|2x-5|$ **79.** $6x$ **81.** $13a$

83. domain $= [-5, \infty)$; $f(-6)$ not a real number, $f(0) = \sqrt{5}$, $f(4) = 3$

85. domain $= \left[-\dfrac{1}{6}, \infty\right)$; $f(-1)$ not a real number, $f\left(-\dfrac{1}{6}\right) = 0$, $f(4) = 5$

87. domain $= (-\infty, \infty)$; $f(-5) = -2, f(0) = -\sqrt[3]{3}, f(3) = 0$

89. 5.64 ft   **91.** $2\pi\sqrt{2}$ sec

93.

x	3	4	7	8
$f(x)$	0	1	2	2.2

95.

x	-2.5	-1	0	2
$f(x)$	0	1.7	2.2	3

97.

x	-6	0	1	2	3	10
$f(x)$	-2	-1.3	-1	0	1	2

99.

x	-8	-1	0	1	8	12
$f(x)$	-4	-3	-2	-1	0	0.3

101. $-20|x^3|$ **103.** $15|h|k^4$ **105.** $|5x-9|$ **107.** $0.7a^2$

109. $5x-6$ **111.** $4a$ **113.** $-2r^2t$ **115.** $\dfrac{5r^6}{7}$

117. $-5r^2|s|$ **119.** $|ab^3|$ **121.** x^3y^2

123. domain $= \left[-\dfrac{4}{3}, \infty\right)$; $f\left(-\dfrac{4}{3}\right) = 0, f(0) = 2, f(2) = \sqrt{10}$

125. domain $= (-\infty, 3]$; $f(-6) = 3, f(0) = \sqrt{3}, f(4)$ not a real number

127. domain $= (-\infty, \infty)$; $f(-4) = 0, f(0) = \sqrt[3]{4}, f(4) = 2$

129. 46.6 mph

131.

x	-3	0	1	1.5
$f(x)$	3	1.7	1	0

133.

x	-8	-1	0	1	8
$f(x)$	3	2	1	0	-1

135. The radicand must be greater than or equal to zero: $\begin{aligned} x - 1 &\geq 0 \\ x &\geq 1 \end{aligned}$

So, the domain of the function is $[1, \infty)$.

137. Since this expression is the square root of a negative value, it is not a real number.

139. 42.83   **141.** 4.31

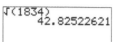

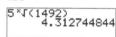

Section 10.2

1. Answers vary; An exponent that is a fraction is called a rational exponent. An example is $x^{2/3}$.

3. Answers vary; Add the exponents: $x^{2/5} \cdot x^{1/3} = x^{2/5+1/3} = x^{6/15+5/15} = x^{11/15}$.

5. Answers vary; Write the rational exponent as 1 divided by the fifth root of 32 raised to the third and then simplify the radical.

$32^{-3/5} = \dfrac{1}{\sqrt[5]{32^3}} = \dfrac{1}{(\sqrt[5]{32})^3} = \dfrac{1}{2^3} = \dfrac{1}{8}$

7. 7   **9.** 11   **11.** -4 **13.** -15 **15.** not a real number

17. $\sqrt[3]{x}$ **19.** $\sqrt[4]{5}$ **21.** 2   **23.** $2x^2$ **25.** $3a^3$ **27.** 8   **29.** 125

31. -32 **33.** -9 **35.** 25   **37.** $\dfrac{1}{9}$ **39.** $\dfrac{8}{125}$ **41.** $16x^6y^4$

43. $16x^{12}y^8$ **45.** $\dfrac{a^4}{16b^2}$ **47.** $\sqrt[5]{(x+8)^4}$ **49.** $\sqrt[3]{(x+3)^2}$

51. $\dfrac{1}{4}$ **53.** $-\dfrac{1}{4}$ **55.** $\dfrac{1}{25}$ **57.** $\dfrac{1}{4x^4}$ **59.** $\dfrac{1}{32a^5b^{10}}$ **61.** $\dfrac{y^{10}}{x^8}$

63. 5   **65.** 4   **67.** $x^{11/10}$ **69.** $\dfrac{1}{a^{2/3}}$ **71.** $\dfrac{1}{5}$ **73.** $x^{1/10}$

75. x^4y^8 **77.** $64x^6y^{18}$ **79.** $\dfrac{8x^2}{y^{3/2}}$ **81.** $x^{1/2}y^{5/3}$ **83.** $\dfrac{x^{5/2}}{y^{2/5}}$ **85.** $x + x^{5/6}$

87. $2a + 5a^{1/2}b^{1/3} - 12b^{2/3}$ **89.** $16r - s^{4/3}$ **91.** a^4 **93.** $\sqrt[3]{x}$

95. $\sqrt{a}$ **97.** $\sqrt{2}$ **99.** $\sqrt{7}$ **101.** $\sqrt{3a}$ **103.** $\sqrt{11x}$

105. $\sqrt[10]{128}$ **107.** $\sqrt[10]{x^{11}}$ **109.** $\sqrt[6]{a^5}$ **111.** \$1003.31   **113.** \$802.66

115. $19.99°F$ **117.** $-2x^3$ **119.** $\dfrac{x^5}{32y^{10}}$ **121.** $25a^2b^4$

123. $\dfrac{27y^3}{x^6}$ **125.** $\sqrt[3]{(y+5)^4}$ **127.** $-\dfrac{1}{4x^2}$ **129.** 4   **131.** $\dfrac{1}{8x^6}$

133. $\dfrac{a^{10}}{b^{20}}$ **135.** $r^{23/12}$ **137.** x^2y^4 **139.** $\dfrac{1}{x^{3/14}}$ **141.** $\dfrac{243x^2}{y^{10/3}}$

143. $\dfrac{y^2}{x^{33/7}}$ **145.** $9x - y^{2/3}$ **147.** $5r + r^{7/6}$

149. $10x + 14x^{1/2}y^{1/3} - 12y^{2/3}$ **151.** $4x^2$

153. $\sqrt{7a}$ **155.** $\dfrac{\sqrt{7}}{x}$ **157.** $\sqrt[6]{x^{13}}$ **159.** \$506.02   **161.** $-63.66°F$

163. When like bases are multiplied, the exponents are added but the bases stay the same. $5^{1/3} \cdot 5^{2/3} = 5^{1/3+2/3} = 5^{3/3} = 5^1 = 5$

165. We can't take the opposite of the base until after we apply the exponent. $-(-x^3y^6)^{2/3} = -(-x^3)^{2/3}(y^6)^{2/3} = -x^2y^4$

167. 392.50   **169.** $\dfrac{1}{16}$

Section 10.3

1. The radicand must not contain any perfect square factors or fractions. There also can't be radicals in the denominators of a fraction.

3. We must rewrite the radical as a product in which one of the factors is a power with an exponent equal to the index. Then we apply the product rule and simplify by extracting any roots.

5. x^2 and y^4 are perfect squares since their exponents are even. Factor 72 as $36 \cdot 2$. Apply the product rule. Group the perfect square factors, x^2, y^4, and 36 in the first square root and the remaining factor 2 in the other square root. Simplify.
$$\sqrt{72x^2y^4} = \sqrt{36x^2y^4} \cdot \sqrt{2} = 6xy^2\sqrt{2}$$

7. $6\sqrt{2}$ 9. $-6\sqrt{10}$ 11. not a real number   13. $60\sqrt{2}$

15. $4a^4\sqrt{3}$ 17. $10x^2\sqrt{3x}$ 19. $3a^3\sqrt[3]{6a}$ 21. $5r^4t^5\sqrt{5rt}$

23. $12x^2y^3\sqrt{x}$ 25. $5\sqrt[3]{2}$ 27. $-10\sqrt[3]{2}$ 29. $4a^2\sqrt[3]{3}$

31. $2x^2y\sqrt[3]{7x^2}$ 33. $2xy^2\sqrt[3]{2x^2y^2}$ 35. $\dfrac{3\sqrt{5}}{7}$ 37. $\dfrac{4x^2\sqrt{2x}}{3}$

39. $\dfrac{5x^3\sqrt{5x}}{3}$ 41. $\dfrac{2r^2\sqrt[3]{2r^2}}{5}$ 43. 15   45. 17   47. $3\sqrt{6}$

49. $2\sqrt{7}$ 51. $\dfrac{17}{5}$ 53. 1.3   55. 284.60 mi

57. $-4a^3b^2\sqrt{3}$ 59. $15h^3k^5\sqrt{2hk}$ 61. $\dfrac{5r^2\sqrt{3r}}{4s^2}$ 63. $2ab^2\sqrt[3]{20a^2b}$

65. $-3rt^2\sqrt[5]{2r^2t^3}$ 67. $\dfrac{4x^2\sqrt{2x^2}}{3y^2}$ 69. $\dfrac{8a^3b\sqrt{a}}{5}$ 71. $\dfrac{2d^3\sqrt{2d}}{5c^2}$

73. 26   75. $2\sqrt{22}$ 77. $\dfrac{5}{6}$ 79. $2\sqrt{6}$ 81. 0.41   83. $\dfrac{5}{4}$

85. The variable in the radicand has an exponent greater than 2, so it is not simplified.
$$\sqrt{63x^7} = \sqrt{9 \cdot 7x^6x} = 3x^3\sqrt{7x}$$

87. The order of operations requires us to simplify the expression inside the radical before we apply the square root.
$$\sqrt{(11-3)^2 + (-16+1)^2} = \sqrt{(8)^2 + (-15)^2}$$
$$= \sqrt{64 + 225}$$
$$= \sqrt{289}$$
$$= 17$$

89.

Since the Y_1 and Y_2 columns agree for each value of x, we can conclude that the two expressions are the same.

91.

Since the Y_1 and Y_2 columns agree for each value of x, we can conclude that the two expressions are the same.

Section 10.4

1. Like radicals are radicals with the same index and same radicand.

3. If the indices are the same, we multiply the radicands of the radicals using the product rule for radicals.

5. We must first simplify each radical.
$\sqrt{18x} + \sqrt{12x} = 3\sqrt{2x} + 2\sqrt{3x}$; The radicals are not alike, so we cannot add the expressions.

7. To multiply, we use the FOIL method.
$$(\sqrt{x} + \sqrt{y})^2 = (\sqrt{x} + \sqrt{y})(\sqrt{x} + \sqrt{y})$$
$$= \sqrt{x^2} + \sqrt{xy} + \sqrt{xy} + \sqrt{y^2}$$
$$= x + 2\sqrt{xy} + y$$

9. $5\sqrt{6}$ 11. $-14\sqrt{7}$ 13. $48\sqrt{6}$ 15. $8x\sqrt{2x}$ 17. $-9\sqrt[3]{5}$

19. $31\sqrt[3]{2}$ 21. $-3a\sqrt[3]{3b} + 2a\sqrt[3]{6b}$ 23. $\dfrac{17\sqrt{5}}{12}$ 25. $3\sqrt{5}$

27. $\dfrac{9\sqrt[3]{3}}{4}$ 29. $\dfrac{26r\sqrt{2r}}{9}$ 31. $4\sqrt{14}$ 33. $48\sqrt{10}$ 35. $8x^2\sqrt{6x}$

37. $4x\sqrt[3]{9x^2}$ 39. $5\sqrt{7} + 7\sqrt{3}$ 41. $-45 - 25\sqrt{6}$

43. $24 - 12\sqrt{3}$ 45. $41 - 4\sqrt{10}$ 47. $4 - 6x$

49. $6x - 3\sqrt{6xy} - 10y$ 51. $36\sqrt{2} + 36$ or 86.9 cm; 324 cm^2

53. $20\sqrt{14} + 20\sqrt{7}$ or 127.7 ft; 700 ft^2

55. $12\sqrt{7}$ cm; $8\sqrt{21}$ cm   57. $9\sqrt{6}$ 59. $27r\sqrt[3]{3t}$

61. $-4a\sqrt[3]{2b} + 10a\sqrt[3]{3b}$ 63. $-10d\sqrt{2cd}$ 65. $7xy\sqrt[4]{y^3}$

67. $25 - 3a$ 69. 0   71. $\dfrac{83r\sqrt{3t}}{14}$ 73. $\dfrac{17\sqrt[3]{3a}}{20}$ 75. $2xy\sqrt[5]{5x^3}$

77. $-2 + 46\sqrt{3}$ 79. $63 - 24\sqrt{5}$ 81. $2\sqrt{6} - 6\sqrt{2}$

83. $5x + 2\sqrt{15xy} - 24y$

85. $42\sqrt{2} + 42$ or 101.40 m; 441 m^2

87. $36\sqrt{6} + 36\sqrt{3}$ or 150.54 in.; 972 in.^{2}

89. $15\sqrt{7}$ cm; $10\sqrt{21}$ cm

91. $\sqrt{28x^2y} + \sqrt{63x^2y} = \sqrt{4 \cdot 7x^2y} + \sqrt{9 \cdot 7x^2y}$
$$= 2x\sqrt{7y} + 3x\sqrt{7y}$$
$$= 5x\sqrt{7y}$$

93. $(2\sqrt{x} - y)^2 = (2\sqrt{x} - y)(2\sqrt{x} - y)$
$$= (2\sqrt{x})^2 - 2(2y\sqrt{x}) + (y)^2$$
$$= 4x - 4y\sqrt{x} + y^2$$

95.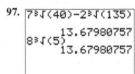

97.

The two radical expressions have the same decimal value. Therefore, they are equivalent.

The two radical expressions have the same decimal value. Therefore, they are equivalent.

Piece It Together Sections 10.1–10.4

1. $7a^6$ 2. $\dfrac{5x^3}{2y^2}$ 3. $-3a^2b^5$ 4. $2ab^3$ 5. $|2x + 1|$ 6. $5b$

7.

x	-8	-1	0	1	8	12
$f(x)$	-4	-3	-2	-1	0	0.3

8. not a real number   9. $\dfrac{1}{9}$ 10. $16x^6y^4$ 11. $\dfrac{1}{4x^4}$

12. $\dfrac{a^6}{b^9}$ 13. $\dfrac{x^{5/2}}{y^{2/5}}$ 14. $\dfrac{a^2}{b^{2/3}}$ 15. $12x^{4/5} + 7x^{2/5}y^{1/2} - 5y$

16. $49r - s^{4/3}$ 17. $15\sqrt{3}$ 18. $12x^2y^3\sqrt{wx}$

19. $2a^2b^4\sqrt[3]{5a}$ 20. $2xy^2\sqrt[3]{3xy^3}$ 21. $\dfrac{4x^2\sqrt{5x}}{3y}$ 22. $2\sqrt{7}$

23. $86\sqrt{2}$ 24. $32\sqrt[3]{2}$ 25. $8x^3\sqrt{6x}$

26. $-45 - 25\sqrt{6}$ 27. $24 - 12\sqrt{3}$

Section 10.5

1. To divide radical expressions with the same index, you divide the radicands and keep the root of the quotient.

3. The conjugate of $a + \sqrt{b}$ is $a - \sqrt{b}$.

5. To rationalize the denominator of $\dfrac{1}{\sqrt[5]{6}}$, you need to multiply the numerator and the denominator by a term that makes the radicand in the denominator a perfect fifth, that is, $\sqrt[5]{6^4}$:
$$\dfrac{1}{\sqrt[5]{6}} \cdot \dfrac{\sqrt[5]{6^4}}{\sqrt[5]{6^4}} = \dfrac{\sqrt[5]{6^4}}{\sqrt[5]{6^5}} = \dfrac{\sqrt[5]{6^4}}{6}.$$

7. 4   9. $10s^4$ 11. $\dfrac{5}{a^5}$ 13. $2c\sqrt{3c}$ 15. $\dfrac{\sqrt{2}}{3}$

17. $\dfrac{\sqrt{2}}{4}$ **19.** $\dfrac{a^2\sqrt{3a}}{9}$ **21.** $12t^3\sqrt[4]{2}$ **23.** $27x^5\sqrt[3]{3}$

25. $14r^2\sqrt[5]{2r}$ **27.** $\dfrac{8a^3\sqrt{a}}{5b^3}$ **29.** $27x^5y\sqrt[3]{3y}$

31. $30a^6b^5\sqrt[3]{ab}$ **33.** $\dfrac{\sqrt{66}}{3}$ **35.** $\dfrac{\sqrt{6a}}{a}$ **37.** $\dfrac{\sqrt{5y}}{2y}$

39. $\dfrac{2\sqrt[3]{5}}{5}$ **41.** $\dfrac{2\sqrt[5]{27a^3}}{a}$ **43.** $\dfrac{3\sqrt[6]{8x^5}}{x}$ **45.** $\dfrac{6\sqrt[4]{27x}}{x}$

47. $4(\sqrt{6}-\sqrt{3})$ **49.** $\dfrac{5(\sqrt{11}+\sqrt{3})}{4}$

51. $-8-4\sqrt{5}+2\sqrt{6}+\sqrt{30}$ **53.** $7-4\sqrt{3}$

55. $\dfrac{45+11\sqrt{15}}{10}$ **57.** $\sqrt{h+100}+10$ **59.** $8x$ **61.** $\dfrac{7}{a^2}$

63. $30x^3\sqrt[3]{2x^2}$ **65.** $12a^3\sqrt[5]{5}$ **67.** $10s^8\sqrt{3s}$ **69.** $\dfrac{x\sqrt{2x}}{4}$

71. $6b^2\sqrt[4]{b^3}$ **73.** $14r\sqrt[5]{5r}$ **75.** $\dfrac{2a^4\sqrt{3a}}{b^2}$ **77.** $15x^2y^2\sqrt[3]{x}$

79. $\dfrac{2a^3\sqrt{3a}}{5b^2}$ **81.** $\dfrac{12r\sqrt[5]{5r^2}}{s^2}$ **83.** $\dfrac{\sqrt{42}}{3}$ **85.** $\dfrac{\sqrt[4]{72a}}{2a}$

87. $\dfrac{\sqrt{21}-\sqrt{5}}{2}$ **89.** $\dfrac{\sqrt{6x}}{x}$ **91.** $\dfrac{3\sqrt[3]{18}}{2}$

93. $\dfrac{-3-\sqrt{5}-3\sqrt{2}-\sqrt{10}}{4}$ **95.** $-9-2\sqrt{21}$

97. $11+3\sqrt{13}$ **99.** $\dfrac{3\sqrt[5]{8x^3}}{x}$

101. $\dfrac{-2+\sqrt{6}+2\sqrt{3}-3\sqrt{2}}{2}$ **103.** $\sqrt{a+36}-6$

105. $\dfrac{\sqrt{28}}{14}=\dfrac{\sqrt{4\cdot7}}{14}=\dfrac{2\sqrt{7}}{14}=\dfrac{\sqrt{7}}{7}$

107. $\dfrac{\sqrt{7}-\sqrt{3}}{\sqrt{7}+\sqrt{3}}=\dfrac{(\sqrt{7}-\sqrt{3})}{(\sqrt{7}+\sqrt{3})}\cdot\dfrac{(\sqrt{7}-\sqrt{3})}{(\sqrt{7}-\sqrt{3})}$

$$=\dfrac{7-2\sqrt{21}+3}{7-3}$$

$$=\dfrac{10-2\sqrt{21}}{4}$$

$$=\dfrac{\overset{5}{\cancel{10}}-\overset{1}{\cancel{2}}\sqrt{21}}{\underset{2}{\cancel{4}}}$$

$$=\dfrac{5-\sqrt{21}}{2}$$

109.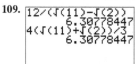
111.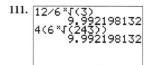

The two radical expressions have the same decimal value. Therefore, they are equivalent.

The two radical expressions have the same decimal value. Therefore, they are equivalent.

Section 10.6

1. You raise both sides of the equation to an exponent equal to the index n and solve the resulting equation.

3. With one radical on one side of the equation, you raise both sides of the equation to the exponent equal to the index.

$$(\sqrt{x-5}+1)^2=(\sqrt{x})^2$$

Next you use the FOIL method to multiply.

$$(\sqrt{x-5}+1)(\sqrt{x-5}+1)=(\sqrt{x})^2$$

$$x-5+2\sqrt{x-5}+1=x$$

$$2\sqrt{x-5}=4$$

$$\sqrt{x-5}=2$$

$$(\sqrt{x-5})^2=(2)^2$$

$$x-5=4$$

$$x=9$$

5. You just need to substitute 655 for x and check the equation.

$$\sqrt[3]{2(655)+21}-9\overset{?}{=}2$$

$$\sqrt[3]{1331}-9\overset{?}{=}2$$

$$11-9\overset{?}{=}2$$

7. $\{10\}$ **9.** $\{20\}$ **11.** $\{7\}$ **13.** $\varnothing$ **15.** $\{5\}$

17. $\{-3\}$ **19.** $\{2\}$ **21.** $\{8\}$ **23.** $\{-1\}$ **25.** $\{1,4\}$

27. $\{1,9\}$ **29.** $\varnothing$ **31.** $\{1,5\}$ **33.** $\{5\}$ **35.** $\varnothing$

37. 35 ft   **39.** 61 cm   **41.** 20 ft   **43.** $5\sqrt{5}$ ft

45. a. 2.30 **b.** 146.79 lb   **47. a.** 1.67 **b.** 148.43 lb   **49.** $\{5\}$

51. $\varnothing$ **53.** $\left\{\dfrac{23}{5}\right\}$ **55.** $\{3\}$ **57.** $\{1,5\}$ **59.** $\{6\}$

61. $\varnothing$ **63.** $\{16\}$ **65.** $\{6\}$ **67.** $\varnothing$ **69.** $\{1,6\}$

71. 7 ft   **73.** (a) 2.39 (b) 163.81 lb   **75.** $\dfrac{2\pi\sqrt{14}}{7}$ sec

77. 67.5 mph

79.
$$\sqrt{4x+25}=9$$
$$\left(\sqrt{4x+25}\right)^2=9^2$$
$$4x+25=81$$
$$4x=56$$
$$x=14 \text{ The solution checks.}$$

81.
$$\sqrt{13-3x}=x-3$$
$$\left(\sqrt{13-3x}\right)^2=(x-3)^2$$
$$13-3x=x^2-6x+9$$
$$0=x^2-3x-4$$
$$0=(x-4)(x+1)$$
$$x=4,-1$$
Since -1 is an extraneous solution, the solution set is $\{4\}$.

83.
$$\sqrt{2x-23}=\sqrt{x}-1$$
$$(\sqrt{2x-23})^2=(\sqrt{x}-1)^2$$
$$2x-23=x-2\sqrt{x}+1$$
$$x-24=-2\sqrt{x}$$
$$(x-24)^2=(-2\sqrt{x})^2$$
$$x^2-48x+576=4x$$
$$x^2-52x+576=0$$
$$(x-16)(x-36)=0$$
$$x=16,36$$
Since 36 is an extraneous solution, the solution set is $\{16\}$.

85.

X	Y₁	Y₂
-9	4.5826	-6
-8	3	-5
-4	6.4031	-1
0	7.5488	3
4	8.544	7
-8	5	-5

X=6

The two radical expressions have the same value at 6 but not at -8. Therefore, the solution set is $\{6\}$.

87.

X	Y₁	Y₂
-6	1	ERROR
-5	3	3
-4	4.1231	4.4142
-3	5	5
-2	5.7446	5.4495
-1	6.4031	5.8284
0	7	6.1623

X=-5

The two radical expressions have the same value at -5 and -3. Therefore, the solution set is $\{-5,-3\}$.

Section 10.7

1. An imaginary number is a number whose value, when squared, is a negative number. We write the imaginary number in the form of $\sqrt{-c}=i\sqrt{c}$, where $c>0$.

3. The conjugate of $a+bi$ is $a-bi$.

5. You multiply complex numbers by applying the distributive property and the FOIL method.

7. $(5-i)^2=25-2(5)(i)+i^2=25-10i-1=24-10i\neq24$

9. $6i$ **11.** $5i\sqrt{3}$ **13.** $1-2i$ **15.** $10+4i$ **17.** $1+\dfrac{9}{2}i$

19. $\dfrac{7}{15}-\dfrac{1}{5}i$ **21.** $\dfrac{5}{7}-\dfrac{11\sqrt{3}}{14}i$ **23.** $\dfrac{5}{11}+\dfrac{4\sqrt{2}}{11}i$ **25.** $10+4i$

27. $-4+5i$ **29.** $-10+24i$ **31.** $7-19i$ **33.** -14

35. $15i$ **37.** $-30+6i$ **39.** $-12+21i$ **41.** $17-19i$

43. $-24+23i$ **45.** $46-3i$ **47.** $13+5i\sqrt{5}$

49. $-40-42i$ **51.** $-13-84i$ **53.** $63+16i$

55. $3 - 4i\sqrt{7}$ **57.** 26   **59.** 29   **61.** $\dfrac{2}{5}$ **63.** $-\dfrac{3i}{5}$

65. $\dfrac{3i}{2}$ **67.** $\dfrac{5}{2} + \dfrac{3}{2}i$ **69.** $\dfrac{5}{17} + \dfrac{3}{17}i$ **71.** $-2 + 3i$

73. $-\dfrac{1}{5} - \dfrac{3}{5}i$ **75.** -1 **77.** i **79.** -1 **81.** 1

83. $\dfrac{1}{2} - \dfrac{7}{4}i$ **85.** $\dfrac{4}{5} + \dfrac{3}{2}i$ **87.** $3 + 3i$ **89.** $77 + 36i$

91. $2 - 7i$ **93.** $-20 - 11i$ **95.** $-\dfrac{2\sqrt{2}}{3}i$ **97.** $-15 + 21i$

99. $5 + 2i$ **101.** $33 - 56i$ **103.** $27 + 72i$

105. $-17 - 19i$ **107.** $-72 + 30i$ **109.** $34 - 6i\sqrt{7}$

111. $-1 + 8i$ **113.** 43   **115.** 100   **117.** $-\dfrac{2}{3}i$

119. $\dfrac{53}{29} - \dfrac{2}{29}i$ **121.** i **123.** 1   **125.** $-128i$ **127.** $-\dfrac{1}{8}i$

129. $\sqrt{-9} \cdot \sqrt{-16} = 3i \cdot 4i = 12i^2 = -12$

131. $(4 - i\sqrt{3})^2 = 16 - 8i\sqrt{3} + 3i^2 = 13 - 8i\sqrt{3}$

133.
```
(3-5i)(7+2i)
          31-29i
```

135.
```
(-1-3i)/(3+4i)
            -.6-.2i
-3/5-1/5i
            -.6-.2i
```

CHAPTER 10 SUMMARY

Section 10.1

1. $b^2 = A$ **2.** $\sqrt{}$; positive   **3.** radicand
4. not; real number   **5.** $b^3 = A$ **6.** $b^n = A$
7. $A \geq 0$ **8.** A can be any real number   **9.** a; $|a|$
10. a **11.** a **12.** $[0, \infty)$ **13.** $(-\infty, \infty)$

Section 10.2

14. rational   **15.** $\sqrt[n]{b}$ **16.** $\sqrt[n]{b^m}$ **17.** $\dfrac{1}{b^{m/n}}$ **18.** rational

Section 10.3

19. $\sqrt[n]{a}\, \sqrt[n]{b}$, where $a, b \geq 0$ **20.** perfect power; index
21. $\dfrac{\sqrt[n]{a}}{\sqrt[n]{b}}$ **22.** $d = \sqrt{(x_2 - x_1)^2 + (y_2 - y_1)^2}$

Section 10.4

23. radicand; index   **24.** coefficients; radicand
25. simplified   **26.** index; radicands
27. distributive; FOIL   **28.** conjugates

Section 10.5

29. index; radicands   **30.** rationalizing the denominator
31. one; index   **32.** conjugate

Section 10.6

33. $a^n = b^n$ **34.** extraneous; check   **35.** isolated
36. radical; twice   **37.** $a^2 + b^2 = c^2$

Section 10.7

38. imaginary unit; i; i; i^2 **39.** $a + bi$; a; b
40. adding   **41.** distributive; FOIL
42. conjugate; $a - bi$ **43.** remainder; 4; i^r

CHAPTER 10 REVIEW EXERCISES

Section 10.1

1. $-14t^{10}$ **3.** $10h^4k$ **5.** $\dfrac{9s^8}{5}$ **7.** $1.9a^3$ **9.** $3x - 1$

11. $-10b$ **13.** $2x^2y\sqrt[4]{4}$ **15.** $4a^2b$

17. domain $= \left[\dfrac{5}{8}, \infty\right)$ or $\left\{x \,\middle|\, x \geq \dfrac{5}{8}\right\}$; not a real number, 3.312

19. domain $= (-\infty, \infty)$ or $\{x \,|\, x \in \mathbb{R}\}$; not a real number, 3, 1.44, -2.35

21.

x	$f(x)$
-3	5
0	4
1	3.61
4	2

Section 10.2

23. $-4x^4$ **25.** $100a^4b^2$ **27.** $\dfrac{8y^3}{x^9}$ **29.** $\sqrt[5]{(2x - 3)^2}$

31. $-\dfrac{1}{5a}$ **33.** $\dfrac{x^4}{y^{16}}$ **35.** 4   **37.** $\dfrac{1}{x^{2/21}}$

39. $-8x^9y^{12}$ **41.** $25x - 9y^{2/3}$ **43.** $2x + 3x^{7/6}$ **45.** $5x^3$

47. $v\sqrt{u}$ **49.** $\sqrt{\dfrac{6}{c^3}}$ **51.** $\sqrt[6]{8192}$ **53.** \$1207.21

Section 10.3

55. $6a^2b^6\sqrt{10a}$ **57.** $5x^2y^5\sqrt[3]{2x^2y}$ **59.** $\dfrac{10x\sqrt[3]{2x^2}}{3y^4}$

61. $\dfrac{d^5\sqrt{3d}}{c^3}$ **63.** $4\sqrt{2}$ **65.** $2\sqrt{5}$

Section 10.4

67. $6\sqrt{6}$ **69.** $63ab\sqrt{6a}$ **71.** $11y\sqrt[3]{3x} - 15y\sqrt[3]{x}$

73. $\dfrac{113a\sqrt{b}}{10}$ **75.** $-\dfrac{3y\sqrt[3]{3x^2}}{5}$ **77.** $2x^2y\sqrt[5]{5x^2}$

79. $5\sqrt{7} - 7\sqrt{2}$ **81.** $9\sqrt{2} - 57$ **83.** $49 - 2a$
85. 62.8, 169   **87.** $36\sqrt{2}$ cm, $24\sqrt{6}$ cm

Section 10.5

89. $5x^4\sqrt{3x}$ **91.** $\dfrac{3y^2\sqrt{2y}}{2}$ **93.** $12c\sqrt[5]{2^2c^3}$

95. $ab^3\sqrt[4]{2ab^2}$ **97.** $\dfrac{2\sqrt{6x}}{3x}$ **99.** $\dfrac{4\sqrt[4]{3a^3}}{a}$

101. $\sqrt{7} - \sqrt{5}$ **103.** $3\sqrt{2} + 2\sqrt{6} - 5\sqrt{3} - 10$
105. $\sqrt{a + 4} + 2$

Section 10.6

107. $\varnothing$ **109.** $\{7\}$ **111.** $\{9\}$ **113.** $\{36, 100\}$
115. $\{-1\}$ **117.** 12 ft   **119.** $3\sqrt{10}$ ft or 9.49 ft

Section 10.7

121. $-8 + 2i$ **123.** $7i$ **125.** -15 **127.** $2 + 23i$

129. $22 - \sqrt{6}i$ **131.** 15   **133.** $\dfrac{6 - 13i}{5}$ **135.** -64 **137.** i

CHAPTER 10 TEST

1. d   **3.** b   **5.** a   **7.** c   **9.** b   **11.** c   **13.** b   **15.** a
17. $5x^2y\sqrt{2y}$ **19.** $2\sqrt[3]{5}$ **21.** $3a\sqrt[4]{2a^3}$ **23.** $2y\sqrt[3]{3y}$

25. $3b$ **27.** $\dfrac{6\sqrt[3]{50}}{5}$ **29.** $-46\sqrt{3}$ **31.** $41 + 12\sqrt{5}$

33. $\{7\}$ **35.** $\{4\}$ **37.** $8i$ **39.** $5 + 7i$ **41.** 97   **43.** $-\dfrac{4}{3}i$

CHAPTERS 1–10 CUMULATIVE REVIEW EXERCISES

1. 32, 34   **3.** \$223,500   **5.** $h = \dfrac{S - \pi r^2}{2\pi r}$ **7.** $\left(\dfrac{3}{4}, 2\right)$

9. $\left[-1, \dfrac{11}{3}\right]$ **11.** $-\dfrac{2}{3}$ **13.** $a^2 + a - 7$

15. $f(x) = \dfrac{6}{5}x - 2, \dfrac{6}{5}, -2$ **17.** $y = 5x + 1$ **19.** $\varnothing$

21. $\{(-5, 3)\}$ **23.** $\left\{\left(-\dfrac{21}{2}, -\dfrac{9}{2}\right)\right\}$

25. 50 L of water and 30 L of 48% acid solution

27. \$6500 at 1.25% in the savings account and \$1500 at 0.9% in the money market fund

29. -9 **31.** $\dfrac{25}{9}$ **33.** $6\pi r^2$ **35.** $16x^8y^{32}$ **37.** $\dfrac{125}{27p^9q^{12}}$

39. $5x^2 - 2x + 26$ **41.** $a^3 + 125$ **43.** $a^2 + 4ab + 4b^2$

45. $(2x + 7)(3x^2 - 5)$ **47.** $(a^3 + 5)(a^3 - 7)$

49. $(x^2 + 9)(x + 3)(x - 3)$ **51.** $(a + 10)(a^2 - 10a + 100)$

53. $\left\{-1, \dfrac{3}{4}\right\}$ **55.** $\dfrac{x + 7}{5x + 6}$ **57.** $\dfrac{2x - 7}{4x}$ **59.** $2x^2 + x + 4$

61. $\dfrac{5x - 16}{(x + 2)(x - 2)(x - 5)}$ **63.** $\dfrac{1}{ab}$ **65.** $\{-18\}$ **67.** 224 mL

69. $x = 2y^3(u + v)$ **71.** $\dfrac{7x^3}{2}$ **73.** $\dfrac{5x^2}{4y}$ **75.** $2x^2y$

77. domain $= (-\infty, 5]; f(-4) = 3, f(1) = 2, f(6)$ undefined

79. $4x^3$ **81.** 16   **83.** $\dfrac{1}{32a^5b^{10}}$ **85.** $x^{5/12}$

87. $6x^{4/5} - 19x^{2/5}y^{1/2} + 3y$ **89.** $\sqrt[4]{a}$ **91.** $\sqrt{2}$ **93.** $6x^2\sqrt{2}$

95. $7a^2b^2\sqrt{2a}$ **97.** $\dfrac{2x\sqrt[3]{x}}{y}$ **99.** $2\sqrt{17}$ **101.** $28x\sqrt{2}$

103. $2\sqrt{3} + 3\sqrt{2}$ **105.** $5x\sqrt{2x}$ **107.** $5cd^3\sqrt{c}$ **109.** $\dfrac{2\sqrt{5a}}{a}$

111. $\dfrac{5(\sqrt{7} - \sqrt{3})}{2}$ **113.** $\{15\}$ **115.** $\{5\}$ **117.** $-2 + 7i$

119. $5 - 12i$ **121.** $3 - \dfrac{1}{2}i$ **123.** $\dfrac{1}{5} + \dfrac{13}{5}i$ **125.** 1   **127.** i

CHAPTER 11

Section 11.1

1. The vertex is the ordered pair $(0, k)$, the parabola opens upward, and the range is $[k, \infty)$.

3. The vertex is the ordered pair (h, k), the parabola opens downward, and the range is $(-\infty, k]$.

5. no x-intercept; y-intercept: $(0, 5)$; vertex: $(0, 5)$; axis of symmetry: $x = 0$; domain: $(-\infty, \infty)$; range: $[5, \infty)$; shifts up 5 units

7. x-intercepts: $(-2, 0)$ and $(2, 0)$; y-intercept: $(0, -4)$; vertex: $(0, -4)$; axis of symmetry: $x = 0$; domain: $(-\infty, \infty)$; range: $[-4, \infty)$; shifts down 4 units

9. x-intercepts: $(-1.5, 0)$ and $(1.5, 0)$; y-intercept: $(0, -2.25)$; vertex: $(0, -2.25)$; axis of symmetry: $x = 0$; domain: $(-\infty, \infty)$; range: $[-2.25, \infty)$; shifts down 2.25 units

11. no x-intercept; y-intercept: $(0, 5.76)$; vertex: $(0, 5.76)$; axis of symmetry: $x = 0$; domain: $(-\infty, \infty)$; range: $[5.76, \infty)$; shifts up 5.76 units

13. x-intercept: $(-3, 0)$; y-intercept: $(0, 9)$; vertex: $(-3, 0)$; axis of symmetry: $x = -3$; domain: $(-\infty, \infty)$; range: $[0, \infty)$; shifts left 3 units

15. x-intercept: $(4, 0)$; y-intercept: $(0, 16)$; vertex: $(4, 0)$; axis of symmetry: $x = 4$; domain: $(-\infty, \infty)$; range: $[0, \infty)$; shifts right 4 units

17. x-intercept: $\left(-\dfrac{3}{2}, 0\right)$; y-intercept: $\left(0, \dfrac{9}{4}\right)$; vertex: $\left(-\dfrac{3}{2}, 0\right)$; axis of symmetry: $x = -\dfrac{3}{2}$; domain: $(-\infty, \infty)$; range: $[0, \infty)$; shifts left $\dfrac{3}{2}$ units

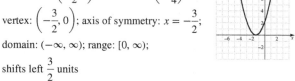

19. x-intercept: $\left(\dfrac{5}{8}, 0\right)$; y-intercept: $\left(0, \dfrac{25}{64}\right)$; vertex: $\left(\dfrac{5}{8}, 0\right)$; axis of symmetry: $x = \dfrac{5}{8}$; domain: $(-\infty, \infty)$; range: $[0, \infty)$; shifts right $\dfrac{5}{8}$ units

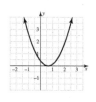

21. x-intercept: $(0, 0)$; y-intercept: $(0, 0)$; vertex: $(0, 0)$; axis of symmetry: $x = 0$; domain: $(-\infty, \infty)$; range: $[0, \infty)$; stretches by a factor of 4

23. x-intercept: $(0, 0)$; y-intercept: $(0, 0)$; vertex: $(0, 0)$; axis of symmetry: $x = 0$; domain: $(-\infty, \infty)$; range: $[0, \infty)$; shrinks by a factor of $\dfrac{1}{4}$

25. x-intercept: $(0, 0)$; y-intercept: $(0, 0)$; vertex: $(0, 0)$; axis of symmetry: $x = 0$; domain: $(-\infty, \infty)$; range: $(-\infty, 0]$; stretches by a factor of 3 and reflects over the x-axis

27. x-intercept: $(0, 0)$; y-intercept: $(0, 0)$; vertex: $(0, 0)$; axis of symmetry: $x = 0$; domain: $(-\infty, \infty)$; range: $(-\infty, 0)$; shrinks by a factor of $\dfrac{1}{5}$ and reflects over the x-axis

29. x-intercept: $(0, 0)$; y-intercept: $(0, 0)$; vertex: $(0, 0)$; axis of symmetry: $x = 0$; domain: $(-\infty, \infty)$; range: $(-\infty, 0]$; stretches by a factor of 2.5 and reflects over the x-axis

31. x-intercepts: $(-2, 0)$ and $(4, 0)$; y-intercept: $(0, -8)$; vertex: $(1, -9)$; axis of symmetry: $x = 1$; domain: $(-\infty, \infty)$; range: $[-9, \infty)$; shifts right 1 unit and down 9 units

51.

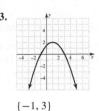

$\{-1, 3\}$

53.

$\{-1, 3\}$

55. x-intercepts: $\left(-\frac{5}{3}, 0\right)$ and $\left(\frac{5}{3}, 0\right)$; y-intercept: $\left(0, -\frac{25}{9}\right)$; vertex: $\left(0, -\frac{25}{9}\right)$; axis of symmetry: $x = 0$; domain: $(-\infty, \infty)$; range: $\left[-\frac{25}{9}, \infty\right)$; shifts down $\frac{25}{9}$ units

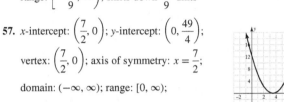

33. no x-intercept; y-intercept: $(0, 13)$; vertex: $(-3, 4)$; axis of symmetry: $x = -3$; domain: $(-\infty, \infty)$; range: $[4, \infty)$; shifts left 3 units and up 4 units

35. no x-intercepts; y-intercept: $(0, -27)$; vertex: $(5, -2)$; axis of symmetry: $x = 5$; domain: $(-\infty, \infty)$; range: $(-\infty, 2]$; reflects over the x-axis, and shifts right 5 units and down 2 units

57. x-intercept: $\left(\frac{7}{2}, 0\right)$; y-intercept: $\left(0, \frac{49}{4}\right)$; vertex: $\left(\frac{7}{2}, 0\right)$; axis of symmetry: $x = \frac{7}{2}$; domain: $(-\infty, \infty)$; range: $[0, \infty)$; shifts right $\frac{7}{2}$ units

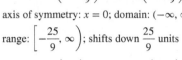

37. x-intercepts: $(-7, 0)$ and $(-1, 0)$; y-intercept: $(0, 14)$; vertex: $(-4, -18)$; axis of symmetry: $x = -4$; domain: $(-\infty, \infty)$; range: $[-18, \infty)$, stretches by a factor of 2, and shifts left 4 units and down 18 units

39. no x-intercepts; y-intercept: $(0, 101)$; vertex: $(5, 1)$; axis of symmetry: $x = 5$; domain: $(-\infty, \infty)$; range: $[1, \infty)$; stretches by a factor of 5, and shifts up 1 unit and right 5 units

59. x-intercepts: $(-1, 0)$ and $(1, 0)$; y-intercept: $(0, 1)$; vertex: $(0, 1)$; axis of symmetry: $x = 0$; domain: $(-\infty, \infty)$; range: $(-\infty, 1]$; reflects over the x-axis and shifts up 1 unit

61. x-intercepts: $(-3, 0)$ and $(1, 0)$; y-intercept: $(0, -6)$; vertex: $(-1, -8)$; axis of symmetry: $x = -1$; domain: $(-\infty, \infty)$; range: $[-8, \infty)$; shifts down 8 units and left 1 unit, and stretches by a factor of 2

41. x-intercepts: $(-10, 0)$ and $(2, 0)$; y-intercept: $(0, 10)$; vertex: $(-4, 18)$; axis of symmetry: $x = -4$; domain: $(-\infty, \infty)$; range: $(-\infty, 18]$; shrinks by a factor of $\frac{1}{2}$, reflects over the x-axis, and shifts left 4 units and up 18 units

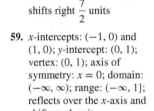

63. x-intercepts: $(4.5, 0)$ and $(5.5, 0)$; y-intercept: $(0, -198)$; vertex: $(5, 2)$; axis of symmetry: $x = 5$; domain: $(-\infty, \infty)$; range: $(-\infty, 2]$; reflects over the x-axis, shifts right 5 units and up 2 units, and stretches by a factor of 8

43. x-intercepts: $(-1, 0)$ and $(5, 0)$; y-intercept: $\left(0, -\frac{10}{3}\right)$; vertex: $(2, -6)$; axis of symmetry: $x = 2$; domain: $(-\infty, \infty)$; range: $[-6, \infty)$; shrinks by a factor of $\frac{2}{3}$, and shifts right 2 units and down 6 units

65. x-intercepts: $(-1.1, 0)$ and $(6.1, 0)$; y-intercept: $(0, -6.71)$; vertex: $(2.5, -12.96)$; axis of symmetry: $x = 2.5$; domain: $(-\infty, \infty)$; range: $[-12.96, \infty)$; shifts right 2.5 units and down 12.96 units

45.

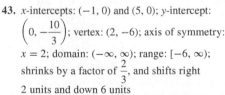

$\{-4, 4\}$

47.

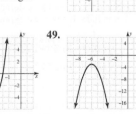

$\{-2.5, -1.5\}$

49.

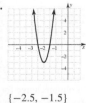

$\varnothing$

67. x-intercepts: $(-3, 0)$ and $(1, 0)$; y-intercept: $\left(0, \frac{3}{2}\right)$; vertex: $(-1, 2)$; axis of symmetry: $x = -1$; domain: $(-\infty, \infty)$; range: $(-\infty, 2]$; reflects over the x-axis, shifts left 1 unit and up 2 units, and shrinks by a factor of $\frac{1}{2}$

69. no x-intercepts; y-intercept: $\left(0, \frac{1}{4}\right)$; vertex: $\left(0, \frac{1}{4}\right)$; axis of symmetry: $x = 0$; domain: $(-\infty, \infty)$; range: $\left[\frac{1}{4}, \infty\right)$; shifts up $\frac{1}{4}$ unit

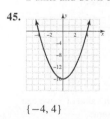

71. no x-intercept; y-intercept: $\left(0, \dfrac{5}{4}\right)$;
vertex: $(1, 1)$; axis of symmetry: $x = 1$;
domain: $(-\infty, \infty)$; range: $[1, \infty)$;
shifts right 1 unit and up 1 unit, and
shrinks by a factor of $\dfrac{1}{4}$

73. x-intercepts: $(-6.3, 0)$ and $(2.7, 0)$;
y-intercept: $(0, 17.01)$; vertex: $(-1.8, 20.25)$;
axis of symmetry: $x = -1.8$;
domain: $(-\infty, \infty)$; range: $(-\infty, 20.25]$;
reflects over the x-axis, and shifts left
1.8 units and up 20.25 units

75. x-intercepts: $(-0.6, 0)$ and $(3, 0)$; y-intercept:
$(0, -3.6)$; vertex: $(1.2, -6.48)$; axis of
symmetry: $x = 1.2$; domain: $(-\infty, \infty)$;
range: $[-6.48, \infty)$; shifts right 1.2 units
and down 6.48 units, and stretches by
a factor of 2

77.   **79.**   **81.**

$\{0.5, 3.5\}$ $\varnothing$ $\{0.5, 3.5\}$

83.   **85.** 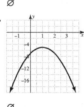 **87.** The graph of $y = x^2$ is shifted right 3 units and up 25 units.

$\{-1, 7\}$ $\varnothing$

89. Reflection over the x-axis and stretching/shrinking must occur before vertical translation. The graph of $y = x^2$ is shifted right 5 units, reflected over the x-axis, shrunk by a factor of $\dfrac{1}{2}$, and shifted up 4 units.

91. $(-0.68, 0)$, $(2.61, 0)$; $(0.96, -7.30)$

Section 11.2

1. Factor $x^2 - a^2$ as a difference of two squares and use the zero products property to solve. To apply the square root property, rewrite the equation as $x^2 = a^2$ and take the square root of both sides of the equation.

3. First take out 2 as a common factor and write the binomial as $2\left(x^2 - \dfrac{3}{2}x\right)$. Add the square of half of the coefficient of x, that is, $\left(\dfrac{1}{2} \cdot \dfrac{3}{2}\right)^2$ or $\dfrac{9}{16}$, to the binomial inside the parentheses and get $2\left(x^2 - \dfrac{3}{2}x + \dfrac{9}{16}\right)$. Write the expression as a perfect square trinomial $2\left(x - \dfrac{3}{4}\right)^2$.

5. The graph of the corresponding quadratic function will not cross the x-axis; that is, there are no x-intercepts.

7. $\{-7, 7\}$ **9.** $\{-8, 8\}$ **11.** $\left\{-\dfrac{1}{7}, \dfrac{1}{7}\right\}$ **13.** $\left\{-\dfrac{2}{9}, \dfrac{2}{9}\right\}$

15. $\{\pm 10\sqrt{21}\}$ **17.** $\{\pm 6i\sqrt{5}\}$ **19.** $\left\{-1, \dfrac{3}{2}\right\}$

21. $\left\{\dfrac{4}{7}, \dfrac{16}{7}\right\}$ **23.** $\left\{\dfrac{1 \pm 4\sqrt{3}}{3}\right\}$ **25.** $\left\{\dfrac{-5 \pm 5i\sqrt{6}}{6}\right\}$

27. $x^2 + 18x + 81 = (x + 9)^2$ **29.** $a^2 - 14a + 49 = (a - 7)^2$

31. $r^2 + r + \dfrac{1}{4} = \left(r + \dfrac{1}{2}\right)^2$ **33.** $a^2 - 9a + \dfrac{81}{4} = \left(a - \dfrac{9}{2}\right)^2$

35. $\{-1, 7\}$ **37.** $\{-10, 5\}$ **39.** $\{3, 7\}$ **41.** $\left\{-\dfrac{1}{2}, \dfrac{3}{4}\right\}$

43. $\left\{-7, -\dfrac{3}{2}\right\}$ **45.** $\left\{\dfrac{-4 \pm \sqrt{11}}{2}\right\}$ **47.** $\left\{\dfrac{-4 \pm i\sqrt{10}}{2}\right\}$

49. 260.77 ft   **51.** 6.18 sec   **53.** 8.09 sec   **55.** 3.1%
57. 3.4%
59. zeros: $x = -13, x = -1$; x-intercepts: $(-13, 0), (-1, 0)$

61. zeros: $x = -1, x = -\dfrac{1}{3}$; x-intercepts: $(-1, 0), \left(-\dfrac{1}{3}, 0\right)$

63. zeros: $x = -1, \pm 2i\sqrt{6}$; no x-intercepts
65. zeros: $x = 3 \pm 5\sqrt{2} \approx 10.07$ or -4.07; x-intercepts: $(-4.07, 0)$, $(10.07, 0)$

67. zeros: $x = \dfrac{-1 \pm \sqrt{3}}{10} \approx 0.07$ or -0.27; x-intercepts: $(-0.27, 0)$, $(0.07, 0)$

69. zeros: $x = \dfrac{-5 \pm 2i\sqrt{21}}{3}$, no x-intercepts   **71.** $\left\{\pm \dfrac{5}{8}\right\}$

73. $\{\pm 10i\}$ **75.** $\{\pm 6\sqrt{7}\}$ **77.** $\left\{-\dfrac{17}{2}, -\dfrac{3}{2}\right\}$

79. $\left\{\dfrac{-2 \pm 5i\sqrt{3}}{9}\right\}$ **81.** $\left\{\dfrac{3 \pm \sqrt{21}}{4}\right\}$ **83.** $\left\{\dfrac{3 \pm i\sqrt{2}}{4}\right\}$

85. $\left\{\dfrac{1 \pm 10\sqrt{3}}{6}\right\}$ **87.** $\{-9, 4\}$ **89.** $\left\{-\dfrac{3}{2}, 2\right\}$

91. $\left\{-\dfrac{7}{2}, 0\right\}$ **93.** $\left\{\dfrac{1 \pm i\sqrt{15}}{4}\right\}$ **95.** $\left\{\dfrac{3 \pm \sqrt{6}}{2}\right\}$

97. $\left\{\dfrac{4 \pm i\sqrt{3}}{3}\right\}$ **99.** $x^2 + 8x + 16 = (x + 4)^2$

101. $a^2 - 1.8a + 0.81 = (a - 0.9)^2$
103. $a^2 + 0.9a + 0.2025 = (a + 0.45)^2$

105. $r^2 + \dfrac{1}{2}r + \dfrac{1}{16} = \left(r + \dfrac{1}{4}\right)^2$ **107.** $a^2 - \dfrac{3}{2}a + \dfrac{9}{16} = \left(a - \dfrac{3}{4}\right)^2$

109. 28.54 ft   **111.** 5.51 sec   **113.** 2.6%

115. $4x^2 - 9 = 0$
$\qquad 4x^2 = 9$
$\qquad x^2 = \dfrac{9}{4}$
$\qquad x = \pm\dfrac{3}{2}$

117. $x^2 + 14x + \left(\dfrac{14}{2}\right)^2 = x^2 + 14x + 49$
$\qquad\qquad\qquad\qquad = (x + 7)^2$

119.
```
Ploti Plot2 Plot3
\Y1=X²-15X+225/4
\Y2=(X-15/2)²
\Y3=
\Y4=
\Y5=
\Y6=
```

X	Y1	Y2
-4	132.25	132.25
-3	110.25	110.25
-2	90.25	90.25
-1	72.25	72.25
0	56.25	56.25
1	42.25	42.25
2	30.25	30.25

Y2=(X-15/2)²

Since the Y_1 and Y_2 columns agree for each value of x, we can conclude that the two expressions are the same.

121.
```
Ploti Plot2 Plot3
\Y1=X²-2.3X+1.32
25
\Y2=(X-1.15)²
\Y3=
\Y4=
\Y5=
\Y6=
```

X	Y1	Y2
0	1.3225	1.3225
1	.0225	.0225
2	.7225	.7225
3	3.4225	3.4225
4	8.1225	8.1225
5	14.823	14.823
6	23.523	23.523

Y2=(X-1.15)²

Since the Y_1 and Y_2 columns agree for each value of x, we can conclude that the two expressions are the same.

Section 11.3

1. I would use the zero products property to solve the equation since the equation is factored and set equal to zero.

3. I would use completing the square since the coefficient of x^2 is 1 and the coefficient of x is even.

5. The advantage is that we can just substitute values into a formula and not have to complete the square when the coefficient of x^2 is not 1 and the coefficient of x is not even.

7. If the discriminant is positive, there are two real solutions. If the discriminant is zero, there is one real solution. If the discriminant is negative, there are no real solutions.

9. $\{3, 4\}$ **11.** $\left\{\dfrac{9 \pm \sqrt{29}}{2}\right\}$ **13.** $\{-4 \pm i\sqrt{2}\}$

15. $\left\{-\dfrac{11}{2}, 0\right\}$ **17.** $\left\{\dfrac{1 \pm i\sqrt{10}}{9}\right\}$ **19.** $\{-10 \pm \sqrt{23}\}$

21. $\left\{-\dfrac{8}{3}, 3\right\}$ **23.** $\left\{\dfrac{5 \pm \sqrt{23}}{6}\right\}$ **25.** two irrational solutions

27. two complex, nonreal solutions   **29.** two rational solutions
31. one repeating rational solution   **33.** 0.39 sec and 2.43 sec
35. 5000 items   **37.** 80 ft by 160 ft   **39.** 1 P.M.   **41.** $\{-8, 2\}$

43. $\left\{-\dfrac{10}{3}, 2\right\}$ **45.** $\left\{\dfrac{8 \pm \sqrt{10}}{2}\right\}$ **47.** $\left\{\dfrac{3 \pm i\sqrt{15}}{2}\right\}$

49. $\left\{-\dfrac{9}{5}, 0\right\}$ **51.** $\left\{\dfrac{5 \pm \sqrt{15}}{2}\right\}$ **53.** $\left\{\dfrac{1 \pm \sqrt{10}}{4}\right\}$

55. $\left\{-\dfrac{3}{2}, \dfrac{8}{3}\right\}$ **57.** $\left\{\dfrac{1 \pm i\sqrt{2}}{3}\right\}$ **59.** two irrational solutions

61. two complex, nonreal solutions   **63.** two rational solutions
65. one repeating rational solution   **67.** 400 items
69. 70 ft by 140 ft   **71.** 10:17 A.M.
73. $x^2 - 3x - 4 = 0$

$$x = \frac{-(-3) \pm \sqrt{(-3)^2 - 4(1)(-4)}}{2(1)}$$
$$= \frac{3 \pm \sqrt{9 + 16}}{2}$$
$$= \frac{3 \pm \sqrt{25}}{2}$$
$$= \frac{3 + 5}{2} \text{ or } \frac{3 - 5}{2}$$
$$= 4 \text{ or } -1$$

75. The quadratic formula was used and simplified correctly. The error was in approximating the solutions.

```
(5+√(89))/4
        3.608495283
(5-√(89))/4
        -1.108495283
```

So, the approximate solutions are 3.61 and -1.11.

Piece It Together Sections 11.1–11.3

1. none; $(0, 10)$; $(2, 6)$; $x = 2$; $(-\infty, \infty)$; $[6, \infty)$; It is the graph of $f(x) = x^2$ shifted right 2 units and up 6 units.

2. $(-7, 0)$ and $(-1, 0)$; $(0, 7)$; $(-4, -9)$; $x = -4$; $(-\infty, \infty)$; $[-9, \infty)$; It is the graph of $f(x) = x^2$ shifted left 4 units and down 9 units.

3. The solution set of $f(x) = 0$ is $\{2, 8\}$. The zeros of $f(x)$ are $x = 2$ and $x = 8$. The x-intercepts of the graph of $f(x)$ are $(2, 0)$ and $(8, 0)$.

4. The solution set of $f(x) = 0$ is $\{-4, -2\}$. The zeros of $f(x)$ are $x = -4$ and $x = -2$. The x-intercepts of the graph of $f(x)$ are $(-4, 0)$ and $(-2, 0)$.

5. $\left\{-\dfrac{3}{7}, \dfrac{1}{7}\right\}$ **6.** $\left\{-\dfrac{7}{3}, \dfrac{7}{3}\right\}$ **7.** $\left\{-2, \dfrac{7}{3}\right\}$ **8.** $\left\{\dfrac{-5 \pm i}{2}\right\}$

9. $y^2 - 22y + 121 = (y - 11)^2$ **10.** $b^2 - 7b + \dfrac{49}{4} = \left(b - \dfrac{7}{2}\right)^2$

11. 28; two irrational solutions   **12.** -28; two complex solutions
13. 256; two rational solutions
14. 0; one repeating rational solution   **15.** 5.92 sec
16. 400 items

Section 11.4

1. Multiply both sides of the equation by the least common denominator, x.
3. Let $u = 2x - 1$. Rewrite the equation as $u^2 - 2u - 8 = 0$.
5. We back-substitute to solve for x.
$$w = 5 \quad \text{or} \quad -3$$
$$x^{1/4} = 5 \quad \text{or} \quad -3$$
$$x = 5^4 \quad \text{or} \quad (-3)^4$$
$$x = 625 \quad \text{or} \quad 81$$

7. $\left\{-\dfrac{3}{2}, \dfrac{5}{3}\right\}$ **9.** $\left\{-\dfrac{3}{2}\right\}$ **11.** $\{5 \pm 2i\sqrt{2}\}$ **13.** $\left\{\dfrac{1}{3}, \dfrac{4}{3}\right\}$

15. $\left\{\dfrac{5}{12}\right\}$ **17.** $\left\{\dfrac{3 \pm \sqrt{6}}{3}\right\}$ **19.** $\left\{\dfrac{3 \pm i\sqrt{3}}{2}\right\}$ **21.** $\{7, 8\}$

23. $\{-2, 6\}$ **25.** $\{2\}$ **27.** $\left\{\dfrac{81}{49}\right\}$ **29.** $\varnothing$ **31.** $\left\{\dfrac{9}{16}, 16\right\}$

33. $\{\pm 5, \pm i\}$ **35.** $\left\{\pm 1, \pm\dfrac{2}{3}\right\}$ **37.** $\{\pm 3i, \pm i\}$

39. $\{\pm 2, \pm 2i\}$ **41.** $\{4, -2 \pm 2i\sqrt{3}\}$ **43.** $\left\{-2, \dfrac{17}{5}\right\}$

45. $\left\{-\dfrac{5}{3}\right\}$ **47.** $\left\{\dfrac{8}{27}, 8\right\}$ **49.** $\left\{-1, -\dfrac{1}{32}\right\}$

51. Marilyn: 9.12 hr; Carolina: 7.12 hr   **53.** 75.89 mph
55. $\{-6, 3\}$ **57.** $\{5, 11\}$ **59.** $\left\{\dfrac{4 \pm 2\sqrt{6}}{5}\right\}$ **61.** $\{9\}$

63. $\left\{\dfrac{-3 \pm i\sqrt{14}}{2}\right\}$ **65.** $\left\{\dfrac{25}{4}, 9\right\}$ **67.** $\left\{-5, -\dfrac{5}{4}\right\}$

69. $\left\{-\dfrac{8}{5}, -\dfrac{4}{5}\right\}$ **71.** $\left\{\dfrac{3}{4}\right\}$ **73.** $\left\{-\dfrac{29}{20}, -\dfrac{16}{15}\right\}$ **75.** $\left\{\dfrac{7}{2}\right\}$

77. $\left\{\pm 1, \pm\dfrac{5}{2}\right\}$ **79.** $\{-1, 64\}$ **81.** $\{-3, 8\}$ **83.** $\{5\}$

85. $\varnothing$ **87.** $\{\pm 5, \pm 5i\}$ **89.** $\left\{-\dfrac{1}{2}, \dfrac{2}{3}\right\}$

91. $\left\{\dfrac{5}{2}, \dfrac{-5 \pm 5i\sqrt{3}}{4}\right\}$ **93.** Ted: 6.54 hr; Ron: 5.54 hr
95. 60 mph

97. After multiplying by the LCD, Jeung didn't distribute the negative sign to the middle product.

$$\frac{2x}{x+4} - \frac{x-6}{x-4} = \frac{32}{x^2-16}$$

$$\frac{2x}{x+4} - \frac{x-6}{x-4} = \frac{32}{(x+4)(x-4)}$$

$$2x(x-4) - (x-6)(x+4) = 32$$

$$2x^2 - 8x - (x^2 - 2x - 24) = 32$$

$$2x^2 - 8x - x^2 + 2x + 24 = 32$$

$$x^2 - 6x - 8 = 0$$

$$x = \frac{-(-6) \pm \sqrt{(-6)^2 - 4(1)(-8)}}{2(1)}$$

$$x = \frac{6 \pm \sqrt{68}}{2}$$

$$x = \frac{6 \pm 2\sqrt{17}}{2}$$

$$x = 3 \pm \sqrt{17}$$

The solution set is $\{3 \pm \sqrt{17}\}$.

99. Ken didn't apply the square root property correctly. When $x^2 = k$, $x = \pm\sqrt{k}$. There should be four solutions of the equation.

$$4x^4 + 7x^2 - 36 = 0$$

$$(4x^2 - 9)(x^2 + 4) = 0$$

$$4x^2 - 9 = 0 \quad \text{or} \quad x^2 + 4 = 0$$

$$x^2 = \frac{9}{4} \quad \text{or} \quad x^2 = -4$$

$$x = \pm\sqrt{\frac{9}{4}} \quad \text{or} \quad x = \pm\sqrt{-4}$$

$$x = \pm\frac{3}{2} \quad \text{or} \quad x = \pm 2i$$

The solution set is $\left\{\pm\frac{3}{2}, \pm 2i\right\}$.

101. Both are solutions to the equation.

103. The solution set is $\left\{-4, \frac{2}{3}\right\}$.

Section 11.5

1. Subtract c from each side, add $\left(\frac{1}{2}b\right)^2$ to both sides of the equation, and complete the square on the right side of the equation.

$$y = x^2 + bx + c$$

$$y - c = x^2 + bx$$

$$y - c + \frac{b^2}{4} = x^2 + bx + \frac{b^2}{4}$$

$$y - c + \frac{b^2}{4} = \left(x + \frac{b}{2}\right)^2$$

$$y = \left(x + \frac{b}{2}\right)^2 + c - \frac{b^2}{4}$$

The vertex form is $y = \left(x + \frac{b}{2}\right)^2 + c - \frac{b^2}{4}$.

3. The standard form, $y = ax^2 + bx + c$, will make it easier to find the y-intercept, $(0, c)$.

5. The x-coordinate of the vertex is between the two x-intercepts.

7. The x-coordinate of the vertex, $x = \frac{-b}{2a}$, is midway between the solutions of the quadratic equation; that is, solutions from the quadratic formula,

$$x = \frac{-b - \sqrt{b^2 - 4ac}}{2a} \quad \text{and} \quad x = \frac{-b + \sqrt{b^2 - 4ac}}{2a}$$

9. $f(x) = (x + 1)^2 - 16$
x-intercepts: $(3, 0)$ and $(-5, 0)$; y-intercept: $(0, -15)$; vertex: $(-1, -16)$

11. $f(x) = (x + 2)^2 - 64$
x-intercepts: $(6, 0)$ and $(-10, 0)$; y-intercept: $(0, -60)$; vertex: $(-2, -64)$

13. $f(x) = \left(x + \frac{3}{2}\right)^2 - \frac{81}{4}$
x-intercepts: $(3, 0)$ and $(-6, 0)$; y-intercept: $(0, -18)$; vertex: $\left(-\frac{3}{2}, -\frac{81}{4}\right)$

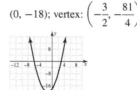

15. $f(x) = -(x + 1)^2 + 9$
x-intercepts: $(2, 0)$ and $(-4, 0)$; y-intercept: $(0, 8)$; vertex: $(-1, 9)$

17. $f(x) = -\left(x + \frac{5}{2}\right)^2 + \frac{169}{4}$
x-intercepts: $(4, 0)$ and $(-9, 0)$; y-intercept: $(0, 36)$; vertex: $\left(-\frac{5}{2}, \frac{169}{4}\right)$

19. $f(x) = 2(x + 3)^2 - 16$
x-intercepts: $\left(-3 \pm 2\sqrt{2}, 0\right) \approx (-0.17, 0)$ and $(-5.83, 0)$; y-intercept: $(0, 2)$; vertex: $(-3, -16)$

21. $f(x) = -5(x + 1)^2 + 12$
x-intercepts: $\left(-1 \pm \frac{2\sqrt{15}}{5}, 0\right)$ $\approx (0.55, 0)$ and $(-2.55, 0)$; y-intercept: $(0, 7)$; vertex: $(-1, 12)$

23. $f(x) = -2(x - 3)^2 - 9$
no x-intercepts; y-intercept: $(0, -27)$; vertex: $(3, -9)$

25. $f(x) = 2(x + 2)^2 + 3$
no x-intercepts; y-intercept: $(0, 11)$; vertex: $(-2, 3)$

27. $\left(\frac{9}{2}, -\frac{73}{4}\right)$ **29.** $\left(\frac{3}{2}, \frac{11}{2}\right)$

31. $\left(-\frac{3}{10}, \frac{11}{20}\right)$

33. x-intercepts: $(\pm 3.16, 0)$; y-intercept: $(0, -10)$; vertex: $(0, -10)$

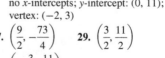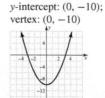

35. no x-intercepts; y-intercept: $(0, -5)$; vertex: $(0, -5)$

37. x-intercepts: $(0, 0)$ and $(-6, 0)$; y-intercept: $(0, 0)$; vertex: $(-3, -9)$

39. x-intercepts: $(0, 0)$ and $(7, 0)$; y-intercept: $(0, 0)$; vertex: $\left(\dfrac{7}{2}, \dfrac{49}{4}\right)$

41. x-intercepts: $(1, 0)$ and $(3, 0)$; y-intercept: $(0, 3)$; vertex: $(2, -1)$

43. x-intercepts: $(-2, 0)$ and $(-4, 0)$; y-intercept: $(0, -8)$; vertex: $(-3, 1)$

45. x-intercepts: $\left(-\dfrac{1}{2}, 0\right)$ and $(5, 0)$; y-intercept: $(0, -5)$; vertex: $\left(\dfrac{9}{4}, -\dfrac{121}{8}\right)$

47. 1000 computers with a maximum monthly revenue of \$435,000
49. 10,500 graphing calculators with a maximum monthly profit of \$347,750
51. 154 ft in 1.5 sec   **53.** 178.84 ft in 3.2 sec
55. x-intercepts: $(-3, 0)$ and $(-8, 0)$; y-intercept: $(0, 24)$; vertex: $\left(-\dfrac{11}{2}, -\dfrac{25}{4}\right)$

57. x-intercepts: $(-5, 0)$ and $(2, 0)$; y-intercept: $(0, 10)$; vertex: $\left(-\dfrac{3}{2}, \dfrac{49}{4}\right)$

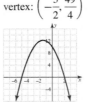

59. x-intercepts: $\left(4 \pm 2\sqrt{5}, 0\right) \approx$ $(8.47, 0)$ and $(-0.47, 0)$; y-intercept: $(0, 4)$ vertex: $(4, 20)$

61. x-intercepts: $(0, 0)$ and $\left(\dfrac{5}{2}, 0\right)$; y-intercept: $(0, 0)$; vertex: $\left(\dfrac{5}{4}, -\dfrac{25}{8}\right)$

63. no x-intercepts; y-intercept: $(0, 4)$; vertex: $\left(-\dfrac{1}{2}, 3\right)$

65. x-intercepts: $(-0.15, 0)$ and $(-2.65, 0)$; y-intercept: $(0, 1.59)$; vertex: $(-1.4, -6.25)$

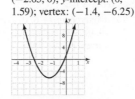

67. x-intercepts: $(-8.2, 0)$ and $(2.6, 0)$; y-intercept: $(0, -21.32)$; vertex: $(-2.8, -29.16)$

69. In about 1.74 years after 2004, or 2005, the unemployment rate reached a minimum of about 3.88%.
71. In month 7, or July, the maximum mean temperature in Topeka reaches a maximum value of about 88°F.
73.
$$y = -x^2 + 6x - 1$$
$$y + 1 = -x^2 + 6x$$
$$-y - 1 = x^2 - 6x$$
$$-y - 1 + 9 = x^2 - 6x + 9$$
$$-y + 8 = (x - 3)^2$$
$$-y = (x - 3)^2 - 8$$
$$y = -(x - 3)^2 + 8$$

75.
$$-x^2 + 5x + 6 = 0$$
$$-(x^2 - 5x - 6) = 0$$
$$-(x - 6)(x + 1) = 0$$
$$x = 6, x = -1$$
The x-intercepts are $(6, 0)$ and $(-1, 0)$.

77.

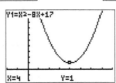

Since the Y_1 and Y_2 columns agree for each value of x, we can conclude that the two forms are the same.

Section 11.6

1. To find the boundary numbers, first find solutions of the associated equation. For rational inequalities, we also find the values that make the function undefined.
3. We always use parentheses on the boundary numbers that make the expression undefined since they are not in the domain of the function.
5. The graph of the quadratic function lies below the x-axis for all values of x.
7. If $c > 0$, then $x = \sqrt{c - b}$. If $c = 0$, then $x = -b$.
9. $(-\infty, -12) \cup (1, \infty)$ **11.** $(-\infty, -5] \cup [-2, \infty)$
13. $(-10, -2)$ **15.** $[-5, 7]$ **17.** $(-\infty, -9) \cup (4, \infty)$
19. $(-\infty, 2) \cup (2, \infty)$ **21.** $(-\infty, \infty)$ **23.** $\left[\dfrac{5}{2}, 4\right]$
25. $\left(-\infty, -\dfrac{1}{3}\right) \cup \left(\dfrac{1}{2}, \infty\right)$ **27.** $\left\{-\dfrac{5}{3}\right\}$ **29.** $\varnothing$ **31.** $\varnothing$
33. $[-4, 0] \cup [3, \infty)$ **35.** $(-\infty, -2) \cup (2, 9)$
37. $\left(-\dfrac{2}{7}, \dfrac{3}{7}\right) \cup (1, \infty)$ **39.** $(-\infty, -1) \cup \left(-\dfrac{1}{3}, \dfrac{1}{3}\right) \cup (1, \infty)$
41. $[-2, 2]$ **43.** $(-\infty, \infty)$ **45.** $(-\infty, -1) \cup [5, \infty)$
47. $\left(-6, \dfrac{8}{3}\right]$ **49.** $(-\infty, -4) \cup (9, \infty)$ **51.** $(-3, 4]$
53. $[-10, -8]$ **55.** $\left[-\dfrac{5}{2}, 3\right]$ **57.** $(-\infty, 2) \cup (7, \infty)$
59. $[0, 3] \cup [6, \infty)$ **61.** $[-4, 2.5]$ **63.** $(-\infty, -1.5) \cup (1, \infty)$
65. $(-\infty, 5) \cup (8, \infty)$ **67.** $[-10, 4]$ **69.** $(-\infty, \infty)$
71. $\left[-3, \dfrac{5}{2}\right]$ **73.** between 5 and 15 units

75. between 2 and 18 sec **77.** [2, 5] **79.** $(-\infty, -3) \cup (8, \infty)$

81. $(-\infty, -5) \cup (12, \infty)$ **83.** $\{1\}$ **85.** $\left(-\dfrac{1}{2}, 1\right)$

87. $(-5, -1)$ **89.** $[-5, 4] \cup [10, \infty)$ **91.** $\left(-\infty, \dfrac{5}{2}\right] \cup (3, \infty)$

93. $(-\infty, 0) \cup \left(\dfrac{2}{3}, \dfrac{6}{5}\right)$ **95.** $(-6, 11]$

97. $(-\infty, -5) \cup (-1, 1) \cup (5, \infty)$ **99.** $\varnothing$ **101.** $(-\infty, 6) \cup [9, \infty)$

103. $\left[-1, \dfrac{5}{2}\right]$ **105.** $\left(-\infty, -\dfrac{7}{2}\right) \cup (3, \infty)$

107. $[-3, 3] \cup [6, \infty)$ **109.** $(-\infty, -2) \cup (2, \infty)$

111.
$$\dfrac{5}{x+3} = 2 \qquad x+3 = 0$$
$$5 = 2x + 6 \qquad x = -3$$
$$2x + 1 = 0$$
$$x = -\dfrac{1}{2}$$

The boundary numbers are -3 and $-\dfrac{1}{2}$. The inequality is true when $x > -3$ and $x \le -\dfrac{1}{2}$. The solution is $\left(-3, -\dfrac{1}{2}\right]$.

113. $(-\infty, -4) \cup (1, \infty)$

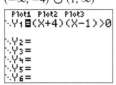 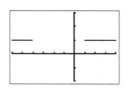

115. $[-1, 3)$

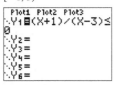 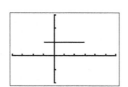

CHAPTER 11 SUMMARY

Section 11.1

1. parabola **2.** vertex **3.** symmetric; axis of symmetry
4. up; down; up; down; $(0, k)$ **5.** left; right; right; left; $(h, 0)$
6. stretched; shrunk; reflected; up; narrow; up; wider; down
7. $a(x - h)^2 + k$; (h, k); $f(x) = 0$; $f(0)$ **8.** $x = r$, zero

Section 11.2

9. $\sqrt{k}; -\sqrt{k}$; **a.** two; **b.** one; **c.** two **10.** isolated; constant
11. perfect square trinomial
12. middle; coefficient; squared; $\left(\dfrac{b}{2}\right)^2$ **13.** completing the square
14. one; divide; constant; completes; square; add; Factor; square root property

Section 11.3

15. quadratic formula; $\dfrac{-b \pm \sqrt{b^2 - 4ac}}{2a}$
16. discriminant; $b^2 - 4ac$; **a.** two rational; **b.** two irrational; **c.** one rational; **d.** two complex, nonreal

Section 11.4

17. rational; LCD; undefined
18. radical; radical; square; extraneous
19. substitution; u; quadratic; substitute

Section 11.5

20. completing; square **21.** $-\dfrac{b}{2a}$; $f\left(-\dfrac{b}{2a}\right)$
22. x-intercepts; y-intercepts; vertex **23.** lowest; highest; k; $x = h$

Section 11.6

24. boundary number method; associated; inequality; equal
25. boundary numbers; tested; one **26.** not included; parentheses
27. included; brackets **28.** associated; undefined
29. x-intercepts; above; below

CHAPTER 11 REVIEW EXERCISES

Section 11.1

1. no x-intercepts; y-intercept: $(0, 9)$; vertex: $(0, 9)$; axis of symmetry: $x = 0$; domain: $(-\infty, \infty)$; range: $(9, \infty)$; shifts up 9 units

3. x-intercepts: $(3, 0)$, $(7, 0)$; y-intercept: $(0, -10.5)$; vertex: $(5, 2)$; axis of symmetry: $x = 5$; domain $(-\infty, \infty)$; range: $(-\infty, 2]$; shifts right 5 units, reflects about the x-axis, shrinks by a factor of $\dfrac{1}{2}$, and shifts up 2 units

5. $\{-10, -6\}$ **7.** $\{0, 4\}$

Section 11.2

9. $\{-3\sqrt{2}, 3\sqrt{2}\}$ **11.** $\left\{\dfrac{8}{3}, \dfrac{22}{3}\right\}$ **13.** $\left\{\dfrac{8}{5}, \dfrac{22}{5}\right\}$

15. $\left\{\dfrac{2 - 8i}{3}, \dfrac{2 + 8i}{3}\right\}$ **17.** $x^2 + 12x + 36 = (x + 6)^2$

19. $r^2 - \dfrac{3}{2}r + \dfrac{9}{16} = \left(r - \dfrac{3}{4}\right)^2$ **21.** $\{-3\sqrt{2} - 3, 3\sqrt{2} - 3\}$

23. $\left\{-\dfrac{1}{3}, \dfrac{7}{3}\right\}$ **25.** $\left\{\dfrac{1 - \sqrt{15}}{2}, \dfrac{1 + \sqrt{15}}{2}\right\}$ **27.** 260 ft

29. 1.5%

Section 11.3

31. $\left\{-\dfrac{1}{7}, 5\right\}$ **33.** $\left\{-\dfrac{3}{4}, \dfrac{4}{5}\right\}$ **35.** $\left\{\dfrac{-2 \pm \sqrt{17}}{3}\right\}$

37. $\{3 \pm i\sqrt{7}\}$ **39.** two complex, nonreal solutions
41. two irrational solutions **43.** two irrational solutions
45. 80 ft by 160 ft **47.** 11:55 A.M.

Section 11.4

49. $\left\{\dfrac{-4 \pm i\sqrt{6}}{4}\right\}$ **51.** $\left\{\dfrac{1 \pm 2\sqrt{6}}{4}\right\}$ **53.** $\{-2, 4\}$

55. $\varnothing$ **57.** $\left\{\dfrac{9}{16}\right\}$ **59.** $\{\pm 2, \pm 4\}$ **61.** $\{\pm 2i, \pm 2.5\}$

63. $\left\{-\dfrac{5}{2}, -\dfrac{9}{8}\right\}$ **65.** $\left\{-\dfrac{9}{4}, -\dfrac{1}{2}\right\}$ **67.** $\{-125, 64\}$

69. $\left\{-32, \dfrac{1}{32}\right\}$ **71.** Jay takes 9.77 hr and Mel takes 6.77 hr.

Section 11.5

73. x-intercepts: $(-6, 0)$ and $(4, 0)$; y-intercept: $(0, -24)$; vertex: $(-1, -25)$

75. 10,000 cameras sold with a maximum profit of $350,000

Section 11.6

77. $[-9, 4]$ **79.** $(-\infty, -4) \cup (8, \infty)$ **81.** $(-\infty, -6) \cup (-6, \infty)$

83. $\varnothing$ **85.** $(-\infty, -5] \cup \left[-\frac{3}{2}, 4\right]$ **87.** $\left(-\frac{12}{5}, 0\right) \cup \left(\frac{3}{7}, \infty\right)$

89. $(-\infty, -1) \cup \left(-\frac{2}{5}, \frac{2}{5}\right) \cup (1, \infty)$ **91.** $\left(-\infty, \frac{3}{2}\right] \cup \left(\frac{5}{2}, \infty\right)$

93. $\left(-4, \frac{7}{2}\right]$ **95.** $(-\infty, -5] \cup \left[\frac{3}{2}, \infty\right)$

97. $(-\infty, -6] \cup [-1, 3]$ **99.** between 4 and 6 sec

CHAPTER 11   TEST

1. The four methods for solving a quadratic equation are applying the zero products property (factoring), applying the square root property, completing the square, and using the quadratic formula. You can complete the square and use the quadratic formula to solve any quadratic equation.

3. d   **5.** c   **7.** b   **9.** a   **11.** $\{4\}$

13. $\left\{\frac{-1 - i\sqrt{19}}{2}, \frac{-1 + i\sqrt{19}}{2}\right\}$ or $\left\{-\frac{1}{2} - \frac{\sqrt{19}}{2}i, -\frac{1}{2} + \frac{\sqrt{19}}{2}i\right\}$

15. $\{-9, -7\}$ **17.** $\{\pm\sqrt{2}, \pm\sqrt{6}\}$

19. x-intercepts: $(-2, 0)$ and $(6, 0)$; y-intercept: $(0, -12)$; vertex: $(2, -16)$; domain: $(-\infty, \infty)$; range: $[-16, \infty)$

21. no x-intercepts; y-intercept: $(0, 5)$; vertex: $\left(\frac{3}{2}, \frac{1}{2}\right)$; domain: $(-\infty, \infty)$; range: $\left[\frac{1}{2}, \infty\right)$

23. 3.3 sec   **25.** $(-\infty, -2] \cup [5, \infty)$ **27.** $(-\infty, 0] \cup (2, \infty)$

CHAPTERS 1–11   CUMULATIVE REVIEW EXERCISES

1. 49   **3.** $4 - 2x > x + 6$ **5.** $\{13\}$ **7.** $\{23, 25\}$

9. $r = \dfrac{3q - 4p}{4s}$ **11.** $[-2, 2]$ **13.** 3

15. can't be written in function notation

17. $m = 4, b = -1; (0, -1)$ **19.** $y = -2x - 1$ **21.** $\{(3, -7)\}$

23. Speed of the plane is 351 mph and speed of the wind is 39 mph.

25. $-65x^7y^5$ **27.** $\frac{4}{3}p^9q^4$ **29.** $\frac{a^2}{b^7}$ **31.** $\frac{y^{19}}{4x^{11}}$

33. $6x^4 - 35x^2 + 50$ **35.** $a^3 + 2a^2 - 25a - 50$

37. $4x(x - 3)(x^2 + 2)$ **39.** $2(x + 2)(2x - 3)$

41. $(2x + 5)(4x^2 - 10x + 25)$ **43.** $\{-2, 2\}$ **45.** $\frac{x - 2}{2x}$

47. 10   **49.** $\dfrac{7x + 14}{(x - 2)^2(x + 5)}$ **51.** 4 and -2 **53.** \$52.50

55. $5x^4$ **57.** $\frac{4a^3}{b^2}$ **59.** $\frac{a^{10}}{b^4}$ **61.** $x^{2/3} - y^{1/2}$

63. $4xy^3\sqrt{5x}$ **65.** $\frac{a\sqrt[3]{a^2}}{b}$ **67.** $14 + 6\sqrt{5}$ **69.** $2c^2d^5\sqrt{10d}$

71. $\{-1, 0\}$ **73.** $-\frac{1}{2} + \frac{3}{2}i$

75. no x-intercepts; y-intercept: $(0, 4)$; vertex: $(1, 3)$; axis of symmetry: $x = 1$; domain: $(-\infty, \infty)$; range: $[3, \infty)$; shifts right 1 unit and up 3 units

77. x-intercepts: $(1, 0), (3, 0)$; y-intercept: $(0, -3)$; vertex: $(2, 1)$; axis of symmetry: $x = 2$; domain: $(-\infty, \infty)$; range: $(-\infty, 1]$; shifts right 2 units, reflects over the x-axis, and shifts up 1 unit

79. $\{-6, 2\}$ **81.** $\{-4 \pm \sqrt{7}\}$ **83.** $\left\{\dfrac{2 \pm 5i}{2}\right\}$

85. $-1, 3; (-1, 0), (3, 0)$ **87.** $\{1 \pm 9i\}$ **89.** $\{-3 \pm \sqrt{17}\}$

91. two complex solutions   **93.** one rational solution

95. 120 ft by 200 ft   **97.** $\left\{-1, \dfrac{5}{6}\right\}$ **99.** $\{6\}$ **101.** $\{-1, 125\}$

103. $f(x) = 2(x - 3)^2 + 1$; no x-intercepts; y-intercept: $(0, 19)$; vertex: $(3, 1)$

105. $\left(-\infty, -\dfrac{5}{3}\right] \cup [4, \infty)$

107. $\left(-\dfrac{5}{2}, 0\right) \cup (3, \infty)$

CHAPTER 12

Section 12.1

1. Answers vary. We add and subtract functions by combining like terms. For example, let $f(x) = 13x + 15$ and $g(x) = 7x - 8$.
$f(x) + g(x) = 13x + 15 + 7x - 8 = 20x + 7$
$f(x) - g(x) = 13x + 15 - 7x + 8 = 6x + 23$

3. Answers vary. When finding the quotient of two functions, the restriction on the domain is the values of x that make the denominator zero. For example, let $f(x) = 2x + 5$ and $g(x) = 3x - 15$.
$$\frac{f(x)}{g(x)} = \frac{2x + 5}{3x - 15}$$
So, $g(x) \neq 0$.
$3x - 15 \neq 0$
$3x \neq 15$
$x \neq 5$
The restriction is $x \neq 5$.

5. Answers vary. Find the difference of the two functions f and g first and evaluate the resulting function at $x = 2$, or find $f(2)$ and $g(2)$ and subtract these values.

7. Answers vary. To find the composite function, $g(f(x))$, substitute the function f for x in the function of g. That is, substitute $2x^2 - x$ in place of x in the function g.
$g(f(x)) = g(2x^2 - x) = 5(2x^2 - x) - 1$

9. $(f + g)(x) = 5x + 19, (f - g)(x) = 3x + 3$,
$(f \cdot g)(x) = 4x^2 + 43x + 88$
domain for $f + g, f - g, f \cdot g$: $(-\infty, \infty)$
$\left(\dfrac{f}{g}\right)(x) = \dfrac{4x + 11}{x + 8}, x \neq -8$
domain for $\dfrac{f}{g}$: $(-\infty, -8) \cup (-8, \infty)$

11. $(f + g)(x) = \sqrt{x} + 11x - 2$, $(f - g)(x) = \sqrt{x} - 11x + 2$,
$(f \cdot g)(x) = 11x\sqrt{x} - 2\sqrt{x}$
domain for $f + g$, $f - g$, $f \cdot g$: $[0, \infty)$
$\left(\dfrac{f}{g}\right)(x) = \dfrac{\sqrt{x}}{11x - 2}$, $x \neq \dfrac{2}{11}$
domain for $\dfrac{f}{g}$: $\left[0, \dfrac{2}{11}\right) \cup \left(\dfrac{2}{11}, \infty\right)$

13. $(f + g)(x) = \sqrt{5x} + x^2 + 9$, $(f - g)(x) = \sqrt{5x} - x^2 - 9$,
$(f \cdot g)(x) = x^2\sqrt{5x} + 9\sqrt{5x}$
domain for $f + g$, $f - g$, $f \cdot g$: $[0, \infty)$
$\left(\dfrac{f}{g}\right)(x) = \dfrac{\sqrt{5x}}{x^2 + 9}$
domain for $\dfrac{f}{g}$: $[0, \infty)$

15. $(f + g)(x) = -x^2 + 9x - 3$, $(f - g)(x) = x^2 - x - 3$,
$(f \cdot g)(x) = -4x^3 + 23x^2 - 15x$
domain for $f + g$, $f - g$, $f \cdot g$: $(-\infty, \infty)$
$\left(\dfrac{f}{g}\right)(x) = \dfrac{4x - 3}{-x^2 + 5x}$, $x \neq 0, 5$
domain for $\dfrac{f}{g}$: $(-\infty, 0) \cup (0, 5) \cup (5, \infty)$

17. -30 **19.** 94 **21.** 18 **23.** $-\dfrac{11}{48}$ **25.** $-\dfrac{1}{4}$ **27.** 8

29. -4 **31.** 16 **33.** -12 **35.** undefined **37.** -6
39. 8 **41.** 33 **43.** 3

45. $(f \circ g)(x) = 12x - 13$ **47.** $(f \circ g)(x) = 5x - 11$
$(g \circ f)(x) = 12x - 35$ $(g \circ f)(x) = 5x + 5$
domain: $(-\infty, \infty)$ domain: $(-\infty, \infty)$

49. $(f \circ g)(2) = 17$ **51.** $(f \circ g)(-2) = 33$
$(g \circ f)(-3) = -28$ $(g \circ f)(3) = \sqrt{34}$

53. $(f \circ g)(-6) = 5$ **55.** -2
$(g \circ f)(-1) = 22$

57. -4 **59.** 0

61. $(f \circ g)(-2) = -10$ **63.** $(f \circ g)(-3) = -7$
$(g \circ f)(-2) = 17$ $(g \circ f)(-3) = -31$

65. $t(x) = 2.7x$ **67.** $P(x) = 37.8x - 421$
69. a. $h(x) = 27x$, $f(x) = 20x$, $p(x) = 4x$
b. $c(x) = 51x$

71. $(f + g)(x) = 3x^2 + 1$ **73.** $(g \cdot h)(x) = -3x^3 + 14x^2 - 8x$
$(f \cdot g)(x) = 6x^3 - x^2 - 2x$ $\left(\dfrac{h}{g}\right)(x) = \dfrac{4 - x}{3x^2 - 2x}$, $x \neq 0, \dfrac{2}{3}$

75. $(f \circ g)(x) = 6x^2 - 4x + 1$ **77.** $(f + g)(x) = 3x^2 - 7x + 9$
$(g \circ f)(x) = 12x^2 + 8x + 1$ $(f - g)(x) = x^2 + 7x - 3$

79. $(f \cdot g)(x) = 2x^4 - 14x^3 + 15x^2 - 21x + 18$
$\left(\dfrac{f}{g}\right)(x) = \dfrac{2x^2 + 3}{x^2 - 7x + 6}$, $x \neq 1, 6$

81. $(f \circ h)(x) = 18x^2 - 60x + 53$, $(h \circ f)(x) = -6x^2 - 4$

83. 13 **85.** -24 **87.** $-\dfrac{2}{3}$ **89.** 83 **91.** -5

93. -1 **95.** 1 **97.** -12 **99.** 0 **101.** -3 **103.** 1

105. -1 **107.** 15 **109.** 16 **111.** $\dfrac{11}{4}$ **113.** 8 **115.** 4

117. $t(x) = (g \circ f)(x) = 1.54x$ **119.** $P(x) = 26.4x - 214$
121. a. $h(x) = 29x$, $f(x) = 15x$, $p(x) = 12x$
b. $c(x) = 56x$

123. $(f - g)(x) = (2x^2 + 4x - 10) - (3x^2 - 2x + 9)$
$= 2x^2 + 4x - 10 - 3x^2 + 2x - 9$
$= -x^2 + 6x - 19$

125. $(f \circ g)(x) = 4(5 - 2x) - 9$
$= 20 - 8x - 9$
$= -8x + 11$

127.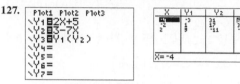

129. $c(x) = (g \circ f)(x)$ **131.** $c(x) = (h \circ g)(x)$

Section 12.2

1. Answers vary. A function is one-to-one if each y-value corresponds to only one x-value.

3. Answers vary. The inverse of a one-to-one function f is a function that when applied to the output of the function f takes us back to the original input value. It is a function that "un-does" the effects of the function f.

5. Answers vary. For each of the ordered pairs of the function f, interchange the x and y. Plot the new points and connect them with a smooth graph. The resulting graph is the graph of the inverse.

7. yes, a one-to-one function **9.** not a one-to-one function
11. not a one-to-one function **13.** not a one-to-one function
15. not a one-to-one function **17.** not a one-to-one function
19. yes, a one-to-one function **21.** not a one-to-one function
23. yes, a one-to-one function
25. one-to-one; $f^{-1}(x) = \{(3, -2), (4, 0), (6, 1), (-1, 2), (1, 3)\}$
27. not a one-to-one function
29. one-to-one;

Mayor in 2011 (x)	City (y)
Michael Bloomberg	New York City
Antonio Villaraigosa	Los Angeles
Rahm Emanuel	Chicago
Annise Parker	Houston
Phil Gordon	Phoenix

31. one-to-one; **33.** $f^{-1}(x) = x + 8$

35. $f^{-1}(x) = \dfrac{1}{11}x + \dfrac{2}{11}$ **37.** $f^{-1}(x) = \dfrac{2}{3}x + \dfrac{5}{3}$

39. $f^{-1}(x) = \dfrac{\sqrt[3]{x} - 9}{2}$ **41.** $f^{-1}(x) = \dfrac{\sqrt[3]{x - 1}}{2}$

43. $f^{-1}(x) = \dfrac{15 - 3x}{x}$ **45.** -10 **47.** $\dfrac{3}{2}$ **49.** 3 **51.** $-\dfrac{1}{6}$

53. **55.**

57. yes **59.** yes **61.** no **63.** yes **65.** no
67. not a one-to-one function **69.** yes, a one-to-one function
71. yes, a one-to-one function **73.** not a one-to-one function
75. yes, a one-to-one function
77. one-to-one; $f^{-1}(x) = \{(-8, -2), (2, 0), (7, 1), (12, 2), (22, 4)\}$
79. not a one-to-one function

81. one-to-one;

Senator in 2011 (y)	State (x)
Joseph Lieberman	Connecticut
Lindsey Graham	South Carolina
Daniel Inouye	Hawaii
Johnny Isakson	Georgia
Mary Landrieu	Louisiana

83. $f^{-1}(x) = \dfrac{1}{7}x - \dfrac{3}{7}$ **85.** $f^{-1}(x) = \dfrac{3x + 5}{8x}$

87. $f^{-1}(x) = \dfrac{\sqrt[3]{2x - 4}}{2}$ **89.** $f^{-1}(x) = \dfrac{5}{6}x + \dfrac{1}{3}$

91. $f^{-1}(x) = \dfrac{\sqrt[3]{x - 4}}{5}$ **93.** -7 **95.** $-\dfrac{4}{5}$

97. 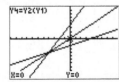   **99.**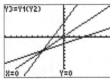

101.
$$f(x) = \frac{3x - 7}{8}$$
$$x = \frac{3y - 7}{8}$$
$$8x = 3y - 7$$
$$8x + 7 = 3y$$
$$y = \frac{8x + 7}{3}$$
$$f^{-1}(x) = \frac{8x + 7}{3}$$

103.

```
Plot1 Plot2 Plot3
\Y1■2X+4
\Y2■1/2X-2
\Y3■Y1(Y2)
\Y4■Y2(Y1)
\Y5=
\Y6=
\Y7=
```

X	Y1	Y2
-5	-6	-4.5
-3	-2	-3.5
-1	2	-2.5
1	6	-1.5
3	10	-.5
5	14	.5
7	18	1.5

Y2■1/2X-2

X	Y3	Y4
-5	-5	-5
-3	-3	-3
-1	-1	-1
1	1	1
3	3	3
5	5	5
7	7	7

Y4■Y2(Y1)

Y3=Y1(Y2)

X=0   Y=0

Y4=Y2(Y1)

X=0   Y=0

105. $f^{-1}(x) = -\dfrac{3}{8}x + \dfrac{3}{4}$ **107.** $f^{-1}(x) = -x + 3$

Section 12.3

1. Answers vary. Since the base $6 > 1$, the function is increasing. At the left end of the x-axis, the graph gets very close to the x-axis but never touches it. At the right end of the x-axis, the graph is increasing to ∞.

3. Answers vary. We must express 625 as a power of 5; that is, $625 = 5^4$. Set the exponents equal and solve the resulting equation for x.

5. Answers vary. We must express $\dfrac{1}{64}$ as a power of 4; that is, $\dfrac{1}{64} = 4^{-3}$. Set the exponents equal and solve the resulting equation for x.

7. Domain: $(-\infty, \infty)$, Range: $(0, \infty)$

9. Domain: $(-\infty, \infty)$, Range: $(0, \infty)$

11. Domain: $(-\infty, \infty)$, Range: $(0, \infty)$

13. $\{3\}$

15. $\{-3\}$ **17.** $\{-4\}$ **19.** $\left\{\dfrac{2}{3}\right\}$ **21.** $\left\{-\dfrac{2}{3}\right\}$

23. $\left\{\dfrac{8}{3}\right\}$ **25.** $\left\{-\dfrac{1}{4}\right\}$

27. $6760.95   **29.** 63.95 million, 67.22 million, 76.14 million
31. 20 g, 6.59 g, 2.17 g
33. Domain: $(-\infty, \infty)$, Range: $(0, \infty)$

35. Domain: $(-\infty, \infty)$, Range: $(0, \infty)$

37. Domain: $(-\infty, \infty)$, Range: $(0, \infty)$

39. $\{-4\}$

41. $\left\{\dfrac{1}{2}\right\}$ **43.** $\left\{-\dfrac{1}{10}\right\}$ **45.** $\left\{-\dfrac{1}{3}\right\}$ **47.** $\{0\}$

49. $\left\{\dfrac{1}{3}\right\}$ **51.** $\{-5\}$

53. $\{-6\}$ **55.** $\left\{\dfrac{5}{6}\right\}$ **57.** $10,817.35

59. 1.34 billion, 1.37 billion, 1.42 billion

61. 30 mg, 15 mg, 2.23 mg
63.
$$16^{3x-2} = 8$$
$$2^{4(3x-2)} = 2^3$$
$$4(3x - 2) = 3$$
$$12x - 8 = 3$$
$$12x = 11$$
$$x = \frac{11}{12}$$

65.

```
Y1=(1/5)^X


X=0        Y=1
```

67. 0.32

```
114.5(5.14)^(-0.
72*5)
         .3157462914
```

Section 12.4

1. Answers vary. A logarithm is an exponent.
3. Answers vary. Since z is the log of y with base w, we say z is the exponent of w that produces y. We write $w^z = y$.
5. Answers vary. Write it in exponential form: $a^1 = a$.
7. Answers vary. Set the expression equal to x, convert to exponential form, and solve the resulting equation for x.

9. $2^5 = 32$ 11. $8^1 = 8$ 13. $7^0 = 1$ 15. $2^{-4} = \dfrac{1}{16}$

17. $\log_6 216 = 3$ 19. $\log_{11} 11 = 1$ 21. $\log_{15} 1 = 0$

23. $\log_7 \dfrac{1}{49} = -2$ 25. $\log_4 8 = \dfrac{3}{2}$ 27. $\log_{144} 12 = \dfrac{1}{2}$

29. $\log_{81} 3 = \dfrac{1}{4}$ 31. 0   33. 1   35. 1   37. -5

39. -2 41. undefined   43. $\{4\}$ 45. $\{3\}$ 47. $\{512\}$
49. $\{7\}$ 51. $\{0\}$ 53. $\{3\}$ 55. $\{-2\}$

57. Domain: $(0, \infty)$,   59. Domain: $(0, \infty)$,
Range: $(-\infty, \infty)$ Range: $(-\infty, \infty)$

61. Domain: $(0, \infty)$,   63. 3
Range: $(-\infty, \infty)$

65. $\{4\}$ 67. $\{625\}$ 69. $\left\{\dfrac{1}{5}\right\}$ 71. $\{27\}$ 73. $\{0\}$

75. $\{-1\}$ 77. $\{-3\}$ 79. $\left\{\dfrac{1}{2}\right\}$

81. Domain: $(0, \infty)$,   83. Domain: $(0, \infty)$,
Range: $(-\infty, \infty)$ Range: $(-\infty, \infty)$

85. $\log_{1/4} 16 = x$
$$\left(\dfrac{1}{4}\right)^x = 16$$
$$4^{-x} = 4^2$$
$$-x = 2$$
$$x = -2$$

Piece It Together Sections 12.1–12.4

1. $(f + g)(x) = 2x + 18$, domain is $(-\infty, \infty)$; $(f - g)(x) = 6x + 12$, domain is $(-\infty, \infty)$; $(f \cdot g)(x) = -8x^2 - 18x + 45$, domain is $(-\infty, \infty)$; $\left(\dfrac{f}{g}\right)(x) = \dfrac{4x + 15}{3 - 2x}$, domain is $\left(-\infty, \dfrac{3}{2}\right) \cup \left(\dfrac{3}{2}, \infty\right)$; $(f \circ g)(x) = -8x + 27$, domain is $(-\infty, \infty)$; $(g \circ f)(x) = -8x - 27$, domain is $(-\infty, \infty)$.

2. 43   3. 91   4. -138 5. $\dfrac{21}{2}$

6. not a one-to-one function   7. one-to-one; $f^{-1}(x) = \dfrac{4x}{x - 1}$

8. Domain: $(-\infty, \infty)$,
Range: $(0, \infty)$

9. Domain: $(0, \infty)$,
Range: $(-\infty, \infty)$

10. $\{-5\}$ 11. $\left\{\dfrac{5}{3}\right\}$ 12. 115 million, 126 million, 141 million

13. -3 14. -3 15. 1   16. undefined   17. $\{1296\}$
18. $\{2\}$

Section 12.5

1. Answers vary. We apply the product rule for logarithms to combine a sum of logarithms with the same base into one logarithm which is written as the logarithm of the product of the arguments of the logarithms.
3. Answers vary. We apply the power rule for logarithms to an expression with a coefficient of a logarithm that becomes the exponent of the argument of the logarithm.
5. Answers vary. Applying the product rule for logarithms, the sum of logarithms with the same base is equal to the logarithm of the product of the arguments. For example,
$$\log_2(2 + 4) = \log_2 6$$
$$\log_2 2 + \log_2 4 = 1 + 2 = 3$$
$$\log_2 6 \neq 3$$

7. $\log_2 15$ 9. $\log_3 30$ 11. $\log_3 8$ 13. 1   15. $\log_4 7x$
17. $\log_2(x^2 - 4x)$ 19. $\log_a(x^2 - x - 6)$ 21. $\log_2 5$ 23. 1

25. $\log_3 \dfrac{x}{8}$ 27. $\log_6 12$ 29. $\log_2\left(\dfrac{x - 2}{x + 3}\right)$ 31. $\log_9(x + 5)$

33. $2\log_3 5$ 35. $4\log_5 x$ 37. $4\log_3(x - 2)$ 39. $\dfrac{1}{5}\log_3 x$

41. $\dfrac{1}{2}$ 43. $2x - 3$ 45. $\log_3 200$ 47. $\log_6 675$

49. $\log_5 x^4 \sqrt[5]{x + 6}$ 51. $\log_{10} \dfrac{x}{y^3}$ 53. $\log_5 \dfrac{a^2}{b^4}$

55. $5\log_{10} a + \dfrac{1}{2}\log_{10}(a - 3)$ 57. $15\log_2 x - 3\log_2 y$

59. $\log_4 3 + \log_4 x - 2\log_4 y$ 61. $2\log_3 x - 3 - \log_3 y$ 63. 6

65. $4x - 5$ 67. 6   69. $x + 5$ 71. $2x$ 73. $\log_3 12$

75. $\log_5\left(\dfrac{xz}{y}\right)$ 77. $\log_{10} 3$ 79. $\log_3 32$ 81. $\log_9 45$

83. $\log_3 \dfrac{x^2 - x}{x^2 + 4}$ 85. $\log_{10} \dfrac{x^{3/2}}{(x + 1)^3}$ 87. $\log_6\left(\dfrac{a^3 b^2}{c^5}\right)$

89. $\log_{10}\left(\dfrac{a}{b^2 \sqrt[3]{c}}\right)$ 91. $\log_6(x^2 + 2x + 4)$

93. $\log_3 5 + 2\log_3 x - \log_3 y$ 95. $\dfrac{1}{2}\log_7 6 + \dfrac{1}{2}\log_7 x - \dfrac{1}{2}\log_7 y$

97. $5\log_2 x + 3\log_2 y + \dfrac{1}{2}\log_2(x + 3)$ 99. $2 + 3\log_{10} a - 5\log_{10} b$

101. $2\log_4 r + \log_4(r + 3) - 3\log_4 s$ 103. $6x + 5$ 105. x
107. $16x$ 109. $2x + 1$ 111. $9a^2$
113. $\log_3 4 + \log_3 x = \log_3(4x)$
115. $(\log_2 8)^2 = (\log_2 2^3)^2 = (3\log_2 2)^2 = 3^2 = 9$

Section 12.6

1. Answers vary. Base is 10. The equivalent exponential form is $x = 10^y$.
3. Answers vary. You rewrite any logarithm by applying the change-of-base formula and using either the common or natural logarithm. For example, $\log_6 20 = \dfrac{\log 20}{\log 6} \approx 1.67$ and $\log_6 20 = \dfrac{\ln 20}{\ln 6} \approx 1.67$

5. Answers vary. The equivalent exponential form is $10^0 = 1$ and $e^0 = 1$.

7. Answers vary. Use the common or natural logarithm to rewrite 64 in the equivalent exponential form.

a. $\log_3 64 = \dfrac{\log 64}{\log 3}$ **b.** $\log_5 64 = \dfrac{\log 64}{\log 5}$ **c.** $\log_{3.75} 64 = \dfrac{\log 64}{\log 3.75}$

9. 5   **11.** -3 **13.** -3 **15.** $\dfrac{1}{3}$ **17.** $\dfrac{2}{5}$ **19.** 4

21. 8   **23.** -4 **25.** $\dfrac{1}{3}$ **27.** $\dfrac{3}{5}$ **29.** 1.30   **31.** -1.07

33. 0.43   **35.** 3.40   **37.** 1.15   **39.** $\{1,000,000\}$
41. $\{e^3\}$ or $\{20.09\}$ **43.** $\{10^{0.6} + 4\}$ or $\{7.98\}$

45. $\{e^{-2} + 4\}$ or $\{4.14\}$ **47.** $\left\{\dfrac{10^{-4.4} - 12}{5}\right\}$ or $\{-2.40\}$

49. $\left\{\dfrac{e^{1.9} + 10}{2}\right\}$ or $\{8.34\}$ **51.** $\left\{\dfrac{e^{-0.9} - 5}{3}\right\}$ or $\{-1.53\}$

53. $\left\{\dfrac{10^{-1}}{2}\right\}$ or $\{0.05\}$ **55.** 1.48   **57.** 0.63   **59.** -0.58

61. 86.34% after 2 days, 70.68% after 7 days, 62.01% after 14 days

63. 9.04   **65.** $\dfrac{5}{2}$ **67.** $\dfrac{5}{3}$ **69.** $\dfrac{1}{5}$ **71.** $\dfrac{4}{7}$ **73.** 1.73

75. 4.20   **77.** 0.77   **79.** -4.81

81. $\{e^{5.5} + 8\}$ or $\{252.69\}$ **83.** $\left\{\dfrac{10^{1.7} + 1}{6}\right\}$ or $\{8.52\}$

85. $\left\{\dfrac{10^{1.2} + 9}{10}\right\}$ or $\{2.48\}$ **87.** $\left\{\dfrac{e^{1.5} - 9}{4}\right\}$ or $\{-1.13\}$

89. $\{2\}$ **91.** $\left\{\dfrac{10^0 - 3}{4}\right\}$ or $\{-0.5\}$ **93.** 7.76   **95.** -0.79

97. 65.60% after 6 days, 56.94% after 12 days, 51.87% after 18 days
99. 180 dB
101. $\log(3x - 5) = 2$
    $3x - 5 = 10^2 = 100$
        $3x = 105$
         $x = 35$

103.
```
(10^1.7+15)/4
        16.27968084
log(4*Ans-15)
              1.7
```

105.
```
log(32)/log(2)
              5
ln(32)/ln(2)
              5
```

Section 12.7

1. Answers vary. When solving for x, you will apply the change-of-base formula to evaluate $x = \log_7 15$ using either the common or natural logarithm.

3. Answers vary. Since the given base is 10, you should use the common logarithm to solve for x.

5. Answers vary. You apply the product rule of logarithms to combine the sum of logarithms on the left side into one logarithm. Then write the resulting equation in exponential form and solve for x.

7. $\{\log 29\}$ or $\{1.46\}$ **9.** $\{\ln 6\}$ or $\{1.79\}$

11. $\left\{\dfrac{\log 90}{7}\right\}$ or $\{0.28\}$ **13.** $\left\{\dfrac{\ln 13}{4}\right\}$ or $\{0.64\}$

15. $\left\{\dfrac{\ln 38}{\ln 7} + 2\right\}$ or $\{3.87\}$ **17.** $\left\{\dfrac{\ln 19}{4\ln 5.75}\right\}$ or $\{0.42\}$

19. $\left\{\dfrac{\ln 94 - \ln 9}{5\ln 0.53}\right\}$ or $\{-0.74\}$

21. $\{23\}$ **23.** $\left\{\dfrac{13}{2}\right\}$ **25.** $\{2\}$

27. $\{4\}$ **29.** $\left\{\dfrac{4}{3}\right\}$ **31.** $\{10\}$ **33.** $\varnothing$ **35.** $\{6\}$

37. $\{3, 4\}$
39. $1107.33 after 4 yr; It will take about 15 yr for his account to grow to $1500.

41. The population of Mexico in 2021 will be 126.97 million. It will take about 4.87 yr to reach 120 million.

43. 9.33 mg   **45.** $\{4\}$ **47.** $\left\{\dfrac{\ln 37.5}{4.6}\right\}$ or $\{0.79\}$ **49.** $\{12\}$

51. $\left\{\dfrac{\ln 2 + 8\ln 12}{6\ln 12}\right\}$ or $\{1.38\}$ **53.** $\left\{\dfrac{9}{4}\right\}$ **55.** $\{7\}$

57. $\left\{\dfrac{-\ln 7}{4\ln 4.08}\right\}$ or $\{-0.35\}$ **59.** $\{40\}$ **61.** $\{2\}$

63. $\left\{\dfrac{\ln 21 - \ln 5}{7\ln 5}\right\}$ or $\{0.13\}$ **65.** $\varnothing$ **67.** $\varnothing$

69. $\left\{\dfrac{\ln 2}{\ln 6.25}\right\}$ or $\{0.38\}$ **71.** $\{5, 20\}$ **73.** $\left\{\dfrac{\log 17.5}{4.7}\right\}$ or $\{0.26\}$

75. $\{9\}$
77. $2247.77 after 4 yr; It will take about 5 yr for her account to grow to $2400.
79. It will take about 16 yr for his account to grow to $23,000.
81. The population of Japan in 2029 will be 120.30 million. It will take about 34.22 yr to reach 115 million.
83. 300.04 mg

85. $8(2)^{3x} = 48$
        $2^{3x} = 6$
     $3x \ln 2 = \ln 6$
            $x = \dfrac{\ln 6}{3\ln 2} \approx 0.86$

87. $\log_6 x + \log_6(x - 35) = 2$
        $\log_6 x(x - 35) = 2$
            $x^2 - 35x = 6^2$
       $x^2 - 35x - 36 = 0$
       $(x - 36)(x + 1) = 0$
                    $x = 36, -1$

Since -1 doesn't check, the solution set is $\{36\}$.

89. $x = \dfrac{\ln 64}{8\ln 4.02} \approx 0.374$ **91.** $x = 4$

```
ln(64)/(8ln(4.02
))
          .37365568
4.02^(8Ans)
               64
```

```
ln(4)/ln(6)+ln(4
+5)/ln(6)
               2
```

CHAPTER 12 SUMMARY

Section 12.1

1. operations, add, subtract, multiply, divide, intersection, zero
2. individually, combined   **3.** output, input, $f(g(x))$, $g(f(x))$

Section 12.2

4. inverse   **5.** one-to-one; different   **6.** horizontal, horizontal
7. interchange, $f^{-1}(x)$, (y, x), $f^{-1}(x)$.
8. y, x and y, y, $f^{-1}(x)$ **9.** $f^{-1}(y) = x$ **10.** symmetric, $y = x$
11. $(f \circ f^{-1})(x) = x$, $(f^{-1} \circ f)(x) = x$

Section 12.3

12. $f(x) = b^x$, $b > 0$, $b \neq 1$, $(-\infty, \infty)$, $(0, \infty)$, $(0, 1)$, $(1, b)$, increasing, decreasing, x-axis, touches
13. $b^x = b^y$, $x = y$ **14.** $A = P\left(1 + \dfrac{r}{n}\right)^{nt}$

Section 12.4

15. b, x, 0, 1   **16.** $b^y = x$
17. $f(x) = \log_b x$, $(0, \infty)$, $(-\infty, \infty)$, $(1, 0)$, no
18. exponential, y, x, y-axis, touches

Section 12.5

19. sum, factors, $\log_b(xy) = \log_b(x) + \log_b(y)$
20. difference, numerator, denominator, $\log_b \dfrac{x}{y} = \log_b x - \log_b y$
21. exponent, base, $\log_b x^n = n\log_b x$ **22.** $\log_b b^x$, $b^{\log_b x}$

Section 12.6

23. $\log x$, 10   **24.** $\ln x$, e **25.** $\dfrac{\log a}{\log b}$, $\dfrac{\ln a}{\ln b}$, 10, e

Section 12.7

26. exponent, log   **27.** single, exponential

CHAPTER 12 REVIEW EXERCISES

Section 12.1

1. $(f+g)(x) = x^2 + x - 9$, $(f-g)(x) = -x^2 + 3x - 5$; domain: $(-\infty, \infty)$
3. $(f \circ h)(x) = -4x + 1$, $(h \circ f)(x) = -4x + 18$; domain: $(-\infty, \infty)$
5. 1   **7.** -21 **9.** 15   **11.** -1 **13.** -12 **15.** 0
17. 6   **19.** 1   **21.** -10 **23.** 9   **25.** $g(x) = 2.25x$

Section 12.2

27. yes, a one-to-one function   **29.** yes, a one-to-one function
31. not a one-to-one function   **33.** yes, a one-to-one function
35. $f^{-1}(x) = \{(-5, -2), (1, 0), (5, 3), (3, 2), (7, 4)\}$

37.

39. $f^{-1}(x) = \dfrac{x + 4}{5}$

41. $f^{-1}(x) = \left(\dfrac{x - 9}{2}\right)^{1/3}$ **43.** -5 **45.** $-\dfrac{1}{6}$

47.

49. yes   **51.** no

Section 12.3

53.

Domain: $(-\infty, \infty)$,
Range: $(0, \infty)$

55. $\{-2\}$ **57.** $\{-3\}$ **59.** $\left\{\dfrac{7}{4}\right\}$ **61.** \$3657.00

Section 12.4

63. $128 = 2^7$ **65.** $\log_9 27 = \dfrac{3}{2}$ **67.** 0   **69.** -3

71. -5 **73.** $\{3\}$ **75.** $\{64\}$ **77.** $\{0\}$
79.
Domain: $(0, \infty)$,
Range: $(-\infty, \infty)$

Section 12.5

81. $\log_3 2$ **83.** $\log_5 \dfrac{x^2 z^5}{y^3}$ **85.** $\log_4 7 + 3\log_4 x - 2\log_4 y$

87. $2\log_{10} x + 5\log_{10} y + \dfrac{1}{2}\log_{10}(x - 1)$ **89.** $5x + 7$ **91.** x

93. $3x - 5$

Section 12.6

95. $\dfrac{5}{3}$ **97.** $\dfrac{2}{3}$ **99.** 0.82   **101.** 4.07

103. $\left\{\dfrac{e^4 + 5}{2}\right\}$ or $\{29.80\}$ **105.** $\{3.5\}$

107. $\left\{\dfrac{e}{5}\right\}$ or $\{0.54\}$ **109.** 1.83

111. 6.90   **113.** 71.88% after 5 days, 63.22% after 10 days

Section 12.7

115. $\left\{\dfrac{\log 32.6}{5.2}\right\}$ or $\{0.29\}$ **117.** $\left\{\dfrac{\ln 28 + 2\ln 7}{3\ln 7}\right\}$ or $\{1.24\}$

119. $\left\{\dfrac{5}{2}\right\}$ **121.** $\left\{\dfrac{83}{5}\right\}$ or $\{16.6\}$ **123.** $\{8\}$

125. $\left\{\dfrac{\ln 0.25}{8\ln 0.54}\right\}$ or $\{0.28\}$ **127.** $\{8\}$

129. a. \$1446.31   **b.** about 30 yr
131. a. 233.08 million   **b.** about 2.80 yr

CHAPTER 12 TEST

1. c   **3.** a   **5.** d   **7.** b
9. a. -4 **b.** $(f \cdot g)(x) = \dfrac{36x + 9}{x - 2}$; domain: $(-\infty, 2) \cup (2, \infty)$
c. $\dfrac{1}{3}$ **d.** $f^{-1}(x) = \dfrac{9 + 2x}{x}$ **11. a.** $\{-3\}$ **b.** $\left\{\dfrac{3}{2}\right\}$
c. $\{-2\}$ **d.** $\{-2\}$ **e.** $\{1\}$ **f.** $\{9\}$ **g.** $\{e^3 - 1\}$
h. $\{20\}$ **i.** $\{\ln 4 - 1\}$
13. a. \$2440.38   **b.** about 40 yr

CHAPTERS 1–12 CUMULATIVE REVIEW EXERCISES

1. 0   **3.** $165 - x$ **5.** $\varnothing$ **7.** $\{-4, 1\}$ **9.** $\dfrac{8}{7}$

11. $(-\infty, \infty)$ **13.** $f(x) = \dfrac{4}{7}x - 4$, $m = \dfrac{4}{7}$, $b = -4$ **15.** neither
17. $\{(4, 4)\}$ **19.** $\{(-2, 1)\}$ **21.** $\{(12, 15)\}$ **23.** $35°, 55°$
25. $\{(1, -2, 3)\}$

27.
29. $-30x^{15}$ **31.** $\dfrac{a^7}{b^6}$

33. $\dfrac{y^{13}}{125x^2}$ **35.** $-2a - 6$ **37.** $14x^2 + 27x - 20$
39. $27x^3 - 54x^2 + 36x - 8$ **41.** $5y^2 + 12y - 3$
43. $(w - 2)(w^2 + 3)$ **45.** $(5x + 4y)^2$
47. $(2x + 5y)(4x^2 - 10xy + 25y^2)$ **49.** $(4x - 1 + 3y)(4x - 1 - 3y)$
51. $\left\{-\dfrac{7}{2}, 1\right\}$ **53.** $\left\{-4, \dfrac{13}{2}\right\}$ **55.** $\dfrac{x + 4}{x^2 + 2x + 4}$
57. $\dfrac{4}{5}(x - 2)$ **59.** $-x^2 + \dfrac{3}{2}x - 2$ **61.** $7x^2 + 3x - 3 - \dfrac{11}{x + 1}$
63. $\dfrac{5x + 11}{(x + 4)(x - 4)(x + 1)}$ **65.** $\dfrac{x - 5}{(x + 1)^2(x - 1)}$ **67.** $\dfrac{a}{bc}$
69. $-\dfrac{1}{xy}$ **71.** $-\dfrac{39}{28}$ **73.** $\dfrac{25}{36}$ **75.** $\{4\}$ **77.** $b = \dfrac{2ac}{3a + c}$
79. $\dfrac{y^6}{x^9}$ **81.** $a + a^{1/2}b^{1/3} - 2b^{2/3}$ **83.** $2x^2y\sqrt{30xy}$

85. $\dfrac{x\sqrt[3]{x}}{2y}$ **87.** $7 - 4\sqrt{3}$ **89.** $\{12\}$

91. x-intercepts: $(-4, 0)$, $(-2, 0)$; y-intercept: $(0, 8)$; vertex: $(-3, -1)$; axis of symmetry: $x = -3$; shift graph of $y = x^2$ left 3 units and down 1 unit; domain: $(-\infty, \infty)$; range: $[-1, \infty)$

93. $\{2 \pm \sqrt{3}\}$ **95.** $1 \pm 7i$ **97.** $\left\{-\dfrac{1}{27}, \dfrac{27}{8}\right\}$

99. $\left(-\infty, -\dfrac{5}{2}\right) \cup (0, 3)$

101. $(f + g)(x) = 8x + 2$ **103.** 10   **105.** 185
$(f - g)(x) = 2x + 6$
$(f \cdot g)(x) = 15x^2 + 2x - 8$
domain: $(-\infty, \infty)$
$\left(\dfrac{f}{g}\right)(x) = \dfrac{5x + 4}{3x - 2}, x \neq \dfrac{2}{3}$
domain: $\left(-\infty, \dfrac{2}{3}\right) \cup \left(\dfrac{2}{3}, \infty\right)$

107. $(f \circ g)(x) = -3x + 7$ **109.** yes, a one-to-one function
$(g \circ f)(x) = -3x + 9$
domain: $(-\infty, \infty)$

111. $f^{-1}(x) = \dfrac{\sqrt[3]{x} + 2}{7}$ **113.** 6.9

115. Domain: $(-\infty, \infty)$,   **117.** 10   **119.** \$1944.55
Range: $(0, \infty)$

121. -4 **123.** 243
125. Domain: $(0, \infty)$,   **127.** $\log_3(x^2 + x - 20)$
Range: $(-\infty, \infty)$

129. $\log_4 \dfrac{x}{x + 6}$ **131.** $5\log(x + 2)$ **133.** $3\ln x + \dfrac{1}{2}\ln(x - 1)$

135. $\log_{10} \dfrac{x^5}{y^2}$ **137.** $3x + 2$ **139.** $2x + 7$ **141.** 9

143. -8 **145.** $\left\{\dfrac{10^{1.5} + 4}{5}\right\}$ or $\{7.125\}$

147. $\left\{\dfrac{\ln 72}{8\ln 0.75}\right\}$ or $\{-1.86\}$ **149.** 6

CHAPTER 13

Section 13.1

1. Answers vary. The vertex is $(1, 3)$. The axis of symmetry is $y = 3$. There is no y-intercept. The x-intercept is $\left(\dfrac{11}{2}, 0\right)$. Since $a = \dfrac{1}{2} > 0$, the parabola opens to the right.

3. Answers vary. We rewrite the equation as $x = -1(y - 0)^2 + 25$. The vertex is $(25, 0)$. The y-intercepts are $(0, 5)$ and $(0, -5)$. Plot the points and connect them to draw a parabola that opens to the left since $a = -1$.

5. Answers vary. You can easily identify the center and radius of the circle.

7. Answers vary. We need to complete the squares for the x-terms and the y-terms. For the x-terms, take one-half of the coefficient of x and square it to complete the square. Add the same number to the right side of the equation. Repeat the process for the y-terms. Write the equation in standard form.
$$x^2 + y^2 - 6x + 10y = 15$$
$$(x^2 - 6x + 9) + (y^2 + 10y + 25) = 15 + 9 + 25$$
$$(x - 3)^2 + (y + 5)^2 = 49$$

9. vertex: $(3, -2)$; axis of symmetry: $y = -2$

11. vertex: $(1, 2)$; axis of symmetry: $y = 2$

13. vertex: $(-4, 3)$; axis of symmetry: $y = 3$

15. vertex: $(4, -2)$; axis of symmetry: $y = -2$

17. vertex: $(-1, 2)$; axis of symmetry: $y = 2$

19. vertex: $(9, 0)$; axis of symmetry: $y = 0$

21. vertex: $(-9, 0)$; axis of symmetry: $y = 0$

23. vertex: $(-9, -1)$; axis of symmetry: $y = -1$

25. vertex: $\left(-\dfrac{1}{4}, -\dfrac{3}{2}\right)$; axis of symmetry: $y = -\dfrac{3}{2}$

27. center: $(0, 0)$; radius: 5; points: $(5, 0)$, $(0, 5)$, $(-5, 0)$, $(0, -5)$

29. center: $(2, 1)$; radius: 3; points: $(2, 4)$, $(2, -2)$, $(5, 1)$, $(-1, 1)$

31. center: $(0, -3)$; radius: 5; points: $(0, 2)$, $(0, -8)$, $(5, -3)$, $(-5, -3)$

33. $(x + 4)^2 + (y - 5)^2 = 1$ **35.** $(x - 4)^2 + (y + 2)^2 = 3$
37. $(x + 2)^2 + (y + 7)^2 = 6$
39. $(x - 6)^2 + (y - 4)^2 = 1$; center: $(6, 4)$, radius = 1
41. $(x + 1)^2 + (y + 1)^2 = 4$; center: $(-1, -1)$, radius = 2
43. $(x - 3)^2 + y^2 = 25$; center: $(3, 0)$, radius = 5
45. $\left(x - \dfrac{1}{2}\right)^2 + \left(y + \dfrac{1}{2}\right)^2 = 1$; center: $\left(\dfrac{1}{2}, -\dfrac{1}{2}\right)$, radius = 1

47. vertex: $(-5, -1)$; axis of symmetry: $y = -1$

49. center: $(-1, 1)$; radius: 1; points: $(-1, 0)$, $(-1, 2)$, $(0, 1)$, $(-2, 1)$

51. vertex: $(4, 3)$; axis of symmetry: $y = 3$

53. vertex: $(1, 1)$; axis of symmetry: $y = 1$

55. center: $(1, 0)$; radius: 4; points: $(1, 4)$, $(1, -4)$, $(5, 0)$, $(-3, 0)$

57. vertex: $(3, 1)$; axis of symmetry: $y = 1$

59. vertex: $(4, 0)$; axis of symmetry: $y = 0$

61. vertex: $(3, 0)$; axis of symmetry: $y = 0$

63. vertex: $(-4, -1)$; axis of symmetry: $y = -1$

65. vertex: $(25, 1)$; axis of symmetry: $y = 1$

67. vertex: $\left(-\dfrac{5}{4}, \dfrac{5}{2}\right)$; axis of symmetry: $y = \dfrac{5}{2}$

69. $(x + 1)^2 + (y + 2)^2 = 16$
71. $(x + 4)^2 + (y - 6)^2 = 6$
73. $(x + 7)^2 + y^2 = 5$
75. $x^2 + \left(y + \dfrac{3}{4}\right)^2 = \dfrac{1}{4}$

77. $(x - 1.2)^2 + (y + 2.5)^2 = 20.25$
79. $(x - 2)^2 + y^2 = 9$
81. $(x + 7)^2 + (y + 3)^2 = 36$; center: $(-7, -3)$, radius = 6
83. $(x + 3)^2 + (y - 6)^2 = 45$; center: $(-3, 6)$, radius = $3\sqrt{5}$

85. $(x + 4)^2 + (y - 3)^2 = 25$; center: $(-4, 3)$, radius = 5
87. $x^2 + (y - 1)^2 = 9$; center: $(0, 1)$, radius = 3
89. $\left(x - \dfrac{3}{2}\right)^2 + \left(y - \dfrac{5}{2}\right)^2 = 9$; center: $\left(\dfrac{3}{2}, \dfrac{5}{2}\right)$, radius = 3
91. $(x - 0.8)^2 + (y + 2.1)^2 = 5.76$; center: $(0.8, -2.1)$, radius = 2.4
93. $(-2, -1)$
95. vertex: $(-1.5, 1.2)$

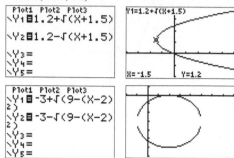

97.

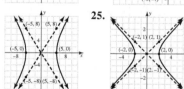

Section 13.2

1. Answers vary. Since the center of the ellipse is $(0, 0)$, the key points will be the two x-intercepts and two y-intercepts. The two x-intercepts are $(3, 0)$ and $(-3, 0)$. The two y-intercepts are $(0, 2)$ and $(0, -2)$.
3. Answers vary. To find the x-intercepts, set y equal to zero and find the corresponding x-values. The x-intercepts are $(5, 0)$ and $(-5, 0)$.
5. Answers vary. The hyperbola is horizontal with x-intercepts of $(5, 0)$ and $(-5, 0)$. Draw the rectangle and asymptotes as described in Exercise 4. Draw two branches of the hyperbola passing through the x-intercepts and approaching the asymptotes.
7. Answers vary. To graph the ellipse, you need to solve for y first and enter the resulting equations as Y_1 and Y_2. The intercepts of the ellipse are $(12, 0)$, $(-12, 0)$, $(0, 15)$, and $(0, -15)$. If the viewing window is set to $[-10, 10]$ by $[-10, 10]$, you will not see the graph.

9.   **11.**   **13.**

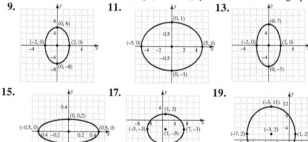

15.   **17.**   **19.**

21.   **23.**   **25.**

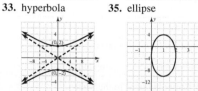

27. ellipse   **29.** parabola   **31.** circle

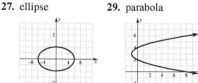

33. hyperbola   **35.** ellipse   **37.** hyperbola

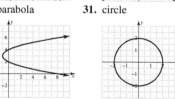

39. hyperbola   **41.** parabola   **43.**

 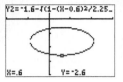

45.

Piece It Together Sections 13.1 and 13.2

1. vertex: $(-36, -3)$; axis of symmetry: $y = -3$

2. vertex: $(25, -2)$; axis of symmetry: $y = -2$

3. center: $(0, 0)$; radius: 10; points: $(10, 0)$, $(0, 10)$, $(-10, 0)$, $(0, -10)$

4. center: $(0, 0)$; radius: 6; points: $(6, 0)$, $(0, 6)$, $(-6, 0)$, $(0, -6)$

5. center: $(0, -1)$; radius: 2; points: $(2, -1)$, $(-2, -1)$, $(0, 1)$, $(0, -3)$

6. center: $(-2, 0)$; radius: 1; points: $(-2, 1)$, $(-2, -1)$, $(-1, 0)$, $(-3, 0)$

7.

8.

9.

10.

11. $(x + 5)^2 + (y - 7)^2 = 16$; center: $(-5, 7)$; radius $= 4$

12. $(x - 1)^2 + (y - 6)^2 = 1$; center: $(1, 6)$; radius $= 1$

Section 13.3

1. Answers vary. The number of possible solutions can be two if the line intersects the parabola at two points, one if the line touches the parabola at one point, and none if the line does not intersect the parabola.

3. Answers vary. Solve the second equation for y and substitute the expression into the first equation.

5. Answers vary. Rewrite the second equation in the same form as in the first equation. Multiply the second equation by -2 and add to the first equation.

7. Answers vary. Since the second equation is linear, use the substitution method. Solve the second equation for one of the two variables and substitute the expression into the first equation.

9. $\{(0, 25), (3, 4)\}$ **11.** $\{(0, 1), (3, 16)\}$ **13.** $\{(4, 2)\}$

15. $\{(-2, \sqrt{3})\}$ **17.** $\{(-1, -4), (2, -1)\}$

19. $\{(-6, -8), (8, 6)\}$ **21.** $\{(3, 1), (-3, 1), (3, -1), (-3, -1)\}$

23. $\{(5, 2), (-5, 2), (5, -2), (-5, -2)\}$

25. $\{(2, 4), (-2, 4), (2, -4), (-2, -4)\}$

27. $\{(2, 5), (-2, 5), (2, -5), (-2, -5)\}$

29. $\{(1, 4), (-1, 4), (1, -4), (-1, -4)\}$

31. $\{(4, 28), (-1, -2)\}$ **33.** $\{(0, 64), (15, 49)\}$ **35.** $\{(2, 4)\}$

37. $\{(-2, \sqrt{3})\}$ **39.** $\{(4, -3), (0, 5)\}$ **41.** $\{(-3, 0), (2, -5)\}$

43. $\{(3, 7), (-2, 2)\}$ **45.** $\{(6, 8), (-6, 8), (6, -8), (-6, -8)\}$

47. $\{(1, 2), (-1, 2), (1, -2), (-1, -2)\}$

49. $\left\{\left(\dfrac{5}{3}, 0\right), \left(-\dfrac{5}{2}, \dfrac{125}{4}\right)\right\}$

51.
$$y = 1 - 8x$$
$$1 - 8x = (x - 1)^2$$
$$1 - 8x = x^2 - 2x + 1$$
$$0 = x^2 + 6x$$
$$0 = x(x + 6)$$
$$x = 0 \quad \text{or} \quad x = -6$$
Substitute the values into the first equation to solve for y.
$$x = 0: y = (0 - 1)^2 = 1$$
$$x = -6: y = (-6 - 1)^2 = 49$$
$$\{(0, 1), (-6, 49)\}$$

53.
$$3x^2 + 2y^2 = 12$$
$$\underline{-3x^2 + y^2 = 3}$$
$$3y^2 = 15$$
$$y^2 = 5$$
$$y = \sqrt{5} \quad \text{or} \quad y = -\sqrt{5}$$
$y = \sqrt{5}$ or $y = -\sqrt{5}$:
$$3x^2 + 10 = 12$$
$$3x^2 = 2$$
$$x^2 = \frac{2}{3}$$
$$x = \sqrt{\frac{2}{3}} \quad \text{or} \quad x = -\sqrt{\frac{2}{3}}$$
$$\left\{\left(\sqrt{\frac{2}{3}}, \sqrt{5}\right), \left(-\sqrt{\frac{2}{3}}, \sqrt{5}\right), \left(\sqrt{\frac{2}{3}}, -\sqrt{5}\right), \left(-\sqrt{\frac{2}{3}}, -\sqrt{5}\right)\right\}$$

55. $\{(-4, 14), (3, 7)\}$

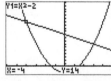  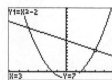

Section 13.4

1. Answers vary. The boundary curve is the graph of the equation formed by replacing the inequality symbol with an equals sign. It is dashed if the inequality symbol is $<$ or $>$. It is solid if the inequality symbol is $\leq$ or $\geq$.

3. Answers vary. The boundary curve for the inequality is $x^2 - 4y^2 = 16$, which is a horizontal hyperbola with x-intercepts, $(4, 0)$ and $(-4, 0)$. Since the inequality symbol is $\leq$, draw the hyperbola with a solid curve.

5. Answers vary. The boundary curve for the inequality, $y > x^2 - 4$, is $y = x^2 - 4$ which is a parabola with x-intercepts, $(2, 0)$ and $(-2, 0)$, and y-intercept $(0, -4)$. Since the inequality symbol is $>$, draw the parabola with a dashed curve. The boundary curve for the inequality, $\frac{x^2}{9} + \frac{y^2}{16} < 1$, is $\frac{x^2}{9} + \frac{y^2}{16} = 1$ which is a vertical ellipse with x-intercepts, $(3, 0)$ and $(-3, 0)$, and y-intercepts, $(0, 4)$ and $(0, -4)$. Since the inequality symbol is $<$, draw the ellipse with a dashed curve.

7.   9.   11.

13.   15.   17.

19. 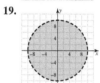   21.   23.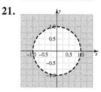

25. 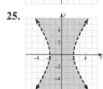   27.   29.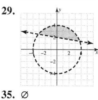

31. ∅   33.    35. ∅

37. 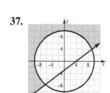   39. 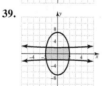   41. ∅

43. 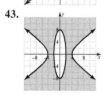   45.   47.

49.   51.   53.

55.   57. 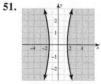   59.

61. 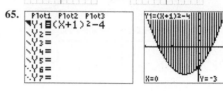   63.

65.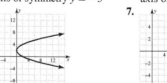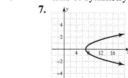

67. Answers vary. For example, $\frac{x^2}{4} + y^2 \le 1$ and $x^2 + \frac{y^2}{9} \le 1$.

69. Answers vary. For example, $\frac{y^2}{9} - x^2 < 1$ and $\frac{x^2}{9} + \frac{y^2}{4} > 1$.

CHAPTER 13 SUMMARY

Section 13.1

1. horizontal; (h, k); right; left   2. circle; (h, k); r

Section 13.2

3. ellipse; (h, k); ellipse; horizontal; ellipse; vertical
4. $(a, 0)$; $(-a, 0)$; $(0, b)$; $(0, -b)$
5. $\frac{x^2}{a^2} - \frac{y^2}{b^2} = 1$; $\frac{y^2}{b^2} - \frac{x^2}{a^2} = 1$; horizontally; vertically
6. rectangle; (a, b); $(a, -b)$; $(-a, b)$; $(-a, -b)$; asymptotes

Section 13.3

7. substitution; elimination   8. algebraically; graphically
9. 0; 1; 2

Section 13.4

10. test points; boundary curve; region
11. inequality; intersection

CHAPTER 13 REVIEW EXERCISES

Section 13.1

1. 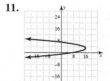   3.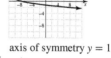
axis of symmetry $y = -3$ axis of symmetry $y = 1$

5.
axis of symmetry $y = 2$ 7.
axis of symmetry $y = 0$

9.
axis of symmetry $y = 2$ 11.
axis of symmetry $y = 3$

13.

15.

The center is (0, 0) and the radius is $r = 7$. Four key points on the circle are (7, 0), (0, 7), (−7, 0), and (0, −7).

The center is (4, −5) and the radius is $r = 3$. Four key points on the circle are (7, −5), (4, −2), (1, −5), and (4, −8).

17. $(x + 1)^2 + (y + 3)^2 = 1$

19. $(x + 6)^2 + (y − 7)^2 = 10$

21. $(x − 3.5)^2 + (y − 1.5)^2 = 20.25$

23. $(x − 1)^2 + (y − 7)^2 = 200$

center: (1, 7); radius $= 10\sqrt{2}$

25. $x^2 + (y − 10)^2 = 216$

center: (0, 10); radius $= 6\sqrt{6}$

27. $\left(x − \dfrac{1}{2}\right)^2 + \left(y + \dfrac{3}{2}\right)^2 = 3$

center: $\left(\dfrac{1}{2}, −\dfrac{3}{2}\right)$; radius $= \sqrt{3}$

Section 13.2

29.

ellipse

31.

circle

33.

hyperbola

35.

hyperbola

Section 13.3

37. $\{(0, 25), (7, 4)\}$

39. $\left\{(4, −3), \left(\dfrac{84}{41}, −\dfrac{187}{41}\right)\right\}$

41. $\{(−3, −3), (1, −11)\}$

43. $\{(−2, −5), (−2, 5), (2, −5), (2, 5)\}$

45. $\left\{\left(\dfrac{3}{4}, \dfrac{51}{16}\right), \left(−\dfrac{1}{2}, −\dfrac{1}{4}\right)\right\}$

47. $\left\{\left(−\dfrac{3}{4}, −\dfrac{4}{3}\right), \left(2, \dfrac{1}{2}\right)\right\}$

Section 13.4

49.

51.

53.

55.

57.   ∅

CHAPTER 13 TEST

1. d   **3.** c

5.

7.

9.

11.

13.

15. $\{(−2, 2)\}$

17.

CHAPTERS 1–13 CUMULATIVE REVIEW EXERCISES

1. −8   **3.** $4x^2 − 13x$ **5.** 34.5°, 55.5°

7. The exact weight of the person is between 104.3 lb and 104.9 lb.

9. $−\dfrac{7}{2}$

11.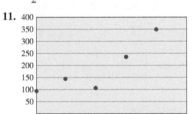

13. 21

15. $m = −\dfrac{2}{5}, (0, 4)$

17. $y = −6$

19.

21. intersecting lines, one solution, consistent system with independent equations

23. 200 adult tickets and 350 student tickets were sold.

25. $\{(−2, 1, 4)\}$ **27.** −1   **29.** −15a   **31.** $−2x^2 − 10x + 4$

33. $75c^2 − 30cd + 3d^2$ **35.** $x(x − 9)$

37. $(3x + 2)(x − 1)(x^2 + x + 1)$ **39.** $\left\{−\dfrac{2}{3}, 0, \dfrac{2}{3}\right\}$

41. $\dfrac{2}{x − 4}$ **43.** $\dfrac{2x^2 + 11x}{(3x − 1)^2(x + 2)}$ **45.** −4   **47.** $\{−7\}$

49. $\dfrac{2y^6\sqrt{7x}}{x}$ **51.** $−1 + 32\sqrt{3}$ **53.** $\dfrac{1}{2} + \dfrac{1}{2}i$

55.

The solution set is $\{-2, 4\}$.

57. $\left\{\dfrac{-1 \pm \sqrt{3}}{2}\right\}$ **59.** two complex, nonreal solutions

61. one rational solution   **63.** $\{\pm 3i, \pm 2\}$ **65.** $\{-1, 243\}$

67. $(-2, 2)$

69. $(f + g)(x) = 2x^2 - 2x + 1$

   $(f - g)(x) = 2x + 7$

   $(fg)(x) = x^4 - 2x^3 + x^2 - 8x - 12$

   domain: $(-\infty, \infty)$

   $\left(\dfrac{f}{g}\right)(x) = \dfrac{x^2 + 4}{(x - 3)(x + 1)}, x \neq 3, -1$

   domain: $(-\infty, -1) \cup (-1, 3) \cup (3, \infty)$

71. 1   **73.** 0

75. $(f \circ g)(x) = -15x + 9$

   $(g \circ f)(x) = -15x + 5$

   domain: $(-\infty, \infty)$

77. $f^{-1}(x) = \dfrac{\sqrt[3]{x + 5}}{2}$ **79.** -1 **81.** 1   **83.** -4 **85.** $\{0.25\}$

87. $\dfrac{1}{2}\ln x + 3\ln(x + 2)$ **89.** $\log_{10}\dfrac{x^2 y^3}{z}$ **91.** $2x + 6$ **93.** $x + 1$

95. $\left\{\dfrac{10^{1.8} - 9}{2}\right\}$ or $\{27.05\}$ **97.** $\left\{\dfrac{\ln 7}{4\ln 1.035}\right\}$ or $\{14.14\}$

99. $\{25\}$ **101.** $\left\{\dfrac{1}{3}\right\}$

103. vertex: $(-36, -3)$; axis of symmetry: $y = -3$

105. center: $(0, 0)$; radius: 10;
   points: $(10, 0)$, $(0, 10)$,
   $(-10, 0)$, $(0, -10)$

107. center: $(0, -1)$; radius: 2;
   points: $(-2, -1)$, $(0, -3)$,
   $(2, -1)$, $(0, 1)$

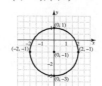

109. $(x - 5)^2 + (y + 6)^2 = 81$

111. $(x - 2)^2 + (y + 1)^2 = 9$; center: $(2, -1)$, radius $= 3$

113.

115.

117. $\{(-\sqrt{3}, -2), (\sqrt{3}, -2)\}$ **119.** $\{(0, 1), (6, 25)\}$

121. $\{(-2\sqrt{5}, -\sqrt{6}), (-2\sqrt{5}, \sqrt{6}), (2\sqrt{5}, -\sqrt{6}), (2\sqrt{5}, \sqrt{6})\}$

123. $\left\{\left(\dfrac{3}{2}, \dfrac{15}{2}\right), (-1, 5)\right\}$

125.

127.

129.

131.

Credits

Photo Credits

Design Icons
Write about it: © Blend Images/Getty RF; **You be the teacher:** © Rubberball/Getty RF; **Practice makes perfect, Think about it, Calculate it, Mix' em up:** © Getty RF; **Piece it together** © PhotoDisc/Getty RF; **Group Activity:** © Banana Stock/PictureQuest RF.

Chapter S
Opener: © Purestock/Punchstock RF; **p. S2(top)** Library of Congress Prints and Photographs Division[LC-USW33-019093-C; **p. S2(right):** © Getty RF; **P. S3:** © Ingram Publishing/Superstock RF; **p. S6(top):** © Janice Ortiz; **p. S6(right), p. S7, p. S8(both), p. S9:** © Getty RF; **p. S10:** © Alamy RF; **p. S11(top):** © Todd Hendricks; **p. S11(right):** © Punchstock RF; **p. S11(bottom):** © Getty RF; **p. S12:** © Ingram Publishing RF; **p. S13:** © Corbis RF; **p. S14:** © Getty RF; **p. S15:** © Ingram Publishing RF; **p. S16, p. S18:** © Getty RF; **p. S19(top):** © Todd Hendricks; **p. S19(right):** © Veer RF; **p. S20:** © Corbis RF; **p. S23, S24:** © Getty RF; **p. S25(top):** Courtesy of Grace Shyu; **p. S25(bottom):** © Veer RF; **p. S25(right)** © Getty RF; **p. S26:** © Veer RF; **p. S27, p. S 28:** © Getty RF.

Chapter 1
Opener: © Stockbyte/Getty RF; **p. 1(bottom):** © Brand X.Punchstock RF; **p. 2(top):** © Najlah Feanny/Corbis; **p. 2(bottom):** © Getty RF; **p. 13:** © The McGraw-Hill Companies, Inc. John Flournoy, photographer; **p. 15:** © Getty RF; **p. 27:** © Punchstock RF; **p. 34, p. 35:** © Getty RF; **p. 39:** © Corbis RF; **p. 40(top):** © Picturequest RF; **p. 40(bottom):** © RubberBall RF; **p. 41:** © Getty RF; **p. 46:** © Alamy RF; **p. 48:** © The McGraw-Hill Companies, Inc. Jill Braaten, photographer; **p. 49:** © Jupiter RF; **p. 50, p. 53:** © Alamy RF; **p. 56(top left):** © Corbis RF; **p. 56(bottom left):** © Getty RF; **p. 56(right):** © Punchstock RF; **p. 57:** © Corbis RF; **p. 64:** © The McGraw-Hill Companies, Inc. Mark Dierker, photographer; **p. 84, p. 90:** © Getty RF.

Chapter 2
Opener: © Corbis RF; **p. 91(bottom):** © Brand X Pictures/PunchStock; **p. 92:** Library of Congress, Prints and Photographs Division. (LC-USZ62-117122); **p. 96:** © Jupiter RF; **p. 99(left):** © Corbis RF; **p. 99(right), p. 101, p. 107(top):** © Getty RF; **p. 107(bottom):** © Comstock/Picturequest RF; **p. 110(top):** © Getty RF; **p. 110(bottom):** © The McGraw-Hill Companies, Inc. Mark Dierker, photographer; **p. 112, p. 118:** © Brand X RF; **p. 122:** © McGraw-Hill Companies, Inc. Gary He, photographer; **p. 136(top left):** © Getty RF; **p. 136(right):** © Brand X Pictures/PunchStock RF; **p. 136(bottom left), p. 137(top, bottom):** © Getty RF; **p. 138, p. 139(top, bottom):** © Brand X Pictures/PunchStock RF; **p. 153:** © Alamy RF; **p. 154:** © The McGraw-Hill Companies, Inc. John Flournoy, photographer; **p. 157(top, middle):** © Getty RF; **p. 158:** © Comstock/Punchstock RF; **p. 162:** © Alamy RF; **p. 163(top, bottom):** © Getty RF; **p. 166:** © Corbis RF; **p. 174:** © Getty RF; **p. 175:** © Brand X/Punchstock RF; **p. 180(left):** © Getty RF; **p. 180(right):** © Copyright 1997 IMS Communications Ltd./Capstone Design. All Rights Reserved; **p. 181, p. 188:** © Corbis RF.

Chapter 3
Opener: U.S. Department of Defense; **p. 193(bottom):** © Alamy RF; **p. 194(left):** © Eric Gay/AP Photo; **p. 194(football):** © Getty RF; **p. 200:** © Rubberball RF; **p. 206:** © Alamy RF; **p. 223:** © The McGraw-Hill Companies, Inc. John Flournoy, photographer; **p. 224(left):** © BananaStock/PunchStock RF; **p. 224(right):** © Creatas Images/PictureQuest RF; **p. 239:** © Getty RF; **p. 241:** © Corbis RF; **p. 247:** © Getty RF; **p. 249, p. 258:** © Corbis RF; **p. 263, p. 264, p. 266(top, bottom), p. 273:** © Getty RF.

Chapter 4
Opener: © Superstock RF; **p. 293(bottom):** © The McGraw-Hill Companies, Inc. Jill Braaten, photographer; **p. 294(top):** Photo courtesy of The Napoleon Hill Foundation; **p. 294(right):** © Rubberball RF; **p. 317, 324:** © The McGraw-Hill Companies, Inc. Jill Braaten, photographer; **p. 329:** © Corbis RF; **p. 330(top):** © Brand X/Punchstock RF; **p. 330(bottom), p. 331:** © Getty RF; **p. 332:** © Corbis RF; **p. 333:** © Stockbyte/Getty RF; **p. 334:** USDA Natural Resources Conservation Services/Photo by Jeff Vanuga; **p. 335, p. 338:** © Corbis RF; **p. 340, p. 341, p. 342:** © Digital Vision RF; **p. 344:** © Corbis RF; **p. 348:** © Corbis RF; **p. 349:** © TongRo Image/Alamy RF; **p. 351, p. 355:** © Getty RF; **p. 360:** Ingram Publishing RF.

Chapter 5
Opener: © Getty RF; **p. 365(bottom):** © Ingram Publishing RF; **p. 366(top):** Library of Congress, Prints and Photographs Division (LC-J601-302); **p. 366(bottom):** Jupiter Images RF; **p. 377(bottom):** © Getty RF; **p. 377(top):** Mark Steinmetz RF; **p. 378:** © Ingram Publishing RF; **p. 379, p. 383:** © Getty RF; **p. 385(top):** © Comstock/Alamy RF; **p. 385(bottom):** © Getty RF; **p. 388:** © Ingram Publishing RF; **p. 389:** © Getty RF; **p. 390, p. 394(top):** © Alasdair Drysdale; **p. 394(middle):** © Corbis RF; **p. 394(bottom):** © Goodshoot/PunchStock RF; **p. 398:** © Getty RF; **p. 401:** © flickr RF/Getty RF; **p. 402:** © Getty RF; **p. 403:** © Alamy RF; **p. 404, p. 407:** © Corbis RF; **p. 410:** © Ingram Publishing RF; **p. 430:** © Goodshoot/PunchStock.

Chapter 6
Opener: © Corbis RF; **p. 439(bottom):** © PictureQuest RF; **p. 440(top):** © Todd Hendricks; **p. 440(bottom), p. 444, p. 450, p. 453:** © Getty RF; **p. 457, p. 460:** © Lars A. Niki; **p. 474:** © Brand X Pictures/Superstock RF; **p. 485:** © Getty RF; **p. 487:** © Photolibrary RF; **p. 489(top):** © Lars A. Niki; **p. 489(bottom), p. 494(left, right), p. 500:** © Getty RF.

Chapter 7
Opener: © Getty RF; **p. 503(bottom):** © Getty RF; **p. 504(top):** Library of Congress Prints and Photographs Division [LC-USZ62-60242]; **p. 504(bottom):** © PictureQuest RF; **p. 511:** © Getty RF; **p. 512(top):** © Corbis RF; **p. 512(bottom):** © Jupiterimages/Getty RF; **p. 518:** © The McGraw-Hill Companies, Inc. John Flournoy, photographer; **p. 521, p. 522:** © Corbis RF; **p. 524(top):** © IMS Communications Ltd./Capstone Design/FlatEarth Images; **p. 524(top left):** © Corbis RF; **p. 524(bottom left):** © Getty RF; **p. 525:** © Alamy RF; **p. 542, p. 546:** © Getty RF; **p. 552:** © Corbis RF; **p. 560, p. 562:** © Getty RF; **p. 563:** NOAA; **p. 565:** © Getty RF; **p. 566:** SuperStock RF.

Chapter 8
Opener: © Ingram Publishing RF; **p. 579:** © Getty RF; **p. 580(top):** Library of Congress Prints and Photographs Division [LC-U9-31687B-13A/14]; **p. 580(bottom):** © The McGraw-Hill Companies, Inc. Jill Braaten, photographer; **p. 590:** © Creatas/Punchstock RF; **p. 591:** © Ingram Publishing RF; **p. 595:** © Blend Images LLC; **p. 601:** © Getty RF; **p. 609:** © Ingram Publishing RF; **610:** © Getty RF; **p. 616, 617:** © Brand X/Fotosearch RF; **p. 618:** © The McGraw-Hill Companies, Inc. Ken Cavanagh photographer; **p. 620:** © Ingram Publishing/SuperStock RF; **p. 621:** © Getty RF; **p. 622:** © Stockbyte/PunchStock RF; **p. 623:** © Rubberball/SuperStock RF.

Chapter 9
Opener: © Ingram Publishing RF; **p. 635:** © GettyRF; **p. 636(top):** © Photo Courtesy Minding International, Ltd.; **p. 636(middle, bottom):** © Getty RF; **p. 649(left):** © Creatas/PunchStock; **p. 649(right), p. 654:** © Corbis RF; **p. 658(top):** © The McGraw-Hill Companies, Inc.; **p. 658(bottom):** © Corbis RF; **p. 676:** © Stockdisc/PunchStock RF; **p. 682, 684, p. 689:** © Ingram Publishing RF.

Chapter 10
Opener: © Getty RF; **p. 699:(bottom):** © Comstock Images/Jupiterimages RF; **p. 700(top):** Used with permission from Jim Rohn International © 2011; **p. 700(bottom):** © Comstock Images/Alamy RF; **p. 713:** © Rubberball/SuperStock RF; **p.714:** © David R. Frazier Photolibrary RF; **p.715:** © Corbis RF; **p. 726:** © Pixtal/SuperStock RF; **p. 737:** © Getty RF; **p. 763:** © Comstock Images/Jupiterimages RF.

Chapter 11

Opener: © Getty RF; **p. 783(bottom):** © Getty RF; **p. 784(top):** Library of Congress, Prints and Photographs Division [LC-USZ62-128302]; **p. 784(bridge):** © Corbis RF; **p. 784(monument):** © Design Pics/PunchStock RF; **p. 784(satellite, fountain):** © Getty RF; **p. 799:** © Brand X Pictures/Jupiter RF; **p. 806(top):** © Alamy RF; **p. 806(bottom):** © imageshop/PunchStock RF; **p. 811, p. 812:** © Corbis RF; **p. 813:** © The McGraw-Hill Companies, Inc. Gerald Wofford, photographer; **p. 818:** © Getty RF; **p. 819(top):** © Valerie Martin RF; **p. 819(bottom):** © Corbis RF; **p. 822, p. 823:** © Getty RF; **p. 825, p. 832:** © Corbis RF; **p. 833:** © Alamy RF; **p. 836, p. 838, p.844:** © Getty RF; **p. 846:** © Comstock Images/Alamy RF; **p. 848:** © Alamy RF; **p. 868:** © Corbis RF.

Chapter 12

Opener: © Corbis RF; **p. 874(top):** Library of Congress Prints and Photographs Division [LC-USZ62-112513] **p. 874(bottom), p. 881, p. 888:** © Getty RF; **p. 906:** © The McGraw-Hill Companies Inc. John Flournoy, photographer; **p. 909(top):** © Getty RF; **p. 909(bottom):** © Imagestate Media RF; **p. 926:** © Getty RF; **p. 931, p. 934:** USGS photo by Walter D. Mooney; **p. 935, p. 940:** © Getty RF; **p. 941:** © Alasdair Drysdale RF. **p. 945:** © Bethany Wheeler

Chapter 13

Opener: © The McGraw-Hill Companies, Inc. Mark Dierker, photographer; **p. 955(bottom):** © The McGraw-Hill Companies, Inc.; **p. 956(bottom):** © EyeWire, Inc. RF; **p. 956(top):** David Shankbone(http://shankbone.org); http://en.wikipedia.org/wiki/File:Joan_Didion_at_the_Brooklyn_Book_Festival.jpg; **p. 965:** © The McGraw-Hill Companies, Inc.

Text Credits

Chapter S

Page S-2: Reproduced with permission of Curtis Brown, London on behalf of the Estate of Sir Winston Churchill. Copyright © Winston S. Churchill.

Chapter 1

Page 2: Stephen Covey. From The 7 habits of highly effective people. © 1989, 2004. Free Press, a division of Simon & Schuster. ISBN 0-7432-7245-5, page 161.

Chapter 2

Page 92: Source: Harry S. Truman, United States President, The Harry S. Truman Library and Museum, Independence, MO.

Chapter 3

Page 194: Source: From Quotable Eddie Robinson, by Aaron S. Lee, page 30 ISBN: 978-1931249218. Used by permission of Taylor Trade Publishing, a member of RLPG.

Chapter 4

Page 294: Source: W. Clement Stone, Philanthropist and author, The W. Clement and Jessie V. Stone Foundation, San Francisco, CA.

Chapter 5

Page 366: Copyright Tuskegee University Archives.

Chapter 6

Page 440: Source: Andrea Hendricks, author.

Chapter 7

Page 504: © 1987-Current Year Hebrew University and Princeton University Press.

Chapter 9

Page 636: Source: Edward de Bono, The World Centre for New Thinking, Villa Bighi Kalkara, Malta.

Chapter 10

Page 700: Source: Article by Jim Rohn, America's Foremost Business Philosopher, reprinted with permission from Jim Rohn International © 2010. As a world-renowned author and success expert, Jim Rohn touched millions of lives during his 46-year career as a motivational speaker and messenger of positive life change. For more information on Jim and his popular personal achievement resources or to subscribe to the weekly Jim Rohn Newsletter, visit www.JimRohn.com.

Chapter 11

Page 784: Source: Leo Tolstoy, Leo Tolstoy Museum-Estate.

Chapter 12

Page 874: "Courtesy of the American Foundation for the Blind, Helen Keller Archives."

Chapter 13

Page 956: Excerpt from SLOUCHING TOWARDS BETHLEHEM by Joan Didion. Copyright © 1966, 1968, renewed 1996 by Joan Didion. Reprinted by permission of Farrar, Straus and Giroux, LLC.

Index

Shape	Formulas for Area (A), and Perimeter (P), Circumference (C)
Triangle	$A = \frac{1}{2}bh = \frac{1}{2} \times$ base $\times$ height $P = a + b + c =$ sum of sides
Rectangle	$A = lw =$ length $\times$ width $P = 2l + 2w =$ sum of twice the length and twice the width
Trapezoid	$A = \frac{1}{2}(b_1 + b_2)h = \frac{1}{2} \times$ sum of bases $\times$ height $P = a + b_1 + c + b_2 =$ sum of sides
Parallelogram	$A = bh =$ base $\times$ height $P = 2a + 2b =$ sum of twice the base and twice the side
Circle	$A = \pi r^2 = \pi \times$ square of radius $C = 2\pi r = 2 \times \pi \times$ radius $C = \pi d = \pi \times$ diameter

Figure	Formulas for Volume (V) and Surface Area (SA)
Rectangular Prism	$V = lwh =$ length $\times$ width $\times$ height $SA = 2lw + 2hw + 2lh$ $= 2($length $\times$ width$) + 2($height $\times$ width$) + 2($length $\times$ height$)$
General Prisms	$V = Bh =$ area of base $\times$ height $SA =$ sum of the areas of the faces
Right Circular Cylinder	$V = Bh =$ area of base $\times$ height $SA = 2B + Ch = (2 \times$ area of base$) + ($circumference $\times$ height$)$
Pyramid	$V = \frac{1}{3}Bh = \frac{1}{3} \times$ area of base $\times$ height $SA = B + \frac{1}{2}P\ell$ $=$ area of base $+ \left(\frac{1}{2} \times$ perimeter of base $\times$ slant height$\right)$
Right Circular Cone	$V = \frac{1}{3}Bh = \frac{1}{3} \times$ area of base $\times$ height $SA = B + \frac{1}{2}C\ell$ $=$ area of base $+ \left(\frac{1}{2} \times$ circumference $\times$ slant height$\right)$
Sphere	$V = \frac{4}{3}\pi r^3 = \frac{4}{3} \times \pi \times$ cube of radius $SA = 4\pi r^2 = 4 \times \pi \times$ square of radius